THE CALCULUS
WITH ANALYTIC GEOMETRY
Sixth Edition

Louis Leithold

1817

HARPER & ROW, PUBLISHERS, New York
Grand Rapids, Philadelphia, St. Louis, San Francisco,
London, Singapore, Sydney, Tokyo

THE COVER AND CHAPTER OPENING ARTIST

David Furman, a sculptor working in Southern California, is currently professor and head of the studio arts program at Pitzer College, Claremont.

The art used for the front and back covers and chapter openings grew out of a body of work exhibited in 1987 at Tortue Gallery, Santa Monica. Realistic in their depiction, these trompe l'oeil wall sculptures were made from a variety of materials including glazed ceramics, styrene, acrylic and oil base paint, underglazes, lusters, and wood. All the objects on the "ersatz blackboard" were made of ceramic.

Sponsoring Editor: Peter Coveney
Development Editor: Jonathan Haber
Project Editor: David Nickol
Text Design: Circa 86
Production: Kewal K. Sharma

The Calculus with Analytic Geometry, Sixth Edition

Library of Congress Cataloging-in-Publication Data

Leithold, Louis.
    The calculus with analytic geometry / Louis Leithold. — 6th ed.
      p. cm.
    ISBN 0-06-043930-0
    1. Calculus. 2. Geometry, Analytic.   I. Title.
    QA303.L428 1990
    515'.15—dc20   89-27810
                                             CIP

90 91 92 93 9 8 7 6 5 4 3 2 1

18. $\displaystyle\int \frac{u^2\,du}{\sqrt{a+bu}} = \frac{2}{15b^3}(3b^2u^2 - 4abu + 8a^2)\sqrt{a+bu} + C$

19. $\displaystyle\int \frac{u^n\,du}{\sqrt{a+bu}} = \frac{2u^n\sqrt{a+bu}}{b(2n+1)} - \frac{2an}{b(2n+1)}\int \frac{u^{n-1}\,du}{\sqrt{a+bu}}$

20. $\displaystyle\int \frac{du}{u\sqrt{a+bu}} = \begin{cases} \dfrac{1}{\sqrt{a}}\ln\left|\dfrac{\sqrt{a+bu}-\sqrt{a}}{\sqrt{a+bu}+\sqrt{a}}\right| + C & \text{if } a > 0 \\[3mm] \dfrac{2}{\sqrt{-a}}\tan^{-1}\sqrt{\dfrac{a+bu}{-a}} + C & \text{if } a < 0 \end{cases}$

21. $\displaystyle\int \frac{du}{u^n\sqrt{a+bu}} = -\frac{\sqrt{a+bu}}{a(n-1)u^{n-1}} - \frac{b(2n-3)}{2a(n-1)}\int \frac{du}{u^{n-1}\sqrt{a+bu}}$

22. $\displaystyle\int \frac{\sqrt{a+bu}\,du}{u} = 2\sqrt{a+bu} + a\int \frac{du}{u\sqrt{a+bu}}$

23. $\displaystyle\int \frac{\sqrt{a+bu}\,du}{u^n} = -\frac{(a+bu)^{3/2}}{a(n-1)u^{n-1}} - \frac{b(2n-5)}{2a(n-1)}\int \frac{\sqrt{a+bu}\,du}{u^{n-1}}$

Forms Containing $a^2 \pm u^2$

24. $\displaystyle\int \frac{du}{a^2+u^2} = \frac{1}{a}\tan^{-1}\frac{u}{a} + C$

25. $\displaystyle\int \frac{du}{a^2-u^2} = \frac{1}{2a}\ln\left|\frac{u+a}{u-a}\right| + C = \begin{cases} \dfrac{1}{a}\tanh^{-1}\dfrac{u}{a} + C & \text{if } |u| < a \\[3mm] \dfrac{1}{a}\coth^{-1}\dfrac{u}{a} + C & \text{if } |u| > a \end{cases}$

26. $\displaystyle\int \frac{du}{u^2-a^2} = \frac{1}{2a}\ln\left|\frac{u-a}{u+a}\right| + C = \begin{cases} -\dfrac{1}{a}\tanh^{-1}\dfrac{u}{a} + C & \text{if } |u| < a \\[3mm] -\dfrac{1}{a}\coth^{-1}\dfrac{u}{a} + C & \text{if } |u| > a \end{cases}$

Forms Containing $\sqrt{u^2 \pm a^2}$

In formulas 27 through 38, we may replace

$\ln(u + \sqrt{u^2+a^2})$ by $\sinh^{-1}\dfrac{u}{a}$

$\ln|u + \sqrt{u^2-a^2}|$ by $\cosh^{-1}\dfrac{u}{a}$

$\ln\left|\dfrac{a + \sqrt{u^2+a^2}}{u}\right|$ by $\sinh^{-1}\dfrac{a}{u}$

27. $\displaystyle\int \frac{du}{\sqrt{u^2 \pm a^2}} = \ln|u + \sqrt{u^2 \pm a^2}| + C$

28. $\displaystyle\int \sqrt{u^2 \pm a^2}\,du = \frac{u}{2}\sqrt{u^2 \pm a^2} \pm \frac{a^2}{2}\ln|u + \sqrt{u^2 \pm a^2}| + C$

29. $\displaystyle\int u^2\sqrt{u^2 \pm a^2}\,du = \frac{u}{8}(2u^2 \pm a^2)\sqrt{u^2 \pm a^2}$
$\displaystyle\qquad\qquad\qquad\qquad\qquad - \frac{a^4}{8}\ln|u + \sqrt{u^2 \pm a^2}| + C$

30. $\displaystyle\int \frac{\sqrt{u^2+a^2}\,du}{u} = \sqrt{u^2+a^2} - a\ln\left|\frac{a + \sqrt{u^2+a^2}}{u}\right| + C$

31. $\displaystyle\int \frac{\sqrt{u^2-a^2}\,du}{u} = \sqrt{u^2-a^2} - a\sec^{-1}\frac{u}{a} + C$

32. $\displaystyle\int \frac{\sqrt{u^2 \pm a^2}\,du}{u^2} = -\frac{\sqrt{u^2 \pm a^2}}{u} + \ln|u + \sqrt{u^2 \pm a^2}| + C$

33. $\displaystyle\int \frac{u^2\,du}{\sqrt{u^2 \pm a^2}} = \frac{u}{2}\sqrt{u^2 \pm a^2} - \frac{\pm a^2}{2}\ln|u + \sqrt{u^2 \pm a^2}| + C$

34. $\displaystyle\int \frac{du}{u\sqrt{u^2+a^2}} = -\frac{1}{a}\ln\left|\frac{a + \sqrt{u^2+a^2}}{u}\right| + C$

35. $\displaystyle\int \frac{du}{u\sqrt{u^2-a^2}} = \frac{1}{a}\sec^{-1}\frac{u}{a} + C$

36. $\displaystyle\int \frac{du}{u^2\sqrt{u^2 \pm a^2}} = -\frac{\sqrt{u^2 \pm a^2}}{\pm a^2 u} + C$

37. $\displaystyle\int (u^2 \pm a^2)^{3/2}\,du = \frac{u}{8}(2u^2 \pm 5a^2)\sqrt{u^2 \pm a^2}$
$\displaystyle\qquad\qquad\qquad\qquad\qquad + \frac{3a^4}{8}\ln|u + \sqrt{u^2 \pm a^2}| + C$

38. $\displaystyle\int \frac{du}{(u^2 \pm a^2)^{3/2}} = \frac{u}{\pm a^2\sqrt{u^2 \pm a^2}} + C$

Forms Containing $\sqrt{a^2 - u^2}$

39. $\displaystyle\int \frac{du}{\sqrt{a^2-u^2}} = \sin^{-1}\frac{u}{a} + C$

40. $\displaystyle\int \sqrt{a^2-u^2}\,du = \frac{u}{2}\sqrt{a^2-u^2} + \frac{a^2}{2}\sin^{-1}\frac{u}{a} + C$

41. $\displaystyle\int u^2\sqrt{a^2-u^2}\,du = \frac{u}{8}(2u^2 - a^2)\sqrt{a^2-u^2} + \frac{a^4}{8}\sin^{-1}\frac{u}{a} + C$

42. $\displaystyle\int \frac{\sqrt{a^2-u^2}\,du}{u} = \sqrt{a^2-u^2} - a\ln\left|\frac{a + \sqrt{a^2-u^2}}{u}\right| + C$
$\displaystyle\qquad\qquad\qquad\quad = \sqrt{a^2-u^2} - a\cosh^{-1}\frac{a}{u} + C$

43. $\displaystyle\int \frac{\sqrt{a^2-u^2}\,du}{u^2} = -\frac{\sqrt{a^2-u^2}}{u} - \sin^{-1}\frac{u}{a} + C$

(This table is continued on the back endpapers)

CONTENTS

To the following teachers who influenced me the most:

Florence Balensiefer; English, Lowell High School, San Francisco
Ivan Barker; Mathematics, Lowell High School, San Francisco
Alan McKeever; Journalism, Lowell High School, San Francisco
Benjamin Bernstein; Mathematics, University of California, Berkeley
Pauline Sperry; Mathematics, University of California, Berkeley
Virginia Wakerling; Mathematics, University of California, Berkeley

And to my former students at

Phoenix College
California State University at Los Angeles
University of Southern California
Open University of Great Britain
Pepperdine University

I have learned from all of them.

PREFACE

> "Everything should be made as simple as possible, but not simpler."
> —ALBERT EINSTEIN

The Calculus with Analytic Geometry (*TCWAG*) is designed both for prospective mathematics majors and for students whose primary interest is in engineering, the physical and social sciences, or nontechnical fields. Its step-by-step explanations, abundant worked examples, and wide variety of exercises continue to be distinctive features of the text in its sixth edition.

Because a textbook should be written for the student, I have endeavored to keep the presentation geared to a beginner's experience and maturity and to leave no step unexplained or omitted. I desire that the reader be aware that proofs of theorems are necessary and that these proofs be well motivated and carefully explained, so that they are understandable to the student who has achieved an average mastery of the preceding sections of the book. If a theorem is stated without proof, the discussion is augmented by both figures and examples, and in such cases I have always stressed that what is presented is an illustration of the statement of the theorem and is not a proof. Some of the theoretical discussions appear in supplementary sections at the end of chapters, where, if so desired, they may be omitted without loss of continuity.

THE SIXTH EDITION OF *TCWAG*

Since the first edition of this book in 1968, there have been significant changes in the content and teaching of the calculus course. With each successive edition, I have attempted to incorporate these changes and to maintain a healthy balance between a rigorous approach and an intuitive point of view.

The nineteen chapters of *TCWAG* form four segments: Chapter 1, review topics in precalculus; Chapters 2–11, functions of a single variable; Chapters 12–13, infinite series; and Chapters 14–19, vectors and functions of more than one variable. The sixth edition incorporates changes in each of these segments, some to reflect the growing importance of computers and programmable calculators and the computations they facilitate.

REVIEW TOPICS IN CALCULUS

CHAPTER 1

This chapter, "Real Numbers, Functions, and Graphs," is less detailed than in previous editions. A section on the basic facts about the real number system is followed by an introduction to analytic geometry that includes the traditional material on lines and circles. The definition of a function, operations with functions, particular kinds of functions, and graphs of functions are discussed. The presentation of the six trigonometric functions here allows their early use in examples of differentiation and integration of nonalgebraic functions.

FUNCTIONS OF A SINGLE VARIABLE

CHAPTER 2   With the section on limits at infinity moved to this chapter, the discussion of limits and continuity is now complete in one chapter. These topics are at the heart of any first course in the calculus. All the limit theorems are stated, and some proofs are presented in the text, while other proofs have been outlined in the exercises. New to this edition are examples and exercises that involve using a calculator to hazard a conjecture about a particular limit.

CHAPTER 3   In Section 3.1, I define the tangent line to a curve to demonstrate in advance the geometrical interpretation of the derivative, which is defined in Section 3.2. The physical application of instantaneous velocity in rectilinear motion is presented after theorems on differentiation are proved. The derivatives of all six trigonometric functions are presented here, and they are then available as examples for the initial presentation of the chain rule. There are some new exercises requiring the use of a calculator to estimate a particular value of the derivative from the definition.

CHAPTER 4   The traditional applications of the derivative to problems involving maxima and minima, as well as to curve sketching are presented in this chapter. The topics of limits at infinity and horizontal and vertical asymptotes have been moved forward to Chapter 2. The separate section on applications in economics and business that appeared here in previous editions has been deleted, but some of the material is discussed elsewhere. The section on the differential has been moved to this chapter so that it appears closer to its reference in the treatment of antidifferentiation.

CHAPTER 5   The definite integral and integration are the subjects of Chapter 5. The first two sections involve antidifferentiation. I use the term "antidifferentiation" instead of "indefinite integration," but the standard notation $\int f(x)\,dx$ is retained. This notation will suggest that some relation must exist between definite integrals and antiderivatives, but I see no harm in this as long as the presentation gives the theoretically proper view of the definite integral as the limit of sums. Two numerical methods for approximating a definite integral are given in the chapter's final section which has been moved forward from the previous edition. These procedures are important now because of their suitability to computers and programmable calculators. The material on the approximation of definite integrals includes the statement of theorems on the bounds of the error involved in these approximations. The chapter also includes a section on separable differential equations, and the complete discussion of area of a plane region appears here.

CHAPTER 6   In this chapter, I have given applications of the definite integral that highlight not only the manipulative techniques but also the fundamental principles involved. In each application, the definitions of the new terms are intuitively motivated and explained. The treatment of volumes of solids, the subject matter of the first two sections, has been revised from earlier editions. Section 6.1 begins with volumes by slicing, and then volumes of solids of revolution by disks and washers are considered as special cases of volumes by slicing. Volumes of solids of revolution by cylindrical shells are discussed in Section 6.2. Another geometrical application of the definite integral is length of arc in Section 6.3. The remaining sections of the chapter are devoted to physical applications including centers of mass of rods and plane regions, work, and liquid pressure.

CHAPTERS 7 AND 8   Inverse functions are covered in the first two sections of Chapter 7, and the next five sections are devoted to logarithmic and exponential functions. The natural logarithmic function is defined first, and then the natural exponential function is defined as its inverse. This procedure allows us to give a precise meaning to an irrational exponent of a positive number. We then define the exponential function to the base a, where a is positive, and its inverse is the logarithmic function to the base a. Applications of these functions include the laws of growth and decay, bounded growth involving the learning curve, and the standardized normal probability density function. Section 7.8, new to this edition, involves the solution of first-order linear differential equations. In Chapter 8 the remaining transcendental (nonalgebraic) functions are introduced. These are the inverse trigonometric functions and the hyperbolic functions.

CHAPTER 9   Techniques of integration involve an important computational aspect of calculus. They are discussed in this chapter, which has been shortened to eight sections in this edition. I have explained the theoretical backgrounds of each different method after an introductory motivation. The mastery of integration techniques depends upon the examples, and I have used as illustrations problems that the student will certainly meet in practice. Two more applications of integration are introduced in Section 9.5: logistic growth, occurring in economics, biology, and sociology; and the law of mass action from chemistry.

CHAPTER 10   The order of analytic geometry topics in this chapter has been changed in this edition. The first four sections pertain to conic sections: the parabola, the ellipse, and the hyperbola. Each of the conics is introduced by indicating how it is formed by intersecting a plane and a cone; then the analytic definition is given and its equation in rectangular coordinates is obtained. Polar coordinates and some of their applications are presented in Sections 10.5 through 10.7. Polar equations of the conics appear in Section 10.8, where they occur as part of a unified treatment of conic sections.

CHAPTER 11   This chapter, "Indeterminate Forms, Improper Integrals, and Taylor's Formula," has been repositioned in this edition so that it immediately precedes the material on infinite series, where many of the results are applied. Applications of improper integrals, appearing in Sections 11.3 and 11.4, include the probability density function as well as some in geometry and economics.

INFINITE SERIES

CHAPTERS 12 AND 13   The study of infinite series in these two chapters is considered as a separate segment of the course to make it more apparent that it is self-contained and can be covered anytime after the completion of the calculus of functions of a single variable. Chapter 12 is devoted to sequences and infinite series of constant terms, with the last section giving a summary of tests for convergence of an infinite series. Chapter 13 is concerned with infinite series of variable terms called power series. The exercise sets have been expanded from earlier editions to include more applications.

VECTORS AND FUNCTIONS OF MORE THAN ONE VARIABLE

CHAPTERS 14 AND 15 These two chapters contain the calculus of vectors as well as a vector approach to solid analytic geometry. The first four sections of Chapter 14 on vectors in the plane can be taken up after Chapter 5 if you wish to study vectors earlier in the course. Chapter 15 treats vectors in three-dimensional space, and if desired, topics in Sections 15.1 and 15.2 may be studied concurrently with the corresponding ones in Chapter 14. Applications of vectors to geometry, physics, and engineering occur in both chapters.

CHAPTERS 16, 17, AND 18 The differential and integral calculus of functions of more than one variable are presented in these three chapters. Limits, continuity, partial differentiation, differentiability, and the total differential are discussed in Chapter 16, where applications include finding rates of change and computing approximations. In Chapter 17, a section on directional derivatives and gradients is followed by a section that shows the application of the gradient to find an equation of the tangent plane to a surface. Additional applications of partial derivatives in Chapter 17 are the solution of extrema problems and Lagrange multipliers. Exact differential equations are solved in Section 17.5. The double integral of a function of two variables and the triple integral of a function of three variables, along with some applications to physics, engineering, and geometry, are given in Chapter 18.

CHAPTER 19 The final chapter, "Introduction to the Calculus of Vector Fields," has an expanded treatment of vector calculus. The coverage includes line and surface integrals, Green's theorem, Gauss's divergence theorem, and Stokes's theorem. The approach in this chapter is intuitive and the applications are to physics and engineering.

SUPPLEMENTARY SECTIONS Ten sections, appearing at the ends of some of the chapters, are designated as supplementary. These self-contained topics can be covered or omitted without affecting the understanding of subsequent material.

The supplementary sections are of two types. Some present additional subject matter that is not necessarily part of the traditional syllabus of a calculus course: Sections 4.10, 6.7, 7.8, 8.5, 9.8, 10.9, and 14.8. Others include theoretical discussions, including proofs of some of the theorems: Sections 2.9, 2.10, and 16.8. Both types increase the flexibility of the text.

EXAMPLES AND ILLUSTRATIONS Examples and illustrations—nearly 1000 in all—appear in every section. The examples, which were carefully chosen to prepare students for the exercises, should be used as models for their solutions. An illustration serves to demonstrate a particular concept, definition, or theorem; it is a prototype of the idea being presented.

EXERCISES There are now over 7400 exercises, which have been revised and graded in difficulty to provide a wide variety of types, ranging from computational to applied and theoretical problems. They occur at the end of sections and as review exercises following the last section of a chapter.

Answers to the odd-numbered exercises are given in the back of the book, and answers to the even-numbered ones are available in a separate booklet. Detailed step-by-step solutions for nearly half the even-numbered exercises (those having numbers divisible by 4) appear in a supplement to this text, *An Outline for the Study of Calculus*, by John H. Minnick and Leon Gerber, published by Harper & Row in three volumes.

ENHANCED THREE-DIMENSIONAL GRAPHICS   In response to the needs of students for a more modern, easier to visualize presentation of three-dimensional graphs, over 200 figures are new to this edition. Many were computer generated to ensure mathematical accuracy. These figures, which instructors should find clearer and move vivid than the airbrushed style of geometric solids in the previous edition and older texts, were created with the assistance of the Mathematica program using Illustrator 88.

Louis Leithold

ACKNOWLEDGMENTS

Reviewers of the Sixth Edition of *TCWAG*:
Peter P. Andre, United States Naval Academy
Leon E. Arnold, Delaware County Community College
Harold R. Bennett, Texas Tech University
Michael L. Berry, West Virginia Wesleyan College
John Broughton, Indiana University of Pennsylvania
Floyd A. Cohen, California State University at Long Beach
Joel Davis, Oregon State University
K. Joe Davis, East Carolina University
N. J. DeLillo, Manhattan College
William A. Echols, Houston Community College
John Garlow, Tarrant County Junior College
Stuart Goldenberg, California Polytechnic State University at San Luis Obispo
Joel K. Haack, Oklahoma State University
Norvin Holm, Charles S. Mott Community College
Roy A. Johnson, Washington State University
Dan Kemp, South Dakota State University
Joh Klippert, James Madison University
Walter F. Martens, University of Alabama at Birmingham
Roger B. Nelsen, Lewis and Clark College
William L. Perry, Texas A&M University
Walter A. Rosenkrantz, University of Massachusetts at Amherst
Daniel B. Shapiro, Ohio State University
Charles R. Stone, Dekalb College
Jere Strickland, James H. Faulkner State Junior College
Richard B. Thompson, University of Arizona
G. B. Turney, University of Texas at Arlington
J. Terry Wilson, San Jacinto College
Richard M. Witt, University of Wisconsin–Eau Claire

Assistants for Solutions to the Exercises:
Leon Gerber, St. John's University
Gloria Langer, University of Colorado

Cover and Chapter Opening Artist
David Furman, Claremont
Courtesy of Tortue Gallery, Santa Monica

To these people, to the staff of Harper & Row, and to all the users of the first five editions who have suggested changes, I express my deep appreciation.

L. L.

SUPPLEMENTS

SUPPLEMENTS FOR THE STUDENT

AN OUTLINE FOR THE STUDY OF CALCULUS

In three volumes, this student outline by John Minnick of DeAnza College and Leon Gerber of St. John's University contains all of the important theorems and definitions from the text and includes sample tests with solutions for each chapter, as well as detailed step-by-step solutions for all exercises having numbers divisible by 4.

MICROCOMPUTER LABORATORY WORKBOOK

This laboratory workbook by Michael Moody of Washington State University introduces students to mathematical computing using applications of the calculus to a variety of problems. Its contents, covering 15 topics from the text, are designed to be worked with the aid of a microcomputer.

SUPPLEMENTS FOR THE INSTRUCTOR

INSTRUCTOR'S SOLUTIONS MANUAL

In two volumes, this manual, written by Leon Gerber of St. John's University, contains complete worked-out solutions for all odd-numbered exercises as well as selected even-numbered ones. Figures appear when required.

EVEN-NUMBERED ANSWER BOOKLET

This booklet, produced by Gloria Langer of the University of Colorado, contains answers for all even-numbered exercises.

COMPUTERIZED TEST BANK

This computerized test bank (IBM only) contains problems and questions keyed to the text.

PRINTED TEST BANK

This printed test bank includes all the problems and questions in the computerized test bank.

PACKAGE OF TRANSPARENCIES

This package contains transparencies of three-dimensional figures from the text. Each figure is reproduced on acetate in two colors.

ADVANCED PLACEMENT BOOKLET

This booklet, containing multiple-choice questions for each chapter, is designed to prepare students for the advanced placement calculus examination.

HISTORICAL BACKGROUND
OF THE CALCULUS

Some of the ideas of calculus can be found in the works of the ancient Greek mathematicians at the time of Archimedes (287–212 B.C.) and in works of the early seventeenth century by René Descartes (1596–1650), Pierre de Fermat (1601–1665), John Wallis (1616–1703), and Isaac Barrow (1630–1677). However, the invention of calculus is often attributed to Sir Isaac Newton (1642–1727) and Gottfried Wilhelm Leibniz (1646–1716) because they began the generalization and unification of the subject. There were other mathematicians of the seventeenth and eighteenth century who joined in the development of the calculus; some of them were Jakob Bernoulli (1654–1705), Johann Bernoulli (1667–1748), Leonhard Euler (1707–1783), and Joseph L. Lagrange (1736–1813). However, it wasn't until the nineteenth century that the processes of calculus were given a sound foundation by such mathematicians as Bernhard Bolzano (1781–1848), Augustin L. Cauchy (1789–1857), Karl Weierstrass (1815–1897), and Richard Dedekind (1831–1916).

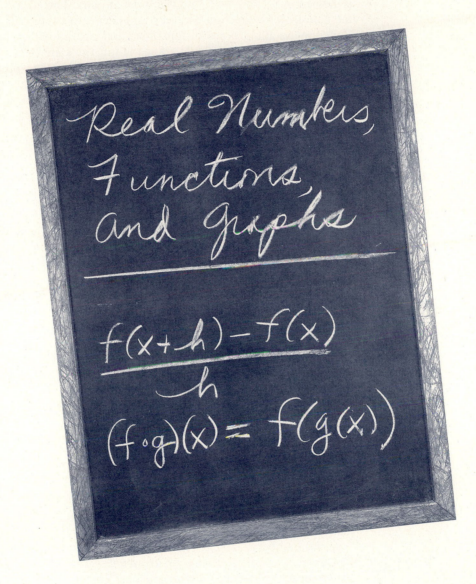

Real Numbers, Functions, and graphs

$$\frac{f(x+h) - f(x)}{h}$$

$$(f \circ g)(x) = f(g(x))$$

Learning calculus can be your most stimulating and exciting educational experience because it is the basis for much of mathematics and for many of the greatest accomplishments of the modern world. You must enter into the study of calculus with a knowledge of certain mathematical concepts. In the first place, it is assumed that you have had courses in high school algebra and geometry. Secondly, there are particular topics that are of special importance. You may have studied these topics in a precalculus course or you may be exposed to them in this chapter for the first time.

You need to be familiar with facts about the *real numbers* and have facility with operations involving inequalities, and this material forms the subject matter of the first section. The next two sections contain an introduction to some of the ideas of *analytic geometry* that are necessary for the sequel.

The notion of a *function* is one of the important concepts in calculus, and it is defined in Section 1.4 as a set of ordered pairs. This idea is used to point up the concept of a function as a correspondence between sets of real numbers. Function notation, types of functions, and operations on functions are also discussed in Section 1.4, while graphs of functions are treated in Section 1.5.

You have probably studied *trigonometric functions* in a previous course, but a review of the basic definitions is presented in Section 1.6. An application of the tangent function to the slope of a line is also given in that section.

Dependent upon your preparation, this chapter may be covered in detail, treated as a review, or omitted.

1.1 REAL NUMBERS AND INEQUALITIES

The **real number system** consists of a set R of elements called **real numbers** and two operations called **addition** and **multiplication**, denoted by the symbols $+$ and $\cdot$, respectively. If a and b are elements of the set R, $a + b$ indicates the **sum** of a and b, and $a \cdot b$ (or ab) indicates their **product**. The operation of subtraction is defined by the equation

$$a - b = a + (-b)$$

where $-b$ denotes the **negative** of b such that $b + (-b) = 0$. The operation of **division** is defined by the equation

$$a \div b = a \cdot b^{-1} \qquad b \neq 0$$

where b^{-1} denotes the **reciprocal** of b such that $b \cdot b^{-1} = 1$.

The real number system can be completely described by a set of axioms (the word **axiom** is used to indicate a formal statement that is assumed to be true without proof). With these axioms we can derive the properties of the real numbers from which follow the familiar algebraic operations of addition, subtraction, multiplication, and division, as well as the algebraic concepts of solving equations, factoring, and so forth.

Properties that can be shown to be logical consequences of axioms are **theorems**. In the statement of most theorems there are two parts: the "if" part, called the **hypothesis**, and the "then" part, called the **conclusion**. The argument verifying a theorem is a **proof**. A proof consists in showing that the conclusion follows from the assumed truth of the hypothesis.

A real number is either positive, negative, or zero, and any real number can be classified as either *rational* or *irrational*. A **rational number** is one that can be expressed as the ratio of two integers. That is, a rational number is of the form p/q, where p and q are integers and $q \neq 0$. The rational numbers consist of the following:

The **integers** (positive, negative, and zero)
$$\ldots, -5, -4, -3, -2, -1, 0, 1, 2, 3, 4, 5, \ldots$$

The positive and negative **fractions**, such as
$$\tfrac{2}{7} \qquad -\tfrac{4}{5} \qquad \tfrac{83}{5}$$

The positive and negative **terminating decimals**, such as
$$2.36 = \frac{236}{100} \qquad -0.003251 = -\frac{3{,}251}{1{,}000{,}000}$$

The positive and negative **nonterminating repeating decimals**, such as

$$0.333\ldots = \tfrac{1}{3} \qquad -0.549549549\ldots = -\tfrac{61}{111}$$

The real numbers that are not rational are called **irrational numbers**. These are positive and negative **nonterminating nonrepeating** decimals, for example,

$$\sqrt{3} = 1.732\ldots \qquad \pi = 3.14159\ldots$$

From time to time we will use some set notation and terminology. The idea of *set* is used extensively in mathematics and is such a basic concept that it is not given a formal definition here. We can say that a **set** is a collection of objects, and the objects in a set are called **elements**. If every element of a set S is also an element of a set T, then S is a **subset** of T. In calculus we are concerned with the set R of real numbers. Two subsets of R are the set N of natural numbers (the positive integers) and the set Z of integers.

We use the symbol $\in$ to indicate that a specific element belongs to a set. Hence we may write $8 \in N$, which is read "8 is an element of N." The notation $a, b \in S$ indicates that both a and b are elements of S. The symbol $\notin$ is read "is not an element of." Thus we read $\tfrac{1}{2} \notin N$ as "$\tfrac{1}{2}$ is not an element of N."

A pair of braces { } used with words or symbols can describe a set. If S is the set of natural numbers less than 6, we can write the set S as

$$\{1, 2, 3, 4, 5\}$$

We can also write the set S as

$$\{x, \text{ such that } x \text{ is a natural number less than 6}\}$$

where the symbol x is called a *variable*. A **variable** is a symbol used to represent any element of a given set.

The set S can be written as follows with **set-builder notation**, where a vertical bar replaces the words *such that*:

$$\{x \,|\, x \text{ is a natural number less than 6}\}$$

which is read "the set of all x such that x is a natural number less than 6."

Two sets A and B are said to be **equal**, written $A = B$, if A and B have identical elements. The **union** of two sets A and B, denoted by $A \cup B$ and read "A union B," is the set of all elements that are in A or in B or in both A and B. The **intersection** of A and B, denoted by $A \cap B$ and read "A intersection B," is the set of only those elements that are in both A and B. The set that contains no elements is called the **empty set** and is denoted by $\varnothing$.

▶ **ILLUSTRATION 1**   Suppose $A = \{2, 4, 6, 8, 10, 12\}$, $B = \{1, 4, 9, 16\}$,   and $C = \{2, 10\}$. Then

$$A \cup B = \{1, 2, 4, 6, 8, 9, 10, 12, 16\} \qquad A \cap B = \{4\}$$
$$B \cup C = \{1, 2, 4, 9, 10, 16\} \qquad\qquad B \cap C = \varnothing \qquad ◀$$

There is an ordering for the set R by means of a relation denoted by the symbols < (read "is less than") and > (read "is greater than").

1.1.1 DEFINITION   If $a, b \in R$,

   (i) $a < b$ if and only if $b - a$ is positive;
  (ii) $a > b$ if and only if $a - b$ is positive.

▶ **ILLUSTRATION 2**

$3 < 5$ because $5 - 3 = 2$, and 2 is positive

$-10 < -6$ because $-6 - (-10) = 4$, and 4 is positive

$7 > 2$ because $7 - 2 = 5$, and 5 is positive

$-2 > -7$ because $-2 - (-7) = 5$, and 5 is positive

$\frac{3}{4} > \frac{2}{3}$ because $\frac{3}{4} - \frac{2}{3} = \frac{1}{12}$, and $\frac{1}{12}$ is positive   ◀

We now define the symbols $\leq$ (read "is less than or equal to") and $\geq$ (read "is greater than or equal to").

1.1.2 DEFINITION

If $a, b \in R$,

(i) $a \leq b$ if and only if either $a < b$ or $a = b$;
(ii) $a \geq b$ if and only if either $a > b$ or $a = b$.

The statements $a < b$, $a > b$, $a \leq b$, and $a \geq b$ are called **inequalities**. In particular, $a < b$ and $a > b$ are called **strict** inequalities, whereas $a \leq b$ and $a \geq b$ are called **nonstrict** inequalities.

The following theorem follows immediately from Definition 1.1.1.

1.1.3 THEOREM

(i) $a > 0$ if and only if a is positive;
(ii) $a < 0$ if and only if a is negative.

A number x is **between** a and b if $a < x$ and $x < b$. We can write this as a **continued inequality** as follows:

$a < x < b$

Another continued inequality is

$a \leq x \leq b$

which means that both $a \leq x$ and $x \leq b$. Other continued inequalities are $a \leq x < b$ and $a < x \leq b$.

The following theorems can be proved by using axioms for the set R and 1.1.1 through 1.1.3.

1.1.4 THEOREM

(i) If $a > 0$ and $b > 0$, then $a + b > 0$.
(ii) If $a > 0$ and $b > 0$, then $ab > 0$.

Part (i) of the above theorem states that the sum of two positive numbers is positive and part (ii) states that the product of two positive numbers is positive.

1.1.5 THEOREM
Transitive Property of Order

If $a, b, c \in R$, and

if $a < b$ and $b < c$, then $a < c$

▶ **ILLUSTRATION 3** If $x < 5$ and $5 < y$, then by the transitive property of order, it follows that $x < y$.   ◀

1.1.6 THEOREM

Suppose $a, b, c \in R$

(i) If $a < b$, then $a + c < b + c$.
(ii) If $a < b$ and $c > 0$, then $ac < bc$.
(iii) If $a < b$ and $c < 0$, then $ac > bc$.

▶ **ILLUSTRATION 4**   (a) If $x < y$, it follows from Theorem 1.1.6(i) that $x + 4 < y + 4$. For instance, $3 < 9$; thus $3 + 4 < 9 + 4$ or, equivalently, $7 < 13$. Furthermore, if $x < y$, then $x - 11 < y - 11$. For instance, $3 < 9$; thus $3 - 11 < 9 - 11$ or, equivalently, $-8 < -2$.

(b) If $x < y$, it follows from Theorem 1.1.6(ii) that $7x < 7y$. For instance, because $5 < 8$, then $7 \cdot 5 < 7 \cdot 8$ or, equivalently, $35 < 56$.

(c) Because $4 < 6$, then if $z < 0$, it follows from Theorem 1.1.6(iii) that $4z > 6z$. For instance, because $4 < 6$, then $4(-3) > 6(-3)$ or, equivalently, $-12 > -18$.   ◀

Part (ii) of Theorem 1.1.6 states that if both sides of an inequality are multiplied by a positive number, the direction of the inequality remains unchanged, whereas part (iii) states that if both sides of an inequality are multiplied by a negative number, the direction of the inequality is reversed. Parts (ii) and (iii) also hold for division because dividing both sides of an inequality by a number d $(d \neq 0)$ is equivalent to multiplying them by $\frac{1}{d}$.

1.1.7 THEOREM   If $a < b$ and $c < d$, then $a + c < b + d$.

▶ **ILLUSTRATION 5**   If $x < 8$ and $y < -3$, then from Theorem 1.1.7, $x + y < 8 + (-3)$; that is, $x + y < 5$.   ◀

We impose upon the set R a condition called the **axiom of completeness** (Axiom 12.2.5). The statement of this axiom is deferred until Section 12.2 because it requires some terminology that is best introduced and discussed later. However, we now give a geometric interpretation to the set of real numbers by associating them with the points on a line, called an **axis**. The axiom of completeness guarantees that there is a one-to-one correspondence between the set R and the set of points on an axis.

Refer to Figure 1, where the axis is a horizontal line. A point on the axis is chosen to represent the number 0. This point is called the **origin**. A unit of distance is selected. Then each positive number x is represented by the point at a distance of x units to the right of the origin, and each negative number x is represented by the point at a distance of $-x$ units to the left of the origin (note that if x is negative, then $-x$ is positive). To each real number there corresponds a unique point on the axis, and with each point on the axis there is associated only one real number; hence there is a one-to-one correspondence between R and the points on the axis. So the points on the axis are identified with the numbers they represent, and the same symbol is used for both the number and the point representing that number on the axis. We identify R with the axis, and we call the axis the **real-number line**.

We see that $a < b$ if and only if the point representing the number a is to the left of the point representing the number b. Similarly, $a > b$ if and only if the point representing a is to the right of the point representing b. For instance, the number 2 is less than the number 5 and the point 2 is to the left of the point 5. We could also write $5 > 2$ and say that the point 5 is to the right of the point 2.

The set of all numbers x satisfying the continued inequality $a < x < b$ is called an **open interval** denoted by (a, b). Therefore

$$(a, b) = \{x \mid a < x < b\}$$

FIGURE 1

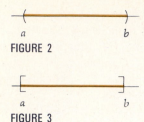

FIGURE 2

FIGURE 3

The **closed interval** from a to b is the open interval (a, b) together with the two endpoints a and b and is denoted by $[a, b]$. Thus

$$[a, b] = \{x \mid a \leq x \leq b\}$$

Figure 2 illustrates the open interval (a, b), and Figure 3 shows the closed interval $[a, b]$.

The **interval half-open on the left** is the open interval (a, b) together with the right endpoint b. It is denoted by $(a, b]$; so

$$(a, b] = \{x \mid a < x \leq b\}$$

We define an **interval half-open on the right** in a similar way and denote it by $[a, b)$. Thus

$$[a, b) = \{x \mid a \leq x < b\}$$

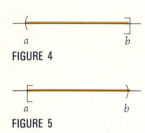

FIGURE 4

FIGURE 5

The interval $(a, b]$ appears in Figure 4 and the interval $[a, b)$ is shown in Figure 5.

We shall use the symbol $+\infty$ (positive infinity) and the symbol $-\infty$ (negative infinity); however, take care not to confuse these symbols with real numbers, for they do not obey the properties of the real numbers. We have the following intervals:

$$(a, +\infty) = \{x \mid x > a\}$$
$$(-\infty, b) = \{x \mid x < b\}$$
$$[a, +\infty) = \{x \mid x \geq a\}$$
$$(-\infty, b] = \{x \mid x \leq b\}$$
$$(-\infty, +\infty) = R$$

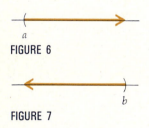

FIGURE 6

FIGURE 7

Figure 6 illustrates the interval $(a, +\infty)$, and the interval $(-\infty, b)$ appears in Figure 7. Note that $(-\infty, +\infty)$ denotes the set of all real numbers.

For each of the intervals (a, b), $[a, b]$, $[a, b)$, and $(a, b]$ the numbers a and b are called the **endpoints** of the interval. The closed interval $[a, b]$ contains both its endpoints, whereas the open interval (a, b) contains neither endpoint. The interval $[a, b)$ contains its left endpoint but not its right one, and the interval $(a, b]$ contains its right endpoint but not its left one. An open interval can be thought of as one that contains none of its endpoints, and a closed interval can be regarded as one that contains all of its endpoints. Consequently, the interval $[a, +\infty)$ is considered to be a closed interval because it contains its only endpoint a. Similarly, $(-\infty, b]$ is a closed interval, whereas $(a, +\infty)$ and $(-\infty, b)$ are open. The intervals $[a, b)$ and $(a, b]$ are neither open nor closed. The interval $(-\infty, +\infty)$ has no endpoints, and it is considered both open and closed.

Intervals are used to represent *solution sets* of inequalities in one variable. The **solution set** of such an inequality is the set of all numbers that satisfy the inequality.

EXAMPLE 1   Find and show on the real-number line the solution set of the inequality

$$2 + 3x < 5x + 8$$

Solution The following inequalities are equivalent:

$$2 + 3x < 5x + 8$$

$$2 + 3x - 2 < 5x + 8 - 2$$

$$3x < 5x + 6$$

$$-2x < 6$$

$$x > -3$$

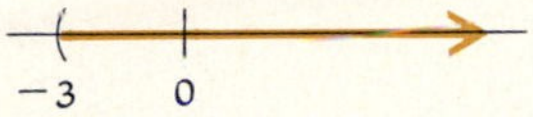

FIGURE 8

Therefore the solution set is the interval $(-3, +\infty)$, which is illustrated in Figure 8.

EXAMPLE 2 Find and show on the real-number line the solution set of the inequality

$$4 < 3x - 2 \leq 10$$

Solution By adding 2 to each member of the given inequality we obtain the equivalent inequalities

$$6 < 3x \leq 12$$

$$2 < x \leq 4$$

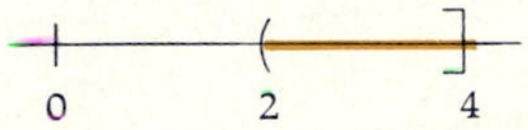

FIGURE 9

Thus the solution set is the interval $(2, 4]$ as shown in Figure 9.

EXAMPLE 3 Find and show on the real-number line the solution set of the inequality

$$\frac{7}{x} > 2$$

Solution We wish to multiply both sides of the inequality by x. However, the direction of the inequality that results will depend on whether x is positive or negative. Observe that if $x < 0$, then

$$\frac{7}{x} < 0$$

which contradicts the given inequality. Therefore we need consider only $x > 0$. Multiplying both sides of the given inequality by x, we obtain the following equivalent inequalities:

$$7 > 2x$$

$$\tfrac{7}{2} > x$$

$$x < \tfrac{7}{2}$$

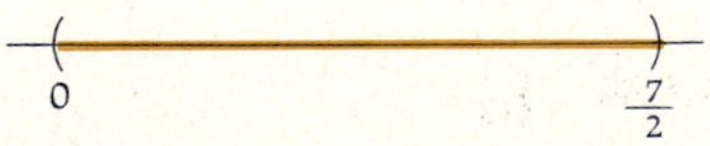

FIGURE 10

Therefore the solution set is $\{x \mid x > 0\} \cap \{x \mid x < \frac{7}{2}\}$ or, equivalently, the set $\{x \mid 0 < x < \frac{7}{2}\}$, which is the interval $(0, \frac{7}{2})$ shown in Figure 10.

EXAMPLE 4 Find and show on the real-number line the solution set of the inequality

$$\frac{x}{x-3} < 4$$

Solution   To multiply both sides of the inequality by $x - 3$ we must consider two cases.

Case 1: $x - 3 > 0$; that is, $x > 3$.
    Multiplying both sides of the inequality by $x - 3$ we get

$$x < 4x - 12$$

$$-3x < -12$$

$$x > 4$$

Thus the solution set of Case 1 is $\{x \mid x > 3\} \cap \{x \mid x > 4\}$ or, equivalently, $\{x \mid x > 4\}$, which is the interval $(4, +\infty)$.

Case 2: $x - 3 < 0$; that is, $x < 3$.
    Multiplying both sides by $x - 3$ and reversing the direction of the inequality we have

$$x > 4x - 12$$

$$-3x > -12$$

$$x < 4$$

Therefore x must be less than 4 and also less than 3. Thus the solution set of Case 2 is the interval $(-\infty, 3)$.
    Combining the solution sets for Cases 1 and 2, we obtain $(-\infty, 3) \cup (4, +\infty)$, which appears in Figure 11.

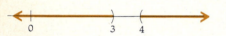

FIGURE 11

The concept of the *absolute value* of a number is used in some important definitions. Furthermore, you will need to work with inequalities involving absolute value.

1.1.8 DEFINITION

The **absolute value** of x, denoted by $|x|$, is defined by

$$|x| = \begin{cases} x & \text{if } x \geq 0 \\ -x & \text{if } x < 0 \end{cases}$$

▶ **ILLUSTRATION 6**

$$|3| = 3 \qquad |-5| = -(-5) \qquad |8 - 14| = |-6|$$
$$= 5 \qquad\qquad\quad = -(-6)$$
$$= 6 \qquad\quad ◀$$

From the definition, the absolute value of a number is either a positive number or zero; that is, it is nonnegative.
    In terms of geometry, the absolute value of a number x is its distance from 0. In general, $|a - b|$ is the distance between a and b, without regard to which is the larger number. Refer to Figure 12.
    The inequality $|x| < a$, where $a > 0$, states that on the real-number line the distance from the origin to the point x is less than a units; that is, $-a < x < a$. Therefore, x is in the open interval $(-a, a)$. See Figure 13. It appears then that the solution set of $|x| < a$ is $\{x \mid -a < x < a\}$. This is indeed the case as stated in the following theorem. The double arrow $\Leftrightarrow$ is used here and throughout the text to indicate that the statement preceding it and the statement following it are *equivalent*.

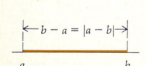

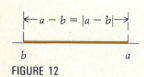

FIGURE 12

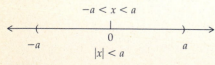

FIGURE 13

1.1.9 THEOREM $|x| < a \iff -a < x < a$ where $a > 0$

Proof   Because $|x| = x$ if $x \geq 0$ and $|x| = -x$ if $x < 0$, it follows that the solution set of the inequality $|x| \leq a$ is the union of the sets

$$\{x \mid x < a \text{ and } x \geq 0\} \quad \text{and} \quad \{x \mid -x < a \text{ and } x < 0\}$$

Observe that the first of these sets is equivalent to $\{x \mid 0 \leq x < a\}$, and the second is equivalent to $\{x \mid -a < x < 0\}$ because $-x < a$ is equivalent to $x > -a$. Thus the solution set of $|x| < a$ is

$$\{x \mid 0 \leq x < a\} \cup \{x \mid -a < x < 0\}$$
$$\iff \{x \mid -a < x < a\}$$

By comparing the given inequality and its solution set we conclude that

$$|x| < a \iff -a < x < a \qquad \blacksquare$$

1.1.10 COROLLARY $|x| \leq a \iff -a \leq x \leq a$ where $a > 0$

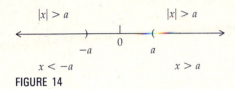

$x < -a$ $x > a$

FIGURE 14

The inequality $|x| > a$, where $a > 0$, states that on the real-number line the distance from the origin to the point x is greater than a units; that is, either $x > a$ or $x < -a$. Therefore x is in $(-\infty, -a) \cup (a, +\infty)$. See Figure 14. Thus it appears that the solution set of $|x| > a$ is $\{x \mid x > a\} \cup \{x \mid x < -a\}$. The next theorem states that this is the situation. You are asked to prove it in Exercise 61.

1.1.11 THEOREM $|x| > a \iff x > a \text{ or } x < -a$ where $a > 0$

1.1.12 COROLLARY $|x| \geq a \iff x \geq a \text{ or } x \leq -a$ where $a > 0$

The following examples illustrate the solution of equations and inequalities involving absolute values.

EXAMPLE 5   Solve each of the equations for x: (a) $|3x + 2| = 5$; (b) $|2x - 1| = |4x + 3|$; (c) $|5x + 4| = -3$.

Solution
(a) $|3x + 2| = 5$

This equation will be satisfied if either

$$3x + 2 = 5 \quad \text{or} \quad -(3x + 2) = 5$$
$$x = 1 \qquad\qquad\qquad x = -\tfrac{7}{3}$$

(b) $|2x - 1| = |4x + 3|$

This equation will be satisfied if either

$$2x - 1 = 4x + 3 \quad \text{or} \quad 2x - 1 = -(4x + 3)$$
$$x = -2 \qquad\qquad\qquad x = -\tfrac{1}{3}$$

(c) $|5x + 4| = -3$

Because the absolute value of a number may never be negative, this equation has no solution.

EXAMPLE 6   Find and show on the real-number line the solution set of the inequality

$$|x - 5| < 4$$

Solution   From Theorem 1.1.9 the following inequalities are equivalent:

$$|x - 5| < 4$$

$$-4 < x - 5 < 4$$

$$1 < x < 9$$

Therefore the solution set is the open interval $(1, 9)$, shown in Figure 15.

FIGURE 15

EXAMPLE 7   Find the solution set of the inequality

$$|3x + 2| > 5$$

Solution   By Theorem 1.1.11, the given inequality is equivalent to

$$3x + 2 > 5 \quad \text{or} \quad 3x + 2 < -5$$

That is, the given inequality will be satisfied if either of these inequalities is satisfied.

Considering the first inequality we have

$$3x + 2 > 5$$

$$x > 1$$

Therefore every number in the interval $(1, +\infty)$ is a solution.

From the second inequality

$$3x + 2 < -5$$

$$x < -\tfrac{7}{3}$$

Hence every number in the interval $(-\infty, -\tfrac{7}{3})$ is a solution.

The solution set of the given inequality is therefore $(-\infty, -\tfrac{7}{3}) \cup (1, +\infty)$.

You may recall from algebra that the symbol $\sqrt{a}$, where $a \geq 0$, is defined as the unique *nonnegative* number x such that $x^2 = a$. We read $\sqrt{a}$ as "the principal square root of a." For example,

$$\sqrt{4} = 2 \qquad \sqrt{0} = 0 \qquad \sqrt{\tfrac{9}{25}} = \tfrac{3}{5}$$

Note: $\sqrt{4} \neq -2$ even though $(-2)^2 = 4$, because $\sqrt{4}$ denotes only the *positive* square root of 4. The *negative* square root of 4 is designated by $-\sqrt{4}$.

Because we are concerned only with real numbers in calculus, $\sqrt{a}$ is not defined if $a < 0$.

EXAMPLE 8   Find all values of x for which $\sqrt{x^2 + 7x + 12}$ is real.

Solution

$$x^2 + 7x + 12 = (x + 3)(x + 4)$$

$$\sqrt{(x + 3)(x + 4)} \quad \text{is real when} \quad (x + 3)(x + 4) \geq 0$$

We find the solution set of this inequality. The inequality will be satisfied when both factors are nonnegative or when both factors are nonpositive, that is, if $x + 3 \geq 0$ and $x + 4 \geq 0$, or if $x + 3 \leq 0$ and $x + 4 \leq 0$. We consider two cases.

Case 1: $x + 3 \geq 0$ and $x + 4 \geq 0$. That is,

$$x \geq -3 \quad \text{and} \quad x \geq -4$$

Both inequalities hold if $x \geq -3$, which is the interval $[-3, +\infty)$.

Case 2: $x + 3 \leq 0$ and $x + 4 \leq 0$. That is,

$$x \leq -3 \quad \text{and} \quad x \leq -4$$

Both inequalities hold if $x \leq -4$, which is the interval $(-\infty, -4]$.

Combining the solution sets of Cases 1 and 2, we have $(-\infty, -4] \cup [-3, +\infty)$.

From the definition of $\sqrt{a}$ it follows that

$$\sqrt{x^2} = |x|$$

▶ **ILLUSTRATION 7**

$$\sqrt{5^2} = |5| \qquad \sqrt{(-3)^2} = |-3|$$
$$\phantom{\sqrt{5^2}} = 5 \qquad\qquad\quad = 3 \qquad ◀$$

The following theorems about absolute value will be useful later.

1.1.13 THEOREM

If $a, b \in R$, then

$$|ab| = |a| \cdot |b|$$

Proof

$$\begin{aligned}
|ab| &= \sqrt{(ab)^2} \\
&= \sqrt{a^2 b^2} \\
&= \sqrt{a^2} \cdot \sqrt{b^2} \\
&= |a| \cdot |b|
\end{aligned}$$

∎

1.1.14 THEOREM

If $a, b \in R$, and $b \neq 0$,

$$\left| \frac{a}{b} \right| = \frac{|a|}{|b|}$$

The proof of Theorem 1.1.14 is left as an exercise (see Exercise 62).

1.1.15 THEOREM
The Triangle Inequality

If $a, b \in R$, then

$$|a + b| \leq |a| + |b|$$

Proof   By Definition 1.1.8, either $a = |a|$ or $a = -|a|$; thus

$$-|a| \leq a \leq |a| \tag{1}$$

Furthermore,

$$-|b| \leq b \leq |b| \tag{2}$$

From inequalities (1) and (2) and Theorem 1.1.7,

$$-(|a| + |b|) \leq a + b \leq |a| + |b|$$

Hence, from Corollary 1.1.10 it follows that

$$|a + b| \leq |a| + |b| \qquad \blacksquare$$

Theorem 1.1.15 has two important corollaries, which we now state and prove.

1.1.16 COROLLARY   If $a, b \in R$, then

$$|a - b| \leq |a| + |b|$$

Proof

$$|a - b| = |a + (-b)| \leq |a| + |-b| = |a| + |b| \qquad \blacksquare$$

1.1.17 COROLLARY   If $a, b \in R$, then

$$|a| - |b| \leq |a - b|$$

Proof

$$|a| = |(a - b) + b| \leq |a - b| + |b|$$

thus, subtracting $|b|$ from both sides of the inequality we have

$$|a| - |b| \leq |a - b| \qquad \blacksquare$$

EXERCISES 1.1

In Exercises 1 through 22, find and show on the real-number line the solution set of the inequality.

1. $5x + 2 > x - 6$
2. $3 - x < 5 + 3x$
3. $\frac{2}{3}x - \frac{1}{2} \leq 0$
4. $3 - 2x \geq 9 + 4x$
5. $13 \geq 2x - 3 \geq 5$
6. $-2 < 6 - 4x \leq 8$
7. $2 > -3 - 3x \geq -7$
8. $2 \leq 5 - 3x < 11$
9. $\dfrac{4}{x} - 3 > \dfrac{2}{x} - 7$
10. $\dfrac{5}{x} < \dfrac{3}{4}$
11. $\dfrac{1}{x + 1} < \dfrac{2}{3x - 1}$
12. $\dfrac{x + 1}{2 - x} < \dfrac{x}{3 + x}$
13. $x^2 > 4$
14. $x^2 \leq 9$
15. $(x - 3)(x + 5) > 0$
16. $x^2 - 3x + 2 > 0$
17. $1 - x - 2x^2 \geq 0$
18. $x^2 + 3x + 1 > 0$
19. $4x^2 + 9x < 9$
20. $2x^2 - 6x + 3 < 0$
21. $\dfrac{1}{3x - 7} \geq \dfrac{4}{3 - 2x}$
22. $x^3 + 1 > x^2 + x$

In Exercises 23 through 30, solve for x.

23. $|4x + 3| = 7$
24. $|3x - 8| = 4$
25. $|5x - 3| = |3x + 5|$
26. $|x - 2| = |3 - 2x|$
27. $|7x| = 4 - x$
28. $2x + 3 = |4x + 5|$

29. $\left| \dfrac{x + 2}{x - 2} \right| = 5$
30. $\left| \dfrac{3x + 8}{2x - 3} \right| = 4$

In Exercises 31 through 36, find all values of x for which the number is real.

31. $\sqrt{8x - 5}$
32. $\sqrt{x^2 - 16}$
33. $\sqrt{x^2 - 3x - 10}$
34. $\sqrt{2x^2 + 5x - 3}$
35. $\sqrt{x^2 - 5x + 4}$
36. $\sqrt{x^2 + 2x - 1}$

In Exercises 37 through 52, find and show on the real-number line the solution set of the inequality.

37. $|x + 4| < 7$
38. $|2x - 5| < 3$
39. $|3x - 4| \leq 2$
40. $|3x + 2| \geq 1$
41. $|5 - x| > 7$
42. $|3 - x| < 5$
43. $|7 - 4x| \leq 9$
44. $|6 - 2x| \geq 7$
45. $|2x - 5| > 3$
46. $|x + 4| \leq |2x - 6|$
47. $|3x| > |6 - 3x|$
48. $|3 + 2x| < |4 - x|$
49. $|9 - 2x| \geq |4x|$
50. $|5 - 2x| \geq 7$
51. $\left| \dfrac{x + 2}{2x - 3} \right| < 4$
52. $\left| \dfrac{6 - 5x}{3 + x} \right| \leq \dfrac{1}{2}$

In Exercises 53 through 56, solve for x, and write the answer with absolute value notation.

53. $\dfrac{x - a}{x + a} > 0$

54. $\dfrac{a - x}{a + x} \geq 0$

55. $\dfrac{x - 2}{x - 4} > \dfrac{x + 2}{x}$

56. $\dfrac{x + 5}{x + 3} < \dfrac{x + 1}{x - 1}$

57. Prove Theorem 1.1.5.
58. Prove Theorem 1.1.6(i).
59. Prove Theorem 1.1.6(ii) and (iii).
60. Prove that if $x < y$, then $x < \frac{1}{2}(x + y) < y$.
61. Prove Theorem 1.1.11.
62. Prove Theorem 1.1.14.

1.2 COORDINATES AND LINES

Ordered pairs of real numbers are important in our discussions. Any two real numbers form a pair, and when the order of appearance of the number is significant, we call it an **ordered pair**. If x is the first real number and y is the second, this ordered pair is denoted by (x, y). Observe that the ordered pair $(3, 7)$ is different from the ordered pair $(7, 3)$.

The set of all ordered pairs of real numbers is called the **number plane**, denoted by R^2, and each ordered pair (x, y) is a **point** in the number plane. Just as R can be identified with points on an axis (a one-dimensional space), we can identify R^2 with points in a geometric plane (a two-dimensional space). The concept is attributed to the French mathematician René Descartes (1596–1650), who is credited with the origination of analytic geometry in 1637. A horizontal line, called the **x axis**, is chosen in the geometric plane. A vertical line is selected and is called the **y axis**. The point of intersection of the x axis and the y axis is called the **origin** and is denoted by the letter O. A unit of length, usually the same on each axis, is chosen. We establish the positive direction on the x axis to the right of the origin, and the positive direction on the y axis above the origin. See Figure 1.

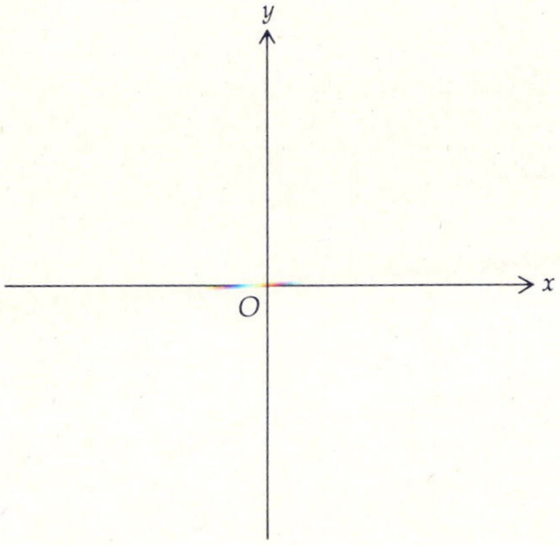

We now associate an ordered pair of real numbers (x, y) with a point in the geometric plane. At the point x on the horizontal axis and the point y on the vertical axis, line segments are drawn perpendicular to the respective axes. The intersection of these two perpendicular line segments is the point P associated with the ordered pair (x, y). Refer to Figure 2. The first number x of the pair is called the **abscissa** (or **x coordinate**) of P, and the second number y is called the **ordinate** (or **y coordinate**) of P. If the abscissa is positive, P is to the right of the y axis; and if it is negative, P is to the left of the y axis. If the ordinate is positive, P is above the x axis; and if it is negative, P is below the x axis.

The abscissa and ordinate of a point are called the **rectangular cartesian coordinates** of the point. The word *cartesian* comes from the name Descartes. There is a one-to-one correspondence between the points in a geometric plane and R^2; that is, with each point there corresponds a unique ordered pair (x, y), and with each ordered pair (x, y) there is associated only one point. This one-to-one correspondence is called a **rectangular cartesian coordinate system**. Figure 3 illustrates a rectangular cartesian coordinate system with some points plotted.

The x and y axes are called the **coordinate axes**. They divide the plane into four parts, called **quadrants**. The first quadrant is the one in which the abscissa and ordinate are both positive, that is, the upper right quadrant. The other quadrants are numbered in the counterclockwise direction, with the fourth being the lower right quadrant. See Figure 4.

Because of the one-to-one correspondence, we identify R^2 with the geometric plane. For this reason we call an ordered pair (x, y) a *point*.

We now discuss the problem of finding the distance between two points in R^2. If A is the point (x_1, y_1) and B is the point (x_2, y_1) (i.e., A and B have the same

FIGURE 1

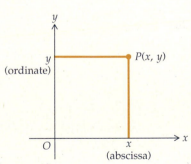

FIGURE 2

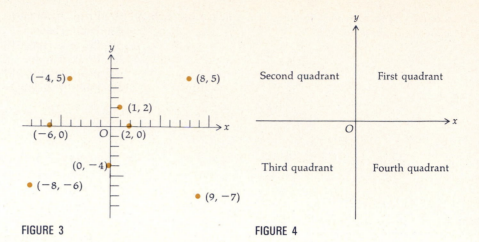

FIGURE 3   FIGURE 4

ordinate but different abscissas), then the **directed distance** from A to B is denoted by $\overline{AB}$, and we define

$$\overline{AB} = x_2 - x_1$$

▶ **ILLUSTRATION 1** Refer to Figure 5(a)–(c). If A is the point $(3, 4)$ and B is the point $(9, 4)$, then $\overline{AB} = 9 - 3$; that is, $\overline{AB} = 6$. If A is the point $(-8, 0)$ and B is the point $(6, 0)$, then $\overline{AB} = 6 - (-8)$; that is, $\overline{AB} = 14$. If A is the point $(4, 2)$ and B is the point $(1, 2)$, then $\overline{AB} = 1 - 4$; that is, $\overline{AB} = -3$. We see that $\overline{AB}$ is positive if B is to the right of A, and $\overline{AB}$ is negative if B is to the left of A. ◀

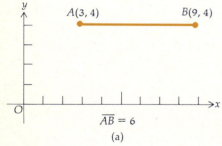
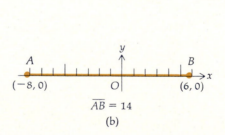
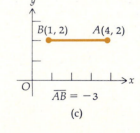

FIGURE 5   (a)   (b)   (c)

If C is the point (x_1, y_1) and D is the point (x_1, y_2), then the directed distance from C to D, denoted by $\overline{CD}$, is defined by

$$\overline{CD} = y_2 - y_1$$

▶ **ILLUSTRATION 2** Refer to Figure 6(a) and (b). If C is the point $(1, -2)$ and D is the point $(1, -8)$, then $\overline{CD} = -8 - (-2)$; that is, $\overline{CD} = -6$. If C is the point $(-2, -3)$ and D is the point $(-2, 4)$, then $\overline{CD} = 4 - (-3)$; that is, $\overline{CD} = 7$. The number $\overline{CD}$ is positive if D is above C, and $\overline{CD}$ is negative if D is below C. ◀

Observe that the terminology *directed distance* indicates both a distance and a direction (positive or negative). If we are concerned only with the length of the line segment between two points P_1 and P_2 (i.e., the distance between the points P_1 and P_2 without regard to direction), then we use the terminology *undirected distance*. We denote the **undirected distance** from P_1 to P_2 by $|\overline{P_1P_2}|$, which is a nonnegative number. If we use the word *distance* without an adjective, *directed* or *undirected*, it is understood that we mean an undirected distance.

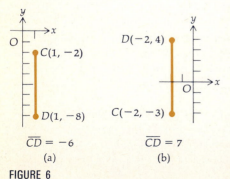

FIGURE 6

We now wish to obtain a formula for computing $|\overline{P_1P_2}|$ if $P_1(x_1, y_1)$ and $P_2(x_2, y_2)$ are any two points in the plane. We use the Pythagorean theorem from plane geometry, which is as follows:

In a right triangle, the sum of the squares of the lengths of the perpendicular sides is equal to the square of the length of the hypotenuse.

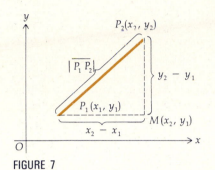

FIGURE 7

Figure 7 shows P_1 and P_2 in the first quadrant and the point $M(x_2, y_1)$. Note that $|\overline{P_1P_2}|$ is the length of the hypotenuse of right triangle P_1MP_2. Using the Pythagorean theorem we have

$$|\overline{P_1P_2}|^2 = |\overline{P_1M}|^2 + |\overline{MP_2}|^2$$
$$|\overline{P_1P_2}| = \sqrt{|\overline{P_1M}|^2 + |\overline{MP_2}|^2}$$
$$|\overline{P_1P_2}| = \sqrt{(x_2 - x_1)^2 + (y_2 - y_1)^2}$$

Observe that in this formula we do not have a $\pm$ symbol in front of the radical in the right member because $|\overline{P_1P_2}|$ is a nonnegative number. The formula holds for all possible positions of P_1 and P_2 in all four quadrants. The length of the hypotenuse is always $|\overline{P_1P_2}|$, and the lengths of the legs are always $|\overline{P_1M}|$ and $|\overline{MP_2}|$. The result is stated as a theorem.

1.2.1 THEOREM   The distance between two points $P_1(x_1, y_1)$ and $P_2(x_2, y_2)$ is given by

$$|\overline{P_1P_2}| = \sqrt{(x_2 - x_1)^2 + (y_2 - y_1)^2}$$

Observe that if P_1 and P_2 are on the same horizontal line, then $y_1 = y_2$, and

$$|\overline{P_1P_2}| = \sqrt{(x_2 - x_1)^2 + 0^2}$$
$$|\overline{P_1P_2}| = |x_2 - x_1| \quad (\text{because } \sqrt{a^2} = |a|)$$

Furthermore, if P_1 and P_2 are on the same vertical line, then $x_1 = x_2$, and

$$|\overline{P_1P_2}| = \sqrt{0^2 + (y_2 - y_1)^2}$$
$$|\overline{P_1P_2}| = |y_2 - y_1|$$

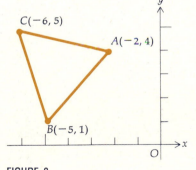

FIGURE 8

EXAMPLE 1   By showing that two sides have the same length, prove that the triangle with vertices at $A(-2, 4)$, $B(-5, 1)$, and $C(-6, 5)$ is isosceles.

Solution   The triangle appears in Figure 8.

$$|\overline{BC}| = \sqrt{(-6 + 5)^2 + (5 - 1)^2} \qquad |\overline{AC}| = \sqrt{(-6 + 2)^2 + (5 - 4)^2}$$
$$= \sqrt{1 + 16} \qquad\qquad\qquad = \sqrt{16 + 1}$$
$$= \sqrt{17} \qquad\qquad\qquad\quad = \sqrt{17}$$

Because $|\overline{BC}| = |\overline{AC}|$ the triangle is isosceles.

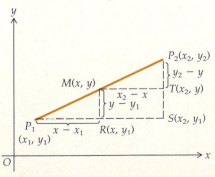

FIGURE 9

If P_1 and P_2 are the endpoints of a line segment, we denote this line segment by P_1P_2. This is not to be confused with the notation $\overline{P_1P_2}$, which denotes the directed distance from P_1 to P_2. That is, $\overline{P_1P_2}$ is a number, whereas P_1P_2 is a line segment. We now obtain the formulas for finding the midpoint of a line segment. Refer to Figure 9, where $M(x, y)$ is the midpoint of the line segment from $P_1(x_1, y_1)$ to $P_2(x_2, y_2)$. Because triangles P_1RM and MTP_2 are congruent

$$|\overline{P_1R}| = |\overline{MT}| \quad \text{and} \quad |\overline{RM}| = |\overline{TP_2}|$$

Thus

$$x - x_1 = x_2 - x \qquad y - y_1 = y_2 - y$$

$$2x = x_1 + x_2 \qquad 2y = y_1 + y_2$$

$$x = \frac{x_1 + x_2}{2} \qquad y = \frac{y_1 + y_2}{2}$$

These are the **midpoint formulas**. In their derivation it was assumed that $x_2 > x_1$ and $y_2 > y_1$. The same formulas are obtained by using any orderings of these numbers.

In analytic geometry the validity of theorems in plane geometry is established by using coordinates and techniques of algebra. The following example demonstrates the procedure.

EXAMPLE 2   Use analytic geometry to prove that the line segments joining the midpoints of the opposite sides of any quadrilateral bisect each other.

Solution   We draw a general quadrilateral. Because the coordinate axes can be chosen anywhere in the plane and because the choice of the position of the axes does not affect the truth of the theorem, we take the origin at one vertex and the x axis along one side. This selection simplifies the coordinates of the two vertices on the x axis. See Figure 10.

The hypothesis and conclusion of the theorem are as follows:

Hypothesis: $OABC$ is a quadrilateral. M is the midpoint of OA, N is the midpoint of CB, R is the midpoint of OC, and S is the midpoint of AB.

Conclusion: MN and RS bisect each other.

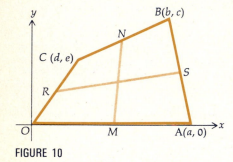

FIGURE 10

Proof   To prove that two line segments bisect each other, we show that they have the same midpoint. From the midpoint formulas we obtain the coordinates of M, N, R, and S. M is the point $(\frac{1}{2}a, 0)$, N is the point $(\frac{1}{2}(b + d), \frac{1}{2}(c + e))$, R is the point $(\frac{1}{2}d, \frac{1}{2}e)$, and S is the point $(\frac{1}{2}(a + b), \frac{1}{2}c)$.

The abscissa of the midpoint of MN is $\frac{1}{2}[\frac{1}{2}a + \frac{1}{2}(b + d)] = \frac{1}{4}(a + b + d)$.
The ordinate of the midpoint of MN is $\frac{1}{2}[0 + \frac{1}{2}(c + e)] = \frac{1}{4}(c + e)$.
Therefore, the midpoint of MN is the point $(\frac{1}{4}(a + b + d), \frac{1}{4}(c + e))$.
The abscissa of the midpoint of RS is $\frac{1}{2}[\frac{1}{2}d + \frac{1}{2}(a + b)] = \frac{1}{4}(a + b + d)$.
The ordinate of the midpoint of RS is $\frac{1}{2}[\frac{1}{2}e + \frac{1}{2}c] = \frac{1}{4}(c + e)$.
Therefore, the midpoint of RS is the point $(\frac{1}{4}(a + b + d), \frac{1}{4}(c + e))$.
Thus, the midpoint of MN is the same point as the midpoint of RS.
Therefore, MN and RS bisect each other. ■

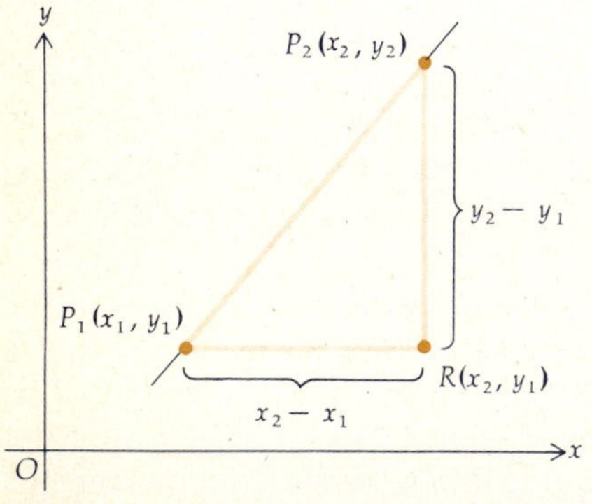

We now discuss *lines* in R^2. Let l be a nonvertical line and $P_1(x_1, y_1)$ and $P_2(x_2, y_2)$ be any two distinct points on l. Figure 11 shows such a line. In the figure, R is the point (x_2, y_1), and the points P_1, P_2, and R are vertices of a right triangle; furthermore, $\overline{P_1R} = x_2 - x_1$ and $\overline{RP_2} = y_2 - y_1$. The number $y_2 - y_1$ gives the measure of the change in the ordinate from P_1 to P_2, and it may be positive, negative, or zero. The number $x_2 - x_1$ gives the measure of the change in the abscissa from P_1 to P_2, and it may be positive or negative.

FIGURE 11

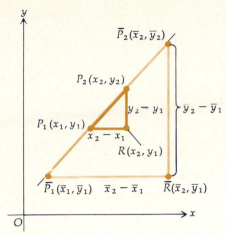

FIGURE 12

Because the line l is not vertical, $x_2 \neq x_1$, and therefore $x_2 - x_1$ is not zero. Let

$$m = \frac{y_2 - y_1}{x_2 - x_1} \qquad (1)$$

The value of m computed from this equation is independent of the choice of the two points P_1 and P_2 on l. To show this, suppose we choose two different points $\bar{P}_1(\bar{x}_1, \bar{y}_1)$ and $\bar{P}_2(\bar{x}_2, \bar{y}_2)$ and compute a number $\bar{m}$ from (1).

$$\bar{m} = \frac{\bar{y}_2 - \bar{y}_1}{\bar{x}_2 - \bar{x}_1}$$

We shall show that $\bar{m} = m$. Refer to Figure 12. Triangles $\bar{P}_1 \bar{R} \bar{P}_2$ and $P_1 R P_2$ are similar; so the lengths of corresponding sides are proportional. Therefore

$$\frac{\bar{y}_2 - \bar{y}_1}{\bar{x}_2 - \bar{x}_1} = \frac{y_2 - y_1}{x_2 - x_1}$$

or

$$\bar{m} = m$$

Thus the value of m computed from (1) is the same number no matter what two points on l are selected. This number m is called the *slope* of the line.

1.2.2 DEFINITION   If $P_1(x_1, y_1)$ and $P_2(x_2, y_2)$ are any two distinct points on line l, which is not parallel to the y axis, then the **slope** of l, denoted by m, is given by

$$m = \frac{y_2 - y_1}{x_2 - x_1}$$

Multiplying both sides of the preceding equation by $x_2 - x_1$, we obtain

$$y_2 - y_1 = m(x_2 - x_1)$$

It follows from this equation that if we consider a particle moving along a line, the change in the ordinate of the particle is equal to the product of the slope and the change in the abscissa.

▶ **ILLUSTRATION 3**   If l is the line through the points $P_1(2, 3)$ and $P_2(4, 7)$, and m is the slope of l, then by Definition 1.2.2,

$$m = \frac{7 - 3}{4 - 2}$$
$$= 2$$

Refer to Figure 13. If a particle is moving along the line l, the change in the ordinate is two times the change in the abscissa. That is, if the particle is at $P_2(4, 7)$ and the abscissa is increased by one unit, then the ordinate is increased by two units, and the particle is at the point $P_3(5, 9)$. Similarly, if the particle is at $P_1(2, 3)$ and the abscissa is decreased by three units, then the ordinate is decreased by six units, and the particle is at $P_4(-1, -3)$.   ◀

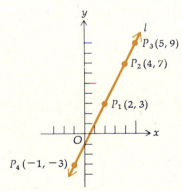

FIGURE 13

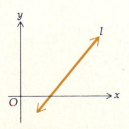

FIGURE 14

If the slope of a line is positive, then as the abscissa of a point on the line increases, the ordinate increases. Such a line is shown in Figure 14. In Figure

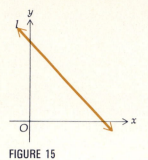

FIGURE 15

15 there is a line whose slope is negative. For this line, as the abscissa of a point on the line increases, the ordinate decreases.

If a line is parallel to the x axis, then $y_2 = y_1$; so the slope of the line is zero.

If a line is parallel to the y axis, $x_2 = x_1$; thus the fraction $\dfrac{y_2 - y_1}{x_2 - x_1}$ is meaningless because we cannot divide by zero. For this reason lines parallel to the y axis are excluded in the definition of slope. Therefore the slope of a vertical line is not defined.

By an *equation of a line* we mean an equation that is satisfied by those, and only those, points on the line. Because a point $P_1(x_1, y_1)$ and a slope m determine a unique line, we should be able to obtain an equation of this line. Let $P(x, y)$ be any point on the line except (x_1, y_1). Then since the slope of the line through P_1 and P is m, we have from the definition of slope

$$\frac{y - y_1}{x - x_1} = m$$

$$y - y_1 = m(x - x_1)$$

This equation is called the **point-slope form** of an equation of the line. It gives an equation of the line if its slope and a point on the line are known.

▶ **ILLUSTRATION 4**   To find an equation of the line through the two points $A(6, -3)$ and $B(-2, 3)$ we first compute m.

$$m = \frac{3 - (-3)}{-2 - 6}$$

$$= \frac{6}{-8}$$

$$= -\tfrac{3}{4}$$

Using the point-slope form of an equation of the line with A as P_1 we have

$$y - (-3) = -\tfrac{3}{4}(x - 6)$$

$$4y + 12 = -3x + 18$$

$$3x + 4y - 6 = 0$$

Of course, the point B may be taken as P_1; in which case we have

$$y - 3 = -\tfrac{3}{4}(x + 2)$$

$$4y - 12 = -3x - 6$$

$$3x + 4y - 6 = 0 \qquad\qquad ◀$$

If in the point-slope form we choose the particular point $(0, b)$ (that is, the point where the line intersects the y axis) for the point (x_1, y_1), we have

$$y - b = m(x - 0)$$

$$y = mx + b$$

The number b, the ordinate of the point where the line intersects the y axis, is the **y intercept** of the line. Consequently, the preceding equation is called the **slope-intercept form** of an equation of the line. This form is especially useful because it enables us to find the slope of a line from its equation. It is also important because it expresses the y coordinate of a point on the line explicitly in terms of its x coordinate.

EXAMPLE 3   Find the slope of the line having the equation

$$6x + 5y - 7 = 0$$

Solution   We solve the equation for y.

$$5y = -6x + 7$$

$$y = -\tfrac{6}{5}x + \tfrac{7}{5}$$

This equation is in the slope-intercept form where $m = -\tfrac{6}{5}$.

Because the slope of a vertical line is undefined, we cannot apply the point-slope form to obtain its equation. We use instead the following theorem involving the **x intercept** of the line (the abscissa of the point at which the line intersects the x axis). The theorem also gives an equation of a horizontal line.

1.2.3 THEOREM

(i) An equation of the vertical line having x intercept a is

$$x = a$$

(ii) An equation of the horizontal line having y intercept b is

$$y = b$$

Proof   (i) Figure 16 shows the vertical line that intersects the x axis at the point $(a, 0)$. This line contains those and only those points on the line having the same abscissa. Thus $P(x, y)$ is any point on the line if and only if

$$x = a$$

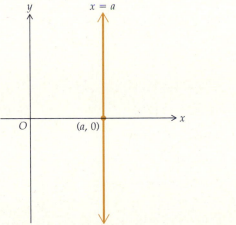

FIGURE 16   FIGURE 17

(ii) The horizontal line that intersects the y axis at the point $(0, b)$ appears in Figure 17. For this line, $m = 0$. Therefore, from the slope-intercept form, an equation of this line is

$$y = b$$ ■

We have shown that an equation of a nonvertical line is of the form $y = mx + b$, and an equation of a vertical line is of the form $x = a$. Because each of these equations is a special case of an equation of the form

$$Ax + By + C = 0 \qquad (2)$$

where A, B, and C are constants and not both A and B are zero, it follows that every line has an equation of the form (2). The converse of this fact is given by Theorem 1.2.5 which follows. But before stating this theorem we define what is meant by the *graph of an equation*.

1.2.4 DEFINITION The **graph of an equation** in R^2 is the set of all points in R^2 whose coordinates are numbers satisfying the equation.

1.2.5 THEOREM The graph of the equation

$$Ax + By + C = 0$$

where A, B, and C are constants and not both A and B are zero, is a line.

The proof of this theorem is left as an exercise. See Exercise 57.

Because the graph of (2) is a line, it is called a **linear equation**; it is the general equation of the first degree in x and y.

Since two points determine a line, to draw a sketch of its graph, we need only determine the coordinates of two points on the line, plot the points, and then draw the line. Any two points will suffice, but it is often convenient to plot the two points where the line intersects the axes.

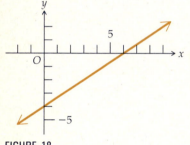

FIGURE 18

▶ **ILLUSTRATION 5** To draw a sketch of the line having the equation

$$2x - 3y = 12$$

we first find the x intercept a and the y intercept b. In the equation we substitute $(a, 0)$ for (x, y) and get $a = 6$. Substituting $(0, b)$ for (x, y), we obtain $b = -4$. Thus we have the line appearing in Figure 18. ◀

An application of slopes is given by the following theorem.

1.2.6 THEOREM If l_1 and l_2 are two distinct nonvertical lines having slopes m_1 and m_2, respectively, then l_1 and l_2 are parallel if and only if $m_1 = m_2$.

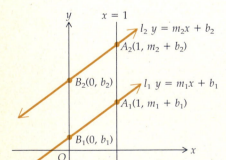

FIGURE 19

Proof Let equations of l_1 and l_2 be, respectively,

$$y = m_1x + b_1 \quad \text{and} \quad y = m_2x + b_2$$

See Figure 19, showing the two lines intersecting the y axis at the points $B_1(0, b_1)$ and $B_2(0, b_2)$. Let the vertical line $x = 1$ intersect l_1 at the point $A_1(1, m_1 + b_1)$ and l_2 at the point $A_2(1, m_2 + b_2)$. Then

$$|\overline{B_1B_2}| = b_2 - b_1 \quad \text{and} \quad |\overline{A_1A_2}| = (m_2 + b_2) - (m_1 + b_1)$$

The two lines are parallel if and only if the vertical distances $|\overline{B_1B_2}|$ and $|\overline{A_1A_2}|$ are equal; that is, l_1 and l_2 are parallel if and only if

$$b_2 - b_1 = (m_2 + b_2) - (m_1 + b_1)$$
$$b_2 - b_1 = m_2 + b_2 - m_1 - b_1$$
$$m_1 = m_2$$

Thus l_1 and l_2 are parallel if and only if $m_1 = m_2$. ■

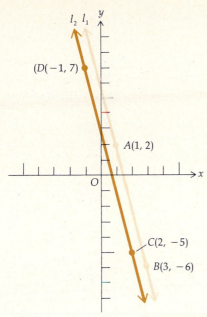

FIGURE 20

▶ **ILLUSTRATION 6**   Let l_1 be the line through the points $A(1, 2)$ and $B(3, -6)$ and m_1 be the slope of l_1; and let l_2 be the line through the points $C(2, -5)$ and $D(-1, 7)$ and m_2 be the slope of l_2. Then

$$m_1 = \frac{-6 - 2}{3 - 1} \qquad m_2 = \frac{7 - (-5)}{-1 - 2}$$

$$= \frac{-8}{2} \qquad\qquad = \frac{12}{-3}$$

$$= -4 \qquad\qquad = -4$$

Because $m_1 = m_2$, it follows that l_1 and l_2 are parallel. See Figure 20.   ◀

Any two distinct points determine a line. Three distinct points may or may not lie on the same line. If three or more points lie on the same line, they are said to be **collinear**. Hence three points A, B, and C are collinear if and only if the line through the points A and B is the same as the line through the points B and C. Because the line through A and B and the line through B and C both contain the point B, they are the same line if and only if their slopes are equal.

EXAMPLE 4   Determine by means of slopes if the points $A(-3, -4)$, $B(2, -1)$, and $C(7, 2)$ are collinear.

Solution   If m_1 is the slope of the line through A and B, and m_2 is the slope of the line through B and C, then

$$m_1 = \frac{-1 - (-4)}{2 - (-3)} \qquad m_2 = \frac{2 - (-1)}{7 - 2}$$

$$= \frac{3}{5} \qquad\qquad\quad = \frac{3}{5}$$

Hence $m_1 = m_2$. Therefore the line through A and B and the line through B and C have the same slope and contain the common point B. Thus they are the same line, and therefore A, B, and C are collinear.

We now state and prove a theorem regarding the slopes of two perpendicular lines.

1.2.7 THEOREM   Two nonvertical lines l_1 and l_2 having slopes m_1 and m_2, respectively, are perpendicular if and only if $m_1 m_2 = -1$.

Proof   Let us choose the coordinate axes so that the origin is at the point of intersection of l_1 and l_2. See Figure 21. Because neither l_1 nor l_2 is vertical, these two lines intersect the line $x = 1$ at points P_1 and P_2, respectively. The abscissa of both P_1 and P_2 is 1. Let $\bar{y}$ be the ordinate of P_1. Since l_1 contains the points $(0, 0)$ and $(1, \bar{y})$ and its slope is m_1, then

$$m_1 = \frac{\bar{y} - 0}{1 - 0}$$

Thus $\bar{y} = m_1$. Similarly, the ordinate of P_2 is m_2. From the Pythagorean theorem and its converse, triangle $P_1 O P_2$ is a right triangle if and only if

$$|\overline{OP_1}|^2 + |\overline{OP_2}|^2 = |\overline{P_1 P_2}|^2 \qquad\qquad (3)$$

FIGURE 21

Applying the distance formula, we obtain

$$|\overline{OP_1}|^2 = (1 - 0)^2 + (m_1 - 0)^2 \qquad |\overline{OP_2}|^2 = (1 - 0)^2 + (m_2 - 0)^2$$
$$= 1 + m_1{}^2 \qquad\qquad\qquad\quad = 1 + m_2{}^2$$
$$|\overline{P_1P_2}|^2 = (1 - 1)^2 + (m_2 - m_1)^2$$
$$= m_2{}^2 - 2m_1m_2 + m_1{}^2$$

Substituting into (3), we can conclude that P_1OP_2 is a right triangle if and only if

$$1 + m_1{}^2 + 1 + m_2{}^2 = m_2{}^2 - 2m_1m_2 + m_1{}^2$$
$$2 = -2m_1m_2$$
$$m_1m_2 = -1$$

Because $m_1m_2 = -1$ is equivalent to

$$m_1 = -\frac{1}{m_2} \quad \text{and} \quad m_2 = -\frac{1}{m_1}$$

Theorem 1.2.7 states that two nonvertical lines are perpendicular if and only if the slope of one of them is the negative reciprocal of the slope of the other.

EXAMPLE 5   Given the line l having the equation

$$5x + 4y - 20 = 0$$

find an equation of the line through the point $(2, -3)$ and (a) parallel to l and (b) perpendicular to l.

Solution   We first determine the slope of l by writing its equation in the slope-intercept form. Solving the equation for y, we have

$$4y = -5x + 20$$
$$y = -\tfrac{5}{4}x + 5$$

The slope of l is the coefficient of x, which is $-\tfrac{5}{4}$.

(a) The slope of a line parallel to l is also $-\tfrac{5}{4}$. Because the required line contains the point $(2, -3)$, we use the point-slope form, which gives

$$y - (-3) = -\tfrac{5}{4}(x - 2)$$
$$4y + 12 = -5x + 10$$
$$5x + 4y + 2 = 0$$

(b) The slope of a line perpendicular to l is the negative reciprocal of $-\tfrac{5}{4}$, which is $\tfrac{4}{5}$. From the point-slope form, an equation of the line through $(2, -3)$ and having slope $\tfrac{4}{5}$ is

$$y - (-3) = \tfrac{4}{5}(x - 2)$$
$$5y + 15 = 4x - 8$$
$$4x - 5y - 23 = 0$$

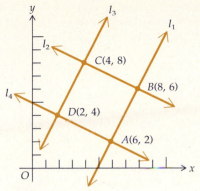

FIGURE 22

EXAMPLE 6   Prove by means of slopes that the four points $A(6, 2)$, $B(8, 6)$, $C(4, 8)$, and $D(2, 4)$ are the vertices of a rectangle.

Solution   See Figure 22, where l_1 is the line through A and B, l_2 is the line through B and C, l_3 is the line through D and C, and l_4 is the line through A and D; m_1, m_2, m_3, and m_4 are their respective slopes.

$$m_1 = \frac{6-2}{8-6} \qquad m_2 = \frac{8-6}{4-8} \qquad m_3 = \frac{8-4}{4-2} \qquad m_4 = \frac{4-2}{2-6}$$

$$= 2 \qquad\qquad = -\tfrac{1}{2} \qquad\qquad = 2 \qquad\qquad = -\tfrac{1}{2}$$

Because $m_1 = m_3$, l_1 is parallel to l_3; and because $m_2 = m_4$, l_2 is parallel to l_4. Because $m_1 m_2 = -1$, l_1 and l_2 are perpendicular. Therefore the quadrilateral has its opposite sides parallel, and a pair of adjacent sides are perpendicular. Thus the quadrilateral is a rectangle.

EXERCISES 1.2

In Exercises 1 through 6, plot the point P and each of the following points as may apply:

(a) *The point Q such that the line segment through Q and P is perpendicular to the x axis and bisected by it. Give the coordinates of Q.*

(b) *The point R such that the line segment through P and R is perpendicular to and bisected by the y axis. Give the coordinates of R.*

(c) *The point S such that the line segment through P and S is bisected by the origin. Give the coordinates of S.*

(d) *The point T such that the line segment through P and T is perpendicular to and bisected by the 45° line through the origin bisecting the first and third quadrants. Give the coordinates of T.*

1. $P(1, -2)$ **2.** $P(-2, 2)$ **3.** $P(2, 2)$
4. $P(-2, -2)$ **5.** $P(-1, -3)$ **6.** $P(0, -3)$

7. Prove that the points $A(-7, 2)$, $B(3, -4)$, and $C(1, 4)$ are the vertices of an isosceles triangle.

8. Prove that the points $A(-4, -1)$, $B(-2, -3)$, $C(4, 3)$, and $D(2, 5)$ are the vertices of a rectangle.

9. A median of a triangle is a line segment from a vertex to the midpoint of the opposite side. Find the length of the medians of the triangle having vertices $A(2, 3)$, $B(3, -3)$, and $C(-1, -1)$.

10. Find the length of the medians of the triangle having vertices $A(-3, 5)$, $B(2, 4)$, and $C(-1, -4)$.

11. Prove that the triangle with vertices $A(3, -6)$, $B(8, -2)$, and $C(-1, -1)$ is a right triangle. Find the area of the triangle. (*Hint: Use the converse of the Pythagorean theorem.*)

12. Find the midpoints of the diagonals of the quadrilateral whose vertices are $(0, 0)$, $(0, 4)$, $(3, 5)$, and $(3, 1)$.

13. Prove that the points $A(6, -13)$, $B(-2, 2)$, $C(13, 10)$, and $D(21, -5)$ are the vertices of a square. Find the length of a diagonal.

14. If one end of a line segment is the point $(-4, 2)$ and the midpoint is $(3, -1)$, find the coordinates of the other end of the line segment.

15. If one end of a line segment is the point $(6, -2)$ and the midpoint is $(-1, 5)$, find the coordinates of the other end of the line segment.

16. The abscissa of a point is -6, and its distance from the point $(1, 3)$ is $\sqrt{74}$. Find the ordinate of the point.

17. By using distance formula (1), prove that the points $(-3, 2)$, $(1, -2)$, and $(9, -10)$ lie on a line.

18. Determine whether the points $(14, 7)$, $(2, 2)$, and $(-4, -1)$ lie on a line by using distance formula (1).

19. If two vertices of an equilateral triangle are $(-4, 3)$ and $(0, 0)$, find the third vertex.

20. Given the two points $A(-3, 4)$ and $B(2, 5)$, find the coordinates of a point P on the line through A and B such that P is (a) twice as far from A as from B, and (b) twice as far from B as from A.

In Exercises 21 through 24, find the slope of the line through the points.

21. $(2, -3), (-4, 3)$
22. $(5, 2), (-2, -3)$
23. $(\tfrac{1}{3}, \tfrac{1}{2}), (-\tfrac{5}{6}, \tfrac{2}{3})$
24. $(-2.1, 0.3), (2.3, 1.4)$

In Exercises 25 through 38, find an equation of the line satisfying the conditions.

25. The slope is 4 and through the point $(2, -3)$.

26. The slope is 3 and through the point $(-4, -1)$.

27. The slope is -2 and through the point $(-3, 5)$.

28. Through the points $(-2, 7)$ and $(6, 0)$.

29. Through the points $(4, 6)$ and $(0, -7)$.

30. Through the points $(3, 1)$ and $(-5, 4)$.

31. The x intercept is -3 and the y intercept is 4.

32. Through the point $(1, 4)$ and parallel to the line whose equation is $2x - 5y + 7 = 0$.

33. Through the point $(-2, 3)$ and perpendicular to the line whose equation is $2x - y - 2 = 0$.

34. Through the point $(-3, -4)$ and parallel to the y axis.

35. Through the point $(1, -7)$ and parallel to the x axis.

36. The slope is -2 and the x intercept is 4.

37. Through the point $(-2, -5)$ and having a slope of $\sqrt{3}$.

38. Through the origin and bisecting the angle between the axes in the first and third quadrants.

In Exercises 39 and 40, find the slope of the line.

39. (a) $x + 3y = 7$; (b) $2y + 9 = 0$

40. (a) $4x - 6y = 5$; (b) $3x - 5 = 0$

In Exercises 41 and 42, determine by means of slopes if the three points are collinear.

41. (a) $(2, 3)$, $(-4, -7)$, $(5, 8)$; (b) $(2, -1)$, $(1, 1)$, $(3, 4)$

42. (a) $(4, 6)$, $(1, 2)$, $(-5, -4)$; (b) $(-3, 6)$, $(3, 2)$, $(9, -2)$

43. (a) Write an equation whose graph is the x axis. (b) Write an equation whose graph is the y axis. (c) Write an equation whose graph is the set of all points on either the x axis or the y axis.

44. (a) Write an equation whose graph consists of all points having an abscissa of 4. (b) Write an equation whose graph consists of all points having an ordinate of -3.

45. Show that the lines having the equations $3x + 5y + 7 = 0$ and $6x + 10y - 5 = 0$ are parallel.

46. Show that the lines having the equations $3x + 5y + 7 = 0$ and $5x - 3y - 2 = 0$ are perpendicular.

47. Given the line l having the equation $2y - 3x = 4$ and the point $P(1, -3)$, find (a) an equation of the line through P and perpendicular to l; (b) the shortest distance from P to line l.

48. Find the value of k such that the lines whose equations are $3kx + 8y = 5$ and $6y - 4kx = -1$ are perpendicular.

49. Show by means of slopes that the points $(-4, -1)$, $(3, \frac{8}{3})$, $(8, -4)$, and $(2, -9)$ are the vertices of a trapezoid.

50. Prove by means of slopes that the three points $A(3, 1)$, $B(6, 0)$, and $C(4, 4)$ are the vertices of a right triangle, and find the area of the triangle.

51. Find the coordinates of the three points that divide the line segment from $A(-5, 3)$ to $B(6, 8)$ into four equal parts.

52. Three consecutive vertices of a parallelogram are $(-4, 1)$, $(2, 3)$, and $(8, 9)$. Find the coordinates of the fourth vertex.

53. Given the line l having the equation $Ax + By + C = 0$, $B \neq 0$, find (a) the slope; (b) the y intercept; (c) the x intercept; (d) an equation of the line through the origin perpendicular to l.

54. If A, B, C, and D are constants, show that (a) the lines $Ax + By + C = 0$ and $Ax + By + D = 0$ are parallel and (b) the lines $Ax + By + C = 0$ and $Bx - Ay + D = 0$ are perpendicular.

55. Find equations of the three medians of the triangle having vertices $A(3, -2)$, $B(3, 4)$, and $C(-1, 1)$, and prove that they meet in a point.

56. Find equations of the perpendicular bisectors of the sides of the triangle having vertices $A(-1, -3)$, $B(5, -3)$, and $C(5, 5)$, and prove that they meet in a point.

57. Prove Theorem 1.2.5: The graph of the equation

$$Ax + By + C = 0$$

where A, B, and C are constants and where not both A and B are zero, is a line. (*Hint:* Consider two cases $B \neq 0$ and $B = 0$. If $B \neq 0$, show that the equation is that of a line having slope $-A/B$ and y intercept $-C/B$. If $B = 0$ show that the equation is that of a vertical line.)

58. Let l_1 be the line $A_1x + B_1y + C_1 = 0$, and let l_2 be the line $A_2x + B_2y + C_2 = 0$. If l_1 is not parallel to l_2 and if k is any constant, the equation

$$A_1x + B_1y + C_1 + k(A_2x + B_2y + C_2) = 0$$

represents an unlimited number of lines. Prove that each of these lines contains the point of intersection of l_1 and l_2.

59. Given that an equation of l_1 is $2x + 3y - 5 = 0$ and that an equation of l_2 is $3x + 5y - 8 = 0$, by using Exercise 58 and without finding the coordinates of the point of intersection of l_1 and l_2, find an equation of the line through this point and (a) containing the point $(1, 3)$; (b) parallel to the x axis; (c) having slope -2.

60. For the lines l_1 and l_2 of Exercise 59, use Exercise 58 and without finding the coordinates of the point of intersection of l_1 and l_2, find an equation of the line through this point and (a) parallel to the y axis; (b) perpendicular to the line having the equation $2x + y = 7$; (c) forming an isosceles triangle with the coordinate axes.

In Exercises 61 through 66, use analytic geometry to prove the given theorem from plane geometry.

61. The sum of the squares of the distances of any point from two opposite vertices of any rectangle is equal to the sum of the squares of its distances from the other two vertices.

62. The midpoint of the hypotenuse of any right triangle is equidistant from each of the three vertices.

63. The line segment joining the midpoints of two opposite sides of any quadrilateral and the line segment joining the midpoints of the diagonals of the quadrilateral bisect each other.

64. The line segments joining consecutive midpoints of the sides of any quadrilateral form a parallelogram.

65. The diagonals of a parallelogram bisect each other.

66. If the diagonals of a quadrilateral bisect each other, then the quadrilateral is a parallelogram.

1.3 CIRCLES AND GRAPHS OF EQUATIONS

An **equation of a graph** is an equation that is satisfied by the coordinates of those, and only those, points on the graph. You learned in Section 1.2 that a first-degree equation in two variables has a line as its graph. One of the simplest curves that is the graph of a second-degree equation in two variables is the *circle*.

1.3.1 DEFINITION

A **circle** is the set of all points in a plane equidistant from a fixed point. The fixed point is called the **center**, and the constant equal distance is called the **radius**.

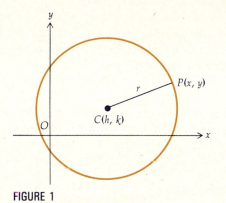

FIGURE 1

To obtain an equation of the circle having center at $C(h, k)$ and radius r, we use the distance formula. Refer to Figure 1. The point $P(x, y)$ is on the circle if and only if $|\overline{PC}| = r$, that is, if and only if

$$\sqrt{(x - h)^2 + (y - k)^2} = r$$

This equation is true if and only if

$$(x - h)^2 + (y - k)^2 = r^2 \qquad (r > 0)$$

This equation is satisfied by the coordinates of those and only those points that lie on the circle, and therefore it is an equation of the circle. We have proved the following theorem.

1.3.2 THEOREM

The circle with center at the point (h, k) and radius r has as an equation

$$(x - h)^2 + (y - k)^2 = r^2$$

If the center of a circle is at the origin, then $h = 0$ and $k = 0$; therefore, its equation is

$$x^2 + y^2 = r^2$$

Such a circle appears in Figure 2. If the radius of a circle is 1, it is called a **unit circle**.

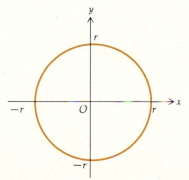

FIGURE 2

EXAMPLE 1　　Find an equation of the circle having a diameter with endpoints at $A(-2, 3)$ and $B(4, 5)$.

Solution　　The midpoint of the line segment from A to B is the center of the circle. See Figure 3. If $C(h, k)$ is the center of the circle, then

$$h = \frac{-2 + 4}{2} \qquad k = \frac{3 + 5}{2}$$
$$= 1 \qquad\qquad = 4$$

The center is at $C(1, 4)$. The radius of the circle can be computed as either $|\overline{CA}|$ or $|\overline{CB}|$. If $r = |\overline{CA}|$, then

$$r = \sqrt{(1 + 2)^2 + (4 - 3)^2}$$
$$= \sqrt{10}$$

An equation of the circle is therefore

$$(x - 1)^2 + (y - 4)^2 = 10$$
$$x^2 + y^2 - 2x - 8y + 7 = 0$$

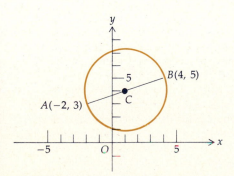

FIGURE 3

The equation $(x - h)^2 + (y - k)^2 = r^2$ is called the **center-radius form** of an equation of a circle. If we remove parentheses and combine like terms, we obtain

$$x^2 + y^2 - 2hx - 2ky + (h^2 + k^2 - r^2) = 0$$

By letting $D = -2h$, $E = -2k$, and $F = h^2 + k^2 - r^2$, this equation becomes

$$x^2 + y^2 + Dx + Ey + F = 0$$

which is called the **general form** of an equation of a circle. Because every circle has a center and radius, its equation can be put in the center-radius form, and hence into the general form, as we did in Example 1. If we start with an equation of a circle in the general form, we can write it in the center-radius form by completing squares. The next example shows the procedure.

EXAMPLE 2   Find the center and radius of the circle having the equation

$$x^2 + y^2 + 6x - 2y - 15 = 0$$

Solution   The given equation may be written as

$$(x^2 + 6x) + (y^2 - 2y) = 15$$

Completing the squares of the terms in parentheses by adding 9 and 1 on both sides of the equation we have

$$(x^2 + 6x + 9) + (y^2 - 2y + 1) = 15 + 9 + 1$$
$$(x + 3)^2 + (y - 1)^2 = 25$$

Because this equation is in the center-radius form, the center is at $(-3, 1)$ and the radius is 5.

We now show that there are equations of the form

$$x^2 + y^2 + Dx + Ey + F = 0 \qquad (1)$$

whose graphs are not circles. Suppose when we complete the squares we obtain

$$(x - h)^2 + (y - k)^2 = d$$

If $d > 0$, we have a circle with center at (h, k) and radius $\sqrt{d}$. However, if $d < 0$, there are no real values of x and y that satisfy the equation; thus there is no graph. In such a case we state that the graph is the empty set. Finally, if $d = 0$, we have

$$(x - h)^2 + (y - k)^2 = 0$$

The only real values of x and y satisfying this equation are $x = h$ and $y = k$. Thus the graph is the point (h, k).

▶ **ILLUSTRATION 1**   Suppose we have the equation

$$x^2 + y^2 - 4x + 10y + 29 = 0$$

which can be written as

$$(x^2 - 4x) + (y^2 + 10y) = -29$$

Completing the squares of the terms in parentheses by adding 4 and 25 on both sides, we have

$$(x^2 - 4x + 4) + (y^2 + 10y + 25) = -29 + 4 + 25$$

$$(x - 2)^2 + (y + 5)^2 = 0$$

Because the only real values of x and y satisfying this equation are $x = 2$ and $y = -5$, the graph is the point $(2, -5)$.   ◄

Observe that an equation of the form

$$Ax^2 + Ay^2 + Dx + Ey + F = 0 \qquad \text{where } A \neq 0 \tag{2}$$

can be written in the form of (1) by dividing by A, thereby obtaining

$$x^2 + y^2 + \frac{D}{A}x + \frac{E}{A}y + \frac{F}{A} = 0$$

Equation (2) is a special case of the general equation of the second degree

$$Ax^2 + Bxy + Cy^2 + Dx + Ey + F = 0$$

in which the coefficients of x^2 and y^2 are equal and which has no xy term. The following theorem is a result of this discussion.

1.3.3 THEOREM   The graph of any second-degree equation in R^2 in x and y, in which the coefficients of x^2 and y^2 are equal and which has no xy term is either a circle, a point, or the empty set.

► **ILLUSTRATION 2**   The equation

$$2x^2 + 2y^2 + 12x - 8y + 31 = 0$$

is of the form (2), and therefore its graph is either a circle, a point, or the empty set. The equation is put in the form of (1):

$$x^2 + y^2 + 6x - 4y + \tfrac{31}{2} = 0$$

$$(x^2 + 6x) + (y^2 - 4y) = -\tfrac{31}{2}$$

$$(x^2 + 6x + 9) + (y^2 - 4y + 4) = -\tfrac{31}{2} + 9 + 4$$

$$(x + 3)^2 + (y - 2)^2 = -\tfrac{5}{2}$$

Therefore the graph is the empty set.   ◄

In Section 1.2 we defined the graph of an equation in R^2 as the set of all points (x, y) whose coordinates are numbers satisfying the equation. The graph of an equation in R^2 is also called a curve. We have already discussed two kinds of curves: lines, which are graphs of first-degree equations, and circles, which are graphs of second-degree equations of form (2). We now examine some graphs of another kind of second-degree equation in x and y:

$$y = ax^2 + bx + c \tag{3}$$

where a, b, and c are constants and $a \neq 0$. In particular, consider

$$y = x^2 - 2 \tag{4}$$

Table 1

x	$y = x^2 - 2$
0	-2
1	-1
2	2
3	7
4	14
-1	-1
-2	2
-3	7
-4	14

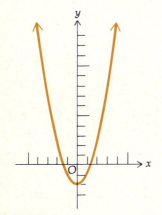

FIGURE 4

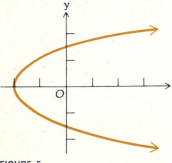

FIGURE 5

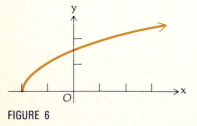

FIGURE 6

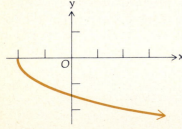

FIGURE 7

A solution of this equation is an ordered pair of real numbers, one for x and one for y, that satisfies the equation. For example, if x is replaced by 3 in the equation, we see that $y = 7$; thus the ordered pair $(3, 7)$ constitutes a solution. If any number is substituted for x in the right side of (4), a corresponding value for y is obtained. Therefore (4) has an unlimited number of solutions. Table 1 gives a few of them.

If we plot the points having as coordinates the number pairs (x, y) from Table 1 and connect them by a smooth curve, we obtain a sketch of the graph of Equation (4), which appears in Figure 4. Any point (x, y) on this curve has coordinates satisfying Equation (4), and the coordinates of any point not on this curve do not satisfy the equation.

The graph in Figure 4 is a *parabola*. The lowest point on the graph is $(0, -2)$ and is the *vertex* of the parabola. The parabola opens upward. A complete treatment of parabolas occurs in Chapter 10, where we show that the graph of an equation of the form (3) is a parabola opening upward if $a > 0$ and downward if $a < 0$. In the following example, the graph is a parabola opening to the right.

EXAMPLE 3   Draw a sketch of the graph of the equation

$$y^2 - x - 2 = 0$$

Solution   Solving the equation for y, we have

$$y^2 = x + 2$$
$$y = \pm\sqrt{x + 2}$$

Thus the given equation is equivalent to the two equations

$$y = \sqrt{x + 2} \quad \text{and} \quad y = -\sqrt{x + 2}$$

The coordinates of any point that satisfy either one of these two equations will satisfy the given equation. Conversely, the coordinates of any point satisfying the given equation will satisfy either $y = \sqrt{x + 2}$ or $y = -\sqrt{x + 2}$. Table 2 gives some of these values of x and y.

Table 2

x	0	0	1	1	2	2	3	3	-1	-1	-2
y	$\sqrt{2}$	$-\sqrt{2}$	$\sqrt{3}$	$-\sqrt{3}$	2	-2	$\sqrt{5}$	$-\sqrt{5}$	1	-1	0

Observe that for any value of $x < -2$ there is no real value for y. Also, for each value of $x > -2$ there are two values for y. A sketch of the graph appears in Figure 5.

EXAMPLE 4   Draw sketches of the graphs of the equations

$$y = \sqrt{x + 2} \quad \text{and} \quad y = -\sqrt{x + 2}$$

Solution   Recall from Example 3 that these two equations together are equivalent to the equation $y^2 - x - 2 = 0$. In the equation $y = \sqrt{x + 2}$, the value of y is nonnegative. Hence the graph of the equation, appearing in Figure 6, is the upper half of the graph in Figure 5.

Similarly, the graph of the equation $y = -\sqrt{x + 2}$, a sketch of which appears in Figure 7, is the lower half of the graph in Figure 5.

In drawing a sketch of the graph of an equation it is often helpful to consider properties of *symmetry* of a graph.

1.3.4 DEFINITION

Two points P and Q are said to be **symmetric with respect to a line** if and only if the line is the perpendicular bisector of the line segment PQ. Two points P and Q are said to be **symmetric with respect to a third point** if and only if the third point is the midpoint of the line segment PQ.

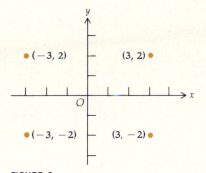

FIGURE 8

▶ **ILLUSTRATION 3** The points $(3, 2)$ and $(3, -2)$ are symmetric with respect to the x axis, the points $(3, 2)$ and $(-3, 2)$ are symmetric with respect to the y axis, and the points $(3, 2)$ and $(-3, -2)$ are symmetric with respect to the origin (see Figure 8). ◀

In general, the points (x, y) and $(x, -y)$ are symmetric with respect to the x axis, (x, y) and $(-x, y)$ are symmetric with respect to the y axis, and (x, y) and $(-x, -y)$ are symmetric with respect to the origin.

1.3.5 DEFINITION

The graph of an equation is symmetric with respect to a line l if and only if for every point P on the graph there is a point Q, also on the graph, such that P and Q are symmetric with respect to l. The graph of an equation is symmetric with respect to a point R if and only if for every point P on the graph there is a point S, also on the graph, such that P and S are symmetric with respect to R.

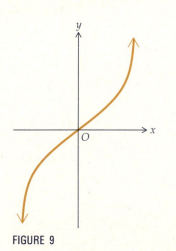

FIGURE 9

In Figure 5 we have a graph symmetric with respect to the x axis, and Figure 4 shows one symmetric with respect to the y axis. We show a graph that is symmetric with respect to the origin in Figure 9. The circle shown in Figure 2 is symmetric with respect to the x axis, the y axis, and the origin.

From Definition 1.3.5 it follows that if a point (x, y) is on a graph symmetric with respect to the x axis, then the point $(x, -y)$ also must be on the graph. And if both the points (x, y) and $(x, -y)$ are on the graph, then the graph is symmetric with respect to the x axis. Therefore the coordinates of the point $(x, -y)$ as well as (x, y) must satisfy an equation of the graph. Hence the graph of an equation in x and y is symmetric with respect to the x axis if and only if an equivalent equation is obtained when y is replaced by $-y$ in the equation. We have thus proved part (i) in the following theorem. The proofs of parts (ii) and (iii) are similar.

1.3.6 THEOREM
Tests for Symmetry

The graph of an equation in x and y is

(i) symmetric with respect to the x axis if and only if an equivalent equation is obtained when y is replaced by $-y$ in the equation;

(ii) symmetric with respect to the y axis if and only if an equivalent equation is obtained when x is replaced by $-x$ in the equation;

(iii) symmetric with respect to the origin if and only if an equivalent equation is obtained when x is replaced by $-x$ and y is replaced by $-y$ in the equation.

Refer back to the graph in Figure 4. It is symmetric with respect to the y axis, and its equation is $y = x^2 - 2$. Observe that an equivalent equation is obtained when x is replaced by $-x$. In Example 3 we have the equation $y^2 - x - 2 = 0$ for which an equivalent equation is obtained when y is replaced by $-y$, and its graph, sketched in Figure *4*, is symmetric with respect to the *x* axis. The following example gives a graph that is symmetric with respect to the origin.

EXAMPLE 5   Draw a sketch of the graph of the equation

$$xy = 1$$

Solution   Observe that if in the given equation *x* is replaced by $-x$ and *y* is replaced by $-y$, an equivalent equation is obtained; hence by Theorem 1.3.6(iii) the graph is symmetric with respect to the origin. Table 3 gives some values of *x* and *y* satisfying the given equation.

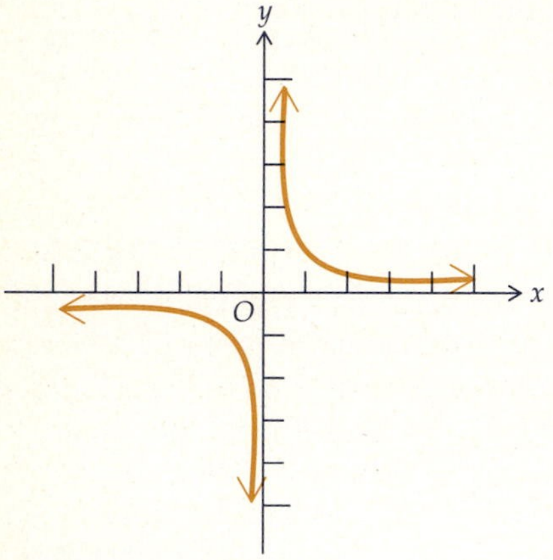

Table 3

x	1	2	3	4	$\frac{1}{2}$	$\frac{1}{3}$	$\frac{1}{4}$	-1	-2	-3	-4	$-\frac{1}{2}$	$-\frac{1}{3}$	$-\frac{1}{4}$
y	1	$\frac{1}{2}$	$\frac{1}{3}$	$\frac{1}{4}$	2	3	4	-1	$-\frac{1}{2}$	$-\frac{1}{3}$	$-\frac{1}{4}$	-2	-3	-4

From the given equation, $y = 1/x$. We see that as *x* increases through positive values, *y* decreases through positive values and gets closer and closer to zero. As *x* decreases through positive values, *y* increases through positive values and gets larger and larger. As *x* increases through negative values (i.e., *x* takes on the values $-4, -3, -2, -1, -\frac{1}{2}$, etc.), *y* takes on negative values having larger and larger absolute values. A sketch of the graph is shown in Figure 10.

FIGURE 10

EXERCISES 1.3

In Exercises 1 through 4, find an equation of the circle with center at C and radius r. Write the equation in both the center-radius form and the general form.

1. $C(4, -3)$, $r = 5$ **2.** $C(0, 0)$, $r = 8$
3. $C(-5, -12)$, $r = 3$ **4.** $C(-1, 1)$, $r = 2$

In Exercises 5 and 6, find an equation of the circle satisfying the conditions.

5. Center is at $(1, 2)$ and through the point $(3, -1)$.
6. Through the three points $(2, 8)$, $(7, 3)$, and $(-2, 0)$.

In Exercises 7 through 10, find the center and radius of the circle, and draw a sketch of the graph.

7. $x^2 + y^2 - 6x - 8y + 9 = 0$
8. $2x^2 + 2y^2 - 2x + 2y + 7 = 0$
9. $3x^2 + 3y^2 + 4y - 7 = 0$
10. $x^2 + y^2 - 10x - 10y + 25 = 0$

In Exercises 11 through 16, determine whether the graph is a circle, a point, or the empty set.

11. $x^2 + y^2 - 2x + 10y + 19 = 0$
12. $4x^2 + 4y^2 + 24x - 4y + 1 = 0$
13. $x^2 + y^2 - 10x + 6y + 36 = 0$
14. $x^2 + y^2 + 2x - 4y + 5 = 0$
15. $36x^2 + 36y^2 - 48x + 36y - 119 = 0$
16. $9x^2 + 9y^2 + 6x - 6y + 5 = 0$

In Exercises 17 through 44, draw a sketch of the graph of the equation.

17. $y = 2x + 5$ **18.** $y = 4x - 3$
19. $y = \sqrt{x + 4}$ **20.** $y = \sqrt{x - 1}$
21. $y = -\sqrt{x + 4}$ **22.** $y = -\sqrt{x - 1}$
23. $y^2 = x + 4$ **24.** $y^2 = x - 1$
25. $y = 3 - x^2$ **26.** $y = x^2 + 2$
27. $y = x^2 - 4$ **28.** $y = 9 - x^2$
29. $y = 4 + x^2$ **30.** $y = x^2 - 9$
31. $xy = 4$ **32.** $xy = -1$
33. $xy = -9$ **34.** $xy = 9$
35. $x = y^2 + 2$ **36.** $x = y^2 - 4$
37. (a) $x + 3y = 0$; (b) $x - 3y = 0$; (c) $x^2 - 9y^2 = 0$
38. (a) $2x - 5y = 0$; (b) $2x + 5y = 0$; (c) $4x^2 - 25y^2 = 0$
39. (a) $y = \sqrt{2x}$; (b) $y = -\sqrt{2x}$; (c) $y^2 = 2x$
40. (a) $y = \sqrt{-2x}$; (b) $y = -\sqrt{-2x}$; (c) $y^2 = -2x$
41. (a) $y = \sqrt{4 - x^2}$; (b) $y = -\sqrt{4 - x^2}$; (c) $x^2 + y^2 = 4$
42. (a) $y = \sqrt{1 - x^2}$; (b) $y = -\sqrt{1 - x^2}$; (c) $x^2 + y^2 = 1$
43. (a) $x = \frac{1}{2}\sqrt{1 - 4y^2}$; (b) $x = -\frac{1}{2}\sqrt{1 - 4y^2}$; (c) $4x^2 + 4y^2 = 1$
44. (a) $xy = 2$; (b) $xy = -2$; (c) $x^2y^2 = 4$

In Exercises 45 through 48, find an equation of the circle satisfying the conditions.

45. Center is at $(-3, -5)$ and tangent to the line $12x + 5y = 4$.
46. Center is at $(-2, 5)$ and tangent to the line $x = 7$.
47. Tangent to the line $3x + y + 2 = 0$ at $(-1, 1)$ and through the point $(3, 5)$.
48. Tangent to the line $3x + 4y - 16 = 0$ at $(4, 1)$ and with a radius of 5 (two possible circles).
49. Find an equation of the line that is tangent to the circle $x^2 + y^2 - 4x + 6y - 12 = 0$ at the point $(5, 1)$.
50. Find an equation of each of the two lines having slope $-\frac{4}{3}$ that are tangent to the circle $x^2 + y^2 + 2x - 8y - 8 = 0$.

51. Prove that a graph that is symmetric with respect to both coordinate axes is also symmetric with respect to the origin.
52. The graph of an equation in x and y is symmetric with respect to the line with equation $y = x$ if and only if an equivalent equation is obtained when x is replaced by y and y is replaced by x. Show that the graph of the equation $ax^2 + by^2 = c$, where a, b, and c are positive, is symmetric with respect to this line if and only if $a = b$.
53. Prove that a graph that is symmetric with respect to any two perpendicular lines is also symmetric with respect to their point of intersection.

1.4 FUNCTIONS

Often in practical applications the value of one quantity depends on the value of another. A person's salary may depend on the number of hours worked; the total production at a factory may depend on the number of machines used; the distance traveled by an object may depend on the time elapsed since it left a specific point; the volume of the space occupied by a gas having a constant pressure depends on the temperature of the gas; the resistance of an electrical cable of fixed length depends on its diameter; and so forth. A relationship between such quantities is often given by means of a *function*. For our purposes we confine the quantities in the relationship to be real numbers. Then

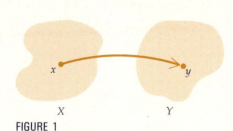

FIGURE 1

> A function can be thought of as a correspondence from a set X of real numbers x to a set Y of real numbers y, where the number y is unique for a specific value of x.

Figure 1 gives a visualization of such a correspondence where the sets X and Y consist of points in a plane region.

Stating the concept of a function another way, we intuitively consider the real number y in set Y to be a *function* of the real number x in set X if there is some rule by which a unique value of y is assigned to a value of x. This rule is often given by an equation. For example, the equation

$$y = x^2$$

Table 1

x	$y = x^2$
1	1
$\frac{3}{2}$	$\frac{9}{4}$
4	16
0	0
-1	1
$-\frac{3}{2}$	$\frac{9}{4}$
-4	16

defines a function for which X is the set of all real numbers and Y is the set of nonnegative numbers. The value of y in Y assigned to the value of x in X is obtained by multiplying x by itself. Table 1 gives the value of y assigned to some particular values of x, and Figure 2 visualizes the correspondence for the numbers in the table.

We use symbols such as f, g, and h to denote a function. The set X of real numbers described above is the *domain* of the function, and the set Y of real numbers assigned to the values of x in X is the *range* of the function.

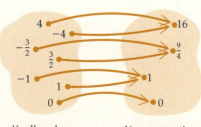

X: all real numbers
Y: nonnegative numbers

FIGURE 2

▶ **ILLUSTRATION 1** Let f be the function defined by the equation

$$y = \sqrt{x - 2}$$

Because the numbers are confined to real numbers, y is a function of x only for $x - 2 \geq 0$ because for any x satisfying this inequality, a unique value of y is determined. However, if $x < 2$, a square root of a negative number is obtained, and hence no real number y exists. Therefore we must restrict x so that $x \geq 2$. Thus the domain of f is the interval $[2, +\infty)$, and the range is $[0, +\infty)$. ◄

▶ **ILLUSTRATION 2**   Let g be the function defined by the equation

$$y = \sqrt{x^2 - 9}$$

We observe that y is a function of x only for $x \geq 3$ or $x \leq -3$ (or simply $|x| \geq 3$); for any x satisfying either of these inequalities, a unique value of y is determined. No real value of y is determined if x is in the open interval $(-3, 3)$, because for these values of x a square root of a negative number is obtained. Hence the domain of g is $(-\infty, -3] \cup [3, +\infty)$, and the range is $[0, +\infty)$. ◄

We can consider a function as a set of ordered pairs. For instance, the function defined by the equation $y = x^2$ consists of all the ordered pairs (x, y) satisfying the equation. The ordered pairs in this function given by Table 1 are $(1, 1)$, $(\frac{3}{2}, \frac{9}{4})$, $(4, 16)$, $(0, 0)$, $(-1, 1)$, $(-\frac{3}{2}, \frac{9}{4})$, and $(-4, 16)$. Of course, there is an unlimited number of ordered pairs in the function. Some others are $(2, 4)$, $(-2, 4)$, $(5, 25)$, $(-5, 25)$, $(\sqrt{3}, 3)$, and so on.

▶ **ILLUSTRATION 3**   The function f of Illustration 1 is the set of ordered pairs (x, y) for which $y = \sqrt{x - 2}$. With symbols we write

$$f = \{(x, y) \mid y = \sqrt{x - 2}\}$$

Some of the ordered pairs in f are $(2, 0)$, $(\frac{9}{4}, \frac{1}{2})$, $(3, 1)$, $(4, \sqrt{2})$, $(5, \sqrt{3})$, $(6, 2)$, $(11, 3)$. ◄

▶ **ILLUSTRATION 4**   The function g of Illustration 2 is the set of ordered pairs (x, y) for which $y = \sqrt{x^2 - 9}$; that is,

$$g = \{(x, y) \mid y = \sqrt{x^2 - 9}\}$$

Some of the ordered pairs in g are $(3, 0)$, $(4, \sqrt{7})$, $(5, 4)$, $(-3, 0)$, $(-\sqrt{13}, 2)$. ◄

We now give the formal definition of a function. Defining a function as a set of ordered pairs rather than as a rule or correspondence makes its meaning precise.

1.4.1 DEFINITION   A **function** is a set of ordered pairs of numbers (x, y) in which no two distinct ordered pairs have the same first number. The set of all admissible values of x is called the **domain** of the function, and the set of all resulting values of y is called the **range** of the function.

In this definition, the restriction that no two distinct ordered pairs can have the same first number ensures that y is unique for a specific value of x. The numbers x and y are **variables**. Because values are assigned to x and because the value of y is dependent on the choice of x, x is the **independent variable** and y is the **dependent variable**.

The concept of a function as a set of ordered pairs permits us to give the following definition of the *graph of a function*.

1.4.2 DEFINITION If f is a function, then the **graph** of f is the set of all points (x, y) in R^2 for which (x, y) is an ordered pair in f.

By comparing this definition with the definition of the graph of an equation (1.2.4), it follows that the graph of a function f is the same as the graph of the equation $y = f(x)$.

▶ **ILLUSTRATION 5** (a) A sketch of the graph of the function f of Illustrations 1 and 3 appears in Figure 3. It is the top half of a parabola.

(b) Figure 4 shows a sketch of the graph of the function g of Illustrations 2 and 4. ◀

Recall that for a function, there must be a unique value of the dependent variable for each value of the independent variable in the domain of the function. In geometric terms this means

The graph of a function can be intersected by a vertical line in at most one point.

Observe that the graphs in Figures 3 and 4 are intersected by a vertical line in at most one point.

▶ **ILLUSTRATION 6** Consider the set

$$\{(x, y) \mid x^2 + y^2 = 25\}$$

A sketch of the graph of this set is shown in Figure 5. This set of ordered pairs is not a function because for any x in the interval $(-5, 5)$ there are two ordered pairs having x as the first number. For example, both $(3, 4)$ and $(3, -4)$ are ordered pairs in the given set. Furthermore, observe that the graph of the given set is a circle with center at the origin and radius 5, and a vertical line having the equation $x = a$, where $-5 < a < 5$, intersects the circle in two points. ◀

The next section deals with graphs of functions, and more examples appear there.

To introduce the notation for a **function value**, let f be the function having as its domain variable x and as its range variable y. Then the symbol $f(x)$ (read "f of x" or "f at x") denotes the particular value of y corresponding to the value of x.

▶ **ILLUSTRATION 7** In Illustration 1, $f = \{(x, y) \mid y = \sqrt{x - 2}\}$. Thus

$$f(x) = \sqrt{x - 2}$$

We compute $f(x)$ for some specific values of x.

$$f(3) = \sqrt{3 - 2} \qquad f(5) = \sqrt{5 - 2} \qquad f(6) = \sqrt{6 - 2} \qquad f(9) = \sqrt{9 - 2}$$
$$= 1 \qquad\qquad\quad = \sqrt{3} \qquad\qquad\quad = 2 \qquad\qquad\quad = \sqrt{7}$$ ◀

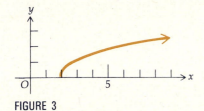

FIGURE 3

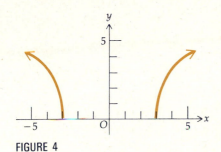

FIGURE 4

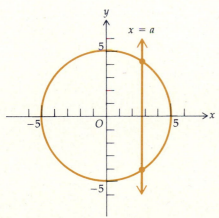

FIGURE 5

When defining a function, the domain of the function must be given either implicitly or explicitly. For instance, if f is defined by

$$f(x) = 3x^2 - 5x + 2$$

it is implied that x can be any real number. However, if f is defined by

$$f(x) = 3x^2 - 5x + 2 \qquad 1 \le x \le 10$$

then the domain of f consists of all real numbers between and including 1 and 10.

Similarly, if g is defined by the equation

$$g(x) = \frac{5x - 2}{x + 4}$$

it is implied that $x \ne -4$, because the quotient is undefined for $x = -4$; hence, the domain of g is the set of all real numbers except -4.

If

$$h(x) = \sqrt{9 - x^2}$$

it is implied that x is in the closed interval $[-3, 3]$ because $\sqrt{9 - x^2}$ is not a real number for $x > 3$ or $x < -3$. Thus the domain of h is $[-3, 3]$ and the range is $[0, 3]$.

EXAMPLE 1   Given that f is the function defined by

$$f(x) = x^2 + 3x - 4$$

find: (a) $f(0)$; (b) $f(2)$; (c) $f(h)$; (d) $f(2h)$; (e) $f(2x)$; (f) $f(x + h)$; (g) $f(x) + f(h)$.

Solution

(a) $f(0) = 0^2 + 3 \cdot 0 - 4$ $\qquad$ (b) $f(2) = 2^2 + 3 \cdot 2 - 4$
$\qquad\quad = 4$ $\qquad\qquad\qquad\qquad\qquad = 6$

(c) $f(h) = h^2 + 3h - 4$ $\qquad$ (d) $f(2h) = (2h)^2 + 3(2h) - 4$
$\qquad\qquad\qquad\qquad\qquad\qquad\qquad = 4h^2 + 6h - 4$

(e) $f(2x) = (2x)^2 + 3(2x) - 4$ $\quad$ (f) $f(x + h) = (x + h)^2 + 3(x + h) - 4$
$\qquad\quad = 4x^2 + 6x - 4$ $\qquad\qquad\qquad\qquad = x^2 + 2hx + h^2 + 3x + 3h - 4$
$\qquad\qquad\qquad\qquad\qquad\qquad\qquad\qquad = x^2 + (2h + 3)x + (h^2 + 3h - 4)$

(g) $f(x) + f(h) = (x^2 + 3x - 4) + (h^2 + 3h - 4)$
$\qquad\qquad\qquad = x^2 + 3x + (h^2 + 3h - 8)$

Compare the computations in parts (f) and (g) of Example 1. In part (f) the computation is for $f(x + h)$, which is the function value at the sum of x and h. In part (g), where $f(x) + f(h)$ is computed, we obtain the sum of the two function values $f(x)$ and $f(h)$.

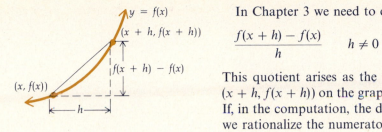

FIGURE 6

In Chapter 3 we need to compute quotients of the form

$$\frac{f(x + h) - f(x)}{h} \qquad h \neq 0$$

This quotient arises as the slope of the line through the points $(x, f(x))$ and $(x + h, f(x + h))$ on the graph of the function defined by $y = f(x)$. See Figure 6. If, in the computation, the difference of two radicals appears in the numerator, we rationalize the numerator as in part (b) of the following example.

EXAMPLE 2 Find

$$\frac{f(x + h) - f(x)}{h}$$

where $h \neq 0$, if (a) $f(x) = 4x^2 - 5x + 7$; (b) $f(x) = \sqrt{x}$.

Solution

(a) $\dfrac{f(x + h) - f(x)}{h} = \dfrac{4(x + h)^2 - 5(x + h) + 7 - (4x^2 - 5x + 7)}{h}$

$$= \frac{4x^2 + 8hx + 4h^2 - 5x - 5h + 7 - 4x^2 + 5x - 7}{h}$$

$$= \frac{8hx - 5h + 4h^2}{h}$$

$$= 8x - 5 + 4h$$

(b) $\dfrac{f(x + h) - f(x)}{h} = \dfrac{\sqrt{x + h} - \sqrt{x}}{h}$

$$= \frac{(\sqrt{x + h} - \sqrt{x})(\sqrt{x + h} + \sqrt{x})}{h(\sqrt{x + h} + \sqrt{x})}$$

$$= \frac{(x + h) - x}{h(\sqrt{x + h} + \sqrt{x})}$$

$$= \frac{h}{h(\sqrt{x + h} + \sqrt{x})}$$

$$= \frac{1}{\sqrt{x + h} + \sqrt{x}}$$

In the second step of this solution the numerator and denominator are multiplied by the conjugate of the numerator in order to rationalize the numerator, and this gives a common factor of h in the numerator and denominator.

We now define some operations on functions. In the definition new functions are formed from given functions by adding, subtracting, multiplying, and dividing function values. Accordingly, these new functions are known as the *sum*, *difference*, *product*, and *quotient* of the original functions.

1.4.3 DEFINITION

Given the two functions f and g:

  (i) their **sum**, denoted by $f + g$, is the function defined by

$$(f + g)(x) = f(x) + g(x)$$

  (ii) their **difference**, denoted by $f - g$, is the function defined by

$$(f - g)(x) = f(x) - g(x)$$

  (iii) their **product**, denoted by $f \cdot g$, is the function defined by

$$(f \cdot g)(x) = f(x) \cdot g(x)$$

  (iv) their **quotient**, denoted by f/g, is the function defined by

$$(f/g)(x) = f(x)/g(x)$$

In each case the *domain* of the resulting function consists of those values of x common to the domains of f and g, with the additional requirement in case (iv) that the values of x for which $g(x) = 0$ are excluded.

EXAMPLE 3   Given that f and g are the functions defined by

$$f(x) = \sqrt{x + 1} \quad \text{and} \quad g(x) = \sqrt{x - 4}$$

find: (a) $(f + g)(x)$; (b) $(f - g)(x)$; (c) $(f \cdot g)(x)$; (d) $(f/g)(x)$.

Solution

(a) $(f + g)(x) = \sqrt{x + 1} + \sqrt{x - 4}$ (b) $(f - g)(x) = \sqrt{x + 1} - \sqrt{x - 4}$

(c) $(f \cdot g)(x) = \sqrt{x + 1} \cdot \sqrt{x - 4}$ (d) $(f/g)(x) = \dfrac{\sqrt{x + 1}}{\sqrt{x - 4}}$

The domain of f is $[-1, +\infty)$, and the domain of g is $[4, +\infty)$. So in parts (a), (b), and (c) the domain of the resulting function is $[4, +\infty)$. In part (d) the denominator is zero when $x = 4$; thus 4 is excluded from the domain, and the domain is therefore $(4, +\infty)$.

Obtaining the *composite function* of two given functions is another operation on functions.

1.4.4 DEFINITION

Given the two functions f and g, the **composite function**, denoted by $f \circ g$, is defined by

$$(f \circ g)(x) = f(g(x))$$

and the domain of $f \circ g$ is the set of all numbers x in the domain of g such that $g(x)$ is in the domain of f.

The definition indicates that when computing $(f \circ g)(x)$, we first apply function g to x and then function f to $g(x)$. The procedure is demonstrated in the following illustration and example.

▶ **ILLUSTRATION 8** If f and g are defined by

$$f(x) = \sqrt{x} \quad \text{and} \quad g(x) = 2x - 3$$

$$(f \circ g)(x) = f(g(x))$$
$$= f(2x - 3)$$
$$= \sqrt{2x - 3}$$

The domain of g is $(-\infty, +\infty)$, and the domain of f is $[0, +\infty)$. Therefore the domain of $f \circ g$ is the set of real numbers for which $2x - 3 \geq 0$ or, equivalently, $[\frac{3}{2}, +\infty)$. ◀

EXAMPLE 4 Given that f and g are defined by

$$f(x) = \sqrt{x} \quad \text{and} \quad g(x) = x^2 - 1$$

find: (a) $f \circ f$; (b) $g \circ g$; (c) $f \circ g$; (d) $g \circ f$. Also determine the domain of the composite function in each part.

Solution

(a) $(f \circ f)(x) = f(f(x))$
$$= f(\sqrt{x})$$
$$= \sqrt{\sqrt{x}}$$
$$= \sqrt[4]{x}$$

The domain is $[0, +\infty)$.

(b) $(g \circ g)(x) = g(g(x))$
$$= g(x^2 - 1)$$
$$= (x^2 - 1)^2 - 1$$
$$= x^4 - 2x^2$$

The domain is $(-\infty, +\infty)$.

(c) $(f \circ g)(x) = f(g(x))$
$$= f(x^2 - 1)$$
$$= \sqrt{x^2 - 1}$$

The domain is $(\infty, -1] \cup [1, +\infty)$.

(d) $(g \circ f)(x) = g(f(x))$
$$= g(\sqrt{x})$$
$$= (\sqrt{x})^2 - 1$$
$$= x - 1$$

The domain is $[0, +\infty)$.

In part (d) note that even though $x - 1$ is defined for all values of x, the domain of $g \circ f$, by the definition of a composite function, is the set of all numbers x in the domain of f such that $f(x)$ is in the domain of g. Thus the domain of $g \circ f$ must be a subset of the domain of f.

Observe from the results of parts (c) and (d) of Example 4 that $(f \circ g)(x)$ and $(g \circ f)(x)$ are not necessarily equal.

1.4.5 DEFINITION (i) A function f is said to be an **even** function if for every x in the domain of f, $f(-x) = f(x)$.
(ii) A function f is said to be an **odd** function if for every x in the domain of f, $f(-x) = -f(x)$.

In both parts (i) and (ii) it is understood that $-x$ is in the domain of f whenever x is.

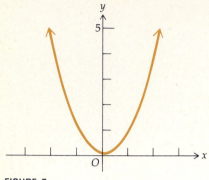

FIGURE 7

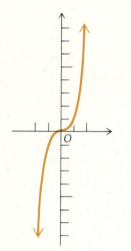

FIGURE 8

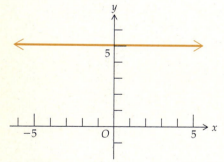

FIGURE 9

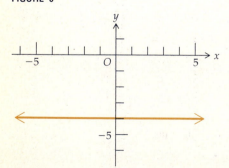

FIGURE 10

▶ **ILLUSTRATION 9**   (a) If $f(x) = 3x^4 - 2x^2 + 7$, then

$$f(-x) = 3(-x)^4 - 2(-x)^2 + 7$$
$$= 3x^4 - 2x^2 + 7$$
$$= f(x)$$

Therefore f is an even function.

(b) If $g(x) = 3x^5 - 4x^3 - 9x$, then

$$g(-x) = 3(-x)^5 - 4(-x)^3 - 9(-x)$$
$$= -3x^5 + 4x^3 + 9x$$
$$= -(3x^5 - 4x^3 - 9x)$$
$$= -g(x)$$

Therefore g is an odd function.

(c) If $h(x) = 2x^4 + 7x^3 - x^2 + 9$, then

$$h(-x) = 2(-x)^4 + 7(-x)^3 - (-x)^2 + 9$$
$$= 2x^4 - 7x^3 - x^2 + 9$$

Because $h(-x) \neq h(x)$ and $h(-x) \neq -h(x)$, h is neither even nor odd.   ◀

From the symmetry tests given in Section 1.3, it follows that the graph of an even function is symmetric with respect to the y axis and the graph of an odd function is symmetric with respect to the origin.

▶ **ILLUSTRATION 10**   (a) If $f(x) = x^2$, f is an even function, and its graph is a parabola that is symmetric with respect to the y axis. See Figure 7.

(b) If $g(x) = x^3$, g is an odd function. The graph of g, shown in Figure 8, is symmetric with respect to the origin.   ◀

A function, whose range consists of only one number, is called a **constant function**. Thus if $f(x) = c$, and if c is any real number, then f is a constant function, and its graph is a horizontal line at a directed distance of c units from the x axis.

▶ **ILLUSTRATION 11**   (a) The function defined by $f(x) = 5$ is a constant function, and its graph, shown in Figure 9, is a horizontal line 5 units above the x axis.

(b) The function defined by $g(x) = -4$ is a constant function whose graph is a horizontal line 4 units below the x axis. See Figure 10.   ◀

A **linear function** is defined by

$$f(x) = mx + b$$

where m and b are constants and $m \neq 0$. Its graph is a line having slope m and y intercept b.

▶ **ILLUSTRATION 12**   The function defined by

$$f(x) = 2x - 6$$

is linear. Its graph is the line appearing in Figure 11.   ◀

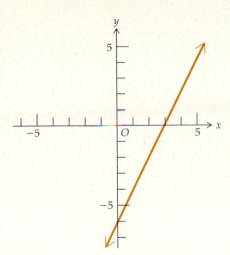

FIGURE 11

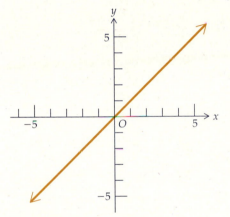

FIGURE 12

The particular linear function defined by

$$f(x) = x$$

is called the **identity function**. Its graph, shown in Figure 12, is the line bisecting the first and third quadrants.

If a function f is defined by

$$f(x) = a_n x^n + a_{n-1} x^{n-1} + a_{n-2} x^{n-2} + \ldots + a_1 x + a_0$$

where $a_0, a_1, \ldots, a_n$ are real numbers ($a_n \neq 0$) and n is a nonnegative integer, then f is called a **polynomial function** of degree n. Thus the function defined by

$$f(x) = 3x^5 - x^2 + 7x - 1$$

is a polynomial function of degree 5.

A linear function is a polynomial function of degree 1. If the degree of a polynomial function is 2, it is called a **quadratic function**, and if the degree is 3, it is called a **cubic function**.

If a function can be expressed as the quotient of two polynomial functions, it is called a **rational function**.

An **algebraic function** is one formed by a finite number of algebraic operations on the identity function and a constant function. These algebraic operations include addition, subtraction, multiplication, division, raising to powers, and extracting roots. Polynomial and rational functions are particular kinds of algebraic functions. A complicated example of an algebraic function is the one defined by

$$f(x) = \frac{(x^2 - 3x + 1)^3}{\sqrt{x^4 + 1}}$$

In addition to algebraic functions, we consider **transcendental functions** in calculus. Examples of transcendental functions are the trigonometric functions discussed in Section 1.6 and the logarithmic and exponential functions introduced in Chapter 7.

EXERCISES 1.4

In Exercises 1 through 4, determine if the set is a function. If it is a function, what is its domain?

1. (a) $\{(x, y) \mid y = \sqrt{x - 4}\}$; (b) $\{(x, y) \mid y = \sqrt{x^2 - 4}\}$;
(c) $\{(x, y) \mid y = \sqrt{4 - x^2}\}$; (d) $\{(x, y) \mid x^2 + y^2 = 4\}$.

2. (a) $\{(x, y) \mid y = \sqrt{x + 1}\}$; (b) $\{(x, y) \mid y = \sqrt{x^2 - 1}\}$;
(c) $\{(x, y) \mid y = \sqrt{1 - x^2}\}$; (d) $\{(x, y) \mid x^2 + y^2 = 1\}$.

3. (a) $\{(x, y) \mid y = x^2\}$; (b) $\{(x, y) \mid x = y^2\}$;
(c) $\{(x, y) \mid y = x^3\}$; (d) $\{(x, y) \mid x = y^3\}$.

4. (a) $\{(x, y) \mid y = (x - 1)^2 + 2\}$; (b) $\{(x, y) \mid x = (y - 2)^2 + 1\}$;
(c) $\{(x, y) \mid y = (x + 2)^3 - 1\}$; (d) $\{(x, y) \mid x = (y + 1)^3 - 2\}$.

5. Given $f(x) = 2x - 1$, find (a) $f(3)$; (b) $f(-2)$; (c) $f(0)$;
(d) $f(a + 1)$; (e) $f(x + 1)$; (f) $f(2x)$; (g) $2f(x)$; (h) $f(x + h)$;
(i) $f(x) + f(h)$; (j) $\dfrac{f(x + h) - f(x)}{h}$, $h \neq 0$.

6. Given $f(x) = \dfrac{3}{x}$, find (a) $f(1)$; (b) $f(-3)$; (c) $f(6)$; (d) $f(\tfrac{1}{3})$;
(e) $f\left(\dfrac{3}{a}\right)$; (f) $f\left(\dfrac{3}{x}\right)$; (g) $\dfrac{f(3)}{f(x)}$; (h) $f(x - 3)$; (i) $f(x) - f(3)$;
(j) $\dfrac{f(x + h) - f(x)}{h}$, $h \neq 0$.

7. Given $f(x) = 2x^2 + 5x - 3$, find (a) $f(-2)$; (b) $f(-1)$; (c) $f(0)$;
(d) $f(3)$; (e) $f(h + 1)$; (f) $f(2x^2)$; (g) $f(x^2 - 3)$; (h) $f(x + h)$;
(i) $f(x) + f(h)$; (j) $\dfrac{f(x + h) - f(x)}{h}$, $h \neq 0$.

8. Given $g(x) = 3x^2 - 4$, find (a) $g(-4)$; (b) $g(\tfrac{1}{2})$; (c) $g(x^2)$;
(d) $g(3x^2 - 4)$; (e) $g(x - h)$; (f) $g(x) - g(h)$; (g) $\dfrac{g(x + h) - g(x)}{h}$,
$h \neq 0$.

9. Given $F(x) = \sqrt{2x + 3}$, find (a) $F(-1)$; (b) $F(4)$; (c) $F(\frac{1}{2})$; (d) $F(11)$; (e) $F(2x + 3)$; (f) $\dfrac{F(x + h) - F(x)}{h}$, $h \neq 0$.

10. Given $G(x) = \sqrt{4 - x}$, find (a) $G(-5)$; (b) $G(0)$; (c) $G(1)$; (d) $G(\frac{11}{9})$; (e) $G(4 - x)$; (f) $\dfrac{G(x + h) - G(x)}{h}$; $h \neq 0$

In Exercises 11 through 20, define the following functions and determine the domain of the resulting function: (a) $f + g$; (b) $f - g$; (c) $f \cdot g$; (d) f/g; (e) g/f.

11. $f(x) = x - 5$; $g(x) = x^2 - 1$ 12. $f(x) = \sqrt{x}$; $g(x) = x^2 + 1$

13. $f(x) = \dfrac{x + 1}{x - 1}$; $g(x) = \dfrac{1}{x}$ 14. $f(x) = \sqrt{x}$; $g(x) = 4 - x^2$

15. $f(x) = \sqrt{x}$; $g(x) = x^2 - 1$ 16. $f(x) = |x|$; $g(x) = |x - 3|$

17. $f(x) = x^2 + 1$; $g(x) = 3x - 2$

18. $f(x) = \sqrt{x + 4}$; $g(x) = x^2 - 4$

19. $f(x) = \dfrac{1}{x + 1}$; $g(x) = \dfrac{x}{x - 2}$ 20. $f(x) = x^2$; $g(x) = \dfrac{1}{\sqrt{x}}$

In Exercises 21 through 30, define the following functions and determine the domain of the composite function: (a) $f \circ g$; (b) $g \circ f$; (c) $f \circ f$; (d) $g \circ g$.

21. $f(x) = x - 2$; $g(x) = x + 7$

22. $f(x) = 3 - 2x$; $g(x) = 6 - 3x$

23. The functions of Exercise 11.

24. The functions of Exercise 12.

25. $f(x) = \sqrt{x - 2}$; $g(x) = x^2 - 2$

26. $f(x) = x^2 - 1$; $g(x) = \dfrac{1}{x}$ 27. $f(x) = \dfrac{1}{x}$; $g(x) = \sqrt{x}$

28. $f(x) = \sqrt{x}$; $g(x) = -\dfrac{1}{x}$ 29. $f(x) = |x|$; $g(x) = |x + 2|$

30. $f(x) = \sqrt{x^2 - 1}$; $g(x) = \sqrt{x - 1}$

In Exercises 31 and 32, define the following functions, and determine the domain of the resulting function: (a) $f(x^2)$; (b) $[f(x)]^2$; (c) $(f \circ f)(x)$.

31. $f(x) = 2x - 3$ 32. $f(x) = \dfrac{2}{x - 1}$

33. Given $G(x) = |x - 2| - |x| + 2$, express $G(x)$ without absolute-value bars if x is in the given interval: (a) $[2, +\infty)$; (b) $(-\infty, 0)$; (c) $[0, 2)$.

34. Given $f(t) = \dfrac{|3 + t| - |t| - 3}{t}$, express $f(t)$ without absolute-value bars if t is in the given interval: (a) $(0, +\infty)$; (b) $[-3, 0)$; (c) $(-\infty, -3)$.

35. Given
$$f(x) = \begin{cases} \dfrac{|x|}{x} & \text{if } x \neq 0 \\ 1 & \text{if } x = 0 \end{cases}$$
find (a) $f(1)$; (b) $f(-1)$; (c) $f(4)$; (d) $f(-4)$; (e) $f(-x)$; (f) $f(x + 1)$; (g) $f(x^2)$; (h) $f(-x^2)$.

In each part of Exercises 36 and 37, determine whether the function is even, odd, or neither.

36. (a) $f(x) = 2x^4 - 3x^2 + 1$ (b) $f(x) = 5x^3 - 7x$
   (c) $f(s) = s^2 + 2s + 2$ (d) $g(x) = x^6 - 1$
   (e) $h(t) = 5t^7 + 1$ (f) $f(x) = |x|$
   (g) $f(y) = \dfrac{y^3 - y}{y^2 + 1}$ (h) $g(z) = \dfrac{z - 1}{z + 1}$

37. (a) $g(x) = 5x^2 - 4$ (b) $f(x) = x^3 + 1$
   (c) $f(t) = 4t^5 + 3t^3 - 2t$ (d) $g(r) = \dfrac{r^2 - 1}{r^2 + 1}$
   (e) $f(x) = \begin{cases} 1 & \text{if } x > 0 \\ -1 & \text{if } x < 0 \end{cases}$ (f) $h(x) = \dfrac{4x^2 - 5}{2x^3 + x}$
   (g) $f(z) = (z - 1)^2$ (h) $g(x) = \dfrac{|x|}{x^2 + 1}$
   (i) $g(x) = \sqrt{x^2 - 1}$ (j) $f(x) = \sqrt[3]{x}$

38. There is one function that is both even and odd. What is it?

39. Determine whether the composite function $f \circ g$ is odd or even in each of the following cases: (a) f and g are both even; (b) f and g are both odd; (c) f is even and g is odd; (d) f is odd and g is even.

If f and g are functions such that $(f \circ g)(x) = x$ and $(g \circ f)(x) = x$, then f and g are inverse functions. In Exercises 40 through 42, show that f and g are inverse functions.

40. $f(x) = 2x - 3$ and $g(x) = \dfrac{x + 3}{2}$

41. $f(x) = \dfrac{1}{x + 1}$ and $g(x) = \dfrac{1 - x}{x}$

42. $f(x) = x^2$, $x \geq 0$ and $g(x) = \sqrt{x}$

1.5 GRAPHS OF FUNCTIONS In preparation for our study of limits and continuity in Chapter 2 we now discuss graphs of functions. Recall from Section 1.4 that the graph of a function f is the same as the graph of the equation $y = f(x)$. While the domain of a function is usually apparent from the function's definition, the range is often determined from the function's graph.

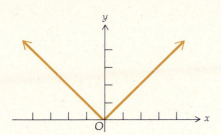

FIGURE 1

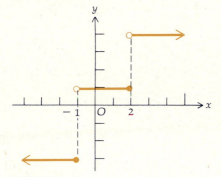

FIGURE 2

FIGURE 3

EXAMPLE 1   The **absolute value function** is defined by

$$f(x) = |x|$$

Determine its domain and range and draw a sketch of its graph.

Solution   From the definition (1.1.8) of $|x|$, we have

$$f(x) = \begin{cases} x & \text{if } x \geq 0 \\ -x & \text{if } x < 0 \end{cases}$$

The domain is $(-\infty, +\infty)$. The graph of f consists of two half lines through the origin and above the x axis; one has slope 1 and the other has slope -1. See Figure 1. The range is $[0, +\infty)$.

EXAMPLE 2   Let f be the function defined by

$$f(x) = \begin{cases} -3 & \text{if } x \leq -1 \\ 1 & \text{if } -1 < x \leq 2 \\ 4 & \text{if } 2 < x \end{cases}$$

Determine the domain and range of f and draw a sketch of its graph.

Solution   The domain is $(-\infty, +\infty)$, and a sketch of the graph appears in Figure 2. The range consists of the three numbers -3, 1, and 4.

EXAMPLE 3   Let g be the function defined by

$$g(x) = \begin{cases} 3x - 2 & \text{if } x < 1 \\ x^2 & \text{if } 1 \leq x \end{cases}$$

Determine the domain and range of g, and draw a sketch of its graph.

Solution   The domain of g is $(-\infty, +\infty)$. The graph contains the portion of the line $y = 3x - 2$ for which $x < 1$ and the portion of the parabola $y = x^2$ for which $1 \leq x$. A sketch of the graph is shown in Figure 3. The range is $(-\infty, +\infty)$.

EXAMPLE 4   The function h is defined by

$$h(x) = \frac{x^2 - 9}{x - 3}$$

Determine the domain and range of h, and draw a sketch of its graph.

Solution   Because $h(x)$ is defined for all x except 3, the domain of h is the set of all real numbers except 3. When $x = 3$, both the numerator and denominator are zero, and $0/0$ is undefined.

Factoring the numerator into $(x - 3)(x + 3)$ we obtain

$$h(x) = \frac{(x - 3)(x + 3)}{x - 3}$$

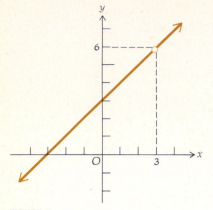

FIGURE 4

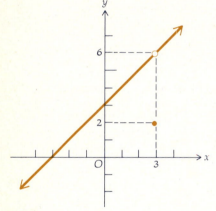

FIGURE 5

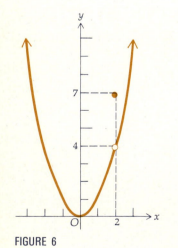

FIGURE 6

or $h(x) = x + 3$, provided that $x \neq 3$. In other words, the function h can be defined by

$$h(x) = x + 3 \quad \text{if } x \neq 3$$

The graph of h consists of all points on the line $y = x + 3$ except the point $(3, 6)$, and it appears in Figure 4. The range of h is the set of all real numbers except 6.

EXAMPLE 5   Let H be the function defined by

$$H(x) = \begin{cases} x + 3 & \text{if } x \neq 3 \\ 2 & \text{if } x = 3 \end{cases}$$

Determine the domain and range of H, and draw a sketch of its graph.

Solution   Because H is defined for all values of x, its domain is $(-\infty, +\infty)$. A sketch of the graph of H is shown in Figure 5. The range is the set of all real numbers except 6.

EXAMPLE 6   The function f is defined by

$$f(x) = \begin{cases} x^2 & \text{if } x \neq 2 \\ 7 & \text{if } x = 2 \end{cases}$$

Determine the domain and range of f, and draw a sketch of its graph.

Solution   Because f is defined for all values of x, the domain is $(-\infty, +\infty)$. The graph, appearing in Figure 6, consists of the point $(2, 7)$ and all points on the parabola $y = x^2$ except $(2, 4)$. The range is $[0, +\infty)$.

EXAMPLE 7   Let F be the function defined by

$$F(x) = \begin{cases} x - 1 & \text{if } x < 3 \\ 5 & \text{if } x = 3 \\ 2x + 1 & \text{if } 3 < x \end{cases}$$

Determine the domain and range of F, and draw a sketch of its graph.

Solution   The domain of F is $(-\infty, +\infty)$. Figure 7 shows a sketch of the graph of F; it consists of the portion of the line $y = x - 1$ for which $x < 3$, the point $(3, 5)$, and the portion of the line $y = 2x + 1$ for which $3 < x$. The function values are either numbers less than 2, the number 5, or numbers greater than 7. Therefore the range of F is the number 5 and those numbers in $(-\infty, 2) \cup (7, +\infty)$.

EXAMPLE 8   The function g is defined by

$$g(x) = \sqrt{x(x - 2)}$$

Determine the domain and range of g, and draw a sketch of its graph.

FIGURE 7

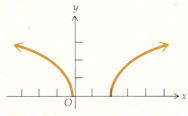

FIGURE 8

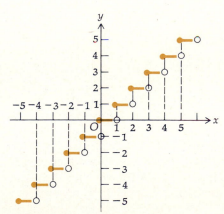

FIGURE 9

Solution   Because $\sqrt{x(x-2)}$ is not a real number when $x(x-2) < 0$, the domain of g consists of the values of x for which $x(x-2) \geq 0$. This inequality will be satisfied when one of the following two cases holds: $x \geq 0$ and $x - 2 \geq 0$; or $x \leq 0$ and $x - 2 \leq 0$.

Case 1: $x \geq 0$ and $x - 2 \geq 0$. That is,

$$x \geq 0 \quad \text{and} \quad x \geq 2$$

Both inequalities hold if $x \geq 2$, which is the interval $[2, +\infty)$.

Case 2: $x \leq 0$ and $x - 2 \leq 0$. That is,

$$x \leq 0 \quad \text{and} \quad x \leq 2$$

Both inequalities hold if $x \leq 0$, which is the interval $(-\infty, 0]$.

The solutions for the two cases are combined to obtain the domain of g. It is $(-\infty, 0] \cup [2, +\infty)$.

Figure 8 shows a sketch of the graph of g. The range is $[0, +\infty)$.

The symbol $[\![x]\!]$ is used to denote the greatest integer less than or equal to x; that is,

$$[\![x]\!] = n \quad \text{if } n \leq x < n + 1, \text{ where } n \text{ is an integer}$$

The function having values $[\![x]\!]$ is called the **greatest integer function**. Therefore, if I is this function

$$I = \{(x, y) \mid y = [\![x]\!]\}$$

and the domain of I is $(-\infty, +\infty)$. To obtain a sketch of the graph of I, we first compute some function values.

$$\text{If} \quad -5 \leq x < -4, \quad [\![x]\!] = -5$$
$$\text{If} \quad -4 \leq x < -3, \quad [\![x]\!] = -4$$
$$\text{If} \quad -3 \leq x < -2, \quad [\![x]\!] = -3$$
$$\text{If} \quad -2 \leq x < -1, \quad [\![x]\!] = -2$$
$$\text{If} \quad -1 \leq x < 0, \quad [\![x]\!] = -1$$
$$\text{If} \quad 0 \leq x < 1, \quad [\![x]\!] = 0$$
$$\text{If} \quad 1 \leq x < 2, \quad [\![x]\!] = 1$$
$$\text{If} \quad 2 \leq x < 3, \quad [\![x]\!] = 2$$
$$\text{If} \quad 3 \leq x < 4, \quad [\![x]\!] = 3$$
$$\text{If} \quad 4 \leq x < 5, \quad [\![x]\!] = 4$$

Figure 9 shows a sketch of the graph of I. The range is the set of all the integers.

EXAMPLE 9   Draw a sketch of the graph of the function defined by

$$G(x) = [\![x]\!] - x$$

State the domain and range of G.

Solution   Because G is defined for all values of x, its domain is $(-\infty, +\infty)$. From the definition of $[\![x]\!]$, we have the following:

If $-2 \le x < -1$, $[\![x]\!] = -2$;   therefore $G(x) = -2 - x$

If $-1 \le x < 0$, $[\![x]\!] = -1$;   therefore $G(x) = -1 - x$

If $0 \le x < 1$, $[\![x]\!] = 0$;   therefore $G(x) = -x$

If $1 \le x < 2$, $[\![x]\!] = 1$;   therefore $G(x) = 1 - x$

If $2 \le x < 3$, $[\![x]\!] = 2$;   therefore $G(x) = 2 - x$

and so on. More generally, if n is any integer, then

If $n \le x < n + 1$, $[\![x]\!] = n$;   therefore $G(x) = n - x$

With these function values we obtain the sketch of the graph of G appearing in Figure 10. From the graph we observe that the range of G is $(-1, 0]$.

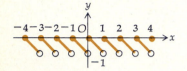

FIGURE 10

EXERCISES 1.5

In each exercise, determine the domain and range of the function and draw a sketch of its graph.

1. $f(x) = 3x - 1$

2. $g(x) = 6 - 2x$

3. $F(x) = x^2 - 1$

4. $G(x) = 5 - x^2$

5. $g(x) = \sqrt{x + 1}$

6. $f(x) = \sqrt{3x - 6}$

7. $f(x) = \sqrt{4 - 2x}$

8. $g(x) = \sqrt{9 - x^2}$

9. $h(x) = \sqrt{-x}$

10. $H(x) = |x - 1|$

11. $f(x) = |4 - x|$

12. $h(x) = |x| - 1$

13. $F(x) = 4 - |x|$

14. $f(x) = 5 - |x + 1|$

15. $g(x) = |x - 2| + 4$

16. $F(x) = \dfrac{4x^2 - 1}{2x + 1}$

17. $f(x) = \dfrac{x^2 - 4x + 3}{x - 1}$

18. $g(x) = \dfrac{x^3 - 3x^2 - 4x + 12}{x^2 - x - 6}$

19. $G(x) = \begin{cases} -2 & \text{if } x \le 3 \\ 2 & \text{if } 3 < x \end{cases}$

20. $h(x) = \begin{cases} -4 & \text{if } x < -2 \\ -1 & \text{if } -2 \le x \le 2 \\ 3 & \text{if } 2 < x \end{cases}$

21. $f(x) = \begin{cases} 2x - 1 & \text{if } x \ne 2 \\ 0 & \text{if } x = 2 \end{cases}$

22. $f(x) = \begin{cases} x^2 - 4 & \text{if } x \ne -3 \\ -2 & \text{if } x = -3 \end{cases}$

23. $H(x) = \begin{cases} x^2 - 4 & \text{if } x < 3 \\ 2x - 1 & \text{if } 3 \le x \end{cases}$

24. $\phi(x) = \begin{cases} x + 5 & \text{if } x < -5 \\ \sqrt{25 - x^2} & \text{if } -5 \le x \le 5 \\ x - 5 & \text{if } 5 < x \end{cases}$

25. $f(x) = \begin{cases} x + 6 & \text{if } x \le -4 \\ \sqrt{16 - x^2} & \text{if } -4 < x < 4 \\ 6 - x & \text{if } 4 \le x \end{cases}$

26. $g(x) = \begin{cases} 6x + 7 & \text{if } x < -2 \\ 3 & \text{if } x = -2 \\ 4 - x & \text{if } -2 < x \end{cases}$

27. $F(x) = \begin{cases} x - 2 & \text{if } x < 0 \\ 0 & \text{if } x = 0 \\ x^2 + 1 & \text{if } 0 < x \end{cases}$

28. $G(x) = \dfrac{(x^2 + 3x - 4)(x^2 - 5x + 6)}{(x^2 - 3x + 2)(x - 3)}$

29. $F(x) = \dfrac{(x + 1)(x^2 + 3x - 10)}{x^2 + 6x + 5}$

30. $h(x) = \sqrt{x^2 - 5x + 6}$ **31.** $f(x) = \sqrt{x^2 - 3x - 4}$

32. $f(x) = \dfrac{x^3 + 3x^2 + x + 3}{x + 3}$ **33.** $g(x) = \dfrac{x^3 - 2x^2}{x - 2}$

34. $F(x) = \dfrac{x^4 + x^3 - 9x^2 - 3x + 18}{x^2 + x - 6}$

35. $h(x) = \dfrac{x^3 + 5x^2 - 6x - 30}{x + 5}$

36. $g(x) = |x| \cdot |x - 1|$

37. $f(x) = |x| + |x - 1|$ **38.** $F(x) = [\![x + 2]\!]$

39. $g(x) = [\![x - 4]\!]$ **40.** $H(x) = |x| + [\![x]\!]$

41. $G(x) = x - [\![x]\!]$ **42.** $h(x) = [\![x^2]\!]$

43. $g(x) = \dfrac{[\![x]\!]}{|x|}$ **44.** $h(x) = \dfrac{|x|}{[\![x]\!]}$

1.6 THE TRIGONOMETRIC FUNCTIONS

We assume that you have studied trigonometry in a previous course. However, because of the importance of the trigonometric functions in calculus, a brief review of them is presented here.

In geometry an **angle** is defined as the union of two rays called the **sides**, having a common endpoint called the **vertex**. Any angle is congruent to some angle having its vertex at the origin and one side, called the **initial side**, lying on the positive side of the x axis. Such an angle is said to be in **standard position**. Figure 1 shows an angle AOB in standard position with OA as the initial side. The other side, OB, is called the **terminal side**. The angle AOB can be formed by rotating the side OA to the side OB, and under such a rotation the point A moves along the circumference of a circle having its center at O and radius $|\overline{OA}|$ to the point B.

In dealing with problems involving angles of triangles, the measurement of an angle is usually given in degrees. However, in calculus we are concerned with trigonometric functions of real numbers, and these functions are defined in terms of *radian measure*.

The length of an arc of a circle is used to define the radian measure of an angle.

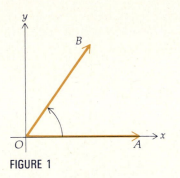

FIGURE 1

1.6.1 DEFINITION

Let AOB be an angle in standard position and $|\overline{OA}| = 1$. If s units is the length of the arc of the circle traveled by point A as the initial side OA is rotated to the terminal side OB, the **radian measure**, t, of angle AOB is given by

$t = s$ if the rotation is counterclockwise

and

$t = -s$ if the rotation is clockwise

▶ **ILLUSTRATION 1**   From the fact that the measure of the length of the unit circle's circumference is 2π, the radian measures of the angles in Figure 2(a)–(f) are determined. They are $\frac{1}{2}\pi$, $\frac{1}{4}\pi$, $-\frac{1}{2}\pi$, $\frac{3}{2}\pi$, $-\frac{3}{4}\pi$, and $\frac{7}{4}\pi$, respectively.   ◀

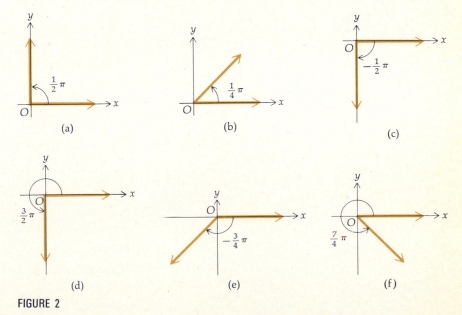

FIGURE 2

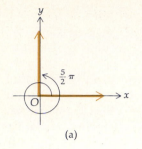

(a)

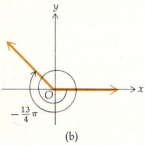

(b)

FIGURE 3

Table 1

Degree Measure	Radian Measure
30	$\frac{1}{6}\pi$
45	$\frac{1}{4}\pi$
60	$\frac{1}{3}\pi$
90	$\frac{1}{2}\pi$
120	$\frac{2}{3}\pi$
135	$\frac{3}{4}\pi$
150	$\frac{5}{6}\pi$
180	π
270	$\frac{3}{2}\pi$
360	2π

In Definition 1.6.1 it is possible that there may be more than one complete revolution in the rotation of OA.

▶ **ILLUSTRATION 2**　　Figure 3(a) shows an angle whose radian measure is $\frac{5}{2}\pi$, and Figure 3(b) shows one whose radian measure is $-\frac{13}{4}\pi$.　　◀

An angle formed by one complete revolution so that OA is coincident with OB has degree measure of 360 and radian measure of 2π. Hence there is the following correspondence between degree measure and radian measure (where the symbol $\sim$ indicates that the given measurements are for the same or congruent angles):

$$360° \sim 2\pi \text{ rad} \qquad 180° \sim \pi \text{ rad}$$

From this it follows that

$$1° \sim \tfrac{1}{180}\pi \text{ rad} \qquad 1 \text{ rad} \sim \frac{180°}{\pi}$$

$$\approx 57°18'$$

Note that the symbol $\approx$ before $57°18'$ indicates that 1 rad and approximately $57°18'$ are measurements for the same or congruent angles.

From this correspondence the measurement of an angle can be converted from one system of units to the other.

EXAMPLE 1　　(a) Find the equivalent radian measurement for $162°$; (b) find the equivalent degree measurement for $\frac{5}{12}\pi$ rad.

Solution

(a) $162° \sim 162 \cdot \tfrac{1}{180}\pi \text{ rad}$　　(b) $\tfrac{5}{12}\pi \text{ rad} \sim \dfrac{5}{12}\pi \cdot \dfrac{180°}{\pi}$

　　$162° \sim \tfrac{9}{10}\pi \text{ rad}$　　　　　　　$\tfrac{5}{12}\pi \text{ rad} \sim 75°$

Table 1 gives the corresponding degree and radian measures of certain angles. We now define the *sine* and *cosine* functions of any real number.

1.6.2 DEFINITION　　Suppose that t is a real number. Place an angle, having radian measure t, in standard position and let point P be at the intersection of the terminal side of the angle with the unit circle having its center at the origin. If P is the point (x, y), then the **sine** function is defined by

$$\sin t = y$$

and the **cosine** function is defined by

$$\cos t = x$$

From this definition, $\sin t$ and $\cos t$ are defined for any value of t. Therefore the domain of the sine and cosine is the set of all real numbers. Figure 4 shows the point $(\cos t, \sin t)$ when $0 < t < \frac{1}{2}\pi$, and Figure 5 shows the point $(\cos t, \sin t)$ when $-\frac{3}{2}\pi < t < -\pi$.

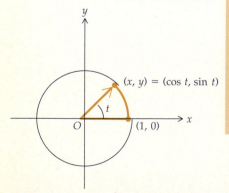

FIGURE 4

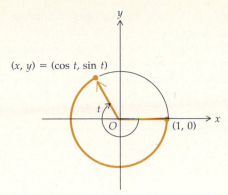

$(x, y) = (\cos t, \sin t)$

FIGURE 5

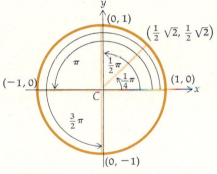

FIGURE 6

Table 2

t	$\sin t$	$\cos t$
0	0	1
$\frac{1}{6}\pi$	$\frac{1}{2}$	$\frac{1}{2}\sqrt{3}$
$\frac{1}{4}\pi$	$\frac{1}{2}\sqrt{2}$	$\frac{1}{2}\sqrt{2}$
$\frac{1}{3}\pi$	$\frac{1}{2}\sqrt{3}$	$\frac{1}{2}$
$\frac{1}{2}\pi$	1	0
$\frac{2}{3}\pi$	$\frac{1}{2}\sqrt{3}$	$-\frac{1}{2}$
$\frac{3}{4}\pi$	$\frac{1}{2}\sqrt{2}$	$-\frac{1}{2}\sqrt{2}$
$\frac{5}{6}\pi$	$\frac{1}{2}$	$-\frac{1}{2}\sqrt{3}$
π	0	-1
$\frac{3}{2}\pi$	-1	0
2π	0	1

The largest value either function may have is 1 and the smallest value is -1. We will show later that the sine and cosine functions assume all values between -1 and 1, and from this fact it follows that the range of the two functions is $[-1, 1]$.

For certain values of t, the sine and cosine are easily obtained from a figure. In Figure 6 we observe that $\sin 0 = 0$ and $\cos 0 = 1$, $\sin \frac{1}{4}\pi = \frac{1}{2}\sqrt{2}$ and $\cos \frac{1}{4}\pi = \frac{1}{2}\sqrt{2}$, $\sin \frac{1}{2}\pi = 1$ and $\cos \frac{1}{2}\pi = 0$, $\sin \pi = 0$ and $\cos \pi = -1$, $\sin \frac{3}{2}\pi = -1$ and $\cos \frac{3}{2}\pi = 0$. Table 2 gives these values and some others that are frequently used.

An equation of the unit circle having its center at the origin is $x^2 + y^2 = 1$. Because $x = \cos t$ and $y = \sin t$, it follows that

$$\sin^2 t + \cos^2 t = 1 \tag{1}$$

Note that $\sin^2 t$ and $\cos^2 t$ stand for $(\sin t)^2$ and $(\cos t)^2$. Equation (1) is an identity because it is valid for any real number t. It is called the **fundamental Pythagorean identity** showing the relationship between the sine and cosine values and can be used to compute one of them when the other is known.

Figures 7 and 8 show angles having a negative radian measure of $-t$ and corresponding angles having a positive radian measure of t. From these figures observe that

$$\sin(-t) = -\sin t \quad \text{and} \quad \cos(-t) = \cos t$$

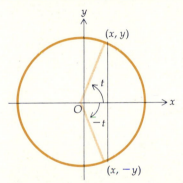

FIGURE 7

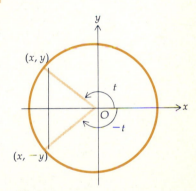

FIGURE 8

These equations hold for any real number t because the points where the terminal sides of the angles (having radian measures t and $-t$) intersect the unit circle have equal abscissas and ordinates that differ only in sign. Hence they are identities. From these identities it follows that the sine is an odd function and the cosine is an even function.

From Definition 1.6.2 we can obtain the following identities:

$$\sin(t + 2\pi) = \sin t \quad \text{and} \quad \cos(t + 2\pi) = \cos t \tag{2}$$

The property of sine and cosine stated by these two equations is called **periodicity**.

1.6.3 DEFINITION   A function f is said to be **periodic** if there exists a positive real number p such that whenever x is in the domain of f, then $x + p$ is also in the domain of f, and

$$f(x + p) = f(x)$$

The smallest such positive real number p is called the **period** of f.

Compare this definition with Equations (2). Because 2π can be shown to be the smallest positive number p having the property that $\sin(t + p) = \sin t$ and $\cos(t + p) = \cos t$, the sine and cosine are periodic with period 2π; that is, whenever the value of the independent variable t is increased by 2π, the value of each of the functions is repeated. It is because of the periodicity of the sine and cosine that these functions have important applications in connection with periodically repetitive phenomena, such as wave motion, alternating electrical current, vibrating strings, oscillating pendulums, business cycles, and biological rhythms.

EXAMPLE 2 Use the periodicity of the sine and cosine functions as well as the values of $\sin t$ and $\cos t$ when $0 \leq t < 2\pi$ to find an exact value of each of the following: (a) $\sin \frac{17}{4}\pi$; (b) $\cos \frac{7}{3}\pi$; (c) $\sin \frac{15}{2}\pi$; (d) $\cos(-\frac{7}{6}\pi)$.

Solution

(a) $\sin \frac{17}{4}\pi = \sin(\frac{1}{4}\pi + 2 \cdot 2\pi)$ (b) $\cos \frac{7}{3}\pi = \cos(\frac{1}{3}\pi + 2\pi)$
    $= \sin \frac{1}{4}\pi$ $= \cos \frac{1}{3}\pi$
    $= \frac{1}{2}\sqrt{2}$ $= \frac{1}{2}$

(c) $\sin \frac{15}{2}\pi = \sin(\frac{3}{2}\pi + 3 \cdot 2\pi)$ (d) $\cos(-\frac{7}{6}\pi) = \cos[\frac{5}{6}\pi + (-1)2\pi]$
    $= \sin \frac{3}{2}\pi$ $= \cos \frac{5}{6}\pi$
    $= -1$ $= -\frac{1}{2}\sqrt{3}$

We now define the other four trigonometric functions in terms of the sine and cosine.

1.6.4 DEFINITION The **tangent** and **secant** functions are defined by

$$\tan t = \frac{\sin t}{\cos t} \qquad \sec t = \frac{1}{\cos t}$$

for all real numbers t for which $\cos t \neq 0$.
  The **cotangent** and **cosecant** functions are defined by

$$\cot t = \frac{\cos t}{\sin t} \qquad \csc t = \frac{1}{\sin t}$$

for all real numbers t for which $\sin t \neq 0$.

The tangent and secant functions are not defined when $\cos t = 0$. Therefore the domain of the tangent and secant functions is the set of all real numbers except numbers of the form $\frac{1}{2}\pi + k\pi$, where k is any integer. Similarly, because $\cot t$ and $\csc t$ are not defined when $\sin t = 0$, the domain of the cotangent and cosecant functions is the set of all real numbers except numbers of the form $k\pi$, where k is any integer.
  It can be shown that the tangent and cotangent are periodic with period π; that is,

$$\tan(t + \pi) = \tan t \quad \text{and} \quad \cot(t + \pi) = \cot t$$

Furthermore, the secant and cosecant are periodic with period 2π; therefore

$$\sec(t + 2\pi) = \sec t \quad \text{and} \quad \csc(t + 2\pi) = \csc t$$

By using the fundamental Pythagorean identity (1) and Definition 1.6.4, we obtain two other important identities. One of these identities is obtained by dividing both sides of (1) by $\cos^2 t$, and the other is obtained by dividing both sides of (1) by $\sin^2 t$. We have

$$\frac{\sin^2 t}{\cos^2 t} + \frac{\cos^2 t}{\cos^2 t} = \frac{1}{\cos^2 t} \quad \text{and} \quad \frac{\sin^2 t}{\sin^2 t} + \frac{\cos^2 t}{\sin^2 t} = \frac{1}{\sin^2 t}$$

$$\tan^2 t + 1 = \sec^2 t \quad \text{and} \quad 1 + \cot^2 t = \csc^2 t$$

These two identities are also called Pythagorean identities.

Three other important identities that follow from Definition 1.6.4 are

$$\sin t \csc t = 1 \qquad \cos t \sec t = 1 \qquad \tan t \cot t = 1$$

These three identities, the three Pythagorean identities, and the two identities in Definition 1.6.4 that define the tangent and cotangent are the *eight fundamental trigonometric identities*. These as well as other formulas from trigonometry appear in the Appendix.

We have defined the trigonometric functions with real-number domains. There are important uses of trigonometric functions for which the domains are sets of angles. For these purposes, we define a trigonometric function of an angle θ as the corresponding function of the real number t, where t is the radian measure of θ.

1.6.5 DEFINITION   If θ is an angle having radian measure t, then

$$\sin \theta = \sin t \qquad \cos \theta = \cos t \qquad \tan \theta = \tan t$$

$$\cot \theta = \cot t \qquad \sec \theta = \sec t \qquad \csc \theta = \csc t$$

When considering a trigonometric function of an angle θ, often the measurement of the angle is used in place of θ. For instance, if the degree measure of an angle θ is 60 (or, equivalently, the radian measure of θ is $\frac{1}{3}\pi$), then in place of $\sin \theta$ we could write $\sin 60°$ or $\sin \frac{1}{3}\pi$. Notice that when the measurement of an angle is in degrees, the degree symbol is written. However, when there is no symbol attached, the measurement of the angle is in radians. For example, $\cos 2°$ means the cosine of an angle having degree measure 2, while $\cos 2$ means the cosine of an angle having radian measure 2. This is consistent with the fact that the cosine of an angle having radian measure 2 is equal to the cosine of the real number 2.

We now show how the tangent function can be used in connection with the slope of a line. We first define the *angle of inclination* of a line.

1.6.6 DEFINITION   The **angle of inclination** of a line not parallel to the x axis is the smallest angle measured counterclockwise from the positive direction of the x axis to the line. The angle of inclination of a line parallel to the x axis is defined to have measure zero.

If α is the angle of inclination of a line, then $0° \leq \alpha < 180°$. Figure 9 shows a line L for which $0° < \alpha < 90°$, and Figure 10 shows one for which $90° < \alpha < 180°$.

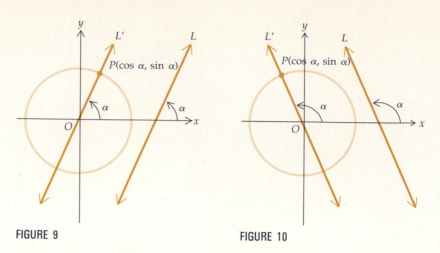

FIGURE 9   FIGURE 10

1.6.7 THEOREM   If α is the angle of inclination of a nonvertical line L, then the slope m of L is given by

$$m = \tan \alpha$$

Proof Refer to Figures 9 and 10, which show the given line L whose angle of inclination is α and whose slope is m. The line L' that passes through the origin and is parallel to L also has slope m and an angle of inclination α. The point $P(\cos \alpha, \sin \alpha)$ at the intersection of L' and the unit circle U lies on L'. And because the point $(0, 0)$ also lies on L', it follows from Definition 1.2.2 that

$$m = \frac{\sin \alpha - 0}{\cos \alpha - 0}$$

$$= \frac{\sin \alpha}{\cos \alpha}$$

$$= \tan \alpha \qquad \blacksquare$$

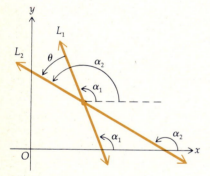

FIGURE 11

If line L is vertical, the angle of inclination of L is $90°$ and $\tan 90°$ does not exist. This is consistent with the fact that the slope of a vertical line is not defined.

Theorem 1.6.7 can be used to obtain a formula for finding the angle between two nonvertical intersecting lines. If two lines intersect, two supplementary angles are formed at their point of intersection. To distinguish these two angles let L_2 be the line with the greater angle of inclination α_2 and let L_1 be the other line whose angle of inclination is α_1. If θ is the angle between the two lines, then we define

$$\theta = \alpha_2 - \alpha_1$$

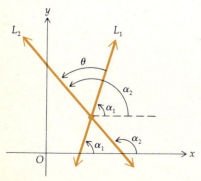

FIGURE 12

If L_1 and L_2 are parallel, then $\alpha_1 = \alpha_2$ and the angle between the two lines is $0°$. Thus if L_1 and L_2 are two distinct lines, then $0° \leq \theta < 180°$. Refer to Figures 11 and 12. The following theorem enables us to find θ when the slopes of L_1 and L_2 are known.

1.6.8 THEOREM

Let L_1 and L_2 be two nonvertical lines that intersect and are not perpendicular, and let L_2 be the line having the greater angle of inclination. Then if m_1 is the slope of L_1, m_2 is the slope of L_2, and θ is the angle between L_1 and L_2.

$$\tan \theta = \frac{m_2 - m_1}{1 + m_1 m_2}$$

The proof of Theorem 1.6.8 is left as an exercise (see Exercise 34).

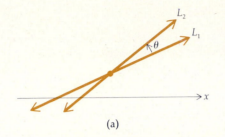

(a)

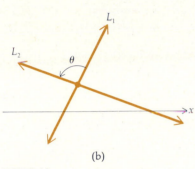

(b)

FIGURE 13

EXAMPLE 3　　Find the angle between the lines having the given slopes: (a) $\frac{3}{4}$ and $\frac{2}{5}$; (b) 2 and $-\frac{2}{5}$.

Solution　　When using the formula of Theorem 1.6.8, m_2 must be the slope of the line having the greater inclination. For part (a) refer to Figure 13(a) where $m_2 = \frac{3}{4}$ and $m_1 = \frac{2}{5}$. For part (b) refer to Figure 13(b) where $m_2 = -\frac{2}{5}$ and $m_1 = 2$. In each part we compute to the nearest degree the angle θ between the lines.

(a) $\tan \theta = \dfrac{m_2 - m_1}{1 + m_1 m_2}$　　　(b) $\tan \theta = \dfrac{m_2 - m_1}{1 + m_1 m_2}$

$\qquad = \dfrac{\frac{3}{4} - \frac{2}{5}}{1 + \frac{3}{4} \cdot \frac{2}{5}}$　　　　　$\qquad = \dfrac{-\frac{2}{5} - 2}{1 + (-\frac{2}{5})(2)}$

$\qquad = \dfrac{15 - 8}{20 + 6}$　　　　　$\qquad = \dfrac{-2 - 10}{5 - 4}$

$\qquad = \dfrac{7}{26}$　　　　　　$\qquad = -12$

$\qquad \theta = 15°$　　　　　　$\qquad \theta = 95°$

The procedure used in Example 3 can be applied to find the angles in a triangle. See Exercises 41 and 42.

The exercises of this section are intended as a review of some of the fundamental concepts of trigonometry. As you do them, you may find it helpful to consult a text in trigonometry.

EXERCISES 1.6

In Exercises 1 and 2, find the equivalent radian measurement.

1. (a) $60°$; (b) $135°$; (c) $210°$; (d) $-150°$; (e) $20°$; (f) $450°$; (g) $-75°$; (h) $100°$

2. (a) $45°$; (b) $120°$; (c) $240°$; (d) $-225°$; (e) $15°$; (f) $540°$; (g) $-48°$; (h) $2°$

In Exercises 3 and 4, find the equivalent degree measurement.

3. (a) $\frac{1}{4}\pi$ rad; (b) $\frac{2}{3}\pi$ rad; (c) $\frac{11}{6}\pi$ rad; (d) $-\frac{1}{2}\pi$ rad; (e) $\frac{1}{2}$ rad; (f) 3π rad; (g) -2 rad; (h) $\frac{1}{12}\pi$ rad

4. (a) $\frac{1}{6}\pi$ rad; (b) $\frac{4}{3}\pi$ rad; (c) $\frac{3}{4}\pi$ rad; (d) -5π rad; (e) $\frac{1}{3}$ rad; (f) -5 rad; (g) $\frac{11}{12}\pi$ rad; (h) 0.2 rad

In Exercises 5 through 12, determine the exact function value.

5. (a) $\sin \frac{1}{6}\pi$; (b) $\cos \frac{1}{4}\pi$; (c) $\sin(-\frac{3}{2}\pi)$; (d) $\cos \frac{1}{3}\pi$

6. (a) $\cos \frac{1}{3}\pi$; (b) $\sin \frac{1}{4}\pi$; (c) $\cos(-\frac{1}{2}\pi)$; (d) $\sin(-2\pi)$

7. (a) $\cos \frac{5}{6}\pi$; (b) $\sin \frac{3}{4}\pi$; (c) $\cos 3\pi$; (d) $\sin(-5\pi)$

8. (a) $\sin \frac{4}{3}\pi$; (b) $\cos(-\frac{1}{6}\pi)$; (c) $\sin 7\pi$; (d) $\cos(-\frac{5}{2}\pi)$

9. (a) $\tan \frac{1}{3}\pi$; (b) $\cot \frac{1}{4}\pi$; (c) $\sec(-\pi)$; (d) $\csc \frac{1}{2}\pi$

10. (a) $\cot \frac{1}{6}\pi$; (b) $\tan \frac{1}{4}\pi$; (c) $\csc(-\frac{3}{2}\pi)$; (d) $\sec \pi$

11. (a) $\sec(-\frac{1}{6}\pi)$; (b) $\csc \frac{3}{4}\pi$; (c) $\tan \frac{5}{6}\pi$; (d) $\cot(-\frac{3}{4}\pi)$

12. (a) $\csc(-\frac{1}{3}\pi)$; (b) $\sec \frac{5}{6}\pi$; (c) $\tan \frac{3}{4}\pi$; (d) $\cot \frac{3}{2}\pi$

In Exercises 13 through 20, use the periodicity of the sine, cosine, secant, and cosecant functions as well as the values of $\sin t$, $\cos t$, $\sec t$, and $\csc t$ when $0 \le t < 2\pi$ to find the exact function value.

13. (a) $\sin \frac{9}{4}\pi$; (b) $\cos \frac{9}{4}\pi$; (c) $\sec \frac{9}{4}\pi$; (d) $\csc \frac{9}{4}\pi$

14. (a) $\sin \frac{17}{6}\pi$; (b) $\cos \frac{17}{6}\pi$; (c) $\sec \frac{17}{6}\pi$; (d) $\csc \frac{17}{6}\pi$

15. (a) $\sin(-\frac{2}{3}\pi)$; (b) $\cos(-\frac{2}{3}\pi)$; (c) $\sec(-\frac{2}{3}\pi)$; (d) $\csc(-\frac{2}{3}\pi)$

16. (a) $\sin(-\frac{5}{4}\pi)$; (b) $\cos(-\frac{5}{4}\pi)$; (c) $\sec(-\frac{5}{4}\pi)$; (d) $\csc(-\frac{5}{4}\pi)$

17. (a) $\sin 8\pi$; (b) $\cos 10\pi$; (c) $\sec 7\pi$; (d) $\csc 9\pi$

18. (a) $\sin \frac{7}{2}\pi$; (b) $\cos \frac{5}{2}\pi$; (c) $\sec \frac{11}{2}\pi$; (d) $\csc \frac{9}{2}\pi$

19. (a) $\sin(-\frac{7}{2}\pi)$; (b) $\cos(-\frac{5}{2}\pi)$; (c) $\sec(-\frac{11}{2}\pi)$; (d) $\csc(-\frac{9}{2}\pi)$

20. (a) $\sin(-8\pi)$; (b) $\cos(-10\pi)$; (c) $\sec(-7\pi)$; (d) $\csc(-9\pi)$

In Exercises 21 through 24, use the periodicity of the tangent and cotangent functions as well as the values of $\tan t$ and $\cot t$ when $0 \le t < \pi$ to find the exact function value.

21. (a) $\tan \frac{7}{4}\pi$; (b) $\cot \frac{7}{4}\pi$; (c) $\tan(-\frac{5}{6}\pi)$; (d) $\cot(-\frac{5}{6}\pi)$

22. (a) $\tan \frac{4}{3}\pi$; (b) $\cot \frac{4}{3}\pi$; (c) $\tan(-\frac{1}{6}\pi)$; (d) $\cot(-\frac{1}{6}\pi)$

23. (a) $\tan \frac{11}{3}\pi$; (b) $\cot \frac{11}{3}\pi$; (c) $\tan(-5\pi)$; (d) $\cot(-\frac{9}{2}\pi)$

24. (a) $\tan(-\frac{11}{4}\pi)$; (b) $\cot(-\frac{11}{4}\pi)$; (c) $\tan 11\pi$; (d) $\cot \frac{15}{2}\pi$

In Exercises 25 through 30, find all values of t in the interval $[0, 2\pi)$ for which the equation is satisfied.

25. (a) $\sin t = 1$; (b) $\cos t = -1$; (c) $\tan t = 1$; (d) $\sec t = 1$

26. (a) $\sin t = -1$; (b) $\cos t = 1$; (c) $\tan t = -1$; (d) $\csc t = 1$

27. (a) $\sin t = 0$; (b) $\cos t = 0$; (c) $\tan t = 0$; (d) $\cot t = 0$

28. (a) $\sin t = \frac{1}{2}$; (b) $\cos t = -\frac{1}{2}$; (c) $\cot t = 1$; (d) $\sec t = 2$

29. (a) $\sin t = -\frac{1}{2}$; (b) $\cos t = \frac{1}{2}$; (c) $\cot t = -1$; (d) $\csc t = 2$

30. (a) $\sin t = -\frac{1}{2}\sqrt{2}$; (b) $\cos t = \frac{1}{2}\sqrt{2}$; (c) $\tan t = -\frac{1}{3}\sqrt{3}$; (d) $\cot t = \frac{1}{3}\sqrt{3}$

31. For what values of t in $[0, 2\pi)$ is (a) $\tan t$ undefined and (b) $\csc t$ undefined?

32. For what values of t in $[0, \pi)$ is (a) $\cot t$ undefined and (b) $\sec t$ undefined?

33. For what values of t in $[\pi, 2\pi)$ is (a) $\cot t$ undefined and (b) $\sec t$ undefined?

34. Prove Theorem 1.6.8 (*Hint:* Use the formula for $\tan(u - v)$ found in the Appendix.)

In Exercises 35 through 38, find $\tan \theta$ if θ is the angle between the lines having the given slopes.

35. (a) 1 and $\frac{1}{4}$; (b) 4 and $-\frac{5}{3}$ **36.** (a) $\frac{1}{2}$ and $-\frac{3}{4}$; (b) $\frac{2}{7}$ and $\frac{7}{2}$

37. (a) $-\frac{1}{2}$ and $\frac{2}{3}$; (b) $-\frac{5}{4}$ and $-\frac{7}{5}$

38. (a) $-\frac{3}{5}$ and 2; (b) $-\frac{1}{3}$ and $-\frac{1}{10}$

In Exercises 39 and 40, find to the nearest degree the measurement of the angle between the lines having the given slopes.

39. (a) 5 and $-\frac{7}{9}$; (b) $-\frac{3}{2}$ and $-\frac{3}{4}$

40. (a) -3 and 2; (b) $\frac{3}{2}$ and $\frac{1}{4}$

41. Find to the nearest degree the measurements of the interior angles of the triangle formed by the lines that have equations $2x + y - 6 = 0$, $3x - y - 4 = 0$, and $3x + 4y + 8 = 0$.

42. Find to the nearest degree the measurements of the interior angles of the triangle having vertices at $(1, 0)$, $(-3, 2)$, and $(2, 3)$.

43. Find an equation of a line through the point $(-1, 4)$ making an angle of radian measure $\frac{1}{4}\pi$ with the line having equation $2x + y - 5 = 0$ (two solutions).

44. Find an equation of a line through the point $(-3, -2)$ making an angle of radian measure $\frac{1}{3}\pi$ with the line having equation $3x - 2y - 7 = 0$ (two solutions).

In Exercises 45 through 48, define the function $f \circ g$ and determine its domain.

45. (a) $f(x) = \sin x$, $g(x) = 3x$; (b) $f(x) = \tan x$, $g(x) = \dfrac{x}{2}$

46. (a) $f(x) = \cos x$, $g(x) = x^2$; (b) $f(x) = \csc x$, $g(x) = 2x$

47. (a) $f(x) = \cot x$, $g(x) = \dfrac{1}{x}$; (b) $f(x) = \sec \dfrac{1}{x}$, $g(x) = \dfrac{1}{x - \pi}$

48. (a) $f(x) = \sin x$, $g(x) = \dfrac{1}{2x}$; (b) $f(x) = \tan x$, $g(x) = x + \pi$

REVIEW EXERCISES FOR CHAPTER 1

In Exercises 1 through 12, find and show on the real-number line the solution set of the inequality.

1. $3x - 7 \le 5x - 17$

2. $8 < 5x + 4 \le 10$

3. $\dfrac{x}{x - 1} > \dfrac{1}{4}$

4. $\dfrac{3}{x + 4} < \dfrac{2}{x - 5}$

5. $2x^2 + x < 3$

6. $|x - 2| < 4$

7. $|3 + 4x| \le 9$

8. $|5 - 2x| \le 3$

9. $|4 - 3x| < 8$

10. $|2x - 5| > 7$

11. $|2x + 7| > 5$

12. $\left| \dfrac{2 - 3x}{3 + x} \right| \ge \dfrac{1}{4}$

In Exercises 13 through 15, solve for x.

13. $|3 - 4x| = 15$

14. $\left| \dfrac{2x - 1}{x + 3} \right| = 4$

15. $|3x - 4| = |6 - 2x|$

16. Find all the values of x for which $\sqrt{x^2 + 2x - 15}$ is real.

17. Find the abscissas of the points having ordinate 4 that are at a distance of $\sqrt{117}$ from the point $(5, -2)$

18. Find an equation that must be satisfied by the coordinates of any point (x, y) whose distance from $(-1, 2)$ is equal to its distance from $(3, -4)$. Draw a sketch of the graph of this equation.

19. Find an equation that must be satisfied by the coordinates of any point whose distance from the point $(-3, 4)$ is 10.

20. Define the following sets of points by either an equation or an inequality: (a) the point $(3, -5)$; (b) the set of all points whose distance from the point $(3, -5)$ is less than 4; (c) the set of all points whose distance from the point $(3, -5)$ is at least 5.

In Exercises 21 through 26, draw a sketch of the graph of the equation.

21. $y^2 = x - 4$

22. $y = |x - 4|$

23. $y = x^2 - 4$

24. $y = \sqrt{4 - x}$

25. $y = \sqrt{16 - x^2}$

26. $xy = 16$

27. Prove that the quadrilateral having vertices at $(1, 2)$, $(5, -1)$, $(11, 7)$, and $(7, 10)$ is a rectangle.

28. Prove that the triangle with vertices at $(-8, 1)$, $(-1, -6)$, and $(2, 4)$ is isosceles, and find its area.

29. Prove that the points $(2, 4)$, $(1, -4)$, and $(5, -2)$ are the vertices of a right triangle, and find the area of the triangle.

30. Prove that the points $(1, -1)$, $(3, 2)$, and $(7, 8)$ are collinear in two ways: (a) by using the distance formula; (b) by using slopes.

31. Two vertices of a parallelogram are at $(-3, 4)$ and $(2, 3)$, and its center is at $(0, -1)$. Find the other two vertices.

32. Two opposite vertices of a square are at $(3, -4)$ and $(9, -4)$. Find the other two vertices.

33. Prove that the points $(2, 13)$, $(-2, 5)$, $(3, -1)$, and $(7, 7)$ are the vertices of a parallelogram.

34. Find an equation of the circle having its center at $(-2, 4)$ and radius $\sqrt{5}$. Write the equation in the general form.

35. Find an equation of the circle having the points $(-3, 2)$ and $(5, 6)$ as endpoints of a diameter.

36. Find an equation of the circle through the points $(3, -1)$, $(2, 2)$, and $(-4, 5)$.

37. Find the center and radius of the circle having the equation $4x^2 + 4y^2 - 12x + 8y + 9 = 0$.

38. Find an equation of the line through the points $(2, -4)$ and $(7, 3)$, and write the equation in slope-intercept form.

39. Find an equation of the line through the point $(-1, 6)$ and having slope 3.

40. Find an equation of the line that is the perpendicular bisector of the line segment from $(-1, 5)$ to $(3, 2)$.

41. Find an equation of the line through the point $(5, -3)$ and perpendicular to the line whose equation is $2x - 5y = 1$.

42. Find an equation of the circle having as its diameter the common chord of the two circles $x^2 + y^2 + 2x - 2y - 14 = 0$ and $x^2 + y^2 - 4x + 4y - 2 = 0$.

43. Find an equation of the circle circumscribed about the triangle having sides on the lines $x - 3y + 2 = 0$, $3x - 2y + 6 = 0$, and $2x + y - 3 = 0$.

44. Find the shortest distance from the point $(2, -5)$ to the line having the equation $3x + y = 2$.

45. Find an equation of the line through the point of intersection of the lines $5x + 6y - 4 = 0$ and $x - 3y + 2 = 0$ and perpendicular to the line $x - 4y - 20 = 0$ without finding the point of intersection of the two lines. (*Hint:* See Exercise 58 in Exercises 1.2.)

46. The sides of a parallelogram are on the line $x + 2y = 10$, $3x - y = -20$, $x + 2y = 15$, and $3x - y = -10$. Find equations of the diagonals without finding the vertices of the parallelogram. (*Hint:* See Exercise 58 in Exercises 1.2.)

47. Determine all values of k for which the graphs of the two equations $x^2 + y^2 = k$ and $x + y = k$ intersect.

48. Determine the values of k and h if $3x + ky + 2 = 0$ and $5x - y + h = 0$ are equations of the same line.

49. Given $f(x) = 3x^2 - x + 5$, find (a) $f(-3)$; (b) $f(-x^2)$; (c) $\dfrac{f(x + h) - f(x)}{h}$, $h \neq 0$.

50. Given $g(x) = \sqrt{x - 1}$, find $\dfrac{g(x + h) - g(x)}{h}$, $h \neq 0$.

51. Determine if the function is even, odd, or neither:
(a) $f(x) = 2x^3 - 3x$; (b) $g(x) = 5x^4 + 2x^2 - 1$;
(c) $h(x) = 3x^5 - 2x^3 + x^2 - x$; (d) $F(x) = \dfrac{x^2 + 1}{x^3 - x}$.

In Exercises 52 through 56, define the following functions and determine the domain of the resulting function: (a) $f + g$; (b) $f - g$; (c) $f \cdot g$; (d) f/g; (e) g/f; (f) $f \circ g$; (g) $g \circ f$.

52. $f(x) = \sqrt{x + 2}$ and $g(x) = x^2 + 4$

53. $f(x) = x^2 - 4$ and $g(x) = 4x - 3$

54. $f(x) = x^2 - 9$ and $g(x) = \sqrt{x + 5}$

55. $f(x) = \dfrac{1}{x - 3}$ and $g(x) = \dfrac{x}{x + 1}$

56. $f(x) = \sqrt{x}$ and $g(x) = \dfrac{1}{x^2}$

In Exercises 57 through 64, determine the domain and range of the function and draw a sketch of its graph.

57. $f(x) = |3 - x|$

58. $f(x) = 3 - |x|$

59. $g(x) = \dfrac{x^2 + 2x - 8}{x - 2}$

60. $G(x) = \begin{cases} x + 4 & \text{if } x \neq 2 \\ 1 & \text{if } x = 2 \end{cases}$

61. $f(x) = \begin{cases} x^2 - 1 & \text{if } x < 1 \\ 2x + 3 & \text{if } 1 \leq x \end{cases}$

62. $g(x) = \dfrac{x^3 + 3x^2 - 4x - 12}{x^2 + x - 6}$

63. $F(x) = \begin{cases} x + 3 & \text{if } x < -2 \\ 4 - x^2 & \text{if } -2 \leq x \leq 2 \\ 3 - x & \text{if } 2 < x \end{cases}$

64. $f(x) = 2x - [\![x]\!]$

In Exercises 65 and 66, determine the exact function value.

65. (a) $\sin \frac{2}{3}\pi$; (b) $\cos \frac{5}{4}\pi$; (c) $\tan(-\frac{5}{6}\pi)$; (d) $\cot \frac{13}{12}\pi$; (e) $\sec \pi$; (f) $\csc(-\frac{1}{8}\pi)$

66. (a) $\cos \frac{7}{6}\pi$; (b) $\sin(-\frac{7}{4}\pi)$; (c) $\cot(-\frac{1}{2}\pi)$; (d) $\tan \frac{5}{8}\pi$; (e) $\sec(-\frac{1}{12}\pi)$; (f) $\csc \frac{5}{3}\pi$

In Exercises 67 and 68, find all values of t in the interval $[0, 2\pi)$ for which the equation is satisfied.

67. (a) $\sin t = \frac{1}{2}$; (b) $\cos t = 1$; (c) $\tan t = -1$; (d) $\cot t = \sqrt{3}$; (e) $\sec t = -2$; (f) $\csc t = \sqrt{2}$

68. (a) $\sin t = \frac{1}{2}\sqrt{3}$; (b) $\cos t = -1$; (c) $\tan t = -\sqrt{3}$; (d) $\cot t = 1$; (e) $\sec t = -\sqrt{2}$; (f) $\csc t = 2$

69. Find to the nearest degree the measurements of the four interior angles of the quadrilateral having vertices at $(5, 6)$, $(-2, 4)$, $(-2, 1)$, and $(3, 1)$, and verify that the sum is $360°$.

70. Find an equation of a line through the point $(2, 5)$ making an angle of radian measure $\frac{1}{4}\pi$ with the line having equation $3x - 4y + 7 = 0$ (two solutions).

71. Find equations of the lines through the origin that are tangent to the circle having its center at $(2, 1)$ and radius 2.

72. Prove that the two lines

$$A_1 x + B_1 y + C_1 = 0 \quad \text{and} \quad A_2 x + B_2 y + C_2 = 0$$

are parallel if and only if $A_1 B_2 - A_2 B_1 = 0$.

73. Use analytic geometry to prove that if the diagonals of a rectangle are perpendicular, then the rectangle is a square.

74. Use analytic geometry to prove that the line segment joining the midpoints of any two sides of a triangle is parallel to the third side and that its length is one-half the length of the third side.

75. In a triangle the point of intersection of the medians, the point of intersection of the altitudes, and the center of the circumscribed circle are collinear. Find these three points and prove that they are collinear for the triangle having vertices at $(2, 8)$, $(5, -1)$, and $(6, 6)$.

76. Use analytic geometry to prove that the set of points equidistant from two given points is the perpendicular bisector of the line segment joining the two points.

77. Prove that if x is any real number, $|x| < x^2 + 1$.

78. Use analytic geometry to prove that the three medians of any triangle meet in a point.

T W O

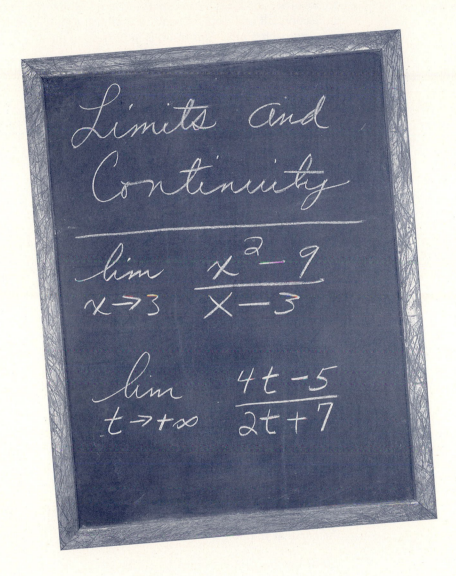

Limits and Continuity

$$\lim_{x \to 3} \frac{x^2 - 9}{x - 3}$$

$$\lim_{t \to +\infty} \frac{4t - 5}{2t + 7}$$

The two fundamental mathematical operations in calculus are *differentiation* and *integration*. These operations involve the computation of the *derivative* and the *definite integral*, each of which is based on the notion of *limit*.

In Section 2.1 the idea of the limit of a function is first given a step-by-step motivation, which brings the discussion from computing the value of a function near a number, through an intuitive treatment of the limiting process, up to a rigorous epsilon-delta definition. Limit theorems are introduced in Section 2.2 to simplify the computation of limits of elementary functions. In Section 2.3 the concept of limit is extended to include additional types of functions. Limits involving infinity are treated in Sections 2.4 and 2.5 and are used to define vertical and horizontal asymptotes of graphs of functions.

55

Probably the most important class of functions studied in calculus are *continuous* functions. Continuity of a function at a number is defined in Section 2.6, while continuity of a composite function and continuity on an interval are the topics of Section 2.7. The *squeeze theorem*, a key theorem in calculus, is presented in Section 2.8 and applied there to establish the limit of the ratio of sin *t* to *t* as *t* approaches zero. This result is important in the discussion of the continuity of the trigonometric functions in Section 2.8.

The final two sections of the chapter, 2.9 and 2.10, are supplementary and contain proofs of some of the theorems on limits of functions.

2.1 THE LIMIT OF A FUNCTION

To begin our discussion of limits, let us consider a particular function:

$$f(x) = \frac{2x^2 + x - 3}{x - 1} \tag{1}$$

Observe that $f(x)$ exists for all x except $x = 1$. We shall investigate the function values when x is close to 1 but not equal to 1. The following illustration shows how the function defined by (1) could arise and why we would wish to consider such function values.

▶ **ILLUSTRATION 1**   The point $P(1, 2)$ is on the curve having the equation

$$y = 2x^2 + x - 1$$

FIGURE 1

Let $Q(x, 2x^2 + x - 1)$ be another point on this curve, distinct from P. Figures 1 and 2 each show a portion of the graph of the equation and the secant line through Q and P where Q is near P. In Figure 1 the x coordinate of Q is less than 1, and in Figure 2 it is greater than 1. Suppose $f(x)$ is the slope of the line PQ. Then

$$f(x) = \frac{(2x^2 + x - 1) - 2}{x - 1}$$

$$f(x) = \frac{2x^2 + x - 3}{x - 1}$$

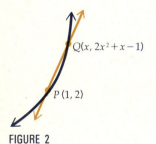

FIGURE 2

which is Equation (1). Furthermore, $x \neq 1$ because P and Q are distinct points. As x gets closer and closer to 1, the values of $f(x)$ get closer and closer to the number that we shall define in Section 3.1 as the slope of the tangent line to the curve at the point P.   ◀

For the function f defined by (1) let x take on the values 0, 0.25, 0.50, 0.75, 0.9, 0.99, 0.999, 0.9999, and so on. We are taking values of x closer and closer to 1 but less than 1; in other words, the variable x is approaching 1 through numbers that are less than 1. Table 1 gives the function values of these numbers obtained from a calculator. Now let the variable x approach 1 through numbers that are greater than 1; that is, let x take on the values 2, 1.75, 1.5, 1.25, 1.1, 1.01, 1.001, 1.0001, 1.00001, and so on. The function values of these numbers are also obtained from a calculator and they appear in Table 2.

Observe from both tables that as x gets closer and closer to 1, $f(x)$ gets closer and closer to 5; and the closer x is to 1, the closer $f(x)$ is to 5. For instance, from Table 1, when $x = 0.9$, $f(x) = 4.8$; that is, when x is 0.1 less than 1, $f(x)$ is 0.2 less than 5. When $x = 0.999$, $f(x) = 4.998$; that is, when x is 0.001 less than 1, $f(x)$ is 0.002 less than 5. Furthermore, when $x = 0.9999$, $f(x) = 0.49998$; that is, when x is 0.0001 less than 1, $f(x)$ is 0.0002 less than 5.

Table 1

x	$f(x) = \dfrac{2x^2 + x - 3}{x - 1}$
0	3
0.25	3.5
0.5	4
0.75	4.5
0.9	4.8
0.99	4.98
0.999	4.998
0.9999	4.9998
0.99999	4.99998

Table 2

x	$f(x) = \dfrac{2x^2 + x - 3}{x - 1}$
2	7
1.75	6.5
1.5	6.0
1.25	5.5
1.1	5.2
1.01	5.02
1.001	5.002
1.0001	5.0002
1.00001	5.00002

Table 2 shows that when $x = 1.1$, $f(x) = 5.2$; that is, when x is 0.1 greater than 1, $f(x)$ is 0.2 greater than 5. When $x = 1.001$, $f(x) = 5.002$; that is, when x is 0.001 greater than 1, $f(x)$ is 0.002 greater than 5. When $x = 1.0001$, $f(x) = 5.0002$; that is, when x is 0.0001 greater than 1, $f(x)$ is 0.0002 greater than 5.

Therefore, from the two tables we see that when x differs from 1 by ± 0.001 (i.e., $x = 0.999$ or $x = 1.001$), $f(x)$ differs from 5 by ± 0.002 (i.e., $f(x) = 4.998$ or $f(x) = 5.002$). And when x differs from 1 by $+0.0001$, $f(x)$ differs from 5 by ± 0.0002.

Now, looking at the situation another way, we consider the values of $f(x)$ first. We see that we can make the value of $f(x)$ as close to 5 as we please by taking x close enough to 1. Another way of saying this is that we can make the absolute value of the difference between $f(x)$ and 5 as small as we please by making the absolute value of the difference between x and 1 small enough. That is, $|f(x) - 5|$ can be made as small as we please by making $|x - 1|$ small enough. But bear in mind that $f(x)$ never takes on the value 5.

A more precise way of noting this is by using two symbols for these small differences. The symbols usually used are the Greek letters ϵ (epsilon) and δ (delta). So we state that for any given positive number ϵ there is an appropriately chosen positive number δ such that, if $|x - 1|$ is less than δ and $|x - 1| \neq 0$ (i.e., $x \neq 1$), then $|f(x) - 5|$ will be less than ϵ. It is important to realize that the size of δ depends on the size of ϵ. Still another way of phrasing this is: Given any positive number ϵ, we can make $|f(x) - 5| < \epsilon$ by taking $|x - 1|$ small enough; that is, there is some sufficiently small positive number δ such that

$$\text{if}\quad 0 < |x - 1| < \delta \quad \text{then} \quad |f(x) - 5| < \epsilon \qquad (2)$$

Observe that the numerator of the fraction in (1) can be factored so that

$$f(x) = \frac{(2x + 3)(x - 1)}{x - 1}$$

If $x \neq 1$, the numerator and denominator can be divided by $x - 1$ to obtain

$$f(x) = 2x + 3 \qquad x \neq 1 \qquad (3)$$

Equation (3) with the stipulation that $x \neq 1$ is just as suitable as (1) for a definition of $f(x)$. From (3) and the two tables notice that if $|x - 1| < 0.1$, then $|f(x) - 5| < 0.2$. Thus given $\epsilon = 0.2$, we take $\delta = 0.1$ and state that

$$\text{if}\quad 0 < |x - 1| < 0.1 \quad \text{then} \quad |f(x) - 5| < 0.2$$

This is statement (2) with $\epsilon = 0.2$ and $\delta = 0.1$.

Also, if $|x - 1| < 0.001$, then $|f(x) - 5| < 0.002$. Hence if $\epsilon = 0.002$, we take $\delta = 0.001$ and state that

$$\text{if}\quad 0 < |x - 1| < 0.001 \quad \text{then} \quad |f(x) - 5| < 0.002$$

This is statement (2) with $\epsilon = 0.002$ and $\delta = 0.001$.

Similarly, if $\epsilon = 0.0002$, we take $\delta = 0.0001$ and state that

$$\text{if}\quad 0 < |x - 1| < 0.0001 \quad \text{then} \quad |f(x) - 5| < 0.0002$$

This is statement (2) with $\epsilon = 0.0002$ and $\delta = 0.0001$.

We could go on and give ϵ any small positive value and find a suitable value for δ such that if $|x - 1|$ is less than δ and $x \neq 1$ (or $|x - 1| > 0$), then $|f(x) - 5|$ will be less than ϵ. Now, because for any $\epsilon > 0$ we can find a $\delta > 0$ such that if $0 < |x - 1| < \delta$ then $|f(x) - 5| < \epsilon$, we state that the limit of $f(x)$ as x approaches

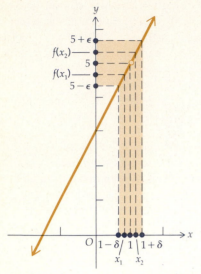

FIGURE 3

1 is equal to 5, or, expressed with symbols,

$$\lim_{x \to 1} f(x) = 5$$

Observe that in this equation we have a new use of the "equals" symbol. Here for no value of x does $f(x)$ have the value 5. The "equals" symbol is appropriate because the left side is written as $\lim_{x \to 1} f(x)$.

From (3) it is apparent that $f(x)$ can be made as close to 5 as we please by taking x sufficiently close to 1, and this property of the function f does not depend on f being defined when $x = 1$. This fact gives the distinction between $\lim_{x \to 1} f(x)$ and the function value at 1; that is, $\lim_{x \to 1} f(x) = 5$, but $f(1)$ does not exist. Consequently, in statement (2), $0 < |x - 1|$ because we are concerned only with values of $f(x)$ for x close to 1, but not for $x = 1$.

Let us see what all this means geometrically for the particular function defined by either (1) or (3). Figure 3 illustrates the geometric significance of ϵ and δ. Observe that if x on the horizontal axis lies between $1 - \delta$ and $1 + \delta$, then $f(x)$ on the vertical axis will lie between $5 - \epsilon$ and $5 + \epsilon$; or

$$\text{if}\quad 0 < |x - 1| < \delta \quad \text{then}\quad |f(x) - 5| < \epsilon$$

Another way of stating this is that $f(x)$ on the vertical axis can be restricted to lie between $5 - \epsilon$ and $5 + \epsilon$ by restricting x on the horizontal axis to lie between $1 - \delta$ and $1 + \delta$.

Note that the values of ϵ are chosen arbitrarily and can be as small as desired, and that the value of δ is dependent on the ϵ chosen. It should also be pointed out that the smaller the value of ϵ, the smaller will be the corresponding value of δ.

Summing up for this example, we state that $\lim_{x \to 1} f(x) = 5$ because for any $\epsilon > 0$, however small, there exists a $\delta > 0$ such that

$$\text{if}\quad 0 < |x - 1| < \delta \quad \text{then}\quad |f(x) - 5| < \epsilon$$

We now define the limit of a function in general.

2.1.1 DEFINITION   Let f be a function that is defined at every number in some open interval containing a, except possibly at the number a itself. The **limit of $f(x)$ as x approaches a is L**, written as

$$\lim_{x \to a} f(x) = L$$

if the following statement is true:
Given any $\epsilon > 0$, however small, there exists a $\delta > 0$ such that

$$\text{if}\quad 0 < |x - a| < \delta \quad \text{then}\quad |f(x) - L| < \epsilon \tag{4}$$

In words, Definition 2.1.1 states that the function values $f(x)$ approach a limit L as x approaches a number a if the absolute value of the difference between $f(x)$ and L can be made as small as we please by taking x sufficiently near a but not equal to a.

It is important to realize that in the above definition nothing is mentioned about the value of the function when $x = a$. That is, it is not necessary that the function be defined for $x = a$ in order for $\lim_{x \to a} f(x)$ to exist. Moreover, even

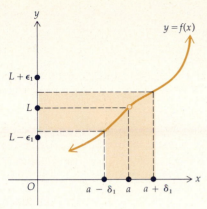

FIGURE 4

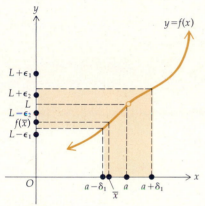

FIGURE 5

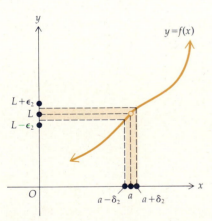

FIGURE 6

if the function is defined for $x = a$, it is possible for $\lim\limits_{x \to a} f(x)$ to exist without having the same value as $f(a)$ (see Example 3 in Section 2.2).

A geometric interpretation of Definition 2.1.1 for a function f is shown in Figure 4. A portion of the graph of f near the point where $x = a$ is shown in the figure. Because f is not necessarily defined at a, there need be no point on the graph with abscissa a. Observe that if x on the horizontal axis lies between $a - \delta_1$ and $a + \delta_1$, then $f(x)$ on the vertical axis will lie between $L - \epsilon_1$ and $L + \epsilon_1$. Stated another way, by restricting x on the horizontal axis to lie between $a - \delta_1$ and $a + \delta_1$, then $f(x)$ on the vertical axis can be restricted to lie between $L - \epsilon_1$ and $L + \epsilon_1$. Thus

$$\text{if} \quad 0 < |x - a| < \delta_1 \quad \text{then} \quad |f(x) - L| < \epsilon_1$$

Figure 5 shows how a smaller value of ϵ can require a different choice for δ. In the figure it is seen that for $\epsilon_2 < \epsilon_1$, the δ_1 value is too large; that is, there are values of x for which $0 < |x - a| < \delta_1$, but $|f(x) - L|$ is not less than ϵ_2. For instance, $0 < |\bar{x} - a| < \delta_1$ but $|f(\bar{x}) - L| > \epsilon_2$. So we must choose a smaller value δ_2 shown in Figure 6 such that

$$\text{if} \quad 0 < |x - a| < \delta_2 \quad \text{then} \quad |f(x) - L| < \epsilon_2$$

However, for any choice of $\epsilon > 0$, no matter how small, there exists a $\delta > 0$ such that statement (4) holds. Therefore $\lim\limits_{x \to a} f(x) = L$.

In the examples of this section we use the symbol $\Rightarrow$ for the first time. The arrow $\Rightarrow$ means *implies*. We also use the double arrow $\Leftrightarrow$, which as previously indicated means the statement preceding it and the statement following it are *equivalent*.

EXAMPLE 1   Let the function f be defined by

$$f(x) = 4x - 7$$

and assume

$$\lim\limits_{x \to 3} f(x) = 5$$

(a) By using a figure similar to Figure 3, for $\epsilon = 0.01$, determine a $\delta > 0$ such that

$$\text{if} \quad 0 < |x - 3| < \delta \quad \text{then} \quad |f(x) - 5| < 0.01$$

(b) By using properties of inequalities, determine a $\delta > 0$ such that the statement in part (a) holds.

Solution

(a) Refer to Figure 7 and observe that the function values increase as the values of x increase. Thus the figure indicates that we need a value of x_1 such that $f(x_1) = 4.99$ and a value of x_2 such that $f(x_2) = 5.01$; that is, we need an x_1 and an x_2 such that

$$4x_1 - 7 = 4.99 \qquad 4x_2 - 7 = 5.01$$

$$x_1 = \frac{11.99}{4} \qquad x_2 = \frac{12.01}{4}$$

$$x_1 = 2.9975 \qquad x_2 = 3.0025$$

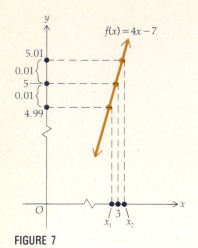

$f(x) = 4x - 7$

FIGURE 7

Because $3 - 2.9975 = 0.0025$ and $3.0025 - 3 = 0.0025$, we choose $\delta = 0.0025$ so that we have the statement

if $0 < |x - 3| < 0.0025$ then $|f(x) - 5| < 0.01$

(b) Because $f(x) = 4x - 7$,

$$|f(x) - 5| = |(4x - 7) - 5|$$
$$= |4x - 12|$$
$$= 4|x - 3|$$

We wish to determine a $\delta > 0$ such that

if $0 < |x - 3| < \delta$ then $|f(x) - 5| < 0.01$

$\Leftrightarrow$ if $0 < |x - 3| < \delta$ then $4|x - 3| < 0.01$

$\Leftrightarrow$ if $0 < |x - 3| < \delta$ then $|x - 3| < 0.0025$

This statement indicates that a suitable choice for δ is 0.0025. Then we have the following argument:

$$0 < |x - 3| < 0.0025$$
$$\Rightarrow \quad 4|x - 3| < 4(0.0025)$$
$$\Rightarrow \quad |4x - 12| < 0.01$$
$$\Rightarrow \quad |(4x - 7) - 5| < 0.01$$
$$\Rightarrow \quad |f(x) - 5| < 0.01$$

We have shown that

if $0 < |x - 3| < 0.0025$ then $|f(x) - 5| < 0.01$ (5)

In this example any positive number less than 0.0025 can be used in place of 0.0025 as the required δ. Observe this fact in Figure 7. Furthermore, if $0 < \gamma < 0.0025$ and statement (5) holds, we have

if $0 < |x - 3| < \gamma$ then $|f(x) - 5| < 0.01$

because every number x satisfying the inequality $0 < |x - 3| < \gamma$ also satisfies the inequality $0 < |x - 3| < 0.0025$.

The solution of Example 1 consisted of finding a δ for a specific ϵ. If for any $\epsilon > 0$ we can find a $\delta > 0$ such that

if $0 < |x - 3| < \delta$ then $|f(x) - 5| < \epsilon$

we shall have established that the limit is 5. This is done in the next example.

EXAMPLE 2   Use Definition 2.1.1 to prove that

$$\lim_{x \to 3} (4x - 7) = 5$$

Solution   The first requirement of Definition 2.1.1 is that $4x - 7$ be defined at every number in some open interval containing 3 except possibly at 3. Because $4x - 7$ is defined for all real numbers, any open interval containing 3 will satisfy this requirement. Now we must show that for any $\epsilon > 0$ there exists a

$\delta > 0$ such that

$$\text{if} \quad 0 < |x - 3| < \delta \quad \text{then} \quad |(4x - 7) - 5| < \epsilon \tag{6}$$

In Example 1 we showed that $|(4x - 7) - 5| = 4|x - 3|$. Therefore (6) is equivalent to the statement

$$\text{if} \quad 0 < |x - 3| < \delta \quad \text{then} \quad 4|x - 3| < \epsilon$$

$$\Leftrightarrow \quad \text{if} \quad 0 < |x - 3| < \delta \quad \text{then} \quad |x - 3| < \tfrac{1}{4}\epsilon$$

This statement indicates that $\tfrac{1}{4}\epsilon$ is a satisfactory δ. With this choice of δ we have the following argument:

$$0 < |x - 3| < \delta$$

$$\Rightarrow \quad 4|x - 3| < 4\delta$$

$$\Rightarrow \quad |4x - 12| < 4\delta$$

$$\Rightarrow \quad |(4x - 7) - 5| < 4\delta$$

$$\Rightarrow \quad |(4x - 7) - 5| < \epsilon \quad (\text{because } \delta = \tfrac{1}{4}\epsilon)$$

Thus we have established that if $\delta = \tfrac{1}{4}\epsilon$, statement (6) holds. This proves that $\lim_{x \to 3} (4x - 7) = 5$.

In particular, if $\epsilon = 0.01$, then we take $\delta = \tfrac{1}{4}(0.01)$, that is, $\delta = 0.0025$. This value of δ corresponds to the one found in Example 1.

Any positive number less than $\tfrac{1}{4}\epsilon$ can be used in place of $\tfrac{1}{4}\epsilon$ as the required δ.

EXAMPLE 3 Let the function f be defined by the equation

$$f(x) = x^2$$

and assume

$$\lim_{x \to 2} f(x) = 4$$

By using a figure for $\epsilon = 0.001$, determine a $\delta > 0$ such that

$$\text{if} \quad 0 < |x - 2| < \delta \quad \text{then} \quad |f(x) - 4| < 0.001$$

Solution Refer to Figure 8 and observe that if $x > 0$, the function values increase as the values of x increase. Therefore the figure indicates that we need a positive value of x_1 such that $f(x_1) = 3.999$ and a positive value of x_2 such that $f(x_2) = 4.001$; that is, we need an x_1 and an x_2 such that

$$x_1{}^2 = 3.999 \qquad x_2{}^2 = 4.001$$

Each of these equations has two solutions. In each case we reject the negative square root because x_1 and x_2 are positive. Thus

$$x_1 = \sqrt{3.999} \qquad x_2 = \sqrt{4.001}$$

$$x_1 \approx 1.9997 \qquad x_2 \approx 2.0002$$

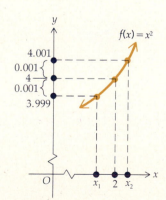

FIGURE 8

Then $2 - 1.9997 = 0.0003$ and $2.0002 - 2 = 0.0002$. Because $0.0002 < 0.0003$, we choose $\delta = 0.0002$ so that we have the statement

$$\text{if} \quad 0 < |x - 2| < 0.0002 \quad \text{then} \quad |f(x) - 4| < 0.001$$

Any positive number less than 0.0002 can be selected as the required δ.

EXAMPLE 4   Use Definition 2.1.1 to prove that

$$\lim_{x \to 2} x^2 = 4$$

Solution   Because x^2 is defined for all real numbers, any open interval containing 2 will satisfy the first requirement of Definition 2.1.1. We must show that for any $\epsilon > 0$ there exists a $\delta > 0$ such that

$$\text{if } \quad 0 < |x - 2| < \delta \quad \text{then} \quad |x^2 - 4| < \epsilon$$

$$\Leftrightarrow \quad \text{if } \quad 0 < |x - 2| < \delta \quad \text{then} \quad |x - 2||x + 2| < \epsilon \tag{7}$$

Observe that on the right end of (7), in addition to the factor $|x - 2|$ we have the factor $|x + 2|$. Thus to prove (7) we wish to place a restriction on δ that will give us an inequality involving $|x + 2|$. Such a restriction is to choose the open interval required by Definition 2.1.1 to be the interval $(1, 3)$, and this implies that $\delta \leq 1$. Then

$$0 < |x - 2| < \delta \quad \text{and} \quad \delta \leq 1$$

$$\Rightarrow \qquad |x - 2| < 1$$

$$\Rightarrow \quad -1 < x - 2 < 1$$

$$\Rightarrow \qquad 3 < x + 2 < 5$$

$$\Rightarrow \qquad |x + 2| < 5 \tag{8}$$

Now

$$0 < |x - 2| < \delta \quad \text{and} \quad |x + 2| < 5$$

$$\Rightarrow \quad |x - 2||x + 2| < \delta \cdot 5 \tag{9}$$

Remember that our goal is to have $|x - 2||x + 2| < \epsilon$. Statement (9) indicates that we should require $\delta \cdot 5 \leq \epsilon$, that is, $\delta \leq \epsilon/5$. This means that we have put two restrictions on δ: $\delta \leq 1$ and $\delta \leq \epsilon/5$. So that both restrictions hold we take δ as the smaller of the two numbers 1 and $\epsilon/5$; with symbols we write this as $\delta = \min(1, \epsilon/5)$. Using this δ we have the following argument:

$$0 < |x - 2| < \delta$$

$$\Rightarrow \quad |x - 2||x + 2| < \delta|x + 2|$$

$$\Rightarrow \quad |(x - 2)(x + 2)| < \delta|x + 2|$$

$$\Rightarrow \qquad |x^2 - 4| < \delta|x + 2| \tag{10}$$

However, we showed in (8) that if $\delta \leq 1$ and $0 < |x - 2| < \delta$, then $|x + 2| < 5$; that is, $\delta|x + 2| < \delta \cdot 5$. Continuing from where we left off at (10) we have

$$|x^2 - 4| < \delta|x + 2| \quad \text{and} \quad \delta|x + 2| < \delta \cdot 5$$

$$\Rightarrow \quad |x^2 - 4| < \delta \cdot 5$$

$$\Rightarrow \quad |x^2 - 4| < \frac{\epsilon}{5} \cdot 5 \quad \text{(because } \delta \leq \epsilon/5)$$

$$\Rightarrow \quad |x^2 - 4| < \epsilon$$

We have demonstrated that for any $\epsilon > 0$ the choice of $\delta = \min(1, \epsilon/5)$ makes the following statement true:

$$\text{if} \quad 0 < |x - 2| < \delta \quad \text{then} \quad |x^2 - 4| < \epsilon$$

This proves that $\lim\limits_{x \to 2} x^2 = 4$.

The following theorem states that a function cannot approach two different limits at the same time. It is called a *uniqueness theorem* because it guarantees that if the limit of a function exists, it is unique. We state the theorem here but defer its proof until Supplementary Section 2.9.

2.1.2 THEOREM   If $\lim\limits_{x \to a} f(x) = L_1$ and $\lim\limits_{x \to a} f(x) = L_2$, then $L_1 = L_2$.

Because of Theorem 2.1.2, we can state that if a function f has a limit L at the number a, then L is *the* limit of f at a.

EXERCISES 2.1

In Exercises 1 through 22, you are given $f(x)$, a, and L, as well as $\lim\limits_{x \to a} f(x) = L$. (a) By using a figure and arguments similar to those in Examples 1 and 3, determine a $\delta > 0$ for the given ϵ such that

$$\text{if} \quad 0 < |x - a| < \delta \quad \text{then} \quad |f(x) - L| < \epsilon \qquad (11)$$

(b) By using properties of inequalities, determine a $\delta > 0$ such that statement (11) holds for the given value of ϵ.

1. $\lim\limits_{x \to 4} (x - 1) = 3$; $\epsilon = 0.2$

2. $\lim\limits_{x \to 3} (x + 2) = 5$; $\epsilon = 0.02$

3. $\lim\limits_{x \to 3} (2x + 4) = 10$; $\epsilon = 0.01$

4. $\lim\limits_{x \to 2} (3x - 1) = 5$; $\epsilon = 0.1$

5. $\lim\limits_{x \to 1} (5x - 3) = 2$; $\epsilon = 0.05$

6. $\lim\limits_{x \to 2} (4x - 5) = 3$; $\epsilon = 0.001$

7. $\lim\limits_{x \to -1} (3 - 4x) = 7$; $\epsilon = 0.02$

8. $\lim\limits_{x \to -2} (2 + 5x) = -8$; $\epsilon = 0.002$

9. $\lim\limits_{x \to 3} x^2 = 9$; $\epsilon = 0.005$

10. $\lim\limits_{x \to -4} x^2 = 16$; $\epsilon = 0.03$

11. $\lim\limits_{x \to -2} x^2 = 4$; $\epsilon = 0.003$

12. $\lim\limits_{x \to 1} (x^2 - 5) = -4$; $\epsilon = 0.01$

13. $\lim\limits_{x \to 2} (x^2 - 2x + 1) = 1$; $\epsilon = 0.001$

14. $\lim\limits_{x \to -1} (x^2 + 4x + 4) = 1$; $\epsilon = 0.002$

15. $\lim\limits_{x \to 0} (x^2 + 3x - 4) = -4$; $\epsilon = 0.03$

16. $\lim\limits_{x \to 3} (x^2 - x - 6) = 0$; $\epsilon = 0.005$

17. $\lim\limits_{x \to -2} (2x^2 + 5x + 3) = 1$; $\epsilon = 0.004$

18. $\lim\limits_{x \to 1} (3x^2 - 7x + 2) = -2$; $\epsilon = 0.02$

19. $\lim\limits_{x \to -2} \dfrac{x^2 - 4}{x + 2} = -4$; $\epsilon = 0.01$

20. $\lim\limits_{x \to 1/3} \dfrac{9x^2 - 1}{3x - 1} = 2$; $\epsilon = 0.01$

21. $\lim\limits_{x \to 1/2} \dfrac{3x^2 - 8x - 3}{x - 3} = \dfrac{5}{2}$; $\epsilon = 0.001$

22. $\lim\limits_{x \to 4} \dfrac{4x^2 - 4x - 3}{2x + 1} = 5$; $\epsilon = 0.003$

In Exercises 23 through 42, prove the limit is the indicated number by applying Definition 2.1.1.

23. $\lim\limits_{x \to 2} 7 = 7$

24. $\lim\limits_{x \to 5} (-4) = -4$

25. $\lim\limits_{x \to 4} (2x + 1) = 9$

26. $\lim\limits_{x \to 1} (4x + 3) = 7$

27. $\lim\limits_{x \to -1} (5x + 8) = 3$

28. $\lim\limits_{x \to 3} (3x - 5) = 4$

29. $\lim\limits_{x \to 3} (7 - 3x) = -2$

30. $\lim\limits_{x \to -4} (2x + 7) = -1$

31. $\lim\limits_{x \to -2} (1 + 3x) = -5$

32. $\lim\limits_{x \to -2} (7 - 2x) = 11$

33. $\lim\limits_{x \to -1} \dfrac{x^2 - 1}{x + 1} = -2$

34. $\lim\limits_{x \to 3} \dfrac{x^2 - 9}{x - 3} = 6$

35. $\lim\limits_{x \to 1} x^2 = 1$

36. $\lim\limits_{x \to -3} x^2 = 9$

37. $\lim\limits_{x \to 5} (x^2 - 3x) = 10$

38. $\lim\limits_{x \to 2} (x^2 + 2x - 1) = 7$

39. $\lim\limits_{x \to -3} (5 - x - x^2) = -1$

40. $\lim\limits_{x \to -1} (3 + 2x - x^2) = 0$

41. $\lim\limits_{x \to 2} (6x^2 - 13x + 5) = 3$

42. $\lim\limits_{x \to 1} (4x^2 - 13x + 12) = 3$

43. Prove that $\lim\limits_{x \to a} x^2 = a^2$ if a is any positive number.

44. Prove that $\lim\limits_{x \to a} x^2 = a^2$ if a is any negative number.

2.2 THEOREMS ON LIMITS OF FUNCTIONS

In Section 2.1 we proved that the limit of a function was a particular number by applying Definition 2.1.1. To compute limits of functions by easier methods we use theorems whose proofs are based on Definition 2.1.1. These theorems, as well as others on limits of functions that appear in later sections of this chapter, are labeled "limit theorems" and are so designated as they are presented.

2.2.1 LIMIT THEOREM 1

If m and b are any constants,

$$\lim_{x \to a} (mx + b) = ma + b$$

Proof   To prove this theorem we use Definition 2.1.1. For any $\epsilon > 0$ we must show that there exists a $\delta > 0$ such that

$$\text{if} \quad 0 < |x - a| < \delta \quad \text{then} \quad |(mx + b) - (ma + b)| < \epsilon \tag{1}$$

Case 1: $m \neq 0$.

Because $|(mx + b) - (ma + b)| = |m| \cdot |x - a|$, we want to find a $\delta > 0$ for any $\epsilon > 0$ such that

$$\text{if} \quad 0 < |x - a| < \delta \quad \text{then} \quad |m| \cdot |x - a| < \epsilon$$

or, because $m \neq 0$,

$$\text{if} \quad 0 < |x - a| < \delta \quad \text{then} \quad |x - a| < \frac{\epsilon}{|m|}$$

This statement will hold if $\delta = \epsilon/|m|$; so we conclude that

$$\text{if} \quad 0 < |x - a| < \delta \text{ and } \delta = \frac{\epsilon}{|m|} \quad \text{then} \quad |(mx + b) - (ma + b)| < \epsilon$$

This proves the theorem for Case 1.

Case 2: $m = 0$.

If $m = 0$, then $|(mx + b) - (ma + b)| = 0$ for all values of x. So take δ to be any positive number, and statement (1) holds. This proves the theorem for Case 2. ■

▶ **ILLUSTRATION 1**   From Limit Theorem 1 it follows that

$$\lim_{x \to 2} (3x + 5) = 3 \cdot 2 + 5$$
$$= 11$$

◀

2.2.2 LIMIT THEOREM 2

If c is a constant, then for any number a,

$$\lim_{x \to a} c = c$$

Proof   This follows immediately from Limit Theorem 1 by taking $m = 0$ and $b = c$. ■

2.2.3 LIMIT THEOREM 3

$$\lim_{x \to a} x = a$$

Proof   This also follows immediately from Limit Theorem 1 by taking $m = 1$ and $b = 0$. ■

▶ **ILLUSTRATION 2** From Limit Theorem 2,

$$\lim_{x \to 5} 7 = 7$$

and from Limit Theorem 3,

$$\lim_{x \to -6} x = -6 \qquad\qquad ◀$$

2.2.4 LIMIT THEOREM 4

If $\lim_{x \to a} f(x) - L$ and $\lim_{x \to a} g(x) - M$, then

$$\lim_{x \to a} [f(x) \pm g(x)] = L \pm M$$

Proof We shall prove this theorem with the plus sign. Given

$$\lim_{x \to a} f(x) = L \tag{2}$$

and

$$\lim_{x \to a} g(x) = M \tag{3}$$

we wish to prove that

$$\lim_{x \to a} [f(x) + g(x)] = L + M$$

We use Definition 2.1.1; that is, for any $\epsilon > 0$ we must show that there exists a $\delta > 0$ such that

$$\text{if} \quad 0 < |x - a| < \delta \quad \text{then} \quad |[f(x) + g(x)] - (L + M)| < \epsilon \tag{4}$$

Because (2) is given, it follows from the definition of a limit that for $\frac{1}{2}\epsilon > 0$ there exists a $\delta_1 > 0$ such that

$$\text{if} \quad 0 < |x - a| < \delta_1 \quad \text{then} \quad |f(x) - L| < \tfrac{1}{2}\epsilon$$

Similarly, from (3), for $\frac{1}{2}\epsilon > 0$ there exists a $\delta_2 > 0$ such that

$$\text{if} \quad 0 < |x - a| < \delta_2 \quad \text{then} \quad |g(x) - M| < \tfrac{1}{2}\epsilon$$

Now let δ be the smaller of the two numbers δ_1 and δ_2. Therefore $\delta \leq \delta_1$ and $\delta \leq \delta_2$. So

$$\text{if} \quad 0 < |x - a| < \delta \quad \text{then} \quad |f(x) - L| < \tfrac{1}{2}\epsilon$$

and

$$\text{if} \quad 0 < |x - a| < \delta \quad \text{then} \quad |g(x) - M| < \tfrac{1}{2}\epsilon$$

Hence if $0 < |x - a| < \delta$, then

$$\begin{aligned}
|[f(x) + g(x)] - (L + M)| &= |(f(x) - L) + (g(x) - M)| \\
&\leq |f(x) - L| + |g(x) - M| \\
&< \tfrac{1}{2}\epsilon + \tfrac{1}{2}\epsilon \\
&= \epsilon
\end{aligned}$$

In this way we have obtained statement (4), thereby proving that

$$\lim_{x \to a} [f(x) + g(x)] = L + M$$

The proof of Limit Theorem 4 using the minus sign is left as an exercise (see Exercise 46). ∎

Limit Theorem 4 can be extended to any finite number of functions.

2.2.5 LIMIT THEOREM 5
If $\lim\limits_{x \to a} f_1(x) = L_1$, $\lim\limits_{x \to a} f_2(x) = L_2, \ldots,$ and $\lim\limits_{x \to a} f_n(x) = L_n$, then

$$\lim_{x \to a} [f_1(x) \pm f_2(x) \pm \cdots \pm f_n(x)] = L_1 \pm L_2 \pm \cdots \pm L_n$$

This theorem may be proved by applying Limit Theorem 4 and mathematical induction (see Exercise 47).

2.2.6 LIMIT THEOREM 6
If $\lim\limits_{x \to a} f(x) = L$ and $\lim\limits_{x \to a} g(x) = M$, then

$$\lim_{x \to a} [f(x) \cdot g(x)] = L \cdot M$$

The proof of this theorem is more sophisticated than those of the preceding theorems. The steps of the proof are indicated in Exercises 49 and 50.

▶ **ILLUSTRATION 3** From Limit Theorem 3, $\lim\limits_{x \to 3} x = 3$, and from Limit Theorem 1, $\lim\limits_{x \to 3} (2x + 1) = 7$. Thus from Limit Theorem 6,

$$\lim_{x \to 3} [x(2x + 1)] = \lim_{x \to 3} x \cdot \lim_{x \to 3} (2x + 1)$$
$$= 3 \cdot 7$$
$$= 21$$
◀

Limit Theorem 6 also can be extended to any finite number of functions by applying mathematical induction.

2.2.7 LIMIT THEOREM 7
If $\lim\limits_{x \to a} f_1(x) = L_1$, $\lim\limits_{x \to a} f_2(x) = L_2, \ldots,$ and $\lim\limits_{x \to a} f_n(x) = L_n$, then

$$\lim_{x \to a} [f_1(x) f_2(x) \ldots f_n(x)] = L_1 L_2 \ldots L_n$$

The proof is left as an exercise (see Exercise 48).

2.2.8 LIMIT THEOREM 8
If $\lim\limits_{x \to a} f(x) = L$ and n is any positive integer, then

$$\lim_{x \to a} [f(x)]^n = L^n$$

The proof follows immediately from Limit Theorem 7 by taking $f_1(x)$, $f_2(x), \ldots, f_n(x)$ all equal to $f(x)$ and $L_1, L_2, \ldots, L_n$ all equal to L.

▶ **ILLUSTRATION 4** From Limit Theorem 1, $\lim\limits_{x \to -2} (5x + 7) = -3$. Therefore, from Limit Theorem 8 it follows that

$$\lim_{x \to -2} (5x + 7)^4 = \left[\lim_{x \to -2} (5x + 7) \right]^4$$
$$= (-3)^4$$
$$= 81$$
◀

2.2.9 LIMIT THEOREM 9   If $\lim\limits_{x \to a} f(x) = L$ and $\lim\limits_{x \to a} g(x) = M$, then

$$\lim_{x \to a} \frac{f(x)}{g(x)} = \frac{L}{M} \qquad \text{if } M \neq 0$$

The proof, based on Definition 2.1.1, is given in Supplementary Section 2.9.

▶ **ILLUSTRATION 5**   From Limit Theorem 3, $\lim\limits_{x \to 4} x = 4$, and from Limit Theorem 1, $\lim\limits_{x \to 4} (-7x + 1) = -27$. Therefore from Limit Theorem 9,

$$\lim_{x \to 4} \frac{x}{-7x + 1} = \frac{\lim\limits_{x \to 4} x}{\lim\limits_{x \to 4} (-7x + 1)}$$

$$= \frac{4}{-27}$$

$$= -\frac{4}{27} \qquad \blacktriangleleft$$

2.2.10 LIMIT THEOREM 10   If n is a positive integer and $\lim\limits_{x \to a} f(x) = L$, then

$$\lim_{x \to a} \sqrt[n]{f(x)} = \sqrt[n]{L}$$

with the restriction that if n is even, $L > 0$.

The proof of this theorem is also given in Supplementary Section 2.9.

▶ **ILLUSTRATION 6**   From the result of Illustration 5 and Limit Theorem 10 it follows that

$$\lim_{x \to 4} \sqrt[3]{\frac{x}{-7x + 1}} = \sqrt[3]{\lim_{x \to 4} \frac{x}{-7x + 1}}$$

$$= \sqrt[3]{-\frac{4}{27}}$$

$$= -\frac{\sqrt[3]{4}}{3} \qquad \blacktriangleleft$$

Following are some examples illustrating the application of the above theorems. To indicate the limit theorem being used we use the abbreviation "L.T." followed by the theorem number; for example, "L.T. 2" refers to Limit Theorem 2.

EXAMPLE 1   Find $\lim\limits_{x \to 3} (x^2 + 7x - 5)$, and, when applicable, indicate the limit theorems being used.

Solution

$$\lim_{x \to 3} (x^2 + 7x - 5) = \lim_{x \to 3} x^2 + \lim_{x \to 3} 7x - \lim_{x \to 3} 5 \qquad \text{(L.T. 5)}$$

$$= \lim_{x \to 3} x \cdot \lim_{x \to 3} x + \lim_{x \to 3} 7 \cdot \lim_{x \to 3} x - \lim_{x \to 3} 5 \qquad \text{(L.T. 6)}$$

$$= 3 \cdot 3 + 7 \cdot 3 - 5 \qquad \text{(L.T. 3 and L.T. 2)}$$

$$= 9 + 21 - 5$$

$$= 25$$

It is important at this point to realize that the limit in Example 1 was evaluated by direct application of the theorems on limits. For the function f defined by $f(x) = x^2 + 7x - 5$, note that $f(3) = 25$, which is the same as $\lim_{x \to 3} (x^2 + 7x - 5)$. It is not always true that $\lim_{x \to a} f(x)$ and $f(a)$ are equal (see Example 3).

EXAMPLE 2   Find

$$\lim_{x \to 2} \sqrt{\frac{x^3 + 2x + 3}{x^2 + 5}}$$

and, when applicable, indicate the limit theorems being used.

Solution

$$\lim_{x \to 2} \sqrt{\frac{x^3 + 2x + 3}{x^2 + 5}} = \sqrt{\lim_{x \to 2} \frac{x^3 + 2x + 3}{x^2 + 5}} \qquad \text{(L.T. 10)}$$

$$= \sqrt{\frac{\lim_{x \to 2} (x^3 + 2x + 3)}{\lim_{x \to 2} (x^2 + 5)}} \qquad \text{(L.T. 9)}$$

$$= \sqrt{\frac{\lim_{x \to 2} x^3 + \lim_{x \to 2} 2x + \lim_{x \to 2} 3}{\lim_{x \to 2} x^2 + \lim_{x \to 2} 5}} \qquad \text{(L.T. 5)}$$

$$= \sqrt{\frac{(\lim_{x \to 2} x)^3 + \lim_{x \to 2} 2 \cdot \lim_{x \to 2} x + \lim_{x \to 2} 3}{(\lim_{x \to 2} x)^2 + \lim_{x \to 2} 5}} \qquad \text{(L.T. 6 and L.T. 8)}$$

$$= \sqrt{\frac{2^3 + 2 \cdot 2 + 3}{2^2 + 5}} \qquad \text{(L.T. 3 and L.T. 2)}$$

$$= \sqrt{\frac{8 + 4 + 3}{9}}$$

$$= \frac{\sqrt{15}}{3}$$

EXAMPLE 3   Find $\lim_{x \to 4} f(x)$ given that

$$f(x) = \begin{cases} x - 3 & \text{if } x \neq 4 \\ 5 & \text{if } x = 4 \end{cases}$$

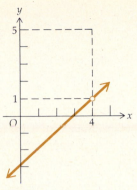

FIGURE 1

Solution   When evaluating $\lim\limits_{x \to 4} f(x)$ we are considering values of x close to 4 but not equal to 4. Thus

$$\lim_{x \to 4} f(x) = \lim_{x \to 4} (x - 3)$$

$$= 1 \qquad \text{(L.T. 1)}$$

In Example 3, $\lim\limits_{x \to 4} f(x) = 1$ but $f(4) = 5$; therefore $\lim\limits_{x \to 4} f(x) \neq f(4)$. In terms of geometry, there is a break in the graph of the function at the point where $x = 4$ (see Figure 1). The graph of the function consists of the point $(4, 5)$ and the line whose equation is $y = x - 3$, with the point $(4, 1)$ deleted.

EXAMPLE 4   Given

$$f(x) = \frac{x^2 - 25}{x - 5}$$

(a) Use a calculator to tabulate values of $f(x)$ when x is 4, 4.5, 4.9, 4.99, 4.999 and when x is 6, 5.5, 5.1, 5.01, 5.001. What does $f(x)$ appear to be approaching as x approaches 5?
(b) Use limit theorems to compute $\lim\limits_{x \to 5} f(x)$.

Solution
(a) Tables 1 and 2 give the values of $f(x)$ for the specified values of x. From the tables, $f(x)$ appears to be approaching 10 as x approaches 5.
(b) Here we have a situation different than in the preceding examples. Limit Theorem 9 cannot be applied to the quotient $\dfrac{x^2 - 25}{x - 5}$ because $\lim\limits_{x \to 5} (x - 5) = 0$. However, factoring the numerator we obtain

$$\frac{x^2 - 25}{x - 5} = \frac{(x - 5)(x + 5)}{x - 5}$$

If $x \neq 5$, the numerator and denominator can be divided by $x - 5$ to obtain $x + 5$. Remember that when computing the limit of a function as x approaches 5, we are considering values of x close to 5 but not equal to 5. Therefore, it is possible to divide the numerator and denominator by $x - 5$. The solution takes the following form:

$$\lim_{x \to 5} \frac{x^2 - 25}{x - 5} = \lim_{x \to 5} \frac{(x - 5)(x + 5)}{x - 5}$$

$$= \lim_{x \to 5} (x + 5)$$

$$= 10 \qquad \text{(L.T. 1)}$$

EXAMPLE 5   Given

$$g(x) = \frac{\sqrt{x} - 2}{x - 4}$$

Table 1

x	$f(x) = \dfrac{x^2 - 25}{x - 5}$
4	9
4.5	9.5
4.9	9.9
4.99	9.99
4.999	9.999

Table 2

x	$f(x) = \dfrac{x^2 - 25}{x - 5}$
6	11
5.5	10.5
5.1	10.1
5.01	10.01
5.001	10.001

(a) Use a calculator to tabulate to four decimal places values of $g(x)$ when x is 3, 3.5, 3.9, 3.99, 3.999 and when x is 5, 4.5, 4.1, 4.01, 4.001. What does $g(x)$ appear to be approaching as x approaches 4?

(b) Find $\lim\limits_{x \to 4} g(x)$ and, when applicable, indicate the limit theorems being used.

Solution

(a) Tables 3 and 4 give the values of $g(x)$ for the specified values of x. From the tables, $g(x)$ appears to be approaching 0.2500 as x approaches 4.

(b) As in Example 4, Limit Theorem 9 cannot be applied to the quotient $\dfrac{\sqrt{x}-2}{x-4}$ because $\lim\limits_{x \to 4} (x-4) = 0$. To simplify the quotient we rationalize the numerator by multiplying the numerator and denominator by $\sqrt{x}+2$.

$$\frac{\sqrt{x}-2}{x-4} = \frac{(\sqrt{x}-2)(\sqrt{x}+2)}{(x-4)(\sqrt{x}+2)}$$

$$= \frac{x-4}{(x-4)(\sqrt{x}+2)}$$

Because we are evaluating the limit as x approaches 4, we are considering values of x close to 4 but not equal to 4. Hence we can divide the numerator and denominator by $x-4$. Therefore

$$\frac{\sqrt{x}-2}{x-4} = \frac{1}{\sqrt{x}+2} \qquad \text{if } x \neq 4$$

The solution is as follows:

$$\lim_{x \to 4} \frac{\sqrt{x}-2}{x-4} = \lim_{x \to 4} \frac{(\sqrt{x}-2)(\sqrt{x}+2)}{(x-4)(\sqrt{x}+2)}$$

$$= \lim_{x \to 4} \frac{x-4}{(x-4)(\sqrt{x}+2)}$$

$$= \lim_{x \to 4} \frac{1}{\sqrt{x}+2}$$

$$= \frac{\lim\limits_{x \to 4} 1}{\lim\limits_{x \to 4} (\sqrt{x}+2)} \qquad\qquad \text{(L.T. 9)}$$

$$= \frac{1}{\lim\limits_{x \to 4} \sqrt{x} + \lim\limits_{x \to 4} 2} \qquad\qquad \text{(L.T. 2 and L.T. 4)}$$

$$= \frac{1}{\sqrt{\lim\limits_{x \to 4} x} + 2} \qquad\qquad \text{(L.T. 10 and L.T. 2)}$$

$$= \frac{1}{\sqrt{4}+2} \qquad\qquad \text{(L.T. 3)}$$

$$= \tfrac{1}{4}$$

Table 3

x	$g(x) = \dfrac{\sqrt{x}-2}{x-4}$
3	0.2679
3.5	0.2583
3.9	0.2516
3.99	0.2502
3.999	0.2500

Table 4

x	$g(x) = \dfrac{\sqrt{x}-2}{x-4}$
5	0.2361
4.5	0.2426
4.1	0.2485
4.01	0.2498
4.001	0.2500

From time to time we will need two other equations that are equivalent to the equation

$$\lim_{x \to a} f(x) = L$$

They are given in the following two theorems.

2.2.11 THEOREM

$$\lim_{x \to a} f(x) = L \quad \text{if and only if} \quad \lim_{x \to a} [f(x) - L] = 0$$

Proof Because the theorem has an *if and only if* qualification, the proof requires two parts.

Part 1: Prove that $\lim_{x \to a} f(x) = L$ if $\lim_{x \to a} [f(x) - L] = 0$.

We start with $\lim_{x \to a} f(x)$, replace $f(x)$ by $[f(x) - L] + L$, and then apply Limit Theorem 4.

$$\lim_{x \to a} f(x) = \lim_{x \to a} ([f(x) - L] + L)$$

$$= \lim_{x \to a} [f(x) - L] + \lim_{x \to a} L$$

$$= 0 + L$$

$$= L$$

Part 2: Prove that $\lim_{x \to a} f(x) = L$ only if $\lim_{x \to a} [f(x) - L] = 0$.

Here we must show that if $\lim_{x \to a} f(x) = L$, then $\lim_{x \to a} [f(x) - L] = 0$. We apply Limit Theorem 4 to $\lim_{x \to a} [f(x) - L]$.

$$\lim_{x \to a} [f(x) - L] = \lim_{x \to a} f(x) - \lim_{x \to a} L$$

$$= L - L$$

$$= 0 \qquad \blacksquare$$

2.2.12 THEOREM

$$\lim_{x \to a} f(x) = L \quad \text{if and only if} \quad \lim_{t \to 0} f(t + a) = L$$

Proof Let $t + a = x$; then $x - a = t$. There are two parts to the proof.

Part 1: Prove that $\lim_{x \to a} f(x) = L$ if $\lim_{t \to 0} f(t + a) = L$.

If $\lim_{t \to 0} f(t + a) = L$, from Definition 2.1.1 it follows that for any $\epsilon > 0$ there exists a $\delta > 0$ such that

$$\text{if} \quad 0 < |t| < \delta \quad \text{then} \quad |f(t + a) - L| < \epsilon \tag{5}$$

or, equivalently, by replacing $t + a$ by x and t by $x - a$,

$$\text{if} \quad 0 < |x - a| < \delta \quad \text{then} \quad |f(x) - L| < \epsilon \tag{6}$$

From Definition 2.1.1, statement (6) implies that

$$\lim_{x \to a} f(x) = L$$

Part 2: Prove that $\lim_{x \to a} f(x) = L$ only if $\lim_{t \to 0} f(t + a) = L$.

If $\lim\limits_{x \to a} f(x) = L$, then by Definition 2.1.1, for any $\epsilon > 0$ there exists a $\delta > 0$ such that statement (6) holds. By replacing x by $t + a$ and $x - a$ by t we have the equivalent statement (5). Thus from Definition 2.1.1 we can conclude that

$$\lim_{t \to 0} f(t + a) = L$$ ■

EXERCISES 2.2

In Exercises 1 through 14, find the limit and, when applicable, indicate the limit theorems being used.

1. $\lim\limits_{x \to 5} (3x - 7)$

2. $\lim\limits_{x \to -4} (5x + 2)$

3. $\lim\limits_{x \to 2} (x^2 + 2x - 1)$

4. $\lim\limits_{x \to 3} (2x^2 - 4x + 5)$

5. $\lim\limits_{z \to -2} (z^3 + 8)$

6. $\lim\limits_{y \to -1} (y^3 - 2y^2 + 3y - 4)$

7. $\lim\limits_{x \to 3} \dfrac{4x - 5}{5x - 1}$

8. $\lim\limits_{x \to 2} \dfrac{3x + 4}{8x - 1}$

9. $\lim\limits_{t \to 2} \dfrac{t^2 - 5}{2t^3 + 6}$

10. $\lim\limits_{x \to -1} \dfrac{2x + 1}{x^2 - 3x + 4}$

11. $\lim\limits_{r \to 1} \sqrt{\dfrac{8r + 1}{r + 3}}$

12. $\lim\limits_{x \to 2} \sqrt{\dfrac{x^2 + 3x + 4}{x^3 + 1}}$

13. $\lim\limits_{x \to 4} \sqrt[3]{\dfrac{x^2 - 3x + 4}{2x^2 - x - 1}}$

14. $\lim\limits_{x \to -3} \sqrt[3]{\dfrac{5 + 2x}{5 - x}}$

In Exercises 15 through 20, do the following: (a) Use a calculator to tabulate to four decimal places values of $f(x)$ for the specified values of x. What does $f(x)$ appear to be approaching as x approaches c? (b) Find $\lim\limits_{x \to c} f(x)$ and, when applicable, indicate the limit theorems being used.

15. $f(x) = \dfrac{x - 2}{x^2 - 4}$; x is 1, 1.5, 1.9, 1.99, 1.999 and x is 3, 2.5, 2.1, 2.01, 2.001; $c = 2$

16. $f(x) = \dfrac{2x^2 + 3x - 2}{x^2 - 6x - 16}$; x is -3, -2.5, -2.1, -2.01, -2.001 and x is -1, -1.5, -1.9, -1.99, -1.999; $c = -2$

17. $f(x) = \dfrac{x^2 + 5x + 6}{x^2 - x - 12}$; x is -4, -3.5, -3.1, -3.01, -3.001, -3.0001 and x is -2, -2.5, -2.9, -2.99, -2.999, -2.9999; $c = -3$

18. $f(x) = \dfrac{2x - 3}{4x^2 - 9}$; x is 1, 1.4, 1.49, 1.499, 1.4999 and x is 2, 1.6, 1.51, 1.501, 1.5001; $c = \frac{3}{2}$

19. $f(x) = \dfrac{3 - \sqrt{x}}{9 - x}$; x is 8, 8.5, 8.9, 8.99, 8.999 and x is 10, 9.5, 9.1, 9.01, 9.001; $c = 9$

20. $f(x) = \dfrac{2 - \sqrt{4 - x}}{x}$; x is -1, -0.5, -0.1, -0.01, -0.001 and x is 1, 0.5, 0.1, 0.01, 0.001; $c = 0$

In Exercises 21 through 39, find the limit and, when applicable, indicate the limit theorems being used.

21. $\lim\limits_{x \to 7} \dfrac{x^2 - 49}{x - 7}$

22. $\lim\limits_{z \to -5} \dfrac{z^2 - 25}{z + 5}$

23. $\lim\limits_{x \to -3/2} \dfrac{4x^2 - 9}{2x + 3}$

24. $\lim\limits_{x \to 1/3} \dfrac{3x - 1}{9x^2 - 1}$

25. $\lim\limits_{s \to 4} \dfrac{3s^2 - 8s - 16}{2s^2 - 9s + 4}$

26. $\lim\limits_{x \to 4} \dfrac{3x^2 - 17x + 20}{4x^2 - 25x + 36}$

27. $\lim\limits_{y \to -2} \dfrac{y^3 + 8}{y + 2}$

28. $\lim\limits_{s \to 1} \dfrac{s^3 - 1}{s - 1}$

29. $\lim\limits_{y \to -3} \sqrt{\dfrac{y^2 - 9}{2y^2 + 7y + 3}}$

30. $\lim\limits_{t \to 3/2} \sqrt{\dfrac{8t^3 - 27}{4t^2 - 9}}$

31. $\lim\limits_{x \to 1} \dfrac{\sqrt{x} - 1}{x - 1}$

32. $\lim\limits_{h \to -1} \dfrac{\sqrt{h + 5} - 2}{h + 1}$

33. $\lim\limits_{x \to 0} \dfrac{\sqrt{x + 2} - \sqrt{2}}{x}$

34. $\lim\limits_{x \to 1} \dfrac{\sqrt[3]{x} - 1}{x - 1}$

35. $\lim\limits_{h \to 0} \dfrac{\sqrt[3]{h + 1} - 1}{h}$

36. $\lim\limits_{x \to -2} \dfrac{x^3 - x^2 - x + 10}{x^2 + 3x + 2}$

37. $\lim\limits_{x \to -1} \dfrac{2x^2 - x - 3}{x^3 + 2x^2 + 6x + 5}$

38. $\lim\limits_{y \to 4} \dfrac{2y^3 - 11y^2 + 10y + 8}{3y^3 - 17y^2 + 16y + 16}$

39. $\lim\limits_{x \to 3} \dfrac{2x^3 - 5x^2 - 2x - 3}{4x^3 - 13x^2 + 4x - 3}$

40. If $f(x) = x^2 + 5x - 3$, show that $\lim\limits_{x \to 2} f(x) = f(2)$.

41. If $F(x) = 2x^3 + 7x - 1$, show that $\lim\limits_{x \to -1} F(x) = F(-1)$.

42. If $g(x) = \dfrac{x^2 - 16}{x - 4}$, show that $\lim\limits_{x \to 4} g(x) = 8$ but that $g(4)$ is not defined.

43. If $h(x) = \dfrac{\sqrt{x + 9} - 3}{x}$, show that $\lim\limits_{x \to 0} h(x) = \frac{1}{6}$, but that $h(0)$ is not defined.

44. Given that f is the function defined by

$$f(x) = \begin{cases} 2x - 1 & \text{if } x \neq 2 \\ 1 & \text{if } x = 2 \end{cases}$$

(a) Find $\lim\limits_{x \to 2} f(x)$, and show that $\lim\limits_{x \to 2} f(x) \neq f(2)$. (b) Draw a sketch of the graph of f.

45. Given that f is the function defined by

$$f(x) = \begin{cases} x^2 - 9 & \text{if } x \neq -3 \\ 4 & \text{if } x = -3 \end{cases}$$

(a) Find $\lim_{x \to -3} f(x)$, and show that $\lim_{x \to -3} f(x) \neq f(-3)$. (b) Draw a sketch of the graph of f.

46. Use Definition 2.1.1 to prove that if

$$\lim_{x \to a} f(x) = L \quad \text{and} \quad \lim_{x \to a} g(x) = M$$

then

$$\lim_{x \to a} [f(x) - g(x)] = L - M$$

47. Prove Limit Theorem 5 by applying Limit Theorem 4 and mathematical induction.

48. Prove Limit Theorem 7 by applying Limit Theorem 6 and mathematical induction.

49. Use Definition 2.1.1 to prove that if

$$\lim_{x \to a} f(x) = L \quad \text{and} \quad \lim_{x \to a} g(x) = 0$$

then

$$\lim_{x \to a} [f(x) \cdot g(x)] = 0$$

(*Hint:* To prove that $\lim_{x \to a} [f(x) \cdot g(x)] = 0$ we must show that for any $\epsilon > 0$ there exists a $\delta > 0$ such that if $0 < |x - a| < \delta$, then $|f(x) \cdot g(x)| < \epsilon$. First show that there is a $\delta_1 > 0$ such that if $0 < |x - a| < \delta_1$, then $|f(x)| < 1 + |L|$, by applying Definition 2.1.1 to $\lim_{x \to a} f(x) = L$ with $\epsilon = 1$ and $\delta = \delta_1$, and then use the triangle inequality. Then show that there is a $\delta_2 > 0$ such that if $0 < |x - a| < \delta_2$, then $|g(x)| < \epsilon/(1 + |L|)$ by applying Definition 2.1.1 to $\lim_{x \to a} g(x) = 0$. By taking δ as the smaller of the two numbers δ_1 and δ_2, the theorem is proved.)

50. Prove Limit Theorem 6: If $\lim_{x \to a} f(x) = L$ and $\lim_{x \to a} g(x) = M$, then

$$\lim_{x \to a} [f(x) \cdot g(x)] = L \cdot M$$

(*Hint:* Let $f(x) \cdot g(x) = [f(x) - L]g(x) + L[g(x) - M] + L \cdot M$. Apply Limit Theorem 5 and the result of Exercise 49.)

2.3 ONE-SIDED LIMITS

When considering $\lim_{x \to a} f(x)$, we are concerned with values of x in an open interval containing a but not at a itself, that is, at values of x close to a and either greater than a or less than a. However, suppose that we have the function f for which, say, $f(x) = \sqrt{x - 4}$. Because $f(x)$ does not exist if $x < 4$, f is not defined on any open interval containing 4. Hence $\lim_{x \to 4} \sqrt{x - 4}$ has no meaning. However, if x is restricted to values greater than 4, the value of $\sqrt{x - 4}$ can be made as close to 0 as we please by taking x sufficiently close to 4 but greater than 4. In such a case we let x approach 4 from the right and consider the **one-sided limit from the right**, or the **right-hand limit**, which is now defined.

2.3.1 DEFINITION

Let f be a function that is defined at every number in some open interval (a, c). Then the **limit of $f(x)$, as x approaches a from the right, is L**, written

$$\lim_{x \to a^+} f(x) = L$$

if for any $\epsilon > 0$, however small, there exists a $\delta > 0$ such that

if $\quad 0 < x - a < \delta \quad$ then $\quad |f(x) - L| < \epsilon$

Note that in the preceding statement there are no absolute-value bars around $x - a$ since the condition $x > a$ is equivalent to $x - a > 0$.

It follows from Definition 2.3.1 that

$$\lim_{x \to 4^+} \sqrt{x - 4} = 0$$

If when considering the limit of a function the independent variable x is restricted to values less than a number a, we say that x approaches a from the left; the limit is called the **one-sided limit from the left**, or the **left-hand limit**.

2.3.2 DEFINITION Let f be a function that is defined at every number in some open interval (d, a). Then the **limit of $f(x)$, as x approaches a from the left, is L**, written

$$\lim_{x \to a^-} f(x) = L$$

if for any $\epsilon > 0$, however small, there exists a $\delta > 0$ such that

$$\text{if} \quad 0 < a - x < \delta \quad \text{then} \quad |f(x) - L| < \epsilon$$

We can refer to $\lim\limits_{x \to a} f(x)$ as the **two-sided limit** to distinguish it from the one-sided limits.

Limit Theorems 1–10 given in Section 2.2 remain unchanged when "$x \to a$" is replaced by "$x \to a^+$" or "$x \to a^-$."

EXAMPLE 1   The **signum function** is defined by

$$\text{sgn } x = \begin{cases} -1 & \text{if } x < 0 \\ 0 & \text{if } x = 0 \\ 1 & \text{if } 0 < x \end{cases}$$

Signum is the latin word for *sign*.

(a) Draw a sketch of the graph of this function. (b) Find $\lim\limits_{x \to 0^-} \text{sgn } x$ and $\lim\limits_{x \to 0^+} \text{sgn } x$, if they exist.

Solution

(a) A sketch of the graph appears in Figure 1.
(b) Because $\text{sgn } x = -1$ if $x < 0$ and $\text{sgn } x = 1$ if $0 < x$, we have

$$\lim_{x \to 0^-} \text{sgn } x = \lim_{x \to 0^-} (-1) \qquad \lim_{x \to 0^+} \text{sgn } x = \lim_{x \to 0^+} 1$$

$$= -1 \qquad\qquad\qquad = 1$$

FIGURE 1

In Example 1, $\lim\limits_{x \to 0^-} \text{sgn } x \neq \lim\limits_{x \to 0^+} \text{sgn } x$. Because the left-hand limit and the right-hand limit are not equal, the two-sided limit $\lim\limits_{x \to 0} \text{sgn } x$ does not exist. The concept of the two-sided limit failing to exist because the two one-sided limits are unequal is stated in the following theorem.

2.3.3 THEOREM $\lim\limits_{x \to a} f(x)$ exists and is equal to L if and only if $\lim\limits_{x \to a^-} f(x)$ and $\lim\limits_{x \to a^+} f(x)$ both exist and both are equal to L.

The proof of Theorem 2.3.3 is left as an exercise (see Exercise 34).

▶ **ILLUSTRATION 1**   A wholesaler sells a product by the pound (or fraction of a pound); if not more than 10 pounds are ordered, the wholesaler charges $1 per pound. However, to invite large orders the wholesaler charges only 90 cents per pound if more than 10 pounds are purchased. Thus if x pounds of the product are purchased and $C(x)$ dollars is the total cost of the order, then

$$C(x) = \begin{cases} x & \text{if } 0 \leq x \leq 10 \\ 0.9x & \text{if } 10 < x \end{cases}$$

A sketch of the graph of C is shown in Figure 2. Observe that $C(x)$ is obtained from the equation $C(x) = x$ when $0 \leq x \leq 10$ and from the equation $C(x) = 0.9x$ when $10 < x$. Because of this situation, when considering the limit of $C(x)$ as x approaches 10, we must distinguish between the left-hand limit at 10 and the right-hand limit at 10. For the function C we have

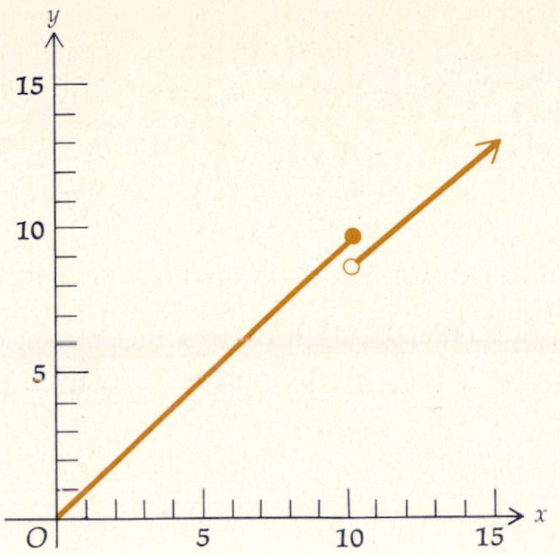

$$\lim_{x \to 10^-} C(x) = \lim_{x \to 10^-} x \qquad \lim_{x \to 10^+} C(x) = \lim_{x \to 10^+} 0.9x$$

$$= 10 \qquad\qquad = 9$$

Because $\lim_{x \to 10^-} C(x) \neq \lim_{x \to 10^+} C(x)$, we conclude from Theorem 2.3.3 that $\lim_{x \to 10} C(x)$ does not exist. Observe in Figure 2 that at $x = 10$ there is a break in the graph of the function C. We return to this function in Section 2.6. ◀

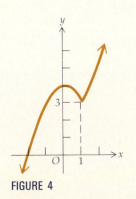

FIGURE 2

EXAMPLE 2   Let g be defined by

$$g(x) = \begin{cases} |x| & \text{if } x \neq 0 \\ 2 & \text{if } x = 0 \end{cases}$$

(a) Draw a sketch of the graph of g. (b) Find $\lim_{x \to 0} g(x)$ if it exists.

Solution
(a) A sketch of the graph is shown in Figure 3.

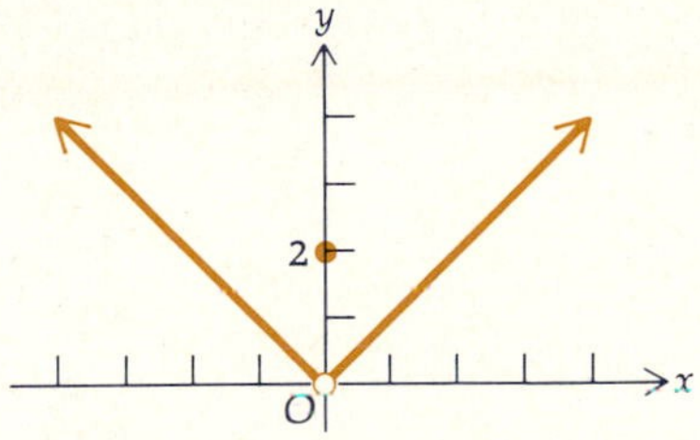

(b) $\lim_{x \to 0^-} g(x) = \lim_{x \to 0^-} (-x) \qquad \lim_{x \to 0^+} g(x) = \lim_{x \to 0^+} x$

$$= 0 \qquad\qquad = 0$$

Because $\lim_{x \to 0^-} g(x) = \lim_{x \to 0^+} g(x)$, it follows from Theorem 2.3.3 that $\lim_{x \to 0} g(x)$ exists and is equal to 0.

FIGURE 3

Observe in Example 2 that $g(0) = 2$, which has no effect on $\lim_{x \to 0} g(x)$. Also observe that there is a break in the graph of g at $x = 0$.

EXAMPLE 3   Let h be defined by

$$h(x) = \begin{cases} 4 - x^2 & \text{if } x \leq 1 \\ 2 + x^2 & \text{if } 1 < x \end{cases}$$

(a) Draw a sketch of the graph of h. (b) Find each of the following limits if they exist: $\lim_{x \to 1^-} h(x)$, $\lim_{x \to 1^+} h(x)$, $\lim_{x \to 1} h(x)$.

Solution
(a) A sketch of the graph appears in Figure 4.

(b) $\lim_{x \to 1^-} h(x) = \lim_{x \to 1^-} (4 - x^2) \qquad \lim_{x \to 1^+} h(x) = \lim_{x \to 1^+} (2 + x^2)$

$$= 3 \qquad\qquad = 3$$

Because $\lim_{x \to 1^-} h(x) = \lim_{x \to 1^+} h(x)$ and both are equal to 3, it follows from Theorem 2.3.3 that $\lim_{x \to 1} h(x) = 3$.

FIGURE 4

EXAMPLE 4   Let f be defined by

$$f(x) = \begin{cases} x + 5 & \text{if } x < -3 \\ \sqrt{9 - x^2} & \text{if } -3 \le x \le 3 \\ 3 - x & \text{if } 3 < x \end{cases}$$

(a) Draw a sketch of the graph of f. (b) Find, if they exist, each of the following limits: $\lim\limits_{x \to -3^-} f(x)$, $\lim\limits_{x \to -3^+} f(x)$, $\lim\limits_{x \to -3} f(x)$, $\lim\limits_{x \to 3^-} f(x)$, $\lim\limits_{x \to 3^+} f(x)$, $\lim\limits_{x \to 3} f(x)$.

Solution

(a) A sketch of the graph of f is shown in Figure 5.

(b) $\lim\limits_{x \to -3^-} f(x) = \lim\limits_{x \to -3^-} (x + 5)$ $\lim\limits_{x \to -3^+} f(x) = \lim\limits_{x \to -3^+} \sqrt{9 - x^2}$

$= 2$ $= 0$

Because $\lim\limits_{x \to -3^-} f(x) \ne \lim\limits_{x \to -3^+} f(x)$, then $\lim\limits_{x \to -3} f(x)$ does not exist.

$\lim\limits_{x \to 3^-} f(x) = \lim\limits_{x \to 3^-} \sqrt{9 - x^2}$ $\lim\limits_{x \to 3^+} f(x) = \lim\limits_{x \to 3^+} (3 - x)$

$= 0$ $= 0$

Because $\lim\limits_{x \to 3^-} f(x) = \lim\limits_{x \to 3^+} f(x)$, then $\lim\limits_{x \to 3} f(x)$ exists and is 0.

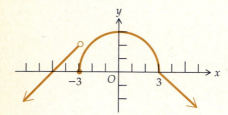

FIGURE 5

EXERCISES 2.3

In Exercises 1 through 22, draw a sketch of the graph, and find the indicated limit if it exists; if the limit does not exist, state the reason.

1. $f(x) = \begin{cases} 2 & \text{if } x < 1 \\ -1 & \text{if } x = 1 \\ -3 & \text{if } 1 < x \end{cases}$

(a) $\lim\limits_{x \to 1^+} f(x)$; (b) $\lim\limits_{x \to 1^-} f(x)$; (c) $\lim\limits_{x \to 1} f(x)$

2. $f(x) = \begin{cases} -2 & \text{if } x < 0 \\ 2 & \text{if } 0 \le x \end{cases}$

(a) $\lim\limits_{x \to 0^+} f(x)$; (b) $\lim\limits_{x \to 0^-} f(x)$; (c) $\lim\limits_{x \to 0} f(x)$

3. $f(t) = \begin{cases} t + 4 & \text{if } t \le -4 \\ 4 - t & \text{if } -4 < t \end{cases}$

(a) $\lim\limits_{t \to -4^+} f(t)$; (b) $\lim\limits_{t \to -4^-} f(t)$; (c) $\lim\limits_{t \to -4} f(t)$

4, $g(s) = \begin{cases} s + 3 & \text{if } s \le -2 \\ 3 - s & \text{if } -2 < s \end{cases}$

(a) $\lim\limits_{s \to -2^+} g(s)$; (b) $\lim\limits_{s \to -2^-} g(s)$; (c) $\lim\limits_{s \to -2} g(s)$

5. $F(x) = \begin{cases} x^2 & \text{if } x \le 2 \\ 8 - 2x & \text{if } 2 < x \end{cases}$

(a) $\lim\limits_{x \to 2^+} F(x)$; (b) $\lim\limits_{x \to 2^-} F(x)$; (c) $\lim\limits_{x \to 2} F(x)$

6. $h(x) = \begin{cases} 2x + 1 & \text{if } x < 3 \\ 10 - x & \text{if } 3 \le x \end{cases}$

(a) $\lim\limits_{x \to 3^+} h(x)$; (b) $\lim\limits_{x \to 3^-} h(x)$; (c) $\lim\limits_{x \to 3} h(x)$

7. $g(r) = \begin{cases} 2r + 3 & \text{if } r < 1 \\ 2 & \text{if } r = 1 \\ 7 - 2r & \text{if } 1 < r \end{cases}$

(a) $\lim\limits_{r \to 1^+} g(r)$; (b) $\lim\limits_{r \to 1^-} g(r)$; (c) $\lim\limits_{r \to 1} g(r)$

8. $g(t) = \begin{cases} 3 + t^2 & \text{if } t < -2 \\ 0 & \text{if } t = -2 \\ 11 - t^2 & \text{if } -2 < t \end{cases}$

(a) $\lim\limits_{t \to -2^+} g(t)$; (b) $\lim\limits_{t \to -2^-} g(t)$; (c) $\lim\limits_{t \to -2} g(t)$

9. $f(x) = \begin{cases} x^2 - 4 & \text{if } x < 2 \\ 4 & \text{if } x = 2 \\ 4 - x^2 & \text{if } 2 < x \end{cases}$

(a) $\lim\limits_{x \to 2^+} f(x)$; (b) $\lim\limits_{x \to 2^-} f(x)$; (c) $\lim\limits_{x \to 2} f(x)$

10. $f(x) = \begin{cases} 2x + 3 & \text{if } x < 1 \\ 4 & \text{if } x = 1 \\ x^2 + 2 & \text{if } 1 < x \end{cases}$

(a) $\lim\limits_{x \to 1^+} f(x)$; (b) $\lim\limits_{x \to 1^-} f(x)$; (c) $\lim\limits_{x \to 1} f(x)$

11. $F(x) = |x - 5|$
   (a) $\lim\limits_{x \to 5^+} F(x)$; (b) $\lim\limits_{x \to 5^-} F(x)$; (c) $\lim\limits_{x \to 5} F(x)$

12. $f(x) = 3 + |2x - 4|$
   (a) $\lim\limits_{x \to 2^+} f(x)$; (b) $\lim\limits_{x \to 2^-} f(x)$; (c) $\lim\limits_{x \to 2} f(x)$

13. $G(x) = |2x - 3| - 4$
   (a) $\lim\limits_{x \to 3/2^+} G(x)$; (b) $\lim\limits_{x \to 3/2^-} G(x)$; (c) $\lim\limits_{x \to 3/2} G(x)$

14. $F(x) = \begin{cases} |x - 1| & \text{if } x < -1 \\ 0 & \text{if } x = -1 \\ |1 - x| & \text{if } -1 < x \end{cases}$
   (a) $\lim\limits_{x \to -1^+} F(x)$; (b) $\lim\limits_{x \to -1^-} F(x)$; (c) $\lim\limits_{x \to -1} F(x)$

15. $f(x) = \dfrac{|x|}{x}$
   (a) $\lim\limits_{x \to 0^+} f(x)$; (b) $\lim\limits_{x \to 0^-} f(x)$; (c) $\lim\limits_{x \to 0} f(x)$

16. $S(x) = |\text{sgn } x|$ (sgn x is defined in Example 1)
   (a) $\lim\limits_{x \to 0^+} S(x)$; (b) $\lim\limits_{x \to 0^-} S(x)$; (c) $\lim\limits_{x \to 0} S(x)$

17. $f(x) = \begin{cases} 2 & \text{if } x < -2 \\ \sqrt{4 - x^2} & \text{if } -2 \le x \le 2 \\ -2 & \text{if } 2 < x \end{cases}$
   (a) $\lim\limits_{x \to -2^-} f(x)$; (b) $\lim\limits_{x \to -2^+} f(x)$; (c) $\lim\limits_{x \to -2} f(x)$; (d) $\lim\limits_{x \to 2^-} f(x)$;
   (e) $\lim\limits_{x \to 2^+} f(x)$; (f) $\lim\limits_{x \to 2} f(x)$

18. $f(x) = \begin{cases} x + 1 & \text{if } x < -1 \\ x^2 & \text{if } -1 \le x \le 1 \\ 2 - x & \text{if } 1 < x \end{cases}$
   (a) $\lim\limits_{x \to -1^-} f(x)$; (b) $\lim\limits_{x \to -1^+} f(x)$; (c) $\lim\limits_{x \to -1} f(x)$; (d) $\lim\limits_{x \to 1^-} f(x)$;
   (e) $\lim\limits_{x \to 1^+} f(x)$; (f) $\lim\limits_{x \to 1} f(x)$

19. $f(t) = \begin{cases} \sqrt[3]{t} & \text{if } t < 0 \\ \sqrt{t} & \text{if } 0 \le t \end{cases}$
   (a) $\lim\limits_{t \to 0^+} f(t)$; (b) $\lim\limits_{t \to 0^-} f(t)$; (c) $\lim\limits_{t \to 0} f(t)$

20. $g(x) = \begin{cases} \sqrt[3]{-x} & \text{if } x \le 0 \\ \sqrt[3]{x} & \text{if } 0 < x \end{cases}$
   (a) $\lim\limits_{x \to 0^+} g(x)$; (b) $\lim\limits_{x \to 0^-} g(x)$; (c) $\lim\limits_{x \to 0} g(x)$

21. $F(x) = \begin{cases} \sqrt{x^2 - 9} & \text{if } x \le -3 \\ \sqrt{9 - x^2} & \text{if } -3 < x < 3 \\ \sqrt{x^2 - 9} & \text{if } 3 \le x \end{cases}$
   (a) $\lim\limits_{x \to -3^-} F(x)$; (b) $\lim\limits_{x \to -3^+} F(x)$; (c) $\lim\limits_{x \to -3} F(x)$; (d) $\lim\limits_{x \to 3^-} F(x)$;
   (e) $\lim\limits_{x \to 3^+} F(x)$; (f) $\lim\limits_{x \to 3} F(x)$

22. $G(t) = \begin{cases} \sqrt[3]{t + 1} & \text{if } t \le -1 \\ \sqrt{1 - t^2} & \text{if } -1 < t < 1 \\ \sqrt[3]{t - 1} & \text{if } 1 \le t \end{cases}$
   (a) $\lim\limits_{t \to -1^-} G(t)$; (b) $\lim\limits_{t \to -1^+} G(t)$; (c) $\lim\limits_{t \to -1} G(t)$; (d) $\lim\limits_{t \to 1^-} G(t)$;
   (e) $\lim\limits_{t \to 1^+} G(t)$; (f) $\lim\limits_{t \to 1} G(t)$

23. $F(x) = x - 2 \text{ sgn } x$, where sgn x is defined in Example 1. Find, if they exist: (a) $\lim\limits_{x \to 0^+} F(x)$; (b) $\lim\limits_{x \to 0^-} F(x)$; (c) $\lim\limits_{x \to 0} F(x)$.

24. $h(x) = \text{sgn } x - U(x)$, where sgn x is defined in Example 1, and U is the unit step function defined by
$$U(x) = \begin{cases} 0 & \text{if } x < 0 \\ 1 & \text{if } 0 \le x \end{cases}$$
   Find, if they exist: (a) $\lim\limits_{x \to 0^+} h(x)$; (b) $\lim\limits_{x \to 0^-} h(x)$; (c) $\lim\limits_{x \to 0} h(x)$.

25. Find, if they exist: (a) $\lim\limits_{x \to 2^+} [\![x]\!]$; (b) $\lim\limits_{x \to 2^-} [\![x]\!]$; (c) $\lim\limits_{x \to 2} [\![x]\!]$.

26. Find, if they exist: (a) $\lim\limits_{x \to 4^+} [\![x - 3]\!]$; (b) $\lim\limits_{x \to 4^-} [\![x - 3]\!]$;
   (c) $\lim\limits_{x \to 4} [\![x - 3]\!]$.

27. Let $h(x) = (x - 1) \text{ sgn } x$. Draw a sketch of the graph of h. Find, if they exist: (a) $\lim\limits_{x \to 0^+} h(x)$; (b) $\lim\limits_{x \to 0^-} h(x)$; (c) $\lim\limits_{x \to 0} h(x)$.

28. Let $G(x) = [\![x]\!] + [\![4 - x]\!]$. Draw a sketch of the graph of G. Find, if they exist: (a) $\lim\limits_{x \to 3^+} G(x)$; (b) $\lim\limits_{x \to 3^-} G(x)$; (c) $\lim\limits_{x \to 3} G(x)$.

29. Given $f(x) = \begin{cases} 3x + 2 & \text{if } x < 4 \\ 5x + k & \text{if } 4 \le x \end{cases}$. Find the value of k such that $\lim\limits_{x \to 4} f(x)$ exists.

30. Given $f(x) = \begin{cases} kx - 3 & \text{if } x \le -1 \\ x^2 + k & \text{if } -1 < x \end{cases}$. Find the value of k such that $\lim\limits_{x \to -1} f(x)$ exists.

31. Given $f(x) = \begin{cases} x^2 & \text{if } x \le -2 \\ ax + b & \text{if } -2 < x < 2 \\ 2x - 6 & \text{if } 2 \le x \end{cases}$. Find the values of a and b such that $\lim\limits_{x \to -2} f(x)$ and $\lim\limits_{x \to 2} f(x)$ both exist.

32. Given $f(x) = \begin{cases} 2x - a & \text{if } x < -3 \\ ax + 2b & \text{if } -3 \le x \le 3 \\ b - 5x & \text{if } 3 < x \end{cases}$. Find the values of a and b such that $\lim\limits_{x \to -3} f(x)$ and $\lim\limits_{x \to 3} f(x)$ both exist.

33. Let $f(x) = \begin{cases} -1 & \text{if } x < 0 \\ 1 & \text{if } 0 < x \end{cases}$. Show that $\lim\limits_{x \to 0} f(x)$ does not exist but that $\lim\limits_{x \to 0} |f(x)|$ does exist.

34. Prove Theorem 2.3.3.

35. Shipping charges are often based on a formula that offers a lower charge per pound as the size of the shipment is increased. Suppose x pounds is the weight of a shipment, $C(x)$ dollars is the total cost of the shipment, and
$$C(x) = \begin{cases} 0.80x & \text{if } 0 < x \le 50 \\ 0.70x & \text{if } 50 < x \le 200 \\ 0.65x & \text{if } 200 < x \end{cases}$$

(a) Draw a sketch of the graph of C. Find each of the following limits: (b) $\lim\limits_{x \to 50^-} C(x)$; (c) $\lim\limits_{x \to 50^+} C(x)$; (d) $\lim\limits_{x \to 200^-} C(x)$; (e) $\lim\limits_{x \to 200^+} C(x)$.

36. Given $f(x) = \begin{cases} x^2 + 3 & \text{if } x \le 1 \\ x + 1 & \text{if } 1 < x \end{cases}$ and $g(x) = \begin{cases} x^2 & \text{if } x \le 1 \\ 2 & \text{if } 1 < x \end{cases}$

(a) Show that $\lim\limits_{x \to 1^-} f(x)$ and $\lim\limits_{x \to 1^+} f(x)$ both exist but are not equal, and hence $\lim\limits_{x \to 1} f(x)$ does not exist.

(b) Show that $\lim\limits_{x \to 1^-} g(x)$ and $\lim\limits_{x \to 1^+} g(x)$ both exist but are not equal, and hence $\lim\limits_{x \to 1} g(x)$ does not exist.

(c) Find formulas for $f(x) \cdot g(x)$.

(d) Prove that $\lim\limits_{x \to 1} [f(x) \cdot g(x)]$ exists by showing that $\lim\limits_{x \to 1^-} [f(x) \cdot g(x)] = \lim\limits_{x \to 1^+} [f(x) \cdot g(x)]$.

2.4 INFINITE LIMITS

In this section we discuss functions whose values increase or decrease without bound as the independent variable gets closer and closer to a fixed number. First let us consider the function defined by

$$f(x) = \frac{3}{(x - 2)^2}$$

The domain of f is the set of all real numbers except 2 and the range is the set of all positive numbers. We investigate the function values of f when x is close to 2. Let x approach 2 from the right; in particular, we let x be 3, 2.5, 2.25, 2.1, 2.01, and 2.001 and use a calculator to compute the corresponding values of $f(x)$ shown in Table 1. From this table you see intuitively that as x gets closer and closer to 2 through values greater than 2, $f(x)$ increases without bound. In other words, we can make $f(x)$ greater than any preassigned positive number (i.e., $f(x)$ can be made as large as we please) for all values of x close enough to 2 and x greater than 2.

To indicate that $f(x)$ increases without bound as x approaches 2 through values greater than 2 we write

$$\lim_{x \to 2^+} \frac{3}{(x - 2)^2} = +\infty$$

Now let x approach 2 from the left; in particular, we let x take on the values 1, 1.5, 1.75, 1.9, 1.99, and 1.999 and use a calculator to compute the corresponding values of $f(x)$ appearing in Table 2. You see intuitively from this table that as x gets closer and closer to 2 through values less than 2, $f(x)$ increases without bound; so we write

$$\lim_{x \to 2^-} \frac{3}{(x - 2)^2} = +\infty$$

Therefore, as x approaches 2 from either the right or the left, $f(x)$ increases without bound, and we write

$$\lim_{x \to 2} \frac{3}{(x - 2)^2} = +\infty$$

From the information in Tables 1 and 2 we obtain the sketch of the graph of f shown in Figure 1. Observe that both "branches" of the curve get closer and closer to the dashed line $x = 2$ as x increases without bound. This dashed line is called a *vertical asymptote*, defined later in this section.

We now state the formal definition of *function values increasing without bound*.

Table 1

x	$f(x) = \dfrac{3}{(x-2)^2}$
3	3
2.5	12
2.25	48
2.1	300
2.01	30,000
2.001	3,000,000

Table 2

x	$f(x) = \dfrac{3}{(x-2)^2}$
1	3
1.5	12
1.75	48
1.9	300
1.99	30,000
1.999	3,000,000

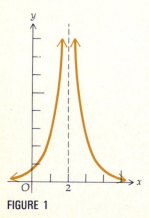

FIGURE 1

2.4.1 DEFINITION   Let f be a function that is defined at every number in some open interval I containing a, except possibly at the number a itself. **As x approaches a, $f(x)$ increases without bound**, which is written

$$\lim_{x \to a} f(x) = +\infty \tag{1}$$

if for any number $N > 0$ there exists a $\delta > 0$ such that

if $0 < |x - a| < \delta$ then $f(x) > N$

Another way of stating Definition 2.4.1 is as follows: "The function values $f(x)$ increase without bound as x approaches a number a if $f(x)$ can be made as large as we please (i.e., greater than any positive number N) for all values of x sufficiently close to a but not equal to a."

We stress again that $+\infty$ is not a symbol for a real number; hence, when we write $\lim_{x \to a} f(x) = +\infty$, it does not have the same meaning as $\lim_{x \to a} f(x) = L$, where L is a real number. Equation (1) can be read as "the limit of $f(x)$ as x approaches a is positive infinity." In such a case the limit does not exist, but the symbol $+\infty$ indicates the behavior of the function values $f(x)$ as x gets closer and closer to a.

In an analogous manner we can indicate the behavior of a function whose function values decrease without bound. To lead up to this, consider the function g defined by the equation

$$g(x) = \frac{-3}{(x - 2)^2}$$

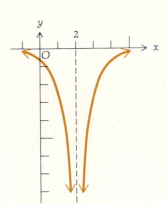

FIGURE 2

A sketch of the graph of this function is in Figure 2.

The function values given by $g(x) = -3/(x - 2)^2$ are the negatives of the function values given by $f(x) = 3/(x - 2)^2$. So for the function g, as x approaches 2, either from the right or the left, $g(x)$ decreases without bound, and we write

$$\lim_{x \to 2} \frac{-3}{(x - 2)^2} = -\infty$$

2.4.2 DEFINITION   Let f be a function that is defined at every number in some open interval I containing a, except possibly at the number a itself. **As x approaches a, $f(x)$ decreases without bound**, which is written

$$\lim_{x \to a} f(x) = -\infty \tag{2}$$

if for any number $N < 0$ there exists a $\delta > 0$ such that

if $0 < |x - a| < \delta$ then $f(x) < N$

Note: Equation (2) can be read as "the limit of $f(x)$ as x approaches a is negative infinity," observing again that the limit does not exist and that the symbol $-\infty$ indicates only the behavior of the function values as x approaches a.

We can consider one-sided limits that are "infinite." In particular, $\lim_{x \to a^+} f(x) = +\infty$ if f is defined at every number in some open interval (a, c) and if for any number $N > 0$ there exists a $\delta > 0$ such that

if $0 < x - a < \delta$, then $f(x) > N$

Similar definitions can be given if $\lim\limits_{x \to a^-} f(x) = +\infty$, $\lim\limits_{x \to a^+} f(x) = -\infty$, and $\lim\limits_{x \to a^-} f(x) = -\infty$.

Now suppose that h is the function defined by the equation

$$h(x) = \frac{2x}{x - 1} \tag{3}$$

A sketch of the graph of this function is in Figure 3. By referring to Figures 1, 2, and 3, note the difference in the behavior of the function whose graph is sketched in Figure 3 from the functions of the other two figures. Observe that

$$\lim_{x \to 1^-} \frac{2x}{x - 1} = -\infty \tag{4}$$

$$\lim_{x \to 1^+} \frac{2x}{x - 1} = +\infty \tag{5}$$

That is, for the function defined by (3), as x approaches 1 through values less than 1, the function values decrease without bound, and as x approaches 1 through values greater than 1, the function values increase without bound.

Before giving some examples, we need two limit theorems involving "infinite" limits.

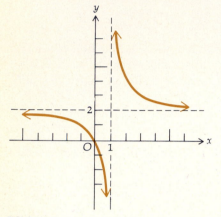

FIGURE 3

2.4.3 LIMIT THEOREM 11   If r is any positive integer, then

(i) $\lim\limits_{x \to 0^+} \dfrac{1}{x^r} = +\infty$;

(ii) $\lim\limits_{x \to 0^-} \dfrac{1}{x^r} = \left\{ \begin{array}{ll} -\infty & \text{if } r \text{ is odd} \\ +\infty & \text{if } r \text{ is even} \end{array} \right\}$.

Proof   We prove part (i). The proof of part (ii) is analogous and is left as an exercise (see Exercise 45). We must show that for any $N > 0$ there exists a $\delta > 0$ such that

$$\text{if}\quad 0 < x < \delta \quad \text{then}\quad \frac{1}{x^r} > N$$

or, equivalently, because $x > 0$ and $N > 0$,

$$\text{if}\quad 0 < x < \delta \quad \text{then}\quad x^r < \frac{1}{N}$$

or, equivalently, because $r > 0$,

$$\text{if}\quad 0 < x < \delta \quad \text{then}\quad x < \left(\frac{1}{N} \right)^{1/r}$$

The above statement holds if $\delta = \left(\dfrac{1}{N} \right)^{1/r}$. Therefore when $\delta = \left(\dfrac{1}{N} \right)^{1/r}$

$$\text{if}\quad 0 < x < \delta \quad \text{then}\quad \frac{1}{x^r} > N \qquad\blacksquare$$

▶ **ILLUSTRATION 1** From Limit Theorem 11(i) it follows that

$$\lim_{x \to 0^+} \frac{1}{x^3} = +\infty \quad \text{and} \quad \lim_{x \to 0^+} \frac{1}{x^4} = +\infty$$

From Limit Theorem 11(ii)

$$\lim_{x \to 0^-} \frac{1}{x^3} = -\infty \quad \text{and} \quad \lim_{x \to 0^-} \frac{1}{x^4} = +\infty$$ ◀

Limit Theorem 12, which follows, involves the limit of a rational function for which the limit of the denominator is zero and the limit of the numerator is a nonzero constant. Such a situation occurs in (4) and (5).

2.4.4 LIMIT THEOREM 12 If a is any real number, and if $\lim_{x \to a} f(x) = 0$ and $\lim_{x \to a} g(x) = c$, where c is a constant not equal to 0, then

(i) if $c > 0$ and if $f(x) \to 0$ through positive values of $f(x)$,

$$\lim_{x \to a} \frac{g(x)}{f(x)} = +\infty$$

(ii) if $c > 0$ and if $f(x) \to 0$ through negative values of $f(x)$,

$$\lim_{x \to a} \frac{g(x)}{f(x)} = -\infty$$

(iii) if $c < 0$ and if $f(x) \to 0$ through positive values of $f(x)$,

$$\lim_{x \to a} \frac{g(x)}{f(x)} = -\infty$$

(iv) if $c < 0$ and if $f(x) \to 0$ through negative values of $f(x)$,

$$\lim_{x \to a} \frac{g(x)}{f(x)} = +\infty$$

The theorem is also valid if "$x \to a$" is replaced by "$x \to a^+$" or "$x \to a^-$."

Proof We shall prove part (i) and leave the proofs of the other parts as exercises (see Exercises 46–48).

To prove that

$$\lim_{x \to a} \frac{g(x)}{f(x)} = +\infty$$

we must show that for any $N > 0$ there exists a $\delta > 0$ such that

$$\text{if} \quad 0 < |x - a| < \delta \quad \text{then} \quad \frac{g(x)}{f(x)} > N \tag{6}$$

Since $\lim_{x \to a} g(x) = c > 0$, by taking $\epsilon = \frac{1}{2}c$ in Definition 2.1.1 it follows that there exists a $\delta_1 > 0$ such that

$$\text{if} \quad 0 < |x - a| < \delta_1 \quad \text{then} \quad |g(x) - c| < \tfrac{1}{2}c$$

By applying Theorem 1.1.10 to the above second inequality it follows that there exists a $\delta_1 > 0$ such that

$$\text{if}\quad 0 < |x - a| < \delta_1 \quad \text{then} \quad -\tfrac{1}{2}c < g(x) - c < \tfrac{1}{2}c$$

$$\Leftrightarrow \quad \text{if}\quad 0 < |x - a| < \delta_1 \quad \text{then} \quad \tfrac{1}{2}c < g(x) < \tfrac{3}{2}c$$

So there exists a $\delta_1 > 0$ such that

$$\text{if}\quad 0 < |x - a| < \delta_1 \quad \text{then} \quad g(x) > \tfrac{1}{2}c \tag{7}$$

Now $\lim\limits_{x \to a} f(x) = 0$. Thus for any $\epsilon > 0$ there exists a $\delta_2 > 0$ such that

$$\text{if}\quad 0 < |x - a| < \delta_2 \quad \text{then} \quad |f(x)| < \epsilon$$

Since $f(x)$ is approaching zero through positive values of $f(x)$, the absolute value bars around $f(x)$ can be removed; hence for any $\epsilon > 0$ there exists a $\delta_2 > 0$ such that

$$\text{if}\quad 0 < |x - a| < \delta_2 \quad \text{then} \quad 0 < f(x) < \epsilon \tag{8}$$

From statements (7) and (8) we can conclude that for any $\epsilon > 0$ there exist a $\delta_1 > 0$ and a $\delta_2 > 0$ such that

$$\text{if}\quad 0 < |x - a| < \delta_1 \quad \text{and} \quad 0 < |x - a| < \delta_2 \quad \text{then} \quad \frac{g(x)}{f(x)} > \frac{\tfrac{1}{2}c}{\epsilon}$$

Hence, if $\epsilon = c/(2N)$ and $\delta = \min(\delta_1, \delta_2)$, then

$$\text{if}\quad 0 < |x - a| < \delta \quad \text{then} \quad \frac{g(x)}{f(x)} > \frac{\tfrac{1}{2}c}{c/(2N)} = N$$

which is statement (6). Hence part (i) is proved.   ■

When Limit Theorem 12 is applied, we can often get an indication of whether the result is $+\infty$ or $-\infty$ by taking a *suitable value* of x near a to ascertain if the quotient is positive or negative, as shown in the following illustration.

▶ **ILLUSTRATION 2**   In (4) we have

$$\lim_{x \to 1^-} \frac{2x}{x - 1}$$

Limit Theorem 12 is applicable because $\lim\limits_{x \to 1^-} 2x = 2$ and $\lim\limits_{x \to 1^-} (x - 1) = 0$. We wish to determine if we have $+\infty$ or $-\infty$. Because $x \to 1^-$, take a value of x near 1 and less than 1; for instance, take $x = 0.9$. Then

$$\frac{2x}{x - 1} = \frac{2(0.9)}{0.9 - 1} \quad \Leftrightarrow \quad \frac{2x}{x - 1} = -18$$

The negative quotient leads us to suspect that

$$\lim_{x \to 1^-} \frac{2x}{x - 1} = -\infty$$

This result follows from part (ii) of Limit Theorem 12, because when $x \to 1^-$, $x - 1$ is approaching 0 through negative values.

For the limit in (5), because $x \to 1^+$, take $x = 1.1$. Then

$$\frac{2x}{x-1} = \frac{2(1.1)}{1.1-1} \quad \Leftrightarrow \quad \frac{2x}{x-1} = 22$$

Because the quotient is positive we suspect that

$$\lim_{x \to 1^+} \frac{2x}{x-1} = +\infty$$

This result follows from part (i) of Limit Theorem 12, because when $x \to 1^+$, $x - 1$ is approaching 0 through positive values.   ◄

When using the procedure described in Illustration 2, be careful that the value of x selected is close enough to a to indicate the true behavior of the quotient. For instance, when computing $\lim\limits_{x \to 1^-} \dfrac{2x}{x-1}$, the value of x selected must not only be less than 1 but also greater than 0.

EXAMPLE 1   Find

(a) $\lim\limits_{x \to 3^+} \dfrac{x^2 + x + 2}{x^2 - 2x - 3}$ (b) $\lim\limits_{x \to 3^-} \dfrac{x^2 + x + 2}{x^2 - 2x - 3}$

Solution

(a) $\lim\limits_{x \to 3^+} \dfrac{x^2 + x + 2}{x^2 - 2x - 3} = \lim\limits_{x \to 3^+} \dfrac{x^2 + x + 2}{(x-3)(x+1)}$

The limit of the numerator is 14, which can be verified easily.

$$\lim_{x \to 3^+} (x-3)(x+1) = \lim_{x \to 3^+} (x-3) \cdot \lim_{x \to 3^+} (x+1)$$
$$= 0 \cdot 4$$
$$= 0$$

The limit of the denominator is 0, and the denominator is approaching 0 through positive values. Then from Limit Theorem 12(i),

$$\lim_{x \to 3^+} \frac{x^2 + x + 2}{x^2 - 2x - 3} = +\infty$$

(b) $\lim\limits_{x \to 3^-} \dfrac{x^2 + x + 2}{x^2 - 2x - 3} = \lim\limits_{x \to 3^-} \dfrac{x^2 + x + 2}{(x-3)(x+1)}$

As in part (a), the limit of the numerator is 14.

$$\lim_{x \to 3^-} (x-3)(x+1) = \lim_{x \to 3^-} (x-3) \cdot \lim_{x \to 3^-} (x+1)$$
$$= 0 \cdot 4$$
$$= 0$$

In this case, the limit of the denominator is zero, but the denominator is approaching zero through negative values. From Limit Theorem 12(ii),

$$\lim_{x \to 3^-} \frac{x^2 + x + 2}{x^2 - 2x - 3} = -\infty$$

EXAMPLE 2   Find

(a) $\displaystyle\lim_{x \to 2^+} \frac{\sqrt{x^2 - 4}}{x - 2}$ (b) $\displaystyle\lim_{x \to 2^-} \frac{\sqrt{4 - x^2}}{x - 2}$

Solution
(a) Because $x \to 2^+$, $x - 2 > 0$; so $x - 2 = \sqrt{(x - 2)^2}$. Thus

$$\lim_{x \to 2^+} \frac{\sqrt{x^2 - 4}}{x - 2} = \lim_{x \to 2^+} \frac{\sqrt{(x - 2)(x + 2)}}{\sqrt{(x - 2)^2}}$$

$$= \lim_{x \to 2^+} \frac{\sqrt{x - 2}\sqrt{x + 2}}{\sqrt{x - 2}\sqrt{x - 2}}$$

$$= \lim_{x \to 2^+} \frac{\sqrt{x + 2}}{\sqrt{x - 2}}$$

The limit of the numerator is 2. The limit of the denominator is 0, and the denominator is approaching 0 through positive values. Therefore, by Limit Theorem 12(i) it follows that

$$\lim_{x \to 2^+} \frac{\sqrt{x^2 - 4}}{x - 2} = +\infty$$

(b) Because $x \to 2^-$, $x - 2 < 0$; so $x - 2 = -\sqrt{(2 - x)^2}$. Therefore

$$\lim_{x \to 2^-} \frac{\sqrt{4 - x^2}}{x - 2} = \lim_{x \to 2^-} \frac{\sqrt{2 - x}\sqrt{2 + x}}{-\sqrt{2 - x}\sqrt{2 - x}}$$

$$= \lim_{x \to 2^-} \frac{\sqrt{2 + x}}{-\sqrt{2 - x}}$$

The limit of the numerator is 2. The limit of the denominator is 0, and the denominator is approaching 0 through negative values. Hence by Limit Theorem 12(ii),

$$\lim_{x \to 2^-} \frac{\sqrt{4 - x^2}}{x - 2} = -\infty$$

EXAMPLE 3   Find

$$\lim_{x \to 4^-} \frac{[\![x]\!] - 4}{x - 4}$$

Solution $\displaystyle\lim_{x \to 4^-} [\![x]\!] = 3$.   Therefore $\displaystyle\lim_{x \to 4^-} ([\![x]\!] - 4) = -1$.   Furthermore, $\displaystyle\lim_{x \to 4^-} (x - 4) = 0$, and $x - 4$ is approaching 0 through negative values. Hence from Limit Theorem 12(iv),

$$\lim_{x \to 4^-} \frac{[\![x]\!] - 4}{x - 4} = +\infty$$

Remember that because $+\infty$ and $-\infty$ are not symbols for real numbers, the Limit Theorems 1–10 of Section 2.2 do not hold for "infinite" limits. However, there are the following properties regarding such limits. The proofs are left as exercises (see Exercises 49–51).

2.4.5 THEOREM

(i) If $\lim\limits_{x \to a} f(x) = +\infty$, and $\lim\limits_{x \to a} g(x) = c$, where c is any constant, then

$$\lim_{x \to a} [f(x) + g(x)] = +\infty$$

(ii) If $\lim\limits_{x \to a} f(x) = -\infty$, and $\lim\limits_{x \to a} g(x) = c$, where c is any constant, then

$$\lim_{x \to a} [f(x) + g(x)] = -\infty$$

The theorem is valid if "$x \to a$" is replaced by "$x \to a^+$" or "$x \to a^-$."

▶ **ILLUSTRATION 3** Because $\lim\limits_{x \to 2^+} \dfrac{1}{x - 2} = +\infty$ and $\lim\limits_{x \to 2^+} \dfrac{1}{x + 2} = \dfrac{1}{4}$, it follows from Theorem 2.4.5(i) that

$$\lim_{x \to 2^+} \left[\frac{1}{x - 2} + \frac{1}{x + 2} \right] = +\infty$$ ◀

2.4.6 THEOREM If $\lim\limits_{x \to a} f(x) = +\infty$ and $\lim\limits_{x \to a} g(x) = c$, where c is any constant except 0, then

(i) if $c > 0$, $\lim\limits_{x \to a} f(x) \cdot g(x) = +\infty$;

(ii) if $c < 0$, $\lim\limits_{x \to a} f(x) \cdot g(x) = -\infty$.

The theorem is valid if "$x \to a$" is replaced by "$x \to a^+$" or "$x \to a^-$."

▶ **ILLUSTRATION 4**

$$\lim_{x \to 3} \frac{5}{(x - 3)^2} = +\infty \quad \text{and} \quad \lim_{x \to 3} \frac{x + 4}{x - 4} = -7$$

Therefore, from Theorem 2.4.6(ii),

$$\lim_{x \to 3} \left[\frac{5}{(x - 3)^2} \cdot \frac{x + 4}{x - 4} \right] = -\infty$$ ◀

2.4.7 THEOREM If $\lim\limits_{x \to a} f(x) = -\infty$ and $\lim\limits_{x \to a} g(x) = c$, where c is any constant except 0, then

(i) if $c > 0$, $\lim\limits_{x \to a} f(x) \cdot g(x) = -\infty$;

(ii) if $c < 0$, $\lim\limits_{x \to a} f(x) \cdot g(x) = +\infty$.

The theorem is valid if "$x \to a$" is replaced by "$x \to a^+$" or "$x \to a^-$."

▶ **ILLUSTRATION 5** In Example 2(b) we showed

$$\lim_{x \to 2^-} \frac{\sqrt{4 - x^2}}{x - 2} = -\infty$$

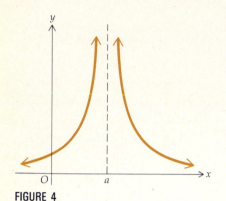

FIGURE 4

Furthermore,

$$\lim_{x \to 2^-} \frac{x-3}{x+2} = -\frac{1}{4}$$

Thus, from Theorem 2.4.7(ii) it follows that

$$\lim_{x \to 2^-} \left[\frac{\sqrt{4-x^2}}{x-2} \cdot \frac{x-3}{x+2} \right] = +\infty \qquad \blacktriangleleft$$

Infinite limits are applied to find *vertical asymptotes* of a graph if there are any. Refer to Figure 4 showing a sketch of the graph of the function defined by

$$f(x) = \frac{1}{(x-a)^2} \tag{9}$$

Any line parallel to and above the x axis will intersect this graph in two points: one point to the left of the line $x = a$ and one point to the right of this line. Thus for any $k > 0$, no matter how large, the line $y = k$ will intersect the graph of f in two points; the distance of these two points from the line $x = a$ gets smaller and smaller as k gets larger and larger. The line $x = a$ is called a *vertical asymptote* of the graph of f.

2.4.8 DEFINITION   The line $x = a$ is said to be a **vertical asymptote** of the graph of the function f if at least one of the following statements is true:

(i) $\lim\limits_{x \to a^+} f(x) = +\infty$

(ii) $\lim\limits_{x \to a^+} f(x) = -\infty$

(iii) $\lim\limits_{x \to a^-} f(x) = +\infty$

(iv) $\lim\limits_{x \to a^-} f(x) = -\infty$

▶ **ILLUSTRATION 6**   Each of the Figures 5 through 8 shows a portion of the graph of a function for which the line $x = a$ is a vertical asymptote. In Figure 5, part (i) of Definition 2.4.8 applies; in Figure 6, part (ii) applies; and in Figures 7 and 8, parts (iii) and (iv), respectively, apply.   ◀

$\lim\limits_{x \to a^+} f(x) = +\infty$

FIGURE 5

$\lim\limits_{x \to a^+} f(x) = -\infty$

FIGURE 6

$\lim\limits_{x \to a^-} f(x) = +\infty$

FIGURE 7

$\lim\limits_{x \to a^-} f(x) = -\infty$

FIGURE 8

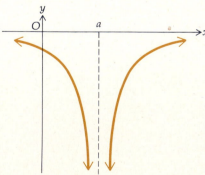

FIGURE 9

For the function defined by (9), both parts (i) and (iii) of the above definition are true. See Figure 4. If g is the function defined by

$$g(x) = -\frac{1}{(x-a)^2}$$

then both parts (ii) and (iv) are true, and the line $x = a$ is a vertical asymptote of the graph of g. This is shown in Figure 9.

EXAMPLE 4   Find the vertical asymptote and draw a sketch of the graph of the function defined by

$$f(x) = \frac{3}{x-3}$$

Solution

$$\lim_{x \to 3^+} \frac{3}{x-3} = +\infty \qquad \lim_{x \to 3^-} \frac{3}{x-3} = -\infty$$

It follows from Definition 2.4.8 that the line $x = 3$ is a vertical asymptote of the graph of f. A sketch of the graph of f appears in Figure 10.

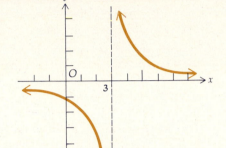

FIGURE 10

EXERCISES 2.4

In Exercises 1 through 12, do the following: (a) Use a calculator to tabulate values of $f(x)$ for the specified values of x, and from these values make a statement regarding the apparent behavior of $f(x)$. (b) Find the indicated limit.

1. (a) $f(x) = \dfrac{1}{x-5}$; x is 6, 5.5, 5.1, 5.01, 5.001, 5.0001;

(b) $\displaystyle\lim_{x \to 5^+} \dfrac{1}{x-5}$

2. (a) $f(x) = \dfrac{1}{x-5}$; x is 4, 4.5, 4.9, 4.99, 4.999, 4.9999;

(b) $\displaystyle\lim_{x \to 5^-} \dfrac{1}{x-5}$

3. (a) $f(x) = \dfrac{1}{(x-5)^2}$; x is 6, 5.5, 5.1, 5.01, 5.001, 5.0001 and x

is 4, 4.5, 4.9, 4.99, 4.999, 4.9999; (b) $\displaystyle\lim_{x \to 5} \dfrac{1}{(x-5)^2}$

4. (a) $f(x) = \dfrac{x+2}{1-x}$; x is 0, 0.5, 0.9, 0.99, 0.999, 0.9999;

(b) $\displaystyle\lim_{x \to 1^-} \dfrac{x+2}{1-x}$

5. (a) $f(x) = \dfrac{x+2}{1-x}$; x is 2, 1.5, 1.1, 1.01, 1.001, 1.0001;

(b) $\displaystyle\lim_{x \to 1^+} \dfrac{x+2}{1-x}$

6. (a) $f(x) = \dfrac{x+2}{(x-1)^2}$; x is 0, 0.5, 0.9, 0.99, 0.999, 0.9999 and x is

2, 1.5, 1.1, 1.01, 1.001, 1.0001; (b) $\displaystyle\lim_{x \to 1} \dfrac{x+2}{(x-1)^2}$

7. (a) $f(x) = \dfrac{x-2}{x+1}$; x is 0, -0.5, -0.9, -0.99, -0.999,

-0.9999; (b) $\displaystyle\lim_{x \to -1^+} \dfrac{x-2}{x+1}$

8. (a) $f(x) = \dfrac{x-2}{x+1}$; x is -2, -1.5, -1.1, -1.01, -1.001,

-1.0001; (b) $\displaystyle\lim_{x \to -1^-} \dfrac{x-2}{x+1}$

9. (a) $f(x) = \dfrac{x}{x+4}$; x is -5, -4.5, -4.1, -4.01, -4.001,

-4.0001; (b) $\displaystyle\lim_{x \to -4^-} \dfrac{x}{x+4}$

10. (a) $f(x) = \dfrac{x}{x-4}$; x is 5, 4.5, 4.1, 4.01, 4.001, 4.0001;

(b) $\displaystyle\lim_{x \to 4^+} \dfrac{x}{x-4}$

11. (a) $f(x) = \dfrac{4x}{9-x^2}$; x is -4, -3.5, -3.1, -3.01, -3.001,

-3.0001; (b) $\displaystyle\lim_{x \to -3^-} \dfrac{4x}{9-x^2}$

12. (a) $f(x) = \dfrac{4x^2}{9 - x^2}$; x is 4, 3.5, 3.1, 3.01, 3.001, 3.0001;

   (b) $\displaystyle\lim_{x \to 3^+} \dfrac{4x^2}{9 - x^2}$

In Exercises 13 through 32, find the limit.

13. $\displaystyle\lim_{t \to 2^+} \dfrac{t + 2}{t^2 - 4}$

14. $\displaystyle\lim_{t \to 2^-} \dfrac{-t + 2}{(t - 2)^2}$

15. $\displaystyle\lim_{t \to 2^-} \dfrac{t + 2}{t^2 - 4}$

16. $\displaystyle\lim_{x \to 0^+} \dfrac{\sqrt{3 + x^2}}{x}$

17. $\displaystyle\lim_{x \to 0^-} \dfrac{\sqrt{3 + x^2}}{x}$

18. $\displaystyle\lim_{x \to 0} \dfrac{\sqrt{3 + x^2}}{x^2}$

19. $\displaystyle\lim_{x \to 3^+} \dfrac{\sqrt{x^2 - 9}}{x - 3}$

20. $\displaystyle\lim_{x \to 4^-} \dfrac{\sqrt{16 - x^2}}{x - 4}$

21. $\displaystyle\lim_{x \to 0^+} \left(\dfrac{1}{x} - \dfrac{1}{x^2} \right)$

22. $\displaystyle\lim_{x \to 0^+} \dfrac{x^2 - 3}{x^3 + x^2}$

23. $\displaystyle\lim_{x \to 0^-} \dfrac{2 - 4x^3}{5x^2 + 3x^3}$

24. $\displaystyle\lim_{s \to 2^-} \left(\dfrac{1}{s - 2} - \dfrac{3}{s^2 - 4} \right)$

25. $\displaystyle\lim_{t \to -4^-} \left(\dfrac{2}{t^2 + 3t - 4} - \dfrac{3}{t + 4} \right)$

26. $\displaystyle\lim_{x \to 1^-} \dfrac{2x^3 - 5x^2}{x^2 - 1}$

27. $\displaystyle\lim_{x \to 3^-} \dfrac{[\![x]\!] - x}{3 - x}$

28. $\displaystyle\lim_{x \to 1^-} \dfrac{[\![x^2]\!] - 1}{x^2 - 1}$

29. $\displaystyle\lim_{x \to 3^-} \dfrac{x^3 + 9x^2 + 20x}{x^2 + x - 12}$

30. $\displaystyle\lim_{x \to -2^+} \dfrac{6x^2 + x - 2}{2x^2 + 3x - 2}$

31. $\displaystyle\lim_{x \to 1^+} \dfrac{x - 1}{\sqrt{2x - x^2} - 1}$

32. $\displaystyle\lim_{x \to 2^-} \dfrac{x - 2}{2 - \sqrt{4x - x^2}}$

33. For each of the following functions, find the vertical asymptote of the graph of the function, and draw a sketch of the

graph: (a) $f(x) = \dfrac{1}{x}$; (b) $g(x) = \dfrac{1}{x^2}$; (c) $h(x) = \dfrac{1}{x^3}$;

   (d) $\phi(x) = \dfrac{1}{x^4}$.

34. For each of the following functions, find the vertical asymptote of the graph of the function, and draw a sketch of the graph: (a) $f(x) = -\dfrac{1}{x}$; (b) $g(x) = -\dfrac{1}{x^2}$; (c) $h(x) = -\dfrac{1}{x^3}$;

   (d) $\phi(x) = -\dfrac{1}{x^4}$.

In Exercises 35 through 42, find the vertical asymptote(s) of the graph of the function, and draw a sketch of the graph.

35. $f(x) = \dfrac{2}{x - 4}$

36. $f(x) = \dfrac{3}{x + 1}$

37. $f(x) = \dfrac{-2}{x + 3}$

38. $f(x) = \dfrac{-4}{x - 5}$

39. $f(x) = \dfrac{-2}{(x + 3)^2}$

40. $f(x) = \dfrac{4}{(x - 5)^2}$

41. $f(x) = \dfrac{5}{x^2 + 8x + 15}$

42. $f(x) = \dfrac{1}{x^2 + 5x - 6}$

43. Prove that $\displaystyle\lim_{x \to 2} \dfrac{3}{(x - 2)^2} = +\infty$ by using Definition 2.4.1.

44. Prove that $\displaystyle\lim_{x \to 4} \dfrac{-2}{(x - 4)^2} = -\infty$ by using Definition 2.4.2.

45. Prove Theorem 2.4.3(ii).   46. Prove Theorem 2.4.4(ii).
47. Prove Theorem 2.4.4(iii).   48. Prove Theorem 2.4.4(iv).
49. Prove Theorem 2.4.5.   50. Prove Theorem 2.4.6.
51. Prove Theorem 2.4.7.

52. Use Definition 2.4.1 to prove that $\displaystyle\lim_{x \to -3} \left| \dfrac{5 - x}{3 + x} \right| = +\infty$.

2.5 LIMITS AT INFINITY

The previous section was devoted to infinite limits where function values either increased or decreased without bound as the independent variable approached a real number. We now consider limits of functions when the independent variable either increases or decreases without bound. We begin with the function defined by

$$f(x) = \dfrac{2x^2}{x^2 + 1}$$

Let x take on the values 0, 1, 2, 3, 4, 5, 10, 100, 1000, and so on, allowing x to increase without bound. The corresponding function values, either exact or approximated by a calculator to six decimal places, are given in Table 1. Observe from the table that as x increases through positive values, the function values get closer and closer to 2.

Table 1

x	$f(x) = \dfrac{2x^2}{x^2 + 1}$
0	0
1	1
2	1.6
3	1.8
4	1.882353
5	1.923077
10	1.980198
100	1.999800
1000	1.999998

In particular, when $x = 4$

$$2 - \frac{2x^2}{x^2 + 1} = 2 - 1.882353$$

$$= 0.117647$$

Therefore the difference between 2 and $f(x)$ is 0.117647 when $x = 4$. When $x = 100$,

$$2 - \frac{2x^2}{x^2 + 1} = 2 - 1.999800$$

$$= 0.000200$$

Hence the difference between 2 and $f(x)$ is 0.000200 when $x = 100$.

Continuing on, we see intuitively that the value of $f(x)$ can be made as close to 2 as we please by taking x large enough. In other words, the difference between 2 and $f(x)$ can be made as small as we please by taking x any number greater than some sufficiently large positive number. Or, going a step further, for any $\epsilon > 0$, however small, we can find a number $N > 0$ such that if $x > N$, then $|f(x) - 2| < \epsilon$.

When an independent variable x is increasing without bound through positive values, we write "$x \to +\infty$." From the illustrative example above, then, we can say that

$$\lim_{x \to +\infty} \frac{2x^2}{x^2 + 1} = 2$$

2.5.1 DEFINITION

Let f be a function that is defined at every number in some interval $(a, +\infty)$. The **limit of $f(x)$, as x increases without bound, is L,** written

$$\lim_{x \to +\infty} f(x) = L$$

if for any $\epsilon > 0$, however small, there exists a number $N > 0$ such that

$$\text{if} \quad x > N \quad \text{then} \quad |f(x) - L| < \epsilon$$

Table 2

x	$f(x) = \dfrac{2x^2}{x^2 + 1}$
-1	1
-2	1.6
-3	1.8
-4	1.882353
-5	1.923077
-10	1.980198
-100	1.999800
-1000	1.999998

Note: When $x \to +\infty$ is written, it does not have the same meaning as, for instance, $x \to 1000$. The symbol $x \to +\infty$ indicates the behavior of the variable x.

Now consider the same function, and let x take on the values -1, -2, -3, -4, -5, -10, -100, -1000, and so on, allowing x to decrease through negative values without bound. Table 2 gives the corresponding function values of $f(x)$.

Observe that the function values are the same for the negative numbers as for the corresponding positive numbers. So we see intuitively that as x decreases without bound, $f(x)$ approaches 2; that is, $|f(x) - 2|$ can be made as small as we please by taking x any number less than some negative number having a sufficiently large absolute value. Formally we say that for any $\epsilon > 0$, however small, we can find a number $N < 0$ such that if $x < N$, then $|f(x) - 2| < \epsilon$. Using the symbol $x \to -\infty$ to denote that the variable x is decreasing without

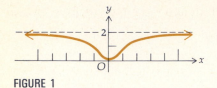

FIGURE 1

bound we write

$$\lim_{x \to -\infty} \frac{2x^2}{x^2 + 1} = 2$$

Figure 1 shows a sketch of the graph of our function. The line $x = 2$ appears as a dashed line, called a *horizontal asymptote*, which we define later in this section.

2.5.2 DEFINITION Let f be a function that is defined at every number in some interval $(-\infty, a)$. The **limit of $f(x)$, as x decreases without bound, is L,** written

$$\lim_{x \to -\infty} f(x) = L$$

if for any $\epsilon > 0$, however small, there exists a number $N < 0$ such that

if $x < N$ then $|f(x) - L| < \epsilon$

Note: As in the note following Definition 2.5.1, the symbol $x \to -\infty$ indicates only the behavior of the variable x.

Limit Theorems 2, 4, 5, 6, 7, 8, 9, and 10 in Section 2.2 and Limit Theorems 11 and 12 in Section 2.4 remain unchanged when "$x \to a$" is replaced by "$x \to +\infty$" or "$x \to \infty$." We have the following additional limit theorem.

2.5.3 LIMIT THEOREM 13 If r is any positive integer, then

(i) $\displaystyle \lim_{x \to +\infty} \frac{1}{x^r} = 0$

(ii) $\displaystyle \lim_{x \to -\infty} \frac{1}{x^r} = 0$

Proof of (i) To prove part (i) we must show that Definition 2.5.1 holds for $f(x) = 1/x^r$ and $L = 0$; that is, we must show that for any $\epsilon > 0$ there exists a number $N > 0$ such that

if $x > N$ then $\left| \dfrac{1}{x^r} - 0 \right| < \epsilon$

$\Leftrightarrow$ if $x > N$ then $|x|^r > \dfrac{1}{\epsilon}$

or equivalently, since $r > 0$,

if $x > N$ then $|x| > \left(\dfrac{1}{\epsilon} \right)^{1/r}$

For the above to hold, take $N = (1/\epsilon)^{1/r}$. Thus

if $N = \left(\dfrac{1}{\epsilon} \right)^{1/r}$ and $x > N$ then $\left| \dfrac{1}{x^r} - 0 \right| < \epsilon$

This proves part (i).   ■

The proof of part (ii) is analogous and is left as an exercise (see Exercise 66).

EXAMPLE 1   Find

$$\lim_{x \to +\infty} \frac{4x - 3}{2x + 5}$$

Solution   To use Limit Theorem 13, divide the numerator and the denominator by x, thus giving

$$\lim_{x \to +\infty} \frac{4x - 3}{2x + 5} = \lim_{x \to +\infty} \frac{4 - \dfrac{3}{x}}{2 + \dfrac{5}{x}}$$

$$= \frac{\displaystyle\lim_{x \to +\infty} 4 - \lim_{x \to +\infty} 3 \cdot \lim_{x \to +\infty} \frac{1}{x}}{\displaystyle\lim_{x \to +\infty} 2 + \lim_{x \to +\infty} 5 \cdot \lim_{x \to +\infty} \frac{1}{x}}$$

$$= \frac{4 - 3 \cdot 0}{2 + 5 \cdot 0}$$

$$= 2$$

EXAMPLE 2   Find

$$\lim_{x \to -\infty} \frac{2x^2 - x + 5}{4x^3 - 1}$$

Solution   To use Limit Theorem 13, divide the numerator and the denominator by the highest power of x occurring in either the numerator or denominator, which in this case is x^3.

$$\lim_{x \to -\infty} \frac{2x^2 - x + 5}{4x^3 - 1} = \lim_{x \to -\infty} \frac{\dfrac{2}{x} - \dfrac{1}{x^2} + \dfrac{5}{x^3}}{4 - \dfrac{1}{x^3}}$$

$$= \frac{\displaystyle\lim_{x \to -\infty} 2 \cdot \lim_{x \to -\infty} \frac{1}{x} - \lim_{x \to -\infty} \frac{1}{x^2} + \lim_{x \to -\infty} 5 \cdot \lim_{x \to -\infty} \frac{1}{x^3}}{\displaystyle\lim_{x \to -\infty} 4 - \lim_{x \to -\infty} \frac{1}{x^3}}$$

$$= \frac{2 \cdot 0 - 0 + 5 \cdot 0}{4 - 0}$$

$$= 0$$

EXAMPLE 3   Find

$$\lim_{x \to +\infty} \frac{3x + 4}{\sqrt{2x^2 - 5}}$$

Solution   Because the highest power of x is 2 and it appears under the radical sign, we divide numerator and denominator by $\sqrt{x^2}$, which is $|x|$. We have, then,

$$\lim_{x \to +\infty} \frac{3x + 4}{\sqrt{2x^2 - 5}} = \lim_{x \to +\infty} \frac{\dfrac{3x}{\sqrt{x^2}} + \dfrac{4}{\sqrt{x^2}}}{\dfrac{\sqrt{2x^2 - 5}}{\sqrt{x^2}}}$$

$$= \lim_{x \to +\infty} \frac{\dfrac{3x}{|x|} + \dfrac{4}{|x|}}{\sqrt{2 - \dfrac{5}{x^2}}}$$

Because $x \to +\infty$, $x > 0$; therefore $|x| = x$. Thus we have

$$\lim_{x \to +\infty} \frac{3x + 4}{\sqrt{2x^2 - 5}} = \lim_{x \to +\infty} \frac{\dfrac{3x}{x} + \dfrac{4}{x}}{\sqrt{2 - \dfrac{5}{x^2}}}$$

$$= \frac{\lim_{x \to +\infty} 3 + \lim_{x \to +\infty} 4 \cdot \lim_{x \to +\infty} \left(\dfrac{1}{x}\right)}{\sqrt{\lim_{x \to +\infty} 2 - \lim_{x \to +\infty} 5 \cdot \lim_{x \to +\infty} \left(\dfrac{1}{x^2}\right)}}$$

$$= \frac{3 + 4 \cdot 0}{\sqrt{2 - 5 \cdot 0}}$$

$$= \frac{3}{\sqrt{2}}$$

EXAMPLE 4   Find

$$\lim_{x \to -\infty} \frac{3x + 4}{\sqrt{2x^2 - 5}}$$

Solution   The function is the same as the one in Example 3. Again we begin by dividing numerator and denominator by $\sqrt{x^2}$ or, equivalently, $|x|$.

$$\lim_{x \to -\infty} \frac{3x + 4}{\sqrt{2x^2 - 5}} = \lim_{x \to -\infty} \frac{\dfrac{3x}{|x|} + \dfrac{4}{|x|}}{\sqrt{2 - \dfrac{5}{x^2}}}$$

Because $x \to -\infty$, $x < 0$; therefore $|x| = -x$. We have, then,

$$\lim_{x \to -\infty} \frac{3x + 4}{\sqrt{2x^2 - 5}} = \lim_{x \to -\infty} \frac{\dfrac{3x}{-x} + \dfrac{4}{-x}}{\sqrt{2 - \dfrac{5}{x^2}}}$$

$$= \frac{\displaystyle\lim_{x \to -\infty} (-3) - \lim_{x \to -\infty} 4 \cdot \lim_{x \to -\infty} \frac{1}{x}}{\sqrt{\displaystyle\lim_{x \to -\infty} 2 - \lim_{x \to -\infty} 5 \cdot \lim_{x \to -\infty} \frac{1}{x^2}}}$$

$$= \frac{-3 - 4 \cdot 0}{\sqrt{2 - 5 \cdot 0}}$$

$$= -\frac{3}{\sqrt{2}}$$

"Infinite" limits at infinity can be considered. There are formal definitions for each of the following.

$$\lim_{x \to +\infty} f(x) = +\infty \qquad \lim_{x \to -\infty} f(x) = +\infty$$

$$\lim_{x \to +\infty} f(x) = -\infty \qquad \lim_{x \to -\infty} f(x) = -\infty$$

For example, $\displaystyle\lim_{x \to +\infty} f(x) = +\infty$ if the function f is defined on some interval $(a, +\infty)$ and if for any number $N > 0$ there exists an $M > 0$ such that if $x > M$, then $f(x) > N$. The other definitions are left as an exercise (see Exercise 63).

EXAMPLE 5   Find

$$\lim_{x \to +\infty} \frac{x^2}{x + 1}$$

Solution   Divide the numerator and denominator by x^2.

$$\lim_{x \to +\infty} \frac{x^2}{x + 1} = \lim_{x \to +\infty} \frac{1}{\dfrac{1}{x} + \dfrac{1}{x^2}}$$

Evaluating the limit of the denominator we have

$$\lim_{x \to +\infty} \left(\frac{1}{x} + \frac{1}{x^2}\right) = \lim_{x \to +\infty} \frac{1}{x} + \lim_{x \to +\infty} \frac{1}{x^2}$$

$$= 0 + 0$$

$$= 0$$

Therefore the limit of the denominator is 0, and the denominator is approaching 0 through positive values.

The limit of the numerator is 1, and so by Limit Theorem 12(i) (2.4.4) it follows that

$$\lim_{x \to +\infty} \frac{x^2}{x+1} = +\infty$$

EXAMPLE 6   Find

$$\lim_{x \to +\infty} \frac{2x - x^2}{3x + 5}$$

Solution

$$\lim_{x \to +\infty} \frac{2x - x^2}{3x + 5} = \lim_{x \to +\infty} \frac{\dfrac{2}{x} - 1}{\dfrac{3}{x} + \dfrac{5}{x^2}}$$

The limits of the numerator and denominator are considered separately.

$$\lim_{x \to +\infty} \left(\frac{2}{x} - 1 \right) = \lim_{x \to +\infty} \frac{2}{x} - \lim_{x \to +\infty} 1 \qquad \lim_{x \to +\infty} \left(\frac{3}{x} + \frac{5}{x^2} \right) = \lim_{x \to +\infty} \frac{3}{x} + \lim_{x \to +\infty} \frac{5}{x^2}$$

$$= 0 - 1 \qquad\qquad\qquad = 0 + 0$$

$$= -1 \qquad\qquad\qquad = 0$$

Therefore we have the limit of a quotient in which the limit of the numerator is -1 and the limit of the denominator is 0, where the denominator is approaching 0 through positive values. By Limit Theorem 12(iii) it follows that

$$\lim_{x \to +\infty} \frac{2x - x^2}{3x + 5} = -\infty$$

In the previous section we discussed vertical asymptotes of a graph as an application of infinite limits. *Horizontal asymptotes* of a graph provide an application of limits at infinity. A horizontal asymptote is a line parallel to the x axis, and following is the formal definition.

2.5.4 DEFINITION   The line $y = b$ is said to be a **horizontal asymptote** of the graph of the function f if at least one of the following statements is true:

(i) $\lim\limits_{x \to +\infty} f(x) = b$, and for some number N, if $x > N$, then $f(x) \neq b$;

(ii) $\lim\limits_{x \to -\infty} f(x) = b$, and for some number N, if $x < N$, then $f(x) \neq b$.

▶ **ILLUSTRATION 1**   In each of Figures 2 through 5 there is a portion of the graph of a function for which the line $y = b$ is a horizontal asymptote. In Figures 2 and 3 part (i) of Definition 2.5.4 applies, and in Figures 4 and 5 part (ii) is true. Both parts (i) and (ii) hold for the function whose graph appears in Figure 6.   ◀

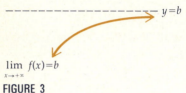

$$\lim_{x \to +\infty} f(x) = b$$

FIGURE 2

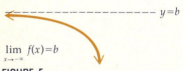

$$\lim_{x \to +\infty} f(x) = b$$

FIGURE 3

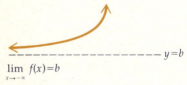

$$\lim_{x \to -\infty} f(x) = b$$

FIGURE 4

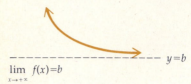

$$\lim_{x \to -\infty} f(x) = b$$

FIGURE 5

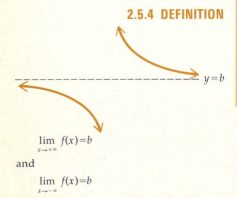

$$\lim_{x \to +\infty} f(x) = b$$

and

$$\lim_{x \to -\infty} f(x) = b$$

FIGURE 6

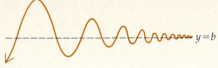

FIGURE 7

In Figure 7 we have a sketch of the graph of a function f for which $\lim\limits_{x \to +\infty} f(x) = b$, but there is no number N such that if $x > N$, then $f(x) \neq b$. Consequently the line $y = b$ is not a horizontal asymptote of the graph of f. An example of such a function is given in Exercise 59 of Exercises 7.5.

EXAMPLE 7   Find the horizontal asymptotes and draw a sketch of the graph of the function defined by

$$f(x) = \frac{x}{\sqrt{x^2 + 1}}$$

Solution   First consider $\lim\limits_{x \to +\infty} f(x)$.

$$\lim_{x \to +\infty} f(x) = \lim_{x \to +\infty} \frac{x}{\sqrt{x^2 + 1}}$$

We divide the numerator and denominator by $\sqrt{x^2}$ and we have

$$\lim_{x \to +\infty} \frac{x}{\sqrt{x^2 + 1}} = \lim_{x \to +\infty} \frac{\dfrac{x}{\sqrt{x^2}}}{\sqrt{\dfrac{x^2}{x^2} + \dfrac{1}{x^2}}}$$

$$= \lim_{x \to +\infty} \frac{\dfrac{x}{|x|}}{\sqrt{1 + \dfrac{1}{x^2}}}$$

Because $x \to +\infty$, $x > 0$; therefore $|x| = x$. Thus

$$\lim_{x \to +\infty} \frac{x}{\sqrt{x^2 + 1}} = \lim_{x \to +\infty} \frac{\dfrac{x}{x}}{\sqrt{1 + \dfrac{1}{x^2}}}$$

$$= \frac{\lim\limits_{x \to +\infty} 1}{\sqrt{\lim\limits_{x \to +\infty} 1 + \lim\limits_{x \to +\infty} \dfrac{1}{x^2}}}$$

$$= \frac{1}{\sqrt{1 + 0}}$$

$$= 1$$

Therefore, by Definition 2.5.4(i), the line $y = 1$ is a horizontal asymptote.

Now consider $\lim\limits_{x \to -\infty} f(x)$. Again we divide numerator and denominator by $\sqrt{x^2}$, which is $|x|$; because $x \to -\infty$, $x < 0$ and so $|x| = -x$. We have

$$\lim_{x \to -\infty} f(x) = \lim_{x \to -\infty} \frac{\dfrac{x}{-x}}{\sqrt{1 + \dfrac{1}{x^2}}}$$

$$= \frac{\lim\limits_{x \to -\infty} (-1)}{\sqrt{\lim\limits_{x \to -\infty} 1 + \lim\limits_{x \to -\infty} \dfrac{1}{x^2}}}$$

$$= \frac{-1}{\sqrt{1 + 0}}$$

$$= -1$$

Accordingly, by Definition 2.5.4(ii), the line $y = -1$ is a horizontal asymptote. A sketch of the graph appears in Figure 8.

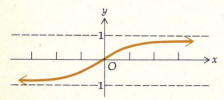

FIGURE 8

EXAMPLE 8   Find the vertical and horizontal asymptotes and draw a sketch of the graph of the equation

$$xy^2 - 2y^2 - 4x = 0$$

Solution   We solve the given equation for y and obtain

$$y = \pm 2 \sqrt{\frac{x}{x - 2}}$$

This equation defines two functions:

$$y = f_1(x) \quad \text{where } f_1 \text{ is defined by} \quad f_1(x) = +2 \sqrt{\frac{x}{x - 2}}$$

and

$$y = f_2(x) \quad \text{where } f_2 \text{ is defined by} \quad f_2(x) = -2 \sqrt{\frac{x}{x - 2}}$$

The graph of the given equation is composed of the graphs of the two functions f_1 and f_2. The domains of the two functions consist of those values of x for which $x/(x - 2) \geq 0$. By using the result of Example 8 in Section 1.5 and excluding $x = 2$ we see that the domain of f_1 and f_2 is $(-\infty, 0] \cup (2, +\infty)$.

Now consider f_1. Because

$$\lim_{x \to 2^+} f_1(x) = \lim_{x \to 2^+} 2 \sqrt{\frac{x}{x - 2}}$$

$$= +\infty$$

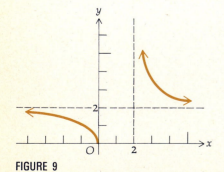

FIGURE 9

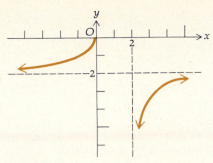

FIGURE 10

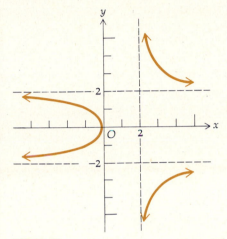

FIGURE 11

by Definition 2.4.8(i) the line $x = 2$ is a vertical asymptote of the graph of f_1.

$$\lim_{x \to +\infty} f_1(x) = \lim_{x \to +\infty} 2\sqrt{\frac{1}{1 - \dfrac{2}{x}}}$$

$$= 2$$

Thus by Definition 2.5.4(i) the line $y = 2$ is a horizontal asymptote of the graph of f_1.

Similarly, $\lim_{x \to -\infty} f_1(x) = 2$. A sketch of the graph of f_1 is in Figure 9.

$$\lim_{x \to 2^+} f_2(x) = \lim_{x \to 2^+} \left[-2\sqrt{\frac{x}{x - 2}} \right]$$

$$= -\infty$$

Hence, by Definition 2.4.8(ii) the line $x = 2$ is a vertical asymptote of the graph of f_2.

$$\lim_{x \to +\infty} f_2(x) = \lim_{x \to +\infty} \left[-2\sqrt{\frac{1}{1 - \dfrac{2}{x}}} \right]$$

$$= -2$$

Thus, by Definition 2.5.4(i) the line $y = -2$ is a horizontal asymptote of the graph of f_2.

Also, $\lim_{x \to -\infty} f_2(x) = -2$. A sketch of the graph of f_2 appears in Figure 10.

The graph of the given equation is the union of the graphs of f_1 and f_2, and a sketch is in Figure 11.

EXERCISES 2.5

In Exercises 1 through 10, do the following: Use a calculator to tabulate the values of $f(x)$ for the specified values of x. (a) What does $f(x)$ appear to be approaching as x increases without bound? (b) What does $f(x)$ appear to be approaching as x decreases without bound. (c) Find $\lim_{x \to +\infty} f(x)$. (d) Find $\lim_{x \to -\infty} f(x)$.

1. $f(x) = \dfrac{4}{x^2}$; x is 1, 2, 4, 6, 8, 10, 100, 1000 and x is $-1, -2,$ $-4, -6, -8, -10, -100, -1000.$

2. $f(x) = \dfrac{3}{x^4}$; x is 1, 2, 4, 6, 8, 10, 100, 1000 and x is $-1, -2,$ $-4, -6, -8, -10, -100, -1000.$

3. $f(x) = \dfrac{1}{x^3}$; x is 1, 2, 4, 6, 8, 10, 100, 1000 and x is $-1, -2,$ $-4, -6, -8, -10, -100, -1000.$

4. $f(x) = -\dfrac{2}{x^3}$; x is 1, 2, 4, 6, 8, 10, 100, 1000 and x is $-1, -2,$ $-4, -6, -8, -10, -100, -1000.$

5. $f(x) = -\dfrac{3x^2}{x^2 + 1}$; x is 0, 1, 2, 4, 6, 8, 10, 100, 1000 and x is $-1, -2, -4, -6, -8, -10, -100, -1000.$

6. $f(x) = \dfrac{x^3}{x^3 + 2}$; x is 2, 4, 6, 8, 10, 100, 1000 and x is $-2, -4,$ $-6, -8, -10, -100, -1000.$

7. $f(x) = \dfrac{4x + 1}{2x - 1}$; x is 2, 6, 10, 100, 1000, 10,000, 100,000 and x is $-2, -6, -10, -100, -1000, -10,000, -100,000.$

8. $f(x) = \dfrac{5x - 3}{10x + 1}$; x is 2, 6, 10, 100, 1000, 10,000, 100,000 and x is $-2, -6, -10, -100, -1000, -10,000, -100,000.$

9. $f(x) = \dfrac{x + 1}{x^2}$; x is 2, 6, 10, 100, 1000, 10,000, 100,000 and x is $-2, -6, -10, -100, -1000, -10,000, -100,000.$

10. $f(x) = \dfrac{x^2}{x + 1}$; x is 2, 6, 10, 100, 1000, 10,000, 100,000 and x is $-2, -6, -10, -100, -1000, -10,000, -100,000.$

In Exercises 11 through 30, find the limit.

11. $\lim\limits_{t \to +\infty} \dfrac{2t + 1}{5t - 2}$

12. $\lim\limits_{x \to -\infty} \dfrac{6x - 4}{3x + 1}$

13. $\lim\limits_{x \to -\infty} \dfrac{2x + 7}{4 - 5x}$

14. $\lim\limits_{x \to +\infty} \dfrac{1 + 5x}{2 - 3x}$

15. $\lim\limits_{x \to +\infty} \dfrac{7x^2 - 2x + 1}{3x^2 + 8x + 5}$

16. $\lim\limits_{s \to -\infty} \dfrac{4s^2 + 3}{2s^2 - 1}$

17. $\lim\limits_{x \to +\infty} \dfrac{x + 4}{3x^2 - 5}$

18. $\lim\limits_{x \to +\infty} \dfrac{x^2 + 5}{x^3}$

19. $\lim\limits_{y \to +\infty} \dfrac{2y^2 - 3y}{y + 1}$

20. $\lim\limits_{x \to +\infty} \dfrac{x^2 - 2x + 5}{7x^3 + x + 1}$

21. $\lim\limits_{x \to -\infty} \dfrac{4x^3 + 2x^2 - 5}{8x^3 + x + 2}$

22. $\lim\limits_{x \to +\infty} \dfrac{3x^4 - 7x^2 + 2}{2x^4 + 1}$

23. $\lim\limits_{y \to +\infty} \dfrac{2y^3 - 4}{5y + 3}$

24. $\lim\limits_{x \to -\infty} \dfrac{5x^3 - 12x + 7}{4x^2 - 1}$

25. $\lim\limits_{x \to -\infty} \left(3x + \dfrac{1}{x^2} \right)$

26. $\lim\limits_{t \to +\infty} \left(\dfrac{2}{t^2} - 4t \right)$

27. $\lim\limits_{x \to +\infty} \dfrac{\sqrt{x^2 + 4}}{x + 4}$

28. $\lim\limits_{x \to -\infty} \dfrac{\sqrt{x^2 + 4}}{x + 4}$

29. $\lim\limits_{w \to -\infty} \dfrac{\sqrt{w^2 - 2w + 3}}{w + 5}$

30. $\lim\limits_{y \to -\infty} \dfrac{\sqrt{y^4 + 1}}{2y^2 - 3}$

In Exercises 31 through 36, find the limit (Hint: First obtain a fraction with a rational numerator.)

31. $\lim\limits_{x \to +\infty} (\sqrt{x^2 + 1} - x)$

32. $\lim\limits_{x \to +\infty} (\sqrt{x^2 + x} - x)$

33. $\lim\limits_{r \to +\infty} (\sqrt{3r^2 + r} - 2r)$

34. $\lim\limits_{x \to +\infty} (\sqrt[3]{x^3 + 1} - x)$

35. $\lim\limits_{x \to -\infty} (\sqrt[3]{x^3 + x} - \sqrt[3]{x^3 + 1})$

36. $\lim\limits_{t \to +\infty} \dfrac{\sqrt{t + \sqrt{t + \sqrt{t}}}}{\sqrt{t + 1}}$

In Exercises 37 through 48, find the horizontal and vertical asymptotes and draw a sketch of the graph of the function.

37. $f(x) = \dfrac{2x + 1}{x - 3}$

38. $f(x) = \dfrac{4 - 3x}{x + 1}$

39. $g(x) = 1 - \dfrac{1}{x}$

40. $h(x) = 1 + \dfrac{1}{x^2}$

41. $f(x) = \dfrac{2}{\sqrt{x^2 - 4}}$

42. $F(x) = \dfrac{-3x}{\sqrt{x^2 + 3}}$

43. $G(x) = \dfrac{4x^2}{x^2 - 9}$

44. $g(x) = \dfrac{x^2}{4 - x^2}$

45. $h(x) = \dfrac{2x}{6x^2 + 11x - 10}$

46. $f(x) = \dfrac{-1}{\sqrt{x^2 + 5x + 6}}$

47. $f(x) = \dfrac{4x^2}{\sqrt{x^2 - 2}}$

48. $h(x) = \dfrac{x}{\sqrt{x^2 - 9}}$

In Exercises 49 through 56, find the horizontal and vertical asymptotes and draw a sketch of the graph of the equation.

49. $3xy - 2x - 4y - 3 = 0$

50. $2xy + 4x - 3y + 6 = 0$

51. $x^2y^2 - x^2 + 4y^2 = 0$

52. $xy^2 + 3y^2 - 9x = 0$

53. $(y^2 - 1)(x - 3) = 6$

54. $2xy^2 + 4y^2 - 3x = 0$

55. $x^2y - 2x^2 - y - 2 = 0$

56. $x^2y + 4xy - x^2 + x + 4y - 6 = 0$

In Exercises 57 through 60, prove that $\lim\limits_{x \to +\infty} f(x) = 1$ by applying Definition 2.5.1; that is, for any $\epsilon > 0$ show that there exists a number $N > 0$ such that if $x > N$, then $|f(x) - 1| < \epsilon$.

57. $f(x) = \dfrac{x}{x - 1}$

58. $f(x) = \dfrac{2x}{2x + 3}$

59. $f(x) = \dfrac{x^2 - 1}{x^2 + 1}$

60. $f(x) = \dfrac{x^2 + 2x}{x^2 - 1}$

61. Prove that $\lim\limits_{x \to -\infty} \dfrac{8x + 3}{2x - 1} = 4$ by showing that for any $\epsilon > 0$ there exists a number $N < 0$ such that if $x < N$ then $\left| \dfrac{8x + 3}{2x - 1} - 4 \right| < \epsilon$.

62. Prove part (i) of Limit Theorem 12 (2.4.4) if "$x \to a$" is replaced by "$x \to +\infty$."

63. Give a definition for each of the following:
    (a) $\lim\limits_{x \to +\infty} f(x) = -\infty$; (b) $\lim\limits_{x \to -\infty} f(x) = +\infty$;
    (c) $\lim\limits_{x \to -\infty} f(x) = -\infty$.

64. Prove that $\lim\limits_{x \to +\infty} (x^2 - 4) = +\infty$ by showing that for any $N > 0$ there exists an $M > 0$ such that if $x > M$ then $x^2 - 4 > N$.

65. Prove that $\lim\limits_{x \to +\infty} (6 - x - x^2) = -\infty$ by applying the definition in Exercise 63(a).

66. Prove part (ii) of Limit Theorem 13 (2.5.3).

2.6 CONTINUITY OF A FUNCTION AT A NUMBER

In Illustration 1 of Section 2.3 we discussed the function C defined by

$$C(x) = \begin{cases} x & \text{if } 0 \leq x \leq 10 \\ 0.9x & \text{if } 10 < x \end{cases} \tag{1}$$

where $C(x)$ dollars is the total cost of x pounds of a product. We showed that $\lim\limits_{x \to 10} C(x)$ does not exist because $\lim\limits_{x \to 10^-} C(x) \neq \lim\limits_{x \to 10^+} C(x)$. A sketch of the graph

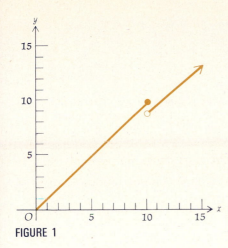

FIGURE 1

of C is shown in Figure 1. Observe that there is a break in the graph of C where $x = 10$. We state that C is *discontinuous* at 10. This discontinuity is caused by the fact that $\lim_{x \to 10} C(x)$ does not exist. We refer to this function again in Illustration 1.

In Section 2.1 we considered the function f defined by

$$f(x) = \frac{(2x + 3)(x - 1)}{x - 1} \qquad (2)$$

We noted that f is defined for all values of x except 1. A sketch of the graph consisting of all points on the line $y = 2x + 3$ except $(1, 5)$ appears in Figure 2. There is a break in the graph at the point $(1, 5)$, and we state that this function f is *discontinuous* at the number 1. This discontinuity occurs because $f(1)$ does not exist.

If f is the function defined by (2) when $x \neq 1$, and if we define $f(1) = 2$, for instance, the function is defined for all values of x, but there is still a break in the graph (see Figure 3), and the function is still *discontinuous* at 1. If we define $f(1) = 5$, however, there is no break in the graph, and the function f is said to be *continuous* at all values of x. We have the following definition.

2.6.1 DEFINITION

The function f is said to be **continuous** at the number a if and only if the following three conditions are satisfied:

   (i) $f(a)$ exists;

   (ii) $\lim_{x \to a} f(x)$ exists;

   (iii) $\lim_{x \to a} f(x) = f(a)$.

If one or more of these three conditions fails to hold at a, the function f is said to be **discontinuous** at a.

FIGURE 2

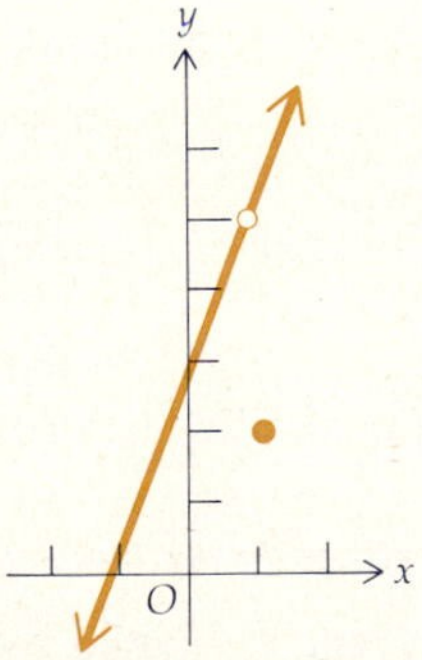

FIGURE 3

▶ **ILLUSTRATION 1** The function C defined by (1) has the graph shown in Figure 1. Because there is a break in the graph at the point where $x = 10$, we will investigate the conditions of Definition 2.6.1 at 10.

Because $C(10) = 10$, condition (i) is satisfied.

Because $\lim_{x \to 10} C(x)$ does not exist, condition (ii) fails to hold at 10. We conclude that C is discontinuous at 10.

Observe that because of the discontinuity of C, it would be advantageous to increase the size of some orders to take advantage of a lower total cost. In particular, it would be unwise to purchase $9\frac{1}{2}$ lb for \$9.50 when $10\frac{1}{2}$ lb can be bought for \$9.45.   ◀

In Illustration 2 there is another situation in which the formula for computing the cost of more than 10 lb of a product is different from the formula for computing the cost of 10 lb or less. However, here the cost function is continuous at 10.

▶ **ILLUSTRATION 2** A wholesaler who sells a product by the pound (or fraction of a pound) charges \$1 per pound if 10 lb or less are ordered. However, if more than 10 lb are ordered, the wholesaler charges \$10 plus 70 cents for each pound in excess of 10 lb. Therefore, if x pounds of the product are purchased and $C(x)$ dollars is the total cost, then $C(x) = x$ if $0 \le x \le 10$, and $C(x) = 10 + 0.7(x - 10)$ if $10 < x$. Therefore,

$$C(x) = \begin{cases} x & \text{if } 0 \le x \le 10 \\ 0.7x + 3 & \text{if } 10 < x \end{cases}$$

A sketch of the graph of C is in Figure 4. For this function $C(10) = 10$, and

$$\lim_{x \to 10^-} C(x) = \lim_{x \to 10^-} x \qquad \lim_{x \to 10^+} C(x) = \lim_{x \to 10^+} (0.7x + 3)$$
$$= 10 \qquad\qquad\qquad = 10$$

Therefore, $\lim_{x \to 10} C(x)$ exists and equals $C(10)$. Thus C is continuous at 10. ◄

We now consider some illustrations of discontinuous functions. For each illustration there is a sketch of the graph of the function. We determine the points where there is a break in the graph and show which of the three conditions in the definition of continuity fails to hold at each discontinuity.

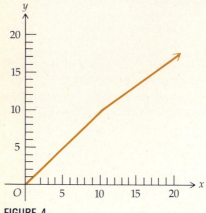

FIGURE 4

► **ILLUSTRATION 3**   Let f be defined by

$$f(x) = \begin{cases} 2x + 3 & \text{if } x \neq 1 \\ 2 & \text{if } x = 1 \end{cases}$$

A sketch of the graph of this function is given in Figure 3. Observe that there is a break in the graph at the point where $x = 1$. So we investigate there the conditions of Definition 2.6.1.

$f(1) = 2$; therefore, condition (i) is satisfied.

$\lim_{x \to 1} f(x) = 5$; therefore, condition (ii) is satisfied.

$\lim_{x \to 1} f(x) = 5$, but $f(1) = 2$; therefore, condition (iii) is not satisfied.

Thus f is discontinuous at 1. ◄

Note that if in Illustration 3 $f(1)$ is defined to be 5, then $\lim_{x \to 1} f(x) = f(1)$ and f would be continuous at 1.

► **ILLUSTRATION 4**   Let f be defined by

$$f(x) = \frac{1}{x - 2}$$

A sketch of the graph of f appears in Figure 5. There is a break in the graph at the point where $x = 2$, and so we investigate there the conditions of Definition 2.6.1.

Because $f(2)$ is not defined, condition (i) is not satisfied. Hence f is discontinuous at 2. ◄

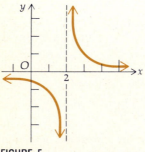

FIGURE 5

► **ILLUSTRATION 5**   Let g be defined by

$$g(x) = \begin{cases} \dfrac{1}{x - 2} & \text{if } x \neq 2 \\ 3 & \text{if } x = 2 \end{cases}$$

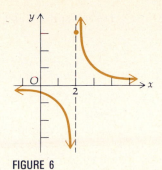

FIGURE 6

A sketch of the graph of g appears in Figure 6. We check the three conditions of Definition 2.6.1 at 2.

Since $g(2) = 3$, condition (i) is satisfied.

$$\lim_{x \to 2^-} g(x) = \lim_{x \to 2^-} \frac{1}{x-2} \qquad \lim_{x \to 2^+} g(x) = \lim_{x \to 2^+} \frac{1}{x-2}$$

$$= -\infty \qquad\qquad\qquad = +\infty$$

Because $\lim_{x \to 2} g(x)$ does not exist, condition (ii) is not satisfied. Therefore g is discontinuous at 2. ◀

▶ **ILLUSTRATION 6**   Let h be defined by

$$h(x) = \begin{cases} 3 + x & \text{if } x \le 1 \\ 3 - x & \text{if } 1 < x \end{cases}$$

Figure 7 shows a sketch of the graph of h. Because there is a break in the graph at the point where $x = 1$, we investigate the conditions of Definition 2.6.1 at 1.

Because $h(1) = 4$, condition (i) is satisfied.

$$\lim_{x \to 1^-} h(x) = \lim_{x \to 1^-} (3 + x) \qquad \lim_{x \to 1^+} h(x) = \lim_{x \to 1^+} (3 - x)$$

$$= 4 \qquad\qquad\qquad = 2$$

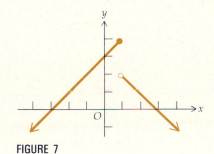

FIGURE 7

Because $\lim_{x \to 1^-} h(x) \ne \lim_{x \to 1^+} h(x)$, then $\lim_{x \to 1} h(x)$ does not exist; therefore condition (ii) fails to hold at 1. Hence h is discontinuous at 1. ◀

▶ **ILLUSTRATION 7**   Let F be defined by

$$F(x) = \begin{cases} |x - 3| & \text{if } x \ne 3 \\ 2 & \text{if } x = 3 \end{cases}$$

A sketch of the graph of F appears in Figure 8. We check the three conditions of Definition 2.6.1 at the point where $x = 3$.

Since $F(3) = 2$, condition (i) is satisfied.

$$\lim_{x \to 3^-} F(x) = \lim_{x \to 3^-} (3 - x) \qquad \lim_{x \to 3^+} F(x) = \lim_{x \to 3^+} (x - 3)$$

$$= 0 \qquad\qquad\qquad = 0$$

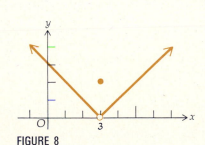

FIGURE 8

Therefore, $\lim_{x \to 3} F(x) = 0$, and so condition (ii) is satisfied. $\lim_{x \to 3} F(x) = 0$, but $F(3) = 2$; therefore condition (iii) is not satisfied. Thus F is discontinuous at 3. ◀

It should be apparent that the geometric notion of a break in the graph at a certain point is synonymous with the concept of a function being discontinuous at a certain value of the independent variable.

After Illustration 3 it was mentioned that if $f(1)$ had been defined to be 5, then f would be continuous at 1. This illustrates the concept of a *removable discontinuity*. In general, suppose that f is a function that is discontinuous at the number a, but for which $\lim_{x \to a} f(x)$ exists. Then either $f(a) \ne \lim_{x \to a} f(x)$ or else $f(a)$ does not exist. Such a discontinuity is called a **removable discontinuity** because if f is redefined at a such that $f(a)$ is equal to $\lim_{x \to a} f(x)$, the new function becomes continuous at a. If the discontinuity is not removable, it is called an **essential discontinuity**.

EXAMPLE 1   In each of the Illustrations 3 through 7, determine if the discontinuity is removable or essential.

Solution   In Illustration 3 the function is discontinuous at 1, but $\lim_{x \to 1} f(x) = 5$. By redefining $f(1) = 5$, then $\lim_{x \to 1} f(x) = f(1)$; so the discontinuity is removable.

In Illustration 4 the function f is discontinuous at 2. $\lim_{x \to 2} f(x)$ does not exist; hence the discontinuity is essential.

In Illustration 5, $\lim_{x \to 2} g(x)$ does not exist; so the discontinuity is essential.

In Illustration 6 the function h is discontinuous because $\lim_{x \to 1} h(x)$ does not exist; so again the discontinuity is essential.

In Illustration 7, $\lim_{x \to 3} F(x) = 0$, but $F(3) = 2$; so F is discontinuous at 3. However, if $F(3)$ is redefined to be 0, then the function is continuous at 3; so the discontinuity is removable.

EXAMPLE 2   The function defined by

$$f(x) = \frac{\sqrt{x} - 2}{x - 4}$$

is discontinuous at 4. Show that the discontinuity is removable and redefine $f(4)$ so that the discontinuity is removed.

Solution   The function f is discontinuous at 4 because $f(4)$ does not exist. If $\lim_{x \to 4} f(x)$ exists, the discontinuity can be removed by redefining $f(4) = \lim_{x \to 4} f(x)$. We compute the limit.

$$\lim_{x \to 4} f(x) = \lim_{x \to 4} \frac{\sqrt{x} - 2}{x - 4}$$

$$= \lim_{x \to 4} \frac{(\sqrt{x} - 2)(\sqrt{x} + 2)}{(x - 4)(\sqrt{x} + 2)}$$

$$= \lim_{x \to 4} \frac{x - 4}{(x - 4)(\sqrt{x} + 2)}$$

$$= \lim_{x \to 4} \frac{1}{\sqrt{x} + 2}$$

$$= \frac{1}{4}$$

Therefore we let $f(4) = \frac{1}{4}$, and we have the new function defined by

$$f(x) = \begin{cases} \dfrac{\sqrt{x} - 2}{x - 4} & \text{if } x \neq 4 \\[2mm] \frac{1}{4} & \text{if } x = 4 \end{cases}$$

and this function is continuous at 4.

The following theorem about functions that are continuous at a number is obtained by applying Definition 2.6.1 and limit theorems.

2.6.2 THEOREM   If f and g are two functions that are continuous at the number a, then

    (i) $f + g$ is continuous at a;
    (ii) $f - g$ is continuous at a;
    (iii) $f \cdot g$ is continuous at a;
    (iv) f/g is continuous at a, provided that $g(a) \neq 0$.

To illustrate the kind of proof required for each part of this theorem, we prove part (i).

Because f and g are continuous at a, from Definition 2.6.1 it follows that

$$\lim_{x \to a} f(x) = f(a) \quad \text{and} \quad \lim_{x \to a} g(x) = g(a)$$

From these two limits and Limit Theorem 4,

$$\lim_{x \to a} [f(x) + g(x)] = f(a) + g(a)$$

which is the condition that $f + g$ is continuous at a. We have proved (i).

The proofs of parts (ii), (iii), and (iv) are similar.

Consider the polynomial function f defined by

$$f(x) = b_0 x^n + b_1 x^{n-1} + b_2 x^{n-2} + \ldots + b_{n-1} x + b_n \qquad b_0 \neq 0$$

where n is a nonnegative integer and $b_0, b_1, \ldots, b_n$ are real numbers. By successive applications of limit theorems we can show that if a is any number,

$$\lim_{x \to a} f(x) = b_0 a^n + b_1 a^{n-1} + b_2 a^{n-2} + \ldots + b_{n-1} a + b_n$$

$$\Rightarrow \quad \lim_{x \to a} f(x) = f(a)$$

thus establishing the following theorem.

2.6.3 THEOREM   A polynomial function is continuous at every number.

▶ **ILLUSTRATION 8**   If $f(x) = x^3 - 2x^2 + 5x + 1$, then f is a polynomial function and therefore, by Theorem 2.6.3, is continuous at every number. In particular, because f is continuous at 3, $\lim_{x \to 3} f(x) = f(3)$. Thus

$$\lim_{x \to 3} (x^3 - 2x^2 + 5x + 1) = 3^3 - 2(3)^2 + 5(3) + 1$$
$$= 27 - 18 + 15 + 1$$
$$= 25 \qquad ◀$$

2.6.4 THEOREM   A rational function is continuous at every number in its domain.

Proof   If f is a rational function, it can be expressed as the quotient of two polynomial functions. So f can be defined by

$$f(x) = \frac{g(x)}{h(x)}$$

where g and h are two polynomial functions, and the domain of f consists of all numbers except those for which $h(x) = 0$.

If a is any number in the domain of f, then $h(a) \neq 0$; so by Limit Theorem 9,

$$\lim_{x \to a} f(x) = \frac{\lim_{x \to a} g(x)}{\lim_{x \to a} h(x)} \tag{3}$$

Because g and h are polynomial functions, by Theorem 2.6.3 they are continuous at a; so $\lim_{x \to a} g(x) = g(a)$ and $\lim_{x \to a} h(x) = h(a)$. Consequently, from (3),

$$\lim_{x \to a} f(x) = \frac{g(a)}{h(a)}$$

Therefore f is continuous at every number in its domain.   ■

EXAMPLE 3   Determine the numbers at which the following function is continuous:

$$f(x) = \frac{x^3 + 1}{x^2 - 9}$$

Solution   The domain of f is the set R of real numbers except those for which $x^2 - 9 = 0$. Because $x^2 - 9 = 0$ when $x = \pm 3$, it follows that the domain of f is the set of all real numbers except 3 and -3.

Because f is a rational function, it follows from Theorem 2.6.4 that f is continuous at all real numbers except 3 and -3.

EXAMPLE 4   Determine the numbers at which the following function is continuous:

$$f(x) = \begin{cases} 2x - 3 & \text{if } x \leq 1 \\ x^2 & \text{if } 1 < x \end{cases}$$

Solution   The functions having values $2x - 3$ and x^2 are polynomials and are therefore continuous everywhere. Thus the only number at which continuity is questionable is 1. We check the three conditions for continuity at 1.

(i) $f(1) = -1$. Thus condition (i) holds.

(ii) $\lim_{x \to 1^-} f(x) = \lim_{x \to 1^-} (2x - 3)$ $\qquad \lim_{x \to 1^+} f(x) = \lim_{x \to 1^+} x^2$

$\qquad \qquad \qquad \quad = -1$ $\qquad \qquad \qquad \qquad \qquad = 1$

Because $\lim_{x \to 1^-} f(x) \neq \lim_{x \to 1^+} f(x)$, the two-sided limit $\lim_{x \to 1} f(x)$ does not exist. Therefore f is discontinuous at 1. However, f is continuous at every real number except 1.

In Supplementary Section 2.9 we prove the following theorem about the continuity of the function defined by $f(x) = \sqrt[n]{x}$.

2.6.5 THEOREM   If n is a positive integer and

$$f(x) = \sqrt[n]{x}$$

then

(i) if n is odd, f is continuous at every number;
(ii) if n is even, f is continuous at every positive number.

▶ **ILLUSTRATION 9**   (a) If $f(x) = \sqrt[3]{x}$, it follows from Theorem 2.6.5(i) that f is continuous at every real number. A sketch of the graph of f appears in Figure 9.
(b) If $g(x) = \sqrt{x}$, then from Theorem 2.6.5(ii) g is continuous at every positive number. In Figure 10 there is a sketch of the graph of g.   ◀

In Section 2.7 we apply a definition of continuity of a function that uses ϵ and δ notation. To obtain this alternate definition we start with Definition 2.6.1, which states that the function f is continuous at the number a if $f(a)$ exists, if $\lim\limits_{x \to a} f(x)$ exists, and if

$$\lim_{x \to a} f(x) = f(a) \qquad (4)$$

Applying Definition 2.1.1 where L is $f(a)$, it follows that (4) will hold if for any $\epsilon > 0$ there exists a $\delta > 0$ such that

$$\text{if} \quad 0 < |x - a| < \delta \quad \text{then} \quad |f(x) - f(a)| < \epsilon \qquad (5)$$

If f is to be continuous at a, $f(a)$ must exist; therefore in statement (5) the condition that $|x - a| > 0$ is not necessary, because when $x = a$, $|f(x) - f(a)|$ will be 0 and thus less than ϵ. We have, then, the following theorem, which serves as our desired alternate definition of continuity.

2.6.6 THEOREM   The function f is continuous at the number a if f is defined on some open interval containing a and if for any $\epsilon > 0$ there exists a $\delta > 0$ such that

$$\text{if} \quad |x - a| < \delta \quad \text{then} \quad |f(x) - f(a)| < \epsilon$$

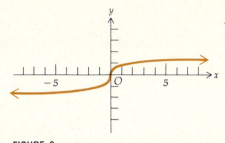

FIGURE 9

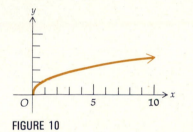

FIGURE 10

EXERCISES 2.6

In Exercises 1 through 22, draw a sketch of the graph of the function; then, by observing where there are breaks in the graph, determine the values of the independent variable at which the function is discontinuous, and show why Definition 2.6.1 is not satisfied at each discontinuity.

1. $f(x) = \dfrac{x^2 + x - 6}{x + 3}$

2. $F(x) = \dfrac{x^2 - 3x - 4}{x - 4}$

3. $g(x) = \begin{cases} \dfrac{x^2 + x - 6}{x + 3} & \text{if } x \neq -3 \\ 1 & \text{if } x = -3 \end{cases}$

4. $G(x) = \begin{cases} \dfrac{x^2 - 3x - 4}{x - 4} & \text{if } x \neq 4 \\ 2 & \text{if } x = 4 \end{cases}$

5. $h(x) = \dfrac{5}{x - 4}$

6. $H(x) = \dfrac{1}{x + 2}$

7. $f(x) = \begin{cases} \dfrac{5}{x - 4} & \text{if } x \neq 4 \\ 2 & \text{if } x = 4 \end{cases}$

8. $g(x) = \begin{cases} \dfrac{1}{x + 2} & \text{if } x \neq -2 \\ 0 & \text{if } x = -2 \end{cases}$

9. $F(x) = \dfrac{x^4 - 16}{x^2 - 4}$

10. $h(x) = \dfrac{(x - 1)(x^2 - x - 12)}{x^2 - 5x + 4}$

11. $G(x) = \dfrac{x^2 - 4}{x^4 - 16}$

12. $H(x) = \dfrac{x^2 - 5x + 4}{(x - 1)(x^2 - x - 12)}$

13. $f(x) = \begin{cases} -1 & \text{if } x < 0 \\ 0 & \text{if } x = 0 \\ \sqrt{x} & \text{if } 0 < x \end{cases}$

14. $f(x) = \begin{cases} x - 1 & \text{if } x < 1 \\ 1 & \text{if } x = 1 \\ 1 - x & \text{if } 1 < x \end{cases}$

15. $g(t) = \begin{cases} t^2 - 4 & \text{if } t < 2 \\ 4 & \text{if } t = 2 \\ 4 - t^2 & \text{if } 2 < t \end{cases}$

16. $H(x) = \begin{cases} 1 + x & \text{if } x \le -2 \\ 2 - x & \text{if } -2 < x \le 2 \\ 2x - 1 & \text{if } 2 < x \end{cases}$

17. $g(x) = \begin{cases} \sqrt{-x} & \text{if } x < 0 \\ \sqrt[3]{x + 1} & \text{if } 0 \le x \end{cases}$

18. $f(t) = \begin{cases} |t + 2| & \text{if } t \ne -2 \\ 3 & \text{if } t = -2 \end{cases}$

19. $f(x) = \dfrac{|x|}{x}$

20. $g(x) = \begin{cases} \dfrac{|x|}{x} & \text{if } x \ne 0 \\ 1 & \text{if } x = 0 \end{cases}$

21. The greatest integer function.

22. The signum function (see Example 1 in Section 2.3).

In Exercises 23 through 32, prove that the function is discontinuous at the number a. Then determine if the discontinuity is removable or essential. If the discontinuity is removable, redefine $f(a)$ so that the discontinuity is removed.

23. $f(x) = \dfrac{9x^2 - 4}{3x - 2}; a = \dfrac{2}{3}$

24. $f(s) = \begin{cases} \dfrac{1}{s + 5} & \text{if } s \ne -5 \\ 0 & \text{if } s = -5 \end{cases}; a = -5$

25. $f(t) = \begin{cases} 9 - t^2 & \text{if } t \le 2 \\ 3t + 2 & \text{if } 2 < t \end{cases}; a = 2$

26. $f(x) = \dfrac{x^2 - x - 12}{x^2 + 2x - 3}; a = -3$

27. $f(x) = \begin{cases} |x - 3| & \text{if } x \ne 3 \\ 2 & \text{if } x = 3 \end{cases}; a = 3$

28. $f(x) = \begin{cases} \dfrac{x^2 - 4x + 3}{x - 3} & \text{if } x \ne 3 \\ 5 & \text{if } x = 3 \end{cases}; a = 3$

29. $f(t) = \begin{cases} t^2 - 4 & \text{if } t \le 2 \\ t & \text{if } 2 < t \end{cases}; a = 2$

30. $f(y) = \dfrac{\sqrt{y + 5} - \sqrt{5}}{y}; a = 0$

31. $f(x) = \dfrac{3 - \sqrt{x + 9}}{x}; a = 0$

32. $f(x) = \dfrac{\sqrt[3]{x + 1} - 1}{x}; a = 0$

In Exercises 33 through 42, determine the numbers at which the function is continuous.

33. $f(x) = x^2(x + 3)^2$

34. $f(x) = (x - 5)^3(x^2 + 4)^5$

35. $g(x) = \dfrac{x}{x - 3}$

36. $h(x) = \dfrac{x + 1}{2x + 5}$

37. $F(x) = \dfrac{x^3 + 7}{x^2 - 4}$

38. $G(x) = \dfrac{x - 2}{x^2 + 2x - 8}$

39. $f(x) = \begin{cases} 3x - 1 & \text{if } x < 2 \\ 4 - x^2 & \text{if } 2 \le x \end{cases}$

40. $f(x) = \begin{cases} (x + 2)^2 & \text{if } x \le 0 \\ x^2 + 2 & \text{if } 0 < x \end{cases}$

41. $f(x) = \begin{cases} \dfrac{1}{x - 2} & \text{if } x \le 1 \\ \dfrac{1}{x} & \text{if } 1 < x \end{cases}$

42. $f(x) = \begin{cases} \dfrac{1}{x + 1} & \text{if } x < 4 \\ \sqrt{x - 4} & \text{if } 4 \le x \end{cases}$

43. The function C of Exercise 35 in Exercises 2.3 is defined by

$$C(x) = \begin{cases} 0.80x & \text{if } 0 < x \le 50 \\ 0.70x & \text{if } 50 < x \le 200 \\ 0.65x & \text{if } 200 < x \end{cases}$$

(a) Draw a sketch of the graph of C. (b) At what numbers is C discontinuous? (c) Show why Definition 2.6.1 is not satisfied at each discontinuity in part (b).

44. Suppose the postage of a letter is computed as follows: 25 cents for the first ounce or less, then 20 cents for each ounce (or fractional part of an ounce) for the next 11 oz. If x ounces is the weight of the letter and $0 < x \le 12$, express the number of cents in the postage as a function of x. (a) Draw a sketch of the graph of this function. (b) At what numbers in the open interval $(0, 12)$ is the function discontinuous? (c) Show why Definition 2.6.1 is not satisfied at each discontinuity in part (b).

45. Let f be the function defined by

$$f(x) = \begin{cases} |x - [\![x]\!]| & \text{if } [\![x]\!] \text{ is even} \\ |x - [\![x + 1]\!]| & \text{if } [\![x]\!] \text{ is odd} \end{cases}$$

Draw a sketch of the graph of f. At what numbers is f discontinuous?

46. The function f defined by

$$f(x) = \dfrac{\sqrt{2 + \sqrt[3]{x}} - 2}{x - 8}$$

is discontinuous at 8. Show that the discontinuity is removable, and redefine $f(8)$ so that the discontinuity is removed.

47. The function g defined by

$$g(x) = \frac{\sqrt[3]{x + a^3} - a}{x}$$

is discontinuous at 0. Show that the discontinuity is removable, and redefine $g(0)$ so that the discontinuity is removed.

48. The function f is defined by

$$f(x) = \lim_{n \to 0} \frac{2nx}{n^2 - nx}$$

Draw a sketch of the graph of f. At what values of x is f discontinuous?

49. If

$$f(x) = \begin{cases} -x & \text{if } x < 0 \\ 1 & \text{if } 0 \le x \end{cases} \quad \text{and} \quad g(x) = \begin{cases} 1 & \text{if } x < 0 \\ x & \text{if } 0 \le x \end{cases}$$

prove that f and g are both discontinuous at 0 but that the product $f \cdot g$ is continuous at 0.

50. Give an example to show that the product of two functions f and g may be continuous at a number a, where f is continuous at a but g is discontinuous at a.

51. Give an example of two functions that are both discontinuous at a number a but whose sum is continuous at a.

52. Prove that if f is continuous at a and g is discontinuous at a, then $f + g$ is discontinuous at a.

2.7 CONTINUITY OF A COMPOSITE FUNCTION AND CONTINUITY ON AN INTERVAL

Recall from Section 1.4 the definition of a composite function:

Given the functions f and g, the composite function, denoted by $f \circ g$, is defined by

$$(f \circ g)(x) = f(g(x))$$

and the domain of $f \circ g$ is the set of all numbers in the domain of g such that $g(x)$ is in the domain of f.

▶ **ILLUSTRATION 1** If $f(x) = \sqrt{x}$ and $g(x) = 4 - x^2$, and if h is the composite function $f \circ g$, then

$$\begin{aligned} h(x) &= f(g(x)) \\ &= f(4 - x^2) \\ &= \sqrt{4 - x^2} \end{aligned}$$

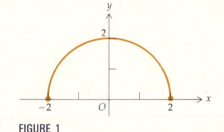

FIGURE 1

Because the domain of g is the set of all real numbers and the domain of f is the set of all nonnegative numbers, the domain of h is the set of all real numbers such that $4 - x^2 \ge 0$, that is, all numbers in the closed interval $[-2, 2]$. A sketch of the graph of h is shown in Figure 1.   ◀

From Figure 1 it appears that h is continuous at every number in the open interval $(-2, 2)$. We show later in Example 1 how this fact can be proved by theorems on continuity. But first we need an important theorem regarding the limit of a composite function. The proof of this theorem makes use of Theorem 2.6.6, which is the definition of continuity involving ϵ and δ notation.

2.7.1 THEOREM   If $\lim\limits_{x \to a} g(x) = b$ and if the function f is continuous at b,

$$\lim_{x \to a} (f \circ g)(x) = f(b)$$

or, equivalently,

$$\lim_{x \to a} f(g(x)) = f(\lim_{x \to a} g(x))$$

Proof Because f is continuous at b, we have the following statement from Theorem 2.6.6. For any $\epsilon_1 > 0$ there exists a $\delta_1 > 0$ such that

$$\text{if } |y - b| < \delta_1 \quad \text{then} \quad |f(y) - f(b)| < \epsilon_1 \tag{1}$$

Because $\lim\limits_{x \to a} g(x) = b$, for any $\delta_1 > 0$ there exists a $\delta_2 > 0$ such that

$$\text{if} \quad 0 < |x - a| < \delta_2 \quad \text{then} \quad |g(x) - b| < \delta_1 \tag{2}$$

If $0 < |x - a| < \delta_2$, we replace y in statement (1) by $g(x)$ and obtain the following: For any $\epsilon_1 > 0$ there exists a $\delta_1 > 0$ such that

$$\text{if} \quad |g(x) - b| < \delta_1 \quad \text{then} \quad |f(g(x)) - f(b)| < \epsilon_1 \tag{3}$$

From statements (3) and (2) we conclude that for any $\epsilon_1 > 0$ there exists a $\delta_2 > 0$ such that

$$\text{if} \quad 0 < |x - a| < \delta_2 \quad \text{then} \quad |f(g(x)) - f(b)| < \epsilon_1$$

from which it follows that

$$\lim_{x \to a} f(g(x)) = f(b)$$

$$\Leftrightarrow \quad \lim_{x \to a} f(g(x)) = f(\lim_{x \to a} g(x))$$ ■

Theorem 2.7.1 is applied in Supplementary Section 2.9 to prove Limit Theorems 9 and 10. Another application of Theorem 2.7.1 is in proving the following theorem about the continuity of a composite function.

2.7.2 THEOREM   If the function g is continuous at a and the function f is continuous at $g(a)$, then the composite function $f \circ g$ is continuous at a.

Proof   Because g is continuous at a,

$$\lim_{x \to a} g(x) = g(a) \tag{4}$$

Now f is continuous at $g(a)$; thus we can apply Theorem 2.7.1 to the composite function $f \circ g$, thereby giving

$$\lim_{x \to a} (f \circ g)(x) = \lim_{x \to a} f(g(x))$$

$$= f(\lim_{x \to a} g(x))$$

$$= f(g(a)) \qquad \text{(by (4))}$$

$$= (f \circ g)(a)$$

which proves that $f \circ g$ is continuous at a.   ■

Theorem 2.7.2 states that *a continuous function of a continuous function is continuous*. The following example shows how it is used to determine the numbers for which a particular function is continuous.

EXAMPLE 1   Determine the numbers at which the following function is continuous:

$$h(x) = \sqrt{4 - x^2}$$

Solution   The function h is the one obtained in Illustration 1 as the composite function $f \circ g$, where $f(x) = \sqrt{x}$ and $g(x) = 4 - x^2$. Because g is a polynomial function, it is continuous everywhere. Furthermore, f is continuous at every positive number by Theorem 2.6.5(ii). Therefore, by Theorem 2.7.2, h is continuous at every number x for which $g(x) > 0$, that is, when $4 - x^2 > 0$. Hence h is continuous at every number in the open interval $(-2, 2)$.

Because the function h of Example 1 is continuous at every number in the open interval $(-2, 2)$, we say that h is *continuous on the open interval* $(-2, 2)$.

2.7.3 DEFINITION A function is said to be **continuous on an open interval** if and only if it is continuous at every number in the open interval.

We refer again to the function h of Example 1. Because h is not defined on any open interval containing either -2 or 2, we cannot consider $\lim\limits_{x \to -2} h(x)$ or $\lim\limits_{x \to 2} h(x)$. Therefore our definition (2.6.1) of continuity at a number does not permit h to be continuous at -2 and 2. Hence, to discuss the question of the continuity of h on the closed interval $[-2, 2]$, we must extend the concept of continuity to include continuity at an endpoint of a closed interval. This is done by first defining *right-hand continuity* and *left-hand continuity*.

2.7.4 DEFINITION The function f is said to be **continuous from the right at the number a** if and only if the following three conditions are satisfied.

  (i) $f(a)$ exists;

  (ii) $\lim\limits_{x \to a^+} f(x)$ exists;

  (iii) $\lim\limits_{x \to a^+} f(x) = f(a)$.

2.7.5 DEFINITION The function f is said to be **continuous from the left at the number a** if and only if the following three conditions are satisfied:

  (i) $f(a)$ exists;

  (ii) $\lim\limits_{x \to a^-} f(x)$ exists;

  (iii) $\lim\limits_{x \to a^-} f(x) = f(a)$.

2.7.6 DEFINITION A function whose domain includes the closed interval $[a, b]$ is said to be **continuous on $[a, b]$** if and only if it is continuous on the open interval (a, b), as well as continuous from the right at a and continuous from the left at b.

EXAMPLE 2   Prove that the function h of Example 1 is continuous on the closed interval $[-2, 2]$.

Solution   The function h is defined by

$$h(x) = \sqrt{4 - x^2}$$

and in Example 1 we showed that h is continuous on the open interval $(-2, 2)$. By applying Theorem 2.7.1 we compute $\lim\limits_{x \to -2^+} h(x)$ and $\lim\limits_{x \to 2^-} h(x)$.

$$\lim_{x \to -2^+} h(x) = \lim_{x \to -2^+} \sqrt{4 - x^2} \qquad \lim_{x \to 2^-} h(x) = \lim_{x \to 2^-} \sqrt{4 - x^2}$$

$$= 0 \qquad\qquad = 0$$

$$= h(-2) \qquad\qquad = h(2)$$

Thus h is continuous from the right at -2 and continuous from the left at 2. Hence by Definition 2.7.6, h is continuous on the closed interval $[-2, 2]$.

A sketch of the graph of h appears in Figure 1.

Observe the difference in terminology we used in Examples 1 and 2. In Example 1 we stated that *h is continuous at every number in the open interval* $(-2, 2)$, while in Example 2 we concluded that *h is continuous on the closed interval* $[-2, 2]$.

2.7.7 DEFINITION
> (i) A function whose domain includes the interval half-open on the right $[a, b)$ is said to be **continuous on** $[a, b)$ if and only if it is continuous on the open interval (a, b) and continuous from the right at a.
> (ii) A function whose domain includes the interval half-open on the left $(a, b]$ is said to be **continuous on** $(a, b]$ if and only if it is continuous on the open interval (a, b) and continuous from the left at b.

Definitions similar to those in Definition 2.7.7 apply to continuity on the intervals $[a, +\infty)$ and $(-\infty, b]$.

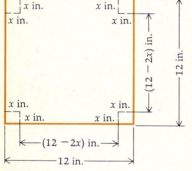

FIGURE 2

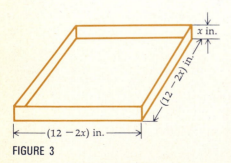

FIGURE 3

EXAMPLE 3   Determine the largest interval (or union of intervals) on which the following function is continuous:

$$f(x) = \frac{\sqrt{25 - x^2}}{x - 3}$$

Solution   We first determine the domain of f. The function is defined everywhere except when $x = 3$ or when $25 - x^2 < 0$ (that is, when $x > 5$ or $x < -5$). Therefore the domain of f is $[-5, 3) \cup (3, 5]$. Because

$$\lim_{x \to -5^+} f(x) = 0 \qquad \text{and} \qquad \lim_{x \to 5^-} f(x) = 0$$
$$= f(-5) \qquad\qquad = f(5)$$

f is continuous from the right at -5 and from the left at 5. Furthermore, f is continuous on the open intervals $(-5, 3)$ and $(3, 5)$. Therefore f is continuous on $[-5, 3) \cup (3, 5]$.

EXAMPLE 4   A cardboard-box manufacturer wishes to make open boxes from square pieces of cardboard of side 12 in. by cutting equal squares from the four corners and turning up the sides. (a) If x inches is the length of the side of the square to be cut out, express the number of cubic inches in the volume of the box as a function of x. (b) What is the domain of the function? (c) Prove that the function is continuous on its domain.

Solution

(a) Figure 2 represents a given piece of cardboard, and Figure 3 represents the box obtained from the cardboard. The numbers of inches in the dimensions of the box are then x, $12 - 2x$, and $12 - 2x$. The volume of the box is the product of the three dimensions. Therefore, if $V(x)$ cubic inches is the volume of the box,

$$V(x) = x(12 - 2x)(12 - 2x)$$

$$= 144x - 48x^2 + 4x^3$$

(b) Observe that $V(0) = 0$ and $V(6) = 0$. From the conditions of the problem we see that x cannot be negative and x cannot be greater than 6. Thus the domain of V is the closed interval $[0, 6]$.

(c) Because V is a polynomial function, V is continuous everywhere. Thus V is continuous on the closed interval $[0, 6]$.

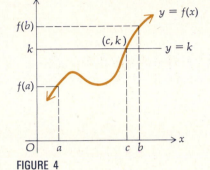

FIGURE 4

The function of Example 4 is discussed again in Section 4.2, where we use the fact that V is continuous on the closed interval $[0, 6]$ to determine the value of x that will give a box having the largest possible volume.

We now discuss an important theorem about a function that is continuous on a closed interval. It is called the **intermediate-value theorem**.

2.7.8 THEOREM

Intermediate-Value Theorem

> If the function f is continuous on the closed interval $[a, b]$, and if $f(a) \neq f(b)$, then for any number k between $f(a)$ and $f(b)$ there exists a number c between a and b such that $f(c) = k$.

The proof of this theorem is beyond the scope of this book; it can be found in an advanced calculus text. However, we discuss the geometric interpretation of the theorem. In Figure 4, $(0, k)$ is any point on the y axis between the points $(0, f(a))$ and $(0, f(b))$. Theorem 2.7.8 states that the line $y = k$ must intersect the curve whose equation is $y = f(x)$ at the point (c, k), where c lies between a and b. Figure 4 shows this intersection.

Note that for some values of k there may be more than one possible value for c. The theorem states that there is always at least one value of c but that it is not necessarily unique. Figure 5 shows three possible values of c (c_1, c_2, and c_3) for a particular k.

Theorem 2.7.8 states that if the function f is continuous on a closed interval $[a, b]$, then f assumes every value between $f(a)$ and $f(b)$ as x assumes all values between a and b. The importance of the continuity of f on $[a, b]$ is demonstrated in the following illustrations.

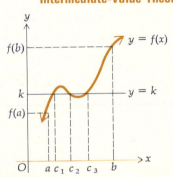

FIGURE 5

▶ **ILLUSTRATION 2** Consider the function f defined by

$$f(x) = \begin{cases} x - 1 & \text{if } 0 \leq x \leq 2 \\ x^2 & \text{if } 2 < x \leq 3 \end{cases}$$

A sketch of the graph of this function appears in Figure 6.

The function f is discontinuous at 2, which is in the closed interval $[0, 3]$; $f(0) = -1$ and $f(3) = 9$. If k is any number between 1 and 4, there is no value of c such that $f(c) = k$ because there are no function values between 1 and 4. ◀

FIGURE 6

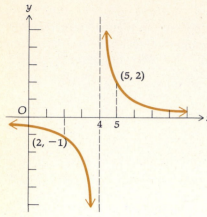

FIGURE 7

▶ **ILLUSTRATION 3**   Let the function g be defined by

$$g(x) = \frac{2}{x - 4}$$

A sketch of the graph of this function is shown in Figure 7.

The function g is discontinuous at 4, which is in the closed interval $[2, 5]$; $g(2) = -1$ and $g(5) = 2$. If k is any number between -1 and 2, there is no value of c between 2 and 5 such that $g(c) = k$. In particular, if $k = 1$, then $g(6) = 1$, but 6 is not in the interval $[2, 5]$.   ◀

EXAMPLE 5   Given the function f defined by

$$f(x) = 4 + 3x - x^2 \qquad 2 \le x \le 5$$

(a) Verify the intermediate-value theorem if $k = 1$; that is, find a number c in the interval $[2, 5]$ such that $f(c) = 1$. (b) Draw a sketch of the graph of f on $[2, 5]$, and show the point $(c, 1)$.

Solution

(a) Because f is a polynomial function, it is continuous everywhere, and thus continuous on $[2, 5]$. Since $f(2) = 6$ and $f(5) = -6$, the intermediate-value theorem guarantees that there is a number c between 2 and 5 such that $f(c) = 1$; that is,

$$4 + 3c - c^2 = 1$$

$$c^2 - 3c - 3 = 0$$

$$c = \frac{3 \pm \sqrt{9 + 12}}{2}$$

$$c = \frac{3 \pm \sqrt{21}}{2}$$

We reject $\frac{1}{2}(3 - \sqrt{21})$ because this number is outside the interval $[2, 5]$. The number $\frac{1}{2}(3 + \sqrt{21})$ is in the interval $[2, 5]$, and

$$f\left(\frac{3 + \sqrt{21}}{2}\right) = 1$$

(b) The required sketch appears in Figure 8.

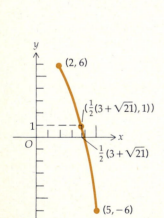

FIGURE 8

EXERCISES 2.7

In Exercises 1 through 14, define $f \circ g$ and determine the numbers at which $f \circ g$ is continuous.

1. $f(x) = \sqrt{x}$; $g(x) = 9 - x^2$

2. $f(x) = \sqrt{x}$; $g(x) = 16 - x^2$

3. $f(x) = \sqrt{x}$; $g(x) = x^2 - 16$

4. $f(x) = \sqrt{x}$; $g(x) = x^2 + 4$

5. $f(x) = x^3$; $g(x) = \sqrt{x}$

6. $f(x) = x^2$; $g(x) = x^2 - 3$

7. $f(x) = \frac{1}{x}$; $g(x) = x - 2$

8. $f(x) = \frac{1}{x^2}$; $g(x) = x + 3$

9. $f(x) = \sqrt{x}$; $g(x) = \frac{1}{x - 2}$

10. $f(x) = \sqrt[3]{x}$; $g(x) = \sqrt{x + 1}$

11. $f(x) = \frac{1}{x - 2}$; $g(x) = \sqrt{x}$

12. $f(x) = \sqrt{x + 1}$; $g(x) = \sqrt[3]{x}$

13. $f(x) = \frac{\sqrt{4 - x^2}}{\sqrt{x - 1}}$; $g(x) = |x|$

14. $f(x) = \frac{\sqrt{x^2 - 1}}{\sqrt{4 - x}}$; $g(x) = |x|$

In Exercises 15 through 24, find the domain of the function, and then determine for each of the indicated intervals whether the function is continuous on that interval.

15. $f(x) = \frac{2}{x + 5}$; $(3, 7)$, $[-6, 4]$, $(-\infty, 0)$, $(-5, +\infty)$, $[-5, +\infty)$, $[-10, -5)$

16. $g(x) = \dfrac{x}{x-2}$; $(-\infty, 0]$, $[0, +\infty)$, $(0, 2)$, $(0, 2]$, $[2, +\infty)$, $(2, +\infty)$

17. $f(t) = \dfrac{t}{t^2 - 1}$; $(0, 1)$, $(-1, 1)$, $[0, 1]$, $(-1, 0]$, $(-\infty, -1]$, $(1, +\infty)$

18. $f(r) = \dfrac{r + 3}{r^2 - 4}$; $(0, 4]$, $(-2, 2)$, $(-\infty, -2]$, $(2, +\infty)$, $[-4, 4]$, $(-2, 2]$

19. $g(x) = \sqrt{x^2 - 9}$; $(-\infty, -3)$, $(-\infty, -3]$, $(3, +\infty)$, $[3, +\infty)$, $(-3, 3)$

20. $f(x) = [\![x]\!]$; $(-\frac{1}{2}, \frac{1}{2})$, $(\frac{1}{4}, \frac{1}{2})$, $(1, 2)$, $[1, 2)$, $(1, 2]$

21. $f(t) = \dfrac{|t - 1|}{t - 1}$; $(-\infty, 1)$ $(-\infty, 1]$, $[-1, 1]$, $(-1, +\infty)$, $(1, +\infty)$

22. $h(x) = \begin{cases} 2x - 3 & \text{if } x < -2 \\ x - 5 & \text{if } -2 \le x \le 1 \\ 3 - x & \text{if } 1 < x \end{cases}$; $(-\infty, 1)$, $(-2, +\infty)$, $(-2, 1)$, $[-2, 1)$, $[-2, 1]$

23. $f(x) = \sqrt{4 - x^2}$; $(-2, 2)$, $[-2, 2]$, $[-2, 2)$, $(-2, 2]$, $(-\infty, -2]$, $(2, +\infty)$

24. $F(y) = \dfrac{1}{3 + 2y - y^2}$; $(-1, 3)$, $[-1, 3]$, $[-1, 3)$, $(-1, 3]$

In Exercises 25 through 34, determine the largest interval (or union of intervals) on which the function f ∘ g of the indicated exercise is continuous.

25. Exercise 1   **26.** Exercise 2   **27.** Exercise 3
28. Exercise 4   **29.** Exercise 9   **30.** Exercise 10
31. Exercise 11   **32.** Exercise 12   **33.** Exercise 13
34. Exercise 14

In Exercises 35 through 38, draw a sketch of the graph of a function f that satisfies the given conditions.

35. f is continuous on $(-\infty, 2]$ and $(2, +\infty)$; $\lim\limits_{x \to 0} f(x) = 4$; $\lim\limits_{x \to 2^-} f(x) = -3$; $\lim\limits_{x \to 2^+} f(x) = +\infty$; $\lim\limits_{x \to 5} f(x) = 0$

36. f is continuous on $(-\infty, 0)$ and $[0, +\infty)$; $\lim\limits_{x \to -4} f(x) = 0$; $\lim\limits_{x \to 0^-} f(x) = 3$; $\lim\limits_{x \to 0^+} f(x) = -3$; $\lim\limits_{x \to 4} f(x) = 2$

37. f is continuous on $(-\infty, -3]$, $(-3, 3)$, and $[3, +\infty)$; $\lim\limits_{x \to -5} f(x) = 2$; $\lim\limits_{x \to -3^-} f(x) = 0$; $\lim\limits_{x \to -3^+} f(x) = 4$; $\lim\limits_{x \to 0} f(x) = 1$; $\lim\limits_{x \to 3^-} f(x) = 0$; $\lim\limits_{x \to 3^+} f(x) = -5$; $\lim\limits_{x \to 4} f(x) = 0$

38. f is continuous on $(-\infty, -2)$, $[-2, 4]$, and $(4, +\infty)$; $\lim\limits_{x \to -5} f(x) = 0$; $\lim\limits_{x \to -2^-} f(x) = -\infty$; $\lim\limits_{x \to -2^+} f(x) = -3$; $\lim\limits_{x \to 0} f(x) = -1$; $\lim\limits_{x \to 4^-} f(x) = 2$; $\lim\limits_{x \to 4^+} f(x) = 5$; $\lim\limits_{x \to 6} f(x) = 0$

39. Determine the largest interval (or union of intervals) on which the function of Exercise 17 in Exercises 2.3 is continuous.

40. Determine the largest interval (or union of intervals) on which the function of Example 4 in Section 2.3 is continuous.

41. A manufacturer of open tin boxes wishes to make use of pieces of tin with dimensions 8 in. by 15 in. by cutting equal squares from the four corners and turning up the sides. (a) If x inches is the length of the side of the square cut out, express the number of cubic inches in the volume of the box as a function of x. (b) What is the domain of the function? (c) Prove that the function is continuous on its domain.

42. Suppose that the manufacturer of Exercise 41 makes the open boxes from square pieces of tin that measure k centimeters on a side. (a) If x centimeters is the length of the side of the square cut out, express the number of cubic centimeters in the volume of the box as a function of x. (b) What is the domain of the function? (c) Prove that the function is continuous on its domain.

43. A rectangular field is to be enclosed with 240 m of fence. (a) If x meters is the length of the field, express the number of square meters in the area of the field as a function of x. (b) What is the domain of the function? (c) Prove that the function is continuous on its domain.

44. A rectangular garden is to be placed so that a side of a house serves as a boundary and 100 ft of fencing material is to be used for the other three sides. (a) If x feet is the length of the side of the garden that is parallel to the house, express the number of square feet in the area of the garden as a function of x. (b) What is the domain of the function? (c) Prove that the function is continuous on its domain.

In Exercises 45 through 48, find the values of the constants c and k that make the function continuous on $(-\infty, +\infty)$, and draw a sketch of the graph of the resulting function.

45. $f(x) = \begin{cases} 3x + 7 & \text{if } x \le 4 \\ kx - 1 & \text{if } 4 < x \end{cases}$

46. $f(x) = \begin{cases} kx - 1 & \text{if } x < 2 \\ kx^2 & \text{if } 2 \le x \end{cases}$

47. $f(x) = \begin{cases} x & \text{if } x \le 1 \\ cx + k & \text{if } 1 < x < 4 \\ -2x & \text{if } 4 \le x \end{cases}$

48. $f(x) = \begin{cases} x + 2c & \text{if } x < -2 \\ 3cx + k & \text{if } -2 \le x \le 1 \\ 3x - 2k & \text{if } 1 < x \end{cases}$

In Exercises 49 through 56, a function f and a closed interval $[a, b]$ are given. Determine if the intermediate-value theorem holds for the given value of k. If the theorem holds, find a number c such that $f(c) = k$. If the theorem does not hold, give the reason. Draw a sketch of the curve and the line $y = k$.

49. $f(x) = 2 + x - x^2$; $[a, b] = [0, 3]$; $k = 1$

50. $f(x) = x^2 + 5x - 6$; $[a, b] = [-1, 2]$; $k = 4$

51. $f(x) = \sqrt{25 - x^2}$; $[a, b] = [-4.5, 3]$; $k = 3$

52. $f(x) = -\sqrt{100 - x^2}$; $[a, b] = [0, 8]$; $k = -8$

53. $f(x) = \dfrac{4}{x + 2}$; $[a, b] = [-3, 1]$; $k = \frac{1}{2}$

54. $f(x) = \begin{cases} 1 + x & \text{if } -4 \le x \le -2 \\ 2 - x & \text{if } -2 < x \le 1 \end{cases}$; $[a, b] = [-4, 1]$; $k = \frac{1}{2}$

55. $f(x) = \begin{cases} x^2 - 4 & \text{if } -2 \le x < 1 \\ x^2 - 1 & \text{if } 1 \le x \le 3 \end{cases}$; $[a, b] = [-2, 3]$; $k = -1$

56. $f(x) = \dfrac{5}{2x - 1}$; $[a, b] = [0, 1]$; $k = 2$

57. Given that f is defined by

$$f(x) = \begin{cases} g(x) & \text{if } a \le x < b \\ h(x) & \text{if } b \le x \le c \end{cases}$$

If g is continuous on $[a, b)$, and h is continuous on $[b, c]$, can we conclude that f is continuous on $[a, c]$? If your answer is yes, prove it. If your answer is no, what additional condition or conditions would assure continuity of f on $[a, c]$?

58. Prove that if the function f is continuous at a, then $\lim\limits_{t \to 0} f(a - t) = f(a)$.

59. Find the largest value of k for which the function defined by $f(x) = [\![x^2 - 2]\!]$ is continuous on the interval $[3, 3 + k]$.

60. Suppose that f is a function for which $0 \le f(x) \le 1$ if $0 \le x \le 1$. Prove that if f is continuous on $[0, 1]$, there is at least one number c in $[0, 1]$ such that $f(c) = c$. (*Hint:* If neither 0 nor 1 qualifies as c, then $f(0) > 0$ and $f(1) < 1$. Consider the function g for which $g(x) = f(x) - x$, and apply the intermediate-value theorem to g on $[0, 1]$.)

61. Show that the intermediate-value theorem guarantees that the equation $x^3 - 4x^2 + x + 3 = 0$ has a root between 1 and 2.

62. Show that the intermediate-value theorem guarantees that the equation $x^3 + x + 3 = 0$ has a root between -2 and -1.

2.8 CONTINUITY OF THE TRIGONOMETRIC FUNCTIONS AND THE SQUEEZE THEOREM

In our discussion of the continuity of the trigonometric functions we make use of the following limit:

$$\lim_{t \to 0} \frac{\sin t}{t} \tag{1}$$

It is apparent that the function given by $\dfrac{\sin t}{t}$ is not defined when $t = 0$. To get an intuitive idea about the existence of the limit in (1), we consider values of $\dfrac{\sin t}{t}$ when t is close to 0. In Table 1 there are function values obtained from a calculator when t is 1.0, 0.9, 0.8, ..., 0.1, and 0.01, and in Table 2 there are values of the function when t is $-1.0, -0.9, -0.8, \ldots, -0.1$, and -0.01.

From the two tables it appears that if the limit in (1) exists, it may be equal to 1. That the limit does exist and is equal to 1 is proved in Theorem 2.8.2, but in the proof of that theorem we need to use the following theorem, sometimes referred to as the **squeeze theorem**. The squeeze theorem not only is important in the proof of Theorem 2.8.2 but also is used to prove some major theorems in later sections.

Table 1

t	$\dfrac{\sin t}{t}$
1.0	0.84147
0.9	0.87036
0.8	0.89670
0.7	0.92031
0.6	0.94107
0.5	0.95885
0.4	0.97355
0.3	0.98507
0.2	0.99335
0.1	0.99833
0.01	0.99998

Table 2

t	$\dfrac{\sin t}{t}$
-1.0	0.84147
-0.9	0.87036
-0.8	0.89670
-0.7	0.92031
-0.6	0.94107
-0.5	0.95885
-0.4	0.97355
-0.3	0.98507
-0.2	0.99335
-0.1	0.99833
-0.01	0.99998

2.8.1 THEOREM
The Squeeze Theorem

Suppose that the functions f, g, and h are defined on some open interval I containing a except possibly at a itself, and that $f(x) \le g(x) \le h(x)$ for all x in I for which $x \ne a$. Also suppose that $\lim\limits_{x \to a} f(x)$ and $\lim\limits_{x \to a} h(x)$ both exist and are equal to L. Then $\lim\limits_{x \to a} g(x)$ exists and is equal to L.

Before Theorem 2.8.1 is proved, consider the following illustration, which interprets the theorem geometrically.

▶ **ILLUSTRATION 1** Let the functions f, g, and h be defined by

$$f(x) = -4(x - 2)^2 + 3$$

$$g(x) = \frac{(x - 2)(x^2 - 4x + 7)}{x - 2}$$

$$h(x) = 4(x - 2)^2 + 3$$

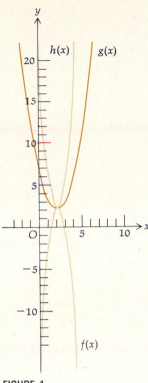

FIGURE 1

The graphs of f and h are parabolas having their vertex at $(2, 3)$. The graph of g is a parabola with its vertex $(2, 3)$ deleted. Sketches of these graphs are shown in Figure 1. The function g is not defined when $x = 2$; however, for all $x \neq 2$, $f(x) \leq g(x) \leq h(x)$. Furthermore, $\lim_{x \to 2} f(x) = 3$ and $\lim_{x \to 2} h(x) = 3$. The hypothesis of Theorem 2.8.1 is therefore satisfied, and it follows that $\lim_{x \to 2} g(x) = 3$.

◀

Proof of Theorem 2.8.1   To prove that $\lim_{x \to a} g(x) = L$ we must show that for any $\epsilon > 0$ there is a $\delta > 0$ such that

$$\text{if} \quad 0 < |x - a| < \delta \quad \text{then} \quad |g(x) - L| < \epsilon \tag{2}$$

We are given that

$$\lim_{x \to a} f(x) = L \quad \text{and} \quad \lim_{x \to a} h(x) = L$$

and so for any $\epsilon > 0$ there is a $\delta_1 > 0$ such that

$$\text{if} \quad 0 < |x - a| < \delta_1 \quad \text{then} \quad |f(x) - L| < \epsilon$$
$$\Leftrightarrow \quad \text{if} \quad 0 < |x - a| < \delta_1 \quad \text{then} \quad L - \epsilon < f(x) < L + \epsilon \tag{3}$$

and a $\delta_2 > 0$ such that

$$\text{if} \quad 0 < |x - a| < \delta_2 \quad \text{then} \quad |h(x) - L| < \epsilon$$
$$\Leftrightarrow \quad \text{if} \quad 0 < |x - a| < \delta_2 \quad \text{then} \quad L - \epsilon < h(x) < L + \epsilon \tag{4}$$

Let $\delta = \min(\delta_1, \delta_2)$, and so $\delta \leq \delta_1$ and $\delta \leq \delta_2$. Therefore it follows from statement (3) that

$$\text{if} \quad 0 < |x - a| < \delta \quad \text{then} \quad L - \epsilon < f(x) \tag{5}$$

and from statement (4) that

$$\text{if} \quad 0 < |x - a| < \delta \quad \text{then} \quad h(x) < L + \epsilon \tag{6}$$

We are given that

$$f(x) \leq g(x) \leq h(x) \tag{7}$$

From statements (5), (6), and (7),

$$\text{if} \quad 0 < |x - a| < \delta \quad \text{then} \quad L - \epsilon < f(x) \leq g(x) \leq h(x) < L + \epsilon$$

Therefore

$$\text{if} \quad 0 < |x - a| < \delta \quad \text{then} \quad L - \epsilon < g(x) < L + \epsilon$$
$$\Leftrightarrow \quad \text{if} \quad 0 < |x - a| < \delta \quad \text{then} \quad |g(x) - L| < \epsilon$$

which is statement (2). Hence

$$\lim_{x \to a} g(x) = L$$

■

EXAMPLE 1   Given $|g(x) - 2| \leq 3(x - 1)^2$ for all x. Use the squeeze theorem to find $\lim_{x \to 1} g(x)$.

Solution   Because $|g(x) - 2| \leq 3(x - 1)^2$ for all x, it follows from Theorem 1.1.10 that

$$-3(x - 1)^2 \leq g(x) - 2 \leq 3(x - 1)^2 \qquad \text{for all } x$$

$$\Leftrightarrow \qquad -3(x - 1)^2 + 2 \leq g(x) \leq 3(x - 1)^2 + 2 \qquad \text{for all } x$$

Let $f(x) = -3(x - 1)^2 + 2$ and $h(x) = 3(x - 1)^2 + 2$. Then

$$\lim_{x \to 1} f(x) = 2 \quad \text{and} \quad \lim_{x \to 1} h(x) = 2 \tag{8}$$

Furthermore, for all x,

$$f(x) \leq g(x) \leq h(x) \tag{9}$$

Thus from (8), (9), and the squeeze theorem

$$\lim_{x \to 1} g(x) = 2$$

EXAMPLE 2   Use the squeeze theorem to prove that

$$\lim_{x \to 0} \left| x \sin \frac{1}{x} \right| = 0$$

Solution   Because $-1 \leq \sin t \leq 1$ for all t, then

$$0 \leq \left| \sin \frac{1}{x} \right| \leq 1 \qquad \text{if } x \neq 0$$

Therefore, if $x \neq 0$,

$$\left| x \sin \frac{1}{x} \right| = |x| \left| \sin \frac{1}{x} \right|$$

$$\leq |x|$$

Hence

$$0 \leq \left| x \sin \frac{1}{x} \right| \leq |x| \qquad \text{if } x \neq 0 \tag{10}$$

Because $\lim_{x \to 0} 0 = 0$ and $\lim_{x \to 0} |x| = 0$, it follows from inequality (10) and the squeeze theorem that

$$\lim_{x \to 0} \left| x \sin \frac{1}{x} \right| = 0$$

2.8.2 THEOREM $\displaystyle \lim_{t \to 0} \frac{\sin t}{t} = 1$

Proof   First assume that $0 < t < \frac{1}{2}\pi$. Refer to Figure 2, which shows the unit circle $x^2 + y^2 = 1$ and the shaded sector BOP, where B is the point $(1, 0)$ and P is the point $(\cos t, \sin t)$. The area of a circular sector of radius r and central angle of radian measure t is determined by $\frac{1}{2}r^2 t$; so if S square units is the area of sector BOP,

$$S = \tfrac{1}{2}t \tag{11}$$

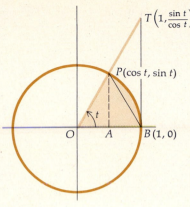

FIGURE 2

Consider now the triangle BOP, and let K_1 square units be the area of this triangle. Because $K_1 = \frac{1}{2}|\overline{AP}| \cdot |\overline{OB}|$, $|\overline{AP}| = \sin t$ and $|\overline{OB}| = 1$, we have

$$K_1 = \tfrac{1}{2}\sin t \tag{12}$$

The line through the points $O(0, 0)$ and $P(\cos t, \sin t)$ has slope $\dfrac{\sin t}{\cos t}$; therefore its equation is

$$y = \frac{\sin t}{\cos t}\,x$$

This line intersects the line $x = 1$ at the point $(1, \sin t/\cos t)$, which is the point T in Figure 2. If K_2 square units is the area of right triangle BOT, then $K_2 = \frac{1}{2}|\overline{BT}| \cdot |\overline{OB}|$. Because $|\overline{BT}| = \sin t/\cos t$ and $|\overline{OB}| = 1$, we have

$$K_2 = \frac{1}{2}\cdot\frac{\sin t}{\cos t} \tag{13}$$

From Figure 2 observe that

$$K_1 < S < K_2$$

Substituting from (11), (12), and (13) into this inequality,

$$\frac{1}{2}\sin t < \frac{1}{2}t < \frac{1}{2}\cdot\frac{\sin t}{\cos t}$$

Multiplying each member of this inequality by $2/\sin t$, which is positive because $0 < t < \frac{1}{2}\pi$,

$$1 < \frac{t}{\sin t} < \frac{1}{\cos t}$$

Taking the reciprocal of each member of this inequality (which reverses the direction of the inequality signs),

$$\cos t < \frac{\sin t}{t} < 1 \tag{14}$$

From the right-hand inequality in the above,

$$\sin t < t \tag{15}$$

and from a half-measure identity in trigonometry,

$$\frac{1 - \cos t}{2} = \sin^2\frac{1}{2}t \tag{16}$$

Replacing t by $\frac{1}{2}t$ in inequality (15) and squaring,

$$\sin^2\tfrac{1}{2}t < \tfrac{1}{4}t^2 \tag{17}$$

Thus from (16) and (17) it follows that

$$\frac{1 - \cos t}{2} < \frac{t^2}{4}$$

$$\Leftrightarrow\quad 1 - \tfrac{1}{2}t^2 < \cos t \tag{18}$$

From (14) and (18) and because $0 < t < \frac{1}{2}\pi$,

$$1 - \frac{1}{2}t^2 < \frac{\sin t}{t} < 1 \qquad \text{if } 0 < t < \tfrac{1}{2}\pi \tag{19}$$

If $-\frac{1}{2}\pi < t < 0$, then $0 < -t < \frac{1}{2}\pi$; and so from (19),

$$1 - \frac{1}{2}(-t)^2 < \frac{\sin(-t)}{-t} < 1 \qquad \text{if } -\frac{1}{2}\pi < t < 0$$

But $\sin(-t) = -\sin t$; thus the above can be written as

$$1 - \frac{1}{2}t^2 < \frac{\sin t}{t} < 1 \qquad \text{if } -\frac{1}{2}\pi < t < 0 \qquad (20)$$

From (19) and (20) we conclude that

$$1 - \frac{1}{2}t^2 < \frac{\sin t}{t} < 1 \qquad \text{if } -\frac{1}{2}\pi < t < \frac{1}{2}\pi \text{ and } t \neq 0 \qquad (21)$$

Because $\lim\limits_{t \to 0} (1 - \frac{1}{2}t^2) = 1$ and $\lim\limits_{t \to 0} 1 = 1$, it follows from (21) and the squeeze theorem that

$$\lim_{t \to 0} \frac{\sin t}{t} = 1 \qquad \blacksquare$$

EXAMPLE 3   Find the limit if it exists:

$$\lim_{x \to 0} \frac{\sin 3x}{\sin 5x}$$

Solution   We wish to write the quotient $\sin 3x/\sin 5x$ in such a way that Theorem 2.8.2 can be applied. If $x \neq 0$,

$$\frac{\sin 3x}{\sin 5x} = \frac{3\left(\dfrac{\sin 3x}{3x}\right)}{5\left(\dfrac{\sin 5x}{5x}\right)}$$

As x approaches zero, so do $3x$ and $5x$. Hence

$$\lim_{x \to 0} \frac{\sin 3x}{3x} = \lim_{3x \to 0} \frac{\sin 3x}{3x} \qquad \lim_{x \to 0} \frac{\sin 5x}{5x} = \lim_{5x \to 0} \frac{\sin 5x}{5x}$$

$$= 1 \qquad\qquad\qquad = 1$$

Therefore

$$\lim_{x \to 0} \frac{\sin 3x}{\sin 5x} = \frac{3 \lim\limits_{x \to 0} \left(\dfrac{\sin 3x}{3x}\right)}{5 \lim\limits_{x \to 0} \left(\dfrac{\sin 5x}{5x}\right)}$$

$$= \frac{3 \cdot 1}{5 \cdot 1}$$

$$= \frac{3}{5}$$

From Theorem 2.8.2 we can prove that the sine function and the cosine function are continuous at 0.

2.8.3 THEOREM   The sine function is continuous at 0.

Proof   The three conditions necessary for continuity at a number are shown to be satisfied.

(i) $\sin 0 = 0$

(ii) $\displaystyle\lim_{t \to 0} \sin t = \lim_{t \to 0} \frac{\sin t}{t} \cdot t$

$$= \lim_{t \to 0} \frac{\sin t}{t} \cdot \lim_{t \to 0} t$$

$$= 1 \cdot 0$$

$$= 0$$

(iii) $\displaystyle\lim_{t \to 0} \sin t = \sin 0$

Therefore the sine function is continuous at 0.   ■

2.8.4 THEOREM   The cosine function is continuous at 0.

Proof   We check the three conditions necessary for continuity at a number In checking condition (ii) we use the fact that the sine function is continuous at 0, and we replace $\cos t$ by $\sqrt{1 - \sin^2 t}$ because $\cos t > 0$ when $0 < t < \frac{1}{2}\pi$ and when $-\frac{1}{2}\pi < t < 0$.

(i) $\cos 0 = 1$

(ii) $\displaystyle\lim_{t \to 0} \cos t = \lim_{t \to 0} \sqrt{1 - \sin^2 t}$

$$= \sqrt{\lim_{t \to 0}(1 - \sin^2 t)}$$

$$= \sqrt{1 - 0}$$

$$= 1$$

(iii) $\displaystyle\lim_{t \to 0} \cos t = \cos 0$

Thus the cosine function is continuous at 0.   ■

We need to use the limit in the following theorem later. It is obtained from the previous three theorems and limit theorems.

2.8.5 THEOREM $\displaystyle\lim_{t \to 0} \frac{1 - \cos t}{t} = 0$

Proof

$$\lim_{t \to 0} \frac{1 - \cos t}{t} = \lim_{t \to 0} \frac{(1 - \cos t)(1 + \cos t)}{t(1 + \cos t)}$$

$$= \lim_{t \to 0} \frac{(1 - \cos^2 t)}{t(1 + \cos t)}$$

$$= \lim_{t \to 0} \frac{\sin^2 t}{t(1 + \cos t)}$$

$$= \lim_{t \to 0} \frac{\sin t}{t} \cdot \lim_{t \to 0} \frac{\sin t}{1 + \cos t}$$

By Theorem 2.8.2,

$$\lim_{t \to 0} \frac{\sin t}{t} = 1$$

and because the sine and cosine functions are continuous at 0 it follows that

$$\lim_{t \to 0} \frac{\sin t}{1 + \cos t} = \frac{0}{1 + 1}$$

$$= 0$$

Therefore

$$\lim_{t \to 0} \frac{1 - \cos t}{t} = 1 \cdot 0$$

$$= 0$$

■

EXAMPLE 4   Find the limit if it exists:

$$\lim_{x \to 0} \frac{1 - \cos x}{\sin x}$$

Solution   Because $\lim_{x \to 0} (1 - \cos x) = 0$ and $\lim_{x \to 0} \sin x = 0$, the limit theorems cannot be applied to the quotient $(1 - \cos x)/\sin x$. However, if the numerator and denominator are divided by x, which is permissible because $x \neq 0$, we are able to apply Theorems 2.8.2 and 2.8.5. Thus

$$\lim_{x \to 0} \frac{1 - \cos x}{\sin x} = \lim_{x \to 0} \frac{\dfrac{1 - \cos x}{x}}{\dfrac{\sin x}{x}}$$

$$= \frac{\lim\limits_{x \to 0} \dfrac{1 - \cos x}{x}}{\lim\limits_{x \to 0} \dfrac{\sin x}{x}}$$

$$= \frac{0}{1}$$

$$= 0$$

EXAMPLE 5   Find the limit if it exists:

$$\lim_{x \to 0} \frac{2 \tan^2 x}{x^2}$$

Solution   We use the trigonometric identity

$$\tan x = \frac{\sin x}{\cos x}$$

and we have

$$\lim_{x \to 0} \frac{2 \tan^2 x}{x^2} = 2 \lim_{x \to 0} \frac{\sin^2 x}{x^2 \cdot \cos^2 x}$$

$$= 2 \lim_{x \to 0} \frac{\sin x}{x} \cdot \lim_{x \to 0} \frac{\sin x}{x} \cdot \lim_{x \to 0} \frac{1}{\cos^2 x}$$

$$= 2 \cdot 1 \cdot 1 \cdot 1$$

$$= 2$$

By using Theorem 2.2.12 and the facts that the sine and cosine functions are continuous at 0 we can prove the following theorem.

2.8.6 THEOREM   The sine and cosine functions are continuous at every real number.

Proof   The set of all real numbers is the domain of both the sine and cosine functions. Therefore we must show that if a is any real number,

$$\lim_{x \to a} \sin x = \sin a \quad \text{and} \quad \lim_{x \to a} \cos x = \cos a$$

or, equivalently, from Theorem 2.2.12,

$$\lim_{t \to 0} \sin(t + a) = \sin a \text{ and } \lim_{t \to 0} \cos(t + a) = \cos a \tag{22}$$

We use the identities

$$\sin(t + a) = \sin t \cos a + \cos t \sin a \tag{23}$$

$$\cos(t + a) = \cos t \cos a - \sin t \sin a \tag{24}$$

From (23),

$$\lim_{t \to 0} \sin(t + a) = \lim_{t \to 0} (\sin t \cos a + \cos t \sin a)$$

$$= \lim_{t \to 0} \sin t \cdot \lim_{t \to 0} \cos a + \lim_{t \to 0} \cos t \cdot \lim_{t \to 0} \sin a$$

$$= 0 \cdot \cos a + 1 \cdot \sin a$$

$$= \sin a$$

Therefore the first equation in (22) holds; so the sine function is continuous at every real number. From (24),

$$\lim_{t \to 0} \cos(t + a) = \lim_{t \to 0} (\cos t \cos a - \sin t \sin a)$$

$$= \lim_{t \to 0} \cos t \cdot \lim_{t \to 0} \cos a - \lim_{t \to 0} \sin t \cdot \lim_{t \to 0} \sin a$$

$$= 1 \cdot \cos a - 0 \cdot \sin a$$

$$= \cos a$$

Thus the second equation in (22) holds; so the cosine function is continuous at every real number.   ∎

By using trigonometric identities, Theorem 2.6.4 about the continuity of a rational function, and Theorem 2.8.6 we can prove that the other four trigonometric functions are continuous on their domains.

2.8.7 THEOREM   The tangent, cotangent, secant, and cosecant functions are continuous on their domains.

The proof of Theorem 2.8.7 is left as exercises (see Exercises 35 through 38).

EXERCISES 2.8

In Exercises 1 through 26, find the limit if it exists.

1. $\lim\limits_{x \to 0} \dfrac{\sin 4x}{x}$

2. $\lim\limits_{x \to 0} \dfrac{2x}{\sin 3x}$

3. $\lim\limits_{x \to 0} \dfrac{\sin 9x}{\sin 7x}$

4. $\lim\limits_{t \to 0} \dfrac{\sin 3t}{\sin 6t}$

5. $\lim\limits_{y \to 0} \dfrac{3y}{\sin 5y}$

6. $\lim\limits_{x \to 0} \dfrac{\sin^3 x}{x^2}$

7. $\lim\limits_{x \to 0} \dfrac{x^2}{\sin^2 3x}$

8. $\lim\limits_{x \to 0} \dfrac{\sin^5 2x}{4x^5}$

9. $\lim\limits_{x \to 0} \dfrac{x}{\cos x}$

10. $\lim\limits_{x \to 0} \dfrac{1 - \cos x}{1 + \sin x}$

11. $\lim\limits_{x \to 0} \dfrac{1 - \cos 4x}{x}$

12. $\lim\limits_{z \to 0} \dfrac{1 - \cos 2z}{4z}$

13. $\lim\limits_{x \to 0} \dfrac{3x^2}{1 - \cos^2 \frac{1}{2}x}$

14. $\lim\limits_{x \to 0} \dfrac{1 - \cos^2 x}{2x^2}$

15. $\lim\limits_{x \to 0} \dfrac{\tan x}{2x}$

16. $\lim\limits_{x \to 0} \dfrac{\tan^4 2x}{4x^4}$

17. $\lim\limits_{t \to 0^+} \dfrac{\sin t}{t^2}$

18. $\lim\limits_{x \to 0} \dfrac{1 - \cos x}{x^2}$

19. $\lim\limits_{x \to 0} \dfrac{1 - \cos 2x}{\sin 3x}$

20. $\lim\limits_{y \to 0^+} \dfrac{\sin 4y}{\cos 3y - 1}$

21. $\lim\limits_{x \to \pi/2} \dfrac{1 - \sin x}{\frac{1}{2}\pi - x}$ (*Hint:* Let $t = \frac{1}{2}\pi - x$.)

22. $\lim\limits_{x \to \pi/2} \dfrac{\frac{1}{2}\pi - x}{\cos x}$ (*Hint:* Let $t = \frac{1}{2}\pi - x$.)

23. $\lim\limits_{x \to \pi^+} \dfrac{\sin x}{x - \pi}$ (*Hint:* Let $t = x - \pi$.)

24. $\lim\limits_{x \to \pi^+} \dfrac{\tan x}{x - \pi}$ (*Hint:* Let $t = x - \pi$.)

25. $\lim\limits_{x \to 0} \dfrac{x^2 + 3x}{\sin x}$

26. $\lim\limits_{x \to 0} \dfrac{\sin x}{3x^2 + 2x}$

In Exercises 27 through 30, use the squeeze theorem to find the limit.

27. $\lim\limits_{x \to 0} x \cos \dfrac{1}{x}$

28. $\lim\limits_{x \to 0} x^2 \sin \dfrac{1}{\sqrt[3]{x}}$

29. $\lim\limits_{x \to 3} g(x)$, if $|g(x) + 4| < 2(3 - x)^4$ for all x

30. $\lim\limits_{x \to -2} g(x)$, if $|g(x) - 3| < 5(x + 2)^2$ for all x

In Exercises 31 and 32, find the limit if it exists.

31. $\lim\limits_{x \to 0} \dfrac{\sin(\sin x)}{x}$

32. $\lim\limits_{x \to 0} \sin x \sin \dfrac{1}{x}$

33. Given: $1 - \cos^2 x \le f(x) \le x^2$ for all x in the open interval $(-\frac{1}{2}\pi, \frac{1}{2}\pi)$. Find $\lim\limits_{x \to 0} f(x)$.

34. Given: $-\sin x \le f(x) \le 2 + \sin x$ for all x in the open interval $(-\pi, 0)$. Find $\lim\limits_{x \to -\pi/2} f(x)$.

In Exercises 35 through 38, prove that the function is continuous on its domain.

35. the tangent function   **36.** the cotangent function
37. the secant function   **38.** the cosecant function

39. If $|f(x)| \le M$ for all x and M is a constant, use the squeeze theorem to prove that $\lim\limits_{x \to 0} x^2 f(x) = 0$.

40. Suppose that $|f(x)| \le M$ for all x, where M is a constant. Furthermore, suppose that $\lim\limits_{x \to a} |g(x)| = 0$. Use the squeeze theorem to prove that $\lim\limits_{x \to a} f(x)g(x) = 0$.

41. If $|f(x)| \le k|x - a|$ for all $x \ne a$, where k is a constant, prove that $\lim\limits_{x \to a} f(x) = 0$.

2.9 PROOFS OF SOME THEOREMS ON LIMITS OF FUNCTIONS *(Supplementary)*

In Section 2.1 there is one theorem (Theorem 2.1.2) and in Section 2.2 there are two theorems (Limit Theorems 9 and 10) whose proofs were deferred to this section.

As stated in Section 2.1, Theorem 2.1.2 is a uniqueness theorem that guarantees that if the limit of a function exists, it is unique. We now restate it and give its proof.

2.1.2 THEOREM   If $\lim\limits_{x \to a} f(x) = L_1$ and $\lim\limits_{x \to a} f(x) = L_2$, then $L_1 = L_2$.

Proof We shall assume that $L_1 \neq L_2$ and show that this assumption leads to a contradiction. Because $\lim_{x \to a} f(x) = L_1$, it follows from Definition 2.1.1 that for any $\epsilon > 0$ there exists a $\delta_1 > 0$ such that

$$\text{if} \quad 0 < |x - a| < \delta_1 \quad \text{then} \quad |f(x) - L_1| < \epsilon \tag{1}$$

Also, because $\lim_{x \to a} f(x) = L_2$, there exists a $\delta_2 > 0$ such that

$$\text{if} \quad 0 < |x - a| < \delta_2 \quad \text{then} \quad |f(x) - L_2| < \epsilon \tag{2}$$

Now, writing $L_1 - L_2$ as $L_1 - f(x) + f(x) - L_2$ and applying the triangle inequality (Theorem 1.1.15) we have

$$|L_1 - L_2| = |[L_1 - f(x)] + [f(x) - L_2]|$$
$$\leq |L_1 - f(x)| + |f(x) - L_2| \tag{3}$$

So from (1), (2), and (3) we may conclude that for any $\epsilon > 0$ there exists a $\delta_1 > 0$ and a $\delta_2 > 0$ such that

$$\text{if} \quad 0 < |x - a| < \delta_1 \quad \text{and} \quad 0 < |x - a| < \delta_2 \quad \text{then} \quad |L_1 - L_2| < \epsilon + \epsilon \tag{4}$$

If δ is the smaller of δ_1 and δ_2, then $\delta \leq \delta_1$ and $\delta \leq \delta_2$, and (4) states that for any $\epsilon > 0$ there exists a $\delta > 0$ such that

$$\text{if} \quad 0 < |x - a| < \delta \quad \text{then} \quad |L_1 - L_2| < 2\epsilon \tag{5}$$

However, if $\epsilon = \frac{1}{2}|L_1 - L_2|$, then (5) states that there exists a $\delta > 0$ such that

$$\text{if} \quad 0 < |x - a| < \delta \quad \text{then} \quad |L_1 - L_2| < |L_1 - L_2|$$

Obviously $|L_1 - L_2|$ is not less than itself. So we have a contradiction and our assumption is false. Thus $L_1 = L_2$, and the theorem is proved. ■

To prove Limit Theorem 9 (the limit of the quotient of two functions) and Limit Theorem 10 (the limit of the nth root of a function) we apply Theorem 2.7.1 regarding the limit of a composite function, which we now restate.

2.7.1 THEOREM If $\lim_{x \to a} g(x) = b$, and if the function f is continuous at b,

$$\lim_{x \to a} (f \circ g)(x) = f(b)$$

$$\Leftrightarrow \quad \lim_{x \to a} f(g(x)) = f(\lim_{x \to a} g(x))$$

Theorem 2.7.1 was proved in Section 2.7. Before this theorem can be applied to prove Limit Theorem 9, we must prove a theorem about the continuity of the function defined by $f(x) = \dfrac{1}{x}$.

2.9.1 THEOREM If a is any real number except zero, and

$$f(x) = \frac{1}{x}$$

then f is continuous at a.

Proof The domain of f is the set of all real numbers except zero. Therefore a is in the domain. The proof then will be complete if we can show that

$$\lim_{x \to a} \frac{1}{x} = \frac{1}{a}$$

To prove this, two cases need to be considered: $a > 0$ and $a < 0$. We prove the case when $a > 0$ and leave the proof for $a < 0$ as an exercise (see Exercise 18).

Because $1/x$ is defined for every x except zero, the open interval required by Definition 2.1.1 can be any open interval containing a but not containing 0.

Considering $a > 0$, we must show that for any $\epsilon > 0$ there exists a $\delta > 0$ such that

$$\text{if} \quad 0 < |x - a| < \delta \quad \text{then} \quad \left| \frac{1}{x} - \frac{1}{a} \right| < \epsilon \tag{6}$$

Because

$$\left| \frac{1}{x} - \frac{1}{a} \right| = \left| \frac{a - x}{ax} \right|$$

$$= \frac{|x - a|}{|a| \, |x|}$$

$$= |x - a| \cdot \frac{1}{a|x|} \qquad \text{(because } a > 0\text{)}$$

statement (6) is equivalent to

$$\text{if} \quad 0 < |x - a| < \delta \quad \text{then} \quad |x - a| \cdot \frac{1}{a|x|} < \epsilon \tag{7}$$

On the right end of (7), in addition to the factor $|x - a|$ we have as another factor the quotient $\dfrac{1}{a|x|}$. Therefore, to prove (7) we need to restrict δ so that we will have an inequality involving $\dfrac{1}{a|x|}$. By choosing the open interval required by Definition 2.1.1 to be the interval $(\frac{1}{2}a, \frac{3}{2}a)$, which contains a but not 0, we are implying that $\delta \leq \frac{1}{2}a$. Then

$$0 < |x - a| < \delta \quad \text{and} \quad \delta \leq \tfrac{1}{2}a$$

$$\Rightarrow \qquad |x - a| < \tfrac{1}{2}a$$

$$\Rightarrow \quad -\tfrac{1}{2}a < x - a < \tfrac{1}{2}a$$

$$\Rightarrow \qquad \tfrac{1}{2}a < x < \tfrac{3}{2}a$$

$$\Rightarrow \qquad \tfrac{1}{2}a < |x| < \tfrac{3}{2}a \qquad \text{(because } a > 0\text{)}$$

$$\Rightarrow \qquad \frac{2}{3a} < \frac{1}{|x|} < \frac{2}{a}$$

$$\Rightarrow \qquad \frac{2}{3a^2} < \frac{1}{a|x|} < \frac{2}{a^2} \tag{8}$$

Now

$$0 < |x - a| < \delta \quad \text{and} \quad \frac{1}{a|x|} < \frac{2}{a^2}$$

$$\Rightarrow \quad |x - a| \cdot \frac{1}{a|x|} < \delta \cdot \frac{2}{a^2} \tag{9}$$

Because our goal is to have $|x - a| \cdot \dfrac{1}{a|x|} < \epsilon$, statement (9) indicates that we should require $\delta \cdot \dfrac{2}{a^2} \leq \epsilon$, that is, $\delta \leq \frac{1}{2}a^2\epsilon$. Thus with the two restrictions on δ we choose $\delta = \min(\frac{1}{2}a, \frac{1}{2}a^2\epsilon)$. With this δ we have the following argument:

$$0 < |x - a| < \delta$$

$$\Rightarrow \quad |x - a| \cdot \frac{1}{a|x|} < \delta \cdot \frac{1}{a|x|}$$

$$\Rightarrow \quad \frac{|x - a|}{|a| \, |x|} < \delta \cdot \frac{1}{a|x|} \quad \text{(because } a > 0\text{)}$$

$$\Rightarrow \quad \left| \frac{a - x}{ax} \right| < \delta \cdot \frac{1}{a|x|}$$

$$\Rightarrow \quad \left| \frac{1}{x} - \frac{1}{a} \right| < \delta \cdot \frac{1}{a|x|} \tag{10}$$

We showed in (8) that if $\delta \leq \frac{1}{2}a$ and $0 < |x - a| < \delta$, then $\dfrac{1}{a|x|} < \dfrac{2}{a^2}$, that is, $\delta \cdot \dfrac{1}{a|x|} < \delta \cdot \dfrac{2}{a^2}$. Continuing from (10) we have

$$\left| \frac{1}{x} - \frac{1}{a} \right| < \delta \cdot \frac{1}{a|x|} \quad \text{and} \quad \delta \cdot \frac{1}{a|x|} < \delta \cdot \frac{2}{a^2}$$

$$\Rightarrow \quad \left| \frac{1}{x} - \frac{1}{a} \right| < \delta \cdot \frac{2}{a^2}$$

$$\Rightarrow \quad \left| \frac{1}{x} - \frac{1}{a} \right| < \frac{1}{2} a^2\epsilon \cdot \frac{2}{a^2} \quad \text{(because } \delta \leq \frac{1}{2}a^2\epsilon\text{)}$$

$$\Rightarrow \quad \left| \frac{1}{x} - \frac{1}{a} \right| < \epsilon$$

Thus we have shown that for any $\epsilon > 0$, if $\delta = \min(\frac{1}{2}a, \frac{1}{2}a^2\epsilon)$, then the following statement is true:

$$\text{if} \quad 0 < |x - a| < \delta \quad \text{then} \quad \left| \frac{1}{x} - \frac{1}{a} \right| < \epsilon$$

This proves that $\lim\limits_{x \to a} \dfrac{1}{x} = \dfrac{1}{a}$, if $a > 0$. Therefore f is continuous at a if $a > 0$.

■

Before restating Limit Theorem 9 and proving it by using Theorems 2.7.1 and 2.9.1, we give an example showing how Theorems 2.7.1 and 2.9.1 can be applied to find the limit of a particular quotient.

EXAMPLE 1   By using Theorems 2.7.1 and 2.9.1, and not Limit Theorem 9 (the limit of a quotient), find

$$\lim_{x \to 2} \frac{4x - 3}{x^2 + 2x + 5}$$

Solution   Let the functions f and g be defined by

$$f(x) = 4x - 3 \quad \text{and} \quad g(x) = x^2 + 2x + 5$$

We consider $f(x)/g(x)$ as the product of $f(x)$ and $1/g(x)$ and apply Limit Theorem 6 (limit of a product). First, though, we must find $\lim\limits_{x \to 2} 1/g(x)$, which is done by considering $1/g(x)$ as a composite function value.

If h is the function defined by $h(x) = 1/x$, then the composite function $h \circ g$ is the function defined by $h(g(x)) = 1/g(x)$. Now

$$\lim_{x \to 2} g(x) = \lim_{x \to 2} (x^2 + 2x + 5)$$

$$= 13$$

From Theorem 2.9.1, h is continuous at 13; thus we can use Theorem 2.7.1, and we have

$$\lim_{x \to 2} \frac{1}{x^2 + 2x + 5} = \lim_{x \to 2} \frac{1}{g(x)}$$

$$= \lim_{x \to 2} h(g(x))$$

$$= h(\lim_{x \to 2} g(x)) \qquad \text{(by Theorem 2.7.1)}$$

$$= h(13)$$

$$= \tfrac{1}{13}$$

Therefore, from Limit Theorem 6,

$$\lim_{x \to 2} \frac{4x - 3}{x^2 + 2x + 5} = \lim_{x \to 2} (4x - 3) \cdot \lim_{x \to 2} \frac{1}{x^2 + 2x + 5}$$

$$= 5 \cdot \tfrac{1}{13}$$

$$= \tfrac{5}{13}$$

2.2.9 LIMIT THEOREM 9   If $\lim\limits_{x \to a} f(x) = L$ and if $\lim\limits_{x \to a} g(x) = M$, then

$$\lim_{x \to a} \frac{f(x)}{g(x)} = \frac{L}{M} \qquad \text{if } M \neq 0$$

Proof   Let h be the function defined by $h(x) = 1/x$. Then the composite function $h \circ g$ is defined by $h(g(x)) = 1/g(x)$. The function h is continuous everywhere except at 0, which follows from Theorem 2.9.1. Hence

$$\lim_{x \to a} \frac{1}{g(x)} = \lim_{x \to a} h(g(x))$$

$$= h(\lim_{x \to a} g(x)) \qquad \text{(by Theorem 2.7.1)}$$

$$= h(M)$$

$$= \frac{1}{M}$$

From Limit Theorem 6 and the above result,

$$\lim_{x \to a} \frac{f(x)}{g(x)} = \lim_{x \to a} f(x) \cdot \lim_{x \to a} \frac{1}{g(x)}$$

$$= L \cdot \frac{1}{M}$$

$$= \frac{L}{M} \qquad\qquad\qquad ∎$$

To prove Limit Theorem 10 by using Theorem 2.7.1 we must also use the theorem about the continuity of the function f defined by $f(x) = \sqrt[n]{x}$. This theorem was stated as Theorem 2.6.5 but not proved in Section 2.6. We now give its proof, which makes use of the following formula, where n is any positive integer:

$$a^n - b^n = (a - b)(a^{n-1} + a^{n-2}b + a^{n-3}b^2 + \ldots + ab^{n-2} + b^{n-1}) \qquad (11)$$

Formula (11) follows from

$$a(a^{n-1} + a^{n-2}b + \ldots + ab^{n-2} + b^{n-1}) = a^n + a^{n-1}b + \ldots + ab^{n-1}$$

and

$$b(a^{n-1} + a^{n-2}b + \ldots + ab^{n-2} + b^{n-1}) = a^{n-1}b + \ldots + ab^{n-1} + b^n$$

by subtracting the terms of the second equation from those of the first.

2.6.5 THEOREM   If n is a positive integer and

$$f(x) = \sqrt[n]{x}$$

then

(i) if n is odd, f is continuous at every real number a;
(ii) if n is even, f is continuous at every positive number a.

Proof   We prove the theorem if a is a positive number, whether n is odd or even. The case when a is negative or zero and n is odd is left as an exercise (see Exercise 19).

We wish to prove that if $a > 0$, and $f(x) = \sqrt[n]{x}$, then f is continuous at a. Because $\sqrt[n]{a}$ exists when $a > 0$, the proof will be complete if we show that

$$\lim_{x \to a} \sqrt[n]{x} = \sqrt[n]{a}$$

Because $\sqrt[n]{x}$ is defined for every nonnegative number, the open interval required by Definition 2.1.1 can be any open interval containing a and having a nonnegative number as its left endpoint. We must show that for any $\epsilon > 0$ there exists a $\delta > 0$ such that

$$\text{if} \quad 0 < |x - a| < \delta \quad \text{then} \quad |\sqrt[n]{x} - \sqrt[n]{a}| < \epsilon \tag{12}$$

To express $|\sqrt[n]{x} - \sqrt[n]{a}|$ in terms of $|x - a|$ we use (11).

$$|\sqrt[n]{x} - \sqrt[n]{a}| = \left| \frac{(x^{1/n} - a^{1/n})[(x^{1/n})^{n-1} + (x^{1/n})^{n-2}a^{1/n} + \ldots + x^{1/n}(a^{1/n})^{n-2} + (a^{1/n})^{n-1}]}{(x^{1/n})^{n-1} + (x^{1/n})^{n-2}a^{1/n} + \ldots + x^{1/n}(a^{1/n})^{n-2} + (a^{1/n})^{n-1}} \right|$$

If (11) is applied to the numerator,

$$|\sqrt[n]{x} - \sqrt[n]{a}| = |x - a| \cdot \frac{1}{|x^{(n-1)/n} + x^{(n-2)/n}a^{1/n} + \ldots + x^{1/n}a^{(n-2)/n} + a^{(n-1)/n}|}$$

In the above equation we let

$$|\phi(x)| = |x^{(n-1)/n} + x^{(n-2)/n}a^{1/n} + \ldots + x^{1/n}a^{(n-2)/n} + a^{(n-1)/n}|$$

and obtain

$$|\sqrt[n]{x} - \sqrt[n]{a}| = |x - a| \cdot \frac{1}{|\phi(x)|} \tag{13}$$

Thus statement (12) is equivalent to

$$\text{if} \quad 0 < |x - a| < \delta \quad \text{then} \quad |x - a| \cdot \frac{1}{|\phi(x)|} < \epsilon \tag{14}$$

On the right end of (14), in addition to the factor $|x - a|$ we have the fraction $\frac{1}{|\phi(x)|}$. Hence to prove (14) we need to restrict δ so that we will have an inequality involving this fraction. If we choose the open interval stipulated in Definition 2.1.1 as the interval $(0, 2a)$, we are requiring that $\delta \leq a$. Then

$$0 < |x - a| < \delta \quad \text{and} \quad \delta \leq a$$

$$\Rightarrow \qquad |x - a| < a$$

$$\Rightarrow \quad -a < x - a < a$$

$$\Rightarrow \qquad 0 < x < 2a$$

$$\Rightarrow \qquad a^{(n-1)/n} < |\phi(x)| \qquad \text{(because } x > 0\text{)}$$

$$\Rightarrow \qquad \frac{1}{|\phi(x)|} < \frac{1}{a^{(n-1)/n}} \tag{15}$$

Now

$$0 < |x - a| < \delta \quad \text{and} \quad \frac{1}{|\phi(x)|} < \frac{1}{a^{(n-1)/n}}$$

$$\Rightarrow \quad |x - a| \cdot \frac{1}{|\phi(x)|} < \delta \cdot \frac{1}{a^{(n-1)/n}} \tag{16}$$

Our goal is to have $|x - a| \cdot \frac{1}{|\phi(x)|} < \epsilon$. Thus statement (16) tells us that we should require $\delta \cdot \dfrac{1}{a^{(n-1)/n}} \leq \epsilon$, that is, $\delta \leq a^{(n-1)/n}\epsilon$. Thus we choose $\delta = \min(a, a^{(n-1)/n}\epsilon)$. With this δ we have the following argument:

$$0 < |x - a| < \delta$$

$$\Rightarrow \quad |x - a| \cdot \frac{1}{|\phi(x)|} < \delta \cdot \frac{1}{|\phi(x)|}$$

$$\Rightarrow \quad |\sqrt[n]{x} - \sqrt[n]{a}| < \delta \cdot \frac{1}{|\phi(x)|} \qquad \text{(from (13))}$$

We showed in (15) that if $\delta \leq a$ and $0 < |x - a| < \delta$, then $\dfrac{1}{|\phi(x)|} < \dfrac{1}{a^{(n-1)/n}}$, that is, $\delta \cdot \dfrac{1}{|\phi(x)|} < \delta \cdot \dfrac{1}{a^{(n-1)/n}}$. Continuing on we have

$$|\sqrt[n]{x} - \sqrt[n]{a}| < \delta \cdot \frac{1}{|\phi(x)|} \quad \text{and} \quad \delta \cdot \frac{1}{|\phi(x)|} < \delta \cdot \frac{1}{a^{(n-1)/n}}$$

$$\Rightarrow \quad |\sqrt[n]{x} - \sqrt[n]{a}| < \delta \cdot \frac{1}{a^{(n-1)/n}}$$

$$\Rightarrow \quad |\sqrt[n]{x} - \sqrt[n]{a}| < a^{(n-1)/n}\epsilon \cdot \frac{1}{a^{(n-1)/n}} \qquad \text{(because } \delta \leq a^{(n-1)/n}\epsilon)$$

$$\Rightarrow \quad |\sqrt[n]{x} - \sqrt[n]{a}| < \epsilon$$

We have demonstrated that for any $\epsilon > 0$, if $\delta = \min(a, a^{(n-1)/n}\epsilon)$, then the following statement is true:

$$\text{if} \quad 0 < |x - a| < \delta \quad \text{then} \quad |\sqrt[n]{x} - \sqrt[n]{a}| < \epsilon$$

Hence we have proved that $\lim\limits_{x \to a} \sqrt[n]{x} = \sqrt[n]{a}$, if a is a positive number. Therefore f is continuous at a if $a > 0$. ∎

As we did with Limit Theorem 9, before restating Limit Theorem 10 and giving its proof using Theorems 2.7.1 and 2.6.5, we give an example showing how Theorems 2.7.1 and 2.6.5 can be used to find the limit of the nth root of a particular function.

EXAMPLE 2   By using Theorems 2.7.1 and 2.6.5, and not Limit Theorem 10 (the limit of the nth root), find

$$\lim_{x \to 7} \sqrt[4]{3x - 5}$$

Solution   Let the functions h and f be defined by

$$h(x) = \sqrt[4]{x} \quad \text{and} \quad f(x) = 3x - 5$$

The composite function $h \circ f$ is defined by $h(f(x)) = \sqrt[4]{f(x)}$. Now

$$\lim_{x \to 7} f(x) = \lim_{x \to 7} (3x - 5)$$

$$= 16$$

By Theorem 2.6.5, the function h is continuous at 16. Therefore we can use Theorem 2.7.1, and we have

$$\lim_{x \to 7} \sqrt[4]{3x - 5} = \lim_{x \to 7} \sqrt[4]{f(x)}$$

$$= \lim_{x \to 7} h(f(x))$$

$$= h(\lim_{x \to 7} f(x)) \qquad \text{(by Theorem 2.7.1)}$$

$$= h(16)$$

$$= 2$$

2.2.10 LIMIT THEOREM 10   If n is a positive integer and $\lim_{x \to a} f(x) = L$, then

$$\lim_{x \to a} \sqrt[n]{f(x)} = \sqrt[n]{L}$$

with the restriction that if n is even, $L > 0$.

Proof   Let h be the function defined by $h(x) = \sqrt[n]{x}$. Then the composite function $h \circ f$ is defined by $h(f(x)) = \sqrt[n]{f(x)}$. From Theorem 2.6.5 it follows that h is continuous at L if n is odd, or if n is even and $L > 0$. Therefore

$$\lim_{x \to a} \sqrt[n]{f(x)} = \lim_{x \to a} h(f(x))$$

$$= h(\lim_{x \to a} f(x)) \qquad \text{(by Theorem 2.7.1)}$$

$$= h(L)$$

$$= \sqrt[n]{L} \qquad \blacksquare$$

EXERCISES 2.9

In Exercises 1 through 6, use Theorems 2.7.1 and 2.9.1, but not Limit Theorem 9, to find the limit.

1. $\lim_{x \to 3} \dfrac{4}{x + 1}$

2. $\lim_{x \to -2} \dfrac{8}{3x + 2}$

3. $\lim_{t \to -1} \dfrac{t + 4}{t - 2}$

4. $\lim_{y \to 3} \dfrac{y - 1}{y + 5}$

5. $\lim_{x \to 2} \dfrac{3x}{2x^2 - 3x + 4}$

6. $\lim_{t \to 4} \dfrac{2t}{t^2 - 4}$

In Exercises 7 through 12, use Theorems 2.7.1 and 2.6.5, but not Limit Theorem 10, to find the limit.

7. $\lim_{x \to 3} \sqrt{x^2 - 5}$

8. $\lim_{x \to 2} \sqrt[3]{x - 1}$

9. $\lim_{y \to -4} \sqrt[3]{3y + 4}$

10. $\lim_{t \to -1/2} \sqrt{8t^3 + 6}$

11. $\lim_{t \to 1} \sqrt[4]{t^2 + 5t + 3}$

12. $\lim_{y \to 0} \sqrt[3]{y^2 - 1}$

13. If the function g is continuous at a number a and the function f is discontinuous at a, is it possible for the quotient of the two functions, f/g, to be continuous at a? Prove your answer.

14. If $f(x)$ is nonnegative for all x in its domain, and $\lim_{x \to a} [f(x)]^2$ exists and is positive, prove that $\lim_{x \to a} f(x) = \sqrt{\lim_{x \to a} [f(x)]^2}$.

15. Prove that if $\lim_{x \to a} f(x)$ exists and is L, then $\lim_{x \to a} |f(x)|$ exists and is $|L|$.

16. Prove that if $f(x) = g(x)$ for all values of x except $x = a$, then $\lim_{x \to a} f(x) = \lim_{x \to a} g(x)$ if the limits exist.

17. Prove that if $f(x) = g(x)$ for all values of x except $x = a$, then if $\lim_{x \to a} g(x)$ does not exist, $\lim_{x \to a} f(x)$ does not exist. (*Hint:* Show that the assumption that $\lim_{x \to a} f(x)$ does exist leads to a contradiction.)

18. Prove Theorem 2.9.1 if $a < 0$.

19. Prove Theorem 2.6.5 for the case when a is negative or zero and n is odd.

2.10 ADDITIONAL THEOREMS ON LIMITS OF FUNCTIONS *(Supplementary)*

We now discuss four theorems that are needed to prove some important theorems in later sections. A graphical illustration is given after the statement of each theorem.

2.10.1 THEOREM If $\lim\limits_{x \to c} f(x)$ exists and is positive, then there is an open interval containing c such that $f(x) > 0$ for every $x \neq c$ in the interval.

▶ **ILLUSTRATION 1** Consider the function f defined by

$$f(x) = \frac{5}{2x - 1}$$

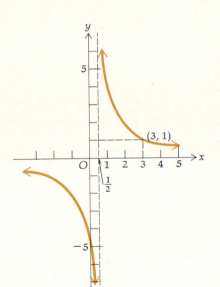

FIGURE 1

A sketch of the graph of f is in Figure 1. Because $\lim\limits_{x \to 3} f(x) = 1$, and $1 > 0$, according to Theorem 2.10.1 there is an open interval containing 3 such that $f(x) > 0$ for every $x \neq 3$ in the interval. Such an interval is $(2, 4)$. Actually, any open interval (a, b) for which $\frac{1}{2} \leq a < 3$ and $b > 3$ will do. ◀

Proof of Theorem 2.10.1 Let $L = \lim\limits_{x \to c} f(x)$. By hypothesis, $L > 0$. Applying Definition 2.1.1 and taking $\epsilon = \frac{1}{2}L$, there is a $\delta > 0$ such that

$$\text{if} \quad 0 < |x - c| < \delta \quad \text{then} \quad |f(x) - L| < \tfrac{1}{2}L \tag{1}$$

Also, $|f(x) - L| < \frac{1}{2}L$ is equivalent to $-\frac{1}{2}L < f(x) - L < \frac{1}{2}L$ (refer to Theorem 1.1.10), which in turn is equivalent to

$$\tfrac{1}{2}L < f(x) < \tfrac{3}{2}L \tag{2}$$

Also, $0 < |x - c| < \delta$ is equivalent to $-\delta < x - c < \delta$ but $x \neq c$, which in turn is equivalent to

$$c - \delta < x < c + \delta \quad \text{but} \quad x \neq c$$

$$\Leftrightarrow \quad x \text{ is in the open interval } (c - \delta, c + \delta) \quad \text{but} \quad x \neq c \tag{3}$$

From statements (2) and (3) we can replace (1) by the statement

if x is in the open interval $(c - \delta, c + \delta)$ but $x \neq c$ then $\quad \frac{1}{2}L < f(x) < \frac{2}{3}L$

Since $L > 0$, it follows that $f(x) > 0$ for every $x \neq c$ in the open interval $(c - \delta, c + \delta)$. ■

EXAMPLE 1 Given

$$f(x) = \frac{-3x}{2x - 7}$$

(a) Show that $\lim\limits_{x \to 1} f(x) > 0$. (b) Verify Theorem 2.10.1 for this function by finding an open interval containing 1 such that $f(x) > 0$ for every $x \neq 1$ in the interval.

Solution

(a) $\lim\limits_{x \to 1} \dfrac{-3x}{2x - 7} = \dfrac{3}{5}$ and $\dfrac{3}{5} > 0$

(b) The values of x in the required open interval must be such that

$$\frac{-3x}{2x - 7} > 0 \qquad (4)$$

Inequality (4) will be satisfied when both the numerator and denominator are positive or when both are negative. Two cases are considered.

Case 1: $-3x > 0$ and $2x - 7 > 0$.
That is, $x < 0$ and $x > \frac{7}{2}$. There is no value of x for which both of these inequalities hold. Therefore the solution set of Case 1 is the empty set.

Case 2: $-3x < 0$ and $2x - 7 < 0$.
That is, $x > 0$ and $x < \frac{7}{2}$. The solution set of Case 2 is the interval $(0, \frac{7}{2})$.

Thus inequality (4) is satisfied if x is in the open interval $(0, \frac{7}{2})$.
We conclude, then, that any open interval (a, b) for which $0 \le a < 1$ and $1 < b \le \frac{7}{2}$ will be such that it contains 1 and $f(x) > 0$ for every $x \ne 1$ in the interval. In particular, the interval $(\frac{1}{2}, \frac{3}{2})$ will do.

2.10.2 THEOREM   If $\lim\limits_{x \to c} f(x)$ exists and is negative, there is an open interval containing c such that $f(x) < 0$ for every $x \ne c$ in the interval.

The proof of this theorem is similar to the proof of Theorem 2.10.1 and is left as an exercise (see Exercise 17).

▶ **ILLUSTRATION 2**   Let

$$g(x) = \frac{6 - x}{3 - 2x}$$

Figure 2 shows a sketch of the graph of g. $\lim\limits_{x \to 2} g(x) = -4$ and $-4 < 0$; hence, by Theorem 2.10.2 there is an open interval containing 2 such that $g(x) < 0$ for every $x \ne 2$ in the interval. Such an interval is $(\frac{3}{2}, 3)$. Any open interval (a, b) for which $\frac{3}{2} \le a < 2$ and $2 < b \le 6$ will suffice.   ◀

The following example also illustrates Theorem 2.10.2.

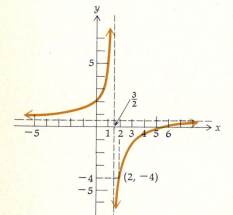

FIGURE 2

EXAMPLE 2   Given

$$g(x) = \frac{1}{1 - 2x}$$

(a) Determine the values of k such that $\lim\limits_{x \to k} g(x) < 0$. (b) Find all open intervals (a, b) containing a particular value of k satisfying part (a) such that $g(x) < 0$ for all $x \ne k$ in (a, b). (c) Draw a sketch of the graph of g, and show on the graph the geometric interpretation of the results of parts (a) and (b).

Solution
(a) If $k \ne \frac{1}{2}$,

$$\lim_{x \to k} \frac{1}{1 - 2x} = \frac{1}{1 - 2k}$$

We wish to find the values of k for which

$$\frac{1}{1 - 2k} < 0$$

This inequality will be satisfied if and only if the denominator is negative, that is, if and only if

$$1 - 2k < 0$$

$$-2k < -1$$

$$k > \tfrac{1}{2}$$

Therefore $\lim\limits_{x \to k} g(x) < 0$ whenever k is in the interval $(\tfrac{1}{2}, +\infty)$.

(b) Because $g(x) < 0$ if and only if $x > \tfrac{1}{2}$, the required open intervals (a, b) containing k such that $g(x) < 0$ are those for which $\tfrac{1}{2} \le a < k$ and $b > k$.

(c) A sketch of the graph of g is shown in Figure 3. In the figure a value of $k > \tfrac{1}{2}$ is selected. Note that $\lim\limits_{x \to k} g(x) = g(k)$ and so g is continuous at k. The point $(k, g(k))$ is shown on the graph, and $g(k) < 0$. Observe that for all values of x in any open interval (a, b) where $\tfrac{1}{2} \le a < k$ and $b > k$, the graph is in the fourth quadrant, and so $g(x) < 0$.

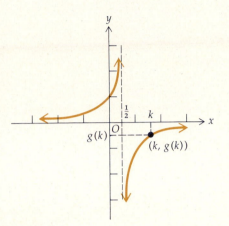

FIGURE 3

2.10.3 THEOREM

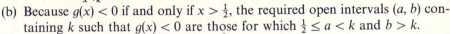
Suppose that the function f is defined on some open interval I containing c, except possibly at c. Also suppose that there is some number M for which there is a $\delta > 0$ such that if $0 < |x - c| < \delta$ then $f(x) \le M$. Then if $\lim\limits_{x \to c} f(x)$ exists and is equal to L, $L \le M$.

▶ **ILLUSTRATION 3** Figure 4 shows a sketch of the graph of a function f satisfying the hypothesis of Theorem 2.10.3. From the figure note that $f(1)$ is not defined, but f is defined on the open interval $(\tfrac{1}{2}, \tfrac{3}{2})$ except at 1. Furthermore, if $0 < |x - 1| < \tfrac{1}{2}$, then $f(x) \le \tfrac{9}{4}$. Thus it follows from Theorem 2.10.3 that if $\lim\limits_{x \to 1} f(x)$ exists and is L, then $L \le \tfrac{9}{4}$. From the figure, observe that there is an L and it is 2. ◀

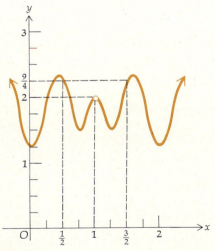

FIGURE 4

Proof of Theorem 2.10.3 We assume that $M < L$ and show that this assumption leads to a contradiction. If $M < L$, there is some $\epsilon > 0$ such that $M + \epsilon = L$. Because $\lim\limits_{x \to c} f(x) = L$, there exists a $\delta_1 > 0$ such that

$$\text{if} \quad 0 < |x - c| < \delta_1 \quad \text{then} \quad |f(x) - L| < \epsilon$$

$$\Leftrightarrow \quad \text{if} \quad 0 < |x - c| < \delta_1 \quad \text{then} \quad L - \epsilon < f(x) < L + \epsilon$$

If we replace L by $M + \epsilon$, it follows that there exists a $\delta_1 > 0$ such that

$$\text{if} \quad 0 < |x - c| < \delta_1 \quad \text{then} \quad (M + \epsilon) - \epsilon < f(x)$$

$$\Leftrightarrow \quad \text{if} \quad 0 < |x - c| < \delta_1 \quad \text{then} \quad M < f(x) \tag{5}$$

But, by hypothesis, there is a δ such that

$$\text{if} \quad 0 < |x - c| < \delta \quad \text{then} \quad f(x) \le M \tag{6}$$

Statements (5) and (6) contradict each other. Hence our assumption that $M < L$ is false. Therefore $L \le M$. ■

2.10.4 THEOREM   Suppose that the function f is defined on some open interval I containing c, except possibly at c. Also suppose that there is some number M for which there is a $\delta > 0$ such that if $0 < |x - c| < \delta$, then $f(x) \ge M$. Then if $\lim\limits_{x \to c} f(x)$ exists and is equal to L, $L \ge M$.

The proof is left as an exercise (see Exercise 18).

▶ **ILLUSTRATION 4**   Figure 4 also illustrates Theorem 2.10.4. From the figure observe that if $0 < |x - 1| < \frac{1}{2}$, then $f(x) \ge \frac{3}{2}$; and, because, as previously stated, f is defined on the open interval $(\frac{1}{2}, \frac{3}{2})$ except at 1, Theorem 2.10.4 states that if $\lim\limits_{x \to 1} f(x)$ exists and is L, then $L \ge \frac{3}{2}$. ◀

EXERCISES 2.10

In Exercises 1 through 4, a function f and a number c arre given. (a) Show that $\lim\limits_{x \to c} f(x) > 0$; *(b) verify Theorem 2.10.1 for the function f by finding an open interval containing c such that f(x) > 0 for every x ≠ c in the interval.*

1. $f(x) = \dfrac{5}{2x + 4}$; $c = 3$ **2.** $f(x) = \dfrac{7}{2x - 1}$; $c = 4$

3. $f(x) = \dfrac{\sqrt{x} - 1}{x - 1}$; $c = 1$ **4.** $f(x) = \dfrac{x^4 - 1}{x^2 - 1}$; $c = -1$

In Exercises 5 through 8, a function g and a number c are given. (a) Show that $\lim\limits_{x \to c} g(x) < 0$; *(b) verify Theorem 2.10.2 for the function g by finding an open interval containing c such that g(x) < 0 for every x ≠ c in the interval.*

5. $g(x) = \dfrac{x^2 - 9}{x + 3}$; $c = -3$ **6.** $g(x) = \dfrac{2 - 7x + 3x^2}{2 - x}$; $c = 2$

7. $g(x) = \dfrac{2 - \sqrt{4 + x}}{x}$; $c = 0$ **8.** $g(x) = \dfrac{\sqrt{x + 1} - 2}{3 - x}$; $c = 3$

In Exercises 9 and 10, a function f is given. (a) Determine the values of k such that $\lim\limits_{x \to k} f(x) > 0$. *(b) Find all open intervals (a, b) containing a particular value of k satisfying part (a) such that f(x) > 0 for all x ≠ k in (a, b). (c) Draw a sketch of the graph of f, and show on the graph the geometric interpretation of the results of parts (a) and (b).*

9. $f(x) = \dfrac{3x}{x + 1}$

10. $f(x) = \dfrac{x^2 - 25}{x^2 - 4x - 5}$

In Exercises 11 and 12, a function g is given. (a) Determine the values of k such that $\lim\limits_{x \to k} g(x) < 0$. *(b) Find all open intervals (a, b) containing a particular value of k satisfying part (a) such that g(x) < 0 for all x ≠ k in (a, b). (c) Draw a sketch of the graph of g, and show on the graph the geometric interpretation of the results of parts (a) and (b).*

11. $g(x) = \dfrac{2x^2 + x - 6}{2x^2 + 11x + 14}$ **12.** $g(x) = 2 - \sqrt{x - 1}$

13. Draw a sketch of the graph of a function f satisfying the hypothesis of Theorem 2.10.3 for which $f(2)$ is not defined but f is defined on the open interval $(1, 3)$ except at 2. Furthermore, let $f(x) \le 5$ if $0 < |x - 2| < 1$, and let $\lim\limits_{x \to 2} f(x)$ be the number L. Show on the figure that $L \le 5$.

14. Draw a sketch of the graph of a function f satisfying the hypothesis of Theorem 2.10.3 for which $f(0)$ is not defined but f is defined on the open interval $(-2, 2)$ except at 0. Furthermore, let $f(x) \le 3$ if $0 < |x| < 2$, and let $\lim\limits_{x \to 0} f(x)$ be the number L. Show on the figure that $L \le 3$.

15. Draw a sketch of the graph of a function satisfying the hypothesis of Theorem 2.10.4 for which $f(0)$ is not defined but f is defined on the open interval $(-\frac{1}{2}, \frac{1}{2})$ except at 0. Furthermore, let $f(x) \ge 2$ if $0 < |x| < \frac{1}{2}$, and let $\lim\limits_{x \to 0} f(x)$ be the number L. Show on the figure that $L \ge 2$.

16. Draw a sketch of the graph of a function f satisfying the hypothesis of Theorem 2.10.4 for which $f(1)$ is not defined but f is defined on the open interval $(\frac{3}{4}, \frac{5}{4})$ except at 1. Furthermore, let $f(x) \ge 0$ if $0 < |x - 1| < \frac{1}{4}$, and let $\lim\limits_{x \to 1} f(x)$ be the number L. Show on the figure that $L \ge 0$.

17. Prove Theorem 2.10.2.   **18.** Prove Theorem 2.10.4.

REVIEW EXERCISES FOR CHAPTER 2

In Exercises 1 through 8, find the limit and, when applicable, indicate the limit theorems used.

1. $\lim\limits_{x \to 2} (3x^2 - 4x + 5)$

2. $\lim\limits_{h \to 1} \dfrac{h^2 - 4}{3h^3 + 6}$

3. $\lim\limits_{z \to -3} \dfrac{z^2 - 9}{z + 3}$

4. $\lim\limits_{x \to -2} \dfrac{x^2 - x - 6}{x^2 - 5x - 14}$

5. $\lim\limits_{x \to 1/2} \sqrt[3]{\dfrac{4x^2 + 4x - 3}{4x^2 - 1}}$

6. $\lim\limits_{y \to -4} \sqrt{\dfrac{5y + 4}{y - 5}}$

7. $\lim\limits_{t \to 0} \dfrac{\sqrt{9 - t} - 3}{t}$

8. $\lim\limits_{t \to 0} \dfrac{1 - \sqrt{1 + t}}{t}$

In Exercises 9 through 18, prove the limit by applying Definition 2.1.1; that is, for any $\epsilon > 0$ find a $\delta > 0$ such that if $0 < |x - a| < \delta$, then $|f(x) - L| < \epsilon$.

9. $\lim\limits_{x \to 3} (2x - 5) = 1$

10. $\lim\limits_{x \to -2} (8 - 3x) = 14$

11. $\lim\limits_{x \to -1} (3x + 8) = 5$

12. $\lim\limits_{x \to 5} (4x - 11) = 9$

13. $\lim\limits_{x \to -4} x^2 = 16$

14. $\lim\limits_{x \to 1/2} x^2 = \tfrac{1}{4}$

15. $\lim\limits_{x \to 2} (x^2 - 3x) = -2$

16. $\lim\limits_{x \to 3} (2x^2 - x - 6) = 9$

17. $\lim\limits_{x \to -3/4} \dfrac{16x^2 - 9}{4x + 3} = -6$

18. $\lim\limits_{x \to 1/3} \dfrac{1 - 9x^2}{1 - 3x} = 2$

In Exercises 19 through 48, find the limit if it exists.

19. $\lim\limits_{x \to -1} \dfrac{2x^2 - x - 3}{3x^2 + 8x + 5}$

20. $\lim\limits_{y \to 3} \sqrt{\dfrac{y - 3}{y^3 - 27}}$

21. $\lim\limits_{x \to 3} f(x)$ if $f(x) = \begin{cases} x^2 - 1 & \text{if } x \le 3 \\ x + 5 & \text{if } 3 < x \end{cases}$

22. $\lim\limits_{x \to 1/3} (|3x - 1| - 5)$

23. $\lim\limits_{x \to 9} \dfrac{2\sqrt{x} - 6}{x - 9}$

24. $\lim\limits_{y \to 5^-} \dfrac{\sqrt{25 - y^2}}{y - 5}$

25. $\lim\limits_{x \to -4^+} \dfrac{2x}{16 - x^2}$

26. $\lim\limits_{s \to 7} \dfrac{5 - \sqrt{4 + 3s}}{7 - s}$

27. $\lim\limits_{t \to 5} \dfrac{\sqrt{t} - 4}{t^2 - 10t + 25}$

28. $\lim\limits_{x \to 0^-} \dfrac{x^2 - 5}{2x^3 - 3x^2}$

29. $\lim\limits_{x \to 2^+} \dfrac{[\![x]\!] - 1}{[\![x]\!] - x}$

30. $\lim\limits_{x \to 5} \dfrac{\sqrt{x - 1} - 2}{x - 5}$

31. $\lim\limits_{x \to +\infty} \dfrac{3x^2 + 2x - 5}{x^2 + 4}$

32. $\lim\limits_{x \to -\infty} \dfrac{4x - 3}{5x^2 - x + 1}$

33. $\lim\limits_{x \to -\infty} \dfrac{x^2 + 5}{2x - 4}$

34. $\lim\limits_{x \to +\infty} \left(\dfrac{8x^3 + 7x - 2}{7x^3 + 3x^2 + 5x} \right)^2$

35. $\lim\limits_{x \to +\infty} (\sqrt{x + 1} - \sqrt{x})$

36. $\lim\limits_{t \to +\infty} (\sqrt{t^2 + t} - \sqrt{t^2 + 4})$

37. $\lim\limits_{x \to 0} \dfrac{x}{\sin 3x}$

38. $\lim\limits_{x \to 0} \dfrac{x^2}{1 - \cos x}$

39. $\lim\limits_{t \to 0} \dfrac{\sin 5t}{\sin 2t}$

40. $\lim\limits_{x \to 0} \dfrac{1 - \cos 3x}{\sin 3x}$

41. $\lim\limits_{x \to 0} \dfrac{1 - \cos^2 x}{x}$

42. $\lim\limits_{t \to 0} \dfrac{4t}{\tan t}$

43. $\lim\limits_{\theta \to 0} \dfrac{\csc 3\theta}{\cot \theta}$

44. $\lim\limits_{x \to 8} \dfrac{\sqrt{7 + \sqrt[3]{x}} - 3}{x - 8}$

45. $\lim\limits_{t \to 0} \dfrac{\sqrt[3]{(t + a)^2} - \sqrt[3]{a^2}}{t}$

46. $\lim\limits_{x \to 0} \dfrac{2x^2 - 3x}{2 \sin x}$

47. $\lim\limits_{x \to 0} \dfrac{\sqrt[4]{x^4 + 1} - \sqrt{x^2 + 1}}{x^2}$ $\left(\text{Hint: Write} \right.$

$\left. \dfrac{\sqrt[4]{x^4 + 1} - \sqrt{x^2 + 1}}{x^2} = \dfrac{1 - \sqrt{x^2 + 1}}{x^2} + \dfrac{\sqrt[4]{x^4 + 1} - 1}{x^2}. \right)$

48. $\lim\limits_{x \to 1^+} \dfrac{[\![x^2]\!] - [\![x]\!]^2}{x^2 - 1}$

In Exercises 49 through 54, find the horizontal and vertical asymptotes and draw a sketch of the graph of the function.

49. $f(x) = \dfrac{x + 8}{x - 4}$

50. $f(x) = \dfrac{3x - 2}{x - 2}$

51. $g(x) = 1 - \dfrac{1}{x^2}$

52. $f(x) = \dfrac{-2}{x^2 - x - 6}$

53. $f(x) = \dfrac{5x^2}{x^2 - 4}$

54. $h(x) = \dfrac{2x^2}{x^2 - 1}$

In Exercises 55 and 56, find the horizontal and vertical asymptotes and draw a sketch of the graph of the equation.

55. $xy - 3x + 4y = 0$

56. $xy - 5x - 3y + 2 = 0$

In Exercises 57 through 62, draw a sketch of the graph of the function; then by observing where there are breaks in the graph, determine the values of the independent variable at which the function is discontinuous, and show why Definition 2.6.1 is not satisfied at each discontinuity.

57. $f(x) = \dfrac{x + 2}{x^2 + x - 2}$

58. $g(x) = \dfrac{x^4 - 1}{x^2 - 1}$

59. $g(x) = \begin{cases} 2x + 1 & \text{if } x \le -2 \\ x - 2 & \text{if } -2 < x \le 2 \\ 2 - x & \text{if } 2 < x \end{cases}$

60. $F(x) = \begin{cases} |4 - x| & \text{if } x \ne 4 \\ -2 & \text{if } x = 4 \end{cases}$

61. $h(x) = \begin{cases} \dfrac{1}{x} & \text{if } x \le 1 \\ x^2 - 1 & \text{if } 1 < x \end{cases}$

62. $f(x) = \begin{cases} x^2 - 9 & \text{if } x < 3 \\ 5 & \text{if } x = 3 \\ 9 - x^2 & \text{if } 3 \le x \end{cases}$

In Exercises 63 through 68, define $f \circ g$ and determine the numbers at which $f \circ g$ is continuous.

63. $f(x) = \sqrt{x}$; $g(x) = 25 - x^2$

64. $f(x) = \sqrt{x}$; $g(x) = x^2 - 25$

65. $f(x) = \dfrac{\sqrt{x^2 - 4}}{\sqrt{3 - x}}$; $g(x) = |x|$

66. $f(x) = \sqrt{x + 1}$; $g(x) = \dfrac{1}{x - 3}$

67. $f(x) = \operatorname{sgn} x$; $g(x) = x^2 - 1$

68. $f(x) = \operatorname{sgn} x$; $g(x) = x^2 - x$

In Exercises 69 through 72, determine the largest interval (or union of intervals) on which the function $f \circ g$ of the indicated exercise is continuous.

69. Exercise 63   **70.** Exercise 64

71. Exercise 65   **72.** Exercise 66

In Exercises 73 and 74, determine the largest interval (or union of intervals) on which the function is continuous.

73. $f(x) = \begin{cases} -x - 6 & \text{if } x < -4 \\ \sqrt{16 - x^2} & \text{if } -4 \le x < 4 \\ x - 6 & \text{if } 4 \le x \end{cases}$

74. $f(x) = \dfrac{\sqrt{9 - x^2}}{x + 2}$

In Exercises 75 through 78, prove that the function is discontinuous at the number a. Then determine if the discontinuity is removable or essential. If the discontinuity is removable, redefine $f(a)$ such that the discontinuity is removed.

75. $f(x) = \dfrac{x^2 + 2x - 8}{x^2 + 3x - 4}$; $a = -4$

76. $f(x) = \begin{cases} 4 - x^2 & \text{if } x < 1 \\ 2x + 3 & \text{if } 1 \le x \end{cases}$; $a = 1$

77. $f(x) = \dfrac{|2x - 1|}{2x - 1}$; $a = \dfrac{1}{2}$

78. $f(x) = \dfrac{2 - \sqrt{x + 4}}{x}$; $a = 0$

In Exercises 79 and 80, use the squeeze theorem to find the limit.

79. $\lim\limits_{x \to 1} \left[(x - 1)^2 \sin \dfrac{1}{\sqrt[3]{x - 1}} \right]$

80. $\lim\limits_{x \to 4} g(x)$ if $|g(x) + 5| < 3(4 - x)^2$ for all x

In Exercises 81 and 82, find the values of the constants a and b that make the function f continuous on $(-\infty, +\infty)$, and draw a sketch of the graph of f.

81. $f(x) = \begin{cases} 2x + 1 & \text{if } x \le 3 \\ ax + b & \text{if } 3 < x < 5 \\ x^2 + 2 & \text{if } 5 \le x \end{cases}$

82. $f(x) = \begin{cases} 3x + 6a & \text{if } x < -3 \\ 3ax - 7b & \text{if } -3 \le x \le 3 \\ x - 12b & \text{if } 3 < x \end{cases}$

83. Let f be the function defined by

$$f(x) = \begin{cases} 1 & \text{if } x \text{ is an integer} \\ 0 & \text{if } x \text{ is not an integer} \end{cases}$$

(a) Draw a sketch of the graph of f. (b) For what values of a does $\lim\limits_{x \to a} f(x)$ exist? (c) At what real numbers is f continuous?

84. Draw a sketch of the graph of a function f that satisfies the following conditions: f is continuous on $(-\infty, -2)$, $[-2, 1)$, $[1, 3]$, and $(3, +\infty)$; $\lim\limits_{x \to -4} f(x) = 0$; $\lim\limits_{x \to -2} f(x) = +\infty$; $\lim\limits_{x \to -2^+} f(x) = 0$; $\lim\limits_{x \to 0} f(x) = -3$; $\lim\limits_{x \to 1^-} f(x) = -\infty$; $\lim\limits_{x \to 1^+} f(x) = 2$; $\lim\limits_{x \to 3^-} f(x) = 4$; $\lim\limits_{x \to 3^+} f(x) = -1$; $\lim\limits_{x \to 5} f(x) = 0$.

85. Draw a sketch of the graph of f if $f(x) = [\![1 - x^2]\!]$ and $-2 \le x \le 2$. (a) Does $\lim\limits_{x \to 0} f(x)$ exist? (b) Is f continuous at 0?

86. Draw a sketch of the graph of g if $g(x) = (x - 1)[\![x]\!]$, and $0 \le x \le 2$. (a) Does $\lim\limits_{x \to 1} g(x)$ exist? (b) Is g continuous at 1?

87. Give an example of a function for which $\lim\limits_{x \to 0} |f(x)|$ exists but $\lim\limits_{x \to 0} f(x)$ does not exist.

88. If the function g is continuous at a and f is continuous at $g(a)$ is the composite function $f \circ g$ continuous at a? Why?

In Exercises 89 and 90, a function f and a closed interval $[a, b]$ are given. Verify that the intermediate-value theorem holds for the given value of k, and find a number c such that $f(c) = k$. Draw a sketch of the curve and the line $y = k$.

89. $f(x) = x^2 + 3$; $[a, b] = [2, 4]$; $k = 10$

90. $f(x) = -\sqrt{16 - x^2}$; $[a, b] = [0, 4]$; $k = -1$

91. (a) Prove that if $\lim\limits_{h \to 0} f(x + h) = f(x)$, then

$$\lim\limits_{h \to 0} f(x + h) = \lim\limits_{h \to 0} f(x - h)$$

(b) Show that the converse of the theorem in (a) is not true by giving an example of a function for which $\lim\limits_{h \to 0} f(x + h) = \lim\limits_{h \to 0} f(x - h)$ but $\lim\limits_{h \to 0} f(x + h) \ne f(x)$.

92. Give an example of a function f that is discontinuous at 1 for which (a) $\lim\limits_{x \to 1} f(x)$ exists but $f(1)$ does not exist; (b) $f(1)$ exists but $\lim\limits_{x \to 1} f(x)$ does not exist; (c) $\lim\limits_{x \to 1} f(x)$ and $f(1)$ both exist but are not equal.

93. If the domain of f is the set of all real numbers and f is continuous at 0, prove that if $f(a + b) = f(a) + f(b)$ for all a and b, then f is continuous at every number.

94. If the domain of f is the set of all real numbers and f is continuous at 0, prove that if $f(a + b) = f(a) \cdot f(b)$ for all a and b, then f is continuous at every number.

Exercises 95 through 98 pertain to Supplementary Section 2.9. In Exercises 95 and 96 use Theorems 2.7.1 and 2.9.1, but not Limit Theorem 9 (the limit of a quotient), to find the limit.

95. $\lim\limits_{x \to -1} \dfrac{5 - 3x}{3 - x}$

96. $\lim\limits_{x \to 2} \dfrac{5x - 3}{x^2 - x + 5}$

In Exercises 97 and 98 use Theorems 2.7.1 and 2.6.5, but not Limit Theorem 10 (the limit of the nth root), to find the limit.

97. $\lim\limits_{x \to 3} \sqrt{2x^2 - 9}$

98. $\lim\limits_{x \to -4} \sqrt[3]{7 - 5x}$

Exercises 99 and 100 pertain to Supplementary Section 2.10.

99. Given $f(x) = \dfrac{4x^2 - 9}{2x^2 + 7x + 6}$. (a) Show that $\lim\limits_{x \to -3/2} f(x) < 0$. (b) Verify Theorem 2.10.2 for the function f by finding an open interval containing $-\frac{3}{2}$ such that $f(x) < 0$ for every $x \neq -\frac{3}{2}$ in the interval.

100. Given $f(x) = \dfrac{x - 2}{x + 5}$. (a) Determine the values of k such that $\lim\limits_{x \to k} f(x) > 0$. (b) Find all open intervals (a, b) containing a particular value of k satisfying part (a) such that $f(x) > 0$ for all $x \neq k$ in (a, b). (c) Draw a sketch of the graph of f, and show on the graph the geometric interpretation of the results of parts (a) and (b).

THREE

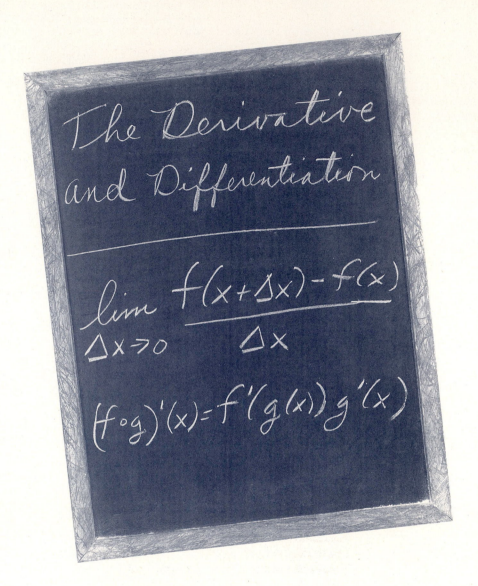

The Derivative and Differentiation

$$\lim_{\Delta x \to 0} \frac{f(x + \Delta x) - f(x)}{\Delta x}$$

$$(f \circ g)'(x) = f'(g(x))g'(x)$$

We introduce the *derivative* in Section 3.1 by first considering its geometrical interpretation as the slope of the tangent line to a curve. A function that has a derivative is said to be *differentiable*, and in Section 3.2 we discuss the relationship between differentiability and continuity. A derivative is computed by the operation of *differentiation* and theorems that help perform this computation on algebraic functions are stated and proved in Section 3.3.

In Section 3.4 we interpret the derivative as a rate of change. This interpretation leads to its importance in many fields. For instance, in physics, velocity in rectilinear motion is defined in terms of a derivative because it is the measure of the rate of change of distance with respect to time. The rate of growth of bacteria gives an application of the derivative in biology. The rate of change of a chemical reaction is of interest to a chemist. Economists are concerned with marginal concepts such as marginal revenue, marginal cost, and marginal profit, which are all rates of change.

Differentiation of trigonometric functions is discussed in Section 3.5, and in Section 3.6 we state and prove the *chain rule*, a powerful tool used to differentiate composite functions. We apply the chain rule in Section 3.7 to obtain the formula giving the derivative of the power function for rational exponents, in Section 3.8 to differentiate functions defined implicitly, and in Section 3.9 to solve problems involving related rates. Higher-order derivatives and their application in physics to rectilinear motion are treated in Section 3.10.

3.1 THE TANGENT LINE AND THE DERIVATIVE

Many important problems in calculus depend on determining the tangent line to a given curve at a specific point on the curve. For a circle, we know from plane geometry that the tangent line at a point on the circle is the line intersecting the circle at only that point. This definition does not suffice for a curve in general. For example, in Figure 1 the line that we wish to be the tangent line to the curve at the point P intersects the curve at another point Q. To arrive at a suitable definition of the tangent line to the graph of a function at a point on the graph, we proceed by considering how to define the slope of the tangent line at the point. Then the tangent line is determined by its slope and the point of tangency.

Consider the function f continuous at x_1. We wish to define the slope of the tangent line to the graph of f at $P(x_1, f(x_1))$. Let I be the open interval that contains x_1 and on which f is defined. Let $Q(x_2, f(x_2))$ be another point on the graph of f such that x_2 is also in I. Draw a line through P and Q. Any line through two points on a curve is called a **secant line**; therefore the line through P and Q is a secant line. In Figure 2 the secant line is shown for various values of x_2. Figure 3 shows one particular secant line. In this figure Q is to the right of P. However, Q may be on either the right or the left side of P, as seen in Figure 2.

Denote the difference of the abscissas of Q and P by Δx (read "delta x") so that

$$\Delta x = x_2 - x_1$$

Observe that Δx denotes a change in the value of x as x changes from x_1 to x_2 and may be either positive or negative. This change is called an *increment of x*. Be careful to note that the symbol Δx for an increment of x does not mean "delta multiplied by x."

Refer back to the secant line PQ in Figure 3; its slope is given by

$$m_{PQ} = \frac{f(x_2) - f(x_1)}{\Delta x}$$

provided that line PQ is not vertical. Because $x_2 = x_1 + \Delta x$, the above equation can be written as

$$m_{PQ} = \frac{f(x_1 + \Delta x) - f(x_1)}{\Delta x}$$

FIGURE 1

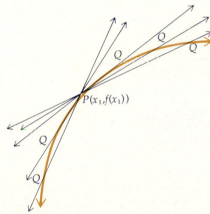

FIGURE 2

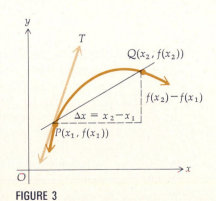

FIGURE 3

Now think of point P as being fixed, and move point Q along the curve toward P; that is, Q approaches P. This is equivalent to stating that Δx approaches zero. As this occurs, the secant line turns about the fixed point P. If this secant line has a limiting position, it is this limiting position that we wish to be the tangent line to the graph at P. So we want the slope of the tangent line to the graph at P to be the limit of m_{PQ} as Δx approaches zero, if this limit exists. If $\lim\limits_{\Delta x \to 0} m_{PQ}$ is $+\infty$ or $-\infty$, then as Δx approaches zero, the line PQ approaches the line through P that is parallel to the y axis. In this case we would want the tangent line to the graph at P to be the line $x = x_1$. The preceding discussion leads to the following definition.

3.1.1 DEFINITION   Suppose the function f is continuous at x_1. The **tangent line** to the graph of f at the point $P(x_1, f(x_1))$ is

(i) the line through P having slope $m(x_1)$, given by

$$m(x_1) = \lim_{\Delta x \to 0} \frac{f(x_1 + \Delta x) - f(x_1)}{\Delta x} \tag{1}$$

if this limit exists;

(ii) the line $x = x_1$ if

$$\lim_{\Delta x \to 0^+} \frac{f(x_1 + \Delta x) - f(x_1)}{\Delta x} \text{ is } +\infty \text{ or } -\infty$$

and

$$\lim_{\Delta x \to 0^-} \frac{f(x_1 + \Delta x) - f(x_1)}{\Delta x} \text{ is } +\infty \text{ or } -\infty$$

If neither (i) nor (ii) of Definition 3.1.1 holds, then there is no tangent line to the graph of f at the point $P(x_1, f(x_1))$.

EXAMPLE 1   Given the parabola $y = x^2$. In parts (a) through (c) find the slope of the secant line through the two points: (a) $(2, 4), (3, 9)$; (b) $(2, 4), (2.1, 4.41)$; (c) $(2, 4), (2.01, 4.0401)$. (d) Find the slope of the tangent line to the parabola at the point $(2, 4)$. (e) Draw a sketch of the graph, and show a segment of the tangent line at $(2, 4)$.

Solution   Let m_a, m_b, and m_c be the slopes of the secant lines for (a), (b), and (c), respectively.

(a) $m_a = \dfrac{9 - 4}{3 - 2}$ (b) $m_b = \dfrac{4.41 - 4}{2.1 - 2}$ (c) $m_c = \dfrac{4.0401 - 4}{2.01 - 2}$

$\qquad = 5$ $= \dfrac{0.41}{0.1}$ $= \dfrac{0.0401}{0.01}$

$\qquad\qquad\qquad\qquad = 4.1$ $= 4.01$

(d) Let $f(x) = x^2$. From (1) we have

$$m(2) = \lim_{\Delta x \to 0} \frac{f(2 + \Delta x) - f(2)}{\Delta x}$$

$$= \lim_{\Delta x \to 0} \frac{(2 + \Delta x)^2 - 4}{\Delta x}$$

$$= \lim_{\Delta x \to 0} \frac{4 + 4\,\Delta x + (\Delta x)^2 - 4}{\Delta x}$$

$$= \lim_{\Delta x \to 0} \frac{4\,\Delta x + (\Delta x)^2}{\Delta x}$$

$$= \lim_{\Delta x \to 0}\ (4 + \Delta x)$$

$$= 4$$

(e) Figure 4 shows a sketch of the graph and a segment of the tangent line at (2, 4).

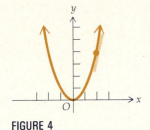

FIGURE 4

EXAMPLE 2   Find the slope of the tangent line to the graph of the function defined by $f(x) = x^3 - 3x + 4$ at the point $P(x_1, f(x_1))$.

Solution

$$f(x_1) = x_1{}^3 - 3x_1 + 4$$

$$f(x_1 + \Delta x) = (x_1 + \Delta x)^3 - 3(x_1 + \Delta x) + 4$$

From (1),

$$m(x_1) = \lim_{\Delta x \to 0} \frac{f(x_1 + \Delta x) - f(x_1)}{\Delta x}$$

$$= \lim_{\Delta x \to 0} \frac{(x_1 + \Delta x)^3 - 3(x_1 + \Delta x) + 4 - (x_1{}^3 - 3x_1 + 4)}{\Delta x}$$

$$= \lim_{\Delta x \to 0} \frac{x_1{}^3 + 3x_1{}^2\,\Delta x + 3x_1(\Delta x)^2 + (\Delta x)^3 - 3x_1 - 3\,\Delta x + 4 - x_1{}^3 + 3x_1 - 4}{\Delta x}$$

$$= \lim_{\Delta x \to 0} \frac{3x_1{}^2\,\Delta x + 3x_1(\Delta x)^2 + (\Delta x)^3 - 3\,\Delta x}{\Delta x}$$

Because $\Delta x \neq 0$, the numerator and denominator can be divided by Δx to obtain

$$m(x_1) = \lim_{\Delta x \to 0}\ [3x_1{}^2 + 3x_1\,\Delta x + (\Delta x)^2 - 3]$$

$$m(x_1) = 3x_1{}^2 - 3 \qquad (2)$$

Table 1

x	y	m
0	4	−3
1	2	0
2	6	9
−1	6	0
−2	2	9

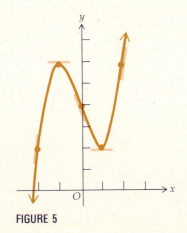

FIGURE 5

To draw a sketch of the graph of the function in Example 2 we plot some points and a segment of the tangent line at some points. Values of x are taken arbitrarily, and the corresponding function value is computed from the given equation; the value of m is found from (2). The results are given in Table 1, and a sketch of the graph appears in Figure 5. It is important to determine the points where the graph has a horizontal tangent. Because a horizontal line has a slope of zero, these points are found by setting $m(x_1) = 0$ and solving for x_1. Doing this calculation for this example we have $3x_1{}^2 - 3 = 0$, that is $x_1 = \pm 1$. Therefore at the points having abscissas of -1 and 1 the tangent line is parallel to the x axis.

EXAMPLE 3   Find an equation of the tangent line to the graph of Example 2 at the point $(2, 6)$.

Solution   Because the slope of the tangent line at any point (x_1, y_1) is

$$m(x_1) = 3x_1{}^2 - 3$$

the slope of the tangent line at the point $(2, 6)$ is $m(2) = 9$. Therefore an equation of the desired line in the point-slope form is

$$y - 6 = 9(x - 2)$$

$$9x - y - 12 = 0$$

3.1.2 DEFINITION   The **normal line** to a graph at a given point is the line perpendicular to the tangent line at that point.

▶ **ILLUSTRATION 1**   The normal line to the graph of Example 2 at the point $(2, 6)$ is perpendicular to the tangent line at that point. From Example 3, the slope of the tangent line at $(2, 6)$ is 9. Therefore the slope of the normal line at $(2, 6)$ is $-\frac{1}{9}$, and an equation of this normal line is

$$y - 6 = -\tfrac{1}{9}(x - 2)$$

$$9y - 54 = -x + 2$$

$$x + 9y - 56 = 0$$

Figure 6 shows the graph and the tangent and normal lines at $(2, 6)$.   ◀

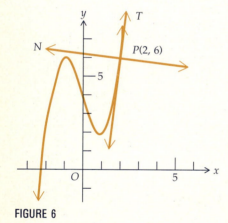

FIGURE 6

The type of limit in (1) used to define the slope of a tangent line is one of the most important in calculus. It occurs often, and it has a specific name.

3.1.3 DEFINITION   The **derivative** of the function f is that function, denoted by f', such that its value at a number x in the domain of f is given by

$$f'(x) = \lim_{\Delta x \to 0} \frac{f(x + \Delta x) - f(x)}{\Delta x} \tag{3}$$

if this limit exists.

If x_1 is a particular number in the domain of f, then

$$f'(x_1) = \lim_{\Delta x \to 0} \frac{f(x_1 + \Delta x) - f(x_1)}{\Delta x} \tag{4}$$

if this limit exists. Comparing formulas (1) and (4), note that the slope of the tangent line to the graph of $y = f(x)$ at the point $(x_1, f(x_1))$ is precisely the derivative of f evaluated at x_1.

EXAMPLE 4   Find the derivative of f if

$$f(x) = 3x^2 + 12$$

Solution   If x is a number in the domain of f, then from (3),

$$f'(x) = \lim_{\Delta x \to 0} \frac{f(x + \Delta x) - f(x)}{\Delta x}$$

$$= \lim_{\Delta x \to 0} \frac{[3(x + \Delta x)^2 + 12] - (3x^2 + 12)}{\Delta x}$$

$$= \lim_{\Delta x \to 0} \frac{3x^2 + 6x\,\Delta x + 3(\Delta x)^2 + 12 - 3x^2 - 12}{\Delta x}$$

$$= \lim_{\Delta x \to 0} \frac{6x\,\Delta x + 3(\Delta x)^2}{\Delta x}$$

$$= \lim_{\Delta x \to 0} (6x + 3\,\Delta x)$$

$$= 6x$$

Therefore the derivative of f is the function f' defined by $f'(x) = 6x$. The domain of f' is the set of all real numbers, which is the same as the domain of f.

Consider now formula (4), which is

$$f'(x_1) = \lim_{\Delta x \to 0} \frac{f(x_1 + \Delta x) - f(x_1)}{\Delta x}$$

In this formula let

$$x_1 + \Delta x = x \tag{5}$$

Then

"$\Delta x \to 0$"   is equivalent to   "$x \to x_1$"   (6)

From (4), (5), and (6) we obtain the following formula for $f'(x_1)$:

$$f'(x_1) = \lim_{x \to x_1} \frac{f(x) - f(x_1)}{x - x_1} \tag{7}$$

if this limit exists. Formula (7) is an alternative formula to (4) for computing $f'(x_1)$.

EXAMPLE 5   For the function f of Example 4 find the derivative of f at 2 in three ways: (a) Apply formula (4); (b) apply formula (7); (c) substitute 2 for x in the expression for $f'(x)$ in Example 4.

Solution

(a) $f(x) = 3x^2 + 12$. From formula (4),

$$f'(2) = \lim_{\Delta x \to 0} \frac{f(2 + \Delta x) - f(2)}{\Delta x}$$

$$= \lim_{\Delta x \to 0} \frac{[3(2 + \Delta x)^2 + 12] - [3(2)^2 + 12]}{\Delta x}$$

$$= \lim_{\Delta x \to 0} \frac{12 + 12\,\Delta x + 3(\Delta x)^2 + 12 - 12 - 12}{\Delta x}$$

$$= \lim_{\Delta x \to 0} \frac{12\,\Delta x + 3(\Delta x)^2}{\Delta x}$$

$$= \lim_{\Delta x \to 0} (12 + 3\,\Delta x)$$

$$= 12$$

(b) From formula (7),

$$f'(2) = \lim_{x \to 2} \frac{f(x) - f(2)}{x - 2}$$

$$= \lim_{x \to 2} \frac{(3x^2 + 12) - 24}{x - 2}$$

$$= \lim_{x \to 2} \frac{3x^2 - 12}{x - 2}$$

$$= 3 \lim_{x \to 2} \frac{(x - 2)(x + 2)}{x - 2}$$

$$= 3 \lim_{x \to 2} (x + 2)$$

$$= 12$$

(c) Because, from Example 4, $f'(x) = 6x$, then $f'(2) = 12$.

The use of the symbol f' for the derivative of the function f was introduced by the French mathematician Joseph Louis Lagrange (1736–1813) in the eighteenth century. This notation emphasizes that the function f' is derived from the function f and its value at x is $f'(x)$.

If (x, y) is a point on the graph of f, then $y = f(x)$, and y' is also used as a notation for the derivative of $f(x)$. With the function f defined by the equation $y = f(x)$, we can let

$$\Delta y = f(x + \Delta x) - f(x) \tag{8}$$

where Δy is called an *increment* of y and denotes a change in the function value as x changes by Δx. By using (8) and writing $\dfrac{dy}{dx}$ in place of $f'(x)$, formula (3) becomes

$$\frac{dy}{dx} = \lim_{\Delta x \to 0} \frac{\Delta y}{\Delta x}$$

The symbol $\dfrac{dy}{dx}$ as a notation for the derivative was first used by the German mathematician Gottfried Wilhelm Leibniz (1646–1716). In the seventeenth century Leibniz and Sir Isaac Newton (1642–1727), working independently, introduced almost simultaneously the derivative. Leibniz probably thought of dx and dy as small changes in the variables x and y and of the derivative of y with respect to x as the ratio of dy to dx as dy and dx become small. The concept of a limit as we know it today was not known to Leibniz.

In the Lagrange notation the value of the derivative at $x = x_1$ is indicated by $f'(x_1)$. With the Leibniz notation we would write

$$\dfrac{dy}{dx}\bigg]_{x=x_1}$$

You must remember that when $\dfrac{dy}{dx}$ is used as a notation for a derivative, dy and dx have so far in this book not been given independent meaning, although later they will be defined separately. So at this time $\dfrac{dy}{dx}$ is a symbol for a derivative and should not be thought of as a ratio. As a matter of fact, $\dfrac{d}{dx}$ can be considered as an operator (a symbol for the operation of computing the derivative), and when we write $\dfrac{dy}{dx}$, it means $\dfrac{d}{dx}(y)$, that is, the derivative of y with respect to x.

EXAMPLE 6   Find $\dfrac{dy}{dx}$ if

$$y = \sqrt{x - 3}$$

Solution   We are given $y = f(x)$, where $f(x) = \sqrt{x - 3}$.

$$\frac{dy}{dx} = \lim_{\Delta x \to 0} \frac{\Delta y}{\Delta x}$$

$$= \lim_{\Delta x \to 0} \frac{f(x + \Delta x) - f(x)}{\Delta x}$$

$$= \lim_{\Delta x \to 0} \frac{\sqrt{x + \Delta x - 3} - \sqrt{x - 3}}{\Delta x}$$

To evaluate this limit we rationalize the numerator.

$$\frac{dy}{dx} = \lim_{\Delta x \to 0} \frac{(\sqrt{x + \Delta x - 3} - \sqrt{x - 3})(\sqrt{x + \Delta x - 3} + \sqrt{x - 3})}{\Delta x(\sqrt{x + \Delta x - 3} + \sqrt{x - 3})}$$

$$= \lim_{\Delta x \to 0} \frac{(x + \Delta x - 3) - (x - 3)}{\Delta x(\sqrt{x + \Delta x - 3} + \sqrt{x - 3})}$$

$$= \lim_{\Delta x \to 0} \frac{\Delta x}{\Delta x(\sqrt{x + \Delta x - 3} + \sqrt{x - 3})}$$

Numerator and denominator are divided by Δx (since $\Delta x \neq 0$) to obtain

$$\frac{dy}{dx} = \lim_{\Delta x \to 0} \frac{1}{\sqrt{x + \Delta x - 3} + \sqrt{x - 3}}$$

$$= \frac{1}{2\sqrt{x - 3}}$$

Two other notations for the derivative of a function f are

$$\frac{d}{dx}[f(x)] \quad \text{and} \quad D_x[f(x)]$$

Each of these notations allows us to indicate the original function in the expression for the derivative. For instance, we can write the result of Example 6 as

$$\frac{d}{dx}(\sqrt{x - 3}) = \frac{1}{2\sqrt{x - 3}} \quad \text{or as} \quad D_x(\sqrt{x - 3}) = \frac{1}{2\sqrt{x - 3}}$$

EXAMPLE 7   Compute

$$\frac{d}{dx}\left(\frac{2 + x}{3 - x}\right)$$

Solution   We wish to find the derivative of $f(x)$ where $f(x) = \dfrac{2 + x}{3 - x}$. Thus

$$\frac{d}{dx}\left(\frac{2 + x}{3 - x}\right) = \lim_{\Delta x \to 0} \frac{f(x + \Delta x) - f(x)}{\Delta x}$$

$$= \lim_{\Delta x \to 0} \frac{\dfrac{2 + x + \Delta x}{3 - x - \Delta x} - \dfrac{2 + x}{3 - x}}{\Delta x}$$

$$= \lim_{\Delta x \to 0} \frac{(3 - x)(2 + x + \Delta x) - (2 + x)(3 - x - \Delta x)}{\Delta x(3 - x - \Delta x)(3 - x)}$$

$$= \lim_{\Delta x \to 0} \frac{(6 + x - x^2 + 3\,\Delta x - x\,\Delta x) - (6 + x - x^2 - 2\,\Delta x - x\,\Delta x)}{\Delta x(3 - x - \Delta x)(3 - x)}$$

$$= \lim_{\Delta x \to 0} \frac{5\,\Delta x}{\Delta x(3 - x - \Delta x)(3 - x)}$$

$$= \lim_{\Delta x \to 0} \frac{5}{(3 - x - \Delta x)(3 - x)}$$

$$= \frac{5}{(3 - x)^2}$$

Of course, if the function and the variables are denoted by letters other than f, x, and y, the notations for the derivative incorporate those letters. For instance, if the function g is defined by the equation $s = g(t)$, then the derivative of g can be indicated in each of the following ways:

$$g'(t) \qquad \frac{ds}{dt} \qquad \frac{d}{dt}[g(t)] \qquad D_t[g(t)]$$

EXERCISES 3.1

In Exercises 1 through 6, find the slope of the tangent line to the graph at the point (x_1, y_1). Make a table of values of x, y, and m over the closed interval $[a, b]$, and include in the table all points where the graph has a horizontal tangent. Draw a sketch of the graph, and show a segment of the tangent line at each of the points given by the table.

1. $y = 9 - x^2$; $[a, b] = [-3, 3]$
2. $y = x^2 + 4$; $[a, b] = [-2, 2]$
3. $y = -2x^2 + 4x$; $[a, b] = [-1, 3]$
4. $y = x^2 - 6x + 9$; $[a, b] = [1, 5]$
5. $y = x^3 + 1$; $[a, b] = [-2, 2]$
6. $y = 1 - x^3$; $[a, b] = [-2, 2]$

In Exercises 7 through 12, find the slope of the tangent line to the graph of the function f at the point $(x_1, f(x_1))$. Make a table of values of x, $f(x)$, and m at various points on the graph, and include in the table all points where the graph has a horizontal tangent. Draw a sketch of the graph.

7. $f(x) = 3x^2 - 12x + 8$
8. $f(x) = 7 - 6x - x^2$
9. $f(x) = \sqrt{4 - x}$
10. $f(x) = \sqrt{x + 1}$
11. $f(x) = x^3 - 6x^2 + 9x - 2$
12. $f(x) = x^3 - x^2 - x + 10$

In Exercises 13 through 20, find equations of the tangent line and normal line to the curve at the indicated point. Draw a sketch of the curve together with the tangent line and normal line.

13. $y = x^2 - 4x - 5$; $(-2, 7)$
14. $y = x^2 - x + 2$; $(2, 4)$
15. $y = \frac{1}{8}x^3$; $(4, 8)$
16. $y = x^2 + 2x + 1$; $(1, 4)$
17. $y = \frac{6}{x}$; $(3, 2)$
18. $y = 2x - x^3$; $(-2, 4)$
19. $y = x^4 - 4x$; $(0, 0)$
20. $y = -\frac{8}{\sqrt{x}}$; $(4, -4)$

21. Find an equation of the tangent line to the curve $y = 2x^2 + 3$ that is parallel to the line $8x - y + 3 = 0$.
22. Find an equation of the tangent line to the curve $y = 3x^2 - 4$ that is parallel to the line $3x + y = 4$.
23. Find an equation of the normal line to the curve $y = 2 - \frac{1}{3}x^2$ that is parallel to the line $x - y = 0$.
24. Find an equation of each normal line to the curve $y = x^3 - 3x$ that is parallel to the line $2x + 18y - 9 = 0$.

In Exercises 25 through 30, find $f'(x)$ by applying formula (3).

25. $f(x) = 7x + 3$
26. $f(x) = 8 - 5x$
27. $f(x) = -4$
28. $f(x) = 3x^2 + 4$
29. $f(x) = 4 - 2x^2$
30. $f(x) = 3x^2 - 2x + 1$

In Exercises 31 through 38, find the indicated derivative.

31. $\frac{d}{dx}(8 - x^3)$
32. $\frac{d}{dx}(x^3)$
33. $\frac{d}{dx}(\sqrt{x})$
34. $D_x\left(\frac{1}{x+1}\right)$
35. $\frac{d}{dx}\left(\frac{2x+3}{3x-2}\right)$
36. $D_x(\sqrt{3x+5})$
37. $D_x\left(\frac{1}{x^2} - x\right)$
38. $D_x\left(\frac{3}{1+x^2}\right)$

39. Given: $f(x) = x^2$. (a) Use a calculator to tabulate values of $\frac{f(3 + \Delta x) - f(3)}{\Delta x}$ when Δx is 1, 0.5, 0.1, 0.01, 0.001 and Δx is -1, 0.5, -0.1, -0.01, -0.001. What does $\frac{f(3 + \Delta x) - f(3)}{\Delta x}$ appear to be approaching as Δx approaches 0? (b) Find $f'(3)$ by applying formula (4). (c) Use a calculator to tabulate values of $\frac{f(x) - f(3)}{x - 3}$ when x is 4, 3.5, 3.1, 3.01, 3.001 and x is 2, 2.5, 2.9, 2.99, 2.999. What does $\frac{f(x) - f(3)}{x - 3}$ appear to be approaching as x approaches 3? (d) Find $f'(3)$ by applying formula (7).

40. Given: $f(x) = x^3$. (a) Use a calculator to tabulate values of $\frac{f(2 + \Delta x) - f(2)}{\Delta x}$ when Δx is 1, 0.5, 0.1, 0.01, 0.001 and Δx is $-1, -0.5, -0.1, -0.01, -0.001$. What does $\frac{f(2 + \Delta x) - f(2)}{\Delta x}$ appear to be approaching as Δx approaches 0? (b) Find $f'(2)$ by applying formula (4). (c) Use a calculator to tabulate values of $\frac{f(x) - f(2)}{x - 2}$ when x is 3, 2.5, 2.1, 2.01, 2.001 and x is 1, 1.5, 1.9, 1.99, 1.999. What does $\frac{f(x) - f(2)}{x - 2}$ appear to be approaching as x approaches 2? (d) Find $f'(2)$ by applying formula (7).

41. Given: $f(x) = \sqrt{x - 3}$. (a) Use a calculator to tabulate values of $\frac{f(7 + \Delta x) - f(7)}{\Delta x}$ when Δx is 1, 0.5, 0.1, 0.01, 0.001 and Δx is $-1, -0.5, -0.1, -0.01, -0.001$. What does $\frac{f(7 + \Delta x) - f(7)}{\Delta x}$ appear to be approaching as Δx approaches 0? (b) Find $f'(7)$ by applying formula (4). (c) Use a calculator to tabulate values of $\frac{f(x) - f(7)}{x - 7}$ when x is 8, 7.5, 7.1, 7.01, 7.001 and x is 6, 6.5, 6.9, 6.99, 6.999. What does $\frac{f(x) - f(7)}{x - 7}$ appear to be approaching as x approaches 7? (d) Find $f'(7)$ by applying formula (7).

42. Given: $f(x) = \frac{10}{x^2}$. (a) Use a calculator to tabulate values of $\frac{f(5 + \Delta x) - f(5)}{\Delta x}$ when Δx is 1, 0.5, 0.1, 0.01, 0.001 and Δx is $-1, -0.5, -0.1, -0.01, -0.001$. What does $\frac{f(5 + \Delta x) - f(5)}{\Delta x}$ appear to be approaching as Δx approaches 0? (b) Find $f'(5)$ by applying formula (4). (c) Use a calculator to tabulate values of $\frac{f(x) - f(5)}{x - 5}$ when x is 6, 5.5, 5.1, 5.01, 5.001 and x is 4, 4.5,

4.9, 4.99, 4.999. What does $\dfrac{f(x) - f(5)}{x - 5}$ appear to be approaching as x approaches 5? (d) Find $f'(5)$ by applying formula (7).

In Exercises 43 through 46, find $f'(a)$ by applying formula (4).

43. $f(x) = 4 - x^2$; $a = 5$ 44. $f(x) = \dfrac{4}{5x}$; $a = 2$

45. $f(x) = \dfrac{2}{x^3}$; $a = 4$ 46. $f(x) = \dfrac{2}{\sqrt{x}} - 1$; $a = 4$

In Exercises 47 through 50, find $f'(a)$ by applying formula (7).

47. $f(x) = 2 - x^3$; $a = -2$ 48. $f(x) = x^2 - x + 4$; $a = 4$

49. $f(x) = \dfrac{1}{\sqrt{2x + 3}}$; $a = 3$ 50. $f(x) = \sqrt{1 + 9x}$; $a = 7$

In Exercises 51 through 56, find $\dfrac{dy}{dx}$.

51. $y = \dfrac{4}{x^2} + 3x$ 52. $y = \dfrac{4}{2x - 5}$ 53. $y = \sqrt{2 - 7x}$

54. $y = \dfrac{1}{\sqrt{x - 1}}$ 55. $y = \sqrt[3]{x}$ 56. $y = \dfrac{1}{\sqrt[3]{x}} - x$

57. If g is continuous at a and $f(x) = (x - a)g(x)$, find $f'(a)$. (*Hint:* use formula (7).)
58. If g is continuous at a and $f(x) = (x^2 - a^2)g(x)$, find $f'(a)$. (*Hint:* use formula (7).)
59. Prove that there is no line through the point (1,5) that is tangent to the curve $y = 4x^2$.
60. Prove that there is no line through the point (1,2) that is tangent to the curve $y = 4 - x^2$.
61. If

$$f''(x) = \lim_{\Delta x \to 0} \frac{f'(x + \Delta x) - f'(x)}{\Delta x}$$

find $f''(x)$ if $f(x) = ax^2 + bx$.
62. Use the formula of Exercise 61 to find $f''(x)$ if $f(x) = a/x$.
63. If $f'(a)$ exists, prove that

$$f'(a) = \lim_{\Delta x \to 0} \frac{f(a + \Delta x) - f(a - \Delta x)}{2 \Delta x}$$

(*Hint:*
$f(a + \Delta x) - f(a - \Delta x) = f(a + \Delta x) - f(a) + f(a) - f(a - \Delta x)$.)
64. Let f be a function whose domain is the set of all real numbers and $f(a + b) = f(a) \cdot f(b)$ for all a and b. Furthermore, suppose that $f(0) = 1$ and $f'(0)$ exists. Prove that $f'(x)$ exists for all x and that $f'(x) = f'(0) \cdot f(x)$.

3.2 DIFFERENTIABILITY AND CONTINUITY

The process of computing the derivative is called **differentiation**. Thus differentiation is the operation of deriving a function f' from a function f.

If a function has a derivative at x_1, the function is said to be **differentiable** at x_1. That is, the function f is differentiable at x_1, if $f'(x_1)$ exists. A function is **differentiable on an open interval** if it is differentiable at every number in the open interval.

▶ **ILLUSTRATION 1**   In Example 4 of Section 3.1, $f(x) = 3x^2 + 12$ and $f'(x) = 6x$. Because the domain of f is the set of all real numbers, and $6x$ exists if x is any real number, f is a differentiable function.   ◀

▶ **ILLUSTRATION 2**   Let g be the function defined by $g(x) = \sqrt{x - 3}$. The domain of g is $[3, +\infty)$. From the result of Example 6 of Section 3.1,

$$g'(x) = \frac{1}{2\sqrt{x - 3}}$$

Because $g'(3)$ does not exist, g is not differentiable at 3. However, g is differentiable at every other number in its domain. Therefore, g is differentiable on the open interval $(3, +\infty)$.   ◀

We lead into a discussion of differentiability and continuity with the following example.

EXAMPLE 1 Let

$$f(x) = x^{1/3}$$

(a) Find $f'(x)$. (b) Show that f is not differentiable at 0 even though it is continuous there. (c) Draw a sketch of the graph of f.

Solution

(a) From Definition 3.1.1

$$f'(x) = \lim_{\Delta x \to 0} \frac{(x + \Delta x)^{1/3} - x^{1/3}}{\Delta x} \qquad (1)$$

We rationalize the numerator to obtain a common factor of Δx in the numerator and the denominator; this yields

$$f'(x) = \lim_{\Delta x \to 0} \frac{[(x + \Delta x)^{1/3} - x^{1/3}][(x + \Delta x)^{2/3} + (x + \Delta x)^{1/3}x^{1/3} + x^{2/3}]}{\Delta x[(x + \Delta x)^{2/3} + (x + \Delta x)^{1/3}x^{1/3} + x^{2/3}]}$$

$$= \lim_{\Delta x \to 0} \frac{(x + \Delta x) - x}{\Delta x[(x + \Delta x)^{2/3} + (x + \Delta x)^{1/3}x^{1/3} + x^{2/3}]}$$

$$= \lim_{\Delta x \to 0} \frac{1}{(x + \Delta x)^{2/3} + (x + \Delta x)^{1/3}x^{1/3} + x^{2/3}}$$

$$= \frac{1}{x^{2/3} + x^{1/3}x^{1/3} + x^{2/3}}$$

$$= \frac{1}{3x^{2/3}}$$

(b) Observe that $\dfrac{1}{3x^{2/3}}$ is not defined when $x = 0$. Also note that if (1) is used to compute $f'(0)$ we have

$$\lim_{\Delta x \to 0} \frac{(0 + \Delta x)^{1/3} - 0^{1/3}}{\Delta x} = \lim_{\Delta x \to 0} \frac{1}{(\Delta x)^{2/3}}$$

and this limit does not exist. Therefore, f is not differentiable at 0. However, f is continuous at 0 because

$$\lim_{x \to 0} f(x) = \lim_{x \to 0} x^{1/3}$$

$$= 0$$

$$= f(0)$$

(c) A sketch of the graph of f appears in Figure 1.

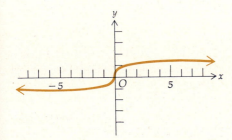

FIGURE 1

▶ **ILLUSTRATION 3** For the function f of Example 1, because

$$\lim_{\Delta x \to 0} \frac{f(0 + \Delta x) - f(0)}{\Delta x} = \lim_{\Delta x \to 0} \frac{1}{(\Delta x)^{2/3}}$$

$$= +\infty$$

it follows from Definition 3.1.1(ii) that $x = 0$ is the tangent line to the graph of f at the origin. ◀

From Example 1 and Illustration 3, the function defined by $f(x) = x^{1/3}$ has the following properties:

1. f is continuous at zero.
2. f is not differentiable at zero.
3. The graph of f has a vertical tangent line at the point where x is zero.

In the following illustration we have another function that is continuous but not differentiable at zero. The graph of this function does not have a tangent line at the point where x is zero.

▶ **ILLUSTRATION 4**   Let f be the absolute value function defined by

$$f(x) = |x|$$

A sketch of the graph of this function is shown in Figure 2. From formula (4) of Section 3.1,

$$f'(0) = \lim_{\Delta x \to 0} \frac{f(0 + \Delta x) - f(0)}{\Delta x}$$

if this limit exists. Because $f(0 + \Delta x) = |\Delta x|$ and $f(0) = 0$,

$$\lim_{\Delta x \to 0} \frac{f(0 + \Delta x) - f(0)}{\Delta x} = \lim_{\Delta x \to 0} \frac{|\Delta x|}{\Delta x}$$

Because $|\Delta x| = \Delta x$ if $\Delta x > 0$ and $|\Delta x| = -\Delta x$ if $\Delta x < 0$, we consider one-sided limits at 0:

$$\lim_{\Delta x \to 0^+} \frac{|\Delta x|}{\Delta x} = \lim_{\Delta x \to 0^+} \frac{\Delta x}{\Delta x} \qquad \lim_{\Delta x \to 0^-} \frac{|\Delta x|}{\Delta x} = \lim_{\Delta x \to 0^-} \frac{-\Delta x}{\Delta x}$$

$$= \lim_{\Delta x \to 0^+} 1 \qquad\qquad\qquad = \lim_{\Delta x \to 0^-} (-1)$$

$$= 1 \qquad\qquad\qquad\qquad = -1$$

Because $\displaystyle\lim_{\Delta x \to 0^+} \frac{|\Delta x|}{\Delta x} \neq \lim_{\Delta x \to 0^-} \frac{|\Delta x|}{\Delta x}$, it follows that the two-sided limit $\displaystyle\lim_{\Delta x \to 0} \frac{|\Delta x|}{\Delta x}$ does not exist. Therefore $f'(0)$ does not exist, and so f is not differentiable at 0.

Because Definition 3.1.1 is not satisfied when $x = 0$, there is no tangent line at the origin for the graph of the absolute-value function. ◀

Because the functions of Illustration 4 and Example 1 are continuous at a number but not differentiable there, we may conclude that continuity of a function at a number does not imply differentiability of the function at that number. However, differentiability *does* imply continuity, which is given by the next theorem.

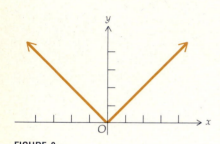

FIGURE 2

3.2.1 THEOREM   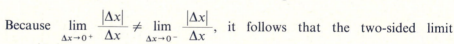 If a function f is differentiable at x_1, then f is continuous at x_1.

Proof   To prove that f is continuous at x_1 we must show that the three conditions of Definition 2.6.1 hold there. That is, we must show that (i) $f(x_1)$ exists; (ii) $\displaystyle\lim_{x \to x_1} f(x)$ exists; and (iii) $\displaystyle\lim_{x \to x_1} f(x) = f(x_1)$.

By hypothesis, f is differentiable at x_1. Therefore $f'(x_1)$ exists. Because by formula (7) of Section 3.1

$$f'(x_1) = \lim_{x \to x_1} \frac{f(x) - f(x_1)}{x - x_1}$$

$f(x_1)$ must exist; otherwise the above limit has no meaning. Therefore condition (i) holds at x_1. Now consider

$$\lim_{x \to x_1} [f(x) - f(x_1)]$$

We can write

$$\lim_{x \to x_1} [f(x) - f(x_1)] = \lim_{x \to x_1} \left[(x - x_1) \cdot \frac{f(x) - f(x_1)}{x - x_1} \right] \tag{2}$$

Because

$$\lim_{x \to x_1} (x - x_1) = 0 \quad \text{and} \quad \lim_{x \to x_1} \frac{f(x) - f(x_1)}{x - x_1} = f'(x_1)$$

we apply the theorem on the limit of a product (Theorem 2.2.6) to the right side of (2) and obtain

$$\lim_{x \to x_1} [f(x) - f(x_1)] = \lim_{x \to x_1} (x - x_1) \cdot \lim_{x \to x_1} \frac{f(x) - f(x_1)}{x - x_1}$$
$$= 0 \cdot f'(x_1)$$
$$= 0$$

Then

$$\lim_{x \to x_1} f(x) = \lim_{x \to x_1} [f(x) - f(x_1) + f(x_1)]$$
$$= \lim_{x \to x_1} [f(x) - f(x_1)] + \lim_{x \to x_1} f(x_1)$$
$$= 0 + f(x_1)$$

which gives

$$\lim_{x \to x_1} f(x) = f(x_1)$$

From this equation it follows that conditions (ii) and (iii) for continuity of f at x_1 hold. Therefore the theorem is proved. ■

A function f can fail to be differentiable at a number c for one of the following reasons:

1. The function f is discontinuous at c. This follows from Theorem 3.2.1. Refer to Figure 3 for a sketch of the graph of such a function.
2. The function f is continuous at c, and the graph of f has a vertical tangent line at the point where $x = c$. See Figure 4 for a sketch of the graph of a function having this property. This situation also occurs in Example 1.
3. The function f is continuous at c, and the graph of f does not have a tangent line at the point where $x = c$. In Figure 5 there is a sketch of the

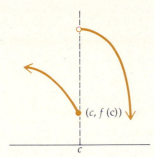

FIGURE 3

$(c, f(c))$

c

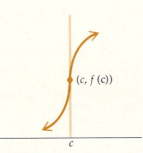

FIGURE 4

$(c, f(c))$

c

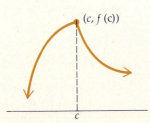

FIGURE 5

$(c, f(c))$

c

graph of a function satisfying this condition. Observe that there is a "sharp turn" in the graph at $x = c$. In Illustration 4 there is another such function.

Before giving an additional example of a function that is continuous at a number but not differentiable there, we introduce the concept of a **one-sided derivative**.

3.2.2 DEFINITION   If the function f is defined at x_1, then the **derivative from the right** of f at x_1, denoted by $f'_+(x_1)$, is defined by

$$f'_+(x_1) = \lim_{\Delta x \to 0^+} \frac{f(x_1 + \Delta x) - f(x_1)}{\Delta x}$$

$$\Leftrightarrow \quad f'_+(x_1) = \lim_{x \to x_1^+} \frac{f(x) - f(x_1)}{x - x_1}$$

if the limit exists.

3.2.3 DEFINITION   If the function f is defined at x_1, then the **derivative from the left** of f at x_1, denoted by $f'_-(x_1)$, is defined by

$$f'_-(x_1) = \lim_{\Delta x \to 0^-} \frac{f(x_1 + \Delta x) - f(x_1)}{\Delta x}$$

$$\Leftrightarrow \quad f'_-(x_1) = \lim_{x \to x_1^-} \frac{f(x) - f(x_1)}{x - x_1}$$

if the limit exists.

From Definitions 3.2.2 and 3.2.3 and Theorem 2.3.3 it follows that a function f defined on an open interval containing x_1 is differentiable at x_1 if and only if $f'_+(x_1)$ and $f'_-(x_1)$ both exist and are equal. Of course, then $f'(x_1)$, $f'_+(x_1)$, and $f'_-(x_1)$ are all equal.

EXAMPLE 2   Let f be defined by

$$f(x) = |1 - x^2|$$

(a) Draw a sketch of the graph of f. (b) Prove that f is continuous at 1.
(c) Determine if f is differentiable at 1.

Solution   By the definition of absolute value, if $x < -1$ or $x > 1$, then $f(x) = -(1 - x^2)$, and if $-1 \le x \le 1$, $f(x) = 1 - x^2$. Therefore f can be defined as follows:

$$f(x) = \begin{cases} x^2 - 1 & \text{if } x < -1 \\ 1 - x^2 & \text{if } -1 \le x \le 1 \\ x^2 - 1 & \text{if } 1 < x \end{cases}$$

(a) A sketch of the graph of f appears in Figure 6.
(b) To prove that f is continuous at 1 we verify the three conditions for continuity.

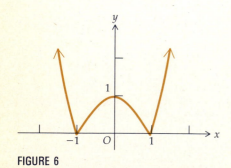

FIGURE 6

(i) $f(1) = 0$

(ii) $\lim\limits_{x\to 1^-} f(x) = \lim\limits_{x\to 1^-} (1 - x^2)$ $\lim\limits_{x\to 1^+} f(x) = \lim\limits_{x\to 1^+} (x^2 - 1)$

$\qquad\qquad\qquad = 0 \qquad\qquad\qquad\qquad\qquad = 0$

Thus $\lim\limits_{x\to 1} f(x) = 0$.

(iii) $\lim\limits_{x\to 1} f(x) = f(1)$

Because conditions (i)–(iii) all hold at 1, f is continuous at 1.

(c) $f'_-(1) = \lim\limits_{x\to 1^-} \dfrac{f(x) - f(1)}{x - 1}$ $f'_+(1) = \lim\limits_{x\to 1^+} \dfrac{f(x) - f(1)}{x - 1}$

$\qquad = \lim\limits_{x\to 1^-} \dfrac{(1 - x^2) - 0}{x - 1}$ $\qquad = \lim\limits_{x\to 1^+} \dfrac{(x^2 - 1) - 0}{x - 1}$

$\qquad = \lim\limits_{x\to 1^-} \dfrac{(1 - x)(1 + x)}{x - 1}$ $\qquad = \lim\limits_{x\to 1^+} \dfrac{(x - 1)(x + 1)}{x - 1}$

$\qquad = \lim\limits_{x\to 1^-} [-(1 + x)]$ $\qquad = \lim\limits_{x\to 1^+} (x + 1)$

$\qquad = -2$ $\qquad\qquad = 2$

Because $f'_-(1) \neq f'_+(1)$, it follows that $f'(1)$ does not exist, and so f is not differentiable at 1.

The function of Example 2 is also not differentiable at $x = -1$, as can be shown by a method similar to that used for $x = 1$ (see Exercise 26).

▶ **ILLUSTRATION 5**   In Illustration 2 of Section 2.6 we had the function C defined by

$$C(x) = \begin{cases} x & \text{if } 0 \le x \le 10 \\ 0.7x + 3 & \text{if } 10 < x \end{cases}$$

where $C(x)$ dollars is the total cost of x pounds of a product. Figure 7 shows a sketch of the graph of C. In Section 2.6 we showed that C is continuous at 10.

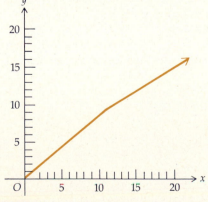

FIGURE 7

$C'_-(10) = \lim\limits_{x\to 10^-} \dfrac{C(x) - C(10)}{x - 10}$ $C'_+(10) = \lim\limits_{x\to 10^+} \dfrac{C(x) - C(10)}{x - 10}$

$\qquad = \lim\limits_{x\to 10^-} \dfrac{x - 10}{x - 10}$ $\qquad = \lim\limits_{x\to 10^+} \dfrac{(0.7x + 3) - 10}{x - 10}$

$\qquad = \lim\limits_{x\to 10^-} 1$ $\qquad = \lim\limits_{x\to 10^+} \dfrac{0.7(x - 10)}{x - 10}$

$\qquad = 1$ $\qquad = \lim\limits_{x\to 10^+} 0.7$

$\qquad\qquad = 0.7$

Because $C'_-(10) \neq C'_+(10)$, C is not differentiable at 10.   ◀

EXAMPLE 3   Given

$$f(x) = \begin{cases} \dfrac{1}{x} & \text{if } 0 < x < b \\[2mm] 1 - \frac{1}{4}x & \text{if } b \leq x \end{cases}$$

(a) Determine a value of b so that f is continuous at b. (b) Is f differentiable at the value of b found in part (a)?

Solution

(a) The function f will be continuous at b if $\lim\limits_{x \to b^-} f(x) = f(b)$ and $\lim\limits_{x \to b^+} f(x) = f(b)$.

$$\lim_{x \to b^-} f(x) = \lim_{x \to b^-} \frac{1}{x} \qquad \lim_{x \to b^+} f(x) = \lim_{x \to b^+} (1 - \tfrac{1}{4}x)$$

$$= \frac{1}{b} \qquad\qquad\qquad = 1 - \tfrac{1}{4}b$$

$f(b) = 1 - \frac{1}{4}b$; therefore f will be continuous at b if

$$\frac{1}{b} = 1 - \frac{1}{4}b$$

$$4 = 4b - b^2$$

$$b^2 - 4b + 4 = 0$$

$$(b - 2)^2 = 0$$

$$b = 2$$

Thus

$$f(x) = \begin{cases} \dfrac{1}{x} & \text{if } 0 < x < 2 \\[2mm] 1 - \frac{1}{4}x & \text{if } 2 \leq x \end{cases}$$

and f is continuous at 2.

(b) To determine if f is differentiable at 2 we compute $f'_-(2)$ and $f'_+(2)$.

$$f'_-(2) = \lim_{x \to 2^-} \frac{f(x) - f(2)}{x - 2} \qquad f'_+(2) = \lim_{x \to 2^+} \frac{f(x) - f(2)}{x - 2}$$

$$= \lim_{x \to 2^-} \frac{\dfrac{1}{x} - \dfrac{1}{2}}{x - 2} \qquad\qquad = \lim_{x \to 2^+} \frac{(1 - \frac{1}{4}x) - \frac{1}{2}}{x - 2}$$

$$\qquad\qquad\qquad\qquad = \lim_{x \to 2^+} \frac{\frac{1}{2} - \frac{1}{4}x}{x - 2}$$

$$= \lim_{x \to 2^-} \frac{2 - x}{2x(x - 2)} \qquad\qquad = \lim_{x \to 2^+} \frac{2 - x}{4(x - 2)}$$

$$= \lim_{x \to 2^-} \frac{-1}{2x} \qquad\qquad\qquad = \lim_{x \to 2^+} \frac{-1}{4}$$

$$= -\tfrac{1}{4} \qquad\qquad\qquad\qquad = -\tfrac{1}{4}$$

Because $f'_-(2) = f'_+(2)$, it follows that $f'(2)$ exists, and hence f is differentiable at 2.

EXERCISES 3.2

In Exercises 1 through 20, do the following: (a) Draw a sketch of the graph of the function; (b) determine if f is continuous at x_1; (c) find $f'_-(x_1)$ and $f'_+(x_1)$ if they exist; (d) determine if f is differentiable at x_1.

1. $f(x) = \begin{cases} x + 2 & \text{if } x \le -4 \\ -x - 6 & \text{if } -4 < x \end{cases}$ $x_1 = -4$

2. $f(x) = \begin{cases} 3 - 2x & \text{if } x < 2 \\ 3x - 7 & \text{if } 2 \le x \end{cases}$ $x_1 = 2$

3. $f(x) = |x - 3|; x_1 = 3$

4. $f(x) = 1 + |x + 2|; x_1 = -2$

5. $f(x) = \begin{cases} -1 & \text{if } x < 0 \\ x - 1 & \text{if } 0 \le x \end{cases}$ $x_1 = 0$

6. $f(x) = \begin{cases} x & \text{if } x \le 0 \\ x^2 & \text{if } 0 < x \end{cases}$ $x_1 = 0$

7. $f(x) = \begin{cases} x^2 & \text{if } x \le 0 \\ -x^2 & \text{if } 0 < x \end{cases}$ $x_1 = 0$

8. $f(x) = \begin{cases} x^2 - 4 & \text{if } x < 2 \\ \sqrt{x - 2} & \text{if } 2 \le x \end{cases}$ $x_1 = 2$

9. $f(x) = \begin{cases} \sqrt{1 - x} & \text{if } x < 1 \\ (1 - x)^2 & \text{if } 1 \le x \end{cases}$ $x_1 = 1$

10. $f(x) = \begin{cases} x^2 & \text{if } x < -1 \\ -1 - 2x & \text{if } -1 \le x \end{cases}$ $x_1 = -1$

11. $f(x) = \begin{cases} 2x^2 - 3 & \text{if } x \le 2 \\ 8x - 11 & \text{if } 2 < x \end{cases}$ $x_1 = 2$

12. $f(x) = \begin{cases} x^2 - 9 & \text{if } x < 3 \\ 6x - 18 & \text{if } 3 \le x \end{cases}$ $x_1 = 3$

13. $f(x) = \sqrt[3]{x + 1}; x_1 = -1$

14. $f(x) = (x - 2)^{-2}; x_1 = 2$

15. $f(x) = \begin{cases} 5 - 6x & \text{if } x \le 3 \\ -4 - x^2 & \text{if } 3 < x \end{cases}$ $x_1 = 3$

16. $f(x) = \begin{cases} -x^{2/3} & \text{if } x \le 0 \\ x^{2/3} & \text{if } 0 < x \end{cases}$ $x_1 = 0$

17. $f(x) = \begin{cases} x - 2 & \text{if } x < 0 \\ x^2 & \text{if } 0 \le x \end{cases}$ $x_1 = 0$

18. $f(x) = \begin{cases} x^3 & \text{if } x \le 1 \\ x + 1 & \text{if } 1 < x \end{cases}$ $x_1 = 1$

19. $f(x) = \begin{cases} 3x^2 & \text{if } x \le 2 \\ x^3 & \text{if } 2 < x \end{cases}$ $x_1 = 2$

20. $f(x) = \begin{cases} x^2 + 1 & \text{if } x < -1 \\ 1 - x^2 & \text{if } -1 \le x \end{cases}$ $x_1 = -1$

21. Given $f(x) = \sqrt{x - 4}$. (a) Prove that f is continuous from the right at 4. (b) Prove that $f'_+(4)$ does not exist. (c) Draw a sketch of the graph of f.

22. Given $f(x) = \sqrt{4 - x^2}$. (a) Prove that f is continuous on the closed interval $[-2, 2]$. (b) Prove that neither $f'_+(-2)$ nor $f'_-(2)$ exist. (c) Draw a sketch of the graph of f.

23. Given $f(x) = \sqrt{x^2 - 9}$. (a) Prove that f is continuous on $(-\infty, -3]$ and $[3, +\infty)$. (b) Prove that neither $f'_-(-3)$ nor $f'_+(3)$ exists. (c) Draw a sketch of the graph of f.

24. Given $f(x) = \sqrt{8 - x}$. (a) Prove that f is continuous from the left at 8. (b) Prove that $f'_-(8)$ does not exist. (c) Draw a sketch of the graph of f.

25. Given $f(x) = \text{sgn } x$. (a) Prove that $f'_+(0)$ and $f'_-(0)$ do not exist. (b) Prove that $\lim_{x \to 0^+} f'(x) = 0$ and $\lim_{x \to 0^-} f'(x) = 0$. (c) Draw a sketch of the graph of f.

26. Prove that the function of Example 2 is continuous at -1 but not differentiable there.

27. Given $f(x) = x^{3/2}$. (a) Prove that f is continuous from the right at 0. (b) Prove that $f'_+(0)$ exists, and find its value. (c) Draw a sketch of the graph of f.

28. Given $f(x) = (1 - x^2)^{3/2}$. (a) Prove that f is continuous on the closed interval $[-1, 1]$. (b) Prove that f is differentiable on the open interval $(-1, 1)$ and that both $f'_+(-1)$ and $f'_-(1)$ exist. (c) Draw a sketch of the graph of f.

29. Given

$$f(x) = \begin{cases} x^2 - 7 & \text{if } 0 < x \le b \\ \dfrac{6}{x} & \text{if } b < x \end{cases}$$

(a) Determine a value of b for which f is continuous at b. (b) Is f differentiable at the value of b found in part (a)?

30. Find the values of a and b such that f is differentiable at 1 if

$$f(x) = \begin{cases} x^2 & \text{if } x < 1 \\ ax + b & \text{if } 1 \le x \end{cases}$$

31. Find the values of a and b such that f is differentiable at 2 if

$$f(x) = \begin{cases} ax + b & \text{if } x < 2 \\ 2x^2 - 1 & \text{if } 2 \le x \end{cases}$$

32. Given $f(x) = [\![x]\!]$, find $f'(x_1)$ if x_1 is not an integer. Prove by applying Theorem 3.2.1 that $f'(x_1)$ does not exist if x_1 is an integer. If x_1 is an integer, what can you say about $f'_-(x_1)$ and $f'_+(x_1)$?

33. Given $f(x) = (x - 1)[\![x]\!]$. Draw a sketch of the graph of f for x in $[0, 2]$. Find, if they exist: (a) $f'_-(1)$; (b) $f'_+(1)$; (c) $f'(1)$.

34. Given $f(x) = (5 - x)[\![x]\!]$. Draw a sketch of the graph of f for x in $[4, 6]$. Find, if they exist: (a) $f'_-(5)$; (b) $f'_+(5)$; (c) $f'(5)$.

35. Given $f(x) = (x - a)[\![x]\!]$ where a is an integer, show that $f'_-(a) + 1 = f'_+(a)$.

36. Let the function f be defined by

$$f(x) = \begin{cases} \dfrac{g(x) - g(a)}{x - a} & \text{if } x \ne a \\ g'(a) & \text{if } x = a \end{cases}$$

Prove that if $g'(a)$ exists, f is continuous at a.

37. In Illustration 4 we have the absolute-value function, defined by $f(x) = |x|$, and this function is not differentiable at 0. Prove that $f'(x) = |x|/x$ for all $x \neq 0$. (*Hint:* Let $|x| = \sqrt{x^2}$.)

38. If $f(x) = |x|$ and $g(x) = -|x|$, (a) find a formula for $(f + g)(x)$. (b) Prove that $f + g$ is differentiable at 0 while neither f nor g is differentiable there.

39. A school-sponsored trip that can accommodate up to 250 students will cost each student \$15 if not more than 150 students make the trip; however, the cost per student will be reduced \$0.05 for each student in excess of 150 until the cost reaches \$10 per student. (a) If x students make the trip, express the number of dollars in the gross income as a function of x. (b) Because x is the number of students, x is a nonnegative integer. To have the continuity requirements to apply the calculus, x must be a nonnegative real number. With this assumption show that the function in part (a) is continuous at 150 but not differentiable there.

40. A manufacturer can make a profit of \$20 on each item if not more than 800 items are produced each week. The profit decreases \$0.02 per item over 800. (a) If x items are produced each week, express the number of dollars in the manufacturer's weekly profit as a function of x. (b) Because x is the number of items, x is a nonnegative integer. To have the continuity requirements to apply the calculus, x must be a nonnegative real number. With this assumption show that the function in part (a) is continuous at 800 but not differentiable there.

41. Given

$$f(x) = \begin{cases} 0 & \text{if } x \leq 0 \\ x^n & \text{if } 0 < x \end{cases}$$

where n is a positive integer. (a) For what values of n is f differentiable for all values of x? (b) For what values of n is f' continuous for all values of x?

3.3 THEOREMS ON DIFFERENTIATION OF ALGEBRAIC FUNCTIONS

Because the process of computing the derivative of a function from the definition (3.1.3) is usually rather lengthy, we now state and prove some theorems that enable us to find derivatives more easily. These theorems are proved by applying Definition 3.1.3. Following the proof of each theorem we state the corresponding formula of differentiation.

3.3.1 THEOREM   If c is a constant and if $f(x) = c$ for all x, then

$$f'(x) = 0$$

Proof

$$f'(x) = \lim_{\Delta x \to 0} \frac{f(x + \Delta x) - f(x)}{\Delta x}$$

$$= \lim_{\Delta x \to 0} \frac{c - c}{\Delta x}$$

$$= \lim_{\Delta x \to 0} 0$$

$$= 0 \qquad \blacksquare$$

$$D_x(c) = 0$$

The derivative of a constant is zero.

▶ **ILLUSTRATION 1**   If $f(x) = 5$, then

$$f'(x) = 0 \qquad ◀$$

3.3.2 THEOREM   If n is a positive integer and if $f(x) = x^n$, then

$$f'(x) = nx^{n-1}$$

Proof

$$f'(x) = \lim_{\Delta x \to 0} \frac{f(x + \Delta x) - f(x)}{\Delta x}$$

$$= \lim_{\Delta x \to 0} \frac{(x + \Delta x)^n - x^n}{\Delta x}$$

Applying the binomial theorem to $(x + \Delta x)^n$ we have

$$f'(x) = \lim_{\Delta x \to 0} \frac{\left[x^n + nx^{n-1}\Delta x + \frac{n(n-1)}{2!} x^{n-2}(\Delta x)^2 + \ldots + nx(\Delta x)^{n-1} + (\Delta x)^n \right] - x^n}{\Delta x}$$

$$= \lim_{\Delta x \to 0} \frac{nx^{n-1}\Delta x + \frac{n(n-1)}{2!} x^{n-2}(\Delta x)^2 + \ldots + nx(\Delta x)^{n-1} + (\Delta x)^n}{\Delta x}$$

We divide the numerator and denominator by Δx to obtain

$$f'(x) = \lim_{\Delta x \to 0} \left[nx^{n-1} + \frac{n(n-1)}{2!} x^{n-2} \Delta x + \ldots + nx(\Delta x)^{n-2} + (\Delta x)^{n-1} \right]$$

Every term except the first has a factor of Δx; therefore every term except the first approaches zero as Δx approaches zero. Thus

$$f'(x) = nx^{n-1}$$ ■

$$D_x(x^n) = nx^{n-1}$$

▶ **ILLUSTRATION 2**   If $f(x) = x^8$, then $f'(x) = 8x^7$.   ◀

▶ **ILLUSTRATION 3**   Let $f(x) = x$.   (a) If $x \neq 0$, by Theorem 3.3.2

$$f'(x) = 1 \cdot x^0$$

$$= 1 \cdot 1$$

$$= 1$$

(b) By formula (7) of Section 3.1

$$f'(0) = \lim_{x \to 0} \frac{f(x) - f(0)}{x - 0}$$

$$= \lim_{x \to 0} \frac{x - 0}{x}$$

$$= \lim_{x \to 0} 1$$

$$= 1$$

Therefore, for all x, $D_x(x) = 1$.   ◀

3.3.3 THEOREM   If f is a function, c is a constant, and g is the function defined by

$$g(x) = c \cdot f(x)$$

then if $f'(x)$ exists,

$$g'(x) = c \cdot f'(x)$$

Proof

$$g'(x) = \lim_{\Delta x \to 0} \frac{g(x + \Delta x) - g(x)}{\Delta x}$$

$$= \lim_{\Delta x \to 0} \frac{cf(x + \Delta x) - cf(x)}{\Delta x}$$

$$= \lim_{\Delta x \to 0} c \cdot \left[\frac{f(x + \Delta x) - f(x)}{\Delta x}\right]$$

$$= c \cdot \lim_{\Delta x \to 0} \frac{f(x + \Delta x) - f(x)}{\Delta x}$$

$$= cf'(x) \qquad\blacksquare$$

$$D_x[c \cdot f(x)] = c \cdot D_x f(x)$$

The derivative of a constant times a function is the constant times the derivative of the function if this derivative exists.

By combining Theorems 3.3.2 and 3.3.3 we obtain the following result: If $f(x) = cx^n$, where n is a positive integer and c is a constant, then

$$f'(x) = cnx^{n-1}$$

$$D_x(cx^n) = cnx^{n-1}$$

▶ **ILLUSTRATION 4**   If $f(x) = 5x^7$, then

$$f'(x) = 5 \cdot 7x^6$$
$$= 35x^6 \qquad\qquad\qquad\qquad\qquad\quad ◀$$

3.3.4 THEOREM   If f and g are functions and if h is the function defined by

$$h(x) = f(x) + g(x)$$

then if $f'(x)$ and $g'(x)$ exist,

$$h'(x) = f'(x) + g'(x)$$

Proof

$$h'(x) = \lim_{\Delta x \to 0} \frac{h(x + \Delta x) - h(x)}{\Delta x}$$

$$= \lim_{\Delta x \to 0} \frac{[f(x + \Delta x) + g(x + \Delta x)] - [f(x) + g(x)]}{\Delta x}$$

$$= \lim_{\Delta x \to 0} \left[\frac{f(x + \Delta x) - f(x)}{\Delta x} + \frac{g(x + \Delta x) - g(x)}{\Delta x}\right]$$

$$= \lim_{\Delta x \to 0} \frac{f(x + \Delta x) - f(x)}{\Delta x} + \lim_{\Delta x \to 0} \frac{g(x + \Delta x) - g(x)}{\Delta x}$$

$$= f'(x) + g'(x) \qquad\qquad\qquad\qquad\qquad\qquad\blacksquare$$

$$D_x[f(x) + g(x)] = D_x f(x) + D_x g(x)$$

The derivative of the sum of two functions is the sum of their derivatives if these derivatives exist.

The result of the preceding theorem can be extended to any finite number of functions by mathematical induction, and this is stated as another theorem.

3.3.5 THEOREM   The derivative of the sum of a finite number of functions is equal to the sum of their derivatives if these derivatives exist.

From the preceding theorems the derivative of any polynomial function can be found easily.

EXAMPLE 1   Find $f'(x)$ if

$$f(x) = 7x^4 - 2x^3 + 8x + 5$$

Solution

$$\begin{aligned}
f'(x) &= D_x(7x^4 - 2x^3 + 8x + 5) \\
&= D_x(7x^4) + D_x(-2x^3) + D_x(8x) + D_x(5) \\
&= 28x^3 - 6x^2 + 8
\end{aligned}$$

3.3.6 THEOREM   If f and g are functions and if h is the function defined by

$$h(x) = f(x)g(x)$$

then if $f'(x)$ and $g'(x)$ exist,

$$h'(x) = f(x)g'(x) + g(x)f'(x)$$

Proof

$$h'(x) = \lim_{\Delta x \to 0} \frac{h(x + \Delta x) - h(x)}{\Delta x}$$

$$= \lim_{\Delta x \to 0} \frac{f(x + \Delta x) \cdot g(x + \Delta x) - f(x) \cdot g(x)}{\Delta x}$$

If $f(x + \Delta x) \cdot g(x)$ is subtracted and added in the numerator, then

$$h'(x) = \lim_{\Delta x \to 0} \frac{f(x + \Delta x) \cdot g(x + \Delta x) - f(x + \Delta x) \cdot g(x) + f(x + \Delta x) \cdot g(x) - f(x) \cdot g(x)}{\Delta x}$$

$$= \lim_{\Delta x \to 0} \left[f(x + \Delta x) \cdot \frac{g(x + \Delta x) - g(x)}{\Delta x} + g(x) \cdot \frac{f(x + \Delta x) - f(x)}{\Delta x} \right]$$

$$= \lim_{\Delta x \to 0} \left[f(x + \Delta x) \cdot \frac{g(x + \Delta x) - g(x)}{\Delta x} \right] + \lim_{\Delta x \to 0} \left[g(x) \cdot \frac{f(x + \Delta x) - f(x)}{\Delta x} \right]$$

$$= \lim_{\Delta x \to 0} f(x + \Delta x) \cdot \lim_{\Delta x \to 0} \frac{g(x + \Delta x) - g(x)}{\Delta x} + \lim_{\Delta x \to 0} g(x) \cdot \lim_{\Delta x \to 0} \frac{f(x + \Delta x) - f(x)}{\Delta x}$$

Because f is differentiable at x, by Theorem 3.2.1 f is continuous at x; therefore $\lim_{\Delta x \to 0} f(x + \Delta x) = f(x)$. Also, $\lim_{\Delta x \to 0} g(x) = g(x)$ and

$$\lim_{\Delta x \to 0} \frac{g(x + \Delta x) - g(x)}{\Delta x} = g'(x) \qquad \lim_{\Delta x \to 0} \frac{f(x + \Delta x) - f(x)}{\Delta x} = f'(x)$$

thus giving

$$h'(x) = f(x)g'(x) + g(x)f'(x)$$

$$D_x[f(x)g(x)] = f(x) \cdot D_x g(x) + g(x) \cdot D_x f(x)$$

The derivative of the product of two functions is the first function times the derivative of the second function plus the second function times the derivative of the first function if these derivatives exist.

EXAMPLE 2   Find $h'(x)$ if

$$h(x) = (2x^3 - 4x^2)(3x^5 + x^2)$$

Solution

$$\begin{aligned}
h'(x) &= (2x^3 - 4x^2)(15x^4 + 2x) + (3x^5 + x^2)(6x^2 - 8x) \\
&= (30x^7 - 60x^6 + 4x^4 - 8x^3) + (18x^7 - 24x^6 + 6x^4 - 8x^3) \\
&= 48x^7 - 84x^6 + 10x^4 - 16x^3
\end{aligned}$$

In Example 2, note that if multiplication is performed before the differentiation, the same result is obtained. Doing this we have

$$h(x) = 6x^8 - 12x^7 + 2x^5 - 4x^4$$

$$h'(x) = 48x^7 - 84x^6 + 10x^4 - 16x^3$$

3.3.7 THEOREM   If f and g are functions and if h is the function defined by

$$h(x) = \frac{f(x)}{g(x)}, \qquad \text{where } g(x) \neq 0$$

then if $f'(x)$ and $g'(x)$ exist,

$$h'(x) = \frac{g(x)f'(x) - f(x)g'(x)}{[g(x)]^2}$$

Proof

$$\begin{aligned}
h'(x) &= \lim_{\Delta x \to 0} \frac{h(x + \Delta x) - h(x)}{\Delta x} \\[2ex]
&= \lim_{\Delta x \to 0} \frac{\dfrac{f(x + \Delta x)}{g(x + \Delta x)} - \dfrac{f(x)}{g(x)}}{\Delta x} \\[2ex]
&= \lim_{\Delta x \to 0} \frac{f(x + \Delta x) \cdot g(x) - f(x) \cdot g(x + \Delta x)}{\Delta x \cdot g(x) \cdot g(x + \Delta x)}
\end{aligned}$$

If $f(x) \cdot g(x)$ is subtracted and added in the numerator, then

$$h'(x) = \lim_{\Delta x \to 0} \frac{f(x + \Delta x) \cdot g(x) - f(x) \cdot g(x) - f(x) \cdot g(x + \Delta x) + f(x) \cdot g(x)}{\Delta x \cdot g(x) \cdot g(x + \Delta x)}$$

$$= \lim_{\Delta x \to 0} \frac{\left[g(x) \cdot \dfrac{f(x + \Delta x) - f(x)}{\Delta x} \right] - \left[f(x) \cdot \dfrac{g(x + \Delta x) - g(x)}{\Delta x} \right]}{g(x) \cdot g(x + \Delta x)}$$

$$= \frac{\displaystyle\lim_{\Delta x \to 0} g(x) \cdot \lim_{\Delta x \to 0} \frac{f(x + \Delta x) - f(x)}{\Delta x} - \lim_{\Delta x \to 0} f(x) \cdot \lim_{\Delta x \to 0} \frac{g(x + \Delta x) - g(x)}{\Delta x}}{\displaystyle\lim_{\Delta x \to 0} g(x) \cdot \lim_{\Delta x \to 0} g(x + \Delta x)}$$

Because g is differentiable at x, then g is continuous at x; thus we have $\lim_{\Delta x \to 0} g(x + \Delta x) = g(x)$. Furthermore, $\lim_{\Delta x \to 0} g(x) = g(x)$ and $\lim_{\Delta x \to 0} f(x) = f(x)$. With these results and the definitions of $f'(x)$ and $g'(x)$ we get

$$h'(x) = \frac{g(x) \cdot f'(x) - f(x) \cdot g'(x)}{g(x) \cdot g(x)}$$

$$= \frac{g(x)f'(x) - f(x)g'(x)}{[g(x)]^2} \qquad \blacksquare$$

$$D_x\left[\frac{f(x)}{g(x)} \right] = \frac{g(x)D_x f(x) - f(x)D_x g(x)}{[g(x)]^2}$$

The derivative of the quotient of two functions is the fraction having as its denominator the square of the original denominator, and as its numerator the denominator times the derivative of the numerator minus the numerator times the derivative of the denominator if these derivatives exist.

EXAMPLE 3   Find

$$D_x\left(\frac{2x^3 + 4}{x^2 - 4x + 1} \right)$$

Solution

$$D_x\left(\frac{2x^3 + 4}{x^2 - 4x + 1} \right) = \frac{(x^2 - 4x + 1)(6x^2) - (2x^3 + 4)(2x - 4)}{(x^2 - 4x + 1)^2}$$

$$= \frac{6x^4 - 24x^3 + 6x^2 - 4x^4 + 8x^3 - 8x + 16}{(x^2 - 4x + 1)^2}$$

$$= \frac{2x^4 - 16x^3 + 6x^2 - 8x + 16}{(x^2 - 4x + 1)^2}$$

3.3.8 THEOREM   If $f(x) = x^{-n}$, where $-n$ is a negative integer and $x \neq 0$, then

$$f'(x) = -nx^{-n-1}$$

Proof   If $-n$ is a negative integer, then n is a positive integer. We write

$$f(x) = \frac{1}{x^n}$$

From Theorem 3.3.7

$$f'(x) = \frac{x^n \cdot 0 - 1 \cdot nx^{n-1}}{(x^n)^2}$$

$$= \frac{-nx^{n-1}}{x^{2n}}$$

$$= -nx^{n-1-2n}$$

$$= -nx^{-n-1}$$ ■

EXAMPLE 4   Find

$$\frac{d}{dx}\left(\frac{3}{x^5}\right)$$

Solution

$$\frac{d}{dx}\left(\frac{3}{x^5}\right) = \frac{d}{dx}(3x^{-5})$$

$$= 3(-5x^{-6})$$

$$= -\frac{15}{x^6}$$

If r is any positive or negative integer, it follows from Theorems 3.3.2 and 3.3.8 that

$$D_x(x^r) = rx^{r-1}$$

and from Theorems 3.3.2, 3.3.3, and 3.3.8 we obtain

$$D_x(cx^r) = crx^{r-1}$$

EXERCISES 3.3

In Exercises 1 through 24, differentiate the function by applying the theorems of this section.

1. $f(x) = 7x - 5$

2. $g(x) = 8 - 3x$

3. $g(x) = 1 - 2x - x^2$

4. $f(x) = 4x^2 + x + 1$

5. $f(x) = x^3 - 3x^2 + 5x - 2$

6. $f(x) = 3x^4 - 5x^2 + 1$

7. $f(x) = \frac{1}{8}x^8 - x^4$

8. $g(x) = x^7 - 2x^5 + 5x^3 - 7x$

9. $F(t) = \frac{1}{4}t^4 - \frac{1}{2}t^2$

10. $H(x) = \frac{1}{3}x^3 - x + 2$

11. $v(r) = \frac{4}{3}\pi r^3$

12. $G(y) = y^{10} + 7y^5 - y^3 + 1$

13. $F(x) = x^2 + 3x + \frac{1}{x^2}$

14. $f(x) = \frac{x^3}{3} + \frac{3}{x^3}$

15. $g(x) = 4x^4 - \frac{1}{4x^4}$

16. $f(x) = x^4 - 5 + x^{-2} + 4x^{-4}$

17. $g(x) = \frac{3}{x^2} + \frac{5}{x^4}$

18. $H(x) = \frac{5}{6x^5}$

19. $f(s) = \sqrt{3}(s^3 - s^2)$

20. $g(x) = (2x^2 + 5)(4x - 1)$

21. $f(x) = (2x^4 - 1)(5x^3 + 6x)$

22. $f(x) = (4x^2 + 3)^2$

23. $G(y) = (7 - 3y^3)^2$

24. $F(t) = (t^3 - 2t + 1)(2t^2 + 3t)$

In Exercises 25 through 36, compute the derivative by applying the theorems of this section.

25. $D_x[(x^2 - 3x + 2)(2x^3 + 1)]$

26. $D_x\left(\frac{2x}{x+3}\right)$

27. $D_x\left(\frac{x}{x-1}\right)$

28. $D_y\left(\frac{2y+1}{3y+4}\right)$

29. $\frac{d}{dx}\left(\frac{x^2 + 2x + 1}{x^2 - 2x + 1}\right)$

30. $\frac{d}{dx}\left(\frac{4 - 3x - x^2}{x - 2}\right)$

31. $\frac{d}{dt}\left(\frac{5t}{1 + 2t^2}\right)$

32. $\frac{d}{dx}\left(\frac{x^4 - 2x^2 + 5x + 1}{x^4}\right)$

33. $\frac{d}{dy}\left(\frac{y^3 - 8}{y^3 + 8}\right)$

34. $\frac{d}{ds}\left(\frac{s^2 - a^2}{s^2 + a^2}\right)$

35. $D_x \left[\dfrac{2x + 1}{x + 5} (3x - 1) \right]$

36. $D_x \left[\dfrac{x^3 + 1}{x^2 + 3} (x^2 - 2x^{-1} + 1) \right]$

37. Find an equation of the tangent line to the curve $y = x^3 - 4$ at the point $(2, 4)$.

38. Find an equation of the normal line to the curve $y = 4x^2 - 8x$ at the point $(1, -4)$.

39. Find an equation of the normal line to the curve $y = 10/(14 - x^2)$ at the point $(4, -5)$.

40. Find an equation of the tangent line to the curve $y = 8/(x^2 + 4)$ at the point $(2, 1)$.

41. Find an equation of the line tangent to the curve $y = 3x^2 - 4x$ and parallel to the line $2x - y + 3 = 0$.

42. Find an equation of the tangent line to the curve $y = x^4 - 6x$ that is perpendicular to the line $x - 2y + 6 = 0$.

43. Find an equation of each of the normal lines to the curve $y = x^3 - 4x$ that is parallel to the line $x + 8y - 8 = 0$.

44. Find an equation of each of the tangent lines to the curve $3y = x^3 - 3x^2 + 6x + 4$ that is parallel to the line $2x - y + 3 = 0$.

45. Find an equation of each of the lines through the point $(4, 13)$ that is tangent to the curve $y = 2x^2 - 1$.

46. Given $f(x) = \frac{1}{3}x^3 + 2x^2 + 5x + 5$. Show that $f'(x) \geq 0$ for all values of x.

47. If f, g, and h are functions and $\phi(x) = f(x) \cdot g(x) \cdot h(x)$, prove that if $f'(x)$, $g'(x)$, and $h'(x)$ exist,

$$\phi'(x) = f(x) \cdot g(x) \cdot h'(x) + f(x) \cdot g'(x) \cdot h(x) + f'(x) \cdot g(x) \cdot h(x)$$

(*Hint:* Apply Theorem 3.3.6 twice.)

Use the result of Exercise 47 to differentiate the functions in Exercises 48 through 51.

48. $f(x) = (x^2 + 3)(2x - 5)(3x + 2)$

49. $h(x) = (3x + 2)^2(x^2 - 1)$

50. $g(x) = (3x^3 + x^{-3})(x + 3)(x^2 - 5)$

51. $\phi(x) = (2x^2 + x + 1)^3$

3.4 RECTILINEAR MOTION AND THE DERIVATIVE AS A RATE OF CHANGE

The derivative of a function f at the number x_1 has a powerful interpretation as the *instantaneous rate of change of f at x_1*. We begin by considering an application in physics: the motion of a particle on a line. Such a motion is called **rectilinear motion**. One direction is chosen arbitrarily as positive, and the opposite direction is negative. For simplicity assume that the motion of the particle is along a horizontal line, with distance to the right as positive and distance to the left as negative. Select some point on the line and denote it by the letter O. Let f be the function determining the directed distance of the particle from O at any particular time.

To be more specific, let s meters (m) be the directed distance of the particle from O at t seconds (sec). Then f is the function defined by the equation

$$s = f(t)$$

which gives the directed distance from the point O to the particle at a particular instant.

▶ **ILLUSTRATION 1** Let

$$s = t^2 + 2t - 3$$

Then when $t = 0$, $s = -3$; therefore the particle is 3 m to the left of point O when $t = 0$. When $t = 1$, $s = 0$; so the particle is at point O at 1 sec. When $t = 2$, $s = 5$; so the particle is 5 m to the right of point O at 2 sec. When $t = 3$, $s = 12$; so the particle is 12 m to the right of point O at 3 sec.

Figure 1 illustrates the various positions of the particle for specific values of t.

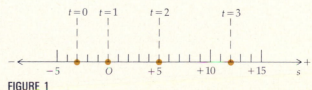

FIGURE 1

Between the time for $t = 1$ and $t = 3$, the particle moves from the point where $s = 0$ to the point where $s = 12$; thus in the 2-second interval the change in the directed distance from O is 12 m. The average velocity of the particle is the ratio of the change in the directed distance from a fixed point to the change in the time. So the number of meters per second in the average velocity of the particle from $t = 1$ to $t = 3$ is $\frac{12}{2} = 6$. From $t = 0$ to $t = 2$, the change in the directed distance from O of the particle is 8 m, and so the number of meters per second in the average velocity of the particle in this 2-second interval is $\frac{8}{2} = 4$.   ◄

In Illustration 1 the average velocity of the particle is obviously not constant; and the average velocity supplies no specific information about the motion of the particle at any particular instant. For example, if a car travels a distance of 100 km in the same direction and it takes 2 hr, we say that the average velocity in traveling that distance is 50 km/hr. However, from this information we cannot determine the speedometer reading of the car at any particular time in the 2-hour period. The speedometer reading at a specific time is referred to as the *instantaneous velocity*. The following discussion enables us to arrive at a definition of what is meant by *instantaneous velocity*.

Let the equation $s = f(t)$ define s (the number of meters in the directed distance of the particle from point O) as a function of t (the number of seconds in the time). When $t = t_1$, $s = s_1$. The change in the directed distance from O is $(s - s_1)$ meters over the interval of time $(t - t_1)$ seconds, and the number of meters per second in the average velocity of the particle over this interval of time is given by

$$\frac{s - s_1}{t - t_1}$$

or, because $s = f(t)$ and $s_1 = f(t_1)$, the average velocity is found from

$$\frac{f(t) - f(t_1)}{t - t_1} \tag{1}$$

Now the shorter the interval is from t_1 to t, the closer the average velocity will be to what we would intuitively think of as the instantaneous velocity at t_1.

For example, if the speedometer reading of a car as it passes a point P_1 is 80 km/hr and if a point P is, for instance, 10 m from P_1, then the average velocity of the car as it travels this 10 m will very likely be close to 80 km/hr because the variation of the velocity of the car along this short stretch is probably slight. Now if the distance from P_1 to P were shortened to 5 m, the average velocity of the car in this interval would be even closer to the speedometer reading of the car as it passes P_1. We can continue this process, and the speedometer reading at P_1 can be represented as the limit of the average velocity between P_1 and P as P approaches P_1. That is, the *instantaneous velocity* can be defined as the limit of quotient (1) as t approaches t_1, provided the limit exists. This limit is the derivative of the function f at t_1. We have, then, the following definition.

3.4.1 DEFINITION

> If f is a function given by the equation
>
> $$s = f(t)$$
>
> and a particle is moving along a straight line such that s is the number of units in the directed distance of the particle from a fixed point on the line at t units of time, then the **instantaneous velocity** of the particle at t units of time is v units of velocity, where
>
> $$v = f'(t) \quad \Leftrightarrow \quad v = \frac{ds}{dt}$$
>
> if it exists.

The instantaneous velocity may be either positive or negative, depending on whether the particle is moving along the line in the positive or the negative direction. When the instantaneous velocity is zero, the particle is at rest.

The **speed** of a particle at any time is defined as the absolute value of the instantaneous velocity. Hence the speed is a nonnegative number. The terms "speed" and "instantaneous velocity" are often confused. Note that the speed indicates only how fast the particle is moving, whereas the instantaneous velocity also tells the direction of motion.

EXAMPLE 1   A particle is moving along a horizontal line according to the equation

$$s = 2t^3 - 4t^2 + 2t - 1$$

Determine the intervals of time when the particle is moving to the right and when it is moving to the left. Also determine the instant when the particle reverses its direction.

Solution

$$v = \frac{ds}{dt}$$

$$= 6t^2 - 8t + 2$$
$$= 2(3t^2 - 4t + 1)$$
$$= 2(3t - 1)(t - 1)$$

The instantaneous velocity is zero when $t = \frac{1}{3}$ and $t = 1$. Therefore the particle is at rest at these two times. The particle is moving to the right when v is positive, and it is moving to the left when v is negative. We determine the sign of v for various intervals of t, and the results are given in Table 1.

Table 1

	$3t - 1$	$t - 1$	Conclusion
$t < \frac{1}{3}$	$-$	$-$	v is positive, and the particle is moving to the right
$t = \frac{1}{3}$	0	$-$	v is zero, and the particle is changing direction from right to left
$\frac{1}{3} < t < 1$	$+$	$-$	v is negative, and the particle is moving to the left
$t = 1$	$+$	0	v is zero, and the particle is changing direction from left to right
$1 < t$	$+$	$+$	v is positive, and the particle is moving to the right

Table 2

t	s	v
-1	-9	16
0	-1	2
$\frac{1}{3}$	$-\frac{19}{27}$	0
1	-1	0
2	3	10

The motion of the particle, indicated in Figure 2, is along the horizontal line; however, the behavior of the motion is indicated above the line. Table 2 gives values of s and v for specific values of t.

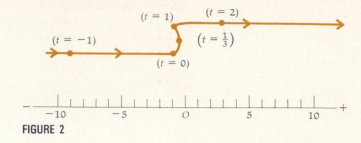

FIGURE 2

EXAMPLE 2   A ball is thrown vertically upward from the ground with an initial velocity of 64 ft/sec. If the positive direction of the distance from the starting point is up, the equation of motion is

$$s = -16t^2 + 64t$$

Let t seconds be the time that has elapsed since the ball was thrown and s feet be the distance of the ball from the starting point at t seconds. (a) Find the instantaneous velocity of the ball at the end of 1 sec. Is the ball rising or falling at the end of 1 sec? (b) Find the instantaneous velocity of the ball at the end of 3 sec. Is the ball rising or falling at the end of 3 sec? (c) How many seconds does it take the ball to reach its highest point? (d) How high will the ball go? (e) Find the speed of the ball at the end of 1 sec and at the end of 3 sec. (f) How many seconds does it take the ball to reach the ground? (g) Find the instantaneous velocity of the ball when it reaches the ground.

Table 3

t	s	v
0	0	64
$\frac{1}{2}$	28	48
1	48	32
2	64	0
3	48	-32
$\frac{7}{2}$	28	-48
4	0	-64

Solution   Let $v(t)$ be the number of feet per second in the instantaneous velocity of the ball at t seconds. Then because $v(t) = ds/dt$,

$$v(t) = -32t + 64$$

(a) $v(1) = -32(1) + 64$; that is, $v(1) = 32$; so at the end of 1 sec the ball is rising with an instantaneous velocity of 32 ft/sec.

(b) $v(3) = -32(3) + 64$; that is, $v(3) = -32$; so at the end of 3 sec the ball is falling with an instantaneous velocity of -32 ft/sec.

(c) The ball reaches its highest point when the direction of motion changes, that is, when $v(t) = 0$. Setting $v(t) = 0$ we obtain $-32t + 64 = 0$. Thus $t = 2$.

(d) When $t = 2$, $s = 64$; therefore the ball reaches a highest point of 64 ft above the starting point.

(e) $|v(t)|$ is the number of feet per second in the speed of the ball at t seconds; $|v(1)| = 32$ and $|v(3)| = 32$.

(f) The ball will reach the ground for $s = 0$. Setting $s = 0$ we have $-16t^2 + 64t = 0$, from which we obtain $t = 0$ and $t = 4$. Therefore the ball will reach the ground in 4 sec.

(g) $v(4) = -64$; when the ball reaches the ground, its instantaneous velocity is -64 ft/sec.

Table 3 gives values of s and v for some specific values of t. The motion of the ball is indicated in Figure 3. The motion is assumed to be in a vertical line, and the behavior of the motion is indicated to the left of the line.

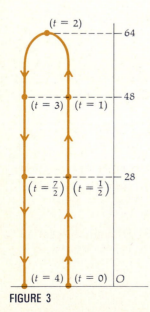

FIGURE 3

The concept of velocity in rectilinear motion corresponds to the more general concept of instantaneous rate of change. For example, if a particle is moving along a line according to the equation $s = f(t)$, the velocity of the particle at t units of time is determined by the derivative of s with respect to t. Because velocity can be interpreted as a rate of change of distance per unit change in time, the derivative of s with respect to t is the rate of change of s per unit change in t.

In a similar way, if a quantity y is a function of a quantity x, we may express the rate of change of y per unit change in x. The discussion is analogous to the discussions of the slope of a tangent line to a graph and the instantaneous velocity of a particle moving along a line.

If the functional relationship between y and x is given by

$$y = f(x)$$

and if x changes from the value x_1 to $x_1 + \Delta x$, then y changes from $f(x_1)$ to $f(x_1 + \Delta x)$. So the change in y, which we denote by Δy, is $f(x_1 + \Delta x) - f(x_1)$ when the change in x is Δx. The average rate of change of y per unit change in x, as x changes from x_1 to $x_1 + \Delta x$, is then

$$\frac{f(x_1 + \Delta x) - f(x_1)}{\Delta x} = \frac{\Delta y}{\Delta x} \tag{2}$$

If the limit of this quotient exists as $\Delta x \to 0$, this limit is what we intuitively think of as the instantaneous rate of change of y per unit change in x at x_1. Accordingly, we have the following definition.

3.4.2 DEFINITION If $y = f(x)$, the **instantaneous rate of change of y per unit change in x at x_1** is $f'(x_1)$ or, equivalently, the derivative of y with respect to x at x_1, if it exists.

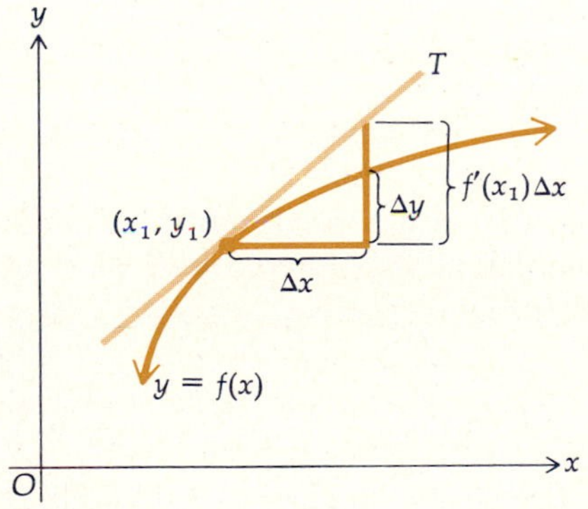

FIGURE 4

To illustrate geometrically, let $f'(x_1)$ be the instantaneous rate of change of y per unit change in x at x_1. Then if $f'(x_1)$ is multiplied by Δx (the change in x), we have the change that would occur in y if the point (x, y) were to move along the tangent line at (x_1, y_1) of the graph of $y = f(x)$. See Figure 4. The average rate of change of y per unit change in x is given by the fraction in (2), and if this fraction is multiplied by Δx, we get Δy, which is the actual change in y caused by a change of Δx in x when the point (x, y) moves along the graph.

EXAMPLE 3 If $V(x)$ cubic centimeters is the volume of a cube having an edge of x centimeters, use a calculator to compute the average rate of change of $V(x)$ with respect to x as x changes from (a) 3.000 to 3.200; (b) 3.000 to 3.100; (c) 3.000 to 3.010; (d) 3.000 to 3.001. (e) What is the instantaneous rate of change of $V(x)$ with respect to x when x is 3?

Solution The average rate of change of $V(x)$ with respect to x as x changes from x_1 to $x_1 + \Delta x$ is

$$\frac{V(x_1 + \Delta x) - V(x_1)}{\Delta x}$$

(a) $x_1 = 3.000$, $\Delta x = 0.200$

$$\frac{V(3.200) - V(3.000)}{0.200} = \frac{(3.200)^3 - (3.000)^3}{0.200}$$

$$= 28.84$$

(b) $x_1 = 3.000$, $\Delta x = 0.100$

$$\frac{V(3.100) - V(3.000)}{0.100} = \frac{(3.100)^3 - (3.000)^3}{0.100}$$

$$= 27.91$$

(c) $x_1 = 3.000$, $\Delta x = 0.010$

$$\frac{V(3.010) - V(3.000)}{0.010} = \frac{(3.010)^3 - (3.000)^3}{0.010}$$

$$= 27.09$$

(d) $x_1 = 3.000$, $\Delta x = 0.001$

$$\frac{V(3.001) - V(3.000)}{0.001} = \frac{(3.001)^3 - (3.000)^3}{0.001}$$

$$= 27.01$$

In part (a) we see that as the length of the edge of the cube changes from 3.000 cm to 3.200 cm, the average rate of change of the volume is 28.84 cm^3 per centimeter change in the length of the edge. There are similar interpretations of parts (b)–(d).

(e) The instantaneous rate of change of $V(x)$ with respect to x when x is 3 is $V'(3)$.

$$V'(x) = 3x^2 \qquad V'(3) = 27$$

Therefore, when the length of the edge of the cube is 3 cm, the instantaneous rate of change of the volume is 27 cm^3 per centimeter change in the length of the edge.

EXAMPLE 4   In an electric circuit, if E volts is the electromotive force, R ohms is the resistance, and I amperes is the current, it follows from Ohm's law that

$$IR = E$$

Assuming that E is a positive constant, show that R decreases at a rate proportional to the inverse square of I.

Solution   Solving the given equation for R, we obtain

$$R = E \cdot I^{-1}$$

Differentiating R with respect to I, we have

$$\frac{dR}{dI} = -E \cdot I^{-2}$$

$$\frac{dR}{dI} = -\frac{E}{I^2}$$

This equation states that the rate of change of R with respect to I is negative and proportional to $\dfrac{1}{I^2}$. Therefore, R decreases at a rate proportional to the inverse square of I.

In economics the variation of one quantity with respect to another may be described by either an *average* concept or a *marginal* concept. The average concept expresses the variation of one quantity over a specified range of values of a second quantity, whereas the marginal concept is the instantaneous change in the first quantity that results from a very small unit change in the second quantity. We begin our examples in economics with the definitions of average cost and marginal cost. To define a marginal concept precisely we use the notion of a limit, and this leads to the derivative.

Suppose $C(x)$ dollars is the total cost of producing x units of a commodity. The function C is called a **total cost function**. In normal circumstances x and $C(x)$ are positive. Since x represents the number of units of a commodity, x is usually a nonnegative integer. However, to apply the calculus we assume that x is a nonnegative real number to give the continuity requirements for the function C.

The **average cost** of producing each unit of a commodity is obtained by dividing the total cost by the number of units produced. If $Q(x)$ dollars is the average cost,

$$Q(x) = \frac{C(x)}{x}$$

and Q is called an **average cost function**.

Now suppose that the number of units in a particular output is x_1, and this is changed by Δx. Then the change in the total cost is given by $C(x_1 + \Delta x) - C(x_1)$, and the average change in the total cost with respect to the change in the number of units produced is given by

$$\frac{C(x_1 + \Delta x) - C(x_1)}{\Delta x}$$

Economists use the term *marginal cost* for the limit of this quotient as Δx approaches zero, provided the limit exists. This limit, being the derivative of C at x_1, states that the **marginal cost**, when $x = x_1$, is given by $C'(x_1)$, if it exists. The function C' is called the **marginal cost function**, and $C'(x_1)$ may be interpreted as the rate of change of the total cost when x_1 units are produced.

▶ **ILLUSTRATION 2**   Suppose that $C(x)$ dollars is the total cost of manufacturing x toys, and

$$C(x) = 110 + 4x + 0.02x^2$$

(a) The marginal cost function is C', and

$$C'(x) = 4 + 0.04x$$

(b) The marginal cost when $x = 50$ is $C'(50)$, and

$$C'(50) = 4 + 0.04(50)$$
$$= 6$$

Therefore the rate of change of the total cost, when 50 toys are manufactured, is $6 per toy.

(c) The number of dollars in the actual cost of manufacturing the fifty-first toy is $C(51) - C(50)$, and

$$C(51) - C(50) = [110 + 4(51) + 0.02(51)^2] - [110 + 4(50) + 0.02(50)^2]$$
$$= 366.02 - 360$$
$$= 6.02$$

Note that the answers in (b) and (c) differ by 0.02. This discrepancy occurs because the marginal cost is the instantaneous rate of change of $C(x)$ with respect to a unit change in x. Hence $C'(50)$ is the approximate number of dollars in the cost of producing the fifty-first toy.   ◀

Observe that the computation of $C'(50)$ in Illustration 2 is simpler than computing $C(51) - C(50)$. Economists frequently approximate the cost of producing one additional unit by using the marginal cost function. Specifically, $C'(k)$ dollars is the approximate cost of the $(k + 1)$st unit after the first k units have been produced.

Another function important in economics is the **total revenue function**, denoted by R, and

$$R(x) = px$$

where $R(x)$ dollars is the total revenue received when x units are sold at p dollars per unit.

The **marginal revenue**, when $x = x_1$, is given by $R'(x_1)$, if it exists. The function R' is called the **marginal revenue function**. $R'(x_1)$ may be positive, negative, or zero, and it may be interpreted as the rate of change of the total revenue when x_1 units are sold. $R'(k)$ dollars is the approximate revenue from the sale of the $(k + 1)$st unit after the first k units have been sold.

EXAMPLE 5   Suppose that $R(x)$ dollars is the total revenue received from the sale of x tables, and

$$R(x) = 300x - \tfrac{1}{2}x^2$$

Find (a) the marginal revenue function; (b) the marginal revenue when $x = 40$; (c) the actual revenue from the sale of the forty-first table.

Solution   (a) The marginal revenue function is R', and

$$R'(x) = 300 - x$$

(b) The marginal revenue when $x = 40$ is given by $R'(40)$, and

$$R'(40) = 300 - 40$$
$$= 260$$

Thus the rate of change of the total revenue when 40 tables are sold is $260 per table.

(c) The number of dollars in the actual revenue from the sale of the forty-first table is $R(41) - R(40)$, and

$$R(41) - R(40) = \left[300(41) - \frac{(41)^2}{2}\right] - \left[300(40) - \frac{(40)^2}{2}\right]$$

$$= [12,300 - 840.50] - [12,000 - 800]$$
$$= 11,459.50 - 11,200$$
$$= 259.50$$

Hence the actual revenue from the sale of the forty-first table is $259.50.

Observe in part (b) of Example 5 we obtained $R'(40) = 260$, and $260 is an approximation of the revenue received from the sale of the forty-first table, which from (c) is $259.50.

EXERCISES 3.4

In Exercises 1 through 8, a particle is moving along a horizontal line according to the given equation, where s meters is the directed distance of the particle from a point O at t seconds. Find the instantaneous velocity v(t) meters per second at t seconds, and then find v(t₁) for the particular value of t_1.

1. $s = 3t^2 + 1$; $t_1 = 3$

2. $s = 8 - t^2$; $t_1 = 5$

3. $s = \dfrac{1}{4t}$; $t_1 = \frac{1}{2}$

4. $s = \dfrac{3}{t^2}$; $t_1 = -2$

5. $s = 2t^3 - t^2 + 5$; $t_1 = -1$

6. $s = 4t^3 + 2t - 1$; $t_1 = \frac{1}{2}$

7. $s = \dfrac{2t}{4 + t}$; $t_1 = 0$

8. $s = \dfrac{1}{t} + \dfrac{3}{t^2}$; $t_1 = 2$

In Exercises 9 through 14, a particle is moving along a horizontal line according to the given equation, where s meters is the directed distance of the particle from a point O at t seconds. The positive direction is to the right. Determine the intervals of time when the particle is moving to the right and when it is moving to the left. Also determine when the particle reverses its direction. Show the behavior of the motion by a figure similar to Figure 2, and choose values of t at random but include the values of t when the particle reverses its direction.

9. $s = t^3 + 3t^2 - 9t + 4$

10. $s = 2t^3 - 3t^2 - 12t + 8$

11. $s = \frac{2}{3}t^3 + \frac{3}{2}t^2 - 2t + 4$

12. $s = \dfrac{t}{1 + t^2}$

13. $s = \dfrac{t}{9 + t^2}$

14. $s = \dfrac{t + 1}{t^2 + 4}$

15. An object falls from rest, and $s = -16t^2$ where s feet is the distance of the object from the starting point at t seconds, and the positive direction is upward. If a stone is dropped from a building 256 ft high, find (a) the instantaneous velocity of the stone 1 sec after it is dropped; (b) the instantaneous velocity of the stone 2 sec after it is dropped; (c) how long it takes the stone to reach the ground; (d) the instantaneous velocity of the stone when it reaches the ground.

16. A stone is dropped from a height of 64 ft. If s feet is the height of the stone t seconds after being dropped, then $s = -16t^2 + 64$. (a) How long does it take the stone to reach the ground? (b) Find the instantaneous velocity of the stone when it reaches the ground.

17. A billiard ball is hit and travels in a straight line. If s centimeters is the distance of the ball from its initial position at t seconds, then $s = 100t^2 + 100t$. If the ball hits a cushion that is 39 cm from its initial position, at what velocity does it hit the cushion?

18. If a stone is thrown vertically upward from the ground with an initial velocity of 32 ft/sec, then $s = -16t^2 + 32t$, where s feet is the distance of the stone from the starting point at t seconds, and the positive direction is upward. Find (a) the average velocity of the stone during the time interval $\frac{3}{4} \leq t \leq \frac{5}{4}$; (b) the instantaneous velocity of the stone at $\frac{3}{4}$ sec and at $\frac{5}{4}$ sec; (c) the speed of the stone at $\frac{3}{4}$ sec and at $\frac{5}{4}$ sec; (d) the average velocity of the stone during the time interval $\frac{1}{2} \leq t \leq \frac{3}{4}$; (e) how many seconds it will take the stone to reach the highest point; (f) how high the stone will go; (g) how many seconds it will take the stone to reach the ground; (h) the instantaneous velocity of the stone when it reaches the ground. Show the behavior of the motion by a figure similar to Figure 3.

19. If a ball is given a push so that it has an initial velocity of 24 ft/sec down a certain inclined plane, then $s = 24t + 10t^2$, where s feet is the distance of the ball from the starting point at t seconds and the positive direction is down the inclined plane. (a) What is the instantaneous velocity of the ball at t_1 seconds? (b) How long does it take for the velocity to increase to 48 ft/sec?

20. A rocket is fired vertically upward, and it is s feet above the ground t seconds after being fired, where $s = 560t - 16t^2$ and the positive direction is upward. Find (a) the velocity of the rocket 2 sec after being fired, and (b) how long it takes for the rocket to reach its maximum height.

21. If $A(x)$ square centimeters is the area of a square having a side of x centimeters, use a calculator to compute the average rate of change of $A(x)$ with respect to x as x changes from (a) 4.000 to 4.600; (b) 4.000 to 4.300; (c) 4.000 to 4.100; (d) 4.000 to 4.050. (e) What is the instantaneous rate of change of $A(x)$ with respect to x when x is 4?

22. The length of a rectangle is 4 in. more than its width, and the 4-in. difference is maintained as the rectangle increases in size. If $A(w)$ square inches is the area of the rectangle having a width of w inches, use a calculator to compute the average rate of change of $A(w)$ with respect to w as w changes from

(a) 3.000 to 3.200; (b) 3.000 to 3.100; (c) 3.000 to 3.010; (d) 3.000 to 3.001. (e) What is the instantaneous rate of change of $A(w)$ with respect to w when w is 3?

23. Boyle's law for the expansion of a gas is $PV = C$, where P is the number of pounds per square unit of pressure, V is the number of cubic units in the volume of the gas, and C is a constant. Show that V decreases at a rate proportional to the inverse square of P.

24. From Boyle's law for the expansion of a gas, given in Exercise 23, find the instantaneous rate of change of V with respect to P when $P = 4$ and $V = 8$.

25. A cold front approaches the college campus. The temperature is T degrees t hours after midnight, and

$$T = 0.1(400 - 40t + t^2) \qquad 0 \le t \le 12$$

(a) Find the average rate of change of T with respect to t between 5 A.M. and 6 A.M. (b) Find the instantaneous rate of change of T with respect to t at 5 A.M.

26. It is estimated that a worker in a shop that makes picture frames can paint y frames x hours after starting work at 8 A.M., and

$$y = 3x + 8x^2 - x^3 \qquad 0 \le x \le 4$$

(a) Find the rate at which the worker is painting at 10 A.M. (b) Find the number of frames that the worker paints between 10 A.M. and 11 A.M.

27. If water is being drained from a swimming pool and V liters is the volume of water in the pool t minutes after the draining starts, where $V = 250(1600 - 80t + t^2)$, find (a) the average rate at which the water leaves the pool during the first 5 min, and (b) how fast the water is flowing out of the pool 5 min after the draining starts.

28. A balloon maintains the shape of a sphere as it is being inflated. Find the rate of change of the surface area with respect to the radius at the instant when the radius is 2 m.

29. The number of dollars in the total cost of manufacturing x watches in a certain plant is given by $C(x) = 1500 + 3x + x^2$. Find (a) the marginal cost function; (b) the marginal cost when $x = 40$; (c) the actual cost of manufacturing the forty-first watch.

30. If $C(x)$ dollars is the total cost of manufacturing x paperweights and

$$C(x) = 200 + \frac{50}{x} + \frac{x^2}{5}$$

find (a) the marginal cost function; (b) the marginal cost when $x = 10$; (c) the actual cost of manufacturing the eleventh paperweight.

31. If $R(x)$ dollars is the total revenue received from the sale of x television sets and $R(x) = 600x - \frac{1}{20}x^3$, find (a) the marginal revenue function; (b) the marginal revenue when $x = 20$; (c) the actual revenue from the sale of the twenty-first television set.

32. The total revenue received from the sale of x desks is $R(x)$ dollars, and $R(x) = 200x - \frac{1}{3}x^2$. Find (a) the marginal reve-

nue function; (b) the marginal revenue when $x = 30$; (c) the actual revenue from the sale of the thirty-first desk.

In Exercises 33 through 35, we use the concept of relative rate, defined as follows: if $y = f(x)$, the **relative rate of change of y** *with respect to x at x_1 is given by* $\dfrac{f'(x_1)}{f(x_1)}$ *or, equivalently,* $\dfrac{dy/dx}{y}$ *evaluated at $x = x_1$.*

33. The annual gross earnings of a particular corporation t years from January 1, 1988 is p millions of dollars and $p = \frac{2}{5}t^2 + 2t + 10$. Find (a) the rate at which the gross earnings were growing on January 1, 1990; (b) the relative rate of growth of the gross earnings on January 1, 1990 to the nearest 0.1 percent; (c) the rate at which the gross earnings should be growing on January 1, 1994; (d) the anticipated relative rate of growth of the gross earnings on January 1, 1994 to the nearest 0.1 percent.

34. A particular company started doing business on April 1, 1987. The annual gross earnings of the company after t years of operation are p dollars, where $p = 50,000 + 18,000t + 600t^2$. Find (a) the rate at which the gross earnings were growing on April 1, 1989; (b) the relative rate of growth of the gross earnings on April 1, 1989 to the nearest 0.1 percent; (c) the rate at which the gross earnings should be growing on April 1, 1997; (d) the anticipated relative rate of growth of the gross earnings on April 1, 1997 to the nearest 0.1 percent.

35. Suppose that the number of people in the population of a particular city t years after January 1, 1986 is expected to be $40t^2 + 200t + 10,000$. Find (a) the rate at which the population is expected to be growing on January 1, 1995; (b) the expected relative rate of growth of the population on January 1, 1995 to the nearest 0.1 percent; (c) the rate at which the population is expected to be growing on January 1, 2001; (d) the expected relative rate of growth of the population on January 1, 2001 to the nearest 0.1 percent.

36. Two particles, A and B, move to the right along a horizontal line. They start at a point O, s meters is the directed distance of the particle from O at t seconds, and the equations of motion are

$$s = 4t^2 + 5t \qquad \text{(for particle } A)$$
$$s = 7t^2 + 3t \qquad \text{(for particle } B)$$

If $t = 0$ at the start, for what values of t will the velocity of particle A exceed the velocity of particle B?

37. The profit of a retail store is $100y$ dollars when x dollars are spent daily on advertising and $y = 2500 + 36x - 0.2x^2$. Use the derivative to determine if it would be profitable for the daily advertising budget to be increased if the daily advertising budget is (a) $60 and (b) $300. (c) What is the maximum value for x below which it is profitable to increase the advertising budget?

38. Show that for any linear function f the average rate of change of $f(x)$ as x changes from x_1 to $x_1 + k$ is the same as the instantaneous rate of change of $f(x)$ at x_1.

3.5   DERIVATIVES OF THE TRIGONOMETRIC FUNCTIONS

To show that the sine function has a derivative we apply the trigonometric identity

$$\sin(a + b) = \sin a \cos b + \cos a \sin b \tag{1}$$

as well as Theorems 2.8.2 and 2.8.5.

Let f be the sine function, so that

$$f(x) = \sin x$$

From the definition of a derivative,

$$f'(x) = \lim_{\Delta x \to 0} \frac{f(x + \Delta x) - f(x)}{\Delta x}$$

$$= \lim_{\Delta x \to 0} \frac{\sin(x + \Delta x) - \sin x}{\Delta x}$$

Formula (1) for $\sin(x + \Delta x)$ is used to obtain

$$f'(x) = \lim_{\Delta x \to 0} \frac{\sin x \cos(\Delta x) + \cos x \sin(\Delta x) - \sin x}{\Delta x}$$

$$= \lim_{\Delta x \to 0} \frac{\sin x[\cos(\Delta x) - 1]}{\Delta x} + \lim_{\Delta x \to 0} \frac{\cos x \sin(\Delta x)}{\Delta x}$$

$$= -\lim_{\Delta x \to 0} \frac{1 - \cos(\Delta x)}{\Delta x}\left(\lim_{\Delta x \to 0} \sin x\right) + \left(\lim_{\Delta x \to 0} \cos x\right) \lim_{\Delta x \to 0} \frac{\sin(\Delta x)}{\Delta x} \tag{2}$$

From Theorem 2.8.5,

$$\lim_{\Delta x \to 0} \frac{1 - \cos(\Delta x)}{\Delta x} = 0 \tag{3}$$

and from Theorem 2.8.2,

$$\lim_{\Delta x \to 0} \frac{\sin(\Delta x)}{\Delta x} = 1 \tag{4}$$

Substituting from (3) and (4) into (2) we get

$$f'(x) = -0 \cdot \sin x + \cos x \cdot 1$$

$$= \cos x$$

We have proved the following theorem.

3.5.1 THEOREM

$$D_x(\sin x) = \cos x$$

EXAMPLE 1   Find $f'(x)$ if

$$f(x) = x^2 \sin x$$

Solution   We find the derivative of a product of two functions by applying Theorem 3.3.6.

$$f'(x) = x^2\, D_x(\sin x) + D_x(x^2)\sin x$$

$$= x^2 \cos x + 2x \sin x$$

To find the derivative of the cosine function we proceed as with the sine function. Here we apply the identity

$$\cos(a + b) = \cos a \cos b - \sin a \sin b \tag{5}$$

If g is the cosine function, then

$$g(x) = \cos x$$

$$g'(x) = \lim_{\Delta x \to 0} \frac{g(x + \Delta x) - g(x)}{\Delta x}$$

$$= \lim_{\Delta x \to 0} \frac{\cos(x + \Delta x) - \cos x}{\Delta x}$$

Formula (5) for $\cos(x + \Delta x)$ is used to obtain

$$g'(x) = \lim_{\Delta x \to 0} \frac{\cos x \cos(\Delta x) - \sin x \sin(\Delta x) - \cos x}{\Delta x}$$

$$= \lim_{\Delta x \to 0} \frac{\cos x [\cos(\Delta x) - 1]}{\Delta x} - \lim_{\Delta x \to 0} \frac{\sin x \sin(\Delta x)}{\Delta x}$$

$$= -\lim_{\Delta x \to 0} \frac{1 - \cos(\Delta x)}{\Delta x} \left(\lim_{\Delta x \to 0} \cos x \right) - \left(\lim_{\Delta x \to 0} \sin x \right) \lim_{\Delta x \to 0} \frac{\sin(\Delta x)}{\Delta x} \tag{6}$$

We substitute from (3) and (4) into (6) and obtain

$$g'(x) = -0 \cdot \cos x - \sin x \cdot 1$$

$$= -\sin x$$

The following theorem has been proved.

3.5.2 THEOREM $D_x(\cos x) = -\sin x$

EXAMPLE 2   Find $\dfrac{dy}{dx}$ if

$$y = \frac{\sin x}{1 - 2 \cos x}$$

Solution   We apply Theorem 3.3.7 (the derivative of a quotient).

$$\frac{dy}{dx} = \frac{(1 - 2 \cos x) D_x(\sin x) - \sin x \cdot D_x(1 - 2 \cos x)}{(1 - 2 \cos x)^2}$$

$$= \frac{(1 - 2 \cos x)(\cos x) - \sin x(2 \sin x)}{(1 - 2 \cos x)^2}$$

$$= \frac{\cos x - 2(\cos^2 x + \sin^2 x)}{(1 - 2 \cos x)^2}$$

$$= \frac{\cos x - 2}{(1 - 2 \cos x)^2}$$

The derivatives of the tangent, cotangent, secant, and cosecant functions are obtained from trigonometric identities involving the sine and cosine as well as the derivatives of the sine and cosine and theorems on differentiation. For the derivative of the tangent we apply the identities

$$\tan x = \frac{\sin x}{\cos x} \qquad \sec x = \frac{1}{\cos x} \qquad \sin^2 x + \cos^2 x = 1$$

3.5.3 THEOREM $D_x(\tan x) = \sec^2 x$

Proof

$$D_x(\tan x) = D_x\left(\frac{\sin x}{\cos x}\right)$$

$$= \frac{\cos x \cdot D_x(\sin x) - \sin x \cdot D_x(\cos x)}{\cos^2 x}$$

$$= \frac{(\cos x)(\cos x) - (\sin x)(-\sin x)}{\cos^2 x}$$

$$= \frac{\cos^2 x + \sin^2 x}{\cos^2 x}$$

$$= \frac{1}{\cos^2 x}$$

$$= \sec^2 x \qquad \blacksquare$$

3.5.4 THEOREM $D_x(\cot x) = -\csc^2 x$

The proof of this theorem is left as an exercise (see Exercise 1). It is analogous to the proof of Theorem 3.5.3. The following identities are used:

$$\cot x = \frac{\cos x}{\sin x} \qquad \csc x = \frac{1}{\sin x}$$

3.5.5 THEOREM $D_x(\sec x) = \sec x \tan x$

Proof

$$D_x(\sec x) = D_x\left(\frac{1}{\cos x}\right)$$

$$= \frac{\cos x \cdot D_x(1) - 1 \cdot D_x(\cos x)}{\cos^2 x}$$

$$= \frac{\cos x \cdot 0 - 1 \cdot (-\sin x)}{\cos^2 x}$$

$$= \frac{\sin x}{\cos^2 x}$$

$$= \frac{1}{\cos x} \cdot \frac{\sin x}{\cos x}$$

$$= \sec x \tan x \qquad \blacksquare$$

EXAMPLE 3   Compute

$$\frac{d}{dx}(\tan x \sec x)$$

Solution

$$\frac{d}{dx}(\tan x \sec x) = \tan x \cdot \frac{d}{dx}(\sec x) + \frac{d}{dx}(\tan x) \cdot \sec x$$

$$= \tan x(\sec x \tan x) + \sec^2 x(\sec x)$$

$$= \sec x \tan^2 x + \sec^3 x$$

3.5.6 THEOREM $D_x(\csc x) = -\csc x \cot x$

The proof of this theorem is also left as an exercise (see Exercise 2).

In a trigonometry course the graphs of the functions are sketched by intuitive considerations. Now we can obtain these graphs in a formal manner by using the derivatives of the trigonometric functions. We first discuss the graphs of the sine and cosine. For each of these functions the domain is the set of all real numbers, and the range is $[-1, 1]$. Let

$$f(x) = \sin x \qquad f'(x) = \cos x$$

To determine where the graph has a horizontal tangent, set $f'(x) = 0$ and get $x = \frac{1}{2}\pi + k\pi$, where k is any integer. At these values of x, $\sin x$ is either $+1$ or -1, and these are the largest and smallest values that $\sin x$ assumes. The graph intersects the x axis at the points where $\sin x = 0$, that is, at the points where $x = k\pi$ and k is any integer. Furthermore, when k is an even integer, $f'(k\pi) = 1$, and when k is an odd integer, $f'(k\pi) = -1$. Thus at the points of intersection of the graph with the x axis the slope of the tangent line is either 1 or -1. From this information we draw the sketch of the graph of the sine function shown in Figure 1.

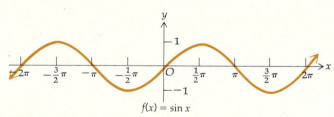

FIGURE 1

For the graph of the cosine function we use the identity

$$\cos x = \sin(x + \tfrac{1}{2}\pi)$$

Thus the graph of the cosine is obtained from the graph of the sine by translating the y axis $\frac{1}{2}\pi$ units to the right. See Figure 2.

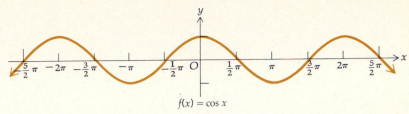

$$f(x) = \cos x$$

FIGURE 2

EXAMPLE 4   Find an equation of the tangent line to the graph of the cosine function at the point $(\frac{3}{2}\pi, 0)$.

Solution   If $f(x) = \cos x$, $f'(x) = -\sin x$. Thus $f'(\frac{3}{2}\pi) = -\sin \frac{3}{2}\pi$. Because $\sin \frac{3}{2}\pi = -1$, $f'(\frac{3}{2}\pi) = 1$. From the point-slope form of an equation of the tangent line having slope 1 and containing the point $(\frac{3}{2}\pi, 0)$, we have

$$y - 0 = 1(x - \tfrac{3}{2}\pi)$$

$$y = x - \tfrac{3}{2}\pi$$

We now consider the graph of the tangent function. Because

$$\tan(-x) = -\tan x$$

the graph is symmetric with respect to the origin. Furthermore,

$$\tan(x + \pi) = \tan x$$

and the tangent function is periodic with period π. The tangent function is continuous at all numbers in its domain, which is the set of all real numbers except those of the form $\frac{1}{2}\pi + k\pi$, where k is any integer. The range is the set of all real numbers. If k is any integer, $\tan k\pi = 0$. Therefore the graph intersects the x axis at the points $(k\pi, 0)$. Let

$$f(x) = \tan x \qquad f'(x) = \sec^2 x$$

Because $f'(k\pi) = \sec^2 k\pi$ and $\sec^2 k\pi = 1$ for k any integer, it follows that where the graph intersects the x axis, the slope of the tangent line is 1. Setting $f'(x) = 0$ gives $\sec^2 x = 0$. Because $\sec^2 x \geq 1$ for all x, we conclude that there are no horizontal tangent lines.

Consider the interval $[0, \frac{1}{2}\pi)$ on which the tangent function is defined everywhere.

$$\lim_{x \to \pi/2^-} \tan x = \lim_{x \to \pi/2^-} \frac{\sin x}{\cos x}$$

Because $\lim\limits_{x \to \pi/2^-} \sin x = 1$ and $\lim\limits_{x \to \pi/2^-} \cos x = 0$, where $\cos x$ is approaching zero through positive values,

$$\lim_{x \to \pi/2^-} \tan x = +\infty$$

Therefore the line $x = \frac{1}{2}\pi$ is a vertical asymptote of the graph. In Table 1 there are some values of x in the interval $[0, \frac{1}{2}\pi)$ and the corresponding values of $\tan x$. By plotting the points having as coordinates the number pairs $(x, \tan x)$ we get the portion of the graph for x in $[0, \frac{1}{2}\pi)$. Because of symmetry with

Table 1

x	$\tan x$
0	0
$\frac{1}{6}\pi$	$\dfrac{1}{\sqrt{3}} \approx 0.58$
$\frac{1}{4}\pi$	1
$\frac{1}{3}\pi$	$\sqrt{3} \approx 1.73$

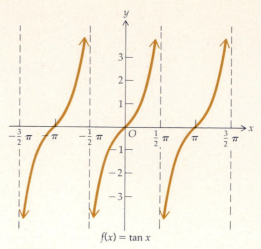

FIGURE 3

respect to the origin, the portion of the graph for x in $(-\frac{1}{2}\pi, 0]$ is obtained. Since the period is π, we complete the sketch of the graph shown in Figure 3.

We can get the graph of the cotangent function from that of the tangent function by using the identity

$$\cot x = -\tan(x + \tfrac{1}{2}\pi)$$

From this identity it follows that the graph of the cotangent is obtained from the graph of the tangent by translating the y axis $\frac{1}{2}\pi$ units to the right and then taking a reflection of the graph with respect to the x axis. A sketch of the graph of the cotangent function appears in Figure 4.

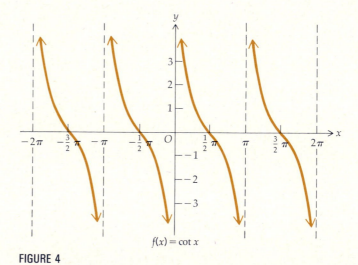

FIGURE 4

Because

$$\sec(x + 2\pi) = \sec x$$

the secant function is periodic with period 2π. The domain of the secant function is the set of all real numbers except those of the form $\frac{1}{2}\pi + k\pi$, where k is any integer. The range is $(-\infty, -1] \cup [1, +\infty)$. The function is continuous at all numbers in its domain. There is no intersection of the graph with the x axis because sec x is never zero.

We use the derivative to determine if the graph has any horizontal tangent lines. Let

$$f(x) = \sec x \qquad f'(x) = \sec x \tan x$$

Setting $f'(x) = 0$ gives sec $x \tan x = 0$. Because sec $x \neq 0$, $f'(x) = 0$ when $\tan x = 0$, which is when $x = k\pi$, where k is any integer.

We first consider the graph for x in $(-\frac{1}{2}\pi, \frac{1}{2}\pi) \cup (\frac{1}{2}\pi, \frac{3}{2}\pi)$. There are horizontal tangent lines at $x = 0$ and $x = \pi$.

$$\lim_{x \to -\pi/2^+} \sec x = \lim_{x \to -\pi/2^+} \frac{1}{\cos x} \qquad\qquad \lim_{x \to \pi/2^-} \sec x = \lim_{x \to \pi/2^-} \frac{1}{\cos x}$$

$$= +\infty \qquad\qquad\qquad\qquad = +\infty$$

$$\lim_{x \to \pi/2^+} \sec x = \lim_{x \to \pi/2^+} \frac{1}{\cos x} \qquad\qquad \lim_{x \to 3\pi/2^-} \sec x = \lim_{x \to 3\pi/2^-} \frac{1}{\cos x}$$

$$= -\infty \qquad\qquad\qquad\qquad = -\infty$$

Therefore the lines $x = -\frac{1}{2}\pi$, $x = \frac{1}{2}\pi$, and $x = \frac{3}{2}\pi$ are vertical asymptotes of the graph.

With the above information and plotting a few points we get a sketch of the graph of the secant function for x in $(-\frac{1}{2}\pi, \frac{1}{2}\pi) \cup (\frac{1}{2}\pi, \frac{3}{2}\pi)$. Because the period is 2π, the sketch of the graph shown in Figure 5 is obtained.

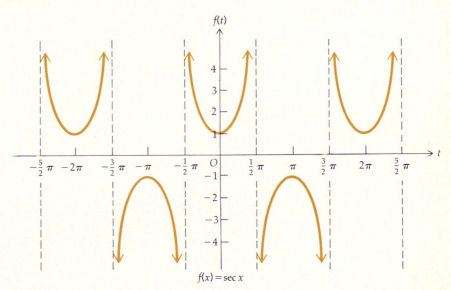

$$f(x) = \sec x$$

FIGURE 5

From the identity

$$\csc x = \sec(x - \tfrac{1}{2}\pi)$$

we get the graph of the cosecant function from that of the secant by translating the y axis $\frac{1}{2}\pi$ units to the left. A sketch of the graph of the cosecant function appears in Figure 6.

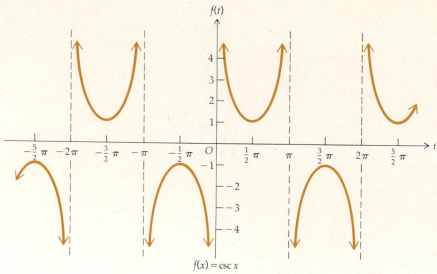

FIGURE 6

$f(x) = \csc x$

EXERCISES 3.5

1. Prove: $D_x(\cot x) = -\csc^2 x$.
2. Prove: $D_x(\csc x) = -\csc x \cot x$.

In Exercises 3 through 16, find the derivative of the function.

3. $f(x) = 3 \sin x$
4. $g(x) = \sin x + \cos x$
5. $g(x) = \tan x + \cot x$
6. $f(x) = 4 \sec x - 2 \csc x$
7. $f(t) = 2t \cos t$
8. $f(x) = 4x^2 \cos x$
9. $g(x) = x \sin x + \cos x$
10. $g(y) = 3 \sin y - y \cos y$
11. $h(x) = 4 \sin x \cos x$
12. $f(x) = x^2 \sin x + 2x \cos x$
13. $f(x) = x^2 \cos x - 2x \sin x - 2 \cos x$
14. $h(y) = y^3 - y^2 \cos y + 2y \sin y + 2 \cos y$
15. $f(x) = 3 \sec x \tan x$
16. $f(t) = \sin t \tan t$

In Exercises 17 through 30, compute the derivative.

17. $D_y(\cot y \csc y)$
18. $D_x(\cos x \cot x)$

19. $D_z\left(\dfrac{2 \cos z}{z + 1}\right)$
20. $D_t\left(\dfrac{\sin t}{t}\right)$

21. $\dfrac{d}{dx}\left(\dfrac{\sin x}{1 - \cos x}\right)$
22. $\dfrac{d}{dx}\left(\dfrac{x + 4}{\cos x}\right)$

23. $\dfrac{d}{dt}\left(\dfrac{\tan t}{\cos t - 4}\right)$
24. $\dfrac{d}{dy}\left(\dfrac{\cot y}{1 - \sin y}\right)$

25. $\dfrac{d}{dy}\left(\dfrac{1 + \sin y}{1 - \sin y}\right)$
26. $\dfrac{d}{dx}\left(\dfrac{\sin x - 1}{\cos x + 1}\right)$

27. $D_x[(x - \sin x)(x + \cos x)]$
28. $D_z[(z^2 + \cos z)(2z - \sin z)]$

29. $D_t\left(\dfrac{2 \csc t - 1}{\csc t + 2}\right)$
30. $D_y\left(\dfrac{\tan y + 1}{\tan y - 1}\right)$

In Exercises 31 through 42, find $f'(a)$.

31. $f(x) = x \cos x; \ a = 0$
32. $f(x) = x \sin x; \ a = \frac{3}{2}\pi$

33. $f(x) = \dfrac{\cos x}{x}; \ a = \frac{1}{2}\pi$
34. $f(x) = \dfrac{\sec x}{x^2}; \ a = \pi$

35. $f(x) = x^2 \tan x; \ a = \pi$
36. $f(x) = x^2 \cos x - \sin x; \ a = 0$
37. $f(x) = \sin x(\cos x - 1); \ a = \pi$
38. $f(x) = (\cos x + 1)(x \sin x - 1); \ a = \frac{1}{2}\pi$
39. $f(x) = x \cos x + x \sin x; \ a = \frac{1}{4}\pi$
40. $f(x) = \tan x + \sec x; \ a = \frac{1}{6}\pi$
41. $f(x) = 2 \cot x - \csc x; \ a = \frac{2}{3}\pi$

42. $f(x) = \dfrac{1}{\cot x - 1}; \ a = \frac{3}{4}\pi$

43. (a) Use a calculator to tabulate to four decimal places values of $\dfrac{\sin(\frac{1}{3}\pi + h) - \sin \frac{1}{3}\pi}{h}$ when h is 1, 0.5, 0.1, 0.01, 0.001 and h is -1, -0.5, -0.1, -0.01, -0.001. What does the quotient appear to be approaching as h approaches 0? (b) Find $\displaystyle\lim_{h \to 0} \dfrac{\sin(\frac{1}{3}\pi + h) - \sin \frac{1}{3}\pi}{h}$ by interpreting it as a derivative.

44. (a) Use a calculator to tabulate to four decimal places values of $\dfrac{\cos(\frac{5}{6}\pi + h) - \cos \frac{5}{6}\pi}{h}$ when h is 1, 0.5, 0.1, 0.01, 0.001 and h is -1, -0.5, -0.1, -0.01, -0.001. What does the quotient appear to be approaching as h approaches 0? (b) Find $\displaystyle\lim_{h \to 0} \dfrac{\cos(\frac{5}{6}\pi + h) - \cos \frac{5}{6}\pi}{h}$ by interpreting it as a derivative.

45. (a) Use a calculator to tabulate to four decimal places values of $\dfrac{\tan(\frac{1}{4}\pi + h) - \tan\frac{1}{4}\pi}{h}$ when h is 0.1, 0.01, 0.001, 0.0001, 0.00001 and h is $-0.1, -0.01, -0.001, -0.0001, -0.00001$. What does the quotient appear to be approaching as h approaches 0? (b) Find $\lim\limits_{h \to 0} \dfrac{\tan(\frac{1}{4}\pi + h) - \tan\frac{1}{4}\pi}{h}$ by interpreting it as a derivative.

46. (a) Use a calculator to tabulate to four decimal places values of $\dfrac{\sec(\frac{1}{6}\pi + h) - \sec\frac{1}{6}\pi}{h}$ when h is 0.1, 0.01, 0.001, 0.0001, 0.00001 and h is $-0.1, -0.01, -0.001, -0.0001, -0.00001$. What does the quotient appear to be approaching as h approaches 0? (b) Find $\lim\limits_{h \to 0} \dfrac{\sec(\frac{1}{6}\pi + h) - \sec\frac{1}{6}\pi}{h}$ by interpreting it as a derivative.

47. (a) Use a calculator to tabulate to four decimal places values of $\dfrac{\cos x - \cos\frac{1}{6}\pi}{x - \frac{1}{6}\pi}$ when x is $\frac{3}{20}\pi, \frac{19}{120}\pi, \frac{33}{200}\pi, \frac{199}{1200}\pi, \frac{333}{2000}\pi$ and x is $\frac{11}{60}\pi, \frac{7}{40}\pi, \frac{101}{600}\pi, \frac{67}{400}\pi, \frac{1001}{6000}\pi$. What does the quotient appear to be approaching as x approaches $\frac{1}{6}\pi$? (b) Find $\lim\limits_{x \to \pi/6} \dfrac{\cos x - \cos\frac{1}{6}\pi}{x - \frac{1}{6}\pi}$ by interpreting it as a derivative.

48. (a) Use a calculator to tabulate to four decimal places values of $\dfrac{\sin x - \sin\frac{1}{3}\pi}{x - \frac{1}{3}\pi}$ when x is $\frac{3}{10}\pi, \frac{19}{60}\pi, \frac{33}{100}\pi, \frac{199}{600}\pi, \frac{333}{1000}\pi$ and x is $\frac{11}{30}\pi, \frac{7}{20}\pi, \frac{101}{300}\pi, \frac{67}{200}\pi, \frac{1001}{3000}\pi$. What does the quotient appear to be approaching as x approaches $\frac{1}{3}\pi$? (b) Find $\lim\limits_{x \to \pi/3} \dfrac{\sin x - \sin\frac{1}{3}\pi}{x - \frac{1}{3}\pi}$ by interpreting it as a derivative.

49. (a) Use a calculator to tabulate to four decimal places values of $\dfrac{\csc x - \csc\frac{2}{3}\pi}{x - \frac{2}{3}\pi}$ when x is $\frac{3}{5}\pi, \frac{19}{30}\pi, \frac{33}{50}\pi, \frac{199}{300}\pi, \frac{333}{500}\pi$ and x is $\frac{11}{15}\pi, \frac{7}{10}\pi, \frac{101}{150}\pi, \frac{67}{100}\pi, \frac{1001}{1500}\pi$. What does the quotient appear to be approaching as x approaches $\frac{2}{3}\pi$? (b) Find $\lim\limits_{x \to 2\pi/3} \dfrac{\csc x - \csc\frac{2}{3}\pi}{x - \frac{2}{3}\pi}$ by interpreting it as a derivative.

50. (a) Use a calculator to tabulate to four decimal places values of $\dfrac{\cot x - \cot\frac{3}{4}\pi}{x - \frac{3}{4}\pi}$ when x is $\frac{29}{40}\pi, \frac{59}{80}\pi, \frac{299}{400}\pi, \frac{599}{800}\pi, \frac{2999}{4000}\pi$ and x is $\frac{31}{40}\pi, \frac{61}{80}\pi, \frac{301}{400}\pi, \frac{601}{800}\pi, \frac{3001}{4000}\pi$. What does the quotient appear to be approaching as x approaches $\frac{3}{4}\pi$? (b) Find $\lim\limits_{x \to 3\pi/4} \dfrac{\cot x - \cot\frac{3}{4}\pi}{x - \frac{3}{4}\pi}$ by interpreting it as a derivative.

51. Find an equation of the tangent line to the graph of the sine function at the point where (a) $x = 0$; (b) $x = \frac{1}{3}\pi$; (c) $x = \pi$.

52. Find an equation of the tangent line to the graph of the cosine function at the point where (a) $x = \frac{1}{2}\pi$; (b) $x = -\frac{1}{2}\pi$; (c) $x = \frac{1}{6}\pi$.

53. Find an equation of the tangent line to the graph of the tangent function at the point where (a) $x = 0$; (b) $x = \frac{1}{4}\pi$; (c) $x = -\frac{1}{4}\pi$.

54. Find an equation of the tangent line to the graph of the secant function at the point where (a) $x = \frac{1}{4}\pi$; (b) $x = -\frac{1}{4}\pi$; (c) $x = \frac{3}{4}\pi$.

In Exercises 55 through 58, a particle is moving along a straight line according to the equation, where s centimeters is the directed distance of the particle from the origin at t seconds. (a) What is the instantaneous velocity of the particle at t seconds? (b) Find the instantaneous velocity of the particle at t_1 seconds for each value of t_1.

55. $s = 4 \sin t$; t_1 is $0, \frac{1}{3}\pi, \frac{1}{2}\pi, \frac{2}{3}\pi$, and π

56. $s = 6 \cos t$; t_1 is $0, \frac{1}{6}\pi, \frac{1}{2}\pi, \frac{5}{6}\pi$, and π

57. $s = -3 \cos t$; t_1 is $0, \frac{1}{6}\pi, \frac{1}{3}\pi, \frac{1}{2}\pi, \frac{2}{3}\pi, \frac{5}{6}\pi$, and π

58. $s = -\frac{1}{2} \sin t$; t_1 is $0, \frac{1}{6}\pi, \frac{1}{3}\pi, \frac{1}{2}\pi, \frac{2}{3}\pi, \frac{5}{6}\pi$, and π

59. If a body of weight W pounds is dragged along a horizontal floor at constant velocity by means of a force of magnitude F pounds and directed at an angle of θ radians with the plane of the floor, then F is given by the equation

$$F = \frac{kW}{k \sin \theta + \cos \theta}$$

where k is a constant called the coefficient of friction. If $k = 0.5$, find the instantaneous rate of change of F with respect to θ when (a) $\theta = \frac{1}{4}\pi$; (b) $\theta = \frac{1}{2}\pi$.

60. A projectile is shot from a gun at an angle of elevation having radian measure $\frac{1}{2}\alpha$ and an initial velocity of v_0 feet per second. If R feet is the range of the projectile, then

$$R = \frac{v_0{}^2}{g} \sin \alpha \qquad 0 \le \alpha \le \pi$$

where g ft/sec^2 is the acceleration due to gravity. (a) If $v_0 = 480$, find the rate of change of R with respect to α when $\alpha = \frac{1}{2}\pi$ (i.e., the angle of elevation has radian measure $\frac{1}{4}\pi$). Take $g = 32$. (b) Find the values of α for which $D_\alpha R > 0$.

3.6   THE DERIVATIVE OF A COMPOSITE FUNCTION AND THE CHAIN RULE

To compute the derivative of a composite function we use one of the important theorems in calculus called the *chain rule*. Before stating this theorem we give three illustrations showing how previous theorems can be used to determine the derivatives of some particular composite functions. In each illustration the final expression for the derivative is written in a form that will seem unusual to you. We are writing it in that form so we can associate it with the chain rule.

▶ **ILLUSTRATION 1**   If

$$F(x) = (4x^2 + 1)^3$$

we can obtain $F'(x)$ by twice applying Theorem 3.3.6 (the derivative of a product). The computation is as follows:

$$F(x) = (4x^2 + 1)^2(4x^2 + 1)$$

$$F'(x) = (4x^2 + 1)^2 \cdot D_x(4x^2 + 1) + (4x^2 + 1) \cdot D_x[(4x^2 + 1)(4x^2 + 1)]$$
$$= (4x^2 + 1)^2(8x) + (4x^2 + 1)[(4x^2 + 1)(8x) + (4x^2 + 1)(8x)]$$
$$= (4x^2 + 1)^2(8x) + (4x^2 + 1)[2(4x^2 + 1)(8x)]$$
$$= (4x^2 + 1)^2(8x) + 2[(4x^2 + 1)^2(8x)]$$

Thus

$$F'(x) = [3(4x^2 + 1)^2](8x) \qquad (1)$$

Observe that F is the composite function $f \circ g$, where $f(x) = x^3$ and $g(x) = 4x^2 + 1$; that is,

$$F(x) = f(g(x))$$
$$= f(4x^2 + 1)$$
$$= (4x^2 + 1)^3$$

Because $f'(x) = 3x^2$ and $g'(x) = 8x$, we have from (1)

$$F'(x) = f'(g(x))g'(x) \qquad (2)$$

◀

▶ **ILLUSTRATION 2**   If

$$G(x) = \sin 2x$$

then to find $G'(x)$ we can use the trigonometric identities

$$\sin 2x = 2 \sin x \cos x \qquad \cos 2x = \cos^2 x - \sin^2 x$$

and Theorem 3.3.6. We have

$$G(x) = 2 \sin x \cos x$$

$$G'(x) = (2 \sin x)D_x(\cos x) + (2 \cos x)D_x(\sin x)$$
$$= (2 \sin x)(-\sin x) + (2 \cos x)(\cos x)$$
$$= 2(\cos^2 x - \sin^2 x)$$

Therefore

$$G'(x) = (\cos 2x)(2) \qquad (3)$$

If we let $f(x) = \sin x$ and $g(x) = 2x$, then G is the composite function $f \circ g$; that is,

$$G(x) = f(g(x))$$
$$= f(2x)$$
$$= \sin 2x$$

Because $f'(x) = \cos x$ and $g'(x) = 2$, we can write (3) in the form

$$G'(x) = f'(g(x))g'(x) \qquad (4)$$

◀

▶ **ILLUSTRATION 3** If

$$H(x) = (\cos x)^{-1}$$

we can compute $H'(x)$ by first using the identity $(\cos x)^{-1} = \sec x$.

$$H(x) = \sec x$$

$$H'(x) = \sec x \tan x$$

$$= \frac{1}{\cos x} \cdot \frac{\sin x}{\cos x}$$

$$= (-1) \frac{1}{\cos^2 x} (-\sin x)$$

Hence

$$H'(x) = [-1(\cos x)^{-2}](-\sin x) \tag{5}$$

◀

With $f(x) = x^{-1}$ and $g(x) = \cos x$, H is the composite function $f \circ g$; that is,

$$H(x) = f(g(x))$$
$$= f(\cos x)$$
$$= (\cos x)^{-1}$$

Since $f'(x) = -1 \cdot x^{-2}$ and $g'(x) = -\sin x$, we can write (5) in the form

$$H'(x) = f'(g(x))g'(x) \tag{6}$$

Observe that the right-hand sides of (2), (4), and (6) are all $f'(g(x))g'(x)$, which is the right-hand side of the chain rule, stated as the following theorem.

3.6.1 THEOREM If the function g is differentiable at x and the function f is differentiable at $g(x)$,
The Chain Rule then the composite function $f \circ g$ is differentiable at x, and

$$(f \circ g)'(x) = f'(g(x))g'(x) \tag{7}$$

The proof of the chain rule is sophisticated and lengthy. So that you can become familiar with the statement of the theorem through illustrations and examples, we defer the proof to the end of this section.

▶ **ILLUSTRATION 4** Let

$$f(x) = x^{10} \qquad g(x) = 2x^3 - 5x^2 + 4$$

Then the composite function $f \circ g$ is defined by

$$(f \circ g)(x) = f(g(x))$$
$$= (2x^3 - 5x^2 + 4)^{10}$$

To apply (7) we need to compute $f'(g(x))$ and $g'(x)$. Because $f(x) = x^{10}$, $f'(x) = 10x^9$; thus

$$f'(g(x)) = 10[g(x)]^9$$
$$f'(g(x)) = 10(2x^3 - 5x^2 + 4)^9 \tag{8}$$

Furthermore, because $g(x) = 2x^3 - 5x^2 + 4$,

$$g'(x) = 6x^2 - 10x \tag{9}$$

Therefore, from (7), (8), and (9) we have

$$(f \circ g)'(x) = f'(g(x))g'(x)$$
$$= 10(2x^3 - 5x^2 + 4)^9(6x^2 - 10x) \qquad \blacktriangleleft$$

▶ **ILLUSTRATION 5**   Let

$$f(x) = \sin x \qquad g(x) = x^2 + 3$$

Then the composite function $f \circ g$ is defined by

$$(f \circ g)(x) = f(g(x))$$
$$= \sin(x^2 + 3)$$

We compute $f'(g(x))$ and $g'(x)$. Because $f(x) = \sin x$, $f'(x) = \cos x$. Hence

$$f'(g(x)) = \cos[g(x)]$$
$$f'(g(x)) = \cos(x^2 + 3) \tag{10}$$

Because $g(x) = x^2 + 3$,

$$g'(x) = 2x \tag{11}$$

Thus from (7), (10), and (11) we obtain

$$(f \circ g)'(x) = f'(g(x))g'(x)$$
$$= [\cos(x^2 + 3)](2x)$$
$$= 2x \cos(x^2 + 3) \qquad \blacktriangleleft$$

▶ **ILLUSTRATION 6**   Suppose

$$h(x) = \left(\frac{2}{x-1}\right)^5$$

To determine $h'(x)$, let

$$f(x) = x^5 \qquad g(x) = \frac{2}{x-1}$$

Then

$$f'(x) = 5x^4 \qquad g'(x) = \frac{-2}{(x-1)^2}$$

Because $h(x) = f(g(x))$, we have from the chain rule

$$h'(x) = f'(g(x)) \cdot g'(x)$$
$$= 5\left(\frac{2}{x-1}\right)^4 \cdot \frac{-2}{(x-1)^2}$$
$$= \frac{-160}{(x-1)^6} \qquad \blacktriangleleft$$

When computing derivatives by the chain rule we don't actually write the functions f and g as we did in Illustrations 4, 5, and 6, but we bear them in mind. For instance, we could write the computation in Illustration 6 as

$$h(x) = \left(\frac{2}{x-1}\right)^5$$

$$h'(x) = 5\left(\frac{2}{x-1}\right)^4 \cdot D_x\left(\frac{2}{x-1}\right)$$

$$= 5\left(\frac{2}{x-1}\right)^4 \cdot \frac{-2}{(x-1)^2}$$

$$= \frac{-160}{(x-1)^6}$$

EXAMPLE 1   Find $f'(x)$ by the chain rule if

$$f(x) = \frac{1}{4x^3 + 5x^2 - 7x + 8}$$

Solution   We write $f(x) = (4x^3 + 5x^2 - 7x + 8)^{-1}$ and apply the chain rule to obtain

$$f'(x) = -1(4x^3 + 5x^2 - 7x + 8)^{-2} \cdot D_x(4x^3 + 5x^2 - 7x + 8)$$
$$= -1(4x^3 + 5x^2 - 7x + 8)^{-2}(12x^2 + 10x - 7)$$

$$= \frac{-12x^2 - 10x + 7}{(4x^3 + 5x^2 - 7x + 8)^2}$$

EXAMPLE 2   Compute

$$\frac{d}{dx}\left[\left(\frac{2x+1}{3x-1}\right)^4\right]$$

Solution   From the chain rule,

$$\frac{d}{dx}\left[\left(\frac{2x+1}{3x-1}\right)^4\right] = 4\left(\frac{2x+1}{3x-1}\right)^3 \cdot \frac{d}{dx}\left(\frac{2x+1}{3x-1}\right)$$

$$= 4\left(\frac{2x+1}{3x-1}\right)^3 \frac{(3x-1)(2) - (2x+1)(3)}{(3x-1)^2}$$

$$= \frac{4(2x+1)^3(-5)}{(3x-1)^5}$$

$$= -\frac{20(2x+1)^3}{(3x-1)^5}$$

If the Leibniz notation is used for the derivative, the chain rule can be stated as follows:

If y is a function of u, defined by $y = f(u)$ and $\dfrac{dy}{du}$ exists, and if u is a function

of x, defined by $u = g(x)$ and $\dfrac{du}{dx}$ exists, then y is a function of x and $\dfrac{dy}{dx}$ exists and is given by

$$\frac{dy}{dx} = \frac{dy}{du} \cdot \frac{du}{dx} \tag{12}$$

Observe from this equation the convenient form for remembering the chain rule. The formal statement suggests a symbolic "division" of du in the numerator and denominator of the right-hand side. However, remember from Section 3.1 that when we introduced the Leibniz notation $\frac{dy}{dx}$, it was emphasized that neither dy nor dx has been given independent meaning. Therefore you should consider (12) as an equation involving formal differentiation notation.

Another way of writing the chain rule is to let $u = g(x)$. Then

$$(f \circ g)(x) = f(u) \qquad (f \circ g)'(x) = D_x f(u) \qquad f'(g(x)) = f'(u) \qquad g'(x) = D_x u$$

With these substitutions (7) becomes

$$D_x[f(u)] = f'(u)D_x u$$

We shall use this form of the chain rule to state important differentiation formulas. In particular, we have from Theorems 3.5.1–3.5.6, the following formulas involving the derivatives of the trigonometric functions: If u is a differentiable function of x

$$D_x(\sin u) = \cos u \, D_x u \qquad\qquad D_x(\cos u) = -\sin u \, D_x u$$

$$D_x(\tan u) = \sec^2 u \, D_x u \qquad\qquad D_x(\cot u) = -\csc^2 u \, D_x u$$

$$D_x(\sec u) = \sec u \tan u \, D_x u \qquad D_x(\csc u) = -\csc u \cot u \, D_x u$$

EXAMPLE 3   Find $F'(t)$ if

$$F(t) = \tan(3t^2 + 2t)$$

Solution   We use the chain rule and obtain

$$F'(t) = \sec^2(3t^2 + 2t) \cdot D_t(3t^2 + 2t)$$
$$= \sec^2(3t^2 + 2t) \cdot (6t + 2)$$
$$= 2(3t + 1)\sec^2(3t^2 + 2t)$$

EXAMPLE 4   Find $\dfrac{dy}{dx}$ if

$$y = \sin(\cos x)$$

Solution   We apply the chain rule.

$$\frac{dy}{dx} = \cos(\cos x)[D_x(\cos x)]$$

$$= \cos(\cos x)[-\sin x]$$
$$= -\sin x \, [\cos(\cos x)]$$

EXAMPLE 5   Find $f'(x)$ if

$$f(x) = (3x^2 + 2)^2(x^2 - 5x)^3$$

Solution   We consider f as the product of the two functions g and h, where

$$g(x) = (3x^2 + 2)^2 \qquad h(x) = (x^2 - 5x)^3$$

From Theorem 3.3.6 for the derivative of the product of two functions,

$$f'(x) = g(x)h'(x) + h(x)g'(x)$$

We find $h'(x)$ and $g'(x)$ by the chain rule.

$$\begin{aligned}
f'(x) &= (3x^2 + 2)^2[3(x^2 - 5x)^2(2x - 5)] + (x^2 - 5x)^3[2(3x^2 + 2)(6x)] \\
&= 3(3x^2 + 2)(x^2 - 5x)^2[(3x^2 + 2)(2x - 5) + 4x(x^2 - 5x)] \\
&= 3(3x^2 + 2)(x^2 - 5x)^2[6x^3 - 15x^2 + 4x - 10 + 4x^3 - 20x^2] \\
&= 3(3x^2 + 2)(x^2 - 5x)^2(10x^3 - 35x^2 + 4x - 10)
\end{aligned}$$

EXAMPLE 6   Compute

$$D_x(\sec^4 2x^2)$$

Solution   We use the chain rule twice.

$$\begin{aligned}
D_x(\sec^4 2x^2) &= 4\sec^3 2x^2[D_x(\sec 2x^2)] \\
&= 4\sec^3 2x^2[(\sec 2x^2 \tan 2x^2)\, D_x(2x^2)] \\
&= (4\sec^4 2x^2 \tan 2x^2)(4x) \\
&= 16x \sec^4 2x^2 \tan 2x^2
\end{aligned}$$

We now prove the chain rule. An important part of the proof consists of introducing the new function F that has useful properties. This device of "making up" a function is a common one for mathematicians.

Proof of the Chain Rule   Let x_1 be any number in the domain of g such that g is differentiable at x_1 and f is differentiable at $g(x_1)$. Form the function F defined by

$$F(t) = \begin{cases} \dfrac{f(t) - f(g(x_1))}{t - g(x_1)} & \text{if } t \neq g(x_1) \\ f'(g(x_1)) & \text{if } t = g(x_1) \end{cases} \qquad (13)$$

Then

$$\lim_{t \to g(x_1)} F(t) = \lim_{t \to g(x_1)} \frac{f(t) - f(g(x_1))}{t - g(x_1)}$$

From (7) in Section 3.1, the function on the right-hand side of this equation is $f'(g(x_1))$. Therefore

$$\lim_{t \to g(x_1)} F(t) = f'(g(x_1)) \qquad (14)$$

But from (13)

$$f'(g(x_1)) = F(g(x_1))$$

Substituting from this equation into (14) we get

$$\lim_{t \to g(x_1)} F(t) = F(g(x_1))$$

Therefore F is continuous at $g(x_1)$. Furthermore, from (13),

$$F(t) = \frac{f(t) - f(g(x_1))}{t - g(x_1)} \qquad \text{if } t \neq g(x_1)$$

Multiplying both sides of this equation by $t - g(x_1)$ gives

$$f(t) - f(g(x_1)) = F(t)[t - g(x_1)] \qquad \text{if } t \neq g(x_1) \tag{15}$$

Observe that (15) holds even if $t = g(x_1)$ because the left-hand side is

$$f(g(x_1)) - f(g(x_1)) = 0$$

and the right-hand side is

$$F(g(x_1))[g(x_1) - g(x_1)] = 0$$

Therefore the stipulation in (15) that $t \neq g(x_1)$ is not necessary, and we write

$$f(t) - f(g(x_1)) = F(t)[t - g(x_1)] \tag{16}$$

Now let h be the composite function $f \circ g$, so that

$$h(x) = f(g(x)) \tag{17}$$

Then from (7) in Section 3.1, if the limit exists,

$$h'(x_1) = \lim_{x \to x_1} \frac{h(x) - h(x_1)}{x - x_1}$$

Substituting from (17) in the right-hand side of this equation we get

$$h'(x_1) = \lim_{x \to x_1} \frac{f(g(x)) - f(g(x_1))}{x - x_1} \tag{18}$$

if the limit exists. Now by letting $t = g(x)$ in (16) it follows that for every x in the domain of g such that $g(x)$ is in the domain of f,

$$f(g(x)) - f(g(x_1)) = F(g(x))[g(x) - g(x_1)]$$

Substituting from this equation into (18) we have

$$h'(x_1) = \lim_{x \to x_1} \frac{F(g(x))[g(x) - g(x_1)]}{x - x_1}$$

Thus if the limits exist,

$$h'(x_1) = \lim_{x \to x_1} F(g(x)) \cdot \lim_{x \to x_1} \frac{g(x) - g(x_1)}{x - x_1} \tag{19}$$

Because F is continuous at $g(x_1)$,

$$\lim_{x \to x_1} F(g(x)) = F(g(x_1)) \tag{20}$$

But from (13),

$$F(g(x_1)) = f'(g(x_1))$$

Substituting from this equation into (20) we get

$$\lim_{x \to x_1} F(g(x)) = f'(g(x_1)) \tag{21}$$

Furthermore, because g is differentiable at x_1,

$$\lim_{x \to x_1} \frac{g(x) - g(x_1)}{x - x_1} = g'(x_1)$$

Substituting from (21) and this equation into (19) and replacing $h'(x_1)$ by $(f \circ g)'(x_1)$ we have

$$(f \circ g)'(x_1) = f'(g(x_1)) \cdot g'(x_1)$$

which is (7) with x replaced by x_1. Thus we have proved the chain rule. ∎

EXERCISES 3.6

In Exercises 1 through 12, find the derivative of the function.

1. $f(x) = (2x + 1)^3$ **2.** $f(x) = (10 - 5x)^4$
3. $F(x) = (x^2 + 4x - 5)^4$ **4.** $g(r) = (2r^4 + 8r^2 + 1)^5$
5. $f(t) = (2t^4 - 7t^3 + 2t - 1)^2$ **6.** $H(z) = (z^3 - 3z^2 + 1)^{-3}$
7. $f(x) = (x^2 + 4)^{-2}$ **8.** $g(x) = \sin x^2$
9. $f(x) = 4 \cos 3x - 3 \sin 4x$ **10.** $G(x) = \sec^2 x$
11. $h(t) = \frac{1}{3} \sec^3 2t - \sec 2t$ **12.** $f(x) = \cos(3x^2 + 1)$

In Exercises 13 through 24, compute the derivative.

13. $\dfrac{d}{dx} (\sec^2 x \tan^2 x)$ **14.** $\dfrac{d}{dt} (2 \sin^3 t \cos^2 t)$

15. $\dfrac{d}{dt} (\cot^4 t - \csc^4 t)$ **16.** $\dfrac{d}{dx} [(4x^2 + 7)^2 (2x^3 + 1)^4]$

17. $D_u[(3u^2 + 5)^3 (3u - 1)^2]$ **18.** $D_x[(x^2 - 4x^{-2})^2 (x^2 + 1)^{-1}]$
19. $D_x[(2x - 5)^{-1}(4x + 3)^{-2}]$ **20.** $D_x[(2x - 9)^2 (x^3 + 4x - 5)^3]$
21. $D_r[(r^2 + 1)^3 (2r^2 + 5r - 3)^2]$
22. $D_y[(y + 3)^3 (5y + 1)^2 (3y^2 - 4)]$

23. $\dfrac{d}{dy} \left[\left(\dfrac{y - 7}{y + 2} \right)^2 \right]$ **24.** $\dfrac{d}{dt} \left[\left(\dfrac{2t^2 + 1}{3t^3 + 1} \right)^2 \right]$

In Exercises 25 through 36, find the derivative of the function.

25. $f(x) = \left(\dfrac{2x - 1}{3x^2 + x - 2} \right)^3$ **26.** $F(x) = \dfrac{(x^2 + 3)^3}{(5x - 8)^2}$

27. $f(z) = \dfrac{(z^2 - 5)^3}{(z^2 + 4)^2}$ **28.** $G(x) = \dfrac{(4x - 1)^3 (x^2 + 2)^4}{(3x^2 + 5)^2}$

29. $g(t) = \sin^2(3t^2 - 1)$ **30.** $f(x) = \tan^2 x^2$
31. $f(x) = (\tan^2 x - x^2)^3$ **32.** $G(x) = (2 \sin x - 3 \cos x)^3$

33. $f(y) = \dfrac{3 \sin 2y}{\cos^2 2y + 1}$ **34.** $g(x) = \dfrac{\cot^2 2x}{1 + x^2}$

35. $F(x) = 4 \cos(\sin 3x)$ **36.** $f(x) = \sin^2(\cos 2x)$

37. Find an equation of the tangent line to the curve $y = (x^2 - 1)^2$ at each of the following points: $(-2, 9)$, $(-1, 0)$, $(0, 1)$, $(1, 0)$, and $(2, 9)$. Draw a sketch of the graph and segments of the tangent lines at the given points.
38. Find an equation of the tangent line to the curve $y = 4 \tan 2x$ at the point where $x = \frac{1}{8}\pi$.

In Exercises 39 through 42, a particle is moving along a straight line according to the given equation, where s meters is the directed distance of the particle from an origin at t seconds. (a) What is the instantaneous velocity of the particle at t seconds? (b) Find the instantaneous velocity of the particle at t_1 seconds for each of the given values of t_1.

39. $s = \dfrac{(t^2 - 1)^2}{(t^2 + 1)^2}$, $t \geq 0$; t_1 is $1, 2$

40. $s = \left(\dfrac{3t}{2t + 1} \right)^4$; $t \geq 0$; t_1 is $\frac{1}{2}, 1$

41. $s = 5 \sin \pi t + 3 \cos \pi t$; t_1 is $\frac{1}{4}, \frac{3}{2}$
42. $s = 2 \cos \pi(t + 1)$; t_1 is $\frac{1}{4}, \frac{3}{2}$

43. The electromotive force for an electric circuit with a simplified generator is $E(t)$ volts at t seconds, where $E(t) = 50 \sin 120\pi t$. Find the instantaneous rate of change of $E(t)$ with respect to t at (a) 0.02 sec and (b) 0.2 sec.
44. A wave produced by a simple sound has the equation $P(t) = 0.003 \sin 1800\pi t$, where $P(t)$ dynes per square centimeter is the difference between the atmospheric pressure and the air pressure at the eardrum at t seconds. Find the instantaneous rate of change of $P(t)$ with respect to t at (a) $\frac{1}{9}$ sec; (b) $\frac{1}{8}$ sec; (c) $\frac{1}{7}$ sec.
45. When a pendulum of length 10 cm has swung so that θ is the radian measure of the angle formed by the pendulum and a vertical line, then if $h(\theta)$ centimeters is the vertical height of the end of the pendulum above its lowest position, $h(\theta) = 20 \sin^2 \frac{1}{2}\theta$. Find the instantaneous rate of change of $h(\theta)$ with respect to θ when (a) $\theta = \frac{1}{3}\pi$; (b) $\theta = \frac{1}{2}\pi$.
46. If K square units is the area of a right triangle, 10 units is the length of the hypotenuse, and α is the radian measure of an

acute angle, then $K = 25 \sin 2\alpha$. Find the instantaneous rate of change of K with respect to α when (a) $\alpha = \frac{1}{6}\pi$; (b) $\alpha = \frac{1}{4}\pi$; (c) $\alpha = \pi$.

47. In a forest a predator feeds on prey, and the predator population at any time is a function of the number of prey in the forest at that time. Suppose that when there are x prey in the forest, the predator population is y, and $y = \frac{1}{6}x^2 + 90$. Furthermore, if t weeks have elapsed since the end of the hunting season, $x = 7t + 85$. At what rate is the population of the predator growing 8 weeks after the close of the hunting season? Do not express y in terms of t, but use the chain rule.

48. The demand equation for a particular toy is $p^2x = 5000$, where x toys are demanded per month when p dollars is the price per toy. It is expected that in t months, where $t \in [0, 6]$, the price of the toy will be p dollars, where $20p = t^2 + 7t + 100$. What is the anticipated rate of change of the demand with respect to time in 5 months? Do not express x in terms of t, but use the chain rule.

49. Given $f(x) = x^3$ and $g(x) = f(x^2)$. Find (a) $f'(x^2)$; (b) $g'(x)$.

50. Given $f(u) = u^2 + 5u + 5$ and $g(x) = (x + 1)/(x - 1)$. Find the derivative of $f \circ g$ in two ways: (a) by first finding $(f \circ g)(x)$ and then finding $(f \circ g)'(x)$; (b) by using the chain rule.

51. Derive the formula for the derivative of the cosine function by using the formula for the derivative of the sine function, the chain rule, and the identities

$$\cos x = \sin(\tfrac{1}{2}\pi - x) \quad \text{and} \quad \sin x = \cos(\tfrac{1}{2}\pi - x)$$

52. Use the chain rule to prove that (a) the derivative of an even function is an odd function, and (b) the derivative of an odd function is an even function, provided that these derivatives exist.

53. Use the result of Exercise 52(a) to prove that if g is an even function and $g'(x)$ exists, then if $h(x) = (f \circ g)(x)$ and f is differentiable everywhere, $h'(0) = 0$.

54. Suppose that f and g are two functions such that (i) $g'(x_1)$ and $f'(g(x_1))$ exist and (ii) for all $x \neq x_1$ in some open interval containing x_1, $g(x) - g(x_1) \neq 0$. Then

$$\frac{(f \circ g)(x) - (f \circ g)(x_1)}{x - x_1} = \frac{(f \circ g)(x) - (f \circ g)(x_1)}{g(x) - g(x_1)} \cdot \frac{g(x) - g(x_1)}{x - x_1}$$

(a) Prove that as $x \to x_1$, $g(x) \to g(x_1)$ and hence that

$$(f \circ g)'(x_1) = f'(g(x_1))g'(x_1)$$

thus simplifying the proof of the chain rule under the additional hypothesis (ii). (b) Show that the proof of the chain rule given in part (a) applies if $f(x) = x^2$ and $g(x) = x^3$, but that it does not apply if $f(x) = x^2$ and $g(x) = \operatorname{sgn} x$.

In Exercises 55 through 58, show that the simplified proof of the chain rule under the additional hypothesis (ii) of Exercise 54 is not valid for the functions f and g. In each exercise draw a sketch of the graph of g.

55. $f(x) = x^4$; $g(x) = [\![x]\!]$

56. $f(x) = x^2 + 1$; $g(x) = |x - 2| + |x + 2|$

57. $f(x) = x^2$; $g(x) = |x| + |x - 1|$

58. $f(x) = \tan x$; $g(x) = \begin{cases} -1 & \text{if } x < 0 \\ 0 & \text{if } x = 0 \\ 1 & \text{if } 0 < x \end{cases}$

59. Suppose that f and g are functions such that $f'(x) = \dfrac{1}{x}$ and $(f \circ g)(x) = x$. Prove that if $g'(x)$ exists, then $g'(x) = g(x)$.

3.7 THE DERIVATIVE OF THE POWER FUNCTION FOR RATIONAL EXPONENTS

The function defined by

$$f(x) = x^r \tag{1}$$

is called the **power function**. In Section 3.3 we obtained the following formula for the derivative of this function when r is a positive or negative integer:

$$f'(x) = rx^{r-1} \tag{2}$$

We now prove that this formula holds when r is a rational number, with certain stipulations when $x = 0$.

First consider $x \neq 0$, and $r = 1/q$, where q is a positive integer. Equation (1) then can be written

$$f(x) = x^{1/q} \tag{3}$$

From Definition 3.1.3,

$$f'(x) = \lim_{\Delta x \to 0} \frac{(x + \Delta x)^{1/q} - x^{1/q}}{\Delta x} \tag{4}$$

To evaluate the limit in (4) we must rationalize the numerator. To do this we use the following formula obtained in Section 2.9:

$$a^n - b^n = (a - b)(a^{n-1} + a^{n-2}b + a^{n-3}b^2 + \ldots + ab^{n-2} + b^{n-1}) \tag{5}$$

The numerator of the fraction in (4) is rationalized by applying (5), where $a = (x + \Delta x)^{1/q}$, $b = x^{1/q}$, and $n = q$. So we multiply the numerator and denominator by

$$[(x + \Delta x)^{1/q}]^{(q-1)} + [(x + \Delta x)^{1/q}]^{(q-2)}x^{1/q} + \ldots + (x^{1/q})^{(q-1)}$$

Then, from (4), $f'(x)$ equals

$$\lim_{\Delta x \to 0} \frac{[(x + \Delta x)^{1/q} - x^{1/q}][(x + \Delta x)^{(q-1)/q} + (x + \Delta x)^{(q-2)/q}x^{1/q} + \ldots + x^{(q-1)/q}]}{\Delta x[(x + \Delta x)^{(q-1)/q} + (x + \Delta x)^{(q-2)/q}x^{1/q} + \ldots + x^{(q-1)/q}]}$$

$$(6)$$

Now if (5) is applied to the numerator, we get $(x + \Delta x)^{q/q} - x^{q/q}$, which is Δx. So from (6),

$$f'(x) = \lim_{\Delta x \to 0} \frac{\Delta x}{\Delta x[(x + \Delta x)^{(q-1)/q} + (x + \Delta x)^{(q-2)/q}x^{1/q} + \ldots + x^{(q-1)/q}]}$$

$$= \lim_{\Delta x \to 0} \frac{1}{(x + \Delta x)^{(q-1)/q} + (x + \Delta x)^{(q-2)/q}x^{1/q} + \ldots + x^{(q-1)/q}}$$

$$= \frac{1}{x^{(q-1)/q} + x^{(q-1)/q} + \ldots + x^{(q-1)/q}}$$

Because there are exactly q terms in the denominator of the above fraction,

$$f'(x) = \frac{1}{qx^{1-(1/q)}}$$

$$f'(x) = \frac{1}{q}x^{1/q-1} \qquad (7)$$

which is formula (2) with $r = 1/q$. We have completed a crucial part of the proof. We have shown that the function defined by (3) is differentiable; furthermore, its derivative is given by (7).

Now, in (1) with $x \neq 0$, let $r = p/q$, where p is any nonzero integer and q is any positive integer; that is, r is any rational number except zero. Then (1) is written as

$$f(x) = x^{p/q} \quad \Leftrightarrow \quad f(x) = (x^{1/q})^p$$

Because p is either a positive or negative integer, it follows from the chain rule and Theorems 3.3.2 and 3.3.8 that

$$f'(x) = p(x^{1/q})^{p-1} \cdot D_x(x^{1/q})$$

Applying formula (7) for $D_x(x^{1/q})$ we get

$$f'(x) = p(x^{1/q})^{p-1} \cdot \frac{1}{q}x^{1/q-1}$$

$$f'(x) = \frac{p}{q}x^{p/q-1/q+1/q-1}$$

$$f'(x) = \frac{p}{q}x^{p/q-1}$$

This formula is the same as formula (2) with $r = p/q$.

If $r = 0$ and $x \neq 0$, (1) becomes $f(x) = x^0$; that is, $f(x) = 1$. Thus $f'(x) = 0$, which can be written as $f'(x) = 0 \cdot x^{0-1}$. Therefore (2) holds if $r = 0$ with $x \neq 0$. We have therefore shown that formula (2) holds when r is any rational number and $x \neq 0$.

Now 0 is in the domain of the power function f if and only if r is a positive number, because when $r \leq 0$, $f(0)$ is not defined. Hence we wish to determine for what positive values of r, $f'(0)$ will be given by formula (2). We must exclude the values of r for which $0 < r \leq 1$ because for those values of r, x^{r-1} is not a real number when $x = 0$. Suppose, then, that $r > 1$. By the definition of a derivative,

$$f'(0) = \lim_{x \to 0} \frac{x^r - 0^r}{x - 0}$$

$$= \lim_{x \to 0} x^{r-1}$$

When $r > 1$, $\lim\limits_{x \to 0} x^{r-1}$ exists and equals 0, provided that r is a number such that x^{r-1} is defined on some open interval containing 0. For example, if $r = \frac{3}{2}$, then $x^{r-1} = x^{1/2}$, which is not defined on any open interval containing 0 (since $x^{1/2}$ does not exist when $x < 0$). However, if $r = \frac{5}{3}$, $x^{r-1} = x^{2/3}$, which is defined on every open interval containing 0. Hence formula (2) gives the derivative of the power function when $x = 0$, provided that r is a number for which x^{r-1} is defined on some open interval containing 0. Thus we have proved the following theorem.

3.7.1 THEOREM   If f is the power function defined by $f(x) = x^r$, where r is any rational number, then f is differentiable and

$$f'(x) = rx^{r-1}$$

For this formula to give $f'(0)$, r must be a number such that x^{r-1} is defined on some open interval containing 0.

EXAMPLE 1   Find $f'(x)$ if

$$f(x) = 4\sqrt[3]{x^2}$$

Solution $f(x) = 4x^{2/3}$. From Theorem 3.7.1

$$f'(x) = 4 \cdot \tfrac{2}{3}(x^{2/3 - 1})$$

$$= \tfrac{8}{3}x^{-1/3}$$

$$= \frac{8}{3x^{1/3}}$$

$$= \frac{8}{3\sqrt[3]{x}}$$

An immediate consequence of Theorem 3.7.1 and the chain rule is the following theorem.

3.7.2 THEOREM   If f and g are functions such that $f(x) = [g(x)]^r$, where r is any rational number, and if $g'(x)$ exists, then f is differentiable, and

$$f'(x) = r[g(x)]^{r-1}g'(x)$$

EXAMPLE 2   Compute

$$D_x(\sqrt{2x^3 - 4x + 5})$$

Solution   We write $\sqrt{2x^3 - 4x + 5}$ as $(2x^3 - 4x + 5)^{1/2}$ and apply Theorem 3.7.2.

$$D_x[(2x^3 - 4x + 5)^{1/2}] = \tfrac{1}{2}(2x^3 - 4x + 5)^{-1/2} \cdot D_x(2x^3 - 4x + 5)$$
$$= \tfrac{1}{2}(2x^3 - 4x + 4)^{-1/2}(6x^2 - 4)$$
$$= \frac{3x^2 - 2}{\sqrt{2x^3 - 4x + 5}}$$

EXAMPLE 3   Find $g'(x)$ if

$$g(x) = \frac{x^3}{\sqrt[3]{3x^2 - 1}}$$

Solution   We write the given fraction as a product.

$$g(x) = x^3(3x^2 - 1)^{-1/3}$$

From Theorems 3.3.6 and 3.7.2,

$$g'(x) = 3x^2(3x^2 - 1)^{-1/3} - \tfrac{1}{3}(3x^2 - 1)^{-4/3}(6x)(x^3)$$
$$= x^2(3x^2 - 1)^{-4/3}[3(3x^2 - 1) - 2x^2]$$
$$= \frac{x^2(7x^2 - 3)}{(3x^2 - 1)^{4/3}}$$

EXAMPLE 4   Find $f'(r)$ if

$$f(r) = \sqrt{4 \sin^2 r + 9 \cos^2 r}$$

Solution $f(r) = (4 \sin^2 r + 9 \cos^2 r)^{1/2}$. We apply Theorem 3.7.2.

$$f'(r) = \tfrac{1}{2}(4 \sin^2 r + 9 \cos^2 r)^{-1/2} \cdot D_r(4 \sin^2 r + 9 \cos^2 r)$$
$$= \frac{8 \sin r \cdot D_r(\sin r) + 18 \cos r \cdot D_r(\cos r)}{2\sqrt{4 \sin^2 r + 9 \cos^2 r}}$$
$$= \frac{8 \sin r \cos r + 18 \cos r(-\sin r)}{2\sqrt{4 \sin^2 r + 9 \cos^2 r}}$$
$$= \frac{-10 \sin r \cos r}{2\sqrt{4 \sin^2 r + 9 \cos^2 r}}$$
$$= -\frac{5 \sin r \cos r}{\sqrt{4 \sin^2 r + 9 \cos^2 r}}$$

EXERCISES 3.7

In Exercises 1 through 24, find the derivative of the function.

1. $f(x) = 4x^{1/2} + 5x^{-1/2}$
2. $f(x) = 3x^{2/3} - 6x^{1/3} + x^{-1/3}$
3. $g(x) = \sqrt{1 + 4x^2}$
4. $f(s) = \sqrt{2 - 3s^2}$
5. $f(x) = (5 - 3x)^{2/3}$
6. $g(x) = \sqrt[3]{4x^2 - 1}$
7. $g(y) = \dfrac{1}{\sqrt{25 - y^2}}$
8. $f(x) = (5 - 2x^2)^{-1/3}$
9. $h(t) = 2 \cos \sqrt{t}$
10. $f(x) = 4 \sec \sqrt{x}$
11. $g(r) = \cot \sqrt{3r}$
12. $g(x) = \sqrt{3 \sin x}$
13. $f(x) = (\sin 3x)^{-1/2}$
14. $f(y) = \sqrt{1 + \csc^2 y}$
15. $f(x) = \tan \sqrt{x^2 + 1}$
16. $f(y) = 3 \cos \sqrt[3]{2y^2}$
17. $g(x) = \sqrt{\dfrac{2x - 5}{3x + 1}}$
18. $h(t) = \dfrac{\sqrt{t} - 1}{\sqrt{t} + 1}$
19. $F(x) = \sqrt[3]{2x^3 - 5x^2 + x}$
20. $G(t) = \sqrt{\dfrac{5t + 6}{5t - 4}}$
21. $g(t) = \sqrt{2t} + \sqrt{\dfrac{2}{t}}$
22. $g(x) = \sqrt[3]{(3x^2 + 5x - 1)^2}$
23. $f(x) = (5 - x^2)^{1/2}(x^3 + 1)^{1/4}$
24. $g(y) = (y^2 + 3)^{1/3}(y^3 - 1)^{1/2}$

In Exercises 25 through 36, compute the derivative.

25. $\dfrac{d}{dx}\left(\dfrac{\sqrt{x^2 - 1}}{x}\right)$
26. $\dfrac{d}{dx}\left(\sqrt{x^2 - 5}\,\sqrt[3]{x^2 + 3}\right)$
27. $\dfrac{d}{dt}\left(\sqrt{\dfrac{\sin t + 1}{1 - \sin t}}\right)$
28. $\dfrac{d}{dz}\left(\sin \sqrt[3]{z} \cos \sqrt[3]{z}\right)$
29. $\dfrac{d}{dy}\left(\tan \sqrt{y} \sec \sqrt{y}\right)$
30. $\dfrac{d}{dx}\left(\sqrt{\dfrac{\cos x - 1}{\sin x}}\right)$
31. $D_x\left(\dfrac{\sqrt{x - 1}}{\sqrt[3]{x + 1}}\right)$
32. $D_x\left(\dfrac{4x + 6}{\sqrt{x^2 + 3x + 4}}\right)$
33. $D_x(\sqrt{9 + \sqrt{9 - x}})$
34. $D_y\left(\sqrt[4]{\dfrac{y^3 + 1}{y^3 - 1}}\right)$
35. $D_z\left(\dfrac{1}{\sqrt{1 + \cos^2 2z}}\right)$
36. $D_x\left(\sqrt{x} \tan \sqrt{\dfrac{1}{x}}\right)$

In Exercises 37 and 38, find an equation of the tangent line to the curve at each of the points. Draw a sketch of the graph and segments of the tangent lines at the points.

37. $y = (2x - 2)^{2/3}$; $(-3, 4)$, $(0, \sqrt[3]{4})$, $(1, 0)$, $(2, \sqrt[3]{4})$, $(5, 4)$
38. $y = (6 - 2x)^{1/3}$; $(-1, 2)$, $(1, \sqrt[3]{4})$, $(3, 0)$, $(5, -\sqrt[3]{4})$, $(7, -2)$
39. Find an equation of the tangent line to the curve $y = \sqrt{x^2 + 9}$ at the point $(4, 5)$.
40. Find an equation of the tangent line to the curve $y = (7x - 6)^{-1/3}$ that is perpendicular to the line $12x - 7y + 2 = 0$.
41. Find an equation of the normal line to the curve $y = x\sqrt{16 + x^2}$ at the origin.

42. Find an equation of the tangent line to the curve $y = \sqrt{\sin x + \cos x}$ at the point where $x = \frac{1}{4}\pi$.
43. An object is moving along a straight line according to the equation of motion $s = \sqrt{4t^2 + 3}$, with $t \geq 0$. Find the value of t for which the measure of the instantaneous velocity is (a) 0; (b) 1; (c) 2.
44. An object is moving along a straight line according to the equation of motion $s = \sqrt{5 + t^2}$, with $t \geq 0$. Find the value of t for which the measure of the instantaneous velocity is (a) 0; (b) 1.
45. Suppose that a liquid is produced by a certain chemical process and that the total cost function C is given by $C(x) = 6 + 4\sqrt{x}$, where $C(x)$ dollars is the total cost of producing x liters of the liquid. Find (a) the marginal cost when 16 liters are produced and (b) the number of liters produced when the marginal cost is \$0.40 per liter.
46. The number of dollars in the total cost of producing x units of a certain commodity is given by $C(x) = 40 + 3x + 9\sqrt{2x}$. Find (a) the marginal cost when 50 units are produced and (b) the number of units produced when the marginal cost is \$4.50.
47. A property development company rents each apartment at p dollars per month when x apartments are rented, and $p = 30\sqrt{300 - 2x}$. If $R(x)$ dollars is the total revenue received from the rental of x apartments, then $R(x) = px$. How many apartments must be rented before the marginal revenue is zero? *Note:* Because x is the number of apartments rented, it is a nonnegative integer. However, to apply calculus, assume that x is a nonnegative real number.
48. The daily production at a particular factory is $f(x)$ units when the capital investment is x thousands of dollars, and $f(x) = 200\sqrt{2x + 1}$. If the current capitalization is \$760,000, use the derivative to estimate the change in the daily production if the capital investment is increased by \$1000.
49. An airplane is flying parallel to the ground at an altitude of 2 km and at a speed of $4\frac{1}{2}$ km/min. If the plane flies directly over the Statue of Liberty, at what rate is the line-of-sight distance between the plane and the statue changing 20 sec later?
50. Given $f(u) = 1/u^2$ and $g(x) = \sqrt{x}/\sqrt{2x^3 - 6x + 1}$, find the derivative of $f \circ g$ in two ways: (a) by first finding $(f \circ g)(x)$ and then finding $(f \circ g)'(x)$; (b) by using the chain rule.

In Exercises 51 through 54, find the derivative of the given function. (Hint: $|a| = \sqrt{a^2}$.)

51. $f(x) = |x^2 - 4|$
52. $g(x) = x|x|$
53. $g(x) = |x|^3$
54. $h(x) = \sqrt[3]{|x| + x}$
55. Suppose $g(x) = |f(x)|$. Prove that if $f'(x)$ and $g'(x)$ exist, then $|g'(x)| = |f'(x)|$.
56. Suppose that $g(x) = \sqrt{9 - x^2}$ and $h(x) = f(g(x))$, where f is differentiable at 3. Prove that $h'(0) = 0$.

3.8 IMPLICIT DIFFERENTIATION

If $f = \{(x, y) \mid y = 3x^2 + 5x + 1\}$, then the equation

$$y = 3x^2 + 5x + 1$$

defines the function f explicitly. However, not all functions are defined by such an equation. For example, if we have the equation

$$x^6 - 2x = 3y^6 + y^5 - y^2 \tag{1}$$

we cannot solve for y in terms of x; however, there may exist one or more functions f such that if $y = f(x)$, Equation (1) is satisfied, that is, such that the equation

$$x^6 - 2x = 3[f(x)]^6 + [f(x)]^5 - [f(x)]^2$$

holds for all values of x in the domain of f. In this case the function f is defined *implicitly* by the given equation.

With the assumption that (1) defines y as at least one differentiable function of x, the derivative of y with respect to x can be found by *implicit differentiation*.

Equation (1) is a special type of equation involving x and y because it can be written so that all the terms involving x are on one side of the equation and all the terms involving y are on the other side. It serves as a first example to illustrate the process of implicit differentiation.

The left side of (1) is a function of x, and the right side is a function of y. Let F be the function defined by the left side, and let G be the function defined by the right side. Thus

$$F(x) = x^6 - 2x \qquad G(y) = 3y^6 + y^5 - y^2$$

where y is a function of x, say $y = f(x)$. So (1) can be written as

$$F(x) = G(f(x))$$

This equation is satisfied by all values of x in the domain of f for which $G(f(x))$ exists.

Then for all values of x for which f is differentiable,

$$D_x(x^6 - 2x) = D_x(3y^6 + y^5 - y^2) \tag{2}$$

The derivative on the left side of (2) is easily found, and

$$D_x(x^6 - 2x) = 6x^5 - 2 \tag{3}$$

We find the derivative on the right side of (2) by the chain rule.

$$D_x(3y^6 + y^5 - y^2) = 18y^5 \cdot \frac{dy}{dx} + 5y^4 \cdot \frac{dy}{dx} - 2y \cdot \frac{dy}{dx} \tag{4}$$

Substituting the values from (3) and (4) into (2) we obtain

$$6x^5 - 2 = (18y^5 + 5y^4 - 2y)\frac{dy}{dx}$$

$$\frac{dy}{dx} = \frac{6x^5 - 2}{18y^5 + 5y^4 - 2y}$$

Observe that by using implicit differentiation we have obtained an expression for $\frac{dy}{dx}$ that involves both variables x and y.

In the following illustration the method of implicit differentiation is used to find $\dfrac{dy}{dx}$ from a more general type of equation.

▶ **ILLUSTRATION 1**   Consider the equation

$$3x^4y^2 - 7xy^3 = 4 - 8y \tag{5}$$

and assume that there exists at least one differentiable function f such that if $y = f(x)$, Equation (5) is satisfied. Differentiating on both sides of (5) (bearing in mind that y is a differentiable function of x) and applying the theorems for the derivative of a product, the derivative of a power, and the chain rule, we obtain

$$12x^3y^2 + 3x^4\left(2y\frac{dy}{dx}\right) - 7y^3 - 7x\left(3y^2\frac{dy}{dx}\right) = 0 - 8\frac{dy}{dx}$$

$$\frac{dy}{dx}(6x^4y - 21xy^2 + 8) = 7y^3 - 12x^3y^2$$

$$\frac{dy}{dx} = \frac{7y^3 - 12x^3y^2}{6x^4y - 21xy^2 + 8} \qquad ◀$$

Remember we assumed that both (1) and (5) define y as at least one differentiable function of x. It may be that an equation in x and y does not imply the existence of any real-valued function, as is the case for the equation

$$x^2 + y^2 + 4 = 0$$

which is not satisfied by any real values of x and y. Furthermore, it is possible that an equation in x and y may be satisfied by many different functions, some of which are differentiable and some of which are not. A general discussion is beyond the scope of this book but can be found in an advanced calculus text. In subsequent discussions when we state that an equation in x and y defines y implicitly as a function of x, it is assumed that one or more of these functions is differentiable. Example 4, which follows, illustrates the fact that implicit differentiation gives the derivative of two differentiable functions defined by the given equation.

EXAMPLE 1   Given $(x + y)^2 - (x - y)^2 = x^4 + y^4$, find $\dfrac{dy}{dx}$.

Solution   Differentiating implicitly with respect to x we have

$$2(x + y)\left(1 + \frac{dy}{dx}\right) - 2(x - y)\left(1 - \frac{dy}{dx}\right) = 4x^3 + 4y^3\frac{dy}{dx}$$

$$2x + 2y + (2x + 2y)\frac{dy}{dx} - 2x + 2y + (2x - 2y)\frac{dy}{dx} = 4x^3 + 4y^3\frac{dy}{dx}$$

$$\frac{dy}{dx}(4x - 4y^3) = 4x^3 - 4y$$

$$\frac{dy}{dx} = \frac{x^3 - y}{x - y^3}$$

EXAMPLE 2   Find an equation of the tangent line to the curve $x^3 + y^3 = 9$ at the point $(1, 2)$.

Solution   We differentiate implicitly with respect to x.

$$3x^2 + 3y^2 \frac{dy}{dx} = 0$$

$$\frac{dy}{dx} = -\frac{x^2}{y^2}$$

Therefore, at the point $(1, 2)$, $\frac{dy}{dx} = -\frac{1}{4}$. An equation of the tangent line is then

$$y - 2 = -\tfrac{1}{4}(x - 1)$$

$$x + 4y - 9 = 0$$

EXAMPLE 3   Given $x \cos y + y \cos x = 1$, find $\frac{dy}{dx}$.

Solution   Differentiating implicitly with respect to x we get

$$1 \cdot \cos y + x(-\sin y)\frac{dy}{dx} + \frac{dy}{dx}(\cos x) + y(-\sin x) = 0$$

$$\frac{dy}{dx}(\cos x - x \sin y) = y \sin x - \cos y$$

$$\frac{dy}{dx} = \frac{y \sin x - \cos y}{\cos x - x \sin y}$$

EXAMPLE 4   Given the equation $x^2 + y^2 = 9$, find (a) $\frac{dy}{dx}$ by implicit differentiation; (b) two functions defined by the equation; (c) the derivative of each of the functions obtained in part (b) by explicit differentiation. (d) Verify that the result obtained in part (a) agrees with the results obtained in part (c).

Solution
(a) We differentiate implicitly.

$$2x + 2y \frac{dy}{dx} = 0$$

$$\frac{dy}{dx} = -\frac{x}{y}$$

(b) If the given equation is solved for y,

$$y = \sqrt{9 - x^2} \quad \text{and} \quad y = -\sqrt{9 - x^2}$$

Let f_1 and f_2 be the two functions for which

$$f_1(x) = \sqrt{9 - x^2} \quad \text{and} \quad f_2(x) = -\sqrt{9 - x^2}$$

(c) Because $f_1(x) = (9 - x^2)^{1/2}$ and $f_2(x) = -(9 - x^2)^{1/2}$ from the chain rule we obtain

$$f_1'(x) = \tfrac{1}{2}(9 - x^2)^{-1/2}(-2x) \qquad f_2'(x) = -\tfrac{1}{2}(9 - x^2)^{-1/2}(-2x)$$

$$= -\frac{x}{\sqrt{9 - x^2}} \qquad\qquad = \frac{x}{\sqrt{9 - x^2}}$$

(d) For $y = f_1(x)$, where $f_1(x) = \sqrt{9 - x^2}$, it follows from part (c) that

$$f_1'(x) = -\frac{x}{\sqrt{9 - x^2}}$$

$$= -\frac{x}{y}$$

which agrees with the answer in part (a).
For $y = f_2(x)$, where $f_2(x) = -\sqrt{9 - x^2}$, we have from part (c)

$$f_2'(x) = \frac{x}{\sqrt{9 - x^2}}$$

$$= -\frac{x}{-\sqrt{9 - x^2}}$$

$$= -\frac{x}{y}$$

which also agrees with the answer in part (a).

EXERCISES 3.8

In Exercises 1 through 28, find $\dfrac{dy}{dx}$ by implicit differentiation.

1. $x^2 + y^2 = 16$

2. $4x^2 - 9y^2 = 1$

3. $x^3 + y^3 = 8xy$

4. $x^2 + y^2 = 7xy$

5. $\dfrac{1}{x} + \dfrac{1}{y} = 1$

6. $\dfrac{3}{x} - \dfrac{3}{y} = 2x$

7. $\sqrt{x} + \sqrt{y} = 4$

8. $2x^3y + 3xy^3 = 5$

9. $x^2y^2 = x^2 + y^2$

10. $(2x + 3)^4 = 3y^4$

11. $x^2 = \dfrac{x + 2y}{x - 2y}$

12. $\dfrac{x}{\sqrt{y}} - 4y = x$

13. $\sqrt[3]{x} + \sqrt[3]{xy} = 4y^2$

14. $\sqrt{y} + \sqrt[3]{y} + \sqrt[4]{y} = x$

15. $\sqrt{xy} + 2x = \sqrt{y}$

16. $y + \sqrt{xy} = 3x^3$

17. $\dfrac{y}{\sqrt{x - y}} = 2 + x^2$

18. $x^2y^3 = x^4 - y^4$

19. $y = \cos(x - y)$

20. $x = \sin(x + y)$

21. $\sec^2 x + \csc^2 y = 4$

22. $\cot xy + xy = 0$

23. $x \sin y + y \cos x = 1$

24. $\cos(x + y) = y \sin x$

25. $\sec^2 y + \cot(x - y) = \tan^2 x$

26. $\csc(x - y) + \sec(x + y) = x$

27. $(x + y)^2 - (x - y)^2 = x^3 + y^3$

28. $y\sqrt{2 + 3x} + x\sqrt{1 + y} = x$

In Exercises 29 through 32, consider y as the independent variable and find $\dfrac{dx}{dy}$.

29. $x^4 + y^4 = 12x^2y$

30. $y = 2x^3 - 5x$

31. $x^3y + 2y^4 - x^4 = 0$

32. $y\sqrt{x} - x\sqrt{y} = 9$

33. Find an equation of the tangent line to the curve $16x^4 + y^4 = 32$ at the point $(1, 2)$.

34. Find an equation of the normal line to the curve $9x^3 - y^3 = 1$ at the point $(1, 2)$.

35. Find an equation of the normal line to the curve $x^2 + xy + y^2 - 3y = 10$ at the point $(2, 3)$.

36. Find an equation of the tangent line to the curve $\sqrt[3]{xy} = 14x + y$ at the point $(2, -32)$.

37. Find the rate of change of y with respect to x at the point $(3, 2)$ if $7y^2 - xy^3 = 4$.

38. For the circle $x^2 + y^2 = r^2$, show that the tangent line at any point (x_1, y_1) on the curve is perpendicular to the line through (x_1, y_1) and the center of the circle.

39. At what point of the curve $xy = (1 - x - y)^2$ is the tangent line parallel to the x axis?

40. There are two lines through the point $(-1, 3)$ that are tangent to the curve $x^2 + 4y^2 - 4x - 8y + 3 = 0$. Find an equation of each of these lines.

In Exercises 41 through 46, do the following: (a) Find two functions defined by the equation, and state their domains. (b) Draw a sketch of the graph of each of the functions obtained in part (a). (c) Draw a sketch of the graph of the equation. (d) Find the derivative of each of the functions obtained in part (a), and state the domains of the derivatives. (e) Find $\dfrac{dy}{dx}$ by implicit differentiation from the given equation, and verify that the result so obtained agrees with the results in part (d). (f) Find an equation of each tangent line at the given value of x_1.

41. $y^2 = 4x - 8;\ x_1 = 3$ **42.** $x^2 + y^2 = 25;\ x_1 = 4$
43. $x^2 - y^2 = 9;\ x_1 = -5$ **44.** $y^2 - x^2 = 16;\ x_1 = -3$
45. $x^2 + y^2 - 2x - 4y - 4 = 0;\ x_1 = 1$
46. $x^2 + 4y^2 + 6x - 40y + 93 = 0;\ x_1 = -2$

47. At 8 A.M. a ship sailing due north at 24 knots (nautical miles per hour) is at a point P. At 10 A.M. a second ship sailing due east at 32 knots is at P. At what rate is the distance between the two ships changing at (a) 9 A.M. and (b) 11 A.M.?

48. If $x^n y^m = (x + y)^{n+m}$, prove that $x \cdot \dfrac{dy}{dx} = y$.

49. Find equations of the tangent lines to the curve $x^{2/3} + y^{2/3} = 1$ at the points where $x = -\frac{1}{8}$.

50. Prove that the sum of the x and y intercepts of any tangent line to the curve $x^{1/2} + y^{1/2} = k^{1/2}$ is constant and equal to k.

51. Let f be the power function defined by $f(x) = x^r$, where r is any rational number. Under the assumption that f is differentiable, use implicit differentiation to show that $f'(x) = rx^{r-1}$. $\left(\textit{Hint: Let } r = \dfrac{p}{q}, \text{ where } p \text{ and } q \text{ are integers} \right.$ and $q > 0$. Then replace $f(x)$ by y and write the equation as $y^q = x^p$. Use implicit differentiation to find $\dfrac{dy}{dx}$.$\Big)$

3.9 RELATED RATES A problem involving rates of change of related variables is called a problem in **related rates**. We begin our discussion with an example that describes a real-world situation.

EXAMPLE 1 A ladder 25 ft long is leaning against a vertical wall. If the bottom of the ladder is pulled horizontally away from the wall at 3 ft/sec, how fast is the top of the ladder sliding down the wall when the bottom is 15 ft from the wall?

Solution Let t seconds be the time that has elapsed since the ladder started to slide down the wall, y feet be the distance from the ground to the top of the ladder at t seconds, and x feet be the distance from the bottom of the ladder to the wall at t seconds. See Figure 1.

Because the bottom of the ladder is pulled horizontally away from the wall at 3 ft/sec, $\dfrac{dx}{dt} = 3$. We wish to find $\dfrac{dy}{dt}$ when $x = 15$. From the Pythagorean theorem,

$$y^2 = 625 - x^2 \tag{1}$$

Because x and y are functions of t, we differentiate both sides of (1) with respect to t and obtain

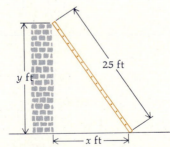

FIGURE 1

$$2y \frac{dy}{dt} = -2x \frac{dx}{dt}$$

$$\frac{dy}{dt} = -\frac{x}{y} \frac{dx}{dt}$$

When $x = 15$, it follows from (1) that $y = 20$. Because $\dfrac{dx}{dt} = 3$, we get from (2)

$$\left.\frac{dy}{dt}\right]_{y=20} = -\tfrac{15}{20} \cdot 3$$
$$= -\tfrac{9}{4}$$

Therefore the top of the ladder is sliding down the wall at the rate of $2\tfrac{1}{4}$ ft/sec when the bottom is 15 ft from the wall. The significance of the minus sign is that y is decreasing as t is increasing.

In related-rate problems the variables have a specific relationship for values of t, where t is a measure of time. This relationship is usually expressed in the form of an equation, as in Example 1 with Equation (1). Values of the variables and rates of change of the variables with respect to t are often given at a particular instant. In Example 1, at the instant when $x = 15$, then $y = 20$ and $\dfrac{dx}{dt} = 3$, and we wish to find $\dfrac{dy}{dt}$.

Before presenting more applications we give another example to demonstrate the computation involved.

EXAMPLE 2   Given

$$x \cos y = 5$$

where x and y are functions of a third variable t. If $\dfrac{dx}{dt} = -4$, find $\dfrac{dy}{dt}$ when $y = \tfrac{1}{3}\pi$.

Solution   Differentiating both sides of the given equation with respect to t, we obtain

$$(\cos y)\frac{dx}{dt} - (x \sin y)\frac{dy}{dt} = 0$$

$$\frac{dy}{dt} = \frac{\cos y}{x \sin y} \cdot \frac{dx}{dt} \tag{3}$$

From the given equation, when $y = \tfrac{1}{3}\pi$, $x = 10$. From (3) with $y = \tfrac{1}{3}\pi$, $x = 10$, and $\dfrac{dx}{dt} = -4$,

$$\left.\frac{dy}{dt}\right]_{y=\pi/3} = \frac{\tfrac{1}{2}}{10(\tfrac{1}{2}\sqrt{3})}(-4)$$
$$= -\tfrac{2}{15}\sqrt{3}$$

The following steps represent a possible procedure for solving a word problem involving related rates.

1. Draw a figure if it is feasible to do so.
2. Define the variables. Generally define t first, because the other variables usually depend on t.
3. Write down any numerical facts known about the variables and their derivatives with respect to t.
4. Write an equation to relate the variables that depend on t.
5. Differentiate with respect to t both sides of the equation found in step 4 to relate the rates of change of the variables.
6. Substitute values of known quantities in the equation of step 5, and solve for the desired quantity.

EXAMPLE 3   A tank is in the form of an inverted cone having an altitude of 16 m and a radius of 4 m. Water is flowing into the tank at the rate of 2 m³/min. How fast is the water level rising when the water is 5 m deep?

Solution   Let t minutes be the time that has elapsed since water started to flow into the tank. At t minutes let h meters be the height of the water level, r meters be the radius of the surface of the water, and V cubic meters be the volume of the water in the tank.

At any time, the volume of water in the tank may be expressed in terms of the volume of a cone. See Figure 2.

$$V = \tfrac{1}{3}\pi r^2 h \qquad\qquad (4)$$

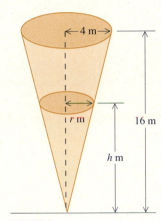

FIGURE 2

V, r, and h are all functions of t. Because water is flowing into the tank at the rate of 2 m³/min, $\dfrac{dV}{dt} = 2$. We wish to find $\dfrac{dh}{dt}$ when $h = 5$. To express r in terms of h we have from similar triangles

$$\frac{r}{h} = \frac{4}{16} \quad\Leftrightarrow\quad r = \tfrac{1}{4}h$$

Substituting this value of r into (4) we obtain

$$V = \tfrac{1}{3}\pi(\tfrac{1}{4}h)^2(h) \quad\Leftrightarrow\quad V = \tfrac{1}{48}\pi h^3$$

By differentiating both sides of this equation with respect to t,

$$\frac{dV}{dt} = \frac{1}{16}\pi h^2 \frac{dh}{dt}$$

Substituting 2 for $\dfrac{dV}{dt}$ and solving for $\dfrac{dh}{dt}$ we get

$$\frac{dh}{dt} = \frac{32}{\pi h^2}$$

Thus

$$\frac{dh}{dt}\bigg]_{h=5} = \frac{32}{25\pi}$$

Therefore the water level is rising at the rate of $\dfrac{32}{25\pi}$ m/min when the water is 5 m deep.

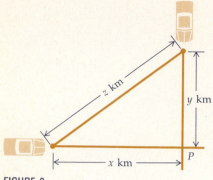

FIGURE 3

EXAMPLE 4   Two cars, one going due east at the rate of 90 km/hr and the other going due south at the rate of 60 km/hr, are traveling toward the intersection of the two roads. At what rate are the two cars approaching each other at the instant when the first car is 0.2 km and the second car is 0.15 km from the intersection?

Solution   Refer to Figure 3, where the point P is the intersection of the two roads. Let t hours be the time elapsed since the cars started to approach P. At t hours, x kilometers is the distance of the first car from P, y kilometers is the distance of the second car from P, and z kilometers is the distance between the two cars. Because the first car is approaching P at the rate of 90 km/hr and x is decreasing as t is increasing, $\dfrac{dx}{dt} = -90$. Similarly, $\dfrac{dy}{dt} = -60$. We wish to find $\dfrac{dz}{dt}$ when $x = 0.2$ and $y = 0.15$. From the Pythagorean theorem

$$z^2 = x^2 + y^2 \tag{5}$$

Differentiating both sides of this equation with respect to t, we obtain

$$2z\frac{dz}{dt} = 2x\frac{dx}{dt} + 2y\frac{dy}{dt}$$

$$\frac{dz}{dt} = \frac{x\dfrac{dx}{dt} + y\dfrac{dy}{dt}}{z} \tag{6}$$

When $x = 0.2$ and $y = 0.15$, it follows from (5) that $z = 0.25$. In (6) we let $\dfrac{dx}{dt} = -90$, $\dfrac{dy}{dt} = -60$, $x = 0.2$, $y = 0.15$, and $z = 0.25$ to get

$$\frac{dz}{dt}\bigg]_{z=0.25} = \frac{(0.2)(-90) + (0.15)(-60)}{0.25}$$

$$= -108$$

Therefore, at the instant in question the cars are approaching each other at the rate of 108 km/hr.

EXAMPLE 5   Suppose in a certain market that x thousands of crates of oranges are supplied daily when p dollars is the price per crate, and the supply equation is

$$px - 20p - 3x + 105 = 0$$

If the daily supply is decreasing at the rate of 250 crates per day, at what rate is the price changing when the daily supply is 5000 crates?

Solution   Let t days be the time that has elapsed since the daily supply of oranges started to decrease. Then p and x are both functions of t. Because the daily supply is decreasing at the rate of 250 crates per day, $\dfrac{dx}{dt} = -\dfrac{250}{1000}$;

that is, $\dfrac{dx}{dt} = -\dfrac{1}{4}$. We wish to find $\dfrac{dp}{dt}$ when $x = 5$. From the given supply equation we differentiate implicitly with respect to t and obtain

$$p\frac{dx}{dt} + x\frac{dp}{dt} - 20\frac{dp}{dt} - 3\frac{dx}{dt} = 0$$

$$\frac{dp}{dt} = \frac{3 - p}{x - 20} \cdot \frac{dx}{dt}$$

When $x = 5$, it follows from the supply equation that $p = 6$. Because $\dfrac{dx}{dt} = -\dfrac{1}{4}$, we have from the preceding equation

$$\frac{dp}{dt}\bigg]_{p=6} = \frac{3 - 6}{5 - 20}\left(-\frac{1}{4}\right)$$

$$= -\frac{1}{20}$$

Thus the price of a crate of oranges is decreasing at the rate of \$0.05 per day when the daily supply is 5000 crates.

EXAMPLE 6   An airplane is flying west at 500 ft/sec at an altitude of 4000 ft and a searchlight on the ground lies directly under the path of the plane. If the light is to be kept on the plane, how fast is the searchlight revolving when the airline distance of the plane from the searchlight is 2000 ft due east?

Solution   Refer to Figure 4. The searchlight is at point L, and at a particular instant the plane is at point P. Let x feet due east be the airline distance of the plane from the searchlight and θ radians be the angle of elevation of the plane at the searchlight at t seconds.

We are given $\dfrac{dx}{dt} = -500$, and we wish to find $\dfrac{d\theta}{dt}$ when $x = 2000$.

$$\tan\theta = \frac{4000}{x}$$

Differentiating both sides of this equation with respect to t we obtain

$$\sec^2\theta\,\frac{d\theta}{dt} = -\frac{4000}{x^2}\frac{dx}{dt}$$

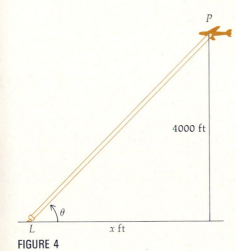

FIGURE 4

Substituting $\dfrac{dx}{dt} = -500$ in the above and dividing by $\sec^2\theta$ gives

$$\frac{d\theta}{dt} = \frac{2{,}000{,}000}{x^2\sec^2\theta} \qquad (7)$$

When $x = 2000$, $\tan\theta = 2$. Because $\sec^2\theta = 1 + \tan^2\theta$, $\sec^2\theta = 5$. Substituting these values into (7) we have, when $x = 2000$,

$$\frac{d\theta}{dt} = \frac{2{,}000{,}000}{4{,}000{,}000(5)}$$

$$= \tfrac{1}{10}$$

We conclude that at the given instant the measurement of the angle is increasing at the rate of $\frac{1}{10}$ rad/sec, and this is how fast the searchlight is revolving.

EXERCISES 3.9

In Exercises 1 through 8, x and y are functions of a third variable t.

1. If $2x + 3y = 8$ and $\frac{dy}{dt} = 2$, find $\frac{dx}{dt}$.

2. If $\frac{x}{y} = 10$ and $\frac{dx}{dt} = -5$, find $\frac{dy}{dt}$.

3. If $xy = 20$ and $\frac{dy}{dt} = 10$, find $\frac{dx}{dt}$ when $x = 2$.

4. If $2 \sin x + 4 \cos y = 3$ and $\frac{dy}{dt} = 3$, find $\frac{dx}{dt}$ at $(\frac{1}{6}\pi, \frac{1}{3}\pi)$.

5. If $\sin^2 x + \cos^2 y = \frac{5}{4}$ and $\frac{dx}{dt} = -1$, find $\frac{dy}{dt}$ at $(\frac{2}{3}\pi, \frac{3}{4}\pi)$.

6. If $x^2 + y^2 = 25$ and $\frac{dx}{dt} = 5$, find $\frac{dy}{dt}$ when $y = 4$.

7. If $\sqrt{x} + \sqrt{y} = 5$ and $\frac{dy}{dt} = 3$, find $\frac{dx}{dt}$ when $x = 1$.

8. If $y(\tan x + 1) = 4$ and $\frac{dy}{dt} = -4$, find $\frac{dx}{dt}$ when $x = \pi$.

9. A kite is flying at a height of 40 ft. A child is flying it so that it is moving horizontally at a rate of 3 ft/sec. If the string is taut, at what rate is the string being paid out when the length of the string released is 50 ft?

10. A spherical balloon is being inflated so that its volume is increasing at the rate of 5 m³/min. At what rate is the diameter increasing when the diameter is 12 m?

11. A spherical snowball is being made so that its volume is increasing at the rate of 8 ft³/min. Find the rate at which the radius is increasing when the snowball is 4 ft in diameter.

12. Suppose that when the diameter is 6 ft, the snowball in Exercise 11 stopped growing and started to melt at the rate of $\frac{1}{4}$ ft³/min. Find the rate at which the radius is changing when the radius is 2 ft.

13. Sand is being dropped at the rate of 10 m³/min onto a conical pile. If the height of the pile is always twice the base radius, at what rate is the height increasing when the pile is 8 m high?

14. A light is hung 15 ft above a straight horizontal path. If a man 6 ft tall is walking away from the light at the rate of 5 ft/sec, how fast is his shadow lengthening?

15. In Exercise 14, at what rate is the tip of the man's shadow moving?

16. A man 6 ft tall is walking toward a building at the rate of 5 ft/sec. If there is a light on the ground 50 ft from the building, how fast is the man's shadow on the building growing shorter when he is 30 ft from the building?

17. Suppose that a tumor in a person's body is spherical in shape. If, when the radius of the tumor is 0.5 cm, the radius is increasing at the rate of 0.001 cm per day, what is the rate of increase of the volume of the tumor at that time?

18. A bacterial cell is spherical in shape. If the radius of the cell is increasing at the rate of 0.01 micrometers per day when it is 1.5 μm, what is the rate of increase of the volume of the cell at that time?

19. For the tumor in Exercise 17, what is the rate of increase of the surface area when its radius is 0.5 cm?

20. For the cell of Exercise 18, what is the rate of increase of the surface area when its radius is 1.5 μm?

21. A water tank in the form of an inverted cone is being emptied at the rate of 6 m³/min. The altitude of the cone is 24 m, and the radius is 12 m. Find how fast the water level is lowering when the water is 10 m deep.

22. A trough is 12 ft long and its ends are in the form of inverted isosceles triangles having an altitude of 3 ft and a base of 3 ft. Water is flowing into the trough at the rate of 2 ft³/min. How fast is the water level rising when the water is 1 ft deep?

23. Boyle's law for the expansion of gas is $PV = C$, where P is the number of pounds per square unit of pressure, V is the number of cubic units of volume of the gas, and C is a constant. At a certain instant the pressure is 3000 lb/ft², the volume is 5 ft³, and the volume is increasing at the rate of 3 ft³/min. Find the rate of change of the pressure at this instant.

24. The adiabatic law (no gain or loss of heat) for the expansion of air is $PV^{1.4} = C$, where P is the number of pounds per square unit of pressure, V is the number of cubic units of volume, and C is a constant. At a specific instant the pressure is 40 lb/in.² and is increasing at the rate of 8 lb/in.² each second. If $C = \frac{5}{16}$, what is the rate of change of volume at this instant?

25. A stone is dropped into a still pond. Concentric circular ripples spread out, and the radius of the disturbed region increases at the rate of 16 cm/sec. At what rate does the area of the disturbed region increase when its radius is 4 cm?

26. Oil is running into an inverted conical tank at the rate of 3π m³/min. If the tank has a radius of 2.5 m at the top and a depth of 10 m, how fast is the depth of the oil changing when it is 8 m?

27. An automobile traveling at a rate of 30 ft/sec is approaching an intersection. When the automobile is 120 ft from the intersection, a truck traveling at the rate of 40 ft/sec crosses the intersection. The automobile and the truck are on roads that are at right angles to each other. How fast are the automobile and the truck separating 2 sec after the truck leaves the intersection?

28. A rope is attached to a boat at water level, and a woman on a dock is pulling on the rope at the rate of 50 ft/min. If her hands are 16 ft above the water level, how fast is the boat approaching the dock when the amount of rope out is 20 ft?

29. This week a factory is producing 50 units of a particular commodity, and the amount being produced is increasing at the rate of 2 units per week. If $C(x)$ dollars is the total cost of producing x units and $C(x) = 0.08x^3 - x^2 + 10x + 48$, find the current rate at which the production cost is increasing.

30. The demand for a particular breakfast cereal is given by the demand equation $px + 50p = 16,000$, where x thousands of boxes are demanded when p cents is the price per box. If the current price of the cereal is $1.60 per box and the price per box is increasing at the rate of 0.4 cent each week, find the rate of change in the demand.

31. The supply equation for a certain commodity is $x = 1000\sqrt{3p^2 + 20p}$, where x units are supplied per month when p dollars is the price per unit. Find the rate of change in the supply if the current price is $20 per unit and the price is increasing at the rate of $0.50 per month.

32. Suppose that y workers are needed to produce x units of a certain commodity, and $x = 4y^2$. If the production of the commodity this year is 250,000 units and the production is increasing at the rate of 18,000 units per year, what is the current rate at which the labor force should be increased?

33. The demand equation for a particular kind of shirt is $2px + 65p - 4950 = 0$, where x hundreds of shirts are demanded per week when p dollars is the price of a shirt. If the shirt is selling this week at $30 and the price is increasing at the rate of $0.20 per week, find the rate of change in the demand.

34. The measure of one of the acute angles of a right triangle is decreasing at the rate of $\frac{1}{36}\pi$ rad/sec. If the length of the hypotenuse is constant and 40 cm, find how fast the area is changing when the measure of the acute angle is $\frac{1}{6}\pi$.

35. Two trucks, one traveling west and the other traveling south, are approaching an intersection. If both trucks are traveling at the rate of k km/hr, show that they are approaching each other at the rate of $k\sqrt{2}$ km/hr when they are each m kilometers from the intersection.

36. A horizontal trough is 16 m long, and its ends are isosceles trapezoids with an altitude of 4 m, a lower base of 4 m, and an upper base of 6 m. Water is being poured into the trough at the rate of 10 m³/min. How fast is the water level rising when the water is 2 m deep?

37. In Exercise 36, if the water level is decreasing at the rate of 25 cm/min when the water is 3 m deep, at what rate is water being drawn from the trough?

38. A ladder 7 m long is leaning against a wall. If the bottom of the ladder is pushed horizontally toward the wall at 1.5 m/sec, how fast is the top of the ladder sliding up the wall when the bottom is 2 m from the wall?

39. A ladder 20 ft long is leaning against an embankment inclined 60° to the horizontal. If the bottom of the ladder is being moved horizontally toward the embankment at 1 ft/sec, how fast is the top of the ladder moving when the bottom is 4 ft from the embankment?

40. The volume of a balloon is decreasing at a rate proportional to its surface area. Show that the radius of the balloon shrinks at a constant rate.

41. An airplane is flying at a constant speed at an altitude of 10,000 ft on a line that will take it directly over an observer on the ground. At a given instant the observer notes that the angle of elevation of the airplane is $\frac{1}{3}\pi$ radians and is increasing at the rate of $\frac{1}{60}$ rad/sec. Find the speed of the airplane.

42. A ship is located 4 mi from a straight shore and has a radar transmitter that rotates 32 times per minute. How fast is the radar beam moving along the shoreline when the beam makes an angle of 45° with the shore?

43. If a ladder of length 30 ft that is leaning against a wall has its upper end sliding down the wall at the rate of $\frac{1}{2}$ ft/sec, what is the rate of change of the measure of the acute angle made by the ladder with the ground when the upper end is 18 ft above the ground?

44. Water is poured at the rate of 8 ft³/min into a tank in the form of a cone. The cone is 20 ft deep and 10 ft in diameter at the top. If there is a leak in the bottom and the water level is rising at the rate of 1 in./min, when the water is 16 ft deep, how fast is the water leaking?

3.10   DERIVATIVES OF HIGHER ORDER

If the function f is differentiable, then its derivative f' is sometimes called the **first derivative** of f or the first derived function. If the function f' is differentiable, then the derivative of f' is called the **second derivative** of f, or the second derived function. The second derivative of f is denoted by f'' (read as "f double prime"). Similarly, the **third derivative** of f, or the third derived function, is defined as the derivative of f'' provided the derivative of f'' exists. The third derivative of f is denoted by f''' (read as "f triple prime").

The **nth derivative** of the function f, where n is a positive integer greater than 1, is the derivative of the $(n-1)$st derivative of f. We denote the nth derivative of f by $f^{(n)}$. Thus if $f^{(n)}$ is the nth derived function, we can write the function f itself as $f^{(0)}$.

EXAMPLE 1   Find all the derivatives of the function f defined by

$$f(x) = 8x^4 + 5x^3 - x^2 + 7$$

Solution

$$f'(x) = 32x^3 + 15x^2 - 2x$$

$$f''(x) = 96x^2 + 30x - 2$$

$$f'''(x) = 192x + 30$$

$$f^{(4)}(x) = 192$$

$$f^{(5)}(x) = 0$$

$$f^{(n)}(x) = 0 \qquad n \geq 5$$

The Leibniz notation for the first derivative is $\dfrac{dy}{dx}$. For the second derivative of y with respect to x the Leibniz notation is $\dfrac{d^2y}{dx^2}$, because it represents $\dfrac{d}{dx}\left[\dfrac{d}{dx}(y)\right]$. The symbol $\dfrac{d^ny}{dx^n}$ is a notation for the nth derivative of y with respect to x.

Other symbols for the nth derivative of f are

$$\frac{d^n}{dx^n}[f(x)] \qquad D_x{}^n[f(x)]$$

EXAMPLE 2   Compute

$$\frac{d^3}{dx^3}(2\sin x + 3\cos x - x^3)$$

Solution

$$\frac{d}{dx}(2\sin x + 3\cos x - x^3) = 2\cos x - 3\sin x - 3x^2$$

$$\frac{d^2}{dx^2}(2\sin x + 3\cos x - x^3) = -2\sin x - 3\cos x - 6x$$

$$\frac{d^3}{dx^3}(2\sin x + 3\cos x - x^3) = -2\cos x + 3\sin x - 6$$

Because $f'(x)$ gives the instantaneous rate of change of $f(x)$ with respect to x, $f''(x)$, being the derivative of $f'(x)$, gives the instantaneous rate of change of $f'(x)$ with respect to x. Furthermore, if (x, y) is any point on the graph of $y = f(x)$, then $\dfrac{dy}{dx}$ gives the slope of the tangent line to the graph at the point (x, y). Thus $\dfrac{d^2y}{dx^2}$ is the instantaneous rate of change of the slope of the tangent line with respect to x at the point (x, y).

EXAMPLE 3   Let $m(x)$ be the slope of the tangent line to the curve

$$y = x^3 - 2x^2 + x$$

at the point (x, y). Find the instantaneous rate of change of $m(x)$ with respect to x at the point $(2, 2)$.

Solution

$$m(x) = \frac{dy}{dx}$$

$$= 3x^2 - 4x + 1$$

The instantaneous rate of change of $m(x)$ with respect to x is given by $m'(x)$ or, equivalently, $\dfrac{d^2y}{dx^2}$.

$$m'(x) = \frac{d^2y}{dx^2}$$

$$= 6x - 4$$

At the point $(2, 2)$, $\dfrac{d^2y}{dx^2} = 8$.

The second derivative $f''(x)$ is expressed in units of $f'(x)$ per unit of x, which is units of $f(x)$ per unit of x, per unit of x. For example, in rectilinear motion, if $f(t)$ meters is the distance of a particle from the origin at t seconds, then $f'(t)$ meters per second is the velocity of the particle at t seconds, and $f''(t)$ meters per second per second is the instantaneous rate of change of the velocity at t seconds. In physics, the instantaneous rate of change of the velocity is called the **instantaneous acceleration**. Therefore, if a particle is moving along a straight line according to the equation of motion $s = f(t)$, where the instantaneous velocity at t seconds is given by v m/sec, and the instantaneous acceleration at t seconds is given by a m/sec^2, then a is the first derivative of v with respect to t or, equivalently, the second derivative of s with respect to t; that is,

$$v = \frac{ds}{dt}$$

$$a = \frac{dv}{dt} \quad \Leftrightarrow \quad a = \frac{d^2s}{dt^2}$$

When $a > 0$, v is increasing, and when $a < 0$, v is decreasing. When $a = 0$, then v is not changing. Because the speed of the particle at t seconds is $|v|$ m/sec, we have the following results:

(i) If $v \geq 0$ and $a > 0$, the speed is increasing.
(ii) If $v \geq 0$ and $a < 0$, the speed is decreasing.
(iii) If $v \leq 0$ and $a > 0$, the speed is decreasing.
(iv) If $v \leq 0$ and $a < 0$, the speed is increasing.

▶ **ILLUSTRATION 1**   A particle is moving along a horizontal line according to the equation

$$s = 3t^2 - t^3 \qquad t \geq 0 \tag{1}$$

where s meters is the distance of the particle from the origin at t seconds. If v meters per second is the instantaneous velocity at t seconds, then $v = \dfrac{ds}{dt}$. Therefore

$$v = 6t - 3t^2 \tag{2}$$

If a m/sec^2 is the instantaneous acceleration at t seconds, then $a = \dfrac{dv}{dt}$. Thus

$$a = 6 - 6t \tag{3}$$

Let us determine the values of t when any one of the quantities s, v, or a is 0. From (1),

$s = 0$ when $t = 0$ or $t = 3$

From (2),

$v = 0$ when $t = 0$ or $t = 2$

From (3),

$a = 0$ when $t = 1$

Table 1

	s	v	a	*Conclusion*
$t = 0$	0	0	6	Particle is at the origin. The velocity is 0 and is increasing. The speed is increasing.
$0 < t < 1$	+	+	+	Particle is at the right of the origin, and it is moving to the right. The velocity is increasing. The speed is increasing.
$t = 1$	2	3	0	Particle is 2 m to the right of the origin, and it is moving to the right at 3 m/sec. The velocity is not changing; so the speed is not changing.
$1 < t < 2$	+	+	−	Particle is at the right of the origin, and it is moving to the right. The velocity is decreasing. The speed is decreasing.
$t = 2$	4	0	−6	Particle is 4 m to the right of the origin, and it is changing its direction of motion from right to left. The velocity is decreasing. The speed is increasing.
$2 < t < 3$	+	−	−	Particle is at the right of the origin, and it is moving to the left. The velocity is decreasing. The speed is increasing.
$t = 3$	0	−9	−12	Particle is at the origin, and it is moving to the left at 9 m/sec. The velocity is decreasing. The speed is increasing.
$3 < t$	−	−	−	Particle is at the left of the origin, and it is moving to the left. The velocity is decreasing. The speed is increasing.

In Table 1 there are values of s, v, and a for these values of t: 0, 1, 2, and 3. Also indicated is the sign of the quantities s, v, and a in the intervals of t excluding 0, 1, 2, and 3. A conclusion is formed regarding the position and motion of the particle for the various values of t.

In Figure 1 the motion of the particle is along the horizontal line, and the behavior of the motion is indicated above the line.   ◄

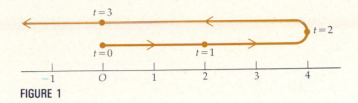

FIGURE 1

EXAMPLE 4   A particle is moving along a straight line according to the equation of motion

$$s = \frac{1}{2}t^2 + \frac{4t}{t + 1}$$

where s meters is the directed distance of the particle from the origin at t seconds. If v m/sec is the instantaneous velocity at t seconds and a m/sec^2 is the instantaneous acceleration at t seconds, find t, s, and v when $a = 0$.

Solution

$$v = \frac{ds}{dt} \qquad\qquad a = \frac{dv}{dt}$$

$$= t + \frac{4}{(t + 1)^2} \qquad = 1 - \frac{8}{(t + 1)^3}$$

Setting $a = 0$ we have

$$\frac{(t + 1)^3 - 8}{(t + 1)^3} = 0$$

$$(t + 1)^3 = 8$$

from which the only real value of t is obtained from the principal cube root of 8, so that $t + 1 = 2$; that is, $t = 1$. When $t = 1$,

$$s = \frac{1}{2}(1)^2 + \frac{4 \cdot 1}{1 + 1} \qquad v = 1 + \frac{4}{(1 + 1)^2}$$

$$= \tfrac{5}{2} \qquad\qquad\qquad = 2$$

Therefore the acceleration is 0 at 1 sec when the particle is $\frac{5}{2}$ m from the origin and moving to the right at a velocity of 2 m/sec.

A particle moving in a straight line is said to have **simple harmonic motion** if the measure of its acceleration is always proportional to the measure of its displacement from a fixed point on the line and its acceleration and displacement are oppositely directed.

EXAMPLE 5   Show that if a particle is moving along a straight line according to the equation of motion

$$s = b \sin(kt + \theta) \tag{4}$$

where b, k, and θ are constants and s meters is the directed distance of the particle from the origin at t seconds, then the motion is simple harmonic.

Solution   We wish to show that if a m/sec^2 is the acceleration of the particle at t seconds, then a is proportional to s, and a and s have opposite signs. To determine a we first find v, where v m/sec is the velocity of the particle at t seconds.

$$v = \frac{ds}{dt}$$

$$= b[\cos(kt + \theta)](k)$$
$$= bk \cos(kt + \theta)$$

$$a = \frac{dv}{dt}$$

$$= bk[-\sin(kt + \theta)](k)$$
$$= -bk^2 \sin(kt + \theta)$$

Substituting from (4) into this equation we have

$$a = -k^2 s$$

Because $-k^2$ is a constant, a is proportional to s. Furthermore, because $-k^2$ is negative, a and s are oppositely directed. Therefore the motion is simple harmonic.

Further applications of the second derivative are its uses in the sketch of the graph of a function (Section 4.5) and the second-derivative test for relative extrema (Section 4.6). An important application of other higher-order derivatives is to determine infinite series, as shown in Chapter 13.

The following example illustrates how the second derivative is found for functions defined implicitly.

EXAMPLE 6   Given

$$4x^2 + 9y^2 = 36$$

find $\dfrac{d^2y}{dx^2}$ by implicit differentiation.

Solution   Differentiating implicitly with respect to x we have

$$8x + 18y \frac{dy}{dx} = 0$$

$$\frac{dy}{dx} = \frac{-4x}{9y} \tag{5}$$

To find $\dfrac{d^2y}{dx^2}$ we compute the derivative of a quotient and keep in mind that y is a function of x. Thus

$$\frac{d^2y}{dx^2} = \frac{9y(-4) - (-4x)\left(9 \cdot \dfrac{dy}{dx}\right)}{81y^2}$$

Substituting the value of $\dfrac{dy}{dx}$ from (5) into this equation we get

$$\frac{d^2y}{dx^2} = \frac{-36y + (36x)\dfrac{-4x}{9y}}{81y^2}$$

$$= \frac{-36y^2 - 16x^2}{81y^3}$$

$$= \frac{-4(9y^2 + 4x^2)}{81y^3}$$

Because any values of x and y satisfying this equation must also satisfy the original equation, we can replace $9y^2 + 4x^2$ by 36 and obtain

$$\frac{d^2y}{dx^2} = \frac{-4(36)}{81y^3}$$

$$= -\frac{16}{9y^3}$$

EXERCISES 3.10

In Exercises 1 through 16, find the first and second derivative of the function.

1. $f(x) = x^5 - 2x^3 + x$
2. $F(x) = 7x^3 - 8x^2$
3. $g(s) = 2s^4 - 4s^3 + 7s - 1$
4. $G(t) = t^3 - t^2 + t$

5. $F(x) = x^2\sqrt{x} - 5x$
6. $g(r) = \sqrt{r} + \dfrac{1}{\sqrt{r}}$

7. $f(x) = \sqrt{x^2 + 1}$
8. $h(y) = \sqrt[3]{2y^3 + 5}$
9. $f(t) = 4\cos t^2$
10. $g(t) = 2\sin^3 t$

11. $G(x) = \cot^2 x$
12. $f(x) = \dfrac{2 - \sqrt{x}}{2 + \sqrt{x}}$

13. $g(x) = \dfrac{x^2}{x^2 + 4}$
14. $g(x) = (2x - 3)^2(x + 4)^3$

15. $f(x) = \sqrt{\sin x + 1}$
16. $f(x) = \sec 2x + \tan 2x$

17. Find $D_x^{\,3}(x^4 - 2x^2 + x - 5)$.
18. Find $D_t^{\,3}(\sqrt{4t + 1})$.
19. Find $\dfrac{d^4}{dx^4}\left(\dfrac{3}{2x - 1}\right)$.

20. Find $f^{(4)}(x)$ if $f(x) = \dfrac{2}{x - 1}$.

21. Find $D_x^{\,3}(2\tan 3x)$.
22. Find $\dfrac{d^4}{dt^4}(3\sin^2 2t)$.

23. Find $f^{(5)}(x)$ if $f(x) = \cos 2x - \sin 2x$.

24. Find $\dfrac{d^3u}{dv^3}$ if $u = v\sqrt{v - 2}$.

25. Given $x^2 + y^2 = 1$, show that $\dfrac{d^2y}{dx^2} = -\dfrac{1}{y^3}$.

26. Given $x^2 + 25y^2 = 100$, show that $\dfrac{d^2y}{dx^2} = -\dfrac{4}{25y^3}$.

27. Given $x^3 + y^3 = 1$, show that $\dfrac{d^2y}{dx^2} = \dfrac{-2x}{y^5}$.

28. Given $x^{1/2} + y^{1/2} = 2$, show that $\dfrac{d^2y}{dx^2} = \dfrac{1}{x^{3/2}}$.

29. Given $x^4 + y^4 = a^4$ (a is a constant), find $\dfrac{d^2y}{dx^2}$ in simplest form.

30. Given $b^2x^2 - a^2y^2 = a^2b^2$ (a and b are constants), find $\dfrac{d^2y}{dx^2}$ in simplest form.

31. Find the slope of the tangent line at each point of the graph of $y = x^4 + x^3 - 3x^2$ where the rate of change of the slope is zero.

32. Find the instantaneous rate of change of the slope of the tangent line to the graph of $y = 2x^3 - 6x^2 - x + 1$ at the point $(3, -2)$.

In Exercises 33 and 34, a particle is moving along a horizontal line according to the given equation, where s meters is the directed distance of the particle from the origin at t seconds, v m/sec is the velocity of the particle at t seconds, and a m/sec² is the acceleration of the particle at t seconds. Find v and a in terms of t. Make a table similar to Table 1 that gives a description of the position and motion of the particle. Include in the table the intervals of time when the particle is moving to the left, when it is moving to the right, when the velocity is increasing, when the velocity is decreasing, when the speed is increasing, when the speed is decreasing, and the position of the particle with respect to the origin during these intervals of time. Show the behavior of the motion by a figure similar to Figure 1.

33. $s = t^3 - 9t^2 + 15t, \ t \geq 0$ **34.** $s = \frac{1}{6}t^3 - 2t^2 + 6t - 2, \ t \geq 0$

In Exercises 35 through 39, a particle is moving along a straight line according to the given equation, where s feet is the directed distance of the particle from the origin at t seconds. Find the time when the instantaneous acceleration is zero, and then find the directed distance of the particle from the origin and the instantaneous velocity at this time.

35. $s = \frac{1}{3}t^3 - \frac{3}{2}t^2 + 2t + 1, \ t \geq 0$

36. $s = 2t^3 - 6t^2 + 3t - 4, \ t \geq 0$ **37.** $s = \dfrac{125}{16t + 32} - \dfrac{2}{5}t^5, \ t \geq 0$

38. $s = 9t^2 + 2\sqrt{2t + 1}, \ t \geq 0$ **39.** $s = \frac{4}{9}t^{3/2} + 2t^{1/2}, \ t \geq 0$

In Exercises 40 through 45, a particle is moving along a straight line according to the given equation of motion, where s meters is the directed distance of the particle from the origin at t seconds. Show that the motion is simple harmonic.

40. $s = A \sin 2\pi kt + B \cos 2\pi kt$, where A, B, and k are constants
41. $s = b \cos(kt + \theta)$, where b, k, and θ are constants
42. $s = 6 \sin(t + \frac{1}{3}\pi) + 4 \sin(t - \frac{1}{6}\pi)$
43. $s = \sin(6t - \frac{1}{3}\pi) + \sin(6t + \frac{1}{6}\pi)$

44. $s = 8 \cos^2 6t - 4$
45. $s = 5 - 10 \sin^2 2t$

In Exercises 46 through 49, find formulas for f'(x) and f''(x), and state the domains of f' and f''.

46. $f(x) = \begin{cases} \dfrac{x^2}{|x|} & \text{if } x \neq 0 \\ 0 & \text{if } x = 0 \end{cases}$ **47.** $f(x) = \begin{cases} -x^2 & \text{if } x < 0 \\ x^2 & \text{if } 0 \leq x \end{cases}$

48. $f(x) = |x|^3$ **49.** $f(x) = \begin{cases} \dfrac{x^5}{|x|} & \text{if } x \neq 0 \\ 0 & \text{if } x = 0 \end{cases}$

50. For the function of Exercise 48, find $f'''(x)$ when it exists.
51. For the function of Exercise 49, find $f'''(x)$ when it exists.
52. Show that if $xy = 1$, then $\dfrac{d^2y}{dx^2} \cdot \dfrac{d^2x}{dy^2} = 4$.
53. If f'' and g'' exist and if $h = f \circ g$, express $h''(x)$ in terms of the derivatives of f and g.
54. If f and g are two functions such that their first and second derivatives exist and h is the function defined by the equation $h(x) = f(x) \cdot g(x)$, prove that

$$h''(x) = f(x) \cdot g''(x) + 2f'(x) \cdot g'(x) + f''(x) \cdot g(x)$$

55. If $y = x^n$, where n is any positive integer, prove by mathematical induction that $\dfrac{d^n y}{dx^n} = n!$

56. If

$$y = \frac{1}{1 - 2x}$$

prove by mathematical induction that

$$\frac{d^n y}{dx^n} = \frac{2^n n!}{(1 - 2x)^{n+1}}$$

57. If k is any positive integer, prove by mathematical induction that

$$D_x{}^n(\sin x) = \begin{cases} \sin x & \text{if } n = 4k \\ \cos x & \text{if } n = 4k + 1 \\ -\sin x & \text{if } n = 4k + 2 \\ -\cos x & \text{if } n = 4k + 3 \end{cases}$$

58. Obtain a formula similar to that in Exercise 57 for $D_x{}^n(\cos x)$.

REVIEW EXERCISES FOR CHAPTER 3

In Exercises 1 through 16, find the derivative of the function.

1. $f(x) = 5x^3 - 7x^2 + 2x - 3$ **2.** $g(x) = 5(x^4 + 3x^7)$

3. $g(x) = \dfrac{x^2}{4} + \dfrac{4}{x^2}$ **4.** $f(x) = \dfrac{4}{x^2} - \dfrac{3}{x^4}$

5. $F(x) = 2x^{1/2} - \dfrac{1}{2}x^{-1/2}$ **6.** $G(x) = \dfrac{x^2 - 4x + 4}{x - 1}$

7. $G(t) = (3t^2 - 4)(4t^3 + t - 1)$

8. $f(x) = (x^4 - 2x)(4x^2 + 2x + 5)$

9. $g(x) = \dfrac{x^3 + 1}{x^3 - 1}$

10. $h(y) = \dfrac{y^2}{y^3 + 8}$

11. $f(s) = (2s^3 - 3s + 7)^4$

12. $F(x) = (4x^4 - 4x^2 + 1)^{-1/3}$

13. $f(x) = \sqrt[3]{\dfrac{x}{x^3 + 1}}$

14. $g(x) = \left(\dfrac{3x^2 + 4}{x^7 + 1}\right)^{10}$

15. $F(x) = (x^2 - 1)^{3/2}(x^2 - 4)^{1/2}$

16. $g(x) = (x^4 - x)^{-3}(5 - x^2)^{-1}$

In Exercises 17 through 24, compute the derivative.

17. $D_x[(x + 1)\sin x - x \cos x]$

18. $D_r(\tan^2 3r)$

19. $\dfrac{d}{dx}\left(x \tan \dfrac{1}{x}\right)$

20. $D_t(\sin^2 3t \sqrt{\cos 2t})$

21. $\dfrac{d}{dt}\left(\dfrac{\sec^2 t}{1 + t^2}\right)$

22. $\dfrac{d}{dx}\left(\dfrac{1 + \sin x}{x \cos x}\right)$

23. $D_w[\sin(\cos 3w) - 3 \cos^2 2w]$

24. $D_x(\tan^3 x \cdot \sec x)$

In Exercises 25 through 34, find $\dfrac{dy}{dx}$.

25. $4x^2 + 4y^2 - y^3 = 0$

26. $y = \sqrt{1 + x} + \sqrt{1 - x}$

27. $y = \dfrac{1}{x - \sqrt{x^2 - 1}}$

28. $xy^2 + 2y^3 = x - 2y$

29. $\sin(x + y) + \sin(x - y) = 1$

30. $x^{2/3} + y^{2/3} = a^{2/3}$

31. $y = x^2 + [x^3 + (x^4 + x)^2]^3$

32. $y = \dfrac{x\sqrt{3 + 2x}}{4x - 1}$

33. $\tan x + \tan y = xy$

34. $\sec(x + y) - \sec(x - y) = 1$

35. Find equations of the tangent lines to the curve $y = 2x^3 + 4x^2 - x$ that have slope $\frac{1}{2}$.

36. Find an equation of the normal line to the curve $x - y = \sqrt{x + y}$ at the point $(3, 1)$.

37. Find equations of the tangent and normal lines to the curve $2x^3 + 2y^3 - 9xy = 0$ at the point $(2, 1)$.

38. Find equations of the tangent and normal lines to the curve $y = 8 \sin^3 2x$ at the point $(\frac{1}{12}\pi, 1)$.

39. Prove that the line tangent to the curve $y = -x^4 + 2x^2 + x$ at the point $(1, 2)$ is also tangent to the curve at another point, and find this point.

40. Prove that the tangent lines to the curves

$$4y^3 - x^2y - x + 5y = 0 \quad \text{and} \quad x^4 - 4y^3 + 5x + y = 0$$

at the origin are perpendicular.

41. Find $\dfrac{d^3y}{dx^3}$ if $y = \sqrt{3 - 2x}$.

42. Given $\dfrac{dy}{dx} = y^k$, where k is a constant and y is a function of x. Express $\dfrac{d^3y}{dx^3}$ in terms of y and k.

43. Given $f(x) = \frac{1}{12}x^4 + \frac{2}{3}x^3 + \frac{3}{2}x^2 + 8x + 2$. For what values of x is $f''(x) > 0$?

44. Show that if $xy = k$, where k is a nonzero constant, then

$$\dfrac{d^2y}{dx^2} \cdot \dfrac{d^2x}{dy^2} = \dfrac{4}{k}.$$

45. A particle is moving in a horizontal line according to the equation $s = t^3 - 11t^2 + 24t + 100$, where s meters is the directed distance of the particle from the origin at t seconds. (a) The particle is at the starting point when $t = 0$. For what other values of t is the particle at the starting point? (b) Determine the velocity of the particle at each instant that it is at the starting point, and interpret the sign of the velocity in each case.

46. A particle is moving in a horizontal line according to the equation $s = t^3 - 3t^2 - 9t + 2$, where s meters is the directed distance of the particle from a point O at t seconds. The positive direction is to the right. Determine the intervals of time when the particle is moving to the right and when it is moving to the left. Also determine when the particle reverses its direction. Show the behavior of the motion by a figure, and choose values of t at random, but include the values of t when the particle reverses its direction.

47. A particle is moving along a straight line according to the equation $s = 5 - 2 \cos^2 t$, where s meters is the directed distance of the particle from the origin at t seconds. If v m/sec and a m/sec^2 are, respectively, the velocity and acceleration of the particle at t seconds, find v and a in terms of s.

48. An object is sliding down an inclined plane according to the equation $s = 12t^2 + 6t$, where s meters is the directed distance of the object from the top t seconds after starting. (a) Find the velocity 3 sec after the start. (b) Find the initial velocity.

49. A ball is thrown vertically upward from the top of a house 112 ft high. Its equation of motion is $s = -16t^2 + 96t$, where s feet is the directed distance of the ball from the starting point at t seconds. Find (a) the instantaneous velocity of the ball at 2 sec; (b) how high the ball will go; (c) how long it takes for the ball to reach the ground; (d) the instantaneous velocity of the ball when it reaches the ground.

50. Stefan's law states that a body emits radiant energy according to the formula $R = kT^4$, where R is the measure of the rate of emission of the radiant energy per square unit of area, T is the measure of the Kelvin temperature of the surface, and k is a constant. Find (a) the average rate of change of R with respect to T as T increases from 200 to 300; (b) the instantaneous rate of change of R with respect to T when T is 200.

51. If A square units is the area of an isosceles right triangle for which each leg has a length of x units, find (a) the average rate of change of A with respect to x as x changes from 8.00 to 8.01; (b) the instantaneous rate of change of A with respect to x when x is 8.00.

52. If $y = x^{2/3}$, find the relative rate of change of y with respect to x when (a) $x = 8$, and (b) $x = c$, where c is a constant.

53. The supply equation for a calculator is $y = m^2 + \sqrt{m}$, where $100y$ calculators are supplied when m dollars is the price per calculator. Find (a) the average rate of change of the supply with respect to the price when the price is increased from $16 to $17; (b) the instantaneous (or marginal) rate of change of the supply with respect to the price when the price is $16.

54. Use the definition of a derivative to find $f'(x)$ if $f(x) = 3x^2 - 5x + 1$.

55. Use the definition of a derivative to find $f'(-5)$ if
$$f(x) = \frac{3}{x + 2}.$$

56. Use the definition of a derivative to find $f'(5)$ if $f(x) = \sqrt{3x + 1}$.

57. Use the definition of a derivative to find $f'(x)$ if $f(x) = \sqrt{4x - 3}$.

58. Find $f''(x)$ if $f(x) = 3 \sin^2 x - 4 \cos^2 x$.

59. Find $f''(\pi)$ if $f(x) = \sqrt{2 + \cos x}$.

60. Find $f'(-3)$ if $f(x) = (|x| - x)\sqrt[3]{9x}$.

61. Find $f'(x)$ if $f(x) = (|x + 1| - |x|)^2$.

62. Given
$$f(x) = \begin{cases} x^2 - 16 & \text{if } x < 4 \\ 8x - 32 & \text{if } 4 \le x \end{cases}$$
(a) Draw a sketch of the graph of f. (b) Determine if f is continuous at 4. (c) Determine if f is differentiable at 4.

63. Given
$$f(x) = \begin{cases} x^2 + 2 & \text{if } x \le 3 \\ 20 - x^2 & \text{if } 3 < x \end{cases}$$
(a) Draw a sketch of the graph of f. (b) Determine if f is continuous at 3. (c) Determine if f is differentiable at 3.

64. The remainder theorem of elementary algebra states that if $P(x)$ is a polynomial in x and r is any real number, then there is a polynomial $Q(x)$ such that $P(x) = Q(x)(x - r) + P(r)$. What is $\lim\limits_{x \to r} Q(x)$?

65. Given $f(x) = |x|^3$. (a) Draw a sketch of the graph of f. (b) Find $\lim\limits_{x \to 0} f(x)$ if it exists. (c) Find $f'(0)$ if it exists.

66. Given $f(x) = x^2 \operatorname{sgn} x$. (a) Where is f differentiable? (b) Is f' continuous on its domain?

67. Given
$$f(x) = \begin{cases} ax^2 + b & \text{if } x \le 1 \\ \dfrac{1}{|x|} & \text{if } 1 < x \end{cases}$$
Find the values of a and b such that $f'(1)$ exists.

68. Suppose
$$f(x) = \begin{cases} x^3 & \text{if } x < 1 \\ ax^2 + bx + c & \text{if } 1 \le x \end{cases}$$
Find the values of a, b, and c such that $f''(1)$ exists.

69. If $C(x)$ dollars is the total cost of manufacturing x chairs, and $C(x) = x^2 + 40x + 800$, find (a) the marginal cost function; (b) the marginal cost when 20 chairs are manufactured; (c) the actual cost of manufacturing the twenty-first chair.

70. The total revenue received from the sale of x lamps is $R(x)$ dollars and $R(x) = 100x - \frac{1}{6}x^2$. Find (a) the marginal revenue function; (b) the marginal revenue when $x = 15$; (c) the actual revenue from the sale of the sixteenth lamp.

71. In a large lake a predator fish feeds on a smaller fish, and the predator population at any time is a function of the num-

ber of small fish in the lake at that time. Suppose that when there are x small fish in the lake, the predator population is y, and $y = \frac{1}{4}x^2 + 80$. If the fishing season ended t weeks ago, $x = 8t + 90$. At what rate is the population of the predator fish growing 9 weeks after the close of the fishing season? Do not express y in terms of t, but use the chain rule.

72. The demand equation for a particular candy bar is
$$px + x + 20p = 3000$$
where $1000x$ candy bars are demanded per week when p cents is the price per bar. If the current price of the candy is 49 cents per bar and the price per bar is increasing at the rate of 0.2 cent each week, find the rate of change in the demand.

73. A particle is moving along a straight line, and
$$s = \sin(4t + \tfrac{1}{3}\pi) + \sin(4t + \tfrac{1}{6}\pi)$$
where s feet is the directed distance of the particle from the origin at t seconds. Prove that the motion is simple harmonic.

74. If an equation of motion is $s = \cos 2t + 2 \sin 2t$, prove that the motion is simple harmonic.

75. If a particle is moving along a straight line such that $s = \cos 2t + \cos t$, prove that the motion is not simple harmonic.

76. A particle is moving in a straight line according to the equation $s = \sqrt{a + bt^2}$, where a and b are positive constants. Prove that the measure of the acceleration of the particle is inversely proportional to s^3 for any t.

77. A ship leaves a port at noon and travels due west at 20 knots. At noon the next day a second ship leaves the same port and travels northwest at 15 knots. How fast are the two ships separating when the second ship has traveled 90 nautical miles?

78. A reservoir is 80 m long and its cross section is an isosceles trapezoid having equal sides of 10 m, an upper base of 17 m, and a lower base of 5 m. At the instant when the water is 5 m deep, find the rate at which the water is leaking out if the water level is falling at the rate of 0.1 m/hr.

79. A funnel in the form of a cone is 10 in. across the top and 8 in. deep. Water is flowing into the funnel at the rate of 12 in.^{3}/sec and out at the rate of 4 in.^{3}/sec. How fast is the surface of the water rising when it is 5 in. deep?

80. As the last car of a train passes under a bridge, an automobile crosses the bridge on a roadway perpendicular to the track and 30 ft above it. The train is traveling at the rate of 80 ft/sec and the automobile is traveling at the rate of 40 ft/sec. How fast are the train and the automobile separating after 2 sec?

81. A man 6 ft tall is walking toward a building at the rate of 4 ft/sec. If there is a light on the ground 40 ft from the building, how fast is the man's shadow on the building growing shorter when he is 30 ft from the building?

82. A burn on a person's skin is in the shape of a circle. If the radius of the burn is decreasing at the rate of 0.05 cm per day when it is 1.0 cm, what is the rate of decrease of the area of the burn at that instant?

83. Suppose $f(x) = 3x + |x|$ and $g(x) = \frac{3}{4}x - \frac{1}{4}|x|$. Prove that neither $f'(0)$ nor $g'(0)$ exists but that $(f \circ g)'(0)$ does exist.

84. Give an example of two functions f and g for which f is differentiable at $g(0)$, g is not differentiable at 0, and $f \circ g$ is differentiable at 0.

85. Give an example of two functions f and g for which f is not differentiable at $g(0)$, g is differentiable at 0, and $f \circ g$ is differentiable at 0.

86. In Exercise 59 of Exercises 3.1, you are to prove that if $f'(a)$ exists, then

$$f'(a) = \lim_{\Delta x \to 0} \frac{f(a + \Delta x) - f(a - \Delta x)}{2 \, \Delta x}$$

Show by using the absolute-value function that it is possible for the limit in the above equation to exist even though $f'(a)$ does not exist.

87. If $f'(x_1)$ exists, prove that

$$\lim_{x \to x_1} \frac{xf(x_1) - x_1 f(x)}{x - x_1} = f(x_1) - x_1 f'(x_1)$$

88. Let f and g be two functions whose domains are the set of all real numbers. Furthermore, suppose that (i) $g(x) = xf(x) + 1$;

(ii) $g(a + b) = g(a) \cdot g(b)$ for all a and b; (iii) $\lim_{x \to 0} f(x) = 1$.
Prove that $g'(x) = g(x)$.

89. If the two functions f and g are differentiable at the number x_1, is the composite function $f \circ g$ necessarily differentiable at x_1? If your answer is yes, prove it. If your answer is no, give a counterexample.

90. Suppose $g(x) = |f(x)|$. If $f^{(n)}(x)$ exists and $f(x) \neq 0$, prove that

$$g^{(n)}(x) = \frac{f(x)}{|f(x)|} f^{(n)}(x)$$

91. Prove that $D_x{}^n(\sin x) = \sin(x + \frac{1}{2}n\pi)$. (*Hint:* Use mathematical induction and the formulas $\sin(x + \frac{1}{2}\pi) = \cos x$ or $\cos(x + \frac{1}{2}\pi) = -\sin x$ after each differentiation.)

92. Suppose the function f is defined on the open interval $(0, 1)$ and

$$f(x) = \frac{\sin \pi x}{x(x - 1)}$$

Define f at 0 and 1 so that f is continuous on the closed interval $[0, 1]$.

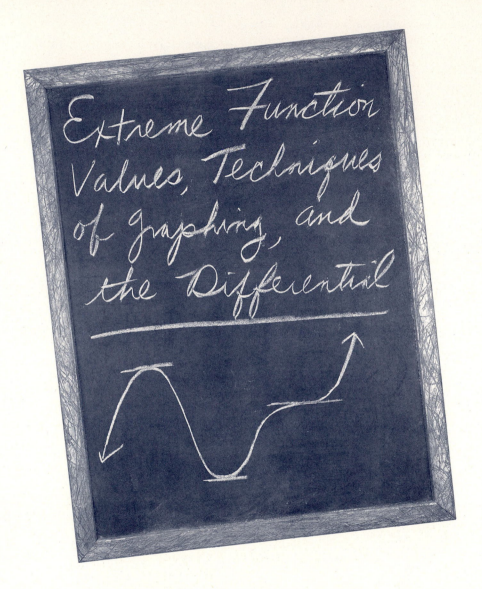

Extreme Function Values, Techniques of graphing, and the Differential

The interpretation of the derivative as the slope of a tangent line gives us information about the behavior of functions and thus it is used in techniques of graphing functions. Sections 4.1 and 4.4–4.7 all pertain to this application. In Section 4.1 we define *extreme function values* and utilize them in Sections 4.2 and 4.8 to solve word problems involving maxima and minima. For example, we determine the strongest rectangular beam that can be cut from a given cylindrical log as well as the dimensions of a box requiring the least amount of material for a specific volume.

One of the most important theorems in calculus is the *mean-value theorem*, discussed in Section 4.3. It is used to prove many theorems of both differential and integral calculus, as well as of other subjects, such as numerical analysis. In Section 4.9 we introduce the concept of the *differential*. Supplementary Section 4.10 is devoted to *Newton's method*, an application of the derivative to numerical processes for approximating solutions of equations.

4.1 MAXIMUM AND MINIMUM FUNCTION VALUES

We have seen that the geometrical interpretation of the derivative of a function is the slope of the tangent line to the graph of the function at a point. This fact enables us to apply derivatives as an aid in sketching graphs. For example, the derivative may be used to determine at what points the tangent line is horizontal; these are the points where the derivative is zero. Also, the derivative may be used to find the intervals for which the graph of a function lies above the tangent line and the intervals for which the graph lies below the tangent line. First we need some definitions and theorems.

4.1.1 DEFINITION

The function f is said to have a **relative maximum value** at c if there exists an open interval containing c, on which f is defined, such that $f(c) \geq f(x)$ for all x in this interval.

Figures 1 and 2 each show a sketch of a portion of the graph of a function having a relative maximum value at c.

4.1.2 DEFINITION

The function f is said to have a **relative minimum value** at c if there exists an open interval containing c, on which f is defined, such that $f(c) \leq f(x)$ for all x in this interval.

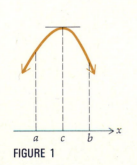

FIGURE 1

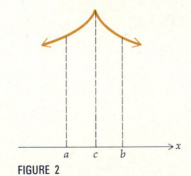

FIGURE 2

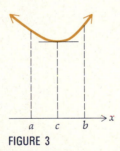

FIGURE 3

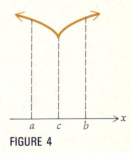

FIGURE 4

Figures 3 and 4 each show a sketch of a portion of the graph of a function having a relative minimum value at c.

If the function f has either a relative maximum or a relative minimum value at c, then f is said to have a **relative extremum** at c.

The following theorem is used to locate the possible values of c for which there is a relative extremum.

4.1.3 THEOREM

If $f(x)$ exists for all values of x in the open interval (a, b), and if f has a relative extremum at c, where $a < c < b$, and if $f'(c)$ exists, then $f'(c) = 0$.

The geometrical interpretation of this theorem is that if f has a relative extremum at c, and if $f'(c)$ exists, then the graph of f must have a horizontal tangent line at the point where $x = c$.

The proof of Theorem 4.1.3 makes use of Theorems 2.10.3 and 2.10.4. You may wish to refer to those theorems at this time and also refer to Illustrations 3 and 4 in Supplementary Section 2.10, which give geometrical interpretations of the theorems.

Proof of Theorem 4.1.3   The proof will be given for the case when f has a relative minimum value at c.

If $f'(c)$ exists, then

$$f'(c) = \lim_{x \to c} \frac{f(x) - f(c)}{x - c} \tag{1}$$

Because f has a relative minimum value at c, by Definition 4.1.2 there exists a $\delta > 0$ such that

$$\text{if} \quad 0 < |x - c| < \delta \quad \text{then} \quad f(x) - f(c) \geq 0$$

If x is approaching c from the right, $x - c > 0$, and therefore

$$\text{if} \quad 0 < x - c < \delta \quad \text{then} \quad \frac{f(x) - f(c)}{x - c} \geq 0$$

By Theorem 2.10.4, if the limit exists,

$$\lim_{x \to c^+} \frac{f(x) - f(c)}{x - c} \geq 0 \tag{2}$$

Similarly, if x is approaching c from the left, $x - c < 0$, and therefore

$$\text{if} \quad -\delta < x - c < 0 \quad \text{then} \quad \frac{f(x) - f(c)}{x - c} \leq 0$$

so that by Theorem 2.10.3, if the limit exists,

$$\lim_{x \to c^-} \frac{f(x) - f(c)}{x - c} \leq 0 \tag{3}$$

Because $f'(c)$ exists, the limits in inequalities (2) and (3) must be equal, and both must be equal to $f'(c)$. So from (2),

$$f'(c) \geq 0$$

and from (3),

$$f'(c) \leq 0$$

Because both of these inequalities are to be true, we conclude that

$$f'(c) = 0$$

which was to be proved.

The proof for the case when f has a relative maximum value at c is similar and is left as an exercise (see Exercise 59).   ∎

If f is a differentiable function, then the only possible values of x for which f can have a relative extremum are those for which $f'(x) = 0$. However, $f'(x)$ can be equal to zero for a specific value of x, and yet f may not have a relative extremum there, as shown in the following illustration.

▶ **ILLUSTRATION 1**   Consider the function f defined by

$$f(x) = (x - 1)^3$$

A sketch of the graph of this function is shown in Figure 5. $f'(x) = 3(x - 1)^2$, and so $f'(1) = 0$. However, $f(x) < 0$ if $x < 1$, and $f(x) > 0$ if $x > 1$. So f does not have a relative extremum at 1.   ◀

A function f may have a relative extremum at a number and f' may fail to exist there. This situation is shown in Illustration 2.

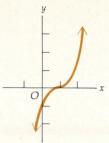

FIGURE 5

▶ **ILLUSTRATION 2**   Let the function f be defined as follows:

$$f(x) = \begin{cases} 2x - 1 & \text{if } x \le 3 \\ 8 - x & \text{if } 3 < x \end{cases}$$

A sketch of the graph of this function appears in Figure 6. The function f has a relative maximum value at 3. The derivative from the left at 3 is given by $f'_-(3) = 2$, and the derivative from the right at 3 is given by $f'_+(3) = -1$. Therefore we conclude that $f'(3)$ does not exist.   ◀

Illustration 2 demonstrates why the condition "$f'(c)$ exists" must be included in the hypothesis of Theorem 4.1.3.

It is possible that a function f can be defined at a number c where $f'(c)$ does not exist and yet f may not have a relative extremum there. The following illustration gives such a function.

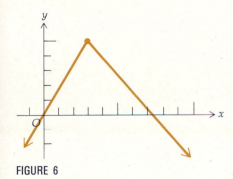

FIGURE 6

▶ **ILLUSTRATION 3**   Let the function f be defined by

$$f(x) = x^{1/3}$$

The domain of f is the set of all real numbers.

$$f'(x) = \frac{1}{3x^{2/3}} \qquad \text{if } x \ne 0$$

Furthermore, $f'(0)$ does not exist. Figure 7 shows a sketch of the graph of f. The function has no relative extrema.   ◀

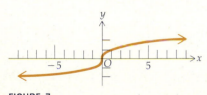

FIGURE 7

In summary, then, if a function f is defined at a number c, a necessary condition for f to have a relative extremum there is that either $f'(c) = 0$ or $f'(c)$ does not exist. But this condition is not sufficient.

4.1.4 DEFINITION   If c is a number in the domain of the function f, and if either $f'(c) = 0$ or $f'(c)$ does not exist, then c is called a **critical number** of f.

Because of this definition and the previous discussion, a necessary (but not sufficient) condition for a function to have a relative extremum at c is for c to be a critical number.

EXAMPLE 1   Find the critical numbers of the function f defined by

$$f(x) = x^{4/3} + 4x^{1/3}$$

Solution

$$f'(x) = \tfrac{4}{3}x^{1/3} + \tfrac{4}{3}x^{-2/3}$$
$$= \tfrac{4}{3}x^{-2/3}(x + 1)$$
$$= \frac{4(x + 1)}{3x^{2/3}}$$

When $x = -1$, $f'(x) = 0$, and when $x = 0$, $f'(x)$ does not exist. Both -1 and 0 are in the domain of f; therefore the critical numbers of f are -1 and 0.

EXAMPLE 2   Find the critical numbers of the function g defined by

$$g(x) = \sin x \cos x$$

Solution   Because $\sin 2x = 2 \sin x \cos x$,

$$g(x) = \tfrac{1}{2}\sin 2x$$
$$g'(x) = \tfrac{1}{2}(\cos 2x)2$$
$$= \cos 2x$$

Since $g'(x)$ exists for all x, the only critical numbers are those for which $g'(x) = 0$. Because $\cos 2x = 0$ when

$$2x = \tfrac{1}{2}\pi + k\pi \qquad \text{where } k \text{ is any integer}$$

the critical numbers of g are $\tfrac{1}{4}\pi + \tfrac{1}{2}k\pi$, where k is any integer.

We are frequently concerned with a function defined on a given interval, and we wish to find the largest or smallest function value on the interval. These intervals can be either closed, open, or closed at one end and open at the other. The greatest function value on an interval is called the *absolute maximum value*, and the smallest function value on an interval is called the *absolute minimum value*. Following are the precise definitions.

4.1.5 DEFINITION   The function f is said to have an **absolute maximum value on an interval** if there is some number c in the interval such that $f(c) \geq f(x)$ for all x in the interval. In such a case, $f(c)$ is the absolute maximum value of f on the interval.

4.1.6 DEFINITION   The function f is said to have an **absolute minimum value on an interval** if there is some number c in the interval such that $f(c) \leq f(x)$ for all x in the interval. In such a case, $f(c)$ is the absolute minimum value of f on the interval.

An **absolute extremum** of a function on an interval is either an absolute maximum value or an absolute minimum value of the function on the interval. A function may or may not have an absolute extremum on a particular interval. In each of the following illustrations, a function and an interval are given, and we find the absolute extrema of the function on the interval if there are any.

▶ **ILLUSTRATION 4**   Suppose f is the function defined by

$$f(x) = 2x$$

FIGURE 8

A sketch of the graph of f on $[1, 4)$ is in Figure 8. This function has an absolute minimum value of 2 on $[1, 4)$. There is no absolute maximum value of f on $[1, 4)$ because $\lim\limits_{x \to 4^-} f(x) = 8$, but $f(x)$ is always less than 8 on the interval. ◀

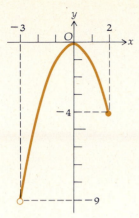

FIGURE 9

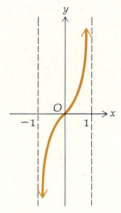

FIGURE 10

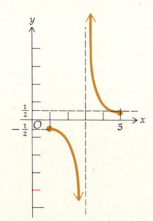

FIGURE 12

▶ **ILLUSTRATION 5**   Consider the function f defined by

$$f(x) = -x^2$$

A sketch of the graph of f on $(-3, 2]$ appears in Figure 9. This function has an absolute maximum value of 0 on $(-3, 2]$. There is no absolute minimum value of f on $(-3, 2]$ because $\lim\limits_{x \to -3^+} f(x) = -9$, but $f(x)$ is always greater than -9 on the given interval.   ◀

▶ **ILLUSTRATION 6**   The function f defined by

$$f(x) = \frac{x}{1 - x^2}$$

has neither an absolute maximum value nor an absolute minimum value on $(-1, 1)$. Figure 10 shows a sketch of the graph of f on $(-1, 1)$. Observe that

$$\lim_{x \to -1^+} f(x) = -\infty \qquad \lim_{x \to 1^-} f(x) = +\infty$$ ◀

▶ **ILLUSTRATION 7**   Let f be the function defined by

$$f(x) = \begin{cases} x + 1 & \text{if } x < 1 \\ x^2 - 6x + 7 & \text{if } 1 \le x \end{cases}$$

There is a sketch of the graph of f on $[-5, 4]$ in Figure 11. The absolute maximum value of f on $[-5, 4]$ occurs at 1, and $f(1) = 2$; the absolute minimum value of f on $[-5, 4]$ occurs at -5, and $f(-5) = -4$. Note that f has a relative maximum value at 1 and a relative minimum value at 3. Also observe that 1 is a critical number of f because $f'(1)$ does not exist, and 3 is a critical number of f because $f'(3) = 0$.   ◀

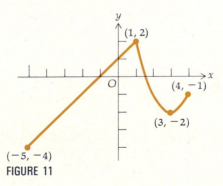

FIGURE 11

▶ **ILLUSTRATION 8**   The function f defined by

$$f(x) = \frac{1}{x - 3}$$

has neither an absolute maximum value nor an absolute minimum value on $[1, 5]$. See Figure 12 for a sketch of the graph of f. $\lim\limits_{x \to 3^-} f(x) = -\infty$; so $f(x)$ can be made less than any negative number by taking $3 - x > 0$ and less than a suitable positive δ. Also, $\lim\limits_{x \to 3^+} f(x) = +\infty$; so $f(x)$ can be made greater than any positive number by taking $x - 3 > 0$ and less than a suitable positive δ.   ◀

We may speak of an absolute extremum of a function when no interval is specified. In such a case we are referring to an absolute extremum of the function on the entire domain of the function.

4.1.7 DEFINITION $f(c)$ is said to be the **absolute maximum value** of the function f if c is in the domain of f and if $f(c) \geq f(x)$ for all values of x in the domain of f.

4.1.8 DEFINITION $f(c)$ is said to be the **absolute minimum value** of the function f if c is in the domain of f and if $f(c) \leq f(x)$ for all values of x in the domain of f.

▶ **ILLUSTRATION 9**   The graph of the function f defined by

$$f(x) = x^2 - 4x + 8$$

is a parabola, and a sketch is shown in Figure 13. The lowest point of the parabola is at $(2, 4)$, and the parabola opens upward. The function has an absolute minimum value of 4 at 2. There is no absolute maximum value of f.   ◀

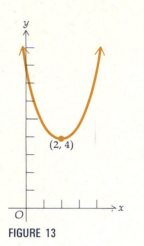

FIGURE 13

Referring back to Illustrations 4–9, we see that the only case in which there are both an absolute maximum function value and an absolute minimum function value is in Illustration 7, where the function is continuous on the closed interval $[-5, 4]$. In the other illustrations, either we do not have a closed interval or we do not have a continuous function. If a function is continuous on a closed interval, there is a theorem, called the *extreme-value theorem*, which assures that the function has both an absolute maximum value and an absolute minimum value on the interval. The proof of this theorem is beyond the scope of this book. You are referred to an advanced calculus text for the proof.

4.1.9 THEOREM
Extreme-Value Theorem

If the function f is continuous on the closed interval $[a, b]$, then f has an absolute maximum value and an absolute minimum value on $[a, b]$.

Theorem 4.1.9 states that continuity of a function on a closed interval is a sufficient condition to guarantee that the function has both an absolute maximum value and an absolute minimum value on the interval. However, it is not a necessary condition. For example, the function whose graph appears in Figure 14 has an absolute maximum value at $x = c$ and an absolute minimum value at $x = d$, even though the function is discontinuous on the open interval $(-1, 1)$.

An absolute extremum of a function continuous on a closed interval must be either a relative extremum or a function value at an endpoint of the interval. Because a necessary condition for a function to have a relative extremum at a number c is for c to be a critical number, the absolute maximum value and the absolute minimum value of a continuous function f on a closed interval $[a, b]$ can be determined by the following procedure:

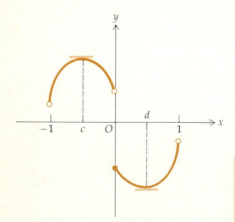

FIGURE 14

1. Find the function values at the critical numbers of f on (a, b).
2. Find the values of $f(a)$ and $f(b)$.
3. The largest of the values from steps 1 and 2 is the absolute maximum value, and the smallest of the values is the absolute minimum value.

Table 1

x	-2	-1	$\frac{1}{3}$	$\frac{1}{2}$
$f(x)$	-1	2	$\frac{22}{27}$	$\frac{7}{8}$

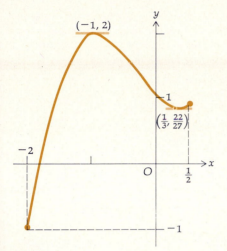

FIGURE 15

Table 2

x	1	2	5
$f(x)$	1	0	$\sqrt[3]{9}$

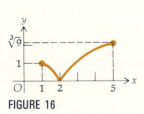

FIGURE 16

EXAMPLE 3 Find the absolute extrema of f on $[-2, \frac{1}{2}]$ if

$$f(x) = x^3 + x^2 - x + 1$$

Solution Because f is continuous on $[-2, \frac{1}{2}]$, the extreme-value theorem applies. To find the critical numbers of f, first find f':

$$f'(x) = 3x^2 + 2x - 1$$

Because $f'(x)$ exists for all real numbers, the only critical numbers of f will be the values of x for which $f'(x) = 0$. Set $f'(x) = 0$.

$$(3x - 1)(x + 1) = 0$$

$$x = \tfrac{1}{3} \qquad x = -1$$

The critical numbers of f are -1 and $\frac{1}{3}$, and each of these numbers is in the given closed interval $[-2, \frac{1}{2}]$. The function values at the critical numbers and at the endpoints of the interval are given in Table 1.

The absolute maximum value of f on $[-2, \frac{1}{2}]$ is therefore 2, which occurs at -1, and the absolute minimum value of f on $[-2, \frac{1}{2}]$ is -1, which occurs at the left endpoint -2. Figure 15 shows a sketch of the graph of f on $[-2, \frac{1}{2}]$.

EXAMPLE 4 Find the absolute extrema of f on $[1, 5]$ if

$$f(x) = (x - 2)^{2/3}$$

Solution Because f is continuous on $[1, 5]$, the extreme-value theorem applies.

$$f'(x) = \frac{2}{3(x - 2)^{1/3}}$$

There is no value of x for which $f'(x) = 0$. However, because $f'(x)$ does not exist at 2, we conclude that 2 is a critical number of f; so the absolute extrema occur either at 2 or at one of the endpoints of the interval. The function values at these numbers are given in Table 2.

From the table we conclude that the absolute minimum value of f on $[1, 5]$ is 0, occurring at 2, and the absolute maximum value of f on $[1, 5]$ is $\sqrt[3]{9}$, occurring at 5. A sketch of the graph of this function on $[1, 5]$ appears in Figure 16.

EXERCISES 4.1

In Exercises 1 through 20, find the critical numbers of the function.

1. $f(x) = x^3 + 7x^2 - 5x$

2. $g(x) = 2x^3 - 2x^2 - 16x + 1$

3. $f(x) = x^4 + 4x^3 - 2x^2 - 12x$

4. $f(x) = x^{7/3} + x^{4/3} - 3x^{1/3}$

5. $g(x) = x^{6/5} - 12x^{1/5}$

6. $f(x) = x^4 + 11x^3 + 34x^2 + 15x - 2$

7. $f(t) = (t^2 - 4)^{2/3}$

8. $f(x) = (x^3 - 3x^2 + 4)^{1/3}$

9. $h(x) = \dfrac{x - 3}{x + 7}$

10. $f(t) = t^{5/3} - 3t^{2/3}$

11. $f(x) = \dfrac{x}{x^2 - 9}$

12. $f(x) = \dfrac{x + 1}{x^2 - 5x + 4}$

13. $f(x) = \sin^2 3x$

14. $f(z) = \cos^2 4z$

15. $g(t) = \sin 2t \cos 2t$

16. $f(x) = \sin 2x + \cos 2x$

17. $f(x) = \tan^2 4x$

18. $g(x) = \sec^2 3x$

19. $G(x) = (x - 2)^3(x + 1)^2$

20. $F(x) = (5 + x)^3(2 - x)^2$

In Exercises 21 through 40, find the absolute extrema of the function on the indicated interval, if there are any, and determine the values of x at which the absolute extrema occur. Draw a sketch of the graph of the function on the interval.

21. $f(x) = 4 - 3x; \ (-1, 2]$

22. $f(x) = x^2 - 2x + 4; \ (-\infty, +\infty)$

23. $g(x) = \dfrac{1}{x}; \ [-2, 3]$

24. $f(x) = \dfrac{1}{x}; \ [2, 3)$

25. $f(x) = 2 \cos x; \ [-\frac{2}{3}\pi, \frac{1}{3}\pi)$

26. $G(x) = -3 \sin x; \ [0, \frac{3}{4}\pi)$

27. $f(x) = \sqrt{3 + x}; \ [-3, +\infty)$

28. $f(x) = \sqrt{4 - x^2}; \ (-2, 2)$

29. $h(x) = \dfrac{4}{(x - 3)^2}; \ [2, 5]$

30. $g(x) = \dfrac{3x}{9 - x^2}; \ (-3, 2)$

31. $F(x) = |x - 4| + 1; \ (0, 6)$

32. $f(x) = |4 - x^2|; \ (-\infty, +\infty)$

33. $g(x) = \sqrt{4 + 7x}; \ [0, 3)$

34. $F(x) = U(x) - U(x - 1)$ where $U(x) = \begin{cases} 0 & \text{if } x < 0 \\ 1 & \text{if } 0 \le x \end{cases}; \ (-1, 1)$

35. $f(x) = \begin{cases} \dfrac{2}{x - 5} & \text{if } x \ne 5 \\ 2 & \text{if } x = 5 \end{cases}; \ [3, 5]$

36. $f(x) = \begin{cases} |x + 1| & \text{if } x \ne -1 \\ 3 & \text{if } x = -1 \end{cases}; \ [-2, 1]$

37. $f(x) = x - [\![x]\!]; \ (1, 3)$

38. $h(x) = 2x + [\![2x - 1]\!]; \ (1, 2]$

39. $g(x) = \sec 3x; \ [-\frac{1}{6}\pi, \frac{1}{6}\pi]$

40. $f(x) = \tan 2x; \ [-\frac{1}{4}\pi, \frac{1}{6}\pi]$

In Exercises 41 through 58, find the absolute maximum value and the absolute minimum value of the function on the indicated interval by the method used in Examples 3 and 4 of this section. Draw a sketch of the graph of the function on the interval.

41. $g(x) = x^3 + 5x - 4; \ [-3, -1]$

42. $f(x) = x^3 + 3x^2 - 9x; \ [-4, 4]$

43. $f(x) = x^4 - 8x^2 + 16; \ [-4, 0]$

44. $f(x) = x^4 - 8x^2 + 16; \ [-3, 2]$

45. $f(x) = x^4 - 8x^2 + 16; \ [0, 3]$

46. $g(x) = x^4 - 8x^2 + 16; \ [-1, 4]$

47. $f(t) = 2 \sin t; \ [-\pi, \pi]$

48. $f(w) = 3 \cos 2w; \ [\frac{1}{6}\pi, \frac{3}{4}\pi]$

49. $f(x) = \frac{1}{2} \csc 2x; \ [-\frac{1}{4}\pi, \frac{1}{6}\pi]$

50. $h(x) = 2 \sec \frac{1}{2}x; \ [-\frac{1}{3}\pi, \frac{1}{2}\pi]$

51. $f(x) = \dfrac{x}{x + 2}; \ [-1, 2]$

52. $f(x) = \dfrac{x + 5}{x - 3}; \ [-5, 2]$

53. $f(x) = \dfrac{x + 1}{2x - 3}; \ [0, 1]$

54. $f(x) = \begin{cases} 2x - 7 & \text{if } -1 \le x \le 2 \\ 1 - x^2 & \text{if } 2 < x \le 4 \end{cases}; \ [-1, 4]$

55. $F(x) = \begin{cases} 3x - 4 & \text{if } -3 \le x < 1 \\ x^2 - 2 & \text{if } 1 \le x \le 3 \end{cases}; \ [-3, 3]$

56. $G(x) = \begin{cases} 4 - (x + 5)^2 & \text{if } -6 \le x \le -4 \\ 12 - (x + 1)^2 & \text{if } -4 < x \le 0 \end{cases}; \ [-6, 0]$

57. $f(x) = (x + 1)^{2/3}; \ [-2, 1]$

58. $g(x) = 1 - (x - 3)^{2/3}; \ [-5, 4]$

59. Prove Theorem 4.1.3 for the case when f has a relative maximum value at c.

4.2 APPLICATIONS INVOLVING AN ABSOLUTE EXTREMUM ON A CLOSED INTERVAL

We now apply the extreme-value theorem to some problems in which the solution is an absolute extremum of a function on a closed interval. The theorem assures us that both an absolute maximum value and an absolute minimum value of a function exist on a closed interval if the function is continuous on that interval. In the following illustration we demonstrate the procedure by considering the problem discussed in Example 4 of Section 2.7.

▶ **ILLUSTRATION 1** A cardboard-box manufacturer wishes to make open boxes from pieces of cardboard 12 in. square by cutting equal squares from the four corners and turning up the sides. We wish to find the length of the side of the square to be cut out to obtain a box of the largest possible volume. Figure 1 represents a given piece of cardboard, and Figure 2 represents the box. We showed in Example 4 of Section 2.7 that if x inches is the length of the side of the square to be cut out and $V(x)$ cubic inches is the volume of the box, then

$$V(x) = 144x - 48x^2 + 4x^3$$

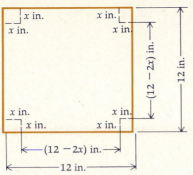

FIGURE 1   FIGURE 2

and the domain of V is the closed interval $[0, 6]$. Because V is continuous on $[0, 6]$, it follows from the extreme-value theorem that V has an absolute maximum value on this interval. We also know that this absolute maximum value of V must occur at either a critical number or at an endpoint of the interval. To find the critical numbers of V we find $V'(x)$ and then find the values of x for which either $V'(x) = 0$ or $V'(x)$ does not exist.

$$V'(x) = 144 - 96x + 12x^2$$

$V'(x)$ exists for all values of x. If $V'(x) = 0$,

$$12(x^2 - 8x + 12) = 0$$

$$x = 6 \qquad x = 2$$

The critical numbers of V are 2 and 6, both of which are in the closed interval $[0, 6]$. The absolute maximum value of V on $[0, 6]$ must occur at either a critical number or at an endpoint of the interval. Because $V(0) = 0$ and $V(6) = 0$, while $V(2) = 128$, the absolute maximum value of V on $[0, 6]$ is 128, occurring when $x = 2$.

Therefore the largest possible volume is 128 in.³, and this is obtained when the length of the side of the square cut out is 2 in. ◀

EXAMPLE 1   Points A and B are opposite each other on shores of a straight river 3 km wide. Point C is on the same shore as B but 2 km down the river from B. A telephone company wishes to lay a cable from A to C. If the cost per kilometer of the cable is 25 percent more under the water than it is on land, what line of cable would be least expensive for the company?

Solution   Refer to Figure 3. Let P be a point on the same shore as B and C and between B and C so that the cable will run from A to P to C. Let x kilometers be the distance from B to P. Then $(2 - x)$ kilometers is the distance from P to C, and $x \in [0, 2]$. Let k dollars be the cost per kilometer on land and $\frac{5}{4}k$ dollars be the cost per kilometer under the water (k is a constant). If $C(x)$ dollars is the total cost of running the cable from A to P and from P to C, then

$$C(x) = \tfrac{5}{4}k\sqrt{3^2 + x^2} + k(2 - x)$$

Because C is continuous on $[0, 2]$, the extreme-value theorem applies; thus C has both an absolute maximum value and an absolute minimum value on $[0, 2]$. We wish to find the absolute minimum value.

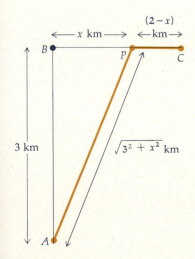

FIGURE 3

$$C'(x) = \frac{5kx}{4\sqrt{9 + x^2}} - k$$

$C'(x)$ exists for all values of x. Setting $C'(x) = 0$ and solving for x we have

$$\frac{5kx}{4\sqrt{9 + x^2}} - k = 0$$

$$5x = 4\sqrt{9 + x^2} \tag{1}$$

$$25x^2 = 16(9 + x^2)$$

$$9x^2 = 16 \cdot 9$$

$$x^2 = 16$$

$$x = \pm 4$$

The number -4 is an extraneous root of (1), and 4 is not in the interval $[0, 2]$. Therefore there are no critical numbers of C in $[0, 2]$. The absolute minimum value of C on $[0, 2]$ must therefore occur at an endpoint of the interval. Computing $C(0)$ and $C(2)$ we get

$$C(0) = \tfrac{23}{4}k \quad \text{and} \quad C(2) = \tfrac{5}{4}k\sqrt{13}$$

Because $\tfrac{5}{4}k\sqrt{13} < \tfrac{23}{4}k$, the absolute minimum value of C on $[0, 2]$ is $\tfrac{5}{4}k\sqrt{13}$, occurring when $x = 2$. Therefore, for the cost of the cable to be the least, the cable should go directly from A to C under the water.

EXAMPLE 2   A rectangular field is to be fenced off along the bank of a river; no fence is required along the river. If the material for the fence costs $8 per running foot for the two ends and $12 per running foot for the side parallel to the river, find the dimensions of the field of largest possible area that can be enclosed with $3600 worth of fence.

Solution   Let x feet be the length of an end of the field, y feet be the length of the side parallel to the river, and A square feet be the area of the field. See Figure 4. Hence

$$A = xy \tag{2}$$

Because the cost of the material for each end is $8 per running foot and the length of an end is x feet, the total cost for the fence for each end is $8x$ dollars. Similarly, the total cost of the fence for the third side is $12y$ dollars. Then

$$8x + 8x + 12y = 3600 \tag{3}$$

To express A in terms of a single variable we solve (3) for y in terms of x and substitute this value into (2), yielding A as a function of x, and

$$A(x) = x(300 - \tfrac{4}{3}x) \tag{4}$$

From (3), if $y = 0$, $x = 225$, and if $x = 0$, $y = 300$. Because both x and y must be nonnegative, the value of x that will make A an absolute maximum is in the closed interval $[0, 225]$. Because A is continuous on the closed interval $[0, 225]$, from the extreme-value theorem A has an absolute maximum value

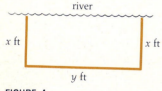

river

x ft x ft

y ft

FIGURE 4

on this interval. From (4)

$$A(x) = 300x - \tfrac{4}{3}x^2$$

$$A'(x) = 300 - \tfrac{8}{3}x$$

Because $A'(x)$ exists for all x, the critical numbers of A are found by setting $A'(x) = 0$, which gives

$$x = 112\tfrac{1}{2}$$

The only critical number of A is $112\tfrac{1}{2}$, which is in the closed interval $[0, 225]$. Thus the absolute maximum value of A must occur at either 0, $112\tfrac{1}{2}$, or 225. Because $A(0) = 0$ and $A(225) = 0$, while $A(112\tfrac{1}{2}) = 16,875$, the absolute maximum value of A on $[0, 225]$ is $16,875$, occurring when $x = 112\tfrac{1}{2}$ and $y = 150$ (obtained from (3) by substituting $112\tfrac{1}{2}$ for x).

Therefore the largest possible area that can be enclosed for \$3600 is $16,875$ ft^2, and this is obtained when the side parallel to the river is 150 ft long and the ends are each $112\tfrac{1}{2}$ ft long.

EXAMPLE 3   In the planning of a coffee shop it is estimated that if there are places for 40 to 80 people, the daily profit will be \$16 per place. However, if the seating capacity is above 80 places, the daily profit on each place will be decreased by \$0.08 times the number of places above 80. What should be the seating capacity to yield the greatest daily profit?

Solution   Let x places be the seating capacity and $P(x)$ dollars be the daily profit. $P(x)$ is obtained by multiplying x by the number of dollars in the profit per place. When $40 \leq x \leq 80$, \$16 is the profit per place; so $P(x) = 16x$. However, when $x > 80$, the number of dollars in the profit per place is $16 - 0.08(x - 80)$, thus giving $P(x) = x[16 - 0.08(x - 80)]$; that is, $P(x) = 22.40x - 0.08x^2$. Therefore

$$P(x) = \begin{cases} 16x & \text{if } 40 \leq x \leq 80 \\ 22.40x - 0.08x^2 & \text{if } 80 < x \leq 280 \end{cases}$$

The upper bound of 280 for x is obtained by noting that $22.40x - 0.08x^2 = 0$ when $x = 280$; and $22.40x - 0.08x^2 < 0$ when $x > 280$.

Even though x, by definition, is a positive integer, to have a continuous function we let x take on all real values in the interval $[40, 280]$. There is continuity at 80 because $P(80) = 1280$ and

$$\lim_{x \to 80^-} P(x) = \lim_{x \to 80^-} 16x \qquad \lim_{x \to 80^+} P(x) = \lim_{x \to 80^+} (22.40x - 0.08x^2)$$

$$= 1280 \qquad\qquad\qquad = 1280$$

Thus P is continuous on the closed interval $[40, 280]$, and the extreme-value theorem guarantees an absolute maximum value of P on this interval.

When $40 < x < 80$, $P'(x) = 16$; when $80 < x < 280$, $P'(x) = 22.40 - 0.16x$. $P'(80)$ does not exist since $P'_-(80) = 16$ and $P'_+(80) = 9.60$. Set $P'(x) = 0$.

$$22.40 - 0.16x = 0$$

$$x = 140$$

The critical numbers of P are then 80 and 140. We evaluate $P(x)$ at the endpoints of the interval $[40, 280]$ and at the critical numbers.

$$P(40) = 640 \qquad P(80) = 1280 \qquad P(140) = 1568 \qquad P(280) = 0$$

The absolute maximum value of P, then, is 1568, occurring when $x = 140$.

The seating capacity should be 140 places, which gives a daily profit of \$1568.

EXAMPLE 4   Find the dimensions of the right-circular cylinder of greatest volume that can be inscribed in a right-circular cone with a radius of 5 cm and a height of 12 cm.

Solution   Let r centimeters be the radius of the cylinder, h centimeters be the height of the cylinder, and V cubic centimeters be the volume of the cylinder.

Figure 5 illustrates the cylinder inscribed in the cone, and Figure 6 shows a plane section through the axis of the cone.

If $r = 0$ and $h = 12$, we have a degenerate cylinder, which is the axis of the cone. If $r = 5$ and $h = 0$, we also have a degenerate cylinder, which is a diameter of the base of the cone. The number r is in the closed interval $[0, 5]$ and h is in the closed interval $[0, 12]$.

The following formula expresses V in terms of r and h:

$$V = \pi r^2 h \tag{5}$$

To express V in terms of a single variable we need another equation involving r and h. From Figure 6, and by similar triangles,

$$\frac{12 - h}{r} = \frac{12}{5}$$

$$h = \frac{60 - 12r}{5} \tag{6}$$

Substituting from (6) into formula (5) we obtain V as a function of r and write

$$V(r) = \tfrac{12}{5}\pi(5r^2 - r^3) \qquad \text{with } r \text{ in } [0, 5] \tag{7}$$

Because V is continuous on the closed interval $[0, 5]$, it follows from the extreme-value theorem that V has an absolute maximum value on this interval. The values of r and h that give this absolute maximum value are the numbers to find.

$$V'(r) = \tfrac{12}{5}\pi(10r - 3r^2)$$

To find the critical numbers of V, we set $V'(r) = 0$ and solve for r:

$$r(10 - 3r) = 0$$

$$r = 0 \qquad r = \tfrac{10}{3}$$

Because $V'(r)$ exists for all values of r, the only critical numbers of V are 0 and $\tfrac{10}{3}$, both of which are in the closed interval $[0, 5]$. The absolute maximum value of V on $[0, 5]$ must occur at either 0, $\tfrac{10}{3}$, or 5. From (7) we obtain

$$V(0) = 0 \qquad V(\tfrac{10}{3}) = \tfrac{400}{9}\pi \qquad V(5) = 0$$

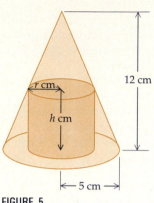

FIGURE 5

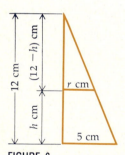

FIGURE 6

Therefore the absolute maximum value of V is $\frac{400}{9}\pi$, and this occurs when $r = \frac{10}{3}$. When $r = \frac{10}{3}$, we find from (6) that $h = 4$.

Thus the greatest volume of an inscribed cylinder in the given cone is $\frac{400}{9}\pi$ cm^3, which occurs when the radius is $\frac{10}{3}$ cm and the height is 4 cm.

EXERCISES 4.2

In some of the exercises, the independent variable, by definition, may represent a nonnegative integer. For instance, in Exercise 17, if x represents the number of students, then x must be a nonnegative integer. In such exercises, to have the continuity requirements necessary to apply the calculus, allow the independent variable to represent a nonnegative real number.

1. Refer to Exercise 41 in Exercises 2.7. A manufacturer of open tin boxes wishes to make use of pieces of tin with dimensions 8 in. by 15 in. by cutting equal squares from the four corners and turning up the sides. Find the length of the side of the square to be cut out if an open box having the largest possible volume is to be obtained from each piece of tin.

2. Refer to Exercise 42 in Exercises 2.7. Suppose the manufacturer of Exercise 1 makes the open boxes from square pieces of tin that measure k cm on a side. Determine the length of the side of the square cut out for the volume of the box to be a maximum.

3. Refer to Exercise 43 in Exercises 2.7. Find the dimensions of the largest rectangular field that can be enclosed with 240 m of fence.

4. Find the dimensions of the largest rectangular garden that can be fenced off with 100 ft of fencing material.

5. If one side of a rectangular field is to have a river as a natural boundary, find the dimensions of the largest rectangular field that can be enclosed by using 240 m of fence for the other three sides.

6. Refer to Exercise 44 in Exercises 2.7. Find the dimensions of the largest rectangular garden that can be placed so that a side of a house serves as a boundary and 100 ft of fencing material is to be used for the other three sides.

7. Find the number in the interval $[0, 1]$ such that the difference between the number and its square is a maximum.

8. Find the number in the interval $[\frac{1}{3}, 2]$ such that the sum of the number and its reciprocal is a maximum.

9. Find the area of the largest rectangle having two vertices on the x axis and two vertices on or above the x axis and on the parabola $y = 9 - x^2$.

10. Find the area of the largest rectangle that can be inscribed in a given circle of radius r.

11. An island is at point A, 6 km offshore from the nearest point B on a straight beach. A woman on the island wishes to go to a point C, 9 km down the beach from B. The woman can rent a boat for \$15 per kilometer and travel by water to a point P between B and C, and then she can hire a car with a driver at a cost of \$12 per kilometer and travel a straight road from P to C. Find the least expensive route from point A to point C.

12. Solve Exercise 11 if point C is only 7 km down the beach from B.

13. Solve Example 1 of this section if point C is 6 km down the river from B.

14. Example 1 and Exercises 11, 12, and 13 are special cases of the following more general problem. Let

$$f(x) = u\sqrt{a^2 + x^2} + v(b - x)$$

where x is in $[0, b]$ and $u > v > 0$. Show that for the absolute minimum value of f to occur at a number in the open interval $(0, b)$, the following inequality must be satisfied: $av < b\sqrt{u^2 - v^2}$.

15. If R feet is the range of a projectile, then

$$R = \frac{v_0{}^2 \sin 2\theta}{g} \qquad 0 \le \theta \le \tfrac{1}{2}\pi$$

where v_0 feet per second is the initial velocity, g ft/sec^2 is the acceleration due to gravity, and θ is the radian measure of the angle that the gun makes with the horizontal. Find the value of θ that makes the range a maximum.

16. If a body of weight W pounds is dragged along a horizontal floor at constant velocity by means of a force of magnitude F pounds and directed at an angle of θ radians with the plane of the floor, then F is given by the equation

$$F = \frac{kW}{k \sin \theta + \cos \theta}$$

where k is a constant called the coefficient of friction and $0 < k < 1$. If $0 \le \theta \le \tfrac{1}{2}\pi$, find $\cos \theta$ when F is least.

17. Refer to Exercise 39 in Exercises 3.2. A school-sponsored trip that can accommodate up to 250 students will cost each student \$15 if not more than 150 students make the trip; however, the cost per student will be reduced \$0.05 for each student in excess of 150 until the cost reaches \$10 per student. How many students should make the trip for the school to receive the largest gross income?

18. Solve Exercise 17 if the reduction per student in excess of 150 is \$0.07.

19. A private club charges annual membership dues of \$100 per member, less \$0.50 for each member over 600 and plus \$0.50 for each member less than 600. How many members will give the club the most revenue from annual dues?

20. Refer to Exercise 40 in Exercises 3.2. A manufacturer can make a profit of $20 on each item if not more than 800 items are produced each week. The profit decreases $0.02 per item over 800. How many items should the manufacturer produce each week to have the greatest profit?

21. Orange trees grown in California produce 600 oranges per year if no more than 20 trees are planted per acre. For each additional tree planted per acre, the yield per tree decreases by 15 oranges. How many trees per acre should be planted to obtain the greatest number of oranges?

22. Show that the largest rectangle having a given perimeter of p units is a square.

23. Find the dimensions of the right-circular cylinder of greatest lateral surface area that can be inscribed in a sphere with a radius of 6 in.

24. Find the dimensions of the right-circular cylinder of greatest volume that can be inscribed in a sphere with a radius of 6 in.

25. For a package to be accepted by a particular mailing service, the sum of the length and girth (the perimeter of a cross section) must not be greater than 100 in. If a package is to be in the shape of a rectangular box with a square cross section, find the dimensions of the package having the greatest possible volume that can be mailed by the service.

26. Two products A and B are manufactured at a particular factory. If C dollars is the total cost of production for an 8-hour day, then $C = 3x^2 + 42y$, where x machines are used to produce product A and y machines are used to produce product B. If during an 8-hour day there are 15 machines working, determine how many of these machines should be used to produce A and how many should be used to produce B for the total cost to be least.

27. Given the circle having the equation $x^2 + y^2 = 9$, find (a) the shortest distance from the point $(4, 5)$ to a point on the circle; (b) the longest distance from the point $(4, 5)$ to a point on the circle.

28. Suppose that a weight is to be held 10 ft below a horizontal line AB by a wire in the shape of a Y. If the points A and B are 8 ft apart, what is the shortest total length of wire that can be used?

29. Assume that the decrease in a person's blood pressure depends on the amount of a particular drug taken by the person. Thus if x milligrams of the drug is taken, the decrease in blood pressure is a function of x. Suppose that $f(x)$ defines this function and

$$f(x) = \tfrac{1}{2}x^2(k - x)$$

and x is in $[0, k]$, where k is a positive constant. Determine the value of x that causes the greatest decrease in blood pressure.

30. During a cough there is a decrease in the radius of a person's trachea (or windpipe). Suppose that the normal radius of the trachea is R centimeters and the radius of the trachea during a cough is r centimeters, where R is a constant and r is a variable. The velocity of air through the trachea can be shown to be a function of r, and if $V(r)$ centimeters per second is this velocity, then

$$V(r) = kr^2(R - r)$$

where k is a positive constant and r is in $[\tfrac{1}{2}R, R]$. Determine the radius of the trachea during a cough for which the velocity of air through the trachea is greatest.

31. In a particular small town the rate at which a rumor spreads is jointly proportional to the number of people who have heard the rumor and the number of people who have not heard it. Show that the rumor is being spread at the greatest rate when half the population of the town knows the rumor.

32. A particular lake can support up to 14,000 fish, and the rate of growth of the fish population is jointly proportional to the number of fish present and the difference between 14,000 and the number present. What should be the size of the fish population for the growth rate to be a maximum?

33. The strength of a rectangular beam is jointly proportional to its breadth and the square of its depth. Find the dimensions of the strongest beam that can be cut from a log in the shape of a right-circular cylinder of radius 72 cm.

34. The stiffness of a rectangular beam is jointly proportional to the breadth and the cube of the depth. Find the dimensions of the stiffest beam that can be cut from a log in the shape of a right-circular cylinder of radius a centimeters.

35. A piece of wire 10 ft long is cut into two pieces. One piece is bent into the shape of a circle and the other into the shape of a square. How should the wire be cut so that (a) the combined area of the two figures is as small as possible; (b) the combined area of the two figures is as large as possible?

36. Solve Exercise 35 if one piece of wire is bent into the shape of an equilateral triangle and the other piece is bent into the shape of a square.

4.3 ROLLE'S THEOREM AND THE MEAN-VALUE THEOREM

We stressed the importance of the mean-value theorem in the introduction to this chapter. The proof of this theorem is based on a special case of it known as *Rolle's theorem*, which we discuss first.

Let f be a function that is continuous on the closed interval $[a, b]$, differentiable on the open interval (a, b), and such that $f(a) = 0$ and $f(b) = 0$. The French mathematician Michel Rolle (1652–1719) proved that if a function f satisfies these conditions, there is at least one number c between a and b for which $f'(c) = 0$.

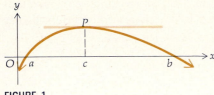

FIGURE 1

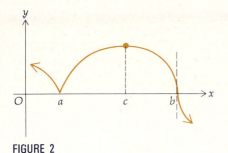

FIGURE 2

Let us see what this means geometrically. Figure 1 shows a sketch of the graph of a function f that satisfies the conditions in the preceding paragraph. We see intuitively that there is at least one point on the curve between the points $(a, 0)$ and $(b, 0)$ where the tangent line is parallel to the x axis; that is, the slope of the tangent line is zero. This situation is illustrated in Figure 1 at the point P. So the abscissa of P is the c such that $f'(c) = 0$.

The function, whose graph is sketched in Figure 1, not only is differentiable on the open interval (a, b) but also is differentiable at the endpoints of the interval. However, the condition that f be differentiable at the endpoints is not necessary for the graph to have a horizontal tangent line at some point in the interval; Figure 2 illustrates this. We see in Figure 2 that the function is not differentiable at a and b; there is, however, a horizontal tangent line at the point where $x = c$, and c is between a and b.

It is necessary, however, that the function be continuous at the endpoints of the interval to guarantee a horizontal tangent line at an interior point. Figure 3 shows a sketch of the graph of a function that is continuous on the interval $[a, b]$ but discontinuous at b; the function is differentiable on the open interval (a, b), and the function values are zero at both a and b. However, there is no point at which the graph has a horizontal tangent line.

We now state and prove Rolle's theorem.

4.3.1 THEOREM
Rolle's Theorem

Let f be a function such that

    (i) it is continuous on the closed interval $[a, b]$;
    (ii) it is differentiable on the open interval (a, b);
    (iii) $f(a) = 0$ and $f(b) = 0$.

Then there is a number c in the open interval (a, b) such that

    $f'(c) = 0$

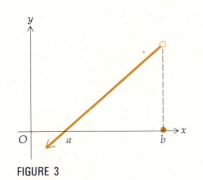

FIGURE 3

Proof   Consider two cases.

Case 1: $f(x) = 0$ for all x in $[a, b]$.
    Then $f'(x) = 0$ for all x in (a, b); therefore any number between a and b can be taken for c.

Case 2: $f(x)$ is not zero for some value of x in the open interval (a, b).
    Because f is continuous on the closed interval $[a, b]$, from the extreme-value theorem, f has an absolute maximum value on $[a, b]$ and an absolute minimum value on $[a, b]$. From (iii), $f(a) = 0$ and $f(b) = 0$. Furthermore, $f(x)$ is not zero for some x in (a, b). Hence f will have either a positive absolute maximum value at some c_1 in (a, b) or a negative absolute minimum value at some c_2 in (a, b), or both. Thus for $c = c_1$, or $c = c_2$ as the case may be, there is an absolute extremum at an interior point of the interval $[a, b]$. Therefore the absolute extremum $f(c)$ is also a relative extremum, and because $f'(c)$ exists by hypothesis, it follows from Theorem 4.1.3 that $f'(c) = 0$. This proves the theorem. ■

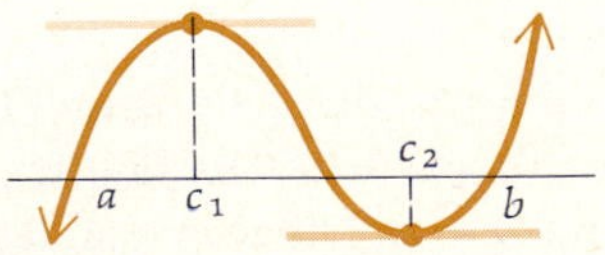

There may be more than one number in the open interval (a, b) for which the derivative of f is zero. This is illustrated geometrically in Figure 4, where there is a horizontal tangent line at the point where $x = c_1$ and also at the point where $x = c_2$, so that both $f'(c_1) = 0$ and $f'(c_2) = 0$.

FIGURE 4

The converse of Rolle's theorem is not true. That is, we cannot conclude that if a function f is such that $f'(c) = 0$, with $a < c < b$, then the conditions (i), (ii), and (iii) must hold. Refer to Exercise 32.

EXAMPLE 1   Given

$$f(x) = 4x^3 - 9x$$

verify that conditions (i), (ii), and (iii) of the hypothesis of Rolle's theorem are satisfied for each of the following intervals: $[-\frac{3}{2}, 0]$, $[0, \frac{3}{2}]$, and $[-\frac{3}{2}, \frac{3}{2}]$. Then find a suitable value for c in each of these intervals for which $f'(c) = 0$.

Solution

$$f'(x) = 12x^2 - 9$$

Because $f'(x)$ exists for all values of x, f is differentiable on $(-\infty, +\infty)$ and therefore continuous on $(-\infty, +\infty)$. Conditions (i) and (ii) of Rolle's theorem thus hold on any interval. To determine on which intervals condition (iii) holds, we find the values of x for which $f(x) = 0$. If $f(x) = 0$,

$$4x(x^2 - \tfrac{9}{4}) = 0$$

$$x = -\tfrac{3}{2} \qquad x = 0 \qquad x = \tfrac{3}{2}$$

With $a = -\frac{3}{2}$ and $b = 0$, Rolle's theorem holds on $[-\frac{3}{2}, 0]$. Similarly, Rolle's theorem holds on $[0, \frac{3}{2}]$ and $[-\frac{3}{2}, \frac{3}{2}]$.

To find the suitable values for c, set $f'(x) = 0$ and get

$$12x^2 - 9 = 0$$

$$x = -\tfrac{1}{2}\sqrt{3} \qquad x = \tfrac{1}{2}\sqrt{3}$$

Therefore, in the interval $[-\frac{3}{2}, 0]$ a suitable choice for c is $-\frac{1}{2}\sqrt{3}$. In the interval $[0, \frac{3}{2}]$, take $c = \frac{1}{2}\sqrt{3}$. In the interval $[-\frac{3}{2}, \frac{3}{2}]$ there are two possibilities for c: either $-\frac{1}{2}\sqrt{3}$ or $\frac{1}{2}\sqrt{3}$.

We now apply Rolle's theorem to prove the mean-value theorem. You should become thoroughly familiar with the content of this theorem.

4.3.2 THEOREM
Mean-Value Theorem

Let f be a function such that

   (i) it is continuous on the closed interval $[a, b]$;
   (ii) it is differentiable on the open interval (a, b).

Then there is a number c in the open interval (a, b) such that

$$f'(c) = \frac{f(b) - f(a)}{b - a}$$

Before proving this theorem, we interpret it geometrically. In a sketch of the graph of the function f, $[f(b) - f(a)]/(b - a)$ is the slope of the line segment joining the points $A(a, f(a))$ and $B(b, f(b))$. The mean-value theorem states that there is some point on the curve between A and B where the tangent line is paral-

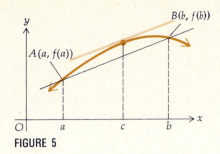

FIGURE 5

lel to the secant line through A and B; that is, there is some number c in (a, b) such that

$$f'(c) = \frac{f(b) - f(a)}{b - a}$$

Refer to Figure 5.

Take the x axis along the line segment AB and observe that the mean-value theorem is a generalization of Rolle's theorem, which is used in its proof.

Proof of Theorem 4.3.2   An equation of the line through A and B in Figure 5 is

$$y - f(a) = \frac{f(b) - f(a)}{b - a}(x - a)$$

$$\Leftrightarrow \qquad y = \frac{f(b) - f(a)}{b - a}(x - a) + f(a)$$

Now if $F(x)$ measures the vertical distance between a point $(x, f(x))$ on the graph of the function f and the corresponding point on the secant line through A and B, then

$$F(x) = f(x) - \frac{f(b) - f(a)}{b - a}(x - a) - f(a) \qquad (1)$$

We show that this function F satisfies the three conditions of the hypothesis of Rolle's theorem.

The function F is continuous on the closed interval $[a, b]$ because it is the sum of f and a linear polynomial function, both of which are continuous there. Therefore condition (i) is satisfied by F. Condition (ii) is satisfied by F because f is differentiable on (a, b). From (1) it follows that $F(a) = 0$ and $F(b) = 0$. Therefore condition (iii) of Rolle's theorem is satisfied by F.

The conclusion of Rolle's theorem states that there is a c in the open interval (a, b) such that $F'(c) = 0$. But

$$F'(x) = f'(x) - \frac{f(b) - f(a)}{b - a}$$

Thus

$$F'(c) = f'(c) - \frac{f(b) - f(a)}{b - a}$$

Therefore there is a number c in (a, b) such that

$$0 = f'(c) - \frac{f(b) - f(a)}{b - a}$$

$$\Leftrightarrow \quad f'(c) = \frac{f(b) - f(a)}{b - a}$$

which was to be proved. ■

EXAMPLE 2   Given

$$f(x) = x^3 - 5x^2 - 3x$$

verify that the hypothesis of the mean-value theorem is satisfied for $a = 1$ and $b = 3$. Then find all numbers c in the open interval $(1, 3)$ such that

$$f'(c) = \frac{f(3) - f(1)}{3 - 1}$$

Solution   Because f is a polynomial function, f is continuous and differentiable for all values of x. Therefore the hypothesis of the mean-value theorem is satisfied for any a and b.

$$f'(x) = 3x^2 - 10x - 3$$

$$f(1) = -7 \quad \text{and} \quad f(3) = -27$$

Hence

$$\frac{f(3) - f(1)}{3 - 1} = \frac{-27 - (-7)}{2}$$

$$= -10$$

Set $f'(c) = -10$ to obtain

$$3c^2 - 10c - 3 = -10$$

$$3c^2 - 10c + 7 = 0$$

$$(3c - 7)(c - 1) = 0$$

$$c = \tfrac{7}{3} \qquad c = 1$$

Because 1 is not in the open interval $(1, 3)$, the only possible value for c is $\tfrac{7}{3}$.

EXAMPLE 3   Given

$$f(x) = x^{2/3}$$

draw a sketch of the graph of f. Show that there is no number c in the open interval $(-2, 2)$ such that

$$f'(c) = \frac{f(2) - f(-2)}{2 - (-2)}$$

Which condition of the hypothesis of the mean-value theorem fails to hold for f when $a = -2$ and $b = 2$?

Solution   A sketch of the graph of f appears in Figure 6.

$$f'(x) = \tfrac{2}{3}x^{-1/3}$$

So

$$f'(c) = \frac{2}{3c^{1/3}}$$

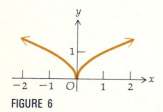

FIGURE 6

$$\frac{f(2) - f(-2)}{2 - (-2)} = \frac{4^{1/3} - 4^{1/3}}{4}$$

$$= 0$$

There is no number c for which $\dfrac{2}{3c^{1/3}} = 0$.

The function f is continuous on the closed interval $[-2, 2]$; however, f is not differentiable on the open interval $(-2, 2)$ because $f'(0)$ does not exist. Therefore condition (ii) of the hypothesis of the mean-value theorem fails to hold for f when $a = -2$ and $b = 2$.

We have indicated that the mean-value theorem is one of the most important theorems in calculus because it is used to prove many other theorems. In such cases it is not necessary to find the value of the number c guaranteed by the theorem. The crucial fact of the theorem is that such a number c exists. To indicate the power of the mean-value theorem we show its use in the proof of the following theorem, which is needed in Chapter 5.

4.3.3 THEOREM   If f is a function such that $f'(x) = 0$ for all values of x in an interval I, then f is constant on I.

Proof   Assume that f is not constant on the interval I. Then there exist two distinct numbers x_1 and x_2 in I, where $x_1 < x_2$, such that $f(x_1) \neq f(x_2)$. Because, by hypothesis, $f'(x) = 0$ for all x in I, then $f'(x) = 0$ for all x in the closed interval $[x_1, x_2]$. Hence f is differentiable at all x in $[x_1, x_2]$ and f is continuous on $[x_1, x_2]$. Therefore the hypothesis of the mean-value theorem is satisfied, and so there is a number c, with $x_1 < c < x_2$, such that

$$f'(c) = \frac{f(x_1) - f(x_2)}{x_1 - x_2} \tag{2}$$

But because $f'(x) = 0$ for all x in the interval $[x_1, x_2]$, then $f'(c) = 0$, and from (2) it follows that $f(x_1) = f(x_2)$. Yet we assumed that $f(x_1) \neq f(x_2)$. Hence there is a contradiction, and so f is constant on I.   ∎

In the next section you will see another application of the mean-value theorem in the proof of Theorem 4.4.3.

EXERCISES 4.3

In Exercises 1 through 4, verify that conditions (i), (ii), and (iii) of the hypothesis of Rolle's theorem are satisfied by the function on the indicated interval. Then find a suitable value for c that satisfies the conclusion of Rolle's theorem.

1. $f(x) = x^2 - 4x + 3$; $[1, 3]$
2. $f(x) = x^3 - 2x^2 - x + 2$; $[1, 2]$
3. $f(x) = \sin 2x$; $[0, \frac{1}{2}\pi]$
4. $f(x) = 3 \cos^2 x$; $[\frac{1}{2}\pi, \frac{3}{2}\pi]$

In Exercises 5 through 10, verify that the hypothesis of the mean-value theorem is satisfied for the function on the indicated interval. Then find a suitable value for c that satisfies the conclusion of the mean-value theorem.

5. $f(x) = x^2 + 2x - 1$; $[0, 1]$
6. $f(x) = x^3 + x^2 - x$; $[-2, 1]$
7. $f(x) = x^{2/3}$; $[0, 1]$
8. $f(x) = \sqrt{1 - \sin x}$; $[0, \frac{1}{2}\pi]$

9. $f(x) = \sqrt{1 + \cos x}$; $[-\frac{1}{2}\pi, \frac{1}{2}\pi]$
10. $f(x) = \dfrac{x^2 + 4x}{x - 7}$; $[2, 6]$

In Exercises 11 through 16, (a) draw a sketch of the graph of the function on the indicated interval; (b) test the three conditions (i), (ii), and (iii) of the hypothesis of Rolle's theorem, and determine which conditions are satisfied and which, if any, are not satisfied; and (c) if the three conditions in part (b) are satisfied, determine a point at which there is a horizontal tangent line.

11. $f(x) = x^{4/3} - 3x^{1/3}$; $[0, 3]$
12. $f(x) = x^{3/4} - 2x^{1/4}$; $[0, 4]$
13. $f(x) = \dfrac{x^2 - x - 12}{x - 3}$; $[-3, 4]$
14. $f(x) = \begin{cases} 3x + 6 & \text{if } x < 1 \\ x - 4 & \text{if } 1 \leq x \end{cases}$; $[-2, 4]$

15. $f(x) = \begin{cases} x^2 - 4 & \text{if } x < 1 \\ 5x - 8 & \text{if } 1 \le x \end{cases}; [-2, \frac{8}{5}]$

16. $f(x) = 1 - |x|; [-1, 1]$

The geometric interpretation of the mean-value theorem is that for a suitable c in the open interval (a, b), the tangent line to the curve $y = f(x)$ at the point $(c, f(c))$ is parallel to the secant line through the points $(a, f(a))$ and $(b, f(b))$. In Exercises 17 through 20, find a value of c satisfying the conclusion of the mean-value theorem, draw a sketch of the graph on the closed interval $[a, b]$, and show the tangent line and secant line.

17. $f(x) = x^2; a = 3, b = 5$ 18. $f(x) = x^2; a = 2, b = 4$

19. $f(x) = \sin x; a = 0, b = \frac{1}{2}\pi$

20. $f(x) = 2 \cos x; a = \frac{1}{3}\pi, b = \frac{2}{3}\pi$

For each of the functions in Exercises 21 through 24, there is no number c in the open interval (a, b) that satisfies the conclusion of the mean-value theorem. In each exercise, determine which part of the hypothesis of the mean-value theorem fails to hold. Draw a sketch of the graph of $y = f(x)$ and the line through the points $(a, f(a))$ and $(b, f(b))$.

21. $f(x) = \dfrac{4}{(x - 3)^2}; a = 1, b = 6$

22. $f(x) = \dfrac{2x - 1}{3x - 4}; a = 1, b = 2$

23. $f(x) = 3(x - 4)^{2/3}; a = -4, b = 5$

24. $f(x) = \begin{cases} 2x + 3 & \text{if } x < 3 \\ 15 - 2x & \text{if } 3 \le x \end{cases}; a = -1, b = 5$

25. If $f(x) = x^4 - 2x^3 + 2x^2 - x$, then $f'(x) = 4x^3 - 6x^2 + 4x - 1$. Prove by Rolle's theorem that the following equation has at least one real root in the open interval (0, 1):
$$4x^3 - 6x^2 + 4x - 1 = 0.$$

26. Prove by Rolle's theorem that the equation $x^3 + 2x + c = 0$, where c is any constant, cannot have more than one real root.

27. Use Rolle's theorem to prove that the equation
$$4x^5 + 3x^3 + 3x - 2 = 0$$
has exactly one root that lies in the interval (0, 1). (*Hint:* First show that there is at least one number in (0, 1) that is a root of the equation. Then assume that there is more than one root of the equation in (0, 1), and show that this leads to a contradiction.)

28. Suppose that $s = f(t)$ is an equation of the motion of a particle moving in a straight line, where f satisfies the hypothesis of the mean-value theorem. Show that the conclusion of the mean-value theorem assures that there will be some instant during any time interval when the instantaneous velocity will equal the average velocity during that time interval.

29. If the equation of the motion in Exercise 28 is $s = t^2 - t + 4$ and $t \in [0, 3]$, find the value of t where the instantaneous velocity equals the average velocity over the interval.

30. If $f(x) = \sin^2 x + \cos^2 x$, use Theorem 4.3.3 to show that $f(x) = 1$ for all x in $[-2\pi, 2\pi]$.

31. Suppose that the function f is continuous on $[a, b]$ and $f'(x) = 1$ for all x in (a, b). Prove that $f(x) = x - a + f(a)$ for all x in $[a, b]$.

32. The converse of Rolle's theorem is not true. Make up an example of a function for which the conclusion of Rolle's theorem is true and for which (a) condition (i) is not satisfied but conditions (ii) and (iii) are satisfied; (b) condition (ii) is not satisfied but conditions (i) and (iii) are satisfied; (c) condition (iii) is not satisfied but conditions (i) and (ii) are satisfied. Draw a sketch of the graph showing the horizontal tangent line for each case.

33. Use Rolle's theorem to prove that if every polynomial of the fourth degree has at most four real roots, then every polynomial of the fifth degree has at most five real roots. (*Hint:* Assume that a polynomial of the fifth degree has six real roots and show that this leads to a contradiction.)

34. Use the method of Exercise 33 and mathematical induction to prove that a polynomial of the nth degree has at most n real roots.

4.4 INCREASING AND DECREASING FUNCTIONS AND THE FIRST-DERIVATIVE TEST

Figure 1 represents a sketch of the graph of a function f for all x in the closed interval $[x_1, x_7]$. In this sketch we have assumed that f is continuous on $[x_1, x_7]$. The figure shows that as a point moves along the curve from A to B, the function values increase as the abscissa increases, and that as a point moves along the curve from B to C, the function values decrease as the abscissa increases. We say, then, that f is *increasing* on the closed interval $[x_1, x_2]$ and that f is *decreasing* on the closed interval $[x_2, x_3]$. Following are the precise definitions of a function increasing or decreasing on an interval.

4.4.1 DEFINITION

A function f defined on an interval is said to be **increasing** on that interval if and only if
$$f(x_1) < f(x_2) \quad \text{whenever} \quad x_1 < x_2$$
where x_1 and x_2 are any numbers in the interval.

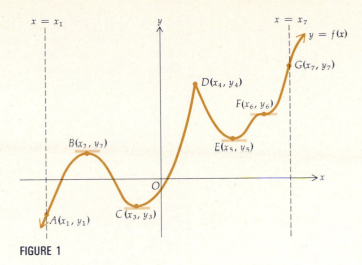

FIGURE 1

The function of Figure 1 is increasing on the following closed intervals: $[x_1, x_2]$; $[x_3, x_4]$; $[x_5, x_6]$; $[x_6, x_7]$; $[x_5, x_7]$.

4.4.2 DEFINITION

A function f defined on an interval is said to be **decreasing** on that interval if and only if

$$f(x_1) > f(x_2) \quad \text{whenever} \quad x_1 < x_2$$

where x_1 and x_2 are any numbers in the interval.

The function of Figure 1 is decreasing on the following closed intervals: $[x_2, x_3]$; $[x_4, x_5]$.

If a function is either increasing on an interval or decreasing on an interval, then it is said to be **monotonic** on the interval.

Before stating a theorem that gives a test for determining if a function is monotonic on an interval, let us see what is happening geometrically. Refer to Figure 1, and observe that when the slope of the tangent line is positive the function is increasing, and when it is negative the function is decreasing. Because $f'(x)$ is the slope of the tangent line to the curve $y = f(x)$, f is increasing when $f'(x) > 0$ and decreasing when $f'(x) < 0$. Also, because $f'(x)$ is the rate of change of the function values $f(x)$ with respect to x, when $f'(x) > 0$, the function values are increasing as x increases; and when $f'(x) < 0$, the function values are decreasing as x increases.

4.4.3 THEOREM

Let the function f be continuous on the closed interval $[a, b]$ and differentiable on the open interval (a, b):

(i) if $f'(x) > 0$ for all x in (a, b), then f is increasing on $[a, b]$;
(ii) if $f'(x) < 0$ for all x in (a, b), then f is decreasing on $[a, b]$.

Proof of (i) Let x_1 and x_2 be any two numbers in $[a, b]$ such that $x_1 < x_2$. Then f is continuous on $[x_1, x_2]$ and differentiable on (x_1, x_2). From the mean-value theorem it follows that there is some number c in (x_1, x_2) such that

$$f'(c) = \frac{f(x_2) - f(x_1)}{x_2 - x_1}$$

Because $x_1 < x_2$, then $x_2 - x_1 > 0$. Also, $f'(c) > 0$ by hypothesis. Therefore $f(x_2) - f(x_1) > 0$, and so $f(x_2) > f(x_1)$. We have shown that $f(x_1) < f(x_2)$ whenever $x_1 < x_2$, where x_1 and x_2 are any numbers in the interval $[a, b]$. Therefore, by Definition 4.4.1 it follows that f is increasing on $[a, b]$.

The proof of part (ii) is similar and is left as an exercise (see Exercise 41).

◼

An application of Theorem 4.4.3 is in the proof of the **first-derivative test for relative extrema** of a function.

4.4.4 THEOREM
First-Derivative Test for Relative Extrema

Let the function f be continuous at all points of the open interval (a, b) containing the number c, and suppose that f' exists at all points of (a, b) except possibly at c:

(i) if $f'(x) > 0$ for all values of x in some open interval having c as its right endpoint, and if $f'(x) < 0$ for all values of x in some open interval having c as its left endpoint, then f has a relative maximum value at c;

(ii) if $f'(x) < 0$ for all values of x in some open interval having c as its right endpoint, and if $f'(x) > 0$ for all values of x in some open interval having c as its left endpoint, then f has a relative minimum value at c.

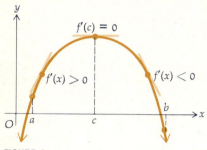

FIGURE 2

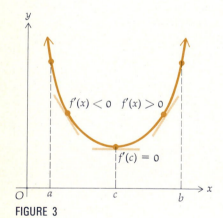

FIGURE 3

Proof of (i) Let (d, c) (where $d > a$) be the interval having c as its right endpoint for which $f'(x) > 0$ for all x in the interval. It follows from Theorem 4.4.3(i) that f is increasing on $[d, c]$. Let (c, e) (where $e < b$) be the interval having c as its left endpoint for which $f'(x) < 0$ for all x in the interval. By Theorem 4.4.3(ii), f is decreasing on $[c, e]$. Because f is increasing on $[d, c]$, it follows from Definition 4.4.1 that if x_1 is in $[d, c]$ and $x_1 \neq c$, then $f(x_1) < f(c)$. Also, because f is decreasing on $[c, e]$, it follows from Definition 4.4.2 that if x_2 is in $[c, e]$ and $x_2 \neq c$, then $f(c) > f(x_2)$. Therefore, from Definition 4.1.1, f has a relative maximum value at c.

The proof of part (ii) is similar to the proof of part (i) and is left as an exercise (see Exercise 42).

◼

The first-derivative test for relative extrema states that if f is continuous at c and $f'(x)$ changes algebraic sign from positive to negative as x increases through the number c, then f has a relative maximum value at c; and if $f'(x)$ changes algebraic sign from negative to positive as x increases through c, then f has a relative minimum value at c.

Figures 2 and 3 illustrate parts (i) and (ii), respectively, of Theorem 4.4.4 when $f'(c)$ exists. Figure 4 shows a sketch of the graph of a function f that has a relative maximum value at a number c, but $f'(c)$ does not exist; however, $f'(x) > 0$ when $x < c$, and $f'(x) < 0$ when $x > c$. In Figure 5 there is a sketch of the graph of a function f for which c is a critical number, and $f'(x) < 0$ when $x < c$, and $f'(x) < 0$ when $x > c$; f does not have a relative extremum at c.

Further illustrations of Theorem 4.4.4 occur in Figure 1. At x_2 and x_4 the function has a relative maximum value, and at x_3 and x_5 the function has a relative minimum value; even though x_6 is a critical number for the function, there is no relative extremum at x_6.

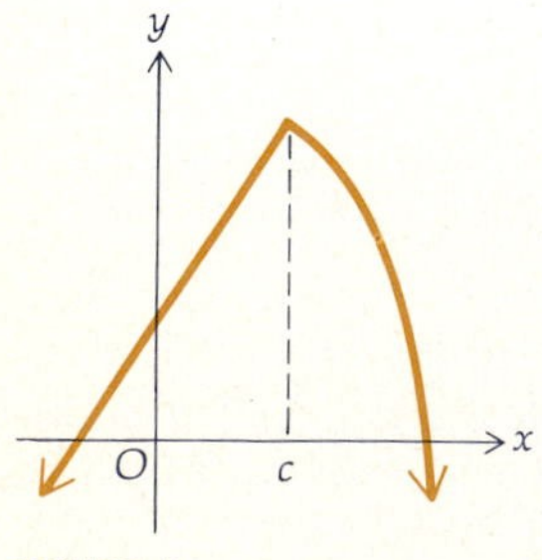

FIGURE 4

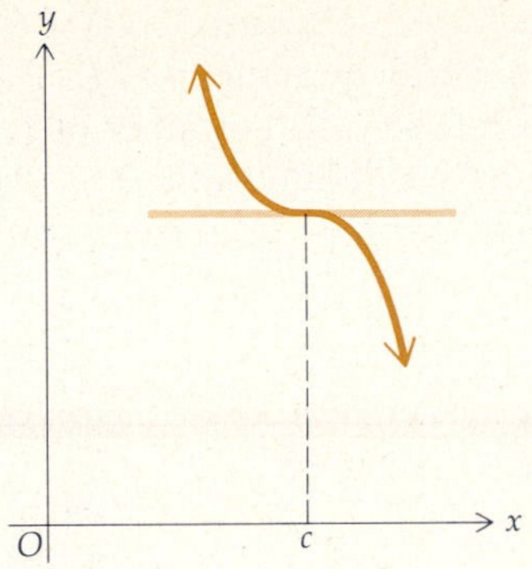

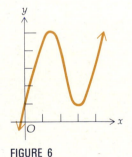

FIGURE 5

In summary, to determine the relative extrema of a function f:

1. Find $f'(x)$.
2. Find the critical numbers of f, that is, the values of x for which $f'(x) = 0$ or for which $f'(x)$ does not exist.
3. Apply the first-derivative test (Theorem 4.4.4).

The following examples illustrate this procedure.

EXAMPLE 1   Given

$$f(x) = x^3 - 6x^2 + 9x + 1$$

find the relative extrema of f by applying the first-derivative test. Determine the values of x at which the relative extrema occur, as well as the intervals on which f is increasing and the intervals on which f is decreasing. Draw a sketch of the graph.

Solution

$$f'(x) = 3x^2 - 12x + 9$$

$f'(x)$ exists for all values of x. Set $f'(x) = 0$.

$$3x^2 - 12x + 9 = 0$$

$$3(x - 3)(x - 1) = 0$$

$$x = 3 \qquad x = 1$$

Thus the critical numbers of f are 1 and 3. To determine whether f has a relative extremum at either of these numbers, apply the first-derivative test. The results are summarized in Table 1.

Table 1

	$f(x)$	$f'(x)$	*Conclusion*
$x < 1$		$+$	f is increasing
$x = 1$	5	0	f has a relative maximum value
$1 < x < 3$		$-$	f is decreasing
$x = 3$	1	0	f has a relative minimum value
$3 < x$		$+$	f is increasing

From the table, 5 is a relative maximum value of f occurring at $x = 1$, and 1 is a relative minimum value of f occurring at $x = 3$. A sketch of the graph is shown in Figure 6.

FIGURE 6

EXAMPLE 2   Given

$$f(x) = \begin{cases} x^2 - 4 & \text{if } x < 3 \\ 8 - x & \text{if } 3 \le x \end{cases}$$

find the relative extrema of f by applying the first-derivative test. Determine the values of x at which the relative extrema occur, as well as the intervals on which f is increasing and the intervals on which f is decreasing. Draw a sketch of the graph.

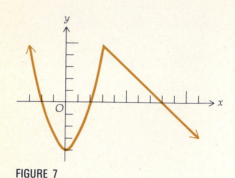

FIGURE 7

Solution   If $x < 3$, $f'(x) = 2x$. If $x > 3$, $f'(x) = -1$. Because $f'_-(3) = 6$ and $f'_+(3) = -1$, $f'(3)$ does not exist. Therefore 3 is a critical number of f.

Because $f'(x) = 0$ when $x = 0$, it follows that 0 is a critical number of f. Applying the first-derivative test, we summarize the results in Table 2. A sketch of the graph appears in Figure 7.

Table 2

	$f(x)$	$f'(x)$	Conclusion
$x < 0$		$-$	f is decreasing
$x = 0$	-4	0	f has a relative minimum value
$0 < x < 3$		$+$	f is increasing
$x = 3$	5	does not exist	f has a relative maximum value
$3 < x$		$-$	f is decreasing

EXAMPLE 3   Given

$$f(x) = x^{4/3} + 4x^{1/3}$$

find the relative extrema of f, determine the values of x at which the relative extrema occur, and determine the intervals on which f is increasing and the intervals on which f is decreasing. Draw a sketch of the graph.

Solution

$$f'(x) = \tfrac{4}{3}x^{1/3} + \tfrac{4}{3}x^{-2/3}$$
$$= \tfrac{4}{3}x^{-2/3}(x + 1)$$

Because $f'(x)$ does not exist when $x = 0$, and $f'(x) = 0$ when $x = -1$, the critical numbers of f are -1 and 0. We apply the first-derivative test and summarize the results in Table 3. Figure 8 shows a sketch of the graph.

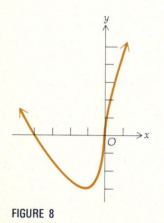

FIGURE 8

Table 3

	$f(x)$	$f'(x)$	Conclusion
$x < -1$		$-$	f is decreasing
$x = -1$	-3	0	f has a relative minimum value
$-1 < x < 0$		$+$	f is increasing
$x = 0$	0	does not exist	f does not have a relative extremum at $x = 0$
$0 < x$		$+$	f is increasing

EXERCISES 4.4

In Exercises 1 through 40, do each of the following: (a) Find the relative extrema of f by applying the first-derivative test; (b) determine the values of x at which the relative extrema occur; (c) determine the intervals on which f is increasing; (d) determine the intervals on which f is decreasing; (e) draw a sketch of the graph.

1. $f(x) = x^2 - 4x - 1$
2. $f(x) = 3x^2 - 3x + 2$
3. $f(x) = x^3 - x^2 - x$
4. $f(x) = x^3 - 9x^2 + 15x - 5$
5. $f(x) = 2x^3 - 9x^2 + 2$
6. $f(x) = x^3 - 3x^2 - 9x$
7. $f(x) = 4 \sin \tfrac{1}{2}x$
8. $f(x) = x^4 + 4x$
9. $f(x) = \tfrac{1}{4}x^4 - x^3 + x^2$
10. $f(x) = 2 \cos 3x$
11. $f(x) = \tfrac{1}{5}x^5 - \tfrac{5}{3}x^3 + 4x + 1$
12. $f(x) = x^5 - 5x^3 - 20x - 2$
13. $f(x) = \sqrt{x} - \dfrac{1}{\sqrt{x}}$
14. $f(x) = \dfrac{x - 2}{x + 2}$
15. $f(x) = x + \dfrac{1}{x^2}$
16. $f(x) = 2x + \dfrac{1}{2x}$
17. $f(x) = (1 - x)^2(1 + x)^3$
18. $f(x) = (x + 2)^2(x - 1)^2$
19. $f(x) = 2x\sqrt{3 - x}$
20. $f(x) = x\sqrt{5 - x^2}$
21. $f(x) = x - 3x^{1/3}$
22. $f(x) = 4x - 6x^{2/3}$
23. $f(x) = 2 - 3(x - 4)^{2/3}$
24. $f(x) = 2 - (x - 1)^{1/3}$
25. $f(x) = x^{2/3} - x^{1/3}$
26. $f(x) = 3 \csc 2x$
27. $f(x) = \tfrac{1}{2} \sec 4x$
28. $f(x) = x^{2/3}(x - 1)^2$

29. $f(x) = (x + 1)^{2/3}(x - 2)^{1/3}$ **30.** $f(x) = \begin{cases} 5 - 2x & \text{if } x < 3 \\ 3x - 10 & \text{if } 3 \leq x \end{cases}$

31. $f(x) = \begin{cases} 2x + 9 & \text{if } x \leq -2 \\ x^2 + 1 & \text{if } -2 < x \end{cases}$

32. $f(x) = \begin{cases} \sqrt{25 - (x + 7)^2} & \text{if } -12 \leq x \leq -3 \\ 12 - x^2 & \text{if } -3 < x \end{cases}$

33. $f(x) = \begin{cases} 3x + 5 & \text{if } x < -1 \\ x^2 + 1 & \text{if } -1 \leq x < 2 \\ 7 - x & \text{if } 2 \leq x \end{cases}$

34. $f(x) = \begin{cases} 4 - (x + 5)^2 & \text{if } x < -4 \\ 12 - (x + 1)^2 & \text{if } -4 \leq x \end{cases}$

35. $f(x) = \begin{cases} (x + 9)^2 - 8 & \text{if } x < -7 \\ -\sqrt{25 - (x + 4)^2} & \text{if } -7 \leq x \leq 0 \\ (x - 2)^2 - 7 & \text{if } 0 < x \end{cases}$

36. $f(x) = \begin{cases} 12 - (x + 5)^2 & \text{if } x \leq -3 \\ 5 - x & \text{if } -3 < x \leq -1 \\ \sqrt{100 - (x - 7)^2} & \text{if } -1 < x \leq 17 \end{cases}$

37. $f(x) = x^{5/4} + 10x^{1/4}$ **38.** $f(x) = (x - a)^{2/5} + 1$

39. $f(x) = x^{1/3}(x + 4)^{-2/3}$ **40.** $f(x) = x^{5/3} - 10x^{2/3}$

41. Prove Theorem 4.4.3(ii).   **42.** Prove Theorem 4.4.4(ii).

43. Find a and b such that the function defined by

$$f(x) = x^3 + ax^2 + b$$

will have a relative extremum at (2, 3).

44. Find a, b, and c such that the function defined by

$$f(x) = ax^2 + bx + c$$

will have a relative maximum value of 7 at 1 and the graph of $y = f(x)$ will go through the point $(2, -2)$.

45. Find a, b, c, and d such that the function defined by

$$f(x) = ax^3 + bx^2 + cx + d$$

will have relative extrema at (1, 2) and (2, 3).

46. Given that the function f is continuous for all values of x, $f(3) = 2$, $f'(x) < 0$ if $x < 3$, and $f'(x) > 0$ if $x > 3$, draw a sketch of a possible graph of f in each of the following cases, where the additional condition is satisfied: (a) f' is continuous at 3; (b) $f'(x) = -1$ if $x < 3$, and $f'(x) = 1$ if $x > 3$; (c) $\lim_{x \to 3^-} f'(x) = -1$, $\lim_{x \to 3^+} f'(x) = 1$, and $f'(a) \neq f'(b)$ if $a \neq b$.

47. Given $f(x) = x^p(1 - x)^q$, where p and q are positive integers greater than 1, prove each of the following: (a) If p is even, f has a relative minimum value at 0; (b) if q is even, f has a relative minimum value at 1; (c) f has a relative maximum value at $p/(p + q)$ whether p and q are odd or even.

48. The function f is differentiable at each number in the closed interval $[a, b]$. Prove that if $f'(a) \cdot f'(b) < 0$, there is a number c in the open interval (a, b) such that $f'(c) = 0$.

49. If $f(x) = x^k$, where k is an odd positive integer, show that f has no relative extrema.

50. Prove that if f is increasing on $[a, b]$ and if g is increasing on $[f(a), f(b)]$, then if $g \circ f$ exists on $[a, b]$, $g \circ f$ is increasing on $[a, b]$.

51. The function f is increasing on the interval I. Prove that (a) if $g(x) = -f(x)$, then g is decreasing on I; (b) if $h(x) = 1/f(x)$ and $f(x) > 0$ on I, then h is decreasing on I.

4.5 CONCAVITY AND POINTS OF INFLECTION

Figure 1 shows a sketch of the graph of a function f whose first and second derivatives exist on the closed interval $[x_1, x_7]$. Because both f and f' are differentiable there, f and f' are continuous on $[x_1, x_7]$.

If we consider a point P moving along the graph of Figure 1 from A to G, then the position of P varies as we increase x from x_1 to x_7. As P moves along the graph from A to B, the slope of the tangent line to the graph is positive and is decreasing; that is, the tangent line is turning clockwise, and the graph lies below the tangent line. When the point P is at B, the slope of the tangent line is zero and is still decreasing. As P moves along the graph from B to C, the slope of the tangent line is negative and is still decreasing; the tangent line is still

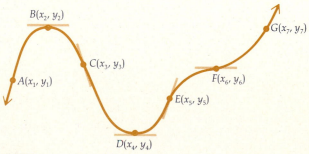

FIGURE 1

turning clockwise, and the graph is below its tangent line. We say that the graph is *concave downward* from A to C. As P moves along the graph from C and D, the slope of the tangent line is negative and is increasing; that is, the tangent line is turning counterclockwise, and the graph is above its tangent line. At D the slope of the tangent line is zero and is still increasing. From D to E, the slope of the tangent line is positive and increasing; the tangent line is still turning counterclockwise, and the graph is above its tangent line. We say that the graph is *concave upward* from C to E. At the point C the graph changes from concave downward to concave upward. Point C is called a *point of inflection*. We have the following definitions.

4.5.1 DEFINITION   The graph of a function f is said to be **concave upward** at the point $(c, f(c))$ if $f'(c)$ exists and if there is an open interval I containing c such that for all values of $x \neq c$ in I the point $(x, f(x))$ on the graph is above the tangent line to the graph at $(c, f(c))$.

4.5.2 DEFINITION   The graph of a function f is said to be **concave downward** at the point $(c, f(c))$ if $f'(c)$ exists and if there is an open interval I containing c such that for all values of $x \neq c$ in I the point $(x, f(x))$ on the graph is below the tangent line to the graph at $(c, f(c))$.

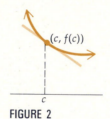

FIGURE 2

▶ **ILLUSTRATION 1**   Figure 2 shows a sketch of a portion of the graph of a function f that is concave upward at the point $(c, f(c))$, and Figure 3 shows a sketch of a portion of the graph of a function f that is concave downward at the point $(c, f(c))$. ◀

The graph in Figure 1 is concave downward at all points $(x, f(x))$ for which x is in either of the following open intervals: (x_1, x_3) or (x_5, x_6). Similarly, the graph in Figure 1 is concave upward at all points $(x, f(x))$ for which x is in either (x_3, x_5) or (x_6, x_7).

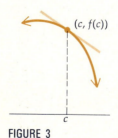

FIGURE 3

▶ **ILLUSTRATION 2**   If f is the function defined by $f(x) = x^2$, then $f'(x) = 2x$ and $f''(x) = 2$. Thus $f''(x) > 0$ for all x. Furthermore, because the graph of f, appearing in Figure 4, is above all of its tangent lines, the graph is concave upward at all of its points.

If g is the function defined by $g(x) = -x^2$, then $g'(x) = -2x$ and $g''(x) = -2$. Hence $g''(x) < 0$ for all x. Also, because the graph of g, shown in Figure 5, is below all of its tangent lines, it is concave downward at all of its points. ◀

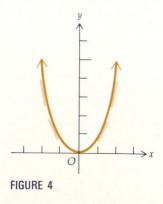

FIGURE 4

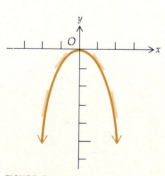

FIGURE 5

The function f of Illustration 2 is such that $f''(x) > 0$ for all x, and the graph of f is concave upward everywhere. For function g of Illustration 2, $g''(x) < 0$ for all x, and the graph of g is concave downward everywhere. These two situations are special cases of the following theorem.

4.5.3 THEOREM Let f be a function that is differentiable on some open interval containing c. Then

(i) if $f''(c) > 0$, the graph of f is concave upward at $(c, f(c))$;
(ii) if $f''(c) < 0$, the graph of f is concave downward at $(c, f(c))$.

Proof of (i)

$$f''(c) = \lim_{x \to c} \frac{f'(x) - f'(c)}{x - c}$$

Because $f''(c) > 0$,

$$\lim_{x \to c} \frac{f'(x) - f'(c)}{x - c} > 0$$

Then, by Theorem 2.10.1 there is an open interval I containing c such that

$$\frac{f'(x) - f'(c)}{x - c} > 0 \tag{1}$$

for every $x \neq c$ in I.

Now consider the tangent line to the graph of f at the point $(c, f(c))$. An equation of this tangent line is

$$y = f(c) + f'(c)(x - c) \tag{2}$$

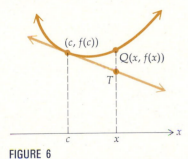

FIGURE 6

Let x be a number in the interval I such that $x \neq c$, and let Q be the point on the graph of f whose abscissa is x. Through Q draw a line parallel to the y axis, and let T be the point of intersection of this line with the tangent line (see Figure 6).

To prove that the graph of f is concave upward at $(c, f(c))$ we must show that the point Q is above the point T or, equivalently, that the directed distance $\overline{TQ} > 0$ for all values of $x \neq c$ in I. $\overline{TQ}$ equals the ordinate of Q minus the ordinate of T. The ordinate of Q is $f(x)$, and the ordinate of T is obtained from (2); so

$$\overline{TQ} = f(x) - [f(c) + f'(c)(x - c)]$$
$$\overline{TQ} = [f(x) - f(c)] - f'(c)(x - c) \tag{3}$$

From the mean-value theorem there exists some number d between x and c such that

$$f'(d) = \frac{f(x) - f(c)}{x - c}$$

That is,

$$f(x) - f(c) = f'(d)(x - c) \qquad \text{for some } d \text{ between } x \text{ and } c$$

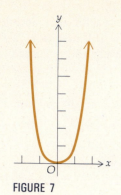

FIGURE 7

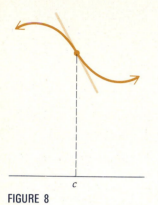

FIGURE 8

Substituting from this equation into (3) we have

$$\overline{TQ} = f'(d)(x - c) - f'(c)(x - c)$$

$$\overline{TQ} = (x - c)[f'(d) - f'(c)] \tag{4}$$

Because d is between x and c, d is in the interval I, and so by taking $x = d$ in inequality (1) we obtain

$$\frac{f'(d) - f'(c)}{d - c} > 0 \tag{5}$$

To prove that $\overline{TQ} > 0$ we show that both of the factors on the right side of (4) have the same sign. If $x - c > 0$, then $x > c$. And because d is between x and c, then $d > c$; therefore, from inequality (5), $f'(d) - f'(c) > 0$. If $x - c < 0$, then $x < c$ and so $d < c$; therefore, from (5), $f'(d) - f'(c) < 0$. We conclude that $x - c$ and $f'(d) - f'(c)$ have the same sign; therefore $\overline{TQ}$ is a positive number. Thus the graph of f is concave upward at $(c, f(c))$.

The proof of part (ii) is similar and is omitted.   ■

The converse of Theorem 4.5.3 is not true. For example, if f is the function defined by $f(x) = x^4$, the graph of f is concave upward at the point $(0, 0)$ but because $f''(x) = 12x^2$, $f''(0) = 0$ (see Figure 7). Accordingly, a sufficient condition for the graph of a function f to be concave upward at the point $(c, f(c))$ is that $f''(c) > 0$, but this is not a necessary condition. Similarly, a sufficient—but not a necessary—condition that the graph of a function f be concave downward at the point $(c, f(c))$ is that $f''(c) < 0$.

If there is a point on the graph of a function at which the sense of concavity changes, and the graph has a tangent line there, then the graph crosses its tangent line at this point, as shown in Figures 8, 9, and 10. Such a point is called a *point of inflection*.

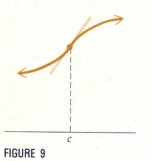

FIGURE 9

4.5.4 DEFINITION   The point $(c, f(c))$ is a **point of inflection** of the graph of the function f if the graph has a tangent line there, and if there exists an open interval I containing c such that if x is in I, then either

     (i) $f''(x) < 0$ if $x < c$, and $f''(x) > 0$ if $x > c$, or
     (ii) $f''(x) > 0$ if $x < c$ and $f''(x) < 0$ if $x > c$

▶ **ILLUSTRATION 3**   Figure 8 illustrates a point of inflection where condition (i) of Definition 4.5.4 holds; in this case the graph is concave downward at points immediately to the left of the point of inflection, and the graph is concave upward at points immediately to the right of the point of inflection. Condition (ii) is illustrated in Figure 9, where the sense of concavity changes from upward to downward at the point of inflection. Figure 10 is another illustration of condition (i), where the sense of concavity changes from downward to upward at the point of inflection. Note that in Figure 10 there is a horizontal tangent line at the point of inflection.   ◀

FIGURE 10

For the graph in Figure 1 there are points of inflection at C, E, and F.

A crucial part of Definition 4.5.4 is that the graph must have a tangent line at a point of inflection. Consider, for instance, the function of Example 3 in

Section 2.3. It is defined by

$$h(x) = \begin{cases} 4 - x^2 & \text{if } x < 1 \\ 2 + x^2 & \text{if } 1 < x \end{cases}$$

A sketch of the graph of h appears in Figure 11. Observe that $h''(x) = -2$ if $x < 1$ and $h''(x) = 2$ if $x > 1$. Thus at the point $(1, 3)$ on the graph the sense of concavity changes from downward to upward. However, $(1, 3)$ is not a point of inflection because the graph does not have a tangent line there.

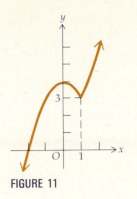

FIGURE 11

▶ **ILLUSTRATION 4** Suppose it is estimated that t hours after starting work at 7 A.M. a factory worker on an assembly line has performed a particular task on $f(t)$ units, and

$$f(t) = 21t + 9t^2 - t^3 \qquad 0 \le t \le 5$$

Table 1

t	1	2	3	4	5
$f(t)$	29	70	117	164	205

In Table 1 there are function values for integer values of t from 1 through 5, and Figure 12 shows a sketch of the graph of f on $[0, 5]$.

$$f'(t) = 21 + 18t - 3t^2 \qquad f''(t) = 18 - 6t$$
$$= 6(3 - t)$$

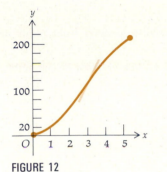

FIGURE 12

Observe that $f''(t) > 0$ if $0 < t < 3$ and $f''(t) < 0$ if $3 < t < 5$. From Definition 4.5.4(ii) it follows that the graph of f has a point of inflection at $t = 3$. From Theorem 4.4.3, because $f''(t) > 0$ when $0 < t < 3$, $f'(t)$ is increasing on $[0, 3]$, and because $f''(t) < 0$ when $3 < t < 5$, $f'(t)$ is decreasing on $[3, 5]$. Therefore, since $f'(t)$ is the rate of change of $f(t)$ with respect to t, we conclude that in the first three hours (from 7 A.M. until 10 A.M.) the worker is performing the task at an increasing rate, and during the remaining two hours (from 10 A.M. until noon) the worker is performing the task at a decreasing rate. At $t = 3$ (10 A.M.) the worker is producing most efficiently, and when $3 < t < 5$ (after 10 A.M.) there is a reduction in the worker's production rate. The point at which the worker is producing most efficiently is called the *point of diminishing returns*; this point is a point of inflection of the graph of f.   ◀

Definition 4.5.4 indicates nothing about the value of the second derivative of f at a point of inflection. The following theorem states that if the second derivative exists at a point of inflection, it must be zero there.

4.5.5 THEOREM   If the function f is differentiable on some open interval containing c, and if $(c, f(c))$ is a point of inflection of the graph of f, then if $f''(c)$ exists, $f''(c) = 0$.

Proof Let g be the function such that $g(x) = f'(x)$; then $g'(x) = f''(x)$. Because $(c, f(c))$ is a point of inflection of the graph of f, then $f''(x)$ changes sign at c and so $g'(x)$ changes sign at c. Therefore, by the first-derivative test (Theorem 4.4.4), g has a relative extremum at c, and c is a critical number of g. Because $g'(c) = f''(c)$, and since by hypothesis $f''(c)$ exists, it follows that $g'(c)$ exists. Therefore, by Theorem 4.1.3, $g'(c) = 0$ and $f''(c) = 0$, which is what we wanted to prove.   ■

The converse of Theorem 4.5.5 is not true. That is, if the second derivative of a function is zero at a number c, it is not necessarily true that the graph of the function has a point of inflection where $x = c$. This fact is shown in the following illustration.

▶ **ILLUSTRATION 5**   Consider the function f defined by $f(x) = x^4$. A sketch of the graph of f is shown in Figure 7.

$$f'(x) = 4x^3 \qquad f''(x) = 12x^2$$

Observe that $f''(0) = 0$; but because $f''(x) > 0$ if $x < 0$ and $f''(x) > 0$ if $x < 0$, the graph is concave upward at points on the graph immediately to the left of $(0, 0)$ and at points immediately to the right of $(0, 0)$. Consequently, $(0, 0)$ is not a point of inflection.   ◀

EXAMPLE 1   The function of Example 1 in Section 4.4 is defined by

$$f(x) = x^3 - 6x^2 + 9x + 1$$

Find the point of inflection of the graph of f, and determine where the graph is concave upward and where it is concave downward. Draw a sketch of the graph and show a segment of the inflectional tangent.

Solution

$$f'(x) = 3x^2 - 12x + 9 \qquad f''(x) = 6x - 12$$

$f''(x)$ exists for all values of x; so the only possible point of inflection is where $f''(x) = 0$, which occurs at $x = 2$. To determine whether there is a point of inflection at $x = 2$, we must check to see if $f''(x)$ changes sign; at the same time we determine the concavity of the graph for the respective intervals. The results are summarized in Table 2.

Table 2

	$f(x)$	$f'(x)$	$f''(x)$	*Conclusion*
$x < 2$			$-$	graph is concave downward
$x = 2$	3	-3	0	graph has a point of inflection
$2 < x$			$+$	graph is concave upward

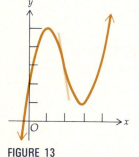

FIGURE 13

In Example 1 of Section 4.4 we showed that f has a relative maximum value at 1 and a relative minimum value at 3. A sketch of the graph showing a segment of the inflectional tangent appears in Figure 13.

The graph of a function may have a point of inflection where the second derivative fails to exist. This situation is shown in the next example.

EXAMPLE 2   Given

$$f(x) = x^{1/3}$$

find the point of inflection of the graph of f, and determine where the graph is concave upward and where it is concave downward. Draw a sketch of the graph.

Solution

$$f'(x) = \tfrac{1}{3}x^{-2/3} \qquad f''(x) = -\tfrac{2}{9}x^{-5/3}$$

Neither $f'(0)$ nor $f''(0)$ exists. In Illustration 3 of Section 3.2 we showed that

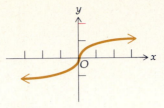

FIGURE 14

the y axis is the tangent line to the graph of this function at $(0, 0)$. Furthermore,

$$f''(x) > 0 \text{ if } x < 0 \quad \text{and} \quad f''(x) < 0 \text{ if } x > 0$$

Therefore, from Definition 4.5.4(ii), f has a point of inflection at $(0, 0)$. The concavity of the graph is determined from the sign of $f''(x)$. The results are summarized in Table 3.

Figure 14 shows a sketch of the graph of f.

Table 3

	$f(x)$	$f'(x)$	$f''(x)$	Conclusion
$x < 0$		$+$	$+$	f is increasing; graph is concave upward
$x = 0$	0	does not exist	does not exist	graph has a point of inflection
$0 < x$		$+$	$-$	f is increasing; graph is concave downward

EXAMPLE 3   If

$$f(x) = (1 - 2x)^3$$

find the point of inflection of the graph of f, and determine where the graph is concave upward and where it is concave downward. Draw a sketch of the graph of f.

Solution

$$f'(x) = -6(1 - 2x)^2 \qquad f''(x) = 24(1 - 2x)$$

Because $f''(x)$ exists for all values of x, the only possible point of inflection is where $f''(x) = 0$, that is, at $x = \frac{1}{2}$. From the results summarized in Table 4, $f''(x)$ changes sign from $+$ to $-$ at $x = \frac{1}{2}$; so the graph has a point of inflection there. Note also that because $f'(\frac{1}{2}) = 0$, the graph has a horizontal tangent line at the point of inflection. A sketch of the graph appears in Figure 15.

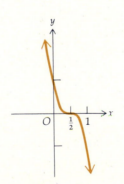

FIGURE 15

Table 4

	$f(x)$	$f'(x)$	$f''(x)$	Conclusion
$x < \frac{1}{2}$			$+$	graph is concave upward
$x = \frac{1}{2}$	0	0	0	graph has a point of inflection
$\frac{1}{2} < x$			$-$	graph is concave downward

EXAMPLE 4   Find the points of inflection of the graph of the sine function. Also find the slopes of the inflectional tangents. Draw the graph of the sine function on an interval of length 2π and containing the point of inflection having the smallest positive abscissa. Show a segment of the inflectional tangent at this point of inflection.

Solution   Let

$$f(x) = \sin x$$

Then

$$f'(x) = \cos x \qquad f''(x) = -\sin x$$

$f''(x)$ exists for all x. To determine the points of inflection set $f''(x) = 0$.

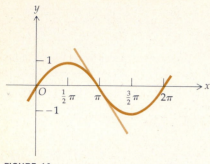

FIGURE 16

$$-\sin x = 0$$

$$x = k\pi \qquad k \text{ is any integer}$$

Because $f''(x)$ changes sign at each of these values of x, the graph has a point of inflection at every point having these abscissas. At each point of inflection,

$$f'(k\pi) = \cos k\pi \qquad k \text{ is any integer}$$

$$= \begin{cases} 1 & \text{if } k \text{ is an even integer} \\ -1 & \text{if } k \text{ is an odd integer} \end{cases}$$

Therefore the slopes of the inflectional tangents are either $+1$ or -1. Figure 16 shows the graph of the sine function on the interval $[0, 2\pi]$. At the point of inflection $(\pi, 0)$ a segment of the inflectional tangent is shown.

EXERCISES 4.5

In Exercises 1 through 16, find the points of inflection of the graph of the function, if there are any. Determine where the graph is concave upward and where it is concave downward. Draw a sketch of the graph, and show a segment of each inflectional tangent.

1. $f(x) = x^3 + 9x$
2. $g(x) = x^3 - 3x^2 + 7x - 3$
3. $g(x) = 2x^3 + 3x^2 - 12x + 1$
4. $f(x) = x^3 - 6x^2 + 20$
5. $F(x) = x^4 - 8x^3$
6. $f(x) = x^4 - 2x^3$
7. $g(x) = (x - 1)^3$
8. $G(x) = (x + 2)^3$
9. $f(x) = (x + 2)^{1/3}$
10. $g(x) = (x - 1)^{1/3}$
11. $G(x) = \dfrac{2}{x^2 + 3}$
12. $F(x) = \dfrac{x}{x^2 + 4}$
13. $g(x) = 2 \sin 3x; \; x \in [-\pi, \pi]$
14. $f(x) = 3 \cos 2x; \; x \in [-\pi, \pi]$
15. $f(x) = \tan \frac{1}{2}x; \; x \in (-\pi, \pi)$
16. $g(x) = \cot 2x; \; x \in (0, \frac{1}{2}\pi)$

In Exercises 17 through 24, find the point of inflection of the graph of the function, if there is one. Determine where the graph is concave upward and where it is concave downward. Draw a sketch of the graph.

17. $f(x) = \begin{cases} x^2 - 1 & \text{if } x < 2 \\ 7 - x^2 & \text{if } 2 \le x \end{cases}$
18. $f(x) = \begin{cases} 2 + x^2 & \text{if } x \le 1 \\ 4 - x^2 & \text{if } 1 < x \end{cases}$
19. $g(x) = \begin{cases} x^2 & \text{if } x \le 0 \\ -x^2 & \text{if } 0 < x \end{cases}$
20. $g(x) = \begin{cases} -x^3 & \text{if } x < 0 \\ x^3 & \text{if } 0 \le x \end{cases}$
21. $F(x) = \begin{cases} x^3 & \text{if } x < 0 \\ x^4 & \text{if } 0 \le x \end{cases}$
22. $G(x) = \begin{cases} x^2 & \text{if } x \le 0 \\ x^4 & \text{if } 0 < x \end{cases}$
23. $f(x) = (x - 2)^{1/5} + 3$
24. $g(x) = (2x - 6)^{3/2} + 1$

25. If $f(x) = ax^3 + bx^2$, determine a and b so that the graph of f will have a point of inflection at $(1, 2)$.
26. If $f(x) = ax^3 + bx^2 + cx$, determine a, b, and c so that the graph of f will have a point of inflection at $(1, 2)$ and so that the slope of the inflectional tangent there will be -2.
27. If $f(x) = ax^3 + bx^2 + cx + d$, determine a, b, c, and d so that f will have a relative extremum at $(0, 3)$ and so that the graph of f will have a point of inflection at $(1, -1)$.

28. If $f(x) = ax^4 + bx^3 + cx^2 + dx + e$, determine the values of a, b, c, d, and e so the graph of f will have a point of inflection at $(1, -1)$, have the origin on it, and be symmetric with respect to the y axis.

In Exercises 29 through 31, (a) find the points of inflection of the graph of the trigonometric function; (b) find the slopes of the inflectional tangents; (c) draw the graph of the function on an interval of length 2π and containing the point of inflection having the smallest positive abscissa, and show a segment of the inflectional tangent at this point of inflection.

29. the cosine function
30. the tangent function
31. the cotangent function

32. Prove that the graphs of the secant and cosecant functions have no points of inflection.

In Exercises 33 through 46, draw a portion of the graph of a function f through the point where $x = c$ if the given conditions are satisfied. Assume that f is continuous on some open interval containing c.

33. $f'(x) > 0$ if $x < c$; $f'(x) < 0$ if $x > c$; $f''(x) < 0$ if $x < c$; $f''(x) < 0$ if $x > c$
34. $f'(x) > 0$ if $x < c$; $f'(x) > 0$ if $x > c$; $f''(x) > 0$ if $x < c$; $f''(x) < 0$ if $x > c$
35. $f'(x) > 0$ if $x < c$; $f'(x) < 0$ if $x > c$; $f''(x) > 0$ if $x < c$; $f''(x) > 0$ if $x > c$
36. $f'(x) < 0$ if $x < c$; $f'(x) > 0$ if $x > c$; $f''(x) > 0$ if $x < c$; $f''(x) < 0$ if $x > c$
37. $f''(c) = 0$; $f'(c) = 0$; $f''(x) > 0$ if $x < c$; $f''(x) < 0$ if $x > c$
38. $f'(c) = 0$; $f'(c) > 0$ if $x < c$; $f''(x) > 0$ if $x > c$
39. $f'(c) = 0$; $f'(c) = 0$; $f''(x) > 0$ if $x < c$; $f''(x) > 0$ if $x > c$
40. $f'(c) = 0$; $f'(x) < 0$ if $x < c$; $f''(x) > 0$ if $x > c$
41. $f''(c) = 0$; $f'(c) = -1$; $f''(x) < 0$ if $x < c$; $f''(x) > 0$ if $x > c$
42. $f''(c) = 0$; $f'(c) = \frac{1}{2}$; $f''(x) > 0$ if $x < c$; $f''(x) < 0$ if $x > c$
43. $f'(c)$ does not exist; $f''(x) > 0$ if $x < c$; $f''(x) > 0$ if $x > c$
44. $f'(c)$ does not exist; $f''(c)$ does not exist; $f''(x) < 0$ if $x < c$; $f''(x) > 0$ if $x > c$

45. $\lim\limits_{x \to c^-} f'(x) = +\infty$; $\lim\limits_{x \to c^+} f'(x) = 0$; $f''(x) > 0$ if $x < c$; $f''(x) < 0$ if $x > c$

46. $\lim\limits_{x \to c^-} f'(x) = +\infty$; $\lim\limits_{x \to c^+} f'(x) = -\infty$; $f''(x) > 0$ if $x < c$; $f''(x) > 0$ if $x > c$

47. Draw a sketch of the graph of a function f for which $f(x)$, $f'(x)$, and $f''(x)$ exist and are positive for all x.

48. Draw a sketch of the graph of a function f for which $f(x)$, $f'(x)$, and $f''(x)$ exist and are negative for all x.

49. If $f(x) = x^5$, show that 0 is a critical number of f but that $f(0)$ is not a relative extremum. Is the origin a point of inflection of the graph of f? Prove your answer.

50. If $f(x) = 3x^2 + x|x|$, prove that $f''(0)$ does not exist but the graph of f is concave upward everywhere.

51. It is estimated that a worker in a shop that makes picture frames can paint y frames x hours after starting work at 8 A.M., and

$$y = 3x + 8x^2 - x^3 \qquad 0 \le x \le 4$$

Find at what time the worker is working most efficiently (that is, at what time does the worker reach the point of diminishing returns?). (*Hint:* See Illustration 4.)

52. Draw a sketch of the graph of the equation $x^{2/3} + y^{2/3} = 1$. (*Hint:* The graph is not that of a function. However, the portion in the first quadrant is the graph of a function. Obtain this portion and then complete the graph by symmetry properties. Concavity plays an important part.)

4.6 THE SECOND-DERIVATIVE TEST FOR RELATIVE EXTREMA

In Section 4.4 you learned how to determine whether a function f has a relative maximum value or a relative minimum value at a critical number c by checking the algebraic sign of f' at numbers in intervals to the left and right of c. Another test for relative extrema is one that involves only the critical number c. Before stating the test in the form of a theorem, we give an informal geometric discussion that should appeal to your intuition.

Suppose that f is a function such that f' and f'' exist on some open interval (a, b) containing c and that $f'(c) = 0$. Also suppose that f'' is negative on (a, b). From Theorem 4.4.3(ii), because $f''(x) < 0$ on (a, b), then f' is decreasing on $[a, b]$. Since the value of f' at a point on the graph of f gives the slope of the tangent line at the point, it follows that the slope of the tangent line is decreasing on $[a, b]$. In Figure 1 there is a sketch of the graph of a function f having these properties. From Theorem 4.5.3(ii), the graph of f is concave downward at all points in the figure, and a segment of the tangent line is shown at some of these points. Observe that the slope of the tangent line is decreasing on $[a, b]$. Note that f has a relative maximum value at c, where $f'(c) = 0$ and $f''(c) < 0$.

Now suppose that f is a function having the properties of the function in the previous paragraph except that f'' is positive on (a, b). Then from Theorem 4.4.3(i), because $f''(x) > 0$ on (a, b), it follows that f' is increasing on $[a, b]$. Thus the slope of the tangent line is increasing on $[a, b]$. Figure 2 shows a sketch of the graph of a function f having these properties. From Theorem 4.5.3(i),

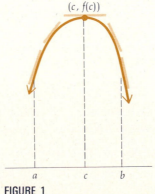

FIGURE 1

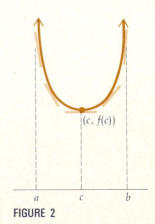

FIGURE 2

the graph of f is concave upward at all points in the figure, and a segment of the tangent line is shown at some of these points. The slopes of these tangent lines are increasing on $[a, b]$. The function f has a relative minimum value at c, where $f'(c) = 0$ and $f''(c) > 0$.

The facts in the preceding two paragraphs are given in the **second-derivative test for relative extrema**, which is now stated.

4.6.1 THEOREM
Second-Derivative Test for Relative Extrema

Let c be a critical number of a function f at which $f'(c) = 0$, and let f' exist for all values of x in some open interval containing c. If $f''(c)$ exists and

(i) if $f''(c) < 0$, then f has a relative maximum value at c;
(ii) if $f''(c) > 0$, then f has a relative minimum value at c.

The proof of part (i) makes use of Theorem 2.10.2 and that of part (ii) utilizes Theorem 2.10.1. You may wish to refer to those theorems now and also refer to Illustrations 1 and 2 in Supplementary Section 2.10, which give geometrical interpretations of the theorems.

Proof of (i) By hypothesis, $f''(c)$ exists and is negative; so

$$f''(c) = \lim_{x \to c} \frac{f'(x) - f'(c)}{x - c} < 0$$

Therefore, by Theorem 2.10.2 there is an open interval I containing c such that

$$\frac{f'(x) - f'(c)}{x - c} < 0 \tag{1}$$

for every $x \neq c$ in the interval.

Let I_1 be the open interval containing all values of x in I for which $x < c$; therefore c is the right endpoint of the open interval I_1. Let I_2 be the open interval containing all values of x in I for which $x > c$; so c is the left endpoint of the open interval I_2.

Then if x is in I_1, $x - c < 0$, and it follows from inequality (1) that $f'(x) - f'(c) > 0$ or, equivalently, $f'(x) > f'(c)$. If x is in I_2, $x - c > 0$, and it follows from (1) that $f'(x) - f'(c) < 0$ or, equivalently, $f'(x) < f'(c)$.

But because $f'(c) = 0$, we conclude that if x is in I_1, $f'(x) > 0$, and if x is in I_2, $f'(x) < 0$. Therefore $f'(x)$ changes algebraic sign from positive to negative as x increases through c, and so, by Theorem 4.4.4, f has a relative maximum value at c.

The proof of part (ii) is similar and is left as an exercise (see Exercise 46). ■

EXAMPLE 1 Given

$$f(x) = x^4 + \tfrac{4}{3}x^3 - 4x^2$$

find the relative maxima and minima of f by applying the second-derivative test. Draw a sketch of the graph of f.

Solution We compute the first and second derivatives of f.

$$f'(x) = 4x^3 + 4x^2 - 8x \qquad f''(x) = 12x^2 + 8x - 8$$

FIGURE 3

Set $f'(x) = 0$.

$$4x(x + 2)(x - 1) = 0$$

$$x = 0 \qquad x = -2 \qquad x = 1$$

Thus the critical numbers of f are -2, 0, and 1. We determine whether or not there is a relative extremum at any of these critical numbers by finding the sign of the second derivative there. The results are summarized in Table 1.

Table 1

	$f(x)$	$f'(x)$	$f''(x)$	Conclusion
$x = -2$	$-\frac{32}{3}$	0	$+$	f has a relative minimum value
$x = 0$	0	0	$-$	f has a relative maximum value
$x = 1$	$-\frac{5}{3}$	0	$+$	f has a relative minimum value

From the facts in the table and by plotting a few more points we obtain the sketch of the graph of f shown in Figure 3.

EXAMPLE 2   Find the relative extrema of the sine function by applying the second-derivative test.

Solution   Let

$$f(x) = \sin x$$

Then

$$f'(x) = \cos x \qquad f''(x) = -\sin x$$

$f'(x)$ exists for all x. The critical numbers are obtained by setting $f'(x) = 0$.

$$\cos x = 0$$

$$x = \tfrac{1}{2}\pi + k\pi \qquad k \text{ is any integer}$$

We determine whether or not there is a relative extremum at any of these critical numbers by finding the sign of the second derivative there.

$$f''(\tfrac{1}{2}\pi + k\pi) = -\sin(\tfrac{1}{2}\pi + k\pi)$$

$$= -\cos k\pi$$

$$= \begin{cases} -1 & \text{if } k \text{ is an even integer} \\ 1 & \text{if } k \text{ is an odd integer} \end{cases}$$

We summarize the results of applying the second-derivative test in Table 2.

Table 2

	$f(x)$	$f'(x)$	$f''(x)$	Conclusion
$x = \tfrac{1}{2}\pi + k\pi$ (k is an even integer)	1	0	$-$	f has a relative maximum value
$x = \tfrac{1}{2}\pi + k\pi$ (k is an odd integer)	-1	0	$+$	f has a relative minimum value

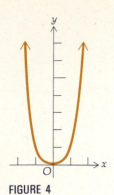

FIGURE 4

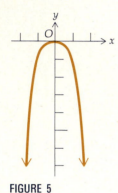

FIGURE 5

FIGURE 6

If $f''(c) = 0$, as well as $f'(c) = 0$, nothing can be concluded regarding a relative extremum of f at c. The following three illustrations justify this statement.

▶ **ILLUSTRATION 1**   If $f(x) = x^4$, then $f'(x) = 4x^3$ and $f''(x) = 12x^2$. Thus $f(0)$, $f'(0)$, and $f''(0)$ all have the value zero. By applying the first-derivative test we see that f has a relative minimum value at 0. A sketch of the graph of f is shown in Figure 4.   ◀

▶ **ILLUSTRATION 2**   If $g(x) = -x^4$, then $g'(x) = -4x^3$ and $g''(x) = -12x^2$. Hence $g(0)$, $g'(0)$, and $g''(0)$ are all zero. In this case g has a relative maximum value at 0, as can be seen by applying the first-derivative test. Figure 5 shows a sketch of the graph of g.   ◀

▶ **ILLUSTRATION 3**   If $h(x) = x^3$, then $h'(x) = 3x^2$ and $h''(x) = 6x$; so $h(0)$, $h'(0)$, and $h''(0)$ are all zero. The function h does not have a relative extremum at 0 because if $x < 0$, $h(x) < h(0)$; and if $x > 0$, $h(x) > h(0)$. A sketch of the graph of h appears in Figure 6.   ◀

In Illustrations 1, 2, and 3 we have examples of three functions, each of which has zero for its second derivative at a number for which its first derivative is zero; yet one function has a relative minimum value at that number, another function has a relative maximum value at that number, and the third function has neither a relative maximum value nor a relative minimum value at that number.

EXAMPLE 3   Given

$$f(x) = x^{2/3} - 2x^{1/3}$$

find the relative extrema of f by applying the second-derivative test when possible. Use the second derivative to find any points of inflection of the graph of f, and determine where the graph is concave upward and where it is concave downward. Draw a sketch of the graph.

Solution

$$f'(x) = \tfrac{2}{3}x^{-1/3} - \tfrac{2}{3}x^{-2/3} \qquad f''(x) = -\tfrac{2}{9}x^{-4/3} + \tfrac{4}{9}x^{-5/3}$$

Because $f'(0)$ does not exist, 0 is a critical number of f. Other critical numbers are found by setting $f'(x) = 0$.

$$\frac{2}{3x^{1/3}} - \frac{2}{3x^{2/3}} = 0$$

$$2x^{1/3} - 2 = 0$$

$$x^{1/3} = 1$$

$$x = 1$$

Thus 1 is also a critical number. We can determine if there is a relative extremum at 1 by applying the second-derivative test. We cannot use the second-derivative test at the critical number 0 because $f'(0)$ does not exist. We apply the first-derivative test at $x = 0$. Table 3 shows the result of these tests.

Because $f''(0)$ does not exist, $(0, 0)$ is a possible point of inflection. To find other possible points of inflection we set $f''(x) = 0$.

$$-\frac{2}{9x^{4/3}} + \frac{4}{9x^{5/3}} = 0$$

$$-2x^{1/3} + 4 = 0$$

$$x^{1/3} = 2$$

$$x = 8$$

To determine if there are points of inflection where x is 0 and 8, we check to see if $f''(x)$ changes sign; at the same time we learn about the concavity of the graph in the respective intervals. At a point of inflection it is necessary that the graph have a tangent line there. At the origin there is a vertical tangent line because

$$\lim_{x \to 0} f'(x) = \lim_{x \to 0} \frac{2x^{1/3} - 2}{3x^{2/3}}$$

$$= -\infty$$

Table 3 summarizes our results and from them we obtain the sketch of the graph shown in Figure 7.

Table 3

	$f(x)$	$f'(x)$	$f''(x)$	Conclusion
$x < 0$		$-$	$-$	f is decreasing; graph is concave downward
$x = 0$	0	does not exist	does not exist	f does not have a relative extremum; graph has a point of inflection
$0 < x < 1$		$-$	$+$	f is decreasing; graph is concave upward
$x = 1$	-1	0	$+$	f has a relative minimum value; graph is concave upward
$1 < x < 8$		$+$	$+$	f is increasing; graph is concave upward
$x = 8$	0	$\frac{1}{6}$	0	f is increasing; graph has a point of inflection
$8 < x$		$+$	$-$	f is increasing; graph is concave downward

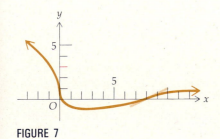

FIGURE 7

EXERCISES 4.6

In Exercises 1 through 26, find the relative extrema of the function by using the second-derivative test if it can be applied. If the second-derivative test cannot be applied, use the first-derivative test. Use the second derivative to find any points of inflection of the graph of the function, and determine where the graph is concave upward and where it is concave downward. Draw a sketch of the graph.

1. $f(x) = 3x^2 - 2x + 1$
2. $g(x) = 7 - 6x - 3x^2$
3. $f(x) = -4x^3 + 3x^2 + 18x$
4. $h(x) = 2x^3 - 9x^2 + 27$
5. $g(x) = \frac{1}{3}x^3 - x^2 + 3$
6. $f(y) = y^3 - 5y + 6$
7. $f(z) = (4 - z)^4$
8. $G(x) = (x + 2)^3$
9. $h(x) = x^4 - \frac{1}{3}x^3 - \frac{3}{2}x^2$
10. $f(x) = \frac{1}{5}x^5 - \frac{2}{3}x^3$
11. $F(x) = \cos 3x$; $x \in \left[-\frac{1}{6}\pi, \frac{1}{2}\pi\right]$
12. $f(x) = 2 \sin 4x$; $x \in \left[0, \frac{1}{2}\pi\right]$
13. $f(x) = x(x + 2)^3$
14. $g(t) = (t - 2)^{7/3}$
15. $f(x) = 4x^{1/2} + 4x^{-1/2}$
16. $f(x) = x(x - 1)^3$
17. $h(x) = x\sqrt{x + 3}$
18. $f(x) = x\sqrt{8 - x^2}$

19. $F(x) = 6x^{1/3} - x^{2/3}$ **20.** $g(x) = \dfrac{9}{x} + \dfrac{x^2}{9}$

21. $f(x) = 5x^3 - 3x^5$ **22.** $G(x) = x^{2/3}(x-4)^2$

23. $f(x) = \tan^2 x; \; x \in (-\frac{1}{2}\pi, \frac{1}{2}\pi)$

24. $g(x) = \sec x \tan x; \; x \in (-\frac{1}{2}\pi, \frac{1}{2}\pi)$

25. $g(x) = x + \cos x; \; x \in [-2\pi, 2\pi]$

26. $f(x) = \sin x - x; \; x \in [-\frac{3}{2}\pi, \frac{3}{2}\pi]$

In Exercises 27 through 30, find the relative extrema of the trigonometric function by applying the second-derivative test.

27. the cosine function   **28.** the secant function

29. the cosecant function   **30.** $f(x) = \sin x + \cos x$

In Exercises 31 through 42, draw a portion of the graph of a function f through the points $(c, f(c))$, $(d, f(d))$, and $(e, f(e))$ if the given conditions are satisfied. Also draw a segment of the tangent line at each of these points, if there is a tangent line. Assume that $c < d < e$ and f is continuous on some open interval containing c, d, and e.

31. $f'(c) = 0; \; f'(d) = 1; \; f''(d) = 0; \; f'(e) = 0; \; f''(x) > 0$ if $x < d$; $f''(x) < 0$ if $x > d$

32. $f'(c) = 0; \; f'(d) = -1; \; f''(d) = 0; \; f'(e) = 0; \; f''(x) < 0$ if $x < d$; $f''(x) > 0$ if $x > d$

33. $f'(c) = 0; \; f'(d) = -1; f''(d) = 0; \; f'(e) = 0; \; f''(e) = 0; \; f''(x) < 0$ if $x < d$; $f''(x) > 0$ if $d < x < e$; $f''(x) < 0$ if $x > e$

34. $f'(c) = 0; \; f'(d) = 1; \; f''(d) = 0; \; f'(e) = 0; \; f''(e) = 0; \; f''(x) > 0$ if $x < d$; $f''(x) < 0$ if $d < x < e$; $f''(x) > 0$ if $x > e$

35. $f'(c) = 0; f''(c) = 0; \; f'(d) = -1; f''(d) = 0; \; f'(e) = 0; f''(x) > 0$ if $x < c$; $f''(x) < 0$ if $c < x < d$; $f''(x) > 0$ if $x > d$

36. $f'(c) = 0; \; f''(c) = 0; \; f'(d) = 1; \; f''(d) = 0; \; f'(e) = 0; \; f''(x) < 0$ if $x < c$; $f''(x) > 0$ if $c < x < d$; $f''(x) < 0$ if $x > d$

37. $f'(c) = 0; \; \lim\limits_{x \to d^-} f'(x) = +\infty; \; \lim\limits_{x \to d^+} f'(x) = +\infty; \; f'(e) = 0;$ $f''(x) > 0$ if $x < d$; $f''(x) < 0$ if $x > d$

38. $f'(c) = 0; \; \lim\limits_{x \to d^-} f'(x) = -\infty; \; \lim\limits_{x \to d^+} f'(x) = -\infty; \; f'(e) = 0;$ $f''(x) < 0$ if $x < d$; $f''(x) > 0$ if $x > d$

39. $f'(c)$ does not exist; $f'(d) = -1; \; f''(d) = 0; \; f'(e) = 0; \; f''(x) > 0$ if $x < c$; $f''(x) < 0$ if $c < x < d$; $f''(x) > 0$ if $x > d$

40. $f'(c) = 0; \; f'(d) = -1; \; f''(d) = 0; \; f'(e)$ does not exist; $f''(x) < 0$ if $x < d$; $f''(x) > 0$ if $d < x < e$; $f''(x) < 0$ if $x > e$

41. $f'(c) = 0; \; f'(d)$ does not exist; $f'(e) = 0; \; f''(e) = 0; \; f''(x) < 0$ if $x < d$; $f''(x) < 0$ if $d < x < e$; $f''(x) > 0$ if $x > e$

42. $f'(c) = 0; \; f''(c) = 0; \; f'(d)$ does not exist; $f'(e) = 0; \; f''(x) < 0$ if $x < c$; $f''(x) > 0$ if $c < x < d$; $f''(x) > 0$ if $x > d$

43. Suppose that $\frac{1}{2}\sqrt{2}$ and $-\frac{1}{2}\sqrt{3}$ are critical numbers of a function f and that $f''(x) = x[\![\frac{1}{2}x^2 + 1]\!]$. At each of these numbers, determine if f has a relative extremum, and if so, whether it is a relative maximum or a relative minimum.

44. Given $f(x) = x^r - rx + k$, where $r > 0$ and $r \neq 1$, prove that (a) if $0 < r < 1$, f has a relative maximum value at 1; (b) if $r > 1$, f has a relative minimum value at 1.

45. Given $f(x) = x^3 + 3rx + 5$, prove that (a) if $r > 0$, f has no relative extrema; (b) if $r < 0$, f has both a relative maximum value and a relative minimum value.

46. Prove Theorem 4.6,1(ii).

47. Given $f(x) = x^2 + rx^{-1}$, prove that regardless of the value of r, f has a relative minimum value and no relative maximum value.

48. If $f(x) = ax^2 + bx + c$, use the second-derivative test to show that f has a relative maximum value if $a < 0$. Find the number at which the relative maximum value occurs.

49. Suppose that f is a function for which $f''(x)$ exists for all values of x in some open interval I and that at a number c in I, $f''(c) = 0$ and $f'''(c)$ exists and is not zero. Prove that the point $(c, f(c))$ is a point of inflection of the graph of f. (*Hint:* The proof is similar to the proof of the second-derivative test.)

4.7   DRAWING A SKETCH OF THE GRAPH OF A FUNCTION

In Sections 2.4 and 2.5 we indicated that an aid in drawing the sketch of the graph of a function is to find, if there are any, the asymptotes of the graph. We discussed vertical asymptotes in Section 2.4 and horizontal asymptotes in Section 2.5. You may wish to review those discussions at this time because we will be applying the concepts in this section.

An asymptote of a graph that is neither horizontal nor vertical is called an **oblique asymptote**. In Section 11.6 we give the formal definition (11.6.3) of an oblique asymptote. At this time we show that the graph of a rational function of the form $f(x)/g(x)$, where the degree of $f(x)$ is one more than the degree of $g(x)$, has the line $y = mx + b$ as an oblique asymptote by proving that

$$\lim_{x \to +\infty} \left| \frac{f(x)}{g(x)} - (mx + b) \right| = 0 \qquad (1)$$

If we divide the polynomial $f(x)$ in the numerator by the polynomial $g(x)$ in the denominator, we obtain the sum of a linear function and a rational function; that is,

$$\frac{f(x)}{g(x)} = mx + b + \frac{h(x)}{g(x)}$$

where the degree of the polynomial $h(x)$ is less than the degree of $g(x)$. Then

$$\lim_{x \to +\infty} \left| \frac{f(x)}{g(x)} - (mx + b) \right| = \lim_{x \to +\infty} \left| \frac{h(x)}{g(x)} \right|$$

When the numerator and denominator of $h(x)/g(x)$ are divided by the highest power of x appearing in $g(x)$, there will be a constant term in the denominator and all other terms in the denominator and every term in the numerator will be of the form k/x^r where k is a constant. Therefore, as $x \to +\infty$, the limit of the numerator will be 0 and the limit of the denominator will be a constant. Thus

$$\lim_{x \to +\infty} \left| \frac{h(x)}{g(x)} \right| = 0$$

and (1) is established.

EXAMPLE 1   Find the asymptotes of the graph of the function h defined by

$$h(x) = \frac{x^2 + 3}{x - 1} \tag{2}$$

and draw a sketch of the graph.

Solution   Because

$$\lim_{x \to 1^-} h(x) = -\infty \quad \text{and} \quad \lim_{x \to 1^+} h(x) = +\infty$$

the line $x = 1$ is a vertical asymptote. There are no horizontal asymptotes, because if the numerator and denominator of $h(x)$ are divided by x^2, we obtain

$$\frac{1 + \dfrac{3}{x^2}}{\dfrac{1}{x} - \dfrac{1}{x^2}}$$

and as $x \to +\infty$ or $x \to -\infty$, the limit of the numerator is 1 and the limit of the denominator is 0. However, the degree of the numerator of $h(x)$ is one more than the degree of the denominator, and when the numerator is divided by the denominator, we obtain

$$h(x) = x + 1 + \frac{4}{x - 1}$$

Therefore the line $y = x + 1$ is an oblique asymptote.
   To draw a sketch of the graph of h we determine if there are any horizontal tangent lines. From (2),

$$h'(x) = \frac{2x(x - 1) - (x^2 + 3)}{(x - 1)^2}$$

$$= \frac{x^2 - 2x - 3}{(x - 1)^2}$$

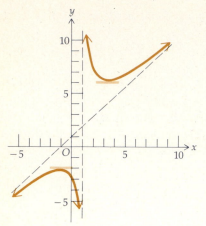

FIGURE 1

Setting $h'(x) = 0$ we get

$$x^2 - 2x - 3 = 0$$

$$(x + 1)(x - 3) = 0$$

$$x = -1 \qquad x = 3$$

Thus there are horizontal tangent lines at the points $(-1, -2)$ and $(3, 6)$.

By indicating the asymptotes and the horizontal tangent lines and plotting a few points we obtain the sketch of the graph of h shown in Figure 1.

To obtain a sketch of the graph of a function f you should apply the properties discussed in this chapter and proceed as follows:

1. Determine the domain of f.
2. Find any y intercepts of the graph. Locate any x intercepts if the resulting equation is easy to solve.
3. Test for symmetry with respect to the y axis and origin.
4. Compute $f'(x)$ and $f''(x)$.
5. Determine the critical numbers of f. These are the values of x in the domain of f for which either $f'(x)$ does not exist or $f'(x) = 0$.
6. Apply either the first-derivative test (Theorem 4.4.4) or the second-derivative test (Theorem 4.6.1) to determine whether at a critical number there is a relative maximum value, a relative minimum value, or neither.
7. Determine the intervals on which f is increasing by finding the values of x for which $f'(x)$ is positive; determine the intervals on which f is decreasing by finding the values of x for which $f'(x)$ is negative. In locating the intervals on which f is monotonic, also check the critical numbers at which f does not have a relative extremum.
8. To obtain possible points of inflection, find the critical numbers of f', that is, the values of x for which $f''(x)$ does not exist or $f''(x) = 0$. At each of these values of x check tó see if $f''(x)$ changes sign and if the graph has a tangent line there to determine if there actually is a point of inflection.
9. Check for concavity of the graph. Find the values of x for which $f''(x)$ is positive to obtain points at which the graph is concave upward; to obtain points at which the graph is concave downward find the values of x for which $f''(x)$ is negative.
10. It is helpful to find the slope of each inflectional tangent.
11. Check for any possible horizontal, vertical, or oblique asymptotes.

It is suggested that all the information obtained be incorporated into a table, as in the following examples.

EXAMPLE 2　　Given

$$f(x) = x^3 - 3x^2 + 3$$

Draw a sketch of the graph of f by first finding the following: the relative extrema of f; the points of inflection of the graph of f; the intervals on which f is increasing; the intervals on which f is decreasing; where the graph is concave upward; where the graph is concave downward; and the slope of any inflectional tangent.

Solution   The domain of f is the set of all real numbers. The y intercept is 3.

$$f'(x) = 3x^2 - 6x \qquad f''(x) = 6x - 6$$

Set $f'(x) = 0$ to obtain $x = 0$ and $x = 2$. From $f''(x) = 0$ we get $x = 1$. In making the table, consider the points at which $x = 0$, $x - 1$, and $x - 2$, and the intervals excluding these values of x:

$$x < 0 \qquad 0 < x < 1 \qquad 1 < x < 2 \qquad 2 < x$$

From the information in Table 1 and by plotting a few points, we obtain the sketch of the graph appearing in Figure 2.

Table 1

	$f(x)$	$f'(x)$	$f''(x)$	Conclusion
$x < 0$		$+$	$-$	f is increasing; graph is concave downward
$x = 0$	3	0	$-$	f has a relative maximum value; graph is concave downward
$0 < x < 1$		$-$	$-$	f is decreasing; graph is concave downward
$x = 1$	1	-3	0	f is decreasing; graph has a point of inflection
$1 < x < 2$		$-$	$+$	f is decreasing; graph is concave upward
$x = 2$	-1	0	$+$	f has a relative minimum value; graph is concave upward
$2 < x$		$+$	$+$	f is increasing; graph is concave upward

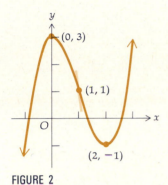

FIGURE 2

EXAMPLE 3   Given

$$f(x) = \frac{x^2}{x^2 - 4}$$

Draw a sketch of the graph of f by following the instructions of Example 2. Also find the horizontal and vertical asymptotes.

Solution   The domain of f is the set of all real numbers except ± 2. The only intersection of the graph with an axis is at the origin. Because $f(-x) = f(x)$, the graph is symmetric with respect to the y axis.

$$f'(x) = \frac{2x(x^2 - 4) - 2x(x^2)}{(x^2 - 4)^2} \qquad f''(x) = \frac{-8(x^2 - 4)^2 + 8x[2(x^2 - 4)(2x)]}{(x^2 - 4)^4}$$

$$= \frac{-8x}{(x^2 - 4)^2} \qquad\qquad = \frac{24x^2 + 32}{(x^2 - 4)^3}$$

Set $f'(x) = 0$ to obtain $x = 0$; $f''(x)$ is never zero. For Table 2 consider the points at which $x = 0$ and $x = \pm 2$, because 2 and -2 are not in the domain of f. Also for the table consider the intervals excluding these values of x:

$$x < -2 \qquad -2 < x < 0 \qquad 0 < x < 2 \qquad 2 < x$$

Because 2 and -2 are excluded from the domain of f, we compute the following limits:

$$\lim_{x \to 2^+} \frac{x^2}{x^2 - 4} = +\infty$$

$$\lim_{x \to 2^-} \frac{x^2}{x^2 - 4} = -\infty$$

$$\lim_{x \to -2^+} \frac{x^2}{x^2 - 4} = -\infty$$

$$\lim_{x \to -2^-} \frac{x^2}{x^2 - 4} = +\infty$$

Therefore $x = 2$ and $x = -2$ are vertical asymptotes of the graph.

$$\lim_{x \to +\infty} \frac{x^2}{x^2 - 4} = \lim_{x \to +\infty} \frac{1}{1 - \dfrac{4}{x^2}} \qquad \lim_{x \to -\infty} \frac{x^2}{x^2 - 4} = \lim_{x \to -\infty} \frac{1}{1 - \dfrac{4}{x^2}}$$

$$= 1 \qquad\qquad = 1$$

Hence $y = 1$ is a horizontal asymptote of the graph.

With the facts from Table 2, the asymptotes as guides, the plotting of a few points, and the symmetry property, we obtain the sketch of the graph of f shown in Figure 3.

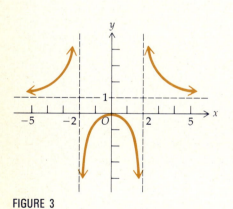

FIGURE 3

Table 2

	$f(x)$	$f'(x)$	$f''(x)$	Conclusion
$x < -2$		$+$	$+$	f is increasing; graph is concave upward
$x = -2$	does not exist	does not exist	does not exist	
$-2 < x < 0$		$+$	$-$	f is increasing; graph is concave downward
$x = 0$	0	0	$-$	f has a relative maximum value
$0 < x < 2$		$-$	$-$	f is decreasing; graph is concave downward
$x = 2$	does not exist	does not exist	does not exist	
$2 < x$		$-$	$+$	f is decreasing; graph is concave upward

Observe in Example 3 that because the graph is symmetric with respect to the y axis, it can be obtained by doing the calculus computations for x in $[0, +\infty)$ and then applying the symmetry property.

EXAMPLE 4   Given

$$f(x) = 5x^{2/3} - x^{5/3}$$

Draw a sketch of the graph of f by following the instructions of Example 2.

Solution   The domain of f is the set of all real numbers. The y intercept is 0. Setting $f(x) = 0$ we get

$$x^{2/3}(5 - x) = 0$$

$$x = 0 \qquad x = 5$$

Therefore the x intercepts are 0 and 5.

$$f'(x) = \tfrac{10}{3}x^{-1/3} - \tfrac{5}{3}x^{2/3} \qquad f''(x) = -\tfrac{10}{9}x^{-4/3} - \tfrac{10}{9}x^{-1/3}$$

$$= \tfrac{5}{3}x^{-1/3}(2 - x) \qquad\qquad = -\tfrac{10}{9}x^{-4/3}(1 + x)$$

When $x = 0$, neither $f'(x)$ nor $f''(x)$ exists. Set $f'(x) = 0$ to obtain $x = 2$. Therefore the critical numbers of f are 0 and 2. From $f''(x) = 0$ we obtain $x = -1$. In making the table, consider the points at which x is -1, 0, and 2, and the following intervals:

$$x < -1 \qquad -1 < x < 0 \qquad 0 < x < 2 \qquad 2 < x$$

A sketch of the graph, drawn from the information in Table 3 and by plotting a few points, appears in Figure 4.

Table 3

	$f(x)$	$f'(x)$	$f''(x)$	*Conclusion*
$x < -1$		$-$	$+$	f is decreasing; graph is concave upward
$x = -1$	6	-5	0	f is decreasing; graph has a point of inflection
$-1 < x < 0$		$-$	$-$	f is decreasing; graph is concave downward
$x = 0$	0	does not exist	does not exist	f has a relative minimum value
$0 < x < 2$		$+$	$-$	f is increasing; graph is concave downward
$x = 2$	$3\sqrt[3]{4} \approx 4.8$	0	$-$	f has a relative maximum value; graph is concave downward
$2 < x$		$-$	$-$	f is decreasing; graph is concave downward

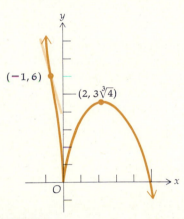

$(-1, 6)$

$(2, 3\sqrt[3]{4})$

FIGURE 4

EXERCISES 4.7

In Exercises 1 through 8, find the vertical and oblique asymptotes of the graph of the function and draw a sketch of the graph.

1. $f(x) = \dfrac{x^2}{x - 1}$

2. $f(x) = \dfrac{x^2 - 3}{x - 2}$

3. $f(x) = \dfrac{x^2 - 8}{x - 3}$

4. $f(x) = \dfrac{x^2 - 3x + 2}{x + 4}$

5. $f(x) = \dfrac{x^2 - 4x - 5}{x + 2}$

6. $f(x) = \dfrac{x^3 - 4}{x^2}$

7. $f(x) = \dfrac{x^3 + 2x^2 + 4}{x^2}$

8. $f(x) = \dfrac{(x + 1)^3}{(x - 1)^2}$

In Exercises 9 through 58 draw a sketch of the graph of f by first finding the following: the relative extrema of f; the points of inflection of the graph of f; the intervals on which f is increasing; the intervals on which f is decreasing; where the graph is concave upward; where the graph is concave downward; the slope of any inflectional tangent; and the horizontal, vertical, and oblique asymptotes, if there are any.

9. $f(x) = 2x^3 - 6x + 1$ **10.** $f(x) = x^3 + x^2 - 5x$

11. $f(x) = x^4 - 2x^3$ **12.** $f(x) = 3x^4 + 2x^3$

13. $f(x) = x^3 + 5x^2 + 3x - 4$

14. $f(x) = 2x^3 - \tfrac{1}{2}x^2 - 12x + 1$

15. $f(x) = x^4 - 3x^3 + 3x^2 + 1$ **16.** $f(x) = x^4 - 4x^3 + 16x$ **36.** $f(x) = 3 \sec \frac{1}{4}x; \; x \in (-2\pi, 2\pi)$

17. $f(x) = \frac{1}{4}x^4 - \frac{1}{3}x^3 - x^2 + 1$ **18.** $f(x) = \frac{1}{4}x^4 - x^3$ **37.** $f(x) = \sin x + \cos x; \; x \in [-2\pi, 2\pi]$

19. $f(x) = \frac{1}{2}x^4 - 2x^3 + 3x^2 + 2$ **20.** $f(x) = 3x^4 + 4x^3 + 6x^2 - 4$ **38.** $f(x) = |\sin x|; \; x \in [-2\pi, 2\pi]$

21. $f(x) = \begin{cases} x^2 & \text{if } x < 0 \\ 2x^2 & \text{if } 0 \le x \end{cases}$ **22.** $f(x) = \begin{cases} -x^3 & \text{if } x < 0 \\ x^3 & \text{if } 0 \le x \end{cases}$ **39.** $f(x) = \dfrac{x^2}{x - 1}$ **40.** $f(x) = \dfrac{x}{x^2 - 4}$

23. $f(x) = \begin{cases} -x^4 & \text{if } x < 0 \\ x^4 & \text{if } 0 \le x \end{cases}$ **24.** $f(x) = \begin{cases} 2(x-1)^3 & \text{if } x < 1 \\ (x-1)^4 & \text{if } 1 \le x \end{cases}$ **41.** $f(x) = \dfrac{x^2 + 1}{x^2 - 1}$ **42.** $f(x) = \dfrac{x^2 + 1}{x - 3}$

25. $f(x) = \begin{cases} 3(x-2)^2 & \text{if } x \le 2 \\ (2-x)^3 & \text{if } 2 < x \end{cases}$ **26.** $f(x) = x^2(x + 4)^3$ **43.** $f(x) = \dfrac{2x}{x^2 + 1}$ **44.** $f(x) = \dfrac{x^2 - 4}{x^2 - 9}$

27. $f(x) = (x + 1)^3(x - 2)^2$ **28.** $f(x) = 3x^5 + 5x^3$ **45.** $f(x) = 3x^{2/3} - 2x$ **46.** $f(x) = x^{1/3} + 2x^{4/3}$

29. $f(x) = 3x^5 + 5x^4$ **30.** $f(x) = |4 - x^2|$ **47.** $f(x) = 3x^{4/3} - 4x$ **48.** $f(x) = 3x^{1/3} - x$

31. $f(x) = 3 \cos 2x; \; x \in [-\pi, \pi]$ **49.** $f(x) = 2 + (x - 3)^{1/3}$ **50.** $f(x) = 2 + (x - 3)^{4/3}$

32. $f(x) = 2 \sin 3x; \; x \in [-\pi, \pi]$ **51.** $f(x) = 2 + (x - 3)^{5/3}$ **52.** $f(x) = 2 + (x - 3)^{2/3}$

33. $f(x) = \begin{cases} \sin x & \text{if } 0 \le x < \frac{1}{2}\pi \\ \sin(x - \frac{1}{2}\pi) & \text{if } \frac{1}{2}\pi \le x \le \pi \end{cases}$ **53.** $f(x) = x^2\sqrt{4 - x}$ **54.** $f(x) = x\sqrt{9 - x^2}$

55. $f(x) = (x + 2)\sqrt{-x}$ **56.** $f(x) = \dfrac{9x}{x^2 + 9}$

34. $f(x) = \begin{cases} \cos x & \text{if } -\pi \le x \le 0 \\ \cos(\pi - x) & \text{if } 0 < x \le \pi \end{cases}$

57. $f(x) = (x + 1)^{2/3}(x - 2)^{1/3}$ **58.** $f(x) = \dfrac{x^3}{x^2 - 1}$

35. $f(x) = 2 \tan \frac{1}{2}x; \; x \in (-\pi, \pi)$

4.8 FURTHER TREATMENT OF ABSOLUTE EXTREMA AND APPLICATIONS

The extreme-value theorem (4.1.9) guarantees an absolute maximum value and an absolute minimum value for a function that is continuous on a closed interval. In this section we consider some functions defined on intervals for which the extreme-value theorem does not apply and which may or may not have absolute extrema.

EXAMPLE 1   Given

$$f(x) = \frac{x^2 - 27}{x - 6}$$

find the absolute extrema of f on the interval $[0, 6)$ if there are any.

Solution   The function f is continuous on the interval $[0, 6)$ because the only discontinuity of f is at 6, which is not in the interval.

$$f'(x) = \frac{2x(x - 6) - (x^2 - 27)}{(x - 6)^2}$$

$$= \frac{x^2 - 12x + 27}{(x - 6)^2}$$

$$= \frac{(x - 3)(x - 9)}{(x - 6)^2}$$

Observe that $f'(x)$ exists for all values of x in $[0, 6)$ and $f'(x) = 0$ when x is 3 or 9; so the only critical number of f in the interval $[0, 6)$ is 3. The first-derivative test is applied to determine if f has a relative extremum at 3, and the results are summarized in Table 1.

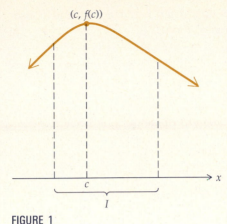

FIGURE 1

Table 1

	$f(x)$	$f'(x)$	Conclusion
$0 \leq x < 3$		+	f is increasing
$x = 3$	6	0	f has a relative maximum value
$3 < x < 6$		−	f is decreasing

Because f has a relative maximum value at 3, and f is increasing on the interval $[0, 3)$ and decreasing on the interval $(3, 6)$, then on $[0, 6)$ f has an absolute maximum value at 3, and it is $f(3)$, which is 6. Note that $\lim\limits_{x \to 6^-} f(x) = -\infty$, and conclude that there is no absolute minimum value of f on $[0, 6)$.

The functions whose graphs appear in Figures 1 and 2 are continuous on an interval I and have only one relative extremum, $f(c)$, on I. Observe that $f(c)$ is also an absolute extremum on I. These functions illustrate the following theorem, which is sometimes useful to determine if a relative extremum is an absolute extremum.

4.8.1 THEOREM

Let the function f be continuous on the interval I containing the number c. If $f(c)$ is a relative extremum of f on I and c is the only number in I for which f has a relative extremum, then $f(c)$ is an absolute extremum of f on I. Furthermore,

(i) if $f(c)$ is a relative maximum value of f on I, then $f(c)$ is an absolute maximum value of f on I;

(ii) if $f(c)$ is a relative minimum value of f on I, then $f(c)$ is an absolute minimum value of f on I.

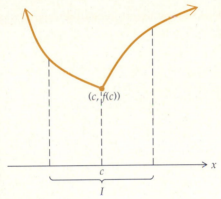

FIGURE 2

Proof   We prove part (i). The proof of part (ii) is similar.

Because $f(c)$ is a relative maximum value of f on I, then by Definition 4.1.1 there is an open interval J, where $J \subset I$, and where J contains c, such that

$$f(c) \geq f(x) \qquad \text{for all } x \in J$$

Because c is the only number in I for which f has a relative maximum value, it follows that

$$f(c) > f(k) \qquad \text{if } k \in J \text{ and } k \neq c \tag{1}$$

To show that $f(c)$ is an absolute maximum value of f on I we show that if d is any number other than c in I, then $f(c) > f(d)$. We assume that

$$f(c) \leq f(d) \tag{2}$$

and show that this assumption leads to a contradiction. Because $d \neq c$, then either $c < d$ or $d < c$. We consider the case that $c < d$ (the proof is similar if $d < c$).

Because f is continuous on I, then f is continuous on the closed interval $[c, d]$. Therefore, by the extreme-value theorem, f has an absolute minimum value on $[c, d]$. Assume this absolute minimum value occurs at e, where $c \leq e \leq d$. From inequality (1) it follows that $e \neq c$, and from inequalities (1) and (2) it follows that $e \neq d$. Therefore $c < e < d$, and hence f has a relative minimum value at e. But this statement contradicts the hypothesis that c is the only number in I for which f has a relative extremum. Thus our assumption that $f(c) \le f(d)$ is false. Therefore $f(c) > f(d)$ if $d \in I$ and $d \ne c$, and consequently $f(c)$ is an absolute maximum value of f on I.   ∎

EXAMPLE 2   Given

$$f(x) = 3x^4 - 8x^3 + 12x^2 - 12x + 3$$

find the absolute extrema of f on $(-\infty, +\infty)$, if there are any.

Solution

$$f'(x) = 12x^3 - 24x^2 + 24x - 12 \qquad f''(x) = 36x^2 - 48x + 24$$

$f'(x)$ exists for all values of x. Set $f'(x) = 0$ and obtain

$$12(x^3 - 2x^2 + 2x - 1) = 0$$

We use synthetic division to determine that $x - 1$ is a factor of the left member of the equation and have

$$12(x - 1)(x^2 - x + 1) = 0$$

Because the equation $x^2 - x + 1 = 0$ has only imaginary roots, the only real solution is 1. Therefore $f'(1) = 0$. To determine if $f(1)$ is a relative extremum we apply the second-derivative test, the results of which are summarized in Table 2.

Table 2

	$f(x)$	$f'(x)$	$f''(x)$	*Conclusion*
$x = 1$	-2	0	$+$	f has a relative minimum value

The function f is continuous on $(-\infty, +\infty)$, and the one and only relative extremum of f on $(-\infty, +\infty)$ is at $x = 1$. Therefore it follows from Theorem 4.8.1(ii) that -2, the relative minimum value of f, is the absolute minimum value of f.

In Section 4.2 the applications involved finding absolute extrema of functions continuous on a closed interval, and the extreme-value theorem was used in the solutions of the problems. We now deal with applications involving absolute extrema for which the extreme-value theorem cannot be applied.

EXAMPLE 3   A closed box with a square base is to have a volume of 2000 in.^{3}. The material for the top and bottom of the box is to cost 3¢ per square inch, and the material for the sides is to cost 1.5¢ per square inch. Find the dimensions of the box so that the total cost of the material is least.

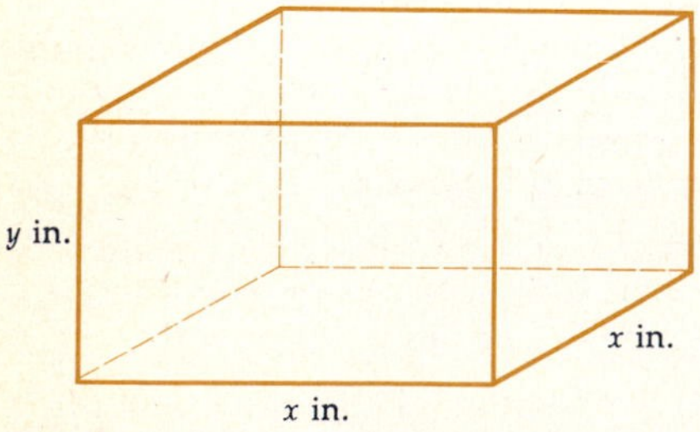

Solution   Let x inches be the length of a side of the square base and $C(x)$ dollars be the total cost of the material. The area of the base is x^2 square inches. Let y inches be the depth of the box. See Figure 3. Because the volume of the box is the product of the area of the base and the depth

$$x^2 y = 2000$$

$$y = \frac{2000}{x^2} \tag{3}$$

y in.

x in.

x in.

FIGURE 3

The total number of square inches in the combined area of the top and bottom is $2x^2$, and for the sides it is $4xy$. Therefore the number of cents in the total cost of the material is

$$3(2x^2) + \tfrac{3}{2}(4xy)$$

Replacing y by its equal from (3) we have

$$C(x) = 6x^2 + 6x\left(\frac{2000}{x^2}\right)$$

$$C(x) = 6x^2 + \frac{12,000}{x}$$

The domain of C is $(0, +\infty)$. Furthermore, C is continuous on its domain.

$$C'(x) = 12x - \frac{12,000}{x^2} \qquad C''(x) = 12 + \frac{24,000}{x^3}$$

Observe that $C'(x)$ does not exist when $x = 0$, but 0 is not in the domain of C. Therefore the only critical numbers will be those obtained by setting $C'(x) = 0$, which gives

$$12x - \frac{12,000}{x^2} = 0$$

$$x^3 = 1000$$

The only real solution is 10. Thus 10 is the only critical number. To determine if $x = 10$ makes C a relative minimum we apply the second-derivative test. The results are summarized in Table 3

Table 3

	$C'(x)$	$C''(x)$	*Conclusion*
$x = 10$	0	+	C has a relative minimum value

Because C is continuous on its domain $(0, +\infty)$ and the one and only relative extremum of C on $(0, +\infty)$ is at $x = 10$, it follows from Theorem 4.8.1(ii) that this relative minimum value of C is the absolute minimum value. Thus the total cost of the material will be least when the side of the square base is 10 in. The depth then will be 20 in., because the area of the base will be 100 in.^{2} and the volume is 2000 in.^{3}.

EXAMPLE 4   If a closed tin can of volume 16π in.^{3} is to be in the form of a right-circular cylinder, find the height and radius if the least amount of material is to be used in its manufacture.

Solution   Let r inches be the base radius of the cylinder, h inches be the height of the cylinder, and S square inches be the total surface area of the cylinder. See Figure 4. The lateral surface area is $2\pi rh$ square inches, the area of the top is πr^2 square inches, and the area of the bottom is πr^2 square inches. Therefore

$$S = 2\pi rh + 2\pi r^2 \tag{4}$$

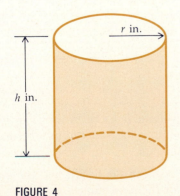

r in.

h in.

FIGURE 4

If V cubic inches is the volume of a right-circular cylinder, then $V = \pi r^2 h$. Thus

$$16\pi = \pi r^2 h \tag{5}$$

Solving (5) for h and substituting into (4) we obtain S as a function of r:

$$S(r) = 2\pi r \left(\frac{16}{r^2}\right) + 2\pi r^2$$

$$S(r) = \frac{32\pi}{r} + 2\pi r^2$$

The domain of S is $(0, +\infty)$, and S is continuous on its domain.

$$S'(r) = -\frac{32\pi}{r^2} + 4\pi r \qquad S''(r) = \frac{64\pi}{r^3} + 4\pi$$

$S'(r)$ does not exist when $r = 0$, but 0 is not in the domain of S. The only critical numbers are those obtained by setting $S'(r) = 0$, from which we have

$$4\pi r^3 = 32\pi$$

$$r^3 = 8$$

$$r = 2$$

The only critical number of S is 2. The results of applying the second-derivative test are summarized in Table 4.

Table 4

	$S'(r)$	$S''(r)$	Conclusion
$r = 2$	0	+	S has a relative minimum value

Because S is continuous on its domain $(0, +\infty)$ and the one and only relative extremum of S on $(0, +\infty)$ is at $r = 2$, it follows from Theorem 4.8.1(ii) that this relative minimum value of S is the absolute minimum value. When $r = 2$, we have, from (5), $h = 4$. Therefore the least amount of material will be used in the manufacture of the can when the radius is 2 in. and the height is 4 in.

In the preceding examples and in the exercises of Section 4.2, the variable for which we wished to find an absolute extremum was expressed as a function of only one variable. Sometimes this procedure is either too difficult or too laborious, or occasionally even impossible. Often the given information enables us to obtain two equations involving three variables. Instead of eliminating one of the variables, it may be more advantageous to differentiate implicitly. The following example illustrates this method. The problem is similar to the one in Example 4, but in this example the volume of the required can is not specified.

EXAMPLE 5   If a closed tin can of fixed volume is to be in the form of a right-circular cylinder, find the ratio of the height to the base radius if the least amount of material is to be used in its manufacture.

Solution   We wish to find a relationship between the height and the base radius of the right-circular cylinder in order for the total surface area to be an absolute minimum for a fixed volume. Therefore we consider the volume of the cylinder a constant.

Let V cubic units be the volume of a cylinder (a constant).

We now define the variables.

Let r units be the base radius of the cylinder; $r > 0$. Let h units be the height of the cylinder; $h > 0$. Let S square units be the total surface area of the cylinder.

We have the following equations:

$$S = 2\pi r^2 + 2\pi rh \tag{6}$$

$$V = \pi r^2 h \tag{7}$$

Because V is a constant, we could solve (7) for either r or h in terms of the other and substitute into (6), which will give S as a function of one variable. The alternative method is to consider S as a function of two variables r and h; however, r and h are not independent of each other. That is, if we choose r as the independent variable, then S depends on r; also, h depends on r.

Differentiating S and V with respect to r and bearing in mind that h is a function of r, we have

$$\frac{dS}{dr} = 4\pi r + 2\pi h + 2\pi r \frac{dh}{dr} \tag{8}$$

$$\frac{dV}{dr} = 2\pi rh + \pi r^2 \frac{dh}{dr}$$

Because V is a constant, $\dfrac{dV}{dr} = 0$; therefore, from the above equation,

$$2\pi rh + \pi r^2 \frac{dh}{dr} = 0$$

with $r \neq 0$. We divide by r and solve for $\dfrac{dh}{dr}$ to obtain

$$\frac{dh}{dr} = -\frac{2h}{r} \tag{9}$$

Substituting from (9) into (8) we obtain

$$\frac{dS}{dr} = 2\pi \left[2r + h + r\left(-\frac{2h}{r} \right) \right]$$

$$\frac{dS}{dr} = 2\pi(2r - h) \tag{10}$$

To find when S has a relative minimum value, set $\dfrac{dS}{dr} = 0$ and obtain $2r - h = 0$, which gives

$$r = \tfrac{1}{2}h$$

To determine if this relationship between r and h makes S a relative minimum we apply the second-derivative test. Then from (10),

$$\frac{d^2S}{dr^2} = 2\pi \left(2 - \frac{dh}{dr} \right)$$

Substituting from (9) into this equation we get

$$\frac{d^2S}{dr^2} = 2\pi\left[2 - \left(\frac{-2h}{r}\right)\right]$$

$$= 2\pi\left(2 + \frac{2h}{r}\right)$$

The results of the second-derivative test are summarized in Table 5.

Table 5

	$\dfrac{dS}{dr}$	$\dfrac{d^2S}{dr^2}$	Conclusion
$r = \frac{1}{2}h$	0	+	S has a relative minimum value

From (6) and (7), S is a continuous function of r on $(0, +\infty)$. Because the one and only relative extremum of S on $(0, +\infty)$ is at $r = \frac{1}{2}h$, we conclude from Theorem 4.8.1(ii) that S has an absolute minimum value at $r = \frac{1}{2}h$. Therefore the total surface area of the tin can will be least for a specific volume when the ratio of the height to the base radius is 2.

Geometric problems involving absolute extrema occasionally are more easily solved by using trigonometric functions. The following example illustrates this fact.

EXAMPLE 6   A right-circular cylinder is to be inscribed in a sphere of given radius. Find the ratio of the altitude to the base radius of the cylinder having the largest lateral surface area.

Solution   Refer to Figure 5. The measure of the constant radius of the sphere is taken as a.

Let θ radians be the angle at the center of the sphere subtended by the radius of the cylinder, r units be the radius of the cylinder, h units be the altitude of the cylinder, and S square units be the lateral surface area of the cylinder. From Figure 5,

$$r = a \sin\theta \quad \text{and} \quad h = 2a \cos\theta$$

Because $S = 2\pi rh$,

$$S = 2\pi(a \sin\theta)(2a \cos\theta)$$

$$= 2\pi a^2(2 \sin\theta \cos\theta)$$

$$= 2\pi a^2 \sin 2\theta$$

Thus S is a function of θ and its domain is $(0, \frac{1}{2}\pi)$.

$$\frac{dS}{d\theta} = 4\pi a^2 \cos 2\theta \quad \text{and} \quad \frac{d^2S}{d\theta^2} = -8\pi a^2 \sin 2\theta$$

Set $\dfrac{dS}{d\theta} = 0$.

$$\cos 2\theta = 0$$

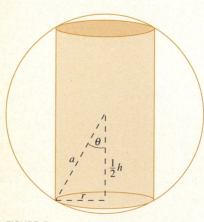

FIGURE 5

Because $0 < \theta < \frac{1}{2}\pi$,

$$\theta = \tfrac{1}{4}\pi$$

We apply the second-derivative test for relative extrema and summarize the results in Table 6.

Table 6

	$\dfrac{dS}{d\theta}$	$\dfrac{d^2S}{d\theta^2}$	Conclusion
$\theta = \tfrac{1}{4}\pi$	0	$-$	S has a relative maximum value

S is continuous on its domain, and because there is only one relative extremum, it follows from Theorem 4.8.1(i) that the relative maximum value of S is the absolute maximum value of S.

When $\theta = \tfrac{1}{4}\pi$,

$$r = a \sin \tfrac{1}{4}\pi \qquad h = 2a \cos \tfrac{1}{4}\pi$$
$$= \tfrac{1}{2}\sqrt{2}a \qquad\qquad = \sqrt{2}a$$

So for the cylinder having the largest lateral surface area, $h/r = 2$.

EXERCISES 4.8

In Exercises 1 through 16, find the absolute extrema of the function on the indicated interval if there are any.

1. $f(x) = x^2$; $(-3, 2]$
2. $g(x) = x^3 + 2x^2 - 4x + 1$; $(-3, 2)$
3. $F(x) = \dfrac{x + 2}{x - 2}$; $[-4, 4]$
4. $f(x) = \dfrac{x^2}{x + 3}$; $[-4, -1]$
5. $g(x) = 4x^2 - 2x + 1$; $(-\infty, +\infty)$
6. $f(x) = x^3 - 3x + 5$; $(-\infty, 0)$
7. $f(x) = (x - 1)^{1/3}$; $(-\infty, +\infty)$
8. $G(x) = (x - 5)^{2/3}$; $(-\infty, +\infty)$
9. $f(x) = x^4 + x^3 - \tfrac{9}{2}x^2 + 2x + 1$; $(-\infty, +\infty)$
10. $h(x) = 3x^4 - 2x^3 - 12x^2 - 6x - 1$; $(-\infty, +\infty)$
11. $g(x) = \dfrac{x^2 + 7}{x - 3}$; $[-3, 3)$ 12. $f(x) = \dfrac{x^2 - 30}{x - 4}$; $(-\infty, 4)$
13. $f(x) = \dfrac{x}{(x^2 + 4)^{3/2}}$; $[0, +\infty)$
14. $g(x) = \sqrt{x^4 - 4x + 7}$; $(-\infty, +\infty)$
15. $f(x) = \tan x + 2 \sec x$; $(-\tfrac{1}{2}\pi, \tfrac{1}{2}\pi)$
16. $h(x) = x + \cot x$; $[\tfrac{1}{6}\pi, \tfrac{2}{3}\pi]$

17. A rectangular field having an area of 2700 m² is to be enclosed by a fence, and an additional fence is to be used to divide the field down the middle. The cost of the fence down the middle is $12 per running meter, and the fence along the

sides costs $18 per running meter. Find the dimensions of the field such that the cost of the fencing will be the least.

18. A rectangular open tank is to have a square base, and its volume is to be 125 m³. The cost per square meter for the bottom is $24 and for the sides is $12. Find the dimensions of the tank for the cost of the material to be the least.

19. A page of print is to contain 24 in.² of printed region, a margin of $1\tfrac{1}{2}$ in. at the top and bottom, and a margin of 1 in. at the sides. What are the dimensions of the smallest page that will fill these requirements?

20. A one-story building having a rectangular floor space of 13,200 ft² is to be constructed where a walkway 22 ft wide is required in the front and back and a walkway 15 ft wide is required on each side. Find the dimensions of the lot having the least area on which this building can be located.

21. A box manufacturer is to produce a closed box of volume 288 in.³, where the base is a rectangle having a length three times its width. Find the dimensions of the box constructed from the least amount of material.

22. Solve Exercise 21 if the box is to have an open top.

23. Find an equation of the tangent line to the curve $y = x^3 - 3x^2 + 5x$ that has the least slope.

24. A cardboard poster containing 32 in.² of printed region is to have a margin of 2 in. at the top and bottom and $\tfrac{4}{3}$ in. at the sides. Determine the dimensions of the smallest piece of cardboard that can be used to make the poster.

25. In a particular community, a certain epidemic spreads in such a way that x months after the start of the epidemic, P

percent of the population is infected, where

$$P = \frac{30x^2}{(1 + x^2)^2}$$

In how many months will the most people be infected, and what percent of the population is this?

26. A direct current generator has an electromotive force of E volts and an internal resistance of r ohms, where E and r are constants. If R ohms is the external resistance, the total resistance is $(r + R)$ ohms, and if P watts is the power, then

$$P = \frac{E^2 R}{(r + R)^2}$$

Show that the most power is consumed when the external resistance is equal to the internal resistance.

27. For the tin can of Example 4, suppose that the cost of material for the top and bottom is twice as much as it is for the sides. Find the height and base radius for the cost of the material to be the least.

28. Solve Example 4 if the tin can is open instead of closed.

In Exercises 29 and 30, we use the economics term perfect competition. When a company is operating under **perfect competition**, there are many small firms; so any one firm cannot affect price by increasing production. Therefore, under perfect competition the price of a commodity is constant, and the company can sell as much as it wishes to sell at this constant price.

29. Under perfect competition a firm can sell at a price of $200 per unit all of a particular commodity it produces. If $C(x)$ dollars is the total cost of each day's production when x units are produced, and $C(x) = 2x^2 + 40x + 1400$, find the number of units that should be produced daily for the firm to have the greatest daily total profit. (Hint: Total profit equals total revenue minus total cost.)

30. A company that builds and sells desks is operating under perfect competition and can sell at a price of $400 per desk all the desks it produces. If x desks are produced and sold each week and $C(x)$ dollars is the total cost of the week's production, then $C(x) = 2x^2 + 80x + 6000$. Determine how many desks should be built each week for the manufacturer to have the greatest weekly total profit. What is the greatest weekly total profit? See the hint for Exercise 29.

31. Under a **monopoly**, which means that there is only one producer of a certain commodity, price and hence demand can be controlled by regulating the quantity of the commodity produced. Suppose that under a monopoly x units are demanded daily when p dollars is the price per unit and $x = 140 - p$. If the number of dollars in the total cost of producing x units is given by $C(x) = x^2 + 20x + 300$, find the maximum daily total profit.

32. Find the shortest distance from the point $P(2, 0)$ to a point on the curve $y^2 - x^2 = 1$, and find the point on the curve that is closest to P.

33. Find the shortest distance from the origin to the line $3x + y = 6$, and find the point P on the line that is closest

to the origin. Then show that the origin lies on the line perpendicular to the given line at P.

34. Find the shortest distance from the point $A(2, \frac{1}{2})$ to a point on the parabola $y = x^2$, and find the point B on the parabola that is closest to A. Then show that A lies on the normal line of the parabola at B.

35. A Norman window consists of a rectangle surmounted by a semicircle. If the perimeter of a Norman window is to be 32 ft, determine what should be the radius of the semicircle and the height of the rectangle such that the window will admit the most light.

36. Solve Exercise 35 if the window is to be such that the semicircle transmits only half as much light per square foot of area as the rectangle.

37. It is determined that if salaries are excluded, the number of dollars in the cost per kilometer for operating a truck is $8 + \frac{1}{300}x$, where x kilometers per hour is the speed of the truck. If the combined salary of the driver and the driver's assistant is $27 per hour, what should be the average speed of the truck for the cost per kilometer to be the least?

38. The number of dollars in the cost per hour of fuel for a cargo ship is $\frac{1}{50}v^3$, where v knots (nautical miles per hour) is the speed of the ship. If there are additional costs of $400 per hour, what should be the average speed of the ship for the cost per nautical mile to be the least?

39. An automobile traveling at the rate of 30 ft/sec is approaching an intersection. When the automobile is 120 ft from the intersection, a truck traveling at the rate of 40 ft/sec crosses the intersection. The automobile and the truck are on roads that are at right angles to each other. How long after the truck leaves the intersection are the two vehicles closest?

40. Two airplanes A and B are flying horizontally at the same altitude. The position of plane B is southwest of plane A and 20 km to the west and 20 km to the south of A. If plane A is flying due west at 16 km/min and plane B is flying due north at $\frac{64}{3}$ km/min, (a) in how many seconds will they be closest, and (b) what will be their closest distance?

41. A steel girder 27 ft long is moved horizontally along a passageway 8 ft wide and into a corridor at right angles to the passageway. How wide must the corridor be for the girder to go around the corner? Neglect the horizontal width of the girder.

42. If two corridors at right angles to each other are 10 ft and 15 ft wide, respectively, what is the length of the longest steel girder that can be moved horizontally around the corner? Neglect the horizontal width of the girder.

43. A funnel of specific volume is to be in the shape of a right-circular cone. Find the ratio of the height to the base radius if the least amount of material is to be used in its manufacture.

44. A right-circular cone is to be inscribed in a sphere of given radius. Find the ratio of the altitude to the base radius of the cone of largest possible volume.

45. A right-circular cone is to be circumscribed about a sphere of given radius. Find the ratio of the altitude to the base radius of the cone of least possible volume.

46. Prove by the method of this section that the shortest distance from the point $P_1(x_1, y_1)$ to the line l having the equation $Ax + By + C = 0$ is $|Ax_1 + By_1 + C|/\sqrt{A^2 + B^2}$. (*Hint:* If s is the number of units from P_1 to a point $P(x, y)$ on l, then s will be an absolute minimum when s^2 is an absolute minimum.)

47. Find the altitude of the right-circular cone of largest possible volume that can be inscribed in a sphere of radius a units. Let 2θ be the radian measure of the vertical angle of the cone.

48. The cross section of a trough has the shape of an inverted isosceles triangle. If the lengths of the equal sides are 15 in., find the size of the vertex angle that will give maximum capacity for the trough.

4.9   THE DIFFERENTIAL

Suppose the function f is defined by the equation

$$y = f(x)$$

We now show how increments Δy can be approximated for this function near points where f is differentiable. At such points

$$f'(x) = \lim_{\Delta x \to 0} \frac{\Delta y}{\Delta x} \tag{1}$$

where

$$\Delta y = f(x + \Delta x) - f(x)$$

From (1) it follows that for any $\epsilon > 0$ there exists a $\delta > 0$ such that

$$\text{if} \quad 0 < |\Delta x| < \delta \quad \text{then} \quad \left| \frac{\Delta y}{\Delta x} - f'(x) \right| < \epsilon$$

$$\Leftrightarrow \quad \text{if} \quad 0 < |\Delta x| < \delta \quad \text{then} \quad \frac{|\Delta y - f'(x)\Delta x|}{|\Delta x|} < \epsilon$$

This means that $|\Delta y - f'(x)\,\Delta x|$ is small compared to $|\Delta x|$. That is, for a sufficiently small $|\Delta x|$, $f'(x)\,\Delta x$ is a good approximation to the value of Δy, and we write

$$\Delta y \approx f'(x)\,\Delta x \tag{2}$$

if $|\Delta x|$ is sufficiently small.

For a graphical interpretation of statement (2), refer to Figure 1. In the figure, an equation of the curve is $y = f(x)$. The line PT is tangent to the curve at $P(x, f(x))$, Q is the point $(x + \Delta x, f(x + \Delta x))$, and the directed distance MQ is $\Delta y = f(x + \Delta x) - f(x)$. In the figure, Δx and Δy are both positive; however, they could be negative. For a small value of Δx, the slope of the secant line PQ and the slope of the tangent line at P are approximately equal; that is,

$$\frac{\Delta y}{\Delta x} \approx f'(x)$$

$$\Delta y \approx f'(x)\,\Delta x$$

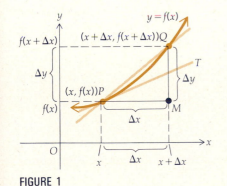

FIGURE 1

which is statement (2).

The right side of statement (2) is defined to be the *differential* of y.

4.9.1 DEFINITION

If the function f is defined by $y = f(x)$, then the **differential of y**, denoted by dy, is given by

$$dy = f'(x)\,\Delta x \tag{3}$$

where x is in the domain of f' and Δx is an arbitrary increment of x.

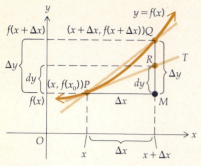

FIGURE 2

Refer now to Figure 2, which is the same as Figure 1 except the vertical line segment MR is shown, where the directed distance $\overline{MR} = dy$. Observe that dy represents the change in y along the tangent line to the graph of the equation $y = f(x)$ at the point $P(x, f(x))$, when x is changed by Δx.

This concept of the differential involves a special type of function of two variables and a detailed study of such functions appears in Chapter 16. The symbol df may be used to represent this function. The variable x can be any number in the domain of f', and Δx can be any number whatsoever. To state that df is a function of the two independent variables x and Δx means that to each ordered pair $(x, \Delta x)$ in the domain of df there corresponds one and only one number in the range of df, and this number can be represented by $df(x, \Delta x)$ so that

$$df(x, \Delta x) = f'(x)\,\Delta x$$

Comparing this equation with (3) we see that when $y = f(x)$, dy and $df(x, \Delta x)$ are two different notations for $f'(x)\,\Delta x$. The dy symbolism is used in subsequent discussions.

▶ **ILLUSTRATION 1**   If $y = 3x^2 - x$, then $f(x) = 3x^2 - x$; so $f'(x) = 6x - 1$. Therefore, from Definition 4.9.1

$$dy = (6x - 1)\,\Delta x$$

In particular, if $x = 2$, then $dy = 11\,\Delta x$.   ◀

When $y = f(x)$, Definition 4.9.1 indicates what is meant by dy, the differential of the dependent variable. We also wish to define the differential of the independent variable, or dx. To arrive at a suitable definition for dx that is consistent with the definition of dy we consider the identity function, which is the function defined by $f(x) = x$. Then $f'(x) = 1$ and $y = x$; thus from (3), $dy = 1 \cdot \Delta x$; that is,

$$\text{if} \quad y = x \quad \text{then} \quad dy = \Delta x \tag{4}$$

For the identity function we would want dx to be equal to dy; that is, because of statement (4) we would want dx to be equal to Δx. It is this reasoning that leads to the following definition.

4.9.2 DEFINITION

If the function f is defined by $y = f(x)$, then the **differential of x**, denoted by dx, is given by

$$dx = \Delta x$$

where Δx is an arbitrary increment of x, and x is any number in the domain of f'.

From Definitions 4.9.1 and 4.9.2

$$dy = f'(x)\,dx \tag{5}$$

Dividing both sides of (5) by dx we have

$$\frac{dy}{dx} = f'(x) \qquad \text{if } dx \neq 0$$

This equation expresses the derivative as the quotient of two differentials. Recall that when the notation $\dfrac{dy}{dx}$ for a derivative was introduced in Section 3.1, dy and dx had not been given independent meaning.

EXAMPLE 1   Given $y = 4x^2 - 3x + 1$, find Δy, dy, and $\Delta y - dy$ for (a) any x and Δx; (b) $x = 2$, $\Delta x = 0.1$; (c) $x = 2$, $\Delta x = 0.01$; (d) $x = 2$, $\Delta x = 0.001$.

Solution
(a) Because $y = 4x^2 - 3x + 1$, let

$$f(x) = 4x^2 - 3x + 1$$

Then,

$$\begin{aligned}
\Delta y &= f(x + \Delta x) - f(x) \\
&= 4(x + \Delta x)^2 - 3(x + \Delta x) + 1 - (4x^2 - 3x + 1) \\
&= 4x^2 + 8x\,\Delta x + 4(\Delta x)^2 - 3x - 3\,\Delta x + 1 - 4x^2 + 3x - 1 \\
&= (8x - 3)\,\Delta x + 4(\Delta x)^2
\end{aligned}$$

From (5),

$$\begin{aligned}
dy &= f'(x)\,dx \\
&= (8x - 3)\,dx \\
&= (8x - 3)\,\Delta x
\end{aligned}$$

Thus

$$\begin{aligned}
\Delta y - dy &= (8x - 3)\,\Delta x + 4(\Delta x)^2 - (8x - 3)\,\Delta x \\
&= 4(\Delta x)^2
\end{aligned}$$

The results for parts (b), (c), and (d) are given in Table 1, where

$$\Delta y = (8x - 3)\,\Delta x + 4(\Delta x)^2 \qquad \text{and} \qquad dy = (8x - 3)\,\Delta x$$

Table 1

	x	Δx	Δy	dy	$\Delta y - dy$
(b)	2	0.1	1.34	1.3	0.04
(c)	2	0.01	0.1304	0.13	0.0004
(d)	2	0.001	0.013004	0.013	0.000004

Note from Table 1 that the closer Δx is to zero, the smaller is the difference between Δy and dy. Furthermore, observe that for each value of Δx, the corresponding value of $\Delta y - dy$ is smaller than the value of Δx. More generally, dy is an approximation of Δy when Δx is small, and the approximation is of better accuracy than the size of Δx.

For a fixed value of x, say x_0,

$$dy = f'(x_0)\,dx$$

That is, dy is a linear function of dx; consequently dy is usually easier to compute than Δy, as was seen in Example 1. Because

$$f(x_0 + \Delta x) - f(x_0) = \Delta y$$

then

$$f(x_0 + \Delta x) = f(x_0) + \Delta y$$

Thus

$$f(x_0 + \Delta x) \approx f(x_0) + dy \tag{6}$$

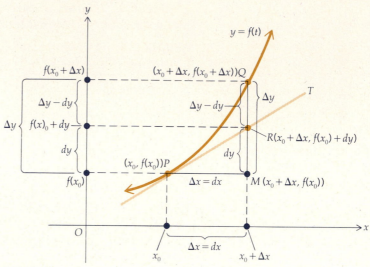

FIGURE 3

Our results are illustrated in Figure 3. The equation of the curve in the figure is $y = f(x)$, and the graph is concave upward. The line PT is tangent to the curve at $P(x_0, f(x_0))$; Δx and dx are equal and are represented by the directed distance $\overline{PM}$, where M is the point $(x_0 + \Delta x, f(x_0))$. We let Q be the point $(x_0 + \Delta x, f(x_0 + \Delta x))$, and the directed distance $\overline{MQ}$ is Δy or, equivalently, $f(x_0 + \Delta x) - f(x_0)$. The slope of PT is $f'(x) = dy/dx$. Also, the slope of PT is $\overline{MR}/\overline{PM}$, and because $\overline{PM} = dx$, we have $dy = \overline{MR}$ and $\overline{RQ} = \Delta y - dy$. Note that the smaller the value of dx (i.e., the closer the point Q is to the point P), then the smaller will be the value of $\Delta y - dy$ (i.e., the smaller will be the length of the line segment RQ). An equation of the tangent line PT is

$$y = f(x_0) + f'(x_0)(x - x_0)$$

Thus, if $\bar{y}$ is the ordinate of R, then

$$\bar{y} = f(x_0) + f'(x_0)[(x_0 + \Delta x) - x_0]$$

$$\bar{y} = f(x_0) + f'(x_0)\,\Delta x$$

$$\bar{y} = f(x_0) + dy$$

By comparing this equation with (6), observe that when using $f(x_0) + dy$ to approximate the value of $f(x_0 + \Delta x)$, we are approximating the ordinate of the point $Q(x_0 + \Delta x, f(x_0 + \Delta x))$ on the curve by the ordinate of the point $R(x_0 + \Delta x, f(x_0) + dy)$ on the line that is tangent to the curve at $P(x_0, f(x_0))$.

In Figure 3, $\Delta y > dy$, so that $\Delta y - dy > 0$. Refer now to Figure 4, where $\Delta y < dy$, so that $\Delta y - dy < 0$. Again observe that the smaller the value of dx, then the smaller will be the values of $|\Delta y - dy|$; that is, the closer the point Q is to the point P, the smaller will be the length of the line segment QR.

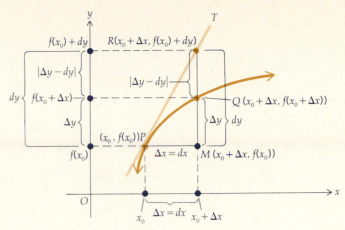

FIGURE 4

EXAMPLE 2   Find the approximate volume of a spherical shell whose inner radius is 4 in. and whose thickness is $\frac{1}{16}$ in.

Solution   We consider the volume of the spherical shell as an increment of the volume of a sphere. Let r inches be the radius of a sphere, V cubic inches be the volume of a sphere, and ΔV cubic inches be the volume of a spherical shell.

$$V = \tfrac{4}{3}\pi r^3 \qquad dV = 4\pi r^2 \, dr$$

Substituting $r = 4$ and $dr = \frac{1}{16}$ into the above we obtain

$$dV = 4\pi(4)^2 \tfrac{1}{16}$$
$$= 4\pi$$

Therefore $\Delta V \approx 4\pi$, and we conclude that the volume of the spherical shell is approximately 4π in.^{3}.

Suppose that y is a function x and that x, in turn, is a function of a third variable t; that is,

$$y = f(x) \quad \text{and} \quad x = g(t)$$

These two equations together define y as a function of t. For example, suppose that $y = x^3$ and $x = 2t^2 - 1$. Combining these two equations we get $y = (2t^2 - 1)^3$. In general, if the two equations $y = f(x)$ and $x = g(t)$ are combined, we obtain

$$y = f(g(t))$$

The derivative of y with respect to t can be found by the chain rule, which yields

$$\frac{dy}{dt} = \left(\frac{dy}{dx}\right)\left(\frac{dx}{dt}\right) \tag{7}$$

Equation (7) expresses $\dfrac{dy}{dt}$ as a function of x and t because $\dfrac{dy}{dx}$ is a function of x and $\dfrac{dx}{dt}$ is a function of t.

▶ **ILLUSTRATION 2**   If $y = x^3$ and $x = 2t^2 - 1$, then

$$\frac{dy}{dt} = \left(\frac{dy}{dx}\right)\left(\frac{dx}{dt}\right)$$

$$= 3x^2(4t)$$

$$= 12x^2 t \qquad\qquad\qquad ◀$$

Because $y = f(g(t))$ defines y as a function of the independent variable t, we obtain the differential of y from (5):

$$dy = \left(\frac{dy}{dt}\right) dt$$

This equation expresses dy as a function of t and dt, and substituting from (7) into it we get

$$dy = \left(\frac{dy}{dx}\right)\left(\frac{dx}{dt}\right) dt \qquad\qquad (8)$$

Now, because x is a function of the independent variable t, (5) can be applied to obtain the differential of x, and we have

$$dx = \left(\frac{dx}{dt}\right) dt$$

This equation expresses dx as a function of t and dt. By using it to replace $\left(\dfrac{dx}{dt}\right) dt$ in (8) by dx, we get

$$dy = \left(\frac{dy}{dx}\right) dx \qquad\qquad (9)$$

You should bear in mind that in (9), dy is a function of t and dt, and that dx is a function of t and dt. If $\dfrac{dy}{dx}$ in (9) is replaced by $f'(x)$, we have

$$dy = f'(x)\, dx \qquad\qquad (10)$$

Equation (10) resembles (5). However, in (5), x is the independent variable, and dy is expressed in terms of x and dx, whereas in (10), t is the independent variable, and both dy and dx are expressed in terms of t and dt. Thus we have the following theorem.

4.9.3 THEOREM   If $y = f(x)$, then when $f'(x)$ exists,

$$dy = f'(x)\, dx$$

whether or not x is an independent variable.

If we divide both sides of (10) by dx (provided $dx \neq 0$), we obtain

$$f'(x) = \frac{dy}{dx} \qquad dx \neq 0$$

This equation states that if $y = f(x)$, then $f'(x)$ is the quotient of the two differentials dy and dx, even though x may not be an independent variable.

▶ **ILLUSTRATION 3**   Suppose $x = \cos t$, $y = \sin t$, and $0 < t < \pi$. Then

$$dx = -\sin t\, dt \quad \text{and} \quad dy = \cos t\, dt$$

Therefore

$$\frac{dy}{dx} = \frac{\cos t \, dt}{-\sin t \, dt}$$

$$\frac{dy}{dx} = -\frac{\cos t}{\sin t}$$

$$\frac{dy}{dx} = -\frac{x}{y} \tag{11}$$

Because $x^2 + y^2 = \cos^2 t + \sin^2 t$, then $x^2 + y^2 = 1$. Thus

$$y^2 = 1 - x^2$$

Furthermore $y > 0$ because $y = \sin t$ and $0 < t < \pi$. Thus from the above equation,

$$y = \sqrt{1 - x^2} \tag{12}$$

Substituting the value of y from (12) into (11) we have

$$\frac{dy}{dx} = -\frac{x}{\sqrt{1 - x^2}}$$

This result can also be obtained by computing $\frac{dy}{dx}$ from (12). ◀

In Section 3.3 we proved theorems for finding derivatives of algebraic functions. The formulas from these theorems are now stated with the Leibniz notation. Along with the formula for the derivative there is a corresponding formula for the differential. In these formulas, u and v are functions of x, and it is understood that the formulas hold providing $\frac{du}{dx}$ and $\frac{dv}{dx}$ exist. When c appears, it is a constant.

$$\text{I} \quad \frac{d(c)}{dx} = 0 \qquad\qquad \text{I}' \quad d(c) = 0$$

$$\text{II} \quad \frac{d(x^n)}{dx} = nx^{n-1} \qquad\qquad \text{II}' \quad d(x^n) = nx^{n-1}\, dx$$

$$\text{III} \quad \frac{d(cu)}{dx} = c\frac{du}{dx} \qquad\qquad \text{III}' \quad d(cu) = c\, du$$

$$\text{IV} \quad \frac{d(u + v)}{dx} = \frac{du}{dx} + \frac{dv}{dx} \qquad\qquad \text{IV}' \quad d(u + v) = du + dv$$

$$\text{V} \quad \frac{d(uv)}{dx} = u\frac{dv}{dx} + v\frac{du}{dx} \qquad\qquad \text{V}' \quad d(uv) = u\, dv + v\, du$$

$$\text{VI} \quad \frac{d\left(\dfrac{u}{v}\right)}{dx} = \frac{v\dfrac{du}{dx} - u\dfrac{dv}{dx}}{v^2} \qquad\qquad \text{VI}' \quad d\left(\frac{u}{v}\right) = \frac{v\, du - u\, dv}{v^2}$$

$$\text{VII} \quad \frac{d(u^n)}{dx} = nu^{n-1}\frac{du}{dx} \qquad\qquad \text{VII}' \quad d(u^n) = nu^{n-1}\, du$$

We extend the operation of differentiation to include the process of computing the differential as well as computing the derivative. If $y = f(x)$, dy can be found either by applying formulas I′–VII′ or by finding $f'(x)$ and multiplying it by dx.

EXAMPLE 3   Given

$$2x^2y^2 - 3x^3 + 5y^3 + 6xy^2 = 5$$

where x and y are functions of a third variable, find $\dfrac{dy}{dx}$ by computing the differential term by term.

Solution   This problem involves implicit differentiation. Taking the differential term by term we get

$$4xy^2\,dx + 4x^2y\,dy - 9x^2\,dx + 15y^2\,dy + 6y^2\,dx + 12xy\,dy = 0$$

Dividing by dx, if $dx \neq 0$, we have

$$(4x^2y + 15y^2 + 12xy)\frac{dy}{dx} = -4xy^2 + 9x^2 - 6y^2$$

$$\frac{dy}{dx} = \frac{9x^2 - 6y^2 - 4xy^2}{4x^2y + 15y^2 + 12xy}$$

EXERCISES 4.9

In Exercises 1 through 4, (a) find dy and Δy for the values of x and Δx. (b) Draw a sketch of the graph, and indicate the line segments whose lengths are dy and Δy.

1. $y = x^2$; $x = 2$ and $\Delta x = 1$ **2.** $y = x^3$; $x = 2$ and $\Delta x = 1$

3. $y = \sqrt[3]{x}$; $x = 8$ and $\Delta x = 2$ **4.** $y = \sqrt{x}$; $x = 4$ and $\Delta x = 3$

In Exercises 5 through 10, find (a) Δy; (b) dy; (c) Δy − dy.

5. $y = 6 - 3x - 2x^2$ **6.** $y = 3x^2 - x$

7. $y = x^3 - x^2$ **8.** $y = \dfrac{1}{x^2 + 1}$

9. $y = \dfrac{2}{x - 1}$ **10.** $y = \dfrac{1}{x} - x^3$

In Exercises 11 through 16, find (a) Δy; (b) dy; (c) Δy − dy.

11. $y = x^2 - 3x$; $x = 2$; $\Delta x = 0.03$

12. $y = x^2 - 3x$; $x = -1$; $\Delta x = 0.02$

13. $y = \dfrac{1}{x}$; $x = -2$; $\Delta x = -0.1$

14. $y = \dfrac{1}{x}$; $x = 3$; $\Delta x = -0.2$

15. $y = x^3 + 1$; $x = 1$; $\Delta x = -0.5$

16. $y = x^3 + 1$; $x = -1$; $\Delta x = 0.1$

In Exercises 17 through 24, find dy.

17. $y = (3x^2 - 2x + 1)^3$ **18.** $y = \sqrt{4 - x^2}$

19. $y = x^2\sqrt[3]{2x + 3}$ **20.** $y = \dfrac{3x}{x^2 + 2}$

21. $y = \dfrac{2 + \cos x}{2 - \sin x}$ **22.** $y = \cot 2x \csc 2x$

23. $y = \tan^2 x \sec^2 x$ **24.** $y = x^2 \sin \dfrac{1}{x} - x \cos \dfrac{1}{x}$

In Exercises 25 through 30, x and y are functions of a third variable. Find $\dfrac{dy}{dx}$ by computing the differential term by term (see Example 3).

25. $3x^2 + 4y^2 = 48$ **26.** $8x^2 - y^2 = 32$

27. $\sqrt{x} + \sqrt{y} = 4$ **28.** $2x^2y - 3xy^3 + 6y^2 = 1$

29. $\sin x \cos y - \cos x \sin y = \frac{1}{2}$

30. $3\tan^2 x + 4\sec^2 y = 1$

In Exercises 31 through 34, do the following: (a) Find dx and dy in terms of t and dt; (b) use the results of part (a) to find $\dfrac{dy}{dx}$; (c) express y in terms of x by eliminating t between the pair of equations, and then find $\dfrac{dy}{dx}$ by using theorems of differentiation.

31. $x = 2t^2$, $y = 3t$ **32.** $x = 1 + t$, $y = 1 - t^2$

33. $x = 2\cos t$, $y = 3\sin t$, $0 < t < \pi$

34. $x = 1 - \cos t$, $y = 2 + \sin t$, $0 < t < \pi$

(*Hint for Exercises 33 and 34: Eliminate t by using the identity $\sin^2 t + \cos^2 t = 1$.*)

35. The measurement of an edge of a cube is found to be 15 cm with a possible error of 0.01 cm. Use differentials to find the approximate error in computing from this measurement (a) the volume; (b) the area of one of the faces.

36. A metal box in the form of a cube is to have an interior volume of 1000 cm³. The six sides are to be made of metal $\frac{1}{2}$ cm thick. If the cost of the metal to be used is $0.20 per cubic centimeter, use differentials to find the approximate cost of the metal to be used in the manufacture of the box.

37. An open cylindrical tank is to have an outside coating of thickness 2 cm. If the inner radius is 6 m and the altitude is 10 m, find by differentials the approximate amount of coating material to be used.

38. The stem of a particular mushroom is cylindrical in shape, and a stem of height 2 cm and radius r centimeters has a volume of V cubic centimeters, where $V = 2\pi r^2$. Use the differential to find the approximate increase in the volume of the stem when the radius increases from 0.4 cm to 0.5 cm.

39. A burn on a person's skin is in the shape of a circle such that if r centimeters is the radius and A square centimeters is the area of the burn, then $A = \pi r^2$. Use the differential to find the approximate decrease in the area of the burn when the radius decreases from 1 cm to 0.8 cm.

40. A certain bacterial cell is spherical in shape such that if r micrometers is its radius and V cubic micrometers is its volume, then $V = \frac{4}{3}\pi r^3$. Use the differential to find the approximate increase in the volume of the cell when the radius increases from 2.2 μm to 2.3 μm.

41. A tumor in a person's body is spherical in shape such that

if r centimeters is the radius and V cubic centimeters is the volume of the tumor, then $V = \frac{4}{3}\pi r^3$. Use the differential to find the approximate increase in the volume of the tumor when the radius increases from 1.5 cm to 1.6 cm.

42. If t seconds is the time for one complete swing of a simple pendulum of length l feet, then $4\pi^2 l = gt^2$, where $g = 32.2$. A clock having a pendulum of length 1 ft gains 5 min each day. Find the approximate amount by which the pendulum should be lengthened to correct the inaccuracy.

43. The measure of the electrical resistance of a wire is proportional to the measure of its length and inversely proportional to the square of the measure of its diameter. Suppose the resistance of a wire of given length is computed from a measurement of the diameter with a possible 2 percent error. Find the possible percent error in the computed value of the resistance.

44. A contractor agrees to paint both sides of 1000 circular signs each of radius 3 m. Upon receiving the signs it is discovered that the radius of each sign is 1 cm too large. Use differentials to find the approximate percent increase of paint that will be needed.

45. If the possible error in the measurement of the volume of a gas is 0.1 ft³ and the allowable error in the pressure is $0.001C$ lb/ft², find the size of the smallest container for which Boyle's law (Exercise 23 in Exercises 3.9) holds.

46. For the adiabatic law for the expansion of air (Exercise 24 in Exercises 3.9), prove that $\dfrac{dP}{P} = -1.4\dfrac{dV}{V}$.

4.10 NUMERICAL SOLUTIONS OF EQUATIONS BY NEWTON'S METHOD (*Supplementary*)

The solutions of an equation of the form

$$f(x) = 0$$

are called the **roots** of the equation or the **zeros** of the function f. If f is a polynomial function of degree less than five, there are formulas for obtaining the zeros. You are familiar with the formulas if f has degree one (a linear function) or two (a quadratic function). For a polynomial function of degree three or four, the general method for obtaining the zeros is complicated. Furthermore, for the zeros of a polynomial function of degree five or more there is a theorem, credited to Niels Abel (1802–1829), that states that there can be no general formula in terms of a finite number of operations on the coefficients. There are, however, numerical processes for approximating solutions of equations, and they are more important now than ever before because of the widespread use of computers and programmable calculators. One of the processes involves an application of the derivative, and it was devised by Sir Isaac Newton in the seventeenth century. It is known as Newton's method and is the subject of this section.

Consider the equation $f(x) = 0$, where f is a differentiable function. Newton's method gives a procedure for approximating a root of this equation or, equivalently, a zero of f, that is, a number r such that $f(r) = 0$. We first give a geometric interpretation of the concepts involved. Refer to Figure 1, which shows a sketch of the graph of $y = f(x)$. The number r is an x intercept of the

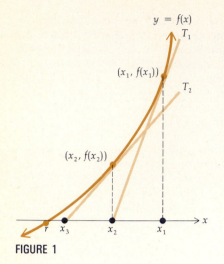

FIGURE 1

graph. To obtain an approximation of r we first select a number x_1. The choice of x_1 can be made by referring to a sketch of the graph, and it should be reasonably close to the number r. We then consider the tangent line to the graph of f at the point $(x_1, f(x_1))$. The tangent line, denoted by T_1, appears in Figure 1, and T_1 has an x intercept x_2. The number x_2 now serves as a second approximation of r. We then repeat the process with the tangent line T_2 at the point $(x_2, f(x_2))$. The x intercept of T_2 is x_3. We continue the process until we have the required degree of accuracy. For this graph it appears that the numbers x_1, x_2, x_3, and so on are getting closer and closer to the number r. This situation occurs for many functions.

To obtain the successive approximations $x_2, x_3, \ldots$ from the first approximation x_1 we use the equations of the tangent lines. The tangent line T_1 at the point $(x_1, f(x_1))$ has a slope of $f'(x_1)$. Thus an equation of T_1 is

$$y - f(x_1) = f'(x_1)(x - x_1)$$

The x intercept of T_1 is x_2, and we determine x_2 by letting $x = x_2$ and $y = 0$ in the above equation. We get

$$0 - f(x_1) = f'(x_1)(x_2 - x_1)$$

$$x_2 = x_1 - \frac{f(x_1)}{f'(x_1)} \qquad \text{if } f'(x_1) \neq 0$$

With this value of x_2 an equation of T_2 is

$$y - f(x_2) = f'(x_2)(x - x_2)$$

Then in this equation we let $x = x_3$ and $y = 0$, and we have

$$0 - f(x_2) = f'(x_2)(x_3 - x_2)$$

$$x_3 = x_2 - \frac{f(x_2)}{f'(x_2)} \qquad \text{if } f'(x_2) \neq 0$$

Continuing on in this manner we obtain the general formula for the approximation x_{n+1} in terms of the preceding approximation x_n:

$$x_{n+1} = x_n - \frac{f(x_n)}{f'(x_n)} \qquad \text{if } f'(x_n) \neq 0 \tag{1}$$

Formula (1) is easily adapted for use on a computer or a programmable calculator.

From formula (1) we can obtain the $(n + 1)$st approximation from the nth approximation, provided $f'(x_n) \neq 0$. When $f'(x_n) = 0$, the tangent line is horizontal, and in such a case, unless the tangent line is the x axis itself, it has no x intercept. Figure 2 shows this happening when $f'(x_2) = 0$. So Newton's method is not applicable if $f'(x_n) = 0$ for some x_n. You should also be aware that the value of x_{n+1} obtained from (1) is not necessarily a better approximation of r than x_n. If, for instance, x_1 is not reasonably close to r, then $|f'(x_1)|$ may be small so that the tangent line T_1 is nearly horizontal. Then x_2, the x intercept of T_1, could be farther away from r than x_1. See Figure 3 where this situation occurs.

In the following illustration we show how Newton's method is applied to an equation for which we know the answer.

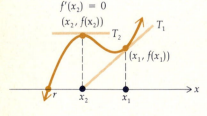

FIGURE 2

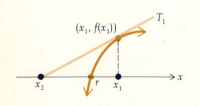

FIGURE 3

▶ **ILLUSTRATION 1** Let us use Newton's method to obtain the positive root of the equation $x^2 = 9$ by starting with a first approximation of 4. We write the equation as $x^2 - 9 = 0$ and let

$$f(x) = x^2 - 9$$

$$f'(x) = 2x$$

From (1) we obtain

$$x_{n+1} = x_n - \frac{f(x_n)}{f'(x_n)}$$

$$x_{n+1} = x_n - \frac{x_n^2 - 9}{2x_n} \qquad (2)$$

We now apply (2) with values of n and corresponding values of x_n to compute by a calculator x_{n+1}. We start with $x_1 = 4$.

$$x_2 = x_1 - \frac{x_1^2 - 9}{2x_1} \qquad\qquad x_3 = x_2 - \frac{x_2^2 - 9}{2x_2}$$

$$= 4 - \frac{16 - 9}{8} \qquad\qquad = 3.125 - \frac{(3.125)^2 - 9}{2(3.125)}$$

$$= 3.125 \qquad\qquad = 3.0025$$

$$x_4 = x_3 - \frac{x_3^2 - 9}{2x_3} \qquad\qquad x_5 = x_4 - \frac{x_4^2 - 9}{2x_4}$$

$$= 3.0025 - \frac{(3.0025)^2 - 9}{2(3.0025)} \qquad = 3.0000 - \frac{(3.0000)^2 - 9}{2(3.0000)}$$

$$= 3.0000 \qquad\qquad = 3.0000$$

Certainly all successive approximations will be 3.0000. Thus the positive root of the equation $x^2 - 9 = 0$ is 3.0000 to four decimal places. ◀

Observe that when x_n is a solution of $f(x) = 0$, $f(x_n) = 0$. Thus from (1),

$$x_{n+1} = x_n - \frac{f(x_n)}{f'(x_n)}$$

$$= x_n - 0$$

$$= x_n$$

Consequently all subsequent approximations are equal to x_n. Note that this situation occurs in Illustration 1, where all approximations after and including x_4 have the same value to four decimal places.

Also observe from (1) that $x_{n+1} = x_n$ implies that $f(x_n) = 0$. Therefore we can conclude that when two successive approximations are equal, we have an approximation for a zero of f.

It is possible, however, that for certain functions, if your initial choice of x_1 is not close enough to the desired zero, you may obtain approximations for a different zero. See Figure 4 for a sketch of the graph of a function where this situation could happen. Note that the indicated choice of x_1 near the desired zero r gives successive approximations $x_2, x_3, x_4, \ldots$ near another zero s. Thus when applying Newton's method you should first draw a rough sketch of the

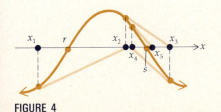

FIGURE 4

graph of the function to obtain your initial approximation. Refer to the graph as you proceed to make sure you are getting successive approximations to the zero you are seeking.

In summary, when using Newton's method to solve an equation of the form $f(x) = 0$, do the following:

> 1. Make a *good guess* for the first approximation x_1. A rough sketch of the graph of f will help to obtain a reasonable choice.
> 2. With the value of x_1 in formula (1), get a second approximation x_2. Then use x_2 in (1) to get a third approximation x_3, and so on until $x_{n+1} = x_n$ to the required degree of accuracy.

EXAMPLE 1   Use Newton's method to find the real root of the equation

$$x^3 - 2x - 2 = 0$$

to four decimal places.

Solution   Let $f(x) = x^3 - 2x - 2$; thus $f'(x) = 3x^2 - 2$. Then from (1) we have

$$x_{n+1} = x_n - \frac{x_n^3 - 2x_n - 2}{3x_n^2 - 2} \qquad (3)$$

To obtain a sketch of the graph of f we plot the points $(-2, -6)$, $(-1, -1)$, $(0, -2)$, $(1, -3)$, and $(2, 2)$. There are relative extrema of f when $f'(x) = 0$, that is, when $x = \pm\frac{1}{3}\sqrt{6}$. The graph appears as in Figure 5. Because the graph intersects the x axis at only one point, there is one real root of the given equation. Because $f(1) = -3$ and $f(2) = 2$, this root lies between 1 and 2. A suitable choice for our first approximation is $x_1 = 1.5$. Table 1 shows the results obtained from a calculator by successive approximations computed from (3) with this x_1. We wish the root to be accurate to four decimal places; thus we use five places in the computations. Because x_5 and x_6 are equal (to five decimal places), we round off that number to four places and obtain 1.7693 as the required root.

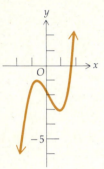

FIGURE 5

Table 1

n	x_n	$x_n^3 - 2x_n - 2$	$3x_n^2 - 2$	$\dfrac{x_n^3 - 2x_n - 2}{3x_n^2 - 2}$	x_{n+1}
1	1.50000	-1.62500	4.75000	-0.34211	1.84211
2	1.84211	0.56674	8.18011	0.06928	1.77283
3	1.77283	0.02621	7.42878	0.00353	1.76930
4	1.76930	0.00006	7.39127	0.00001	1.76929
5	1.76929	-0.00002	7.39116	0.00000	1.76929

EXAMPLE 2   Use Newton's method to find to three decimal places the x coordinate of the point of intersection in the first quadrant of the line $y = \frac{1}{3}x$ and the curve $y = \sin x$.

Solution   Figure 6 shows the line and the curve. We wish to find the positive value of x for which

$$\sin x = \tfrac{1}{3}x$$

$$3 \sin x - x = 0$$

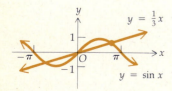

FIGURE 6

Let

$$f(x) = 3 \sin x - x$$

$$f'(x) = 3 \cos x - 1$$

From formula (1),

$$x_{n+1} = x_n - \frac{f(x_n)}{f'(x_n)}$$

$$x_{n+1} = x_n - \frac{3 \sin x_n - x_n}{3 \cos x_n - 1} \tag{4}$$

From Figure 6 it appears that a reasonable choice of x_1 is 2. We use a calculator to compute the successive approximations from formula (4); these appear in Table 2. The results are expressed to four decimal places. Observe that to four decimal places x_4 and x_5 are both equal to 2.2789. Thus to three decimal places the positive value of x for which $\sin x = \frac{1}{3}x$ is 2.279.

Table 2

n	x_n	$3 \sin x_n - x_n$	$3 \cos x_n - 1$	$\dfrac{3 \sin x_n - x_n}{3 \cos x_n - 1}$	x_{n+1}
1	2.0000	0.7279	-2.2484	-0.3237	2.3237
2	2.3237	-0.1346	-3.0513	0.0441	2.2796
3	2.2796	-0.0022	-2.9528	0.0007	2.2789
4	2.2789	-0.0001	-2.9512	0.0000	2.2789

There are theorems that state conditions for which Newton's method is applicable, as well as theorems relating to its accuracy. Such theorems can be found in texts on numerical analysis.

EXERCISES 4.10

In Exercises 1 through 4, use Newton's method to find the real root of the equation to four decimal places.

1. $x^3 - 4x^2 - 2 = 0$ **2.** $6x^3 + 9x + 1 = 0$
3. $x^5 - x + 1 = 0$ **4.** $x^5 + x - 1 = 0$

In Exercises 5 through 10, use Newton's method to find, to the nearest thousandth, the approximate value of the indicated root.

5. $x^3 - 4x - 8 = 0$; the positive root
6. $x^3 - 2x + 7 = 0$; the negative root
7. $x^4 - 10x + 5 = 0$; the smallest positive root
8. $x^4 - 10x + 5 = 0$; the largest positive root
9. $2x^4 - 2x^3 + x^2 + 3x - 4 = 0$; the negative root
10. $x^4 + x^3 - 3x^2 - x - 4 = 0$; the positive root

In Exercises 11 through 14, use Newton's method to find the value of the radical to five decimal places.

11. $\sqrt{3}$ by solving the equation $x^2 - 3 = 0$
12. $\sqrt{10}$ by solving the equation $x^2 - 10 = 0$

13. $\sqrt[3]{6}$ by solving the equation $x^3 - 6 = 0$
14. $\sqrt[3]{7}$ by solving the equation $x^3 - 7 = 0$

In Exercises 15 through 18, use Newton's method to find to four decimal places the x coordinate of the point of intersection in the first quadrant of the graphs of the two equations.

15. $y = x$; $y = \cos x$ **16.** $y = \frac{1}{2}x$; $y = \sin x$
17. $y = x^2$; $y = \sin x$ **18.** $y = x^2$; $y = \cos x$

19. Equations of the form $\tan x + ax = 0$, where a is an integer, arise in heat conduction problems. The positive roots of the equation in increasing order are $\alpha_1, \alpha_2, \alpha_3, \ldots$. If $a = 1$, find α_1 and α_2 to four decimal places.
20. Follow the instructions of Exercise 19 if $a = -2$.

In Exercises 21 and 22, obtain an approximation for π to five decimal places by using Newton's method to solve the equation.

21. $\tan x = 0$ **22.** $\cos x + 1 = 0$

REVIEW EXERCISES FOR CHAPTER 4

In Exercises 1 through 12, find the absolute extrema of the function on the indicated interval, if there are any, and find the values of x at which the absolute extrema occur. Draw a sketch of the graph of the function on the interval.

1. $f(x) = \sqrt{5+x}; [-5, +\infty)$

2. $f(x) = \sqrt{9-x^2}; (-3, 3)$

3. $f(x) = \frac{5}{2}x^6 - 3x^5; [-1, 2]$

4. $f(x) = x^4 - 12x^2 + 36; [-2, 6]$

5. $f(x) = |9 - x^2|; [-2, 3]$

6. $f(x) = \frac{3}{x-2}; [0, 4]$

7. $f(x) \neq x^4 - 12x^2 + 36; [0, 5]$

8. $f(x) = \begin{cases} 2x+3 & \text{if } -2 \leq x < 1 \\ x^2+4 & \text{if } 1 \leq x \leq 2 \end{cases}; [-2, 2]$

9. $f(x) = 2\sin 3x; [-\frac{1}{3}\pi, \frac{1}{3}\pi]$

10. $f(x) = 4\cos^2 2x; [0, \frac{3}{4}\pi]$

11. $f(x) = \tan 4x; [-\frac{1}{8}\pi, \frac{1}{12}\pi]$

12. $f(x) = \csc 3x; [0, \frac{1}{3}\pi]$

In Exercises 13 through 32, draw a sketch of the graph of the function f by first finding the following: the relative extrema of f; the points of inflection of the graph of f; the intervals on which f is increasing; the intervals on which f is decreasing; where the graph is concave upward; where the graph is concave downward; the slope of any inflectional tangent; the horizontal, vertical, and oblique asymptotes, if there are any.

13. $f(x) = x^3 + 3x^2 - 4$

14. $f(x) = x^3 + 2x^2 + x - 5$

15. $f(x) = (x-4)^2(x+2)^3$

16. $f(x) = (x-1)^3(x-3)$

17. $f(x) = (x-4)^{1/3} - 3$

18. $f(x) = (x+2)^3 + 2$

19. $f(x) = \frac{4x^2}{3x^2+1}$

20. $f(x) = \frac{x^2}{x^2-9}$

21. $f(x) = \frac{5x^2}{x^2-4}$

22. $f(x) = \frac{x}{x^2-1}$

23. $f(x) = \frac{x^2}{x-3}$

24. $f(x) = \frac{x^2+9}{x}$

25. $f(x) = \begin{cases} (1-x)^3 & \text{if } x \leq 1 \\ (x-1)^3 & \text{if } 1 < x \end{cases}$

26. $f(x) = \begin{cases} x^3 - 3x & \text{if } x < 2 \\ 6 - x^2 & \text{if } 2 \leq x \end{cases}$

27. $f(x) = (x-3)^{5/3} + 1$

28. $f(x) = (x+2)^{4/3}$

29. $f(x) = (x+1)^{2/3}(x-3)^2$

30. $f(x) = x\sqrt{25-x^2}$

31. $f(x) = \sin 2x - \cos 2x; x \in [-\frac{3}{8}\pi, \frac{5}{8}\pi]$

32. $f(x) = x - \tan x; x \in (-\frac{1}{2}\pi, \frac{1}{2}\pi)$

In Exercises 33 and 34, verify that the three conditions of the hypothesis of Rolle's theorem are satisfied by the function on the indicated interval. Then find a suitable value for c that satisfies the conclusion of Rolle's theorem. Draw a sketch of the graph of f on the interval, and use the sketch to give the geometric interpretation of Rolle's theorem.

33. $f(x) = x^3 - x^2 - 4x + 4; [-2, 1]$

34. $f(x) = 2\sin 3x; [0, \frac{1}{3}\pi]$

In Exercises 35 and 36, verify that the hypothesis of the mean-value theorem is satisfied for the function on the indicated interval. Then find a suitable value for c that satisfies the conclusion of the mean-value theorem. Draw a sketch of the graph of f on the interval, and use the sketch to give the geometric interpretation of the mean-value theorem.

35. $f(x) = \sqrt{3-x}; [-6, -1]$ **36.** $f(x) = x^3; [-2, 2]$

37. (a) If f is a polynomial function and $f(a)$, $f(b)$, $f'(a)$, and $f'(b)$ are zero, prove that there are at least two numbers in the open interval (a, b) that are roots of the equation $f''(x) = 0$. (b) Show that the function defined by $f(x) = (x^2 - 4)^2$ satisfies part (a) for the open interval $(-2, 2)$.

38. If f is the function defined by $f(x) = |2x - 4| - 6$, then $f(-1) = 0$ and $f(5) = 0$. However, f' never has the value 0. Show why Rolle's theorem does not apply.

39. Suppose that f and g are two functions that satisfy the hypothesis of the mean-value theorem on $[a, b]$. Furthermore, suppose that $f'(x) = g'(x)$ for all x in the open interval (a, b). Prove that $f(x) - g(x) = f(a) - g(a)$ for all x in the closed interval $[a, b]$.

40. If f is a polynomial function, show that between any two consecutive roots of the equation $f'(x) = 0$ there is at most one root of the equation $f(x) = 0$.

41. Find the absolute maximum value attained by the function f if $f(x) = A\sin kx + B\cos kx$, where A, B, and k are positive constants.

42. Find the point (or points) of inflection of the graph of the function defined by $f(x) = x^{1/5}$, and determine where the graph is concave upward and where it is concave downward.

43. Find the point (or points) of inflection of the graph of the function defined by

$$f(x) = \begin{cases} 2 - x^2 & \text{if } x < 0 \\ x^2 + 2 & \text{if } 0 \leq x \end{cases}$$

and determine where the graph is concave upward and where it is concave downward.

44. If $f(x) = ax^3 + bx^2$, determine a and b so that the graph of f will have a point of inflection at $(2, 16)$.

45. If $f(x) = ax^3 + bx^2 + cx$, determine a, b, and c so that the graph of f will have a point of inflection at $(1, -1)$ and so that the slope of the inflectional tangent there will be -3.

46. If $f(x) = \frac{x+1}{x^2+1}$, prove that the graph of f has three points of inflection that are collinear. Draw a sketch of the graph.

47. If $f(x) = x|x|$, show that the graph of f has a point of inflection at the origin.

48. Let $f(x) = x^n$, where n is a positive integer. Prove that the graph of f has a point of inflection at the origin if and only if n is an odd integer and $n > 1$. Furthermore show that if n is even, f has a relative minimum value at 0.

49. If $f(x) = (x^2 + a^2)^p$, where p is a rational number and $p \neq 0$, prove that if $p < \frac{1}{2}$, the graph of f has two points of inflection, and if $p \geq \frac{1}{2}$, the graph of f has no points of inflection.

50. Suppose that the graph of a function has a point of inflection at $x = c$. What can you conclude, if anything, about (a) the continuity of f at c; (b) the continuity of f' at c; (c) the continuity of f'' at c?

51. Find two nonnegative numbers whose sum is 12 such that the sum of their squares is an absolute minimum.

52. Find two nonnegative numbers whose sum is 12 such that their product is an absolute maximum.

53. Show that among all the rectangles having a perimeter of 36 in., the square of side 9 in. has the greatest area.

54. Show that among all the rectangles having an area of 81 in.^{2}, the square of side 9 in. has the least perimeter.

55. An open box having a square base is to have a volume of 4000 in.^{3}. Find the dimensions of the box that can be constructed with the least amount of material.

56. Solve Exercise 55 if the box is to be closed.

57. Solve Exercise 55 if the open box is to have a volume of k in.^{3}.

58. In a town of population 11,000 the rate of growth of an epidemic is jointly proportional to the number of people infected and the number of people not infected. Determine the number of people infected when the epidemic is growing at a maximum rate.

59. Because of various restrictions, the size of a particular community is limited to 3000 inhabitants, and the rate of increase of the population is jointly proportional to its size and the difference between 3000 and its size. Determine the size of the population for which the rate of growth of the population is a maximum.

60. A manufacturer offers to deliver to a dealer 300 chairs at $90 per chair and to reduce the price per chair on the entire order by $0.25 for each additional chair over 300. Find the dollar total involved in the largest possible transaction between the manufacturer and the dealer under these circumstances.

61. Two towns A and B are to get their water supply from the same pumping station to be located on the bank of a straight river that is 15 km from town A and 10 km from town B. If the points on the river nearest to A and B are 20 km apart and A and B are on the same side of the river, where should the pumping station be located so that the least amount of piping is required?

62. One of the acute angles of a triangle is to have a radian measure of $\frac{1}{6}\pi$, and the side opposite this angle is to have a length of 10 in. Prove that of all the triangles satisfying these requirements, the one having the maximum area is isosceles. (*Hint:* Express the measure of the area of the triangle in terms of trigonometric functions of one of the other acute angles.)

63. In a warehouse, goods weighing 1000 lb are transported along a level floor by securing a heavy rope under a low mobile platform and pulling it with a motorized vehicle. If the rope is directed at an angle of θ radians with the plane of the floor, then the force of magnitude F pounds along the rope

is given by

$$F = \frac{1000k}{k \sin \theta + \cos \theta}$$

where k is the constant coefficient of friction and $0 < k < 1$. If $0 \leq \theta \leq \frac{1}{2}\pi$, show that F is least when $\tan \theta = k$.

64. A closed tin can having a volume of 27 in.^{3} is to be in the form of a right-circular cylinder. If the circular top and bottom are cut from square pieces of tin, find the radius and height of the can if the least amount of tin is to be used in its manufacture. Include the tin that is wasted when obtaining the top and bottom.

65. If $100x$ units of a particular commodity are demanded when p dollars is the price per unit, $x^2 + p^2 = 36$. Find the absolute maximum total revenue.

66. Under a monopoly (see Exercise 31 in Exercises 4.8), x units are demanded daily when p dollars is the price per unit and $x^2 + p = 320$. If the number of dollars in the total cost of producing x units is given by $C(x) = 20x$, find the maximum daily total profit.

67. A firm operating under perfect competition (see the instructions for Exercises 29 and 30 in Exercises 4.8) manufactures and sells portable radios. The firm can sell at a price of $75 per radio all the radios it produces. If x radios are manufactured each day and $C(x)$ dollars is the daily total cost of production, then $C(x) = x^2 + 25x + 100$. How many radios should be produced each day for the firm to have the greatest daily total profit? What is the greatest daily total profit?

68. Find the shortest distance from the point $P(0, 4)$ to a point on the curve $x^2 - y^2 = 16$, and find the point on the curve that is closest to P.

69. Two particles start their motion at the same time. One particle is moving along a horizontal line and its equation of motion is $x = t^2 - 2t$, where x centimeters is the directed distance of the particle from the origin at t seconds. The other particle is moving along a vertical line that intersects the horizontal line at the origin, and its equation of motion is $y = t^2 - 2$, where y centimeters is the directed distance of the particle from the origin at t seconds. Find when the directed distance between the two particles is least, and their velocities at that time.

70. A ladder is to reach over a fence $\frac{27}{8}$ m high to a wall 8 m behind the fence. Find the length of the shortest ladder that may be used.

71. Solve Exercise 70 if the fence is h meters high and the wall is w meters behind the fence.

72. A tent is to be in the shape of a cone. Find the ratio of the measure of the radius to the measure of the altitude for a tent of given volume to require the least material.

73. A sign is to contain 50 m^2 of printed material. If margins of 4 m at the top and bottom and 2 m on each side are required, find the dimensions of the smallest sign that will meet these specifications.

74. Find the volume of the largest right-circular cylinder that can be inscribed in a right-circular cone having a radius of 4 in. and a height of 8 in.

75. Find the dimensions of the right-circular cone of least volume that can be circumscribed about a right-circular cylinder of radius r centimeters and height h centimeters.

76. A piece of wire 80 cm long is bent to form a rectangle. Find the dimensions of the rectangle so that its area is as large as possible.

77. A piece of wire 20 cm long is cut into two pieces, and each piece is bent into the shape of a square. How should the wire be cut so that the total area of the two squares is as small as possible?

78. If $f(x) = x + \sin x$, prove that f has no relative extrema but that the graph of f has horizontal tangent lines. Draw a sketch of the graph for $x \in [-2\pi, 2\pi]$.

79. For a certain commodity, where x units are demanded weekly when p dollars is the price of each unit,

$$10^6 px = 10^9 - 2 \cdot 10^6 x + 18 \cdot 10^3 x^2 - 6x^3$$

The number of dollars in the average cost of producing each unit is given by $Q(x) = \frac{1}{50}x - 24 + 11 \cdot 10^3 x^{-1}$ and $x \geq 100$. Find the number of units that should be produced each week and the price of each unit for the total weekly profit to be maximized.

80. If $y = 2x^2 - 3$, (a) find dy and Δy for $x = 2$ and $\Delta x = 0.5$. (b) Draw a sketch of the graph, and indicate the line segments whose lengths are dy and Δy.

81. If $y = 80x - 16x^2$, find the difference $\Delta y - dy$ if (a) $x = 2$ and $\Delta x = 0.1$; (b) $x = 4$ and $\Delta x = -0.2$.

82. If $x^3 + y^3 - 3xy^2 + 1 = 0$, find dy at the point $(1, 1)$ if $dx = 0.1$.

83. Use differentials to approximate the volume of material needed to make a rubber ball if the radius of the hollow inner core is 2 in. and the thickness of the rubber is $\frac{1}{8}$ in.

84. A container in the form of a cube having a volume of 1000 in.^{3} is to be made by using six equal squares of material costing 20 cents per square inch. How accurately must the side of each square be made so that the total cost of the material shall be correct to within $5.00?

85. If t seconds is the time for one complete swing of a simple pendulum of length l feet, then $4\pi^2 l = gt^2$, where $g = 32.2$.

What is the effect upon the time if an error of 0.01 ft is made in measuring the length of the pendulum?

86. The measure of the radius of a right-circular cone is $\frac{4}{3}$ times the measure of the altitude. How accurately must the altitude be measured if the error in the computed volume is not to exceed 3 percent?

87. (a) If $f(x) = 3|x| + 4|x - 1|$, prove that f has an absolute minimum value of 3. (b) If $g(x) = 4|x| + 3|x - 1|$, prove that g has an absolute minimum value of 3. (c) If $a > 0$ and $b > 0$, and if $h(x) = a|x| + b|x - 1|$, prove that h has an absolute minimum value that is the smaller of the two numbers a and b.

88. If $f(x) = |x|^a \cdot |x - 1|^b$, where a and b are positive rational numbers, prove that f has a relative maximum value of $a^a b^b / (a + b)^{a+b}$.

89. Let f and g be two functions that are differentiable at every number in the closed interval $[a, b]$. Suppose further that $f(a) = g(a)$ and $f(b) = g(b)$. Prove that there exists a number c in the open interval (a, b) such that $f'(c) = g'(c)$. (*Hint:* Consider the function $f - g$.)

90. Draw a sketch of the graph of a function f on the interval I in each case: (a) I is the open interval $(0, 2)$ and f is continuous on I. At 1, f has a relative maximum value but $f'(1)$ does not exist. (b) I is the closed interval $[0, 2]$. The function f has a relative minimum value at 1, but the absolute minimum value of f is at 0. (c) I is the open interval $(0, 2)$, and f' has a relative maximum value at 1.

Exercises 91 through 94 pertain to Supplementary Section 4.10.

91. Use Newton's method to find to three decimal places the positive root of the equation $4x^4 - 3x^3 + 2x - 5 = 0$.

92. Use Newton's method to find to three decimal places the negative root of the equation $3x^4 - 4x^3 + 36x^2 + 2x - 8 = 0$.

93. Find to four decimal places by Newton's method the x coordinate of the point of intersection of the curve $y = \sin x$ and the line $y = 2x - 3$.

94. Find to four decimal places the value of x in the interval $(\frac{1}{2}\pi, \frac{3}{2}\pi)$ for which $\tan x = x$ by applying Newton's method.

F I V E

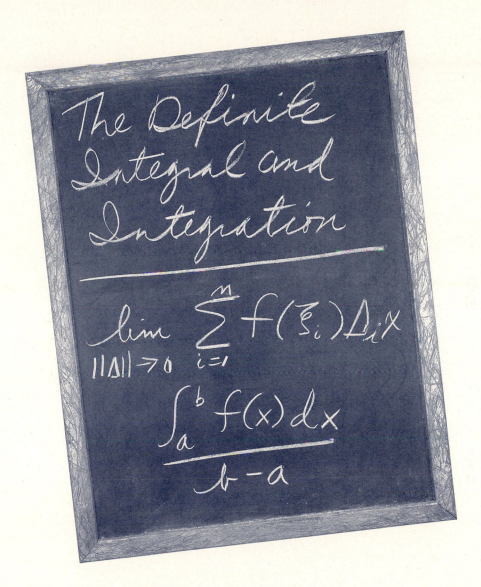

The Definite
Integral and
Integration

$$\lim_{\|\Delta\| \to 0} \sum_{i=1}^{n} f(\xi_i) \Delta_i x$$

$$\frac{\int_a^b f(x)\,dx}{b-a}$$

Antidifferentiation, the inverse operation of differentiation, is treated in the first two sections of this chapter in preparation for its use later in integration. In Section 5.3 we use antidifferentiation to solve *separable differential equations* with applications to rectilinear motion.

We define the area of a plane region in Section 5.4 as a new kind of limit, and in Section 5.5 a *definite integral* is defined in terms of this limit. Sections 5.6 and 5.7 are devoted to theorems giving properties of definite integrals. These properties are used in Section 5.8 to prove the *fundamental theorems of the calculus* that enable us to evaluate a definite integral by computing an antiderivative. In Section 5.9 we apply this powerful tool to compute areas of plane regions.

In the last section of the chapter we discuss numerical methods for finding an approximate value of a definite integral. Computations by these processes are easily done by programmable calculators and computers.

5.1 ANTIDIFFERENTIATION

You are already familiar with *inverse operations*. Addition and subtraction are inverse operations; multiplication and division are inverse operations, as are raising to powers and extracting roots. In this section we develop the inverse operation of differentiation called *antidifferentiation*.

5.1.1 DEFINITION

A function F is called an **antiderivative** of a function f on an interval I if $F'(x) = f(x)$ for every value of x in I.

▶ **ILLUSTRATION 1**   If F is defined by

$$F(x) = 4x^3 + x^2 + 5$$

then $F'(x) = 12x^2 + 2x$. Thus if f is the function defined by

$$f(x) = 12x^2 + 2x$$

we state that f is the derivative of F and that F is an antiderivative of f. If G is the function defined by

$$G(x) = 4x^3 + x^2 - 17$$

then G is also an antiderivative of f, because $G'(x) = 12x^2 + 2x$. Actually, any function whose function value is given by $4x^3 + x^2 + C$, where C is any constant, is an antiderivative of f.   ◀

In general, if a function F is an antiderivative of a function f on an interval I, and if G is defined by

$$G(x) = F(x) + C$$

where C is an arbitrary constant, then

$$G'(x) = F'(x)$$
$$= f(x)$$

and G is also an antiderivative of f on the interval I.

We now proceed to prove that if F is any particular antiderivative of f on an interval I, then every antiderivative of f on I is given by $F(x) + C$, where C is an arbitrary constant. First, a preliminary theorem is needed.

5.1.2 THEOREM

If f and g are two functions such that $f'(x) = g'(x)$ for all x in the interval I, then there is a constant K such that

$$f(x) = g(x) + K \qquad \text{for all } x \text{ in } I$$

Proof   Let h be the function defined on I by

$$h(x) = f(x) - g(x)$$

so that for all x in I,

$$h'(x) = f'(x) - g'(x)$$

But, by hypothesis, $f'(x) = g'(x)$ for all x in I. Therefore

$$h'(x) = 0 \qquad \text{for all } x \text{ in } I$$

Thus Theorem 4.3.3 applies to the function h, and there is a constant K such that

$$h(x) = K \qquad \text{for all } x \text{ in } I$$

Replacing $h(x)$ by $f(x) - g(x)$ we have

$$f(x) = g(x) + K \qquad \text{for all } x \text{ in } I$$

and the theorem is proved. ∎

The next theorem follows immediately from Theorem 5.1.2.

5.1.3 THEOREM If F is a particular antiderivative of f on an interval I, then every antiderivative of f on I is given by

$$F(x) + C \tag{1}$$

where C is an arbitrary constant, and all antiderivatives of f on I can be obtained from (1) by assigning particular values to C.

Proof Let G represent any antiderivative of f on I; then

$$G'(x) = f(x) \qquad \text{for all } x \text{ in } I \tag{2}$$

Because F is a particular antiderivative of f on I,

$$F'(x) = f(x) \qquad \text{for all } x \text{ in } I \tag{3}$$

From (2) and (3) it follows that

$$G'(x) = F'(x) \qquad \text{for all } x \text{ in } I$$

Therefore, from Theorem 5.1.2 there is a constant K such that

$$G(x) = F(x) + K \qquad \text{for all } x \text{ in } I$$

Because G represents any antiderivative of f on I, it follows that all antiderivatives of f can be obtained from $F(x) + C$, where C is an arbitrary constant. Hence the theorem is proved. ∎

Antidifferentiation is the process of finding the set of all antiderivatives of a given function. The symbol $\int$ denotes the operation of antidifferentiation, and we write

$$\int f(x)\, dx = F(x) + C \tag{4}$$

where

$$F'(x) = f(x)$$

and

$$d(F(x)) = f(x)\, dx \tag{5}$$

Leibniz introduced the convention of writing the differential of a function after the antidifferentiation symbol. The advantage of using the differential in this manner will be apparent to you when we compute antiderivatives by changing the variable in Section 5.2. From (4) and (5) we can write

$$\int d(F(x)) = F(x) + C$$

This equation will be used to obtain formulas for antidifferentiation in the sections that follow; the equation states that when we antidifferentiate the differential of a function, we obtain that function plus an arbitrary constant. So we can think of the $\int$ symbol for antidifferentiation as meaning that operation which is the inverse of the operation denoted by d for computing a differential.

If $\{F(x) + C\}$ is the set of all functions whose differentials are $f(x)\, dx$, it is also the set of all functions whose derivatives are $f(x)$. Therefore antidifferentiation is considered as the operation of finding the set of all functions having a given derivative.

Because antidifferentiation is the inverse operation of differentiation, antidifferentiation theorems can be obtained from differentiation theorems. Thus the following theorems can be proved from the corresponding ones for differentiation.

5.1.4 THEOREM

$$\int dx = x + C$$

5.1.5 THEOREM

$$\int af(x)\, dx = a \int f(x)\, dx$$

where a is a constant.

Theorem 5.1.5 states that to find an antiderivative of a constant times a function, first find an antiderivative of the function and then multiply it by the constant.

5.1.6 THEOREM   If f_1 and f_2 are defined on the same interval, then

$$\int [f_1(x) + f_2(x)]\, dx = \int f_1(x)\, dx + \int f_2(x)\, dx$$

Theorem 5.1.6 states that to find an antiderivative of the sum of two functions, find an antiderivative of each of the functions separately and then add the results, it being understood that both functions are defined on the same interval. Theorem 5.1.6 can be extended to any finite number of functions. Combining Theorem 5.1.6 with Theorem 5.1.5 we have the following one.

5.1.7 THEOREM   If $f_1, f_2, \ldots, f_n$ are defined on the same interval,

$$\int [c_1 f_1(x) + c_2 f_2(x) + \ldots + c_n f_n(x)]\, dx$$
$$= c_1 \int f_1(x)\, dx + c_2 \int f_2(x)\, dx + \ldots + c_n \int f_n(x)\, dx$$

where $c_1, c_2, \ldots, c_n$ are constants.

5.1.8 THEOREM If n is a rational number,

$$\int x^n \, dx = \frac{x^{n+1}}{n+1} + C \qquad n \neq -1$$

Proof

$$D_x\left(\frac{x^{n+1}}{n+1}\right) = \frac{(n+1)x^n}{n+1}$$

$$= x^n \qquad \blacksquare$$

▶ **ILLUSTRATION 2** Applying Theorem 5.1.8 for particular values of n we have

$$\int x^2 \, dx = \frac{x^3}{3} + C \qquad \qquad \int x^3 \, dx = \frac{x^4}{4} + C$$

$$\int \frac{1}{x^2} \, dx = \int x^{-2} \, dx \qquad \int \sqrt[3]{x} \, dx = \int x^{1/3} \, dx$$

$$= \frac{x^{-2+1}}{-2+1} + C \qquad = \frac{x^{1/3+1}}{\frac{1}{3}+1} + C$$

$$= \frac{x^{-1}}{-1} + C \qquad = \frac{x^{4/3}}{\frac{4}{3}} + C$$

$$= -\frac{1}{x} + C \qquad = \tfrac{3}{4}x^{4/3} + C$$

◀

The next illustration shows how Theorems 5.1.4 through 5.1.8 are used to antidifferentiate.

▶ **ILLUSTRATION 3**

$$\int (3x + 5) \, dx = \int 3x \, dx + \int 5 \, dx \qquad \text{(by Theorem 5.1.6)}$$

$$= 3 \int x \, dx + 5 \int dx \qquad \text{(by Theorem 5.1.5)}$$

$$= 3\left(\frac{x^2}{2} + C_1\right) + 5(x + C_2) \qquad \text{(by Theorems 5.1.8 and 5.1.4)}$$

$$= \tfrac{3}{2}x^2 + 5x + (3C_1 + 5C_2)$$

Because $3C_1 + 5C_2$ is an arbitrary constant, it may be denoted by C; so the result can be written as

$$\tfrac{3}{2}x^2 + 5x + C$$

The answer can be checked by finding its derivative.

$$D_x(\tfrac{3}{2}x^2 + 5x + C) = 3x + 5$$

◀

EXAMPLE 1　　Evaluate

$$\int (5x^4 - 8x^3 + 9x^2 - 2x + 7)\, dx$$

Solution

$$\int (5x^4 - 8x^3 + 9x^2 - 2x + 7)\, dx$$

$$= 5 \int x^4\, dx - 8 \int x^3\, dx + 9 \int x^2\, dx - 2 \int x\, dx + 7 \int dx$$

$$= 5 \cdot \frac{x^5}{5} - 8 \cdot \frac{x^4}{4} + 9 \cdot \frac{x^3}{3} - 2 \cdot \frac{x^2}{2} + 7x + C$$

$$= x^5 - 2x^4 + 3x^3 - x^2 + 7x + C$$

EXAMPLE 2　　Evaluate

$$\int \sqrt{x}\left(x + \frac{1}{x}\right) dx$$

Solution

$$\int \sqrt{x}\left(x + \frac{1}{x}\right) dx = \int x^{1/2}(x + x^{-1})\, dx$$

$$= \int (x^{3/2} + x^{-1/2})\, dx$$

$$= \frac{x^{5/2}}{\frac{5}{2}} + \frac{x^{1/2}}{\frac{1}{2}} + C$$

$$= \tfrac{2}{5}x^{5/2} + 2x^{1/2} + C$$

EXAMPLE 3　　Evaluate

$$\int \frac{5t^2 + 7}{t^{4/3}}\, dt$$

Solution

$$\int \frac{5t^2 + 7}{t^{4/3}}\, dt = 5 \int \frac{t^2}{t^{4/3}}\, dt + 7 \int \frac{1}{t^{4/3}}\, dt$$

$$= 5 \int t^{2/3}\, dt + 7 \int t^{-4/3}\, dt$$

$$= 5\left(\frac{t^{5/3}}{\frac{5}{3}}\right) + 7\left(\frac{t^{-1/3}}{-\frac{1}{3}}\right) + C$$

$$= 5(\tfrac{3}{5}t^{5/3}) + 7(-3t^{-1/3}) + C$$

$$= 3t^{5/3} - \frac{21}{t^{1/3}} + C$$

The theorems for the antiderivative of the sine and cosine functions follow immediately from the corresponding ones for differentiation.

5.1.9 THEOREM

$$\int \sin x \, dx = -\cos x + C$$

Proof $D_x(-\cos x) = -(-\sin x)$
$$= \sin x$$ ∎

5.1.10 THEOREM

$$\int \cos x \, dx = \sin x + C$$

Proof $D_x(\sin x) = \cos x$ ∎

The following theorems are consequences of the theorems for the derivatives of the tangent, cotangent, secant, and cosecant functions. The proofs are again immediate by finding the derivatives of the right-hand side of the equations.

5.1.11 THEOREM

$$\int \sec^2 x \, dx = \tan x + C$$

5.1.12 THEOREM

$$\int \csc^2 x \, dx = -\cot x + C$$

5.1.13 THEOREM

$$\int \sec x \tan x \, dx = \sec x + C$$

5.1.14 THEOREM

$$\int \csc x \cot x \, dx = -\csc x + C$$

EXAMPLE 4 Evaluate

$$\int (3 \sec x \tan x - 5 \csc^2 x) \, dx$$

Solution We apply Theorems 5.1.13 and 5.1.12.

$$\int (3 \sec x \tan x - 5 \csc^2 x) \, dx = 3 \int \sec x \tan x \, dx - 5 \int \csc^2 x \, dx$$
$$= 3 \sec x - 5(-\cot x) + C$$
$$= 3 \sec x + 5 \cot x + C$$

Trigonometric identities are often used when computing antiderivatives involving trigonometric functions. The following eight fundamental identities are crucial:

$$\sin x \csc x = 1 \qquad \cos x \sec x = 1 \qquad \tan x \cot x = 1$$

$$\tan x = \frac{\sin x}{\cos x} \qquad \cot x = \frac{\cos x}{\sin x}$$

$$\sin^2 x + \cos^2 x = 1 \qquad \tan^2 x + 1 = \sec^2 x \qquad \cot^2 x + 1 = \csc^2 x$$

EXAMPLE 5   Evaluate

$$\int \frac{2 \cot x - 3 \sin^2 x}{\sin x} \, dx$$

Solution

$$\int \frac{2 \cot x - 3 \sin^2 x}{\sin x} \, dx$$

$$= 2 \int \frac{1}{\sin x} \cdot \cot x \, dx - 3 \int \frac{\sin^2 x}{\sin x} \, dx$$

$$= 2 \int \csc x \cot x \, dx - 3 \int \sin x \, dx$$

$$= 2(-\csc x) - 3(-\cos x) + C \qquad \text{(from Theorems 5.1.14 and 5.1.9)}$$

$$= -2 \csc x + 3 \cos x + C$$

EXAMPLE 6   Evaluate

$$\int (\tan^2 x + \cot^2 x + 4) \, dx$$

Solution

$$\int (\tan^2 x + \cot^2 x + 4) \, dx$$

$$= \int [(\sec^2 x - 1) + (\csc^2 x - 1) + 4] \, dx$$

$$= \int \sec^2 x \, dx + \int \csc^2 x \, dx + 2 \int dx$$

$$= \tan x - \cot x + 2x + C \qquad \text{(from Theorems 5.1.11 and 5.1.12)}$$

Often in applications involving antidifferentiation it is desired to find a particular antiderivative that satisfies certain conditions called **initial** or **boundary conditions**, depending on whether they occur at one or more than one point. For example, if an equation involving $\dfrac{dy}{dx}$ is given as well as the initial condition that $y = y_1$ when $x = x_1$, then after the set of all antiderivatives is found, if x and y are replaced by x_1 and y_1, a particular value of the arbitrary constant C is determined. With this value of C a particular antiderivative is obtained.

▶ **ILLUSTRATION 4**   Suppose we wish to find the particular antiderivative satisfying the equation

$$\frac{dy}{dx} = 2x$$

and the initial condition that $y = 6$ when $x = 2$. From the equation

$$y = \int 2x \, dx$$

$$y = x^2 + C \tag{6}$$

In (6) we substitute 2 for x and 6 for y and get

$$6 = 4 + C$$

$$C = 2$$

When this value of C is substituted back in (6), we obtain

$$y = x^2 + 2$$

which gives the particular antiderivative desired.   ◀

EXAMPLE 7   At any point (x, y) on a particular curve the tangent line has a slope equal to $4x - 5$. If the curve contains the point $(3, 7)$, find its equation.

Solution   Because the slope of the tangent line to a curve at any point (x, y) is the value of the derivative at that point, we have

$$\frac{dy}{dx} = 4x - 5$$

$$y = \int (4x - 5)\, dx$$

$$y = 4\left(\frac{x^2}{2}\right) - 5x + C$$

$$y = 2x^2 - 5x + C \tag{7}$$

Equation (7) represents a *family* of curves. Because we wish to determine the particular curve of this family that contains the point $(3, 7)$, we substitute 3 for x and 7 for y in (7) and get

$$7 = 2(9) - 5(3) + C$$

$$7 = 18 - 15 + C$$

$$C = 4$$

Replacing C by 4 in (7) we get the required equation, which is

$$y = 2x^2 - 5x + 4$$

In Section 3.4 we introduced the marginal cost and marginal revenue functions from economics. They are the first derivatives, C' and R' of the total cost function C and the total revenue function R, respectively. Thus C and R can be obtained from C' and R' by antidifferentiation. When finding the function C from C', the arbitrary constant can be evaluated if we know the overhead cost (i.e., the cost when no units are produced) or the cost of production of a specific number of units of the commodity. Because it is generally true that the total revenue function is zero when the number of units produced is zero, this fact may be used to evaluate the arbitrary constant when finding the function R from R'.

EXAMPLE 8   The marginal cost function C' is given by

$$C'(x) = 4x - 8$$

where $C(x)$ dollars is the total cost of producing x units. If the cost of producing 5 units is \$20, find the total cost function.

Solution   Because $C'(x) = 4x - 8$

$$C(x) = \int (4x - 8)\, dx$$

$$= 2x^2 - 8x + k$$

Because $C(5) = 20$, we obtain $k = 10$. Hence

$$C(x) = 2x^2 - 8x + 10$$

Observe that because the marginal cost must be nonnegative, $4x - 8 \geq 0$ or, equivalently, $x \geq 2$. Therefore the domain of C is $[2, +\infty)$; remember that even though x represents the number of units of a commodity, we assume that x is a real number to give the continuity requirements for the functions C and C'.

EXERCISES 5.1

In Exercises 1 through 36, perform the antidifferentiation. In Exercises 1 through 8, 15 through 18, and 31 through 34, check by finding the derivative of your answer.

1. $\int 3x^4\, dx$

2. $\int 2x^7\, dx$

3. $\int \dfrac{1}{x^3}\, dx$

4. $\int \dfrac{3}{t^5}\, dt$

5. $\int 5u^{3/2}\, du$

6. $\int 10\sqrt[3]{x^2}\, dx$

7. $\int \dfrac{2}{\sqrt[3]{x}}\, dx$

8. $\int \dfrac{3}{\sqrt{y}}\, dy$

9. $\int 6t^2\sqrt[3]{t}\, dt$

10. $\int 7x^3\sqrt{x}\, dx$

11. $\int (4x^3 + x^2)\, dx$

12. $\int (3u^5 - 2u^3)\, du$

13. $\int y^3(2y^2 - 3)\, dy$

14. $\int x^4(5 - x^2)\, dx$

15. $\int (3 - 2t + t^2)\, dt$

16. $\int (4x^3 - 3x^2 + 6x - 1)\, dx$

17. $\int (8x^4 + 4x^3 - 6x^2 - 4x + 5)\, dx$

18. $\int (2 + 3x^2 - 8x^3)\, dx$

19. $\int \sqrt{x}(x + 1)\, dx$

20. $\int (ax^2 + bx + c)\, dx$

21. $\int (x^{3/2} - x)\, dx$

22. $\int \left(\sqrt{x} - \dfrac{1}{\sqrt{x}}\right) dx$

23. $\int \left(\dfrac{2}{x^3} + \dfrac{3}{x^2} + 5\right) dx$

24. $\int \left(3 - \dfrac{1}{x^4} + \dfrac{1}{x^2}\right) dx$

25. $\int \dfrac{x^2 + 4x - 4}{\sqrt{x}}\, dx$

26. $\int \dfrac{y^4 + 2y^2 - 1}{\sqrt{y}}\, dy$

27. $\int \left(\sqrt[3]{x} + \dfrac{1}{\sqrt[3]{x}}\right) dx$

28. $\int \dfrac{27t^3 - 1}{\sqrt[3]{t}}\, dt$

29. $\int (3 \sin t - 2 \cos t)\, dt$

30. $\int (5 \cos x - 4 \sin x)\, dx$

31. $\int \dfrac{\sin x}{\cos^2 x}\, dx$

32. $\int \dfrac{\cos x}{\sin^2 x}\, dx$

33. $\int (4 \csc x \cot x + 2 \sec^2 x)\, dx$

34. $\int (3 \csc^2 t - 5 \sec t \tan t)\, dt$

35. $\int (2 \cot^2 \theta - 3 \tan^2 \theta)\, d\theta$

36. $\int \dfrac{3 \tan \theta - 4 \cos^2 \theta}{\cos \theta}\, d\theta$

37. The point $(3, 2)$ is on a curve, and at any point (x, y) on the curve the tangent line has a slope equal to $2x - 3$. Find an equation of the curve.

38. The slope of the tangent line at any point (x, y) on a curve is $3\sqrt{x}$. If the point $(9, 4)$ is on the curve, find an equation of the curve.

39. The points $(-1, 3)$ and $(0, 2)$ are on a curve, and at any point (x, y) on the curve $\dfrac{d^2y}{dx^2} = 2 - 4x$. Find an equation of the curve. $\left(\text{Hint: Let } \dfrac{d^2y}{dx^2} = \dfrac{dy'}{dx}, \text{ and obtain an equation involving } y', x, \text{ and an arbitrary constant } C_1. \text{ From this equation obtain another equation involving } y, x, C_1, \text{ and } C_2.\right.$ Compute C_1 and C_2 from the conditions.$\Big)$

40. An equation of the tangent line to a curve at the point $(1, 3)$ is $y = x + 2$. If at any point (x, y) on the curve, $\dfrac{d^2 y}{dx^2} = 6x$, find an equation of the curve. See the hint for Exercise 39.

41. At any point (x, y) on a curve, $\dfrac{d^2 y}{dx^2} = 1 - x^2$, and an equation of the tangent line to the curve at the point $(1, 1)$ is $y = 2 - x$. Find an equation of the curve. See the hint for Exercise 39.

42. At any point (x, y) on a curve, $\dfrac{d^3 y}{dx^3} = 2$, and $(1, 3)$ is a point of inflection at which the slope of the inflectional tangent is -2. Find an equation of the curve.

43. The marginal cost function is given by $C'(x) = 3x^2 + 8x + 4$, and the overhead cost is $6. Find the total cost function.

44. A company has determined that the marginal cost function for the production of a particular commodity is given by $C'(x) = 125 + 10x + \frac{1}{9}x^2$, where $C(x)$ dollars is the total cost of producing x units of the commodity. If the overhead cost is $250, what is the cost of producing 15 units?

45. The marginal cost function is defined by $C'(x) = 6x$, where $C(x)$ is the number of hundreds of dollars in the total cost of x hundred units of a certain commodity. If the cost of 200 units is $2000, find (a) the total cost function and (b) the overhead cost.

46. The marginal revenue function for a certain commodity is $R'(x) = 12 - 3x$. If x units are demanded when p dollars is the price per unit, find (a) the total revenue function and (b) an equation involving p and x (the demand equation).

47. For a particular article of merchandise, the marginal revenue function is given by $R'(x) = 15 - 4x$. If x units are demanded when p dollars is the price per unit, find (a) the total revenue function and (b) an equation involving p and x (the demand equation).

48. The efficiency of a factory worker is expressed as a percent. For instance, if the worker's efficiency at a particular time is given as 70 percent, then the worker is performing at 70 percent of her full potential. Suppose that E percent is a factory worker's efficiency t hours after beginning work, and the rate at which E is changing is $(35 - 8t)$ percent per hour. If the worker's efficiency is 81 percent after working 3 hr, find her efficiency after working (a) 4 hr and (b) 8 hr.

49. The volume of water in a tank is V cubic meters when the depth of the water is h meters. If the rate of change of V with respect to h is $\pi(4h^2 + 12h + 9)$, find the volume of water in the tank when the depth is 3 m.

50. An art collector purchased for $1000 a painting by an artist whose works are currently increasing in value with respect to time according to the formula $\dfrac{dV}{dt} = 5t^{3/2} + 10t + 50$, where V dollars is the anticipated value of a painting t years after its purchase. If this formula were valid for the next 6 years, what would be the anticipated value of the painting 4 years from now?

51. Let $f(x) = 1$ for all x in $(-1, 1)$, and let

$$g(x) = \begin{cases} -1 & \text{if } -1 < x \le 0 \\ 1 & \text{if } 0 < x < 1 \end{cases}$$

Then $f'(x) = 0$ for all x in $(-1, 1)$ and $g'(x) = 0$ whenever g' exists in $(-1, 1)$. However, $f(x) \ne g(x) + K$ for x in $(-1, 1)$. Why doesn't Theorem 5.1.2 apply?

52. Let

$$f(x) = \begin{cases} -1 & \text{if } x < 0 \\ 0 & \text{if } x = 0 \\ 1 & \text{if } 0 < x \end{cases}$$

and $F(x) = |x|$. Show that $F'(x) = f(x)$ if $x \ne 0$. Is F an antiderivative of f on $(-\infty, +\infty)$? Explain.

53. Let $f(x) = |x|$ and F be defined by

$$F(x) = \begin{cases} -\frac{1}{2}x^2 & \text{if } x < 0 \\ \frac{1}{2}x^2 & \text{if } 0 \le x \end{cases}$$

Show that F is an antiderivative of f on $(-\infty, +\infty)$.

54. Let

$$U(x) = \begin{cases} 0 & \text{if } x < 0 \\ 1 & \text{if } 0 \le x \end{cases}$$

Show that U does not have an antiderivative on $(-\infty, +\infty)$. (*Hint:* Assume that U has an antiderivative F on $(-\infty, +\infty)$, and a contradiction is obtained by showing that it follows from the mean-value theorem that there exists a number k such that $F(x) = x + k$ if $x > 0$, and $F(x) = k$ if $x < 0$.)

5.2 SOME TECHNIQUES OF ANTIDIFFERENTIATION

Many antiderivatives cannot be found directly by applying the theorems of Section 5.1. It is therefore necessary to learn certain techniques that can be used in the computation of such antiderivatives. In this section we discuss techniques requiring the *chain rule for antidifferentiation* and those involving a change of variable.

▶ **ILLUSTRATION 1** To differentiate $\frac{1}{10}(1 + x^2)^{10}$ we apply the chain rule for differentiation and obtain

$$D_x\left[\tfrac{1}{10}(1 + x^2)^{10}\right] = (1 + x^2)^9(2x)$$

Suppose we wish to antidifferentiate $(1 + x^2)^9(2x)$. Then we need to compute

$$\int (1 + x^2)^9(2x \, dx) \qquad (1)$$

To lead up to a procedure that can be used in such a situation, let

$$g(x) = 1 + x^2 \qquad g'(x) \, dx = 2x \, dx \qquad (2)$$

Then (1) can be written as

$$\int [g(x)]^9[g'(x) \, dx] \qquad (3)$$

From Theorem 5.1.8,

$$\int u^9 \, du = \tfrac{1}{10}u^{10} + C \qquad (4)$$

Observe that (3) is of the same form as the left-hand side of (4). Thus

$$\int [g(x)]^9[g'(x) \, dx] = \tfrac{1}{10}[g(x)]^{10} + C$$

and with $g(x)$ and $g'(x) \, dx$ given in (2) we have

$$\int (1 + x^2)^9(2x \, dx) = \tfrac{1}{10}(1 + x^2)^{10} + C \qquad \blacktriangleleft$$

Justification of the procedure used to obtain the result of Illustration 1 is provided by the following theorem, which is analogous to the chain rule for differentiation and is called the *chain rule for antidifferentiation*.

5.2.1 THEOREM
Chain Rule for
Antidifferentiation

Let g be a differentiable function, and let the range of g be an interval I. Suppose that f is a function defined on I and that F is an antiderivative of f on I. Then

$$\int f(g(x))[g'(x) \, dx] = F(g(x)) + C$$

Proof By hypothesis,

$$F'(g(x)) = f(g(x)) \qquad (5)$$

By the chain rule for differentiation,

$$D_x[F(g(x))] = F'(g(x))[g'(x)]$$

Substituting from (5) in this equation we get

$$D_x[F(g(x))] = f(g(x))[g'(x)]$$

from which it follows that

$$\int f(g(x))[g'(x) \, dx] = F(g(x)) + C$$

which is what we wished to prove. ∎

As a particular case of Theorem 5.2.1, from Theorem 5.1.8, we have the generalized power formula for antiderivatives, which we now state.

5.2.2 THEOREM

If g is a differentiable function, and n is a rational number,

$$\int [g(x)]^n[g'(x) \, dx] = \frac{[g(x)]^{n+1}}{n+1} + C \qquad n \neq -1$$

EXAMPLE 1   Evaluate

$$\int \sqrt{3x + 4}\, dx$$

Solution   To apply Theorem 5.2.2 we first write

$$\int \sqrt{3x + 4}\, dx = \int (3x + 4)^{1/2}\, dx$$

We observe that if

$$g(x) = 3x + 4 \quad \text{then} \quad g'(x)\, dx = 3\, dx \tag{6}$$

Therefore we need a factor of 3 to go with dx to give $g'(x)\, dx$. Hence we write

$$\int (3x + 4)^{1/2}\, dx = \int (3x + 4)^{1/2}\tfrac{1}{3}(3\, dx)$$

$$= \tfrac{1}{3} \int (3x + 4)^{1/2}(3\, dx)$$

Thus from Theorem 5.2.2, with $g(x)$ and $g'(x)\, dx$ given in (6), we have

$$\frac{1}{3} \int (3x + 4)^{1/2}(3\, dx) = \frac{1}{3} \cdot \frac{(3x + 4)^{3/2}}{\frac{3}{2}} + C$$

$$= \tfrac{2}{9}(3x + 4)^{3/2} + C$$

EXAMPLE 2   Evaluate

$$\int x^2(5 + 2x^3)^8\, dx$$

Solution   Observe that if

$$g(x) = 5 + 2x^3 \quad \text{then} \quad g'(x)\, dx = 6x^2\, dx \tag{7}$$

Because

$$\int x^2(5 + 2x^3)^8\, dx = \int (5 + 2x^3)^8(x^2\, dx)$$

we need a factor of 6 to go with $x^2\, dx$ to give $g'(x)\, dx$. Therefore we write

$$\int x^2(5 + 2x^3)^8\, dx = \tfrac{1}{6} \int (5 + 2x^3)^8(6x^2\, dx)$$

Applying Theorem 5.2.2 with $g(x)$ and $g'(x)\, dx$ given in (7) we get

$$\frac{1}{6} \int (5 + 2x^3)^8(6x^2\, dx) = \frac{1}{6} \cdot \frac{(5 + 2x^3)^9}{9} + C$$

$$= \tfrac{1}{54}(5 + 2x^3)^9 + C$$

The chain rule for antidifferentiation (Theorem 5.2.1) is

$$\int f(g(x))[g'(x)\, dx] = F(g(x)) + C$$

where F is an antiderivative of f. If in this formula f is the cosine function, then F is the sine function and we have

$$\int \cos(g(x))[g'(x)\, dx] = \sin(g(x)) + C \tag{8}$$

We use this formula in the next example.

EXAMPLE 3   Evaluate

$$\int x \cos x^2 \, dx$$

Solution   If

$$g(x) = x^2 \quad \text{then} \quad g'(x) \, dx = 2x \, dx \tag{9}$$

Because

$$\int x \cos x^2 \, dx = \int (\cos x^2)(x \, dx)$$

we need a factor of 2 to go with $x \, dx$ to give $g'(x) \, dx$. Thus we write

$$\int x \cos x^2 \, dx = \tfrac{1}{2} \int (\cos x^2)(2x \, dx)$$

We apply (8) with $g(x)$ and $g'(x) \, dx$ given in (9), and we obtain

$$\tfrac{1}{2} \int (\cos x^2)(2x \, dx) = \tfrac{1}{2} \sin x^2 + C$$

The details of the solutions of Examples 1, 2, and 3 can be shortened by not specifically stating $g(x)$ and $g'(x) \, dx$. The solution of Example 1 then takes the following form:

$$\int \sqrt{3x + 4} \, dx = \frac{1}{3} \int (3x + 4)^{1/2} (3 \, dx)$$

$$= \frac{1}{3} \cdot \frac{(3x + 4)^{3/2}}{\frac{3}{2}} + C$$

$$= \tfrac{2}{9}(3x + 4)^{3/2} + C$$

The solution of Example 2 can be written as

$$\int x^2(5 + 2x^3)^8 \, dx = \frac{1}{6} \int (5 + 2x^3)^8 (6x^2 \, dx)$$

$$= \frac{1}{6} \cdot \frac{(5 + 2x^3)^9}{9} + C$$

$$= \tfrac{1}{54}(5 + 2x^3)^9 + C$$

and the solution of Example 3 can be shortened as follows:

$$\int x \cos x^2 \, dx = \tfrac{1}{2} \int (\cos x^2)(2x \, dx)$$

$$= \tfrac{1}{2} \sin x^2 + C$$

EXAMPLE 4   Evaluate

$$\int \frac{4x^2 \, dx}{(1 - 8x^3)^4}$$

Solution Because $d(1 - 8x^3) = -24x^2 \, dx$, we write

$$\int \frac{4x^2 \, dx}{(1 - 8x^3)^4} = 4 \int (1 - 8x^3)^{-4}(x^2 \, dx)$$

$$= 4\left(-\frac{1}{24}\right) \int (1 - 8x^3)^{-4}(-24x^2 \, dx)$$

$$= -\frac{1}{6} \cdot \frac{(1 - 8x^3)^{-3}}{-3} + C$$

$$= \frac{1}{18(1 - 8x^3)^3} + C$$

The results of each of the above examples can be checked by finding the derivative of the answer.

▶ **ILLUSTRATION 2** In Example 2 we have

$$\int x^2(5 + 2x^3)^8 \, dx = \tfrac{1}{54}(5 + 2x^3)^9 + C$$

Checking by differentiation gives

$$D_x[\tfrac{1}{54}(5 + 2x^3)^9] = \tfrac{1}{54} \cdot 9(5 + 2x^3)^8(6x^2)$$

$$= x^2(5 + 2x^3)^8 \qquad\qquad ◀$$

Sometimes it is possible to compute an antiderivative after a change of variable, as shown in the following example.

EXAMPLE 5 Evaluate

$$\int x^2 \sqrt{1 + x} \, dx$$

Solution Let

$$u = 1 + x \qquad du = dx \qquad x = u - 1$$

We have

$$\int x^2 \sqrt{1 + x} \, dx = \int (u - 1)^2 u^{1/2} \, du$$

$$= \int (u^2 - 2u + 1)u^{1/2} \, du$$

$$= \int u^{5/2} \, du - 2 \int u^{3/2} \, du + \int u^{1/2} \, du$$

$$= \frac{u^{7/2}}{\frac{7}{2}} - 2 \cdot \frac{u^{5/2}}{\frac{5}{2}} + \frac{u^{3/2}}{\frac{3}{2}} + C$$

$$= \tfrac{2}{7}(1 + x)^{7/2} - \tfrac{4}{5}(1 + x)^{5/2} + \tfrac{2}{3}(1 + x)^{3/2} + C$$

▶ **ILLUSTRATION 3** An alternate method for the solution of Example 5 is to let

$$v = \sqrt{1 + x} \qquad v^2 = 1 + x$$

$$x = v^2 - 1 \qquad dx = 2v \, dv$$

The Definite Integral and Integration

The computation then takes the following form:

$$\int x^2 \sqrt{1+x}\,dx = \int (v^2 - 1)^2 \cdot v \cdot (2v\,dv)$$

$$= 2\int v^6\,dv - 4\int v^4\,dv + 2\int v^2\,dv$$

$$= \tfrac{2}{7}v^7 - \tfrac{4}{5}v^5 + \tfrac{2}{3}v^3 + C$$

$$= \tfrac{2}{7}(1+x)^{7/2} - \tfrac{4}{5}(1+x)^{5/2} + \tfrac{2}{3}(1+x)^{3/2} + C$$

Checking by differentiation gives

$$D_x\left[\tfrac{2}{7}(1+x)^{7/2} - \tfrac{4}{5}(1+x)^{5/2} + \tfrac{2}{3}(1+x)^{3/2}\right]$$

$$= (1+x)^{5/2} - 2(1+x)^{3/2} + (1+x)^{1/2}$$

$$= (1+x)^{1/2}\left[(1+x)^2 - 2(1+x) + 1\right]$$

$$= (1+x)^{1/2}\left[1 + 2x + x^2 - 2 - 2x + 1\right]$$

$$= x^2\sqrt{1+x}$$

◀

EXAMPLE 6   Evaluate

$$\int \frac{\sin\sqrt{x}}{\sqrt{x}}\,dx$$

Solution   Let

$$u = \sqrt{x} \qquad du = \frac{1}{2\sqrt{x}}\,dx$$

Therefore

$$\int \frac{\sin\sqrt{x}}{\sqrt{x}} = 2\int \sin\sqrt{x}\left(\frac{1}{2\sqrt{x}}\,dx\right)$$

$$= 2\int \sin u\,du$$

$$= -2\cos u + C$$

$$= -2\cos\sqrt{x} + C$$

EXAMPLE 7   Evaluate

$$\int \sin x\sqrt{1 - \cos x}\,dx$$

Solution   Let

$$u = 1 - \cos x \qquad\qquad du = \sin x\,dx$$

Thus

$$\int \sin x\sqrt{1 - \cos x}\,dx = \int u^{1/2}\,du$$

$$= \tfrac{2}{3}u^{3/2} + C$$

$$= \tfrac{2}{3}(1 - \cos x)^{3/2} + C$$

EXAMPLE 8   Evaluate $\int \tan x \sec^2 x\, dx$ by two methods: (a) Let $u = \tan x$; (b) Let $v = \sec x$. (c) Explain the difference in appearance of the answers in (a) and (b).

Solution

(a) If $u = \tan x$, then $du = \sec^2 x\, dx$. We have

$$\int \tan x \sec^2 x\, dx = \int u\, du$$

$$= \frac{u^2}{2} + C$$

$$= \tfrac{1}{2}\tan^2 x + C$$

(b) If $v = \sec x$, then $dv = \sec x \tan x\, dx$. Thus

$$\int \tan x \sec^2 x\, dx = \int \sec x(\sec x \tan x\, dx)$$

$$= \int v\, dv$$

$$= \frac{v^2}{2} + C$$

$$= \tfrac{1}{2}\sec^2 x + C$$

(c) Because $\sec^2 x = 1 + \tan^2 x$, the functions defined by $\tfrac{1}{2}\tan^2 x$ and $\tfrac{1}{2}\sec^2 x$ differ by a constant; so each serves as an antiderivative of $\tan x \sec^2 x$. Furthermore we can write

$$\tfrac{1}{2}\sec^2 x + C = \tfrac{1}{2}(\tan^2 x + 1) + C$$

$$= \tfrac{1}{2}\tan^2 x + \tfrac{1}{2} + C$$

$$= \tfrac{1}{2}\tan^2 x + K \qquad \text{where } K = \tfrac{1}{2} + C$$

EXAMPLE 9   A wound is healing in such a way that t days since Monday the area of the wound has been decreasing at a rate of $-3(t + 2)^{-2}$ square centimeters per day. If on Tuesday the area of the wound was 2 cm^2, (a) what was the area of the wound on Monday, and (b) what is the anticipated area of the wound on Friday if it continues to heal at the same rate?

Solution   Let A square centimeters be the area of the wound t days since Monday. Then

$$\frac{dA}{dt} = -3(t + 2)^{-2}$$

$$A = -3 \int (t + 2)^{-2}\, dt$$

Because $d(t + 2) = dt$ we obtain

$$A = -3 \cdot \frac{(t + 2)^{-1}}{-1} + C$$

$$A = \frac{3}{t + 2} + C \tag{10}$$

Because on Tuesday the area of the wound was 2 cm^2, we know that $A = 2$ when $t = 1$. Substituting these values in (10) we obtain

$$2 = 1 + C$$

$$C = 1$$

Therefore from (10),

$$A = \frac{3}{t + 2} + 1 \tag{11}$$

(a) On Monday, $t = 0$. Let A_0 be the value of A when $t = 0$. From (11),

$$A_0 = \tfrac{3}{2} + 1$$
$$= \tfrac{5}{2}$$

Thus on Monday the area of the wound was 2.5 cm^2.

(b) On Friday, $t = 4$. Let A_4 be the value of A when $t = 4$. From (11),

$$A_4 = \tfrac{3}{6} + 1$$
$$= \tfrac{3}{2}$$

Hence on Friday the anticipated area of the wound is 1.5 cm^2.

EXERCISES 5.2

In Exercises 1 through 52, perform the antidifferentiation.

1. $\displaystyle\int \sqrt{1 - 4y}\, dy$

2. $\displaystyle\int \sqrt[3]{3x - 4}\, dx$

3. $\displaystyle\int \sqrt[3]{6 - 2x}\, dx$

4. $\displaystyle\int \sqrt{5r + 1}\, dr$

5. $\displaystyle\int x\sqrt{x^2 - 9}\, dx$

6. $\displaystyle\int 3x\sqrt{4 - x^2}\, dx$

7. $\displaystyle\int x^2(x^3 - 1)^{10}\, dx$

8. $\displaystyle\int x(2x^2 + 1)^6\, dx$

9. $\displaystyle\int 5x\sqrt[3]{(9 - 4x^2)^2}\, dx$

10. $\displaystyle\int \frac{x\, dx}{(x^2 + 1)^3}$

11. $\displaystyle\int \frac{y^3\, dy}{(1 - 2y^4)^5}$

12. $\displaystyle\int \frac{s\, ds}{\sqrt{3s^2 + 1}}$

13. $\displaystyle\int (x^2 - 4x + 4)^{4/3}\, dx$

14. $\displaystyle\int x^4\sqrt{3x^5 - 5}\, dx$

15. $\displaystyle\int x\sqrt{x + 2}\, dx$

16. $\displaystyle\int \frac{t\, dt}{\sqrt{t + 3}}$

17. $\displaystyle\int \frac{2r\, dr}{(1 - r)^7}$

18. $\displaystyle\int x^3(2 - x^2)^{12}\, dx$

19. $\displaystyle\int \sqrt{3 - 2x}\, x^2\, dx$

20. $\displaystyle\int (x^3 + 3)^{1/4} x^5\, dx$

21. $\displaystyle\int \cos 4\theta\, d\theta$

22. $\displaystyle\int \sin \tfrac{1}{3}x\, dx$

23. $\displaystyle\int 6x^2 \sin x^3\, dx$

24. $\displaystyle\int \tfrac{1}{2}t \cos 4t^2\, dt$

25. $\displaystyle\int \sec^2 5x\, dx$

26. $\displaystyle\int \csc^2 2\theta\, d\theta$

27. $\displaystyle\int y \csc 3y^2 \cot 3y^2\, dy$

28. $\displaystyle\int r^2 \sec^2 r^3\, dr$

29. $\displaystyle\int \cos x(2 + \sin x)^5\, dx$

30. $\displaystyle\int \frac{4 \sin x\, dx}{(1 + \cos x)^2}$

31. $\displaystyle\int \sqrt{1 + \frac{1}{3x}}\, \frac{dx}{x^2}$

32. $\displaystyle\int \sqrt{\frac{1}{t} - 1}\, \frac{dt}{t^2}$

33. $\displaystyle\int 2 \sin x \sqrt[3]{1 + \cos x}\, dx$

34. $\displaystyle\int \sin 2x\sqrt{2 - \cos 2x}\, dx$

35. $\displaystyle\int \cos^2 t \sin t\, dt$

36. $\displaystyle\int \sin^3 \theta \cos \theta\, d\theta$

37. $\displaystyle\int (\tan 2x + \cot 2x)^2\, dx$

38. $\displaystyle\int \frac{\tfrac{1}{2}\cos \tfrac{1}{4}x}{\sqrt{\sin \tfrac{1}{4}x}}\, dx$

39. $\displaystyle\int \frac{\cos 3x}{\sqrt{1 - 2 \sin 3x}}\, dx$

40. $\displaystyle\int \frac{\sec^2 3\sqrt{t}}{\sqrt{t}}\, dt$

41. $\displaystyle\int \frac{(x^2 + 2x)\, dx}{\sqrt{x^3 + 3x^2 + 1}}$

42. $\displaystyle\int x(x^2 + 1)\sqrt{4 - 2x^2 - x^4}\, dx$

43. $\displaystyle\int \frac{x(3x^2 + 1)\, dx}{(3x^4 + 2x^2 + 1)^2}$

44. $\displaystyle\int \sqrt{3 + s}(s + 1)^2\, ds$

45. $\displaystyle\int \frac{(y + 3)\, dy}{(3 - y)^{2/3}}$

46. $\displaystyle\int (2t^2 + 1)^{1/3} t^3\, dt$

47. $\displaystyle\int \frac{(r^{1/3} + 2)^4\, dr}{\sqrt[3]{r^2}}$

48. $\displaystyle\int \left(t + \frac{1}{t}\right)^{3/2} \left(\frac{t^2 - 1}{t^2}\right) dt$

49. $\displaystyle\int \frac{x^3\, dx}{(x^2 + 4)^{3/2}}$

50. $\displaystyle\int \frac{x^3\, dx}{\sqrt{1 - 2x^2}}$

51. $\displaystyle\int \sin x \sin(\cos x)\, dx$

52. $\displaystyle\int \sec x \tan x \cos(\sec x)\, dx$

53. Evaluate $\int (2x + 1)^3 \, dx$ by two methods: (a) Expand $(2x + 1)^3$ by the binomial theorem; (b) let $u = 2x + 1$. (c) Explain the difference in appearance of the answers obtained in (a) and (b).

54. Evaluate $\int x(x^2 + 2)^2 \, dx$ by two methods: (a) Expand $(x^2 + 2)^2$ and multiply the result by x; (b) let $u = x^2 + 2$. (c) Explain the difference in appearance of the answers obtained in (a) and (b).

55. Evaluate $\int \dfrac{(\sqrt{x} - 1)^2}{\sqrt{x}} \, dx$ by two methods: (a) Expand $(\sqrt{x} - 1)^2$ and multiply the result by $x^{-1/2}$; (b) let $u = \sqrt{x} - 1$. (c) Explain the difference in appearance of the answers obtained in (a) and (b).

56. Evaluate $\int \sqrt{x - 1} \, x^2 \, dx$ by two methods: (a) Let $u = x - 1$; (b) let $v = \sqrt{x - 1}$.

57. Evaluate $\int 2 \sin x \cos x \, dx$ by three methods: (a) Let $u = \sin x$; (b) let $v = \cos x$; (c) use the identity $2 \sin x \cos x = \sin 2x$. (d) Explain the difference in appearance of the answers obtained in (a), (b), and (c).

58. Evaluate $\int \csc^2 x \cot x \, dx$ by two methods: (a) Let $u = \cot x$; (b) let $v = \csc x$. (c) Explain the difference in appearance of the answers obtained in (a) and (b).

59. The marginal cost function for a particular article of merchandise is given by $C'(x) = 3(5x + 4)^{-1/2}$. If the overhead cost is $10, find the total cost function.

60. For a certain commodity the marginal cost function is given by $C'(x) = 3\sqrt{2x + 4}$. If the overhead cost is zero, find the total cost function.

61. If x units are demanded when p dollars is the price per unit, find an equation involving p and x (the demand equation) of a commodity for which the marginal revenue function is given by $R'(x) = 4 + 10(x + 5)^{-2}$.

62. The marginal revenue function for a particular article of merchandise is given by $R'(x) = ab(x + b)^{-2} - c$. Find (a) the total revenue function and (b) an equation involving p and x (the demand equation) where x units are demanded when p dollars is the price per unit.

63. If q coulombs is the charge of electricity received by a condenser from an electric circuit of i amperes at t seconds then $i = \dfrac{dq}{dt}$. If $i = 5 \sin 60t$ and $q = 0$ when $t = \frac{1}{2}\pi$, find the greatest positive charge on the condenser.

64. Do Exercise 63 if $i = 4 \cos 120t$ and $q = 0$ when $t = 0$.

65. The cost of a certain piece of machinery is $700, and its value is depreciating with time according to the formula $\dfrac{dV}{dt} = -500(t + 1)^{-2}$, where V dollars is its value t years after its purchase. What is its value 3 years after its purchase?

66. The volume of a balloon is increasing according to the formula $\dfrac{dV}{dt} = \sqrt{t + 1} + \frac{2}{3}t$, where V cubic centimeters is the volume of the balloon at t seconds. If $V = 33$ when $t = 3$, find (a) a formula for V in terms of t; (b) the volume of the balloon at 8 sec.

67. For the first 10 days in December a plant cell grew in such a way that t days after December 1 the volume of the cell was increasing at a rate of $(12 - t)^{-2}$ cubic micrometers per day. If on December 3 the volume of the cell was 3 μm^3, what was the volume on December 8?

68. The volume of water in a tank is V cubic meters when the depth of the water is h meters. If the rate of change of V with respect to h is given by $\dfrac{dV}{dh} = \pi(2h + 3)^2$, find the volume of water in the tank when the depth is 3 m.

5.3 DIFFERENTIAL EQUATIONS AND RECTILINEAR MOTION

An equation containing derivatives is called a **differential equation**. Some simple differential equations are

$$\frac{dy}{dx} = 2x \tag{1}$$

$$\frac{dy}{dx} = \frac{2x^2}{3y^3} \tag{2}$$

$$\frac{d^2y}{dx^2} = 4x + 3 \tag{3}$$

The **order** of a differential equation is the order of the derivative of highest order that appears in the equation. Therefore (1) and (2) are first-order differential equations and (3) is of the second order. The simplest type of differential equation is a first-order equation of the form

$$\frac{dy}{dx} = f(x)$$

for which (1) is a particular example. Writing this equation with differentials we have

$$dy = f(x)\, dx \tag{4}$$

Another type of differential equation of the first order is one of the form

$$\frac{dy}{dx} = \frac{g(x)}{h(y)}$$

Equation (2) is a particular example of an equation of this type. If this equation is written with differentials, we have

$$h(y)\, dy = g(x)\, dx \tag{5}$$

In both (4) and (5), the left side involves only the variable y and the right side involves only the variable x. Thus the variables are separated, and we say that these are **separable differential equations**.

Consider Equation (4), which is

$$dy = f(x)\, dx$$

To solve this equation we must find all functions G for which $y = G(x)$ such that the equation is satisfied. So if F is an antiderivative of f, all functions G are defined by $G(x) = F(x) + C$, where C is an arbitrary constant. That is, if

$$\begin{aligned} d(G(x)) &= d(F(x) + C) \\ &= f(x)\, dx \end{aligned}$$

then what is called the **complete solution** of (4) is given by

$$y = F(x) + C$$

This equation represents a family of functions depending on an arbitrary constant C and is called a **one-parameter family**. The graphs of these functions form a one-parameter family of curves in the plane, and through any particular point (x_1, y_1) there passes just one curve of the family.

▶ **ILLUSTRATION 1**   Suppose we wish to find the complete solution of the differential equation

$$\frac{dy}{dx} = 2x \tag{6}$$

We separate the variables by writing the equation with differentials as

$$dy = 2x\, dx$$

We antidifferentiate on both sides of the equation and obtain

$$\int dy = \int 2x\, dx$$

$$y + C_1 = x^2 + C_2$$

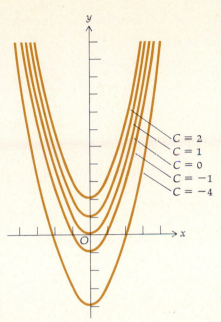

FIGURE 1

Because $C_2 - C_1$ is an arbitrary constant if C_2 and C_1 are arbitrary, we can replace $C_2 - C_1$ by C, thereby obtaining

$$y = x^2 + C \tag{7}$$

which is the complete solution of differential equation (6).

Equation (7) represents a one-parameter family of functions. Figure 1 shows sketches of the graphs of the functions corresponding to $C = -4$, $C = -1$, $C = 0$, $C = 1$, and $C = 2$.   ◀

Now consider (5), which is

$$h(y)\, dy = g(x)\, dx$$

If we antidifferentiate on both sides of the equation, we write

$$\int h(y)\, dy = \int g(x)\, dx$$

If H is an antiderivative of h, and G is an antiderivative of g, the complete solution of (5) is given by

$$H(y) = G(x) + C$$

EXAMPLE 1   Find the complete solution of the differential equation

$$\frac{dy}{dx} = \frac{2x^2}{3y^3}$$

Solution   If the given equation is written with differentials, we have

$$3y^3\, dy = 2x^2\, dx$$

and the variables are separated. We antidifferentiate on both sides of the equation and obtain

$$\int 3y^3\, dy = \int 2x^2\, dx$$

$$\frac{3y^4}{4} = \frac{2x^3}{3} + \frac{C}{12}$$

$$9y^4 = 8x^3 + C$$

which is the complete solution.

At first the arbitrary constant was written as $C/12$ so that when both sides of the equation are multiplied by 12, the arbitrary constant becomes C.

In the following illustration we show how a particular solution of a first-order differential equation is obtained when initial conditions are given.

▶ **ILLUSTRATION 2**   To find the particular solution of differential equation (6) satisfying the initial condition that $y = 6$ when $x = 2$, we substitute these values in (7) and solve for C, giving $6 = 4 + C$, or $C = 2$. Substituting this value of C in (7) we obtain

$$y = x^2 + 2$$

which is the particular solution desired.   ◀

Equation (3) is an example of a particular type of differential equation of the second order

$$\frac{d^2y}{dx^2} = f(x)$$

Two successive antidifferentiations are necessary to solve this equation, and two arbitrary constants occur in the complete solution. The complete solution therefore represents a **two-parameter family** of functions, and the graphs of these functions form a two-parameter family of curves in the plane. The following example shows the method of obtaining the complete solution of an equation of this kind.

EXAMPLE 2   Find the complete solution of the differential equation

$$\frac{d^2y}{dx^2} = 4x + 3$$

Solution   Because

$$\frac{d^2y}{dx^2} = \frac{d}{dx}\left(\frac{dy}{dx}\right)$$

and letting $y' = \dfrac{dy}{dx}$, we can write the given equation as

$$\frac{dy'}{dx} = 4x + 3$$

Thus we have, with differentials,

$$dy' = (4x + 3)\, dx$$

Antidifferentiating, we obtain

$$\int dy' = \int (4x + 3)\, dx$$
$$y' = 2x^2 + 3x + C_1$$

Because $y' = \dfrac{dy}{dx}$, we make this substitution in the above equation and get

$$\frac{dy}{dx} = 2x^2 + 3x + C_1$$
$$dy = (2x^2 + 3x + C_1)\, dx$$
$$\int dy = \int (2x^2 + 3x + C_1)\, dx$$
$$y = \tfrac{2}{3}x^3 + \tfrac{3}{2}x^2 + C_1x + C_2$$

which is the complete solution.

EXAMPLE 3   Find the particular solution of the differential equation in Example 2 for which $y = 2$ and $y' = -3$ when $x = 1$.

Solution   Because $y' = 2x^2 + 3x + C_1$, we substitute -3 for y' and 1 for x, giving $-3 = 2 + 3 + C_1$, or $C_1 = -8$. Substituting this value of C_1 into the complete solution gives

$$y = \tfrac{2}{3}x^3 + \tfrac{3}{2}x^2 - 8x + C_2$$

Because $y = 2$ when $x = 1$, we substitute these values in the above equation and get $2 = \tfrac{2}{3} + \tfrac{3}{2} - 8 + C_2$, from which we obtain $C_2 = \tfrac{47}{6}$. The particular solution desired, then, is

$$y = \tfrac{2}{3}x^3 + \tfrac{3}{2}x^2 - 8x + \tfrac{47}{6}$$

We learned in Sections 3.4 and 3.10 that in considering the motion of a particle along a line, when an equation of motion, $s = f(t)$, is given, then the instantaneous velocity and the instantaneous acceleration can be determined from the equations

$$v = \frac{ds}{dt} \qquad a = \frac{dv}{dt}$$

Therefore, if we are given v or a as a function of t, as well as some boundary conditions, it is possible to determine the equation of motion by solving a differential equation. This procedure is illustrated in the following example.

EXAMPLE 4   A particle is moving on a line; at t seconds, s feet is the directed distance of the particle from the origin, v feet per second is the velocity of the particle, and a feet per second per second is the acceleration of the particle. If

$$a = 2t - 1$$

and $v = 3$ and $s = 4$ when $t = 1$, express v and s as functions of t.

Solution   Because $a = \dfrac{dv}{dt}$, we have the differential equation

$$\frac{dv}{dt} = 2t - 1$$

$$dv = (2t - 1)\,dt$$

$$\int dv = \int (2t - 1)\,dt$$

$$v = t^2 - t + C_1 \tag{8}$$

Substituting $v = 3$ and $t = 1$ in (8) we have

$$3 = 1 - 1 + C_1$$

$$C_1 = 3$$

Substituting this value of C_1 in (8) we obtain

$$v = t^2 - t + 3$$

which expresses v as a function of t. Now, letting $v = \dfrac{ds}{dt}$ in this equation we have

$$\frac{ds}{dt} = t^2 - t + 3$$

$$ds = (t^2 - t + 3)\,dt$$

$$\int ds = \int (t^2 - t + 3)\,dt$$

$$s = \tfrac{1}{3}t^3 - \tfrac{1}{2}t^2 + 3t + C_2 \tag{9}$$

We substitute $s = 4$ and $t = 1$ in (9) and get

$$4 = \tfrac{1}{3} - \tfrac{1}{2} + 3 + C_2$$

$$C_2 = \tfrac{7}{6}$$

Replacing C_2 by $\tfrac{7}{6}$ in (9) we have expressed s as a function of t:

$$s = \tfrac{1}{3}t^3 - \tfrac{1}{2}t^2 + 3t + \tfrac{7}{6}$$

EXAMPLE 5   A particle is moving on a line where v centimeters per second is the velocity of the particle at t seconds and

$$v = \cos 2\pi t$$

If the positive direction is to the right of the origin and the particle is 5 cm to the right of the origin at the start of the motion, find its position $\tfrac{1}{3}$ sec later.

Solution   Let s centimeters be the directed distance of the particle from the origin at t seconds. Because $v = \dfrac{ds}{dt}$,

$$\frac{ds}{dt} = \cos 2\pi t$$

$$ds = \cos 2\pi t\,dt$$

$$\int ds = \int \cos 2\pi t\,dt$$

$$s = \frac{1}{2\pi} \int \cos 2\pi t(2\pi\,dt)$$

$$s = \frac{1}{2\pi} \sin 2\pi t + C$$

Because $s = 5$ when $t = 0$,

$$5 = \frac{1}{2\pi} \sin 0 + C$$

$$C = 5$$

Therefore the equation of motion is

$$s = \frac{1}{2\pi} \sin 2\pi t + 5$$

Let $s = \bar{s}$ when $t = \frac{1}{3}$. Then

$$\bar{s} = \frac{1}{2\pi} \sin \frac{2}{3} \pi + 5$$

$$= \frac{1}{2\pi} \cdot \frac{\sqrt{3}}{2} + 5$$

$$\approx 5.14$$

Thus the particle is 5.14 cm to the right of the origin $\frac{1}{3}$ sec after the start of the motion.

If an object is moving freely in a vertical line and is being pulled toward the earth by a force of gravity, the acceleration due to *gravity* varies with the distance of the object from the center of the earth. However, for small changes of distances the acceleration due to gravity is almost constant. If the object is near sea level, an approximate value of the acceleration due to gravity is 32 ft/sec² or 9.8 m/sec².

EXAMPLE 6   A stone is thrown vertically upward from the ground with an initial velocity of 128 ft/sec. If the only force considered is that attributed to the acceleration due to gravity, find (a) how long it will take for the stone to strike the ground, (b) the speed with which it will strike the ground, and (c) how high the stone will rise.

Solution   The motion of the stone is illustrated in Figure 2. The positive direction is upward.

Let t seconds be the time that has elapsed since the stone was thrown, s feet be the distance of the stone from the ground at t seconds, v feet per second be the velocity of the stone at t seconds, and $|v|$ feet per second be the speed of the stone at t seconds.

When the stone strikes the ground, $s = 0$. Let $\bar{t}$ and $\bar{v}$ be the particular values of t and v when $s = 0$ and $t \neq 0$. The stone will be at its highest point when the velocity is zero. Let $\bar{s}$ be the particular value of s when $v = 0$. Table 1 shows the boundary conditions.

Because the only acceleration is due to gravity, which is the downward direction, the acceleration has an approximate constant value of -32 ft/sec². Because the acceleration is given by $\dfrac{dv}{dt}$ we have

$$\frac{dv}{dt} = -32$$

$$dv = -32 \, dt$$

$$\int dv = -32 \int dt$$

$$v = -32t + C_1$$

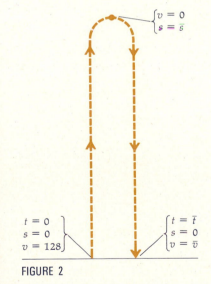

FIGURE 2

Table 1

t	s	v
0	0	128
	$\bar{s}$	0
$\bar{t}$	0	$\bar{v}$

Since $v = 128$ when $t = 0$, we substitute these values in the above equation and get $C_1 = 128$. Therefore

$$v = -32t + 128 \qquad (10)$$

Because $v = \dfrac{ds}{dt}$,

$$\frac{ds}{dt} = -32t + 128$$

$$ds = (-32t + 128)\,dt$$

$$\int ds = \int(-32t + 128)\,dt$$

$$s = -16t^2 + 128t + C_2$$

Since $s = 0$ when $t = 0$, then $C_2 = 0$, and substituting 0 for C_2 in the above equation gives

$$s = -16t^2 + 128t \qquad (11)$$

(a) In (11) we substitute $\bar{t}$ for t and 0 for s and get

$$0 = -16\bar{t}(\bar{t} - 8)$$

from which $\bar{t} = 0$ or $\bar{t} = 8$. However, the value 0 occurs when the stone is thrown; so it takes 8 sec for the stone to strike the ground.

(b) To obtain $\bar{v}$ we use (10) and substitute 8 for t and $\bar{v}$ for v to obtain

$$\bar{v} = -32(8) + 128$$

$$= -128$$

Thus $|\bar{v}| = 128$; so the stone strikes the ground with a speed of 128 ft/sec.

(c) To find $\bar{s}$ we first find the value of t for which $v = 0$. From (10), $t = 4$ when $v = 0$. In (11) we substitute 4 for t and $\bar{s}$ for s and get

$$\bar{s} = -16(16) + 128(4)$$

$$= 256$$

Therefore the stone will rise 256 ft.

EXERCISES 5.3

In Exercises 1 through 14, find the complete solution of the differential equation.

1. $\dfrac{dy}{dx} = 4x - 5$

2. $\dfrac{dy}{dx} = 6 - 3x^2$

3. $\dfrac{dy}{dx} = 3x^2 + 2x - 7$

4. $\dfrac{ds}{dt} = 5\sqrt{s}$

5. $\dfrac{dy}{dx} = 3xy^2$

6. $\dfrac{dy}{dx} = \dfrac{\sqrt{x} + x}{\sqrt{y} - y}$

7. $\dfrac{du}{dv} = \dfrac{3v\sqrt{1 + u^2}}{u}$

8. $\dfrac{dy}{dx} = \dfrac{x^2\sqrt{x^3 - 3}}{y^2}$

9. $\dfrac{dy}{dx} = \dfrac{\sec^2 x}{\tan^2 y}$

10. $\dfrac{du}{dv} = \dfrac{\cos 2v}{\sin 3u}$

11. $\dfrac{d^2y}{dx^2} = 5x^2 + 1$

12. $\dfrac{d^2y}{dx^2} = \sqrt{2x - 3}$

13. $\dfrac{d^2s}{dt^2} = \sin 3t + \cos 3t$

14. $\dfrac{d^2u}{dv^2} = \tan v \sec^2 v$

In Exercises 15 through 20, find the particular solution of the differential equation determined by the initial conditions.

15. $\dfrac{dy}{dx} = x^2 - 2x - 4$; $y = -6$ when $x = 3$

16. $\dfrac{dy}{dx} = (x + 1)(x + 2)$; $y = -\frac{3}{2}$ when $x = -3$

17. $\dfrac{dy}{dx} = \dfrac{\cos 3x}{\sin 2y}$; $y = \frac{1}{3}\pi$ when $x = \frac{1}{2}\pi$

18. $\dfrac{ds}{dt} = \cos \frac{1}{2}t$; $s = 3$ when $t = \frac{1}{3}\pi$

19. $\dfrac{d^2u}{dv^2} = 4(1 + 3v)^2$; $u = -1$ and $\dfrac{du}{dv} = -2$ when $v = -1$

20. $\dfrac{d^2y}{dx^2} = -\dfrac{3}{x^4}$; $y = \dfrac{1}{2}$ and $\dfrac{dy}{dx} = -1$ when $x = 1$

In Exercises 21 through 32, a particle is moving on a line; at t
seconds, s feet is the directed distance of the particle from the origin,
v feet per second is the velocity of the particle, and a feet per second
per second is the acceleration of the particle.

$\left(\text{Hint for Exercises 29 through 32: } a = \dfrac{dv}{dt} = \dfrac{dv}{ds}\dfrac{ds}{dt} = v\dfrac{dv}{ds}. \right)$

21. $v = \sqrt{2t + 4}$; $s = 0$ when $t = 0$. Express s in terms of t.
22. $v = 4 - t$; $s = 0$ when $t = 2$. Express s in terms of t.
23. $a = 5 - 2t$; $v = 2$ and $s = 0$ when $t = 0$. Express v and s in terms of t.
24. $a = 17$; $v = 0$ and $s = 0$ when $t = 0$. Express v and s in terms of t.
25. $a = t^2 + 2t$; $s = 1$ when $t = 0$ and $s = -3$ when $t = 2$. Express v and s in terms of t.
26. $a = 3t - t^2$; $v = \frac{7}{6}$ and $s = 1$ when $t = 1$. Express v and s in terms of t.
27. $a = -4\sqrt{2}\cos(2t - \frac{1}{4}\pi)$; $v = 2$ and $s = 1$ when $t = 0$. Express v and s in terms of t.
28. $a = 18\sin 3t$; $v = -6$ and $s = 4$ when $t = 0$. Express v and s in terms of t.
29. $a = 800$; $v = 20$ when $s = 1$. Find an equation involving v and s.
30. $a = 500$; $v = 10$ when $s = 5$. Find an equation involving v and s.
31. $a = 5s + 2$; $v = 4$ when $s = 2$. Find an equation involving v and s.
32. $a = 2s + 1$; $v = 2$ when $s = 1$. Find an equation involving v and s.

In Exercises 33 through 40, the only force considered is that due
to the acceleration of gravity, which we take as 32 ft/sec² in the
downward direction.

33. A stone is thrown vertically upward from the ground with
an initial velocity of 20 ft/sec. (a) How long will it take the
stone to strike the ground? (b) With what speed will the stone
strike the ground? (c) How long will the stone be going upward? (d) How high will the stone go?

34. A stone is thrown vertically upward from the ground with
an initial velocity of 24 ft/sec. (a) How long will it take the
stone to strike the ground, and (b) with what speed will the
stone strike the ground?

35. A ball is dropped from the top of the Washington Monument,
which is 555 ft high. (a) How long will it take the ball to
reach the ground, and (b) with what speed will it strike the
ground?

36. A stone is thrown vertically upward from the ground with
an initial velocity of 16 ft/sec. (a) How long will the stone be
going upward, and (b) how high will the stone go?

37. A woman in a balloon drops her binoculars when the balloon
is 150 ft above the ground and rising at the rate of 10 ft/sec.
(a) How long will it take the binoculars to strike the ground,
and (b) what is their speed on impact?

38. A ball is thrown downward from a window that is 80 ft above
the ground with an initial velocity of -64 ft/sec. (a) When
does the ball strike the ground, and (b) with what speed will
it strike the ground?

39. A ball is thrown vertically upward with an initial velocity of
40 ft/sec from a point 20 ft above the ground. (a) If v feet per
second is the velocity of the ball when it is s feet from the
starting point, express v in terms of s. (b) What is the velocity
of the ball when it is 36 ft from the ground and rising? (*Hint:*
See hint for Exercises 29 through 32.)

40. A stone is thrown vertically upward from the top of a house
60 ft above the ground with an initial velocity of 40 ft/sec.
(a) How long will it take the stone to reach its greatest height,
and (b) what is its greatest height? (c) How long will it take
the stone to pass the top of the house on its way down, and
(d) what is its velocity at that instant? (e) How long will it
take the stone to strike the ground, and (f) with what velocity
will it strike the ground?

41. A particle is moving along a straight line in such a way that
if v centimeters per second is the velocity of the particle at
t seconds, then $v = \sin \pi t$, where the positive direction is to
the right of the origin. If the particle is at the origin at the
start of the motion, find its position $\frac{2}{3}$ sec later.

42. If a ball is rolled across level ground with an initial velocity
of 20 ft/sec, and if the speed of the ball is decreasing at the
rate of 6 ft/sec² due to friction, how far will the ball roll?

43. If the driver of an automobile wishes to increase the speed
from 40 km/hr to 100 km/hr while traveling a distance of
200 m, what constant acceleration should be maintained?

44. What constant negative acceleration will enable a driver
to decrease the speed from 120 km/hr to 60 km/hr while
traveling a distance of 100 m?

45. If the brakes are applied on a car traveling 100 km/hr and
the brakes can give the car a constant negative acceleration
of 8 m/sec², (a) how long will it take the car to come to a
stop, and (b) how far will the car travel before stopping?

46. A ball started upward from the bottom of an inclined plane
with an initial velocity of 6 ft/sec. If there was a downward

acceleration of 4 ft/sec², how far up the plane did the ball go before starting to roll down?

47. If the brakes on a car can give the car a constant negative acceleration of 8 m/sec², what is the greatest speed it may be going if it is necessary to be able to stop the car within 25 m after the brake is applied?

48. A block of ice slides down a chute with a constant acceleration of 3 m/sec². The chute is 36 m long and it takes 4 sec for the ice to reach the bottom. (a) What is the initial velocity of the ice? (b) What is the speed of the ice after it has traveled 12 m? (c) How long does it take the ice to go the 12 m?

49. The equation $x^2 = 4ay$ represents a one-parameter family of parabolas. Find an equation of another one-parameter family of curves such that at any point (x, y) there is a curve of each family through it and the tangent lines to the two curves at this point are perpendicular. (*Hint:* First show that the slope of the tangent line at any point (x, y), not on the y axis, of the parabola of the given family through that point is $2y/x$.)

50. Solve Exercise 49 if the given one-parameter family of curves has the equation $x^3 + y^3 = a^3$.

5.4 AREA

You probably have an intuitive idea of what is meant by the *area* of certain geometrical figures. It is a measurement that in some way gives the size of the region enclosed by the figure. The area of a rectangle is the product of its length and width, and the area of a triangle is half the product of the lengths of the base and the altitude.

The area of a polygon can be defined as the sum of the areas of triangles into which it is decomposed, and it can be proved that the area thus obtained is independent of how the polygon is decomposed into triangles (see Figure 1). However, how do we define the area of a region in a plane if the region is bounded by a curve? In this section we give the definition of the area of such a region, and in Section 5.5 we use this definition to motivate the definition of the *definite integral*.

In our treatment of area we shall be concerned with the sums of many terms, and so a notation called the **sigma notation** is introduced to facilitate writing these sums. This notation involves the use of the symbol $\sum$, the capital sigma of the Greek alphabet. Some examples of the sigma notation are given in the following illustration.

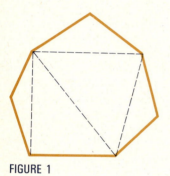

FIGURE 1

▶ **ILLUSTRATION 1**

$$\sum_{i=1}^{5} i^2 = 1^2 + 2^2 + 3^2 + 4^2 + 5^2$$

$$\sum_{i=-2}^{2} (3i + 2) = [3(-2) + 2] + [3(-1) + 2] + [3 \cdot 0 + 2] + [3 \cdot 1 + 2] + [3 \cdot 2 + 2]$$
$$= (-4) + (-1) + 2 + 5 + 8$$

$$\sum_{j=1}^{n} j^3 = 1^3 + 2^3 + 3^3 + \ldots + n^3$$

$$\sum_{k=3}^{8} \frac{1}{k} = \frac{1}{3} + \frac{1}{4} + \frac{1}{5} + \frac{1}{6} + \frac{1}{7} + \frac{1}{8} \qquad \blacktriangleleft$$

We have the formal definition of the sigma notation.

5.4.1 DEFINITION

$$\sum_{i=m}^{n} F(i) = F(m) + F(m + 1) + F(m + 2) + \cdots + F(n - 1) + F(n)$$

where m and n are integers, and $m \leq n$.

The right side of the equation in the definition consists of the sum of $(n - m + 1)$ terms, the first of which is obtained by replacing i by m in $F(i)$, the second by replacing i by $m + 1$ in $F(i)$, and so on, until the last term is obtained by replacing i by n in $F(i)$.

The number m is called the **lower limit** of the sum, and n is called the **upper limit**. The symbol i is called the **index of summation**. It is a "dummy" symbol because any other letter can be used for this purpose. For example,

$$\sum_{k=3}^{5} k^2 = 3^2 + 4^2 + 5^2$$

is equivalent to

$$\sum_{i=3}^{5} i^2 = 3^2 + 4^2 + 5^2$$

▶ **ILLUSTRATION 2** From Definition 5.4.1,

$$\sum_{i=3}^{6} \frac{i^2}{i + 1} = \frac{3^2}{3 + 1} + \frac{4^2}{4 + 1} + \frac{5^2}{5 + 1} + \frac{6^2}{6 + 1} \qquad \blacktriangleleft$$

Sometimes the terms of a sum involve subscripts, as shown in the next illustration.

▶ **ILLUSTRATION 3**

$$\sum_{i=1}^{n} A_i = A_1 + A_2 + \cdots + A_n$$

$$\sum_{k=4}^{9} k b_k = 4b_4 + 5b_5 + 6b_6 + 7b_7 + 8b_8 + 9b_9$$

$$\sum_{i=1}^{5} f(x_i)\,\Delta x = f(x_1)\,\Delta x + f(x_2)\,\Delta x + f(x_3)\,\Delta x + f(x_4)\,\Delta x + f(x_5)\,\Delta x \qquad \blacktriangleleft$$

The following theorems involving the sigma notation are useful for computation and are easily proved.

5.4.2 THEOREM

$$\sum_{i=1}^{n} c = cn, \text{ where } c \text{ is any constant}$$

Proof

$$\sum_{i=1}^{n} c = c + c + \cdots + c \qquad (n \text{ terms})$$
$$= cn \qquad \blacksquare$$

5.4.3 THEOREM

$$\sum_{i=1}^{n} c \cdot F(i) = c \sum_{i=1}^{n} F(i), \text{ where } c \text{ is any constant}$$

Proof

$$\sum_{i=1}^{n} c \cdot F(i) = c \cdot F(1) + c \cdot F(2) + c \cdot F(3) + \ldots + c \cdot F(n)$$
$$= c[F(1) + F(2) + F(3) + \ldots + F(n)]$$

$$= c \sum_{i=1}^{n} F(i) \qquad \blacksquare$$

5.4.4 THEOREM

$$\sum_{i=1}^{n} [F(i) + G(i)] = \sum_{i=1}^{n} F(i) + \sum_{i=1}^{n} G(i)$$

The proof is left as an exercise (see Exercise 59). Theorem 5.4.4 can be extended to the sum of any number of functions.

5.4.5 THEOREM

$$\sum_{i=a}^{b} F(i) = \sum_{i=a+c}^{b+c} F(i - c) \tag{1}$$

and

$$\sum_{i=a}^{b} F(i) = \sum_{i=a-c}^{b-c} F(i + c) \tag{2}$$

The proof of Theorem 5.4.5 is left as an exercise (see Exercise 60). The following illustration shows the application of this theorem.

▶ **ILLUSTRATION 4**   From Equation (1) of Theorem 5.4.5,

$$\sum_{i=3}^{10} F(i) = \sum_{i=5}^{12} F(i - 2) \quad \text{and} \quad \sum_{i=6}^{11} i^2 = \sum_{i=7}^{12} (i - 1)^2$$

From Equation (2) of Theorem 5.4.5,

$$\sum_{i=3}^{10} F(i) = \sum_{i=1}^{8} F(i + 2) \quad \text{and} \quad \sum_{i=6}^{11} i^2 = \sum_{i=1}^{6} (i + 5)^2 \qquad ◀$$

5.4.6 THEOREM

$$\sum_{i=1}^{n} [F(i) - F(i - 1)] = F(n) - F(0)$$

Proof

$$\sum_{i=1}^{n} [F(i) - F(i - 1)] = \sum_{i=1}^{n} F(i) - \sum_{i=1}^{n} F(i - 1)$$

On the right side of this equation we write the first summation in another form and apply Equation (2) of Theorem 5.4.5 with $c = 1$ to the second summation. Then

$$\sum_{i=1}^{n} [F(i) - F(i - 1)] = \left(\sum_{i=1}^{n-1} F(i) + F(n) \right) - \sum_{i=1-1}^{n-1} F[(i + 1) - 1]$$

$$= \sum_{i=1}^{n-1} F(i) + F(n) - \sum_{i=0}^{n-1} F(i)$$

$$= \sum_{i=1}^{n-1} F(i) + F(n) - \left(F(0) + \sum_{i=1}^{n-1} F(i) \right)$$

$$= F(n) - F(0) \qquad \blacksquare$$

EXAMPLE 1   Evaluate

$$\sum_{i=1}^{n} (4^i - 4^{i-1})$$

Solution   From Theorem 5.4.6, where $F(i) = 4^i$, it follows that

$$\sum_{i=1}^{n} (4^i - 4^{i-1}) = 4^n - 4^0$$
$$= 4^n - 1$$

In the following theorem there are four formulas that are useful for computation with sigma notation. They are numbered for future reference.

5.4.7 THEOREM

If n is a positive integer, then

$$\sum_{i=1}^{n} i = \frac{n(n+1)}{2} \qquad \text{(Formula 1)}$$

$$\sum_{i=1}^{n} i^2 = \frac{n(n+1)(2n+1)}{6} \qquad \text{(Formula 2)}$$

$$\sum_{i=1}^{n} i^3 = \frac{n^2(n+1)^2}{4} \qquad \text{(Formula 3)}$$

$$\sum_{i=1}^{n} i^4 = \frac{n(n+1)(6n^3 + 9n^2 + n - 1)}{30} \qquad \text{(Formula 4)}$$

Proof   We prove Formula 1 first.

$$\sum_{i=1}^{n} i = 1 + 2 + 3 + \ldots + (n-1) + n$$

and

$$\sum_{i=1}^{n} i = n + (n-1) + (n-2) + \ldots + 2 + 1$$

If the left sides of these two equations are added, the result is

$$2 \sum_{i=1}^{n} i$$

and if the right sides are added term by term, we obtain n terms, each having the value $(n+1)$. Hence

$$2 \sum_{i=1}^{n} i = (n+1) + (n+1) + (n+1) + \ldots + (n+1) \qquad n \text{ terms}$$
$$= n(n+1)$$

Therefore

$$\sum_{i=1}^{n} i = \frac{n(n+1)}{2}$$

which is Formula 1. Formula 1, as well as the other three formulas, can be proved by mathematical induction. We now give the proof of Formula 2 by mathematical induction. The proof consists of two parts and a conclusion.

Part 1: First Formula 2 is verified for $n = 1$. When $n = 1$, the left side of Formula 2 is

$$\sum_{i=1}^{1} i^2 = 1$$

and the right side is

$$\frac{1(1 + 1)(2 + 1)}{6} = \frac{1 \cdot 2 \cdot 3}{6}$$

$$= 1$$

Hence Formula 2 is true when $n = 1$.

Part 2: We assume that Formula 2 is true when $n = k$, where k is any positive integer; that is, we assume

$$\sum_{i=1}^{k} i^2 = \frac{k(k + 1)(2k + 1)}{6} \tag{3}$$

With this assumption we wish to prove that the formula is also true when $n = k + 1$; that is, we wish to prove

$$\sum_{i=1}^{k+1} i^2 = \frac{(k + 1)[(k + 1) + 1][2(k + 1) + 1]}{6} \tag{4}$$

When $n = k + 1$, we have

$$\sum_{i=1}^{k+1} i^2 = 1^2 + 2^2 + 3^2 + \ldots + k^2 + (k + 1)^2$$

$$= \sum_{i=1}^{k} i^2 + (k + 1)^2$$

$$= \frac{k(k + 1)(2k + 1)}{6} + (k + 1)^2 \qquad \text{(by applying (3))}$$

$$= \frac{k(k + 1)(2k + 1) + 6(k + 1)^2}{6}$$

$$= \frac{(k + 1)[k(2k + 1) + 6(k + 1)]}{6}$$

$$= \frac{(k + 1)(2k^2 + 7k + 6)}{6}$$

$$= \frac{(k + 1)(k + 2)(2k + 3)}{6}$$

$$= \frac{(k + 1)[(k + 1) + 1][2(k + 1) + 1]}{6}$$

which is (4).

Conclusion: From part 1 we know that Formula 2 is true for $n = 1$. Because it is true for $n = 1$, it follows from part 2 that the formula is true for $n = 1 + 1$, or 2; because it is true for $n = 2$, it is true for $n = 2 + 1$, or 3; and so on. By the principle of mathematical induction the formula is true for all positive-integer values of n. ■

The proofs of Formulas 3 and 4 are left as exercises (see Exercises 53 through 58).

EXAMPLE 2   Evaluate

$$\sum_{i=1}^{n} i(3i - 2)$$

Solution

$$\sum_{i=1}^{n} i(3i - 2) = \sum_{i=1}^{n} (3i^2 - 2i)$$

$$= \sum_{i=1}^{n} (3i^2) + \sum_{i=1}^{n} (-2i) \qquad \text{(by Theorem 5.4.4)}$$

$$= 3 \sum_{i=1}^{n} i^2 - 2 \sum_{i=1}^{n} i \qquad \text{(by Theorem 5.4.3)}$$

$$= 3 \cdot \frac{n(n + 1)(2n + 1)}{6} - 2 \cdot \frac{n(n + 1)}{2} \qquad \text{(by Formulas 2 and 1)}$$

$$= \frac{2n^3 + 3n^2 + n - 2n^2 - 2n}{2}$$

$$= \frac{2n^3 + n^2 - n}{2}$$

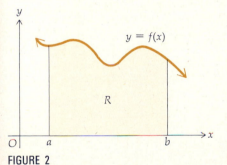

FIGURE 2

Before discussing the area of a plane region, we indicate why we use the terminology "*measure* of the area." The word *measure* refers to a number (no units are included). For example, if the area of a triangle is 20 cm², we say that the square-centimeter measure of the area of the triangle is 20. When the word *measurement* is applied, the units are included. Thus the measurement of the area of the triangle is 20 cm².

Consider now a region R in the plane as shown in Figure 2. The region R is bounded by the x axis, the lines $x = a$ and $x = b$, and the curve having the equation $y = f(x)$, where f is a function continuous on the closed interval $[a, b]$. For simplicity, take $f(x) \geq 0$ for all x in $[a, b]$. We wish to assign a number A to be the measure of the area of R, and we use a limiting process similar to the one used in defining the area of a circle: The area of a circle is defined as the limit of the areas of inscribed regular polygons as the number of sides increases without bound. We realize intuitively that, whatever number is chosen to represent A, that number must be at least as great as the measure of the area of any polygonal region contained in R, and it must be no greater than the measure of the area of any polygonal region containing R.

We first define a polygonal region contained in R. Divide the closed interval $[a, b]$ into n subintervals. For simplicity, we now take each of these subintervals as being of equal length, for instance, Δx. Therefore $\Delta x = (b - a)/n$. Denote the endpoints of these subintervals by $x_0, x_1, x_2, \ldots, x_{n-1}, x_n$, where $x_0 = a$, $x_1 = a + \Delta x, \ldots, x_i = a + i \Delta x, \ldots, x_{n-1} = a + (n - 1) \Delta x, x_n = b$. Let the ith subinterval be denoted by $[x_{i-1}, x_i]$. Because f is continuous on the closed interval $[a, b]$, it is continuous on each closed subinterval. By the extreme-value theorem there is a number in each subinterval for which f has an absolute minimum value. In the ith subinterval let this number be c_i, so that $f(c_i)$ is the absolute minimum value of f on the subinterval $[x_{i-1}, x_i]$. Consider n rectangles, each having a width Δx units and an altitude $f(c_i)$ units (see Figure 3). Let the sum of the areas of these n rectangles be given by S_n square units; then

$$S_n = f(c_1)\,\Delta x + f(c_2)\,\Delta x + \ldots + f(c_i)\,\Delta x + \ldots + f(c_n)\,\Delta x$$

or, with the sigma notation,

$$S_n = \sum_{i=1}^{n} f(c_i)\,\Delta x \tag{5}$$

The summation on the right side of (5) gives the sum of the measures of the areas of n inscribed rectangles. Thus, however we define A, it must be such that

$$A \geq S_n$$

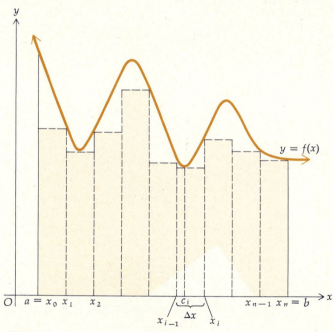

FIGURE 3

In Figure 3 the shaded region has an area of S_n square units. Now let n increase. Specifically, multiply n by 2; then the number of rectangles is doubled, and the width of each rectangle is halved. This is illustrated in Figure 4, showing twice as many rectangles as Figure 3. By comparing the two figures, notice that the shaded region in Figure 4 appears to approximate the region R more nearly than that of Figure 3. So the sum of the measures of the areas of the rectangles in Figure 4 is closer to the number we wish to represent the measure of the area of R.

As n increases, the values of S_n found from Equation (5) increase, and successive values of S_n differ from each other by amounts that become arbitrarily

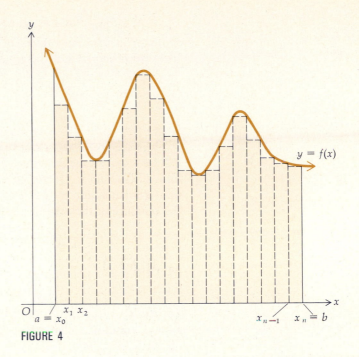

FIGURE 4

small. This is proved in advanced calculus by a theorem that states that if f is continuous on $[a, b]$, then as n increases without bound, the value of S_n given by (5) approaches a limit. It is this limit that we take as the definition of the measure of the area of region R.

5.4.8 DEFINITION Suppose that the function f is continuous on the closed interval $[a, b]$, with $f(x) \geq 0$ for all x in $[a, b]$, and that R is the region bounded by the curve $y = f(x)$, the x axis, and the lines $x = a$ and $x = b$. Divide the interval $[a, b]$ into n subintervals, each of length $\Delta x = (b - a)/n$, and denote the ith subinterval by $[x_{i-1}, x_i]$. Then if $f(c_i)$ is the absolute minimum function value on the ith subinterval, the measure of the area of region R is given by

$$A = \lim_{n \to +\infty} \sum_{i=1}^{n} f(c_i) \, \Delta x \tag{6}$$

This equation means that for any $\epsilon > 0$ there is a number $N > 0$ such that if n is a positive integer and

$$\text{if} \quad n > N \quad \text{then} \quad \left| \sum_{i=1}^{n} f(c_i) \, \Delta x - A \right| < \epsilon$$

We could take circumscribed rectangles instead of inscribed rectangles. In this case we take as the measures of the altitudes of the rectangles the absolute maximum value of f on each subinterval. The existence of an absolute maximum value of f on each subinterval is guaranteed by the extreme-value theorem. The corresponding sums of the measures of the areas of the circumscribed rectangles are at least as great as the measure of the area of the region R, and it can be shown that the limit of these sums as n increases without bound is exactly the same as the limit of the sum of the measures of the areas of the inscribed rectangles. This is also proved in advanced calculus. Thus we could define the measure of the area of the region R by

$$A = \lim_{n \to +\infty} \sum_{i=1}^{n} f(d_i) \, \Delta x \tag{7}$$

where $f(d_i)$ is the absolute maximum value of f in $[x_{i-1}, x_i]$.

The measure of the altitude of the rectangle in the ith subinterval actually can be taken as the function value of any number in that subinterval, and the limit of the sum of the measures of the areas of the rectangles is the same no matter what numbers are selected. In Section 5.5 we extend the definition of the measure of the area of a region to be the limit of such a sum.

EXAMPLE 3   Find the area of the region bounded by the curve $y = x^2$, the x axis, and the line $x = 3$ by taking inscribed rectangles.

Solution   Figure 5 shows the region and the ith inscribed rectangle. We apply Definition 5.4.8. Divide the closed interval $[0, 3]$ into n subintervals, each of length Δx: $x_0 = 0, x_1 = \Delta x, x_2 = 2 \, \Delta x, \ldots, x_i = i \, \Delta x, \ldots, x_{n-1} = (n-1) \, \Delta x$, $x_n = 3$.

$$\Delta x = \frac{3 - 0}{n} \qquad f(x) = x^2$$

$$= \frac{3}{n}$$

Because f is increasing on $[0, 3]$, the absolute minimum value of f on the ith subinterval $[x_{i-1}, x_i]$ is $f(x_{i-1})$. Therefore, from (6)

$$A = \lim_{n \to +\infty} \sum_{i=1}^{n} f(x_{i-1}) \, \Delta x \tag{8}$$

Because $x_{i-1} = (i - 1) \, \Delta x$ and $f(x) = x^2$,

$$f(x_{i-1}) = [(i - 1) \, \Delta x]^2$$

Therefore

$$\sum_{i=1}^{n} f(x_{i-1}) \, \Delta x = \sum_{i=1}^{n} (i - 1)^2 (\Delta x)^3$$

But $\Delta x = 3/n$; so

$$\sum_{i=1}^{n} f(x_{i-1}) \, \Delta x = \sum_{i=1}^{n} (i - 1)^2 \frac{27}{n^3}$$

$$= \frac{27}{n^3} \sum_{i=1}^{n} (i - 1)^2$$

$$= \frac{27}{n^3} \left[\sum_{i=1}^{n} i^2 - 2 \sum_{i=1}^{n} i + \sum_{i=1}^{n} 1 \right]$$

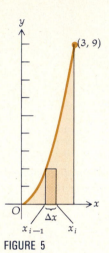

FIGURE 5

and using Formulas 2 and 1 and Theorem 5.4.2 we get

$$\sum_{i=1}^{n} f(x_{i-1})\, \Delta x = \frac{27}{n^3} \left[\frac{n(n + 1)(2n + 1)}{6} - 2 \cdot \frac{n(n + 1)}{2} + n \right]$$

$$= \frac{27}{n^3} \cdot \frac{2n^3 + 3n^2 + n - 6n^2 - 6n + 6n}{6}$$

$$= \frac{9}{2} \cdot \frac{2n^2 - 3n + 1}{n^2}$$

Then, from (8),

$$A = \lim_{n \to +\infty} \left[\frac{9}{2} \cdot \frac{2n^2 - 3n + 1}{n^2} \right]$$

$$= \frac{9}{2} \cdot \lim_{n \to +\infty} \left(2 - \frac{3}{n} + \frac{1}{n^2} \right)$$

$$= \tfrac{9}{2}(2 - 0 + 0)$$

$$= 9$$

Therefore the area of the region is 9 square units.

EXAMPLE 4   Find the area of the region in Example 3 by taking circumscribed rectangles.

Solution   With circumscribed rectangles the measure of the altitude of the ith rectangle is the absolute maximum value of f on the ith subinterval $[x_{i-1}, x_i]$, which is $f(x_i)$. From (7),

$$A = \lim_{n \to +\infty} \sum_{i=1}^{n} f(x_i)\, \Delta x \tag{9}$$

Because $x_i = i\, \Delta x$, then $f(x_i) = (i\, \Delta x)^2$, and so

$$\sum_{i=1}^{n} f(x_i)\, \Delta x = \sum_{i=1}^{n} i^2 (\Delta x)^3$$

$$= \frac{27}{n^3} \sum_{i=1}^{n} i^2$$

$$= \frac{27}{n^3} \left[\frac{n(n + 1)(2n + 1)}{6} \right]$$

$$= \frac{9}{2} \cdot \frac{2n^2 + 3n + 1}{n^2}$$

Therefore, from (9),

$$A = \lim_{n \to +\infty} \frac{9}{2} \cdot \left(2 + \frac{3}{n} + \frac{1}{n^2} \right)$$

$$= 9 \qquad \text{(as in Example 3)}$$

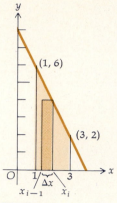

FIGURE 6

EXAMPLE 5   Find the area of the trapezoid that is the region bounded by the lines $x = 1$ and $x = 3$, the x axis, and the line $2x + y = 8$. Take inscribed rectangles.

Solution   The region and the ith inscribed rectangle are shown in Figure 6. The closed interval $[1, 3]$ is divided into n subintervals, each of length Δx: $x_0 = 1$, $x_1 = 1 + \Delta x$, $x_2 = 1 + 2 \Delta x, \ldots, x_i = 1 + i \Delta x, \ldots, x_{n-1} = 1 + (n-1) \Delta x$, $x_n = 3$.

$$\Delta x = \frac{3-1}{n}$$

$$= \frac{2}{n}$$

Solving the equation of the line for y we obtain $y = -2x + 8$. Therefore $f(x) = -2x + 8$, and because f is decreasing on $[1, 3]$, the absolute minimum value of f on the ith subinterval $[x_{i-1}, x_i]$ is $f(x_i)$. Because $x_i = 1 + i \Delta x$ and $f(x) = -2x + 8$, then $f(x_i) = -2(1 + i \Delta x) + 8$, that is, $f(x_i) = 6 - 2i \Delta x$. From (6),

$$A = \lim_{n \to +\infty} \sum_{i=1}^{n} f(x_i) \, \Delta x$$

$$= \lim_{n \to +\infty} \sum_{i=1}^{n} (6 - 2i \, \Delta x) \, \Delta x$$

$$= \lim_{n \to +\infty} \sum_{i=1}^{n} [6 \, \Delta x - 2i(\Delta x)^2]$$

$$= \lim_{n \to +\infty} \sum_{i=1}^{n} \left[6\left(\frac{2}{n}\right) - 2i\left(\frac{2}{n}\right)^2 \right]$$

$$= \lim_{n \to +\infty} \left[\frac{12}{n} \sum_{i=1}^{n} 1 - \frac{8}{n^2} \sum_{i=1}^{n} i \right]$$

From Theorem 5.4.2 and Formula 1,

$$A = \lim_{n \to +\infty} \left[\frac{12}{n} \cdot n - \frac{8}{n^2} \cdot \frac{n(n+1)}{2} \right]$$

$$= \lim_{n \to +\infty} \left(8 - \frac{4}{n} \right)$$

$$= 8$$

Therefore the area is 8 square units. Using the formula from plane geometry for the area of a trapezoid, $A = \frac{1}{2}h(b_1 + b_2)$, where h, b_1, and b_2 are, respectively, the number of units in the lengths of the altitude and the two bases, we get $A = \frac{1}{2}(2)(6 + 2)$; that is, $A = 8$, which agrees with the result.

EXERCISES 5.4

In Exercises 1 through 12, find the sum.

1. $\displaystyle\sum_{i=1}^{6} (3i - 2)$ **2.** $\displaystyle\sum_{i=1}^{20} (5i + 4)$ **3.** $\displaystyle\sum_{i=1}^{7} (i^2 + 1)$ **4.** $\displaystyle\sum_{i=1}^{7} (i + 1)^2$ **5.** $\displaystyle\sum_{i=1}^{10} (i - 1)^3$ **6.** $\displaystyle\sum_{i=1}^{10} (i^3 - 1)$

7. $\displaystyle\sum_{i=2}^{5} \frac{i}{i-1}$

8. $\displaystyle\sum_{j=3}^{6} \frac{2}{j(j-2)}$

9. $\displaystyle\sum_{i=-2}^{3} 2^i$

10. $\displaystyle\sum_{i=0}^{3} \frac{1}{1+i^2}$

11. $\displaystyle\sum_{k=1}^{4} \frac{(-1)^{k+1}}{k}$

12. $\displaystyle\sum_{k=-2}^{3} \frac{k}{k+3}$

In Exercises 13 through 20, evaluate the sum by using Theorems 5.4.2 through 5.4.7.

13. $\displaystyle\sum_{i=1}^{25} 2i(i-1)$

14. $\displaystyle\sum_{i=1}^{20} 3i(i^2+2)$

15. $\displaystyle\sum_{k=1}^{n} (2^k - 2^{k-1})$

16. $\displaystyle\sum_{i=1}^{n} (10^{i+1} - 10^i)$

17. $\displaystyle\sum_{k=1}^{100} \left[\frac{1}{k} - \frac{1}{k+1}\right]$

18. $\displaystyle\sum_{i=1}^{n} 2i(1+i^2)$

19. $\displaystyle\sum_{i=1}^{n} 4i^2(i-2)$

20. $\displaystyle\sum_{k=1}^{n} [(3^{-k} - 3^k)^2 - (3^{k-1} + 3^{-k-1})^2]$

In Exercises 21 through 36, use the method of this section to find the area of the region; use inscribed or circumscribed rectangles as indicated. For each exercise draw a figure showing the region and the ith rectangle.

21. The region bounded by $y = x^2$, the x axis, and the line $x = 2$; inscribed rectangles.
22. The region of Exercise 21; circumscribed rectangles.
23. The region bounded by $y = 2x$, the x axis, and the lines $x = 1$ and $x = 4$; circumscribed rectangles.
24. The region of Exercise 23; inscribed rectangles.
25. The region above the x axis and to the right of the line $x = 1$ bounded by the x axis, the line $x = 1$, and the curve $y = 4 - x^2$; inscribed rectangles.
26. The region of Exercise 25; circumscribed rectangles.
27. The region lying to the left of the line $x = 1$ bounded by the curve and lines of Exercise 25; circumscribed rectangles.
28. The region of Exercise 27; inscribed rectangles.
29. The region bounded by $y = 3x^4$, the x axis, and the line $x = 1$; inscribed rectangles.
30. The region of Exercise 29; circumscribed rectangles.
31. The region bounded by $y = x^3$, the x axis, and the lines $x = -1$ and $x = 2$; inscribed rectangles.
32. The region of Exercise 31; circumscribed rectangles.
33. The region bounded by $y = x^3 + x$, the x axis, and the lines $x = -2$ and $x = 1$; circumscribed rectangles.
34. The region of Exercise 33; inscribed rectangles.
35. The region bounded by $y = mx$, with $m > 0$, the x axis, and the lines $x = a$ and $x = b$, with $b > a > 0$; circumscribed rectangles.
36. The region of Exercise 35; inscribed rectangles.
37. Use the method of this section to find the area of an isosceles trapezoid whose bases have measures b_1 and b_2 and whose altitude has measure h.

38. The graph of $y = 4 - |x|$ and the x axis from $x = -4$ to $x = 4$ form a triangle. Use the method of this section to find the area of this triangle.

In Exercises 39 through 44, find the area of the region by taking as the measure of the altitude of the ith rectangle $f(m_i)$, where m_i is the midpoint of the ith subinterval. (Hint: $m_i = \frac{1}{2}(x_i + x_{i-1})$.)

39. The region of Example 3.
40. The region of Exercise 21.
41. The region of Exercise 23.
42. The region of Exercise 25.
43. The region of Exercise 27.
44. The region of Exercise 29.

In Exercises 45 through 52, a function f and numbers, n, a, and b are given. Approximate to four decimal places the area of the region bounded by the curve $y = f(x)$, the x axis, and the lines $x = a$ and $x = b$ by doing the following: Divide the interval $[a, b]$ into n subintervals of equal length Δx units and use a calculator to compute the sum of the areas of n inscribed or circumscribed (as indicated) rectangles each having a width of Δx units.

45. $f(x) = \dfrac{1}{x}$, $a = 1$, $b = 3$, $n = 10$, inscribed rectangles.

46. $f(x) = \dfrac{1}{x^2}$, $a = 1$, $b = 2$, $n = 12$, circumscribed rectangles.

47. $f(x) = \dfrac{1}{x}$, $a = 1$, $b = 3$, $n = 10$, circumscribed rectangles.

48. $f(x) = \dfrac{1}{x^2}$, $a = 1$, $b = 2$, $n = 12$, inscribed rectangles.

49. $f(x) = \sin x$, $a = \frac{1}{6}\pi$, $b = \frac{5}{6}\pi$, $n = 8$, circumscribed rectangles.
50. $f(x) = \cos x$, $a = 0$, $b = \frac{1}{2}\pi$, $n = 6$, inscribed rectangles.
51. $f(x) = \sin x$, $a = \frac{1}{6}\pi$, $b = \frac{5}{6}\pi$, $n = 8$, inscribed rectangles.
52. $f(x) = \cos x$, $a = 0$, $b = \frac{1}{2}\pi$, $n = 6$, circumscribed rectangles.

53. Prove Formula 2 of Theorem 5.4.7 without mathematical induction. (*Hint:* $i^3 - (i-1)^3 = 3i^2 - 3i + 1$; so $\displaystyle\sum_{i=1}^{n} [i^3 - (i-1)^3] = \sum_{i=1}^{n} (3i^2 - 3i + 1)$. On the left side of this equation use Theorem 5.4.6; on the right side use Theorems 5.4.2, 5.4.3, and 5.4.4 and Formula 1.)

54. Prove Formula 3 of Theorem 5.4.7.
(*Hint:* $i^4 - (i-1)^4 = 4i^3 - 6i^2 + 4i - 1$, and use a method similar to the one for Exercise 53.)

55. Prove Formula 4 of Theorem 5.4.7. (See the hints for Exercises 53 and 54.)
56. Prove Formula 1 of Theorem 5.4.7 by mathematical induction.
57. Prove Formula 3 of Theorem 5.4.7 by mathematical induction.
58. Prove Formula 4 of Theorem 5.4.7 by mathematical induction.
59. Prove Theorem 5.4.4.
60. Prove Theorem 5.4.5.

5.5 THE DEFINITE INTEGRAL

In Section 5.4 the measure of the area of a region was defined as the following limit:

$$\lim_{n \to +\infty} \sum_{i=1}^{n} f(c_i)\, \Delta x \tag{1}$$

To lead up to this definition we divided the closed interval $[a, b]$ into subintervals of equal length and then took c_i as the point in the ith subinterval for which f has an absolute minimum value. We also restricted the function values $f(x)$ to be nonnegative on $[a, b]$ and further required f to be continuous on $[a, b]$.

The limit in (1) is a special case of a new kind of limiting process that leads to the definition of the *definite integral*. We now discuss this "new kind of limit."

Let f be a function defined on the closed interval $[a, b]$. Divide this interval into n subintervals by choosing *any* $n - 1$ intermediate points between a and b. Let $x_0 = a$ and $x_n = b$, and let $x_1, x_2, \ldots, x_{n-1}$ be the intermediate points so that

$$x_0 < x_1 < x_2 < \ldots < x_{n-1} < x_n$$

The points $x_0, x_1, x_2, \ldots, x_{n-1}, x_n$ are not necessarily equidistant. Let $\Delta_1 x$ be the length of the first subinterval so that $\Delta_1 x = x_1 - x_0$; let $\Delta_2 x$ be the length of the second subinterval so that $\Delta_2 x = x_2 - x_1$; and so forth so that the length of the ith subinterval is $\Delta_i x$, and

$$\Delta_i x = x_i - x_{i-1}$$

A set of all such subintervals of the interval $[a, b]$ is called a **partition** of the interval $[a, b]$. Let Δ be such a partition. Figure 1 illustrates one such partition Δ of $[a, b]$.

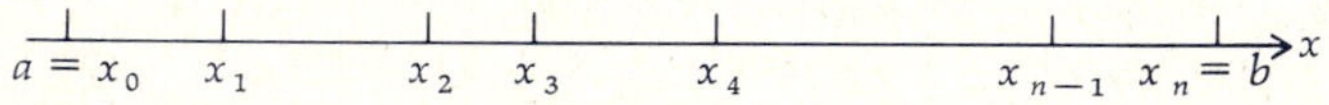

FIGURE 1

The partition Δ contains n subintervals. One of these subintervals is longest; however, there may be more than one such subinterval. The length of the longest subinterval of the partition Δ, called the **norm** of the partition, is denoted by $\|\Delta\|$.

Choose a point in each subinterval of the partition Δ: Let ξ_1 be the point chosen in $[x_0, x_1]$ so that $x_0 \le \xi_1 \le x_1$. Let ξ_2 be the point chosen in $[x_1, x_2]$, so that $x_1 \le \xi_2 \le x_2$, and so forth, so that ξ_i is the point chosen in $[x_{i-1}, x_i]$, and $x_{i-1} \le \xi_i \le x_i$. Form the sum

$$f(\xi_1)\, \Delta_1 x + f(\xi_2)\, \Delta_2 x + \ldots + f(\xi_i)\, \Delta_i x + \ldots + f(\xi_n)\, \Delta_n x$$

or

$$\sum_{i=1}^{n} f(\xi_i)\, \Delta_i x$$

Such a sum is called a **Riemann sum**, named for the mathematician Georg Friedrich Bernhard Riemann (1826–1866).

▶ **ILLUSTRATION 1** Suppose that $f(x) = 10 - x^2$, with $\frac{1}{4} \le x \le 3$. We will find the Riemann sum for the function f on $[\frac{1}{4}, 3]$ for the partition Δ: $x_0 = \frac{1}{4}$, $x_1 = 1$, $x_2 = 1\frac{1}{2}$, $x_3 = 1\frac{3}{4}$, $x_4 = 2\frac{1}{4}$, $x_5 = 3$, and $\xi_1 = \frac{1}{2}$, $\xi_2 = 1\frac{1}{4}$, $\xi_3 = 1\frac{3}{4}$, $\xi_4 = 2$, $\xi_5 = 2\frac{3}{4}$.

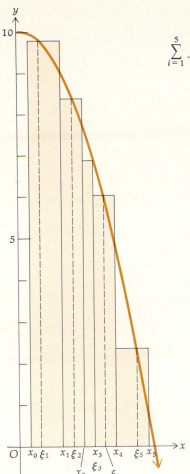

FIGURE 2

Figure 2 shows a sketch of the graph of f on $[\frac{1}{4}, 3]$ and the five rectangles, the measures of whose areas are the terms of the Riemann sum.

$$\sum_{i=1}^{5} f(\xi_i) \Delta_i x = f(\xi_1) \Delta_1 x + f(\xi_2) \Delta_2 x + f(\xi_3) \Delta_3 x + f(\xi_4) \Delta_4 x + f(\xi_5) \Delta_5 x$$

$$= f(\tfrac{1}{2})(1 - \tfrac{1}{4}) + f(\tfrac{5}{4})(1\tfrac{1}{2} - 1) + f(\tfrac{7}{4})(1\tfrac{3}{4} - 1\tfrac{1}{2}) + f(2)(2\tfrac{1}{4} - 1\tfrac{3}{4}) + f(\tfrac{11}{4})(3 - 2\tfrac{1}{4})$$

$$= (9\tfrac{3}{4})(\tfrac{3}{4}) + (8\tfrac{7}{16})(\tfrac{1}{2}) + (6\tfrac{15}{16})(\tfrac{1}{4}) + (6)(\tfrac{1}{2}) + (2\tfrac{7}{16})(\tfrac{3}{4})$$

$$= 18\tfrac{3}{32}$$

The norm of Δ is the length of the longest subinterval. Hence $\|\Delta\| = \tfrac{3}{4}$. ◀

Because the function values $f(x)$ are not restricted to nonnegative values, some of the $f(\xi_i)$ could be negative. In such a case the geometric interpretation of the Riemann sum would be the sum of the measures of the areas of the rectangles lying above the x axis and the negatives of the measures of the areas of the rectangles lying below the x axis. This situation is illustrated in Figure 3. Here

$$\sum_{i=1}^{10} f(\xi_i) \Delta_i x = A_1 + A_2 - A_3 - A_4 - A_5 + A_6 + A_7 - A_8 - A_9 - A_{10}$$

because $f(\xi_3)$, $f(\xi_4)$, $f(\xi_5)$, $f(\xi_8)$, $f(\xi_9)$, and $f(\xi_{10})$ are negative numbers.

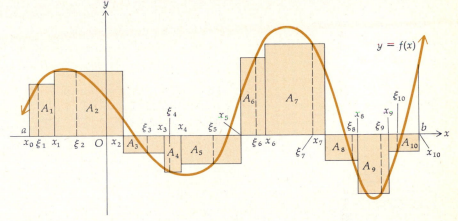

FIGURE 3

Let f be a function whose domain includes the closed interval $[a, b]$. Suppose there is a number L such that $\left| \sum_{i=1}^{n} f(\xi_i) \Delta_i x - L \right|$ can be made as small as we please for all partitions Δ whose norms are sufficiently small, and for any ξ_i in the closed interval $[x_{i-1}, x_i]$, $i = 1, 2, \ldots, n$. In such a case f is said to be *integrable* on $[a, b]$.

5.5.1 DEFINITION   Let f be a function whose domain includes the closed interval $[a, b]$. Then f is said to be **integrable** on $[a, b]$ if there is a number L satisfying the condition that, for any $\epsilon > 0$, there exists a $\delta > 0$ such that for every partition Δ for which $\|\Delta\| < \delta$, and for any ξ_i in the closed interval $[x_{i-1}, x_i]$, $i = 1, 2, \ldots, n$, then

$$\left| \sum_{i=1}^{n} f(\xi_i) \Delta_i x - L \right| < \epsilon \qquad (2)$$

For such a situation we write

$$\lim_{\|\Delta\| \to 0} \sum_{i=1}^{n} f(\xi_i) \Delta_i x = L \qquad (3)$$

This definition states that, for a given function f defined on the closed interval $[a, b]$, we can make the values of the Riemann sums as close to L as we please by taking the norms $\|\Delta\|$ of all partitions Δ of $[a, b]$ sufficiently small for all possible choices of the numbers ξ_i for which $x_{i-1} \le \xi_i \le x_i$, $i = 1, 2, \ldots, n$.

The limiting process given by (3) is different from that discussed in Chapter 2. From Definition 5.5.1 the number L in (3) exists if for any $\epsilon > 0$ there exists a $\delta > 0$ such that for every partition Δ for which $\|\Delta\| < \delta$, and for any ξ_i in the closed interval $[x_{i-1}, x_i]$, $i = 1, 2, \ldots, n$, then inequality (2) holds.

In Definition 2.1.1 we had

$$\lim_{x \to a} f(x) = L \tag{4}$$

if for any $\epsilon > 0$ there exists a $\delta > 0$ such that

$$\text{if} \quad 0 < |x - a| < \delta \quad \text{then} \quad |f(x) - L| < \epsilon$$

In limiting process (3), for a particular $\delta > 0$ there are infinitely many partitions Δ having norm $\|\Delta\| < \delta$. This is analogous to the fact that in limiting process (4), for a given $\delta > 0$ there are infinitely many values of x for which $0 < |x - a| < \delta$. However, in limiting process (3), for each partition Δ there are infinitely many choices of ξ_i. It is in this respect that the two limiting processes differ.

Theorem 2.1.2 proved in Supplementary Section 2.9 states that if the number L in limiting process (4) exists, it is unique. In a similar manner we can show that if there is a number L satisfying Definition 5.5.1, then it is unique. Now we can define the *definite integral*.

5.5.2 DEFINITION

If f is a function defined on the closed interval $[a, b]$, then the **definite integral** of f from a to b, denoted by $\int_a^b f(x)\, dx$, is given by

$$\int_a^b f(x)\, dx = \lim_{\|\Delta\| \to 0} \sum_{i=1}^n f(\xi_i)\, \Delta_i x \tag{5}$$

if the limit exists.

Note that the statement "the function f is integrable on the closed interval $[a, b]$" is synonymous with the statement "the definite integral of f from a to b exists."

In the notation for the definite integral $\int_a^b f(x)\, dx$, $f(x)$ is the **integrand**, a is the **lower limit**, and b is the **upper limit**. The symbol $\int$ is an **integral sign**. The integral sign resembles a capital S, which is appropriate because the definite integral is the limit of a sum. It is the same symbol we have been using to indicate the operation of antidifferentiation. The reason for the common symbol is that a theorem (5.8.2), called the second fundamental theorem of the calculus, enables us to evaluate a definite integral by finding an antiderivative (also called an **indefinite integral**).

The following question now arises: Under what conditions does a number L satisfying Definition 5.5.2 exist; that is, under what conditions is a function integrable? An answer to this question is given by the next theorem.

5.5.3 THEOREM

If a function is continuous on the closed interval $[a, b]$, then it is integrable on $[a, b]$.

The proof of this theorem is beyond the scope of this book and is given in advanced calculus texts. The condition that f is continuous on $[a, b]$, while being sufficient to guarantee that f is integrable on $[a, b]$, is not a necessary condition for the existence of $\int_a^b f(x)\,dx$. That is, if f is continuous on $[a, b]$, then Theorem 5.5.3 assures us that $\int_a^b f(x)\,dx$ exists; however, there are functions that are discontinuous yet integrable on a closed interval. Such a function is given in Example 2 at the end of this section.

At the beginning of this section we stated that the limit used in Definition 5.4.8 to define the measure of the area of a region is a special case of the limit used in Definition 5.5.2 to define the definite integral. In the discussion of area, the interval $[a, b]$ was divided into n subintervals of equal length. Such a partition of the interval $[a, b]$ is called a **regular partition**. If Δx is the length of each subinterval in a regular partition, then each $\Delta_i x = \Delta x$, and the norm of the partition is Δx. Making these substitutions in (5) we have

$$\int_a^b f(x)\,dx = \lim_{\Delta x \to 0} \sum_{i=1}^n f(\xi_i)\,\Delta x \tag{6}$$

Furthermore,

$$\Delta x = \frac{b - a}{n} \quad \text{and} \quad n = \frac{b - a}{\Delta x}$$

Thus

$$\lim_{n \to +\infty} \Delta x = 0 \quad \text{and} \quad \lim_{\Delta x \to 0} n = +\infty$$

The reason that $\lim\limits_{\Delta x \to 0} n = +\infty$ is that $b > a$ and Δx approaches zero through positive values (because $\Delta x > 0$). From these limits we conclude that

$$\Delta x \to 0 \quad \text{is equivalent to} \quad n \to +\infty$$

Thus from this statement and (6), we have

$$\int_a^b f(x)\,dx = \lim_{n \to +\infty} \sum_{i=1}^n f(\xi_i)\,\Delta x \tag{7}$$

You should remember that ξ_i can be any point in the ith subinterval $[x_{i-1}, x_i]$.

Comparing the limit used in Definition 5.4.8, which gives the measure of the area of a region, with the limit on the right side of (7), we have in the first case

$$\lim_{n \to +\infty} \sum_{i=1}^n f(c_i)\,\Delta x \tag{8}$$

where $f(c_i)$ is the absolute minimum function value on $[x_{i-1}, x_i]$. In the second case we have

$$\lim_{n \to +\infty} \sum_{i=1}^n f(\xi_i)\,\Delta x \tag{9}$$

where ξ_i is any number in $[x_{i-1}, x_i]$.

Because the function f is continuous on $[a, b]$, by Theorem 5.5.3, $\int_a^b f(x)\,dx$ exists; therefore this definite integral is the limit of all Riemann sums of f on $[a, b]$ including those in (8) and (9). Because of this, we redefine the area of a region in a more general way.

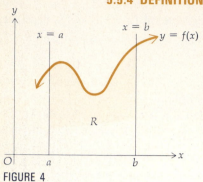

FIGURE 4

5.5.4 DEFINITION Let the function f be continuous on $[a, b]$ and $f(x) \geq 0$ for all x in $[a, b]$. Let R be the region bounded by the curve $y = f(x)$, the x axis, and the lines $x = a$ and $x = b$. Then the measure A of the area of region R is given by

$$A = \lim_{\|\Delta\| \to 0} \sum_{i=1}^{n} f(\xi_i) \, \Delta_i x$$

$$\Leftrightarrow \quad A = \int_a^b f(x) \, dx$$

This definition states that if $f(x) \geq 0$ for all x in $[a, b]$, the definite integral $\int_a^b f(x) \, dx$ can be interpreted geometrically as the measure of the area of the region R shown in Figure 4.

Equation (7) can be used to find the exact value of a definite integral, as illustrated in the following example.

EXAMPLE 1 Find the exact value of the definite integral $\int_1^3 x^2 \, dx$. Interpret the result geometrically.

Solution   Consider a regular partition of the closed interval $[1, 3]$ into n subintervals. Then $\Delta x = 2/n$.

If we choose ξ_i as the right endpoint of each subinterval, we have

$$\xi_1 = 1 + \frac{2}{n}, \; \xi_2 = 1 + 2\left(\frac{2}{n}\right), \; \xi_3 = 1 + 3\left(\frac{2}{n}\right), \ldots, \xi_i = 1 + i\left(\frac{2}{n}\right), \ldots, \xi_n = 1 + n\left(\frac{2}{n}\right)$$

Because $f(x) = x^2$,

$$f(\xi_i) = \left(1 + \frac{2i}{n}\right)^2$$

$$= \left(\frac{n + 2i}{n}\right)^2$$

Therefore, by using (7) and applying theorems from Section 5.4, we get

$$\int_1^3 x^2 \, dx = \lim_{n \to +\infty} \sum_{i=1}^{n} \left(\frac{n + 2i}{n}\right)^2 \frac{2}{n}$$

$$= \lim_{n \to +\infty} \frac{2}{n^3} \sum_{i=1}^{n} (n^2 + 4ni + 4i^2)$$

$$= \lim_{n \to +\infty} \frac{2}{n^3} \left[n^2 \sum_{i=1}^{n} 1 + 4n \sum_{i=1}^{n} i + 4 \sum_{i=1}^{n} i^2\right]$$

$$= \lim_{n \to +\infty} \frac{2}{n^3} \left[n^2 n + 4n \cdot \frac{n(n + 1)}{2} + \frac{4n(n + 1)(2n + 1)}{6}\right]$$

$$= \lim_{n \to +\infty} \frac{2}{n^3} \left[n^3 + 2n^3 + 2n^2 + \frac{2n(2n^2 + 3n + 1)}{3}\right]$$

$$= \lim_{n \to +\infty} \left[6 + \frac{4}{n} + \frac{8n^2 + 12n + 4}{3n^2}\right]$$

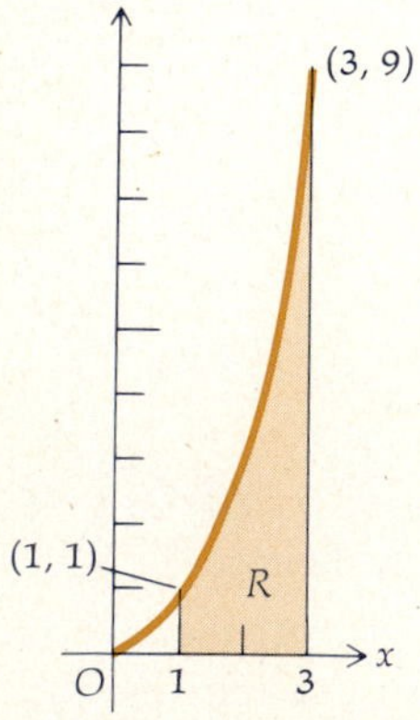

FIGURE 5

$$= \lim_{n \to +\infty} \left[6 + \frac{4}{n} + \frac{8}{3} + \frac{4}{n} + \frac{4}{3n^2} \right]$$

$$= 6 + 0 + \tfrac{8}{3} + 0 + 0$$

$$= \tfrac{26}{3}$$

We interpret the result geometrically. Because $x^2 \geq 0$ for all x in $[1, 3]$, the region bounded by the curve $y = x^2$, the x axis, and the lines $x - 1$ and $x - 3$ has an area of $\tfrac{26}{3}$ square units. The region appears in Figure 5.

In Definition 5.5.2 the closed interval $[a, b]$ is given, and so we assume that $a < b$. To consider the definite integral of a function f from a to b when $a > b$, or when $a = b$, we have the following definitions.

5.5.5 DEFINITION If $a > b$, then

$$\int_a^b f(x)\, dx = - \int_b^a f(x)\, dx$$

if $\int_b^a f(x)\, dx$ exists.

▶ **ILLUSTRATION 2** In Example 1 we showed that $\int_1^3 x^2\, dx = \tfrac{26}{3}$. Therefore, from Definition 5.5.5,

$$\int_3^1 x^2\, dx = - \int_1^3 x^2\, dx$$

$$= -\tfrac{26}{3}$$ ◀

5.5.6 DEFINITION If $f(a)$ exists, then

$$\int_a^a f(x)\, dx = 0$$

▶ **ILLUSTRATION 3** From Definition 5.5.6,

$$\int_1^1 x^2\, dx = 0$$ ◀

As stated earlier, it is possible for a function to be integrable on a closed interval even though it is discontinuous on that interval. This situation occurs in the following example.

EXAMPLE 2 Let f be the function defined by

$$f(x) = \begin{cases} 0 & \text{if } x \neq 0 \\ 1 & \text{if } x = 0 \end{cases}$$

Let $[a, b]$ be any interval for which $a < 0 < b$. Show that f is discontinuous on $[a, b]$, yet integrable on $[a, b]$.

Solution Because $\lim_{x \to 0} f(x) = 0$, but $f(0) = 1$, f is discontinuous at 0 and hence discontinuous on $[a, b]$.

To prove that f is integrable on $[a, b]$ we show that Definition 5.5.1 is satisfied. Consider the Riemann sum

$$\sum_{i=1}^{n} f(\xi_1) \, \Delta_i x$$

If none of the numbers $\xi_1, \xi_2, \ldots, \xi_n$ is zero, then the Riemann sum is zero. Suppose that $\xi_j = 0$. Then

$$\sum_{i=1}^{n} f(\xi_i) \, \Delta_i x = 1 \cdot \Delta_j x$$

In either case,

$$\left| \sum_{i=1}^{n} f(\xi_i) \, \Delta_i x \right| \leq \|\Delta\|$$

Hence

$$\text{if} \quad \|\Delta\| < \epsilon \quad \text{then} \quad \left| \sum_{i=1}^{n} f(\xi_i) \, \Delta_i x - 0 \right| < \epsilon$$

Comparing this statement with Definition 5.5.1 where $\delta = \epsilon$ and $L = 0$, we see that f is integrable on $[a, b]$.

Observe that the function f of Example 2 has a finite number (only 1) of discontinuities on $[a, b]$ and $|f(x)| \leq 1$ for all x in $[a, b]$. This function is a member of a special set of functions that are integrable on a closed interval. Three other functions in this set appear in Exercises 36—38.

EXERCISES 5.5

In Exercises 1 through 9, find the Riemann sum for the function on the interval, using the partition Δ and the given values of ξ_i. Draw a sketch of the graph of the function on the interval, and show the rectangles the measure of whose areas are the terms of the Riemann sum. (See Illustration 1 and Figure 2.)

1. $f(x) = x^2$, $0 \leq x \leq 3$; for Δ: $x_0 = 0$, $x_1 = \frac{1}{2}$, $x_2 = 1\frac{1}{4}$, $x_3 = 2\frac{1}{4}$, $x_4 = 3$; $\xi_1 = \frac{1}{4}$, $\xi_2 = 1$, $\xi_3 = 1\frac{1}{2}$, $\xi_4 = 2\frac{1}{2}$

2. $f(x) = x^2$, $0 \leq x \leq 3$; for Δ: $x_0 = 0$, $x_1 = \frac{3}{4}$, $x_2 = 1\frac{1}{4}$, $x_3 = 2$, $x_4 = 2\frac{3}{4}$, $x_5 = 3$; $\xi_1 = \frac{1}{2}$, $\xi_2 = 1$, $\xi_3 = 1\frac{3}{4}$, $\xi_4 = 2\frac{1}{4}$, $\xi_5 = 2\frac{3}{4}$

3. $f(x) = 1/x$, $1 \leq x \leq 3$; for Δ: $x_0 = 1$, $x_1 = 1\frac{1}{3}$, $x_2 = 2\frac{1}{4}$, $x_3 = 2\frac{2}{3}$, $x_4 = 3$; $\xi_1 = 1\frac{1}{4}$, $\xi_2 = 2$, $\xi_3 = 2\frac{1}{2}$, $\xi_4 = 2\frac{3}{4}$

4. $f(x) = x^3$, $-1 \leq x \leq 2$; for Δ: $x_0 = -1$, $x_1 = -\frac{1}{3}$, $x_2 = \frac{1}{2}$, $x_3 = 1$, $x_4 = 1\frac{1}{4}$, $x_5 = 2$; $\xi_1 = -\frac{1}{2}$, $\xi_2 = 0$, $\xi_3 = \frac{2}{3}$, $\xi_4 = 1$, $\xi_5 = 1\frac{1}{2}$

5. $f(x) = x^2 - x + 1$, $0 \leq x \leq 1$; for Δ: $x_0 = 0$, $x_1 = 0.2$, $x_2 = 0.5$, $x_3 = 0.7$, $x_4 = 1$; $\xi_1 = 0.1$, $\xi_2 = 0.4$, $\xi_3 = 0.6$, $\xi_4 = 0.9$

6. $f(x) = 1/(x + 2)$, $-1 \leq x \leq 3$; for Δ: $x_0 = -1$, $x_1 = -\frac{1}{4}$, $x_2 = 0$, $x_3 = \frac{1}{2}$, $x_4 = 1\frac{1}{4}$, $x_5 = 2$, $x_6 = 2\frac{1}{4}$, $x_7 = 2\frac{3}{4}$, $x_8 = 3$; $\xi_1 = -\frac{3}{4}$, $\xi_2 = 0$, $\xi_3 = \frac{1}{4}$, $\xi_4 = 1$, $\xi_5 = 1\frac{1}{2}$, $\xi_6 = 2$, $\xi_7 = 2\frac{1}{2}$, $\xi_8 = 3$

7. $f(x) = \sin x$, $0 \leq x \leq \pi$; for Δ: $x_0 = 0$, $x_1 = \frac{1}{4}\pi$, $x_2 = \frac{1}{2}\pi$, $x_3 = \frac{2}{3}\pi$, $x_4 = \frac{3}{4}\pi$, $x_5 = \pi$; $\xi_1 = \frac{1}{6}\pi$, $\xi_2 = \frac{1}{3}\pi$, $\xi_3 = \frac{1}{2}\pi$, $\xi_4 = \frac{3}{4}\pi$, $\xi_5 = \frac{5}{6}\pi$.

8. $f(x) = 3 \cos \frac{1}{2}x$, $-\pi \leq x \leq \pi$; for Δ: $x_0 = -\pi$, $x_1 = -\frac{1}{2}\pi$, $x_2 = -\frac{1}{3}\pi$, $x_3 = \frac{1}{3}\pi$, $x_4 = \frac{7}{12}\pi$, $x_5 = \pi$; $\xi_1 = -\frac{2}{3}\pi$, $\xi_2 = -\frac{1}{3}\pi$, $\xi_3 = 0$, $\xi_4 = \frac{1}{2}\pi$, $\xi_5 = \frac{2}{3}\pi$

9. $f(x) = [\![x]\!] + 2$, $-3 \leq x \leq 3$; for Δ: $x_0 = -3$, $x_1 = -1$, $x_2 = 0$, $x_3 = 2$, $x_4 = 3$; $\xi_1 = -2.5$, $\xi_2 = -0.5$, $\xi_3 = 1$, $\xi_4 = 2.5$

In Exercises 10 through 18, find the exact value of the definite integral. Use the method of Example 1 of this section.

10. $\int_2^7 3x \, dx$ 11. $\int_0^2 x^2 \, dx$ 12. $\int_2^4 x^2 \, dx$

13. $\int_1^2 x^3 \, dx$ 14. $\int_0^5 (x^3 - 1) \, dx$

15. $\int_1^4 (x^2 + 4x + 5) \, dx$ 16. $\int_0^4 (x^2 + x - 6) \, dx$

17. $\int_{-2}^2 (x^3 + 1) \, dx$ 18. $\int_{-2}^1 x^4 \, dx$

In Exercises 19 through 28, find the exact area of the region in the following way: (a) Express the measure of the area as the limit of a Riemann sum with regular partitions; (b) express this limit with *definite integral notation; (c) evaluate the definite integral by the method of this section and a suitable choice of ξ_i. Draw a figure showing the region.*

19. Bounded by the line $y = 2x - 1$, the x axis, and the lines $x = 1$ and $x = 5$.

20. Bounded by the line $y = 2x - 6$, the x axis, and the lines $x = 4$ and $x = 7$.

21. Bounded by the line $y = -3x + 2$, the x axis, and the lines $x = -5$ and $x = -1$.

22. Bounded by the line $y = 3x + 2$, the x axis, and the lines $x = -4$ and $x = -1$.

23. Bounded by the curve $y = 4 - x^2$, the x axis, and the lines $x = 1$ and $x = 2$.

24. Bounded by the curve $y = (x + 3)^2$, the x axis, and the lines $x = -3$ and $x = 0$.

25. Bounded by the curve $y = 12 - x - x^2$, the x axis, and the lines $x = -3$ and $x = 2$.

26. Bounded by the curve $y = 10 + x - x^2$, the x axis, and the lines $x = -2$ and $x = 3$.

27. Bounded by the curve $y = x^3 - 4$, the x axis, and the lines $x = -2$ and $x = -1$.

28. Bounded by the curve $y = 6x + x^2 - x^3$, the x axis, and the lines $x = -1$ and $x = 3$.

In Exercises 29 through 32, approximate the value of the definite integral by using a calculator to compute to four decimal places the corresponding Riemann sum with a regular partition of n subintervals and ξ_i as the left or right (as indicated) endpoint of each subinterval.

29. $\int_2^5 \dfrac{1}{x^2} \, dx$, $n = 9$, ξ_i is the right endpoint.

30. $\int_3^4 \dfrac{1}{x} \, dx$, $n = 10$, ξ_i is the left endpoint.

31. $\int_{-\pi/3}^{\pi/3} \sec x \, dx$, $n = 8$, ξ_i is the left endpoint.

32. $\int_{-\pi/6}^{\pi/3} \tan x \, dx$, $n = 6$, ξ_i is the right endpoint.

33. Express as a definite integral: $\displaystyle\lim_{n \to +\infty} \sum_{i=1}^{n} \frac{8i^2}{n^3}$. (*Hint:* Consider the function f for which $f(x) = x^2$.)

34. Express as a definite integral: $\displaystyle\lim_{n \to +\infty} \sum_{i=1}^{n} \frac{1}{n + i}$. $\left(\textit{Hint:}\right.$ Consider the function f for which $f(x) = \dfrac{1}{x}$ on $[1, 2]$. $\left.\right)$

35. Express as a definite integral: $\displaystyle\lim_{n \to +\infty} \sum_{i=1}^{n} \frac{n}{(i + n)^2}$. $\left(\textit{Hint:}\right.$ Consider the function f for which $f(x) = \dfrac{1}{x^2}$ on $[1, 2]$. $\left.\right)$

36. Let $[a, b]$ be any interval such that $a < 0 < b$. Prove that even though the unit step function (Exercise 24 in Exercises 2.3) is discontinuous on $[a, b]$, it is integrable on $[a, b]$ and $\int_a^b U(x) \, dx = b$.

37. Prove that even though the signum function is discontinuous on $[-1, 1]$, it is integrable on $[-1, 1]$. Furthermore show that $\int_{-1}^1 \operatorname{sgn} x \, dx = 0$.

38. Prove that even though the greatest integer function is discontinuous on $\left[0, \frac{3}{2}\right]$, it is integrable on $\left[0, \frac{3}{2}\right]$. Furthermore show that $\int_0^{3/2} [\![x]\!] \, dx = \frac{1}{2}$.

5.6   PROPERTIES OF THE DEFINITE INTEGRAL

Evaluating a definite integral from the definition, by actually finding the limit of a sum as we did in Section 5.5, is usually quite tedious and frequently almost impossible. To establish a much simpler method we first need to develop some properties of the definite integral. First the following two theorems about Riemann sums are needed.

5.6.1 THEOREM

If Δ is any partition of the closed interval $[a, b]$, then

$$\lim_{\|\Delta\| \to 0} \sum_{i=1}^{n} \Delta_i x = b - a$$

Proof

$$\sum_{i=1}^{n} \Delta_i x - (b - a) = (b - a) - (b - a)$$
$$= 0$$

Hence, for any $\epsilon > 0$, any choice of $\delta > 0$ guarantees that

$$\text{if} \quad \|\Delta\| < \delta \quad \text{then} \quad \left| \sum_{i=1}^{n} \Delta_i x - (b - a) \right| < \epsilon$$

Thus by Definition 5.5.1,

$$\lim_{||\Delta|| \to 0} \sum_{i=1}^{n} \Delta_i x = b - a$$

■

5.6.2 THEOREM   If f is defined on the closed interval $[a, b]$, and if

$$\lim_{||\Delta|| \to 0} \sum_{i=1}^{n} f(\xi_i) \Delta_i x$$

exists, where Δ is any partition of $[a, b]$, then if k is any constant,

$$\lim_{||\Delta|| \to 0} \sum_{i=1}^{n} k f(\xi_i) \Delta_i x = k \lim_{||\Delta|| \to 0} \sum_{i=1}^{n} f(\xi_i) \Delta_i x$$

The proof of this theorem is left as an exercise (see Exercise 43).

▶ **ILLUSTRATION 1**   Refer to Figure 1. If $k > 0$, the definite integral $\int_a^b k \, dx$ gives the measure of the area of the shaded region, which is a rectangle whose dimensions are k units and $(b - a)$ units. This fact is a geometric interpretation of the following theorem when $k > 0$.   ◀

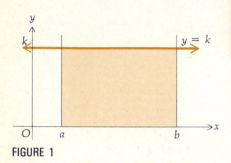

FIGURE 1

5.6.3 THEOREM   If k is any constant, then

$$\int_a^b k \, dx = k(b - a)$$

Proof   From Definition 5.5.2,

$$\int_a^b f(x) \, dx = \lim_{||\Delta|| \to 0} \sum_{i=1}^{n} f(\xi_i) \Delta_i x$$

If $f(x) = k$ for all x in $[a, b]$, we have from this equation

$$\int_a^b k \, dx = \lim_{||\Delta|| \to 0} \sum_{i=1}^{n} k \, \Delta_i x$$

$$= k \lim_{||\Delta|| \to 0} \sum_{i=1}^{n} \Delta_i x \qquad \text{(by Theorem 5.6.2)}$$

$$= k(b - a) \qquad \text{(by Theorem 5.6.1)}$$

■

EXAMPLE 1   Evaluate

$$\int_{-3}^{5} 4 \, dx$$

Solution   We apply Theorem 5.6.3.

$$\int_{-3}^{5} 4 \, dx = 4[5 - (-3)]$$

$$= 4(8)$$

$$= 32$$

5.6.4 THEOREM   If the function f is integrable on the closed interval $[a, b]$, and if k is any constant, then

$$\int_a^b k f(x) \, dx = k \int_a^b f(x) \, dx$$

Proof   Because f is integrable on $[a, b]$, $\displaystyle\lim_{||\Delta|| \to 0} \sum_{i=1}^{n} f(\xi_i) \Delta_i x$ exists; so by Theorem 5.6.2,

$$\lim_{||\Delta|| \to 0} \sum_{i=1}^{n} k f(\xi_i)\Delta_i x = k \lim_{||\Delta|| \to 0} \sum_{i=1}^{n} f(\xi_i) \Delta_i x$$

Therefore

$$\int_a^b k f(x) \, dx = k \int_a^b f(x) \, dx$$

∎

Observe the similarity of the following theorem to Limit Theorem 4 (2.2.4) about the limit of the sum of two functions. Also notice that the proofs of the two theorems are alike.

5.6.5 THEOREM   If the functions f and g are integrable on $[a, b]$, then $f + g$ is integrable on $[a, b]$ and

$$\int_a^b [f(x) + g(x)] \, dx = \int_a^b f(x) \, dx + \int_a^b g(x) \, dx$$

Proof   The functions f and g are integrable on $[a, b]$; thus let

$$\int_a^b f(x) \, dx = M \quad \text{and} \quad \int_a^b g(x) \, dx = N$$

To prove that $f + g$ is integrable on $[a, b]$ and $\displaystyle\int_a^b [f(x) + g(x)] \, dx = M + N$, we must show that for any $\epsilon > 0$ there exists a $\delta > 0$ such that for all partitions Δ and for any ξ_i in $[x_{i-1}, x_i]$, if $||\Delta|| < \delta$, then

$$\left| \sum_{i=1}^{n} [f(\xi_i) + g(\xi_i)] \Delta_i x - (M + N) \right| < \epsilon$$

Because

$$M = \lim_{||\Delta|| \to 0} \sum_{i=1}^{n} f(\xi_i) \Delta_i x \quad \text{and} \quad N = \lim_{||\Delta|| \to 0} \sum_{i=1}^{n} g(\xi_i) \Delta_i x$$

it follows that for any $\epsilon > 0$ there exist a $\delta_1 > 0$ and a $\delta_2 > 0$ such that for all partitions Δ and for any ξ_i in $[x_{i-1}, x_i]$, if $||\Delta|| < \delta_1$ and $||\Delta|| < \delta_2$, then

$$\left| \sum_{i=1}^{n} f(\xi_i) \Delta_i x - M \right| < \frac{\epsilon}{2} \quad \text{and} \quad \left| \sum_{i=1}^{n} g(\xi_i) \Delta_i x - N \right| < \frac{\epsilon}{2}$$

Therefore, if $\delta = \min(\delta_1, \delta_2)$, then for any $\epsilon > 0$, for all partitions Δ and for any ξ_i in $[x_{i-1}, x_i]$, if $||\Delta|| < \delta$,

$$\left| \sum_{i=1}^{n} f(\xi_i) \Delta_i x - M \right| + \left| \sum_{i=1}^{n} g(\xi_i) \Delta_i x - N \right| < \frac{\epsilon}{2} + \frac{\epsilon}{2} = \epsilon \tag{1}$$

By the triangle inequality we have

$$\left| \left(\sum_{i=1}^{n} f(\xi_i) \Delta_i x - M \right) + \left(\sum_{i=1}^{n} g(\xi_i) \Delta_i x - N \right) \right|$$

$$\leq \left| \sum_{i=1}^{n} f(\xi_i) \Delta_i x - M \right| + \left| \sum_{i=1}^{n} g(\xi_i) \Delta_i x - N \right| \tag{2}$$

From inequalities (1) and (2) we have

$$\left|\left(\sum_{i=1}^{n} f(\xi_i)\,\Delta_i x + \sum_{i=1}^{n} g(\xi_i)\,\Delta_i x\right) - (M + N)\right| < \epsilon \tag{3}$$

From Theorem 5.4.4,

$$\sum_{i=1}^{n} f(\xi_i)\,\Delta_i x + \sum_{i=1}^{n} g(\xi_i)\,\Delta_i x = \sum_{i=1}^{n} [f(\xi_i) + g(\xi_i)]\,\Delta_i x$$

So by substituting from this equation into (3) we are able to conclude that for any $\epsilon > 0$, for all partitions Δ and for any ξ_i in $[x_{i-1}, x_i]$, if $\|\Delta\| < \delta$ where $\delta = \min(\delta_1, \delta_2)$, then

$$\left|\sum_{i=1}^{n} [f(\xi_i) + g(\xi_i)]\,\Delta_i x - (M + N)\right| < \epsilon$$

This proves that $f + g$ is integrable on $[a, b]$ and that

$$\int_a^b [f(x) + g(x)]\,dx = \int_a^b f(x)\,dx + \int_a^b g(x)\,dx \qquad\blacksquare$$

The plus sign in the statement of Theorem 5.6.5 can be replaced by a minus sign as a result of applying Theorem 5.6.4, where $k = -1$.

Theorem 5.6.5 can be extended to any number of functions. That is, if the functions $f_1, f_2, \ldots, f_n$ are all integrable on $[a, b]$, then $(f_1 \pm f_2 \pm \ldots \pm f_n)$ is integrable on $[a, b]$ and

$$\int_a^b [f_1(x) \pm f_2(x) \pm \ldots \pm f_n(x)]\,dx = \int_a^b f_1(x)\,dx \pm \int_a^b f_2(x)\,dx \pm \ldots \pm \int_a^b f_n(x)\,dx$$

EXAMPLE 2   Use the result of Example 1, Section 5.5, and the fact that $\int_1^3 x\,dx = 4$ to evaluate $\int_1^3 (3x^2 - 5x + 2)\,dx$.

Solution   In Example 1, Section 5.5, we had the result

$$\int_1^3 x^2\,dx = \tfrac{26}{3}$$

From properties of the definite integral,

$$\int_1^3 (3x^2 - 5x + 2)\,dx = \int_1^3 3x^2\,dx - \int_1^3 5x\,dx + \int_1^3 2\,dx$$
$$= 3\int_1^3 x^2\,dx - 5\int_1^3 x\,dx + 2(3 - 1)$$
$$= 3(\tfrac{26}{3}) - 5(4) + 4$$
$$= 26 - 20 + 4$$
$$= 10$$

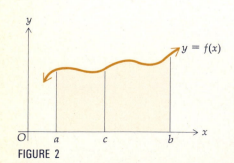

FIGURE 2

▶ **ILLUSTRATION 2**   A geometric interpretation of Theorem 5.6.6, which follows, is shown in Figure 2 where $f(x) \geq 0$. For all x in $[a, b]$, the measure of the area of the region bounded by the curve $y = f(x)$ and the x axis from a to b is equal to the sum of the measures of the areas of the regions from a to c and from c to b. ◀

5.6.6 THEOREM If the function f is integrable on the closed intervals $[a, b]$, $[a, c]$, and $[c, b]$, then

$$\int_a^b f(x)\, dx = \int_a^c f(x)\, dx + \int_c^b f(x)\, dx$$

where $a < c < b$.

Proof Let Δ be a partition of $[a, b]$. Form the partition Δ' of $[a, b]$ in the following way. If c is one of the partitioning points of Δ (i.e., $c = x_i$ for some i), then Δ' is exactly the same as Δ. If c is not one of the partitioning points of Δ but is contained in the subinterval $[x_{i-1}, x_i]$, then the partition Δ' has as its partitioning points all the partitioning points of Δ and, in addition, the point c. Therefore the subintervals of the partition Δ' are the same as the subintervals of Δ, with the exception that the subinterval $[x_{i-1}, x_i]$ of Δ is divided into the two subintervals $[x_{i-1}, c]$ and $[c, x_i]$.

If $\|\Delta'\|$ is the norm of Δ' and if $\|\Delta\|$ is the norm of Δ, then

$$\|\Delta'\| \leq \|\Delta\|$$

If in the partition Δ' the interval $[a, c]$ is divided into r subintervals and the interval $[c, b]$ is divided into $(n - r)$ subintervals, then the part of the partition Δ' from a to c gives a Riemann sum of the form

$$\sum_{i=1}^{r} f(\xi_i)\, \Delta_i x$$

and the other part of the partition Δ', from c to b, gives a Riemann sum of the form

$$\sum_{i=r+1}^{n} f(\xi_i)\, \Delta_i x$$

Using the definition of the definite integral and properties of sigma notation we have

$$\int_a^b f(x)\, dx = \lim_{\|\Delta\| \to 0} \sum_{i=1}^{n} f(\xi_i)\, \Delta_i x$$

$$= \lim_{\|\Delta\| \to 0} \left[\sum_{i=1}^{r} f(\xi_i)\, \Delta_i x + \sum_{i=r+1}^{n} f(\xi_i)\, \Delta_i x \right]$$

$$= \lim_{\|\Delta\| \to 0} \sum_{i=1}^{r} f(\xi_i)\, \Delta_i x + \lim_{\|\Delta\| \to 0} \sum_{i=r+1}^{n} f(\xi_i)\, \Delta_i x$$

Because $0 < \|\Delta'\| \leq \|\Delta\|$, we can replace $\|\Delta\| \to 0$ by $\|\Delta'\| \to 0$, giving

$$\int_a^b f(x)\, dx = \lim_{\|\Delta'\| \to 0} \sum_{i=1}^{r} f(\xi_i)\, \Delta_i x + \lim_{\|\Delta'\| \to 0} \sum_{i=r+1}^{n} f(\xi_i)\, \Delta_i x$$

Applying the definition of the definite integral to the right side of the above equation we have

$$\int_a^b f(x)\, dx = \int_a^c f(x)\, dx + \int_c^b f(x)\, dx \qquad \blacksquare$$

The result of Theorem 5.6.6 is true for any ordering of the numbers a, b, and c. This is stated as another theorem.

5.6.7 THEOREM

If f is integrable on a closed interval containing the three numbers a, b, and c, then

$$\int_a^b f(x)\,dx = \int_a^c f(x)\,dx + \int_c^b f(x)\,dx \tag{4}$$

regardless of the order of a, b, and c.

Proof If a, b, and c are distinct, there are six possible orderings of these three numbers: $a < b < c$, $a < c < b$, $b < a < c$, $b < c < a$, $c < a < b$, and $c < b < a$. The second ordering, $a < c < b$, is Theorem 5.6.6. We make use of Theorem 5.6.6 in proving that Equation (4) holds for the other orderings.

Suppose that $a < b < c$; then from Theorem 5.6.6 we have

$$\int_a^b f(x)\,dx + \int_b^c f(x)\,dx = \int_a^c f(x)\,dx \tag{5}$$

From Definition 5.5.5,

$$\int_b^c f(x)\,dx = -\int_c^b f(x)\,dx$$

Substituting from this equation into (5) we obtain

$$\int_a^b f(x)\,dx - \int_c^b f(x)\,dx = \int_a^c f(x)\,dx$$

Thus

$$\int_a^b f(x)\,dx = \int_a^c f(x)\,dx + \int_c^b f(x)\,dx$$

which is the desired result.

The proofs for the other four orderings are similar and left as exercises (see Exercises 44 through 47).

There is also the possibility that two of the three numbers are equal; for example, $a = c < b$. Then

$$\int_a^c f(x)\,dx = \int_a^a f(x)\,dx$$

$$= 0 \qquad \text{(by Definition 5.5.6)}$$

Also, because $a = c$,

$$\int_c^b f(x)\,dx = \int_a^b f(x)\,dx$$

Therefore

$$\int_a^c f(x)\,dx + \int_c^b f(x)\,dx = 0 + \int_a^b f(x)\,dx$$

which is the desired result. ■

▶ **ILLUSTRATION 3** In Figure 3, $f(x) \geq g(x) \geq 0$ for all x in $[a, b]$. The definite integral $\int_a^b f(x)\,dx$ gives the measure of the area of the region bounded by the curve $y = f(x)$, the x axis, and the lines $x = a$ and $x = b$; $\int_a^b g(x)\,dx$ gives the measure of the area of the region bounded by the curve $y = g(x)$, the x axis, and the lines $x = a$ and $x = b$. In the figure we see that the first area is greater

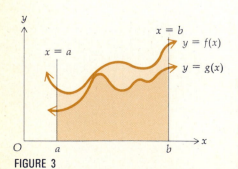

FIGURE 3

than the second area. We have given a geometric interpretation of the following theorem when $f(x)$ and $g(x)$ are nonnegative on $[a, b]$. ◀

5.6.8 THEOREM If the functions f and g are integrable on the closed interval $[a, b]$, and if $f(x) \geq g(x)$ for all x in $[a, b]$, then

$$\int_a^b f(x)\, dx \geq \int_a^b g(x)\, dx$$

Proof Because f and g are integrable on $[a, b]$, $\int_a^b f(x)\, dx$ and $\int_a^b g(x)\, dx$ both exist. Therefore,

$$\int_a^b f(x)\, dx - \int_a^b g(x)\, dx = \int_a^b f(x)\, dx + \int_a^b [-g(x)]\, dx \qquad \text{(by Theorem 5.6.4)}$$

$$= \int_a^b [f(x) - g(x)]\, dx \qquad \text{(by Theorem 5.6.5)}$$

Let h be the function defined by

$$h(x) = f(x) - g(x)$$

Then $h(x) \geq 0$ for all x in $[a, b]$ because $f(x) \geq g(x)$ for all x in $[a, b]$.

We wish to prove that $\int_a^b h(x) \geq 0$. Because

$$\int_a^b h(x)\, dx = \lim_{||\Delta|| \to 0} \sum_{i=1}^n h(\xi_i)\, \Delta_i x$$

we assume that

$$\lim_{||\Delta|| \to 0} \sum_{i=1}^n h(\xi_i)\, \Delta_i x = L < 0 \tag{6}$$

Then by Definition 5.5.1, with $\epsilon = -L$, there exists a $\delta > 0$ such that

$$\text{if} \quad ||\Delta|| < \delta \quad \text{then} \quad \left| \sum_{i=1}^n h(\xi_i)\, \Delta_i x - L \right| < -L \tag{7}$$

But because

$$\sum_{i=1}^n h(\xi_i)\, \Delta_i x - L \leq \left| \sum_{i=1}^n h(\xi_i)\, \Delta_i x - L \right|$$

from (7) we have

$$\text{if} \quad ||\Delta|| < \delta \quad \text{then} \quad \sum_{i=1}^n h(\xi_i)\, \Delta_i x - L < -L$$

$$\Leftrightarrow \quad \text{if} \quad ||\Delta|| < \delta \quad \text{then} \quad \sum_{i=1}^n h(\xi_i)\, \Delta_i x < 0$$

But this statement is impossible, because every $h(\xi_i)$ is nonnegative and every $\Delta_i x > 0$; thus we have a contradiction to our assumption (6). Therefore (6) is false, and

$$\lim_{||\Delta|| \to 0} \sum_{i=1}^n h(\xi_i)\, \Delta_i x \geq 0$$

$$\int_a^b h(x) \geq 0$$

Because $h(x) = f(x) - g(x)$, we have

$$\int_a^b [f(x) - g(x)] \, dx \geq 0$$

$$\int_a^b f(x) \, dx - \int_a^b g(x) \, dx \geq 0$$

$$\int_a^b f(x) \, dx \geq \int_a^b g(x) \, dx$$

$\blacksquare$

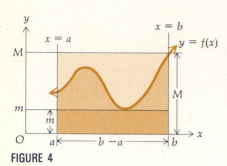

FIGURE 4

▶ **ILLUSTRATION 4**   In Figure 4, $f(x) \geq 0$ for all x in $[a, b]$, and m and M are, respectively, the absolute minimum and absolute maximum function values of f on $[a, b]$. The integral $\int_a^b f(x) \, dx$ gives the measure of the area of the region bounded by the curve $y = f(x)$, the x axis, and the lines $x = a$ and $x = b$. This area is greater than that of the rectangle whose dimensions are m and $b - a$ and less than that of the rectangle whose dimensions are M and $b - a$. Thus we have a geometric interpretation of the next theorem if $f(x) \geq 0$ for all x in $[a, b]$.   ◀

5.6.9 THEOREM   Suppose that the function f is continuous on the closed interval $[a, b]$. If m and M are, respectively, the absolute minimum and absolute maximum function values of f on $[a, b]$ so that

$$m \leq f(x) \leq M \qquad \text{for } a \leq x \leq b$$

then

$$m(b - a) \leq \int_a^b f(x) \, dx \leq M(b - a)$$

Proof   Because f is continuous on $[a, b]$, the extreme-value theorem guarantees the existence of m and M.

By Theorem 5.6.3,

$$\int_a^b m \, dx = m(b - a) \tag{8}$$

and

$$\int_a^b M \, dx = M(b - a) \tag{9}$$

Because f is continuous on $[a, b]$, it follows from Theorem 5.5.3 that f is integrable on $[a, b]$. Then because $f(x) \geq m$ for all x in $[a, b]$, we have from Theorem 5.6.8

$$\int_a^b f(x) \, dx \geq \int_a^b m \, dx$$

which from (8) gives

$$\int_a^b f(x) \, dx \geq m(b - a) \tag{10}$$

Similarly, because $M \geq f(x)$ for all x in $[a, b]$, it follows from Theorem 5.6.8 that

$$\int_a^b M \, dx \geq \int_a^b f(x) \, dx$$

which from (9) gives

$$M(b - a) \geq \int_a^b f(x)\, dx$$

Combining this inequality with (10) we have

$$m(b - a) \leq \int_a^b f(x)\, dx \leq M(b - a) \qquad \blacksquare$$

EXAMPLE 3   Apply Theorem 5.6.9 to find a closed interval containing the value of $\int_{1/2}^4 (x^3 - 6x^2 + 9x + 1)\, dx$. Use the results of Example 1, Section 4.4.

Solution   If

$$f(x) = x^3 - 6x^2 + 9x + 1$$

from Example 1 in Section 4.4, f has a relative minimum value of 1 at $x = 3$ and a relative maximum value of 5 at $x = 1$. $f(\frac{1}{2}) = \frac{33}{8}$ and $f(4) = 5$. Hence the absolute minimum value of f on $[\frac{1}{2}, 4]$ is 1, and the absolute maximum value is 5. Taking $m = 1$ and $M = 5$ in Theorem 5.6.9 we have

$$1(4 - \tfrac{1}{2}) \leq \int_{1/2}^4 (x^3 - 6x^2 + 9x + 1)\, dx \leq 5(4 - \tfrac{1}{2})$$

$$\tfrac{7}{2} \leq \int_{1/2}^4 (x^3 - 6x^2 + 9x + 1)\, dx \leq \tfrac{35}{2}$$

Therefore the closed interval $[\frac{7}{2}, \frac{35}{2}]$ contains the value of the definite integral. In Example 2, Section 5.8, we show that the exact value of the definite integral is $\frac{679}{64}$.

EXAMPLE 4   Apply Theorem 5.6.9 to find a closed interval containing the value of $\int_{\pi/4}^{3\pi/4} \sqrt{\sin x}\, dx$.

Solution   If $f(x) = \sqrt{\sin x}$, then

$$f'(x) = \frac{\cos x}{2\sqrt{\sin x}}$$

For x in $[\frac{1}{4}\pi, \frac{3}{4}\pi]$, $f'(x) = 0$ when $x = \frac{1}{2}\pi$. Since $f'(x) > 0$ when $\frac{1}{4}\pi < x < \frac{1}{2}\pi$, and $f'(x) < 0$ when $\frac{1}{2}\pi < x < \frac{3}{4}\pi$, it follows that f has a relative maximum value at $\frac{1}{2}\pi$; and $f(\frac{1}{2}\pi) = 1$. Furthermore, $f(\frac{1}{4}\pi) = \sqrt[4]{2}/\sqrt{2} \approx 0.841$, and $f(\frac{3}{4}\pi) \approx 0.841$. Thus, on $[\frac{1}{4}\pi, \frac{3}{4}\pi]$ the absolute minimum value of f is 0.841 and the absolute maximum value is 1. So, with $m = 0.841$ and $M = 1$ in Theorem 5.6.9

$$0.841[\tfrac{3}{4}\pi - \tfrac{1}{4}\pi] \leq \int_{\pi/4}^{3\pi/4} \sqrt{\sin x}\, dx \leq 1[\tfrac{3}{4}\pi - \tfrac{1}{4}\pi]$$

$$0.420\pi \leq \int_{\pi/4}^{3\pi/4} \sqrt{\sin x}\, dx \leq 0.5\pi$$

$$1.32 \leq \int_{\pi/4}^{3\pi/4} \sqrt{\sin x}\, dx \leq 1.57$$

The value of the definite integral is therefore in the closed interval $[1.32, 1.57]$.

EXERCISES 5.6

In Exercises 1 through 6, evaluate the definite integral.

1. $\int_2^5 4\,dx$　　　**2.** $\int_{-3}^4 7\,dx$　　　**3.** $\int_{-2}^2 \sqrt{5}\,dx$

4. $\int_5^{-1} 6\,dx$　　　**5.** $\int_{-5}^{-10} dx$　　　**6.** $\int_3^3 dx$

In Exercises 7 through 18, evaluate the definite integral by using the following results:

$$\int_{-1}^2 x^2\,dx = 3 \qquad \int_{-1}^2 x\,dx = \tfrac{3}{2} \qquad \int_0^\pi \sin x\,dx = 2$$

$$\int_0^\pi \cos x\,dx = 0 \qquad \int_0^\pi \sin^2 x\,dx = \tfrac{1}{2}\pi$$

7. $\int_{-1}^2 (2x^2 - 4x + 5)\,dx$　　　**8.** $\int_{-1}^2 (8 - x^2)\,dx$

9. $\int_{-1}^2 (2 - 5x + \tfrac{1}{2}x^2)\,dx$　　　**10.** $\int_{-1}^2 (3x^2 - 4x - 1)\,dx$

11. $\int_2^{-1} (2x + 1)^2\,dx$　　　**12.** $\int_{-1}^2 (5x^2 + \tfrac{1}{3}x - \tfrac{1}{2})\,dx$

13. $\int_{-1}^2 (x - 1)(2x + 3)\,dx$　　　**14.** $\int_2^{-1} 3x(x - 4)\,dx$

15. $\int_0^\pi (2\sin x + 3\cos x + 1)\,dx$　**16.** $\int_0^\pi 3\cos^2 x\,dx$

17. $\int_0^\pi (\cos x + 4)^2\,dx$　　　**18.** $\int_\pi^0 (\sin x - 2)^2\,dx$

In Exercises 19 through 34, apply Theorem 5.6.9 to find a closed interval containing the value of the definite integral.

19. $\int_3^7 2x\,dx$　**20.** $\int_2^5 3x\,dx$　**21.** $\int_0^4 x^2\,dx$　**22.** $\int_{-2}^1 x^3\,dx$

23. $\int_{-3}^6 \sqrt{3 + x}\,dx$　　　**24.** $\int_{-2}^1 (x + 1)^{2/3}\,dx$

25. $\int_{-4}^0 (x^4 - 8x^2 + 16)\,dx$　　**26.** $\int_{-1}^4 (x^4 - 8x^2 + 16)\,dx$

27. $\int_{\pi/6}^{\pi/3} \sin x\,dx$　　　**28.** $\int_{-\pi/3}^{2\pi/3} \cos x\,dx$

29. $\int_1^4 |x - 2|\,dx$　　　**30.** $\int_{-1}^2 \sqrt{x^2 + 5}\,dx$

31. $\int_{-1}^2 \frac{x}{x + 2}\,dx$　　　**32.** $\int_{-5}^2 \frac{x + 5}{x - 3}\,dx$

33. $\int_{-\pi/2}^{\pi/2} (4\cos^3 x - 9\cos x)\,dx$　**34.** $\int_{-\pi/2}^{\pi/2} 3\sin^3 x\,dx$

In Exercises 35 through 38, use Theorem 5.6.8 to determine which of the symbols $\geq$ or $\leq$ should be inserted in the blank to make the inequality correct.

35. $\int_{-1}^3 (2x^2 - 4)\,dx$ _____ $\int_{-1}^3 (x^2 - 6)\,dx$

36. $\int_4^5 \sqrt{6 - x}\,dx$ _____ $\int_4^5 \sqrt{x - 2}\,dx$

37. $\int_{3\pi/4}^{5\pi/4} \sin^2 x\,dx$ _____ $\int_{3\pi/4}^{5\pi/4} \cos^2 x\,dx$

38. $\int_0^{\pi/4} \cos x\,dx$ _____ $\int_0^{\pi/4} \sin x\,dx$

39. Show that if f is continuous on $[-3, 4]$, then

$$\int_3^{-1} f(x)\,dx + \int_4^3 f(x)\,dx + \int_{-3}^4 f(x)\,dx + \int_{-1}^{-3} f(x)\,dx = 0$$

40. Show that if f is continuous on $[-1, 2]$, then

$$\int_{-1}^2 f(x)\,dx + \int_2^0 f(x)\,dx + \int_0^1 f(x)\,dx + \int_1^{-1} f(x)\,dx = 0$$

41. Show that $\int_0^1 x\,dx \geq \int_0^1 x^2\,dx$ but $\int_1^2 x\,dx \leq \int_1^2 x^2\,dx$. Do not evaluate the definite integrals.

42. If f is continuous on $[a, b]$, prove that

$$\left| \int_a^b f(x)\,dx \right| \leq \int_a^b |f(x)|\,dx$$

(*Hint:* $-|f(x)| \leq f(x) \leq |f(x)|$.)

43. Prove Theorem 5.6.2.

In Exercises 44 through 49, prove that Theorem 5.6.7 is valid for the given ordering of a, b, and c; in each case use Theorem 5.6.6.

44. $b < a < c$　　**45.** $b < c < a$　　**46.** $c < a < b$
47. $c < b < a$　　**48.** $a < c = b$　　**49.** $a = b < c$

5.7 THE MEAN-VALUE THEOREM FOR INTEGRALS

FIGURE 1

We continue our development of properties of the definite integral in this section with the *mean-value theorem for integrals*. This theorem is prominent in the proof of the first fundamental theorem of the calculus (Theorem 5.8.1), which is important because it leads to a method for evaluating definite integrals. We begin with an illustration that gives a geometric interpretation of the mean-value theorem for integrals.

▶ **ILLUSTRATION 1**　Consider $f(x) \geq 0$ for all values of x in $[a, b]$. Then $\int_a^b f(x)\,dx$ gives the measure of the area of the region bounded by the curve whose equation is $y = f(x)$, the x axis, and the lines $x = a$ and $x = b$. See Figure 1. The mean-value theorem for integrals states that there is a number χ in $[a, b]$ such that the area of the rectangle $AEFB$ of height $f(\chi)$ units and width $(b - a)$ units is equal to the area of the region $ADCB$. ◀

5.7.1 THEOREM
Mean-Value Theorem for Integrals

If the function f is continuous on the closed interval $[a, b]$, there exists a number χ in $[a, b]$ such that

$$\int_a^b f(x)\,dx = f(\chi)(b - a)$$

Proof Because f is continuous on $[a, b]$, from the extreme-value theorem f has an absolute maximum value and an absolute minimum value on $[a, b]$.
Let m be the absolute minimum value occurring at $x = x_m$. Thus

$$f(x_m) = m \qquad a \le x_m \le b \tag{1}$$

Let M be the absolute maximum value, occurring at $x = x_M$. Thus

$$f(x_M) = M \qquad a \le x_M \le b \tag{2}$$

We have, then,

$$m \le f(x) \le M \qquad \text{for all } x \text{ in } [a, b]$$

From Theorem 5.6.9 it follows that

$$m(b - a) \le \int_a^b f(x)\,dx \le M(b - a)$$

Dividing by $b - a$ and noting that $b - a$ is positive because $b > a$ we get

$$m \le \frac{\int_a^b f(x)\,dx}{b - a} \le M$$

But from (1) and (2), $m = f(x_m)$ and $M = f(x_M)$; so we have

$$f(x_m) \le \frac{\int_a^b f(x)\,dx}{b - a} \le f(x_M)$$

From this inequality and the intermediate-value theorem there is some number χ in a closed interval containing x_m and x_M such that

$$f(\chi) = \frac{\int_a^b f(x)\,dx}{b - a}$$

$$\Leftrightarrow \int_a^b f(x)\,dx = f(\chi)(b - a) \qquad a \le \chi \le b \qquad \blacksquare$$

The value of χ in Theorem 5.7.1 is not necessarily unique. The theorem does not provide a method for finding χ, but it states that a value of χ exists, and this fact is used to prove other theorems. In some particular cases we can find the value of χ guaranteed by the theorem, as we illustrate in the following example.

EXAMPLE 1 Find the value of χ such that $\int_1^3 f(x)\,dx = f(\chi)(3 - 1)$ if $f(x) = x^2$. Use the result of Example 1 in Section 5.5.

Solution In Example 1 of Section 5.5 we obtained

$$\int_1^3 x^2\,dx = \frac{26}{3}$$

Therefore we wish to find χ such that

$$f(\chi) \cdot (2) = \tfrac{26}{3}$$

that is,

$$\chi^2 = \tfrac{13}{3}$$

$$\chi = \pm\tfrac{1}{3}\sqrt{39}$$

We reject $-\tfrac{1}{3}\sqrt{39}$ because it is not in the interval $[1, 3]$, and we have

$$\int_1^3 f(x)\,dx = f(\tfrac{1}{3}\sqrt{39})(3 - 1)$$

The value $f(\chi)$ given by Theorem 5.7.1 is called the *mean value* (or *average value*) of f on the interval $[a, b]$. It is a generalization of the arithmetic mean of a finite set of numbers. That is, if $\{f(x_1), f(x_2), \ldots, f(x_n)\}$ is a set of n numbers, then the arithmetic mean is given by

$$\frac{\displaystyle\sum_{i=1}^{n} f(x_i)}{n}$$

To generalize this definition, consider a regular partition of the closed interval $[a, b]$, which is divided into n subintervals of equal length $\Delta x = (b - a)/n$. Let ξ_i be any point in the ith subinterval. Form the sum:

$$\frac{\displaystyle\sum_{i=1}^{n} f(\xi_i)}{n} \tag{3}$$

This quotient corresponds to the arithmetic mean of n numbers. Because $\Delta x = (b - a)/n$ we have

$$n = \frac{b - a}{\Delta x} \tag{4}$$

Substituting from (4) into (3) we obtain

$$\frac{\displaystyle\sum_{i=1}^{n} f(\xi_i)}{\dfrac{b - a}{\Delta x}} = \frac{\displaystyle\sum_{i=1}^{n} f(\xi_i)\,\Delta x}{b - a}$$

Taking the limit as $n \to +\infty$ (or $\Delta x \to 0$) we have, if the limit exists,

$$\lim_{n \to +\infty} \frac{\displaystyle\sum_{i=1}^{n} f(\xi_i)\,\Delta x}{b - a} = \frac{\displaystyle\int_a^b f(x)\,dx}{b - a}$$

This result leads to the following definition.

5.7.2 DEFINITION   If the function f is integrable on the closed interval $[a, b]$, then the **average value** of f on $[a, b]$ is

$$\frac{\displaystyle\int_a^b f(x)\,dx}{b - a}$$

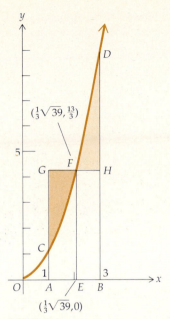

$(\frac{1}{3}\sqrt{39}, \frac{13}{3})$

$(\frac{1}{3}\sqrt{39}, 0)$

FIGURE 2

EXAMPLE 2   (a) If $f(x) = x^2$, find the average value of f on the interval $[1, 3]$. (b) Interpret the result of part (a) geometrically.

Solution
(a) In Example 1, Section 5.5, we obtained

$$\int_1^3 x^2 \, dx = \tfrac{26}{3}$$

So if A.V. is the average value of f on $[1, 3]$, we have

$$\text{A.V.} = \frac{\tfrac{26}{3}}{3 - 1}$$

$$= \tfrac{13}{3}$$

(b) In Example 1 we found for this function

$$f(\tfrac{1}{3}\sqrt{39}) = \tfrac{13}{3}$$

Therefore the average value of f occurs at $\tfrac{1}{3}\sqrt{39}$. In Figure 2 there is a sketch of the graph of f on $[1, 3]$ and the line segment from the point $E(\tfrac{1}{3}\sqrt{39}, 0)$ on the x axis to the point $F(\tfrac{1}{3}\sqrt{39}, \tfrac{13}{3})$ on the graph of f. The area of rectangle $AGHB$ having height $\tfrac{13}{3}$ and width 2 is equal to the area of the region $ACDB$. Consequently, the area of the shaded region CGF is equal to the area of the shaded region FDH.

An application of the average value of a function occurs in physics and engineering in connection with the concept of center of mass. This topic is discussed in Chapter 6.

EXERCISES 5.7

In Exercises 1 through 8, find the value of χ satisfying the mean-value theorem for integrals. For the value of the definite integral use the result of the indicated exercise in Exercises 5.5. Draw a figure illustrating the application of the theorem.

1. $\int_0^2 x^2 \, dx$; Exercise 11   2. $\int_2^4 x^2 \, dx$; Exercise 12

3. $\int_1^2 x^3 \, dx$; Exercise 13   4. $\int_0^5 (x^3 - 1) \, dx$; Exercise 14

5. $\int_1^4 (x^2 + 4x + 5) \, dx$; Exercise 15

6. $\int_0^4 (x^2 + x - 6) \, dx$; Exercise 16

7. $\int_{-2}^2 (x^3 + 1) \, dx$; Exercise 17

8. $\int_{-2}^1 x^4 \, dx$; Exercise 18

In Exercises 9 through 12, use the mean-value theorem for integrals to prove the inequality.

9. $\int_0^2 \dfrac{1}{x^2 + 4} \, dx \le \dfrac{1}{2}$ 10. $\int_{-3}^3 \dfrac{1}{x^2 + 6} \, dx \le 1$

11. $\int_{-\pi/6}^{\pi/6} \cos x^2 \, dx \le \dfrac{\pi}{3}$ 12. $\int_0^\pi \sin\sqrt{x} \, dx \le \pi$

In Exercises 13 through 16, use the mean-value theorem for integrals and Theorem 5.6.8 to prove the inequality.

13. $0 \le \int_2^5 \dfrac{1}{x^3 + 1} \, dx \le \dfrac{1}{3}$ 14. $0 \le \int_5^9 \dfrac{1}{\sqrt{x - 1}} \, dx \le 2$

15. $0 \le \int_0^2 \sin\tfrac{1}{2}\pi x \, dx \le 2$ 16. $0 \le \int_{-1/2}^{1/2} \cos \pi x \le 1$

17. Given that $\int_{-1}^2 x \, dx = \tfrac{3}{2}$, find the average value of the identity function on the interval $[-1, 2]$. Also find the value of x at which the average value occurs. Give a geometric interpretation of the results.

18. Find the average value of the function f defined by $f(x) = x^2$ on the interval $[-1, 2]$ given that $\int_{-1}^2 x^2 \, dx = 3$. Also find the value of x at which the average value occurs. Give a geometric interpretation of the results.

19. Given that $\int_0^\pi \sin x \, dx = 2$, find the average value of the sine function on the interval $[0, \pi]$. Also find the smallest value of x at which the average value occurs. Give a geometric interpretation of the results.

20. Find the average value of the function f where $f(x) = \sec^2 x$ on the interval $[0, \tfrac{1}{4}\pi]$ given that $\int_0^{\pi/4} \sec^2 x \, dx = 1$. Also

find the value of x at which the average value occurs. Give a geometric interpretation of the results.

21. Suppose a ball is dropped from rest and after t seconds its velocity is v feet per second. Neglecting air resistance, express v in terms of t as $v = f(t)$, and find the average value of f on $[0, 2]$. (*Hint:* Find the value of the definite integral by interpreting it as the measure of the area of a region enclosed by a triangle.)

22. Find the average value of the function f defined by $f(x) = \sqrt{49 - x^2}$ on the interval $[0, 7]$. Draw a figure. (*Hint:* Find the value of the definite integral by interpreting it as the measure of the area of a region enclosed by a quarter-circle.)

23. Find the average value of the function f defined by $f(x) = \sqrt{16 - x^2}$ on the interval $[-4, 4]$. Draw a figure. (*Hint:* Find the value of the definite integral by interpreting it as the measure of the area of a region enclosed by a semicircle.)

24. Suppose that f is integrable on $[-4, 7]$. If the average value of f on the interval $[-4, 7]$ is $\frac{17}{4}$, find $\int_{-4}^{7} f(x)\, dx$.

25. If f is continuous on $[a, b]$ and $\int_a^b f(x)\, dx = 0$, prove that there is at least one number c in $[a, b]$ such that $f(c) = 0$.

26. The following theorem is a generalization of the mean-value theorem for integrals: If f and g are two functions that are continuous on the closed interval $[a, b]$ and $g(x) > 0$ for all x in the open interval (a, b), then there exists a number χ in $[a, b]$ such that

$$\int_a^b f(x)g(x)\, dx = f(\chi) \int_a^b g(x)\, dx$$

Prove this theorem by a method similar to that for Theorem 5.7.1: Obtain the inequality $m \le f(x) \le M$ and then conclude that $mg(x) \le f(x)g(x) \le Mg(x)$; apply Theorem 5.6.8 and proceed as in the proof of Theorem 5.7.1.

27. Show that when $g(x) = 1$, the theorem of Exercise 26 becomes the mean-value theorem for integrals.

In Exercises 28 through 32, use the theorem of Exercise 26 to prove the inequality.

28. $\displaystyle\int_0^4 \frac{x\, dx}{x^3 + 2} < \int_0^4 x\, dx$

29. $\displaystyle\int_{-1}^1 \frac{x^2\, dx}{\sqrt{x^2 + 4}} < \int_{-1}^1 x^2\, dx$

30. $\displaystyle\int_0^\pi x \sin x\, dx \le \int_0^\pi x\, dx$

31. $\displaystyle\int_{-1/2}^{1/2} \sin^2 \pi x \cos \pi x\, dx \le \int_{-1/2}^{1/2} \cos \pi x\, dx$

32. $\displaystyle\int_0^1 \frac{x \cos x}{x^2 + 1}\, dx \le \int_0^1 x\, dx$

5.8 THE FUNDAMENTAL THEOREMS OF THE CALCULUS

Historically, the basic concepts of the definite integral were used by the ancient Greeks, principally Archimedes (287–212 B.C.), more than 2000 years ago, which was many years before the differential calculus was discovered.

In the seventeenth century, almost simultaneously but working independently, Newton and Leibniz showed how the calculus could be used to find the area of a region bounded by a curve or a set of curves, by evaluating a definite integral by antidifferentiation. The procedure involves what are known as the *fundamental theorems of the calculus.* Before we state and prove these important theorems, we discuss definite integrals having a variable upper limit.

Let f be a function continuous on the closed interval $[a, b]$. Then the value of the definite integral $\int_a^b f(x)\, dx$ depends only on the function f and the numbers a and b, and not on the symbol x, which is used here as the independent variable. In Example 1, Section 5.5, we found the value of $\int_1^3 x^2\, dx$ to be $\frac{26}{3}$. Any other symbol instead of x could have been used; for example,

$$\int_1^3 t^2\, dt = \frac{26}{3} \qquad \int_1^3 u^2\, du = \frac{26}{3} \qquad \int_1^3 r^2\, dr = \frac{26}{3}$$

If f is continuous on the closed interval $[a, b]$, then by Theorem 5.5.3 the definite integral $\int_a^b f(t)\, dt$ exists. We previously stated that if the definite integral exists, it is a unique number. If x is a number in $[a, b]$, then f is continuous on $[a, x]$ because it is continuous on $[a, b]$. Consequently, $\int_a^x f(t)\, dt$ exists and is a unique number whose value depends on x. Therefore $\int_a^x f(t)\, dt$ defines a function F having as its domain all numbers in the closed interval [a, b] and whose function value at any number x in $[a, b]$ is given by

$$F(x) = \int_a^x f(t)\, dt \tag{1}$$

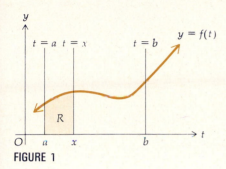

FIGURE 1

As a notational observation, if the limits of the definite integral are variables, different symbols are used for these limits and for the independent variable in the integrand. Hence, in (1), because x is the upper limit, we use the letter t as the independent variable in the integrand.

If, in (1), $f(t) \geq 0$ for all values of t in $[a, b]$, then the function value $F(x)$ can be interpreted geometrically as the measure of the area of the region bounded by the curve whose equation is $y = f(t)$, the t axis, and the lines $t = a$ and $t = x$. (See Figure 1.) Note that $F(a) = \int_a^a f(t)\, dt$, which by Definition 5.5.6 equals 0.

We now state and prove an important theorem giving the derivative of a function F defined as a definite integral having a variable upper limit. This theorem is called *the first fundamental theorem of the calculus.*

5.8.1 THEOREM
First Fundamental Theorem of the Calculus

Let the function f be continuous on the closed interval $[a, b]$ and let x be any number in $[a, b]$. If F is the function defined by

$$F(x) = \int_a^x f(t)\, dt$$

then

$$F'(x) = f(x) \tag{2}$$

(If $x = a$, the derivative in (2) may be a derivative from the right, and if $x = b$, the derivative in (2) may be a derivative from the left.)

Proof   Consider two numbers x_1 and $x_1 + \Delta x$ in $[a, b]$. Then

$$F(x_1) = \int_a^{x_1} f(t)\, dt$$

and

$$F(x_1 + \Delta x) = \int_a^{x_1 + \Delta x} f(t)\, dt$$

so that

$$F(x_1 + \Delta x) - F(x_1) = \int_a^{x_1 + \Delta x} f(t)\, dt - \int_a^{x_1} f(t)\, dt \tag{3}$$

By Theorem 5.6.7,

$$\int_a^{x_1} f(t)\, dt + \int_{x_1}^{x_1 + \Delta x} f(t)\, dt = \int_a^{x_1 + \Delta x} f(t)\, dt$$

$$\int_a^{x_1 + \Delta x} f(t)\, dt - \int_a^{x_1} f(t)\, dt = \int_{x_1}^{x_1 + \Delta x} f(t)\, dt$$

Substituting from this equation into (3) we get

$$F(x_1 + \Delta x) - F(x_1) = \int_{x_1}^{x_1 + \Delta x} f(t)\, dt \tag{4}$$

By the mean-value theorem for integrals (5.7.1) there is some number χ in the closed interval bounded by x_1 and $x_1 + \Delta x$ such that

$$\int_{x_1}^{x_1 + \Delta x} f(t)\, dt = f(\chi)\, \Delta x$$

From this equation and (4) we obtain

$$F(x_1 + \Delta x) - F(x_1) = f(\chi) \, \Delta x$$

$$\frac{F(x_1 + \Delta x) - F(x_1)}{\Delta x} = f(\chi)$$

Taking the limit as Δx approaches zero we have

$$\lim_{\Delta x \to 0} \frac{F(x_1 + \Delta x) - F(x_1)}{\Delta x} = \lim_{\Delta x \to 0} f(\chi) \tag{5}$$

The left side of (5) is $F'(x_1)$. To determine $\lim\limits_{\Delta x \to 0} f(\chi)$, recall that χ is in the closed interval bounded by x_1 and $x_1 + \Delta x$, and because

$$\lim_{\Delta x \to 0} x_1 = x_1 \quad \text{and} \quad \lim_{\Delta x \to 0} (x_1 + \Delta x) = x_1$$

it follows from the squeeze theorem (2.8.1) that $\lim\limits_{\Delta x \to 0} \chi = x_1$. Thus we have $\lim\limits_{\Delta x \to 0} f(\chi) = \lim\limits_{\chi \to x_1} f(\chi)$. Because f is continuous at x_1, $\lim\limits_{\chi \to x_1} f(\chi) = f(x_1)$; thus $\lim\limits_{\Delta x \to 0} f(\chi) = f(x_1)$, and from (5) we get

$$F'(x_1) = f(x_1) \tag{6}$$

If the function f is not defined for values of x less than a but is continuous from the right at a, then in the above argument, if $x_1 = a$ in (5), Δx must approach 0 from the right. Hence the left side of (6) will be $F'_+(x_1)$. Similarly, if f is not defined for values of x greater than b but is continuous from the left at b, then if $x_1 = b$ in (5), Δx must approach 0 from the left. Hence we have $F'_-(x_1)$ on the left side of (6).

Because x_1 is any number in $[a, b]$, Equation (6) states what we wished to prove.   ■

Theorem 5.8.1 states that the definite integral $\int_a^x f(t) \, dt$, with variable upper limit x, is an antiderivative of f.

Equation (2) of the theorem can be written in the following way by replacing $F'(x)$ by $\dfrac{d}{dx} \displaystyle\int_a^x f(t) \, dt$:

$$\frac{d}{dx} \int_a^x f(t) \, dt = f(x) \tag{7}$$

EXAMPLE 1   Compute the following derivatives:

(a) $\dfrac{d}{dx} \displaystyle\int_1^x \frac{1}{t^3 + 1} \, dt$ (b) $\dfrac{d}{dx} \displaystyle\int_3^{x^2} \sqrt{\cos t} \, dt$

Solution

(a) From (7) with $f(t) = \dfrac{1}{t^3 + 1}$, we have

$$\frac{d}{dx} \int_1^x \frac{1}{t^3 + 1} \, dt = \frac{1}{x^3 + 1}$$

(b) We use the chain rule with $u = x^2$, and we have

$$\frac{d}{dx} \int_3^{x^2} \sqrt{\cos t}\; dt = \frac{d}{du} \int_3^{u} \sqrt{\cos t}\; dt \cdot \frac{du}{dx}$$

From (7) with $f(t) = \sqrt{\cos t}$ and because $\dfrac{du}{dx} = 2x$, we have

$$\frac{d}{dx} \int_3^{x^2} \sqrt{\cos t}\; dt = \sqrt{\cos u}\;(2x)$$
$$= 2x\sqrt{\cos x^2}$$

We now use Theorem 5.8.1 to prove the *second fundamental theorem of the calculus*.

5.8.2 THEOREM
Second Fundamental Theorem
of the Calculus

Let the function f be continuous on the closed interval $[a, b]$ and let g be a function such that

$$g'(x) = f(x) \tag{8}$$

for all x in $[a, b]$. Then

$$\int_a^b f(t)\; dt = g(b) - g(a)$$

(If $x = a$, the derivative in (8) may be a derivative from the right, and if $x = b$, the derivative in (8) may be a derivative from the left.)

Proof   If f is continuous at all numbers in $[a, b]$, we know from Theorem 5.8.1 that the definite integral $\int_a^x f(t)\; dt$, with variable upper limit x, defines a function F whose derivative on $[a, b]$ is f. Because by hypothesis $g'(x) = f(x)$, it follows from Theorem 5.1.2 that

$$g(x) = \int_a^x f(t)\; dt + k$$

where k is some constant. Letting $x = b$ and $x = a$, successively, in this equation we get

$$g(b) = \int_a^b f(t)\; dt + k \tag{9}$$

and

$$g(a) = \int_a^a f(t)\; dt + k \tag{10}$$

From (9) and (10),

$$g(b) - g(a) = \int_a^b f(t)\; dt - \int_a^a f(t)\; dt$$

But, by Definition 5.5.6, $\int_a^a f(t)\; dt = 0$; so

$$g(b) - g(a) = \int_a^b f(t)\; dt$$

which is what we wished to prove.

If f is not defined for values of x greater than b but is continuous from the left at b, the derivative in (8) is a derivative from the left, and we have $g'_-(b) = F'_-(b)$, from which (9) follows. Similarly, if f is not defined for values of x less than a but is continuous from the right at a, then the derivative in (8) is a derivative from the right, and we have $g'_+(a) = F'_+(a)$, from which (10) follows. ∎

We are now able to find the exact value of a definite integral by applying Theorem 5.8.2. In the computation we denote

$$[g(b) - g(a)] \quad \text{by} \quad g(x)\Big]_a^b$$

▶ **ILLUSTRATION 1**   We apply the second fundamental theorem of the calculus to evaluate

$$\int_1^3 x^2 \, dx$$

Here $f(x) = x^2$. An antiderivative of x^2 is $\frac{1}{3}x^3$. From this we choose

$$g(x) = \frac{x^3}{3}$$

Therefore, from Theorem 5.8.2,

$$\int_1^3 x^2 \, dx = \frac{x^3}{3}\Big]_1^3$$

$$= 9 - \tfrac{1}{3}$$

$$= \tfrac{26}{3}$$

Compare this result with that of Example 1, Section 5.5.   ◀

Because of the connection between definite integrals and antiderivatives, we used the integral sign $\int$ for the notation $\int f(x) \, dx$ for an antiderivative. We now dispense with the terminology of antiderivatives and antidifferentiation and begin to call $\int f(x) \, dx$ an **indefinite integral**. The process of evaluating an indefinite integral or a definite integral is called **integration**.

The difference between an indefinite integral and a definite integral should be emphasized. The indefinite integral $\int f(x) \, dx$ is a function g such that its derivative $D_x[g(x)] = f(x)$. However, the definite integral $\int_a^b f(x) \, dx$ is a number whose value depends on the function f and the numbers a and b, and it is defined as the limit of a Riemann sum. The definition of the definite integral makes no reference to differentiation.

The indefinite integral involves an arbitrary constant; for instance,

$$\int x^2 \, dx = \frac{x^3}{3} + C$$

This arbitrary constant C is called a **constant of integration**. In applying the second fundamental theorem to evaluate a definite integral, we do not need to include the arbitrary constant C in the expression for $g(x)$ because the theorem permits us to select *any* antiderivative, including the one for which $C = 0$.

EXAMPLE 2   Evaluate

$$\int_{1/2}^{4} (x^3 - 6x^2 + 9x + 1) \, dx$$

Solution

$$\int_{1/2}^{4} (x^3 - 6x^2 + 9x + 1) \, dx = \int_{1/2}^{4} x^3 \, dx - 6\int_{1/2}^{4} x^2 \, dx + 9\int_{1/2}^{4} x \, dx + \int_{1/2}^{4} dx$$

$$= \frac{x^4}{4} - 6 \cdot \frac{x^3}{3} + 9 \cdot \frac{x^2}{2} + x \Big]_{1/2}^{4}$$

$$= (64 - 128 + 72 + 4) - (\tfrac{1}{64} - \tfrac{1}{4} + \tfrac{9}{8} + \tfrac{1}{2})$$

$$= \tfrac{679}{64}$$

EXAMPLE 3   Evaluate

$$\int_{-1}^{1} (x^{4/3} + 4x^{1/3}) \, dx$$

Solution

$$\int_{-1}^{1} (x^{4/3} + 4x^{1/3}) \, dx = \tfrac{3}{7}x^{7/3} + 4 \cdot \tfrac{3}{4}x^{4/3} \Big]_{-1}^{1}$$

$$= \tfrac{3}{7} + 3 - (-\tfrac{3}{7} + 3)$$

$$= \tfrac{6}{7}$$

EXAMPLE 4   Evaluate

$$\int_{0}^{2} 2x^2 \sqrt{x^3 + 1} \, dx$$

Solution

$$\int_{0}^{2} 2x^2 \sqrt{x^3 + 1} \, dx = \frac{2}{3} \int_{0}^{2} \sqrt{x^3 + 1} \, (3x^2 \, dx)$$

$$= \frac{2}{3} \cdot \frac{(x^3 + 1)^{3/2}}{\frac{3}{2}} \Big]_{0}^{2}$$

$$= \tfrac{4}{9}(8 + 1)^{3/2} - \tfrac{4}{9}(0 + 1)^{3/2}$$

$$= \tfrac{4}{9}(27 - 1)$$

$$= \tfrac{104}{9}$$

EXAMPLE 5   Evaluate

$$\int_{0}^{3} x\sqrt{1 + x} \, dx$$

Solution   To evaluate the indefinite integral $\int x\sqrt{1 + x} \, dx$ we let

$$u = \sqrt{1 + x} \qquad u^2 = 1 + x \qquad x = u^2 - 1 \qquad dx = 2u \, du$$

Substituting we have

$$\int x\sqrt{1 + x}\, dx = \int (u^2 - 1)u(2u\, du)$$

$$= 2\int (u^4 - u^2)\, du$$

$$= \tfrac{2}{5}u^5 - \tfrac{2}{3}u^3 + C$$

$$= \tfrac{2}{5}(1 + x)^{5/2} - \tfrac{2}{3}(1 + x)^{3/2} + C$$

Therefore the definite integral

$$\int_0^3 x\sqrt{1 + x}\, dx = \tfrac{2}{5}(1 + x)^{5/2} - \tfrac{2}{3}(1 + x)^{3/2}\Big]_0^3$$

$$= \tfrac{2}{5}(4)^{5/2} - \tfrac{2}{3}(4)^{3/2} - \tfrac{2}{3}(1)^{5/2} + \tfrac{2}{3}(1)^{3/2}$$

$$= \tfrac{64}{5} - \tfrac{16}{3} - \tfrac{2}{5} + \tfrac{2}{3}$$

$$= \tfrac{116}{15}$$

Another method for evaluating the definite integral in Example 5 is provided by a formula that follows from Theorems 5.8.2 and 5.2.1 (the chain rule for antidifferentiation). From these theorems, if F is an antiderivative of f,

$$\int_a^b f(g(x))g'(x)\, dx = F(g(x))\Big]_a^b$$

$$\Leftrightarrow \quad \int_a^b f(g(x))g'(x)\, dx = F(g(b)) - F(g(a))$$

Thus

$$\int_a^b f(g(x))g'(x)\, dx = F(u)\Big]_{g(a)}^{g(b)}$$

$$\Leftrightarrow \quad \int_a^b f(g(x))g'(x)\, dx = \int_{g(a)}^{g(b)} f(u)\, du \tag{11}$$

To apply (11), change the variables in the given integral by letting $u = g(x)$. Then $du = g'(x)\, dx$. Then change the x-limits of integration a and b to u-limits, which are $g(a)$ and $g(b)$.

▶ **ILLUSTRATION 2** To evaluate the integral of Example 5, let $u = \sqrt{1 + x}$, $x = u^2 - 1$, and $dx = 2u\, du$. Furthermore, when $x = 0$, $u = 1$, and when $x = 3$, $u = 2$. Thus, from (11) we have

$$\int_0^3 x\sqrt{1 + x}\, dx = 2\int_1^2 (u^4 - u^2)\, du$$

$$= \tfrac{2}{5}u^5 - \tfrac{2}{3}u^3\Big]_1^2$$

$$= \tfrac{64}{5} - \tfrac{16}{3} - \tfrac{2}{5} + \tfrac{2}{3}$$

$$= \tfrac{116}{15}$$

◀

EXAMPLE 6   Evaluate

$$\int_0^{\pi/2} \sin^3 x \cos x\, dx$$

Solution   Let

$$u = \sin x \qquad du = \cos x\, dx$$

When $x = 0$, $u = 0$; when $x = \frac{1}{2}\pi$, $u = 1$. Therefore

$$\int_0^{\pi/2} \sin^3 x \cos x \, dx = \int_0^1 u^3 \, du$$

$$= \frac{u^4}{4}\Bigg]_0^1$$

$$= \frac{1}{4}$$

EXAMPLE 7   Evaluate

$$\int_{-3}^4 |x + 2| \, dx$$

Solution

$$|x + 2| = \begin{cases} -x - 2 & \text{if} \quad x \le -2 \\ x + 2 & \text{if} \; -2 \le x \end{cases}$$

From Theorem 5.6.7,

$$\int_{-3}^4 |x + 2| \, dx = \int_{-3}^{-2} (-x - 2) \, dx + \int_{-2}^4 (x + 2) \, dx$$

$$= \left[-\frac{x^2}{2} - 2x \right]_{-3}^{-2} + \left[\frac{x^2}{2} + 2x \right]_{-2}^4$$

$$= [(-2 + 4) - (-\tfrac{9}{2} + 6)] + [(8 + 8) - (2 - 4)]$$

$$= \tfrac{1}{2} + 18$$

$$= \tfrac{37}{2}$$

EXAMPLE 8   Given $f(x) = \sec^2 x$. Find the average value of f on the interval $[-\frac{1}{4}\pi, \frac{1}{4}\pi]$.

Solution   Let A.V. be the average value of f on $[-\frac{1}{4}\pi, \frac{1}{4}\pi]$. From Definition 5.7.2,

$$\text{A.V.} = \frac{1}{\frac{1}{4}\pi - (-\frac{1}{4}\pi)} \int_{-\pi/4}^{\pi/4} \sec^2 x \, dx$$

$$= \frac{2}{\pi} \Big[\tan x \Big]_{-\pi/4}^{\pi/4}$$

$$= \frac{2}{\pi} [1 - (-1)]$$

$$= \frac{4}{\pi}$$

EXERCISES 5.8

In Exercises 1 through 36, evaluate the definite integral.

1. $\int_0^3 (3x^2 - 4x + 1) \, dx$

2. $\int_0^4 (x^3 - x^2 + 1) \, dx$

3. $\int_3^6 (x^2 - 2x) \, dx$

4. $\int_{-1}^3 (3x^2 + 5x - 1) \, dx$

5. $\int_1^2 \frac{x^2 + 1}{x^2} \, dx$

6. $\int_{-3}^5 (y^3 - 4y) \, dy$

7. $\int_0^1 \frac{z}{(z^2 + 1)^3} \, dz$

8. $\int_1^4 \sqrt{x}(2 + x) \, dx$

9. $\int_1^{10} \sqrt{5x - 1} \, dx$

10. $\int_0^{\sqrt{5}} t \sqrt{t^2 + 1} \, dt$

11. $\int_{-2}^0 3w \sqrt{4 - w^2} \, dw$

12. $\int_{-1}^3 \frac{dy}{(y + 2)^3}$

13. $\int_0^{\pi/2} \sin 2x \, dx$

14. $\int_0^\pi \cos \tfrac{1}{2} x \, dx$

15. $\int_1^2 t^2 \sqrt{t^3 + 1} \, dx$

16. $\int_1^3 \frac{x \, dx}{(3x^2 - 1)^3}$

17. $\int_0^1 \frac{(y^2 + 2y) \, dy}{\sqrt[3]{y^3 + 3y^2 + 4}}$

18. $\int_2^4 \frac{w^4 - w}{w^3} \, dw$

19. $\int_0^{15} \frac{w \, dw}{(1 + w)^{3/4}}$

20. $\int_4^5 x^2 \sqrt{x - 4} \, dx$

21. $\int_{-2}^5 |x - 3| \, dx$

22. $\int_{-4}^4 |x - 2| \, dx$

23. $\int_{-1}^1 \sqrt{|x| - x} \, dx$

24. $\int_{-3}^3 \sqrt{3 + |x|} \, dx$

25. $\int_0^3 (x + 2)\sqrt{x + 1} \, dx$

26. $\int_{-2}^1 (x + 1)\sqrt{x + 3} \, dx$

27. $\int_0^1 \frac{x^3 + 1}{x + 1} \, dx$

(*Hint:* Divide the numerator by the denominator.)

28. $\int_1^4 \frac{x^5 - x}{3x^3} \, dx$

29. $\int_1^{64} \left(\sqrt{t} - \frac{1}{\sqrt{t}} + \sqrt[3]{t} \right) dt$

30. $\int_0^1 \sqrt{x} \sqrt{1 + x\sqrt{x}} \, dx$

31. $\int_1^2 \frac{x^3 + 2x^2 + x + 2}{(x + 1)^2} \, dx$

32. $\int_{-3}^2 \frac{3x^3 - 24x^2 + 48x + 5}{x^2 - 8x + 16} \, dx$

(*Hint* for Exercises 31 and 32: Divide the numerator by the denominator.)

33. $\int_0^1 \sin \pi x \cos \pi x \, dx$

34. $\int_0^{1/2} \sec^2 \tfrac{1}{2}\pi t \tan \tfrac{1}{2}\pi t \, dt$

35. $\int_{\pi/8}^{\pi/4} 3 \csc^2 2x \, dx$

36. $\int_0^{\pi/6} (\sin 2x + \cos 3x) \, dx$

In Exercises 37 through 46, compute the derivative.

37. $\dfrac{d}{dx} \int_0^x \sqrt{4 + t^6} \, dt$

38. $\dfrac{d}{dx} \int_2^x \dfrac{1}{t^4 + 4} \, dt$

39. $\dfrac{d}{dx} \int_x^3 \sqrt{\sin t} \, dt$

40. $\dfrac{d}{dx} \int_x^3 \sqrt{1 + t^4} \, dt$

41. $\dfrac{d}{dx} \int_{-x}^x \dfrac{1}{3 + t^2} \, dt$

42. $\dfrac{d}{dx} \int_{-x}^x \cos(t^2 + 1) \, dt$

43. $\dfrac{d}{dx} \int_1^{x^3} \sqrt[3]{t^2 + 1} \, dt$

44. $\dfrac{d}{dx} \int_0^{x^2} \dfrac{1}{\sqrt{t^2 + 1}} \, dt$

45. $\dfrac{d}{dx} \int_2^{\tan x} \dfrac{1}{1 + t^2} \, dt$

46. $\dfrac{d}{dx} \int_3^{\sin x} \dfrac{1}{1 - t^2} \, dt$

In Exercises 47 through 50, find the average value of the function f on the interval [a, b]. In Exercises 47 and 48, find the value of x at which the average value of f occurs, and make a sketch.

47. $f(x) = 9 - x^2$; $[a, b] = [0, 3]$

48. $f(x) = 8x - x^2$; $[a, b] = [0, 4]$

49. $f(x) = 3x\sqrt{x^2 - 16}$; $[a, b] = [4, 5]$

50. $f(x) = x^2\sqrt{x - 3}$; $[a, b] = [7, 12]$

51. In an electric circuit, suppose that E volts is the electromotive force at t seconds and that $E = 2 \sin 3t$. Find the average value of E from $t = 0$ to $t = \tfrac{1}{3}\pi$.

52. For the electric circuit of Exercise 51, find the square root of the average value of E^2 from $t = 0$ to $t = \tfrac{1}{3}\pi$. (*Hint:* Use the identity $\sin^2 x = \tfrac{1}{2}(1 - \cos 2x)$.)

53. A ball is dropped from rest, and after t seconds its velocity is v feet per second. Neglecting air resistance, show that the average velocity during the first $\tfrac{1}{2}T$ seconds is one-third of the average velocity during the next $\tfrac{1}{2}T$ seconds.

54. A stone is thrown downward with an initial velocity of v_0 feet per second. Neglect air resistance. (a) Show that if v feet per second is the velocity of the stone after falling s feet, then $v = \sqrt{v_0^2 + 2gs}$. (b) Find the average velocity during the first 100 ft of fall if the initial velocity is 60 ft/sec. (Take $g = 32$ and downward as the positive direction.)

55. Find $\int_4^{16} \left[D_x \int_5^x (2\sqrt{t} - 1) \, dt \right] dx$.

56. Let f be a function whose derivative f' is continuous on $[a, b]$. Find the average value of the slope of the tangent line of the graph of f on $[a, b]$, and give a geometric interpretation of the result.

5.9   AREA OF A PLANE REGION

In Section 5.4 we defined the area of a plane region as the limit of a Riemann sum, and in Section 5.5 we learned that such a limit is a definite integral. Now that we have techniques for computing definite integrals we consider more problems involving areas of plane regions.

In the examples that follow we first express the measure of the required area as the limit of a Riemann sum to reinforce the procedure for setting up such sums for future applications.

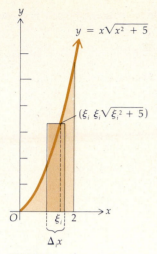

$y = x\sqrt{x^2 + 5}$

$(\xi_i, \xi_i\sqrt{\xi_i^2 + 5})$

ξ_i 2

$\Delta_i x$

FIGURE 1

EXAMPLE 1 Find the area of the region in the first quadrant bounded by the curve

$$y = x\sqrt{x^2 + 5}$$

the x axis, and the line $x = 2$. Draw a sketch.

Solution See Figure 1. The region is shown together with one of the rectangular elements of area.

We take a partition of the interval $[0, 2]$. The width of the ith rectangle is $\Delta_i x$ units, and the altitude is $\xi_i\sqrt{\xi_i^2 + 5}$ units, where ξ_i is any number in the ith subinterval. Therefore the measure of the area of the rectangular element is $\xi_i\sqrt{\xi_i^2 + 5}\,\Delta_i x$. The sum of the measures of the areas of n such rectangles is

$$\sum_{i=1}^{n} \xi_i\sqrt{\xi_i^2 + 5}\,\Delta_i x$$

which is a Riemann sum. The limit of this sum as $\|\Delta\|$ approaches 0 gives the measure of the desired area. The limit of the Riemann sum is a definite integral that we evaluate by the second fundamental theorem of the calculus. Let A square units be the area of the region. Then

$$A = \lim_{\|\Delta\| \to 0} \sum_{i=1}^{n} \xi_i\sqrt{\xi_i^2 + 5}\,\Delta_i x$$

$$= \int_0^2 x\sqrt{x^2 + 5}\,dx$$

$$= \tfrac{1}{2}\int_0^2 \sqrt{x^2 + 5}\,(2x\,dx)$$

$$= \tfrac{1}{2}\cdot\tfrac{2}{3}(x^2 + 5)^{3/2}\Big]_0^2$$

$$= \tfrac{1}{3}[(9)^{3/2} - (5)^{3/2}]$$

$$= \tfrac{1}{3}(27 - 5\sqrt{5})$$

$$\approx 5.27$$

Hence the area is $\tfrac{1}{3}(27 - 5\sqrt{5})$ square units, or approximately 5.27 square units.

So far we have considered the area of a region for which the function values are nonnegative on $[a, b]$. Suppose now that $f(x) < 0$ for all x in $[a, b]$. Then each $f(\xi_i)$ is a negative number; so we define the number of square units in the area of the region bounded by $y = f(x)$, the x axis, and the lines $x = a$ and $x = b$ to be

$$\lim_{\|\Delta\| \to 0} \sum_{i=1}^{n} [-f(\xi_i)]\,\Delta_i x$$

which equals

$$-\int_a^b f(x)\,dx$$

EXAMPLE 2 Find the area of the region bounded by the curve

$$y = x^2 - 4x$$

the x axis, and the lines $x = 1$ and $x = 3$.

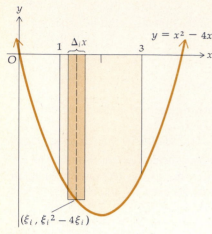

$y = x^2 - 4x$

$(\xi_i, \xi_i{}^2 - 4\xi_i)$

FIGURE 2

Solution   The region, together with a rectangular element of area, is shown in Figure 2.

We take a partition of the interval $[1, 3]$; the width of the ith rectangle is $\Delta_i x$. Because $x^2 - 4x < 0$ on $[1, 3]$, the altitude of the ith rectangle is $-(\xi_i{}^2 - 4\xi_i) = 4\xi_i - \xi_i{}^2$. Hence the sum of the measures of the areas of n rectangles is given by

$$\sum_{i=1}^{n} (4\xi_i - \xi_i{}^2) \, \Delta_i x$$

The measure of the desired area is given by the limit of this sum as $\|\Delta\|$ approaches 0; so if A square units is the area of the region,

$$A = \lim_{\|\Delta\| \to 0} \sum_{i=1}^{n} (4\xi_i - \xi_i{}^2) \, \Delta_i x$$

$$= \int_1^3 (4x - x^2) \, dx$$

$$= 2x^2 - \tfrac{1}{3}x^3 \Big]_1^3$$

$$= \tfrac{22}{3}$$

Thus the area of the region is $\tfrac{22}{3}$ square units.

EXAMPLE 3   Find the area of the region bounded by the curve

$$y = x^3 - 2x^2 - 5x + 6$$

the x axis, and the lines $x = -1$ and $x = 2$.

Solution   The region appears in Figure 3. Let

$$f(x) = x^3 - 2x^2 - 5x + 6$$

Because $f(x) \geq 0$ when x is in the closed interval $[-1, 1]$ and $f(x) \leq 0$ when x is in the closed interval $[1, 2]$, we separate the region into two parts. Let A_1 be the number of square units in the area of the region when x is in $[-1, 1]$, and let A_2 be the number of square units in the area of the region when x is in $[1, 2]$. Then

$$A_1 = \lim_{\|\Delta\| \to 0} \sum_{i=1}^{n} f(\xi_i) \, \Delta_i x$$

$$= \int_{-1}^{1} f(x) \, dx$$

$$= \int_{-1}^{1} (x^3 - 2x^2 - 5x + 6) \, dx$$

and

$$A_2 = \lim_{\|\Delta\| \to 0} \sum_{i=1}^{n} [-f(\xi_i)] \, \Delta_i x$$

$$= \int_{1}^{2} -(x^3 - 2x^2 - 5x + 6) \, dx$$

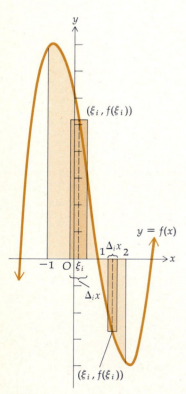

$(\xi_i, f(\xi_i))$

$y = f(x)$

$\Delta_i x$

$(\xi_i, f(\xi_i))$

FIGURE 3

If A square units is the area of the entire region, then

$$A = A_1 + A_2$$

$$= \int_{-1}^{1} (x^3 - 2x^2 - 5x + 6)\, dx - \int_{1}^{2} (x^3 - 2x^2 - 5x + 6)\, dx$$

$$= \left[\tfrac{1}{4}x^4 - \tfrac{2}{3}x^3 - \tfrac{5}{2}x^2 + 6x \right]_{-1}^{1} - \left[\tfrac{1}{4}x^4 - \tfrac{2}{3}x^3 - \tfrac{5}{2}x^2 + 6x \right]_{1}^{2}$$

$$- \left[(\tfrac{1}{4} - \tfrac{2}{3} - \tfrac{5}{2} + 6) - (\tfrac{1}{4} + \tfrac{2}{3} - \tfrac{5}{2} - 6) \right] - \left[(4 - \tfrac{16}{3} - 10 + 12) - (\tfrac{1}{4} - \tfrac{2}{3} - \tfrac{5}{2} + 6) \right]$$

$$= \tfrac{32}{3} - (-\tfrac{29}{12})$$

$$= \tfrac{157}{12}$$

The area of the region is therefore $\tfrac{157}{12}$ square units.

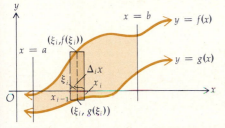

Now consider two functions f and g continuous on the closed interval $[a, b]$ and such that $f(x) \geq g(x)$ for all x in $[a, b]$. We wish to find the area of the region bounded by the two curves $y = f(x)$ and $y = g(x)$ and the two lines $x = a$ and $x = b$. Such a situation is shown in Figure 4.

Take a partition of the interval $[a, b]$, with the ith subinterval having a length of $\Delta_i x$. In each subinterval choose a point ξ_i. Consider the rectangle having altitude $[f(\xi_i) - g(\xi_i)]$ units and width $\Delta_i x$ units. A rectangle is shown in Figure 4. There are n such rectangles, one associated with each subinterval. The sum of the measures of the areas of these n rectangles is given by the following Riemann sum:

$$\sum_{i=1}^{n} [f(\xi_i) - g(\xi_i)]\, \Delta_i x$$

FIGURE 4

This Riemann sum is an approximation to what we intuitively think of as the number representing the "measure of the area" of the region. The smaller the value of $\|\Delta\|$, the better is this approximation. If A square units is the area of the region, we define

$$A = \lim_{\|\Delta\| \to 0} \sum_{i=1}^{n} [f(\xi_i) - g(\xi_i)]\, \Delta_i x \tag{1}$$

Because f and g are continuous on $[a, b]$, so too is $f - g$; therefore the limit in (1) exists and is equal to the definite integral

$$\int_{a}^{b} [f(x) - g(x)]\, dx$$

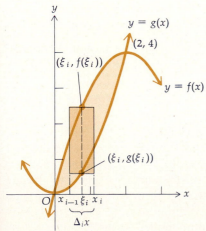

FIGURE 5

EXAMPLE 4 Find the area of the region bounded by the curves $y = x^2$ and $y = -x^2 + 4x$.

Solution To find the points of intersection of the two curves we solve the equations simultaneously and obtain the points $(0, 0)$ and $(2, 4)$. Figure 5 shows the region.

Let

$$f(x) = -x^2 + 4x \qquad g(x) = x^2$$

Therefore, in the interval $[0, 2]$ the curve $y = f(x)$ is above the curve $y = g(x)$. We draw a vertical rectangular element of area, having altitude $[f(\xi_i) - g(\xi_i)]$ units and width $\Delta_i x$ units. The measure of the area of this rectangle then is given by $[f(\xi_i) - g(\xi_i)] \Delta_i x$. The sum of the measures of the areas of n such rectangles is given by the Riemann sum

$$\sum_{i=1}^{n} [f(\xi_i) - g(\xi_i)] \Delta_i x$$

If A square units is the area of the region, then

$$A = \lim_{\|\Delta\| \to 0} \sum_{i=1}^{n} [f(\xi_i) - g(\xi_i)] \Delta_i x$$

and the limit of the Riemann sum is a definite integral. Hence

$$
\begin{aligned}
A &= \int_0^2 [f(x) - g(x)] \, dx \\
&= \int_0^2 [(-x^2 + 4x) - x^2] \, dx \\
&= \int_0^2 (-2x^2 + 4x) \, dx \\
&= \left[-\tfrac{2}{3}x^3 + 2x^2 \right]_0^2 \\
&= -\tfrac{16}{3} + 8 - 0 \\
&= \tfrac{8}{3}
\end{aligned}
$$

The area of the region is $\tfrac{8}{3}$ square units.

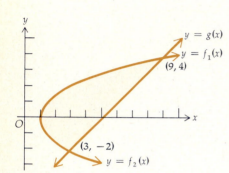

FIGURE 6

EXAMPLE 5   Find the area of the region bounded by the parabola $y^2 = 2x - 2$ and the line $y = x - 5$.

Solution   The two curves intersect at the points $(3, -2)$ and $(9, 4)$. The region is shown in Figure 6.

The equation $y^2 = 2x - 2$ is equivalent to the two equations

$$y = \sqrt{2x - 2} \quad \text{and} \quad y = -\sqrt{2x - 2}$$

with the first equation giving the upper half of the parabola and the second equation giving the bottom half. If $f_1(x) = \sqrt{2x - 2}$ and $f_2(x) = -\sqrt{2x - 2}$, the equation of the top half of the parabola is $y = f_1(x)$, and the equation of the bottom half is $y = f_2(x)$. If we let $g(x) = x - 5$, the equation of the line is $y = g(x)$.

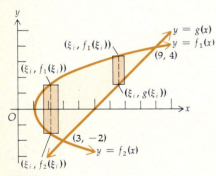

FIGURE 7

In Figure 7 we see two vertical rectangular elements of area. Each rectangle has the upper base on the curve $y = f_1(x)$. Because the lower base of the first rectangle is on the curve $y = f_2(x)$, the altitude is $[f_1(\xi_i) - f_2(\xi_i)]$ units. Because the lower base of the second rectangle lies on the curve $y = g(x)$, its altitude is $[f_1(\xi_i) - g(\xi_i)]$ units. If we wish to solve this problem by using vertical rectangular elements of area, we must divide the region into two separate regions, for instance R_1 and R_2, where R_1 is the region bounded by the curves $y = f_1(x)$ and $y = f_2(x)$ and the line $x = 3$, and where R_2 is the region bounded by the curves $y = f_1(x)$ and $y = g(x)$ and the line $x = 3$ (see Figure 8).

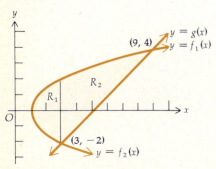

FIGURE 8

If A_1 is the number of square units in the area of region R_1, we have

$$A_1 = \lim_{||\Delta|| \to 0} \sum_{i=1}^{n} [f_1(\xi_i) - f_2(\xi_i)] \, \Delta_i x$$

$$= \int_1^3 [f_1(x) - f_2(x)] \, dx$$

$$= \int_1^3 [\sqrt{2x - 2} + \sqrt{2x - 2}] \, dx$$

$$= 2 \int_1^3 \sqrt{2x - 2} \, dx$$

$$= \tfrac{2}{3}(2x - 2)^{3/2} \Big]_1^3$$

$$= \tfrac{16}{3}$$

If A_2 is the number of square units in the area of region R_2, we have

$$A_2 = \lim_{||\Delta|| \to 0} \sum_{i=1}^{n} [f_1(\xi_i) - g(\xi_i)] \, \Delta_i x$$

$$= \int_3^9 [f_1(x) - g(x)] \, dx$$

$$= \int_3^9 [\sqrt{2x - 2} - (x - 5)] \, dx$$

$$= \left[\tfrac{1}{3}(2x - 2)^{3/2} - \tfrac{1}{2}x^2 + 5x \right]_3^9$$

$$= \left[\tfrac{64}{3} - \tfrac{81}{2} + 45 \right] - \left[\tfrac{8}{3} - \tfrac{9}{2} + 15 \right]$$

$$= \tfrac{38}{3}$$

Hence $A_1 + A_2 = \tfrac{16}{3} + \tfrac{38}{3}$. Therefore the area of the entire region is 18 square units.

EXAMPLE 6   Find the area of the region in Example 5 by taking horizontal rectangular elements of area.

Solution   Figure 9 illustrates the region with a horizontal rectangular element of area.

If in the equations of the parabola and the line we solve for x,

$$x = \tfrac{1}{2}(y^2 + 2) \qquad x = y + 5$$

Letting $\phi(y) = \tfrac{1}{2}(y^2 + 2)$ and $\lambda(y) = y + 5$, the equation of the parabola may be written as $x = \phi(y)$ and the equation of the line as $x = \lambda(y)$. Consider the closed interval $[-2, 4]$ on the y axis, and take a partition of this interval. The ith subinterval will have a length of $\Delta_i y$. In the ith subinterval $[y_{i-1}, y_i]$ choose a point ξ_i. Then the length of the ith rectangular element is $[\lambda(\xi_i) - \phi(\xi_i)]$ units and the width is $\Delta_i y$ units. The measure of the area of the region can be approximated by the Riemann sum

$$\sum_{i=1}^{n} [\lambda(\xi_i) - \phi(\xi_i)] \, \Delta_i y$$

If A square units is the area of the region, then

$$A = \lim_{||\Delta|| \to 0} \sum_{i=1}^{n} [\lambda(\xi_i) - \phi(\xi_i)] \, \Delta_i y$$

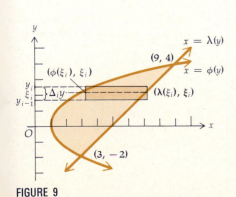

FIGURE 9

Because λ and ϕ are continuous on $[-2, 4]$, so too is $\lambda - \phi$, and the limit of the Riemann sum is a definite integral:

$$A = \int_{-2}^{4} [\lambda(y) - \phi(y)] \, dy$$

$$= \int_{-2}^{4} [(y + 5) - \tfrac{1}{2}(y^2 + 2)] \, dy$$

$$= \tfrac{1}{2} \int_{-2}^{4} (-y^2 + 2y + 8) \, dy$$

$$= \tfrac{1}{2} \left[-\tfrac{1}{3}y^3 + y^2 + 8y \right]_{-2}^{4}$$

$$= \tfrac{1}{2} [(-\tfrac{64}{3} + 16 + 32) - (\tfrac{8}{3} + 4 - 16)]$$

$$= 18$$

Comparing the solutions in Examples 5 and 6 we see that in the first case there are two definite integrals to evaluate, whereas in the second case there is only one. In general, if possible, the rectangular elements of area should be constructed so that a single definite integral is obtained. The following example illustrates a situation where two definite integrals are necessary.

EXAMPLE 7   Find the area of the region bounded by the two curves $y = x^3 - 6x^2 + 8x$ and $y = x^2 - 4x$.

Solution   The points of intersection of the two curves are $(0, 0)$, $(3, -3)$, and $(4, 0)$. The region is shown in Figure 10.
   Let

$$f(x) = x^3 - 6x^2 + 8x \qquad g(x) = x^2 - 4x$$

In the interval $[0, 3]$ the curve $y = f(x)$ is above the curve $y = g(x)$, and in the interval $[3, 4]$ the curve $y = g(x)$ is above the curve $y = f(x)$. So the region must be divided into two separate regions R_1 and R_2, where R_1 is the region bounded by the two curves in the interval $[0, 3]$ and R_2 is the region bounded by the two curves in the interval $[3, 4]$. If A_1 square units is the area of R_1 and A_2 square units is the area of R_2,

$$A_1 = \lim_{||\Delta|| \to 0} \sum_{i=1}^{n} [f(\xi_i) - g(\xi_i)] \, \Delta_i x \qquad A_2 = \lim_{||\Delta|| \to 0} \sum_{i=1}^{n} [g(\xi_i) - f(\xi_i)] \, \Delta_i x$$

so that

$$A_1 + A_2 = \int_{0}^{3} [(x^3 - 6x^2 + 8x) - (x^2 - 4x)] \, dx + \int_{3}^{4} [(x^2 - 4x) - (x^3 - 6x^2 + 8x)] \, dx$$

$$= \int_{0}^{3} (x^3 - 7x^2 + 12x) \, dx + \int_{3}^{4} (-x^3 + 7x^2 - 12x) \, dx$$

$$= \left[\tfrac{1}{4}x^4 - \tfrac{7}{3}x^3 + 6x^2 \right]_{0}^{3} + \left[-\tfrac{1}{4}x^4 + \tfrac{7}{3}x^3 - 6x^2 \right]_{3}^{4}$$

$$= \tfrac{45}{4} + \tfrac{7}{12}$$

$$= \tfrac{71}{6}$$

Therefore the required area is $\tfrac{71}{6}$ square units.

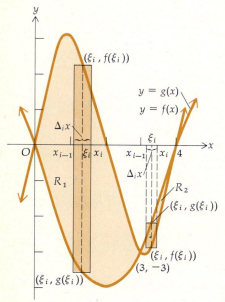

FIGURE 10

EXERCISES 5.9

In Exercises 1 through 38, find the area of the region bounded by the curves. In each problem do the following: (a) Draw a figure showing the region and a rectangular element of area; (b) express the area of the region as the limit of a Riemann sum; (c) find the limit in part (b) by evaluating a definite integral by the second fundamental theorem of the calculus.

1. $y = 4 - x^2$; x axis
2. $y = x^2 - 2x + 3$; x axis; $x = -2$; $x = 1$
3. $y = 4x - x^2$; x axis; $x = 1$; $x = 3$
4. $y = 6 - x - x^2$; x axis
5. $y = \sqrt{x + 1}$; x axis; y axis; $x = 8$
6. $y = \dfrac{1}{x^2} - x$; x axis; $x = 2$; $x = 3$
7. $y = x^2 + x - 12$; x axis
8. $y = x^2 - 6x + 5$; x axis
9. $y = \sin x$; x axis; $x = \frac{1}{3}\pi$; $x = \frac{2}{3}\pi$
10. $y = \cos x$; x axis; y axis; $x = \frac{1}{6}\pi$
11. $y = \sec^2 x$; x axis; y axis; $x = \frac{1}{4}\pi$
12. $y = \csc^2 x$; x axis; $x = \frac{1}{4}\pi$; $x = \frac{1}{3}\pi$
13. $x^2 = -y$; $y = -4$
14. $y^2 = -x$; $x = -2$; $x = -4$
15. $x^2 + y + 4 = 0$; $y = -8$. Take the elements of area perpendicular to the y axis.
16. The same region as in Exercise 15. Take the elements of area parallel to the y axis.
17. $x^2 - y + 1 = 0$; $x - y + 1 = 0$. Take the elements of area perpendicular to the x axis.
18. The same region as in Exercise 17. Take the elements of area parallel to the x axis.
19. $x^3 = 2y^2$; $x = 0$; $y = -2$
20. $y^3 = 4x$; $x = 0$; $y = -2$
21. $y = 2 - x^2$; $y = -x$
22. $y = x^2$; $y = x^4$
23. $y^2 = x - 1$; $x = 3$
24. $y = x^2$; $x^2 = 18 - y$
25. $y = \sqrt{x}$; $y = x^3$
26. $x = 4 - y^2$; $x = 4 - 4y$
27. $y^3 = x^2$; $x - 3y + 4 = 0$
28. $xy^2 = y^2 - 1$; $x = 1$; $y = 1$; $y = 4$
29. $x = y^2 - 2$; $x = 6 - y^2$
30. $x = y^2 - y$; $x = y - y^2$
31. $y = 2x^3 - 3x^2 - 9x$; $y = x^3 - 2x^2 - 3x$
32. $3y = x^3 - 2x^2 - 15x$; $y = x^3 - 4x^2 - 11x + 30$
33. $y = x^3 + 3x^2 + 2x$; $y = 2x^2 + 4x$
34. $y = |x - 1| + 3$; $y = 0$; $x = -2$; $x = 4$
35. $y = \cos x - \sin x$; $x = 0$; $y = 0$
36. $y = \sin x$; $y = -\sin x$; $x = -\frac{1}{2}\pi$; $x = \frac{1}{2}\pi$
37. $y = |x|$; $y = x^2 - 1$; $x = -1$; $x = 1$
38. $y = |x + 1| + |x|$; $y = 0$; $x = -2$; $x = 3$

39. Find by integration the area of the triangle having vertices at $(5, 1)$, $(1, 3)$, and $(-1, -2)$.
40. Find by integration the area of the triangle having vertices at $(3, 4)$, $(2, 0)$, and $(0, 1)$.
41. Find the area of the region bounded by the line $x = 4$ and the curve $x^3 - x^2 + 2xy - y^2 = 0$. (*Hint:* Solve the cubic equation for y in terms of x, and express y as two functions of x).
42. Find the area of the region bounded by the three curves $y = x^2$, $x = y^3$, and $x + y = 2$.
43. Find the area of the region bounded by the three curves $y = x^2$, $y = 8 - x^2$, and $4x - y + 12 = 0$.
44. Find by integration the area of the trapezoid having vertices at $(-1, -1)$, $(2, 2)$, $(6, 2)$, and $(7, -1)$.
45. Find the area of the region bounded by the curve $y = \sin x$, the line $y = 1$, and the y axis to the right of the y axis.
46. Find the area of the region bounded by the two curves $y = \sin x$ and $y = \cos x$ between two consecutive points of intersection.
47. Find the area of the region bounded by the curve $y = \tan^2 x$, the x axis, and the line $x = \frac{1}{4}\pi$.
48. Find the area of the region above the parabola $x^2 = 4py$ and inside the triangle formed by the x axis and the lines $y = x + 8p$ and $y = -x + 8p$.
49. Find the area of the region bounded by the two parabolas $y^2 = 4px$ and $x^2 = 4py$.
50. Find the rate of change of the measure of the area of Exercise 48 with respect to p when $p = \frac{3}{8}$.
51. Find the rate of change of the measure of the area of Exercise 49 with respect to p when $p = 3$.
52. Determine m so that the region above the line $y = mx$ and below the parabola $y = 2x - x^2$ has an area of 36 square units.
53. Determine m so that the region above the curve $y = mx^2$ $(m > 0)$, to the right of the y axis, and below the line $y = m$ has an area of K square units, where $K > 0$.
54. If A square units is the area of the region bounded by the parabola $y^2 = 4x$ and the line $y = mx$ $(m > 0)$, find the rate of change of A with respect to m.

5.10 NUMERICAL INTEGRATION

When applying the second fundamental theorem of the calculus to evaluate a definite integral it is necessary to obtain an indefinite integral. In Section 5.8 we computed the indefinite integral by techniques presented in Sections 5.1 and 5.2. Even though you will learn in Chapter 9 additional methods for obtaining indefinite integrals, there will still remain many functions for which a definite integral cannot be evaluated exactly by using elementary functions. Examples of such definite integrals are

$$\int_0^1 \sqrt{1 + x^4}\, dx \qquad \int_0^{\pi/2} \sin x^2\, dx$$

Other examples appear in the exercises. In this section we learn two methods for computing an approximate value of a definite integral of a function that is continuous on a closed interval. These methods often give fairly good accuracy, and variations of them are used for evaluating a definite integral on programmable calculators and computers. The first method is known as the *trapezoidal rule.*

Let f be a function continuous on a closed interval $[a, b]$. The definite integral $\int_a^b f(x)\, dx$ is the limit of a Riemann sum; that is,

$$\int_a^b f(x)\, dx = \lim_{\|\Delta\| \to 0} \sum_{i=1}^{n} f(\xi_i)\, \Delta_i x$$

The geometric interpretation of the Riemann sum

$$\sum_{i=1}^{n} f(\xi_i)\, \Delta_i x$$

is that it is equal to the sum of the measures of the areas of the rectangles lying above the x axis plus the negative of the measures of the areas of the rectangles lying below the x axis (see Figure 3 in Section 5.5).

To approximate the measure of the area of a region we shall use trapezoids instead of rectangles. Let us also use regular partitions and function values at equally spaced points.

Thus for the definite integral $\int_a^b f(x)\, dx$ we divide the interval $[a, b]$ into n subintervals, each of length $\Delta x = (b - a)/n$. This gives the following $n + 1$ points: $x_0 = a$, $x_1 = a + \Delta x$, $x_2 = a + 2\Delta x$, ..., $x_i = a + i\Delta x$, ..., $x_{n-1} = a + (n - 1)\Delta x$, $x_n = b$. Then the definite integral $\int_a^b f(x)\, dx$ may be expressed as the sum of n definite integrals as follows:

$$\int_a^b f(x)\, dx = \int_a^{x_1} f(x)\, dx + \int_{x_1}^{x_2} f(x)\, dx + \ldots + \int_{x_{i-1}}^{x_i} f(x)\, dx + \ldots + \int_{x_{n-1}}^{b} f(x)\, dx \quad (1)$$

To interpret (1) geometrically, refer to Figure 1, in which $f(x) \geq 0$ for all x in $[a, b]$; however, (1) holds for any function that is continuous on $[a, b]$.

Then the integral $\int_a^{x_1} f(x)\, dx$ is the measure of the area of the region bounded by the x axis, the lines $x = a$ and $x = x_1$, and the portion of the curve from P_0 to

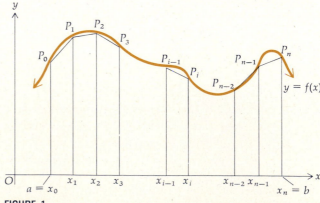

FIGURE 1

P_1. This integral may be approximated by the measure of the area of the trapezoid formed by the lines $x = a$, $x = x_1$, $P_0 P_1$, and the x axis. By a formula from geometry the measure of the area of this trapezoid is

$$\tfrac{1}{2}[f(x_0) + f(x_1)]\,\Delta x$$

Similarly, the other integrals on the right side of (1) may be approximated by the measure of the area of a trapezoid. For the ith integral,

$$\int_{x_{i-1}}^{x_i} f(x)\,dx \approx \tfrac{1}{2}[f(x_{i-1}) + f(x_i)]\,\Delta x$$

Using this for each of the integrals on the right side of (1) we have

$$\int_a^b f(x)\,dx \approx \tfrac{1}{2}[f(x_0) + f(x_1)]\,\Delta x + \tfrac{1}{2}[f(x_1) + f(x_2)]\,\Delta x + \ldots + \tfrac{1}{2}[f(x_{n-2}) + f(x_{n-1})]\,\Delta x + \tfrac{1}{2}[f(x_{n-1}) + f(x_n)]\,\Delta x$$

Thus

$$\int_a^b f(x)\,dx \approx \tfrac{1}{2}\Delta x[f(x_0) + 2f(x_1) + 2f(x_2) + \ldots + 2f(x_{n-1}) + f(x_n)]$$

This formula is known as the *trapezoidal rule* and is stated in the following theorem.

5.10.1 THEOREM
The Trapezoidal Rule

If the function f is continuous on the closed interval $[a, b]$ and the numbers $a = x_0, x_1, x_2, \ldots, x_n = b$ form a regular partition of $[a, b]$, then

$$\int_a^b f(x)\,dx \approx \frac{b-a}{2n}\left[f(x_0) + 2f(x_1) + 2f(x_2) + \ldots + 2f(x_{n-1}) + f(x_n)\right]$$

EXAMPLE 1   Find an approximation for

$$\int_0^3 \frac{dx}{16 + x^2}$$

by using the trapezoidal rule with $n = 6$. Express the result to three decimal places.

Solution   Because $[a, b] = [0, 3]$ and $n = 6$,

$$\Delta x = \frac{b-a}{n} \qquad \frac{b-a}{2n} = \frac{3-0}{12}$$

$$= \frac{3-0}{6} \qquad\qquad = 0.25$$

$$= 0.5$$

Table 1

i	x_i	$f(x_i)$	k_i	$k_i \cdot f(x_i)$
0	0	0.0625	1	0.0625
1	0.5	0.0615	2	0.1230
2	1	0.0588	2	0.1176
3	1.5	0.0548	2	0.1096
4	2	0.0500	2	0.1000
5	2.5	0.0450	2	0.0900
6	3	0.0400	1	0.0400

$$\sum_{i=0}^{6} k_i f(x_i) = 0.6427$$

Therefore

$$\int_0^3 \frac{dx}{16 + x^2} \approx 0.25[f(x_0) + 2f(x_1) + 2f(x_2) + 2f(x_3) + 2f(x_4) + 2f(x_5) + f(x_6)]$$

where $f(x) = 1/(16 + x^2)$. The computation of the sum in brackets in the above is shown in Table 1, where the entries are obtained from a calculator. Thus

$$\int_0^3 \frac{dx}{16 + x^2} \approx (0.25)(0.6427)$$

$$\approx 0.1607$$

$$\approx 0.161$$

To obtain an exact value of this definite integral requires a technique you will learn in Section 8.3. The exact value to four decimal places is 0.1609.

To consider the accuracy of the approximation of a definite integral by the trapezoidal rule, we refer to two kinds of errors that are introduced. One is the error due to the approximation of the graph of the function by segments of straight lines. The term **truncation error** refers to this kind of error. The other kind of error, which is unavoidable, is called the **round-off error**. It arises because numbers having a finite number of digits are used to approximate numbers. As the value of n (the number of subintervals) increases, the accuracy of the approximation of the area of the region by areas of trapezoids improves; thus the truncation error is reduced. However, as n increases, more computations are necessary; hence there is an increase in the round-off error. By methods discussed in *numerical analysis*, it is possible in a particular problem to determine the value of n that minimizes the combined errors. Obviously the round-off error is affected by how the calculations are performed. The truncation error can be estimated by a theorem. We prove first that as Δx approaches zero and n increases without bound, the limit of the approximation by the trapezoidal rule is the exact value of the definite integral. Let

$$T = \tfrac{1}{2}\Delta x[f(x_0) + 2f(x_1) + \ldots + 2f(x_{n-1}) + f(x_n)]$$

Then

$$T = [f(x_1) + f(x_2) + \ldots + f(x_n)]\,\Delta x + \tfrac{1}{2}[f(x_0) - f(x_n)]\,\Delta x$$

$$\Leftrightarrow \quad T = \sum_{i=1}^{n} f(x_i)\,\Delta x + \tfrac{1}{2}[f(a) - f(b)]\,\Delta x$$

Therefore, if $n \to +\infty$ and $\Delta x \to 0$,

$$\lim_{\Delta x \to 0} T = \lim_{\Delta x \to 0} \sum_{i=1}^{n} f(x_i)\,\Delta x + \lim_{\Delta x \to 0} \tfrac{1}{2}[f(a) - f(b)]\,\Delta x$$

$$= \int_{a}^{b} f(x)\,dx + 0$$

Thus we can make the difference between T and the value of the definite integral as small as we please by taking n sufficiently large (and consequently Δx sufficiently small).

The following theorem, which is proved in numerical analysis, gives a method for estimating the truncation error obtained when using the trapezoidal rule. The truncation error is denoted by ϵ_T.

5.10.2 THEOREM   Let the function f be continuous on the closed interval $[a, b]$, and f' and f'' both exist on $[a, b]$. If

$$\epsilon_T = \int_{a}^{b} f(x)\,dx - T$$

where T is the approximate value of $\int_{a}^{b} f(x)\,dx$ found by the trapezoidal rule, then there is some number η in $[a, b]$ such that

$$\epsilon_T = -\tfrac{1}{12}(b - a)f''(\eta)(\Delta x)^2 \tag{2}$$

EXAMPLE 2   Find the bounds for the truncation error in the result of Example 1.

Solution   We first find the absolute minimum and absolute maximum values of $f''(x)$ on $[0, 3]$.

$$f(x) = (16 + x^2)^{-1}$$

$$f'(x) = -2x(16 + x^2)^{-2}$$

$$f''(x) = 8x^2(16 + x^2)^{-3} - 2(16 + x^2)^{-2}$$

$$= (6x^2 - 32)(16 + x^2)^{-3}$$

$$f'''(x) = -6x(6x^2 - 32)(16 + x^2)^{-4} + 12x(16 + x^2)^{-3}$$

$$= 24x(16 - x^2)(16 + x^2)^{-4}$$

Because $f'''(x) > 0$ for all x in the open interval $(0, 3)$, then f'' is increasing on the open interval $(0, 3)$. Therefore the absolute minimum value of f'' on $[0, 3]$ is $f''(0)$, and the absolute maximum value of f'' on $[0, 3]$ is $f''(3)$.

$$f''(0) = -\frac{1}{128} \qquad f''(3) = \frac{22}{15,625}$$

Taking $\eta = 0$ on the right side of (2) we get

$$-\frac{3}{12}\left(-\frac{1}{128}\right)\frac{1}{4} = \frac{1}{2048}$$

Taking $\eta = 3$ on the right side of (2) we have

$$-\frac{3}{12}\left(\frac{22}{15,625}\right)\frac{1}{4} = -\frac{11}{125,000}$$

Therefore if ϵ_T is the truncation error in the result of Example 1,

$$-\frac{11}{125,000} \le \epsilon_T \le \frac{1}{2048}$$

$$-0.0001 \le \epsilon_T \le 0.0005$$

If in Theorem 5.10.2 $f(x) = mx + b$, then $f''(x) = 0$ for all x. Therefore $\epsilon_T = 0$; so the trapezoidal rule gives the exact value of the definite integral of a linear function.

Another method for approximating the value of a definite integral is provided by *Simpson's rule* (sometimes referred to as the *parabolic rule*), named after the British mathematician Thomas Simpson (1710–1761). For a given partition of the closed interval $[a, b]$, Simpson's rule usually gives a better approximation than the trapezoidal rule. In the trapezoidal rule, successive points on the graph of $y = f(x)$ are connected by segments of straight lines, whereas in Simpson's rule the points are connected by segments of parabolas. Before Simpson's rule is developed, we state and prove a theorem that will be needed.

5.10.3 THEOREM   If $P_0(x_0, y_0)$, $P_1(x_1, y_1)$, and $P_2(x_2, y_2)$ are three noncollinear points on the parabola having the equation $y = Ax^2 + Bx + C$, where $y_0 \ge 0$, $y_1 \ge 0$, $y_2 \ge 0$, $x_1 = x_0 + h$, and $x_2 = x_0 + 2h$, then the measure of the area of the region bounded by the parabola, the x axis, and the lines $x = x_0$ and $x = x_2$ is given by

$$\tfrac{1}{3}h(y_0 + 4y_1 + y_2)$$

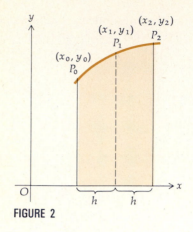

FIGURE 2

Proof The parabola whose equation is $y = Ax^2 + Bx + C$ has a vertical axis. Refer to Figure 2, which shows the region bounded by the parabola, the x axis, and the lines $x = x_0$ and $x = x_2$.

Because P_0, P_1, and P_2 are points on the parabola, their coordinates satisfy the equation of the parabola. So when we replace x_1 by $x_0 + h$, and x_2 by $x_0 + 2h$, we have

$$y_0 = Ax_0^2 + Bx_0 + C$$

$$y_1 = A(x_0 + h)^2 + B(x_0 + h) + C$$
$$= A(x_0^2 + 2hx_0 + h^2) + B(x_0 + h) + C$$

$$y_2 = A(x_0 + 2h)^2 + B(x_0 + 2h) + C$$
$$= A(x_0^2 + 4hx_0 + 4h^2) + B(x_0 + 2h) + C$$

Therefore

$$y_0 + 4y_1 + y_2 = A(6x_0^2 + 12hx_0 + 8h^2) + B(6x_0 + 6h) + 6C \tag{3}$$

Now if K square units is the area of the region, then K can be computed by the limit of a Riemann sum, and we have

$$K = \lim_{\|\Delta\| \to 0} \sum_{i=1}^{n} (A\xi_i^2 + B\xi_i + C) \, \Delta x$$

$$= \int_{x_0}^{x_0 + 2h} (Ax^2 + Bx + C) \, dx$$

$$= \tfrac{1}{3}Ax^3 + \tfrac{1}{2}Bx^2 + Cx \Big]_{x_0}^{x_0 + 2h}$$

$$= \tfrac{1}{3}A(x_0 + 2h)^3 + \tfrac{1}{2}B(x_0 + 2h)^2 + C(x_0 + 2h) - (\tfrac{1}{3}Ax_0^3 + \tfrac{1}{2}Bx_0^2 + Cx_0)$$

$$= \tfrac{1}{3}h[A(6x_0^2 + 12hx_0 + 8h^2) + B(6x_0 + 6h) + 6C]$$

Substituting from (3) in this expression for K we get

$$K = \tfrac{1}{3}h(y_0 + 4y_1 + y_2) \qquad \blacksquare$$

Let the function f be continuous on the closed interval $[a, b]$. Consider a regular partition of the interval $[a, b]$ of n subintervals, where n is even. The length of each subinterval is given by $\Delta x = (b - a)/n$. Let the points on the curve $y = f(x)$ having these partitioning points as abscissas be denoted by $P_0(x_0, y_0)$, $P_1(x_1, y_1), \ldots, P_n(x_n, y_n)$; see Figure 3, where $f(x) \geq 0$ for all x in $[a, b]$.

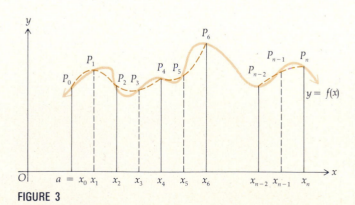

FIGURE 3

We approximate the segment of the curve $y = f(x)$ from P_0 to P_2 by the segment of the parabola with a vertical axis and through P_0, P_1, and P_2. Then by Theorem 5.10.3 the measure of the area of the region bounded by this parabola, the x axis, and the lines $x = x_0$ and $x = x_2$, with $h = \Delta x$, is given by

$$\tfrac{1}{3} \Delta x(y_0 + 4y_1 + y_2) \quad \text{or} \quad \tfrac{1}{3} \Delta x[f(x_0) + 4f(x_1) + f(x_2)]$$

In a similar manner we approximate the segment of the curve $y = f(x)$ from P_2 to P_4 by the segment of the parabola with a vertical axis and through P_2, P_3, and P_4. The measure of the area of the region bounded by this parabola, the x axis, and the lines $x = x_2$ and $x = x_4$ is given by

$$\tfrac{1}{3} \Delta x(y_2 + 4y_3 + y_4) \quad \text{or} \quad \tfrac{1}{3} \Delta x[f(x_2) + 4f(x_3) + f(x_4)]$$

This process is continued until there are $\tfrac{1}{2}n$ such regions, and the measure of the area of the last region is given by

$$\tfrac{1}{3} \Delta x(y_{n-2} + 4y_{n-1} + y_n) \quad \text{or} \quad \tfrac{1}{3} \Delta x[f(x_{n-2}) + 4f(x_{n-1}) + f(x_n)]$$

The sum of the measures of the areas of these regions approximates the measure of the area of the region bounded by the curve whose equation is $y = f(x)$, the x axis, and the lines $x = a$ and $x = b$ The measure of the area of this region is given by the definite integral $\int_a^b f(x)\,dx$. So we have as an approximation to the definite integral

$$\tfrac{1}{3} \Delta x[f(x_0) + 4f(x_1) + f(x_2)] + \tfrac{1}{3} \Delta x[f(x_2) + 4f(x_3) + f(x_4)] + \cdots$$
$$+ \tfrac{1}{3} \Delta x[f(x_{n-4}) + 4f(x_{n-3}) + f(x_{n-2})] + \tfrac{1}{3} \Delta x[f(x_{n-2}) + 4f(x_{n-1}) + f(x_n)]$$

Thus

$$\int_a^b f(x)\,dx \approx \tfrac{1}{3} \Delta x[f(x_0) + 4f(x_1) + 2f(x_2) + 4f(x_3) + 2f(x_4) + \cdots + 2f(x_{n-2}) + 4f(x_{n-1}) + f(x_n)]$$

where $\Delta x = (b - a)/n$.

This formula is called *Simpson's rule* and is given in the next theorem.

5.10.4   THEOREM
Simpson's Rule

If the function f is continuous on the closed interval $[a, b]$, n is an even integer, and the numbers $a = x_0, x_1, x_2, \ldots, x_{n-1}, x_n = b$ form a regular partition of $[a, b]$, then

$$\int_a^b f(x)\,dx \approx \frac{b - a}{3n} [f(x_0) + 4f(x_1) + 2f(x_2) + 4f(x_3) + 2f(x_4) + \cdots$$
$$+ 2f(x_{n-2}) + 4f(x_{n-1}) + f(x_n)]$$

EXAMPLE 3   Use Simpson's rule with $n = 4$ to approximate the value of

$$\int_0^2 \frac{dx}{x + 1}$$

Solution   Applying Simpson's rule with $n = 4$, we have

$$\Delta x = \frac{b - a}{n} \qquad \frac{b - a}{3n} = \frac{1 - 0}{3(4)}$$

$$= \frac{1 - 0}{4} \qquad\qquad = \tfrac{1}{12}$$

$$= \tfrac{1}{4}$$

Table 2

i	x_i	$f(x_i)$	k_i	$k_i \cdot f(x_i)$
0	0	1.00000	1	1.00000
1	0.25	0.80000	4	3.20000
2	0.5	0.66667	2	1.33334
3	0.75	0.57143	4	2.28572
4	1	0.50000	1	0.50000

$$\sum_{i=0}^{4} k_i f(x_i) = 8.31906$$

Therefore, if $f(x) = 1/(x + 1)$,

$$\int_0^1 \frac{dx}{x + 1} \approx \tfrac{1}{12}[f(x_0) + 4f(x_1) + 2f(x_2) + 4f(x_3) + f(x_4)]$$

In Table 2, where the entries are obtained from a calculator, we have the computation of the above sum in brackets. Therefore

$$\int_0^1 \frac{dx}{x + 1} \approx \frac{1}{12}(8.31906)$$

$$\approx 0.69325^+$$

Rounding off the result to four decimal places gives

$$\int_0^1 \frac{dx}{x + 1} \approx 0.6933$$

In Section 7.1 you will learn the procedure for computing the exact value of this definite integral, which to four decimal places is 0.6931. Our approximation agrees with the exact value in the first three places and the error in the approximation is -0.0002.

In applying Simpson's rule, the larger we take the value of n, the smaller will be the value of Δx, and so geometrically it seems evident that the smaller will be the truncation error of the approximation, because a parabola passing through three points of a curve that are close to each other will be close to the curve throughout the subinterval of width $2\Delta x$.

The following theorem that is proved in numerical analysis gives a method for determining the truncation error in applying Simpson's rule. This error is denoted by ϵ_S.

5.10.5 THEOREM   Let the function f be continuous on the closed interval $[a, b]$, and f', f'', f''', and $f^{(iv)}$ all exist on $[a, b]$. If

$$\epsilon_S = \int_a^b f(x)\, dx - S$$

where S is the approximate value of $\int_a^b f(x)\, dx$ found by Simpson's rule, then there is some number η in $[a, b]$ such that

$$\epsilon_S = -\tfrac{1}{180}(b - a)f^{(iv)}(\eta)(\Delta x)^4 \tag{4}$$

EXAMPLE 4   Find the bounds for the truncation error in Example 3.

Solution

$$f(x) = (x + 1)^{-1}$$
$$f'(x) = -1(x + 1)^{-2}$$
$$f''(x) = 2(x + 1)^{-3}$$
$$f'''(x) = -6(x + 1)^{-4}$$
$$f^{(iv)}(x) = 24(x + 1)^{-5}$$
$$f^{(v)}(x) = -120(x + 1)^{-6}$$

Because $f^{(v)}(x) < 0$ for all x in $[0, 1]$, $f^{(iv)}$ is decreasing on $[0, 1]$. Thus the absolute minimum value of $f^{(iv)}$ is at the right endpoint 1, and the absolute maximum value of $f^{(iv)}$ on $[0, 1]$ is at the left endpoint 0.

$$f^{(iv)}(0) = 24 \quad \text{and} \quad f^{(iv)}(1) = \tfrac{3}{4}$$

Substituting 0 for η in the right side of (4) we get

$$-\tfrac{1}{180}(b - a)f^{(iv)}(0)(\Delta x)^4 = -\tfrac{1}{180}(24)(\tfrac{1}{4})^4$$

$$\approx -0.00052$$

Substituting 1 for η in the right side of (4) we have

$$-\frac{1}{180}(b - a)f^{(iv)}(1)(\Delta x)^4 = -\frac{1}{180} \cdot \frac{3}{4}\left(\frac{1}{4}\right)^4$$

$$\approx -0.00002$$

So

$$-0.00052 \le \epsilon_S \le -0.00002$$

This inequality agrees with the discussion in Example 3 regarding the error in the approximation of $\int_0^1 dx/(x + 1)$ by Simpson's rule because $-0.00052 < -0.0002 < -0.00002$.

If $f(x)$ is a polynomial of degree three or less, then $f^{(iv)}(x) \equiv 0$ and therefore $\epsilon_S = 0$. In other words, Simpson's rule gives an exact result for a polynomial of the third degree or lower. This statement is geometrically obvious if $f(x)$ is of the second or first degree because in the first case the graph of $y = f(x)$ is a parabola, and in the second case the graph is a line.

Numerical methods can be applied to approximate $\int_a^b f(x)\, dx$ even when we do not know a formula for $f(x)$ provided, of course, we have access to some function values. Such function values are often obtained experimentally. The following example involves such a situation.

EXAMPLE 5   A particle moving along a horizontal line has a velocity of $v(t)$ meters per second at t seconds. Table 3 gives values of $v(t)$ for $\tfrac{1}{2}$-sec intervals of time for a period of 4 sec. Use these values and Simpson's rule to approximate the distance the particle travels during the 4 sec.

Table 3

t	0	0.5	1.0	1.5	2.0	2.5	3.0	3.5	4.0
$v(t)$	0	0.15	0.35	0.55	0.78	1.02	1.27	1.57	1.90

Solution   The number of meters the particle travels during the 4 sec is $\int_0^4 v(t)\, dt$. From Simpson's rule with $n = 8$, we have

$$\Delta x = \frac{b - a}{n} \qquad \frac{b - a}{3n} = \frac{4 - 0}{24}$$

$$= \frac{4 - 0}{8} \qquad = \frac{1}{6}$$

$$= \frac{1}{2}$$

Therefore

$$\int_0^4 v(t)\, dt \approx \tfrac{1}{6}[v(0) + 4v(1) + 2v(2) + 4v(3) + 2v(4) + 4v(5) + 2v(6) + 4v(7) + v(8)]$$
$$= \tfrac{1}{6}[0 + 4(0.15) + 2(0.35) + 4(0.55) + 2(0.78) + 4(1.02) + 2(1.27) + 4(1.57) + 1.90]$$
$$\approx \tfrac{1}{6}[19.86]$$
$$= 3.31$$

Thus the particle travels approximately 3.31 meters during the 4 sec.

EXERCISES 5.10

In Exercises 1 through 12, compute the approximate value of the definite integral by the trapezoidal rule for the indicated value of n. Express the result to three decimal places. In Exercises 1 through 4, find the exact value of the definite integral, and compare the result with the approximation.

1. $\int_0^2 x^3\, dx$; $n = 4$

2. $\int_0^2 x\sqrt{4 - x^2}\, dx$; $n = 8$

3. $\int_0^\pi \cos x\, dx$; $n = 4$

4. $\int_0^\pi \sin x\, dx$; $n = 6$

5. $\int_1^2 \dfrac{dx}{x}$; $n = 5$

6. $\int_2^{10} \dfrac{dx}{1 + x}$; $n = 8$

7. $\int_0^1 \dfrac{dx}{\sqrt{1 + x^2}}$; $n = 5$

8. $\int_2^3 \sqrt{1 + x^2}\, dx$; $n = 6$

9. $\int_{\pi/2}^{3\pi/2} \dfrac{\sin x}{x}\, dx$; $n = 6$

10. $\int_0^1 \sqrt{1 + x^3}\, dx$; $n = 4$

11. $\int_0^2 \sqrt{1 + x^4}\, dx$; $n = 6$

12. $\int_0^\pi \dfrac{\sin x}{1 + x}\, dx$; $n = 6$

In Exercises 13 through 18, find the bounds for the truncation error in the approximation of the indicated exercise.

13. Exercise 1   14. Exercise 4   15. Exercise 3
16. Exercise 6   17. Exercise 5   18. Exercise 8

19. Approximate $\int_0^2 x^3\, dx$ to three decimal places by Simpson's rule with $n = 4$. Compare the result with those obtained in Exercise 1, and observe that Simpson's rule gives better accuracy than the trapezoidal rule with the same number of subintervals.

20. Approximate $\int_0^\pi \sin x\, dx$ to three decimal places by Simpson's rule with $n = 6$. Compare the result with those obtained in Exercise 4, and observe that Simpson's rule gives better accuracy than the trapezoidal rule with the same number of subintervals.

In Exercises 21 through 24, approximate the definite integral by Simpson's rule with the indicated value of n. Express the answer to three decimal places.

21. $\int_{-1}^0 \dfrac{dx}{1 - x}$; $n = 4$

22. $\int_{-1/2}^0 \dfrac{dx}{\sqrt{1 - x^2}}$; $n = 4$

23. $\int_0^1 \dfrac{dx}{x^2 + x + 1}$; $n = 4$

24. $\int_1^2 \dfrac{dx}{x + 1}$; $n = 8$

In Exercises 25 through 28, find bounds for the truncation error in the approximation of the indicated exercise.

25. Exercise 19   26. Exercise 20
27. Exercise 21   28. Exercise 24

Each of the definite integrals in Exercises 29 through 34 cannot be evaluated exactly in terms of elementary functions. Use Simpson's rule, with the indicated value of n, to find an approximate value of the definite integral. Express the result to three decimal places.

29. $\int_{\pi/2}^{3\pi/2} \dfrac{\sin x}{x}\, dx$; $n = 6$

30. $\int_0^2 \sqrt{1 + x^4}\, dx$; $n = 6$

31. $\int_1^{1.8} \sqrt{1 + x^3}\, dx$; $n = 4$

32. $\int_0^1 \sqrt[3]{1 - x^2}\, dx$; $n = 4$

33. $\int_0^2 \dfrac{dx}{\sqrt{1 + x^3}}$; $n = 8$

34. $\int_0^{\pi/2} \sqrt{\sin x}\, dx$; $n = 6$

35. Show that the exact value of the integral $\int_0^2 \sqrt{4 - x^2}\, dx$ is π by interpreting it as the measure of the area of a region. Approximate the definite integral by the trapezoidal rule with $n = 8$. Give the result to three decimal places, and compare the value so obtained with the exact value.

36. Show that the exact value of the integral $\int_0^1 4\sqrt{1 - x^2}\, dx$ is π by interpreting it as the measure of the area of a region. Use Simpson's rule with $n = 6$ to get an approximate value of the definite integral to three decimal places. Compare the results.

In Exercises 37 and 38, the function values $f(x)$ were obtained experimentally. With the assumption that f is continuous on $[0, 4]$ approximate $\int_0^4 f(x)\, dx$ by (a) the trapezoidal rule and (b) Simpson's rule.

37.

x	0	0.50	1.00	1.50	2.00	2.50	3.00	3.50	4.00
$f(x)$	3.25	4.17	4.60	3.84	3.59	4.23	4.01	3.96	3.75

38.

x	0	0.4	0.8	1.2	1.6	2.0	2.4	2.8	3.2	3.6	4.0
$f(x)$	8.4	8.1	7.9	7.5	7.6	7.2	6.8	6.3	6.5	6.0	5.7

In Exercises 39 and 40, use Simpson's rule to approximate the area of the shaded region in the figure.

39.

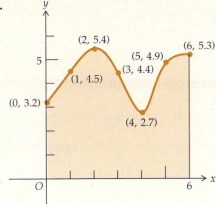

40.

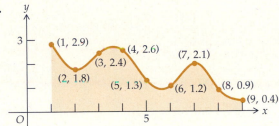

41. A woman took 10 minutes to drive from her home to the supermarket. At every 1-minute interval she observed the speedometer reading given in the following table, where $v(t)$ miles per hour was the speedometer reading t minutes after she left her home. Use Simpson's rule to approximate the distance from the woman's home to the supermarket.

t	0	1	2	3	4	5	6	7	8	9	10
$v(t)$	0	30	33	41	38	32	42	45	41	37	22

42. The shape of a parking lot is irregular and the length of the lot from west to east is 240 ft. At the west end of the lot the width is 150 ft and at the east end it is 175 ft. At 40, 80, 120, 160, and 200 ft from the west end, the widths are 154, 158, 165, 163, and 172 ft, respectively. Use Simpson's rule to approximate the area of the parking lot.

43. Find the area of the region enclosed by the loop of the curve whose equation is $y^2 = 8x^2 - x^5$. Evaluate the definite integral by Simpson's rule with $n = 8$ and express the result to three decimal places.

44. Apply Simpson's rule to the definite integral $\int_a^b f(x)\,dx$ where $f(x)$ is a third-degree polynomial to prove the **prismoidal formula**:

$$\int_a^b f(x)\,dx = \frac{b-a}{6}\left[f(a) + 4f\left(\frac{a+b}{2}\right) + f(b)\right]$$

In Exercises 45 through 48, evaluate the definite integral by two methods: (a) use the prismoidal formula given in Exercise 44; (b) use the second fundamental theorem of the calculus.

45. $\int_1^3 (4x^3 - 3x^2 + 1)\,dx$ **46.** $\int_2^6 (2x^3 - 2x - 3)\,dx$

47. $\int_{-1}^5 (x^3 + 3x^2 - 2x - 6)\,dx$ **48.** $\int_{-2}^2 (x^3 + x^2 - 4x - 2)\,dx$

49. Suppose that f is a function continuous on the closed interval $[a, b]$. Let T be the approximate value of $\int_a^b f(x)\,dx$ by using the trapezoidal rule, and let S be the approximate value of $\int_a^b f(x)\,dx$ by using Simpson's rule, where the same partition of the interval $[a, b]$ is used for both approximations. Show that

$$S = \tfrac{2}{3}[T + \Delta x(f(x_1) + f(x_3) + f(x_5) + \cdots + f(x_{n-1}))]$$

where n is even.

REVIEW EXERCISES FOR CHAPTER 5

In Exercises 1 through 24, perform the antidifferentiation; that is, evaluate the indefinite integral.

1. $\int (2x^3 - x^2 + 3)\,dx$ **2.** $\int (5x^4 + 3x - 1)\,dx$

3. $\int (4y + 6\sqrt{y})\,dy$ **4.** $\int 3z^{-2}\,dz$

5. $\int \sin 3t\,dt$ **6.** $\int \frac{\sin \sqrt{w}}{\sqrt{w}}\,dw$

7. $\int \cos^2 x \sin x\,dx$ **8.** $\int \sqrt{x}(1 + x^2)\,dx$

9. $\int \left(\frac{2}{x^4} - \frac{5}{x^2}\right)dx$ **10.** $\int \left(\sqrt[3]{t} - \frac{1}{\sqrt[3]{t}}\right)dt$

11. $\int 5x(2 + 3x^2)^8\,dx$ **12.** $\int x^4\sqrt{x^5 - 1}\,dx$

13. $\int \left(\sqrt{3x} + \frac{1}{\sqrt{5x}}\right)dx$ **14.** $\int \sqrt[3]{7w + 3}\,dw$

15. $\int \frac{x^3 + x}{(x^4 + 2x^2)^7}\,dx$ **16.** $\int (x^3 + x)\sqrt{x^2 + 3}\,dx$

17. $\int \frac{s}{\sqrt{2s + 3}}\,ds$ **18.** $\int t \sec^2 t^2\,dt$

19. $\int \tan^2 3\theta\,d\theta$ **20.** $\int (3 \cot^2 2x - 2 \csc^2 3x)\,dx$

21. $\int \frac{5 \cos^2 x - 3 \tan x}{\cos x}\,dx$ **22.** $\int \sin^3 2\theta \cot 2\theta\,d\theta$

23. $\int \sqrt{4x + 3}\,(x^2 + 1)\,dx$ **24.** $\int \left(\frac{x^3 + 2}{x^3}\right)\sqrt{x - \frac{1}{x^2}}\,dx$

In Exercises 25 through 28, find the exact value of the definite integral by using the definition of the definite integral; do not use the fundamental theorems of the calculus.

25. $\int_0^2 (3x^2 - 1)\, dx$ **26.** $\int_0^1 (2x^2 + 4x + 1)\, dx$

27. $\int_{-2}^1 (x^3 + 2x)\, dx$ **28.** $\int_{-1}^3 (x^2 - 1)^2\, dx$

In Exercises 29 through 38, evaluate the definite integral by using the second fundamental theorem of the calculus.

29. $\int_{-2}^2 (t^3 - 3t)\, dt$ **30.** $\int_1^5 \dfrac{dy}{\sqrt{2y - 1}}$

31. $\int_2^3 \dfrac{12x\, dx}{(x^2 - 1)^2}$ **32.** $\int_{-5}^5 2x\sqrt[3]{x^2 + 2}\, dx$

33. $\int_0^{\pi/6} \dfrac{\sin 2\theta\, d\theta}{\cos^2 2\theta}$ **34.** $\int_{\pi/3}^{\pi} \sin^2 \tfrac{1}{2}t \cos \tfrac{1}{2}t\, dt$

35. $\int_{-1}^7 \dfrac{x^2\, dx}{\sqrt{x + 2}}$ **36.** $\int_1^2 \dfrac{y\, dy}{\sqrt{5 - y}}$

37. $\int_0^{\pi/2} (\tan^2 \tfrac{1}{2}x + \sec^2 \tfrac{1}{2}x)\, dx$ **38.** $\int_{\pi/6}^{\pi/3} (1 - \cos \theta) \csc^2 \theta\, d\theta$

In Exercises 39 through 42, find the complete solution of the differential equation.

39. $x^2 y \dfrac{dy}{dx} = (y^2 - 1)^2$ **40.** $\dfrac{d^2 y}{dx^2} = 12x^2 - 30x$

41. $\dfrac{d^2 y}{dx^2} = \sqrt{2x - 1}$ **42.** $\dfrac{dy}{dx} = \dfrac{x\sqrt{1 - y^2}}{y\sqrt{2x^2 + 1}}$

43. Evaluate $\int (x^3 + 1)^2 x^2\, dx$ by two methods: (a) Let $u = x^3 + 1$; (b) first expand $(x^3 + 1)^2$. Compare the results, and explain the difference in the appearance of the answers obtained in (a) and (b).

44. Evaluate $\int (x^4)^6 4x^3\, dx$ as $\int u^6\, du$ and as $\int 4x^{27}\, dx$, and compare the results.

45. The slope of the tangent line at any point (x, y) on a curve is $10 - 4x$, and the point $(1, -1)$ is on the curve. Find an equation of the curve.

46. The marginal cost function for a particular commodity is given by $C'(x) = 6x - 17$. If the cost of producing 2 units is \$25, find the total cost function.

47. The marginal revenue function for a certain article of merchandise is given by $R'(x) = \tfrac{3}{4}x^2 - 10x + 12$. Find (a) the total revenue function and (b) an equation involving p and x (the demand equation) where x units are demanded when p dollars is the price per unit.

48. The enrollment at a certain college has been increasing at the rate of $1000(t + 1)^{-1/2}$ students per year since 1985. If the enrollment in 1988 was 10,000, (a) what was the enrollment in 1985, and (b) what is the anticipated enrollment in 1993 if it is expected to be increasing at the same rate?

49. The volume of a balloon is increasing according to the formula $\dfrac{dV}{dt} = \sqrt{t + 1} + \tfrac{2}{3}t$, where V cubic centimeters is the

volume of the balloon at t seconds. If $V = 33$ when $t = 3$, find (a) a formula for V in terms of t; (b) the volume of the balloon at 8 sec.

50. Suppose that a particular company estimates its growth income from sales by the formula $\dfrac{dS}{dt} = 2(t - 1)^{2/3}$, where S millions of dollars is the gross income from sales t years hence. If the gross income from the current year's sales is \$8 million, what should be the expected gross income from sales 2 years from now?

51. It is July 31 and a tumor has been growing inside a person's body in such a way that t days since July 1 the volume of the tumor has been increasing at a rate of $\tfrac{1}{100}(t + 6)^{1/2}$ cubic centimeters per day. If the volume of the tumor on July 4 was 0.20 cm^3, what is the volume today?

52. After experimentation, a certain manufacturer determined that if x units of a certain article of merchandise are produced per day, the marginal cost is given by $C'(x) = 0.3x - 11$ where $C(x)$ dollars is the total cost of producing x units. If the selling price of the article is fixed at \$19 per unit and the overhead cost is \$100 per day, find the maximum daily profit that can be obtained.

53. A manufacturer of children's toys has a new toy coming on the market and wishes to determine a selling price for the toy such that the total profit will be a maximum. From analyzing the price and demand of another similar toy, it is anticipated that if x toys are demanded when p dollars is the price per toy, then $\dfrac{dp}{dx} = -\dfrac{p^2}{30,000}$, and the demand should be 1800 when the price is \$10. If $C(x)$ dollars is the total cost of producing x toys, then $C(x) = x + 7500$. Find the price that should be charged for the manufacturer's total profit to be a maximum.

In Exercises 54 and 55, a particle is moving on a straight line, s feet is the directed distance of the particle from the origin at t seconds, v feet per second is the velocity of the particle at t seconds, and a feet per second per second is the acceleration of the particle at t seconds.

54. $a = 3t + 4$; $v = 5$ and $s = 0$, when $t = 0$. Express v and s in terms of t.

55. $a = 6 \cos 2t$; $v = 3$ and $s = 4$ when $t = \tfrac{1}{2}\pi$. Express v and s in terms of t.

56. Neglecting air resistance, if an object is dropped from an airplane at a height of 30,000 ft above the ocean, how long will it take the object to strike the water?

57. Suppose a bullet is fired directly downward from the airplane in Exercise 56 with a muzzle velocity of 2500 ft/sec. If air resistance is neglected, how long will it take the bullet to reach the ocean?

58. A ball is thrown vertically upward from the top of a house 64 ft above the ground, and the initial velocity is 48 ft/sec. (a) How long will it take the ball to reach its greatest height, and (b) what is its greatest height? (c) How long will it take the ball to strike the ground, and (d) with what velocity will it strike the ground?

59. Suppose the ball in Exercise 58 is thrown downward with an initial velocity of 48 ft/sec. (a) How long will it take the ball to strike the ground, and (b) with what velocity will it strike the ground?

60. Suppose the ball in Exercise 58 is dropped from the top of the house. (a) How long will it take for the ball to strike the ground, and (b) with what velocity will it strike the ground?

In Exercises 61 and 62, find the sum.

61. $\sum_{i=1}^{100} 2i(i^3 - 1)$

62. $\sum_{i=1}^{41} (\sqrt[3]{3i - 1} - \sqrt[3]{3i + 2})$

63. Prove that $\sum_{i=1}^{n} i^3 = \left(\sum_{i=1}^{n} i\right)^2$, and verify the formula for $n = 1, 2,$ and 3.

64. Express as a definite integral and evaluate the definite integral: $\lim_{n \to +\infty} \sum_{i=1}^{n} (8\sqrt{i}/n^{3/2})$ (*Hint:* Consider the function f for which $f(x) = \sqrt{x}$.)

65. Show that each of the following inequalities holds:

(a) $\int_{-2}^{-1} \frac{dx}{x - 3} \geq \int_{-2}^{-1} \frac{dx}{x}$; (b) $\int_{1}^{2} \frac{dx}{x} \geq \int_{1}^{2} \frac{dx}{x - 3}$;

(c) $\int_{4}^{5} \frac{dx}{x - 3} \geq \int_{4}^{5} \frac{dx}{x}$.

66. If

$$f(x) = \begin{cases} 0 & \text{if } a \leq x < c \\ k & \text{if } x = c \\ 1 & \text{if } c < x \leq b \end{cases}$$

prove that f is integrable on $[a, b]$ and $\int_{a}^{b} f(x)\, dx = b - c$ regardless of the value of k.

In Exercises 67 through 68, apply Theorem 5.6.9 to find a closed interval containing the value of the given definite integral.

67. $\int_{-\pi/2}^{\pi/2} \sqrt{\cos t}\, dt$

68. $\int_{-2}^{1} \sqrt[3]{2x^3 - 3x^2 + 1}\, dx$

In Exercises 69 through 70, evaluate the definite integral.

69. $\int_{-3}^{3} |x - 2|^3\, dx$

70. $\int_{-2}^{2} x|x - 3|\, dx$

In Exercises 71 through 74, compute the derivative.

71. $\dfrac{d}{dx} \int_{x}^{4} (3t^2 - 4)^{3/2}\, dt$

72. $\dfrac{d}{dx} \int_{-x}^{x} \dfrac{4}{1 + t^2}\, dt$

73. $\dfrac{d}{dx} \int_{x}^{x^2} \dfrac{1}{t}\, dt \qquad x > 0$

74. $\dfrac{d}{dx} \int_{1}^{\sec x} \sqrt{t^2 - 1}\, dt \qquad 0 < x < \frac{1}{2}\pi$

75. Find the average value of the cosine function on the closed interval $[a, a + 2\pi]$.

76. Interpret the mean-value theorem for integrals (5.7.1) in terms of an average function value.

77. If $f(x) = x^2 \sqrt{x - 3}$, find the average value of f on $[7, 12]$.

78. (a) Find the average value of the function f defined by $f(x) = 1/x^2$ on the interval $[1, r]$. (b) If A is the average value found in part (a), find $\lim_{r \to +\infty} A$.

79. A body falls from rest and travels a distance of s feet before striking the ground. If the only force acting is that of gravity, which gives the body an acceleration of g feet per second squared toward the ground, show that the average value of the velocity, expressed as a function of distance, while traveling this distance is $\frac{2}{3}\sqrt{2gs}$ feet per second, and that this average velocity is two-thirds of the final velocity.

80. Suppose a ball is dropped from rest and after t seconds its directed distance from the starting point is s feet and its velocity is v feet per second. Neglect air resistance. When $t = t_1, s = s_1,$ and $v = v_1$. (a) Express v as a function of t as $v = f(t)$, and find the average value of f on $[0, t_1]$. (b) Express v as a function of s as $v = h(s)$, and find the average value of h on $[0, s_1]$. (c) Write the results of parts (a) and (b) in terms of t_1, and determine which average velocity is larger.

In Exercises 81 through 88, find the area of the region bounded by the curve and lines. Draw a figure showing the region and a rectangular element of area. Express the measure of the area as the limit of a Riemann sum, and then with definite integral notation. Evaluate the definite integral by the second fundamental theorem of the calculus.

81. $y = 9 - x^2$; x axis; y axis; $x = 3$
82. $y = 3 \cos \frac{1}{2}x$; x axis; $x = -\frac{1}{2}\pi$; $x = \frac{1}{2}\pi$
83. $y = 2\sqrt{x - 1}$; x axis; $x = 5$; $x = 17$
84. $y = 16 - x^2$; x axis
85. $y = x\sqrt{x + 5}$; x axis, $x = -1$; $x = 4$
86. $y = \dfrac{4}{x^2} - x$; x axis; $x = -2$; $x = -1$
87. $x^2 + y - 5 = 0$; $y = -4$
88. $y = x^2 - 7x$; x axis; $x = 2$; $x = 4$

89. Find the area of the region bounded by the curves $x = y^2$ and $x = y^3$.

90. Find the area of the region bounded by the curves $y = \sin 2x$ and $y = \sin x$ from $x = 0$ to $x = \frac{1}{3}\pi$.

91. Find the area of the region bounded by the curves $y = \cos x$ and $y = \sin x$ from $x = \frac{1}{4}\pi$ to $x = \frac{5}{4}\pi$.

92. Find the area of the region bounded by the loop of the curve $y^2 = x^2(4 - x)$.

93. Find the area of the region in the first quadrant bounded by the y axis and the curves $y = \sec^2 x$ and $y = 2 \tan^2 x$.

94. An automobile traveling at a constant speed of 60 mi/hr along a straight highway fails to stop at a stop sign. If 3 sec later a highway patrol car starts from rest from the stop sign and maintains a constant acceleration of 8 ft/sec², how long will it take the patrol car to overtake the automobile, and how far from the stop sign will this occur? Also determine the speed of the patrol car when it overtakes the automobile.

95. Suppose that on a particular day in a certain city the Fahrenheit temperature is $f(t)$ degrees t hours since midnight, where

$$f(t) = 60 - 15 \sin \tfrac{1}{12}\pi(8 - t) \qquad 0 \le t \le 24$$

(a) Draw a sketch of the graph of f. Find the temperature at (b) 12 midnight; (c) 8 A.M.; (d) 12 noon; (e) 2 P.M.; and (f) 6 P.M. (g) Find the average temperature between 8 A.M. and 6 P.M.

96. If n is any positive integer, prove that $\int_0^\pi \sin^2 nx\, dx = \tfrac{1}{2}\pi$.

In Exercises 97 and 98, find an approximate value for the integral by using the trapezoidal rule with $n = 4$. Express the result to three decimal places.

97. $\int_0^2 \sqrt{1 + x^2}\, dx$

98. $\int_1^{9/5} \sqrt{1 + x^3}\, dx$

In Exercises 99 and 100, find an approximate value for the integral of the indicated exercise by using Simpson's rule with $n = 4$. Express the result to three decimal places.

99. Exercise 97

100. Exercise 98

101. Find an approximate value of the following integral to three decimal places by two methods: (a) use the trapezoidal rule with $n = 4$; (b) use Simpson's rule with $n = 4$:

$$\int_{1/10}^{1/2} \frac{\cos x}{x}\, dx$$

102. The function values $f(x)$ in the following table were obtained experimentally. With the assumption that f is continuous on $[1, 3]$ approximate $\int_1^3 f(x)\, dx$ by (a) the trapezoidal rule and (b) Simpson's rule.

x	1	1.2	1.4	1.6	1.8	2.0	2.2	2.4	2.6	2.8	3.0
$f(x)$	5.2	5.7	5.8	6.3	6.1	6.4	6.0	6.5	6.8	6.7	6.4

103. Evaluate $\int_0^\pi |\cos x + \tfrac{1}{2}|\, dx$.

104. Make up an example of a discontinuous function for which the mean-value theorem for integrals (a) does not apply and (b) does apply.

In Exercises 105 and 106, use the second fundamental theorem of the calculus to evaluate the definite integral. Then find the value of χ satisfying the mean-value theorem for integrals.

105. $\int_0^3 (x^2 + 1)\, dx$

106. $\int_1^4 \sqrt{x}\, dx$

107. Let f be continuous on $[a, b]$ and $\int_a^b f(t)\, dt \ne 0$. Show that for any number k in $(0, 1)$ there is a number c in (a, b) such that $\int_a^c f(t)\, dx = k \int_a^b f(t)\, dt$. $\Big($ *Hint:* Consider the function F for which $F(x) = \int_a^x f(t)\, dt \Big/ \int_a^b f(t)\, dt$, and apply the intermediate-value theorem. $\Big)$

108. Given $F(x) = \int_x^{2x} \dfrac{1}{t}\, dt$ and $x > 0$. Prove that F is a constant function by showing that $F'(x) = 0$. (*Hint:* Use the first fundamental theorem of the calculus (5.9.1) after writing the given integral as the difference of two integrals.)

109. If $f(x) = x + |x - 1|$ and

$$F(x) = \begin{cases} x & \text{if } x < 1 \\ x^2 - x + 1 & \text{if } 1 \le x \end{cases}$$

show that F is an antiderivative of f on $(-\infty, +\infty)$.

110. Let f and g be two functions such that for all x in $(-\infty, +\infty)$, $f'(x) = g(x)$ and $g'(x) = -f(x)$. Further suppose that $f(0) = 0$ and $g(0) = 1$. Prove that

$$[f(x)]^2 + [g(x)]^2 = 1$$

(*Hint:* Consider the functions F and G where $F(x) = [f(x)]^2$ and $G(x) = -[g(x)]^2$, and show that $F'(x) = G'(x)$ for all x.)

111. Given the integral $\int_d^{d+6} (x^2 + bx + c)\, dx$, where b, c, and d are constants. Suppose this integral is approximated by using the trapezoidal rule with $n = k$. (a) Show that the error in the approximation is exactly $-36/k^2$. (b) What is the smallest value of k such that the approximation is accurate to one decimal place?

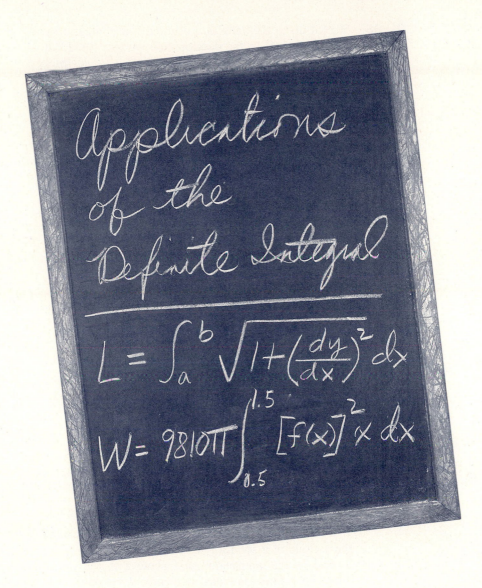

Applications
of the
Definite Integral

$$L = \int_a^b \sqrt{1 + \left(\frac{dy}{dx}\right)^2}\, dx$$

$$W = 9810\pi \int_{0.5}^{1.5} [f(x)]^2 x\, dx$$

The power of integral calculus in geometry, physics, and engineering is demonstrated in this chapter.

In Sections 6.1 and 6.2 we apply the definite integral to compute the volumes of various kinds of solids. We use *slicing*, *disks*, and *washers* in Section 6.1 and *cylindrical shells* in Section 6.2. We show in Section 6.3 how the definite integral can be used to compute the *length of arc* of the graph of a function between two points.

Physical applications of integration appear in the other four sections. We determine *centers of mass of rods* in Section 6.4 and *centers of mass of plane regions* in Section 6.5. The *work* done by a variable force acting on an object is calculated in Section 6.6. Supplementary Section 6.7 deals with applying the definite integral to find the force caused by *liquid pressure*, such as water pressure against the side of a container.

6.1 VOLUMES OF SOLIDS BY SLICING, DISKS, AND WASHERS

We led up to the definition of the definite integral by first defining the area of a plane region. In the development we used the formula from plane geometry for the area of a rectangle. We now use a similar process to obtain volumes of particular kinds of solids. One such solid is a *right cylinder*. Note that in this chapter a cylinder is considered to be a solid, whereas later, in Chapter 15, we define a cylinder as a surface.

A solid is a **right cylinder** if it is bounded by two congruent plane regions R_1 and R_2 lying in parallel planes and by a lateral surface generated by a line segment, having its endpoints on the boundaries of R_1 and R_2, which moves so that it is always perpendicular to the planes of R_1 and R_2. Figure 1 shows a right cylinder. The height of the cylinder is the perpendicular distance between the planes of R_1 and R_2 and the base is either R_1 or R_2. If the base of the right cylinder is a region enclosed by a rectangle, we have a **rectangular parallelepiped**, appearing in Figure 2, and if the base is a region enclosed by a circle, we have a **right-circular cylinder** as shown in Figure 3.

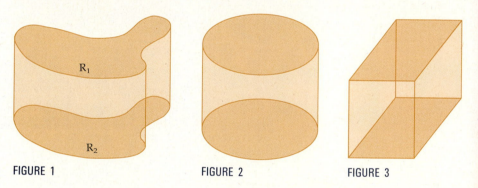

FIGURE 1   FIGURE 2   FIGURE 3

If the area of the base of a right cylinder is A square units and the height is h units, then from solid geometry if V cubic units is the volume

$$V = Ah$$

We shall use this formula to obtain a method of computing the measure of the volume of a solid for which the area of any plane section (a plane region formed by the intersection of a plane with the solid) that is perpendicular to an axis is a function of the perpendicular distance of the plane section from a fixed point on the axis. Figure 4 shows such a solid S that lies between the planes perpendicular to the x axis at a and b. We let $A(x)$ square units be the area of the plane section of S that is perpendicular to the x axis at x. We require A to be continuous on $[a, b]$.

Let Δ be a partition of the closed interval $[a, b]$ given by

$$a = x_0 < x_1 < x_2 < \ldots < x_n = b$$

There are, then, n subintervals of the form $[x_{i-1}, x_i]$, where $i = 1, 2, \ldots, n$, with the length of the ith subinterval being $\Delta_i x = x_i - x_{i-1}$. Choose any number ξ_i, with $x_{i-1} \leq \xi_i \leq x_i$, in each subinterval, and construct the right cylinders of heights $\Delta_i x$ units and plane section areas $A(\xi_i)$ square units. Figure 5 shows the

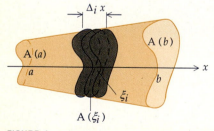

FIGURE 4

FIGURE 5

ith right cylinder, which we call an element of volume. If $\Delta_i V$ cubic units is the volume of the ith element then

$$\Delta_i V = A(\xi_i)\,\Delta_i x$$

The sum of the measures of the volumes of the n elements is

$$\sum_{i=1}^{n} \Delta_i V = \sum_{i=1}^{n} A(\xi_i)\,\Delta_i x \qquad (1)$$

which is a Riemann sum. This Riemann sum is an approximation of what we intuitively think of as the number of cubic units in the volume of the solid. The smaller we take the norm $\|\Delta\|$ of the partition, the larger will be n, and the closer this approximation will be to the number V we wish to assign to the measure of the volume. We therefore define V to be the limit of the Riemann sum in (1) as $\|\Delta\|$ approaches zero. This limit exists because A is continuous on $[a, b]$. We have, then, the following definition.

6.1.1 DEFINITION   Let S be a solid such that S lies between planes drawn perpendicular to the x axis at a and b. If the measure of the area of the plane section of S drawn perpendicular to the x axis at x is given by $A(x)$, where A is continuous on $[a, b]$, then the measure of the volume of S is given by

$$V = \lim_{\|\Delta\| \to 0} \sum_{i=1}^{n} A(\xi_i)\,\Delta_i x$$

$$= \int_a^b A(x)\,dx$$

The terminology **slicing** is used when applying Definition 6.1.1 to find the volume of a solid. The process is similar to slicing a loaf of bread into very thin pieces where all the pieces together make up the whole loaf. In the following illustration we show that Definition 6.1.1 is consistent with the formula from solid geometry for the volume of a right-circular cylinder.

▶ **ILLUSTRATION 1**   In Figure 6 there is a right-circular cylinder having a height h units and a base radius r units with the coordinate axes chosen so that the origin is at the center of one base and the height is measured along the positive x axis. A plane section at a distance of x units from the origin has an area of $A(x)$ square units where

$$A(x) = \pi r^2$$

An element of volume, shown in Figure 6, is a right cylinder of base area $A(\xi_i)$ square units and a thickness of $\Delta_i x$ units. Thus if V cubic units is the volume of the right-circular cylinder

$$V = \lim_{\|\Delta\| \to 0} \sum_{i=1}^{n} A(\xi_i)\,\Delta_i x$$

$$= \int_0^h A(x)\,dx$$

$$= \int_0^h \pi r^2\,dx$$

$$= \pi r^2 x \Big]_0^h$$

$$= \pi r^2 h$$ ◀

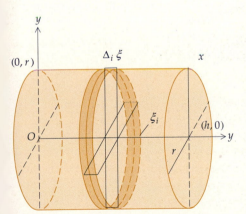

FIGURE 6

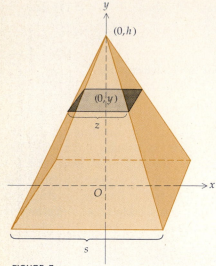

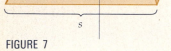

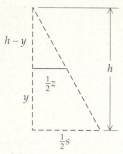

FIGURE 7

In Definition 6.1.1 we can replace x by y. In such a situation S is a solid lying between planes drawn perpendicular to the y axis at c and d, and the measure of the area of the plane section of S drawn perpendicular to the y axis at y is given by $A(y)$, where A is continuous on $[c, d]$. Then the measure of the volume of S is given by

$$V = \lim_{||\Delta|| \to 0} \sum_{i=1}^{n} A(\xi_i)\, \Delta_i y$$

$$= \int_c^d A(y)\, dy$$

EXAMPLE 1   Use slicing to find the volume of a right pyramid whose altitude is h units and whose base is a square of side s units.

Solution   Figure 7 shows the right pyramid and the coordinate axes chosen so that the center of the base is at the origin and the altitude is measured along the positive side of the y axis. The plane section of the pyramid drawn perpendicular to the y axis at $(0, y)$ is a square. If the length of a side of this square is z units, then by similar triangles (see Figure 8)

$$\frac{\frac{1}{2}z}{h - y} = \frac{\frac{1}{2}s}{h}$$

$$z = \frac{s}{h}(h - y)$$

Therefore, if $A(y)$ square units is the area of the plane section

$$A(y) = \frac{s^2}{h^2}(h - y)^2$$

FIGURE 8

Figure 9 shows an element of volume which is a right cylinder of area $A(\xi_i)$ square units and a thickness of $\Delta_i y$ units. Thus if V cubic units is the volume of the right pyramid

$$V = \lim_{||\Delta|| \to 0} \sum_{i=1}^{n} A(\xi_i)\, \Delta_i y$$

$$= \int_0^h A(y)\, dy$$

$$= \int_0^h \frac{s^2}{h^2}(h - y)^2\, dy$$

$$= \frac{s^2}{h^2}\left[-\frac{(h - y)^3}{3} \right]_0^h$$

$$= \frac{s^2}{h^2}\left[0 + \frac{h^3}{3} \right]$$

$$= \tfrac{1}{3}s^2 h$$

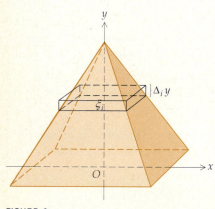

FIGURE 9

We now show how Definition 6.1.1 can be applied to find the volume of a **solid of revolution**, which is a solid obtained by revolving a region in a plane about a line in the plane, called the **axis of revolution**, which may or may not

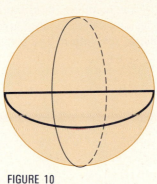

FIGURE 10

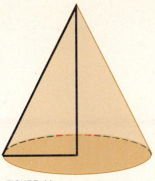

FIGURE 11

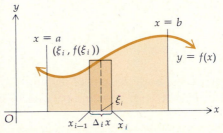

FIGURE 12

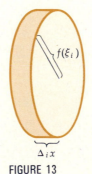

FIGURE 13

intersect the region. For example, if the region bounded by a semicircle and its diameter is revolved about the diameter, a sphere is generated (see Figure 10). A right-circular cone is generated if the region bounded by a right triangle is revolved about one of its legs (see Figure 11).

Consider first the case where the axis of revolution is a boundary of the region that is revolved. Let the function f be continuous on the closed interval $[a, b]$, and assume that $f(x) \geq 0$ for all x in $[a, b]$. Let R be the region bounded by the curve $y = f(x)$, the x axis, and the lines $x = a$ and $x = b$. Figure 12 shows the region R and the ith rectangle. When the ith rectangle is revolved about the x axis we obtain an element of volume which is a disk whose base is a circle of radius $f(\xi_1)$ units and whose altitude is $\Delta_i x$ units as shown in Figure 13. If $\Delta_i V$ cubic units is the volume of this disk,

$$\Delta_i V = \pi [f(\xi_i)]^2 \, \Delta_i x$$

Because there are n rectangles, n disks are obtained in this way, and the sum of the measures of the volumes of these n disks is

$$\sum_{i=1}^{n} \Delta_i V = \sum_{i=1}^{n} \pi [f(\xi_i)]^2 \, \Delta_i x$$

This is a Riemann sum of the form (1) where $A(\xi_i) = \pi [f(\xi_i)]^2$. Therefore if V cubic units is the volume of the solid of revolution, it follows from Definition 6.1.1 that V is the limit of this Riemann sum as $\|\Delta\|$ approaches zero. This limit exists because f^2 is continuous on $[a, b]$, since we assumed that f is continuous there. We have then the following theorem.

6.1.2 THEOREM   Let the function f be continuous on the closed interval $[a, b]$, and assume that $f(x) \geq 0$ for all x in $[a, b]$. If S is the solid of revolution obtained by revolving about the x axis the region bounded by the curve $y = f(x)$, the x axis, and the lines $x = a$ and $x = b$, and if V cubic units is the volume of S, then

$$V = \lim_{\|\Delta\| \to 0} \sum_{i=1}^{n} \pi [f(\xi_i)]^2 \, \Delta_i x$$

$$= \pi \int_{a}^{b} [f(x)]^2 \, dx$$

▶ **ILLUSTRATION 2**   We find the volume of the solid of revolution generated when the region bounded by the curve $y = x^2$, the x axis, and the lines $x = 1$

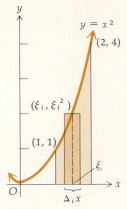

FIGURE 14

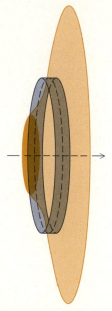

FIGURE 15

and $x = 2$ is revolved about the x axis. Refer to Figure 14, showing the region and a rectangular element of area. Figure 15 shows an element of volume and the solid of revolution. The measure of the volume of the disk is given by

$$\Delta_i V = \pi(\xi_i^2)^2 \, \Delta_i x$$
$$= \pi \xi_i^4 \, \Delta_i x$$

Then

$$V = \lim_{||\Delta|| \to 0} \sum_{i=1}^{n} \pi \xi_i^4 \, \Delta_i x$$
$$= \pi \int_1^2 x^4 \, dx$$
$$= \pi \left(\tfrac{1}{5} x^5\right)\Big]_1^2$$
$$= \tfrac{31}{5} \pi$$

Therefore the volume of the solid of revolution is $\tfrac{31}{5}\pi$ cubic units. ◀

A theorem similar to Theorem 6.1.2 applies when both the axis of revolution and a boundary of the revolved region are the y axis or any line parallel to either the x axis or the y axis.

EXAMPLE 2 Find the volume of the solid generated by revolving about the line $x = 1$ the region bounded by the curve

$$(x - 1)^2 = 20 - 4y$$

and the lines $x = 1$, $y = 1$, and $y = 3$ and to the right of $x = 1$.

Solution The region and a rectangular element of area are shown in Figure 16. An element of volume and the solid of revolution appear in Figure 17.

In the equation of the curve we solve for x and have

$$x = \sqrt{20 - 4y} + 1$$

Let $g(y) = \sqrt{20 - 4y} + 1$. We take a partition of the interval $[1, 3]$ on the y

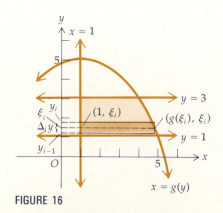

FIGURE 16

FIGURE 17

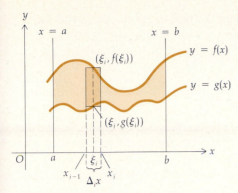

FIGURE 18

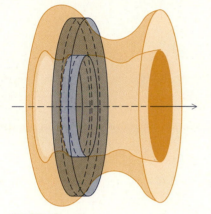

FIGURE 19

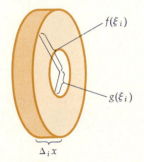

FIGURE 20

axis. Then if $\Delta_i V$ cubic units is the volume of the ith disk,

$$\Delta_i V = \pi[g(\xi_i) - 1]^2 \, \Delta_i y$$
$$= \pi[(\sqrt{20 - 4\xi_1} + 1) - 1]^2 \, \Delta_i y$$
$$= \pi(20 - 4\xi_i) \, \Delta_i y$$

If V cubic units is the volume of the solid of revolution,

$$V = \lim_{\|\Delta\| \to 0} \sum_{i=1}^{n} \pi(20 - 4\xi_i) \, \Delta_i y$$
$$= \pi \int_{1}^{3} (20 - 4y) \, dy$$
$$= \pi \left[20y - 2y^2 \right]_{1}^{3}$$
$$= \pi[(60 - 18) - (20 - 2)]$$
$$= 24\pi$$

The volume of the solid of revolution is therefore 24π cubic units.

Now suppose that the axis of revolution is not a boundary of the region being revolved. Let f and g be two continuous functions on the closed interval $[a, b]$, and assume that $f(x) \geq g(x) \geq 0$ for all x in $[a, b]$. Let R be the region bounded by the curves $y = f(x)$ and $y = g(x)$ and the lines $x = a$ and $x = b$. The region R and the ith rectangle are shown in Figure 18, and the solid of revolution appears in Figure 19. When the ith rectangle is revolved about the x axis, we obtain a washer (or circular ring) as in Figure 20. The difference of the areas of the two circular regions is $\pi[f(\xi_i)]^2 - \pi[g(\xi_i)]^2$ units and the thickness is $\Delta_i x$ units. If $\Delta_i V$ cubic units is the volume of the washer,

$$\Delta_i V = \pi([f(\xi_i)]^2 - [g(\xi_i)]^2) \, \Delta_i x$$

The sum of the measures of the volumes of the n washers formed by revolving the n rectangular elements of area about the x axis is

$$\sum_{i=1}^{n} \Delta_i V = \sum_{i=1}^{n} \pi([f(\xi_i)]^2 - [g(\xi_i)]^2) \, \Delta_i x$$

This is a Riemann sum of the form (1) where $A(\xi_i) = \pi[f(\xi_i)]^2 - \pi[g(\xi_i)]^2$. From Definition 6.1.1, the number of cubic units in the volume of the solid of revolution is the limit of this Riemann sum as $\|\Delta\|$ approaches zero. The limit exists since $f^2 - g^2$ is continuous on $[a, b]$ because f and g are continuous there. We have then the following theorem.

6.1.3 THEOREM

Let the functions f and g be continuous on the closed interval $[a, b]$, and assume that $f(x) \geq g(x) \geq 0$ for all x in $[a, b]$. Then if V cubic units is the volume of the solid of revolution generated by revolving about the x axis the region bounded by the curves $y = f(x)$ and $y = g(x)$ and the lines $x = a$ and $x = b$,

$$V = \lim_{\|\Delta\| \to 0} \sum_{i=1}^{n} \pi([f(\xi_i)]^2 - [g(\xi_i)]^2) \, \Delta_i x$$
$$= \pi \int_{a}^{b} ([f(x)]^2 - [g(x)]^2) \, dx$$

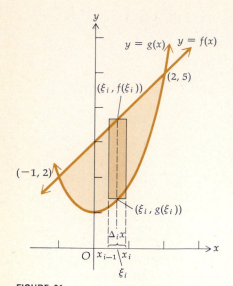

FIGURE 21

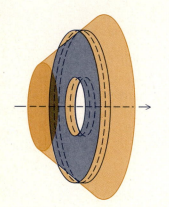

FIGURE 22

As before, a similar theorem applies when the axis of revolution is the y axis or any line parallel to either the x or y axis.

EXAMPLE 3   Find the volume of the solid generated by revolving about the x axis the region bounded by the parabola $y = x^2 + 1$ and the line $y = x + 3$.

Solution   The points of intersection are $(-1, 2)$ and $(2, 5)$. Figure 21 shows the region and a rectangular element of area. An element of volume and the solid of revolution are shown in Figure 22.

If $f(x) = x + 3$ and $g(x) = x^2 + 1$, the measure of the volume of the circular ring is

$$\Delta_i V = \pi([f(\xi_i)]^2 - [g(\xi_i)]^2)\,\Delta_i x$$

If V cubic units is the volume of the solid, then

$$V = \lim_{\|\Delta\| \to 0} \sum_{i=1}^{n} \pi([f(\xi_i)]^2 - [g(\xi_i)]^2)\,\Delta_i x$$

$$= \pi \int_{-1}^{2} ([f(x)]^2 - [g(x)]^2)\,dx$$

$$= \pi \int_{-1}^{2} [(x + 3)^2 - (x^2 + 1)^2]\,dx$$

$$= \pi \int_{-1}^{2} [-x^4 - x^2 + 6x + 8]\,dx$$

$$= \pi \left[-\tfrac{1}{5}x^5 - \tfrac{1}{3}x^3 + 3x^2 + 8x \right]_{-1}^{2}$$

$$= [(-\tfrac{32}{5} - \tfrac{8}{3} + 12 + 16) - (\tfrac{1}{5} + \tfrac{1}{3} + 3 - 8)]$$

$$= \tfrac{117}{5}\pi$$

Therefore the volume of the solid of revolution is $\tfrac{117}{5}\pi$ cubic units.

EXAMPLE 4   Find the volume of the solid generated by revolving about the line $x = -4$ the region bounded by the two parabolas $x = y - y^2$ and $x = y^2 - 3$.

Solution   The curves intersect at the points $(-2, -1)$ and $(-\tfrac{3}{4}, \tfrac{3}{2})$. The region and a rectangular element of area are shown in Figure 23. Figure 24 shows the solid of revolution as well as an element of volume, which is a circular ring.

Let $F(y) = y - y^2$ and $G(y) = y^2 - 3$. The number of cubic units in the volume of the circular ring is

$$\Delta_i V = \pi([4 + F(\xi_i)]^2 - [4 + G(\xi_i)]^2)\,\Delta_i y$$

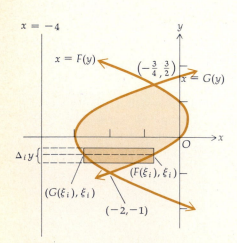

FIGURE 23

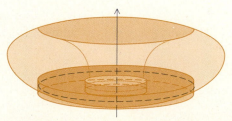

FIGURE 24

Thus

$$V = \lim_{||\Delta|| \to 0} \sum_{i=1}^{n} \pi([4 + F(\xi_i)]^2 - [4 + G(\xi_i)]^2) \, \Delta_i y$$

$$= \pi \int_{-1}^{3/2} [(4 + y - y^2)^2 - (4 + y^2 - 3)^2] \, dy$$

$$= \pi \int_{-1}^{3/2} (-2y^3 - 9y^2 + 8y + 15) \, dy$$

$$= \pi \left[-\tfrac{1}{2} y^4 - 3y^3 + 4y^2 + 15y \right]_{-1}^{3/2}$$

$$= \tfrac{875}{32} \pi$$

The volume of the solid of revolution is then $\frac{875}{32}\pi$ cubic units.

As we have seen, computing volumes by disks and washers is a special case of computing volumes by slicing. We now give another example of finding a volume by slicing.

EXAMPLE 5   A wedge is cut from a right-circular cylinder with a radius of r centimeters by two planes, one perpendicular to the axis of the cylinder and the other intersecting the first along a diameter of the circular plane section at an angle of measurement 60°. Find the volume of the wedge.

Solution   The wedge is shown in Figure 25. The xy plane is taken as the plane perpendicular to the axis of the cylinder, and the origin is at the point of perpendicularity. An equation of the circular plane section is then $x^2 + y^2 = r^2$. Every plane section of the wedge perpendicular to the x axis is a right triangle. An element of volume is a right cylinder having altitude $\Delta_i x$ centimeters, and area of the base given by $\frac{1}{2}\sqrt{3}[f(\xi_i)]^2$ square centimeters, where $f(x)$ is obtained by solving the equation of the circle for y and setting $y = f(x)$. Therefore, we have $f(x) = \sqrt{r^2 - x^2}$. Thus, if V cubic centimeters is the volume of the wedge,

$$V = \lim_{||\Delta|| \to 0} \sum_{i=1}^{n} \tfrac{1}{2}\sqrt{3}(r^2 - \xi_i^2) \, \Delta_i x$$

$$= \tfrac{1}{2}\sqrt{3} \int_{-r}^{r} (r^2 - x^2) \, dx$$

$$= \tfrac{1}{2}\sqrt{3} \left[r^2 x - \tfrac{1}{3} x^3 \right]_{-r}^{r}$$

$$= \tfrac{2}{3}\sqrt{3} r^3$$

Hence the volume of the wedge is $\frac{2}{3}\sqrt{3}r^3$ cm³.

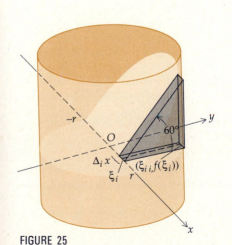

FIGURE 25

EXERCISES 6.1

1. Derive the formula for the volume of a sphere of radius r units by slicing.
2. Derive the formula for the volume of a right-circular cone of altitude h units and base radius a units by slicing.
3. Find the volume of the solid of revolution generated when the

region bounded by the curve $y = x^3$, the x axis, and the lines $x = 1$ and $x = 2$ is revolved about the x axis.
4. Find the volume of the solid of revolution generated when the region bounded by the curve $y = x^2 + 1$, the x axis, and the lines $x = 2$ and $x = 3$ is revolved about the x axis.

In Exercises 5 through 12, find the volume of the solid of revolution generated when the given region of the figure is revolved about the indicated line. An equation of the curve in the figure is $y^2 = x^3$.

5. OAC about the x axis
6. OAC about the line AC
7. OAC about the line BC
8. OAC about the y axis
9. OBC about the y axis
10. OBC about the line BC
11. OBC about the line AC
12. OBC about the x axis

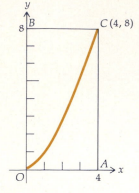

In Exercises 13 through 16, find the volume of the solid of revolution generated by revolving about the indicated line the region bounded by the curve $y = \sqrt{x}$, the x axis, and the line $x = 4$.

13. the line $x = 4$ 14. the x axis
15. the y axis   16. the line $y = 2$

17. Derive the formula for the volume of a sphere by revolving the region bounded by the circle $x^2 + y^2 = r^2$ and the x axis about the x axis.

18. Derive the formula for the volume of a right-circular cone of altitude h units and base radius a units by revolving the region bounded by a right triangle about one of its legs.

19. Derive the formula for the volume of the frustum of a right-circular cone by revolving the line segment from $(0, b)$ to (h, a) about the x axis.

20. Find by slicing the volume of a tetrahedron having three mutually perpendicular faces and three mutually perpendicular edges whose lengths are 3 in., 4 in., and 7 in.

21. The region bounded by the curve $y = \sec x$, the x axis, the y axis, and the line $x = \frac{1}{4}\pi$ is revolved about the x axis. Find the volume of the solid generated.

22. Find the volume of the solid generated when the region bounded by the curve $y = \csc x$, the x axis, and the lines $x = \frac{1}{6}\pi$ and $x = \frac{1}{3}\pi$ is revolved about the x axis.

23. Find the volume of the solid of revolution generated if the region bounded by one arch of the sine curve is revolved about the x axis. (*Hint:* Use the identity $\sin^2 x = \frac{1}{2}(1 - \cos 2x)$.)

24. The region bounded by the y axis and the curves $y = \sin x$ and $y = \cos x$ for $0 \leq x \leq \frac{1}{4}\pi$ is revolved about the x axis. Find the volume of the solid of revolution generated. (*Hint:* Use the following identities: $\sin^2 x = \frac{1}{2}(1 - \cos 2x)$ and $\cos^2 x = \frac{1}{2}(1 + \cos 2x)$.)

25. Find the volume of the solid generated if the region of Exercise 23 is revolved about the line $y = 1$.

26. Find the volume of the solid generated if the region of Exercise 24 is revolved about the line $y = 1$.

27. The region bounded by the curve $y = \cot x$, the line $x = \frac{1}{6}\pi$, and the x axis is revolved about the x axis. Find the volume of the solid generated.

28. The region bounded by the curve $y = \tan x$, the line $x = \frac{1}{3}\pi$, and the x axis is revolved about the x axis. Find the volume of the solid generated.

29. Find the volume of the solid generated by revolving about the line $x = -4$ the region bounded by that line and the parabola $x = 4 + 6y - 2y^2$.

30. Find the volume of the solid generated by revolving about the x axis the region bounded by the parabola $y^2 = 4x$ and the line $y = x$.

31. Find the volume of the solid generated by revolving the region of Exercise 30 about the line $x = 4$.

32. Find the volume of the solid generated by revolving about the y axis the region bounded by the line through $(1, 3)$ and $(3, 7)$, and the lines $y = 3$, $y = 7$, and $x = 0$.

33. Find the volume of the solid generated by revolving about the line $y = -3$ the region bounded by the two parabolas $y = x^2$ and $y = 1 + x - x^2$.

34. Find the volume of the solid generated by revolving about the x axis the region bounded by the loop of the curve whose equation is $2y^2 = x(x^2 - 4)$.

35. Find the volume of the solid generated when the region bounded by one loop of the curve, having the equation $x^2y^2 = (x^2 - 9)(1 - x^2)$, is revolved about the x axis.

36. An oil tank in the shape of a sphere has a diameter of 60 ft. How much oil does the tank contain if the depth of the oil is 25 ft?

37. A paraboloid of revolution is obtained by revolving the parabola $y^2 = 4px$ about the x axis. Find the volume bounded by a paraboloid of revolution and a plane perpendicular to its axis if the plane is 10 cm from the vertex, and if the plane section of intersection is a circle having a radius of 6 cm.

38. The region in the first quadrant bounded by the curve $y = \sec x$, the y axis, and the line $y = 2$ is revolved about the x axis. Find the volume of the solid generated.

39. The region bounded by the curve $y = \csc x$ and the lines $y = 2$, $x = \frac{1}{6}\pi$, and $x = \frac{5}{6}\pi$ is revolved about the x axis. Find the volume of the solid generated.

40. The region in the first quadrant bounded by the coordinate axes, the line $y = 1$, and the curve $y = \cot x$ is revolved about the x axis. Find the volume of the solid generated.

41. A solid of revolution is formed by revolving about the x axis the region bounded by the curve $y = \sqrt{2x + 4}$, the x axis, the y axis, and the line $x = c$ ($c > 0$). For what value of c will the volume be 12π cubic units?

42. The base of a solid is the region enclosed by a circle with a radius of 2 units. Find the volume of the solid if all plane sections perpendicular to a fixed diameter of the base are squares.

43. The base of a solid is the region enclosed by a circle having a radius of 7 cm. Find the volume of the solid if all plane sections perpendicular to a fixed diameter of the base are equilateral triangles.

44. The base of a solid is the region enclosed by a circle having a radius of 4 in., and each plane section perpendicular to a fixed diameter of the base is an isosceles triangle having an altitude of 10 in. and a chord of the circle as a base. Find the volume of the solid.

45. The base of a solid is the region of Exercise 43. Find the volume of the solid if all plane sections perpendicular to a fixed diameter of the base are isosceles triangles of height equal to the distance of the plane section from the center of the circle. The side of the triangle lying in the base of the solid is not one of the two sides of equal length.

46. The base of a solid is the region enclosed by a circle with a radius of r units, and all plane sections perpendicular to a fixed diameter of the base are isosceles right triangles having the hypotenuse in the plane of the base. Find the volume of the solid.

47. Solve Exercise 46 if the isosceles right triangles have one leg in the plane of the base.

48. Two right-circular cylinders, each having a radius of r units, have axes that intersect at right angles. Find the volume of the solid common to the two cylinders.

49. A wedge is cut from a solid in the shape of a right-circular cylinder with a radius of r centimeters by a plane through a diameter of the base and inclined to the plane of the base at an angle of measurement 45°. Find the volume of the wedge.

50. A wedge is cut from a solid in the shape of a right-circular cone having a base radius of 5 ft and an altitude of 20 ft by two half planes containing the axis of the cone. The angle between the two planes has a measurement of 30°. Find the volume of the wedge cut out.

6.2   VOLUMES OF SOLIDS BY CYLINDRICAL SHELLS

In the preceding section we found the volume of a solid of revolution by taking the rectangular elements of area perpendicular to the axis of revolution, and the element of volume was either a disk or a washer. For some solids of revolution this method may not be feasible. For example, suppose we wish to find the volume of the solid of revolution obtained by revolving about the y axis the region bounded by the graph of $y = 3x - x^3$, the y axis, and the line $y = 2$. Figure 1 shows the region. If an element of area is perpendicular to the y axis, as shown in the figure, the element of volume is a disk, and to determine the volume of the solid of revolution involves an integral of the form $\int_0^2 A(y)\,dy$. But to obtain a formula for $A(y)$ requires solving the cubic equation $y = 3x - x^3$ for x in terms of y, which is a laborious undertaking. So we now discuss an alternative procedure for computing the volume of a solid of revolution, which is easier to apply in this and some other situations.

The method involves taking the rectangular elements of area parallel to the axis of revolution. Then when an element of area is revolved about the axis of revolution, a *cylindrical shell* is obtained. A **cylindrical shell** is a solid contained between two cylinders having the same center and axis. Such a cylindrical shell is shown in Figure 2.

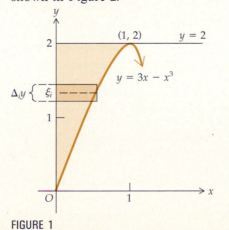

FIGURE 1

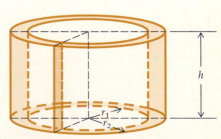

FIGURE 2

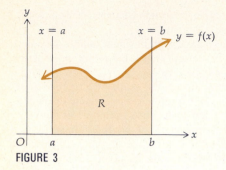

FIGURE 3

If the cylindrical shell has an inner radius r_1 units, outer radius r_2 units, and altitude h units, then its volume V cubic units is given by

$$V = \pi r_2^2 h - \pi r_1^2 h \tag{1}$$

Let R be the region bounded by the curve $y = f(x)$, the x axis, and the lines $x = a$ and $x = b$, where f is continuous on the closed interval $[a, b]$ and $f(x) \geq 0$ for all x in $[a, b]$; further assume that $a \geq 0$. Such a region is shown in Figure 3. If R is revolved about the y axis, a solid of revolution S is generated. Such a solid appears in Figure 4. To find the volume of S when the rectangular elements of area are taken parallel to the y axis we proceed in the following manner.

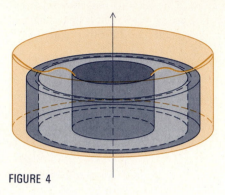

FIGURE 4

Let Δ be a partition of the closed interval $[a, b]$ given by

$$a = x_0 < x_1 < x_2 < \ldots < x_{n-1} < x_n = b$$

Let m_i be the midpoint of the ith subinterval $[x_{i-1}, x_i]$. Then we have $m_i = \frac{1}{2}(x_{i-1} + x_i)$. Consider the rectangle having altitude $f(m_i)$ units and width $\Delta_i x$ units. If this rectangle is revolved about the y axis, a cylindrical shell is obtained. Figure 4 shows the cylindrical shell generated by the rectangular element of area.

If $\Delta_i V$ gives the measure of the volume of this cylindrical shell, we have, from formula (1), where $r_1 = x_{i-1}$, $r_2 = x_i$, and $h = f(m_i)$,

$$\Delta_i V = \pi x_i^2 f(m_i) - \pi x_{i-1}^2 f(m_i)$$

$$\Delta_i V = \pi(x_i^2 - x_{i-1}^2) f(m_i)$$

$$\Delta_i V = \pi(x_i - x_{i-1})(x_i + x_{i-1}) f(m_i)$$

Because $x_i - x_{i-1} = \Delta_i x$, and because $x_i + x_{i-1} = 2m_i$, then from this equation

$$\Delta_i V = 2\pi m_i f(m_i) \Delta_i x$$

If n rectangular elements of area are revolved about the y axis, n cylindrical shells are obtained. The sum of the measures of their volumes is

$$\sum_{i=1}^{n} \Delta_i V = \sum_{i=1}^{n} 2\pi m_i f(m_i) \Delta_i x$$

which is a Riemann sum. The limit of this Riemann sum as $\|\Delta\|$ approaches zero exists because if f is continuous on $[a, b]$ so is the function having values $2\pi x f(x)$. The limit is the definite integral $\int_a^b 2\pi x f(x) \, dx$, and it gives the volume of the solid of revolution. This result is summarized in the following theorem.

6.2.1 THEOREM   Let the function f be continuous on the closed interval $[a, b]$, where $a \geq 0$. Assume that $f(x) \geq 0$ for all x in $[a, b]$. If R is the region bounded by the curve $y = f(x)$, the x axis, and the lines $x = a$ and $x = b$, if S is the solid of revolution obtained by revolving R about the y axis, and if V cubic units is the volume of S, then

$$V = \lim_{\|\Delta\| \to 0} \sum_{i=1}^{n} 2\pi m_i f(m_i) \, \Delta_i x$$

$$= 2\pi \int_a^b x f(x) \, dx$$

While the truth of Theorem 6.2.1 should seem plausible because of the discussion preceding its statement, a proof requires showing that the same volume is obtained by the disk method of Theorem 6.1.2. In the February 1984 issue of the *American Mathematical Monthly* (Vol. 91, No. 2) Charles A. Cable of Allegheny College gave such a proof using integration by parts (discussed in Section 9.1).

The formula for the measure of the volume of the cylindrical shell is easily remembered by noticing that $2\pi m_i$, $f(m_i)$, and $\Delta_i x$ are, respectively, the measures of the circumference of the circle having as radius the mean of the inner and outer radii of the shell, the altitude of the shell, and the thickness. Thus the volume of the shell is

$$2\pi(\text{mean radius})(\text{altitude})(\text{thickness})$$

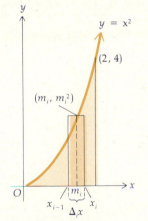

FIGURE 5

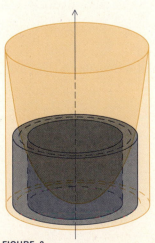

FIGURE 6

EXAMPLE 1   The region bounded by the curve $y = x^2$, the x axis, and the line $x = 2$ is revolved about the y axis. Find the volume of the solid generated. Take the elements of area parallel to the axis of revolution.

Solution   Figure 5 shows the region and a rectangular element of area. Figure 6 shows the solid of revolution and the cylindrical shell obtained by revolving the rectangular element of area about the y axis.

The element of volume is a cylindrical shell the measure of whose volume is

$$\Delta_i V = 2\pi m_i (m_i^2) \, \Delta_i x$$

$$= 2\pi m_i^3 \, \Delta_i x$$

Thus

$$V = \lim_{\|\Delta\| \to 0} \sum_{i=1}^{n} 2\pi m_i^3 \, \Delta_i x$$

$$= 2\pi \int_0^2 x^3 \, dx$$

$$= 2\pi \left(\tfrac{1}{4} x^4 \right) \Big]_0^2$$

$$= 8\pi$$

Therefore the volume of the solid of revolution is 8π cubic units.

In the next example we compute the volume of the solid of revolution discussed at the beginning of this section.

EXAMPLE 2   Find the volume of the solid generated by revolving about the y axis the region bounded by the graph of $y = 3x - x^3$, the y axis, and the line $y = 2$.

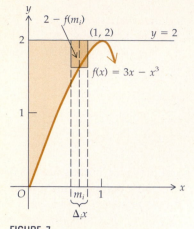

FIGURE 7

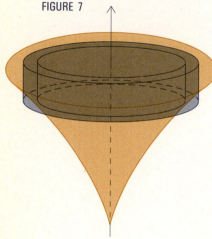

FIGURE 8

Solution   Let $f(x) = 3x - x^3$. Figure 7 shows the region and a rectangular element of area parallel to the y axis. The solid of revolution and a cylindrical shell element of volume appear in Figure 8. The mean radius of the cylindrical shell is m_i units, the altitude is $[2 - f(m_i)]$ units, and the thickness is $\Delta_i x$ units. Therefore, if $\Delta_i V$ cubic units is the volume of the shell

$$\Delta_i V = 2\pi m_i [2 - f(m_i)] \Delta_i x$$

Thus if V cubic units is the volume of the solid of revolution

$$V = \lim_{||\Delta|| \to 0} \sum_{i=1}^{n} 2\pi m_i [2 - f(m_i)] \Delta_i x$$

$$= 2\pi \int_0^1 x[2 - f(x)] \, dx$$

$$= 2\pi \int_0^1 x(2 - 3x + x^3) \, dx$$

$$= 2\pi \int_0^1 (2x - 3x^2 + x^4) \, dx$$

$$= 2\pi \left[x^2 - x^3 + \frac{x^5}{5} \right]_0^1$$

$$= 2\pi (1 - 1 + \tfrac{1}{5})$$

$$= \tfrac{2}{5}\pi$$

Therefore the volume is $\tfrac{2}{5}\pi$ cubic units.

EXAMPLE 3   The region bounded by the curve $y = x^2$ and the lines $y = 1$ and $x = 2$ is revolved about the line $y = -3$. Find the volume of the solid generated by taking the rectangular elements of area parallel to the axis of revolution.

Solution   The region and a rectangular element of area appear in Figure 9.

The equation of the curve is $y = x^2$. Solving for x we obtain $x = \pm\sqrt{y}$. Because $x > 0$ for the given region, $x = \sqrt{y}$.

The solid of revolution and a cylindrical shell element of volume are shown in Figure 10. The outer radius of the cylindrical shell is $(y_i + 3)$ units and the inner radius is $(y_{i-1} + 3)$ units. Hence the mean of the inner and outer radii is $(m_i + 3)$ units. Because the altitude and thickness of the cylindrical shell are, respectively, $(2 - \sqrt{m_i})$ units and $\Delta_i y$ units,

$$\Delta_i V = 2\pi(m_i + 3)(2 - \sqrt{m_i}) \Delta_i y$$

Hence, if V cubic units is the volume of the solid of revolution,

$$V = \lim_{||\Delta|| \to 0} \sum_{i=1}^{n} 2\pi(m_i + 3)(2 - \sqrt{m_i}) \Delta_i y$$

$$= \int_1^4 2\pi(y + 3)(2 - \sqrt{y}) \, dy$$

$$= 2\pi \int_1^4 (-y^{3/2} + 2y - 3y^{1/2} + 6) \, dy$$

$$= 2\pi \left[-\tfrac{2}{5}y^{5/2} + y^2 - 2y^{3/2} + 6y \right]_1^4$$

$$= \tfrac{66}{5}\pi$$

Therefore the volume is $\tfrac{66}{5}\pi$ cubic units.

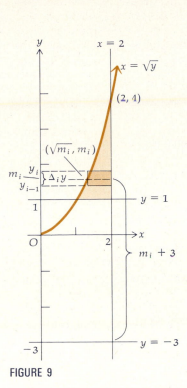

FIGURE 9

FIGURE 10

EXERCISES 6.2

1–12. Solve Exercises 5 through 16 in Section 6.1 by the cylindrical-shell method.

In the figure below, the region bounded by the x axis, the line $x = 1$, and the curve $y = x^2$ is denoted by R_1; the region bounded by the two curves $y = x^2$ and $y^2 = x$ is denoted by R_2; the region bounded by the y axis, the line $y = 1$, and the curve $y^2 = x$ is denoted by R_3. In Exercises 13 through 20, find the volume of the solid generated when the indicated region is revolved about the given line.

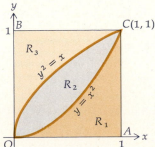

13. R_1 is revolved about the y axis; the rectangular elements are parallel to the axis of revolution.

14. Same as Exercise 13, but the rectangular elements are perpendicular to the axis of revolution.

15. R_2 is revolved about the x axis; the rectangular elements are parallel to the axis of revolution.

16. Same as Exercise 15, but the rectangular elements are perpendicular to the axis of revolution.

17. R_3 is revolved about the line $y = 2$; the rectangular elements are parallel to the axis of revolution.

18. Same as Exercise 17, but the rectangular elements are perpendicular to the axis of revolution.

19. R_2 is revolved about the line $x = -2$; the rectangular elements are parallel to the axis of revolution.

20. Same as Exercise 19, but the rectangular elements are perpendicular to the axis of revolution.

In Exercises 21 through 24, the region bounded by the curves $x = y^2 - 2$ and $x = 6 - y^2$ is revolved about the indicated axis. Find the volume of the solid generated.

21. the x axis   **22.** the y axis
23. the line $x = 2$ **24.** the line $y = 2$

25. Find the volume of the solid generated if the region bounded by the parabola $y^2 = 4px$ $(p > 0)$ and the line $x = p$ is revolved about $x = p$.

26. Find the volume of the solid generated if the region of Exercise 25 is revolved about the y axis.

27. Find the volume of the solid generated by revolving about the y axis the region bounded by the graph of $y = 3x - x^3$, the x axis, and the line $x = 1$.

28. Find the volume of the solid generated by revolving the region of Exercise 27 about the line $x = 1$.

29. Find the volume of the solid generated by revolving the region of Example 2 about the line $x = 1$.

30. Find the volume of the solid generated by revolving about the y axis the region bounded by the graph of $y = 4x - \frac{1}{8}x^4$, the x axis, and the line $x = 2$.

31. Find the volume of the solid generated by revolving the region of Exercise 30 about the line $x = 2$.

32. Find the volume of the solid generated by revolving the region bounded by the graph of $y = 4x - \frac{1}{8}x^4$, the y axis, and the line $y = 6$ about the line $x = 2$.

33. Find the volume of the solid generated by revolving the region of Exercise 32 about the y axis.

34. Find the volume of the solid generated by revolving about the x axis the region bounded by the curves $y = x^3$ and $x = y^3$. Take the rectangular elements of area parallel to the axis of revolution.

35. Find the volume of the solid generated by revolving about the line $y = 1$ the region bounded by that line and the parabola $x^2 = 4y$. Take the rectangular elements of area parallel to the axis of revolution.

36. Find the volume of the solid generated by revolving about the y axis the region bounded by the curve $x^{2/3} + y^{2/3} = a^{2/3}$.

37. Find the volume of the solid generated by revolving about the y axis the region bounded by the curve $y = \sin x^2$, the x axis, and the lines $x = \frac{1}{2}\sqrt{\pi}$ and $x = \sqrt{\pi}$.

38. Find the volume of the solid generated by revolving about the y axis the region in the first quadrant bounded by the curve $y = \cos x^2$ and the coordinate axes.

39. The region in the first quadrant bounded by the curve $x = \cos y^2$, the y axis, and the x axis, where $0 \le x \le 1$, is revolved about the x axis. Find the volume of the solid of revolution generated.

40. A hole of radius 2 cm is bored through a spherical shaped solid of radius 6 cm, and the axis of the hole is a diameter of the sphere. Find the volume of the part of the solid that remains.

41. A hole of radius $2\sqrt{3}$ in. is bored through the center of a spherical shaped solid of radius 4 in. Find the volume of the portion of the solid cut out.

42. Find the volume of the solid generated by revolving about the y axis the region bounded by the graph of $y = |x - 3|$, and the lines $x = 1$, $x = 5$, and $y = 0$. Take the rectangular elements of area parallel to the axis of revolution.

43. A solid of revolution is formed by revolving about the y axis the region bounded by the curve $y = \sqrt[3]{x}$, the x axis, and the line $x = c$ ($c > 0$). Take the rectangular elements of area parallel to the axis of revolution to determine the value of c that will give a volume of 12π cubic units.

44. Find the volume of the solid generated by revolving about the y axis the region outside the curve $y = x^2$ and between the lines $y = 2x - 1$ and $y = x + 2$.

6.3 LENGTH OF ARC OF THE GRAPH OF A FUNCTION

Another geometric application of the definite integral is to calculate the length of arc of the graph of a function. In our discussions of areas and volumes, we used the words "measure of the area" and "measure of the volume" to indicate a number without any units of measurement included. In our treatment of length of arc, we shall use the word "length" in place of the words "measure of the length." Thus it is understood that the length of an arc is a number without any units of measurement attached to it.

Let the function f be continuous on the closed interval $[a, b]$. Consider the graph of this function defined by the equation $y = f(x)$, of which a sketch is shown in Figure 1. The portion of the curve from the point $A(a, f(a))$ to the point $B(b, f(b))$ is called an *arc*. We wish to assign a number to what we intuitively think of as the length of such an arc. If the arc is a line segment from the point (x_1, y_1) to the point (x_2, y_2), we know from the formula for the distance between two points that its length is given by $\sqrt{(x_1 - x_2)^2 + (y_1 - y_2)^2}$. We use this formula for defining the length of an arc in general. Recall from geometry that the circumference of a circle is defined as the limit of the perimeters of regular polygons inscribed in the circle. For other curves we proceed in a similar way.

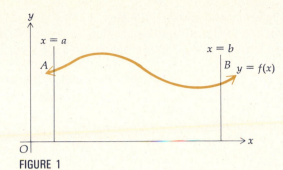

FIGURE 1

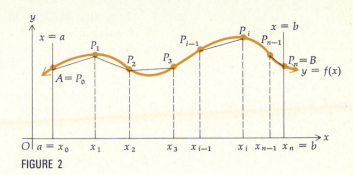

FIGURE 2

Let Δ be a partition of the closed interval $[a, b]$ formed by dividing the interval into n subintervals by choosing any $n - 1$ intermediate numbers between a and b. Let $x_0 = a$ and $x_n = b$, and let $x_1, x_2, x_3, \ldots, x_{n-1}$ be the intermediate numbers so that $x_0 < x_1 < x_2 < \ldots < x_{n-1} < x_n$. Then the ith subinterval is $[x_{i-1}, x_i]$; its length, denoted by $\Delta_i x$, is $x_i - x_{i-1}$, where $i = 1, 2, 3, \ldots, n$. Then if $\|\Delta\|$ is the norm of the partition Δ, each $\Delta_i x \leq \|\Delta\|$.

Associated with each point $(x_i, 0)$ on the x axis is a point $P_i(x_i, f(x_i))$ on the curve. Draw a line segment from each point P_{i-1} to the next point P_i, as shown in Figure 2. The length of the line segment from P_{i-1} to P_i is denoted by $|\overline{P_{i-1}P_i}|$ and is given by the distance formula

$$|\overline{P_{i-1}P_i}| = \sqrt{(x_i - x_{i-1})^2 + (y_i - y_{i-1})^2} \tag{1}$$

The sum of the lengths of the line segments is

$$|\overline{P_0P_1}| + |\overline{P_1P_2}| + |\overline{P_2P_3}| + \ldots + |\overline{P_{i-1}P_i}| + \ldots + |\overline{P_{n-1}P_n}|$$

which can be written with sigma notation as

$$\sum_{i=1}^{n} |\overline{P_{i-1}P_i}| \tag{2}$$

It seems plausible that if n is sufficiently large, the sum in (2) will be "close to" what we would intuitively think of as the length of the arc AB. So we define the length of the arc as the limit of the sum in (2) as the norm of Δ approaches zero, in which case n increases without bound. We have, then, the following definition.

6.3.1 DEFINITION   Suppose the function f is continuous on the closed interval $[a, b]$. Further suppose that there exists a number L having the following property:

For any $\epsilon > 0$ there is a $\delta > 0$ such that for every partition Δ of the interval $[a, b]$ it is true that

$$\text{if} \quad \|\Delta\| < \delta \quad \text{then} \quad \left| \sum_{i=1}^{n} |\overline{P_{i-1}P_i}| - L \right| < \epsilon$$

Then we write

$$L = \lim_{\|\Delta\| \to 0} \sum_{i=1}^{n} |\overline{P_{i-1}P_i}| \tag{3}$$

and L is called the **length of the arc** of the curve $y = f(x)$ from the point $A(a, f(a))$ to the point $B(b, f(b))$.

If the limit in (3) exists, the arc is said to be **rectifiable**.

We derive a formula for finding the length L of an arc that is rectifiable. The derivation requires that the derivative of f be continuous on $[a, b]$; such a function is said to be **smooth** on $[a, b]$.

Refer now to Figure 3. If P_{i-1} has coordinates (x_{i-1}, y_{i-1}) and P_i has coordinates (x_i, y_i), then the length of the chord $P_{i-1}P_i$ is given by formula (1).

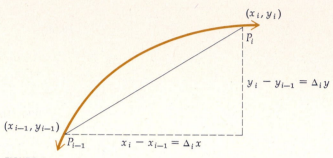

FIGURE 3

Letting $x_i - x_{i-1} = \Delta_i x$ and $y_i - y_{i-1} = \Delta_i y$ we have

$$|\overline{P_{i-1}P_i}| = \sqrt{(\Delta_i x)^2 + (\Delta_i y)^2}$$

or, equivalently, because $\Delta_i x \neq 0$,

$$|\overline{P_{i-1}P_i}| = \sqrt{1 + \left(\frac{\Delta_i y}{\Delta_i x}\right)^2} \, (\Delta_i x) \tag{4}$$

Because we required that f' be continuous on $[a, b]$, the hypothesis of the mean-value theorem (4.3.2) is satisfied by f; so there is a number z_i in the open interval (x_{i-1}, x_i) such that

$$f(x_i) - f(x_{i-1}) = f'(z_i)(x_i - x_{i-1})$$

Because $\Delta_i y = f(x_i) - f(x_{i-1})$ and $\Delta_i x = x_i - x_{i-1}$, from the above equation we have

$$\frac{\Delta_i y}{\Delta_i x} = f'(z_i)$$

Substituting from this equation into (4) we get

$$|\overline{P_{i-1}P_i}| = \sqrt{1 + [f'(z_i)]^2} \, \Delta_i x$$

For each i from 1 to n there is an equation of this form, so that

$$\sum_{i=1}^{n} |\overline{P_{i-1}P_i}| = \sum_{i=1}^{n} \sqrt{1 + [f'(z_i)]^2} \, \Delta_i x$$

Taking the limit on both sides of this equation as $\|\Delta\|$ approaches zero we obtain

$$\lim_{\|\Delta\| \to 0} \sum_{i=1}^{n} |\overline{P_{i-1}P_i}| = \lim_{\|\Delta\| \to 0} \sum_{i=1}^{n} \sqrt{1 + [f'(z_i)]^2} \, \Delta_i x \tag{5}$$

if this limit exists.

To show that the limit on the right side of (5) exists, let F be the function defined by

$$F(x) = \sqrt{1 + [f'(x)]^2}$$

Because we are requiring f' to be continuous on $[a, b]$, F is continuous on $[a, b]$. Since $x_{i-1} < z_i < x_i$, for $i = 1, 2, \ldots, n$, on the right side of (5) we have the limit of a Riemann sum which is a definite integral. Therefore from (5)

$$\lim_{\|\Delta\| \to 0} \sum_{i=1}^{n} |\overline{P_{i-1}P_i}| = \int_a^b \sqrt{1 + [f'(x)]^2} \, dx$$

From (3) the left side is L; therefore

$$L = \int_a^b \sqrt{1 + [f'(x)]^2} \, dx$$

In this way we have proved the following theorem.

6.3.2 THEOREM If the function f and its derivative f' are continuous on the closed interval $[a, b]$, then the length of arc of the curve $y = f(x)$ from the point $(a, f(a))$ to the point $(b, f(b))$ is given by

$$L = \int_a^b \sqrt{1 + [f'(x)]^2} \, dx$$

We also have the following theorem, which gives the length of the arc of a curve when x is expressed as a function of y.

6.3.3 THEOREM If the function g and its derivative g' are continuous on the closed interval $[c, d]$, then the length of arc of the curve $x = g(y)$ from the point $(g(c), c)$ to the point $(g(d), d)$ is given by

$$L = \int_c^d \sqrt{1 + [g'(y)]^2} \, dy$$

The proof of Theorem 6.3.3 is identical to that of Theorem 6.3.2; here we interchange x and y as well as the functions f and g.

The definite integral obtained when applying Theorem 6.3.2 or Theorem 6.3.3 is often difficult to evaluate. Because our techniques of integration have so far been limited to integration of powers and some trigonometric functions, we are further restricted in finding equations of curves for which we can evaluate the resulting definite integrals to find the length of an arc.

EXAMPLE 1 Find the length of the arc of the curve $y = x^{2/3}$ from the point $(1, 1)$ to $(8, 4)$ by using Theorem 6.3.2.

Solution See Figure 4. Because $f(x) = x^{2/3}$, $f'(x) = \frac{2}{3}x^{-1/3}$. From Theorem 6.3.2,

$$L = \int_1^8 \sqrt{1 + \frac{4}{9x^{2/3}}} \, dx$$

$$= \frac{1}{3} \int_1^8 \frac{\sqrt{9x^{2/3} + 4}}{x^{1/3}} \, dx$$

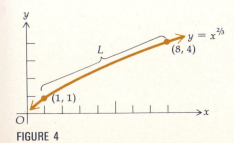

FIGURE 4

To evaluate this definite integral let $u = 9x^{2/3} + 4$; then $du = 6x^{-1/3} \, dx$. When $x = 1$, $u = 13$; when $x = 8$, $u = 40$. Therefore

$$L = \tfrac{1}{18} \int_{13}^{40} u^{1/2} \, du$$

$$= \tfrac{1}{18} \left[\tfrac{2}{3} u^{3/2} \right]_{13}^{40}$$

$$= \tfrac{1}{27} (40^{3/2} - 13^{3/2})$$

$$\approx 7.6$$

EXAMPLE 2   Find the length of the arc in Example 1 by using Theorem 6.3.3.

Solution   Because $y = x^{2/3}$ and $x > 0$, we solve for x and get $x = y^{3/2}$. Letting $g(y) = y^{3/2}$ we have $g'(y) = \tfrac{3}{2} y^{1/2}$. Then, from Theorem 6.3.3,

$$L = \int_{1}^{4} \sqrt{1 + \tfrac{9}{4} y} \; dy$$

$$= \tfrac{1}{2} \int_{1}^{4} \sqrt{4 + 9y} \; dy$$

$$= \tfrac{1}{18} \left[\tfrac{2}{3} (4 + 9y)^{3/2} \right]_{1}^{4}$$

$$= \tfrac{1}{27} (40^{3/2} - 13^{3/2})$$

$$\approx 7.6$$

Using the Leibniz notation for derivatives, the formulas of Theorems 6.3.2 and 6.3.3 can be written as

$$L = \int_{a}^{b} \sqrt{1 + \left(\frac{dy}{dx} \right)^2} \; dx \quad \text{and} \quad L = \int_{c}^{d} \sqrt{1 + \left(\frac{dx}{dy} \right)^2} \; dy \tag{6}$$

We now introduce the *arc length function* and the differential of arc length which provide a mnemonic device for remembering these formulas.

If f' is continuous on $[a, b]$, the definite integral $\int_{a}^{x} \sqrt{1 + [f'(t)]^2} \; dt$ is a function of x; and it gives the length of the arc of the curve $y = f(x)$ from the point $(a, f(a))$ to the point $(x, f(x))$, where x is any number in the closed interval $[a, b]$. Let $s(x)$ denote the length of this arc; s is called the **arc length function** and

$$s(x) = \int_{a}^{x} \sqrt{1 + [f'(t)]^2} \; dt$$

From Theorem 5.8.1,

$$s'(x) = \sqrt{1 + [f'(x)]^2}$$

or, because $s'(x) = ds/dx$ and $f'(x) = dy/dx$,

$$\frac{ds}{dx} = \sqrt{1 + \left(\frac{dy}{dx} \right)^2}$$

Multiplying by dx we obtain

$$ds = \sqrt{1 + \left(\frac{dy}{dx} \right)^2} \; dx \tag{7}$$

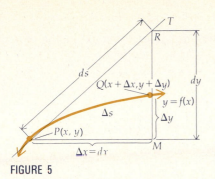

FIGURE 5

Similarly, for the length of arc of the curve $x = g(y)$ from $(g(c), c)$ to $(g(y), y)$,

$$ds = \sqrt{1 + \left(\frac{dx}{dy}\right)^2}\, dy \tag{8}$$

Observe that ds (the differential of arc length) is the integrand in formulas (6). Squaring on both sides of either (7) or (8) gives

$$(ds)^2 = (dx)^2 + (dy)^2 \tag{9}$$

From this equation we get the geometric interpretation of ds, which is shown in Figure 5. In the figure, line T is tangent to the curve $y = f(x)$ at the point P. $|\overline{PM}| = \Delta x = dx$; $|\overline{MQ}| = \Delta y$; $|\overline{MR}| = dy$; $|\overline{PR}| = ds$; the length of arc PQ is Δs. Figure 5 provides an easy way to remember (9) from which formulas (6) can be obtained.

EXERCISES 6.3

1. Compute the length of the segment of the line $y = 3x$ from the point $(1, 3)$ to the point $(2, 6)$ by three methods: (a) Use the distance formula; (b) use Theorem 6.3.2; (c) use Theorem 6.3.3.

2. Compute the length of the segment of the line $x + 3y = 4$ from the point $(-2, 2)$ to the point $(4, 0)$ by three methods: (a) Use the distance formula; (b) use Theorem 6.3.2; (c) use Theorem 6.3.3.

3. Compute the length of the segment of the line $4x + 9y = 36$ between its x and y intercepts by three methods: (a) Use the Pythagorean theorem; (b) use Theorem 6.3.2; (c) use Theorem 6.3.3.

4. Follow instructions of Exercise 3 for the line $5x - 2y = 10$.

5. Find the length of the arc of the curve $9y^2 = 4x^3$ from the origin to the point $(3, 2\sqrt{3})$.

6. Find the length of the arc of the curve $x^2 = (2y + 3)^3$ from $(1, -1)$ to $(7\sqrt{7}, 2)$.

7. Find the length of the arc of the curve $8y = x^4 + 2x^{-2}$ from the point where $x = 1$ to the point where $x = 2$.

8. Use Theorem 6.3.2 to find the length of the arc of the curve $y^3 = 8x^2$ from the point $(1, 2)$ to the point $(27, 18)$.

9. Solve Exercise 8 by using Theorem 6.3.3.

10. Find the length of the arc of the curve $y = \frac{2}{3}(x - 5)^{3/2}$ from the point where $x = 6$ to the point where $x = 8$.

11. Find the length of the arc of the curve $y = \frac{1}{3}(x^2 + 2)^{3/2}$ from the point where $x = 0$ to the point where $x = 3$.

12. Find the length of the arc of the curve $6xy = y^4 + 3$ from the point where $y = 1$ to the point where $y = 2$.

13. Find the length of the arc of the curve $y = \frac{1}{3}\sqrt{x}(3x - 1)$ from the point where $x = 1$ to the point where $x = 4$.

14. Find the length of the arc of the curve $y = \frac{1}{6}x^3 + \frac{1}{2}x^{-1}$ from the point $(2, \frac{19}{12})$ to the point $(5, \frac{314}{15})$.

15. Find the length of the arc of the curve $x^{2/3} + y^{2/3} = 1$ in the first quadrant from the point where $x = \frac{1}{8}$ to the point where $x = 1$.

16. Find the length of the arc of the curve $x^{2/3} + y^{2/3} = a^{2/3}$ (a is a constant, $a > 1$) in the first quadrant from the point where $x = 1$ to the point where $x = a$.

17. Find the length of the curve $(x/a)^{2/3} + (y/b)^{2/3} = 1$ in the first quadrant from the point where $x = \frac{1}{8}a$ to the point where $x = a$.

18. Find the length of the curve $9y^2 = x^2(2x + 3)$ in the second quadrant from the point where $x = -1$ to the point where $x = 0$.

19. Find the length of the curve $9y^2 = x(x - 3)^2$ in the first quadrant from the point where $x = 1$ to the point where $x = 3$.

20. Find the length of the curve $9y^2 = 4(1 + x^2)^3$ in the first quadrant from the point where $x = 0$ to the point where $x = 2\sqrt{2}$.

21. If $f(x) = \int_0^x \sqrt{\cos t}\, dt$, find the length of the arc of the graph of f from the point where $x = 0$ to the point where $x = \frac{1}{2}\pi$. (*Hint:* Use the identity $\cos^2 \frac{1}{2}x = \frac{1}{2}(1 + \cos x)$ and Theorem 5.8.1.)

22. If $f(x) = \int_0^x \sqrt{\sin t}\, dt$, find the length of the arc of the graph of f from the point where $x = 0$ to the point where $x = \frac{1}{2}\pi$. (*Hint:* Use the hint for Exercise 21 and the identity $\sin x = \cos(\frac{1}{2}\pi - x)$.)

In Exercises 23 through 26, use Simpson's rule with $n = 8$ to approximate to four decimal places the length of arc.

23. The arc of the sine curve from the origin to the point $(\pi, 0)$.

24. The arc of the cosine curve from the origin to the point $(\frac{1}{3}\pi, \frac{1}{2})$.

25. The arc of the curve $y = \frac{1}{3}x^3$ from the origin to the point $(1, \frac{1}{3})$.

26. The arc of the curve $y = \tan x$ from the origin to the point $(\frac{1}{4}\pi, 1)$.

6.4 CENTER OF MASS OF A ROD

In Section 5.7 you learned that if a function f is continuous on the closed interval $[a, b]$, the average value of f on $[a, b]$ is given by

$$\frac{\int_a^b f(x)\,dx}{b - a}$$

An important application of the average value of a function occurs in physics in connection with the concept of *center of mass*.

To arrive at a definition of *mass*, consider a particle that is set into motion along an axis by a force exerted on the particle. Assume that all velocities are small compared with the velocity of light. So long as the force is acting on the particle, the velocity of the particle is increasing; that is, the particle is accelerating. The ratio of the force to the acceleration is constant regardless of the magnitude of the force, and this constant ratio is called the **mass** of the particle. Therefore, if the force is F units, the acceleration is a units, and the mass is M units, then

$$M = \frac{F}{a}$$

We shall be measuring force, mass, and acceleration in units of the British and metric systems. We now discuss these units.

In the British system the unit of force is 1 lb and the unit of acceleration is 1 ft/sec². The unit of mass is defined as the mass of a particle whose acceleration is 1 ft/sec² when the particle is subjected to a force of 1 lb. This unit of mass is called 1 *slug*. Hence

$$1 \text{ slug} = \frac{1 \text{ lb}}{1 \text{ ft/sec}^2}$$

▶ **ILLUSTRATION 1** The acceleration of a certain particle moving on a horizontal line is 10 ft/sec² when the force is 30 lb. Therefore the mass of the particle is

$$\frac{30 \text{ lb}}{10 \text{ ft/sec}^2} = \frac{3 \text{ lb}}{1 \text{ ft/sec}^2}$$

Thus for every 1 ft/sec² of acceleration, a force of 3 lb must be exerted on the particle. The mass of the particle is then 3 slugs. ◀

The metric system that has been officially adopted by every major country except the United States is the International System of Units, abbreviated SI for its French name, Système International d'Unités. In the SI system the unit of mass is 1 kilogram (kg) and the unit of acceleration is 1 meter per second squared (m/sec²). The unit of force in the SI system is 1 *newton* (N), which is the force that provides a mass of 1 kg an acceleration of 1 m/sec².

▶ **ILLUSTRATION 2** A particle of mass 6 kg is subjected to a constant horizontal force of 3 N. The acceleration of the particle is obtained by dividing the force by the mass, and so it is

$$\frac{3 \text{ N}}{6 \text{ kg}} = 0.5 \text{ m/sec}^2$$ ◀

In the SI system, the acceleration due to gravity near the surface of the earth is 9.81 m/sec². If M kilograms is the mass of an object, and if F newtons is the force on the object due to gravity near the surface of the earth, then

$$F = 9.81M$$

Another metric system is the centimeter-gram-second system, abbreviated CGS. In the CGS system, the standard unit of mass is the gram, and 1 g = 0.001 kg. The unit of acceleration in the CGS system is 1 cm/sec². Therefore the unit of force, called 1 *dyne*, is the force that gives a particle of mass 1 g an acceleration of 1 cm/sec². Because

$$1 \text{ kg} = 10^3 \text{ g} \quad \text{and} \quad 1 \text{ m/sec}^2 = 10^2 \text{ cm/sec}^2$$

then

$$1 \text{ N} = 10^5 \text{ dynes}$$

We summarize the discussion of units in the British, SI, and CGS systems in Table 1.

Table 1

System of Units	Force	Mass	Acceleration
British	pound (lb)	slug	ft/sec²
SI	newton (N)	kilogram (kg)	m/sec²
CGS	dyne	gram (g)	cm/sec²

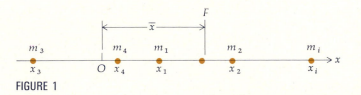

FIGURE 1

Consider now a horizontal rod, of negligible weight and thickness, placed on the x axis. On the rod is a system of n particles located at points $x_1, x_2, \ldots, x_n$. The ith particle ($i = 1, 2, \ldots, n$) is at a directed distance x_i meters from the origin and its mass is m_i kilograms. See Figure 1. The number of kilograms in the total mass of the system is $\sum_{i=1}^{n} m_i$. We define the number of kilogram-meters in the *moment of mass* of the ith particle with respect to the origin as $m_i x_i$. The **moment of mass** for the system is defined as the sum of the moments of mass of all the particles. Hence, if M_0 kilogram-meters is the moment of mass of the system with respect to the origin, then

$$M_0 = \sum_{i=1}^{n} m_i x_i$$

If the measurement of the distance is in feet and of the mass in slugs, then the moment of mass is measured in slug-feet.

We wish to find a point $\bar{x}$ such that if the total mass of the system were concentrated there, its moment of mass with respect to the origin would be equal to the moment of mass of the system with respect to the origin. Then

$\bar{x}$ must satisfy the equation

$$\bar{x} \sum_{i=1}^{n} m_i = \sum_{i=1}^{n} m_i x_i$$

$$\bar{x} = \frac{\displaystyle\sum_{i=1}^{n} m_i x_i}{\displaystyle\sum_{i=1}^{n} m_i} \tag{1}$$

The point $\bar{x}$ is called the **center of mass** of the system, and it is the point where the system will balance. The position of the center of mass is independent of the position of the origin; that is, the location of the center of mass relative to the positions of the particles does not change when the origin is changed. The center of mass is important because the behavior of an entire system of particles can be described by the behavior of the center of mass of the system.

EXAMPLE 1   Given four particles of masses 2, 3, 1, and 5 kg located on the x axis at the points having coordinates 5, 2, -3, and -4, respectively, where distance measurement is in meters, find the center of mass of this system.

Solution   If $\bar{x}$ is the coordinate of the center of mass, we have from formula (1)

$$\bar{x} = \frac{2(5) + 3(2) + (1)(-3) + 5(-4)}{2 + 3 + 1 + 5}$$

$$= -\frac{7}{11}$$

Thus the center of mass is $\frac{7}{11}$ m to the left of the origin.

We now extend the preceding discussion to a rigid horizontal rod having a continuously distributed mass. The rod is said to be **homogeneous** if it has constant linear density, that is, if its mass is directly proportional to its length. In other words, if the segment of the rod whose length is $\Delta_i x$ meters has a mass of $\Delta_i m$ kilograms, and $\Delta_i m = k \Delta_i x$, then the rod is homogeneous. The number k is a constant, and k kilograms per meter is called the **linear density** of the rod.

Suppose that we have a nonhomogeneous rod, in which case the linear density varies along the rod. Let L meters be the length of the rod, and place the rod on the x axis so the left endpoint of the rod is at the origin and the right endpoint is at L. See Figure 2. The linear density at any point x on the rod is $\rho(x)$ kilograms per meter, where ρ is continuous on $[0, L]$. To find the total mass of the rod we consider a partition Δ of the closed interval $[0, L]$ into n subintervals. The ith subinterval is $[x_{i-1}, x_i]$, and its length is $\Delta_i x$ meters. If ξ_i is any point in $[x_{i-1}, x_i]$, an approximation to the mass of the part of the rod

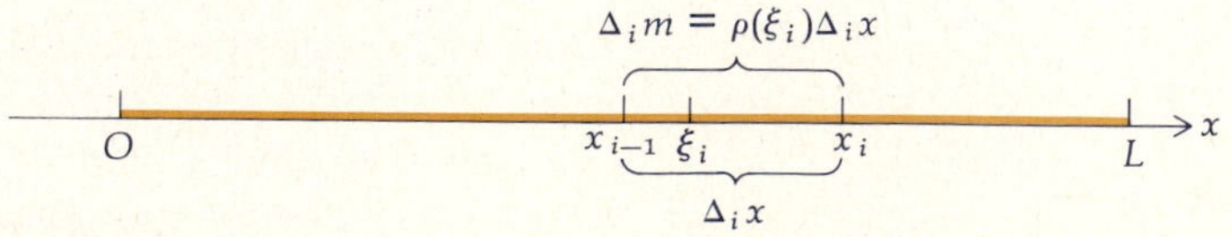

$$\Delta_i m = \rho(\xi_i) \Delta_i x$$

FIGURE 2

contained in the ith subinterval is $\Delta_i m$ kilograms, where

$$\Delta_i m = \rho(\xi_i)\, \Delta_i x$$

The number of kilograms in the total mass of the rod is approximated by

$$\sum_{i=1}^{n} \Delta_i m = \sum_{i=1}^{n} \rho(\xi_i)\, \Delta_i x$$

The smaller we take the norm of the partition Δ, the closer this Riemann sum will be to what we intuitively think of as the measure of the mass of the rod, and so we define the measure of the mass as the limit of the Riemann sum.

6.4.1 DEFINITION   A rod of length L meters has its left endpoint at the origin. If $\rho(x)$ kilograms per meter is the linear density at a point x meters from the origin, where ρ is continuous on $[0, L]$, then the total **mass** of the rod is M kilograms, where

$$M = \lim_{\|\Delta\| \to 0} \sum_{i=1}^{n} \rho(\xi_i)\, \Delta_i x$$

$$= \int_0^L \rho(x)\, dx \tag{2}$$

In Definition 6.4.1, if the distance is measured in feet and the mass is measured in slugs, then the density is measured in slugs per foot.

EXAMPLE 2   The linear density at any point of a rod 4 m long varies directly as the distance from the point to an external point in the line of the rod and 2 m from an end, where the linear density is 5 kg/m. Find the total mass of the rod.

Solution   Figure 3 shows the rod placed on the x axis. If $\rho(x)$ kilograms per meter is the linear density of the rod at the point x meters from the end having the greater density, then

$$\rho(x) = c(6 - x)$$

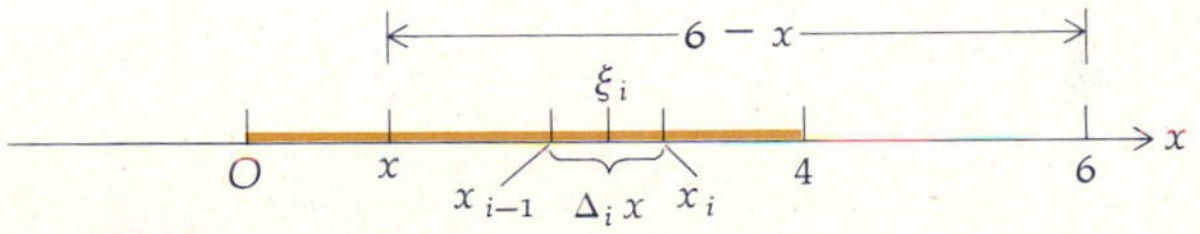

FIGURE 3

where c is the constant of proportionality. Because $\rho(4) = 5$, then $5 = 2c$ or $c = \frac{5}{2}$. Hence $\rho(x) = \frac{5}{2}(6 - x)$. Therefore, if M kilograms is the total mass of the rod, we have from Definition 6.4.1

$$M = \lim_{\|\Delta\| \to 0} \sum_{i=1}^{n} \tfrac{5}{2}(6 - \xi_i)\, \Delta_i x$$

$$= \int_0^4 \tfrac{5}{2}(6 - x)\, dx$$

$$= \tfrac{5}{2}\left[6x - \tfrac{1}{2}x^2\right]_0^4$$

$$= 40$$

The total mass of the rod is therefore 40 kg.

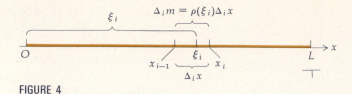

FIGURE 4

We now proceed to define the center of mass of the rod of Definition 6.4.1. However, first we must define the moment of mass of the rod with respect to the origin.

As before, place the rod on the x axis with the left endpoint at the origin and the right endpoint at L. See Figure 4. Let Δ be a partition of $[0, L]$ into n subintervals, with the ith subinterval $[x_{i-1}, x_i]$ having length $\Delta_i x$ meters. If ξ_i is any point in $[x_{i-1}, x_i]$, an approximation to the moment of mass with respect to the origin of the part of the rod contained in the ith subinterval is $\xi_i \Delta_i m$ kilogram-meters, where $\Delta_i m = \rho(\xi_i) \Delta_i x$. The number of kilogram-meters in the moment of mass of the entire rod is approximated by

$$\sum_{i=1}^{n} \xi_1 \Delta_i m = \sum_{i=1}^{n} \xi_i \rho(\xi_i) \Delta_i x$$

The smaller we take the norm of the partition Δ, the closer this Riemann sum will be to what we intuitively think of as the measure of the moment of mass of the rod with respect to the origin. We have, then, the following definition.

6.4.2 DEFINITION   A rod of length L meters has its left endpoint at the origin and $\rho(x)$ kilograms per meter is the linear density at a point x meters from the origin, where ρ is continuous on $[0, L]$. The **moment of mass** of the rod with respect to the origin is M_0 kilogram-meters, where

$$M_0 = \lim_{\|\Delta\| \to 0} \sum_{i=1}^{n} \xi_i \rho(\xi_i) \Delta_i x$$

$$= \int_0^L x\rho(x) \, dx \tag{3}$$

The center of mass of the rod is at the point $\bar{x}$ such that if M kilograms is the total mass of the rod, $\bar{x}M = M_0$. Thus, from (2) and (3),

$$\bar{x} = \frac{\int_0^L x\rho(x) \, dx}{\int_0^L \rho(x) \, dx} \tag{4}$$

EXAMPLE 3   Find the center of mass of the rod in Example 2.

Solution   In Example 2, $M = 40$. From (4) with $\rho(x) = \frac{5}{2}(6 - x)$,

$$\bar{x} = \frac{\int_0^4 \frac{5}{2}x(6 - x) \, dx}{40}$$

$$= \frac{1}{16}\left[3x^2 - \frac{1}{3}x^3\right]_0^4$$

$$= \frac{5}{3}$$

Therefore the center of mass is $\frac{5}{3}$ m from the end having the greater density.

▶ **ILLUSTRATION 3**   If a rod is of uniform linear density k kilograms per meter, where k is a constant, then from formula (4),

$$\bar{x} = \frac{\displaystyle\int_0^L xk\,dx}{\displaystyle\int_0^L k\,dx}$$

$$= \frac{\dfrac{kx^2}{2}\bigg]_0^L}{kx\bigg]_0^L}$$

$$= \frac{\dfrac{kL^2}{2}}{kL}$$

$$= \frac{L}{2}$$

Thus the center of mass is at the center of the rod, as is to be expected.   ◀

EXERCISES 6.4

In Exercises 1 through 4, a particle is moving on a horizontal line. Find the force exerted on the particle if it has the given mass and acceleration.

1. Mass is 50 slugs; acceleration is 5 ft/sec^2.
2. Mass is 10 kg; acceleration is 6 m/sec^2.
3. Mass is 80 g; acceleration is 50 cm/sec^2.
4. Mass is 22 slugs; acceleration is 4 ft/sec^2.

In Exercises 5 through 8, a particle is subjected to the given horizontal force, and either the mass or acceleration of the particle is given. Find the other quantity.

5. Force is 6 N; mass is 4 kg.
6. Force is 32 lb; mass is 8 slugs.
7. Force is 24 lb; acceleration is 9 ft/sec^2.
8. Force is 700 dynes; acceleration is 80 cm/sec^2.

In Exercises 9 through 12, a system of particles is located on the x axis. The number of kilograms in the mass of each particle and the coordinate of its position are given. Distance is measured in meters. Find the center of mass of each system.

9. $m_1 = 5$ at 2; $m_2 = 6$ at 3; $m_3 = 4$ at 5; $m_4 = 3$ at 8
10. $m_1 = 2$ at -4; $m_2 = 8$ at -1; $m_3 = 4$ at 2; $m_4 = 2$ at 3
11. $m_1 = 2$ at -3; $m_2 = 4$ at -2; $m_3 = 20$ at 4; $m_4 = 10$ at 6; $m_5 = 30$ at 9
12. $m_1 = 5$ at -7; $m_2 = 3$ at -2; $m_3 = 5$ at 0; $m_4 = 1$ at 2; $m_5 = 8$ at 10

In Exercises 13 through 21, find the total mass of the given rod and the center of mass.

13. The length of a rod is 6 m and the linear density of the rod at a point x meters from one end is $(2x + 3)$ kg/m.

14. The length of a rod is 20 cm and the linear density of the rod at a point x centimeters from one end is $(3x + 2)$ g/cm.
15. The length of a rod is 9 in. and the linear density of the rod at a point x inches from one end is $(4x + 1)$ slugs/in.
16. The length of a rod is 3 ft, and the linear density of the rod at a point x feet from one end is $(5 + 2x)$ slugs/ft.
17. The length of a rod is 12 cm, and the measure of the linear density at a point is a linear function of the measure of the distance from the left end of the rod. The linear density at the left end is 3 g/cm and at the right end is 4 g/cm.
18. The length of a rod is 10 m and the measure of the linear density at a point is a linear function of the measure of the distance of the point from the left end of the rod. The linear density at the left end is 2 kg/m and at the right end is 3 kg/m.
19. The measure of the linear density at any point of a rod 6 m long varies directly as the distance from the point to an external point in the line of the rod and 4 m from an end, where the density is 3 kg/m.
20. A rod is 10 ft long, and the measure of the linear density at a point is a linear function of the measure of the distance from the center of the rod. The linear density at each end of the rod is 5 slugs/ft and at the center the linear density is $3\frac{1}{2}$ slugs/ft.
21. The measure of the linear density at a point of a rod varies directly as the third power of the measure of the distance of the point from one end. The length of the rod is 4 ft and the linear density is 2 slugs/ft at the center.
22. The linear density at any point of a rod 5 m long varies directly as the distance from the point to an external point in the line of the rod and 2 m from an end, where the linear density is K kg/m. Find K if the total mass of the rod is 135 kg.

23. The linear density at any point of a rod 3 m long varies directly as the distance from a point to an external point in the line of the rod and 1 m from an end where the linear density is 2 kg/m. If the total mass of the rod is 15 kg, find the center of mass of the rod.

24. The measure of the linear density at a point on a rod varies directly as the fourth power of the measure of the distance of the point from one end. The length of the rod is 2 m and the linear density is 2 kg/m at the center. If the total mass of the rod is $\frac{64}{5}$ kg, find the center of mass of the rod.

25. A rod is L meters long and the center of mass of the rod is at the point $\frac{3}{4}L$ meters from the left end. If the measure of the linear density at a point is proportional to a power of the measure of the distance of the point from the left end and

the linear density at the right end is 20 kg/m, find the linear density at a point x meters from the left end. The mass is in kilograms.

26. The total mass of a rod of length L meters is M kilograms, and the measure of the linear density at a point x meters from the left end is proportional to the measure of the distance of the point from the right end. Show that the linear density at a point on the rod x meters from the left end is $2M(L - x)/L^2$ kilograms per meter.

27. A rod is 6 m long and its mass is 24 kg. If the measure of the linear density at any point of the rod varies directly as the square of the distance of the point from one end, find the largest value of the linear density.

6.5 CENTROID OF A PLANE REGION

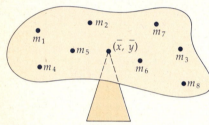

FIGURE 1

Let the masses of n particles located at the points $(x_1, y_1), (x_2, y_2), \ldots, (x_n, y_n)$ in the xy plane be measured by $m_1, m_2, \ldots, m_n$, and consider the problem of finding the center of mass of this system. We may imagine the particles being supported by a sheet of negligible weight and negligible thickness and may assume that each particle has its position at exactly one point. The center of mass is the point where the sheet will balance. Refer to Figure 1, which shows eight particles placed on the sheet. The label of the ith particle in the figure is m_i, which is the measure of its mass. The sheet will balance on a fulcrum located at the center of mass denoted by $(\bar{x}, \bar{y})$. To determine the center of mass we first define the moment of mass of a system of particles with respect to an axis.

Suppose a particle at a distance of d meters from an axis has a mass of m kilograms. If M_1 kilogram-meters is the moment of mass of the particle with respect to the axis, then

$$M_1 = md \tag{1}$$

If the ith particle, having mass m_i kilograms, is located at the point (x_i, y_i), its distance from the y axis is x_i meters; thus, from formula (1), the moment of mass of this particle with respect to the y axis is $m_i x_i$ kilogram-meters. Similarly, the moment of mass of the particle with respect to the x axis is $m_i y_i$ kilogram-meters. The moment of the system of n particles with respect to the y axis is M_y kilogram-meters, where

$$M_y = \sum_{i=1}^{n} m_i x_i$$

and the moment of the system with respect to the x axis is M_x kilogram-meters, where

$$M_x = \sum_{i=1}^{n} m_i y_i$$

The total mass of the system is M kilograms, where

$$M = \sum_{i=1}^{n} m_i$$

The center of mass of the system is at the point $(\bar{x}, \bar{y})$, where

$$\bar{x} = \frac{M_y}{M} \quad \text{and} \quad \bar{y} = \frac{M_x}{M}$$

The point $(\bar{x}, \bar{y})$ can be interpreted as the point such that, if the total mass M kilograms of the system were concentrated there, its moment of mass with respect to the y axis, M_y kilogram-meters, would be determined by $M_y = M\bar{x}$, and its moment of mass with respect to the x axis, M_x kilogram-meters, would be determined by $M_x = M\bar{y}$.

EXAMPLE 1   Find the center of mass of the four particles having masses 2, 6, 4, and 1 kg located at the points $(5, -2), (-2, 1), (0, 3)$, and $(4, -1)$, respectively.

Solution

$$M_y = \sum_{i=1}^{4} m_i x_i \qquad\qquad M_x = \sum_{i=1}^{4} m_i y_i$$

$$= 2(5) + 6(-2) + 4(0) + 1(4) \qquad = 2(-2) + 6(1) + 4(3) + 1(-1)$$

$$= 2 \qquad\qquad\qquad\qquad\qquad = 13$$

$$M = \sum_{i=1}^{4} m_i$$

$$= 2 + 6 + 4 + 1$$

$$= 13$$

Therefore

$$\bar{x} = \frac{M_y}{M} \qquad \bar{y} = \frac{M_x}{M}$$

$$= \tfrac{2}{13} \qquad\quad = \tfrac{13}{13}$$

$$\qquad\qquad\quad = 1$$

The center of mass is at $(\tfrac{2}{13}, 1)$.

Consider now a thin sheet of continuously distributed mass, for example, a piece of paper or a flat strip of tin. We regard such sheets as being two-dimensional and call such a plane region a *lamina*. In this section we confine our discussion to homogeneous laminae, that is, laminae having constant area density. Laminae of variable area density are considered in Chapter 18 in connection with applications of multiple integrals.

Let a homogeneous lamina of area A square meters have a mass of M kilograms. Then if the constant area density is k kilograms per square meter, $M = kA$. If the homogeneous lamina is a rectangle, its center of mass is defined to be at the center of the rectangle. We use this definition to define the center of mass of a more general homogeneous lamina.

Let L be the homogeneous lamina whose constant area density is k kg/m² and which is bounded by the curve $y = f(x)$, the x axis, and the lines $x = a$ and $x = b$. The function f is continuous on the closed interval $[a, b]$, and $f(x) \geq 0$

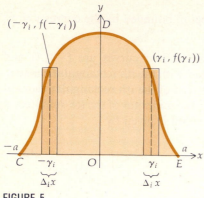

FIGURE 5

Proof   Choose the coordinate axes so that L is on the y axis and the origin is in the region R. Figure 5 illustrates an example of this situation. In the figure, R is the region CDE, C is the point $(-a, 0)$, E is the point $(a, 0)$, and an equation of the curve CDE is $y = f(x)$.

Consider a partition of the interval $[0, a]$. Let γ_i be the midpoint of the ith subinterval $[x_{i-1}, x_i]$. The moment with respect to the y axis of the rectangular element having an altitude $f(\gamma_i)$ and a width $\Delta_i x$ is $\gamma_i[f(\gamma_i)\Delta_i x]$. Because of symmetry, for a similar partition of the interval $[-a, 0]$ there is a corresponding element having as its moment with respect to the y axis $-\gamma_i f(\gamma_i)\Delta_i x$. The sum of these two moments is 0; therefore $M_y = 0$. Because $\bar{x} = M_y/A$, we conclude that $\bar{x} = 0$. Thus the centroid of the region R lies on the y axis, which is what was to be proved.   ∎

By applying the preceding theorem we can simplify the problem of finding the centroid of a plane region that can be divided into regions having axes of symmetry.

EXAMPLE 4   Find the centroid of the region bounded by the semicircle $y = \sqrt{4 - x^2}$ and the x axis.

Solution   The region is shown in Figure 6.

Because the y axis is an axis of symmetry, the centroid lies on the y axis; so $\bar{x} = 0$.

The moment of the region with respect to the x axis is given by

$$M_x = \lim_{\|\Delta\| \to 0} \sum_{i=1}^{n} \tfrac{1}{2}[\sqrt{4 - \gamma_i^2}]^2 \, \Delta_i x$$

$$= 2 \cdot \tfrac{1}{2} \int_0^2 (4 - x^2) \, dx$$

$$= 4x - \tfrac{1}{3}x^3 \Big]_0^2$$

$$= \tfrac{16}{3}$$

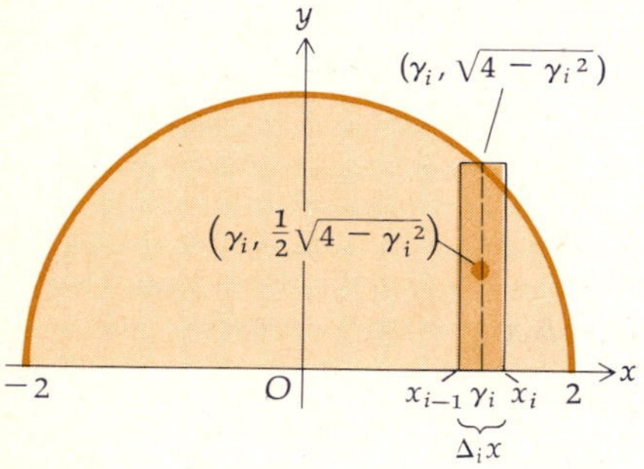

FIGURE 6

The area of the region is 2π square units; so

$$\bar{y} = \frac{\frac{16}{3}}{2\pi}$$

$$= \frac{8}{3\pi}$$

The centroid is at the point $\left(0, \dfrac{8}{3\pi}\right)$.

EXERCISES 6.5

1. Find the center of mass of the three particles having masses of 1, 2, and 3 kg and located at the points $(-1, 3)$, $(2, 1)$, and $(3, -1)$, respectively.

2. Find the center of mass of the four particles having masses of 2, 3, 3, and 4 kg and located at the points $(-1, -2)$, $(1, 3)$, $(0, 5)$, and $(2, 1)$, respectively.

3. The y coordinate of the center of mass of four particles is 5. The particles have masses 2, 5, 4, and m kg and are located

at the points $(3, 2)$, $(-1, 0)$, $(0, 20)$, and $(2, -2)$, respectively. Find m.

4. Find the center of mass of the three particles having masses 3, 7, and 2 kg located at the points $(2, 3)$, $(-1, 4)$, and $(0, 2)$, respectively.
5. Find the center of mass of three particles of equal mass located at the points $(4, -2)$, $(-3, 0)$, and $(1, 5)$.
6. Prove that the center of mass of three particles of equal mass in a plane lies at the point of intersection of the medians of the triangle having as vertices the points at which the particles are located.

In Exercises 7 through 14, find the centroid of the region with the indicated boundaries.

7. The parabola $y = 4 - x^2$ and the x axis.
8. The parabola $x = 2y - y^2$ and the y axis.
9. The parabola $y = x^2$ and the line $y = 4$.
10. The parabola $y^2 = 4x$, the y axis, and the line $y = 4$.
11. The curves $y = x^3$ and $y = 4x$ in the first quadrant.
12. The lines $y = 2x + 1$, $x + y = 7$, and $x = 8$.
13. The curves $y = x^2 - 4$ and $y = 2x - x^2$.
14. The curves $y = x^2$ and $y = x^3$.

15. Find the center of mass of the lamina bounded by the parabola $2y^2 = 18 - 3x$ and the y axis if the area density at any point (x, y) is $\sqrt{6 - x}$ kilograms per square meter.
16. Solve Exercise 15 if the area density at any point (x, y) is x kilograms per square meter.
17. If the centroid of the region bounded by the parabola $y^2 = 4px$ and the line $x = a$ is to be at the point $(p, 0)$, find the value of a.
18. Prove that the distance from the centroid of a triangle to any side of the triangle is equal to one-third of the length of the altitude to that side.
19. Let R be the region bounded by the curves $y = f_1(x)$ and $y = f_2(x)$ (see the figure). If A is the measure of the area of R and if $\bar{y}$ is the ordinate of the centroid of R, prove that the measure of the volume V of the solid of revolution obtained by revolving R about the x axis is given by $V = 2\pi\bar{y}A$. Stating this equation in words we have:

If a plane region is revolved about a line in its plane that does not cut the region, then the measure of the volume of the solid of revolution generated is equal to the product of the measure of the area of the region and the measure of the distance traveled by the centroid of the region.

The above statement is known as the *theorem of Pappus* for volumes of solids of revolution.

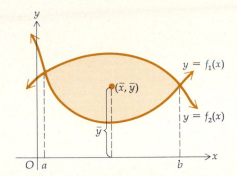

20. Use the theorem of Pappus to find the volume of the torus (doughnut shape) generated by revolving a circle with a radius of r units about a line in its plane at a distance of b units from its center, where $b > r$.
21. Use the theorem of Pappus to find the centroid of the region bounded by a semicircle and its diameter.
22. Use the theorem of Pappus to find the volume of a sphere with a radius of r units.
23. Use the theorem of Pappus to find the volume of a right-circular cone with base radius r units and height h units.
24. Let R be the region bounded by the semicircle $y = \sqrt{r^2 - x^2}$ and the x axis. Use the theorem of Pappus to find the moment of R with respect to the line $y = -r$.
25. If R is the region of Exercise 24, use the theorem of Pappus to find the volume of the solid of revolution generated by revolving R about the line $x - y = r$. (*Hint:* Use the result of Exercise 46 in Exercises 4.8.)
26. Give an example to show that the centroid of a plane region is not necessarily a point within the region.

6.6 WORK

The *work* done by a force acting on an object is defined in physics as "force times displacement." For example, suppose that an object is moving to the right along the x axis from a point a to a point b, and a constant force of F lb is acting on the object in the direction of motion. Then if the displacement is measured in feet, $b - a$ is the number of feet in the displacement. And if W is the number of foot-pounds of work done by the force, W is defined by

$$W = F(b - a) \qquad (1)$$

▶ **ILLUSTRATION 1** If W foot-pounds is the work necessary to lift a 70-lb weight to a height of 3 ft, then

$$W = 70 \cdot 3$$
$$= 210$$

◀

The unit of measurement for work depends on the units of force and distance. In the British system, where the force is measured in pounds and the distance is measured in feet, work is measured in foot-pounds. In the SI system, the unit of force is a newton, the unit of distance is a meter, and the unit of work is a newton-meter, called a *joule*. In the CGS system, the unit of force is a dyne, the unit of distance is a centimeter, and the unit of work is a dyne-centimeter, called an *erg*. For conversion purposes, 1 newton is 10^5 dynes and 1 joule is 10^7 ergs.

The following illustration shows a computation for work using units in the SI system.

▶ **ILLUSTRATION 2**   We wish to find the work done in lifting a rock of mass 8 kg a distance of 4 m. We use the formula $F = Ma$, where F newtons is the force required to give a mass of M kilograms an acceleration of a meters per second squared. The force in this case is the force of gravity and the acceleration is that due to gravity, which is 9.81 m/sec². The mass is 8 kg. Therefore $M = 8$, $a = 9.81$, and

$$F = 8(9.81)$$
$$= 78.48$$

Thus we wish to find the work done by a force of 78.48 N and a displacement of 4 m. If W joules is the work,

$$W = (78.48)(4)$$
$$= 313.92$$

Hence the work done is 313.92 joules.   ◀

Consider now the work done by a variable force acting along a line in the direction of the displacement. We wish to define what is meant by the term "work" in such a case.

Suppose that f is continuous on the closed interval $[a, b]$ and $f(x)$ units is the force acting in the direction of motion on an object as it moves to the right along the x axis from point a to point b. Let Δ be a partition of $[a, b]$:

$$a = x_0 < x_1 < x_2 < \ldots < x_{n-1} < x_n = b$$

The ith subinterval is $[x_{i-1}, x_i]$; and if x_{i-1} is close to x_i, the force is almost constant in this subinterval. If we assume that the force is constant in the ith subinterval and if ξ_i is any point such that $x_{i-1} \le \xi_i \le x_i$, then if $\Delta_i W$ units of work is done on the object as it moves from the point x_{i-1} to the point x_i, from formula (1) we have

$$\Delta_i W = f(\xi_i)(x_i - x_{i-1})$$

Replacing $x_i - x_{i-1}$ by $\Delta_i x$ we have

$$\Delta_i W = f(\xi_i) \, \Delta_i x$$

$$\sum_{i=1}^{n} \Delta_i W = \sum_{i=1}^{n} f(\xi_i) \, \Delta_i x$$

The smaller we take the norm of the partition Δ, the larger n will be and the closer the Riemann sum will be to what we intuitively think of as the measure of the total work done. We therefore define the measure of the total work as the limit of the Riemann sum.

6.6.1 DEFINITION

Let the function f be continuous on the closed interval $[a, b]$ and $f(x)$ units be the force acting on an object at the point x on the x axis. Then if W units is the **work** done by the force as the object moves from a to b,

$$W = \lim_{||\Delta|| \to 0} \sum_{i=1}^{n} f(\xi_i)\, \Delta_i x$$

$$= \int_a^b f(x)\, dx$$

EXAMPLE 1   A particle is moving along the x axis under the action of a force of $f(x)$ pounds when the particle is x feet from the origin. If $f(x) = x^2 + 4$, find the work done as the particle moves from the point where $x = 2$ to the point where $x = 4$.

Solution   We take a partition of the closed interval $[2, 4]$. If W foot-pounds is the work done as the particle moves from the point where $x = 2$ to the point where $x = 4$, then from Definition 6.6.1,

$$W = \lim_{||\Delta|| \to 0} \sum_{i=1}^{n} f(\xi_i)\, \Delta_i x$$

$$= \int_2^4 f(x)\, dx$$

$$= \int_2^4 (x^2 + 4)\, dx$$

$$= \frac{x^3}{3} + 4x \bigg]_2^4$$

$$= \tfrac{64}{3} + 16 - (\tfrac{8}{3} + 8)$$

$$= 26\tfrac{2}{3}$$

Therefore the work done is $26\tfrac{2}{3}$ ft-lb.

In the following example we use **Hooke's law**, which states that if a spring is stretched x units beyond its natural length, it is pulled back with a force equal to kx units, where k is a constant depending on the material and size of the spring.

EXAMPLE 2   A spring has a natural length of 14 cm. If a force of 500 dynes is required to keep the spring stretched 2 cm, how much work is done in stretching the spring from its natural length to a length of 18 cm?

Solution   Place the spring along the x axis with the origin at the point where the stretching starts. See Figure 1. Let $f(x)$ dynes be the force required

FIGURE 1

to stretch the spring x centimeters beyond its natural length. Then, by Hooke's law

$$f(x) = kx$$

Because $f(2) = 500$, we have

$$500 = k \cdot 2$$

$$k = 250$$

Thus

$$f(x) = 250x$$

Because the spring is being stretched from 14 cm to 18 cm, we consider a partition of the closed interval $[0, 4]$ on the x axis. Let $\Delta_i x$ centimeters be the length of the ith subinterval and let ξ_i be any point in that subinterval. If W ergs is the work done in stretching the spring from 14 cm to 18 cm, then

$$
\begin{aligned}
W &= \lim_{\|\Delta\| \to 0} \sum_{i=1}^{n} f(\xi_i)\, \Delta_i x \\
&= \int_0^4 f(x)\, dx \\
&= \int_0^4 250x\, dx \\
&= \frac{250}{2} x^2 \bigg]_0^4 \\
&= 2000
\end{aligned}
$$

Therefore the work done in stretching the spring is 2000 ergs.

In Examples 3 and 4 we are concerned with the weight of water. In the SI system the mass density of water is 1000 kg/m^3 and the acceleration due to gravity is 9.81 m/sec^2. Therefore the weight density of water is $(1000)(9.81)$ N/m^3 = 9810 N/m^3. In the British system the mass density of water is 1.94 slugs/ft^3 and the acceleration due to gravity is 32.2 ft/sec^2. Therefore the weight density of water is $(1.94)(32.2)$ lb/ft^3 = 62.5 lb/ft^3.

EXAMPLE 3   A water tank in the form of an inverted right-circular cone is 2 m across the top and 1.5 m deep. If the surface of the water is 0.5 m below the top of the tank, find the work done in pumping the water to the top of the tank.

Solution   Refer to Figure 2. The positive x axis is chosen in the downward direction because the motion is vertical. Take the origin at the top of the tank. We consider a partition of the closed interval $[0.5, 1.5]$ on the x axis and let ξ_i be any point in the ith subinterval $[x_{i-1}, x_i]$. An element of volume is a circular disk having thickness $\Delta_i x$ meters and radius $f(\xi_i)$ meters, where the function f is determined by an equation of the line through the points $(0, 1)$ and $(1.5, 0)$ in the form $y = f(x)$. The volume of this element is $\pi[f(\xi_i)]^2\, \Delta_i x$ cubic meters. With the weight of 1 m^3 of water as 9810 N, the weight of the element is $9810\pi[f(\xi_i)]^2\, \Delta_i x$ newtons, which is the force acting on the element. If x_{i-1} is close to x_i, then the distance the element moves is approximately ξ_i

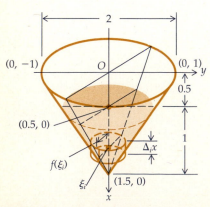

FIGURE 2

meters. Thus the work done in pumping the element to the top of the tank is approximately $(9810\pi[f(\xi_i)]^2 \, \Delta_i x) \cdot \xi_i$ joules. So if W joules is the total work done

$$W = \lim_{||\Delta|| \to 0} \sum_{i=1}^{n} 9810\pi[f(\xi_i)]^2 \cdot \xi_i \, \Delta_i x$$

$$= 9810\pi \int_{0.5}^{1.5} [f(x)]^2 x \, dx$$

To determine $f(x)$ we find an equation of the line through $(0, 1)$ and $(1.5, 0)$ by using the slope-intercept form:

$$y = \frac{0 - 1}{1.5 - 0} x + 1 \quad \Leftrightarrow \quad y = -\tfrac{2}{3}x + 1$$

Therefore, $f(x) = -\tfrac{2}{3}x + 1$, and

$$W = 9810\pi \int_{0.5}^{1.5} (-\tfrac{2}{3}x + 1)^2 x \, dx$$

$$= 9810\pi \int_{0.5}^{1.5} (\tfrac{4}{9}x^3 - \tfrac{4}{3}x^2 + x) \, dx$$

$$= 9810\pi \left[\tfrac{1}{9}x^4 - \tfrac{4}{9}x^3 + \tfrac{1}{2}x^2 \right]_{0.5}^{1.5}$$

$$= 1090\pi$$

Therefore the work done is 1090π joules.

EXAMPLE 4   As a water tank is being raised, water spills out at a constant rate of 2 ft^3 per foot of rise. If the weight of the tank is 200 lb and it originally contains 1000 ft^3 of water, find the work done in raising the tank 20 ft.

Solution   Refer to Figure 3. Here we are taking the origin at the starting point of the bottom of the tank and the positive x axis is in the upward direction because the motion is vertically upward from O. We consider a partition of the closed interval $[0, 20]$ on the x axis. Let ξ_i be any point in the ith subinterval $[x_{i-1}, x_i]$. When the bottom of the tank is at ξ_i, there is $(1000 - 2\xi_i)$ cubic feet of water in the tank. With the weight of 1 ft^3 of water as 62.5 lb the weight of the tank and its contents when it is at ξ_i is $[200 + 62.5(1000 - 2\xi_i)]$ pounds or, equivalently, $(62{,}700 - 125\xi_i)$ pounds, which is the force acting on the tank. The work done in raising the tank through the ith subinterval is approximately $(62{,}700 - 125\xi_i) \, \Delta_i x$ foot-pounds. We use the terminology "approximately" because we are assuming that the amount of water in the tank is constant throughout the subinterval. If W foot-pounds is the total work done in raising the tank 20 ft,

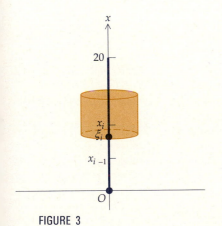

FIGURE 3

$$W = \lim_{||\Delta|| \to 0} \sum_{i=1}^{n} (62{,}700 - 125\xi_i) \, \Delta_i x$$

$$= \int_{0}^{20} (62{,}700 - 125x) \, dx$$

$$= 62{,}700x - \tfrac{125}{2}x^2 \Big]_{0}^{20}$$

$$= 1{,}229{,}000$$

Therefore the work done is 1,229,000 ft-lb.

EXERCISES 6.6

In Exercises 1 and 2, a particle is moving along the x axis under the action of a force of f(x) pounds when the particle is x feet from the origin. Find the work done as the particle moves from the point where x = a to the point where x = b.

1. $f(x) = (2x + 1)^2$; $a = 1$; $b = 3$
2. $f(x) = x^2\sqrt{x^3 + 1}$; $a = 0$; $b = 2$

In Exercises 3 and 4, a particle is moving along the x axis under the action of a force of f(x) newtons when the particle is x meters from the origin. Find the work done as the particle moves from the point where x = a to the point where x = b.

3. $f(x) = x\sqrt{x + 1}$; $a = 3$; $b = 8$
4. $f(x) = (4x - 1)^2$; $a = 1$; $b = 4$

5. An object is moving along the x axis under the action of a force of $f(x)$ dynes when the object is x centimeters from the origin. If 96 ergs is the work done in moving the object from the origin to the point where $x = K$, and $f(x) = 2x - 3$, find K if $K > 0$.
6. Solve Exercise 5 if 90 ergs is the work done, and $f(x) = 4x - 3$.
7. A spring has a natural length of 8 in. If a force of 20 lb stretches the spring $\frac{1}{2}$ in., find the work done in stretching the spring from 8 in. to 11 in.
8. A spring has a natural length of 10 in., and a 30-lb force stretches it to $11\frac{1}{2}$ in. (a) Find the work done in stretching the spring from 10 in. to 12 in. (b) Find the work done in stretching the spring from 12 in. to 14 in.
9. A force of 8 N stretches a spring of natural length 4 m to an additional 50 cm. Find the work done in stretching the spring from its natural length to 5 m.
10. A force of 500 dynes stretches a spring from its natural length of 20 cm to a length of 24 cm. Find the work done in stretching the spring from its natural length to a length of 28 cm.
11. A spring has a natural length of 12 cm. A force of 600 dynes compresses the spring to 10 cm. Find the work done in compressing the spring from 12 cm to 9 cm. Hooke's law holds for compression as well as for extension.
12. A spring has a natural length of 6 in. A 1200-lb force compresses it to $5\frac{1}{2}$ in. Find the work done in compressing it from 6 in. to $4\frac{1}{2}$ in.
13. A tank full of water is in the form of a rectangular parallelepiped 5 ft deep, 15 ft wide, and 25 ft long. Find the work required to pump the water in the tank up to a level 1 ft above the surface of the tank.
14. A trough full of water is 10 ft long, and its cross section is in the shape of an isosceles triangle 2 ft wide across the top and 2 ft high. How much work is done in pumping all the water out of the trough over the top?
15. A hemispherical tank, placed so that the top is a circular region of radius 6 ft, is filled with water to a depth of 4 ft. Find the work done in pumping the water to the top of the tank.

16. A right-circular cylindrical tank with a depth of 12 ft and a radius of 4 ft is half full of oil weighing 60 lb/ft³. Find the work done in pumping the oil to a height 6 ft above the tank.
17. A cable 200 ft long and weighing 4 lb/ft is hanging vertically down a well. If a weight of 100 lb is suspended from the lower end of the cable, find the work done in pulling the cable and weight to the top of the well.
18. A bucket weighing 20 lb containing 60 lb of sand is attached to the lower end of a chain 100 ft long and weighing 10 lb that is hanging in a deep well. Find the work done in raising the bucket to the top of the well.
19. Solve Exercise 18 if the sand is leaking out of the bucket at a constant rate and has all leaked out just as soon as the bucket is at the top of the well.
20. As a flour sack is being raised a distance of 9 ft, flour leaks out at such a rate that the number of pounds lost is directly proportional to the square root of the distance traveled. If the sack originally contained 60 lb of flour and it loses a total of 12 lb while being raised the 9 ft, find the work done in raising the sack.
21. A right-circular cylindrical tank with a depth of 10 m and a radius of 5 m is half filled with water. Find the work necessary to pump the water to the top of the tank.
22. A tank in the form of an inverted right-circular cone is 8 m across the top and 10 m deep. If the tank is filled to a height of 9 m with water, find the work done in pumping the water to the top of the tank.
23. If the tank of Exercise 22 is filled to a height of 8 m with oil weighing 950 kg/m³, find the work done in pumping the oil to the top of the tank. (*Hint:* The number of newtons of force necessary to lift an element is the product of the number of kilograms of mass (the same as the number of kilograms of weight) and 9.81, the number of meters per second squared in the acceleration due to gravity.)
24. If in Exercise 22, only half of the water is to be pumped to the top of the tank, find the work.
25. A 1-horsepower motor can do 550 ft-lb of work per second. If a 0.1 hp motor is used to pump water from a full tank in the shape of a rectangular parallelepiped 2 ft deep, 2 ft wide, and 6 ft long to a point 5 ft above the top of the tank, how long will it take?
26. A meteorite is a miles from the center of the earth and falls to the surface of the earth. The force of gravity is inversely proportional to the square of the distance of a body from the center of the earth. Find the work done by gravity if the weight of the meteorite is w pounds at the surface of the earth. Let R miles be the radius of the earth.
27. A tank in the form of a rectangular parallelepiped 6 ft deep, 4 ft wide, and 12 ft long is full of oil weighing 50 lb/ft³. When one-third of the work necessary to pump the oil to the top of the tank has been done, find by how much the surface of the oil is lowered.

28. A cylindrical tank 10 ft high and 5 ft in radius is standing on a platform 50 ft high. Find the depth of the water when one-half of the work required to fill the tank from the ground level through a pipe in the bottom has been done.

6.7 LIQUID PRESSURE (*Supplementary*)

Another application of the definite integral in physics is to find the force caused by liquid pressure on a plate submerged in the liquid or on a side of a container holding the liquid. First suppose that a flat plate is inserted horizontally into a liquid in a container. The weight of the liquid exerts a force on the plate. The force per square unit of area exerted by the liquid on the plate is called the **pressure** of the liquid.

Let ρ be the measure of the mass density of the liquid and h units be the depth of a point below the surface of the liquid. If P is the measure of the pressure exerted by the liquid at the point, then

$$P = \rho g h \tag{1}$$

where g is the measure of the acceleration due to gravity.

If A square units is the area of a flat plate that is submerged horizontally in the liquid, and F is the measure of the force caused by liquid pressure acting on the upper face of the plate, then

$$F = PA$$

Substituting from (1) into this equation gives

$$F = \rho g h A$$

From (1) it follows that the size of the container is immaterial so far as liquid pressure is concerned. For example, at a depth of 5 ft in a swimming pool filled with salt water the pressure is the same as at a depth of 5 ft in the Pacific Ocean, assuming the mass density of the water is the same.

▶ **ILLUSTRATION 1** A rectangular sheet of tin 8 ft × 12 ft is submerged in a tank of water at a depth of 10 ft. If P lb/ft² is the pressure exerted by the water at a point on the upper face of the sheet of tin,

$$P = 10\rho g$$

The area of the piece of tin is 96 ft². So if F lb is the force caused by liquid pressure acting on the upper face of the sheet of tin,

$$F = 96P$$

Substituting $10\rho g$ for P, we get

$$F = 960\rho g$$

In the British system $g = 32.2$ and for water $\rho = 1.94$; thus $\rho g = 62.5$. Therefore

$$F = 960(62.5)$$
$$= 60{,}000$$

Hence the force due to water pressure on the upper face of the sheet of tin is 60,000 lb. ◀

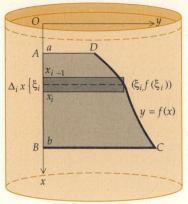

FIGURE 1

Now suppose that the plate is submerged vertically in the liquid. Then at points on the plate at different depths the pressure, computed from (1), will be different and will be greater at the bottom of the plate than at the top. We now proceed to define the force caused by liquid pressure when the plate is submerged vertically in the liquid. We use Pascal's principle: At any point in a liquid, the pressure is the same in all directions.

In Figure 1 let $ABCD$ be the region bounded by the x axis, the lines $x = a$ and $x = b$, and the curve $y = f(x)$, where the function f is continuous and $f(x) \geq 0$ on the closed interval $[a, b]$. Choose the coordinate axes so the y axis lies along the line of the surface of the liquid. Take the x axis vertical with the positive direction downward. The length of the plate at a depth x units is given by $f(x)$ units.

Let Δ be a partition of the closed interval $[a, b]$ that divides the interval into n subintervals. Choose a point ξ_i in the ith subinterval, with $x_{i-1} \leq \xi_i \leq x_i$. Draw n horizontal rectangles. The ith rectangle has a length of $f(\xi_i)$ units and a width of $\Delta_i x$ units (see Figure 1).

If we rotate each rectangular element through an angle of $90°$, each element becomes a plate submerged in the liquid at a depth of ξ_i units below the surface of the liquid and perpendicular to the region $ABCD$. Then the measure of the force on the ith rectangular element is $\rho g \xi_i f(\xi_i) \Delta_i x$. An approximation to F, the measure of the total force on the vertical plate, is

$$\sum_{i=1}^{n} \rho g \xi_i f(\xi_i) \Delta_i x$$

which is a Riemann sum. The smaller we take $\|\Delta\|$, the larger n will be and the closer the approximation of this Riemann sum will be to what we wish to be the measure of the total force. We have, then, the following definition.

6.7.1 DEFINITION   Suppose that a flat plate is submerged vertically in a liquid for which a measure of its mass density is ρ. The length of the plate at a depth of x units below the surface of the liquid is $f(x)$ units, where f is continuous on the closed interval $[a, b]$ and $f(x) \geq 0$ on $[a, b]$. Then if F is the measure of the **force caused by liquid pressure** on the plate

$$F = \lim_{\|\Delta\| \to 0} \sum_{i=1}^{n} \rho g \xi_i f(\xi_i) \Delta_i x$$

$$= \int_a^b \rho g x f(x)\, dx \qquad (2)$$

EXAMPLE 1   A trough having a trapezoidal cross section is full of water. If the trapezoid is 3 ft wide at the top, 2 ft wide at the bottom, and 2 ft deep, find the total force owing to water pressure on one end of the trough.

Solution   Figure 2 illustrates one end of the trough together with a rectangular element of area. Because an equation of the line AB is $y = \frac{3}{2} - \frac{1}{4}x$, $f(x) = \frac{3}{2} - \frac{1}{4}x$. If we rotate the rectangular element through $90°$, the force on the element is $2\rho g \xi_i f(\xi_i) \Delta_i x$ pounds. If F pounds is the total force on the side

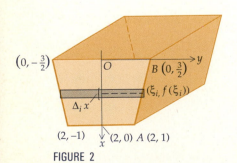

FIGURE 2

of the trough,

$$F = \lim_{||\Delta|| \to 0} \sum_{i=1}^{n} 2\rho g \xi_i f(\xi_i) \, \Delta_i x$$

$$= 2\rho g \int_0^2 x f(x) \, dx$$

$$= 2\rho g \int_0^2 x(\tfrac{3}{2} - \tfrac{1}{4}x) \, dx$$

$$= 2\rho g \left[\tfrac{3}{4}x^2 - \tfrac{1}{12}x^3 \right]_0^2$$

$$= \tfrac{14}{3}\rho g$$

Taking $\rho g = 62.5$, we find that the total force is 292 lb.

EXAMPLE 2   The ends of a trough are semicircular regions, each with a radius of 2 ft. Find the force caused by water pressure on one end if the trough is full of water.

Solution   Figure 3 shows one end of the trough together with a rectangular element of area. An equation of the semicircle is $x^2 + y^2 = 4$. Solving for y gives $y = \sqrt{4 - x^2}$. The force on the rectangular element is $2\rho g \xi_i \sqrt{4 - \xi_i^2} \, \Delta_i x$ pounds. Therefore, if F pounds is the total force on the side of the trough,

$$F = \lim_{||\Delta|| \to 0} \sum_{i=1}^{n} 2\rho g \xi_i \sqrt{4 - \xi_i^2} \, \Delta_i x$$

$$= 2\rho g \int_0^2 x \sqrt{4 - x^2} \, dx$$

$$= -\tfrac{2}{3}\rho g (4 - x^2)^{3/2} \Big]_0^2$$

$$= \tfrac{16}{3}\rho g$$

With $\rho g = 62.5$, the total force is 333 lb.

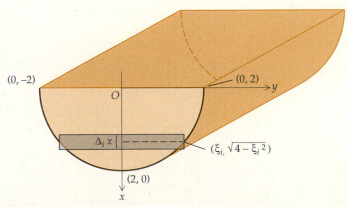

FIGURE 3

There is a useful relation between the force caused by liquid pressure on a plane region and the location of the centroid of the region. Refer to Figure 1 where $ABCD$ is the region bounded by the x axis, the lines $x = a$ and $x = b$, and the curve $y = f(x)$, where f is continuous and $f(x) \geq 0$ on the closed interval $[a, b]$. We consider the region $ABCD$ as a vertical plate immersed in a liquid having mass density of measure ρ. If F is the measure of the force owing to liquid pressure on the vertical plate, from (2),

$$F = \rho g \int_a^b x f(x)\, dx \tag{3}$$

If $\bar{x}$ is the abscissa of the centroid of the region $ABCD$, then $\bar{x} = M_y/A$. Because $M_y = \int_a^b x f(x)\, dx$,

$$\bar{x} = \frac{\int_a^b x f(x)\, dx}{A}$$

$$\int_a^b x f(x)\, dx = \bar{x} A$$

Substituting from this equation into (3) we obtain

$$F = \rho g \bar{x} A \tag{4}$$

From (4) it follows that the total force owing to liquid pressure against a vertical plane region is the same as it would be if the region were horizontal at a depth $\bar{x}$ units below the surface of a liquid.

▶ **ILLUSTRATION 2**   Consider a trough full of water having as ends semicircular regions each with a radius of 2 ft. Using the result of Example 4 in Section 6.5 we find that the centroid of the region is at a depth of $8/(3\pi)$ ft. Therefore, using (4) we see that if F pounds is the force on one end of the trough,

$$F = \rho g \frac{8}{3\pi} \cdot 2\pi$$

$$= \frac{16}{3} \rho g$$

This agrees with the result found in Example 2.   ◀

For various simple plane regions the centroid may be found in a table. When both the area of the region and the centroid of the region may be obtained directly, (4) is easy to apply and is used in such cases by engineers to find the force caused by liquid pressure.

In the next example we use SI units, where $g = 9.81$ and for water $\rho = 1000$; thus $\rho g = 9810$.

EXAMPLE 3   A container in the shape of a right-circular cylinder having a base of radius 3 m is on its side at the bottom of a tank full of water. The depth of the tank is 13 m. Find the total force due to water pressure on one end of the container.

Solution   Figure 4 shows one end of the container in the tank and a rectangular element of area. The coordinate system is chosen so that the origin is at the center of the circle. An equation of the circle is $x^2 + y^2 = 9$. Solving for x gives $x = \sqrt{9 - y^2}$. The number of newtons in the force on the rectangular

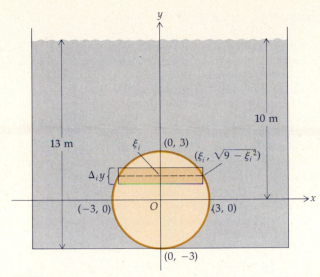

FIGURE 4

element is

$$\rho g(10 - \xi_i)[2\sqrt{9 - \xi_i^2}]\Delta_i y$$

So if F newtons is the total force on the end of the container,

$$F = \lim_{\|\Delta\| \to 0} \sum_{i=1}^{n} \rho g(10 - \xi_i)[2\sqrt{9 - \xi_i^2}]\Delta_i y$$

$$= 2\rho g \int_{-3}^{3} (10 - y)\sqrt{9 - y^2}\, dy$$

Because $\rho g = 9810$, we have

$$F = 196{,}200 \int_{-3}^{3} \sqrt{9 - y^2}\, dy - 19{,}620 \int_{-3}^{3} y\sqrt{9 - y^2}\, dy \qquad (5)$$

To evaluate $\int_{-3}^{3} \sqrt{9 - y^2}\, dy$ requires a technique of integration you will learn in Section 9.4. At present we determine its value by considering it as the measure of the area of the region enclosed by a semicircle of radius 3. Therefore

$$\int_{-3}^{3} \sqrt{9 - y^2}\, dy = \tfrac{9}{2}\pi$$

Substituting this value into (5) and evaluating the second integral we have

$$F = 196{,}200(\tfrac{9}{2}\pi) + 19{,}620\left[\tfrac{1}{3}(9 - y^2)^{3/2}\right]_{-3}^{3}$$

$$= 882{,}900\pi$$

Thus the total force is $882{,}900\pi$ N.

EXERCISES 6.7

1. A plate in the shape of a rectangle is submerged vertically in a tank of water, with the upper edge lying in the surface. If the width of the plate is 10 ft and the depth is 8 ft, find the force due to liquid pressure on one side of the plate.

2. A square plate of side 4 ft is submerged vertically in a tank of water and its center is 2 ft below the surface. Find the force due to liquid pressure on one side of the plate.

3. Solve Exercise 2 if the center of the plate is 4 ft below the surface.

4. A plate in the shape of an isosceles right triangle is submerged vertically in a tank of water, with one leg lying in the surface. The legs are each 6 ft long. Find the force due to liquid pressure on one side of the plate.

5. A rectangular tank full of water is 2 ft wide and 18 in. deep. Find the force due to liquid pressure on one end of the tank.

6. The ends of a trough are equilateral triangles having sides with lengths of 2 ft. If the water in the trough is 1 ft deep, find the force due to liquid pressure on one end.

7. The face of the gate of a dam is in the shape of an isosceles triangle 4 m wide at the top and 3 m high. If the upper edge of the face of the gate is 15 m below the surface of the water, find the total force due to water pressure on the gate.

8. The face of a gate of a dam is vertical and in the shape of an isosceles trapezoid 3 m wide at the top, 4 m wide at the bottom, and 3 m high. If the upper base is 20 m below the surface of the water, find the total force due to water pressure on the gate.

9. The face of a dam adjacent to the water is vertical, and its shape is in the form of an isosceles triangle 250 m wide across the top and 100 m high in the center. If the water is 10 m deep in the center, find the total force on the dam due to water pressure.

10. An oil tank is in the shape of a right-circular cylinder 4 m in diameter, and its axis is horizontal. If the tank is half full of oil having mass density 750 kg/m^3, find the total force on one end due to liquid pressure.

11. An oil tank is in the shape of a right-circular cylinder with a radius of r meters, and its axis is horizontal. If the tank is half full of oil having mass density 750 kg/m^3, find r if the total force on one end of the tank due to liquid pressure is 80,000 N.

12. Solve Exercise 4 by using Equation (4).

13. Solve Exercise 5 by using Equation (4).

14. Solve Exercise 6 by using Equation (4).

15. The face of a dam adjacent to the water is vertical and is in the shape of an isosceles trapezoid 90 ft wide at the top, 60 ft wide at the bottom, and 20 ft high. Use Equation (4) to find the total force due to water pressure on the face of the dam.

16. A semicircular plate with a radius of 3 ft is submerged vertically in a tank of water, with its diameter lying in the surface. Use Equation (4) to find the total force due to water pressure on one side of the plate.

17. Find the moment about the lower base of the trapezoid of the force in Exercise 15.

18. A plate in the shape of a region bounded by the parabola $x^2 = 6y$ and the line $2y = 3$ is placed in a water tank with its vertex downward and the line in the surface of the water. Find the total force due to water pressure on one side of the plate if distance is measured in meters.

19. A cylindrical tank is half full of gasoline having weight density 42 lb/ft^3. If the axis is horizontal and the diameter is 6 ft, find the force on an end due to liquid pressure.

20. If the end of a water tank is in the shape of a rectangle and the tank is full, show that the measure of the force due to water pressure on the end is the product of the measure of the area of the end and the measure of the force at the geometrical center.

21. The bottom of a swimming pool is an inclined plane. The pool is 2 ft deep at one end and 8 ft deep at the other. If the width of the pool is 25 ft and the length is 40 ft, find the total force due to water pressure on the bottom.

22. The face of a dam adjacent to the water is inclined at an angle of 45° from the vertical. The face is a rectangle of width 80 ft and slant height 40 ft. If the dam is full of water, find the total force due to water pressure on the face.

23. The face of a dam adjacent to the water is inclined at an angle of 30° from the vertical. The shape of the face is a rectangle of width 50 ft and slant height of 30 ft. If the dam is full of water, find the total force due to water pressure on the face.

24. Solve Exercise 23 if the face of the dam is an isosceles trapezoid 120 ft wide at the top, 80 ft wide at the bottom, and with a slant height of 40 ft.

REVIEW EXERCISES FOR CHAPTER 6

1. Find the volume of the solid generated by revolving about the x axis the region bounded by the curve $y = x^4$, the line $x = 1$, and the x axis.

2. Find the volume of the solid generated if the region of Exercise 1 is revolved about the y axis.

3. The region bounded by the curve $y = \sqrt{\sin x}$, the x axis, and the line $x = \frac{1}{2}\pi$ is revolved about the x axis. Find the volume of the solid generated.

4. The region bounded by the curve $x = \sqrt{\cos y}$, the line $y = \frac{1}{6}\pi$, and the y axis, where $\frac{1}{6}\pi \leq y \leq \frac{1}{2}\pi$, is revolved about the y axis. Find the volume of the solid generated.

5. The region bounded by the curve $y = \csc x$, the x axis, and the lines $x = \frac{1}{4}\pi$ and $x = \frac{1}{2}\pi$ is revolved about the x axis. Find the volume of the solid generated.

6. The region in the first quadrant bounded by the curves $x = y^2$ and $x = y^4$ is revolved about the y axis. Find the volume of the solid generated.

7. Find the volume of the solid generated by revolving about the y axis the region bounded by the parabola $x = y^2 + 2$ and the line $x = y + 8$.

8. Find the volume of the solid of revolution generated when the region bounded by the parabola $y^2 = x$, the x axis, and the line $x = 4$ is revolved about the line $x = 4$. Take the elements of area parallel to the axis of revolution.

9. The base of a solid is the region bounded by the parabola $y^2 = 8x$ and the line $x = 8$. Find the volume of the solid if every plane section perpendicular to the axis of the base is a square.

10. The base of a solid is the region enclosed by a circle having a radius of r units, and every plane section perpendicular to a fixed diameter of the base is a square for which a chord of the circle is a diagonal. Find the volume of the solid.

11. Find the volume of the solid generated by revolving the region bounded by the curve $y = |x - 2|$, the x axis, and the lines $x = 1$ and $x = 4$ about the x axis.

12. Use integration to find the volume of a segment of a sphere if the sphere has a radius of r units and the altitude of the segment is h units.

13. Find the volume of the solid generated by revolving about the line $y = -1$ the region above the x axis bounded by the line $2y = x + 3$ and the curves $y^2 + x = 0$ and $y^2 - 4x = 0$ from $x = -1$ to $x = 1$.

14. Find the length of the arc of the curve $ay^2 = x^3$ from the origin to $(4a, 8a)$.

15. Find the length of the arc of the curve $6y^2 = x(x - 2)^2$ from $(2, 0)$ to $(8, 4\sqrt{3})$.

16. A sphere of radius 10 cm is intersected by two parallel planes on the same side of the center of the sphere. The distance from the center of the sphere to one of the planes is 1 cm, and the distance between the two planes is 6 cm. Find the volume of the solid portion of the sphere between the two planes.

17. Solve Exercise 16 if the two planes lie on opposite sides of the center of the sphere but the other facts are the same.

18. A solid is formed by revolving about the y axis the region bounded by the curve $y^3 = x$, the x axis, and the line $x = c$, where $c > 0$. For what value of c will the volume of the solid be 12π cubic units?

19. Find the length of the arc of the curve $9x^{2/3} + 4y^{2/3} = 36$ in the second quadrant from the point where $x = -1$ to the point where $x = -\frac{1}{8}$.

20. Find the length of the arc of the curve $3y = (x^2 - 2)^{3/2}$ from the point where $x = 3$ to the point where $x = 6$.

21. Three particles of masses 4, 2, and 7 kg are located on the x axis at the points having coordinates -5, 4, and 2, respectively, where the distance is measured in meters. Find the center of mass of the system.

22. Three particles having masses 5, 2, and 8 slugs are located, respectively, at the points $(-1, 3)$, $(2, -1)$, and $(5, 2)$. Find the center of mass, where distance is measured in feet.

23. Find the coordinates of the center of mass of the four particles having equal masses located at the points $(3, 0)$, $(2, 2)$, $(2, 4)$, and $(-1, 2)$.

24. Three particles, each having the same mass, are located on the x axis at the points having coordinates -4, 1, and 5, where the distance is measured in meters. Find the coordinates of the center of mass of the system.

25. The length of a rod is 8 in. and the linear density of the rod at a point x inches from the left end is $2\sqrt{x + 1}$ slugs per inch. Find the total mass of the rod and the center of mass.

26. The length of a rod is 4 m and the linear density of the rod at a point x meters from the left end is $(3x + 1)$ kilograms per meter. Find the total mass of the rod and the center of mass.

27. Find the centroid of the region in the first quadrant bounded by the coordinate axes and the parabola $y = 9 - x^2$.

28. Find the centroid of the region bounded by the parabola $y^2 = x$ and the line $y = x - 2$.

29. Find the centroid of the region bounded by the curves $y = \sqrt{x}$ and $y = x^2$.

30. Find the centroid of the region bounded above by the parabola $4x^2 = 36 - 9y$ and below by the x axis.

31. Use the theorem of Pappus to find the volume of a sphere of radius 4 m.

32. Use the theorem of Pappus to find the volume of a right-circular cone with base radius 2 m and height 3 m.

33. Find the volume of the solid generated by revolving about the y axis the region bounded by the graph of $y = 2x - \frac{1}{6}x^3$, the x axis, and the line $x = 2$.

34. Find the volume of the solid generated by revolving the region of Exercise 33 about the line $x = 2$.

35. Find the volume of the solid generated by revolving about the line $x = 3$ the region bounded by that line, the x axis, and the graph of $y = 9x - \frac{1}{12}x^4$.

36. A force of 500 lb is required to compress a spring whose natural length is 10 in. to a length of 9 in. Find the work done to compress the spring to a length of 8 in.

37. A force of 600 dynes stretches a spring from its natural length of 30 cm to a length of 35 cm. Find the work done in stretching the spring from its natural length to a length of 40 cm.

38. A trough full of water is 6 ft long, and its cross section is in the shape of a semicircle with a diameter of 2 ft at the top. How much work is required to pump the water out over the top?

39. A cable 20 ft long and weighing 2 lb/ft is hanging vertically from the top of a pole. Find the work done in raising the entire cable to the top of the pole.

40. The work necessary to stretch a spring from 9 in. to 10 in. is $\frac{3}{2}$ times the work necessary to stretch it from 8 in. to 9 in. What is the natural length of the spring?

41. A tank full of water is in the form of a rectangular parallelepiped 4 m deep, 15 m wide, and 30 m long. Find the work required to pump the water in the tank up to a level 50 cm above the top of the tank.

42. A hemispherical tank having a diameter of 10 m is filled with water to a depth of 3 m. Find the work done in pumping the water to the top of the tank.

43. A water tank is in the shape of a hemisphere surmounted by a right-circular cylinder. The radius of both the hemisphere and the cylinder is 4 ft, and the altitude of the cylinder is 8 ft. If the tank is full of water, find the work necessary to empty the tank by pumping it through an outlet at the top of the tank.

44. A container has the same shape and dimensions as a solid of revolution formed by revolving about the y axis the region in the first quadrant bounded by the parabola $x^2 = 4py$, the y axis, and the line $y = p$. If the container is full of water, find the work done in pumping all the water up to a point $3p$ feet above the top of the container.

45. The surface of a tank is the same as that of a paraboloid of revolution obtained by revolving the parabola $y = x^2$ about the y axis. The vertex of the parabola is at the bottom of the tank, and the tank is 36 ft high. If the tank is filled with water to a depth of 20 ft, find the work done in pumping all of the water out over the top.

46. A wedge is cut from a right-circular cylinder with a radius of r units by two planes, one perpendicular to the axis of the cylinder and the other intersecting the first along a diameter of the circular plane section at an angle of measurement $30°$. Find the volume of the wedge.

47. A church steeple is 30 ft high, and every horizontal plane section is a square having sides of length one-tenth of the distance of the plane section from the top of the steeple. Find the volume of the steeple.

48. Find by slicing the volume of a tetrahedron having three mutually perpendicular faces and three mutually perpendicular edges whose lengths are a, b, and c units.

49. The region bounded by a pentagon having vertices at $(-4, 4)$, $(-2, 0)$, $(0, 8)$, $(2, 0)$, and $(4, 4)$ is revolved about the x axis. Find the volume of the solid generated.

50. The region bounded by the curves $y = \tan x$ and $y = \cot x$ and the x axis, where $0 \le x \le \frac{1}{4}\pi$, is revolved about the x axis. Find the volume of the solid generated.

51. The region from $x = 0$ to $x = \frac{1}{2}\pi$ bounded by the curve $y = \sin x$, the line $y = 1$, and the y axis is revolved about the

x axis. Find the volume of the solid generated. (*Hint:* Use the identity $\sin^2 x = \frac{1}{2}(1 - \cos 2x)$.)

52. If $f(x) = \int_0^x \sqrt{\cos t}\, dt$, find the length of the arc of the graph of f from the point where $x = \frac{1}{3}\pi$ to the point where $x = \frac{1}{2}\pi$. (*Hint:* Use the identity $\cos^2 \frac{1}{2}x = \frac{1}{2}(1 + \cos x)$ and Theorem 5.8.1.)

Exercises 53 through 56 pertain to Supplementary Section 6.7.

53. A plate in the shape of the region bounded by the parabola $x^2 = 6y$ and the line $2y = 3$ is placed in a water tank with its vertex downward and the line in the surface of the water. Find the force due to water pressure on one side of the plate if distance is measured in feet.

54. The face of a dam adjacent to the water is inclined at an angle of $45°$ from the vertical. The face is a rectangle of width 80 ft and slant height 40 ft. If the dam is full of water, find the total force due to water pressure on the face.

55. A cylindrical tank is half full of gasoline having weight density 40 lb/ft³. If the axis is horizontal and the diameter is 8 ft, find the force on an end due to liquid pressure.

56. A semicircular plate with a radius of 4 ft is submerged vertically in a tank of water, with its diameter lying in the surface. Use Equation (4) of Section 6.7 to find the force due to water pressure on one side of the plate.

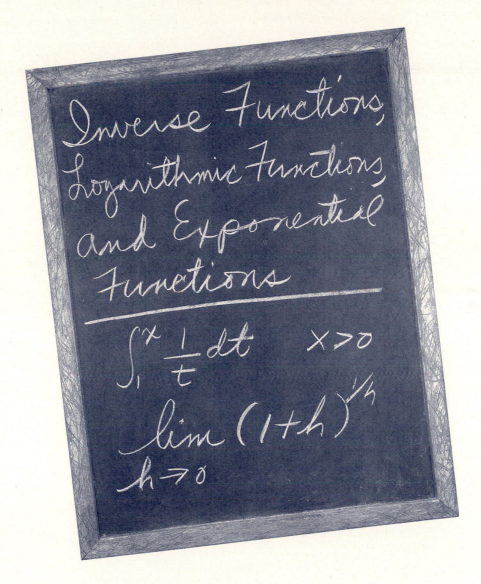

Inverse Functions, Logarithmic Functions, and Exponential Functions

$$\int_1^x \frac{1}{t}\,dt \qquad x>0$$

$$\lim_{h \to 0} (1+h)^{1/h}$$

The first two sections of this chapter pertain to *inverse functions* and their properties. In Section 7.1 the inverse of a function is defined and sufficient conditions for a function to have an inverse are presented. The inverse function theorems and the derivative of the inverse of a function form the subject matter of Section 7.2.

We previously stated that functions that are not algebraic are called transcendental, examples of which are the six trigonometric functions. The *natural logarithmic and exponential functions* are also transcendental and they are treated in Sections 7.3 through 7.6. The natural logarithm is defined as an integral in Section 7.3 and the natural exponential function *exp* is defined as the inverse of the natural logarithmic function *ln* in Section 7.5. With this background, we are able to define an irrational power of a real number. The functions *ln* and *exp* are applied in Section 7.7 to problems involving *laws of growth and decay*. Another application of these functions is to solve *first-order linear differential equations* as shown in Supplementary Section 7.8.

7.1 INVERSE FUNCTIONS

One of a pair of inverse operations essentially "undoes" the other. For instance, addition and subtraction are inverse operations: if 4 is added to x, the sum is $x + 4$; if then 4 is subtracted from this sum the difference is x. In the following illustration we use pairs of functions associated with inverse operations.

▶ **ILLUSTRATION 1**   (a) Let $f(x) = x + 4$ and $g(x) = x - 4$. Then

$$f(g(x)) = f(x - 4) \qquad g(f(x)) = g(x + 4)$$
$$= (x - 4) + 4 \qquad\qquad = (x + 4) - 4$$
$$= x \qquad\qquad\qquad = x$$

(b) Let $f(x) = 2x$ and $g(x) = \dfrac{x}{2}$. Then

$$f(g(x)) = f\left(\frac{x}{2}\right) \qquad g(f(x)) = g(2x)$$

$$= 2\left(\frac{x}{2}\right) \qquad\qquad = \frac{2x}{2}$$

$$= x \qquad\qquad\qquad = x$$

(c) Let $f(x) = x^3$ and $g(x) = \sqrt[3]{x}$. Then

$$f(g(x)) = f(\sqrt[3]{x}) \qquad g(f(x)) = g(x^3)$$
$$= (\sqrt[3]{x})^3 \qquad\qquad = \sqrt[3]{x^3}$$
$$= x \qquad\qquad\qquad = x \qquad\qquad ◀$$

Each pair of functions f and g in Illustration 1 satisfies the following two statements:

$$f(g(x)) = x \qquad \text{for } x \text{ in the domain of } g$$

and

$$g(f(x)) = x \qquad \text{for } x \text{ in the domain of } f$$

Observe that for the functions f and g in these two equations the composite functions $f(g(x))$ and $g(f(x))$ are equal, a relationship that is not generally true for arbitrary functions f and g. You will learn subsequently (in Illustration 5) that each pair of functions in Illustration 1 is a set of *inverse functions*, and that is the reason the two equations are satisfied.

We lead up to the formal definition of the *inverse of a function* by considering some more particular functions. Figure 1 shows a sketch of the graph of the function defined by

$$f(x) = x^2$$

The domain of f is the set of real numbers and the range is the interval $[0, +\infty)$. Observe that because $f(2) = 4$ and $f(-2) = 4$, the number 4 is the

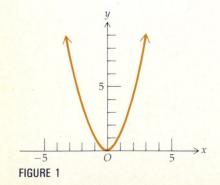

FIGURE 1

function value of two distinct numbers in the domain. Furthermore, every number except 0 in the range of this function is the function value of two distinct numbers in the domain. In particular, $\frac{25}{4}$ is the function value of both $\frac{5}{2}$ and $-\frac{5}{2}$, 1 is the function value of both 1 and -1, and 9 is the function value of both 3 and -3.

A different situation occurs with the function g defined by

$$g(x) = x^3 \qquad -2 \le x \le 2$$

The domain of g is the closed interval $[-2, 2]$, and the range is $[-8, 8]$. A sketch of the graph of g is shown in Figure 2. This function is one for which a number in its range is the function value of one and only one number in the domain. Such a function is called *one-to-one*.

7.1.1 DEFINITION   A function f is said to be **one-to-one** if every number in its range corresponds to exactly one number in its domain; that is, for all x_1 and x_2 in the domain of f

$$\text{if} \quad x_1 \ne x_2, \quad \text{then} \quad f(x_1) \ne f(x_2)$$

$$\Leftrightarrow \quad f(x_1) = f(x_2) \quad \text{only when} \quad x_1 = x_2$$

▶ **ILLUSTRATION 2**   As discussed above, for the function f defined by

$$f(x) = x^2$$

every number except 0 in the range of f is the function value of two distinct numbers in the domain. Therefore, by Definition 7.1.1, this function is not one-to-one.   ◀

We know that a vertical line can intersect the graph of a function in only one point. For a one-to-one function, it is also true that a horizontal line can intersect the graph in at most one point. Notice that this is the situation for the one-to-one function defined by $g(x) = x^3$, where $-2 \le x \le 2$, whose graph appears in Figure 2. Furthermore, observe in Figure 1 that for the function defined by $f(x) = x^2$, which is not one-to-one, any horizontal line above the x axis intersects the graph in two points.

FIGURE 2

EXAMPLE 1   Given

$$f(x) = 4x - 3 \qquad h(x) = 4 - x^2$$

(a) Prove that f is one-to-one and draw a sketch of its graph.
(b) Prove that h is not one-to-one, and draw a sketch of its graph.

Solution
(a) We are given $f(x) = 4x - 3$. The domain of f is the set of real numbers. Suppose x_1 and x_2 are two real numbers and $f(x_1) = f(x_2)$; then

$$4x_1 - 3 = 4x_2 - 3$$

$$4x_1 = 4x_2$$

$$x_1 = x_2$$

Because $f(x_1) = f(x_2)$ implies $x_1 = x_2$, it follows from Definition 7.1.1 that f is one-to-one. A sketch of the graph of f is shown in Figure 3. Observe that a horizontal line intersects the graph in only one point.

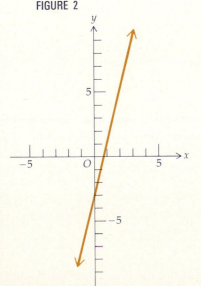

FIGURE 3

(b) We are given $h(x) = 4 - x^2$. The domain of h is the set of real numbers. The numbers 1 and -1 are in the domain of h, and

$$h(1) = 3 \qquad h(-1) = 3$$

Because $1 \neq -1$ and $h(1) = h(-1)$, it follows from Definition 7.1.1 that h is not one-to-one. A sketch of the graph of h appears in Figure 4. Note that a horizontal line $y = b$, with $b < 4$, intersects the graph in two points.

It is not always a simple matter to apply Definition 7.1.1 to prove that a function is one-to-one. The following theorem gives a test that can sometimes be used.

7.1.2 THEOREM   A function that is monotonic on an interval is one-to-one on the interval.

Proof   Assume that the function f is increasing on an interval. If x_1 and x_2 are two numbers in the interval and $x_1 \neq x_2$, then either $x_1 < x_2$ or $x_2 < x_1$. If $x_1 < x_2$, it follows from the definition of an increasing function (Definition 4.4.1) that $f(x_1) < f(x_2)$; so $f(x_1) \neq f(x_2)$. If $x_2 < x_1$, then $f(x_2) < f(x_1)$, and so again $f(x_1) \neq f(x_2)$. Therefore from Definition 7.1.1 f is one-to-one on the interval. The proof is similar if f is decreasing on an interval.   ■

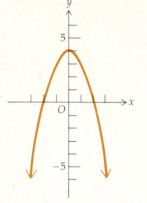

FIGURE 4

To apply Theorem 7.1.2 it is necessary first to determine if the function is increasing or decreasing on an interval. Theorem 4.4.3 can usually be used for this purpose.

EXAMPLE 2   If

$$f(x) = \frac{2x + 3}{x - 1}$$

prove that f is one-to-one on each of the intervals $(-\infty, 1)$ and $(1, +\infty)$. Draw a sketch of the graph of f to show that f is one-to-one on its entire domain.

Solution   The domain of f is the set of all real numbers except 1, or, equivalently, the set $(-\infty, 1) \cup (1, +\infty)$. We compute $f'(x)$.

$$f'(x) = \frac{2(x - 1) - (2x + 3)}{(x - 1)^2}$$

$$= -\frac{5}{(x - 1)^2}$$

Because $f'(x) < 0$ for all $x \neq 1$, we can conclude from Theorem 4.4.3 that f is decreasing on each of the intervals $(-\infty, 1)$ and $(1, +\infty)$. Therefore, from Theorem 7.1.2, f is one-to-one on each of these intervals.
Furthermore,

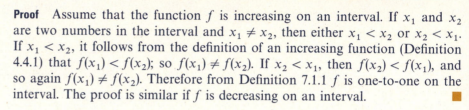

$$\lim_{x \to -\infty} f(x) = \lim_{x \to -\infty} \frac{2 + \dfrac{3}{x}}{1 - \dfrac{1}{x}} \qquad \lim_{x \to +\infty} f(x) = \lim_{x \to +\infty} \frac{2 + \dfrac{3}{x}}{1 - \dfrac{1}{x}}$$

$$= 2 \qquad\qquad\qquad = 2$$

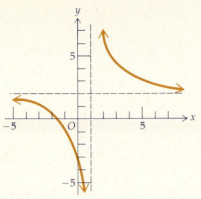

FIGURE 5

Thus the line $y = 2$ is a horizontal asymptote of the graph of f. Also

$$\lim_{x \to 1^-} f(x) = \lim_{x \to 1^-} \frac{2x + 3}{x - 1} \qquad \lim_{x \to 1^+} f(x) = \lim_{x \to 1^+} \frac{2x + 3}{x - 1}$$
$$= -\infty \qquad\qquad\qquad = +\infty$$

Hence the line $x = 1$ is a vertical asymptote of the graph of f. With this information we obtain the sketch of the graph of f shown in Figure 5. From the graph we observe that f is one-to-one on its domain.

▶ **ILLUSTRATION 3** Consider the equation

$$y = x^3 \qquad -2 \le x \le 2 \tag{1}$$

This equation defines the one-to-one function g discussed prior to Definition 7.1.1, where

$$g(x) = x^3 \qquad -2 \le x \le 2$$

Function g is the set of ordered pairs (x, y) satisfying (1). If we solve (1) for x, we obtain

$$x = \sqrt[3]{y} \qquad -8 \le y \le 8 \tag{2}$$

which defines a function G where

$$G(y) = \sqrt[3]{y} \qquad -8 \le y \le 8$$

The function G is the set of ordered pairs (y, x) satisfying (2). ◀

The function G of Illustration 3 is called the *inverse* of the function g. In the following formal definition of the inverse of a function, we use the notation f^{-1} to denote the inverse of f. This notation is read "f inverse," and it should not be confused with the use of -1 as an exponent.

7.1.3 DEFINITION   If f is a one-to-one function, then there is a function f^{-1}, called the **inverse** of f such that

$$x = f^{-1}(y) \quad \text{if and only if} \quad y = f(x)$$

The domain of f^{-1} is the range of f and the range of f^{-1} is the domain of f.

It is crucial in Definition 7.1.3 that f be a one-to-one function. This requirement ensures that $f^{-1}(y)$ is unique for each value of y.

We eliminate y from the equations of the definition by replacing y by $f(x)$ in the equation

$$f^{-1}(y) = x$$

to obtain

$$f^{-1}(f(x)) = x \tag{3}$$

where x is in the domain of f.

We eliminate x between the same pair of equations by replacing x by $f^{-1}(y)$ in the equation

$$f(x) = y$$

We get

$$f(f^{-1}(y)) = y$$

where y is in the domain of f^{-1}. Because the symbol used for the independent variable is arbitrary, we can replace y by x to obtain

$$f(f^{-1}(x)) = x \tag{4}$$

where x is in the domain of f^{-1}.

From (3) and (4) we see that if the inverse of the function f is the function f^{-1}, then the inverse of f^{-1} is f. We state these results formally as the following theorem.

7.1.4 THEOREM　　If f is a one-to-one function having f^{-1} as its inverse, then f^{-1} is a one-to-one function having f as its inverse. Furthermore,

$$f^{-1}(f(x)) = x \qquad \text{for } x \text{ in the domain of } f$$

and

$$f(f^{-1}(x)) = x \qquad \text{for } x \text{ in the domain of } f^{-1}$$

We use the terminology *inverse functions* when referring to a function and its inverse.

▶ **ILLUSTRATION 4**　　In Illustration 3 the function G defined by

$$G(y) = \sqrt[3]{y} \qquad -8 \le y \le 8$$

is the inverse of the function g defined by

$$g(x) = x^3 \qquad -2 \le x \le 2$$

Therefore g^{-1} can be written in place of G, and we have

$$g^{-1}(y) = \sqrt[3]{y} \qquad -8 \le y \le 8$$

or, equivalently, if we replace y by x,

$$g^{-1}(x) = \sqrt[3]{x} \qquad -8 \le x \le 8$$

Observe that the domain of g is $[-2, 2]$, which is the range of g^{-1}; also the range of g is $[-8, 8]$, which is the domain of g^{-1}.　　◀

If a function f has an inverse, then $f^{-1}(x)$ can be found by the method used in the following illustration.

▶ **ILLUSTRATION 5**　　Each of the functions f in Illustration 1 is one-to-one. Therefore, $f^{-1}(x)$ exists. For each function we compute $f^{-1}(x)$ from the definition of $f(x)$ by substituting y for $f(x)$ and solving the resulting equation for x. This procedure gives the equation $x = f^{-1}(y)$. We then have the definition of $f^{-1}(y)$, from which we obtain $f^{-1}(x)$.

(a) $f(x) = x + 4$ (b) $f(x) = 2x$ (c) $f(x) = x^3$

$y = x + 4$ $y = 2x$ $y = x^3$

$x = y - 4$ $x = \dfrac{y}{2}$ $x = \sqrt[3]{y}$

$f^{-1}(y) = y - 4$ $f^{-1}(y) = \dfrac{y}{2}$ $f^{-1}(y) = \sqrt[3]{y}$

$f^{-1}(x) = x - 4$ $f^{-1}(y) = \dfrac{y}{2}$ $f^{-1}(x) = \sqrt[3]{x}$

$f^{-1}(x) = \dfrac{x}{2}$

Observe that the function f^{-1} in each part is the function g in the corresponding part of Illustration 1.   ◄

EXAMPLE 3   Find $f^{-1}(x)$ for the function f of Example 1. Verify the equations of Theorem 7.1.4 for f and f^{-1}, and draw a sketch of the graph of f^{-1}.

Solution   In Example 1 we showed that the function f defined by

$$f(x) = 4x - 3$$

is one-to-one. Therefore f^{-1} exists. To find $f^{-1}(x)$, we write the equation

$$y = 4x - 3$$

and solve for x. We obtain

$$x = \frac{y + 3}{4}$$

Therefore

$$f^{-1}(y) = \frac{y + 3}{4} \quad \Leftrightarrow \quad f^{-1}(x) = \frac{x + 3}{4}$$

We verify the equations of Theorem 7.1.4.

$$f^{-1}(f(x)) = f^{-1}(4x - 3) \qquad f(f^{-1}(x)) = f\left(\frac{x + 3}{4}\right)$$

$$= \frac{(4x - 3) + 3}{4} \qquad\qquad = 4\left(\frac{x + 3}{4}\right) - 3$$

$$= \frac{4x}{4} \qquad\qquad\qquad\quad = (x + 3) - 3$$

$$\qquad\qquad\qquad\qquad = x$$

$$= x$$

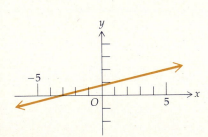

FIGURE 6

Figure 6 shows a sketch of the graph of f^{-1}.

Figure 3 shows a sketch of the graph of the function f of Example 3. In Figure 6 there is a sketch of the graph of f^{-1}. When the graphs of f and f^{-1}

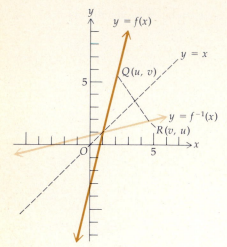

FIGURE 7

are shown on the same set of axes, as in Figure 7, it appears that if $Q(u, v)$ is on the graph of f, then the point $R(v, u)$ is on the graph of f^{-1}.

In general, if Q is the point (u, v) and R is the point (v, u), the line segment QR is perpendicular to the line $y = x$ and is bisected by it. The point Q is a *reflection of the point R* with respect to the line $y = x$, and the point R is a *reflection of the point Q* with respect to the line $y = x$. If x and y are interchanged in the equation $y = f(x)$, we obtain the equation $x = f(y)$, and the graph of the equation $x = f(y)$ is a *reflection of the graph* of the equation $y = f(x)$ with respect to the line $y = x$. Because the equation $x = f(y)$ is equivalent to the equation $y = f^{-1}(x)$, the graph of the equation $y = f^{-1}(x)$ is a reflection of the graph of the equation $y = f(x)$ with respect to the line $y = x$. Therefore, if a function has an inverse, the graphs of the functions are reflections of each other with respect to the line $y = x$.

▶ **ILLUSTRATION 6**   Functions g and g^{-1} of Illustration 4 are defined by

$$g(x) = x^3 \qquad -2 \le x \le 2$$

and

$$g^{-1}(x) = \sqrt[3]{x} \qquad -8 \le x \le 8$$

Sketches of the graphs of g and g^{-1} appear in Figure 8. Observe that these graphs are reflections of each other with respect to the line $y = x$.   ◀

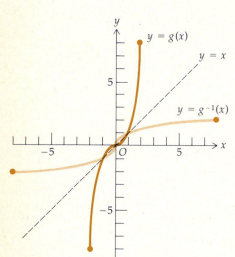

FIGURE 8

EXAMPLE 4   Find $f^{-1}(x)$ for the function f of Example 2. Verify the equations of Theorem 7.1.4 and draw sketches of the graphs of f and f^{-1} on the same set of axes.

Solution   The function f of Example 2 is defined by

$$f(x) = \frac{2x + 3}{x - 1}$$

Because f is one-to-one, as shown in Example 2, f has an inverse f^{-1}. To find $f^{-1}(x)$, we let $y = f(x)$ and solve for x, giving us $x = f^{-1}(y)$. So we have

$$y = \frac{2x + 3}{x - 1}$$

$$xy - y = 2x + 3$$

$$x(y - 2) = y + 3$$

$$x = \frac{y + 3}{y - 2}$$

Therefore

$$f^{-1}(y) = \frac{y + 3}{y - 2} \quad \Leftrightarrow \quad f^{-1}(x) = \frac{x + 3}{x - 2}$$

The domain of f^{-1} is the set of all real numbers except 2. We verify the equations of Theorem 7.1.4.

$$f^{-1}(f(x)) = f^{-1}\left(\frac{2x+3}{x-1}\right) \qquad f(f^{-1}(x)) = f\left(\frac{x+3}{x-2}\right)$$

$$= \frac{\left(\dfrac{2x+3}{x-1}\right)+3}{\left(\dfrac{2x+3}{x-1}\right)-2} \qquad\qquad = \frac{2\left(\dfrac{x+3}{x-2}\right)+3}{\left(\dfrac{x+3}{x-2}\right)-1}$$

$$= \frac{(2x+3)+3(x-1)}{(2x+3)-2(x-1)} \qquad = \frac{2(x+3)+3(x-2)}{(x+3)-(x-2)}$$

$$= \frac{5x}{5} \qquad\qquad\qquad = \frac{5x}{5}$$

$$= x \qquad\qquad\qquad\qquad = x$$

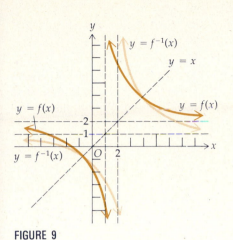

FIGURE 9

Sketches of the graphs of f and f^{-1} on the same set of axes appear in Figure 9.

We have the following theorem about the inverse of a function that is continuous and monotonic on a closed interval.

7.1.5 THEOREM Suppose the function f has the closed interval $[a, b]$ as its domain. Then

(i) if f is continuous and increasing on $[a, b]$, f has an inverse f^{-1} that is defined on $[f(a), f(b)]$;

(ii) if f is continuous and decreasing on $[a, b]$, f has an inverse f^{-1} that is defined on $[f(b), f(a)]$.

Proof For part (i), if f is continuous on $[a, b]$ and if k is any number such that $f(a) < k < f(b)$, then by the intermediate-value theorem (2.7.8) there exists a number c in (a, b) such that $f(c) = k$. Therefore the range of f is the closed interval $[f(a), f(b)]$. Because f is increasing on $[a, b]$, f is one-to-one and so f has an inverse f^{-1}. Because the domain of f^{-1} is the range of f, f^{-1} is defined on $[f(a), f(b)]$.

The proof of part (ii) is similar. However, because f is decreasing on $[a, b]$, $f(a) > f(b)$; so the range of f is $[f(b), f(a)]$. Thus f^{-1} is defined on $[f(b), f(a)]$. ∎

▶ **ILLUSTRATION 7** Let f be the function defined by

$$f(x) = x^2 - 1$$

and a sketch of whose graph appears in Figure 10. Because f is not one-to-one, f does not have an inverse.

However, let us restrict the domain to the closed interval $[0, c]$ and consider the function f_1 defined by

$$f_1(x) = x^2 - 1 \qquad x \in [0, c] \tag{5}$$

Also consider the function f_2 for which the domain is restricted to the closed interval $[-c, 0]$; that is

FIGURE 10

$$f_2(x) = x^2 - 1 \qquad x \in [-c, 0] \tag{6}$$

The functions f_1 and f_2 are distinct functions because their domains are different. We find the derivatives of f_1 and f_2 and obtain

$$f_1'(x) = 2x \qquad x \in [0, c]$$

$$f_2'(x) = 2x \qquad x \in [-c, 0]$$

Because f_1 is continuous on $[0, c]$ and $f_1'(x) > 0$ for all x in $(0, c)$, by Theorem 4.4.3(i) f_1 is increasing on $[0, c]$. Therefore, by Theorem 7.1.5(i), f_1 has an inverse f_1^{-1} that is defined on $[f_1(0), f_1(c)]$ which is $[-1, c^2 - 1]$. We obtain $f_1^{-1}(x)$ by replacing $f_1(x)$ by y in (5) and solving for x, where $x \geq 0$. We get

$$y = x^2 - 1$$

$$x^2 = y + 1$$

$$x = \sqrt{y + 1}$$

Therefore $f_1^{-1}(y) = \sqrt{y + 1}$, and if we replace y by x,

$$f_1^{-1}(x) = \sqrt{x + 1} \qquad x \in [-1, c^2 - 1]$$

Because f_2 is continuous on $[-c, 0]$ and because $f_2'(x) < 0$ for all x in $(-c, 0)$, by Theorem 4.4.3(ii) f_2 is decreasing on $[-c, 0]$. Thus, by Theorem 7.1.5(ii), f_2 has an inverse f_2^{-1} that is defined on $[f_2(0), f_2(-c)]$ which is $[-1, c^2 - 1]$. If in (6) we replace $f_2(x)$ by y and solve for x, where $x \leq 0$, we obtain

$$x = -\sqrt{y + 1}$$

Therefore $f_2^{-1}(y) = -\sqrt{y + 1}$, and

$$f_2^{-1}(x) = -\sqrt{x + 1} \qquad x \in [-1, c^2 - 1]$$

In Figure 11 there are sketches of the graphs of f_1 and its inverse f_1^{-1} plotted on the same set of axes. Sketches of the graphs of f_2 and its inverse f_2^{-1} are plotted on the same set of axes in Figure 12. In both figures observe that the graphs of the function and its inverse are reflections of each other with respect to the line $y = x$. ◀

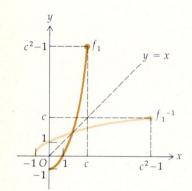

FIGURE 11

There are functions that have an inverse but for which we cannot obtain an equation defining the inverse function explicitly. For example, let

$$f(x) = x^5 + 5x^3 + 2x - 4 \tag{7}$$

$$f'(x) = 5x^4 + 15x^2 + 2$$

Because $f'(x) > 0$ for all x, f is an increasing function and therefore has an inverse. However, if we replace $f(x)$ by y in (7), we obtain a fifth-degree equation in x. Equations of the fifth degree and higher in general do not have solutions that can be expressed by a formula. Nevertheless we can determine some properties of the inverse function. These properties are discussed in Section 7.2.

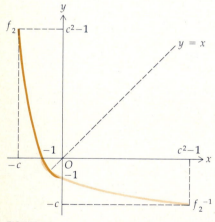

FIGURE 12

EXERCISES 7.1

In Exercises 1 through 18, determine if the function is one-to-one and draw a sketch of its graph.

1. $f(x) = 2x + 3$ **2.** $g(x) = 8 - 4x$ **3.** $f(x) = \frac{1}{2}x^2 - 2$

4. $f(x) = 3 - x^2$ **5.** $g(x) = 4 - x^3$ **6.** $h(x) = \frac{1}{2}x^3 + 1$

7. $f(x) = \sqrt{x + 3}$ **8.** $g(x) = \sqrt{1 - x^2}$

9. $h(x) = 2 \sin x, \ -\frac{1}{2}\pi \leq x \leq \frac{1}{2}\pi$

10. $f(x) = 1 - \cos x, \ 0 \leq x \leq \pi$

11. $f(x) = \frac{1}{2} \tan x, \ -\frac{1}{2}\pi < x < \frac{1}{2}\pi$

12. $F(x) = \cot \frac{1}{2}x, \ 0 < x < 2\pi$

13. $G(x) = \sec x,\; x \in [0, \tfrac{1}{2}\pi) \cup [\pi, \tfrac{3}{2}\pi)$
14. $g(x) = \csc x,\; x \in (0, \tfrac{1}{2}\pi] \cup (\pi, \tfrac{3}{2}\pi]$

15. $f(x) = \dfrac{2}{x+3}$ **16.** $g(x) = 5$

17. $g(x) = |x - 2|$ **18.** $f(x) = \dfrac{1}{2x-4}$

In Exercises 19 through 40, determine if the function has an inverse. If the inverse exists, do the following: (a) Find it and state its domain and range; (b) draw sketches of the graphs of the function and its inverse on the same set of axes. If the function does not have an inverse, show that a horizontal line intersects the graph of the function in more than one point.

19. $f(x) = 5x - 7$ **20.** $f(x) = 3x + 6$
21. $g(x) = 1 - x^2$ **22.** $g(x) = x^5$
23. $f(x) = (4 - x)^3$ **24.** $F(x) = 3(x^2 + 1)$
25. $h(x) = \sqrt{2x - 6}$ **26.** $g(x) = \sqrt{1 - x^2}$
27. $F(x) = \sqrt[3]{x + 1}$ **28.** $f(x) = |x| + x$
29. $f(x) = (x + 2)^4$ **30.** $g(x) = 3\sqrt[3]{x} + 1$
31. $f(x) = 2\sqrt[5]{x}$ **32.** $f(x) = \dfrac{2x - 1}{x}$
33. $f(x) = \dfrac{x - 3}{x + 1}$ **34.** $g(x) = \dfrac{8}{x^3 + 1}$
35. $g(x) = x^2 + 5,\; x \geq 0$ **36.** $f(x) = (2x - 1)^2,\; x \leq \tfrac{1}{2}$
37. $f(x) = (2x + 1)^3,\; -\tfrac{1}{2} \leq x \leq \tfrac{1}{2}$
38. $f(x) = \tfrac{1}{8}x^3,\; -1 \leq x \leq 1$
39. $F(x) = \sqrt{9 - x^2},\; 0 \leq x \leq 3$
40. $G(x) = \sqrt{4x^2 - 9},\; x \geq \tfrac{3}{2}$

In Exercises 41 through 46, (a) prove that the function f has an inverse, (b) find $f^{-1}(x)$, and (c) verify the equations of Theorem 7.1.4 for f and f^{-1}.

41. $f(x) = 4x - 3$ **42.** $f(x) = 5x + 2$ **43.** $f(x) = x^3 + 2$
44. $f(x) = (x + 2)^3$ **45.** $f(x) = \dfrac{3x + 1}{2x + 4}$ **46.** $f(x) = \dfrac{x - 3}{3x - 6}$

47. If x degrees is the Celsius temperature, then the number of degrees in the Fahrenheit temperature can be expressed as a function of x. If f is this function, then $f(x)$ degrees is the Fahrenheit temperature, and $f(x) = 32 + \tfrac{9}{5}x$. Determine the inverse function f^{-1} that expresses the number of degrees in the Celsius temperature as a function of the number of degrees in the Fahrenheit temperature.

48. If $f(t)$ dollars is the amount in t years of an investment of $1000 at 12 percent simple interest, then

$$f(t) = 1000(1 + 0.12t)$$

Determine the inverse function f^{-1} that expresses the number of years that $1000 has been invested at 12 percent simple interest as a function of the amount of the investment.

49. If $f(x) = \sqrt{16 - x^2},\; 0 \leq x \leq 4$, show that f is its own inverse function.

50. Determine the value of the constant k so that the function defined by

$$f(x) = \frac{x + 5}{x + k}$$

will be its own inverse.

In Exercises 51 through 54, do the following: (a) Show that the function f is not one-to-one and hence does not have an inverse; (b) restrict the domain and obtain two one-to-one functions, f_1 and f_2, having the same range as f; (c) find $f_1^{-1}(x)$ and $f_2^{-1}(x)$ and state the domains of f_1^{-1} and f_2^{-1}; (d) draw sketches of the graphs of f_1 and f_1^{-1} on the same set of axes; (e) draw sketches of the graphs of f_2 and f_2^{-1} on the same set of axes.

51. $f(x) = x^2 + 4$ **52.** $f(x) = 2x^2 - 6$
53. $f(x) = \sqrt{9 - x^2}$ **54.** $f(x) = \tfrac{1}{2}\sqrt{x^2 - 16}$

55. Given

$$f(x) = \begin{cases} x & \text{if } x < 1 \\ x^2 & \text{if } 1 \leq x \leq 9 \\ 27\sqrt{x} & \text{if } 9 < x \end{cases}$$

Prove that f has an inverse function and find $f^{-1}(x)$.

7.2 INVERSE FUNCTION THEOREMS AND THE DERIVATIVE OF THE INVERSE OF A FUNCTION

Information about the continuity and differentiability of the inverse f^{-1} of a function f can be acquired from properties of f even though $f^{-1}(x)$ is not defined explicitly by an equation. The *inverse function theorems* of this section provide the means of obtaining this information. Before stating the inverse function theorem for increasing functions, we present two illustrations giving examples of a function and its inverse that satisfy the conditions of the theorem.

▶ **ILLUSTRATION 1** In Example 3 of Section 7.1 we had the function f and its inverse f^{-1} defined by

$$f(x) = 4x - 3 \qquad f^{-1}(x) = \frac{x + 3}{4}$$

In Figure 7 of Section 7.1 the graphs of f and f^{-1} are shown on the same set of axes. We see that both f and f^{-1} are continuous and increasing on their domains. ◀

▶ **ILLUSTRATION 2**   In Illustration 6 of Section 7.1 the function g and its inverse g^{-1} are defined by

$$g(x) = x^3 \qquad -2 \le x \le 2$$
$$g^{-1}(x) = \sqrt[3]{x} \qquad -8 \le x \le 8$$

and the graphs of g and g^{-1} are shown in Figure 8 of Section 7.1. Each of these functions is continuous and increasing on its domain.   ◀

7.2.1 THEOREM
Inverse Function Theorem

Suppose that the function f is continuous and increasing on the closed interval $[a, b]$. Then if f^{-1} is its inverse, which is defined on $[f(a), f(b)]$,

(i) f^{-1} is increasing on $[f(a), f(b)]$.
(ii) f^{-1} is continuous on $[f(a), f(b)]$.

Proof of (i)   The existence of f^{-1} is guaranteed by Theorem 7.1.5. To prove that f^{-1} is increasing on $[f(a), f(b)]$ we must show that

$$\text{if} \quad y_1 < y_2 \quad \text{then} \quad f^{-1}(y_1) < f^{-1}(y_2)$$

where y_1 and y_2 are two numbers in $[f(a), f(b)]$. Because f^{-1} is defined on $[f(a), f(b)]$, there exist numbers x_1 and x_2 in $[a, b]$ such that $y_1 = f(x_1)$ and $y_2 = f(x_2)$. Therefore

$$f^{-1}(y_1) = f^{-1}(f(x_1)) \quad \text{and} \quad f^{-1}(y_2) = f^{-1}(f(x_2))$$
$$\Leftrightarrow \quad f^{-1}(y_1) = x_1 \qquad\qquad \text{and} \quad f^{-1}(y_2) = x_2 \qquad\qquad (1)$$

If $x_2 < x_1$, then because f is increasing on $[a, b]$, $f(x_2) < f(x_1)$ or, equivalently, $y_2 < y_1$. But $y_1 < y_2$; therefore x_2 cannot be less than x_1.

If $x_2 = x_1$, then because f is a function, $f(x_1) = f(x_2)$ or, equivalently, $y_1 = y_2$, but this also contradicts the fact that $y_1 < y_2$. Therefore $x_2 \neq x_1$.

So if x_2 is not less than x_1 and $x_2 \neq x_1$, it follows that $x_1 < x_2$; hence, from the two equations in (1), $f^{-1}(y_1) < f^{-1}(y_2)$. Thus we have proved that f^{-1} is increasing on $[f(a), f(b)]$.

Proof of (ii)   To prove that f^{-1} is continuous on the closed interval $[f(a), f(b)]$ we must show that if r is any number in the open interval $(f(a), f(b))$, then f^{-1} is continuous at r, and f^{-1} is continuous from the right at $f(a)$, and f^{-1} is continuous from the left at $f(b)$.

We prove that f^{-1} is continuous at any r in the open interval $(f(a), f(b))$ by showing that Theorem 2.6.6 holds at r. We wish to show that, for any $\epsilon > 0$ small enough so that $f^{-1}(r) - \epsilon$ and $f^{-1}(r) + \epsilon$ are both in $[a, b]$, there exists a $\delta > 0$ such that

$$\text{if} \quad |y - r| < \delta \quad \text{then} \quad |f^{-1}(y) - f^{-1}(r)| < \epsilon$$

Let $f^{-1}(r) = s$. Then $f(s) = r$. Because, from (i), f^{-1} is increasing on $[f(a), f(b)]$, we conclude that $a < s < b$. Therefore

$$a \le s - \epsilon < s < s + \epsilon \le b$$

Because f is increasing on $[a, b]$,

$$f(a) \le f(s - \epsilon) < r < f(s + \epsilon) \le f(b) \qquad\qquad (2)$$

Let δ be the smaller of the two numbers $r - f(s - \epsilon)$ and $f(s + \epsilon) - r$; so $\delta \leq r - f(s - \epsilon)$ and $\delta \leq f(s + \epsilon) - r$ or, equivalently,

$$f(s - \epsilon) \leq r - \delta \quad \text{and} \quad r + \delta \leq f(s + \epsilon) \tag{3}$$

If $|y - r| < \delta$, then $-\delta < y - r < \delta$ or, equivalently,

$$r - \delta < y < r + \delta$$

From this inequality and (2) and (3), we have

$$\text{if} \quad |y - r| < \delta \quad \text{then} \quad f(a) \leq f(s - \epsilon) < y < f(s + \epsilon) \leq f(b)$$

Because f^{-1} is increasing on $[f(a), f(b)]$, it follows from the above that

$$\text{if} \quad |y - r| < \delta \quad \text{then} \quad f^{-1}(f(s - \epsilon)) < f^{-1}(y) < f^{-1}(f(s + \epsilon))$$

$$\Leftrightarrow \quad \text{if} \quad |y - r| < \delta \quad \text{then} \quad s - \epsilon < f^{-1}(y) < s + \epsilon$$

$$\Leftrightarrow \quad \text{if} \quad |y - r| < \delta \quad \text{then} \quad -\epsilon < f^{-1}(y) - s < \epsilon$$

$$\Leftrightarrow \quad \text{if} \quad |y - r| < \delta \quad \text{then} \quad |f^{-1}(y) - f^{-1}(r)| < \epsilon$$

So f^{-1} is continuous on the open interval $(f(a), f(b))$.

The proofs that f^{-1} is continuous from the right at $f(a)$ and continuous from the left at $f(b)$ appear as an exercise (see Exercise 40). ∎

We now state the inverse function theorem for decreasing functions. The proof is similar to that for Theorem 7.2.1 and is left for the exercises (see Exercises 41 and 42).

7.2.2 THEOREM
Inverse Function Theorem

Suppose that the function f is continuous and decreasing on the closed interval $[a, b]$. Then if f^{-1} is its inverse, which is defined on $[f(b), f(a)]$,

(i) f^{-1} is decreasing on $[f(b), f(a)]$;
(ii) f^{-1} is continuous on $[f(b), f(a)]$.

The inverse function theorems are used to prove the following theorem, which expresses a relationship between the derivative of a function and the derivative of its inverse if the function has an inverse. In the statement and proof of the theorem we use the Leibniz notation for a derivative. Notice how this notation makes the equation easy to remember.

7.2.3 THEOREM

Suppose that the function f is continuous and monotonic on the closed interval $[a, b]$, and let $y = f(x)$. If f is differentiable on $[a, b]$ and $f'(x) \neq 0$ for any x in $[a, b]$, then the derivative of the inverse function f^{-1}, defined by $x = f^{-1}(y)$, is given by

$$\frac{dx}{dy} = \frac{1}{\dfrac{dy}{dx}}$$

Proof Because f is continuous and monotonic on $[a, b]$, then by Theorems 7.2.1 and 7.2.2, f has an inverse that is continuous and monotonic on $[f(a), f(b)]$ (or $[f(b), f(a)]$ if $f(b) < f(a)$).

If x is a number in $[a, b]$, let Δx be an increment of x, $\Delta x \neq 0$, such that $x + \Delta x$ is also in $[a, b]$. Then the corresponding increment of y is given by

$$\Delta y = f(x + \Delta x) - f(x) \tag{4}$$

$\Delta y \neq 0$ since $\Delta x \neq 0$ and f is monotonic on $[a, b]$; that is, either

$$f(x + \Delta x) < f(x) \quad \text{or} \quad f(x + \Delta x) > f(x) \qquad \text{on } [a, b]$$

If x is in $[a, b]$ and $y = f(x)$, then y is in $[f(a), f(b)]$ (or $[f(b), f(a)]$). Also, if $x + \Delta x$ is in $[a, b]$, then $y + \Delta y$ is in $[f(a), f(b)]$ (or $[f(b), f(a)]$) because $y + \Delta y = f(x + \Delta x)$ by (4). So

$$x = f^{-1}(y) \quad \text{and} \quad x + \Delta x = f^{-1}(y + \Delta y)$$

From these two equations, we have

$$\Delta x = f^{-1}(y + \Delta y) - f^{-1}(y) \tag{5}$$

From the definition of a derivative,

$$\frac{dx}{dy} = \lim_{\Delta y \to 0} \frac{f^{-1}(y + \Delta y) - f^{-1}(y)}{\Delta y}$$

Substituting from (4) and (5) into the above equation we get

$$\frac{dx}{dy} = \lim_{\Delta y \to 0} \frac{\Delta x}{f(x + \Delta x) - f(x)}$$

and because $\Delta x \neq 0$,

$$\frac{dx}{dy} = \lim_{\Delta y \to 0} \frac{1}{\dfrac{f(x + \Delta x) - f(x)}{\Delta x}} \tag{6}$$

Before we find the limit in (6) we show that under the hypothesis of this theorem $\Delta x \to 0$ is equivalent to $\Delta y \to 0$. First we show that $\lim\limits_{\Delta y \to 0} \Delta x = 0$. From (5),

$$\lim_{\Delta y \to 0} \Delta x = \lim_{\Delta y \to 0} \left[f^{-1}(y + \Delta y) - f^{-1}(y) \right]$$

Because f^{-1} is continuous on $[f(a), f(b)]$ (or $[f(b), f(a)]$), the limit on the right side of the above equation is zero. So

$$\lim_{\Delta y \to 0} \Delta x = 0 \tag{7}$$

Now we demonstrate that $\lim\limits_{\Delta x \to 0} \Delta y = 0$. From (4),

$$\lim_{\Delta x \to 0} \Delta y = \lim_{\Delta x \to 0} \left[f(x + \Delta x) - f(x) \right]$$

Because f is continuous on $[a, b]$, the limit on the right side of the above equation is zero, and therefore

$$\lim_{\Delta x \to 0} \Delta y = 0$$

From this equation and (7) it follows that

$$\Delta x \to 0 \quad \text{if and only if} \quad \Delta y \to 0$$

From this statement and by applying the theorem regarding the limit of a quotient to (6) we have

$$\frac{dx}{dy} = \frac{1}{\lim\limits_{\Delta x \to 0} \dfrac{f(x + \Delta x) - f(x)}{\Delta x}}$$

Because f is differentiable on $[a, b]$, the limit in the denominator of the above is $f'(x)$ or, equivalently, $\dfrac{dy}{dx}$. Thus

$$\frac{dx}{dy} = \frac{1}{\dfrac{dy}{dx}} \qquad\qquad \blacksquare$$

▶ **ILLUSTRATION 3**   We verify Theorem 7.2.3 for the function f defined by $f(x) = \sqrt{x}$. If we let $y = f(x)$, we have the equation

$$y = \sqrt{x} \qquad x \geq 0, y \geq 0$$

Because f is one-to-one, f^{-1} exists and is defined by $f^{-1}(y) = y^2$. If we let $x = f^{-1}(y)$, we have the equation

$$x = y^2 \qquad y \geq 0, x \geq 0$$

Because $y = \sqrt{x}$

$$\frac{dy}{dx} = \frac{1}{2\sqrt{x}}$$

and because $x = y^2$

$$\frac{dx}{dy} = 2y$$

Replacing y by $\sqrt{x}$, we get

$$\frac{dx}{dy} = 2\sqrt{x}$$

$$= \frac{1}{\dfrac{1}{2\sqrt{x}}}$$

$$= \frac{1}{\dfrac{dy}{dx}}$$

When $x = 0$, $\dfrac{dy}{dx}$ does not exist; thus the above equation is not satisfied for this value of x. Because the domain of f is the closed interval $[0, +\infty)$, the theorem is valid for this function when x is in the open interval $(0, +\infty)$.   ◀

EXAMPLE 1   Show that Theorem 7.2.3 holds for the function f of Examples 2 and 4 in Section 7.1.

Solution   The function f is defined by $(2x + 3)/(x - 1)$. If we let $y = f(x)$, we have

$$y = \frac{2x + 3}{x - 1} \tag{8}$$

$$\frac{dy}{dx} = -\frac{5}{(x - 1)^2}$$

In the solution of Example 4 in Section 7.1 we showed that

$$x = \frac{y + 3}{y - 2}$$

We compute $\dfrac{dx}{dy}$ from this equation and get

$$\frac{dx}{dy} = -\frac{5}{(y - 2)^2}$$

In the above equation we substitute the value of y from (8) and obtain

$$\frac{dx}{dy} = -\frac{5}{\left(\dfrac{2x + 3}{x - 1} - 2\right)^2}$$

$$= -\frac{5(x - 1)^2}{(2x + 3 - 2x + 2)^2}$$

$$= -\tfrac{5}{25}(x - 1)^2$$

$$= -\tfrac{1}{5}(x - 1)^2$$

$$= \frac{1}{\dfrac{dy}{dx}}$$

EXAMPLE 2   Determine if the function f defined by

$$f(x) = x^3 + x$$

has an inverse. If it does, find the derivative of the inverse function.

Solution

$$f'(x) = 3x^2 + 1$$

Therefore $f'(x) > 0$ for all real numbers; so f is increasing on its domain. Thus f is a one-to-one function, and hence it has an inverse f^{-1}. Let $y = f(x)$, and then $x = f^{-1}(y)$. By Theorem 7.2.3,

$$\frac{dx}{dy} = \frac{1}{\dfrac{dy}{dx}}$$

$$= \frac{1}{3x^2 + 1}$$

When using Theorem 7.2.3 to compute the value of the derivative of the inverse of a function at a particular number, it is more convenient to have the statement of the theorem with f' and $(f^{-1})'$ notations for the derivatives. With this notation we restate the theorem as Theorem 7.2.4.

7.2.4 THEOREM
Suppose the function f is continuous and monotonic on a closed interval $[a, b]$ containing the number c, and let $f(c) = d$. If $f'(c)$ exists and $f'(c) \neq 0$, then $(f^{-1})'(d)$ exists and

$$(f^{-1})'(d) = \frac{1}{f'(c)}$$

▶ **ILLUSTRATION 4** We show that Theorem 7.2.4 holds for a particular function and particular values of c and d. If f is the function of Illustration 3,

$$f(x) = \sqrt{x} \qquad f'(x) = \frac{1}{2\sqrt{x}}$$

$$f^{-1}(x) = x^2 \qquad (f^{-1})'(x) = 2x \qquad x \geq 0$$

The function f is continuous and monotonic on any closed interval $[a, b]$ for which $0 \leq a < b$. Let $c = 9$, and then $d = f(9)$; that is, $d = 3$. Theorem 7.2.4 states that

$$(f^{-1})'(3) = \frac{1}{f'(9)}$$

This equation is valid because $(f^{-1})'(3) = 6$ and $f'(9) = \frac{1}{6}$. ◀

▶ **ILLUSTRATION 5** The function f mentioned at the conclusion of Section 7.1 is defined by

$$f(x) = x^5 + 5x^3 + 2x - 4$$

and as stated there, f has an inverse f^{-1}. But we do not have an equation defining explicitly the function value of f^{-1}. Nevertheless it is possible to compute the derivative of f^{-1} at a particular point on the graph of $y = f^{-1}(x)$. For instance, because $(1, 4)$ is on the graph of $y = f(x)$, the point $(4, 1)$ is on the graph of $y = f^{-1}(x)$. We compute the value of $(f^{-1})'(4)$ by using Theorem 7.2.4, which states that

$$(f^{-1})'(4) = \frac{1}{f'(1)}$$

We first find $f'(x)$ and then compute $f'(1)$.

$$f'(x) = 5x^4 + 15x^2 + 2 \qquad f'(1) = 22$$

Thus

$$(f^{-1})'(4) = \frac{1}{22}$$ ◀

EXAMPLE 3 (a) On the same set of axes draw sketches of the graphs of the functions f and f^{-1} of Example 2. (b) Find the slope of the tangent line to the graph of f at the point $(1, 2)$. (c) Find the slope of the tangent line to the graph of f^{-1} at the point $(2, 1)$.

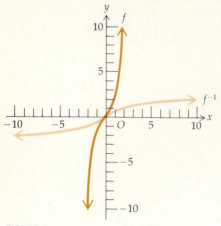

FIGURE 1

Solution

$$f(x) = x^3 + x$$

$$f'(x) = 3x^2 + 1$$

$$f''(x) = 6x$$

(a) The function is increasing on its domain. Because $f''(0) = 0$ and $f''(x)$ changes sign at $x = 0$, the graph of f has a point of inflection at the origin. The slope of the inflectional tangent at the origin is $f'(0) = 1$. We use this information to draw the sketch of the graph of f shown in Figure 1. This figure also shows a sketch of the graph of f^{-1}, which is obtained from the fact that it is the reflection of the graph of f with respect to the line $y = x$. Observe that to find an equation defining $f^{-1}(x)$ explicitly it is necessary to solve the third degree equation $y = x^3 + x$.

(b) The slope of the tangent line to the graph of f at the point $(1, 2)$ is $f'(1) = 4$.

(c) The slope of the tangent line to the graph of f^{-1} at the point $(2, 1)$ is $(f^{-1})'(2)$. From Theorem 7.2.4,

$$(f^{-1})'(2) = \frac{1}{f'(1)}$$

$$= \tfrac{1}{4}$$

EXERCISES 7.2

In Exercises 1 through 10, let $y = f(x)$ and $x = f^{-1}(y)$, and verify that $\dfrac{dx}{dy} = \dfrac{1}{\dfrac{dy}{dx}}$.

1. $f(x) = 4x - 3$ **2.** $f(x) = 7 - 2x$ **3.** $f(x) = \sqrt{x + 1}$
4. $f(x) = 8x^3$ **5.** $f(x) = \tfrac{1}{5}x^5$ **6.** $f(x) = \sqrt{4 - 3x}$

7. $f(x) = \sqrt[3]{x - 8}$ **8.** $f(x) = \sqrt[5]{x}$ **9.** $f(x) = \dfrac{2x - 3}{x + 2}$

10. $f(x) = \dfrac{3x + 4}{2x + 6}$

In Exercises 11 through 28, find $(f^{-1})'(d)$.

11. $f(x) = \sqrt{3x + 1}; d = 1$ **12.** $f(x) = x^5 + 2; d = 1$
13. $f(x) = x^2 - 16, x \geq 0; d = 9$ **14.** $f(x) = \sqrt{4 - x}; d = 3$
15. $f(x) = x^3 + 5; d = -3$ **16.** $f(x) = 4x^3 + 2x; d = 6$
17. $f(x) = 3x^5 + 2x^3; d = 5$
18. $f(x) = \sin x; -\tfrac{1}{2}\pi \leq x \leq \tfrac{1}{2}\pi; d = \tfrac{1}{2}$
19. $f(x) = \tfrac{1}{2}\cos^2 x; 0 \leq x \leq \tfrac{1}{2}\pi; d = \tfrac{1}{4}$
20. $f(x) = \tan 2x; -\tfrac{1}{4}\pi < x < \tfrac{1}{4}\pi; d = 1$
21. $f(x) = 2 \cot x; 0 < x < \pi; d = 2$
22. $f(x) = \sec \tfrac{1}{2}x; 0 \leq x < \pi; d = 2$
23. $f(x) = \tfrac{1}{2} \csc x; 0 < x < \tfrac{1}{2}\pi; d = 1$
24. $f(x) = x^2 - 6x + 7, x \leq 3; d = 0$
25. $f(x) = 2x^2 + 8x + 7, x \leq -2; d = 1$
26. $f(x) = 2x^3 + x + 20; d = 2$

27. $f(x) = \int_{-3}^{x} \sqrt{t + 3}\, dt, x > -3; d = 18$
28. $f(x) = \int_{x}^{2} t\, dt, x < 0; d = -6$

In Exercises 29 through 34, do the following: (a) Solve the equation for y in terms of x, and express y as one or more functions of x; (b) for each of the functions obtained in (a) determine if the function has an inverse, and, if it does, determine the domain of the inverse function; (c) use implicit differentiation to find $\dfrac{dy}{dx}$ and $\dfrac{dx}{dy}$, and determine the values of x and y for which $\dfrac{dy}{dx}$ and $\dfrac{dx}{dy}$ are reciprocals.

29. $x^2 + y^2 = 9$ **30.** $x^2 - 4y^2 = 16$
31. $xy = 4$ **32.** $9y^2 - 8x^3 = 0$
33. $2x^2 - 3xy + 1 = 0$ **34.** $2x^2 + 2y + 1 = 0$

35. Given $f(x) = x^3 + 3x - 1$. (a) On the same set of axes draw sketches of the graphs of the functions f and f^{-1}. (b) Find the slope of the tangent line to the graph of f at the point $(1, 3)$. (c) Find the slope of the tangent line to the graph of f^{-1} at the point $(3, 1)$.

36. Given $f(x) = 6 - x - x^3$. (a) On the same set of axes draw sketches of the graphs of the functions f and f^{-1}. (b) Find the slope of the tangent line to the graph of f at the point $(2, -4)$. (c) Find the slope of the tangent line to the graph of f^{-1} at the point $(-4, 2)$.

37. Given $f(x) = \int_1^x \sqrt{16 - t^4}\, dt$, $-2 \leq x \leq 2$. Prove that f has an inverse f^{-1}, and compute $(f^{-1})'(0)$.

38. Given $f(x) = \int_2^x \sqrt{9 + t^4}\, dt$. Prove that f has an inverse f^{-1}, and compute $(f^{-1})'(0)$.

39. Given $f(x) = \int_1^{2x} \dfrac{dt}{\sqrt{1 + t^4}}$. Prove that f has an inverse and compute $(f^{-1})'(0)$.

40. Given that the function f is continuous and increasing on the closed interval $[a, b]$, by assuming Theorem 7.2.1(i) prove that f^{-1} is continuous from the right at $f(a)$ and continuous from the left at $f(b)$.

41. Prove Theorem 7.2.2(i).

42. Prove Theorem 7.2.2(ii).

43. Show that the formula of Theorem 7.2.3 can be written as

$$(f^{-1})'(x) = \frac{1}{f'(f^{-1}(x))}$$

44. Use the formula of Exercise 43 to show that

$$(f^{-1})''(x) = -\frac{f''(f^{-1}(x))}{[f'(f^{-1}(x))]^3}$$

7.3 THE NATURAL LOGARITHMIC FUNCTION

The definition of the logarithmic function that you encountered in algebra is based on exponents. The properties of logarithms are then proved from corresponding properties of exponents. One such property of exponents is

$$a^x \cdot a^y = a^{x+y} \tag{1}$$

If the exponents x and y are positive integers and if a is any real number, (1) follows from the definition of a positive integer exponent and mathematical induction. If the exponents are allowed to be any integers, either positive, negative, or zero, and $a \neq 0$, then (1) will hold if a zero exponent and a negative integer exponent are defined by

$$a^0 = 1 \quad \text{and} \quad a^{-n} = \frac{1}{a^n} \quad n > 0$$

If the exponents are rational numbers and $a \geq 0$, then (1) holds when $a^{m/n}$ is defined by

$$a^{m/n} = (\sqrt[n]{a})^m$$

It is not quite so simple to define a^x when x is an irrational number. For example, what is meant by $2^{\sqrt{3}}$? The definition of the logarithmic function, as given in elementary algebra, is based on the assumption that a^x exists if a is any positive number and x is any real number.

This definition states that the equation

$$a^x = N$$

where a is any positive number except 1 and N is any positive number, can be solved for x, and x is uniquely determined by

$$x = \log_a N$$

The following properties of logarithms are proved from those of exponents:

$$\log_a 1 = 0 \tag{2}$$

$$\log_a MN = \log_a M + \log_a N \tag{3}$$

$$\log_a \frac{M}{N} = \log_a M - \log_a N \tag{4}$$

$$\log_a M^n = n \log_a M \tag{5}$$

$$\log_a a = 1 \tag{6}$$

In this chapter we define the logarithmic function by using calculus and prove the properties of logarithms by means of this definition. Then the exponential function is defined in terms of the logarithmic function. This definition enables us to define a^x when x is any real number and $a > 0$. The properties of exponents are then proved if the exponent is any real number.

Recall the formula

$$\int t^n \, dt = \frac{t^{n+1}}{n+1} + C \qquad n \neq -1$$

This formula does not hold when $n = -1$.

To evaluate $\int t^n \, dt$ for $n = -1$ we need a function whose derivative is $\frac{1}{x}$. The first fundamental theorem of the calculus (5.8.1) gives us one; it is

$$\int_a^x \frac{1}{t} \, dt$$

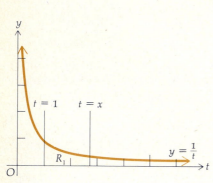

FIGURE 1

where a can be any real number having the same sign as x. To interpret such a function let R_1 be the region bounded by the curve $y = 1/t$, by the t axis, on the left by the line $t = 1$, and on the right by the line $t = x$, where $x > 1$. This region R_1 is shown in Figure 1. The measure of the area of R_1 is a function of x; call it $A(x)$ and define it as a definite integral by

$$A(x) = \int_1^x \frac{1}{t} \, dt$$

Now consider this integral if $0 < x < 1$. From Definition 5.5.5,

$$\int_1^x \frac{1}{t} \, dt = -\int_x^1 \frac{1}{t} \, dt$$

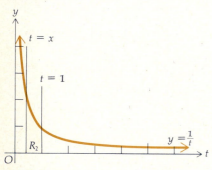

FIGURE 2

Then the integral $\int_x^1 (1/t) \, dt$ represents the measure of the area of the region R_2 bounded by the curve $y = 1/t$, by the t axis, on the left by the line $t = x$, and on the right by the line $t = 1$. So the integral $\int_1^x (1/t) \, dt$ is then the negative of the measure of the area of the region R_2 shown in Figure 2.

If $x = 1$, the integral $\int_1^x (1/t) \, dt$ becomes $\int_1^1 (1/t) \, dt$, which equals 0 by Definition 5.5.6. In this case the left and right boundaries of the region are the same and the measure of the area is 0.

Thus the integral $\int_1^x (1/t) \, dt$ for $x > 0$ can be interpreted in terms of the measure of the area of a region. Its value depends on x and is used to define the *natural logarithmic function*.

7.3.1 DEFINITION   The **natural logarithmic function** is the function defined by

$$\ln x = \int_1^x \frac{1}{t} \, dt \qquad x > 0$$

The domain of the natural logarithmic function is the set of all positive numbers. We read $\ln x$ as "the natural logarithm of x."

The natural logarithmic function is differentiable because by applying the first fundamental theorem of the calculus (5.8.1)

$$D_x(\ln x) = D_x\left(\int_1^x \frac{1}{t}\, dt\right)$$

$$= \frac{1}{x}$$

From this result and the chain rule we have the following theorem.

7.3.2 THEOREM   If u is a differentiable function of x and $u(x) > 0$, then

$$D_x(\ln u) = \frac{1}{u} \cdot D_x u$$

EXAMPLE 1   Find $f'(x)$ if

$$f(x) = \ln(3x^2 - 6x + 8)$$

Solution   From Theorem 7.3.2,

$$f'(x) = \frac{1}{3x^2 - 6x + 8}(6x - 6)$$

$$= \frac{6x - 6}{3x^2 - 6x + 8}$$

EXAMPLE 2   Find $\dfrac{dy}{dx}$ if

$$y = \ln[(4x^2 + 3)(2x - 1)]$$

Solution   Applying Theorem 7.3.2 we get

$$\frac{dy}{dx} = \frac{1}{(4x^2 + 3)(2x - 1)} \cdot [8x(2x - 1) + 2(4x^2 + 3)]$$

$$= \frac{24x^2 - 8x + 6}{(4x^2 + 3)(2x - 1)} \tag{7}$$

EXAMPLE 3   Find $\dfrac{dy}{dx}$ if

$$y = \ln\left(\frac{x}{x + 1}\right)$$

Solution   From Theorem 7.3.2,

$$\frac{dy}{dx} = \frac{1}{\dfrac{x}{x + 1}} \cdot \frac{(x + 1) - x}{(x + 1)^2}$$

$$= \frac{x + 1}{x} \cdot \frac{1}{(x + 1)^2}$$

$$= \frac{1}{x(x + 1)}$$

Observe that when applying Theorem 7.3.2, $u(x)$ must be positive; that is, a number in the domain of the derivative must be in the domain of the given function, ln u.

▶ **ILLUSTRATION 1**   In Example 1 the domain of the given function is the set of all real numbers, because $3x^2 - 6x + 8 > 0$ for all x. This can be seen from the fact that the parabola having equation $y = 3x^2 - 6x + 8$ has its vertex at $(1, 5)$ and opens upward. Hence $(6x - 6)/(3x^2 - 6x + 8)$ is the derivative for all values of x.

In Example 2, because $(4x^2 + 3)(2x - 1) > 0$ only when $x > \frac{1}{2}$, the domain of the given function is the interval $(\frac{1}{2}, +\infty)$. Therefore it is understood that fraction (7) is the derivative only if $x > \frac{1}{2}$.

Because $x/(x + 1) > 0$ when either $x < -1$ or $x > 0$, the domain of the function in Example 3 is $(-\infty, -1) \cup (0, +\infty)$; so $1/[x(x + 1)]$ is the derivative if either $x < -1$ or $x > 0$.   ◀

We show that the natural logarithmic function obeys the properties of logarithms given earlier.

7.3.3 THEOREM   ln $1 = 0$

Proof   If $x = 1$ in Definition 7.3.1,

$$\ln 1 = \int_1^1 \frac{1}{t}\, dt$$

The right side of the above is zero by Definition 5.5.6. Thus

$$\ln 1 = 0 \qquad \blacksquare$$

Theorem 7.3.3 corresponds to the property of logarithms given by (2). The next three theorems correspond to the properties of logarithms given by (3), (4), and (5). The discussion of property (6) is postponed until Section 7.5.

7.3.4 THEOREM   If a and b are any positive numbers, then

$$\ln(ab) = \ln a + \ln b$$

Proof   Consider the function f defined by

$$f(x) = \ln(ax)$$

where $x > 0$. Then

$$f'(x) = \frac{1}{ax}\,(a)$$

$$= \frac{1}{x}$$

Therefore the derivatives of $\ln(ax)$ and ln x are equal. Thus from Theorem 5.1.2 there is a constant K such that

$$\ln(ax) = \ln x + K \quad \text{for all } x > 0 \tag{8}$$

To determine K, let $x = 1$ in this equation and we have

$\ln a = \ln 1 + K$

Because $\ln 1 = 0$, we obtain $K = \ln a$. Replacing K by $\ln a$ in (8), we obtain

$\ln(ax) = \ln x + \ln a$ for all $x > 0$

Now, letting $x = b$, we have

$\ln(ab) = \ln a + \ln b$ ∎

7.3.5 THEOREM   If a and b are any positive numbers, then

$$\ln \frac{a}{b} = \ln a - \ln b$$

Proof Because $a = (a/b) \cdot b$,

$$\ln a = \ln \left(\frac{a}{b} \cdot b \right)$$

Applying Theorem 7.3.4 to the right side of this equation we get

$$\ln a = \ln \frac{a}{b} + \ln b$$

$$\ln \frac{a}{b} = \ln a - \ln b$$ ∎

7.3.6 THEOREM   If a is any positive number and r is any rational number, then

$\ln a^r = r \ln a$

Proof   From Theorem 7.3.2, if r is any rational number and $x > 0$,

$$D_x(\ln x^r) = \frac{1}{x^r} \cdot rx^{r-1}$$

$$= \frac{r}{x}$$

and

$$D_x(r \ln x) = \frac{r}{x}$$

Therefore

$D_x(\ln x^r) = D_x(r \ln x)$

From this equation the derivatives of $\ln x^r$ and $r \ln x$ are equal; so it follows from Theorem 5.1.2 that there is a constant K such that

$\ln x^r = r \ln x + K$ for all $x > 0$ (9)

To determine K, we substitute 1 for x in (9) and get

$\ln 1^r = r \ln 1 + K$

But $\ln 1 = 0$; hence $K = 0$. Replacing K by 0 in (9) gives

$$\ln x^r = r \ln x \quad \text{for all } x > 0$$

from which it follows that if $x = a$, where a is any positive number, then

$$\ln a^r = r \ln a$$ ∎

▶ **ILLUSTRATION 2**   In Example 2, if Theorem 7.3.4 is applied before finding the derivative, we have

$$y = \ln(4x^2 + 3) + \ln(2x - 1) \tag{10}$$

The domain of the function defined by this equation is the interval $(\frac{1}{2}, +\infty)$, which is the same as the domain of the given function. From (10),

$$\frac{dy}{dx} = \frac{8x}{4x^2 + 3} + \frac{2}{2x - 1}$$

and combining the fractions gives

$$\frac{dy}{dx} = \frac{8x(2x - 1) + 2(4x^2 + 3)}{(4x^2 + 3)(2x - 1)}$$

which is equivalent to the first line of the solution of Example 2.   ◀

▶ **ILLUSTRATION 3**   If we apply Theorem 7.3.5 before finding the derivative in Example 3, we have

$$y = \ln x - \ln(x + 1) \tag{11}$$

Because $\ln x$ is defined only when $x > 0$, and $\ln(x + 1)$ is defined only when $x > -1$, the domain of the function defined by (11) is the interval $(0, +\infty)$. But the domain of the function given in Example 3 consists of the two intervals $(-\infty, -1)$ and $(0, +\infty)$. Computing the derivative from (11) we have

$$\frac{dy}{dx} = \frac{1}{x} - \frac{1}{x + 1}$$

$$= \frac{1}{x(x + 1)}$$

but remember here that x must be greater than 0, whereas in the solution of Example 3 values of x less than -1 are also included.   ◀

Illustration 3 shows the care that must be taken when applying Theorems 7.3.4, 7.3.5, and 7.3.6 to functions involving the natural logarithm.

EXAMPLE 4   Find $f'(x)$ if

$$f(x) = \ln(2x - 1)^3$$

Solution   From Theorem 7.3.6,

$$f(x) = 3 \ln(2x - 1)$$

Observe that $\ln(2x - 1)^3$ and $3\ln(2x - 1)$ both have the same domain: $x > \frac{1}{2}$. Applying Theorem 7.3.2 gives

$$f'(x) = 3 \cdot \frac{1}{2x - 1} \cdot 2$$

$$= \frac{6}{2x - 1}$$

To draw a sketch of the graph of the natural logarithmic function we must first consider some properties of this function.

Let f be the function defined by $f(x) = \ln x$; that is,

$$f(x) = \int_1^x \frac{1}{t}\, dt \qquad x > 0$$

The domain of f is the set of all positive numbers. The function f is differentiable for all x in its domain, and

$$f'(x) = \frac{1}{x}$$

Because f is differentiable for all $x > 0$, f is continuous for all $x > 0$. From the above equation, $f'(x) > 0$ for all $x > 0$, and therefore f is an increasing function.

$$f''(x) = -\frac{1}{x^2}$$

Observe that $f''(x) < 0$ when $x > 0$. Therefore the graph of $y = f(x)$ is concave downward at every point. Because $f(1) = \ln 1$ and $\ln 1 = 0$, the x intercept of the graph is 1.

We can determine an approximate numerical value for $\ln 2$ by using the equation

$$\ln 2 = \int_1^2 \frac{1}{t}\, dt$$

and interpreting the definite integral as the measure of the area of the region appearing in Figure 3. From this figure we observe that $\ln 2$ is between the measures of the areas of the rectangles, each having a base of length 1 unit and altitudes of lengths $\frac{1}{2}$ unit and 1 unit; that is,

$$0.5 < \ln 2 < 1$$

An inequality can be obtained analytically, using Theorem 5.6.8, by proceeding as follows. Let $f(t) = 1/t$ and $g(t) = \frac{1}{2}$. Then $f(t) \geq g(t)$ for all t in $[1, 2]$. Because f and g are continuous on $[1, 2]$, they are integrable on $[1, 2]$, and from Theorem 5.6.8,

$$\int_1^2 \frac{1}{t}\, dt \geq \int_1^2 \frac{1}{2}\, dt$$

$$\ln 2 \geq \tfrac{1}{2} \tag{12}$$

Similarly, if $f(t) = 1/t$ and $h(t) = 1$, then $h(t) \geq f(t)$ for all t in $[1, 2]$. Because h and f are continuous on $[1, 2]$, they are integrable there; and again using

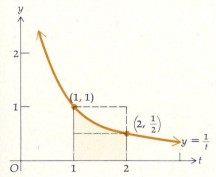

FIGURE 3

Theorem 5.6.8 we obtain

$$\int_1^2 1 \, dt \geq \int_1^2 \frac{1}{t} \, dt$$

$$1 \geq \ln 2$$

Combining this inequality with (12) we get

$$0.5 \leq \ln 2 \leq 1 \tag{13}$$

The number 0.5 is a lower bound of $\ln 2$ and 1 is an upper bound. In a similar manner we can obtain a lower and upper bound for the natural logarithm of any positive real number. Later we learn, by applying infinite series, how to compute the natural logarithm of any positive real number to any desired number of decimal places.

The value of $\ln 2$ to five decimal places is given by

$$\ln 2 \approx 0.69315$$

Using this value of $\ln 2$ and applying Theorem 7.3.6 we can find an approximate value for the natural logarithm of any power of 2. In particular,

$\ln 4 = \ln 2^2$	$\ln 8 = \ln 2^3$	$\ln \frac{1}{2} = \ln 2^{-1}$	$\ln \frac{1}{4} = \ln 2^{-2}$
$= 2 \ln 2$	$= 3 \ln 2$	$= -1 \cdot \ln 2$	$= -2 \ln 2$
≈ 1.3863	≈ 2.0795	≈ -0.69315	≈ -1.3863

A calculator with an $\boxed{\ln}$ key can be used to obtain values of the natural logarithmic function.

We now determine the behavior of the natural logarithmic function for large values of x by considering $\lim\limits_{x \to +\infty} \ln x$.

Because the natural logarithmic function is an increasing function, if we take p as any positive number, we have

$$\text{if} \quad x > 2^p \quad \text{then} \quad \ln x > \ln 2^p \tag{14}$$

From Theorem 7.3.6,

$$\ln 2^p = p \ln 2$$

Substituting from this equation into (14) we get

$$\text{if} \quad x > 2^p \quad \text{then} \quad \ln x > p \ln 2$$

Because from (12) $\ln 2 \geq \frac{1}{2}$, we have from the above

$$\text{if} \quad x > 2^p \quad \text{then} \quad \ln x > \tfrac{1}{2} p$$

Letting $p = 2n$, where $n > 0$, we have

$$\text{if} \quad x > 2^{2n} \quad \text{then} \quad \ln x > n$$

It follows from this statement, by taking $N = 2^{2n}$, that for any $n > 0$

$$\text{if} \quad x > N \quad \text{then} \quad \ln x > n$$

So we may conclude that

$$\lim_{x \to +\infty} \ln x = +\infty \tag{15}$$

To determine the behavior of the natural logarithmic function for positive values of x near zero we investigate $\lim\limits_{x \to 0^+} \ln x$. Because $\ln x = \ln(x^{-1})^{-1}$,

$$\ln x = -\ln \frac{1}{x}$$

The expression "$x \to 0^+$" is equivalent to "$1/x \to +\infty$"; so from this equation we write

$$\lim_{x \to 0^+} \ln x = - \lim_{1/x \to +\infty} \ln \frac{1}{x} \tag{16}$$

From (15) we have

$$\lim_{1/x \to +\infty} \ln \frac{1}{x} = +\infty$$

Therefore, from this result and (16) we get

$$\lim_{x \to 0^+} \ln x = -\infty \tag{17}$$

From (17), (15), and the intermediate-value theorem (2.7.8) it follows that the range of the natural logarithmic function is the set of all real numbers. From (17) we conclude that the graph of the natural logarithmic function is asymptotic to the negative part of the y axis through the fourth quadrant. The properties of the natural logarithmic function are summarized as follows:

(i) The domain is the set of all positive numbers.

(ii) The range is the set of all real numbers.

(iii) The function is increasing on its entire domain.

(iv) The function is continuous at all numbers in its domain.

(v) The graph of the function is concave downward at all points.

(vi) The graph of the function is asymptotic to the negative part of the y axis through the fourth quadrant.

By using these properties and plotting a few points with a segment of the tangent line at the points, we can draw a sketch of the graph of the natural logarithmic function. In Figure 4 we have plotted the points having abscissas of $\frac{1}{4}$, $\frac{1}{2}$, 1, 2, 4, and 8. The slope of the tangent line is found from the formula $D_x(\ln x) = 1/x$.

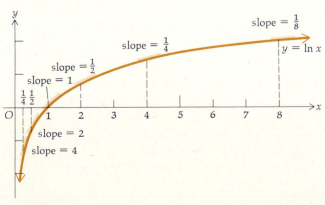

FIGURE 4

EXERCISES 7.3

In Exercises 1 through 26, differentiate the function and simplify the result.

1. $f(x) = \ln(4 + 5x)$

2. $g(x) = \ln(1 + 4x^2)$

3. $h(x) = \ln\sqrt{4 + 5x}$

4. $f(x) = \ln(8 - 2x)$

5. $f(t) = \ln(3t + 1)^2$

6. $h(x) = \ln(8 - 2x)^5$

7. $g(t) = \ln^2(3t + 1)$

8. $G(x) = \ln\sqrt{1 + 4x^2}$

9. $f(x) = \ln\sqrt[3]{4 - x^2}$

10. $g(y) = \ln(\ln y)$

11. $F(y) = \ln(\sin 5y)$

12. $f(x) = x \ln x$

13. $f(x) = \cos(\ln x)$

14. $g(x) = \ln \cos\sqrt{x}$

15. $G(x) = \ln(\sec 2x + \tan 2x)$

16. $h(y) = \csc(\ln y)$

17. $f(x) = \ln\sqrt{\tan x}$

18. $f(t) = \ln\sqrt[4]{\dfrac{t^2 - 1}{t^2 + 1}}$

19. $f(w) = \ln\sqrt{\dfrac{3w + 1}{2w - 5}}$

20. $f(x) = \ln[(5x - 3)^4(2x^2 + 7)^3]$

21. $h(x) = \dfrac{x}{\ln x}$

22. $g(x) = \ln(\cos 2x + \sin 2x)$

23. $g(x) = \ln\sqrt[3]{\dfrac{x + 1}{x^2 + 1}}$

24. $f(x) = \sqrt[3]{\ln x^3}$

25. $F(x) = \sqrt{x + 1} - \ln(1 + \sqrt{x + 1})$

26. $G(x) = x \ln(x + \sqrt{1 + x^2}) - \sqrt{1 + x^2}$

In Exercises 27 through 32, find $\dfrac{dy}{dx}$ by implicit differentiation.

27. $\ln xy + x + y = 2$

28. $\ln\dfrac{y}{x} + xy = 1$

29. $x = \ln(x + y + 1)$

30. $\ln(x + y) - \ln(x - y) = 4$

31. $x + \ln x^2 y + 3y^2 = 2x^2 - 1$

32. $x \ln y + y \ln x = xy$

33. Draw a sketch of the graph of $y = \ln x$ by plotting the points having the abscissas $\frac{1}{9}$, $\frac{1}{3}$, 1, 3, and 9, and use $\ln 3 \approx 1.1$. At each of the five points find the slope of the tangent line and draw a segment of the tangent line.

In Exercises 34 through 41, draw a sketch of the graph of the equation.

34. $x = \ln y$

35. $y = \ln(-x)$

36. $y = \ln(x + 1)$

37. $y = \ln|x|$

38. $y = \ln\dfrac{1}{x - 1}$

39. $y = x - \ln x$

40. $y = x + 2 \ln x$

41. $y = \ln \sin x$

42. Find an equation of the tangent line to the curve $y = \ln x$ at the point whose abscissa is 2.

43. Find an equation of the normal line to the curve $y = \ln x$ that is parallel to the line $x + 2y - 1 = 0$.

44. Find an equation of the normal line to the graph of $y = x \ln x$ that is perpendicular to the line having the equation $x - y + 7 = 0$.

45. Find an equation of the tangent line to the graph of $y = \ln(4x^2 - 3)^5$ at the point whose abscissa is 1.

46. A particle is moving on a straight line according to the equation of motion $s = (t + 1)^2 \ln(t + 1)$, where s is the directed distance of the particle from the starting point at t seconds. Find the velocity and acceleration when $t = 3$.

47. In a telegraph cable, the measure of the speed of the signal is proportional to $x^2 \ln(1/x)$, where x is the ratio of the measure of the radius of the core of the cable to the measure of the thickness of the cable's winding. Find the value of $\ln x$ for which the speed of the signal is greatest.

48. A manufacturer of electric generators began operations on January 1, 1980. During the first year there were no sales because the company concentrated on product development and research. After the first year the sales increased steadily according to the equation $y = x \ln x$, where x is the number of years during which the company has been operating and y is the number of millions of dollars in the sales volume. (a) Draw a sketch of the graph of the equation. Determine the rate at which the sales were increasing on (b) January 1, 1985, and (c) January 1, 1990.

49. A particular company has determined that when its weekly advertising expense is x dollars, then if S dollars is its total weekly income from sales, $S = 4000 \ln x$. (a) Determine the rate of change of sales income with respect to advertising expense when \$800 is the weekly advertising budget. (b) If the weekly advertising budget is increased to \$950, what is the approximate increase in the total weekly income from sales?

50. The linear density of a rod at a point x centimeters from one end is $2/(1 + x)$ grams per centimeter. If the rod is 15 cm long, find the mass and center of mass of the rod.

51. Prove that $x - 1 - \ln x > 0$ and $1 - \ln x - 1/x < 0$ for all $x > 0$ and $x \neq 1$, thus establishing the inequality

$$1 - \frac{1}{x} < \ln x < x - 1$$

for all $x > 0$ and $x \neq 1$. (*Hint:* Let $f(x) = x - 1 - \ln x$ and $g(x) = 1 - \ln x - 1/x$, and determine the signs of $f'(x)$ and $g'(x)$ on the intervals $(0, 1)$ and $(1, +\infty)$.)

52. Show the geometric interpretation of the inequality in Exercise 51 by drawing on the same pair of axes sketches of the graphs of the functions f, g, and h defined by the equations

$$f(x) = 1 - \frac{1}{x} \qquad g(x) = \ln x \qquad h(x) = x - 1$$

53. Use the result of Exercise 51 to prove that

$$\lim_{x \to 0} \frac{\ln(1 + x)}{x} = 1$$

54. Establish the limit of Exercise 53 by using the definition of the derivative to find $F'(0)$ for the function F for which $F(x) = \ln(1 + x)$.

55. Prove that $\lim\limits_{x \to 0^+} x \ln x = 0$. (*Hint:* First show that if $x > 0$, $x > \ln x$, and use this result to show that if $0 < x < 1$, $-2\sqrt{x} < x \ln x < 0$. Then use the squeeze theorem.)

56. Use the result of Exercise 55 to draw a sketch of the graph of $f(x) = x \ln x$.

7.4 LOGARITHMIC DIFFERENTIATION AND INTEGRALS YIELDING THE NATURAL LOGARITHMIC FUNCTION

For the discussion of both of the topics, logarithmic differentiation and integrals yielding the natural logarithmic function, we need a formula for $D_x(\ln|x|)$. To find such a formula by using Theorem 7.3.2 we substitute $\sqrt{x^2}$ for $|x|$ and use the chain rule. Thus

$$D_x(\ln|x|) = D_x(\ln\sqrt{x^2})$$

$$= \frac{1}{\sqrt{x^2}} \cdot D_x(\sqrt{x^2})$$

$$= \frac{1}{\sqrt{x^2}} \cdot \frac{x}{\sqrt{x^2}}$$

$$= \frac{x}{x^2}$$

$$= \frac{1}{x}$$

From this formula and the chain rule we obtain the following theorem.

7.4.1 THEOREM   If u is a differentiable function of x,

$$D_x(\ln|u|) = \frac{1}{u} \cdot D_x u$$

In Exercise 37 of Exercises 7.3 you were asked to draw a sketch of the graph of $y = \ln|x|$. This graph appears in Figure 1. The slope of the tangent line at each point (x, y) of the graph is $\dfrac{1}{x}$.

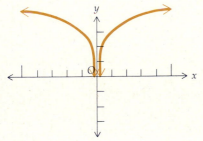

FIGURE 1

EXAMPLE 1   Find $f'(x)$ if

$$f(x) = \ln|x^4 + x^3|$$

Solution   From Theorem 7.4.1

$$f'(x) = \frac{1}{x^4 + x^3}(4x^3 + 3x^2)$$

$$= \frac{x^2(4x + 3)}{x^4 + x^3}$$

$$= \frac{4x + 3}{x^2 + x}$$

The following example illustrates how the properties of the natural logarithmic function, given in Theorems 7.3.4, 7.3.5, and 7.3.6, can simplify the work involved in differentiating complicated expressions involving products, quotients, and powers.

EXAMPLE 2   Find $\dfrac{dy}{dx}$ if

$$y = \frac{\sqrt[3]{x+1}}{(x+2)\sqrt{x+3}}$$

Solution   From the given equation,

$$|y| = \left| \frac{\sqrt[3]{x+1}}{(x+2)\sqrt{x+3}} \right|$$

$$= \frac{|\sqrt[3]{x+1}|}{|x+2||\sqrt{x+3}|}$$

Taking the natural logarithm and applying the properties of logarithms we obtain

$$\ln|y| = \tfrac{1}{3}\ln|x+1| - \ln|x+2| - \tfrac{1}{2}\ln|x+3|$$

Differentiating on both sides implicitly with respect to x and applying Theorem 7.4.1 we get

$$\frac{1}{y} \cdot \frac{dy}{dx} = \frac{1}{3(x+1)} - \frac{1}{x+2} - \frac{1}{2(x+3)}$$

Multiplying on both sides by y we have

$$\frac{dy}{dx} = y \cdot \frac{2(x+2)(x+3) - 6(x+1)(x+3) - 3(x+1)(x+2)}{6(x+1)(x+2)(x+3)}$$

Replacing y by its given value we obtain

$$\frac{dy}{dx} = \frac{(x+1)^{1/3}}{(x+2)(x+3)^{1/2}} \cdot \frac{2x^2 + 10x + 12 - 6x^2 - 24x - 18 - 3x^2 - 9x - 6}{6(x+1)(x+2)(x+3)}$$

$$= \frac{-7x^2 - 23x - 12}{6(x+1)^{2/3}(x+2)^2(x+3)^{3/2}}$$

The process illustrated in Example 2 is called **logarithmic differentiation**, which was developed in 1697 by Johann Bernoulli (1667–1748).

From Theorem 7.4.1 we obtain the following one for indefinite integration.

7.4.2 THEOREM $\displaystyle\int \frac{1}{u}\, du = \ln|u| + C$

From Theorems 7.4.2 and 5.1.8, for n any rational number,

$$\int u^n\, du = \begin{cases} \dfrac{u^{n+1}}{n+1} + C & \text{if } n \neq -1 \\[2mm] \ln|u| + C & \text{if } n = -1 \end{cases}$$

EXAMPLE 3   Evaluate

$$\int \frac{x^2\, dx}{x^3 + 1}$$

Solution

$$\int \frac{x^2\, dx}{x^3 + 1} = \frac{1}{3} \int \frac{3x^2\, dx}{x^3 + 1}$$

$$= \tfrac{1}{3}\ln|x^3 + 1| + C$$

EXAMPLE 4   Evaluate

$$\int_0^2 \frac{x^2 + 2}{x + 1}\, dx$$

Solution   Because $(x^2 + 2)/(x + 1)$ is an improper fraction, we divide the numerator by the denominator and obtain

$$\frac{x^2 + 2}{x + 1} = x - 1 + \frac{3}{x + 1}$$

Therefore

$$\int_0^2 \frac{x^2 + 2}{x + 1}\, dx = \int_0^2 \left(x - 1 + \frac{3}{x + 1} \right) dx$$

$$= \tfrac{1}{2}x^2 - x + 3\ln|x + 1| \Big]_0^2$$

$$= 2 - 2 + 3\ln 3 - 3\ln 1$$

$$= 3\ln 3 - 3\cdot 0$$

$$= 3\ln 3$$

Because $3\ln 3 = \ln 3^3$, the answer can be written as $\ln 27$.

EXAMPLE 5   Evaluate

$$\int \frac{\ln x}{x}\, dx$$

Solution   Let

$$u = \ln x \qquad du = \frac{1}{x}\, dx$$

Therefore

$$\int \frac{\ln x}{x} \, dx = \int u \, du$$

$$= \frac{u^2}{2} + C$$

$$= \tfrac{1}{2}(\ln x)^2 + C$$

We have delayed obtaining the formulas for the indefinite integrals of the tangent, cotangent, secant, and cosecant functions until now because the natural logarithmic function is needed.

A formula for the indefinite integral of the tangent function is derived as follows: Because

$$\int \tan u \, du = \int \frac{\sin u}{\cos u} \, du$$

we let

$$v = \cos u \qquad dv = -\sin u \, du$$

and obtain

$$\int \tan u \, du = -\int \frac{dv}{v}$$

$$= -\ln|v| + C$$
$$= -\ln|\cos u| + C$$
$$= \ln|(\cos u)^{-1}| + C$$
$$= \ln|\sec u| + C$$

We have proved the following theorem.

7.4.3 THEOREM $\displaystyle\int \tan u \, du = \ln|\sec u| + C$

▶ **ILLUSTRATION 1**

$$\int \tan 3x \, dx = \tfrac{1}{3} \int \tan 3x (3 \, dx)$$

$$= \tfrac{1}{3} \ln|\sec 3x| + C \qquad\qquad ◀$$

The theorem giving the indefinite integral of the cotangent function is proved in a way similar to that of Theorem 7.4.3. See Exercise 45.

7.4.4 THEOREM $\displaystyle\int \cot u \, du = \ln|\sin u| + C$

To obtain the formula for $\int \sec u \, du$ we multiply the numerator and denominator of the integrand by $\sec u + \tan u$, and we have

$$\int \sec u \, du = \int \frac{\sec u(\sec u + \tan u)}{\sec u + \tan u} \, du$$

$$= \int \frac{(\sec^2 u + \sec u \tan u)}{\sec u + \tan u} \, du$$

Let

$$v = \sec u + \tan u \qquad dv = (\sec u \tan u + \sec^2 u) \, du$$

Therefore we have

$$\int \sec u \, du = \int \frac{dv}{v}$$

$$= \ln|v| + C$$

$$= \ln|\sec u + \tan u| + C$$

We have proved the following theorem.

7.4.5 THEOREM $\int \sec u \, du = \ln|\sec u + \tan u| + C$

In Section 9.7 we obtain a formula for the integral of the secant function by a method that does not depend on the "trick" of multiplying the numerator and denominator by $\sec u + \tan u$ used to prove Theorem 7.4.5.

A formula for $\int \csc u \, du$ can be derived by multiplying the numerator and denominator of the integrand by $\csc u - \cot u$ and proceeding as we did for Theorem 7.4.5. Another procedure is to let $\int \csc u \, du = \int \sec(u - \frac{1}{2}\pi) \, du$ and use Theorem 7.4.5 and trigonometric identities. You are asked to provide these derivations in Exercise 45. The formula obtained is given in the next theorem.

7.4.6 THEOREM $\int \csc u \, du = \ln|\csc u - \cot u| + C$

▶ **ILLUSTRATION 2**

$$\int \frac{dx}{\sin 2x} = \int \csc 2x \, dx$$

$$= \frac{1}{2} \int \csc 2x(2 \, dx)$$

$$= \frac{1}{2} \ln|\csc 2x - \cot 2x| + C \qquad ◀$$

EXAMPLE 6   Evaluate

$$\int_{\pi/8}^{\pi/6} (\csc 4x - \cot 4x) \, dx$$

Solution

$$\int_{\pi/8}^{\pi/6} (\csc 4x - \cot 4x) \, dx$$

$$= \tfrac{1}{4} \int_{\pi/8}^{\pi/6} (\csc 4x - \cot 4x)(4 \, dx)$$

$$= \tfrac{1}{4} \Big[\ln|\csc 4x - \cot 4x| - \ln|\sin 4x| \Big]_{\pi/8}^{\pi/6}$$

$$= \tfrac{1}{4} \big[(\ln|\csc \tfrac{2}{3}\pi - \cot \tfrac{2}{3}\pi| - \ln|\sin \tfrac{2}{3}\pi|) - (\ln|\csc \tfrac{1}{2}\pi - \cot \tfrac{1}{2}\pi| - \ln|\sin \tfrac{1}{2}\pi|) \big]$$

$$= \frac{1}{4} \left[\left(\ln \left| \frac{2}{\sqrt{3}} - \left(-\frac{1}{\sqrt{3}} \right) \right| - \ln \left| \frac{\sqrt{3}}{2} \right| \right) - (\ln|1 - 0| - \ln|1|) \right]$$

$$= \frac{1}{4} \left[\left(\ln \frac{3}{\sqrt{3}} - \ln \frac{\sqrt{3}}{2} \right) - (0 - 0) \right]$$

$$= \frac{1}{4} \left[\ln \sqrt{3} - \ln \frac{\sqrt{3}}{2} \right]$$

$$= \frac{1}{4} \left[\ln \frac{\sqrt{3}}{\frac{\sqrt{3}}{2}} \right] \qquad \text{(by Theorem 7.3.5)}$$

$$= \tfrac{1}{4} \ln 2$$

EXERCISES 7.4

In Exercises 1 through 8, use Theorem 7.4.1 to find $\dfrac{dy}{dx}$.

1. $y = \ln|x^3 + 1|$ **2.** $y = \ln|x^2 - 1|$

3. $y = \ln|\cos 3x|$ **4.** $y = \ln|\sec 2x|$

5. $y = \ln|\tan 4x + \sec 4x|$ **6.** $y = \ln|\cot 3x - \csc 3x|$

7. $y = \ln \left| \dfrac{3x}{x^2 + 4} \right|$ **8.** $y = \sin(\ln|2x + 1|)$

In Exercises 9 through 14, find $\dfrac{dy}{dx}$ by logarithmic differentiation.

9. $y = x^2(x^2 - 1)^3(x + 1)^4$

10. $y = (5x - 4)(x^2 + 3)(3x^3 - 5)$

11. $y = \dfrac{x^2(x - 1)^2(x + 2)^3}{(x - 4)^5}$

12. $y = \dfrac{x^5(x + 2)}{x - 3}$ **13.** $y = \dfrac{x^3 + 2x}{\sqrt[5]{x^7 + 1}}$ **14.** $y = \dfrac{\sqrt{1 - x^2}}{(x + 1)^{2/3}}$

In Exercises 15 through 36, evaluate the indefinite integral.

15. $\displaystyle\int \frac{dx}{3 - 2x}$ **16.** $\displaystyle\int \frac{dx}{7x + 10}$ **17.** $\displaystyle\int \frac{3x}{x^2 + 4} \, dx$

18. $\displaystyle\int \frac{x}{2 - x^2} \, dx$ **19.** $\displaystyle\int \frac{3x^2}{5x^3 - 1} \, dx$ **20.** $\displaystyle\int \frac{2x - 1}{x(x - 1)} \, dx$

21. $\displaystyle\int \frac{\cos t}{1 + 2 \sin t} \, dt$ **22.** $\displaystyle\int \frac{\sin 3t}{\cos 3t - 1} \, dt$

23. $\displaystyle\int (\cot 5x + \csc 5x) \, dx$ **24.** $\displaystyle\int (\tan 2x - \sec 2x) \, dx$

25. $\displaystyle\int \frac{2 - 3 \sin 2x}{\cos 2x} \, dx$ **26.** $\displaystyle\int \frac{\cos 3x + 3}{\sin 3x} \, dx$

27. $\displaystyle\int \frac{2x^3}{x^2 - 4} \, dx$ **28.** $\displaystyle\int \frac{5 - 4y^2}{3 + 2y} \, dy$

29. $\displaystyle\int \frac{dx}{x \ln x}$ **30.** $\displaystyle\int \frac{dx}{\sqrt{x}(1 + \sqrt{x})}$

31. $\displaystyle\int \frac{\ln^2 3x}{x} \, dx$ **32.** $\displaystyle\int \frac{(2 + \ln^2 x)}{x(1 - \ln x)} \, dx$

33. $\displaystyle\int \frac{2 \ln x + 1}{x[(\ln x)^2 + \ln x]} \, dx$ **34.** $\displaystyle\int \frac{3x^5 - 2x^3 + 5x^2 - 2}{x^3 + 1} \, dx$

35. $\displaystyle\int \frac{\tan(\ln x)}{x} \, dx$ **36.** $\displaystyle\int \frac{\cot \sqrt{t}}{\sqrt{t}} \, dt$

In Exercises 37 through 44, evaluate the definite integral.

37. $\displaystyle\int_3^5 \frac{2x}{x^2 - 5} \, dx$ **38.** $\displaystyle\int_4^5 \frac{x}{4 - x^2} \, dx$

39. $\displaystyle\int_1^3 \frac{2t + 3}{t + 1} \, dt$ **40.** $\displaystyle\int_1^5 \frac{4z^3 - 1}{2z - 1} \, dz$

41. $\int_0^{\pi/6} (\tan 2x + \sec 2x)\,dx$ **42.** $\int_{\pi/12}^{\pi/6} (\cot 3x + \csc 3x)\,dx$

43. $\int_2^4 \dfrac{dx}{x\ln^2 x}$ **44.** $\int_2^4 \dfrac{\ln x}{x}\,dx$

45. (a) Prove Theorem 7.4.4. (b) Prove Theorem 7.4.6 by multiplying the numerator and denominator of the integrand by $\csc u - \cot u$. (c) Prove Theorem 7.4.6 by letting

$$\int \csc u\,du = \int \sec(u - \tfrac{1}{2}\pi)\,du$$ and using Theorem 7.4.5 and trigonometric identities.

46. Prove that $\int \csc u\,du = -\ln|\csc u + \cot u| + C$ in two ways: (a) Use Theorem 7.4.6; (b) multiply the numerator and denominator of the integrand by $\csc u + \cot u$.

In Exercises 47 through 55, give the exact value of the number to be found, and then obtain an approximation of this number to three decimal places by using a calculator.

47. If $f(x) = 1/x$, find the average value of f on the interval $[1, 5]$.
48. If $f(x) = (x + 2)/(x - 3)$, find the average value of f on the interval $[4, 6]$.
49. Use Boyle's law for the expansion of a gas (see Exercise 23 in Exercises 3.4) to find the average pressure with respect to the volume as the volume increases from 4 ft³ to 8 ft³ and the pressure is 2000 lb/ft² when the volume is 4 ft³.

50. Find the area of the region bounded by the curve $y = x/(2x^2 + 4)$, the x axis, the y axis, and the line $x = 4$.
51. Find the area of the region bounded by the curve $y = 2/(x - 3)$, the x axis, and the lines $x = 4$ and $x = 5$.
52. Find the volume of the solid of revolution generated when the region bounded by the curve $y = 1 - 3/x$, the x axis, and the line $x = 1$ is revolved about the x axis.
53. Find the volume of the solid of revolution generated when the region bounded by the curve $y = 1 + 2/\sqrt{x}$, the x axis, and the lines $x = 1$ and $x = 4$ is revolved about the x axis.
54. Find the length of the arc of the curve $y = \ln \sec x$ from $x = 0$ to $x = \tfrac{1}{4}\pi$.
55. Find the length of the arc of the curve $y = \ln \csc x$ from $x = \tfrac{1}{6}\pi$ to $x = \tfrac{1}{2}\pi$.
56. Prove that $\lim\limits_{x \to +\infty} \dfrac{\ln x}{x} = 0$ by two methods. (a) Let $x = \dfrac{1}{t}$ and use the result of Exercise 55 in Exercises 7.3. (b) First prove that $\int_1^x \dfrac{1}{\sqrt{t}}\,dt \geq \int_1^x \dfrac{1}{t}\,dt$ by applying Theorem 5.6.8. Then use the squeeze theorem.

7.5 THE NATURAL EXPONENTIAL FUNCTION

Because the natural logarithmic function is increasing on its entire domain, then by Theorems 7.1.5 and 7.2.1 it has an inverse that is also an increasing function. The inverse of the natural logarithmic function is called the *natural exponential function*, which we now define.

7.5.1 DEFINITION

> The **natural exponential function** is the inverse of the natural logarithmic function; thus it is defined by
>
> $$\exp(x) = y \quad \text{if and only if} \quad x = \ln y$$

The notation $\exp(x)$ is read as "the value of the natural exponential function at x."

Because the range of the natural logarithmic function is the set of all real numbers, the domain of the natural exponential function is the set of all real numbers. The range of the natural exponential function is the set of positive numbers, because this is the domain of the natural logarithmic function.

Because the natural logarithmic function and the natural exponential function are inverses of each other, it follows from Theorem 7.1.4 that

$$\ln(\exp(x)) = x \quad \text{and} \quad \exp(\ln x) = x \tag{1}$$

We now wish to define what is meant by a^x, where a is a positive number and x is an irrational number. To arrive at a reasonable definition consider the case a^r, where $a > 0$ and r is a rational number. Replacing x by a^r in the equation $x = \exp(\ln x)$ we have

$$a^r = \exp[\ln(a^r)]$$

But by Theorem 7.3.6, $\ln a^r = r \ln a$; so from the above

$$a^r = \exp(r \ln a)$$

Because the right side of this equation has a meaning if r is any real number, we use it for our definition.

7.5.2 DEFINITION   If a is any positive number and x is any real number, we define
$$a^x = \exp(x \ln a)$$

Theorem 7.3.6 states that if a is any positive number and r is any rational number, then $\ln a^r = r \ln a$. Because of Definition 7.5.2, this equation is also valid if r is any real number, and we now state this fact as a theorem.

7.5.3 THEOREM   If a is any positive number and x is any real number,
$$\ln a^x = x \ln a$$

Proof   From Definition 7.5.2,
$$a^x = \exp(x \ln a)$$
Hence, from Definition 7.5.1,
$$\ln a^x = x \ln a$$
∎

Following is the definition of one of the most important numbers in mathematics.

7.5.4 DEFINITION   The number e is defined by the formula
$$e = \exp 1$$

The letter e was chosen because of the Swiss mathematician and physicist Leonhard Euler (1707–1783).

The number e is a *transcendental* number; that is, it cannot be expressed as the root of any polynomial with integer coefficients. The number π is another example of a transcendental number. The proof that e is transcendental was first given in 1873 by Charles Hermite, and its value can be expressed to any required degree of accuracy. In Chapter 13 you will learn a method for doing this. The value of e to seven decimal places is 2.7182818. Thus

$$e \approx 2.7182818$$

The importance of the number e will become apparent as you proceed through this chapter.

7.5.5 THEOREM $\ln e = 1$

Proof   By Definition 7.5.4
$$e = \exp 1$$
Therefore
$$\ln e = \ln(\exp 1)$$
Because the natural logarithmic function and the natural exponential function are inverses, it follows that the right side of this equation is 1. Thus
$$\ln e = 1$$
∎

Observe that Theorem 7.5.5 corresponds to property (6) in Section 7.3. We have now completed showing that the function ln satisfies properties (2)–(6) of logarithms, given in Section 7.3.

7.5.6 THEOREM For all values of x,

$$\exp(x) = e^x$$

Proof By Definition 7.5.2, with $a = e$,

$$e^x = \exp(x \ln e)$$

But by Theorem 7.5.5, $\ln e = 1$, and substituting in the above equation we obtain

$$e^x = \exp(x) \qquad \blacksquare$$

From now on we write e^x in place of $\exp(x)$, and so from Definition 7.5.1,

$$e^x = y \qquad \text{if and only if} \qquad x = \ln y \tag{2}$$

With e^x in place of $\exp(x)$, (1) becomes

$$\ln e^x = x \quad \text{and} \quad e^{\ln x} = x$$

If we replace $\exp(x \ln a)$ by $e^{x \ln a}$ in the equation of Definition 7.5.2, we have

$$a^x = e^{x \ln a} \qquad \text{for every } a > 0$$

This equation can be used to compute a^x if x is any real number. For values of powers of e you can use a calculator with an $\boxed{e^x}$ key. If your calculator has no $\boxed{e^x}$ key, powers of e can be obtained by pressing the $\boxed{\text{INV}}$ key followed by the $\boxed{\ln}$ key. The reason that these two successive operations give a value for e^x is that the natural exponential function and the natural logarithmic function are inverses of each other.

EXAMPLE 1 Compute the value of $2^{\sqrt{3}}$ to two decimal places.

Solution Because $a^x = e^{x \ln a}$ if $a > 0$,

$$2^{\sqrt{3}} = e^{\sqrt{3} \ln 2}$$
$$\approx e^{1.732(0.6931)}$$
$$\approx e^{1.200}$$
$$\approx 3.32$$

Because $0 = \ln 1$, we have from statement (2)

$$e^0 = 1$$

We now state some properties of the natural exponential function as theorems.

7.5.7 THEOREM If a and b are any real numbers, then

$$e^a \cdot e^b = e^{a+b}$$

Proof   Let $A = e^a$ and $B = e^b$. Then from statement (2),

$$\ln A = a \quad \text{and} \quad \ln B = b \tag{3}$$

From Theorem 7.3.4,

$$\ln AB = \ln A + \ln B$$

Substituting from (3) into this equation we obtain

$$\ln AB = a + b$$

Thus

$$e^{\ln AB} = e^{a+b}$$

Because $e^{\ln x} = x$ the left side of the above equation is AB; so

$$AB = e^{a+b}$$

Replacing A and B by their values we get

$$e^a \cdot e^b = e^{a+b} \qquad \blacksquare$$

7.5.8 THEOREM   If a and b are any real numbers, then

$$e^a \div e^b = e^{a-b}$$

The proof is analogous to the proof of Theorem 7.5.7, where Theorem 7.3.4 is replaced by Theorem 7.3.5.

7.5.9 THEOREM   If a and b are any real numbers, then

$$(e^a)^b = e^{ab}$$

Proof   If in the equation $x = e^{\ln x}$ we let x be $(e^a)^b$, we have

$$(e^a)^b = e^{\ln(e^a)^b}$$

Applying Theorem 7.5.3 to the exponent in the right side of this equation we obtain

$$(e^a)^b = e^{b \ln e^a}$$

But $\ln e^a = a$, and therefore

$$(e^a)^b = e^{ab} \qquad \blacksquare$$

Because the natural exponential function is the inverse of the natural logarithmic function, it follows from Theorem 7.2.3 that it is differentiable. We obtain the theorem for the derivative of the natural exponential function by implicit differentiation. Let

$$y = e^x$$

Then from statement (2),

$$x = \ln y$$

On both sides of this equation we differentiate implicitly with respect to x to get

$$1 = \frac{1}{y} \cdot \frac{dy}{dx}$$

$$\frac{dy}{dx} = y$$

Replacing y by e^x we obtain

$$\frac{d}{dx}(e^x) = e^x$$

The next theorem follows from this equation and the chain rule.

7.5.10 THEOREM

If u is a differentiable function of x,

$$D_x(e^u) = e^u \, D_x u$$

Observe tnat the derivative of the function defined by $f(x) = ke^x$, where k is a constant, is itself. The only other function we have previously encountered that has this property is the constant function zero; actually, this is the special case of $f(x) = ke^x$ when $k = 0$. It can be proved that the most general function that is its own derivative is given by $f(x) = ke^x$ (see Exercise 63).

EXAMPLE 2   Find $\dfrac{dy}{dx}$ if

$$y = e^{1/x^2}$$

Solution   From Theorem 7.5.10,

$$\frac{dy}{dx} = e^{1/x^2}\left(-\frac{2}{x^3}\right)$$

$$= -\frac{2e^{1/x^2}}{x^3}$$

EXAMPLE 3   Find $\dfrac{dy}{dx}$ if

$$y = e^{2x + \ln x}$$

Solution   Because $e^{2x + \ln x} = e^{2x}e^{\ln x}$ and $e^{\ln x} = x$, then

$$y = xe^{2x}$$

Therefore

$$\frac{dy}{dx} = e^{2x} + 2xe^{2x}$$

The indefinite integration formula given in the following theorem is a consequence of Theorem 7.5.10.

7.5.11 THEOREM

$$\int e^u \, du = e^u + C$$

EXAMPLE 4 Evaluate

$$\int \frac{e^{\sqrt{x}}}{\sqrt{x}} \, dx$$

Solution Let

$$u = \sqrt{x} \qquad du = \frac{1}{2\sqrt{x}} \, dx$$

Therefore

$$\int \frac{e^{\sqrt{x}}}{\sqrt{x}} \, dx = 2 \int e^u \, du$$

$$= 2e^u + C$$

$$= 2e^{\sqrt{x}} + C$$

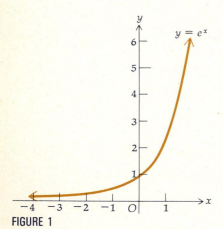

FIGURE 1

Because from (2) $e^x = y$ if and only if $x = \ln y$, the graph of $y = e^x$ is identical to the graph of $x = \ln y$. So we can obtain the graph of $y = e^x$, shown in Figure 1, by interchanging the x and y axes in Figure 4 of Section 7.3.

The graph of $y = e^x$ can be obtained without referring to the graph of the natural logarithmic function. Because the range of the exponential function is the set of all positive numbers, it follows that $e^x > 0$ for all values of x. Therefore the graph lies entirely above the x axis. Because $\frac{dy}{dx} = e^x > 0$ for all x the function is increasing for all x. Because $\frac{d^2y}{dx^2} = e^x > 0$ for all x the graph is concave upward at all points.

We have the following two limits:

$$\lim_{x \to +\infty} e^x = +\infty \quad \text{and} \quad \lim_{x \to -\infty} e^x = 0$$

The proofs of these limits are left as exercises (see Exercises 57 and 58). To plot some specific points use a calculator for powers of e.

Functions having values of the form Ce^{kx}, where C and k are constants, occur frequently in various fields. Some of these applications occur in the exercises of this section and others are discussed in Sections 7.7 and 7.8.

In Definition 7.5.4 the number e was defined as the value of the natural exponential function at 1; that is, $e = \exp 1$. To arrive at another way of defining e we consider the natural logarithmic function

$$f(x) = \ln x$$

We know that the derivative of f is given by $f'(x) = 1/x$; hence $f'(1) = 1$. However, let us apply the definition of the derivative to find $f'(1)$. We have

$$f'(1) = \lim_{\Delta x \to 0} \frac{f(1 + \Delta x) - f(1)}{\Delta x}$$

$$= \lim_{\Delta x \to 0} \frac{\ln(1 + \Delta x) - \ln 1}{\Delta x}$$

$$= \lim_{\Delta x \to 0} \frac{1}{\Delta x} \ln(1 + \Delta x)$$

Therefore

$$\lim_{\Delta x \to 0} \frac{1}{\Delta x} \ln(1 + \Delta x) = 1$$

Replacing Δx by h we have from the above equation and Theorem 7.5.3

$$\lim_{h \to 0} \ln(1 + h)^{1/h} = 1 \tag{4}$$

Now, because the natural exponential function and the natural logarithmic function are inverse functions, we have

$$\lim_{h \to 0} (1 + h)^{1/h} = \lim_{h \to 0} \exp[\ln(1 + h)^{1/h}] \tag{5}$$

Because the natural exponential function is continuous and $\lim_{h \to 0} \ln(1 + h)^{1/h}$ exists and equals 1, as shown in Equation (4), we can apply Theorem 2.7.1 to the right side of (5) and get

$$\lim_{h \to 0} (1 + h)^{1/h} = \exp\left[\lim_{h \to 0} \ln(1 + h)^{1/h}\right]$$

$$= \exp 1$$

Hence

$$\lim_{h \to 0} (1 + h)^{1/h} = e \tag{6}$$

Equation (6) is sometimes given as the definition of e; however, to use this as a definition it is necessary to prove that the limit exists.

Let us consider the function F defined by

$$F(h) = (1 + h)^{1/h}$$

and determine the function values for some values of h close to zero. These values are obtained from a calculator. When h is positive, the values appear in Table 1, and when h is negative they appear in Table 2.

The two tables lead us to suspect that $\lim_{h \to 0} (1 + h)^{1/h}$ is probably a number that lies between 2.7181 and 2.7184. As previously mentioned, in Chapter 13 we learn a method for finding the value of e to any desired number of decimal places.

Table 1

h	$F(h) = (1 + h)^{1/h}$
1	2
0.5	2.25
0.05	2.6533
0.01	2.7048
0.001	2.7169
0.0001	2.7181

Table 2

h	$F(h) = (1 + h)^{1/h}$
-0.5	4
-0.05	2.7895
-0.01	2.7320
-0.001	2.7196
-0.0001	2.7184

EXERCISES 7.5

In Exercises 1 through 16, find $\dfrac{dy}{dx}$.

1. $y = e^{5x}$ **2.** $y = e^{-7x}$ **3.** $y = e^{-3x^2}$

4. $y = e^{x^2-3}$ **5.** $y = e^{\cos x}$ **6.** $y = e^{2\sin 3x}$

7. $y = e^x \sin e^x$ **8.** $y = \dfrac{e^x}{x}$ **9.** $y = \tan e^{\sqrt{x}}$

10. $y = e^{e^x}$ **11.** $y = \dfrac{e^x - e^{-x}}{e^x + e^{-x}}$ **12.** $y = \ln \dfrac{e^{4x} - 1}{e^{4x} + 1}$

13. $y = x^5 e^{-3\ln x}$ **14.** $y = \ln(e^x + e^{-x})$

15. $y = \sec e^{2x} + e^{2\sec x}$ **16.** $y = \tan e^{3x} + e^{\tan 3x}$

In Exercises 17 through 20, find $\dfrac{dy}{dx}$ by implicit differentiation.

17. $e^x + e^y = e^{x+y}$ **18.** $e^y = \ln(x^3 + 3y)$

19. $y^2 e^{2x} + xy^3 = 1$ **20.** $ye^{2x} + xe^{2y} = 1$

In Exercises 21 through 28, evaluate the indefinite integral.

21. $\displaystyle\int e^{2-5x}\, dx$ **22.** $\displaystyle\int e^{2x+1}\, dx$ **23.** $\displaystyle\int \dfrac{1 + e^{2x}}{e^x}\, dx$

24. $\displaystyle\int e^{3x} e^{2x}\, dx$ **25.** $\displaystyle\int \dfrac{e^{3x}}{(1 - 2e^{3x})^2}\, dx$

26. $\displaystyle\int x^2 e^{2x^3}\, dx$ **27.** $\displaystyle\int \dfrac{e^{2x}}{e^x + 3}\, dx$ **28.** $\displaystyle\int \dfrac{dx}{1 + e^x}$

In Exercises 29 through 36, evaluate the definite integral.

29. $\displaystyle\int_0^1 e^2\, dx$ **30.** $\displaystyle\int_1^{e^2} \dfrac{dx}{x}$ **31.** $\displaystyle\int_e^{e^3} \dfrac{dx}{x}$

32. $\displaystyle\int_1^e \dfrac{\ln x}{x}\, dx$ **33.** $\displaystyle\int_e^{e^2} \dfrac{dx}{x(\ln x)^2}$ **34.** $\displaystyle\int_0^3 \dfrac{e^x + e^{-x}}{2}\, dx$

35. $\displaystyle\int_0^2 xe^{4-x^2}\, dx$ **36.** $\displaystyle\int_1^2 \dfrac{e^x}{e^x + e}\, dx$

In Exercises 37 through 40, compute the value to two decimal places.

37. $2^{\sqrt{2}}$ **38.** $(\sqrt{2})^{\sqrt{2}}$

39. $(\sqrt{2})^e$ **40.** e^e

41. Draw a sketch of the graph of $y = e^{-x}$.

42. Draw a sketch of the graph of $y = e^{|x|}$.

43. Find the area of the region bounded by the curve $y = e^x$, the coordinate axes, and the line $x = 2$.

44. Find the area of the region bounded by the curve $y = e^x$ and the line through the points $(0, 1)$ and $(1, e)$.

45. Find an equation of the tangent line to the curve $y = e^{-x}$ that is perpendicular to the line $2x - y = 5$.

46. Find an equation of the normal line to the curve $y = e^{2x}$ at the point where $x = \ln 2$.

47. A body is moving along a straight line and at t seconds the velocity is v feet per second, where $v = e^3 - e^{2t}$. Find the distance traveled by the particle while $v > 0$ after $t = 0$.

48. A particle is moving along a straight line, where s feet is the directed distance of the particle from the origin, v feet per second is the velocity of the particle, and a feet per second squared is the acceleration of the particle at t seconds. If $a = e^t + e^{-t}$ and $v = 1$ and $s = 2$ when $t = 0$, find v and s in terms of t.

49. If p pounds per square foot is the atmospheric pressure at a height of h feet above sea level, then $p = 2116e^{-0.0000318h}$. Find the time rate of change of the atmospheric pressure outside an airplane that is 5000 ft high and rising at the rate of 160 ft/sec.

50. At a certain height the gauge on an airplane indicates that the atmospheric pressure is 1500 lb/ft². Applying the formula of Exercise 49, approximate by differentials how much higher the airplane must rise so that the pressure will be 1480 lb/ft².

51. If l feet is the length of an iron rod when t degrees is its temperature, then $l = 60e^{0.00001t}$. Use differentials to find the approximate increase in l when t increases from 0 to 10.

52. A simple electric circuit containing no condensers, a resistance of R ohms, and an inductance of L henrys has the electromotive force cut off when the current is I_0 amperes. The current dies down so that at t seconds the current is i amperes, and $i = I_0 e^{-(R/L)t}$. Show that the rate of change of the current is proportional to the current.

53. A tank is in the shape of the solid of revolution formed by rotating about the x axis (with the positive x axis downward) the region bounded by the curve $y^2 x = e^{-2x}$ and the lines $x = 1$ and $x = 4$. If the tank is full of water, find the work done in pumping all the water to a point 1 ft above the top of the tank. Distance is measured in feet.

54. An advertising agency determined statistically that if a breakfast food manufacturer increases its budget for television commercials by x thousand dollars, there will be an increase in the total profit of $25x^2 e^{-0.2x}$ hundred dollars. What should be the advertising budget increase in order for the manufacturer to have the greatest profit? What will be the corresponding increase in the company's profit?

55. (a) By letting $h = \dfrac{1}{z}$ in Equation (6) show that

$$\lim_{z \to +\infty} \left(1 + \dfrac{1}{z}\right)^z = e \quad \text{and} \quad \lim_{z \to -\infty} \left(1 + \dfrac{1}{z}\right)^z = e$$

(b) Use a calculator to compute the values of $\left(1 + \dfrac{1}{z}\right)^z$ when $z = 10,000$ and $z = -10,000$. Then obtain an approximation of the number e by using these values to find the average value of $(1.0001)^{10,000}$ and $(0.9999)^{-10,000}$.

56. Draw a sketch of the graph of the function defined by

$$f(x) = \begin{cases} (1 + x)^{1/x} & \text{if } x \neq 0 \\ e & \text{if } x = 0 \end{cases}$$

on the interval $[-0.5, 0.5]$ and plotting the points for which x has the values $-0.5, -0.1, -0.15, -0.01, 0.01, 0.05, 0.1,$ and 0.5. Use a calculator for the function values, and take different scales on the two axes. The y coordinate of the point at which the graph intersects the y axis is the number e.

57. Prove that $\lim\limits_{x \to +\infty} e^x = +\infty$ by showing that for any $N > 0$ there exists an $M > 0$ such that if $x > M$, then $e^x > N$.

58. Prove that $\lim\limits_{x \to -\infty} e^x = 0$ by showing that for any $\epsilon > 0$ there exists an $N < 0$ such that if $x < N$, then $e^x < \epsilon$.

59. Prove that $\lim\limits_{x \to +\infty} \dfrac{x}{e^x} = 0$. (*Hint:* Use the result of Exercise 56 in Exercises 7.4.)

In Exercises 60 and 61, find the relative extrema of f, the points of inflection of the graph of f, the intervals on which f is increasing, the intervals on which f is decreasing, where the graph is concave upward, where the graph is concave downward, and the slope of any inflectional tangent. Draw a sketch of the graph of f. In Exercise 61 use the result of Exercise 59 to draw the sketch.

60. $f(x) = e^{-x^2}$

61. $f(x) = xe^{-x}$

62. Draw a sketch of the graph of the function f for which $f(x) = e^{-x} \sin x$. Show that $\lim\limits_{x \to +\infty} f(x) = 0$ but that the line $y = 0$ is not a horizontal asymptote of the graph of f. Why does Definition 2.5.4(i) not hold for this function?

63. Prove that the most general function that is equal to its derivative is given by $f(x) = ke^x$. $\left(\textit{Hint: Let } y = f(x), \text{ and solve the differential equation } \dfrac{dy}{dx} = y.\right)$

7.6 OTHER EXPONENTIAL AND LOGARITHMIC FUNCTIONS

From Definition 7.5.2, if $a > 0$,

$$a^x = e^{x \ln a} \tag{1}$$

The expression on the left side of this equation is called the *exponential function to the base a*.

7.6.1 DEFINITION
If a is any positive number and x is any real number, then the function f defined by

$$f(x) = a^x$$

is called the **exponential function to the base a**.

The exponential function to the base a satisfies the same properties as the natural exponential function.

▶ **ILLUSTRATION 1** If x and y are any real numbers and a is positive, then from (1),

$$a^x a^y = e^{x \ln a} e^{y \ln a}$$
$$= e^{x \ln a + y \ln a}$$
$$= e^{(x+y)\ln a}$$
$$= a^{x+y} \qquad ◀$$

From Illustration 1 we have the property

$$a^x a^y = a^{x+y}$$

We also have the following properties:

$$a^x \div a^y = a^{x-y}$$

$$(a^x)^y = a^{xy}$$

$$(ab)^x = a^x b^x$$

$$a^0 = 1$$

The proofs of these properties are left as exercises (see Exercises 37 to 40).

To find the derivative of the exponential function to the base a we set $a^x = e^{x \ln a}$ and apply the chain rule.

$$a^x = e^{x \ln a}$$

$$D_x(a^x) = e^{x \ln a} D_x(x \ln a)$$

$$= e^{x \ln a}(\ln a)$$

$$= a^x \ln a$$

Therefore we have the following theorem.

7.6.2 THEOREM   If a is any positive number and u is a differentiable function of x,

$$D_x(a^u) = a^u \ln a \, D_x u$$

▶ **ILLUSTRATION 2**   If $y = 3^{x^2}$, then from Theorem 7.6.2,

$$\frac{dy}{dx} = 3^{x^2}(\ln 3)(2x)$$

$$= 2(\ln 3)x3^{x^2}$$ ◀

The next theorem, giving the indefinite integration formula for the exponential function to the base a, follows from Theorem 7.6.2.

7.6.3 THEOREM   If a is any positive number other than 1,

$$\int a^u \, du = \frac{a^u}{\ln a} + C$$

EXAMPLE 1   Evaluate
$$\int \sqrt{10^{3x}} \, dx$$

Solution   Because $\sqrt{10^{3x}} = 10^{3x/2}$, we apply Theorem 7.6.3 with $u = \frac{3}{2}x$. We have, then,

$$\int \sqrt{10^{3x}} \, dx = \int 10^{3x/2} \, dx$$

$$= \frac{2}{3} \int 10^{3x/2}(\tfrac{3}{2} \, dx)$$

$$= \frac{2}{3} \cdot \frac{10^{3x/2}}{\ln 10} + C$$

$$= \frac{2\sqrt{10^{3x}}}{3 \ln 10} + C$$

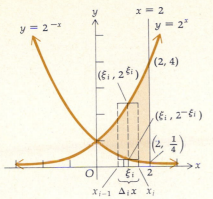

FIGURE 1

EXAMPLE 2   Draw sketches of the graphs of $y = 2^x$ and $y = 2^{-x}$ on the same set of axes. Find the area of the region bounded by these two graphs and the line $x = 2$.

Solution   The required sketches are shown in Figure 1. The region is shaded in the figure. If A square units is the desired area,

$$A = \lim_{\|\Delta\| \to 0} \sum_{i=1}^{n} \left[2^{\xi_i} - 2^{-\xi_i} \right] \Delta_i x$$

$$= \int_0^2 (2^x - 2^{-x}) \, dx$$

$$= \frac{2^x}{\ln 2} + \frac{2^{-x}}{\ln 2} \Bigg]_0^2$$

$$= \frac{4}{\ln 2} + \frac{\frac{1}{4}}{\ln 2} - \frac{1}{\ln 2} - \frac{1}{\ln 2}$$

$$= \frac{9}{4 \ln 2}$$

$$\approx 3.25$$

We can now define the *logarithmic function to the base a* if a is any positive number other than 1.

7.6.4 DEFINITION   If a is any positive number except 1, the **logarithmic function to the base a** is the inverse of the exponential function to the base a; we write

$$y = \log_a x \quad \text{if and only if} \quad a^y = x \tag{2}$$

The above is the definition of the logarithmic function to the base a usually given in elementary algebra; however, (2) has meaning for y any real number because a^y has been precisely defined. If $a = e$, we have the logarithmic function to the base e, which is the natural logarithmic function.

$\log_a x$ is read as "the logarithm of x to the base a."

The logarithmic function to the base a obeys the same laws as the natural logarithmic function. We list them:

$$\log_a(xy) = \log_a x + \log_a y$$

$$\log_a(x \div y) = \log_a x - \log_a y$$

$$\log_a 1 = 0$$

$$\log_a x^y = y \log_a x$$

The proofs of these properties are left as exercises (see Exercises 41 to 44).

A relationship between logarithms to the base a and natural logarithms follows easily. Let

$$y = \log_a x$$

Then

$$a^y = x$$

$$\ln a^y = \ln x$$

$$y \ln a = \ln x$$

$$y = \frac{\ln x}{\ln a}$$

Replacing y by $\log_a x$ we obtain

$$\log_a x = \frac{\ln x}{\ln a} \tag{3}$$

Most calculators do not have a $\boxed{\log_a x}$ key. Thus (3) is a convenient formula to apply to compute values of $\log_a x$ on a calculator.

Equation (3) sometimes is used as the definition of the logarithmic function to the base a. Because the natural logarithmic function is continuous at all $x > 0$, it follows from (3) that the logarithmic function to the base a is continuous at all $x > 0$.

If in (3) $x = e$, we have

$$\log_a e = \frac{\ln e}{\ln a}$$

$$\log_a e = \frac{1}{\ln a} \tag{4}$$

We now find the derivative of the logarithmic function to the base a by differentiating both sides of (3) with respect to x.

$$D_x(\log_a x) = \frac{1}{\ln a} D_x(\ln x)$$

$$D_x(\log_a x) = \frac{1}{\ln a} \cdot \frac{1}{x} \tag{5}$$

Substituting from (4) into this equation we get

$$D_x(\log_a x) = \frac{\log_a e}{x}$$

By applying the chain rule to this formula and (5) we have the following theorem.

7.6.5 THEOREM   If u is a differentiable function of x,

$$D_x(\log_a u) = \frac{\log_a e}{u} \cdot D_x u$$

$$\Leftrightarrow \quad D_x(\log_a u) = \frac{1}{(\ln a)u} \cdot D_x u$$

If in Theorem 7.6.5 $a = e$, then we have

$$D_x(\log_e u) = \frac{\log_e e}{u} D_x u$$

$$\Leftrightarrow \quad D_x(\ln u) = \frac{1}{u} D_x u$$

which is Theorem 7.3.2 for the derivative of the natural logarithmic function.

EXAMPLE 3   Find $\dfrac{dy}{dx}$ if

$$y = \log_{10} \frac{x + 1}{x^2 + 1}$$

Solution   Using a property of logarithms, we write

$$y = \log_{10}(x + 1) - \log_{10}(x^2 + 1)$$

From Theorem 7.6.5

$$\frac{dy}{dx} = \frac{\log_{10} e}{x + 1} - \frac{\log_{10} e}{x^2 + 1} \cdot 2x$$

$$= \log_{10} e \left(\frac{1}{x + 1} - \frac{2x}{x^2 + 1} \right)$$

$$= \frac{\log_{10} e(1 - 2x - x^2)}{(x + 1)(x^2 + 1)}$$

Because x^n has been defined for any real number n, we can now prove the theorem for the derivative of the power function if the exponent is any real number.

7.6.6 THEOREM   If n is any real number and the function f is defined by

$$f(x) = x^n \qquad \text{where } x > 0$$

then

$$f'(x) = nx^{n-1}$$

Proof   From Definition 7.5.2,

$$f(x) = e^{n \ln x}$$

Thus

$$f'(x) = e^{n \ln x} D_x(n \ln x)$$

$$= e^{n \ln x} \left(\frac{n}{x} \right)$$

$$= x^n \cdot \frac{n}{x}$$

$$= nx^{n-1}$$ ∎

Theorem 7.6.6 enables us to find the derivative of a variable to a constant power. Previously in this section we learned how to differentiate a constant to a variable power. We now consider the derivative of a function whose value is a variable to a variable power.

EXAMPLE 4   If $y = x^x$, where $x > 0$, find $\dfrac{dy}{dx}$.

Solution   From Definition 7.5.2, if $x > 0$, $x^x = e^{x \ln x}$. Therefore

$$y = e^{x \ln x}$$

$$\frac{dy}{dx} = e^{x \ln x} D_x(x \ln x)$$

$$= e^{x \ln x}\left(x \cdot \frac{1}{x} + \ln x\right)$$

$$= x^x(1 + \ln x)$$

The method of logarithmic differentiation can also be used to find the derivative of a function whose value is a variable to a variable power, as shown in the following example.

EXAMPLE 5   Find $\dfrac{dy}{dx}$ in Example 4 by using logarithmic differentiation.

Solution   We are given $y = x^x$ with $x > 0$. We take the natural logarithm on both sides of the equation and obtain

$$\ln y = \ln x^x$$

$$\ln y = x \ln x$$

Differentiating on both sides of the above equation with respect to x gives

$$\frac{1}{y} \cdot \frac{dy}{dx} = x \cdot \frac{1}{x} + \ln x$$

$$\frac{dy}{dx} = y(1 + \ln x)$$

$$= x^x(1 + \ln x)$$

EXERCISES 7.6

In Exercises 1 through 24, find the derivative of the function.

1. $f(x) = 3^{5x}$

2. $f(x) = 6^{-3x}$

3. $f(t) = 4^{3t^2}$

4. $g(x) = 10^{x^2 - 2x}$

5. $f(x) = 4^{\sin 2x}$

6. $f(z) = 2^{\csc 3z}$

7. $g(x) = 2^{5x}3^{4x^2}$

8. $f(x) = (x^3 + 3)2^{-7x}$

9. $h(x) = \dfrac{\log_{10} x}{x}$

10. $f(t) = \log_{10} \dfrac{t}{t + 1}$

11. $f(x) = \sqrt{\log_a x}$

12. $f(x) = \log_a[\log_a(\log_a x)]$

13. $f(x) = \log_{10}[\log_{10}(x + 1)]$

14. $g(w) = \tan 2^{3w}$

15. $f(t) = \sec 3^{t^2}$

16. $f(x) = x^{x^2}; x > 0$

17. $f(x) = x^{\sqrt{x}}; x > 0$

18. $f(x) = x^{\ln x}; x > 0$

19. $g(z) = z^{\cos z}; z > 0$

20. $g(t) = (\cos t)^t; \cos t > 0$

21. $h(x) = (\sin x)^{\tan x}; \sin x > 0$

22. $f(x) = (x)^{x^x}; x > 0$

23. $f(x) = x^{e^x}; x > 0$

24. $f(x) = (\ln x)^{\ln x}; x > 1$

In Exercises 25 through 32, evaluate the indefinite integral.

25. $\int 3^{2x} dx$

26. $\int a^{nx} dx$

27. $\int a^t e^t dt$

28. $\int 5^{x^4 + 2x}(2x^3 + 1) dx$

29. $\int x^2 10^{x^3} dx$

30. $\int a^{z \ln z}(\ln z + 1) dz$

31. $\int e^y 2^{e^y} 3^{e^y} dy$

32. $\int \frac{4^{\ln(1/x)}}{x} dx$

In Exercises 33 and 34, compute the value of the logarithm.

33. $\log_5 e$

34. $\log_2 10$

In Exercises 35 and 36, use differentials to find an approximate value of the logarithm, and express the answer to three decimal places.

35. $\log_{10} 997$

36. $\log_{10} 1.015$

In Exercises 37 through 40, prove the property if a and b are any positive numbers and x and y are real numbers.

37. $a^x \div a^y = a^{x-y}$

38. $(a^x)^y = a^{xy}$

39. $(ab)^x = a^x b^x$

40. $a^0 = 1$

In Exercises 41 through 44, prove the property if a is any positive number except 1, and x and y are any positive numbers.

41. $\log_a(xy) = \log_a x + \log_a y$

42. $\log_a(x \div y) = \log_a x - \log_a y$

43. $\log_a 1 = 0$

44. $\log_a x^y = y \log_a x$

45. A company has learned that when it initiates a new sales campaign, the number of sales per day increases. However, the number of extra daily sales per day decreases as the impact of the campaign wears off. For a specific campaign the company has determined that if there are $S(t)$ extra daily sales as a result of the campaign and t days have elapsed since the campaign ended, then $S(t) = 1000(3^{-t/2})$. Find the rate at which the extra daily sales are decreasing when (a) $t = 4$ and (b) $t = 10$.

46. A company estimates that in t years the number of its employees will be $N(t)$, where $N(t) = 1000(0.8)^{t/2}$. (a) How many employees does the company expect to have in 4 years? (b) At what rate is the number of employees expected to be changing in 4 years?

47. A particle is moving along a straight line according to the equation of motion $s = A \cdot 2^{kt} + B \cdot 2^{-kt}$, where A, B, and k are constants and s feet is the directed distance of the particle from the origin at t seconds. Prove that the motion is simple harmonic by showing that if a feet per second squared is the acceleration at t seconds, then a is proportional to s.

48. A particle moves along a straight line according to the equation of motion $s = t^{1/t}$, where s feet is the directed distance of the particle from the origin at t seconds. Find the velocity and acceleration at 2 sec.

49. An historically important abstract painting was purchased in 1924 for $200, and its value has doubled every 10 years since its purchase. If y dollars is the value of the painting t years after its purchase, (a) define y in terms of t. (b) What was the value of the painting in 1984? (c) Find the rate at which the value of the painting was increasing in 1984.

50. Draw sketches of the graphs of $y = \log_{10} x$ and $y = \ln x$ on the same set of axes.

51. Draw sketches of the graphs of $y = e^x$ and $y = 2^x$ on the same set of axes. Find the area of the region bounded by these two graphs and the line $x = 1$.

52. Find the area of the region bounded by the graph of $y = 3^x$ and the lines $x = 1$ and $y = 1$.

53. Find the volume of the solid generated if the region of Exercise 52 is revolved about the x axis.

54. Given $f(x) = \frac{1}{2}(a^x + a^{-x})$. Prove that

$$f(b + c) + f(b - c) = 2f(b)f(c)$$

7.7 APPLICATIONS OF THE NATURAL EXPONENTIAL FUNCTION

Mathematical models involving differential equations, having solutions containing powers of e, occur in many fields such as chemistry, physics, biology, psychology, sociology, business, and economics. We begin by discussing models involving the *laws of growth and decay*. These laws arise when the rate of change of the amount of a quantity with respect to time is proportional to the amount of the quantity present at a given instant. For example, it may be that the rate of growth of the population of a community is proportional to the actual population at any given instant. In biology, under certain circumstances, the rate of growth of a culture of bacteria is proportional to the amount of bacteria present at any specific time. In a chemical reaction it is often the case that the rate of the reaction is proportional to the quantity of the substance present; for instance, it is known from experiments that the rate of decay of radium is proportional to the amount of radium present at a given moment. An application in business occurs when interest is compounded continuously.

In such cases, if the time is represented by t units, and if y units represents the amount of the quantity present at any time, then

$$\frac{dy}{dt} = ky$$

where k is a constant and $y > 0$ for all $t \geq 0$. If y increases as t increases, then $k > 0$, and we have the **law of natural growth**. If y decreases as t increases, then $k < 0$, and we have the **law of natural decay**.

If by definition y is a positive integer (for instance, if y is the population of a certain community), we assume that y can be any positive real number in order for y to be a continuous function of t.

Suppose that we have a mathematical model involving the law of natural growth or decay and the initial condition that $y = y_0$ when $t = 0$. The differential equation is

$$\frac{dy}{dt} = ky$$

Separating the variables we obtain

$$\frac{dy}{y} = k\, dt$$

Integrating we have

$$\int \frac{dy}{y} = k \int dt$$

$$\ln|y| = kt + \bar{c}$$

$$|y| = e^{kt + \bar{c}}$$

$$|y| = e^{\bar{c}} \cdot e^{kt}$$

Letting $e^{\bar{c}} = C$ we have $|y| = Ce^{kt}$, and because y is positive we can omit the absolute-value bars, thereby giving

$$y = Ce^{kt}$$

Because $y = y_0$ when $t = 0$, we obtain $C = y_0$. Thus

$$y = y_0 e^{kt}$$

We have proved the following theorem.

7.7.1 THEOREM   Suppose that y is a continuous function of t with $y > 0$ for all $t \geq 0$. Furthermore,

$$\frac{dy}{dt} = ky$$

where k is a constant and $y = y_0$ when $t = 0$. Then

$$y = y_0 e^{kt}$$

EXAMPLE 1   In a certain culture the rate of growth of bacteria is proportional to the amount present. If 1000 bacteria are present initially and the amount doubles in 12 min, how long will it take before there will be 1,000,000 bacteria present?

Table 1

t	0	12	T
y	1000	2000	1,000,000

Solution   Let t minutes be the time from now and let y bacteria be present at t minutes. Table 1 gives the boundary conditions, where T minutes is the time it takes until there are 1,000,000 bacteria present.

The differential equation is

$$\frac{dy}{dt} = ky$$

where k is a constant and $y = 1000$ when $t = 0$. We have the law of natural growth, and from Theorem 7.7.1,

$$y = 1000e^{kt} \tag{1}$$

Because $y = 2000$ when $t = 12$ we obtain from (1)

$$e^{12k} = 2$$

Thus

$$12k = \ln 2$$

$$k = \tfrac{1}{12} \ln 2$$

$$k = 0.05776$$

Hence from (1) with this value of k we get

$$y = 1000e^{0.05776t}$$

Replacing t by T and y by 1,000,000 we have

$$1,000,000 = 1000e^{0.05776T}$$

$$e^{0.05776T} = 1000$$

$$0.5776T = \ln 1000$$

$$T = \frac{\ln 1000}{0.05776}$$

$$T = 119.6$$

Therefore there will be 1,000,000 bacteria present in 119.6 min, which is 1 hr, 59 min, 36 sec.

EXAMPLE 2   The rate of increase of the population of a certain city is proportional to the population. If the population in 1950 was 50,000 and in 1980 it was 75,000, what will be the population in 2010?

Table 2

t	0	30	60
y	50,000	75,000	y_{60}

Solution   Let t years be the time since 1950. Let y be the population in t years. We have the boundary conditions given in Table 2, where y_{60} is the population in 2010.

The differential equation is

$$\frac{dy}{dt} = ky$$

where k is a constant and $y = 50,000$ when $t = 0$. We have the law of natural growth, and from Theorem 7.7.1,

$$y = 50,000e^{kt} \tag{2}$$

Because $y = 75,000$ when $t = 30$, we get from (2)

$$75,000 = 50,000e^{30k}$$

$$e^{30k} = \tfrac{3}{2} \tag{3}$$

When $t = 60$, $y = y_{60}$. Therefore from (2),

$$y_{60} = 50,000e^{60k}$$

$$y_{60} = 50,000(e^{30k})^2$$

Substituting from (3) into this equation we obtain

$$y_{60} = 50,000(\tfrac{3}{2})^2$$

$$= 112,500$$

Therefore the population in 2010 will be 112,500.

In the above example, because the population is increasing with time we have a case of the law of natural growth. If a population decreases with time, which can occur if the death rate is greater than the birth rate, then we have a case of the law of natural decay (see Exercise 3). In the next example there is another situation involving the law of natural decay. In problems involving the law of natural decay, the **half-life** of a substance is the time for half of it to decay.

EXAMPLE 3　　The rate of decay of radium is proportional to the amount present at any time. If 60 mg of radium are present now and its half-life is 1690 years, how much radium will be present 100 years from now?

Solution　　Let t be the number of years in the time from now. Let y milligrams of radium be present at t years. We have the boundary conditions given in Table 3, where y_{100} milligrams of radium will be present 100 years from now.
　　The differential equation is

$$\frac{dy}{dt} = ky$$

Table 3

t	0	1690	100
y	60	30	y_{100}

where k is a constant and $y = 60$ when $t = 0$. We have the law of natural decay, and from Theorem 7.7.1,

$$y = 60e^{kt} \tag{4}$$

Because $y = 30$ when $t = 1690$, from (4) we get $30 = 60e^{1690k}$; that is,

$$e^{1690k} = 0.5$$

$$1690k = \ln 0.5$$

$$k = \frac{\ln 0.5}{1690}$$

$$k = -0.000410$$

Substituting this value of k in (4) we obtain

$$y = 60e^{-0.000410t}$$

When $t = 100$, $y = y_{100}$; thus from the above equation

$$y_{100} = 60e^{-0.0410}$$
$$= 57.6$$

Therefore 57.6 mg of radium will be present 100 years from now.

EXAMPLE 4   There are 100 million liters of fluoridated water in the reservoir containing a city's water supply, and the water contains 700 kg of fluoride. To decrease the fluoride content, fresh water runs into the reservoir at the rate of 3 million liters per day, and the mixture of water and fluoride, kept uniform, runs out of the reservoir at the same rate. How many kilograms of fluoride are in the reservoir 60 days after the pure water started to flow into the reservoir?

Solution   Let t days be the time elapsed since the pure water started to flow into the reservoir. Let x kilograms be the weight of the fluoride in the reservoir at t days.

Because 100 million liters of fluoridated water are in the tank at all times, at t days the weight of fluoride per million liters is $x/100$ kilograms. Three million liters of the fluoridated water run out of the reservoir each day; so the reservoir loses $3(x/100)$ kilograms of fluoride per day. Because $\dfrac{dx}{dt}$ is the rate of change of x with respect to t, and x is decreasing as t increases, we have the differential equation

$$\frac{dx}{dt} = -\frac{3x}{100}$$

This equation is of the form $\dfrac{dx}{dt} = kx$, where k is -0.03. Therefore from Theorem 7.7.1

$$x = Ce^{-0.03t}$$

Table 4

t	0	60
x	700	x_{60}

We have the initial conditions given in Table 4 where x_{60} kilograms is the weight of the fluoride in the reservoir after 60 days. When $t = 0$, $x = 700$; thus $C = 700$. Letting $t = 60$ and $x = x_{60}$ we have

$$x_{60} = 700e^{-1.8}$$
$$= 700(0.1653)$$
$$= 115.7$$

Hence there are 115.7 kg of fluoride in the reservoir 60 days after the pure water started to flow into the reservoir.

The calculus is often very useful to the economist for evaluating certain business decisions. However, to use the calculus we must deal with continuous functions. Consider, for example, the following formula, which gives A, the number of dollars in the amount after t years, if P dollars is invested at an annual rate of $100i$ percent, compounded m times per year:

$$A = P\left(1 + \frac{i}{m}\right)^{mt} \tag{5}$$

Let us conceive of a situation in which the interest is continuously compounding; that is, consider formula (5), where we let the number of interest periods per year increase without bound. Then going to the limit in formula (5) we have

$$A = P \lim_{m \to +\infty} \left(1 + \frac{i}{m}\right)^{mt}$$

which can be written as

$$A = P \lim_{m \to +\infty} \left[\left(1 + \frac{i}{m}\right)^{m/i}\right]^{it} \tag{6}$$

To compute this limit by Theorem 2.7.1 we must first determine if

$$\lim_{m \to +\infty} \left(1 + \frac{i}{m}\right)^{m/i}$$

exists. Letting $h = i/m$ we have $m/i = 1/h$; and because $m \to +\infty$ is equivalent to $h \to 0^+$,

$$\lim_{m \to +\infty} \left(1 + \frac{i}{m}\right)^{m/i} = \lim_{h \to 0^+} (1 + h)^{1/h}$$
$$= e$$

Hence, using Theorem 2.7.1 we obtain

$$\lim_{m \to +\infty} \left[\left(1 + \frac{i}{m}\right)^{m/i}\right]^{it} = \left[\lim_{m \to +\infty} \left(1 + \frac{i}{m}\right)^{m/i}\right]^{it}$$
$$= e^{it}$$

and so (6) becomes

$$A = Pe^{it} \tag{7}$$

By letting t vary through the set of nonnegative real numbers we see that (7) expresses A as a continuous function of t.

Another way of looking at the same situation is to consider an investment of P dollars, which increases at a rate proportional to its size. This is the law of natural growth. Then if A dollars is the amount at t years, we have

$$\frac{dA}{dt} = kA$$

where k is a constant and $A = P$ when $t = 0$. From Theorem 7.7.1

$$A = Pe^{kt}$$

Comparing this equation with (7) we see that they are the same if $k = i$. So if an investment increases at a rate proportional to its size, we say that the interest is **compounded continuously**, and the annual interest rate is the constant of proportionality.

▶ **ILLUSTRATION 1** If P dollars is invested at a rate of 8% per year compounded continuously, and A dollars is the amount of the investment at t years,

$$\frac{dA}{dt} = 0.08A$$

and

$$A = Pe^{0.08t}$$ ◀

If in (7) we take $P = 1$, $i = 1$, and $t = 1$, we get $A = e$, which gives a justification for the economist's interpretation of the number e as the yield on an investment of \$1 for a year at an interest rate of 100% compounded continuously.

In the following example we use the terminology *effective annual rate* of interest, which is the rate that gives the same amount of interest compounded once a year.

EXAMPLE 5   If \$5000 is borrowed at an interest rate of 12% per year, compounded continuously, and the loan is to be repaid in one payment at the end of a year, how much must the borrower repay? Also, find the effective annual rate of interest.

Solution   Let A dollars be the amount to be repaid, and because $P = 5000$, $i = 0.12$, and $t = 1$, we have from (7)

$$A = 5000e^{0.12}$$
$$= 5000(1.1275)$$
$$= 5637.50$$

Hence the borrower must repay \$5637.50. Letting j be the effective annual rate of interest we have

$$5000(1 + j) = 5000e^{0.12}$$
$$1 + j = e^{0.12}$$
$$j = 1.1275 - 1$$
$$j = 0.1275$$
$$j = 12.75\%$$

The effective annual interest rate is 12.75%.

If the function of Theorem 7.7.1 is denoted by f, then $y = f(t)$. Furthermore, if $f(0) = B$ $(B > 0)$, then from the theorem, when

$$f'(t) = kt \qquad t \geq 0 \tag{8}$$

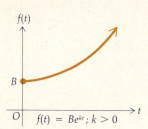

FIGURE 1

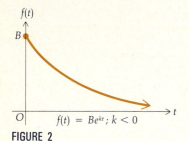

FIGURE 2

we have

$$f(t) = Be^{kt} \qquad t \geq 0 \tag{9}$$

If $k > 0$, then (8) is the law of natural growth, and (9) defines a function that is said to have **exponential growth**. Because when $k > 0$,

$$\lim_{t \to +\infty} f(t) = B \lim_{t \to +\infty} e^{kt}$$
$$= +\infty$$

then $f(t)$ increases without bound as t increases without bound. A sketch of the graph of (9) when $k > 0$ appears in Figure 1.

If $k < 0$, then (8) is the law of natural decay and (9) defines a function that is said to have **exponential decay**. Observe from (9) that if $k < 0$,

$$\lim_{t \to +\infty} f(t) = B \lim_{t \to +\infty} e^{kt}$$
$$= 0$$

and $f(t)$ is approaching 0 through positive values. A sketch of the graph of (9) when $k < 0$ is shown in Figure 2.

Suppose now that a quantity increases at a rate proportional to the difference between a fixed positive number A and its size. Then if time is represented by t units and y units is the amount of the quantity present at any time,

$$\frac{dy}{dt} = k(A - y) \tag{10}$$

where k is a positive constant, and $y < A$ for all $t \geq 0$. Separating the variables in (10) we obtain

$$\frac{dy}{A - y} = kt$$

Integrating we have

$$\int \frac{dy}{A - y} = k \int dt$$
$$-\ln|A - y| = kt + C$$
$$\ln|A - y| = -kt - C$$
$$|A - y| = e^{-C}e^{-kt}$$

Let $e^{-C} = B$. Because $A - y$ is positive, we can omit the absolute-value bars. Thus we have

$$A - y = Be^{-kt}$$
$$y = A - Be^{-kt} \tag{11}$$

If in this equation we let $y = f(t)$, it becomes

$$f(t) = A - Be^{-kt}$$

where A, B, and k are positive constants. This equation describes **bounded growth**.

$$\lim_{t \to +\infty} f(t) = \lim_{t \to +\infty} (A - Be^{-kt})$$

$$= A - B \lim_{t \to +\infty} e^{-kt}$$

$$= A - B \cdot 0$$

$$= A$$

and $f(t)$ is approaching A through values less than A. Therefore, from Definition 2.5.4(i) the graph of f has as a horizontal asymptote the line A units above the t axis. Also note that

$$f(0) = A - Be^{-k \cdot 0}$$

$$= A - B$$

From this information we obtain the sketch of the graph of f in Figure 3. This graph is sometimes called a **learning curve**. The name appears appropriate when $f(t)$ represents the competence at which a person performs a job. As a person's experience increases, the competence increases rapidly at first and then slows down as additional experience has little effect on the skill at which the task is performed.

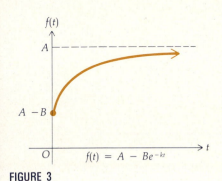

FIGURE 3

EXAMPLE 6   A new employee is performing a job more efficiently each day in such a way that if y units are produced per day after t days on the job,

$$\frac{dy}{dt} = k(80 - y)$$

where k is a positive constant and $y < 80$ for all $t \geq 0$. The employee produces 20 units the first day of work and 50 units per day after being on the job 10 days. (a) How many units per day does the employee produce after being on the job 30 days? (b) Show that after being on the job 60 days the employee is producing just 1 unit less than full potential.

Solution   The given differential equation is the same as (10) with $A = 80$. Therefore the general solution is of the form of (11) with $A = 80$; thus it is

$$y = 80 - Be^{-kt} \tag{12}$$

Table 5 shows the boundary conditions, where y_{30} units are produced per day after being on the job 30 days and y_{60} units are produced per day after being on the job 60 days. Because $y = 20$ when $t = 0$, we obtain from (12)

$$20 = 80 - Be^0$$

$$20 = 80 - B$$

$$B = 60$$

Replacing B by 60 in (12) we get

$$y = 80 - 60e^{-kt} \tag{13}$$

Table 5

t	0	10	30	60
y	20	50	y_{30}	y_{60}

From (13), with $y = 50$ when $t = 10$, we have

$$50 = 80 - 60e^{-10k}$$

$$e^{-10k} = 0.5$$

Substituting this value of e^{-10k} into (13) we have

$$y = 80 - 60(e^{-10k})^{t/10}$$
$$= 80 - 60(0.5)^{t/10} \qquad (14)$$

(a) Because $y = y_{30}$ when $t = 30$, we get from (14)

$$y_{30} = 80 - 60(0.5)^3$$
$$= 80 - 60(0.125)$$
$$= 80 - 7.5$$
$$= 72.5$$

Therefore the employee is producing 72 units per day after being on the job 30 days.

(b) From (14), with $y = y_{60}$, when $t = 60$, we obtain

$$y_{60} = 80 - 60(0.5)^6$$
$$= 80 - 60(0.0156)$$
$$= 80 - 0.938$$
$$= 79.062$$

After being on the job 60 days, the employee is producing 79 units per day. Because

$$\lim_{t \to +\infty} [80 - 60(0.5)^{t/10}] = 80 - 60 \lim_{t \to +\infty} \frac{1}{2^{t/10}}$$

$$= 80 - 60 \cdot 0$$
$$= 80$$

full potential is 80 units per day. Thus after 60 days the employee is producing just 1 unit less than full potential.

In Section 9.6 we consider another type of bounded growth, one that gives a model of population growth that takes into account environmental factors.

There is an important function in statistics called the **standardized normal probability density function** that is defined by

$$N(x) = \frac{1}{\sqrt{2\pi}} e^{-x^2/2}$$

A sketch of the graph of N appears in Figure 4. It is the bell-shaped curve well known to statisticians. The curve is symmetric with respect to the y axis. Table 6 gives values of $N(x)$ for some values of x.

Observe that as either x increases without bound or decreases without bound, then $N(x)$ rapidly approaches 0. For instance, $N(3) = 0.004$ and $N(4) = 0.0001$. The probability that a random choice of x will be in the closed interval $[a, b]$

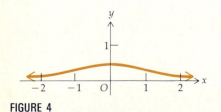

FIGURE 4

Table 6

x	$N(x)$
0	0.40
±0.5	0.35
±1.0	0.24
±1.5	0.13
±2.0	0.05
±2.5	0.02
±3.0	0.004
±3.5	0.001
±4.0	0.0001

is denoted by $P([a, b])$, and

$$P([a, b]) = \frac{1}{\sqrt{2\pi}} \int_a^b e^{-x^2/2} \, dx \tag{15}$$

The definite integral in (15) is the measure of the area of the region bounded above by the curve $y = N(x)$, below by the x axis, and on the sides by the lines $x = a$ and $x = b$. This definite integral cannot be evaluated by the second fundamental theorem of the calculus because we cannot find an antiderivative of the integrand by using elementary functions. However, we can compute an approximate value of such a definite integral by either the trapezoidal rule or Simpson's rule, as shown in the following example.

EXAMPLE 7   For the standardized normal probability density function, determine the probability that a random choice of x will be in the interval $[0, 2]$. Approximate the value of the definite integral by (a) the trapezoidal rule with $n = 4$ and (b) Simpson's rule with $n = 4$.

Solution   The probability that a random choice of x will be in the interval $[0, 2]$ is $P([0, 2])$, and from (15)

$$P([0, 2]) = \frac{1}{\sqrt{2\pi}} \int_0^2 e^{-x^2/2} \, dx \tag{16}$$

(a) We approximate the integral in (16) by the trapezoidal rule with $n = 4$. Because $[a, b] = [0, 2]$, $\Delta x = \frac{1}{2}$. Therefore, with $f(x) = e^{-x^2/2}$,

$$\begin{aligned}
\int_0^2 e^{-x^2/2} \, dx &\approx \tfrac{1}{4}[f(0) + 2f(\tfrac{1}{2}) + 2f(1) + 2f(\tfrac{3}{2}) + f(2)] \\
&= \tfrac{1}{4}[e^0 + 2e^{-1/8} + 2e^{-1/2} + 2e^{-9/8} + e^{-2}] \\
&\approx \tfrac{1}{4}[1 + 2(0.8825) + 2(0.6065) + 2(0.3246) + 0.1353] \\
&= \tfrac{1}{4}(4.7625) \\
&\approx 1.191
\end{aligned}$$

Thus

$$P([0, 2]) \approx \frac{1}{\sqrt{2\pi}} (1.191)$$

$$\approx 0.475$$

(b) If Simpson's rule with $n = 4$ is used to approximate the integral in (16), we have

$$\begin{aligned}
\int_0^2 e^{-x^2/2} \, dx &\approx \tfrac{1}{6}[f(0) + 4f(\tfrac{1}{2}) + 2f(1) + 4f(\tfrac{3}{2}) + f(2)] \\
&= \tfrac{1}{6}[e^0 + 4e^{-1/8} + 2e^{-1/2} + 4e^{-9/8} + e^{-2}] \\
&\approx \tfrac{1}{6}[1 + 4(0.8825) + 2(0.6065) + 4(0.3246) + 0.1353] \\
&= \tfrac{1}{6}(7.1767) \\
&\approx 1.196
\end{aligned}$$

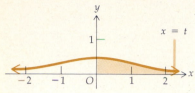

FIGURE 5

Therefore

$$P([0, 2]) \approx \frac{1}{\sqrt{2\pi}} (1.196)$$

$$\approx 0.477$$

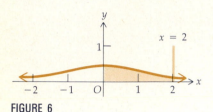

FIGURE 6

In Figure 5 the shaded region is in the first quadrant and is bounded by the graph of the standardized normal probability density function, the coordinate axes, and the line $x = t$, where $t > 0$. If $A(t)$ square units is the area of this region, it can be shown, although it is difficult to do so, that $\lim_{t \to +\infty} A(t) = 0.5$. Thus the exact value of $P([0, 2])$ is less than 0.5, because it is the measure of the area of the shaded region in Figure 6. This fact agrees with the results in Example 7.

EXERCISES 7.7

1. The rate of natural increase of the population of a certain city is proportional to the population. If the population increases from 40,000 to 60,000 in 40 years, when will the population be 80,000?

2. The population of a particular city doubled in 60 years from 1890 to 1950. If the rate of natural increase of the population at any time is proportional to the population at the time, and the population in 1950 was 60,000, estimate the population in the year 2000.

3. The population of a town is decreasing at a rate proportional to its size. In 1975 the population was 50,000 and in 1985 it was 44,000. What will be the population in 1995?

4. After the pre-opening and opening day publicity of a certain exploitation film stopped, the attendance decreased at a rate proportional to its size. If the opening day's attendance at a specific theater was 5000 and the attendance on the third day was 2000, what is the expected attendance on the sixth day?

5. After an automobile is 1 year old, its rate of depreciation at any time is proportional to its value at that time. If an automobile was purchased on June 1, 1988, and its values on June 1, 1989, and June 1, 1990, were, respectively, $7000 and $5800, what is its expected value on June 1, 1994?

6. Suppose that the value of a certain antique collection increases with age and its rate of appreciation at any time is proportional to its value at that time. If the value of the collection was $25,000 10 years ago and its present value is $35,000, in how many years is its value expected to be $50,000?

7. Bacteria grown in a certain culture increase at a rate proportional to the number present. If 1000 bacteria are present initially and the number doubles in 30 min, how many bacteria will there be in 2 hours?

8. In a certain bacterial culture where the rate of growth of bacteria is proportional to the number present, the number triples in 1 hour. If at the end of 4 hours there were 10 million bacteria, how many bacteria were present initially?

9. If the half-life of radium is 1690 years, what percent of the amount present now will be remaining after (a) 100 years and (b) 1000 years.

10. Thirty percent of a radioactive substance disappears in 15 years. Find the half-life of the substance.

11. The winter mortality rate of a certain species of wildlife in a particular geographical region is proportional to the number of the species present at any time. There were 2400 of the species present in the region on December 21, the first day of winter; 30 days later there were 2000. How many of the species were expected to survive the winter? That is, how many will be living 90 days after December 21?

12. When a simple electric circuit, containing no capacitors but having inductance and resistance, has the electromotive force removed, the rate of decrease of the current is proportional to the current. The current is i amperes t seconds after the cutoff, and $i = 40$ when $t = 0$. If the current dies down to 15 amperes in 0.01 sec, find i in terms of t.

13. (a) Find the interest earned on an investment of $500 at the end of one year if the annual interest rate is 10% compounded continuously. (b) What is the effective annual rate of interest?

14. A loan of $1000 is repaid in one payment at the end of a year. If the interest rate is 8% compounded continuously, determine (a) the total amount repaid and (b) the effective annual rate of interest.

15. If an amount of money invested doubles itself in 10 years at interest compounded continuously, how long will it take for the original amount to triple itself?

16. If the purchasing power of a dollar is decreasing at the rate of 10% annually, compounded continuously, how long will it take for the purchasing power to be $0.50?

17. How long will it take for $500 to accumulate to $1000 if

money is invested at 8 percent per year compounded continuously?

18. In biology an equation sometimes used to describe the restricted growth of a population is the Gompertz growth equation,

$$\frac{dy}{dt} = ky \ln \frac{a}{y}$$

where a and k are positive constants. Find the general solution of this differential equation.

19. In a certain chemical reaction the rate of conversion of a substance is proportional to the amount of the substance still unreacted at that time. After 10 min one-third of the original amount of the substance has been reacted and 20 g has been reacted after 15 min. What was the original amount of the substance?

20. Sugar decomposes in water at a rate proportional to the amount still unchanged. If 50 kg of sugar were present initially and at the end of 5 hr this is reduced to 20 kg, how long will it take until 90 percent of the sugar is decomposed?

21. There are 100 liters of brine in a tank, and the brine contains 70 kg of dissolved salt. Fresh water runs into the tank at the rate of 3 liters/min, and the mixture, kept uniform by stirring, runs out at the same rate. How many kilograms of salt are there in the tank at the end of 1 hour?

22. A tank contains 200 liters of brine in which there are 3 kg of salt per liter. It is desired to dilute this solution by adding brine containing 1 kg of salt per liter, which flows into the tank at the rate of 4 liters/min and runs out at the same rate. When will the tank contain $1\frac{1}{2}$ kg of salt per liter?

23. Professor Willard Libby of the University of California at Los Angeles was awarded the Nobel prize in chemistry for discovering a method of determining the date of death of a once-living object. Professor Libby made use of the fact that the tissue of a living organism is composed of two kinds of carbons, a radioactive carbon-14 (commonly written ^{14}C) and a stable carbon-12 (^{12}C), in which the ratio of the amount of ^{14}C to the amount of ^{12}C is approximately constant. When the organism dies, the law of natural decay applies to ^{14}C. If it is determined that the amount of ^{14}C in a piece of charcoal is only 45 percent of its original amount and the half-life of ^{14}C is 5600 years, when did the tree from which the charcoal came die?

24. Refer to Exercise 23. Suppose that after finding a fossil an archaeologist determines that the amount of ^{14}C present in the fossil is 25 percent of its original amount. Using the fact

that the half-life of ^{14}C is 5600 years, what is the age of the fossil?

25. Suppose that a student has 3 hours to cram for an examination and during this time wishes to memorize a set of 60 facts. According to psychologists, the rate at which a person can memorize a set of facts is proportional to the number of facts remaining to be memorized. Thus if the student memorizes y facts in t minutes,

$$\frac{dy}{dt} = k(60 - y)$$

where k is a positive constant and $y < 60$ for all $t \geq 0$. It is assumed that initially zero facts are memorized. If the student memorizes 15 facts in the first 20 min, how many facts will the student memorize in (a) 1 hour and (b) 3 hours?

26. A new worker on an assembly line can do a particular task in such a way that if y units are completed per day after t days on the assembly line, then

$$\frac{dy}{dt} = k(90 - y)$$

where k is a positive constant and $y < 90$ for all $t \geq 0$. On the day the worker starts, 60 units are completed, and after being on the job 5 days, the worker does 75 units per day. (a) How many units per day are completed after the worker is on the job 9 days? (a) Show that the worker is producing at almost full potential after 30 days.

27. For the standardized normal probability density function determine the probability that a random choice of x will be in the interval $[0, 1]$. Approximate the value of the definite integral by (a) the trapezoidal rule with $n = 4$ and (b) Simpson's rule with $n = 4$.

28. For the standardized normal probability density function, determine the probability that a random choice of x will be in the interval $[-3, 3]$. Approximate the value of the definite integral by (a) the trapezoidal rule with $n = 6$ and (b) Simpson's rule with $n = 6$.

29. The error function, denoted by erf, is defined by

$$\text{erf}(x) = \frac{2}{\sqrt{\pi}} \int_0^x e^{-t^2} \, dt$$

Find an approximate value of erf(1) to four decimal places by using Simpson's rule with $n = 10$.

30. For the error function defined in Exercise 29, find an approximate value of erf(10) to four decimal places by using Simpson's rule with $n = 10$.

7.8 FIRST-ORDER LINEAR DIFFERENTIAL EQUATIONS (*Supplementary*)

We introduced separable differential equations in Section 5.3, and additional applications of such equations appeared in Section 7.7. We now discuss **first-order linear differential equations**, which are equations of the form

$$\frac{dy}{dx} + P(x)y = Q(x) \tag{1}$$

where P and Q are continuous and $y = F(x)$ where F is differentiable. A function F that satisfies this equation is called a **solution** of the equation. The **complete solution** involves an arbitrary constant.

To arrive at a method of finding the complete solution of (1) we begin by taking $Q(x) = 0$ to obtain the equation

$$\frac{dy}{dx} + P(x)y = 0 \tag{2}$$

We can separate the variables and get

$$\frac{dy}{y} = -P(x)\,dx$$

Integrating both sides, we have

$$\ln|y| + \bar{C} = -\int P(x)\,dx$$

If we let $\bar{C} = -\ln|C|$, we have

$$\ln|y| - \ln|C| = -\int P(x)\,dx$$

$$\ln\left|\frac{y}{C}\right| = -\int P(x)\,dx$$

$$\frac{y}{C} = e^{-\int P(x)\,dx}$$

$$ye^{\int P(x)\,dx} = C \tag{3}$$

The complete solution of (2) is given by (3). But remember our goal is to solve equation (1). Observe that if we compute the differential of the left side of (3), we have

$$d(ye^{\int P(x)\,dx}) = dye^{\int P(x)\,dx} + ye^{\int P(x)\,dx}P(x)\,dx$$

$$d(ye^{\int P(x)\,dx}) = e^{\int P(x)\,dx}(dy + P(x)y\,dx) \tag{4}$$

Now we can show that if we multiply both sides of (1) by $e^{\int P(x)\,dx}\,dx$, the left side becomes the right side of (4). We do this by first writing (1):

$$\frac{dy}{dx} + P(x)y = Q(x)$$

$$e^{\int P(x)\,dx}\,dx\left[\frac{dy}{dx} + P(x)y\right] = e^{\int P(x)\,dx}\,dx[Q(x)]$$

$$e^{\int P(x)\,dx}(dy + P(x)y\,dx) = Q(x)e^{\int P(x)\,dx}\,dx$$

Substituting from (4) into this equation, we get

$$d(ye^{\int P(x)\,dx}) = Q(x)e^{\int P(x)\,dx}\,dx$$

Integrating both sides gives

$$ye^{\int P(x)\,dx} = \int Q(x)e^{\int P(x)\,dx}\,dx$$

If G is any antiderivative of $Q(x)e^{\int P(x)\,dx}$, we have

$$ye^{\int P(x)\,dx} = G(x) + C$$

$$y = G(x)e^{-\int P(x)\,dx} + Ce^{-\int P(x)\,dx}$$

$$\Leftrightarrow \quad y = e^{-\int P(x)\,dx}\int Q(x)e^{\int P(x)\,dx}\,dx + Ce^{-\int P(x)\,dx} \tag{5}$$

We have shown that if Equation (1) has a solution it is of the form (5). To verify that (5) is the complete solution of (1), we must show that any y of the form (5) is a solution of (1). To do this, we first multiply both sides of (5) by $e^{\int P(x)\,dx}$ to obtain

$$ye^{\int P(x)\,dx} = \int Q(x)e^{\int P(x)\,dx}\,dx + C$$

Differentiating both sides of this equation with respect to x gives

$$\frac{dy}{dx}e^{\int P(x)\,dx} + ye^{\int P(x)\,dx}\,P(x) = Q(x)e^{\int P(x)\,dx}$$

We now multiply both sides by $e^{-\int P(x)\,dx}$ and get

$$\frac{dy}{dx} + P(x)y = Q(x)$$

We have proved the following theorem.

7.8.1 THEOREM

The complete solution of the differential equation

$$\frac{dy}{dx} + P(x)y = Q(x)$$

where P and Q are continuous, is given by

$$y = e^{-\int P(x)\,dx}\int Q(x)e^{\int P(x)\,dx}\,dx + Ce^{-\int P(x)\,dx}$$

To apply Theorem 7.8.1 requires memorizing the formula for y. Instead, we can solve a first-order linear differential equation by first computing $e^{\int P(x)\,dx}$, which is called an **integrating factor** of the equation. Then we proceed in the same manner as we did to obtain (5); that is, we multiply both sides of the equation by the integrating factor and then integrate both sides of the resulting equation. When computing the integrating factor, we take the arbitrary constant of integration to be zero because any antiderivative of P will give an integrating factor. We demonstrate the procedure in the following illustration.

▶ **ILLUSTRATION 1** The differential equation

$$\frac{dy}{dx} - 2xy = 3x$$

is of the form of (1) where $P(x) = -2x$ and $Q(x) = 3x$. To solve this equation we first compute $e^{\int P(x)\,dx}$.

$$e^{\int -2x\,dx} = e^{-x^2}$$

We multiply both sides of the given equation by this integrating factor to obtain

$$e^{-x^2}\left(\frac{dy}{dx}\right) + e^{-x^2}(-2xy) = e^{-x^2}(3x)$$

$$\frac{d}{dx}(e^{-x^2}y) = 3xe^{-x^2}$$

Integrating both sides gives

$$e^{-x^2}y = \int 3xe^{-x^2}\,dx$$

$$e^{-x^2}y = -\tfrac{3}{2}e^{-x^2} + C$$

We multiply both sides by e^{x^2} and get

$$y = -\tfrac{3}{2} + Ce^{x^2}$$

which is the complete solution of the given differential equation. ◀

Observe that the differential equation in Illustration 1 can also be solved by separating the variables after writing the equation in the form

$$\frac{dy}{dx} = x(3 + 2y)$$

A linear differential equation of the form (1) for which $Q(x) = 0$ can also be solved by separating the variables. In the following illustration we solve such an equation by two methods.

▶ **ILLUSTRATION 2**   (a) We solve the equation

$$\frac{dy}{dx} + 2xy = 0$$

as a first-order linear differential equation of the form (1) where $P(x) = 2x$ and $Q(x) = 0$. An integrating factor is

$$e^{\int 2x\,dx} = e^{x^2}$$

Multiplying both sides of the differential equation by e^{x^2} yields

$$e^{x^2}\frac{dy}{dx} + 2xe^{x^2}y = 0$$

$$\frac{d}{dx}(e^{x^2}y) = 0$$

Integrating both sides, we get

$$e^{x^2}y = C$$

$$y = Ce^{-x^2}$$

(b) We now solve the same equation by separating the variables.

$$\frac{dy}{dx} = -2xy$$

$$\frac{dy}{y} = -2x\, dx$$

$$\int \frac{dy}{y} = -2 \int x\, dx$$

$$\ln|y| = -x^2 + K$$

$$|y| = e^{-x^2}e^K$$

$$y = \pm e^K e^{-x^2}$$

$$y = Ce^{-x^2}$$

where $C = \pm e^K$.   ◄

EXAMPLE 1   Find the complete solution of the differential equation

$$x\frac{dy}{dx} - 2y = x^2$$

where $x \neq 0$.

Solution   To write the given differential equation in the form (1) we divide both sides of the equation by x so that the coefficient of dy/dx is 1. We have

$$\frac{dy}{dx} - \frac{2}{x}y = x \tag{6}$$

Because $P(x) = -\dfrac{2}{x}$ an integrating factor is

$$e^{\int (-2/x)\, dx} = e^{-\ln x^2}$$

$$= \frac{1}{x^2}$$

Multiplying both sides of (6) by $1/x^2$ yields

$$\frac{1}{x^2}\frac{dy}{dx} - \frac{2}{x^3}y = \frac{1}{x}$$

$$\frac{d}{dx}\left(\frac{1}{x^2}y\right) = \frac{1}{x}$$

Integrating both sides gives

$$\frac{1}{x^2}y = \int \frac{1}{x}\, dx$$

$$\frac{1}{x^2}y = \ln|x| + C$$

$$y = x^2 \ln|x| + Cx^2$$

which is the complete solution.

EXAMPLE 2   Find the particular solution of the differential equation

$$\frac{dy}{dx} + y \cot x = \cos x$$

if $y = 1$ when $x = \frac{1}{6}\pi$.

Solution   We have a first-order linear differential equation for which $P(x) = \cot x$. Therefore, an integrating factor is

$$e^{\int \cot x \, dx} = e^{\ln|\sin x|}$$
$$= |\sin x|$$

Multiplying both sides of the given differential equation by $\sin x$ (we delete the absolute value bars because $\sin x > 0$ when $x = \frac{1}{6}\pi$), we get

$$\sin x \frac{dy}{dx} + (\cos x)y = \sin x \cos x$$

$$\frac{d}{dx}(y \sin x) = \sin x \cos x$$

Integrating both sides yields

$$y \sin x = \int \sin x \cos x \, dx$$
$$y \sin x = \tfrac{1}{2}\sin^2 x + C \tag{7}$$

Because $y = 1$ when $x = \frac{1}{6}\pi$, we substitute these values for x and y in this equation and obtain

$$1 \sin \tfrac{1}{6}\pi = \tfrac{1}{2}\sin^2 \tfrac{1}{6}\pi + C$$
$$\tfrac{1}{2} = \tfrac{1}{2} \cdot \tfrac{1}{4} + C$$
$$C = \tfrac{3}{8}$$

Replacing C by $\frac{3}{8}$ in (7) we get the desired particular solution:

$$y \sin x = \tfrac{1}{2}\sin^2 x + \tfrac{3}{8}$$
$$8y \sin x = 4\sin^2 x + 3$$

EXAMPLE 3   Find the complete solution of the differential equation

$$\frac{dy}{dx} = \frac{y}{4x + y^2}$$

where $y \neq 0$.

Solution   This equation is not linear in y. However,

$$\frac{dx}{dy} = \frac{4x + y^2}{y}$$

$$\frac{dx}{dy} - \frac{4}{y}x = y \tag{8}$$

This is a first-order differential equation that is linear in x; that is, it is of the form

$$\frac{dx}{dy} + P(y)x = Q(y)$$

where $P(y) = -4/y$ and $Q(y) = y$. Therefore, an integrating factor is

$$e^{\int (-4/y)\, dy} = e^{-\ln y^4}$$

$$= \frac{1}{y^4}$$

Multiplying both sides of (8) by $1/y^4$ yields

$$\frac{1}{y^4}\frac{dx}{dy} - \frac{4}{y^5}x = \frac{1}{y^3}$$

$$\frac{d}{dy}\left(\frac{1}{y^4}x\right) = \frac{1}{y^3}$$

Integrating both sides, we obtain

$$\frac{x}{y^4} = \int \frac{1}{y^3}\, dy$$

$$\frac{x}{y^4} = -\frac{1}{2y^2} + C$$

$$x = -\tfrac{1}{2}y^2 + Cy^4$$

which is the complete solution.

An application of linear first-order differential equations in physics is afforded by *Newton's law of cooling*, which states that the rate at which a body changes temperature is proportional to the difference between its temperature and that of the surrounding medium.

EXAMPLE 4   If a body is in air of temperature $35°$ and the body cools from $120°$ to $60°$ in 40 min, use Newton's law of cooling to find the temperature of the body after 100 min.

Solution   Let t minutes be the time that has elapsed since the body started to cool. Let y degrees be the temperature of the body at t minutes. Table 1 gives the boundary conditions, where y_{100} degrees is the temperature of the body after 100 min.

From Newton's law of cooling we have

$$\frac{dy}{dt} = k(y - 35)$$

where k is a constant and $y > 35$ for all $t \geq 0$.

Writing the equation in the form

$$\frac{dy}{dt} - ky = -35k \tag{9}$$

Table 1

t	0	40	100
y	120	60	y_{100}

we observe that it is a first-order linear differential equation for which an integrating factor is

$$e^{\int -k\,dt} = e^{-kt}$$

Multiplying both sides of (9) by e^{-kt} yields

$$e^{-kt}\frac{dy}{dt} - ke^{-kt}y = -35ke^{-kt}$$

$$\frac{d}{dt}(e^{-kt}y) = -35ke^{-kt}$$

Integrating both sides, we obtain

$$e^{-kt}y = \int -35ke^{-kt}\,dt$$

$$e^{-kt}y = 35e^{-kt} + C$$

$$y = 35 + Ce^{kt}$$

When $t = 0$, $y = 120$; so $C = 85$. Hence

$$y = 35 + 85e^{kt} \tag{10}$$

When $t = 40$, $y = 60$, and we obtain

$$60 = 35 + 85e^{40k}$$

$$e^{40k} = \tfrac{5}{17}$$

Substituting this value of e^{40k} in (10), we have

$$y = 35 + 85(e^{40k})^{t/40}$$

$$y = 35 + 85(\tfrac{5}{17})^{t/40}$$

Because $y = y_{100}$ when $t = 100$, we obtain from the above equation

$$y_{100} = 35 + 85(\tfrac{5}{17})^{5/2}$$
$$= 39$$

Therefore after 100 min the temperature of the body is 39°.

Another application of linear first-order differential equations arises in connection with electricity. Suppose a simple electrical circuit, shown in Figure 1, has a voltage supply of E volts from a generator or battery. The current flows when the switch at S in the figure is closed. The German physicist Gustav Kirchhoff (1824–1887) formulated a law, called *Kirchhoff's second law*, stating that at each instant the sum of the voltage drops around an electrical loop or circuit is equal to the electromotive force at that instant. In the circuit of Figure 1 let i amperes be the current at t seconds. If L henrys is the inductance, then the voltage drop across an inductor is $L\dfrac{di}{dt}$, and if R ohms is the resistance, the voltage drop across a resistor is Ri. Therefore by Kirchhoff's second law

$$L\frac{di}{dt} + Ri = E \tag{11}$$

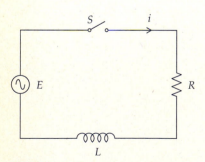

FIGURE 1

▶ **ILLUSTRATION 3** Suppose an electrical circuit has a generator supplying 150 volts and a resistor of 15 ohms and an inductor of 5 henrys connected in series. Furthermore, assume that the switch is closed when $t = 0$; that is, $i = 0$ when $t = 0$. From Equation (11) with $E = 150$, $R = 15$, and $L = 5$, we have

$$5 \frac{di}{dt} + 15i = 150$$

$$\frac{di}{dt} + 3i = 30 \tag{12}$$

An integrating factor for (12) is $e^{\int 3\,dt} = e^{3t}$. Multiplying both sides of (12) by e^{3t}, we have

$$e^{3t} \frac{di}{dt} + 3e^{3t}i = 30e^{3t}$$

$$\frac{d}{dt}(e^{3t}i) = 30e^{3t}$$

Integrating both sides, we obtain

$$e^{3t}i = \int 30e^{3t}\,dt$$

$$e^{3t}i = 10e^{3t} + C$$

$$i = 10 + Ce^{-3t}$$

Because $i = 0$ when $t = 0$, $C = -10$. Therefore

$$i = 10 - 10e^{-3t}$$

$$i = 10(1 - e^{-3t})$$

Because

$$\lim_{t \to +\infty} i = \lim_{t \to +\infty} 10(1 - e^{-3t})$$

$$= 10$$

i is always less than 10 but approaches 10 as t increases without bound. Even though the current never actually reaches 10 amperes, we state that the theoretically maximum current is 10 amperes. ◀

The electrical circuit shown in Figure 2 indicates that an electromotive force of E volts is connected in series with a resistor of R ohms and a capacitor of C farads. For this circuit, the voltage drop across a resistor is Ri and the voltage drop across a capacitor is q/C where q coulombs is the instantaneous charge on the capacitor at t seconds. By Kirchhoff's second law

$$Ri + \frac{q}{C} = E \tag{13}$$

Because the current is the rate of change of the charge with respect to time

$$i = \frac{dq}{dt}$$

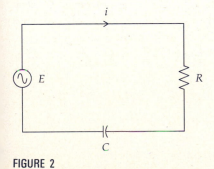

FIGURE 2

Substituting from this equation into (13), we obtain the first-order linear differential equation

$$R\frac{dq}{dt} + \frac{q}{C} = E \tag{14}$$

EXAMPLE 5   Suppose a resistor of $(1 + 0.1t)$ ohms at t seconds is connected in series with a capacitor of $\frac{1}{5}$ farad and an electromotive force of 100 volts. The initial charge on the capacitor is 6 coulombs. If at t seconds, the charge is q coulombs and the current is i amperes, (a) express q as a function of t, and (b) express i as a function of t. (c) Find the theoretically maximum charge.

Solution

(a) From (14) with $E = 100$, $C = \frac{1}{5}$, and $R = 1 + 0.1t$, we have

$$(1 + 0.1t)\frac{dq}{dt} + 5q = 100$$

$$\frac{dq}{dt} + \frac{5}{1 + 0.1t}q = \frac{100}{1 + 0.1t}$$

An integrating factor of this differential equation is

$$e^{\int 5\,dt/(1+0.1t)} = e^{50\ln(1+0.1t)}$$

$$= (1 + 0.1t)^{50}$$

Multiplying both sides of the differential equation by $(1 + 0.1t)^{50}$ yields

$$(1 + 0.1t)^{50}\frac{dq}{dt} + 5(1 + 0.1t)^{49}q = 100(1 + 0.1t)^{49}$$

$$\frac{d}{dt}[(1 + 0.1t)^{50}q] = 100(1 + 0.1t)^{49}$$

$$(1 + 0.1t)^{50}q = \int 100(1 + 0.1t)^{49}\,dt$$

$$(1 + 0.1t)^{50}q = 20(1 + 0.1t)^{50} + C$$

$$q = 20 + C(1 + 0.1t)^{-50}$$

Because the initial charge is 6, $q = 6$ when $t = 0$. Substituting these values for q and t in the above equation, we get $C = -14$. Therefore,

$$q = 20 - 14(1 + 0.1t)^{-50}$$

(b) Because $i = \dfrac{dq}{dt}$

$$i = \frac{d}{dt}[20 - 14(1 + 0.1t)^{-50}]$$

$$= 70(1 + 0.1t)^{-51}$$

(c) $\lim\limits_{t\to+\infty} q = \lim\limits_{t\to+\infty}[20 - 14(1 + 0.1t)^{-50}]$

$$= 20$$

Therefore the theoretically maximum charge is 20 coulombs.

EXERCISES 7.8

In Exercises 1 through 4, solve the differential equation by two methods: (a) use an integrating factor; (b) separate the variables.

1. $\dfrac{dy}{dx} + \dfrac{y}{x} = 0$

2. $\dfrac{dy}{dx} + y \cos x = 0$

3. $\dfrac{dy}{dx} + y \tan x = 0$

4. $\left(x + \dfrac{1}{x} \right) \dfrac{dy}{dx} + y = 0$

In Exercises 5 through 16, find the complete solution of the differential equation.

5. $\dfrac{dy}{dx} - y = e^x$

6. $x \dfrac{dy}{dx} + 4y = 5x$

7. $\dfrac{dy}{dx} + y = \dfrac{1}{1 + e^x}$

8. $x^2 \dfrac{dy}{dx} + 2xy = e^x$

9. $(3y - 2x) \dfrac{dy}{dx} = y$

10. $\dfrac{dy}{dx} = \dfrac{y}{2x + y^4}$

11. $\dfrac{dy}{dx} + 2y \cot x = \cos x$

12. $\cos x \dfrac{dy}{dx} = 2 + 2y \sin x$

13. $\dfrac{dy}{dx} = \dfrac{1 - 2xy}{1 + x^2}$

14. $\dfrac{dy}{dx} = \dfrac{1 - 2y}{2 \sec x}$

15. $\dfrac{dy}{dx} = \dfrac{y \ln y}{x + \ln y}$

16. $2y \dfrac{dy}{dx} = \dfrac{1}{e^{y^2} + x}$

In Exercises 17 through 22, find the particular solution of the differential equation which satisfies the given condition.

17. $x \dfrac{dy}{dx} - 2y = x^3 + 1; \; y = 1$ when $x = 1$

18. $(x^2 + 1) \dfrac{dy}{dx} + 2xy = x^2; \; y = 0$ when $x = -1$

19. $\dfrac{dy}{dx} - \dfrac{y}{x} = y - x; \; y = 2$ when $x = 1$

20. $\dfrac{dy}{dx} + y \tan x = \sec x; \; y = 1$ when $x = 0$

21. $\dfrac{dy}{dx} = 1 + 2y \cot x; \; y = 2$ when $x = \frac{1}{4}\pi$

22. $x \dfrac{dy}{dx} = 2y + 4x^2 \ln x; \; y = 4$ when $x = 1$

23. Find an equation of the curve through the point $(\frac{1}{6}\pi, 0)$ and for which the slope at any point (x, y) on it is $(2y + 4)/\tan x$.

24. Find an equation of the curve through the point $(1, 2)$ and for which the slope at any point (x, y) on it is $(3 - 2xy)/x^2$.

25. Under the conditions of Example 4, after how many minutes will the temperature of the body be $45°$?

26. If a body in air at a temperature of $0°$ cools from $200°$ to $100°$ in 40 min, how many more minutes will it take for the body to cool to $50°$? Use Newton's law of cooling.

27. If a thermometer is taken from a room in which the temperature is $75°$ into the open, where the temperature is $35°$ and the reading of the thermometer is $65°$ after 30 sec, (a) how long after the removal will the reading be $50°$? (b) What is the thermometer reading 3 min after the removal? Use Newton's law of cooling.

28. A pot of water was initially boiling at $100°$ and was cooling in air at a temperature of $0°$. After 20 min the temperature of the water was $90°$. (a) After how many minutes was the temperature of the water $80°$? (b) What was the temperature of the water after 1 hr? Use Newton's law of cooling.

29. An electromotive force of 60 volts is applied to a simple electrical circuit consisting of a resistor of 20 ohms and an inductor of 4 henrys. If i amperes is the current at t seconds and $i = 0$ when $t = 0$, (a) express i as a function of t. (b) Find the theoretically maximum current.

30. Do Exercise 29 if, instead of a constant inductance, an inductor of L henrys varies with time so that at t seconds, $L = 0.04 + 0.01t$ where $0 < t \leq 500$.

31. A simple electrical circuit consists of an inductor of 5 henrys in series with a resistor of 40 ohms, and at t seconds the electromotive force is e^t volts. If the current is i amperes at t seconds and $i = 0$ when $t = 0$, express i as a function of t.

32. Do Exercise 31 if the inductance is L henrys and the resistance is R ohms where L and R are positive constants. The electromotive force is still e^t volts at t seconds.

33. A simple electrical circuit has a resistor of R ohms where $R = 1 + 0.01t$ at t seconds and $0 \leq t \leq 500$. The resistor is connected in series with a capacitor of 0.1 farad and an electromotive force of 80 volts. The charge is q coulombs and the current is i amperes at t seconds. If the initial charge on the capacitor is 4 coulombs, (a) express q as a function of t, and (b) express i as a function of t. (c) Find the theoretically maximum charge.

34. A first-order differential equation of the form

$$\frac{dy}{dx} + P(x)y = Q(x)y^n$$

where n is a constant, is called a *Bernoulli differential equation*, named for James Bernoulli (1654–1705). Show that a Bernoulli equation can be made linear in u by the substitution $u = y^{1-n}$.

In Exercises 35 through 38, find the complete solution of the Bernoulli differential equation by the result of Exercise 34.

35. $\dfrac{dy}{dx} + \dfrac{2}{x} y = \dfrac{y^4}{x^4}$

36. $x^2 \dfrac{dy}{dx} - y^2 = 2xy$

37. $e^{x^2} \dfrac{dy}{dx} + 2xe^{x^2}y = xy^3$

38. $\cos x \dfrac{dy}{dx} + y^2 = y \sin x; \; -\frac{1}{2}\pi < x < \frac{1}{2}\pi$

REVIEW EXERCISES FOR CHAPTER 7

In Exercises 1 through 6, determine if the function has an inverse. If the inverse exists, do the following: (a) Define it and state its domain and range; (b) draw sketches of the graphs of the function and its inverse on the same set of axes. If the function does not have an inverse, show that a horizontal line intersects the graph of the function in more than one point.

1. $f(x) = x^3 - 4$ **2.** $f(x) = 2\sqrt[3]{x} - 1$ **3.** $f(x) = 9 - x^2$

4. $f(x) = \sqrt{4 - x^2}$ **5.** $f(x) = \dfrac{3x - 4}{x}$ **6.** $f(x) = |2x - 3|$

In Exercises 7 and 8, (a) prove that the function f has an inverse, (b) find $f^{-1}(x)$, and (c) verify the equations of Theorem 7.1.4 for f and f^{-1}.

7. $f(x) = \sqrt[3]{x + 1}$ **8.** $f(x) = \dfrac{2x - 1}{2x + 1}$

In Exercises 9 through 12, find $(f^{-1})'(d)$.

9. $f(x) = x^2 - 6x + 8$, $x \geq 3$; $d = 3$
10. $f(x) = \sqrt{3x + 4}$; $d = 5$
11. $f(x) = 8x^3 + 6x$; $d = 4$
12. $f(x) = x^5 + x - 22$; $d = 12$

In Exercises 13 through 24, differentiate the function.

13. $f(x) = (\ln x^2)^2$

14. $f(t) = 3 \sin(e^{4t})$

15. $g(x) = 2^{\cos 4x}$

16. $f(x) = \dfrac{e^x}{(e^x + e^{-x})^2}$

17. $f(x) = e^{x/(4 + x^2)}$

18. $f(x) = 10^{-5x^2}$

19. $f(x) = \sqrt{\log_{10} \dfrac{1 + x}{1 - x}}$

20. $g(x) = \ln \sqrt{\dfrac{2x + 1}{x - 3}}$

21. $f(t) = t^{3/\ln t}$

22. $F(x) = \log_{10}(x^2 2^{x^2 + 1})$

23. $f(x) = (x)^{xe^x}$; $x > 0$

24. $f(x) = 3^{x^{x^2}}$

In Exercises 25 through 32, evaluate the indefinite integral.

25. $\displaystyle\int \dfrac{3e^{2x}}{1 + e^{2x}}\, dx$

26. $\displaystyle\int e^{2x^2 - 4x}(x - 1)\, dx$

27. $\displaystyle\int (e^{3x} + 2^{3x})\, dx$

28. $\displaystyle\int \dfrac{10^{\ln x^2}}{x}\, dx$

29. $\displaystyle\int \dfrac{xe^{6x^2}}{\sqrt{1 + e^{6x^2}}}\, dx$

30. $\displaystyle\int (x + 1)e^x 7^{xe^x}\, dx$

31. $\displaystyle\int \dfrac{2^x\, dx}{\sqrt{3 \cdot 2^x + 4}}$

32. $\displaystyle\int \dfrac{10^x + 1}{10^x - 1}\, dx$

In Exercises 33 through 38, evaluate the definite integral.

33. $\displaystyle\int_0^2 x^2 e^{x^3}\, dx$

34. $\displaystyle\int_0^1 (e^{2x} + 1)^2\, dx$

35. $\displaystyle\int_1^8 \dfrac{x^{1/3}}{x^{4/3} + 4}\, dx$

36. $\displaystyle\int_{1/3}^{1/2} \dfrac{4x^{-3} + 2}{x^{-2} - x}\, dx$

37. $\displaystyle\int_0^{\ln 2} \dfrac{e^{2x}}{e^x - 5}\, dx$

38. $\displaystyle\int_e^{e^2} \dfrac{dx}{x(\ln x)}$

39. Find $\dfrac{dy}{dx}$ if $ye^x + xe^y + x + y = 0$.

40. If $f(x) = \log_{(e^x)}(x + 1)$, find $f'(x)$.

41. Use differentials to find an approximate value to three decimal places of $\log_{10} 100937$. Use the fact that $\log_{10} e = 0.434$, with accuracy to three decimal places.

42. Find an equation of the tangent line to the curve $y = x^{x-1}$ at $(2, 2)$.

43. A particle is moving on a straight line, where s feet is the directed distance of the particle from the origin, v feet per second is the velocity of the particle, and a feet per second squared is the acceleration of the particle at t seconds. If $a = e^t + e^{-t}$ and $v = 1$ and $s = 2$ when $t = 0$, find v and s in terms of t.

44. The area of the region bounded by the curve $y = e^{-x}$, the coordinate axes, and the line $x = b(b > 0)$ is a function of b. If f is this function, find $f(b)$. Also find $\lim\limits_{b \to +\infty} f(b)$.

45. The volume of the solid of revolution obtained by revolving the region in Exercise 44 about the x axis is a function of b. If g is this function, find $g(b)$. Also find $\lim\limits_{b \to +\infty} g(b)$.

46. The rate of natural increase of the population of a certain city is proportional to the population. If the population doubles in 60 years and if the population in 1950 was 60,000, estimate the population in the year 2000.

47. The rate of decay of a radioactive substance is proportional to the amount present. If half of a given deposit of the substance disappears in 1900 years, how long will it take for 95% of the deposit to disappear?

48. Prove that if a rectangle is to have its base on the x axis and two of its vertices on the curve $y = e^{-x^2}$, then the rectangle will have the largest possible area if the two vertices are at the points of inflection of the graph.

49. Given $f(x) = \ln|x|$ and $x < 0$. Show that f has an inverse function. If g is the inverse function, find $g(x)$ and the domain of g.

50. Prove that if $x < 1$, $\ln x < x$. (*Hint:* Let $f(x) = x - \ln x$, and show that f is decreasing on $(0, 1)$ and find $f(1)$.)

51. When a gas undergoes an adiabatic (no gain or loss of heat) expansion or compression, then the rate of change of the pressure with respect to the volume varies directly as the pressure and inversely as the volume. If the pressure is p pounds per square inch when the volume is v cubic inches, and the initial pressure and volume are p_0 pounds per square inch and v_0 cubic inches, show that $pv^k = p_0 v_0{}^k$.

52. If W inch-pounds is the work done by a gas expanding against a piston in a cylinder and P pounds per square inch is the pressure of the gas when the volume of the gas is V cubic inches, show that if V_1 cubic inches and V_2 cubic inches are the initial and final volumes, respectively, then

$$W = \int_{V_1}^{V_2} P \, dV$$

53. Suppose that a piston compresses a gas in a cylinder from an initial volume of 60 in.^{3} to a volume of 40 in.^{3}. If Boyle's law (Exercise 23 in Exercises 3.4) holds, and the initial pressure is 50 lb/in.^{2}, find the work done by the piston. (Use the result of Exercise 52.)

54. The charge of electricity on a spherical surface leaks off at a rate proportional to the charge. Initially, the charge of electricity was 8 coulombs and one-fourth leaks off in 15 min. When will there be only 2 coulombs remaining?

55. How long will it take for an investment to double itself if interest is paid at the rate of 8% per year compounded continuously?

56. Interest on a savings account is computed at 10% per year compounded continuously. If one wishes to have $1000 in the account at the end of a year by making a single deposit now, what should be the amount of the deposit?

57. The rate of bacterial growth in a certain culture is proportional to the number present, and the number doubles in 20 min. If at the end of 1 hour there were 1,500,000 bacteria, how many bacteria were present initially?

58. Refer to Exercise 23 in Exercises 7.7. A paleontologist discovered an insect preserved inside a transparent amber, which is hardened tree pitch, and the amount of ^{14}C present in the insect was determined to be 2 percent of its original amount. Use the fact that the half-life of ^{14}C is 5600 years to determine the age of the insect at the time of discovery.

59. A student studying a foreign language has 50 verbs to memorize. The rate at which the student can memorize these verbs is proportional to the number of verbs remaining to be memorized; that is, if the student memorizes y verbs in t minutes,

$$\frac{dy}{dt} = k(50 - y)$$

Assume that initially no verbs are memorized, and suppose that 20 verbs are memorized in the first 30 minutes. How many verbs will the student memorize in (a) 1 hour and (b) 2 hours? (c) After how many hours will the student have only one verb left to memorize?

60. A tank contains 100 liters of fresh water, and brine containing 2 kg of salt per liter flows into the tank at the rate of 3 liters/min. If the mixture, kept uniform by stirring, flows out at the same rate, how many kilograms of salt are there in the tank at the end of 30 min?

61. A tank contains 60 gal of salt water with 120 lb of dissolved salt. Salt water with 3 lb of salt per gallon flows into the tank at the rate of 2 gal/min, and the mixture, kept uniform by stirring, flows out at the same rate. How long will it be before there are 135 lb of salt in the tank?

62. If the population of a particular country doubles every 25

years, at what percent is it growing per year?

63. Find the point on the graph of $f(x) = e^x$ for which the tangent line to the graph there passes through the origin.

64. Find the volume of the solid of revolution generated if the region bounded by the curve $y = 2^{-x}$ and the lines $x = 1$ and $x = 4$ is revolved about the x axis.

65. Prove by using the definition of a derivative that

$$\lim_{x \to 0} \frac{\log_a(1 + x)}{x} = \log_a e$$

(*Note:* Compare with Exercise 53 in Exercises 7.3.)

66. Prove without using the definition of a derivative that

$$\lim_{x \to 0} \frac{a^x - 1}{x} = \ln a$$

(*Hint:* Let $y = a^x - 1$ and express $(a^x - 1)/x$ as a function of y, say $g(y)$. Then show that $y \to 0$ as $x \to 0$, and find $\lim_{y \to 0} g(y)$.)

67. Use the results of Exercises 65 and 66 to prove that

$$\lim_{x \to 1} \frac{x^b - 1}{x - 1} = b$$

(*Hint:* Write

$$\frac{x^b - 1}{x - 1} = \frac{e^{b \ln x} - 1}{b \ln x} \cdot \frac{b \ln x}{x - 1}$$

Then let $s = b \ln x$ and $t = x - 1$.)

68. Prove by using the definition of a derivative that

$$\lim_{x \to 0} \frac{e^{ax} - 1}{x} = a$$

69. If the domain of f is the set of real numbers and $f'(x) = cf(x)$ for all x where c is a constant, prove that there is a constant k for which $f(x) = ke^{cx}$ for all x. (*Hint:* Consider the function g for which $g(x) = f(x)e^{-cx}$, and find $g'(x)$.)

70. Prove that

$$D_x{}^n(\ln x) = (-1)^{n-1} \frac{(n-1)!}{x^n}$$

(*Hint:* Use mathematical induction.)

71. Find $\int_0^t e^{-|x|} \, dx$ if t is any real number.

72. Prove that if $x > 0$, and $\int_1^x t^{h-1} \, dt = 1$, then

$$\lim_{h \to 0} x = \lim_{h \to 0} (1 + h)^{1/h}$$

73. Do Exercise 65 in the Review Exercises for Chapter 5 by evaluating each integral.

74. An important function in statistics is the probability density function defined by

$$f(x) = \frac{1}{\sigma\sqrt{2\pi}} \exp\left[-\frac{(x-\mu)^2}{2\sigma^2}\right]$$

where σ and μ are constants such that $\sigma > 0$ and μ is any real number. Find (a) the relative extrema of f; (b) the points of inflection of the graph of f; (c) $\lim\limits_{x \to +\infty} f(x)$; (d) $\lim\limits_{x \to -\infty} f(x)$. (e) Draw a sketch of the graph of f.

Exercises 75 through 81 pertain to Supplementary Section 7.8. In Exercises 75 and 76, find the complete solution of the differential equation.

75. $x^2 \dfrac{dy}{dx} + 2xy = 3x^2 + 1$ **76.** $\dfrac{dx}{dy} + 2x \cot y = \sec^2 y$

In Exercises 77 and 78, find the particular solution of the differential equation which satisfies the given condition.

77. $\cos x \dfrac{dy}{dx} = 1 - y - \sin x$; $y = 2$ when $x = 0$

78. $x \dfrac{dy}{dx} + y = e^x$; $y = 1$ when $x = \ln 2$

79. An electromotive force of 100 volts is applied to a simple electrical circuit consisting of a resistor of 10 ohms and an inductor of L henrys where L varies with time so that at t seconds $L = 0.03 + 0.01t$ where $0 \le t \le 1000$. If i amperes is the current at t seconds and $i = 0$ when $t = 0$, (a) express i as a function of t. (b) Find the theoretically maximum current.

80. Find an equation of the curve through the point $(0, -1)$ and for which the slope at any point (x, y) on it is $\cos x - y \tan x$.

81. Use Newton's law of cooling, to determine the current temperature of a body in air of temperature $40°$ if 30 min ago the body's temperature was $150°$ and 10 min ago it was $90°$.

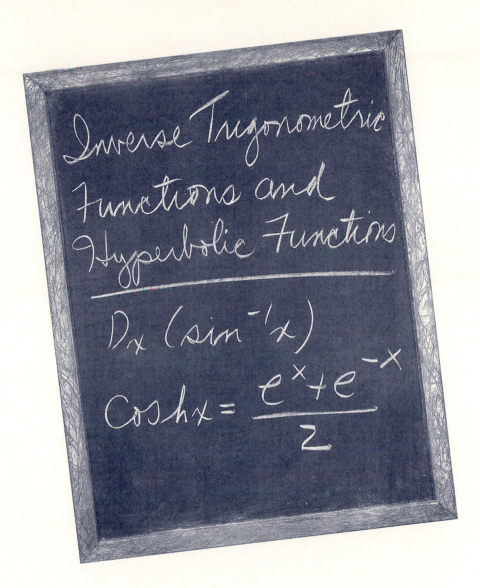

EIGHT

Inverse Trigonometric Functions and Hyperbolic Functions

$$D_x\left(\sin^{-1}x\right)$$

$$\cosh x = \frac{e^x + e^{-x}}{2}$$

More transcendental functions appear in this chapter. We define the *inverse trigonometric functions* in Section 8.1. These functions play an important part in our discussion of integration by trigonometric substitution in the next chapter. Sections 8.2 and 8.3 are devoted to derivatives and integrals involving the inverse trigonometric functions.

Hyperbolic functions involve powers of *e*, and they are introduced in Section 8.4. These functions have properties similar to those of the trigonometric functions. Supplementary Section 8.5 deals with the *inverse hyperbolic functions*, which are primarily used in connection with integration later in Supplementary Section 9.8.

8.1 THE INVERSE TRIGONOMETRIC FUNCTIONS

Recall that it is necessary for a function to be one-to-one in order for it to have an inverse. A sketch of the graph of the sine function appears in Figure 1 showing that every number in its range is the function value of more than one number in its domain. Thus the sine function is not one-to-one and therefore does not have an inverse. However, observe from Figure 1 that the sine is increasing on the interval $[-\frac{1}{2}\pi, \frac{1}{2}\pi]$. This fact follows from Theorem 4.4.3, because if $f(x) = \sin x$, then $f'(x) = \cos x > 0$ for all x in $(-\frac{1}{2}\pi, \frac{1}{2}\pi)$. Therefore,

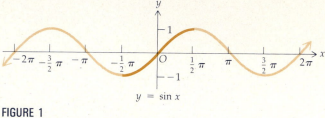

$$y = \sin x$$

FIGURE 1

even though the sine function does not have an inverse, it follows from Theorem 7.1.5 that the function F for which

$$F(x) = \sin x \quad \text{and} \quad -\tfrac{1}{2}\pi \le x \le \tfrac{1}{2}\pi \tag{1}$$

does have an inverse. The domain of F is $[-\frac{1}{2}\pi, \frac{1}{2}\pi]$, and its range is $[-1, 1]$. A sketch of the graph of F is shown in Figure 2. The inverse of the function defined by (1) is called the *inverse sine function*.

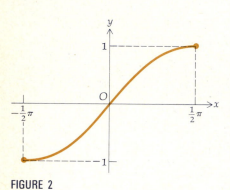

FIGURE 2

8.1.1 DEFINITION

The **inverse sine function**, denoted by $\sin^{-1}$, is defined as follows:

$$y = \sin^{-1} x \quad \text{if and only if} \quad x = \sin y \text{ and } -\tfrac{1}{2}\pi \le y \le \tfrac{1}{2}\pi$$

The domain of $\sin^{-1}$ is the closed interval $[-1, 1]$, and the range is the closed interval $[-\frac{1}{2}\pi, \frac{1}{2}\pi]$. A sketch of the graph is shown in Figure 3.

The use of the symbol -1 to represent the inverse sine function makes it necessary to denote the reciprocal of $\sin x$ by $(\sin x)^{-1}$ to avoid confusion. A similar convention is applied when using any negative exponent with a trigonometric function. For instance,

$$\frac{1}{\sin x} = (\sin x)^{-1} \qquad \frac{1}{\sin^2 x} = (\sin x)^{-2} \qquad \frac{1}{\cos^3 x} = (\cos x)^{-3}$$

and so on.

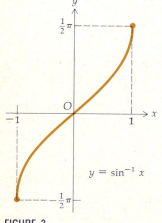

$$y = \sin^{-1} x$$

FIGURE 3

▶ **ILLUSTRATION 1**

(a) $\sin^{-1} \dfrac{1}{\sqrt{2}} = \dfrac{1}{4}\pi$ (b) $\sin^{-1}\left(-\dfrac{1}{\sqrt{2}}\right) = -\dfrac{1}{4}\pi$ ◀

In (1) the domain of F is restricted to the closed interval $[-\frac{1}{2}\pi, \frac{1}{2}\pi]$ so that the function is monotonic on its domain and therefore has an inverse function. However, the sine function has period 2π and is increasing on other intervals, for instance, $[-\frac{5}{2}\pi, -\frac{3}{2}\pi]$ and $[\frac{3}{2}\pi, \frac{5}{2}\pi]$. Also, the function is decreasing on certain closed intervals, in particular the intervals $[-\frac{3}{2}\pi, -\frac{1}{2}\pi]$ and $[\frac{1}{2}\pi, \frac{3}{2}\pi]$. Any one of these intervals could just as well be chosen for the domain of the function F of Equation (1). The choice of the interval $\left[-\frac{1}{2}\pi, \frac{1}{2}\pi\right]$, however, is customary because it is the largest interval containing the number 0 on which the function is monotonic.

▶ **ILLUSTRATION 2**   To determine $\sin^{-1}(0.4695)$ on a calculator, first set the calculator in the radian mode. Obtain 0.4695 in the display, and then press the $\boxed{\sin^{-1}}$ key (or the $\boxed{\text{INV}}$ key followed by the $\boxed{\sin}$ key) and read 0.4887.   ◀

It follows from Definition 8.1.1 that

$$\sin(\sin^{-1} x) = x \quad \text{for } x \text{ in } [-1, 1]$$

$$\sin^{-1}(\sin y) = y \quad \text{for } y \text{ in } \left[-\tfrac{1}{2}\pi, \tfrac{1}{2}\pi\right]$$

Observe that $\sin^{-1}(\sin y) \neq y$ if y is not in the interval $\left[-\frac{1}{2}\pi, \frac{1}{2}\pi\right]$. For example,

$$\sin^{-1}(\sin \tfrac{3}{4}\pi) = \sin^{-1}\frac{1}{\sqrt{2}} \quad \text{and} \quad \sin^{-1}(\sin \tfrac{7}{4}\pi) = \sin^{-1}\left(-\frac{1}{\sqrt{2}}\right)$$

$$= \tfrac{1}{4}\pi \qquad\qquad\qquad\qquad\qquad = -\tfrac{1}{4}\pi$$

EXAMPLE 1   Find (a) $\cos[\sin^{-1}(-\tfrac{1}{2})]$; (b) $\sin^{-1}(\cos \tfrac{2}{3}\pi)$

Solution   Because the range of the inverse sine function is $\left[-\frac{1}{2}\pi, \frac{1}{2}\pi\right]$, $\sin^{-1}(-\tfrac{1}{2}) = -\tfrac{1}{6}\pi$.

(a) $\cos[\sin^{-1}(-\tfrac{1}{2})] = \cos(-\tfrac{1}{6}\pi)$ (b) $\sin^{-1}(\cos \tfrac{2}{3}\pi) = \sin^{-1}(-\tfrac{1}{2})$

$$\qquad\qquad\qquad\quad = \frac{\sqrt{3}}{2} \qquad\qquad\qquad\qquad\qquad = -\tfrac{1}{6}\pi$$

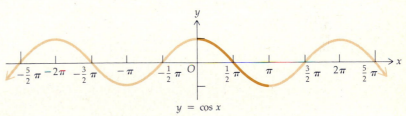

$$y = \cos x$$

FIGURE 4

A sketch of the graph of the cosine function appears in Figure 4. The cosine function does not have an inverse because it also is not one-to-one. To define the inverse cosine function we restrict the cosine to an interval on which the function is monotonic. We choose the interval $[0, \pi]$ on which the cosine is decreasing, as shown in Figure 4. So consider the function G defined by

$$G(x) = \cos x \quad \text{and} \quad 0 \le x \le \pi$$

The domain of G is the closed interval $[0, \pi]$ and the range is the closed interval $[-1, 1]$. A sketch of the graph of G is in Figure 5. Because G is continuous and decreasing on its domain, it has an inverse function, called the *inverse cosine function*.

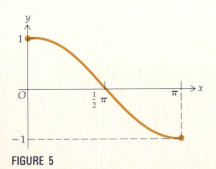

FIGURE 5

8.1.2 DEFINITION   The **inverse cosine function**, denoted by $\cos^{-1}$, is defined as follows:

$$y = \cos^{-1} x \quad \text{if and only if} \quad x = \cos y \text{ and } 0 \le y \le \pi$$

The domain of $\cos^{-1}$ is the closed interval $[-1, 1]$, and the range is the closed interval $[0, \pi]$. A sketch of its graph is shown in Figure 6.

▶ **ILLUSTRATION 3**

(a) $\cos^{-1} \dfrac{1}{\sqrt{2}} = \dfrac{1}{4}\pi$ (b) $\cos^{-1}\left(-\dfrac{1}{\sqrt{2}}\right) = \dfrac{3}{4}\pi$ ◀

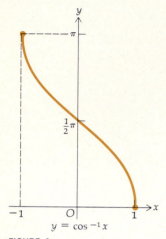

$y = \cos^{-1} x$

FIGURE 6

From Definition 8.1.2 it follows that

$$\cos(\cos^{-1} x) = x \quad \text{for } x \text{ in } [-1, 1]$$

$$\cos^{-1}(\cos y) = y \quad \text{for } y \text{ in } [0, \pi]$$

Notice there is again a restriction on y in order to have the equality $\cos^{-1}(\cos y) = y$. For example, because $\frac{3}{4}\pi$ is in $[0, \pi]$

$$\cos^{-1}(\cos \tfrac{3}{4}\pi) = \tfrac{3}{4}\pi$$

However,

$$\cos^{-1}(\cos \tfrac{5}{4}\pi) = \cos^{-1}\left(-\frac{1}{\sqrt{2}}\right) \quad \text{and} \quad \cos^{-1}(\tfrac{7}{4}\pi) = \cos^{-1}\left(\frac{1}{\sqrt{2}}\right)$$

$$= \tfrac{3}{4}\pi \qquad\qquad\qquad\qquad = \tfrac{1}{4}\pi$$

EXAMPLE 2   Find the exact value of $\sin[2 \cos^{-1}(-\tfrac{3}{5})]$.

Solution   Because we wish to obtain trigonometric functions of the number $\cos^{-1}(-\tfrac{3}{5})$, we shall let t represent this number.

$$t = \cos^{-1}(-\tfrac{3}{5})$$

Because the range of the inverse cosine function is $[0, \pi]$ and $\cos t$ is negative, t is in the second quadrant. Thus

$$\cos t = -\tfrac{3}{5} \quad \text{and} \quad \tfrac{1}{2}\pi < t < \pi$$

We wish to find the exact value of $\sin 2t$. From the sine double-measure identity, $\sin 2t = 2 \sin t \cos t$. Thus we need to compute $\sin t$. From the identity $\sin^2 t + \cos^2 t = 1$, and because $\sin t > 0$ since t is in $(\tfrac{1}{2}\pi, \pi)$, $\sin t = \sqrt{1 - \cos^2 t}$. Thus

$$\sin t = \sqrt{1 - (-\tfrac{3}{5})^2}$$
$$= \tfrac{4}{5}$$

Therefore

$$\sin 2t = 2 \sin t \cos t$$
$$= 2(\tfrac{4}{5})(-\tfrac{3}{5})$$
$$= -\tfrac{24}{25}$$

from which we conclude that

$$\sin[2\cos^{-1}(-\tfrac{3}{5})] = -\tfrac{24}{25}$$

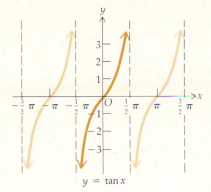

FIGURE 7

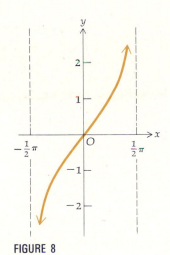

FIGURE 8

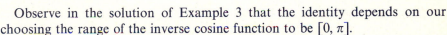

FIGURE 9

EXAMPLE 3   Prove

$$\cos^{-1} x = \tfrac{1}{2}\pi - \sin^{-1} x \quad \text{for } |x| \leq 1 \tag{2}$$

Solution   Let x be in $[-1, 1]$, and let

$$t = \cos(\tfrac{1}{2}\pi - \sin^{-1} x) \tag{3}$$

Applying the reduction formula $\cos(\tfrac{1}{2}\pi - v) = \sin v$ with $v = \sin^{-1} x$ on the right side of (3), we get

$$t = \sin(\sin^{-1} x)$$

Because $\sin(\sin^{-1} x) = x$,

$$t = x$$

Replacing t by x in (3) gives

$$x = \cos(\tfrac{1}{2}\pi - \sin^{-1} x) \tag{4}$$

Because $-\tfrac{1}{2}\pi \leq \sin^{-1} x \leq \tfrac{1}{2}\pi$, by adding $-\tfrac{1}{2}\pi$ to each member we have

$$-\pi \leq -\tfrac{1}{2}\pi + \sin^{-1} x \leq 0$$

Multiplying each member of this inequality by -1 and reversing the direction of the inequality signs gives

$$0 \leq \tfrac{1}{2}\pi - \sin^{-1} x \leq \pi$$

From this inequality, (4), and Definition 8.1.2 it follows that

$$\cos^{-1} x = \tfrac{1}{2}\pi - \sin^{-1} x \quad \text{for } |x| \leq 1$$

which is (2).

Observe in the solution of Example 3 that the identity depends on our choosing the range of the inverse cosine function to be $[0, \pi]$.

To obtain the inverse tangent function we first consider the graph of the tangent function shown in Figure 7. The function is continuous and increasing on the open interval $(-\tfrac{1}{2}\pi, \tfrac{1}{2}\pi)$ as indicated in the figure. We restrict the tangent function to this interval and let H be the function defined by

$$H(x) = \tan x \quad \text{and} \quad -\tfrac{1}{2}\pi < x < \tfrac{1}{2}\pi$$

The domain of H is the open interval $(-\tfrac{1}{2}\pi, \tfrac{1}{2}\pi)$, and the range is the set R of real numbers. A sketch of its graph is shown in Figure 8. Because the function H is continuous and increasing on its domain, it has an inverse function, called the *inverse tangent function*.

8.1.3 DEFINITION   The **inverse tangent function**, denoted by $\tan^{-1}$, is defined as follows:

$$y = \tan^{-1} x \quad \text{if and only if} \quad x = \tan y \text{ and } -\tfrac{1}{2}\pi < y < \tfrac{1}{2}\pi$$

The domain of $\tan^{-1}$ is the set R of real numbers, and the range is the open interval $(-\tfrac{1}{2}\pi, \tfrac{1}{2}\pi)$. Figure 9 shows a sketch of its graph.

▶ **ILLUSTRATION 4**

(a) $\tan^{-1}\sqrt{3} = \frac{1}{3}\pi$ (b) $\tan^{-1}\left(-\frac{1}{\sqrt{3}}\right) = -\frac{1}{6}\pi$ (c) $\tan^{-1}0 = 0$ ◀

From Definition 8.1.3 we have

$$\tan(\tan^{-1}x) = x \qquad \text{for } x \text{ in } (-\infty, +\infty)$$

$$\tan^{-1}(\tan y) = y \qquad \text{for } y \text{ in } (-\tfrac{1}{2}\pi, \tfrac{1}{2}\pi)$$

▶ **ILLUSTRATION 5**

$$\tan^{-1}(\tan\tfrac{1}{4}\pi) = \tfrac{1}{4}\pi \quad \text{and} \quad \tan^{-1}[\tan(-\tfrac{1}{4}\pi)] = -\tfrac{1}{4}\pi$$

However

$$\tan^{-1}(\tan\tfrac{3}{4}\pi) = \tan^{-1}(-1) \quad \text{and} \quad \tan^{-1}(\tan\tfrac{5}{4}\pi) = \tan^{-1}1$$
$$= -\tfrac{1}{4}\pi \qquad\qquad\qquad\qquad = \tfrac{1}{4}\pi \quad ◀$$

EXAMPLE 4   Find the exact value of

$$\sec[\tan^{-1}(-3)]$$

Solution   We shall do this problem by letting $\tan^{-1}(-3)$ be an angle. Let

$$\theta = \tan^{-1}(-3)$$

Because the range of the inverse tangent function is $(-\tfrac{1}{2}\pi, \tfrac{1}{2}\pi)$, and because $\tan\theta$ is negative, $-\tfrac{1}{2}\pi < \theta < 0$. Thus

$$\tan\theta = -3 \quad \text{and} \quad -\tfrac{1}{2}\pi < \theta < 0$$

Figure 10 shows an angle θ that satisfies these requirements. Observe that the point P selected on the terminal side of θ is $(1, -3)$. From the Pythagorean theorem r is $\sqrt{1^2 + (-3)^2} = \sqrt{10}$. Therefore $\sec\theta = \sqrt{10}$. Hence

$$\sec[\tan^{-1}(-3)] = \sqrt{10}$$

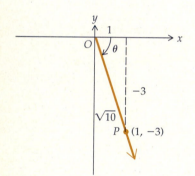

FIGURE 10

Before defining the inverse cotangent function we refer back to Example 3, in which we proved identity (2) involving $\cos^{-1}$ and $\sin^{-1}$. This identity can be used to define the inverse cosine function, and then it can be proved that the range of $\cos^{-1}$ is $[0, \pi]$. We use this kind of procedure in discussing the inverse cotangent function.

8.1.4 DEFINITION   The **inverse cotangent function**, denoted by $\cot^{-1}$, is defined by

$$\cot^{-1}x = \tfrac{1}{2}\pi - \tan^{-1}x \qquad \text{where } x \text{ is any real number}$$

It follows from the definition that the domain of $\cot^{-1}$ is the set R of real numbers. To obtain the range we write the equation in the definition as

$$\tan^{-1}x = \tfrac{1}{2}\pi - \cot^{-1}x \tag{5}$$

Because

$$-\tfrac{1}{2}\pi < \tan^{-1}x < \tfrac{1}{2}\pi$$

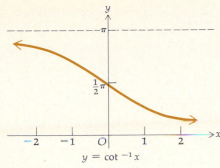

FIGURE 11

by substituting from (5) into this inequality we get

$$-\tfrac{1}{2}\pi < \tfrac{1}{2}\pi - \cot^{-1} x < \tfrac{1}{2}\pi$$

Subtracting $\tfrac{1}{2}\pi$ from each member we get

$$-\pi < -\cot^{-1} x < 0$$

Now multiplying each member by -1 and reversing the direction of the inequality signs we obtain

$$0 < \cot^{-1} x < \pi$$

Therefore the range of the inverse cotangent function is the open interval $(0, \pi)$. A sketch of its graph is in Figure 11.

▶ **ILLUSTRATION 6**

(a) $\tan^{-1} 1 = \tfrac{1}{4}\pi$

(b) $\tan^{-1}(-1) = -\tfrac{1}{4}\pi$

(c) $\cot^{-1} 1 = \tfrac{1}{2}\pi - \tan^{-1} 1$
   $= \tfrac{1}{2}\pi - \tfrac{1}{4}\pi$
   $= \tfrac{1}{4}\pi$

(d) $\cot^{-1}(-1) = \tfrac{1}{2}\pi - \tan^{-1}(-1)$
   $= \tfrac{1}{2}\pi - (-\tfrac{1}{4}\pi)$
   $= \tfrac{3}{4}\pi$ ◀

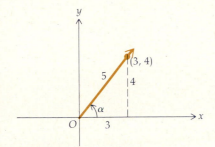

FIGURE 12

EXAMPLE 5   Find the exact value of $\cos[\cot^{-1}\tfrac{3}{4} + \cot^{-1}(-\tfrac{5}{12})]$.

Solution   Let $\alpha = \cot^{-1}\tfrac{3}{4}$ and $\beta = \cot^{-1}(-\tfrac{5}{12})$. Then

$$\cot \alpha = \tfrac{3}{4} \quad\text{and}\quad 0 < \alpha < \tfrac{1}{2}\pi$$

$$\cot \beta = -\tfrac{5}{12} \quad\text{and}\quad \tfrac{1}{2}\pi < \beta < \pi$$

We wish to find $\cos(\alpha + \beta)$. From the cosine sum identity,

$$\cos(\alpha + \beta) = \cos \alpha \cos \beta - \sin \alpha \sin \beta \tag{6}$$

To determine $\cos \alpha$ and $\sin \alpha$, refer to Figure 12, which shows a first-quadrant angle α for which $\cot \alpha = \tfrac{3}{4}$. From the figure,

$$\sin \alpha = \tfrac{4}{5} \qquad \cos \alpha = \tfrac{3}{5} \tag{7}$$

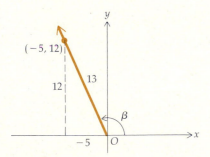

FIGURE 13

Figure 13 shows a second-quadrant angle β for which $\cot \beta = -\tfrac{5}{12}$. From the figure,

$$\sin \beta = \tfrac{12}{13} \qquad \cos \beta = -\tfrac{5}{13} \tag{8}$$

Substituting from (7) and (8) into (6) we have

$$\cos(\alpha + \beta) = \tfrac{3}{5}(-\tfrac{5}{13}) - \tfrac{4}{5}(\tfrac{12}{13})$$
$$= -\tfrac{63}{65}$$

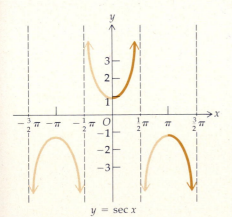

$y = \sec x$

FIGURE 14

To lead up to the definition of the *inverse secant function* we first consider a sketch of the graph of the secant function shown in Figure 14. Observe that the secant function is increasing on the interval $[0, \tfrac{1}{2}\pi)$ and decreasing on the interval $[\pi, \tfrac{3}{2}\pi)$. Furthermore, if $x \in [0, \tfrac{1}{2}\pi) \cup [\pi, \tfrac{3}{2}\pi)$, then $\sec x \in (-\infty, -1] \cup [1, +\infty)$. We choose the interval $[\pi, \tfrac{3}{2}\pi)$ because differentiation of the inverse secant function, as well as some other computations with it (in Section 9.4, for instance), is simplified if the inverse secant function is defined so that when $x < 0$, then $\pi \leq \sec^{-1} x < \frac{3}{2}\pi$. Thus we make the following definition.

8.1.5 DEFINITION

The **inverse secant function**, denoted by $\sec^{-1}$, is defined as follows:

$$y = \sec^{-1} x \quad \text{if and only if} \quad x = \sec y \text{ and } \begin{cases} 0 \leq y < \frac{1}{2}\pi & \text{if } x \geq 1 \\ \pi \leq y < \frac{3}{2}\pi & \text{if } x \leq -1 \end{cases}$$

The domain of $\sec^{-1}$ is $(-\infty, -1) \cup [1, +\infty)$. The range of $\sec^{-1}$ is $[0, \frac{1}{2}\pi) \cup [\pi, \frac{3}{2}\pi)$. A sketch of the graph of $\sec^{-1}$ appears in Figure 15.
From Definition 8.1.5 we have

$$\sec(\sec^{-1} x) = x \qquad \text{for } x \text{ in } (-\infty, -1] \cup [1, +\infty)$$

$$\sec^{-1}(\sec y) = y \qquad \text{for } y \text{ in } [0, \frac{1}{2}\pi) \cup [\pi, \frac{3}{2}\pi)$$

We now define the *inverse cosecant function* in terms of the inverse secant function.

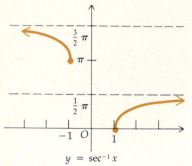

$y = \sec^{-1} x$

FIGURE 15

8.1.6 DEFINITION

The **inverse cosecant function**, denoted by $\csc^{-1}$, is defined by

$$\csc^{-1} x = \frac{1}{2}\pi - \sec^{-1} x \qquad \text{for } |x| \geq 1$$

From the definition the domain of $\csc^{-1}$ is $(-\infty, -1] \cup [1, +\infty)$. The range of $\csc^{-1}$ can be found in a manner similar to that used to determine the range of $\cot^{-1}$. The range of $\csc^{-1}$ is $(-\pi, -\frac{1}{2}\pi] \cup (0, \frac{1}{2}\pi]$, and you are asked to show this in Exercise 53. A sketch of the graph of $\csc^{-1}$ appears in Figure 16.

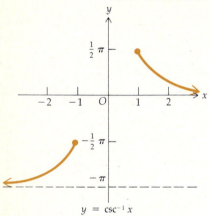

$y = \csc^{-1} x$

FIGURE 16

▶ **ILLUSTRATION 7**

(a) $\sec^{-1} 2 = \frac{1}{3}\pi$

(b) $\sec^{-1}(-2) = \frac{4}{3}\pi$

(c) $\csc^{-1} 2 = \frac{1}{2}\pi - \sec^{-1} 2$
$\phantom{\csc^{-1} 2} = \frac{1}{2}\pi - \frac{1}{3}\pi$
$\phantom{\csc^{-1} 2} = \frac{1}{6}\pi$

(d) $\csc^{-1}(-2) = \frac{1}{2}\pi - \sec^{-1}(-2)$
$\phantom{\csc^{-1}(-2)} = \frac{1}{2}\pi - \frac{4}{3}\pi$
$\phantom{\csc^{-1}(-2)} = -\frac{5}{6}\pi$ ◀

EXERCISES 8.1

In Exercises 1 through 6, determine the exact function value.

1. (a) $\sin^{-1} \frac{1}{2}$; (b) $\sin^{-1}(-\frac{1}{2})$; (c) $\cos^{-1} \frac{1}{2}$; (d) $\cos^{-1}(-\frac{1}{2})$

2. (a) $\sin^{-1} \frac{\sqrt{3}}{2}$; (b) $\sin^{-1}\left(-\frac{\sqrt{3}}{2}\right)$; (c) $\cos^{-1} \frac{\sqrt{3}}{2}$;

(d) $\cos^{-1}\left(-\frac{\sqrt{3}}{2}\right)$

3. (a) $\tan^{-1} \frac{1}{\sqrt{3}}$; (b) $\tan^{-1}(-\sqrt{3})$; (c) $\sec^{-1} \frac{2}{\sqrt{3}}$;

(d) $\sec^{-1}\left(-\frac{2}{\sqrt{3}}\right)$

4. (a) $\cot^{-1} \frac{1}{\sqrt{3}}$; (b) $\cot^{-1}(-\sqrt{3})$; (c) $\csc^{-1} \frac{2}{\sqrt{3}}$; (d) $\csc^{-1}\left(-\frac{2}{\sqrt{3}}\right)$

5. (a) $\sin^{-1} 1$; (b) $\sin^{-1}(-1)$; (c) $\csc^{-1} 1$; (d) $\csc^{-1}(-1)$; (e) $\sin^{-1} 0$

6. (a) $\cos^{-1} 1$; (b) $\cos^{-1}(-1)$; (c) $\sec^{-1} 1$; (d) $\sec^{-1}(-1)$; (e) $\cos^{-1} 0$

7. Given $x = \sin^{-1} \frac{1}{3}$, find the exact value of each of the following: (a) $\cos x$; (b) $\tan x$; (c) $\cot x$; (d) $\sec x$; (e) $\csc x$.

8. Given $x = \cos^{-1} \frac{2}{3}$, find the exact value of each of the following: (a) $\sin x$; (b) $\tan x$; (c) $\cot x$; (d) $\sec x$; (e) $\csc x$.

9. Do Exercise 7 if $x = \sin^{-1}(-\frac{1}{3})$.

10. Do Exercise 8 if $x = \cos^{-1}(-\frac{2}{3})$.

11. Given $y = \tan^{-1}(-2)$, find the exact value of each of the following: (a) $\sin y$; (b) $\cos y$; (c) $\cot y$; (d) $\sec y$; (e) $\csc y$.

12. Given $y = \cot^{-1}(-\frac{1}{2})$, find the exact value of each of the following: (a) $\sin y$; (b) $\cos y$; (c) $\tan y$; (d) $\sec y$; (e) $\csc y$.

13. Given $t = \csc^{-1}(-\frac{3}{2})$, find the exact value of each of the following: (a) $\sin t$; (b) $\cos t$; (c) $\tan t$; (d) $\cot t$; (e) $\sec t$.

14. Given $t = \sec^{-1}(-3)$, find the exact value of each of the following: (a) $\sin t$; (b) $\cos t$; (c) $\tan t$; (d) $\cot t$; (e) $\csc t$.

In Exercises 15 through 40, find the exact value of the quantity.

15. (a) $\sin^{-1}(\sin \frac{1}{6}\pi)$; (b) $\sin^{-1}[\sin(-\frac{1}{6}\pi)]$; (c) $\sin^{-1}(\sin \frac{5}{6}\pi)$;
   (d) $\sin^{-1}(\sin \frac{11}{6}\pi)$

16. (a) $\sin^{-1}(\sin \frac{1}{3}\pi)$; (b) $\sin^{-1}[\sin(-\frac{1}{3}\pi)]$; (c) $\sin^{-1}(\sin \frac{2}{3}\pi)$;
   (d) $\sin^{-1}(\sin \frac{5}{3}\pi)$

17. (a) $\cos^{-1}(\cos \frac{1}{3}\pi)$; (h) $\cos^{-1}[\cos(-\frac{1}{3}\pi)]$; (c) $\cos^{-1}(\cos \frac{2}{3}\pi)$;
   (d) $\cos^{-1}(\cos \frac{4}{3}\pi)$

18. (a) $\cos^{-1}(\cos \frac{1}{4}\pi)$; (b) $\cos^{-1}[\cos(-\frac{1}{4}\pi)]$; (c) $\cos^{-1}(\cos \frac{3}{4}\pi)$;
   (d) $\cos^{-1}(\cos \frac{5}{4}\pi)$

19. (a) $\tan^{-1}(\tan \frac{1}{6}\pi)$; (b) $\tan^{-1}[\tan(-\frac{1}{3}\pi)]$; (c) $\tan^{-1}(\tan \frac{7}{6}\pi)$;
   (d) $\tan^{-1}[\tan(-\frac{4}{3}\pi)]$

20. (a) $\tan^{-1}(\tan \frac{1}{3}\pi)$; (b) $\tan^{-1}[\tan(-\frac{1}{6}\pi)]$; (c) $\tan^{-1}(\tan \frac{4}{3}\pi)$;
   (d) $\tan^{-1}[\tan(-\frac{7}{6}\pi)]$

21. (a) $\cot^{-1}(\cot \frac{1}{6}\pi)$; (b) $\cot^{-1}[\cot(-\frac{1}{3}\pi)]$; (c) $\cot^{-1}(\cot \frac{7}{6}\pi)$;
   (d) $\cot^{-1}[\cot(-\frac{4}{3}\pi)]$

22. (a) $\cot^{-1}(\cot \frac{1}{3}\pi)$; (b) $\cot^{-1}[\cot(-\frac{1}{6}\pi)]$; (c) $\cot^{-1}(\cot \frac{4}{3}\pi)$;
   (d) $\cot^{-1}[\cot(-\frac{7}{6}\pi)]$

23. (a) $\sec^{-1}(\sec \frac{1}{3}\pi)$; (b) $\sec^{-1}[\sec(-\frac{1}{3}\pi)]$; (c) $\sec^{-1}(\sec \frac{2}{3}\pi)$;
   (d) $\sec^{-1}(\sec \frac{4}{3}\pi)$

24. (a) $\sec^{-1}(\sec \frac{1}{4}\pi)$; (b) $\sec^{-1}[\sec(-\frac{1}{4}\pi)]$; (c) $\sec^{-1}(\sec \frac{3}{4}\pi)$;
   (d) $\sec^{-1}(\sec \frac{5}{4}\pi)$

25. $\csc^{-1}(\csc \frac{1}{6}\pi)$; (b) $\csc^{-1}[\csc(-\frac{1}{6}\pi)]$; (c) $\csc^{-1}(\csc \frac{5}{6}\pi)$;
   (d) $\csc^{-1}(\csc \frac{11}{6}\pi)$

26. (a) $\csc^{-1}(\csc \frac{1}{3}\pi)$; (b) $\csc^{-1}[\csc(-\frac{1}{3}\pi)]$; (c) $\csc^{-1}(\csc \frac{2}{3}\pi)$;
   (d) $\csc^{-1}(\csc \frac{5}{3}\pi)$

27. (a) $\tan[\sin^{-1} \frac{1}{2}\sqrt{3}]$; (b) $\sin[\tan^{-1} \frac{1}{2}\sqrt{3}]$

28. (a) $\cos[\tan^{-1}(-3)]$; (b) $\tan[\sec^{-1}(-3)]$

29. (a) $\cos[\sin^{-1}(-\frac{1}{2})]$; (b) $\sin[\cos^{-1}(-\frac{1}{2})]$

30. (a) $\tan[\cot^{-1}(-1)]$; (b) $\cot[\tan^{-1}(-1)]$

31. $\cos[2 \sin^{-1}(-\frac{5}{13})]$

32. $\tan[2 \sec^{-1}(-\frac{5}{4})]$

33. $\sin(\sin^{-1} \frac{2}{3} + \cos^{-1} \frac{1}{3})$

34. $\cos[\sin^{-1}(-\frac{1}{2}) + \sin^{-1} \frac{1}{4}]$

35. $\cos[\sin^{-1} \frac{2}{3} + 2 \sin^{-1}(-\frac{1}{3})]$

36. $\tan[\tan^{-1}(-\frac{2}{5}) - \cos^{-1}(-\frac{1}{2}\sqrt{2})]$

37. $\tan(\tan^{-1} \frac{3}{4} - \sin^{-1} \frac{1}{2})$

38. $\tan[\sec^{-1} \frac{5}{3} + \csc^{-1}(-\frac{13}{12})]$

39. $\cos(\sin^{-1} \frac{1}{3} - \tan^{-1} \frac{1}{2})$

40. $\sin[\cos^{-1}(-\frac{2}{3}) + 2 \sin^{-1}(-\frac{1}{3})]$

41. Prove: $\cos^{-1} \dfrac{3}{\sqrt{10}} + \cos^{-1} \dfrac{2}{\sqrt{5}} = \dfrac{1}{4}\pi$.

42. Prove: $2 \tan^{-1} \frac{1}{3} - \tan^{-1}(-\frac{1}{7}) = \frac{1}{4}\pi$.

In Exercises 43 through 50, draw a sketch of the graph of the function.

43. $f(x) = \frac{1}{2} \sin^{-1} x$ **44.** $g(x) = \sin^{-1} \frac{1}{2}x$

45. $g(x) = \tan^{-1} 2x$ **46.** $f(x) = 2 \tan^{-1} x$

47. $F(x) = \cos^{-1} 3x$ **48.** $G(x) = \frac{1}{2} \sec^{-1} 2x$

49. $f(x) = 2 \cot^{-1} \frac{1}{2}x$ **50.** $g(x) = \frac{1}{2} \csc^{-1} \frac{1}{2}x$

51. A weight is suspended from a spring and vibrating vertically according to the equation $y = 2 \sin 4\pi(t + \frac{1}{8})$, where y centimeters is the directed distance of the weight from its central position t seconds after the start of the motion and the positive direction is upward. (a) Solve the equation for t. (b) Use the equation in part (a) to determine the smallest three positive values of t for which the weight is 1 cm above its central position.

52. A 60-cycle alternating current is described by the equation $x = 20 \sin 120\pi(t - \frac{11}{720})$, where x amperes is the current at t seconds. (a) Solve the equation for t. (b) Use the equation in part (a) to determine the smallest three positive values of t for which the current is 10 amperes.

53. Prove that the range of $\csc^{-1}$ is $(-\pi, -\frac{1}{2}\pi] \cup (0, \frac{1}{2}\pi]$.

54. Prove that $\tan^{-1} x + \tan^{-1}\left(\dfrac{1}{x}\right) = \dfrac{1}{2}\pi$, if $x > 0$. (*Hint:* Use the identity $\cot A = \tan(\frac{1}{2}\pi - A)$).

8.2 DERIVATIVES OF THE INVERSE TRIGONOMETRIC FUNCTIONS

Because the inverse sine function is continuous and increasing on its domain, it follows from Theorem 7.2.3 that it has a derivative. To derive the formula for this derivative let

$$y = \sin^{-1} x$$

which is equivalent to

$$x = \sin y \quad \text{and} \quad y \text{ is in } [-\tfrac{1}{2}\pi, \tfrac{1}{2}\pi]$$

Differentiating on both sides of this equation with respect to y we obtain

$$\frac{dx}{dy} = \cos y \quad \text{and} \quad y \text{ is in } [-\tfrac{1}{2}\pi, \tfrac{1}{2}\pi] \tag{1}$$

From the identity $\sin^2 y + \cos^2 y = 1$, and replacing $\sin y$ by x, we obtain

$$\cos^2 y = 1 - x^2$$

If y is in $[-\frac{1}{2}\pi, \frac{1}{2}\pi]$, $\cos y$ is nonnegative; thus

$$\cos y = \sqrt{1 - x^2} \qquad \text{if } y \text{ is in } [-\tfrac{1}{2}\pi, \tfrac{1}{2}\pi]$$

Substituting from this equation into (1) we get

$$\frac{dx}{dy} = \sqrt{1 - x^2}$$

From Theorem 7.2.3, $\dfrac{dy}{dx}$ is the reciprocal of $\dfrac{dx}{dy}$; hence

$$D_x(\sin^{-1} x) = \frac{1}{\sqrt{1 - x^2}} \tag{2}$$

The domain of the derivative of the inverse sine function is the open interval $(-1, 1)$.

From (2) and the chain rule we have the following theorem.

8.2.1 THEOREM   If u is a differentiable function of x,

$$D_x(\sin^{-1} u) = \frac{1}{\sqrt{1 - u^2}} \, D_x u$$

▶ **ILLUSTRATION 1**   If $y = \sin^{-1} x^2$, then

$$\frac{dy}{dx} = \frac{1}{\sqrt{1 - (x^2)^2}} \, (2x)$$

$$= \frac{2x}{\sqrt{1 - x^4}} \qquad\blacktriangleleft$$

To derive the formula for the derivative of the inverse cosine function we use (2) of Section 8.1, which is

$$\cos^{-1} x = \tfrac{1}{2}\pi - \sin^{-1} x$$

Differentiating with respect to x we have

$$D_x(\cos^{-1} x) = D_x(\tfrac{1}{2}\pi - \sin^{-1} x)$$

Thus

$$D_x(\cos^{-1} x) = -\frac{1}{\sqrt{1 - x^2}} \tag{3}$$

where x is in $(-1, 1)$.

The next theorem follows from (3) and the chain rule.

8.2.2 THEOREM   If u is a differentiable function of x,

$$D_x(\cos^{-1} u) = -\frac{1}{\sqrt{1 - u^2}} \, D_x u$$

We now derive the formula for the derivative of the inverse tangent function. If

$$y = \tan^{-1} x$$

then

$$x = \tan y \quad \text{and} \quad y \text{ is in } (-\tfrac{1}{2}\pi, \tfrac{1}{2}\pi)$$

Differentiating on both sides of this equation with respect to y we obtain

$$\frac{dx}{dy} = \sec^2 y \quad \text{and} \quad y \text{ is in } (-\tfrac{1}{2}\pi, \tfrac{1}{2}\pi) \tag{4}$$

From the identity $\sec^2 y = 1 + \tan^2 y$, and replacing $\tan y$ by x, we have

$$\sec^2 y = 1 + x^2$$

Substituting from this equation into (4) we get

$$\frac{dx}{dy} = 1 + x^2$$

So from Theorem 7.2.3,

$$D_x(\tan^{-1} x) = \frac{1}{1 + x^2} \tag{5}$$

The domain of the derivative of the inverse tangent function is the set R of real numbers.

From (5) and the chain rule we obtain the following theorem.

8.2.3 THEOREM

If u is a differentiable function of x,

$$D_x(\tan^{-1} u) = \frac{1}{1 + u^2} D_x u$$

▶ **ILLUSTRATION 2** If $f(x) = \tan^{-1} \dfrac{1}{x + 1}$, then

$$f'(x) = \frac{1}{1 + \dfrac{1}{(x + 1)^2}} \cdot \frac{-1}{(x + 1)^2}$$

$$= \frac{-1}{(x + 1)^2 + 1}$$

$$= \frac{-1}{x^2 + 2x + 2} \qquad ◀$$

From Definition 8.1.4,

$$\cot^{-1} x = \tfrac{1}{2}\pi - \tan^{-1} x$$

Differentiating with respect to x we obtain the formula

$$D_x(\cot^{-1} x) = -\frac{1}{1 + x^2}$$

From this formula and the chain rule we have the following theorem.

8.2.4 THEOREM   If u is a differentiable function of x,

$$D_x(\cot^{-1} u) = -\frac{1}{1 + u^2} D_x u$$

EXAMPLE 1   Find $\dfrac{dy}{dx}$ if

$$y = x^3 \cot^{-1} \tfrac{1}{3}x$$

Solution

$$\frac{dy}{dx} = 3x^2 \cot^{-1} \tfrac{1}{3}x + x^3 \cdot \frac{-1}{1 + \frac{1}{9}x^2} \cdot \frac{1}{3}$$

$$= 3x^2 \cot^{-1} \tfrac{1}{3}x - \frac{3x^3}{9 + x^2}$$

EXAMPLE 2   Find $\dfrac{dy}{dx}$ if

$$\ln(x + y) = \tan^{-1}\left(\frac{x}{y}\right)$$

Solution   Differentiating implicitly on both sides of the given equation with respect to x we get

$$\frac{1}{x + y}\left(1 + \frac{dy}{dx}\right) = \frac{1}{1 + \dfrac{x^2}{y^2}} \cdot \frac{y - x\dfrac{dy}{dx}}{y^2}$$

$$\frac{1 + \dfrac{dy}{dx}}{x + y} = \frac{y - x\dfrac{dy}{dx}}{y^2 + x^2}$$

$$y^2 + x^2 + (y^2 + x^2)\frac{dy}{dx} = xy + y^2 - (x^2 + xy)\frac{dy}{dx}$$

$$\frac{dy}{dx} = \frac{xy - x^2}{2x^2 + xy + y^2}$$

To obtain the formula for the derivative of the inverse secant function let

$$y = \sec^{-1} x \qquad \text{and} \quad |x| \geq 1$$

Then

$$x = \sec y \qquad \text{and } y \text{ is in } [0, \tfrac{1}{2}\pi) \cup [\pi, \tfrac{3}{2}\pi) \tag{6}$$

Differentiating on both sides of (6) with respect to y we get

$$\frac{dx}{dy} = \sec y \tan y \qquad \text{and } y \text{ is in } [0, \tfrac{1}{2}\pi) \cup [\pi, \tfrac{3}{2}\pi) \tag{7}$$

From the identity $\tan^2 y = \sec^2 y - 1$, with $\sec y = x$, we get

$$\tan^2 y = x^2 - 1$$

Because y is in $[0, \tfrac{1}{2}\pi) \cup [\pi, \tfrac{3}{2}\pi)$, $\tan y$ is nonnegative. Thus

$$\tan y = \sqrt{x^2 - 1} \qquad \text{if } y \text{ is in } [0, \tfrac{1}{2}\pi) \cup [\pi, \tfrac{3}{2}\pi)$$

Substituting from (6) and this equation in (7) we have

$$\frac{dx}{dy} = x\sqrt{x^2 - 1}$$

Thus from Theorem 7.2.3,

$$D_x(\sec^{-1} x) = \frac{1}{x\sqrt{x^2 - 1}} \tag{8}$$

where $|x| > 1$. From (8) and the chain rule the next theorem follows.

8.2.5 THEOREM   If u is a differentiable function of x,

$$D_x(\sec^{-1} u) = \frac{1}{u\sqrt{u^2 - 1}} D_x u$$

▶ **ILLUSTRATION 3**   If $f(x) = \sec^{-1}(3e^x)$, then

$$f'(x) = \frac{1}{3e^x\sqrt{(3e^x)^2 - 1}} (3e^x)$$

$$= \frac{1}{\sqrt{9e^{2x} - 1}} \qquad\qquad ◀$$

From Definition 8.1.6,

$$\csc^{-1} x = \tfrac{1}{2}\pi - \sec^{-1} x \qquad \text{for } |x| \geq 1$$

Differentiating with respect to x we obtain

$$D_x(\csc^{-1} x) = -\frac{1}{x\sqrt{x^2 - 1}} \tag{9}$$

where $|x| > 1$. From (9) and the chain rule we have the following theorem.

8.2.6 THEOREM   If u is a differentiable function of x,

$$D_x(\csc^{-1} u) = -\frac{1}{u\sqrt{u^2 - 1}} D_x u$$

EXAMPLE 3   Find $f'(x)$ if

$$f(x) = x \csc^{-1} \frac{1}{x}$$

Solution

$$f'(x) = \csc^{-1} \frac{1}{x} + x \left[-\frac{1}{\frac{1}{x} \sqrt{\left(\frac{1}{x}\right)^2 - 1}} \left(-\frac{1}{x^2}\right) \right]$$

$$= \csc^{-1} \frac{1}{x} + \left[\frac{-x^2}{\frac{\sqrt{1-x^2}}{\sqrt{x^2}}} \left(-\frac{1}{x^2}\right) \right]$$

$$= \csc^{-1} \frac{1}{x} + \frac{1}{\frac{\sqrt{1-x^2}}{|x|}}$$

$$= \csc^{-1} \frac{1}{x} + \frac{|x|}{\sqrt{1-x^2}}$$

In the following example, an observer is looking at a picture placed high on a wall. See Figure 1. When the observer is far away from the wall, the angle subtended at the observer's eye by the picture is small. As the observer gets closer to the wall, that angle increases until it reaches a maximum value. Then as the observer gets even closer to the wall, the angle gets smaller. When the angle is a maximum, we say that the observer has the "best view" of the picture.

EXAMPLE 4   A picture 7 ft high is placed on a wall with its base 9 ft above the eye level of an observer. How far from the wall should the observer stand to get the best view of the picture, that is, so that the angle subtended at the observer's eye by the picture is a maximum?

Solution   Let x feet be the distance of the observer from the wall, θ be the radian measure of the angle subtended at the observer's eye by the picture, α be the radian measure of the angle subtended at the observer's eye by the portion of the wall above eye level and below the picture, and $\beta = \alpha + \theta$. Refer to Figure 1.

We wish to find the value of x that will make θ an absolute maximum. Because x is in the interval $(0, +\infty)$, the absolute maximum value of θ will be a relative maximum value. We see from the figure that

$$\cot \beta = \frac{x}{16} \quad \text{and} \quad \cot \alpha = \frac{x}{9}$$

Because $0 < \beta < \frac{1}{2}\pi$ and $0 < \alpha < \frac{1}{2}\pi$,

$$\beta = \cot^{-1} \frac{x}{16} \quad \text{and} \quad \alpha = \cot^{-1} \frac{x}{9}$$

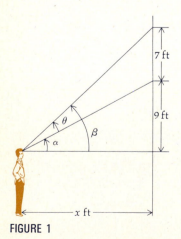

FIGURE 1

Substituting these values of β and α in the equation $\theta = \beta - \alpha$ we get

$$\theta = \cot^{-1}\frac{x}{16} - \cot^{-1}\frac{x}{9}$$

Differentiating with respect to x gives

$$\frac{d\theta}{dx} = -\frac{\frac{1}{16}}{1 + \left(\frac{x}{16}\right)^2} + \frac{\frac{1}{9}}{1 + \left(\frac{x}{9}\right)^2}$$

$$= -\frac{16}{16^2 + x^2} + \frac{9}{9^2 + x^2}$$

Setting $\dfrac{d\theta}{dx} = 0$ we obtain

$$9(16^2 + x^2) - 16(9^2 + x^2) = 0$$

$$-7x^2 + 9 \cdot 16(16 - 9) = 0$$

$$x^2 = 9 \cdot 16$$

$$x = 12$$

The -12 is rejected because it is not in the interval $(0, +\infty)$. The results of the first-derivative test are shown in Table 1. Because the relative maximum value of θ is an absolute maximum value, we conclude that the observer should stand 12 ft from the wall.

Table 1

	$\dfrac{d\theta}{dx}$	Conclusion
$0 < x < 12$	$+$	
$x = 12$	0	θ has a relative maximum value
$12 < x < +\infty$	$-$	

EXERCISES 8.2

In Exercises 1 through 26, find the derivative of the function.

1. $f(x) = \sin^{-1}\frac{1}{2}x$

2. $f(x) = \cos^{-1} 3x$

3. $g(x) = \tan^{-1} 2x$

4. $f(x) = \csc^{-1} 2x$

5. $F(x) = 2\cos^{-1}\sqrt{x}$

6. $g(x) = \frac{1}{2}\sin^{-1} x^2$

7. $g(t) = \sec^{-1} 5t + \csc^{-1} 5t$

8. $f(y) = \cot^{-1} e^y$

9. $f(x) = \sin^{-1}\sqrt{1 - x^2}$

10. $f(w) = 2\tan^{-1}\frac{1}{w}$

11. $F(x) = \cot^{-1}\frac{2}{x} + \tan^{-1}\frac{x}{2}$

12. $f(x) = \ln(\tan^{-1} 3x)$

13. $h(y) = y\sin^{-1} 2y$

14. $f(t) = t^2\cos^{-1} t$

15. $g(x) = x^2\sec^{-1}\frac{1}{x}$

16. $g(s) = \cos^{-1} s + \frac{s}{1 - s^2}$

17. $f(x) = \cos^{-1}(\sin x)$

18. $h(x) = \tan^{-1}\frac{2x}{1 - x^2}$

19. $F(x) = \ln(\tan^{-1} x^2)$

20. $F(x) = \sec^{-1}\sqrt{x^2 + 4}$

21. $f(x) = 4\sin^{-1}\frac{1}{2}x + x\sqrt{4 - x^2}$

22. $f(t) = a\sin^{-1}\frac{t}{a} + \sqrt{a^2 - t^2}$

23. $h(x) = \csc^{-1}(2e^{3x})$

24. $f(x) = \sin^{-1} x + \cos^{-1} x$

25. $G(x) = x\cot^{-1} x + \ln\sqrt{1 + x^2}$

26. $f(t) = \sin^{-1}\frac{t - 1}{t + 1}$

In Exercises 27 through 30, find $\dfrac{dy}{dx}$.

27. $e^{x+y} = \cos^{-1} x$ **28.** $\ln(\sin^2 3x) = e^x + \cot^{-1} y$

29. $x \sin y + x^3 = \tan^{-1} y$ **30.** $\sin^{-1}(xy) = \cos^{-1}(x + y)$

31. Find equations of the tangent line and normal line to the graph of the equation $y = \sec^{-1}(2x + 1)$ at the point $(\frac{1}{2}, \frac{1}{3}\pi)$.

32. In Example 4 show that another equation defining θ in terms of x is

$$\theta = \tan^{-1} \frac{7x}{x^2 + 144}$$

Use this equation to determine how far from the wall the observer should stand to get the best view of the picture.

33. A sign 3 ft high is placed on a wall with its base 2 ft above the eye level of a woman attempting to read it. How far from the wall should she stand to get the best view of the sign, that is, so that the angle subtended at the woman's eye by the sign is a maximum?

34. Example 4 and Exercise 33 are particular cases of the following more general situation: An object (for instance, a picture or a sign) a feet high is placed on a wall with its base b feet above the eye level of an observer. Show that the observer gets the best view of the object when the distance of the observer from the wall is $\sqrt{b(a + b)}$ feet.

35. A picture 40 cm high is placed on a wall with its base 30 cm above the level of the eye of an observer. If the observer is approaching the wall at the rate of 40 cm/sec, how fast is the measure of the angle subtended at the observer's eye by the picture changing when the observer is 1 m from the wall?

36. A ladder 25 ft long is leaning against a vertical wall. If the bottom of the ladder is pulled horizontally away from the wall so that the top is sliding down at 3 ft/sec, how fast is the measure of the angle between the ladder and the ground changing when the bottom of the ladder is 15 ft from the wall?

37. A light is 3 km from a straight beach. If the light revolves and makes 2 rpm, find the speed of light along the beach when the spot is 2 km from the point on the beach nearest the light.

38. A man on a dock is pulling in at the rate of 2 ft/sec a rowboat by means of a rope. The man's hands are 20 ft above the level of the point where the rope is attached to the boat. How fast is the measure of the angle of depression of the rope changing when there are 52 ft of rope out?

39. A woman is walking at the rate of 5 ft/sec along the diameter of a circular courtyard. A light at one end of a diameter perpendicular to her path casts a shadow on the circular wall. How fast is the shadow moving along the wall when the distance from the woman to the center of the courtyard is $\frac{1}{2}r$, where r feet is the radius of the courtyard?

40. In Exercise 39, how far is the woman from the center of the courtyard when the speed of her shadow along the wall is 9 ft/sec?

41. A rope is attached to a weight and it passes over a hook that is 8 ft above the ground. The rope is pulled over the hook at the rate of $\frac{3}{4}$ ft/sec and drags the weight along level ground. If the length of the rope between the weight and the hook is x feet when the radian measure of the angle between the rope and the floor is θ, find the time rate of change of θ in terms of x.

42. Use differentials to find an approximate value of $\sin^{-1} 0.52$ to three decimal places.

43. Do Example 3 of Section 8.1 by showing that $\sin^{-1} x$ and $-\cos^{-1} x$ differ by a constant and then evaluating the constant.

44. Do Exercise 54 of Exercises 8.1 by showing that $\tan^{-1} x$ and $-\tan^{-1}\left(\dfrac{1}{x}\right)$ differ by a constant and then evaluating the constant.

45. Given: $f(x) = \tan^{-1}(1/x) - \cot^{-1} x$. (a) Show that $f'(x) = 0$ for all x in the domain of f. (b) Prove that there is no constant C for which $f(x) = C$ for all x in its domain. (c) Why doesn't the statement in part (b) contradict Theorem 5.1.2?

8.3 INTEGRALS YIELDING INVERSE TRIGONOMETRIC FUNCTIONS

From the theorems for the derivatives of the inverse trigonometric functions we obtain the following one giving some indefinite integral formulas.

8.3.1 THEOREM

$$\int \frac{du}{\sqrt{1 - u^2}} = \sin^{-1} u + C \tag{1}$$

$$\int \frac{du}{1 + u^2} = \tan^{-1} u + C \tag{2}$$

$$\int \frac{du}{u\sqrt{u^2 - 1}} = \sec^{-1} u + C \tag{3}$$

The proof of each formula is immediate by taking the derivative of the right-hand side. We also have the formulas given in the next theorem.

8.3.2 THEOREM

$$\int \frac{du}{\sqrt{a^2 - u^2}} = \sin^{-1} \frac{u}{a} + C \qquad \text{where } a > 0 \tag{4}$$

$$\int \frac{du}{a^2 + u^2} = \frac{1}{a} \tan^{-1} \frac{u}{a} + C \qquad \text{where } a \neq 0 \tag{5}$$

$$\int \frac{du}{u\sqrt{u^2 - a^2}} = \frac{1}{a} \sec^{-1} \frac{u}{a} + C \qquad \text{where } a > 0 \tag{6}$$

Proof   These formulas can be proved by finding the derivative of the right side and obtaining the integrand. We prove formula (4).

$$D_u\left(\sin^{-1}\frac{u}{a}\right) = \frac{1}{\sqrt{1 - \left(\frac{u}{a}\right)^2}} D_u\left(\frac{u}{a}\right)$$

$$= \frac{\sqrt{a^2}}{\sqrt{a^2 - u^2}} \cdot \frac{1}{a}$$

$$= \frac{a}{\sqrt{a^2 - u^2}} \cdot \frac{1}{a} \qquad \text{if } a > 0$$

$$= \frac{1}{\sqrt{a^2 - u^2}} \qquad \text{if } a > 0$$

The proofs of (5) and (6) are left as exercises (see Exercises 38 and 39).   ∎

The formulas of Theorem 8.3.2 can also be proved by making a suitable change of variable and then using Theorem 8.3.1 (see Exercises 40–42). Observe that the formulas of Theorem 8.3.2 include those of Theorem 8.3.1 by taking $a = 1$.

EXAMPLE 1   Evaluate

$$\int \frac{dx}{\sqrt{4 - 9x^2}}$$

Solution

$$\int \frac{dx}{\sqrt{4 - 9x^2}} = \frac{1}{3} \int \frac{d(3x)}{\sqrt{4 - (3x)^2}}$$

$$= \frac{1}{3} \sin^{-1} \frac{3x}{2} + C$$

In the next three examples we complete the square of a quadratic expression to express the integrand in a form that enables us to use Theorem 8.3.2.

EXAMPLE 2   Evaluate

$$\int \frac{dx}{3x^2 - 2x + 5}$$

Solution

$$\int \frac{dx}{3x^2 - 2x + 5} = \int \frac{dx}{3(x^2 - \frac{2}{3}x) + 5}$$

To complete the square of $x^2 - \frac{2}{3}x$ we add $\frac{1}{9}$, and because $\frac{1}{9}$ is multiplied by 3, we actually add $\frac{1}{3}$ to the denominator, and so we also subtract $\frac{1}{3}$ from the denominator. Therefore we have

$$\int \frac{dx}{3x^2 - 2x + 5} = \int \frac{dx}{3(x^2 - \frac{2}{3}x + \frac{1}{9}) + 5 - \frac{1}{3}}$$

$$= \int \frac{dx}{3(x - \frac{1}{3})^2 + \frac{14}{3}}$$

$$= \frac{1}{3} \int \frac{dx}{(x - \frac{1}{3})^2 + \frac{14}{9}}$$

$$= \frac{1}{3} \cdot \frac{3}{\sqrt{14}} \tan^{-1}\left(\frac{x - \frac{1}{3}}{\frac{1}{3}\sqrt{14}}\right) + C$$

$$= \frac{1}{\sqrt{14}} \tan^{-1}\left(\frac{3x - 1}{\sqrt{14}}\right) + C$$

EXAMPLE 3   Evaluate

$$\int \frac{(2x + 7)\,dx}{x^2 + 2x + 5}$$

Solution   Because $d(x^2 + 2x + 5) = (2x + 2)\,dx$, we write the numerator as $(2x + 2)\,dx + 5\,dx$ and express the original integral as the sum of two integrals.

$$\int \frac{(2x + 7)\,dx}{x^2 + 2x + 5} = \int \frac{(2x + 2)\,dx}{x^2 + 2x + 5} + 5 \int \frac{dx}{x^2 + 2x + 5}$$

$$= \ln|x^2 + 2x + 5| + 5 \int \frac{dx}{(x + 1)^2 + 4}$$

$$= \ln(x^2 + 2x + 5) + \frac{5}{2} \tan^{-1}\frac{x + 1}{2} + C$$

Note: Because $x^2 + 2x + 5 > 0$, for all x, $|x^2 + 2x + 5| = x^2 + 2x + 5$.

EXAMPLE 4   Evaluate

$$\int \frac{3\,dx}{(x + 2)\sqrt{x^2 + 4x + 3}}$$

Solution

$$\int \frac{3\,dx}{(x+2)\sqrt{x^2+4x+3}} = \int \frac{3\,dx}{(x+2)\sqrt{(x^2+4x+4)-1}}$$

$$= 3\int \frac{d(x+2)}{(x+2)\sqrt{(x+2)^2-1}}$$

$$= 3\sec^{-1}(x+2) + C$$

EXAMPLE 5 Find the area of the region in the first quadrant bounded by the curve

$$y = \frac{1}{1+x^2}$$

the x axis, the y axis, and the line $x = 1$.

Solution Figure 1 shows the region and a rectangular element of area. If A square units is the area of the region,

$$A = \lim_{||\Delta|| \to 0} \sum_{i=1}^{n} \frac{1}{1+\xi_i^2}\,\Delta_i x$$

$$= \int_0^1 \frac{dx}{1+x^2}$$

$$= \tan^{-1}x \Big]_0^1$$

$$= \tan^{-1}1 - \tan^{-1}0$$

$$= \tfrac{1}{4}\pi - 0$$

$$= \tfrac{1}{4}\pi$$

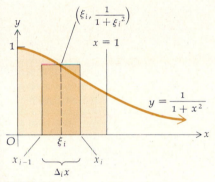

FIGURE 1

EXERCISES 8.3

In Exercises 1 through 25, evaluate the indefinite integral.

1. $\displaystyle\int \frac{dx}{\sqrt{1-4x^2}}$

2. $\displaystyle\int \frac{dx}{x^2+25}$

3. $\displaystyle\int \frac{dx}{9x^2+16}$

4. $\displaystyle\int \frac{dt}{\sqrt{1-16t^2}}$

5. $\displaystyle\int \frac{dx}{4+(x-1)^2}$

6. $\displaystyle\int \frac{dx}{9+(3-x)^2}$

7. $\displaystyle\int \frac{dx}{4x\sqrt{x^2-16}}$

8. $\displaystyle\int \frac{x\,dx}{x^4+16}$

9. $\displaystyle\int \frac{dx}{\sqrt{2-5x^2}}$

10. $\displaystyle\int \frac{3\,dx}{x\sqrt{x^2-9}}$

11. $\displaystyle\int \frac{r\,dr}{\sqrt{16-9r^4}}$

12. $\displaystyle\int \frac{du}{u\sqrt{16u^2-9}}$

13. $\displaystyle\int \frac{e^x\,dx}{7+e^{2x}}$

14. $\displaystyle\int \frac{\sin x\,dx}{\sqrt{2-\cos^2 x}}$

15. $\displaystyle\int \frac{dx}{(1+x)\sqrt{x}}$

16. $\displaystyle\int \frac{ds}{\sqrt{2s-s^2}}$

17. $\displaystyle\int \frac{dx}{x^2-x+2}$

18. $\displaystyle\int \frac{dx}{\sqrt{3x-x^2-2}}$

19. $\displaystyle\int \frac{dx}{\sqrt{15+2x-x^2}}$

20. $\displaystyle\int \frac{dx}{2x^2+2x+3}$

21. $\displaystyle\int \frac{x\,dx}{\sqrt{3-2x-x^2}}$

22. $\displaystyle\int \frac{x\,dx}{x^2+x+1}$

23. $\displaystyle\int \frac{(2+x)\,dx}{\sqrt{4-2x-x^2}}$

24. $\displaystyle\int \frac{2\,dt}{(t-3)\sqrt{t^2-6t+5}}$

25. $\displaystyle\int \frac{2x^3\,dx}{2x^2-4x+3}$

In Exercises 26 through 34, evaluate the definite integral.

26. $\displaystyle\int_0^{1/2} \frac{dx}{\sqrt{1-x^2}}$

27. $\displaystyle\int_0^1 \frac{1+x}{1+x^2}\,dx$

28. $\displaystyle\int_0^{\sqrt{3}} \frac{x\,dx}{\sqrt{1+x^2}}$

29. $\displaystyle\int_{-4}^{-2} \frac{dt}{\sqrt{-t^2-6t-5}}$

30. $\displaystyle\int_2^5 \frac{dx}{x^2-4x+13}$

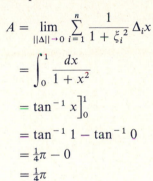

31. $\displaystyle\int_0^1 \frac{dx}{e^x + e^{-x}}$

32. $\displaystyle\int_{1/\sqrt{2}}^1 \frac{dx}{x\sqrt{4x^2 - 1}}$

33. $\displaystyle\int_1^e \frac{dx}{x[1 + (\ln x)^2]}$

34. $\displaystyle\int_0^{\pi/6} \frac{\sec^2 x \, dx}{1 + 9 \tan^2 x}$

35. Find the area of the region bounded by the curve $y = 8/(x^2 + 4)$, the x axis, the y axis, and the line $x = 2$.

36. Find the area of the region bounded by the curves $x^2 = 4ay$ and $y = 8a^3/(x^2 + 4a^2)$.

37. Find the area of the region bounded by the curve $y = 1/\sqrt{5 - 4x - x^2}$, the x axis, and the lines $x = -\frac{7}{2}$ and $x = -\frac{1}{2}$.

In Exercises 38 and 39, prove the formula by showing that the derivative of the right side is equal to the integrand.

38. $\displaystyle\int \frac{du}{a^2 + u^2} = \frac{1}{a} \tan^{-1} \frac{u}{a} + C$

39. $\displaystyle\int \frac{du}{u\sqrt{u^2 - a^2}} = \frac{1}{a} \sec^{-1} \frac{u}{a} + C$ if $a > 0$

In Exercises 40 through 42, prove the indicated formula of Theorem 8.3.2 by making a suitable change of variable and then using Theorem 8.3.1.

40. Formula (4)   **41.** Formula (5)   **42.** Formula (6)

43. In Section 3.10 we stated that a particle moving in a straight line is said to have *simple harmonic motion* if the measure of its acceleration is always proportional to the measure of its displacement from a fixed point on the line and its acceleration and displacement are oppositely directed. Therefore, if at t seconds s centimeters is the directed distance of the particle from the origin and v centimeters per second is the velocity of the particle, then a differential equation for simple harmonic motion is

$$\frac{dv}{dt} = -k^2 s \qquad (7)$$

where k^2 is the constant of proportionality and the minus sign indicates that the acceleration is opposite in direction from the displacement. Because $\dfrac{dv}{dt} = \dfrac{dv}{ds} \cdot \dfrac{ds}{dt}$, it follows that $\dfrac{dv}{dt} = v \dfrac{dv}{ds}$. Thus (7) may be written as

$$v \frac{dv}{ds} = -k^2 s \qquad (8)$$

(a) Solve (8) for v to get $v = \pm k\sqrt{a^2 - s^2}$. *Note:* Take $a^2 k^2$ as the arbitrary constant of integration, and justify this choice. (b) Letting $v = ds/dt$ in the solution of part (a) we obtain the differential equation

$$\frac{ds}{dt} = \pm k\sqrt{a^2 - s^2} \qquad (9)$$

Taking $t = 0$ at the instant when $v = 0$ (and hence $s = a$), solve (9) to obtain

$$s = a \cos kt \qquad (10)$$

(c) Show that the largest value for $|s|$ is a. The number a is called the *amplitude* of the motion. (d) The particle will oscillate between the points where $s = a$ and $s = -a$. If T seconds is the time for the particle to go from a to $-a$ and return, show that $T = 2\pi/k$. The number T is called the *period* of the motion.

44. A particle is moving in a straight line according to the equation of motion $s = 5 - 10 \sin^2 2t$, where s centimeters is the directed distance of the particle from the origin at t seconds. Use the result of part (b) of Exercise 43 to show that the motion is simple harmonic. Find the amplitude and period of this motion.

45. Show that the motion of Exercise 44 is simple harmonic by showing that differential equation (7) is satisfied.

In Exercises 46 through 48, prove the formula by showing the derivative of the right side is equal to the integrand.

46. $\displaystyle\int \frac{du}{\sqrt{1 - u^2}} = -\cos^{-1} u + C$

Is this formula equivalent to formula (1)? Why?

47. $\displaystyle\int \frac{du}{1 + u^2} = -\cot^{-1} u + C$

Is this formula equivalent to formula (2)? Why?

48. $\displaystyle\int \frac{du}{u\sqrt{u^2 - 1}} = -\csc^{-1} u + C$

Is this formula equivalent to formula (3)? Why?

8.4 THE HYPERBOLIC FUNCTIONS

Certain combinations of e^x and e^{-x} appear so frequently in applications of mathematics that they are given special names. Two of these functions are the *hyperbolic sine function* and the *hyperbolic cosine function*. At the end of this section we show that the function values are related to the coordinates of the points on an equilateral hyperbola in a manner similar to that in which the values of the corresponding trigonometric functions are related to the coordinates of points on a circle. Following are the definitions of the hyperbolic sine function and the hyperbolic cosine function.

8.4.1 DEFINITION | The **hyperbolic sine function** is defined by

$$\sinh x = \frac{e^x - e^{-x}}{2}$$

The domain and range are the set R of real numbers.

8.4.2 DEFINITION | The **hyperbolic cosine function** is defined by

$$\cosh x = \frac{e^x + e^{-x}}{2}$$

The domain is the set R of real numbers and the range is the set of numbers in the interval $[1, +\infty)$.

Because

$$\sinh(-x) = \frac{e^{-x} - e^x}{2} \qquad \cosh(-x) = \frac{e^{-x} + e^x}{2}$$

$$= -\frac{e^x - e^{-x}}{2} \qquad\qquad\quad = \cosh x$$

$$= -\sinh x$$

the hyperbolic sine is an odd function and the hyperbolic cosine is an even function.

The formulas for the derivatives of the hyperbolic sine and hyperbolic cosine functions are obtained by applying Definitions 8.4.1 and 8.4.2 and differentiating the resulting expressions involving exponential functions. Thus

$$D_x(\sinh x) = D_x\left(\frac{e^x - e^{-x}}{2}\right) \qquad D_x(\cosh x) = D_x\left(\frac{e^x + e^{-x}}{2}\right)$$

$$= \frac{e^x + e^{-x}}{2} \qquad\qquad\qquad = \frac{e^x - e^{-x}}{2}$$

$$= \cosh x \qquad\qquad\qquad\quad = \sinh x$$

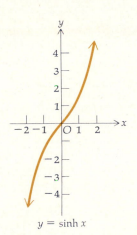

$y = \sinh x$

FIGURE 1

From these formulas and the chain rule we have the following theorem.

8.4.3 THEOREM | If u is a differentiable function of x,

$$D_x(\sinh u) = \cosh u \, D_x u$$

$$D_x(\cosh u) = \sinh u \, D_x u$$

Because $D_x(\sinh x) > 0$ for all x, the hyperbolic sine function is increasing on its entire domain. With this information, the knowledge that it is an odd function, and some values obtained from a calculator, we have the graph of the hyperbolic sine function shown in Figure 1.

The hyperbolic cosine function is decreasing on the interval $(-\infty, 0]$ because $D_x(\cosh x) < 0$ if $x < 0$, and it is increasing on the interval $[0, +\infty)$ because $D_x(\cosh x) > 0$ if $x > 0$. Furthermore, the hyperbolic cosine is an even function. With these facts and some values of $\cosh x$ from a calculator, we obtain the graph of the hyperbolic cosine function shown in Figure 2.

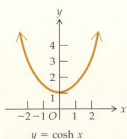

$y = \cosh x$

FIGURE 2

The remaining four hyperbolic functions are defined in terms of the hyperbolic sine and hyperbolic cosine functions. Observe that each satisfies an identity analogous to one satisfied by corresponding trigonometric functions.

8.4.4 DEFINITION   The **hyperbolic tangent function, hyperbolic cotangent function, hyperbolic secant function**, and **hyperbolic cosecant function** are defined, respectively, as follows:

$$\tanh x = \frac{\sinh x}{\cosh x} \tag{1}$$

$$\coth x = \frac{\cosh x}{\sinh x} \tag{2}$$

$$\operatorname{sech} x = \frac{1}{\cosh x} \tag{3}$$

$$\operatorname{csch} x = \frac{1}{\sinh x} \tag{4}$$

The hyperbolic functions in Definition 8.4.4 can be expressed in terms of exponential functions by using Definitions 8.4.1 and 8.4.2. We have

$$\tanh x = \frac{e^x - e^{-x}}{e^x + e^{-x}} \qquad \coth x = \frac{e^x + e^{-x}}{e^x - e^{-x}}$$

$$\operatorname{sech} x = \frac{2}{e^x + e^{-x}} \qquad \operatorname{csch} x = \frac{2}{e^x - e^{-x}}$$

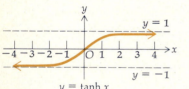

$y = \tanh x$

FIGURE 3

A sketch of the graph of the hyperbolic tangent function is shown in Figure 3. Notice from Figures 1 through 3 that none of these functions is periodic, while the corresponding trigonometric functions are periodic.

There are identities that are satisfied by the hyperbolic functions which are similar to those satisfied by the trigonometric functions. Four of the fundamental identities are given in Definition 8.4.4. The other four fundamental identities are as follows:

$$\tanh x = \frac{1}{\coth x} \tag{5}$$

$$\cosh^2 x - \sinh^2 x = 1 \tag{6}$$

$$1 - \tanh^2 x = \operatorname{sech}^2 x \tag{7}$$

$$1 - \coth^2 x = -\operatorname{csch}^2 x \tag{8}$$

Equation (5) follows immediately from (1) and (2). Following is the proof of (6).

$$\cosh^2 x - \sinh^2 x = \left(\frac{e^x + e^{-x}}{2}\right)^2 - \left(\frac{e^x - e^{-x}}{2}\right)^2$$

$$= \tfrac{1}{4}(e^{2x} + 2e^0 + e^{-2x} - e^{2x} + 2e^0 - e^{-2x})$$

$$= 1$$

Equation (7) can be proved by using the formulas for $\tanh x$ and $\operatorname{sech} x$ in terms of e^x and e^{-x} as in the proof above, or an alternate proof is obtained by using other identities as follows:

$$1 - \tanh^2 x = 1 - \frac{\sinh^2 x}{\cosh^2 x}$$

$$= \frac{\cosh^2 x - \sinh^2 x}{\cosh^2 x}$$

$$= \frac{1}{\cosh^2 x}$$

$$= \operatorname{sech}^2 x$$

The proof of (8) is left as an exercise (see Exercise 1). From the eight fundamental identities it is possible to prove others. Some of these appear in the exercises. Other identities, such as the hyperbolic functions of the sum or difference of two numbers and the hyperbolic functions of twice a number or one-half a number, are similar to the corresponding trigonometric identities. Sometimes it is helpful to make use of the following two relations that follow from Definitions 8.4.1 and 8.4.2.

$$\cosh x + \sinh x = e^x \tag{9}$$

$$\cosh x - \sinh x = e^{-x} \tag{10}$$

We use them in proving the following identity:

$$\sinh(x + y) = \sinh x \cosh y + \cosh x \sinh y \tag{11}$$

From Definition 8.4.1,

$$\sinh(x + y) = \frac{e^{x+y} - e^{-(x+y)}}{2}$$

$$= \frac{e^x e^y - e^{-x} e^{-y}}{2}$$

Applying (9) and (10) to the right side of this equation we obtain

$$\sinh(x + y) = \tfrac{1}{2}[(\cosh x + \sinh x)(\cosh y + \sinh y) - (\cosh x - \sinh x)(\cosh y - \sinh y)]$$

Expanding the right side of the above equation and combining terms we have formula (11).

In a similar manner we can prove

$$\cosh(x + y) = \cosh x \cosh y + \sinh x \sinh y \tag{12}$$

If in (11) and (12) y is replaced by x, the following two formulas are obtained:

$$\sinh 2x = 2 \sinh x \cosh x \tag{13}$$

$$\cosh 2x = \cosh^2 x + \sinh^2 x \tag{14}$$

Formula (14) combined with identity (6) gives two alternate formulas for $\cosh 2x$, which are

$$\cosh 2x = 2 \sinh^2 x + 1 \tag{15}$$

$$\cosh 2x = 2 \cosh^2 x - 1 \tag{16}$$

Solving (15) and (16) for sinh x and cosh x, respectively, and replacing x by $\frac{1}{2}x$ we get

$$\sinh \frac{x}{2} = \begin{cases} \sqrt{\dfrac{\cosh x - 1}{2}} & \text{if } x \geq 0 \\[3mm] -\sqrt{\dfrac{\cosh x - 1}{2}} & \text{if } x < 0 \end{cases} \tag{17}$$

$$\cosh \frac{x}{2} = \sqrt{\frac{\cosh x + 1}{2}} \tag{18}$$

We do not have a $\pm$ symbol on the right side of (18) because the range of the hyperbolic cosine function is $[1, +\infty)$. The details of the proofs of (12) through (18) are left as exercises (see Exercises 2, 4, and 5).

To find the derivative of the hyperbolic tangent function we use some of the identities.

$$D_x(\tanh x) = D_x\left(\frac{\sinh x}{\cosh x}\right)$$

$$= \frac{\cosh^2 x - \sinh^2 x}{\cosh^2 x}$$

$$= \frac{1}{\cosh^2 x}$$

$$= \operatorname{sech}^2 x$$

The formulas for the derivatives of the remaining three hyperbolic functions are as follows: $D_x(\coth x) = -\operatorname{csch}^2 x$; $D_x(\operatorname{sech} x) = -\operatorname{sech} x \tanh x$; $D_x(\operatorname{csch} x) = -\operatorname{csch} x \coth x$. The proofs of these formulas are left as exercises (see Exercises 15 and 16).

From these formulas and the chain rule we have the following theorem.

8.4.5 THEOREM   If u is a differentiable function of x,

$$D_x(\tanh u) = \operatorname{sech}^2 u D_x u$$

$$D_x(\coth u) = -\operatorname{csch}^2 u D_x u$$

$$D_x(\operatorname{sech} u) = -\operatorname{sech} u \tanh u D_x u$$

$$D_x(\operatorname{csch} u) = -\operatorname{csch} u \coth u D_x u$$

Observe that the formulas for the derivatives of the hyperbolic sine, cosine, and tangent all have a plus sign, whereas those for the derivatives of the hyperbolic cotangent, secant, and cosecant all have a minus sign. Otherwise the formulas are similar to the corresponding ones for the derivatives of the trigonometric functions.

EXAMPLE 1   Find $\dfrac{dy}{dx}$ if

$$y = \tanh(1 - x^2)$$

Solution

$$\frac{dy}{dx} = \text{sech}^2(1 - x^2) \cdot D_x(1 - x^2)$$
$$= -2x \, \text{sech}^2(1 - x^2)$$

EXAMPLE 2   Find $f'(x)$ if

$$f(x) = \ln \sinh x$$

Solution

$$f'(x) = \frac{1}{\sinh x} \cdot \cosh x$$

$$= \frac{\cosh x}{\sinh x}$$

$$= \coth x$$

The indefinite integration formulas in the next theorem follow from the differentiation formulas in Theorems 8.4.3 and 8.4.5.

8.4.6 THEOREM

$$\int \sinh u \, du = \cosh u + C$$

$$\int \cosh u \, du = \sinh u + C$$

$$\int \text{sech}^2 u \, du = \tanh u + C$$

$$\int \text{csch}^2 u \, du = -\coth u + C$$

$$\int \text{sech } u \tanh u \, du = -\text{sech } u + C$$

$$\int \text{csch } u \coth u \, du = -\text{csch } u + C$$

The techniques used to integrate the hyperbolic functions are similar to those used for the trigonometric functions. The following two examples illustrate the method.

EXAMPLE 3   Evaluate

$$\int \sinh x \cosh^2 x \, dx$$

Solution

$$\int \sinh x \cosh^2 x \, dx = \int \cosh^2 x \, (\sinh x \, dx)$$
$$= \tfrac{1}{3} \cosh^3 x + C$$

EXAMPLE 4   Evaluate

$$\int \tanh^2 x \, dx$$

Solution

$$\int \tanh^2 x \, dx = \int (1 - \text{sech}^2 \, x) \, dx$$

$$= \int dx - \int \text{sech}^2 \, x \, dx$$

$$= x - \tanh x + C$$

A **catenary** is the curve formed by a flexible cable of uniform density hanging from two points under its own weight. Some cables of suspension bridges and some attached to telephone poles hang in this shape. If the lowest point of the catenary is at $(0, a)$, it can be shown that an equation of it is

$$y = a \cosh\left(\frac{x}{a}\right) \qquad a > 0 \tag{19}$$

A sketch of a catenary appears in Figure 4.

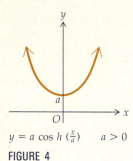

$y = a \cos h\left(\frac{x}{a}\right) \qquad a > 0$

FIGURE 4

EXAMPLE 5   Find the length of the arc of the catenary defined by (19) from the point $(0, a)$ to the point (x_1, y_1), where $x_1 > 0$.

Solution   If

$$f(x) = a \cosh\left(\frac{x}{a}\right) \quad \text{then} \quad f'(x) = a \sinh\left(\frac{x}{a}\right) \cdot \left(\frac{1}{a}\right)$$

$$= \sinh\left(\frac{x}{a}\right)$$

From Theorem 6.3.3, if L units is the length of the given arc,

$$L = \int_0^{x_1} \sqrt{1 + [f'(x)]^2} \, dx$$

$$= \int_0^{x_1} \sqrt{1 + \sinh^2\left(\frac{x}{a}\right)} \, dx$$

$$= \int_0^{x_1} \sqrt{\cosh^2\left(\frac{x}{a}\right)} \, dx \qquad \text{(from (6))}$$

$$= \int_0^{x_1} \cosh\left(\frac{x}{a}\right) dx \qquad \left(\text{because } \cosh\left(\frac{x}{a}\right) \geq 1\right)$$

$$= a \sinh\left(\frac{x}{a}\right)\Bigg]_0^{x_1}$$

$$= a \sinh\left(\frac{x_1}{a}\right) - a \sinh 0$$

$$= a \sinh\left(\frac{x_1}{a}\right)$$

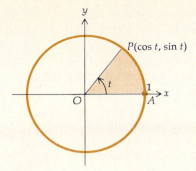

FIGURE 5

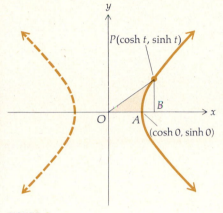

FIGURE 6

In Section 1.6 the sine and cosine functions were defined as coordinates of a point on the unit circle

$$x^2 + y^2 = 1$$

Recall from Definition 1.6.2 that if t is the radian measure of the angle between the x axis and a line from the origin to the point $P(x, y)$ on the unit circle then

$$\sin t = y \qquad \cos t = x$$

See Figure 5. By substituting these values of x and y in the equation of the unit circle we obtain the identity

$$\cos^2 t + \sin^2 t = 1$$

In Section 10.3 you will learn that the curve

$$x^2 - y^2 = 1$$

is an equilateral hyperbola, called the unit hyperbola. Refer to Figure 6. If t is any real number, then the point $(\cosh t, \sinh t)$ is on this hyperbola because

$$\cosh^2 t - \sinh^2 t = 1$$

Observe that because $\cosh t$ is never less than 1, all points $(\cosh t, \sinh t)$ are on the right branch of the hyperbola.

We now show how the areas of the shaded regions in Figures 5 and 6 are related. Because the area of a circular sector of radius r units and a central angle of radian measure t is given by $\frac{1}{2}r^2 t$ square units, the area of the circular sector in Figure 5 is $\frac{1}{2}t$ square units, since $r = 1$. The sector AOP in Figure 6 is the region bounded by the x axis, the line OP, and the arc AP of the unit hyperbola. If A_1 square units is the area of sector AOP, A_2 square units is the area of triangle OBP, and A_3 square units is the area of region ABP,

$$A_1 = A_2 - A_3 \tag{20}$$

From the formula for determining the area of a triangle,

$$A_2 = \tfrac{1}{2} \cosh t \sinh t \tag{21}$$

We find A_3 by integration:

$$A_3 = \int_0^t \sinh u \, d(\cosh u)$$

$$= \int_0^t \sinh^2 u \, du$$

$$= \tfrac{1}{2} \int_0^t (\cosh 2u - 1) \, du$$

$$= \tfrac{1}{4} \sinh 2u - \tfrac{1}{2} u \Big]_0^t$$

Therefore

$$A_3 = \tfrac{1}{2} \cosh t \sinh t - \tfrac{1}{2} t$$

Substituting from this equation and (21) into (20) we have

$$A_1 = \tfrac{1}{2} \cosh t \sinh t - (\tfrac{1}{2} \cosh t \sinh t - \tfrac{1}{2} t)$$
$$= \tfrac{1}{2} t$$

Thus the number of square units in the area of circular sector AOP of Figure 5 and the number of square units in the area of sector AOP of Figure 6 is, in each case, one-half the value of the parameter associated with the point P. For the unit circle, the parameter t is the radian measure of the angle AOP. The parameter t for the unit hyperbola is not interpreted as the measure of an angle; however, sometimes the term *hyperbolic radian* is used in connection with t.

We have therefore shown that the function values of sinh and cosh have the same relationship to the unit hyperbola as the function values of sine and cosine have to the unit circle. So just as the sine and cosine are called *circular functions*, sinh and cosh are called hyperbolic functions.

EXERCISES 8.4

In Exercises 1 through 10, prove the identity.

1. $1 - \coth^2 x = -\csch^2 x$

2. $\cosh(x + y) = \cosh x \cosh y + \sinh x \sinh y$

3. $\tanh(x + y) = \dfrac{\tanh x + \tanh y}{1 + \tanh x \tanh y}$

4. (a) $\sinh 2x = 2 \sinh x \cosh x$; (b) $\cosh 2x = \cosh^2 x + \sinh^2 x$; (c) $\cosh 2x = 2 \sinh^2 x + 1$; (d) $\cosh 2x = 2 \cosh^2 x - 1$

5. (a) $\cosh \tfrac{1}{2}x = \sqrt{\dfrac{\cosh x + 1}{2}}$;

(b) $\sinh \tfrac{1}{2}x = \pm\sqrt{\dfrac{\cosh x - 1}{2}}$

6. $\csch 2x = \tfrac{1}{2} \sech x \csch x$

7. (a) $\tanh(-x) = -\tanh x$; (b) $\sech(-x) = \sech x$

8. $\sinh 3x = 3 \sinh x + 4 \sinh^3 x$

9. $\cosh 3x = 4 \cosh^3 x - 3 \cosh x$

10. $\sinh^2 x - \sinh^2 y = \sinh(x + y)\sinh(x - y)$

11. Prove: $(\sinh x + \cosh x)^n = \cosh nx + \sinh nx$, if n is any positive integer. (*Hint:* Use formula (9).)

12. Prove that the hyperbolic tangent function is an odd function and the hyperbolic secant function is an even function.

13. Prove: $\tanh(\ln x) = \dfrac{x^2 - 1}{x^2 + 1}$ **14.** Prove: $\dfrac{1 + \tanh x}{1 - \tanh x} = e^{2x}$

15. Prove: (a) $D_x(\sinh x) = \cosh x$; (b) $D_x(\coth x) = -\csch^2 x$.

16. Prove: (a) $D_x(\sech x) = -\sech x \tanh x$;
(b) $D_x(\csch x) = -\csch x \coth x$.

In Exercises 17 through 30, find the derivative of the function.

17. $f(x) = \tanh \dfrac{4x + 1}{5}$ **18.** $g(t) = \cosh t^3$

19. $f(y) = \sinh e^{2y}$ **20.** $f(x) = \coth(\ln x)$

21. $f(w) = \sech^2 4w$ **22.** $g(x) = \ln(\sinh x^3)$

23. $f(x) = e^x \cosh x$ **24.** $h(x) = \coth \dfrac{1}{x}$

25. $h(t) = \ln(\tanh t)$ **26.** $F(x) = \tan^{-1}(\sinh x^2)$

27. $G(x) = \sin^{-1}(\tanh x^2)$

28. $f(x) = \ln(\coth 3x - \csch 3x)$

29. $f(x) = x^{\sinh x}$; $x > 0$ **30.** $g(x) = (\cosh x)^x$

31. Prove: (a) $\displaystyle\int \tanh u \, du = \ln|\cosh u| + C$;

(b) $\displaystyle\int \coth u \, du = \ln|\sinh u| + C$.

32. Prove: $\displaystyle\int \sech u \, du = 2 \tan^{-1} e^u + C$.

33. Prove: $\displaystyle\int \csch u \, du = \ln|\tanh \tfrac{1}{2}u| + C$.

In Exercises 34 through 43, evaluate the indefinite integral.

34. $\displaystyle\int \sinh^4 x \cosh x \, dx$ **35.** $\displaystyle\int e^t \cosh(e^t) \sinh(e^t) \, dt$

36. $\displaystyle\int e^x \csch^2 e^x \, dx$ **37.** $\displaystyle\int \dfrac{\sinh \sqrt{x}}{\sqrt{x}} \, dx$

38. $\displaystyle\int x \sech^2 x^2 \, dx$ **39.** $\displaystyle\int \coth^2 3x \, dx$

40. $\displaystyle\int \sech^3 x \tanh x \, dx$ **41.** $\displaystyle\int \sech^2 x \tanh^5 x \, dx$

42. $\displaystyle\int \dfrac{e^{\cosh t}}{\csch t} \, dt$ **43.** $\displaystyle\int \tanh x \ln(\cosh x) \, dx$

In Exercises 44 through 47, evaluate the definite integral.

44. $\displaystyle\int_0^{\ln 2} \tanh z \, dz$ **45.** $\displaystyle\int_0^{\ln 3} \sech^2 t \, dt$

46. $\displaystyle\int_{-1}^{1} (\cosh x - \sinh x) \, dx$ **47.** $\displaystyle\int_0^{2} \sinh^3 x \cosh x \, dx$

48. Draw a sketch of the graph of the hyperbolic cotangent function. Find equations of the asymptotes.

49. Draw a sketch of the graph of the hyperbolic secant function. Find the relative extrema of the function, the points of inflection of the graph, the intervals on which the graph is concave upward, and the intervals on which the graph is concave downward.

50. Prove that a catenary is concave upward at each point.

51. Find the area of the region bounded by the catenary of Example 5, the y axis, the x axis, and the line $x = x_1$, where $x_1 > 0$.

52. Find the volume of the solid of revolution if the region of Exercise 51 is revolved about the x axis.

53. A particle is moving along a straight line according to the equation of motion $s = e^{-ct/2}(A \sinh t + B \cosh t)$, where s centimeters is the directed distance of the particle from the origin at t seconds. If v centimeters per second and a centimeters per second squared are the velocity and acceleration, respectively, of the particle at t seconds, find v and a. Also show that a is the sum of two numbers, one of which is proportional to s and the other proportional to v.

54. Prove that the hyperbolic sine function is continuous and increasing on its entire domain.

55. Prove that the hyperbolic tangent function is continuous and increasing on its entire domain.

56. Prove that the hyperbolic cosine function is continuous on its entire domain but is not monotonic on its entire domain. Find the intervals on which the function is increasing and the intervals on which it is decreasing.

8.5   THE INVERSE HYPERBOLIC FUNCTIONS (*Supplementary*)

From the graph of the hyperbolic sine function (Figure 1 in Section 8.4) we see that a horizontal line intersects the graph in one and only one point; therefore the hyperbolic sine function is one-to-one. Furthermore, the hyperbolic sine function is continuous and increasing on its domain (see Exercise 54 in Exercises 8.4). Therefore, by Theorem 7.1.5, the function has an inverse.

8.5.1 DEFINITION

The **inverse hyperbolic sine function**, denoted by $\sinh^{-1}$, is defined as follows:

$$y = \sinh^{-1} x \quad \text{if and only if} \quad x = \sinh y$$

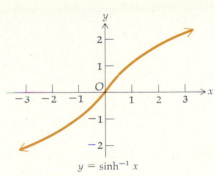

$$y = \sinh^{-1} x$$

FIGURE 1

A sketch of the graph of the inverse hyperbolic sine function appears in Figure 1. Its domain is the set R of real numbers, and its range is also the set R. It follows from Definition 8.5.1 that

$$\sinh(\sinh^{-1} x) = x \quad \text{and} \quad \sinh^{-1}(\sinh y) = y$$

From Figure 2 in Section 8.4 we notice that a horizontal line, $y = k$, where $k > 1$, intersects the graph of the hyperbolic cosine function in two points. Therefore, for each number greater than 1 in the range of this function there correspond two numbers in the domain. So the hyperbolic cosine function is not one-to-one and does not have an inverse. However, we define a function F as follows:

$$F(x) = \cosh x \qquad \text{for } x \geq 0$$

The domain of F is the interval $[0, +\infty)$, and the range is the interval $[1, +\infty)$. A sketch of the graph of F is shown in Figure 2. The function F is continuous

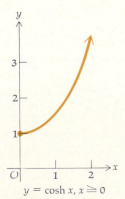

$$y = \cosh x, x \geq 0$$

FIGURE 2

and increasing on its domain and so by Theorem 7.1.5 F has an inverse, which is called the *inverse hyperbolic cosine function.*

8.5.2 DEFINITION

The **inverse hyperbolic cosine function**, denoted by $\cosh^{-1}$, is defined as follows:

$$y = \cosh^{-1} x \quad \text{if and only if} \quad x = \cosh y \text{ and } y \geq 0$$

A sketch of the graph of the inverse hyperbolic cosine function is in Figure 3. The domain of this function is the interval $[1, +\infty)$, and the range is the interval $[0, +\infty)$. From Definition 8.5.2,

$$\cosh(\cosh^{-1} x) = x \qquad \text{if } x \geq 1$$

and

$$\cosh^{-1}(\cosh y) = y \qquad \text{if } y \geq 0$$

As with the hyperbolic sine function, a horizontal line intersects each of the graphs of the hyperbolic tangent function, the hyperbolic cotangent function, and the hyperbolic cosecant function at one and only one point. For the hyperbolic tangent function this may be seen in Figure 3 of Section 8.4. Each of these three functions then is one-to-one; furthermore, each is continuous and monotonic on its domain. Thus each has an inverse function.

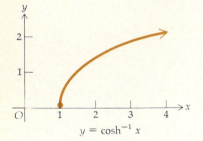

$y = \cosh^{-1} x$

FIGURE 3

8.5.3 DEFINITION

The **inverse hyperbolic tangent function**, the **inverse hyperbolic cotangent function**, and the **inverse hyperbolic cosecant function**, denoted respectively by $\tanh^{-1}$, $\coth^{-1}$, and csch^{-1}, are defined as follows:

$$y = \tanh^{-1} x \quad \text{if and only if} \quad x = \tanh y$$

$$y = \coth^{-1} x \quad \text{if and only if} \quad x = \coth y$$

$$y = \operatorname{csch}^{-1} x \quad \text{if and only if} \quad x = \operatorname{csch} y$$

A sketch of the graph of the inverse hyperbolic tangent function is in Figure 4.

Because $\operatorname{sech} x = 1/\cosh x$ and the hyperbolic cosine function is not one-to-one, it follows that the hyperbolic secant function is not one-to-one. Therefore it does not have an inverse. However, we proceed as we did with the hyperbolic cosine function and define a new function that has an inverse function. Let the function G be defined by

$$G(x) = \operatorname{sech} x \qquad \text{for } x \geq 0$$

It can be shown that G is continuous and monotonic on its domain; therefore G has an inverse, which is called the *inverse hyperbolic secant function.*

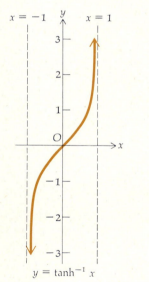

$y = \tanh^{-1} x$

FIGURE 4

8.5.4 DEFINITION

The **inverse hyperbolic secant function**, denoted by sech^{-1}, is defined as follows:

$$y = \operatorname{sech}^{-1} x \quad \text{if and only if} \quad x = \operatorname{sech} y \text{ and } y \geq 0$$

The inverse hyperbolic functions can be expressed in terms of natural logarithms. This should not be surprising, because the hyperbolic functions were defined in terms of the exponential function, and the natural logarithmic function is the inverse of the exponential function.

Following are these expressions for $\sinh^{-1}$, $\cosh^{-1}$, $\tanh^{-1}$, and $\coth^{-1}$.

$$\sinh^{-1} x = \ln(x + \sqrt{x^2 + 1}) \qquad x \text{ any real number} \tag{1}$$

$$\cosh^{-1} x = \ln(x + \sqrt{x^2 - 1}) \qquad x \geq 1 \tag{2}$$

$$\tanh^{-1} x = \frac{1}{2} \ln \frac{1 + x}{1 - x} \qquad |x| < 1 \tag{3}$$

$$\coth^{-1} x = \frac{1}{2} \ln \frac{x + 1}{x - 1} \qquad |x| > 1 \tag{4}$$

We prove (2) and leave the proofs of the other three formulas as exercises (see Exercises 1 through 3).

To prove (2), let $y = \cosh^{-1} x$, $x \geq 1$. Then from Definition 8.5.2, we have $x = \cosh y$, $y \geq 0$. Applying Definition 8.4.2 to $\cosh y$ we get

$$x = \frac{e^y + e^{-y}}{2} \qquad y \geq 0$$

from which we have

$$e^{2y} - 2xe^y + 1 = 0 \qquad y \geq 0$$

Solving this equation for e^y by the quadratic formula we obtain

$$e^y = \frac{2x \pm \sqrt{4x^2 - 4}}{2}$$

$$= x \pm \sqrt{x^2 - 1} \qquad y \geq 0 \tag{5}$$

We know that $y \geq 0$ and $x \geq 1$. Therefore $e^y \geq 1$. When $x = 1$, $x + \sqrt{x^2 - 1} = 1$ and $x - \sqrt{x^2 - 1} = 1$. Furthermore, when $x > 1$, we have $0 < x - 1 < x + 1$; so $\sqrt{x - 1} < \sqrt{x + 1}$. Therefore

$$\sqrt{x - 1}\sqrt{x - 1} < \sqrt{x - 1}\sqrt{x + 1}$$

giving $x - 1 < \sqrt{x^2 - 1}$. Hence when $x > 1$, $x - \sqrt{x^2 - 1} < 1$. Consequently, we can reject the minus sign in (5). And since $y = \cosh^{-1} x$, we have

$$\cosh^{-1} x = \ln(x + \sqrt{x^2 - 1}) \qquad x \geq 1$$

which is (2).

EXAMPLE 1   Express each of the following in terms of a natural logarithm: (a) $\sinh^{-1} 2$; (b) $\tanh^{-1}(-\frac{4}{5})$.

Solution
(a) From (1),   (b) From (3),

$$\sinh^{-1} 2 = \ln(2 + \sqrt{5})$$

$$\tanh^{-1}(-\tfrac{4}{5}) = \frac{1}{2} \ln \frac{\frac{1}{5}}{\frac{9}{5}}$$

$$= \tfrac{1}{2} \ln 3^{-2}$$

$$= -\ln 3$$

To obtain a formula for the derivative of the inverse hyperbolic sine function let $y = \sinh^{-1} x$, and then $x = \sinh y$. Thus, because $\dfrac{dx}{dy} = \cosh y$ and $\dfrac{dy}{dx}$ is the reciprocal of $\dfrac{dx}{dy}$,

$$\frac{dy}{dx} = \frac{1}{\cosh y} \tag{6}$$

From the identity $\cosh^2 y - \sinh^2 y = 1$ we get $\cosh y = \sqrt{\sinh^2 y + 1}$, where when we take the square root, the minus sign is rejected because $\cosh y \geq 1$. Therefore, because $x = \sinh y$, $\cosh y = \sqrt{x^2 + 1}$. Substituting this value of $\cosh y$ in (6) we obtain

$$D_x(\sinh^{-1} x) = \frac{1}{\sqrt{x^2 + 1}}$$

This formula can also be derived by using (1), as follows:

$$D_x(\sinh^{-1} x) = D_x \ln(x + \sqrt{x^2 + 1})$$

$$= \frac{1 + \dfrac{x}{\sqrt{x^2 + 1}}}{x + \sqrt{x^2 + 1}}$$

$$= \frac{\sqrt{x^2 + 1} + x}{\sqrt{x^2 + 1}(x + \sqrt{x^2 + 1})}$$

$$= \frac{1}{\sqrt{x^2 + 1}}$$

The formulas for the derivatives of the other five inverse hyperbolic functions are obtained in a way analogous to that for the derivative of the inverse hyperbolic sine function. Their derivations are left as exercises (see Exercises 4 through 8). From these formulas and the chain rule the next theorem follows.

8.5.5 THEOREM   If u is a differentiable function of x,

$$D_x(\sinh^{-1} u) = \frac{1}{\sqrt{u^2 + 1}} D_x u \tag{7}$$

$$D_x(\cosh^{-1} u) = \frac{1}{\sqrt{u^2 - 1}} D_x u \qquad u > 1 \tag{8}$$

$$D_x(\tanh^{-1} u) = \frac{1}{1 - u^2} D_x u \qquad |u| < 1 \tag{9}$$

$$D_x(\coth^{-1} u) = \frac{1}{1 - u^2} D_x u \qquad |u| > 1 \tag{10}$$

$$D_x(\operatorname{sech}^{-1} u) = -\frac{1}{u\sqrt{1 - u^2}} D_x u \qquad 0 < u < 1 \tag{11}$$

$$D_x(\operatorname{csch}^{-1} u) = -\frac{1}{|u|\sqrt{1 + u^2}} D_x u \qquad u \neq 0 \tag{12}$$

EXAMPLE 2 Find $f'(x)$ if

$$f(x) = \tanh^{-1}(\cos 2x)$$

Solution From (9),

$$f'(x) = \frac{1}{1 - \cos^2 2x}(-2\sin 2x)$$

$$= \frac{-2\sin 2x}{\sin^2 2x}$$

$$= \frac{-2}{\sin 2x}$$

$$= -2\csc 2x$$

EXAMPLE 3 Verify that $D_x(x\cosh^{-1} x - \sqrt{x^2 - 1}) = \cosh^{-1} x$.

Solution

$$D_x(x\cosh^{-1} x - \sqrt{x^2 - 1}) = \cosh^{-1} x + x \cdot \frac{1}{\sqrt{x^2-1}} - \frac{x}{\sqrt{x^2-1}}$$

$$= \cosh^{-1} x$$

The main application of the inverse hyperbolic functions is in connection with integration. This topic is the subject of Supplementary Section 9.8.

EXERCISES 8.5

In Exercises 1 through 8, prove the indicated formula.

1. Formula (1) **2.** Formula (3) **3.** Formula (4)
4. Formula (8) **5.** Formula (9) **6.** Formula (10)
7. Formula (11) **8.** Formula (12)

In Exercises 9 and 10, express the quantity in terms of a natural logarithm.

9. (a) $\sinh^{-1}\frac{1}{4}$; (b) $\tanh^{-1}\frac{1}{2}$ **10.** (a) $\cosh^{-1} 3$; (b) $\coth^{-1}(-2)$

In Exercises 11 through 28, find the derivative of the function.

11. $f(x) = \sinh^{-1} x^2$ **12.** $G(x) = \cosh^{-1}\frac{1}{3}x$
13. $F(x) = \tanh^{-1} 4x$ **14.** $h(w) = \tanh^{-1} w^3$
15. $g(x) = \coth^{-1}(3x + 1)$ **16.** $f(r) = \csch^{-1}\frac{1}{2}r^2$

17. $f(x) = x^2\cosh^{-1} x^2$ **18.** $h(x) = (\sech^{-1} x)^2$
19. $f(x) = \sinh^{-1}(\tan x)$ **20.** $g(x) = \tanh^{-1}(\sin 3x)$
21. $h(x) = \cosh^{-1}(\csc x)$ **22.** $F(x) = \coth^{-1}(\cosh x)$
23. $f(z) = (\coth^{-1} z^2)^3$ **24.** $f(t) = \sinh^{-1} e^{2t}$
25. $g(x) = \tanh^{-1}(\cos e^x)$ **26.** $h(x) = \cosh^{-1}(\ln x)$
27. $G(x) = x\sinh^{-1} x - \sqrt{1 + x^2}$
28. $H(x) = \ln\sqrt{x^2 + 1} - x\tanh^{-1} x$

In Exercises 29 through 31 draw a sketch of the graph of the indicated function.

29. The inverse hyperbolic cotangent function.
30. The inverse hyperbolic secant function.
31. The inverse hyperbolic cosecant function.

REVIEW EXERCISES FOR CHAPTER 8

1. Given $x = \sin^{-1}\frac{3}{5}$ and $y = \cos^{-1}(-\frac{4}{5})$, find the exact value of each of the following: (a) $\cos x$; (b) $\sin y$; (c) $\tan x$; (d) $\tan y$.
2. Given $x = \cos^{-1}\frac{7}{25}$ and $y = \tan^{-1}(-\frac{7}{24})$, find the exact value of each of the following: (a) $\sin x$; (b) $\sin y$; (c) $\cos y$; (d) $\tan x$.

In Exercises 3 and 4, find the exact value of the quantity.

3. (a) $\sin[2\cos^{-1}(-\frac{12}{13})]$; (b) $\tan[\cos^{-1}\frac{3}{5} + \sin^{-1}(-\frac{7}{25})]$
4. (a) $\tan[2\sin^{-1}(-\frac{24}{25})]$; (b) $\cos[\tan^{-1}\frac{4}{3} - \sin^{-1}(-\frac{5}{13})]$

In Exercises 5 through 16, find the derivative of the function.

5. $f(x) = \tan^{-1} 2^x$ **6.** $g(x) = x^2 \sin^{-1} 2x$

7. $F(x) = \sinh^3 2x$ **8.** $h(x) = e^x(\cosh x + \sinh x)$

9. $f(x) = \sec^{-1} \sqrt{x^2 + 1}$ **10.** $G(w) = \tan^{-1}\left(\dfrac{w - 3}{3w + 1}\right)$

11. $g(x) = \sin^{-1}(\tanh \sqrt{x})$ **12.** $f(x) = \text{sech}(\tan 3x)$

13. $h(x) = \text{csch}^2 e^{2x}$ **14.** $g(x) = \cos^{-1}(\tanh 2x)$

15. $f(x) = (\cosh x)^{1/x}$ **16.** $f(x) = \sec^{-1}(\cosh x)$

In Exercises 17 and 18, draw a sketch of the graph of the function.

17. $f(x) = 2 \tanh \frac{1}{2}x$ **18.** $f(x) = \frac{1}{2} \sin^{-1} 3x$

In Exercises 19 and 20, find $\dfrac{dy}{dx}$ by implicit differentiation.

19. $\cot^{-1} \dfrac{x^2}{3y} + xy^2 = 0$ **20.** $\tan^{-1}(x + 3y) = \ln y$

In Exercises 21 through 28, evaluate the indefinite integral.

21. $\displaystyle\int \frac{3x\, dx}{\sqrt{1 - x^4}}$ **22.** $\displaystyle\int x \coth \frac{1}{2}x^2\, dx$

23. $\displaystyle\int \tanh^2 3x\, dx$ **24.** $\displaystyle\int \frac{dx}{\sqrt{7 + 5x - 2x^2}}$

25. $\displaystyle\int \frac{dt}{2t^2 + 3t + 5}$ **26.** $\displaystyle\int \frac{dy}{9e^y + e^{-y}}$

27. $\displaystyle\int \frac{dx}{\sqrt{e^{2x} - 8}}$ **28.** $\displaystyle\int \frac{\cosh t\, dt}{\sqrt{\sinh t}}$

In Exercises 29 through 32, evaluate the definite integral.

29. $\displaystyle\int_1^2 \frac{(t + 2)\, dt}{\sqrt{4t - t^2}}$ **30.** $\displaystyle\int_0^2 \text{sech}^2 \frac{1}{2}x\, dx$

31. $\displaystyle\int_0^1 \sqrt{1 + \cosh y}\, dy$ **32.** $\displaystyle\int_{-1}^1 \frac{(2x + 6)\, dx}{x^2 + 2x + 5}$

33. In an electric circuit the electromotive force is E volts at t seconds, where $E = 20 \cos 120\pi t$. (a) Solve the equation for t. Use the equation in part (a) to find the smallest positive value of t for which the electromotive force is (b) 10 volts; (c) 5 volts; (d) -10 volts; (e) -5 volts.

34. A weight is suspended from a spring and vibrating vertically according to the equation $y = 4 \sin 2\pi(t + \frac{1}{6})$, where y centimeters is the directed distance of the weight from its central position t seconds after the start of the motion, and the positive direction is upward. (a) Solve the equation for t. Use the equation in part (a) to determine the smallest positive value of t for which the displacement of the weight above its central position is (b) 2 cm and (c) 3 cm.

35. Show that $\cosh(\ln x) = \dfrac{x^2 + 1}{2x}$.

36. Prove: (a) $\displaystyle\lim_{x \to +\infty} \coth x = 1$; (b) $\displaystyle\lim_{x \to +\infty} \text{csch } x = 0$.

37. Find the area of the region bounded by the line $x = 2\sqrt{2}$, the curve $y = 9/\sqrt{9 - x^2}$, and the two coordinate axes.

38. Find the length of the catenary $y = \cosh x$ from $(\ln 2, \frac{5}{4})$ to $(\ln 3, \frac{5}{3})$.

39. Find the volume of the solid of revolution generated when the region bounded by the curve $y = \sqrt{\sinh x}$, the x axis, and the lines $x = 0$ and $x = \ln 2$ is revolved about the x axis.

40. The *gudermannian*, named after the German mathematician Christoph Gudermann (1798–1852), is the function defined by gd $x = \tan^{-1}(\sinh x)$. Show that $D_x(\text{gd } x) = \text{sech } x$.

41. A searchlight is $\frac{1}{2}$ km from a straight road and it keeps a light trained on an automobile that is traveling at a constant speed of 60 km/hr. Find the rate at which the light beam is changing direction (a) when the car is at the point on the road nearest the searchlight and (b) when the car is $\frac{1}{2}$ km down the road from this point.

42. A picture 5 ft high is placed on a wall with its base 7 ft above the level of the eye of an observer. If the observer is approaching the wall at the rate of 3 ft/sec, how fast is the measure of the angle subtended at her eye by the picture changing when the observer is 10 ft from the wall?

43. An airplane is flying at a speed of 300 mi/hr at an altitude of 4 mi. If an observer is on the ground, find the time rate of change of the measure of the observer's angle of elevation of the airplane when the airplane is directly over a point on the ground 2 mi from the observer.

44. A helicopter leaves the ground at a point 800 ft from an observer and rises vertically at 25 ft/sec. Find the time rate of change of the measure of the observer's angle of elevation of the helicopter when the helicopter is 600 ft above the ground.

45. Two points A and B are diametrically opposite each other on the shores of a circular lake 1 km in diameter. A man desires to go from point A to point B. He can row at the rate of $1\frac{1}{2}$ km/hr and walk at the rate of 5 km/hr. Find the least amount of time it can take him to get from point A to point B.

46. Solve Exercise 45 if the rates of rowing and walking are, respectively, 2 km/hr and 4 km/hr.

Exercises 47 through 55, pertain to Supplementary Section 8.5. In Exercises 47 and 48, express the quantity in terms of a natural logarithm.

47. $\cosh^{-1} 2$ **48.** $\tanh^{-1} \frac{1}{4}$

In Exercises 49 through 52, find the derivative of the function.

49. $h(w) = w^2 \sinh^{-1} 2w$ **50.** $f(x) = \tanh^{-1}(\sin \sqrt{x})$

51. $f(x) = \text{sech}^{-1} e^{-2x}$ **52.** $g(t) = \cosh^{-1} \sqrt{t^2 + 1}$

53. The graph of the equation

$$x = a \sinh^{-1} \sqrt{\frac{a^2}{y^2} - 1} - \sqrt{a^2 - y^2}$$

is called a *tractrix*. Prove that the slope of the curve at any point (x, y) is $-y/\sqrt{a^2 - y^2}$.

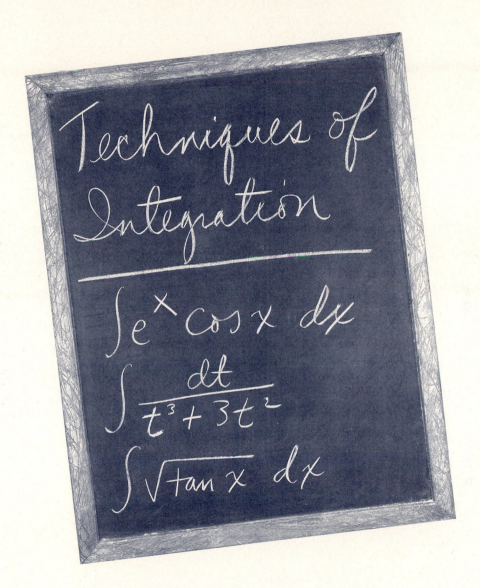

Techniques of Integration

$$\int e^x \cos x \, dx$$

$$\int \frac{dt}{t^3 + 3t^2}$$

$$\int \sqrt{\tan x} \, dx$$

The exact value of a definite integral may be calculated by the second fundamental theorem of the calculus, provided that an antiderivative (or indefinite integral) of the integrand can be found. Some methods for evaluating indefinite integrals have already been given, and additional ones are presented in this chapter.

In Section 9.1, we discuss the very widely used technique of integration, called *integration by parts*, based on the formula for the derivative of a product. The next two sections pertain to integrals of powers of trigonometric functions. Integrands involving powers of the sine and cosine appear in Section 9.2 and those containing powers of the other four trigonometric functions appear in

Section 9.3. Trigonometric substitutions are applied in Section 9.4 to simplify integrands involving $\sqrt{a^2 - x^2}$, $\sqrt{a^2 + x^2}$, and $\sqrt{x^2 - a^2}$. Partial fractions are used to integrate rational functions in Sections 9.5 and 9.6. More substitutions for evaluating indefinite integrals are given in Section 9.7. The integrands in that section either involve fractional powers of a variable or are rational functions of sine and cosine. In Supplementary Section 9.8 the inverse hyperbolic functions are applied in integration to give new forms of results obtained earlier by other methods.

From time to time you may find it desirable to resort to a table of integrals instead of performing a complicated integration. A short table of integrals can be found on the front and back endpapers, and Section A.1 in the appendix provides a discussion of the use of such a table. Sometimes it may be necessary to employ some of the techniques of integration to express the integrand in a form that is found in a table. Therefore you should acquire proficiency in recognizing which technique to apply to a given integral. Furthermore, development of computational skills is important in all branches of mathematics, and the exercises in this chapter provide a good training ground. For this reason you are advised to use a table of integrals only after you have mastered integration.

In practice, it is not always possible to evaluate a definite integral by computing an indefinite integral. That is, we may have a definite integral that exists, but the integrand has no antiderivative that can be expressed in terms of elementary functions. An example of such a definite integral is

$$\int_0^{1/2} e^{-t^2}\, dt$$

A programmable calculator or computer utilizing the numerical methods discussed in Section 5.10 can be applied to approximate the value of this definite integral. Another procedure is given in Example 2 of Section 13.3, where an infinite series is used.

Standard indefinite integration formulas that you learned in previous chapters and that are used frequently are listed below and numbered for reference.

1. $\displaystyle\int du = u + C$

2. $\displaystyle\int a\, du = au + C$ where a is any constant

3. $\displaystyle\int [f(u) + g(u)]\, du = \int f(u)\, du + \int g(u)\, du$

4. $\displaystyle\int u^n\, du = \frac{u^{n+1}}{n+1} + C$ $n \neq -1$

5. $\displaystyle\int \frac{du}{u} = \ln |u| + C$

6. $\displaystyle\int a^u\, du = \frac{a^u}{\ln a} + C$ where $a > 0$ and $a \neq 1$

7. $\displaystyle\int e^u\, du = e^u + C$

8. $\displaystyle\int \sin u\, du = -\cos u + C$

9. $\displaystyle\int \cos u\, du = \sin u + C$

10. $\displaystyle\int \sec^2 u\, du = \tan u + C$

11. $\displaystyle\int \csc^2 u\, du = -\cot u + C$

12. $\displaystyle\int \sec u \tan u\, du = \sec u + C$

13. $\displaystyle\int \csc u \cot u\, du = -\csc u + C$

14. $\displaystyle\int \tan u\, du = \ln|\sec u| + C$

15. $\displaystyle\int \cot u\, du = \ln|\sin u| + C$

16. $\displaystyle\int \sec u\, du = \ln|\sec u + \tan u| + C$

17. $\displaystyle\int \csc u\, du = \ln|\csc u - \cot u| + C$

18. $\displaystyle\int \frac{du}{\sqrt{a^2 - u^2}} = \sin^{-1}\frac{u}{a} + C \qquad$ where $a > 0$

19. $\displaystyle\int \frac{du}{a^2 + u^2} = \frac{1}{a}\tan^{-1}\frac{u}{a} + C \qquad$ where $a \neq 0$

20. $\displaystyle\int \frac{du}{u\sqrt{u^2 - a^2}} = \frac{1}{a}\sec^{-1}\frac{u}{a} + C \qquad$ where $a > 0$

21. $\displaystyle\int \sinh u\, du = \cosh u + C$

22. $\displaystyle\int \cosh u\, du = \sinh u + C$

23. $\displaystyle\int \operatorname{sech}^2 u\, du = \tanh u + C$

24. $\displaystyle\int \operatorname{csch}^2 u\, du = -\coth u + C$

25. $\displaystyle\int \operatorname{sech} u \tanh u\, du = -\operatorname{sech} u + C$

26. $\displaystyle\int \operatorname{csch} u \coth u\, du = -\operatorname{csch} u + C$

9.1 INTEGRATION BY PARTS

From the formula for the derivative of the product of two functions we obtain a very useful method of integration called *integration by parts*. If f and g are differentiable functions, then

$$D_x[f(x)g(x)] = f(x)g'(x) + g(x)f'(x)$$

$$\Leftrightarrow \qquad f(x)g'(x) = D_x[f(x)g(x)] - g(x)f'(x)$$

Integrating on each side of this equation we obtain

$$\int f(x)g'(x)\, dx = \int D_x[f(x)g(x)]\, dx - \int g(x)f'(x)\, dx$$

$$\int f(x)g'(x)\, dx = f(x)g(x) - \int g(x)f'(x)\, dx \tag{1}$$

We call (1) the **formula for integration by parts**. For computational purposes a more convenient way of writing this formula is obtained by letting

$$u = f(x) \quad \text{and} \quad v = g(x)$$

Then

$$du = f'(x)\, dx \quad \text{and} \quad dv = g'(x)\, dx$$

so that (1) becomes

$$\int u \, dv = uv - \int v \, du \tag{2}$$

This formula expresses the integral $\int u \, dv$ in terms of another integral, $\int v \, du$. By a suitable choice of u and dv, it may be easier to evaluate the second integral than the first. When choosing the substitutions for u and dv, we usually want dv to be the most complicated factor of the integrand that can be integrated directly and u to be a function whose derivative is a simpler function. The method is shown by the following illustrations and examples.

▶ **ILLUSTRATION 1** We wish to evaluate

$$\int x \ln x \, dx$$

To determine the substitutions for u and dv, bear in mind that to find v we must be able to integrate dv. This suggests letting $dv = x \, dx$ and $u = \ln x$. Then

$$v = \frac{x^2}{2} + C_1 \quad \text{and} \quad du = \frac{dx}{x}$$

From formula (2)

$$\int x \ln x \, dx = \ln x \left(\frac{x^2}{2} + C_1 \right) - \int \left(\frac{x^2}{2} + C_1 \right) \frac{dx}{x}$$

$$= \frac{x^2}{2} \ln x + C_1 \ln x - \frac{1}{2} \int x \, dx - C_1 \int \frac{dx}{x}$$

$$= \frac{x^2}{2} \ln x + C_1 \ln x - \frac{x^2}{4} - C_1 \ln x + C_2$$

$$= \tfrac{1}{2} x^2 \ln x - \tfrac{1}{4} x^2 + C_2 \qquad \blacktriangleleft$$

In Illustration 1 observe that the first constant of integration C_1 does not appear in the final answer. The C_1 is used only to show that all choices for v of the form $\tfrac{1}{2} x^2 + C_1$ produce the same result for $\int x \ln x \, dx$. This situation is true in general, and we prove it as follows: By writing $v + C_1$ in formula (2) we have

$$\int u \, dv = u(v + C_1) - \int (v + C_1) \, du$$

$$= uv + C_1 u - \int v \, du - C_1 \int du$$

$$= uv + C_1 u - \int v \, du - C_1 u$$

$$= uv - \int v \, du$$

Therefore it is not necessary to write C_1 when finding v from dv.

▶ **ILLUSTRATION 2** The answer in Illustration 1 can be written as

$$\tfrac{1}{2} x^2 (\ln x - \tfrac{1}{2}) + C$$

We check this result by computing the derivative of a product.

$$D_x\left[\frac{1}{2}x^2(\ln x - 1)\right] = \frac{1}{2}x^2\left(\frac{1}{x}\right) + x\left(\ln x - \frac{1}{2}\right)$$

$$= \tfrac{1}{2}x + x\ln x - \tfrac{1}{2}x$$

$$= x\ln x \qquad \blacktriangleleft$$

▶ **ILLUSTRATION 3**   To evaluate

$$\int x^3 e^{x^2}\, dx$$

we use integration by parts with $dv = xe^{x^2}\, dx$ and $u = x^2$. Then

$$v = \tfrac{1}{2}e^{x^2} \quad \text{and} \quad du = 2x\, dx$$

From formula (2)

$$\int x^3 e^{x^2}\, dx = x^2(\tfrac{1}{2}e^{x^2}) - \int (\tfrac{1}{2}e^{x^2})2x\, dx$$

$$= \tfrac{1}{2}x^2 e^{x^2} - \int xe^{x^2}\, dx$$

$$= \tfrac{1}{2}x^2 e^{x^2} - \tfrac{1}{2}e^{x^2} + C \qquad \blacktriangleleft$$

EXAMPLE 1   Evaluate

$$\int x\cos x\, dx$$

Solution   Let $u = x$ and $dv = \cos x\, dx$. Then

$$du = dx \quad \text{and} \quad v = \sin x$$

Therefore

$$\int x\cos x\, dx = x\sin x - \int \sin x\, dx$$

$$= x\sin x + \cos x + C$$

▶ **ILLUSTRATION 4**   In Example 1, if instead of our choices of u and dv as above we let

$$u = \cos x \quad \text{and} \quad dv = x\, dx$$

then

$$du = -\sin x\, dx \quad \text{and} \quad v = \tfrac{1}{2}x^2$$

Thus

$$\int x\cos x\, dx = \frac{x^2}{2}\cos x + \frac{1}{2}\int x^2 \sin x\, dx$$

The integral on the right is more complicated than the one with which we started, thereby indicating that these are not desirable choices for u and dv. $\qquad \blacktriangleleft$

It may happen that a particular integral requires repeated applications of integration by parts. This is illustrated in the following example.

EXAMPLE 2   Evaluate

$$\int x^2 e^x \, dx$$

Solution   Let $u = x^2$ and $dv = e^x \, dx$. Then

$$du = 2x \, dx \quad \text{and} \quad v = e^x$$

We have, then,

$$\int x^2 e^x \, dx = x^2 e^x - 2 \int x e^x \, dx$$

We now apply integration by parts to the integral on the right. Let $\bar{u} = x$ and $d\bar{v} = e^x \, dx$. Then

$$d\bar{u} = dx \quad \text{and} \quad \bar{v} = e^x$$

So we obtain

$$\int x e^x \, dx = x e^x - \int e^x \, dx$$
$$= x e^x - e^x + \bar{C}$$

Therefore

$$\int x^2 e^x \, dx = x^2 e^x - 2(x e^x - e^x + \bar{C})$$
$$= x^2 e^x - 2x e^x + 2 e^x + C \qquad \text{where } C = -2\bar{C}$$

Integration by parts often is used when the integrand involves logarithms, inverse trigonometric functions, and products of functions.

EXAMPLE 3   Evaluate

$$\int \tan^{-1} x \, dx$$

Solution   Let $u = \tan^{-1} x$ and $dv = dx$. Then

$$du = \frac{dx}{1 + x^2} \quad \text{and} \quad v = x$$

Thus

$$\int \tan^{-1} x \, dx = x \tan^{-1} x - \int \frac{x \, dx}{1 + x^2}$$
$$= x \tan^{-1} x - \tfrac{1}{2} \ln(1 + x^2) + C$$

A situation that sometimes occurs when using integration by parts is shown in Example 4.

EXAMPLE 4   Evaluate

$$\int e^x \sin x \, dx$$

Solution   Let $u = e^x$ and $dv = \sin x \, dx$. Then

$$du = e^x \, dx \quad \text{and} \quad v = -\cos x$$

Therefore

$$\int e^x \sin x \, dx = -e^x \cos x + \int e^x \cos x \, dx$$

The integral on the right is similar to the first integral except that it has $\cos x$ in place of $\sin x$. We apply integration by parts again by letting $\bar{u} = e^x$ and $d\bar{v} = \cos x \, dx$. So

$$d\bar{u} = e^x \, dx \quad \text{and} \quad \bar{v} = \sin x$$

Thus

$$\int e^x \sin x \, dx = -e^x \cos x + \left(e^x \sin x - \int e^x \sin x \, dx \right)$$

On the right there is the same integral as on the left. So we add $\int e^x \sin x \, dx$ to both sides of the equation and obtain

$$2 \int e^x \sin x \, dx = -e^x \cos x + e^x \sin x + 2C$$

Observe that the right side of the above equation contains an arbitrary constant because on the left side we have an indefinite integral. This arbitrary constant is written as $2C$ so that when we divide on both sides of the equation by 2, the arbitrary constant in the answer becomes C. Thus we have

$$\int e^x \sin x \, dx = \tfrac{1}{2} e^x (\sin x - \cos x) + C$$

In applying integration by parts to a specific integral, one pair of choices for u and dv may work while another pair may not. We saw this in Illustration 4, and another case occurs in Illustration 5.

▶ **ILLUSTRATION 5**   In Example 4, in the step where we have

$$\int e^x \sin x \, dx = -e^x \cos x + \int e^x \cos x \, dx$$

if we evaluate the integral on the right by letting $\bar{u} = \cos x$ and $d\bar{v} = e^x \, dx$, we have

$$d\bar{u} = -\sin x \, dx \quad \text{and} \quad \bar{v} = e^x$$

Thus we get

$$\int e^x \sin x \, dx = -e^x \cos x + \left(e^x \cos x + \int e^x \sin x \, dx \right)$$

$$= \int e^x \sin x \, dx$$

◀

In Exercises 53 and 54 you are asked to derive the following formulas, where a and n are nonzero real numbers.

$$\int e^{au} \sin nu \, du = \frac{e^{au}}{a^2 + n^2} (a \sin nu - n \cos nu) + C \tag{3}$$

$$\int e^{au} \cos nu \, du = \frac{e^{au}}{a^2 + n^2} (a \cos nu + n \sin nu) + C \tag{4}$$

Integrals of the form of those in (3) and (4) often occur in problems involving electrical circuits as in Exercises 55–57 pertaining to Supplementary Section 7.8.

EXERCISES 9.1

In Exercises 1 through 24, evaluate the indefinite integral.

1. $\int x e^{3x} \, dx$

2. $\int x \cos 2x \, dx$

3. $\int x \sec x \tan x \, dx$

4. $\int x \, 3^x \, dx$

5. $\int \ln x \, dx$

6. $\int \sin^{-1} w \, dw$

7. $\int (\ln x)^2 \, dx$

8. $\int x \sec^2 x \, dx$

9. $\int x \tan^{-1} x \, dx$

10. $\int x^2 \ln x \, dx$

11. $\int \frac{x e^x}{(x + 1)^2} \, dx$

12. $\int x^2 \sin 3x \, dx$

13. $\int \sin x \ln(\cos x) \, dx$

14. $\int \sin(\ln x) \, dx$

15. $\int e^x \cos x \, dx$

16. $\int x^5 e^{x^2} \, dx$

17. $\int \frac{x^3 \, dx}{\sqrt{1 - x^2}}$

18. $\int \frac{\sin 2x}{e^x} \, dx$

19. $\int x^2 \sinh x \, dx$

20. $\int \frac{e^{2x}}{\sqrt{1 - e^x}} \, dx$

21. $\int \frac{\cot^{-1} \sqrt{z}}{\sqrt{z}} \, dz$

22. $\int \cos^{-1} 2x \, dx$

23. $\int \cos \sqrt{x} \, dx$

24. $\int \tan^{-1} \sqrt{x} \, dx$

In Exercises 25 through 34, evaluate the definite integral.

25. $\int_0^2 x^2 3^x \, dx$

26. $\int_{-1}^2 \ln(x + 2) \, dx$

27. $\int_0^{\pi/3} \sin 3x \cos x \, dx$

28. $\int_0^{\pi^2/2} \cos \sqrt{2x} \, dx$

29. $\int_0^2 x e^{2x} \, dx$

30. $\int_{-\pi}^{\pi} z^2 \cos 2z \, dz$

31. $\int_0^{\pi/4} e^{3x} \sin 4x \, dx$

32. $\int_0^1 x \sin^{-1} x \, dx$

33. $\int_2^4 \sec^{-1} \sqrt{t} \, dt$

34. $\int_{\pi/4}^{3\pi/4} x \cot x \csc x \, dx$

35. Find the area of the region bounded by the curve $y = \ln x$, the x axis, and the line $x = e^2$.

36. Find the volume of the solid generated by revolving the region in Exercise 35 about the x axis.

37. Find the volume of the solid generated by revolving the region in Exercise 35 about the y axis.

38. Find the area of the region bounded by the curve $y = x \csc^2 x$, the x axis, and the lines $x = \frac{1}{6}\pi$ and $x = \frac{1}{4}\pi$.

39. Find the area of the region bounded by the curve $y = 2x e^{-x/2}$, the x axis, and the line $x = 4$.

40. Find the volume of the solid of revolution generated by revolving about the x axis the region of Exercise 39.

41. The linear density of a rod at a point x meters from one end is $2e^{-x}$ kilograms per meter. If the rod is 6 m long, find the mass and center of mass of the rod.

42. Find the centroid of the region bounded by the curve $y = e^x$, the coordinate axes, and the line $x = 3$.

43. Find the centroid of the region in the first quadrant bounded by the curves $y = \sin x$ and $y = \cos x$, and the y axis.

44. The region in the first quadrant bounded by the curve $y = \cos x$ and the lines $y = 1$ and $x = \frac{1}{2}\pi$ is revolved about the line $x = \frac{1}{2}\pi$. Find the volume of the solid generated.

45. A water tank full of water is in the shape of the solid of revolution formed by rotating about the x axis the region bounded by the curve $y = e^{-x}$, the coordinate axes, and the line $x = 4$. Find the work done in pumping all the water to the top of the tank. Distance is measured in feet. Take the positive x axis vertically downward.

46. A particle is moving along a straight line, and s feet is the directed distance of the particle from the origin at t seconds. If v feet per second is the velocity at t seconds, $s = 0$ when $t = 0$, and $v \cdot s = t \sin t$, find s in terms of t and also s when $t = \frac{1}{2}\pi$.

47. The marginal cost function is C' and $C'(x) = \ln x$, where $x > 1$. Find the total cost function if $C(x)$ dollars is the total cost of producing x units and $C(1) = 5$.

48. A manufacturer has discovered that if $100x$ units of a particular commodity are produced per week, the marginal cost is determined by $x2^{x/2}$ and the marginal revenue is determined by $8 \cdot 2^{-x/2}$, where both the production cost and the revenue are in thousands of dollars. If the weekly fixed costs amount to $2000, find the maximum weekly profit that can be obtained.

49. (a) Derive the following formula, where r is any real number:

$$\int x^r \ln x \, dx = \begin{cases} \dfrac{x^{r+1}}{r+1} \ln x - \dfrac{x^{r+1}}{(r+1)^2} + C & \text{if } r \neq -1 \\ \frac{1}{2}(\ln x)^2 + C & \text{if } r = -1 \end{cases}$$

(b) Use the formula derived in (a) to find $\int x^3 \ln x \, dx$.

50. (a) Derive the following formula, where r and q are any real numbers:

$$\int x^r (\ln x)^q \, dx = \begin{cases} \dfrac{x^{r+1}(\ln x)^q}{r+1} - \dfrac{q}{r+1} \int x^r (\ln x)^{q-1} \, dx & \text{if } r \neq -1 \\ \dfrac{(\ln x)^{q+1}}{q+1} + C & \text{if } r = -1 \text{ and } q \neq -1 \end{cases}$$

(b) Use the formula derived in (a) to find $\int x^4 (\ln x)^2 \, dx$.

51. (a) Show that if $m \geq 2$ and m is an integer,

$$\int \sec^m x \, dx = \frac{\sec^{m-2} x \tan x}{m-1} + \frac{m-2}{m-1} \int \sec^{m-2} x \, dx$$

(b) Use the formula in part (a) to find $\int \sec^6 x \, dx$.

52. (a) Derive the following formula, where r is any real number:

$$\int x^r e^x \, dx = x^r e^x - r \int x^{r-1} e^x \, dx$$

(b) Use the formula derived in (a) to find $\int x^4 e^x \, dx$.

53. Derive formula (3).

54. Derive formula (4).

Exercises 55 through 57 pertain to Supplementary Section 7.8.

55. An electrical circuit has an electromotive force of $10 \sin 10t$ volts at t seconds and a resistor of 3 ohms and an inductor of 1 henry connected in series. If the current is i amperes at t seconds and $i = 0$ when $t = 0$, find (a) i when $t = 0.1$ and (b) i when $t = 3$.

56. An electrical circuit has an electromotive force of $100 \sin 200t$ volts at t seconds and a resistor of 10 ohms and an inductor of 0.1 henry connected in series. (a) If the current is i amperes at t seconds and $i = 0$ when $t = 0$, express i as a function of t. (b) Give an approximate value of i for large values of t in terms of only sine and cosine functions.

57. Do Exercise 56 if the electrical circuit has an electromotive force of $120 \sin 120\pi t$ volts and a resistor of 100 ohms and the other conditions are the same.

9.2 INTEGRATION OF POWERS OF SINE AND COSINE

There are four cases of indefinite integrals involving powers of sine and cosine, dependent on whether the powers are odd or even.

Case 1: $\int \sin^n u \, du$ or $\int \cos^n u \, du$, where n is an odd integer.

▶ **ILLUSTRATION 1**

$$\int \cos^3 x \, dx = \int \cos^2 x (\cos x \, dx)$$

$$= \int (1 - \sin^2 x)(\cos x \, dx)$$

$$= \int \cos x \, dx - \int \sin^2 x \cos x \, dx \tag{1}$$

For the second integral in (1) observe that because $d(\sin x) = \cos x \, dx$, we have

$$\int \sin^2 x (\cos x \, dx) = \tfrac{1}{3} \sin^3 x + C_1$$

Because the first integral in (1) is $\sin x + C_2$,

$$\int \cos^3 x \, dx = \sin x - \tfrac{1}{3} \sin^3 x + C$$

◀

EXAMPLE 1   Evaluate

$$\int \sin^5 x \, dx$$

Solution

$$
\begin{aligned}
\int \sin^5 x \, dx &= \int (\sin^2 x)^2 \sin x \, dx \\
&= \int (1 - \cos^2 x)^2 \sin x \, dx \\
&= \int (1 - 2\cos^2 x + \cos^4 x) \sin x \, dx \\
&= \int \sin x \, dx - 2 \int \cos^2 x \sin x \, dx + \int \cos^4 x \sin x \, dx \\
&= -\cos x + 2 \int \cos^2 x(-\sin x \, dx) - \int \cos^4 x(-\sin x \, dx) \\
&= -\cos x + \tfrac{2}{3} \cos^3 x - \tfrac{1}{5} \cos^5 x + C
\end{aligned}
$$

Case 2: $\int \sin^n x \cos^m x \, dx$, where at least one of the exponents is odd.
The solution of this case is similar to the method used for Case 1.

▶ **ILLUSTRATION 2**

$$
\begin{aligned}
\int \sin^3 x \cos^4 x \, dx &= \int \sin^2 x \cos^4 x(\sin x \, dx) \\
&= \int (1 - \cos^2 x) \cos^4 x(\sin x \, dx) \\
&= \int \cos^4 x \sin x \, dx - \int \cos^6 x \sin x \, dx \\
&= -\tfrac{1}{5} \cos^5 x + \tfrac{1}{7} \cos^7 x + C
\end{aligned}
$$
◀

Case 3: $\int \sin^n u \, du$ and $\int \cos^n u \, du$, where n is an even integer.
The method used in Cases 1 and 2 does not work for this case. We use the following identities from trigonometry:

$$\sin^2 x = \frac{1 - \cos 2x}{2} \qquad \cos^2 x = \frac{1 + \cos 2x}{2}$$

▶ **ILLUSTRATION 3**

$$
\begin{aligned}
\int \sin^2 x \, dx &= \int \frac{1 - \cos 2x}{2} \, dx \\
&= \tfrac{1}{2}x - \tfrac{1}{4} \sin 2x + C
\end{aligned}
$$
◀

Case 4: $\int \sin^n x \cos^m x \, dx$, where both m and n are even.
The solution of this case is similar to the method used for Case 3.

EXAMPLE 2   Evaluate

$$\int \sin^2 x \cos^4 x \, dx$$

Solution

$$\int \sin^2 x \cos^4 x \, dx$$

$$= \int \left(\frac{1 - \cos 2x}{2}\right)\left(\frac{1 + \cos 2x}{2}\right)^2 dx$$

$$= \frac{1}{8}\int dx + \frac{1}{8}\int \cos 2x \, dx - \frac{1}{8}\int \cos^2 2x \, dx - \frac{1}{8}\int \cos^3 2x \, dx$$

$$= \frac{1}{8}x + \frac{1}{16}\sin 2x - \frac{1}{8}\int \frac{1 + \cos 4x}{2} dx - \frac{1}{8}\int (1 - \sin^2 2x)\cos 2x \, dx$$

$$= \frac{x}{8} + \frac{\sin 2x}{16} - \frac{x}{16} - \frac{\sin 4x}{64} - \frac{1}{8}\int \cos 2x \, dx + \frac{1}{8}\int \sin^2 2x \cos 2x \, dx$$

$$= \frac{x}{16} + \frac{\sin 2x}{16} - \frac{\sin 4x}{64} - \frac{\sin 2x}{16} + \frac{\sin^3 2x}{48} + C$$

$$= \frac{x}{16} + \frac{\sin^3 2x}{48} - \frac{\sin 4x}{64} + C$$

EXAMPLE 3   Evaluate

$$\int \sin^4 x \cos^4 x \, dx$$

Solution   Using the identity $\sin x \cos x = \frac{1}{2}\sin 2x$, we have

$$\int \sin^4 x \cos^4 x \, dx = \frac{1}{16}\int \sin^4 2x \, dx$$

$$= \frac{1}{16}\int \left(\frac{1 - \cos 4x}{2}\right)^2 dx$$

$$= \frac{1}{64}\int dx - \frac{1}{32}\int \cos 4x \, dx + \frac{1}{64}\int \cos^2 4x \, dx$$

$$= \frac{x}{64} - \frac{\sin 4x}{128} + \frac{1}{64}\int \frac{1 + \cos 8x}{2} dx$$

$$= \frac{x}{64} - \frac{\sin 4x}{128} + \frac{x}{128} + \frac{\sin 8x}{1024} + C$$

$$= \frac{3x}{128} - \frac{\sin 4x}{128} + \frac{\sin 8x}{1024} + C$$

EXAMPLE 4   Find the centroid of the region in the first quadrant to the left of the line $x = \frac{1}{2}\pi$ and bounded by the curve $y = \sin x$, the x axis, and the line $x = \frac{1}{2}\pi$.

Solution   We use the symbols A, M_x, M_y, $\bar{x}$, and $\bar{y}$, as defined in Section 6.4. The region and the ith rectangular element are shown in Figure 1. We first compute the area of the region.

$$A = \lim_{\|\Delta\| \to 0} \sum_{i=1}^{n} \sin \gamma_i \, \Delta_i x$$

$$= \int_0^{\pi/2} \sin x \, dx$$

$$= -\cos x \Big]_0^{\pi/2}$$

$$= 1$$

We apply Definition 6.4.2 to compute M_x and M_y. To evaluate the integral for M_y we use integration by parts with $u = x$ and $dv = \sin x \, dx$.

$$M_x = \lim_{\|\Delta\| \to 0} \sum_{i=1}^{n} \tfrac{1}{2}[\sin \gamma_i]^2 \, \Delta_i x \qquad M_y = \lim_{\|\Delta\| \to 0} \sum_{i=1}^{n} \gamma_i \sin \gamma_i \, \Delta_i x$$

$$= \frac{1}{2} \int_0^{\pi/2} \sin^2 x \, dx \qquad\qquad = \int_0^{\pi/2} x \sin x \, dx$$

$$= \frac{1}{2} \int_0^{\pi/2} \frac{1 - \cos 2x}{2} \, dx \qquad = -x \cos x \Big]_0^{\pi/2} + \int_0^{\pi/2} \cos x \, dx$$

$$= \tfrac{1}{4}\left[x - \tfrac{1}{2}\sin 2x\right]_0^{\pi/2} \qquad = -x \cos x + \sin x \Big]_0^{\pi/2}$$

$$= \tfrac{1}{4}\left[\tfrac{1}{2}\pi\right] \qquad\qquad\qquad = -\tfrac{1}{2}\pi \cos \tfrac{1}{2}\pi + \sin \tfrac{1}{2}\pi$$

$$= \tfrac{1}{8}\pi \qquad\qquad\qquad\qquad = 1$$

Therefore

$$\bar{x} = \frac{M_y}{A} \qquad \bar{y} = \frac{M_x}{A}$$

$$= \frac{1}{1} \qquad\qquad = \frac{\tfrac{1}{8}\pi}{1}$$

$$= 1 \qquad\qquad = \tfrac{1}{8}\pi$$

Thus the centroid is at the point $(1, \tfrac{1}{8}\pi)$.

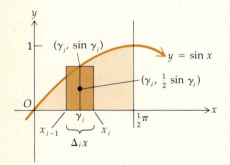

FIGURE 1

The next example involves another type of integral containing a product of a sine and a cosine.

EXAMPLE 5   Evaluate

$$\int \sin 3x \cos 2x \, dx$$

Solution   We use the following identity from trigonometry:

$$\sin mx \cos nx = \tfrac{1}{2} \sin(m - n)x + \tfrac{1}{2} \sin(m + n)x$$

$$\int \sin 3x \cos 2x \, dx = \int (\tfrac{1}{2} \sin x + \tfrac{1}{2} \sin 5x) \, dx$$

$$= \tfrac{1}{2} \int \sin x \, dx + \tfrac{1}{2} \int \sin 5x \, dx$$

$$= -\tfrac{1}{2} \cos x - \tfrac{1}{10} \cos 5x + C$$

EXERCISES 9.2

In Exercises 1 through 24, evaluate the indefinite integral.

1. $\int \sin^4 x \cos x \, dx$

2. $\int \sin^5 x \cos x \, dx$

3. $\int \cos^3 4x \sin 4x \, dx$

4. $\int \cos^6 \tfrac{1}{2}x \sin \tfrac{1}{2}x \, dx$

5. $\int \sin^3 x \, dx$

6. $\int \sin^2 3x \, dx$

7. $\int \sin^4 z \, dz$

8. $\int \cos^5 x \, dx$

9. $\int \cos^2 \tfrac{1}{2}x \, dx$

10. $\int \sin^3 x \cos^3 x \, dx$

11. $\int \sin^2 x \cos^3 x \, dx$

12. $\int \cos^6 x \, dx$

13. $\int \sin^5 x \cos^2 x \, dx$

14. $\int \sin^2 2t \cos^4 2t \, dt$

15. $\int \sin^2 3t \cos^2 3t \, dt$

16. $\int \sqrt{\cos z} \sin^3 z \, dz$

17. $\int \dfrac{\cos^3 3x}{\sqrt[3]{\sin 3x}} \, dx$

18. $\int \sin^3 \tfrac{1}{2}y \cos^2 \tfrac{1}{2}y \, dy$

19. $\int \cos 4x \cos 3x \, dx$

20. $\int \sin 2x \cos 4x \, dx$

21. $\int \sin 3y \cos 5y \, dy$

22. $\int \cos t \cos 3t \, dt$

23. $\int (\sin 3t - \sin 2t)^2 \, dt$

24. $\int \sin x \sin 3x \sin 5x \, dx$

In Exercises 25 through 32, evaluate the definite integral.

25. $\int_0^{\pi/2} \cos^3 x \, dx$

26. $\int_0^1 \sin^3 \tfrac{1}{2}\pi t \, dt$

27. $\int_0^1 \sin^4 \tfrac{1}{2}\pi x \, dx$

28. $\int_0^{\pi/3} \sin^3 t \cos^2 t \, dt$

29. $\int_0^1 \sin^2 \pi t \cos^2 \pi t \, dt$

30. $\int_0^{\pi/6} \sin 2x \cos 4x \, dx$

31. $\int_0^{\pi/8} \sin 3x \cos 5x \, dx$

32. $\int_0^{\pi/2} \sin^2 \tfrac{1}{2}x \cos^2 \tfrac{1}{2}x \, dx$

33. Evaluate $\int 2 \sin x \cos x \, dx$ by three methods: (a) Make the substitution $u = \sin x$; (b) make the substitution $u = \cos x$; (c) use the identity $2 \sin x \cos x = \sin 2x$. Explain the difference in appearance of the answers obtained in (a), (b), and (c).

34. If n is any positive integer, prove that

$$\int_0^\pi \sin^2 nx \, dx = \tfrac{1}{2}\pi$$

35. If n is a positive odd integer, prove that

$$\int_0^\pi \cos^n x \, dx = 0$$

In Exercises 36 through 38, m and n are any positive integers; show that the formula is true.

36. $\int_{-1}^1 \cos n\pi x \cos m\pi x \, dx = \begin{cases} 0 & \text{if } m \neq n \\ 1 & \text{if } m = n \end{cases}$

37. $\int_{-1}^1 \cos n\pi x \sin m\pi x \, dx = 0$

38. $\int_{-1}^1 \sin n\pi x \sin m\pi x \, dx = \begin{cases} 0 & \text{if } m \neq n \\ 1 & \text{if } m = n \end{cases}$

39. Find the area of the region bounded by the curve $y = \sin^2 x$ and the x axis from $x = 0$ to $x = \pi$.

40. Find the volume of the solid of revolution generated by revolving one arch of the sine curve about the x axis.

41. Find the volume of the solid of revolution generated if the region of Exercise 39 is revolved about the x axis.

42. The region in the first quadrant bounded by the curve $y = \cos x$ and the lines $y = 1$ and $x = \tfrac{1}{2}\pi$ is revolved about the x axis. Find the volume of the solid generated.

43. Find the volume of the solid of revolution generated if the region of Exercise 39 is revolved about the line $y = 1$.

44. The region bounded by the y axis and the curves $y = \sin x$ and $y = \cos x$ for $0 \leq x \leq \tfrac{1}{4}\pi$ is revolved about the x axis. Find the volume of the solid of revolution generated.

45. Find the centroid of the region from $x = 1$ to $x = \tfrac{1}{2}\pi$ bounded by the curve $y = \cos x$ and the x axis.

46. Find the centroid of the region described in Exercise 44.

47. Prove:

$$\int \sin^n u \, du = -\frac{1}{n} \sin^{n-1} u \cos u + \frac{n-1}{n} \int \sin^{n-2} u \, du$$

if n is a positive integer greater than 1.

48. Prove:

$$\int \cos^n u\, du = \frac{1}{n} \cos^{n-1} u \sin u + \frac{n-1}{n} \int \cos^{n-2} u\, du$$

if n is a positive integer greater than 1.

Exercise 49 pertains to Supplementary Section 6.7.

49. The face of a dam is in the shape of one arch of the curve $y = -100 \cos \frac{1}{200}\pi x$, $x \in [-100, 100]$, and the surface of the water is at the top of the dam. Find the force due to water pressure on the face of the dam if distance is measured in feet.

9.3 INTEGRATION OF POWERS OF TANGENT, COTANGENT, SECANT, AND COSECANT

Recall the following integration formulas involving tangent, cotangent, secant, and cosecant:

$$\int \tan u\, du = \ln|\sec u| + C \qquad \int \cot u\, du = \ln|\sin u| + C$$

$$\int \sec u\, du = \ln|\sec u + \tan u| + C \qquad \int \csc u\, du = \ln|\csc u - \cot u| + C$$

$$\int \sec^2 u\, du = \tan u + C \qquad \int \csc^2 u\, du = -\cot u + C$$

$$\int \sec u \tan u\, du = \sec u + C \qquad \int \csc u \cot u\, du = -\csc u + C$$

With these formulas and the trigonometric identities

$$1 + \tan^2 u = \sec^2 u \qquad 1 + \cot^2 u = \csc^2 u$$

we can evaluate integrals of the form

$$\int \tan^m u \sec^n u\, du \quad \text{and} \quad \int \cot^m u \csc^n u\, du \tag{1}$$

where m and n are nonnegative integers.

▶ **ILLUSTRATION 1**

(a) $\displaystyle \int \tan^2 x\, dx = \int (\sec^2 x - 1)\, dx$

$\displaystyle \qquad\qquad = \int \sec^2 x\, dx - \int dx$

$\displaystyle \qquad\qquad = \tan x - x + C$

(b) $\displaystyle \int \cot^2 x\, dx = \int (\csc^2 x - 1)\, dx$

$\displaystyle \qquad\qquad = \int \csc^2 x\, dx - \int dx$

$\displaystyle \qquad\qquad = -\cot x - x + C$ ◀

We now distinguish various cases of integrals of the form (1).

Case 1: $\displaystyle \int \tan^n u\, du$ or $\displaystyle \int \cot^n u\, du$, where n is a positive integer.
We write

$$\tan^n u = \tan^{n-2} u \tan^2 u \qquad\qquad \cot^n u = \cot^{n-2} u \cot^2 u$$

$$= \tan^{n-2} u(\sec^2 u - 1) \qquad\qquad = \cot^{n-2} u(\csc^2 u - 1)$$

EXAMPLE 1   Evaluate

$$\int \tan^3 x\, dx$$

Solution

$$\int \tan^3 x \, dx = \int \tan x (\sec^2 x - 1) \, dx$$

$$= \int \tan x \sec^2 x \, dx - \int \tan x \, dx$$

$$= \tfrac{1}{2} \tan^2 x + \ln|\cos x| + C$$

EXAMPLE 2   Evaluate

$$\int \cot^4 3x \, dx$$

Solution

$$\int \cot^4 3x \, dx = \int \cot^2 3x (\csc^2 3x - 1) \, dx$$

$$= \int \cot^2 3x \csc^2 3x \, dx - \int \cot^2 3x \, dx$$

$$= \tfrac{1}{9}(-\cot^3 3x) - \int (\csc^2 3x - 1) \, dx$$

$$= -\tfrac{1}{9} \cot^3 3x + \tfrac{1}{3} \cot 3x + x + C$$

Case 2: $\int \sec^n u \, du$ or $\int \csc^n u \, du$, where n is a positive even integer.
We write

$$\sec^n u = \sec^{n-2} u \sec^2 u \qquad\qquad \csc^n u = \csc^{n-2} u \csc^2 u$$
$$= (\tan^2 u + 1)^{(n-2)/2} \sec^2 u \qquad = (\cot^2 u + 1)^{(n-2)/2} \csc^2 u$$

EXAMPLE 3   Evaluate

$$\int \csc^6 x \, dx$$

Solution

$$\int \csc^6 x \, dx = \int (\cot^2 x + 1)^2 \csc^2 x \, dx$$

$$= \int \cot^4 x \csc^2 x \, dx + 2 \int \cot^2 x \csc^2 x \, dx + \int \csc^2 x \, dx$$

$$= -\tfrac{1}{5} \cot^5 x - \tfrac{2}{3} \cot^3 x - \cot x + C$$

Case 3: $\int \sec^n u \, du$ or $\int \csc^n u \, du$, where n is a positive odd integer.
   To integrate odd powers of the secant and cosecant we use integration by parts. The process is illustrated in the following example.

EXAMPLE 4   Evaluate

$$\int \sec^3 x \, dx$$

Solution   Let $u = \sec x$ and $dv = \sec^2 x \, dx$. Then

$$du = \sec x \tan x \, dx \quad \text{and} \quad v = \tan x$$

Therefore

$$\int \sec^3 x \, dx = \sec x \tan x - \int \sec x \tan^2 x \, dx$$

$$\int \sec^3 x \, dx = \sec x \tan x - \int \sec x(\sec^2 x - 1) \, dx$$

$$\int \sec^3 x \, dx = \sec x \tan x - \int \sec^3 x \, dx + \int \sec x \, dx$$

Adding $\int \sec^3 x \, dx$ to both sides we get

$$2 \int \sec^3 x \, dx = \sec x \tan x + \ln|\sec x + \tan x| + 2C$$

$$\int \sec^3 x \, dx = \tfrac{1}{2} \sec x \tan x + \tfrac{1}{2} \ln|\sec x + \tan x| + C$$

Case 4: $\int \tan^m u \sec^n u \, du$ or $\int \cot^m u \csc^n u \, du$, where n is a positive even integer.

This case is illustrated by the following example.

EXAMPLE 5   Evaluate

$$\int \tan^5 x \sec^4 x \, dx$$

Solution

$$\int \tan^5 x \sec^4 x \, dx = \int \tan^5 x(\tan^2 x + 1) \sec^2 x \, dx$$

$$= \int \tan^7 x \sec^2 x \, dx + \int \tan^5 x \sec^2 x \, dx$$

$$= \tfrac{1}{8} \tan^8 x + \tfrac{1}{6} \tan^6 x + C$$

Case 5: $\int \tan^m u \sec^n u \, du$ or $\int \cot^m u \csc^n u \, du$, where m is a positive odd integer.

The next example illustrates this case.

EXAMPLE 6   Evaluate

$$\int \tan^5 x \sec^7 x \, dx$$

Solution

$$\int \tan^5 x \sec^7 x \, dx = \int \tan^4 x \sec^6 x \sec x \tan x \, dx$$

$$= \int (\sec^2 x - 1)^2 \sec^6 x(\sec x \tan x \, dx)$$

$$= \int \sec^{10} x(\sec x \tan x \, dx) - 2 \int \sec^8 x(\sec x \tan x \, dx) + \int \sec^6 x(\sec x \tan x \, dx)$$

$$= \tfrac{1}{11} \sec^{11} x - \tfrac{2}{9} \sec^9 x + \tfrac{1}{7} \sec^7 x + C$$

Case 6: $\int \tan^m u \sec^n u \, du$ or $\int \cot^m u \csc^n u \, du$, where m is a positive even integer and n is a positive odd integer.

The integrand can be expressed in terms of odd powers of secant or cosecant. For example,

$$\int \tan^2 x \sec^3 x \, dx = \int (\sec^2 x - 1) \sec^3 x \, dx$$

$$= \int \sec^5 x \, dx - \int \sec^3 x \, dx$$

To evaluate each of these integrals we use integration by parts, as indicated in Case 3.

EXERCISES 9.3

In Exercises 1 through 30, evaluate the indefinite integral.

1. $\int \tan^2 5x \, dx$ **2.** $\int \cot^2 4t \, dt$ **3.** $\int x \cot^2 2x^2 \, dx$

4. $\int e^x \tan^2(e^x) \, dx$ **5.** $\int \cot^3 t \, dt$ **6.** $\int \tan^4 x \, dx$

7. $\int \tan^6 3x \, dx$ **8.** $\int \cot^5 2x \, dx$ **9.** $\int \sec^4 x \, dx$

10. $\int \csc^4 x \, dx$ **11.** $\int \csc^3 x \, dx$ **12.** $\int \sec^5 x \, dx$

13. $\int e^x \tan^4(e^x) \, dx$

14. $\int \dfrac{\sec^4(\ln x)}{x} \, dx$

15. $\int \tan^6 x \sec^4 x \, dx$

16. $\int \tan^5 x \sec^3 x \, dx$

17. $\int \cot^2 3x \csc^4 3x \, dx$

18. $\int (\sec 5x + \csc 5x)^2 \, dx$

19. $\int (\tan 2x + \cot 2x)^2 \, dx$

20. $\int \dfrac{dx}{1 + \cos x}$

21. $\int \dfrac{2 \sin w - 1}{\cos^2 w} \, dw$

22. $\int \dfrac{\tan^3 \sqrt{x}}{\sqrt{x}} \, dx$

23. $\int \tan^5 3x \, dx$

24. $\int \dfrac{\tan^4 y}{\sec^5 y} \, dy$

25. $\int \dfrac{du}{1 + \sec \frac{1}{2}u}$

26. $\int \dfrac{\csc^4 x}{\cot^2 x} \, dx$

27. $\int \dfrac{\sec^3 x}{\tan^4 x} \, dx$

28. $\int \dfrac{\sin^2 \pi x}{\cos^6 \pi x} \, dx$

29. $\int \dfrac{\tan^3(\ln x) \sec^6(\ln x)}{x} \, dx$

30. $\int \dfrac{\sec^4 w}{\sqrt{\tan w}} \, dw$

In Exercises 31 through 36, evaluate the definite integral.

31. $\int_{\pi/16}^{\pi/12} \tan^3 4x \, dx$

32. $\int_{\pi/8}^{\pi/6} 3 \sec^4 2t \, dt$

33. $\int_{-\pi/4}^{\pi/4} \sec^6 x \, dx$

34. $\int_0^{\pi/3} \dfrac{\tan^3 x}{\sec x} \, dx$

35. $\int_{\pi/4}^{\pi/2} \dfrac{\cos^4 t}{\sin^6 t} \, dt$

36. $\int_{\pi/6}^{\pi/4} \cot^3 w \, dw$

37. Find the area of the region bounded by the curve $y = \tan^2 x$, the x axis, and the line $x = \frac{1}{4}\pi$.

38. Find the volume of the solid of revolution generated if the region bounded by the curve $y = 3 \csc^3 x$, the x axis, and the lines $x = \frac{1}{6}\pi$ and $x = \frac{1}{2}\pi$ is revolved about the x axis.

39. Find the volume of the solid of revolution generated if the region bounded by the curve $y = \sec^2 x$, the axes, and the line $x = \frac{1}{4}\pi$ is revolved about the x axis.

40. Prove: $\int \cot x \csc^n x \, dx = -\dfrac{\csc^n x}{n} + C$ if $n \neq 0$.

41. Prove: $\int \tan^n x \, dx = \dfrac{\tan^{n-1} x}{n-1} - \int \tan^{n-2} x \, dx$ if n is a positive integer greater than 1.

42. Derive a formula similar to that in Exercise 40 for $\int \tan x \sec^n x \, dx$ if $n \neq 0$.

43. Derive a formula similar to that in Exercise 41 for $\int \cot^n x \, dx$, if n is a positive integer greater than 1.

9.4   INTEGRATION BY TRIGONOMETRIC SUBSTITUTION

If the integrand contains an expression of the form $\sqrt{a^2 - u^2}$, $\sqrt{a^2 + u^2}$, or $\sqrt{u^2 - a^2}$, where $a > 0$, it is often possible to perform the integration by making a trigonometric substitution that results in an integral involving trigonometric functions. We consider each form as a separate case.

Case 1: The integrand contains an expression of the form $\sqrt{a^2 - u^2}$, where $a > 0$.

We introduce a new variable θ by letting $u = a \sin \theta$, where

$$0 \le \theta \le \tfrac{1}{2}\pi \quad \text{if } u \ge 0 \quad \text{and} \quad -\tfrac{1}{2}\pi \le \theta < 0 \quad \text{if } u < 0$$

Then $du = a \cos \theta \, d\theta$, and

$$\sqrt{a^2 - u^2} = \sqrt{a^2 - a^2 \sin^2 \theta}$$
$$= \sqrt{a^2(1 - \sin^2 \theta)}$$
$$= a\sqrt{\cos^2 \theta}$$

Because $-\tfrac{1}{2}\pi \le \theta \le \tfrac{1}{2}\pi$, $\cos \theta \ge 0$. Hence $\sqrt{\cos^2 \theta} = \cos \theta$, and

$$\sqrt{a^2 - u^2} = a \cos \theta$$

Because $\sin \theta = u/a$ and $-\tfrac{1}{2}\pi \le \theta \le \tfrac{1}{2}\pi$,

$$\theta = \sin^{-1} \frac{u}{a}$$

EXAMPLE 1   Evaluate

$$\int \frac{\sqrt{9 - x^2}}{x^2} \, dx$$

Solution   Let $x = 3 \sin \theta$, where $0 < \theta \le \tfrac{1}{2}\pi$ if $x > 0$ and $-\tfrac{1}{2}\pi \le \theta < 0$ if $x < 0$. Then $dx = 3 \cos \theta \, d\theta$ and

$$\sqrt{9 - x^2} = \sqrt{9 - 9 \sin^2 \theta}$$
$$= 3\sqrt{\cos^2 \theta}$$
$$= 3 \cos \theta$$

Therefore

$$\int \frac{\sqrt{9 - x^2}}{x^2} \, dx = \int \frac{3 \cos \theta}{9 \sin^2 \theta} (3 \cos \theta \, d\theta)$$

$$= \int \cot^2 \theta \, d\theta$$

$$= \int (\csc^2 \theta - 1) \, d\theta$$

$$= -\cot \theta - \theta + C$$

Because $\sin \theta = \tfrac{1}{3}x$ and $-\tfrac{1}{2}\pi \le \theta \le \tfrac{1}{2}\pi$, $\theta = \sin^{-1} \tfrac{1}{3}x$. To find $\cot \theta$, refer to Figures 1 (for $x > 0$) and 2 (for $x < 0$). Observe that in either case $\cot \theta = \sqrt{9 - x^2}/x$. Therefore

$$\int \frac{\sqrt{9 - x^2}}{x^2} \, dx = -\frac{\sqrt{9 - x^2}}{x} - \sin^{-1} \frac{x}{3} + C$$

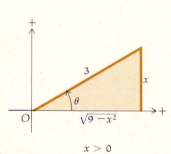

$x > 0$

FIGURE 1

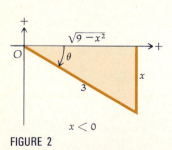

$x < 0$

FIGURE 2

Case 2. The integrand contains an expression of the form $\sqrt{a^2 + u^2}$, where $a > 0$.

We introduce a new variable θ by letting $u = a \tan \theta$, where

$$0 \le \theta < \tfrac{1}{2}\pi \quad \text{if } u \ge 0 \quad \text{and} \quad -\tfrac{1}{2}\pi < \theta < 0 \quad \text{if } u < 0$$

Then $du = a \sec^2 \theta \, d\theta$, and

$$\sqrt{a^2 + u^2} = \sqrt{a^2 + a^2 \tan^2 \theta}$$
$$= a\sqrt{1 + \tan^2 \theta}$$
$$= a\sqrt{\sec^2 \theta}$$

Because $-\frac{1}{2}\pi < \theta < \frac{1}{2}\pi$, $\sec \theta \geq 1$. Thus $\sqrt{\sec^2 \theta} = \sec \theta$, and

$$\sqrt{a^2 + u^2} = a \sec \theta$$

Because $\tan \theta = u/a$ and $-\frac{1}{2}\pi < \theta < \frac{1}{2}\pi$,

$$\theta = \tan^{-1} \frac{u}{a}$$

EXAMPLE 2 Evaluate

$$\int \sqrt{x^2 + 5} \, dx$$

Solution Substitute $x = \sqrt{5} \tan \theta$, where $0 \leq \theta < \frac{1}{2}\pi$ if $x \geq 0$ and $-\frac{1}{2}\pi < \theta < 0$ if $x < 0$. Then $dx = \sqrt{5} \sec^2 \theta \, d\theta$ and

$$\sqrt{x^2 + 5} = \sqrt{5 \tan^2 \theta + 5}$$
$$= \sqrt{5}\sqrt{\sec^2 \theta}$$
$$= \sqrt{5} \sec \theta$$

Therefore

$$\int \sqrt{x^2 + 5} \, dx = \int \sqrt{5} \sec \theta (\sqrt{5} \sec^2 \theta \, d\theta)$$
$$= 5 \int \sec^3 \theta \, d\theta$$

Using the result of Example 4 of Section 9.3 we have

$$\int \sqrt{x^2 + 5} \, dx = \frac{5}{2} \sec \theta \tan \theta + \frac{5}{2} \ln|\sec \theta + \tan \theta| + C$$

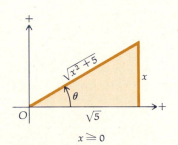

FIGURE 3

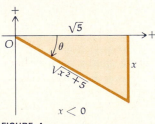

FIGURE 4

We determine $\sec \theta$ from Figures 3 (for $x \geq 0$) and 4 (for $x < 0$), where $\tan \theta = x/\sqrt{5}$. We see in either case that $\sec \theta = \sqrt{x^2 + 5}/\sqrt{5}$. Hence

$$\int \sqrt{x^2 + 5} \, dx = \frac{5}{2} \cdot \frac{\sqrt{x^2 + 5}}{\sqrt{5}} \cdot \frac{x}{\sqrt{5}} + \frac{5}{2} \ln \left| \frac{\sqrt{x^2 + 5}}{\sqrt{5}} + \frac{x}{\sqrt{5}} \right| + C$$
$$= \frac{1}{2}x\sqrt{x^2 + 5} + \frac{5}{2} \ln|\sqrt{x^2 + 5} + x| - \frac{5}{2} \ln \sqrt{5} + C$$
$$= \frac{1}{2}x\sqrt{x^2 + 5} + \frac{5}{2} \ln(\sqrt{x^2 + 5} + x) + C_1$$

Observe that we replaced $-\frac{5}{2} \ln \sqrt{5} + C$ by the arbitrary constant C_1. Furthermore, because $\sqrt{x^2 + 5} + x > 0$, we delete the absolute-value bars.

Case 3: The integrand contains an expression of the form $\sqrt{u^2 - a^2}$, where $a > 0$.

We introduce a new variable by letting $u = a \sec \theta$, where

$$0 \leq \theta < \frac{1}{2}\pi \quad \text{if } u \geq a \quad \text{and} \quad \pi \leq \theta < \frac{3}{2}\pi \quad \text{if } u \leq -a$$

Then $du = a \sec \theta \tan \theta \, d\theta$ and

$$\sqrt{u^2 - a^2} = \sqrt{a^2 \sec^2 \theta - a^2}$$
$$= \sqrt{a^2(\sec^2 \theta - 1)}$$
$$= a\sqrt{\tan^2 \theta}$$

Because either $0 \le \theta < \frac{1}{2}\pi$ or $\pi \le \theta < \frac{3}{2}\pi$, $\tan \theta \ge 0$. Thus $\sqrt{\tan^2 \theta} = \tan \theta$, and we have

$$\sqrt{u^2 - a^2} = a \tan \theta$$

Because $\sec \theta = u/a$ and θ is in $[0, \frac{1}{2}\pi) \cup [\pi, \frac{3}{2}\pi)$,

$$\theta = \sec^{-1} \frac{u}{a}$$

EXAMPLE 3   Evaluate

$$\int \frac{dx}{x^3\sqrt{x^2 - 9}}$$

Solution   Let $x = 3 \sec \theta$, where $0 < \theta < \frac{1}{2}\pi$ if $x > 3$ and $\pi < \theta < \frac{3}{2}\pi$ if $x < -3$. Then $dx = 3 \sec \theta \tan \theta \, d\theta$ and

$$\sqrt{x^2 - 9} = \sqrt{9 \sec^2 \theta - 9}$$
$$= 3\sqrt{\tan^2 \theta}$$
$$= 3 \tan \theta$$

Hence

$$\int \frac{dx}{x^3\sqrt{x^2 - 9}} = \int \frac{3 \sec \theta \tan \theta \, d\theta}{27 \sec^3 \theta \cdot 3 \tan \theta}$$

$$= \tfrac{1}{27} \int \cos^2 \theta \, d\theta$$

$$= \tfrac{1}{54} \int (1 + \cos 2\theta) \, d\theta$$

$$= \tfrac{1}{54}(\theta + \tfrac{1}{2} \sin 2\theta) + C$$
$$= \tfrac{1}{54}(\theta + \sin \theta \cos \theta) + C$$

Because $\sec \theta = \frac{1}{3}x$ and θ is in $(0, \frac{1}{2}\pi) \cup (\pi, \frac{3}{2}\pi)$, $\theta = \sec^{-1} \frac{1}{3}x$. When $x > 3$, $0 < \theta < \frac{1}{2}\pi$, and we obtain $\sin \theta$ and $\cos \theta$ from Figure 5. When $x < -3$, $\pi < \theta < \frac{3}{2}\pi$, and we obtain $\sin \theta$ and $\cos \theta$ from Figure 6. In either case $\sin \theta = \sqrt{x^2 - 9}/x$ and $\cos \theta = 3/x$. Therefore

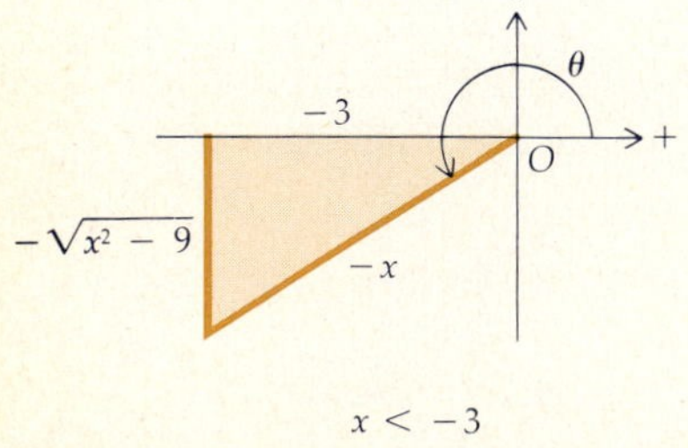

$$\int \frac{dx}{x^3\sqrt{x^2 - 9}} = \frac{1}{54}\left(\sec^{-1} \frac{x}{3} + \frac{\sqrt{x^2 - 9}}{x} \cdot \frac{3}{x}\right) + C$$

$$= \frac{1}{54} \sec^{-1} \frac{x}{3} + \frac{\sqrt{x^2 - 9}}{18x^2} + C$$

EXAMPLE 4   Evaluate

$$\int \frac{dx}{\sqrt{x^2 - 25}}$$

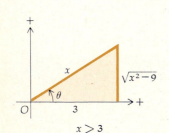

$x > 3$

FIGURE 5

$x < -3$

FIGURE 6

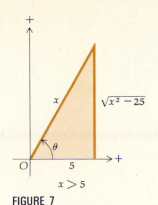

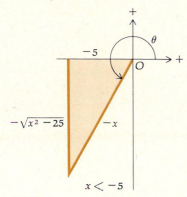

FIGURE 7

$x > 5$

FIGURE 8

$x < -5$

Solution   Let $x = 5 \sec \theta$, where $0 < \theta < \frac{1}{2}\pi$ if $x > 5$ and $\pi < \theta < \frac{3}{2}\pi$ if $x < -5$. Then $dx = 5 \sec \theta \tan \theta \, d\theta$ and

$$\sqrt{x^2 - 25} = \sqrt{25 \sec^2 \theta - 25}$$
$$= 5\sqrt{\tan^2 \theta}$$
$$= 5 \tan \theta$$

Therefore

$$\int \frac{dx}{\sqrt{x^2 - 25}} = \int \frac{5 \sec \theta \tan \theta \, d\theta}{5 \tan \theta}$$

$$= \int \sec \theta \, d\theta$$

$$= \ln|\sec \theta + \tan \theta| + C$$

To find $\tan \theta$ refer to Figure 7 (for $x > 5$) and Figure 8 (for $x < -5$). In either case, $\sec \theta = \frac{1}{5}x$ and $\tan \theta = \frac{1}{5}\sqrt{x^2 - 25}$. We have, then,

$$\int \frac{dx}{\sqrt{x^2 - 25}} = \ln\left| \frac{x}{5} + \frac{\sqrt{x^2 - 25}}{5} \right| + C$$

$$= \ln|x + \sqrt{x^2 - 25}| - \ln 5 + C$$
$$= \ln|x + \sqrt{x^2 - 25}| + C_1$$

EXAMPLE 5   Evaluate

$$\int_1^2 \frac{dx}{(6 - x^2)^{3/2}}$$

Solution   To compute the indefinite integral $\int dx/(6 - x^2)^{3/2}$ we make the substitution $x = \sqrt{6} \sin \theta$. In this case we can restrict θ to the interval $0 < \theta < \frac{1}{2}\pi$ because we are evaluating a definite integral for which $x > 0$ since x is in $[1, 2]$. So $x = \sqrt{6} \sin \theta$, $0 < \theta < \frac{1}{2}\pi$, and $dx = \sqrt{6} \cos \theta \, d\theta$. Furthermore,

$$(6 - x^2)^{3/2} = (6 - 6 \sin^2 \theta)^{3/2}$$
$$= 6\sqrt{6}(1 - \sin^2 \theta)^{3/2}$$
$$= 6\sqrt{6}(\cos^2 \theta)^{3/2}$$
$$= 6\sqrt{6} \cos^3 \theta$$

Hence

$$\int \frac{dx}{(6 - x^2)^{3/2}} = \int \frac{\sqrt{6} \cos \theta \, d\theta}{6\sqrt{6} \cos^3 \theta}$$

$$= \frac{1}{6} \int \frac{d\theta}{\cos^2 \theta}$$

$$= \frac{1}{6} \int \sec^2 \theta \, d\theta$$

$$= \frac{1}{6} \tan \theta + C$$

We find $\tan \theta$ from Figure 9, in which $\sin \theta = x/\sqrt{6}$ and $0 < \theta < \frac{1}{2}\pi$. Thus $\tan \theta = x/\sqrt{6 - x^2}$, and so

$$\int \frac{dx}{(6 - x^2)^{3/2}} = \frac{x}{6\sqrt{6 - x^2}} + C$$

Therefore

$$\int_1^2 \frac{dx}{(6 - x^2)^{3/2}} = \frac{x}{6\sqrt{6 - x^2}}\Bigg]_1^2$$

$$= \frac{1}{3\sqrt{2}} - \frac{1}{6\sqrt{5}}$$

$$= \frac{\sqrt{2}}{6} - \frac{\sqrt{5}}{30}$$

$$= \frac{5\sqrt{2} - \sqrt{5}}{30}$$

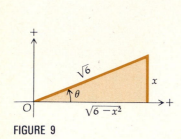

FIGURE 9

EXERCISES 9.4

In Exercises 1 through 24, evaluate the indefinite integral.

1. $\int \dfrac{dx}{x^2\sqrt{4 - x^2}}$

2. $\int \dfrac{\sqrt{4 - x^2}}{x^2}\, dx$

3. $\int \dfrac{dx}{x\sqrt{x^2 + 4}}$

4. $\int \dfrac{x^2 dx}{\sqrt{x^2 + 6}}$

5. $\int \dfrac{dx}{x\sqrt{25 - x^2}}$

6. $\int \sqrt{1 - u^2}\, du$

7. $\int \dfrac{dx}{\sqrt{x^2 - a^2}}$

8. $\int \dfrac{dw}{w^2\sqrt{w^2 - 7}}$

9. $\int \dfrac{x^2\, dx}{(x^2 + 4)^2}$

10. $\int \dfrac{dx}{(4 + x^2)^{3/2}}$

11. $\int \dfrac{dx}{(4x^2 - 9)^{3/2}}$

12. $\int \dfrac{dx}{x^4\sqrt{16 + x^2}}$

13. $\int \dfrac{2\, dt}{t\sqrt{t^4 + 25}}$

14. $\int \dfrac{x^3\, dx}{(25 - x^2)^2}$

15. $\int \dfrac{dx}{\sqrt{4x + x^2}}$

16. $\int \dfrac{dx}{\sqrt{4x - x^2}}$

17. $\int \dfrac{dx}{(5 - 4x - x^2)^{3/2}}$

18. $\int \dfrac{dx}{x\sqrt{x^4 - 4}}$

19. $\int \dfrac{\sec^2 x\, dx}{(4 - \tan^2 x)^{3/2}}$

20. $\int \dfrac{e^{-x}\, dx}{(9e^{-2x} + 1)^{3/2}}$

21. $\int \dfrac{\ln^3 w\, dw}{w\sqrt{\ln^2 w - 4}}$

22. $\int \dfrac{dz}{(z^2 - 6z + 18)^{3/2}}$

23. $\int \dfrac{e^t\, dt}{(e^{2t} + 8e^t + 7)^{3/2}}$

24. $\int \dfrac{\sqrt{16 - e^{2x}}}{e^x}\, dx$

In Exercises 25 through 32, evaluate the definite integral.

25. $\int_0^2 \dfrac{x^3\, dx}{\sqrt{16 - x^2}}$

26. $\int_0^4 \dfrac{dx}{(16 + x^2)^{3/2}}$

27. $\int_{\sqrt{3}}^{3\sqrt{3}} \dfrac{dx}{x^2\sqrt{x^2 + 9}}$

28. $\int_0^1 \dfrac{x^2\, dx}{\sqrt{4 - x^2}}$

29. $\int_4^6 \dfrac{dx}{x\sqrt{x^2 - 4}}$

30. $\int_1^3 \dfrac{dx}{x^4\sqrt{x^2 + 3}}$

31. $\int_0^5 x^2\sqrt{25 - x^2}\, dx$

32. $\int_4^8 \dfrac{dw}{(w^2 - 4)^{3/2}}$

33. Use methods previous to this section (i.e., without a trigonometric substitution) to evaluate the integrals:

(a) $\int \dfrac{3\, dx}{x\sqrt{4x^2 - 9}}$ (b) $\int \dfrac{5x\, dx}{\sqrt{3 - 2x^2}}$

34. Find the area of the region bounded by the curve $y = \sqrt{x^2 - 9}/x^2$, the x axis, and the line $x = 5$.

35. Find the length of the arc of the curve $y = \ln x$ from $x = 1$ to $x = 3$.

36. Find the volume of the solid of revolution generated by revolving the region of Exercise 34 about the y axis.

37. Find the volume of the solid of revolution generated when the region to the right of the y axis bounded by the curve $y = x\sqrt[4]{9 - x^2}$ and the x axis is revolved about the x axis.

38. Find the length of the arc of the parabola $y = x^2$ from $(0, 0)$ to $(1, 1)$.

39. Find the center of mass of a rod 8 cm long if the linear density at a point x centimeters from the left end is $\rho(x)$ grams per centimeter, where $\rho(x) = \sqrt{x^2 + 36}$.

40. The linear density of a rod at a point x meters from one end is $\sqrt{9 + x^2}$ kilograms per meter. Find the mass and center of mass of the rod if it is 3 m long.

41. Find the centroid of the region bounded by the curve $yx^2 = \sqrt{x^2 - 9}$, the x axis, and the line $x = 5$.

42. Use integration to obtain πr^2 square units as the area of the region enclosed by a circle of radius r units.

Exercises 43 through 45 pertain to Supplementary Section 6.7.

43. A horizontal cylindrical pipe has a 4-ft inner diameter and is closed at one end by a circular gate that just fits over the pipe. If the pipe contains water at a depth of 3 ft, find the force on the gate due to water pressure.

44. A gate in an irrigation ditch is in the shape of a segment of a circle of radius 4 ft. The top of the gate is horizontal and 3 ft above the lowest point on the gate. If the water level is 2 ft above the top of the gate, find the force on the gate due to the water pressure.

45. An automobile's gasoline tank is in the shape of a right-circular cylinder of radius 8 in. with a horizontal axis. Find the total force on one end when the gasoline is 12 in. deep and ρ ounces per cubic inch is the weight density of gasoline.

9.5 INTEGRATION OF RATIONAL FUNCTIONS BY PARTIAL FRACTIONS WHEN THE DENOMINATOR HAS ONLY LINEAR FACTORS

From the definition of a rational function, H is rational if $H(x) = P(x)/Q(x)$, where $P(x)$ and $Q(x)$ are polynomials. We have seen that if the degree of the numerator is not less than the degree of the denominator, we have an improper fraction, and in that case we divide the numerator by the denominator until we obtain a proper fraction, one in which the degree of the numerator is less than the degree of the denominator. For example,

$$\frac{x^4 - 10x^2 + 3x + 1}{x^2 - 4} = x^2 - 6 + \frac{3x - 23}{x^2 - 4}$$

So if we wish to integrate

$$\int \frac{x^4 - 10x^2 + 3x + 1}{x^2 - 4}\, dx$$

the problem is reduced to integrating

$$\int (x^2 - 6)\, dx + \int \frac{3x - 23}{x^2 - 4}\, dx$$

In general, then, we are concerned with the integration of expressions of the form

$$\int \frac{P(x)}{Q(x)}\, dx$$

where the degree of $P(x)$ is less than the degree of $Q(x)$.

To do this it is often necessary to write $P(x)/Q(x)$ as the sum of *partial fractions*. The denominators of the partial fractions are obtained by factoring $Q(x)$ into a product of linear and quadratic factors where the quadratic factors have no real zeros. Sometimes it may be difficult to find these factors of $Q(x)$; however, a theorem from algebra states that theoretically this can always be done.

After $Q(x)$ has been factored into products of linear and quadratic factors, the method of determining the partial fractions depends on the nature of these factors. We consider various cases separately. The results of algebra, which are not proved here, provide us with the form of the partial fractions in each case.

Case 1: The factors of $Q(x)$ are all linear, and none is repeated. That is,

$$Q(x) = (a_1 x + b_1)(a_2 x + b_2) \ldots (a_n x + b_n)$$

where no two of the factors are identical. In this case we write

$$\frac{P(x)}{Q(x)} \equiv \frac{A_1}{a_1 x + b_1} + \frac{A_2}{a_2 x + b_2} + \ldots + \frac{A_n}{a_n x + b_n}$$

where $A_1, A_2, \ldots, A_n$ are constants to be determined.

Note that we used $\equiv$ (read as "identically equal") instead of $=$ in the above equality because it is an identity.

The following illustration shows how the values of A_i are found.

▶ **ILLUSTRATION 1** To evaluate

$$\int \frac{(x-1)\,dx}{x^3 - x^2 - 2x}$$

we factor the denominator and have

$$\frac{x-1}{x^3 - x^2 - 2x} \equiv \frac{x-1}{x(x-2)(x+1)}$$

So

$$\frac{x-1}{x(x-2)(x+1)} \equiv \frac{A}{x} + \frac{B}{x-2} + \frac{C}{x+1} \tag{1}$$

Because (1) is an identity it must hold for all x except 0, 2, and -1. From (1),

$$x - 1 \equiv A(x-2)(x+1) + Bx(x+1) + Cx(x-2) \tag{2}$$

Equation (2) is an identity that is true for all values of x including 0, 2, and -1. We wish to find the constants A, B, and C. Substituting 0 for x in (2) we obtain

$$-1 = -2A \quad \Leftrightarrow \quad A = \tfrac{1}{2}$$

Substituting 2 for x in (2) we get

$$1 = 6B \quad \Leftrightarrow \quad B = \tfrac{1}{6}$$

Substituting -1 for x in (2) we obtain

$$-2 = 3C \quad \Leftrightarrow \quad C = -\tfrac{2}{3}$$

There is another method for finding the values of A, B, and C. If on the right side of (2) we combine terms,

$$x - 1 \equiv (A + B + C)x^2 + (-A + B - 2C)x - 2A$$

Because we have an identity the coefficients on the left must equal the corresponding coefficients on the right. Hence

$$A + B + C = 0$$

$$-A + B - 2C = 1$$

$$-2A = -1$$

Solving these equations simultaneously we get $A = \tfrac{1}{2}$, $B = \tfrac{1}{6}$, and $C = -\tfrac{2}{3}$. Substituting these values in (1) we get

$$\frac{x-1}{x(x-2)(x+1)} \equiv \frac{\tfrac{1}{2}}{x} + \frac{\tfrac{1}{6}}{x-2} + \frac{-\tfrac{2}{3}}{x+1}$$

So the given integral can be expressed as follows:

$$\int \frac{x-1}{x^3 - x^2 - 2x}\, dx = \frac{1}{2}\int \frac{dx}{x} + \frac{1}{6}\int \frac{dx}{x-2} - \frac{2}{3}\int \frac{dx}{x+1}$$

$$= \tfrac{1}{2}\ln|x| + \tfrac{1}{6}\ln|x-2| - \tfrac{2}{3}\ln|x+1| + \tfrac{1}{6}\ln C$$

$$= \tfrac{1}{6}(3\ln|x| + \ln|x-2| - 4\ln|x+1| + \ln C)$$

$$= \frac{1}{6}\ln\left|\frac{Cx^3(x-2)}{(x+1)^4}\right| \qquad \blacktriangleleft$$

Case 2: The factors of $Q(x)$ are all linear, and some are repeated.

Suppose that $(a_i x + b_i)$ is a p-fold factor. Then, corresponding to this factor there will be the sum of p partial fractions

$$\frac{A_1}{(a_i x + b_i)^p} + \frac{A_2}{(a_i x + b_i)^{p-1}} + \cdots + \frac{A_{p-1}}{(a_i x + b_i)^2} + \frac{A_p}{a_i x + b_i}$$

where $A_1, A_2, \ldots, A_p$ are constants to be determined.

Example 1 following illustrates this case and the method of determining each A_i.

EXAMPLE 1   Evaluate

$$\int \frac{(x^3 - 1)\, dx}{x^2(x-2)^3}$$

Solution   The fraction in the integrand is written as a sum of partial fractions as follows:

$$\frac{x^3 - 1}{x^2(x-2)^3} \equiv \frac{A}{x^2} + \frac{B}{x} + \frac{C}{(x-2)^3} + \frac{D}{(x-2)^2} + \frac{E}{x-2} \qquad (3)$$

Multiplying on both sides of (3) by the lowest common denominator we get

$$x^3 - 1 \equiv A(x-2)^3 + Bx(x-2)^3 + Cx^2 + Dx^2(x-2) + Ex^2(x-2)^2 \qquad (4)$$

We substitute 2 for x in (4) and obtain

$$7 = 4C \quad \Leftrightarrow \quad C = \tfrac{7}{4}$$

Substituting 0 for x in (4) we get

$$-1 = -8A \quad \Leftrightarrow \quad A = \tfrac{1}{8}$$

We substitute these values for A and C in (4) and expand the powers of the binomials, and we have

$$x^3 - 1 \equiv \tfrac{1}{8}(x^3 - 6x^2 + 12x - 8) + Bx(x^3 - 6x^2 + 12x - 8) + \tfrac{7}{4}x^2 + Dx^3 - 2Dx^2 + Ex^2(x^2 - 4x + 4)$$

$$x^3 - 1 \equiv (B + E)x^4 + (\tfrac{1}{8} - 6B + D - 4E)x^3 + (-\tfrac{3}{4} + 12B + \tfrac{7}{4} - 2D + 4E)x^2 + (\tfrac{3}{2} - 8B)x - 1$$

Equating the coefficients of like powers of x we obtain

$$B + E = 0$$

$$\tfrac{1}{8} - 6B + D - 4E = 1$$

$$-\tfrac{3}{4} + 12B + \tfrac{7}{4} - 2D + 4E = 0$$

$$\tfrac{3}{2} - 8B = 0$$

Solving, we get

$$B = \tfrac{3}{16} \qquad D = \tfrac{5}{4} \qquad E = -\tfrac{3}{16}$$

Therefore, from (3)

$$\frac{x^3 - 1}{x^2(x - 2)^3} \equiv \frac{\tfrac{1}{8}}{x^2} + \frac{\tfrac{3}{16}}{x} + \frac{\tfrac{7}{4}}{(x - 2)^3} + \frac{\tfrac{5}{4}}{(x - 2)^2} + \frac{-\tfrac{3}{16}}{x - 2}$$

Thus

$$\int \frac{x^3 - 1}{x^2(x - 2)^3}\, dx$$

$$= \frac{1}{8} \int \frac{dx}{x^2} + \frac{3}{16} \int \frac{dx}{x} + \frac{7}{4} \int \frac{dx}{(x - 2)^3} + \frac{5}{4} \int \frac{dx}{(x - 2)^2} - \frac{3}{16} \int \frac{dx}{x - 2}$$

$$= -\frac{1}{8x} + \frac{3}{16} \ln|x| - \frac{7}{8(x - 2)^2} - \frac{5}{4(x - 2)} - \frac{3}{16} \ln|x - 2| + C$$

$$= \frac{-11x^2 + 17x - 4}{8x(x - 2)^2} + \frac{3}{16} \ln \left| \frac{x}{x - 2} \right| + C$$

EXAMPLE 2   Evaluate

$$\int \frac{du}{u^2 - a^2}$$

Solution

$$\frac{1}{u^2 - a^2} \equiv \frac{A}{u - a} + \frac{B}{u + a}$$

Multiplying by $(u - a)(u + a)$ we get

$$1 \equiv A(u + a) + B(u - a)$$

$$1 \equiv (A + B)u + Aa - Ba$$

Equating coefficients we have

$$A + B = 0$$

$$Aa - Ba = 1$$

Solving simultaneously we get

$$A = \frac{1}{2a} \qquad B = -\frac{1}{2a}$$

Therefore

$$\int \frac{du}{u^2 - a^2} = \frac{1}{2a} \int \frac{du}{u - a} - \frac{1}{2a} \int \frac{du}{u + a}$$

$$= \frac{1}{2a} \ln|u - a| - \frac{1}{2a} \ln|u + a| + C$$

$$= \frac{1}{2a} \ln \left| \frac{u - a}{u + a} \right| + C$$

The type of integral of the above example occurs frequently enough for it to be listed as a formula. It is not necessary to memorize it because an integration by partial fractions is fairly simple.

$$\int \frac{du}{u^2 - a^2} = \frac{1}{2a} \ln \left| \frac{u - a}{u + a} \right| + C$$

If we have $\int du/(a^2 - u^2)$, we write

$$\int \frac{du}{a^2 - u^2} = -\int \frac{du}{u^2 - a^2}$$

$$= -\frac{1}{2a} \ln \left| \frac{u - a}{u + a} \right| + C$$

$$= \frac{1}{2a} \ln \left| \frac{u + a}{u - a} \right| + C$$

This is also listed as a formula.

$$\int \frac{du}{a^2 - u^2} = \frac{1}{2a} \ln \left| \frac{u + a}{u - a} \right| + C$$

In Section 7.7 we discussed exponential growth that occurs when the rate of increase of the amount of a quantity is proportional to the amount present at a given instant. It has the mathematical model

$$f(t) = Be^{kt} \tag{5}$$

where k is a positive constant, B units is the amount present initially, and $f(t)$ units is the amount present at t units of time, where $f(t) \geq B$ for $t \geq 0$. We also discussed in Section 7.7 bounded growth that happens when a quantity increases at a rate proportional to the difference between a fixed positive number A and its size. A mathematical model for this bounded growth is

$$f(t) = A - Be^{-kt} \tag{6}$$

where B and k are positive constants and $A - B \leq f(t) < A$ for $t \geq 0$. Sketches of the graphs of (5) and (6) appear in Figures 1 and 2, respectively.

Consider now the growth of a population that is affected by the environment imposing an upper bound on its size. For instance, space or reproduction may be factors limited by the environment. In such cases a mathematical model of the form (5) does not apply because the population does not increase beyond a certain point. A model that takes into account environmental factors is obtained when a quantity is increasing at a rate that is jointly proportional to its size and the difference between a fixed positive number A and its size. Thus if y units is the amount of the quantity present at t units of time,

$$\frac{dy}{dt} = ky(A - y) \tag{7}$$

where k is a positive constant, and $0 < y < A$ for $t \geq 0$. To solve (7) we first separate the variables and obtain

$$\frac{dy}{y(A - y)} = k \, dt$$

$$\int \frac{dy}{y(A - y)} = k \int dt \tag{8}$$

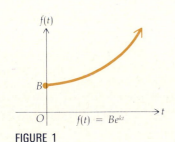

FIGURE 1

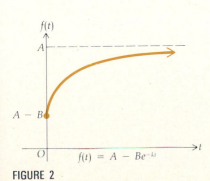

FIGURE 2

Writing the integrand on the left as the sum of partial fractions gives

$$\frac{1}{y(A - y)} = \frac{1}{A}\left(\frac{1}{y} + \frac{1}{A - y}\right)$$

Thus

$$\int \frac{dy}{y(A - y)} = \frac{1}{A}\int \left(\frac{1}{y} + \frac{1}{A - y}\right)dy$$

$$= \frac{1}{A}\left(\ln|y| - \ln|A - y|\right) + C_1$$

Therefore, from (8), we have

$$\frac{1}{A}\left(\ln|y| - \ln|A - y|\right) = kt + C_2$$

$$\ln|A - y| - \ln|y| = -Akt - AC_2$$

$$\ln\left|\frac{A - y}{y}\right| = -Akt - AC_2$$

$$\left|\frac{A - y}{y}\right| = e^{-Akt}e^{-AC_2}$$

Because $0 < y < A$, $(A - y)/y > 0$. Therefore we can omit the absolute-value bars and with $B = e^{-AC_2}$ we have

$$A - y = Bye^{-Akt}$$

$$y(1 + Be^{-Akt}) = A$$

$$y = \frac{A}{1 + Be^{-Akt}} \qquad (9)$$

Letting $y = f(t)$, we write this equation as

$$f(t) = \frac{A}{1 + Be^{-Akt}} \qquad t \geq 0 \qquad (10)$$

where A, B, and k are positive constants. To draw a sketch of the graph of f, first consider $\lim_{t \to +\infty} f(t)$. From (10),

$$\lim_{t \to +\infty} f(t) = \frac{A}{1 + B \lim_{t \to +\infty} e^{-Akt}}$$

$$= \frac{A}{1 + B \cdot 0}$$

$$= A$$

and $f(t)$ is approaching A through values less than A. Therefore, the line A units above the t axis is a horizontal asymptote of the graph of f. Because

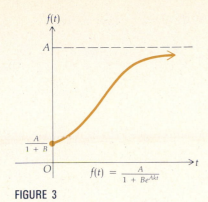

$f(t)$

$f(t) = \dfrac{A}{1 + Be^{Akt}}$

FIGURE 3

$f(0) = \dfrac{A}{1 + B}$, the graph intersects the $f(t)$ axis at $\dfrac{A}{1 + B}$. In Exercise 38 you are

asked to show that the graph of f has a point of inflection at $t = \dfrac{1}{Ak} \ln B$. With

this information we draw a sketch of the graph as shown in Figure 3. It is called a curve of *logistic growth*. Observe that when t is small, the graph is similar to the one for exponential growth in Figure 1, and as t increases the curve is analogous to that shown in Figure 2 for bounded growth.

An application of logistic growth in economics is the distribution of information about a particular product. Logistic growth is used by biologists to describe the spread of a disease and by sociologists to describe the spread of a rumor or joke.

EXAMPLE 3   In a community of 45,000 people, the rate of growth of a flu epidemic is jointly proportional to the number of people who have contracted the flu and the number of people who have not contracted it. (a) If 200 people had the flu at the outbreak of the epidemic and 2800 people have it now after 3 weeks, find a mathematical model describing the epidemic. How many people are expected to have the flu (b) after 5 weeks, and (c) after 10 weeks? (d) If the epidemic continues indefinitely, how many people will contract the flu?

Solution

(a) If t weeks have elapsed since the outbreak of the epidemic and y people have the flu after t weeks, then

$$\frac{dy}{dt} = ky(45{,}000 - y) \tag{11}$$

where k is a constant and $0 < y < 45{,}000$ for $t \geq 0$. We have the boundary conditions given in Table 1, where y_5 and y_{10} are the number of people having the flu after 5 weeks and 10 weeks, respectively.

Differential equation (11) is of the form of (7), and its general solution is of the form of (9). Therefore the general solution of (11) is

$$y = \frac{45{,}000}{1 + Be^{-45{,}000kt}} \tag{12}$$

Because $y = 200$ when $t = 0$, from (12) we get

$$200 = \frac{45{,}000}{1 + Be^0}$$

$$1 + B = 225$$

$$B = 224$$

Substituting this value of B in (12) we obtain

$$y = \frac{45{,}000}{1 + 224e^{-45{,}000kt}} \tag{13}$$

Table 1

t	0	3	5	10
y	200	2800	y_5	y_{10}

When $t = 3$, $y = 2800$. Thus from (13) we get

$$2800 = \frac{45,000}{1 + 224e^{-135,000k}}$$

$$1 + 224e^{-135,000k} = \frac{45,000}{2800}$$

$$1 + 224e^{-135,000k} = 16.0714$$

$$e^{-135,000k} = \frac{15.0714}{224}$$

$$-135,000k = \ln 0.0672830$$

$$k = -\frac{\ln 0.0672830}{135,000}$$

$$k = 0.0000199915$$

Substituting this value of k in (13) we obtain

$$y = \frac{45,000}{1 + 224e^{-0.899616t}}$$

which is the mathematical model desired.

(b) Because $y = y_5$ when $t = 5$,   (c) Because $y = y_{10}$ when $t = 10$,

$$y_5 = \frac{45,000}{1 + 224e^{-4.49808}}$$ $$y_{10} = \frac{45,000}{1 + 224e^{-8.99616}}$$

$$= \frac{45,000}{1 + 224(0.0111303)}$$ $$= \frac{45,000}{1 + 224(0.000123885)}$$

$$= 12,882.2$$ $$= 43,785.0$$

Therefore 12,882 people are expected to have the flu after 5 weeks and 43,785 are expected to have it after 10 weeks.

(d) Because

$$\lim_{t \to +\infty} \frac{45,000}{1 + 224e^{-0.899616t}} = \frac{45,000}{1 + 224 \cdot 0}$$

$$= 45,000$$

the entire community of 45,000 people will contract the flu if the epidemic continues indefinitely.

In chemistry, the *law of mass action* affords an application of integration that leads to the use of partial fractions. Under certain conditions it is found that a substance A reacts with a substance B to form a third substance C in such a way that the rate of change of the amount of C is proportional to the product of the amounts of A and B remaining at any given time.

Suppose that initially there are α grams of A and β grams of B and that r grams of A combine with s grams of B to form $(r + s)$ grams of C. If x grams of substance C is present at t units of time, then C contains $rx/(r + s)$ grams of A and $sx/(r + s)$ grams of B. The number of grams of substance A remaining is then $\alpha - rx/(r + s)$, and the number of grams of substance B remaining is $\beta - sx/(r + s)$. Therefore the law of mass action gives

$$\frac{dx}{dt} = K\left(\alpha - \frac{rx}{r + s}\right)\left(\beta - \frac{sx}{r + s}\right)$$

where K is the constant of proportionality. This equation can be written as

$$\frac{dx}{dt} = \frac{Krs}{(r + s)^2}\left(\frac{r + s}{r}\alpha - x\right)\left(\frac{r + s}{s}\beta - x\right)$$

Letting

$$k = \frac{Krs}{(r + s)^2} \qquad a = \frac{r + s}{r}\alpha \qquad b = \frac{r + s}{s}\beta$$

this equation becomes

$$\frac{dx}{dt} = k(a - x)(b - x) \tag{14}$$

We can separate the variables in (14) and get

$$\frac{dx}{(a - x)(b - x)} = k\,dt$$

If $a = b$, then the left side of the above equation can be integrated by the power formula. If $a \neq b$, partial fractions can be used for the integration.

EXAMPLE 4   A chemical reaction causes a substance A to combine with a substance B to form a substance C so that the law of mass action is obeyed. If in Equation (14) $a = 8$ and $b = 6$, and 2 g of substance C are formed in 10 min, how many grams of C are formed in 15 min?

Solution   If x grams of substance C is present at t minutes, we have the boundary conditions shown in Table 2, where x_{15} grams of substance C is present at 15 min. Equation (14) becomes

Table 2

t	0	10	15
x	0	2	x_{15}

$$\frac{dx}{dt} = k(8 - x)(6 - x)$$

Separating the variables we have

$$\int \frac{dx}{(8 - x)(6 - x)} = k \int dt \tag{15}$$

Writing the integrand as the sum of partial fractions gives

$$\frac{1}{(8 - x)(6 - x)} \equiv \frac{A}{8 - x} + \frac{B}{6 - x}$$

$$1 \equiv A(6 - x) + B(8 - x)$$

Substituting 6 for x gives $B = \frac{1}{2}$, and substituting 8 for x gives $A = -\frac{1}{2}$. Hence (15) is written as

$$-\frac{1}{2}\int \frac{dx}{8 - x} + \frac{1}{2}\int \frac{dx}{6 - x} = k \int dt$$

Integrating we have

$$\frac{1}{2}\ln|8 - x| - \frac{1}{2}\ln|6 - x| + \frac{1}{2}\ln|C| = kt$$

$$\ln\left|\frac{6 - x}{C(8 - x)}\right| = -2kt$$

$$\frac{6 - x}{8 - x} = Ce^{-2kt}$$

Substituting $x = 0$, $t = 0$ in this equation gives $C = \frac{3}{4}$. Hence

$$\frac{6 - x}{8 - x} = \frac{3}{4}e^{-2kt} \tag{16}$$

Substituting $x = 2$, $t = 10$ in (16) we have

$$\frac{4}{6} = \frac{3}{4}e^{-20k}$$

$$e^{-20k} = \frac{8}{9}$$

Substituting $x = x_{15}$, $t = 15$ into (16) we get

$$\frac{6 - x_{15}}{8 - x_{15}} = \frac{3}{4}e^{-30k}$$

$$4(6 - x_{15}) = 3(e^{-20k})^{3/2}(8 - x_{15})$$

$$24 - 4x_{15} = 3(\tfrac{8}{9})^{3/2}(8 - x_{15})$$

$$24 - 4x_{15} = \frac{16\sqrt{2}}{9}(8 - x_{15})$$

$$x_{15} = \frac{54 - 32\sqrt{2}}{9 - 4\sqrt{2}}$$

$$x_{15} \approx 2.6$$

Therefore 2.6 g of substance C will be formed in 15 min.

EXERCISES 9.5

In Exercises 1 through 20, evaluate the indefinite integral.

1. $\displaystyle\int \frac{dx}{x^2 - 4}$

2. $\displaystyle\int \frac{x^2\,dx}{x^2 + x - 6}$

3. $\displaystyle\int \frac{5x - 2}{x^2 - 4}\,dx$

4. $\displaystyle\int \frac{(4x - 2)\,dx}{x^3 - x^2 - 2x}$

5. $\displaystyle\int \frac{4w - 11}{2w^2 + 7w - 4}\,dw$

6. $\displaystyle\int \frac{9t^2 - 26t - 5}{3t^2 - 5t - 2}\,dt$

7. $\displaystyle\int \frac{6x^2 - 2x - 1}{4x^3 - x}\,dx$

8. $\displaystyle\int \frac{x^2 + x + 2}{x^2 - 1}\,dx$

9. $\displaystyle\int \frac{dx}{x^3 + 3x^2}$

10. $\displaystyle\int \frac{x^2 + 4x - 1}{x^3 - x}\,dx$

11. $\displaystyle\int \frac{dx}{x^2(x + 1)^2}$

12. $\displaystyle\int \frac{3x^2 - x + 1}{x^3 - x^2}\,dx$

13. $\displaystyle\int \frac{x^2 - 3x - 7}{(2x + 3)(x + 1)^2}\,dx$

14. $\displaystyle\int \frac{dt}{(t + 2)^2(t + 1)}$

15. $\displaystyle\int \frac{3z + 1}{(z^2 - 4)^2}\,dz$

16. $\displaystyle\int \frac{(5x^2 - 11x + 5)\,dx}{x^3 - 4x^2 + 5x - 2}$

17. $\displaystyle\int \frac{x^4 + 3x^3 - 5x^2 - 4x + 17}{x^3 + x^2 - 5x + 3}\,dx$

18. $\displaystyle\int \frac{2x^4 - 2x + 1}{2x^5 - x^4}\,dx$

19. $\displaystyle\int \frac{-24x^3 + 30x^2 + 52x + 17}{9x^4 - 6x^3 - 11x^2 + 4x + 4}\,dx$

20. $\displaystyle\int \frac{dx}{16x^4 - 8x^2 + 1}$

In Exercises 21 through 28, evaluate the definite integral.

21. $\int_1^2 \dfrac{x-3}{x^3+x^2}\,dx$

22. $\int_0^4 \dfrac{(x-2)\,dx}{2x^2+7x+3}$

23. $\int_1^3 \dfrac{x^2-4x+3}{x(x+1)^2}\,dx$

24. $\int_1^4 \dfrac{(2x^2+13x+18)\,dx}{x^3+6x^2+9x}$

25. $\int_1^2 \dfrac{5x^2-3x+18}{9x-x^3}\,dx$

26. $\int_0^1 \dfrac{(3x^2+7x)\,dx}{x^3+6x^2+11x+6}$

27. $\int_0^5 \dfrac{(x^2-3)\,dx}{x^3+4x^2+5x+2}$

28. $\int_0^4 \dfrac{x^2\,dx}{2x^3+9x^2+12x+4}$

29. Find the area of the region bounded by the curve $y=(x-1)/(x^2-5x+6)$, the x axis, and the lines $x=4$ and $x=6$.

30. Find the area of the region in the first quadrant bounded by the curve $(x+2)^2 y=4-x$.

31. Find the volume of the solid of revolution generated by revolving the region in Exercise 29 about the y axis.

32. Find the volume of the solid of revolution generated if the region in Exercise 30 is revolved about the x axis.

33. Find the centroid of the region bounded by the curve $y=(x-1)/(x^2-5x+6)$, the x axis, and the lines $x=4$ and $x=6$.

34. Find the centroid of the region in the first quadrant bounded by the curve $(x+2)^2 y=4-x$.

35. One day on a college campus, when there were 5000 people in attendance, a particular student heard that a certain controversial speaker was going to make an unscheduled appearance. This information was told to friends who in turn related it to others, and the rate of growth of the spread of this information was jointly proportional to the number of people who had heard it and the number of people who had not heard it. (a) If after 10 min 144 people had heard the rumor, find a mathematical model describing the spread of information. How many people had heard the rumor (b) after 15 min and (c) after 20 min? (d) How many people will eventually hear the rumor?

36. In a particular town of population A, 20 percent of the residents heard a radio announcement about a local political scandal. The rate of growth of the spread of information about the scandal was jointly proportional to the number of people who had heard it and the number of people who had not heard it. If 50 percent of the population heard about the scandal after 1 hour, how long was it until 80 percent of the population heard it?

37. In a community in which A people are susceptible to a particular virus, the rate of growth of the spread of the virus was jointly proportional to the number of people who had caught the virus and the number of susceptible people who had not caught it. If 10 percent of those susceptible had the virus initially and 25 percent had been infected after 3 weeks, what percent of those susceptible had been infected after 6 weeks?

38. Show that the graph of the function f defined by (10) has a point of inflection at $t=\dfrac{1}{Ak}\ln B$.

39. A manufacturer who began operations four years ago has determined that income from sales has increased steadily at the rate of $\dfrac{t^3+3t^2+6t+7}{t^2+3t+2}$ millions of dollars per year, where t is the number of years that the company has been operating. It is estimated that the total income from sales will increase at the same rate for the next 2 years. If the total income from sales for the year just ended was \$6 million, what is the total income from sales expected for the period ending 1 year from now? Give the answer to the nearest \$100.

40. A particle is moving along a straight line so that if v feet per second is the velocity of the particle at t seconds, then

$$v=\frac{t+3}{t^2+3t+2}$$

Find the distance traveled by the particle from the time when $t=0$ to the time when $t=2$.

41. Suppose in Example 4 that $a=5$ and $b=4$ and 1 g of substance C is formed in 5 min. How many grams of C are formed in 10 min?

42. Suppose in Example 4 that $a=6$ and $b=3$ and 1 g of substance C is formed in 4 min. How long will it take 2 g of substance C to be formed?

43. At any instant the rate at which a substance dissolves is proportional to the product of the amount of the substance present at that instant and the difference between the concentration of the substance in solution at that instant and the concentration of the substance in a saturated solution. A quantity of insoluble material is mixed with 10 lb of salt initially, and the salt is dissolving in a tank containing 20 gal of water. If 5 lb of salt dissolves in 10 min and the concentration of salt in a saturated solution is 3 lb/gal, how much salt will dissolve in 20 min?

9.6 INTEGRATION OF RATIONAL FUNCTIONS BY PARTIAL FRACTIONS WHEN THE DENOMINATOR CONTAINS QUADRATIC FACTORS

We continue our discussion of integration of rational functions by partial fractions with the two cases where the denominator contains *irreducible* quadratic factors. Recall from algebra that a factor ax^2+bx+c is irreducible if the equation $ax^2+bx+c=0$ has no real roots, that is, $b^2-4ac<0$.

Case 3: The factors of $Q(x)$ are linear and quadratic, and none of the quadratic factors is repeated.

Corresponding to the quadratic factor $ax^2 + bx + c$ in the denominator is the partial fraction of the form

$$\frac{Ax + B}{ax^2 + bx + C}$$

EXAMPLE 1   Evaluate

$$\int \frac{(x^2 - 2x - 3)\,dx}{(x - 1)(x^2 + 2x + 2)}$$

Solution   We write the fraction in the integrand as a sum of partial fractions as follows:

$$\frac{x^2 - 2x - 3}{(x - 1)(x^2 + 2x + 2)} \equiv \frac{Ax + B}{x^2 + 2x + 2} + \frac{C}{x - 1} \tag{1}$$

Multiplying on both sides of (1) by the lowest common denominator we have

$$x^2 - 2x - 3 \equiv (Ax + B)(x - 1) + C(x^2 + 2x + 2) \tag{2}$$

We compute C by substituting 1 for x in (2), and get

$$-4 = 5C \quad \Leftrightarrow \quad C = -\tfrac{4}{5}$$

We replace C by $-\tfrac{4}{5}$ in (2) and multiply on the right side to obtain

$$x^2 - 2x - 3 \equiv (A - \tfrac{4}{5})x^2 + (B - A - \tfrac{8}{5})x + (-\tfrac{8}{5} - B)$$

Equating coefficients of like powers of x gives

$$A - \tfrac{4}{5} = 1$$
$$B - A - \tfrac{8}{5} = -2$$
$$-\tfrac{8}{5} - B = -3$$

Therefore

$$A = \tfrac{9}{5} \qquad B = \tfrac{7}{5}$$

Substituting the values of A, B, and C in (1) we have

$$\frac{x^2 - 2x - 3}{(x - 1)(x^2 + 2x + 2)} \equiv \frac{\tfrac{9}{5}x + \tfrac{7}{5}}{x^2 + 2x + 2} + \frac{-\tfrac{4}{5}}{x - 1}$$

So

$$\int \frac{x^2 - 2x - 3}{(x - 1)(x^2 + 2x + 2)}\,dx$$

$$= \frac{9}{5}\int \frac{x\,dx}{x^2 + 2x + 2} + \frac{7}{5}\int \frac{dx}{x^2 + 2x + 2} - \frac{4}{5}\int \frac{dx}{x - 1} \tag{3}$$

To integrate $\int (x\,dx)/(x^2 + 2x + 2)$ we see that the differential of the denominator is $2(x + 1)\,dx$; so we add and subtract 1 in the numerator, thereby giving

$$\frac{9}{5}\int \frac{x\,dx}{x^2 + 2x + 2} = \frac{9}{5}\int \frac{(x + 1)\,dx}{x^2 + 2x + 2} - \frac{9}{5}\int \frac{dx}{x^2 + 2x + 2}$$

Substituting from this equation into (3) and combining terms we get

$$\int \frac{x^2 - 2x - 3}{(x - 1)(x^2 + 2x + 2)}\, dx$$

$$= \frac{9}{5} \cdot \frac{1}{2} \int \frac{2(x + 1)\, dx}{x^2 + 2x + 2} - \frac{2}{5} \int \frac{dx}{x^2 + 2x + 2} - \frac{4}{5} \int \frac{dx}{x - 1} \qquad (4)$$

$$= \frac{9}{10} \ln|x^2 + 2x + 2| - \frac{2}{5} \int \frac{dx}{(x + 1)^2 + 1} - \frac{4}{5} \ln|x - 1|$$

$$= \tfrac{9}{10} \ln|x^2 + 2x + 2| - \tfrac{2}{5} \tan^{-1}(x + 1) - \tfrac{8}{10} \ln|x - 1| + \tfrac{1}{10} \ln C$$

$$= \frac{1}{10} \ln \left| \frac{C(x^2 + 2x + 2)^9}{(x - 1)^8} \right| - \frac{2}{5} \tan^{-1}(x + 1)$$

▶ **ILLUSTRATION 1** In Example 1 some steps are saved if instead of (1) the original fraction is expressed as

$$\frac{x^2 - 2x - 3}{(x - 1)(x^2 + 2x + 2)} \equiv \frac{D(2x + 2) + E}{x^2 + 2x + 2} + \frac{F}{x - 1}$$

Note: We write $D(2x + 2) + E$ instead of $Ax + B$ because

$$2x + 2 = D_x(x^2 + 2x + 2)$$

Then, solving for D, E, and F we obtain

$$D = \tfrac{9}{10} \qquad E = -\tfrac{2}{5} \qquad F = -\tfrac{4}{5}$$

giving (4) directly. ◀

Case 4: The factors of $Q(x)$ are linear and quadratic, and some of the quadratic factors are repeated.

If $ax^2 + bx + c$ is a p-fold quadratic factor of $Q(x)$, then corresponding to this factor $(ax^2 + bx + c)^p$ we have the sum of the following p partial fractions:

$$\frac{A_1 x + B_1}{(ax^2 + bx + c)^p} + \frac{A_2 x + B_2}{(ax^2 + bx + c)^{p-1}} + \cdots + \frac{A_p x + B_p}{ax^2 + bx + c}$$

▶ **ILLUSTRATION 2** If the denominator contains the factor $(x^2 - 5x + 2)^3$, we have, corresponding to this factor,

$$\frac{Ax + B}{(x^2 - 5x + 2)^3} + \frac{Cx + D}{(x^2 - 5x + 2)^2} + \frac{Ex + F}{x^2 - 5x + 2}$$

or, more conveniently,

$$\frac{A(2x - 5) + B}{(x^2 - 5x + 2)^3} + \frac{C(2x - 5) + D}{(x^2 - 5x + 2)^2} + \frac{E(2x - 5) + F}{x^2 - 5x + 2}$$ ◀

EXAMPLE 2   Evaluate

$$\int \frac{(x - 2)\, dx}{x(x^2 - 4x + 5)^2}$$

Solution

$$\frac{x - 2}{x(x^2 - 4x + 5)^2} \equiv \frac{A}{x} + \frac{B(2x - 4) + C}{(x^2 - 4x + 5)^2} + \frac{D(2x - 4) + E}{x^2 - 4x + 5}$$

Multiplying on both sides of this equation by the lowest common denominator we have

$$x - 2 \equiv A(x^2 - 4x + 5)^2 + x(2Bx - 4B + C)$$
$$+ x(x^2 - 4x + 5)(2Dx - 4D + E) \quad (5)$$

$$x - 2 \equiv Ax^4 + 16Ax^2 + 25A - 8Ax^3 + 10Ax^2 - 40Ax + 2Bx^2 - 4Bx$$
$$+ Cx + 2Dx^4 - 12Dx^3 + Ex^3 + 26Dx^2 - 4Ex^2 - 20Dx + 5Ex$$

$$x - 2 \equiv (A + 2D)x^4 + (-8A - 12D + E)x^3 + (26A + 2B + 26D - 4E)x^2$$
$$+ (-40A - 4B + C - 20D + 5E)x + 25A \quad (6)$$

The value of A can be computed from (5) by substituting 0 for x. If we equate coefficients in (6) and solve the resulting equations simultaneously, we obtain

$$A = -\tfrac{2}{25} \qquad B = \tfrac{1}{5} \qquad C = \tfrac{1}{5} \qquad D = \tfrac{1}{25} \qquad E = -\tfrac{4}{25}$$

Therefore

$$\int \frac{(x-2)\,dx}{x(x^2 - 4x + 5)^2} = -\frac{2}{25} \int \frac{dx}{x} + \frac{1}{5} \int \frac{(2x-4)\,dx}{(x^2 - 4x + 5)^2} + \frac{1}{5} \int \frac{dx}{(x^2 - 4x + 5)^2}$$
$$+ \frac{1}{25} \int \frac{(2x-4)\,dx}{x^2 - 4x + 5} - \frac{4}{25} \int \frac{dx}{x^2 - 4x + 5}$$

$$= -\frac{2}{25} \ln|x| - \frac{1}{5(x^2 - 4x + 5)} + \frac{1}{5} \int \frac{dx}{[(x^2 - 4x + 4) + 1]^2}$$
$$+ \frac{1}{25} \ln|x^2 - 4x + 5| - \frac{4}{25} \int \frac{dx}{(x^2 - 4x + 4) + 1} \quad (7)$$

We evaluate separately the integrals in the third and fifth terms on the right side of (7),

$$\int \frac{dx}{[(x^2 - 4x + 4) + 1]^2} = \int \frac{dx}{[(x - 2)^2 + 1]^2}$$

Let $x - 2 = \tan\theta$, where $0 \le \theta < \tfrac{1}{2}\pi$ if $x \ge 2$, and $-\tfrac{1}{2}\pi < \theta < 0$ if $x < 2$. Then $dx = \sec^2\theta\,d\theta$ and $(x-2)^2 + 1 = \tan^2\theta + 1$. Hence

$$\int \frac{dx}{[(x-2)^2 + 1]^2} = \int \frac{\sec^2\theta\,d\theta}{(\tan^2\theta + 1)^2}$$

$$= \int \frac{\sec^2\theta\,d\theta}{\sec^4\theta}$$

$$= \int \frac{d\theta}{\sec^2\theta}$$

$$= \int \cos^2\theta\,d\theta$$

$$= \int \frac{1 + \cos 2\theta}{2}\,d\theta$$

$$= \frac{\theta}{2} + \frac{1}{4}\sin 2\theta + C_1$$

$$= \frac{\theta}{2} + \frac{1}{2}\sin\theta\cos\theta + C_1$$

Because $\tan\theta = x - 2$ and $-\frac{1}{2}\pi < \theta < \frac{1}{2}\pi$, $\theta = \tan^{-1}(x-2)$. We find $\sin\theta$ and $\cos\theta$ from Figures 1 (if $x \geq 2$) and 2 (if $x < 2$). In either case

$$\sin\theta = \frac{x-2}{\sqrt{x^2-4x+5}} \qquad \cos\theta = \frac{1}{\sqrt{x^2-4x+5}}$$

Thus

$$\int \frac{dx}{[(x-2)^2+1]^2} = \frac{1}{2}\tan^{-1}(x-2) + \frac{1}{2}\cdot\frac{x-2}{\sqrt{x^2-4x+5}}\cdot\frac{1}{\sqrt{x^2-4x+5}} + C_1$$

$$\int \frac{dx}{[(x-2)^2+1]^2} = \frac{1}{2}\tan^{-1}(x-2) + \frac{x-2}{2(x^2-4x+5)} + C_1 \tag{8}$$

Now, considering the other integral on the right side of (7), we have

$$\int \frac{dx}{(x^2-4x+4)+1} = \int \frac{dx}{(x-2)^2+1}$$

$$\int \frac{dx}{(x^2-4x+4)+1} = \tan^{-1}(x-2) + C_2$$

Substituting from this equation and (8) into (7) we get

$$\int \frac{(x-2)\,dx}{x(x^2-4x+5)^2}$$

$$= -\frac{2}{25}\ln|x| - \frac{1}{5(x^2-4x+5)} + \frac{1}{10}\tan^{-1}(x-2) + \frac{x-2}{10(x^2-4x+5)}$$

$$+ \frac{1}{25}\ln|x^2-4x+5| - \frac{4}{25}\tan^{-1}(x-2) + C$$

$$= \frac{1}{25}\ln\left|\frac{x^2-4x+5}{x^2}\right| - \frac{3}{50}\tan^{-1}(x-2) + \frac{x-4}{10(x^2-4x+5)} + C$$

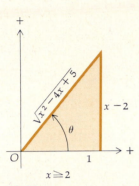

FIGURE 1

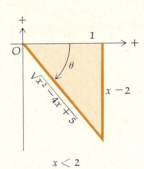

FIGURE 2

EXERCISES 9.6

In Exercises 1 through 20, evaluate the indefinite integral.

1. $\displaystyle\int \frac{dx}{2x^3+x}$

2. $\displaystyle\int \frac{(x+4)\,dx}{x(x^2+4)}$

3. $\displaystyle\int \frac{dx}{16x^4-1}$

4. $\displaystyle\int \frac{(x^2-4x-4)\,dx}{x^3-2x^2+4x-8}$

5. $\displaystyle\int \frac{(t^2+t+1)\,dt}{(2t+1)(t^2+1)}$

6. $\displaystyle\int \frac{3w^3+13w+4}{w^3+4w}\,dw$

7. $\displaystyle\int \frac{(x^2+x)\,dx}{x^3-x^2+x-1}$

8. $\displaystyle\int \frac{dx}{9x^4+x^2}$

9. $\displaystyle\int \frac{dx}{x^3+x^2+x}$

10. $\displaystyle\int \frac{(x+3)\,dx}{4x^4+4x^3+x^2}$

11. $\displaystyle\int \frac{(2x^2-x+2)\,dx}{x^5+2x^3+x}$

12. $\displaystyle\int \frac{(2x^3+9x)\,dx}{(x^2+3)(x^2-2x+3)}$

13. $\displaystyle\int \frac{(5z^3-z^2+15z-10)\,dz}{(z^2-2z+5)^2}$

14. $\displaystyle\int \frac{dt}{(t^2+1)^3}$

15. $\displaystyle\int \frac{(x^2+2x-1)\,dx}{27x^3-1}$

16. $\displaystyle\int \frac{e^{5x}\,dx}{(e^{2x}+1)^2}$

17. $\displaystyle\int \frac{18\,dx}{(4x^2+9)^2}$

18. $\displaystyle\int \frac{(2x^2+3x+2)\,dx}{x^3+4x^2+6x+4}$

19. $\displaystyle\int \frac{(\sec^2 x+1)\sec^2 x\,dx}{1+\tan^3 x}$

20. $\displaystyle\int \frac{(6w^4+4w^3+9w^2+24w+32)\,dw}{(w^3+8)(w^2+3)}$

In Exercises 21 through 29, evaluate the definite integral.

21. $\displaystyle\int_1^4 \frac{(4+5x^2)\,dx}{x^3+4x}$

22. $\displaystyle\int_0^1 \frac{x\,dx}{x^3+2x^2+x+2}$

23. $\int_3^4 \dfrac{(5x^3 - 4x)\,dx}{x^4 - 16}$

24. $\int_0^1 \dfrac{9\,dx}{8x^3 + 1}$

25. $\int_{-1}^0 \dfrac{x^2\,dx}{(2x^2 + 2x + 1)^2}$

26. $\int_0^{1/2} \dfrac{(x + 1)\,dx}{x^3 - 1}$

27. $\int_0^1 \dfrac{(x^2 + 3x + 3)\,dx}{x^3 + x^2 + x + 1}$

28. $\int_{\pi/6}^{\pi/2} \dfrac{\cos x\,dx}{\sin x + \sin^3 x}$

29. $\int_{\ln 2}^{\ln 5} \dfrac{12\,dt}{e^{2t} + 16}$

30. Use methods previous to those in this section (i.e., without partial fractions) to evaluate the integrals:

(a) $\int \dfrac{(x^2 - 4x + 6)\,dx}{x^3 - 6x^2 + 18x}$ (b) $\int \dfrac{3x + 1}{(x + 2)^4}\,dx$

31. Find the area of the region bounded by the x axis, the y axis, the curve $y(x^2 + 1)^3 = x^3$, and the line $x = 1$.

32. Find the area of the region bounded by the x axis, the y axis, the curve $y(x^3 + 8) = 4$, and the line $x = 1$.

33. Find the volume of the solid of revolution generated by revolving the region of Exercise 32 about the y axis.

34. Find the abscissa of the centroid of the region of Exercise 32.

35. A particle is moving along a straight line so that if v centimeters per second is the velocity of the particle at t seconds, then

$$v = \frac{t^2 - t + 1}{(t + 2)^2(t^2 + 1)}$$

Find a formula for the distance traveled by the particle from the time when $t = 0$ to the time when $t = t_1$.

9.7 MISCELLANEOUS SUBSTITUTIONS

If an integrand involves fractional powers of a variable x, the integrand can be simplified by the substitution

$$x = z^n$$

where n is the lowest common denominator of the denominators of the exponents. This substitution is illustrated in the following example.

EXAMPLE 1   Evaluate

$$\int \frac{\sqrt{x}\,dx}{1 + \sqrt[3]{x}}$$

Solution   We let $x = z^6$; then $dx = 6z^5\,dz$. So

$$\int \frac{x^{1/2}\,dx}{1 + x^{1/3}} = \int \frac{z^3(6z^5\,dz)}{1 + z^2}$$

$$= 6\int \frac{z^8}{z^2 + 1}\,dz$$

Dividing the numerator by the denominator we have

$$\int \frac{x^{1/2}\,dx}{1 + x^{1/3}} = 6\int \left(z^6 - z^4 + z^2 - 1 + \frac{1}{z^2 + 1}\right)dz$$

$$= 6(\tfrac{1}{7}z^7 - \tfrac{1}{5}z^5 + \tfrac{1}{3}z^3 - z + \tan^{-1} z) + C$$

$$= \tfrac{6}{7}x^{7/6} - \tfrac{6}{5}x^{5/6} + 2x^{1/2} - 6x^{1/6} + 6\tan^{-1} x^{1/6} + C$$

No general rule can be given to determine a substitution that will result in a simpler integrand. The following example shows another situation where we rationalize the given integrand.

EXAMPLE 2   Evaluate

$$\int x^5\sqrt{x^2 + 4}\,dx$$

Solution   Let $z = \sqrt{x^2 + 4}$. Then $z^2 = x^2 + 4$, and $2z\,dz = 2x\,dx$. So

$$\int x^5 \sqrt{x^2 + 4}\,dx = \int (x^2)^2 \sqrt{x^2 + 4}\; x\,dx$$

$$= \int (z^2 - 4)^2 z(z\,dz)$$

$$= \int (z^6 - 8z^4 + 16z^2)\,dz$$

$$= \tfrac{1}{7}z^7 - \tfrac{8}{5}z^5 + \tfrac{16}{3}z^3 + C$$

$$= \tfrac{1}{105}z^3[15z^4 - 168z^2 + 560] + C$$

$$= \tfrac{1}{105}(x^2 + 4)^{3/2}[15(x^2 + 4)^2 - 168(x^2 + 4) + 560] + C$$

$$= \tfrac{1}{105}(x^2 + 4)^{3/2}(15x^4 - 48x^2 + 128) + C$$

If an integrand is a rational function of $\sin x$ and $\cos x$, it can be reduced to a rational function of z by the substitution

$$z = \tan \tfrac{1}{2}x$$

as we will show by an example. To obtain the formulas for $\sin x$ and $\cos x$ in terms of z we use the following identities: $\sin 2y = 2 \sin y \cos y$ and $\cos 2y = 2 \cos^2 y - 1$ with $y = \tfrac{1}{2}x$. We have, then,

$$\sin x = 2 \sin \tfrac{1}{2}x \cos \tfrac{1}{2}x \qquad\qquad \cos x = 2 \cos^2 \tfrac{1}{2}x - 1$$

$$= 2 \cdot \frac{\sin \tfrac{1}{2}x \cos^2 \tfrac{1}{2}x}{\cos \tfrac{1}{2}x} \qquad\qquad = \frac{2}{\sec^2 \tfrac{1}{2}x} - 1$$

$$= 2 \tan \tfrac{1}{2}x \cdot \frac{1}{\sec^2 \tfrac{1}{2}x} \qquad\qquad = \frac{2}{1 + \tan^2 \tfrac{1}{2}x} - 1$$

$$= \frac{2 \tan \tfrac{1}{2}x}{1 + \tan^2 \tfrac{1}{2}x} \qquad\qquad = \frac{2}{1 + z^2} - 1$$

$$= \frac{2z}{1 + z^2} \qquad\qquad\qquad = \frac{1 - z^2}{1 + z^2}$$

Because $z = \tan \tfrac{1}{2}x$,

$$dz = \tfrac{1}{2} \sec^2 \tfrac{1}{2}x\,dx$$

$$= \tfrac{1}{2}(1 + \tan^2 \tfrac{1}{2}x)\,dx$$

Thus

$$dx = \frac{2\,dz}{1 + z^2}$$

We state these results as a theorem.

9.7.1 THEOREM   If $z = \tan \tfrac{1}{2}x$, then

$$\sin x = \frac{2z}{1 + z^2} \qquad \cos x = \frac{1 - z^2}{1 + z^2} \qquad dx = \frac{2\,dz}{1 + z^2}$$

EXAMPLE 3   Evaluate

$$\int \frac{dx}{1 - \sin x + \cos x}$$

Solution   Let $z = \tan \frac{1}{2}x$. Then from the formulas of Theorem 9.7.1 we have

$$\int \frac{dx}{1 - \sin x + \cos x} = \int \frac{\dfrac{2\, dz}{1 + z^2}}{1 - \dfrac{2z}{1 + z^2} + \dfrac{1 - z^2}{1 + z^2}}$$

$$= 2 \int \frac{dz}{(1 + z^2) - 2z + (1 - z^2)}$$

$$= 2 \int \frac{dz}{2 - 2z}$$

$$= \int \frac{dz}{1 - z}$$

$$= -\ln|1 - z| + C$$

$$= -\ln|1 - \tan \tfrac{1}{2}x| + C$$

EXAMPLE 4   Let $z = \tan \frac{1}{2}x$ to evaluate

$$\int \sec x \, dx$$

Solution   With $z = \tan \frac{1}{2}x$ and the formulas of Theorem 9.7.1 we have

$$\int \sec x \, dx = \int \frac{dx}{\cos x}$$

$$= \int \frac{2\, dz}{1 + z^2} \cdot \frac{1 + z^2}{1 - z^2}$$

$$= 2 \int \frac{dz}{1 - z^2}$$

$$= \ln \left| \frac{1 + z}{1 - z} \right| + C \qquad \text{(from Section 9.5)}$$

$$= \ln \left| \frac{1 + \tan \frac{1}{2}x}{1 - \tan \frac{1}{2}x} \right| + C$$

The value of $\int \sec x \, dx$ from Example 4 can be written in another form by letting $1 = \tan \frac{1}{4}\pi$ and using the trigonometric identity

$$\tan(a + b) = \frac{\tan a + \tan b}{1 - \tan a \tan b}$$

Thus

$$\int \sec x \, dx = \ln \left| \frac{\tan \frac{1}{4}\pi + \tan \frac{1}{2}x}{1 - \tan \frac{1}{4}\pi \cdot \tan \frac{1}{2}x} \right| + C$$

$$\int \sec x \, dx = \ln \left| \tan(\tfrac{1}{4}\pi + \tfrac{1}{2}x) \right| + C \tag{1}$$

In Theorem 5.4.5 we have the formula

$$\int \sec x \, dx = \ln|\sec x + \tan x| + C$$

which is obtained by the trick of multiplying the numerator and denominator of the integrand by $\sec x + \tan x$. Still another form for $\int \sec x \, dx$ is obtained as follows:

$$\int \sec x \, dx = \int \frac{dx}{\cos x}$$

$$= \int \frac{\cos x \, dx}{\cos^2 x}$$

$$= \int \frac{\cos x \, dx}{1 - \sin^2 x}$$

$$= \int \frac{du}{1 - u^2} \qquad \text{(by letting } u = \sin x \text{ and } du = \cos x \, dx)$$

$$= \frac{1}{2} \ln \left| \frac{1 + u}{1 - u} \right| + C$$

$$= \ln \left| \frac{1 + \sin x}{1 - \sin x} \right|^{1/2} + C$$

Because $-1 \le \sin x \le 1$ for all x, $1 + \sin x$ and $1 - \sin x$ are nonnegative. Hence the absolute-value bars can be removed, and we have

$$\int \sec x \, dx = \ln \sqrt{\frac{1 + \sin x}{1 - \sin x}} + C \tag{2}$$

EXERCISES 9.7

In Exercises 1 through 31, evaluate the indefinite integral.

1. $\displaystyle\int \frac{x \, dx}{3 + \sqrt{x}}$

2. $\displaystyle\int \frac{dx}{\sqrt[3]{x} - x}$

3. $\displaystyle\int \frac{dx}{x\sqrt{1 + 4x}}$

4. $\displaystyle\int x(1 + x)^{2/3} \, dx$

5. $\displaystyle\int \frac{\sqrt{1 + x}}{1 - x} \, dx$

6. $\displaystyle\int \frac{dx}{3 + \sqrt{x + 2}}$

7. $\displaystyle\int \frac{dx}{1 + \sqrt[3]{x - 2}}$

8. $\displaystyle\int \frac{dx}{2\sqrt[3]{x} + \sqrt{x}}$

9. $\displaystyle\int \frac{dx}{\sqrt{2x} - \sqrt{x + 4}}$

10. $\displaystyle\int \frac{x^3 \, dx}{\sqrt{5x^2 + 4}}$

11. $\displaystyle\int \frac{(2x^5 + 3x^2) \, dx}{\sqrt{1 + 2x^3}}$

12. $\displaystyle\int \frac{dx}{\sqrt{\sqrt{x} + 1}}$

13. $\displaystyle\int \frac{3 \, dx}{8 + 7 \cos x}$

14. $\displaystyle\int \frac{dx}{1 + \sin x}$

15. $\displaystyle\int \frac{3 \, dx}{7 + 8 \cos x}$

16. $\displaystyle\int \frac{dx}{\sin x - \cos x + 2}$

17. $\displaystyle\int \frac{dx}{\sin x + \tan x}$

18. $\displaystyle\int \frac{dx}{5 + 4 \cos x}$

19. $\displaystyle\int \frac{dx}{3 - 5 \sin x}$

20. $\displaystyle\int \frac{dx}{\tan x - 1}$

21. $\displaystyle\int \frac{dx}{4 \sin x - 3 \cos x}$

22. $\displaystyle\int \frac{dx}{3 \cos x - 2 \sin x + 3}$

23. $\int \dfrac{8\,dx}{3\cos 2x + 1}$

24. $\int \dfrac{\cos x\,dx}{3\cos x - 5}$

25. $\int \dfrac{\cos x\,dx}{1 + 2\cos x}$

26. $\int \dfrac{dx}{\sin x - \tan x}$

27. $\int \dfrac{dx}{\cot x(6 + 7\cos 2x)}$

28. $\int \dfrac{dx}{\cot 2x(1 - \cos 2x)}$

29. $\int \dfrac{dx}{2\sin x + 2\cos x + 3}$

30. $\int \dfrac{5\,dx}{6 + 4\sec x}$

31. $\int \dfrac{dx}{\sqrt{x}\,\sqrt[3]{x}(1 + \sqrt[3]{x})^2}$

32. Evaluate the indefinite integral $\int \dfrac{dx}{x\sqrt{x^2 + 2x - 1}}$ by two methods: (a) use the substitution $x = 1/z$; (b) use the substitution $\sqrt{x^2 + 2x - 1} = z - x$.

In Exercises 33 through 44, evaluate the definite integral.

33. $\int_0^4 \dfrac{dx}{1 + \sqrt{x}}$

34. $\int_0^1 \dfrac{x^{3/2}}{x + 1}\,dx$

35. $\int_{1/2}^2 \dfrac{dx}{\sqrt{2x}(\sqrt{2x} + 9)}$

36. $\int_{16}^{18} \dfrac{dx}{\sqrt{x} - \sqrt[4]{x^3}}$

37. $\int_0^{\pi/2} \dfrac{dx}{5\sin x + 3}$

38. $\int_0^{\pi/2} \dfrac{dx}{3 + \cos 2x}$

39. $\int_{\pi/6}^{\pi/3} \dfrac{3\,dx}{2\sin 2x + 1}$

40. $\int_0^{\pi/4} \dfrac{8\,dx}{\tan x + 1}$

41. $\int_{-\pi/3}^{\pi/2} \dfrac{3\,dx}{2\cos x + 1}$

42. $\int_0^{\pi/2} \dfrac{\sin 2x\,dx}{2 + \cos x}$

43. $\int_0^1 \dfrac{\sqrt{x}}{1 + \sqrt[3]{x}}\,dx$

44. $\int_2^{11} \dfrac{x^3\,dx}{\sqrt[3]{x^2 + 4}}$

45. Evaluate $\int \dfrac{dx}{x - \sqrt{x}}$ by two methods: (a) Let $x = z^2$; (b) write $x - \sqrt{x} = \sqrt{x}(\sqrt{x} - 1)$ and let $u = \sqrt{x} - 1$.

46. Use the substitution of this section, $z = \tan \frac{1}{2}x$, to show that $\int \sin x\,dx = -\cos x + C$.

47. Show that formula (2) of this section is equivalent to the formula $\int \sec x\,dx = \ln|\sec x + \tan x| + C$. (*Hint:* Multiply the numerator and denominator under the radical sign by $(1 + \sin x)$.)

48. By using the substitution $z = \tan \frac{1}{2}x$, prove that
$$\int \csc x\,dx = \ln \sqrt{\dfrac{1 - \cos x}{1 + \cos x}} + C$$

49. Show that the result in Exercise 48 is equivalent to the formula $\int \csc x\,dx = \ln|\csc x - \cot x| + C$. (*Hint:* Use a method similar to that suggested in the hint for Exercise 47.)

50. Evaluate the integral
$$\int \dfrac{\tan \frac{1}{2}x}{\sin x}\,dx$$
by two methods: (a) Let $z = \tan \frac{1}{2}x$; (b) let $u = \frac{1}{2}x$ and obtain an integral involving trigonometric functions of u.

9.8 INTEGRALS YIELDING INVERSE HYPERBOLIC FUNCTIONS (*Supplementary*)

Sometimes the use of inverse hyperbolic functions in integration shortens the computation considerably. However, no new types of integrals are evaluated by this procedure. Only new forms of the results are obtained.

From formula (7) in Section 8.5

$$D_x(\sinh^{-1} u) = \dfrac{1}{\sqrt{u^2 + 1}}\,D_x u$$

from which we obtain the integration formula

$$\int \dfrac{du}{\sqrt{u^2 + 1}} = \sinh^{-1} u + C$$

If formula (1) of Section 8.5 is used to express $\sinh^{-1} u$ as a natural logarithm, the following theorem is obtained.

9.8.1 THEOREM

$$\int \dfrac{du}{\sqrt{u^2 + 1}} = \sinh^{-1} u + C$$
$$= \ln(u + \sqrt{u^2 + 1}) + C$$

Formula (8) in Section 8.5 is

$$D_x(\cosh^{-1} u) = \dfrac{1}{\sqrt{u^2 - 1}}\,D_x u \qquad \text{where } u > 1$$

from which it follows that

$$\int \frac{du}{\sqrt{u^2 - 1}} = \cosh^{-1} u + C \qquad \text{if } u > 1$$

By combining this with formula (2) in Section 8.5, we obtain the following theorem.

9.8.2 THEOREM

$$\int \frac{du}{\sqrt{u^2 - 1}} = \cosh^{-1} u + C$$

$$= \ln(u + \sqrt{u^2 - 1}) + C \qquad \text{if } u > 1$$

Formulas (9) and (10) of Section 8.5 are, respectively,

$$D_x(\tanh^{-1} u) = \frac{1}{1 - u^2} D_x u \qquad \text{where } |u| < 1$$

$$D_x(\coth^{-1} u) = \frac{1}{1 - u^2} D_x u \qquad \text{where } |u| > 1$$

From the above two formulas we get

$$\int \frac{du}{1 - u^2} = \begin{cases} \tanh^{-1} u + C & \text{if } |u| < 1 \\ \coth^{-1} u + C & \text{if } |u| > 1 \end{cases}$$

With this formula and (3) and (4) of Section 8.5 we have the next theorem.

9.8.3 THEOREM

$$\int \frac{du}{1 - u^2} = \begin{cases} \tanh^{-1} u + C & \text{if } |u| < 1 \\ \coth^{-1} u + C & \text{if } |u| > 1 \end{cases}$$

$$= \frac{1}{2} \ln \left| \frac{1 + u}{1 - u} \right| + C \qquad \text{if } u \neq 1$$

We also have the three formulas given in the following theorem.

9.8.4 THEOREM

$$\int \frac{du}{\sqrt{u^2 + a^2}} = \sinh^{-1} \frac{u}{a} + C$$

$$= \ln(u + \sqrt{u^2 + a^2}) + C \qquad \text{if } a > 0 \tag{1}$$

$$\int \frac{du}{\sqrt{u^2 - a^2}} = \cosh^{-1} \frac{u}{a} + C$$

$$= \ln(u + \sqrt{u^2 - a^2}) + C \qquad \text{if } u > a > 0 \tag{2}$$

$$\int \frac{du}{a^2 - u^2} = \begin{cases} \dfrac{1}{a} \tanh^{-1} \dfrac{u}{a} + C & \text{if } |u| < a \\ \dfrac{1}{a} \coth^{-1} \dfrac{u}{a} + C & \text{if } |u| > a \end{cases}$$

$$= \frac{1}{2a} \ln \left| \frac{a + u}{a - u} \right| + C \qquad \text{if } u \neq a \text{ and } a \neq 0 \tag{3}$$

Proof   The formulas can be proved by finding the derivative of the right side and obtaining the integrand, or else more directly by using a hyperbolic function substitution. The proof of (1) by differentiating the right side is as follows:

$$D_u\left(\sinh^{-1}\frac{u}{a}\right) = \frac{1}{\sqrt{\left(\frac{u}{a}\right)^2 + 1}} \cdot \frac{1}{a}$$

$$= \frac{\sqrt{a^2}}{\sqrt{u^2 + a^2}} \cdot \frac{1}{a}$$

and because $a > 0$, $\sqrt{a^2} = a$; thus

$$D_u\left(\sinh^{-1}\frac{u}{a}\right) = \frac{1}{\sqrt{u^2 + a^2}}$$

To obtain the natural logarithm representation we use formula (1) of Section 8.5, and we have

$$\sinh^{-1}\frac{u}{a} = \ln\left(\frac{u}{a} + \sqrt{\left(\frac{u}{a}\right)^2 + 1}\right)$$

$$= \ln\left(\frac{u}{a} + \frac{\sqrt{u^2 + a^2}}{a}\right)$$

$$= \ln(u + \sqrt{u^2 + a^2}) - \ln a$$

Therefore

$$\sinh^{-1}\frac{u}{a} + C = \ln(u + \sqrt{u^2 + a^2}) - \ln a + C$$

$$= \ln(u + \sqrt{u^2 + a^2}) + C_1$$

where $C_1 = C - \ln a$.

The proof of (2) by a hyperbolic function substitution makes use of the identity $\cosh^2 x - \sinh^2 x = 1$. Let $u = a \cosh x$, where $x > 0$; then $du = a \sinh x \, dx$, and $x = \cosh^{-1}(u/a)$. Substituting in the integrand we get

$$\int \frac{du}{\sqrt{u^2 - a^2}} = \int \frac{a \sinh x \, dx}{\sqrt{a^2 \cosh^2 x - a^2}}$$

$$= \int \frac{a \sinh x \, dx}{\sqrt{a^2}\sqrt{\cosh^2 x - 1}}$$

$$= \int \frac{a \sinh x \, dx}{\sqrt{a^2}\sqrt{\sinh^2 x}}$$

Because $a > 0$, $\sqrt{a^2} = a$. Furthermore, because $x > 0$, $\sinh x > 0$; therefore, $\sqrt{\sinh^2 x} = \sinh x$. Then

$$\int \frac{du}{\sqrt{u^2 - a^2}} = \int \frac{a \sinh x \, dx}{a \sinh x}$$

$$= \int dx$$

$$= x + C$$

$$= \cosh^{-1}\frac{u}{a} + C$$

The natural logarithm representation may be obtained in a way similar to the way we obtained the one for formula (1). The proof of formula (3) is left as an exercise (see Exercise 23). ∎

The formulas of Theorems 9.8.1 through 9.8.4 give alternate representations of the integral in question. In evaluating an integral in which one of these forms occurs, the inverse hyperbolic function representation may be easier to use and is sometimes less cumbersome to write. Observe that the natural logarithm form of formula (3) was obtained in Section 9.5 by using partial fractions.

EXAMPLE 1   Evaluate

$$\int \frac{dx}{\sqrt{x^2 - 6x + 13}}$$

Solution   We apply formula (1) after completing the square.

$$\int \frac{dx}{\sqrt{x^2 - 6x + 13}} = \int \frac{dx}{\sqrt{(x^2 - 6x + 9) + 4}}$$

$$= \int \frac{dx}{\sqrt{(x - 3)^2 + 4}}$$

$$= \sinh^{-1}\left(\frac{x - 3}{2}\right) + C$$

$$= \ln(x - 3 + \sqrt{x^2 - 6x + 13}) + C$$

EXAMPLE 2   Evaluate

$$\int_6^{10} \frac{dx}{\sqrt{x^2 - 5}}$$

Solution   From (2),

$$\int_6^{10} \frac{dx}{\sqrt{x^2 - 25}} = \cosh^{-1}\frac{x}{5}\bigg]_6^{10}$$

$$= \cosh^{-1} 2 - \cosh^{-1} 1.2$$

From a calculator, we obtain $\cosh^{-1} 2 \approx 1.32$ and $\cosh^{-1} 1.2 \approx 0.62$. Thus

$$\int_6^{10} \frac{dx}{\sqrt{x^2 - 25}} \approx 0.70$$

Instead of applying the formulas, integrals of the forms of those in Theorems 9.8.1 through 9.8.4 can be obtained by using a hyperbolic function substitution and proceeding in a manner similar to that using a trigonometric substitution.

EXAMPLE 3   Evaluate the integral of Example 1 without using a formula but by using a hyperbolic function substitution.

Solution   In the solution of Example 1 the given integral was rewritten as

$$\int \frac{dx}{\sqrt{(x-3)^2 + 4}}$$

If we let $x - 3 = 2 \sinh u$, then $dx = 2 \cosh u\, du$ and $u = \sinh^{-1} \frac{1}{2}(x-3)$. Therefore

$$\int \frac{dx}{\sqrt{(x-3)^2 + 4}} = \int \frac{2 \cosh u\, du}{\sqrt{4 \sinh^2 u + 4}}$$

$$= \int \frac{2 \cosh u\, du}{2\sqrt{\sinh^2 u + 1}}$$

$$= \int \frac{\cosh u\, du}{\sqrt{\cosh^2 u}}$$

$$= \int \frac{\cosh u\, du}{\cosh u}$$

$$= \int du$$

$$= u + C$$

$$= \sinh^{-1}\left(\frac{x-3}{2}\right) + C$$

which agrees with the result of Example 1.

EXERCISES 9.8

In Exercises 1 through 16, express the indefinite integral in terms of an inverse hyperbolic function and as a natural logarithm.

In Exercises 17 through 22, evaluate the definite integral and express the answer in terms of a natural logarithm.

1. $\displaystyle\int \frac{dx}{\sqrt{4 + x^2}}$

2. $\displaystyle\int \frac{dx}{\sqrt{4x^2 - 9}}$

3. $\displaystyle\int \frac{x\, dx}{\sqrt{x^4 - 1}}$

4. $\displaystyle\int \frac{dx}{25 - x^2}$

5. $\displaystyle\int \frac{dx}{9x^2 - 16}$

6. $\displaystyle\int \frac{dx}{4e^x - e^{-x}}$

7. $\displaystyle\int \frac{\cos x\, dx}{\sqrt{4 - \cos^2 x}}$

8. $\displaystyle\int \frac{dx}{\sqrt{x^2 - 4x + 1}}$

9. $\displaystyle\int \frac{dt}{\sqrt{5 - e^{-2t}}}$

10. $\displaystyle\int \frac{dw}{4w - w^2 - 3}$

11. $\displaystyle\int \frac{dx}{2 - 4x - x^2}$

12. $\displaystyle\int \frac{dx}{\sqrt{25x^2 + 9}}$

13. $\displaystyle\int \frac{dx}{x^2 + 10x + 24}$

14. $\displaystyle\int \frac{dz}{\sqrt{9z^2 - 6z - 8}}$

15. $\displaystyle\int \frac{3\, dw}{w\sqrt{4 \ln^2 w + 9}}$

16. $\displaystyle\int \frac{3x\, dx}{\sqrt{x^4 + 6x^2 + 5}}$

17. $\displaystyle\int_3^5 \frac{dx}{\sqrt{x^2 - 4}}$

18. $\displaystyle\int_{-4}^{-3} \frac{dx}{1 - x^2}$

19. $\displaystyle\int_{-1/2}^{1/2} \frac{dx}{1 - x^2}$

20. $\displaystyle\int_1^2 \frac{dx}{\sqrt{x^2 + 2x}}$

21. $\displaystyle\int_2^3 \frac{dx}{\sqrt{9x^2 - 12x - 5}}$

22. $\displaystyle\int_{-2}^2 \frac{dx}{\sqrt{16 + x^2}}$

23. Prove formula (3) by using a hyperbolic function substitution.

24. A curve goes through the point $(0, a)$, $a > 0$, and the slope at any point is $\sqrt{y^2/a^2 - 1}$. Prove that the curve is a catenary.

25. A man wearing a parachute falls out of an airplane, and when the parachute opens his velocity is 200 ft/sec. If v feet per second is his velocity t seconds after the parachute opens,

$$\frac{324}{g} \cdot \frac{dv}{dt} = 324 - v^2$$

Solve this differential equation to obtain

$$t = \frac{18}{g}\left(\coth^{-1}\frac{v}{18} - \coth^{-1}\frac{100}{9}\right)$$

26. Show that $\sinh^{-1} u = \ln(u + \sqrt{u^2 + 1})$ by using a trigonometric substitution to evaluate $\int \dfrac{du}{\sqrt{u^2 + 1}}$.

27. The region bounded by the curve $y = (16 - x^2)^{-1/2}$, the x

axis, and the lines $x = -2$ and $x = 3$ is revolved about the x axis. Show that the volume of the solid generated is $\frac{1}{4}\pi[\tanh^{-1}\frac{3}{4} - \tanh^{-1}(-\frac{1}{2})]$.

REVIEW EXERCISES FOR CHAPTER 9 AND REVIEW OF INTEGRATION

In Exercises 1 through 62, evaluate the indefinite integral.

1. $\int \tan^2 4x \cos^4 4x \, dx$

2. $\int \dfrac{5x^2 - 3}{x^3 - x} \, dx$

3. $\int \dfrac{e^x \, dx}{\sqrt{4 - e^x}}$

4. $\int \dfrac{dx}{x^2\sqrt{a^2 + x^2}}$

5. $\int \tan^{-1}\sqrt{x} \, dx$

6. $\int \dfrac{dt}{t^4 + 1}$

7. $\int \cos^2 \frac{1}{3}x \, dx$

8. $\int \dfrac{\sqrt{x+1}+1}{\sqrt{x+1}-1} \, dx$

9. $\int \dfrac{x^2 + 1}{(x-1)^3} \, dx$

10. $\int \dfrac{dy}{\sqrt{y}+1}$

11. $\int \sin x \sin 3x \, dx$

12. $\int \cos\theta \cos 2\theta \, d\theta$

13. $\int \dfrac{dx}{x + x^{4/3}}$

14. $\int t\sqrt{2t - t^2} \, dt$

15. $\int (\sec 3x + \csc 3x)^2 \, dx$

16. $\int \dfrac{dx}{\sqrt{e^x - 1}}$

17. $\int \dfrac{2t^3 + 11t + 8}{t^3 + 4t^2 + 4t} \, dt$

18. $\int x^3 e^{3x} \, dx$

19. $\int \dfrac{x^4 + 1}{x^4 - 1} \, dx$

20. $\int \dfrac{\sqrt{x^2 - 4}}{x^2} \, dx$

21. $\int \sin^4 3x \cos^2 3x \, dx$

22. $\int t \sin^2 2t \, dt$

23. $\int \dfrac{dr}{\sqrt{3 - 4r - r^2}}$

24. $\int \dfrac{4x^2 + x - 2}{x^3 - 5x^2 + 8x - 4} \, dx$

25. $\int x^3 \cos x^2 \, dx$

26. $\int \dfrac{y \, dy}{9 + 16y^4}$

27. $\int e^{t/2} \cos 2t \, dt$

28. $\int \dfrac{du}{u^{5/8} - u^{1/8}}$

29. $\int \dfrac{\sin x \cos x}{4 + \sin^4 x} \, dx$

30. $\int \dfrac{\sqrt{w - a}}{w} \, dw, \ a > 0$

31. $\int \sin^5 nx \, dx$

32. $\int \dfrac{dx}{x \ln x(\ln x - 1)}$

33. $\int \csc^5 x \, dx$

34. $\int \dfrac{dx}{5 + 4\sec x}$

35. $\int \dfrac{2y^2 + 1}{y^3 - 6y^2 + 12y - 8} \, dy$

36. $\int \dfrac{x^5 \, dx}{(x^2 - a^2)^3}$

37. $\int \dfrac{\sin x \, dx}{1 + \cos^2 x}$

38. $\int \dfrac{dx}{x\sqrt{x^2 + x + 1}}$

39. $\int \sqrt{4t - t^2} \, dt$

40. $\int \dfrac{dx}{\sqrt{1 - x + 3x^2}}$

41. $\int \dfrac{dx}{x^4 - x}$

42. $\int \dfrac{\sqrt{t} - 1}{\sqrt{t} + 1} \, dt$

43. $\int \dfrac{e^x \, dx}{\sqrt{4 - 9e^{2x}}}$

44. $\int \dfrac{dx}{5 + 4\cos 2x}$

45. $\int \cot^2 3x \csc^4 3x \, dx$

46. $\int \dfrac{\cot x \, dx}{3 + 2\sin x}$

47. $\int x^2 \sin^{-1} x \, dx$

48. $\int \dfrac{dx}{x\sqrt{5x - 6 - x^2}}$

49. $\int \dfrac{dx}{\sin x - 2\csc x}$

50. $\int \cos x \ln(\sin x) \, dx$

51. $\int \dfrac{\cos 3t \, dt}{\sin 3t\sqrt{\sin^2 3t - \frac{1}{4}}}$

52. $\int \dfrac{dx}{(x^2 + 6x + 34)^2}$

53. $\int \dfrac{\sqrt{x^2 + a^2}}{x^4} \, dx$

54. $\int \tan x \sin x \, dx$

55. $\int \dfrac{\sin^{-1}\sqrt{2t}}{\sqrt{1 - 2t}} \, dt$

56. $\int \ln(x^2 + 1) \, dx$

57. $\int \dfrac{dx}{\sqrt{2 + \sqrt{x - 1}}}$

58. $\int \dfrac{dx}{2 + 2\sin x + \cos x}$

59. $\int \sqrt{\tan x} \, dx$

60. $\int \dfrac{dx}{\sqrt{1 + \sqrt[3]{x}}}$

61. $\int x^n \ln x \, dx$

62. $\int \tan^n x \sec^4 x \, dx, \ n > 0$

In Exercises 63 through 94, evaluate the definite integral.

63. $\int_0^\pi \sqrt{2 + 2\cos x} \, dx$

64. $\int_{1/2}^1 \sqrt{\dfrac{1 - x}{x}} \, dx$

65. $\int_1^2 \dfrac{2x^2 + x + 4}{x^3 + 4x^2} \, dx$

66. $\int_0^1 \dfrac{dx}{e^x + e^{-x}}$

67. $\int_0^2 \dfrac{t^3 \, dt}{\sqrt{4 + t^2}}$

68. $\int_0^{\pi/2} \sin^3 t \cos^3 t \, dt$

69. $\int_{-2}^{2\sqrt{3}} \dfrac{x^2 \, dx}{(16 - x^2)^{3/2}}$

70. $\int_0^1 \dfrac{xe^x \, dx}{(1 + x)^2}$

71. $\int_0^{\pi/4} \sec^4 x\, dx$

72. $\int_0^2 \dfrac{(1-x)\, dx}{x^2 + 3x + 2}$

73. $\int_{\pi/12}^{\pi/8} \cot^3 2y\, dy$

74. $\int_0^2 (2^x + x^2)\, dx$

75. $\int_a^{a/2} \dfrac{\sqrt{(a^2 - x^2)^3}}{x^2}\, dx$

76. $\int_1^2 (\ln x)^2\, dx$

77. $\int_{\sqrt{3}/3}^1 \dfrac{(2x^2 - 2x + 1)\, dx}{x^3 + x}$

78. $\int_{\sqrt{2}/2}^1 \dfrac{x^3\, dx}{\sqrt{2 - x^2}}$

79. $\int_1^{10} \log_{10} \sqrt{ex}\, dx$

80. $\int_0^{2\pi} |\sin x - \cos x|\, dx$

81. $\int_1^2 \dfrac{x + 2}{(x + 1)^2}\, dx$

82. $\int_0^{\sqrt{\pi/2}} xe^{x^2} \cos x^2\, dx$

83. $\int_0^{\pi} |\cos^3 x|\, dx$

84. $\int_{-\pi/4}^{\pi/4} |\tan^5 x|\, dx$

85. $\int_0^{1/2} \dfrac{2x\, dx}{x^3 - x^2 - x + 1}$

86. $\int_0^1 x^3 \sqrt{1 + x^2}\, dx$

87. $\int_0^{1/2} \dfrac{x\, dx}{\sqrt{1 - 4x^4}}$

88. $\int_0^{\pi/12} \dfrac{dx}{\cos^4 3x}$

89. $\int_0^1 \sqrt{2y + y^2}\, dy$

90. $\int_0^4 \dfrac{x^2\, dx}{x^3 + 4x^2 + 5x + 2}$

91. $\int_0^{\pi/2} \dfrac{dt}{12 + 13 \cos t}$

92. $\int_0^3 \dfrac{dr}{(r + 2)\sqrt{r + 1}}$

93. $\int_0^{16} \sqrt{4 - \sqrt{x}}\, dx$

94. $\int_{2\pi/3}^{\pi} \dfrac{\sin \frac{1}{2}t}{1 + \cos \frac{1}{2}t}\, dt$

95. The linear density of a rod 3 m long at a point x meters from one end is ke^{-3x} kilograms per meter. Find the mass and center of mass of the rod.

96. Find the center of mass of a rod 4 m long if the linear density at the point x meters from the left end is $\sqrt{9 + x^2}$ kilograms per meter.

97. Find the length of the arc of the parabola $y^2 = 6x$ from $x = 6$ to $x = 12$.

98. Find the area of the region bounded by the curve $y = \sin^{-1} 2x$, the line $x = \frac{1}{4}\sqrt{3}$, and the x axis.

99. Find the area of the region enclosed by one loop of the curve $x^2 = y^4(1 - y^2)$.

100. Find the length of the arc of the curve $y = \ln x$ from $x = 1$ to $x = e$.

101. Find the volume of the solid of revolution generated by revolving about the y axis the region bounded by the curve $y = \ln 2x$, the x axis, and the line $x = e$.

102. The region in the first quadrant bounded by the curve $y = \dfrac{5 - x}{(x + 1)^2}$, the x axis, and the y axis is revolved about the x axis. Find the volume of the solid generated.

103. Two chemicals A and B react to form a chemical C, and the rate of change of the amount of C is proportional to the product of the amounts of A and B remaining at any given time. Initially there are 60 lb of chemical A and 60 lb of chemical B, and to form 5 lb of C, 3 lb of A and 2 lb of

B are required. After 1 hour, 15 lb of C are formed. (a) If x pounds of C are formed at t hours, find an expression for x in terms of t. (b) Find the amount of C after 3 hours.

104. A tank is in the shape of the solid of revolution formed by rotating about the x axis the region bounded by the curve $y = \ln x$, the x axis, and the lines $x = e$ and $x = e^2$. If the tank is full of water, find the work done in pumping all the water to the top of the tank. Distance is measured in feet. Take the positive x axis vertically downward.

105. Find the centroid of the region of Exercise 99.

106. Find the centroid of the region enclosed by the loop of the curve $y^2 = x^2 - x^3$.

107. Find the centroid of the region bounded by the y axis and the curves $y = \sin x - \cos x$ and $y = \sin x + \cos x$ from $x = 0$ to $x = \frac{1}{2}\pi$.

108. Find the centroid of the region in the first quadrant bounded by the coordinate axes and the curve $y = \cos x$.

109. A pond can support a maximum of 10,000 fish so that the rate of growth of the fish population is jointly proportional to the number of fish present and the difference between 10,000 and the number present. The pond initially contained 400 fish, and 6 weeks later there were 3000 fish. (a) How many fish did the pond contain after 8 weeks? (b) When was the growth rate greatest? That is, after how many weeks did the pond contain 5000 fish?

110. In a town of population 12,000 the growth rate of a flu epidemic is jointly proportional to the number of people who have the flu and the number of people who do not have it. Five days ago 400 people in the town had the flu, and today 1000 people have it. How many people are expected to have the flu tomorrow? (b) In how many days from now will the epidemic be spreading the fastest? That is, when will half the population have the flu?

Exercises 111 and 112 pertain to Supplementary Section 6.7.

111. The vertical end of a water trough is 3 ft wide at the top and 2 ft deep, and it has the form of the region bounded by the x axis and one arch of the curve $y = 2 \sin \frac{1}{3}\pi x$. If the trough is full of water, find the force due to water pressure on the end.

112. A board is in the shape of a region bounded by a straight line and one arch of the sine curve. If the board is submerged vertically in water so that the straight line is the lower boundary 2 ft below the surface of the water, find the force on the board due to water pressure.

Exercises 113 and 114 pertain to Supplementary Section 9.8. Obtain the result by a hyperbolic function substitution.

113. $\int \dfrac{dx}{x^2\sqrt{a^2 + x^2}} = -\dfrac{1}{a^2} \coth\left(\sinh^{-1} \dfrac{x}{a}\right) + C, a > 0$

114. $\int \dfrac{dx}{(a^2 - x^2)^{3/2}} = \dfrac{1}{a^2} \sinh\left(\tanh^{-1} \dfrac{x}{a}\right) + C, a > 0$

The Conic
Sections and
Polar Coordinates

$$\frac{(x-h)^2}{a^2} - \frac{(y-k)^2}{b^2} = 1$$

$$R = 2 \sin 3\theta$$

This chapter is concerned with subject matter from analytic geometry. You may have studied some of this material in a precalculus course, and if so those topics can be treated as a review or omitted.

In the first four sections we discuss *conic sections* (or *conics*), which are curves of intersection of a plane with a right-circular cone. There are three types of curves that occur in this way: the *parabola*, the *ellipse* (including the circle as a special case), and the *hyperbola*. The curve obtained depends on the inclination of the axis of the cone to the cutting plane. The Greek mathematician Apollonius studied conic sections about 225 B.C., in terms of geometry, by using this concept.

In Section 10.1 we discuss the parabola, and the following two sections are devoted to the ellipse and the hyperbola. In our treatment of each of these curves we first show how the cone and the cutting plane are taken to obtain the particular conic section. We then define the curve as a set of points in a plane. It was proved that such a definition is a consequence of the definition of the curve as a conic section by the Belgian mathematician G. P. Dandelin (1794–1847) in 1822. Dandelin's proof for an ellipse appears in my text *Before Calculus* (second edition published in 1989 by Harper & Row Publishers, Inc.). Variations of that proof can be given for a parabola and a hyperbola. We introduce *rotation of axes* in Section 10.4 to enable us to consider conics whose axes are neither horizontal nor vertical.

Sections 10.5 through 10.7 are devoted to a discussion of *polar coordinates* and some of their applications. There is a unified treatment of conic sections that utilizes their polar equations in Section 10.8. In Supplementary Section 10.9 we present tangent lines of polar curves.

10.1 THE PARABOLA AND TRANSLATION OF AXES

When considering the geometry of conic sections, a cone is regarded as having two nappes, extending indefinitely far in both directions. A portion of a right-circular cone of two nappes is shown in Figure 1. A **generator** (or **element**) of the cone is a line lying in the cone, and all the generators of a cone contain the point *V*, called the **vertex**. In Figure 2 we have a cone and a cutting plane that is parallel to one and only one generator. The conic is a *parabola*.

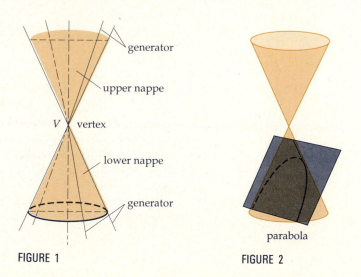

FIGURE 1

parabola

FIGURE 2

10.1.1 DEFINITION   A **parabola** is the set of points in a plane equidistant from a fixed point and a fixed line. The fixed point is called the **focus** and the fixed line is called the **directrix**.

We now derive an equation of a parabola from the definition. For this equation to be as simple as possible, we choose the *x* axis as perpendicular to the directrix and containing the focus. The origin is taken as the point on the *x* axis midway between the focus and the directrix. It should be stressed that we are choosing the axes (*not* the parabola) in a special way. See Figure 3.

Let *p* be the directed distance $\overline{OF}$. The focus is the point $F(p, 0)$, and the directrix is the line having the equation $x = -p$. A point $P(x, y)$ is on the

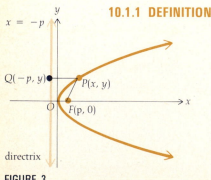

FIGURE 3

parabola if and only if P is equidistant from F and the directrix. That is, if $Q(-p, y)$ is the foot of the perpendicular line from P to the directrix, then P is on the parabola if and only if

$$|\overline{FP}| = |\overline{QP}|$$

Because

$$|\overline{FP}| = \sqrt{(x - p)^2 + y^2}$$

and

$$|\overline{QP}| = \sqrt{(x + p)^2 + (y - y)^2}$$

P is on the parabola if and only if

$$\sqrt{(x - p)^2 + y^2} = \sqrt{(x + p)^2}$$

By squaring on both sides of the equation we obtain

$$x^2 - 2px + p^2 + y^2 = x^2 + 2px + p^2$$

$$y^2 = 4px$$

This result is stated as a theorem.

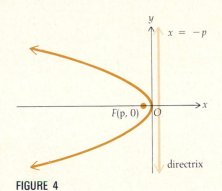

FIGURE 4

10.1.2 THEOREM An equation of the parabola having its focus at $(p, 0)$ and as its directrix the line $x = -p$ is

$$y^2 = 4px$$

In Figure 3, p is positive; p may be negative, however, because it is the directed distance $\overline{OF}$. Figure 4 shows a parabola for $p < 0$.

From Figures 3 and 4 we see that for the equation $y^2 = 4px$ the parabola opens to the right if $p > 0$ and to the left if $p < 0$. The point midway between the focus and the directrix on the parabola is called the **vertex**. The vertex of the parabolas in Figures 3 and 4 is the origin. The line through the vertex and the focus is called the **axis** of the parabola. The axis of the parabolas in Figures 3 and 4 is the x axis.

In the above derivation, if the x axis and the y axis are interchanged, then the focus is at the point $F(0, p)$, and the directrix is the line having the equation $y = -p$. An equation of this parabola is $x^2 = 4py$.

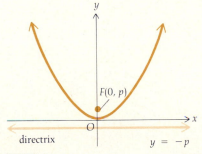

FIGURE 5

10.1.3 THEOREM An equation of the parabola having its focus at $(0, p)$ and as its directrix the line $y = -p$ is

$$x^2 = 4py$$

If $p > 0$, the parabola opens upward, as shown in Figure 5; if $p < 0$, it opens downward, as shown in Figure 6. In each case the vertex is at the origin, and the y axis is the axis of the parabola.

When drawing a sketch of a parabola, it is helpful to consider the chord through the focus, perpendicular to the axis of the parabola, because the endpoints of this chord give two points on the parabola. This chord is called the **latus rectum** of the parabola. The length of the latus rectum is $|4p|$. (See Exercise 20.)

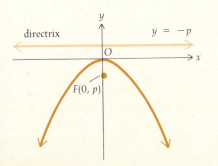

FIGURE 6

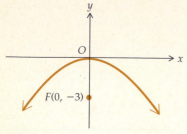

FIGURE 7

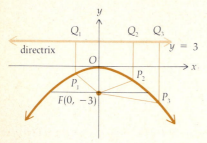

FIGURE 8

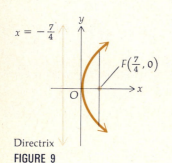

FIGURE 9

EXAMPLE 1   Find an equation of the parabola having its focus at $(0, -3)$ and as its directrix the line $y = 3$. Draw a sketch of the parabola.

Solution   Because the focus is on the y axis and is also below the directrix, the parabola opens downward, and $p = -3$. Hence an equation of the parabola is

$$x^2 = -12y$$

The length of the latus rectum is

$$|4(-3)| = 12$$

A sketch of the parabola appears in Figure 7.

Any point on the parabola of Figure 7 is equidistant from the focus and the directrix. In Figure 8 three such points (P_1, P_2, and P_3) are shown, and

$$|\overline{FP_1}| = |\overline{P_1Q_1}| \qquad |\overline{FP_2}| = |\overline{P_2Q_2}| \qquad |\overline{FP_3}| = |\overline{P_3Q_3}|$$

EXAMPLE 2   Given the parabola having the equation $y^2 = 7x$, find the co-ordinates of the focus, an equation of the directrix, and the length of the latus rectum. Draw a sketch of the parabola.

Solution   The given equation is of the form $y^2 = 4px$; so

$$4p = 7$$
$$p = \tfrac{7}{4}$$

Because $p > 0$, the parabola opens to the right. The focus is at the point $F(\tfrac{7}{4}, 0)$. An equation of the directrix is $x = -\tfrac{7}{4}$. The length of the latus rectum is 7. A sketch of the parabola is shown in Figure 9.

To find the general equation of a parabola having its vertex at a point other than the origin and its directrix parallel to a coordinate axis, we first consider the concept of *translation of axes*.

Note that the shape of a curve is not affected by the position of the coordinate axes but that an equation of the curve is affected.

▶ **ILLUSTRATION 1**   If a circle with a radius of 3 has its center at the point $(4, -1)$, then an equation of this circle is

$$(x - 4)^2 + (y + 1)^2 = 9$$
$$x^2 + y^2 - 8x + 2y + 8 = 0$$

However, if the origin is at the center, the circle has a simpler equation, namely,

$$x^2 + y^2 = 9 \qquad \qquad ◀$$

If the coordinate axes may be taken as we please, they are generally chosen in such a way that the equations will be as simple as possible. If the axes are given, however, we often wish to find a simpler equation of a given curve referred to another set of axes.

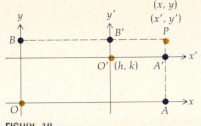

FIGURE 10

In general, if in the plane with given x and y axes new coordinate axes are chosen parallel to the given ones, we say that there has been a **translation of axes** in the plane.

In particular, let the given x and y axes be translated to the x' and y' axes, having origin (h, k) with respect to the given axes. Also, assume that the positive numbers are on the same side of the origin on the x' and y' axes as they are on the x and y axes. See Figure 10.

A point P in the plane, having coordinates (x, y) with respect to the given coordinate axes, will have coordinates (x', y') with respect to the new axes. To obtain relationships between these two sets of coordinates, draw a line through P parallel to the y axis and the y' axis, and also a line through P parallel to the x axis and the x' axis. Let the first line intersect the x axis at the point A and the x' axis at the point A', and the second line intersect the y axis at the point B and the y' axis at the point B'.

With respect to the x and y axes, the coordinates of P are (x, y), the coordinates of A are $(x, 0)$, and the coordinates of A' are (x, k). Because $\overline{A'P} = \overline{AP} - \overline{AA'}$,

$$y' = y - k \quad \text{and} \quad y = y' + k$$

With respect to the x and y axes, the coordinates of B are $(0, y)$, and the coordinates of B' are (h, y). Because $\overline{B'P} = \overline{BP} - \overline{BB'}$,

$$x' = x - h \quad \text{and} \quad x = x' + h$$

We have proved the following theorem.

10.1.4 THEOREM   If (x, y) represents a point P with respect to a given set of axes, and (x', y') is a representation of P after the axes are translated to a new origin having coordinates (h, k) with respect to the given axes, then

$$x = x' + h \quad \text{and} \quad y = y' + k$$
$$\Leftrightarrow \quad x' = x - h \quad \text{and} \quad y' = y - k$$

If an equation of a curve is given in x and y then an equation in x' and y' is obtained by replacing x by $x' + h$ and y by $y' + k$. The graph of the equation in x and y, with respect to the x and y axes, is exactly the same set of points as the graph of the corresponding equation in x' and y' with respect to the x' and y' axes.

EXAMPLE 3   Given the equation

$$x^2 + 10x + 6y + 19 = 0$$

find an equation of the graph with respect to the x' and y' axes after a translation of axes to the new origin $(-5, 1)$. Draw a sketch of the graph and show both sets of axes.

Solution   A point P, represented by (x, y) with respect to the old axes, has the representation (x', y') with respect to the new axes. Then from Theorem 10.1.4, with $h = -5$ and $k = 1$,

$$x = x' - 5 \quad \text{and} \quad y = y' + 1$$

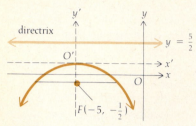

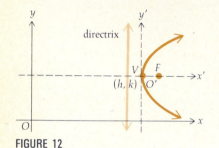

FIGURE 11

FIGURE 12

Substituting these values of x and y into the given equation we obtain

$$(x' - 5)^2 + 10(x' - 5) + 6(y' + 1) + 19 = 0$$

$$x'^2 - 10x' + 25 + 10x' - 50 + 6y' + 6 + 19 = 0$$

$$x'^2 = -6y'$$

The graph of this equation with respect to the x' and y' axes is a parabola with its vertex at the origin, opening downward, and with $4p = -6$. The graph with respect to the x and y axes is, then, a parabola having its vertex at $(-5, 1)$, its focus at $(-5, -\frac{1}{2})$, and as its directrix the line $y = \frac{5}{2}$. A sketch of the graph and both sets of axes appear in Figure 11.

The above example illustrates how an equation can be reduced to a simpler form by a suitable translation of axes.

We now apply the translation of axes to finding the general equation of a parabola having its directrix parallel to a coordinate axis and its vertex at the point (h, k). In particular, let the directrix be parallel to the y axis. If the vertex is at the point $V(h, k)$, then the directrix has the equation $x = h - p$, and the focus is at the point $F(h + p, k)$. Let the x' and y' axes be such that the origin O' is at $V(h, k)$. See Figure 12.

An equation of the parabola in Figure 12 with respect to the x' and y' axes is

$$y'^2 = 4px'$$

To obtain an equation of this parabola with respect to the x and y axes, we replace x' by $x - h$ and y' by $y - k$, which gives

$$(y - k)^2 = 4p(x - h)$$

The axis of this parabola is parallel to the x axis.

Similarly, if the directrix of a parabola is parallel to the x axis and the vertex is at $V(h, k)$, then its focus is at $F(h, k + p)$ and the directrix has the equation $y = k - p$, and an equation of the parabola with respect to the x and y axes is

$$(x - h)^2 = 4p(y - k)$$

The axis of this parabola is parallel to the y axis. We have proved, then, the following theorem.

10.1.5 THEOREM   If p is the directed distance from the vertex to the focus, an equation of the parabola with its vertex at (h, k) and with its axis parallel to the x axis is

$$(y - k)^2 = 4p(x - h) \tag{1}$$

A parabola with the same vertex and with its axis parallel to the y axis has for an equation

$$(x - h)^2 = 4p(y - k) \tag{2}$$

EXAMPLE 4   Find an equation of the parabola having as its directrix the line $y = 1$ and as its focus the point $F(-3, 7)$. Draw a sketch of this parabola.

Solution   Because the directrix is parallel to the x axis, the axis will be parallel to the y axis, and the equation will have the form (2).

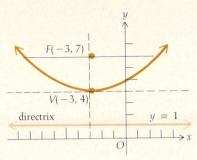

FIGURE 13

Because the vertex V is halfway between the directrix and the focus, V has coordinates $(-3, 4)$. The directed distance from the vertex to the focus is p; so

$$p = 7 - 4 \quad \Leftrightarrow \quad p = 3$$

Therefore an equation is

$$(x + 3)^2 = 12(y - 4)$$

Squaring and simplifying we have

$$x^2 + 6x - 12y + 57 = 0$$

A sketch of the parabola is shown in Figure 13.

EXAMPLE 5   Given the parabola having the equation

$$y^2 + 6x + 8y + 1 = 0$$

find the vertex, the focus, an equation of the directrix, an equation of the axis, and the length of the latus rectum. Draw a sketch of the parabola.

Solution   We rewrite the given equation as

$$y^2 + 8y = -6x - 1$$

Completing the square of the terms involving y on the left side of this equation by adding 16 on both sides we obtain

$$y^2 + 8y + 16 = -6x + 15$$

$$(y + 4)^2 = -6(x - \tfrac{5}{2})$$

Comparing this equation with (1) we let

$$k = -4 \qquad h = \tfrac{5}{2}$$

and

$$4p = -6 \quad \Leftrightarrow \quad p = -\tfrac{3}{2}$$

Therefore the vertex is at $(\tfrac{5}{2}, -4)$, an equation of the axis is $y = -4$, the focus is at $(1, -4)$, an equation of the directrix is $x = 4$, and the length of the latus rectum is 6. A sketch of the parabola appears in Figure 14.

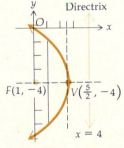

FIGURE 14

In Section 1.3 there is a discussion of the general equation of the second degree in two variables:

$$Ax^2 + Bxy + Cy^2 + Dx + Ey + F = 0 \tag{3}$$

where $B = 0$ and $A = C$. Then the graph of (3) is either a circle, a point, or the empty set. When the graph is a point or the empty set, we say it is a *degenerate circle*. We now consider (3) where $B = 0$ and $AC = 0$. In such a case either $A = 0$ or $C = 0$, but not both are zero because if the three numbers A, B, and C are zero, (3) is not a second-degree equation. Suppose in (3) $B = 0$, $A = 0$, and $C \neq 0$; we then have the equation

$$Cy^2 + Dx + Ey + F = 0 \tag{4}$$

If $D \neq 0$, this is an equation of a parabola because it can be obtained from (1) by squaring and combining terms. If in (4) $D = 0$, then the equation is

$$Cy^2 + Ey + F = 0$$

The graph of this equation may be two parallel lines, one line, or the empty set; any one of these graphs is then referred to as a *degenerate parabola*.

▶ **ILLUSTRATION 2**   The graph of the equation $4y^2 - 9 = 0$ is two parallel lines; $9y^2 + 6y + 1 = 0$ is an equation of one line; and $2y^2 + y + 1 = 0$ is satisfied by no real values of y.   ◀

A similar discussion holds if in (3) $B = 0$, $C = 0$, and $A \neq 0$. The results are summarized in the following theorem.

10.1.6 THEOREM

If in the general second-degree equation

$$Ax^2 + Bxy + Cy^2 + Dx + Ey + F = 0$$

$B = 0$, and either $A = 0$ and $C \neq 0$ or $C = 0$ and $A \neq 0$, then the graph is one of the following: a parabola, two parallel lines, one line, or the empty set.

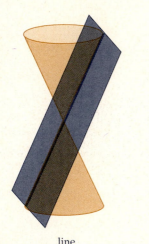

line
FIGURE 15

At the beginning of this section we indicated that a parabola is obtained as a conic section when the cutting plane is parallel to one and only one generator of the cone. If the cutting plane contains the vertex of the cone and only one generator, as in Figure 15, then a straight line is obtained, and this is a degenerate parabola. The degenerate parabola consisting of two parallel lines cannot be obtained as a plane section of a cone unless we consider a circular cylinder as a degenerate cone (one with its vertex at infinity). Then a plane parallel to the elements of the cylinder and cutting two distinct elements produces the degenerate parabola consisting of two parallel lines.

There is an interesting property of parabolas that has applications to the construction of searchlights, automobile headlights, and telescopes. In Figure 16 the line PT is the tangent line at point P to the graph of a parabola. The point F is the focus of the parabola, and α is the measure of the angle between the line segment FP and the tangent line PT. Line PR is parallel to the axis of the parabola, and β is the measure of the angle between PR and PT. In Exercise 48 you are asked to prove that $\alpha = \beta$. Because of this equality, for a parabolic mirror in a searchlight, rays of light from a source at the focus are reflected along a line parallel to the axis. A similar principle involves parabolic reflectors in automobile headlights. For a parabolic mirror in a reflecting telescope, a converse situation occur, where rays of light from an object in the sky, which go toward the mirror and are parallel to the axis, are reflected by the mirror and go through the focus.

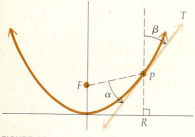

FIGURE 16

EXAMPLE 6   A parabolic mirror has a depth of 12 cm at the center, and the distance across the top of the mirror is 32 cm. Find the distance from the vertex to the focus.

Solution   See Figure 17. The coordinate axes are chosen so that the parabola has its vertex at the origin and its axis along the y axis, and it opens upward. Therefore an equation of the parabola is of the form

$$x^2 = 4py$$

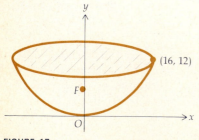

FIGURE 17

where p centimeters is the distance from the vertex to the focus. Because the point (16, 12) is on the parabola, its coordinates satisfy the equation, and we have

$$16^2 = 4p(12)$$

$$p = \tfrac{16}{3}$$

Therefore the distance from the vertex to the focus is $5\tfrac{1}{3}$ cm.

There are other practical applications of parabolas. The path of a projectile is a parabola if motion is considered to be in a plane and air resistance is neglected. Arches are sometimes parabolic in shape, and the cable of a suspension bridge could hang in the shape of a parabola. Disk antennas for receiving satellite television signals are also parabolic in shape.

EXERCISES 10.1

For each of the parabolas in Exercises 1 through 8, find the coordinates of the focus, an equation of the directrix, and the length of the latus rectum. Draw a sketch of the parabola.

1. $x^2 = 4y$ **2.** $y^2 = 6x$ **3.** $y^2 = -8x$
4. $x^2 = -16y$ **5.** $x^2 + y = 0$ **6.** $y^2 + 5x = 0$
7. $2y^2 - 9x = 0$ **8.** $3x^2 + 4y = 0$

In Exercises 9 through 17, find an equation of the parabola having the given properties.

9. Focus (5, 0); directrix $x = -5$.
10. Focus (0, 4); directrix $y = -4$.
11. Focus (0, -2); directrix $y - 2 = 0$.
12. Focus ($-\tfrac{5}{3}$, 0); directrix $5 - 3x = 0$.
13. Focus ($\tfrac{1}{2}$, 0); directrix $2x + 1 = 0$.
14. Focus (0, $\tfrac{2}{3}$); directrix $3y + 2 = 0$.
15. Vertex (0, 0); opens to the left; length of latus rectum is 6.
16. Vertex (0, 0); opens upward; length of latus rectum is 3.
17. Vertex (0, 0); directrix $2x = -5$.
18. Find an equation of the parabola having its vertex at the origin, the x axis as its axis, and passing through the point (2, -4).
19. Find an equation of the parabola having its vertex at the origin, the y axis as its axis, and passing through the point (-2, -4).
20. Prove that the length of the latus rectum of a parabola is $|4p|$.

In Exercises 21 through 26, find a new equation of the graph of the equation after a translation of axes to the new origin as indicated. Draw the original and the new axes and a sketch of the graph.

21. $x^2 + y^2 + 6x + 4y = 0$; (-3, -2)
22. $x^2 + y^2 - 6x - 10y + 18 = 0$; (3, 5)
23. $y^2 - 6x + 9 = 0$; ($\tfrac{3}{2}$, 0)
24. $y^2 + 3x - 2y + 7 = 0$; (-2, 1)
25. $y - 4 = 2(x - 1)^3$; (1, 4)
26. $(y + 1)^2 = 4(x - 2)^3$; (2, -1)

In Exercises 27 through 32, find the vertex, the focus, an equation of the axis, and an equation of the directrix of the parabola. Draw a sketch of the parabola.

27. $x^2 + 6x + 4y + 8 = 0$ **28.** $4x^2 - 8x + 3y - 2 = 0$
29. $y^2 + 6x + 10y + 19 = 0$ **30.** $3y^2 - 8x - 12y - 4 = 0$
31. $2y^2 = 4y - 3x$ **32.** $y = 3x^2 - 3x + 3$

In Exercises 33 through 40, find an equation of the parabola having the given properties. Draw a sketch of the parabola.

33. Vertex at (2, 4); focus at (-3, 4).
34. Vertex at (1, -3); directrix $y = 1$.
35. Focus at (-1, 7); directrix $y = 3$.
36. Focus at ($-\tfrac{3}{4}$, 4); directrix $x = -\tfrac{5}{4}$.
37. Vertex at (3, -2); axis $x = 3$; length of the latus rectum is 6.
38. Directrix $x = -2$; axis $y = 4$; length of the latus rectum is 8.
39. Vertex at (-4, 2); axis $y = 2$; through the point (0, 6).
40. Endpoints of the latus rectum are (1, 3) and (7, 3).
41. The endpoints of the latus rectum of a parabola are (5, k) and (-5, k). If the vertex of the parabola is at the origin and the parabola opens downward, find (a) the value of k; (b) an equation of the parabola.
42. Assume that water issuing from the end of a horizontal pipe 25 ft above the ground describes a parabolic curve, the vertex of the parabola being at the end of the pipe. If at a point 8 ft below the line of the pipe the flow of water has curved outward 10 ft beyond a vertical line through the end of the pipe, how far beyond this vertical line will the water strike the ground?
43. The cable of a suspension bridge hangs in the form of a parabola when the load is uniformly distributed horizontally. The distance between two towers is 150 m, the points of support of the cable on the towers are 22 m above the roadway, and the lowest point on the cable is 7 m above the roadway. Find the vertical distance to the cable from a point in the roadway 15 m from the foot of a tower.

44. A parabolic arch has a height of 20 m and a width of 36 m at the base. If the vertex of the parabola is at the top of the arch, at which height above the base is it 18 m wide?

45. Prove that on a parabola, the point closest to the focus is the vertex.

46. Suppose the orbit of a particular comet is a parabola having the sun at the focus. When the comet is 100 million km from the sun, the angle between the axis of the parabola and the line from the sun to the comet is 45°. Use the result of Exercise 45 to determine the shortest distance from the comet to the sun.

47. A reflecting telescope has a parabolic mirror for which the distance from the vertex to the focus is 30 ft. If the distance across the top of the mirror is 64 in., how deep is the mirror at the center?

48. In Figure 16, prove that $\alpha = \beta$. (*Hint:* Choose the coordinate axes so that the parabola has its vertex at the origin and its axis along the y axis and opens upward. Let Q be the point of intersection of the tangent line PT with the y axis. Prove that $\alpha = \beta$ by showing that $\triangle QPF$ is isosceles.)

49. An equation of the directrix of a parabola is $x + y = 0$ and its focus is at the point $(1, 1)$. Find (a) an equation of the axis of the parabola, (b) the coordinates of the vertex, and (c) the length of the latus rectum.

50. If a parabola has its focus at the origin and the x axis is its axis, prove that it must have an equation of the form $y^2 = 4kx + 4k^2$, $k \neq 0$.

51. If the axes are translated to a new origin having coordinates (h, k), show that after the translation the equation $y = \sin x$ becomes $y' = A \sin x' + B \cos x' + C$, where A, B, and C are constants. Find A, B, and C in terms of h and k. Show that $A^2 + B^2 = 1$.

52. Show that after a translation of axes to the new origin $(-\frac{1}{4}\pi, 1)$ the equation $y = \dfrac{1}{\sqrt{2}} (\sin x + \cos x) + 1$ becomes $y' = \sin x'$.

53. Refer to the results of Exercises 51 and 52, and determine a translation of axes so that the equation

$$y = \frac{1}{\sqrt{2}} (\sin x - \cos x) - 3$$

becomes $y' = \sin x'$.

54. Refer to the results of Exercises 51 and 52, and determine a translation of axes so that the equation

$$y = \tfrac{1}{2}(\sqrt{3} \sin x + \cos x) + 2$$

becomes $y' = \sin x'$.

10.2 THE ELLIPSE

An ellipse is obtained as a conic section if the cutting plane is parallel to no generator, in which case the cutting plane intersects each generator as in Figure 1. A special case of the ellipse is a circle, as shown in Figure 2, and it is formed if the cutting plane that intersects each generator is also perpendicular to the axis of the cone. We now define an ellipse as a set of points in a plane.

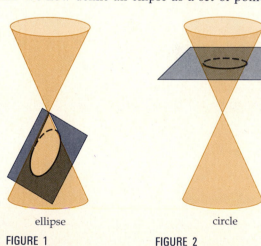

ellipse   circle

FIGURE 1   FIGURE 2

10.2.1 DEFINITION An **ellipse** is the set of points in a plane, the sum of whose distances from two fixed points is a constant. Each fixed point is called a **focus**.

Let the undirected distance between the foci (the plural of focus) be $2c$, where $c > 0$. To obtain an equation of an ellipse we select the x axis as the line through the foci F and F' and we choose the origin as the midpoint of the segment FF'.

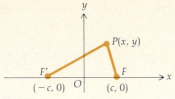

FIGURE 3

See Figure 3. The foci F and F' have coordinates $(c, 0)$ and $(-c, 0)$, respectively. Let the constant sum referred to in Definition 10.2.1 be $2a$. Then $a > c$ and the point $P(x, y)$ in Figure 3 is any point on the ellipse if and only if

$$|\overline{FP}| + |\overline{F'P}| = 2a \qquad (1)$$

Because

$$|\overline{FP}| = \sqrt{(x - c)^2 + y^2} \quad \text{and} \quad |\overline{F'P}| = \sqrt{(x + c)^2 + y^2}$$

P is on the ellipse if and only if

$$\sqrt{(x - c)^2 + y^2} + \sqrt{(x + c)^2 + y^2} = 2a$$

We simplify this equation by first writing it so that one radical is on the left side and the other is on the right and then squaring on both sides of the equation. Thus we have

$$\sqrt{(x - c)^2 + y^2} = 2a - \sqrt{(x + c)^2 + y^2} \qquad (2)$$

$$(x - c)^2 + y^2 = 4a^2 - 4a\sqrt{(x + c)^2 + y^2} + (x + c)^2 + y^2 \qquad (3)$$

$$x^2 - 2cx + c^2 + y^2 = 4a^2 - 4a\sqrt{(x + c)^2 + y^2} + x^2 + 2cx + c^2 + y^2$$

$$4a\sqrt{(x + c)^2 + y^2} = 4a^2 + 4cx$$

$$\sqrt{(x + c)^2 + y^2} = a + \frac{c}{a}x \qquad (4)$$

$$x^2 + 2cx + c^2 + y^2 = a^2 + 2cx + \frac{c^2}{a^2}x^2 \qquad (5)$$

$$x^2\left(1 - \frac{c^2}{a^2}\right) + y^2 = a^2 - c^2$$

$$(a^2 - c^2)x^2 + a^2y^2 = a^2(a^2 - c^2)$$

$$\frac{x^2}{a^2} + \frac{y^2}{a^2 - c^2} = 1 \qquad (6)$$

Because $a > c$, $a^2 - c^2 > 0$, and we can let

$$b^2 = a^2 - c^2 \qquad (7)$$

Substituting from this equation in (6) we get

$$\frac{x^2}{a^2} + \frac{y^2}{b^2} = 1 \qquad (8)$$

We have shown that the coordinates (x, y) of any point P on the ellipse satisfies Equation (8). To prove that (8) is an equation of the ellipse we must also show that any point P whose coordinates (x, y) satisfy (8) is on the ellipse. To do this we start with (8) and reverse the steps to obtain (1). When reversing the steps to go from (5) to (4) we must verify that

$$a + \frac{c}{a}x \geq 0 \qquad (9)$$

and to go from (3) to (2) we must verify that

$$2a - \sqrt{(x + c)^2 + y^2} \geq 0 \qquad (10)$$

You are asked to do this in Exercise 42. An alternate method for obtaining

(1) from (8) is outlined in Exercise 43. As a result of this discussion, we have the following theorem.

10.2.2 THEOREM

If $2a$ is the constant referred to in Definition 10.2.1 and an ellipse has its foci at $(c, 0)$ and $(-c, 0)$, then if $b^2 = a^2 - c^2$, an equation of the ellipse is

$$\frac{x^2}{a^2} + \frac{y^2}{b^2} = 1 \tag{11}$$

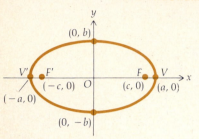

FIGURE 4

To obtain a sketch of the graph of the ellipse having Equation (11) first observe from the equation that the graph is symmetric with respect to both the x and y axes. Furthermore, the graph intersects the x axis at the points $(a, 0)$ and $(-a, 0)$ and it intersects the y axis at the points $(0, b)$ and $(0, -b)$. See Figure 4 and refer to it as you read the next paragraph.

The line through the foci is called the **principal axis** of the ellipse. For the ellipse of Theorem 10.2.2 the x axis is the principal axis. The points of intersection of the ellipse and its principal axis are called the **vertices**. Thus for this ellipse the vertices are at $V(a, 0)$ and $V'(-a, 0)$. The point on the principal axis that lies halfway between the two vertices is called the **center** of the ellipse. The origin is the center of this ellipse. The segment of the principal axis between the two vertices is called the **major axis** of the ellipse and its length is $2a$ units. Then we state that a is the number of units in the length of the semimajor axis. For this ellipse the segment of the y axis between the points $(0, b)$ and $(0, -b)$ is called the **minor axis** of the ellipse. Its length is $2b$ units. Hence b is the number of units in the length of the semiminor axis. Observe from (7) that $a > b$.

The ellipse is called a **central conic** in contrast to the parabola, which has no center because it has only one vertex.

EXAMPLE 1   Given the ellipse having the equation

$$\frac{x^2}{25} + \frac{y^2}{16} = 1$$

find the vertices, foci, and extremities of the minor axis. Draw a sketch of the ellipse, and show the foci.

Solution   From the equation of the ellipse, $a^2 = 25$ and $b^2 = 16$; thus $a = 5$ and $b = 4$. Therefore the vertices are at the points $V(5, 0)$ and $V'(-5, 0)$ and the extremities of the minor axis are at the points $B(0, 4)$ and $B'(0, -4)$. From (7),

$$c^2 = 25 - 16$$
$$= 9$$

Therefore $c = 3$ and so the foci are at $F(3, 0)$ and $F'(-3, 0)$. A sketch of the ellipse and its foci appear in Figure 5.

From the definition of an ellipse it follows that if P is any point on the ellipse $|\overline{FP}| + |\overline{F'P}| = 10$.

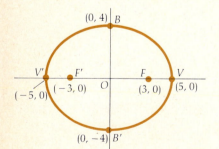

FIGURE 5

EXAMPLE 2   An arch is in the form of a semiellipse. It is 48 ft wide at the base and has a height of 20 ft. How wide is the arch at a height of 10 ft above the base?

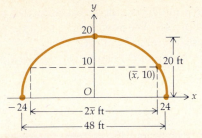

FIGURE 6

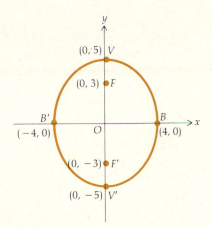

FIGURE 7

Solution   Figure 6 shows a sketch of the arch and the coordinate axes chosen so that the x axis is along the base and the origin is at the midpoint of the base. Then the ellipse has its principal axis on the x axis, its center at the origin, $a = 24$, and $b = 20$. An equation of the ellipse is of the form of (11):

$$\frac{x^2}{576} + \frac{y^2}{400} = 1$$

Let $2\bar{x}$ be the number of feet in the width of the arch at a height of 10 ft above the base. Therefore the point $(\bar{x}, 10)$ lies on the ellipse. Thus

$$\frac{\bar{x}^2}{576} + \frac{100}{400} = 1$$

$$\bar{x}^2 = 432$$

$$\bar{x} = 12\sqrt{3}$$

Hence, at a height of 10 ft above the base the width of the arch is $24\sqrt{3}$ ft.

If an ellipse has its center at the origin and its principal axis on the y axis, then an equation of the ellipse is of the form

$$\frac{y^2}{a^2} + \frac{x^2}{b^2} = 1 \tag{12}$$

which is obtained from (11) by interchanging x and y.

▶ **ILLUSTRATION 1**   Because for an ellipse $a > b$, it follows that the ellipse having the equation

$$\frac{x^2}{16} + \frac{y^2}{25} = 1$$

has its foci on the y axis. This ellipse has the same shape as the ellipse of Example 1. The vertices are at $(0, 5)$ and $(0, -5)$ and the foci are at $(0, 3)$ and $(0, -3)$. A sketch of the graph of this ellipse appears in Figure 7.   ◀

If the center of an ellipse is at the point (h, k) rather than at the origin, and if the principal axis is parallel to one of the coordinate axes, then by a translation of axes so that the point (h, k) is the new origin, an equation of the ellipse is $\bar{x}^2/a^2 + \bar{y}^2/b^2 = 1$ if the principal axis is horizontal, and $\bar{y}^2/a^2 + \bar{x}^2/b^2 = 1$ if the principal axis is vertical. Because $\bar{x} = x - h$ and $\bar{y} = y - k$, these equations become the following in x and y:

$$\frac{(x - h)^2}{a^2} + \frac{(y - k)^2}{b^2} = 1 \tag{13}$$

if the principal axis is horizontal, and

$$\frac{(y - k)^2}{a^2} + \frac{(x - h)^2}{b^2} = 1 \tag{14}$$

if the principal axis is vertical.

Recall that for the general equation of the second degree in two variables

$$Ax^2 + Bxy + Cy^2 + Dx + Ey + F = 0$$

when $B = 0$ and $A = C$, the graph is either a circle or a degenerate case of a circle, which is either a point-circle or the empty set. We now discuss this equation when $B = 0$ and A and C are not necessarily equal but $AC > 0$.

If we eliminate the fractions and combine terms in (13) and (14), we obtain an equation of the form

$$Ax^2 + Cy^2 + Dx + Ey + F = 0 \tag{15}$$

where $A \neq C$ if $a \neq b$ and $AC > 0$. It can be shown by completing the squares in x and y that an equation of the form (15) can be put in the form

$$\frac{(x-h)^2}{\dfrac{1}{A}} + \frac{(y-k)^2}{\dfrac{1}{C}} = G \tag{16}$$

If $AC > 0$, then A and C have the same sign. If G has the same sign as A and C, then (16) can be written in the form of (13) or (14). Thus the graph of (15) is an ellipse.

▶ **ILLUSTRATION 2** Suppose we have the equation

$$6x^2 + 9y^2 - 24x - 54y + 51 = 0$$

which can be written as

$$6(x^2 - 4x) + 9(y^2 - 6y) = -51$$

Completing the squares in x and y we get

$$6(x^2 - 4x + 4) + 9(y^2 - 6y + 9) = -51 + 24 + 81$$
$$6(x-2)^2 + 9(y-3)^2 = 54$$
$$\frac{(x-2)^2}{\frac{1}{6}} + \frac{(y-3)^2}{\frac{1}{9}} = 54$$

This is an equation of the form of (16). By dividing on both sides by 54 we have

$$\frac{(x-2)^2}{9} + \frac{(y-3)^2}{6} = 1$$

which has the form of (13). ◀

If in (16) G has a sign opposite to that of A and C, then (16) is not satisfied by any real values of x and y. Hence the graph of (15) is the empty set.

▶ **ILLUSTRATION 3** Suppose that (15) is

$$6x^2 + 9y^2 - 24x - 54y + 115 = 0$$

Then, upon completing the squares in x and y, we get

$$6(x-2)^2 + 9(y-3)^2 = -115 + 24 + 81$$
$$\frac{(x-2)^2}{\frac{1}{6}} + \frac{(y-3)^2}{\frac{1}{9}} = -10 \tag{17}$$

This is of the form of (16), where $G = -10$, $A = 6$, and $C = 9$. For all values of x and y the left side of (17) is nonnegative; hence the graph of (17) is the empty set. ◀

If $G = 0$ in (16), then the equation is satisfied by only the ordered pair (h, k). Therefore the graph of (15) is a point.

▶ **ILLUSTRATION 4**   Because the equation

$$6x^2 + 9y^2 - 24x - 54y + 105 = 0$$

can be written as

$$\frac{(x - 2)^2}{\frac{1}{6}} + \frac{(y - 3)^2}{\frac{1}{9}} = 0$$

its graph is the point $(2, 3)$.   ◀

If the graph of (15) is a point or the empty set, the graph is said to be degenerate.

If $A = C$ in (15), we have either a circle or a degenerate case of a circle, as mentioned above. A circle is a limiting form of an ellipse. This fact can be shown by considering the equation relating a, b, and c for an ellipse:

$$b^2 = a^2 - c^2$$

From this equation we see that as c approaches zero, b^2 approaches a^2. If $b^2 = a^2$, (13) and (14) become

$$(x - h)^2 + (y - k)^2 = a^2$$

which is an equation of a circle having its center at (h, k) and radius a. The results of Section 1.3 for a circle are the same as those obtained for (15) applied to an ellipse.

The results of the preceding discussion are summarized in the following theorem.

10.2.3 THEOREM   If in the general second-degree equation

$$Ax^2 + Bxy + Cy^2 + Dx + Ey + F = 0$$

$B = 0$ and $AC > 0$, then the graph is either an ellipse, a point, or the empty set. In addition, if $A = C$, the graph is either a circle, a point, or the empty set.

The degenerate case of an ellipse, a point, is obtained as a conic section if the cutting plane contains the vertex of the cone but does not contain a generator. See Figure 8.

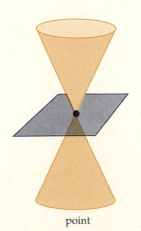

point

FIGURE 8

EXAMPLE 3   Determine the graph of the equation

$$25x^2 + 16y^2 + 150x - 128y - 1119 = 0$$

Solution   From Theorem 10.2.3, because $B = 0$ and $AC > 0$, the graph is either an ellipse or is degenerate. Completing the squares in x and y we have

$$25(x^2 + 6x + 9) + 16(y^2 - 8y + 16) = 1119 + 225 + 256$$

$$25(x + 3)^2 + 16(y - 4)^2 = 1600$$

$$\frac{(x + 3)^2}{64} + \frac{(y - 4)^2}{100} = 1 \tag{18}$$

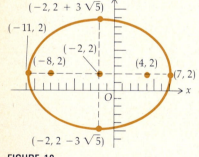

FIGURE 9

This equation is of the form of (14); so the graph is an ellipse having its principal axis parallel to the y axis and its center at $(-3, 4)$.

EXAMPLE 4   For the ellipse of Example 3, find the vertices, foci, and extremities of the minor axis. Draw a sketch of the ellipse, and show the foci.

Solution   From (18), $a = 10$ and $b = 8$. Because the center of the ellipse is at $(-3, 4)$ and the principal axis is vertical, the vertices are at the points $V(-3, 14)$ and $V'(-3, -6)$. The extremities of the minor axis are at the points $B(5, 4)$ and $B'(-11, 4)$. Because $b^2 = a^2 - c^2$,

$$64 = 100 - c^2$$

$$c^2 = 36$$

$$c = 6$$

Thus the distance from the center to a focus is 6, and so the foci are at the points $F(-3, 10)$ and $F'(-3, -2)$. A sketch of the ellipse and the foci appears in Figure 9.

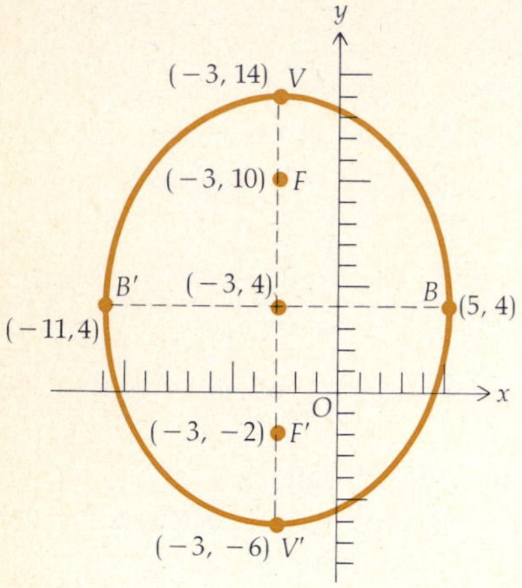

EXAMPLE 5   Find an equation of the ellipse having foci at $(-8, 2)$ and $(4, 2)$ and for which the constant referred to in Definition 10.2.1 is 18. Draw a sketch of the ellipse.

Solution   The center of the ellipse is halfway between the foci and is the point $(-2, 2)$. The distance between the foci of an ellipse is $2c$ and the distance between $(-8, 2)$ and $(4, 2)$ is 12. Therefore $c = 6$. The constant referred to in Definition 10.2.1 is $2a$; thus $2a = 18$ and $a = 9$. Because $b^2 = a^2 - c^2$,

$$b^2 = 81 - 36$$

$$b^2 = 45$$

$$b = 3\sqrt{5}$$

The principal axis is parallel to the x axis; hence an equation of the ellipse is of the form of (13). Because (h, k) is the point $(-2, 2)$, $a = 9$ and $b = 3\sqrt{5}$, the required equation is

$$\frac{(x + 2)^2}{81} + \frac{(y - 2)^2}{45} = 1$$

A sketch of this ellipse is in Figure 10.

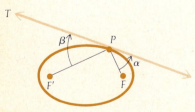

FIGURE 10

FIGURE 11

There are applications of ellipses in astronomy because orbits of planets and satellites are ellipses. Ellipses are used in making machine gears. Arches of bridges are sometimes elliptical in shape.

There is a reflective property of the ellipse that is analogous to the one shown for the parabola in Figure 16 of Section 10.1. For the ellipse refer to Figure 11, where the line PT is the tangent line at P to the graph of the ellipse having foci at F and F'. The measure of the angle between the line segment FP and the tangent line PT is α, and the measure of the angle between the line segment $F'P$ and the tangent line PT is β. In Exercise 31 you are asked to prove that

$\alpha = \beta$. Therefore a light ray from a source at one focus of an elliptical mirror that hits the mirror is reflected along a line through the other focus. This property of ellipses is used in so-called whispering galleries, where the ceilings have cross sections that are arcs of ellipses with common foci. A person located at one focus F can hear another person whispering at the other focus F' because the sound waves originating from the whisperer at F' hit the ceiling and are reflected by the ceiling to the listener at F. A famous example of a whispering gallery is under the dome of the Capitol in Washington, D.C. Another is at the Mormon Tabernacle in Salt Lake City.

EXERCISES 10.2

In Exercises 1 through 16, find the center, vertices, foci, and extremities of the minor axis of the ellipse. Draw a sketch of the ellipse, and show the foci.

1. $4x^2 + 9y^2 = 36$
2. $4x^2 + 9y^2 = 4$
3. $25x^2 + 4y^2 = 100$
4. $16x^2 + 9y^2 = 144$
5. $2x^2 + 3y^2 = 18$
6. $64x^2 + y^2 = 16$
7. $16x^2 + 4y^2 = 1$
8. $3x^2 + 4y^2 = 9$
9. $6x^2 + 9y^2 - 24x - 54y + 51 = 0$
10. $9x^2 + 4y^2 - 18x + 16y - 11 = 0$
11. $5x^2 + 3y^2 - 3y - 12 = 0$
12. $2x^2 + 2y^2 - 2x + 18y + 33 = 0$
13. $4x^2 + 4y^2 + 20x - 32y + 89 = 0$
14. $3x^2 + 4y^2 - 30x + 16y + 100 = 0$
15. $3x^2 + 5y^2 - 6x - 12 = 0$
16. $2x^3 + 3y^2 - 4x + 12y + 2 = 0$

In Exercises 17 and 18, determine whether the graph of the equation is an ellipse, a point, or the empty set.

17. $4x^2 + y^2 - 8x + 2y + 5 = 0$
18. $2x^2 + 3y^2 + 8x - 6y + 20 = 0$

In Exercises 19 through 28, find an equation of the ellipse having the given properties, and draw a sketch of the ellipse.

19. Vertices at $(-\tfrac{5}{2}, 0)$ and $(\tfrac{5}{2}, 0)$ and one focus at $(\tfrac{3}{2}, 0)$.
20. Foci at $(-5, 0)$ and $(5, 0)$ and for which the constant referred to in Definition 10.2.1 is 20.
21. Foci at $(0, 3)$ and $(0, -3)$ and for which the constant referred to in Definition 10.2.1 is $6\sqrt{3}$.
22. Center at the origin, its foci on the x axis, the length of the major axis equal to three times the length of the minor axis, and passing through the point $(3, 3)$.
23. Vertices at $(2, 0)$ and $(-2, 0)$ and through the point $(-1, \tfrac{1}{2}\sqrt{3})$.
24. Vertices at $(0, 5)$ and $(0, -5)$ and through the point $(2, -\tfrac{5}{3}\sqrt{5})$.
25. Center at $(4, -2)$, a vertex at $(9, -2)$, and one focus at $(0, -2)$.
26. A focus at $(2, -3)$, a vertex at $(2, 4)$, and center on the x axis.
27. Foci at $(-1, -1)$ and $(-1, 7)$ and the semimajor axis of length 8 units.
28. Foci at $(2, 3)$ and $(2, -7)$ and the length of the semiminor axis is two-thirds of the length of the semimajor axis.

29. Find an equation of the tangent line to the ellipse $4x^2 + 9y^2 = 72$ at the point $(3, 2)$.
30. Show that an equation of the tangent line to the ellipse $x^2/a^2 + y^2/b^2 = 1$ at the point (x_0, y_0) on the ellipse is $x_0 x/a^2 + y_0 y/b^2 = 1$.
31. In Figure 11 prove that $\alpha = \beta$. (*Hint:* Choose the coordinate axes so that the center of the ellipse is at the origin and the axes of the ellipse are along the coordinate axes. Then use Theorem 1.6.8.)
32. The orbit of the earth around the sun is elliptical in shape with the sun at one focus and a semimajor axis of length 92.9 million miles. If the distance between the foci is 3.16 million miles, find (a) how close the earth gets to the sun and (b) the greatest possible distance between the earth and the sun.
33. The ceiling in a hallway 10 m wide is in the shape of a semiellipse and is 9 m high in the center and 6 m high at the side walls. Find the height of the ceiling 2 m from either wall.
34. The arch of a bridge is in the shape of a semiellipse having a horizontal span of 40 m and a height of 16 m at its center. How high is the arch 9 m to the right or left of the center?
35. Suppose that the orbit of a planet is in the shape of an ellipse with a major axis whose length is 500 million km. If the distance between the foci is 400 million km, find an equation of the orbit.
36. A football is 12 in. long, and a plane section containing a seam is an ellipse, of which the length of the minor axis is 7 in. Find the volume of the football if the leather is so stiff that every cross section is a square.
37. Solve Exercise 36 if every cross section is a circle.
38. Definition 10.2.1 gives a procedure for drawing the graph of an ellipse. To apply the method to the ellipse $4x^2 + 9y^2 = 36$, first determine the points of intersection with the coordinate axes. Obtain the foci on the x axis by using a compass with its center at one of the points of intersection with the y axis and with radius of 3. Then fasten thumbtacks at each focus. Take a piece of string of length 6, which is $2a$, and attach an end at one thumbtack and an end at the other thumbtack. Place a pencil against the string and make it tight. Move the pencil against the string and trace a curve. This curve is an ellipse because the pencil traces a set of points the sum of whose distances from the two tacks is the constant 6.

39. For the ellipse whose equation is

$$\frac{(x-h)^2}{a^2} + \frac{(y-k)^2}{b^2} = 1$$

where $a > b > 0$, find the coordinates of the foci in terms of h, k, a, and b.

Exercises 40 and 41 pertain to Supplementary Section 6.7.

40. A plate is in the shape of the region bounded by the ellipse having a semimajor axis of length 3 ft and a semiminor axis of length 2 ft. If the plate is lowered vertically in a tank of water until the minor axis lies in the surface of the water, find the force due to water pressure on one side of the submerged portion of the plate.

41. If the plate of Exercise 40 is lowered until the center is 3 ft below the surface of the water, find the force due to water pressure on one side of the plate. The minor axis is still horizontal.

42. By reversing the steps used to obtain Equation (8) from Equation (1), prove that if $P(x, y)$ is any point whose coordinates satisfy (8), then

$$|\overline{FP}| + |\overline{F'P}| = 2a$$

where F and F' are the foci of the ellipse in Figure 3. (*Hint:* To show inequality (9) use the fact that $a > c > 0$ and that for every point (x, y) satisfying (8), $-a \le x \le a$. To show inequality (10) first show that $\sqrt{(x+c)^2 + y^2} = a + \frac{c}{a}x$ by replacing y^2 by $b^2\left(1 - \frac{x^2}{a^2}\right)$ and using (7).)

43. Instead of reversing the steps used to obtain Equation (8) from Equation (1), an alternate method for obtaining (1) from (8) is as follows:

$$|\overline{FP}| + |\overline{F'P}| = \sqrt{(x-c)^2 + y^2} + \sqrt{(x+c)^2 + y^2}$$

$$= \frac{1}{a}\left(\sqrt{a^2(x-c)^2 + a^2 y^2} + \sqrt{a^2(x+c)^2 + a^2 y^2}\right)$$

Replace $a^2 y^2$ by $a^2 b^2 - b^2 x^2$ and use (7) to obtain

$$|\overline{FP}| + |\overline{F'P}| = \left| a - \frac{c}{a}x \right| + \left| a + \frac{c}{a}x \right|$$

Show that the right-hand side of the above equation is $2a$ by using the fact that $a > c > 0$ and that for every point (x, y) satisfying (8), $-a \le x \le a$.

10.3 THE HYPERBOLA

When a cutting plane of a cone is parallel to two generators, it intersects both nappes of the cone, and the conic section obtained is a *hyperbola*, shown in Figure 1. Following is the definition of a hyperbola as a set of points in a plane.

10.3.1 DEFINITION

A **hyperbola** is the set of points in a plane, the absolute value of the difference of whose distances from two fixed points is a constant. The two fixed points are called the **foci**.

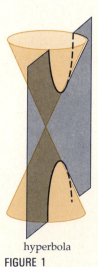

hyperbola

FIGURE 1

To obtain an equation of a hyperbola we begin as we did with the ellipse by letting the undirected distance between the foci be $2c$, where $c > 0$. Then we choose the x axis as the line through the foci F and F', and we take the origin as the midpoint of the segment FF'. Refer to Figure 2. The points $(c, 0)$ and $(-c, 0)$ are the foci F and F', respectively. Let $2a$ be the constant referred to in Definition 10.3.1. In Figure 2 the point $P(x, y)$ represents any point on the hyperbola. Then from Definition 10.3.1,

$$\left| |\overline{FP}| - |\overline{F'P}| \right| = 2a \tag{1}$$

To determine the relationship between a and c, we use the fact that the sum of the lengths of any two sides of a triangle is greater than the length of the third side and write the two inequalities

$$|\overline{F'F}| + |\overline{FP}| > |\overline{F'P}| \qquad\qquad |\overline{F'F}| + |\overline{F'P}| > |\overline{FP}|$$

$$|\overline{F'F}| > |\overline{F'P}| - |\overline{FP}| \qquad\qquad |\overline{F'F}| > |\overline{FP}| - |\overline{F'P}|$$

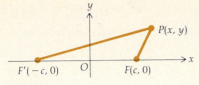

FIGURE 2

By using absolute value bars these two inequalities can be written as the inequality

$$|\overline{F'F}| > \left||\overline{FP}| - |\overline{F'P}|\right|$$

Because $|\overline{F'F}| = 2c$ and $\left||\overline{FP}| - |\overline{F'P}|\right| = 2a$, we have

$$2c > 2a$$

$$c > a \qquad (2)$$

Because

$$|\overline{FP}| = \sqrt{(x-c)^2 + y^2} \quad \text{and} \quad |\overline{F'P}| = \sqrt{(x+c)^2 + y^2}$$

then from (1), P is on the hyperbola if and only if

$$\left|\sqrt{(x-c)^2 + y^2} - \sqrt{(x+c)^2 + y^2}\right| = 2a$$

or, equivalently, without absolute value bars,

$$\sqrt{(x-c)^2 + y^2} - \sqrt{(x+c)^2 + y^2} = \pm 2a$$

$$\sqrt{(x-c)^2 + y^2} = \pm 2a + \sqrt{(x+c)^2 + y^2}$$

$$x^2 - 2cx + c^2 + y^2 = 4a^2 \pm 4a\sqrt{(x+c)^2 + y^2} + x^2 + 2cx + c^2 + y^2$$

$$\pm 4a\sqrt{(x+c)^2 + y^2} = 4a^2 + 4cx$$

$$\pm\sqrt{(x+c)^2 + y^2} = a + \frac{c}{a}x$$

$$x^2 + 2cx + c^2 + y^2 = a^2 + 2cx + \frac{c^2}{a^2}x^2$$

$$x^2\left(\frac{c^2}{a^2} - 1\right) - y^2 = c^2 - a^2$$

$$(c^2 - a^2)x^2 - a^2y^2 = a^2(c^2 - a^2)$$

$$\frac{x^2}{a^2} - \frac{y^2}{c^2 - a^2} = 1 \qquad (3)$$

Because, from (2), $c > a$, we can let

$$\qquad (4)$$

$$b^2 = c^2 - a^2$$

Substituting from (4) in (3) we get

$$\frac{x^2}{a^2} - \frac{y^2}{b^2} = 1 \qquad (5)$$

We have proved that the coordinates (x, y) of any point P on the hyperbola satisfies Equation (5). To prove that (5) is an equation of the hyperbola we must also show that any point P whose coordinates (x, y) satisfy (5) is on the hyperbola. You are asked to do this in Exercise 41. The procedure is similar to what we indicated in Section 11.2 for the ellipse as outlined in Exercises 42 and 43 of that section. This discussion gives us the following theorem.

10.3.2 THEOREM

If $2a$ is the constant referred to in Definition 10.3.1 and a hyperbola has its foci at $(c, 0)$ and $(-c, 0)$, then if $b^2 = c^2 - a^2$, an equation of the hyperbola is

$$\frac{x^2}{a^2} - \frac{y^2}{b^2} = 1 \qquad (6)$$

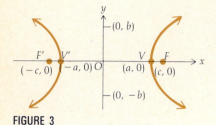

FIGURE 3

A sketch of the hyperbola of Theorem 10.3.2 appears in Figure 3. We now show how this graph is obtained. From the equation we observe that the graph is symmetric with respect to both the x and y axes. As with the ellipse, the line through the foci is called the **principal axis** of the hyperbola. Thus for this hyperbola the x axis is the principal axis. The points where the graph intersects the principal axis are called the **vertices**, and the point that is halfway between the vertices is called the **center** of the hyperbola. For this hyperbola the vertices are at $V(a, 0)$ and $V'(-a, 0)$ and the center is at the origin. The segment $V'V$ of the principal axis is called the **transverse axis** of the hyperbola, and its length is $2a$ units; thus a units is the length of the semitransverse axis.

Substituting 0 for x in (6) we obtain $y^2 = -b^2$, which has no real solutions. Consequently the hyperbola does not intersect the y axis. However, the line segment having extremities at the points $(0, -b)$ and $(0, b)$ is called the **conjugate axis** of the hyperbola and its length is $2b$ units. Thus b is the number of units in the length of the semiconjugate axis.

Solving (6) for y in terms of x we obtain

$$y = \pm \frac{b}{a} \sqrt{x^2 - a^2} \qquad (7)$$

We conclude from (7) that if $|x| < a$, there is no real value of y. Therefore there are no points (x, y) on the hyperbola for which $-a < x < a$. We also see from (7) that if $|x| > a$, then y has two real values. Thus the hyperbola has two *branches*. One branch contains the vertex $V(a, 0)$ and extends indefinitely far to the right of V. The other branch contains the vertex $V'(-a, 0)$ and extends indefinitely far to the left of V'.

As was the case with an ellipse, because the hyperbola has a center it is called a **central conic**.

EXAMPLE 1   Given the hyperbola

$$\frac{x^2}{9} - \frac{y^2}{16} = 1$$

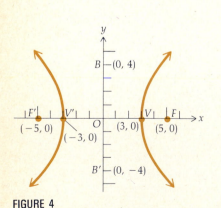

FIGURE 4

find the vertices, foci, and lengths of the transverse and conjugate axes. Draw a sketch of the hyperbola, and show the foci.

Solution   The given equation is of the form of (6); thus $a = 3$ and $b = 4$. The vertices are therefore the points $V(3, 0)$ and $V'(-3, 0)$. The number of units in the length of the transverse axis is $2a$ or 6, and the number of units in the length of the conjugate axis is $2b$ or 8. Because from (5), $b^2 = c^2 - a^2$ we have $16 = c^2 - 9$; so $c = 5$. Hence the foci are at $F(5, 0)$ and $F'(-5, 0)$. A sketch of the hyperbola and its foci are in Figure 4.

From Definition 10.3.1 it follows that if P is any point on this hyperbola, $\left| |\overline{FP}| - |\overline{F'P}| \right| = 6$.

EXAMPLE 2   Find an equation of the hyperbola having a focus at $(5, 0)$ and the ends of its conjugate axis at $(0, 2)$ and $(0, -2)$.

Solution   Because the ends of the conjugate axis are at $(0, 2)$ and $(0, -2)$, then $b = 2$, the principal axis is on the x axis, and the center is at the origin. Hence an equation is of the form

$$\frac{x^2}{a^2} - \frac{y^2}{b^2} = 1$$

Since a focus is at $(5, 0)$, $c = 5$, and because $b^2 = c^2 - a^2$, $a^2 = 25 - 4$. Thus $a = \sqrt{21}$, and an equation of the hyperbola is

$$\frac{x^2}{21} - \frac{y^2}{4} = 1$$

If in Equation (6), x and y are interchanged we obtain

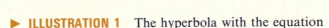

$$\frac{y^2}{a^2} - \frac{x^2}{b^2} = 1 \qquad\qquad (8)$$

which is an equation of a hyperbola having its center at the origin and its principal axis on the y axis.

▶ **ILLUSTRATION 1**   The hyperbola with the equation

$$\frac{y^2}{9} - \frac{x^2}{16} = 1$$

has its foci and vertices on the y axis because the equation is of the form (8). A sketch of this hyperbola appears in Figure 5.   ◀

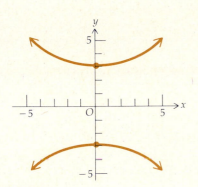

FIGURE 5

There is no general inequality involving a and b corresponding to the inequality $a > b$ for an ellipse. That is, for a hyperbola it is possible to have $a < b$, as in Illustration 1, where $a = 3$ and $b = 4$; or it is possible to have $a > b$, as for the hyperbola of Example 2, where $a = \sqrt{21}$ and $b = 2$. If, for a hyperbola, $a = b$, then the hyperbola is said to be **equilateral**.

We will now prove that a hyperbola has asymptotes and show how to obtain equations of these asymptotes. In Sections 2.4 and 2.5 vertical and horizontal asymptotes of the graph of a function were defined. In Section 4.7 we also discussed oblique asymptotes of a rational function. What follows is a more general definition, of which the previous definitions of asymptotes are special cases.

10.3.3 DEFINITION   The graph of the equation $y = f(x)$ has the line $y = mx + b$ as an **asymptote** if either of the following statements is true:

(i) $\displaystyle\lim_{x \to +\infty} [f(x) - (mx + b)] = 0$, and for some number $M > 0$, $f(x) \neq mx + b$ whenever $x > M$.

(ii) $\displaystyle\lim_{x \to -\infty} [f(x) - (mx + b)] = 0$, and for some number $M < 0$, $f(x) \neq mx + b$ whenever $x < M$.

Statement (i) indicates that for any $\epsilon > 0$ there exists a number $N > 0$ such that

if $\quad x > N \quad$ then $\quad 0 < |f(x) - (mx + b)| < \epsilon$

that is, we can make the function value $f(x)$ as close to the value of $mx + b$ as we please by taking x large enough. This is consistent with our intuitive notion of an asymptote of a graph. A similar statement may be made for part (ii) of Definition 10.3.3.

For the hyperbola $x^2/a^2 - y^2/b^2 = 1$, upon solving for y we get

$$y = \pm \frac{b}{a}\sqrt{x^2 - a^2}$$

So if

$$f(x) = \frac{b}{a}\sqrt{x^2 - a^2}$$

then

$$\lim_{x \to +\infty}\left[f(x) - \frac{b}{a}x \right] = \lim_{x \to +\infty}\left[\frac{b}{a}\sqrt{x^2 - a^2} - \frac{b}{a}x \right]$$

$$= \frac{b}{a}\lim_{x \to +\infty}\frac{(\sqrt{x^2 - a^2} - x)(\sqrt{x^2 - a^2} + x)}{\sqrt{x^2 - a^2} + x}$$

$$= \frac{b}{a}\lim_{x \to +\infty}\frac{-a^2}{\sqrt{x^2 - a^2} + x}$$

$$= 0$$

Therefore, by Definition 10.3.3, the line $y = \frac{b}{a}x$ is an asymptote of the graph of $y = \frac{b}{a}\sqrt{x^2 - a^2}$. Similarly, it can be shown that the line $y = \frac{b}{a}x$ is an asymptote of the graph of $y = -\frac{b}{a}\sqrt{x^2 - a^2}$. Consequently, the line $y = \frac{b}{a}x$ is an asymptote of the hyperbola $\frac{x^2}{a^2} - \frac{y^2}{b^2} = 1$. In an analogous manner we can demonstrate that the line $y = -\frac{b}{a}x$ is an asymptote of this same hyperbola. We have, then, the following theorem.

10.3.4 THEOREM

The lines

$$y = \frac{b}{a}x \quad \text{and} \quad y = -\frac{b}{a}x$$

are asymptotes of the hyperbola

$$\frac{x^2}{a^2} - \frac{y^2}{b^2} = 1$$

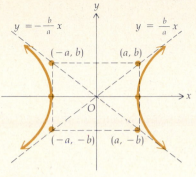

$$y = -\frac{b}{a}x \qquad y = \frac{b}{a}x$$

$(-a, b)$ (a, b)

$(-a, -b)$ $(a, -b)$

FIGURE 6

Figure 6 shows a sketch of the hyperbola of Theorem 10.3.4 together with its asymptotes. In the figure note that the diagonals of the rectangle having vertices at (a, b), $(a, -b)$, $(-a, b)$, and $(-a, -b)$ are on the asymptotes of the hyperbola. This rectangle is called the **auxiliary rectangle** of the hyperbola. The vertices of the hyperbola are the points of intersection of the principal axis and the auxiliary rectangle. A fairly good sketch of a hyperbola can be made by first drawing the auxiliary rectangle and then drawing the branch of the hyperbola through each vertex tangent to the side of the auxiliary rectangle there and approaching asymptotically the lines on which the diagonals of the rectangle lie. Observe that because $a^2 + b^2 = c^2$, the circle having its center at the origin and passing through the vertices of the auxiliary rectangle also passes through the foci of the hyperbola.

There is a mnemonic device for obtaining equations of the asymptotes of a hyperbola. For example, for the hyperbola having the equation $x^2/a^2 - y^2/b^2 = 1$, if the right side is replaced by zero, we obtain

$$\frac{x^2}{a^2} - \frac{y^2}{b^2} = 0$$

Upon factoring, this equation becomes

$$\left(\frac{x}{a} - \frac{y}{b}\right)\left(\frac{x}{a} + \frac{y}{b}\right) = 0$$

which is equivalent to the two equations

$$\frac{x}{a} - \frac{y}{b} = 0 \quad \text{and} \quad \frac{x}{a} + \frac{y}{b} = 0$$

which, by Theorem 10.3.4, are equations of the asymptotes of the given hyperbola. Using this device for the hyperbola having Equation (8) we see that the asymptotes are the lines having equations

$$\frac{y}{a} - \frac{x}{b} = 0 \quad \text{and} \quad \frac{y}{a} + \frac{x}{b} = 0$$

which are the same lines as the asymptotes of the hyperbola with the equation $x^2/b^2 - y^2/a^2 = 1$. This hyperbola and the one having Equation (8) are called **conjugate hyperbolas**.

The asymptotes of an equilateral hyperbola ($a = b$) are perpendicular to each other. The auxiliary rectangle for such a hyperbola is a square, and the transverse and conjugate axes have equal lengths.

If the center of a hyperbola is at (h, k) and its principal axis is parallel to the x axis, then if the axes are translated so that the point (h, k) is the new origin, an equation of the hyperbola relative to this new coordinate system is $\bar{x}^2/a^2 - \bar{y}^2/b^2 = 1$. If we replace $\bar{x}$ by $x - h$ and $\bar{y}$ by $y - k$, this equation becomes

$$\frac{(x - h)^2}{a^2} - \frac{(y - k)^2}{b^2} = 1 \tag{9}$$

Similarly, an equation of a hyperbola having its center at (h, k) and its principal axis parallel to the y axis is

$$\frac{(y - k)^2}{a^2} - \frac{(x - h)^2}{b^2} = 1 \tag{10}$$

EXAMPLE 3   The vertices of a hyperbola are at $(-5, -3)$ and $(-5, -1)$ and the extremities of the conjugate axis are at $(-7, -2)$ and $(-3, -2)$. Find an equation of the hyperbola and equations of the asymptotes. Draw a sketch of the hyperbola and the asymptotes.

Solution   The distance between the vertices is $2a$; so $2a = 2$ and $a = 1$. The length of the conjugate axis is $2b$; thus $2b = 4$ and $b = 2$. Because the principal axis is parallel to the y axis, an equation of the hyperbola is of the form (10). The center (h, k) is halfway between the vertices and is therefore at the point $(-5, -2)$. Hence an equation of the hyperbola is

$$\frac{(y + 2)^2}{1} - \frac{(x + 5)^2}{4} = 1$$

Using the mnemonic device to obtain equations of the asymptotes we have

$$\left(\frac{y + 2}{1} - \frac{x + 5}{2}\right)\left(\frac{y + 2}{1} + \frac{x + 5}{2}\right) = 0$$

which gives

$$y + 2 = \tfrac{1}{2}(x + 5) \quad \text{and} \quad y + 2 = -\tfrac{1}{2}(x + 5)$$

A sketch of the hyperbola and the asymptotes are in Figure 7.

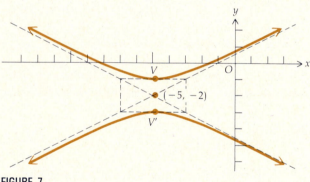

FIGURE 7

If in (9) and (10) we eliminate fractions and combine terms, the resulting equations are of the form

$$Ax^2 + Cy^2 + Dx + Ey + F = 0 \tag{11}$$

where A and C have different signs; that is, $AC < 0$. We now wish to show that the graph of an equation of the form (11), where $AC < 0$, is either a hyperbola or it degenerates. Completing the squares in x and y in (11), where $AC < 0$, the resulting equation has the form

$$\alpha^2(x - h)^2 - \beta^2(y - k)^2 = H \tag{12}$$

If $H > 0$, (12) can be written as

$$\frac{(x - h)^2}{\dfrac{H}{\alpha^2}} - \frac{(y - k)^2}{\dfrac{H}{\beta^2}} = 1$$

which has the form of (9).

▶ **ILLUSTRATION 2**   The equation

$$4x^2 - 12y^2 + 24x + 96y - 181 = 0$$

can be written as

$$4(x^2 + 6x) - 12(y^2 - 8y) = 181$$

and upon completing the squares in x and y we have

$$4(x^2 + 6x + 9) - 12(y^2 - 8y + 16) = 181 + 36 - 192$$
$$4(x + 3)^2 - 12(y - 4)^2 = 25$$

This has the form of (12), where $H = 25 > 0$. It may be written as

$$\frac{(x + 3)^2}{\frac{25}{4}} - \frac{(y - 4)^2}{\frac{25}{12}} = 1$$

which has the form of (9).   ◀

If $H < 0$, then (12) may be written as

$$\frac{(y - k)^2}{\frac{|H|}{\alpha^2}} - \frac{(x - h)^2}{\frac{|H|}{\beta^2}} = 1$$

which has the form of (10).

▶ **ILLUSTRATION 3**   Suppose that (11) is

$$4x^2 - 12y^2 + 24x + 96y - 131 = 0$$

Upon completing the squares in x and y we get

$$4(x + 3)^2 - 12(y - 4)^2 = -25$$

This has the form of (12), where $H = -25 < 0$, and it may be written as

$$\frac{(y - 4)^2}{\frac{25}{12}} - \frac{(x + 3)^2}{\frac{25}{4}} = 1$$

which has the form of (10).   ◀

If $H = 0$, then (12) is equivalent to the two equations

$$\alpha(x - h) - \beta(y - k) = 0 \quad \text{and} \quad \alpha(x - h) + \beta(y - k) = 0$$

which are equations of two lines through the point (h, k). This is the degenerate case of the hyperbola.

The following theorem summarizes the results of the preceding discussion.

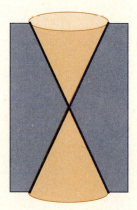

two intersecting lines

FIGURE 8

10.3.5 THEOREM   If in the general second-degree equation

$$Ax^2 + Bxy + Cy^2 + Dx + Ey + F = 0$$

$B = 0$ and $AC < 0$, then the graph is either a hyperbola or two intersecting lines.

The degenerate case of a hyperbola, two intersecting lines, is obtained as a conic section if the cutting plane contains the vertex of the cone and two generators, as shown in Figure 8.

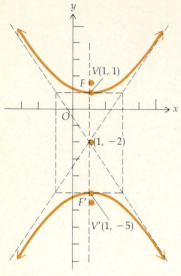

$V(1, 1)$
F
O
x
$(1, -2)$
F'
$V'(1, -5)$

FIGURE 9

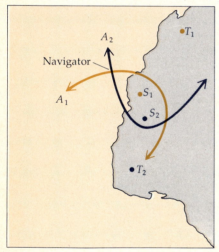

A_2
T_1
Navigator
A_1
S_1
S_2
T_2

FIGURE 10

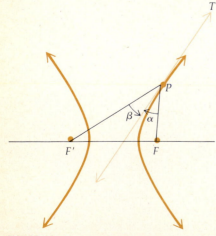

T
P
β α
F'
F

FIGURE 11

EXAMPLE 4   Determine the graph of the equation

$$9x^2 - 4y^2 - 18x - 16y + 29 = 0$$

Solution   From Theorem 10.3.5, because $B = 0$ and $AC = -36 < 0$, the graph is either a hyperbola or two intersecting lines.

Completing the squares in x and y we get

$$9(x^2 - 2x + 1) - 4(y^2 + 4y + 4) = -29 + 9 - 16$$

$$9(x - 1)^2 - 4(y + 2)^2 = -36$$

$$\frac{(y + 2)^2}{9} - \frac{(x - 1)^2}{4} = 1 \tag{13}$$

This equation has the form of (10); so the graph is a hyperbola whose principal axis is parallel to the y axis and whose center is at $(1, -2)$.

EXAMPLE 5   Find the vertices and foci of the hyperbola of Example 4. Draw a sketch showing the hyperbola, its asymptotes, and the foci.

Solution   From (13) we observe that $a = 3$ and $b = 2$. Because the principal axis is vertical, the center is at $(1, -2)$, and $a = 3$, it follows that the vertices are at the points $V(1, 1)$ and $V'(1, -5)$. For a hyperbola, $c^2 = a^2 + b^2$; thus $c^2 = 9 + 4$ and $c = \sqrt{13}$. Therefore the foci are at $F(1, -2 + \sqrt{13})$ and $F'(1, -2 - \sqrt{13})$. Figure 9 shows the hyperbola, its asymptotes, and the foci.

The property of the hyperbola given in Definition 10.3.1 forms the basis of several important navigational systems. These systems involve a network of pairs of radio transmitters at fixed positions at a known distance from one another. The transmitters send out radio signals that are received by a navigator. The difference in arrival time of the two signals determines the difference $2a$ of the distances from the navigator. Thus the navigator's position is known to be somewhere along one arc of a hyperbola having foci at the locations of the two transmitters. One arc, rather than both, is determined because of the signal delay between the two transmitters that is built into the system. The procedure is then repeated for a different pair of radio transmitters, and another arc of a hyperbola that contains the navigator's position is determined. The point of intersection of the two hyperbolic arcs is the actual position. For example, in Figure 10 suppose a pair of transmitters is located at points T_1 and S_1 and the signals from this pair determine the hyperbolic arc A_1. Another pair of transmitters is located at points T_2 and S_2 and hyperbolic arc A_2 is determined from their signals. Then the intersection of A_1 and A_2 is the position of the navigator.

The hyperbola has a reflective property that is used in the design of certain telescopes. In Figure 11 the line PT is the tangent line at P to the graph of the hyperbola having foci at F and F'. The measure of the angle between the line segment FP and the tangent line PT is α and the measure of the angle between the line segment $F'P$ and the tangent line PT is β. You are asked to prove that $\alpha = \beta$ in Exercise 39. From this equality it follows that a ray of light from a source at one focus of a hyperbolic mirror (one with hyperbolic cross section) is reflected along the line through the other focus.

Hyperbolas are also used in combat in *sound ranging* to locate the position of enemy guns by the sound of the firing of those guns. Some comets move in hyperbolic orbits. If a quantity varies inversely as another quantity, such as pressure and volume in Boyle's law for a perfect gas ($PV = k$), the graph is a hyperbola, as you will learn in Section 10.4.

From Theorems 10.1.6, 10.2.3, and 10.3.5 we may conclude that the graph of the general quadratic equation in two unknowns when $B = 0$ is either a conic or a degenerate conic. The type of conic can be determined from the product of A and C as stated in the following theorem.

10.3.6 THEOREM

The graph of the equation

$$Ax^2 + Cy^2 + Dx + Ey + F = 0$$

where A and C are not both zero, is either a conic or a degenerate conic; if it is a conic, then the graph is

(i) a *parabola* if either $A = 0$ or $C = 0$, that is, if $AC = 0$;
(ii) an *ellipse* if A and C have the same sign, that is, if $AC > 0$;
(iii) a *hyperbola* if A and C have opposite signs, that is, if $AC < 0$.

In the next section we discuss the graph of the general quadratic equation where $B \neq 0$.

EXERCISES 10.3

In Exercises 1 through 16, find the center, vertices, foci, and equations of the asymptotes of the hyperbola. Draw a sketch of the curve and its asymptotes, and show the foci.

1. $9x^2 - 4y^2 = 36$
2. $x^2 - 9y^2 = 9$
3. $4x^2 - 25y^2 = 100$
4. $16y^2 - 9x^2 = 144$
5. $4y^2 - x^2 = 16$
6. $4y^2 - 7x^2 = 56$
7. $9y^2 - 16x^2 = 1$
8. $25x^2 - 25y^2 = 1$
9. $x^2 - y^2 + 8x - 2y - 21 = 0$
10. $4x^2 - y^2 - 8x - 12 = 0$
11. $9x^2 - 18y^2 + 54x - 36y + 79 = 0$
12. $x^2 - y^2 + 6x + 10y - 4 = 0$
13. $3y^2 - 4x^2 - 8x - 24y - 40 = 0$
14. $4x^2 - y^2 + 56x + 2y + 195 = 0$
15. $4y^2 - 9x^2 + 16y + 18x = 29$
16. $y^2 - x^2 + 2y - 2x - 1 = 0$

In Exercises 17 through 26, find an equation of the hyperbola satisfying the given conditions, and draw a sketch of the hyperbola.

17. Vertices at $(-2, 0)$ and $(2, 0)$ and a conjugate axis of length 6.
18. Foci at $(0, 5)$ and $(0, -5)$ and a vertex at $(0, 4)$.
19. Center at the origin, its foci on the y axis, and passing through the points $(-2, 4)$ and $(-6, 7)$.
20. Extremities of its conjugate axis at $(0, -3)$ and $(0, 3)$ and one focus at $(5, 0)$.
21. One focus at $(26, 0)$ and asymptotes the lines $12y = \pm 5x$.
22. Center at $(3, -5)$, a vertex at $(7, -5)$, and a focus at $(8, -5)$.
23. Center at $(-2, -1)$, a vertex at $(-2, 11)$, and a focus at $(-2, 14)$.
24. Foci at $(3, 6)$ and $(3, 0)$ and passing through the point $(5, 3 + \frac{6}{5}\sqrt{5})$.

25. Foci at $(-1, 4)$ and $(7, 4)$ and length of the transverse axis is $\frac{8}{3}$.
26. One focus at $(-3 - 3\sqrt{13}, 1)$, asymptotes intersecting at $(-3, 1)$ and one asymptote passing through the point $(1, 7)$.
27. The vertices of a hyperbola are at $(-3, -1)$ and $(-1, -1)$ and the distance between the foci is $2\sqrt{5}$. Find (a) an equation of the hyperbola and (b) equations of the asymptotes.
28. The foci of a hyperbola are at $(2, 7)$ and $(2, -7)$ and the distance between the vertices is $8\sqrt{3}$. Find (a) an equation of the hyperbola and (b) equations of the asymptotes.
29. Find an equation of the tangent line to the hyperbola $4x^2 - y^2 = -1$ at the point $(\frac{1}{2}, \sqrt{2})$.
30. Find an equation of the normal line to the hyperbola $4x^2 - 3y^2 = 24$ at the point $(3, 2)$.
31. Find an equation of the hyperbola whose foci are the vertices of the ellipse $7x^2 + 11y^2 = 77$ and whose vertices are the foci of this ellipse.
32. Find an equation of the ellipse whose foci are the vertices of the hyperbola $11x^2 - 7y^2 = 77$ and whose vertices are the foci of this hyperbola.
33. The cost of production of a commodity is $12 less per unit at a point A than it is at a point B, and the distance between A and B is 100 km. Assuming that the route of delivery of the commodity is along a straight line, and that the delivery cost is 20 cents per unit per kilometer, find the curve at any point of which the commodity can be supplied from either A or B at the same total cost. (*Hint:* Take points A and B at $(-50, 0)$ and $(50, 0)$, respectively.)
34. Prove that there is no tangent line to the hyperbola $x^2 - y^2 = 1$ that passes through the origin.

35. Find the volume of the solid of revolution generated by revolving about the x axis the region bounded by the hyperbola $x^2/a^2 - y^2/b^2 = 1$ and the line $x = 2a$.

36. Find the centroid of the solid of revolution of Exercise 35.

37. Three listening posts are located at the points $A(0, 0)$, $B(0, \frac{21}{4})$, and $C(\frac{25}{3}, 0)$, the unit being 1 km. Microphones located at these points show that a gun is $\frac{5}{3}$ km closer to A than to C and $\frac{7}{4}$ km closer to B than to A. Determine the position of the gun by use of Definition 10.3.1.

38. Two LORAN (short for Long-Range Navigation) stations A and B lie on a line running east and west, and A is 80 mi due east of B. An airplane is traveling east on a straight line course that is 60 mi north of the line through A and B. Signals are sent at the same time from A and B, and the signal from A reaches the plane 350 μsec before the one from B. If the signals travel at the rate of 0.2 mi/μsec, locate the position of the plane by use of Definition 10.3.1.

39. In Figure 11 prove that $\alpha = \beta$. (*Hint:* Choose the coordinate axes so that the center of the hyperbola is at the origin and the axes of the hyperbola are along the coordinate axes. Then prove that the tangent line PT bisects the angle formed by the line segments FP and $F'P$.)

40. For an equilateral hyperbola having its center at (h, k) and principal axis parallel to the x axis, prove that the asymptotes are perpendicular.

41. Prove that if $P(x, y)$ is any point whose coordinates satisfy Equation (5), then

$$\left\| \overline{FP} \right| - \left| \overline{F'P} \right\| = 2a$$

where F and F' are the foci of the hyperbola in Figure 2. (*Hint:* Use a method similar to that suggested for the ellipse in either Exercise 42 or 43 of Exercises 10.2.)

10.4 ROTATION OF AXES

We have shown how a translation of coordinate axes can simplify the form of certain equations. We now use a rotation of coordinate axes to transform a second-degree equation having an xy term into one having no such term. A translation of axes gives a new coordinate system whose axes are parallel to the original x and y axes. For a rotation the new coordinate system will in general have axes that are *not* parallel to the original ones.

Suppose that we have two rectangular cartesian coordinate systems with the same origin. Let one system be the xy system and the other the $\bar{x}\bar{y}$ system. Suppose further that the $\bar{x}$ axis makes an angle α with the x axis. See Figure 1. Of course, the $\bar{y}$ axis then makes an angle α with the y axis. In such a case we state that the xy system of coordinates is *rotated* through an angle α to form the $\bar{x}\bar{y}$ system of coordinates. A point P having coordinates (x, y) with respect to the original coordinate system will have coordinates $(\bar{x}, \bar{y})$ with respect to the new one. We now obtain relationships between these two sets of coordinates.

In Figure 1, let r denote the undirected distance $|\overline{OP}|$ and let θ be the angle measured from the x axis to the line segment OP. From the figure we observe that

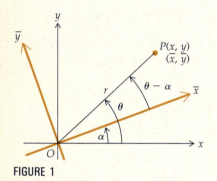

FIGURE 1

$$x = r \cos \theta \quad \text{and} \quad y = r \sin \theta \tag{1}$$

Also from Figure 1

$$\bar{x} = r \cos(\theta - \alpha) \quad \text{and} \quad \bar{y} = r \sin(\theta - \alpha)$$

With the cosine and sine difference identities these two equations become

$$\bar{x} = r \cos \theta \cos \alpha + r \sin \theta \sin \alpha \quad \text{and} \quad \bar{y} = r \sin \theta \cos \alpha - r \cos \theta \sin \alpha$$

Substituting from Equations (1) into the preceding equations, we get

$$\bar{x} = x \cos \alpha + y \sin \alpha \quad \text{and} \quad \bar{y} = -x \sin \alpha + y \cos \alpha \tag{2}$$

Solving Equations (2) simultaneously for x and y in terms of $\bar{x}$ and $\bar{y}$ (see Exercise 26), we obtain

$$x = \bar{x} \cos \alpha - \bar{y} \sin \alpha \quad \text{and} \quad y = \bar{x} \sin \alpha + \bar{y} \cos \alpha \tag{3}$$

We state these results formally in the following theorem.

10.4.1 THEOREM If (x, y) represents a point P with respect to a given set of axes and $(\bar{x}, \bar{y})$ is a representation of P after the axes have been rotated through an angle α, then

(i) $x = \bar{x} \cos \alpha - \bar{y} \sin \alpha$ and $y = \bar{x} \sin \alpha + \bar{y} \cos \alpha$
(ii) $\bar{x} = x \cos \alpha + y \sin \alpha$ and $\bar{y} = -x \sin \alpha + y \cos \alpha$

EXAMPLE 1   Given the equation

$$xy = 1$$

(a) Find an equation of the graph with respect to the $\bar{x}$ and $\bar{y}$ axes after a rotation of axes through an angle of radian measure $\frac{1}{4}\pi$, and (b) draw a sketch of the graph and show both sets of axes.

Solution
(a) With $\alpha = \frac{1}{4}\pi$ in Theorem 10.4.1(i) we obtain

$$x = \frac{1}{\sqrt{2}}\bar{x} - \frac{1}{\sqrt{2}}\bar{y} \quad \text{and} \quad y = \frac{1}{\sqrt{2}}\bar{x} + \frac{1}{\sqrt{2}}\bar{y}$$

Substituting these expressions for x and y in the equation $xy = 1$ we get

$$\left(\frac{1}{\sqrt{2}}\bar{x} - \frac{1}{\sqrt{2}}\bar{y}\right)\left(\frac{1}{\sqrt{2}}\bar{x} + \frac{1}{\sqrt{2}}\bar{y}\right) = 1$$

$$\frac{\bar{x}^2}{2} - \frac{\bar{y}^2}{2} = 1$$

(b) This is an equation of an equilateral hyperbola whose asymptotes are the bisectors of the quadrants in the $\bar{x}\bar{y}$ system. Thus the graph of the equation $xy = 1$ is an equilateral hyperbola lying in the first and third quadrants and the asymptotes are the x and y axes. See Figure 2 for the required graph.

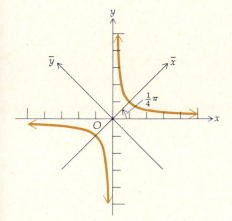

FIGURE 2

In Section 10.3 we showed that when $B = 0$ and A and C are not both zero, the graph of the general second-degree equation in two unknowns,

$$Ax^2 + Bxy + Cy^2 + Dx + Ey + F = 0 \tag{4}$$

is either a conic or a degenerate conic. We now show that if $B \neq 0$, then any equation of the form (4) can be transformed by a suitable rotation of axes into an equation of the form

$$\bar{A}\bar{x}^2 + \bar{C}\bar{y}^2 + \bar{D}\bar{x} + \bar{E}\bar{y} + \bar{F} = 0 \tag{5}$$

where $\bar{A}$ and $\bar{C}$ are not both zero.

If the xy system is rotated through an angle α, then to obtain an equation of the graph of (4) with respect to the $\bar{x}\bar{y}$ system we replace x by $\bar{x} \cos \alpha - \bar{y} \sin \alpha$ and y by $\bar{x} \sin \alpha + \bar{y} \cos \alpha$. We get

$$\bar{A}\bar{x}^2 + \bar{B}\bar{x}\bar{y} + \bar{C}\bar{y}^2 + \bar{D}\bar{x} + \bar{E}\bar{y} + \bar{F} = 0 \tag{6}$$

where

$$\bar{A} = A\cos^2\alpha + B\sin\alpha\cos\alpha + C\sin^2\alpha$$

$$\bar{B} = -2A\sin\alpha\cos\alpha + B(\cos^2\alpha - \sin^2\alpha) + 2C\sin\alpha\cos\alpha$$

$$\bar{C} = A\sin^2\alpha - B\sin\alpha\cos\alpha + C\cos^2\alpha$$

We wish to find an α so that the rotation transforms (4) into an equation of the form (5). Setting the expression for $\bar{B}$ equal to zero, we have

$$B(\cos^2 \alpha - \sin^2 \alpha) + (C - A)(2 \sin \alpha \cos \alpha) = 0$$

or, equivalently, with trigonometric identities,

$$B \cos 2\alpha + (C - A) \sin 2\alpha = 0$$

Because $B \neq 0$, this gives

$$\cot 2\alpha = \frac{A - C}{B}$$

We have shown that a rotation of axes through an angle α satisfying this equation will transform an equation of the form (4) where $B \neq 0$ to an equation of the form (5). We now wish to show that $\bar{A}$ and $\bar{C}$ in (5) are not both zero. To prove this, notice that (6) is obtained from (4) by rotating the axes through the angle α. Also, (4) can be obtained from (6) by rotating the axes back through the angle $-\alpha$. If $\bar{A}$ and $\bar{C}$ in (6) are both zero, then the substitutions

$$\bar{x} = x \cos \alpha + y \sin \alpha \quad \text{and} \quad \bar{y} = -x \sin \alpha + y \cos \alpha$$

in (6) would result in the equation

$$\bar{D}(x \cos \alpha + y \sin \alpha) + \bar{E}(-x \sin \alpha + y \cos \alpha) + \bar{F} = 0$$

which is an equation of the first degree and hence different from (4) because we have assumed that at least $B \neq 0$. The following theorem has, therefore, been proved.

10.4.2 THEOREM   If $B \neq 0$, the equation

$$Ax^2 + Bxy + Cy^2 + Dx + Ey + F = 0$$

can be transformed into the equation

$$\bar{A}\bar{x}^2 + \bar{C}\bar{y}^2 + \bar{D}\bar{x} + \bar{E}\bar{y} + \bar{F} = 0$$

where $\bar{A}$ and $\bar{C}$ are not both zero, by a rotation of axes through an angle α for which

$$\cot 2\alpha = \frac{A - C}{B}$$

By Theorems 10.4.2 and 10.3.6 it follows that the graph of an equation of the form (4) is either a conic or a degenerate conic. To determine which type of conic is the graph of a particular equation, we examine the expression $B^2 - 4AC$.

We use the fact that A, B, and C of (4) and $\bar{A}$, $\bar{B}$, and $\bar{C}$ of (6) satisfy the relation

$$B^2 - 4AC = \bar{B}^2 - 4\bar{A}\bar{C} \tag{7}$$

which can be proved by substituting the expressions for $\bar{A}$, $\bar{B}$, and $\bar{C}$ given after Equation (6) in the right side of (7). This is left as an exercise. See Exercise 25. The expression $B^2 - 4AC$ is called the **discriminant** of (4). Equation (7) states that the discriminant of the general quadratic equation in two variables is **invariant** under a rotation of axes.

If the angle of rotation is chosen so that $\bar{B} = 0$, then (7) becomes

$$B^2 - 4AC = -4\bar{A}\bar{C} \tag{8}$$

From Theorem 10.3.6 it follows that if the graph of (5) is not degenerate, then it is a parabola if $\bar{A}\bar{C} = 0$, an ellipse if $\bar{A}\bar{C} > 0$, and a hyperbola if $\bar{A}\bar{C} < 0$. So the graph of (5) is a parabola, an ellipse, or a hyperbola depending on whether $-4\bar{A}\bar{C}$ is zero, negative, or positive. Because the graph of (4) is the same as the graph of (5), we conclude from (8) that if the graph of (4) is not degenerate, then it is a parabola, an ellipse, or a hyerbola depending on whether the discriminant $B^2 - 4AC$ is zero, negative, or positive. We have proved the following theorem.

10.4.3 THEOREM

The graph of the equation

$$Ax^2 + Bxy + Cy^2 + Dx + Ey + F = 0$$

is either a conic or a degenerate conic. If it is a conic, then it is

(i) a *parabola* if $B^2 - 4AC = 0$;
(ii) an *ellipse* if $B^2 - 4AC < 0$;
(iii) a *hyperbola* if $B^2 - 4AC > 0$.

EXAMPLE 2 Simplify the equation

$$17x^2 - 12xy + 8y^2 - 80 = 0$$

by a rotation of axes. Draw a sketch of the graph of the equation and show both sets of axes.

Solution

$$B^2 - 4AC = (-12)^2 - 4(17)(8)$$
$$= -400$$

Because $B^2 - 4AC < 0$, by Theorem 10.4.3 the graph is an ellipse or else it is degenerate. To eliminate the xy term by a rotation of axes we must choose an α such that

$$\cot 2\alpha = \frac{A - C}{B}$$

$$= \frac{17 - 8}{-12}$$

$$= -\tfrac{3}{4}$$

There is a 2α in the inverval $(0, \pi)$ for which $\cot 2\alpha = -\tfrac{3}{4}$. Therefore α is in the interval $(0, \tfrac{1}{2}\pi)$. To apply Theorem 10.4.1 it is not necessary to find α so long as we find $\cos \alpha$ and $\sin \alpha$. These functions can be found from the value of $\cot 2\alpha$ by the trigonometric identities

$$\cos \alpha = \sqrt{\frac{1 + \cos 2\alpha}{2}} \quad \text{and} \quad \sin \alpha = \sqrt{\frac{1 - \cos 2\alpha}{2}} \qquad 0 < \alpha < \tfrac{1}{2}\pi$$

Because $\cot 2\alpha = -\tfrac{3}{4}$ and $0 < \alpha < \tfrac{1}{2}\pi$, it follows that $\cos 2\alpha = -\tfrac{3}{5}$. So

$$\cos \alpha = \sqrt{\frac{1 - \tfrac{3}{5}}{2}} \quad \text{and} \quad \sin \alpha = \sqrt{\frac{1 + \tfrac{3}{5}}{2}}$$

$$= \frac{1}{\sqrt{5}} \qquad\qquad\qquad = \frac{2}{\sqrt{5}}$$

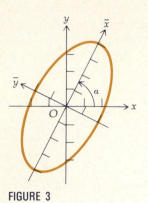

FIGURE 3

Substituting $x = \bar{x}/\sqrt{5} - 2\bar{y}/\sqrt{5}$ and $y = 2\bar{x}/\sqrt{5} + \bar{y}/\sqrt{5}$ in the given equation we obtain

$$17\left(\frac{\bar{x}^2 - 4\bar{x}\bar{y} + 4\bar{y}^2}{5}\right) - 12\left(\frac{2\bar{x}^2 - 3\bar{x}\bar{y} - 2\bar{y}^2}{5}\right) + 8\left(\frac{4\bar{x}^2 + 4\bar{x}\bar{y} + \bar{y}^2}{5}\right) - 80 = 0$$

Upon simplification this equation becomes

$$\bar{x}^2 + 4\bar{y}^2 = 16$$

$$\frac{\bar{x}^2}{16} + \frac{\bar{y}^2}{4} = 1$$

So the graph is an ellipse for which $a = 4$ and $b = 2$. Therefore the length of the major axis is 8 units and the length of the minor axis is 4 units. A sketch of the ellipse with both sets of axes appears in Figure 3.

EXERCISES 10.4

In Exercises 1 through 4, for the given equation, (a) find an equation of the graph with respect to the $\bar{x}$ and $\bar{y}$ axes after a rotation of axes through an angle of radian measure $\frac{1}{4}\pi$, and (b) draw a sketch of the graph and show both sets of axes.

1. $xy = 8$ **2.** $xy = -4$
3. $x^2 - y^2 = 8$ **4.** $y^2 - x^2 = 16$

In Exercises 5 through 12, remove the xy term from the equation by a rotation of axes. Draw a sketch of the graph and show both sets of axes.

5. $24xy - 7y^2 + 36 = 0$ **6.** $4xy + 3x^2 = 4$
7. $x^2 + 2xy + y^2 - 8x + 8y = 0$
8. $x^2 + xy + y^2 = 3$ **9.** $xy + 16 = 0$
10. $5x^2 + 6xy + 5y^2 = 9$
11. $31x^2 + 10\sqrt{3}xy + 21y^2 = 144$
12. $6x^2 + 20\sqrt{3}xy + 26y^2 = 324$

In Exercises 13 through 22, simplify the equation by a rotation and translation of axes. Draw a sketch of the graph and show the three sets of axes.

13. $x^2 + xy + y^2 - 3y - 6 = 0$
14. $19x^2 + 6xy + 11y^2 - 26x + 38y + 31 = 0$
15. $17x^2 - 12xy + 8y^2 - 68x + 24y - 12 = 0$
16. $x^2 - 10xy + y^2 + x + y + 1 = 0$

17. $x^2 + 2xy + y^2 + x - y - 4 = 0$
18. $16x^2 - 24xy + 9y^2 - 60x - 80y + 400 = 0$
19. $11x^2 - 24xy + 4y^2 + 30x + 40y - 45 = 0$
20. $3x^2 - 4xy + 8x - 1 = 0$
21. $4x^2 + 4xy + y^2 - 6x + 12y = 0$
22. $x^2 + 2xy + y^2 - x - 3y = 0$

23. Show that the graph of $\sqrt{x} + \sqrt{y} = 1$ is a segment of a parabola by rotating the axes through an angle of radian measure $\frac{1}{4}\pi$. *Hint:* Eliminate the radicals in the equation before applying Theorem 10.4.1.

24. Given the equation $(a^2 + b^2)xy = 1$, where $a > 0$ and $b > 0$, find an equation of the graph with respect to the $\bar{x}$ and $\bar{y}$ axes after a rotation of the axes through an angle of radian measure $\tan^{-1}(b/a)$.

25. Show that for the general second-degree equation in two variables, the discriminant $B^2 - 4AC$ is invariant under a rotation of axes.

26. Derive Equations (3) from Equations (2) by solving for x and y in terms of $\bar{x}$ and $\bar{y}$. (*Hint:* To solve for x, multiply each member of the first equation in (2) by $\cos \alpha$ and each member of the second equation by $\sin \alpha$, and then subtract corresponding members of the resulting equations. Use a similar procedure to solve for y.)

10.5 POLAR COORDINATES

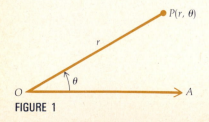

FIGURE 1

Until now, we have located a point in a plane by its rectangular cartesian coordinates. There are other coordinate systems that give the position of a point in a plane. The **polar coordinate system** is one of them, and it is important because certain curves have simpler equations in that system. In polar coordinates all three conics (the parabola, ellipse, and hyperbola) have one equation. This equation is applied in the derivation of Kepler's laws in physics and in the study of the motion of planets in astronomy.

Cartesian coordinates are numbers, the abscissa and ordinate, and these numbers are directed distances from two fixed lines. Polar coordinates consist

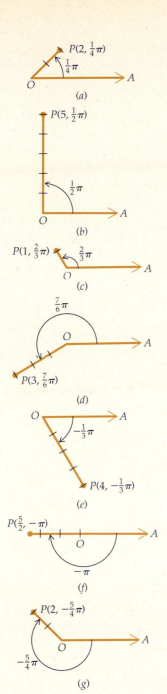

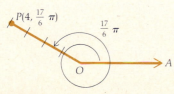

FIGURE 2

of a directed distance and the measure of an angle, which is taken relative to a fixed point and a fixed ray (or half line). The fixed point is called the **pole** (or origin), designed by the letter O. The fixed ray is called the **polar axis** (or polar line), which we label OA. The ray OA is usually drawn horizontally and to the right, and it extends indefinitely. See Figure 1.

Let P be any point in the plane distinct from O. Let θ be the radian measure of a directed angle AOP, positive when measured counterclockwise and negative when measured clockwise, having as its initial side the ray OA and as its terminal side the ray OP. Then if r is the undirected distance from O to P (that is, $r = |\overline{OP}|$), one set of polar coordinates of P is given by r and θ, and we write these coordinates as (r, θ).

EXAMPLE 1   Plot each of the following points having the given set of polar coordinates: (a) $(2, \frac{1}{4}\pi)$; (b) $(5, \frac{1}{2}\pi)$; (c) $(1, \frac{2}{3}\pi)$; (d) $(3, \frac{7}{6}\pi)$; (e) $(4, -\frac{1}{3}\pi)$; (f) $(\frac{5}{2}, -\pi)$; (g) $(2, -\frac{5}{4}\pi)$

Solution
(a) The point $(2, \frac{1}{4}\pi)$ is determined by first drawing the angle with radian measure $\frac{1}{4}\pi$, having its vertex at the pole and its initial side along the polar axis. The point on the terminal side that is 2 units from the pole is the point $(2, \frac{1}{4}\pi)$. See Figure 2(a).

In a similar manner we obtain the points appearing in Figure 2(b)–(g).

▶ **ILLUSTRATION 1**   Figure 3 shows the point $(4, \frac{5}{6}\pi)$. Another set of polar coordinates for this point is $(4, -\frac{7}{6}\pi)$; see Figure 4. Furthermore the polar coordinates $(4, \frac{17}{6}\pi)$ also yield the same point, as shown in Figure 5.   ◀

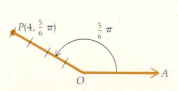

FIGURE 3

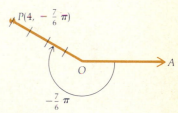

FIGURE 4

Actually the coordinates $(4, \frac{5}{6}\pi + 2n\pi)$, where n is any integer, give the same point as $(4, \frac{5}{6}\pi)$. So a given point has an unlimited number of sets of polar coordinates. This is unlike the rectangular cartesian coordinate system because there is a one-to-one correspondence between the rectangular cartesian coordinates and the position of points in the plane. There is no such one-to-one correspondence between the polar coordinates and the position of points in the plane. A further example is obtained by considering sets of polar coordinates for the pole. If $r = 0$ and θ is any real number, we have the pole, which is designated by $(0, \theta)$.

We now consider polar coordinates for which r is negative. In this case, instead of the point being on the terminal side of the angle, it is on the extension of the terminal side, which is the ray from the pole extending in the direction

FIGURE 5

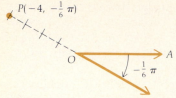

FIGURE 6

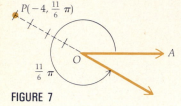

FIGURE 7

opposite to the terminal side. So if P is on the extension of the terminal side of the angle of radian measure θ, a set of polar coordinates of P is (r, θ), where $r = -|\overline{OP}|$.

▶ **ILLUSTRATION 2** The point $(-4, -\frac{1}{6}\pi)$ shown in Figure 6 is the same point as $(4, \frac{5}{6}\pi)$, $(4, -\frac{7}{6}\pi)$, and $(4, \frac{17}{6}\pi)$ in Illustration 1. Still another set of polar coordinates for this point is $(-4, \frac{11}{6}\pi)$; see Figure 7. ◀

The angle is usually measured in radians: thus a set of polar coordinates of a point is an ordered pair of real numbers. For each ordered pair of real numbers there is a unique point having this set of polar coordinates. However, we have seen that a particular point can be given by an unlimited number of ordered pairs of real numbers. If the point P is not the pole, and r and θ are restricted so that $r > 0$ and $0 \le \theta < 2\pi$, then there is a unique set of polar coordinates for P.

EXAMPLE 2 (a) Plot the point having polar coordinates $(3, -\frac{2}{3}\pi)$. Find another set of polar coordinates of this point for which (b) $r < 0$ and $0 < \theta < 2\pi$; (c) $r > 0$ and $0 < \theta < 2\pi$; (d) $r < 0$ and $-2\pi < \theta < 0$.

Solution
(a) The point is plotted by drawing the angle of radian measure $-\frac{2}{3}\pi$ in a clockwise direction from the polar axis. Because $r > 0$, P is on the terminal side of the angle, three units from the pole; see Figure 8(a).
   The answers to (b), (c), and (d) are, respectively, $(-3, \frac{1}{3}\pi)$, $(3, \frac{4}{3}\pi)$, and $(-3, -\frac{5}{3}\pi)$. They are illustrated in Figure 8(b)–(d).

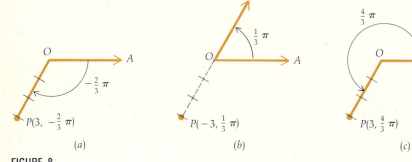

FIGURE 8

Sometimes we wish to refer to both the rectangular cartesian coordinates and the polar coordinates of a point. To do this we take the origin of the first system and the pole of the second system coincident, the polar axis as the positive side of the x axis, and the ray for which $\theta = \frac{1}{2}\pi$ as the positive side of the y axis.

Suppose that P is a point whose representation in the rectangular cartesian coordinate system is (x, y) and (r, θ) is a polar coodinate representation of P. We distinguish two cases: $r > 0$ and $r < 0$. In the first case, if $r > 0$, then the point P is on the terminal side of the angle of radian measure θ, and $r = |\overline{OP}|$.

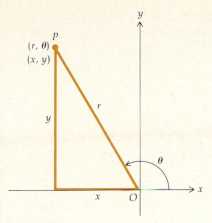

FIGURE 9

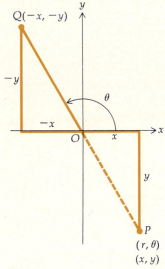

FIGURE 10

Such a case is shown in Figure 9. Then

$$\cos\theta = \frac{x}{|\overline{OP}|} \qquad \sin\theta = \frac{-y}{|\overline{OP}|}$$

$$= \frac{x}{r} \qquad\qquad = \frac{y}{r}$$

Thus

$$x = r\cos\theta \quad \text{and} \quad y = r\sin\theta \tag{1}$$

In the second case, if $r < 0$, then the point P is on the extension of the terminal side and $r = -|\overline{OP}|$. See Figure 10. Then if Q is the point $(-x, -y)$,

$$\cos\theta = \frac{-x}{|\overline{OQ}|} \qquad \sin\theta = \frac{-y}{|\overline{OQ}|}$$

$$= \frac{-x}{|\overline{OP}|} \qquad\qquad = \frac{-y}{|\overline{OP}|}$$

$$= \frac{-x}{-r} \qquad\qquad = \frac{-y}{-r}$$

$$= \frac{x}{r} \qquad\qquad = \frac{y}{r}$$

Hence

$$x = r\cos\theta \quad \text{and} \quad y = r\sin\theta$$

These equations are the same as Equations (1); therefore they hold in all cases.

From Equations (1) we can obtain the rectangular cartesian coordinates of a point when its polar coordinates are known. Also, from the equations we can obtain a polar equation of a curve if a rectangular cartesian equation is known.

To obtain equations that give a set of polar coordinates of a point when its rectangular cartesian coordinates are known, we square on both sides of each equation in (1) and obtain

$$x^2 = r^2\cos^2\theta \quad \text{and} \quad y^2 = r^2\sin^2\theta$$

Equating the sum of the left members of the above to the sum of the right members we have

$$x^2 + y^2 = r^2\cos^2\theta + r^2\sin^2\theta$$

$$x^2 + y^2 = r^2(\cos^2\theta + \sin^2\theta)$$

$$x^2 + y^2 = r^2$$

$$r = \pm\sqrt{x^2 + y^2} \tag{2}$$

From the equations in (1) and dividing we have

$$\frac{r\sin\theta}{r\cos\theta} = \frac{y}{x}$$

$$\tan\theta = \frac{y}{x} \tag{3}$$

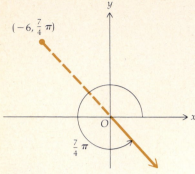

FIGURE 11

▶ **ILLUSTRATION 3**   The point whose polar coordinates are $(-6, \frac{7}{4}\pi)$ is plotted in Figure 11. We find its rectangular cartesian coordinates. From (1),

$$x = r \cos \theta \qquad\qquad y = r \sin \theta$$
$$= -6 \cos \tfrac{7}{4}\pi \qquad\qquad = -6 \sin \tfrac{7}{4}\pi$$
$$= -6 \cdot \frac{\sqrt{2}}{2} \qquad\qquad = -6\left(-\frac{\sqrt{2}}{2}\right)$$
$$= -3\sqrt{2} \qquad\qquad = 3\sqrt{2}$$

So the point is $(-3\sqrt{2}, 3\sqrt{2})$.   ◀

The graph of an equation in polar coordinates r and θ consists of all those points and only those points P having at least one pair of coordinates that satisfy the equation. If an equation of a graph is given in polar coordinates, it is called a **polar equation** to distinguish it from a **cartesian equation**, which is the term used when an equation is given in rectangular cartesian coordinates. In Section 11.6 we discuss methods of obtaining the graph of a polar equation.

EXAMPLE 3   Given that a polar equation of a graph is

$$r^2 = 4 \sin 2\theta$$

find a cartesian equation.

Solution   Because $\sin 2\theta = 2 \sin \theta \cos \theta$ we have $\sin 2\theta = 2(y/r)(x/r)$. With this substitution and $r^2 = x^2 + y^2$, we obtain from the given polar equation

$$x^2 + y^2 = 4(2)\frac{y}{r} \cdot \frac{x}{r}$$

$$x^2 + y^2 = \frac{8xy}{r^2}$$

$$x^2 + y^2 = \frac{8xy}{x^2 + y^2}$$

$$(x^2 + y^2)^2 = 8xy$$

EXAMPLE 4   Find (r, θ) if $r > 0$ and $0 \le \theta < 2\pi$ for the point whose rectangular cartesian coordinate representation is $(-\sqrt{3}, -1)$.

Solution   The point $(-\sqrt{3}, -1)$ is plotted in Figure 12. From (2), because $r > 0$,

$$r = \sqrt{3 + 1}$$
$$= 2$$

From (3), $\tan \theta = -1/(-\sqrt{3})$, and since $\pi < \theta < \frac{3}{2}\pi$,

$$\theta = \tfrac{7}{6}\pi$$

So the point is $(2, \frac{7}{6}\pi)$.

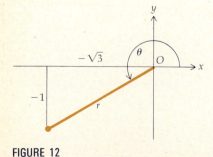

FIGURE 12

EXAMPLE 5   Find a polar equation of the graph whose cartesian equation is

$$x^2 + y^2 - 4x = 0$$

Solution   Substituting $x = r \cos \theta$ and $y = r \sin \theta$ in

$$x^2 + y^2 - 4x = 0$$

we have

$$r^2 \cos^2 \theta + r^2 \sin^2 \theta - 4r \cos \theta = 0$$
$$r^2 - 4r \cos \theta = 0$$
$$r(r - 4 \cos \theta) = 0$$

Therefore

$$r = 0 \quad \text{or} \quad r - 4 \cos \theta = 0$$

The graph of $r = 0$ is the pole. However, the pole is a point on the graph of $r - 4 \cos \theta = 0$ because $r = 0$ when $\theta = \frac{1}{2}\pi$. Therefore a polar equation of the graph is

$$r = 4 \cos \theta$$

The graph of $x^2 + y^2 - 4x = 0$ is a circle. The equation may be written in the form

$$(x - 2)^2 + y^2 = 4$$

which is an equation of the circle with center at $(2, 0)$ and radius 2.

EXERCISES 10.5

In Exercises 1 through 4, plot the point having the given set of polar coordinates.

1. (a) $(3, \frac{1}{6}\pi)$; (b) $(2, \frac{2}{3}\pi)$; (c) $(1, \pi)$; (d) $(4, \frac{5}{4}\pi)$; (e) $(5, \frac{11}{6}\pi)$

2. (a) $(4, \frac{1}{3}\pi)$ (b) $(3, \frac{3}{4}\pi)$; (c) $(1, \frac{7}{6}\pi)$; (d) $(2, \frac{3}{2}\pi)$; (e) $(5, \frac{5}{3}\pi)$

3. (a) $(1, -\frac{1}{4}\pi)$; (b) $(3, -\frac{5}{6}\pi)$; (c) $(-1, \frac{1}{4}\pi)$; (d) $(-3, \frac{5}{6}\pi)$; (e) $(-2, -\frac{1}{2}\pi)$

4. (a) $(5, -\frac{2}{3}\pi)$; (b) $(2, -\frac{7}{6}\pi)$; (c) $(-5, \frac{2}{3}\pi)$; (d) $(-2, \frac{7}{6}\pi)$; (e) $(-4, -\frac{5}{4}\pi)$

In Exercises 5 through 10, plot the point having the given set of polar coordinates; then find another set of polar coordinates for the same point for which (a) $r < 0$ and $0 \le \theta < 2\pi$; (b) $r > 0$ and $-2\pi < \theta \le 0$; (c) $r < 0$ and $-2\pi < \theta \le 0$.

5. $(4, \frac{1}{4}\pi)$ **6.** $(3, \frac{5}{6}\pi)$ **7.** $(2, \frac{1}{2}\pi)$

8. $(3, \frac{3}{2}\pi)$ **9.** $(\sqrt{2}, \frac{7}{4}\pi)$ **10.** $(2, \frac{4}{3}\pi)$

11. Plot the point having the polar coordinates $(2, -\frac{1}{4}\pi)$. Find another set of polar coordinates for this point for which (a) $r < 0$ and $0 \le \theta < 2\pi$; (b) $r < 0$ and $-2\pi < \theta \le 0$; (c) $r > 0$ and $2\pi \le \theta < 4\pi$.

12. Plot the point having the polar coordinates $(-3, -\frac{2}{3}\pi)$. Find another set of polar coordinates for this point for which (a)

$r > 0$ and $0 \le \theta < 2\pi$; (b) $r > 0$ and $-2\pi < \theta \le 0$; (c) $r < 0$ and $2\pi \le \theta < 4\pi$.

In Exercises 13 through 20, plot the point having the given set of polar coordinates; then give two other sets of polar coordinates of the same point, one with the same value of r and one with an r having opposite sign.

13. $(3, -\frac{2}{3}\pi)$ **14.** $(\sqrt{2}, -\frac{1}{4}\pi)$ **15.** $(-4, \frac{5}{6}\pi)$

16. $(-2, \frac{4}{3}\pi)$ **17.** $(-2, -\frac{5}{4}\pi)$ **18.** $(-3, -\pi)$

19. $(2, 6)$ **20.** $(5, \frac{1}{6}\pi)$

In Exercises 21 and 22, find the rectangular cartesian coordinates of the points whose polar coordinates are given.

21. (a) $(3, \pi)$: (b) $(\sqrt{2}, -\frac{3}{4}\pi)$; (c) $(-4, \frac{2}{3}\pi)$; (d) $(-1, -\frac{7}{6}\pi)$

22. (a) $(-2, -\frac{1}{2}\pi)$; (b) $(-1, \frac{1}{4}\pi)$; (c) $(2, -\frac{7}{6}\pi)$; (d) $(2, \frac{7}{4}\pi)$

In Exercises 23 and 24, find a set of polar coordinates of the points whose rectangular cartesian coordinates are given. Take $r > 0$ and $0 \le \theta < 2\pi$.

23. (a) $(1, -1)$; (b) $(-\sqrt{3}, 1)$; (c) $(2, 2)$; (d) $(-5, 0)$

24. (a) $(3, -3)$; (b) $(-1, \sqrt{3})$; (c) $(0, -2)$; (d) $(-2, -2\sqrt{3})$

In Exercises 25 through 34, find a polar equation of the graph having the cartesian equation.

25. $x^2 + y^2 = a^2$ **26.** $x + y = 1$
27. $y^2 = 4(x + 1)$ **28.** $x^3 = 4y^2$
29. $x^2 = 6y - y^2$ **30.** $x^2 - y^2 = 16$
31. $(x^2 + y^2)^2 = 4(x^2 - y^2)$ **32.** $2xy = a^2$
33. $x^3 + y^3 - 3axy = 0$ **34.** $y = \dfrac{2x}{x^2 + 1}$

In Exercises 35 through 44, find a cartesian equation of the graph having the polar equation.

35. $r^2 = 2 \sin 2\theta$ **36.** $r^2 \cos 2\theta = 10$
37. $r^2 = \cos \theta$ **38.** $r^2 = 4 \cos 2\theta$
39. $r^2 = \theta$ **40.** $r = 2 \sin 3\theta$
41. $r \cos \theta = -1$ **42.** $r^6 = r^2 \cos^2 \theta$
43. $r = \dfrac{6}{2 - 3 \sin \theta}$ **44.** $r = \dfrac{4}{3 - 2 \cos \theta}$

10.6 GRAPHS OF EQUATIONS IN POLAR COORDINATES

In Section 10.5 we stated that the graph of a polar equation consists of those points and only those points having at least one pair of polar coordinates that satisfy the equation. In this section we show how to obtain a sketch of such a graph.

The equation

$$\theta = C$$

where C is a constant, is satisfied by all points having polar coordinates (r, C) whatever the value of r. Therefore, the graph of this equation is a line containing the pole and making an angle of radian measure C with the polar axis. See Figure 1. The same line is given by the equation

$$\theta = C \pm k\pi$$

where k is any integer.

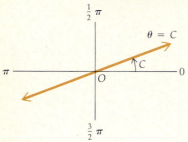

FIGURE 1

▶ **ILLUSTRATION 1**   (a) The graph of the equation

$$\theta = \tfrac{1}{4}\pi$$

appears in Figure 2. It is the line passing through the pole and making an angle of radian measure $\tfrac{1}{4}\pi$ with the polar axis. The same line is given by the equations

$$\theta = \tfrac{5}{4}\pi \qquad \theta = \tfrac{9}{4}\pi \qquad \theta = -\tfrac{3}{4}\pi \qquad \theta = -\tfrac{7}{4}\pi$$

and so on.

(b) The graph of the equation

$$\theta = \tfrac{2}{3}\pi$$

is shown in Figure 3. It is the line passing through the pole and making an angle of radian measure $\tfrac{2}{3}\pi$ with the polar axis. Other equations of this line are

$$\theta = \tfrac{5}{3}\pi \qquad \theta = \tfrac{8}{3}\pi \qquad \theta = -\tfrac{1}{3}\pi \qquad \theta = -\tfrac{4}{3}\pi$$

and so on.   ◀

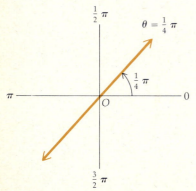

FIGURE 2

In general, the polar form of an equation of a line is not as simple as the cartesian form. However, if the line is parallel to either the polar axis or the $\tfrac{1}{2}\pi$ axis, the equation is fairly simple.

If a line is parallel to the polar axis and contains the point B whose cartesian coordinates are $(0, b)$ and polar coordinates are $(b, \tfrac{1}{2}\pi)$, then a cartesian equation is $y = b$. If we replace y by $r \sin \theta$, we have

$$r \sin \theta = b$$

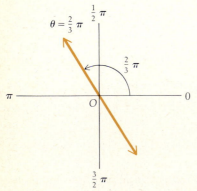

FIGURE 3

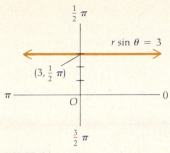

FIGURE 4

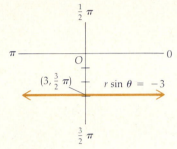

FIGURE 5

which is a polar equation of any line parallel to the polar axis. If b is positive, the line is above the polar axis. If b is negative, it is below the polar axis.

▶ **ILLUSTRATION 2** In Figure 4 we have a sketch of the graph of the equation

$$r \sin \theta = 3$$

and in Figure 5 we have a sketch of the graph of the equation

$$r \sin \theta = -3$$ ◀

Now consider a line parallel to the $\frac{1}{2}\pi$ axis or, equivalently, perpendicular to the polar axis. If the line goes through the point A whose cartesian coordinates are $(a, 0)$ and polar coordinates are $(a, 0)$, a cartesian equation is $x = a$. Replacing x by $r \cos \theta$ we obtain

$$r \cos \theta = a$$

which is an equation of any line perpendicular to the polar axis. If a is positive, the line is to the right of the $\frac{1}{2}\pi$ axis. If a is negative, the line is to the left of the $\frac{1}{2}\pi$ axis.

▶ **ILLUSTRATION 3** Figure 6 shows a sketch of the graph of the equation

$$r \cos \theta = 3$$

and Figure 7 shows a sketch of the graph of the equation

$$r \cos \theta = -3$$ ◀

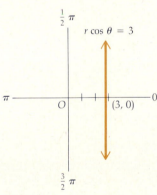

FIGURE 7

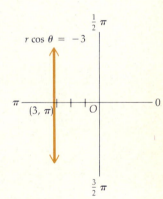

FIGURE 8

The graph of the equation

$$r = C$$

where C is any constant, is a circle whose center is at the pole and radius is $|C|$. The same circle is given by the equation

$$r = -C$$

▶ **ILLUSTRATION 4** In Figure 8 there is a sketch of the graph of the equation

$$r = 4$$

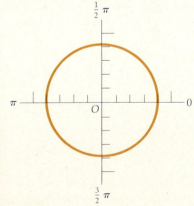

FIGURE 6

It is a circle with center at the pole and radius 4. The same circle is given by the equation

$$r = -4$$

although the use of such an equation is uncommon.   ◀

As was the case with the line, the general polar equation of a circle is not as simple as the cartesian form. However, there are special cases of an equation of a circle that are worth considering in polar form.

If a circle contains the origin (the pole) and has its center at the point having cartesian coordinates (a, b), then a cartesian equation of the circle is

$$x^2 + y^2 - 2ax - 2by = 0$$

A polar equation of this circle is

$$(r\cos\theta)^2 + (r\sin\theta)^2 - 2a(r\cos\theta) - 2b(r\sin\theta) = 0$$
$$r^2(\cos^2\theta + \sin^2\theta) - 2ar\cos\theta - 2br\sin\theta = 0$$
$$r^2 - 2ar\cos\theta - 2br\sin\theta = 0$$
$$r(r - 2a\cos\theta - 2b\sin\theta) = 0$$
$$r = 0 \qquad r - 2a\cos\theta - 2b\sin\theta = 0$$

Because the graph of the equation $r = 0$ is the pole and the pole ($r = 0$ when $\theta = \tan^{-1}\left(-\frac{a}{b}\right)$) is on the graph of $r - 2a\cos\theta - 2b\sin\theta = 0$, a polar equation of the circle is

$$r = 2a\cos\theta + 2b\sin\theta$$

When $b = 0$ in this equation, we have

$$r = 2a\cos\theta$$

This is a polar equation of the circle of radius $|a|$ units, tangent to the $\frac{1}{2}\pi$ axis, and with its center on the polar axis or its extension. If $a > 0$, the circle is to the right of the pole as in Figure 9, and if $a < 0$, the circle is to the left of the pole.

If $a = 0$ in the equation $r = 2a\cos\theta + 2b\sin\theta$, we have

$$r = 2b\sin\theta$$

which is a polar equation of the circle of radius $|b|$ units, with its center on the $\frac{1}{2}\pi$ axis or its extension and tangent to the polar axis. If $b > 0$, the circle is above the pole, and if $b < 0$, the circle is below the pole.

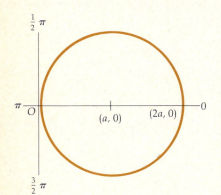

FIGURE 9

EXAMPLE 1   Draw a sketch of the graph of each of the following equations:

(a) $r = 5\cos\theta$ (b) $r = -6\sin\theta$

Solution

(a) The equation

$$r = 5\cos\theta$$

is of the form $r = 2a\cos\theta$ with $a = \frac{5}{2}$. Thus the graph is a circle with center at the point having polar coordinates $(\frac{5}{2}, 0)$ and tangent to the $\frac{1}{2}\pi$ axis. A sketch of the graph appears in Figure 10.

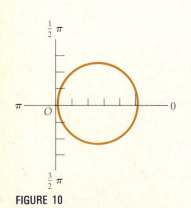

FIGURE 10

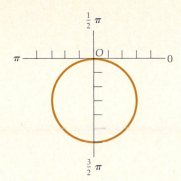

FIGURE 11

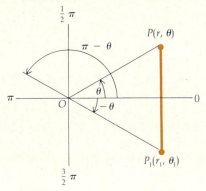

FIGURE 12

(b) The equation

$$r = -6 \sin \theta$$

is of the form $r = 2b \sin \theta$ with $b = -3$. The graph is the circle with center at the point having polar coordinates $(3, \frac{3}{2}\pi)$ and tangent to the polar axis. Figure 11 shows a sketch of the graph.

We have presented important special cases (lines and circles) of graphs of polar equations. Before discussing more general curves having polar equations, we consider properties of symmetry.

In Section 1.2 (Definition 1.2.4) we stated that two points P and Q are symmetric with respect to a line if and only if the line is the perpendicular bisector of the line segment PQ, and that two points P and Q are symmetric with respect to a third point if and only if the third point is the midpoint of the line segment PQ. Therefore the points $(2, \frac{1}{3}\pi)$ and $(2, \frac{2}{3}\pi)$ are symmetric with respect to the $\frac{1}{2}\pi$ axis, and the points $(2, \frac{1}{3}\pi)$ and $(2, -\frac{2}{3}\pi)$ are symmetric with respect to the pole. We also stated (Definition 1.2.5) that the graph of an equation is symmetric with respect to a line l if and only if for every point P on the graph there is a point Q, also on the graph, such that P and Q are symmetric with respect to l. Similarly, the graph of an equation is symmetric with respect to a point R if and only if for every point P on the graph there is a point S, also on the graph, such that P and S are symmetric with respect to R. We have three theorems giving tests for symmetry of graphs of polar equations.

10.6.1 THEOREM If for an equation in polar coordinates an equivalent equation is obtained when (r, θ) is replaced by either $(r, -\theta + 2n\pi)$ or $(-r, \pi - \theta + 2n\pi)$, where n is any integer, the graph of the equation is symmetric with respect to the polar axis.

Proof If the point $P(r, \theta)$ is a point on the graph of an equation, then the graph is symmetric with respect to the polar axis if there is a point $P_1(r_1, \theta_1)$ on the graph such that the polar axis is the perpendicular bisector of the line segment P_1P (see Figure 12). So if $r_1 = r$, then θ_1 must equal $-\theta + 2n\pi$, where n is an integer. And if $r_1 = -r$, then θ_1 must be $\pi - \theta + 2n\pi$. ∎

10.6.2 THEOREM If for an equation in polar coordinates an equivalent equation is obtained when (r, θ) is replaced by either $(r, \pi - \theta + 2n\pi)$ or $(-r, -\theta + 2n\pi)$, where n is any integer, the graph of the equation is symmetric with respect to the $\frac{1}{2}\pi$ axis.

10.6.3 THEOREM If for an equation in polar coordinates an equivalent equation is obtained when (r, θ) is replaced by either $(-r, \theta + 2n\pi)$ or $(r, \pi + \theta + 2n\pi)$, where n is any integer, the graph of the equation is symmetric with respect to the pole.

The proofs of Theorems 10.6.2 and 10.6.3 are similar to the proof of Theorem 10.6.1 and are left as exercises (see Exercises 53 and 54).

▶ **ILLUSTRATION 5** For the graph of the equation

$$r = 4 \cos 2\theta$$

we test for symmetry with respect to the polar axis, the $\frac{1}{2}\pi$ axis, and the pole.

Table 1

θ	r
0	-1
$\frac{1}{6}\pi$	$1 - \sqrt{3}$
$\frac{1}{3}\pi$	0
$\frac{1}{2}\pi$	1
$\frac{2}{3}\pi$	2
$\frac{5}{6}\pi$	$1 + \sqrt{3}$
π	3

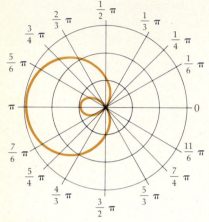

FIGURE 13

Using Theorem 10.6.1 to test for symmetry with respect to the polar axis, we replace (r, θ) by $(r, -\theta)$ and obtain $r = 4 \cos(-2\theta)$, which is equivalent to $r = 4 \cos 2\theta$. So the graph is symmetric with respect to the polar axis.

Using Theorem 10.6.2 to test for symmetry with respect to the $\frac{1}{2}\pi$ axis, we replace (r, θ) by $(r, \pi - \theta)$ and get $r = 4 \cos(2(\pi - \theta))$ or, equivalently, $r = 4 \cos(2\pi - 2\theta)$, which is equivalent to the equation $r = 4 \cos 2\theta$. Therefore the graph is symmetric with respect to the $\frac{1}{2}\pi$ axis.

To test for symmetry with respect to the pole, we replace (r, θ) by $(-r, \theta)$ and obtain the equation $-r = 4 \cos 2\theta$, which is not equivalent to the given equation. But we must also determine if the other set of coordinates works. We replace (r, θ) by $(r, \pi + \theta)$ and obtain $r = 4 \cos 2(\pi + \theta)$ or, equivalently, $r = 4 \cos(2\pi + 2\theta)$, which is equivalent to the equation $r = 4 \cos 2\theta$. Therefore the graph is symmetric with respect to the pole. ◀

When drawing a sketch of a graph, it is desirable to determine if the pole is on the graph. This is done by substituting 0 for r and solving for θ. Also, it is advantageous to plot the points for which r has a relative maximum or relative minimum value. As a further aid in plotting, if a curve contains the pole, it is sometimes desirable to consider the tangent lines there. When helpful we use the fact shown in Supplementary Section 10.9 that if θ_1 is a value of θ that satisfies a polar equation of the curve when $r = 0$, then the line $\theta = \theta_1$ is tangent to the curve at the pole.

EXAMPLE 2   Draw a sketch of the graph of the equation

$$r = 1 - 2 \cos \theta$$

Solution   Replacing (r, θ) by $(r, -\theta)$, we obtain an equivalent equation. Therefore the graph is symmetric with respect to the polar axis.

Table 1 gives the coordinates of some points on the graph. From these points we draw half of the graph; the remainder is drawn from its symmetry with respect to the polar axis.

If $r = 0$, we obtain $\cos \theta = \frac{1}{2}$, and if $0 \leq \theta \leq \pi$, then $\theta = \frac{1}{3}\pi$. Thus the point $(0, \frac{1}{3}\pi)$ is on the graph. Furthermore, an equation of the tangent line at the pole is $\theta = \frac{1}{3}\pi$. A sketch of the graph appears in Figure 13.

The curve in Example 2 is called a *limaçon*. The graph of an equation of the form

$$r = a \pm b \cos \theta \quad \text{or} \quad r = a \pm b \sin \theta$$

is a **limaçon**. There are four types of limaçons, and the particular type depends on the ratio a/b, where a and b are positive. We show these four types obtained from the equation

$$r = a + b \cos \theta \qquad a > 0 \text{ and } b > 0$$

1. $0 < \dfrac{a}{b} < 1$ **Limaçon with a loop**. See Figure 14(a).

2. $\dfrac{a}{b} = 1$ **Cardioid** (heart-shaped). See Figure 14(b).

3. $1 < \dfrac{a}{b} < 2$ **Limaçon with a dent**. See Figure 14(c).

4. $2 \le \dfrac{a}{b}$ **Convex limaçon** (no dent). See Figure 14(d).

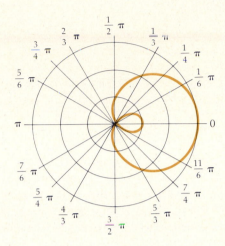

$0 < \dfrac{a}{b} < 1$
Limaçon with a loop
(a)

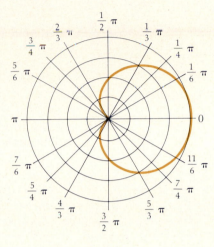

$\dfrac{a}{b} = 1$
Cardioid
(b)

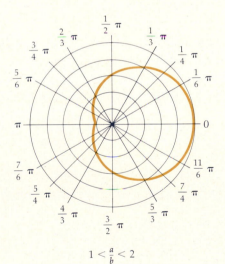

$1 < \dfrac{a}{b} < 2$
Limaçon with a dent
(c)

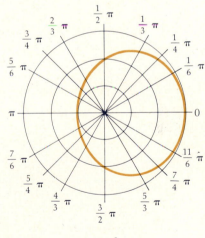

$2 \le \dfrac{a}{b}$
Convex Limaçon
(d)

FIGURE 14

If you study Supplementary Section 10.9, where horizontal and vertical tangent lines of polar curves are discussed, the distinction between limaçons of type 3 (with a dent) and those of type 4 (no dent) will be apparent.

The limaçons obtained from the equation

$$r = a + b \sin \theta \qquad a > 0 \text{ and } b > 0$$

have the $\frac{1}{2}\pi$ axis as the axis of symmetry. If a limaçon has the equation

$$r = a - b \cos \theta \qquad a > 0 \text{ and } b > 0$$

the limaçon points in the direction of π, and if it has the equation

$$r = a - b \sin \theta \qquad a > 0 \text{ and } b > 0$$

it points in the direction of $\frac{3}{2}\pi$.

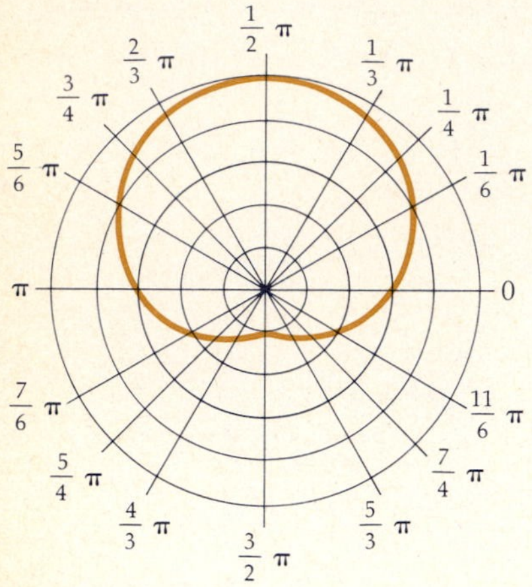

EXAMPLE 3   Draw a sketch of the graph of each of the following limaçons:

(a) $r = 3 + 2 \sin \theta$ (b) $r = 2 + 2 \cos \theta$ (c) $r = 2 - \sin \theta$

Solution

Table 2

θ	r
0	3
$\frac{1}{6}\pi$	4
$\frac{1}{3}\pi$	$3 + \sqrt{3}$
$\frac{1}{2}\pi$	5
π	3
$\frac{7}{6}\pi$	2
$\frac{4}{3}\pi$	$3 - \sqrt{3}$
$\frac{3}{2}\pi$	1

(a) The equation

$$r = 3 + 2 \sin \theta$$

is of the form of $r = a + b \sin \theta$ with $a = 3$ and $b = 2$. Because $a/b = 3/2$, and $1 < \frac{3}{2} < 2$, the graph is a limaçon with a dent. It is symmetric with respect to the $\frac{1}{2}\pi$ axis. Table 2 gives the coordinates of some of the points on the graph. A sketch of the graph shown in Figure 15 is drawn by plotting the points whose coordinates are given in Table 2 and using the symmetry property.

Table 3

θ	r
0	4
$\frac{1}{6}\pi$	$2 + \sqrt{3}$
$\frac{1}{3}\pi$	3
$\frac{1}{2}\pi$	2
$\frac{2}{3}\pi$	1
$\frac{5}{6}\pi$	$2 - \sqrt{3}$
π	0

(b) The equation

$$r = 2 + 2 \cos \theta$$

is of the form of $r = a + b \cos \theta$ with $a = 2$ and $b = 2$. Because $a/b = 1$, the graph is a cardioid. It is symmetric with respect to the polar axis. The coordinates of some of the points on the graph are given in Table 3, and the sketch of the graph appearing in Figure 16 is drawn by plotting these points and using the symmetry property.

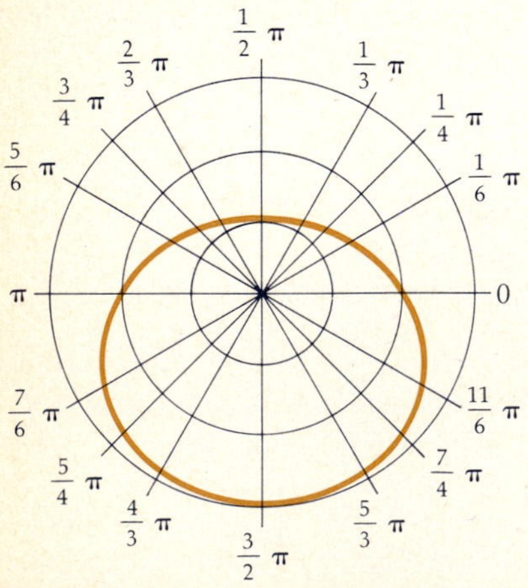

Table 4

θ	r
0	2
$\frac{1}{6}\pi$	$\frac{3}{2}$
$\frac{1}{3}\pi$	$2 - \frac{1}{2}\sqrt{3}$
$\frac{1}{2}\pi$	1
π	2
$\frac{7}{6}\pi$	$\frac{5}{2}$
$\frac{4}{3}\pi$	$2 + \frac{1}{2}\sqrt{3}$
$\frac{3}{2}\pi$	3

(c) The equation

$$r = 2 - \sin \theta$$

is of the form of $r = a - b \sin \theta$ with $a = 2$ and $b = 1$. Because $a/b = 2$, the graph is a convex limaçon. It is symmetric with respect to the $\frac{1}{2}\pi$ axis and points in the direction of $\frac{3}{2}\pi$. Figure 17 shows a sketch of the graph obtained by plotting the points whose coordinates are given in Table 4 and using the symmetry property.

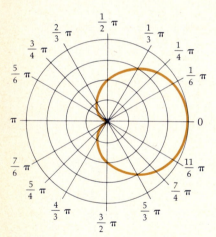

FIGURE 15

FIGURE 16

FIGURE 17

The graph of an equation of the form

$$r = a \cos n\theta \quad \text{or} \quad r = a \sin n\theta$$

is a **rose**, having n leaves if n is odd and $2n$ leaves if n is even.

Table 5

θ	r
0	4
$\frac{1}{12}\pi$	$2\sqrt{3}$
$\frac{1}{6}\pi$	2
$\frac{1}{4}\pi$	0
$\frac{1}{3}\pi$	-2
$\frac{5}{12}\pi$	$-2\sqrt{3}$
$\frac{1}{2}\pi$	-4

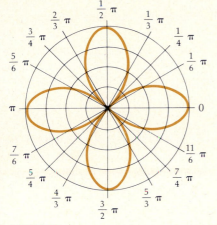

FIGURE 18

EXAMPLE 4   Draw a sketch of the four-leafed rose

$$r = 4\cos 2\theta$$

Solution   In Illustration 5 we proved that the graph is symmetric with respect to the polar axis, the $\frac{1}{2}\pi$ axis, and the pole. Substituting 0 for r in the given equation we get

$$\cos 2\theta - 0$$

from which we obtain, for $0 \le \theta < 2\pi$,

$$\theta = \tfrac{1}{4}\pi \qquad \theta = \tfrac{3}{4}\pi \qquad \theta = \tfrac{5}{4}\pi \qquad \theta = \tfrac{7}{4}\pi$$

The lines having these equations are tangent to the graph at the pole.

Table 5 gives values of r for some values of θ from 0 to $\frac{1}{2}\pi$. From these values and the symmetry properties we draw a sketch of the graph shown in Figure 18.

Observe that if in the equations for a rose we take $n = 1$, we get

$$r = a\cos\theta \quad \text{or} \quad r = a\sin\theta$$

which are equations for a circle. Thus a circle can be considered as a one-leafed rose.

Other polar curves that occur frequently are *lemniscates* (see Exercises 29 through 32) and *spirals* (see Exercises 25 through 28). The curve in the next example is called a *spiral of Archimedes*.

EXAMPLE 5   Draw a sketch of the graph of

$$r = \theta \qquad \theta \ge 0$$

Solution   When $\theta = n\pi$, where n is any integer, the graph intersects the polar axis or its extension, and when $\theta = \frac{1}{2}n\pi$, where n is any odd integer, the graph intersects the $\frac{1}{2}\pi$ axis or its extension. When $r = 0$, $\theta = 0$; thus the tangent line to the curve at the pole is the polar axis. A sketch of the graph appears in Figure 19.

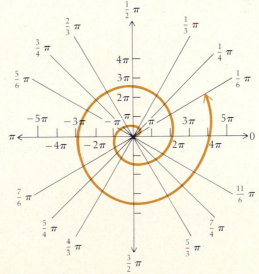

FIGURE 19

To find the points of intersection of two curves whose equations are in cartesian coordinates, we solve the two equations simultaneously. The common solutions give all the points of intersection. However, because a point has an unlimited number of sets of polar coordinates, it is possible to have as the intersection of two curves a point for which no single pair of polar coordinates satisfies both equations. This is illustrated in the following example.

EXAMPLE 6   Draw sketches of the graphs of

$$r = 2 \sin 2\theta \quad \text{and} \quad r = 1$$

with the same pole and polar axis, and find the points of intersection.

Solution   The graph of $r = 2 \sin 2\theta$ is a four-leafed rose, and the graph of $r = 1$ is the circle with its center at the pole and radius 1. Sketches of the graphs appear in Figure 20. Solving the two equations simultaneously we have

$$2 \sin 2\theta = 1$$

$$\sin 2\theta = \tfrac{1}{2}$$

Therefore

$$2\theta = \tfrac{1}{6}\pi \qquad 2\theta = \tfrac{5}{6}\pi \qquad 2\theta = \tfrac{13}{6}\pi \qquad 2\theta = \tfrac{17}{6}\pi$$

$$\theta = \tfrac{1}{12}\pi \qquad \theta = \tfrac{5}{12}\pi \qquad \theta = \tfrac{13}{12}\pi \qquad \theta = \tfrac{17}{12}\pi$$

Hence we obtain the points of intersection $(1, \tfrac{1}{12}\pi)$, $(1, \tfrac{5}{12}\pi)$, $(1, \tfrac{13}{12}\pi)$, and $(1, \tfrac{17}{12}\pi)$. We notice in Figure 20 that eight points of intersection are shown. The other four points are obtained if we take another form of the equation of the circle $r = 1$; that is, consider the equation $r = -1$, which is the same circle. Solving this equation simultaneously with the equation of the four-leafed rose we have

$$\sin 2\theta = -\tfrac{1}{2}$$

Then we get

$$2\theta = \tfrac{7}{6}\pi \qquad 2\theta = \tfrac{11}{6}\pi \qquad 2\theta = \tfrac{19}{6}\pi \qquad 2\theta = \tfrac{23}{6}\pi$$

$$\theta = \tfrac{7}{12}\pi \qquad \theta = \tfrac{11}{12}\pi \qquad \theta = \tfrac{19}{12}\pi \qquad \theta = \tfrac{23}{12}\pi$$

Thus we have the four points $(-1, \tfrac{7}{12}\pi)$, $(-1, \tfrac{11}{12}\pi)$, $(-1, \tfrac{19}{12}\pi)$, and $(-1, \tfrac{23}{12}\pi)$. Incidentally, $(-1, \tfrac{7}{12}\pi)$ also can be written as $(1, \tfrac{19}{12}\pi)$, $(-1, \tfrac{11}{12}\pi)$ can be written as $(1, \tfrac{23}{12}\pi)$, $(-1, \tfrac{19}{12}\pi)$ can be written as $(1, \tfrac{7}{12}\pi)$, and $(-1, \tfrac{23}{12}\pi)$ can be written as $(1, \tfrac{11}{12}\pi)$.

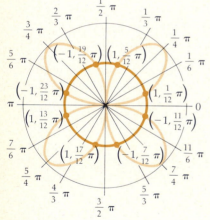

FIGURE 20

Because $(0, \theta)$ represents the pole for any θ, we determine if the pole is a point of intersection by setting $r = 0$ in each equation and solving for θ.

Often the coordinates of the points of intersection of two curves can be found directly from their graphs. However, the following is a general method.

If an equation of a curve in polar coordinates is given by $r = f(\theta)$, then the same curve is given by

$$(-1)^n r = f(\theta + n\pi) \tag{1}$$

where n is any integer.

▶ **ILLUSTRATION 6**   Consider the curves of Example 6. The graph of the equation $r = 2 \sin 2\theta$ also has the equation (by taking $n = 1$ in (1))

$$(-1)r = 2 \sin 2(\theta + \pi) \quad \Leftrightarrow \quad -r = 2 \sin 2\theta$$

If we take $n = 2$ in (1), the graph of $r = 2 \sin 2\theta$ also has the equation

$$(-1)^2 r = 2 \sin 2(\theta + 2\pi) \quad \Leftrightarrow \quad r = 2 \sin 2\theta$$

which is the same as the original equation. Taking n any other integer, we get either $r = 2 \sin 2\theta$ or $r = -2 \sin 2\theta$. The graph of the equation $r = 1$ also has the equation (by taking $n = 1$ in (1))

$$r = -1$$

Other integer values of n in (1) applied to the equation $r = 1$ give either $r = 1$ or $r = -1$.   ◀

If we are given the two equations $r = f(\theta)$ and $r = g(\theta)$, we obtain all the points of intersection of the graphs of the equations by doing the following:
  (a) Use (1) to determine all the distinct equations of the two curves:

$$r = f_1(\theta), r = f_2(\theta), r = f_3(\theta), \ldots \tag{2}$$

$$r = g_1(\theta), r = g_2(\theta), r = g_3(\theta), \ldots \tag{3}$$

  (b) Solve each equation in (2) simultaneously with each equation in (3).
  (c) Check to see if the pole is a point of intersection by setting $r = 0$ in each equation, thereby giving

$$f(\theta) = 0 \quad \text{and} \quad g(\theta) = 0$$

If these equations each have a solution for θ, not necessarily the same, then the pole lies on both curves.

EXAMPLE 7   Find the points of intersection of the two curves

$$r = 2 - 2 \cos \theta \quad \text{and} \quad r = 2 \cos \theta$$

Draw sketches of their graphs.

Solution   To find other equations of the curve represented by

$$r = 2 - 2 \cos \theta$$

we have

$$(-1)r = 2 - 2 \cos(\theta + \pi)$$

$$-r = 2 + 2 \cos \theta$$

and

$$(-1)^2 r = 2 - 2 \cos(\theta + 2\pi)$$

$$r = 2 - 2 \cos \theta$$

which is the same as the original equation.

In a similar manner we find other equations of the curve given by $r = 2 \cos \theta$:

$$(-1)r = 2 \cos(\theta + \pi)$$

$$-r = -2 \cos \theta$$

$$r = 2 \cos \theta$$

which is the same as the original equation.

So there are two possible equations for the first curve, $r = 2 - 2 \cos \theta$ and $-r = 2 + 2 \cos \theta$, and one equation for the second curve, $r = 2 \cos \theta$. Solving simultaneously $r = 2 - 2 \cos \theta$ and $r = 2 \cos \theta$ gives

$$2 \cos \theta = 2 - 2 \cos \theta$$

$$4 \cos \theta = 2$$

$$\cos \theta = \tfrac{1}{2}$$

Thus $\theta = \tfrac{1}{3}\pi$ and $\theta = \tfrac{5}{3}\pi$, giving the points $(1, \tfrac{1}{3}\pi)$ and $(1, \tfrac{5}{3}\pi)$. Solving simultaneously $-r = 2 + 2 \cos \theta$ and $r = 2 \cos \theta$ yields

$$2 + 2 \cos \theta = -2 \cos \theta$$

$$4 \cos \theta = -2$$

$$\cos \theta = -\tfrac{1}{2}$$

Hence $\theta = \tfrac{2}{3}\pi$ and $\theta = \tfrac{4}{3}\pi$, giving the points $(-1, \tfrac{2}{3}\pi)$ and $(-1, \tfrac{4}{3}\pi)$. However, $(-1, \tfrac{2}{3}\pi)$ is the same point as $(1, \tfrac{5}{3}\pi)$, and $(-1, \tfrac{4}{3}\pi)$ is the same point as $(1, \tfrac{1}{3}\pi)$.

Checking to see if the pole is on the first curve by substituting $r = 0$ in the equation $r = 2 - 2 \cos \theta$, we have

$$0 = 2 - 2 \cos \theta$$

$$\cos \theta = 1$$

$$\theta = 0$$

Therefore the pole lies on the first curve. In a similar fashion, by substituting $r = 0$ in $r = 2 \cos \theta$ we get

$$0 = 2 \cos \theta$$

$$\cos \theta = 0$$

$$\theta = \tfrac{1}{2}\pi \qquad \theta = \tfrac{3}{2}\pi$$

So the pole lies on the second curve.

Therefore the points of intersection of the two curves are $(1, \tfrac{1}{3}\pi)$, $(1, \tfrac{5}{3}\pi)$, and the pole. Sketches of the two curves are shown in Figure 21.

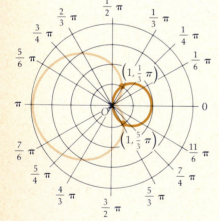

FIGURE 21

EXERCISES 10.6

In Exercises 1 through 36, draw a sketch of the graph of the equation.

1. (a) $\theta = \tfrac{1}{3}\pi$; (b) $r = \tfrac{1}{3}\pi$

2. (a) $\theta = \tfrac{3}{4}\pi$; (b) $r = \tfrac{3}{4}\pi$

3. (a) $\theta = 2$; (b) $r = 2$

4. (a) $\theta = -3$; (b) $r = -3$

5. (a) $r \cos \theta = 4$; (b) $r = 4 \cos \theta$

6. (a) $r \sin \theta = 2$; (b) $r = 2 \sin \theta$

7. (a) $r \sin \theta = -4$; (b) $r = -4 \sin \theta$

8. (a) $r \cos \theta = -5$; (b) $r = -5 \cos \theta$

9. $r = 4 - 4 \cos \theta$

10. $r = 3 - 3 \sin \theta$

11. $r = 2 + 2 \sin \theta$

12. $r = 3 + 3 \cos \theta$

13. $r = 2 - 3 \sin \theta$

14. $r = 4 - 3 \sin \theta$

15. $r = 3 - 2 \cos \theta$

16. $r = 3 - 4 \cos \theta$

17. $r = 4 + 2 \sin \theta$ 18. $r = 6 + 2 \cos \theta$

19. $r = 2 \sin 3\theta$ 20. $r = 4 \sin 5\theta$ 21. $r = 2 \cos 4\theta$

22. $r = 3 \cos 2\theta$ 23. $r = 4 \sin 2\theta$ 24. $r = 3 \cos 3\theta$

25. $r = e^\theta$ (logarithmic spiral)

26. $r = e^{\theta/3}$ (logarithmic spiral)

27. $r = \dfrac{1}{\theta}$ (reciprocal spiral)

28. $r = 2\theta$ (spiral of Archimedes)
29. $r^2 = 9 \sin 2\theta$ (lemniscate)
30. $r^2 = 16 \cos 2\theta$ (lemniscate)
31. $r^2 = -25 \cos 2\theta$ (lemniscate)
32. $r^2 = -4 \sin 2\theta$ (lemniscate)
33. $r = 2 \sin \theta \tan \theta$ (cissoid)
34. $r = 2 \sec \theta - 1$ (conchoid of Nicomedes)
35. $r = |\sin 2\theta|$ 36. $r = 2|\cos \theta|$

In Exercises 37 through 52, find the points of intersection of the graphs of the pair of equations. Draw a sketch of each pair of graphs with the same pole and polar axis.

37. $\begin{cases} 2r = 3 \\ r = 3 \sin \theta \end{cases}$ 38. $\begin{cases} 2r = 3 \\ r = 1 + \cos \theta \end{cases}$

39. $\begin{cases} r = 2 \cos \theta \\ r = 2 \sin \theta \end{cases}$ 40. $\begin{cases} r = 2 \cos 2\theta \\ r = 2 \sin \theta \end{cases}$

41. $\begin{cases} r = 4\theta \\ r = \frac{1}{2}\pi \end{cases}$ 42. $\begin{cases} r \sin \theta = 4 \\ r \cos \theta = 4 \end{cases}$

43. $\begin{cases} r = \tan \theta \\ r = 4 \sin \theta \end{cases}$ 44. $\begin{cases} r = 2 \cos \theta \\ r = 2\sqrt{3} \sin \theta \end{cases}$

45. $\begin{cases} r = 3 \\ r = 2(1 + \cos \theta) \end{cases}$ 46. $\begin{cases} r = \sin \theta \\ r = \sin 2\theta \end{cases}$

47. $\begin{cases} r^2 \sin 2\theta = 8 \\ r \cos \theta = 2 \end{cases}$ 48. $\begin{cases} r = 4(1 + \sin \theta) \\ r(1 - \sin \theta) = 3 \end{cases}$

49. $\begin{cases} r = \cos \theta - 1 \\ r = \cos 2\theta \end{cases}$ 50. $\begin{cases} r = 1 - \sin \theta \\ r = \cos 2\theta \end{cases}$

51. $\begin{cases} r = \sin 2\theta \\ r = \cos 2\theta \end{cases}$ 52. $\begin{cases} r = 4 \tan \theta \sin \theta \\ r = 4 \cos \theta \end{cases}$

53. Prove Theorem 10.6.2.   54. Prove Theorem 10.6.3.

In Exercises 55 and 56, the graph of the equation intersects itself. Find the points at which this occurs.

55. $r = \sin \frac{3}{2}\theta$ 56. $r = 1 + 2 \cos 2\theta$

10.7 AREA OF A REGION IN POLAR COORDINATES

We now develop a method for finding the area of a region bounded by two lines through the pole and a curve whose equation is given in polar coordinates.

Let the function f be continuous and nonnegative on the closed interval $[\alpha, \beta]$. Let R be the region bounded by the curve whose equation is $r = f(\theta)$ and by the lines $\theta = \alpha$ and $\theta = \beta$. Then the region R is the region AOB shown in Figure 1.

Consider a partition Δ of $[\alpha, \beta]$ defined by

$$\alpha = \theta_0 < \theta_1 < \theta_2 < \ldots < \theta_{i-1} < \theta_i < \ldots < \theta_{n-1} < \theta_n = \beta$$

Therefore we have n subintervals of the form $[\theta_{i-1}, \theta_i]$, where $i = 1, 2, \ldots, n$. Let ξ_i be a value of θ in the ith subinterval $[\theta_{i-1}, \theta_i]$. See Figure 2, where the ith subinterval is shown together with $\theta = \xi_i$. The radian measure of the angle between the lines $\theta = \theta_{i-1}$ and $\theta = \theta_i$ is denoted by $\Delta_i\theta$. The number of square units in the area of the circular sector of radius $f(\xi_i)$ units and central angle of radian measure $\Delta_i\theta$ is given by

$$\tfrac{1}{2}[f(\xi_i)]^2 \, \Delta_i\theta$$

There is such a circular sector for each of the n subintervals. The sum of the measures of the areas of these n circular sectors is

$$\tfrac{1}{2}[f(\xi_1)]^2 \, \Delta_1\theta + \tfrac{1}{2}[f(\xi_2)]^2 \, \Delta_2\theta + \ldots + \tfrac{1}{2}[f(\xi_i)]^2 \, \Delta_i\theta + \ldots + \tfrac{1}{2}[f(\xi_n)]^2 \, \Delta_n\theta$$

which can be written, using sigma notation, as

$$\sum_{i=1}^n \tfrac{1}{2}[f(\xi_2)]^2 \, \Delta_2\theta \qquad (1)$$

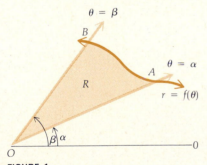

FIGURE 1

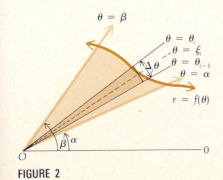

FIGURE 2

Let $\|\Delta\|$ be the norm of the partition Δ; that is, $\|\Delta\|$ is the measure of the largest $\Delta_i\theta$. Then if we let A square units be the area of the region R, we define A as the limit of Riemann sum (1) as $\|\Delta\|$ approaches 0, which is a definite integral.

10.7.1 DEFINITION   Let R be the region bounded by the lines $\theta = \alpha$ and $\theta = \beta$ and the curve whose equation is $r = f(\theta)$, where f is continuous and nonnegative on the closed interval $[\alpha, \beta]$. Then if A square units is the area of region R,

$$A = \lim_{\|\Delta\| \to 0} \sum_{i=1}^{n} \tfrac{1}{2}[f(\xi_i)]^2 \, \Delta_i\theta$$

$$= \frac{1}{2} \int_{\alpha}^{\beta} [f(\theta)]^2 \, d\theta$$

EXAMPLE 1   Find the area of the region bounded by the graph of

$$r = 2 + 2\cos\theta$$

Solution   The region together with an element of area is shown in Figure 3. Because the curve is symmetric with respect to the polar axis, we take the θ limits from 0 to π that determine the area of the region bounded by the curve above the polar axis. Then the area of the entire region is determined by multiplying that area by 2. Thus, if A square units is the required area,

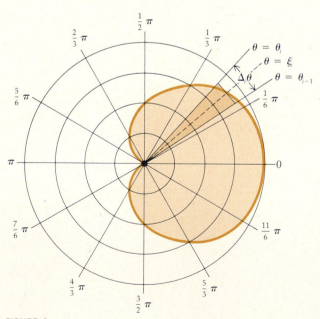

FIGURE 3

$$A = 2 \lim_{\|\Delta\| \to 0} \sum_{i=1}^{n} \tfrac{1}{2}(2 + 2 \cos \xi_i)^2 \, \Delta_i \theta$$

$$= 2 \int_0^\pi \tfrac{1}{2}(2 + 2 \cos \theta)^2 \, d\theta$$

$$= 4 \int_0^\pi (1 + 2 \cos \theta + \cos^2 \theta) \, d\theta$$

$$= 4 \left[\theta + 2 \sin \theta + \tfrac{1}{2}\theta + \tfrac{1}{4} \sin 2\theta \right]_0^\pi$$

$$= 4(\pi + 0 + \tfrac{1}{2}\pi + 0 - 0)$$

$$= 6\pi$$

Therefore the area is 6π square units.

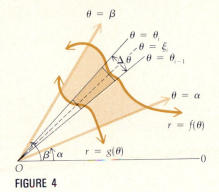

FIGURE 4

Consider now the region bounded by the lines $\theta = \alpha$ and $\theta = \beta$ and the two curves whose equations are $r = f(\theta)$ and $r = g(\theta)$, where f and g are continuous on the closed interval $[\alpha, \beta]$ and $f(\theta) \geq g(\theta)$ on $[\alpha, \beta]$. See Figure 4. We wish to find the area of this region. We take a partition of the interval $[\alpha, \beta]$ with ξ_i a value of θ in the ith subinterval $[\theta_{i-1}, \theta_i]$. The measure of the area of an element of area is the difference of the measures of the areas of two circular sectors:

$$\tfrac{1}{2}[f(\xi_i)]^2 \, \Delta_i \theta - \tfrac{1}{2}[g(\xi_i)]^2 \, \Delta_i \theta = \tfrac{1}{2}([f(\xi_i)]^2 - [g(\xi_i)]^2) \, \Delta_i \theta$$

The sum of the measures of the areas of n such elements is given by

$$\sum_{i=1}^{n} \tfrac{1}{2}([f(\xi_i)]^2 - [g(\xi_i)]^2) \, \Delta_i \theta$$

Hence, if A square units is the area of the region desired, we have

$$A = \lim_{\|\Delta\| \to 0} \sum_{i=1}^{n} \tfrac{1}{2}([f(\xi_i)]^2 - [g(\xi_i)]^2) \, \Delta_i \theta$$

Because f and g are continuous on $[\alpha, \beta]$, so also is $f - g$; therefore the limit exists and is equal to a definite integral. Thus

$$A = \frac{1}{2} \int_\alpha^\beta ([f(\theta)]^2 - [g(\theta)]^2) \, d\theta$$

EXAMPLE 2   Find the area of the region inside the circle $r = 3 \sin \theta$ and outside the limaçon $r = 2 - \sin \theta$.

Solution   To find the points of intersection we set

$$3 \sin \theta = 2 - \sin \theta$$

$$\sin \theta = \tfrac{1}{2}$$

$$\theta = \tfrac{1}{6}\pi \qquad \theta = \tfrac{5}{6}\pi$$

The curves are sketched and the region is shown together with an element of area in Figure 5.

If we let $f(\theta) = 3 \sin \theta$ and $g(\theta) = 2 - \sin \theta$, then the equation of the circle is $r = f(\theta)$, and the equation of the limaçon is $r = g(\theta)$.

Instead of taking the limits $\tfrac{1}{6}\pi$ to $\tfrac{5}{6}\pi$ we use the property of symmetry with respect to the $\tfrac{1}{2}\pi$ axis and take the limits from $\tfrac{1}{6}\pi$ to $\tfrac{1}{2}\pi$ and multiply by 2.

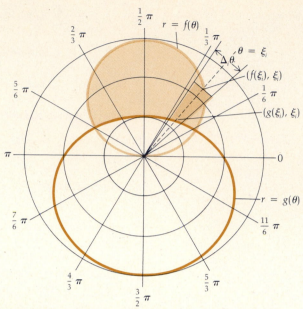

FIGURE 5

Then, if A square units is the area of the given region,

$$A = 2 \lim_{||\Delta|| \to 0} \sum_{i=1}^{n} \tfrac{1}{2}([f(\xi_i)]^2 - [g(\xi_i)]^2) \Delta_i \theta$$

$$= 2 \cdot \frac{1}{2} \int_{\pi/6}^{\pi/2} ([f(\theta)]^2 - [g(\theta)]^2) \, d\theta$$

$$= \int_{\pi/6}^{\pi/2} [9 \sin^2 \theta - (2 - \sin \theta)^2] \, d\theta$$

$$= 8 \int_{\pi/6}^{\pi/2} \sin^2 \theta \, d\theta + 4 \int_{\pi/6}^{\pi/2} \sin \theta \, d\theta - 4 \int_{\pi/6}^{\pi/2} d\theta$$

$$= 4 \int_{\pi/6}^{\pi/2} (1 - \cos 2\theta) \, d\theta + \Big[-4 \cos \theta - 4\theta \Big]_{\pi/6}^{\pi/2}$$

$$= 4\theta - 2 \sin 2\theta - 4 \cos \theta - 4\theta \Big]_{\pi/6}^{\pi/2}$$

$$= -2 \sin 2\theta - 4 \cos \theta \Big]_{\pi/6}^{\pi/2}$$

$$= (-2 \sin \pi - 4 \cos \tfrac{1}{2}\pi) - (-2 \sin \tfrac{1}{3}\pi - 4 \cos \tfrac{1}{6}\pi)$$

$$= 2 \cdot \tfrac{1}{2}\sqrt{3} + 4 \cdot \tfrac{1}{2}\sqrt{3}$$

$$= 3\sqrt{3}$$

Therefore the area is $3\sqrt{3}$ square units.

EXERCISES 10.7

In Exercises 1 through 6, find the area of the region enclosed by the graph of the equation.

1. $r = 3 \cos \theta$

2. $r = 2 - \sin \theta$

3. $r = 4 \cos 3\theta$

4. $r = 4 \sin^2 \tfrac{1}{2}\theta$

5. $r^2 = 4 \sin 2\theta$

6. $r = 4 \sin^2 \theta \cos \theta$

7. Find the area of the region enclosed by the graph of the equation $r = \theta$ from $\theta = 0$ to $\theta = \tfrac{3}{2}\pi$.

8. Find the area of the region enclosed by the graph of $r = e^\theta$ and the lines $\theta = 0$ and $\theta = 1$.

In Exercises 9 through 12, find the area of the region enclosed by one loop of the graph of the equation.

9. $r = 3 \cos 2\theta$

10. $r = a(1 - 2 \cos \theta)$

11. $r = 1 + 3 \sin \theta$

12. $r = a \sin 3\theta$

In Exercises 13 through 16, find the area of the intersection of the regions enclosed by the graphs of the two equations.

13. $\begin{cases} r = 2 \\ r = 3 - 2 \cos \theta \end{cases}$

14. $\begin{cases} r = 4 \sin \theta \\ r = 4 \cos \theta \end{cases}$

15. $\begin{cases} r = 3 \sin 2\theta \\ r = 3 \cos 2\theta \end{cases}$

16. $\begin{cases} r^2 = 2 \cos 2\theta \\ r = 1 \end{cases}$

In Exercises 15 through 21, find the area of the region that is inside the graph of the first equation and outside the graph of the second equation.

17. $\begin{cases} r = a \\ r = a(1 - \cos \theta) \end{cases}$

18. $\begin{cases} r^2 = 4 \sin 2\theta \\ r = \sqrt{2} \end{cases}$

19. $\begin{cases} r = 2 \sin \theta \\ r = \sin \theta + \cos \theta \end{cases}$

20. $\begin{cases} r = 2a \sin \theta \\ r = a \end{cases}$

21. $\begin{cases} r = a(1 + \cos \theta) \\ r = 2a \cos \theta \end{cases}$

22. Determine the value of a for which the area of the region enclosed by the cardioid $r = a(1 - \cos \theta)$ is 9π square units.

23. The face of a bow tie is the region enclosed by the graph of the equation $r^2 = 4 \cos 2\theta$. How much material is necessary to cover the face of the tie?

24. Find the area of the region swept out by the radius vector of the spiral $r = a\theta$ during its second revolution that was not swept out during its first revolution.

25. Find the area of the region swept out by the radius vector of the curve of Exercise 24 during its third revolution that was not swept out during its second revolution.

10.8 A UNIFIED TREATMENT OF CONIC SECTIONS AND POLAR EQUATIONS OF CONICS

In Sections 10.1 through 10.3 we defined each of the three types of conic sections separately. An alternative approach is to start with a definition that gives a common property of conics and then introduce each of the conics as a special case of the general definition. We state this definition in the following theorem. The positive constant e in the statement of the theorem is called the **eccentricity** of the conic.

10.8.1 THEOREM

A conic section can be defined as the set of all points P in a plane such that the ratio of the undirected distance of P from a fixed point to the undirected distance of P from a fixed line that does not contain the fixed point is a positive constant e. Furthermore, if $e = 1$, the conic is a parabola; if $0 < e < 1$, it is an ellipse; and if $e > 1$, it is a hyperbola.

Proof If $e = 1$, we see by comparing Definition 10.1.1 and the statement of the theorem that the set is a parabola having the fixed point as its focus and the fixed line as its directrix.

Suppose now that $e \neq 1$. We first obtain a polar equation of the set of points described. Let F denote the fixed point and l denote the fixed line. We take the pole at F and the polar axis and its extension perpendicular to l. We first consider the situation when the line l is to the left of the point F. Let D be the point of intersection of l with the extension of the polar axis, and let d denote the undirected distance from F to l. Refer to Figure 1. Let $P(r, \theta)$ be any point in the set to the right of l and on the terminal side of the angle of measure θ. Draw perpendiculars PQ and PR to the polar axis and line l, respectively. The point P is in the set described if and only if

$$|\overline{FP}| = e|\overline{RP}| \tag{1}$$

Because P is to the right of l, $\overline{RP} > 0$; thus $|\overline{RP}| = \overline{RP}$. Furthermore, $|\overline{FP}| = r$ because $r > 0$. Thus from (1),

$$r = e(\overline{RP}) \tag{2}$$

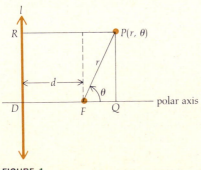

FIGURE 1

is the directrix. We now show that the point F is one of the foci when we have a central conic. Because (4) is an equation of a central conic for which the fixed point F is at the origin (pole) and (7) is obtained from (4) by translating the origin to the point $(e^2 d/(1 - e^2), 0)$, it follows that for the conic having Equation (7) the point F is at $(-e^2 d/(1 - e^2), 0)$. We wish to show that this point is a focus of the conic. From (6),

$$a = \begin{cases} \dfrac{ed}{1 - e^2} & \text{if } 0 < e < 1 \\[2mm] \dfrac{ed}{e^2 - 1} & \text{if } e > 1 \end{cases}$$

$$\Leftrightarrow \quad \frac{-e^2 d}{1 - e^2} = \begin{cases} -ae & \text{if } 0 < e < 1 \\ ae & \text{if } e > 1 \end{cases} \tag{12}$$

If (7) is an equation of an ellipse ($0 < e < 1$), we know that the foci are at $(-c, 0)$ and $(c, 0)$, where

$$c^2 = a^2 - b^2 \qquad c > 0$$

Substituting from (8) in this equation we get

$$c^2 = a^2 - a^2(1 - e^2)$$

$$c^2 = a^2 e^2$$

$$c = ae \tag{13}$$

By comparing (13) and (12) we conclude that if (7) is an equation of an ellipse, the point F is the left-hand focus.

If (7) is an equation of a hyperbola ($e > 1$), again the foci are at $(-c, 0)$ and $(c, 0)$, but for a hyperbola,

$$c^2 = a^2 + b^2 \qquad c > 0$$

Substituting from (9) in this equation we have

$$c^2 = a^2 + a^2(e^2 - 1)$$

$$c^2 = a^2 e^2$$

$$c = ae \tag{14}$$

By comparing (14) and (12) it follows that if (7) is an equation of a hyperbola, the point F is the right-hand focus.

The fixed line l mentioned in Theorem 10.8.1 is called the **directrix** corresponding to the focus at F. When the graph of (7) is an ellipse, the directrix corresponding to the focus at $(-c, 0)$ or, equivalently, $(-ae, 0)$ has the equation

$$x = -ae - d$$

Because when $0 < e < 1$, $d = a(1 - e^2)/e$, this equation becomes

$$x = -ae - \frac{a(1 - e^2)}{e}$$

$$x = -\frac{a}{e}$$

Similarly, if the graph of (7) is a hyperbola, the directrix corresponding to the focus at $(c, 0)$ or, equivalently, $(ae, 0)$ has the equation

$$x = ae - d$$

When $e > 1$, $d = a(e^2 - 1)/e$; thus the above equation of the directrix can be written as

$$x = \frac{a}{e}$$

Hence we have shown that if (7) is an equation of an ellipse, a focus and its corresponding directrix are $(-ae, 0)$ and $x = -a/e$; and if (7) is an equation of a hyperbola, a focus and its corresponding directrix are $(ae, 0)$ and $x = a/e$.

Because (7) contains only even powers of x and y, its graph is symmetric with respect to both the x and y axes. Therefore, if there is a focus at $(-ae, 0)$ having a corresponding directrix of $x = -a/e$, by symmetry there is also a focus at $(ae, 0)$ having a corresponding directrix of $x = a/e$. Similarly, for a focus at $(ae, 0)$ and a corresponding directrix of $x = a/e$, there is also a focus at $(-ae, 0)$ and a corresponding directrix of $x = -a/e$. These results are summarized in the following theorem.

10.8.2 THEOREM

> The central conic having the equation
>
> $$\frac{x^2}{a^2} + \frac{y^2}{a^2(1 - e^2)} = 1 \tag{15}$$
>
> where $a > 0$, has a focus at $(-ae, 0)$, whose corresponding directrix is $x = -a/e$, and a focus at $(ae, 0)$, whose corresponding directrix is $x = a/e$.

Figures 2 and 3 show sketches of the graph of (15) together with the foci and directrices in the respective cases of an ellipse and a hyperbola.

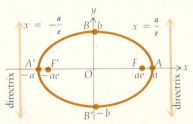

FIGURE 2   FIGURE 3

From both Equations (13) and (14)

$$e = \frac{c}{a} \tag{16}$$

Because, also $e = 2c/(2a)$, the eccentricity of both the ellipse and the hyperbola is the ratio of the undirected distance between the foci to the undirected distance between the vertices. Thus the eccentricity gives an indication of the shape of a central conic.

For an ellipse the eccentricity takes on values between 0 and 1. When the foci are close together, e is close to zero, and the shape of the ellipse is close to that of a circle. See Figure 4(a) showing an ellipse for which $e = 0.3$. If a remains fixed, then as e increases the "flatness" of the ellipse increases. Figures 4(b) and (c) show ellipses with eccentricities of 0.7 and 0.95, respectively, each with the same value of a as in Figure 4(a). The limiting forms of the ellipse are a circle of diameter $2a$ and a line segment of length $2a$.

For a hyperbola, the eccentricity is greater than 1. The eccentricity of an equilateral hyperbola is $\sqrt{2}$, which is obtained from (9) by taking $a = b$. See Figure 5(b). If e approaches 1 and a remains fixed, then c approaches a, b approaches 0, and the shape of the hyperbola becomes "thin" around its principal axis. In Figure 5(a) there is a hyperbola with $e = 1.05$ and the same value of a as in Figure 5(b). If e increases as a remains fixed, then c increases and b increases, and the hyperbola becomes "fat" around its principal axis. See Figure 5(c) for a hyperbola with $e = 2$ and the same value of a as in Figures 5(a) and (b).

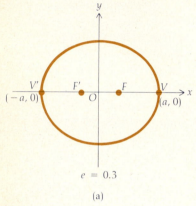

$e = 0.3$

(a)

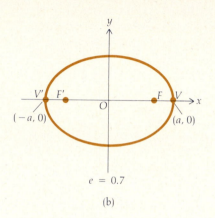

$e = 0.7$

(b)

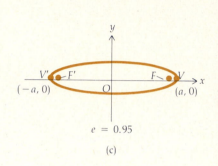

$e = 0.95$

(c)

FIGURE 4

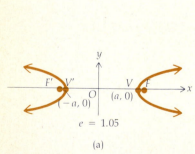

$e = 1.05$

(a)

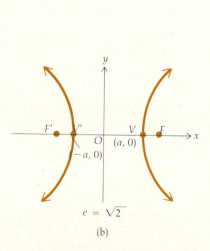

$e = \sqrt{2}$

(b)

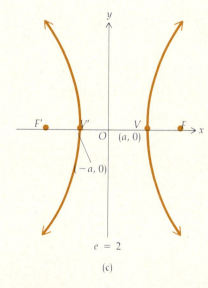

$e = 2$

(c)

FIGURE 5

EXAMPLE 1   The ellipse of Example 1 in Section 10.2 has the equation

$$\frac{x^2}{25} + \frac{y^2}{16} = 1$$

(a) Find the eccentricity and directrices of this ellipse. (b) Draw a sketch showing the ellipse, the directrices, and the foci. Also choose any three points P on the ellipse and draw the line segments whose lengths are the undirected distances from P to a focus and its corresponding directrix. Observe that the ratio of these distances is e.

Solution

(a) From the equation of the ellipse, $a = 5$ and $b = 4$. We showed in Example 1 of Section 10.2 that $c = 3$. Therefore, from (16), $e = \frac{3}{5}$. Because $a/e = 25/3$, it follows from Theorem 10.8.2 that the directrix corresponding to the focus at $(3, 0)$ has the equation $x = \frac{25}{3}$, and the directrix corresponding to the focus at $(-3, 0)$ has the equation $x = -\frac{25}{3}$.

(b) Figure 6 shows the ellipse, the directrices, and the foci, as well as three points P_1, P_2, and P_3 on the ellipse. For each of these points,

$$\frac{|\overline{FP}|}{|\overline{RP}|} = \frac{3}{5}$$

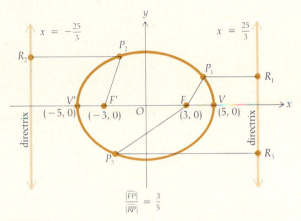

FIGURE 6

EXAMPLE 2   The hyperbola of Example 1 in Section 10.3 has the equation

$$\frac{x^2}{9} - \frac{y^2}{16} = 1$$

(a) Find the eccentricity and directrices of this hyperbola. (b) Draw a sketch showing the hyperbola, the directrices, and the foci. Also choose any three points P on the hyperbola and draw the line segments whose lengths are the undirected distances from P to a focus and its corresponding directrix. Observe that the ratio of these distances is e.

Solution

(a) From the equation of the hyperbola, $a = 3$, and $b = 4$. In Example 1 of Section 10.3 we showed that $c = 5$. From (16), $e = \frac{5}{3}$. Because $a/e = 9/5$, we conclude from Theorem 10.8.2 that the directrix corresponding to the focus at $(5, 0)$ has the equation $x = \frac{9}{5}$, and the directrix corresponding to the focus at $(-5, 0)$ has the equation $x = -\frac{9}{5}$.

(b) Figure 7 shows the hyperbola, the directrices, and the foci, as well as three points P_1, P_2, and P_3 on the hyperbola. For each of these points

$$\frac{|\overline{FP}|}{|\overline{RP}|} = \frac{5}{3}$$

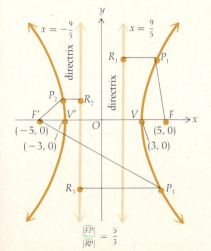

FIGURE 7

In the proof of Theorem 10.8.1 we learned that all three types of conics have polar equations of the same form. When a focus is at the pole and the corresponding directrix is either perpendicular or parallel to the polar axis, an equation of the conic has the form of (3), (10), or (11). Thus we have the following theorem.

10.8.3 THEOREM

Suppose we have a conic for which e and d are, respectively, the eccentricity and the undirected distance between the focus and the corresponding directrix.

(i) If a focus of the conic is at the pole and the corresponding directrix is perpendicular to the polar axis, then an equation of the conic is

$$r = \frac{ed}{1 \pm e \cos \theta} \tag{17}$$

where the plus sign is taken when the directrix corresponding to the focus at the pole is to the right of the focus and the minus sign is taken when it is to the left of the focus.

(ii) If a focus of the conic is at the pole and the corresponding directrix is parallel to the polar axis, then an equation of the conic is

$$r = \frac{ed}{1 \pm e \sin \theta} \tag{18}$$

where the plus sign is taken when the directrix corresponding to the focus at the pole is above the focus, and the minus sign is taken when it is below the focus.

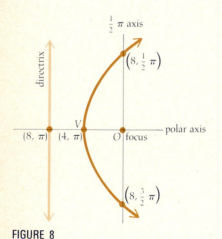

FIGURE 8

EXAMPLE 3   A parabola has its focus at the pole and its vertex at $(4, \pi)$. Find an equation of the parabola and an equation of the directrix. Draw a sketch of the parabola and the directrix.

Solution   Because the focus is at the pole and the vertex is at $(4, \pi)$, the polar axis and its extension are along the axis of the parabola. Furthermore, the vertex is to the left of the focus; so the directrix is also to the left of the focus. Hence an equation of the parabola is of the form of (17) with the minus sign. Because the vertex is at $(4, \pi)$, $\frac{1}{2}d = 4$; thus $d = 8$. The eccentricity $e = 1$, and therefore we obtain the equation

$$r = \frac{8}{1 - \cos \theta}$$

An equation of the directrix is given by $r \cos \theta = -d$, and because $d = 8$, then $r \cos \theta = -8$. Figure 8 shows a sketch of the parabola and the directrix.

EXAMPLE 4   An equation of a conic is

$$r = \frac{5}{3 + 2 \sin \theta}$$

Find the eccentricity, identify the conic, write an equation of the directrix corresponding to the focus at the pole, find the vertices, and draw a sketch of the curve.

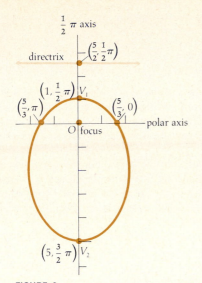

directrix

$\left(\frac{5}{2}, \frac{1}{2}\pi\right)$

$\left(1, \frac{1}{2}\pi\right) V_1$

$\left(\frac{5}{3}, \pi\right)$

$\left(\frac{5}{3}, 0\right)$

O focus — polar axis

$\frac{1}{2}\pi$ axis

$\left(5, \frac{3}{2}\pi\right) V_2$

FIGURE 9

Solution   Dividing the numerator and denominator of the fraction in the given equation by 3 we obtain

$$r = \frac{\frac{5}{3}}{1 + \frac{2}{3}\sin\theta}$$

which is of the form of (18) with the plus sign. The eccentricity $e = \frac{2}{3}$. Because $e < 1$, the conic is an ellipse. Because $ed = \frac{5}{3}$, $d = \frac{5}{3} \div \frac{2}{3}$; thus $d = \frac{5}{2}$. The $\frac{1}{2}\pi$ axis and its extension are along the principal axis. The directrix corresponding to the focus at the pole is above the focus, and an equation of it is $r\sin\theta = \frac{5}{2}$. When $\theta = \frac{1}{2}\pi$, $r = 1$; and when $\theta = \frac{3}{2}\pi$, $r = 5$. The vertices are therefore at $(1, \frac{1}{2}\pi)$ and $(5, \frac{3}{2}\pi)$. A sketch of the ellipse appears in Figure 9.

EXAMPLE 5   The polar axis and its extension are along the principal axis of a hyperbola having a focus at the pole. The corresponding directrix is to the left of the focus. If the hyperbola contains the point $(1, \frac{2}{3}\pi)$ and $e = 2$, find (a) an equation of the hyperbola; (b) the vertices; (c) the center; (d) an equation of the directrix corresponding to the focus at the pole. (e) Draw a sketch of the hyperbola.

Solution   An equation of the hyperbola is of the form of (17) with the minus sign, where $e = 2$. We have, then,

$$r = \frac{2d}{1 - 2\cos\theta}$$

(a) Because the point $(1, \frac{2}{3}\pi)$ lies on the hyperbola, its coordinates satisfy the equation. Therefore

$$1 = \frac{2d}{1 - 2(-\frac{1}{2})}$$

from which we obtain $d = 1$. Hence an equation of the hyperbola is

$$r = \frac{2}{1 - 2\cos\theta} \qquad (19)$$

(b) The vertices are the points on the hyperbola for which $\theta = 0$ and $\theta = \pi$. From (19), when $\theta = 0$, $r = -2$; and when $\theta = \pi$, $r = \frac{2}{3}$. Consequently, the left vertex V_1 is at the point $(-2, 0)$, and the right vertex V_2 is at the point $(\frac{2}{3}, \pi)$.
(c) The center of the hyperbola is the point on the principal axis halfway between the two vertices. This is the point $(\frac{4}{3}, \pi)$.
(d) An equation of the directrix corresponding to the focus at the pole is given by $r\cos\theta = -d$. Because $d = 1$, this equation is $r\cos\theta = -1$.
(e) As an aid in drawing a sketch of the hyperbola, we first draw the two asymptotes. These are lines through the center of the hyperbola that are parallel to the lines $\theta = \theta_1$ and $\theta = \theta_2$, where θ_1 and θ_2 are the values of θ in the interval $[0, 2\pi)$ for which r is not defined. From (19), r is not defined when $1 - 2\cos\theta = 0$. Therefore $\theta_1 = \frac{1}{3}\pi$ and $\theta_2 = \frac{5}{3}\pi$. Figure 10 shows a sketch of the hyperbola, as well as the two asymptotes and the directrix corresponding to the focus at the pole.

$\theta = \frac{5}{3}\pi$

$\theta = \frac{1}{3}\pi$

$\frac{1}{2}\pi$ axis

directrix

$\left(2, \frac{1}{2}\pi\right)$

V_1 C V_2 O — polar axis

$(-2, 0)$

focus

$\left(\frac{2}{3}, \pi\right)$

$\left(2, \frac{3}{2}\pi\right)$

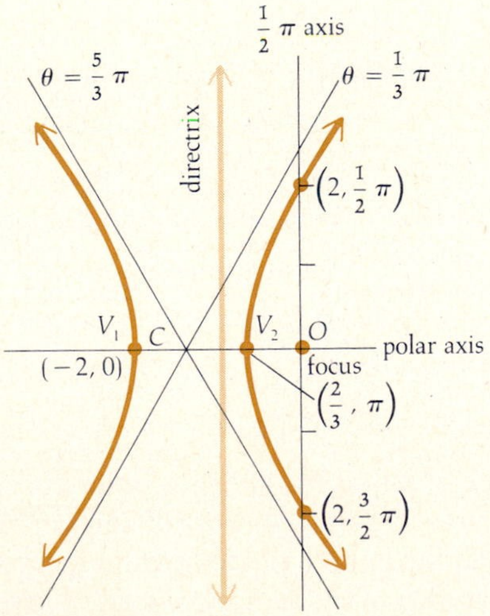

FIGURE 10

EXERCISES 10.8

In Exercises 1 through 8, (a) find the eccentricity, foci, and directrices of the central conic. (b) Draw a sketch showing the conic, the foci, and the directrices. Also choose any four points P (one in each quadrant) on the conic and draw the line segments whose lengths are the undirected distances from P to a focus and its corresponding directrix. Observe that the ratio of these distances is e.

1. $4x^2 + 9y^2 = 36$ 2. $4x^2 + 9y^2 = 4$
3. $25x^2 + 4y^2 = 100$ 4. $16x^2 + 9y^2 = 144$
5. $4x^2 - 25y^2 = 100$ 6. $x^2 - 9y^2 = 9$
7. $16x^2 - 9y^2 = 144$ 8. $4y^2 - x^2 = 16$

In Exercises 9 and 10, the polar equation represents a conic having a focus at the pole. Identify the conic.

9. (a) $r = \dfrac{3}{1 - \cos\theta}$;   (b) $r = \dfrac{6}{4 + 5\sin\theta}$;   (c) $r = \dfrac{5}{4 - \cos\theta}$;

     (d) $r = \dfrac{4}{1 + \sin\pi}$

10. (a) $r = \dfrac{1}{1 - \sin\theta}$;   (b) $r = \dfrac{2}{3 + \sin\theta}$;   (c) $r = \dfrac{3}{2 + 4\cos\theta}$;

     (d) $r = \dfrac{5}{1 - \cos\pi}$

In Exercises 11 through 22, the equation is that of a conic having a focus at the pole. (a) Find the eccentricity, (b) identify the conic, (c) write an equation of the directrix that corresponds to the focus at the pole, (d) draw a sketch of the curve.

11. $r = \dfrac{2}{1 - \cos\theta}$ 12. $r = \dfrac{4}{1 + \cos\theta}$ 13. $r = \dfrac{5}{2 + \sin\theta}$

14. $r = \dfrac{4}{1 - 3\cos\theta}$ 15. $r = \dfrac{6}{3 - 2\cos\theta}$ 16. $r = \dfrac{1}{2 + \sin\theta}$

17. $r = \dfrac{9}{5 - 6\sin\theta}$ 18. $r = \dfrac{1}{1 - 2\sin\theta}$ 19. $r = \dfrac{10}{7 - 2\sin\theta}$

20. $r = \dfrac{7}{3 + 4\cos\theta}$ 21. $r = \dfrac{10}{4 + 5\cos\theta}$ 22. $r = \dfrac{1}{5 - 3\sin\theta}$

In Exercises 23 through 28, find a polar equation of the conic having a focus at the pole and satisfying the given conditions.

23. Parabola; vertex at $(4, \frac{3}{2}\pi)$.
24. Ellipse; $e = \frac{1}{2}$; a vertex at $(4, \pi)$.
25. Hyperbola; $e = \frac{4}{3}$; $r\cos\theta = 9$ is the directrix corresponding to the focus at the pole.
26. Hyperbola; vertices at $(1, \frac{1}{2}\pi)$ and $(3, \frac{1}{2}\pi)$.

27. Ellipse; vertices at $(3, 0)$ and $(1, \pi)$.
28. Parabola; vertex at $(6, \frac{1}{2}\pi)$.
29. (a) Find a polar equation of the hyperbola having a focus at the pole and the corresponding directrix to the left of the focus if the point $(2, \frac{4}{3}\pi)$ is on the hyperbola and $e = 3$. (b) Write an equation of the directrix that corresponds to the focus at the pole.
30. (a) Find a polar equation of the hyperbola for which $e = 3$ and which has the line $r\sin\theta = 3$ as the directrix corresponding to a focus at the pole. (b) Find the polar equations of the two lines through the pole that are parallel to the asymptotes of the hyperbola.
31. Find the area of the region inside the ellipse $r = 6/(2 - \sin\theta)$ and above the parabola $r = 3/(1 + \sin\theta)$.
32. For the ellipse and parabola of Exercise 31, find the area of the region inside the ellipse and below the parabola.
33. Show that an equation of a conic having its principal axis along the polar axis and its extension, a focus at the pole, and the corresponding directrix to the right of the focus is $r = ed/(1 + e\cos\theta)$.
34. Show that an equation of a conic having its principal axis along the $\frac{1}{2}\pi$ axis and its extension, a focus at the pole, and the corresponding directrix above the focus is $r = ed/(1 + e\sin\theta)$.
35. Show that an equation of a conic having its principal axis along the $\frac{1}{2}\pi$ axis and its extension, a focus at the pole, and the corresponding directrix below the focus is $r = ed/(1 - e\sin\theta)$.
36. Show that the equation $r = k\csc^2\frac{1}{2}\theta$, where k is a constant, is a polar equation of a parabola.
37. A comet is moving in a parabolic orbit around the sun at the focus of the parabola. When the comet is 80 million miles from the sun, the line segment from the sun to the comet makes an angle of $\frac{1}{3}\pi$ radians with the axis of the orbit. (a) Find an equation of the comet's orbit. (b) How close does the comet come to the sun?
38. The orbit of a planet is in the form of an ellipse having the equation $r = p/(1 + e\cos\theta)$ where the pole is at the sun. Find the average measure of the distance of the planet from the sun with respect to θ.
39. Use Theorem 10.8.1 to find a polar equation of a central conic for which the center is at the pole, the principal axis is along the polar axis and its extension, and the distance from the pole to a directrix is a/e.
40. Show that the tangent lines at the points of intersection of the parabolas $r = a/(1 + \cos\theta)$ and $r = b/(1 - \cos\theta)$ are perpendicular.

10.9   TANGENT LINES OF POLAR CURVES
(Supplementary)

We first derive a formula for finding the slope of a tangent line to a polar curve at a point (r, θ) on the curve. Let $r = f(\theta)$ be a polar equation of the curve. Consider a rectangular cartesian coordinate system and a polar coordinate system in the same plane and having the positive side of the x axis coincident with the polar axis. In Section 10.5 we learned that the two sets of coordinates are

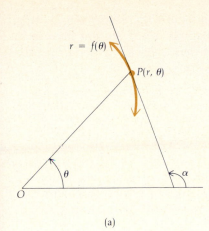

(a)

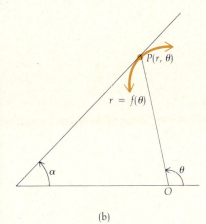

(b)

FIGURE 1

related by the equations

$$x = r \cos \theta \quad \text{and} \quad y = r \sin \theta$$

The variables x and y can be considered as functions of θ because $r = f(\theta)$. If we differentiate with respect to θ on both sides of these equations, we get, by applying the chain rule,

$$\frac{dx}{d\theta} = \cos \theta \frac{dr}{d\theta} - r \sin \theta \tag{1}$$

and

$$\frac{dy}{d\theta} = \sin \theta \frac{dr}{d\theta} + r \cos \theta \tag{2}$$

Let α be the radian measure of the inclination of the tangent line to the curve at (r, θ). See Figure 1(a) and (b), where in (a) $\alpha \geq \theta$ and in (b) $\alpha < \theta$. In either case

$$\tan \alpha = \frac{dy}{dx}$$

and if $\dfrac{dx}{d\theta} \neq 0$,

$$\frac{dy}{dx} = \frac{\dfrac{dy}{d\theta}}{\dfrac{dx}{d\theta}}$$

Substituting from (1) and (2) into this equation we obtain

$$\frac{dy}{dx} = \frac{\sin \theta \dfrac{dr}{d\theta} + r \cos \theta}{\cos \theta \dfrac{dr}{d\theta} - r \sin \theta} \tag{3}$$

If $\cos \theta \neq 0$, we divide the numerator and the denominator of the right side of (3) by $\cos \theta$ and replace $\dfrac{dy}{dx}$ by $\tan \alpha$, thereby giving

$$\tan \alpha = \frac{\tan \theta \dfrac{dr}{d\theta} + r}{\dfrac{dr}{d\theta} - r \tan \theta} \tag{4}$$

If in (3), $\cos \theta = 0$, we obtain

$$\tan \alpha = -\frac{1}{r} \frac{dr}{d\theta} \tag{5}$$

which holds for points on the $\frac{1}{2}\pi$ axis or its extension.

Observe from (4) that if $r = 0$ and $\dfrac{dr}{d\theta} \neq 0$,

$$\tan \alpha = \tan \theta$$

From this equation we conclude that the values of θ that satisfy a polar equation of the curve when $r = 0$ are the inclinations of the tangent lines to the curve at the pole. Thus if $\theta_1, \theta_2, \ldots, \theta_k$ are these values of θ, then equations of the tangent lines to the curve at the pole are

$$\theta = \theta_1, \ \theta = \theta_2, \ldots, \theta = \theta_k$$

In Section 10.6 we used this information as an aid in drawing sketches of graphs of polar curves.

EXAMPLE 1   Find the slope of the tangent line to the curve in Example 3(c) of Section 10.6 at the point where (a) $\theta = 0$ and (b) $\theta = \frac{5}{6}\pi$.

Solution   An equation of the curve is

$$r = 2 - \sin \theta \tag{6}$$

Thus

$$\frac{dr}{d\theta} = -\cos \theta$$

If α is the radian measure of the angle of inclination of the tangent line at (r, θ), then from (4),

$$\tan \alpha = \frac{\dfrac{\sin \theta}{\cos \theta}(-\cos \theta) + (2 - \sin \theta)}{-\cos \theta - (2 - \sin \theta)\dfrac{\sin \theta}{\cos \theta}}$$

$$= \frac{-\sin \theta \cos \theta + 2\cos \theta - \sin \theta \cos \theta}{-\cos^2 \theta - 2 \sin \theta + \sin^2 \theta}$$

$$= \frac{-2 \sin \theta \cos \theta + 2 \cos \theta}{-1 + \sin^2 \theta - 2 \sin \theta + \sin^2 \theta}$$

$$= \frac{2 \cos \theta(1 - \sin \theta)}{2 \sin^2 \theta - 2 \sin \theta - 1} \tag{7}$$

(a) At $\theta = 0$, $r = 2$, and from (7),

$$\tan \alpha = \frac{2(1)(1 - 0)}{2(0) - 2(0) - 1}$$

$$= -2$$

(b) At $\theta = \frac{5}{6}\pi$, $r = \frac{3}{2}$, and

$$\tan \alpha = \frac{2\left(-\dfrac{\sqrt{3}}{2}\right)\left(1 - \dfrac{1}{2}\right)}{2(\frac{1}{4}) - 2(\frac{1}{2}) - 1}$$

$$= \frac{-\sqrt{3}(\frac{1}{2})}{\frac{1}{2} - 1 - 1}$$

$$= \frac{\sqrt{3}}{3}$$

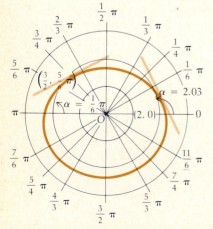

FIGURE 2

Figure 2 shows a sketch of the graph of (6) and the tangent lines at $\theta = 0$ and $\theta = \frac{5}{6}\pi$. At $\theta = 0$, because $\tan \alpha = -2$, $\alpha \approx 2.03$; at $\theta = \frac{5}{6}\pi$, because $\tan \alpha = \frac{1}{3}\sqrt{3}$, $\alpha = \frac{1}{6}\pi$.

Formula (4) can be used to determine where a polar curve has horizontal and vertical tangent lines. This information is helpful when sketching graphs of polar equations. The procedure is shown in the following illustration and next example applied to two of the limaçons in Example 3 of Section 10.6.

▶ **ILLUSTRATION 1**   The curve in Example 3(c) of Section 10.6 and Example 1 of this section has the equation

$$r = 2 - \sin \theta$$

The horizontal tangent lines of this curve occur when $\tan \alpha = 0$. Thus we equate the numerator of (7) to zero and solve for θ.

$$2 \cos \theta(1 - \sin \theta) = 0$$

$$\cos \theta = 0 \qquad\qquad 1 - \sin \theta = 0$$

$$\theta = \tfrac{1}{2}\pi \qquad \theta = \tfrac{3}{2}\pi \qquad\qquad \sin \theta = 1$$

$$\theta = \tfrac{1}{2}\pi$$

Therefore the curve has a horizontal tangent line at the points $(1, \tfrac{1}{2}\pi)$ and $(3, \tfrac{3}{2}\pi)$. The vertical tangent lines occur when the denominator of (7) is zero and the numerator is not zero. We solve the resulting equation.

$$2 \sin^2 \theta - 2 \sin \theta - 1 = 0$$

$$\sin \theta = \frac{-b \pm \sqrt{b^2 - 4ac}}{2a}$$

$$= \frac{2 \pm \sqrt{4 + 8}}{4}$$

$$= \frac{2 \pm 2\sqrt{3}}{4}$$

$$= \frac{1 \pm \sqrt{3}}{2}$$

$$\sin \theta = 1.3660 \qquad\qquad \sin \theta = -0.3660$$

$$\text{no solution} \qquad\qquad \theta \approx 3.51 \qquad \theta \approx 5.91$$

Hence the curve has a vertical tangent line at the points having polar coordinates (2.37, 3.51) and (2.37, 5.91). Figure 3 shows a sketch of the graph of the curve and the horizontal and vertical tangent lines.   ◀

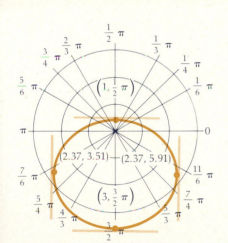

FIGURE 3

EXAMPLE 2   Determine the points at which the curve in Example 3(a) of Section 10.6 has horizontal and vertical tangent lines. Draw a sketch of the graph and show these tangent lines.

Solution   The curve has the equation

$$r = 3 + 2 \sin \theta$$

Therefore

$$\frac{dr}{d\theta} = 2 \cos \theta$$

If α is the radian measure of the angle of inclination of the tangent line at (r, θ), we have, from (4),

$$\tan \alpha = \frac{\dfrac{\sin \theta}{\cos \theta}(2 \cos \theta) + (3 + 2 \sin \theta)}{2 \cos \theta - (3 + 2 \sin \theta)\dfrac{\sin \theta}{\cos \theta}}$$

$$= \frac{2 \sin \theta \cos \theta + 3 \cos \theta + 2 \sin \theta \cos \theta}{2 \cos^2 \theta - 3 \sin \theta - 2 \sin^2 \theta}$$

$$= \frac{4 \sin \theta \cos \theta + 3 \cos \theta}{2 - 2 \sin^2 \theta - 3 \sin \theta - 2 \sin^2 \theta}$$

$$= -\frac{\cos \theta(4 \sin \theta + 3)}{4 \sin^2 \theta + 3 \sin \theta - 2} \tag{8}$$

To determine the horizontal tangent lines we equate the numerator of (8) to zero.

$$\cos \theta(4 \sin \theta + 3) = 0$$

$$\cos \theta = 0 \qquad\qquad 4 \sin \theta + 3 = 0$$

$$\theta = \tfrac{1}{2}\pi \qquad \theta = \tfrac{3}{2}\pi \qquad\qquad \sin \theta = -\tfrac{3}{4}$$

$$\theta \approx 3.99 \qquad \theta \approx 5.44$$

Thus the curve has a horizontal tangent line at the points $(5, \tfrac{1}{2}\pi)$, $(1, \tfrac{3}{2}\pi)$, $(\tfrac{3}{2}, 3.99)$ and $(\tfrac{3}{2}, 5.44)$.

The vertical tangent lines are found by equating the denominator of (8) to zero.

$$4 \sin^2 \theta + 3 \sin \theta - 2 = 0$$

$$\sin \theta = \frac{-b \pm \sqrt{b^2 - 4ac}}{2a}$$

$$= \frac{-3 \pm \sqrt{41}}{8}$$

$$\sin \theta = 0.4254 \qquad\qquad \sin \theta = -1.1754$$

$$\theta \approx 0.44 \qquad \theta \approx 2.70 \qquad \text{no solution}$$

Therefore the curve has a vertical tangent line at the points with polar coordinates (3.85, 0.44) and (3.85, 2.70).

Refer to Figure 4, which shows a sketch of the graph of the curve and the horizontal and vertical tangent lines.

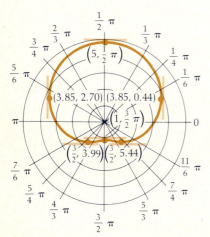

FIGURE 4

Observe from Figure 4 that the limaçon has a dent. This fact is apparent because of the four horizontal tangent lines. The limaçon of Figure 3 has no dent; it has only two horizontal tangent lines.

Formula (4) is generally complicated to apply. A simpler formula is obtained by considering the angle between the line OP and the tangent line. This angle will have radian measure χ and is measured from the line OP counterclockwise to the tangent line, $0 \leq \chi < \pi$.

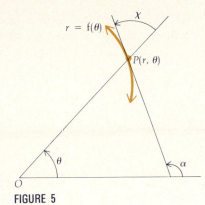

FIGURE 5

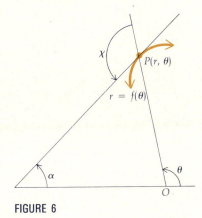

FIGURE 6

There are two possible cases: $\alpha \geq \theta$ and $\alpha < \theta$. These two cases are illustrated in Figures 5 and 6. In Figure 5, $\alpha > \theta$ and $\chi = \alpha - \theta$. In Figure 6, $\alpha < \theta$ and $\chi = \pi - (\theta - \alpha)$. In each case

$$\tan \chi = \tan(\alpha - \theta)$$

$$\tan \chi = \frac{\tan \alpha - \tan \theta}{1 + \tan \alpha \tan \theta}$$

Substituting the value of $\tan \alpha$ from (4) into this equation we get

$$\tan \chi = \frac{\dfrac{\tan \theta \dfrac{dr}{d\theta} + r}{\dfrac{dr}{d\theta} - r \tan \theta} - \tan \theta}{1 + \left(\dfrac{\tan \theta \dfrac{dr}{d\theta} + r}{\dfrac{dr}{d\theta} - r \tan \theta}\right) \tan \theta}$$

$$= \frac{\tan \theta \dfrac{dr}{d\theta} + r - \tan \theta \dfrac{dr}{d\theta} + r \tan^2 \theta}{\dfrac{dr}{d\theta} - r \tan \theta + \tan^2 \theta \dfrac{dr}{d\theta} + r \tan \theta}$$

$$= \frac{r(1 + \tan^2 \theta)}{(1 + \tan^2 \theta) \dfrac{dr}{d\theta}}$$

$$= \frac{r}{\dfrac{dr}{d\theta}}$$

We have proved the following theorem.

10.9.1 THEOREM Let χ be the radian measure of the angle between the line OP and the tangent line to the graph of $r = f(\theta)$ at the point $P(r, \theta)$, where P is not on the $\frac{1}{2}\pi$ axis or its extension. The angle χ is measured from OP counterclockwise to the tangent line, and $0 \leq \chi < \pi$. Then

$$\tan \chi = \frac{r}{\dfrac{dr}{d\theta}} \tag{9}$$

Comparing the equation in Theorem 10.9.1 with (4) you can see why it is more desirable to consider χ instead of α when working with polar coordinates. Theorem 10.9.1 cannot be used if P is on the $\frac{1}{2}\pi$ axis or its extension. In that case use (5) to compute $\tan \alpha$.

EXAMPLE 3   Given the cardioid having the equation

$$r = 2 + 2 \sin \theta$$

Find χ and α at the points where (a) $\theta = \frac{1}{6}\pi$, (b) $\theta = \frac{1}{3}\pi$, and (c) $\theta = \frac{2}{3}\pi$. (d) Draw a sketch of the cardioid, and show χ and α at each of these points.

Solution   From the given equation

$$\frac{dr}{d\theta} = 2 \cos \theta$$

From Theorem 10.9.1

$$\tan \chi = \frac{2 + 2 \sin \theta}{2 \cos \theta}$$

$$= \frac{1 + \sin \theta}{\cos \theta} \tag{9}$$

(a) When $\theta = \frac{1}{6}\pi$, from (9) we get

$$\tan \chi = \frac{1 + \frac{1}{2}}{\frac{\sqrt{3}}{2}}$$

$$= \sqrt{3}$$

Thus $\chi = \frac{1}{3}\pi$. Because $\alpha > \theta$, $\chi = \alpha - \theta$. Therefore

$$\alpha = \frac{1}{6}\pi + \frac{1}{3}\pi$$

$$= \frac{1}{2}\pi$$

(b) When $\theta = \frac{1}{3}\pi$, from (9)

$$\tan \chi = \frac{1 + \frac{\sqrt{3}}{2}}{\frac{1}{2}}$$

$$= 2 + \sqrt{3}$$

Therefore $\chi \approx 1.31$. Because $\alpha > \theta$, $\chi = \alpha - \theta$. Thus

$$\alpha \approx \frac{1}{3}\pi + 1.31$$

$$\approx 2.36$$

(c) When $\theta = \frac{2}{3}\pi$, from (9)

$$\tan \chi = \frac{1 + \frac{\sqrt{3}}{2}}{-\frac{1}{2}}$$

$$= -2 - \sqrt{3}$$

Hence $\chi \approx 1.83$. Because $\alpha < \theta$, $\chi = \pi - (\theta - \alpha)$. Therefore

$$\alpha \approx (\tfrac{2}{3}\pi + 1.83) - \pi$$

$$= 1.83 - \tfrac{1}{3}\pi$$

$$\approx 0.78$$

(d) Figure 7 shows a sketch of the cardioid and χ and α at each of the points $(3, \frac{1}{6}\pi)$, $(2 + \sqrt{3}, \frac{1}{3}\pi)$ and $(2 + \sqrt{3}, \frac{2}{3}\pi)$.

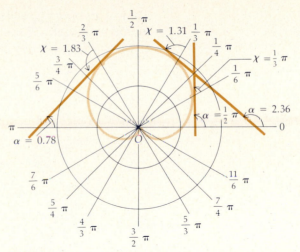

FIGURE 7

EXAMPLE 4   Find to the nearest 10 min the measurement of the smaller angle between the tangent lines to the curves

$$r = 3\cos 2\theta \quad \text{and} \quad r = 3\sin 2\theta$$

at the point $P(\frac{3}{2}\sqrt{2}, \frac{1}{8}\pi)$.

Solution   The graph of each equation is a four-leafed rose. Refer to Figure 8, which shows a portion of each of the curves at the point P; χ_1 is the radian measure of the angle between the line OP and the tangent line to the curve $r = 3\cos 2\theta$, and χ_2 is the radian measure of the angle between the line OP and the tangent line to the curve $r = 3\sin 2\theta$. If β is the angle between the tangent lines at P, then

$$\beta = \chi_1 - \chi_2$$

$$\tan \beta = \tan(\chi_1 - \chi_2)$$

$$\tan \beta = \frac{\tan \chi_1 - \tan \chi_2}{1 + \tan \chi_1 \tan \chi_2} \tag{10}$$

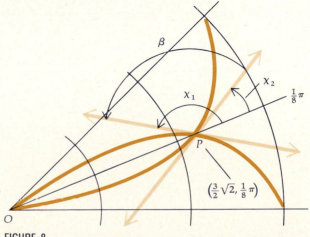

FIGURE 8

The values of $\tan \chi_1$ and $\tan \chi_2$ are found from Theorem 11.9.1. When computing $\tan \chi_1$, $r = 3 \cos 2\theta$ and $dr/d\theta = -6 \sin 2\theta$. For $\tan \chi_2$ we have $r = 3 \sin 2\theta$ and $dr/d\theta = 6 \cos 2\theta$. Thus

$$\tan \chi_1 = \frac{3 \cos 2\theta}{-6 \sin 2\theta} \qquad \tan \chi_2 = \frac{3 \sin 2\theta}{6 \cos 2\theta}$$

$$= -\tfrac{1}{2} \cot 2\theta \qquad\qquad = \tfrac{1}{2} \tan 2\theta$$

When $\theta = \tfrac{1}{8}\pi$,

$$\tan \chi_1 = -\tfrac{1}{2} \cot \tfrac{1}{4}\pi \qquad \tan \chi_2 = \tfrac{1}{2} \tan \tfrac{1}{4}\pi$$

$$= -\tfrac{1}{2} \qquad\qquad\qquad = \tfrac{1}{2}$$

Substituting these values in (10) we obtain

$$\tan \beta = \frac{-\tfrac{1}{2} - \tfrac{1}{2}}{1 + (-\tfrac{1}{2})(\tfrac{1}{2})}$$

$$= -\tfrac{4}{3}$$

The angle β has a measurement of $126°50'$ to the nearest 10 min. So, to the nearest 10 min, the measurement of the smaller angle between the two tangent lines is $180° - 126°50' = 53°10'$.

EXERCISES 10.9

In Exercises 1 through 8, find the slope of the tangent line to the curve at the point having the indicated value of θ. Also draw a sketch of the graph and the tangent line.

1. $r = 2 \sin \theta$; $\theta = \tfrac{1}{6}\pi$ **2.** $r = 4 \cos \theta$; $\theta = \tfrac{1}{3}\pi$
3. $r = 1 - \cos \theta$; $\theta = \tfrac{1}{4}\pi$ **4.** $r = 2 + 2 \sin \theta$; $\theta = \tfrac{1}{4}\pi$
5. $r = 6 + 2 \sin \theta$; $\theta = \tfrac{5}{6}\pi$ **6.** $r = 3 - 2 \cos \theta$; $\theta = \tfrac{2}{3}\pi$
7. $r = 3 \cos 2\theta$; $\theta = \tfrac{2}{3}\pi$ **8.** $r = 2 \sin 3\theta$; $\theta = \tfrac{5}{6}\pi$

In Exercises 9 through 20, determine the points at which the curve has horizontal and vertical tangent lines. Draw a sketch of the graph and show these tangent lines.

9. $r = 4 + 3 \sin \theta$ **10.** $r = 2 + \cos \theta$
11. $r = 4 - 2 \cos \theta$ **12.** $r = 3 - 2 \sin \theta$
13. $r = 1 - \sin \theta$ **14.** $r = 3 - 3 \cos \theta$
15. $r = 2 + 3 \cos \theta$ **16.** $r = 1 - 2 \sin \theta$
17. $r = \cos 2\theta$ **18.** $r = 2 \sin 3\theta$
19. $r^2 = 4 \sin 2\theta$ **20.** $r^2 = 9 \cos 2\theta$

In Exercises 21 through 28, find χ and α at the point having the indicated value of θ. Draw a sketch of the curve and show χ and α.

21. $r\theta = 4$; $\theta = 1$ **22.** $r = 2\theta$; $\theta = \tfrac{5}{2}\pi$
23. $r = \sec 2\theta$; $\theta = -\tfrac{1}{8}\pi$ **24.** $r = 4 \sin \tfrac{1}{2}\theta$; $\theta = \tfrac{1}{3}\pi$
25. $r = \theta^2$; $\theta = \tfrac{1}{2}\pi$ **26.** $r = \cos 2\theta$; $\theta = \tfrac{1}{12}\pi$
27. $r = 3\sqrt{\cos 2\theta}$; $\theta = \tfrac{1}{6}\pi$ **28.** $r = 2(1 - \sin \theta)$; $\theta = \pi$

In Exercises 29 through 32, find the measurement, in degrees, of the smaller angle between the tangent lines of the pair of curves at the indicated point of intersection.

29. $\begin{cases} r = 2 \cos \theta \\ r = 2 \sin \theta \end{cases}$; $(\sqrt{2}, \tfrac{1}{4}\pi)$ **30.** $\begin{cases} r = 3 \\ r = 6 \sin \theta \end{cases}$; $(3, \tfrac{1}{6}\pi)$

31. $\begin{cases} r = 4 \cos \theta \\ r = 4 \cos^2 \theta - 3 \end{cases}$; $(-2, \tfrac{2}{3}\pi)$ **32.** $\begin{cases} r = -4 \sin \theta \\ r = 4 \cos 2\theta \end{cases}$; the pole

In Exercises 33 through 36, find the measurement, in degrees, of the angle between the tangent lines of the pair of curves at all points of intersection.

33. $\begin{cases} r = 1 - \sin \theta \\ r = 1 + \sin \theta \end{cases}$ **34.** $\begin{cases} r = 2 \sec \theta \\ r = \csc^2 \tfrac{1}{2}\theta \end{cases}$

35. $\begin{cases} r = \cos \theta \\ r = \sin 2\theta \end{cases}$ **36.** $\begin{cases} r = 3 \cos \theta \\ r = 1 + \cos \theta \end{cases}$

37. Given the curve $r = k$, where k is a nonzero constant and α is the radian measure of the inclination of the tangent line to the curve at the point (r, θ). Show that $\tan \alpha \tan \theta = -1$ for all θ for which $\tan \theta \neq 0$.

38. Prove that at each point of the logarithmic spiral $r = be^{a\theta}$, χ is the same.

39. Prove that $\tan \chi = \tan \tfrac{1}{2}\theta$ at all points of the cardioid $r = 2(1 - \cos \theta)$.

40. For the curve $r \cos \theta = 4$ find $\lim\limits_{\theta \to \pi/2} \chi$.

41. Prove that at the points of intersection of the cardioids $r = a(1 + \sin \theta)$ and $r = b(1 - \sin \theta)$ their tangent lines are perpendicular for all values of a and b.

42. Prove that at the points of intersection of the two curves $r = a \sec^2 \frac{1}{2}\theta$ and $r = b \csc^2 \frac{1}{2}\theta$ their tangent lines are perpendicular.

REVIEW EXERCISES FOR CHAPTER 10

In Exercises 1 through 4, (a) find an equation of the parabola having the given focus and directrix, (b) find the length of the latus rectum, and (c) draw a sketch of the parabola.

1. Focus at $(0, -3)$; directrix, $y = 3$
2. Focus at $(0, 4)$; directrix, $y = -4$
3. Focus at $(1, 0)$; directrix, $x = -1$
4. Focus at $(6, 0)$; directrix, $x = -6$

5. Draw a sketch of the parabola having the equation $x^2 = 6y$, and find (a) the focus and (b) an equation of the directrix.
6. Draw a sketch of the parabola having the equation $y = x^2 - 8x$. Find (a) the vertex, (b) an equation of the axis, (c) the focus, and (d) an equation of the directrix.
7. Given the parabola having the equation $y^2 + 8x - 6y = 7$, find (a) the vertex, (b) an equation of the axis, (c) the focus, and (d) an equation of the directrix. Draw a sketch of the curve.
8. Find an equation of the parabola having its vertex at $(-3, 5)$ and its focus at $(-3, -1)$.
9. Find an equation of the parabola having vertex at $(5, 1)$, axis parallel to the y axis, and through the point $(9, 3)$.
10. Show that any equation of the form $xy + ax + by + c = 0$ can always be written in the form $x'y' = k$ by a translation of the axes, and determine the value of k.
11. Any section of a parabolic mirror made by passing a plane through the axis of the mirror is a segment of a parabola. The altitude of the segment is 12 cm and the length of the base is 18 cm. A section of the mirror made by a plane perpendicular to its axis is a circle. Find the circumference of the circular plane section if the plane perpendicular to the axis is 3 cm from the vertex.
12. The directrix of the parabola $y^2 = 4px$ is tangent to a circle having the focus of the parabola as its center. Find an equation of the circle and the points of intersection of the two curves.

In Exercises 13 through 20, the graph of the equation is either an ellipse or a hyperbola. Find the center, vertices, and foci of the conic. If the graph is an ellipse, also find the extremities of the minor axis. If the graph is a hyperbola, also find equations of the asymptotes. Draw a sketch of the curve and show the foci. If the curve is a hyperbola, also draw the asymptotes.

13. $4x^2 + 25y^2 = 100$
14. $25x^2 - 4y^2 = 100$
15. $16x^2 - 9y^2 = 144$
16. $9x^2 + 16y^2 = 144$
17. $4x^2 + y^2 + 24x - 16y + 84 = 0$
18. $4x^2 + 9y^2 + 32x - 18y + 37 = 0$
19. $25x^2 - y^2 + 50x + 6y - 9 = 0$
20. $3x^2 - 2y^2 + 6x - 8y + 11 = 0$

In Exercises 21 and 22, find an equation of the ellipse having the given properties and draw a sketch of the graph.

21. Vertices at $(0, 8)$ and $(0, -4)$ and one focus at $(0, 6)$.
22. Foci at $(-3, -2)$ and $(-3, 6)$ and the length of the semiminor axis is three-fourths the length of the semimajor axis.

In Exercises 23 and 24, find an equation of the hyperbola having the given properties, and draw a sketch of the graph.

23. Foci at $(-5, 1)$ and $(1, 1)$ and one vertex at $(-4, 1)$.
24. Center at $(-5, 2)$, a vertex at $(-1, 2)$, and a focus at $(0, 2)$.

In Exercises 25 through 28, find a polar equation of the graph having the cartesian equation.

25. $4x^2 - 9y^2 = 36$
26. $2xy = 1$
27. $x^2 + y^2 - 9x + 8y = 0$
28. $y^4 = x^2(a^2 - y^2)$

In Exercises 29 through 32, find a cartesian equation of the graph having the polar equation.

29. $r = 9 \sin^2 \frac{1}{2}\theta$
30. $r(1 - \cos \theta) = 2$
31. $r = 4 \cos 3\theta$
32. $r = a \tan^2 \theta$

33. Draw a sketch of the graph of (a) $r\theta = 3$ (reciprocal spiral) and (b) $3r = \theta$ (spiral of Archimedes).
34. Show that the equations $r = 1 + \sin \theta$ and $r = \sin \theta - 1$ have the same graph.
35. Draw a sketch of the graph of $r = \sqrt{|\cos \theta|}$.
36. Draw a sketch of the graph of $r = \sqrt{|\cos 2\theta|}$.
37. Draw a sketch of the graph of $r = 2(1 + \cos \theta)$.
38. Draw a sketch of the graph of $r^2 = 16 \cos \theta$.

In Exercises 39 and 40, find all the points of intersection of the graphs of the two equations.

39. $\begin{cases} r = 2(1 + \cos \theta) \\ r^2 = 2 \cos \theta \end{cases}$
40. $\begin{cases} r \cos \theta = 1 \\ r = 1 + 2 \cos \theta \end{cases}$

41. Find the area of the region enclosed by (a) the loop of the limaçon $r = 4(1 + 2 \cos \theta)$ and (b) the outer part of the limaçon.
42. Find the area of one leaf of the rose $r = 2 \sin 3\theta$.
43. Find the area of the region inside the graph of $r = 2a \sin \theta$ and outside the graph of $r = a$.
44. Find the area of the region inside the graph of the lemniscate $r^2 = 2 \sin 2\theta$ and outside the graph of the circle $r = 1$.
45. Find the area of the region swept out by the radius vector of the logarithmic spiral $r = e^{k\theta}$ ($k > 0$) as θ varies from 0 to 2π.

46. Find the area of the intersection of the regions enclosed by the graphs of the two equations $r = a \cos \theta$ and $r = a(1 - \cos \theta)$, $a > 0$.

47. Prove that the distance between the two points $P_1(r_1, \theta_1)$ and $P_2(r_2, \theta_2)$ is $\sqrt{r_1^2 + r_2^2 - 2r_1 r_2 \cos(\theta_2 - \theta_1)}$.

48. Find the points of intersection of the graphs of the equations $r = \tan \theta$ and $r = \cot \theta$.

49. Find a polar equation of the circle having its center at (r_0, θ_0) and a radius of a units. (*Hint:* Apply the law of cosines to the triangle having vertices at the pole, (r_0, θ_0) and (r, θ).)

50. Find the area enclosed by one loop of the curve $r = a \sin n\theta$, where n is a positive integer.

In Exercises 51 through 54, the equation is that of a conic having a focus at the pole. (a) Find the eccentricity; (b) identify the conic; (c) write an equation of the directrix that corresponds to the focus at the pole; (d) draw a sketch of the curve.

51. $r = \dfrac{2}{2 - \sin \theta}$ **52.** $r = \dfrac{5}{3 + 3 \sin \theta}$

53. $r = \dfrac{4}{2 + 3 \cos \theta}$ **54.** $r = \dfrac{4}{3 - 2 \cos \theta}$

In Exercises 55 through 58, find a polar equation of the conic satisfying the conditions, and draw a sketch of the graph.

55. A focus at the pole; vertices at $(2, \pi)$ and $(4, \pi)$.

56. A focus at the pole; a vertex at $(6, \frac{1}{2}\pi)$; $e = \frac{3}{4}$.

57. A focus at the pole; a vertex at $(3, \frac{3}{2}\pi)$; $e = 1$.

58. The line $r \sin \theta = 6$ is the directrix corresponding to the focus at the pole and $e = \frac{5}{3}$.

In Exercises 59 and 60, simplify the equation by a rotation and translation of axes. Draw a sketch of the graph and show the three sets of axes.

59. $3x^2 - 3xy - y^2 - 6y = 0$

60. $4x^2 + 3xy + y^2 - 6x + 12y = 0$

In Exercises 61 and 62, (a) find the eccentricity, foci, and directrices of the central conic. (b) Draw a sketch showing the conic, the foci, and the directrices.

61. $25x^2 - 9y^2 = 225$ **62.** $4x^2 + y^2 = 64$

63. Find the volume of the solid of revolution generated if the region bounded by the hyperbola $x^2/a^2 - y^2/b^2 = 1$ and the line $x = 2a$ is revolved about the y axis.

64. Show that the hyperbola $x^2 - y^2 = 4$ has the same foci as the ellipse $x^2 + 9y^2 = 9$.

65. A satellite is traveling around the earth in an elliptical orbit having the earth at one focus and an eccentricity of $\frac{1}{3}$. The closest distance that the satellite gets to the earth is 300 mi. Find the farthest distance that the satellite gets from the earth.

66. The orbit of the planet Mercury around the sun is elliptical in shape with the sun at one focus, a semimajor axis of length 36 million miles, and an eccentricity of 0.206. Find

(a) how close Mercury gets to the sun and (b) the greatest possible distance between Mercury and the sun.

67. A comet is moving in a parabolic orbit around the sun at the focus F of the parabola. An observation of the comet is made when it is at point P_1, 15 million miles from the sun, and a second observation is made when it is at point P_2, 5 million miles from the sun. The line segments FP_1 and FP_2 are perpendicular. With this information there are two possible orbits for the comet. Find how close the comet comes to the sun for each orbit.

68. The arch of a bridge is in the shape of a semiellipse having a horizontal span of 60 m and a height of 20 m at its center. How high is the arch 10 m to the right or left of the center?

69. Find a polar equation of the parabola containing the point $(2, \frac{1}{3}\pi)$, whose focus is at the pole and whose vertex is on the extension of the polar axis.

70. If the distance between the two directrices of an ellipse is three times the distance between the foci, find the eccentricity.

71. Points A and B are 1000 m apart, and it is determined from the sound of an explosion heard at these points at different times that the location of the explosion is 600 m closer to A than to B. Show that the location of the explosion is restricted to a particular curve, and find an equation of it.

72. A focal chord of a conic is divided into two segments by the focus. Prove that the sum of the reciprocals of the measures of the lengths of the two segments is the same, regardless of what chord is taken. (*Hint:* Use polar coordinates.)

73. A focal chord of a conic is a line segment passing through a focus and having its endpoints on the conic. Prove that if two focal chords of a parabola are perpendicular, the sum of the reciprocals of the measures of their lengths is a constant. (*Hint:* Use polar coordinates.)

74. Find the area of the region bounded by the two parabolas $r = 2/(1 - \cos \theta)$ and $r = 2/(1 + \cos \theta)$.

75. Prove that the line $y = 2x$ is an asymptote of the graph of the function f for which $f(x) = 2\sqrt{x^2 - 1}$.

76. Prove that the midpoints of all chords parallel to a fixed chord of a parabola lie on a line that is parallel to the axis of the parabola.

77. Show that $\sqrt{x} \pm \sqrt{y} = \pm\sqrt{a}$ represents a one-parameter family of conics, and determine the type of conic. Draw sketches of the conics for $a = 1$, $a = 2$, and $a = 4$.

78. The graph of the equation $(1 - e^2)x^2 + y^2 - 2px + p^2 = 0$ is a central conic having eccentricity e and a focus at $(p, 0)$. Simplify the equation by a translation of axes. Find the new origin with respect to the x and y axes, and determine the type of conic.

Exercises 79 through 86 pertain to Supplementary Section 11.9.

79. Determine the points at which the curve $r = 3 - 3 \sin \theta$ has horizontal and vertical tangent lines. Draw a sketch of the graph and show these tangent lines.

80. Determine the points at which the curve $r = 4 + 3 \cos \theta$ has horizontal and vertical tangent lines. Draw a sketch of the graph and show these tangent lines.

81. Find the slope of the tangent line to the graph of $r = \frac{1}{2} + \sin\theta$ at the point $(1, \frac{1}{6}\pi)$. Draw a sketch of the graph and the tangent line.

82. Find the slope of the tangent line to the graph of $r = 6\cos\theta - 2$ at the point $(1, \frac{5}{3}\pi)$. Draw a sketch of the graph and the tangent line.

83. Find the radian measure of the angle between the tangent lines at each point of intersection of the graphs of the equations $r = 6\cos\theta$ and $r = 2(1 - \cos\theta)$.

84. Prove that the graphs of $r = a\theta$ and $r\theta = a$ have an unlimited number of points of intersection. Also prove that the tangent lines are perpendicular at only two of these points of intersection, and find these points.

85. For the curve $r = 2\sqrt{\cos\frac{1}{2}\theta}$, find χ and α at the point $(1, \frac{2}{3}\pi)$. Draw a sketch of the curve and show χ and α.

86. For the curve $r = 2(1 + \sin\theta)$, find χ and α at the point $(3, \frac{1}{6}\pi)$. Draw a sketch of the curve and show χ and α.

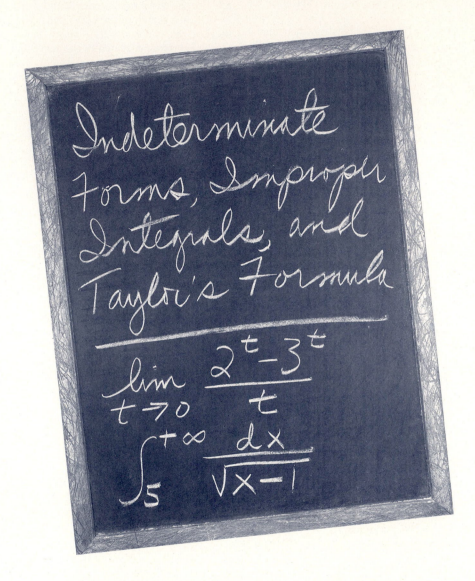

ELEVEN

On the chalkboard:

Indeterminate Forms, Improper Integrals, and Taylor's Formula

$$\lim_{t \to 0} \frac{2^t - 3^t}{t}$$

$$\int_5^{+\infty} \frac{dx}{\sqrt{x-1}}$$

Methods for computing certain limits involving *indeterminate forms* are treated in the first two sections of this chapter. The technique used is called *L'Hôpital's rule*, named for the French mathematician Guillaume François de *L'Hôpital* (1661–1707), who wrote the first calculus textbook, published in 1696.

Sections 11.3 and 11.4 pertain to *improper integrals*. Those in Section 11.3 are integrals over unbounded intervals; that is, they have infinite limits of integration. Each of the improper integrals in Section 11.4 involves an unbounded function on a closed interval [a, b]; that is, the integrand has an infinite discontinuity on [a, b].

In Section 11.5, we discuss *Taylor's formula*, which gives a procedure for approximating functions by means of polynomials.

11.1 THE INDETERMINATE FORM 0/0

Limit Theorem 9 (2.2.9) states that if $\lim\limits_{x \to a} f(x)$ and $\lim\limits_{x \to a} g(x)$ both exist, then

$$\lim_{x \to a} \frac{f(x)}{g(x)} = \frac{\lim\limits_{x \to a} f(x)}{\lim\limits_{x \to a} g(x)}$$

provided that $\lim\limits_{x \to a} g(x) \neq 0$.

There are various situations for which this theorem cannot be used. In particular, if $\lim\limits_{x \to a} g(x) = 0$ and $\lim\limits_{x \to a} f(x) = k$, where k is a constant not equal to 0, then Limit Theorem 12 (2.4.4) can be applied. Consider the case when both $\lim\limits_{x \to a} f(x) = 0$ and $\lim\limits_{x \to a} g(x) = 0$. Some limits of this type have previously been discussed.

▶ **ILLUSTRATION 1** We wish to find

$$\lim_{x \to 4} \frac{x^2 - x - 12}{x^2 - 3x - 4}$$

Here, $\lim\limits_{x \to 4} (x^2 - x - 12) = 0$ and $\lim\limits_{x \to 4} (x^2 - 3x - 4) = 0$. However, the numerator and denominator can be factored, which gives

$$\lim_{x \to 4} \frac{x^2 - x - 12}{x^2 - 3x - 4} = \lim_{x \to 4} \frac{(x - 4)(x + 3)}{(x - 4)(x + 1)}$$

$$= \lim_{x \to 4} \frac{x + 3}{x + 1}$$

$$= \tfrac{7}{5}$$ ◀

▶ **ILLUSTRATION 2** $\lim\limits_{x \to 0} \sin x = 0$ and $\lim\limits_{x \to 0} x = 0$; and by Theorem 2.8.2,

$$\lim_{x \to 0} \frac{\sin x}{x} = 1$$ ◀

11.1.1 DEFINITION If f and g are two functions such that $\lim\limits_{x \to a} f(x) = 0$ and $\lim\limits_{x \to a} g(x) = 0$, then the function f/g has the **indeterminate form 0/0 at a**.

▶ **ILLUSTRATION 3** From Definition 11.1.1, $(x^2 - x - 12)/(x^2 - 3x - 4)$ has the indeterminate form 0/0 at 4; however, we saw in Illustration 1 that

$$\lim_{x \to 4} \frac{x^2 - x - 12}{x^2 - 3x - 4} = \frac{7}{5}$$

Also, $\sin x/x$ has the indeterminate form 0/0 at 0, while $\lim\limits_{x \to 0} (\sin x/x) = 1$, as indicated in Illustration 2. ◀

We now consider a general method for finding the limit, if it exists, of a function at a number where it has the indeterminate form 0/0. It is known as *L'Hôpital's rule*.

11.1.2 THEOREM
L'Hôpital's Rule

Let f and g be functions that are differentiable on an open interval I, except possibly at the number a in I. Suppose that for all $x \neq a$ in I, $g'(x) \neq 0$. Then if $\lim\limits_{x \to a} f(x) = 0$ and $\lim\limits_{x \to a} g(x) = 0$, and if

$$\lim_{x \to a} \frac{f'(x)}{g'(x)} = L$$

it follows that

$$\lim_{x \to a} \frac{f(x)}{g(x)} = L$$

The theorem is valid if all the limits are right-hand limits or all the limits are left-hand limits.

Before giving the proof of Theorem 11.1.2, we show the use of the theorem by an illustration and examples.

▶ **ILLUSTRATION 4**   We use L'Hôpital's rule to evaluate the limits in Illustrations 1 and 2. In Illustration 1, because $\lim\limits_{x \to 4} (x^2 - x - 12) = 0$ and $\lim\limits_{x \to 4} (x^2 - 3x - 4) = 0$, we can apply L'Hôpital's rule and obtain

$$\lim_{x \to 4} \frac{x^2 - x - 12}{x^2 - 3x - 4} = \lim_{x \to 4} \frac{2x - 1}{2x - 3}$$

$$= \tfrac{7}{5}$$

L'Hôpital's rule can be applied in Illustration 2 because $\lim\limits_{x \to 0} \sin x = 0$ and $\lim\limits_{x \to 0} x = 0$. Then

$$\lim_{x \to 0} \frac{\sin x}{x} = \lim_{x \to 0} \frac{\cos x}{1}$$

$$= 1$$ ◀

EXAMPLE 1   Evaluate the limit, if it exists:

$$\lim_{x \to 0} \frac{x}{1 - e^x}$$

Solution   Because $\lim\limits_{x \to 0} x = 0$ and $\lim\limits_{x \to 0} (1 - e^x) = 0$, L'Hôpital's rule can be applied. Thus

$$\lim_{x \to 0} \frac{x}{1 - e^x} = \lim_{x \to 0} \frac{1}{-e^x}$$

$$= \frac{1}{-1}$$

$$= -1$$

EXAMPLE 2   Evaluate the limit, if it exists:

$$\lim_{x \to 1} \frac{1 - x + \ln x}{x^3 - 3x + 2}$$

Solution

$$\lim_{x \to 1} (1 - x + \ln x) = 1 - 1 + 0 \qquad \lim_{x \to 1} (x^3 - 3x + 2) = 1 - 3 + 2$$
$$= 0 \qquad\qquad\qquad\qquad\qquad = 0$$

Therefore, applying L'Hôpital's rule we have

$$\lim_{x \to 1} \frac{1 - x + \ln x}{x^3 - 3x + 2} = \lim_{x \to 1} \frac{-1 + \dfrac{1}{x}}{3x^2 - 3}$$

Now, because $\lim_{x \to 1} (-1 + 1/x) = 0$ and $\lim_{x \to 1} (3x^2 - 3) = 0$, we apply L'Hôpital's rule again, giving

$$\lim_{x \to 1} \frac{-1 + \dfrac{1}{x}}{3x^2 - 3} = \lim_{x \to 1} \frac{-\dfrac{1}{x^2}}{6x}$$
$$= -\tfrac{1}{6}$$

Therefore

$$\lim_{x \to 1} \frac{1 - x + \ln x}{x^3 - 3x + 2} = -\frac{1}{6}$$

To prove Theorem 11.1.2 we need to use the theorem known as *Cauchy's mean-value theorem*, which extends to two functions the mean-value theorem (4.3.2) for a single function. This theorem is attributed to the French mathematician Augustin L. Cauchy (1789–1857).

11.1.3 THEOREM
Cauchy's Mean-Value Theorem

If f and g are two functions such that

(i) f and g are continuous on the closed interval $[a, b]$;
(ii) f and g are differentiable on the open interval (a, b);
(iii) for all x in the open interval (a, b), $g'(x) \neq 0$,

then there exists a number z in the open interval (a, b) such that

$$\frac{f(b) - f(a)}{g(b) - g(a)} = \frac{f'(z)}{g'(z)}$$

Proof   We first show that $g(b) \neq g(a)$. Assume $g(b) = g(a)$. Because g satisfies the two conditions in the hypothesis of the mean-value theorem, there is some number c in (a, b) such that $g'(c) = [g(b) - g(a)]/(b - a)$. But if $g(b) = g(a)$, then there is some number c in (a, b) such that $g'(c) = 0$. But condition (iii) of the hypothesis of this theorem states that for all x in (a, b), $g'(x) \neq 0$. Therefore there is a contradiction. Hence the assumption that $g(b) = g(a)$ is false. So $g(b) \neq g(a)$, and consequently $g(b) - g(a) \neq 0$.

Now consider the function h defined by

$$h(x) = f(x) - f(a) - \left[\frac{f(b) - f(a)}{g(b) - g(a)}\right][g(x) - g(a)]$$

Then

$$h'(x) = f'(x) - \left[\frac{f(b) - f(a)}{g(b) - g(a)}\right]g'(x) \qquad (1)$$

Therefore h is differentiable on (a, b) because f and g are differentiable there, and h is continuous on $[a, b]$ because f and g are continuous there.

$$h(a) = f(a) - f(a) - \left[\frac{f(b) - f(a)}{g(b) - g(a)}\right][g(a) - g(a)]$$

$$= 0$$

$$h(b) = f(b) - f(a) - \left[\frac{f(b) - f(a)}{g(b) - g(a)}\right][g(b) - g(a)]$$

$$= 0$$

Hence the three conditions of the hypothesis of Rolle's theorem are satisfied by the function h. So there exists a number z in the open interval (a, b) such that $h'(z) = 0$. Thus, from (1),

$$f'(z) - \frac{f(b) - f(a)}{g(b) - g(a)} g'(z) = 0$$

Because $g'(z) \neq 0$ on (a, b), we have from the above equation

$$\frac{f(b) - f(a)}{g(b) - g(a)} = \frac{f'(z)}{g'(z)}$$

where z is some number in (a, b). This proves the theorem. ■

If g is the function such that $g(x) = x$, then the conclusion of Cauchy's mean-value theorem becomes the conclusion of the former mean-value theorem because then $g'(z) = 1$. So the former mean-value theorem is a special case of Cauchy's mean-value theorem.

A geometric interpretation of Cauchy's mean-value theorem involves parametric equations, and is therefore postponed until Section 14.3.

▶ **ILLUSTRATION 5** Suppose $f(x) = 3x^2 + 3x - 1$ and $g(x) = x^3 - 4x + 2$. We find a number z in $(0, 1)$, predicted by Theorem 11.1.3.

$$f'(x) = 6x + 3 \qquad g'(x) = 3x^2 - 4$$

Thus f and g are differentiable and continuous everywhere, and for all x in $(0, 1)$, $g'(x) \neq 0$. Hence, by Theorem 11.1.3 there exists a z in $(0, 1)$ such that

$$\frac{f(1) - f(0)}{g(1) - g(0)} = \frac{6z + 3}{3z^2 - 4}$$

Substituting $f(1) = 5$, $g(1) = -1$, $f(0) = -1$, and $g(0) = 2$ and solving for z we have

$$\frac{5 - (-1)}{-1 - 2} = \frac{6z + 3}{3z^2 - 4}$$

$$6z^2 + 6z - 5 = 0$$

$$z = \frac{-6 \pm \sqrt{36 + 120}}{12}$$

$$= \frac{-6 \pm 2\sqrt{39}}{12}$$

$$= \frac{-3 \pm \sqrt{39}}{6}$$

Only one of these numbers is in $(0, 1)$, namely $z = \frac{1}{6}(-3 + \sqrt{39})$. ◀

We are now in a position to prove Theorem 11.1.2. We distinguish three cases: (i) $x \to a^+$; (ii) $x \to a^-$; (iii) $x \to a$.

Proof of Theorem 11.1.2(i) Because in the hypothesis it is not assumed that f and g are defined at a, we consider two new functions F and G for which

$$F(x) = f(x) \qquad \text{if } x \neq a \text{ and } F(a) = 0$$

$$G(x) = g(x) \qquad \text{if } x \neq a \text{ and } G(a) = 0 \tag{2}$$

Let b be the right endpoint of the open interval I given in the hypothesis. Because f and g are both differentiable on I, except possibly at a, we conclude that F and G are both differentiable on the interval $(a, x]$, where $a < x < b$. Therefore F and G are both continuous on $(a, x]$. The functions F and G are also both continuous from the right at a because $\lim\limits_{x \to a^+} F(x) = \lim\limits_{x \to a^+} f(x)$ and $\lim\limits_{x \to a^+} f(x) = 0$, which is $F(a)$; similarly, $\lim\limits_{x \to a^+} G(x) = G(a)$. Therefore F and G are continuous on the closed interval $[a, x]$. So F and G satisfy the three conditions of the hypothesis of Cauchy's mean-value theorem (Theorem 11.1.3) on the interval $[a, x]$. Hence

$$\frac{F(x) - F(a)}{G(x) - G(a)} = \frac{F'(z)}{G'(z)}$$

where z is some number such that $a < z < x$. From (2) and the above equation we have

$$\frac{f(x)}{g(x)} = \frac{f'(z)}{g'(z)}$$

Because $a < z < x$, it follows that as $x \to a^+$, $z \to a^+$; therefore

$$\lim_{x \to a^+} \frac{f(x)}{g(x)} = \lim_{x \to a^+} \frac{f'(z)}{g'(z)}$$

$$= \lim_{z \to a^+} \frac{f'(z)}{g'(z)}$$

But by hypothesis, this limit is L. Therefore

$$\lim_{x \to a^+} \frac{f(x)}{g(x)} = L$$

■

The proof of case (ii) is similar to the proof of case (i) and is left as an exercise (see Exercise 34). The proof of case (iii) is based on the results of cases (i) and (ii) and is also left as an exercise (see Exercise 35).

L'Hôpital's rule also holds if either x increases without bound or x decreases without bound, as given in the next theorem.

11.1.4 THEOREM
L'Hôpital's Rule

Let f and g be functions that are differentiable for all $x > N$, where N is a positive constant, and suppose that for all $x > N$, $g'(x) \neq 0$. Then if $\lim\limits_{x \to +\infty} f(x) = 0$ and $\lim\limits_{x \to +\infty} g(x) = 0$, and if

$$\lim_{x \to +\infty} \frac{f'(x)}{g'(x)} = L$$

it follows that

$$\lim_{x \to +\infty} \frac{f(x)}{g(x)} = L$$

The theorem is also valid if $x \to +\infty$ is replaced by $x \to -\infty$.

Proof We prove the theorem for $x \to +\infty$. The proof for $x \to -\infty$ is left as an exercise (see Exercise 38).

For all $x > N$, let $x = 1/t$; then $t = 1/x$. Let F and G be the functions defined by $F(t) = f(1/t)$ and $G(t) = g(1/t)$, if $t \neq 0$. Then $f(x) = F(t)$ and $g(x) = G(t)$, where $x > N$ and $0 < t < 1/N$. From Definitions 2.5.1 and 2.3.1 it may be shown that the statements

$$\lim_{x \to +\infty} f(x) = M \quad \text{and} \quad \lim_{t \to 0^+} F(t) = M$$

have the same meaning. It is left as an exercise to prove this (see Exercise 36). Because by hypothesis $\lim\limits_{x \to +\infty} f(x) = 0$ and $\lim\limits_{x \to +\infty} g(x) = 0$, we can conclude that

$$\lim_{t \to 0^+} F(t) = 0 \quad \text{and} \quad \lim_{t \to 0^+} G(t) = 0 \tag{3}$$

Using the chain rule in the quotient $F'(t)/G'(t)$, we have

$$\frac{F'(t)}{G'(t)} = \frac{-\dfrac{1}{t^2} f'\left(\dfrac{1}{t}\right)}{-\dfrac{1}{t^2} g'\left(\dfrac{1}{t}\right)}$$

$$= \frac{f'\left(\dfrac{1}{t}\right)}{g'\left(\dfrac{1}{t}\right)}$$

$$= \frac{f'(x)}{g'(x)}$$

Because by hypothesis $\lim\limits_{x \to +\infty} f'(x)/g'(x) = L$, it follows from the above that

$$\lim_{t \to 0^+} \frac{F'(t)}{G'(t)} = L \tag{4}$$

Because for all $x > N$, $g'(x) \neq 0$,

$$G'(t) \neq 0 \qquad \text{for all } 0 < t < \frac{1}{N}$$

From this statement, (3), and (4) it follows from Theorem 11.1.2 that

$$\lim_{t \to 0^+} \frac{F(t)}{G(t)} = L$$

But because $F(t)/G(t) = f(x)/g(x)$ for all $x > N$ and $t \neq 0$, then

$$\lim_{x \to +\infty} \frac{f(x)}{g(x)} = L$$

and so the theorem is proved. ∎

Theorems 11.1.2 and 11.1.4 also hold if L is replaced by $+\infty$ or $-\infty$. The proofs for these cases are omitted.

EXAMPLE 3 Evaluate the limit, if it exists:

$$\lim_{x \to +\infty} \frac{\sin \dfrac{1}{x}}{\tan^{-1}\left(\dfrac{1}{x}\right)}$$

Solution $\lim\limits_{x \to +\infty} \sin(1/x) = 0$ and $\lim\limits_{x \to +\infty} \tan^{-1}(1/x) = 0$. So from L'Hôpital's rule,

$$\lim_{x \to +\infty} \frac{\sin \dfrac{1}{x}}{\tan^{-1}\left(\dfrac{1}{x}\right)} = \lim_{x \to +\infty} \frac{\cos \dfrac{1}{x} \cdot \left(-\dfrac{1}{x^2}\right)}{\dfrac{1}{1 + \dfrac{1}{x^2}} \cdot \left(-\dfrac{1}{x^2}\right)}$$

$$= \lim_{x \to +\infty} \frac{\cos \dfrac{1}{x}}{\dfrac{x^2}{x^2 + 1}}$$

Because

$$\lim_{x \to +\infty} \cos \frac{1}{x} = 1 \quad \text{and} \quad \lim_{x \to +\infty} \frac{x^2}{x^2 + 1} = \lim_{x \to +\infty} \frac{1}{1 + \dfrac{1}{x^2}}$$

$$= 1$$

it follows that

$$\lim_{x \to +\infty} \frac{\cos \dfrac{1}{x}}{\dfrac{x^2}{x^2 + 1}} = 1$$

Therefore the given limit is 1.

EXAMPLE 4   Given

$$f(x) = \begin{cases} \dfrac{e^x - 1}{x} & \text{if } x \neq 0 \\ 1 & \text{if } x = 0 \end{cases}$$

(a) Prove that f is continuous at 0 by using Definition 2.6.1. (b) Prove that f is differentiable at 0 by computing $f'(0)$.

Solution
(a) We check the three conditions for continuity.

(i) $f(0) = 1$

(ii) $\lim_{x \to 0} f(x) = \lim_{x \to 0} \dfrac{e^x - 1}{x}$

To compute the limit we apply L'Hôpital's rule because $\lim_{x \to 0} (e^x - 1) = 0$ and $\lim_{x \to 0} x = 0$. We have

$$\lim_{x \to 0} f(x) = \lim_{x \to 0} \frac{e^x}{1}$$

$$= 1$$

(iii) $\lim_{x \to 0} f(x) = f(0)$

Therefore f is continuous at 0.

(b) $f'(0) = \lim_{x \to 0} \dfrac{f(x) - f(0)}{x - 0}$

$$= \lim_{x \to 0} \frac{\dfrac{e^x - 1}{x} - 1}{x}$$

$$= \lim_{x \to 0} \frac{e^x - 1 - x}{x^2}$$

Because $\lim_{x \to 0} (e^x - 1 - x) = 0$ and $\lim_{x \to 0} x^2 = 0$, we apply L'Hôpital's rule and get

$$f'(0) = \lim_{x \to 0} \frac{e^x - 1}{2x}$$

Because $\lim\limits_{x\to0} (e^x - 1) = 0$ and $\lim\limits_{x\to0} 2x = 0$, we apply L'Hôpital's rule again and obtain

$$f'(0) = \lim_{x\to0} \frac{e^x}{2}$$

$$= \tfrac{1}{2}$$

EXERCISES 11.1

In Exercises 1 through 5, find all values of z in the interval (a, b) satisfying the conclusion of Theorem 11.1.3 for the given pair of functions.

1. $f(x) = x^3$, $g(x) = x^2$; $(a, b) = (0, 2)$

2. $f(x) = \dfrac{2x}{1 + x^2}$, $g(x) = \dfrac{1 - x^2}{1 + x^2}$; $(a, b) = (0, 2)$

3. $f(x) = \sin x$, $g(x) = \cos x$; $(a, b) = (0, \pi)$

4. $f(x) = \cos 2x$, $g(x) = \sin x$; $(a, b) = (0, \tfrac{1}{2}\pi)$

5. $f(x) = \sqrt{x + 5}$, $g(x) = x + 3$; $(a, b) = (-4, -1)$

In Exercises 6 through 30, evaluate the limit, if it exists.

6. $\lim\limits_{x\to\pi/2} \dfrac{3\cos x}{2x - \pi}$

7. $\lim\limits_{x\to0} \dfrac{x}{\tan x}$

8. $\lim\limits_{x\to0} \dfrac{\tan x - x}{x - \sin x}$

9. $\lim\limits_{x\to+\infty} \dfrac{\sin\dfrac{2}{x}}{\dfrac{1}{x}}$

10. $\lim\limits_{x\to2} \dfrac{\sin \pi x}{2 - x}$

11. $\lim\limits_{x\to-\infty} \dfrac{2^x}{e^x}$

12. $\lim\limits_{x\to0} \dfrac{e^x - \cos x}{x \sin x}$

13. $\lim\limits_{x\to\pi/2} \dfrac{\ln(\sin x)}{(\pi - 2x)^2}$

14. $\lim\limits_{x\to0} \dfrac{\sin^{-1} x}{x}$

15. $\lim\limits_{x\to0} \dfrac{2^x - 3^x}{x}$

16. $\lim\limits_{t\to2} \dfrac{t^n - 2^n}{t - 2}$

17. $\lim\limits_{\theta\to0} \dfrac{\theta - \sin \theta}{\tan^3 \theta}$

18. $\lim\limits_{x\to0} \dfrac{e^{2x^2} - 1}{\sin^2 x}$

19. $\lim\limits_{z\to+\infty} \dfrac{1 - e^{1/z}}{-\dfrac{3}{z}}$

20. $\lim\limits_{y\to0} \dfrac{y^2}{1 - \cosh y}$

21. $\lim\limits_{t\to0} \dfrac{\sin t}{\ln(2e^t - 1)}$

22. $\lim\limits_{z\to0} \dfrac{5z}{5^z - e^z}$

23. $\lim\limits_{t\to0} \dfrac{\sin^2 t}{\sin t^2}$

24. $\lim\limits_{x\to0} \dfrac{\tanh 2x}{\tanh x}$

25. $\lim\limits_{x\to0} \dfrac{(1 + x)^{1/5} - (1 - x)^{1/5}}{(1 + x)^{1/3} - (1 - x)^{1/3}}$

26. $\lim\limits_{x\to0} \dfrac{e^x - 10^x}{x}$

27. $\lim\limits_{x\to\pi} \dfrac{1 + \cos 2x}{1 - \sin x}$

28. $\lim\limits_{x\to+\infty} \dfrac{\dfrac{1}{x^2} - 2\tan^{-1}\dfrac{1}{x}}{\dfrac{1}{x}}$

29. $\lim\limits_{x\to0} \dfrac{\cos x - \cosh x}{x^2}$

30. $\lim\limits_{x\to0} \dfrac{\sinh x - \sin x}{\sin^3 x}$

31. An electrical circuit has a resistance of R ohms, an inductance of L henrys, and an electromotive force of E volts, where R, L, and E are positive. If i amperes is the current flowing in the circuit t seconds after a switch is turned on, then

$$i = \frac{E}{R} (1 - e^{-Rt/L})$$

If t, E, and L are constants, find $\lim\limits_{R\to0^+} i$.

32. In a geometric progression, if a is the first term, r is the common ratio of two successive terms, and S is the sum of the first n terms, then if $r \neq 1$,

$$S = \frac{a(r^n - 1)}{r - 1}$$

Find $\lim\limits_{r\to1} S$. Is the result consistent with the sum of the first n terms if $r = 1$?

33. Find values for a and b such that

$$\lim_{x\to0} \frac{\sin 3x + ax + bx^3}{x^3} = 0$$

34. Prove Theorem 11.1.2(ii).   35. Prove Theorem 11.1.2(iii).

36. Suppose that f is a function defined for all $x > N$, where N is a positive constant. If $t = 1/x$ and $F(t) = f(1/t)$, where $t \neq 0$, prove that the statements $\lim\limits_{x\to+\infty} f(x) = M$ and $\lim\limits_{t\to0^+} F(t) = M$ have the same meaning.

37. Given:

$$f(x) = \begin{cases} \dfrac{\cos x - 1}{x} & \text{if } x \neq 0 \\ 0 & \text{if } x = 0 \end{cases}$$

(a) Prove that f is continuous at 0 by using Definition 2.6.1.
(b) Prove that f is differentiable at 0 by computing $f'(0)$.

38. Prove Theorem 11.1.4 for $x \to -\infty$.

39. Prove that $x^{10} < e^x$ for x sufficiently large by evaluating $\lim\limits_{x\to+\infty} \dfrac{x^{10}}{e^x}$.

40. Prove that $\ln x < x^p$ for all $p > 0$ and x sufficiently large by evaluating $\lim\limits_{x\to+\infty} \dfrac{\ln x}{x^p}$.

11.2 OTHER INDETERMINATE FORMS

Suppose that we wish to determine whether

$$\lim_{x \to 0^+} \frac{\ln x}{\dfrac{1}{x}}$$

exists. We cannot apply the theorem involving the limit of a quotient because $\lim_{x \to 0^+} \ln x = -\infty$ and $\lim_{x \to 0^+} (1/x) = +\infty$. In this case we say that the function defined by $\ln x/(1/x)$ has the indeterminate form $(-\infty)/(+\infty)$ at $x = 0$. L'Hôpital's rule also applies to an indeterminate form of this type as well as to $(+\infty)/(+\infty)$, $(-\infty)/(-\infty)$, and $(+\infty)/(-\infty)$. This is given by the following theorems, for which the proofs are omitted because they are beyond the scope of this book.

11.2.1 THEOREM L'Hôpital's Rule

Let f and g be functions that are differentiable on an open interval I, except possibly at the number a in I, and suppose that for all $x \neq a$ in I, $g'(x) \neq 0$. Then if $\lim_{x \to a} f(x)$ is $+\infty$ or $-\infty$, and $\lim_{x \to a} g(x)$ is $+\infty$ or $-\infty$, and if

$$\lim_{x \to a} \frac{f'(x)}{g'(x)} = L$$

it follows that

$$\lim_{x \to a} \frac{f(x)}{g(x)} = L$$

The theorem is valid if all the limits are right-hand limits or if all the limits are left-hand limits.

EXAMPLE 1   Evaluate the limit, if it exists:

$$\lim_{x \to 0^+} \frac{\ln x}{\dfrac{1}{x}}$$

Solution   Because $\lim_{x \to 0^+} \ln x = -\infty$ and $\lim_{x \to 0^+} (1/x) = +\infty$, we apply L'Hôpital's rule and get

$$\lim_{x \to 0^+} \frac{\ln x}{\dfrac{1}{x}} = \lim_{x \to 0^+} \frac{\dfrac{1}{x}}{-\dfrac{1}{x^2}}$$

$$= \lim_{x \to 0^+} (-x)$$

$$= 0$$

11.2.2 THEOREM
L'Hôpital's Rule

Let f and g be functions that are differentiable for all $x > N$, where N is a positive constant, and suppose that for all $x > N$, $g(x) \neq 0$. Then if $\lim\limits_{x \to +\infty} f(x)$ is $+\infty$ or $-\infty$, and $\lim\limits_{x \to +\infty} g(x)$ is $+\infty$ or $-\infty$, and if

$$\lim_{x \to +\infty} \frac{f'(x)}{g'(x)} = L$$

if follows that

$$\lim_{x \to +\infty} \frac{f(x)}{g(x)} = L$$

The theorem is also valid if $x \to +\infty$ is replaced by $x \to -\infty$.

Theorems 11.2.1 and 11.2.2 also hold if L is replaced by $+\infty$ or $-\infty$, and the proofs for these cases are also omitted.

EXAMPLE 2   Evaluate the limit, if it exists:

$$\lim_{x \to +\infty} \frac{\ln(2 + e^x)}{3x}$$

Solution   Because $\lim\limits_{x \to +\infty} \ln(2 + e^x) = +\infty$ and $\lim\limits_{x \to +\infty} 3x = +\infty$, by applying L'Hôpital's rule we obtain

$$\lim_{x \to +\infty} \frac{\ln(2 + e^x)}{3x} = \lim_{x \to +\infty} \frac{\dfrac{1}{2 + e^x} \cdot e^x}{3}$$

$$= \lim_{x \to +\infty} \frac{e^x}{6 + 3e^x}$$

Now because $\lim\limits_{x \to +\infty} e^x = +\infty$ and $\lim\limits_{x \to +\infty} (6 + 3e^x) = +\infty$, we apply L'Hôpital's rule again and get

$$\lim_{x \to +\infty} \frac{\ln(2 + e^x)}{3x} = \lim_{x \to +\infty} \frac{e^x}{3e^x}$$

$$= \lim_{x \to +\infty} \frac{1}{3}$$

$$= \tfrac{1}{3}$$

EXAMPLE 3   Evaluate the limit, if it exists:

$$\lim_{x \to \pi/2^-} \frac{\sec^2 x}{\sec^2 3x}$$

Solution $\lim\limits_{x\to\pi/2^-} \sec^2 x = +\infty$ and $\lim\limits_{x\to\pi/2^-} \sec^2 3x = +\infty$. So from L'Hôpital's rule,

$$\lim_{x\to\pi/2^-} \frac{\sec^2 x}{\sec^2 3x} = \lim_{x\to\pi/2^-} \frac{2\sec^2 x \tan x}{6\sec^2 3x \tan 3x}$$

$$\lim_{x\to\pi/2^-} 2\sec^2 x \tan x = +\infty \quad \text{and} \quad \lim_{x\to\pi/2^-} 6\sec^2 3x \tan 3x = +\infty$$

Observe that further applications of L'Hôpital's rule will not help us. However, the original quotient may be rewritten, and we have

$$\lim_{x\to\pi/2^-} \frac{\sec^2 x}{\sec^2 3x} = \lim_{x\to\pi/2^-} \frac{\cos^2 3x}{\cos^2 x}$$

Now, because $\lim\limits_{x\to\pi/2^-} \cos^2 3x = 0$ and $\lim\limits_{x\to\pi/2^-} \cos^2 x = 0$, we may apply L'Hôpital's rule, giving

$$\lim_{x\to\pi/2^-} \frac{\cos^2 3x}{\cos^2 x} = \lim_{x\to\pi/2^-} \frac{-6\cos 3x \sin 3x}{-2\cos x \sin x}$$

$$= \lim_{x\to\pi/2^-} \frac{3(2\cos 3x \sin 3x)}{(2\cos x \sin x)}$$

$$= \lim_{x\to\pi/2^-} \frac{3\sin 6x}{\sin 2x}$$

Because $\lim\limits_{x\to\pi/2^-} 3\sin 6x = 0$ and $\lim\limits_{x\to\pi/2^-} \sin 2x = 0$, we use L'Hôpital's rule again and have

$$\lim_{x\to\pi/2^-} \frac{3\sin 6x}{\sin 2x} = \lim_{x\to\pi/2^-} \frac{18\cos 6x}{2\cos 2x}$$

$$= \frac{18(-1)}{2(-1)}$$

$$= 9$$

Therefore

$$\lim_{x\to\pi/2^-} \frac{\sec^2 x}{\sec^2 3x} = 9$$

The limit in Example 3 can be evaluated without L'Hôpital's rule by using Theorem 2.8.2. This is left as an exercise (see Exercise 38).

In addition to $0/0$ and $\pm\infty/\pm\infty$, other indeterminate forms are $0\cdot(+\infty)$, $+\infty-(+\infty)$, 0^0, $(\pm\infty)^0$, and $1^{\pm\infty}$. These indeterminate forms are defined analogously to the other two. For instance, if $\lim\limits_{x\to a} f(x) = +\infty$ and $\lim\limits_{x\to a} g(x) = 0$, then the function defined by $f(x)^{g(x)}$ has the indeterminate form $(+\infty)^0$ at a. To find the limit of a function having one of these indeterminate forms, it must be changed to either the form $0/0$ or $\pm\infty/\pm\infty$ before L'Hôpital's rule can be applied. The following examples illustrate the method.

EXAMPLE 4   Evaluate the limit if it exists:

$$\lim_{x\to 0^+} \sin^{-1} x \csc x$$

Solution   Because $\lim\limits_{x\to0^+} \sin^{-1} x = 0$ and $\lim\limits_{x\to0^+} \csc x = +\infty$, the function defined by $\sin^{-1} x \csc x$ has the indeterminate form $0 \cdot (+\infty)$ at 0. Before we can apply L'Hôpital's rule we rewrite $\sin^{-1} x \csc x$ as $\sin^{-1} x/\sin x$, and consider $\lim\limits_{x\to0^+} (\sin^{-1} x/\sin x)$. Now $\lim\limits_{x\to0^+} \sin^{-1} x = 0$ and $\lim\limits_{x\to0^+} \sin x = 0$; so we have the indeterminate form $0/0$. Therefore, from L'Hôpital's rule we obtain

$$\lim_{x\to0^+} \frac{\sin^{-1} x}{\sin x} = \lim_{x\to0^+} \frac{\dfrac{1}{\sqrt{1-x^2}}}{\cos x}$$

$$= \frac{1}{1}$$

$$= 1$$

EXAMPLE 5   Evaluate the limit, if it exists:

$$\lim_{x\to0} \left(\frac{1}{x^2} - \frac{1}{x^2 \sec x} \right)$$

Solution   Because

$$\lim_{x\to0} \frac{1}{x^2} = +\infty \quad \text{and} \quad \lim_{x\to0} \frac{1}{x^2 \sec x} = +\infty$$

we have the indeterminate form $+\infty - (+\infty)$. Rewriting the expression we have

$$\lim_{x\to0} \left(\frac{1}{x^2} - \frac{1}{x^2 \sec x} \right) = \lim_{x\to0} \frac{\sec x - 1}{x^2 \sec x}$$

$\lim\limits_{x\to0} (\sec x - 1) = 0$ and $\lim\limits_{x\to0} (x^2 \sec x) = 0$; so we apply L'Hôpital's rule and obtain

$$\lim_{x\to0} \frac{\sec x - 1}{x^2 \sec x} = \lim_{x\to0} \frac{\sec x \tan x}{2x \sec x + x^2 \sec x \tan x}$$

$$= \lim_{x\to0} \frac{\tan x}{2x + x^2 \tan x}$$

$\lim\limits_{x\to0} \tan x = 0$ and $\lim\limits_{x\to0} (2x + x^2 \tan x) = 0$

Thus we apply the rule again and obtain

$$\lim_{x\to0} \frac{\tan x}{2x + x^2 \tan x} = \lim_{x\to0} \frac{\sec^2 x}{2 + 2x \tan x + x^2 \sec^2 x}$$

$$= \tfrac{1}{2}$$

Therefore

$$\lim_{x\to0} \left(\frac{1}{x^2} - \frac{1}{x^2 \sec x} \right) = \frac{1}{2}$$

For any one of the indeterminate forms 0^0, $(\pm\infty)^0$, $1^{\pm\infty}$, the procedure for evaluating the limit is illustrated in Example 6.

EXAMPLE 6   Evaluate the limit, if it exists:

$$\lim_{x \to 0^+} (x + 1)^{\cot x}$$

Solution   Because $\lim_{x \to 0^+} (x + 1) = 1$ and $\lim_{x \to 0^+} \cot x = +\infty$, we have the indeterminate form $1^{+\infty}$. Let

$$y = (x + 1)^{\cot x} \tag{1}$$

Then

$$\ln y = \cot x \ln(x + 1)$$

$$= \frac{\ln(x + 1)}{\tan x}$$

So

$$\lim_{x \to 0^+} \ln y = \lim_{x \to 0^+} \frac{\ln(x + 1)}{\tan x} \tag{2}$$

Because $\lim_{x \to 0^+} \ln(x + 1) = 0$ and $\lim_{x \to 0^+} \tan x = 0$, we may apply L'Hôpital's rule to the right side of (2) and obtain

$$\lim_{x \to 0^+} \frac{\ln(x + 1)}{\tan x} = \lim_{x \to 0^+} \frac{\dfrac{1}{x + 1}}{\sec^2 x}$$

$$= 1$$

Therefore substituting 1 on the right side of (2) we have

$$\lim_{x \to 0^+} \ln y = 1 \tag{3}$$

Because the exponential function is continuous on its entire domain, which is the set of all real numbers, we may apply Theorem 2.7.1, and so

$$\lim_{x \to 0^+} \exp(\ln y) = \exp\left(\lim_{x \to 0^+} \ln y \right)$$

Therefore it follows from (3) and this equation that

$$\lim_{x \to 0^+} y = e^1$$

But from (1), $y = (x + 1)^{\cot x}$, and therefore

$$\lim_{x \to 0^+} (x + 1)^{\cot x} = e$$

EXERCISES 11.2

In Exercises 1 through 34, evaluate the limit, if it exists.

1. $\displaystyle\lim_{x \to +\infty} \frac{x^2}{e^x}$

2. $\displaystyle\lim_{x \to \pi/2^-} \frac{\ln(\cos x)}{\ln(\tan x)}$

3. $\displaystyle\lim_{x \to 1/2^-} \frac{\ln(1 - 2x)}{\tan \pi x}$

4. $\displaystyle\lim_{x \to +\infty} \frac{(\ln x)^3}{x}$

5. $\displaystyle\lim_{x \to 0^+} x \csc x$

6. $\displaystyle\lim_{x \to 1/2^+} (2x - 1) \tan \pi x$

7. $\lim\limits_{x \to +\infty} \dfrac{\ln x}{\sqrt{x}}$

8. $\lim\limits_{x \to 0^+} \tan^{-1} x \cot x$

9. $\lim\limits_{x \to 0^+} \tan x(\ln x)$

10. $\lim\limits_{x \to 0^+} \left(\dfrac{1}{\sin x} - \dfrac{1}{x} \right)$

11. $\lim\limits_{x \to 1} \left(\dfrac{1}{\ln x} - \dfrac{1}{x-1} \right)$

12. $\lim\limits_{x \to 0^+} (\sinh x)^{\tan x}$

13. $\lim\limits_{x \to 0^+} x^{\sin x}$

14. $\lim\limits_{x \to 0} (x + e^{2x})^{1/x}$

15. $\lim\limits_{x \to +\infty} (x^2 - \sqrt{x^4 - x^2 + 2})$

16. $\lim\limits_{x \to 2} \left(\dfrac{5}{x^2 + x - 6} - \dfrac{1}{x-2} \right)$

17. $\lim\limits_{x \to +\infty} x^{1/x}$

18. $\lim\limits_{x \to 0^+} (1 + x)^{\ln x}$

19. $\lim\limits_{x \to +\infty} (e^x + x)^{2/x}$

20. $\lim\limits_{x \to 0^+} x^{1/\ln x}$

21. $\lim\limits_{x \to 0} (1 + ax)^{1/x}; \; a \neq 0$

22. $\lim\limits_{x \to +\infty} \dfrac{x^2 + 2x}{e^{3x} - 1}$

23. $\lim\limits_{x \to 0^+} (\sin x)^{x^2}$

24. $\lim\limits_{x \to +\infty} \left(1 + \dfrac{1}{2x} \right)^{x^2}$

25. $\lim\limits_{x \to 0} (1 + \sinh x)^{2/x}$

26. $\lim\limits_{x \to 2} (x - 2) \tan \tfrac{1}{4}\pi x$

27. $\lim\limits_{x \to 0} [(\cos x)e^{x^2/2}]^{4/x^4}$

28. $\lim\limits_{x \to 0} (\cos x)^{1/x^2}$

29. $\lim\limits_{x \to +\infty} [(x^6 + 3x^5 + 4)^{1/6} - x]$

30. $\lim\limits_{x \to +\infty} \dfrac{\ln(x + e^x)}{3x}$

31. $\lim\limits_{x \to 0^+} \dfrac{e^{-1/x}}{x}$

32. $\lim\limits_{x \to 0^+} x^{x^x}$

33. $\lim\limits_{x \to +\infty} \dfrac{x}{\sqrt{1 + x^2}}$

34. $\lim\limits_{x \to +\infty} (x - \sqrt{x^2 + x})$

35. If $f(x) = \begin{cases} (1 - e^{4x})^x & \text{if } x < 0 \\ k & \text{if } 0 \leq x \end{cases}$, find k so that f is continuous at $x = 0$.

36. If $f(x) = \begin{cases} (x + 1)^{(\ln k)/x} & \text{if } x \neq 0 \\ 5 & \text{if } x = 0 \end{cases}$, find k so that f is continuous at $x = 0$.

37. If $\lim\limits_{x \to +\infty} \left(\dfrac{nx + 1}{nx - 1} \right)^x = 9$, find n.

38. Evaluate the limit in Example 3 without using L'Hôpital's rule but by using Theorem 2.8.2 and the identities $\cos(\tfrac{1}{2}\pi - t) = \sin t$ and $\sin(\tfrac{1}{2}\pi - t) = \cos t$.

39. (a) Prove that $\lim\limits_{x \to 0}(e^{-1/x^2}/x^n) = 0$ for any positive integer n.
(b) If $f(x) = e^{-1/x^2}$, use the result of (a) to prove that the limits of f and all of its derivatives, as x approaches 0, are 0.

40. Suppose $f(x) = \int_1^x e^{3t} \sqrt{9t^4 + 1} \, dt$ and $g(x) = x^n e^{3x}$. If
$$\lim\limits_{x \to +\infty} \left[\dfrac{f'(x)}{g'(x)} \right] = 1, \text{ find } n.$$

In Exercises 41 and 42, draw a sketch of the graph of f by first finding the relative extrema of f and the horizontal asymptotes of the graph, if there are any.

41. $f(x) = x^{1/x}, \; x > 0$

42. $f(x) = x^x, \; x > 0$

43. Determine each of the following limits: (a) $\lim\limits_{x \to 0^+} (\sin x)^{\csc x}$;
(b) $\lim\limits_{x \to +\infty} \left(\sin \dfrac{1}{x} \right)^x$. Is $0^{+\infty}$ an indeterminate form?

11.3 IMPROPER INTEGRALS WITH INFINITE LIMITS OF INTEGRATION

In defining the definite integral $\int_a^b f(x) \, dx$, we assumed that the function f was defined on the closed interval $[a, b]$. We now extend the definition of the definite integral to consider an infinite interval of integration and call such an integral an **improper integral**. In Section 11.4 we discuss another kind of improper integral.

▶ **ILLUSTRATION 1** Consider the problem of finding the area of the region bounded by the curve $y = e^{-x}$, the x axis, the y axis, and the line $x = b$, where $b > 0$. This region appears in Figure 1. If A square units is the area of the region,

FIGURE 1

$$A = \lim\limits_{||\Delta|| \to 0} \sum_{i=1}^{n} e^{-\xi_i} \Delta_i x$$
$$= \int_0^b e^{-x} \, dx$$
$$= -e^{-x} \Big]_0^b$$
$$= 1 - e^{-b}$$

If we let b increase without bound, then

$$\lim_{b \to +\infty} \int_0^b e^{-x}\, dx = \lim_{b \to +\infty} (1 - e^{-b})$$

$$\lim_{b \to +\infty} \int_0^b e^{-x}\, dx = 1 \tag{1}$$

From (1) it follows that no matter how large a value we take for b, the area of the region shown in Figure 1 will always be less than 1 square unit.   ◄

Equation (1) states that if $b > 0$, for any $\epsilon > 0$ there exists an $N > 0$ such that

$$\text{if} \quad b > N \quad \text{then} \quad \left| \int_0^b e^{-x}\, dx - 1 \right| < \epsilon$$

In place of (1) we write

$$\int_0^{+\infty} e^{-x}\, dx = 1$$

In general, we have the following definition.

11.3.1 DEFINITION   If f is continuous for all $x \geq a$, then

$$\int_a^{+\infty} f(x)\, dx = \lim_{b \to +\infty} \int_a^b f(x)\, dx$$

if this limit exists.

If the lower limit of integration is infinite, we have the following definition.

11.3.2 DEFINITION   If f is continuous for all $x \leq b$, then

$$\int_{-\infty}^b f(x)\, dx = \lim_{a \to -\infty} \int_a^b f(x)\, dx$$

if this limit exists.

Finally, we have the case when both limits of integration are infinite.

11.3.3 DEFINITION   If f is continuous for all values of x, and c is any real number, then

$$\int_{-\infty}^{+\infty} f(x)\, dx = \lim_{a \to -\infty} \int_a^c f(x)\, dx + \lim_{b \to +\infty} \int_c^b f(x)\, dx \tag{2}$$

if both these limits exist.

In Exercise 40 you are asked to prove that when the limits exist, the right side of (2) is independent of the choice of c. When Definition 11.3.3 is applied, c is usually taken as 0.

In the above three definitions, if the limits exist, we say that the improper integral is **convergent**. If the limits do not exist, we say that the improper integral is **divergent**.

EXAMPLE 1   Evaluate the integral, if it converges:

$$\int_{-\infty}^{2} \frac{dx}{(4 - x)^2}$$

Solution

$$\int_{-\infty}^{2} \frac{dx}{(4 - x)^2} = \lim_{a \to -\infty} \int_{a}^{2} \frac{dx}{(4 - x)^2}$$

$$= \lim_{a \to -\infty} \left[\frac{1}{4 - x} \right]_{a}^{2}$$

$$= \lim_{a \to -\infty} \left(\frac{1}{2} - \frac{1}{4 - a} \right)$$

$$= \tfrac{1}{2} - 0$$

$$= \tfrac{1}{2}$$

EXAMPLE 2   Evaluate the integral, if it converges:

$$\int_{0}^{+\infty} xe^{-x} \, dx$$

Solution

$$\int_{0}^{+\infty} xe^{-x} \, dx = \lim_{b \to +\infty} \int_{0}^{b} xe^{-x} \, dx$$

To evaluate the integral we use integration by parts with $u = x$, $dv = e^{-x} \, dx$, $du = dx$, and $v = -e^{-x}$. Thus

$$\int_{0}^{+\infty} xe^{-x} \, dx = \lim_{b \to +\infty} \left[-xe^{-x} - e^{-x} \right]_{0}^{b}$$

$$= \lim_{b \to +\infty} (-be^{-b} - e^{-b} + 1)$$

$$= -\lim_{b \to +\infty} \frac{b}{e^b} - 0 + 1 \qquad\qquad (3)$$

To evaluate $\lim_{b \to +\infty} \dfrac{b}{e^b}$, we apply L'Hôpital's rule because $\lim_{b \to +\infty} b = +\infty$ and $\lim_{b \to +\infty} e^b = +\infty$. We have

$$\lim_{b \to +\infty} \frac{b}{e^b} = \lim_{b \to +\infty} \frac{1}{e^b}$$

$$= 0$$

Therefore, from (3),

$$\int_{0}^{+\infty} xe^{-x} \, dx = 1$$

EXAMPLE 3   Evaluate if they exist:

(a) $\displaystyle\int_{-\infty}^{+\infty} x\,dx$ (b) $\displaystyle\lim_{r \to +\infty} \int_{-r}^{r} x\,dx$

Solution

(a) From Definition 11.3.3 with $c = 0$ we have

$$\int_{-\infty}^{+\infty} x\,dx = \lim_{a \to -\infty} \int_{a}^{0} x\,dx + \lim_{b \to +\infty} \int_{0}^{b} x\,dx$$

$$= \lim_{a \to -\infty} \left[\tfrac{1}{2}x^2\right]_{a}^{0} + \lim_{b \to +\infty} \left[\tfrac{1}{2}x^2\right]_{0}^{b}$$

$$= \lim_{a \to -\infty} (-\tfrac{1}{2}a^2) + \lim_{b \to +\infty} \tfrac{1}{2}b^2$$

Because neither of these two limits exists, the improper integral diverges.

(b) $\displaystyle\lim_{r \to +\infty} \int_{-r}^{r} x\,dx = \lim_{r \to +\infty} \left[\tfrac{1}{2}x^2\right]_{-r}^{r}$

$$= \lim_{r \to +\infty} (\tfrac{1}{2}r^2 - \tfrac{1}{2}r^2)$$

$$= \lim_{r \to +\infty} 0$$

$$= 0$$

Example 3 illustrates why we do not use the limit in (b) to determine the convergence of an improper integral where both limits of integration are infinite. That is, the improper integral in (a) is divergent, but the limit in (b) exists and is zero.

EXAMPLE 4   Evaluate the integral, if it converges:

$$\int_{-\infty}^{+\infty} \frac{dx}{x^2 + 6x + 12}$$

Solution

$$\int_{-\infty}^{+\infty} \frac{dx}{x^2 + 6x + 12} = \lim_{a \to -\infty} \int_{a}^{0} \frac{dx}{(x+3)^2 + 3} + \lim_{b \to +\infty} \int_{0}^{b} \frac{dx}{(x+3)^2 + 3}$$

$$= \lim_{a \to -\infty} \left[\frac{1}{\sqrt{3}} \tan^{-1} \frac{x+3}{\sqrt{3}}\right]_{a}^{0} + \lim_{b \to +\infty} \left[\frac{1}{\sqrt{3}} \tan^{-1} \frac{x+3}{\sqrt{3}}\right]_{0}^{b}$$

$$= \lim_{a \to -\infty} \left(\frac{1}{\sqrt{3}} \tan^{-1} \sqrt{3} - \frac{1}{\sqrt{3}} \tan^{-1} \frac{a+3}{\sqrt{3}}\right) + \lim_{b \to +\infty} \left(\frac{1}{\sqrt{3}} \tan^{-1} \frac{b+3}{\sqrt{3}} - \frac{1}{\sqrt{3}} \tan^{-1} \sqrt{3}\right)$$

$$= \frac{1}{\sqrt{3}} \left[\lim_{a \to -\infty} \left(-\tan^{-1} \frac{a+3}{\sqrt{3}}\right) + \lim_{b \to +\infty} \tan^{-1} \frac{b+3}{\sqrt{3}}\right]$$

$$= \frac{1}{\sqrt{3}} \left[-\left(-\frac{\pi}{2}\right) + \frac{\pi}{2}\right]$$

$$= \frac{\pi}{\sqrt{3}}$$

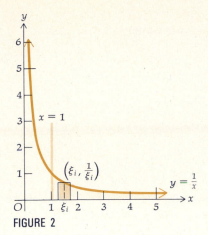

FIGURE 2

EXAMPLE 5   Is it possible to assign a finite number to represent the measure of the area of the region to the right of the line $x = 1$, below the graph of $y = 1/x$, and above the x axis?

Solution   The region is shown in Figure 2. Let L be the number we wish to assign to the measure of the area, if possible. Let A be the measure of the area of the region bounded by the graphs of the equations $y = 1/x$, $y = 0$, $x = 1$, and $x = b$, where $b > 1$. Then

$$A = \lim_{||\Delta|| \to 0} \sum_{i=1}^{n} \frac{1}{\xi_i} \Delta_i x$$

$$= \int_1^b \frac{1}{x} \, dx$$

So we shall let $L = \lim_{b \to +\infty} A$ if this limit exists. But

$$\lim_{b \to +\infty} A = \lim_{b \to +\infty} \int_1^b \frac{1}{x} \, dx$$

$$= \lim_{b \to +\infty} \left[\ln b - \ln 1 \right]$$

$$= +\infty$$

Therefore it is not possible to assign a finite number to represent the measure of the area of the region.

EXAMPLE 6   Is it possible to assign a finite number to represent the measure of the volume of the solid formed by revolving the region in Example 5 about the x axis?

Solution   The element of volume is a circular disk having a thickness of $\Delta_i x$ and a base radius of $1/\xi_i$. Let L be the number we wish to assign to the measure of the volume, and let V be the measure of the volume of the solid formed by revolving about the x axis the region bounded by the graphs of the equations $y = 1/x$, $y = 0$, $x = 1$, and $x = b$, where $b > 1$. Then

$$V = \lim_{||\Delta|| \to 0} \sum_{i=1}^{n} \pi \left(\frac{1}{\xi_i} \right)^2 \Delta_i x$$

$$= \pi \int_1^b \frac{1}{x^2} \, dx$$

We shall let $L = \lim_{b \to +\infty} V$, if the limit exists.

$$\lim_{b \to +\infty} V = \lim_{b \to +\infty} \pi \int_1^b \frac{dx}{x^2}$$

$$= \pi \lim_{b \to +\infty} \left[-\frac{1}{x} \right]_1^b$$

$$= \pi \lim_{b \to +\infty} \left(-\frac{1}{b} + 1 \right)$$

Therefore we assign π to represent the measure of the volume of the solid.

EXAMPLE 7   Determine if $\int_0^{+\infty} \sin x \, dx$ is convergent or divergent.

Solution

$$\int_0^{+\infty} \sin x \, dx = \lim_{b \to +\infty} \int_0^b \sin x \, dx$$

$$= \lim_{b \to +\infty} \left[-\cos x \right]_0^b$$

$$= \lim_{b \to +\infty} (-\cos b + 1)$$

For any integer n, as b takes on all values from $n\pi$ to $2n\pi$, $\cos b$ takes on all values from -1 to 1. Hence $\lim_{b \to +\infty} \cos b$ does not exist. Therefore the improper integral is divergent.

Example 7 illustrates the case for which an improper integral is divergent where the limit is not infinite.

An application of an improper integral with an infinite limit of integration involves probability. The probability of a particular event occurring is a number in the closed interval $[0, 1]$. If an event is certain to occur, then the probability of its happening is 1; if the event will never occur, then the probability is 0. The surer it is that an event will occur, the closer its probability is to 1.

Suppose that the set of all possible outcomes of a particular situation is the set of all numbers x in some interval I. For instance, x may be the number of minutes in the waiting time for a table at a particular restaurant, the number of hours in the life of a tube for a television set, or the number of inches in a person's height. It is sometimes necessary to determine the probability of x being in some subinterval of I. For example, one may wish to find the probability that a person will have to wait between 20 and 30 minutes for a table at a restaurant, or the probability of a television tube lasting more than 2000 hours, or the probability that someone chosen at random will have a height between 66 and 72 inches. Such problems involve evaluating an integral of a function called a *probability density function*. Probability density functions are obtained from statistical experiments. We give a brief informal discussion of them here to show how improper integrals arise.

A **probability density function** is a function f having as its domain the set R of real numbers and that satisfies the following two conditions:

1. $f(x) \geq 0$ for all x in R.

2. $\int_{-\infty}^{+\infty} f(x) \, dx = 1$

We shall consider here the *exponential density function* defined by

$$f(x) = \begin{cases} ke^{-kx} & \text{if } x \geq 0 \\ 0 & \text{if } x < 0 \end{cases} \tag{4}$$

where $k > 0$. To verify that this function qualifies as a probability density function we show that the two properties hold.

1. If $x < 0$, $f(x) = 0$; if $x \geq 0$, $f(x) = ke^{-kx}$, and because $k > 0$, $ke^{-kx} > 0$.

2. $\displaystyle\int_{-\infty}^{+\infty} f(x)\,dx = \int_{-\infty}^{0} f(x)\,dx + \int_{0}^{+\infty} f(x)\,dx$

$\displaystyle\qquad\qquad\quad = \int_{-\infty}^{0} 0\,dx + \int_{0}^{+\infty} ke^{-kx}\,dx$

$\displaystyle\qquad\qquad\quad = 0 + \lim_{b \to +\infty}\left[-\int_{0}^{b} e^{-kx}(-k\,dx) \right]$

$\displaystyle\qquad\qquad\quad = \lim_{b \to +\infty}\left[-e^{-kx} \right]_{0}^{b}$

$\displaystyle\qquad\qquad\quad = \lim_{b \to +\infty} (-e^{-kb} + 1)$

$\displaystyle\qquad\qquad\quad = 1$

If f is a probability density function for a particular event occurring, then the **probability that the event will occur over the closed interval $[a, b]$** is denoted by $P([a, b])$ and

$$P([a, b]) = \int_{a}^{b} f(x)\,dx$$

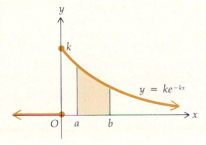

$y = ke^{-kx}$

FIGURE 3

Figure 3 shows a sketch of the graph of the exponential density function. Because $\int_{0}^{+\infty} ke^{-kx}\,dx = 1$, the measure of the area of the region bounded by $y = ke^{-kx}$, the x axis, and the y axis is 1. The measure of the area of the shaded region in the figure is $P([a, b])$.

EXAMPLE 8   For a particular kind of battery the probability density function for x hours to be the life of a battery selected at random is given by

$$f(x) = \begin{cases} \frac{1}{60}e^{-x/60} & \text{if } x \geq 0 \\ 0 & \text{if } x < 0 \end{cases} \qquad\qquad (5)$$

Find the probability that the life of a battery selected at random will be (a) between 15 and 25 hours and (b) at least 50 hours.

Solution   The function defined by (5) is of the form of (4) with $k = \frac{1}{60}$. (a) The probability that the life of a battery selected at random will be between 15 and 25 hours is $P([15, 25])$, and (b) the probability that it will be at least 50 hours is $P([50, +\infty))$.

(a) $\displaystyle P([15, 25]) = \int_{15}^{25} \frac{1}{60}e^{-x/60}\,dx$

$\displaystyle\qquad\qquad\quad = -\int_{15}^{25} e^{-x/60}\left(-\tfrac{1}{60}dx\right)$

$\displaystyle\qquad\qquad\quad = -e^{-x/60}\Big]_{15}^{25}$

$\displaystyle\qquad\qquad\quad = -e^{-25/60} + e^{-15/60}$

$\displaystyle\qquad\qquad\quad = -e^{-0.417} + e^{-0.250}$

$\displaystyle\qquad\qquad\quad = -0.659 + 0.779$

$\displaystyle\qquad\qquad\quad = 0.120$

$$(b) \quad P([50, +\infty)) = \lim_{b \to +\infty} \int_{50}^{b} \tfrac{1}{60} e^{-x/60}\, dx$$

$$= \lim_{b \to +\infty} \left[-e^{-x/60} \right]_{50}^{b}$$

$$= \lim_{b \to +\infty} (-e^{-b/60} + e^{-50/60})$$

$$= 0 + e^{-0.833}$$

$$= 0.435$$

EXERCISES 11.3

In Exercises 1 through 18, determine whether the improper integral is convergent or divergent. If it is convergent, evaluate it.

1. $\displaystyle\int_{0}^{+\infty} e^{-x}\, dx$

2. $\displaystyle\int_{-\infty}^{1} e^{x}\, dx$

3. $\displaystyle\int_{-\infty}^{0} x5^{-x^2}\, dx$

4. $\displaystyle\int_{1}^{+\infty} 2^{-x}\, dx$

5. $\displaystyle\int_{0}^{+\infty} x2^{-x}\, dx$

6. $\displaystyle\int_{5}^{+\infty} \frac{dx}{\sqrt{x-1}}$

7. $\displaystyle\int_{-\infty}^{+\infty} x \cosh x\, dx$

8. $\displaystyle\int_{-\infty}^{0} x^2 e^{x}\, dx$

9. $\displaystyle\int_{5}^{+\infty} \frac{x\, dx}{\sqrt[3]{9-x^2}}$

10. $\displaystyle\int_{-\infty}^{+\infty} \frac{3x\, dx}{(3x^2+2)^3}$

11. $\displaystyle\int_{\sqrt{3}}^{+\infty} \frac{3\, dx}{x^2+9}$

12. $\displaystyle\int_{e}^{+\infty} \frac{dx}{x \ln x}$

13. $\displaystyle\int_{-\infty}^{+\infty} e^{-|x|}\, dx$

14. $\displaystyle\int_{-\infty}^{+\infty} xe^{-x^2}\, dx$

15. $\displaystyle\int_{e}^{+\infty} \frac{dx}{x(\ln x)^2}$

16. $\displaystyle\int_{-\infty}^{+\infty} \frac{dx}{16+x^2}$

17. $\displaystyle\int_{1}^{+\infty} \ln x\, dx$

18. $\displaystyle\int_{0}^{+\infty} e^{-x} \cos x\, dx$

19. Evaluate if they exist:

(a) $\displaystyle\int_{-\infty}^{+\infty} \sin x\, dx$ (b) $\displaystyle\lim_{r \to +\infty} \int_{-r}^{r} \sin x\, dx$

20. Prove that if $\displaystyle\int_{-\infty}^{b} f(x)\, dx$ is convergent, then $\displaystyle\int_{-b}^{+\infty} f(-x)\, dx$ is also convergent and has the same value.

21. Show that the improper integral $\displaystyle\int_{-\infty}^{+\infty} x(1+x^2)^{-2}\, dx$ is convergent and the improper integral $\displaystyle\int_{-\infty}^{+\infty} x(1+x^2)^{-1}\, dx$ is divergent.

22. Prove that the improper integral $\displaystyle\int_{1}^{+\infty} \frac{dx}{x^n}$ is convergent if and only if $n > 1$.

23. Determine if it is possible to assign a finite number to represent the measure of the area of the region bounded by the curve whose equation is $y = 1/(e^x + e^{-x})$ and the x axis. If a finite number can be assigned, find it.

24. Determine if it is possible to assign a finite number to represent the measure of the area of the region bounded by the x axis, the line $x = 2$, and the curve whose equation is $y = 1/(x^2 - 1)$. If a finite number can be assigned, find it.

25. Determine if it is possible to assign a finite number to represent the measure of the volume of the solid formed by revolving about the x axis the region to the right of the line $x = 1$ and bounded by the curve whose equation is $y = 1/x^{3/2}$ and the x axis. If a finite number can be assigned, find it.

26. Determine if it is possible to assign a finite number to represent the measure of the volume of the solid formed by revolving about the x axis the region bounded by the x axis, the y axis, and the curve whose equation is $y = e^{-2x}$. If a finite number can be assigned, find it.

27. For the battery of Example 8, find the probability that the life of a battery selected at random will be (a) not more than 50 hours and (b) at least 75 hours.

28. For a certain type of light bulb, the probability density function that x hours will be the life of a bulb selected at random is given by

$$f(x) = \begin{cases} \tfrac{1}{40} e^{-x/40} & \text{if } x \ge 0 \\ 0 & \text{if } x < 0 \end{cases}$$

Find the probability that the life of a bulb selected at random will be (a) between 40 and 60 hours and (b) at least 60 hours.

29. In a certain city, the probability density function for x minutes to be the length of a telephone call selected at random is given by

$$f(x) = \begin{cases} \tfrac{1}{3} e^{-x/3} & \text{if } x \ge 0 \\ 0 & \text{if } x < 0 \end{cases}$$

Find the probability that a telephone call selected at random will last (a) between 1 min and 2 min, and (b) at least 5 min.

30. For a particular appliance, the probability density function that it will need servicing x months after it is purchased is given by

$$f(x) = \begin{cases} 0.02 e^{-0.02x} & \text{if } x \ge 0 \\ 0 & \text{if } x < 0 \end{cases}$$

If the appliance is guaranteed for a year, what is the probability that a customer selected at random will not need servicing during the 1-year warranty period?

31. If f is a probability density function, then the *mean* (or *average value*) of the probabilities is given by $\int_{-\infty}^{+\infty} xf(x)\, dx$. Find the mean of the probabilities obtained from the exponential density function (4).

32. A *uniform* probability density function is defined by

$$f(x) = \begin{cases} 0 & \text{if } x < c \\ \dfrac{1}{d-c} & \text{if } c \le x \le d \\ 0 & \text{if } d < x \end{cases}$$

Show that this function satisfies the conditions necessary for it to qualify as a probability density function.

Exercises 33 through 36 show an application of improper integrals in the field of economics. Suppose there is a continuous flow of income for which interest is compounded continuously at the annual rate of 100i percent and $f(t)$ dollars is the income per year at any time t years. If the income continues indefinitely, the present value, V dollars, of all future income is defined by

$$V = \int_0^{+\infty} f(t)e^{-it}\, dt$$

33. A continuous flow of income is decreasing with time, and at t years the number of dollars in the annual income is $1000 \cdot 2^{-t}$. Find the present value of this income if it continues indefinitely using an interest rate of 8 percent compounded continuously.

34. Suppose that the owner of a piece of business property holds a permanent lease on the property so that the rent is paid perpetually. If the annual rent is \$12,000 and money is worth 10 percent compounded continuously, find the present value of all future rent payments.

35. The British Consol is a bond with no maturity (i.e., it never comes due), and it affords the holder an annual lump-sum payment. By finding the present value of a flow of payments of R dollars annually and using the current interest rate 100i percent, compounded continuously, show that the fair selling price of a British Consol is R/i dollars.

36. The continuous flow of profit for a company is increasing with time, and at t years the number of dollars in the profit per year is proportional to t. Show that the present value of the company is inversely proportional to i^2, where 100i percent is the interest rate compounded continuously.

37. Determine the values of n for which the following improper integral is convergent: $\int_e^{+\infty} \dfrac{dx}{x(\ln x)^n}$.

38. Determine a value of n for which the improper integral $\int_1^{+\infty} \left(\dfrac{n}{x+1} - \dfrac{3x}{2x^2+n} \right) dx$ is convergent, and evaluate the integral for this value of n.

39. Determine a value of n for which the improper integral $\int_1^{+\infty} \left(\dfrac{nx^2}{x^3+1} - \dfrac{1}{3x+1} \right) dx$ is convergent, and evaluate the integral for this value of n.

40. Suppose f is continuous for all values of x. Prove that if

$$\lim_{a \to -\infty} \int_a^c f(x)\, dx = L \quad \text{and} \quad \lim_{b \to +\infty} \int_c^b f(x)\, dx = M$$

then if d is any real number,

$$\lim_{a \to -\infty} \int_a^d f(x)\, dx + \lim_{b \to +\infty} \int_d^b f(x)\, dx = L + M$$

Hint: $\int_a^d f(x)\, dx = \int_a^c f(x)\, dx + \int_c^d f(x)\, dx$
$\int_d^b f(x)\, dx = \int_d^c f(x)\, dx + \int_c^b f(x)\, dx$

11.4 OTHER IMPROPER INTEGRALS

Figure 1 shows the region bounded by the curve whose equation is $y = 1/\sqrt{x}$, the x axis, the y axis, and the line $x = 4$. If it is possible to assign a finite number to the measure of the area of this region, it would be given by

$$\lim_{\|\Delta\| \to 0} \sum_{i=1}^n \frac{1}{\sqrt{\xi_i}} \Delta_i x$$

If this limit exists, it is the definite integral denoted by

$$\int_0^4 \frac{dx}{\sqrt{x}} \tag{1}$$

However, the integrand is discontinuous at the lower limit zero. Furthermore, $\lim\limits_{x \to 0^+} 1/\sqrt{x} = +\infty$, and so we state that the integrand has an infinite discontinuity at the lower limit. Such an integral is improper, and its existence can be determined from the following definition.

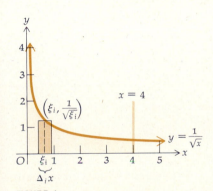

FIGURE 1

11.4.1 DEFINITION   If f is continuous at all x in the interval half open on the left $(a, b]$, and if $\lim\limits_{x \to a^+} f(x) = \pm\infty$, then

$$\int_a^b f(x)\,dx = \lim_{t \to a^+} \int_t^b f(x)\,dx$$

if this limit exists.

▶ **ILLUSTRATION 1**   We determine if a finite number can be assigned to the measure of the area of the region in Figure 1. From the discussion preceding Definition 11.4.1, the measure of the area of the given region will be the improper integral (1) if it exists. By Definition 11.4.1,

$$\int_0^4 \frac{dx}{\sqrt{x}} = \lim_{t \to 0^+} \int_t^4 \frac{dx}{\sqrt{x}}$$

$$= \lim_{t \to 0^+} 2x^{1/2} \Big]_t^4$$

$$= \lim_{t \to 0^+} (4 - 2\sqrt{t})$$

$$= 4 - 0$$

$$= 4$$

Therefore we assign 4 to the measure of the area of the given region.   ◀

If the integrand has an infinite discontinuity at the upper limit of integration, we use the following definition to determine the existence of the improper integral.

11.4.2 DEFINITION   If f is continuous at all x in the interval half open on the right $[a, b)$, and if $\lim\limits_{x \to b^-} f(x) = \pm\infty$, then

$$\int_a^b f(x)\,dx = \lim_{t \to b^-} \int_a^t f(x)\,dx$$

if this limit exists.

If there is an infinite discontinuity at an interior point of the interval of integration, the existence of the improper integral is determined from the following definition.

11.4.3 DEFINITION   If f is continuous at all x in the interval $[a, b]$ except c, where $a < c < b$, and if $\lim\limits_{x \to c} |f(x)| = +\infty$, then

$$\int_a^b f(x)\,dx = \lim_{t \to c^-} \int_a^t f(x)\,dx + \lim_{s \to c^+} \int_s^b f(x)\,dx$$

if both these limits exist.

If $\int_a^b f(x)\,dx$ is an improper integral, it is convergent if the corresponding limit (or limits) exists; otherwise it is divergent.

EXAMPLE 1   Evaluate the integral, if it is convergent:

$$\int_0^2 \frac{dx}{(x-1)^2}$$

Solution   The integrand has an infinite discontinuity at 1. Applying Definition 11.4.3 we have

$$\int_0^2 \frac{dx}{(x-1)^2} = \lim_{t \to 1^-} \int_0^t \frac{dx}{(x-1)^2} + \lim_{s \to 1^+} \int_s^2 \frac{dx}{(x-1)^2}$$

$$= \lim_{t \to 1^-} \left[-\frac{1}{x-1} \right]_0^t + \lim_{s \to 1^+} \left[-\frac{1}{x-1} \right]_s^2$$

$$= \lim_{t \to 1^-} \left[-\frac{1}{t-1} - 1 \right] + \lim_{s \to 1^+} \left[-1 + \frac{1}{s-1} \right]$$

Because neither of these limits exist, the improper integral is divergent.

▶ **ILLUSTRATION 2**   Suppose that in evaluating the integral in Example 1 we had failed to note the infinite discontinuity of the integrand at 1. We would have obtained

$$-\frac{1}{x-1} \Bigg]_0^2 = -\frac{1}{1} + \frac{1}{-1}$$

$$= -2$$

This is obviously an incorrect result. Because $1/(x-1)^2$ is never negative, the integral from 0 to 2 could not possibly be a negative number.   ◀

EXAMPLE 2   Evaluate the integral, if it is convergent:

$$\int_0^1 x \ln x \, dx$$

Solution   The integrand has a discontinuity at the lower limit. From Definition 11.4.1,

$$\int_0^1 x \ln x \, dx = \lim_{t \to 0^+} \int_t^1 x \ln x \, dx$$

$$= \lim_{t \to 0^+} \left[\tfrac{1}{2}x^2 \ln x - \tfrac{1}{4}x^2 \right]_t^1$$

$$= \lim_{t \to 0^+} \left[\tfrac{1}{2} \ln 1 - \tfrac{1}{4} - \tfrac{1}{2}t^2 \ln t + \tfrac{1}{4}t^2 \right]$$

Hence

$$\int_0^1 x \ln x \, dx = 0 - \tfrac{1}{4} - \tfrac{1}{2} \lim_{t \to 0^+} t^2 \ln t + 0 \qquad (2)$$

To evaluate

$$\lim_{t \to 0^+} t^2 \ln t = \lim_{t \to 0^+} \frac{\ln t}{\dfrac{1}{t^2}}$$

we apply L'Hôpital's rule, because $\lim\limits_{t\to 0^+} \ln t = -\infty$ and $\lim\limits_{t\to 0^+} 1/t^2 = +\infty$. We have

$$\lim_{t\to 0^+} \frac{\ln t}{\dfrac{1}{t^2}} = \lim_{t\to 0^+} \frac{\dfrac{1}{t}}{-\dfrac{2}{t^3}}$$

$$= \lim_{t\to 0^+} \left[-\frac{t^2}{2} \right]$$

$$= 0$$

Therefore, from (2),

$$\int_0^1 x \ln x \, dx = -\tfrac{1}{4}$$

EXAMPLE 3   Evaluate the integral, if it is convergent:

$$\int_1^{+\infty} \frac{dx}{x\sqrt{x^2 - 1}}$$

Solution   For this integral there is both an infinite upper limit and an infinite discontinuity of the integrand at the lower limit. We proceed as follows.

$$\int_1^{+\infty} \frac{dx}{x\sqrt{x^2 - 1}} = \lim_{t\to 1^+} \int_t^2 \frac{dx}{x\sqrt{x^2 - 1}} + \lim_{b\to +\infty} \int_2^b \frac{dx}{x\sqrt{x^2 - 1}}$$

$$= \lim_{t\to 1^+} \left[\sec^{-1} x \right]_t^2 + \lim_{b\to +\infty} \left[\sec^{-1} x \right]_2^b$$

$$= \lim_{t\to 1^+} (\sec^{-1} 2 - \sec^{-1} t) + \lim_{b\to +\infty} (\sec^{-1} b - \sec^{-1} 2)$$

$$= \tfrac{1}{3}\pi - \lim_{t\to 1^+} \sec^{-1} t + \lim_{b\to +\infty} \sec^{-1} b - \tfrac{1}{3}\pi$$

$$= -0 + \tfrac{1}{2}\pi$$

$$= \tfrac{1}{2}\pi$$

The integral in Example 3 is called an *improper integral of mixed type.*

EXERCISES 11.4

In Exercises 1 through 25, determine whether the improper integral is convergent or divergent. If it is convergent, evaluate it.

1. $\int_0^1 \dfrac{dx}{\sqrt{1 - x}}$

2. $\int_0^{16} \dfrac{dx}{x^{3/4}}$

3. $\int_{-5}^{-3} \dfrac{x \, dx}{\sqrt{x^2 - 9}}$

4. $\int_0^4 \dfrac{x \, dx}{\sqrt{16 - x^2}}$

5. $\int_2^4 \dfrac{dt}{\sqrt{16 - t^2}}$

6. $\int_{-4}^1 \dfrac{dz}{(z + 3)^3}$

7. $\int_{\pi/4}^{\pi/2} \sec \theta \, d\theta$

8. $\int_{-2}^0 \dfrac{dx}{\sqrt{4 - x^2}}$

9. $\int_0^{+\infty} \dfrac{dx}{x^3}$

10. $\int_0^{\pi/2} \tan \theta \, d\theta$

11. $\int_0^{\pi/2} \dfrac{dy}{1 - \sin y}$

12. $\int_0^2 \dfrac{dx}{(x - 1)^{2/3}}$

13. $\int_0^4 \dfrac{dx}{x^2 - 2x - 3}$

14. $\int_2^{+\infty} \dfrac{dx}{x\sqrt{x^2 - 4}}$

15. $\int_0^{+\infty} \ln x \, dx$

16. $\int_0^2 \dfrac{dx}{\sqrt{2x - x^2}}$

17. $\int_{-2}^0 \dfrac{dw}{(w + 1)^{1/3}}$

18. $\int_{-1}^1 \dfrac{dx}{x^2}$

19. $\int_{-2}^2 \dfrac{dx}{x^3}$

20. $\int_0^{+\infty} \dfrac{e^{-\sqrt{x}}}{\sqrt{x}} \, dx$

21. $\int_{1/2}^2 \dfrac{dz}{z(\ln z)^{1/5}}$

22. $\int_0^2 \dfrac{x\,dx}{1-x}$ 23. $\int_1^2 \dfrac{dx}{x\sqrt{x^2-1}}$ 24. $\int_0^1 \dfrac{dx}{x\sqrt{4-x^2}}$

25. $\int_1^3 \dfrac{dy}{\sqrt[3]{y-2}}$

26. Evaluate, if they exist: (a) $\int_{-1}^1 \dfrac{dx}{x}$; (b) $\lim\limits_{r\to 0^+}\left[\int_{-1}^{-r}\dfrac{dx}{x}+\int_r^1\dfrac{dx}{x}\right]$.

In Exercises 27 through 29, find the values of n for which the improper integral converges, and evaluate the integral for these values of n.

27. $\int_0^1 x^n\,dx$ 28. $\int_0^1 x^n\ln x\,dx$ 29. $\int_0^1 x^n\ln^2 x\,dx$

30. Show that it is possible to assign a finite number to represent the measure of the area of the region bounded by the curve

whose equation is $y=1/\sqrt{x}$, the line $x=1$, and the x and y axes, but that it is not possible to assign a finite number to represent the measure of the volume of the solid of revolution generated if this region is revolved about the x axis.

31. Determine if it is possible to assign a finite number to represent the measure of the volume of the solid formed by revolving about the x axis the region bounded by the curve whose equation is $y=x^{-1/3}$, the line $x=8$, and the x and y axes.

32. Given the improper integral $\int_a^b \dfrac{dx}{(x-a)^n}$ where $b>a$. Determine if the integral is convergent or divergent in each case: (a) $0<n<1$; (b) $n=1$; (c) $n>1$. If the integral is convergent, evaluate it.

33. Use integration to verify that the circumference of the circle $x^2+y^2=a^2$ is $2\pi a$.

11.5 TAYLOR'S FORMULA

Values of polynomial functions can be found by performing a finite number of additions and multiplications. However, there are other functions, such as the logarithmic, exponential, and trigonometric functions, that cannot be evaluated as easily. We show in this section that many functions can be approximated by polynomials and that the polynomial, instead of the original function, can be used for computations when the difference between the actual function value and the polynomial approximation is sufficiently small.

There are various methods of approximating a given function by polynomials. One of the most widely used is that involving **Taylor's formula**, named in honor of the English mathematician Brook Taylor (1685–1731). The following theorem, which can be considered as a generalization of the mean-value theorem (4.3.2), gives Taylor's formula.

11.5.1 THEOREM   Let f be a function such that f and its first n derivatives are continuous on the closed interval $[a, b]$. Furthermore, let $f^{(n+1)}(x)$ exist for all x in the open interval (a, b). Then there is a number ξ in the open interval (a, b) such that

$$f(b)=f(a)+\frac{f'(a)}{1!}(b-a)+\frac{f''(a)}{2!}(b-a)^2+\cdots$$

$$+\frac{f^{(n)}(a)}{n!}(b-a)^n+\frac{f^{(n+1)}(\xi)}{(n+1)!}(b-a)^{n+1} \quad (1)$$

Equation (1) also holds if $b<a$; in such a case $[a, b]$ is replaced by $[b, a]$, and (a, b) is replaced by (b, a).

Note that when $n=0$, (1) becomes

$$f(b)=f(a)+f'(\xi)(b-a)$$

where ξ is between a and b. This is the mean-value theorem.

We defer the proof of Theorem 11.5.1 until later in this section.

If in (1) b is replaced by x, Taylor's formula is obtained. It is

$$f(x) = f(a) + \frac{f'(a)}{1!}(x - a) + \frac{f''(a)}{2!}(x - a)^2 + \ldots$$

$$+ \frac{f^{(n)}(a)}{n!}(x - a)^n + \frac{f^{(n+1)}(\xi)}{(n+1)!}(x - a)^{n+1} \qquad (2)$$

where ξ is between a and x.

The condition under which (2) holds is that f and its first n derivatives must be continuous on a closed interval containing a and x, and the $(n + 1)$st derivative of f must exist at all points of the corresponding open interval. Formula (2) may be written as

$$f(x) = P_n(x) + R_n(x) \qquad (3)$$

where

$$P_n(x) = f(a) + \frac{f'(a)}{1!}(x - a) + \frac{f''(a)}{2!}(x - a)^2 + \ldots + \frac{f^{(n)}(a)}{n!}(x - a)^n \qquad (4)$$

and

$$R_n(x) = \frac{f^{(n+1)}(\xi)}{(n+1)!}(x - a)^{n+1} \qquad \text{where } \xi \text{ is between } a \text{ and } x \qquad (5)$$

$P_n(x)$ is called the nth-degree **Taylor polynomial** of the function f at the number a, and $R_n(x)$ is called the *remainder*. The term $R_n(x)$ as given in (5) is called the *Lagrange form* of the remainder, named in honor of the French mathematician Joseph L. Lagrange (1736–1813).

The special case of Taylor's formula obtained by taking $a = 0$ in (2) is

$$f(x) = f(0) + \frac{f'(0)}{1!}x + \frac{f''(0)}{2!}x^2 + \ldots + \frac{f^{(n)}(0)}{n!}x^n + \frac{f^{(n+1)}(\xi)}{(n+1)!}x^{n+1}$$

where ξ is between 0 and x. This formula is called **Maclaurin's formula**, named in honor of the Scottish mathematician Colin Maclaurin (1698–1746). However, the formula was obtained earlier by Taylor and by another British mathematician, James Stirling (1692–1770). The nth degree **Maclaurin polynomial** for a function f, obtained from (4) with $a = 0$, is

$$P_n(x) = f(0) + \frac{f'(0)}{1!}x + \frac{f''(0)}{2!}x^2 + \ldots + \frac{f^{(n)}(0)}{n!}x^n \qquad (6)$$

We may approximate a function by means of a Taylor polynomial at a number a or by a Maclaurin polynomial.

▶ **ILLUSTRATION 1**   We compute the nth degree Maclaurin polynomial for the natural exponential function. If $f(x) = e^x$, all the derivatives of f at x are e^x and the derivatives at zero are 1. Therefore, from (6),

$$P_n(x) = 1 + x + \frac{x^2}{2!} + \frac{x^3}{3!} + \ldots + \frac{x^n}{n!} \qquad (7)$$

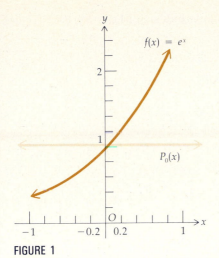

FIGURE 1

Thus the first four Maclaurin polynomials of the natural exponential function are

$$P_0(x) = 1$$

$$P_1(x) = 1 + x$$

$$P_2(x) = 1 + x + \tfrac{1}{2}x^2$$

$$P_3(x) = 1 + x + \tfrac{1}{2}x^2 + \tfrac{1}{6}x^3$$

In Figures 1 through 4 the graph of $f(x) = e^x$ appears along with the graphs of $P_0(x)$, $P_1(x)$, $P_2(x)$, and $P_3(x)$, respectively. In Figure 5 the graphs of the four Maclaurin polynomials and the graph of $f(x) = e^x$ are shown on one coordinate system. Observe how the polynomials approximate e^x for values of x near zero, and notice that as n increases, the approximation improves. Tables 1 and 2 give values of e^x, $P_n(x)$ (when n is 0, 1, 2, and 3) and $e^x - P_n(x)$ for $x = 0.4$ and $x = 0.2$, respectively. Observe that from these two values of x it appears that the closer x is to zero, the better is the approximation for a specific $P_n(x)$. ◀

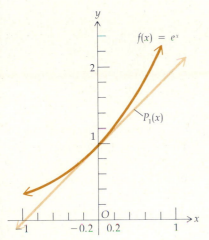

FIGURE 2

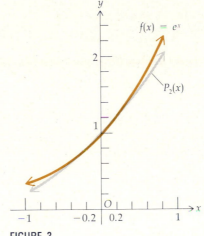

FIGURE 3

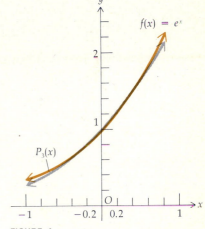

FIGURE 4

Table 1

n	$e^{0.4}$	$P_n(0.4)$	$e^{0.4} - P_n(0.4)$
0	1.4918	1	0.4918
1	1.4918	1.4	0.0918
2	1.4918	1.48	0.0118
3	1.4918	1.4907	0.0011

Table 2

n	$e^{0.2}$	$P_n(0.2)$	$e^{0.2} - P_n(0.2)$
0	1.2214	1	0.2214
1	1.2214	1.2	0.0214
2	1.2214	1.22	0.0014
3	1.2214	1.2213	0.0001

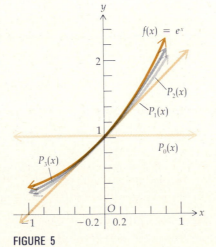

FIGURE 5

▶ **ILLUSTRATION 2**   We now determine the nth degree Maclaurin polynomial for the sine function. If $f(x) = \sin x$, then

$$f'(x) = \cos x \qquad f''(x) = -\sin x \qquad f'''(x) = -\cos x$$

$$f^{(iv)}(x) = \sin x \qquad f^{(v)}(x) = \cos x \qquad f^{(vi)}(x) = -\sin x$$

and so on. Thus $f(0) = 0$,

$$f'(0) = 1 \qquad f''(0) = 0 \qquad f'''(0) = -1 \qquad f^{(iv)}(0) = 0 \qquad f^{(v)}(0) = 1$$

and so on. From (6),

$$P_n(x) = x - \frac{x^3}{3!} + \frac{x^5}{5!} - \frac{x^7}{7!} + \ldots + (-1)^{n-1} \frac{x^{2n-1}}{(2n-1)!}$$

Thus $P_0(x) = 0$,

$$P_1(x) = x \qquad\qquad\qquad\qquad P_2(x) = x$$

$$P_3(x) = x - \frac{x^3}{6} \qquad\qquad\qquad P_4(x) = x - \frac{x^3}{6}$$

$$P_5(x) = x - \frac{x^3}{6} + \frac{x^5}{120} \qquad\qquad P_6(x) = x - \frac{x^3}{6} + \frac{x^5}{120}$$

$$P_7(x) = x - \frac{x^3}{6} + \frac{x^5}{120} - \frac{x^7}{5040} \qquad P_8(x) = x - \frac{x^3}{6} + \frac{x^5}{120} - \frac{x^7}{5040}$$

and so on.

Figure 6 shows the graph of the sine function along with the graphs of its Maclaurin polynomials of degrees 1, 3, 5, and 7. Notice that the polynomial approximations improve as n increases.   ◄

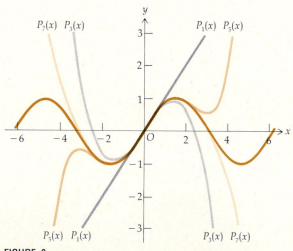

FIGURE 6

EXAMPLE 1   Find the third-degree Taylor polynomial of the cosine function at $\frac{1}{4}\pi$ and the Lagrange form of the remainder.

Solution   Let $f(x) = \cos x$. Then from (4),

$$P_3(x) = f\left(\frac{\pi}{4}\right) + f'\left(\frac{\pi}{4}\right)\left(x - \frac{\pi}{4}\right) + \frac{f''(\frac{1}{4}\pi)}{2!}\left(x - \frac{\pi}{4}\right)^2 + \frac{f'''(\frac{1}{4}\pi)}{3!}\left(x - \frac{\pi}{4}\right)^3$$

Because $f(x) = \cos x$, $f(\frac{1}{4}\pi) = \frac{1}{2}\sqrt{2}$; $f'(x) = -\sin x$, $f'(\frac{1}{4}\pi) = -\frac{1}{2}\sqrt{2}$; $f''(x) = -\cos x$, $f''(\frac{1}{4}\pi) = -\frac{1}{2}\sqrt{2}$; $f'''(x) = \sin x$, $f'''(\frac{1}{4}\pi) = \frac{1}{2}\sqrt{2}$. Therefore

$$P_3(x) = \tfrac{1}{2}\sqrt{2} - \tfrac{1}{2}\sqrt{2}(x - \tfrac{1}{4}\pi) - \tfrac{1}{4}\sqrt{2}(x - \tfrac{1}{4}\pi)^2 + \tfrac{1}{12}\sqrt{2}(x - \tfrac{1}{4}\pi)^3$$

Because $f^{(iv)}(x) = \cos x$, we obtain from (5)

$$R_3(x) = \tfrac{1}{24}(\cos \xi)(x - \tfrac{1}{4}\pi)^4 \qquad \text{where } \xi \text{ is between } \tfrac{1}{4}\pi \text{ and } x$$

Because $|\cos \xi| \le 1$, we may conclude that for all x, $|R_3(x)| \le \tfrac{1}{24}(x - \tfrac{1}{4}\pi)^4$.

From (3),

$$|R_n(x)| = |f(x) - P_n(x)| \tag{8}$$

If $P_n(x)$ is used to approximate $f(x)$, we can obtain an upper bound for the error of this approximation if we can find a number $E > 0$ such that $|R_n(x)| \le E$ or, because of (8), such that $|f(x) - P_n(x)| \le E$ or, equivalently,

$$P_n(x) - E \le f(x) \le P_n(x) + E$$

EXAMPLE 2   Use the result of Example 1 to compute an approximate value of $\cos 47°$, and determine the accuracy of the result.

Solution $47° \sim \frac{47}{180}\pi$ radians. Thus in the solution of Example 1, take $x = \frac{47}{180}\pi$ and $x - \frac{1}{4}\pi = \frac{1}{90}\pi$, and

$$\cos 47° = \tfrac{1}{2}\sqrt{2}\left[1 - \tfrac{1}{90}\pi - \tfrac{1}{2}(\tfrac{1}{90}\pi)^2 + \tfrac{1}{6}(\tfrac{1}{90}\pi)^3\right] + R_3(\tfrac{47}{180}\pi)$$

where

$$R_3(\tfrac{47}{180}\pi) = \tfrac{1}{24}\cos \xi (\tfrac{1}{90}\pi)^4 \qquad \text{with } \tfrac{1}{4}\pi < \xi < \tfrac{47}{180}\pi$$

Because $0 < \cos \xi < 1$,

$$0 < R_3(\tfrac{47}{180}\pi) < \tfrac{1}{24}(\tfrac{1}{90}\pi)^4 < 0.00000007$$

Taking $\frac{1}{90}\pi \approx 0.0349066$ we obtain

$$\cos 47° \approx 0.681998$$

which is accurate to six decimal places.

EXAMPLE 3   Use a Maclaurin polynomial to find the value of $\sqrt{e}$ accurate to four decimal places.

Solution   If $f(x) = e^x$, then from (7), the nth degree Maclaurin polynomial of f is

$$P_n(x) = 1 + x + \frac{x^2}{2!} + \frac{x^3}{3!} + \ldots + \frac{x^n}{n!}$$

and from (5),

$$R_n(x) = \frac{e^\xi}{(n+1)!} x^{n+1} \qquad \text{where } \xi \text{ is between 0 and } x$$

We want $|R_n(\tfrac{1}{2})|$ to be less than 0.00005. If $x = \tfrac{1}{2}$ in the above, and because $e^{1/2} < 2$,

$$|R_n(\tfrac{1}{2})| < \frac{e^{1/2}}{2^{n+1}(n+1)!} < \frac{2}{2^{n+1}(n+1)!} = \frac{1}{2^n(n+1)!}$$

$|R_n(\tfrac{1}{2})|$ will be less than 0.00005 if $1/2^n(n+1)! < 0.00005$. When $n = 5$,

$$\frac{1}{2^n(n+1)!} = \frac{1}{(32)(720)}$$

$$= 0.00004$$

Because $0.00004 < 0.00005$, we take $P_5(\tfrac{1}{2})$ as the approximation of $\sqrt{e}$ accurate to four decimal places. Because

$$P_5(\tfrac{1}{2}) = 1 + \tfrac{1}{2} + \tfrac{1}{8} + \tfrac{1}{48} + \tfrac{1}{384} + \tfrac{1}{3840}$$

we obtain $\sqrt{e} \approx 1.6487$.

We now prove Theorem 11.5.1. There are several known proofs of this theorem, although none is very well motivated. The one following makes use of Cauchy's mean-value theorem (11.1.3).

Proof of Theorem 11.5.1 Let F and G be two functions defined by

$$F(x) = f(b) - f(x) - f'(x)(b-x) - \frac{f''(x)}{2!}(b-x)^2 - \ldots - \frac{f^{(n-1)}(x)}{(n-1)!}(b-x)^{n-1} - \frac{f^{(n)}(x)}{n!}(b-x)^n \quad (9)$$

and

$$G(x) = \frac{(b-x)^{n+1}}{(n+1)!} \tag{10}$$

It follows that $F(b) = 0$ and $G(b) = 0$. Differentiating in (9) we get

$$F'(x) = -f'(x) + f'(x) - f''(x)(b-x) + \frac{2f''(x)(b-x)}{2!}$$

$$- \frac{f'''(x)(b-x)^2}{2!} + \frac{3f'''(x)(b-x)^2}{3!} - \frac{f^{(iv)}(x)(b-x)^3}{3!} + \ldots$$

$$+ \frac{(n-1)f^{(n-1)}(x)(b-x)^{n-2}}{(n-1)!} - \frac{f^{(n)}(x)(b-x)^{n-1}}{(n-1)!}$$

$$+ \frac{nf^{(n)}(x)(b-x)^{n-1}}{n!} - \frac{f^{(n+1)}(x)(b-x)^n}{n!}$$

Combining terms we see that the sum of every odd-numbered term with the following even-numbered term is zero; so only the last term remains. Therefore

$$F'(x) = -\frac{f^{(n+1)}(x)}{n!}(b-x)^n \tag{11}$$

Differentiating in (10) we obtain

$$G'(x) = -\frac{1}{n!}(b - x)^n \qquad (12)$$

Checking the hypothesis of Theorem 11.1.3 we see that

  (i) F and G are continuous on $[a, b]$;
  (ii) F and G are differentiable on (a, b);
  (iii) for all x in (a, b), $G'(x) \neq 0$.

So by the conclusion of Theorem 11.1.3,

$$\frac{F(b) - F(a)}{G(b) - G(a)} = \frac{F'(\xi)}{G'(\xi)}$$

where ξ is in (a, b). But $F(b) = 0$ and $G(b) = 0$. So

$$F(a) = \frac{F'(\xi)}{G'(\xi)} G(a) \qquad (13)$$

for some ξ in (a, b).

Letting $x = a$ in (10), $x = \xi$ in (11), and $x = \xi$ in (12) and substituting into (13) we obtain

$$F(a) = -\frac{f^{(n+1)}(\xi)}{n!}(b - \xi)^n \left[-\frac{n!}{(b - \xi)^n} \right] \frac{(b - a)^{n+1}}{(n + 1)!}$$

$$F(a) = \frac{f^{(n+1)}(\xi)}{(n + 1)!}(b - a)^{n+1} \qquad (14)$$

If $x = a$ in (9), we obtain

$$F(a) = f(b) - f(a) - f'(a)(b - a) - \frac{f''(a)}{2!}(b - a)^2 - \ldots - \frac{f^{(n-1)}(a)}{(n - 1)!}(b - a)^{n-1} - \frac{f^{(n)}(a)}{n!}(b - a)^n$$

Substituting from (14) into the above equation, we get

$$f(b) = f(a) + f'(a)(b - a) + \frac{f''(a)}{2!}(b - a)^2 + \ldots + \frac{f^{(n)}(a)}{n!}(b - a)^n + \frac{f^{(n+1)}(\xi)}{(n + 1)!}(b - a)^{n+1}$$

which is the desired result. The theorem holds if $b < a$ because the conclusion of Theorem 11.1.3 is unaffected if a and b are interchanged. ■

There are other forms of the remainder in Taylor's formula. Depending on the function, one form of the remainder may be more desirable to use than another. The following theorem expresses the remainder as an integral, and it is known as *Taylor's formula with integral form of the remainder.*

11.5.2 THEOREM   If f is a function whose first $n + 1$ derivatives are continuous on a closed interval containing a and x, then $f(x) = P_n(x) + R_n(x)$, where $P_n(x)$ is the nth-degree Taylor polynomial of f at a and $R_n(x)$ is the remainder given by

$$R_n(x) = \frac{1}{n!} \int_a^x (x - t)^n f^{(n+1)}(t)\, dt$$

The proof of this theorem is left as an exercise (see Exercise 30).

EXERCISES 11.5

In Exercises 1 through 14, find the Taylor polynomial of degree n with the Lagrange form of the remainder at the number a for the function defined by the equation.

1. $f(x) = \dfrac{1}{x-2}$; $a = 1$; $n = 3$ **2.** $f(x) = e^{-x}$; $a = 0$; $n = 4$

3. $f(x) = x^{3/2}$; $a = 4$; $n = 3$ **4.** $f(x) = \tan x$; $a = 0$; $n = 3$

5. $f(x) = \sin x$; $a = \frac{1}{6}\pi$; $n = 3$ **6.** $f(x) = \cosh x$; $a = 0$; $n = 4$

7. $f(x) = \sinh x$; $a = 0$; $n = 4$ **8.** $f(x) = \sqrt{x}$; $a = 4$; $n = 4$

9. $f(x) = \ln x$; $a = 1$; $n = 3$

10. $f(x) = \ln(x+2)$; $a = -1$; $n = 3$

11. $f(x) = \ln \cos x$; $a = \frac{1}{3}\pi$; $n = 3$

12. $f(x) = e^{-x^2}$; $a = 0$; $n = 3$

13. $f(x) = (1+x)^{3/2}$; $a = 0$; $n = 3$

14. $f(x) = (1-x)^{-1/2}$; $a = 0$; $n = 3$

15. Compute the value of e correct to five decimal places, and prove that your answer has the required accuracy.

16. Use the Taylor polynomial in Exercise 8 to compute $\sqrt{5}$ accurate to as many places as is justified when R_4 is neglected.

17. Estimate the error that results when $\cos x$ is replaced by $1 - \frac{1}{2}x^2$ if $|x| < 0.1$.

18. Estimate the error that results when $\sqrt{1+x}$ is replaced by $1 + \frac{1}{2}x$ if $0 < x < 0.01$.

19. Compute $\sin 31°$ accurate to three decimal places by using the Taylor polynomial in Exercise 5 at $\frac{1}{6}\pi$. (Use the approximation $\frac{1}{180}\pi \approx 0.0175$.)

20. Use the Maclaurin polynomial for the function defined by $f(x) = \ln(1+x)$ to compute the value of $\ln 1.2$ accurate to four decimal places.

21. Use the Maclaurin polynomial for the function defined by

$$f(x) = \ln \frac{1+x}{1-x}$$

to compute the value of $\ln 1.2$ accurate to four decimal places. Compare the computation with that of Exercise 20.

22. Show that if $0 \le x \le \frac{1}{2}$,

$$\sin x = x - \frac{x^3}{3!} + R(x)$$

where $|R(x)| < \frac{1}{3840}$.

23. Use the result of Exercise 22 to find an approximate value of $\int_0^{1/\sqrt{2}} \sin x^2 \, dx$, and estimate the error.

24. Show that the formula $(1+x)^{3/2} \approx 1 + \frac{3}{2}x$ is accurate to three decimal places if $-0.03 \le x \le 0$.

25. Show that the formula $(1+x)^{-1/2} \approx 1 - \frac{1}{2}x$ is accurate to two decimal places if $-0.1 \le x \le 0$.

26. Draw sketches of the graphs of $y = \sin x$ and $y = mx$ on the same set of axes. Note that if m is positive and close to zero, then the graphs intersect at a point whose abscissa is close to π. By finding the second-degree Taylor polynomial at π for the function f defined by $f(x) = \sin x - mx$, show that an approximate solution of the equation $\sin x = mx$, when m is positive and close to zero, is given by $x \approx \pi/(1+m)$.

27. Use the method described in Exercise 26 to find an approximate solution of the equation $\cot x = mx$ when m is positive and close to zero.

28. (a) Use the first-degree Maclaurin polynomial to approximate e^k if $0 < k < 0.01$. (b) Estimate the error in terms of k.

29. Apply Taylor's formula to express the polynomial

$$P(x) = x^4 - x^3 + 2x^2 - 3x + 1$$

as a polynomial in powers of $x - 1$.

30. Prove Theorem 11.5.2. $\left(\text{Hint: Let } \int_a^x f'(t)\, dt = f(x) - f(a).\right.$ Solve for $f(x)$ and integrate $\int_a^x f'(t)\, dt$ by parts by letting $u = f'(t)$ and $dv = dt$. Repeat this process, and the desired result follows by mathematical induction.$\Big)$

REVIEW EXERCISES FOR CHAPTER 11

In Exercises 1 through 18, evaluate the limit, if it exists.

1. $\lim\limits_{x \to 0^+} \dfrac{\tanh 2x}{\sinh^2 x}$

2. $\lim\limits_{x \to \pi/2} \left(\dfrac{1}{1 - \sin x} - \dfrac{2}{\cos^2 x} \right)$

3. $\lim\limits_{x \to 0} (\csc^2 x - x^{-2})$

4. $\lim\limits_{x \to 0} \dfrac{e - (1+x)^{1/x}}{x}$

5. $\lim\limits_{t \to +\infty} \dfrac{\ln(1 + e^{2t}/t)}{t^{1/2}}$

6. $\lim\limits_{x \to +\infty} x \ln \dfrac{x+1}{x-1}$

7. $\lim\limits_{x \to \pi/2} (\sin^2 x)^{\tan x}$

8. $\lim\limits_{t \to +\infty} \dfrac{t^{100}}{e^t}$

9. $\lim\limits_{y \to 0} \dfrac{\sin 2y}{y - \sin 5y}$

10. $\lim\limits_{z \to \pi/2^-} \dfrac{z \cos 3z}{\cos^2 z}$

11. $\lim\limits_{t \to 0^-} (1 + 4t)^{3/t}$

12. $\lim\limits_{y \to +\infty} (1 + e^{2y})^{-2/y}$

13. $\lim\limits_{x \to +\infty} \dfrac{\ln(\ln x)}{\ln(x - \ln x)}$

14. $\lim\limits_{x \to 0} \dfrac{x - \tan^{-1} x}{4x^3}$

15. $\lim\limits_{\theta \to \pi/2} \dfrac{\tan \theta + 3}{\sec \theta - 1}$

16. $\lim\limits_{x \to 0} \left(\dfrac{\sin x}{x} \right)^{1/x}$

17. $\lim\limits_{x \to +\infty} (e^x - x)^{1/x}$

18. $\lim\limits_{x \to 0^+} \left(\dfrac{1}{x} \right)^{\tan x}$

In Exercises 19 through 32, determine whether the improper integral is convergent or divergent. If it is convergent, evaluate it.

19. $\int_{-2}^{0} \dfrac{dx}{2x + 3}$

20. $\int_{0}^{+\infty} \dfrac{dx}{\sqrt{e^x}}$

21. $\int_{-\infty}^{0} \dfrac{dx}{(x - 2)^2}$

22. $\int_{2}^{4} \dfrac{x\,dx}{\sqrt{x - 2}}$

23. $\int_{0}^{\pi/4} \cot^2 \theta\,d\theta$

24. $\int_{1}^{+\infty} \dfrac{dt}{t^4 + t^2}$

25. $\int_{-\infty}^{3} 4^x\,dx$

26. $\int_{-\infty}^{0} xe^x\,dx$

27. $\int_{0}^{1} \dfrac{(\ln x)^2}{x}\,dx$

28. $\int_{0}^{+\infty} \dfrac{3^{-\sqrt{x}}}{\sqrt{x}}\,dx$

29. $\int_{-\infty}^{+\infty} \dfrac{dx}{4x^2 + 4x + 5}$

30. $\int_{0}^{1} \dfrac{\ln x}{x}\,dx$

31. $\int_{0}^{1} \dfrac{dx}{x + x^3}$

32. $\int_{-3}^{0} \dfrac{dx}{\sqrt{3 - 2x - x^2}}$

33. Given

$$f(x) = \begin{cases} \dfrac{1 - e^{2x}}{2x} & \text{if } x \neq 0 \\ -1 & \text{if } x = 0 \end{cases}$$

(a) Prove that f is continuous at 0 by using Definition 2.6.1.
(b) Prove that f is differentiable at 0 by computing $f'(0)$.

34. For the function of Exercise 33, find, if they exist: (a) $\lim\limits_{x \to +\infty} f(x)$; (b) $\lim\limits_{x \to -\infty} f(x)$.

35. Given:

$$f(x) = \begin{cases} \dfrac{e^x - e^{-x}}{e^{2x} - e^{-2x}} & \text{if } x \neq 0 \\ 1 & \text{if } x = 0 \end{cases}$$

(a) Is f continuous at 0? (b) Find $\lim\limits_{x \to +\infty} f(x)$, if it exists.

36. If the normal line to the curve $y = \ln x$ at the point $(x_1, \ln x_1)$ intersects the x axis at the point having an abscissa of a, prove that $\lim\limits_{x_1 \to +\infty} (a - x_1) = 0$.

37. Evaluate if they exist:

(a) $\int_{-\infty}^{+\infty} \sinh x\,dx$ (b) $\lim\limits_{r \to +\infty} \int_{-r}^{r} \sinh x\,dx$

38. (a) Prove that $\lim\limits_{x \to +\infty} (x^n/e^x) = 0$ for n any positive integer. (b) Find $\lim\limits_{x \to 0} (e^{-1/x}/x^n)$, where $x > 0$ and n is any positive integer, by letting $x = 1/t$ and using the result of part (a).

In Exercises 39 through 41, find the Taylor polynomial of degree n with the Lagrange form of the remainder at the number a for the function defined by the equation.

39. $f(x) = \cos x$; $a = 0$; $n = 6$

40. $f(x) = (1 + x^2)^{-1}$; $a = 1$; $n = 3$
41. $f(x) = x^{-1/2}$; $a = 9$; $n = 4$

42. Express both e^{x^2} and $\cos x$ as a Maclaurin polynomial of degree 4 and use them to evaluate

$$\lim_{x \to 0} \frac{e^{x^2} - \cos x}{x^4}$$

43. Evaluate the limit in Exercise 42 by L'Hôpital's rule.
44. Apply Taylor's formula to express the polynomial

$$P(x) = 4x^3 + 5x^2 - 2x + 1$$

as a polynomial in powers of $x + 2$.

45. Determine if it is possible to assign a finite number to represent the measure of the area of the region to the right of the y axis bounded by the curve $4y^2 - xy^2 - x^2 = 0$ and its asymptote. If a finite number can be assigned, find it.

46. Determine if it is possible to assign a finite number to represent the measure of the area of the region bounded by the curve whose equation is $2xy - y = 1$, the x axis, and the line $x = 1$. If a finite number can be assigned, find it.

47. Determine if it is possible to assign a finite number to represent the measure of the area of the region in the first quadrant and below the curve having the equation $y = e^{-x}$. If a finite number can be assigned, find it.

48. Find the values of n for which the improper integral

$$\int_{1}^{+\infty} \frac{\ln x}{x^n}\,dx$$

converges, and evaluate the integral for those values of n.

49. On a certain college campus the probability density function for the length of a telephone call to be x minutes is given by

$$f(x) = \begin{cases} 0.4e^{-0.4x} & \text{if } x \geq 0 \\ 0 & \text{if } x < 0 \end{cases}$$

What is the probability that a telephone call selected at random will last (a) between 2 and 3 min; (b) at most 3 min; (c) at least 10 min?

50. If the first four terms of the Maclaurin polynomial are used to approximate the value of $\sqrt{e}$, how accurate is the result?

51. Assume a continuous flow of income for a particular business and that at t years from now the number of dollars in the income per year is given by $1000t - 300$. What is the present value of all future income if 8% is the interest rate compounded continuously? (*Hint:* See the paragraph preceding Exercise 33 in Exercises 11.3.)

52. Show that $\lim\limits_{x \to 0} \dfrac{e^x - 1}{x} = 1$ without using L'Hôpital's rule. (*Hint:* Compute $f'(0)$ for $f(x) = e^x$ by using the definition of a derivative.)

TWELVE

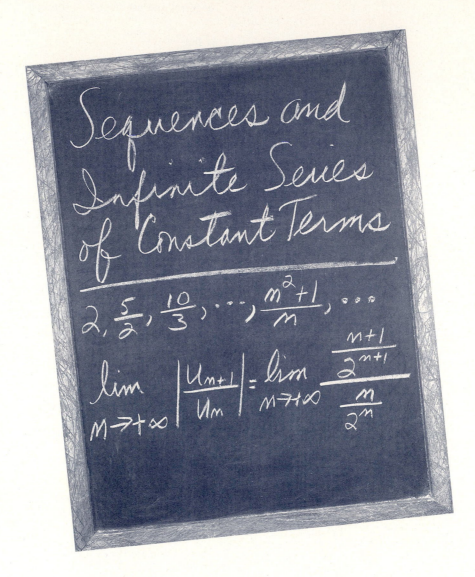

Sequences and Infinite Series of Constant Terms

$$2, \frac{5}{2}, \frac{10}{3}, \ldots, \frac{n^2+1}{n}, \ldots$$

$$\lim_{n \to +\infty} \left| \frac{U_{n+1}}{U_n} \right| = \lim_{n \to +\infty} \frac{\frac{n+1}{2^{n+1}}}{\frac{n}{2^n}}$$

The discussion of *infinite series* in this chapter and Chapter 13 utilizes concepts of calculus you have learned earlier. However, no subject matter in subsequent chapters has a knowledge of infinite series as a prerequisite. Consequently, if so desired Chapters 14 through 19 can be studied prior to Chapters 12 and 13.

In this chapter we study another kind of function having a set of real numbers as its domain and range. In Section 12.1, a *sequence* is defined as a function whose domain is a set of integers. In Section 12.2 you will find the proof of the equivalence of convergence and boundedness of monotonic sequences (Theorems 12.2.6 and 12.2.9) based on the completeness property of the real numbers. In Section 12.3, an *infinite series* is defined as a particular kind of sequence, and four theorems about infinite series are discussed in Section 12.4. The next four sections deal with *tests for convergence* of infinite series. In Sections 12.5 and 12.6 we consider series of positive terms, and then in Section 12.7 we are concerned with series whose terms are alternately positive and negative. Series of arbitrary terms are treated in Section 12.8. A summary of tests for convergence or divergence of an infinite series is given in Section 12.9.

12.1   SEQUENCES

Sequences of numbers are often encountered in mathematics. For instance, the numbers

$$2, 4, 6, 8, 10$$

form a sequence. This sequence is said to be **finite** because there is a last number. If the set of numbers that forms a sequence does not have a last number, the sequence is said to be **infinite**. For instance, the sequence

$$\frac{1}{3}, \frac{2}{5}, \frac{3}{7}, \frac{4}{9}, \ldots \tag{1}$$

is infinite because the three dots with no number following indicate that there is no last number. We are concerned here with infinite sequences, and when the word "sequence" is used, it is understood that we are referring to an infinite sequence. We now give a formal definition of a sequence; it is a particular kind of function.

12.1.1 DEFINITION

A **sequence** is a function whose domain is the set

$$\{1, 2, 3, \ldots, n, \ldots\}$$

of all positive integers.

The numbers in the range of a sequence are called the **elements** of the sequence.

If the nth element is given by $f(n)$, then the sequence is the set of ordered pairs of the form $(n, f(n))$, where n is a positive integer.

▶ **ILLUSTRATION 1**   If $f(n) = n/(2n + 1)$, then

$$f(1) = \tfrac{1}{3} \qquad f(2) = \tfrac{2}{5} \qquad f(3) = \tfrac{3}{7} \qquad f(4) = \tfrac{4}{9}$$

and so on. The range of f consists of the elements of sequence (1). Some of the ordered pairs in the sequence f are $(1, \tfrac{1}{3})$, $(2, \tfrac{2}{5})$, $(3, \tfrac{3}{7})$, $(4, \tfrac{4}{9})$, and $(5, \tfrac{5}{11})$. A sketch of the graph of this sequence is shown in Figure 1. ◀

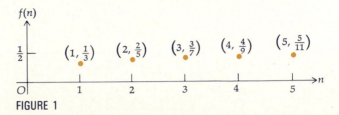

FIGURE 1

Usually the nth element $f(n)$ of the sequence is stated when the elements are listed in order. Thus the elements of sequence (1) can be written as

$$\frac{1}{3}, \frac{2}{5}, \frac{3}{7}, \frac{4}{9}, \ldots, \frac{n}{2n + 1}, \ldots$$

Because the domain of every sequence is the same, the notation $\{f(n)\}$ may be used to denote a sequence. So the sequence (1) can be denoted by $\{n/(2n + 1)\}$. The subscript notation $\{a_n\}$ is also used to denote the sequence for which $f(n) = a_n$.

A sequence

$$a_1, a_2, a_3, \ldots, a_n, \ldots$$

is said to be **equal** to a sequence

$$b_1, b_2, b_3, \ldots, b_n, \ldots$$

if and only if $a_i = b_i$ for every positive integer i. Remember that a sequence consists of an ordering of elements. Therefore it is possible for two sequences to have the same elements and be unequal. This situation is shown in the following illustration.

▶ **ILLUSTRATION 2**   The sequence $\{1/n\}$ has as its elements the reciprocals of the positive integers

$$1, \frac{1}{2}, \frac{1}{3}, \frac{1}{4}, \ldots, \frac{1}{n}, \ldots \tag{2}$$

The sequence for which

$$f(n) = \begin{cases} 1 & \text{if } n \text{ is odd} \\ \dfrac{2}{n + 2} & \text{if } n \text{ is even} \end{cases}$$

has as its elements

$$1, \tfrac{1}{2}, 1, \tfrac{1}{3}, 1, \tfrac{1}{4}, \ldots \tag{3}$$

The elements of sequences (2) and (3) are the same; however, the sequences are different. Sketches of the graphs of sequences (2) and (3) are shown in Figures 2 and 3, respectively.   ◀

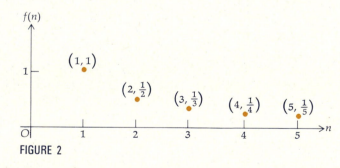

FIGURE 2

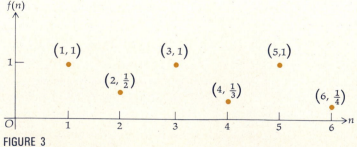

FIGURE 3

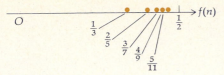

FIGURE 4

We now plot on a horizontal axis the points corresponding to successive elements of a sequence. This is done in Figure 4 for sequence (1), which is $\{n/(2n+1)\}$. Observe that the successive elements of the sequence get closer and closer to $\frac{1}{2}$, even though no element in the sequence has the value $\frac{1}{2}$. Intuitively we see that the element will be as close to $\frac{1}{2}$ as we please by taking the number of the element sufficiently large. Or stating this another way, $|n/(2n+1) - \frac{1}{2}|$ can be made less than any given positive ϵ by taking n large enough. Because of this we state that the limit of the sequence $\{n/(2n+1)\}$ is $\frac{1}{2}$.

In general, if there is a number L such that $|a_n - L|$ is arbitrarily small for n sufficiently large, the sequence $\{a_n\}$ is said to have the limit L. Following is the precise definition of the limit of a sequence.

12.1.2 DEFINITION

A sequence $\{a_n\}$ has the limit L if for any $\epsilon > 0$ there exists a number $N > 0$ such that if n is an integer and

if $n > N$ then $|a_n - L| < \epsilon$

and we write

$$\lim_{n \to +\infty} a_n = L$$

EXAMPLE 1 Use Definition 12.1.2 to prove that the following sequence has the limit $\frac{1}{2}$:

$$\left\{ \frac{n}{2n+1} \right\}$$

Solution We must show that for any $\epsilon > 0$ there exists a number $N > 0$ such that if n is an integer and

if $n > N$ then $\left| \dfrac{n}{2n+1} - \dfrac{1}{2} \right| < \epsilon$

$\Leftrightarrow$ if $n > N$ then $\left| \dfrac{2n - 2n - 1}{2(2n+1)} \right| < \epsilon$

$\Leftrightarrow$ if $n > N$ then $\left| \dfrac{-1}{2(2n+1)} \right| < \epsilon$

$\Leftrightarrow$ if $n > N$ then $\dfrac{1}{2(2n+1)} < \epsilon$

$\Leftrightarrow$ if $n > N$ then $2n + 1 > \dfrac{1}{2\epsilon}$

$\Leftrightarrow$ if $n > N$ then $n > \dfrac{1 - 2\epsilon}{4\epsilon}$

For this statement to hold take $N = (1 - 2\epsilon)/(4\epsilon)$, and we have if n is an integer and

if $n > \dfrac{1 - 2\epsilon}{4\epsilon}$ then $\left| \dfrac{n}{2n+1} - \dfrac{1}{2} \right| < \epsilon$ (4)

Observe that, if in particular $\epsilon = \frac{1}{8}$, then $N = \frac{3}{2}$ and (4) becomes

$$\text{if } n > \tfrac{3}{2} \quad \text{then} \quad \left| \frac{n}{2n+1} - \frac{1}{2} \right| < \frac{1}{8}$$

For instance, if $n = 4$,

$$\left| \frac{n}{2n+1} - \frac{1}{2} \right| = \left| \frac{4}{9} - \frac{1}{2} \right|$$

$$= \frac{1}{18}$$

and $\frac{1}{18} < \frac{1}{8}$. Statement (4) proves that the given sequence has the limit $\frac{1}{2}$.

▶ **ILLUSTRATION 3**   Consider the sequence $\{(-1)^{n+1}/n\}$. Note that the nth element of this sequence is $(-1)^{n+1}/n$, and $(-1)^{n+1}$ is equal to $+1$ when n is odd and to -1 when n is even. Hence the elements of the sequence can be written

$$1, \ -\frac{1}{2}, \ \frac{1}{3}, \ -\frac{1}{4}, \ \frac{1}{5}, \ \ldots, \ \frac{(-1)^{n+1}}{n}, \ \ldots$$

In Figure 5, points corresponding to successive elements of this sequence are plotted. In the figure, $a_1 = 1$, $a_2 = -\frac{1}{2}$, $a_3 = \frac{1}{3}$, $a_4 = -\frac{1}{4}$, $a_5 = \frac{1}{5}$, $a_6 = -\frac{1}{6}$, $a_7 = \frac{1}{7}$, $a_8 = -\frac{1}{8}$, $a_9 = \frac{1}{9}$, $a_{10} = -\frac{1}{10}$. The limit of the sequence is 0, and the elements oscillate about 0.   ◀

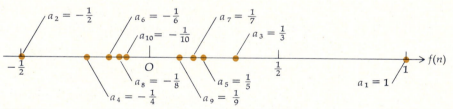

FIGURE 5

Compare Definition 12.1.2 with Definition 2.5.1 of the limit of $f(x)$ as x increases without bound. The two definitions are almost identical; however, when we state that $\lim\limits_{x \to +\infty} f(x) = L$, the function f is defined for all real numbers greater than some real number r, while when we consider $\lim\limits_{n \to +\infty} a_n$, n is restricted to positive integers. There is, however, the following theorem that follows immediately from Definition 2.5.1.

12.1.3 THEOREM   If $\lim\limits_{x \to +\infty} f(x) = L$, and f is defined for every positive integer, then also $\lim\limits_{n \to +\infty} f(n) = L$ when n is any positive integer.

The proof is left as an exercise (see Exercise 26).

▶ **ILLUSTRATION 4**   We verify Theorem 12.1.3 for the sequence of Example 1, for which $f(n) = n/(2n + 1)$. Hence $f(x) = x/(2x + 1)$ and

$$\lim_{x \to +\infty} \frac{x}{2x + 1} = \lim_{x \to +\infty} \frac{1}{2 + \dfrac{1}{x}}$$

$$= \tfrac{1}{2}$$

It follows, then, from Theorem 12.1.3 that $\lim\limits_{n \to +\infty} f(n) = \tfrac{1}{2}$ when n is any positive integer. This agrees with the solution of Example 1.   ◀

12.1.4 DEFINITION   If a sequence $\{a_n\}$ has a limit, the sequence is said to be **convergent**, and a_n **converges** to that limit. If the sequence is not convergent, it is **divergent**.

EXAMPLE 2   Determine whether the sequence is convergent or divergent:

$$\left\{ \frac{4n^2}{2n^2 + 1} \right\}$$

Solution   We wish to determine if $\lim\limits_{n \to +\infty} 4n^2/(2n^2 + 1)$ exists. We let $f(x) = 4x^2/(2x^2 + 1)$, and investigate $\lim\limits_{x \to +\infty} f(x)$.

$$\lim_{x \to +\infty} \frac{4x^2}{2x^2 + 1} = \lim_{x \to +\infty} \frac{4}{2 + \dfrac{1}{x^2}}$$

$$= 2$$

Therefore, by Theorem 12.1.3, $\lim\limits_{n \to +\infty} f(n) = 2$. Thus the given sequence is convergent and $4n^2/(2n^2 + 1)$ converges to 2.

EXAMPLE 3   Prove that if $|r| < 1$, the sequence $\{r^n\}$ is convergent and r^n converges to zero.

Solution   First we consider $r = 0$. Then the sequence is $\{0\}$ and $\lim\limits_{n \to +\infty} 0 = 0$. Thus the sequence is convergent and the nth element converges to zero.

If $0 < |r| < 1$, we wish to show that Definition 12.1.2 holds with $L = 0$. Therefore, we must show that for any $\epsilon > 0$ there exists a number $N > 0$ such that if n is an integer and

$$\text{if } n > N \quad \text{then} \quad |r^n - 0| < \epsilon \tag{5}$$

$\Leftrightarrow \quad$ if $n > N$ then $|r|^n < \epsilon$

$\Leftrightarrow \quad$ if $n > N$ then $\ln|r|^n < \ln \epsilon$

$\Leftrightarrow \quad$ if $n > N$ then $n \ln|r| < \ln \epsilon$

Because $0 < |r| < 1$, $\ln|r| < 0$. The above statement is equivalent to

$$\text{if } n > N \quad \text{then} \quad n > \frac{\ln \epsilon}{\ln|r|}$$

Therefore, if $N = \ln \epsilon / \ln|r|$, we may conclude (5). Consequently, $\lim\limits_{n \to +\infty} r^n = 0$. Hence, by Definitions 12.1.2 and 12.1.4, $\{r^n\}$ is convergent and r^n converges to zero.

EXAMPLE 4   Determine whether the sequence $\{(-1)^n + 1\}$ is convergent or divergent.

Solution   The elements of this sequence are $0, 2, 0, 2, 0, 2, \ldots, (-1)^n + 1, \ldots$. Because $a_n = 0$ if n is odd and $a_n = 2$ if n is even, it appears that the sequence is divergent. To prove this, let us assume that the sequence is convergent and show that this assumption leads to a contradiction. If the sequence has the limit L, then by Definition 12.1.2 for any $\epsilon > 0$ there exists a number $N > 0$ such that if n is an integer and

$$\text{if } n > N \quad \text{then} \quad |a_n - L| < \epsilon$$

In particular, when $\epsilon = \frac{1}{2}$, there exists a number $N > 0$ such that if n is an integer and

$$\text{if } n > N \quad \text{then} \quad |a_n - L| < \tfrac{1}{2}$$

$$\Leftrightarrow \quad \text{if } n > N \quad \text{then} \quad -\tfrac{1}{2} < a_n - L < \tfrac{1}{2}$$

Because $a_n = 0$ if n is odd and $a_n = 2$ if n is even, it follows from this statement that

$$-\tfrac{1}{2} < -L < \tfrac{1}{2} \quad \text{and} \quad -\tfrac{1}{2} < 2 - L < \tfrac{1}{2}$$

But if $-L > -\frac{1}{2}$, then $2 - L > \frac{3}{2}$; hence $2 - L$ cannot be less than $\frac{1}{2}$. So there is a contradiction, and therefore the given sequence is divergent.

EXAMPLE 5   Determine whether the sequence is convergent or divergent:

$$\left\{ n \sin \frac{\pi}{n} \right\}$$

Solution   We wish to determine whether $\lim\limits_{n \to +\infty} n \sin(\pi/n)$ exists. We let $f(x) = x \sin(\pi/x)$ and investigate $\lim\limits_{x \to +\infty} f(x)$. Because $f(x)$ can be written as $[\sin(\pi/x)]/(1/x)$ and $\lim\limits_{x \to +\infty} \sin(\pi/x) = 0$ and $\lim\limits_{x \to +\infty} (1/x) = 0$, L'Hôpital's rule can be applied to obtain

$$\lim_{x \to +\infty} f(x) = \lim_{x \to +\infty} \frac{-\dfrac{\pi}{x^2} \cos \dfrac{\pi}{x}}{-\dfrac{1}{x^2}}$$

$$= \lim_{x \to +\infty} \pi \cos \frac{\pi}{x}$$

$$= \pi$$

Therefore $\lim\limits_{n \to +\infty} f(n) = \pi$ when n is a positive integer. So the given sequence is convergent and $n \sin(\pi/n)$ converges to π.

There are limit theorems for sequences that are analogous to limit theorems for functions given in Chapter 2. The statement of these theorems uses the terminology of sequences. The proofs are omitted because they are similar to the proofs of the corresponding theorems of Chapter 2.

12.1.5 THEOREM If $\{a_n\}$ and $\{b_n\}$ are convergent sequences and c is a constant, then

(i) the constant sequence $\{c\}$ has c as its limit;

(ii) $\lim\limits_{n \to +\infty} ca_n = c \lim\limits_{n \to +\infty} a_n$;

(iii) $\lim\limits_{n \to +\infty} (a_n \pm b_n) = \lim\limits_{n \to +\infty} a_n \pm \lim\limits_{n \to +\infty} b_n$;

(iv) $\lim\limits_{n \to +\infty} a_n b_n = \left(\lim\limits_{n \to +\infty} a_n \right) \left(\lim\limits_{n \to +\infty} b_n \right)$;

(v) $\lim\limits_{n \to +\infty} \dfrac{a_n}{b_n} = \dfrac{\lim\limits_{n \to +\infty} a_n}{\lim\limits_{n \to +\infty} b_n}$ if $\lim\limits_{n \to +\infty} b_n \neq 0$, and all $b_n \neq 0$.

EXAMPLE 6 Use Theorem 12.1.5 to prove that the sequence

$$\left\{ \frac{n^2}{2n + 1} \sin \frac{\pi}{n} \right\}$$

is convergent, and find its limit.

Solution

$$\frac{n^2}{2n + 1} \sin \frac{\pi}{n} = \frac{n}{2n + 1} \cdot n \sin \frac{\pi}{n}$$

In Example 1 the sequence $\{n/(2n + 1)\}$ was shown to be convergent and $\lim\limits_{n \to +\infty} [n/(2n + 1)] = \frac{1}{2}$. In Example 5 we showed that the sequence $\{n \sin(\pi/n)\}$ is convergent and $\lim\limits_{n \to +\infty} [n \sin(\pi/n)] = \pi$. Hence, by Theorem 12.1.5(iv),

$$\lim_{n \to +\infty} \left[\frac{n}{2n + 1} \cdot n \sin \frac{\pi}{n} \right] = \lim_{n \to +\infty} \frac{n}{2n + 1} \cdot \lim_{n \to +\infty} n \sin \frac{\pi}{n}$$

$$= \tfrac{1}{2} \cdot \pi$$

Thus the given sequence is convergent, and its limit is $\frac{1}{2}\pi$.

EXERCISES 12.1

In Exercises 1 through 19, write the first four elements of the sequence and determine whether it is convergent or divergent. If the sequence converges, find its limit.

1. $\left\{ \dfrac{n + 1}{2n - 1} \right\}$

2. $\left\{ \dfrac{2n^2 + 1}{3n^2 - n} \right\}$

3. $\left\{ \dfrac{n^2 + 1}{n} \right\}$

4. $\left\{ \dfrac{3n^3 + 1}{2n^2 + n} \right\}$

5. $\left\{ \dfrac{3 - 2n^2}{n^2 - 1} \right\}$

6. $\left\{ \dfrac{e^n}{n} \right\}$

7. $\left\{ \dfrac{\ln n}{n^2} \right\}$

8. $\left\{ \dfrac{\log_b n}{n} \right\}, b > 1$

9. $\{\tanh n\}$

10. $\{\sinh n\}$

11. $\left\{ \dfrac{n}{n + 1} \sin \dfrac{n\pi}{2} \right\}$

12. $\left\{ \dfrac{\sinh n}{\sin n} \right\}$

13. $\left\{ \dfrac{1}{\sqrt{n^2 + 1} - n} \right\}$

14. $\{\sqrt{n + 1} - \sqrt{n}\}$

15. $\left\{\left(1 + \dfrac{1}{3n}\right)^n\right\}$ $\left(\textit{Hint: } \text{Use } \lim\limits_{x \to 0} (1 + x)^{1/x} = e.\right)$

16. $\left\{\left(1 + \dfrac{2}{n}\right)^n\right\}$ See Hint for Exercise 15.

17. $\{r^{1/n}\}$ and $r > 0$. (*Hint:* Consider two cases: $r \leq 1$ and $r > 1$.)

18. $\{\cos n\pi\}$ **19.** $\left\{\dfrac{n}{c^n}\right\}, c > 1$

In Exercises 20 through 25, use Definition 12.1.2 to prove that the sequence has the limit L.

20. $\left\{\dfrac{4}{2n - 1}\right\}; L = 0$ **21.** $\left\{\dfrac{3}{n - 1}\right\}; L = 0$

22. $\left\{\dfrac{1}{\sqrt{n}}\right\}; L = 0$ **23.** $\left\{\dfrac{8n}{2n + 3}\right\}; L = 4$

24. $\left\{\dfrac{5 - n}{2 + 3n}\right\}; L = -\dfrac{1}{3}$ **25.** $\left\{\dfrac{2n^2}{5n^2 + 1}\right\}; L = \dfrac{2}{5}$

26. Prove Theorem 12.1.3.

27. Show that the sequences $\left\{\dfrac{n^2}{n - 3}\right\}$ and $\left\{\dfrac{n^2}{n + 4}\right\}$ are both divergent, but that the sequence $\left\{\dfrac{n^2}{n - 3} - \dfrac{n^2}{n + 4}\right\}$ is convergent.

28. Given the sequence

$$\left\{\dfrac{1 - \left(1 - \dfrac{1}{n}\right)^a}{1 - \left(1 - \dfrac{1}{n}\right)^b}\right\}, a \text{ and } b \text{ are constants and } b \neq 0$$

Determine whether the sequence is convergent or divergent. If the sequence converges, find its limit.

29. Prove that if $|r| < 1$, the sequence $\{nr^n\}$ is convergent and nr^n converges to zero.

30. Prove that if the sequence $\{a_n\}$ converges, then $\lim\limits_{n \to +\infty} a_n$ is unique. (*Hint:* Assume that $\lim\limits_{n \to +\infty} a_n$ has two different values, L and M, and show that this is impossible by taking $\epsilon = \dfrac{1}{2}|L - M|$ in Definition 12.1.2.)

31. Prove that if the sequence $\{a_n\}$ is convergent and $\lim\limits_{n \to +\infty} a_n = L$, then the sequence $\{a_n{}^2\}$ is also convergent and $\lim\limits_{n \to +\infty} a_n{}^2 = L^2$.

32. Prove that if the sequence $\{a_n\}$ is convergent and $\lim\limits_{n \to +\infty} a_n = L$, then the sequence $\{|a_n|\}$ is also convergent and $\lim\limits_{n \to +\infty} |a_n| = |L|$.

12.2 MONOTONIC AND BOUNDED SEQUENCES

Certain kinds of sequences are given special names.

12.2.1 DEFINITION

A sequence $\{a_n\}$ is said to be

  (i) **increasing** if $a_n \leq a_{n+1}$ for all n;
  (ii) **decreasing** if $a_n \geq a_{n+1}$ for all n.

If a sequence is increasing or if it is decreasing, it is called **monotonic**.

If $a_n < a_{n+1}$ (a special case of $a_n \leq a_{n+1}$), the sequence is **strictly increasing**; if $a_n > a_{n+1}$, the sequence is **strictly decreasing**.

EXAMPLE 1   For each of the following sequences, determine if it is increasing, decreasing, or not monotonic: (a) $\{n/(2n + 1)\}$; (b) $\{1/n\}$; (c) $\{(-1)^{n+1}/n\}$.

Solution
(a) The elements of the sequence can be written

$$\dfrac{1}{3}, \dfrac{2}{5}, \dfrac{3}{7}, \dfrac{4}{9}, \ldots, \dfrac{n}{2n + 1}, \dfrac{n + 1}{2n + 3}, \ldots$$

Note that a_{n+1} is obtained from a_n by replacing n by $n + 1$. Therefore, because $a_n = n/(2n + 1)$,

$$a_{n+1} = \frac{n + 1}{2(n + 1) + 1}$$

$$= \frac{n + 1}{2n + 3}$$

Look at the first four elements of the sequence and observe that the elements increase as n increases. Thus we suspect in general that

$$\frac{n}{2n + 1} \le \frac{n + 1}{2n + 3} \tag{1}$$

Inequality (1) can be verified if an equivalent inequality can be found that we know is valid. Multiplying each member of (1) by $(2n + 1)(2n + 3)$ we obtain the following equivalent inequalities:

$$n(2n + 3) \le (n + 1)(2n + 1)$$

$$2n^2 + 3n \le 2n^2 + 3n + 1 \tag{2}$$

Inequality (2) obviously holds because the right member is 1 greater than the left member. Therefore inequality (1) holds; so the given sequence is increasing.

(b) The elements of the sequence can be written

$$1, \frac{1}{2}, \frac{1}{3}, \frac{1}{4}, \ldots, \frac{1}{n}, \frac{1}{n + 1}, \ldots$$

Because

$$\frac{1}{n} > \frac{1}{n + 1}$$

for all n, the sequence is decreasing.

(c) The elements of the sequence can be written

$$1, -\frac{1}{2}, \frac{1}{3}, -\frac{1}{4}, \ldots, \frac{(-1)^{n+1}}{n}, \frac{(-1)^{n+2}}{n + 1}, \ldots$$

Because $a_1 = 1$ and $a_2 = -\frac{1}{2}$, $a_1 > a_2$. But $a_3 = \frac{1}{3}$; thus $a_2 < a_3$. In a more general sense, consider three consecutive elements

$$a_n = \frac{(-1)^{n+1}}{n} \qquad a_{n+1} = \frac{(-1)^{n+2}}{n + 1} \qquad a_{n+2} = \frac{(-1)^{n+3}}{n + 2}$$

If n is odd, $a_n > a_{n+1}$ and $a_{n+1} < a_{n+2}$; for instance, $a_1 > a_2$ and $a_2 < a_3$. If n is even, $a_n < a_{n+1}$ and $a_{n+1} > a_{n+2}$; for instance, $a_2 < a_3$ and $a_3 > a_4$. Hence the sequence is neither increasing nor decreasing; thus it is not monotonic.

12.2.2 DEFINITION   The number C is called a **lower bound** of the sequence $\{a_n\}$ if $C \le a_n$ for all positive integers n, and the number D is called an **upper bound** of the sequence $\{a_n\}$ if $a_n \le D$ for all positive integers n.

▶ **ILLUSTRATION 1**   The number zero is a lower bound of the sequence $\{n/(2n + 1)\}$ whose elements are

$$\frac{1}{3}, \frac{2}{5}, \frac{3}{7}, \frac{4}{9}, \cdots, \frac{n}{2n + 1}, \cdots$$

Another lower bound of this sequence is $\frac{1}{3}$. Actually any number that is less than or equal to $\frac{1}{3}$ is a lower bound of this sequence.   ◀

▶ **ILLUSTRATION 2**   For the sequence $\{1/n\}$ whose elements are

$$1, \frac{1}{2}, \frac{1}{3}, \frac{1}{4}, \cdots, \frac{1}{n}, \cdots$$

1 is an upper bound; 26 is also an upper bound. Any number that is greater than or equal to 1 is an upper bound of this sequence, and any nonpositive number will serve as a lower bound.   ◀

From Illustrations 1 and 2 we see that a sequence may have many upper and lower bounds.

12.2.3 DEFINITION   If A is a lower bound of a sequence $\{a_n\}$ and if A has the property that for every lower bound C of $\{a_n\}$, $C \leq A$, then A is called the **greatest lower bound** of the sequence. Similarly, if B is an upper bound of a sequence $\{a_n\}$ and if B has the property that for every upper bound D of $\{a_n\}$, $B \leq D$, then B is called the **least upper bound** of the sequence.

▶ **ILLUSTRATION 3**   For the sequence $\{n/(2n + 1)\}$ of Illustration 1, the greatest lower bound is $\frac{1}{3}$ because every lower bound of the sequence is less than or equal to $\frac{1}{3}$. Furthermore,

$$\frac{n}{2n + 1} = \frac{1}{2 + \dfrac{1}{n}} < \frac{1}{2}$$

for all n, and $\frac{1}{2}$ is the least upper bound of the sequence.

In Illustration 2 there is the sequence $\{1/n\}$ whose least upper bound is 1 because every upper bound of the sequence is greater than or equal to 1. The greatest lower bound of this sequence is 0.   ◀

12.2.4 DEFINITION   A sequence $\{a_n\}$ is said to be **bounded** if and only if it has an upper bound and a lower bound.

Because the sequence $\{1/n\}$ has an upper bound and a lower bound, it is bounded. This sequence is also a decreasing sequence and hence is a bounded monotonic sequence. There is a theorem (12.2.6) that guarantees that a bounded monotonic sequence is convergent. In particular, the sequence $\{1/n\}$ is convergent because $\lim\limits_{n \to +\infty} (1/n) = 0$. The sequence $\{n\}$ whose elements are

$$1, 2, 3, \ldots, n, \ldots$$

is monotonic (because it is increasing) but is not bounded (because there is no upper bound). It is not convergent because $\lim\limits_{n \to +\infty} n = +\infty$.

For the proof of Theorem 12.2.6 we need a very important property of the real-number system that is now stated.

12.2.5 THE AXIOM OF COMPLETENESS

Every nonempty set of real numbers that has a lower bound has a greatest lower bound. Also, every nonempty set of real numbers that has an upper bound has a least upper bound.

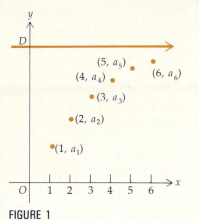

FIGURE 1

The second sentence in the statement of the axiom of completeness is unnecessary because it can be proved from the first sentence. It is included in the axiom here to expedite the discussion.

Suppose $\{a_n\}$ is an increasing sequence that is bounded. Let D be an upper bound of the sequence. Then if the points (n, a_n) are plotted on a rectangular cartesian coordinate system, these points will all lie below the line $y = D$. Furthermore, because the sequence is increasing, then as n increases, the points will get closer and closer to the line $y = D$. See Figure 1. Hence, as n increases, the elements a_n increase toward D. Intuitively, it appears that the sequence $\{a_n\}$ has a limit that is either D or some number less than D. This is indeed the case and is proved in the following theorem.

12.2.6 THEOREM

A bounded monotonic sequence is convergent.

Proof We prove the theorem for the case when the monotonic sequence is increasing. Let the sequence be $\{a_n\}$.

Because $\{a_n\}$ is bounded, there is an upper bound for the sequence. By the axiom of completeness, $\{a_n\}$ has a least upper bound that we call B. Then if ϵ is a positive number, $B - \epsilon$ cannot be an upper bound of the sequence because $B - \epsilon < B$ and B is the least upper bound of the sequence. So for some positive integer N,

$$B - \epsilon < a_N \tag{3}$$

Because B is the least upper bound of $\{a_n\}$, by Definition 12.2.2 it follows that

$$a_n \leq B \qquad \text{for every positive integer } n \tag{4}$$

Because $\{a_n\}$ is an increasing sequence, we have from Definition 12.2.1(i)

$$a_n \leq a_{n+1} \qquad \text{for every positive integer } n$$

and so

$$\text{if } n \geq N \quad \text{then} \quad a_N \leq a_n$$

From this statement and (3) and (4) it follows that

$$\text{if } n \geq N \quad \text{then} \quad B - \epsilon < a_N \leq a_n \leq B < B + \epsilon$$

from which we get

$$\text{if } n \geq N \quad \text{then} \quad B - \epsilon < a_n < B + \epsilon$$

$$\Leftrightarrow \quad \text{if } n \geq N \quad \text{then} \quad -\epsilon < a_n - B < \epsilon$$

$$\Leftrightarrow \quad \text{if } n \geq N \quad \text{then} \quad |a_n - B| < \epsilon$$

But by Definition 12.1.2, this statement is the condition that $\lim\limits_{n \to +\infty} a_n = B$.
Therefore the sequence $\{a_n\}$ is convergent.

To prove the theorem when $\{a_n\}$ is a decreasing sequence, consider the sequence $\{-a_n\}$, which will be increasing, and apply the above results. We leave it as an exercise to fill in the steps (see Exercise 17). ∎

Theorem 12.2.6 states that if $\{a_n\}$ is a bounded monotonic sequence, there exists a number L such that $\lim\limits_{n \to +\infty} a_n = L$, but it does not state how to find L. For this reason Theorem 12.2.6 is called an *existence theorem*. Many important concepts in mathematics are based on existence theorems. In particular, there are many sequences for which the limit cannot be found by direct use of the definition or by use of limit theorems, but the knowledge that such a limit exists can be of value to a mathematician.

In the proof of Theorem 12.2.6 the limit of the bounded increasing sequence is the least upper bound B of the sequence. Hence, if D is an upper bound of the sequence, $\lim\limits_{n \to +\infty} a_n = B \leq D$. We have, then, the following theorem.

12.2.7 THEOREM

> Let $\{a_n\}$ be an increasing sequence, and suppose that D is an upper bound of this sequence. Then $\{a_n\}$ is convergent, and
>
> $$\lim_{n \to +\infty} a_n \leq D$$

In proving Theorem 12.2.6 for the case when the bounded monotonic sequence is decreasing, the limit of the sequence is the greatest lower bound. The next theorem follows in a way similar to that of Theorem 12.2.7.

12.2.8 THEOREM

> Let $\{a_n\}$ be a decreasing sequence, and suppose that C is a lower bound of this sequence. Then $\{a_n\}$ is convergent, and
>
> $$\lim_{n \to +\infty} a_n \geq C$$

EXAMPLE 2   Use Theorem 12.2.6 to prove that the sequence is convergent:

$$\left\{ \frac{2^n}{n!} \right\}$$

Solution   The elements of the given sequence are

$$\frac{2^1}{1!}, \frac{2^2}{2!}, \frac{2^3}{3!}, \frac{2^4}{4!}, \ldots, \frac{2^n}{n!}, \frac{2^{n+1}}{(n+1)!}, \ldots$$

$1! = 1, 2! = 2, 3! = 6, 4! = 24$. Hence the elements of the sequence can be written as

$$2, 2, \frac{4}{3}, \frac{2}{3}, \ldots, \frac{2^n}{n!}, \frac{2^{n+1}}{(n+1)!}, \ldots$$

Then $a_1 = a_2 > a_3 > a_4$; so the given sequence may be decreasing. We must check to see if $a_n \geq a_{n+1}$; that is, it must be determined if

$$\frac{2^n}{n!} \geq \frac{2^{n+1}}{(n+1)!} \tag{5}$$

$$\Leftrightarrow \quad 2^n(n+1)! \geq 2^{n+1}n!$$

$$\Leftrightarrow \quad 2^n n!(n+1) \geq 2 \cdot 2^n n!$$

$$\Leftrightarrow \quad n+1 \geq 2 \tag{6}$$

When $n = 1$, inequality (6) becomes $2 = 2$, and (6) obviously holds when $n > 2$. Because inequality (5) is equivalent to (6), it follows that the given sequence is decreasing and hence monotonic. An upper bound for the given sequence is 2, and a lower bound is 0. Therefore the sequence is bounded.

The sequence $\{2^n/n!\}$ is therefore a bounded monotonic sequence, and by Theorem 12.2.6 it is convergent.

Theorem 12.2.6 states that a sufficient condition for a monotonic sequence to be convergent is that it be bounded. This is also a necessary condition and is given in the following theorem.

12.2.9 THEOREM   A convergent monotonic sequence is bounded.

Proof We prove the theorem for the case when the monotonic sequence is increasing. Let the sequence be $\{a_n\}$.

To prove that $\{a_n\}$ is bounded, it must be shown that it has a lower bound and an upper bound. Because $\{a_n\}$ is an increasing sequence, its first element serves as a lower bound. We must now find an upper bound.

Because $\{a_n\}$ is convergent, the sequence has a limit; call this limit L. Therefore $\lim\limits_{n \to +\infty} a_n = L$, and so by Definition 12.1.2, for any $\epsilon > 0$ there exists a number $N > 0$ such that if n is an integer and

$$\text{if } n > N \quad \text{then} \quad |a_n - L| < \epsilon$$

$$\Leftrightarrow \quad \text{if } n > N \quad \text{then} \quad -\epsilon < a_n - L < \epsilon$$

$$\Leftrightarrow \quad \text{if } n > N \quad \text{then} \quad L - \epsilon < a_n < L + \epsilon$$

Because $\{a_n\}$ is increasing, it follows from this statement that

$$a_n < L + \epsilon \quad \text{for all positive integers } n$$

Therefore $L + \epsilon$ will serve as an upper bound of the sequence $\{a_n\}$.

To prove the theorem when $\{a_n\}$ is a decreasing sequence, do as suggested in the proof of Theorem 12.2.6: Consider the sequence $\{-a_n\}$, which will be increasing, and apply the above results. You are asked to provide this proof in Exercise 18. ∎

EXERCISES 12.2

In Exercises 1 through 16, determine if the sequence is increasing, decreasing, or not monotonic.

1. $\left\{\dfrac{3n-1}{4n+5}\right\}$

2. $\left\{\dfrac{2n-1}{4n-1}\right\}$

3. $\left\{\dfrac{1-2n^2}{n^2}\right\}$

4. $\{\sin n\pi\}$

5. $\{\cos \frac{1}{3}n\pi\}$

6. $\left\{\dfrac{n^3-1}{n}\right\}$

7. $\left\{\dfrac{1}{n+\sin n^2}\right\}$

8. $\left\{\dfrac{2^n}{1+2^n}\right\}$

9. $\left\{\dfrac{5^n}{1+5^{2n}}\right\}$

10. $\left\{\dfrac{(2n)!}{5^n}\right\}$

11. $\left\{\dfrac{n!}{3^n}\right\}$

12. $\left\{\dfrac{n}{2^n}\right\}$

13. $\left\{\dfrac{n^n}{n!}\right\}$

14. $\{n^2 + (-1)^n n\}$

15. $\left\{\dfrac{n!}{1 \cdot 3 \cdot 5 \cdot \ldots \cdot (2n-1)}\right\}$

16. $\left\{\dfrac{1 \cdot 3 \cdot 5 \cdot \ldots \cdot (2n-1)}{2^n \cdot n!}\right\}$

17. Use the fact that Theorem 12.2.6 holds for an increasing sequence to prove that the theorem holds when $\{a_n\}$ is a decreasing sequence. (*Hint:* Consider the sequence $\{-a_n\}$.)

18. Prove Theorem 12.2.9 when $\{a_n\}$ is a decreasing sequence by a method similar to that used in Exercise 17.

In Exercises 19 through 20, determine if the sequence is bounded.

19. $\left\{\dfrac{n^2+3}{n+1}\right\}$

20. $\{3 - (-1)^{n-1}\}$

In Exercises 21 through 30, prove that the sequence is convergent by using Theorem 12.2.6.

21. The sequence of Exercise 1. **22.** $\left\{\dfrac{n}{3^{n+1}}\right\}$

23. $\left\{\dfrac{1 \cdot 3 \cdot 5 \cdot \ldots \cdot (2n-1)}{2 \cdot 4 \cdot 6 \cdot \ldots \cdot (2n)}\right\}$ **24.** The sequence of Exercise 8.

25. The sequence of Exercise 9.
26. The sequence of Exercise 12.

27. The sequence of Exercise 15.
28. The sequence of Exercise 16.

29. $\left\{\dfrac{n^2}{2^n}\right\}$ **30.** $\{k^{1/n}\}, k > 1$

31. Give an example of a sequence that is bounded and convergent but not monotonic.

32. Given the sequence $\{a_n\}$ where $a_n > 0$ for all n, and $a_{n+1} < ka_n$ with $0 < k < 1$. Prove that $\{a_n\}$ is convergent.

12.3 INFINITE SERIES OF CONSTANT TERMS

An important part of the study of calculus involves representing functions as "infinite sums." To do this requires extending the familiar operation of addition of a finite set of numbers to addition of infinitely many numbers. To carry this out we deal with a limiting process by considering sequences.

Let us associate with the sequence

$$u_1, u_2, u_3, \ldots, u_n, \ldots$$

an "infinite sum" denoted by

$$u_1 + u_2 + u_3 + \ldots + u_n + \ldots$$

But what is the meaning of such an expression? That is, what do we mean by the "sum" of an infinite number of terms, and under what circumstances does such a sum exist? To get an intuitive idea of the concept of such a sum, suppose a piece of string of length 2 ft is cut in half. One of these halves of length 1 ft is set aside and the other piece is cut in half again. One of the resulting pieces of length $\frac{1}{2}$ ft is set aside and the other piece is cut in half so that two pieces, each of length $\frac{1}{4}$ ft, are obtained. One of the pieces of length $\frac{1}{4}$ ft is set aside and then the other piece is cut in half; so two pieces, each of length $\frac{1}{8}$ ft, are obtained. Again one of the pieces is set aside and the other is cut in half. If this procedure is continued indefinitely, the number of feet in the sum of the lengths of the pieces set aside can be considered as the infinite sum

$$1 + \frac{1}{2} + \frac{1}{4} + \frac{1}{8} + \frac{1}{16} + \ldots + \frac{1}{2^{n-1}} + \ldots \tag{1}$$

Because we started with a piece of string 2 ft in length, our intuition indicates that the infinite sum (1) should be 2. We demonstrate that this is indeed the case in Illustration 2. However, we first need some preliminary definitions.

From the sequence

$$u_1, u_2, u_3, \ldots, u_n, \ldots$$

we form a new sequence $\{s_n\}$ by adding successive elements of $\{u_n\}$:

$$s_1 = u_1$$

$$s_2 = u_1 + u_2$$

$$s_3 = u_1 + u_2 + u_3$$

$$s_4 = u_1 + u_2 + u_3 + u_4$$

$$\vdots$$

$$s_n = u_1 + u_2 + u_3 + u_4 + \ldots + u_n$$

The sequence $\{s_n\}$ obtained in this manner from the sequence $\{u_n\}$ is called an *infinite series*.

12.3.1 DEFINITION

If $\{u_n\}$ is a sequence and

$$s_n = u_1 + u_2 + u_3 + \ldots + u_n$$

then the sequence $\{s_n\}$ is called an **infinite series**. This infinite series is denoted by

$$\sum_{n=1}^{+\infty} u_n = u_1 + u_2 + u_3 + \ldots + u_n + \ldots$$

The numbers $u_1, u_2, u_3, \ldots, u_n, \ldots$ are called the **terms** of the infinite series. The numbers $s_1, s_2, s_3, \ldots, s_n, \ldots$ are called the **partial sums** of the infinite series.

Observe that Definition 12.3.1 states that an infinite series is a sequence of partial sums.

▶ **ILLUSTRATION 1** Consider the sequence $\{u_n\}$ where $u_n = \dfrac{1}{2^{n-1}}$:

$$1, \frac{1}{2}, \frac{1}{4}, \frac{1}{8}, \frac{1}{16}, \ldots, \frac{1}{2^{n-1}}, \ldots$$

From this sequence let us form a sequence of partial sums:

$$s_1 = 1 \qquad\qquad\qquad s_1 = 1$$

$$s_2 = 1 + \frac{1}{2} \qquad\qquad \Leftrightarrow \quad s_2 = \frac{3}{2}$$

$$s_3 = 1 + \frac{1}{2} + \frac{1}{4} \qquad \Leftrightarrow \quad s_3 = \frac{7}{4}$$

$$s_4 = 1 + \frac{1}{2} + \frac{1}{4} + \frac{1}{8} \qquad \Leftrightarrow \quad s_4 = \frac{15}{8}$$

$$s_5 = 1 + \frac{1}{2} + \frac{1}{4} + \frac{1}{8} + \frac{1}{16} \quad \Leftrightarrow \quad s_5 = \frac{31}{16}$$

$$\vdots$$

$$s_n = 1 + \frac{1}{2} + \frac{1}{4} + \frac{1}{8} + \frac{1}{16} + \ldots + \frac{1}{2^{n-1}}$$

This sequence of partial sums $\{s_n\}$ is the infinite series denoted by

$$\sum_{n=1}^{+\infty} \frac{1}{2^{n-1}} = 1 + \frac{1}{2} + \frac{1}{4} + \frac{1}{8} + \frac{1}{16} + \ldots + \frac{1}{2^{n-1}} + \ldots$$

Observe that this is the infinite sum (1) obtained at the beginning of this section in the discussion of repeatedly cutting the string of length 2 ft. It is an example of a *geometric series* discussed later in this section. ◀

When $\{s_n\}$ is a sequence of partial sums,

$$s_{n-1} = u_1 + u_2 + u_3 + \ldots + u_{n-1}$$

Thus

$$s_n = s_{n-1} + u_n$$

We use this formula in the following example.

EXAMPLE 1   Given the infinite series

$$\sum_{n=1}^{+\infty} u_n = \sum_{n=1}^{+\infty} \frac{1}{n(n+1)}$$

(a) find the first four elements of the sequence of partial sums $\{s_n\}$, and
(b) find a formula for s_n in terms of n.

Solution
(a) Because $s_n = s_{n-1} + u_n$

$$s_1 = u_1 \qquad\qquad s_2 = s_1 + u_2$$

$$= \frac{1}{1 \cdot 2} \qquad\qquad = \frac{1}{2} + \frac{1}{2 \cdot 3}$$

$$= \tfrac{1}{2} \qquad\qquad\qquad = \tfrac{2}{3}$$

$$s_3 = s_2 + u_3 \qquad\quad s_4 = s_3 + u_4$$

$$= \frac{2}{3} + \frac{1}{3 \cdot 4} \qquad\quad = \frac{3}{4} + \frac{1}{4 \cdot 5}$$

$$= \tfrac{3}{4} \qquad\qquad\qquad = \tfrac{4}{5}$$

(b) Because $u_k = \dfrac{1}{k(k+1)}$ we have, by partial fractions,

$$u_k = \frac{1}{k} - \frac{1}{k+1}$$

Therefore,

$$u_1 = 1 - \tfrac{1}{2} \qquad u_2 = \tfrac{1}{2} - \tfrac{1}{3} \qquad u_3 = \tfrac{1}{3} - \tfrac{1}{4}$$

$$\cdots$$

$$u_{n-1} = \frac{1}{n-1} - \frac{1}{n} \qquad u_n = \frac{1}{n} - \frac{1}{n+1}$$

Thus, because $s_n = u_1 + u_2 + \ldots + u_{n-1} + u_n$,

$$s_n = \left(1 - \frac{1}{2}\right) + \left(\frac{1}{2} - \frac{1}{3}\right) + \left(\frac{1}{3} - \frac{1}{4}\right) + \ldots + \left(\frac{1}{n-1} - \frac{1}{n}\right) + \left(\frac{1}{n} - \frac{1}{n+1}\right)$$

Upon removing parentheses and combining terms we obtain

$$s_n = 1 - \frac{1}{n+1}$$

$$= \frac{n}{n+1}$$

By taking n as 1, 2, 3, and 4, we see that the previous results agree.

The method of solution of the above example applies only to a special case. In general, it is not possible to obtain such an expression for s_n.

We now define the *sum* of an infinite series.

12.3.2 DEFINITION Let $\sum\limits_{n=1}^{+\infty} u_n$ be a given infinite series, and let $\{s_n\}$ be the sequence of partial sums defining this infinite series. Then if $\lim\limits_{n \to +\infty} s_n$ exists and is equal to S, the given series is said to be **convergent** and S is the **sum** of the given infinite series. If $\lim\limits_{n \to +\infty} s_n$ does not exist, the series is **divergent** and the series does not have a sum.

Essentially Definition 12.3.2 states that an infinite series is convergent if and only if the corresponding sequence of partial sums is convergent.

If an infinite series has a sum S, we also say that the series converges to S.

Observe that the sum of a convergent series is the limit of a sequence of partial sums and is not obtained by ordinary addition. For a convergent series the symbolism

$$\sum_{n=1}^{+\infty} u_n$$

is used to denote both the series and the sum of the series. The use of the same symbol should not be confusing because the correct interpretation is apparent from the context in which it is employed.

▶ **ILLUSTRATION 2** The infinite series of Illustration 1 is

$$\sum_{n=1}^{+\infty} \frac{1}{2^{n-1}} = 1 + \frac{1}{2} + \frac{1}{4} + \frac{1}{8} + \frac{1}{16} + \ldots + \frac{1}{2^{n-1}} + \ldots \tag{2}$$

and the sequence of partial sums is $\{s_n\}$, where

$$s_n = 1 + \frac{1}{2} + \frac{1}{4} + \frac{1}{8} + \ldots + \frac{1}{2^{n-1}} \tag{3}$$

To determine if infinite series (2) has a sum we must compute $\lim\limits_{n \to +\infty} s_n$. To find a formula for s_n we use the identity from algebra:

$$a^n - b^n = (a - b)(a^{n-1} + a^{n-2}b + a^{n-3}b^2 + \ldots + ab^{n-2} + b^{n-1})$$

Applying this identity with $a = 1$ and $b = \frac{1}{2}$ we have

$$1 - \frac{1}{2^n} = \left(1 - \frac{1}{2}\right)\left(1 + \frac{1}{2} + \frac{1}{2^2} + \frac{1}{2^3} + \ldots + \frac{1}{2^{n-1}}\right)$$

$$\Leftrightarrow \quad 1 + \frac{1}{2} + \frac{1}{4} + \frac{1}{8} + \ldots + \frac{1}{2^{n-1}} = \frac{1 - \dfrac{1}{2^n}}{\frac{1}{2}}$$

Comparing this equation and (3) we obtain

$$s_n = 2\left(1 - \frac{1}{2^n}\right)$$

Because $\lim\limits_{n \to +\infty} \dfrac{1}{2^n} = 0$ we have

$$\lim_{n \to +\infty} s_n = 2$$

Therefore infinite series (2) has the sum 2.   ◀

EXAMPLE 2   Determine if the infinite series of Example 1 has a sum.

Solution   In the solution of Example 1 we showed that the sequence of partial sums for the given series is $\{s_n\} = \{n/(n + 1)\}$. Therefore

$$\lim_{n \to +\infty} s_n = \lim_{n \to +\infty} \frac{n}{n + 1}$$

$$= \lim_{n \to +\infty} \frac{1}{1 + \dfrac{1}{n}}$$

$$= 1$$

So the infinite series has a sum equal to 1, and we write

$$\sum_{n=1}^{+\infty} \frac{1}{n(n + 1)} = \frac{1}{2} + \frac{1}{6} + \frac{1}{12} + \frac{1}{20} + \ldots + \frac{1}{n(n + 1)} + \ldots$$

$$= 1$$

EXAMPLE 3   Find the infinite series that has the following sequence of partial sums:

$$\{s_n\} = \left\{ \frac{1}{2^n} \right\}$$

Also determine if the infinite series is convergent or divergent; if it is convergent, find its sum.

Solution   Because $s_1 = \frac{1}{2}$, then $u_1 = \frac{1}{2}$. If $n > 1$,

$$u_n = s_n - s_{n-1}$$

$$= \frac{1}{2^n} - \frac{1}{2^{n-1}}$$

$$= -\frac{1}{2^n}$$

Therefore the infinite series is

$$\frac{1}{2} - \sum_{n=2}^{+\infty} \frac{1}{2^n}$$

Because

$$\lim_{n \to +\infty} s_n = \lim_{n \to +\infty} \frac{1}{2^n}$$

$$= 0$$

the series is convergent and its sum is 0.

As mentioned above, in most cases it is not possible to obtain an expression for s_n in terms of n; so we must have other methods for determining whether or not a given infinite series has a sum or, equivalently, whether a given infinite series is convergent or divergent.

12.3.3 THEOREM If the infinite series $\sum_{n=1}^{+\infty} u_n$ is convergent, then $\lim_{n \to +\infty} u_n = 0$.

Proof Let $\{s_n\}$ be the sequence of partial sums for the given series, and denote the sum of the series by S. From Definition 12.3.2, $\lim_{n \to +\infty} s_n = S$. Thus for any $\epsilon > 0$ there exists a number $N > 0$ such that

if $n > N$ then $|S - s_n| < \frac{1}{2}\epsilon$

Also

if $n > N$ then $|S - s_{n+1}| < \frac{1}{2}\epsilon$

Then

$$|u_{n+1}| = |s_{n+1} - s_n|$$
$$= |S - s_n + s_{n+1} - S|$$
$$\leq |S - s_n| + |s_{n+1} - S|$$

Therefore

if $n > N$ then $|u_{n+1}| < \frac{1}{2}\epsilon + \frac{1}{2}\epsilon = \epsilon$

Hence

$$\lim_{n \to +\infty} u_n = 0$$ ∎

Theorem 12.3.3 provides a simple test for divergence because if $\lim_{n \to +\infty} u_n \neq 0$, we can conclude that $\sum_{n=1}^{+\infty} u_n$ is divergent.

EXAMPLE 4   Prove that the following two series are divergent:

(a) $\sum_{n=1}^{+\infty} \frac{n^2 + 1}{n^2} = 2 + \frac{5}{4} + \frac{10}{9} + \frac{17}{16} + \ldots$

(b) $\sum_{n=1}^{+\infty} (-1)^{n+1} 3 = 3 - 3 + 3 - 3 + \ldots$

Solution

(a)
$$\lim_{n \to +\infty} u_n = \lim_{n \to +\infty} \frac{n^2 + 1}{n^2}$$

$$= \lim_{n \to +\infty} \frac{1 + \dfrac{1}{n^2}}{1}$$

$$= 1$$

$$\neq 0$$

Therefore, by Theorem 12.3.3, the series is divergent.

(b) $\lim\limits_{n \to +\infty} u_n = \lim\limits_{n \to +\infty} (-1)^{n+1}3$, which does not exist. Therefore, by Theorem 12.3.3, the series is divergent.

The converse of Theorem 12.3.3 is false. That is, if $\lim\limits_{n \to +\infty} u_n = 0$, it does not follow that the series is necessarily convergent. In other words, it is possible to have a divergent series for which $\lim\limits_{n \to +\infty} u_n = 0$. An example of such a series is the **harmonic series**, which is

$$\sum_{n=1}^{+\infty} \frac{1}{n} = 1 + \frac{1}{2} + \frac{1}{3} + \frac{1}{4} + \ldots + \frac{1}{n} + \ldots \tag{4}$$

Clearly, $\lim\limits_{n \to +\infty} 1/n = 0$. In Illustration 3 the harmonic series is proved to diverge and we use the following theorem, which states that the difference between two partial sums s_R and s_T of a convergent series can be made as small as we please by taking R and T sufficiently large.

12.3.4 THEOREM   Let $\{s_n\}$ be the sequence of partial sums for a given convergent series $\sum\limits_{n=1}^{+\infty} u_n$. Then for any $\epsilon > 0$ there exists a number N such that

if $R > N$ and $T > N$ then $|s_R - s_T| < \epsilon$

Proof   Because the series $\sum\limits_{n=1}^{+\infty} u_n$ is convergent, call its sum S. Then for any $\epsilon > 0$ there exists an $N > 0$ such that if $n > N$, then $|S - s_n| < \frac{1}{2}\epsilon$. Therefore, if $R > N$ and $T > N$,

$$|s_R - s_T| = |s_R - S + S - s_T| \leq |s_R - S| + |S - s_T| < \tfrac{1}{2}\epsilon + \tfrac{1}{2}\epsilon$$

So

if $R > N$ and $T > N$ then $|s_R - s_T| < \epsilon$ ■

▶ **ILLUSTRATION 3**   We prove that the harmonic series (4) is divergent. For this series,

$$s_n = 1 + \frac{1}{2} + \ldots + \frac{1}{n}$$

and

$$s_{2n} = 1 + \frac{1}{2} + \ldots + \frac{1}{n} + \frac{1}{n+1} + \ldots + \frac{1}{2n}$$

So

$$s_{2n} - s_n = \frac{1}{n+1} + \frac{1}{n+2} + \frac{1}{n+3} + \ldots + \frac{1}{2n} \tag{5}$$

If $n > 1$,

$$\frac{1}{n+1} + \frac{1}{n+2} + \frac{1}{n+3} + \ldots + \frac{1}{2n} > \frac{1}{2n} + \frac{1}{2n} + \frac{1}{2n} + \ldots + \frac{1}{2n}$$

There are n terms on each side of the inequality sign; so the right side is $n(1/2n) = \frac{1}{2}$. Therefore, from (5) and the above inequality,

$$\text{if } n > 1 \quad \text{then} \quad s_{2n} - s_n > \tfrac{1}{2} \tag{6}$$

But Theorem 12.3.4 states that if the given series is convergent, then $s_{2n} - s_n$ may be made as small as we please by taking n large enough; that is, if $\epsilon = \frac{1}{2}$, there exists an N such that

$$\text{if } 2n > N \text{ and } n > N \quad \text{then} \quad s_{2n} - s_n < \tfrac{1}{2}$$

But this statement contradicts (6). Therefore the harmonic series is divergent even though $\lim\limits_{n \to +\infty} 1/n = 0$. ◀

A **geometric series** is a series of the form

$$\sum_{n=1}^{+\infty} ar^{n-1} = a + ar + ar^2 + \ldots + ar^{n-1} + \ldots$$

Infinite series (2) discussed in Illustrations 1 and 2 is a geometric series with $a = 1$ and $r = \frac{1}{2}$. The nth partial sum of the above geometric series is given by

$$s_n = a(1 + r + r^2 + \ldots + r^{n-1}) \tag{7}$$

From the identity

$$1 - r^n = (1 - r)(1 + r + r^2 + \ldots + r^{n-1})$$

(7) can be written as

$$s_n = \frac{a(1 - r^n)}{1 - r} \quad \text{if } r \neq 1 \tag{8}$$

12.3.5 THEOREM The geometric series converges to the sum $a/(1 - r)$ if $|r| < 1$, and the geometric series diverges if $|r| \geq 1$.

Proof In Example 3, Section 12.1, we showed that $\lim\limits_{n \to +\infty} r^n = 0$ if $|r| < 1$. Therefore, from (8) we can conclude that if $|r| < 1$,

$$\lim_{n \to +\infty} s_n = \frac{a}{1 - r}$$

So if $|r| < 1$, the geometric series converges and its sum is $a/(1 - r)$.

If $r = 1$, $s_n = na$. Then $\lim\limits_{n \to +\infty} s_n = +\infty$ if $a > 0$, and $\lim\limits_{n \to +\infty} s_n = -\infty$ if $a < 0$.

If $r = -1$, then the geometric series becomes

$$a - a + a - \ldots + (-1)^{n-1}a + \ldots$$

Thus $s_n = 0$ if n is even, and $s_n = a$ if n is odd. Therefore $\lim\limits_{n \to +\infty} s_n$ does not exist.

Hence the geometric series diverges when $|r| = 1$.

If $|r| > 1$, $\lim\limits_{n \to +\infty} ar^{n-1} = a \lim\limits_{n \to +\infty} r^{n-1}$. Clearly, $\lim\limits_{n \to +\infty} r^{n-1} \neq 0$ because $|r^{n-1}|$ can be made as large as we please by taking n large enough. Therefore, by Theorem 12.3.3, the series is divergent. This completes the proof. ■

The following example illustrates how Theorem 12.3.5 can be used to express a nonterminating repeating decimal as a common fraction.

EXAMPLE 5   Express the decimal $0.3333 \ldots$ as a common fraction.

Solution

$$0.3333 \ldots = \frac{3}{10} + \frac{3}{100} + \frac{3}{1000} + \frac{3}{10,000} + \ldots + \frac{3}{10^n} + \ldots$$

This is a geometric series in which $a = \frac{3}{10}$ and $r = \frac{1}{10}$. Because $|r| < 1$, it follows from Theorem 12.3.5 that the series converges and its sum is $a/(1 - r)$. Therefore

$$0.3333 \ldots = \frac{\frac{3}{10}}{1 - \frac{1}{10}}$$

$$= \tfrac{1}{3}$$

EXERCISES 12.3

In Exercises 1 through 8, find the first four elements of the sequence of partial sums $\{s_n\}$, and find a formula for s_n in terms of n. Also determine whether the infinite series is convergent or divergent; if it is convergent, find its sum.

1. $\displaystyle\sum_{n=1}^{+\infty} \frac{1}{(2n-1)(2n+1)}$ **2.** $\displaystyle\sum_{n=1}^{+\infty} n$

3. $\displaystyle\sum_{n=1}^{+\infty} \frac{5}{(3n+1)(3n-2)}$ **4.** $\displaystyle\sum_{n=1}^{+\infty} \frac{2}{(4n-3)(4n+1)}$

5. $\displaystyle\sum_{n=1}^{+\infty} \ln\frac{n}{n+1}$ **6.** $\displaystyle\sum_{n=1}^{+\infty} \frac{2n+1}{n^2(n+1)^2}$

7. $\displaystyle\sum_{n=1}^{+\infty} \frac{2}{5^{n-1}}$ **8.** $\displaystyle\sum_{n=1}^{+\infty} \frac{2^{n-1}}{3^n}$

In Exercises 9 through 13, find the infinite series that is the given sequence of partial sums. Also determine whether the infinite series is convergent or divergent; if it is convergent, find its sum.

9. $\{s_n\} = \left\{\dfrac{2n}{3n+1}\right\}$ **10.** $\{s_n\} = \left\{\dfrac{n^2}{n+1}\right\}$

11. $\{s_n\} = \left\{\dfrac{1}{3^n}\right\}$ **12.** $\{s_n\} = \{3^n\}$

13. $\{s_n\} = \{\ln(2n+1)\}$

In Exercises 14 through 26, write the first four terms of the infinite series and determine whether the series is convergent or divergent. If the series is convergent, find its sum.

14. $\displaystyle\sum_{n=1}^{+\infty} \frac{n}{n+1}$ **15.** $\displaystyle\sum_{n=1}^{+\infty} \frac{2n+1}{3n+2}$

16. $\displaystyle\sum_{n=1}^{+\infty} [1 + (-1)^n]$ **17.** $\displaystyle\sum_{n=1}^{+\infty} \left(\frac{2}{3}\right)^n$

18. $\displaystyle\sum_{n=1}^{+\infty} \frac{3n^2}{n^2+1}$ **19.** $\displaystyle\sum_{n=1}^{+\infty} \ln\frac{1}{n}$ **20.** $\displaystyle\sum_{n=1}^{+\infty} \frac{2}{3^{n-1}}$

21. $\displaystyle\sum_{n=1}^{+\infty} (-1)^{n+1}\frac{3}{2^n}$ **22.** $\displaystyle\sum_{n=1}^{+\infty} \tan^n\frac{\pi}{6}$ **23.** $\displaystyle\sum_{n=1}^{+\infty} e^{-n}$

24. $\displaystyle\sum_{n=1}^{+\infty} \frac{\sinh n}{n}$ **25.** $\displaystyle\sum_{n=1}^{+\infty} \cos \pi n$ **26.** $\displaystyle\sum_{n=1}^{+\infty} \sin \pi n$

In Exercises 27 through 30, express the nonterminating repeating decimal as a common fraction.

27. $0.27\ 27\ 27\ldots$

28. $2.045\ 45\ 45\ldots$

29. $1.234\ 234\ 234\ldots$

30. $0.4653\ 4653\ 4653\ldots$

31. The path of each swing, after the first, of a pendulum bob is 0.93 as long as the path of the previous swing (from one side to the other side). If the path of the first swing is 56 cm long, and air resistance eventually brings the pendulum to rest, how far does the bob travel before it comes to rest?

32. A ball is dropped from a height of 12 m. Each time it strikes the ground, it bounces back to a height of three-fourths the distance from which it fell. Find the total distance traveled by the ball before it comes to rest.

33. What is the total distance traveled by a tennis ball before coming to rest if it is dropped from a height of 100 m and if, after each fall, it rebounds eleven-twentieths of the distance from which it fell?

34. An equilateral triangle has sides of length 4 units; therefore its perimeter is 12 units. Another equilateral triangle is constructed by drawing line segments through the midpoints of the sides of the first triangle. This triangle has sides of length 2 units, and its perimeter is 6 units. If this procedure can be repeated an unlimited number of times, what is the total perimeter of all the triangles that are formed?

35. After a woman riding a bicycle removes her feet from the pedals, the front wheel rotates 200 times during the first 10 sec. Then in each succeeding 10-sec time period the wheel rotates four-fifths as many times as it did the previous period. Determine the number of rotations of the wheel before the bicycle stops.

36. Find an infinite geometric series whose sum is 6 and such that each term is four times the sum of all the terms that follow it.

12.4 FOUR THEOREMS ABOUT INFINITE SERIES

The first theorem of this section states that the convergence or divergence of an infinite series is not affected by changing a finite number of terms.

12.4.1 THEOREM

If $\displaystyle\sum_{n=1}^{+\infty} a_n$ and $\displaystyle\sum_{n=1}^{+\infty} b_n$ are two infinite series, differing only in their first m terms (i.e., $a_k = b_k$ if $k > m$), then either both series converge or both series diverge.

Proof Let $\{s_n\}$ and $\{t_n\}$ be the sequences of partial sums of the series $\displaystyle\sum_{n=1}^{+\infty} a_n$ and $\displaystyle\sum_{n=1}^{+\infty} b_n$, respectively. Then

$$s_n = a_1 + a_2 + \ldots + a_m + a_{m+1} + a_{m+2} + \ldots + a_n$$

and

$$t_n = b_1 + b_2 + \ldots + b_m + b_{m+1} + b_{m+2} + \ldots + b_n$$

Because $a_k = b_k$ if $k > m$, then if $n \geq m$,

$$s_n - t_n = (a_1 + a_2 + \ldots + a_m) - (b_1 + b_2 + \ldots + b_m)$$

So

$$\text{if } n \geq m \quad \text{then} \quad s_n - t_n = s_m - t_m \tag{1}$$

We wish to show that either both $\displaystyle\lim_{n \to +\infty} s_n$ and $\displaystyle\lim_{n \to +\infty} t_n$ exist or do not exist. Suppose that $\displaystyle\lim_{n \to +\infty} t_n$ exists. Then from (1),

$$\text{if } n \geq m \quad \text{then} \quad s_n = t_n + (s_m - t_m)$$

Thus

$$\lim_{n \to +\infty} s_n = \lim_{n \to +\infty} t_n + (s_m - t_m)$$

Hence, when $\displaystyle\lim_{n \to +\infty} t_n$ exists, $\displaystyle\lim_{n \to +\infty} s_n$ also exists and both series converge.

Now suppose that $\lim\limits_{n \to +\infty} t_n$ does not exist and $\lim\limits_{n \to +\infty} s_n$ exists. From (1),

if $n \geq m$ then $t_n = s_n + (t_m - s_m)$

Because $\lim\limits_{n \to +\infty} s_n$ exists, it follows that

$$\lim_{n \to +\infty} t_n = \lim_{n \to +\infty} s_n + (t_m - s_m)$$

and so $\lim\limits_{n \to +\infty} t_n$ has to exist, which is a contradiction. Hence, if $\lim\limits_{n \to +\infty} t_n$ does not exist, then $\lim\limits_{n \to +\infty} s_n$ does not exist, and both series diverge. ■

EXAMPLE 1   Determine whether the series is convergent or divergent:

$$\sum_{n=1}^{+\infty} \frac{1}{n+4}$$

Solution   The given series is

$$\frac{1}{5} + \frac{1}{6} + \frac{1}{7} + \ldots + \frac{1}{n+4} + \ldots$$

which can be written as

$$0 + 0 + 0 + 0 + \frac{1}{5} + \frac{1}{6} + \frac{1}{7} + \ldots + \frac{1}{n} + \ldots \tag{2}$$

Now the harmonic series, which is known to be divergent, is

$$1 + \frac{1}{2} + \frac{1}{3} + \frac{1}{4} + \frac{1}{5} + \frac{1}{6} + \frac{1}{7} + \ldots + \frac{1}{n} + \ldots$$

Series (2) differs from the harmonic series only in the first four terms. Hence, by Theorem 12.4.1, series (2) is also divergent.

EXAMPLE 2   Determine whether the series is convergent or divergent:

$$\sum_{n=1}^{+\infty} \frac{\left[\!\left[\cos \dfrac{3}{n}\pi + 2\right]\!\right]}{3^n}$$

Solution   The given series can be written as

$$\frac{[\![\cos 3\pi + 2]\!]}{3} + \frac{[\![\cos \frac{3}{2}\pi + 2]\!]}{3^2} + \frac{[\![\cos \pi + 2]\!]}{3^3} + \frac{[\![\cos \frac{3}{4}\pi + 2]\!]}{3^4}$$

$$+ \frac{[\![\cos \frac{3}{5}\pi + 2]\!]}{3^5} + \frac{[\![\cos \frac{1}{2}\pi + 2]\!]}{3^6} + \frac{[\![\cos \frac{3}{7}\pi + 2]\!]}{3^7} + \ldots$$

$$= \frac{1}{3} + \frac{2}{3^2} + \frac{1}{3^3} + \frac{1}{3^4} + \frac{1}{3^5} + \frac{2}{3^6} + \frac{2}{3^7} + \frac{2}{3^8} + \ldots \tag{3}$$

Consider the geometric series with $a = \frac{2}{3}$ and $r = \frac{1}{3}$:

$$\frac{2}{3} + \frac{2}{3^2} + \frac{2}{3^3} + \frac{2}{3^4} + \frac{2}{3^5} + \frac{2}{3^6} + \frac{2}{3^7} + \frac{2}{3^8} + \ldots \tag{4}$$

This series is convergent by Theorem 11.3.5. Because series (3) differs from series (4) only in the first five terms, it follows from Theorem 12.4.1 that series (3) is also convergent.

As a consequence of Theorem 12.4.1, for a given infinite series a finite number of terms can be added or subtracted without affecting its convergence or divergence. For instance, in Example 1 the given series may be thought of as being obtained from the harmonic series by subtracting the first four terms. And because the harmonic series is divergent, the given series is divergent. In Example 2 we could consider the convergent geometric series

$$\frac{2}{3^6} + \frac{2}{3^7} + \frac{2}{3^8} + \ldots \tag{5}$$

and obtain the given series (3) by adding five terms. Because series (5) is convergent, it follows that series (3) is convergent.

The following theorem states that if an infinite series is multiplied term by term by a nonzero constant, its convergence or divergence is not affected.

12.4.2 THEOREM   Let c be any nonzero constant.

(i) If the series $\sum\limits_{n=1}^{+\infty} u_n$ is convergent and its sum is S, then the series $\sum\limits_{n=1}^{+\infty} cu_n$ is also convergent and its sum is $c \cdot S$.

(ii) If the series $\sum\limits_{n=1}^{+\infty} u_n$ is divergent, then the series $\sum\limits_{n=1}^{+\infty} cu_n$ is also divergent.

Proof   Let the nth partial sum of the series $\sum\limits_{n=1}^{+\infty} u_n$ be s_n. Therefore $s_n = u_1 + u_2 + \ldots + u_n$. The nth partial sum of the series $\sum\limits_{n=1}^{+\infty} cu_n$ is $c(u_1 + u_2 + \ldots + u_n) = cs_n$.

Proof of (i)   If the series $\sum\limits_{n=1}^{+\infty} u_n$ is convergent, then $\lim\limits_{n \to +\infty} s_n$ exists and is S. Therefore

$$\lim_{n \to +\infty} cs_n = c \lim_{n \to +\infty} s_n$$
$$= c \cdot S$$

Hence the series $\sum\limits_{n=1}^{+\infty} cu_n$ is convergent and its sum is $c \cdot S$.

Proof of (ii)   If the series $\sum\limits_{n=1}^{+\infty} u_n$ is divergent, then $\lim\limits_{n \to +\infty} s_n$ does not exist. Now suppose that the series $\sum\limits_{n=1}^{+\infty} cu_n$ is convergent. Then $\lim\limits_{n \to +\infty} cs_n$ exists. But $s_n = cs_n/c$; so

$$\lim_{n \to +\infty} s_n = \lim_{n \to +\infty} \frac{1}{c}(cs_n)$$
$$= \frac{1}{c} \lim_{n \to +\infty} cs_n$$

Thus $\lim\limits_{n \to +\infty} s_n$ must exist, which is a contradiction. Therefore the series $\sum\limits_{n=1}^{+\infty} cu_n$ is divergent. ∎

EXAMPLE 3   Determine whether the series is convergent or divergent:

$$\sum_{n=1}^{+\infty} \frac{1}{4n}$$

Solution

$$\sum_{n=1}^{+\infty} \frac{1}{4n} = \frac{1}{4} + \frac{1}{8} + \frac{1}{12} + \frac{1}{16} + \ldots + \frac{1}{4n} + \ldots$$

Because $\sum\limits_{n=1}^{+\infty} \frac{1}{n}$ is the harmonic series that is divergent, then by Theorem 12.4.2(ii) with $c = \frac{1}{4}$, the given series is divergent.

Theorem 12.4.2(i) is an extension to convergent infinite series of the following property of finite sums:

$$\sum_{k=1}^{n} ca_k = c \sum_{k=1}^{n} a_k$$

Another property of finite sums is

$$\sum_{k=1}^{n} (a_k \pm b_k) = \sum_{k=1}^{n} a_k \pm \sum_{k=1}^{n} b_k$$

and its extension to convergent infinite series is given by the following theorem.

12.4.3 THEOREM   If $\sum\limits_{n=1}^{+\infty} a_n$ and $\sum\limits_{n=1}^{+\infty} b_n$ are convergent infinite series whose sums are S and R, respectively, then

(i) $\sum\limits_{n=1}^{+\infty} (a_n + b_n)$ is a convergent series and its sum is $S + R$;

(ii) $\sum\limits_{n=1}^{+\infty} (a_n - b_n)$ is a convergent series and its sum is $S - R$.

The proof of this theorem is left as an exercise (see Exercise 24).

The next theorem is a corollary of the above theorem and is sometimes used to prove that a series is divergent.

12.4.4 THEOREM   If the series $\sum\limits_{n=1}^{+\infty} a_n$ is convergent and the series $\sum\limits_{n=1}^{+\infty} b_n$ is divergent, then the series $\sum\limits_{n=1}^{+\infty} (a_n + b_n)$ is divergent.

Proof   Assume that $\sum\limits_{n=1}^{+\infty} (a_n + b_n)$ is convergent and its sum is S. Let the sum of the series $\sum\limits_{n=1}^{+\infty} a_n$ be R. Then because

$$\sum_{n=1}^{+\infty} b_n = \sum_{n=1}^{+\infty} [(a_n + b_n) - a_n]$$

it follows from Theorem 12.4.3(ii) that $\sum\limits_{n=1}^{+\infty} b_n$ is convergent and its sum is $S - R$.

But this is a contradiction to the hypothesis that $\sum\limits_{n=1}^{+\infty} b_n$ is divergent. Hence $\sum\limits_{n=1}^{+\infty} (a_n + b_n)$ is divergent. ∎

EXAMPLE 4   Determine whether the series is convergent or divergent:

$$\sum_{n=1}^{+\infty} \left(\frac{1}{4n} + \frac{1}{4^n} \right)$$

Solution   In Example 3 the series $\sum\limits_{n=1}^{+\infty} \frac{1}{4n}$ was proved to be divergent.

Because the series $\sum\limits_{n=1}^{+\infty} \frac{1}{4^n}$ is a geometric series with $|r| = \frac{1}{4} < 1$, it is convergent. Hence, by Theorem 12.4.4, the given series is divergent.

If both series $\sum\limits_{n=1}^{+\infty} a_n$ and $\sum\limits_{n=1}^{+\infty} b_n$ are divergent, the series $\sum\limits_{n=1}^{+\infty} (a_n + b_n)$ may or may not be convergent. For example, if $a_n = \frac{1}{n}$ and $b_n = \frac{1}{n}$, then $a_n + b_n = \frac{2}{n}$ and $\sum\limits_{n=1}^{+\infty} \frac{2}{n}$ is divergent. But if $a_n = \frac{1}{n}$ and $b = -\frac{1}{n}$, then $a_n + b_n = 0$ and $\sum\limits_{n=1}^{+\infty} 0$ is convergent.

EXERCISES 12.4

In Exercises 1 through 22, determine whether the series is convergent or divergent. If the series is convergent, find its sum.

1. $\sum\limits_{n=1}^{+\infty} \frac{1}{n+2}$ **2.** $\sum\limits_{n=3}^{+\infty} \frac{1}{n-1}$ **3.** $\sum\limits_{n=1}^{+\infty} \frac{3}{2n}$

4. $\sum\limits_{n=1}^{+\infty} \frac{2}{3n}$ **5.** $\sum\limits_{n=1}^{+\infty} \frac{3}{2^n}$ **6.** $\sum\limits_{n=1}^{+\infty} \frac{2}{3^n}$

7. $\sum\limits_{n=1}^{+\infty} \frac{4}{3} \left(\frac{5}{7} \right)^n$ **8.** $\sum\limits_{n=1}^{+\infty} \frac{7}{5} \left(\frac{3}{4} \right)^n$

9. $\sum\limits_{n=1}^{+\infty} \frac{\left[\sin \dfrac{4}{n} \pi + 3 \right]}{4^n}$ **10.** $\sum\limits_{n=1}^{+\infty} \frac{\left[\cos \dfrac{1}{n} \pi + 1 \right]}{2^n}$

11. $\sum\limits_{n=1}^{+\infty} \left(\frac{1}{2n} + \frac{1}{2^n} \right)$ **12.** $\sum\limits_{n=1}^{+\infty} \left(\frac{1}{3^n} + \frac{1}{3n} \right)$ **13.** $\sum\limits_{n=1}^{+\infty} \left(\frac{1}{2^n} + \frac{1}{3^n} \right)$

14. $\sum\limits_{n=1}^{+\infty} \left(\frac{1}{3^n} - \frac{1}{4^n} \right)$ **15.** $\sum\limits_{n=1}^{+\infty} (e^{-n} + e^n)$ **16.** $\sum\limits_{n=1}^{+\infty} (2^{-n} + 3^n)$

17. $\sum\limits_{n=1}^{+\infty} \left(\frac{1}{2n} - \frac{1}{3n} \right)$ **18.** $\sum\limits_{n=1}^{+\infty} \left(\frac{3}{2n} - \frac{2}{3n} \right)$ **19.** $\sum\limits_{n=1}^{+\infty} \left(\frac{3}{2^n} - \frac{2}{3^n} \right)$

20. $\sum\limits_{n=1}^{+\infty} \left(\frac{5}{4^n} + \frac{4}{5^n} \right)$ **21.** $\sum\limits_{n=1}^{+\infty} \left(\frac{1}{n^2} + 2 \right)$ **22.** $\sum\limits_{n=1}^{+\infty} \frac{n!}{5^n}$

23. Give an example to show that even if each of the series $\sum\limits_{n=1}^{+\infty} a_n$ and $\sum\limits_{n=1}^{+\infty} b_n$ is divergent, it is possible for the series $\sum\limits_{n=1}^{+\infty} a_n b_n$ to be convergent.

24. Prove Theorem 12.4.3.

12.5 INFINITE SERIES OF POSITIVE TERMS

If all the terms of an infinite series are positive, the sequence of partial sums is increasing. Thus the following theorem follows immediately from Theorems 12.2.6 and 12.2.9.

12.5.1 THEOREM   An infinite series of positive terms is convergent if and only if its sequence of partial sums has an upper bound.

Proof For an infinite series of positive terms, the sequence of partial sums has a lower bound of 0. If the sequence of partial sums also has an upper bound, it is bounded. Furthermore, the sequence of partial sums of an infinite series of positive terms is increasing. It follows, then, from Theorem 12.2.6 that the sequence of partial sums is convergent, and therefore the infinite series is convergent.

Suppose now that an infinite series of positive terms is convergent. Then the sequence of partial sums is also convergent. It follows from Theorem 12.2.9 that the sequence of partial sums is bounded, and so it has an upper bound.

     ■

EXAMPLE 1   Prove that the series is convergent by using Theorem 12.5.1:

$$\sum_{n=1}^{+\infty} \frac{1}{n!}$$

Solution We must find an upper bound for the sequence of partial sums of the series $\sum_{n=1}^{+\infty} \frac{1}{n!}$.

$$s_1 = 1, s_2 = 1 + \frac{1}{1 \cdot 2}, s_3 = 1 + \frac{1}{1 \cdot 2} + \frac{1}{1 \cdot 2 \cdot 3},$$

$$s_4 = 1 + \frac{1}{1 \cdot 2} + \frac{1}{1 \cdot 2 \cdot 3} + \frac{1}{1 \cdot 2 \cdot 3 \cdot 4},$$

$$\vdots$$

$$s_n = 1 + \frac{1}{1 \cdot 2} + \frac{1}{1 \cdot 2 \cdot 3} + \frac{1}{1 \cdot 2 \cdot 3 \cdot 4} + \ldots + \frac{1}{1 \cdot 2 \cdot 3 \cdot \ldots \cdot n} \tag{1}$$

Now consider the first n terms of the geometric series with $a = 1$ and $r = \frac{1}{2}$:

$$\sum_{k=1}^{n} \frac{1}{2^{k-1}} = 1 + \frac{1}{2} + \frac{1}{2^2} + \frac{1}{2^3} + \ldots + \frac{1}{2^{n-1}} \tag{2}$$

By Theorem 12.3.5 the geometric series with $a = 1$ and $r = \frac{1}{2}$ has the sum $a/(1 - r) = 2$. Hence summation (2) is less than 2. Observe that each term of summation (1) is less than or equal to the corresponding term of summation (2); that is,

$$\frac{1}{k!} \le \frac{1}{2^{k-1}}$$

This is true because $k! = 1 \cdot 2 \cdot 3 \cdot \ldots \cdot k$, which in addition to the factor 1 contains $k - 1$ factors each greater than or equal to 2. Hence

$$s_n = \sum_{k=1}^{n} \frac{1}{k!} \le \sum_{k=1}^{n} \frac{1}{2^{k-1}} < 2$$

From the above, s_n has an upper bound of 2. Therefore, by Theorem 12.5.1, the given series is convergent.

In the above example the terms of the given series were compared with those of a known convergent series. This is a particular case of the following theorem known as the *comparison test*.

12.5.2 THEOREM
Comparison Test

Let the series $\sum\limits_{n=1}^{+\infty} u_n$ be a series of positive terms.

(i) If $\sum\limits_{n=1}^{+\infty} v_n$ is a series of positive terms that is known to be convergent, and $u_n \le v_n$ for all positive integers n, then $\sum\limits_{n=1}^{+\infty} u_n$ is convergent.

(ii) If $\sum\limits_{n=1}^{+\infty} w_n$ is a series of positive terms that is known to be divergent, and $u_n \ge w_n$ for all positive integers n, then $\sum\limits_{n=1}^{+\infty} u_n$ is divergent.

Proof of (i)   Let $\{s_n\}$ be the sequence of partial sums for the series $\sum\limits_{n=1}^{+\infty} u_n$ and $\{t_n\}$ be the sequence of partial sums for the series $\sum\limits_{n=1}^{+\infty} v_n$. Because $\sum\limits_{n=1}^{+\infty} v_n$ is a series of positive terms that is convergent, it follows from Theorem 12.5.1 that the sequence $\{t_n\}$ has an upper bound; call it B. Because $u_n \le v_n$ for all positive integers n, we can conclude that $s_n \le t_n \le B$ for all positive integers n. Therefore B is an upper bound of the sequence $\{s_n\}$. And because the terms of the series $\sum\limits_{n=1}^{+\infty} u_n$ are all positive, it follows from Theorem 12.5.1 that $\sum\limits_{n=1}^{+\infty} u_n$ is convergent.

Proof of (ii)   Assume that $\sum\limits_{n=1}^{+\infty} u_n$ is convergent. Then because both $\sum\limits_{n=1}^{+\infty} u_n$ and $\sum\limits_{n=1}^{+\infty} w_n$ are infinite series of positive terms and $w_n \le u_n$ for all positive integers n, it follows from part (i) that $\sum\limits_{n=1}^{+\infty} w_n$ is convergent. However, this contradicts the hypothesis; so our assumption is false. Therefore $\sum\limits_{n=1}^{+\infty} u_n$ is divergent.   ■

As we learned in Section 12.4, the convergence or divergence of an infinite series is not affected by discarding a finite number of terms. Therefore, when applying the comparison test, if $u_i \le v_i$ or $u_i \ge w_i$ when $i > m$, the test is valid regardless of how the first m terms of the two series compare.

EXAMPLE 2   Determine whether the series is convergent or divergent:

$$\sum_{n=1}^{+\infty} \frac{4}{3^n + 1}$$

Solution   The given series is

$$\frac{4}{4} + \frac{4}{10} + \frac{4}{28} + \frac{4}{82} + \ldots + \frac{4}{3^n + 1} + \ldots$$

Comparing the nth term of this series with the nth term of the convergent geometric series

$$\frac{4}{3} + \frac{4}{9} + \frac{4}{27} + \frac{4}{81} + \ldots + \frac{4}{3^n} + \ldots \qquad r = \tfrac{1}{3} < 1$$

we have

$$\frac{4}{3^n + 1} < \frac{4}{3^n}$$

for every positive integer n. Therefore, by the comparison test, Theorem 12.5.2(i), the given series is convergent.

EXAMPLE 3   Determine whether the series is convergent or divergent:

$$\sum_{n=1}^{+\infty} \frac{1}{\sqrt{n}}$$

Solution   The given series is

$$\sum_{n=1}^{+\infty} \frac{1}{\sqrt{n}} = \frac{1}{\sqrt{1}} + \frac{1}{\sqrt{2}} + \frac{1}{\sqrt{3}} + \ldots + \frac{1}{\sqrt{n}} + \ldots$$

Comparing the nth term of this series with the nth term of the divergent harmonic series we have

$$\frac{1}{\sqrt{n}} \geq \frac{1}{n}   \text{for every positive integer } n$$

So by Theorem 12.5.2(ii) the given series is divergent.

The following theorem, known as the *limit comparison test*, is a consequence of Theorem 12.5.2 and is often easier to apply.

12.5.3 THEOREM
Limit Comparison Test

Let $\displaystyle\sum_{n=1}^{+\infty} u_n$ and $\displaystyle\sum_{n=1}^{+\infty} v_n$ be two series of positive terms.

(i) If $\displaystyle\lim_{n \to +\infty} \frac{u_n}{v_n} = c > 0$, then the two series either both converge or both diverge.

(ii) If $\displaystyle\lim_{n \to +\infty} \frac{u_n}{v_n} = 0$, and if $\displaystyle\sum_{n=1}^{+\infty} v_n$ converges, then $\displaystyle\sum_{n=1}^{+\infty} u_n$ converges.

(iii) If $\displaystyle\lim_{n \to +\infty} \frac{u_n}{v_n} = +\infty$, and if $\displaystyle\sum_{n=1}^{+\infty} v_n$ diverges, then $\displaystyle\sum_{n=1}^{+\infty} u_n$ diverges.

Proof of (i)   Because $\displaystyle\lim_{n \to +\infty} (u_n/v_n) = c$, it follows that there exists an $N > 0$ such that

$$\text{if } n > N   \text{ then }   \left| \frac{u_n}{v_n} - c \right| < \frac{c}{2}$$

$$\Leftrightarrow   \text{if } n > N   \text{ then }   -\frac{c}{2} < \frac{u_n}{v_n} - c < \frac{c}{2}$$

$$\Leftrightarrow   \text{if } n > N   \text{ then }   \frac{c}{2} < \frac{u_n}{v_n} < \frac{3c}{2} \tag{3}$$

From the right-hand inequality (3),

$$u_n < \tfrac{3}{2}cv_n \tag{4}$$

If $\sum\limits_{n=1}^{+\infty} v_n$ is convergent, so is $\sum\limits_{n=1}^{+\infty} \tfrac{3}{2}cv_n$. It follows from inequality (4) and the comparison test that $\sum\limits_{n=1}^{+\infty} u_n$ is convergent.

From the left-hand inequality (3),

$$v_n < \frac{2}{c}u_n \tag{5}$$

If $\sum\limits_{n=1}^{+\infty} u_n$ is convergent, so is $\sum\limits_{n=1}^{+\infty} \frac{2}{c}u_n$. From inequality (5) and the comparison test it follows that $\sum\limits_{n=1}^{+\infty} v_n$ is convergent.

If $\sum\limits_{n=1}^{+\infty} v_n$ is divergent, $\sum\limits_{n=1}^{+\infty} u_n$ can be shown to be divergent by assuming that $\sum\limits_{n=1}^{+\infty} u_n$ is convergent and getting a contradiction by applying inequality (5) and the comparison test.

In a similar manner, if $\sum\limits_{n=1}^{+\infty} u_n$ is divergent, it follows that $\sum\limits_{n=1}^{+\infty} v_n$ is divergent because a contradiction is obtained from inequality (4) and the comparison test if $\sum\limits_{n=1}^{+\infty} v_n$ is assumed to be convergent.

We have therefore proved part (i). The proofs of parts (ii) and (iii) are left as exercises (see Exercises 30 and 31). ■

A word of caution is in order regarding part (ii) of Theorem 12.5.3. Note that when $\lim\limits_{n \to +\infty} \dfrac{u_n}{v_n} = 0$, the divergence of the series $\sum\limits_{n=1}^{+\infty} v_n$ does *not* imply that the series $\sum\limits_{n=1}^{+\infty} u_n$ diverges.

EXAMPLE 4   Solve Example 2 by using the limit comparison test.

Solution   Let u_n be the nth term of the given series $\sum\limits_{n=1}^{+\infty} \dfrac{4}{3^n + 1}$ and v_n be the nth term of the convergent geometric series $\sum\limits_{n=1}^{+\infty} \dfrac{4}{3^n}$. Therefore

$$\lim_{n \to +\infty} \frac{u_n}{v_n} = \lim_{n \to +\infty} \frac{\dfrac{4}{3^n + 1}}{\dfrac{4}{3^n}}$$

$$= \lim_{n \to +\infty} \frac{3^n}{3^n + 1}$$

$$= \lim_{n \to +\infty} \frac{1}{1 + 3^{-n}}$$

$$= 1$$

Hence, by part (i) of the limit comparison test it follows that the given series is convergent.

EXAMPLE 5   Solve Example 3 by using the limit comparison test.

Solution   Let u_n be the nth term of the given series $\sum_{n=1}^{+\infty} \dfrac{1}{\sqrt{n}}$ and v_n be the nth term of the divergent harmonic series. Then

$$\lim_{n \to +\infty} \frac{u_n}{v_n} = \lim_{n \to +\infty} \frac{\dfrac{1}{\sqrt{n}}}{\dfrac{1}{n}}$$

$$= \lim_{n \to +\infty} \sqrt{n}$$

$$= +\infty$$

Therefore, by part (iii) of the limit comparison test the given series is divergent.

EXAMPLE 6   Determine whether the series is convergent or divergent:

$$\sum_{n=1}^{+\infty} \frac{n^3}{n!}$$

Solution   In Example 1 we proved that the series $\sum_{n=1}^{+\infty} \dfrac{1}{n!}$ is convergent. By the limit comparison test with $u_n = \dfrac{n^3}{n!}$ and $v_n = \dfrac{1}{n!}$,

$$\lim_{n \to +\infty} \frac{u_n}{v_n} = \lim_{n \to +\infty} \frac{\dfrac{n^3}{n!}}{\dfrac{1}{n!}}$$

$$= \lim_{n \to +\infty} n^3$$

$$= +\infty$$

Part (iii) of the limit comparison test is not applicable because $\sum_{n=1}^{+\infty} v_n$ converges. However, there is a way that the limit comparison test can be used. The given series can be written as

$$\frac{1^3}{1!} + \frac{2^3}{2!} + \frac{3^3}{3!} + \frac{4^3}{4!} + \frac{5^3}{5!} + \ldots + \frac{n^3}{n!} + \ldots$$

Because Theorem 12.4.1 allows us to subtract a finite number of terms without affecting the behavior (convergence or divergence) of a series, we discard the first three terms and obtain

$$\frac{4^3}{4!} + \frac{5^3}{5!} + \frac{6^3}{6!} + \ldots + \frac{(n+3)^3}{(n+3)!} + \ldots$$

Now let $u_n = \dfrac{(n+3)^3}{(n+3)!}$ and, as before, let $v_n = \dfrac{1}{n!}$. Then

$$\lim_{n \to +\infty} \frac{u_n}{v_n} = \lim_{n \to +\infty} \frac{\dfrac{(n+3)^3}{(n+3)!}}{\dfrac{1}{n!}}$$

$$= \lim_{n \to +\infty} \frac{(n+3)^3 n!}{(n+3)!}$$

$$= \lim_{n \to +\infty} \frac{(n+3)^3 n!}{n!(n+1)(n+2)(n+3)}$$

$$= \lim_{n \to +\infty} \frac{(n+3)^2}{(n+1)(n+2)}$$

$$= \lim_{n \to +\infty} \frac{n^2 + 6n + 9}{n^2 + 3n + 2}$$

$$= \lim_{n \to +\infty} \frac{1 + \dfrac{6}{n} + \dfrac{9}{n^2}}{1 + \dfrac{3}{n} + \dfrac{2}{n^2}}$$

$$= 1$$

It follows from part (i) of the limit comparison test that the given series is convergent.

▶ **ILLUSTRATION 1**   Consider the geometric series

$$1 + \frac{1}{2} + \frac{1}{4} + \frac{1}{8} + \frac{1}{16} + \frac{1}{32} + \ldots + \frac{1}{2^{n-1}} + \ldots \tag{6}$$

which converges to 2 as shown in Illustration 2 of Section 12.3. Regroup the terms of this series to obtain

$$\left(1 + \frac{1}{2}\right) + \left(\frac{1}{4} + \frac{1}{8}\right) + \left(\frac{1}{16} + \frac{1}{32}\right) + \ldots + \left(\frac{1}{4^{n-1}} + \frac{1}{2 \cdot 4^{n-1}}\right) + \ldots$$

which is the series

$$\frac{3}{2} + \frac{3}{8} + \frac{3}{32} + \ldots + \frac{3}{2 \cdot 4^{n-1}} + \ldots \tag{7}$$

Series (7) is the geometric series with $a = \frac{3}{2}$ and $r = \frac{1}{4}$. Thus by Theorem 12.3.5 it is convergent, and its sum is

$$\frac{a}{1 - r} = \frac{\frac{3}{2}}{1 - \frac{1}{4}}$$

$$= 2$$

Therefore series (7), which is obtained from the convergent series (6) by regrouping the terms, is also convergent. Its sum is the same as that of series (6). ◀

Illustration 1 gives a particular case of the following theorem.

12.5.4 THEOREM   If $\displaystyle\sum_{n=1}^{+\infty} u_n$ is a given convergent series of positive terms, its terms can be grouped in any manner, and the resulting series also will be convergent and will have the same sum as the given series.

Proof   Let $\{s_n\}$ be the sequence of partial sums for the given convergent series of positive terms. Then $\displaystyle\lim_{n \to +\infty} s_n$ exists; let this limit be S. Consider a series $\displaystyle\sum_{n=1}^{+\infty} v_n$ whose terms are obtained by grouping the terms of $\displaystyle\sum_{n=1}^{+\infty} u_n$ in some manner. For example, $\displaystyle\sum_{n=1}^{+\infty} v_n$ may be the series

$$u_1 + (u_2 + u_3) + (u_4 + u_5 + u_6) + (u_7 + u_8 + u_9 + u_{10}) + \ldots$$

or it may be the series

$$(u_1 + u_2) + (u_3 + u_4) + (u_5 + u_6) + (u_7 + u_8) + \ldots$$

and so forth. Let $\{t_m\}$ be the sequence of partial sums for the series $\displaystyle\sum_{n=1}^{+\infty} v_n$. Each partial sum of the sequence $\{t_m\}$ is also a partial sum of the sequence $\{s_n\}$. Therefore, as m increases without bound, so does n. Because $\displaystyle\lim_{n \to +\infty} s_n = S$, we conclude that $\displaystyle\lim_{m \to +\infty} t_m = S$. This proves the theorem.   ■

Theorem 12.5.4 and the next theorem state properties of the sum of a convergent series of positive terms that are similar to properties that hold for the sum of a finite number of terms.

12.5.5 THEOREM   If $\displaystyle\sum_{n=1}^{+\infty} u_n$ is a given convergent series of positive terms, the order of the terms can be rearranged, and the resulting series also will be convergent and will have the same sum as the given series.

Proof   Let $\{s_n\}$ be the sequence of partial sums for the given convergent series of positive terms, and let $\displaystyle\lim_{n \to +\infty} s_n = S$. Let $\displaystyle\sum_{n=1}^{+\infty} v_n$ be a series formed by rearranging the order of the terms of $\displaystyle\sum_{n=1}^{+\infty} u_n$. For example, $\displaystyle\sum_{n=1}^{+\infty} v_n$ may be the series

$$u_4 + u_3 + u_7 + u_1 + u_9 + u_5 + \ldots$$

Let $\{t_n\}$ be the sequence of partial sums for the series $\displaystyle\sum_{n=1}^{+\infty} v_n$. Each partial sum of the sequence $\{t_n\}$ will be less than S because it is the sum of n terms of the infinite series $\displaystyle\sum_{n=1}^{+\infty} u_n$. Therefore S is an upper bound of the sequence $\{t_n\}$. Furthermore, because all the terms of the series $\displaystyle\sum_{n=1}^{+\infty} v_n$ are positive, $\{t_n\}$ is a monotonic increasing sequence. Hence, by Theorem 12.2.7 the sequence $\{t_n\}$ is convergent, and $\lim\limits_{n \to +\infty} t_n = T \leq S$. Now because the given series $\sum\limits_{n=1}^{+\infty} u_n$ can be obtained from the series $\sum\limits_{n=1}^{+\infty} v_n$ by rearranging the order of the terms, we can use the same argument and conclude that $S \leq T$. If both inequalities, $T \leq S$ and $S \leq T$, must hold, it follows that $S = T$. This proves the theorem. ■

A series that is often used in the comparison test is the one known as the **p series**, or the **hyperharmonic series**. It is

$$\frac{1}{1^p} + \frac{1}{2^p} + \frac{1}{3^p} + \ldots + \frac{1}{n^p} + \ldots \qquad \text{where } p \text{ is a constant} \tag{8}$$

In the following illustration we prove that the p series diverges if $p \leq 1$ and converges if $p > 1$.

▶ **ILLUSTRATION 2** If $p = 1$, the p series is the harmonic series, which diverges. If $p < 1$, then $n^p \leq n$; so

$$\frac{1}{n^p} \geq \frac{1}{n} \qquad \text{for every positive integer } n$$

Hence, by Theorem 12.5.2(ii) the p series is divergent if $p < 1$.

If $p > 1$, group the terms as follows:

$$\frac{1}{1^p} + \left(\frac{1}{2^p} + \frac{1}{3^p}\right) + \left(\frac{1}{4^p} + \frac{1}{5^p} + \frac{1}{6^p} + \frac{1}{7^p}\right) + \left(\frac{1}{8^p} + \frac{1}{9^p} + \ldots + \frac{1}{15^p}\right) + \ldots \tag{9}$$

Consider the series

$$\frac{1}{1^p} + \frac{2}{2^p} + \frac{4}{4^p} + \frac{8}{8^p} + \ldots + \frac{2^{n-1}}{(2^{n-1})^p} + \ldots \tag{10}$$

This is a geometric series whose ratio is $2/2^p = 1/2^{p-1}$, which is a positive number less than 1. Hence series (10) is convergent. Rewrite the terms of series (10) to get

$$\frac{1}{1^p} + \left(\frac{1}{2^p} + \frac{1}{2^p}\right) + \left(\frac{1}{4^p} + \frac{1}{4^p} + \frac{1}{4^p} + \frac{1}{4^p}\right) + \left(\frac{1}{8^p} + \frac{1}{8^p} + \ldots + \frac{1}{8^p}\right) + \ldots \tag{11}$$

By comparing series (9) and series (11) we see that the group of terms in each set of parentheses after the first group is less in sum for (9) than it is for (11). Therefore, by the comparison test, series (9) is convergent. Because (9) is merely a regrouping of the terms of the p series when $p > 1$, it follows from Theorem 12.5.4 that the p series is convergent if $p > 1$. ◀

Note that the series in Example 3 is the p series where $p = \frac{1}{2} < 1$; therefore it is divergent.

EXAMPLE 7   Determine whether the series is convergent or divergent:

$$\sum_{n=1}^{+\infty} \frac{1}{(n^2 + 2)^{1/3}}$$

Solution   Because for large values of n the number $n^2 + 2$ is close to the number n^2, so is the number $1/(n^2 + 2)^{1/3}$ close to the number $1/n^{2/3}$. The series $\sum\limits_{n=1}^{+\infty} \dfrac{1}{n^{2/3}}$ is divergent because it is the p series with $p = \frac{2}{3} < 1$. From the limit comparison test with $u_n = \dfrac{1}{(n^2 + 2)^{1/3}}$ and $v_n = \dfrac{1}{n^{2/3}}$,

$$\lim_{n \to +\infty} \frac{u_n}{v_n} = \lim_{n \to +\infty} \frac{\dfrac{1}{(n^2 + 2)^{1/3}}}{\dfrac{1}{n^{2/3}}}$$

$$= \lim_{n \to +\infty} \frac{n^{2/3}}{(n^2 + 2)^{1/3}}$$

$$= \lim_{n \to +\infty} \left(\frac{n^2}{n^2 + 2}\right)^{1/3}$$

$$= \lim_{n \to +\infty} \left(\frac{1}{1 + \dfrac{2}{n^2}}\right)^{1/3}$$

$$= 1$$

Therefore the given series is divergent.

EXERCISES 12.5

In Exercises 1 through 26, determine whether the series is convergent or divergent.

1. $\sum\limits_{n=1}^{+\infty} \dfrac{1}{n2^n}$

2. $\sum\limits_{n=1}^{+\infty} \dfrac{1}{\sqrt{2n + 1}}$

3. $\sum\limits_{n=1}^{+\infty} \dfrac{1}{n^n}$

4. $\sum\limits_{n=1}^{+\infty} \dfrac{n^2}{4n^3 + 1}$

5. $\sum\limits_{n=1}^{+\infty} \dfrac{3n + 1}{2n^2 + 5}$

6. $\sum\limits_{n=1}^{+\infty} \dfrac{3}{\sqrt{n^3 + n}}$

7. $\sum\limits_{n=1}^{+\infty} \dfrac{\cos^2 n}{3^n}$

8. $\sum\limits_{n=1}^{+\infty} \dfrac{1}{\ln(n + 1)}$

9. $\sum\limits_{n=1}^{+\infty} \dfrac{1}{\sqrt{n^2 + 4n}}$

10. $\sum\limits_{n=1}^{+\infty} \dfrac{|\sin n|}{n^2}$

11. $\sum\limits_{n=1}^{+\infty} \dfrac{n!}{(n + 2)!}$

12. $\sum\limits_{n=1}^{+\infty} \dfrac{1}{\sqrt{n^3 + 1}}$

13. $\sum\limits_{n=1}^{+\infty} \dfrac{n}{5n^2 + 3}$

14. $\sum\limits_{n=1}^{+\infty} \dfrac{(n - 1)!}{(n + 1)!}$

15. $\sum\limits_{n=1}^{+\infty} \dfrac{n!}{(2n)!}$

16. $\sum\limits_{n=1}^{+\infty} \sin \dfrac{1}{n}$

17. $\sum\limits_{n=1}^{+\infty} \dfrac{|\csc n|}{n}$

18. $\sum\limits_{n=1}^{+\infty} \dfrac{1}{n + \sqrt{n}}$

19. $\sum\limits_{n=2}^{+\infty} \dfrac{1}{n\sqrt{n^2 - 1}}$

20. $\sum\limits_{n=1}^{+\infty} \dfrac{2^n}{n!}$

21. $\sum\limits_{n=1}^{+\infty} \dfrac{3}{2n - \sqrt{n}}$

22. $\sum\limits_{n=1}^{+\infty} \dfrac{\sqrt{n}}{n^2 + 1}$

23. $\sum\limits_{n=1}^{+\infty} \dfrac{\ln n}{n^2 + 2}$

24. $\sum\limits_{n=1}^{+\infty} \dfrac{1}{3^n - \cos n}$

25. $\sum\limits_{n=1}^{+\infty} \dfrac{(n + 1)^2}{(n + 2)!}$

26. $\sum\limits_{n=1}^{+\infty} \dfrac{1}{(n + 2)(n + 4)}$

27. Use the series $\sum\limits_{n=1}^{+\infty} (-1)^{n+1}$ to show that Theorem 12.5.1 does not apply to an infinite series of both positive and negative terms; that is, show that the sequence of partial sums has an upper bound but the series is not convergent.

28. Suppose that f is a function such that $f(n) > 0$ for n any positive integer. Furthermore suppose that if p is any positive number, $\lim\limits_{n \to +\infty} n^p f(n)$ exists and is positive. Prove that the series $\sum\limits_{n=1}^{+\infty} f(n)$ is convergent if $p > 1$ and divergent if $0 < p \le 1$.

29. If $\sum\limits_{n=1}^{+\infty} a_n$ and $\sum\limits_{n=1}^{+\infty} b_n$ are two convergent series of positive terms, use the limit comparison test to prove that the series $\sum\limits_{n=1}^{+\infty} a_n b_n$ is also convergent.

30. Prove Theorem 12.5.3(ii).

31. Prove Theorem 12.5.3(iii).

12.6 THE INTEGRAL TEST

The theorem known as the *integral test* makes use of the theory of improper integrals to test an infinite series of positive terms for convergence.

12.6.1 THEOREM
Integral Test

Let f be a function that is continuous, decreasing, and positive-valued for all $x \geq 1$. Then the infinite series

$$\sum_{n=1}^{+\infty} f(n) = f(1) + f(2) + f(3) + \ldots + f(n) + \ldots$$

is convergent if the improper integral

$$\int_1^{+\infty} f(x)\, dx$$

exists, and it is divergent if $\lim\limits_{b \to +\infty} \int_1^b f(x)\, dx = +\infty$.

Proof If i is a positive integer and $i \geq 2$, then by the mean-value theorem for integrals (5.7.1) there exists a number X such that $i - 1 \leq X \leq i$ and

$$\int_{i-1}^i f(x)\, dx = f(X) \cdot 1 \tag{1}$$

Because f is a decreasing function,

$$f(i - 1) \geq f(X) \geq f(i)$$

and so from (1),

$$f(i - 1) \geq \int_{i-1}^i f(x)\, dx \geq f(i)$$

Therefore, if n is a positive integer and $n \geq 2$,

$$\sum_{i=2}^n f(i - 1) \geq \sum_{i=2}^n \int_{i-1}^i f(x)\, dx \geq \sum_{i=2}^n f(i)$$

$$\Leftrightarrow \sum_{i=1}^{n-1} f(i) \geq \int_1^n f(x)\, dx \geq \sum_{i=1}^n f(i) - f(1) \tag{2}$$

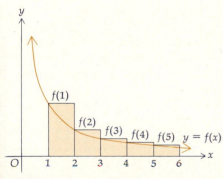

FIGURE 1

Figures 1 and 2 show the geometric interpretation of the above discussion for $n = 6$. In Figure 1 there is a sketch of the graph of a function f satisfying the hypothesis. The sum of the measures of the areas of the shaded rectangles is $f(1) + f(2) + f(3) + f(4) + f(5)$, which is the left member of inequality (2) when $n = 6$. Clearly, the sum of the measures of the areas of these rectangles is greater than the measure of the area given by the definite integral when $n = 6$. In Figure 2 the sum of the measures of the areas of the shaded rectangles is $f(2) + f(3) + f(4) + f(5) + f(6)$, which is the right member of the inequality (2) when $n = 6$. This sum is less than the value of the definite integral when $n = 6$.

If the given improper integral exists, let L be its value. Then

$$\int_1^n f(x)\, dx \leq L \tag{3}$$

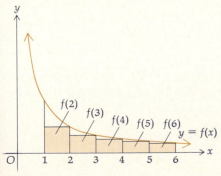

FIGURE 2

From the second and third members of the inequality (2) and from (3),

$$\sum_{i=1}^n f(i) \leq f(1) + \int_1^n f(x)\, dx \leq f(1) + L \tag{4}$$

Consider now the infinite series $\sum_{n=1}^{+\infty} f(n)$. Let the sequence of partial sums of this series be $\{s_n\}$, where $s_n = \sum_{i=1}^{n} f(i)$. From (4), $\{s_n\}$ has an upper bound of $f(1) + L$. Hence, by Theorem 12.5.1, $\sum_{n=1}^{+\infty} f(n)$ is convergent.

Suppose that $\lim_{b \to +\infty} \int_1^b f(x)\, dx = +\infty$. From (2)

$$\sum_{i=1}^{n-1} f(i) \geq \int_1^n f(x)\, dx$$

for all positive integers n. Therefore

$$\lim_{n \to +\infty} s_n = \lim_{n \to +\infty} \sum_{i=1}^{n} f(i)$$

$$= +\infty$$

Hence $\sum_{n=1}^{+\infty} f(n)$ is divergent. ∎

EXAMPLE 1   Use the integral test to show that the p series diverges if $p \leq 1$ and converges if $p > 1$.

Solution   The p series is $\sum_{n=1}^{+\infty} \frac{1}{n^p}$. If $f(x) = \frac{1}{x^p}$, then f is continuous and positive valued for all $x \geq 1$. Furthermore, if $1 \leq x_1 < x_2$, then $\frac{1}{x_1^p} > \frac{1}{x_2^p}$, and so f is decreasing for all $x \geq 1$. Therefore the hypothesis of Theorem 12.6.1 is satisfied by the function f. We consider the improper integral and have

$$\int_1^{+\infty} \frac{dx}{x^p} = \lim_{b \to +\infty} \int_1^b \frac{dx}{x^p}$$

If $p = 1$, the above integral gives

$$\lim_{b \to +\infty} \ln x \Big]_1^b = \lim_{b \to +\infty} \ln b$$

$$= +\infty$$

If $p \neq 1$, the integral gives

$$\lim_{b \to +\infty} \frac{x^{1-p}}{1-p} \Big]_1^b = \lim_{b \to +\infty} \frac{b^{1-p} - 1}{1-p}$$

This limit is $+\infty$ when $p < 1$; it is $-1/(1-p)$ if $p > 1$. Therefore, by the integral test it follows that the p series converges for $p > 1$ and diverges for $p \leq 1$.

EXAMPLE 2   Determine whether the series is convergent or divergent:

$$\sum_{n=1}^{+\infty} n e^{-n}$$

Solution   Let $f(x) = xe^{-x}$. Then

$$f'(x) = e^{-x} - xe^{-x}$$
$$= e^{-x}(1 - x)$$

Because $f'(x) < 0$ if $x > 1$, it follows from Theorem 4.4.3 that f is decreasing if $x \geq 1$. Furthermore, f is continuous and positive valued for all $x \geq 1$. Thus the hypothesis of the integral test is satisfied. By applying integration by parts,

$$\int xe^{-x}\, dx = -e^{-x}(x + 1) + C$$

Hence

$$\int_1^{+\infty} xe^{-x}\, dx = \lim_{b \to +\infty} \left[-e^{-x}(x + 1) \right]_1^b$$

$$= \lim_{b \to +\infty} \left[-\frac{b + 1}{e^b} + \frac{2}{e} \right]$$

Because $\lim\limits_{b \to +\infty} (b + 1) = +\infty$ and $\lim\limits_{b \to +\infty} e^b = +\infty$, L'Hôpital's rule can be used to obtain

$$\lim_{b \to +\infty} \frac{b + 1}{e^b} = \lim_{b \to +\infty} \frac{1}{e^b}$$

$$= 0$$

Therefore

$$\int_1^{+\infty} xe^{-x}\, dx = \frac{2}{e}$$

Thus the given series is convergent.

If for an infinite series the summation index starts with $n = k$ rather than $n = 1$, we have the following modification of the integral test:

If f is a function that is continuous, decreasing, and positive-valued for all $x \geq k$, then the infinite series $\sum\limits_{n=k}^{+\infty} f(n)$ is convergent if the improper integral

$$\int_k^{+\infty} f(x)\, dx$$

exists and is divergent if $\lim\limits_{b \to +\infty} \int_k^b f(x)\, dx = +\infty$.

The proof is identical to that of Theorem 12.6.1.

EXAMPLE 3   Determine whether the series is convergent or divergent:

$$\sum_{n=2}^{+\infty} \frac{1}{n\sqrt{\ln n}}$$

Solution   The function f defined by

$$f(x) = \frac{1}{x\sqrt{\ln x}}$$

is continuous and positive-valued for all $x \geq 2$. Also, if $2 \leq x_1 < x_2$, then $f(x_1) > f(x_2)$; so f is decreasing for all $x \geq 2$. Therefore the integral test can be applied.

$$\int_2^{+\infty} \frac{dx}{x\sqrt{\ln x}} = \lim_{b \to +\infty} \int_2^b (\ln x)^{-1/2} \frac{dx}{x}$$

$$= \lim_{b \to +\infty} \left[2\sqrt{\ln x} \right]_2^b$$

$$= \lim_{b \to +\infty} \left[2\sqrt{\ln b} - 2\sqrt{\ln 2} \right]$$

$$= +\infty$$

Thus the given series is divergent.

EXERCISES 12.6

In Exercises 1 through 8, use the integral test to determine whether the series is convergent or divergent.

1. $\displaystyle\sum_{n=1}^{+\infty} \frac{1}{2n+1}$ **2.** $\displaystyle\sum_{n=1}^{+\infty} \frac{2}{(3n+5)^2}$ **3.** $\displaystyle\sum_{n=1}^{+\infty} \frac{1}{(n+2)^{3/2}}$

4. $\displaystyle\sum_{n=2}^{+\infty} \frac{n}{n^2-2}$ **5.** $\displaystyle\sum_{n=3}^{+\infty} \frac{4}{n^2-4}$ **6.** $\displaystyle\sum_{n=1}^{+\infty} \frac{2n+3}{(n^2+3n)^2}$

7. $\displaystyle\sum_{n=1}^{+\infty} e^{-5n}$ **8.** $\displaystyle\sum_{n=1}^{+\infty} \frac{2n}{n^4+1}$

In Exercises 9 through 22, determine whether the series is convergent or divergent.

9. $\displaystyle\sum_{n=1}^{+\infty} \frac{\ln n}{n}$ **10.** $\displaystyle\sum_{n=2}^{+\infty} \frac{1}{n \ln n}$ **11.** $\displaystyle\sum_{n=1}^{+\infty} \frac{\tan^{-1} n}{n^2+1}$

12. $\displaystyle\sum_{n=1}^{+\infty} ne^{-n^2}$ **13.** $\displaystyle\sum_{n=1}^{+\infty} n^2 e^{-n}$ **14.** $\displaystyle\sum_{n=1}^{+\infty} ne^{-n}$

15. $\displaystyle\sum_{n=2}^{+\infty} \frac{\ln n}{n^3}$ **16.** $\displaystyle\sum_{n=1}^{+\infty} \cot^{-1} n$ **17.** $\displaystyle\sum_{n=1}^{+\infty} \operatorname{csch} n$

18. $\displaystyle\sum_{n=1}^{+\infty} \frac{e^{\tan^{-1} n}}{n^2+1}$ **19.** $\displaystyle\sum_{n=1}^{+\infty} \frac{e^{1/n}}{n^2}$ **20.** $\displaystyle\sum_{n=1}^{+\infty} \operatorname{sech}^2 n$

21. $\displaystyle\sum_{n=1}^{+\infty} \ln\left(\frac{n+3}{n}\right)$ **22.** $\displaystyle\sum_{n=2}^{+\infty} \frac{1}{n(\ln n)^3}$

23. Prove that the series $\displaystyle\sum_{n=2}^{+\infty} \frac{1}{n(\ln n)^p}$ is convergent if and only if $p > 1$.

24. Prove that the series $\displaystyle\sum_{n=3}^{+\infty} \frac{1}{n(\ln n)[\ln(\ln n)]^p}$ is convergent if and only if $p > 1$.

25. Prove that the series $\displaystyle\sum_{n=1}^{+\infty} \frac{\ln n}{n^p}$ is convergent if and only if $p > 1$.

26. If s_k is the kth partial sum of the harmonic series, prove that $\ln(k+1) < s_k < 1 + \ln k$. (*Hint:* $\dfrac{1}{m+1} \leq \dfrac{1}{x} \leq \dfrac{1}{m}$ if $0 < m \leq x \leq m+1$. Integrate each member of the inequality from m to $m+1$; let m take on successively the values $1, 2, \ldots, n-1$, and add the results.)

27. Use the method of Exercise 26 to estimate the sum

$$\sum_{m=50}^{100} \frac{1}{m} = \frac{1}{50} + \frac{1}{51} + \ldots + \frac{1}{100}$$

12.7   ALTERNATING SERIES

In this section and the next we consider infinite series having both positive and negative terms. The first type of such a series that we discuss is one whose terms are alternately positive and negative—an *alternating series*.

12.7.1 DEFINITION   If $a_n > 0$ for all positive integers n, then the series

$$\sum_{n=1}^{+\infty} (-1)^{n+1} a_n = a_1 - a_2 + a_3 - a_4 + \ldots + (-1)^{n+1} a_n + \ldots \tag{1}$$

and the series

$$\sum_{n=1}^{+\infty} (-1)^n a_n = -a_1 + a_2 - a_3 + a_4 - \ldots + (-1)^n a_n + \ldots \tag{2}$$

are called **alternating series**.

▶ **ILLUSTRATION 1**   An example of an alternating series of the form (1), where the first term is positive, is

$$\sum_{n=1}^{+\infty} (-1)^{n+1} \frac{1}{n} = 1 - \frac{1}{2} + \frac{1}{3} - \frac{1}{4} + \ldots + (-1)^{n+1} \frac{1}{n} + \ldots$$

An alternating series of the form (2), where the first term is negative, is

$$\sum_{n=1}^{+\infty} (-1)^n \frac{1}{n!} = -1 + \frac{1}{2!} - \frac{1}{3!} + \frac{1}{4!} - \ldots + (-1)^n \frac{1}{n!} + \ldots \qquad ◀$$

The following theorem gives a test for the convergence of an alternating series. It is called the alternating-series test; it is also known as Leibniz's test for alternating series because Leibniz formulated it in 1705.

12.7.2 THEOREM
Alternating-Series Test

Suppose we have the alternating series $\displaystyle\sum_{n=1}^{+\infty} (-1)^{n+1} a_n \left[\text{ or } \sum_{n=1}^{+\infty} (-1)^n a_n \right]$, where $a_n > 0$ and $a_{n+1} < a_n$ for all positive integers n. If $\displaystyle\lim_{n \to +\infty} a_n = 0$, the alternating series is convergent.

Proof   Assume that the first term of the alternating series is positive. This assumption is not a loss of generality because if this is not the case, then we discard the first term, which does not affect the convergence of the series. Thus we have the alternating series $\displaystyle\sum_{n=1}^{+\infty} (-1)^{n+1} a_n$. Consider the partial sum

$$s_{2n} = (a_1 - a_2) + (a_3 - a_4) + \ldots + (a_{2n-1} - a_{2n})$$

Because by hypothesis $a_{n+1} < a_n$, each quantity in parentheses is positive. Therefore

$$0 < s_2 < s_4 < s_6 < \ldots < s_{2n} < \ldots \tag{3}$$

We can also write s_{2n} as

$$s_{2n} = a_1 - (a_2 - a_3) - (a_4 - a_5) - \ldots - (a_{2n-2} - a_{2n-1}) - a_{2n}$$

Because $a_{n+1} < a_n$, again each quantity in parentheses is positive. Therefore

$$s_{2n} < a_1 \qquad \text{for every positive integer } n \tag{4}$$

From (3) and (4),

$$0 < s_{2n} < a_1 \qquad \text{for every positive integer } n$$

Thus the sequence $\{s_{2n}\}$ is bounded. Furthermore, from (3), the sequence $\{s_{2n}\}$ is increasing. Therefore, by Theorem 12.2.6 the sequence $\{s_{2n}\}$ is convergent. Let $\lim\limits_{n \to +\infty} s_{2n} = S$, and from Theorem 12.2.7, $S \leq a_1$. Because $s_{2n+1} = s_{2n} + a_{2n+1}$,

$$\lim_{n \to +\infty} s_{2n+1} = \lim_{n \to +\infty} s_{2n} + \lim_{n \to +\infty} a_{2n+1}$$

But, by hypothesis, $\lim\limits_{n \to +\infty} a_{2n+1} = 0$; so $\lim\limits_{n \to +\infty} s_{2n+1} = \lim\limits_{n \to +\infty} s_{2n}$. Therefore the sequence of partial sums of the even-numbered terms and the sequence of partial sums of the odd-numbered terms have the same limit S.

We now show that $\lim\limits_{n \to +\infty} s_n = S$. Because $\lim\limits_{n \to +\infty} s_{2n} = S$, then for any $\epsilon > 0$ there exists an integer $N_1 > 0$ such that

$$\text{if } 2n \geq N_1 \quad \text{then} \quad |s_{2n} - S| < \epsilon$$

And because $\lim\limits_{n \to +\infty} s_{2n+1} = S$, there exists an integer $N_2 > 0$ such that

$$\text{if } 2n + 1 \geq N_2 \quad \text{then} \quad |s_{2n+1} - S| < \epsilon$$

If N is the larger of the two integers N_1 and N_2, it follows that if n is any integer, either odd or even, and

$$\text{if } n \geq N \quad \text{then} \quad |s_n - S| < \epsilon$$

Therefore $\lim\limits_{n \to +\infty} s_n = S$; so the alternating series is convergent. ∎

EXAMPLE 1   Prove that the following alternating series is convergent:

$$\sum_{n=1}^{+\infty} (-1)^{n+1} \frac{1}{n}$$

Solution   The given series is

$$1 - \frac{1}{2} + \frac{1}{3} - \frac{1}{4} + \ldots + (-1)^{n+1} \frac{1}{n} + (-1)^{n+2} \frac{1}{n+1} + \ldots$$

Because $\dfrac{1}{n+1} < \dfrac{1}{n}$ for all positive integers n, and $\lim\limits_{n \to +\infty} \dfrac{1}{n} = 0$, it follows from Theorem 12.7.2 that the given alternating series is convergent.

EXAMPLE 2   Determine whether the series is convergent or divergent:

$$\sum_{n=1}^{+\infty} (-1)^n \frac{n+2}{n(n+1)}$$

Solution   The given series is an alternating series.

$$\lim_{n \to +\infty} a_n = \lim_{n \to +\infty} \frac{n+2}{n(n+1)}$$

$$= \lim_{n \to +\infty} \frac{\dfrac{1}{n} + \dfrac{2}{n^2}}{1 + \dfrac{1}{n}}$$

$$= 0$$

Before the alternating-series test can be applied, we must also show that $a_{n+1} < a_n$ or, equivalently, $\dfrac{a_{n+1}}{a_n} < 1$.

$$\frac{a_{n+1}}{a_n} = \frac{\dfrac{n+3}{(n+1)(n+2)}}{\dfrac{n+2}{n(n+1)}}$$

$$= \frac{n(n+3)}{(n+2)^2}$$

$$= \frac{n^2 + 3n}{n^2 + 4n + 4}$$

$$< 1$$

Then it follows from Theorem 12.7.2 that the given series is convergent.

12.7.3 DEFINITION

> If an infinite series is convergent and its sum is S, then the **remainder** obtained by approximating the sum of the series by the kth partial sum s_k is denoted by R_k, and
>
> $$R_k = S - s_k$$

12.7.4 THEOREM

> Suppose we have the alternating series $\displaystyle\sum_{n=1}^{+\infty} (-1)^{n+1} a_n \left[\text{or} \displaystyle\sum_{n=1}^{+\infty} (-1)^n a_n \right]$, where $a_n > 0$ and $a_{n+1} < a_n$ for all positive integers n, and $\displaystyle\lim_{n \to +\infty} a_n = 0$. Then if R_k is the remainder obtained by approximating the sum of the series by the sum of the first k terms, $|R_k| < a_{k+1}$.

Proof The given series converges by the alternating-series test. Assume that the odd-numbered terms of the given series are positive and the even-numbered terms are negative. Then from (3) in the proof of Theorem 12.7.2, the sequence $\{s_{2n}\}$ is increasing. So if S is the sum of the given series,

$$s_{2k} < s_{2k+2} < S \qquad \text{for all } k \geq 1 \tag{5}$$

To show that the sequence $\{s_{2n-1}\}$ is decreasing we write

$$s_{2n-1} = a_1 - (a_2 - a_3) - (a_4 - a_5) - \ldots - (a_{2n-2} - a_{2n-1})$$

Because $a_{n+1} < a_n$, it follows that each quantity in parentheses is positive. Therefore, because $a_1 > 0$

$$s_1 > s_3 > s_5 > \ldots > s_{2n-1} > \ldots$$

Hence the sequence $\{s_{2n-1}\}$ is decreasing. Thus

$$S < s_{2k+1} < s_{2k-1} \qquad \text{for all } k \geq 1 \tag{6}$$

Because $S < s_{2k+1}$

$$S - s_{2k} < s_{2k+1} - s_{2k} = a_{2k+1} \qquad \text{for all } k \geq 1 \tag{7}$$

From (5), $s_{2k} < S$. Hence

$$0 < S - s_{2k} \qquad \text{for all } k \geq 1$$

Therefore, from this inequality and (7),

$$0 < S - s_{2k} < a_{2k+1} \qquad \text{for all } k \geq 1 \tag{8}$$

From (5), $-S < -s_{2k}$. Hence

$$s_{2k-1} - S < s_{2k-1} - s_{2k} = a_{2k} \qquad \text{for all } k \geq 1 \tag{9}$$

From (6),

$$0 < s_{2k-1} - S \qquad \text{for all } k \geq 1$$

Then from this inequality and (9),

$$0 < s_{2k-1} - S < a_{2k} \qquad \text{for all } k \geq 1 \tag{10}$$

Because from Definition 12.7.3, $R_k = S - s_k$, then (8) can be written as

$$0 < R_{2k} < a_{2k+1} \qquad \text{for all } k \geq 1 \tag{11}$$

and (10) can be written as

$$0 < -R_{2k-1} < a_{2k} \qquad \text{for all } k \geq 1$$

Combining this inequality and (11) we have

$$|R_k| < a_{k+1} \qquad \text{for all } k \geq 1$$

and the theorem is proved.   ■

EXAMPLE 3   A series for computing $\ln(1 + x)$ if x is in the open interval $(-1, 1)$ is

$$\ln(1 + x) = \sum_{n=1}^{+\infty} (-1)^{n+1} \frac{x^n}{n}$$

Find an upper bound for the error when the first three terms of this series are used to approximate the value of $\ln 1.1$.

Solution   We use the given series with $x = 0.1$ to obtain

$$\ln 1.1 = 0.1 - \frac{(0.1)^2}{2} + \frac{(0.1)^3}{3} - \frac{(0.1)^4}{4} + \cdots$$

This series satisfies the conditions of Theorem 12.7.4; so if R_3 is the difference between the actual value of $\ln 1.1$ and the sum of the first three terms, then

$$|R_3| < 0.000025$$

Thus the sum of the first three terms will yield a value of $\ln 1.1$ accurate to at least four decimal places. Using the first three terms we get

$$\ln 1.1 \approx 0.0953$$

EXERCISES 12.7

In Exercises 1 through 14, determine whether the alternating series is convergent or divergent.

1. $\sum_{n=1}^{+\infty} (-1)^{n+1} \frac{1}{2n}$

2. $\sum_{n=1}^{+\infty} (-1)^n \frac{1}{n^2}$

3. $\sum_{n=1}^{+\infty} (-1)^n \frac{3}{n^2 + 1}$

4. $\sum_{n=1}^{+\infty} (-1)^{n+1} \frac{4}{3n - 2}$

5. $\sum_{n=2}^{+\infty} (-1)^n \frac{1}{\ln n}$

6. $\sum_{n=1}^{+\infty} (-1)^{n+1} \sin \frac{\pi}{n}$

7. $\displaystyle\sum_{n=1}^{+\infty} (-1)^{n+1} \frac{n^2}{n^3+2}$

8. $\displaystyle\sum_{n=1}^{+\infty} (-1)^{n+1} \frac{\ln n}{n}$

9. $\displaystyle\sum_{n=1}^{+\infty} (-1)^{n+1} \frac{\ln n}{n^2}$

10. $\displaystyle\sum_{n=1}^{+\infty} (-1)^{n} \frac{e^n}{n}$

11. $\displaystyle\sum_{n=1}^{+\infty} (-1)^{n} \frac{3^n}{n^2}$

12. $\displaystyle\sum_{n=1}^{+\infty} (-1)^{n} \frac{\sqrt{n}}{3n-1}$

13. $\displaystyle\sum_{n=1}^{+\infty} (-1)^{n} \frac{n}{2^n}$

14. $\displaystyle\sum_{n=1}^{+\infty} (-1)^{n+1} \frac{3^n}{1+3^{2n}}$

19. $\displaystyle\sum_{n=1}^{+\infty} (-1)^{n} \frac{1}{n^2}$

20. $\displaystyle\sum_{n=1}^{+\infty} (-1)^{n+1} \frac{1}{n^n}$

21. $\displaystyle\sum_{n=1}^{+\infty} (-1)^{n+1} \frac{1}{(n+1)\ln(n+1)}$

22. $\displaystyle\sum_{n=1}^{+\infty} (-1)^{n} \frac{1}{n!}$

In Exercises 23 through 30, find the sum of the infinite series, accurate to three decimal places.

23. $\displaystyle\sum_{n=1}^{+\infty} (-1)^{n+1} \frac{1}{2^n}$

24. $\displaystyle\sum_{n=1}^{+\infty} (-1)^{n+1} \frac{1}{n^4}$

25. $\displaystyle\sum_{n=1}^{+\infty} (-1)^{n+1} \frac{1}{n!}$

26. $\displaystyle\sum_{n=1}^{+\infty} (-1)^{n+1} \frac{2}{3^n}$

27. $\displaystyle\sum_{n=1}^{+\infty} (-1)^{n+1} \frac{1}{(2n)^3}$

28. $\displaystyle\sum_{n=1}^{+\infty} (-1)^{n} \frac{1}{(2n+1)^3}$

29. $\displaystyle\sum_{n=1}^{+\infty} (-1)^{n+1} \frac{1}{n2^n}$

30. $\displaystyle\sum_{n=1}^{+\infty} (-1)^{n+1} \frac{1}{(2n)!}$

In Exercises 15 through 22, find an upper bound for the error if the sum of the first four terms is used as an approximation to the sum of the infinite series.

15. $\displaystyle\sum_{n=1}^{+\infty} (-1)^{n+1} \frac{1}{n}$

16. $\displaystyle\sum_{n=1}^{+\infty} (-1)^{n} \frac{2}{n^2}$

17. $\displaystyle\sum_{n=1}^{+\infty} (-1)^{n+1} \frac{1}{(2n-1)^2}$

18. $\displaystyle\sum_{n=1}^{+\infty} (-1)^{n+1} \frac{n}{(n+1)^2}$

12.8 ABSOLUTE AND CONDITIONAL CONVERGENCE, THE RATIO TEST, AND THE ROOT TEST

If all the terms of a given infinite series are replaced by their absolute values and the resulting series is convergent, then the given series is said to be *absolutely convergent*.

12.8.1 DEFINITION

The infinite series $\displaystyle\sum_{n=1}^{+\infty} u_n$ is said to be **absolutely convergent** if the series $\displaystyle\sum_{n=1}^{+\infty} |u_n|$ is convergent.

▶ **ILLUSTRATION 1**   Consider the series

$$\sum_{n=1}^{+\infty} (-1)^{n+1} \frac{2}{3^n} = \frac{2}{3} - \frac{2}{3^2} + \frac{2}{3^3} - \frac{2}{3^4} + \ldots + (-1)^{n+1} \frac{2}{3^n} + \ldots \qquad (1)$$

This series will be absolutely convergent if the series

$$\sum_{n=1}^{+\infty} \frac{2}{3^n} = \frac{2}{3} + \frac{2}{3^2} + \frac{2}{3^3} + \frac{2}{3^4} + \ldots + \frac{2}{3^n} + \ldots$$

is convergent. Because this is the geometric series with $r = \frac{1}{3} < 1$, it is convergent. Therefore series (1) is absolutely convergent.   ◀

▶ **ILLUSTRATION 2**   A convergent series that is not absolutely convergent is the series

$$\sum_{n=1}^{+\infty} \frac{(-1)^{n+1}}{n}$$

In Example 1 of Section 12.7 this series was proved to be convergent. The series is not absolutely convergent because the series of absolute values is the harmonic series, which is divergent.   ◀

The series of Illustration 2 is an example of a *conditionally convergent* series.

12.8.2 DEFINITION A series that is convergent, but not absolutely convergent, is said to be **conditionally convergent**.

It is possible, then, for a series to be convergent but not absolutely convergent. If a series is absolutely convergent, it must be convergent, however, and this is given by the next theorem.

12.8.3 THEOREM If the infinite series $\sum\limits_{n=1}^{+\infty} u_n$ is absolutely convergent, it is convergent and

$$\left| \sum_{n=1}^{+\infty} u_n \right| \le \sum_{n=1}^{+\infty} |u_n|$$

Proof   Consider the three infinite series

$$\sum_{n=1}^{+\infty} u_n \qquad \sum_{n=1}^{+\infty} |u_n| \qquad \sum_{n=1}^{+\infty} (u_n + |u_n|)$$

and let their sequences of partial sums be $\{s_n\}$, $\{t_n\}$, and $\{r_n\}$, respectively. For every positive integer n, $u_n + |u_n|$ is either 0 or $2|u_n|$; so we have the inequality

$$0 \le u_n + |u_n| \le 2|u_n| \qquad (2)$$

Because $\sum\limits_{n=1}^{+\infty} |u_n|$ is convergent, it has a sum, which we denote by T. $\{t_n\}$ is an increasing sequence of positive numbers; so $t_n \le T$ for all positive integers n. From (2) it follows that

$$0 \le r_n \le 2t_n \le 2T$$

Therefore the sequence $\{r_n\}$ has an upper bound of $2T$. Thus, by Theorem 12.5.1 the series $\sum\limits_{n=1}^{+\infty} (u_n + |u_n|)$ is convergent. Let its sum be R. Because from (2), $\{r_n\}$ is an increasing sequence, we may conclude from Theorem 12.2.7 that $R \le 2T$.

Each of the series $\sum\limits_{n=1}^{+\infty} (u_n + |u_n|)$ and $\sum\limits_{n=1}^{+\infty} |u_n|$ is convergent; hence, from Theorem 12.4.3 the series

$$\sum_{n=1}^{+\infty} [(u_n + |u_n|) - |u_n|] = \sum_{n=1}^{+\infty} u_n$$

is also convergent.

Let the sum of the series $\sum\limits_{n=1}^{+\infty} u_n$ be S. Then, also from Theorem 12.4.3, $S = R - T$. And because $R \le 2T$, $S \le 2T - T = T$.

Because $\sum\limits_{n=1}^{+\infty} u_n$ is convergent and has the sum S, it follows from Theorem 12.4.2 that $\sum\limits_{n=1}^{+\infty} (-u_n)$ is convergent and has the sum $-S$. Because $\sum\limits_{n=1}^{+\infty} |-u_n|$ and $\sum\limits_{n=1}^{+\infty} |u_n|$ both equal T, we can replace $\sum\limits_{n=1}^{+\infty} u_n$ by $\sum\limits_{n=1}^{+\infty} (-u_n)$ in the above discussion and show that $-S \le T$. Because $S \le T$ and $-S \le T$, we have $|S| \le T$; therefore $\left| \sum\limits_{n=1}^{+\infty} u_n \right| \le \sum\limits_{n=1}^{+\infty} |u_n|$, and the theorem is proved. ∎

EXAMPLE 1   Determine whether the series is convergent or divergent:

$$\sum_{n=1}^{+\infty} \frac{\cos \frac{1}{3}n\pi}{n^2}$$

Solution   Denoting the given series by $\sum_{n=1}^{+\infty} u_n$, we have

$$\sum_{n=1}^{+\infty} u_n = \frac{\frac{1}{2}}{1^2} - \frac{\frac{1}{2}}{2^2} - \frac{1}{3^2} - \frac{\frac{1}{2}}{4^2} + \frac{\frac{1}{2}}{5^2} + \frac{1}{6^2} + \frac{\frac{1}{2}}{7^2} - \cdots + \frac{\cos \frac{1}{3}n\pi}{n^2} + \cdots$$

$$= \frac{1}{2} - \frac{1}{8} - \frac{1}{9} - \frac{1}{32} + \frac{1}{50} + \frac{1}{36} + \frac{1}{98} - \cdots$$

This is a series of positive and negative terms. We can prove this series is convergent if we can show that it is absolutely convergent.

$$\sum_{n=1}^{+\infty} |u_n| = \sum_{n=1}^{+\infty} \frac{|\cos \frac{1}{3}n\pi|}{n^2}$$

Because

$$|\cos \tfrac{1}{3}n\pi| \le 1 \qquad \text{for all } n$$

$$\frac{|\cos \frac{1}{3}n\pi|}{n^2} \le \frac{1}{n^2} \qquad \text{for all positive integers } n$$

The series $\sum_{n=1}^{+\infty} \frac{1}{n^2}$ is the p series, with $p = 2$, and is therefore convergent. So by the comparison test $\sum_{n=1}^{+\infty} |u_n|$ is convergent. The given series is therefore absolutely convergent; hence, by Theorem 12.8.3 it is convergent.

Observe that the terms of the series $\sum_{n=1}^{+\infty} |u_n|$ neither increase monotonically nor decrease monotonically. For example, $|u_4| = \frac{1}{32}$, $|u_5| = \frac{1}{50}$, $|u_6| = \frac{1}{36}$; and so $|u_5| < |u_4|$, but $|u_6| > |u_5|$.

The *ratio test*, given in the next theorem, is used frequently to determine whether a given series is absolutely convergent.

12.8.4 THEOREM
Ratio Test

Let $\sum_{n=1}^{+\infty} u_n$ be a given infinite series for which every u_n is nonzero. Then

(i) if $\displaystyle\lim_{n \to +\infty} \left| \frac{u_{n+1}}{u_n} \right| = L < 1$, the series is absolutely convergent;

(ii) if $\displaystyle\lim_{n \to +\infty} \left| \frac{u_{n+1}}{u_n} \right| = L > 1$ or if $\displaystyle\lim_{n \to +\infty} \left| \frac{u_{n+1}}{u_n} \right| = +\infty$, the series is divergent;

(iii) if $\displaystyle\lim_{n \to +\infty} \left| \frac{u_{n+1}}{u_n} \right| = 1$, no conclusion regarding convergence may be made from this test.

Proof of (i) It is given that $L < 1$. Let R be a number such that $L < R < 1$. Let $R - L = \epsilon < 1$. Because $\lim\limits_{n \to +\infty} \left| \dfrac{u_{n+1}}{u_n} \right| = L$, there exists an integer $N > 0$ such that

if $n \geq N$ then $\left| \left| \dfrac{u_{n+1}}{u_n} \right| - L \right| < \epsilon$

Therefore

if $n \geq N$ then $0 < \left| \dfrac{u_{n+1}}{u_n} \right| < L + \epsilon = R$ $\qquad\qquad$ (3)

Let n take on the successive values $N, N + 1, N + 2, \ldots$, and so forth. We obtain, from (3),

$$|u_{N+1}| < R|u_N|$$
$$|u_{N+2}| < R|u_{N+1}| < R^2|u_N|$$
$$|u_{N+3}| < R|u_{N+2}| < R^3|u_N|$$
$$\cdots$$

In general,

$$|u_{N+k}| < R^k|u_N| \qquad \text{for every positive integer } k \qquad\qquad (4)$$

The series

$$\sum_{k=1}^{+\infty} |u_N|R^k = |u_N|R + |u_N|R^2 + \ldots + |u_N|R^n + \ldots$$

is convergent because it is a geometric series whose ratio is less than 1. So from (4) and the comparison test it follows that the series $\sum\limits_{k=1}^{+\infty} |u_{N+k}|$ is convergent. The series $\sum\limits_{k=1}^{+\infty} |u_{N+k}|$ differs from the series $\sum\limits_{n=1}^{+\infty} |u_n|$ in only the first N terms. Therefore $\sum\limits_{n=1}^{+\infty} |u_n|$ is convergent; so the given series is absolutely convergent.

Proof of (ii) If $\lim\limits_{n \to +\infty} \left| \dfrac{u_{n+1}}{u_n} \right| = L > 1$ or $\lim\limits_{n \to +\infty} \left| \dfrac{u_{n+1}}{u_n} \right| = +\infty$, then in either case there is an integer $N > 0$ such that if $n \geq N$, then $\left| \dfrac{u_{n+1}}{u_n} \right| > 1$. Let n take on the successive values $N, N + 1, N + 2, \ldots$, and so on. We obtain

$$|u_{N+1}| > |u_N|$$
$$|u_{N+2}| > |u_{N+1}| > |u_N|$$
$$|u_{N+3}| > |u_{N+2}| > |u_N|$$
$$\cdots$$

Thus if $n > N$, then $|u_n| > |u_N|$. Hence $\lim\limits_{n \to +\infty} u_n \neq 0$; so the given series is divergent.

Proof of (iii) If the ratio test is applied to the p series, we have

$$\lim_{n \to +\infty} \left| \frac{u_{n+1}}{u_n} \right| = \lim_{n \to +\infty} \left| \frac{\dfrac{1}{(n+1)^p}}{\dfrac{1}{n^p}} \right|$$

$$= \lim_{n \to +\infty} \left| \left(\frac{n}{n+1} \right)^p \right|$$

$$= 1$$

Because the p series diverges if $p \leq 1$ and converges if $p > 1$, we have shown that it is possible to have both convergent and divergent series for which $\lim\limits_{n \to +\infty} \left| \dfrac{u_{n+1}}{u_n} \right| = 1$. This proves part (iii). ■

EXAMPLE 2   Determine whether the series is convergent or divergent:

$$\sum_{n=1}^{+\infty} (-1)^{n+1} \frac{n}{2^n}$$

Solution $u_n = (-1)^{n+1} \dfrac{n}{2^n}$ and $u_{n+1} = (-1)^{n+2} \dfrac{n+1}{2^{n+1}}$. Therefore

$$\left| \frac{u_{n+1}}{u_n} \right| = \frac{n+1}{2^{n+1}} \cdot \frac{2^n}{n}$$

$$= \frac{n+1}{2n}$$

So

$$\lim_{n \to +\infty} \left| \frac{u_{n+1}}{u_n} \right| = \lim_{n \to +\infty} \frac{1 + \dfrac{1}{n}}{2}$$

$$= \frac{1}{2}$$

$$< 1$$

Therefore, by the ratio test, the given series is absolutely convergent and hence, by Theorem 12.8.3, it is convergent.

EXAMPLE 3   In Example 2 of Section 12.7 we showed that the series

$$\sum_{n=1}^{+\infty} (-1)^n \frac{n+2}{n(n+1)}$$

is convergent. Is this series absolutely convergent or conditionally convergent?

Solution   To test for absolute convergence we apply the ratio test. In the solution of Example 2 of Section 12.7 we showed that the ratio

$$\frac{|u_{n+1}|}{|u_n|} = \frac{n^2 + 3n}{n^2 + 4n + 4}$$

Hence

$$\lim_{n \to +\infty} \left| \frac{u_{n+1}}{u_n} \right| = \lim_{n \to +\infty} \frac{1 + \dfrac{3}{n}}{1 + \dfrac{4}{n} + \dfrac{4}{n^2}}$$

$$= 1$$

So the ratio test fails. Because

$$|u_n| = \frac{n + 2}{n(n + 1)}$$

$$= \frac{n + 2}{n + 1} \cdot \frac{1}{n}$$

$$> \frac{1}{n}$$

the comparison test can be applied. And because the series $\sum\limits_{n=1}^{+\infty} \dfrac{1}{n}$ is the harmonic series, which diverges, we conclude that the series $\sum\limits_{n=1}^{+\infty} |u_n|$ is divergent and hence $\sum\limits_{n=1}^{+\infty} u_n$ is not absolutely convergent. Therefore the series is conditionally convergent.

Note that the ratio test does not include all possibilities for $\lim\limits_{n \to +\infty} \left| \dfrac{u_{n+1}}{u_n} \right|$ because it is possible that the limit does not exist and is not $+\infty$. The discussion of such cases is beyond the scope of this book.

The proof of the ratio test was based on using the comparison test with the geometric series. Another test whose proof is similar is the *root test*.

12.8.5 THEOREM
Root Test

Let $\sum\limits_{n=1}^{+\infty} u_n$ be a given infinite series for which every u_n is nonzero. Then

(i) if $\lim\limits_{n \to +\infty} \sqrt[n]{|u_n|} = L < 1$, the series is absolutely convergent;

(ii) if $\lim\limits_{n \to +\infty} \sqrt[n]{|u_n|} = L > 1$, or if $\lim\limits_{n \to +\infty} \sqrt[n]{|u_n|} = +\infty$, the series is divergent;

(iii) if $\lim\limits_{n \to +\infty} \sqrt[n]{|u_n|} = 1$, no conclusion regarding convergence may be made from this test.

Because of the similarity of the proof of the root test with that of the ratio test it is left for you to do as an exercise (see Exercises 26 through 28).

EXAMPLE 4   Use the root test to determine whether the series is convergent or divergent:

$$\sum_{n=1}^{+\infty} (-1)^n \frac{3^{2n+1}}{n^{2n}}$$

Solution   By applying the root test we have

$$\lim_{n \to +\infty} \sqrt[n]{|u_n|} = \lim_{n \to +\infty} \left(\frac{3^{2n+1}}{n^{2n}} \right)^{1/n}$$

$$= \lim_{n \to +\infty} \frac{3^{2+(1/n)}}{n^2}$$

$$= 0$$

$$< 1$$

Therefore, by the root test, the given series is absolutely convergent. Hence by Theorem 12.8.3 it is convergent.

The ratio test and the root test are closely related; however, the ratio test is usually easier to apply. If the terms of the series contain factorials, this is certainly the case. If the terms contain powers, as in Example 4, it may be advantageous to use the root test. The next example provides a series for which the root test is the better one to apply.

EXAMPLE 5   Determine whether the series is convergent or divergent:

$$\sum_{n=1}^{+\infty} \frac{1}{[\ln(n+1)]^n}$$

Solution

$$\lim_{n \to +\infty} \sqrt[n]{|u_n|} = \lim_{n \to +\infty} \sqrt[n]{\left| \frac{1}{[\ln(n+1)]^n} \right|}$$

$$= \lim_{n \to +\infty} \left| \frac{1}{\ln(n+1)} \right|$$

$$= 0$$

$$< 1$$

From the root test, the given series is absolutely convergent. Therefore, by Theorem 12.8.3 it is convergent.

There are series for which the root test can be used to show convergence but for which the ratio test fails. Such a series appears in Exercise 25.

EXERCISES 12.8

In Exercises 1 through 20, determine if the series is absolutely convergent, conditionally convergent, or divergent. Prove your answer.

1. $\displaystyle\sum_{n=1}^{+\infty} \left(-\frac{2}{3}\right)^n$

2. $\displaystyle\sum_{n=1}^{+\infty} (-1)^n \frac{2^n}{n^3}$

3. $\displaystyle\sum_{n=1}^{+\infty} (-1)^{n+1} \frac{2^n}{n!}$

4. $\displaystyle\sum_{n=1}^{+\infty} n\left(\frac{2}{3}\right)^n$

5. $\displaystyle\sum_{n=1}^{+\infty} \frac{n^2}{n!}$

6. $\displaystyle\sum_{n=1}^{+\infty} (-1)^{n+1} \frac{1}{(2n-1)!}$

7. $\displaystyle\sum_{n=1}^{+\infty} (-1)^n \frac{n!}{2^{n+1}}$

8. $\displaystyle\sum_{n=1}^{+\infty} (-1)^{n+1} \frac{1}{n(n+2)}$

9. $\displaystyle\sum_{n=1}^{+\infty} \frac{1-2\sin n}{n^3}$

10. $\displaystyle\sum_{n=1}^{+\infty} (-1)^n \frac{1}{(n+1)^3}$

11. $\displaystyle\sum_{n=1}^{+\infty} (-1)^{n+1} \frac{3^n}{n!}$

12. $\displaystyle\sum_{n=1}^{+\infty} (-1)^n \frac{n^2+1}{n^3}$

13. $\displaystyle\sum_{n=2}^{+\infty} (-1)^{n+1} \frac{1}{n(\ln n)^2}$

14. $\displaystyle\sum_{n=1}^{+\infty} \frac{\cos n}{n^2}$

15. $\displaystyle\sum_{n=1}^{+\infty} \frac{\sin \pi n}{n}$

16. $\displaystyle\sum_{n=2}^{+\infty} (-1)^{n+1} \frac{n}{\ln n}$

17. $\displaystyle\sum_{n=2}^{+\infty} \frac{1}{(\ln n)^n}$

18. $\displaystyle\sum_{n=1}^{+\infty} \frac{\left(1+\dfrac{1}{n}\right)^{2n}}{e^n}$

19. $\displaystyle\sum_{n=1}^{+\infty} \frac{n^n}{n!}$

20. $\displaystyle\sum_{n=1}^{+\infty} \frac{1\cdot3\cdot5\cdot\ldots\cdot(2n-1)}{1\cdot4\cdot7\cdot\ldots\cdot(3n-2)}$

21. If $|r| < 1$, prove that the series $\displaystyle\sum_{n=1}^{+\infty} r^n \sin nt$ is absolutely convergent for all values of t.

22. Prove that if $\displaystyle\sum_{n=1}^{+\infty} u_n$ is absolutely convergent and $u_n \neq 0$ for all n, then $\displaystyle\sum_{n=1}^{+\infty} \frac{1}{|u_n|}$ is divergent.

23. Prove that if $\displaystyle\sum_{n=1}^{+\infty} u_n$ is absolutely convergent, then $\displaystyle\sum_{n=1}^{+\infty} u_n^2$ is convergent.

24. Show by means of an example that the converse of Exercise 23 is not true.

25. Given the series $\displaystyle\sum_{n=1}^{+\infty} \frac{1}{2^{n+1+(-1)^n}}$. (a) Show that the ratio test fails for this series. (b) Use the root test to determine whether the series is convergent or divergent.

26. Prove part (i) of the root test (Theorem 12.8.5). (*Hint:* Because $L < 1$, let R be a number such that $L < R < 1$, and let $R - L = \epsilon < 1$. Show that there is an integer N such that if $n > N$, then $|u_n| < R^n$. Then use the comparison test.)

27. Prove part (ii) of the root test. See the hint for Exercise 26.

28. Prove part (iii) of the root test by applying it to the two series $\displaystyle\sum_{n=1}^{+\infty} \frac{1}{n}$ and $\displaystyle\sum_{n=1}^{+\infty} \frac{1}{n^2}$. $\left(\textit{Hint: } \text{Determine } \lim_{n \to +\infty} \sqrt[n]{n} \text{ by letting } \sqrt[n]{n} = e^{(\ln n)/n} \text{ and using L'Hôpital's rule to find } \lim_{n \to +\infty} \frac{\ln n}{n}.\right)$

12.9 A SUMMARY OF TESTS FOR CONVERGENCE OR DIVERGENCE OF AN INFINITE SERIES

To conclude the discussion of infinite series of constant terms we summarize the various tests that can be used to determine the convergence or divergence of a given series. In Sections 12.3 through 12.8 we had a number of these tests, and to gain proficiency in recognizing and applying the appropriate one requires considerable practice. You will gain this practice by doing Exercises 11 through 48 in the Review Exercises that follow. Here is a list of these tests. You might attempt them in the indicated order. If a particular step is not applicable or no conclusion can be made, continue on to the next one. Of course, sometimes more than one test can be used, but you should select the most efficient one.

1. Compute $\lim_{n \to +\infty} u_n$. If $\lim_{n \to +\infty} u_n \neq 0$, then the series diverges. If $\lim_{n \to +\infty} u_n = 0$, no conclusion can be made.

2. Examine the series to determine if it is one of the special types:

   (i) A geometric series: $\displaystyle\sum_{n=1}^{+\infty} ar^{n-1}$. It converges to the sum $\dfrac{a}{1-r}$ if $|r| < 1$; it diverges if $|r| \geq 1$.

(ii) A p series: $\sum_{n=1}^{+\infty} \dfrac{1}{n^p}$ (where p is a constant). It converges if $p > 1$; it diverges if $p \leq 1$.

(iii) An alternating series: $\sum_{n=1}^{+\infty} (-1)^{n+1} a_n$ or $\sum_{n=1}^{+\infty} (-1)^n a_n$. Apply the alternating-series test (Theorem 12.7.2): If $a_n > 0$ and $a_{n+1} < a_n$ for all positive integers n, and $\lim\limits_{n \to +\infty} a_n = 0$, then the alternating series is convergent.

3. Try the ratio test (Theorem 12.8.4): Let $\sum_{n=1}^{+\infty} u_n$ be a given infinite series for which every u_n is nonzero. Then

   (i) if $\lim\limits_{n \to +\infty} \left| \dfrac{u_{n+1}}{u_n} \right| = L < 1$, the series is absolutely convergent;

   (ii) if $\lim\limits_{n \to +\infty} \left| \dfrac{u_{n+1}}{u_n} \right| = L > 1$ or if $\lim\limits_{n \to +\infty} \left| \dfrac{u_{n+1}}{u_n} \right| = +\infty$, the series is divergent;

   (iii) if $\lim\limits_{n \to +\infty} \left| \dfrac{u_{n+1}}{u_n} \right| = 1$, no conclusion regarding convergence may be made from this test.

4. Try the root test (Theorem 12.8.5): Let $\sum_{n=1}^{+\infty} u_n$ be a given infinite series for which every u_n is nonzero. Then

   (i) if $\lim\limits_{n \to +\infty} \sqrt[n]{|u_n|} = L < 1$, the series is absolutely convergent;

   (ii) if $\lim\limits_{n \to +\infty} \sqrt[n]{|u_n|} = L > 1$, or if $\lim\limits_{n \to +\infty} \sqrt[n]{|u_n|} = +\infty$, the series is divergent;

   (iii) if $\lim\limits_{n \to +\infty} \sqrt[n]{|u_n|} = 1$, no conclusion regarding convergence may be made from this test.

5. Try the integral test (Theorem 12.6.1): Let f be a function that is continuous, decreasing, and positive valued for all $x \geq 1$. Then the infinite series

$$\sum_{n=1}^{+\infty} f(n) = f(1) + f(2) + f(3) + \ldots + f(n) + \ldots$$

is convergent if the improper integral

$$\int_1^{+\infty} f(x)\, dx$$

exists, and it is divergent if $\lim\limits_{b \to +\infty} \int_1^b f(x)\, dx = +\infty$.

6. Try the comparison test (Theorem 12.5.2): Let the series $\sum_{n=1}^{+\infty} u_n$ be a series of positive terms.

   (i) If $\sum_{n=1}^{+\infty} v_n$ is a series of positive terms that is known to be convergent, and $u_n \leq v_n$ for all positive integers n, then $\sum_{n=1}^{+\infty} u_n$ is convergent.

(ii) If $\sum\limits_{n=1}^{+\infty} w_n$ is a series of positive terms that is known to be divergent, and $u_n \geq w_n$ for all positive integers n, then $\sum\limits_{n=1}^{+\infty} u_n$ is divergent.

or the limit comparison test (Theorem 12.5.3): Let $\sum\limits_{n=1}^{+\infty} u_n$ and $\sum\limits_{n=1}^{+\infty} v_n$ be two series of positive terms.

(i) If $\lim\limits_{n \to +\infty} \dfrac{u_n}{v_n} = c > 0$, then the two series either both converge or both diverge.

(ii) If $\lim\limits_{n \to +\infty} \dfrac{u_n}{v_n} = 0$, and if $\sum\limits_{n=1}^{+\infty} v_n$ converges, then $\sum\limits_{n=1}^{+\infty} u_n$ converges.

(iii) If $\lim\limits_{n \to +\infty} \dfrac{u_n}{v_n} = +\infty$, and if $\sum\limits_{n=1}^{+\infty} v_n$ diverges, then $\sum\limits_{n=1}^{+\infty} u_n$ diverges.

REVIEW EXERCISES FOR CHAPTER 12

In Exercises 1 through 8, write the first four numbers of the sequence and find its limit, if it exists.

1. $\left\{ \dfrac{3n}{n+2} \right\}$ **2.** $\left\{ \dfrac{(-1)^{n-1}}{(n+1)^2} \right\}$ **3.** $\left\{ \dfrac{n^2-1}{n^2+1} \right\}$

4. $\left\{ \dfrac{n+3n^2}{4+2n^3} \right\}$ **5.** $\{2 + (-1)^n\}$ **6.** $\left\{ \dfrac{n^2}{\ln(n+1)} \right\}$

7. $\left\{ \left(1 + \dfrac{1}{n}\right)^{2n} \right\}$ **8.** $\left\{ \dfrac{(n+2)^2}{n+4} - \dfrac{(n+2)^2}{n} \right\}$

In Exercises 9 and 10, find the first four elements of the sequence of partial sums $\{s_n\}$, and find a formula for s_n in terms of n. Also determine whether the infinite series is convergent or divergent; if it is convergent, find its sum.

9. $\sum\limits_{n=1}^{+\infty} \dfrac{3}{4^{n+1}}$ **10.** $\sum\limits_{n=1}^{+\infty} \ln\left(\dfrac{2n-1}{2n+1} \right)$

In Exercises 11 through 20, determine whether the series is convergent or divergent. If the series is convergent, find its sum.

11. $\sum\limits_{n=1}^{+\infty} \left(\dfrac{3}{4} \right)^n$ **12.** $\sum\limits_{n=1}^{+\infty} e^{-2n}$

13. $\sum\limits_{n=1}^{+\infty} \dfrac{n-1}{n+1}$ **14.** $\sum\limits_{n=0}^{+\infty} [(-1)^n + (-1)^{n+1}]$

15. $\sum\limits_{n=0}^{+\infty} \sin^n \tfrac{1}{3}\pi$ **16.** $\sum\limits_{n=0}^{+\infty} \cos^n \tfrac{1}{3}\pi$

17. $\sum\limits_{n=1}^{+\infty} \dfrac{1}{(3n-1)(3n+2)}$ (Hint: To find the sum, first find the sequence of partial sums.)

18. $\sum\limits_{n=1}^{+\infty} \dfrac{3}{2} \left(\dfrac{1}{5} \right)^n$ **19.** $\sum\limits_{n=1}^{+\infty} \dfrac{\left[\sin\dfrac{3}{n}\pi + 2 \right]}{3^n}$

20. $\sum\limits_{n=1}^{+\infty} \left(\dfrac{1}{4^n} + \dfrac{1}{3^n} \right)$

In Exercises 21 through 38, determine whether the series is convergent or divergent.

21. $\sum\limits_{n=1}^{+\infty} \dfrac{2}{n^2 + 6n}$ **22.** $\sum\limits_{n=1}^{+\infty} \dfrac{1}{(2n+1)^3}$

23. $\sum\limits_{n=1}^{+\infty} \cos\left(\dfrac{\pi}{2n^2-1} \right)$ **24.** $\sum\limits_{n=1}^{+\infty} \dfrac{3 + \sin n}{n^2}$

25. $\sum\limits_{n=1}^{+\infty} \dfrac{(n!)^2}{(2n)!}$ **26.** $\sum\limits_{n=1}^{+\infty} \dfrac{n}{\sqrt{3n+2}}$ **27.** $\sum\limits_{n=1}^{+\infty} (-1)^n \ln\dfrac{1}{n}$

28. $\sum\limits_{n=1}^{+\infty} \dfrac{(-1)^{n+1}}{1 + \sqrt{n}}$ **29.** $\sum\limits_{n=2}^{+\infty} \dfrac{1}{n(\ln n)^2}$ **30.** $\sum\limits_{n=1}^{+\infty} \dfrac{\ln n}{n^2}$

31. $\sum\limits_{n=1}^{+\infty} \left(\dfrac{2}{5n} - \dfrac{3}{2n} \right)$ **32.** $\sum\limits_{n=0}^{+\infty} \dfrac{n!}{10^n}$ **33.** $\sum\limits_{n=1}^{+\infty} \dfrac{1}{1 + 2\ln n}$

34. $\sum\limits_{n=1}^{+\infty} \dfrac{|\sec n|}{n^{3/4}}$ **35.** $\sum\limits_{n=1}^{+\infty} \dfrac{\cos n}{n^3}$ **36.** $\sum\limits_{n=1}^{+\infty} n3^{-n^2}$

37. $\sum\limits_{n=1}^{+\infty} \dfrac{1}{2^n + \sin n}$ **38.** $\sum\limits_{n=1}^{+\infty} \dfrac{(n+2)^2}{(n+3)!}$

In Exercises 39 through 48, determine if the series is absolutely convergent, conditionally convergent, or divergent. Prove your answer.

39. $\sum\limits_{n=0}^{+\infty} (-1)^n \dfrac{n^2}{3^n}$ **40.** $\sum\limits_{n=0}^{+\infty} (-1)^n \dfrac{5^{2n+1}}{(2n+1)!}$

41. $\sum\limits_{n=1}^{+\infty} (-1)^{n-1} \dfrac{1}{(n+1)^{3/4}}$ **42.** $\sum\limits_{n=1}^{+\infty} (-1)^{n-1} \dfrac{6^n}{5^{n+1}}$

43. $\sum\limits_{n=1}^{+\infty} (-1)^n \dfrac{n!}{10n}$ **44.** $\sum\limits_{n=1}^{+\infty} (-1)^n \dfrac{\sqrt{2n-1}}{n}$

45. $\displaystyle\sum_{n=1}^{+\infty} (-1)^{n-1} \frac{2^{3n}}{n^n}$

46. $\displaystyle\sum_{n=1}^{+\infty} (-1)^{n-1} \frac{1}{[\ln(n+2)]^n}$

47. $\displaystyle\sum_{n=1}^{+\infty} c_n$, where $c_n = \begin{cases} -\dfrac{1}{n} & \text{if } n \text{ is a perfect square} \\ \dfrac{1}{n^2} & \text{if } n \text{ is not a perfect square} \end{cases}$

48. $\displaystyle\sum_{n=1}^{+\infty} c_n$, where $c_n = \begin{cases} -\dfrac{1}{n} & \text{if } \frac{1}{4}n \text{ is an integer} \\ \dfrac{1}{n^2} & \text{if } \frac{1}{4}n \text{ is not an integer} \end{cases}$

49. Express the nonterminating repeating decimal 1.324 24 24 . . . as a common fraction.

50. A ball is dropped from a height of 18 m. Each time it strikes the ground, it bounces back to a height of two-thirds the distance from which it fell. Find the total distance traveled by the ball before it comes to rest.

51. The path of each swing, after the first, of a pendulum bob is 80 percent as long as the path of the previous swing from one side to the other side. If the path of the first swing is 18 in. long, and air resistance eventually brings the pendulum to rest, how far does the bob travel before it comes to rest?

THIRTEEN

$$\cos x = \sum_{n=0}^{+\infty} \frac{(-1)^n x^{2n}}{(2n)!}$$

$$\int_0^{1/2} e^{-t^2}\, dt$$

$$\int_{1/2}^1 \frac{\sin x}{x}\, dx$$

Our treatment of infinite series is as complete as is feasible in a beginning calculus text. In Section 13.1 we introduce *power series* and in Sections 13.2 through 13.5 you will learn how to use power series to express as an infinite series many functions such as rational, trigonometric, exponential, and logarithmic functions. The proofs (Theorems 13.2.3 and 13.3.1) of the computational processes involving differentiation and integration of power series are included.

An application of power series is to find approximations of irrational numbers such as $\sqrt{2}$, π, e, ln 5, and sin 0.3. Another application is to approximate definite integrals for which the integrand has no antiderivative that can be expressed in terms of elementary functions. For example, you will learn how to use power series to compute values of integrals such as $\int_0^{1/2} e^{-t^2}\,dt$, $\int_0^1 \cos\sqrt{x}\,dx$, and $\int_0^{0.1} \ln(1 + \sin x)\,dx$ to any required accuracy. Furthermore, solutions of many differential equations can be expressed as power series.

13.1 INTRODUCTION TO POWER SERIES

The infinite series in Chapter 12 involved constant terms. We now discuss an important type of series of variable terms called *power series*, which can be considered as a generalization of a polynomial function. You will learn in this chapter how power series can be used to calculate values of such functions as $\sin x$, e^x, $\ln x$, and $\sqrt{x}$, which cannot be evaluated by the familiar operations of arithmetic used for determining rational function values.

13.1.1 DEFINITION

A **power series** in $x - a$ is a series of the form

$$c_0 + c_1(x - a) + c_2(x - a)^2 + \ldots + c_n(x - a)^n + \ldots \tag{1}$$

We use the notation $\sum\limits_{n=0}^{+\infty} c_n(x - a)^n$ to represent series (1). (Note that we take $(x - a)^0 = 1$, even when $x = a$, for convenience in writing the general term.) If x is a particular number, the power series (1) becomes an infinite series of constant terms. A special case of (1) is obtained when $a = 0$, and the series becomes a power series in x, which is

$$\sum_{n=0}^{+\infty} c_n x^n = c_0 + c_1 x + c_2 x^2 + \ldots + c_n x^n + \ldots \tag{2}$$

In addition to power series in $x - a$ and x, there are power series of the form

$$\sum_{n=0}^{+\infty} c_n[\phi(x)]^n = c_0 + c_1\phi(x) + c_2[\phi(x)]^2 + \ldots + c_n[\phi(x)]^n + \ldots$$

where ϕ is a function of x. Such a series is called a power series in $\phi(x)$. In this book we are concerned exclusively with power series of the forms (1) and (2), and when the term "power series" is used, we mean either of these forms. The discussion of the theory of power series is confined to series (2). The more general power series (1) can be obtained from (2) by the translation $x = \bar{x} - a$; therefore our results can be applied to series (1) as well.

In dealing with an infinite series of constant terms we were concerned with the question of convergence or divergence of the series. In considering a power series we ask, For what values of x does the power series converge? For each value of x for which the power series converges, the series represents the number that is the sum of the series. Therefore, a power series defines a function. The function f, with function values

$$f(x) = \sum_{n=0}^{+\infty} c_n x^n$$

has as its domain all values of x for which the power series converges. It is apparent that every power series (2) is convergent for $x = 0$. There are some series (see Example 3) that are convergent for no other value of x, and there are also series that converge for every value of x (see Example 2).

The following three examples illustrate how the ratio test can be used to determine the values of x for which a power series is convergent. When $n!$ is used in representing the nth term of a power series (as in Example 2), we take $0! = 1$ so that the expression for the nth term will hold when $n = 0$.

EXAMPLE 1   Find the values of x for which the power series is convergent:

$$\sum_{n=1}^{+\infty} (-1)^{n+1} \frac{2^n x^n}{n3^n}$$

Solution   For the given series,

$$u_n = (-1)^{n+1} \frac{2^n x^n}{n3^n} \quad \text{and} \quad u_{n+1} = (-1)^{n+2} \frac{2^{n+1} x^{n+1}}{(n+1)3^{n+1}}$$

So

$$\lim_{n \to +\infty} \left| \frac{u_{n+1}}{u_n} \right| = \lim_{n \to +\infty} \left| \frac{2^{n+1} x^{n+1}}{(n+1)3^{n+1}} \cdot \frac{n3^n}{2^n x^n} \right|$$

$$= \lim_{n \to +\infty} \frac{2}{3} |x| \frac{n}{n+1}$$

$$= \tfrac{2}{3} |x|$$

Therefore the power series is absolutely convergent when $\tfrac{2}{3}|x| < 1$ or, equivalently, when $|x| < \tfrac{3}{2}$. The series is divergent when $\tfrac{2}{3}|x| > 1$ or, equivalently, when $|x| > \tfrac{3}{2}$. When $\tfrac{2}{3}|x| = 1$ (i.e., when $x = \pm\tfrac{3}{2}$), the ratio test fails. When $x = \tfrac{3}{2}$, the given power series becomes

$$\frac{1}{1} - \frac{1}{2} + \frac{1}{3} - \frac{1}{4} + \ldots + (-1)^{n+1}\frac{1}{n} + \ldots$$

which is convergent, as was shown in Example 1 of Section 12.7. When $x = -\tfrac{3}{2}$, we have

$$-\frac{1}{1} - \frac{1}{2} - \frac{1}{3} - \frac{1}{4} - \ldots - \frac{1}{n} - \ldots$$

which by Theorem 12.4.2 is divergent. We conclude, then, that the given power series is convergent when $-\tfrac{3}{2} < x \leq \tfrac{3}{2}$. The series is absolutely convergent when $-\tfrac{3}{2} < x < \tfrac{3}{2}$ and is conditionally convergent when $x = \tfrac{3}{2}$. If $x \leq -\tfrac{3}{2}$ or $x > \tfrac{3}{2}$, the series is divergent.

EXAMPLE 2   Find the values of x for which the power series is convergent:

$$\sum_{n=0}^{+\infty} \frac{x^n}{n!}$$

Solution   For the given series,

$$u_n = \frac{x^n}{n!} \quad \text{and} \quad u_{n+1} = \frac{x^{n+1}}{(n+1)!}.$$

So by applying the ratio test,

$$\lim_{n \to +\infty} \left| \frac{u_{n+1}}{u_n} \right| = \lim_{n \to +\infty} \left| \frac{x^{n+1}}{(n+1)!} \cdot \frac{n!}{x^n} \right|$$

$$= |x| \lim_{n \to +\infty} \frac{1}{n+1}$$

$$= 0$$

$$< 1$$

Therefore the given power series is absolutely convergent for all values of x.

EXAMPLE 3   Find the values of x for which the power series is convergent:

$$\sum_{n=0}^{+\infty} n! x^n$$

Solution   For the given series, $u_n = n! x^n$ and $u_{n+1} = (n+1)! x^{n+1}$. Applying the ratio test we have

$$\lim_{n \to +\infty} \left| \frac{u_{n+1}}{u_n} \right| = \lim_{n \to +\infty} \left| \frac{(n+1)! x^{n+1}}{n! x^n} \right|$$

$$= \lim_{n \to +\infty} |(n+1)x|$$

$$= \begin{cases} 0 & \text{if } x = 0 \\ +\infty & \text{if } x \neq 0 \end{cases}$$

It follows that the series is divergent for all values of x except 0.

In the next example the root test is used to determine when a power series is convergent.

EXAMPLE 4   Find the values of x for which the power series is convergent:

$$\sum_{n=1}^{+\infty} n^3 x^n$$

Solution   We use the root test and compute $\lim_{n \to +\infty} \sqrt[n]{|u_n|}$.

$$\lim_{n \to +\infty} \sqrt[n]{|n^3 x^n|} = \lim_{n \to +\infty} n^{3/n} |x| \tag{3}$$

To determine $\lim_{n \to +\infty} n^{3/n}$, let $y = n^{3/n}$. Then $\ln y = \frac{3}{n} \ln n$. Thus

$$\lim_{n \to +\infty} \ln y = \lim_{n \to +\infty} \frac{3 \ln n}{n} \tag{4}$$

To compute the limit on the right-hand side of (4) we first find $\lim_{z \to +\infty} \frac{3 \ln z}{z}$, where z is a real number. Because $\lim_{z \to +\infty} \ln z = +\infty$ and $\lim_{z \to +\infty} z = +\infty$, we apply L'Hôpital's rule and get

$$\lim_{z \to +\infty} \frac{3 \ln z}{z} = \lim_{z \to +\infty} \frac{3}{z}$$

$$= 0$$

Thus from Theorem 12.1.3, $\lim_{n \to +\infty} \dfrac{3 \ln n}{n} = 0$; hence, from (4),

$$\lim_{n \to +\infty} \ln y = 0$$

$$\lim_{n \to +\infty} y = 1$$

Substituting this result in (3) we have

$$\lim_{n \to +\infty} \sqrt[n]{|n^3 x^n|} = |x|$$

Therefore the power series is absolutely convergent when $|x| < 1$. The series is divergent when $|x| > 1$. When $x = 1$, the given power series becomes $\sum\limits_{n=1}^{+\infty} n^3$, which is divergent because $\lim\limits_{n \to +\infty} n^3 \neq 0$. Similarly, the power series is divergent when $x = -1$.

13.1.2 THEOREM   If the power series $\sum\limits_{n=0}^{+\infty} c_n x^n$ is convergent for $x = x_1$ ($x_1 \neq 0$), then it is absolutely convergent for all values of x for which $|x| < |x_1|$.

Proof   If $\sum\limits_{n=0}^{+\infty} c_n x_1{}^n$ is convergent, then $\lim\limits_{n \to +\infty} c_n x_1{}^n = 0$. Therefore, if we take $\epsilon = 1$ in Definition 2.5.1, there exists an integer $N > 0$ such that

$$\text{if } n \geq N \quad \text{then} \quad |c_n x_1{}^n| < 1$$

Now if x is any number such that $|x| < |x_1|$, then if $n \geq N$

$$|c_n x^n| = \left| c_n x_1{}^n \, \frac{x^n}{x_1{}^n} \right|$$

$$|c_n x^n| = |c_n x_1{}^n| \left| \frac{x}{x_1} \right|^n$$

$$|c_n x^n| < \left| \frac{x}{x_1} \right|^n \tag{5}$$

The series

$$\sum_{n=N}^{+\infty} \left| \frac{x}{x_1} \right|^n \tag{6}$$

is convergent because it is a geometric series with $r = |x/x_1| < 1$ (because $|x| < |x_1|$). Compare the series $\sum\limits_{n=N}^{+\infty} |c_n x^n|$, where $|x| < |x_1|$, with series (6). From

(5) and the comparison test, $\sum\limits_{n=N}^{+\infty} |c_n x^n|$ is convergent for $|x| < |x_1|$. So the given power series is absolutely convergent for all values of x for which $|x| < |x_1|$. ∎

▶ **ILLUSTRATION 1**   An illustration of Theorem 13.1.2 is given in Example 1. The power series is convergent for $x = \frac{3}{2}$ and is absolutely convergent for all values of x for which $|x| < \frac{3}{2}$. ◀

The following theorem is a corollary of Theorem 13.1.2.

13.1.3 THEOREM   If the power series $\sum\limits_{n=0}^{+\infty} c_n x^n$ is divergent for $x = x_2$, it is divergent for all values of x for which $|x| > |x_2|$.

Proof   Suppose that the given power series is convergent for some number x for which $|x| > |x_2|$. Then by Theorem 13.1.2 the series must converge when $x = x_2$. However, this contradicts the hypothesis. Therefore the given power series is divergent for all values of x for which $|x| > |x_2|$. ∎

▶ **ILLUSTRATION 2**   To illustrate Theorem 13.1.3, consider again the power series of Example 1. It is divergent for $x = -\frac{3}{2}$ and is also divergent for all values of x for which $|x| > |-\frac{3}{2}|$. ◀

From Theorems 13.1.2 and 13.1.3 the following important theorem can be proved.

13.1.4 THEOREM   Let $\sum\limits_{n=0}^{+\infty} c_n x^n$ be a given power series. Then exactly one of the following conditions holds:

   (i) the series converges only when $x = 0$;
   (ii) the series is absolutely convergent for all values of x;
   (iii) there exists a number $R > 0$ such that the series is absolutely convergent for all values of x for which $|x| < R$ and is divergent for all values of x for which $|x| > R$.

Proof   If x is replaced by zero in the given power series, we have $c_0 + 0 + 0 + \ldots$, which is obviously convergent. Therefore every power series of the form $\sum\limits_{n=0}^{+\infty} c_n x^n$ is convergent when $x = 0$. If this is the only value of x for which the series converges, then condition (i) holds.

Suppose that the given series is convergent for $x = x_1$ where $x_1 \neq 0$. Then it follows from Theorem 13.1.2 that the series is absolutely convergent for all values of x for which $|x| < |x_1|$. Now if in addition there is no value of x for which the given series is divergent, then the series is absolutely convergent for all values of x. This is condition (ii).

If the given series is convergent for $x = x_1$ where $x_1 \neq 0$ and is divergent for $x = x_2$ where $|x_2| > |x_1|$, it follows from Theorem 13.1.3 that the series is divergent for all values of x for which $|x| > |x_2|$. Hence $|x_2|$ is an upper bound of the set of values of $|x|$ for which the series is absolutely convergent. Therefore, by the axiom of completeness (12.2.5) this set of numbers has a least upper bound, which is the number R of condition (iii). This proves that exactly one of the three conditions holds. ∎

Theorem 13.1.4(iii) can be illustrated on the number line. See Figure 1.

If instead of the power series $\sum\limits_{n=1}^{+\infty} c_n x^n$ we have the series $\sum\limits_{n=0}^{+\infty} c_n (x - a)^n$, then in conditions (i) and (iii) of Theorem 13.1.4 x is replaced by $x - a$. The conditions become

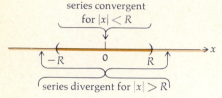

FIGURE 1

    (i) the series converges only when $x = a$;
    (iii) there exists a number $R > 0$ such that the series is absolutely convergent for all values of x for which $|x - a| < R$ and is divergent for all values of x for which $|x - a| > R$. (See Figure 2 for an illustration of this on the number line.)

The set of all values of x for which a given power series is convergent is called the **interval of convergence** of the power series. The number R of condition (iii) of Theorem 13.1.4 is called the **radius of convergence** of the power series. If condition (i) holds, $R = 0$; if condition (ii) holds, we write $R = +\infty$.

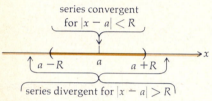

FIGURE 2

▶ **ILLUSTRATION 3**   For the power series of Example 1, $R = \frac{3}{2}$ and the interval of convergence is $(-\frac{3}{2}, \frac{3}{2}]$. In Example 2, $R = +\infty$, and the interval of convergence is written as $(-\infty, +\infty)$. ◀

If R is the radius of convergence of the power series $\sum\limits_{n=0}^{+\infty} c_n x^n$, the interval of convergence is one of the following intervals: $(-R, R)$, $[-R, R]$, $(-R, R]$, or $[-R, R)$. For the more general power series $\sum\limits_{n=0}^{+\infty} c_n (x - a)^n$, the interval of convergence is one of the following:

$$(a - R, a + R) \qquad [a - R, a + R] \qquad (a - R, a + R] \qquad [a - R, a + R)$$

A given power series defines a function having the interval of convergence as its domain. The most useful method at our disposal for determining the interval of convergence of a power series is the ratio test. However, the ratio test will not reveal anything about the convergence or divergence of the power series at the endpoints of the interval of convergence. At an endpoint, a power series may be either absolutely convergent, conditionally convergent, or divergent. If a power series converges absolutely at one endpoint, it follows from the definition of absolute convergence that the series is absolutely convergent at each endpoint (see Exercise 33). If a power series converges at one endpoint and diverges at the other, the series is conditionally convergent at the endpoint at which it converges (see Exercise 34). There are cases for which the convergence or divergence of a power series at the endpoints cannot be determined by the methods of elementary calculus.

EXAMPLE 5   Determine the interval of convergence of the power series:

$$\sum_{n=1}^{+\infty} n(x-2)^n$$

Solution   The given series is

$$(x-2) + 2(x-2)^2 + \ldots + n(x-2)^n + (n+1)(x-2)^{n+1} + \ldots$$

Applying the ratio test we have

$$\lim_{n \to +\infty} \left| \frac{u_{n+1}}{u_n} \right| = \lim_{n \to +\infty} \left| \frac{(n+1)(x-2)^{n+1}}{n(x-2)^n} \right|$$

$$= |x-2| \lim_{n \to +\infty} \frac{n+1}{n}$$

$$= |x-2|$$

The given series then will be absolutely convergent if $|x-2| < 1$ or, equivalently, $-1 < x - 2 < 1$ or, equivalently, $1 < x < 3$.

When $x = 1$, the series is $\sum_{n=1}^{+\infty} (-1)^n n$, which is divergent because $\lim_{n \to +\infty} u_n \neq 0$.

When $x = 3$, the series is $\sum_{n=1}^{+\infty} n$, which is also divergent because $\lim_{n \to +\infty} u_n \neq 0$.

Therefore the interval of convergence is $(1, 3)$. So the given power series defines a function having the interval $(1, 3)$ as its domain.

EXAMPLE 6   Determine the interval of convergence of the power series:

$$\sum_{n=1}^{+\infty} \frac{x^n}{2 + n^2}$$

Solution   The given series is

$$\frac{x}{2 + 1^2} + \frac{x^2}{2 + 2^2} + \frac{x^3}{2 + 3^2} + \ldots + \frac{x^n}{2 + n^2} + \frac{x^{n+1}}{2 + (n+1)^2} + \ldots$$

Applying the ratio test we have

$$\lim_{n \to +\infty} \left| \frac{u_{n+1}}{u_n} \right| = \lim_{n \to +\infty} \left| \frac{x^{n+1}}{2 + (n+1)^2} \cdot \frac{2 + n^2}{x^n} \right|$$

$$= |x| \lim_{n \to +\infty} \frac{2 + n^2}{2 + n^2 + 2n + 1}$$

$$= |x|$$

So the given series will be absolutely convergent if $|x| < 1$ or, equivalently, $-1 < x < 1$. When $x = 1$, the series is

$$\frac{1}{2 + 1^2} + \frac{1}{2 + 2^2} + \frac{1}{2 + 3^2} + \ldots + \frac{1}{2 + n^2} + \ldots$$

Because $\frac{1}{2 + n^2} < \frac{1}{n^2}$ for all positive integers n, and because $\sum_{n=1}^{+\infty} \frac{1}{n^2}$ is a convergent p series, it follows from the comparison test that the given power series is convergent when $x = 1$. When $x = -1$, the series is $\sum\limits_{n=1}^{+\infty} \dfrac{(-1)^n}{2 + n^2}$, which is convergent because we have just seen that it is absolutely convergent. Hence the interval of convergence of the given power series is $[-1, 1]$.

EXERCISES 13.1

In Exercises 1 through 28, determine the interval of convergence of the power series.

1. $\sum\limits_{n=0}^{+\infty} \dfrac{x^n}{n + 1}$

2. $\sum\limits_{n=0}^{+\infty} \dfrac{x^n}{n^2 + 1}$

3. $\sum\limits_{n=0}^{+\infty} \dfrac{x^n}{n^2 - 3}$

4. $\sum\limits_{n=0}^{+\infty} \dfrac{n^2 x^n}{2^n}$

5. $\sum\limits_{n=1}^{+\infty} \dfrac{2^n x^n}{n^2}$

6. $\sum\limits_{n=1}^{+\infty} \dfrac{x^n}{2^n \sqrt{n}}$

7. $\sum\limits_{n=1}^{+\infty} \dfrac{n x^n}{3^n}$

8. $\sum\limits_{n=1}^{+\infty} (-1)^n \dfrac{x^{2n}}{(2n)!}$

9. $\sum\limits_{n=1}^{+\infty} (-1)^{n+1} \dfrac{x^{2n-1}}{(2n - 1)!}$

10. $\sum\limits_{n=1}^{+\infty} \dfrac{n + 1}{n^{2n}} x^n$

11. $\sum\limits_{n=0}^{+\infty} \dfrac{(x + 3)^n}{2^n}$

12. $\sum\limits_{n=0}^{+\infty} \dfrac{x^n}{(n + 1)5^n}$

13. $\sum\limits_{n=1}^{+\infty} (-1)^n \dfrac{x^n}{(2n - 1)3^{2n-1}}$

14. $\sum\limits_{n=1}^{+\infty} (-1)^{n+1} \dfrac{(n + 1)x^n}{n!}$

15. $\sum\limits_{n=1}^{+\infty} (-1)^{n+1} \dfrac{(x - 1)^n}{n}$

16. $\sum\limits_{n=1}^{+\infty} \dfrac{(x + 2)^n}{(n + 1)2^n}$

17. $\sum\limits_{n=0}^{+\infty} (\sinh 2n)x^n$

18. $\sum\limits_{n=1}^{+\infty} \dfrac{x^n}{\ln(n + 1)}$

19. $\sum\limits_{n=2}^{+\infty} (-1)^{n+1} \dfrac{x^n}{n(\ln n)^2}$

20. $\sum\limits_{n=1}^{+\infty} \dfrac{(x + 5)^{n-1}}{n^2}$

21. $\sum\limits_{n=1}^{+\infty} \dfrac{n^2}{5^n} (x - 1)^n$

22. $\sum\limits_{n=0}^{+\infty} \dfrac{4^{n+1} x^{2n}}{n + 3}$

23. $\sum\limits_{n=1}^{+\infty} \dfrac{\ln n(x - 5)^n}{n + 1}$

24. $\sum\limits_{n=1}^{+\infty} \dfrac{x^n}{n^n}$

25. $\sum\limits_{n=1}^{+\infty} (-1)^n \dfrac{1 \cdot 3 \cdot 5 \cdot \ldots \cdot (2n - 1)}{2 \cdot 4 \cdot 6 \cdot \ldots \cdot 2n} x^{2n+1}$

26. $\sum\limits_{n=1}^{+\infty} n^n(x - 3)^n$

27. $\sum\limits_{n=1}^{+\infty} \dfrac{n! x^n}{n^n}$

28. $\sum\limits_{n=1}^{+\infty} \dfrac{(-1)^{n+1} 1 \cdot 3 \cdot 5 \cdot \ldots \cdot (2n - 1)}{2 \cdot 4 \cdot 6 \cdot \ldots \cdot 2n} x^n$

29. If a and b are positive integers, find the radius of convergence of the power series $\sum\limits_{n=1}^{+\infty} \dfrac{(n + a)!}{n!(n + b)!} x^n$.

30. If $\sum\limits_{n=1}^{+\infty} a_n$ is an absolutely convergent series, prove that the power series $\sum\limits_{n=1}^{+\infty} a_n x^n$ is absolutely convergent when $|x| \leq 1$.

31. Prove that if the radius of convergence of the power series $\sum\limits_{n=1}^{+\infty} u_n x^n$ is r, then the radius of convergence of the series $\sum\limits_{n=1}^{+\infty} u_n x^{2n}$ is $\sqrt{r}$.

32. Prove that if $\lim\limits_{n \to +\infty} \sqrt[n]{|u_n|} = L$ $(L \neq 0)$, then the radius of convergence of the power series $\sum\limits_{n=1}^{+\infty} u_n x^n$ is $\dfrac{1}{L}$.

33. Prove that if a power series converges absolutely at one endpoint of its interval of convergence, then the power series is absolutely convergent at each endpoint.

34. Prove that if a power series converges at one endpoint of its interval of convergence and diverges at the other endpoint, then the power series is conditionally convergent at the endpoint at which it converges.

13.2 DIFFERENTIATION OF POWER SERIES

You learned in Section 13.1 that a power series $\sum\limits_{n=0}^{+\infty} c_n x^n$ defines a function whose domain is the interval of convergence of the series.

▶ **ILLUSTRATION 1** Consider the geometric series with $a = 1$ and $r = x$, which is $\sum\limits_{n=0}^{+\infty} x^n$. By Theorem 12.3.5 this series converges to the sum $1/(1 - x)$ if $|x| < 1$. Therefore the power series $\sum\limits_{n=0}^{+\infty} x^n$ defines the function f for which $f(x) = 1/(1 - x)$

From this inequality, statement (9), and the comparison test it follows that $\sum\limits_{n=1}^{+\infty} |c_n x_2{}^n|$ is convergent. Therefore the series $\sum\limits_{n=0}^{+\infty} c_n x_2{}^n$ is convergent, which contradicts statement (8). Hence the assumption that $R' > R$ is false. Therefore R' cannot be greater than R; and because it was shown that $R' \geq R$, it follows that $R' = R$, which proves the theorem.   ∎

▶ **ILLUSTRATION 3**   We verify Theorem 13.2.1 for the power series

$$\sum_{n=0}^{+\infty} \frac{x^{n+1}}{(n+1)^2} = x + \frac{x^2}{4} + \frac{x^3}{9} + \ldots + \frac{x^{n+1}}{(n+1)^2} + \frac{x^{n+2}}{(n+2)^2} + \ldots$$

The radius of convergence is found by applying the ratio test.

$$\lim_{n \to +\infty} \left| \frac{u_{n+1}}{u_n} \right| = \lim_{n \to +\infty} \left| \frac{(n+1)^2 x^{n+2}}{(n+2)^2 x^{n+1}} \right|$$

$$= |x| \lim_{n \to +\infty} \left| \frac{n^2 + 2n + 1}{n^2 + 4n + 4} \right|$$

$$= |x|$$

Hence the power series is convergent when $|x| < 1$; so its radius of convergence $R = 1$.

The power series obtained from the given series by differentiating term by term is

$$\sum_{n=0}^{+\infty} \frac{(n+1)x^n}{(n+1)^2} = \sum_{n=0}^{+\infty} \frac{x^n}{n+1}$$

$$= 1 + \frac{x}{2} + \frac{x^2}{3} + \frac{x^3}{4} + \ldots + \frac{x^n}{n+1} + \frac{x^{n+1}}{n+2} + \ldots$$

Applying the ratio test for this power series we have

$$\lim_{n \to +\infty} \left| \frac{u_{n+1}}{u_n} \right| = \lim_{n \to +\infty} \left| \frac{(n+1)x^{n+1}}{(n+2)x^n} \right|$$

$$= |x| \lim_{n \to +\infty} \left| \frac{n+1}{n+2} \right|$$

$$= |x|$$

This power series is convergent when $|x| < 1$; thus its radius of convergence $R' = 1$. Because $R = R'$, Theorem 13.2.1 is verified.   ◀

13.2.2 THEOREM   If the radius of convergence of the power series $\sum\limits_{n=0}^{+\infty} c_n x^n$ is $R > 0$, then R is also the radius of convergence of the series $\sum\limits_{n=2}^{+\infty} n(n-1)c_n x^{n-2}$.

Proof   If Theorem 13.2.1 is applied to the series $\sum\limits_{n=1}^{+\infty} nc_n x^{n-1}$, the desired result follows.   ∎

We are now in a position to prove the theorem regarding term-by-term differentiation of a power series.

13.2.3 THEOREM   Let $\sum\limits_{n=0}^{+\infty} c_n x^n$ be a power series whose radius of convergence is $R > 0$. Then if f is the function defined by

$$f(x) = \sum_{n=0}^{+\infty} c_n x^n \tag{10}$$

$f'(x)$ exists for every x in the open interval $(-R, R)$ and it is given by

$$f'(x) = \sum_{n=1}^{+\infty} n c_n x^{n-1}$$

Proof   Let x and a be two distinct numbers in the open interval $(-R, R)$. Taylor's formula (formula (2) in Section 11.5), with $n = 1$, is

$$f(x) = f(a) + \frac{f'(a)}{1!}(x - a) + \frac{f''(\xi)}{2!}(x - a)^2$$

From this formula with $f(x) = x^n$ it follows that for every positive integer n,

$$x^n = a^n + na^{n-1}(x - a) + \tfrac{1}{2}n(n-1)(\xi_n)^{n-2}(x - a)^2 \tag{11}$$

where ξ_n is between a and x for every positive integer n. From (10),

$$f(x) - f(a) = \sum_{n=0}^{+\infty} c_n x^n - \sum_{n=0}^{+\infty} c_n a^n$$

$$= c_0 + \sum_{n=1}^{+\infty} c_n x^n - c_0 - \sum_{n=1}^{+\infty} c_n a^n$$

$$= \sum_{n=1}^{+\infty} c_n (x^n - a^n)$$

Dividing by $x - a$ (because $x \neq a$) and using (11) we have from the above equation

$$\frac{f(x) - f(a)}{x - a} = \frac{1}{x - a} \sum_{n=1}^{+\infty} c_n \left[na^{n-1}(x - a) + \tfrac{1}{2}n(n-1)(\xi_n)^{n-2}(x - a)^2 \right]$$

So

$$\frac{f(x) - f(a)}{x - a} = \sum_{n=1}^{+\infty} n c_n a^{n-1} + \tfrac{1}{2}(x - a) \sum_{n=2}^{+\infty} n(n-1)c_n(\xi_n)^{n-2} \tag{12}$$

Because a is in $(-R, R)$, we conclude from Theorem 13.2.1 that $\sum\limits_{n=1}^{+\infty} n c_n a^{n-1}$ is absolutely convergent.

Because both a and x are in $(-R, R)$, there is some number $K > 0$ such that $|a| < K < R$ and $|x| < K < R$. It follows from Theorem 13.2.2 that

$$\sum_{n=2}^{+\infty} n(n-1)c_n K^{n-2}$$

is absolutely convergent. Then because

$$\left| n(n-1)c_n(\xi_n)^{n-2} \right| < \left| n(n-1)c_n K^{n-2} \right| \tag{13}$$

for each ξ_n, we can conclude from the comparison test that

$$\sum_{n=2}^{+\infty} n(n-1)c_n(\xi_n)^{n-2}$$

is absolutely convergent.

From (12),

$$\left| \frac{f(x)-f(a)}{x-a} - \sum_{n=1}^{+\infty} nc_n a^{n-1} \right| = \left| \tfrac{1}{2}(x-a) \sum_{n=2}^{+\infty} n(n-1)c_n(\xi_n)^{n-2} \right| \tag{14}$$

However, from Theorem 12.8.3, if $\sum_{n=1}^{+\infty} u_n$ is absolutely convergent, then

$$\left| \sum_{n=1}^{+\infty} u_n \right| \le \sum_{n=1}^{+\infty} |u_n|$$

Applying this to the right side of (14) we obtain

$$\left| \frac{f(x)-f(a)}{x-a} - \sum_{n=1}^{+\infty} nc_n a^{n-1} \right| \le \tfrac{1}{2}|x-a| \sum_{n=2}^{+\infty} n(n-1)|c_n| \, |\xi_n|^{n-2}$$

From this inequality and (13),

$$\left| \frac{f(x)-f(a)}{x-a} - \sum_{n=1}^{+\infty} nc_n a^{n-1} \right| \le \tfrac{1}{2}|x-a| \sum_{n=2}^{+\infty} n(n-1)|c_n| K^{n-2} \tag{15}$$

where $0 < K < R$. Because the series on the right side of (15) is absolutely convergent, the limit of the right side, as x approaches a, is zero. Therefore, from (15) and the squeeze theorem,

$$\lim_{x \to a} \frac{f(x)-f(a)}{x-a} = \sum_{n=1}^{+\infty} nc_n a^{n-1}$$

$$\Leftrightarrow \quad f'(a) = \sum_{n=1}^{+\infty} nc_n a^{n-1}$$

and because a may be any number in the open interval $(-R, R)$, the theorem is proved. ∎

EXAMPLE 1   Let f be the function defined by the power series of Illustration 3. (a) Find the domain of f; (b) write the power series that defines the function f' and find the domain of f'.

Solution

(a) $f(x) = \sum_{n=0}^{+\infty} \frac{x^{n+1}}{(n+1)^2}$

The domain of f is the interval of convergence of the power series. In Illustration 3 we showed that the radius of convergence of the power series is

1; that is, the series converges when $|x| < 1$. Consider now the power series when $|x| = 1$. When $x = 1$, the series is

$$1 + \frac{1}{4} + \frac{1}{9} + \ldots + \frac{1}{(n+1)^2} + \ldots$$

which is convergent because it is the p series with $p = 2$. When $x = -1$, we have the series $\sum_{n=0}^{+\infty} \frac{(-1)^{n+1}}{(n+1)^2}$, which is convergent because it is absolutely convergent. Hence the domain of f is the interval $[-1, 1]$.

(b) From Theorem 13.2.3 it follows that f' is defined by

$$f'(x) = \sum_{n=0}^{+\infty} \frac{x^n}{n+1} \tag{16}$$

and that $f'(x)$ exists for every x in the open interval $(-1, 1)$. In Illustration 3 we showed that the radius of convergence of the power series in (16) is 1. We now consider the power series in (16) when $x = \pm 1$. When $x = 1$, the series is

$$1 + \frac{1}{2} + \frac{1}{3} + \frac{1}{4} + \ldots + \frac{1}{n+1} + \ldots$$

which is the harmonic series and hence is divergent. When $x = -1$, the series is

$$1 - \frac{1}{2} + \frac{1}{3} - \frac{1}{4} + \ldots + (-1)^n \frac{1}{n+1} + \ldots$$

which is a convergent alternating series. Therefore the domain of f' is the interval $[-1, 1)$.

Example 1 illustrates the fact that if a function f is defined by a power series and this power series is differentiated term by term, the resulting power series, which defines f', has the same radius of convergence but not necessarily the same interval of convergence.

EXAMPLE 2   Obtain a power-series representation of

$$\frac{1}{(1 - x)^2}$$

Solution   From (1),

$$\frac{1}{1 - x} = 1 + x + x^2 + x^3 + \ldots + x^n + \ldots \qquad \text{if } |x| < 1$$

Using Theorem 13.2.3 and differentiating on both sides of the above we get

$$\frac{1}{(1 - x)^2} = 1 + 2x + 3x^2 + \ldots + nx^{n-1} + \ldots \qquad \text{if } |x| < 1$$

EXAMPLE 3   Show that for all real values of x

$$e^x = \sum_{n=0}^{+\infty} \frac{x^n}{n!}$$

$$= 1 + x + \frac{x^2}{2!} + \frac{x^3}{3!} + \ldots + \frac{x^n}{n!} + \ldots$$

Solution   In Example 2 of Section 13.1 we showed that the power series $\sum_{n=0}^{+\infty} \frac{x^n}{n!}$ is absolutely convergent for all real values of x. Therefore, if f is the function defined by

$$f(x) = \sum_{n=0}^{+\infty} \frac{x^n}{n!} \tag{17}$$

the domain of f is the set of all real numbers; that is, the interval of convergence is $(-\infty, +\infty)$. It follows from Theorem 13.2.3 that for all real values of x,

$$f'(x) = \sum_{n=1}^{+\infty} \frac{nx^{n-1}}{n!}$$

Because $\frac{n}{n!} = \frac{1}{(n-1)!}$, the above can be written as

$$f'(x) = \sum_{n=1}^{+\infty} \frac{x^{n-1}}{(n-1)!}$$

$$\Leftrightarrow \quad f'(x) = \sum_{n=0}^{+\infty} \frac{x^n}{n!}$$

From this equality and (17), $f'(x) = f(x)$ for all real values of x. Therefore the function f satisfies the differential equation

$$\frac{dy}{dx} = y$$

for which, from Theorem 7.7.1, the general solution is $y = Ce^x$. Hence for some constant C, $f(x) = Ce^x$. From (17), $f(0) = 1$. (Remember that we take $x^0 = 1$ even when $x = 0$ for convenience in writing the general term.) Therefore $C = 1$; so $f(x) = e^x$, and we have the desired result.

EXAMPLE 4   Use the result of Example 3 to find a power-series representation of e^{-x}.

Solution   If x is replaced by $-x$ in the series for e^x, it follows that

$$e^{-x} = 1 - x + \frac{x^2}{2!} - \frac{x^3}{3!} + \ldots + (-1)^n \frac{x^n}{n!} + \ldots$$

for all real values of x.

EXAMPLE 5   Use the series of Example 4 to find the value of e^{-1} correct to five decimal places.

Solution   If $x = 1$ in the series for e^{-x},

$$e^{-1} = 1 - 1 + \frac{1}{2!} - \frac{1}{3!} + \frac{1}{4!} - \frac{1}{5!} + \frac{1}{6!} - \frac{1}{7!} + \frac{1}{8!} - \frac{1}{9!} + \frac{1}{10!} - \cdots$$

$$= 1 - 1 + \frac{1}{2} - \frac{1}{6} + \frac{1}{24} - \frac{1}{120} + \frac{1}{720} - \frac{1}{5040} + \frac{1}{40,320}$$

$$- \frac{1}{362,880} + \frac{1}{3,628,800} - \cdots$$

$$\approx 1 - 1 + 0.5 - 0.166667 + 0.041667 - 0.008333 + 0.001389$$

$$- 0.000198 + 0.000025 - 0.000003 + 0.0000003 - \cdots$$

This is a convergent alternating series for which $|u_{n+1}| < |u_n|$. So if the first ten terms are used to approximate the sum, by Theorem 12.7.4 the error is less than the absolute value of the eleventh term. Adding the first ten terms we obtain 0.367880. Rounding off to five decimal places gives

$$e^{-1} \approx 0.36788$$

In computation with infinite series two kinds of errors occur. One is the error given by the remainder after the first n terms. The other is the round-off error that occurs when each term of the series is approximated by a decimal with a finite number of places. In particular, in Example 5 we wanted the result accurate to five decimal places; so each term was rounded off to six decimal places. After computing the sum, we rounded off this result to five decimal places. Of course, the error given by the remainder can be reduced by considering additional terms of the series, while the round-off error can be reduced by using more decimal places.

If you take a course in differential equations, you will learn that it is possible to express solutions of many differential equations as power series. In the following example we have this situation.

EXAMPLE 6   Show that

$$y = x + \sum_{n=0}^{+\infty} \frac{x^n}{n!} \tag{18}$$

is a solution of the differential equation $\dfrac{d^2y}{dx^2} - y + x = 0$.

Solution   The power series in (18) is convergent for all values of x. Therefore, from Theorem 13.2.3, for all x,

$$\frac{dy}{dx} = 1 + \sum_{n=1}^{+\infty} \frac{nx^{n-1}}{n!} \qquad \frac{d^2y}{dx^2} = \sum_{n=2}^{+\infty} \frac{(n-1)x^{n-2}}{(n-1)!}$$

$$= 1 + \sum_{n=1}^{+\infty} \frac{x^{n-1}}{(n-1)!} \qquad = \sum_{n=2}^{+\infty} \frac{x^{n-2}}{(n-2)!}$$

$$= \sum_{n=0}^{+\infty} \frac{x^n}{n!}$$

Thus

$$\frac{d^2y}{dx^2} - y + x = \sum_{n=0}^{+\infty} \frac{x^n}{n!} - \left(x + \sum_{n=0}^{+\infty} \frac{x^n}{n!}\right) + x$$

$$= 0$$

Hence the differential equation is satisfied; so (18) is a solution.

EXERCISES 13.2

In Exercises 1 through 10, do the following: (a) Find the radius of convergence of the power series and the domain of f; (b) write the power series that defines the function f′ and find its radius of convergence by using methods of Section 13.1 (thus verifying Theorem 13.2.1); (c) find the domain of f′.

1. $f(x) = \sum_{n=1}^{+\infty} \frac{x^n}{n^2}$

2. $f(x) = \sum_{n=1}^{+\infty} (-1)^{n-1} \frac{x^n}{n}$

3. $f(x) = \sum_{n=1}^{+\infty} \frac{x^n}{\sqrt{n}}$

4. $f(x) = \sum_{n=2}^{+\infty} \frac{(x-2)^n}{\sqrt{n-1}}$

5. $f(x) = \sum_{n=1}^{+\infty} (-1)^{n-1} \frac{x^{2n-1}}{(2n-1)!}$

6. $f(x) = \sum_{n=0}^{+\infty} \frac{x^{2n}}{(n!)^2}$

7. $f(x) = \sum_{n=1}^{+\infty} (n+1)(3x-1)^n$

8. $f(x) = \sum_{n=1}^{+\infty} \frac{x^{2n-2}}{(2n-2)!}$

9. $f(x) = \sum_{n=1}^{+\infty} \frac{(x-1)^n}{n3^n}$

10. $f(x) = \sum_{n=2}^{+\infty} (-1)^n \frac{(x-3)^n}{n(n-1)}$

11. Use the result of Example 2 to find a power-series representation of $\frac{1}{(1-x)^3}$.

12. Use the result of Example 3 to find a power-series representation of $e^{\sqrt{x}}$.

13. Obtain a power-series representation of $\frac{1}{(1+x)^2}$ if $|x| < 1$ by differentiating series (2) term by term.

14. Obtain a power-series representation of $\frac{x}{(1+x^2)^2}$ if $|x| < 1$ by differentiating series (4) term by term.

15. (a) Use series (1) to find a power-series representation for $\frac{1}{1-2x}$. (b) Differentiate term by term the series found in part (a) to find a power-series representation for $\frac{2}{(1-2x)^2}$.

16. (a) Use series (2) to find a power-series representation for $\frac{1}{1+x^3}$. (b) Differentiate term by term the series found in part (a) to find a power-series representation for $\frac{-3x^2}{(1+x^3)^2}$.

17. (a) Use the result of Example 3 to find a power-series representation for e^{x^2}. (b) Differentiate term by term the series found in part (a) to find a power-series representation for xe^{x^2}.

18. Let f be the function defined by $f(x) = \sum_{n=0}^{+\infty} (-1)^n \frac{x^n}{3^n(n+2)}$. (a) Find the domain of f. (b) Find $f'(x)$ and find the domain of f'.

19. Use the result of Example 4 to find the value of $\frac{1}{\sqrt{e}}$ correct to five decimal places.

20. If $f(x) = \sum_{n=0}^{+\infty} (-1)^n \frac{x^{2n}}{3^n}$, find $f'(\frac{1}{2})$ correct to four decimal places.

21. Use the results of Examples 3 and 4 to find a power-series representation of (a) sinh x and (b) cosh x.

22. Show that each of the power series in parts (a) and (b) of Exercise 21 can be obtained from the other by term-by-term differentiation.

23. Use the result of Example 2 to find the sum of the series $\sum_{n=1}^{+\infty} \frac{n}{2^n}$.

24. (a) Find a power-series representation for $\frac{e^x - 1}{x}$. (b) By differentiating term by term the power series in part (a), show that $\sum_{n=1}^{+\infty} \frac{n}{(n+1)!} = 1$.

25. (a) Find a power-series representation for x^2e^{-x}. (b) By differentiating term by term the power series in part (a), show that $\sum_{n=1}^{+\infty} (-2)^{n+1} \frac{n+2}{n!} = 4$.

26. (a) Find a power-series representation for e^{-x^2}. (b) By differentiating the power series in part (a) twice term by term, show that $\sum_{n=1}^{+\infty} (-1)^{n+1} \frac{2n+1}{2^n n!} = 1$.

27. Suppose a function f has the power-series representation $\sum_{n=0}^{+\infty} c_n x^n$, where if R is the radius of convergence, $R > 0$. If $f'(x) = f(x)$ and $f(0) = 1$, find the power series by using only properties of power series and nothing about the exponential function.

28. (a) Use only properties of power series to find a power-series representation of the function f if $f(x) > 0$ and $f'(x) = 2xf(x)$ for all x, and $f(0) = 1$. (b) Verify your result in part (a) by solving the differential equation $\dfrac{dy}{dx} = 2xy$ with the initial condition $y = 1$ when $x = 0$.

29. Suppose a function f has the power-series representation $\displaystyle\sum_{n=0}^{+\infty} c_n x^n$. If f is an even function, show that $c_n = 0$ when n is odd.

30. Assume that the constant 0 has a power-series representation $\displaystyle\sum_{n=0}^{+\infty} c_n x^n$, where the radius of convergence $R > 0$. Prove that $c_n = 0$ for all n.

In Exercises 31 through 35, show that the power series is a solution of the differential equation.

31. $y = \displaystyle\sum_{n=0}^{+\infty} \dfrac{2^n}{n!} x^n;\ \dfrac{dy}{dx} - 2y = 0$

32. $y = \displaystyle\sum_{n=0}^{+\infty} \dfrac{1}{2^n n!} x^{2n};\ \dfrac{dy}{dx} - xy = 0$

33. $y = \displaystyle\sum_{n=1}^{+\infty} \dfrac{(-1)^{n+1}}{(2n-1)!} x^{2n-1};\ \dfrac{d^2y}{dx^2} + y = 0$

34. $y = x + \displaystyle\sum_{n=0}^{+\infty} (-1)^n \dfrac{x^{2n}}{(2n)!};\ \dfrac{d^2y}{dx^2} + y - x = 0$

35. $y = \displaystyle\sum_{n=0}^{+\infty} (-1)^n \dfrac{2^n n!}{(2n+1)!} x^{2n+1};\ \dfrac{d^2y}{dx^2} + x\dfrac{dy}{dx} + y = 0$

13.3 INTEGRATION OF POWER SERIES

The theorem regarding the term-by-term integration of a power series is a consequence of Theorem 13.2.3.

13.3.1 THEOREM

Let $\displaystyle\sum_{n=0}^{+\infty} c_n x^n$ be a power series whose radius of convergence is $R > 0$. Then if f is the function defined by

$$f(x) = \sum_{n=0}^{+\infty} c_n x^n$$

f is integrable on every closed subinterval of $(-R, R)$, and the integral of f is evaluated by integrating the given power series term by term; that is, if x is in $(-R, R)$, then

$$\int_0^x f(t)\, dt = \sum_{n=0}^{+\infty} \dfrac{c_n}{n+1} x^{n+1}$$

Furthermore, R is the radius of convergence of the resulting series.

Proof   Let g be the function defined by

$$g(x) = \sum_{n=0}^{+\infty} \dfrac{c_n}{n+1} x^{n+1}$$

Because the terms of the power-series representation of $f(x)$ are the derivatives of the terms of the power-series representation of $g(x)$, the two series have, by Theorem 13.2.1, the same radius of convergence. By Theorem 13.2.3

$$g'(x) = f(x) \qquad \text{for every } x \text{ in } (-R, R)$$

By Theorem 13.2.2 it follows that $f'(x) = g''(x)$ for every x in $(-R, R)$. Because f is differentiable on $(-R, R)$, f is continuous there; consequently f is continuous on every closed subinterval of $(-R, R)$. From Theorem 5.8.2 we conclude that if x is in $(-R, R)$, then

$$\int_0^x f(t)\, dt = g(x) - g(0)$$

$$= g(x)$$

$$\Leftrightarrow \int_0^x f(t)\, dt = \sum_{n=0}^{+\infty} \dfrac{c_n}{n+1} x^{n+1} \qquad \blacksquare$$

Theorem 13.3.1 often is used to compute a definite integral that cannot be evaluated directly by finding an antiderivative of the integrand. Examples 1 and 2 illustrate the technique. The definite integral $\int_0^x e^{-t^2}\,dt$ appearing in these two examples is similar to the one that represents the measure of the area of a region under the "normal probability curve."

EXAMPLE 1   Find a power-series representation of $\int_0^x e^{-t^2}\,dt$.

Solution   From Example 4 of Section 13.2,

$$e^{-x} = \sum_{n=0}^{+\infty} \frac{(-1)^n x^n}{n!}$$

for all values of x. If x is replaced by t^2,

$$e^{-t^2} = 1 - t^2 + \frac{t^4}{2!} - \frac{t^6}{3!} + \ldots + (-1)^n \frac{t^{2n}}{n!} + \ldots \qquad \text{for all values of } t$$

Applying Theorem 13.3.1 we integrate term by term and obtain

$$\int_0^x e^{-t^2}\,dt = \sum_{n=0}^{+\infty} \int_0^x (-1)^n \frac{t^{2n}}{n!}\,dt$$

$$= x - \frac{x^3}{3} + \frac{x^5}{2! \cdot 5} - \frac{x^7}{3! \cdot 7} + \ldots + (-1)^n \frac{x^{2n+1}}{n!(2n+1)} + \ldots$$

The power series represents the integral for all values of x.

EXAMPLE 2   Use the result of Example 1 to compute accurate to three decimal places the value of $\int_0^{1/2} e^{-t^2}\,dt$.

Solution   We replace x by $\frac{1}{2}$ in the power series obtained in Example 1 to obtain

$$\int_0^{1/2} e^{-t^2}\,dt = \frac{1}{2} - \frac{1}{24} + \frac{1}{320} - \frac{1}{5376} + \ldots$$

$$\approx 0.5 - 0.0417 + 0.0031 - 0.0002 + \ldots$$

This is a convergent alternating series with $|u_{n+1}| < |u_n|$. Thus if we use the first three terms to approximate the sum, by Theorem 12.7.4 the error is less than the absolute value of the fourth term. From the first three terms,

$$\int_0^{1/2} e^{-t^2}\,dt \approx 0.461$$

EXAMPLE 3   Obtain a power-series representation of $\ln(1 + x)$.

Solution   We consider the function f defined by $f(t) = \dfrac{1}{1 + t}$. A power-series representation of this function is given by series (2) in Section 13.2, which is

$$\frac{1}{1 + t} = 1 - t + t^2 - t^3 + \ldots + (-1)^n t^n + \ldots \qquad \text{if } |t| < 1$$

We apply Theorem 13.3.1 and integrate term by term to obtain

$$\int_0^x \frac{dt}{1+t} = \sum_{n=0}^{+\infty} \int_0^x (-1)^n t^n \, dt \qquad \text{if } |x| < 1$$

Therefore

$$\ln(1+x) = x - \frac{x^2}{2} + \frac{x^3}{3} - \frac{x^4}{4} + \ldots + (-1)^n \frac{x^{n+1}}{n+1} + \ldots \qquad \text{if } |x| < 1$$

$$\Leftrightarrow \quad \ln(1+x) = \sum_{n=1}^{+\infty} (-1)^{n-1} \frac{x^n}{n} \qquad \text{if } |x| < 1 \tag{1}$$

Because $|x| < 1$, $|1 + x| = 1 + x$. Thus the absolute-value bars are not needed when writing $\ln(1 + x)$.

In Example 3, Theorem 13.3.1 allows us to conclude that the power series in (1) represents the function only for values of x in the open interval $(-1, 1)$. However, the power series is convergent at the right endpoint 1, as was shown in Example 1 of Section 12.7. When $x = -1$, the power series becomes the negative of the harmonic series and is divergent. Hence the interval of convergence of the power series in (1) is $(-1, 1]$.

In the following illustration we show that the power series in (1) represents $\ln(1 + x)$ at $x = 1$ by proving that the sum of the series $\sum_{n=1}^{+\infty} \frac{(-1)^{n-1}}{n}$ is $\ln 2$.

▶ **ILLUSTRATION 1** For the infinite series $\sum_{n=1}^{+\infty} \frac{(-1)^{n-1}}{n}$, the nth partial sum is

$$s_n = 1 - \frac{1}{2} + \frac{1}{3} - \frac{1}{4} + \ldots + (-1)^{n-1} \frac{1}{n} \tag{2}$$

It follows from Definition 12.3.2 that if we show $\lim_{n \to +\infty} s_n = \ln 2$, we will have proved that the sum of the series is $\ln 2$.

From algebra comes the following formula for the sum of a finite geometric series:

$$a + ar + ar^2 + ar^3 + \ldots + ar^{n-1} = \frac{a - ar^n}{1 - r}$$

From this formula with $a = 1$ and $r = -t$,

$$1 - t + t^2 - t^3 + \ldots + (-t)^{n-1} = \frac{1 - (-t)^n}{1 + t}$$

which may be written as

$$1 - t + t^2 - t^3 + \ldots + (-1)^{n-1} t^{n-1} = \frac{1}{1+t} + (-1)^{n+1} \frac{t^n}{1+t}$$

Integrating from 0 to 1 we get

$$\int_0^1 \left[1 - t + t^2 - t^3 + \ldots + (-1)^{n-1} t^{n-1} \right] dt = \int_0^1 \frac{dt}{1+t} + (-1)^{n+1} \int_0^1 \frac{t^n}{1+t} \, dt$$

which gives

$$1 - \frac{1}{2} + \frac{1}{3} - \frac{1}{4} + \ldots + (-1)^{n-1}\frac{1}{n} = \ln 2 + (-1)^{n+1}\int_0^1 \frac{t^n}{1+t}\,dt \tag{3}$$

Referring to (2), we see that the left side of (3) is s_n. Letting

$$R_n = (-1)^{n+1}\int_0^1 \frac{t^n}{1+t}\,dt$$

(3) may be written as

$$s_n = \ln 2 + R_n \tag{4}$$

Because $\dfrac{t^n}{1+t} \le t^n$ for all t in $[0, 1]$ it follows from Theorem 5.6.8 that

$$\int_0^1 \frac{t^n}{1+t}\,dt \le \int_0^1 t^n\,dt$$

Hence

$$0 \le |R_n| = \int_0^1 \frac{t^n}{1+t}\,dt \le \int_0^1 t^n\,dt = \frac{1}{n+1}$$

Because $\lim\limits_{n \to +\infty} \dfrac{1}{n+1} = 0$, it follows from the above inequality and the squeeze theorem that $\lim\limits_{n \to +\infty} R_n = 0$. Therefore, from (4),

$$\lim_{n \to +\infty} s_n = \ln 2 + \lim_{n \to +\infty} R_n$$

$$= \ln 2$$

Thus

$$\sum_{n=1}^{+\infty}(-1)^{n-1}\frac{1}{n} = 1 - \frac{1}{2} + \frac{1}{3} - \frac{1}{4} + \ldots \tag{5}$$

$$= \ln 2 \qquad \blacktriangleleft$$

The solution of Example 3 shows that the power series in (1) represents $\ln(x + 1)$ if $|x| < 1$. Hence, with the result of Illustration 1 we can conclude that the power series in (1) represents $\ln(x + 1)$ for all x in its interval of convergence $(-1, 1]$.

Although it is interesting that the sum of the series in (5) is $\ln 2$, this series converges too slowly to use it to calculate $\ln 2$. We now proceed to obtain a power series for computation of natural logarithms.

From (1),

$$\ln(1 + x) = x - \frac{x^2}{2} + \frac{x^3}{3} - \ldots + (-1)^{n-1}\frac{x^n}{n} + \ldots \qquad \text{for } x \text{ in } (-1, 1] \tag{6}$$

If x is replaced by $-x$ in this series,

$$\ln(1 - x) = -x - \frac{x^2}{2} - \frac{x^3}{3} - \frac{x^4}{4} - \ldots - \frac{x^n}{n} - \ldots \qquad \text{for } x \text{ in } [-1, 1) \tag{7}$$

Subtracting term by term (7) from (6) we obtain

$$\ln \frac{1+x}{1-x} = 2\left(x + \frac{x^3}{3} + \frac{x^5}{5} + \ldots + \frac{x^{2n-1}}{2n-1} + \ldots\right) \qquad \text{if } |x| < 1 \tag{8}$$

This series can be used to compute the value of the natural logarithm of any positive number.

▶ **ILLUSTRATION 2**   If y is any positive number, let

$$y = \frac{1+x}{1-x} \quad \text{and then} \quad x = \frac{y-1}{y+1} \quad \text{and} \quad |x| < 1$$

For instance, if $y = 2$, then $x = \frac{1}{3}$. From (8),

$$\ln 2 = 2\left(\frac{1}{3} + \frac{1}{3^4} + \frac{1}{5 \cdot 3^5} + \frac{1}{7 \cdot 3^7} + \frac{1}{9 \cdot 3^9} + \frac{1}{11 \cdot 3^{11}} + \ldots\right)$$

$$= 2\left(\frac{1}{3} + \frac{1}{81} + \frac{1}{1215} + \frac{1}{15,309} + \frac{1}{177,147} + \frac{1}{1,948,617} + \ldots\right)$$

$$\approx 2(0.333333 + 0.012346 + 0.000823 + 0.000065 + 0.000006 + 0.000001 + \ldots)$$

Using the first six terms in parentheses, multiplying by 2, and rounding off to five decimal places we get

$$\ln 2 \approx 0.69315 \qquad \blacktriangleleft$$

EXAMPLE 4   Obtain a power-series representation of $\tan^{-1} x$.

Solution   From series (4) in Section 13.2,

$$\frac{1}{1+x^2} = 1 - x^2 + x^4 - x^6 + \ldots + (-1)^n x^{2n} + \ldots \qquad \text{if } |x| < 1$$

We apply Theorem 13.3.1 and integrate term by term to obtain

$$\int_0^x \frac{1}{1+t^2}\, dt = x - \frac{x^3}{3} + \frac{x^5}{5} - \ldots + (-1)^n \frac{x^{2n+1}}{2n+1} + \ldots$$

Therefore

$$\tan^{-1} x = \sum_{n=0}^{+\infty} (-1)^n \frac{x^{2n+1}}{2n+1} \qquad \text{if } |x| < 1 \tag{9}$$

Although Theorem 13.3.1 allows us to conclude that the power series in (9) represents $\tan^{-1} x$ only for values of x such that $|x| < 1$, it can be shown that the interval of convergence of the power series is $[-1, 1]$ and that the power series is a representation of $\tan^{-1} x$ for all x in its interval of convergence. (You are asked to do this in Exercise 36.) Therefore

$$\tan^{-1} x = \sum_{n=0}^{+\infty} (-1)^n \frac{x^{2n+1}}{2n+1}$$

$$= x - \frac{x^3}{3} + \frac{x^5}{5} - \cdots \qquad \text{if } |x| \le 1 \tag{10}$$

▶ **ILLUSTRATION 3**   If $x = 1$ in (10),

$$\frac{\pi}{4} = 1 - \frac{1}{3} + \frac{1}{5} - \frac{1}{7} + \cdots + (-1)^n \frac{1}{2n+1} + \cdots \qquad ◀$$

The series in Illustration 3 is not suitable for computing π because it converges too slowly. The following example gives a better method.

EXAMPLE 5   Prove that $\frac{1}{4}\pi = \tan^{-1}\frac{1}{2} + \tan^{-1}\frac{1}{3}$. Use this formula and the power series for $\tan^{-1} x$ of Example 4 to compute accurate to five significant figures the value of π.

Solution   Let $\alpha = \tan^{-1}\frac{1}{2}$ and $\beta = \tan^{-1}\frac{1}{3}$. Then

$$\tan(\alpha + \beta) = \frac{\tan \alpha + \tan \beta}{1 - \tan \alpha \tan \beta}$$

$$= \frac{\frac{1}{2} + \frac{1}{3}}{1 - \frac{1}{2} \cdot \frac{1}{3}}$$

$$= \frac{3 + 2}{6 - 1}$$

$$= 1$$

$$= \tan \frac{1}{4}\pi$$

Therefore, since $0 < \alpha + \beta < \frac{1}{2}\pi$,

$$\frac{1}{4}\pi = \alpha + \beta$$

$$\frac{1}{4}\pi = \tan^{-1}\frac{1}{2} + \tan^{-1}\frac{1}{3} \tag{11}$$

From formula (10) with $x = \frac{1}{2}$,

$$\tan^{-1}\frac{1}{2} = \frac{1}{2} - \frac{1}{3}\left(\frac{1}{2}\right)^3 + \frac{1}{5}\left(\frac{1}{2}\right)^5 - \frac{1}{7}\left(\frac{1}{2}\right)^7 + \frac{1}{9}\left(\frac{1}{2}\right)^9 - \frac{1}{11}\left(\frac{1}{2}\right)^{11} + \frac{1}{13}\left(\frac{1}{2}\right)^{13} - \frac{1}{15}\left(\frac{1}{2}\right)^{15} + \cdots$$

$$= \frac{1}{2} - \frac{1}{24} + \frac{1}{160} - \frac{1}{896} + \frac{1}{4608} - \frac{1}{22,528} + \frac{1}{106,492} - \frac{1}{491,520} + \cdots$$

$$\approx 0.500000 - 0.041667 + 0.006250 - 0.001116 + 0.000217 - 0.000044 + 0.000009 - 0.000002 + \cdots$$

Because the series is alternating and $|u_{n+1}| < |u_n|$, it follows from Theorem 12.7.4 that if the first seven terms are used to approximate the sum of the series, the error is less than the absolute value of the eighth term. Therefore

$$\tan^{-1}\tfrac{1}{2} \approx 0.463648$$

From formula (10) with $x = \tfrac{1}{3}$,

$$\tan^{-1}\frac{1}{3} = \frac{1}{3} - \frac{1}{3}\left(\frac{1}{3}\right)^3 + \frac{1}{5}\left(\frac{1}{3}\right)^5 - \frac{1}{7}\left(\frac{1}{3}\right)^7 + \frac{1}{9}\left(\frac{1}{3}\right)^9 - \frac{1}{11}\left(\frac{1}{3}\right)^{11} + \cdots$$

$$= \frac{1}{3} - \frac{1}{81} + \frac{1}{1215} - \frac{1}{15,309} + \frac{1}{177,147} - \frac{1}{1,948,617} + \cdots$$

$$\approx 0.333333 - 0.012346 + 0.000823 - 0.000065 + 0.000006 - 0.0000005 + \cdots$$

If the first five terms are used to approximate the sum,

$$\tan^{-1}\tfrac{1}{3} \approx 0.321751$$

By substituting the values of $\tan^{-1}\tfrac{1}{2}$ and $\tan^{-1}\tfrac{1}{3}$ into (11),

$$\tfrac{1}{4}\pi \approx 0.463648 + 0.321751$$

$$\approx 0.78540$$

We multiply by 4 and the result to five significant figures is $\pi \approx 3.1416$.

EXERCISES 13.3

In Exercises 1 through 4, find a power-series representation of the integral and determine its radius of convergence.

1. $\int_0^x e^t \, dt$

2. $\int_0^x \dfrac{dt}{t^2 + 4}$

3. $\int_2^x \dfrac{dt}{4 - t}$

4. $\int_0^x \ln(1 + t) \, dt$

In Exercises 5 through 8, compute accurate to three decimal places the value of the integral by two methods: (a) Use the second fundamental theorem of the calculus; (b) use the result of the indicated exercise.

5. $\int_0^1 e^t \, dt$; Exercise 1

6. $\int_0^1 \dfrac{dt}{t^2 + 4}$; Exercise 2

7. $\int_2^3 \dfrac{dt}{4 - t}$; Exercise 3

8. $\int_0^{1/3} \ln(1 + t) \, dt$; Exercise 4

In Exercises 9 through 12, find a power-series representation of the integral and determine its radius of convergence.

9. $\int_0^x f(t) \, dt$, where $f(t) = \begin{cases} \dfrac{e^t - 1}{t} & \text{if } t \neq 0 \\ 1 & \text{if } t = 0 \end{cases}$

10. $\int_0^x f(t) \, dt$, where $f(t) = \begin{cases} \dfrac{\ln(1 + t)}{t} & \text{if } t \neq 0 \\ 1 & \text{if } t = 0 \end{cases}$

11. $\int_0^x h(t) \, dt$, where $h(t) = \begin{cases} \dfrac{\sinh t}{t} & \text{if } t \neq 0 \\ 1 & \text{if } t = 0 \end{cases}$

12. $\int_0^x g(t) \, dt$, where $g(t) = \begin{cases} \dfrac{\tan^{-1} t}{t} & \text{if } t \neq 0 \\ 1 & \text{if } t = 0 \end{cases}$

In Exercises 13 through 16, compute accurate to three decimal places the value of the definite integral obtained by replacing x by the given number in the indicated exercise.

13. $x = 1$; Exercise 9

14. $x = \tfrac{1}{2}$; Exercise 10

15. $x = 1$; Exercise 11

16. $x = \tfrac{1}{4}$; Exercise 12

In Exercises 17 through 24, compute accurate to three decimal places the value of the integral by using series.

17. $\int_0^{1/2} \dfrac{dx}{1 + x^3}$

18. $\int_0^{1/2} e^{-x^3} \, dx$

19. $\int_0^1 e^{-x^2} \, dx$

20. $\int_0^{1/3} \dfrac{dx}{1 + x^4}$

21. $\int_0^{1/2} \tan^{-1} x^2 \, dx$

22. $\int_0^{1/2} \cosh x^2 \, dx$

23. $\int_0^1 x \sinh \sqrt{x} \, dx$

24. $\int_0^1 g(x) \, dx$, where $g(x) = \begin{cases} \dfrac{\cosh x - 1}{x} & \text{if } x \neq 0 \\ 0 & \text{if } x = 0 \end{cases}$

25. Given $\cosh x = \sum\limits_{n=0}^{+\infty} \dfrac{x^{2n}}{(2n)!}$ for all x. Obtain a power-series representation for $\sinh x$ by integrating term by term from 0 to x the given series.

26. Find a power-series representation for $\ln(1 + ax)$ by integrating term by term from 0 to x a power-series representation for $\dfrac{1}{1 + at}$.

27. Use the power series in (9) to compute $\tan^{-1} \frac{1}{4}$ accurate to four decimal places.

28. Use the power series in (8) to compute $\ln 3$ accurate to four decimal places.

29. If $f'(x) = \sum\limits_{n=0}^{+\infty} (-1)^n \dfrac{(x-1)^n}{n!}$, find $f(\frac{5}{4})$ accurate to three decimal places.

30. If $g'(x) = \sum\limits_{n=0}^{+\infty} (-1)^n \dfrac{x^n}{n^2 + 3}$, find $g(1)$ accurate to two decimal places.

31. Find a power-series representation for $\tanh^{-1} x$ by integrating term by term from 0 to x a power-series representation for $(1 - t^2)^{-1}$.

32. Find a power series for xe^x by multiplying the series for e^x by x, and then integrate the resulting series term by term from 0 to 1 and show that $\sum\limits_{n=1}^{+\infty} \dfrac{1}{n!(n+2)} = \dfrac{1}{2}$.

33. By integrating term by term from 0 to x a power-series representation for $\ln(1 - t)$, show that
$$\sum_{n=2}^{+\infty} \frac{x^n}{(n-1)n} = x + (1 - x)\ln(1 - x)$$

34. By integrating term by term from 0 to x a power-series representation for $\ln(1 + t)$, show that
$$\sum_{n=0}^{+\infty} \frac{(-1)^n}{(n+1)(n+2)} = 2\ln 2 - 1$$

35. By integrating term by term from 0 to x a power-series representation for $t \tan^{-1} t$, show that
$$\sum_{n=1}^{+\infty} (-1)^{n+1} \frac{x^{2n+1}}{(2n-1)(2n+1)} = \frac{1}{2}[(x^2 + 1)\tan^{-1} x - x]$$

36. Show that the interval of convergence of the power series in (9) is $[-1, 1]$ and that the power series is a representation of $\tan^{-1} x$ for all x in its interval of convergence.

37. Find the power series in x of $f(x)$ if $f''(x) = -f(x)$, $f(0) = 0$, and $f'(0) = 1$. Also, find the radius of convergence of the resulting series.

38. Integrate term by term from 0 to x a power-series representation for $(1 - t^2)^{-1}$ to obtain the power series in (8) for $\ln \dfrac{1 + x}{1 - x}$.

13.4 TAYLOR SERIES

If f is the function defined by

$$f(x) = \sum_{n=0}^{+\infty} c_n x^n$$
$$= c_0 + c_1 x + c_2 x^2 + c_3 x^3 + \ldots + c_n x^n + \ldots \qquad (1)$$

whose radius of convergence is $R > 0$, it follows from successive applications of Theorem 13.2.3 that f has derivatives of all orders on $(-R, R)$. Such a function is said to be *infinitely differentiable* on $(-R, R)$. Successive differentiations of the function in (1) give

$$f'(x) = c_1 + 2c_2 x + 3c_3 x^2 + 4c_4 x^3 + \ldots + nc_n x^{n-1} + \ldots \qquad (2)$$

$$f''(x) = 2c_2 + 2 \cdot 3c_3 x + 3 \cdot 4c_4 x^2 + \ldots + (n-1)nc_n x^{n-2} + \ldots \qquad (3)$$

$$f'''(x) = 2 \cdot 3c_3 + 2 \cdot 3 \cdot 4c_4 x + \ldots + (n-2)(n-1)nc_n x^{n-3} + \ldots \qquad (4)$$

$$f^{(iv)}(x) = 2 \cdot 3 \cdot 4c_4 + \ldots + (n-3)(n-2)(n-1)nc_n x^{n-4} + \ldots \qquad (5)$$

etc. If $x = 0$ in (1),

$$f(0) = c_0$$

If $x = 0$ in (2),

$$f'(0) = c_1$$

If in (3) $x = 0$,

$$f''(0) = 2c_2 \quad \Leftrightarrow \quad c_2 = \frac{f''(0)}{2!}$$

From (4), if $x = 0$,

$$f'''(0) = 2 \cdot 3c_3 \quad \Leftrightarrow \quad c_3 = \frac{f'''(0)}{3!}$$

In a similar manner, from (5), if $x = 0$,

$$f^{(iv)}(0) = 2 \cdot 3 \cdot 4c_4 \quad \Leftrightarrow \quad c_4 = \frac{f^{(iv)}(0)}{4!}$$

In general,

$$c_n = \frac{f^{(n)}(0)}{n!} \qquad \text{for every positive integer } n$$

This formula also holds when $n = 0$ if we take $f^{(0)}(0)$ to be $f(0)$ and $0! = 1$. So from this formula and (1) the power series of f in x can be written as

$$\sum_{n=0}^{+\infty} \frac{f^{(n)}(0)}{n!} x^n = f(0) + f'(0)x + \frac{f''(0)}{2!} x^2 + \ldots + \frac{f^{(n)}(0)}{n!} x^n + \ldots \tag{6}$$

In a more general sense, consider the function f as a power series in $x - a$; that is,

$$f(x) = \sum_{n=0}^{+\infty} c_n(x - a)^n$$

$$= c_0 + c_1(x - a) + c_2(x - a)^2 + \ldots + c_n(x - a)^n + \ldots \tag{7}$$

If the radius of convergence of this series is R, then f is infinitely differentiable on $(a - R, a + R)$. Successive differentiations of the function in (7) give

$$f'(x) = c_1 + 2c_2(x - a) + 3c_3(x - a)^2 + 4c_4(x - a)^3 + \ldots + nc_n(x - a)^{n-1} + \ldots$$

$$f''(x) = 2c_2 + 2 \cdot 3c_3(x - a) + 3 \cdot 4c_4(x - a)^2 + \ldots + (n - 1)nc_n(x - a)^{n-2} + \ldots$$

$$f'''(x) = 2 \cdot 3c_3 + 2 \cdot 3 \cdot 4c_4(x - a) + \ldots + (n - 2)(n - 1)nc_n(x - a)^{n-3} + \ldots$$

etc. Letting $x = a$ in the power-series representations of f and its derivatives we get

$$c_0 = f(a) \qquad c_1 = f'(a) \qquad c_2 = \frac{f''(a)}{2!} \qquad c_3 = \frac{f'''(a)}{3!}$$

and in general

$$c_n = \frac{f^{(n)}(a)}{n!} \tag{8}$$

From this formula and (7) the power series of f in $x - a$ can be written as

$$\sum_{n=0}^{+\infty} \frac{f^{(n)}(a)}{n!} (x - a)^n = f(a) + f'(a)(x - a) + \frac{f''(a)}{2!} (x - a)^2 + \ldots + \frac{f^{(n)}(a)}{n!} (x - a)^n + \ldots \tag{9}$$

The series in (9) is called the **Taylor series** of f at a. The special case of (9), when $a = 0$, is (6), which is called the **Maclaurin series**.

Observe that the nth partial sum of infinite series (9) is the nth degree Taylor polynomial of the function f at the number a, discussed in Section 11.5.

EXAMPLE 1   Find the Maclaurin series for e^x.

Solution   If $f(x) = e^x$, $f^{(n)}(x) = e^x$ for all x; therefore $f^{(n)}(0) = 1$ for all n. So from (6) we have the Maclaurin series:

$$e^x = 1 + x + \frac{x^2}{2!} + \frac{x^3}{3!} + \ldots + \frac{x^n}{n!} + \ldots \tag{10}$$

This series is the same as the one obtained in Example 3 of Section 13.2.

EXAMPLE 2   Find the Taylor series for $\sin x$ at a.

Solution   If $f(x) = \sin x$, then $f'(x) = \cos x$, $f''(x) = -\sin x$, $f'''(x) = -\cos x$, $f^{(iv)}(x) = \sin x$, and so forth. Thus, from formula (8), $c_0 = \sin a$, $c_1 = \cos a$, $c_2 = (-\sin a)/2!$, $c_3 = (-\cos a)/3!$, $c_4 = (\sin a)/4!$, and so on. The required Taylor series is obtained from (9), and it is

$$\sin x = \sin a + (\cos a)(x - a) - (\sin a)\frac{(x - a)^2}{2!} - (\cos a)\frac{(x - a)^3}{3!} + (\sin a)\frac{(x - a)^4}{4!} + \ldots$$

We can deduce that a power-series representation of a function is unique. That is, if two functions have the same function values in some interval containing the number a, and if both functions have a power-series representation in $x - a$, then these series must be the same because the coefficients in the series are obtained from the values of the functions and their derivatives at a. Therefore, if a function has a power-series representation in $x - a$, this series must be its Taylor series at a. Hence the Taylor series for a given function does not have to be obtained by using formula (9). Any method that gives a power series in $x - a$ representing the function will be the Taylor series of the function at a.

▶ **ILLUSTRATION 1**   To find the Taylor series for e^x at a, write $e^x = e^a e^{x-a}$ and then use series (10), where x is replaced by $x - a$. Then

$$e^x = e^a\left[1 + (x - a) + \frac{(x - a)^2}{2!} + \frac{(x - a)^3}{3!} + \ldots + \frac{(x - a)^n}{n!} + \ldots\right] \qquad ◀$$

▶ **ILLUSTRATION 2**   The series for $\ln(1 + x)$ found in Example 3 of Section 13.3 can be used to find the Taylor series for $\ln x$ at a $(a > 0)$ by writing

$$\ln x = \ln[a + (x - a)]$$

$$\ln x = \ln a + \ln\left(1 + \frac{x - a}{a}\right) \tag{11}$$

Because

$$\ln(1 + t) = \sum_{n=1}^{+\infty} (-1)^{n-1}\frac{t^n}{n} \qquad \text{if } -1 < t \leq 1$$

then

$$\ln\left(1 + \frac{x - a}{a}\right) = \frac{x - a}{a} - \frac{(x - a)^2}{2a^2} + \frac{(x - a)^3}{3a^3} - \cdots$$

Therefore, from (11),

$$\ln x = \ln a + \frac{x - a}{a} - \frac{(x - a)^2}{2a^2} + \frac{(x - a)^3}{3a^3} - \cdots$$

and the series represents $\ln x$ if $-1 < \dfrac{x - a}{a} \leq 1$ or, equivalently, $0 < x \leq 2a$.

◀

A natural question that arises is: If a function has a Taylor series in $x - a$ having radius of convergence $R > 0$, does this series represent the function for all values of x in the interval $(a - R, a + R)$? For most elementary functions the answer is yes. However, there are functions for which the answer is no. The following example shows this.

EXAMPLE 3   Let f be the function defined by

$$f(x) = \begin{cases} e^{-1/x^2} & \text{if } x \neq 0 \\ 0 & \text{if } x = 0 \end{cases}$$

Find the Maclaurin series for f, and show that it converges for all values of x but that it represents $f(x)$ only when $x = 0$.

Solution   To find $f'(0)$, we use the definition of a derivative.

$$f'(0) = \lim_{x \to 0} \frac{e^{-1/x^2} - 0}{x - 0}$$

$$= \lim_{x \to 0} \frac{\dfrac{1}{x}}{e^{1/x^2}}$$

Because $\lim\limits_{x \to 0} (1/x) = +\infty$ and $\lim\limits_{x \to 0} e^{1/x^2} = +\infty$, we can use L'Hôpital's rule. Therefore

$$f'(0) = \lim_{x \to 0} \frac{-\dfrac{1}{x^2}}{e^{1/x^2}\left(-\dfrac{2}{x^3}\right)}$$

$$= \lim_{x \to 0} \frac{x}{2e^{1/x^2}}$$

$$= 0$$

By a similar method, using the definition of a derivative and L'Hôpital's rule, we get 0 for every derivative. So $f^{(n)}(0) = 0$ for all n. Therefore the Maclaurin series for the given function is $0 + 0 + 0 + \ldots + 0 + \ldots$. This series converges to 0 for all x; however, if $x \neq 0$, $f(x) = e^{-1/x^2}$ and $e^{-1/x^2} \neq 0$.

The following theorem gives a test for determining whether a function is represented by its Taylor series.

13.4.1 THEOREM

Let f be a function such that f and all of its derivatives exist in some interval $(a - r, a + r)$. Then the function is represented by its Taylor series

$$\sum_{n=0}^{+\infty} \frac{f^{(n)}(a)}{n!}(x - a)^n$$

for all x such that $|x - a| < r$ if and only if

$$\lim_{n \to +\infty} R_n(x) = \lim_{n \to +\infty} \frac{f^{(n+1)}(\xi_n)}{(n + 1)!}(x - a)^{n+1}$$

$$= 0$$

where each ξ_n is between x and a.

Proof In the interval $(a - r, a + r)$, the function f satisfies the hypothesis of Theorem 11.5.1 for which

$$f(x) = P_n(x) + R_n(x) \tag{12}$$

where $P_n(x)$ is the nth-degree Taylor polynomial of f at a and $R_n(x)$ is the remainder, given by

$$R_n(x) = \frac{f^{(n+1)}(\xi_n)}{(n + 1)!}(x - a)^{n+1} \tag{13}$$

where each ξ_n is between x and a.

Now $P_n(x)$ is the nth partial sum of the Taylor series of f at a. So if we show that $\lim_{n \to +\infty} P_n(x)$ exists and equals $f(x)$ if and only if $\lim_{n \to +\infty} R_n(x) = 0$, the theorem will be proved. From (12),

$$P_n(x) = f(x) - R_n(x)$$

If $\lim_{n \to +\infty} R_n(x) = 0$, it follows from this equation that

$$\lim_{n \to +\infty} P_n(x) = f(x) - \lim_{n \to +\infty} R_n(x)$$

$$= f(x) - 0$$

$$= f(x)$$

Now under the hypothesis that $\lim_{n \to +\infty} P_n(x) = f(x)$ we wish to show that $\lim_{n \to +\infty} R_n(x) = 0$. From (12),

$$R_n(x) = f(x) - P_n(x)$$

Thus

$$\lim_{n \to +\infty} R_n(x) = f(x) - \lim_{n \to +\infty} P_n(x)$$

$$= f(x) - f(x)$$

$$= 0$$

This proves the theorem. ∎

Theorem 13.4.1 also holds for other forms of the remainder $R_n(x)$ besides the Lagrange form.

It is often difficult to apply Theorem 13.4.1 in practice because the values of ξ_n are arbitrary. However, sometimes an upper bound for $R_n(x)$ can be found, and it may be possible to prove that the limit of the upper bound is zero as $n \to +\infty$. The following limit is helpful in some cases:

$$\lim_{n \to +\infty} \frac{x^n}{n!} = 0 \qquad \text{for all } x \tag{14}$$

This follows from Example 2 of Section 13.1, where we showed that the power series $\sum_{n=0}^{+\infty} \frac{x^n}{n!}$ is convergent for all values of x and hence the limit of its nth term must be zero. In a similar manner, because $\sum_{n=0}^{+\infty} \frac{(x-a)^n}{n!}$ is convergent for all values of x,

$$\lim_{n \to +\infty} \frac{(x-a)^n}{n!} = 0 \qquad \text{for all } x \tag{15}$$

EXAMPLE 4   Use Theorem 13.4.1 to show that the Maclaurin series for e^x, found in Example 1, represents the function for all values of x.

Solution   The Maclaurin series for e^x is series (10) and

$$R_n(x) = \frac{e^{\xi_n}}{(n+1)!} x^{n+1}$$

where each ξ_n is between 0 and x.

We must show that $\lim_{n \to +\infty} R_n = 0$ for all x. There are three cases: $x > 0$, $x < 0$, and $x = 0$.

If $x > 0$, then $0 < \xi_n < x$; hence $e^{\xi_n} < e^x$. So

$$0 < \frac{e^{\xi_n}}{(n+1)!} x^{n+1} < e^x \frac{x^{n+1}}{(n+1)!} \tag{16}$$

From (14) it follows that $\lim_{n \to +\infty} \frac{x^{n+1}}{(n+1)!} = 0$, and so

$$\lim_{n \to +\infty} e^x \frac{x^{n+1}}{(n+1)!} = 0$$

Therefore, from (16) and the squeeze theorem it follows that $\lim_{n \to +\infty} R_n(x) = 0$.

If $x < 0$, then $x < \xi_n < 0$ and $0 < e^{\xi_n} < 1$. Therefore, if $x^{n+1} > 0$,

$$0 < \frac{e^{\xi_n}}{(n+1)!} x^{n+1} < \frac{x^{n+1}}{(n+1)!}$$

and if $x^{n+1} < 0$,

$$\frac{x^{n+1}}{(n+1)!} < \frac{e^{\xi_n}}{(n+1)!} x^{n+1} < 0$$

In either case, because $\lim_{n \to +\infty} \frac{x^{n+1}}{(n+1)!} = 0$, we conclude that $\lim_{n \to +\infty} R_n = 0$.

Finally, if $x = 0$, the series has the sum of 1, which is e^0. Hence series (10) represents e^x for all values of x.

From the results of the above example we can write

$$e^x = \sum_{n=0}^{+\infty} \frac{x^n}{n!}$$

$$= 1 + x + \frac{x^2}{2!} + \frac{x^3}{3!} + \dots \qquad \text{for all } x$$

and this agrees with Example 3 of Section 13.2.

EXAMPLE 5 Show that the Taylor series for $\sin x$ at a, found in Example 2, represents the function for all values of x.

Solution We use Theorem 13.4.1; so we must show that

$$\lim_{n \to +\infty} R_n(x) = \lim_{n \to +\infty} \frac{f^{(n+1)}(\xi_n)}{(n+1)!}(x-a)^{n+1}$$

$$= 0$$

Because $f(x) = \sin x$, $f^{(n+1)}(\xi_n)$ will be one of the following numbers: $\cos \xi_n$, $\sin \xi_n$, $-\cos \xi_n$, or $-\sin \xi_n$. In any case, $|f^{(n+1)}(\xi_n)| \leq 1$. Hence

$$0 < |R_n(x)| \leq \frac{|x-a|^{n+1}}{(n+1)!} \tag{17}$$

From (15), $\displaystyle\lim_{n \to +\infty} \frac{|x-a|^{n+1}}{(n+1)!} = 0$. Thus, by the squeeze theorem and (17) it follows that $\displaystyle\lim_{n \to +\infty} R_n(x) = 0$.

EXAMPLE 6 Compute the value of $\sin 47°$ accurate to four decimal places.

Solution In Examples 2 and 5 we obtained the Taylor series for $\sin x$ at a. It represents the function for all values of x. The series is

$$\sin x = (\sin a) + (\cos a)(x-a) - (\sin a)\frac{(x-a)^2}{2!} - (\cos a)\frac{(x-a)^3}{3!} + \dots$$

To make $x - a$ small we must choose a value of a near the value of x for which the function value is being computed. The sine and cosine of a must also be known. We therefore choose $a = \frac{1}{4}\pi$ and have

$$\sin x = \sin \tfrac{1}{4}\pi + (\cos \tfrac{1}{4}\pi)(x - \tfrac{1}{4}\pi) - (\sin \tfrac{1}{4}\pi)\frac{(x - \tfrac{1}{4}\pi)^2}{2!} - (\cos \tfrac{1}{4}\pi)\frac{(x - \tfrac{1}{4}\pi)^3}{3!} + \dots$$

Because $47°$ is equivalent to $\frac{47}{180}\pi$ radians or $(\frac{1}{4}\pi + \frac{1}{90}\pi)$ radians, from this series with $x = \frac{47}{180}\pi$, we have

$$\sin \tfrac{47}{180}\pi = \tfrac{1}{2}\sqrt{2} + \tfrac{1}{2}\sqrt{2} \cdot \tfrac{1}{90}\pi - \tfrac{1}{2}\sqrt{2} \cdot \tfrac{1}{2}(\tfrac{1}{90}\pi)^2 - \tfrac{1}{2}\sqrt{2} \cdot \tfrac{1}{6}(\tfrac{1}{90}\pi)^3 + \dots$$

$$\approx \tfrac{1}{2}\sqrt{2}(1 + 0.03490 - 0.00061 - 0.000002 + \dots)$$

Taking $\sqrt{2} \approx 1.41421$ and using the first three terms of the series we get

$$\sin \tfrac{47}{180}\pi \approx (0.70711)(1.03429)$$

$$\approx 0.73136$$

Rounding off to four decimal places gives $\sin 47° \approx 0.7314$. The error introduced by using the first three terms is $R_2(\tfrac{47}{180}\pi)$, and from (17),

$$\left| R_2\left(\frac{47}{180}\pi\right) \right| \leq \frac{(\tfrac{1}{90}\pi)^3}{3!} \approx 0.00001$$

The result, then, is between $0.73136 - 0.00001$ and $0.73136 + 0.00001$; that is, the result is between 0.73135 and 0.73137. Thus to four decimal place accuracy, we have

$$\sin \tfrac{47}{180}\pi \approx 0.7314$$

The following Maclaurin series represent the given function for all values of x:

$$\sin x = \sum_{n=0}^{+\infty} \frac{(-1)^n x^{2n+1}}{(2n+1)!}$$

$$= x - \frac{x^3}{3!} + \frac{x^5}{5!} - \frac{x^7}{7!} + \cdots$$

$$\cos x = \sum_{n=0}^{+\infty} \frac{(-1)^n x^{2n}}{(2n)!}$$

$$= 1 - \frac{x^2}{2!} + \frac{x^4}{4!} - \frac{x^6}{6!} + \cdots$$

$$\sinh x = \sum_{n=0}^{+\infty} \frac{x^{2n+1}}{(2n+1)!}$$

$$= x + \frac{x^3}{3!} + \frac{x^5}{5!} + \frac{x^7}{7!} + \cdots$$

$$\cosh x = \sum_{n=0}^{+\infty} \frac{x^{2n}}{(2n)!}$$

$$= 1 + \frac{x^2}{2!} + \frac{x^4}{4!} + \frac{x^6}{6!} + \cdots$$

The series for $\sin x$ is a direct result of Examples 2 and 5 with $a = 0$. You are asked to verify the other series in Exercises 1, 2, and 3.

EXAMPLE 7   Evaluate to five decimal places

$$\int_{1/2}^{1} \frac{\sin x}{x}\, dx$$

Solution An antiderivative of the integrand in terms of elementary functions cannot be found. However, from the Maclaurin series for $\sin x$,

$$\frac{\sin x}{x} = \frac{1}{x} \cdot \sin x$$

$$= \frac{1}{x}\left(x - \frac{x^3}{3!} + \frac{x^5}{5!} - \frac{x^7}{7!} + \frac{x^9}{9!} - \cdots\right)$$

$$= 1 - \frac{x^2}{3!} + \frac{x^4}{5!} - \frac{x^6}{7!} + \frac{x^8}{9!} - \cdots$$

which is true for all $x \neq 0$. Using term-by-term integration we get

$$\int_{1/2}^{1} \frac{\sin x}{x}\, dx = x - \frac{x^3}{3 \cdot 3!} + \frac{x^5}{5 \cdot 5!} - \frac{x^7}{7 \cdot 7!} + \frac{x^9}{9 \cdot 9!} - \cdots \Bigg]_{1/2}^{1}$$

$$\approx (1 - 0.0555555 + 0.0016667 - 0.0000283 + 0.0000003 - \cdots)$$

$$- (0.5 - 0.0069444 + 0.0000521 - 0.0000002 + \cdots)$$

In each set of parentheses there is a convergent alternating series with $|u_{n+1}| < |u_n|$. In the first set of parentheses we use the first four terms because the error obtained is less than 0.0000003. In the second set of parentheses we use the first three terms where the error obtained is less than 0.0000002. Doing the arithmetic and rounding off to five decimal places we get

$$\int_{1/2}^{1} \frac{\sin x}{x}\, dx \approx 0.45298$$

EXERCISES 13.4

1. Prove that the series $\displaystyle\sum_{n=0}^{+\infty} \frac{(-1)^n x^{2n}}{(2n)!}$ represents $\cos x$ for all values of x.

2. Prove that the series $\displaystyle\sum_{n=0}^{+\infty} \frac{x^{2n+1}}{(2n+1)!}$ represents $\sinh x$ for all values of x.

3. Prove that the series $\displaystyle\sum_{n=0}^{+\infty} \frac{x^{2n}}{(2n)!}$ represents $\cosh x$ for all values of x.

4. Obtain the Maclaurin series for the cosine function by differentiating the Maclaurin series for the sine function. Also obtain the Maclaurin series for the sine function by differentiating the one for the cosine function.

5. Obtain the Maclaurin series for the hyperbolic sine function by differentiating the Maclaurin series for the hyperbolic cosine function. Also differentiate the Maclaurin series for the hyperbolic sine function to obtain the one for the hyperbolic cosine function.

6. Find the Taylor series for e^x at 3 by using the Maclaurin series for e^x.

7. Use the Maclaurin series for $\ln(1 + x)$ to find the Taylor series for $\ln x$ at 2.

8. Given $\ln 2 = 0.6931$, use the series obtained in Exercise 7 to find $\ln 3$ accurate to four decimal places.

In Exercises 9 through 14, find a power-series representation for the function at the number a, and determine its radius of convergence.

9. $f(x) = \ln(x + 1); a = 1$ **10.** $f(x) = \sqrt[3]{x}; a = 1$

11. $f(x) = \sqrt{x}; a = 4$ **12.** $f(x) = \dfrac{1}{x}; a = 1$

13. $f(x) = \cos x; a = \frac{1}{3}\pi$ **14.** $f(x) = 2^x; a = 0$

15. Find the Maclaurin series for $\sin^2 x$.
(*Hint:* Use $\sin^2 x = \frac{1}{2}(1 - \cos 2x)$.)

16. Find the Maclaurin series for $\cos^2 x$.
(*Hint:* Use $\cos^2 x = \frac{1}{2}(1 + \cos 2x)$.)

17. (a) Find the first three nonzero terms of the Maclaurin series for $\tan x$. (b) Use the result of part (a) and term-by-term differentiation to find the first three nonzero terms of the Maclaurin series for $\sec^2 x$. (c) Use the result of part (a) and term-by-term integration to find the first three nonzero terms of the Maclaurin series for $\ln|\sec x|$.

18. (a) Find the first three nonzero terms of the Taylor series for $\cot x$ at $\frac{1}{2}\pi$. (b) Use the result of part (a) and term-by-term integration to find the first three nonzero terms of the Taylor series for ln sin x at $\frac{1}{2}\pi$.

In Exercises 19 through 24, use a power series to compute to the indicated accuracy the value of the quantity.

19. cos 58°; four decimal places

20. $\sqrt[5]{e}$; four decimal places   **21.** $\sqrt[5]{30}$; five decimal places

22. sinh $\frac{1}{2}$; five decimal places

23. ln (0.8); four decimal places

24. $\sqrt[3]{29}$; three decimal places

25. Compute the value of e correct to seven decimal places, and prove that your answer has the required accuracy.

In Exercises 26 through 31, compute accurate to three decimal places the value of the definite integral.

26. $\int_0^1 \sqrt{x}\, e^{-x^2}\, dx$ **27.** $\int_0^{1/2} \sin x^2\, dx$

28. $\int_0^1 \cos \sqrt{x}\, dx$ **29.** $\int_0^{0.1} \ln(1 + \sin x)\, dx$

30. $\int_0^{1/3} f(x)\, dx$, where $f(x) = \begin{cases} \dfrac{\sin x}{x} & \text{if } x \neq 0 \\ 1 & \text{if } x = 0 \end{cases}$

31. $\int_0^1 g(x)\, dx$, where $g(x) = \begin{cases} \dfrac{1 - \cos x}{x} & \text{if } x \neq 0 \\ 0 & \text{if } x = 0 \end{cases}$

32. The function E defined by

$$E(x) = \frac{2}{\sqrt{\pi}} \int_0^x e^{-t^2}\, dt$$

is called the *error function*, and it is important in mathematical statistics. Find the Maclaurin series for the error function.

33. Determine a_n ($n = 0, 1, 2, 3, 4$) so that the polynomial

$$f(x) = 3x^4 - 17x^3 + 35x^2 - 32x + 17$$

is written in the form

$$f(x) = a_4(x - 1)^4 + a_3(x - 1)^3 + a_2(x - 1)^2 + a_1(x - 1) + a_0$$

34. Determine a_n ($n = 0, 1, 2, 3$) so that the polynomial

$$f(x) = 4x^3 - 5x^2 + 2x - 3$$

is written in the form

$$f(x) = a_3(x + 2)^3 + a_2(x + 2)^2 + a_1(x + 2) + a_0$$

13.5 THE BINOMIAL SERIES

In algebra you learned that the binomial theorem expresses $(a + b)^m$ as a sum of powers of a and b, where m is a positive integer, as follows:

$$(a + b)^m = a^m + ma^{m-1}b + \frac{m(m - 1)}{2!} a^{m-2}b^2 + \ldots + \frac{m(m - 1) \cdot \ldots \cdot (m - k + 1)}{k!} a^{m-k}b^k + \ldots + b^m$$

We now take $a = 1$ and $b = x$ and apply the binomial theorem to the expression $(1 + x)^m$, where m is not a positive integer. We obtain the power series

$$1 + mx + \frac{m(m - 1)}{2!} x^2 + \frac{m(m - 1)(m - 2)}{3!} x^3 + \ldots + \frac{m(m - 1)(m - 2) \cdot \ldots \cdot (m - n + 1)}{n!} x^n + \ldots \quad (1)$$

This is the Maclaurin series for $(1 + x)^m$. It is called a **binomial series**. To find the radius of convergence of series (1), apply the ratio test and get

$$\lim_{n \to +\infty} \left| \frac{u_{n+1}}{u_n} \right| = \lim_{n \to +\infty} \left| \frac{\dfrac{m(m - 1) \cdot \ldots \cdot (m - n + 1)(m - n)}{(n + 1)!} x^{n+1}}{\dfrac{m(m - 1) \cdot \ldots \cdot (m - n + 1)}{n!} x^n} \right|$$

$$= \lim_{n \to +\infty} \left| \frac{m - n}{n + 1} \right| |x|$$

$$= \lim_{n \to +\infty} \left| \frac{\dfrac{m}{n} - 1}{1 + \dfrac{1}{n}} \right| |x|$$

$$= |x|$$

So the series is convergent if $|x| < 1$. We now prove that series (1) represents $(1 + x)^m$ for all real numbers m if x is in the open interval $(-1, 1)$. This is not done by calculating $R_n(x)$ and showing that its limit is zero because that is quite difficult, as you will soon see if you attempt to do so. Instead we use the following method. Let

$$f(x) = 1 + \sum_{n=1}^{+\infty} \frac{m(m - 1) \cdot \ldots \cdot (m - n + 1)}{n!} x^n \qquad |x| < 1 \tag{2}$$

We wish to show that $f(x) = (1 + x)^m$, where $|x| < 1$. By Theorem 13.2.3,

$$f'(x) = \sum_{n=1}^{+\infty} \frac{m(m - 1) \cdot \ldots \cdot (m - n + 1)}{(n - 1)!} x^{n-1} \qquad |x| < 1 \tag{3}$$

Multiplying on both sides of (3) by x we get, from Theorem 12.4.2,

$$xf'(x) = \sum_{n=1}^{+\infty} \frac{m(m - 1) \cdot \ldots \cdot (m - n + 1)}{(n - 1)!} x^n \tag{4}$$

By rewriting the right side of (3),

$$f'(x) = m + \sum_{n=2}^{+\infty} \frac{m(m - 1) \cdot \ldots \cdot (m - n + 1)}{(n - 1)!} x^{n-1}$$

Rewriting this summation with the lower limit decreased by 1 and n replaced by $n + 1$, we have

$$f'(x) = m + \sum_{n=1}^{+\infty} (m - n) \frac{m(m - 1) \cdot \ldots \cdot (m - n + 1)}{n!} x^n$$

In (4) we multiply the numerator and the denominator by n to obtain

$$xf'(x) = \sum_{n=1}^{+\infty} n \frac{m(m - 1) \cdot \ldots \cdot (m - n + 1)}{n!} x^n$$

The series for $f'(x)$ and $xf'(x)$ are absolutely convergent for $|x| < 1$. So by Theorem 12.4.3 they can be added term by term, and the resulting series will be absolutely convergent for $|x| < 1$. Therefore, from the addition,

$$(1 + x)f'(x) = m\left[1 + \sum_{n=1}^{+\infty} \frac{m(m - 1) \cdot \ldots \cdot (m - n + 1)}{n!} x^n \right]$$

Because by (2) the expression in brackets is $f(x)$, we have

$$(1 + x)f'(x) = mf(x)$$

$$\frac{f'(x)}{f(x)} = \frac{m}{1 + x}$$

The left side of the above equation is $D_x[\ln f(x)]$; thus

$$\frac{d}{dx}[\ln f(x)] = \frac{m}{1 + x}$$

However, we also know that

$$\frac{d}{dx}[\ln(1 + x)^m] = \frac{m}{1 + x}$$

Because $\ln f(x)$ and $\ln(1 + x)^m$ have the same derivative, they differ by a constant. Hence

$$\ln f(x) = \ln(1 + x)^m + C$$

From (2), $f(0) = 1$. Therefore $C = 0$; so

$$f(x) = (1 + x)^m$$

We have proved the following general binomial theorem.

13.5.1 THEOREM
Binomial Theorem

If m is any real number, then

$$(1 + x)^m = 1 + \sum_{n=1}^{+\infty} \frac{m(m - 1)(m - 2) \cdot \ldots \cdot (m - n + 1)}{n!} x^n$$

for all values of x such that $|x| < 1$.

If m is a positive integer, the binomial series will terminate after a finite number of terms.

EXAMPLE 1   Express as a power series in x:

$$\frac{1}{\sqrt{1 + x}}$$

Solution   From Theorem 13.5.1, when $|x| < 1$,

$$(1 + x)^{-1/2} = 1 - \frac{1}{2} x + \frac{(-\frac{1}{2})(-\frac{1}{2} - 1)}{2!} x^2 + \frac{(-\frac{1}{2})(-\frac{1}{2} - 1)(-\frac{1}{2} - 2)}{3!} x^3$$

$$+ \ldots + \frac{(-\frac{1}{2})(-\frac{3}{2})(-\frac{5}{2}) \cdot \ldots \cdot (-\frac{1}{2} - n + 1)}{n!} x^n + \ldots$$

$$= 1 - \frac{1}{2} x + \frac{1 \cdot 3}{2^2 \cdot 2!} x^2 - \frac{1 \cdot 3 \cdot 5}{2^3 \cdot 3!} x^3 + \ldots$$

$$+ (-1)^n \frac{1 \cdot 3 \cdot 5 \cdot \ldots \cdot (2n - 1)}{2^n n!} x^n + \ldots$$

EXAMPLE 2   From the result of Example 1 obtain a binomial series for $(1 - x^2)^{-1/2}$, and use it to find a power series for $\sin^{-1} x$.

Solution   We replace x by $-x^2$ in the series for $(1 + x)^{-1/2}$ and get for $|x| < 1$

$$(1 - x^2)^{-1/2} = 1 + \frac{1}{2} x^2 + \frac{1 \cdot 3}{2^2 \cdot 2!} x^4 + \frac{1 \cdot 3 \cdot 5}{2^3 \cdot 3!} x^6 + \ldots + \frac{1 \cdot 3 \cdot 5 \cdot \ldots \cdot (2n - 1)}{2^n n!} x^{2n} + \ldots$$

We now apply Theorem 13.3.1 and integrate term by term to obtain

$$\int_0^x \frac{dt}{\sqrt{1 - t^2}} = x + \frac{1}{2} \cdot \frac{x^3}{3} + \frac{1 \cdot 3}{2^2 \cdot 2!} \cdot \frac{x^5}{5} + \frac{1 \cdot 3 \cdot 5}{2^3 \cdot 3!} \cdot \frac{x^7}{7} + \ldots + \frac{1 \cdot 3 \cdot 5 \cdot \ldots \cdot (2n - 1)}{2^n n!} \cdot \frac{x^{2n+1}}{2n + 1} + \ldots$$

Therefore

$$\sin^{-1} x = x + \sum_{n=1}^{+\infty} \frac{1 \cdot 3 \cdot 5 \cdot \ldots \cdot (2n - 1)}{2^n n!} \cdot \frac{x^{2n+1}}{2n + 1} \qquad \text{for } |x| < 1$$

The next example shows how the binomial series is used to estimate roots.

EXAMPLE 3   Compute the value of $\sqrt[3]{25}$ accurate to three decimal places by using the binomial series for $(1 + x)^{1/3}$.

Solution   From Theorem 13.5.1,

$$(1 + x)^{1/3} = 1 + \frac{1}{3}x + \left(\frac{1}{3}\right)\left(-\frac{2}{3}\right)\frac{x^2}{2!} + \left(\frac{1}{3}\right)\left(-\frac{2}{3}\right)\left(-\frac{5}{3}\right)\frac{x^3}{3!} + \ldots \quad \text{if } |x| < 1$$

Because $\sqrt[3]{25} = \sqrt[3]{27}\sqrt[3]{\frac{25}{27}}$, we can write

$$\sqrt[3]{25} = 3\left(1 - \frac{2}{27}\right)^{1/3} \tag{5}$$

From the series for $(1 + x)^{1/3}$ with $x = -\frac{2}{27}$,

$$\left(1 - \frac{2}{27}\right)^{1/3} = 1 + \frac{1}{3}\left(-\frac{2}{27}\right) - \frac{2}{3^2 \cdot 2!}\left(-\frac{2}{27}\right)^2 + \frac{2 \cdot 5}{3^3 \cdot 3!}\left(-\frac{2}{27}\right)^3 + \ldots$$

$$\approx 1 - 0.0247 - 0.0006 - 0.00003 - \ldots \tag{6}$$

If the first three terms of the above series are used, it follows from (13) of Section 13.4 that the remainder is

$$R_2\left(-\frac{2}{27}\right) = \frac{f'''(\xi_2)}{3!}\left(-\frac{2}{27}\right)^3$$

$$= \left(\frac{1}{3}\right)\left(-\frac{2}{3}\right)\left(-\frac{5}{3}\right)\left(\frac{1}{3!}\right)(1 + \xi_2)^{(1/3)-3}\left(-\frac{2}{27}\right)^3$$

where $-\frac{2}{27} < \xi_2 < 0$. Therefore

$$\left|R_2\left(-\frac{2}{27}\right)\right| = \left(\frac{2 \cdot 5}{3^3 \cdot 3!}\right)\frac{(1 + \xi_2)^{1/3}}{(1 + \xi_2)^3}\left(\frac{2}{27}\right)^3 \tag{7}$$

Because $-\frac{2}{27} < \xi_2 < 0$, it follows that

$$(1 + \xi_2)^{1/3} < 1^{1/3} = 1 \quad \text{and} \quad \frac{1}{(1 + \xi_2)^3} < \frac{1}{\left(\frac{25}{27}\right)^3} \tag{8}$$

Furthermore,

$$\frac{2 \cdot 5}{3^3 \cdot 3!} = \frac{2 \cdot 5}{3 \cdot 3 \cdot 3 \cdot 1 \cdot 2 \cdot 3}$$

$$= \frac{2}{3} \cdot \frac{5}{6} \cdot \frac{1}{9} < \frac{1}{9}$$

Therefore from this inequality and (8) in (7),

$$\left|R_2\left(-\frac{2}{27}\right)\right| < \frac{1}{9} \cdot \frac{1}{\left(\frac{25}{27}\right)^3} \cdot \left(\frac{2}{27}\right)^3 = \frac{8}{140,625} < 0.00006$$

Using, then, the first three terms of series (6) gives

$$\left(1 - \frac{2}{27}\right)^{1/3} \approx 0.9747$$

with an error less than 0.00006. From (5) we obtain

$$\sqrt[3]{25} \approx 3(0.9747)$$
$$= 2.9241$$

with an error less than $3(0.00006) = 0.00018$. Rounding off to three decimal places gives $\sqrt[3]{25} \approx 2.924$.

EXERCISES 13.5

In Exercises 1 through 10, use a binomial series to find the Maclaurin series for the function and determine its radius of convergence.

1. $f(x) = \sqrt{1 + x}$
2. $f(x) = (3 - x)^{-2}$
3. $f(x) = (4 + x)^{-1/2}$
4. $f(x) = \sqrt[3]{8 + x}$
5. $f(x) = \sqrt[3]{1 - x^3}$
6. $f(x) = (4 + x^2)^{-1}$

7. $f(x) = (9 + x^4)^{-1/2}$
8. $f(x) = \dfrac{x}{\sqrt{1 - x}}$

9. $f(x) = \dfrac{x^2}{\sqrt{1 + x}}$
10. $f(x) = \dfrac{x}{\sqrt[3]{1 + x^2}}$

In Exercises 11 through 16, compute the value of the quantity accurate to three decimal places by using a binomial series.

11. $\sqrt{24}$
12. $\sqrt{51}$
13. $\sqrt[4]{630}$
14. $\sqrt[3]{66}$
15. $\dfrac{1}{\sqrt[3]{128}}$
16. $\dfrac{1}{\sqrt[5]{31}}$

17. Integrate term by term from 0 to x the binomial series for $(1 + t^2)^{-1/2}$ to obtain the Maclaurin series for $\sinh^{-1} x$. Determine the radius of convergence.

18. Use a method similar to that of Exercise 17 to find the Maclaurin series for $\tanh^{-1} x$. Determine the radius of convergence.

19. (a) Express $\sqrt[4]{1 + x}$ as a power series in x. (b) Use the result

of part (a) to express $\sqrt[4]{1 + x^2}$ as a power series in x. (c) Use the result of part (b) to compute accurate to three decimal places the value of $\int_0^{1/2} \sqrt[4]{1 + x^2}\, dx$.

20. Use the procedure indicated in Exercise 19 to compute accurate to three decimal places the value of $\int_0^{1/4} (1 - \sqrt{x})^{2/3}\, dx$.

In Exercises 21 through 28, compute accurate to three decimal places the value of the definite integral.

21. $\int_0^{1/3} \sqrt{1 + x^3}\, dx$
22. $\int_0^{2/5} \sqrt[3]{1 + x^4}\, dx$
23. $\int_0^{1} \sqrt[3]{8 + x^2}\, dx$
24. $\int_0^{1/2} \sqrt{1 - x^3}\, dx$
25. $\int_0^{1/2} \dfrac{dx}{\sqrt{1 + x^4}}$
26. $\int_0^{1/3} \dfrac{dx}{\sqrt[3]{x^2 + 1}}$
27. $\int_0^{1/2} \dfrac{dx}{\sqrt{1 - x^3}}$

28. $\int_0^{1/2} f(x)\, dx$, where $f(x) = \begin{cases} \dfrac{\sin^{-1} x}{x} & \text{if } x \neq 0 \\ 1 & \text{if } x = 0 \end{cases}$

29. Find the Maclaurin series for $\int_0^x \dfrac{t^p}{\sqrt{1 - t^2}}\, dt$ if p is a nonnegative integer. Determine the radius of convergence of the series.

REVIEW EXERCISES FOR CHAPTER 13

In Exercises 1 through 12, find the interval of convergence of the power series.

1. $\displaystyle\sum_{n=1}^{+\infty} \dfrac{x^n}{\sqrt{n}}$
2. $\displaystyle\sum_{n=1}^{+\infty} \dfrac{(x - 2)^n}{n}$
3. $\displaystyle\sum_{n=1}^{+\infty} \dfrac{x^n}{3^n(n^2 + n)}$

4. $\displaystyle\sum_{n=0}^{+\infty} \dfrac{x^n}{2^n}$
5. $\displaystyle\sum_{n=0}^{+\infty} \dfrac{n!}{2^n}(x - 3)^n$

6. $\displaystyle\sum_{n=1}^{+\infty} \dfrac{(-1)^{n-1}x^{2n-1}}{(2n - 1)!}$
7. $\displaystyle\sum_{n=1}^{+\infty} \dfrac{n^2}{6^n}(x + 1)^n$

8. $\displaystyle\sum_{n=1}^{+\infty} n(2x - 1)^n$
9. $\displaystyle\sum_{n=1}^{+\infty} \dfrac{(-1)^{n-1}(x - 1)^n}{n2^n}$

10. $\displaystyle\sum_{n=1}^{+\infty} n^n x^n$
11. $\displaystyle\sum_{n=0}^{+\infty} (\sin 2n)x^n$

12. $\displaystyle\sum_{n=1}^{+\infty} (-1)^{n+1} \dfrac{x^n}{(n + 1)\ln(n + 1)}$

In Exercises 13 through 16 do the following: (a) Find the radius of convergence of the power series and the domain of f; (b) write the power series that defines the function f' and state its radius of convergence; (c) find the domain of f'.

13. $f(x) = \displaystyle\sum_{n=1}^{+\infty} (-1)^n \dfrac{x^{2n}}{2n}$
14. $f(x) = \displaystyle\sum_{n=1}^{+\infty} \dfrac{x^n}{n^3}$

15. $f(x) = \sum\limits_{n=0}^{+\infty} \dfrac{x^n}{(n!)^2}$

16. $f(x) = \sum\limits_{n=1}^{+\infty} \dfrac{(x+1)^n}{n2^n}$

In Exercises 17 through 20, find a power-series representation of the integral and determine its radius of convergence.

17. $\displaystyle\int_0^x \dfrac{dt}{t^2 + 16}$

18. $\displaystyle\int_3^x \dfrac{dt}{t - 1}$

19. $\displaystyle\int_0^x f(t)\,dt$, where $f(t) = \begin{cases} \dfrac{\sin t}{t} & \text{if } t \neq 0 \\ 1 & \text{if } t = 0 \end{cases}$

20. $\displaystyle\int_0^x f(t)\,dt$, where $f(t) = \begin{cases} \dfrac{1 - \cos t}{t} & \text{if } t \neq 0 \\ 0 & \text{if } t = 0 \end{cases}$

In Exercises 21 through 24, compute accurate to three decimal places the value of the definite integral obtained by replacing x by the given number in the indicated exercise.

21. $x = 3$; Exercise 17

22. $x = 4$; Exercise 18

23. $x = \frac{1}{2}$; Exercise 19

24. $x = \frac{1}{4}$; Exercise 20

In Exercises 25 through 34, use a power series to compute accurate to four decimal places the value of the quantity.

25. $\tan^{-1}\frac{1}{5}$

26. $\sin 0.3$

27. $\sqrt[3]{130}$

28. $\sin^{-1} 1$

29. $\cos 3°$

30. $\sqrt[4]{e}$

31. $\ln 5$

32. $\displaystyle\int_0^1 \cos x^3\,dx$

33. $\displaystyle\int_0^{1/4} \sqrt{x}\,\sin x\,dx$

34. $\displaystyle\int_0^{1/2} \dfrac{dx}{1 + x^5}$

In Exercises 35 through 38, find the Maclaurin series for the function, and find its interval of convergence.

35. $f(x) = a^x \;(a > 0)$

36. $f(x) = \dfrac{1}{2 - x}$

37. $f(x) = \sin^3 x$

38. $f(x) = \sqrt{x + 1}$

In Exercises 39 through 42, find the Taylor series for the function at the indicated number.

39. $f(x) = \sin 3x$; at $-\frac{1}{3}\pi$

40. $f(x) = \dfrac{1}{x}$; at 2

41. $f(x) = \ln|x|$; at -1

42. $f(x) = e^{x-2}$; at 2

Exercises 43 through 47 pertain to the functions J_0 and J_1 defined by power series as follows:

$$J_0(x) = \sum\limits_{n=0}^{+\infty} (-1)^n \dfrac{x^{2n}}{n!n!2^{2n}} \qquad J_1(x) = \sum\limits_{n=0}^{+\infty} (-1)^n \dfrac{x^{2n+1}}{n!(n+1)!2^{2n+1}}$$

The functions J_0 and J_1 are called Bessel functions of the first kind of orders zero and one, respectively.

43. Show that both J_0 and J_1 converge for all real values of x.

44. Show that $y = J_0(x)$ is a solution of the differential equation

$$x\dfrac{d^2y}{dx^2} + \dfrac{dy}{dx} + xy = 0$$

45. Show that $J_0'(x) = -J_1(x)$.

46. Show that $D_x(xJ_1(x)) = xJ_0(x)$.

47. Show that $y = J_1(x)$ is a solution of the differential equation

$$x^2\dfrac{d^2y}{dx^2} + x\dfrac{dy}{dx} + (x^2 - 1)y = 0$$

FOURTEEN

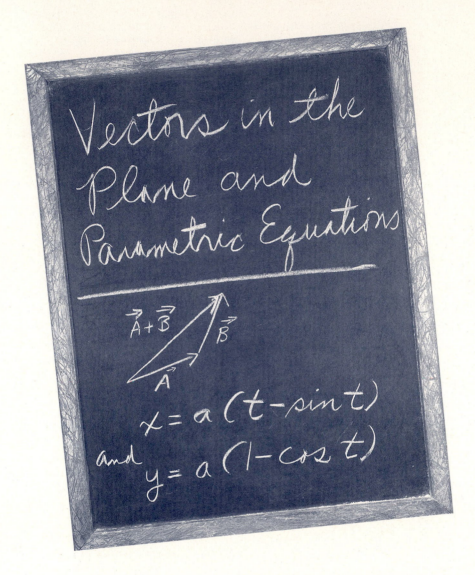

The approach to vectors in this chapter and Chapter 15 is modern, and it serves as an introduction both to the viewpoint of linear algebra and to that of classical vector analysis. In Section 14.1 we define a *vector in the plane* as an ordered pair of real numbers, and in that section as well as in Section 14.2 we perform operations on vectors by applying algebraic operations on their real coordinates. In Section 14.3 we introduce a new type of function whose domain is a set of real numbers and whose range is a set of vectors. These functions are called *vector-valued functions*. The graph of a vector-valued function is a curve, which can also be represented by *parametric equations*. These equations express the *x* and *y* coordinates of points on a plane curve as functions of a third variable *t*, which often represents time. The calculus of vector-valued functions, presented in Section 14.3, involves the calculus of the real-valued functions defined by the corresponding parametric equations.

The remaining sections of the chapter pertain to applications of vectors to geometry, physics, and engineering. The geometrical applications include *length of arc, tangent and normal vectors* to curves, and *curvature*. For applications in physics and engineering, we use vectors to calculate *work* and to discuss *motion along a curve*.

14.1   VECTORS IN THE PLANE

The applications of mathematics are often concerned with quantities that possess both magnitude and direction. An example of such a quantity is *velocity*. For instance, an airplane's velocity has magnitude (the speed of the airplane) and direction (which determines the course of the airplane). Other examples of such quantities are *force, displacement,* and *acceleration*. Physicists and engineers refer to a directed line segment as a *vector*, and the quantities that have both magnitude and direction are called **vector quantities**. In contrast, a quantity that has magnitude but not direction is called a **scalar quantity**. Examples of scalar quantities are length, area, volume, and speed. The study of vectors is called **vector analysis**.

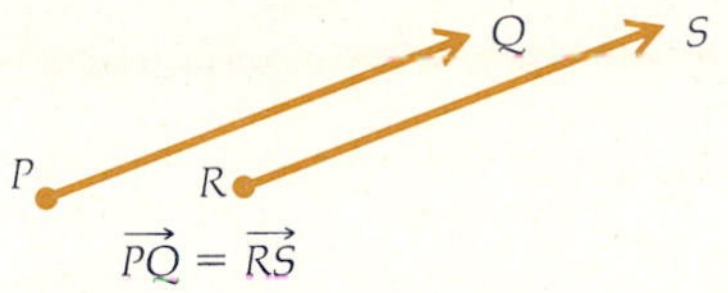

The approach to vector analysis can be on either a geometric or an analytic basis. If the geometric approach is taken, we first define a directed line segment as a line segment from a point P to a point Q and denote this directed line segment by $\overrightarrow{PQ}$. The point P is called the **initial point**, and the point Q is called the **terminal point**. Then two directed line segments $\overrightarrow{PQ}$ and $\overrightarrow{RS}$ are said to be equal if they have the same *length* and *direction*, and we write $\overrightarrow{PQ} = \overrightarrow{RS}$ (see Figure 1). The directed line segment $\overrightarrow{PQ}$ is called the **vector** from P to Q. A vector is denoted by a single letter, set in boldface type, such as **A**. In some books, a letter in lightface type, with an arrow above it, is used to indicate a vector, for example $\vec{A}$. When doing your work, you may use that notation or $\underline{A}$ to distinguish the symbol for a vector from the symbol for a real number.

Continuing with the geometric approach to vector analysis, note that if the directed line segment $\overrightarrow{PQ}$ is the vector **A**, and $\overrightarrow{PQ} = \overrightarrow{RS}$, the directed line segment $\overrightarrow{RS}$ is also the vector **A**. Then a vector is considered to remain unchanged if it is moved parallel to itself. With this interpretation of a vector, we can assume for convenience that every vector has its initial point at some fixed reference point. By taking this point as the origin of a rectangular cartesian coordinate system, a vector can be defined analytically in terms of real numbers. Such a definition permits the study of vector analysis from a purely mathematical viewpoint.

In this book we use the analytic approach; however, the geometric interpretation is used for illustrative purposes. A vector in the plane is denoted by an ordered pair of real numbers and the notation $\langle x, y \rangle$ is used instead of (x, y) to avoid confusing the notation for a vector with the notation for a point. V_2 is the set of all such ordered pairs.

$\overrightarrow{PQ} = \overrightarrow{RS}$

FIGURE 1

14.1.1 DEFINITION

A **vector in the plane** is an ordered pair of real numbers $\langle x, y \rangle$. The numbers x and y are called the **components** of the vector $\langle x, y \rangle$.

There is a one-to-one correspondence between the vectors $\langle x, y \rangle$ in the plane and the points (x, y) in the plane. Let the vector **A** be the ordered pair of real

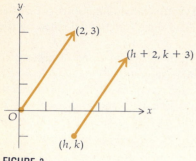

FIGURE 2

numbers $\langle a_1, a_2 \rangle$. If A is the point (a_1, a_2), then the vector $\mathbf{A}$ may be represented geometrically by the directed line segment $\overrightarrow{OA}$. Such a directed line segment is called a **representation** of vector $\mathbf{A}$. Any directed line segment that is equal to $\overrightarrow{OA}$ is also a representation of vector $\mathbf{A}$. The particular representation of a vector that has its initial point at the origin is called the **position representation** of the vector.

▶ **ILLUSTRATION 1**   The vector $\langle 2, 3 \rangle$ has as its position representation the directed line segment from the origin to the point $(2, 3)$. The representation of the vector $\langle 2, 3 \rangle$ whose initial point is (h, k) has as its terminal point $(h + 2, k + 3)$; refer to Figure 2. ◀

The vector $\langle 0, 0 \rangle$ is called the **zero vector**, and it is denoted by $\mathbf{0}$; that is,

$$\mathbf{0} = \langle 0, 0 \rangle$$

Any point is a representation of the zero vector.

14.1.2 DEFINITION

> The **magnitude** of a vector is the length of any of its representations, and the **direction** of a nonzero vector is the direction of any of its representations.

The magnitude of the vector $\mathbf{A}$ is denoted by $\|\mathbf{A}\|$.

14.1.3 THEOREM

> If $\mathbf{A}$ is the vector $\langle a_1, a_2 \rangle$, then $\|\mathbf{A}\| = \sqrt{a_1{}^2 + a_2{}^2}$.

Proof   Because by Definition 14.1.2 $\|\mathbf{A}\|$ is the length of any of the representations of $\mathbf{A}$, then $\|\mathbf{A}\|$ will be the length of the position representation of $\mathbf{A}$, which is the distance from the origin to the point (a_1, a_2). So from the formula for the distance between two points,

$$\|\mathbf{A}\| = \sqrt{(a_1 - 0)^2 + (a_2 - 0)^2}$$
$$= \sqrt{a_1{}^2 + a_2{}^2}$$
■

Observe that $\|\mathbf{A}\|$ is a nonnegative number and not a vector. From Theorem 14.1.3 it follows that $\|\mathbf{0}\| = 0$.

▶ **ILLUSTRATION 2**   If $\mathbf{A} = \langle -3, 5 \rangle$, then

$$\|\mathbf{A}\| = \sqrt{(-3)^2 + 5^2}$$
$$= \sqrt{34}$$
◀

EXAMPLE 1   Let the vector $\mathbf{A}$ be $\langle -4, 5 \rangle$ and the point P be $(6, -2)$. (a) Draw the position representation of $\mathbf{A}$ and also the particular representation of $\mathbf{A}$ having P as its initial point. (b) Find the magnitude of $\mathbf{A}$.

Solution

(a) Let A be the point $(-4, 5)$. Figure 3 shows $\overrightarrow{OA}$, which is the position representation of vector $\mathbf{A}$. Let $\overrightarrow{PQ}$ be the particular representation of vector $\mathbf{A}$ having P as its initial point. If $Q = (x, y)$, then

$$x - 6 = -4 \qquad y + 2 = 5$$
$$x = 2 \qquad\qquad y = 3$$

Therefore $Q = (2, 3)$ and $\overrightarrow{PQ}$ appears in Figure 3.

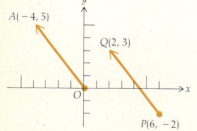

FIGURE 3

(b) From Theorem 14.1.3,

$$\|\mathbf{A}\| = \sqrt{(-4)^2 + 5^2}$$
$$= \sqrt{41}$$

The **direction angle** of any nonzero vector is the angle θ measured from the positive side of the x axis counterclockwise to the position representation of the vector. If θ is measured in radians, $0 \le \theta < 2\pi$. If $\mathbf{A} = \langle a_1, a_2 \rangle$, then

$$\tan \theta = \frac{a_2}{a_1} \qquad \text{if } a_1 \ne 0 \tag{1}$$

If $a_1 = 0$ and $a_2 > 0$, then $\theta = \frac{1}{2}\pi$; if $a_1 = 0$ and $a_2 < 0$, then $\theta = \frac{3}{2}\pi$. Figures 4 through 6 show the direction angle θ for specific vectors whose position representations are drawn.

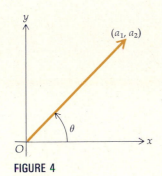

FIGURE 4

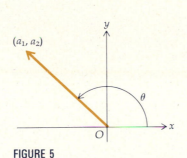

FIGURE 5

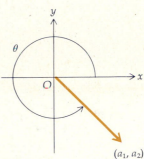

FIGURE 6

EXAMPLE 2 Find the radian measure of the direction angle of each of the following vectors: (a) $\langle -1, 1 \rangle$; (b) $\langle 0, -5 \rangle$; (c) $\langle 1, -2 \rangle$.

Solution The position representation of each of the vectors in (a), (b), and (c) is shown in Figures 7, 8, and 9, respectively. (a) $\tan \theta = -1$, and $\frac{1}{2}\pi < \theta < \pi$; so $\theta = \frac{3}{4}\pi$. (b) $\tan \theta$ does not exist, and $a_2 < 0$; thus $\theta = \frac{3}{2}\pi$. (c) $\tan \theta = -2$, and $\frac{3}{2}\pi < \theta < 2\pi$; therefore $\theta = \tan^{-1}(-2) + 2\pi$; that is, $\theta \approx 5.176$.

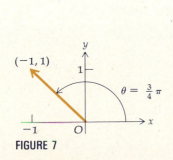

FIGURE 7

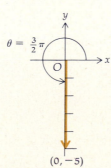

FIGURE 8

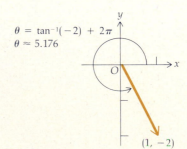

FIGURE 9

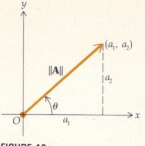

FIGURE 10

Observe that if $\mathbf{A} = \langle a_1, a_2 \rangle$ and θ is the direction angle of $\mathbf{A}$, then

$$a_1 = \|\mathbf{A}\| \cos \theta \qquad a_2 = \|\mathbf{A}\| \sin \theta \qquad (2)$$

See Figure 10, where the point (a_1, a_2) is in the first quadrant.

If the vector $\mathbf{A} = \langle a_1, a_2 \rangle$, then the representation of $\mathbf{A}$ whose initial point is (x, y) has as its endpoint $(x + a_1, y + a_2)$. In this way a vector may be thought of as a translation of the plane into itself. Figure 11 illustrates five representations of the vector $\mathbf{A} = \langle a_1, a_2 \rangle$. In each case $\mathbf{A}$ translates the point (x_i, y_i) into the point $(x_i + a_1, y_i + a_2)$.

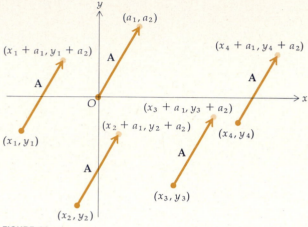

FIGURE 11

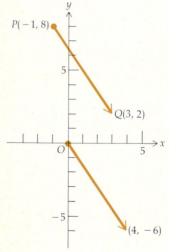

FIGURE 12

EXAMPLE 3   Suppose P is the point $(-1, 8)$ and Q is the point $(3, 2)$. Find the vector $\mathbf{A}$ having $\overrightarrow{PQ}$ as a representation. Draw $\overrightarrow{PQ}$ and the position representation of $\mathbf{A}$.

Solution   Figure 12 shows the directed line segment $\overrightarrow{PQ}$. Let $\mathbf{A} = \langle a_1, a_2 \rangle$. Because $\overrightarrow{PQ}$ is a representation of vector $\mathbf{A}$, the vector $\mathbf{A}$ translates the point $P(-1, 8)$ into the point $Q(3, 2)$. But vector $\langle a_1, a_2 \rangle$ translates the point $(-1, 8)$ into the point $\langle -1 + a_1, 8 + a_2 \rangle$. Thus

$$-1 + a_1 = 3 \qquad 8 + a_2 = 2$$
$$a_1 = 4 \qquad\qquad a_2 = -6$$

Therefore $\mathbf{A} = \langle 4, -6 \rangle$. Figure 12 also shows the position representation of $\mathbf{A}$.

The following definition gives the method for adding two vectors.

14.1.4 DEFINITION   The **sum** of two vectors $\mathbf{A} = \langle a_1, a_2 \rangle$ and $\mathbf{B} = \langle b_1, b_2 \rangle$ is the vector $\mathbf{A} + \mathbf{B}$ defined by

$$\mathbf{A} + \mathbf{B} = \langle a_1 + b_1, a_2 + b_2 \rangle$$

▶ **ILLUSTRATION 3**   If $\mathbf{A} = \langle 3, -1 \rangle$ and $\mathbf{B} = \langle -4, 5 \rangle$, then

$$\mathbf{A} + \mathbf{B} = \langle 3 + (-4), -1 + 5 \rangle$$
$$= \langle -1, 4 \rangle \qquad\qquad\qquad ◀$$

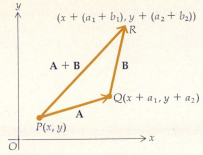

FIGURE 13

The geometric interpretation of the sum of two vectors is shown in Figure 13. Let $\mathbf{A} = \langle a_1, a_2 \rangle$ and $\mathbf{B} = \langle b_1, b_2 \rangle$, and let P be the point (x, y). Then $\mathbf{A}$ translates the point P into the point $(x + a_1, y + a_2) = Q$. The vector $\mathbf{B}$ translates the point Q into the point $((x + a_1) + b_1, (y + a_2) + b_2)$ or, equivalently, $(x + (a_1 + b_1), y + (a_2 + b_2)) = R$. Furthermore,

$$\mathbf{A} + \mathbf{B} = \langle a_1 + b_1, a_2 + b_2 \rangle$$

Therefore the vector $\mathbf{A} + \mathbf{B}$ translates the point P into the point $(x + (a_1 + b_1), y + (a_2 + b_2)) = R$. Thus, in Figure 13 $\overrightarrow{PQ}$ is a representation of the vector $\mathbf{A}$, $\overrightarrow{QR}$ is a representation of the vector $\mathbf{B}$, and $\overrightarrow{PR}$ is a representation of the vector $\mathbf{A} + \mathbf{B}$. The representations of the vectors $\mathbf{A}$ and $\mathbf{B}$ are adjacent sides of a parallelogram, and the representation of the vector $\mathbf{A} + \mathbf{B}$ is a diagonal of the parallelogram. This diagonal is called the **resultant** of the vectors $\mathbf{A}$ and $\mathbf{B}$. The rule for the addition of vectors is sometimes referred to as the **parallelogram law**.

Force is a vector quantity where the magnitude is expressed in force units and the direction angle is determined by the direction of the force. It is shown in physics that two forces applied to an object at a particular point can be replaced by an equivalent force that is their resultant.

EXAMPLE 4   Two forces of magnitudes 200 lb and 250 lb make an angle of $\frac{1}{3}\pi$ with each other and are applied to an object at the same point. Find (a) the magnitude of the resultant force and (b) the angle it makes with the force of 200 lb.

Solution   Refer to Figure 14, where the axes are chosen so that the position representation of the force of 200 lb is along the positive side of the x axis. The vector $\mathbf{A}$ represents this force, and $\mathbf{A} = \langle 200, 0 \rangle$. The vector $\mathbf{B}$ represents the force of 250 lb. From formulas (2), if $\mathbf{B} = \langle b_1, b_2 \rangle$, then

$$b_1 = 250 \cos \tfrac{1}{3}\pi \qquad b_2 = 250 \sin \tfrac{1}{3}\pi$$
$$= 125 \qquad\qquad\quad \approx 216.5$$

Thus $\mathbf{B} = \langle 125, 216.5 \rangle$. The resultant force is $\mathbf{A} + \mathbf{B}$, and

$$\mathbf{A} + \mathbf{B} = \langle 200, 0 \rangle + \langle 125, 216.5 \rangle$$
$$= \langle 325, 216.5 \rangle$$

(a) $\|\mathbf{A} + \mathbf{B}\| = \sqrt{(325)^2 + (216.5)^2}$
$$\approx 390.5$$

(b) If θ is the angle the vector $\mathbf{A} + \mathbf{B}$ makes with $\mathbf{A}$, then

$$\tan \theta = \frac{216.5}{325}$$

$$\tan \theta \approx 0.6662$$

$$\theta \approx 0.5877$$

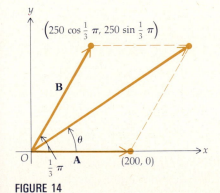

FIGURE 14

14.1.5 DEFINITION   If $\mathbf{A} = \langle a_1, a_2 \rangle$, then the vector $\langle -a_1, -a_2 \rangle$ is defined to be the **negative** of $\mathbf{A}$, denoted by $-\mathbf{A}$.

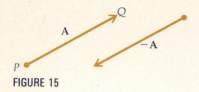

FIGURE 15

If the directed line segment $\overrightarrow{PQ}$ is a representation of the vector **A**, then the directed line segment $\overrightarrow{QP}$ is a representation of $-\mathbf{A}$. Any directed line segment that is parallel to $\overrightarrow{PQ}$, has the same length as $\overrightarrow{PQ}$, and has a direction opposite to that of $\overrightarrow{PQ}$ is also a representation of $-\mathbf{A}$. See Figure 15. We now define subtraction of two vectors.

14.1.6 DEFINITION

The **difference** of the two vectors **A** and **B**, denoted by $\mathbf{A} - \mathbf{B}$, is the vector obtained by adding **A** to the negative of **B**; that is,

$$\mathbf{A} - \mathbf{B} = \mathbf{A} + (-\mathbf{B})$$

Thus if $\mathbf{A} = \langle a_1, a_2 \rangle$ and $\mathbf{B} = \langle b_1, b_2 \rangle$, then $-\mathbf{B} = \langle -b_1, -b_2 \rangle$, and

$$\mathbf{A} - \mathbf{B} = \langle a_1 - b_1, a_2 - b_2 \rangle$$

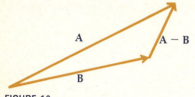

FIGURE 16

▶ **ILLUSTRATION 4**   If $\mathbf{A} = \langle 4, -2 \rangle$ and $\mathbf{B} = \langle 6, -3 \rangle$, then

$$\mathbf{A} - \mathbf{B} = \langle 4, -2 \rangle - \langle 6, -3 \rangle$$
$$= \langle 4, -2 \rangle + \langle -6, 3 \rangle$$
$$= \langle -2, 1 \rangle$$

◀

To interpret the difference of two vectors geometrically, let the representations of the vectors **A** and **B** have the same initial point. Then the directed line segment from the endpoint of the representation of **B** to the endpoint of the representation of **A** is a representation of the vector $\mathbf{A} - \mathbf{B}$. This obeys the parallelogram law $\mathbf{B} + (\mathbf{A} - \mathbf{B}) = \mathbf{A}$. See Figure 16.

The following example, involving the difference of two vectors, is concerned with air navigation. The *air speed* of a plane refers to its speed relative to the air, and the *ground speed* is its speed relative to the ground. When there is a wind, the velocity of the plane relative to the ground is the resultant of the vector representing the wind's velocity and the vector representing the velocity of the plane relative to the air. In navigation, the *course* of a ship or airplane is the angle measured in degrees clockwise from the north to the direction in which the carrier is traveling. The angle is considered positive even though it is in the clockwise sense.

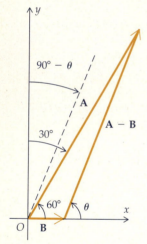

FIGURE 17

EXAMPLE 5   An airplane can fly at an air speed of 300 mi/hr. If there is a wind blowing toward the east at 50 mi/hr, what should be the plane's compass heading in order for its course to be 30°? What will be the plane's ground speed if it flies this course?

Solution   Refer to Figure 17, showing position representations of the vectors **A** and **B** as well as a representation of $\mathbf{A} - \mathbf{B}$. The vector **A** represents the velocity of the plane relative to the ground on a course of 30°. The direction angle of **A** is 60°. The vector **B** represents the velocity of the wind. Because **B** has a magnitude of 50 and a direction angle of 0°, $\mathbf{B} = \langle 50, 0 \rangle$. The vector $\mathbf{A} - \mathbf{B}$ represents the velocity of the plane relative to the air; thus $\|\mathbf{A} - \mathbf{B}\| = 300$. Let θ be the direction angle of $\mathbf{A} - \mathbf{B}$. From Figure 17 we obtain the triangle shown in Figure 18.

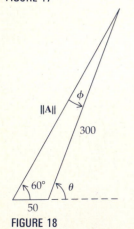

FIGURE 18

Applying the law of sines to this triangle, we get

$$\frac{\sin \phi}{50} = \frac{\sin 60°}{300}$$

$$\sin \phi = \frac{50 \sin 60°}{300}$$

$$\sin \phi = 0.1433$$

$$\phi = 8.3°$$

Therefore

$$\theta = 60° + 8.3°$$
$$= 68.3°$$

Again applying the law of sines to the triangle in Figure 18, we have

$$\frac{\|\mathbf{A}\|}{\sin(180° - \theta)} = \frac{300}{\sin 60°}$$

$$\|\mathbf{A}\| = \frac{300 \sin 111.7°}{\sin 60°}$$

$$\|\mathbf{A}\| = 322$$

The plane's compass heading should be $90° - \theta$, which is $21.7°$, and if the plane flies this course, its ground speed will be 322 mi/hr.

Suppose that P is the point (a_1, a_2) and Q is the point (b_1, b_2). We shall use the notation $\mathbf{V}(\overrightarrow{PQ})$ to denote the vector having the directed line segment $\overrightarrow{PQ}$ as a representation. See Figure 19, showing representations of the vectors $\mathbf{V}(\overrightarrow{PQ})$, $\mathbf{V}(\overrightarrow{OP})$, and $\mathbf{V}(\overrightarrow{OQ})$. Observe that

$$\mathbf{V}(\overrightarrow{PQ}) = \mathbf{V}(\overrightarrow{OQ}) - \mathbf{V}(\overrightarrow{OP})$$

$$\mathbf{V}(\overrightarrow{PQ}) = \langle b_1, b_2 \rangle - \langle a_1, a_2 \rangle$$

$$\mathbf{V}(\overrightarrow{PQ}) = \langle b_1 - a_1, b_2 - a_2 \rangle$$

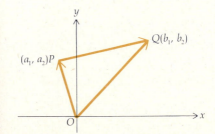

FIGURE 19

▶ **ILLUSTRATION 5**   If P is the point $(-6, 7)$ and Q is the point $(2, 9)$, then

$$\mathbf{V}(\overrightarrow{PQ}) = \langle 2 - (-6), 9 - 7 \rangle$$
$$= \langle 8, 2 \rangle \qquad \qquad ◀$$

Another operation with vectors is *scalar multiplication*. Following is the definition of the multiplication of a vector by a scalar (a real number).

14.1.7 DEFINITION   If c is a scalar and $\mathbf{A}$ is the vector $\langle a_1, a_2 \rangle$, then the product of c and $\mathbf{A}$, denoted by $c\mathbf{A}$, is a vector and is given by

$$c\mathbf{A} = c\langle a_1, a_2 \rangle$$
$$= \langle ca_1, ca_2 \rangle$$

▶ **ILLUSTRATION 6**   If $\mathbf{A} = \langle 4, -5 \rangle$, then

$$3\mathbf{A} = 3\langle 4, -5 \rangle$$
$$= \langle 12, -15 \rangle \qquad \qquad ◀$$

In Exercise 45 you are asked to show that if $\mathbf{A}$ is any vector and c is any scalar

$$0(\mathbf{A}) = \mathbf{0} \quad \text{and} \quad c(\mathbf{0}) = \mathbf{0}$$

The magnitude of the vector $c\mathbf{A}$ is computed as follows:

$$\begin{aligned}
\|c\mathbf{A}\| &= \sqrt{(ca_1)^2 + (ca_2)^2} \\
&= \sqrt{c^2(a_1{}^2 + a_2{}^2)} \\
&= \sqrt{c^2}\,\sqrt{a_1{}^2 + a_2{}^2} \\
&= |c|\,\|\mathbf{A}\|
\end{aligned}$$

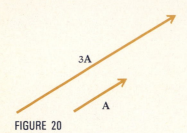

FIGURE 20

Therefore the magnitude of $c\mathbf{A}$ is the absolute value of c times the magnitude of $\mathbf{A}$.

The geometric interpretation of the vector $c\mathbf{A}$ is given in Figures 20 and 21. If $c > 0$, then $c\mathbf{A}$ is a vector whose representation has a length c times the magnitude of $\mathbf{A}$ and the same direction as $\mathbf{A}$; an example of this appears in Figure 20, where $c = 3$. If $c < 0$, then $c\mathbf{A}$ is a vector whose representation has a length that is $|c|$ times the magnitude of $\mathbf{A}$ and a direction opposite to that of $\mathbf{A}$. This is shown in Figure 21, where $c = -\frac{1}{2}$.

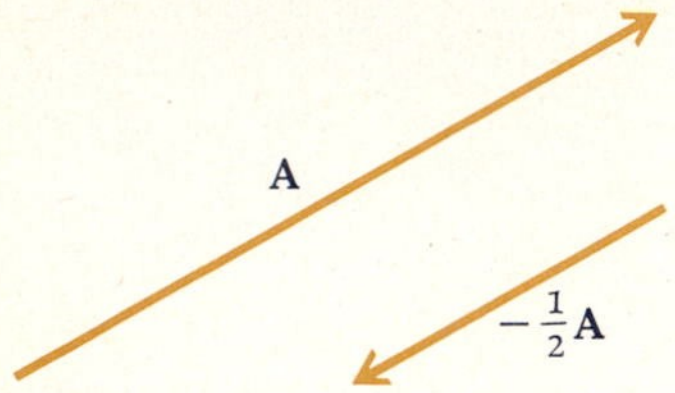

FIGURE 21

The following theorem gives laws satisfied by the operations of vector addition and scalar multiplication of any vectors in V_2.

14.1.8 THEOREM

If $\mathbf{A}$, $\mathbf{B}$, and $\mathbf{C}$ are any vectors in V_2, and c and d are any scalars, then vector addition and scalar multiplication satisfy the following properties:

(i) $\mathbf{A} + \mathbf{B} = \mathbf{B} + \mathbf{A}$ (commutative law)
(ii) $\mathbf{A} + (\mathbf{B} + \mathbf{C}) = (\mathbf{A} + \mathbf{B}) + \mathbf{C}$ (associative law)
(iii) There is a vector $\mathbf{0}$ in V_2 for which $\mathbf{A} + \mathbf{0} = \mathbf{A}$
      (existence of additive identity)
(iv) There is a vector $-\mathbf{A}$ in V_2 such that
      $\mathbf{A} + (-\mathbf{A}) = \mathbf{0}$ (existence of negative)
(v) $(cd)\mathbf{A} = c(d\mathbf{A})$ (associative law)
(vi) $c(\mathbf{A} + \mathbf{B}) = c\mathbf{A} + c\mathbf{B}$ (distributive law)
(vii) $(c + d)\mathbf{A} = c\mathbf{A} + d\mathbf{A}$ (distributive law)
(viii) $1(\mathbf{A}) = \mathbf{A}$ (existence of scalar multiplicative identity)

Proof   We give the proofs of (i) and (vi) and leave the others as exercises (see Exercises 46 through 50). In the proof of (i) we use the commutative law for real numbers and in the proof of (vi) we use the distributive law for real numbers. Let $\mathbf{A} = \langle a_1, a_2 \rangle$ and $\mathbf{B} = \langle b_1, b_2 \rangle$.

Proof of (i)

$$\begin{aligned}
\mathbf{A} + \mathbf{B} &= \langle a_1, a_2 \rangle + \langle b_1, b_2 \rangle \\
&= \langle a_1 + b_1, a_2 + b_2 \rangle \\
&= \langle b_1 + a_1, b_2 + a_2 \rangle \\
&= \langle b_1, b_2 \rangle + \langle a_1, a_2 \rangle \\
&= \mathbf{B} + \mathbf{A}
\end{aligned}$$

Proof of (vi)

$$\begin{aligned}
c(\mathbf{A} + \mathbf{B}) &= c(\langle a_1, a_2 \rangle + \langle b_1, b_2 \rangle) \\
&= c(\langle a_1 + b_1, a_2 + b_2 \rangle) \\
&= \langle c(a_1 + b_1), c(a_2 + b_2) \rangle \\
&= \langle ca_1 + cb_1, ca_2 + cb_2 \rangle \\
&= \langle ca_1, ca_2 \rangle + \langle cb_1, cb_2 \rangle \\
&= c\langle a_1, a_2 \rangle + c\langle b_1, b_2 \rangle \\
&= c\mathbf{A} + c\mathbf{B} \qquad \blacksquare
\end{aligned}$$

Theorem 14.1.8 is important because every algebraic law for the operations of vector addition and scalar multiplication of vectors in V_2 can be derived from the eight properties stated in the theorem. These laws are similar to the laws of arithmetic of real numbers. Furthermore, in linear algebra, a *real vector space* is defined as a set of vectors together with a set of real numbers (*scalars*) and the two operations of vector addition and scalar multiplication that satisfy the eight properties given in Theorem 14.1.8.

14.1.9 DEFINITION A **real vector space** V is a set of elements, called *vectors*, together with a set of real numbers, called *scalars*, with two operations called *vector addition* and *scalar multiplication* such that for every pair of vectors **A** and **B** in V and for every scalar c, a vector $\mathbf{A} + \mathbf{B}$ and a vector $c\mathbf{A}$ are defined so that properties (i)–(viii) of Theorem 14.1.8 are satisfied.

From Definition 14.1.9 and Theorem 14.1.8 it follows that V_2 is a real vector space.

We now take an arbitrary vector in V_2 and write it in a special form.

$$\langle a_1, a_2 \rangle = \langle a_1, 0 \rangle + \langle 0, a_2 \rangle$$

$$\langle a_1, a_2 \rangle = a_1 \langle 1, 0 \rangle + a_2 \langle 0, 1 \rangle \tag{3}$$

Because the magnitude of each of the two vectors $\langle 1, 0 \rangle$ and $\langle 0, 1 \rangle$ is one unit, they are called **unit vectors**. We introduce the following notations for these two unit vectors:

$$\mathbf{i} = \langle 1, 0 \rangle \qquad \mathbf{j} = \langle 0, 1 \rangle$$

With these notations, we have from (3)

$$\langle a_1, a_2 \rangle = a_1 \mathbf{i} + a_2 \mathbf{j} \tag{4}$$

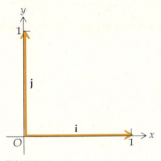

FIGURE 22

The position representation of each of the vectors **i** and **j** is shown in Figure 22. Equation (4) states that any vector in V_2 can be written as a linear combination of **i** and **j**. Because of this statement and the fact that **i** and **j** are independent (their position representations are not collinear), the vectors **i** and **j** are said to form a **basis** for the vector space V_2. See Exercise 52 for an example of a basis consisting of nonunit vectors. The number of elements in a basis of a vector space is called the **dimension** of the vector space. Therefore V_2 is a two-dimensional vector space.

▶ **ILLUSTRATION 7** From (4),

$$\langle 3, -4 \rangle = 3\mathbf{i} - 4\mathbf{j}$$ ◀

Let **A** be the vector $\langle a_1, a_2 \rangle$ and θ the direction angle of **A**. See Figure 23, where the point (a_1, a_2) is in the second quadrant and the position representation of **A** is shown. Because $\mathbf{A} = a_1\mathbf{i} + a_2\mathbf{j}$, $a_1 = \|\mathbf{A}\| \cos \theta$, and $a_2 = \|\mathbf{A}\| \sin \theta$, we can write

$$\mathbf{A} = \|\mathbf{A}\| \cos \theta \mathbf{i} + \|\mathbf{A}\| \sin \theta \mathbf{j}$$

$$\mathbf{A} = \|\mathbf{A}\|(\cos \theta \mathbf{i} + \sin \theta \mathbf{j}) \tag{5}$$

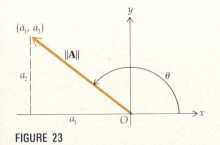

FIGURE 23

This equation expresses the vector **A** in terms of its magnitude, the cosine and sine of its direction angle, and the unit vectors **i** and **j**.

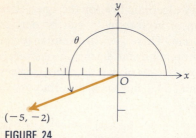

FIGURE 24

EXAMPLE 6   Express the vector $\langle -5, -2 \rangle$ in the form of (5).

Solution   Refer to Figure 24, which shows the position representation of the vector $\langle -5, -2 \rangle$.

$$\|\langle -5, -2 \rangle\| = \sqrt{(-5)^2 + (-2)^2} \qquad \cos \theta = -\frac{5}{\sqrt{29}} \qquad \sin \theta = -\frac{2}{\sqrt{29}}$$
$$= \sqrt{29}$$

Therefore from (5) we have

$$\langle -5, -2 \rangle = \sqrt{29} \left(-\frac{5}{\sqrt{29}} \mathbf{i} - \frac{2}{\sqrt{29}} \mathbf{j} \right)$$

14.1.10 THEOREM   If the nonzero vector $\mathbf{A} = a_1 \mathbf{i} + a_2 \mathbf{j}$, then the unit vector $\mathbf{U}$ having the same direction as $\mathbf{A}$ is given by

$$\mathbf{U} = \frac{a_1}{\|\mathbf{A}\|} \mathbf{i} + \frac{a_2}{\|\mathbf{A}\|} \mathbf{j}$$

Proof   We must show that the vector $\mathbf{U}$ is a unit vector having the same direction as $\mathbf{A}$.

$$\|\mathbf{U}\| = \sqrt{\left(\frac{a_1}{\|\mathbf{A}\|} \right)^2 + \left(\frac{a_2}{\|\mathbf{A}\|} \right)^2} \qquad \mathbf{U} = \frac{1}{\|\mathbf{A}\|} (a_1 \mathbf{i} + a_2 \mathbf{j})$$

$$= \frac{\sqrt{a_1^2 + a_2^2}}{\|\mathbf{A}\|} \qquad\qquad = \frac{1}{\|\mathbf{A}\|} (\mathbf{A})$$

$$= \frac{\|\mathbf{A}\|}{\|\mathbf{A}\|}$$

$$= 1$$

Because $\|\mathbf{U}\| = 1$, $\mathbf{U}$ is a unit vector, and because $\mathbf{U}$ is a positive scalar times the vector $\mathbf{A}$, the direction of $\mathbf{U}$ is the same as the direction of $\mathbf{A}$.   ∎

EXAMPLE 7   Given $\mathbf{A} = 3\mathbf{i} + \mathbf{j}$ and $\mathbf{B} = -2\mathbf{i} + 4\mathbf{j}$, find the unit vector having the same direction as $\mathbf{A} - \mathbf{B}$.

Solution

$$\mathbf{A} - \mathbf{B} = (3\mathbf{i} + \mathbf{j}) - (-2\mathbf{i} + 4\mathbf{j})$$
$$= 5\mathbf{i} - 3\mathbf{j}$$

Thus

$$\|\mathbf{A} - \mathbf{B}\| = \sqrt{5^2 + (-3)^2}$$
$$= \sqrt{34}$$

By Theorem 14.1.10, the desired unit vector is

$$\mathbf{U} = \frac{5}{\sqrt{34}} \mathbf{i} - \frac{3}{\sqrt{34}} \mathbf{j}$$

EXERCISES 14.1

In Exercises 1 through 4, (a) draw the position representation of the vector $\mathbf{A}$ and also the particular representation through the point P. (b) Find the magnitude of $\mathbf{A}$.

1. $\mathbf{A} = \langle 3, 4 \rangle$; $P = (2, 1)$
2. $\mathbf{A} = \langle -2, 5 \rangle$; $P = (-3, 4)$
3. $\mathbf{A} = \langle e, -\frac{1}{2} \rangle$; $P = (-2, -e)$
4. $\mathbf{A} = \langle 4, 0 \rangle$; $P = (2, 6)$

In Exercises 5 and 6, find the exact radian measure of the direction angle of the vector. In part (c) also approximate the radian measure to the nearest hundredth.

5. (a) $\langle 1, -1 \rangle$; (b) $\langle -3, 0 \rangle$; (c) $\langle 5, 2 \rangle$
6. (a) $\langle \sqrt{3}, 1 \rangle$; (b) $\langle 0, 4 \rangle$; (c) $\langle -3, 2 \rangle$

In Exercises 7 through 10, find the vector $\mathbf{A}$ having $\overrightarrow{PQ}$ as a representation. Draw $\overrightarrow{PQ}$ and the position representation of $\mathbf{A}$.

7. $P = (3, 7)$; $Q = (5, 4)$
8. $P = (5, 4)$; $Q = (3, 7)$
9. $P = (-5, -3)$; $Q = (0, 3)$
10. $P = (-\sqrt{2}, 0)$; $Q = (0, 0)$

In Exercises 11 through 14, find the point S so that $\overrightarrow{PQ}$ and $\overrightarrow{RS}$ are each representations of the same vector.

11. $P = (2, 5)$; $Q = (1, 6)$; $R = (-3, 2)$
12. $P = (-2, 0)$; $Q = (-3, -4)$; $R = (4, 2)$
13. $P = (0, 3)$; $Q = (5, -2)$; $R = (7, 0)$
14. $P = (-1, 4)$; $Q = (2, -3)$; $R = (-5, -2)$

In Exercises 15 and 16, find the sum of the pairs of vectors and illustrate geometrically.

15. (a) $\langle 2, 4 \rangle$, $\langle -3, 5 \rangle$; (b) $\langle -3, 0 \rangle$, $\langle 4, -5 \rangle$
16. (a) $\langle 0, 3 \rangle$, $\langle -2, 3 \rangle$; (b) $\langle 2, 3 \rangle$, $\langle -\sqrt{2}, -1 \rangle$

In Exercises 17 and 18, subtract the second vector from the first and illustrate geometrically.

17. (a) $\langle -3, -4 \rangle$, $\langle 6, 0 \rangle$; (b) $\langle 1, e \rangle$, $\langle -3, 2e \rangle$
18. (a) $\langle 0, 5 \rangle$, $\langle 2, 8 \rangle$; (b) $\langle 3, 7 \rangle$, $\langle 3, 7 \rangle$

In Exercises 19 and 20, find the vector or scalar if $\mathbf{A} = \langle 2, 4 \rangle$, $\mathbf{B} = \langle 4, -3 \rangle$, and $\mathbf{C} = \langle -3, 2 \rangle$.

19. (a) $\mathbf{A} + \mathbf{B}$; (b) $\|\mathbf{C} - \mathbf{B}\|$; (c) $\|7\mathbf{A} - \mathbf{B}\|$
20. (a) $\mathbf{A} - \mathbf{B}$; (b) $\|\mathbf{C}\|$; (c) $\|2\mathbf{A} + 3\mathbf{B}\|$

In Exercises 21 through 24, find the given vector or scalar if $\mathbf{A} = 2\mathbf{i} + 3\mathbf{j}$ and $\mathbf{B} = 4\mathbf{i} - \mathbf{j}$.

21. (a) $5\mathbf{A}$; (b) $-6\mathbf{B}$; (c) $\mathbf{A} + \mathbf{B}$; (d) $\|\mathbf{A} + \mathbf{B}\|$
22. (a) $-2\mathbf{A}$; (b) $3\mathbf{B}$; (c) $\mathbf{A} - \mathbf{B}$; (d) $\|\mathbf{A} - \mathbf{B}\|$
23. (a) $\|\mathbf{A}\| + \|\mathbf{B}\|$; (b) $5\mathbf{A} - 6\mathbf{B}$; (c) $\|5\mathbf{A} - 6\mathbf{B}\|$; (d) $\|5\mathbf{A}\| - \|6\mathbf{B}\|$
24. (a) $\|\mathbf{A}\| - \|\mathbf{B}\|$; (b) $3\mathbf{B} - 2\mathbf{A}$; (c) $\|3\mathbf{B} - 2\mathbf{A}\|$; (d) $\|3\mathbf{B}\| - \|2\mathbf{A}\|$

In Exercises 25 and 26, $\mathbf{A} = -4\mathbf{i} + 2\mathbf{j}$, $\mathbf{B} = -\mathbf{i} + 3\mathbf{j}$, and $\mathbf{C} = 5\mathbf{i} - \mathbf{j}$.

25. Find: (a) $5\mathbf{A} - 2\mathbf{B} - 2\mathbf{C}$ and (b) $\|5\mathbf{A} - 2\mathbf{B} - 2\mathbf{C}\|$.
26. Find: (a) $3\mathbf{B} - 2\mathbf{A} - \mathbf{C}$ and (b) $\|3\mathbf{B} - 2\mathbf{A} - \mathbf{C}\|$.

In Exercises 27 and 28, $\mathbf{A} = 8\mathbf{i} + 5\mathbf{j}$ and $\mathbf{B} = 3\mathbf{i} - \mathbf{j}$.

27. Find a unit vector having the same direction as $\mathbf{A} + \mathbf{B}$.
28. Find a unit vector having the same direction as $\mathbf{A} - \mathbf{B}$.

In Exercises 29 through 32, write the given vector in the form $r(\cos\theta\mathbf{i} + \sin\theta\mathbf{j})$, where r is the magnitude and θ is the direction angle. Also find a unit vector having the same direction.

29. (a) $3\mathbf{i} - 4\mathbf{j}$; (b) $2\mathbf{i} + 2\mathbf{j}$
30. (a) $8\mathbf{i} + 6\mathbf{j}$; (b) $2\sqrt{5}\mathbf{i} + 4\mathbf{j}$
31. (a) $-4\mathbf{i} + 4\sqrt{3}\mathbf{j}$; (b) $-16\mathbf{i}$
32. (a) $3\mathbf{i} - 3\mathbf{j}$; (b) $2\mathbf{j}$

33. If $\mathbf{A} = -2\mathbf{i} + \mathbf{j}$, $\mathbf{B} = 3\mathbf{i} - 2\mathbf{j}$, and $\mathbf{C} = 5\mathbf{i} - 4\mathbf{j}$, find scalars h and k such that $\mathbf{C} = h\mathbf{A} + k\mathbf{B}$.

34. If $\mathbf{A} = 5\mathbf{i} - 2\mathbf{j}$, $\mathbf{B} = -4\mathbf{i} + 3\mathbf{j}$, and $\mathbf{C} = -6\mathbf{i} + 8\mathbf{j}$, find scalars h and k such that $\mathbf{B} = h\mathbf{C} - k\mathbf{A}$.

35. If $\mathbf{A} = \mathbf{i} - 2\mathbf{j}$, $\mathbf{B} = -2\mathbf{i} + 4\mathbf{j}$, and $\mathbf{C} = 7\mathbf{i} - 5\mathbf{j}$, show that $\mathbf{C}$ cannot be written in the form $h\mathbf{A} + k\mathbf{B}$, where h and k are scalars.

36. Two forces of magnitudes 340 lb and 475 lb make an angle of 34.6° with each other and are applied to an object at the same point. Find (a) the magnitude of the resultant force and (b) to the nearest tenth of a degree the angle it makes with the force of 475 lb.

37. Two forces of magnitudes 60 lb and 80 lb make an angle of 30° with each other and are applied to an object at the same point. Find (a) the magnitude of the resultant force and (b) to the nearest degree the angle it makes with the force of 60 lb.

38. A force of magnitude 22 lb and one of magnitude 34 lb are applied to an object at the same point and make an angle of θ with each other. If the resultant force has a magnitude of 46 lb, find θ to the nearest degree.

39. A force of magnitude 112 lb and one of 84 lb are applied to an object at the same point, and the resultant force has a magnitude of 162 lb. Find to the nearest tenth of a degree the angle made by the resultant force with the force of 112 lb.

40. A plane has an air speed of 350 mi/hr. In order for the actual course of the plane to be due north, the compass heading is 340°. If the wind is blowing from the west, (a) what is the magnitude of its velocity? (b) What is the plane's ground speed?

41. In an airplane that has an air speed of 250 mi/hr, a pilot wishes to fly due north. If there is a wind blowing at 60 mi/hr toward the east, (a) what should be the plane's compass heading? (b) What will be the plane's ground speed if it flies this course?

42. A boat can travel 15 knots relative to the water. On a river whose current is 3 knots toward the west the boat has a compass heading of south. What is the speed of the boat relative to the land and what is its course?

43. A swimmer who can swim at a speed of 1.5 mi/hr relative to the water leaves the south bank of a river and is headed north directly across the river. If the river's current is toward the east at 0.8 mi/hr, (a) in what direction is the swimmer going? (b) What is the swimmer's speed relative to the land?

(c) If the distance across the river is 1 mile, how far down the river does the swimmer reach the north bank?

44. Suppose the swimmer in Exercise 43 wishes to reach the point directly north across the river. (a) In what direction should the swimmer head? (b) What will be the swimmer's speed relative to the land if this direction is taken?

45. Prove that if **A** is any vector and c is any scalar then $0(\mathbf{A}) = \mathbf{0}$ and $c(\mathbf{0}) = \mathbf{0}$.

46. Prove Theorem 14.1.8(ii).

47. Prove Theorem 14.1.8(iii) and (viii).

48. Prove Theorem 14.1.8(iv).

49. Prove Theorem 14.1.8(v).

50. Prove Theorem 14.1.8(vii).

51. Given $\mathbf{A} = \langle 2, -5 \rangle$; $\mathbf{B} = \langle 3, 1 \rangle$; $\mathbf{C} = \langle -4, 2 \rangle$. (a) Find $\mathbf{A} + (\mathbf{B} + \mathbf{C})$ and illustrate geometrically. (b) Find $(\mathbf{A} + \mathbf{B}) + \mathbf{C}$ and illustrate geometrically.

52. Two vectors are said to be *independent* if and only if their position representations are not collinear. Furthermore, two vectors **A** and **B** are said to form a *basis* for the vector space V_2 if and only if any vector in V_2 can be written as a linear combination of **A** and **B**. A theorem can be proved which states that two vectors form a basis for the vector space V_2 if they are independent. Show that this theorem holds for the two vectors $\langle 2, 5 \rangle$ and $\langle 3, -1 \rangle$ by doing the following: (a) Verify that the vectors are independent by showing that their position representations are not collinear; (b) verify that

the vectors form a basis by showing that any vector $a_1\mathbf{i} + a_2\mathbf{j}$ can be written as $c(2\mathbf{i} + 5\mathbf{j}) + d(3\mathbf{i} - \mathbf{j})$, where c and d are scalars. (*Hint:* Find c and d in terms of a_1 and a_2.)

53. Refer to the first two sentences of Exercise 52. A theorem can be proved which states that two vectors form a basis for the vector space V_2 only if they are independent. Show that this theorem holds for the two vectors $\langle 3, -2 \rangle$ and $\langle -6, 4 \rangle$ by doing the following: (a) Verify that the vectors are dependent (not independent) by showing that their position representations are collinear; (b) verify that the vectors do not form a basis by taking a particular vector and showing that it cannot be written as $c(3\mathbf{i} - 2\mathbf{j}) + d(-6\mathbf{i} + 4\mathbf{j})$, where c and d are scalars.

54. A set of vectors $\mathbf{V}_1, \mathbf{V}_2, \mathbf{V}_3, \ldots, \mathbf{V}_n$ is said to be *linearly dependent* if and only if there are scalars $k_1, k_2, k_3, \ldots, k_n$, not all zero, such that

$$k_1\mathbf{V}_1 + k_2\mathbf{V}_2 + k_3\mathbf{V}_3 + \ldots + k_n\mathbf{V}_n = \mathbf{0}$$

Show that if $\mathbf{V}_1 = 3\mathbf{i} - 2\mathbf{j}$, $\mathbf{V}_2 = \mathbf{i} + 4\mathbf{j}$, and $\mathbf{V}_3 = 2\mathbf{i} + 5\mathbf{j}$, then $\mathbf{V}_1, \mathbf{V}_2$, and $\mathbf{V}_3$ are linearly dependent.

55. Let $\overrightarrow{PQ}$ be a representation of vector **A**, $\overrightarrow{QR}$ be a representation of vector **B**, and $\overrightarrow{RS}$ be a representation of vector **C**. Prove that if $\overrightarrow{PQ}, \overrightarrow{QR}$, and $\overrightarrow{RS}$ are sides of a triangle, then $\mathbf{A} + \mathbf{B} + \mathbf{C} = \mathbf{0}$.

56. Prove analytically the triangle inequality for vectors

$$\|\mathbf{A} + \mathbf{B}\| \le \|\mathbf{A}\| + \|\mathbf{B}\|$$

14.2 DOT PRODUCT

In Section 14.1 we defined addition and subtraction of vectors and multiplication of a vector by a scalar. We now define a multiplication operation on two vectors that gives the *dot product*.

14.2.1 DEFINITION

If $\mathbf{A} = \langle a_1, a_2 \rangle$ and $\mathbf{B} = \langle b_1, b_2 \rangle$ are two vectors in V_2, then the **dot product** of **A** and **B**, denoted by $\mathbf{A} \cdot \mathbf{B}$, is given by

$$\mathbf{A} \cdot \mathbf{B} = \langle a_1, a_2 \rangle \cdot \langle b_1, b_2 \rangle$$
$$= a_1 b_1 + a_2 b_2$$

The dot product of two vectors is a real number (or scalar) and not a vector. It is sometimes called the **scalar product** or **inner product**.

▶ **ILLUSTRATION 1**   If $\mathbf{A} = \langle 2, -3 \rangle$ and $\mathbf{B} = \langle -\frac{1}{2}, 4 \rangle$, then

$$\mathbf{A} \cdot \mathbf{B} = \langle 2, -3 \rangle \cdot \langle -\tfrac{1}{2}, 4 \rangle$$
$$= (2)(-\tfrac{1}{2}) + (-3)(4)$$
$$= -13$$ ◀

The following dot products are useful and are easily verified (see Exercise 5).

$$\mathbf{i} \cdot \mathbf{i} = 1 \qquad \mathbf{j} \cdot \mathbf{j} = 1 \qquad \mathbf{i} \cdot \mathbf{j} = 0$$

The next theorem states that dot multiplication is commutative and distributive with respect to vector addition.

14.2.2 THEOREM   If **A**, **B**, and **C** are any vectors in V_2, then

(i) $\mathbf{A} \cdot \mathbf{B} = \mathbf{B} \cdot \mathbf{A}$ (commutative law)
(ii) $\mathbf{A} \cdot (\mathbf{B} + \mathbf{C}) = \mathbf{A} \cdot \mathbf{B} + \mathbf{A} \cdot \mathbf{C}$ (distributive law)

The proofs are left as exercises (see Exercises 6 and 7).
Because $\mathbf{A} \cdot \mathbf{B}$ is a scalar, the expression $(\mathbf{A} \cdot \mathbf{B}) \cdot \mathbf{C}$ is meaningless. Hence we do not consider associativity of dot multiplication.
Some other laws of dot multiplication are given in the following theorem.

14.2.3 THEOREM   If **A** and **B** are any vectors in V_2 and c is any scalar, then

(i) $c(\mathbf{A} \cdot \mathbf{B}) = (c\mathbf{A}) \cdot \mathbf{B}$
(ii) $\mathbf{0} \cdot \mathbf{A} = 0$
(iii) $\mathbf{A} \cdot \mathbf{A} = \|\mathbf{A}\|^2$

The proofs are left as exercises (see Exercises 8 through 10).
We now consider what is meant by the angle between two vectors, and this leads to another expression for the dot product of two vectors.

14.2.4 DEFINITION   Let **A** and **B** be two nonzero vectors such that **A** is not a scalar multiple of **B**. If $\overrightarrow{OP}$ is the position representation of **A** and $\overrightarrow{OQ}$ is the position representation of **B**, then the **angle between the vectors A and B** is defined to be the angle of positive measure between $\overrightarrow{OP}$ and $\overrightarrow{OQ}$ interior to the triangle determined by the points O, P, and Q. If $\mathbf{A} = c\mathbf{B}$, where c is a scalar, then if $c > 0$ the angle between the vectors has radian measure 0; if $c < 0$, the angle between the vectors has radian measure π.

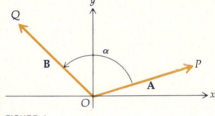

FIGURE 1

The symbol used to denote the angle between two vectors is also used to denote the measure of that angle. From Definition 14.2.4, if α is the radian measure of the angle between two vectors, then $0 \leq \alpha \leq \pi$. Figure 1 shows the angle between two vectors if **A** is not a scalar multiple of **B**.

14.2.5 THEOREM   If α is the angle between the two nonzero vectors **A** and **B**, then

$$\mathbf{A} \cdot \mathbf{B} = \|\mathbf{A}\| \, \|\mathbf{B}\| \cos \alpha$$

Proof   Let $\mathbf{A} = a_1 \mathbf{i} + a_2 \mathbf{j}$ and $\mathbf{B} = b_1 \mathbf{i} + b_2 \mathbf{j}$. Let $\overrightarrow{OP}$ be the position representation of **A** and $\overrightarrow{OQ}$ be the position representation of **B**. Then the angle between the vectors **A** and **B** is the angle at the origin in triangle POQ (see Figure 2); P is the point (a_1, a_2) and Q is the point (b_1, b_2). In triangle POQ, $\|\mathbf{A}\|$ is the length of OP and $\|\mathbf{B}\|$ is the length of OQ. So from the law of cosines,

$$\cos \alpha = \frac{\|\mathbf{A}\|^2 + \|\mathbf{B}\|^2 - |\overrightarrow{PQ}|^2}{2\|\mathbf{A}\| \, \|\mathbf{B}\|}$$

$$= \frac{(a_1^2 + a_2^2) + (b_1^2 + b_2^2) - [(a_1 - b_1)^2 + (a_2 - b_2)^2]}{2\|\mathbf{A}\| \, \|\mathbf{B}\|}$$

$$= \frac{2a_1 b_1 + 2a_2 b_2}{2\|\mathbf{A}\| \, \|\mathbf{B}\|}$$

$$= \frac{a_1 b_1 + a_2 b_2}{\|\mathbf{A}\| \, \|\mathbf{B}\|}$$

FIGURE 2

Hence

$$\cos \alpha = \frac{\mathbf{A} \cdot \mathbf{B}}{\|\mathbf{A}\| \, \|\mathbf{B}\|}$$

$$\mathbf{A} \cdot \mathbf{B} = \|\mathbf{A}\| \, \|\mathbf{B}\| \cos \alpha \qquad \blacksquare$$

Theorem 14.2.5 states that the dot product of two vectors is the product of the magnitudes of the vectors and the cosine of the angle between them.

▶ **ILLUSTRATION 2**   If $\mathbf{A} = 3\mathbf{i} - 2\mathbf{j}$, $\mathbf{B} = 2\mathbf{i} + \mathbf{j}$, and α is the angle between $\mathbf{A}$ and $\mathbf{B}$, then from Theorem 14.2.5,

$$\begin{aligned}
\cos \alpha &= \frac{\mathbf{A} \cdot \mathbf{B}}{\|\mathbf{A}\| \, \|\mathbf{B}\|} \\
&= \frac{(3)(2) + (-2)(1)}{\sqrt{9+4}\sqrt{4+1}} \\
&= \frac{6-2}{\sqrt{13}\sqrt{5}} \\
&= \frac{4}{\sqrt{65}}
\end{aligned}$$
◀

You learned in Section 14.1 that if two nonzero vectors are scalar multiples of each other, they have either the same or opposite directions. We have, then, the following definition.

14.2.6 DEFINITION   Two vectors are said to be **parallel** if and only if one of the vectors is a scalar multiple of the other.

▶ **ILLUSTRATION 3**   The vectors $\langle 3, -4 \rangle$ and $\langle \frac{3}{4}, -1 \rangle$ are parallel because $\langle 3, -4 \rangle = 4\langle \frac{3}{4}, -1 \rangle$. ◀

If $\mathbf{A}$ is any vector, $\mathbf{0} = 0\mathbf{A}$; thus, from Definition 14.2.6 it follows that the zero vector is parallel to any vector.

It is left as an exercise for you to show that two nonzero vectors are parallel if and only if the radian measure of the angle between them is 0 or π (see Exercise 37).

If $\mathbf{A}$ and $\mathbf{B}$ are nonzero vectors, then from Theorem 14.2.5 it follows that

$$\cos \alpha = 0 \quad \text{if and only if} \quad \mathbf{A} \cdot \mathbf{B} = 0$$

Because $0 \le \alpha \le \pi$, it follows from this statement that

$$\alpha = \tfrac{1}{2}\pi \quad \text{if and only if} \quad \mathbf{A} \cdot \mathbf{B} = 0$$

We have, then, the following definition.

14.2.7 DEFINITION   Two vectors $\mathbf{A}$ and $\mathbf{B}$ are said to be **orthogonal (perpendicular)** if and only if $\mathbf{A} \cdot \mathbf{B} = 0$.

▶ **ILLUSTRATION 4**   The vectors $\langle -4, 5 \rangle$ and $\langle 10, 8 \rangle$ are orthogonal because

$$\langle -4, 5 \rangle \cdot \langle 10, 8 \rangle = (-4)(10) + (5)(8)$$
$$= 0 \qquad \blacktriangleleft$$

If **A** is any vector, $\mathbf{0} \cdot \mathbf{A} = 0$, and therefore it follows from Definition 14.2.7 that the zero vector is orthogonal to any vector.

EXAMPLE 1   Given $\mathbf{A} = 3\mathbf{i} + 2\mathbf{j}$ and $\mathbf{B} = 2\mathbf{i} + k\mathbf{j}$, where k is a scalar, find (a) k such that **A** and **B** are orthogonal; (b) k such that **A** and **B** are parallel.

Solution
(a) By Definition 14.2.7, **A** and **B** are orthogonal if and only if $\mathbf{A} \cdot \mathbf{B} = 0$; that is,

$$(3)(2) + 2(k) = 0$$
$$k = -3$$

(b) From Definition 14.2.6, **A** and **B** are parallel if and only if there is some scalar c such that $\langle 3, 2 \rangle = c \langle 2, k \rangle$; that is,

$$3 = 2c \quad \text{and} \quad 2 = ck$$

Solving these two equations simultaneously we obtain $k = \frac{4}{3}$.

A geometric interpretation of the dot product is obtained by considering the *scalar projection* of a vector onto another vector. See Figure 3, where $\overrightarrow{OP}$ and $\overrightarrow{OQ}$ are the position representations of vectors **A** and **B**, respectively. Point R is the foot of the perpendicular from Q to the line containing $\overrightarrow{OP}$. The scalar projection of **B** onto **A** is the magnitude of the vector having $\overrightarrow{OR}$ as its position representation.

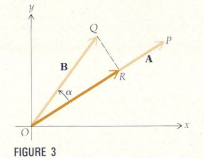

FIGURE 3

14.2.8 DEFINITION   If **A** and **B** are nonzero vectors, the **scalar projection** of **B** onto **A** is defined to be $\|\mathbf{B}\| \cos \alpha$, where α is the angle between **A** and **B**.

Observe that the scalar projection may be either positive or negative, depending on the sign of $\cos \alpha$.

From Theorem 14.2.5,

$$\mathbf{A} \cdot \mathbf{B} = \|\mathbf{A}\|(\|\mathbf{B}\| \cos \alpha) \qquad (1)$$

Thus the dot product of **A** and **B** is the magnitude of **A** multiplied by the scalar projection of **B** onto **A**. See Figure 4(a) and (b). Because dot multiplication is commutative, $\mathbf{A} \cdot \mathbf{B}$ is also the magnitude of **B** multiplied by the scalar projection of **A** onto **B**.

If $\mathbf{B} = b_1 \mathbf{i} + b_2 \mathbf{j}$, then

$$\mathbf{i} \cdot \mathbf{B} = b_1 \quad \text{and} \quad \mathbf{j} \cdot \mathbf{B} = b_2$$

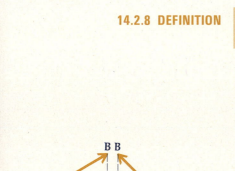

$$\|\mathbf{B}\| \cos \alpha > 0 \qquad \|\mathbf{B}\| \cos \alpha < 0$$

(a)   (b)

FIGURE 4

Hence the dot product of **i** and **B** gives the component of **B** in the direction of **i** and the dot product of **j** and **B** gives the component of **B** in the direction of **j**. To generalize this result, let **U** be any unit vector. Then from (1), if α is the angle between **U** and **B**,

$$\mathbf{U} \cdot \mathbf{B} = \|\mathbf{U}\| \|\mathbf{B}\| \cos \alpha$$
$$= \|\mathbf{B}\| \cos \alpha$$

Therefore $\mathbf{U} \cdot \mathbf{B}$ is the scalar projection of $\mathbf{B}$ onto $\mathbf{U}$, which is called the *component* of the vector $\mathbf{B}$ in the direction of $\mathbf{U}$. More generally, the **component** of a vector $\mathbf{B}$ in the direction of a vector $\mathbf{A}$ is the scalar projection of $\mathbf{B}$ onto a unit vector in the direction of $\mathbf{A}$.

The following theorem can be used to compute the scalar projection of one vector onto another.

14.2.9 THEOREM

The scalar projection of the vector $\mathbf{B}$ onto the vector $\mathbf{A}$ is

$$\frac{\mathbf{A} \cdot \mathbf{B}}{\|\mathbf{A}\|}$$

Proof   From Definition 14.2.8, the scalar projection of $\mathbf{B}$ onto $\mathbf{A}$ is $\|\mathbf{B}\| \cos \alpha$, where α is the angle between $\mathbf{A}$ and $\mathbf{B}$. From Theorem 14.2.5,

$$\|\mathbf{A}\| \|\mathbf{B}\| \cos \alpha = \mathbf{A} \cdot \mathbf{B}$$

$$\|\mathbf{B}\| \cos \alpha = \frac{\mathbf{A} \cdot \mathbf{B}}{\|\mathbf{A}\|} \qquad \blacksquare$$

Refer again to Figure 3. If $\mathbf{C}$ is the vector having $\overrightarrow{OR}$ as its position representation, then $\mathbf{C}$ is called the **vector projection** of $\mathbf{B}$ onto $\mathbf{A}$. To determine $\mathbf{C}$, we multiply $\|\mathbf{B}\| \cos \alpha$ by the unit vector having the same direction as $\mathbf{A}$. Thus

$$\mathbf{C} = (\|\mathbf{B}\| \cos \alpha) \frac{\mathbf{A}}{\|\mathbf{A}\|}$$

$$= \frac{\|\mathbf{A}\| (\|\mathbf{B}\| \cos \alpha)}{\|\mathbf{A}\|^2} \mathbf{A}$$

$$= \left(\frac{\mathbf{A} \cdot \mathbf{B}}{\|\mathbf{A}\|^2} \right) \mathbf{A} \qquad \text{(from Theorem 14.2.5)}$$

We state this result as a theorem.

14.2.10 THEOREM

The vector projection of the vector $\mathbf{B}$ onto the vector $\mathbf{A}$ is

$$\left(\frac{\mathbf{A} \cdot \mathbf{B}}{\|\mathbf{A}\|^2} \right) \mathbf{A}$$

EXAMPLE 2   Given the vectors

$$\mathbf{A} = -5\mathbf{i} + \mathbf{j} \qquad \mathbf{B} = 4\mathbf{i} + 2\mathbf{j}$$

Find: (a) the scalar projection of $\mathbf{B}$ onto $\mathbf{A}$; (b) the vector projection of $\mathbf{B}$ onto $\mathbf{A}$. (c) Show on a figure the position representations of $\mathbf{A}$, $\mathbf{B}$, and the vector projection of $\mathbf{B}$ onto $\mathbf{A}$.

Solution   We first compute $\mathbf{A} \cdot \mathbf{B}$ and $\|\mathbf{A}\|$.

$$\mathbf{A} \cdot \mathbf{B} = \langle -5, 1 \rangle \cdot \langle 4, 2 \rangle \qquad \|\mathbf{A}\| = \sqrt{(-5)^2 + 1^2}$$
$$= -20 + 2 \qquad\qquad\qquad = \sqrt{26}$$
$$= -18$$

(a) From Theorem 14.2.9, the scalar projection of **B** onto **A** is

$$\frac{\mathbf{A} \cdot \mathbf{B}}{\|\mathbf{A}\|} = -\frac{18}{\sqrt{26}}$$

(b) From Theorem 14.2.10, the vector projection of **B** onto **A** is

$$\begin{aligned}
\left(\frac{\mathbf{A} \cdot \mathbf{B}}{\|\mathbf{A}\|^2}\right)\mathbf{A} &= -\tfrac{18}{26}(-5\mathbf{i} + \mathbf{j}) \\
&= -\tfrac{9}{13}(-5\mathbf{i} + \mathbf{j}) \\
&= \tfrac{45}{13}\mathbf{i} - \tfrac{9}{13}\mathbf{j}
\end{aligned}$$

(c) Figure 5 shows the position representations of **A**, **B**, and **C**, where **C** is the vector projection of **B** onto **A**.

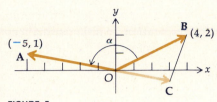

FIGURE 5

In Section 6.6 we stated that if a constant force of F pounds moves an object a distance d feet along a straight line and the force is acting in the direction of motion, then if W is the number of foot-pounds in the work done by the force, $W = Fd$. Suppose, however, that the constant force is not directed along the line of motion. In this case the physicist defines the **work** done as the *product of the component of the force along the line of motion times the displacement*. If the object moves from the point A to the point B, we call the vector, having $\overrightarrow{AB}$ as a representation, the **displacement vector** and denote it by $\mathbf{V}(\overrightarrow{AB})$. So if the magnitude of a constant force vector **F** is expressed in pounds and the distance from A to B is expressed in feet, and α is the radian measure of the angle between the vectors **F** and $\mathbf{V}(\overrightarrow{AB})$, then if W is the number of foot-pounds in the work done by the force **F** in moving an object from A to B,

$$\begin{aligned}
W &= (\|\mathbf{F}\| \cos \alpha)\|\mathbf{V}(\overrightarrow{AB})\| \\
&= \|\mathbf{F}\|\,\|\mathbf{V}(\overrightarrow{AB})\| \cos \alpha \\
&= \mathbf{F} \cdot \mathbf{V}(\overrightarrow{AB})
\end{aligned}$$

EXAMPLE 3 Suppose that a force **F** has a magnitude of 6 lb and $\tfrac{1}{6}\pi$ is the radian measure of the angle giving its direction. Find the work done by **F** in moving an object along a straight line from the origin to the point $P(7, 1)$, where distance is measured in feet.

Solution Figure 6 shows the position representations of **F** and $\mathbf{V}(\overrightarrow{OP})$. Because $\mathbf{F} = \langle 6 \cos \tfrac{1}{6}\pi, 6 \sin \tfrac{1}{6}\pi \rangle$, and $\mathbf{V}(\overrightarrow{OP}) = \langle 7, 1 \rangle$, then if W ft-lb is the work done, we have

$$\begin{aligned}
\mathbf{W} &= \mathbf{F} \cdot \mathbf{V}(\overrightarrow{OP}) \\
&= \langle 6 \cos \tfrac{1}{6}\pi, 6 \sin \tfrac{1}{6}\pi \rangle \cdot \langle 7, 1 \rangle \\
&= \langle 3\sqrt{3}, 3 \rangle \cdot \langle 7, 1 \rangle \\
&= 21\sqrt{3} + 3 \\
&\approx 39.37
\end{aligned}$$

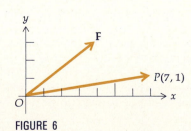

FIGURE 6

Therefore the work done is approximately 39.37 ft-lb.

Vectors have geometric representations that are independent of the coordinate system used. Because of this, vector analysis can be used to prove certain theorems of plane geometry. This is illustrated in the following example.

EXAMPLE 4   Prove by vector analysis that the altitudes of a triangle meet in a point.

Solution   Let ABC be a triangle having altitudes AP and BQ intersecting at point S. Draw a line through C and S intersecting AB at point R. We wish to prove that RC is perpendicular to AB (see Figure 7).

Let $\overrightarrow{AB}$, $\overrightarrow{BC}$, $\overrightarrow{AC}$, $\overrightarrow{AS}$, $\overrightarrow{BS}$, $\overrightarrow{CS}$ be representations of vectors. Let $\mathbf{V}(\overrightarrow{AB})$ be the vector having directed line segment $\overrightarrow{AB}$ as a representation. In a similar manner let $\mathbf{V}(\overrightarrow{BC})$, $\mathbf{V}(\overrightarrow{AC})$, $\mathbf{V}(\overrightarrow{AS})$, $\mathbf{V}(\overrightarrow{BS})$, and $\mathbf{V}(\overrightarrow{CS})$ be the vectors having the directed line segment in parentheses as a representation.

Because AP is an altitude of the triangle,

$$\mathbf{V}(\overrightarrow{AS}) \cdot \mathbf{V}(\overrightarrow{BC}) = 0 \qquad (2)$$

Also, because BQ is an altitude of the triangle,

$$\mathbf{V}(\overrightarrow{BS}) \cdot \mathbf{V}(\overrightarrow{AC}) = 0 \qquad (3)$$

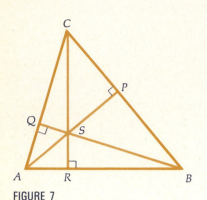

FIGURE 7

To prove that RC is perpendicular to AB we shall show $\mathbf{V}(\overrightarrow{CS}) \cdot \mathbf{V}(\overrightarrow{AB}) = 0$.

$$\begin{aligned}
\mathbf{V}(\overrightarrow{CS}) \cdot \mathbf{V}(\overrightarrow{AB}) &= \mathbf{V}(\overrightarrow{CS}) \cdot [\mathbf{V}(\overrightarrow{AC}) + \mathbf{V}(\overrightarrow{CB})] \\
&= \mathbf{V}(\overrightarrow{CS}) \cdot \mathbf{V}(\overrightarrow{AC}) + \mathbf{V}(\overrightarrow{CS}) \cdot \mathbf{V}(\overrightarrow{CB}) \\
&= [\mathbf{V}(\overrightarrow{CB}) + \mathbf{V}(\overrightarrow{BS})] \cdot \mathbf{V}(\overrightarrow{AC}) + [\mathbf{V}(\overrightarrow{CA}) + \mathbf{V}(\overrightarrow{AS})] \cdot \mathbf{V}(\overrightarrow{CB}) \\
&= \mathbf{V}(\overrightarrow{CB}) \cdot \mathbf{V}(\overrightarrow{AC}) + \mathbf{V}(\overrightarrow{BS}) \cdot \mathbf{V}(\overrightarrow{AC}) + \mathbf{V}(\overrightarrow{CA}) \cdot \mathbf{V}(\overrightarrow{CB}) + \mathbf{V}(\overrightarrow{AS}) \cdot \mathbf{V}(\overrightarrow{CB})
\end{aligned}$$

Replacing $\mathbf{V}(\overrightarrow{CA})$ by $-\mathbf{V}(\overrightarrow{AC})$ and using (2) and (3) we obtain

$$\begin{aligned}
\mathbf{V}(\overrightarrow{CS}) \cdot \mathbf{V}(\overrightarrow{AB}) &= \mathbf{V}(\overrightarrow{CB}) \cdot \mathbf{V}(\overrightarrow{AC}) + 0 + [-\mathbf{V}(\overrightarrow{AC})] \cdot \mathbf{V}(\overrightarrow{CB}) + 0 \\
&= 0
\end{aligned}$$

Therefore altitudes AP, BQ, and RC meet in a point.

EXERCISES 14.2

In Exercises 1 through 4, find $\mathbf{A} \cdot \mathbf{B}$.

1. $\mathbf{A} = \langle -1, 2 \rangle$; $\mathbf{B} = \langle -4, 3 \rangle$

2. $\mathbf{A} = \langle \frac{1}{3}, -\frac{1}{2} \rangle$; $\mathbf{B} = \langle \frac{5}{2}, \frac{4}{3} \rangle$ **3.** $\mathbf{A} = 2\mathbf{i} - \mathbf{j}$; $\mathbf{B} = \mathbf{i} + 3\mathbf{j}$

4. $\mathbf{A} = -2\mathbf{i}$; $\mathbf{B} = -\mathbf{i} + \mathbf{j}$

5. Show that $\mathbf{i} \cdot \mathbf{i} = 1$; $\mathbf{j} \cdot \mathbf{j} = 1$; $\mathbf{i} \cdot \mathbf{j} = 0$.

6. Prove Theorem 14.2.2(i).   **7.** Prove Theorem 14.2.2(ii).

8. Prove Theorem 14.2.3(i).   **9.** Prove Theorem 14.2.3(ii).

10. Prove Theorem 14.2.3(iii).

In Exercises 11 through 14, if α *is the angle between* $\mathbf{A}$ *and* $\mathbf{B}$, *find* $\cos \alpha$.

11. $\mathbf{A} = \langle 4, 3 \rangle$; $\mathbf{B} = \langle 1, -1 \rangle$

12. $\mathbf{A} = \langle -2, -3 \rangle$; $\mathbf{B} = \langle 3, 2 \rangle$

13. $\mathbf{A} = 5\mathbf{i} - 12\mathbf{j}$; $\mathbf{B} = 4\mathbf{i} + 3\mathbf{j}$

14. $\mathbf{A} = 2\mathbf{i} + 4\mathbf{j}$; $\mathbf{B} = -5\mathbf{j}$

15. Find k such that the radian measure of the angle between the vectors in Example 1 is $\frac{1}{4}\pi$.

16. Given $\mathbf{A} = k\mathbf{i} - 2\mathbf{j}$ and $\mathbf{B} = k\mathbf{i} + 6\mathbf{j}$, where k is a scalar. Find k such that $\mathbf{A}$ and $\mathbf{B}$ are orthogonal.

17. Given $\mathbf{A} = 5\mathbf{i} - k\mathbf{j}$ and $\mathbf{B} = k\mathbf{i} + 6\mathbf{j}$, where k is a scalar. Find (a) k such that $\mathbf{A}$ and $\mathbf{B}$ are orthogonal; (b) k such that $\mathbf{A}$ and $\mathbf{B}$ are parallel.

18. Find k such that the vectors in Exercise 16 have opposite directions.

19. Given $\mathbf{A} = 5\mathbf{i} + 12\mathbf{j}$; $\mathbf{B} = \mathbf{i} + k\mathbf{j}$, where k is a scalar. Find k such that the radian measure of the angle between $\mathbf{A}$ and $\mathbf{B}$ is $\frac{1}{3}\pi$.

20. Find two unit vectors each having a representation whose initial point is $(2, 4)$ and which is tangent to the parabola $y = x^2$ there.

21. Find two unit vectors each having a representation whose initial point is $(2, 4)$ and which is normal to the parabola $y = x^2$ there.

22. If $\mathbf{A} = 2\mathbf{i} - 7\mathbf{j}$, find the unit vectors that are orthogonal to $\mathbf{A}$.

23. If $\mathbf{A}$ is the vector $a_1\mathbf{i} + a_2\mathbf{j}$, find the unit vectors that are orthogonal to $\mathbf{A}$.

24. If $\mathbf{A} = 5\mathbf{i} - 9c\mathbf{j}$ and $\mathbf{B} = 7\mathbf{i} - 4c\mathbf{j}$, show that there is no real value for c such that $\mathbf{A}$ and $\mathbf{B}$ are orthogonal.

25. If $\mathbf{A} = -8\mathbf{i} + 4\mathbf{j}$ and $\mathbf{B} = 7\mathbf{i} - 6\mathbf{j}$, find (a) the scalar projection of $\mathbf{A}$ onto $\mathbf{B}$ and (b) the vector projection of $\mathbf{A}$ onto $\mathbf{B}$.

26. For the vectors of Exercise 25, find (a) the scalar projection of $\mathbf{B}$ onto $\mathbf{A}$ and (b) the vector projection of $\mathbf{B}$ onto $\mathbf{A}$.

27. Find the component of the vector $\mathbf{A} = 5\mathbf{i} - 6\mathbf{j}$ in the direction of the vector $\mathbf{B} = 7\mathbf{i} + \mathbf{j}$.

28. For the vectors $\mathbf{A}$ and $\mathbf{B}$ of Exercise 27, find the component of the vector $\mathbf{B}$ in the direction of vector $\mathbf{A}$.

29. A vector $\mathbf{F}$ represents a force that has a magnitude of 8 lb and $\frac{1}{3}\pi$ is the radian measure of its direction angle. Find the work done by the force in moving an object (a) along the x axis from the origin to the point $(6, 0)$ and (b) along the y axis from the origin to the point $(0, 6)$. Distance is measured in feet.

30. A vector $\mathbf{F}$ represents a force that has a magnitude of 10 lb and $\frac{1}{4}\pi$ is the radian measure of its direction angle. Find the work done by the force in moving an object along the y axis from the point $(0, -2)$ to the point $(0, 5)$. Distance is measured in feet.

31. A vector $\mathbf{F}$ represents a force that has a magnitude of 9 lb and $\frac{2}{3}\pi$ is the radian measure of its direction angle. Find the work done by the force in moving an object from the origin to the point $(-4, -2)$. Distance is measured in feet.

32. Two forces represented by the vectors $\mathbf{F}_1$ and $\mathbf{F}_2$ act on a particle and cause it to move along a line from the point $(2, 5)$ to the point $(7, 3)$. If $\mathbf{F}_1 = 3\mathbf{i} - \mathbf{j}$ and $\mathbf{F}_2 = -4\mathbf{i} + 5\mathbf{j}$, the magnitudes of the forces are measured in pounds, and distance is measured in feet, find the work done by the two forces acting together.

33. If $\mathbf{A}$ and $\mathbf{B}$ are vectors, prove that

$$(\mathbf{A} + \mathbf{B}) \cdot (\mathbf{A} + \mathbf{B}) = \mathbf{A} \cdot \mathbf{A} + 2\mathbf{A} \cdot \mathbf{B} + \mathbf{B} \cdot \mathbf{B}$$

34. Prove by vector analysis that the medians of a triangle meet in a point.

35. Prove by vector analysis that the line segment joining the midpoints of two sides of a triangle is parallel to the third side and its length is one-half the length of the third side.

36. Prove by vector analysis that the line segment joining the midpoints of the nonparallel sides of a trapezoid is parallel to the parallel sides and its length is one-half the sum of the lengths of the parallel sides.

37. Prove that two nonzero vectors are parallel if and only if the radian measure of the angle between them is 0 or π.

38. If $\mathbf{A}$ and $\mathbf{B}$ are vectors, prove that $\|\mathbf{A} \cdot \mathbf{B}\| \leq \|\mathbf{A}\| \|\mathbf{B}\|$, where equality pertains if and only if there exists a scalar c such that $\mathbf{A} = c\mathbf{B}$.

14.3 VECTOR-VALUED FUNCTIONS AND PARAMETRIC EQUATIONS

Suppose a particle moves so that the coordinates (x, y) of its position at any time t are given by the equations $x = f(t)$ and $y = g(t)$. Then for every number t in the domain common to f and g there is a vector $f(t)\mathbf{i} + g(t)\mathbf{j}$, and the endpoints of the position representations of these vectors trace a curve C traveled by the particle. This leads us to consider a function whose domain is a set of real numbers and whose range is a set of vectors. Such a function is called a *vector-valued function*.

14.3.1 DEFINITION

Let f and g be two real-valued functions of a real variable t. Then for every number t in the domain common to f and g there is a vector $\mathbf{R}$ defined by

$$\mathbf{R}(t) = f(t)\mathbf{i} + g(t)\mathbf{j}$$

and $\mathbf{R}$ is called a **vector-valued function**.

▶ **ILLUSTRATION 1** Let $\mathbf{R}$ be the vector-valued function defined by

$$\mathbf{R}(t) = \sqrt{t - 2}\,\mathbf{i} + (t - 3)^{-1}\mathbf{j}$$

If $f(t) = \sqrt{t - 2}$ and $g(t) = (t - 3)^{-1}$, the domain of $\mathbf{R}$ is the set of values of t for which both $f(t)$ and $g(t)$ are defined. Because $f(t)$ is defined for $t \geq 2$ and $g(t)$ is defined for all real numbers except 3, the domain of $\mathbf{R}$ is $\{t \,|\, t \geq 2, t \neq 3\}$. ◀

The equation $\mathbf{R}(t) = f(t)\mathbf{i} + g(t)\mathbf{j}$ is called a **vector equation**, and it defines a curve C. The same curve C is also defined by the equations

$$x = f(t) \quad \text{and} \quad y = g(t) \tag{1}$$

which are called **parametric equations** of C. The variable t is a parameter. The curve C is also called a **graph**; that is, the set of all points (x, y) satisfying (1) is the graph of the vector-valued function **R**.

A vector equation of a curve, as well as parametric equations of a curve, gives the curve a direction at each point. That is, if we think of the curve as being traced by a particle, we can consider the positive direction along a curve as the direction in which the particle moves as the parameter t increases. In such a case as this, t may be taken to be the measure of the time, and the vector $\mathbf{R}(t)$ is called the **position vector**. Sometimes $\mathbf{R}(t)$ is referred to as the **radius vector**.

If the parameter t is eliminated from the pair of Equations (1), we obtain one equation in x and y, which is called a **cartesian equation** of C. It may happen that elimination of the parameter leads to a cartesian equation whose graph contains more points than the graph defined by either the vector equation or the parametric equations. This situation occurs in Example 2.

EXAMPLE 1   Given the vector equation

$$\mathbf{R}(t) = 2 \cos t\mathbf{i} + 2 \sin t\mathbf{j}$$

(a) Draw a sketch of the graph of this equation, and (b) find a cartesian equation of the graph.

Solution

(a) The domain of **R** is the set of all real numbers. The values of x and y for particular values of t can be tabulated. We find the magnitude of the position vector. For every t,

$$\|\mathbf{R}(t)\| = \sqrt{4 \cos^2 t + 4 \sin^2 t}$$
$$= 2\sqrt{\cos^2 t + \sin^2 t}$$
$$= 2$$

Therefore the endpoint of the position representation of each vector $\mathbf{R}(t)$ is two units from the origin. By letting t take on all numbers in the closed interval $[0, 2\pi]$, we obtain a circle having its center at the origin and radius 2. This is the entire graph because any value of t will give a point on this circle. A sketch of the circle is shown in Figure 1. Parametric equations of the graph are

$$x = 2 \cos t \quad \text{and} \quad y = 2 \sin t$$

(b) We can find a cartesian equation of the graph by eliminating t from the two parametric equations, which when squaring on both sides of each equation and adding gives

$$x^2 + y^2 = 4$$

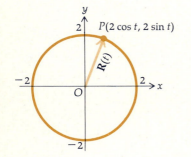

FIGURE 1

EXAMPLE 2   Given the parametric equations

$$x = \cosh t \quad \text{and} \quad y = \sinh t$$

(a) Draw a sketch of the graph defined by these equations, and (b) find a cartesian equation of the graph.

Solution

(a) Squaring on both sides of the given equations and subtracting we have

$$x^2 - y^2 = \cosh^2 t - \sinh^2 t$$

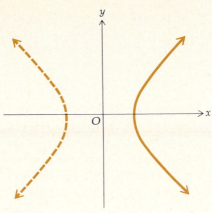

FIGURE 2

From the identity $\cosh^2 t - \sinh^2 t = 1$, this equation becomes

$$x^2 - y^2 = 1$$

This is an equation of an equilateral hyperbola. However, observe that for t any real number, $\cosh t$ is never less than 1. Thus the curve defined by the given parametric equations consists of only the points on the right branch of the hyperbola. A sketch of this curve appears in Figure 2. The "dashed" curve in the figure is the left branch of the equilateral hyperbola.

(b) A cartesian equation is

$$x^2 - y^2 = 1 \qquad x \geq 1$$

We have seen that by eliminating t from parametric equations (1) we obtain a cartesian equation. The cartesian equation either implicitly or explicitly defines y as one or more functions of x. That is, if $x = f(t)$ and $y = g(t)$, then $y = h(x)$. If h is a differentiable function of x and f is a differentiable function of t, then from the chain rule,

$$\frac{dy}{dt} = \frac{dy}{dx} \cdot \frac{dx}{dt}$$

If $\dfrac{dx}{dt} \neq 0$, we can divide on both sides of the above equation by $\dfrac{dx}{dt}$ and obtain

$$\frac{dy}{dx} = \frac{\dfrac{dy}{dt}}{\dfrac{dx}{dt}} \tag{2}$$

Equation (2) enables us to find the derivative of y with respect to x directly from the parametric equations.

Because $\dfrac{d^2y}{dx^2} = \dfrac{d}{dx}\left(\dfrac{dy}{dx}\right)$, then $\dfrac{d^2y}{dx^2} = \dfrac{d(y')}{dx}$. Thus

$$\frac{d^2y}{dx^2} = \frac{\dfrac{d(y')}{dt}}{\dfrac{dx}{dt}} \tag{3}$$

EXAMPLE 3   Given the parametric equations

$$x = 3t^2 \quad \text{and} \quad y = 4t^3$$

find $\dfrac{dy}{dx}$ and $\dfrac{d^2y}{dx^2}$ without eliminating t.

Solution   Because $\dfrac{dy}{dt} = 12t^2$ and $\dfrac{dx}{dt} = 6t$, we have, from (2),

$$\frac{dy}{dx} = \frac{12t^2}{6t}$$

$$= 2t$$

Since $y' = 2t$, $\dfrac{d(y')}{dt} = 2$. Then from (3),

$$\frac{d^2y}{dx^2} = \frac{\dfrac{d(y')}{dt}}{\dfrac{dx}{dt}}$$

$$= \frac{2}{6t}$$

$$= \frac{1}{3t}$$

Table 1

t	x	y
0	0	0
$\frac{1}{2}$	$\frac{3}{4}$	$\frac{1}{2}$
1	3	4
2	12	32
$-\frac{1}{2}$	$\frac{3}{4}$	$-\frac{1}{2}$
-1	3	-4
-2	12	-32

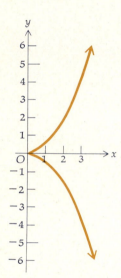

FIGURE 3

EXAMPLE 4   (a) Draw a sketch of the graph defined by the parametric equations of Example 3, and (b) find a cartesian equation of the graph.

Solution   The parametric equations are

$$x = 3t^2 \quad \text{and} \quad y = 4t^3$$

(a) We observe that x is nonnegative. Thus the graph is restricted to the first and fourth quadrants. Table 1 gives values of x and y for particular values of t. Because $\dfrac{dy}{dx} = 2t$, then when $t = 0$, $\dfrac{dy}{dx} = 0$. Hence at the point $(0, 0)$ the tangent line is horizontal. With this fact and points obtained from Table 1, we obtain the sketch of the graph appearing in Figure 3.

(b) From the two parametric equations, we get $x^3 = 27t^6$ and $y^2 = 16t^6$. Solving each of these equations for t^6 and eliminating t^6 we have

$$\frac{x^3}{27} = \frac{y^2}{16}$$

$$16x^3 = 27y^2 \tag{4}$$

which is the cartesian equation desired.

▶ **ILLUSTRATION 2**   If in (4) we differentiate implicitly,

$$48x^2 = 54y\frac{dy}{dx}$$

$$\frac{dy}{dx} = \frac{8x^2}{9y}$$

Substituting for x and y in terms of t from the given parametric equations we obtain

$$\frac{dy}{dx} = \frac{8(3t^2)^2}{9(4t^3)}$$

$$= 2t$$

which agrees with the value of $\dfrac{dy}{dx}$ found in Example 3.   ◀

It follows from (2) that if at a particular point, $\dfrac{dy}{dt} = 0$ and $\dfrac{dx}{dt} \neq 0$, then $\dfrac{dy}{dx} = 0$, and the graph of the pair of parametric equations has a horizontal tangent line at the point. Furthermore, if at a particular point, $\dfrac{dx}{dt} = 0$ and $\dfrac{dy}{dt} \neq 0$, then $\dfrac{dy}{dx}$ does not exist at the point, and the graph may have a vertical tangent line there. These tangent lines can be helpful when drawing a sketch of the graph of the parametric equations.

Table 2

t	x	y
-4	-12	0
-3	-5	-3
-2	0	-4
-1	3	-3
0	4	0
1	3	5
2	0	12
3	-5	21

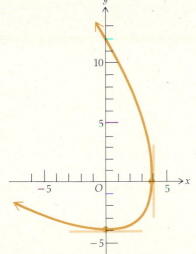

FIGURE 4

EXAMPLE 5   Given the parametric equations

$$x = 4 - t^2 \quad \text{and} \quad y = t^2 + 4t$$

Find the horizontal and vertical tangent lines of the graph of this pair of equations and draw a sketch of the graph.

Solution

$$\frac{dx}{dt} = -2t \qquad \frac{dy}{dt} = 2t + 4$$

Therefore $\dfrac{dy}{dx} = \dfrac{2t + 4}{-2t}$; that is,

$$\frac{dy}{dx} = \frac{t + 2}{-t}$$

When $t = -2$, $\dfrac{dy}{dx} = 0$, $x = 0$, and $y = -4$. Thus the graph has a horizontal tangent line at $(0, -4)$. When $t = 0$, $\dfrac{dy}{dx}$ does not exist, $x = 4$, and $y = 0$; the graph has a vertical tangent line at $(4, 0)$. Table 2 gives values of x and y for particular values of t. With points obtained from these values and knowing the horizontal and vertical tangent lines, we have the sketch of the graph shown in Figure 4.

We now show how parametric equations can be used to define a curve that is described by a physical motion. The curve we consider is a **cycloid**, which is the curve traced by a point on the circumference of a circle as the circle rolls along a straight line. Suppose the circle has radius a. Let the fixed straight line on which the circle rolls be the x axis, and let the origin be one of the points at which the given point P comes in contact with the x axis. See Figure 5, showing the circle after it has rolled through an angle of t radians. From Figure 5,

$$\mathbf{V}(\overrightarrow{OT}) + \mathbf{V}(\overrightarrow{TA}) + \mathbf{V}(\overrightarrow{AP}) = \mathbf{V}(\overrightarrow{OP}) \tag{5}$$

The length of the arc PT is $\|\mathbf{V}(\overrightarrow{OT})\| = at$. Because the direction of $\mathbf{V}(\overrightarrow{OT})$ is along the positive x axis,

$$\mathbf{V}(\overrightarrow{OT}) = at\mathbf{i} \tag{6}$$

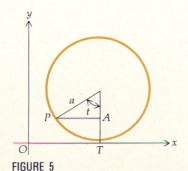

FIGURE 5

Also, $\|\mathbf{V}(\overrightarrow{TA})\| = a - a\cos t$. And because the direction of $\mathbf{V}(\overrightarrow{TA})$ is the same as the direction of $\mathbf{j}$,

$$\mathbf{V}(\overrightarrow{TA}) = a(1 - \cos t)\mathbf{j} \tag{7}$$

$\|\mathbf{V}(\overrightarrow{AP})\| = a\sin t$, and the direction of $\mathbf{V}(\overrightarrow{AP})$ is the same as the direction of $-\mathbf{i}$; thus

$$\mathbf{V}(\overrightarrow{AP}) = -a\sin t\mathbf{i}$$

Substituting from this equation, (6), and (7) into (5) we obtain

$$at\mathbf{i} + a(1 - \cos t)\mathbf{j} - a\sin t\mathbf{i} = \mathbf{V}(\overrightarrow{OP})$$

$$\Leftrightarrow \quad \mathbf{V}(\overrightarrow{OP}) = a(t - \sin t)\mathbf{i} + a(1 - \cos t)\mathbf{j}$$

This is a vector equation of the cycloid. So parametric equations of the cycloid are

$$x = a(t - \sin t) \quad \text{and} \quad y = a(1 - \cos t) \tag{8}$$

where t is any real number. A sketch of a portion of the cycloid is shown in Figure 6.

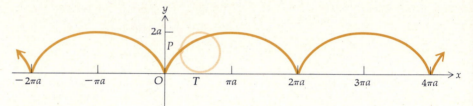

FIGURE 6

In Section 11.1 where we stated and proved Cauchy's mean-value theorem (11.1.3), we indicated that a geometric interpretation would be given in this section because parametric equations are needed. Recall that the theorem states that if f and g are two functions such that (i) f and g are continuous on $[a, b]$, (ii) f and g are differentiable on (a, b), and (iii) for all x in (a, b) $g'(x) \neq 0$, then there exists a number z in the open interval (a, b) such that

$$\frac{f(b) - f(a)}{g(b) - g(a)} = \frac{f'(z)}{g'(z)}$$

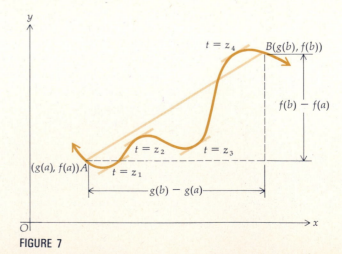

FIGURE 7

Figure 7 shows a curve having the parametric equations $x = g(t)$ and $y = f(t)$, where $a \leq t \leq b$. The slope of the curve in the figure at a particular point is given by

$$\frac{dy}{dx} = \frac{f'(t)}{g'(t)}$$

and the slope of the line segment through the points $A(g(a), f(a))$ and $B(g(b), f(b))$ is given by

$$\frac{f(b) - f(a)}{g(b) - g(a)}$$

Cauchy's mean-value theorem states that the slopes are equal for at least one value of t between a and b. For the curve shown in Figure 7 there are four values of t satisfying the conclusion of the theorem: $t = z_1$, $t = z_2$, $t = z_3$, and $t = z_4$.

EXERCISES 14.3

In Exercises 1 through 6, find the domain of the vector-valued function.

1. $\mathbf{R}(t) = (1/t)\mathbf{i} + \sqrt{4 - t}\,\mathbf{j}$ **2.** $\mathbf{R}(t) = (t^2 + 3)\mathbf{i} + (t - 1)\mathbf{j}$

3. $\mathbf{R}(t) = (\sin^{-1} t)\mathbf{i} + (\cos^{-1} t)\mathbf{j}$

4. $\mathbf{R}(t) = \ln(t + 1)\mathbf{i} + (\tan^{-1} t)\mathbf{j}$

5. $\mathbf{R}(t) = \sqrt{t^2 - 9}\,\mathbf{i} + \sqrt{t^2 + 2t - 8}\,\mathbf{j}$

6. $\mathbf{R}(t) = \sqrt{t - 4}\,\mathbf{i} + \sqrt{4 - t}\,\mathbf{j}$

In Exercises 7 through 12, find $\dfrac{dy}{dx}$ and $\dfrac{d^2y}{dx^2}$ without eliminating the parameter.

7. $x = 3t$, $y = 2t^2$ **8.** $x = 1 - t^2$, $y = 1 + t$

9. $x = t^2 e^t$, $y = t \ln t$ **10.** $x = e^{2t}$, $y = 1 + \cos t$

11. $x = a \cos t$, $y = b \sin t$ **12.** $x = a \cosh t$, $y = b \sinh t$

In Exercises 13 through 19, draw a sketch of the graph of the vector equation, and find a cartesian equation of the graph.

13. $\mathbf{R}(t) = t^2\mathbf{i} + (t + 1)\mathbf{j}$ **14.** $\mathbf{R}(t) = (t - 2)\mathbf{i} + (t^2 + 4)\mathbf{j}$

15. $\mathbf{R}(t) = 3 \cosh t\mathbf{i} + 5 \sinh t\mathbf{j}$ **16.** $\mathbf{R}(t) = \dfrac{4}{t^2}\mathbf{i} + \dfrac{4}{t}\mathbf{j}$

17. $\mathbf{R}(t) = \sec t\mathbf{i} + \tan t\mathbf{j}$; t in $(-\tfrac{1}{2}\pi, \tfrac{1}{2}\pi)$

18. $\mathbf{R}(t) = \cos t\mathbf{i} + \cos t\mathbf{j}$; t in $[0, \tfrac{1}{2}\pi]$

19. $\mathbf{R}(t) = 4 \cos t\mathbf{i} + 3 \sin t\mathbf{j}$; t in $[0, 2\pi]$

20. Find an equation of the tangent line to the curve $x = 1 + 3 \sin t$, $y = 2 - 5 \cos t$, at the point where $t = \tfrac{1}{6}\pi$.

21. Find an equation of the tangent line to the curve $x = 2 \sin t$, $y = 5 \cos t$, at the point where $t = \tfrac{1}{3}\pi$.

In Exercises 22 through 24, find the horizontal and vertical tangent lines of the graph of the pair of parametric equations, and draw a sketch of the graph.

22. $x = t^2 + t$, $y = t^2 - t$ **23.** $x = 4t^2 - 4t$, $y = 1 - 4t^2$

24. $x = \dfrac{3at}{1 + t^3}$, $y = \dfrac{3at^2}{1 + t^3}$

25. Parametric equations for the *trochoid* are

$$x = at - b \sin t \quad \text{and} \quad y = a - b \cos t$$

Show that the trochoid has no vertical tangent line if $a > b > 0$.

26. A projectile moves so that the coordinates of its position at any time t are given by the parametric equations $x = 60t$ and $y = 80t - 16t^2$. Draw a sketch of the path of the projectile.

27. Find $\dfrac{dy}{dx}$, $\dfrac{d^2y}{dx^2}$, and $\dfrac{d^3y}{dx^3}$ at the point on the cycloid having Equations (8) for which y has its largest value when x is in the closed interval $[0, 2\pi a]$.

28. Show that the slope of the tangent line at $t = t_1$ to the cycloid having Equations (8) is $\cot \tfrac{1}{2}t_1$. Deduce, then, that the tangent line is vertical when $t = 2n\pi$, where n is any integer.

29. A *hypocycloid* is the curve traced by a point P on a circle of radius b that is rolling inside a fixed circle of radius a, $a > b$. If the origin is at the center of the fixed circle, $A(a, 0)$ is one of the points at which the point P comes in contact with the fixed circle, B is the moving point of tangency of the two circles, and the parameter t is the number of radians in the angle AOB, prove that parametric equations of the hypocycloid are

$$x = (a - b) \cos t + b \cos \frac{a - b}{b} t$$

and

$$y = (a - b) \sin t - b \sin \frac{a - b}{b} t$$

30. If $a = 4b$ in Exercise 29, we have a *hypocycloid of four cusps*. Show that parametric equations of this curve are $x = a \cos^3 t$ and $y = a \sin^3 t$.

31. Use the parametric equations of Exercise 30 to find a cartesian equation of the hypocycloid of four cusps, and draw a sketch of the graph of the resulting equation.

32. Parametric equations for the *tractrix* are

$$x = t - a \tanh \frac{t}{a} \quad \text{and} \quad y = a \operatorname{sech} \frac{t}{a}$$

Draw a sketch of the curve for $a = 4$.

33. Prove that the parameter t in the parametric equations of a tractrix (see Exercise 32) is the x intercept of the tangent line.

34. Show that the tractrix of Exercise 32 is a curve such that the length of the segment of every tangent line from the point of tangency to the point of intersection with the x axis is constant and equal to a.

35. Find the area of the region bounded by the x axis and one arch of the cycloid having Equations (8).

36. Find the centroid of the region of Exercise 35.

14.4 CALCULUS OF VECTOR-VALUED FUNCTIONS

The definitions of limits, continuity, derivatives, and indefinite integrals of vector-valued functions involve the corresponding definitions for real-valued functions.

14.4.1 DEFINITION

Let $\mathbf{R}$ be a vector-valued function whose function values are given by

$$\mathbf{R}(t) = f(t)\mathbf{i} + g(t)\mathbf{j}$$

Then the **limit of $\mathbf{R}(t)$ as t approaches t_1** is defined by

$$\lim_{t \to t_1} \mathbf{R}(t) = \left[\lim_{t \to t_1} f(t) \right]\mathbf{i} + \left[\lim_{t \to t_1} g(t) \right]\mathbf{j}$$

if $\lim_{t \to t_1} f(t)$ and $\lim_{t \to t_1} g(t)$ both exist.

▶ **ILLUSTRATION 1**   If $\mathbf{R}(t) = \cos t\mathbf{i} + 2e^t\mathbf{j}$, then

$$\lim_{t \to 0} \mathbf{R}(t) = (\lim_{t \to 0} \cos t)\mathbf{i} + (\lim_{t \to 0} 2e^t)\mathbf{j}$$

$$= \mathbf{i} + 2\mathbf{j} \qquad \blacktriangleleft$$

14.4.2 DEFINITION

The vector-valued function $\mathbf{R}$ is **continuous** at t_1 if and only if the following three conditions are satisfied:

(i) $\mathbf{R}(t_1)$ exists;

(ii) $\lim_{t \to t_1} \mathbf{R}(t)$ exists;

(iii) $\lim_{t \to t_1} \mathbf{R}(t) = \mathbf{R}(t_1)$.

From Definitions 14.4.1 and 14.4.2 it follows that the vector-valued function $\mathbf{R}$, defined by $\mathbf{R}(t) = f(t)\mathbf{i} + g(t)\mathbf{j}$, is continuous at t_1 if and only if f and g are continuous there.

In the following definition the expression

$$\frac{\mathbf{R}(t + \Delta t) - \mathbf{R}(t)}{\Delta t}$$

is used to indicate the division of a vector by a scalar. This expression means

$$\frac{1}{\Delta t}\left[\mathbf{R}(t + \Delta t) - \mathbf{R}(t)\right]$$

14.4.3 DEFINITION

If $\mathbf{R}$ is a vector-valued function, then the **derivative** of $\mathbf{R}$ is a vector-valued function, denoted by $\mathbf{R}'$ and defined by

$$\mathbf{R}'(t) = \lim_{\Delta t \to 0} \frac{\mathbf{R}(t + \Delta t) - \mathbf{R}(t)}{\Delta t}$$

if this limit exists.

The notation $D_t \mathbf{R}(t)$ is sometimes used in place of $\mathbf{R}'(t)$.

The following theorem follows from Definition 14.4.3 and the definition of the derivative of a real-valued function.

14.4.4 THEOREM   If $\mathbf{R}$ is a vector-valued function defined by

$$\mathbf{R}(t) = f(t)\mathbf{i} + g(t)\mathbf{j}$$

then

$$\mathbf{R}'(t) = f'(t)\mathbf{i} + g'(t)\mathbf{j}$$

if $f'(t)$ and $g'(t)$ exist.

Proof   From Definition 14.4.3,

$$\mathbf{R}'(t) = \lim_{\Delta t \to 0} \frac{\mathbf{R}(t + \Delta t) - \mathbf{R}(t)}{\Delta t}$$

$$= \lim_{\Delta t \to 0} \frac{[f(t + \Delta t)\mathbf{i} + g(t + \Delta t)\mathbf{j}] - [f(t)\mathbf{i} + g(t)\mathbf{j}]}{\Delta t}$$

$$= \lim_{\Delta t \to 0} \frac{[f(t + \Delta t) - f(t)]}{\Delta t}\mathbf{i} + \lim_{\Delta t \to 0} \frac{[g(t + \Delta t) - g(t)]}{\Delta t}\mathbf{j}$$

$$= f'(t)\mathbf{i} + g'(t)\mathbf{j} \qquad\qquad \blacksquare$$

The direction of $\mathbf{R}'(t)$ is along the tangent line at the point $(f(t), g(t))$ to the graph of $\mathbf{R}(t) = f(t)\mathbf{i} + g(t)\mathbf{j}$. That is, the direction of $\mathbf{R}'(t)$ is given by θ $(0 \le \theta < 2\pi)$, where $\tan \theta = g'(t)/f'(t)$; that is,

$$\tan \theta = \frac{\dfrac{dy}{dt}}{\dfrac{dx}{dt}}$$

$$= \frac{dy}{dx}$$

A geometric interpretation of Definition 14.4.3 is obtained by considering representations of the vectors $\mathbf{R}(t)$, $\mathbf{R}(t + \Delta t)$, and $\mathbf{R}'(t)$. Refer to Figure 1. The curve C is traced by the endpoint of the position representation of $\mathbf{R}(t)$ as t assumes all values in the domain of $\mathbf{R}$. Let $\overrightarrow{OP}$ be the position representation of $\mathbf{R}(t)$ and $\overrightarrow{OQ}$ be the position representation of $\mathbf{R}(t + \Delta T)$. Then $\mathbf{R}(t + \Delta t) - \mathbf{R}(t)$ is a vector for which $\overrightarrow{PQ}$ is a representation. If the vector $\mathbf{R}(t + \Delta t) - \mathbf{R}(t)$ is multiplied by the scalar $1/\Delta t$, we obtain a vector having the same direction and whose magnitude is $1/|\Delta t|$ times the magnitude of $\mathbf{R}(t + \Delta t) - \mathbf{R}(t)$. As Δt approaches zero, the vector $[\mathbf{R}(t + \Delta t) - \mathbf{R}(t)]/\Delta t$ approaches a vector having one of its representations tangent to the curve C at the point P.

▶ **ILLUSTRATION 2**   If $\mathbf{R}(t) = (2 + \sin t)\mathbf{i} + \cos t\mathbf{j}$, then

$$\mathbf{R}'(t) = \cos t\mathbf{i} - \sin t\mathbf{j} \qquad\qquad ◀$$

Higher order derivatives of vector-valued functions are defined as for higher order derivatives of real-valued functions. So if $\mathbf{R}$ is a vector-valued function defined by $\mathbf{R}(t) = f(t)\mathbf{i} + g(t)\mathbf{j}$, the second derivative of $\mathbf{R}$, denoted by $\mathbf{R}''(t)$, is

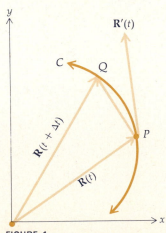

FIGURE 1

given by

$$\mathbf{R}''(t) = D_t\big[\mathbf{R}'(t)\big]$$

The notation $D_t{}^2\mathbf{R}(t)$ can be used in place of $\mathbf{R}''(t)$. By applying Theorem 14.4.4 to $\mathbf{R}'(t)$,

$$\mathbf{R}''(t) = f''(t)\mathbf{i} + g''(t)\mathbf{j}$$

if $f''(t)$ and $g''(t)$ exist.

▶ **ILLUSTRATION 3**   If $\mathbf{R}(t) = (\ln t)\mathbf{i} + \left(\dfrac{1}{t}\right)\mathbf{j}$, then

$$\mathbf{R}'(t) = \frac{1}{t}\mathbf{i} - \frac{1}{t^2}\mathbf{j} \qquad \mathbf{R}''(t) = -\frac{1}{t^2}\mathbf{i} + \frac{2}{t^3}\mathbf{j} \qquad ◀$$

14.4.5 DEFINITION   A vector-valued function $\mathbf{R}$ is said to be **differentiable** on an interval if $\mathbf{R}'(t)$ exists for all values of t in the interval.

The following theorems give differentiation formulas for vector-valued functions. The proofs are based on Theorem 14.4.4 and theorems on differentiation of real-valued functions.

14.4.6 THEOREM   If $\mathbf{R}$ and $\mathbf{Q}$ are differentiable vector-valued functions on an interval, then $\mathbf{R} + \mathbf{Q}$ is differentiable on the interval, and

$$D_t\big[\mathbf{R}(t) + \mathbf{Q}(t)\big] = D_t\mathbf{R}(t) + D_t\mathbf{Q}(t)$$

The proof of this theorem is left as an exercise (see Exercise 25).

EXAMPLE 1   Verify Theorem 14.4.6 if

$$\mathbf{R}(t) = t^2\mathbf{i} + (t - 1)\mathbf{j} \quad \text{and} \quad \mathbf{Q}(t) = \sin t\mathbf{i} + \cos t\mathbf{j}$$

Solution

$$
\begin{aligned}
D_t\big[\mathbf{R}(t) + \mathbf{Q}(t)\big] &= D_t\big([t^2\mathbf{i} + (t - 1)\mathbf{j}] + [\sin t\mathbf{i} + \cos t\mathbf{j}]\big) \\
&= D_t\big[(t^2 + \sin t)\mathbf{i} + (t - 1 + \cos t)\mathbf{j}\big] \\
&= (2t + \cos t)\mathbf{i} + (1 - \sin t)\mathbf{j}
\end{aligned}
$$

$$
\begin{aligned}
D_t\mathbf{R}(t) + D_t\mathbf{Q}(t) &= D_t\big[t^2\mathbf{i} + (t - 1)\mathbf{j}\big] + D_t(\sin t\mathbf{i} + \cos t\mathbf{j}) \\
&= (2t\mathbf{i} + \mathbf{j}) + (\cos t\mathbf{i} - \sin t\mathbf{j}) \\
&= (2t + \cos t)\mathbf{i} + (1 - \sin t)\mathbf{j}
\end{aligned}
$$

Hence $D_t\big[\mathbf{R}(t) + \mathbf{Q}(t)\big] = D_t\mathbf{R}(t) + D_t\mathbf{Q}(t)$.

14.4.7 THEOREM   If $\mathbf{R}$ and $\mathbf{Q}$ are differentiable vector-valued functions on an interval, then $\mathbf{R} \cdot \mathbf{Q}$ is differentiable on the interval, and

$$D_t\big[\mathbf{R}(t) \cdot \mathbf{Q}(t)\big] = \big[D_t\mathbf{R}(t)\big] \cdot \mathbf{Q}(t) + \mathbf{R}(t) \cdot \big[D_t\mathbf{Q}(t)\big]$$

Proof   Let $\mathbf{R}(t) = f_1(t)\mathbf{i} + g_1(t)\mathbf{j}$ and $\mathbf{Q}(t) = f_2(t)\mathbf{i} + g_2(t)\mathbf{j}$. Then by Theorem 14.4.4,

$$D_t\mathbf{R}(t) = f_1'(t)\mathbf{i} + g_1'(t)\mathbf{j} \qquad D_t\mathbf{Q}(t) = f_2'(t)\mathbf{i} + g_2'(t)\mathbf{j}$$

$$\mathbf{R}(t) \cdot \mathbf{Q}(t) = \big[f_1(t)\big]\big[f_2(t)\big] + \big[g_1(t)\big]\big[g_2(t)\big]$$

Therefore

$$D_t[\mathbf{R}(t) \cdot \mathbf{Q}(t)]$$
$$= [f_1{}'(t)][f_2(t)] + [f_1(t)][f_2{}'(t)] + [g_1{}'(t)][g_2(t)] + [g_1(t)][g_2{}'(t)]$$
$$= \{[f_1{}'(t)][f_2(t)] + [g_1{}'(t)][g_2(t)]\} + \{[f_1(t)][f_2{}'(t)] + [g_1(t)][g_2{}'(t)]\}$$
$$[D_t\mathbf{R}(t)] \cdot \mathbf{Q}(t) + \mathbf{R}(t) \cdot [D_t\mathbf{Q}(t)]$$
∎

EXAMPLE 2 Verify Theorem 14.4.7 for the vectors of Example 1.

Solution The vectors are

$$\mathbf{R}(t) = t^2\mathbf{i} + (t - 1)\mathbf{j} \qquad \mathbf{Q}(t) = \sin t\mathbf{i} + \cos t\mathbf{j}$$

Thus $\mathbf{R}(t) \cdot \mathbf{Q}(t) = t^2 \sin t + (t - 1) \cos t$. Therefore

$$D_t[\mathbf{R}(t) \cdot \mathbf{Q}(t)] = 2t \sin t + t^2 \cos t + \cos t + (t - 1)(- \sin t)$$
$$= (t + 1) \sin t + (t^2 + 1) \cos t \tag{1}$$

Because $D_t\mathbf{R}(t) = 2t\mathbf{i} + \mathbf{j}$ and $D_t\mathbf{Q}(t) = \cos t\mathbf{i} - \sin t\mathbf{j}$, we have

$$[D_t\mathbf{R}(t)] \cdot \mathbf{Q}(t) + \mathbf{R}(t) \cdot [D_t\mathbf{Q}(t)]$$
$$= (2t\mathbf{i} + \mathbf{j}) \cdot (\sin t\mathbf{i} + \cos t\mathbf{j}) + [t^2\mathbf{i} + (t - 1)\mathbf{j}] \cdot (\cos t\mathbf{i} - \sin t\mathbf{j})$$
$$= (2t \sin t + \cos t) + [t^2 \cos t - (t - 1) \sin t]$$
$$= (t + 1) \sin t + (t^2 + 1) \cos t \tag{2}$$

Comparing (1) and (2), we see that Theorem 14.4.7 holds for these vectors.

14.4.8 THEOREM If $\mathbf{R}$ is a differentiable vector-valued function on an interval and f is a differentiable real-valued function on the interval, then

$$D_t\{[f(t)][\mathbf{R}(t)]\} = [D_tf(t)]\mathbf{R}(t) + f(t) D_t\mathbf{R}(t)$$

The proof is left as an exercise (see Exercise 26).

We need to apply the following theorem in some future discussions. It is the chain-rule for vector-valued functions. The proof, which is left as an exercise (see Exercise 27), is based on Theorem 14.4.4 and the chain rule for real-valued functions.

14.4.9 THEOREM Suppose that $\mathbf{F}$ is a vector-valued function, h is a real-valued function, and $\mathbf{G}$ is the vector-valued function defined by $\mathbf{G}(t) = \mathbf{F}(h(t))$. If $\phi = h(t)$ and both $\dfrac{d\phi}{dt}$ and $D_\phi\mathbf{G}(t)$ exist, then $D_t\mathbf{G}(t)$ exists and is given by

$$D_t\mathbf{G}(t) = [D_\phi\mathbf{G}(t)]\frac{d\phi}{dt}$$

▶ **ILLUSTRATION 4** Let the functions $\mathbf{F}$ and h of Theorem 14.4.9 be defined by

$$\mathbf{F}(\phi) = \phi^2\mathbf{i} + e^\phi\mathbf{j} \quad \text{and} \quad h(t) = \sin t$$

If $\phi = h(t)$ and $\mathbf{G}(t) = \mathbf{F}(h(t))$, we have

$$\phi = \sin t \quad \text{and} \quad \mathbf{G}(t) = \sin^2 t\mathbf{i} + e^{\sin t}\mathbf{j}$$

Computing $D_t\mathbf{G}(t)$ by Theorem 14.4.4 we have

$$D_t\mathbf{G}(t) = 2 \sin t \cos t\mathbf{i} + e^{\sin t} \cos t\mathbf{j} \tag{3}$$

We show that we get the same result if we apply Theorem 14.4.9. Because $\mathbf{G}(t)$ can also be written as $\phi^2\mathbf{i} + e^\phi\mathbf{j}$, we have

$$\begin{aligned} D_\phi\mathbf{G}(t) &= D_\phi[\phi^2\mathbf{i} + e^\phi\mathbf{j}] \\ &= 2\phi\mathbf{i} + e^\phi\mathbf{j} \end{aligned}$$

But $\phi = \sin t$; thus

$$D_\phi\mathbf{G}(t) = 2 \sin t\mathbf{i} + e^{\sin t}\mathbf{j} \quad \text{and} \quad \frac{d\phi}{dt} = \cos t$$

Substituting these values into the right side of the formula of Theorem 14.4.9 we have

$$\begin{aligned} D_t\mathbf{G}(t) &= [2 \sin t\mathbf{i} + e^{\sin t}\mathbf{j}] \cos t \\ &= 2 \sin t \cos t\mathbf{i} + e^{\sin t} \cos t\mathbf{j} \end{aligned}$$

which agrees with (3).   ◀

We now define an indefinite integral (or antiderivative) of a vector-valued function.

14.4.10 DEFINITION

> If $\mathbf{Q}$ is the vector-valued function given by
>
> $$\mathbf{Q}(t) = f(t)\mathbf{i} + g(t)\mathbf{j}$$
>
> then the **indefinite integral of $\mathbf{Q}(t)$** is defined by
>
> $$\int \mathbf{Q}(t)\, dt = \mathbf{i} \int f(t)\, dt + \mathbf{j} \int g(t)\, dt \tag{4}$$

This definition is consistent with the definition of an indefinite integral of a real-valued function because if we take the derivative on both sides of (4) with respect to t,

$$D_t \int \mathbf{Q}(t)\, dt = \mathbf{i}D_t \int f(t)\, dt + \mathbf{j}D_t \int g(t)\, dt$$

$$D_t \int \mathbf{Q}(t)\, dt = \mathbf{i}f(t) + \mathbf{j}g(t)$$

For each of the indefinite integrals on the right side of (4) there occurs an arbitrary scalar constant. When each of these scalars is multiplied by either $\mathbf{i}$ or $\mathbf{j}$, there occurs an arbitrary constant vector in the sum. So

$$\int \mathbf{Q}(t)\, dt = \mathbf{R}(t) + \mathbf{C}$$

where $D_t\mathbf{R}(t) = \mathbf{Q}(t)$ and $\mathbf{C}$ is an arbitrary constant vector.

EXAMPLE 3   Find the most general vector-valued function whose derivative is

$$\mathbf{Q}(t) = \sin t\mathbf{i} - 3 \cos t\mathbf{j}$$

Solution   If $D_t\mathbf{R}(t) = \mathbf{Q}(t)$, then $\mathbf{R}(t) = \int \mathbf{Q}(t)\, dt$; that is,

$$\mathbf{R}(t) = \mathbf{i} \int \sin t\, dt - 3\mathbf{j} \int \cos t\, dt$$

$$= \mathbf{i}(-\cos t + C_1) - 3\mathbf{j}(\sin t + C_2)$$

$$= -\cos t\mathbf{i} - 3\sin t\mathbf{j} + (C_1\mathbf{i} - 3C_2\mathbf{j})$$

$$= -\cos t\mathbf{i} - 3\sin t\mathbf{j} + \mathbf{C}$$

EXAMPLE 4   Find the vector $\mathbf{R}(t)$ for which

$$D_t\mathbf{R}(t) = e^{-t}\mathbf{i} + e^t\mathbf{j} \quad \text{and} \quad \mathbf{R}(0) = \mathbf{i} + \mathbf{j}$$

Solution

$$\mathbf{R}(t) = \mathbf{i} \int e^{-t}\, dt + \mathbf{j} \int e^t\, dt$$

$$= \mathbf{i}(-e^{-t} + C_1) + \mathbf{j}(e^t + C_2)$$

Because $\mathbf{R}(0) = \mathbf{i} + \mathbf{j}$,

$$\mathbf{i} + \mathbf{j} = \mathbf{i}(-1 + C_1) + \mathbf{j}(1 + C_2)$$

Therefore

$$C_1 - 1 = 1 \qquad C_2 + 1 = 1$$

$$C_1 = 2 \qquad\qquad C_2 = 0$$

Hence

$$\mathbf{R}(t) = (-e^{-t} + 2)\mathbf{i} + e^t\mathbf{j}$$

The following theorem will be useful later.

14.4.11 THEOREM   If $\mathbf{R}$ is a differentiable vector-valued function on an interval and $\|\mathbf{R}(t)\|$ is constant for all t in the interval, then the vectors $\mathbf{R}(t)$ and $D_t\mathbf{R}(t)$ are orthogonal.

Proof   Let $\|\mathbf{R}(t)\| = k$. Then by Theorem 14.2.3(iii),

$$\mathbf{R}(t) \cdot \mathbf{R}(t) = k^2$$

Differentiating on both sides with respect to t and using Theorem 14.4.7 we obtain

$$[D_t\mathbf{R}(t)] \cdot \mathbf{R}(t) + \mathbf{R}(t) \cdot [D_t\mathbf{R}(t)] = 0$$

$$2\mathbf{R}(t) \cdot D_t\mathbf{R}(t) = 0$$

Because the dot product of $\mathbf{R}(t)$ and $D_t\mathbf{R}(t)$ is zero, it follows from Definition 14.2.7 that $\mathbf{R}(t)$ and $D_t\mathbf{R}(t)$ are orthogonal.   ∎

The geometric interpretation of Theorem 14.4.11 is evident. If the vector $\mathbf{R}(t)$ has constant magnitude, then the position representation $\overrightarrow{OP}$ of $\mathbf{R}(t)$ has its terminal point P on the circle with its center at the origin and radius k. So the graph of $\mathbf{R}$ is this circle. Because $D_t\mathbf{R}(t)$ and $\mathbf{R}(t)$ are orthogonal, $\overrightarrow{OP}$ is perpendicular to a representation of $D_t\mathbf{R}(t)$. Figure 2 shows a sketch of a quarter circle, the position representation $\overrightarrow{OP}$ of $\mathbf{R}(t)$, and the representation $\overrightarrow{PB}$ of $D_t\mathbf{R}(t)$.

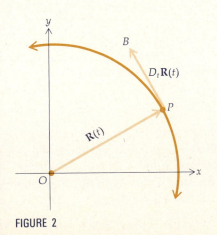

FIGURE 2

EXERCISES 14.4

In Exercises 1 through 5, find the indicated limit, if it exists.

1. $\mathbf{R}(t) = (3t - 2)\mathbf{i} + t^2\mathbf{j}$; $\lim\limits_{t \to 2} \mathbf{R}(t)$

2. $\mathbf{R}(t) = (t - 2)\mathbf{i} + \dfrac{t^2 - 4}{t - 2}\mathbf{j}$; $\lim\limits_{t \to 2} \mathbf{R}(t)$

3. $\mathbf{R}(t) = 2\sin t\mathbf{i} + \cos t\mathbf{j}$; $\lim\limits_{t \to \pi/2} \mathbf{R}(t)$

4. $\mathbf{R}(t) = \dfrac{t^2 - 2t - 3}{t - 3}\mathbf{i} + \dfrac{t^2 - 5t + 6}{t - 3}\mathbf{j}$; $\lim\limits_{t \to 3} \mathbf{R}(t)$

5. $\mathbf{R}(t) = e^{t+1}\mathbf{i} + |t + 1|\mathbf{j}$; $\lim\limits_{t \to -1} \mathbf{R}(t)$

In Exercises 6 through 14, find $\mathbf{R}'(t)$ and $\mathbf{R}''(t)$.

6. $\mathbf{R}(t) = (t^2 - 3)\mathbf{i} + (2t + 1)\mathbf{j}$ **7.** $\mathbf{R}(t) = e^{2t}\mathbf{i} + \ln t\mathbf{j}$

8. $\mathbf{R}(t) = \cos 2t\mathbf{i} + \tan t\mathbf{j}$ **9.** $\mathbf{R}(t) = \tan^{-1} t\mathbf{i} + 2^t\mathbf{j}$

10. $\mathbf{R}(t) = \dfrac{t - 1}{t + 1}\mathbf{i} + \dfrac{t - 2}{t}\mathbf{j}$

11. $\mathbf{R}(t) = (t^2 + 4)^{-1}\mathbf{i} + \sqrt{1 - 5t}\mathbf{j}$
12. $\mathbf{R}(t) = \sqrt{2t + 1}\mathbf{i} + (t - 1)^2\mathbf{j}$
13. $\mathbf{R}(t) = 5\sin 2t\mathbf{i} - \sec 4t\mathbf{j}$
14. $\mathbf{R}(t) = (e^{3t} + 2)\mathbf{i} + 2e^{3t}\mathbf{j}$

In Exercises 15 and 16, find $D_t\|\mathbf{R}(t)\|$.

15. $\mathbf{R}(t) = (t - 1)\mathbf{i} + (2 - t)\mathbf{j}$
16. $\mathbf{R}(t) = (e^t + 1)\mathbf{i} + (e^t - 1)\mathbf{j}$

In Exercises 17 and 18, verify Theorem 14.4.6 for the given vectors.

17. $\mathbf{R}(t) = (t^2 + e^t)\mathbf{i} + (t - e^{2t})\mathbf{j}$; $\mathbf{Q}(t) = (t^3 + 2e^t)\mathbf{i} - (3t + e^{2t})\mathbf{j}$
18. $\mathbf{R}(t) = \cos 2t\mathbf{i} - \sin 2t\mathbf{j}$; $\mathbf{Q}(t) = \sin^2 t\mathbf{i} + \cos 2t\mathbf{j}$

In Exercises 19 and 20, verify Theorem 14.4.7 for the vectors of the indicated exercise.

19. Exercise 17   **20.** Exercise 18

In Exercises 21 through 24, find $\mathbf{R}'(t) \cdot \mathbf{R}''(t)$.

21. $\mathbf{R}(t) = (2t^2 - 1)\mathbf{i} + (t^2 + 3)\mathbf{j}$ **22.** $\mathbf{R}(t) = \ln(t - 1)\mathbf{i} - 3t^{-1}\mathbf{j}$
23. $\mathbf{R}(t) = e^{2t}\mathbf{i} + e^{-2t}\mathbf{j}$ **24.** $\mathbf{R}(t) = -\cos 2t\mathbf{i} + \sin 2t\mathbf{j}$

25. Prove Theorem 14.4.6.
26. Prove Theorem 14.4.8.
27. Prove Theorem 14.4.9.

In Exercises 28 through 33, find the most general vector whose derivative has the given function value.

28. $(t^2 - 9)\mathbf{i} + (2t - 5)\mathbf{j}$ **29.** $\tan t\mathbf{i} - \dfrac{1}{t}\mathbf{j}$

30. $3^t\mathbf{i} - 2^t\mathbf{j}$ **31.** $e^{3t}\mathbf{i} + \dfrac{1}{t - 1}\mathbf{j}$

32. $\dfrac{1}{4 + t^2}\mathbf{i} - \dfrac{4}{1 - t^2}\mathbf{j}$ **33.** $\ln t\mathbf{i} + t^2\mathbf{j}$

34. If $\mathbf{R}'(t) = t^2\mathbf{i} + \dfrac{1}{t - 2}\mathbf{j}$, and $\mathbf{R}(3) = 2\mathbf{i} - 5\mathbf{j}$, find $\mathbf{R}(t)$.

35. If $\mathbf{R}'(t) = \sin^2 t\mathbf{i} + 2\cos^2 t\mathbf{j}$, and $\mathbf{R}(\pi) = \mathbf{0}$, find $\mathbf{R}(t)$.
36. If $\mathbf{R}'(t) = e^t\sin t\mathbf{i} + e^t\cos t\mathbf{j}$, and $\mathbf{R}(0) = \mathbf{i} - \mathbf{j}$, find $\mathbf{R}(t)$.

In Exercises 37 and 38, find a cartesian equation of the curve that is traced by the endpoint of the position representation of $\mathbf{R}'(t)$. Find $\mathbf{R}(t) \cdot \mathbf{R}'(t)$. Interpret the result geometrically.

37. $\mathbf{R}(t) = \cos t\mathbf{i} + \sin t\mathbf{j}$ **38.** $\mathbf{R}(t) = \cosh t\mathbf{i} - \sinh t\mathbf{j}$

In Exercises 39 and 40, if $\alpha(t)$ is the radian measure of the angle between $\mathbf{R}(t)$ and $\mathbf{Q}(t)$, find $D_t\alpha(t)$.

39. $\mathbf{R}(t) = 3e^{2t}\mathbf{i} - 4e^{2t}\mathbf{j}$ and $\mathbf{Q}(t) = 6e^{3t}\mathbf{j}$
40. $\mathbf{R}(t) = 2t\mathbf{i} + (t^2 - 1)\mathbf{j}$ and $\mathbf{Q}(t) = 3t\mathbf{i}$

41. Suppose that $\mathbf{R}$ and $\mathbf{R}'$ are vector-valued functions defined on an interval and $\mathbf{R}'$ is differentiable on the interval. Prove

$$D_t[\mathbf{R}'(t) \cdot \mathbf{R}(t)] = \|\mathbf{R}'(t)\|^2 + \mathbf{R}(t) \cdot \mathbf{R}''(t)$$

42. If $\|\mathbf{R}(t)\| = h(t)$, prove that $\mathbf{R}(t) \cdot \mathbf{R}'(t) = [h(t)][h'(t)]$.
43. If the vector-valued function $\mathbf{R}$ and the real-valued function f are both differentiable on an interval and $f(t) \neq 0$ on the interval, prove that $\mathbf{R}/f$ is also differentiable on the interval and

$$D_t\left[\frac{\mathbf{R}(t)}{f(t)}\right] = \frac{f(t)\mathbf{R}'(t) - f'(t)\mathbf{R}(t)}{[f(t)]^2}$$

44. Prove that if $\mathbf{A}$ and $\mathbf{B}$ are constant vectors and f and g are integrable functions, then

$$\int [\mathbf{A}f(t) + \mathbf{B}g(t)]\, dt = \mathbf{A}\int f(t)\, dt + \mathbf{B}\int g(t)\, dt$$

(*Hint:* Express $\mathbf{A}$ and $\mathbf{B}$ in terms of $\mathbf{i}$ and $\mathbf{j}$.)

14.5 LENGTH OF ARC

In Section 6.3 we obtained a formula for finding the length of arc of the graph of a function. Such a graph is a special kind of curve because the graph of a function cannot be intersected by a vertical line in more than one point.

We now develop a method for finding the length of arc of some other kinds of curves. Let C be the curve having parametric equations

$$x = f(t) \quad \text{and} \quad y = g(t)$$

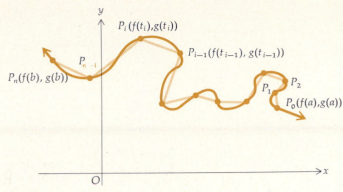

FIGURE 1

and suppose that f and g are continuous on the closed interval $[a, b]$. We wish to assign a number L to represent the number of units in the length of arc of C from $t = a$ to $t = b$. We proceed as in Section 6.3.

Let Δ be a partition of the closed interval $[a, b]$ formed by dividing the interval into n subintervals by choosing $n - 1$ numbers between a and b. Let $t_0 = a$ and $t_n = b$, and let $t_1, t_2, \ldots, t_{n-1}$ be intermediate numbers:

$$t_0 < t_1 < \ldots < t_{n-1} < t_n$$

The ith subinterval is $[t_{i-1}, t_i]$ and the number of units in its length, denoted by $\Delta_i t$, is $t_i - t_{i-1}$, where $i = 1, 2, \ldots, n$. Let $\|\Delta\|$ be the norm of the partition; so each $\Delta_i t \leq \|\Delta\|$.

Associated with each number t_i is a point $P_i(f(t_i), g(t_i))$ on C. From each point P_{i-1} draw a line segment to the next point P_i. See Figure 1. The number of units in the length of the line segment from P_{i-1} to P_i is denoted by $|\overline{P_{i-1}P_i}|$. From the distance formula we have

$$|\overline{P_{i-1}P_i}| = \sqrt{[f(t_i) - f(t_{i-1})]^2 + [g(t_i) - g(t_{i-1})]^2} \tag{1}$$

The sum of the numbers of units of lengths of the n line segments is

$$\sum_{i=1}^{n} |\overline{P_{i-1}P_i}|$$

Our intuitive notion of the length of the arc from $t = a$ to $t = b$ leads us to define the number of units of the length of arc as the limit of this sum as $\|\Delta\|$ approaches zero.

14.5.1 DEFINITION Let the curve C have parametric equations $x = f(t)$ and $y = g(t)$. Suppose there exists a number L having the following property: For any $\epsilon > 0$ there is a $\delta > 0$ such that for every partition Δ of the interval $[a, b]$ for which $\|\Delta\| < \delta$, then

$$\left| \sum_{i=1}^{n} |\overline{P_{i-1}P_i}| - L \right| < \epsilon$$

Then we write

$$L = \lim_{\|\Delta\| \to 0} \sum_{i=1}^{n} |\overline{P_{i-1}P_i}|$$

and L units is called the **length of arc** of the curve C from the point $(f(a), g(a))$ to the point $(f(b), g(b))$.

The arc of the curve is rectifiable if the limit in Definition 14.5.1 exists. If f' and g' are continuous on $[a, b]$, we proceed as follows to find a formula for evaluating this limit.

Because f' and g' are continuous on $[a, b]$, they are continuous on each subinterval of the partition Δ. So the hypothesis of the mean-value theorem (Theorem 4.3.2) is satisfied by f and g on each $[t_{i-1}, t_i]$; therefore there are numbers z_i and w_i in the open interval (t_{i-1}, t_i) such that

$$f(t_i) - f(t_{i-1}) = f'(z_i)\,\Delta_i t \quad \text{and} \quad g(t_i) - g(t_{i-1}) = g'(w_i)\,\Delta_i t$$

Substituting from these equations into (1) we obtain

$$|\overline{P_{i-1}P_i}| = \sqrt{[f'(z_i)\,\Delta_i t]^2 + [g'(w_i)\,\Delta_i t]^2}$$

$$|\overline{P_{i-1}P_i}| = \sqrt{[f'(z_i)]^2 + [g'(w_i)]^2}\,\Delta_i t \tag{2}$$

where z_i and w_i are in the open interval (t_{i-1}, t_i). Then from Definition 14.5.1 and (2), if the limit exists,

$$L = \lim_{\|\Delta\| \to 0} \sum_{i=1}^{n} \sqrt{[f'(z_i)]^2 + [g'(w_i)]^2}\,\Delta_i t \tag{3}$$

The sum in (3) is not a Riemann sum because z_i and w_i are not necessarily the same numbers. So we cannot apply the definition of a definite integral to evaluate the limit in (3). However, there is a theorem that can be applied to evaluate this limit. We state the theorem, but a proof is not given because it is beyond the scope of this book. You can find a proof in an advanced calculus text.

14.5.2 THEOREM   If the functions F and G are continuous on the closed interval $[a, b]$, then the function $\sqrt{F^2 + G^2}$ is also continuous on $[a, b]$, and if Δ is a partition of the interval $[a, b]$ ($\Delta: a = t_0 < t_1 < \ldots < t_{i-1} < t_i < \ldots < t_n = b$), and z_i and w_i are any numbers in (t_{i-1}, t_i), then

$$\lim_{\|\Delta\| \to 0} \sum_{i=1}^{n} \sqrt{[F(z_i)]^2 + [G(w_i)]^2}\,\Delta_i t = \int_a^b \sqrt{[F(t)]^2 + [G(t)]^2}\,dt$$

Applying Theorem 14.5.2 to (3), where F is f' and G is g', we have

$$L = \int_a^b \sqrt{[f'(t)]^2 + [g'(t)]^2}\,dt$$

This result is stated as a theorem.

14.5.3 THEOREM   Let the curve C have parametric equations $x = f(t)$ and $y = g(t)$, and suppose that f' and g' are continuous on the closed interval $[a, b]$. Then if L units is the length of arc of the curve C from the point $(f(a), g(a))$ to the point $(f(b), g(b))$,

$$L = \int_a^b \sqrt{[f'(t)]^2 + [g'(t)]^2}\,dt$$

EXAMPLE 1   Find the length of the arc of the curve having parametric equations

$$x = t^3 \quad \text{and} \quad y = 2t^2$$

in each of the following cases: (a) from $t = 0$ to $t = 1$; (b) from $t = -2$ to $t = 0$.

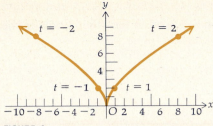

FIGURE 2

Solution   A sketch of the curve appears in Figure 2. Let

$$f(t) = t^3 \qquad g(t) = 2t^2$$
$$f'(t) = 3t^2 \qquad g'(t) = 4t$$

The curve has parametric equations $x = f(t)$ and $y = g(t)$. We apply Theorem 14.5.3 in parts (a) and (b) where L_a units is the length of the arc from $t = 0$ to $t = 1$ and L_b units is the length of the arc from $t = -2$ to $t = 0$.

(a) $L_a = \displaystyle\int_0^1 \sqrt{9t^4 + 16t^2}\, dt$

$= \displaystyle\int_0^1 \sqrt{t^2}\, \sqrt{9t^2 + 16}\, dt$

$= \displaystyle\int_0^1 t\sqrt{9t^2 + 16}\, dt$

$= \frac{1}{18} \cdot \frac{2}{3}(9t^2 + 16)^{3/2}\Big]_0^1$

$= \frac{1}{27}[(25)^{3/2} - (16)^{3/2}]$

$= \frac{1}{27}(125 - 64)$

$= \frac{61}{27}$

(b) $L_b = \displaystyle\int_{-2}^0 \sqrt{9t^4 + 16t^2}\, dt$

$= \displaystyle\int_{-2}^0 \sqrt{t^2}\, \sqrt{9t^2 + 16}\, dt$

$= \displaystyle\int_{-2}^0 -t\sqrt{9t^2 + 16}\, dt$

$= -\frac{1}{18} \cdot \frac{2}{3}(9t^2 + 16)^{3/2}\Big]_{-2}^0$

$= -\frac{1}{27}[(16)^{3/2} - (52)^{3/2}]$

$= \frac{1}{27}(104\sqrt{13} - 64)$

≈ 11.5

Observe in the third integral in part (a) we replaced $\sqrt{t^2}$ by t because $0 \leq t \leq 1$. However, in the third integral in part (b) we replaced $\sqrt{t^2}$ by $-t$ because $-2 \leq t \leq 0$.

For the curve C having parametric equations $x = f(t)$ and $y = g(t)$, let s units be the length of arc of C from the point $(f(t_0), g(t_0))$ to the point $(f(t), g(t))$, and let s increase as t increases. Then s is a function of t and is given by

$$s = \int_{t_0}^t \sqrt{[f'(u)]^2 + [g'(u)]^2}\, du$$

From the first fundamental theorem of the calculus (Theorem 5.8.1),

$$\frac{ds}{dt} = \sqrt{[f'(t)]^2 + [g'(t)]^2} \tag{4}$$

A vector equation of C is

$$\mathbf{R}(t) = f(t)\mathbf{i} + g(t)\mathbf{j} \tag{5}$$

Because

$$\mathbf{R}'(t) = f'(t)\mathbf{i} + g'(t)\mathbf{j}$$

then

$$\|\mathbf{R}'(t)\| = \sqrt{[f'(t)]^2 + [g'(t)]^2} \tag{6}$$

Substituting from (6) into (4) we obtain

$$\|\mathbf{R}'(t)\| = \frac{ds}{dt}$$

From this equation it follows that if s units is the length of arc of curve C having vector equation (5) measured from some fixed point to the point $(f(t), g(t))$ where s increases as t increases, then the derivative of s with respect to t is the magnitude of the derivative of the position vector at the point $(f(t), g(t))$.

We substitute from (6) into the formula of Theorem 14.5.3 and obtain $L = \int_a^b \|\mathbf{R}'(t)\| \, dt$. Thus Theorem 14.5.3 can be stated in terms of vectors in the following way.

14.5.4 THEOREM   Let the curve C have the vector equation $\mathbf{R}(t) = f(t)\mathbf{i} + g(t)\mathbf{j}$, and suppose that f' and g' are continuous on the closed interval $[a, b]$. Then the length of arc of C, traced by the terminal point of the position representation of $\mathbf{R}(t)$ as t increases from a to b, is determined by

$$L = \int_a^b \|\mathbf{R}'(t)\| \, dt$$

EXAMPLE 2   Find the length of the arc traced by the terminal point of the position representation of $\mathbf{R}(t)$ as t increases from 1 to 4 if

$$\mathbf{R}(t) = e^t \sin t\mathbf{i} + e^t \cos t\mathbf{j}$$

Solution

$$\mathbf{R}'(t) = (e^t \sin t + e^t \cos t)\mathbf{i} + (e^t \cos t - e^t \sin t)\mathbf{j}$$

$$\|\mathbf{R}'(t)\| = \sqrt{(e^t \sin t + e^t \cos t)^2 + (e^t \cos t - e^t \sin t)^2}$$
$$= \sqrt{e^{2t}}\sqrt{\sin^2 t + 2\sin t \cos t + \cos^2 t + \cos^2 t - 2\sin t \cos t + \sin^2 t}$$
$$= e^t\sqrt{2}$$

From Theorem 14.5.4,

$$L = \int_1^4 \sqrt{2}\,e^t \, dt$$

$$= \sqrt{2}\,e^t \Big]_1^4$$

$$= \sqrt{2}(e^4 - e)$$

An alternate form of the formula of Theorem 14.5.3 for the length of arc of a curve C, having parametric equations $x = f(t)$ and $y = g(t)$, is obtained by replacing $f'(t)$ by $\dfrac{dx}{dt}$ and $g'(t)$ by $\dfrac{dy}{dt}$. This form is

$$L = \int_a^b \sqrt{\left(\frac{dx}{dt}\right)^2 + \left(\frac{dy}{dt}\right)^2} \, dt \qquad (7)$$

Now suppose that we wish to find the length of arc of a curve C whose polar equation is $r = F(\theta)$. If (x, y) is the cartesian representation of a point P on C and (r, θ) is a polar representation of P, then

$$x = r \cos \theta \quad \text{and} \quad y = r \sin \theta$$

Replacing r by $F(\theta)$ in these two equations we have

$$x = F(\theta) \cos \theta \quad \text{and} \quad y = F(\theta) \sin \theta$$

These equations can be considered as parametric equations of C where θ is the parameter instead of t. Therefore, if F' is continuous on the closed interval $[\alpha, \beta]$, the formula for the length of arc of the curve C whose polar equation is $r = F(\theta)$ is obtained from (7) by taking $t = \theta$. So

$$L = \int_{\alpha}^{\beta} \sqrt{\left(\frac{dx}{d\theta}\right)^2 + \left(\frac{dy}{d\theta}\right)^2} \, d\theta \tag{8}$$

Because $x = r \cos \theta$ and $y = r \sin \theta$,

$$\frac{dx}{d\theta} = \cos \theta \frac{dr}{d\theta} - r \sin \theta \quad \text{and} \quad \frac{dy}{d\theta} = \sin \theta \frac{dr}{d\theta} + r \cos \theta$$

Therefore

$$\sqrt{\left(\frac{dx}{d\theta}\right)^2 + \left(\frac{dy}{d\theta}\right)^2} = \sqrt{\left(\cos \theta \frac{dr}{d\theta} - r \sin \theta\right)^2 + \left(\sin \theta \frac{dr}{d\theta} + r \cos \theta\right)^2}$$

$$= \sqrt{\cos^2 \theta \left(\frac{dr}{d\theta}\right)^2 - 2r \sin \theta \cos \theta \frac{dr}{d\theta} + r^2 \sin^2 \theta + \sin^2 \theta \left(\frac{dr}{d\theta}\right)^2 + 2r \sin \theta \cos \theta \frac{dr}{d\theta} + r^2 \cos^2 \theta}$$

$$= \sqrt{(\cos^2 \theta + \sin^2 \theta)\left(\frac{dr}{d\theta}\right)^2 + (\sin^2 \theta + \cos^2 \theta)r^2}$$

$$= \sqrt{\left(\frac{dr}{d\theta}\right)^2 + r^2}$$

Substituting this into (8) we obtain

$$L = \int_{\alpha}^{\beta} \sqrt{\left(\frac{dr}{d\theta}\right)^2 + r^2} \, d\theta \tag{9}$$

EXAMPLE 3   Find the length of the cardioid $r = 2(1 + \cos \theta)$.

Solution   A sketch of the curve appears in Figure 3. To obtain the length of the entire curve we can let θ take on values from 0 to 2π or we can make use of the symmetry of the curve and find half the length by letting θ take on values from 0 to π.

Because $r = 2(1 + \cos \theta)$, $\dfrac{dr}{d\theta} = -2 \sin \theta$. Substituting into (9), integrating from 0 to π, and multiplying by 2 we have

$$L = 2 \int_0^{\pi} \sqrt{(-2 \sin \theta)^2 + 4(1 + \cos \theta)^2} \, d\theta$$

$$= 4 \int_0^{\pi} \sqrt{\sin^2 \theta + 1 + 2 \cos \theta + \cos^2 \theta} \, d\theta$$

$$= 4\sqrt{2} \int_0^{\pi} \sqrt{1 + \cos \theta} \, d\theta$$

To evaluate this integral, we use the identity $\cos^2 \frac{1}{2}\theta = \frac{1}{2}(1 + \cos \theta)$, which gives $\sqrt{1 + \cos \theta} = \sqrt{2} \left|\cos \frac{1}{2}\theta\right|$. Because $0 \leq \theta \leq \pi$, $0 \leq \frac{1}{2}\theta \leq \frac{1}{2}\pi$; thus $\cos \frac{1}{2}\theta \geq 0$. Therefore $\sqrt{1 + \cos \theta} = \sqrt{2} \cos \frac{1}{2}\theta$. So

$$L = 4\sqrt{2} \int_0^{\pi} \sqrt{2} \cos \frac{1}{2}\theta \, d\theta$$

$$= 16 \sin \frac{1}{2}\theta \Big]_0^{\pi}$$

$$= 16$$

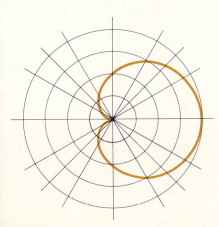

FIGURE 3

EXERCISES 14.5

In Exercises 1 through 28, find the length of the arc. When a appears, $a > 0$.

1. $x = \frac{1}{2}t^2 + t$, $y = \frac{1}{2}t^2 - t$; from $t = 0$ to $t = 1$.
2. $x = 3t^2$, $y = 2t^3$; from $t = 0$ to $t = 3$.
3. $x = t^2 + 2t$, $y = t^2 - 2t$; from $t = 0$ to $t = 2$.
4. $x = t^3$, $y = 3t^2$; from $t = -2$ to $t = 0$.
5. $\mathbf{R}(t) = 2t^2\mathbf{i} + 2t^3\mathbf{j}$; from $t = 1$ to $t = 2$.
6. $\mathbf{R}(t) = t\mathbf{i} + \cosh t\mathbf{j}$; from $t = 0$ to $t = 3$.
7. $\mathbf{R}(t) = 3e^{2t}\mathbf{i} - 4e^{2t}\mathbf{j}$; from $t = 0$ to $t = \ln 5$.
8. $\mathbf{R}(t) = (t^2 + 3)\mathbf{i} + 3t^2\mathbf{j}$; from $t = 1$ to $t = 4$.
9. $\mathbf{R}(t) = e^t \cos t\mathbf{i} + e^t \sin t\mathbf{j}$; from $t = 0$ to $t = 1$.
10. $\mathbf{R}(t) = \ln \sin t\mathbf{i} + (t + 1)\mathbf{j}$; from $t = \frac{1}{6}\pi$ to $t = \frac{1}{2}\pi$.
11. $\mathbf{R}(t) = \tan^{-1} t\mathbf{i} + \frac{1}{2}\ln(t^2 + 1)\mathbf{j}$; from $t = 0$ to $t = 1$.
12. $\mathbf{R}(t) = a(\cos t + t \sin t)\mathbf{i} + a(\sin t - t \cos t)\mathbf{j}$; from $t = 0$ to $t = \frac{1}{3}\pi$.

13. The entire hypocycloid of four cusps:

$$x = a \cos^3 t \qquad y = a \sin^3 t$$

14. $x = e^{-t} \cos t$, $y = e^{-t} \sin t$; from $t = 0$ to $t = \pi$.
15. $x = 4 \sin 2t$, $y = 4 \cos 2t$; from $t = 0$ to $t = \pi$.

16. One arch of the cycloid: $x = a(t - \sin t)$, $y = a(1 - \cos t)$.
17. The tractrix

$$x = t - a \tanh \frac{t}{a} \qquad y = a \operatorname{sech} \frac{t}{a}$$

from $t = -a$ to $t = 2a$.

18. The circumference of the circle: $\mathbf{R}(t) = a \cos t\mathbf{i} + a \sin t\mathbf{j}$.
19. The circumference of the circle: $r = 5 \cos \theta$.
20. The circumference of the circle: $r = a \sin \theta$.
21. The circumference of the circle: $r = a$.
22. The entire curve: $r = 1 - \sin \theta$.
23. The entire curve: $r = 3 \cos^2 \frac{1}{2}\theta$.
24. $r = a\theta$; from $\theta = 0$ to $\theta = 2\pi$.
25. $r = e^{2\theta}$; from $\theta = 0$ to $\theta = 4$.
26. $r = a\theta^2$; from $\theta = 0$ to $\theta = \pi$.
27. $r = a \sin^3 \frac{1}{3}\theta$; from $\theta = 0$ to $\theta = \theta_1$.
28. $r = \sin^2 \frac{1}{2}\theta$; from $\theta = 0$ to $\theta = \frac{1}{2}\pi$.

29. Find the distance traveled by a thumbtack in the tread of a bicycle tire if the radius of the tire is 40 cm and the bicycle goes a distance of 50π m. (*Hint:* The path of the thumbtack is a cycloid.)

14.6 THE UNIT TANGENT AND UNIT NORMAL VECTORS AND ARC LENGTH AS PARAMETER

With each point on a curve in the plane we now associate two unit vectors, the *unit tangent vector* and the *unit normal vector*. These vectors occur in many applications of vector-valued functions.

14.6.1 DEFINITION

If $\mathbf{R}(t)$ is the position vector of curve C at a point P on C, then the **unit tangent vector** of C at P, denoted by $\mathbf{T}(t)$, is the unit vector in the direction of $D_t\mathbf{R}(t)$ if $D_t\mathbf{R}(t) \neq \mathbf{0}$.

The unit vector in the direction of $D_t\mathbf{R}(t)$ is given by $D_t\mathbf{R}(t)/\|D_t\mathbf{R}(t)\|$; thus

$$\mathbf{T}(t) = \frac{D_t\mathbf{R}(t)}{\|D_t\mathbf{R}(t)\|} \tag{1}$$

Because $\mathbf{T}(t)$ is a unit vector, it follows from Theorem 14.4.11 that $D_t\mathbf{T}(t)$ must be orthogonal to $\mathbf{T}(t)$. $D_t\mathbf{T}(t)$ is not necessarily a unit vector. However, the vector $D_t\mathbf{T}(t)/\|D_t\mathbf{T}(t)\|$ is of unit magnitude and has the same direction as $D_t\mathbf{T}(t)$. Therefore $D_t\mathbf{T}(t)/\|D_t\mathbf{T}(t)\|$ is a unit vector that is orthogonal to $\mathbf{T}(t)$, and it is called the *unit normal vector*.

14.6.2 DEFINITION

If $\mathbf{T}(t)$ is the unit tangent vector of curve C at a point P on C, then the **unit normal vector**, denoted by $\mathbf{N}(t)$, is the unit vector in the direction of $D_t\mathbf{T}(t)$.

From Definition 14.6.2 and the previous discussion,

$$\mathbf{N}(t) = \frac{D_t\mathbf{T}(t)}{\|D_t\mathbf{T}(t)\|} \tag{2}$$

EXAMPLE 1   Given the curve having parametric equations

$$x = t^3 - 3t \quad \text{and} \quad y = 3t^2$$

find $\mathbf{T}(t)$ and $\mathbf{N}(t)$. Draw a sketch of a portion of the curve at $t = 2$ and draw the representations of $\mathbf{T}(2)$ and $\mathbf{N}(2)$ having their initial point at $t = 2$.

Solution   A vector equation of the curve is

$$\mathbf{R}(t) = (t^3 - 3t)\mathbf{i} + 3t^2\mathbf{j}$$

Thus

$$D_t\mathbf{R}(t) = (3t^2 - 3)\mathbf{i} + 6t\mathbf{j} \qquad \|D_t\mathbf{R}(t)\| = \sqrt{(3t^2 - 3)^2 + 36t^2}$$
$$= \sqrt{9(t^4 + 2t^2 + 1)}$$
$$= 3(t^2 + 1)$$

From (1),

$$\mathbf{T}(t) = \frac{D_t\mathbf{R}(t)}{\|D_t\mathbf{R}(t)\|}$$

$$= \frac{t^2 - 1}{t^2 + 1}\mathbf{i} + \frac{2t}{t^2 + 1}\mathbf{j}$$

Differentiating $\mathbf{T}(t)$ with respect to t we obtain

$$D_t\mathbf{T}(t) = \frac{4t}{(t^2 + 1)^2}\mathbf{i} + \frac{2 - 2t^2}{(t^2 + 1)^2}\mathbf{j}$$

Therefore

$$\|D_t\mathbf{T}(t)\| = \sqrt{\frac{16t^2}{(t^2 + 1)^4} + \frac{4 - 8t^2 + 4t^4}{(t^2 + 1)^4}}$$

$$= \sqrt{\frac{4 + 8t^2 + 4t^4}{(t^2 + 1)^4}}$$

$$= \sqrt{\frac{4(t^2 + 1)^2}{(t^2 + 1)^4}}$$

$$= \frac{2}{t^2 + 1}$$

From (2),

$$\mathbf{N}(t) = \frac{D_t\mathbf{T}(t)}{\|D_t\mathbf{T}(t)\|}$$

$$= \frac{2t}{t^2 + 1}\mathbf{i} + \frac{1 - t^2}{t^2 + 1}\mathbf{j}$$

We find $\mathbf{R}(t)$, $\mathbf{T}(t)$, and $\mathbf{N}(t)$ when $t = 2$.

$$\mathbf{R}(2) = 2\mathbf{i} + 12\mathbf{j} \qquad \mathbf{T}(2) = \tfrac{3}{5}\mathbf{i} + \tfrac{4}{5}\mathbf{j} \qquad \mathbf{N}(2) = \tfrac{4}{5}\mathbf{i} - \tfrac{3}{5}\mathbf{j}$$

The required sketch is shown in Figure 1.

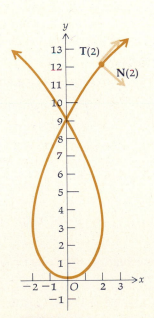

FIGURE 1

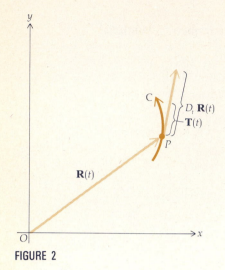

FIGURE 2

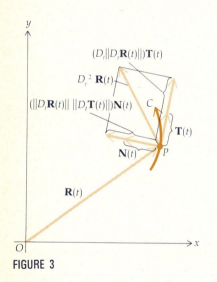

FIGURE 3

From (1),

$$D_t\mathbf{R}(t) = \|D_t\mathbf{R}(t)\|\mathbf{T}(t) \tag{3}$$

This equation expresses the vector $D_t\mathbf{R}(t)$ as a scalar times the unit tangent vector. Figure 2 shows a portion of a curve C with the position representation of $\mathbf{R}(t)$ and the representations of $\mathbf{T}(t)$ and $D_t\mathbf{R}(t)$ whose initial points are at the point P on C.

We now use (3) to compute $D_t{}^2\mathbf{R}(t)$ by applying Theorem 14.4.8.

$$D_t{}^2\mathbf{R}(t) = (D_t\|D_t\mathbf{R}(t)\|)\mathbf{T}(t) + \|D_t\mathbf{R}(t)\|(D_t\mathbf{T}(t)) \tag{4}$$

From (2),

$$D_t\mathbf{T}(t) = \|D_t\mathbf{T}(t)\|\mathbf{N}(t)$$

Substituting from this equation into (4) we get

$$D_t{}^2\mathbf{R}(t) = (D_t\|D_t\mathbf{R}(t)\|)\mathbf{T}(t) + (\|D_t\mathbf{R}(t)\|\,\|D_t\mathbf{T}(t)\|)\mathbf{N}(t) \tag{5}$$

This equation expresses the vector $D_t{}^2\mathbf{R}(t)$ as a scalar times the unit tangent vector plus a scalar times the unit normal vector. The coefficient of $\mathbf{T}(t)$ on the right side of (5) is the component of the vector $D_t{}^2\mathbf{R}(t)$ in the direction of the unit tangent vector. The coefficient of $\mathbf{N}(t)$ on the right side of (5) is the component of $D_t{}^2\mathbf{R}(t)$ in the direction of the unit normal vector.

Figure 3 shows the position representation of $\mathbf{R}(t)$ and the same portion of the curve C as shown in Figure 2. Also shown in Figure 3 are the representations of the following vectors, all of whose initial points are at the point P on C:

$$D_t{}^2\mathbf{R}(t) \qquad \mathbf{T}(t) \qquad (D_t\|D_t\mathbf{R}(t)\|)\mathbf{T}(t) \qquad \mathbf{N}(t) \qquad (\|D_t\mathbf{R}(t)\|\,\|D_t\mathbf{T}(t)\|)\mathbf{N}(t)$$

Observe that the representation of the unit normal vector $\mathbf{N}(t)$ is on the concave side of the curve. This fact is proved in general in Section 14.7.

Sometimes instead of a parameter t, we wish to use as a parameter the number of units of arc length s from an arbitrarily chosen point $P_0(x_0, y_0)$ on curve C to the point $P(x, y)$ on C. Let s increase as t increases so that s is positive if the length of arc is measured in the direction of increasing t and s is negative if the length of arc is measured in the opposite direction. Therefore s units is a directed distance. Also, $\dfrac{ds}{dt} > 0$. To each value of s there corresponds a unique point P on the curve C. Consequently, the coordinates of P are functions of s, and s is a function of t. From Section 14.5 we have

$$\|D_t\mathbf{R}(t)\| = \frac{ds}{dt}$$

Substituting from this equation into (3) we get

$$D_t\mathbf{R}(t) = \frac{ds}{dt}\mathbf{T}(t)$$

If the parameter is s instead of t, we have from this equation, by taking $t = s$ and noting that $\dfrac{ds}{ds} = 1$,

$$D_s\mathbf{R}(s) = \mathbf{T}(s)$$

We state this result as a theorem.

14.6.3 THEOREM   If the vector equation of a curve C is $\mathbf{R}(s) = f(s)\mathbf{i} + g(s)\mathbf{j}$, where s units is the length of arc measured from a particular point P_0 on C to the point P, then the unit tangent vector of C at P is given by

$$\mathbf{T}(s) = D_s\mathbf{R}(s)$$

if it exists.

Now suppose that the parametric equations of a curve C involve a parameter t, and we wish to find parametric equations of C, with s, the number of units of arc length measured from some fixed point, as the parameter. Often the operations involved are quite complicated. However, the method used is illustrated in the following example.

EXAMPLE 2   Suppose that parametric equations of the curve C are

$$x = t^3 \quad \text{and} \quad y = t^2 \qquad t \geq 0$$

Find parametric equations of C having s as a parameter, where s units is the arc length measured from the point where $t = 0$.

Solution   The point where $t = 0$ is the origin. A vector equation of C is

$$\mathbf{R}(t) = t^3\mathbf{i} + t^2\mathbf{j}$$

Because $\dfrac{ds}{dt} = \|D_t R(t)\|$, we differentiate the above vector and get

$$D_t\mathbf{R}(t) = 3t^2\mathbf{i} + 2t\mathbf{j}$$

$$\|D_t\mathbf{R}(t)\| = \sqrt{9t^4 + 4t^2}$$

$$= \sqrt{t^2}\sqrt{9t^2 + 4}$$

$$= t\sqrt{9t^2 + 4} \qquad \text{(because } t \geq 0\text{)}$$

Therefore

$$\frac{ds}{dt} = t\sqrt{9t^2 + 4}$$

$$s = \int t\sqrt{9t^2 + 4}\ dt$$

$$= \tfrac{1}{18}\int \sqrt{9t^2 + 4}(18t\ dt)$$

$$= \tfrac{1}{27}(9t^2 + 4)^{3/2} + C$$

Because $s = 0$ when $t = 0$, we obtain $C = -\tfrac{8}{27}$. Therefore

$$s = \tfrac{1}{27}(9t^2 + 4)^{3/2} - \tfrac{8}{27}$$

Solving this equation for t in terms of s we have

$$(9t^2 + 4)^{3/2} = 27s + 8$$

$$9t^2 + 4 = (27s + 8)^{2/3}$$

Because $t \geq 0$,

$$t = \tfrac{1}{3}\sqrt{(27s + 8)^{2/3} - 4}$$

Substituting this value of t into the given parametric equations for C we obtain

$$x = \tfrac{1}{27}\left[(27s + 8)^{2/3} - 4\right]^{3/2} \quad \text{and} \quad y = \tfrac{1}{9}\left[(27s + 8)^{2/3} - 4\right] \qquad (6)$$

Since $D_s\mathbf{R}(s) = \mathbf{T}(s)$, then if $\mathbf{R}(s) = x(s)\mathbf{i} + y(s)\mathbf{j}$, $\mathbf{T}(s) = \left(\dfrac{dx}{ds}\right)\mathbf{i} + \left(\dfrac{dy}{ds}\right)\mathbf{j}$. Thus, because $\mathbf{T}(s)$ is a unit vector,

$$\left(\frac{dx}{ds}\right)^2 + \left(\frac{dy}{ds}\right)^2 = 1 \tag{7}$$

Equation (7) can be used to check Equations (6). This check is left as an exercise (see Exercise 17).

EXERCISES 14.6

In Exercises 1 through 10, find $\mathbf{T}(t)$ *and* $\mathbf{N}(t)$, *and at* $t = t_1$ *draw a sketch of a portion of the curve and draw the representations of* $\mathbf{T}(t_1)$ *and* $\mathbf{N}(t_1)$ *having initial point at* $t = t_1$.

1. $x = \frac{1}{3}t^3 - t, y = t^2; t_1 = 2$ **2.** $x = \frac{1}{2}t^2, y = \frac{1}{3}t^3; t_1 = 1$
3. $\mathbf{R}(t) = e^t\mathbf{i} + e^{-t}\mathbf{j}; t_1 = 0$ **4.** $\mathbf{R}(t) = e^{-2t}\mathbf{i} + e^{2t}\mathbf{j}; t_1 = 0$
5. $\mathbf{R}(t) = 3 \cos t\mathbf{i} + 3 \sin t\mathbf{j}; t_1 = \frac{1}{2}\pi$
6. $x = e^t \sin t, y = e^t \cos t; t_1 = 0$
7. $x = \cos kt, y = \sin kt, k > 0; t_1 = \pi/k$
8. $x = t - \sin t, y = 1 - \cos t; t_1 = \pi$
9. $\mathbf{R}(t) = \ln \cos t\mathbf{i} + \ln \sin t\mathbf{j}, 0 < t < \frac{1}{2}\pi; t_1 = \frac{1}{4}\pi$
10. $\mathbf{R}(t) = t \cos t\mathbf{i} + t \sin t\mathbf{j}; t_1 = 0$

In Exercises 11 through 16, find $\mathbf{T}(t)$ *and* $\mathbf{N}(t)$.

11. $\mathbf{R}(t) = t^2\mathbf{i} + e^t\mathbf{j}$
12. $\mathbf{R}(t) = t\mathbf{i} + \cosh t\mathbf{j}; t \geq 0$
13. $\mathbf{R}(t) = \sin^3 t\mathbf{i} + \cos^3 t\mathbf{j}; 0 \leq t \leq \frac{1}{2}\pi$
14. $\mathbf{R}(t) = \sqrt{1 + t^2}\mathbf{i} + t\mathbf{j}$
15. $\mathbf{R}(t) = 4t\mathbf{i} + 2t^2\mathbf{j}$
16. $\mathbf{R}(t) = (1 + t)\mathbf{i} + \ln \cos t\mathbf{j}$

17. Check Equations (6) of the solution of Example 2 by using Equation (7).

In Exercises 18 through 23, find parametric equations of the curve having arc length s as a parameter, where s is measured from the point where $t = 0$. *Check your result by using Equation (7).*

18. $x = a \cos t, y = a \sin t$ **19.** $x = 2 + \cos t, y = 3 + \sin t$
20. $x = 2(\cos t + t \sin t), y = 2(\sin t - t \cos t)$
21. $y = x^{3/2}$ **22.** $x = t^2 - 1, y = \frac{1}{3}t^3$
23. One cusp of the hypocycloid of four cusps:
$$\mathbf{R}(t) = a \cos^3 t\mathbf{i} + a \sin^3 t\mathbf{j} \qquad 0 \leq t \leq \frac{1}{2}\pi.$$

24. Given the cycloid $x = 2(t - \sin t)$, $y = 2(1 - \cos t)$, express the arc length s as a function of t, where s is measured from the point where $t = 0$.

25. For the curve having parametric equations $x = e^t \cos t$ and $y = e^t \sin t$, express the arc length s as a function of t, where s is measured from the point where $t = 0$.

26. If the vector equation of curve C is $\mathbf{R}(t) = 2 \sin t\mathbf{i} + \sin 2t\mathbf{j}$, find the cosine of the angle between the vectors $\mathbf{R}(\frac{1}{6}\pi)$ and $\mathbf{T}(\frac{1}{6}\pi)$.

27. If the vector equation of curve C is $\mathbf{R}(t) = 3t^2\mathbf{i} + (t^3 - 3t)\mathbf{j}$, find the cosine of the angle between the vectors $\mathbf{R}(2)$ and $\mathbf{T}(2)$.

28. The vector equation of curve C is $\mathbf{R}(t) = (4 - 3t^2)\mathbf{i} + (t^3 - 3t)\mathbf{j}$. Find the radian measure of the angle between the vectors $\mathbf{N}(1)$ and $D_t^2\mathbf{R}(1)$.

14.7 CURVATURE

Differential geometry and *curvilinear motion* (motion along a curved path) involve a study of curves by means of the calculus of vector-valued functions. One of the important concepts in this subject is *curvature*, which gives the rate of change of the direction of a curve with respect to a change in its length. Related to this idea is the angle giving the direction of the unit tangent vector associated with a curve C. Thus we let ϕ be the radian measure of the angle from the direction of the positive x axis counterclockwise to the direction of the unit tangent vector $\mathbf{T}(t)$. See Figure 1. We compute $D_\phi\mathbf{T}(t)$. Because $\|\mathbf{T}(t)\| = 1$, it follows from Equation (5) in Section 14.1 that

$$\mathbf{T}(t) = \cos \phi\mathbf{i} + \sin \phi\mathbf{j}$$

Differentiating with respect to ϕ we obtain

$$D_\phi\mathbf{T}(t) = -\sin \phi\mathbf{i} + \cos \phi\mathbf{j}$$

Since $\|D_\phi\mathbf{T}(t)\| = \sqrt{(-\sin \phi)^2 + (\cos \phi)^2}$, which is 1, $D_\phi\mathbf{T}(t)$ is a unit vector. Later in this section we show the relationship of the vector $D_\phi\mathbf{T}(t)$ to the unit normal vector $\mathbf{N}(t)$.

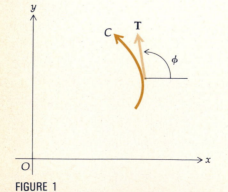

FIGURE 1

To lead up to the definition of curvature, consider the vector $D_s\mathbf{T}(t)$, where s units is the arc length measured from an arbitrarily chosen point on C to point P and s increases as t increases. By the chain rule (Theorem 14.4.9),

$$D_s\mathbf{T}(t) = D_\phi\mathbf{T}(t)\frac{d\phi}{ds}$$

Hence

$$\left\|D_s\mathbf{T}(t)\right\| = \left\|D_\phi\mathbf{T}(t)\frac{d\phi}{ds}\right\|$$

$$= \left\|D_\phi\mathbf{T}(t)\right\|\left|\frac{d\phi}{ds}\right|$$

But because $D_\phi\mathbf{T}(t)$ is a unit vector, $\left\|D_\phi\mathbf{T}(t)\right\| = 1$; thus

$$\left\|D_s\mathbf{T}(t)\right\| = \left|\frac{d\phi}{ds}\right| \tag{1}$$

The number $\left|\dfrac{d\phi}{ds}\right|$ is the absolute value of the rate of change of the measure of the angle giving the direction of the unit tangent vector $\mathbf{T}(t)$ at a point on a curve with respect to the measure of arc length along the curve. This number is what we shall define to be the curvature of the curve at the point; however, before giving the formal definition, we show that taking it as this number is consistent with our intuitive notion of curvature. For example, at point P on C, ϕ is the radian measure of the angle giving the direction of the vector $\mathbf{T}(t)$, and s units is the arc length from a point P_0 on C to P. Let Q be a point on C for which the radian measure of the angle giving the direction of $\mathbf{T}(t + \Delta t)$ at Q is $\phi + \Delta\phi$, and $s + \Delta s$ units is the arc length from P_0 to Q. Then the arc length from P to Q is Δs units, and the ratio $\Delta\phi/\Delta s$ seems like a good measure of what we would intuitively think of as the *average curvature* along arc PQ.

▶ **ILLUSTRATION 1** See Figure 2(a), (b), (c) and (d): In (a), $\Delta\phi > 0$ and $\Delta s > 0$; in (b), $\Delta\phi > 0$ and $\Delta s < 0$; in (c), $\Delta\phi < 0$ and $\Delta s > 0$; and in (d), $\Delta\phi < 0$ and $\Delta s < 0$. ◀

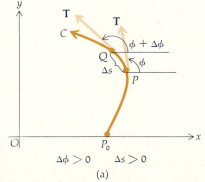

(a)

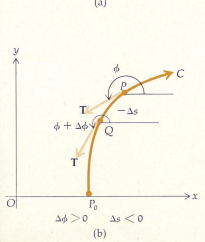

(b)

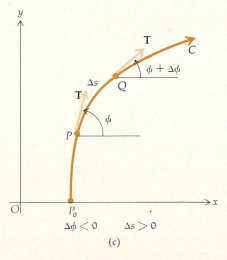

(c)

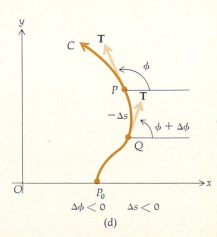

(d)

FIGURE 2

We now define the *curvature vector* of a curve at a point, and its magnitude, the *curvature* of the curve at the point.

14.7.1 DEFINITION

If $\mathbf{T}(t)$ is the unit tangent vector to a curve C at a point P, s is the arc length measured from an arbitrarily chosen point on C to P, and s increases as t increases, then the **curvature vector** of C at P, denoted by $\mathbf{K}(t)$, is given by

$$\mathbf{K}(t) = D_s\mathbf{T}(t)$$

The **curvature** of C at P, denoted by $K(t)$, is the magnitude of the curvature vector; that is,

$$K(t) = \|D_s\mathbf{T}(t)\|$$

Observe that by substituting from (1) in the formula for $K(t)$ in Definition 14.7.1, the curvature $K(t)$ can also be defined by

$$K(t) = \left|\frac{d\phi}{ds}\right|$$

To find the curvature vector and the curvature for a particular curve it is convenient to have a formula expressing the curvature vector in terms of derivatives with respect to t. By the chain rule,

$$D_t\mathbf{T}(t) = D_s\mathbf{T}(t)\frac{ds}{dt}$$

From Section 14.5, $\dfrac{ds}{dt} = \|D_t\mathbf{R}(t)\|$. Thus

$$D_t\mathbf{T}(t) = [D_s\mathbf{T}(t)]\|D_t\mathbf{R}(t)\|$$

$$D_s\mathbf{T}(t) = \frac{D_t\mathbf{T}(t)}{\|D_t\mathbf{R}(t)\|}$$

Substituting from this equation in the formula for $\mathbf{K}(t)$ in Definition 14.7.1 we obtain

$$\mathbf{K}(t) = \frac{D_t\mathbf{T}(t)}{\|D_t\mathbf{R}(t)\|} \tag{2}$$

Because $K(t) = \|\mathbf{K}(t)\|$, the curvature is given by

$$K(t) = \left\|\frac{D_t\mathbf{T}(t)}{\|D_t\mathbf{R}(t)\|}\right\| \tag{3}$$

EXAMPLE 1   Given the circle with radius a:

$$x = a\cos t \qquad y = a\sin t \qquad a > 0$$

find the curvature vector and the curvature at any t.

Solution   The vector equation of the circle is

$$\mathbf{R}(t) = a\cos t\mathbf{i} + a\sin t\mathbf{j}$$

Thus

$$D_t\mathbf{R}(t) = -a\sin t\mathbf{i} + a\cos t\mathbf{j} \qquad \|D_t\mathbf{R}(t)\| = \sqrt{(-a\sin t)^2 + (a\cos t)^2}$$

$$= a$$

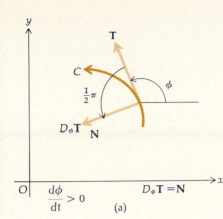

$\dfrac{d\phi}{dt} > 0$ $D_\phi \mathbf{T} = \mathbf{N}$

(a)

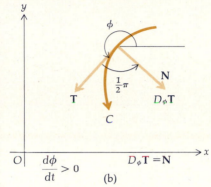

$\dfrac{d\phi}{dt} > 0$ $D_\phi \mathbf{T} = \mathbf{N}$

(b)

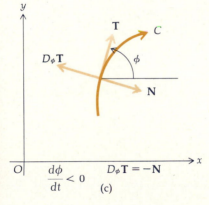

$\dfrac{d\phi}{dt} < 0$ $D_\phi \mathbf{T} = -\mathbf{N}$

(c)

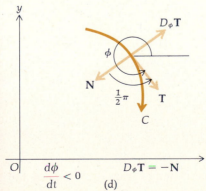

$\dfrac{d\phi}{dt} < 0$ $D_\phi \mathbf{T} = -\mathbf{N}$

(d)

FIGURE 3

Therefore

$$\mathbf{T}(t) = \frac{D_t \mathbf{R}(t)}{\|D_t \mathbf{R}(t)\|} \qquad\qquad D_t \mathbf{T}(t) = -\cos t\,\mathbf{i} - \sin t\,\mathbf{j}$$

$$= -\sin t\,\mathbf{i} + \cos t\,\mathbf{j}$$

$$\frac{D_t \mathbf{T}(t)}{\|D_t \mathbf{R}(t)\|} = -\frac{\cos t}{a}\mathbf{i} - \frac{\sin t}{a}\mathbf{j}$$

Hence the curvature vector and the curvature are given by

$$\mathbf{K}(t) = -\frac{1}{a}\cos t\,\mathbf{i} - \frac{1}{a}\sin t\,\mathbf{j} \qquad K(t) = \|\mathbf{K}(t)\|$$

$$= \frac{1}{a}$$

The result of Example 1 states that the curvature of a circle is constant, which is what you would expect. Furthermore, it is the reciprocal of the radius.

Let us return now to the unit vector $D_\phi \mathbf{T}(t)$ defined by

$$D_\phi \mathbf{T}(t) = -\sin \phi\,\mathbf{i} + \cos \phi\,\mathbf{j} \tag{4}$$

Because $\mathbf{T}(t)$ has constant magnitude, it follows from Theorem 14.4.11 that $D_\phi \mathbf{T}(t)$ is orthogonal to $\mathbf{T}(t)$. Replacing $-\sin \phi$ by $\cos(\tfrac{1}{2}\pi + \phi)$ and $\cos \phi$ by $\sin(\tfrac{1}{2}\pi + \phi)$, we write (4) as

$$D_\phi \mathbf{T}(t) = \cos(\tfrac{1}{2}\pi + \phi)\mathbf{i} + \sin(\tfrac{1}{2}\pi + \phi)\mathbf{j}$$

Thus the vector $D_\phi \mathbf{T}(t)$ is a unit vector orthogonal to $\mathbf{T}(t)$ in the direction $\tfrac{1}{2}\pi$ counterclockwise from the direction of $\mathbf{T}(t)$. The unit normal vector $\mathbf{N}(t)$ is also orthogonal to $\mathbf{T}(t)$. By the chain rule,

$$D_t \mathbf{T}(t) = D_\phi \mathbf{T}(t)\frac{d\phi}{dt}$$

Because the direction of $\mathbf{N}(t)$ is the same as the direction of $D_t \mathbf{T}(t)$, it follows from this equation that the direction of $\mathbf{N}(t)$ is the same as the direction of $D_\phi \mathbf{T}(t)$ if $\dfrac{d\phi}{dt} > 0$ (i.e., if $\mathbf{T}(t)$ turns counterclockwise as t increases), and the direction of $\mathbf{N}(t)$ is opposite that of $D_\phi \mathbf{T}(t)$ if $\dfrac{d\phi}{dt} < 0$ (i.e., if $\mathbf{T}(t)$ turns clockwise as t increases). Because both $D_\phi \mathbf{T}(t)$ and $\mathbf{N}(t)$ are unit vectors, we conclude that

$$D_\phi \mathbf{T}(t) = \begin{cases} \mathbf{N}(t) & \text{if } \dfrac{d\phi}{dt} > 0 \\[2mm] -\mathbf{N}(t) & \text{if } \dfrac{d\phi}{dt} < 0 \end{cases}$$

▶ **ILLUSTRATION 2**   In Figure 3(a), (b), (c), and (d) various situations are shown; in (a) and (b), $\dfrac{d\phi}{dt} > 0$, and in (c) and (d), $\dfrac{d\phi}{dt} < 0$. The positive direction along the curve C is indicated by the tip of the arrow on C. In each figure are shown the angle of radian measure ϕ and representations of the vectors $\mathbf{T}(t)$, $D_\phi \mathbf{T}(t)$, and $\mathbf{N}(t)$.   ◀

Observe in Figure 3 that *the representation of the unit normal vector* $\mathbf{N}(t)$ *is always on the concave side of the curve.*

Suppose we are given a curve C and at a particular point P the curvature exists and is $K(t)$, where $K(t) \neq 0$. Consider the circle that is tangent to curve C at P and has curvature $K(t)$ at P. From Example 1, the radius of this circle is $1/K(t)$ and its center is on a line perpendicular to the tangent line in the direction of $\mathbf{N}(t)$. This circle is called the **circle of curvature**, and its radius is the *radius of curvature* of C at P. The circle of curvature is sometimes referred to as the **osculating circle**.

14.7.2 DEFINITION   If $K(t)$ is the curvature of a curve C at point P and $K(t) \neq 0$, then the **radius of curvature** of C at P, denoted by $\rho(t)$, is defined by

$$\rho(t) = \frac{1}{K(t)}$$

EXAMPLE 2   Given that a vector equation of a curve C is
$$\mathbf{R}(t) = 2t\mathbf{i} + (t^2 - 1)\mathbf{j}$$

(a) Find the unit tangent vector, the curvature, and the radius of curvature at $t = 1$. (b) Draw a sketch of a portion of the curve, the unit tangent vector, and the circle of curvature at $t = 1$.

Solution

$$D_t\mathbf{R}(t) = 2\mathbf{i} + 2t\mathbf{j} \qquad \|D_t\mathbf{R}(t)\| = 2\sqrt{1 + t^2}$$

$$\mathbf{T}(t) = \frac{D_t\mathbf{R}(t)}{\|D_t\mathbf{R}(t)\|}$$

$$= \frac{1}{\sqrt{1 + t^2}}\mathbf{i} + \frac{t}{\sqrt{1 + t^2}}\mathbf{j}$$

$$D_t\mathbf{T}(t) = -\frac{t}{(1 + t^2)^{3/2}}\mathbf{i} + \frac{1}{(1 + t^2)^{3/2}}\mathbf{j}$$

$$\mathbf{K}(t) = \frac{D_t\mathbf{T}(t)}{\|D_t\mathbf{R}(t)\|}$$

$$= -\frac{t}{2(1 + t^2)^2}\mathbf{i} + \frac{1}{2(1 + t^2)^2}\mathbf{j}$$

$$K(t) = \|\mathbf{K}(t)\|$$

$$= \sqrt{\frac{t^2}{4(1 + t^2)^4} + \frac{1}{4(1 + t^2)^4}}$$

$$= \frac{1}{2(1 + t^2)^{3/2}}$$

(a) $\mathbf{T}(1) = \dfrac{1}{\sqrt{2}}\mathbf{i} + \dfrac{1}{\sqrt{2}}\mathbf{j}$ $K(1) = \dfrac{1}{4\sqrt{2}}$ $\rho(1) = 4\sqrt{2}$

(b) Figure 4 shows the required sketch. The accompanying Table 1 gives the corresponding values of x and y when t is $-2, -1, 0, 1$, and 2.

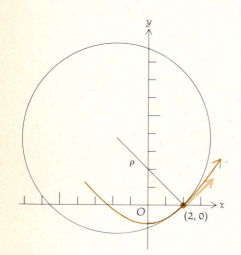

FIGURE 4

Table 1

t	x	y
-2	-4	3
-1	-2	0
0	0	-1
1	2	0
2	4	3

We now find a formula for computing the curvature directly from parametric equations of the curve, $x = f(t)$ and $y = g(t)$. Because $K(t) = \left| \dfrac{d\phi}{ds} \right|$, we first compute $\dfrac{d\phi}{ds}$.

$$\frac{d\phi}{ds} = \frac{\dfrac{d\phi}{dt}}{\dfrac{ds}{dt}}$$

With the assumption that s and t increase together, $\dfrac{ds}{dt} > 0$. Thus

$$\frac{d\phi}{ds} = \frac{\dfrac{d\phi}{dt}}{\sqrt{\left(\dfrac{dx}{dt}\right)^2 + \left(\dfrac{dy}{dt}\right)^2}} \tag{5}$$

To find $\dfrac{d\phi}{dt}$ we observe that because ϕ is the radian measure of the angle giving the direction of the unit tangent vector, $\tan \phi = \dfrac{dy}{dx}$. Therefore

$$\tan \phi = \frac{\dfrac{dy}{dt}}{\dfrac{dx}{dt}}$$

Differentiating implicitly with respect to t the left and right members of this equation, we obtain

$$\sec^2 \phi \, \frac{d\phi}{dt} = \frac{\left(\dfrac{dx}{dt}\right)\left(\dfrac{d^2y}{dt^2}\right) - \left(\dfrac{dy}{dt}\right)\left(\dfrac{d^2x}{dt^2}\right)}{\left(\dfrac{dx}{dt}\right)^2}$$

$$\frac{d\phi}{dt} = \frac{\left(\dfrac{dx}{dt}\right)\left(\dfrac{d^2y}{dt^2}\right) - \left(\dfrac{dy}{dt}\right)\left(\dfrac{d^2x}{dt^2}\right)}{\sec^2 \phi \left(\dfrac{dx}{dt}\right)^2} \tag{6}$$

Because $\sec^2 \phi = 1 + \tan^2 \phi$, we have

$$\sec^2 \phi = 1 + \frac{\left(\dfrac{dy}{dt}\right)^2}{\left(\dfrac{dx}{dt}\right)^2}$$

Substituting this expression for $\sec^2 \phi$ in (6) we get

$$\frac{d\phi}{dt} = \frac{\left(\dfrac{dx}{dt}\right)\left(\dfrac{d^2y}{dt^2}\right) - \left(\dfrac{dy}{dt}\right)\left(\dfrac{d^2x}{dt^2}\right)}{\left(\dfrac{dx}{dt}\right)^2 + \left(\dfrac{dy}{dt}\right)^2}$$

Substituting from this equation into (5), and because $K(t) = \left| \dfrac{d\phi}{ds} \right|$, we have

$$K(t) = \frac{\left| \left(\dfrac{dx}{dt} \right) \left(\dfrac{d^2 y}{dt^2} \right) - \left(\dfrac{dy}{dt} \right) \left(\dfrac{d^2 x}{dt^2} \right) \right|}{\left[\left(\dfrac{dx}{dt} \right)^2 + \left(\dfrac{dy}{dt} \right)^2 \right]^{3/2}} \tag{7}$$

EXAMPLE 3   Find the curvature of the curve in Example 2 by using formula (7).

Solution   Parametric equations of C are $x = 2t$ and $y = t^2 - 1$. Hence

$$\frac{dx}{dt} = 2 \qquad \frac{d^2 x}{dt^2} = 0 \qquad \frac{dy}{dt} = 2t \qquad \frac{d^2 y}{dt^2} = 2$$

Therefore, from (7),

$$K(t) = \frac{|2(2) - 2t(0)|}{[(2)^2 + (2t)^2]^{3/2}}$$

$$= \frac{4}{(4 + 4t^2)^{3/2}}$$

$$= \frac{1}{2(1 + t^2)^{3/2}}$$

Suppose a cartesian equation of a curve is given in either of the forms $y = F(x)$ or $x = G(y)$. Special cases of formula (7) can be used to find the curvature of a curve in such situations.

If $y = F(x)$ is an equation of a curve C, a set of parametric equations of C is $x = t$ and $y = F(t)$. Then

$$\frac{dx}{dt} = 1 \qquad \frac{d^2 x}{dt^2} = 0 \qquad \frac{dy}{dt} = \frac{dy}{dx} \qquad \frac{d^2 y}{dt^2} = \frac{d^2 y}{dx^2}$$

Substituting into (7) we obtain

$$K = \frac{\left| \dfrac{d^2 y}{dx^2} \right|}{\left[1 + \left(\dfrac{dy}{dx} \right)^2 \right]^{3/2}} \tag{8}$$

Similarly, if an equation of a curve C is $x = G(y)$,

$$K = \frac{\left| \dfrac{d^2 x}{dy^2} \right|}{\left[1 + \left(\dfrac{dx}{dy} \right)^2 \right]^{3/2}}$$

EXAMPLE 4　　If an equation of curve C is

$$y = \frac{1}{x}$$

find the radius of curvature of C at the point $(1, 1)$, and draw a sketch of the curve and the circle of curvature at $(1, 1)$.

Solution

$$\frac{dy}{dx} = -\frac{1}{x^2} \qquad \frac{d^2y}{dx^2} = \frac{2}{x^3}$$

We compute K from (8) and then $\rho = 1/K$.

$$K = \frac{\left|\dfrac{2}{x^3}\right|}{\left[1 + \dfrac{1}{x^4}\right]^{3/2}} \qquad \rho = \frac{(x^4 + 1)^{3/2}}{2|x^3|}$$

$$= \frac{2|x^3|}{(x^4 + 1)^{3/2}}$$

Therefore at $(1, 1)$, $\rho = \sqrt{2}$. The required sketch appears in Figure 5.

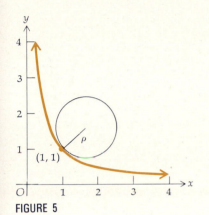

FIGURE 5

EXERCISES 14.7

In Exercises 1 through 4, find the curvature K and the radius of curvature ρ at the point where $t = t_1$. Use formula (3) to find K. Draw a sketch showing a portion of the curve, the unit tangent vector, and the circle of curvature at $t = t_1$.

1. $\mathbf{R}(t) = t^2\mathbf{i} + (2t + 1)\mathbf{j}$; $t_1 = 1$
2. $\mathbf{R}(t) = (t^2 - 2t)\mathbf{i} + (t^3 - t)\mathbf{j}$; $t_1 = 1$
3. $\mathbf{R}(t) = 2e^t\mathbf{i} + 2e^{-t}\mathbf{j}$; $t_1 = 0$
4. $\mathbf{R}(t) = \sin t\mathbf{i} + \sin 2t\mathbf{j}$; $t_1 = \frac{1}{2}\pi$

In Exercises 5 and 6, find the curvature K by using formula (7). Then find K and ρ at the point where $t = t_1$, and draw a sketch showing a portion of the curve, the unit tangent vector, and the circle of curvature at $t = t_1$.

5. $x = \dfrac{1}{1 + t}$, $y = \dfrac{1}{1 - t}$; $t_1 = 0$
6. $x = e^t + e^{-t}$, $y = e^t - e^{-t}$; $t_1 = 0$

In Exercises 7 through 14, find the curvature K and the radius of curvature ρ at the given point. Draw a sketch showing a portion of the curve, a piece of the tangent line, and the circle of curvature at the given point.

7. $y = 2\sqrt{x}$; $(0, 0)$
8. $y^2 = x^3$; $(\frac{1}{4}, \frac{1}{8})$
9. $y = e^x$; $(0, 1)$
10. $y = \ln x$; $(e, 1)$
11. $x = \sin y$; $(\frac{1}{2}, \frac{1}{6}\pi)$
12. $4x^2 + 9y^2 = 36$; $(0, 2)$
13. $x = \sqrt{y - 1}$; $(2, 5)$
14. $x = \tan y$; $(1, \frac{1}{4}\pi)$

In Exercises 15 through 22, find the radius of curvature at any point on the given curve.

15. $y = \sin^{-1} x$
16. $y = \ln \sec x$
17. $4x^2 - 9y^2 = 16$
18. $x = \tan^{-1} y$
19. $x^{1/2} + y^{1/2} = a^{1/2}$
20. $\mathbf{R}(t) = e^t \sin t\mathbf{i} + e^t \cos t\mathbf{j}$
21. The cycloid $x = a(t - \sin t)$, $y = a(1 - \cos t)$
22. The tractrix $x = t - a \tanh \dfrac{t}{a}$, $y = a \operatorname{sech} \dfrac{t}{a}$

23. Show that the curvature of the catenary $y = a \cosh(x/a)$ at any point (x, y) on the curve is a/y^2. Draw the circle of curvature at $(0, a)$. Show that the curvature K is an absolute maximum at the point $(0, a)$ without referring to $K'(x)$.

In Exercises 24 through 28, find a point on the given curve at which the curvature is an absolute maximum.

24. $y = e^x$
25. $y = 6x - x^2$
26. $y = \sin x$
27. $\mathbf{R}(t) = (2t - 3)\mathbf{i} + (t^2 - 1)\mathbf{j}$
28. $y = x^2 - 2x + 3$

29. Find an equation of the circle of curvature for $y = e^x$ at the point $(0, 1)$.

30. If a polar equation of a curve is $r = F(\theta)$, prove that the curvature K is given by the formula

$$K = \frac{\left| r^2 + 2\left(\dfrac{dr}{d\theta}\right)^2 - r\left(\dfrac{d^2r}{d\theta^2}\right) \right|}{\left[r^2 + \left(\dfrac{dr}{d\theta}\right)^2 \right]^{3/2}}$$

In Exercises 31 through 34, find the curvature K and the radius of curvature ρ at the indicated point. Use the formula of Exercise 30 to find K.

31. $r = 4 \cos 2\theta;\ \theta = \frac{1}{12}\pi$　　　**32.** $r = 1 - \sin \theta;\ \theta = 0$
33. $r = a \sec^2 \frac{1}{2}\theta;\ \theta = \frac{2}{3}\pi$　　　**34.** $r = a\theta;\ \theta = 1$

35. The center of the circle of curvature of a curve C at a point P is called the *center of curvature at P*. Prove that the coordinates of the center of curvature of a curve at $P(x, y)$ are given

by

$$x_c = x - \frac{\left(\dfrac{dy}{dx}\right)\left[1 + \left(\dfrac{dy}{dx}\right)^2\right]}{\dfrac{d^2y}{dx^2}} \qquad y_c = y + \frac{\left(\dfrac{dy}{dx}\right)^2 + 1}{\dfrac{d^2y}{dx^2}}$$

In Exercises 36 through 38, find the curvature K, the radius of curvature ρ, and the center of curvature at the given point. Draw a sketch of the curve and the circle of curvature.

36. $y = \ln x;\ (1, 0)$　　　**37.** $y = x^4 - x^2;\ (0, 0)$
38. $y = \cos x;\ (\frac{1}{3}\pi, \frac{1}{2})$

In Exercises 39 through 42, find the coordinates of the center of curvature at any point.

39. $y^2 = 4px$　　　**40.** $y^3 = a^2x$
41. $\mathbf{R}(t) = a \cos t\mathbf{i} + b \sin t\mathbf{j}$　　　**42.** $\mathbf{R}(t) = a \cos^3 t\mathbf{i} + a \sin^3 t\mathbf{j}$

43. Show that the curvature of a line is zero at every point.

14.8 PLANE MOTION

Our previous discussions of the motion of a particle were confined to rectilinear motion. In this connection we defined the velocity and acceleration of a particle moving along a straight line. Consider now the motion of a particle along a curve in the plane. This is called **curvilinear motion**.

Suppose that C is the plane curve having parametric equations $x = f(t)$ and $y = g(t)$, where t units denote time. Then

$$\mathbf{R}(t) = f(t)\mathbf{i} + g(t)\mathbf{j}$$

is a vector equation of C. As t varies, the endpoint $P(f(t), g(t))$ of $\overrightarrow{OP}$ moves along the curve C. The position at time t units of a particle moving along C is the point $P(f(t), g(t))$. The **velocity vector** of the particle at time t units is defined to be $\mathbf{R}'(t)$ and is denoted by the symbol $\mathbf{V}(t)$.

14.8.1 DEFINITION

Let C be the curve having parametric equations $x = f(t)$ and $y = g(t)$. If a particle is moving along C so that its position at any time t units is the point (x, y), then the **instantaneous velocity** of the particle at time t units is determined by the velocity vector

$$\mathbf{V}(t) = f'(t)\mathbf{i} + g'(t)\mathbf{j}$$

if $f'(t)$ and $g'(t)$ exist.

Because the direction of $\mathbf{R}'(t)$ at the point $P(f(t), g(t))$ is along the tangent line to the curve C at P, the velocity vector $\mathbf{V}(t)$ has this direction at P.

The magnitude of the velocity vector is a measure of the **speed** of the particle at time t and is given by

$$\|\mathbf{V}(t)\| = \sqrt{[f'(t)]^2 + [g'(t)]^2} \tag{1}$$

Note that the velocity is a vector and the speed is a scalar. We showed in Section 14.5, the expression on the right side of (1) is ds/dt. So the speed is the rate of change of s with respect to t, and we write

$$\|\mathbf{V}(t)\| = \frac{ds}{dt}$$

The **acceleration vector** of the particle at time t units is denoted by $\mathbf{A}(t)$ and is defined to be the derivative of the velocity vector or, equivalently, the second derivative of the position vector.

14.8.2 DEFINITION

The **instantaneous acceleration** at time t units of a particle moving along a curve C, having parametric equations $x = f(t)$ and $y = g(t)$, is determined by the acceleration vector

$$\mathbf{A}(t) = \mathbf{V}'(t) \quad \Leftrightarrow \quad \mathbf{A}(t) = \mathbf{R}''(t)$$

where $\mathbf{R}(t) = f(t)\mathbf{i} + g(t)\mathbf{j}$ and $\mathbf{R}''(t)$ exists.

Figure 1 shows the representations of the velocity vector and the acceleration vector whose initial point is the point P on C.

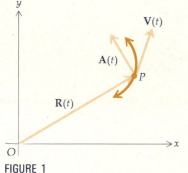

FIGURE 1

EXAMPLE 1   A particle is moving along the curve having parametric equations

$$x = 4 \cos \tfrac{1}{2}t \quad \text{and} \quad y = 4 \sin \tfrac{1}{2}t$$

If x and y are centimeter measures, find the speed and the magnitude of the particle's acceleration vector at t seconds. Draw a sketch of the particle's path, and also draw the representations of the velocity and acceleration vectors having initial point where $t = \tfrac{1}{3}\pi$.

Solution   A vector equation of C is

$$\mathbf{R}(t) = 4 \cos \tfrac{1}{2}t\,\mathbf{i} + 4 \sin \tfrac{1}{2}t\,\mathbf{j}$$

$$\mathbf{V}(t) = \mathbf{R}'(t) \qquad\qquad\qquad \mathbf{A}(t) = \mathbf{V}'(t)$$
$$\quad = -2 \sin \tfrac{1}{2}t\,\mathbf{i} + 2 \cos \tfrac{1}{2}t\,\mathbf{j} \qquad = -\cos \tfrac{1}{2}t\,\mathbf{i} - \sin \tfrac{1}{2}t\,\mathbf{j}$$

$$\|\mathbf{V}(t)\| = \sqrt{(-2 \sin \tfrac{1}{2}t)^2 + (2 \cos \tfrac{1}{2}t)^2} \qquad \|\mathbf{A}(t)\| = \sqrt{(-\cos \tfrac{1}{2}t)^2 + (-\sin \tfrac{1}{2}t)^2}$$
$$= \sqrt{4 \sin^2 \tfrac{1}{2}t + 4 \cos^2 \tfrac{1}{2}t} \qquad\qquad = 1$$
$$= 2$$

Therefore, the speed of the particle is constant and is 2 cm/sec. The magnitude of the acceleration vector is also constant and is 1 cm/sec^2.

Eliminating t between the parametric equations of C, we obtain the cartesian equation

$$x^2 + y^2 = 16$$

which is a circle with its center at the origin and radius 4. We now find the velocity and acceleration vectors at $t = \tfrac{1}{3}\pi$.

$$\mathbf{V}(\tfrac{1}{3}\pi) = -2 \sin \tfrac{1}{6}\pi\,\mathbf{i} + 2 \cos \tfrac{1}{6}\pi\,\mathbf{j} \qquad \mathbf{A}(\tfrac{1}{3}\pi) = -\cos \tfrac{1}{6}\pi\,\mathbf{i} - \sin \tfrac{1}{6}\pi\,\mathbf{j}$$
$$= -\mathbf{i} + \sqrt{3}\,\mathbf{j} \qquad\qquad\qquad = -\tfrac{1}{2}\sqrt{3}\,\mathbf{i} - \tfrac{1}{2}\mathbf{j}$$

The direction of $\mathbf{V}(\tfrac{1}{3}\pi)$ is given by

$$\tan \theta_1 = -\sqrt{3} \qquad \tfrac{1}{2}\pi < \theta_1 < \pi$$

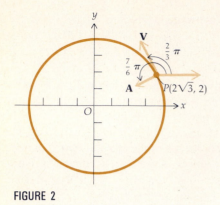

FIGURE 2

and the direction of $\mathbf{A}(\tfrac{1}{3}\pi)$ is given by

$$\tan \theta_2 = \frac{1}{\sqrt{3}} \qquad \pi < \theta_2 < \tfrac{3}{2}\pi$$

Thus $\theta_1 = \tfrac{2}{3}\pi$ and $\theta_2 = \tfrac{7}{6}\pi$. Figure 2 shows the particle's path and representations of the velocity and acceleration vectors having initial point where $t = \tfrac{1}{3}\pi$.

EXAMPLE 2   The position of a moving particle at time t units is given by the vector equation

$$\mathbf{R}(t) = e^{-2t}\mathbf{i} + 3e^t\mathbf{j}$$

Find $\mathbf{V}(t)$, $\mathbf{A}(t)$, $\|\mathbf{V}(t)\|$, $\|\mathbf{A}(t)\|$. Draw a sketch of the path of the particle and the representations of the velocity and acceleration vectors having initial point where $t = \tfrac{1}{2}$.

Solution

$$\mathbf{V}(t) = \mathbf{R}'(t) \qquad\qquad \mathbf{A}(t) = \mathbf{V}'(t)$$
$$= -2e^{-2t}\mathbf{i} + 3e^t\mathbf{j} \qquad = 4e^{-2t}\mathbf{i} + 3e^t\mathbf{j}$$
$$\|\mathbf{V}(t)\| = \sqrt{4e^{-4t} + 9e^{2t}} \quad \|\mathbf{A}(t)\| = \sqrt{16e^{-4t} + 9e^{2t}}$$
$$\|\mathbf{V}(\tfrac{1}{2})\| = \sqrt{4e^{-2} + 9e} \quad \|\mathbf{A}(\tfrac{1}{2})\| = \sqrt{16e^{-2} + 9e}$$
$$\approx 5.00 \qquad\qquad\qquad \approx 5.16$$

Parametric equations of the path of the particle are

$$x = e^{-2t} \quad\text{and}\quad y = 3e^t$$

We eliminate t between these two equations and obtain

$$e^{2t} = \frac{1}{x} \quad\text{and}\quad e^{2t} = \frac{y^2}{9}$$

$$\frac{1}{x} = \frac{y^2}{9}$$

$$xy^2 = 9$$

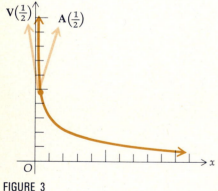

FIGURE 3

Because $x > 0$ and $y > 0$, the path of the particle is the portion of the curve $xy^2 = 9$ in the first quadrant. Figure 3 shows the path of the particle and representations of the velocity and acceleration vectors when $t = \tfrac{1}{2}$. The slope of the representation of $\mathbf{V}(\tfrac{1}{2})$ is $-\tfrac{3}{2}e^{3/2} \approx -6.7$, and the slope of the representation of $\mathbf{A}(\tfrac{1}{2})$ is $\tfrac{3}{4}e^{3/2} \approx 3.4$.

We now derive the equations of motion of a projectile by assuming that the projectile is moving in a vertical plane. We also assume that the only force acting on the projectile is its weight, which has a downward direction and a magnitude of mg pounds, where m slugs is its mass and g feet per second squared is the constant of acceleration caused by gravity. We are neglecting the force attributed to air resistance (which for heavy bodies traveling at small speeds has no noticeable effect). The positive direction is taken as vertically upward and horizontally to the right.

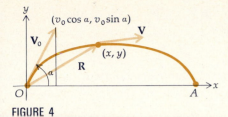

$(v_0 \cos a, v_0 \sin a)$

$\mathbf{V_0}$

$\mathbf{V}$

(x, y)

$\mathbf{R}$

α

FIGURE 4

Suppose, then, that a projectile is shot from a gun having an angle of elevation of radian measure α. Let the number of feet per second in the initial speed, or *muzzle speed*, be denoted by v_0. The coordinate axes are set up so that the gun is located at the origin. Refer to Figure 4. The initial velocity vector, $\mathbf{V_0}$, of the projectile is given by

$$\mathbf{V_0} = v_0 \cos \alpha \mathbf{i} + v_0 \sin \alpha \mathbf{j} \tag{2}$$

Let t seconds be the time that has elapsed since the gun was fired, x feet be the horizontal distance of the projectile from the starting point at t seconds and y feet be the vertical distance of the projectile from the starting point at t seconds. $\mathbf{R}(t)$ is the position vector of the projectile at t seconds, $\mathbf{V}(t)$ is the velocity vector of the projectile at t seconds, and $\mathbf{A}(t)$ is the acceleration vector of the projectile at t seconds.

Because x is a function of t, we write $x(t)$. Similarly, y is a function of t and we write $y(t)$. Then

$$\mathbf{R}(t) = x(t)\mathbf{i} + y(t)\mathbf{j}$$

$$\mathbf{V}(t) = \mathbf{R}'(t)$$

$$\mathbf{A}(t) = \mathbf{V}'(t)$$

Because the only force acting on the projectile has a magnitude of mg lb and is in the downward direction, then if $\mathbf{F}$ denotes this force,

$$\mathbf{F} = -mg\mathbf{j} \tag{3}$$

Newton's second law of motion states that the net force acting on a body is its "mass times acceleration." So

$$\mathbf{F} = m\mathbf{A}$$

From this equation and (3),

$$m\mathbf{A} = -mg\mathbf{j}$$

$$\mathbf{A} = -g\mathbf{j}$$

Because $\mathbf{A}(t) = \mathbf{V}'(t)$, we have from the above

$$\mathbf{V}'(t) = -g\mathbf{j}$$

Integrating on both sides of this equation with respect to t we obtain

$$\mathbf{V}(t) = -gt\mathbf{j} + \mathbf{C}_1 \tag{4}$$

where $\mathbf{C}_1$ is a vector constant of integration.

When $t = 0$, $\mathbf{V} = \mathbf{V_0}$. So $\mathbf{C}_1 = \mathbf{V_0}$. Therefore, from (4),

$$\mathbf{V}(t) = -gt\mathbf{j} + \mathbf{V_0}$$

or, because $\mathbf{V}(t) = \mathbf{R}'(t)$,

$$\mathbf{R}'(t) = -gt\mathbf{j} + \mathbf{V_0}$$

Integrating on both sides of this vector equation with respect to t we obtain

$$\mathbf{R}(t) = -\tfrac{1}{2}gt^2\mathbf{j} + \mathbf{V_0}t + \mathbf{C}_2$$

where $\mathbf{C}_2$ is a vector constant of integration.

When $t = 0$, $\mathbf{R} = \mathbf{0}$ because the projectile is at the origin at the start. So $\mathbf{C}_2 = \mathbf{0}$. Therefore

$$\mathbf{R}(t) = -\tfrac{1}{2}gt^2\mathbf{j} + \mathbf{V}_0 t$$

Substituting the value of $\mathbf{V}_0$ from (2) into the above we obtain

$$\mathbf{R}(t) = -\tfrac{1}{2}gt^2\mathbf{j} + (v_0 \cos \alpha\mathbf{i} + v_0 \sin \alpha\mathbf{j})t$$

$$\mathbf{R}(t) = tv_0 \cos \alpha\mathbf{i} + (tv_0 \sin \alpha - \tfrac{1}{2}gt^2)\mathbf{j} \tag{5}$$

Equation (5) gives the position vector of the projectile at time t seconds. From this equation we can discuss the motion of the projectile. We are usually concerned with the following questions:

1. What is the range of the projectile? The range is the distance $|OA|$ along the x axis (see Figure 4).
2. What is the total time of flight, that is, the time it takes the projectile to go from O to A?
3. What is the maximum height of the projectile?
4. What is a cartesian equation of the curve traveled by the projectile?
5. What is the velocity vector of the projectile at impact?

These questions are answered in the following example.

EXAMPLE 3   A projectile is shot from a gun at an angle of elevation of radian measure $\tfrac{1}{6}\pi$. Its muzzle speed is 480 ft/sec. Find (a) the position vector of the projectile at any time; (b) the time of flight; (c) the range; (d) the maximum height; (e) the velocity vector of the projectile at impact; (f) the position vector and the velocity vector at 2 sec; (g) the speed at 2 sec; (h) a cartesian equation of the curve traveled by the projectile.

Solution   From (2) with $v_0 = 480$ and $\alpha = \tfrac{1}{6}\pi$, the initial velocity vector is

$$\mathbf{V}_0 = 480 \cos \tfrac{1}{6}\pi\mathbf{i} + 480 \sin \tfrac{1}{6}\pi\mathbf{j}$$
$$= 240\sqrt{3}\mathbf{i} + 240\mathbf{j}$$

(a) We can obtain the position vector at t seconds by applying (5); we get

$$\mathbf{R}(t) = 240\sqrt{3}t\mathbf{i} + (240t - \tfrac{1}{2}gt^2)\mathbf{j}$$

By letting $g = 32$ we have

$$\mathbf{R}(t) = 240\sqrt{3}t\mathbf{i} + (240t - 16t^2)\mathbf{j} \tag{6}$$

Thus if (x, y) is the position of the projectile at t seconds,

$$x = 240\sqrt{3}t \quad \text{and} \quad y = 240t - 16t^2 \tag{7}$$

(b) To determine the time of flight, we must find t when $y = 0$. We set $y = 0$ in the second equation of (7), and we have

$$240t - 16t^2 = 0$$
$$t(240 - 16t) = 0$$
$$t = 0 \qquad t = 15$$

The value $t = 0$ occurs when the projectile is fired. The value $t = 15$ gives the time of flight. Thus the time of flight is 15 sec.

(c) To find the range, we determine x when $t = 15$. From the first equation of (7) with $t = 15$, we obtain $x = 3600\sqrt{3}$. Hence the range is $3600\sqrt{3}$ ft $\approx$ 6235 ft.

(d) The maximum height is attained when the vertical component of the velocity vector is 0, that is, when $\frac{dy}{dt} = 0$. We compute $\frac{dy}{dt}$ from the second equation of (7) and get

$$\frac{dy}{dt} = 240 - 32t$$

Setting $\frac{dy}{dt} = 0$ we obtain $t = \frac{15}{2}$, which is half the total time of flight. When $t = \frac{15}{2}$, $y = 900$. So the maximum height attained is 900 ft.

(e) Because the time of flight is 15 sec, the velocity vector at impact is $\mathbf{V}(15)$. Because $\mathbf{V}(t) = \mathbf{R}'(t)$, we get, from (6),

$$\mathbf{V}(t) = 240\sqrt{3}\mathbf{i} + (240 - 32t)\mathbf{j} \tag{8}$$

$$\mathbf{V}(15) = 240\sqrt{3}\mathbf{i} - 240\mathbf{j}$$

(f) If $t = 2$ in (6) and (8), we have

$$\mathbf{R}(2) = 480\sqrt{3}\mathbf{i} + 416\mathbf{j} \qquad \mathbf{V}(2) = 240\sqrt{3}\mathbf{i} + 176\mathbf{j}$$

(g) $\|\mathbf{V}(2)\| = \sqrt{(240\sqrt{3})^2 + (176)^2}$

$\qquad\quad = 32\sqrt{199}$

Therefore at 2 sec the speed is $32\sqrt{199}$ ft/sec ≈ 451.4 ft/sec.

(h) To find a cartesian equation of the curve traveled by the projectile, we eliminate t between the parametric equations (7). Substituting the value of t from the first equation into the second, we have

$$y = 240\left(\frac{x}{240\sqrt{3}}\right) - 16\left(\frac{x}{240\sqrt{3}}\right)^2$$

$$y = \frac{1}{\sqrt{3}}x - \frac{1}{10,800}x^2$$

which is an equation of a parabola.

EXERCISES 14.8

In Exercises 1 through 8, a particle is moving along the curve having the given parametric equations, where t seconds is the time. Find: (a) the velocity vector $\mathbf{V}(t)$; (b) the acceleration vector $\mathbf{A}(t)$; (c) the speed at $t = t_1$; (d) the magnitude of the acceleration vector at $t = t_1$. Draw a sketch of the path of the particle and the representations of the velocity vector and the acceleration vector at $t = t_1$.

1. $x = t^2 + 4$, $y = t - 2$; $t_1 = 3$
2. $x = \ln(t - 2)$, $y = t^3 - 1$; $t_1 = 3$
3. $x = 5 \cos 2t$, $y = 3 \sin 2t$; $t_1 = \frac{1}{4}\pi$
4. $x = 2/t$, $y = -\frac{1}{4}t$; $t_1 = 4$
5. $x = t$, $y = \ln \sec t$; $t_1 = \frac{1}{4}\pi$

6. $x = 2 \cos t$, $y = 3 \sin t$; $t_1 = \frac{1}{3}\pi$
7. $x = \sin t$, $y = \tan t$; $t_1 = \frac{1}{6}\pi$
8. $x = e^{2t}$, $y = e^{3t}$; $t_1 = 0$

In Exercises 9 through 16, the position of a moving particle at t seconds is determined from a vector equation. Find: (a) $\mathbf{V}(t_1)$; (b) $\mathbf{A}(t_1)$; (c) $\|\mathbf{V}(t_1)\|$; (d) $\|\mathbf{A}(t_1)\|$. Draw a sketch of a portion of the path of the particle containing the position of the particle at $t = t_1$, and draw the representations of $\mathbf{V}(t_1)$ and $\mathbf{A}(t_1)$ having initial point where $t = t_1$.

9. $\mathbf{R}(t) = (2t - 1)\mathbf{i} + (t^2 + 1)\mathbf{j}$; $t_1 = 3$
10. $\mathbf{R}(t) = (1 - t)\mathbf{i} + (t^2 - 1)\mathbf{j}$; $t_1 = -1$

11. $R(t) = e^t i + e^{2t} j$; $t_1 = \ln 2$
12. $R(t) = (t^2 + 3t)i + (1 - 3t^2)j$; $t_1 = \frac{1}{2}$
13. $R(t) = \cos 2t i - 3 \sin t j$; $t_1 = \pi$
14. $R(t) = e^{-t} i + e^{2t} j$; $t_1 = \ln 2$
15. $R(t) = 2(1 - \cos t)i + 2(1 - \sin t)j$; $t_1 = \frac{5}{6}\pi$
16. $R(t) = \ln(t + 2)i + \frac{1}{3}t^2 j$; $t_1 = 1$

In Exercises 17 through 20, find the position vector $R(t)$.

17. $V(t) = \dfrac{1}{(t - 1)^2} i - (t + 1)j$, and $R(0) = 3i + 2j$

18. $V(t) = (2t - 1)i + 3t^{-2}j$, and $R(1) = 4i - 3j$
19. $A(t) = e^{-t}i + 2e^{2t}j$, $V(0) = 2i + j$, and $R(0) = 3j$
20. $A(t) = 2 \cos 2t i + 2 \sin 2t j$, $V(0) = i + j$, and $R(0) = \frac{1}{2}i - \frac{1}{2}j$

21. A projectile is shot from a gun at an angle of elevation of 45° with a muzzle speed of 2500 ft/sec. Find (a) the range of the projectile; (b) the maximum height reached; (c) the velocity at impact.
22. A projectile is shot from a gun at an angle of elevation of 60°. The muzzle speed is 160 ft/sec. Find (a) the position vector of the projectile at t seconds; (b) the time of flight; (c) the range; (d) the maximum height reached; (e) the velocity at impact; (f) the speed at 4 sec.
23. A projectile is shot from the top of a building 96 ft high from a gun at an angle of 30° with the horizontal. If the muzzle speed is 1600 ft/sec, find the time of flight and the distance

from the base of the building to the point where the projectile lands.
24. The muzzle speed of a gun is 160 ft/sec. At what angle of elevation should the gun be fired so that a projectile will hit an object on the same level as the gun and a distance of 400 ft from it?
25. What is the muzzle speed of a gun if a projectile fired from it has a range of 2000 ft and reaches a maximum height of 1000 ft?
26. A ball is thrown horizontally from the top of a cliff 256 ft high with an initial speed of 50 ft/sec. Find the time of flight of the ball and the distance from the base of the cliff to the point where the ball lands.
27. A person throws a ball with an initial speed of 60 ft/sec at an angle of elevation of 60° toward a tall building that is 25 ft from the person. If the person's hand is 5 ft from the ground, show that the ball hits the building, and find the direction of the ball when it hits the building.
28. From the top of a building 60 ft high, a girl tosses a rock toward the ground at an angle of 45° with the horizontal at an initial speed of 15 ft/sec. Determine the distance on the ground from the base of the building to the landing position of the rock.
29. Solve Exercise 28 if the girl tosses the rock horizontally with an initial speed of 15 ft/sec.
30. At what angle of elevation should a gun be fired to obtain the maximum range for a given muzzle speed?

14.9 TANGENTIAL AND NORMAL COMPONENTS OF ACCELERATION (*Supplementary*)

If a particle is moving along a curve C having the vector equation

$$R(t) = f(t)i + g(t)j$$

the velocity vector at a point P is given by

$$V(t) = D_t R(t) \tag{1}$$

From Section 14.6, if $T(t)$ is the unit tangent vector at P, s is the length of arc of C from a fixed point P_0 to P, and s increases as t increases,

$$D_t R(t) = \frac{ds}{dt} T(t)$$

Substituting from this equation into (1) we have

$$V(t) = \frac{ds}{dt} T(t)$$

This equation expresses the velocity vector at a point as a scalar times the unit tangent vector at the point. The coefficient of $T(t)$ is called the **tangential component of the velocity vector** and it is $\dfrac{ds}{dt}$. We now proceed to express the acceleration vector at a point in terms of a vector tangent to the direction of motion and a vector normal to the direction of motion.

The acceleration vector at P is given by

$$A(t) = D_t^2 R(t) \tag{2}$$

From (5) in Section 14.6,

$$D_t^2\mathbf{R}(t) = (D_t\|D_t\mathbf{R}(t)\|)\mathbf{T}(t) + (\|D_t\mathbf{R}(t)\|\,\|D_t\mathbf{T}(t)\|)\mathbf{N}(t) \tag{3}$$

From Section 14.5,

$$\frac{ds}{dt} = \|D_t\mathbf{R}(t)\| \tag{4}$$

Differentiating with respect to t on both sides of (4) we obtain

$$\frac{d^2s}{dt^2} = D_t\|D_t\mathbf{R}(t)\| \tag{5}$$

Furthermore,

$$\|D_t\mathbf{R}(t)\|\,\|D_t\mathbf{T}(t)\| = \|D_t\mathbf{R}(t)\|^2 \left\|\frac{D_t\mathbf{T}(t)}{\|D_t\mathbf{R}(t)\|}\right\| \tag{6}$$

Applying (4) above and Equation (3) of Section 14.7 to the right side of (6) we have

$$\|D_t\mathbf{R}(t)\|\,\|D_t\mathbf{T}(t)\| = \left(\frac{ds}{dt}\right)^2 K(t) \tag{7}$$

Substituting from (2), (5), and (7) into (3) we obtain

$$\mathbf{A}(t) = \frac{d^2s}{dt^2}\mathbf{T}(t) + \left(\frac{ds}{dt}\right)^2 K(t)\mathbf{N}(t) \tag{8}$$

Equation (8) expresses the acceleration vector as the sum of a scalar times the unit tangent vector and a scalar times the unit normal vector; that is, it converts $\mathbf{A}(t)$ into the sum of a vector tangent to the direction of motion and a vector normal to the direction of motion. The coefficient of $\mathbf{T}(t)$ is called the **tangential component of the acceleration vector** and is denoted by $A_T(t)$, whereas the coefficient of $\mathbf{N}(t)$ is called the **normal component of the acceleration vector** and is denoted by $A_N(t)$. Thus

$$A_T(t) = \frac{d^2s}{dt^2} \tag{9}$$

and

$$A_N(t) = \left(\frac{ds}{dt}\right)^2 K(t) \quad \Leftrightarrow \quad A_N(t) = \frac{\left(\dfrac{ds}{dt}\right)^2}{\rho(t)} \tag{10}$$

Because $\mathbf{A}(t) = D_t\mathbf{V}(t)$, $\mathbf{A}(t)$ is the rate of change of $\mathbf{V}(t)$. A change in $\mathbf{V}(t)$ can be caused by either a change in its magnitude or a change in its direction. Since $\|\mathbf{V}(t)\|$ is the measure of the speed of the particle at time t units and $\dfrac{ds}{dt} = \|\mathbf{V}(t)\|$, then $A_T(t)$ is the rate of change of the measure of the speed of the particle; that is, $A_T(t)$ is related to the change in the magnitude of $\mathbf{V}(t)$. Because $A_N(t)$ involves the curvature $K(t)$, $A_N(t)$ is related to the change in the direction of $\mathbf{V}(t)$. These results are important in mechanics. From Newton's second law of motion,

$$\mathbf{F} = m\mathbf{A} \tag{11}$$

where $\mathbf{F}$ is the force vector applied to a moving object, m is the constant measure of the mass of the object and $\mathbf{A}$ is the acceleration vector of the object. Substituting from (8) in (11) and letting $v = \dfrac{ds}{dt}$ we have

$$\mathbf{F}(t) = m\frac{dv}{dt}\mathbf{T}(t) + mv^2 K(t)\mathbf{N}(t)$$

Thus in curvilinear motion the normal component of $\mathbf{F}$ is

$$mv^2 K(t) \quad \Leftrightarrow \quad \frac{mv^2}{\rho(t)}$$

which is the magnitude of the force normal to the curve necessary to keep the object on the curve. For example, if an automobile is going around a curve at a high speed, then the normal force must have a large magnitude to keep the car on the road. Also, if the curve is sharp, the radius of curvature is a small number; so the magnitude of the normal force must be a large number.

Substituting from (9) and (10) into (8) we have

$$\mathbf{A}(t) = A_T(t)\mathbf{T}(t) + A_N(t)\mathbf{N}(t)$$

from which it follows that

$$\|\mathbf{A}(t)\| = \sqrt{[A_T(t)]^2 + [A_N(t)]^2}$$

Solving this equation for $A_N(t)$, and noting from (10) that $A_N(t)$ is nonnegative, we have

$$A_N(t) = \sqrt{\|\mathbf{A}(t)\|^2 - [A_T(t)]^2}$$

EXAMPLE 1   A particle is moving along the curve having the vector equation

$$\mathbf{R}(t) = t\mathbf{i} + e^t\mathbf{j}$$

Find the tangential and normal components of the acceleration vector.

Solution

$$\begin{aligned} \mathbf{V}(t) &= D_t\mathbf{R}(t) & \mathbf{A}(t) &= D_t\mathbf{V}(t) \\ &= \mathbf{i} + e^t\mathbf{j} & &= e^t\mathbf{j} \\ \|\mathbf{V}(t)\| &= \sqrt{1 + e^{2t}} & \|\mathbf{A}(t)\| &= e^t \end{aligned}$$

Because $\dfrac{ds}{dt} = \|\mathbf{V}(t)\|$, $\dfrac{ds}{dt} = \sqrt{1 + e^{2t}}$, and $\dfrac{d^2s}{dt^2} = \dfrac{e^{2t}}{\sqrt{1 + e^{2t}}}$. Hence

$$A_T(t) = \frac{e^{2t}}{\sqrt{1 + e^{2t}}} \qquad A_N(t) = \sqrt{\|\mathbf{A}(t)\|^2 - [A_T(t)]^2}$$

$$= \sqrt{e^{2t} - \frac{e^{4t}}{1 + e^{2t}}}$$

$$= \frac{e^t}{\sqrt{1 + e^{2t}}}$$

EXAMPLE 2   A particle is moving along the curve having the vector equation

$$\mathbf{R}(t) = (t^2 - 1)\mathbf{i} + (\tfrac{1}{3}t^3 - t)\mathbf{j}$$

Find each of the following vectors: $\mathbf{V}(t)$, $\mathbf{A}(t)$, $\mathbf{T}(t)$, and $\mathbf{N}(t)$. Also find the following scalars: $\|\mathbf{V}(t)\|$, $A_T(t)$, $A_N(t)$, and $K(t)$. Find the particular values when $t = 2$. Draw a sketch showing a portion of the curve at the point where $t = 2$, and representations of $\mathbf{V}(2)$, $\mathbf{A}(2)$, $A_T(2)\mathbf{T}(2)$, and $A_N(2)\mathbf{N}(2)$, having their initial point at $t = 2$.

Solution   Because $\mathbf{V}(t) = D_t\mathbf{R}(t)$ and $\mathbf{A}(t) = D_t\mathbf{V}(t)$, we have

$$\mathbf{V}(t) = 2t\mathbf{i} + (t^2 - 1)\mathbf{j} \qquad\qquad \mathbf{A}(t) = 2\mathbf{i} + 2t\mathbf{j}$$

$$\|\mathbf{V}(t)\| = \sqrt{4t^2 + (t^2 - 1)^2} \qquad \|\mathbf{A}(t)\| = \sqrt{4 + 4t^2}$$
$$= \sqrt{t^4 + 2t^2 + 1} \qquad\qquad\qquad = 2\sqrt{1 + t^2}$$
$$= t^2 + 1$$

Therefore, $\dfrac{ds}{dt} = t^2 + 1$. Hence

$$A_T(t) = \frac{d^2s}{dt^2} \qquad A_N(t) = \sqrt{\|\mathbf{A}(t)\|^2 - [A_T(t)]^2}$$
$$= \sqrt{4 + 4t^2 - 4t^2}$$
$$= 2t \qquad\qquad\qquad = 2$$

$$\mathbf{T}(t) = \frac{\mathbf{V}(t)}{\|\mathbf{V}(t)\|}$$

$$= \frac{2t}{t^2 + 1}\mathbf{i} + \frac{t^2 - 1}{t^2 + 1}\mathbf{j}$$

To compute $\mathbf{N}(t)$ we use the following formula that comes from (8):

$$\mathbf{N}(t) = \frac{1}{(D_t s)^2 K(t)}\left[\mathbf{A}(t) - (D_t{}^2 s)\mathbf{T}(t)\right] \tag{12}$$

$$\mathbf{A}(t) - (D_t{}^2 s)\mathbf{T}(t) = 2\mathbf{i} + 2t\mathbf{j} - 2t\left(\frac{2t}{t^2 + 1}\mathbf{i} + \frac{t^2 - 1}{t^2 + 1}\mathbf{j}\right)$$

$$\mathbf{A}(t) - (D_t{}^2 s)\mathbf{T}(t) = \frac{2}{t^2 + 1}\left[(1 - t^2)\mathbf{i} + 2t\mathbf{j}\right] \tag{13}$$

From (12), $\mathbf{N}(t)$ is a scalar times the vector in (13). Because $\mathbf{N}(t)$ is a unit vector, $\mathbf{N}(t)$ can be obtained by dividing the vector in (13) by its magnitude. Thus

$$\mathbf{N}(t) = \frac{(1 - t^2)\mathbf{i} + 2t\mathbf{j}}{\sqrt{(1 - t^2)^2 + (2t)^2}}$$

$$= \frac{1 - t^2}{1 + t^2}\mathbf{i} + \frac{2t}{1 + t^2}\mathbf{j}$$

We find the curvature $K(t)$ from the first equation in (10). Because $\dfrac{ds}{dt} = t^2 + 1$, we get

$$K(t) = \frac{2}{(t^2 + 1)^2}$$

The required vectors and scalars at $t = 2$ are as follows:

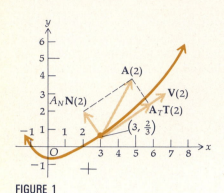

FIGURE 1

$\mathbf{V}(2) = 4\mathbf{i} + 3\mathbf{j}$	$\mathbf{A}(2) = 2\mathbf{i} + 4\mathbf{j}$
$\|\mathbf{V}(2)\| = 5$	$A_T(2) = 4$
$\mathbf{T}(2) = \frac{4}{5}\mathbf{i} + \frac{3}{5}\mathbf{j}$	$\mathbf{N}(2) = -\frac{3}{5}\mathbf{i} + \frac{4}{5}\mathbf{j}$
$A_N(2) = 2$	$K(2) = \frac{2}{25}$

The sketch appears in Figure 1.

EXERCISES 14.9

In Exercises 1 through 4, a particle is moving along the curve having the vector equation. Find the vectors $\mathbf{V}(t)$ and $\mathbf{A}(t)$ and the scalars $A_T(t)$ and $A_N(t)$.

1. $\mathbf{R}(t) = t\mathbf{i} + t^2\mathbf{j}$ **2.** $\mathbf{R}(t) = 2\sin 4t\mathbf{i} + 2\cos 4t\mathbf{j}$
3. $\mathbf{R}(t) = (\cos t + t\sin t)\mathbf{i} + (\sin t - t\cos t)\mathbf{j}$; $t \geq 0$
4. $\mathbf{R}(t) = (t^3 - 3t)\mathbf{i} + 3t^2\mathbf{j}$

In Exercises 5 through 8, a particle is moving along the curve having the vector equation. Find $\mathbf{V}(t_1)$, $\mathbf{A}(t_1)$, $A_T(t_1)$, and $A_N(t_1)$ for the given value of t_1.

5. $\mathbf{R}(t) = e^{-t}\mathbf{i} + e^t\mathbf{j}$; $t_1 = 0$
6. $\mathbf{R}(t) = \cos^2 t\mathbf{i} + \sin^2 t\mathbf{j}$; $t_1 = \frac{1}{6}\pi$
7. $\mathbf{R}(t) = \sin^3 t\mathbf{i} + \cos^3 t\mathbf{j}$; $t_1 = \frac{1}{4}\pi$
8. $\mathbf{R}(t) = e^{-2t}\mathbf{i} + e^{2t}\mathbf{j}$; $t_1 = \ln 2$

In Exercises 9 and 10, find the following scalars: $\|\mathbf{V}(t)\|$, $A_T(t)$, $A_N(t)$, and $K(t)$.

9. $\mathbf{R}(t) = t^2\mathbf{i} + t^3\mathbf{j}$; $t \geq 0$ **10.** $\mathbf{R}(t) = (t^2 + 4)\mathbf{i} + (2t - 5)\mathbf{j}$

In Exercises 11 through 16, a particle is moving along the curve having the vector equation. In each exercise, find the vectors $\mathbf{V}(t)$, $\mathbf{A}(t)$, $\mathbf{T}(t)$, and $\mathbf{N}(t)$, and the following scalars for an arbitrary value of t: $\|\mathbf{V}(t)\|$, $A_T(t)$, $A_N(t)$, and $K(t)$. Also find the particular values when $t = t_1$. At $t = t_1$, draw a sketch of a portion of the curve

and representations of the vectors $\mathbf{V}(t_1)$, $\mathbf{A}(t_1)$, $A_T(t_1)\mathbf{T}(t_1)$, and $A_N(t_1)\mathbf{N}(t_1)$.

11. $\mathbf{R}(t) = (2t + 3)\mathbf{i} + (t^2 - 1)\mathbf{j}$; $t_1 = 2$
12. $\mathbf{R}(t) = (t - 1)\mathbf{i} + t^2\mathbf{j}$; $t_1 = 1$
13. $\mathbf{R}(t) = 5\cos 3t\mathbf{i} + 5\sin 3t\mathbf{j}$; $t_1 = \frac{1}{3}\pi$
14. $\mathbf{R}(t) = 3t^2\mathbf{i} + 2t^3\mathbf{j}$; $t_1 = 1$ **15.** $\mathbf{R}(t) = e^t\mathbf{i} + e^{-t}\mathbf{j}$; $t_1 = 0$
16. $\mathbf{R}(t) = \cos t^2\mathbf{i} + \sin t^2\mathbf{j}$; $t_1 = \frac{1}{2}\sqrt{\pi}$

In Exercises 17 and 18, a particle is moving along the curve having the cartesian equation. At the point find (a) the position vector, (b) the velocity vector, (c) the acceleration vector, (d) A_T, and (e) A_N.

17. $y = 4x^2$; $(1, 4)$ **18.** $y^2 = x^3$; $(4, 8)$

19. A particle is moving along the parabola $y^2 = 8x$ and its speed is constant. Find each of the following when the particle is at $(2, 4)$: the position vector, the velocity vector, the acceleration vector, the unit tangent vector, the unit normal vector, A_T, and A_N.

20. A particle is moving along the top branch of the hyperbola $y^2 - x^2 = 9$, such that $\dfrac{dx}{dt}$ is a positive constant. Find each of the following when the particle is at $(4, 5)$: the position vector, the velocity vector, the acceleration vector, the unit tangent vector, the unit normal vector, A_T, and A_N.

REVIEW EXERCISES FOR CHAPTER 14

In Exercises 1 through 18, $\mathbf{A} = 4\mathbf{i} - 6\mathbf{j}$, $\mathbf{B} = \mathbf{i} + 7\mathbf{j}$, and $\mathbf{C} = 9\mathbf{i} - 5\mathbf{j}$.

1. Find $3\mathbf{B} - 7\mathbf{A}$.   **2.** Find $5\mathbf{B} - 3\mathbf{C}$.
3. Find $\|3\mathbf{B} - 7\mathbf{A}\|$.   **4.** Find $\|5\mathbf{B} - 3\mathbf{C}\|$.
5. Find $\|3\mathbf{B}\| - \|7\mathbf{A}\|$.   **6.** Find $\|5\mathbf{B}\| - \|3\mathbf{C}\|$.

7. Find $(\mathbf{A} - \mathbf{B}) \cdot \mathbf{C}$.   **8.** Find $(\mathbf{A} \cdot \mathbf{B})\mathbf{C}$.
9. Find a unit vector having the same direction as $2\mathbf{A} + \mathbf{B}$.
10. Find the unit vectors that are orthogonal to $\mathbf{B}$.
11. Find scalars h and k such that $\mathbf{A} = h\mathbf{B} + k\mathbf{C}$.
12. Find scalars h and k such that $h\mathbf{A} + k\mathbf{B} = -\mathbf{C}$.
13. Find the scalar projection of $\mathbf{A}$ onto $\mathbf{B}$.

14. Find the scalar projection of **C** onto **A**.
15. Find the vector projection of **A** onto **B**.
16. Find the vector projection of **C** onto **A**.
17. Find the component of **B** in the direction of **A**.
18. Find $\cos \alpha$ if α is the angle between **A** and **C**.

19. Two forces of magnitudes 50 lb and 70 lb make an angle of 60° with each other and are applied to an object at the same point. Find (a) the magnitude of the resultant force and (b) to the nearest degree the angle it makes with the force of 50 lb.
20. Determine the angle between two forces of 112 lb and 136 lb applied to an object at the same point if the resultant force has a magnitude of 168 lb.
21. A force is represented by a vector **F** having a magnitude of 30 lb and a direction angle of radian measure $\frac{3}{4}\pi$. If distance is measured in feet find the work done by the force in moving a particle along a line from the point $(3, 6)$ to the point $(-2, 7)$.
22. The compass heading of an airplane is 107° and its air speed is 210 mi/hr. If there is a wind blowing from the west at 36 mi/hr, what are (a) the plane's ground speed and (b) its course?

In Exercises 23 and 24, for the vector-valued function, find (a) the domain of **R***; (b) $\lim\limits_{t \to 1}$* **R**(t); *(c) D_t* **R**(t).

23. $\mathbf{R}(t) = \dfrac{1}{t+1}\mathbf{i} + \dfrac{\sqrt{t}-1}{t-1}\mathbf{j}$ 24. $\mathbf{R}(t) = |t-1|\mathbf{i} + \ln t\,\mathbf{j}$

In Exercises 25 and 26, find $\dfrac{dy}{dx}$ and $\dfrac{d^2y}{dx^2}$ without eliminating the parameter.

25. $x = 9t^2 - 1$, $y = 3t + 1$ 26. $x = e^{2t}$, $y = e^{-3t}$

In Exercises 27 and 28, find equations of the horizontal and vertical tangent lines, and then draw a sketch of the graph of the given pair of parametric equations.

27. $x = 12 - t^2$, $y = 12t - t^3$
28. $x = \dfrac{2at^2}{1+t^2}$, $y = \dfrac{2at^3}{1+t^2}$, $a > 0$ (the cissoid of Diocles)

29. If $\mathbf{R}(t) = \ln(t^2 - 1)\mathbf{i} - 2t^{-3}\mathbf{j}$, find $\mathbf{R}'(t) \cdot \mathbf{R}''(t)$.
30. Find the length of the arc of the curve having parametric equations $x = t^2$, $y = t^3$, from $t = 1$ to $t = 2$.
31. Find the length of the arc of the curve $\mathbf{R}(t) = (2-t)\mathbf{i} + t^2\mathbf{j}$ from $t = 0$ to $t = 3$.
32. Find the length of the arc of the curve $r = 3 \sec \theta$ from $\theta = 0$ to $\theta = \frac{1}{4}\pi$.
33. (a) Show that the curve defined by the parametric equations $x = a \sin t$ and $y = b \cos t$ is an ellipse. (b) If s is the measure of the length of arc of the ellipse of part (a), show that

$$s = 4 \int_0^{\pi/2} a\sqrt{1 - k^2 \sin^2 t}\; dt$$

where $k^2 = (a^2 - b^2)/a^2 < 1$. This integral is called an *elliptic*

integral and cannot be evaluated exactly in terms of elementary functions.

34. Draw a sketch of the graph of the vector equation $\mathbf{R}(t) = e^t\mathbf{i} + e^{-t}\mathbf{j}$, and find a cartesian equation of the graph.
35. Show that the curvature of the curve $y = \ln x$ at any point (x, y) is $x/(x^2 + 1)^{3/2}$. Also show that the absolute maximum curvature is $\frac{2}{9}\sqrt{3}$, which occurs at the point $(\frac{1}{2}\sqrt{2}, -\frac{1}{2} \ln 2)$.
36. Find the curvature at any point of the branch of the hyperbola defined by $x = a \cosh t$, $y = b \sinh t$. Also show that the curvature is an absolute maximum at the vertex.
37. Find the radius of curvature at any point on the curve $x = a(\cos t + t \sin t)$, $y = a(\sin t - t \cos t)$.
38. Find the curvature, the radius of curvature, and the center of curvature of the curve $y = e^{-x}$ at the point $(0, 1)$.
39. Find the curvature and radius of curvature of the curve $\mathbf{R}(t) = 3t^2\mathbf{i} + (t^3 - 3t)\mathbf{j}$ at the point where $t = 2$.
40. If $\mathbf{R}(t) = e^{\lambda t}\mathbf{i} + e^{-\lambda t}\mathbf{j}$, where λ is a constant, show that $\mathbf{R}(t)$ satisfies the equation $\mathbf{R}''(t) - \lambda^2\mathbf{R}(t) = \mathbf{0}$.
41. A particle is moving along a curve having the vector equation $\mathbf{R}(t) = 3t\mathbf{i} + (4t - t^2)\mathbf{j}$. (a) Find a cartesian equation of the path of the particle. (b) Find the velocity vector and the acceleration vector. (c) Find **V**(1) and **A**(1).
42. Follow the instructions of Exercise 41 if $\mathbf{R}(t) = 2e^t\mathbf{i} + 3e^{-t}\mathbf{j}$.
43. For the hypocycloid of four cusps,

$$x = a \cos^3 t \quad \text{and} \quad y = a \sin^3 t$$

find $\dfrac{dy}{dx}$ and $\dfrac{d^2y}{dx^2}$ without eliminating the parameter.

44. If a particle is moving along a curve, under what conditions will the acceleration vector and the unit tangent vector have the same or opposite directions?

In Exercises 45 and 46, find **T**(t) *and* **N**(t), *and at $t = t_1$ draw a sketch of a portion of the curve and draw the representations of* **T**(t_1) *and* **N**(t_1) *having initial point at $t = t_1$.*

45. $\mathbf{R}(t) = (e^t + e^{-t})\mathbf{i} + 2t\mathbf{j}$; $t_1 = 2$
46. $\mathbf{R}(t) = 3(\cos t + t \sin t)\mathbf{i} + 3(\sin t - t \cos t)\mathbf{j}$, $t > 0$; $t_1 = \frac{1}{2}\pi$

47. Given the curve having parametric equations $x = 4t$, $y = \frac{1}{3}(2t + 1)^{3/2}$, $t \geq 0$, find parametric equations having the measure of arc length s as a parameter, where arc length is measured from the point where $t = 0$. Check your result by using Equation (7) of Section 14.6.
48. Find the radian measure of the angle of elevation at which a gun should be fired in order to obtain the maximum range for a given muzzle speed.
49. Find a formula for obtaining the maximum height reached by a projectile fired from a gun having a given muzzle speed of v_0 feet per second and an angle of elevation of radian measure α.
50. A girl throws a ball horizontally from the top of a cliff 288 ft high at an initial speed of 32 ft/sec. Find (a) the time of flight, and (b) the distance from the base of the cliff to the point where the ball hits the ground.
51. Prove by vector analysis that the diagonals of a parallelogram bisect each other.

52. Find the position vector $\mathbf{R}(t)$ if the acceleration vector $\mathbf{A}(t) = t^2\mathbf{i} - \dfrac{1}{t^2}\mathbf{j}$ and $\mathbf{V}(1) = \mathbf{j}$, and $\mathbf{R}(1) = \frac{1}{4}\mathbf{i} + \frac{1}{2}\mathbf{j}$.

53. An *epicycloid* is the curve traced by a point P on the circumference of a circle of radius b which is rolling externally on a fixed circle of radius a. If the origin is at the center of the fixed circle, $A(a, 0)$ is one of the points at which the given point P comes in contact with the fixed circle, B is the moving point of tangency of the two circles, and the parameter t is the radian measure of the angle AOB, prove that parametric equations of the epicycloid are

$$x = (a + b)\cos t - b\cos\frac{a + b}{b}t$$

and

$$y = (a + b)\sin t - b\sin\frac{a + b}{b}t$$

54. Given triangle ABC, points D, E, and F are on the sides AB, BC, and AC, respectively, and

$$\mathbf{V}(\overrightarrow{AD}) = \tfrac{1}{3}\mathbf{V}(\overrightarrow{AB}) \qquad \mathbf{V}(\overrightarrow{BE}) = \tfrac{1}{3}\mathbf{V}(\overrightarrow{BC}) \qquad \mathbf{V}(\overrightarrow{CF}) = \tfrac{1}{3}\mathbf{V}(\overrightarrow{CA})$$

Prove $\mathbf{V}(\overrightarrow{AE}) + \mathbf{V}(\overrightarrow{BF}) + \mathbf{V}(\overrightarrow{CD}) = \mathbf{0}$.

Exercises 55 through 58 pertain to Supplementary Section 14.9. In Exercises 55 and 56, find (a) the velocity and acceleration vectors, (b) the speed, and (c) the tangential and normal components of acceleration.

55. $\mathbf{R}(t) = \cosh 2t\mathbf{i} + \sinh 2t\mathbf{j}$
56. $\mathbf{R}(t) = (2\tan^{-1}t - t)\mathbf{i} + \ln(1 + t^2)\mathbf{j}$
57. Find the tangential and normal components of the acceleration vector for the particle of Exercise 41.
58. Find the tangential and normal components of the acceleration vector for the particle of Exercise 42.

FIFTEEN

Vectors in Three-Dimensional Space and Solid Analytic Geometry

$$\cos^2\alpha + \cos^2\beta + \cos^2\gamma = 1$$

$$\vec{A} \cdot \vec{B} = \|\vec{A}\| \|\vec{B}\| \cos\theta$$

In this chapter we treat vectors in three-dimensional space. Solid geometry is also included because its discussion is simplified by the use of vectors. We begin in Section 15.1 by setting up a coordinate system in space similar to the one in the plane. In Section 15.2 we extend to three dimensions the definitions and theorems given in Section 14.1 for vectors in the plane. Solid geometry topics appear in Sections 15.3, 15.4, 15.6, and 15.7. These topics include *planes, lines, cylinders, surfaces of revolution,* and *quadric surfaces. Cross product,* discussed in Section 15.5, is a vector operation for three-dimensional vectors that we did not have for vectors in the plane.

Vector-valued functions in three-dimensional space along with a brief introduction to the *differential geometry* of curves and surfaces are presented in Section 15.8. *Cylindrical and spherical coordinates* are generalizations of polar coordinates to three-dimensional space. They are discussed in Section 15.9 so they will be available for engineering applications in Chapter 18.

15.1 THE THREE-DIMENSIONAL NUMBER SPACE

Until now we have been concerned with the number line R (the one-dimensional number space) and the number plane R^2 (the two-dimensional number space). We identified the real numbers in R with points on a horizontal axis and the real number pairs in R^2 with points in a geometric plane. We now introduce the set of all ordered triples of real numbers.

15.1.1 DEFINITION

> The set of all ordered triples of real numbers is called the **three-dimensional number space** and is denoted by R^3. Each ordered triple (x, y, z) is called a **point** in the three-dimensional number space.

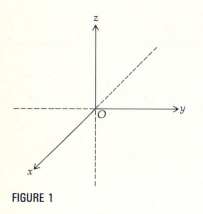

FIGURE 1

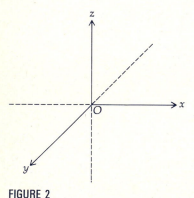

FIGURE 2

To represent R^3 in a geometric three-dimensional space, consider the directed distances of a point from three mutually perpendicular planes. The planes are formed by first considering three mutually perpendicular lines that intersect at a point that we call the **origin** and denote by the letter O. These lines, called the coordinate axes, are designated as the x axis, the y axis, and the z axis. Usually the x axis and the y axis are taken in a horizontal plane, and the z axis is vertical. A positive direction is selected on each axis. If the positive directions are chosen as in Figure 1, the coordinate system is called a **right-handed system**. This terminology follows from the fact that if the right hand is placed so the thumb is pointed in the positive direction of the x axis and the index finger is pointed in the positive direction of the y axis, then the middle finger is pointed in the positive direction of the z axis. If the middle finger is pointed in the negative direction of the z axis, then the coordinate system is called **left handed**. A left-handed system is shown in Figure 2. In general, we use a right-handed system. The three axes determine three coordinate planes: the xy plane containing the x and y axes, the xz plane containing the x and z axes, and the yz plane containing the y and z axes.

An ordered triple of real numbers (x, y, z) is associated with each point P in a geometric three-dimensional space. The directed distance of P from the yz plane is called the **x coordinate**, the directed distance of P from the xz plane is called the **y coordinate**, and the **z coordinate** is the directed distance of P from the xy plane. These three coordinates are called the **rectangular cartesian coordinates** of the point, and there is a one-to-one correspondence (called a **rectangular cartesian coordinate system**) between all such ordered triples of real numbers and the points in a geometric three-dimensional space. Hence we identify R^3 with the geometric three-dimensional space, and we call an ordered triple (x, y, z) a point. The point $(3, 2, 4)$ appears in Figure 3, and the point $(4, -2, -5)$ is shown in Figure 4. The three coordinate planes divide the space into eight parts, called **octants**. The first octant is the one in which all three coordinates are positive.

A line is parallel to a plane if and only if the distance from any point on the line to the plane is the same.

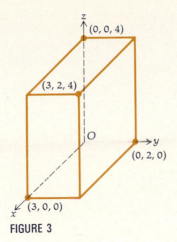

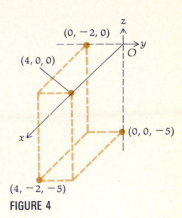

FIGURE 3

FIGURE 4

▶ **ILLUSTRATION 1**   A line parallel to the *yz* plane, one parallel to the *xz* plane, and one parallel to the *xy* plane are shown in Figures 5, 6, and 7, respectively.   ◀

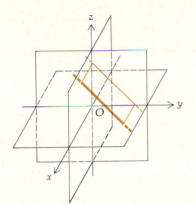

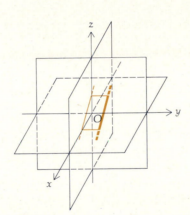

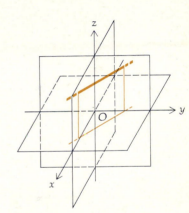

FIGURE 5

FIGURE 6

FIGURE 7

We consider all lines lying in a given plane as being parallel to the plane, in which case the distance from any point on the line to the plane is zero. The following theorem follows immediately.

15.1.2 THEOREM

  (i) A line is parallel to the *yz* plane if and only if all points on the line have equal *x* coordinates.
 (ii) A line is parallel to the *xz* plane if and only if all points on the line have equal *y* coordinates.
(iii) A line is parallel to the *xy* plane if and only if all points on the line have equal *z* coordinates.

In three-dimensional space, if a line is parallel to each of two intersecting planes, it is parallel to the line of intersection of the two planes. Also, if a given line is parallel to a second line, then the given line is parallel to any plane containing the second line. Theorem 15.1.3 follows from these two geometrical facts and from Theorem 15.1.2.

15.1.3 THEOREM

(i) A line is parallel to the x axis if and only if all points on the line have equal y coordinates and equal z coordinates.

(ii) A line is parallel to the y axis if and only if all points on the line have equal x coordinates and equal z coordinates.

(iii) A line is parallel to the z axis if and only if all points on the line have equal x coordinates and equal y coordinates.

▶ **ILLUSTRATION 2**   A line parallel to the x axis, a line parallel to the y axis, and a line parallel to the z axis appear in Figures 8, 9, and 10, respectively.

◀

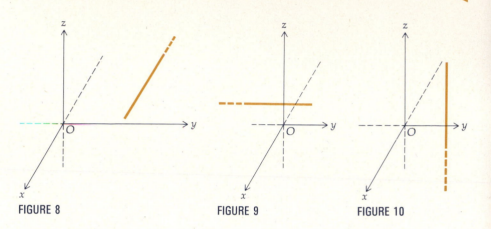

FIGURE 8   FIGURE 9   FIGURE 10

The formulas for finding the directed distance from one point to another on a line parallel to a coordinate axis follow from the definition of directed distance given in Section 1.2 and are stated in the following theorem.

15.1.4 THEOREM

(i) If $A(x_1, y, z)$ and $B(x_2, y, z)$ are two points on a line parallel to the x axis, then the directed distance from A to B, denoted by $\overline{AB}$, is given by

$$\overline{AB} = x_2 - x_1$$

(ii) If $C(x, y_1, z)$ and $D(x, y_2, z)$ are two points on a line parallel to the y axis, then the directed distance from C to D, denoted by $\overline{CD}$, is given by

$$\overline{CD} = y_2 - y_1$$

(iii) If $E(x, y, z_1)$ and $F(x, y, z_2)$ are two points on a line parallel to the z axis, then the directed distance from E to F, denoted by $\overline{EF}$, is given by

$$\overline{EF} = z_2 - z_1$$

▶ **ILLUSTRATION 3**   The directed distance $\overline{PQ}$ from the point $P(2, -5, -4)$ to the point $Q(2, -3, -4)$ is given by Theorem 15.1.4(ii).

$$\overline{PQ} = (-3) - (-5)$$
$$= 2$$

◀

The following theorem gives a formula for finding the undirected distance between any two points in three-dimensional space.

15.1.5 THEOREM The undirected distance between the two points $P_1(x_1, y_1, z_1)$ and $P_2(x_2, y_2, z_2)$ is given by

$$|\overline{P_1P_2}| = \sqrt{(x_2 - x_1)^2 + (y_2 - y_1)^2 + (z_2 - z_1)^2}$$

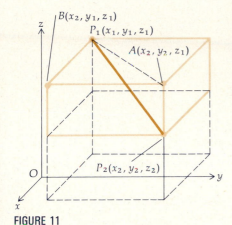

FIGURE 11

Proof We construct a rectangular parallelepiped having P_1 and P_2 as opposite vertices and faces parallel to the coordinate planes (see Figure 11).

By the Pythagorean theorem,

$$|\overline{P_1P_2}|^2 = |\overline{P_1A}|^2 + |\overline{AP_2}|^2 \tag{1}$$

Because

$$|\overline{P_1A}|^2 = |\overline{P_1B}|^2 + |\overline{BA}|^2$$

we obtain, by substituting from this equation into (1),

$$|\overline{P_1P_2}|^2 = |\overline{P_1B}|^2 + |\overline{BA}|^2 + |\overline{AP_2}|^2$$

Applying Theorem 15.1.4(i), (ii), and (iii) to the right side we obtain

$$|\overline{P_1P_2}|^2 = (x_2 - x_1)^2 + (y_2 - y_1)^2 + (z_2 - z_1)^2$$
$$|\overline{P_1P_2}| = \sqrt{(x_2 - x_1)^2 + (y_2 - y_1)^2 + (z_2 - z_1)^2}$$

∎

EXAMPLE 1 Find the undirected distance between the points $P(-3, 4, -1)$ and $Q(2, 5, -4)$.

Solution From Theorem 15.1.5,

$$|\overline{PQ}| = \sqrt{(2 + 3)^2 + (5 - 4)^2 + (-4 + 1)^2}$$
$$= \sqrt{35}$$

The formula for the distance between two points in R^3 is merely an extension of the corresponding formula for the distance between two points in R^2. It is noteworthy that the undirected distance between two points x_2 and x_1 in R is given by

$$|x_2 - x_1| = \sqrt{(x_2 - x_1)^2}$$

The formulas for the coordinates of the midpoint of a line segment are derived by forming congruent triangles and proceeding in a manner analogous to the two-dimensional case. These formulas are given in the following theorem and the proof is left as an exercise (see Exercise 18).

15.1.6 THEOREM The coordinates of the midpoint of the line segment having endpoints $P_1(x_1, y_1, z_1)$ and $P_2(x_2, y_2, z_2)$ are given by

$$\bar{x} = \frac{x_1 + x_2}{2} \qquad \bar{y} = \frac{y_1 + y_2}{2} \qquad \bar{z} = \frac{z_1 + z_2}{2}$$

15.1.7 DEFINITION The **graph of an equation** in R^3 is the set of all points (x, y, z) whose coordinates are numbers satisfying the equation.

The graph of an equation in R^3 is called a **surface**. One particular surface is the *sphere*.

15.1.8 DEFINITION

A **sphere** is the set of all points in three-dimensional space equidistant from a fixed point. The fixed point is called the **center** of the sphere and the measure of the constant distance is called the **radius** of the sphere.

15.1.9 THEOREM

An equation of the sphere of radius r and center at (h, k, l) is

$$(x - h)^2 + (y - k)^2 + (z - l)^2 = r^2 \qquad (2)$$

Proof Let the point (h, k, l) be denoted by C (see Figure 12). The point $P(x, y, z)$ is a point on the sphere if and only if

$$|\overline{CP}| = r$$
$$\Leftrightarrow \quad \sqrt{(x - h)^2 + (y - k)^2 + (z - l)^2} = r$$

Squaring on both sides of the above equation we obtain the desired result. ∎

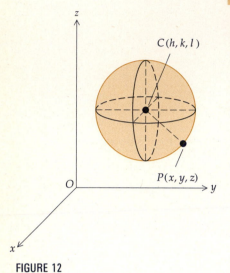

FIGURE 12

If the center of the sphere is at the origin, then $h = 0$, $k = 0$, $l = 0$; so an equation of this sphere is

$$x^2 + y^2 + z^2 = r^2$$

If we expand the terms of (2) and regroup the terms, we have

$$x^2 + y^2 + z^2 - 2hx - 2ky - 2lz + (h^2 + k^2 + l^2 - r^2) = 0$$

This equation is of the form

$$x^2 + y^2 + z^2 + Gx + Hy + Iz + J = 0 \qquad (3)$$

where G, H, I, and J are constants. Equation (3) is called the **general form** of an equation of a sphere, whereas (2) is called the **center-radius form**. Because every sphere has a center and a radius, its equation can be put in the center-radius form and hence the general form.

It can be shown that any equation of the form (3) can be put in the form

$$(x - h)^2 + (y - k)^2 + (z - l)^2 = K \qquad (4)$$

where

$$h = -\tfrac{1}{2}G \qquad k = -\tfrac{1}{2}H \qquad l = -\tfrac{1}{2}I \qquad K = \tfrac{1}{4}(G^2 + H^2 + I^2 - 4J)$$

It is left as an exercise to show this (see Exercise 19).

If $K > 0$, then (4) is of the form of Equation (2); so the graph of the equation is a sphere having its center at (h, k, l) and radius $\sqrt{K}$. If $K = 0$, the graph of the equation is the point (h, k, l). If $K < 0$, the graph is the empty set because the sum of the squares of three real numbers is nonnegative. We state this result as a theorem.

15.1.10 THEOREM

The graph of any second-degree equation in x, y, and z, of the form

$$x^2 + y^2 + z^2 + Gx + Hy + Iz + J = 0$$

is either a sphere, a point, or the empty set.

EXAMPLE 2   Draw a sketch of the graph of the equation

$$x^2 + y^2 + z^2 - 6x - 4y + 2z = 2$$

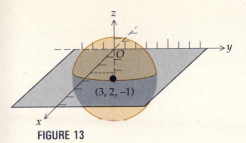

FIGURE 13

Solution   Regrouping terms and completing the squares we have

$$x^2 - 6x + 9 + y^2 - 4y + 4 + z^2 + 2z + 1 = 2 + 9 + 4 + 1$$

$$(x - 3)^2 + (y - 2)^2 + (z + 1)^2 = 16$$

So the graph is a sphere having its center at $(3, 2, -1)$ and radius 4. A sketch of the graph appears in Figure 13.

EXAMPLE 3   Find an equation of the sphere having the points $A(-5, 6, -2)$ and $B(9, -4, 0)$ as endpoints of a diameter.

Solution   The center of the sphere is the midpoint of the line segment AB. Let this point be $C(\bar{x}, \bar{y}, \bar{z})$. By Theorem 15.1.6 we get

$$\bar{x} = \frac{9 - 5}{2} \qquad \bar{y} = \frac{-4 + 6}{2} \qquad \bar{z} = \frac{0 - 2}{2}$$

$$= 2 \qquad\qquad = 1 \qquad\qquad = -1$$

Thus C is the point $(2, 1, -1)$. The radius of the sphere is $|\overline{CB}|$. Hence

$$r = \sqrt{(9 - 2)^2 + (-4 - 1)^2 + (0 + 1)^2}$$

$$= \sqrt{75}$$

Therefore, from Theorem 15.1.9 an equation of the sphere is

$$(x - 2)^2 + (y - 1)^2 + (z + 1)^2 = 75$$

$$x^2 + y^2 + z^2 - 4x - 2y + 2z - 69 = 0$$

EXERCISES 15.1

In Exercises 1 through 5, points A and B are opposite vertices of a rectangular parallelepiped having its faces parallel to the coordinate planes. In each exercise, (a) draw a sketch of the figure, (b) find the coordinates of the other six vertices, (c) find the length of the diagonal AB.

1. $A(0, 0, 0)$; $B(7, 2, 3)$
2. $A(1, 1, 1)$; $B(3, 4, 2)$
3. $A(-1, 1, 2)$; $B(2, 3, 5)$
4. $A(2, -1, -3)$; $B(4, 0, -1)$
5. $A(1, -1, 0)$; $B(3, 3, 5)$

6. The vertex opposite one corner of a room is 18 ft east, 15 ft south, and 12 ft up from the first corner. (a) Draw a sketch of the figure; (b) determine the length of the diagonal joining two opposite vertices; (c) find the coordinates of all eight vertices of the room.

In Exercises 7 through 11, find (a) the undirected distance between the points A and B and (b) the midpoint of the line segment joining A and B.

7. $A(3, 4, 2)$; $B(1, 6, 3)$
8. $A(4, -3, 2)$; $B(-2, 3, -5)$
9. $A(2, -4, 1)$; $B(\frac{1}{2}, 2, 3)$
10. $A(-2, -\frac{1}{2}, 5)$; $B(5, 1, -4)$
11. $A(-5, 2, 1)$; $B(3, 7, -2)$

12. Prove that the three points $(1, -1, 3)$, $(2, 1, 7)$, and $(4, 2, 6)$ are the vertices of a right triangle, and find its area.

13. A line is drawn through the point $(6, 4, 2)$ perpendicular to the yz plane. Find the coordinates of the points on this line at a distance of 10 units from the point $(0, 4, 0)$.

14. Solve Exercise 13 if the line is drawn perpendicular to the xy plane.

15. Prove that the three points $(-3, 2, 4)$, $(6, 1, 2)$, and $(-12, 3, 6)$ are collinear by using the distance formula.

16. Find the vertices of the triangle whose sides have midpoints at $(3, 2, 3)$, $(-1, 1, 5)$ and $(0, 3, 4)$.

17. For the triangle having vertices at $A(2, -5, 3)$, $B(-1, 7, 0)$, and $C(-4, 9, 7)$ find (a) the length of each side and (b) the midpoint of each side.

18. Prove Theorem 15.1.6.

19. Show that any equation of the form

$$x^2 + y^2 + z^2 + Gx + Hy + Iz + J = 0$$

can be put in the form

$$(x - h)^2 + (y - k)^2 + (z - l)^2 = K$$

In Exercises 20 through 25, determine the graph of the equation.

20. $x^2 + y^2 + z^2 - 8y + 6z - 25 = 0$
21. $x^2 + y^2 + z^2 - 8x + 4y + 2z - 4 = 0$
22. $x^2 + y^2 + z^2 - x - y - 3z + 2 = 0$
23. $x^2 + y^2 + z^2 - 6z + 9 = 0$
24. $x^2 + y^2 + z^2 - 8x + 10y - 4z + 13 = 0$
25. $x^2 + y^2 + z^2 - 6x + 2y - 4z + 19 = 0$

In Exercises 26 through 28, find an equation of the sphere satisfying the conditions.

26. A diameter is the line segment having endpoints at $(6, 2, -5)$ and $(-4, 0, 7)$.

27. It is concentric with the sphere having the equation $x^2 + y^2 + z^2 - 2y + 8z - 9 = 0$ and has radius 3.

28. It contains the points $(0, 0, 4)$, $(2, 1, 3)$, and $(0, 2, 6)$ and has its center in the yz plane.

29. Prove by analytic geometry that the four diagonals joining opposite vertices of a rectangular parallelepiped bisect each other.

30. If P, Q, R, and S are four points in three-dimensional space and A, B, C, and D are the midpoints of PQ, QR, RS, and SP, respectively, prove by analytic geometry that $ABCD$ is a parallelogram.

31. Prove by analytic geometry that the four diagonals of a rectangular parallelepiped have the same length.

15.2 VECTORS IN THREE-DIMENSIONAL SPACE

We defined a vector in the plane as an ordered pair of real numbers. We now extend this definition to a vector in three-dimensional space.

15.2.1 DEFINITION

A vector in three-dimensional space is an ordered triple of real numbers $\langle x, y, z \rangle$. The numbers x, y, and z are called the **components** of the vector $\langle x, y, z \rangle$.

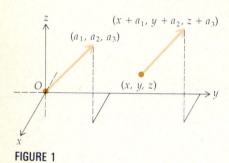

FIGURE 1

We let V_3 be the set of all ordered triples $\langle x, y, z \rangle$ for which x, y, and z are real numbers. In this chapter a vector is always in V_3 unless otherwise stated.

Just as for vectors in V_2, a vector in V_3 can be represented by a directed line segment. If $\mathbf{A} = \langle a_1, a_2, a_3 \rangle$, then the directed line segment having its initial point at the origin and its terminal point at the point (a_1, a_2, a_3) is called the **position representation** of $\mathbf{A}$. A directed line segment having its initial point at (x, y, z) and its terminal point at $(x + a_1, y + a_2, z + a_3)$ is also a representation of the vector $\mathbf{A}$. See Figure 1.

The **zero vector** is the vector $\langle 0, 0, 0 \rangle$ and is denoted by $\mathbf{0}$. Any point is a representation of the zero vector.

The **magnitude** of a vector is the length of any of its representations. If the vector $\mathbf{A} = \langle a_1, a_2, a_3 \rangle$, the magnitude of $\mathbf{A}$ is denoted by $\|\mathbf{A}\|$, and it follows that

$$\|\mathbf{A}\| = \sqrt{a_1{}^2 + a_2{}^2 + a_3{}^2}$$

The **direction** of a nonzero vector in V_3 is given by three angles, called the *direction angles* of the vector.

15.2.2 DEFINITION

The **direction angles** of a nonzero vector are the three angles that have the smallest nonnegative radian measures α, β, and γ measured from the positive x, y, and z axes, respectively, to the position representation of the vector.

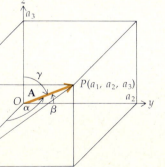

FIGURE 2

The radian measure of each direction angle of a vector is greater than or equal to 0 and less than or equal to π. The direction angles having radian measures α, β, and γ of the vector $\mathbf{A} = \langle a_1, a_2, a_3 \rangle$ are shown in Figure 2. In this figure the components of $\mathbf{A}$ are all positive numbers, and the direction angles of this vector all have positive radian measure less than $\frac{1}{2}\pi$. From the figure we see that triangle POR is a right triangle and

$$\cos \alpha = \frac{a_1}{\|\mathbf{A}\|}$$

It can be shown that the same formula holds if $\frac{1}{2}\pi \leq \alpha \leq \pi$. Similar formulas can be found for $\cos \beta$ and $\cos \gamma$, and we have

$$\cos \alpha = \frac{a_1}{\|\mathbf{A}\|} \qquad \cos \beta = \frac{a_2}{\|\mathbf{A}\|} \qquad \cos \gamma = \frac{a_3}{\|\mathbf{A}\|} \tag{1}$$

The three numbers $\cos \alpha$, $\cos \beta$, and $\cos \gamma$ are called the **direction cosines** of vector **A**. The zero vector has no direction angles and hence no direction cosines.

▶ **ILLUSTRATION 1**   We find the magnitude and direction cosines of the vector $\mathbf{A} = \langle 3, 2, -6 \rangle$.

$$\|\mathbf{A}\| = \sqrt{(3)^2 + (2)^2 + (-6)^2}$$
$$= 7$$

From Equations (1),

$$\cos \alpha = \tfrac{3}{7} \qquad \cos \beta = \tfrac{2}{7} \qquad \cos \gamma = -\tfrac{6}{7} \qquad ◀$$

If the magnitude of a vector and its direction cosines are known, the vector is uniquely determined because from (1) it follows that

$$a_1 = \|\mathbf{A}\| \cos \alpha \qquad a_2 = \|\mathbf{A}\| \cos \beta \qquad a_3 = \|\mathbf{A}\| \cos \gamma \qquad (2)$$

The three direction cosines of a vector are not independent of each other, as we see by the following theorem.

15.2.3 THEOREM   If $\cos \alpha$, $\cos \beta$, and $\cos \gamma$ are the direction cosines of a vector, then

$$\cos^2 \alpha + \cos^2 \beta + \cos^2 \gamma = 1$$

Proof   If $\mathbf{A} = \langle a_1, a_2, a_3 \rangle$, then the direction cosines of **A** are given by (1) and

$$\cos^2 \alpha + \cos^2 \beta + \cos^2 \gamma = \frac{a_1{}^2}{\|\mathbf{A}\|^2} + \frac{a_2{}^2}{\|\mathbf{A}\|^2} + \frac{a_3{}^2}{\|\mathbf{A}\|^2}$$

$$= \frac{a_1{}^2 + a_2{}^2 + a_3{}^2}{\|\mathbf{A}\|^2}$$

$$= \frac{\|\mathbf{A}\|^2}{\|\mathbf{A}\|^2}$$

$$= 1 \qquad ■$$

▶ **ILLUSTRATION 2**   We verify Theorem 15.2.3 for the vector of Illustration 1.

$$\cos^2 \alpha + \cos^2 \beta + \cos^2 \gamma = (\tfrac{3}{7})^2 + (\tfrac{2}{7})^2 + (-\tfrac{6}{7})^2$$

$$= \tfrac{9}{49} + \tfrac{4}{49} + \tfrac{36}{49}$$

$$= \tfrac{49}{49}$$

$$= 1 \qquad ◀$$

The vector $\mathbf{A} = \langle a_1, a_2, a_3 \rangle$ is a unit vector if $\|\mathbf{A}\| = 1$, and from Equations (1) the components of a unit vector are its direction cosines.

The operations of addition, subtraction, and scalar multiplication of vectors in V_3 are given definitions analogous to the corresponding definitions for vectors in V_2.

15.2.4 DEFINITION   If $\mathbf{A} = \langle a_1, a_2, a_3 \rangle$ and $\mathbf{B} = \langle b_1, b_2, b_3 \rangle$, then the **sum** of these vectors is given by

$$\mathbf{A} + \mathbf{B} = \langle a_1 + b_1, a_2 + b_2, a_3 + b_3 \rangle$$

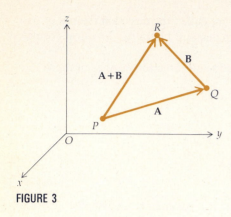

FIGURE 3

EXAMPLE 1   Given $\mathbf{A} = \langle 5, -2, 6 \rangle$ and $\mathbf{B} = \langle 8, -5, -4 \rangle$, find $\mathbf{A} + \mathbf{B}$.

Solution

$$\mathbf{A} + \mathbf{B} = \langle 5 + 8, (-2) + (-5), 6 + (-4) \rangle$$
$$= \langle 13, -7, 2 \rangle$$

The geometric interpretation of the sum of two vectors in V_3 is similar to that for vectors in V_2. See Figure 3. If P is the point (x, y, z), and $\mathbf{A} = \langle a_1, a_2, a_3 \rangle$, and $\overrightarrow{PQ}$ is a representation of $\mathbf{A}$, then Q is the point $(x + a_1, y + a_2, z + a_3)$. Let $\mathbf{B} = \langle b_1, b_2, b_3 \rangle$ and let $\overrightarrow{QR}$ be a representation of $\mathbf{B}$. Then R is the point $(x + (a_1 + b_1), y + (a_2 + b_2), z + (a_3 + b_3))$. Therefore $\overrightarrow{PR}$ is a representation of the vector $\mathbf{A} + \mathbf{B}$, and the parallelogram law holds.

15.2.5 DEFINITION   If $\mathbf{A} = \langle a_1, a_2, a_3 \rangle$, then the vector $\langle -a_1, -a_2, -a_3 \rangle$ is defined to be the **negative** of $\mathbf{A}$, denoted by $-\mathbf{A}$.

15.2.6 DEFINITION   The **difference** of the two vectors $\mathbf{A}$ and $\mathbf{B}$, denoted by $\mathbf{A} - \mathbf{B}$, is defined by

$$\mathbf{A} - \mathbf{B} = \mathbf{A} + (-\mathbf{B})$$

From Definitions 15.2.5 and 15.2.6, it follows that if $\mathbf{A} = \langle a_1, a_2, a_3 \rangle$ and $\mathbf{B} = \langle b_1, b_2, b_3 \rangle$, then $-\mathbf{B} = \langle -b_1, -b_2, -b_3 \rangle$ and

$$\mathbf{A} - \mathbf{B} = \langle a_1 - b_1, a_2 - b_2, a_3 - b_3 \rangle$$

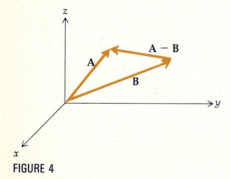

FIGURE 4

EXAMPLE 2   For the vectors $\mathbf{A}$ and $\mathbf{B}$ of Example 1 find $\mathbf{A} - \mathbf{B}$.

Solution

$$\mathbf{A} - \mathbf{B} = \langle 5, -2, 6 \rangle - \langle 8, -5, -4 \rangle$$
$$= \langle 5, -2, 6 \rangle + \langle -8, 5, 4 \rangle$$
$$= \langle -3, 3, 10 \rangle$$

The difference of two vectors in V_3 is also interpreted geometrically as it is in V_2. See Figure 4. A representation of the vector $\mathbf{A} - \mathbf{B}$ is obtained by choosing representations of $\mathbf{A}$ and $\mathbf{B}$ having the same initial point. Then a representation of the vector $\mathbf{A} - \mathbf{B}$ is the directed line segment from the terminal point of the representation of $\mathbf{B}$ to the terminal point of the representation of $\mathbf{A}$.

Figure 5 shows the points $P(a_1, a_2, a_3)$ and $Q(b_1, b_2, b_3)$, and the directed line segments $\overrightarrow{PQ}$, $\overrightarrow{OP}$, and $\overrightarrow{OQ}$. Observe that

$$\mathbf{V}(\overrightarrow{PQ}) = \mathbf{V}(\overrightarrow{OQ}) - \mathbf{V}(\overrightarrow{OP})$$
$$= \langle b_1, b_2, b_3 \rangle - \langle a_1, a_2, a_3 \rangle$$

Therefore

$$\mathbf{V}(\overrightarrow{PQ}) = \langle b_1 - a_1, b_2 - a_2, b_3 - a_3 \rangle$$

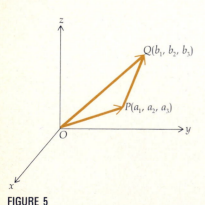

FIGURE 5

▶ **ILLUSTRATION 3** Figure 6 shows the directed line segment $\overrightarrow{PQ}$, where P is the point $(1, 3, 5)$ and Q is the point $(2, -1, 4)$.

$$\mathbf{V}(\overrightarrow{PQ}) = \langle 2 - 1, -1 - 3, 4 - 5 \rangle$$
$$= \langle 1, -4, -1 \rangle \qquad \blacktriangleleft$$

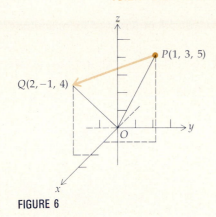

FIGURE 6

15.2.7 DEFINITION

If c is a scalar and $\mathbf{A}$ is the vector $\langle a_1, a_2, a_3 \rangle$, then the product of c and $\mathbf{A}$, denoted by $c\mathbf{A}$, is a vector and is given by

$$c\mathbf{A} = c\langle a_1, a_2, a_3 \rangle$$
$$= \langle ca_1, ca_2, ca_3 \rangle$$

EXAMPLE 3 Given $\mathbf{A} = \langle -4, 7, -2 \rangle$, find $3\mathbf{A}$ and $-5\mathbf{A}$.

Solution

$$3\mathbf{A} = 3\langle -4, 7, -2 \rangle \qquad -5\mathbf{A} = (-5)\langle -4, 7, -2 \rangle$$
$$= \langle -12, 21, -6 \rangle \qquad\qquad = \langle 20, -35, 10 \rangle$$

Suppose that $\mathbf{A} = \langle a_1, a_2, a_3 \rangle$ is a nonzero vector having direction cosines $\cos \alpha$, $\cos \beta$, and $\cos \gamma$, and let c be any nonzero scalar. Then $c\mathbf{A} = \langle ca_1, ca_2, ca_3 \rangle$; and if $\cos \alpha_1$, $\cos \beta_1$, and $\cos \gamma_1$ are the direction cosines of $c\mathbf{A}$, we have, from Equations (1),

$$\cos \alpha_1 = \frac{ca_1}{\|c\mathbf{A}\|} \qquad \cos \beta_1 = \frac{ca_2}{\|c\mathbf{A}\|} \qquad \cos \gamma_1 = \frac{ca_3}{\|c\mathbf{A}\|}$$

$$\cos \alpha_1 = \frac{c}{|c|} \frac{a_1}{\|\mathbf{A}\|} \qquad \cos \beta_1 = \frac{c}{|c|} \frac{a_2}{\|\mathbf{A}\|} \qquad \cos \gamma_1 = \frac{c}{|c|} \frac{a_3}{\|\mathbf{A}\|}$$

$$\cos \alpha_1 = \frac{c}{|c|} \cos \alpha \qquad \cos \beta_1 = \frac{c}{|c|} \cos \beta \qquad \cos \gamma_1 = \frac{c}{|c|} \cos \gamma \qquad (3)$$

Thus if $c > 0$, it follows from Equations (3) that the direction cosines of vector $c\mathbf{A}$ are the same as the direction cosines of $\mathbf{A}$. And if $c < 0$, the direction cosines of $c\mathbf{A}$ are the negatives of the direction cosines of $\mathbf{A}$. Therefore, if c is a nonzero scalar, then the vector $c\mathbf{A}$ is a vector whose magnitude is $|c|$ times the magnitude of $\mathbf{A}$. If $c > 0$, $c\mathbf{A}$ has the same direction as $\mathbf{A}$, whereas if $c < 0$, the direction of $c\mathbf{A}$ is opposite that of $\mathbf{A}$.

The operations of vector addition and scalar multiplication of any vectors in V_3 satisfy the properties given in Theorem 14.1.8 (you are asked to prove them in Exercises 19 and 20). From this fact and Definition 14.1.9 it follows that V_3 is a real vector space. The three unit vectors

$$\mathbf{i} = \langle 1, 0, 0 \rangle \qquad \mathbf{j} = \langle 0, 1, 0 \rangle \qquad \mathbf{k} = \langle 0, 0, 1 \rangle$$

form a basis for the vector space V_3 because any vector $\langle a_1, a_2, a_3 \rangle$ can be written in terms of them as follows:

$$\langle a_1, a_2, a_3 \rangle = a_1 \langle 1, 0, 0 \rangle + a_2 \langle 0, 1, 0 \rangle + a_3 \langle 0, 0, 1 \rangle$$

Hence, if $\mathbf{A} = \langle a_1, a_2, a_3 \rangle$, we also can write

$$\mathbf{A} = a_1 \mathbf{i} + a_2 \mathbf{j} + a_3 \mathbf{k} \qquad (4)$$

Because there are three elements in a basis, V_3 is a three-dimensional vector space.

Substituting from (2) into (4) we have

$$\mathbf{A} = \|\mathbf{A}\| \cos \alpha \mathbf{i} + \|\mathbf{A}\| \cos \beta \mathbf{j} + \|\mathbf{A}\| \cos \gamma \mathbf{k}$$

$$\mathbf{A} = \|\mathbf{A}\|(\cos \alpha \mathbf{i} + \cos \beta \mathbf{j} + \cos \gamma \mathbf{k}) \tag{5}$$

This equation enables us to express any nonzero vector in terms of its magnitude and direction cosines.

EXAMPLE 4   Express the vector of Illustration 1 in terms of its magnitude and direction cosines.

Solution   In Illustration 1, we have $\mathbf{A} = \langle 3, 2, -6 \rangle$, $\|\mathbf{A}\| = 7$, $\cos \alpha = \frac{3}{7}$, $\cos \beta = \frac{2}{7}$, and $\cos \gamma = -\frac{6}{7}$. Hence, from (5),

$$\mathbf{A} = 7(\tfrac{3}{7}\mathbf{i} + \tfrac{2}{7}\mathbf{j} - \tfrac{6}{7}\mathbf{k})$$

15.2.8 THEOREM   If the nonzero vector $\mathbf{A} = a_1\mathbf{i} + a_2\mathbf{j} + a_3\mathbf{k}$, then the unit vector $\mathbf{U}$ having the same direction as $\mathbf{A}$ is given by

$$\mathbf{U} = \frac{a_1}{\|\mathbf{A}\|}\mathbf{i} + \frac{a_2}{\|\mathbf{A}\|}\mathbf{j} + \frac{a_3}{\|\mathbf{A}\|}\mathbf{k}$$

The proof of Theorem 15.2.8 is analogous to the proof of Theorem 14.1.10 for a vector in V_2 and is left as an exercise (see Exercise 46).

EXAMPLE 5   Given the points $R(2, -1, 3)$ and $S(3, 4, 6)$, find the unit vector having the same direction as $\mathbf{V}(\overrightarrow{RS})$.

Solution

$$\mathbf{V}(\overrightarrow{RS}) = \langle 3, 4, 6 \rangle - \langle 2, -1, 3 \rangle \qquad \|\mathbf{V}(\overrightarrow{RS})\| = \sqrt{1^2 + 5^2 + 3^2}$$
$$= \mathbf{i} + 5\mathbf{j} + 3\mathbf{k} \qquad\qquad = \sqrt{35}$$

Therefore, by Theorem 15.2.9 the desired unit vector is

$$\mathbf{U} = \frac{1}{\sqrt{35}}\mathbf{i} + \frac{5}{\sqrt{35}}\mathbf{j} + \frac{3}{\sqrt{35}}\mathbf{k}$$

The definition of the dot product of two vectors in V_3 is an extension of the definition for vectors in V_2.

15.2.9 DEFINITION   If $\mathbf{A} = \langle a_1, a_2, a_3 \rangle$ and $\mathbf{B} = \langle b_1, b_2, b_3 \rangle$, then the **dot product** of $\mathbf{A}$ and $\mathbf{B}$, denoted by $\mathbf{A} \cdot \mathbf{B}$, is given by

$$\mathbf{A} \cdot \mathbf{B} = \langle a_1, a_2, a_3 \rangle \cdot \langle b_1, b_2, b_3 \rangle$$
$$= a_1 b_1 + a_2 b_2 + a_3 b_3$$

▶ **ILLUSTRATION 4**   If $\mathbf{A} = \langle 4, 2, -6 \rangle$ and $\mathbf{B} = \langle -5, 3, -2 \rangle$, then

$$\mathbf{A} \cdot \mathbf{B} = \langle 4, 2, -6 \rangle \cdot \langle -5, 3, -2 \rangle$$
$$= 4(-5) + 2(3) + (-6)(-2)$$
$$= -20 + 6 + 12$$
$$= -2$$

◀

For the unit vectors **i**, **j**, and **k**,

$$\mathbf{i} \cdot \mathbf{i} = 1 \qquad \mathbf{j} \cdot \mathbf{j} = 1 \qquad \mathbf{k} \cdot \mathbf{k} = 1$$

$$\mathbf{i} \cdot \mathbf{j} = 0 \qquad \mathbf{i} \cdot \mathbf{k} = 0 \qquad \mathbf{j} \cdot \mathbf{k} = 0$$

Laws of dot multiplication are the same as those in Theorems 14.2.2 and 14.2.3 for vectors in V_2. The proofs are left as exercises (see Exercises 27 and 28).

We now define the angle between two vectors and then express the dot product in terms of the cosine of the radian measure of this angle.

15.2.10 DEFINITION

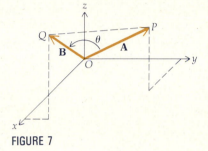

FIGURE 7

Let **A** and **B** be two nonzero vectors in V_3 such that **A** is not a scalar multiple of **B**. If $\overrightarrow{OP}$ is the position representation of **A** and $\overrightarrow{OQ}$ is the position representation of **B**, then the **angle between the vectors A** and **B** is defined to be the angle of positive measure between $\overrightarrow{OP}$ and $\overrightarrow{OQ}$ interior to the triangle POQ. If $\mathbf{A} = c\mathbf{B}$, where c is a scalar, then if $c > 0$, the angle between the vectors has radian measure 0, and if $c < 0$, the angle between the vectors has radian measure π.

Figure 7 shows the angle θ between the two vectors if **A** is not a scalar multiple of **B**.

15.2.11 THEOREM

If θ is the angle between the two nonzero vectors **A** and **B** in V_3, then

$$\mathbf{A} \cdot \mathbf{B} = \|\mathbf{A}\| \, \|\mathbf{B}\| \cos \theta$$

The proof of Theorem 15.2.11 is comparable to the proof of Theorem 14.2.5 for vectors in V_2 and is left as an exercise (see Exercise 47).

The definition of *parallel* vectors in V_3 is analogous to Definition 14.2.6 for vectors in V_2; that is, two vectors in V_3 are **parallel** if and only if one of the vectors is a scalar multiple of the other. As with vectors in V_2, we can prove from this definition and Theorem 15.2.11 that two nonzero vectors in V_3 are parallel if and only if the radian measure of the angle between them is 0 or π.

The definition of *orthogonal* vectors in V_3 corresponds to Definition 14.2.7 for vectors in V_2; that is, if **A** and **B** are two vectors in V_3, **A** and **B** are said to be **orthogonal** if and only if $\mathbf{A} \cdot \mathbf{B} = 0$.

EXAMPLE 6 Prove by using vectors that the points $A(4, 9, 1)$, $B(-2, 6, 3)$, and $C(6, 3, -2)$ are the vertices of a right triangle.

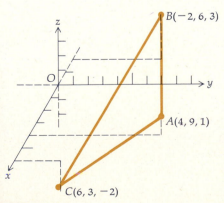

FIGURE 8

Solution Triangle CAB appears in Figure 8. From the figure it looks as if the angle at A is the one that may be a right angle. We shall find $\mathbf{V}(\overrightarrow{AB})$ and $\mathbf{V}(\overrightarrow{AC})$ and if the dot product of these two vectors is zero, the angle is a right angle.

$$\mathbf{V}(\overrightarrow{AB}) = \langle -2 - 4, 6 - 9, 3 - 1 \rangle \qquad \mathbf{V}(\overrightarrow{AC}) = \langle 6 - 4, 3 - 9, -2 - 1 \rangle$$

$$= \langle -6, -3, 2 \rangle \qquad\qquad\qquad = \langle 2, -6, -3 \rangle$$

$$\mathbf{V}(\overrightarrow{AB}) \cdot \mathbf{V}(\overrightarrow{AC}) = \langle -6, -3, 2 \rangle \cdot \langle 2, -6, -3 \rangle$$

$$= -12 + 18 - 6$$

$$= 0$$

Therefore $\mathbf{V}(\overrightarrow{AB})$ and $\mathbf{V}(\overrightarrow{AC})$ are orthogonal; thus the angle at A in triangle CAB is a right angle, and CAB is a right triangle.

If **U** is a unit vector in the direction of **A**, and θ is the angle between **A** (or **U**) and a vector **B**, then, from Theorem 15.2.11,

$$\mathbf{U} \cdot \mathbf{B} = \|\mathbf{U}\| \|\mathbf{B}\| \cos \theta$$
$$= \|\mathbf{B}\| \cos \theta$$

As with vectors in V_2, $\|\mathbf{B}\| \cos \theta$ is the **scalar projection** of **B** onto **A** and the **component** of **B** in the direction of **A**.

The formula of Theorem 14.2.9 giving the scalar projection of **B** onto **A** holds for vectors in V_3, and the proof is identical to that for vectors in V_2. Thus the scalar projection of **B** onto **A** is

$$\frac{\mathbf{A} \cdot \mathbf{B}}{\|\mathbf{A}\|} \tag{6}$$

Theorem 14.2.10, for vectors in V_2, gives a formula for computing the **vector projection** of the vector **B** onto the vector **A**. This theorem and its proof are identical for vectors in V_3, and the vector is

$$\left(\frac{\mathbf{A} \cdot \mathbf{B}}{\|\mathbf{A}\|^2} \right) \mathbf{A} \tag{7}$$

EXAMPLE 7   Given the vectors

$$\mathbf{A} = 6\mathbf{i} - 3\mathbf{j} + 2\mathbf{k} \qquad \mathbf{B} = 2\mathbf{i} + \mathbf{j} - 3\mathbf{k}$$

Find: (a) $\cos \theta$ if θ is the angle between **A** and **B**; (b) the component of **B** in the direction of **A**; (c) the vector projection of **B** onto **A**.

Solution   We first compute $\mathbf{A} \cdot \mathbf{B}$, $\|\mathbf{A}\|$, and $\|\mathbf{B}\|$.

$$\mathbf{A} \cdot \mathbf{B} = \langle 6, -3, 2 \rangle \cdot \langle 2, 1, -3 \rangle \qquad \|\mathbf{A}\| = \sqrt{36 + 9 + 4} \qquad \|\mathbf{B}\| = \sqrt{4 + 1 + 9}$$
$$= 12 - 3 - 6 \qquad\qquad\qquad = \sqrt{49} \qquad\qquad = \sqrt{14}$$
$$= 3 \qquad\qquad\qquad\qquad = 7$$

(a) From Theorem 15.2.11,

$$\cos \theta = \frac{\mathbf{A} \cdot \mathbf{B}}{\|\mathbf{A}\| \|\mathbf{B}\|}$$

$$= \frac{3}{7\sqrt{14}}$$

(b) The component of **B** in the direction of **A** is the scalar projection of **B** onto **A**, which is

$$\|\mathbf{B}\| \cos \theta = \sqrt{14} \left(\frac{3}{7\sqrt{14}} \right)$$

$$= \tfrac{3}{7}$$

(c) From (7), the vector projection of **B** onto **A** is

$$\left(\frac{\mathbf{A} \cdot \mathbf{B}}{\|\mathbf{A}\|^2} \right) \mathbf{A} = \tfrac{3}{49}(6\mathbf{i} - 3\mathbf{j} + 2\mathbf{k})$$

$$= \tfrac{18}{49}\mathbf{i} - \tfrac{9}{49}\mathbf{j} + \tfrac{6}{49}\mathbf{k}$$

EXAMPLE 8   Find the distance from the point $P(4, 1, 6)$ to the line through the points $A(8, 3, 2)$ and $B(2, -3, 5)$.

Solution   Figure 9 shows the point P and a sketch of the line through A and B. The point M is the foot of the perpendicular line from P to the line through A and B. Let d units be the distance $|\overline{PM}|$. Thus from the Pythagorean theorem,

$$d = \sqrt{|\overline{AP}|^2 - |\overline{AM}|^2} \tag{8}$$

To apply (8) we need to compute $|\overline{AP}|$, which is the magnitude of $\mathbf{V}(\overrightarrow{AP})$, and $|\overline{AM}|$, which is the scalar projection of $\mathbf{V}(\overrightarrow{AP})$ onto $\mathbf{V}(\overrightarrow{AB})$. We first find $\mathbf{V}(\overrightarrow{AP})$ and $\mathbf{V}(\overrightarrow{AB})$.

$$\mathbf{V}(\overrightarrow{AP}) = \langle 4 - 8, 1 - 3, 6 - 2 \rangle \qquad \mathbf{V}(\overrightarrow{AB}) = \langle 2 - 8, -3 - 3, 5 - 2 \rangle$$
$$= \langle -4, -2, 4 \rangle \qquad\qquad\qquad = \langle -6, -6, 3 \rangle$$

We compute $|\overline{AP}|$ by finding $\|\mathbf{V}(\overrightarrow{AP})\|$, and we compute $|\overline{AM}|$ by using formula (6) with $\mathbf{A} = \mathbf{V}(\overrightarrow{AB})$ and $\mathbf{B} = \mathbf{V}(\overrightarrow{AP})$.

$$|\overline{AP}| = \|\mathbf{V}(\overrightarrow{AP})\| \qquad\qquad |\overline{AM}| = \frac{\mathbf{V}(\overrightarrow{AB}) \cdot \mathbf{V}(\overrightarrow{AP})}{\|\mathbf{V}(\overrightarrow{AB})\|}$$
$$= \sqrt{(-4)^2 + (-2)^2 + 4^2} \qquad\qquad = \frac{\langle -6, -6, 3 \rangle \cdot \langle -4, -2, 4 \rangle}{\sqrt{(-6)^2 + (-6)^2 + 3^2}}$$
$$= \sqrt{36} \qquad\qquad\qquad\qquad = \frac{24 + 12 + 12}{\sqrt{81}}$$
$$= 6 \qquad\qquad\qquad\qquad\qquad = \tfrac{48}{9}$$

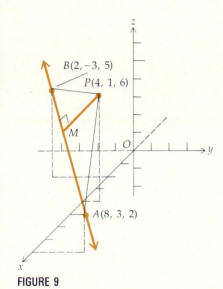

$B(2, -3, 5)$

$P(4, 1, 6)$

M

O

$A(8, 3, 2)$

FIGURE 9

Substituting these values of $|\overline{AP}|$ and $|\overline{AM}|$ in (8) we get

$$d = \sqrt{6^2 - \left(\tfrac{48}{9}\right)^2}$$
$$= 6\sqrt{1 - \tfrac{64}{81}}$$
$$= \tfrac{2}{3}\sqrt{17}$$

EXERCISES 15.2

In Exercises 1 through 6, $\mathbf{A} = \langle 1, 2, 3 \rangle$, $\mathbf{B} = \langle 4, -3, -1 \rangle$, $\mathbf{C} = \langle -5, -3, 5 \rangle$, *and* $\mathbf{D} = \langle -2, 1, 6 \rangle$.

1. Find: (a) $\mathbf{A} + 5\mathbf{B}$; (b) $7\mathbf{C} - 5\mathbf{D}$; (c) $\|7\mathbf{C}\| - \|5\mathbf{D}\|$; (d) $\|7\mathbf{C} - 5\mathbf{D}\|$.
2. Find: (a) $2\mathbf{A} - \mathbf{C}$; (b) $\|2\mathbf{A}\| - \|\mathbf{C}\|$; (c) $4\mathbf{B} + 6\mathbf{C} - 2\mathbf{D}$; (d) $\|4\mathbf{B}\| + \|6\mathbf{C}\| - \|2\mathbf{D}\|$.
3. Find: (a) $\mathbf{C} + 3\mathbf{D} - 8\mathbf{A}$; (b) $\|\mathbf{A}\|\,\|\mathbf{B}\|(\mathbf{C} - \mathbf{D})$.
4. Find: (a) $3\mathbf{A} - 2\mathbf{B} + \mathbf{C} - 12\mathbf{D}$; (b) $\|\mathbf{A}\|\mathbf{C} - \|\mathbf{B}\|\mathbf{D}$.
5. Find scalars a and b such that $a(\mathbf{A} + \mathbf{B}) + b(\mathbf{C} + \mathbf{D}) = \mathbf{0}$.
6. Find scalars a, b, and c that $a\mathbf{A} + b\mathbf{B} + c\mathbf{C} = \mathbf{D}$.

In Exercises 7 through 10, find the direction cosines of the vector $\mathbf{V}(\overrightarrow{P_1 P_2})$ *and check the answers by verifying that the sum of their squares is 1.*

7. $P_1(3, -1, -4)$; $P_2(7, 2, 4)$ 8. $P_1(-2, 6, 5)$; $P_2(2, 4, 1)$
9. $P_1(4, -3, -1)$; $P_2(-2, -4, -8)$
10. $P_1(1, 3, 5)$; $P_2(2, -1, 4)$

11. Use the points P_1 and P_2 of Exercise 7 and find the point Q such that $\mathbf{V}(\overrightarrow{P_1 P_2}) = 3\mathbf{V}(\overrightarrow{P_1 Q})$.
12. Use the points P_1 and P_2 of Exercise 10 and find the point R such that $\mathbf{V}(\overrightarrow{P_1 R}) = -2\mathbf{V}(\overrightarrow{P_2 R})$.
13. Given $P_1(3, 2, -4)$ and $P_2(-5, 4, 2)$, find the point P_3 such that $4\mathbf{V}(\overrightarrow{P_1 P_2}) = -3\mathbf{V}(\overrightarrow{P_2 P_3})$.
14. Given $P_1(7, 0, -2)$ and $P_2(2, -3, 5)$, find the point P_3 such that $\mathbf{V}(\overrightarrow{P_1 P_3}) = 5\mathbf{V}(\overrightarrow{P_2 P_3})$.

In Exercises 15 and 16, express the vector in terms of its magnitude and direction cosines.

15. (a) $-6\mathbf{i} + 2\mathbf{j} + 3\mathbf{k}$; (b) $-2\mathbf{i} + \mathbf{j} - 3\mathbf{k}$
16. (a) $2\mathbf{i} - 2\mathbf{j} + \mathbf{k}$; (b) $3\mathbf{i} + 4\mathbf{j} - 5\mathbf{k}$

In Exercises 17 and 18, find the unit vector having the same direction as $\mathbf{V}(\overrightarrow{P_1 P_2})$.

17. (a) $P_1(4, -1, -6)$ and $P_2(5, 7, -2)$;
    (b) $P_1(-2, 5, 3)$ and $P_2(-4, 7, 5)$

18. (a) $P_1(3, 0, -1)$ and $P_2(-3, 8, -1)$;
    (b) $P_1(-8, -5, 2)$ and $P_2(-3, -9, 4)$

In Exercises 19 and 20, prove the property if **A**, **B**, *and* **C** *are any vectors in* V_3 *and* c *and* d *are any scalars.*

19. (a) $\mathbf{A} + \mathbf{B} = \mathbf{B} + \mathbf{A}$ (commutative law)
    (b) There is a vector **0** in V_3 for which $\mathbf{A} + \mathbf{0} = \mathbf{A}$ (existence of additive identity)
    (c) There is a vector $-\mathbf{A}$ in V_3 such that $\mathbf{A} + (-\mathbf{A}) = \mathbf{0}$ (existence of negative)
    (d) $c(\mathbf{A} + \mathbf{B}) = c\mathbf{A} + c\mathbf{B}$ (distributive law)
20. (a) $\mathbf{A} + (\mathbf{B} + \mathbf{C}) = (\mathbf{A} + \mathbf{B}) + \mathbf{C}$ (associative law)
    (b) $(cd)\mathbf{A} = c(d\mathbf{A})$ (associative law)
    (c) $(c + d)\mathbf{A} = c\mathbf{A} + d\mathbf{A}$ (distributive law)

In Exercises 21 through 26, $\mathbf{A} = \langle -4, -2, 4 \rangle$; $\mathbf{B} = \langle 2, 7, -1 \rangle$; $\mathbf{C} = \langle 6, -3, 0 \rangle$, *and* $\mathbf{D} = \langle 5, 4, -3 \rangle$.

21. Find: (a) $\mathbf{A} \cdot (\mathbf{B} + \mathbf{C})$; (b) $(\mathbf{A} \cdot \mathbf{B})(\mathbf{C} \cdot \mathbf{D})$; (c) $\mathbf{A} \cdot \mathbf{D} - \mathbf{B} \cdot \mathbf{C}$; (d) $(\mathbf{D} \cdot \mathbf{B})\mathbf{A} - (\mathbf{D} \cdot \mathbf{A})\mathbf{B}$.
22. Find: (a) $\mathbf{A} \cdot \mathbf{B} + \mathbf{A} \cdot \mathbf{C}$; (b) $(\mathbf{A} \cdot \mathbf{B})(\mathbf{B} \cdot \mathbf{C})$; (c) $(\mathbf{A} \cdot \mathbf{B})\mathbf{C} + (\mathbf{B} \cdot \mathbf{C})\mathbf{D}$; (d) $(2\mathbf{A} + 3\mathbf{B}) \cdot (4\mathbf{C} - \mathbf{D})$.
23. Find: (a) $\cos \theta$ if θ is the angle between **A** and **C**; (b) the component of **C** in the direction of **A**; (c) the vector projection of **C** onto **A**.
24. Find: (a) $\cos \theta$ if θ is the angle between **B** and **D**; (b) the component of **B** in the direction of **D**; (c) the vector projection of **B** onto **D**.
25. Find: (a) the scalar projection of **A** onto **B**; (b) the vector projection of **A** onto **B**.
26. Find: (a) the scalar projection of **D** onto **C**; (b) the vector projection of **D** onto **C**.

In Exercises 27 and 28, prove the law of dot multiplication if **A**, **B**, *and* **C** *are any vectors in* V_3 *and* c *is a scalar.*

27. (a) $\mathbf{A} \cdot \mathbf{B} = \mathbf{B} \cdot \mathbf{A}$ (commutative law)
    (b) $\mathbf{A} \cdot (\mathbf{B} + \mathbf{C}) = \mathbf{A} \cdot \mathbf{B} + \mathbf{A} \cdot \mathbf{C}$ (distributive law)
28. (a) $c(\mathbf{A} \cdot \mathbf{B}) = (c\mathbf{A}) \cdot \mathbf{B}$; (b) $\mathbf{0} \cdot \mathbf{A} = 0$; (c) $\mathbf{A} \cdot \mathbf{A} = \|\mathbf{A}\|^2$

29. Find the distance from the point $(2, -1, -4)$ to the line through the points $(3, -2, 2)$ and $(-9, -6, 6)$.
30. Find the distance from the point $(3, 2, 1)$ to the line through the points $(1, 2, 9)$ and $(-3, -6, -3)$.
31. Prove by using vectors that the points $(2, 2, 2)$, $(2, 0, 1)$, $(4, 1, -1)$, and $(4, 3, 0)$ are the vertices of a rectangle.
32. Prove by using vectors that the points $(2, 2, 2)$, $(0, 1, 2)$, $(-1, 3, 3)$, and $(3, 0, 1)$ are the vertices of a parallelogram.
33. Find the area of the triangle having vertices at $(-2, 3, 1)$, $(1, 2, 3)$, and $(3, -1, 2)$.
34. Prove by using vectors that the points $(-2, 1, 6)$, $(2, 4, 5)$, and $(-1, -2, 1)$ are the vertices of a right triangle, and find the area of the triangle.
35. If $\mathbf{A} = 3\mathbf{i} + 5\mathbf{j} - 3\mathbf{k}$, $\mathbf{B} = -\mathbf{i} - 2\mathbf{j} + 3\mathbf{k}$, and $\mathbf{C} = 2\mathbf{i} - \mathbf{j} + 4\mathbf{k}$, find the component of **B** in the direction of $\mathbf{A} - 2\mathbf{C}$.
36. Find the cosines of the angles of the triangle having vertices at $A(0, 0, 0)$, $B(4, -1, 3)$, and $C(1, 2, 3)$.
37. If a force has the vector representation $\mathbf{F} = 3\mathbf{i} - 2\mathbf{j} + \mathbf{k}$, find the work done by the force in moving an object from

the point $P_1(-2, 4, 3)$ along a straight line to the point $P_2(1, -3, 5)$. The magnitude of the force is measured in pounds and distance is measured in feet. (*Hint:* Review Section 14.2.)

38. If a force has the vector representation $\mathbf{F} = 5\mathbf{i} - 3\mathbf{k}$, find the work done by the force in moving an object from the point $P_1(4, 1, 3)$ along a straight line to the point $P_2(-5, 6, 2)$. The magnitude of the force is measured in pounds and distance is measured in feet. (See hint for Exercise 37.)
39. A force is represented by the vector **F**, it has a magnitude of 10 lb, and direction cosines of **F** are $\cos \alpha = \frac{1}{6}\sqrt{6}$ and $\cos \beta = \frac{1}{3}\sqrt{6}$. If the force moves an object from the origin along a straight line to the point $(7, -4, 2)$, find the work done. Distance is measured in feet. (See hint for Exercise 37.)
40. If **A** and **B** are nonzero vectors, prove that the vector $\mathbf{A} - c\mathbf{B}$ is orthogonal to **B** if $c = \mathbf{A} \cdot \mathbf{B}/\|\mathbf{B}\|^2$.
41. If $\mathbf{A} = 12\mathbf{i} + 9\mathbf{j} - 5\mathbf{k}$ and $\mathbf{B} = 4\mathbf{i} + 3\mathbf{j} - 5\mathbf{k}$, use the result of Exercise 40 to find the value of the scalar c so that the vector $\mathbf{B} - c\mathbf{A}$ is orthogonal to **A**.
42. For the vectors of Exercise 41, use the result of Exercise 40, to find the value of the scalar d so that the vector $\mathbf{A} - d\mathbf{B}$ is orthogonal to **B**.
43. Prove that if **A** and **B** are any vectors, then the vectors $\|\mathbf{B}\|\mathbf{A} + \|\mathbf{A}\|\mathbf{B}$ and $\|\mathbf{B}\|\mathbf{A} - \|\mathbf{A}\|\mathbf{B}$ are orthogonal.
44. Prove that if **A** and **B** are any nonzero vectors and $\mathbf{C} = \|\mathbf{B}\|\mathbf{A} + \|\mathbf{A}\|\mathbf{B}$, then the angle between **A** and **C** has the same measure as the angle between **B** and **C**.
45. If the radian measure of each direction angle of a vector is the same, what is it? Prove your answer.
46. Prove Theorem 15.2.8.   47. Prove Theorem 15.2.11.
48. Three vectors in V_3 are said to be *independent* if and only if their position representations do not lie in a plane, and three vectors $\mathbf{E}_1$, $\mathbf{E}_2$, and $\mathbf{E}_3$ are said to form a *basis* for the vector space V_3 if and only if any vector in V_3 can be written as a linear combination of $\mathbf{E}_1$, $\mathbf{E}_2$, and $\mathbf{E}_3$. A theorem can be proved which states that three vectors form a basis for the vector space V_3 if they are independent. Show that this theorem holds for the three vectors $\langle 1, 0, 0 \rangle$, $\langle 1, 1, 0 \rangle$, and $\langle 1, 1, 1 \rangle$ by doing the following: (a) Verify that the vectors are independent by showing that their position representations are not coplanar; (b) verify that the vectors form a basis by showing that any vector **A** can be written

$$\mathbf{A} = r\langle 1, 0, 0 \rangle + s\langle 1, 1, 0 \rangle + t\langle 1, 1, 1 \rangle \qquad (9)$$

where r, s, and t are scalars. (c) If $\mathbf{A} = \langle 6, -2, 5 \rangle$, find the particular values of r, s, and t such that (9) holds.

49. See Exercise 48. (a) Verify that the vectors $\langle 2, 0, 1 \rangle$, $\langle 0, -1, 0 \rangle$, and $\langle 1, -1, 0 \rangle$ form a basis for V_3 by showing that any vector **A** can be written

$$\mathbf{A} = r\langle 2, 0, 1 \rangle + s\langle 0, -1, 0 \rangle + t\langle 1, -1, 0 \rangle \qquad (10)$$

where r, s, and t are scalars. (b) If $\mathbf{A} = \langle -2, 3, 5 \rangle$, find the particular values of r, s, and t such that (10) holds.

50. Refer to the first sentence of Exercise 48. A theorem can be proved which states that three vectors form a basis for

the vector space V_3 only if they are independent. Show that this theorem is valid for the three vectors $F_1 = \langle 1, 0, 1 \rangle$, $F_2 = \langle 1, 1, 1 \rangle$, and $F_3 = \langle 2, 1, 2 \rangle$ by doing the following: (a) Verify that F_1, F_2, and F_3 are not independent by show-ing that their position representations are coplanar; (b) verify that the vectors do not form a basis by showing that every vector in V_3 cannot be written as a linear combination of F_1, F_2, and F_3.

15.3 PLANES

The graph of an equation in two variables, x and y, is a curve in the xy plane. The simplest kind of curve in two-dimensional space is a line, and the general equation of a line is of the form $Ax + By + C = 0$, which is an equation of the first degree. In three-dimensional space the graph of an equation in three variables x, y, and z, is a surface. The simplest kind of surface is a *plane*, and we shall see that an equation of a plane is an equation of the first degree in three variables.

15.3.1 DEFINITION

If N is a given nonzero vector and P_0 is a given point, then the set of all points P for which $V(\overrightarrow{P_0P})$ and N are orthogonal is defined to be a **plane** through P_0 having N as a **normal vector**.

Figure 1 shows a portion of a plane through the point $P_0(x_0, y_0, z_0)$ and the representation of the normal vector N having its initial point at P_0.

In plane analytic geometry we can obtain an equation of a line if a point on the line and its direction (slope) are given. In an analogous manner, in solid analytic geometry an equation of a plane can be determined by knowing a point in the plane and the direction of a normal vector.

15.3.2 THEOREM

If $P_0(x_0, y_0, z_0)$ is a point in a plane and $\langle a, b, c \rangle$ is a normal vector to the plane then an equation of the plane is

$$a(x - x_0) + b(y - y_0) + c(z - z_0) = 0$$

Proof Refer to Figure 1 where $N = \langle a, b, c \rangle$. Let $P(x, y, z)$ be any point in the plane. $V(\overrightarrow{P_0P})$ is the vector having $\overrightarrow{P_0P}$ as a representation; so

$$V(\overrightarrow{P_0P}) = \langle x - x_0, y - y_0, z - z_0 \rangle \tag{1}$$

From Definition 15.3.1 and the fact that the dot product of two orthogonal vectors is zero we have

$$V(\overrightarrow{P_0P}) \cdot \langle a, b, c \rangle = 0$$

From (1) and the above equation,

$$a(x - x_0) + b(y - y_0) + c(z - z_0) = 0$$

which is the desired equation. ∎

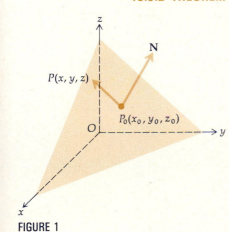

FIGURE 1

EXAMPLE 1   Find an equation of the plane containing the point $(2, 1, 3)$ and having $3i - 4j + k$ as a normal vector.

Solution   Using Theorem 15.3.2 where the point (x_0, y_0, z_0) is $(2, 1, 3)$ and the vector $\langle a, b, c \rangle$ is $\langle 3, -4, 1 \rangle$, we have as an equation of the required plane

$$3(x - 2) - 4(y - 1) + (z - 3) = 0$$

$$3x - 4y + z - 5 = 0$$

15.3.3 THEOREM   If a, b, and c are not all zero, the graph of an equation of the form

$$ax + by + cz + d = 0$$

is a plane and $\langle a, b, c \rangle$ is a normal vector to the plane.

Proof   Suppose that $b \neq 0$. Then the point $(0, -d/b, 0)$ is on the graph of the equation because its coordinates satisfy the equation. The given equation can be written as

$$a(x - 0) + b\left(y + \frac{d}{b}\right) + c(z - 0) = 0$$

which from Theorem 15.3.2 is an equation of a plane through the point $(0, -d/b, 0)$ and for which $\langle a, b, c \rangle$ is a normal vector. This proves the theorem if $b \neq 0$. A similar argument holds if $b = 0$ and either $a \neq 0$ or $c \neq 0$. ∎

The equations of Theorems 15.3.2 and 15.3.3 are called *cartesian* equations of a plane. The equation of Theorem 15.3.2 is analogous to the point-slope form of an equation of a line in two dimensions. The equation of Theorem 15.3.3 is the general first-degree equation in three variables and is called a *linear equation*.

A plane is determined by three noncollinear points, by a line and a point not on the line, by two intersecting lines, or by two parallel lines.

EXAMPLE 2   Find an equation of the plane through the points $P(1, 3, 2)$, $Q(3, -2, 2)$, and $R(2, 1, 3)$.

Solution   From Theorem 15.3.3 the graph of the linear equation

$$ax + by + cz + d = 0 \tag{2}$$

is a plane. If this equation is satisfied by the coordinates of points P, Q, and R, the plane will contain the points. Replacing x, y, and z in (2) by the coordinates of the three points we have the equations

$$a + 3b + 2c + d = 0$$
$$3a - 2b + 2c + d = 0$$
$$2a + b + 3c + d = 0$$

We solve this system of equations for a, b, and c in terms of d, and we obtain

$$a = -\tfrac{5}{9}d \qquad b = -\tfrac{2}{9}d \qquad c = \tfrac{1}{9}d$$

Replacing a, b, and c in (2) by these values we have

$$-\tfrac{5}{9}dx - \tfrac{2}{9}dy + \tfrac{1}{9}dz + d = 0$$

Multiplying on both sides of this equation by $-9/d$ we get

$$5x + 2y - z - 9 = 0$$

which is the required equation.

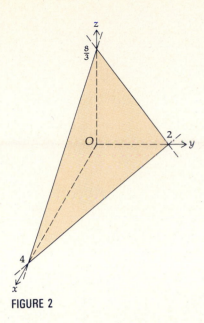

FIGURE 2

To draw a sketch of a plane from its equation it is convenient to find the points at which the plane intersects each of the coordinate axes. The x coordinate of the point at which the plane intersects the x axis is called the x *intercept* of the plane; the y coordinate of the point at which the plane intersects the y axis is called the y *intercept* of the plane; and the z *intercept* of the plane is the z coordinate of the point at which the plane intersects the z axis.

▶ **ILLUSTRATION 1**   We wish to draw a sketch of the plane having the equation

$$2x + 4y + 3z = 8$$

By substituting zero for y and z we obtain $x = 4$; so the x intercept of the plane is 4. The y intercept and the z intercept are obtained in a similar manner; they are 2 and $\frac{8}{3}$, respectively. Plotting the points corresponding to these intercepts and connecting them with lines we have the sketch of the plane shown in Figure 2. Note that only a portion of the plane is shown in the figure.   ◀

▶ **ILLUSTRATION 2**   To draw a sketch of the plane having the equation

$$3x + 2y - 6z = 0$$

first notice that because the equation is satisfied when x, y, and z are all zero, the plane intersects each of the axes at the origin. If $x = 0$ in the given equation, we obtain $y - 3z = 0$, which is a line in the yz plane; this is the line of intersection of the yz plane with the given plane. Similarly, the line of intersection of the xz plane with the given plane is obtained by setting $y = 0$, and we get $x - 2z = 0$. Drawing a sketch of each of these two lines and drawing a line segment from a point on one of the lines to a point on the other line we obtain Figure 3.   ◀

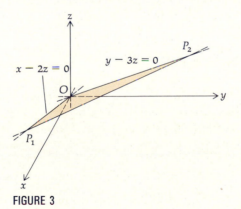

FIGURE 3

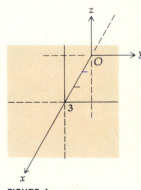

FIGURE 4

In Illustration 2 the line in the yz plane and the line in the xz plane used to draw the sketch of the plane are called the **traces** of the given plane in the yz plane and the xz plane, respectively. The equation $x = 0$ is an equation of the yz plane because the point (x, y, z) is in the yz plane if and only if $x = 0$. Similarly, the equations $y = 0$ and $z = 0$ are equations of the xz plane and the xy plane, respectively.

A plane parallel to the yz plane has an equation of the form $x = k$, where k is a constant. Figure 4 shows a sketch of the plane having the equation $x = 3$. A plane parallel to the xz plane has an equation of the form $y = k$, and

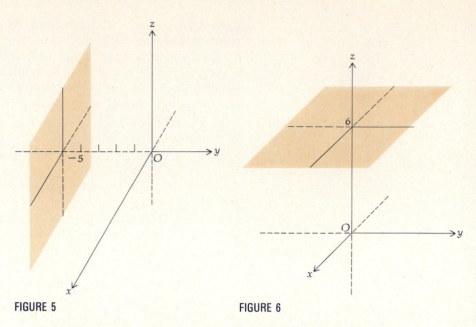

FIGURE 5   FIGURE 6

a plane parallel to the xy plane has an equation of the form $z = k$. Figures 5 and 6 show sketches of the planes having the equations $y = -5$ and $z = 6$, respectively.

15.3.4 DEFINITION   An **angle between two planes** is defined to be the angle between normal vectors of the planes.

There are two angles between two planes. If one of these angles is θ, the other is the supplement of θ.

EXAMPLE 3   Find the radian measure of the acute angle between the planes

$$5x - 2y + 5z - 12 = 0 \quad \text{and} \quad 2x + y - 7z + 11 = 0$$

Solution   Let $\mathbf{N}_1$ be a normal vector to the first plane and $\mathbf{N}_2$ be a normal vector to the second plane. Then

$$\mathbf{N}_1 = 5\mathbf{i} - 2\mathbf{j} + 5\mathbf{k} \qquad \mathbf{N}_2 = 2\mathbf{i} + \mathbf{j} - 7\mathbf{k}$$

From Definition 15.3.4, an angle between the two planes is the angle between $\mathbf{N}_1$ and $\mathbf{N}_2$. Thus from Theorem 15.2.11, if θ is the radian measure of this angle,

$$
\begin{aligned}
\cos \theta &= \frac{\mathbf{N}_1 \cdot \mathbf{N}_2}{\|\mathbf{N}_1\|\,\|\mathbf{N}_2\|} \\[2mm]
&= \frac{\langle 5, -2, 5 \rangle \cdot \langle 2, 1, -7 \rangle}{\sqrt{25 + 4 + 25}\,\sqrt{4 + 1 + 49}} \\[2mm]
&= \frac{10 - 2 - 35}{\sqrt{54}\,\sqrt{54}} \\[2mm]
&= -\tfrac{27}{54} \\[2mm]
&= -\tfrac{1}{2}
\end{aligned}
$$

Therefore $\theta = \frac{2}{3}\pi$. The acute angle between the two planes is the supplement of θ which is $\frac{1}{3}\pi$.

15.3.5 DEFINITION Two planes are **parallel** if and only if their normal vectors are parallel.

Recall that two vectors are parallel if and only if one of the vectors is a scalar multiple of the other. Thus from Definition 15.3.5 it follows that if we have one plane with a normal vector $\mathbf{N}_1$ and another plane with a normal vector $\mathbf{N}_2$, then the two planes are parallel if and only if

$$\mathbf{N}_1 = k\mathbf{N}_2$$

where k is a constant. Figure 7 shows sketches of two parallel planes and representations of some of their normal vectors.

15.3.6 DEFINITION Two planes are **perpendicular** if and only if their normal vectors are orthogonal.

From Definition 15.3.6 and the fact that two vectors are orthogonal if and only if their dot product is zero it follows that two planes having normal vectors $\mathbf{N}_1$ and $\mathbf{N}_2$ are perpendicular if and only if

$$\mathbf{N}_1 \cdot \mathbf{N}_2 = 0 \tag{3}$$

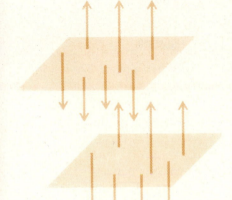

FIGURE 7

EXAMPLE 4 Find an equation of the plane containing the point $(4, 0, -2)$ and perpendicular to each of the planes

$$x - y + z = 0 \quad \text{and} \quad 2x + y - 4z - 5 = 0$$

Solution Let M be the required plane and $\langle a, b, c \rangle$, $a \neq 0$, be a normal vector of M. Let M_1 be the plane having the equation $x - y + z = 0$. By Theorem 15.3.3, a normal vector of M_1 is $\langle 1, -1, 1 \rangle$. Because M and M_1 are perpendicular, it follows from (3) that

$$\langle a, b, c \rangle \cdot \langle 1, -1, 1 \rangle = 0$$
$$a - b + c = 0 \tag{4}$$

Let M_2 be the plane having the equation $2x + y - 4z - 5 = 0$. A normal vector of M_2 is $\langle 2, 1, -4 \rangle$. Because M and M_2 are perpendicular,

$$\langle a, b, c \rangle \cdot \langle 2, 1, -4 \rangle = 0$$
$$2a + b - 4c = 0$$

Solving this equation and (4) simultaneously for b and c in terms of a we get $b = 2a$ and $c = a$. Therefore a normal vector of M is $\langle a, 2a, a \rangle$. Because $(4, 0, -2)$ is a point in M, it follows from Theorem 15.3.2 that an equation of M is

$$a(x - 4) + 2a(y - 0) + a(z + 2) = 0$$

Because $a \neq 0$, we divide by a and combine terms to obtain

$$x + 2y + z - 2 = 0$$

Consider now the plane having the equation $ax + by + d = 0$ and the xy plane whose equation is $z = 0$. Normal vectors to these planes are $\langle a, b, 0 \rangle$ and $\langle 0, 0, 1 \rangle$, respectively. Because $\langle a, b, 0 \rangle \cdot \langle 0, 0, 1 \rangle = 0$, the two planes are perpendicular. This means that a plane having an equation with no z term is perpendicular to the xy plane. Figure 8 illustrates this. In a similar manner we can conclude that a plane having an equation with no x term is perpendicular to the yz plane (see Figure 9), and a plane having an equation with no y term is perpendicular to the xz plane (see Figure 10).

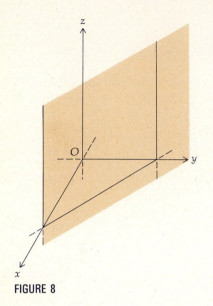

FIGURE 8

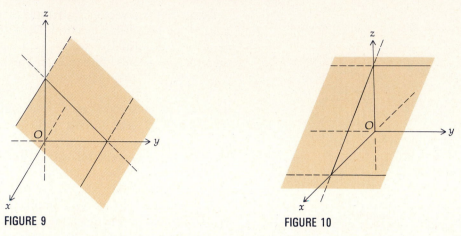

FIGURE 9

FIGURE 10

Vectors can be used to find the distance from a point to a plane. The following example illustrates the procedure.

EXAMPLE 5   Find the distance from the point $(1, 4, 6)$ to the plane

$$2x - y + 2z + 10 = 0$$

Solution   Let P be the point $(1, 4, 6)$ and choose any point Q in the plane. For simplicity choose the point Q as the point where the plane intersects the x axis, that is, the point $(-5, 0, 0)$. The vector having $\overrightarrow{PQ}$ as a representation is given by

$$\mathbf{V}(\overrightarrow{PQ}) = -6\mathbf{i} - 4\mathbf{j} - 6\mathbf{k}$$

A normal vector to the given plane is

$$\mathbf{N} = 2\mathbf{i} - \mathbf{j} + 2\mathbf{k}$$

The negative of $\mathbf{N}$ is also a normal vector to the given plane and

$$-\mathbf{N} = -2\mathbf{i} + \mathbf{j} - 2\mathbf{k}$$

We are not certain which of the two vectors, $\mathbf{N}$ or $-\mathbf{N}$, makes the smaller angle with vector $\mathbf{V}(\overrightarrow{PQ})$. Let $\mathbf{N}'$ be the one of the two vectors $\mathbf{N}$ or $-\mathbf{N}$ that makes an angle of radian measure $\theta < \frac{1}{2}\pi$ with $\mathbf{V}(\overrightarrow{PQ})$. In Figure 11 there is a portion of the given plane containing the point $Q(-5, 0, 0)$, the representation of the vector $\mathbf{N}'$ having its initial point at Q, the point $P(1, 4, 6)$, the directed line segment $\overrightarrow{PQ}$, and the point R, which is the foot of the perpendicular from P to the plane. For simplicity the coordinate axes are not included in this figure. The distance $|\overline{RP}|$ is the required distance, which we call d. Because d is an undirected distance it is nonnegative. We see from Figure 11 that d is the absolute value of the scalar projection of $\mathbf{V}(\overrightarrow{PQ})$ onto $\mathbf{N}'$. Thus from (6) in Section 15.2 we get

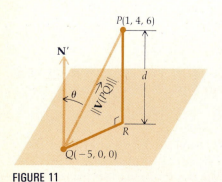

FIGURE 11

$$d = \frac{|\mathbf{N}' \cdot \mathbf{V}(\overrightarrow{PQ})|}{\|\mathbf{N}'\|}$$

Because we have the absolute value of the dot product in the numerator and the magnitude of $\mathbf{N}'$ in the denominator we can replace $\mathbf{N}'$ by $\mathbf{N}$, and we have

$$d = \frac{|\mathbf{N} \cdot \mathbf{V}(\overrightarrow{PQ})|}{\|\mathbf{N}\|}$$

$$= \frac{|\langle 2, -1, 2 \rangle \cdot \langle -6, -4, -6 \rangle|}{\sqrt{4 + 1 + 4}}$$

$$= \frac{|-12 + 4 - 12|}{\sqrt{9}}$$

$$= \tfrac{20}{3}$$

EXERCISES 15.3

In Exercises 1 through 6, find an equation of the plane containing the point P and having the vector $\mathbf{N}$ as a normal vector.

1. $P(3, 1, 2)$; $\mathbf{N} = \langle 1, 2, -3 \rangle$
2. $P(-3, 2, 5)$; $\mathbf{N} = \langle 6, -3, -2 \rangle$
3. $P(0, -1, 2)$; $\mathbf{N} = \langle 0, 1, -1 \rangle$
4. $P(-1, 8, 3)$; $\mathbf{N} = \langle -7, -1, 1 \rangle$
5. $P(2, 1, -1)$; $\mathbf{N} = -\mathbf{i} + 3\mathbf{j} + 4\mathbf{k}$
6. $P(1, 0, 0)$; $\mathbf{N} = \mathbf{i} + \mathbf{k}$

In Exercises 7 and 8, find an equation of the plane containing the three points.

7. $(3, 4, 1)$, $(1, 7, 1)$, $(-1, -2, 5)$
8. $(0, 0, 2)$, $(2, 4, 1)$, $(-2, 3, 3)$

In Exercises 9 through 14, draw a sketch of the plane and find two unit vectors that are normal to the plane.

9. $2x - y + 2z - 6 = 0$
10. $4x - 4y + 2z - 9 = 0$
11. $4x + 3y - 12z = 0$
12. $y + 2z - 4 = 0$
13. $3x + 2z - 6 = 0$
14. $z = 5$

In Exercises 15 through 20, find an equation of the plane satisfying the conditions.

15. Perpendicular to the line through the points $(2, 2, -4)$ and $(7, -1, 3)$ and containing the point $(-5, 1, 2)$.
16. Parallel to the plane $4x - 2y + z - 1 = 0$ and containing the point $(2, 6, -1)$.
17. Perpendicular to the plane $x + 3y - z - 7 = 0$ and containing the points $(2, 0, 5)$ and $(0, 2, -1)$.
18. Perpendicular to each of the planes $x - y + z = 0$ and $2x + y - 4z - 5 = 0$ and containing the point $(4, 0, -2)$.
19. Perpendicular to the yz plane, containing the point $(2, 1, 1)$, and making an angle of radian measure $\cos^{-1}\tfrac{2}{3}$ with the plane $2x - y + 2z - 3 = 0$.
20. Containing the point $P(-3, 5, -2)$ and perpendicular to the representations of the vector $\mathbf{V}(\overrightarrow{OP})$.

In Exercises 21 through 23, find the acute angle between the two planes.

21. $2x - y - 2z - 5 = 0$ and $6x - 2y + 3z + 8 = 0$
22. $2x - 5y + 3z - 1 = 0$ and $y - 5z + 3 = 0$
23. $3x + 4y = 0$ and $4x - 7y + 4z - 6 = 0$

24. Find the distance from the plane $2x + 2y - z - 6 = 0$ to the point $(2, 2, -4)$.
25. Find the distance from the plane $5x + 11y + 2z - 30 = 0$ to the point $(-2, 6, 3)$.
26. Find the perpendicular distance between the parallel planes $4x - 8y - z + 9 = 0$ and $4x - 8y - z - 6 = 0$.
27. Find the perpendicular distance between the parallel planes $4y - 3z - 6 = 0$ and $8y - 6z - 27 = 0$.
28. Prove that the undirected distance from the plane $ax + by + cz + d = 0$ to the point (x_0, y_0, z_0) is given by

$$\frac{|ax_0 + by_0 + cz_0 + d|}{\sqrt{a^2 + b^2 + c^2}}$$

29. Prove that the perpendicular distance between the parallel planes $ax + by + cz + d_1 = 0$ and $ax + by + cz + d_2 = 0$ is given by

$$\frac{|d_1 - d_2|}{\sqrt{a^2 + b^2 + c^2}}$$

30. If a, b, and c are nonzero and are the x intercept, y intercept, and z intercept, respectively, of a plane, prove that an equation of the plane is

$$\frac{x}{a} + \frac{y}{b} + \frac{z}{c} = 1$$

This is called the *intercept form* of an equation of a plane.

15.4 LINES IN R^3

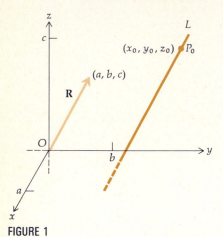

FIGURE 1

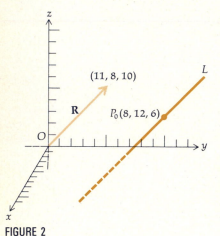

FIGURE 2

Let L be a line in R^3 that contains a given point $P_0(x_0, y_0, z_0)$ and is parallel to the representations of a given vector $\mathbf{R} = \langle a, b, c \rangle$. Figure 1 shows a sketch of L and the position representation of $\mathbf{R}$. Line L is the set of points $P(x, y, z)$ such that $\mathbf{V}(\overrightarrow{P_0P})$ is parallel to $\mathbf{R}$. So P is on L if and only if there is a non-zero scalar t such that

$$\mathbf{V}(\overrightarrow{P_0P}) = t\mathbf{R}$$

Because $\mathbf{V}(\overrightarrow{P_0P}) = \langle x - x_0, y - y_0, z - z_0 \rangle$, we obtain, from this equation

$$\langle x - x_0, y - y_0, z - z_0 \rangle = t\langle a, b, c \rangle$$

from which it follows that

$$x - x_0 = ta \qquad y - y_0 = tb \qquad z - z_0 = tc$$

$$x = x_0 + ta \qquad y = y_0 + tb \qquad z = z_0 + tc \tag{1}$$

Letting the parameter t be any real number (i.e., t takes on all values in the interval $(-\infty, +\infty)$), P may be any point on L. Therefore Equations (1) represent the line L; these equations are called **parametric equations** of the line.

▶ **ILLUSTRATION 1**   From Equations (1), parametric equations of the line L that is parallel to the representations of the vector $\mathbf{R} = \langle 11, 8, 10 \rangle$ and that contains the point $(8, 12, 6)$ are

$$x = 8 + 11t \qquad y = 12 + 8t \qquad z = 6 + 10t$$

Figure 2 shows a sketch of the line and the position representation of $\mathbf{R}$.   ◀

If none of the numbers a, b, or c is zero, we can eliminate t from Equations (1) and obtain

$$\frac{x - x_0}{a} = \frac{y - y_0}{b} = \frac{z - z_0}{c} \tag{2}$$

These equations are called **symmetric equations** of the line. Equations (2) are equivalent to the system of three equations

$$b(x - x_0) = a(y - y_0)$$

$$c(x - x_0) = a(z - z_0)$$

$$c(y - y_0) = b(z - z_0)$$

Actually, these three equations are not independent because any one of them can be derived from the other two. Each of the equations is an equation of a plane containing the line L represented by Equations (2). Any two of these planes have as their intersection the line L; hence any two of the equations define the line. However, there is an unlimited number of planes containing a given line, and because any two of them will determine the line, there is an unlimited number of pairs of equations that represent a line.

The vector $\mathbf{R} = \langle a, b, c \rangle$ determines the direction of the line having symmetric equations (2), and the numbers a, b, and c are called **direction numbers** of the line. Any vector parallel to $\mathbf{R}$ has either the same or the opposite direction as $\mathbf{R}$; hence such a vector can be used in place of $\mathbf{R}$ in the above discussion. Because the components of any vector parallel to $\mathbf{R}$ are proportional to the components of **R**, any set of three numbers proportional to a, b, and c also can serve as a set of direction numbers of the line. So a line has an unlimited number of sets of direction numbers. A set of direction numbers of a line is written in brackets as $[a, b, c]$.

▶ **ILLUSTRATION 2**   If $[2, 3, -4]$ represents a set of direction numbers of a line, other sets of direction numbers of the same line can be represented as $[4, 6, -8]$, $[1, \frac{3}{2}, -2]$, and $[2/\sqrt{29}, 3/\sqrt{29}, -4/\sqrt{29}]$.   ◀

▶ **ILLUSTRATION 3**   A set of direction numbers of the line of Illustration 1 is $[11, 8, 10]$, and the line contains the point $(8, 12, 6)$. Thus, from (2), symmetric equations of this line are

$$\frac{x - 8}{11} = \frac{y - 12}{8} = \frac{z - 6}{10}$$ ◀

EXAMPLE 1 .   Find two sets of symmetric equations of the line through the points $(-3, 2, 4)$ and $(6, 1, 2)$.

Solution   Let P_1 be the point $(-3, 2, 4)$ and P_2 be the point $(6, 1, 2)$. Then the required line is parallel to the representations of the vector $\mathbf{V}(\overrightarrow{P_1P_2})$, and so the components of this vector constitute a set of direction numbers of the line. $\mathbf{V}(\overrightarrow{P_1P_2}) = \langle 9, -1, -2 \rangle$. Taking P_0 as the point $(-3, 2, 4)$ we have, from (2), the equations

$$\frac{x + 3}{9} = \frac{y - 2}{-1} = \frac{z - 4}{-2}$$

Another set of symmetric equations of this line, obtained by taking P_0 as the point $(6, 1, 2)$, is

$$\frac{x - 6}{9} = \frac{y - 1}{-1} = \frac{z - 2}{-2}$$

If one of the numbers a, b, or c is zero, we do not use symmetric equations (2). However, suppose, for example, that $b = 0$ and neither a nor c is zero. Then equations of the line are

$$\frac{x - x_0}{a} = \frac{z - z_0}{c} \quad \text{and} \quad y = y_0$$

A line having these symmetric equations lies in the plane $y = y_0$ and hence is parallel to the xz plane. Figure 3 shows such a line.

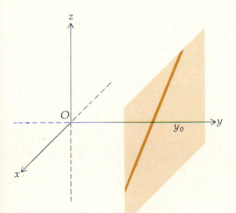

FIGURE 3

EXAMPLE 2   Given the two planes

$$x + 3y - z - 9 = 0 \quad \text{and} \quad 2x - 3y + 4z + 3 = 0$$

For the line of intersection of these two planes, find (a) a set of symmetric equations, (b) a set of parametric equations, and (c) the direction cosines of a vector whose representations are parallel to it.

Solution

(a) A set of symmetric equations is of the form (2). To obtain this form we solve the pair of given equations for x and y in terms of z. The computation is as follows:

$$
\begin{array}{ll}
x + 3y - z - 9 = 0 & 2x + 6y - 2z - 18 = 0 \\
2x - 3y + 4z + 3 = 0 \ (+) & 2x - 3y + 4z + 3 = 0 \ (-) \\
\hline
3x \quad\quad + 3z - 6 = 0 & 9y - 6z - 21 = 0 \\
\end{array}
$$

$$
x = -z + 2 \qquad\qquad y = \tfrac{2}{3}z + \tfrac{7}{3}
$$

We now solve each equation for z and obtain

$$
\frac{x - 2}{-1} = z \qquad \frac{y - \tfrac{7}{3}}{\tfrac{2}{3}} = z
$$

Thus a set of symmetric equations is

$$
\frac{x - 2}{-1} = \frac{y - \tfrac{7}{3}}{\tfrac{2}{3}} = \frac{z - 0}{1}
$$

$$
\Leftrightarrow \quad \frac{x - 2}{-3} = \frac{y - \tfrac{7}{3}}{2} = \frac{z - 0}{3}
$$

(b) A set of parametric equations is obtained by setting each of the ratios in part (a) equal to t, and we have

$$
\frac{x - 2}{-3} = t \qquad \frac{y - \tfrac{7}{3}}{2} = t \qquad \frac{z - 0}{3} = t
$$

$$
x = 2 - 3t \qquad y = \tfrac{7}{3} + 2t \qquad z = 3t
$$

(c) From the symmetric equations in part (a), a set of direction numbers of the line is $[-3, 2, 3]$. Therefore the vector $\langle -3, 2, 3 \rangle$ has its representations parallel to the line. Because $\sqrt{(-3)^2 + 2^2 + 3^2} = \sqrt{22}$, the direction cosines of this vector are

$$
\cos \alpha = -\frac{3}{\sqrt{22}} \qquad \cos \beta = \frac{2}{\sqrt{22}} \qquad \cos \gamma = \frac{3}{\sqrt{22}}
$$

EXAMPLE 3   Find equations of the line through the point $(1, -1, 1)$ perpendicular to the line

$$
3x = 2y = z \tag{3}
$$

and parallel to the plane

$$
x + y - z = 0 \tag{4}
$$

Solution   Let $[a, b, c]$ be a set of direction numbers of the required line. Equations (3) can be written as

$$
\frac{x - 0}{\tfrac{1}{3}} = \frac{y - 0}{\tfrac{1}{2}} = \frac{z - 0}{1}
$$

which are symmetric equations of a line. A set of direction numbers of this line is $[\tfrac{1}{3}, \tfrac{1}{2}, 1]$. Because the required line is perpendicular to this line, it follows that the vectors $\langle a, b, c \rangle$ and $\langle \frac{1}{3}, \frac{1}{2}, 1 \rangle$ are orthogonal. So

$$\langle a, b, c \rangle \cdot \langle \tfrac{1}{3}, \tfrac{1}{2}, 1 \rangle = 0$$

$$\tfrac{1}{3}a + \tfrac{1}{2}b + c = 0 \tag{5}$$

A normal vector to the plane (4) is $\langle 1, 1, -1 \rangle$. Because the required line is parallel to this plane, it is perpendicular to representations of the normal vector. Hence the vectors $\langle a, b, c \rangle$ and $\langle 1, 1, -1 \rangle$ are orthogonal; so

$$\langle a, b, c \rangle \cdot \langle 1, 1, -1 \rangle = 0$$

$$a + b - c = 0$$

Assuming $c \neq 0$, we solve this equation and (5) simultaneously for a and b in terms of c and get $a = 9c$ and $b = -8c$. The required line then has the set of direction numbers $[9c, -8c, c]$ and contains the point $(1, -1, 1)$. Therefore symmetric equations of the line are

$$\frac{x-1}{9c} = \frac{y+1}{-8c} = \frac{z-1}{c}$$

$$\Leftrightarrow \quad \frac{x-1}{9} = \frac{y+1}{-8} = \frac{z-1}{1}$$

In the following example we use the concept of **skew lines**, which are two lines that do not lie in one plane.

EXAMPLE 4   If l_1 is the line through $A(1, 2, 7)$ and $B(-2, 3, -4)$ and l_2 is the line through $C(2, -1, 4)$ and $D(5, 7, -3)$, prove that l_1 and l_2 are skew lines.

Solution   To show that two lines do not lie in one plane we demonstrate that they do not intersect and are not parallel. Parametric equations of a line are

$$x = x_0 + ta \qquad y = y_0 + tb \qquad z = z_0 + tc$$

where $[a, b, c]$ is a set of direction numbers of the line and (x_0, y_0, z_0) is any point on the line. Because $\mathbf{V}(\overrightarrow{AB}) = \langle -3, 1, -11 \rangle$, a set of direction numbers of l_1 is $[-3, 1, -11]$. Taking A as the point P_0 we have as parametric equations of l_1

$$x = 1 - 3t \qquad y = 2 + t \qquad z = 7 - 11t \tag{6}$$

Because $\mathbf{V}(\overrightarrow{CD}) = \langle 3, 8, -7 \rangle$ and l_2 contains the point C, parametric equations of l_2 are

$$x = 2 + 3s \qquad y = -1 + 8s \qquad z = 4 - 7s \tag{7}$$

Because the sets of direction numbers are not proportional, l_1 and l_2 are not parallel. For the lines to intersect, there have to be a value of t and a value of s that give the same point (x_1, y_1, z_1) in both sets of Equations (6) and (7). Therefore we equate the right sides of the respective equations and obtain

$$1 - 3t = 2 + 3s$$

$$2 + t = -1 + 8s$$

$$7 - 11t = 4 - 7s$$

Solving the first two equations simultaneously we obtain $s = \frac{8}{27}$ and $t = -\frac{17}{27}$. This set of values does not satisfy the third equation; hence the two lines do not intersect. Thus l_1 and l_2 are skew lines.

EXERCISES 15.4

In Exercises 1 through 8, find parametric and symmetric equations for the line satisfying the conditions.

1. Through the two points $(1, 2, 1)$ and $(5, -1, 1)$.
2. Through the point $(5, 3, 2)$ with direction numbers $[4, 1, -1]$.
3. Through the origin and perpendicular to the line
$$\tfrac{1}{4}(x - 10) = \tfrac{1}{3}y = \tfrac{1}{2}z$$
   at their intersection.
4. Through the origin and perpendicular to the lines having direction numbers $[4, 2, 1]$ and $[-3, -2, 1]$.
5. Perpendicular to the lines having direction numbers $[-5, 1, 2]$ and $[2, -3, -4]$ at the point $(-2, 0, 3)$.
6. Through the point $(-3, 1, -5)$ and perpendicular to the plane $4x - 2y + z - 7 = 0$.
7. Through the point $(4, -5, 20)$ and perpendicular to the plane $x + 3y - 6z - 8 = 0$.
8. Through the point $(2, 0, -4)$ and parallel to each of the planes $2x + y - z = 0$ and $x + 3y + 5z = 0$.
9. Find a set of symmetric equations for the line
$$\begin{cases} 4x - 3y + z - 2 = 0 \\ 2x + 5y - 3z + 4 = 0 \end{cases}$$

10. Show that the lines
$$\frac{x + 1}{2} = \frac{y + 4}{-5} = \frac{z - 2}{3} \quad \text{and} \quad \frac{x - 3}{-2} = \frac{y + 14}{5} = \frac{z - 8}{-3}$$
   are coincident.
11. Prove that the line $\tfrac{1}{2}(x - 3) = \tfrac{1}{3}(y + 2) = \tfrac{1}{4}(z + 1)$ lies in the plane $x - 2y + z = 6$.
12. Prove that the line $x + 1 = -\tfrac{1}{2}(y - 6) = z$ lies in the plane $3x + y - z = 3$.

*The planes through a line that are perpendicular to the coordinate planes are called the **projecting planes** of the line. In Exercises 13 through 16, find equations of the projecting planes of the line and draw a sketch of the line.*

13. $\begin{cases} 3x - 2y + 5z - 30 = 0 \\ 2x + 3y - 10z - 6 = 0 \end{cases}$ 14. $\begin{cases} x + y - 3z + 1 = 0 \\ 2x - y - 3z + 14 = 0 \end{cases}$

15. $\begin{cases} x - 2y - 3z + 6 = 0 \\ x + y + z - 1 = 0 \end{cases}$ 16. $\begin{cases} 2x - y + z - 7 = 0 \\ 4x - y + 3z - 13 = 0 \end{cases}$

17. Find the cosine of the smallest angle between the vector whose representations are parallel to the line $x = 2y + 4$, $z = -y + 4$, and the vector whose representations are parallel to the line $x = y + 7$, $2z = y + 2$.

18. Find an equation of the plane containing the point $(6, 2, 4)$ and the line $\tfrac{1}{5}(x - 1) = \tfrac{1}{6}(y + 2) = \tfrac{1}{7}(z - 3)$.

In Exercises 19 and 20, find an equation of the plane containing the given intersecting lines.

19. $\dfrac{x - 2}{4} = \dfrac{y + 3}{-1} = \dfrac{z + 2}{3}$ and $\begin{cases} 3x + 2y + z + 2 = 0 \\ x - y + 2z - 1 = 0 \end{cases}$

20. $\dfrac{x}{2} = \dfrac{y - 2}{3} = \dfrac{z - 1}{1}$ and $\dfrac{x}{1} = \dfrac{y - 2}{-1} = \dfrac{z - 1}{1}$

21. Show that the lines
$$\begin{cases} 3x - y - z = 0 \\ 8x - 2y - 3z + 1 = 0 \end{cases} \text{and} \begin{cases} x - 3y + z + 3 = 0 \\ 3x - y - z + 5 = 0 \end{cases}$$
   are parallel, and find an equation of the plane determined by these lines.
22. Show that the lines
$$\frac{x + 2}{5} = \frac{y - 1}{-2} = z + 4 \quad \text{and} \quad \frac{x - 3}{-5} = \frac{y + 4}{2} = \frac{z - 3}{-1}$$
   are parallel, and find an equation of the plane determined by these lines.
23. Find the coordinates of the point of intersection of the line $\tfrac{1}{4}(x - 2) = -\tfrac{1}{2}(y + 3) = \tfrac{1}{7}(z - 1)$ and the plane $5x - y + 2z - 12 = 0$.
24. Find equations of the line through the point $(1, -1, 1)$, perpendicular to the line $3x = 2y = z$, and parallel to the plane $x + y - z = 0$.
25. Find equations of the line through the point $(3, 6, 4)$, intersecting the z axis, and parallel to the plane $x - 3y + 5z - 6 = 0$.
26. Find the perpendicular distance from the origin to the line $x = -2 + \tfrac{6}{7}t$, $y = 7 - \tfrac{2}{7}t$, $z = 4 + \tfrac{3}{7}t$.
27. Find the perpendicular distance from the point $(-1, 3, -1)$ to the line $x - 2z = 7$, $y = 1$.
28. Find equations of the line through the origin, perpendicular to the line $x = y - 5$, $z = 2y - 3$, and intersecting the line $y = 2x + 1$, $z = x + 2$.
29. Prove that the lines
$$\frac{x - 1}{5} = \frac{y - 2}{-2} = \frac{z + 1}{-3} \quad \text{and} \quad \frac{x - 2}{1} = \frac{y + 1}{-3} = \frac{z + 3}{2}$$
   are skew lines.
30. Find equations of the line through the point $(3, -4, -5)$ that intersects each of the skew lines of Exercise 29.
31. What are the symmetric equations of a line if the two direction numbers a and b are zero?

15.5   CROSS PRODUCT   If **A** and **B** are two nonparallel vectors, representations of the two vectors with the same initial point determine a plane as indicated in Figure 1. We show that a vector whose representations are perpendicular to this plane is given by the vector operation called the *cross product* of **A** and **B**. We first define this operation and then consider its algebraic and geometric properties.

15.5.1 DEFINITION   If $\mathbf{A} = \langle a_1, a_2, a_3 \rangle$ and $\mathbf{B} = \langle b_1, b_2, b_3 \rangle$, then the **cross product** of **A** and **B**, denoted by **A** × **B**, is given by

$$\mathbf{A} \times \mathbf{B} = \langle a_2 b_3 - a_3 b_2, a_3 b_1 - a_1 b_3, a_1 b_2 - a_2 b_1 \rangle$$

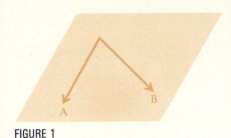

FIGURE 1

Because the cross product of two vectors is a vector, the cross product also is called the **vector product**. The operation of obtaining the cross product is called **vector multiplication**.

▶ **ILLUSTRATION 1**   If $\mathbf{A} = \langle 2, 1, -3 \rangle$ and $\mathbf{B} = \langle 3, -1, 4 \rangle$, then from Definition 15.5.1,

$$\begin{aligned}
\mathbf{A} \times \mathbf{B} &= \langle 2, 1, -3 \rangle \times \langle 3, -1, 4 \rangle \\
&= \langle (1)(4) - (-3)(-1), (-3)(3) - (2)(4), (2)(-1) - (1)(3) \rangle \\
&= \langle 4 - 3, -9 - 8, -2 - 3 \rangle \\
&= \langle 1, -17, -5 \rangle \\
&= \mathbf{i} - 17\mathbf{j} - 5\mathbf{k}
\end{aligned}$$ ◀

There is a mnemonic device for remembering the cross product formula that makes use of determinant notation. A second-order determinant is defined by the equation

$$\begin{vmatrix} a & b \\ c & d \end{vmatrix} = ad - bc$$

where a, b, and c are real numbers. For example,

$$\begin{vmatrix} 3 & 6 \\ -2 & 5 \end{vmatrix} = 3(5) - (6)(-2)$$
$$= 27$$

Therefore the cross product formula can be written as

$$\mathbf{A} \times \mathbf{B} = \begin{vmatrix} a_2 & a_3 \\ b_2 & b_3 \end{vmatrix} \mathbf{i} - \begin{vmatrix} a_1 & a_3 \\ b_1 & b_3 \end{vmatrix} \mathbf{j} + \begin{vmatrix} a_1 & a_2 \\ b_1 & b_2 \end{vmatrix} \mathbf{k}$$

The right side of the above expression can be written symbolically as

$$\begin{vmatrix} \mathbf{i} & \mathbf{j} & \mathbf{k} \\ a_1 & a_2 & a_3 \\ b_1 & b_2 & b_3 \end{vmatrix}$$

which is the notation for a third-order determinant. However, observe that the first row contains vectors and not real numbers as is customary with determinant notation.

▶ **ILLUSTRATION 2**   We use the mnemonic device employing determinant notation to find the cross product of the vectors of Illustration 1.

$$\mathbf{A} \times \mathbf{B} = \begin{vmatrix} \mathbf{i} & \mathbf{j} & \mathbf{k} \\ 2 & 1 & -3 \\ 3 & -1 & 4 \end{vmatrix}$$

$$= \begin{vmatrix} 1 & -3 \\ -1 & 4 \end{vmatrix} \mathbf{i} - \begin{vmatrix} 2 & -3 \\ 3 & 4 \end{vmatrix} \mathbf{j} + \begin{vmatrix} 2 & 1 \\ 3 & -1 \end{vmatrix} \mathbf{k}$$

$$= [(1)(4) - (-3)(-1)]\mathbf{i} - [(2)(4) - (-3)(3)]\mathbf{j} + [(2)(-1) - (1)(3)]\mathbf{k}$$

$$= \mathbf{i} - 17\mathbf{j} - 5\mathbf{k}$$ ◀

15.5.2 THEOREM   If $\mathbf{A}$ is any vector in V_3, then

   (i) $\mathbf{A} \times \mathbf{A} = \mathbf{0}$
  (ii) $\mathbf{0} \times \mathbf{A} = \mathbf{0}$
 (iii) $\mathbf{A} \times \mathbf{0} = \mathbf{0}$

Proof of (i)   If $\mathbf{A} = \langle a_1, a_2, a_3 \rangle$, then by Definition 15.5.1,

$$\mathbf{A} \times \mathbf{A} = \langle a_2 a_3 - a_3 a_2, a_3 a_1 - a_1 a_3, a_1 a_2 - a_2 a_1 \rangle$$

$$= \langle 0, 0, 0 \rangle$$

$$= \mathbf{0}$$

The proofs of (ii) and (iii) are left as exercises (see Exercise 13).   ■

By applying Definition 15.5.1 to pairs of unit vectors $\mathbf{i}$, $\mathbf{j}$, and $\mathbf{k}$ we obtain the following:

$$\mathbf{i} \times \mathbf{i} = \mathbf{0} \qquad \mathbf{j} \times \mathbf{j} = \mathbf{0} \qquad \mathbf{k} \times \mathbf{k} = \mathbf{0}$$

$$\mathbf{i} \times \mathbf{j} = \mathbf{k} \qquad \mathbf{j} \times \mathbf{k} = \mathbf{i} \qquad \mathbf{k} \times \mathbf{i} = \mathbf{j}$$

$$\mathbf{j} \times \mathbf{i} = -\mathbf{k} \qquad \mathbf{k} \times \mathbf{j} = -\mathbf{i} \qquad \mathbf{i} \times \mathbf{k} = -\mathbf{j}$$

As an aid in remembering the above cross products, first notice that the cross product of any one of the unit vectors $\mathbf{i}$, $\mathbf{j}$, or $\mathbf{k}$ with itself is the zero vector. The other six cross products can be obtained from Figure 2 by applying the following rule: The cross product of two consecutive vectors in the clockwise direction is the next vector; and the cross product of two consecutive vectors in the counterclockwise direction is the negative of the next vector.

We can easily show that cross multiplication of two vectors is not commutative because in particular $\mathbf{i} \times \mathbf{j} \neq \mathbf{j} \times \mathbf{i}$. However, we have $\mathbf{i} \times \mathbf{j} = \mathbf{k}$ and $\mathbf{j} \times \mathbf{i} = -\mathbf{k}$; so $\mathbf{i} \times \mathbf{j} = -(\mathbf{j} \times \mathbf{i})$. In general, if $\mathbf{A}$ and $\mathbf{B}$ are any vectors in V_3, $\mathbf{A} \times \mathbf{B} = -(\mathbf{B} \times \mathbf{A})$, which we state and prove as a theorem.

FIGURE 2

15.5.3 THEOREM   If $\mathbf{A}$ and $\mathbf{B}$ are any vectors in V_3,

$$\mathbf{A} \times \mathbf{B} = -(\mathbf{B} \times \mathbf{A})$$

Proof   If $\mathbf{A} = \langle a_1, a_2, a_3 \rangle$ and $\mathbf{B} = \langle b_1, b_2, b_3 \rangle$, then by Definition 15.5.1,

$$\mathbf{A} \times \mathbf{B} = \langle a_2 b_3 - a_3 b_2, a_3 b_1 - a_1 b_3, a_1 b_2 - a_2 b_1 \rangle$$

$$= -1 \langle a_3 b_2 - a_2 b_3, a_1 b_3 - a_3 b_1, a_2 b_1 - a_1 b_2 \rangle$$

$$= -(\mathbf{B} \times \mathbf{A})$$ ■

Cross multiplication of vectors is not associative. We show this by the following example:

$$\mathbf{i} \times (\mathbf{i} \times \mathbf{j}) = \mathbf{i} \times \mathbf{k} \qquad (\mathbf{i} \times \mathbf{i}) \times \mathbf{j} = \mathbf{0} \times \mathbf{j}$$
$$= -\mathbf{j} \qquad\qquad\qquad = \mathbf{0}$$

Thus

$$\mathbf{i} \times (\mathbf{i} \times \mathbf{j}) \neq (\mathbf{i} \times \mathbf{i}) \times \mathbf{j}$$

Cross multiplication of vectors is distributive with respect to vector addition, as given by the following theorem.

15.5.4 THEOREM   If $\mathbf{A}$, $\mathbf{B}$, and $\mathbf{C}$ are any vectors in V_3, then

$$\mathbf{A} \times (\mathbf{B} + \mathbf{C}) = \mathbf{A} \times \mathbf{B} + \mathbf{A} \times \mathbf{C}$$

To prove Theorem 15.5.4 let $\mathbf{A} = \langle a_1, a_2, a_3 \rangle$, $\mathbf{B} = \langle b_1, b_2, b_3 \rangle$, and $\mathbf{C} = \langle c_1, c_2, c_3 \rangle$, and then show that the components of the vector on the left side of the equation are the same as the components of the vector on the right side. The details are left as an exercise (see Exercise 35).

15.5.5 THEOREM   If $\mathbf{A}$ and $\mathbf{B}$ are any two vectors in V_3 and c is a scalar, then

(i) $(c\mathbf{A}) \times \mathbf{B} = \mathbf{A} \times (c\mathbf{B})$;
(ii) $(c\mathbf{A}) \times \mathbf{B} = c(\mathbf{A} \times \mathbf{B})$.

The proof of Theorem 15.5.5 is left as an exercise (see Exercise 36).

Theorems 15.5.4 and 15.5.5 can be applied to compute the cross product of two vectors by using laws of algebra, provided the order of the vectors in cross multiplication is not changed, because it is prohibited by Theorem 15.5.3. The following illustration demonstrates this procedure.

▶ **ILLUSTRATION 3**   We find the cross product of the vectors in Illustration 1 by applying Theorems 15.5.4 and 15.5.5.

$$\mathbf{A} \times \mathbf{B} = (2\mathbf{i} + \mathbf{j} - 3\mathbf{k}) \times (3\mathbf{i} - \mathbf{j} + 4\mathbf{k})$$
$$= 6(\mathbf{i} \times \mathbf{i}) - 2(\mathbf{i} \times \mathbf{j}) + 8(\mathbf{i} \times \mathbf{k}) + 3(\mathbf{j} \times \mathbf{i}) - 1(\mathbf{j} \times \mathbf{j})$$
$$\qquad\qquad + 4(\mathbf{j} \times \mathbf{k}) - 9(\mathbf{k} \times \mathbf{i}) + 3(\mathbf{k} \times \mathbf{j}) - 12(\mathbf{k} \times \mathbf{k})$$
$$= 6(\mathbf{0}) - 2(\mathbf{k}) + 8(-\mathbf{j}) + 3(-\mathbf{k}) - 1(\mathbf{0}) + 4(\mathbf{i}) - 9(\mathbf{j}) + 3(-\mathbf{i}) - 12(\mathbf{0})$$
$$= -2\mathbf{k} - 8\mathbf{j} - 3\mathbf{k} + 4\mathbf{i} - 9\mathbf{j} - 3\mathbf{i}$$
$$= \mathbf{i} - 17\mathbf{j} - 5\mathbf{k} \qquad\qquad\qquad ◀$$

The method used in Illustration 3 gives a way of finding the cross product without having to remember the formula of Definition 15.5.1 or to use determinant notation. Actually all the steps shown in the solution need not be included because the various cross products of the unit vectors can be obtained immediately by using Figure 2 and the corresponding rule.

There are two *triple products* that we shall consider. One is the product $\mathbf{A} \cdot (\mathbf{B} \times \mathbf{C})$, called the **triple scalar product** of the vectors $\mathbf{A}$, $\mathbf{B}$, and $\mathbf{C}$. Actually, the parentheses are not needed because $\mathbf{A} \cdot \mathbf{B}$ is a scalar, and therefore $\mathbf{A} \cdot \mathbf{B} \times \mathbf{C}$ can be interpreted in only one way.

15.5.6 THEOREM   If $\mathbf{A}$, $\mathbf{B}$, and $\mathbf{C}$ are vectors in V_3, then

$$\mathbf{A} \cdot \mathbf{B} \times \mathbf{C} = \mathbf{A} \times \mathbf{B} \cdot \mathbf{C}$$

Theorem 15.5.6 can be proved by letting

$$\mathbf{A} = \langle a_1, a_2, a_3 \rangle \qquad \mathbf{B} = \langle b_1, b_2, b_3 \rangle \qquad \mathbf{C} = \langle c_1, c_2, c_3 \rangle$$

and then by showing that the scalar on the left side of the equation is equal to the scalar on the right. The details are left as an exercise (see Exercise 37).

▶ **ILLUSTRATION 4**   We verify Theorem 15.5.6 if

$$\mathbf{A} = \langle 1, -1, 2 \rangle \qquad \mathbf{B} = \langle 3, 4, -2 \rangle \qquad \mathbf{C} = \langle -5, 1, -4 \rangle$$

$$\begin{aligned} \mathbf{B} \times \mathbf{C} &= (3\mathbf{i} + 4\mathbf{j} - 2\mathbf{k}) \times (-5\mathbf{i} + \mathbf{j} - 4\mathbf{k}) \\ &= 3\mathbf{k} - 12(-\mathbf{j}) - 20(-\mathbf{k}) - 16\mathbf{i} + 10\mathbf{j} - 2(-\mathbf{i}) \\ &= -14\mathbf{i} + 22\mathbf{j} + 23\mathbf{k} \end{aligned}$$

$$\begin{aligned} \mathbf{A} \cdot (\mathbf{B} \times \mathbf{C}) &= \langle 1, -1, 2 \rangle \cdot \langle -14, 22, 23 \rangle \\ &= -14 - 22 + 46 \\ &= 10 \end{aligned}$$

$$\begin{aligned} \mathbf{A} \times \mathbf{B} &= (\mathbf{i} - \mathbf{j} + 2\mathbf{k}) \times (3\mathbf{i} + 4\mathbf{j} - 2\mathbf{k}) \\ &= 4\mathbf{k} - 2(-\mathbf{j}) - 3(-\mathbf{k}) + 2\mathbf{i} + 6\mathbf{j} + 8(-\mathbf{i}) \\ &= -6\mathbf{i} + 8\mathbf{j} + 7\mathbf{k} \end{aligned}$$

$$\begin{aligned} (\mathbf{A} \times \mathbf{B}) \cdot \mathbf{C} &= \langle -6, 8, 7 \rangle \cdot \langle -5, 1, -4 \rangle \\ &= 30 + 8 - 28 \\ &= 10 \end{aligned}$$

This verifies the theorem for these three vectors.   ◀

The other triple product is $\mathbf{A} \times (\mathbf{B} \times \mathbf{C})$, called the **triple vector product**.

15.5.7 THEOREM   If $\mathbf{A}$, $\mathbf{B}$, and $\mathbf{C}$ are vectors in V_3, then

$$\mathbf{A} \times (\mathbf{B} \times \mathbf{C}) = (\mathbf{A} \cdot \mathbf{C})\mathbf{B} - (\mathbf{A} \cdot \mathbf{B})\mathbf{C}$$

The proof of Theorem 15.5.7 is similar to the proof of Theorem 15.5.6. By using components of the vectors $\mathbf{A}$, $\mathbf{B}$, and $\mathbf{C}$ we can show that the vector on the left side of the equation is the same as the vector on the right. The computation is left as an exercise (see Exercise 38).

▶ **ILLUSTRATION 5**   We verify Theorem 15.5.7 for the vectors $\mathbf{A}$, $\mathbf{B}$, and $\mathbf{C}$ of Illustration 4. Because $\mathbf{B} \times \mathbf{C} = -14\mathbf{i} + 22\mathbf{j} + 23\mathbf{k}$

$$\begin{aligned} \mathbf{A} \times (\mathbf{B} \times \mathbf{C}) &= \begin{vmatrix} \mathbf{i} & \mathbf{j} & \mathbf{k} \\ 1 & -1 & 2 \\ -14 & 22 & 23 \end{vmatrix} \\ &= -23\mathbf{i} - 28\mathbf{j} + 22\mathbf{k} - 14\mathbf{k} - 44\mathbf{i} - 23\mathbf{j} \\ &= -67\mathbf{i} - 51\mathbf{j} + 8\mathbf{k} \end{aligned}$$

(1)

$$\mathbf{A} \cdot \mathbf{C} = \langle 1, -1, 2 \rangle \cdot \langle -5, 1, -4 \rangle \qquad \mathbf{A} \cdot \mathbf{B} = \langle 1, -1, 2 \rangle \cdot \langle 3, 4, -2 \rangle$$
$$= -5 - 1 - 8 \qquad\qquad\qquad\qquad = 3 - 4 - 4$$
$$= -14 \qquad\qquad\qquad\qquad\qquad\quad = -5$$

Thus

$$(\mathbf{A} \cdot \mathbf{C})\mathbf{B} - (\mathbf{A} \cdot \mathbf{B})\mathbf{C} = -14\langle 3, 4, -2 \rangle - (-5)\langle -5, 1, -4 \rangle$$
$$= \langle -42, -56, 28 \rangle - \langle 25, -5, 20 \rangle$$
$$= \langle -67, -51, 8 \rangle$$
$$= -67\mathbf{i} - 51\mathbf{j} + 8\mathbf{k}$$

By comparing this result with (1), Theorem 15.5.7 is verified for these three vectors. ◄

The following theorem is used for a geometric interpretation of the cross product.

15.5.8 THEOREM   If $\mathbf{A}$ and $\mathbf{B}$ are two vectors in V_3 and θ is the radian measure of the angle between $\mathbf{A}$ and $\mathbf{B}$, then

$$\|\mathbf{A} \times \mathbf{B}\| = \|\mathbf{A}\| \, \|\mathbf{B}\| \sin \theta$$

Proof   From Theorem 14.2.3(iii) applied to vectors in V_3

$$\|\mathbf{A} \times \mathbf{B}\|^2 = (\mathbf{A} \times \mathbf{B}) \cdot (\mathbf{A} \times \mathbf{B}) \tag{2}$$

With the notation $\mathbf{U}$, $\mathbf{V}$, and $\mathbf{W}$ as the vectors, we have from Theorem 15.5.6

$$(\mathbf{U} \times \mathbf{V}) \cdot \mathbf{W} = \mathbf{U} \cdot (\mathbf{V} \times \mathbf{W})$$

If in this equation we let $\mathbf{U} = \mathbf{A}$, $\mathbf{V} = \mathbf{B}$, and $\mathbf{W} = \mathbf{A} \times \mathbf{B}$, we have

$$(\mathbf{A} \times \mathbf{B}) \cdot (\mathbf{A} \times \mathbf{B}) = \mathbf{A} \cdot [\mathbf{B} \times (\mathbf{A} \times \mathbf{B})]$$

Applying Theorem 15.5.7 to the vector in brackets on the right side we get

$$(\mathbf{A} \times \mathbf{B}) \cdot (\mathbf{A} \times \mathbf{B}) = \mathbf{A} \cdot [(\mathbf{B} \cdot \mathbf{B})\mathbf{A} - (\mathbf{B} \cdot \mathbf{A})\mathbf{B}]$$
$$= (\mathbf{A} \cdot \mathbf{A})(\mathbf{B} \cdot \mathbf{B}) - (\mathbf{A} \cdot \mathbf{B})(\mathbf{A} \cdot \mathbf{B})$$
$$= \|\mathbf{A}\|^2 \|\mathbf{B}\|^2 - (\mathbf{A} \cdot \mathbf{B})^2 \tag{3}$$

From Theorem 15.2.11, if θ is the angle between $\mathbf{A}$ and $\mathbf{B}$,

$$\mathbf{A} \cdot \mathbf{B} = \|\mathbf{A}\| \, \|\mathbf{B}\| \cos \theta$$

Substituting from this equation into (3) we have

$$(\mathbf{A} \times \mathbf{B}) \cdot (\mathbf{A} \times \mathbf{B}) = \|\mathbf{A}\|^2 \|\mathbf{B}\|^2 - \|\mathbf{A}\|^2 \|\mathbf{B}\|^2 \cos^2 \theta$$
$$= \|\mathbf{A}\|^2 \|\mathbf{B}\|^2 (1 - \cos^2 \theta)$$

Substituting from (2) into the above, and because $1 - \cos^2 \theta = \sin^2 \theta$, we have

$$\|\mathbf{A} \times \mathbf{B}\|^2 = \|\mathbf{A}\|^2 \|\mathbf{B}\|^2 \sin^2 \theta$$

Because $0 \le \theta \le \pi$, $\sin \theta \ge 0$. Therefore, we take the square root of both sides and get

$$\|\mathbf{A} \times \mathbf{B}\| = \|\mathbf{A}\| \, \|\mathbf{B}\| \sin \theta$$

■

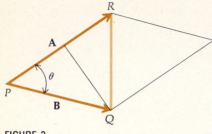

FIGURE 3

We consider now a geometric interpretation of $\|\mathbf{A} \times \mathbf{B}\|$. Let $\overline{PR}$ be a representation of $\mathbf{A}$ and let $\overline{PQ}$ be a representation of $\mathbf{B}$. Then the angle between the vectors $\mathbf{A}$ and $\mathbf{B}$ is the angle at P in triangle RPQ (see Figure 3). Let the radian measure of this angle be θ. Therefore the area of the parallelogram having $\overline{PR}$ and $\overline{PQ}$ as adjacent sides is $\|\mathbf{A}\| \|\mathbf{B}\| \sin \theta$ square units because the altitude of the parallelogram has length $\|\mathbf{B}\| \sin \theta$ units and the length of the base is $\|\mathbf{A}\|$ units. So from Theorem 15.5.8 it follows that $\|\mathbf{A} \times \mathbf{B}\|$ square units is the area of this parallelogram.

EXAMPLE 1   Show that the quadrilateral having vertices at $P(1, -2, 3)$, $Q(4, 3, -1)$, $R(2, 2, 1)$, and $S(5, 7, -3)$ is a parallelogram, and find its area.

Solution   Figure 4 shows the quadrilateral $PQSR$.

$$\mathbf{V}(\overrightarrow{PQ}) = \langle 4 - 1, 3 + 2, -1 - 3 \rangle \qquad \mathbf{V}(\overrightarrow{PR}) = \langle 2 - 1, 2 + 2, 1 - 3 \rangle$$
$$= \langle 3, 5, -4 \rangle \qquad\qquad\qquad = \langle 1, 4, -2 \rangle$$
$$\mathbf{V}(\overrightarrow{RS}) = \langle 5 - 2, 7 - 2, -3 - 1 \rangle \qquad \mathbf{V}(\overrightarrow{QS}) = \langle 5 - 4, 7 - 3, -3 + 1 \rangle$$
$$= \langle 3, 5, -4 \rangle \qquad\qquad\qquad = \langle 1, 4, -2 \rangle$$

Because $\mathbf{V}(\overrightarrow{PQ}) = \mathbf{V}(\overrightarrow{RS})$ and $\mathbf{V}(\overrightarrow{PR}) = \mathbf{V}(\overrightarrow{QS})$, it follows that $\overline{PQ}$ is parallel to $\overline{RS}$ and $\overline{PR}$ is parallel to $\overline{QS}$. Therefore $PQSR$ is a parallelogram.

Let $\mathbf{A} = \mathbf{V}(\overrightarrow{PR})$ and $\mathbf{B} = \mathbf{V}(\overrightarrow{PQ})$; then

$$\mathbf{A} \times \mathbf{B} = (\mathbf{i} + 4\mathbf{j} - 2\mathbf{k}) \times (3\mathbf{i} + 5\mathbf{j} - 4\mathbf{k})$$
$$= 3(\mathbf{i} \times \mathbf{i}) + 5(\mathbf{i} \times \mathbf{j}) - 4(\mathbf{i} \times \mathbf{k}) + 12(\mathbf{j} \times \mathbf{i}) + 20(\mathbf{j} \times \mathbf{j}) - 16(\mathbf{j} \times \mathbf{k})$$
$$-6(\mathbf{k} \times \mathbf{i}) - 10(\mathbf{k} \times \mathbf{j}) + 8(\mathbf{k} \times \mathbf{k})$$
$$= 3(\mathbf{0}) + 5(\mathbf{k}) - 4(-\mathbf{j}) + 12(-\mathbf{k}) + 20(\mathbf{0}) - 16(\mathbf{i}) - 6(\mathbf{j}) - 10(-\mathbf{i}) + 8(\mathbf{0})$$
$$= -6\mathbf{i} - 2\mathbf{j} - 7\mathbf{k}$$

Thus

$$\|\mathbf{A} \times \mathbf{B}\| = \sqrt{36 + 4 + 49}$$
$$= \sqrt{89}$$

Therefore the area of the parallelogram is $\sqrt{89}$ square units.

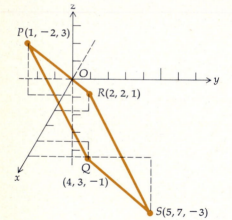

FIGURE 4

The following theorem, which gives a method for determining if two vectors in V_3 are parallel, follows from Theorem 15.5.8.

15.5.9 THEOREM   If $\mathbf{A}$ and $\mathbf{B}$ are two vectors in V_3, $\mathbf{A}$ and $\mathbf{B}$ are parallel if and only if $\mathbf{A} \times \mathbf{B} = \mathbf{0}$.

Proof   If either $\mathbf{A}$ or $\mathbf{B}$ is the zero vector, then from Theorem 15.5.2, $\mathbf{A} \times \mathbf{B} = \mathbf{0}$. Because the zero vector is parallel to any vector, the theorem holds.

If neither $\mathbf{A}$ nor $\mathbf{B}$ is the zero vector, $\|\mathbf{A}\| \neq 0$ and $\|\mathbf{B}\| \neq 0$. Therefore by Theorem 15.5.8, $\|\mathbf{A} \times \mathbf{B}\| = 0$ if and only if $\sin \theta = 0$. Because $\|\mathbf{A} \times \mathbf{B}\| = 0$ if and only if $\mathbf{A} \times \mathbf{B} = \mathbf{0}$ and $\sin \theta = 0$ $(0 \leq \theta \leq \pi)$ if and only if $\theta = 0$ or $\theta = \pi$, we can conclude that

$$\mathbf{A} \times \mathbf{B} = \mathbf{0} \quad \text{if and only if} \quad \theta = 0 \text{ or } \theta = \pi$$

However, two nonzero vectors are parallel if and only if the radian measure of the angle between the two vectors is 0 or π. Thus the theorem follows. ■

15.5.10 THEOREM   If **A** and **B** are two vectors in V_3, then the vector **A** × **B** is orthogonal to both **A** and **B**.

Proof   From Theorem 15.5.6,

$$\mathbf{A} \cdot \mathbf{A} \times \mathbf{B} = \mathbf{A} \times \mathbf{A} \cdot \mathbf{B}$$

From Theorem 15.5.2(i), **A** × **A** = **0**. Therefore, from the above equation,

$$\mathbf{A} \cdot \mathbf{A} \times \mathbf{B} = \mathbf{0} \cdot \mathbf{B}$$
$$= 0$$

Because the dot product of **A** and **A** × **B** is zero, it follows that **A** and **A** × **B** are orthogonal.

Also from Theorem 15.5.6,

$$\mathbf{A} \times \mathbf{B} \cdot \mathbf{B} = \mathbf{A} \cdot \mathbf{B} \times \mathbf{B}$$

Again applying Theorem 15.5.2(i) we get **B** × **B** = **0**; thus from the above equation,

$$\mathbf{A} \times \mathbf{B} \cdot \mathbf{B} = \mathbf{A} \cdot \mathbf{0}$$
$$= 0$$

Therefore, because the dot product of **A** × **B** and **B** is zero, **A** × **B** and **B** are orthogonal and the theorem is proved. ■

From Theorem 15.5.10 we can conclude that if representations of the vectors **A**, **B**, and **A** × **B** have the same initial point, then the representation of **A** × **B** is perpendicular to the plane formed by the representations of **A** and **B**.

EXAMPLE 2   Given the points $P(-1, -2, -3)$, $Q(-2, 1, 0)$, and $R(0, 5, 1)$, find a unit vector whose representations are perpendicular to the plane through the points P, Q, and R.

Solution   Let $\mathbf{A} = \mathbf{V}(\overrightarrow{PQ})$ and $\mathbf{B} = \mathbf{V}(\overrightarrow{PR})$. Then

$$\mathbf{A} = \langle -2 + 1, 1 + 2, 0 + 3 \rangle \qquad \mathbf{B} = \langle 0 + 1, 5 + 2, 1 + 3 \rangle$$
$$= \langle -1, 3, 3 \rangle \qquad\qquad\qquad = \langle 1, 7, 4 \rangle$$

The plane through P, Q, and R is the plane formed by $\overrightarrow{PQ}$ and $\overrightarrow{PR}$, which are, respectively, representations of vectors **A** and **B**. Therefore any representation of the vector **A** × **B** is perpendicular to this plane.

$$\mathbf{A} \times \mathbf{B} = (-\mathbf{i} + 3\mathbf{j} + 3\mathbf{k}) \times (\mathbf{i} + 7\mathbf{j} + 4\mathbf{k})$$
$$= -9\mathbf{i} + 7\mathbf{j} - 10\mathbf{k}$$

The desired vector is a unit vector parallel to **A** × **B**. To find this unit vector we apply Theorem 15.2.8 and divide **A** × **B** by $\|\mathbf{A} \times \mathbf{B}\|$ and obtain

$$\frac{\mathbf{A} \times \mathbf{B}}{\|\mathbf{A} \times \mathbf{B}\|} = -\frac{9}{\sqrt{230}}\mathbf{i} + \frac{7}{\sqrt{230}}\mathbf{j} - \frac{10}{\sqrt{230}}\mathbf{k}$$

The following two examples show the use of the cross product to find an equation of a plane. These examples involve the use of the same information as in Examples 2 and 4 of Section 15.3, respectively.

EXAMPLE 3   Find an equation of the plane through the points $P(1, 3, 2)$, $Q(3, -2, 2)$, and $R(2, 1, 3)$.

Solution $\mathbf{V}(\overrightarrow{QR}) = -\mathbf{i} + 3\mathbf{j} + \mathbf{k}$ and $\mathbf{V}(\overrightarrow{PR}) = \mathbf{i} - 2\mathbf{j} + \mathbf{k}$. A normal vector to the required plane is the cross product $\mathbf{V}(\overrightarrow{QR}) \times \mathbf{V}(\overrightarrow{PR})$, which is

$$(-\mathbf{i} + 3\mathbf{j} + \mathbf{k}) \times (\mathbf{i} - 2\mathbf{j} + \mathbf{k}) = 5\mathbf{i} + 2\mathbf{j} - \mathbf{k}$$

So if $P_0 = (1, 3, 2)$ and $\mathbf{N} = \langle 5, 2, -1 \rangle$, from Theorem 15.3.2 an equation of the required plane is

$$5(x - 1) + 2(y - 3) - (z - 2) = 0$$
$$5x + 2y - z - 9 = 0$$

This result agrees with that of Example 2 in Section 15.3.

EXAMPLE 4   Find an equation of the plane containing the point $(4, 0, -2)$ and perpendicular to each of the planes

$$x - y + z = 0 \quad \text{and} \quad 2x + y - 4z - 5 = 0$$

Solution   By Theorem 15.3.3, a normal vector to the plane $x - y + z = 0$ is $\langle 1, -1, 1 \rangle$ and a normal vector to the plane $2x + y - 4z - 5 = 0$ is $\langle 2, 1, -4 \rangle$. Thus a normal vector to the required plane is orthogonal to both $\langle 1, -1, 1 \rangle$ and $\langle 2, 1, -4 \rangle$. By Theorem 15.5.10, such a vector is

$$\langle 1, -1, 1 \rangle \times \langle 2, 1, -4 \rangle = \begin{vmatrix} \mathbf{i} & \mathbf{j} & \mathbf{k} \\ 1 & -1 & 1 \\ 2 & 1 & -4 \end{vmatrix}$$
$$= 3\mathbf{i} + 6\mathbf{j} + 3\mathbf{k}$$

The required plane contains the point $(4, 0, -2)$ and has $\langle 3, 6, 3 \rangle$ as a normal vector. From Theorem 15.3.2, an equation of this plane is

$$3(x - 4) + 6(y - 0) + 3(z + 2) = 0$$
$$x + 2y + z - 2 = 0$$

This equation is the same as the one obtained in Example 4 of Section 15.3.

A geometric interpretation of the triple scalar product is obtained by considering a parallelepiped having the edges $\overrightarrow{PQ}$, $\overrightarrow{PR}$, and $\overrightarrow{PS}$ and letting $\mathbf{A} = \mathbf{V}(\overrightarrow{PQ})$, $\mathbf{B} = \mathbf{V}(\overrightarrow{PR})$, and $\mathbf{C} = \mathbf{V}(\overrightarrow{PS})$. See Figure 5. The vector $\mathbf{A} \times \mathbf{B}$ is a normal vector to the plane of $\overrightarrow{PQ}$ and $\overrightarrow{PR}$. The vector $-(\mathbf{A} \times \mathbf{B})$ is also a normal vector to this plane. We are not certain which of the two vectors, $\mathbf{A} \times \mathbf{B}$ or $-(\mathbf{A} \times \mathbf{B})$, makes the smaller angle with $\mathbf{C}$. Let $\mathbf{N}$ be the one of the two vectors $\mathbf{A} \times \mathbf{B}$ or $-(\mathbf{A} \times \mathbf{B})$ that makes an angle of radian measure $\theta < \frac{1}{2}\pi$ with $\mathbf{C}$. Then the representations of $\mathbf{N}$ and $\mathbf{C}$ having their initial points at P are on the same side of the plane of $\overrightarrow{PQ}$ and $\overrightarrow{PR}$ as shown in Figure 5. The area of the base of the parallelepiped is $\|\mathbf{A} \times \mathbf{B}\|$ square units. If h units is the length of the altitude of the parallelepiped, and if V cubic units is the volume of the parallelepiped,

$$V = \|\mathbf{A} \times \mathbf{B}\|h \qquad (4)$$

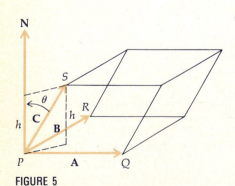

FIGURE 5

Consider now the dot product $\mathbf{N} \cdot \mathbf{C}$. By Theorem 15.2.11,

$$\mathbf{N} \cdot \mathbf{C} = \|\mathbf{N}\| \, \|\mathbf{C}\| \cos \theta$$

But $h = \|\mathbf{C}\| \cos \theta$; thus

$$\mathbf{N} \cdot \mathbf{C} = \|\mathbf{N}\| h \tag{5}$$

Because $\mathbf{N}$ is either $\mathbf{A} \times \mathbf{B}$ or $-(\mathbf{A} \times \mathbf{B})$, it follows that $\|\mathbf{N}\| = \|\mathbf{A} \times \mathbf{B}\|$. Thus, from (5),

$$\mathbf{N} \cdot \mathbf{C} = \|\mathbf{A} \times \mathbf{B}\| h$$

Comparing this equation and (4) we have

$$V = \mathbf{N} \cdot \mathbf{C}$$

It follows that the measure of the volume of the parallelepiped is either $(\mathbf{A} \times \mathbf{B}) \cdot \mathbf{C}$ or $-(\mathbf{A} \times \mathbf{B}) \cdot \mathbf{C}$; that is, the measure of the volume of the parallelepiped is the absolute value of the triple scalar product $\mathbf{A} \times \mathbf{B} \cdot \mathbf{C}$.

EXAMPLE 5 Find the volume of the parallelepiped having vertices $P(5, 4, 5)$, $Q(4, 10, 6)$, $R(1, 8, 7)$, and $S(2, 6, 9)$ and edges $\overrightarrow{PQ}$, $\overrightarrow{PR}$, and $\overrightarrow{PS}$.

Solution Figure 6 shows the parallelepiped. Let $\mathbf{A} = \mathbf{V}(\overrightarrow{PQ})$; then $\mathbf{A} = \langle -1, 6, 1 \rangle$. Let $\mathbf{B} = \mathbf{V}(\overrightarrow{PR})$; then $\mathbf{B} = \langle -4, 4, 2 \rangle$. Let $\mathbf{C} = \mathbf{V}(\overrightarrow{PS})$; then $\mathbf{C} = \langle -3, 2, 4 \rangle$. Thus

$$\begin{aligned} \mathbf{A} \times \mathbf{B} &= (-\mathbf{i} + 6\mathbf{j} + \mathbf{k}) \times (-4\mathbf{i} + 4\mathbf{j} + 2\mathbf{k}) \\ &= 8\mathbf{i} - 2\mathbf{j} + 20\mathbf{k} \end{aligned}$$

Therefore

$$\begin{aligned} \mathbf{A} \times \mathbf{B} \cdot \mathbf{C} &= \langle 8, -2, 20 \rangle \cdot \langle -3, 2, 4 \rangle \\ &= -24 - 4 + 80 \\ &= 52 \end{aligned}$$

Hence the volume is 52 cubic units.

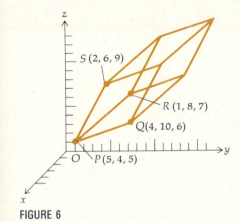

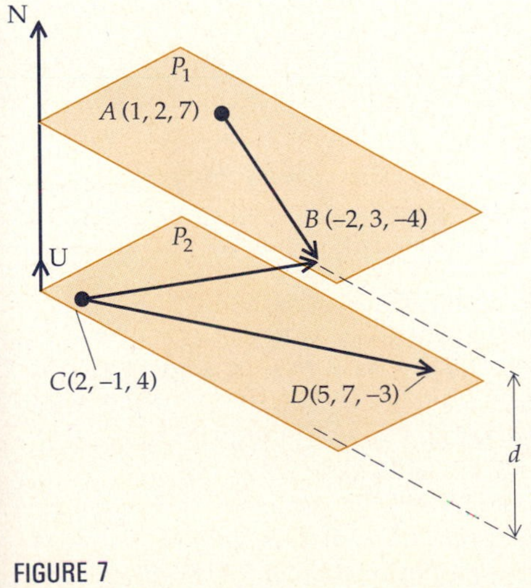

FIGURE 6

EXAMPLE 6 Find the distance between the two skew lines l_1 and l_2 of Example 4 in Section 15.4.

Solution The line l_1 contains the points $A(1, 2, 7)$ and $B(-2, 3, -4)$. The line l_2 contains the points $C(2, -1, 4)$ and $D(5, 7, -3)$. Because l_1 and l_2 are skew lines, there are parallel planes P_1 and P_2 containing the lines l_1 and l_2, respectively. See Figure 7. Let d units be the distance between planes P_1 and P_2. The distance between l_1 and l_2 is also d units. A normal vector to the two planes is

$$\mathbf{N} = \mathbf{V}(\overrightarrow{AB}) \times \mathbf{V}(\overrightarrow{CD})$$

Let $\mathbf{U}$ be a normal vector in the direction of $\mathbf{N}$. Then

$$\mathbf{U} = \frac{\mathbf{V}(\overrightarrow{AB}) \times \mathbf{V}(\overrightarrow{CD})}{\|\mathbf{V}(\overrightarrow{AB}) \times \mathbf{V}(\overrightarrow{CD})\|} \tag{6}$$

FIGURE 7

Now we take two points, one in each plane (for instance, B and C). Then the scalar projection of $\mathbf{V}(\overrightarrow{CB})$ onto $\mathbf{N}$ is $\mathbf{V}(\overrightarrow{CB}) \cdot \mathbf{U}$, and

$$d = |\mathbf{V}(\overrightarrow{CB}) \cdot \mathbf{U}| \tag{7}$$

We now perform the computations.

$$\mathbf{V}(\overrightarrow{AB}) = \langle -2 - 1, 3 - 2, -4 - 7 \rangle \qquad \mathbf{V}(\overrightarrow{CD}) = \langle 5 - 2, 7 + 1, -3 - 4 \rangle$$
$$= \langle -3, 1, -11 \rangle \qquad\qquad\qquad = \langle 3, 8, -7 \rangle$$

Thus

$$\mathbf{V}(\overrightarrow{AB}) \times \mathbf{V}(\overrightarrow{CD}) = \begin{vmatrix} \mathbf{i} & \mathbf{j} & \mathbf{k} \\ -3 & 1 & -11 \\ 3 & 8 & -7 \end{vmatrix}$$

$$= 27(3\mathbf{i} - 2\mathbf{j} - \mathbf{k})$$

Therefore, from (6),

$$\mathbf{U} = \frac{27(3\mathbf{i} - 2\mathbf{j} - \mathbf{k})}{\sqrt{27^2(3^2 + 2^2 + 1^2)}}$$

$$\mathbf{U} = \frac{1}{\sqrt{14}}(3\mathbf{i} - 2\mathbf{j} - \mathbf{k}) \tag{8}$$

Furthermore

$$\mathbf{V}(\overrightarrow{CB}) = \langle -2 - 2, 3 + 1, -4 - 4 \rangle$$
$$\mathbf{V}(\overrightarrow{CB}) = \langle -4, 4, -8 \rangle$$

Substituting from this equation and (8) into (7) we get

$$d = \left| \langle -4, 4, -8 \rangle \cdot \frac{1}{\sqrt{14}} \langle 3, -2, -1 \rangle \right|$$

$$= \frac{1}{\sqrt{14}} |-12 - 8 + 8|$$

$$= \frac{12}{\sqrt{14}}$$

$$\approx 3.21$$

EXERCISES 15.5

In Exercises 1 through 12, $\mathbf{A} = \langle 1, 2, 3 \rangle$, $\mathbf{B} = \langle 4, -3, -1 \rangle$, $\mathbf{C} = \langle -5, -3, 5 \rangle$, $\mathbf{D} = \langle -2, 1, 6 \rangle$, $\mathbf{E} = \langle 4, 0, -7 \rangle$, *and* $\mathbf{F} = \langle 0, 2, 1 \rangle$.

1. Find $\mathbf{A} \times \mathbf{B}$.
2. Find $\mathbf{D} \times \mathbf{E}$.
3. Find $(\mathbf{C} \times \mathbf{D}) \cdot (\mathbf{E} \times \mathbf{F})$.
4. Find $(\mathbf{C} \times \mathbf{E}) \cdot (\mathbf{D} \times \mathbf{F})$.
5. Verify Theorem 15.5.3 for vectors $\mathbf{A}$ and $\mathbf{B}$.
6. Verify Theorem 15.5.4 for vectors $\mathbf{A}$, $\mathbf{B}$, and $\mathbf{C}$.
7. Verify Theorem 15.5.5(i) for vectors $\mathbf{A}$ and $\mathbf{B}$ and $c = 3$.
8. Verify Theorem 15.5.5(ii) for vectors $\mathbf{A}$ and $\mathbf{B}$ and $c = 3$.

9. Verify Theorem 15.5.6 for vectors $\mathbf{A}$, $\mathbf{B}$, and $\mathbf{C}$.
10. Verify Theorem 15.5.7 for vectors $\mathbf{A}$, $\mathbf{B}$, and $\mathbf{C}$.
11. Find $(\mathbf{A} + \mathbf{B}) \times (\mathbf{C} - \mathbf{D})$ and $(\mathbf{D} - \mathbf{C}) \times (\mathbf{A} + \mathbf{B})$, and verify that they are equal.
12. Find $\|\mathbf{A} \times \mathbf{B}\| \, \|\mathbf{C} \times \mathbf{D}\|$.
13. Prove Theorem 15.5.2(ii) and (iii).
14. Given the unit vectors $\mathbf{A} = \frac{4}{9}\mathbf{i} + \frac{7}{9}\mathbf{j} - \frac{4}{9}\mathbf{k}$ and $\mathbf{B} = -\frac{2}{3}\mathbf{i} + \frac{2}{3}\mathbf{j} + \frac{1}{3}\mathbf{k}$. If θ is the angle between $\mathbf{A}$ and $\mathbf{B}$, find $\sin \theta$ in two ways: (a) by using the cross product (Theorem 15.5.8); (b) by using the dot product and a trigonometric identity.

15. Follow the instructions of Exercise 14 for the two unit vectors

$$\mathbf{A} = \frac{1}{\sqrt{3}}\mathbf{i} - \frac{1}{\sqrt{3}}\mathbf{j} + \frac{1}{\sqrt{3}}\mathbf{k} \qquad \mathbf{B} = \frac{1}{3\sqrt{3}}\mathbf{i} + \frac{5}{3\sqrt{3}}\mathbf{j} + \frac{1}{3\sqrt{3}}\mathbf{k}$$

16. Show that the quadrilateral having vertices at $(-2, 1, -1)$, $(1, 1, 3)$, $(-5, 4, 0)$ and $(8, 4, -4)$ is a parallelogram, and find its area.

17. Show that the quadrilateral having vertices at $(1, -2, 3)$, $(4, 3, -1)$, $(2, 2, 1)$, and $(5, 7, -3)$ is a parallelogram, and find its area.

18. Find the area of the parallelogram $PQRS$ if $\mathbf{V}(\overrightarrow{PQ}) = 3\mathbf{i} - 2\mathbf{j}$ and $\mathbf{V}(\overrightarrow{PS}) = 3\mathbf{j} + 4\mathbf{k}$.

19. Find the area of the triangle having vertices at $(0, 2, 2)$, $(8, 8, -2)$, and $(9, 12, 6)$.

20. Find the area of the triangle having vertices at $(4, 5, 6)$, $(4, 4, 5)$, and $(3, 5, 5)$.

In Exercises 21 and 22, use the cross product to find an equation of the plane containing the three points.

21. $(-2, 2, 2)$, $(-8, 1, 6)$, $(3, 4, -1)$
22. $(2, 3, 0)$, $(2, 0, 4)$, $(0, 3, 4)$

23. Do Exercise 18 in Exercises 15.3 by using the cross product.

24. Find a unit vector whose representations are perpendicular to the plane containing $\overrightarrow{PQ}$ and $\overrightarrow{PR}$ if $\overrightarrow{PQ}$ is a representation of the vector $\mathbf{i} + 3\mathbf{j} - 2\mathbf{k}$ and $\overrightarrow{PR}$ is a representation of the vector $2\mathbf{i} - \mathbf{j} - \mathbf{k}$.

In Exercises 25 through 27, find a unit vector whose representations are perpendicular to the plane through the points P, Q, and R.

25. $P(5, 2, -1)$, $Q(2, 4, -2)$, $R(11, 1, 4)$
26. $P(-2, 1, 0)$, $Q(2, -2, -1)$, $R(-5, 0, 2)$

27. $P(1, 4, 2)$, $Q(3, 2, 4)$, $R(4, 3, 1)$

28. Find the volume of the parallelepiped having edges $\overrightarrow{PQ}$, $\overrightarrow{PR}$, and $\overrightarrow{PS}$ if the points P, Q, R, and S are, respectively, $(1, 3, 4)$, $(3, 5, 3)$, $(2, 1, 6)$, and $(2, 2, 5)$.

29. Find the volume of the parallelepiped $PQRS$ if the vectors $\mathbf{V}(\overrightarrow{PQ})$, $\mathbf{V}(\overrightarrow{PR})$, and $\mathbf{V}(PS)$ are, respectively, $\mathbf{i} + 3\mathbf{j} + 2\mathbf{k}$, $2\mathbf{i} + \mathbf{j} - \mathbf{k}$, and $\mathbf{i} - 2\mathbf{j} + \mathbf{k}$.

30. If $\mathbf{A}$ and $\mathbf{B}$ are any two vectors in V_3, prove that

$$(\mathbf{A} - \mathbf{B}) \times (\mathbf{A} + \mathbf{B}) = 2(\mathbf{A} \times \mathbf{B})$$

In Exercises 31 and 32, find the perpendicular distance between the two skew lines.

31. $\dfrac{x - 1}{5} = \dfrac{y - 2}{3} = \dfrac{z + 1}{2}$ and $\dfrac{x + 2}{4} = \dfrac{y + 1}{2} = \dfrac{z - 3}{-3}$

32. $\dfrac{x + 1}{2} = \dfrac{y + 2}{-4} = \dfrac{z - 1}{-3}$ and $\dfrac{x - 1}{5} = \dfrac{y - 1}{3} = \dfrac{z + 1}{2}$

33. Let P, Q, and R be three noncollinear points in R^3 and let $\overrightarrow{OP}$, $\overrightarrow{OQ}$, and $\overrightarrow{OR}$ be the position representations of vectors $\mathbf{A}$, $\mathbf{B}$, and $\mathbf{C}$, respectively. Prove that the representations of the vector $\mathbf{A} \times \mathbf{B} + \mathbf{B} \times \mathbf{C} + \mathbf{C} \times \mathbf{A}$ are perpendicular to the plane containing the points P, Q, and R.

34. Find an equation of the plane containing the endpoints of the position representations of the vectors $2\mathbf{i} - \mathbf{j} + 3\mathbf{k}$, $-\mathbf{i} + \mathbf{j} + 2\mathbf{k}$, and $5\mathbf{i} + \mathbf{j} - \mathbf{k}$.

35. Prove Theorem 15.5.4.   **36.** Prove Theorem 15.5.5.
37. Prove Theorem 15.5.6.   **38.** Prove Theorem 15.5.7.

39. Let $\overrightarrow{OP}$ be the position representation of vector $\mathbf{A}$, $\overrightarrow{OQ}$ be the position representation of vector $\mathbf{B}$, and $\overrightarrow{OR}$ be the position representation of vector $\mathbf{C}$. Prove that the area of triangle PQR is $\frac{1}{2}\|(\mathbf{B} - \mathbf{A}) \times (\mathbf{C} - \mathbf{A})\|$.

15.6 CYLINDERS AND SURFACES OF REVOLUTION

As mentioned previously, the graph of an equation in three variables is a **surface**. A surface is represented by an equation if the coordinates of every point on the surface satisfy the equation and if every point whose coordinates satisfy the equation lies on the surface. We have already discussed two kinds of surfaces, a plane and a sphere. Another kind of surface that is fairly simple is a cylinder. You are probably familiar with right-circular cylinders from previous experience. We now consider a more general cylindrical surface.

15.6.1 DEFINITION   A **cylinder** is a surface that is generated by a line moving along a given plane curve in such a way that it always remains parallel to a fixed line not lying in the plane of the given curve. The moving line is called a **generator** of the cylinder and the given plane curve is called a **directrix** of the cylinder. Any position of a generator is called a **ruling** of the cylinder.

We confine this discussion to cylinders having a directrix in a coordinate plane and rulings perpendicular to that plane. If the rulings of a cylinder are perpendicular to the plane of a directrix, the cylinder is said to be perpendicular to the plane.

The familiar right-circular cylinder is one for which a directrix is a circle in a plane perpendicular to the cylinder.

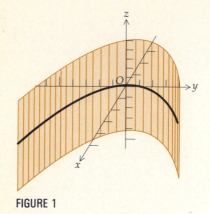

FIGURE 1

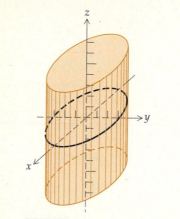

FIGURE 2

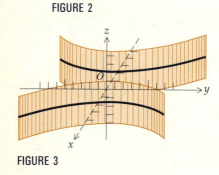

FIGURE 3

▶ **ILLUSTRATION 1** In Figure 1 there is a cylinder whose directrix is the parabola $y^2 = 8x$ in the xy plane and whose rulings are parallel to the z axis. The cylinder is called a **parabolic cylinder**. An **elliptic cylinder** is shown in Figure 2; its directrix is the ellipse $9x^2 + 16y^2 = 144$ in the xy plane, and its rulings are parallel to the z axis. Figure 3 shows a **hyperbolic cylinder** having as a directrix the hyperbola $25x^2 - 4y^2 = 100$ in the xy plane and rulings parallel to the z axis. ◀

Consider the problem of finding an equation of a cylinder having a directrix in a coordinate plane and rulings parallel to the coordinate axis not in that plane. To be specific, take the directrix in the xy plane and the rulings parallel to the z axis. Refer to Figure 4. Suppose that an equation of the directrix in the xy plane is $y = f(x)$. If the point $(x_0, y_0, 0)$ in the xy plane satisfies this equation, any point (x_0, y_0, z) in three-dimensional space, where z is any real number, will satisfy the same equation because z does not appear in the equation. The points having representations (x_0, y_0, z) all lie on the line parallel to the z axis through the point $(x_0, y_0, 0)$. This line is a ruling of the cylinder. Hence any point whose x and y coordinates satisfy the equation $y = f(x)$ lies on the cylinder. Conversely, if the point $P(x, y, z)$ lies on the cylinder (see Figure 5), then the point $(x, y, 0)$ lies on the directrix of the cylinder in the xy plane, and hence the x and y coordinates of P satisfy the equation $y = f(x)$. Therefore, if $y = f(x)$ is considered as an equation of a graph in three-dimensional space, the graph is a cylinder whose rulings are parallel to the z axis and which has as a directrix the curve $y = f(x)$ in the plane $z = 0$. A similar discussion pertains when the directrix is in either of the other coordinate planes. The results are summarized in the following theorem.

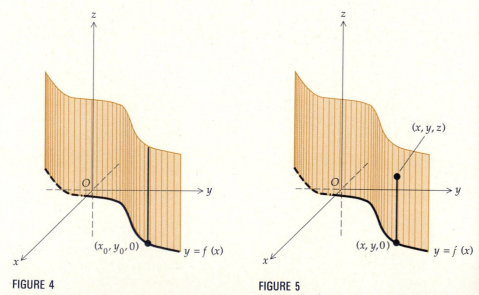

FIGURE 4

FIGURE 5

15.6.2 THEOREM

In three-dimensional space, the graph of an equation in two of the three variables x, y, and z is a cylinder whose rulings are parallel to the axis associated with the missing variable and whose directrix is a curve in the plane associated with the two variables appearing in the equation.

▶ **ILLUSTRATION 2**   It follows from Theorem 15.6.2 that an equation of the parabolic cylinder of Figure 1 is $y^2 = 8x$, considered as an equation in R^3. Similarly, equations of the elliptic cylinder of Figure 2 and the hyperbolic cylinder of Figure 3 are, respectively,

$$9x^2 + 16y^2 = 144 \quad \text{and} \quad 25x^2 - 4y^2 = 100$$

both considered as equations in R^3.   ◀

A **cross section** of a surface in a plane is the set of all points of the surface that lie in the given plane. If a plane is parallel to the plane of the directrix of a cylinder, the cross section of the cylinder is the same as the directrix. For example, the cross section of the elliptic cylinder of Figure 2 in any plane parallel to the xy plane is an ellipse.

EXAMPLE 1   Draw a sketch of the graph of each of the following equations: (a) $y = \ln z$; (b) $z^2 = x^3$.

Solution
(a) The graph is a cylinder whose directrix lying in the yz plane is the curve $y = \ln z$ and whose rulings are parallel to the x axis. A sketch of the graph appears in Figure 6.
(b) The graph is a cylinder whose directrix is in the xz plane and whose rulings are parallel to the y axis. An equation of the directrix is the curve $z^2 = x^3$ in the xz plane. A sketch of the graph is shown in Figure 7.

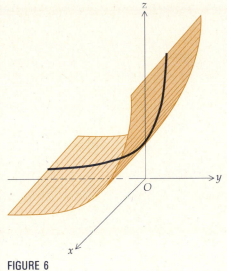

FIGURE 6

15.6.3 DEFINITION   If a plane curve is revolved about a fixed line lying in the plane of the curve, the surface generated is called a **surface of revolution**. The fixed line is called the **axis** of the surface of revolution, and the plane curve is called the **generating curve**.

Figure 8 shows a surface of revolution whose generating curve is the curve C in the yz plane and whose axis is the z axis. A sphere is a particular example of a surface of revolution because a sphere can be generated by revolving a semicircle about a diameter.

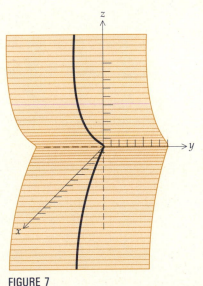

FIGURE 7

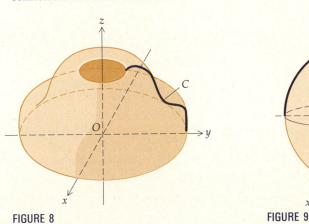

FIGURE 8   FIGURE 9

▶ **ILLUSTRATION 3**   Figure 9 shows a sphere that can be generated by revolving the semicircle $y^2 + z^2 = r^2$, $z \geq 0$, about the y axis. Another example

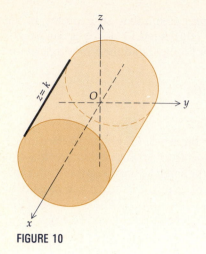

FIGURE 10

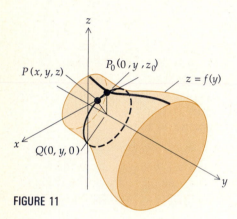

FIGURE 11

of a surface of revolution is a right-circular cylinder for which the generating curve and the axis are parallel straight lines. If the generating curve is the line $z = k$ in the xz plane and the axis is the x axis, we obtain the right-circular cylinder shown in Figure 10. ◀

We now find an equation of the surface generated by revolving about the y axis the curve in the yz plane having the two-dimensional equation

$$z = f(y) \tag{1}$$

Refer to Figure 11. Let $P(x, y, z)$ be any point on the surface of revolution. Through P, we pass a plane perpendicular to the y axis, and denote the point of intersection of this plane with the y axis by $Q(0, y, 0)$. We let $P_0(0, y, z_0)$ be the point of intersection of the plane with the generating curve. Because the cross section of the surface with the plane through P is a circle, P is on the surface if and only if

$$|\overline{QP}|^2 = |\overline{QP_0}|^2$$

Because $|\overline{QP}| = \sqrt{x^2 + z^2}$ and $|\overline{QP_0}| = z_0$, we obtain from this equation

$$x^2 + z^2 = z_0{}^2 \tag{2}$$

The point P_0 is on the generating curve; so its coordinates must satisfy (1). Therefore

$$z_0 = f(y)$$

From this equation and (2), the point P is on the surface of revolution if and only if

$$x^2 + z^2 = [f(y)]^2 \tag{3}$$

This is the desired equation of the surface of revolution. Because (3) is equivalent to

$$\pm \sqrt{x^2 + z^2} = f(y)$$

we can obtain (3) by replacing z in (1) by $\pm \sqrt{x^2 + z^2}$.

In a similar manner we can show that if the curve in the yz plane having the two-dimensional equation

$$y = g(z) \tag{4}$$

is revolved about the z axis, an equation of the surface of revolution generated is obtained by replacing y in (4) by $\pm \sqrt{x^2 + y^2}$. Analogous remarks hold when a curve in any coordinate plane is revolved about either one of the coordinate axes in that plane. In summary, the graphs of the following equations are surfaces of revolution having the indicated axis: $x^2 + y^2 = [F(z)]^2$—z axis; $x^2 + z^2 = [F(y)]^2$—y axis; $y^2 + z^2 = [F(x)]^2$—x axis. In each case, cross sections of the surface in planes perpendicular to the axis are circles having centers on the axis.

EXAMPLE 2   Find an equation of the surface of revolution generated by revolving the parabola $y^2 = 4x$ in the xy plane about the x axis. Draw a sketch of the graph of the surface.

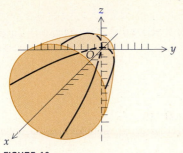

FIGURE 12

FIGURE 13

Solution   In the equation of the parabola we replace y by $\pm\sqrt{y^2 + z^2}$ and obtain

$$y^2 + z^2 = 4x$$

Figure 12 shows a sketch of the graph. The same surface is generated if the parabola $z^2 = 4x$ in the xz plane is revolved about the x axis.

The surface obtained in Example 2 is called a **paraboloid of revolution**. If an ellipse is revolved about one of its axes, the surface obtained is called an **ellipsoid of revolution. A hyperboloid of revolution** is obtained when a hyperbola is revolved about an axis.

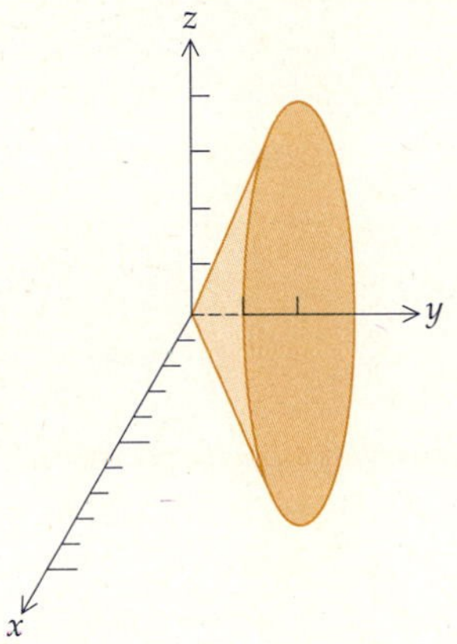

EXAMPLE 3   Draw a sketch of the surface $x^2 + z^2 - 4y^2 = 0$, if $y \geq 0$.

Solution   The given equation is of the form $x^2 + z^2 = [F(y)]^2$; so its graph is a surface of revolution having the y axis as axis. Solving the given equation for y, we obtain

$$2y = \pm\sqrt{x^2 + z^2}$$

Hence the generating curve can be either the line $2y = x$ in the xy plane or the line $2y = z$ in the yz plane. By drawing sketches of the two possible generating curves and using the fact that cross sections of the surface in planes perpendicular to the y axis are circles having centers on the y axis, we obtain the surface shown in Figure 13 (note that because $y \geq 0$ there is only one nappe of the cone).

The surface obtained in Example 3 is called a **right-circular cone**.

EXERCISES 15.6

In Exercises 1 through 4, draw a sketch of the cross section of the given cylinder in the indicated plane.

1. $4x^2 + y^2 = 16$; xy plane   **2.** $4z^2 - y^2 = 4$; yz plane
3. $z = e^x$; xz plane   **4.** $x = |y|$; xy plane

In Exercises 5 through 12, draw a sketch of the cylinder having the given equation.

5. $4x^2 + 9y^2 = 36$ **6.** $z = \sin y$
7. $y = |z|$ **8.** $x^2 - z^2 = 4$
9. $z = 2x^2$ **10.** $z^2 = 4y^2$
11. $y = \cosh x$ **12.** $x^2 = y^3$

In Exercises 13 through 20, find an equation of the surface of revolution generated by revolving the plane curve about the indicated axis. Draw a sketch of the surface.

13. $x^2 = 4y$ in the xy plane, about the y axis.
14. $x^2 + 4z^2 = 16$ in the xz plane, about the z axis.
15. $x^2 + 4z^2 = 16$ in the xz plane, about the x axis.

16. $x^2 = 4y$ in the xy plane, about the x axis.
17. $y = 3z$ in the yz plane, about the y axis.
18. $9y^2 - 4z^2 = 144$ in the yz plane, about the z axis.
19. $y = \sin x$ in the xy plane, about the x axis.
20. $y^2 = z^3$ in the yz plane, about the z axis.

In Exercises 21 through 28, find a generating curve and the axis for the surface of revolution. Draw a sketch of the surface.

21. $x^2 + y^2 + z^2 = 16$ **22.** $x^2 + z^2 = y$
23. $x^2 + y^2 - z^2 = 4$ **24.** $y^2 + z^2 = e^{2x}$
25. $x^2 + z^2 = |y|$ **26.** $4x^2 + 9y^2 + 4z^2 = 36$
27. $9x^2 - y^2 + 9z^2 = 0$ **28.** $4x^2 + 4y^2 - z = 9$
29. The tractrix

$$x = t - a\tanh\frac{t}{a} \qquad y = a\operatorname{sech}\frac{t}{a}$$

from $x = -a$ to $x = 2a$ is revolved about the x axis. Draw a sketch of the surface of revolution.

15.7 QUADRIC SURFACES

You learned in Chapter 10 that the graph of a second-degree equation in two variables x and y,

$$Ax^2 + Bxy + Cy^2 + Dx + Ey + F = 0$$

is a conic section. The graph of a second-degree equation in three variables x, y, and z,

$$Ax^2 + By^2 + Cz^2 + Dxy + Exz + Fyz + Gx + Hy + Iz + J = 0 \qquad (1)$$

is called a **quadric surface**. The simplest types of quadric surfaces are the parabolic, elliptic, and hyperbolic cylinders that were discussed in the preceding section. There are six other types of quadric surfaces that we now consider. We choose the coordinate axes so the equations are in their simplest form. In the discussion of each of these surfaces we refer to the cross sections of the surfaces in planes parallel to the coordinate planes. These cross sections help to visualize the surface.

The ellipsoid

$$\frac{x^2}{a^2} + \frac{y^2}{b^2} + \frac{z^2}{c^2} = 1 \qquad (2)$$

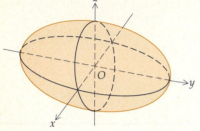

FIGURE 1

where a, b, and c are positive (see Figure 1).

If in (2) z is replaced by zero, we obtain the cross section of the ellipsoid in the xy plane, which is the ellipse

$$\frac{x^2}{a^2} + \frac{y^2}{b^2} = 1$$

To obtain the cross sections of the surface with the planes $z = k$, we replace z by k in the equation of the ellipsoid and get

$$\frac{x^2}{a^2} + \frac{y^2}{b^2} = 1 - \frac{k^2}{c^2}$$

If $|k| < c$, the cross section is an ellipse and the lengths of the semiaxes decrease to zero as $|k|$ increases to the value c. If $|k| = c$, the intersection of a plane $z = k$ with the ellipsoid is the single point $(0, 0, k)$. If $|k| > c$, there is no intersection. The discussion is similar if we consider cross sections formed by planes parallel to either of the other coordinate planes.

The numbers a, b, and c are the lengths of the semiaxes of the ellipsoid. If any two of these three numbers are equal, we have an ellipsoid of revolution, which is also called a **spheroid**. A spheroid for which the third number is greater than the two equal numbers is said to be **prolate**. A prolate spheroid is shaped like a football. An **oblate** spheroid is obtained if the third number is less than the two equal numbers. If all three numbers, a, b, and c in the equation of an ellipsoid are equal, the ellipsoid is a **sphere**.

The elliptic hyperboloid of one sheet

$$\frac{x^2}{a^2} + \frac{y^2}{b^2} - \frac{z^2}{c^2} = 1 \qquad (3)$$

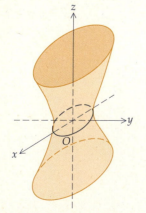

FIGURE 2

where a, b, and c are positive (see Figure 2).

The cross sections in the planes $z = k$ are the ellipses

$$\frac{x^2}{a^2} + \frac{y^2}{b^2} = 1 + \frac{k^2}{c^2}$$

When $k = 0$, the lengths of the semiaxes of the ellipse are smallest, and these lengths increase as $|k|$ increases. The cross sections in the planes $x = k$ are the hyperbolas

$$\frac{y^2}{b^2} - \frac{z^2}{c^2} = 1 - \frac{k^2}{a^2}$$

If $|k| < a$, the transverse axis of the hyperbola is parallel to the y axis, and if $|k| > a$, the transverse axis is parallel to the z axis. If $k = a$, the hyperbola degenerates into two lines:

$$\frac{y}{b} - \frac{z}{c} = 0 \quad \text{and} \quad \frac{y}{b} + \frac{z}{c} = 0$$

In an analogous manner, the cross sections in the planes $y = k$ are also hyperbolas. The axis of this hyperboloid is the z axis.

If $a = b$, the surface is a hyperboloid of revolution for which the axis is the line containing the conjugate axis.

The elliptic hyperboloid of two sheets

$$-\frac{x^2}{a^2} - \frac{y^2}{b^2} + \frac{z^2}{c^2} = 1 \tag{4}$$

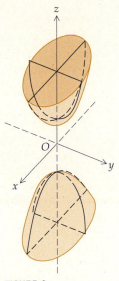

FIGURE 3

where a, b, and c are positive (see Figure 3).

Replacing z by k in (4) we obtain

$$\frac{x^2}{a^2} + \frac{y^2}{b^2} = \frac{k^2}{c^2} - 1$$

If $|k| < c$, there is no intersection of the plane $z = k$ with the surface; hence there are no points of the surface between the planes $z = -c$ and $z = c$. If $|k| = c$, the intersection of the plane $z = k$ with the surface is the single point $(0, 0, k)$. When $|k| > c$, the cross section of the surface in the plane $z = k$ is an ellipse, and the lengths of the semiaxes of the ellipse increase as $|k|$ increases.

The cross sections of the surface in the planes $x = k$ are the hyperbolas

$$\frac{z^2}{c^2} - \frac{y^2}{b^2} = 1 + \frac{k^2}{a^2}$$

whose transverse axes are parallel to the z axis. In a similar fashion, the cross sections in the planes $y = k$ are the hyperbolas

$$\frac{z^2}{c^2} - \frac{x^2}{a^2} = 1 + \frac{k^2}{b^2}$$

for which the transverse axes are also parallel to the z axis.

If $a = b$, the surface is a hyperboloid of revolution in which the axis is the line containing the transverse axis of the hyperbola.

Each of the above three quadric surfaces is symmetric with respect to each of the coordinate planes and symmetric with respect to the origin. Their graphs are called **central quadrics** and their center is at the origin. The graph of any equation of the form

$$\pm\frac{x^2}{a^2} \pm \frac{y^2}{b^2} \pm \frac{z^2}{c^2} = 1$$

where a, b, and c are positive, is a central quadric.

EXAMPLE 1   Draw a sketch of the graph of the equation

$$4x^2 - y^2 + 25z^2 = 100$$

and name the surface.

Solution   We divide both sides of the equation by 100 and obtain

$$\frac{x^2}{25} - \frac{y^2}{100} + \frac{z^2}{4} = 1$$

which is of the form of (3) with y and z interchanged. Hence the surface is an elliptic hyperboloid of one sheet whose axis is the y axis. The cross sections in the planes $y = k$ are the ellipses

$$\frac{x^2}{25} + \frac{z^2}{4} = 1 + \frac{k^2}{100}$$

The cross sections in the planes $x = k$ are the hyperbolas

$$\frac{z^2}{4} - \frac{y^2}{100} = 1 - \frac{k^2}{25}$$

and the cross sections in the planes $z = k$ are the hyperbolas

$$\frac{x^2}{25} - \frac{y^2}{100} = 1 - \frac{k^2}{4}$$

A sketch of the surface appears in Figure 4.

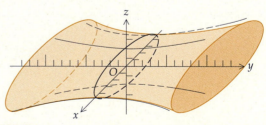

FIGURE 4

EXAMPLE 2   Draw a sketch of the graph of the equation

$$4x^2 - 25y^2 - z^2 = 100$$

and name the surface.

Solution   By dividing on both sides by 100 we can write the given equation as

$$\frac{x^2}{25} - \frac{y^2}{4} - \frac{z^2}{100} = 1$$

which is of the form of (4) with x and z interchanged; thus the surface is an elliptic hyperboloid of two sheets whose axis is the x axis. The cross sections in the planes $x = k$, where $|k| > 5$, are the ellipses

$$\frac{y^2}{4} + \frac{z^2}{100} = \frac{k^2}{25} - 1$$

The planes $x = k$, where $|k| < 5$, do not intersect the surface. The cross sections in the planes $y = k$ are the hyperbolas

$$\frac{x^2}{25} - \frac{z^2}{100} = 1 + \frac{k^2}{4}$$

and the cross sections in the planes $z = k$ are the hyperbolas

$$\frac{x^2}{25} - \frac{y^2}{4} = 1 + \frac{k^2}{100}$$

The required sketch is shown in Figure 5.

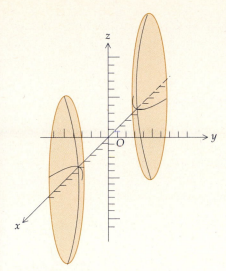

FIGURE 5

The next two surfaces are called noncentral quadrics.

The elliptic paraboloid

$$\frac{x^2}{a^2} + \frac{y^2}{b^2} = \frac{z}{c} \qquad (5)$$

where a and b are positive and $c \neq 0$. Figure 6 shows the surface if $c > 0$.

Substituting k for z in (5) we obtain

$$\frac{x^2}{a^2} + \frac{y^2}{b^2} = \frac{k}{c}$$

When $k = 0$, this equation becomes $b^2 x^2 + a^2 y^2 = 0$, which represents a single point, the origin. If $k \neq 0$ and k and c have the same sign, the equation is that of an ellipse. So we conclude that cross sections of the surface in the planes $z = k$, where k and c have the same sign, are ellipses and the lengths of the semiaxes increase as $|k|$ increases. If k and c have opposite signs, the planes $z = k$ do not intersect the surface. The cross sections of the surface with the planes $x = k$ and $y = k$ are parabolas. When $c > 0$, the parabolas open upward, as shown in Figure 6; when $c < 0$, the parabolas open downward.

If $a = b$, the surface is a paraboloid of revolution.

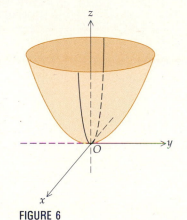

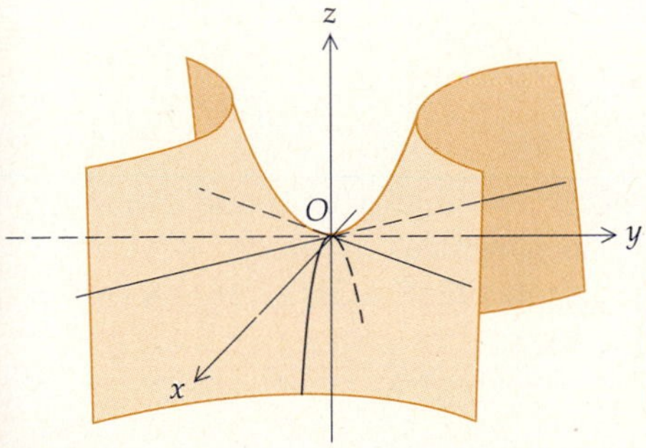

FIGURE 6

The hyperbolic paraboloid

$$\frac{y^2}{b^2} - \frac{x^2}{a^2} = \frac{z}{c} \qquad (6)$$

where a and b are positive and $c \neq 0$. The surface appears in Figure 7 for $c > 0$.

The cross sections of the surface in the planes $z = k$, where $k \neq 0$, are hyperbolas having their transverse axes parallel to the y axis if k and c have the same sign and parallel to the x axis if k and c have opposite signs. The cross section of the surface in the plane $z = 0$ consists of two straight lines through the origin. The cross sections in the planes $x = k$ are parabolas opening upward if $c > 0$ and opening downward if $c < 0$. The cross sections in the planes $y = k$ are parabolas opening downward if $c > 0$ and opening upward if $c < 0$.

FIGURE 7

EXAMPLE 3   Draw a sketch of the graph of the equation

$$3y^2 + 12z^2 = 16x$$

and name the surface.

Solution   The given equation can be written as

$$\frac{y^2}{16} + \frac{z^2}{4} = \frac{x}{3}$$

which is of the form of (5) with x and z interchanged. Hence the graph of the equation is an elliptic paraboloid whose axis is the x axis. The cross sections in the planes $x = k$, where $k > 0$, are the ellipses

$$\frac{y^2}{16} + \frac{z^2}{4} = \frac{k}{3}$$

The planes $x = k$, where $k < 0$, do not intersect the surface. The cross sections in the planes $y = k$ are the parabolas $12z^2 = 16x - 3k^2$, and the cross sections in the planes $z = k$ are the parabolas $3y^2 = 16x - 12k^2$. Figure 8 shows a sketch of the elliptic paraboloid.

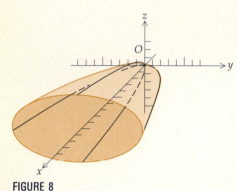

FIGURE 8

EXAMPLE 4   Draw a sketch of the graph of the equation

$$3y^2 - 12z^2 = 16x$$

and name the surface.

Solution   If the given equation is written as

$$\frac{y^2}{16} - \frac{z^2}{4} = \frac{x}{3}$$

it is of the form of (6) with x and z interchanged. The surface is therefore a hyperbolic paraboloid. The cross sections in the planes $x = k$, where $k \neq 0$, are the hyperbolas

$$\frac{y^2}{16} - \frac{z^2}{4} = \frac{k}{3}$$

The cross section in the yz plane consists of the two lines $y = 2z$ and $y = -2z$. In the planes $z = k$ the cross sections are the parabolas $3y^2 = 16x + 12k^2$; in the planes $y = k$ the cross sections are the parabolas $12z^2 = 3k^2 - 16x$. A sketch of the hyperbolic paraboloid appears in Figure 9.

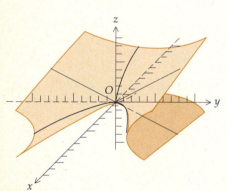

FIGURE 9

The elliptic cone

$$\frac{x^2}{a^2} + \frac{y^2}{b^2} - \frac{z^2}{c^2} = 0 \qquad\qquad (7)$$

where a, b, and c are positive (see Figure 10).

The intersection of the plane $z = 0$ with the surface is a point, the origin. The cross sections of the surface in the planes $z = k$, where $k \neq 0$, are ellipses, and the lengths of the semiaxes increase as k increases. Cross sections in the

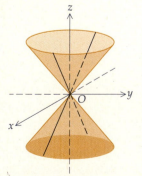

FIGURE 10

planes $x = 0$ and $y = 0$ are pairs of intersecting lines. In the planes $x = k$ and $y = k$, where $k \neq 0$, the cross sections are hyperbolas.

EXAMPLE 5   Draw a sketch of the graph of the equation

$$4x^2 - y^2 + 25z^2 = 0$$

and name the surface.

Solution   We can write the given equation as

$$\frac{x^2}{25} - \frac{y^2}{100} + \frac{z^2}{4} = 0$$

which is of the form of (7) with y and z interchanged. Therefore the surface is an elliptic cone having the y axis as its axis. The surface intersects the plane $y = 0$ at the origin only. The intersection of the surface with the plane $x = 0$ is the pair of intersecting lines $y = \pm 5z$, and the intersection with the plane $z = 0$ is the pair of intersecting lines $y = \pm 2x$. The cross sections in the planes $y = k$, where $k \neq 0$, are the ellipses

$$\frac{x^2}{25} + \frac{z^2}{4} = \frac{k^2}{100}$$

In the planes $x = k$ and $z = k$, where $k \neq 0$, the cross sections are, respectively, the hyperbolas

$$\frac{y^2}{100} - \frac{z^2}{4} = \frac{k^2}{25} \quad \text{and} \quad \frac{y^2}{100} - \frac{x^2}{25} = \frac{k^2}{4}$$

Figure 11 shows a sketch of the surface.

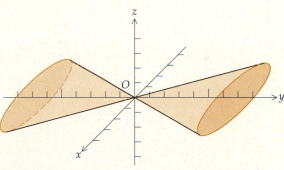

FIGURE 11

Equation (1) is the general equation of the second degree in x, y, and z. It can be shown that by translation and rotation of the three-dimensional coordinate axes (the study of which is beyond the scope of this book) this equation can be reduced to one of the following two forms:

$$Ax^2 + By^2 + Cz^2 + J = 0 \tag{8}$$

$$Ax^2 + By^2 + Iz = 0 \tag{9}$$

Graphs of the equations of the second degree will either be one of the above six types of quadrics or else will degenerate into a cylinder, plane, line, point, or the empty set.

The nondegenerate curves associated with equations of the form (8) are the central quadrics and the elliptic cone, whereas those associated with equations of the form (9) are the noncentral quadrics. Following are examples of some degenerate cases:

$$x^2 - y^2 = 0; \text{ two planes, } x - y = 0 \text{ and } x + y = 0$$
$$z^2 = 0; \text{ one plane, the } xy \text{ plane}$$
$$x^2 + y^2 = 0; \text{ one line, the } z \text{ axis}$$
$$x^2 + y^2 + z^2 = 0; \text{ a point, the origin}$$
$$x^2 + y^2 + z^2 + 1 = 0; \text{ the empty set}$$

EXERCISES 15.7

In Exercises 1 through 6, name the surface having the equation.

1. $9x^2 - 4y^2 + 36z^2 = 36$
2. $4x^2 - 16y^2 + 9z^2 = 0$
3. $5x^2 - 2z^2 - 3y = 0$
4. $25x^2 = 4y^2 + z^2 + 100$
5. $4y^2 - 25x^2 = 100$
6. $3y^2 + 7z^2 = 6x$

In Exercises 7 through 18, draw a sketch of the graph of the equation and name the surface.

7. $4x^2 + 9y^2 + z^2 = 36$
8. $4x^2 - 9y^2 - z^2 = 36$
9. $4x^2 + 9y^2 - z^2 = 36$
10. $4x^2 - 9y^2 + z^2 = 36$
11. $x^2 = y^2 - z^2$
12. $x^2 = y^2 + z^2$
13. $\dfrac{x^2}{36} + \dfrac{z^2}{25} = 4y$
14. $\dfrac{y^2}{25} + \dfrac{x^2}{36} = 4$
15. $\dfrac{x^2}{36} - \dfrac{z^2}{25} = 9y$
16. $x^2 = 2y + 4z$
17. $x^2 + 16z^2 = 4y^2 - 16$
18. $9y^2 - 4z^2 + 18x = 0$

19. Find the values of k for which the intersection of the plane $x + ky = 1$ and the elliptic hyperboloid of two sheets $y^2 - x^2 - z^2 = 1$ is (a) an ellipse and (b) a hyperbola.

20. Find the vertex and focus of the parabola that is the intersection of the plane $y = 2$ with the hyperbolic paraboloid $\dfrac{y^2}{16} - \dfrac{x^2}{4} = \dfrac{z}{9}$.

21. Find the vertex and focus of the parabola that is the intersection of the plane $x = 1$ with the hyperbolic paraboloid $\dfrac{z^2}{4} - \dfrac{x^2}{9} = \dfrac{y}{3}$.

22. Find the area of the plane section formed by the intersection of the plane $y = 3$ with the solid bounded by the ellipsoid $\dfrac{x^2}{9} + \dfrac{y^2}{25} + \dfrac{z^2}{4} = 1$.

23. Show that the intersection of the surface $x^2 - 4y^2 - 9z^2 = 36$ and the plane $x + z = 9$ is a circle.

24. Show that the intersection of the hyperbolic paraboloid $\dfrac{y^2}{b^2} - \dfrac{x^2}{a^2} = \dfrac{z}{c}$ and the plane $z = bx + ay$ consists of two intersecting lines.

In Exercises 25 through 27, use the method of slicing (Section 6.1) to find the volume of the solid. The measure of the area of the region enclosed by the ellipse having semiaxes a and b is πab.

25. The solid bounded by the ellipsoid $36x^2 + 9y^2 + 4z^2 = 36$.

26. The solid bounded by the ellipsoid $\dfrac{x^2}{a^2} + \dfrac{y^2}{b^2} + \dfrac{z^2}{c^2} = 1$.

27. The solid bounded by the plane $z = h$, where $h > 0$, and the elliptic paraboloid $\dfrac{x^2}{a^2} + \dfrac{y^2}{b^2} = \dfrac{z}{c}$, where $c > 0$.

15.8 CURVES IN R^3

We now consider vector-valued functions in three-dimensional space.

15.8.1 DEFINITION

Let f_1, f_2, and f_3 be three real-valued functions of a real variable t. Then for every number t in the domain common to f_1, f_2, and f_3 there is a vector **R** defined by

$$\mathbf{R}(t) = f_1(t)\mathbf{i} + f_2(t)\mathbf{j} + f_3(t)\mathbf{k}$$

and **R** is called a **vector-valued function**.

The graph of a vector-valued function in three-dimensional space is obtained analogously to the way the graph of a vector-valued function in two dimensions was obtained in Section 14.3. As t assumes all values in the domain of **R**, the

Table 1

t	x	y	z
0	a	0	0
$\dfrac{\pi}{4}$	$\dfrac{a}{\sqrt{2}}$	$\dfrac{b}{\sqrt{2}}$	$\dfrac{\pi}{4}$
$\dfrac{\pi}{2}$	0	b	$\dfrac{\pi}{2}$
$\dfrac{3\pi}{4}$	$-\dfrac{a}{\sqrt{2}}$	$\dfrac{b}{\sqrt{2}}$	$\dfrac{3\pi}{4}$
π	$-a$	0	π
$\dfrac{3\pi}{2}$	0	$-b$	$\dfrac{3\pi}{2}$

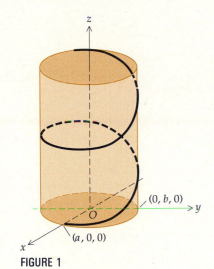

FIGURE 1

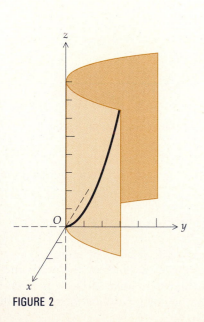

FIGURE 2

terminal point of the position representation of the vector $\mathbf{R}(t)$ traces a curve C, and this curve is called the graph of $\mathbf{R}$. A point on the curve C has the cartesian representation (x, y, z), where

$$x = f_1(t) \qquad y = f_2(t) \qquad z = f_3(t) \tag{1}$$

These equations are called **parametric equations** of C, whereas the equation of Definition 15.8.1 is called a **vector equation** of C. By eliminating t from Equations (1) we obtain two equations in x, y, and z. These equations are called **cartesian equations** of C. Each cartesian equation is an equation of a surface, and curve C is the intersection of the two surfaces. The equations of any two surfaces containing C may be taken as cartesian equations defining C.

▶ **ILLUSTRATION 1**   We draw a sketch of the curve having the vector equation

$$\mathbf{R}(t) = a \cos t\,\mathbf{i} + b \sin t\,\mathbf{j} + t\mathbf{k}$$

Parametric equations of the given curve are

$$x = a \cos t \qquad y = b \sin t \qquad z = t$$

To eliminate t from the first two equations we write them as

$$\frac{x^2}{a^2} = \cos^2 t \quad \text{and} \quad \frac{y^2}{b^2} = \sin^2 t$$

Adding corresponding members of these two equations we obtain

$$\frac{x^2}{a^2} + \frac{y^2}{b^2} = 1$$

Therefore the curve lies entirely on the elliptical cylinder whose directrix is an ellipse in the xy plane and whose rulings are parallel to the z axis. Table 1 gives sets of values of x, y, and z for specific values of t. A sketch of the curve appears in Figure 1.   ◀

The curve of Illustration 1 is called a **helix**. If $a = b$, the helix is a **circular helix** and it lies on the right-circular cylinder $x^2 + y^2 = a^2$.

▶ **ILLUSTRATION 2**   The curve having the vector equation

$$\mathbf{R}(t) = t\mathbf{i} + t^2\mathbf{j} + t^3\mathbf{k}$$

is called a **twisted cubic**. Parametric equations of the twisted cubic are

$$x = t \qquad y = t^2 \qquad z = t^3$$

Eliminating t from the first two of these equations yields $y = x^2$, which is a cylinder whose directrix in the xy plane is a parabola. The twisted cubic lies on this cylinder. Figure 2 shows a sketch of the cylinder and the portion of the twisted cubic from $t = 0$ to $t = 2$.   ◀

Definition 14.4.1 pertaining to the limit of a vector-valued function in two dimensions can be extended to vector-valued functions in three dimensions as follows: If

$$\mathbf{R}(t) = f_1(t)\mathbf{i} + f_2(t)\mathbf{j} + f_3(t)\mathbf{k}$$

then

$$\lim_{t \to t_1} \mathbf{R}(t) = \lim_{t \to t_1} f_1(t)\mathbf{i} + \lim_{t \to t_1} f_2(t)\mathbf{j} + \lim_{t \to t_1} f_3(t)\mathbf{k}$$

if $\lim_{t \to t_1} f_1(t)$, $\lim_{t \to t_1} f_2(t)$, and $\lim_{t \to t_1} f_3(t)$ all exist.

The definitions of continuity (14.4.2) and the derivative (14.4.3) of vector-valued functions in V_2 are the same for those in V_3. The proof of the following theorem is similar to the proof of Theorem 14.4.4 and is left as an exercise (see Exercise 13).

15.8.2 THEOREM   If $\mathbf{R}$ is the vector-valued function defined by

$$\mathbf{R}(t) = f_1(t)\mathbf{i} + f_2(t)\mathbf{j} + f_3(t)\mathbf{k}$$

and $\mathbf{R}'(t)$ exists, then

$$\mathbf{R}'(t) = f_1{}'(t)\mathbf{i} + f_2{}'(t)\mathbf{j} + f_3{}'(t)\mathbf{k}$$

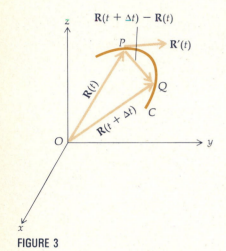

The geometric interpretation for the derivative of $\mathbf{R}$ is the same as that for the derivative of a vector-valued function in R^2. Figure 3 shows a portion of the curve C, which is the graph of $\mathbf{R}$. In the figure, $\overrightarrow{OP}$ is the position representation of $\mathbf{R}(t)$, $\overrightarrow{OQ}$ is the position representation of $\mathbf{R}(t + \Delta t)$, and so $\overrightarrow{PQ}$ is a representation of the vector $\mathbf{R}(t + \Delta t) - \mathbf{R}(t)$. As Δt approaches zero, the vector $[\mathbf{R}(t + \Delta t) - \mathbf{R}(t)]/\Delta t$ has a representation approaching a directed line segment tangent to the curve C at P.

The definition of the **unit tangent vector** is analogous to Definition 14.6.1 for vectors in the plane. So if $\mathbf{T}(t)$ denotes the unit tangent vector to the graph of $\mathbf{R}$, then

$$\mathbf{T}(t) = \frac{D_t\mathbf{R}(t)}{\|D_t\mathbf{R}(t)\|} \tag{2}$$

FIGURE 3

▶ **ILLUSTRATION 3**   We find the unit tangent vector for the twisted cubic of Illustration 2.

Because $\mathbf{R}(t) = t\mathbf{i} + t^2\mathbf{j} + t^3\mathbf{k}$,

$$D_t\mathbf{R}(t) = \mathbf{i} + 2t\mathbf{j} + 3t^2\mathbf{k} \qquad \|D_t\mathbf{R}(t)\| = \sqrt{1 + 4t^2 + 9t^4}$$

From (2),

$$\mathbf{T}(t) = \frac{1}{\sqrt{1 + 4t^2 + 9t^4}}(\mathbf{i} + 2t\mathbf{j} + 3t^2\mathbf{k})$$

Therefore, in particular,

$$\mathbf{T}(1) = \frac{1}{\sqrt{14}}\mathbf{i} + \frac{2}{\sqrt{14}}\mathbf{j} + \frac{3}{\sqrt{14}}\mathbf{k}$$

Figure 4 shows the representation of $\mathbf{T}(1)$ at the point $(1, 1, 1)$.   ◀

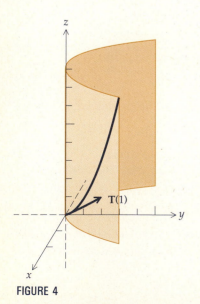

FIGURE 4

Theorems 14.4.6, 14.4.7, and 14.4.8 regarding derivatives of sums and products of two-dimensional vector-valued functions also hold for vectors in three dimensions. The following theorem regarding the derivative of the cross product of two vector-valued functions is similar to the corresponding formula for the

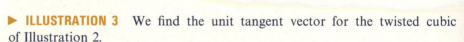

derivative of the product of real-valued functions; however, it is important to maintain the correct order of the vector-valued functions because the cross product is not commutative.

15.8.3 THEOREM If **R** and **Q** are vector-valued functions, then

$$D_t[\mathbf{R}(t) \times \mathbf{Q}(t)] = \mathbf{R}(t) \times \mathbf{Q}'(t) + \mathbf{R}'(t) \times \mathbf{Q}(t)$$

for all values of t for which $\mathbf{R}'(t)$ and $\mathbf{Q}'(t)$ exist.

The proof of Theorem 15.8.3 is left as an exercise (see Exercise 14).

The length of an arc of a curve C in three-dimensional space can be defined in exactly the same way as we defined the length of an arc of a curve in the plane (see Definition 14.5.1). If C is the curve having parametric equations (1), f_1', f_2', f_3' are continuous on the closed interval $[a, b]$, and no two values of t give the same point (x, y, z) on C, then we can prove (as we did for the plane) a theorem similar to Theorem 14.5.3, which states that the length of arc, L units, of the curve C from the point $(f_1(a), f_2(a), f_3(a))$ to the point $(f_1(b), f_2(b), f_3(b))$ is determined by

$$L = \int_a^b \sqrt{[f_1'(t)]^2 + [f_2'(t)]^2 + [f_3'(t)]^2} \, dt \tag{3}$$

If s is the measure of the length of arc of C from the fixed point $(f_1(t_0), f_2(t_0), f_3(t_0))$ to the variable point $(f_1(t), f_2(t), f_3(t))$ and s increases as t increases, then s is a function of t and is given by

$$s = \int_{t_0}^t \sqrt{[f_1'(u)]^2 + [f_2'(u)]^2 + [f_3'(u)]^2} \, du$$

Furthermore, as shown in Section 14.5 for plane curves

$$\frac{ds}{dt} = \|D_t\mathbf{R}(t)\|$$

and the length of arc, L units, given by (3), also can be determined by

$$L = \int_a^b \|D_t\mathbf{R}(t)\| \, dt \tag{4}$$

EXAMPLE 1   Given the circular helix

$$\mathbf{R}(t) = a \cos t\mathbf{i} + a \sin t\mathbf{j} + t\mathbf{k}$$

where $a > 0$, find the length of arc from $t = 0$ to $t = 2\pi$.

Solution

$$D_t\mathbf{R}(t) = -a \sin t\mathbf{i} + a \cos t\mathbf{j} + \mathbf{k}$$

So from (4),

$$L = \int_0^{2\pi} \sqrt{(-a \sin t)^2 + (a \cos t)^2 + 1} \, dt$$

$$= \int_0^{2\pi} \sqrt{a^2 + 1} \, dt$$

$$= 2\pi\sqrt{a^2 + 1}$$

Thus the length of arc is $2\pi\sqrt{a^2 + 1}$ units.

The definitions of the **curvature vector** $\mathbf{K}(t)$ and the **curvature** $K(t)$ at a point P on a curve C in R^3 are the same as for plane curves given in Definition 14.7.1. Hence, if $\mathbf{T}(t)$ is the unit tangent vector to C at P and s is the measure of the arc length from an arbitrarily chosen point on C to P, where s increases as t increases, then

$$\mathbf{K}(t) = D_s \mathbf{T}(t)$$

$$\Leftrightarrow \quad \mathbf{K}(t) = \frac{D_t \mathbf{T}(t)}{\|D_t \mathbf{R}(t)\|} \tag{5}$$

and

$$K(t) = \|D_s \mathbf{T}(t)\|$$

$$\Leftrightarrow \quad K(t) = \left\| \frac{D_t \mathbf{T}(t)}{\|D_t \mathbf{R}(t)\|} \right\|$$

Taking the dot product of $\mathbf{K}(t)$ and $\mathbf{T}(t)$ and using (5) we get

$$\mathbf{K}(t) \cdot \mathbf{T}(t) = \frac{D_t \mathbf{T}(t)}{\|D_t \mathbf{R}(t)\|} \cdot \mathbf{T}(t)$$

$$= \frac{1}{\|D_t \mathbf{R}(t)\|} D_t \mathbf{T}(t) \cdot \mathbf{T}(t) \tag{6}$$

Theorem 14.4.11 states that if a vector-valued function in a plane has a constant magnitude, it is orthogonal to its derivative. This theorem and its proof also hold for vectors in three dimensions. Therefore, because $\|\mathbf{T}(t)\| = 1$, we can conclude from (6) that $\mathbf{K}(t) \cdot \mathbf{T}(t) = 0$. And so the curvature vector and the unit tangent vector of a curve at a point are orthogonal.

The **unit normal vector** is defined as the unit vector having the same direction as the curvature vector, provided that the curvature vector is not the zero vector. So if $\mathbf{N}(t)$ denotes the unit normal vector to a curve C at a point P, then if $\mathbf{K}(t) \neq \mathbf{0}$,

$$\mathbf{N}(t) = \frac{\mathbf{K}(t)}{\|\mathbf{K}(t)\|} \tag{7}$$

From (7) and the previous discussion it follows that the unit normal vector and the unit tangent vector are orthogonal. Thus the angle between these two vectors has a radian measure of $\frac{1}{2}\pi$, and from Theorem 15.5.8,

$$\|\mathbf{T}(t) \times \mathbf{N}(t)\| = \|\mathbf{T}(t)\| \|\mathbf{N}(t)\| \sin \tfrac{1}{2}\pi$$

$$= 1$$

Therefore the cross product of $\mathbf{T}(t)$ and $\mathbf{N}(t)$ is a unit vector. By Theorem 15.5.10, $\mathbf{T}(t) \times \mathbf{N}(t)$ is orthogonal to both $\mathbf{T}(t)$ and $\mathbf{N}(t)$; hence the vector $\mathbf{B}(t)$, defined by

$$\mathbf{B}(t) = \mathbf{T}(t) \times \mathbf{N}(t) \tag{8}$$

is a unit vector orthogonal to $\mathbf{T}(t)$ and $\mathbf{N}(t)$ and is called the **unit binormal vector** to the curve C at P.

The three mutually orthogonal unit vectors, $\mathbf{T}(t)$, $\mathbf{N}(t)$, and $\mathbf{B}(t)$, of a curve C are called the **moving trihedral** of C (see Figure 5).

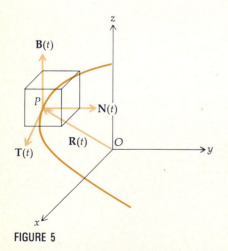

FIGURE 5

EXAMPLE 2   Find the moving trihedral and the curvature at any point of the circular helix of Example 1.

Solution   A vector equation of the circular helix is

$$\mathbf{R}(t) = a \cos t\mathbf{i} + a \sin t\mathbf{j} + t\mathbf{k}$$

So $D_t\mathbf{R}(t) = -a \sin t\mathbf{i} + a \cos t\mathbf{j} + \mathbf{k}$ and $\|D_t\mathbf{R}(t)\| = \sqrt{a^2 + 1}$. From (2),

$$\mathbf{T}(t) = \frac{1}{\sqrt{a^2 + 1}} (-a \sin t\mathbf{i} + a \cos t\mathbf{j} + \mathbf{k})$$

Thus

$$D_t\mathbf{T}(t) = \frac{1}{\sqrt{a^2 + 1}} (-a \cos t\mathbf{i} - a \sin t\mathbf{j})$$

Applying (5) we obtain

$$\mathbf{K}(t) = \frac{1}{a^2 + 1} (-a \cos t\mathbf{i} - a \sin t\mathbf{j})$$

Because the curvature $K(t) = \|\mathbf{K}(t)\|$, we have

$$K(t) = \frac{a}{a^2 + 1}$$

Thus the curvature of the circular helix is constant. From (7),

$$\mathbf{N}(t) = -\cos t\mathbf{i} - \sin t\mathbf{j}$$

Applying (8) we have

$$\mathbf{B}(t) = \frac{1}{\sqrt{a^2 + 1}} (-a \sin t\mathbf{i} + a \cos t\mathbf{j} + \mathbf{k}) \times (-\cos t\mathbf{i} - \sin t\mathbf{j})$$

$$= \frac{1}{\sqrt{a^2 + 1}} (\sin t\mathbf{i} - \cos t\mathbf{j} + a\mathbf{k})$$

A thorough treatment of curves and surfaces by means of calculus is presented in a course in differential geometry. The use of calculus of vectors further enhances this subject. The previous discussion has been but a short introduction.

We now consider briefly the motion of a particle along a curve in three-dimensional space. If the parameter t in the vector equation

$$\mathbf{R}(t) = f_1(t)\mathbf{i} + f_2(t)\mathbf{j} + f_3(t)\mathbf{k} \tag{9}$$

measures time, then the position at t of a particle moving along the curve having vector equation (9) is the point $P(f_1(t), f_2(t), f_3(t))$. The **velocity vector**, $\mathbf{V}(t)$, and the **acceleration vector**, $\mathbf{A}(t)$, are defined as in the plane. The vector $\mathbf{R}(t)$ is called the **position vector**, and

$$\mathbf{V}(t) = D_t\mathbf{R}(t) \qquad \mathbf{A}(t) = D_t\mathbf{V}(t)$$

The speed of the particle at t is the magnitude of the velocity vector. Because $\|D_t\mathbf{R}(t)\| = ds/dt$

$$\|\mathbf{V}(t)\| = \frac{ds}{dt}$$

EXAMPLE 3   A particle is moving along the curve having parametric equations

$$x = 3t \qquad y = t^2 \qquad z = \tfrac{2}{3}t^3$$

Find the velocity and acceleration vectors and the speed of the particle at $t = 1$. Draw a sketch of a portion of the curve at $t = 1$, and draw representations of the velocity and acceleration vectors there.

Solution   A vector equation of the curve is

$$\mathbf{R}(t) = 3t\mathbf{i} + t^2\mathbf{j} + \tfrac{2}{3}t^3\mathbf{k}$$

Therefore

$$\mathbf{V}(t) = D_t\mathbf{R}(t) \qquad\qquad \mathbf{A}(t) = D_t\mathbf{V}(t)$$
$$= 3\mathbf{i} + 2t\mathbf{j} + 2t^2\mathbf{k} \qquad = 2\mathbf{j} + 4t\mathbf{k}$$
$$\|\mathbf{V}(t)\| = \sqrt{9 + 4t^2 + 4t^4}$$

Thus

$$\mathbf{V}(1) = 3\mathbf{i} + 2\mathbf{j} + 2\mathbf{k} \qquad \mathbf{A}(1) = 2\mathbf{j} + 4\mathbf{k} \qquad \|\mathbf{V}(1)\| = \sqrt{17}$$

The required sketch is shown in Figure 6.

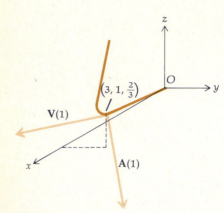

FIGURE 6

EXERCISES 15.8

In Exercises 1 through 5, find the unit tangent vector for the curve having the vector equation.

1. $\mathbf{R}(t) = (t + 1)\mathbf{i} - t^2\mathbf{j} + (1 - 2t)\mathbf{k}$
2. $\mathbf{R}(t) = \sin 2t\mathbf{i} + \cos 2t\mathbf{j} + 2t^{3/2}\mathbf{k}$
3. $\mathbf{R}(t) = e^t \cos t\mathbf{i} + e^t \sin t\mathbf{j} + e^t\mathbf{k}$
4. $\mathbf{R}(t) = t^2\mathbf{i} + (t + \tfrac{1}{3}t^3)\mathbf{j} + (t - \tfrac{1}{3}t^3)\mathbf{k}$
5. $\mathbf{R}(t) = 2t \cos t\mathbf{i} + 5t\mathbf{j} + 2t \sin t\mathbf{k}$

6. Find the unit tangent vector for the curve having the vector equation $\mathbf{R}(t) = 4 \cosh 2t\mathbf{i} + 4 \sinh 2t\mathbf{j} + 6t\mathbf{k}$ at the point where $t = 0$.

In Exercises 7 through 11, find the length of arc of the curve from t_1 to t_2.

7. The curve of Exercise 1; $t_1 = -1$; $t_2 = 2$.
8. The curve of Exercise 2; $t_1 = 0$; $t_2 = 1$.
9. The curve of Exercise 3; $t_1 = 0$; $t_2 = 3$.
10. The curve of Exercise 4; $t_1 = 0$; $t_2 = 1$.
11. $\mathbf{R}(t) = 4t^{3/2}\mathbf{i} - 3 \sin t\mathbf{j} + 3 \cos t\mathbf{k}$; $t_1 = 0$; $t_2 = 2$.

12. Prove that the unit tangent vector of the circular helix of Example 1 makes an angle of constant radian measure with the unit vector $\mathbf{k}$.
13. Prove Theorem 15.8.2.   14. Prove Theorem 15.8.3.
15. Write a vector equation of the curve of intersection of the surfaces $y = e^x$ and $z = xy$.

16. Write a vector equation of the curve of intersection of the surfaces $x = \ln(z^2 + 2)$ and $y = xz^3$.
17. Find at the point where $t = \pi$ the cosine of the angle between the vector $\mathbf{j}$ and the unit tangent vector to the curve $\mathbf{R}(t) = \cos 2t\mathbf{i} - 3t\mathbf{j} + 2 \sin 2t\mathbf{k}$.
18. Find the curvature at the point where $t = 1$ of the curve $\mathbf{R}(t) = t^2\mathbf{i} + (4 + t)\mathbf{j} + (3 - 2t)\mathbf{k}$.
19. Find the moving trihedral and the curvature at the point where $t = 1$ of the twisted cubic of Illustration 2.
20. Find the moving trihedral and the curvature at any point of the curve $\mathbf{R}(t) = \cosh t\mathbf{i} + \sinh t\mathbf{j} + t\mathbf{k}$.

In Exercises 21 through 24, find the moving trihedral and the curvature of the curve at $t = t_1$, if they exist.

21. The curve of Exercise 1; $t_1 = -1$.
22. The curve of Exercise 2; $t_1 = 0$.
23. The curve of Exercise 3; $t_1 = 0$.
24. The curve of Exercise 4; $t_1 = 1$.

In Exercises 25 through 28, a particle is moving along the curve. Find the velocity vector, the acceleration vector, and the speed at $t = t_1$. Draw a sketch of a portion of the curve at $t = t_1$, and draw the velocity and acceleration vectors there.

25. The circular helix of Example 1; $t_1 = \tfrac{1}{2}\pi$.
26. $x = t$, $y = \tfrac{1}{2}t^2$, $z = \tfrac{1}{3}t^3$; $t_1 = 2$.
27. $x = e^{2t}$, $y = e^{-2t}$, $z = te^{2t}$; $t_1 = 1$.
28. $x = \tfrac{1}{2}(t^2 + 1)^{-1}$, $y = \ln(1 + t^2)$, $z = \tan^{-1} t$; $t_1 = 1$.

29. Prove that if $\mathbf{R}(t) = f_1(t)\mathbf{i} + f_2(t)\mathbf{j} + f_3(t)\mathbf{k}$ is a vector equation of curve C, and $K(t)$ is the curvature of C, then

$$K(t) = \frac{\|D_t\mathbf{R}(t) \times D_t^2\mathbf{R}(t)\|}{\|D_t\mathbf{R}(t)\|^3}$$

30. Use the formula of Exercise 29 to show that the curvature of the circular helix of Example 1 is $a/(a^2 + 1)$.

In Exercises 31 and 32, find the curvature of the curve at the indicated point.

31. $x = t$, $y = t^2$, $z = t^3$; the origin.
32. $x = e^t$, $y = e^{-t}$, $z = t$; $t = 0$.

33. Prove that if $\mathbf{R}(t) = f_1(t)\mathbf{i} + f_2(t)\mathbf{j} + f_3(t)\mathbf{k}$ is a vector equation of curve C, $K(t)$ is the curvature of C at a point P, and s units is the arc length measured from an arbitrarily chosen point on C to P, then $D_s\mathbf{R}(t) \cdot D_s^3\mathbf{R}(t) = -[K(t)]^2$.
34. A particle is moving along a curve with vector equation $\mathbf{R}(t) = \tan t\mathbf{i} + \sinh 2t\mathbf{j} + \operatorname{sech} t\mathbf{k}$. Prove that the velocity vector and the acceleration vector are orthogonal at $t = 0$.
35. Prove that if the speed of a moving particle is constant, its acceleration vector is always orthogonal to its velocity vector.
36. Prove that for the twisted cubic of Illustration 2, if $t \neq 0$, no two of the vectors $\mathbf{R}(t)$, $\mathbf{V}(t)$, and $\mathbf{A}(t)$ are orthogonal.

15.9 CYLINDRICAL AND SPHERICAL COORDINATES

The **cylindrical coordinate** representation of a point P is (r, θ, z), where r and θ are the polar coordinates of the projection of P on a polar plane and z is the directed distance from this polar plane to P. See Figure 1.

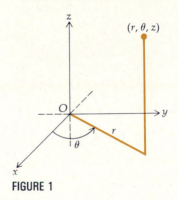

FIGURE 1

EXAMPLE 1 Draw a sketch of the graph of each of the following equations where c is a constant: (a) $r = c$; (b) $\theta = c$; (c) $z = c$.

Solution

(a) For a point $P(r, \theta, z)$ on the graph of $r = c$, θ and z can have any values and r is a constant. The graph is a right-circular cylinder having radius $|c|$ and the z axis as its axis. A sketch of the graph appears in Figure 2.
(b) For all points $P(r, \theta, z)$ on the graph of $\theta = c$, r and z can assume any value while θ remains constant. The graph is a plane through the z axis. See Figure 3 for a sketch of the graph.
(c) The graph of $z = c$ is a plane parallel to the polar plane at a directed distance of c units from it. Figure 4 shows a sketch of the graph.

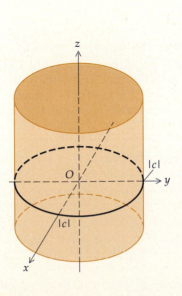

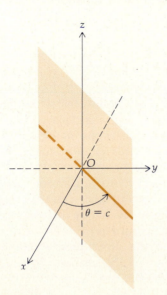

FIGURE 3

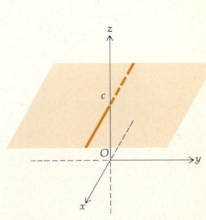

FIGURE 4

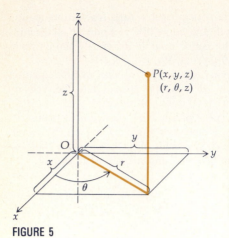

FIGURE 5

The name "cylindrical coordinates" comes from the fact that the graph of $r = c$ is a right-circular cylinder as in Example 1(a). Cylindrical coordinates are often used in a physical problem when there is an axis of symmetry.

Suppose that a cartesian coordinate system and a cylindrical coordinate system are placed so the xy plane is the polar plane of the cylindrical coordinate system and the positive side of the x axis is the polar axis as shown in Figure 5. Then the point P has (x, y, z) and (r, θ, z) as two sets of coordinates that are related by the equations

$$x = r \cos \theta \qquad y = r \sin \theta \qquad z = z \qquad (1)$$

$$r^2 = x^2 + y^2 \qquad \tan \theta = \frac{y}{x} \text{ if } x \neq 0 \qquad z = z \qquad (2)$$

EXAMPLE 2   Find an equation in cartesian coordinates of the following surfaces whose equations are expressed in cylindrical coordinates, and identify the surface: (a) $r = 6 \sin \theta$; (b) $r(3 \cos \theta + 2 \sin \theta) + 6z = 0$.

Solution
(a) Multiplying on both sides of the equation by r we get $r^2 = 6r \sin \theta$. Because $r^2 = x^2 + y^2$ and $r \sin \theta = y$, then $x^2 + y^2 = 6y$. This equation can be written in the form $x^2 + (y - 3)^2 = 9$, which shows that its graph is a right-circular cylinder whose cross section in the xy plane is the circle with its center at $(0, 3)$ and radius 3.
(b) Replacing $r \cos \theta$ with x and $r \sin \theta$ with y we obtain the equation $3x + 2y + 6z = 0$. Hence the graph is a plane through the origin and has $\langle 3, 2, 6 \rangle$ as a normal vector.

EXAMPLE 3   Find an equation in cylindrical coordinates for each of the following surfaces whose equations are given in cartesian coordinates, and identify the surface: (a) $x^2 + y^2 = z$; (b) $x^2 - y^2 = z$.

Solution
(a) The equation is similar to Equation (5) of Section 15.7; so the graph is an elliptic paraboloid. If $x^2 + y^2$ is replaced by r^2, the equation becomes $r^2 = z$.
(b) The equation is similar to Equation (6) of Section 15.7 with x and y interchanged. The graph is therefore a hyperbolic paraboloid having the z axis as its axis. When x is replaced by $r \cos \theta$ and y is replaced by $r \sin \theta$, we get the equation $r^2 \cos^2 \theta - r^2 \sin^2 \theta = z$; because $\cos^2 \theta - \sin^2 \theta = \cos 2\theta$, we can write this as $z = r^2 \cos 2\theta$.

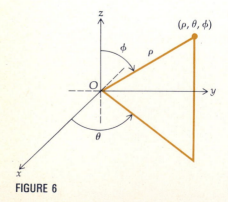

FIGURE 6

In a spherical coordinate system there is a polar plane and an axis perpendicular to the polar plane, with the origin of the z axis at the pole of the polar plane. A point is located by three numbers, and the **spherical coordinate** representation of a point P is (ρ, θ, ϕ), where $\rho = |\overline{OP}|$, θ is the radian measure of the polar angle of the projection of P on the polar plane, and ϕ is the nonnegative radian measure of the smallest angle measured from the positive side of the z axis to the line OP. See Figure 6. The origin has the spherical coordinate representation $(0, \theta, \phi)$, where θ and ϕ may have any values. If the point $P(\rho, \theta, \phi)$ is not the origin, then $\rho > 0$ and $0 \leq \phi \leq \pi$, where $\phi = 0$ if P is on the positive side of the z axis and $\phi = \pi$ if P is on the negative side of the z axis.

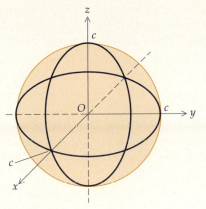

FIGURE 7

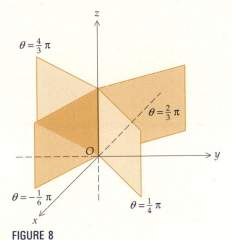

FIGURE 8

EXAMPLE 4 Draw a sketch of the graph of each of the following equations where c is a constant: (a) $\rho = c$, and $c > 0$; (b) $\theta = c$; (c) $\phi = c$, and $0 < c < \pi$.

Solution

(a) Every point $P(\rho, \theta, \phi)$ on the graph of $\rho = c$ has the same value of ρ, θ may be any number, and $0 \leq \phi \leq \pi$. It follows that the graph is a sphere of radius c and has its center at the pole. Figure 7 shows a sketch of the sphere.

(b) For any point $P(\rho, \theta, \phi)$ on the graph of $\theta = c$, ρ may be any nonnegative number, ϕ may be any number in the closed interval $[0, \pi]$, and θ is constant. The graph is a half plane containing the z axis and is obtained by rotating about the z axis through an angle of c radians that half of the xz plane for which $x \geq 0$. Figure 8 shows sketches of the half planes for $\theta = \frac{1}{4}\pi$, $\theta = \frac{2}{3}\pi$, $\theta = \frac{4}{3}\pi$, and $\theta = -\frac{1}{6}\pi$.

(c) The graph of $\phi = c$ contains all the points $P(\rho, \theta, \phi)$ for which ρ is any nonnegative number, θ is any number, and ϕ is the constant c. The graph is half of a cone having its vertex at the origin and the z axis as its axis. Figures 9(a) and (b) each show a sketch of the half cone for $0 < c < \frac{1}{2}\pi$ and $\frac{1}{2}\pi < c < \pi$, respectively.

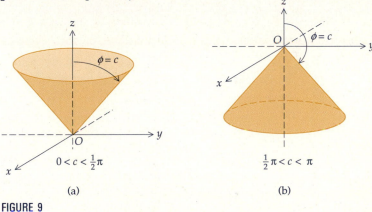

(a)   (b)

FIGURE 9

Because the graph of $\rho = c$ is a sphere as seen in Example 4(a), we have the name "spherical coordinates." In a physical problem when there is a point that is a center of symmetry, spherical coordinates are often used.

By placing a spherical coordinate system and a cartesian coordinate system together as shown in Figure 10, we obtain relationships between the spherical coordinates and the cartesian coordinates of a point P from

$$x = |\overline{OQ}| \cos \theta \qquad y = |\overline{OQ}| \sin \theta \qquad z = |\overline{QP}|$$

Because $|\overline{OQ}| = \rho \sin \phi$ and $\overline{QP} = \rho \cos \phi$, these equations become

$$x = \rho \sin \phi \cos \theta \qquad y = \rho \sin \phi \sin \theta \qquad z = \rho \cos \phi \qquad (3)$$

By squaring each of the equations in (3) and adding,

$$x^2 + y^2 + z^2 = \rho^2 \sin^2 \phi \cos^2 \theta + \rho^2 \sin^2 \phi \sin^2 \theta + \rho^2 \cos^2 \phi$$

$$x^2 + y^2 + z^2 = \rho^2 \sin^2 \phi (\cos^2 \theta + \sin^2 \theta) + \rho^2 \cos^2 \phi$$

$$x^2 + y^2 + z^2 = \rho^2 (\sin^2 \phi + \cos^2 \phi)$$

$$x^2 + y^2 + z^2 = \rho^2$$

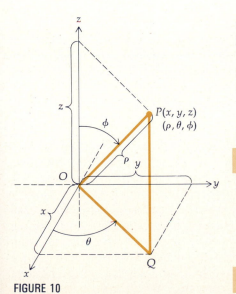

FIGURE 10

EXAMPLE 5   Find an equation in cartesian coordinates of the following surfaces whose equations are expressed in spherical coordinates, and identify the surface: (a) $\rho \cos \phi = 4$; (b) $\rho \sin \phi = 4$.

Solution

(a) Because $z = \rho \cos \phi$, the equation becomes $z = 4$. Hence the graph is a plane parallel to the xy plane and 4 units above it.

(b) For spherical coordinates $\rho \geq 0$ and $\sin \phi \geq 0$ (because $0 \leq \phi \leq \pi$); therefore, by squaring on both sides of the given equation we obtain the equivalent equation $\rho^2 \sin^2 \phi = 16$, which in turn is equivalent to

$$\rho^2(1 - \cos^2 \phi) = 16$$

$$\rho^2 - \rho^2 \cos^2 \phi = 16$$

Replacing ρ^2 by $x^2 + y^2 + z^2$ and $\rho \cos \phi$ by z we get

$$x^2 + y^2 + z^2 - z^2 = 16$$

$$x^2 + y^2 = 16$$

Therefore the graph is the right-circular cylinder having the z axis as its axis and radius 4.

EXAMPLE 6   Find an equation in spherical coordinates for (a) the elliptic paraboloid of Example 3(a); (b) the plane of Example 2(b).

Solution

(a) A cartesian equation of the elliptic paraboloid of Example 3(a) is $x^2 + y^2 = z$. Replacing x by $\rho \sin \phi \cos \theta$, y by $\rho \sin \phi \sin \theta$, and z by $\rho \cos \phi$ we get

$$\rho^2 \sin^2 \phi \cos^2 \theta + \rho^2 \sin^2 \phi \sin^2 \theta = \rho \cos \phi$$

$$\rho^2 \sin^2 \phi(\cos^2 \theta + \sin^2 \theta) = \rho \cos \phi$$

which is equivalent to the two equations

$$\rho = 0 \quad \text{and} \quad \rho \sin^2 \phi = \cos \phi$$

The origin is the only point whose coordinates satisfy $\rho = 0$. Because the origin $(0, \theta, \frac{1}{2}\pi)$ lies on $\rho \sin^2 \phi = \cos \phi$, we can disregard the equation $\rho = 0$. Furthermore, $\sin \phi \neq 0$ because there is no value of ϕ for which both $\sin \phi$ and $\cos \phi$ are 0. Therefore the equation $\rho \sin^2 \phi = \cos \phi$ can be written as $\rho = \csc^2 \phi \cos \phi$, or, equivalently, $\rho = \csc \phi \cot \phi$.

(b) A cartesian equation of the plane of Example 2(b) is $3x + 2y + 6z = 0$. By using Equations (3) this equation becomes

$$3\rho \sin \phi \cos \theta + 2\rho \sin \phi \sin \theta + 6\rho \cos \phi = 0$$

EXERCISES 15.9

1. Find the cartesian coordinates of the point having the given cylindrical coordinates: (a) $(3, \frac{1}{2}\pi, 5$; (b) $(7, \frac{2}{3}\pi, -4)$; (c) $(1, 1, 1)$.

2. Find a set of cylindrical coordinates of the point having the given cartesian coordinates: (a) $(4, 4, -2)$; (b) $(-3\sqrt{3}, 3, 6)$; (c) $(1, 1, 1)$.

3. Find the cartesian coordinates of the point having the given spherical coordinates: (a) $(4, \frac{1}{6}\pi. \frac{1}{4}\pi)$; (b) $(4, \frac{1}{2}\pi, \frac{1}{3}\pi)$; (c) $(\sqrt{6}, \frac{1}{3}\pi, \frac{3}{4}\pi)$.

4. Find a set of spherical coordinates of the point having the given cartesian coordinates: (a) $(1, -1, -\sqrt{2})$; (b) $(-1, \sqrt{3}, 2)$; (c) $(2, 2, 2)$.

SIXTEEN

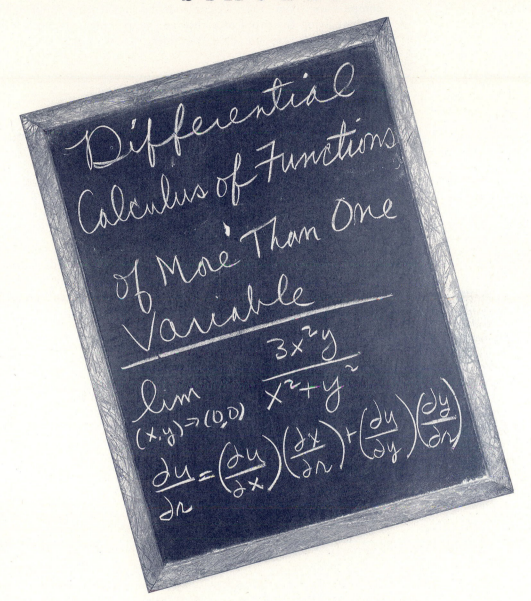

Differential Calculus of Functions of More Than One Variable

$$\lim_{(x,y)\to(0,0)} \frac{3x^2 y}{x^2+y^2}$$

$$\frac{dy}{dx} = \left(\frac{dy}{dx}\right)\left(\frac{dx}{dx}\right) + \left(\frac{dy}{dy}\right)\left(\frac{dy}{dx}\right)$$

In Section 16.1 we extend the concept of a function to a *function of n variables*, and in the next two sections we extend to functions of *n* variables the concepts of *limit* and *continuity*. Most of our discussion is confined to functions of two and three variables; however, we make the definitions for functions of *n* variables and then show the applications of these definitions to functions of two and three variables. We also show that when each of these definitions is applied to a function of one variable, we have the definition previously given.

Our treatment of differentiation of functions of several variables begins in Section 16.4, where we define the *partial derivatives* of such a function. Then in Section 16.5 we discuss *differentiability* of these functions as well as the *total differential.* The several-variable version of the *chain rule* is presented in Section 16.6, and *higher-order partial derivatives* are treated in Section 16.7. The applications of differentiation in this chapter are to find rates of change and to compute approximations. We conclude the chapter with Supplementary Section 16.8 involving the proof of a theorem giving sufficient conditions for differentiability of a function of two variables.

16.1 FUNCTIONS OF MORE THAN ONE VARIABLE

We now generalize the notion of a function to functions of more than one independent variable. Such functions often occur in practical situations. For example, a person's approximate body surface area depends on the person's weight and height. The volume of a right-circular cylinder depends on its radius and height. According to the ideal gas law, the volume occupied by a confined gas is directly proportional to its temperature and inversely proportional to its pressure. The cost of a particular product may be dependent upon the cost of labor, the price of materials, and overhead expenses.

To extend the concept of a function to functions of any number of variables we must first consider points in n-dimensional number space. Just as we denoted a point in R by a real number x, a point in R^2 by an ordered pair of real numbers (x, y), and a point in R^3 by an ordered triple of real numbers (x, y, z), a point in n-dimensional number space, R^n, is represented by an ordered n-tuple of real numbers customarily denoted by $P = (x_1, x_2, \ldots, x_n)$. In particular, if $n = 1$, let $P = x$; if $n = 2, P = (x, y)$; if $n = 3, P = (x, y, z)$; if $n = 6$, $P = (x_1, x_2, x_3, x_4, x_5, x_6)$.

16.1.1 DEFINITION

The set of all ordered n-tuples of real numbers is called the **n-dimensional number space** and is denoted by R^n. Each ordered n-tuple $(x_1, x_2, \ldots, x_n)$ is called a **point** in the n-dimensional number space.

16.1.2 DEFINITION

A **function of n variables** is a set of ordered pairs of the form (P, w) in which no two distinct ordered pairs have the same first element. P is a point in n-dimensional number space and w is a real number. The set of all admissible values of P is called the **domain** of the function, and the set of all resulting values of w is called the **range** of the function.

From this definition, the domain of a function of n variables is a set of points in R^n and the range is a set of real numbers or, equivalently, a set of points in R. When $n = 1$, we have a function of one variable; thus the domain is a set of points in R or, equivalently, a set of real numbers. Hence Definition 1.4.1 is a special case of Definition 16.1.2. If $n = 2$, we have a function of two variables, and the domain is a set of points in R^2 or, equivalently, a set of ordered pairs of real numbers (x, y).

▶ **ILLUSTRATION 1** Let the function f of two variables x and y be the set of all ordered pairs of the form (P, z) such that

$$z = \sqrt{25 - x^2 - y^2}$$

The domain of f is the set $\{(x, y) \mid x^2 + y^2 \le 25\}$. This is the set of points in the xy plane on the circle $x^2 + y^2 = 25$ and in the interior region bounded by the

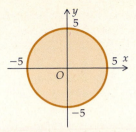

FIGURE 1

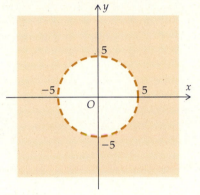

FIGURE 2

circle. In Figure 1 there is a sketch showing as a shaded region in R^2 the set of points in the domain of f.

Because $z = \sqrt{25 - (x^2 + y^2)}$, then $0 \leq z \leq 5$; therefore the range of f is the set of all real numbers in the closed interval $[0, 5]$. ◀

▶ **ILLUSTRATION 2** The function g of two variables x and y is the set of all ordered pairs of the form (P, z) such that

$$z = \frac{1}{\sqrt{x^2 + y^2 - 25}}$$

The domain of g is the set $\{(x, y) | x^2 + y^2 > 25\}$. This is the set of points in the exterior region bounded by the circle $x^2 + y^2 = 25$. In Figure 2 there is a sketch showing as a shaded region in R^2 the set of points in the domain of g. ◀

If f is a function of n variables, then according to Definition 16.1.2, f is a set of ordered pairs of the form (P, w), where $P = (x_1, x_2, \ldots, x_n)$ is a point in R^n and w is a real number. The particular value of w that corresponds to a point P is denoted by the symbol $f(P)$ or $f(x_1, x_2, \ldots, x_n)$. In particular, if $n = 2$ and $P = (x, y)$, we can represent the function value by either $f(P)$ or $f(x, y)$. Similarly, if $n = 3$ and $P = (x, y, z)$, we denote the function value by either $f(P)$ or $f(x, y, z)$. Note that if $n = 1$, $P = x$; hence if f is a function of one variable, $f(P) = f(x)$. Therefore this notation is consistent with the notation for function values of one variable.

A function f of n variables can be defined by the equation

$$w = f(x_1, x_2, \ldots, x_n)$$

The variables $x_1, x_2, \ldots, x_n$ are called the *independent variables*, and w is called the *dependent variable*.

▶ **ILLUSTRATION 3** Let f be the function of Illustration 1; that is,

$$f(x, y) = \sqrt{25 - x^2 - y^2}$$

Then

$$f(3, -4) = \sqrt{25 - 3^2 - (-4)^2} \qquad f(-2, 1) = \sqrt{25 - (-2)^2 - 1^2}$$
$$= \sqrt{25 - 9 - 16} \qquad\qquad\quad = \sqrt{25 - 4 - 1}$$
$$= 0 \qquad\qquad\qquad\qquad\quad = 2\sqrt{5}$$

$$f(u, 3v) = \sqrt{25 - u^2 - (3v)^2}$$
$$= \sqrt{25 - u^2 - 9v^2}$$ ◀

EXAMPLE 1   The function g is defined by

$$g(x, y, z) = x^3 - 4yz^2$$

Find: (a) $g(1, 3, -2)$; (b) $g(2a, -4b, 3c)$; (c) $g(x^2, y^2, z^2)$; (d) $g(y, z, -x)$.

Solution
(a) $g(1, 3, -2) = 1^3 - 4(3)(-2)^2$ (b) $g(2a, -4b, 3c) = (2a)^3 - 4(-4b)(3c)^2$
$$= 1 - 48 \qquad\qquad\qquad\qquad\qquad = 8a^3 + 144bc^2$$
$$= -47$$

(c) $g(x^2, y^2, z^2) = (x^2)^3 - 4y^2(z^2)^2$ (d) $g(y, z, -x) = y^3 - 4z(-x)^2$
    $= x^6 - 4y^2z^4$ $= y^3 - 4x^2z$

16.1.3 DEFINITION

If f is a function of a single variable and g is a function of two variables, then the **composite function** $f \circ g$ is the function of two variables defined by

$$(f \circ g)(x, y) = f(g(x, y))$$

and the domain of $f \circ g$ is the set of all points (x, y) in the domain of g such that $g(x, y)$ is in the domain of f.

EXAMPLE 2   Given $f(t) = \ln t$ and $g(x, y) = x^2 + y$, find $h(x, y)$ if $h = f \circ g$, and determine the domain of h.

Solution

$$\begin{aligned} h(x, y) &= (f \circ g)(x, y) \\ &= f(g(x, y)) \\ &= f(x^2 + y) \\ &= \ln(x^2 + y) \end{aligned}$$

The domain of g is the set of all points in R^2, and the domain of f is $(0, +\infty)$. Therefore the domain of h is the set $\{(x, y) | x^2 + y > 0\}$.

Definition 16.1.3 can be extended to a composite function of n variables as follows.

16.1.4 DEFINITION

If f is a function of a single variable and g is a function of n variables, then the **composite function** $f \circ g$ is the function of n variables defined by

$$(f \circ g)(x_1, x_2, \ldots, x_n) = f(g(x_1, x_2, \ldots, x_n))$$

and the domain of $f \circ g$ is the set of all points $(x_1, x_2, \ldots, x_n)$ in the domain of g such that $g(x_1, x_2, \ldots, x_n)$ is in the domain of f.

EXAMPLE 3   Given $F(x) = \sin^{-1} x$ and $G(x, y, z) = \sqrt{x^2 + y^2 + z^2 - 4}$ find the function $F \circ G$ and its domain.

Solution

$$\begin{aligned} (F \circ G)(x, y, z) &= F(G(x, y, z)) \\ &= F(\sqrt{x^2 + y^2 + z^2 - 4}) \\ &= \sin^{-1}\sqrt{x^2 + y^2 + z^2 - 4} \end{aligned}$$

The domain of G is the set $\{(x, y, z) | x^2 + y^2 + z^2 - 4 \geq 0\}$, and the domain of F is $[-1, 1]$. So the domain of $F \circ G$ is the set of all points (x, y, z) in R^3 such that $0 \leq x^2 + y^2 + z^2 - 4 \leq 1$ or, equivalently $4 \leq x^2 + y^2 + z^2 \leq 5$.

A **polynomial function** of two variables x and y is a function f such that $f(x, y)$ is the sum of terms of the form $cx^n y^m$, where c is a real number and n and m are nonnegative integers. The **degree** of the polynomial function is determined by the largest sum of the exponents of x and y appearing in any one term.

▶ **ILLUSTRATION 4**   (a) The function f defined by

$$f(x, y) = x^3 + 2x^2 y^2 - y^3$$

is a polynomial function of degree 4 because the term of highest degree is $2x^2 y^2$.
(b) If

$$g(x, y) = 6x^3 y^2 - 5xy^3 + 7x^2 y - 2x^2 + y + 4$$

g is a polynomial function of degree 5.   ◀

The graph of a function f of a single variable consists of the set of points (x, y) in R^2 for which $y = f(x)$. Similarly, the graph of a function of two variables is a set of points in R^3.

16.1.5 DEFINITION   If f is a function of two variables, then the **graph** of f is the set of all points (x, y, z) in R^3 for which (x, y) is a point in the domain of f and $z = f(x, y)$.

Hence the graph of a **function** f of two variables is a surface that is the set of all points in three-dimensional space whose cartesian coordinates are given by the ordered triples of real numbers (x, y, z). Because the domain of f is a set of points in the xy plane, and because for each ordered pair (x, y) in the domain of f there corresponds a unique value of z, no line perpendicular to the xy plane can intersect the graph of f in more than one point.

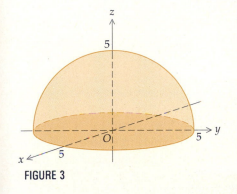

FIGURE 3

▶ **ILLUSTRATION 5**   The function of Illustration 1 is the function f that is the set of all ordered pairs of the form (P, z) such that

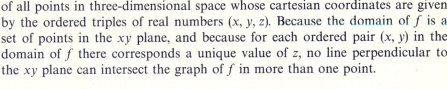
$$z = \sqrt{25 - x^2 - y^2}$$

So the graph of f is the hemisphere on and above the xy plane having a radius of 5 and its center at the origin. A sketch of the graph of this hemisphere appears in Figure 3.   ◀

EXAMPLE 4   Draw a sketch of the graph of the function f having function values $f(x, y) = x^2 + y^2$.

Solution   The graph of f is the surface having the equation $z = x^2 + y^2$. The trace of the surface in the xy plane is found by using the equation $z = 0$ simultaneously with the equation of the surface. We obtain $x^2 + y^2 = 0$, which is the origin. The traces in the xz and yz planes are found by using the equations $y = 0$ and $x = 0$, respectively, with the equation $z = x^2 + y^2$. These traces are the parabolas $z = x^2$ and $z = y^2$. The cross section of the surface in a plane $z = k$, parallel to the xy plane, is a circle with its center on the z axis and radius $\sqrt{k}$. With this information we have the required sketch shown in Figure 4.

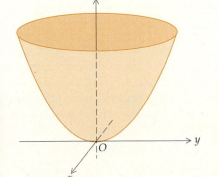

FIGURE 4

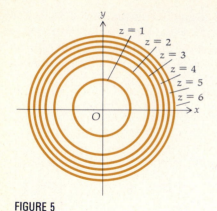

FIGURE 5

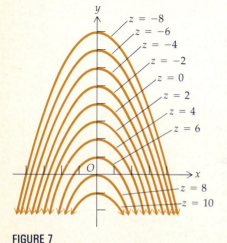

FIGURE 6

Another useful method of representing a function of two variables geometrically is similar to that of representing a three-dimensional landscape by a two-dimensional topographical map. Suppose that the surface $z = f(x, y)$ is intersected by the plane $z = k$, and the curve of intersection is projected onto the xy plane. This projected curve has $f(x, y) = k$ as an equation, and the curve is called the **level curve** (or **contour curve**) of the function f at k. Each point on the level curve corresponds to the unique point on the surface that is k units above it if k is positive, or k units below it if k is negative. By considering different values for the constant k we obtain a set of level curves called a **contour map**. The set of all possible values of k is the range of the function f, and each level curve, $f(x, y) = k$, in the contour map consists of the points (x, y) in the domain of f having equal function values of k. For example, for the function f of Example 4 the level curves are circles with the center at the origin. The particular level curves for z equals 1, 2, 3, 4, 5, and 6 are shown in Figure 5.

A contour map shows the variation of z with x and y. The level curves are usually shown for values of z at constant intervals, and the values of z are changing more rapidly when the level curves are close together than when they are far apart; that is, when the level curves are close together the surface is steep, and when the level curves are far apart the elevation of the surface is changing slowly. On a two-dimensional topographical map of a landscape, a general notion of its steepness is obtained by considering the spacing of its level curves. Also on a topographical map if the path of a level curve is followed, the elevation remains constant.

EXAMPLE 5   Let f be the function for which $f(x, y) = 8 - x^2 - 2y$. Draw a sketch of the graph of f and a contour map of f showing the level curves of f at 10, 8, 6, 4, 2, 0, −2, −4, −6, and −8.

Solution   A sketch of the graph of f appears in Figure 6. This is the surface $z = 8 - x^2 - 2y$. The trace in the xy plane is obtained by setting $z = 0$, which gives the parabola $x^2 = -2(y - 4)$. Setting $y = 0$ and $x = 0$, we obtain the traces in the xz and yz planes, which are, respectively, the parabola $x^2 = -(z - 8)$ and the line $2y + z = 8$. The cross section of the surface made by the plane $z = k$ is a parabola having its vertex on the line $2y + z = 8$ in the yz plane and opening to the left. The cross sections for z equals 8, 6, 4, 2, −2, −4, −6, and −8 are shown in the figure.

The level curves of f are the parabolas $x^2 = -2(y - 4 + \frac{1}{2}k)$. The contour map of f with sketches of the required level curves is shown in Figure 7.

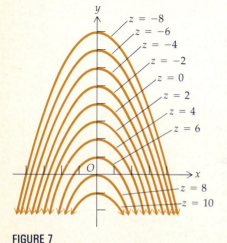

FIGURE 7

To illustrate a use of level curves, suppose that the temperature at any point of a flat metal plate is given by the function f; that is, if t degrees is the temperature, then at the point (x, y), $t = f(x, y)$. Then the curves having equations of the form $f(x, y) = k$, where k is a constant, are curves on which the temperature is constant. These are the level curves of f and are called **isothermals**. Furthermore, if V volts gives the electric potential at any point (x, y) of the xy plane, and $V = f(x, y)$, then the level curves of f are called **equipotential curves** because the electric potential at each point of such a curve is the same.

For an application of level curves in economics, consider the productivity (or output) of an industry that is dependent on several inputs. Among the inputs may be the number of machines used in production, the number of person-hours available, the amount of working capital to be had, the quantity of material used, and the amount of land available. Suppose the amounts of the inputs are given by x and y, the amount of the output is given by z, and $z = f(x, y)$. Such a function is called a **production function**, and the level curves of f, having equations of the form $f(x, y) = k$, where k is a constant, are called **constant product curves**.

EXAMPLE 6   Let f be the production function for which

$$f(x, y) = 2x^{1/2}y^{1/2}$$

Draw a contour map of f showing the constant product curves at 8, 6, 4, and 2.

Solution   The contour map consists of the curves that are the intersection of the surface

$$z = 2x^{1/2}y^{1/2} \tag{1}$$

with the planes $z = k$, where k equals 8, 6, 4, and 2. Substituting $z = 8$ in (1) we obtain $4 = x^{1/2}y^{1/2}$ or, equivalently,

$$xy = 16 \qquad x > 0 \quad \text{and} \quad y > 0 \tag{2}$$

The curve in the xy plane represented by (2) is a branch of a hyperbola lying in the first quadrant. With each of the numbers, 6, 4, and 2 we also obtain a branch of a hyperbola in the first quadrant. These are the constant product curves, and they are shown in Figure 8.

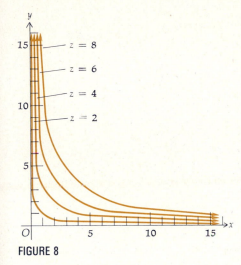

FIGURE 8

The following definition extends the notion of the graph of a function to a function of n variables.

16.1.6 DEFINITION   If f is a function of n variables, then the **graph** of f is the set of all points $(x_1, x_2, \ldots, x_n, w)$ in R^{n+1} for which $(x_1, x_2, \ldots, x_n)$ is a point in the domain of f and $w = f(x_1, x_2, \ldots, x_n)$.

Analogous to level curves for a function of two variables is a similar situation for functions of three variables. If f is a function whose domain is a set of points in R^3, then if k is a number in the range of f, the graph of the equation $f(x, y, z) = k$ is a surface. This surface is called the **level surface** of f at k. Every surface in three-dimensional space can be considered as a level surface of some function of three variables. For example, if the function g is defined by the equation $g(x, y, z) = x^2 + y^2 - z$, then the surface shown in Figure 4 is the level surface of g at 0. Similarly, the surface having equation $z - x^2 - y^2 + 5 = 0$ is the level surface of g at 5.

Computer programs can be used to generate surfaces that would be difficult or impossible to draw otherwise. These computer-generated plots are called **computer graphics**. Many of them show cross sections made by planes $x = k$ and $y = k$ for equally spaced values of k. To demonstrate the diverse and intricate surfaces that arise, we show the computer graphics associated with particular functions in Figures 9 through 16.

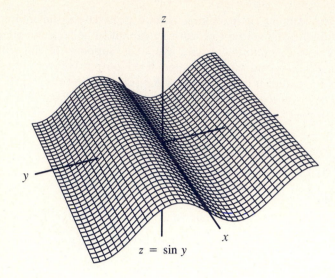

$z = \sin y$

Figure 9

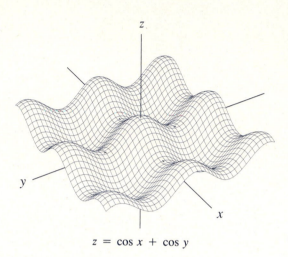

$z = \cos x + \cos y$

Figure 10

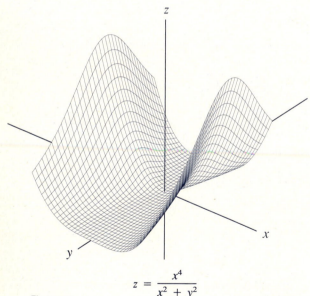

$z = \dfrac{x^4}{x^2 + y^2}$

Figure 11

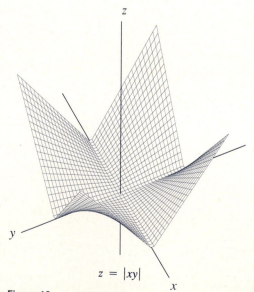

$z = |xy|$

Figure 12

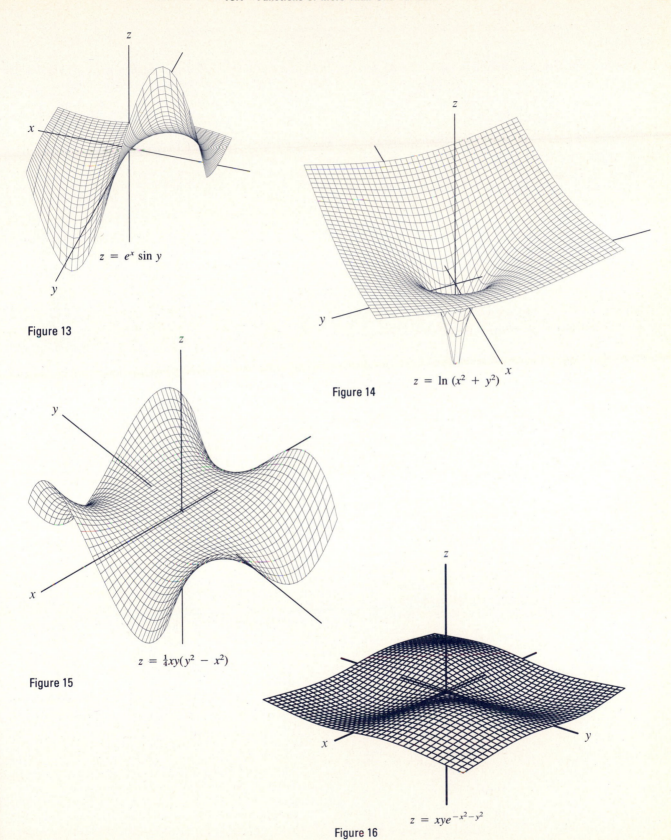

$z = e^x \sin y$

Figure 13

$z = \ln(x^2 + y^2)$

Figure 14

$z = \frac{1}{4}xy(y^2 - x^2)$

Figure 15

$z = xye^{-x^2-y^2}$

Figure 16

EXERCISES 16.1

1. Let the function f of two variables x and y be the set of all ordered pairs of the form (P, z) such that

$$z = \frac{x+y}{x-y} \quad \Leftrightarrow \quad f(x, y) = \frac{x+y}{x-y}$$

Find: (a) $f(-3, 4)$; (b) $f(\frac{1}{2}, \frac{1}{3})$; (c) $f(x + 1, y - 1)$; (d) $f(-x, y) - f(x, -y)$.

2. Let the function g of two variables x and y be the set of all ordered pairs of the form (P, z) such that

$$z = \sqrt{x^2 - y} \quad \Leftrightarrow \quad g(x, y) = \sqrt{x^2 - y}$$

Find: (a) $g(3, 5)$; (b) $g(-4, -9)$; (c) $g(x + 2, 4x + 4)$;
(d) $g\left(\frac{1}{x}, \frac{-3}{x^2}\right)$.

3. Let the function g of three variables x, y, and z be the set of all ordered pairs of the form (P, w) such that

$$w = \sqrt{4 - x^2 - y^2 - z^2} \quad \Leftrightarrow \quad g(x, y, z) = \sqrt{4 - x^2 - y^2 - z^2}$$

Find: (a) $g(1, -1, -1)$; (b) $g(-1, \frac{1}{2}, \frac{3}{2})$; (c) $g(\frac{1}{2}x, \frac{1}{2}y, \frac{1}{2}z)$;
(d) $[g(x, y, z)]^2 - [g(x + 2, y + 2, z)]^2$.

4. Let the function f of three variables x, y, and z be the set of all ordered pairs of the form (P, w) such that

$$w = \frac{4}{x^2 + y^2 + z^2 - 9} \quad \Leftrightarrow \quad f(x, y, z) = \frac{4}{x^2 + y^2 + z^2 - 9}$$

Find: (a) $f(1, 2, 3)$; (b) $f(2, -\frac{1}{2}, \frac{3}{2})$; (c) $f\left(-\frac{2}{x}, \frac{2}{x}, -\frac{1}{x}\right)$;
(d) $f(x + 2, 1, x - 2)$.

In Exercises 5 through 20, determine the domain of f and draw a sketch showing as a region in R^2 the set of points in the domain. Use dashed curves to indicate any part of the boundary not in the domain and solid curves to indicate parts of the boundary in the domain.

5. $f(x, y) = \dfrac{1}{x^2 + y^2 - 1}$ 6. $f(x, y) = \dfrac{4}{4 - x^2 - y^2}$

7. $f(x, y) = \sqrt{1 - x^2 - y^2}$ 8. $f(x, y) = \sqrt{16 - x^2 - 4y^2}$

9. $f(x, y) = \sqrt{x^2 - y^2 - 1}$ 10. $f(x, y) = \sqrt{x^2 - 4y^2 + 16}$

11. $f(x, y) = \sqrt{x^2 + y^2 - 1}$ 12. $f(x, y) = \sqrt{x^2 + 4y^2 - 16}$

13. $f(x, y) = \dfrac{1}{\sqrt{1 - x^2 - y^2}}$ 14. $f(x, y) = \dfrac{1}{\sqrt{16 - x^2 - 4y^2}}$

15. $f(x, y) = \dfrac{x^4 - y^4}{x^2 - y^2}$ 16. $f(x, y) = \dfrac{x - y}{x + y}$

17. $f(x, y) = \cos^{-1}(x - y)$ 18. $f(x, y) = \ln(x^2 + y)$

19. $f(x, y) = \ln(xy - 1)$ 20. $f(x, y) = \sin^{-1}(x + y)$

In Exercises 21 through 28, determine the domain of f and describe the region in R^3 that is the set of points in the domain.

21. $f(x, y, z) = \dfrac{x + y + z}{x - y - z}$ 22. $f(x, y, z) = \dfrac{z}{x^2 - y}$

23. $f(x, y, z) = \sqrt{16 - x^2 - 4y^2 - z^2}$

24. $f(x, y, z) = \sqrt{9 - x^2 - y^2 - z^2}$

25. $f(x, y, z) = \sin^{-1} x + \sin^{-1} y + \sin^{-1} z$

26. $f(x, y, z) = \ln x + \ln y + \ln z$

27. $f(x, y, z) = \ln(4 - x^2 - y^2) + |z|$

28. $f(x, y, z) = xz \cos^{-1}(y^2 - 1)$

In Exercises 29 through 36, determine the domain of f and draw a sketch of the graph of f.

29. $f(x, y) = \sqrt{16 - x^2 - y^2}$ 30. $f(x, y) = 6 - 2x + 2y$

31. $f(x, y) = 16 - x^2 - y^2$ 32. $f(x, y) = \sqrt{100 - 25x^2 - 4y^2}$

33. $f(x, y) = x^2 - y^2$ 34. $f(x, y) = 144 - 9x^2 - 16y^2$

35. $f(x, y) = 4x^2 + 9y^2$ 36. $f(x, y) = \sqrt{x + y}$

In Exercises 37 through 44, draw a sketch of a contour map of f showing the level curves at the given numbers.

37. The function of Exercise 29 at 0, 1, 2, 3, and 4.

38. The function of Exercise 30 at 10, 6, 2, 0, -2, -6, and -10.

39. The function of Exercise 31 at 16, 12, 7, 0, -9, and -20.

40. The function of Exercise 32 at 0, 2, 4, 6, 8, and 10.

41. The function of Exercise 33 at 16, 9, 4, 0, -4, -9, and -16.

42. The function of Exercise 36 at 10, 8, 6, 5, and 0.

43. The function f for which $f(x, y) = \frac{1}{2}(x^2 + y^2)$ at 8, 6, 4, 2, and 0.

44. The function f for which $f(x, y) = (x - 3)/(y + 2)$ at 4, 2, 1, $\frac{1}{2}$, $\frac{1}{4}$, 0, $-\frac{1}{4}$, $-\frac{1}{2}$, -1, -2, and -4.

In Exercises 45 and 46, find $h(x, y)$ if $h = f \circ g$; also find the domain of h.

45. $f(t) = \sin^{-1} t$; $g(x, y) = \sqrt{1 - x^2 - y^2}$

46. $f(t) = e^t$; $g(x, y) = y \ln x$

47. Given $f(x, y) = x - y$, $g(t) = \sqrt{t}$, $h(s) = s^2$. Find (a) $(g \circ f)(5, 1)$; (b) $f(h(3), g(9))$; (c) $f(g(x), h(y))$; (d) $g((h \circ f)(x, y))$; (e) $(g \circ h)(f(x, y))$.

48. Given $f(x, y) = x/y^2$, $g(x) = x^2$, $h(x) = \sqrt{x}$. Find (a) $(h \circ f)(2, 1)$; (b) $f(g(2), h(4))$; (c) $f(g(\sqrt{x}), h(x^2))$; (d) $h((g \circ f)(x, y))$; (e) $(h \circ g)(f(x, y))$.

49. The electric potential at a point (x, y) is $V(x, y)$ volts and $V(x, y) = 4/\sqrt{9 - x^2 - y^2}$. Draw the equipotential curves for V at 16, 12, 8, 4, 1, $\frac{1}{2}$, and $\frac{1}{4}$.

50. The production function f for a certain commodity has function values $f(x, y) = 4x^{1/3}y^{2/3}$, where x and y give the amounts of two inputs. Draw a contour map of f showing the constant product curves at 16, 12, 8, 4, and 2.

51. Suppose the number of units of a commodity produced is z, and $z = 6xy$, where x is the number of machines used in production and y is the number of person-hours available. Then

the function f defined by $f(x, y) = 6xy$ is a production function. Draw a contour map of f showing the constant product curves for z at 30, 24, 18, 12, and 6.

52. The temperature at a point (x, y) of a flat metal plate is $t(x, y)$ degrees and $t(x, y) = 4x^2 + 2y^2$. Draw the isothermals for t at 12, 8, 4, 1, and 0.

In Exercises 53 and 54, draw sketches of the level surfaces of f at the given numbers.

53. $f(x, y, z) = x^2 + y^2 - 4z$ at 8, 4, 0, -4, and -8.
54. $f(x, y, z) = x^2 + y^2 + z^2$ at 9, 4, 1, and 0.

16.2 LIMITS OF FUNCTIONS OF MORE THAN ONE VARIABLE

In R the distance between two points is the absolute value of the difference of two real numbers. That is, $|x - a|$ is the distance between the points x and a. In R^2 the distance between the two points $P(x, y)$ and $P_0(x_0, y_0)$ is given by $\sqrt{(x - x_0)^2 + (y - y_0)^2}$. In R^3 the distance between the two points $P(x, y, z)$ and $P_0(x_0, y_0, z_0)$ is given by $\sqrt{(x - x_0)^2 + (y - y_0)^2 + (z - z_0)^2}$. In R^n the distance between two points is defined analogously.

16.2.1 DEFINITION

If $P(x_1, x_2, \ldots, x_n)$ and $A(a_1, a_2, \ldots, a_n)$ are two points in R^n, then the distance between P and A, denoted by $\|P - A\|$, is given by

$$\|P - A\| = \sqrt{(x_1 - a_1)^2 + (x_2 - a_2)^2 + \ldots + (x_n - a_n)^2}$$

The symbol $\|P - A\|$ represents a nonnegative number and is read as "the distance between P and A."

In R, R^2, and R^3, the formula in Definition 16.2.1 becomes, respectively,

$$\|x - a\| = |x - a|$$
$$\|(x, y) - (x_0, y_0)\| = \sqrt{(x - x_0)^2 + (y - y_0)^2}$$
$$\|(x, y, z) - (x_0, y_0, z_0)\| = \sqrt{(x - x_0)^2 + (y - y_0)^2 + (z - z_0)^2}$$

16.2.2 DEFINITION

If A is a point in R^n and r is a positive number, then the **open ball** $B(A; r)$ is the set of all points P in R^n such that $\|P - A\| < r$.

16.2.3 DEFINITION

If A is a point in R^n and r is a positive number, then the **closed ball** $B[A; r]$ is the set of all points P in R^n such that $\|P - A\| \leq r$.

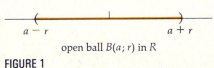

open ball $B(a; r)$ in R

FIGURE 1

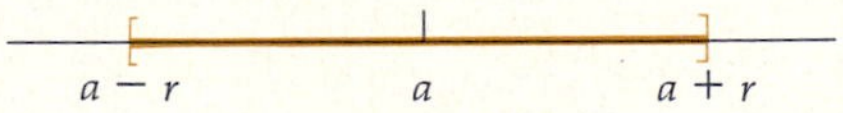

closed ball $B[a; r]$ in R

FIGURE 2

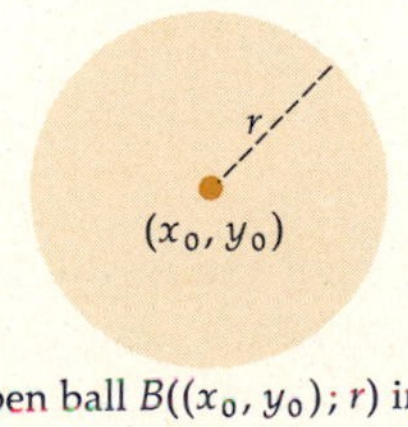

open ball $B((x_0, y_0); r)$ in R^2

FIGURE 3

To illustrate these definitions, we show what they mean in R, R^2, and R^3. First of all, if a is a point in R, then the open ball $B(a; r)$ is the set of all points x in R such that

$$|x - a| < r$$

The set of all points x satisfying this inequality is the set of all points in the open interval $(a - r, a + r)$; so the open ball $B(a; r)$ in R (see Figure 1) is simply an open interval having its midpoint at a and endpoints at $a - r$ and $a + r$. The closed ball $B[a; r]$ in R (Figure 2) is the closed interval $[a - r, a + r]$.

If (x_0, y_0) is a point in R^2, then the open ball $B((x_0, y_0); r)$ is the set of all points (x, y) in R^2 such that

$$\sqrt{(x - x_0)^2 + (y - y_0)^2} < r$$

So the open ball $B((x_0, y_0); r)$ in R^2 (Figure 3) consists of all points in the interior region bounded by the circle having its center at (x_0, y_0) and radius r. An open ball in R^2 is sometimes called an *open disk*. The closed ball, or closed

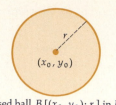

closed ball $B[(x_0, y_0); r]$ in R^2

FIGURE 4

disk, $B[(x_0, y_0); r]$ in R^2 (Figure 4) is the set of all points in the open ball $B((x_0, y_0); r)$ and on the circle having its center at (x_0, y_0) and radius r.

If (x_0, y_0, z_0) is a point in R^3, then the open ball $B((x_0, y_0, z_0); r)$ is the set of all points (x, y, z) in R^3 such that

$$\sqrt{(x - x_0)^2 + (y - y_0)^2 + (z - z_0)^2} < r$$

Therefore the open ball $B((x_0, y_0, z_0); r)$ in R^3 (Figure 5) consists of all points in the interior region bounded by the sphere having its center at (x_0, y_0, z_0) and radius r. Similarly, the closed ball $B[(x_0, y_0, z_0); r]$ in R^3 (Figure 6) consists of all points in the open ball $B((x_0, y_0, z_0); r)$ and on the sphere having its center at (x_0, y_0, z_0) and radius r.

We are now in a position to define the *limit of a function of n variables*.

16.2.4 DEFINITION

Let f be a function of n variables that is defined on some open ball $B(A; r)$, except possibly at the point A itself. Then the **limit of $f(P)$ as P approaches A is L**, written as

$$\lim_{P \to A} f(P) = L$$

if for any $\epsilon > 0$, however small, there exists a $\delta > 0$ such that

$$\text{if} \quad 0 < \|P - A\| < \delta \quad \text{then} \quad |f(P) - L| < \epsilon$$

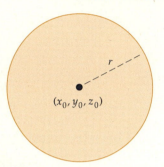

FIGURE 5

If in the above definition, f is a function of one variable, $A = a$ in R, and $P = x$, then the definition states: if f is defined on some open interval centered at a, except possibly at a itself,

$$\lim_{x \to a} f(x) = L$$

if for any $\epsilon > 0$, however small, there exists a $\delta > 0$ such that

$$\text{if} \quad 0 < |x - a| < \delta \quad \text{then} \quad |f(x) - L| < \epsilon$$

So the definition (2.1.1) of the limit of a function of one variable is a special case of Definition 16.2.4.

We now state the definition of the limit of a function of two variables. It is the special case of Definition 16.2.4, where A is the point (x_0, y_0) and P is the point (x, y).

16.2.5 DEFINITION

Let f be a function of two variables that is defined on some open disk $B((x_0, y_0); r)$, except possibly at the point (x_0, y_0) itself. Then the **limit of $f(x, y)$ as (x, y) approaches (x_0, y_0) is L**, written as

$$\lim_{(x,y) \to (x_0,y_0)} f(x, y) = L$$

if for any $\epsilon > 0$, however small, there exists a $\delta > 0$ such that

$$\text{if} \quad 0 < \sqrt{(x - x_0)^2 + (y - y_0)^2} < \delta \quad \text{then} \quad |f(x, y) - L| < \epsilon$$

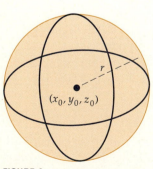

FIGURE 6

In words, Definition 16.2.5 states that the function values $f(x, y)$ approach a limit L as the point (x, y) approaches the point (x_0, y_0) if the absolute value of the difference between $f(x, y)$ and L can be made arbitrarily small by taking the point (x, y) sufficiently close to (x_0, y_0) but not equal to (x_0, y_0). In the definition nothing is said about the function value at the point (x_0, y_0); that is,

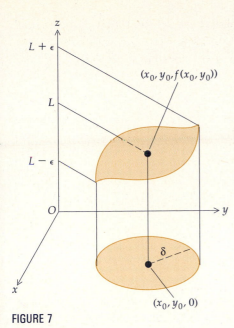

FIGURE 7

it is not necessary that the function be defined at (x_0, y_0) for $\lim\limits_{(x,y)\to(x_0,y_0)} f(x, y)$ to exist.

A geometric interpretation of Definition 16.2.5 is illustrated in Figure 7. The portion, above the open disk $B((x_0, y_0); \delta)$, of the surface having equation $z = f(x, y)$ is shown. We see that $f(x, y)$ on the z axis will lie between $L - \epsilon$ and $L + \epsilon$ whenever the point (x, y) in the xy plane is in the open disk $B((x_0, y_0); \delta)$. Another way of stating this is that $f(x, y)$ on the z axis can be restricted to lie between $L - \epsilon$ and $L + \epsilon$ by restricting the point (x, y) in the xy plane to be in the open disk $B((x_0, y_0); \delta)$.

▶ **ILLUSTRATION 1**   We apply Definition 16.2.5 to prove that

$$\lim_{(x,y)\to(1,3)} (2x + 3y) = 11$$

The first requirement of the definition is that $2x + 3y$ must be defined on some open disk having its center at the point $(1, 3)$, except possibly at $(1, 3)$. Because $2x + 3y$ is defined at every point (x, y), any open disk having its center at $(1, 3)$ will satisfy this requirement. Now, we must show that for any $\epsilon > 0$ there exists a $\delta > 0$ such that

$$\text{if} \quad 0 < \sqrt{(x-1)^2 + (y-3)^2} < \delta \quad \text{then} \quad |(2x + 3y) - 11| < \epsilon \tag{1}$$

From the triangle inequality,

$$|2x + 3y - 11| = |2x - 2 + 3y - 9|$$
$$\leq 2|x - 1| + 3|y - 3|$$

Because

$$|x - 1| \leq \sqrt{(x-1)^2 + (y-3)^2} \quad \text{and} \quad |y - 3| \leq \sqrt{(x-1)^2 + (y-3)^2}$$

it follows that

$$\text{if} \quad 0 < \sqrt{(x-1)^2 + (y-3)^2} < \delta \quad \text{then} \quad 2|x - 1| + 3|y - 3| < 2\delta + 3\delta$$

This statement indicates that a suitable choice for δ is $5\delta = \epsilon$, that is, $\delta = \frac{1}{5}\epsilon$. With this δ we have the following argument:

$$0 < \sqrt{(x-1)^2 + (y-3)^2} < \delta$$
$$\Rightarrow \quad |x - 1| < \delta \quad \text{and} \quad |y - 3| < \delta$$
$$\Rightarrow \quad 2|x - 1| + 3|y - 3| < 5\delta$$
$$\Rightarrow \quad |2(x - 1) + 3(y - 3)| < 5(\tfrac{1}{5}\epsilon)$$
$$\Rightarrow \quad |2x + 3y - 11| < \epsilon$$

We have demonstrated that for any $\epsilon > 0$ choose $\delta = \frac{1}{5}\epsilon$ and statement (1) is true. This proves that $\lim\limits_{(x,y)\to(1,3)} (2x + 3y) = 11$. ◀

EXAMPLE 1   Use Definition 16.2.5 to prove

$$\lim_{(x,y)\to(1,2)} (3x^2 + y) = 5$$

Solution   Because $3x^2 + y$ is defined at every point (x, y), any open disk having its center at $(1, 2)$ will satisfy the first requirement of Definition 16.2.5.

We must show that for any $\epsilon > 0$ there exists a $\delta > 0$ such that

$$\text{if} \quad 0 < \sqrt{(x-1)^2 + (y-2)^2} < \delta \quad \text{then} \quad |(3x^2 + y) - 5| < \epsilon \tag{2}$$

From the triangle inequality,

$$|3x^2 + y - 5| = |3x^2 - 3 + y - 2|$$
$$\leq 3|x^2 - 1| + |y - 2|$$

Thus

$$|3x^2 + y - 5| \leq 3|x-1||x+1| + |y-2| \tag{3}$$

Because

$$|x-1| \leq \sqrt{(x-1)^2 + (y-2)^2} \quad \text{and} \quad |y-2| \leq \sqrt{(x-1)^2 + (y-2)^2}$$

it follows that

$$\text{if} \quad 0 < \sqrt{(x-1)^2 + (y-2)^2} < \delta \quad \text{then} \quad |x-1| < \delta \quad \text{and} \quad |y-2| < \delta$$

Observe that on the right side of inequality (3), in addition to the expressions $|x-1|$ and $|y-2|$, we have the expression $|x+1|$. Thus to prove (2) we wish to place a restriction on δ that will give us an inequality involving $|x+1|$. One such restriction is to choose the radius of the open disk required by Definition 16.2.5 to be less than or equal to 1. Then

$$0 < \sqrt{(x-1)^2 + (y-2)^2} < \delta \quad \text{and} \quad \delta \leq 1$$

$$\Rightarrow \qquad |x-1| < 1$$

$$\Rightarrow \quad -1 < x - 1 < 1$$

$$\Rightarrow \qquad 1 < x + 1 < 3$$

$$\Rightarrow \qquad |x+1| < 3$$

Now

$$|x-1| < \delta \quad \text{and} \quad |x+1| < 3 \quad \text{and} \quad |y-2| < \delta$$

$$\Rightarrow \quad 3|x-1||x+1| + |y-2| < 3 \cdot \delta \cdot 3 + \delta$$

Because our goal is to have $3|x-1||x+1| + |y-2| < \epsilon$, we should require $10\delta \leq \epsilon$, that is $\delta \leq \frac{1}{10}\epsilon$. This means that we have put two restrictions on δ: $\delta \leq 1$ and $\delta \leq \frac{1}{10}\epsilon$. For both restrictions to hold, we take $\delta = \min(1, \frac{1}{10}\epsilon)$. With this δ we have the following argument:

$$0 < \sqrt{(x-1)^2 + (y-2)^2} < \delta \quad \text{and} \quad \delta = \min(1, \tfrac{1}{10}\epsilon)$$

$$\Rightarrow \quad |x-1| < \tfrac{1}{10}\epsilon \quad \text{and} \quad |x-1| < 1 \quad \text{and} \quad |y-2| < \tfrac{1}{10}\epsilon$$

$$\Rightarrow \quad |x-1| < \tfrac{1}{10}\epsilon \quad \text{and} \quad |x+1| < 3 \quad \text{and} \quad |y-2| < \tfrac{1}{10}\epsilon$$

$$\Rightarrow \quad 3|x-1||x+1| + |y-2| < 3 \cdot \tfrac{1}{10}\epsilon \cdot 3 + \tfrac{1}{10}\epsilon$$

$$\Rightarrow \quad |3(x-1)(x+1) + y - 2| < \epsilon$$

$$\Rightarrow \qquad |(3x^2 + y) - 5| < \epsilon$$

We have demonstrated that for any $\epsilon > 0$, choose $\delta = \min(1, \frac{1}{10}\epsilon)$ and statement (2) is true. This proves that $\displaystyle\lim_{(x,y) \to (1,2)} (3x^2 + y) = 5$.

The limit theorems of Section 2.2 and their proofs, with minor modifications, apply to functions of more than one variable. For example, corresponding to Limit Theorem 1 in Section 2.2 we have

$$\lim_{(x,y)\to(a,b)} (mx + ny + d) = ma + nb + d$$

and the proof is a generalization of the proof in Illustration 1. We use the limit theorems without restating them and their proofs.

▶ **ILLUSTRATION 2** By applying the limit theorems on sums and products,

$$\lim_{(x,y)\to(-2,1)} (x^3 + 2x^2y - y^2 + 2) = (-2)^3 + 2(-2)^2(1) - (1)^2 + 2$$
$$= 1 \qquad ◀$$

EXAMPLE 2 Find

$$\lim_{(x,y)\to(0,0)} \frac{x^4 - y^4}{x^2 + y^2}$$

Solution

$$\lim_{(x,y)\to(0,0)} \frac{x^4 - y^4}{x^2 + y^2} = \lim_{(x,y)\to(0,0)} \frac{(x^2 - y^2)(x^2 + y^2)}{x^2 + y^2}$$
$$= \lim_{(x,y)\to(0,0)} (x^2 - y^2)$$
$$= 0 - 0$$
$$= 0$$

Analogous to Theorem 2.7.1 for functions of a single variable is the following theorem regarding the limit of a composite function of two variables.

16.2.6 THEOREM   If g is a function of two variables and $\lim\limits_{(x,y)\to(x_0,y_0)} g(x, y) = b$, and f is a function of a single variable continuous at b, then

$$\lim_{(x,y)\to(x_0,y_0)} (f \circ g)(x, y) = f(b)$$

$$\Leftrightarrow \quad \lim_{(x,y)\to(x_0,y_0)} f(g(x, y)) = f\left(\lim_{(x,y)\to(x_0,y_0)} g(x, y) \right)$$

The proof of this theorem is similar to the proof of Theorem 2.7.1 and is left as an exercise (see Exercise 31).

EXAMPLE 3 Use Theorem 16.2.6 to find $\lim\limits_{(x,y)\to(2,1)} \ln(xy - 1)$.

Solution Let g be the function such that $g(x, y) = xy - 1$, and let f be the function such that $f(t) = \ln t$.

$$\lim_{(x,y)\to(2,1)} (xy - 1) = 1$$

and because f is continuous at 1, from Theorem 16.2.6,

$$\lim_{(x,y)\to(2,1)} \ln(xy - 1) = \ln\left(\lim_{(x,y)\to(2,1)} (xy - 1)\right)$$

$$= \ln 1$$

$$= 0$$

We now introduce the concept of an *accumulation point*, which is needed to continue the discussion of limits of functions of two variables.

16.2.7 DEFINITION

A point P_0 is said to be an **accumulation point** of a set S of points in R^n if every open ball $B(P_0; r)$ contains infinitely many points of S.

▶ **ILLUSTRATION 3** If S is the set of all points in R^2 on the positive side of the x axis, the origin will be an accumulation point of S because no matter how small we take the value of r, every open disk having its center at the origin and radius r will contain infinitely many points of S. This is an example of a set having an accumulation point for which the accumulation point is not a point of the set. Any point of this set S also will be an accumulation point of S. ◀

▶ **ILLUSTRATION 4** If S is the set of all points in R^2 for which the cartesian coordinates are positive integers, then this set has no accumulation point. This can be seen by considering the point (m, n), where m and n are positive integers. Then an open disk having its center at (m, n) and radius less than 1 will contain no points of S other than (m, n); therefore Definition 16.2.7 will not be satisfied (see Figure 8). ◀

FIGURE 8

We now consider the limit of a function of two variables as a point (x, y) approaches a point (x_0, y_0), where (x, y) is restricted to a specific set of points.

16.2.8 DEFINITION

Let f be a function defined on a set of points S in R^2, and let (x_0, y_0) be an accumulation point of S. Then the **limit of $f(x, y)$ as (x, y) approaches (x_0, y_0) in S is L**, written as

$$\lim_{\substack{(x,y)\to(x_0,y_0)\\(P \text{ in } S)}} f(x, y) = L$$

if for any $\epsilon > 0$, however small, there exists a $\delta > 0$ such that

$$\text{if}\quad 0 < \|(x, y) - (x_0, y_0)\| < \delta \quad \text{then}\quad |f(x, y) - L| < \epsilon$$

and (x, y) is in S.

In some cases the limit in the above definition becomes the limit of a function of one variable. For example, consider $\lim_{(x,y)\to(0,0)} f(x, y)$. Then if S_1 is the set of all points on the positive side of the x axis,

$$\lim_{\substack{(x,y)\to(0,0)\\(P \text{ in } S_1)}} f(x, y) = \lim_{x\to 0^+} f(x, 0)$$

If S_2 is the set of all points on the negative side of the y axis,

$$\lim_{\substack{(x,y)\to(0,0)\\(P \text{ in } S_2)}} f(x, y) = \lim_{y\to 0^-} f(0, y)$$

If S_3 is the set of all points on the x axis,

$$\lim_{\substack{(x,y)\to(0,0) \\ (P \text{ in } S_3)}} f(x, y) = \lim_{x\to 0} f(x, 0)$$

If S_4 is the set of all points on the parabola $y = x^2$,

$$\lim_{\substack{(x,y)\to(0,0) \\ (P \text{ in } S_4)}} f(x, y) = \lim_{x\to 0} f(x, x^2)$$

16.2.9 THEOREM

Suppose that the function f is defined for all points on an open disk having its center at (x_0, y_0), except possibly at (x_0, y_0) itself, and

$$\lim_{(x,y)\to(x_0,y_0)} f(x, y) = L$$

Then if S is any set of points in R^2 having (x_0, y_0) as an accumulation point,

$$\lim_{\substack{(x,y)\to(x_0,y_0) \\ (P \text{ in } S)}} f(x, y)$$

exists and always has the value L.

Proof   Because $\lim\limits_{(x,y)\to(x_0,y_0)} f(x, y) = L$, then by Definition 16.2.5, for any $\epsilon > 0$ there exists a $\delta > 0$ such that

if $0 < \|(x, y) - (x_0, y_0)\| < \delta$ then $|f(x, y) - L| < \epsilon$

The above will be true if we further restrict (x, y) by the requirement that (x, y) be in a set S, where S is any set of points having (x_0, y_0) as an accumulation point. Therefore, by Definition 16.2.8,

$$\lim_{\substack{(x,y)\to(x_0,y_0) \\ (P \text{ in } S)}} f(x, y) = L$$

and L does not depend on the set S through which (x, y) is approaching (x_0, y_0). This proves the theorem.   ∎

The next theorem is an immediate consequence of Theorem 16.2.9.

16.2.10 THEOREM

If the function f has different limits as (x, y) approaches (x_0, y_0) through two distinct sets of points having (x_0, y_0) as an accumulation point, then $\lim\limits_{(x,y)\to(x_0,y_0)} f(x, y)$ does not exist.

Proof   Suppose S_1 and S_2 are two distinct sets of points in R^2 having (x_0, y_0) as an accumulation point, and let

$$\lim_{\substack{(x,y)\to(x_0,y_0) \\ (P \text{ in } S_1)}} f(x, y) = L_1 \quad \text{and} \quad \lim_{\substack{(x,y)\to(x_0,y_0) \\ (P \text{ in } S_2)}} f(x, y) = L_2$$

Now assume that $\lim\limits_{(x,y)\to(x_0,y_0)} f(x, y)$ exists. Then by Theorem 16.2.9 L_1 must equal L_2, but by hypothesis $L_1 \neq L_2$, and so we have a contradiction. Therefore $\lim\limits_{(x,y)\to(x_0,y_0)} f(x, y)$ does not exist.   ∎

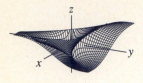

FIGURE 9

EXAMPLE 4   Given

$$f(x, y) = \frac{xy}{x^2 + y^2}$$

find $\lim\limits_{(x,y)\to(0,0)} f(x, y)$ if it exists.

Solution   The function f is defined at all points in R^2 except $(0, 0)$. Figure 9 shows a computer-generated graph of f. Let S_1 be the set of all points on the x axis, and S_2 be the set of all points on the line $y = x$. Then

$$\lim_{\substack{(x,y)\to(0,0)\\(P \text{ in } S_1)}} f(x, y) = \lim_{x\to0} f(x, 0) \qquad \lim_{\substack{(x,y)\to(0,0)\\(P \text{ in } S_2)}} f(x, y) = \lim_{x\to0} f(x, x)$$

$$= \lim_{x\to0} \frac{0}{x^2 + 0} \qquad\qquad = \lim_{x\to0} \frac{x^2}{x^2 + x^2}$$

$$= \lim_{x\to0} 0 \qquad\qquad = \lim_{x\to0} \tfrac{1}{2}$$

$$= 0 \qquad\qquad = \tfrac{1}{2}$$

Because

$$\lim_{\substack{(x,y)\to(0,0)\\(P \text{ in } S_1)}} f(x, y) \neq \lim_{\substack{(x,y)\to(0,0)\\(P \text{ in } S_2)}} f(x, y)$$

it follows from Theorem 16.2.10 that $\lim\limits_{(x,y)\to(0,0)} f(x, y)$ does not exist.

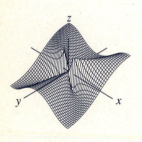

FIGURE 10

EXAMPLE 5   Given

$$f(x, y) = \frac{x^2 y}{x^4 + y^2}$$

find $\lim\limits_{(x,y)\to(0,0)} f(x, y)$ if it exists.

Solution   The function f is defined at all points in R^2 except $(0, 0)$. Figure 10 shows a computer-generated graph of f. Let S_1 be the set of all points on a line through the origin; that is, for any point (x, y) in S_1, $y = mx$. Let S_2 be the set of all points on the parabola $y = x^2$. Then

$$\lim_{\substack{(x,y)\to(0,0)\\(P \text{ in } S_1)}} f(x, y) = \lim_{x\to0} f(x, mx) \qquad \lim_{\substack{(x,y)\to(0,0)\\(P \text{ in } S_2)}} f(x, y) = \lim_{x\to0} f(x, x^2)$$

$$= \lim_{x\to0} \frac{mx^3}{x^4 + m^2 x^2} \qquad\qquad = \lim_{x\to0} \frac{x^4}{x^4 + x^4}$$

$$= \lim_{x\to0} \frac{mx}{x^2 + m^2} \qquad\qquad = \lim_{x\to0} \tfrac{1}{2}$$

$$= 0 \qquad\qquad = \tfrac{1}{2}$$

Because

$$\lim_{\substack{(x,y)\to(0,0)\\(P \text{ in } S_2)}} f(x, y) \neq \lim_{\substack{(x,y)\to(0,0)\\(P \text{ in } S_1)}} f(x, y)$$

it follows that $\lim\limits_{(x,y)\to(0,0)} f(x, y)$ does not exist.

EXAMPLE 6   Given

$$f(x, y) = \frac{3x^2 y}{x^2 + y^2}$$

find $\lim\limits_{(x,y) \to (0,0)} f(x, y)$ if it exists.

Solution   The function f is defined at all points in R^2 except $(0, 0)$. Let S_1 be the set of all points on a line through the origin; so if (x, y) is a point in S_1, $y = mx$. Let S_2 be the set of all points on the parabola $y = x^2$. Then

$$\lim_{\substack{(x,y) \to (0,0) \\ (P \text{ in } S_1)}} f(x, y) = \lim_{x \to 0} \frac{3x^2(mx)}{x^2 + m^2 x^2} \qquad \lim_{\substack{(x,y) \to (0,0) \\ (P \text{ in } S_2)}} f(x, y) = \lim_{x \to 0} \frac{3x^2(x^2)}{x^2 + (x^2)^2}$$

$$= \lim_{x \to 0} \frac{3mx}{1 + m^2} \qquad\qquad\qquad = \lim_{x \to 0} \frac{3x^4}{x^2 + x^4}$$

$$= 0 \qquad\qquad\qquad\qquad\qquad = \lim_{x \to 0} \frac{3x^2}{1 + x^2}$$

$$\qquad\qquad\qquad\qquad\qquad\qquad = 0$$

Even though the same limit of 0 is obtained if (x, y) approaches $(0, 0)$ through a set of points on a line through the origin as well as on the parabola $y = x^2$, we cannot conclude that $\lim\limits_{(x,y) \to (0,0)} f(x, y)$ exists and is zero, even though we may expect that is the case. So, let us attempt to prove that $\lim\limits_{(x,y) \to (0,0)} f(x, y) = 0$.

Any open disk having its center at the origin will satisfy the first requirement of Definition 16.2.5. If we can show that for any $\epsilon > 0$ there exists a $\delta > 0$ such that

$$\text{if} \quad 0 < \sqrt{x^2 + y^2} < \delta \quad \text{then} \quad \left| \frac{3x^2 y}{x^2 + y^2} \right| < \epsilon \tag{4}$$

then we have proved that $\lim\limits_{(x,y) \to (0,0)} f(x, y) = 0$.

Because $x^2 \leq x^2 + y^2$ and $|y| \leq \sqrt{x^2 + y^2}$

$$\left| \frac{3x^2 y}{x^2 + y^2} \right| = \frac{3x^2 |y|}{x^2 + y^2}$$

$$\leq \frac{3(x^2 + y^2)\sqrt{x^2 + y^2}}{x^2 + y^2}$$

$$= 3\sqrt{x^2 + y^2}$$

Thus a suitable choice for δ is $3\delta = \epsilon$, that is, $\delta = \frac{1}{3}\epsilon$. With this δ we have the following argument:

$$0 < \sqrt{x^2 + y^2} < \delta$$

$$\Rightarrow \quad \frac{3(x^2 + y^2)\sqrt{x^2 + y^2}}{x^2 + y^2} < 3\delta$$

$$\Rightarrow \quad \frac{3x^2 |y|}{x^2 + y^2} < 3(\tfrac{1}{3}\epsilon)$$

$$\Rightarrow \quad \left| \frac{3x^2 y}{x^2 + y^2} \right| < \epsilon$$

Thus if $\delta = \frac{1}{3}\epsilon$, statement (4) holds. Hence we have proved that

$$\lim_{(x,y)\to(0,0)} f(x, y) = 0.$$

EXERCISES 16.2

In Exercises 1 through 8, evaluate the limit by use of limit theorems.

1. $\lim\limits_{(x,y)\to(2,3)} (3x^2 + xy - 2y^2)$ **2.** $\lim\limits_{(x,y)\to(-1,4)} (5x^2 - 2xy + y^2)$

3. $\lim\limits_{(x,y)\to(2,-1)} \dfrac{3x - 2y}{x + 4y}$ **4.** $\lim\limits_{(x,y)\to(-2,4)} y\sqrt[3]{x^3 + 2y}$

5. $\lim\limits_{(x,y)\to(0,0)} \dfrac{e^x + e^y}{\cos x + \sin y}$ **6.** $\lim\limits_{(x,y)\to(0,0)} \dfrac{\sin^2 x + \cos^2 y}{e^{2x} + e^{2y}}$

7. $\lim\limits_{(x,y)\to(0,1)} \dfrac{x^4 - (y-1)^4}{x^2 + (y-1)^2}$

8. $\lim\limits_{(x,y)\to(1,1)} \dfrac{(x-1)^{4/3} - (y-1)^{4/3}}{(x-1)^{2/3} + (y-1)^{2/3}}$

In Exercises 9 through 16, establish the limit by finding a $\delta > 0$ for any $\epsilon > 0$ such that Definition 16.2.5 holds.

9. $\lim\limits_{(x,y)\to(3,2)} (3x - 4y) = 1$ **10.** $\lim\limits_{(x,y)\to(2,4)} (5x - 3y) = -2$

11. $\lim\limits_{(x,y)\to(-1,3)} (3x - 2y) = -9$ **12.** $\lim\limits_{(x,y)\to(-2,1)} (5x + 4y) = -6$

13. $\lim\limits_{(x,y)\to(1,1)} (x^2 + y^2) = 2$ **14.** $\lim\limits_{(x,y)\to(2,3)} (2x^2 - y^2) = -1$

15. $\lim\limits_{(x,y)\to(2,4)} (x^2 + 2x - y) = 4$

16. $\lim\limits_{(x,y)\to(3,-1)} (x^2 + y^2 - 4x + 2y) = -4$

In Exercises 17 through 22, prove that $\lim\limits_{(x,y)\to(0,0)} f(x, y)$ does not exist.

17. $f(x, y) = \dfrac{x^2 - y^2}{x^2 + y^2}$ **18.** $f(x, y) = \dfrac{x^2}{x^2 + y^2}$

19. $f(x, y) = \dfrac{x^4 y^4}{(x^2 + y^4)^3}$ **20.** $f(x, y) = \dfrac{x^4 + 3x^2y^2 + 2xy^3}{(x^2 + y^2)^2}$

21. $f(x, y) = \dfrac{x^9 y}{(x^6 + y^2)^2}$ **22.** $f(x, y) = \dfrac{x^2 y^2}{x^4 + y^4}$

In Exercises 23 through 26, prove that $\lim\limits_{(x,y)\to(0,0)} f(x, y)$ exists.

23. $f(x, y) = \dfrac{x^2 y + xy^2}{x^2 + y^2}$ **24.** $f(x, y) = \dfrac{x^3 + y^3}{x^2 + y^2}$

25. $f(x, y) = \dfrac{xy}{\sqrt{x^2 + y^2}}$ **26.** $f(x, y) = \dfrac{x^2 + 2xy}{\sqrt{x^2 + y^2}}$

In Exercises 27 through 30, determine if the limit exists.

27. $\lim\limits_{(x,y)\to(0,0)} \dfrac{x^2 y^2}{x^2 + y^2}$ **28.** $\lim\limits_{(x,y)\to(0,0)} \dfrac{x^2 y^4}{x^4 + y^4}$

29. $\lim\limits_{(x,y)\to(0,0)} \dfrac{x^2 + y}{x^2 + y^2}$ **30.** $\lim\limits_{(x,y)\to(0,0)} \dfrac{x^2 y^2}{x^4 + y^4}$

31. Prove Theorem 16.2.6.

In Exercises 32 through 35, show the application of Theorem 16.2.6 to find the limit.

32. $\lim\limits_{(x,y)\to(\ln 3, \ln 2)} e^{x-y}$ **33.** $\lim\limits_{(x,y)\to(2,2)} \tan^{-1}\dfrac{y}{x}$

34. $\lim\limits_{(x,y)\to(-2,3)} [\![5x + \frac{1}{2}y^2]\!]$ **35.** $\lim\limits_{(x,y)\to(4,2)} \sqrt{\dfrac{1}{3x - 4y}}$

36. (a) Give a definition, similar to Definition 16.2.5, of the limit of a function of three variables as a point (x, y, z) approaches a point (x_0, y_0, z_0). (b) Give a definition, similar to Definition 16.2.8, of the limit of a function of three variables as a point (x, y, z) approaches a point (x_0, y_0, z_0) in a specific set of points S in R^3.

37. (a) State and prove a theorem similar to Theorem 16.2.9 for a function f of three variables. (b) State and prove a theorem similar to Theorem 16.2.10 for a function f of three variables.

In Exercises 38 through 41, evaluate the limit by the use of limit theorems.

38. $\lim\limits_{(x,y,z)\to(-2,1,4)} (4x^2 y - 3xyz^2 + 7y^2 z^3)$

39. $\lim\limits_{(x,y,z)\to(\pi/3,1,\pi)} \dfrac{\sec xy + \sec yz}{y - \sec z}$

40. $\lim\limits_{(x,y,z)\to(0,2,0)} \dfrac{x^2 y^2 + y^2 z^2}{x^2 + z^2}$

41. $\lim\limits_{(x,y,z)\to(0,0,0)} \dfrac{(e^x + e^y + e^z)^2}{e^{2x} + e^{2y} + e^{2z}}$

In Exercises 42 through 45, use the definitions and theorems of Exercises 36 and 37 to prove that $\lim\limits_{(x,y,z)\to(0,0,0)} f(x, y, z)$ does not exist.

42. $f(x, y, z) = \dfrac{x^3 + yz^2}{x^4 + y^2 + z^4}$ **43.** $f(x, y, z) = \dfrac{x^2 + y^2 - z^2}{x^2 + y^2 + z^2}$

44. $f(x, y, z) = \dfrac{x^4 + yx^3 + z^2 x^2}{x^4 + y^4 + z^4}$ **45.** $f(x, y, z) = \dfrac{x^2 y^2 z^2}{x^6 + y^6 + z^6}$

In Exercises 46 and 47, use the definition in Exercise 36(a) to prove that $\lim\limits_{(x,y,z)\to(0,0,0)} f(x, y, z)$ exists.

46. $f(x, y, z) = \dfrac{y^3 + xz^2}{x^2 + y^2 + z^2}$ **47.** $f(x, y, z) = \dfrac{xy + xz + yz}{\sqrt{x^2 + y^2 + z^2}}$

48. Suppose that f and g are functions of two variables satisfying the following conditions:

(i) $f(tx, ty) = t^n f(x, y)$; $g(tx, ty) = t^n g(x, y)$ for some n and for all t;

(ii) $g(1, 1) \neq 0$ and $g(1, 0) \neq 0$;

(iii) $g(1, 1) \cdot f(1, 0) \neq g(1, 0) \cdot f(1, 1)$.

Show that $\lim\limits_{(x,y) \to (0,0)} \dfrac{f(x, y)}{g(x, y)}$ does not exist.

16.3 CONTINUITY OF FUNCTIONS OF MORE THAN ONE VARIABLE

We now define continuity of a function of n variables at a point in R^n. Observe that Definition 2.6.1 of continuity of a function of one variable at a number a is a special case of this definition.

16.3.1 DEFINITION

Suppose that f is a function of n variables and A is a point in R^n. Then f is said to be **continuous** at the point A if and only if the following three conditions are satisfied:

(i) $f(A)$ exists;

(ii) $\lim\limits_{P \to A} f(P)$ exists;

(iii) $\lim\limits_{P \to A} f(P) = f(A)$.

If one or more of these three conditions fails to hold at the point A, then f is said to be **discontinuous** at A.

If f is a function of two variables, A is the point (x_0, y_0), and P is a point (x, y), then Definition 16.3.1 becomes the following.

16.3.2 DEFINITION

The function f of two variables x and y is said to be **continuous** at the point (x_0, y_0) if and only if the following three conditions are satisfied:

(i) $f(x_0, y_0)$ exists;

(ii) $\lim\limits_{(x,y) \to (x_0, y_0)} f(x, y)$ exists;

(iii) $\lim\limits_{(x,y) \to (x_0, y_0)} f(x, y) = f(x_0, y_0)$.

EXAMPLE 1   Determine whether f is continuous at $(0, 0)$ if

$$f(x, y) = \begin{cases} \dfrac{3x^2 y}{x^2 + y^2} & \text{if } (x, y) \neq (0, 0) \\ 0 & \text{if } (x, y) = (0, 0) \end{cases}$$

Solution   We check the three conditions of Definition 16.3.2 at the point $(0, 0)$.

(i) $f(0, 0) = 0$. Therefore condition (i) holds.

(ii) $\lim\limits_{(x,y) \to (0,0)} f(x, y) = \lim\limits_{(x,y) \to (0,0)} \dfrac{3x^2 y}{x^2 + y^2}$

$= 0$

This fact was proved in Example 6, Section 16.2.

(iii) $\lim\limits_{(x,y) \to (0,0)} f(x, y) = f(0, 0)$

Therefore f is continuous at $(0, 0)$.

EXAMPLE 2   Determine whether f is continuous at $(0, 0)$ if

$$f(x, y) = \begin{cases} \dfrac{xy}{x^2 + y^2} & \text{if } (x, y) \neq (0, 0) \\ 0 & \text{if } (x, y) = (0, 0) \end{cases}$$

Solution   Checking the conditions of Definition 16.3.2, we have the following:

(i) $f(0, 0) = 0$; so condition (i) holds.
(ii) When $(x, y) \neq (0, 0)$, $f(x, y) = xy/(x^2 + y^2)$. In Example 4, Section 16.2, we showed that $\lim\limits_{(x,y) \to (0,0)} xy/(x^2 + y^2)$ does not exist; so $\lim\limits_{(x,y) \to (0,0)} f(x, y)$ does not exist. Therefore condition (ii) fails to hold.

Thus f is discontinuous at $(0, 0)$.

If a function f of two variables is discontinuous at the point (x_0, y_0) but $\lim\limits_{(x,y) \to (x_0,y_0)} f(x, y)$ exists, then f is said to have a **removable discontinuity** at (x_0, y_0) because if f is redefined at (x_0, y_0) so that

$$f(x_0, y_0) = \lim_{(x,y) \to (x_0,y_0)} f(x, y)$$

then the new function is continuous at (x_0, y_0). If the discontinuity is not removable, it is called an **essential discontinuity**.

▶ **ILLUSTRATION 1**   (a) If $g(x, y) = 3x^2y/(x^2 + y^2)$, then g is discontinuous at the origin because $g(0, 0)$ is not defined. However, in Example 6, Section 16.2, we showed that $\lim\limits_{(x,y) \to (0,0)} 3x^2y/(x^2 + y^2) = 0$. Therefore the discontinuity is removable if $g(0, 0)$ is redefined to be 0. (Refer to Example 1.)

(b) Let $h(x, y) = xy/(x^2 + y^2)$. Then h is discontinuous at the origin because $h(0, 0)$ is not defined. In Example 4, Section 16.2, we showed that $\lim\limits_{(x,y) \to (0,0)} xy/(x^2 + y^2)$ does not exist. Therefore the discontinuity is essential. (Refer to Example 2.)   ◀

The theorems about continuity for functions of a single variable can be extended to functions of two variables.

16.3.3 THEOREM   If f and g are two functions that are continuous at the point (x_0, y_0), then

    (i) $f + g$ is continuous at (x_0, y_0);
    (ii) $f - g$ is continuous at (x_0, y_0);
    (iii) fg is continuous at (x_0, y_0);
    (iv) f/g is continuous at (x_0, y_0), provided that $g(x_0, y_0) \neq 0$.

The proof of this theorem is analogous to the proof of the corresponding theorem (2.6.2) for functions of one variable, and hence it is omitted.

16.3.4 THEOREM   A polynomial function of two variables is continuous at every point in R^2.

Proof   Every polynomial function is the sum of products of the functions defined by $f(x, y) = x$, $g(x, y) = y$, and $h(x, y) = c$, where c is a real number. Because f, g, and h are continuous at every point in R^2, the theorem follows by repeated applications of Theorem 16.3.3, parts (i) and (iii). ∎

16.3.5 THEOREM   A rational function of two variables is continuous at every point in its domain.

Proof   A rational function is the quotient of two polynomial functions f and g that are continuous at every point in R^2, by Theorem 16.3.4. If (x_0, y_0) is any point in the domain of f/g, then $g(x_0, y_0) \neq 0$; so by Theorem 16.3.3(iv) f/g is continuous there. ∎

EXAMPLE 3   Determine all points at which f is continuous if

$$f(x, y) = \begin{cases} x^2 + y^2 & \text{if } x^2 + y^2 \leq 1 \\ 0 & \text{if } x^2 + y^2 > 1 \end{cases}$$

Solution   The function f is defined at all points in R^2. Therefore condition (i) of Definition 16.3.2 holds for every point (x_0, y_0).

Consider the points (x_0, y_0) if $x_0{}^2 + y_0{}^2 \neq 1$.

If $x_0{}^2 + y_0{}^2 < 1$,

$$\lim_{(x,y) \to (x_0, y_0)} f(x, y) = \lim_{(x,y) \to (x_0, y_0)} (x^2 + y^2)$$
$$= x_0{}^2 + y_0{}^2$$
$$= f(x_0, y_0)$$

If $x_0{}^2 + y_0{}^2 > 1$,

$$\lim_{(x,y) \to (x_0, y_0)} f(x, y) = \lim_{(x,y) \to (x_0, y_0)} 0$$
$$= 0$$
$$= f(x_0, y_0)$$

Thus f is continuous at all points (x_0, y_0) for which $x_0{}^2 + y_0{}^2 \neq 1$.

To determine the continuity of f at points (x_0, y_0) for which $x_0{}^2 + y_0{}^2 = 1$, we determine if $\lim_{(x,y) \to (x_0, y_0)} f(x, y)$ exists and equals 1.

Let S_1 be the set of all points (x, y) such that $x^2 + y^2 \leq 1$, and S_2 be the set of all points (x, y) such that $x^2 + y^2 > 1$. Then

$$\lim_{\substack{(x,y) \to (x_0, y_0) \\ (P \text{ in } S_1)}} f(x, y) = \lim_{\substack{(x,y) \to (x_0, y_0) \\ (P \text{ in } S_1)}} (x^2 + y^2)$$
$$= x_0{}^2 + y_0{}^2$$
$$= 1$$

$$\lim_{\substack{(x,y) \to (x_0, y_0) \\ (P \text{ in } S_2)}} f(x, y) = \lim_{\substack{(x,y) \to (x_0, y_0) \\ (P \text{ in } S_2)}} 0$$
$$= 0$$

Because

$$\lim_{\substack{(x,y) \to (x_0, y_0) \\ (P \text{ in } S_1)}} f(x, y) \neq \lim_{\substack{(x,y) \to (x_0, y_0) \\ (P \text{ in } S_2)}} f(x, y)$$

we conclude that $\lim_{(x,y) \to (x_0, y_0)} f(x, y)$ does not exist. Hence f is discontinuous at all points (x_0, y_0) for which $x_0{}^2 + y_0{}^2 = 1$.

We have proved that f is continuous at all points in R^2 except those on the circle $x^2 + y^2 = 1$.

16.3.6 DEFINITION   The function f of n variables is said to be **continuous on an open ball** if it is continuous at every point of the open ball.

As an illustration of the above definition, the function of Example 3 is continuous on every open disk that does not contain a point of the circle $x^2 + y^2 = 1$.

The following theorem states that a continuous function of a continuous function is continuous. It is analogous to Theorem 2.7.2 and its proof is similar.

16.3.7 THEOREM   Suppose that f is a function of a single variable and g is a function of two variables. Suppose further that g is continuous at (x_0, y_0) and f is continuous at $g(x_0, y_0)$. Then the composite function $f \circ g$ is continuous at (x_0, y_0).

▶ **ILLUSTRATION 2**   Let

$$h(x, y) = \ln(xy - 1)$$

If $g(x, y) = xy - 1$, g is continuous at all points in R^2. The natural logarithmic function is continuous on its entire domain, which is the set of all positive numbers. So if f is the function defined by $f(t) = \ln t$, f is continuous for all $t > 0$. Then the function h is the composite function $f \circ g$ and, by Theorem 16.3.7, is continuous at all points (x, y) in R^2 for which $xy - 1 > 0$.   ◀

EXAMPLE 4   Determine all points at which f is continuous if

$$f(x, y) = \frac{1}{\sqrt{x^2 + y^2 - 25}}$$

Solution   The domain of f is the set of all points (x, y) in R^2 for which $x^2 + y^2 - 25 > 0$. These are the points in the exterior region bounded by the circle $x^2 + y^2 = 25$. Function f is the quotient of functions g and h for which

$$g(x, y) = 1 \qquad h(x, y) = \sqrt{x^2 + y^2 - 25}$$

The function g is a constant function and is therefore continuous everywhere. It follows from Theorem 16.3.7 that h is continuous at all points in R^2 for which $x^2 + y^2 > 25$. Therefore by Theorem 16.3.3(iv), f is continuous at all points in its domain.

EXERCISES 16.3

In Exercises 1 through 24, determine all points at which the function is continuous.

1. $f(x, y) = \dfrac{x^2}{y - 1}$

2. $F(x, y) = \dfrac{1}{x - y}$

3. $h(x, y) = \sin \dfrac{y}{x}$

4. $f(x, y) = \ln xy^2$

5. $f(x, y) = \dfrac{4x^2y + 3y^2}{2x - y}$

6. $g(x, y) = \dfrac{5xy^2 + 2y}{16 - x^2 - 4y^2}$

7. $g(x, y) = \ln(25 - x^2 - y^2)$

8. $f(x, y) = \cos^{-1}(x + y)$

9. $f(x, y) = \begin{cases} \dfrac{xy}{\sqrt{x^2 + y^2}} & \text{if } (x, y) \neq (0, 0) \\ 0 & \text{if } (x, y) = (0, 0) \end{cases}$
(*Hint:* See Exercise 25, Exercises 16.2.)

10. $h(x, y) = \begin{cases} \dfrac{x^2y}{x^4 + y^2} & \text{if } (x, y) \neq (0, 0) \\ 0 & \text{if } (x, y) = (0, 0) \end{cases}$
(*Hint:* See Example 5, Section 16.2.)

11. $f(x, y) = \begin{cases} \dfrac{x + y}{x^2 + y^2} & \text{if } (x, y) \neq (0, 0) \\ 0 & \text{if } (x, y) = (0, 0) \end{cases}$

12. $f(x, y) = \begin{cases} \dfrac{x^3 + y^3}{x^2 + y^2} & \text{if } (x, y) \neq (0, 0) \\ 0 & \text{if } (x, y) = (0, 0) \end{cases}$

13. $G(x, y) = \begin{cases} \dfrac{xy}{|x| + |y|} & \text{if } (x, y) \neq (0, 0) \\ 0 & \text{if } (x, y) = (0, 0) \end{cases}$

14. $F(x, y) = \begin{cases} \dfrac{x^2y^2}{|x^3| + |y^3|} & \text{if } (x, y) \neq (0, 0) \\ 0 & \text{if } (x, y) = (0, 0) \end{cases}$

15. $f(x, y) = \dfrac{xy}{\sqrt{16 - x^2 - y^2}}$

16. $f(x, y) = \dfrac{y}{\sqrt{x^2 - y^2 - 4}}$

17. $f(x, y) = \dfrac{x}{\sqrt{4x^2 + 9y^2 - 36}}$

18. $f(x, y) = \dfrac{x^2 + y^2}{\sqrt{9 - x^2 - y^2}}$

19. $f(x, y) = \sec^{-1}(xy)$

20. $f(x, y) = \ln(x^2 + y^2 - 9) - \ln(1 - x^2 - y^2)$
21. $f(x, y) = \sin^{-1}(x + y) + \ln(xy)$
22. $f(x, y) = \sin^{-1}(xy)$

23. $f(x, y) = \begin{cases} \dfrac{\sin(x + y)}{x + y} & \text{if } x + y \neq 0 \\ 1 & \text{if } x + y = 0 \end{cases}$

24. $f(x, y) = \begin{cases} \dfrac{x^2 - y^2}{x - y} & \text{if } x \neq y \\ x - y & \text{if } x = y \end{cases}$

In Exercises 25 through 31, the function is discontinuous at the origin because $f(0, 0)$ does not exist. Determine if the discontinuity is removable or essential. If the discontinuity is removable, redefine $f(0, 0)$ so that the new function is continuous at $(0, 0)$.

25. $f(x, y) = \dfrac{xy}{x^2 + xy + y^2}$ **26.** $f(x, y) = \dfrac{x}{x^2 + y^2}$

27. $f(x, y) = (x + y) \sin \dfrac{x}{x^2 + y^2}$

28. $f(x, y) = \dfrac{x^2 y^2}{x^2 + y^2}$ **29.** $f(x, y) = \dfrac{x^3 y^2}{x^6 + y^4}$

30. $f(x, y) = \dfrac{2y^2 - 3xy}{\sqrt{x^2 + y^2}}$ **31.** $f(x, y) = \dfrac{x^3 - 4xy^2}{x^2 + y^2}$

32. The function F is defined by

$$F(x, y) = \begin{cases} x^2 - 3y^2 & \text{if } x^2 - 3y^2 \leq 1 \\ 2 & \text{if } x^2 - 3y^2 > 1 \end{cases}$$

Show that F is continuous at all points (x, y) in R^2 except those on the hyperbola $x^2 - 3y^2 = 1$.

33. The function G is defined by

$$G(x, y) = \begin{cases} x^2 + 4y^2 & \text{if } x^2 + 4y^2 \leq 5 \\ 3 & \text{if } x^2 + 4y^2 > 5 \end{cases}$$

Show that G is continuous at all points (x, y) in R^2 except those on the ellipse $x^2 + 4y^2 = 5$.

34. (a) Give a definition of continuity at a point for a function of three variables, similar to Definition 16.3.2. (b) State theorems for functions of three variables similar to Theorem 16.3.3 and 16.3.7.

In Exercises 35 through 38, use the definitions and theorems of Exercise 34 to determine all points at which the function is continuous.

35. $f(x, y, z) = \dfrac{xz}{\sqrt{x^2 + y^2 + z^2 - 1}}$

36. $f(x, y, z) = \ln(36 - 4x^2 - y^2 - 9z^2)$

37. $f(x, y, z) = \begin{cases} \dfrac{3xyz}{x^2 + y^2 + z^2} & \text{if } (x, y, z) \neq (0, 0, 0) \\ 0 & \text{if } (x, y, z) = (0, 0, 0) \end{cases}$

38. $f(x, y, z) = \begin{cases} \dfrac{xz - y^2}{x^2 + y^2 + z^2} & \text{if } (x, y, z) \neq (0, 0, 0) \\ 0 & \text{if } (x, y, z) = (0, 0, 0) \end{cases}$

16.4 PARTIAL DERIVATIVES

The discussion of the differentiation of real-valued functions of n variables is reduced to the one-dimensional case by treating a function of n variables as a function of one variable at a time and holding the others fixed. This leads to the concept of a *partial derivative*. We first define the partial derivative of a function of two variables.

16.4.1 DEFINITION

Let f be a function of two variables, x and y. The **partial derivative of f with respect to x** is that function, denoted by $D_1 f$, such that its function value at any point (x, y) in the domain of f is given by

$$D_1 f(x, y) = \lim_{\Delta x \to 0} \frac{f(x + \Delta x, y) - f(x, y)}{\Delta x}$$

if this limit exists. Similarly, the **partial derivative of f with respect to y** is that function, denoted by $D_2 f$, such that its function value at any point (x, y) in the domain of f is given by

$$D_2 f(x, y) = \lim_{\Delta y \to 0} \frac{f(x, y + \Delta y) - f(x, y)}{\Delta y}$$

if this limit exists.

The process of finding a partial derivative is called **partial differentiation**.

$D_1 f$ is read as "D sub 1 of f," and this denotes the function that is the partial derivative of f with respect to the first variable. $D_1 f(x, y)$ is read as "D sub 1 of f of x and y," and this denotes the function value of $D_1 f$ at the point (x, y). Other notations for $D_1 f$ are f_1, f_x, and $\dfrac{\partial f}{\partial x}$. Other notations for $D_1 f(x, y)$ are $f_1(x, y), f_x(x, y)$, and $\dfrac{\partial f(x, y)}{\partial x}$. Similarly, other notations for $D_2 f$ are f_2, f_y, and $\dfrac{\partial f}{\partial y}$; other notations for $D_2 f(x, y)$ are $f_2(x, y), f_y(x, y)$, and $\dfrac{\partial f(x, y)}{\partial y}$. If $z = f(x, y)$, we can write $\dfrac{\partial z}{\partial x}$ for $D_1 f(x, y)$. A partial derivative cannot be thought of as a ratio of ∂z and ∂x because neither of these symbols has a separate meaning. The notation $\dfrac{dy}{dx}$ can be regarded as the quotient of two differentials when y is a function of the single variable x, but there is not a similar interpretation for $\dfrac{\partial z}{\partial x}$.

EXAMPLE 1   Apply Definition 16.4.1 to find $D_1 f(x, y)$ and $D_2 f(x, y)$ if

$$f(x, y) = 3x^2 - 2xy + y^2$$

Solution

$$D_1 f(x, y) = \lim_{\Delta x \to 0} \frac{f(x + \Delta x, y) - f(x, y)}{\Delta x}$$

$$= \lim_{\Delta x \to 0} \frac{3(x + \Delta x)^2 - 2(x + \Delta x)y + y^2 - (3x^2 - 2xy + y^2)}{\Delta x}$$

$$= \lim_{\Delta x \to 0} \frac{3x^2 + 6x\,\Delta x + 3(\Delta x)^2 - 2xy - 2y\,\Delta x + y^2 - 3x^2 + 2xy - y^2}{\Delta x}$$

$$= \lim_{\Delta x \to 0} \frac{6x\,\Delta x + 3(\Delta x)^2 - 2y\,\Delta x}{\Delta x}$$

$$= \lim_{\Delta x \to 0} (6x + 3\,\Delta x - 2y)$$

$$= 6x - 2y$$

$$D_2 f(x, y) = \lim_{\Delta y \to 0} \frac{f(x, y + \Delta y) - f(x, y)}{\Delta y}$$

$$= \lim_{\Delta y \to 0} \frac{3x^2 - 2x(y + \Delta y) + (y + \Delta y)^2 - (3x^2 - 2xy + y^2)}{\Delta y}$$

$$= \lim_{\Delta y \to 0} \frac{3x^2 - 2xy - 2x\,\Delta y + y^2 + 2y\,\Delta y + (\Delta y)^2 - 3x^2 + 2xy - y^2}{\Delta y}$$

$$= \lim_{\Delta y \to 0} \frac{-2x\Delta y + 2y\,\Delta y + (\Delta y)^2}{\Delta y}$$

$$= \lim_{\Delta y \to 0} (-2x + 2y + \Delta y)$$

$$= -2x + 2y$$

If (x_0, y_0) is a particular point in the domain of f, then

$$D_1 f(x_0, y_0) = \lim_{\Delta x \to 0} \frac{f(x_0 + \Delta x, y_0) - f(x_0, y_0)}{\Delta x} \tag{1}$$

if this limit exists, and

$$D_2 f(x_0, y_0) = \lim_{\Delta y \to 0} \frac{f(x_0, y_0 + \Delta y) - f(x_0, y_0)}{\Delta y} \tag{2}$$

if this limit exists.

▶ **ILLUSTRATION 1**   We apply formula (1) to find $D_1 f(3, -2)$ for the function f of Example 1.

$$\begin{aligned}
D_1 f(3, -2) &= \lim_{\Delta x \to 0} \frac{f(3 + \Delta x, -2) - f(3, -2)}{\Delta x} \\
&= \lim_{\Delta x \to 0} \frac{3(3 + \Delta x)^2 - 2(3 + \Delta x)(-2) + (-2)^2 - (27 + 12 + 4)}{\Delta x} \\
&= \lim_{\Delta x \to 0} \frac{27 + 18\,\Delta x + 3(\Delta x)^2 + 12 + 4\,\Delta x + 4 - 43}{\Delta x} \\
&= \lim_{\Delta x \to 0} (18 + 3\,\Delta x + 4) \\
&= 22
\end{aligned}$$

◀

Alternate formulas to (1) and (2) for $D_1 f(x_0, y_0)$ and $D_2 f(x_0, y_0)$ are given by

$$D_1 f(x_0, y_0) = \lim_{x \to x_0} \frac{f(x, y_0) - f(x_0, y_0)}{x - x_0} \tag{3}$$

if this limit exists, and

$$D_2 f(x_0, y_0) = \lim_{y \to y_0} \frac{f(x_0, y) - f(x_0, y_0)}{y - y_0} \tag{4}$$

if this limit exists.

▶ **ILLUSTRATION 2**   We apply formula (3) to find $D_1 f(3, -2)$ for the function f of Example 1.

$$\begin{aligned}
D_1 f(3, -2) &= \lim_{x \to 3} \frac{f(x, -2) - f(3, -2)}{x - 3} \\
&= \lim_{x \to 3} \frac{3x^2 + 4x + 4 - 43}{x - 3} \\
&= \lim_{x \to 3} \frac{3x^2 + 4x - 39}{x - 3} \\
&= \lim_{x \to 3} \frac{(3x + 13)(x - 3)}{x - 3} \\
&= \lim_{x \to 3} (3x + 13) \\
&= 22
\end{aligned}$$

◀

▶ **ILLUSTRATION 3** In Example 1 we showed that

$$D_1 f(x, y) = 6x - 2y$$

Therefore

$$\begin{aligned} D_1 f(3, -2) &= 18 + 4 \\ &= 22 \end{aligned}$$

This result agrees with those of Illustrations 1 and 2. ◀

Comparing Definition 16.4.1 with the definition of an ordinary derivative (3.1.3), we see that $D_1 f(x, y)$ is the ordinary derivative of f if f is considered as a function of one variable x (i.e., y is held constant), and $D_2 f(x, y)$ is the ordinary derivative of f if f is considered as a function of one variable y (and x is held constant). So the results in Example 1 can be obtained more easily by applying the theorems for ordinary differentiation if y is considered constant when finding $D_1 f(x, y)$ and if x is considered constant when finding $D_2 f(x, y)$. The following example illustrates this.

EXAMPLE 2 Find $f_x(x, y)$ and $f_y(x, y)$ if

$$f(x, y) = 3x^3 - 4x^2y + 3xy^2 + \sin xy^2$$

Solution Treating f as a function of x and holding y constant we have

$$f_x(x, y) = 9x^2 - 8xy + 3y^2 + y^2 \cos xy^2$$

Considering f as a function of y and holding x constant we have

$$f_y(x, y) = -4x^2 + 6xy + 2xy \cos xy^2$$

EXAMPLE 3 Given

$$f(x, y) = \begin{cases} \dfrac{xy(x^2 - y^2)}{x^2 + y^2} & \text{if } (x, y) \neq (0, 0) \\ 0 & \text{if } (x, y) = (0, 0) \end{cases}$$

Show that (a) $f_1(0, y) = -y$ for all y and (b) $f_2(x, 0) = x$ for all x.

Solution

(a) If $y \neq 0$, from (3),

$$\begin{aligned} f_1(0, y) &= \lim_{x \to 0} \frac{f(x, y) - f(0, y)}{x - 0} \\ &= \lim_{x \to 0} \frac{\dfrac{xy(x^2 - y^2)}{x^2 + y^2} - 0}{x} \\ &= \lim_{x \to 0} \frac{y(x^2 - y^2)}{x^2 + y^2} \\ &= -\frac{y^3}{y^2} \\ &= -y \end{aligned}$$

If $y = 0$, from (3)

$$\begin{aligned} f_1(0, 0) &= \lim_{x \to 0} \frac{f(x, 0) - f(0, 0)}{x - 0} \\ &= \lim_{x \to 0} \frac{0 - 0}{x} \\ &= \lim_{x \to 0} 0 \\ &= 0 \end{aligned}$$

Because $f_1(0, y) = -y$ if $y \neq 0$ and $f_1(0, 0) = 0$, we can conclude that $f_1(0, y) = -y$ for all y.

(b) If $x \neq 0$, from (4),   If $x = 0$, from (4),

$$f_2(x, 0) = \lim_{y \to 0} \frac{f(x, y) - f(x, 0)}{y - 0} \qquad f_2(0, 0) = \lim_{y \to 0} \frac{f(0, y) - f(0, 0)}{y - 0}$$

$$= \lim_{y \to 0} \frac{\dfrac{xy(x^2 - y^2)}{x^2 + y^2} - 0}{y} \qquad = \lim_{y \to 0} \frac{0 - 0}{y}$$

$$= \lim_{y \to 0} \frac{x(x^2 - y^2)}{x^2 + y^2} \qquad = \lim_{y \to 0} 0$$

$$= \frac{x^3}{x^2} \qquad\qquad = 0$$

$$= x$$

Because $f_2(x, 0) = x$ if $x \neq 0$ and $f_2(0, 0) = 0$, then $f_2(x, 0) = x$ for all x.

Geometric interpretations of the partial derivatives of a function of two variables are similar to those of a function of one variable. The graph of a function f of two variables is a surface having equation $z = f(x, y)$. If y is held constant (say, $y = y_0$), then $z = f(x, y_0)$ is an equation of the trace of this surface in the plane $y = y_0$. The curve can be represented by the two equations

$$y = y_0 \quad \text{and} \quad z = f(x, y) \tag{5}$$

because the curve is the intersection of these two surfaces.

Then $D_1 f(x_0, y_0)$ is the slope of the tangent line to the curve given by (5) at the point $P_0(x_0, y_0, f(x_0, y_0))$ in the plane $y = y_0$. In an analogous fashion, $D_2 f(x_0, y_0)$ represents the slope of the tangent line to the curve having equations

$$x = x_0 \quad \text{and} \quad z = f(x, y)$$

at the point P_0 in the plane $x = x_0$. Figure 1(a) and (b) shows the portions of the curves and the tangent lines.

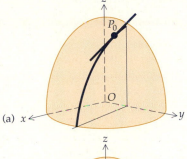

(a)

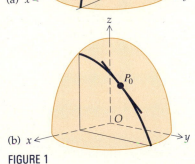

(b)

FIGURE 1

EXAMPLE 4   Find the slope of the tangent line to the curve of intersection of the surface

$$z = \tfrac{1}{2}\sqrt{24 - x^2 - 2y^2}$$

with the plane $y = 2$ at the point $(2, 2, \sqrt{3})$.

Solution   The required slope is the value of $\dfrac{\partial z}{\partial x}$ at the point $(2, 2, \sqrt{3})$.

$$\frac{\partial z}{\partial x} = \frac{-x}{2\sqrt{24 - x^2 - 2y^2}}$$

So at $(2, 2, \sqrt{3})$,

$$\frac{\partial z}{\partial x} = \frac{-2}{2\sqrt{12}}$$

$$= -\frac{1}{2\sqrt{3}}$$

Because every derivative is a measure of a rate of change, a partial derivative can be so interpreted. If f is a function of the two variables x and y, the partial derivative of f with respect to x at the point $P_0(x_0, y_0)$ gives the instantaneous rate of change, at P_0, of $f(x, y)$ per unit change in x (x alone varies and y is held fixed at y_0). Similarly, the partial derivative of f with respect to y at P_0 gives the instantaneous rate of change, at P_0, of $f(x, y)$ per unit change in y.

EXAMPLE 5   According to the *ideal gas law* for a confined gas, if P atmospheres is the pressure, V liters is the volume, and T is the absolute temperature on the Kelvin scale, we have the formula

$$PV = kT \tag{6}$$

where k is a constant of proportionality. Suppose that the volume of a gas in a certain container is 12 liters and the temperature is 290 K with $k = 0.6$. (a) Find the instantaneous rate of change of P per unit change in T if V remains fixed at 12. (b) Use the result of part (a) to approximate the change in the pressure if the temperature is increased to 295 K. (c) Find the instantaneous rate of change of V per unit change in P if T remains fixed at 290. (d) Suppose that the temperature is held constant. Use the result of part (c) to find the approximate change in the volume necessary to produce the same change in the pressure as obtained in part (b).

Solution   Substituting $V = 12$, $T = 290$, and $k = 0.6$ in (6), we obtain $P = 14.5$.

(a) Solving (6) for P when $k = 0.6$ we get

$$P = \frac{0.6T}{V}$$

The instantaneous rate of change of P per unit change in T if V remains constant is $\frac{\partial P}{\partial T}$, and

$$\frac{\partial P}{\partial T} = \frac{0.6}{V}$$

When $T = 290$ and $V = 12$, $\frac{\partial P}{\partial T} = 0.05$, which is the answer required.

(b) From the result of part (a) when T is increased by 5 (290 K to 295 K) and V remains fixed, an approximate increase in P is $5(0.05) = 0.25$. We conclude then that if the temperature is increased from 290 K to 295 K the increase in the pressure is approximately 0.25 atm.

(c) Solving (6) for V when $k = 0.6$, we obtain

$$V = \frac{0.6T}{P}$$

The instantaneous rate of change of V per unit change in P if T remains fixed is $\dfrac{\partial V}{\partial P}$, and

$$\frac{\partial V}{\partial P} = -\frac{0.6T}{P^2}$$

When $T = 290$ and $P = 14.5$,

$$\frac{\partial V}{\partial P} = -\frac{0.6(290)}{(14.5)^2}$$

$$= -0.83$$

which is the instantaneous rate of change of V per unit change in P when $T = 290$ and $P = 14.5$ if T remains fixed at 290.

(d) If P is to be increased by 0.25 and T is held fixed, then from the result of part (c) the change in V should be approximately $(0.25)(-0.83) = -0.21$. Hence the volume should be decreased by approximately 0.21 liter if the pressure is to be increased from 14.5 atm to 14.75 atm.

We now extend the concept of partial derivative to functions of n variables.

16.4.2 DEFINITION Let $P(x_1, x_2, \ldots, x_n)$ be a point in R^n, and let f be a function of the n variables $x_1, x_2, \ldots, x_n$. Then the partial derivative of f with respect to x_k is that function, denoted by $D_k f$, such that its function value at any point P in the domain of f is given by

$$D_k f(x_1, x_2, \ldots, x_n)$$
$$= \lim_{\Delta x_k \to 0} \frac{f(x_1, x_2, \ldots, x_{k-1}, x_k + \Delta x_k, x_{k+1}, \ldots, x_n) - f(x_1, x_2, \ldots, x_n)}{\Delta x_k}$$

if this limit exists.

In particular, if f is a function of the three variables x, y, and z, then the partial derivatives of f are given by

$$D_1 f(x, y, z) = \lim_{\Delta x \to 0} \frac{f(x + \Delta x, y, z) - f(x, y, z)}{\Delta x}$$

$$D_2 f(x, y, z) = \lim_{\Delta y \to 0} \frac{f(x, y + \Delta y, z) - f(x, y, z)}{\Delta y}$$

$$D_3 f(x, y, z) = \lim_{\Delta z \to 0} \frac{f(x, y, z + \Delta z) - f(x, y, z)}{\Delta z}$$

if these limits exist.

EXAMPLE 6   Given $f(x, y, z) = x^2y + yz^2 + z^3$, verify that

$$xf_1(x, y, z) + yf_2(x, y, z) + zf_3(x, y, z) = 3f(x, y, z)$$

Solution   Holding y and z constant we get

$$f_1(x, y, z) = 2xy$$

Holding x and z constant we obtain

$$f_2(x, y, z) = x^2 + z^2$$

Holding x and y constant we get

$$f_3(x, y, z) = 2yz + 3z^2$$

Therefore

$$\begin{aligned} xf_1(x, y, z) + yf_2(x, y, z) + zf_3(x, y, z) &= x(2xy) + y(x^2 + z^2) + z(2yz + 3z^2) \\ &= 2x^2y + x^2y + yz^2 + 2yz^2 + 3z^3 \\ &= 3(x^2y + yz^2 + z^3) \\ &= 3f(x, y, z) \end{aligned}$$

EXERCISES 16.4

In Exercises 1 through 6, apply Definition 16.4.1 to find the partial derivative.

1. $f(x, y) = 6x + 3y - 7$; $D_1 f(x, y)$
2. $f(x, y) = 4x^2 - 3xy$; $D_1 f(x, y)$
3. $f(x, y) = 3xy + 6x - y^2$; $D_2 f(x, y)$
4. $f(x, y) = xy^2 - 5y + 6$; $D_2 f(x, y)$
5. $f(x, y) = \sqrt{x^2 + y^2}$; $f_x(x, y)$
6. $f(x, y) = \dfrac{x + 2y}{x^2 - y}$; $f_y(x, y)$

In Exercises 7 through 10, apply Definition 16.4.2 to find the partial derivative.

7. $f(x, y, z) = x^2y - 3xy^2 + 2yz$; $D_2 f(x, y, z)$
8. $f(x, y, z) = x^2 + 4y^2 + 9z^2$; $D_1 f(x, y, z)$
9. $f(x, y, z, r, t) = xyr + yzt + yrt + zrt$; $f_r(x, y, z, r, t)$
10. $f(r, s, t, u, v, w) = 3r^2st + st^2v - 2tuv^2 - tvw + 3uw^2$; $f_v(r, s, t, u, v, w)$

11. Given $f(x, y) = x^2 - 9y^2$. Find $D_1 f(2, 1)$ by (a) applying formula (1); (b) applying formula (3); (c) applying Definition 16.4.1 and then replacing x and y by 2 and 1, respectively.
12. For the function in Exercise 11, find $D_2 f(2, 1)$ by (a) applying formula (2); (b) applying formula (4); (c) applying Definition 16.4.1 and then replacing x and y by 2 and 1, respectively.

In Exercises 13 through 24, find the partial derivative by holding all but one of the variables constant and applying theorems for ordinary differentiation.

13. $f(x, y) = 4y^3 + \sqrt{x^2 + y^2}$; $D_1 f(x, y)$
14. $f(x, y) = \dfrac{x + y}{\sqrt{y^2 - x^2}}$; $D_2 f(x, y)$

15. $f(\theta, \phi) = \sin 3\theta \cos 2\phi$; $f_\phi(\theta, \phi)$
16. $f(r, \theta) = r^2 \cos \theta - 2r \tan \theta$; $f_\theta(r, \theta)$
17. $z = e^{y/x} \ln \dfrac{x^2}{y}$; $\dfrac{\partial z}{\partial y}$
18. $r = e^{-\theta} \cos(\theta + \phi)$; $\dfrac{\partial r}{\partial \theta}$
19. $u = (x^2 + y^2 + z^2)^{-1/2}$; $\dfrac{\partial u}{\partial z}$
20. $u = \tan^{-1}(xyzw)$; $\dfrac{\partial u}{\partial w}$
21. $f(x, y, z) = 4xyz + \ln(2xyz)$; $f_3(x, y, z)$
22. $f(x, y, z) = e^{xy} \sinh 2z - e^{xy} \cosh 2z$; $f_z(x, y, z)$
23. $f(x, y, z) = e^{xyz} + \tan^{-1} \dfrac{3xy}{z^2}$; $f_y(x, y, z)$
24. $f(r, \theta, \phi) = 4r^2 \sin \theta + 5e^r \cos \theta \sin \phi - 2 \cos \phi$; $f_2(r, \theta, \phi)$
25. If $f(r, \theta) = r \tan \theta - r^2 \sin \theta$, find (a) $f_1(\sqrt{2}, \frac{1}{4}\pi)$; (b) $f_2(3, \pi)$.
26. If $f(x, y, z) = e^{xy^2} + \ln(y + z)$, find (a) $f_1(3, 0, 17)$; (b) $f_2(1, 0, 2)$; (c) $f_3(0, 0, 1)$.

In Exercises 27 and 28, find $f_x(x, y)$ and $f_y(x, y)$.

27. $f(x, y) = \displaystyle\int_x^y \ln \sin t \, dt$ **28.** $f(x, y) = \displaystyle\int_x^y e^{\cos t} \, dt$

29. Given $u = \sin \dfrac{r}{t} + \ln \dfrac{t}{r}$. Verify $t \dfrac{\partial u}{\partial t} + r \dfrac{\partial u}{\partial r} = 0$.
30. Given $w = x^2y + y^2z + z^2x$. Verify

$$\dfrac{\partial w}{\partial x} + \dfrac{\partial w}{\partial y} + \dfrac{\partial w}{\partial z} = (x + y + z)^2$$

31. Given $f(x, y) = \begin{cases} \dfrac{x^3 + y^3}{x^2 + y^2} & \text{if } (x, y) \neq (0, 0) \\ 0 & \text{if } (x, y) = (0, 0) \end{cases}$

Find (a) $f_1(0, 0)$; (b) $f_2(0, 0)$.

32. Given $f(x, y) = \begin{cases} \dfrac{x^2 - xy}{x + y} & \text{if } (x, y) \neq (0, 0) \\ 0 & \text{if } (x, y) = (0, 0) \end{cases}$

Find (a) $f_1(0, y)$ if $y \neq 0$; (b) $f_1(0, 0)$.

33. For the function of Exercise 32 find (a) $f_2(x, 0)$ if $x \neq 0$; (b) $f_2(0, 0)$.

34. Find the slope of the tangent line to the curve of intersection of the surface $36x^2 - 9y^2 + 4z^2 + 36 = 0$ with the plane $x = 1$ at the point $(1, \sqrt{12}, -3)$. Interpret this slope as a partial derivative.

35. Find the slope of the tangent line to the curve of intersection of the surface $z = x^2 + y^2$ with the plane $y = 1$ at the point $(2, 1, 5)$. Draw a sketch. Interpret this slope as a partial derivative.

36. Find equations of the tangent line to the curve of intersection of the surface $x^2 + y^2 + z^2 = 9$ with the plane $y = 2$ at the point $(1, 2, 2)$.

37. The temperature at any point (x, y) of a flat plate is T degrees and $T = 54 - \frac{2}{3}x^2 - 4y^2$. If distance is measured in centimeters, find the rate of change of the temperature with respect to the distance moved along the plate in the directions of the positive x and y axes, respectively, at the point $(3, 1)$.

38. Use the ideal gas law for a confined gas (see Example 5) to show that

$$\frac{\partial V}{\partial T} \cdot \frac{\partial T}{\partial P} \cdot \frac{\partial P}{\partial V} = -1$$

39. If V dollars is the present value of an ordinary annuity of equal payments of \$100 per year for t years at an interest rate of $100i$ percent per year, then

$$V = 100 \left[\frac{1 - (1 + i)^{-t}}{i} \right]$$

(a) Find the instantaneous rate of change of V per unit change in i if t remains fixed at 8. (b) Use the result of part (a) to find the approximate change in the present value if the interest rate changes from 6% to 7% and the time remains fixed at 8 years. (c) Find the instantaneous rate of change of V per unit change in t if i remains fixed at 0.06. (d) Use the result of part (c) to find the approximate change in the present value if the time is decreased from 8 to 7 years and the interest rate remains fixed at 6%.

40. Suppose that $10,000x$ dollars is the inventory carried in a store employing y clerks, P dollars is the weekly profit of the store, and

$$P = 3000 + 240y + 20y(x - 2y) - 10(x - 12)^2$$

where $15 \leq x \leq 25$ and $5 \leq y \leq 12$. At present the inventory is \$180,000 and there are 8 clerks. (a) Find the instantaneous rate of change of P per unit change in x if y remains fixed at 8. (b) Use the result of part (a) to find the approximate change in the weekly profit if the inventory changes from \$180,000 to \$200,000 and the number of clerks remains fixed at 8. (c) Find the instantaneous rate of change of P per unit change in y if x remains fixed at 18. (d) Use the result of part (c) to find the approximate change in the weekly profit if the number of clerks is increased from 8 to 10 and the inventory remains fixed at \$180,000.

41. If S square meters is a person's body surface area, then a formula giving an approximate value of S is

$$S = 2W^{0.4}H^{0.7}$$

where W kilograms is the person's weight and H meters is the person's height. When $W = 70$ and $H = 1.8$, find $\dfrac{\partial S}{\partial W}$ and $\dfrac{\partial S}{\partial H}$, and interpret the results.

16.5 DIFFERENTIABILITY AND THE TOTAL DIFFERENTIAL

We shall define *differentiability* of functions of more than one variable by means of an equation involving the increment of a function. To motivate this definition we first obtain a representation for the increment of a function of a single variable that is similar to what will appear in our Definition 16.5.2 of differentiability. We discussed the increment of a function of a single variable in Section 3.1, and recall from that section that if f is a differentiable function of x, and $y = f(x)$, then

$$f'(x) = \lim_{\Delta x \to 0} \frac{\Delta y}{\Delta x}$$

where Δx and Δy are increments of x and y and

$$\Delta y = f(x + \Delta x) - f(x)$$

When $|\Delta x|$ is small and $\Delta x \neq 0$, $\Delta y/\Delta x$ differs from $f'(x)$ by a small number that depends on Δx, which we shall denote by ϵ. Thus

$$\epsilon = \frac{\Delta y}{\Delta x} - f'(x) \qquad \text{if } \Delta x \neq 0$$

where ϵ is a function of Δx. From this equation we obtain

$$\Delta y = f'(x)\,\Delta x + \epsilon\,\Delta x$$

where ϵ is a function of Δx and $\epsilon \to 0$ as $\Delta x \to 0$.

From the above it follows that if the function f is differentiable at x_0, the increment of f at x_0, denoted by $\Delta f(x_0)$, is given by

$$\Delta f(x_0) = f'(x_0)\,\Delta x + \epsilon\,\Delta x \qquad \text{where} \quad \lim_{\Delta x \to 0} \epsilon = 0$$

For functions of two or more variables an equation corresponding to this one is used to define differentiability. And from the definition we determine criteria for a function to be differentiable at a point. We give the details for a function of two variables and begin by defining the *increment* of such a function.

16.5.1 DEFINITION If f is a function of two variables x and y, then the **increment of f** at the point (x_0, y_0), denoted by $\Delta f(x_0, y_0)$, is given by

$$\Delta f(x_0, y_0) = f(x_0 + \Delta x, y_0 + \Delta y) - f(x_0, y_0)$$

Figure 1 illustrates this definition for a function that is continuous on an open disk containing the points (x_0, y_0) and $(x_0 + \Delta x, y_0 + \Delta y)$. The figure shows a portion of the surface $z = f(x, y)$. $\Delta f(x_0, y_0) = \overline{QR}$, where Q is the point $(x_0 + \Delta x, y_0 + \Delta y, f(x_0, y_0))$ and R is the point having coordinates $(x_0 + \Delta x, y_0 + \Delta y, f(x_0 + \Delta x, y_0 + \Delta y))$.

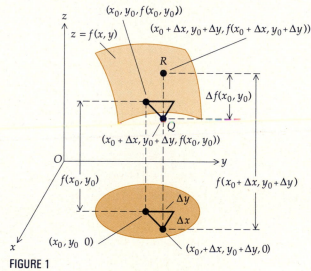

FIGURE 1

▶ **ILLUSTRATION 1** For the function f defined by

$$f(x, y) = 3x - xy^2$$

we find the increment of f at any point (x_0, y_0).

$$\begin{aligned}
\Delta f(x_0, y_0) &= f(x_0 + \Delta x, y_0 + \Delta y) - f(x_0, y_0) \\
&= 3(x_0 + \Delta x) - (x_0 + \Delta x)(y_0 + \Delta y)^2 - (3x_0 - x_0 y_0{}^2) \\
&= 3x_0 + 3\,\Delta x - x_0 y_0{}^2 - y_0{}^2\,\Delta x - 2x_0 y_0\,\Delta y - 2y_0\,\Delta x\,\Delta y \\
&\qquad - x_0(\Delta y)^2 - \Delta x(\Delta y)^2 - 3x_0 + x_0 y_0{}^2 \\
&= 3\,\Delta x - y_0{}^2\,\Delta x - 2x_0 y_0\,\Delta y - 2y_0\,\Delta x\,\Delta y - x_0(\Delta y)^2 - \Delta x(\Delta y)^2
\end{aligned}$$

◀

16.5.2 DEFINITION

If f is a function of two variables x and y and the increment of f at (x_0, y_0) can be written as

$$\Delta f(x_0, y_0) = D_1 f(x_0, y_0)\, \Delta x + D_2 f(x_0, y_0)\, \Delta y + \epsilon_1\, \Delta x + \epsilon_2\, \Delta y$$

where ϵ_1 and ϵ_2 are functions of Δx and Δy such that $\epsilon_1 \to 0$ and $\epsilon_2 \to 0$ as $(\Delta x, \Delta y) \to (0, 0)$, then f is said to be **differentiable** at (x_0, y_0).

▶ **ILLUSTRATION 2**   We use Definition 16.5.2 to prove that the function of Illustration 1 is differentiable at all points in R^2. We must show that for all points (x_0, y_0) in R^2 we can find an ϵ_1 and an ϵ_2 such that

$$\Delta f(x_0, y_0) - D_1 f(x_0, y_0)\, \Delta x - D_2 f(x_0, y_0)\, \Delta y = \epsilon_1\, \Delta x + \epsilon_2\, \Delta y$$

and $\epsilon_1 \to 0$ and $\epsilon_2 \to 0$ as $(\Delta x, \Delta y) \to (0, 0)$.

Because $f(x, y) = 3x - xy^2$,

$$D_1 f(x_0, y_0) = 3 - y_0{}^2 \quad \text{and} \quad D_2 f(x_0, y_0) = -2x_0 y_0$$

With these values and the value of $\Delta f(x_0, y_0)$ from Illustration 1,

$$\Delta f(x_0, y_0) - D_1 f(x_0, y_0)\, \Delta x - D_2 f(x_0, y_0)\, \Delta y = -x_0(\Delta y)^2 - 2y_0\, \Delta x\, \Delta y - \Delta x(\Delta y)^2$$

The right side of the above equation can be written in the following ways:

$$[-2y_0\, \Delta y - (\Delta y)^2]\, \Delta x + (-x_0\, \Delta y)\, \Delta y$$

$$(-2y_0\, \Delta y)\, \Delta x + (-\Delta x\, \Delta y - x_0\, \Delta y)\, \Delta y$$

$$[-(\Delta y)^2]\, \Delta x + (-2y_0\, \Delta x - x_0\, \Delta y)\, \Delta y$$

$$0 \cdot \Delta x + [-2y_0\, \Delta x - \Delta x\, \Delta y - x_0\, \Delta y]\, \Delta y$$

So there are at least four possible pairs of values for ϵ_1 and ϵ_2:

$$\epsilon_1 = -2y_0\, \Delta y - (\Delta y)^2 \quad \text{and} \quad \epsilon_2 = -x_0\, \Delta y$$

$$\epsilon_1 = -2y_0\, \Delta y \qquad\qquad \text{and} \quad \epsilon_2 = -\Delta x\, \Delta y - x_0\, \Delta y$$

$$\epsilon_1 = -(\Delta y)^2 \qquad\qquad \text{and} \quad \epsilon_2 = -2y_0\, \Delta x - x_0\, \Delta y$$

$$\epsilon_1 = 0 \qquad\qquad\qquad \text{and} \quad \epsilon_2 = -2y_0\, \Delta x - \Delta x\, \Delta y - x_0\, \Delta y$$

For each pair,

$$\lim_{(\Delta x, \Delta y) \to (0,0)} \epsilon_1 = 0 \quad \text{and} \quad \lim_{(\Delta x, \Delta y) \to (0,0)} \epsilon_2 = 0$$

It should be noted that it is only necessary to find one pair of values for ϵ_1 and ϵ_2.   ◀

16.5.3 THEOREM

If a function f of two variables is differentiable at a point, it is continuous at that point.

Proof   If f is differentiable at the point (x_0, y_0), it follows from Definition 16.5.2 that

$$f(x_0 + \Delta x, y_0 + \Delta y) - f(x_0, y_0) = D_1 f(x_0, y_0)\, \Delta x + D_2 f(x_0, y_0)\, \Delta y + \epsilon_1\, \Delta x + \epsilon_2\, \Delta y$$

where $\epsilon_1 \to 0$ and $\epsilon_2 \to 0$ as $(\Delta x, \Delta y) \to (0, 0)$. Therefore

$$f(x_0 + \Delta x, y_0 + \Delta y) = f(x_0, y_0) + D_1 f(x_0, y_0)\, \Delta x + D_2 f(x_0, y_0)\, \Delta y + \epsilon_1\, \Delta x + \epsilon_2\, \Delta y$$

Taking the limit on both sides of the above as $(\Delta x, \Delta y) \to (0, 0)$ we obtain

$$\lim_{(\Delta x, \Delta y) \to (0,0)} f(x_0 + \Delta x, y_0 + \Delta y) = f(x_0, y_0) \tag{1}$$

If we let $x_0 + \Delta x = x$ and $y_0 + \Delta y = y$, then "$(\Delta x, \Delta y) \to (0, 0)$" is equivalent to "$(x, y) \to (x_0, y_0)$." Thus, from (1),

$$\lim_{(x,y) \to (x_0,y_0)} f(x, y) = f(x_0, y_0)$$

which proves that f is continuous at (x_0, y_0),   ■

Theorem 16.5.3 states that for a function of two variables *differentiability implies continuity*. However, the mere existence of the partial derivatives $D_1 f$ and $D_2 f$ at a point does not imply differentiability at that point. The following example illustrates this.

EXAMPLE 1   Given

$$f(x, y) = \begin{cases} \dfrac{xy}{x^2 + y^2} & \text{if } (x, y) \neq (0, 0) \\ 0 & \text{if } (x, y) = (0, 0) \end{cases}$$

prove that $D_1 f(0, 0)$ and $D_2 f(0, 0)$ exist but that f is not differentiable at $(0, 0)$.

Solution

$$D_1 f(0, 0) = \lim_{x \to 0} \frac{f(x, 0) - f(0, 0)}{x - 0} \qquad D_2 f(0, 0) = \lim_{y \to 0} \frac{f(0, y) - f(0, 0)}{y - 0}$$

$$= \lim_{x \to 0} \frac{0 - 0}{x} \qquad\qquad\qquad = \lim_{y \to 0} \frac{0 - 0}{y}$$

$$= \lim_{x \to 0} 0 \qquad\qquad\qquad\qquad = \lim_{y \to 0} 0$$

$$= 0 \qquad\qquad\qquad\qquad\qquad = 0$$

Therefore both $D_1 f(0, 0)$ and $D_2 f(0, 0)$ exist.

In Example 4 of Section 16.2 we demonstrated that for this function $\lim_{(x,y) \to (0,0)} f(x, y)$ does not exist; hence f is not continuous at $(0, 0)$. Because f is not continuous at $(0, 0)$, it follows from Theorem 16.5.3 that f is not differentiable there.

The following theorem, which is easier to apply than Definition 16.5.2, gives conditions guaranteeing that a function is differentiable at a point. Its proof appears in Supplementary Section 16.8.

16.5.4 THEOREM   Let f be a function of two variables x and y, and suppose that $D_1 f$ and $D_2 f$ exist on an open disk $B(P_0; r)$, where P_0 is the point (x_0, y_0). Then if $D_1 f$ and $D_2 f$ are continuous at P_0, f is differentiable at P_0.

EXAMPLE 2   Use Theorem 16.5.4 to prove that the function defined by

$$f(x, y) = x^3 + 3xy - 5y^3$$

is differentiable everywhere.

Solution   We compute the partial derivatives:

$$D_1 f(x, y) = 3x^2 + 3y \qquad D_2 f(x, y) - 3x - 15y^2$$

Because $D_1 f$ and $D_2 f$ are continuous everywhere, it follows from Theorem 16.5.4 that f is differentiable everywhere.

The argument given for the polynomial function f in Example 2 can be applied to any polynomial function. Thus all polynomial functions are differentiable everywhere.

Observe that the conditions given in Theorem 16.5.4 are sufficient to prove that a function is differentiable at a point. However, they are not necessary conditions. That is, it is possible for a function to be differentiable at a point even if its partial derivatives are not continuous there. An example of such a function appears in Exercises 42 through 45. A function satisfying the hypothesis of Theorem 16.5.4 is said to be **continuously differentiable** at the point P_0. Thus continuous differentiability at a point is a sufficient condition, but not a necessary one, for differentiability at the point.

EXAMPLE 3   Given

$$f(x, y) = \begin{cases} \dfrac{x^2 y^2}{x^2 + y^2} & \text{if } (x, y) \neq (0, 0) \\ 0 & \text{if } (x, y) = (0, 0) \end{cases}$$

use Theorem 16.5.4 to prove that f is differentiable at $(0, 0)$.

Solution   To find $D_1 f$ consider two cases: $(x, y) = (0, 0)$ and $(x, y) \neq (0, 0)$. If $(x, y) = (0, 0)$, we have

$$D_1 f(0, 0) = \lim_{x \to 0} \frac{f(x, 0) - f(0, 0)}{x - 0}$$

$$= \lim_{x \to 0} \frac{0 - 0}{x}$$

$$= 0$$

If $(x, y) \neq (0, 0)$, $f(x, y) = x^2 y^2/(x^2 + y^2)$. To find $D_1 f(x, y)$ we use the theorem for the ordinary derivative of a quotient and consider y as a constant.

$$D_1 f(x, y) = \frac{2xy^2(x^2 + y^2) - 2x(x^2 y^2)}{(x^2 + y^2)^2}$$

$$= \frac{2xy^4}{(x^2 + y^2)^2}$$

The function $D_1 f$ is therefore defined by

$$D_1 f(x, y) = \begin{cases} \dfrac{2xy^4}{(x^2 + y^2)^2} & \text{if } (x, y) \neq (0, 0) \\ 0 & \text{if } (x, y) = (0, 0) \end{cases}$$

In the same manner we obtain the function $D_2 f$ defined by

$$D_2 f(x, y) = \begin{cases} \dfrac{2x^4 y}{(x^2 + y^2)^2} & \text{if } (x, y) \neq (0, 0) \\ 0 & \text{if } (x, y) = (0, 0) \end{cases}$$

Both $D_1 f$ and $D_2 f$ exist on every open disk having its center at the origin. It remains to show that $D_1 f$ and $D_2 f$ are continuous at $(0, 0)$.

Because $D_1 f(0, 0) = 0$, $D_1 f$ will be continuous at $(0, 0)$ if

$$\lim_{(x,y) \to (0,0)} D_1 f(x, y) = 0$$

Therefore we must show that for any $\epsilon > 0$ there exists a $\delta > 0$ such that

$$\text{if} \quad 0 < \sqrt{x^2 + y^2} < \delta \quad \text{then} \quad \left| \frac{2xy^4}{(x^2 + y^2)^2} \right| < \epsilon \tag{2}$$

$$\left| \frac{2xy^4}{(x^2 + y^2)^2} \right| = \frac{2|x| y^4}{(x^2 + y^2)^2}$$

$$\leq \frac{2\sqrt{x^2 + y^2} \, (\sqrt{x^2 + y^2})^4}{(x^2 + y^2)^2}$$

$$= 2\sqrt{x^2 + y^2}$$

Thus a suitable choice for δ is $2\delta = \epsilon$, that is, $\delta = \tfrac{1}{2}\epsilon$. With this δ we have the following argument:

$$0 < \sqrt{x^2 + y^2} < \delta \quad \text{and} \quad \delta = \tfrac{1}{2}\epsilon$$

$$\Rightarrow \qquad\qquad 2\sqrt{x^2 + y^2} < 2(\tfrac{1}{2}\epsilon)$$

$$\Rightarrow \quad \frac{2\sqrt{x^2 + y^2}(\sqrt{x^2 + y^2})^4}{(x^2 + y^2)^2} < \epsilon$$

$$\Rightarrow \qquad\qquad \frac{2|x| y^4}{(x^2 + y^2)^2} < \epsilon$$

$$\Rightarrow \qquad\qquad \left| \frac{2xy^4}{(x^2 + y^2)^2} \right| < \epsilon$$

We have therefore shown that (2) holds. Hence $D_1 f$ is continuous at $(0, 0)$. In the same way we can show that $D_2 f$ is continuous at $(0, 0)$. It follows from Theorem 16.5.4 that f is differentiable at $(0, 0)$.

The equation in Definition 16.5.2 is

$$\Delta f(x_0, y_0) = D_1 f(x_0, y_0) \Delta x + D_2 f(x_0, y_0) \Delta y + \epsilon_1 \Delta x + \epsilon_2 \Delta y \tag{3}$$

The expression involving the first two terms on the right side of this equation is called the *principal part* of $\Delta f(x_0, y_0)$ or the *total differential* of f at (x_0, y_0).

16.5.5 DEFINITION If f is a function of two variables x and y, and f is differentiable at (x, y), then the **total differential** of f is the function df having function values given by

$$df(x, y, \Delta x, \Delta y) = D_1 f(x, y)\, \Delta x + D_2 f(x, y)\, \Delta y$$

Note that df is a function of the four variables x, y, Δx, and Δy. If $z = f(x, y)$, we sometimes use dz in place of $df(x, y, \Delta x, \Delta y)$, and write

$$dz = D_1 f(x, y)\, \Delta x + D_2 f(x, y)\, \Delta y \tag{4}$$

If in particular $f(x, y) = x$, then $z = x$, $D_1 f(x, y) = 1$, and $D_2 f(x, y) = 0$; so (4) gives $dz = \Delta x$. Because $z = x$, for this function $dx = \Delta x$. In a similar fashion, if we take $f(x, y) = y$, then $z = y$, $D_1 f(x, y) = 0$, and $D_2 f(x, y) = 1$; thus (4) gives $dz = \Delta y$. Because $z = y$, then for this function $dy = \Delta y$. Hence we define the differentials of the independent variables as $dx = \Delta x$ and $dy = \Delta y$. Then (4) can be written as

$$dz = D_1 f(x, y)\, dx + D_2 f(x, y)\, dy \tag{5}$$

and at the point (x_0, y_0),

$$dz = D_1 f(x_0, y_0)\, dx + D_2 f(x_0, y_0)\, dy \tag{6}$$

In (3) let $\Delta z = \Delta f(x_0, y_0)$, $dx = \Delta x$, and $dy = \Delta y$. Then

$$\Delta z = D_1 f(x_0, y_0)\, dx + D_2 f(x_0, y_0)\, dy + \epsilon_1\, dx + \epsilon_2\, dy$$

By comparing this equation and (6), observe that when dx (i.e., Δx) and dy (i.e., Δy) are close to zero, and because then ϵ_1 and ϵ_2 also will be close to zero, dz is an approximation to Δz. Before giving an example, we write (5) with the notation $\dfrac{\partial z}{\partial x}$ and $\dfrac{\partial z}{\partial y}$ instead of $D_1 f(x, y)$ and $D_2 f(x, y)$, respectively:

$$dz = \frac{\partial z}{\partial x}\, dx + \frac{\partial z}{\partial y}\, dy \tag{7}$$

EXAMPLE 4 A closed metal container in the shape of a right-circular cylinder is to have an inside height of 6 in., an inside radius of 2 in., and a thickness of 0.1 in. If the cost of the metal to be used is 20 cents per cubic inch, find by differentials the approximate cost of the metal to be used in the manufacture of the container.

Solution If V cubic inches is the volume of a right-circular cylinder of radius r inches and height h inches, then

$$V = \pi r^2 h$$

The exact volume of metal in the container is the difference between the volumes of two right-circular cylinders for which $r = 2.1$, $h = 6.2$, and $r = 2$, $h = 6$, respectively. The increment ΔV gives the exact volume of metal, but because only an approximate value is wanted, we find dV instead. From (7)

$$dV = \frac{\partial V}{\partial r}\, dr + \frac{\partial V}{\partial h}\, dh$$

$$= 2\pi r h\, dr + \pi r^2\, dh$$

With $r = 2$, $h = 6$, $dr = 0.1$, and $dh = 0.2$, we obtain

$$dV = 2\pi(2)(6)(0.1) + \pi(2)^2(0.2)$$
$$= 3.2\pi$$

Hence $\Delta V \approx 3.2\pi$; therefore there is approximately 3.2π in.^{3} of metal in the container. Because the cost of the metal is 20 cents per cubic inch, and $20 \cdot 3.2\pi = 64\pi$ and $64\pi \approx 201$, the approximate cost of the metal is $2.01.

We now extend the concepts of differentiability and the total differential to a function of n variables.

16.5.6 DEFINITION   If f is a function of the n variables $x_1, x_2, \ldots, x_n$, and $\bar{P}$ is the point $(\bar{x}_1, \bar{x}_2, \ldots, \bar{x}_n)$, then the **increment of f** at $\bar{P}$ is given by

$$\Delta f(\bar{P}) = f(\bar{x}_1 + \Delta x_1, \bar{x}_2 + \Delta x_2, \ldots, \bar{x}_n + \Delta x_n) - f(\bar{P})$$

16.5.7 DEFINITION   If f is a function of the n variables $x_1, x_2, \ldots, x_n$, and the increment of f at the point $\bar{P}$ can be written as

$$\Delta f(\bar{P}) = D_1 f(\bar{P}) \, \Delta x_1 + D_2 f(\bar{P}) \, \Delta x_2 + \ldots + D_n f(\bar{P}) \, \Delta x_n +$$
$$\epsilon_1 \, \Delta x_1 + \epsilon_2 \, \Delta x_2 + \ldots + \epsilon_n \, \Delta x_n$$

where $\epsilon_1 \to 0$, $\epsilon_2 \to 0$, $\ldots$, $\epsilon_n \to 0$, as

$$(\Delta x_1, \Delta x_2, \ldots, \Delta x_n) \to (0, 0, \ldots, 0)$$

then f is said to be **differentiable** at $\bar{P}$.

Analogously to Theorem 16.5.4 it can be proved that sufficient conditions for a function f of n variables to be differentiable at a point $\bar{P}$ are that $D_1 f$, $D_2 f, \ldots, D_n f$ all exist on an open ball $B(\bar{P}; r)$ and that $D_1 f, D_2 f, \ldots, D_n f$ are all continuous at $\bar{P}$. As was the case for functions of two variables, it follows that for functions of n variables differentiability implies continuity. However, the existence of the partial derivatives $D_1 f, D_2 f, \ldots, D_n f$ at a point does not imply differentiability of the function at the point.

16.5.8 DEFINITION   If f is a function of the n variables $x_1, x_2, \ldots, x_n$ and f is differentiable at P, then the **total differential** of f is the function df having function values given by

$$df(P, \Delta x_1, \Delta x_2, \ldots, \Delta x_n) = D_1 f(P) \, \Delta x_1 + D_2 f(P) \, \Delta x_2 + \ldots + D_n f(P) \, \Delta x_n$$

Letting $w = f(x_1, x_2, \ldots, x_n)$, defining $dx_1 = \Delta x_1$, $dx_2 = \Delta x_2, \ldots$, $dx_n = \Delta x_n$, and using the notation $\dfrac{\partial w}{\partial x_i}$ instead of $D_i f(P)$, we can write the equation of Definition 16.5.8 as

$$dw = \frac{\partial w}{\partial x_1} dx_1 + \frac{\partial w}{\partial x_2} dx_2 + \ldots + \frac{\partial w}{\partial x_n} dx_n \tag{8}$$

EXAMPLE 5   The dimensions of a box are measured to be 10 cm, 12 cm, and 15 cm, and the measurements are correct to 0.02 cm. Find approximately the greatest error if the volume of the box is calculated from the given measurements. Also find the approximate percent error.

Solution If V cubic centimeters is the volume of a box whose dimensions are x, y, and z centimeters,

$$V = xyz$$

The exact value of the error is found from ΔV; however, we use dV as an approximation to ΔV. From (8) for three independent variables,

$$dV = \frac{\partial V}{\partial x}\,dx + \frac{\partial V}{\partial y}\,dy + \frac{\partial V}{\partial z}\,dz$$

$$= yz\,dx + xz\,dy + xy\,dz$$

From the given information $|\Delta x| \leq 0.02$, $|\Delta y| \leq 0.02$, and $|\Delta z| \leq 0.02$. To find the greatest error in the volume we take the greatest error in the measurements of the three dimensions. So taking $dx = 0.02$, $dy = 0.02$, $dz = 0.02$, and $x = 10$, $y = 12$, $z = 15$, we have,

$$dV = (12)(15)(0.02) + (10)(15)(0.02) + (10)(12)(0.02)$$

$$= 9$$

Thus $\Delta V \approx 9$, and therefore the greatest possible error in the calculation of the volume from the given measurements is approximately 9 cm^3.

The relative error is found by dividing the error by the actual value. Hence the relative error in computing the volume from the given measurements is $\frac{\Delta V}{V} \approx \frac{dV}{V}$. Because $\frac{dV}{V} = \frac{9}{1800}$, $\frac{\Delta V}{V} \approx 0.005$. So the approximate percent error is 0.5%.

EXERCISES 16.5

1. If $f(x, y) = 3x^2 + 2xy - y^2$, find:
   (a) $\Delta f(1, 4)$, the increment of f at $(1, 4)$;
   (b) $\Delta f(1, 4)$ when $\Delta x = 0.03$ and $\Delta y = -0.02$;
   (c) $df(1, 4, \Delta x, \Delta y)$, the total differential of f at $(1, 4)$;
   (d) $df(1, 4. 0.03, -0.02)$.
2. If $f(x, y) = 2x^2 + 5xy + 4y^2$, find:
   (a) $\Delta f(2, -1)$, the increment of f at $(2, -1)$;
   (b) $\Delta f(2, -1)$ when $\Delta x = -0.01$ and $\Delta y = 0.02$;
   (c) $df(2, -1, \Delta x, \Delta y)$, the total differential of f at $(2, -1)$;
   (d) $df(2, -1, -0.01, 0.02)$.
3. If $g(x, y) = xye^{xy}$, find:
   (a) $\Delta g(2, -4)$, the increment of g at $(2, -4)$;
   (b) $\Delta g(2, -4)$ when $\Delta x = -0.1$ and $\Delta y = 0.2$;
   (c) $dg(2, -4, \Delta x, \Delta y)$, the total differential of g at $(2, -4)$;
   (d) $dg(2, -4, -0.1, 0.2)$.
4. If $h(x, y) = (x + y)/(x - y)$, find:
   (a) $\Delta h(3, 0)$, the increment of h at $(3, 0)$;
   (b) $\Delta h(3, 0)$ when $\Delta x = 0.04$ and $\Delta y = 0.03$;
   (c) $dh(3, 0, \Delta x, \Delta y)$, the total differential of h at $(3, 0)$;
   (d) $dh(3, 0, 0.04, 0.03)$.
5. If $F(x, y, z) = xy + \ln(yz)$, find:
   (a) $\Delta F(4, 1, 5)$, the increment of F at $(4, 1, 5)$;
   (b) $\Delta F(4, 1, 5)$ when $\Delta x = 0.02$, $\Delta y = 0.04$, and $\Delta z = -0.03$;
   (c) $dF(4, 1, 5, \Delta x, \Delta y, \Delta z)$, the total differential of F at $(4, 1, 5)$;
   (d) $dF(4, 1, 5, 0.02, 0.04, -0.03)$.

6. If $G(x, y, z) = x^2y + 2xyz - z^3$, find:
   (a) $\Delta G(-3, 0, 2)$, the increment of G at $(-3, 0, 2)$;
   (b) $\Delta G(-3, 0, 2)$ when $\Delta x = 0.01$, $\Delta y = 0.03$, $\Delta z = -0.01$;
   (c) $dG(-3, 0, 2, \Delta x, \Delta y, \Delta z)$, the total differential of G at $(-3, 0, 2)$;
   (d) $dG(-3, 0, 2, 0.01, 0.03, -0.01)$.

In Exercises 7 through 14, find the total differential dw.

7. $w = 4x^3 - xy^2 + 3y - 7$
8. $w = y \tan x^2 - 2xy$
9. $w = x \cos y - y \sin x$
10. $w = xe^{2y} + e^{-y}$
11. $w = \ln(x^2 + y^2 + z^2)$
12. $w = \dfrac{xyz}{x + y + z}$
13. $w = x \tan^{-1} z - \dfrac{y^2}{z}$
14. $w = e^{yz} - \cos xz$

In Exercises 15 through 18, prove that f is differentiable at all points in its domain by doing each of the following: (a) Find $\Delta f(x_0, y_0)$; (b) find an ϵ_1 and an ϵ_2 so that Equation (3) holds; (c) show that the ϵ_1 and ϵ_2 found in part (b) both approach zero as $(\Delta x, \Delta y) \to (0, 0)$.

15. $f(x, y) = x^2y - 2xy$
16. $f(x, y) = 2x^2 + 3y^2$
17. $f(x, y) = \dfrac{x^2}{y}$
18. $f(x, y) = \dfrac{y}{x}$

19. Given $f(x, y) = \begin{cases} x + y - 2 & \text{if } x = 1 \text{ or } y = 1 \\ 2 & \text{if } x \neq 1 \text{ and } y \neq 1 \end{cases}$

Prove that $D_1 f(1, 1)$ and $D_2 f(1, 1)$ exist but f is not differentiable at $(1, 1)$.

20. Given $f(x, y) = \begin{cases} \dfrac{3x^2 y^2}{x^4 + y^4} & \text{if } (x, y) \neq (0, 0) \\ 0 & \text{if } (x, y) = (0, 0) \end{cases}$

Prove that $D_1 f(0, 0)$ and $D_2 f(0, 0)$ exist but f is not differentiable at $(0, 0)$.

In Exercises 21 through 27, use Theorem 16.5.4 to prove that the function is differentiable at all points in its domain.

21. $g(x, y) = 2x^4 - 3x^2 y^2 + x^{-2} y^{-2}$

22. $f(x, y) = \dfrac{3x - 4y}{x^2 + 8y}$

23. $f(x, y) = 3 \ln xy + 5 \sin x$

24. $g(x, y) = y \ln x - \dfrac{x}{y}$

25. $h(x, y) = \tan^{-1}(x + y) + \dfrac{1}{x - y}$

26. $f(x, y) = e^{2x} \sin y + e^{-2x} \cos y$

27. $f(x, y) = ye^{3x} - xe^{-3y}$

28. Given $f(x, y) = \begin{cases} \dfrac{3x^2 y}{x^2 + y^2} & \text{if } (x, y) \neq (0, 0) \\ 0 & \text{if } (x, y) = (0, 0) \end{cases}$

This function is continuous at $(0, 0)$ (see Example 1, Section 16.3). Prove that $D_1 f(0, 0)$ and $D_2 f(0, 0)$ exist but $D_1 f$ and $D_2 f$ are not continuous at $(0, 0)$.

29. Given $f(x, y) = \begin{cases} \dfrac{xy(x^2 - y^2)}{x^2 + y^2} & \text{if } (x, y) \neq (0, 0) \\ 0 & \text{if } (x, y) = (0, 0) \end{cases}$

Prove that f is differentiable at $(0, 0)$ by using Theorem 16.5.4.

In Exercises 30 through 32, prove that f is differentiable at all points in R^3 by doing each of the following: (a) Find $\Delta f(x_0, y_0, z_0)$; (b) find an $\epsilon_1, \epsilon_2,$ and ϵ_3 such that the equation of Definition 16.5.7 holds; (c) show that the $\epsilon_1, \epsilon_2,$ and ϵ_3 found in (b) all approach zero as $(\Delta x, \Delta y, \Delta z)$ approaches $(0, 0, 0)$.

30. $f(x, y, z) = 3x + 2y - 4z$ **31.** $f(x, y, z) = xy - xz + z^2$

32. $f(x, y, z) = 2x^2 z - 3yz^2$

33. Given $f(x, y, z) = \begin{cases} \dfrac{xy^2 z}{x^4 + y^4 + z^4} & \text{if } (x, y, z) \neq (0, 0, 0) \\ 0 & \text{if } (x, y, z) = (0, 0, 0) \end{cases}$

(a) Show that $D_1 f(0, 0, 0)$, $D_2 f(0, 0, 0)$, and $D_3 f(0, 0, 0)$ exist; (b) make use of the fact that differentiability implies continuity to prove that f is not differentiable at $(0, 0, 0)$.

34. Given $f(x, y, z) = \begin{cases} \dfrac{xyz^2}{x^2 + y^2 + z^2} & \text{if } (x, y, z) \neq (0, 0, 0) \\ 0 & \text{if } (x, y, z) = (0, 0, 0) \end{cases}$

Prove that f is differentiable at $(0, 0, 0)$.

35. A closed container in the shape of a rectangular solid is to have an inside length of 8 m, an inside width of 5 m, an inside height of 4 m, and a thickness of 4 cm. Use differentials to approximate the amount of material needed to construct the container.

36. Use the total differential to find approximately the greatest error in calculating the area of a right triangle from the lengths of the legs if they are measured to be 6 cm and 8 cm, respectively, with a possible error of 0.1 cm for each measurement. Also find the approximate percent error.

37. Find approximately, by using the total differential, the greatest error in calculating the length of the hypotenuse of the right triangle from the measurements of Exercise 36. Also find the approximate percent error.

38. If the ideal gas law (see Example 5, Section 16.4) is used to find P when T and V are given, but there is an error of 0.3% in measuring T and an error of 0.8% in measuring V, find approximately the greatest percent error in P.

39. The specific gravity s of an object is given by the formula

$$s = \frac{A}{A - W}$$

where A pounds is the weight of the object in air and W pounds is the weight of the object in water. If the weight of an object in air is read as 20 lb with a possible error of 0.01 lb and its weight in water is read as 12 lb with a possible error of 0.02 lb, find approximately the largest possible error in calculating s from these measurements. Also find the largest possible relative error.

40. A wooden box is to be made of lumber that is $\frac{2}{3}$ in. thick. The inside length is to be 6 ft, the inside width is to be 3 ft, the inside depth is to be 4 ft, and the box is to have no top. Use the total differential to find the approximate amount of lumber to be used in the box.

41. A company has contracted to manufacture 10,000 closed wooden crates having dimensions 3 m, 4 m, and 5 m. The cost of the wood to be used is $3 per square meter. If the machines that are used to cut the pieces of wood have a possible error of 0.5 cm in each dimension, find approximately, by using the total differential, the greatest possible error in the estimate of the cost of the wood.

In Exercises 42 through 45, we show that a function may be differentiable at a point even though it is not continuously differentiable there. Hence the conditions of Theorem 16.5.4 are sufficient but not necessary for differentiability. The function f in these exercises is defined by

$$f(x, y) = \begin{cases} (x^2 + y^2) \sin \dfrac{1}{\sqrt{x^2 + y^2}} & \text{if } (x, y) \neq (0, 0) \\ 0 & \text{if } (x, y) = (0, 0) \end{cases}$$

42. Find $\Delta f(0, 0)$.

43. Find $D_1 f(x, y)$ and $D_2 f(x, y)$.

44. Prove that f is differentiable at $(0, 0)$ by using Definition 16.5.2 and the results of Exercises 42 and 43.

45. Prove that $D_1 f$ and $D_2 f$ are not continuous at $(0, 0)$.

16.6   THE CHAIN RULE   Recall that with Leibniz notation the chain rule for a function of a single variable is as follows: If y is a function of u and $\dfrac{dy}{du}$ exists, and u is a function of x and $\dfrac{du}{dx}$ exists, then y is a function of x and $\dfrac{dy}{dx}$ exists and is given by

$$\frac{dy}{dx} = \frac{dy}{du} \cdot \frac{du}{dx}$$

We now consider the chain rule for a function of two variables, where each of these variables is also a function of two variables.

16.6.1   THEOREM
The Chain Rule

If u is a differentiable function of x and y, defined by $u = f(x, y)$, where $x = F(r, s)$, $y = G(r, s)$, and $\dfrac{\partial x}{\partial r}, \dfrac{\partial x}{\partial s}, \dfrac{\partial y}{\partial r}$, and $\dfrac{\partial y}{\partial s}$ all exist, then u is a function of r and s and

$$\frac{\partial u}{\partial r} = \left(\frac{\partial u}{\partial x}\right)\left(\frac{\partial x}{\partial r}\right) + \left(\frac{\partial u}{\partial y}\right)\left(\frac{\partial y}{\partial r}\right)$$

$$\frac{\partial u}{\partial s} = \left(\frac{\partial u}{\partial x}\right)\left(\frac{\partial x}{\partial s}\right) + \left(\frac{\partial u}{\partial y}\right)\left(\frac{\partial y}{\partial s}\right)$$

Proof   We prove the chain rule for $\dfrac{\partial u}{\partial r}$. The proof for $\dfrac{\partial u}{\partial s}$ is similar.

If s is held fixed and r is changed by an amount Δr, then x is changed by an amount Δx and y is changed by an amount Δy. Thus

$$\Delta x = F(r + \Delta r, s) - F(r, s) \tag{1}$$

and

$$\Delta y = G(r + \Delta r, s) - G(r, s) \tag{2}$$

Because f is differentiable,

$$\Delta f(x, y) = D_1 f(x, y)\, \Delta x + D_2 f(x, y)\, \Delta y + \epsilon_1\, \Delta x + \epsilon_2\, \Delta y \tag{3}$$

where ϵ_1 and ϵ_2 both approach zero as $(\Delta x, \Delta y)$ approaches $(0, 0)$. Furthermore, we require that $\epsilon_1 = 0$ and $\epsilon_2 = 0$ when $(\Delta x, \Delta y) = (0, 0)$. We make this requirement so that ϵ_1 and ϵ_2, which are functions of Δx and Δy, will be continuous at $(\Delta x, \Delta y) = (0, 0)$.

If in (3) we replace $\Delta f(x, y)$ by Δu, $D_1 f(x, y)$ by $\dfrac{\partial u}{\partial x}$, and $D_2 f(x, y)$ by $\dfrac{\partial u}{\partial y}$ and divide on both sides by Δr ($\Delta r \neq 0$), we obtain

$$\frac{\Delta u}{\Delta r} = \frac{\partial u}{\partial x}\frac{\Delta x}{\Delta r} + \frac{\partial u}{\partial y}\frac{\Delta y}{\Delta r} + \epsilon_1\frac{\Delta x}{\Delta r} + \epsilon_2\frac{\Delta y}{\Delta r}$$

Taking the limit on both sides of the above as Δr approaches zero we get

$$\lim_{\Delta r \to 0}\frac{\Delta u}{\Delta r} = \frac{\partial u}{\partial x}\lim_{\Delta r \to 0}\frac{\Delta x}{\Delta r} + \frac{\partial u}{\partial y}\lim_{\Delta r \to 0}\frac{\Delta y}{\Delta r} + \left(\lim_{\Delta r \to 0}\epsilon_1\right)\lim_{\Delta r \to 0}\frac{\Delta x}{\Delta r} + \left(\lim_{\Delta r \to 0}\epsilon_2\right)\lim_{\Delta r \to 0}\frac{\Delta y}{\Delta r} \tag{4}$$

Because u is a function of x and y and both x and y are functions of r and s, u is a function of r and s. Because s is held fixed and r is changed by an amount Δr,

$$\lim_{\Delta r \to 0} \frac{\Delta u}{\Delta r} = \lim_{\Delta r \to 0} \frac{u(r + \Delta r, s) - u(r, s)}{\Delta r}$$

$$= \frac{\partial u}{\partial r} \tag{5}$$

Also

$$\lim_{\Delta r \to 0} \frac{\Delta x}{\Delta r} = \frac{\partial x}{\partial r} \quad \text{and} \quad \lim_{\Delta r \to 0} \frac{\Delta y}{\Delta r} = \frac{\partial y}{\partial r} \tag{6}$$

Because $\dfrac{\partial x}{\partial r}$ and $\dfrac{\partial y}{\partial r}$ exist, F and G are each continuous with respect to the variable r. (*Note:* The existence of the partial derivatives of a function does not imply continuity with respect to all of the variables simultaneously, as we saw in the preceding section, but as with functions of a single variable it does imply continuity of the function with respect to each variable separately.) Hence, from (1),

$$\lim_{\Delta r \to 0} \Delta x = \lim_{\Delta r \to 0} \left[F(r + \Delta r, s) - F(r, s) \right]$$

$$= F(r, s) - F(r, s)$$

$$= 0$$

and from (2),

$$\lim_{\Delta r \to 0} \Delta y = \lim_{\Delta r \to 0} \left[G(r + \Delta r, s) - G(r, s) \right]$$

$$= G(r, s) - G(r, s)$$

$$= 0$$

Therefore, as Δr approaches zero, both Δx and Δy approach zero. And because both ϵ_1 and ϵ_2 approach zero as $(\Delta x, \Delta y)$ approaches $(0, 0)$, we can conclude that

$$\lim_{\Delta r \to 0} \epsilon_1 = 0 \quad \text{and} \quad \lim_{\Delta r \to 0} \epsilon_2 = 0 \tag{7}$$

Now, it is possible that for certain values of Δr, $\Delta x = 0$ and $\Delta y = 0$. Because we required in such a case that $\epsilon_1 = 0$ and $\epsilon_2 = 0$, the limits in (7) are still zero. Substituting from (5), (6), and (7) into (4) we obtain

$$\frac{\partial u}{\partial r} = \left(\frac{\partial u}{\partial x} \right) \left(\frac{\partial x}{\partial r} \right) + \left(\frac{\partial u}{\partial y} \right) \left(\frac{\partial y}{\partial r} \right)$$

which we wished to prove.   ■

EXAMPLE 1   Given

$$u = \ln \sqrt{x^2 + y^2} \qquad x = re^s \qquad y = re^{-s}$$

find $\dfrac{\partial u}{\partial r}$ and $\dfrac{\partial u}{\partial s}$.

Solution

$$\frac{\partial u}{\partial x} = \frac{x}{x^2 + y^2} \qquad \frac{\partial u}{\partial y} = \frac{y}{x^2 + y^2} \qquad \frac{\partial x}{\partial r} = e^s$$

$$\frac{\partial x}{\partial s} = re^s \qquad \frac{\partial y}{\partial r} = e^{-s} \qquad \frac{\partial y}{\partial s} = -re^{-s}$$

From the chain rule we get

$$\frac{\partial u}{\partial r} = \frac{x}{x^2 + y^2}\,(e^s) + \frac{y}{x^2 + y^2}\,(e^{-s}) \qquad \frac{\partial u}{\partial s} = \frac{x}{x^2 + y^2}\,(re^s) + \frac{y}{x^2 + y^2}\,(-re^{-s})$$

$$= \frac{xe^s + ye^{-s}}{x^2 + y^2} \qquad\qquad\qquad = \frac{r(xe^s - ye^{-s})}{x^2 + y^2}$$

As mentioned earlier, the symbols $\dfrac{\partial u}{\partial r}, \dfrac{\partial u}{\partial s}, \dfrac{\partial u}{\partial x}, \dfrac{\partial u}{\partial y}$, and so forth, must not be considered as fractions. The symbols ∂u, ∂x, and so on, have no meaning by themselves. For functions of one variable, the chain rule is easily remembered by thinking of an ordinary derivative as the quotient of two differentials, but there is no similar interpretation for partial derivatives.

Another troublesome notational problem arises when considering u as a function of x and y and then as a function of r and s. If $u = f(x, y)$, $x = F(r, s)$, and $y = G(r, s)$, then $u = f(F(r, s), G(r, s))$. (It is incorrect to write $u = f(r, s)$.)

▶ **ILLUSTRATION 1**   In Example 1,

$$u = f(x, y) \qquad\quad x = F(r, s) \qquad y = G(r, s)$$
$$= \ln \sqrt{x^2 + y^2} \qquad\quad = re^s \qquad\qquad = re^{-s}$$

Thus

$$u = f(F(r, s), G(r, s))$$
$$= \ln \sqrt{r^2 e^{2s} + r^2 e^{-2s}}$$

Observe that

$$f(r, s) = \ln \sqrt{r^2 + s^2}$$

That is, $f(r, s) \neq u$.   ◀

If we let $f(F(r, s), G(r, s)) = h(r, s)$, then the equations of Theorem 16.6.1 can be written respectively as

$$h_1(r, s) = f_1(x, y)F_1(r, s) + f_2(x, y)G_1(r, s)$$

$$h_2(r, s) = f_1(x, y)F_2(r, s) + f_2(x, y)G_2(r, s)$$

In the statement of Theorem 16.6.1 the independent variables are r and s, and u is the dependent variable. The variables x and y can be called the intermediate variables. We now extend the chain rule to n intermediate variables and m independent variables.

16.6.2 THEOREM
The General Chain Rule

Suppose that u is a differentiable function of the n variables $x_1, x_2, \ldots, x_n$, and each of these variables is in turn a function of the m variables $y_1, y_2, \ldots, y_m$. Suppose further that each of the partial derivatives $\dfrac{\partial x_i}{\partial y_j}$ ($i = 1, 2, \ldots, n$; $j = 1, 2, \ldots, m$) exists. Then u is a function of $y_1, y_2, \ldots, y_m$, and

$$\frac{\partial u}{\partial y_1} = \left(\frac{\partial u}{\partial x_1}\right)\left(\frac{\partial x_1}{\partial y_1}\right) + \left(\frac{\partial u}{\partial x_2}\right)\left(\frac{\partial x_2}{\partial y_1}\right) + \cdots + \left(\frac{\partial u}{\partial x_n}\right)\left(\frac{\partial x_n}{\partial y_1}\right)$$

$$\frac{\partial u}{\partial y_2} = \left(\frac{\partial u}{\partial x_1}\right)\left(\frac{\partial x_1}{\partial y_2}\right) + \left(\frac{\partial u}{\partial x_2}\right)\left(\frac{\partial x_2}{\partial y_2}\right) + \cdots + \left(\frac{\partial u}{\partial x_n}\right)\left(\frac{\partial x_n}{\partial y_2}\right)$$

$$\vdots$$

$$\frac{\partial u}{\partial y_m} = \left(\frac{\partial u}{\partial x_1}\right)\left(\frac{\partial x_1}{\partial y_m}\right) + \left(\frac{\partial u}{\partial x_2}\right)\left(\frac{\partial x_2}{\partial y_m}\right) + \cdots + \left(\frac{\partial u}{\partial x_n}\right)\left(\frac{\partial x_n}{\partial y_m}\right)$$

The proof is an extension of the proof of Theorem 16.6.1.

Observe that in the general chain rule there are as many terms on the right side of each equation as there are intermediate variables.

EXAMPLE 2   Given

$$u = xy + xz + yz \qquad x = r \qquad y = r \cos t \qquad z = r \sin t$$

find $\dfrac{\partial u}{\partial r}$ and $\dfrac{\partial u}{\partial t}$.

Solution   From the chain rule,

$$\frac{\partial u}{\partial r} = \left(\frac{\partial u}{\partial x}\right)\left(\frac{\partial x}{\partial r}\right) + \left(\frac{\partial u}{\partial y}\right)\left(\frac{\partial y}{\partial r}\right) + \left(\frac{\partial u}{\partial z}\right)\left(\frac{\partial z}{\partial r}\right)$$

$$= (y + z)(1) + (x + z)(\cos t) + (x + y)(\sin t)$$
$$= y + z + x \cos t + z \cos t + x \sin t + y \sin t$$
$$= r \cos t + r \sin t + r \cos t + (r \sin t)(\cos t) + r \sin t + (r \cos t)(\sin t)$$
$$= 2r(\cos t + \sin t) + r(2 \sin t \cos t)$$
$$= 2r(\cos t + \sin t) + r \sin 2t$$

$$\frac{\partial u}{\partial t} = \left(\frac{\partial u}{\partial x}\right)\left(\frac{\partial x}{\partial t}\right) + \left(\frac{\partial u}{\partial y}\right)\left(\frac{\partial y}{\partial t}\right) + \left(\frac{\partial u}{\partial z}\right)\left(\frac{\partial z}{\partial t}\right)$$

$$= (y + z)(0) + (x + z)(-r \sin t) + (x + y)(r \cos t)$$
$$= (r + r \sin t)(-r \sin t) + (r + r \cos t)(r \cos t)$$
$$= -r^2 \sin t - r^2 \sin^2 t + r^2 \cos t + r^2 \cos^2 t$$
$$= r^2(\cos t - \sin t) + r^2(\cos^2 t - \sin^2 t)$$
$$= r^2(\cos t - \sin t) + r^2 \cos 2t$$

Now suppose that u is a differentiable function of the two variables x and y, and both x and y are differentiable functions of the single variable t. Then u is a function of the single variable t, and so instead of the partial derivative of u with respect to t we have the ordinary derivative of u with respect to t, which is given by

$$\frac{du}{dt} = \left(\frac{\partial u}{\partial x}\right)\left(\frac{dx}{dt}\right) + \left(\frac{\partial u}{\partial y}\right)\left(\frac{dy}{dt}\right) \tag{8}$$

We call $\dfrac{du}{dt}$ given by (8) the **total derivative** of u with respect to t.

If u is a differentiable function of the n variables $x_1, x_2, \ldots, x_n$ and each x_i is a differentiable function of the single variable t, then u is a function of t and the total derivative of u with respect to t is given by

$$\frac{du}{dt} = \left(\frac{\partial u}{\partial x_1}\right)\left(\frac{dx_1}{dt}\right) + \left(\frac{\partial u}{\partial x_2}\right)\left(\frac{dx_2}{dt}\right) + \cdots + \left(\frac{\partial u}{\partial x_n}\right)\left(\frac{dx_n}{dt}\right)$$

EXAMPLE 3   Given

$$u = x^2 + 2xy + y^2 \qquad x = t\cos t \qquad y = t\sin t$$

find $\dfrac{du}{dt}$ by two methods: (a) Use the chain rule; (b) express u in terms of t before differentiating.

Solution

(a) We compute the derivatives:

$$\frac{\partial u}{\partial x} = 2x + 2y \qquad\qquad \frac{\partial u}{\partial y} = 2x + 2y$$

$$\frac{dx}{dt} = \cos t - t\sin t \qquad \frac{dy}{dt} = \sin t + t\cos t$$

Thus from (8),

$$
\begin{aligned}
\frac{du}{dt} &= (2x + 2y)(\cos t - t\sin t) + (2x + 2y)(\sin t + t\cos t)\\
&= 2(x + y)(\cos t - t\sin t + \sin t + t\cos t)\\
&= 2(t\cos t + t\sin t)(\cos t - t\sin t + \sin t + t\cos t)\\
&= 2t(\cos^2 t - t\sin t\cos t + \sin t\cos t + t\cos^2 t + \sin t\cos t - t\sin^2 t + \sin^2 t + t\sin t\cos t)\\
&= 2t[1 + 2\sin t\cos t + t(\cos^2 t - \sin^2 t)]\\
&= 2t(1 + \sin 2t + t\cos 2t)
\end{aligned}
$$

$$
\begin{aligned}
\text{(b)} \quad u &= (t\cos t)^2 + 2(t\cos t)(t\sin t) + (t\sin t)^2\\
&= t^2\cos^2 t + t^2(2\sin t\cos t) + t^2\sin^2 t\\
&= t^2 + t^2\sin 2t
\end{aligned}
$$

Therefore

$$\frac{du}{dt} = 2t + 2t\sin 2t + 2t^2\cos 2t$$

EXAMPLE 4   If f is a differentiable function and a and b are constant, prove that $z = f(\frac{1}{2}bx^2 - \frac{1}{3}ay^3)$ satisfies the partial differential equation

$$ay^2 \frac{\partial z}{\partial x} + bx \frac{\partial z}{\partial y} = 0$$

Solution   Let $u = \frac{1}{2}bx^2 - \frac{1}{3}ay^3$. We wish to show that $z = f(u)$ satisfies the given equation. By the chain rule,

$$\frac{\partial z}{\partial x} = \left(\frac{dz}{du}\right)\left(\frac{\partial u}{\partial x}\right) \qquad \frac{\partial z}{\partial y} = \left(\frac{dz}{du}\right)\left(\frac{\partial u}{\partial y}\right)$$

$$= f'(u)(bx) \qquad\qquad = f'(u)(-ay^2)$$

Therefore

$$ay^2 \frac{\partial z}{\partial x} + bx \frac{\partial z}{\partial y} = ay^2[f'(u)(bx)] + bx[f'(u)(-ay^2)]$$

$$= 0$$

which we wished to prove.

EXAMPLE 5   Use the ideal gas law (see Example 5, Section 16.4) with $k = 0.8$ to find the rate at which the temperature is changing at the instant when the volume of the gas is 15 liters and the gas is under a pressure of 12 atm if the volume is increasing at the rate of 0.1 liter/min and the pressure is decreasing at the rate of 0.2 atm/min.

Solution   Let t minutes be the time that has elapsed since the volume of the gas started to increase. At t minutes let the Kelvin temperature be T, the pressure be P atmospheres, and the volume be V liters. From the ideal gas law

$$PV = 0.8T$$

$$T = 1.25\,PV$$

At the given instant, $P = 12$, $V = 15$, $\dfrac{dP}{dt} = -0.2$, and $\dfrac{dV}{dt} = 0.1$. From the chain rule

$$\frac{dT}{dt} = \frac{\partial T}{\partial P}\frac{dP}{dt} + \frac{\partial T}{\partial V}\frac{dV}{dt}$$

$$= 1.25V\frac{dP}{dt} + 1.25P\frac{dV}{dt}$$

$$= 1.25(15)(-0.2) + 1.25(12)(0.1)$$

$$= -2.25$$

Therefore the temperature is decreasing at the rate of 2.25 K/min at the given instant.

EXERCISES 16.6

In Exercises 1 through 6, find the indicated partial derivative by two methods: (a) Use the chain rule; (b) make the substitutions for x and y before differentiating.

1. $u - x^2 - y^2$; $x = 3r - s$; $y = r + 2s$; $\dfrac{\partial u}{\partial r}$; $\dfrac{\partial u}{\partial s}$

2. $u = 3x - 4y^2$; $x = 5pq$; $y = 3p^2 - 2q$; $\dfrac{\partial u}{\partial p}$; $\dfrac{\partial u}{\partial q}$

3. $u = 3x^2 + xy - 2y^2 + 3x - y$; $x = 2r - 3s$; $y = r + s$; $\dfrac{\partial u}{\partial r}$; $\dfrac{\partial u}{\partial s}$

4. $u = x^2 + y^2$; $x = \cosh r \cos t$; $y = \sinh r \sin t$; $\dfrac{\partial u}{\partial r}$; $\dfrac{\partial u}{\partial t}$

5. $u = e^{y/x}$; $x = 2r \cos t$; $y = 4r \sin t$; $\dfrac{\partial u}{\partial r}$; $\dfrac{\partial u}{\partial t}$

6. $V = \pi x^2 y$; $x = \cos z \sin t$; $y = z^2 e^t$; $\dfrac{\partial V}{\partial z}$; $\dfrac{\partial V}{\partial t}$

In Exercises 7 through 14, find the indicated partial derivative by using the chain rule.

7. $u = x^2 + xy$; $x = r^2 + s^2$; $y = 3r - 2s$; $\dfrac{\partial u}{\partial r}$; $\dfrac{\partial u}{\partial s}$

8. $u = xy + xz + yz$; $x = rs$; $y = r^2 - s^2$; $z = (r - s)^2$; $\dfrac{\partial u}{\partial r}$; $\dfrac{\partial u}{\partial s}$

9. $u = \sin^{-1}(3x + y)$; $x = r^2 e^s$; $y = \sin rs$; $\dfrac{\partial u}{\partial r}$; $\dfrac{\partial u}{\partial s}$

10. $u = \sin(xy)$; $x = 2ze^t$; $y = t^2 e^{-z}$; $\dfrac{\partial u}{\partial t}$; $\dfrac{\partial u}{\partial z}$

11. $u = \cosh \dfrac{y}{x}$; $x = 3r^2 s$; $y = 6se^r$; $\dfrac{\partial u}{\partial r}$; $\dfrac{\partial u}{\partial s}$

12. $u = xe^{-y}$; $x = \tan^{-1}(rst)$; $y = \ln(3rs + 5st)$; $\dfrac{\partial u}{\partial r}$; $\dfrac{\partial u}{\partial s}$; $\dfrac{\partial u}{\partial t}$

13. $u = x^2 + y^2 + z^2$; $x = r \sin \phi \cos \theta$; $y = r \sin \phi \sin \theta$; $z = r \cos \phi$; $\dfrac{\partial u}{\partial r}$; $\dfrac{\partial u}{\partial \phi}$; $\dfrac{\partial u}{\partial \theta}$

14. $u = x^2 yz$; $x = \dfrac{r}{s}$; $y = re^s$; $z = re^{-s}$; $\dfrac{\partial u}{\partial r}$; $\dfrac{\partial u}{\partial s}$

In Exercises 15 through 18, find the total derivative $\dfrac{du}{dt}$ by two methods: (a) Use the chain rule; (b) make the substitutions for x and y or for x, y, and z before differentiating.

15. $u = ye^x + xe^y$; $x = \cos t$; $y = \sin t$

16. $u = \ln xy + y^2$; $x = e^t$; $y = e^{-t}$

17. $u = \sqrt{x^2 + y^2 + z^2}$; $x = \tan t$; $y = \cos t$; $z = \sin t$; $0 < t < \tfrac{1}{2}\pi$

18. $u = \dfrac{t + e^x}{y - e^t}$; $x = 3 \sin t$; $y = \ln t$

In Exercises 19 through 22, find the total derivative $\dfrac{du}{dt}$ by using the chain rule; do not express u as a function of t before differentiating.

19. $u = \tan^{-1}\left(\dfrac{y}{x}\right)$; $x = \ln t$; $y = e^t$

20. $u = xy + xz + yz$; $x = t \cos t$; $y = t \sin t$; $z = t$

21. $u = \dfrac{x + t}{y + t}$; $x = \ln t$; $y = \ln \dfrac{1}{t}$

22. $u = \ln(x^2 + y^2 + t^2)$; $x = t \sin t$; $y = \cos t$

In Exercises 23 through 26, assume that the equation defines z as a function of x and y. Differentiate implicitly to find $\dfrac{\partial z}{\partial x}$ and $\dfrac{\partial z}{\partial y}$.

23. $3x^2 + y^2 + z^2 - 3xy + 4xz - 15 = 0$

24. $z = (x^2 + y^2) \sin xz$ **25.** $ye^{xyz} \cos 3xz = 5$

26. $ze^{yz} + 2xe^{xz} - 4e^{xy} = 3$

27. If f is a differentiable function of the variable u, let $u = bx - ay$ and prove that $z = f(bx - ay)$ satisfies the equation $a\left(\dfrac{\partial z}{\partial x}\right) + b\left(\dfrac{\partial z}{\partial y}\right) = 0$, where a and b are constants.

28. If f is a differentiable function of two variables u and v, let $u = x - y$ and $v = y - x$; prove that $z = f(x - y, y - x)$ satisfies the equation $\dfrac{\partial z}{\partial x} + \dfrac{\partial z}{\partial y} = 0$.

29. At a given instant, the length of one leg of a right triangle is 10 cm and it is increasing at the rate of 1 cm/min, and the length of the other leg is 12 cm and it is decreasing at the rate of 2 cm/min. Find the rate of change of the measure of the acute angle opposite the leg of length 12 cm at the given instant.

30. The height of a right-circular cone is increasing at the rate of 40 cm/min and the radius is decreasing at the rate of 15 cm/min. Find the rate of change of the volume at the instant when the height is 200 cm and the radius is 60 cm.

31. The height of a right-circular cylinder is decreasing at the rate of 10 cm/min and the radius is increasing at the rate of 4 cm/min. Find the rate of change of the volume at the instant when the height is 50 cm and the radius is 16 cm.

32. Water is flowing into a tank in the form of a right-circular cylinder at the rate of $\tfrac{4}{5}\pi$ m³/min. The tank is stretching in such a way that even though it remains cylindrical, its radius is increasing at the rate of 0.2 cm/min. How fast is the surface of the water rising when the radius is 2 m and the volume of water in the tank is 20π m³?

33. A quantity of gas obeys the ideal gas law (see Example 5, Section 16.4) with $k = 1.2$, and the gas is in a container that is being heated at a rate of 3 K/min. If at the instant when the temperature is 300 K, the pressure is 6 atm and is decreasing at the rate of 0.1 atm/min, find the rate of change of the volume at that instant.

34. A vertical wall makes an angle of radian measure $\frac{2}{3}\pi$ with the ground. A ladder of length 20 ft is leaning against the wall and its top is sliding down the wall at the rate of 3 ft/sec. How fast is the area of the triangle formed by the ladder, the wall, and the ground changing when the ladder makes an angle of $\frac{1}{6}\pi$ radians with the ground?

35. Suppose that f is a differentiable function of x and y, and $u = f(x, y)$. Then if $x = \cosh v \cos w$ and $y = \sinh v \sin w$, express $\dfrac{\partial u}{\partial v}$ and $\dfrac{\partial u}{\partial w}$ in terms of $\dfrac{\partial u}{\partial x}$ and $\dfrac{\partial u}{\partial y}$.

36. Suppose that f is a differentiable finction of x, y, and z, and $u = f(x, y, z)$. Then if $x = r \sin \phi \cos \theta$, $y = r \sin \phi \sin \theta$, and $z = r \cos \phi$, express $\dfrac{\partial u}{\partial r}, \dfrac{\partial u}{\partial \phi}$, and $\dfrac{\partial u}{\partial \theta}$ in terms of $\dfrac{\partial u}{\partial x}, \dfrac{\partial u}{\partial y}$, and $\dfrac{\partial u}{\partial z}$.

37. If $u = f(x, y)$ and $v = g(x, y)$, then the equations

$$\frac{\partial u}{\partial x} = \frac{\partial v}{\partial y} \quad \text{and} \quad \frac{\partial v}{\partial x} = -\frac{\partial u}{\partial y}$$

are called the *Cauchy-Riemann* equations. Show that the Cauchy-Riemann equations are satisfied if $u = \frac{1}{2}\ln(x^2 + y^2)$ and $v = \tan^{-1}\dfrac{y}{x}$.

38. Suppose that f and g are differentiable functions of x and y, and $u = f(x, y)$ and $v = g(x, y)$. Show that if the Cauchy-Riemann equations (see Exercise 37) hold and if $x = r \cos \theta$ and $y = r \sin \theta$, then

$$\frac{\partial u}{\partial r} = \frac{1}{r}\frac{\partial v}{\partial \theta} \quad \text{and} \quad \frac{\partial v}{\partial r} = -\frac{1}{r}\frac{\partial u}{\partial \theta}$$

39. If f is a differentiable function of x and y, and $u = f(x, y)$, $x = r \cos \theta$, and $y = r \sin \theta$, show that

$$\frac{\partial u}{\partial x} = \frac{\partial u}{\partial r}\cos \theta - \frac{\partial u}{\partial \theta}\frac{\sin \theta}{r}$$

$$\frac{\partial u}{\partial y} = \frac{\partial u}{\partial r}\sin \theta + \frac{\partial u}{\partial \theta}\frac{\cos \theta}{r}$$

16.7 HIGHER-ORDER PARTIAL DERIVATIVES

If f is a function of two variables, then in general $D_1 f$ and $D_2 f$ are also functions of two variables. And if the partial derivatives of these functions exist, they are called second partial derivatives of f. In contrast, $D_1 f$ and $D_2 f$ are called first partial derivatives of f. There are four second partial derivatives of a function of two variables. If f is a function of the two variables x and y, the notations

$$D_2(D_1 f) \qquad D_{12}f \qquad f_{12} \qquad f_{xy} \qquad \frac{\partial^2 f}{\partial y \, \partial x}$$

all denote the second partial derivative of f, which is obtained by first partial-differentiating f with respect to x and then partial-differentiating the result with respect to y. This second partial derivative is defined by

$$f_{12}(x, y) = \lim_{\Delta y \to 0} \frac{f_1(x, y + \Delta y) - f_1(x, y)}{\Delta y} \tag{1}$$

if this limit exists. The notations

$$D_1(D_1 f) \qquad D_{11}f \qquad f_{11} \qquad f_{xx} \qquad \frac{\partial^2 f}{\partial x^2}$$

all denote the second partial derivative of f, which is obtained by partial-differentiating twice with respect to x. We have the definition

$$f_{11}(x, y) = \lim_{\Delta x \to 0} \frac{f_1(x + \Delta x, y) - f_1(x, y)}{\Delta x} \tag{2}$$

if this limit exists. The other two second partial derivatives are defined in an analogous way.

$$f_{21}(x, y) = \lim_{\Delta x \to 0} \frac{f_2(x + \Delta x, y) - f_2(x, y)}{\Delta x} \tag{3}$$

$$f_{22}(x, y) = \lim_{\Delta y \to 0} \frac{f_2(x, y + \Delta y) - f_2(x, y)}{\Delta y} \tag{4}$$

if these limits exist.

The definitions of higher-order partial derivatives are similar. Again there are various notations for a specific derivative. For example,

$$D_{112}f \qquad f_{112} \qquad f_{xxy} \qquad \frac{\partial^3 f}{\partial y\, \partial x\, \partial x} \qquad \frac{\partial^3 f}{\partial y\, \partial x^2}$$

all stand for the third partial derivative of f, which is obtained by partial-differentiating twice with respect to x and then once with respect to y. In the subscript notation, the order of partial differentiation is from left to right; in the notation $\dfrac{\partial^3 f}{\partial y\, \partial x\, \partial x}$, the order is from right to left.

EXAMPLE 1   Given

$$f(x, y) = e^x \sin y + \ln xy$$

Find: (a) $D_{11}f(x, y)$; (b) $D_{12}f(x, y)$; (c) $\dfrac{\partial^3 f}{\partial x\, \partial y^2}$.

Solution

$$D_1 f(x, y) = e^x \sin y + \frac{1}{xy}\,(y)$$

$$= e^x \sin y + \frac{1}{x}$$

(a) $D_{11}f(x, y) = e^x \sin y - \dfrac{1}{x^2}$ (b) $D_{12}f(x, y) = e^x \cos y$

(c) To find $\dfrac{\partial^3 f}{\partial x\, \partial y^2}$ we partial-differentiate twice with respect to y and then once with respect to x. We have, then,

$$\frac{\partial f}{\partial y} = e^x \cos y + \frac{1}{y} \qquad \frac{\partial^2 f}{\partial y^2} = -e^x \sin y - \frac{1}{y^2} \qquad \frac{\partial^3 f}{\partial x\, \partial y^2} = -e^x \sin y$$

Higher-order partial derivatives of a function of n variables have definitions that are analogous to the definitions of higher-order partial derivatives of a function of two variables. If f is a function of n variables, there may be n^2 second partial derivatives of f at a particular point. That is, for a function of three variables, if all the second-order partial derivatives exist, there are nine of them: $f_{11}, f_{12}, f_{13}, f_{21}, f_{22}, f_{23}, f_{31}, f_{32},$ and f_{33}.

EXAMPLE 2   Find $D_{132}f(x, y, z)$ if

$$f(x, y, z) = \sin(xy + 2z)$$

Solution

$$D_1 f(x, y, z) = y \cos(xy + 2z)$$

$$D_{13} f(x, y, z) = -2y \sin(xy + 2z)$$

$$D_{132} f(x, y, z) = -2 \sin(xy + 2z) - 2xy \cos(xy + 2z)$$

EXAMPLE 3   Given

$$f(x, y) = x^3 y - y \cosh xy$$

Find: (a) $f_{xy}(x, y)$; (b) $f_{yx}(x, y)$.

Solution

(a) $f_x(x, y) = 3x^2 y - y^2 \sinh xy$

 $f_{xy}(x, y) = 3x^2 - 2y \sinh xy - xy^2 \cosh xy$

(b) $f_y(x, y) = x^3 - \cosh xy - xy \sinh xy$

 $f_{yx}(x, y) = 3x^2 - y \sinh xy - y \sinh xy - xy^2 \cosh xy$

 $= 3x^2 - 2y \sinh xy - xy^2 \cosh xy$

Observe from the above results that for the function of Example 3 the "mixed" partial derivatives $f_{xy}(x, y)$ and $f_{yx}(x, y)$ are equal. So for this particular function, when finding the second partial derivative with respect to x and then y, the order of differentiation is immaterial. This condition holds for many functions. However, the following example shows that it is not always true.

EXAMPLE 4   Find $f_{12}(0, 0)$ and $f_{21}(0, 0)$ if

$$f(x, y) = \begin{cases} (xy) \dfrac{x^2 - y^2}{x^2 + y^2} & \text{if } (x, y) \neq (0, 0) \\ 0 & \text{if } (x, y) = (0, 0) \end{cases}$$

Solution   In Example 3, Section 16.4, we showed that for this function

$$f_1(0, y) = -y \qquad \text{for all } y \tag{5}$$

and

$$f_2(x, 0) = x \qquad \text{for all } x \tag{6}$$

From (1),

$$f_{12}(0, 0) = \lim_{\Delta y \to 0} \frac{f_1(0, 0 + \Delta y) - f_1(0, 0)}{\Delta y}$$

But from (5), $f_1(0, \Delta y) = -\Delta y$ and $f_1(0, 0) = 0$, and so

$$f_{12}(0, 0) = \lim_{\Delta y \to 0} \frac{-\Delta y - 0}{\Delta y}$$

$$= \lim_{\Delta y \to 0} (-1)$$

$$= -1$$

From (3),

$$f_{21}(0, 0) = \lim_{\Delta x \to 0} \frac{f_2(0 + \Delta x, 0) - f_2(0, 0)}{\Delta x}$$

From (6), $f_2(\Delta x, 0) = \Delta x$ and $f_2(0, 0) = 0$. Therefore

$$f_{21}(0, 0) = \lim_{\Delta x \to 0} \frac{\Delta x - 0}{\Delta x}$$

$$= \lim_{\Delta x \to 0} 1$$

$$= 1$$

For the function in Example 4 the mixed partial derivatives $f_{12}(x, y)$ and $f_{21}(x, y)$ are not equal at $(0, 0)$. A set of conditions for which $f_{12}(x_0, y_0)$ and $f_{21}(x_0, y_0)$ are equal is given by Theorem 16.7.1, which follows. The function of Example 4 does not satisfy the hypothesis of this theorem because both f_{12} and f_{21} are discontinuous at $(0, 0)$. It is left as an exercise to show this (see Exercise 24).

16.7.1 THEOREM

Suppose that f is a function of two variables x and y defined on an open disk $B((x_0, y_0); r)$ and f_x, f_y, f_{xy}, and f_{yx} also are defined on B. Furthermore, suppose that f_{xy} and f_{yx} are continuous on B. Then

$$f_{xy}(x_0, y_0) = f_{yx}(x_0, y_0)$$

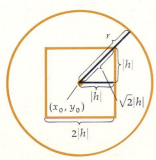

FIGURE 1

Proof Consider a square having its center at (x_0, y_0) and the length of its side $2|h|$ such that $0 < \sqrt{2}|h| < r$. Then all the points in the interior of the square and on the sides of the square are in the open disk B (see Figure 1). So the points $(x_0 + h, y_0 + h)$, $(x_0 + h, y_0)$, and $(x_0, y_0 + h)$ are in B. Let Δ be defined by

$$\Delta = f(x_0 + h, y_0 + h) - f(x_0 + h, y_0) - f(x_0, y_0 + h) + f(x_0, y_0) \qquad (7)$$

Consider the function G defined by

$$G(x) = f(x, y_0 + h) - f(x, y_0) \qquad (8)$$

Then

$$G(x + h) = f(x + h, y_0 + h) - f(x + h, y_0)$$

So (7) can be written as

$$\Delta = G(x_0 + h) - G(x_0) \qquad (9)$$

From (8),

$$G'(x) = f_x(x, y_0 + h) - f_x(x, y_0) \qquad (10)$$

Now, because $f_x(x, y_0 + h)$ and $f_x(x, y_0)$ are defined on B, $G'(x)$ exists if x is in the closed interval having endpoints at x_0 and $x_0 + h$. Hence G is continuous if x is in this closed interval. By the mean-value theorem (4.3.2) there is a number c_1 between x_0 and $x_0 + h$ such that

$$G(x_0 + h) - G(x_0) = hG'(c_1)$$

Substituting from this equation into (9) we get

$$\Delta = hG'(c_1)$$

From this equation and replacing x by c_1 in (10), we have

$$\Delta = h[f_x(c_1, y_0 + h) - f_x(c_1, y_0)] \tag{11}$$

Now if g is the function defined by

$$g(y) = f_x(c_1, y) \tag{12}$$

we can write (11) as

$$\Delta = h[g(y_0 + h) - g(y_0)] \tag{13}$$

From (12),

$$g'(y) = f_{xy}(c_1, y) \tag{14}$$

Because $f_{xy}(c_1, y)$ is defined on B, $g'(y)$ exists if y is in the closed interval having endpoints at y_0 and $y_0 + h$; hence g is continuous if y is in this closed interval. Therefore, by the mean-value theorem there is a number d_1 between y_0 and $y_0 + h$ such that

$$g(y_0 + h) - g(y_0) = hg'(d_1)$$

Substituting from this equation into (13) we get $\Delta = h^2 g'(d_1)$; so from (14) it follows that

$$\Delta = h^2 f_{xy}(c_1, d_1) \tag{15}$$

for some point (c_1, d_1) in the open disk B. We define a function ϕ by

$$\phi(y) = f(x_0 + h, y) - f(x_0, y) \tag{16}$$

and so $\phi(y + h) = f(x_0 + h, y + h) - f(x_0, y + h)$. Therefore (7) can be written as

$$\Delta = \phi(y_0 + h) - \phi(y_0) \tag{17}$$

From (16),

$$\phi'(y) = f_y(x_0 + h, y) - f_y(x_0, y) \tag{18}$$

Because, by hypothesis, each term on the right side of (18) exists on B, ϕ' exists if y is in the closed interval having y_0 and $y_0 + h$ as endpoints. Therefore ϕ is continuous on this closed interval. So by the mean-value theorem there is a number d_2 between y_0 and $y_0 + h$ such that

$$\phi(y_0 + h) - \phi(y_0) = h\phi'(d_2)$$

From this equation, (17), and (18) it follows that

$$\Delta = h[f_y(x_0 + h, d_2) - f_y(x_0, d_2)] \tag{19}$$

We define the function χ by

$$\chi(x) = f_y(x, d_2) \tag{20}$$

and write (19) as

$$\Delta = h[\chi(x_0 + h) - \chi(x_0)] \tag{21}$$

From (20),

$$\chi'(x) = f_{yx}(x, d_2) \tag{22}$$

and by the mean-value theorem there is a number c_2 between x_0 and $x_0 + h$ such that

$$\chi(x_0 + h) - \chi(x_0) = h\chi'(c_2)$$

From this equation, (21), and (22)

$$\Delta = h^2 f_{yx}(c_2, d_2)$$

With this expression for Δ, and (15) we get

$$h^2 f_{xy}(c_1, d_1) = h^2 f_{yx}(c_2, d_2)$$

and because $h \neq 0$, we can divide by h^2, which gives

$$f_{xy}(c_1, d_1) = f_{yx}(c_2, d_2) \tag{23}$$

where (c_1, d_1) and (c_2, d_2) are in B.

Because c_1 and c_2 are each between x_0 and $x_0 + h$, we have $c_1 = x_0 + \epsilon_1 h$, where $0 < \epsilon_1 < 1$, and $c_2 = x_0 + \epsilon_2 h$, where $0 < \epsilon_2 < 1$. Similarly, because both d_1 and d_2 are between y_0 and $y_0 + h$, we have $d_1 = y_0 + \epsilon_3 h$, where $0 < \epsilon_3 < 1$, and $d_2 = y_0 + \epsilon_4 h$, where $0 < \epsilon_4 < 1$. Making these substitutions in (23) gives

$$f_{xy}(x_0 + \epsilon_1 h, y_0 + \epsilon_3 h) = f_{yx}(x_0 + \epsilon_2 h, y_0 + \epsilon_4 h)$$

Because f_{xy} and f_{yx} are continuous on B, upon taking the limit of both sides of this equation as h approaches zero we obtain

$$f_{xy}(x_0, y_0) = f_{yx}(x_0, y_0) \qquad \blacksquare$$

EXAMPLE 5   Suppose $u = f(x, y)$, $x = F(r, s)$, and $y = G(r, s)$, and assume $f_{xy} = f_{yx}$. Prove by using the chain rule that

$$\frac{\partial^2 u}{\partial r^2} = f_{xx}(x, y)[F_r(r, s)]^2 + 2f_{xy}(x, y)F_r(r, s)G_r(r, s) + f_{yy}(x, y)[G_r(r, s)]^2$$
$$+ f_x(x, y)F_{rr}(r, s) + f_y(x, y)G_{rr}(r, s)$$

Solution   From the chain rule

$$\frac{\partial u}{\partial r} = f_x(x, y)F_r(r, s) + f_y(x, y)G_r(r, s)$$

Taking the partial derivative again with respect to r, and using the formula for the derivative of a product and the chain rule, we obtain

$$\frac{\partial^2 u}{\partial r^2} = [f_{xx}(x, y)F_r(r, s) + f_{xy}(x, y)G_r(r, s)]F_r(r, s) + f_x(x, y)F_{rr}(r, s)$$
$$+ [f_{yx}(x, y)F_r(r, s) + f_{yy}(x, y)G_r(r, s)]G_r(r, s) + f_y(x, y)G_{rr}(r, s)$$

Multiplying and combining terms, and using the fact that $f_{xy}(x, y)$ and $f_{yx}(x, y)$ are equal, we get

$$\frac{\partial^2 u}{\partial r^2} = f_{xx}(x, y)[F_r(r, s)]^2 + 2f_{xy}(x, y)F_r(r, s)G_r(r, s) + f_{yy}(x, y)[G_r(r, s)]^2$$
$$+ f_x(x, y)F_{rr}(r, s) + f_y(x, y)G_{rr}(r, s)$$

which is what we wished to prove.

As a result of Theorem 16.7.1, if the function f of two variables has continuous partial derivatives on some open disk, then the order of partial differentiation can be changed without affecting the result; that is,

$$D_{112}f = D_{121}f = D_{211}f$$

$$D_{1122}f = D_{1212}f = D_{1221}f = D_{2112}f = D_{2121}f = D_{2211}f$$

and so forth. In particular, assuming that all of the partial derivatives are continuous on some open disk, we can prove that $D_{211}f = D_{112}f$ by applying Theorem 16.7.1 repeatedly. Doing this we have

$$D_{211}f = D_1(D_{21}f) = D_1(D_{12}f) = D_1[D_2(D_1 f)] = D_2[D_1(D_1 f)]$$
$$= D_2(D_{11}f) = D_{112}f$$

EXERCISES 16.7

In Exercises 1 through 10, do each of the following: (a) Find $D_{11}f(x, y)$; (b) find $D_{22}f(x, y)$; (c) show that $D_{12}f(x, y)$ and $D_{21}f(x, y)$ are equal.

1. $f(x, y) = \dfrac{x^2}{y} - \dfrac{y}{x^2}$

2. $f(x, y) = 2x^3 - 3x^2y + xy^2$

3. $f(x, y) = e^{2x} \sin y$ **4.** $f(x, y) = e^{-x/y} + \ln \dfrac{y}{x}$

5. $f(x, y) = (x^2 + y^2) \tan^{-1} \dfrac{y}{x}$

6. $f(x, y) = \sin^{-1} \dfrac{3y}{x^2}$

7. $f(x, y) = 4x \sinh y + 3y \cosh x$
8. $f(x, y) = x \cos y - ye^x$
9. $f(x, y) = e^x \cos y + \tan^{-1} x \cdot \ln y$
10. $f(x, y) = 3x \cosh y - y \sin^{-1} e^x$

In Exercises 11 through 18, find the indicated partial derivatives.

11. $f(x, y) = 2x^3y + 5x^2y^2 - 3xy^2$; (a) $f_{121}(x, y)$; (b) $f_{211}(x, y)$
12. $G(x, y) = 3x^3y^2 + 5x^2y^3 + 2x$; (a) $G_{yyx}(x, y)$; (b) $G_{yxy}(x, y)$
13. $f(x, y, z) = ye^x + ze^y + e^z$; (a) $f_{xz}(x, y, z)$; (b) $f_{yz}(x, y, z)$
14. $g(x, y, z) = \sin(xyz)$; (a) $g_{23}(x, y, z)$; (b) $g_{12}(x, y, z)$
15. $f(w, z) = w^2 \cos e^z$; (a) $f_{121}(w, z)$; (b) $f_{212}(w, z)$
16. $f(u, v) = \ln \cos(u - v)$; (a) $f_{uuv}(u, v)$; (b) $f_{vuv}(u, v)$
17. $g(r, s, t) = \ln(r^2 + 4s^2 - 5t^2)$; (a) $g_{132}(r, s, t)$; (b) $g_{122}(r, s, t)$
18. $f(x, y, z) = \tan^{-1}(3xyz)$; (a) $f_{113}(x, y, z)$; (b) $f_{123}(x, y, z)$

In Exercises 19 through 22, show that $u(x, y)$ satisfies the equation $\dfrac{\partial^2 u}{\partial x^2} + \dfrac{\partial^2 u}{\partial y^2} = 0$ which is known as Laplace's equation in R^2.

19. $u(x, y) = \ln(x^2 + y^2)$
20. $u(x, y) = e^x \sin y + e^y \cos x$

21. $u(x, y) = \tan^{-1} \dfrac{y}{x} + \dfrac{x}{x^2 + y^2}$

22. $u(x, y) = \tan^{-1} \dfrac{2xy}{x^2 - y^2}$

23. Laplace's equation in R^3 is

$$\frac{\partial^2 u}{\partial x^2} + \frac{\partial^2 u}{\partial y^2} + \frac{\partial^2 u}{\partial z^2} = 0$$

Show that $u(x, y, z) = (x^2 + y^2 + z^2)^{-1/2}$ satisfies this equation.

24. For the function of Example 4, show that f_{12} is discontinuous at $(0, 0)$ and hence that the hypothesis of Theorem 16.7.1 is not satisfied if $(x_0, y_0) = (0, 0)$.

In Exercises 25 through 27, find $f_{12}(0, 0)$ and $f_{21}(0, 0)$, if they exist.

25. $f(x, y) = \begin{cases} \dfrac{2xy}{x^2 + y^2} & \text{if } (x, y) \neq (0, 0) \\ 0 & \text{if } (x, y) = (0, 0) \end{cases}$

26. $f(x, y) = \begin{cases} \dfrac{x^2y^2}{x^4 + y^4} & \text{if } (x, y) \neq (0, 0) \\ 0 & \text{if } (x, y) = (0, 0) \end{cases}$

27. $f(x, y) = \begin{cases} x^2 \tan^{-1} \dfrac{y}{x} - y^2 \tan^{-1} \dfrac{x}{y} & \text{if } x \neq 0 \text{ and } y \neq 0 \\ 0 & \text{if either } x = 0 \text{ or } y = 0 \end{cases}$

28. Given that $u = f(x, y)$, $x = F(t)$, and $y = G(t)$, and assuming that $f_{xy} = f_{yx}$, prove by using the chain rule that

$$\frac{d^2 u}{dt^2} = f_{xx}(x, y)[F'(t)]^2 + 2f_{xy}(x, y)F'(t)G'(t) + f_{yy}(x, y)[G'(t)]^2$$
$$+ f_x(x, y)F''(t) + f_y(x, y)G''(t)$$

29. Given that $u = f(x, y)$, $x = F(r, s)$, and $y = G(r, s)$, and assuming that $f_{xy} = f_{yx}$, prove by using the chain rule that

$$\frac{\partial^2 u}{\partial r \partial s} = f_{xx}(x, y)F_r(r, s)F_s(r, s) + f_{xy}(x, y)[F_s(r, s)G_r(r, s) + F_r(r, s)G_s(r, s)]$$
$$+ f_{yy}(x, y)G_r(r, s)G_s(r, s) + f_x(x, y)F_{sr}(r, s) + f_y(x, y)G_{sr}(r, s)$$

30. Given $u = e^y \cos x$, $x = 2t$, $y = t^2$. Find $\dfrac{d^2u}{dt^2}$ in three ways: (a) first express u in terms of t; (b) use the formula of Exercise 28; (c) use the chain rule.

31. Given $u = 3xy - 4y^2$, $x = 2se^r$, $y = re^{-s}$. Find $\dfrac{\partial^2 u}{\partial r^2}$ in three ways: (a) first express u in terms of r and s; (b) use the formula of Example 5; (c) use the chain rule.

32. For u, x, and y as given in Exercise 31, find $\dfrac{\partial^2 u}{\partial s \, \partial r}$ in three ways: (a) first express u in terms of r and s; (b) use the formula of Exercise 29; (c) use the chain rule.

33. Given $u = 9x^2 + 4y^2$, $x = r \cos \theta$, $y = r \sin \theta$. Find $\dfrac{\partial^2 u}{\partial r^2}$ in three ways: (a) first express u in terms of r and θ; (b) use the formula of Example 5; (c) use the chain rule.

34. For u, x, and y as given in Exercise 33, find $\dfrac{\partial^2 u}{\partial \theta^2}$ in three ways: (a) first express u in terms of r and θ; (b) use the formula of Example 5; (c) use the chain rule.

35. For u, x, and y as given in Exercise 33, find $\dfrac{\partial^2 u}{\partial r \, \partial \theta}$ in three ways: (a) first express u in terms of r and θ; (b) use the formula of Exercise 29; (c) use the chain rule.

36. Suppose that $u = f(x, y)$ and $v = g(x, y)$, and that f and g and their first and second partial derivatives are continuous. Prove that if u and v satisfy the Cauchy-Riemann equations (see Exercise 37 in Exercises 16.6), they also satisfy Laplace's equation (see Exercises 19 through 22).

37. The one-dimensional heat-conduction partial differential equation is

$$\frac{\partial u}{\partial t} = k^2 \frac{\partial^2 u}{\partial x^2}$$

Show that if f is a function of x satisfying the equation

$$\frac{d^2 f}{dx^2} + \lambda^2 f(x) = 0$$

and g is a function of t satisfying the equation

$$\frac{dg}{dt} + k^2 \lambda^2 g(t) = 0,$$

then if $u = f(x)g(t)$ and k and λ are constants, the partial differential equation is satisfied.

38. The partial differential equation for a vibrating string is

$$\frac{\partial^2 u}{\partial t^2} = a^2 \frac{\partial^2 u}{\partial x^2}$$

Show that if f is a function of x satisfying the equation $\dfrac{d^2 f}{dx^2} + \lambda^2 f(x) = 0$ and g is a function of t satisfying the equation $\dfrac{d^2 g}{dt^2} + a^2 \lambda^2 g(t) = 0$, then if $u = f(x)g(t)$ and a and λ are constants, the partial differential equation is satisfied.

39. Prove that if f and g are two arbitrary functions of a real variable having continuous second derivatives and

$$u = f(x + at) + g(x - at)$$

then u satisfies the partial differential equation of the vibrating string given in Exercise 38. (*Hint:* Let $v = x + at$ and $w = x - at$; then u is a function of v and w, and v and w are in turn functions of x and t.)

40. Prove that if f is a function of two variables and all the partial derivatives of f up to the fourth order are continuous on some open disk, then

$$D_{1122} f = D_{2121} f$$

16.8 SUFFICIENT CONDITIONS FOR DIFFERENTIABILITY (*Supplementary*)

The proof of Theorem 16.5.4, which gives sufficient conditions for a function of two variables to be differentiable at a point, was delayed until this section. We first give a theorem that is needed for the proof. It is the mean-value theorem for a function of a single variable applied to a function of two variables.

16.8.1 THEOREM

Let f be a function of two variables defined for all x in the closed interval $[a, b]$ and all y in the closed interval $[c, d]$.

(i) If $D_1 f(x, y_0)$ exists for some y_0 in $[c, d]$ and for all x in $[a, b]$, then there is a number ξ_1 in the open interval (a, b) such that

$$f(b, y_0) - f(a, y_0) = (b - a)D_1 f(\xi_1, y_0) \tag{1}$$

(ii) If $D_2 f(x_0, y)$ exists for some x_0 in $[a, b]$ and for all y in $[c, d]$, then there is a number ξ_2 in the open interval (c, d) such that

$$f(x_0, d) - f(x_0, c) = (d - c)D_2 f(x_0, \xi_2) \tag{2}$$

Before proving this theorem, we interpret it geometrically. For part (i) refer to Figure 1, which shows the portion of the surface $z = f(x, y)$ above the rectangular region in the xy plane bounded by the lines $x = a$, $x = b$, $y = c$, and $y = d$. The plane $y = y_0$ intersects the surface in the curve represented by the two equations $y = y_0$ and $z = f(x, y)$. The slope of the line through the points $A(a, y_0, f(a, y_0))$ and $B(b, y_0, f(b, y_0))$ is $[f(b, y_0) - f(a, y_0)]/(b - a)$. Theorem 16.8.1(i) states that there is some point $(\xi_1, y_0, f(\xi_1, y_0))$ on the curve between the points A and B where the tangent line is parallel to the secant line through A and B; that is, there is some number ξ_1 in (a, b) such that

$$D_1 f(\xi_1, y_0) = \frac{f(b, y_0) - f(a, y_0)}{b - a}$$

and this is illustrated in the figure, for which $D_1 f(\xi_1, y_0) < 0$.

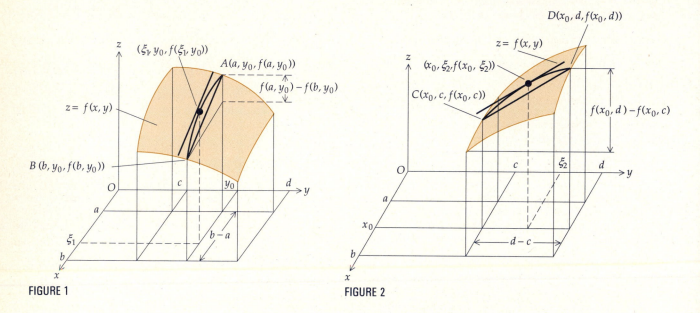

FIGURE 1   **FIGURE 2**

Figure 2 illustrates part (ii) of Theorem 16.8.1. The plane $x = x_0$ intersects the surface $z = f(x, y)$ in the curve represented by the two equations $x = x_0$ and $z = f(x, y)$. The slope of the line through the points $C(x_0, c, f(x_0, c))$ and $D(x_0, d, f(x_0, d))$ is $[f(x_0, d) - f(x_0, c)]/(d - c)$, and Theorem 16.8.1(ii) states that there is some point $(x_0, \xi_2, f(x_0, \xi_2))$ on the curve between the points C and D where the tangent line is parallel to the secant line through C and D; that is, there is some number ξ_2 in (c, d) such that

$$D_2 f(x_0, \xi_2) = \frac{f(x_0, d) - f(x_0, c)}{d - c}$$

Proof of Theorem 16.8.1(i)   Let g be the function of one variable x defined by

$$g(x) = f(x, y_0)$$

Then

$$g'(x) = D_1 f(x, y_0)$$

Because $D_1 f(x, y_0)$ exists for all x in $[a, b]$, then $g'(x)$ exists for all x in $[a, b]$, and therefore g is continuous on $[a, b]$. So by the mean-value theorem (4.3.2) for ordinary derivatives there exists a number ξ_1 in (a, b) such that

$$g'(\xi_1) = \frac{g(b) - g(a)}{b - a}$$

$$\Leftrightarrow \quad D_1 f(\xi_1, y_0) = \frac{f(b, y_0) - f(a, y_0)}{b - a}$$

from which we obtain

$$f(b, y_0) - f(a, y_0) = (b - a)D_1 f(\xi_1, y_0)$$

The proof of part (ii) is similar to the proof of part (i) and is left as an exercise (see Exercise 17).   ∎

Equation (1) can be written in the form

$$f(x_0 + h, y_0) - f(x_0, y_0) = hD_1 f(\xi_1, y_0) \tag{3}$$

where ξ_1 is between x_0 and $x_0 + h$ and h is either positive or negative (see Exercise 1).

Equation (2) can be written in the form

$$f(x_0, y_0 + k) - f(x_0, y_0) = kD_2 f(x_0, \xi_2) \tag{4}$$

where ξ_2 is between y_0 and $y_0 + k$ and k is either positive or negative (see Exercise 2).

EXAMPLE 1   Given

$$f(x, y) = \frac{2xy}{3 + x}$$

find a ξ_1 required by Theorem 16.8.1 if x is in $[2, 5]$ and $y = 4$.

Solution

$$D_1 f(x, y) = \frac{6y}{(3 + x)^2}$$

By Theorem 16.8.1(i) there is a number ξ_1 in the open interval $(2, 5)$ such that

$$f(5, 4) - f(2, 4) = (5 - 2)D_1 f(\xi_1, 4)$$

$$5 - \frac{16}{5} = 3 \cdot \frac{24}{(3 + \xi_1)^2}$$

$$\frac{9}{5} = \frac{72}{(3 + \xi_1)^2}$$

$$(3 + \xi_1)^2 = 40$$

$$3 + \xi_1 = \pm 2\sqrt{10}$$

But because $2 < \xi_1 < 5$, we take only the $+$ sign and obtain

$$\xi_1 = 2\sqrt{10} - 3$$

EXAMPLE 2 Given

$$f(x, y) = 3xe^y - 2ye^x$$

find a ξ_2 required by Theorem 16.8.1 if y is in $[0, 4]$ and $x = 3$.

Solution

$$D_2 f(x, y) = 3xe^y - 2e^x$$

By Theorem 16.8.1(ii) there is a number ξ_2 in the open interval $(0, 4)$ such that

$$f(3, 4) - f(3, 0) = (4 - 0)D_2 f(3, \xi_2)$$

$$(9e^4 - 8e^3) - 9 = 4(9e^{\xi_2} - 2e^3)$$

$$9e^4 - 8e^3 - 9 = 36e^{\xi_2} - 8e^3$$

$$36e^{\xi_2} = 9e^4 - 9$$

$$e^{\xi_2} = \tfrac{1}{4}(e^4 - 1)$$

$$e^{\xi_2} \approx 13.40$$

$$\xi_2 \approx \ln 13.40$$

$$\xi_2 \approx 2.60$$

Following is a restatement of Theorem 16.5.4 along with its proof.

16.5.4 THEOREM   Let f be a function of two variables x and y. Suppose that $D_1 f$ and $D_2 f$ exist on an open disk $B(P_0; r)$, where P_0 is the point (x_0, y_0). Then if $D_1 f$ and $D_2 f$ are continuous at P_0, f is differentiable at P_0.

Proof   We choose the point $(x_0 + \Delta x, y_0 + \Delta y)$ so that it is in $B(P_0; r)$. Then

$$\Delta f(x_0, y_0) = f(x_0 + \Delta x, y_0 + \Delta y) - f(x_0, y_0)$$

Subtracting and adding $f(x_0 + \Delta x, y_0)$ to the right side of the above equation we get

$$\Delta f(x_0, y_0) = [f(x_0 + \Delta x, y_0 + \Delta y) - f(x_0 + \Delta x, y_0)] + [f(x_0 + \Delta x, y_0) - f(x_0, y_0)] \quad (5)$$

Because $D_1 f$ and $D_2 f$ exist on $B(P_0; r)$ and $(x_0 + \Delta x, y_0 + \Delta y)$ is in $B(P_0; r)$, it follows from (4) that

$$f(x_0 + \Delta x, y_0 + \Delta y) - f(x_0 + \Delta x, y_0) = (\Delta y)D_2 f(x_0 + \Delta x, \xi_2) \quad (6)$$

where ξ_2 is between y_0 and $y_0 + \Delta y$.
    From (3)

$$f(x_0 + \Delta x, y_0) - f(x_0, y_0) = (\Delta x)D_1 f(\xi_1, y_0)$$

where ξ_1 is between x_0 and $x_0 + \Delta x$. Substituting from this equation and (6) in (5) we obtain

$$\Delta f(x_0, y_0) = (\Delta y)D_2 f(x_0 + \Delta x, \xi_2) + (\Delta x)D_1 f(\xi_1, y_0) \quad (7)$$

Because $(x_0 + \Delta x, y_0 + \Delta y)$ is in $B(P_0; r)$, ξ_2 is between y_0 and $y_0 + \Delta y$, and $D_2 f$ is continuous at P_0, it follows that

$$\lim_{(\Delta x, \Delta y) \to (0,0)} D_2 f(x_0 + \Delta x, \xi_2) = D_2 f(x_0, y_0) \tag{8}$$

and because ξ_1 is between x_0 and $x_0 + \Delta x$ and $D_1 f$ is continuous at P_0,

$$\lim_{(\Delta x, \Delta y) \to (0,0)} D_1 f(\xi_1, y_0) = D_1 f(x_0, y_0) \tag{9}$$

If

$$\epsilon_1 = D_1 f(\xi_1, y_0) - D_1 f(x_0, y_0) \tag{10}$$

then from (9)

$$\lim_{(\Delta x, \Delta y) \to (0,0)} \epsilon_1 = 0 \tag{11}$$

and if

$$\epsilon_2 = D_2 f(x_0 + \Delta x, \xi_2) - D_2 f(x_0, y_0) \tag{12}$$

then from (8)

$$\lim_{(\Delta x, \Delta y) \to (0,0)} \epsilon_2 = 0 \tag{13}$$

Substituting from (10) and (12) into (7) we get

$$\Delta f(x_0, y_0) = \Delta y [D_2 f(x_0, y_0) + \epsilon_2] + \Delta x [D_1 f(x_0, y_0) + \epsilon_1]$$

$$\Leftrightarrow \quad \Delta f(x_0, y_0) = D_1 f(x_0, y_0) \, \Delta x + D_2 f(x_0, y_0) \, \Delta y + \epsilon_1 \, \Delta x + \epsilon_2 \, \Delta y$$

From this equation, (11), and (13), Definition 16.5.2 holds; so f is differentiable at (x_0, y_0). ∎

EXERCISES 16.8

1. Show that Equation (1) may be written in the form (3), where ξ_1 is between x_0 and $x_0 + h$.
2. Show that Equation (2) may be written in the form (4), where ξ_2 is between y_0 and $y_0 + k$.

In Exercises 3 through 8, apply Theorem 16.8.1(i) to find a ξ_1.

3. $f(x, y) = x^2 + 3xy - y^2$; x is in $[1, 3]$; $y = 4$
4. $f(x, y) = x^3 - y^2$; x is in $[2, 6]$; $y = 3$
5. $f(x, y) = \dfrac{4x}{x + y}$; x is in $[0, 4]$; $y = -6$
6. $f(x, y) = \dfrac{2x - y}{2y + x}$; x is in $[-3, 3]$; $y = 5$
7. $f(x, y) = \cos x + y$; x is in $[-\pi, \pi]$; $y = 4$
8. $f(x, y) = \ln(x + y)$; x is in $[0, 2]$; $y = 1$

In Exercises 9 through 14, apply Theorem 16.8.1(ii) to find a ξ_2.

9. The function of Exercise 3; y is in $[-2, 2]$; $x = 0$

10. The function of Exercise 4; y is in $[-3, -1]$; $x = 7$
11. The function of Exercise 5; y is in $[-2, 2]$; $x = 4$
12. The function of Exercise 6; y is in $[0, 4]$; $x = 2$
13. $f(x, y) = e^y \tan x$; y is in $[3, 5]$; $x = \frac{1}{4}\pi$
14. $f(x, y) = \cos x + \sin y$; y is in $[-\frac{1}{6}\pi, \frac{1}{6}\pi]$; $x = \frac{1}{3}\pi$

In Exercises 15 and 16, use Theorem 16.5.4 to prove that f is differentiable at (0, 0).

15. $f(x, y) = \begin{cases} \dfrac{x^3 y^3}{x^4 + y^4} & \text{if } (x, y) \neq (0, 0) \\ 0 & \text{if } (x, y) = (0, 0) \end{cases}$

16. $f(x, y) = \begin{cases} \dfrac{x^4 + y^4}{x^2 + y^2} & \text{if } (x, y) \neq (0, 0) \\ 0 & \text{if } (x, y) = (0, 0) \end{cases}$

17. Prove Theorem 16.8.1(ii).

REVIEW EXERCISES FOR CHAPTER 16

In Exercises 1 through 4, determine the domain of f and draw a sketch showing as a region in R^2 the set of points in the domain.

1. $f(x, y) = \sqrt{x^2 + 4y^2 - 16}$ **2.** $f(x, y) = \dfrac{6}{\sqrt{36 - x^2 - y^2}}$

3. $f(x, y) = \ln(y - x^2)$
4. $f(x, y) = \sin^{-1}(5 - x^2 - y^2)$

In Exercises 5 and 6, determine the domain of f and describe the region in R^3 that is the set of points in the domain.

5. $f(x, y, z) = \dfrac{x}{|y| - |z|}$

6. $f(x, y, z) = \ln(x^2 + y^2 + z^2 - 4)$

In Exercises 7 and 8, determine the domain of f and draw a sketch of the graph of f.

7. $f(x, y) = \sqrt{36 - 4x^2 - 9y^2}$ **8.** $f(x, y) = 16x^2 - y^2$

9. The production function for a certain commodity is f, where $f(x, y) = 4x^{1/2}y$, and x and y give the amounts of two inputs. Draw a contour map of f showing the constant product curves at 16, 8, 4, and 2.

10. The temperature at a point (x, y) of a flat metal plate is $t(x, y)$ degrees, and $t(x, y) = x^2 + 2y$. Draw the isothermals for t at 0, 2, 4, 6, and 8.

In Exercises 11 through 24, find the indicated partial derivatives.

11. $f(x, y) = 2x^2y - 3xy^2 + 4x - 2y$; (a) $D_1 f(x, y)$; (b) $D_2 f(x, y)$; (c) $D_{11} f(x, y)$; (d) $D_{22} f(x, y)$; (e) $D_{12} f(x, y)$; (f) $D_{21} f(x, y)$.

12. $f(x, y) = (4x^2 - 2y)^3$; (a) $f_1(x, y)$; (b) $f_2(x, y)$; (c) $f_{11}(x, y)$; (d) $f_{22}(x, y)$; (e) $f_{12}(x, y)$; (f) $f_{21}(x, y)$.

13. $f(x, y) = \dfrac{x^2 - y}{3y^2}$; (a) $f_x(x, y)$; (b) $f_y(x, y)$; (c) $f_{xy}(x, y)$; (d) $f_{yx}(x, y)$.

14. $f(r, s) = re^{2rs}$; (a) $D_r f(r, s)$; (b) $D_s f(r, s)$; (c) $D_{rs} f(r, s)$; (d) $D_{sr} f(r, s)$.

15. $g(s, t) = \sin(st^2) + te^s$; (a) $D_s g(s, t)$; (b) $D_t g(s, t)$; (c) $D_{st} g(s, t)$; (d) $D_{ts} g(s, t)$.

16. $h(x, y) = \tan^{-1} \dfrac{x^3}{y^2}$; (a) $D_1 h(x, y)$; (b) $D_2 h(x, y)$; (c) $D_{11} h(x, y)$; (d) $D_{22} h(x, y)$.

17. $f(x, y) = e^{x/y} + \ln \dfrac{x}{y}$; (a) $f_x(x, y)$; (b) $f_y(x, y)$; (c) $f_{xx}(x, y)$; (d) $f_{yy}(x, y)$.

18. $f(x, y) = \ln\sqrt{x^2 + y^2}$; (a) $f_1(x, y)$; (b) $f_{11}(x, y)$; (c) $f_{12}(x, y)$; (d) $f_{121}(x, y)$.

19. $f(x, y, z) = \dfrac{x}{x^2 + y^2 + z^2}$; (a) $D_1 f(x, y, z)$; (b) $D_2 f(x, y, z)$; (c) $D_3 f(x, y, z)$.

20. $f(x, y, z) = \sqrt{x^2 + 3yz - z^2}$; (a) $f_x(x, y, z)$; (b) $f_y(x, y, z)$; (c) $f_z(x, y, z)$.

21. $f(u, v, w) = \ln(u^2 + 4v^2 - 5w^2)$; (a) $f_{uwv}(u, v, w)$; (b) $f_{uvv}(u, v, w)$.

22. $f(r, s, t) = t^2 e^{4rst}$; (a) $f_r(r, s, t)$; (b) $f_{rt}(r, s, t)$; (c) $f_{rts}(r, s, t)$.

23. $f(r, s, t) = \dfrac{\ln 4rs}{t^2}$; (a) $D_1 f(r, s, t)$; (b) $D_{13} f(r, s, t)$; (c) $D_{131} f(r, s, t)$.

24. $f(u, v, w) = w \cos 2v + 3v \sin u - 2uv \tan w$; (a) $D_2 f(u, v, w)$; (b) $D_1 f(u, v, w)$; (c) $D_{131} f(u, v, w)$.

25. If $w = x^2y - y^2x + y^2z - z^2y + z^2x - x^2z$, show that
$$\frac{\partial w}{\partial x} + \frac{\partial w}{\partial y} + \frac{\partial w}{\partial z} = 0$$

26. If $u = (x^2 + y^2 + z^2)^{-1/2}$, show that
$$\frac{\partial^2 u}{\partial x^2} + \frac{\partial^2 u}{\partial y^2} + \frac{\partial^2 u}{\partial z^2} = 0$$

In Exercises 27 and 28, find $\dfrac{\partial u}{\partial t}$ and $\dfrac{\partial u}{\partial s}$ by two methods.

27. $u = y \ln(x^2 + y^2)$, $x = 2s + 3t$, $y = 3t - 2s$

28. $u = e^{2x+y} \cos(2y - x)$, $x = 2s^2 - t^2$, $y = s^2 + 2t^2$

29. If $u = 3x^2y + 2xy - 3yz - 2z^2$, $x = e^{3rs}$, $y = r^3s^2$, and $z = \ln 4$, find $\dfrac{\partial u}{\partial r}$ by two methods: (a) Use the chain rule; (b) make the substitutions for x, y, and z before differentiating.

30. If $u = e^{x^2+y^2} - \dfrac{3x}{y} + 3z$, $x = \sin \theta$, $y = \cos \theta$, and $z = \tan \theta$, find the total derivative $\dfrac{du}{d\theta}$ by two methods: (a) Do not express u in terms of θ before differentiating; (b) express u in terms of θ before differentiating.

31. If $u = xy + x^2$, $x = 4 \cos t$, and $y = 3 \sin t$, find the value of the total derivative $\dfrac{du}{dt}$ at $t = \frac{1}{4}\pi$ by two methods: (a) Do not express u in terms of t before differentiating; (b) express u in terms of t before differentiating.

32. If $f(x, y) = x^2 + ye^x$, find:
(a) $\Delta f(0, 2)$, the increment of f at $(0, 2)$;
(b) $\Delta f(0, 2)$ when $\Delta x = -0.1$ and $\Delta y = 0.2$;
(c) $df(0, 2, \Delta x, \Delta y)$, the total differential of f at $(0, 2)$;
(d) $df(0, 2, -0.1, 0.2)$.

33. If $f(x, y, z) = 3xy^2 - 5xz^2 - 2xyz$, find:
(a) $\Delta f(-1, 3, 2)$, the increment of f at $(-1, 3, 2)$;
(b) $\Delta f(-1, 3, 2)$ when $\Delta x = 0.02$, $\Delta y = -0.01$, and $\Delta z = -0.02$;
(c) $df(-1, 3, 2, \Delta x, \Delta y, \Delta z)$, the total differential of f at $(-1, 3, 2)$;
(d) $df(-1, 3, 2, 0.02, -0.01, -0.02)$.

34. Given $f(x) = x^2 + 1$, $g(x, y) = \dfrac{2x}{3y}$, and $h(x) = \dfrac{1}{x}$, find:
(a) $(h \circ g)(-3, 4)$; (b) $g(f(3), h(\frac{1}{4}))$; (c) $g(f(x), h(y))$; (d) $f((h \circ g)(x, y))$.

In Exercises 35 through 37, evaluate the limit by the use of limit theorems.

35. $\lim_{(x,y) \to (e,0)} \ln\left(\frac{x^2}{y+1}\right)$

36. $\lim_{(x,y) \to (0,\pi/2)} \frac{xy^2 + e^x}{\cos x + \sin y}$

37. $\lim_{(x,y) \to (1,3)} \sin^{-1}\left(\frac{3x}{2y}\right)$

In Exercises 38 through 40, establish the limit by finding a $\delta > 0$ for any $\epsilon > 0$ such that Definition 16.2.5 holds.

38. $\lim_{(x,y) \to (4,-1)} (4x - 5y) = 21$

39. $\lim_{(x,y) \to (2,-2)} (3x^2 - 4y^2) = -4$

40. $\lim_{(x,y) \to (3,1)} (x^2 - y^2 + 2x - 4y) = 10$

In Exercises 41 through 44, determine if the limit exists.

41. $\lim_{(x,y) \to (0,0)} \frac{x^3 y^3}{x^2 + y^2}$

42. $\lim_{(x,y) \to (0,0)} \frac{x^4 - y^4}{x^4 + y^4}$

43. $\lim_{(x,y) \to (0,0)} \frac{x^9 y}{(x^6 + y^2)^2}$

44. $\lim_{(x,y) \to (0,0)} \frac{2x^3 + 4x^2 y}{x^2 + y^2}$

In Exercises 45 through 49, determine all points at which f is continuous.

45. $f(x, y) = \dfrac{x^2 + 4y^2}{x^2 - 4y^2}$

46. $f(x, y) = \dfrac{1}{\cos \frac{1}{2}\pi x} + \dfrac{1}{\cos \frac{1}{2}\pi y}$

47. $f(x, y) = \dfrac{1}{\cos^2 \frac{1}{2}\pi x + \cos^2 \frac{1}{2}\pi y}$

48. $f(x, y) = \begin{cases} \dfrac{x^4 - y^4}{x^4 + y^4} & \text{if } (x, y) \neq (0, 0) \\ 0 & \text{if } (x, y) = (0, 0) \end{cases}$

(*Hint:* See Exercise 42.)

49. $f(x, y) = \begin{cases} \dfrac{x^3 y^3}{x^2 + y^2} & \text{if } (x, y) \neq (0, 0) \\ 0 & \text{if } (x, y) = (0, 0) \end{cases}$

(*Hint:* See Exercise 41.)

50. Suppose α is the radian measure of an acute angle of a right triangle and $\sin \alpha$ is determined by a/c, where a centimeters is the length of the side opposite the angle and c centimeters is the length of the hypotenuse. If by measurement a is found to be 3.52 and c is found to be 7.14, and there is a possible error of 0.01 in each, find the possible error in the computation of $\sin \alpha$ from these measurements.

51. A painting contractor charges \$4 per square meter for painting the four walls and ceiling of a room. If the dimensions of the ceiling are measured to be 4 m and 5 m, the height of the room is measured to be 3 m, and these measurements are correct to 0.5 cm, find approximately, by using the total differential, the greatest error in estimating the cost of the job from these measurements.

52. At a given instant, the length of one side of a rectangle is 6 cm and it is increasing at the rate of 1 cm/sec, and the length of another side of the rectangle is 10 cm and it is decreasing at the rate of 2 cm/sec. Find the rate of change of the area of the rectangle at the given instant.

53. The radius of a right-circular cylinder is decreasing at the rate of 5 cm/min and the height is increasing at the rate of 12 cm/min. Find the rate of change of the volume at the instant when the radius is 20 cm and the height is 40 cm.

54. Find the slope of the tangent line to the curve of intersection of the surface $25x^2 - 16y^2 + 9z^2 - 4 = 0$ with the plane $x = 4$ at the point $(4, 9, 10)$.

55. Use the ideal gas law (see Example 5, Section 16.4) with $k = 1.4$ to find the rate of change of the pressure at the instant when the Kelvin temperature is 75 and the volume of the gas is 20 liters if the temperature is increasing at the rate of 0.5 K/min and the volume is increasing at the rate of 0.3 liter/min.

In Exercises 56 and 57, prove that f is differentiable at all points in its domain by showing that Definition 16.6.2 holds.

56. $f(x, y) = 3xy^2 - 4x^2 + y^2$ **57.** $f(x, y) = \dfrac{2x + y}{y^2}$

58. If $f(x, y)$ units are produced by x workers and y machines, then $D_x f(x, y)$ is called the *marginal productivity of labor* and $D_y f(x, y)$ is called the *marginal productivity of machines*. Suppose that

$$f(x, y) = x^2 + 6xy + 3y^2$$

where $5 \leq x \leq 30$ and $4 \leq y \leq 12$. (a) Find the number of units produced in a day when the labor force for that day consists of 15 workers, and 8 machines are used. (b) Use the marginal productivity of labor to determine the approximate number of additional units that can be produced in 1 day if the labor force is increased from 15 to 16 and the number of machines remains fixed at 8. (c) Use the marginal productivity of machines to determine the approximate number of additional units that can be produced in 1 day if the number of machines is increased from 8 to 9 and the number of workers remains fixed at 15.

59. Find (a) $f_2(x, 0)$ if $x \neq 0$ and (b) $f_2(0, 0)$ if

$$f(x, y) = \begin{cases} \dfrac{12x^2 y - 3y^2}{x^2 + y} & \text{if } (x, y) \neq (0, 0) \\ 0 & \text{if } (x, y) = (0, 0) \end{cases}$$

60. Verify that $u(x, y) = (\sinh x)(\sin y)$ satisfies Laplace's equation in R^2:

$$\frac{\partial^2 u}{\partial x^2} + \frac{\partial^2 u}{\partial y^2} = 0$$

61. If f is a differentiable function of u, let $u = x^2 + y^2$ and prove that $z = xy + f(x^2 + y^2)$ satisfies the equation

$$y\frac{\partial z}{\partial x} - x\frac{\partial z}{\partial y} = y^2 - x^2$$

62. Laplace's equation in polar coordinates is

$$r^2\frac{\partial^2 u}{\partial r^2} + r\frac{\partial u}{\partial r} + \frac{\partial^2 u}{\partial \theta^2} = 0$$

Verify that $u(r, \theta) = r^n \sin n\theta$, where n is a constant, satisfies this equation.

63. Verify that $u(x, y, z) = e^{3x+4y} \sin 5z$ satisfies Laplace's equation in R^3:

$$\frac{\partial^2 u}{\partial x^2} + \frac{\partial^2 u}{\partial y^2} + \frac{\partial^2 u}{\partial z^2} = 0$$

64. Verify that $u(x, t) = A \cos(kat) \sin(kx)$, where A and k are arbitrary constants, satisfies the partial differential equation for a vibrating string:

$$\frac{\partial^2 u}{\partial t^2} = a^2 \frac{\partial^2 u}{\partial x^2}$$

65. Verify that

$$u(x, t) = \sin\frac{n\pi x}{L} e^{(-n^2\pi^2 k^2/L^2)t}$$

satisfies the one-dimensional heat-conduction partial differential equation:

$$\frac{\partial u}{\partial t} = k^2 \frac{\partial^2 u}{\partial x^2}$$

66. Given

$$f(x, y) = \begin{cases} \dfrac{x^2 y}{x^4 + y^2} & \text{if } (x, y) \neq (0, 0) \\ 0 & \text{if } (x, y) = (0, 0) \end{cases}$$

Prove that $D_1 f(0, 0)$ and $D_2 f(0, 0)$ exist but that f is not differentiable at $(0, 0)$. (*Hint:* See Example 5, Section 16.2 and Exercise 10 in Exercises 16.3).

67. Given

$$f(x, y, z) = \begin{cases} \dfrac{x^2 y^2 z^2}{(x^2 + y^2 + z^2)^2} & \text{if } (x, y, z) \neq (0, 0, 0) \\ 0 & \text{if } (x, y, z) = (0, 0, 0) \end{cases}$$

Prove that f is differentiable at $(0, 0, 0)$.

68. Let f be the function defined by

$$f(x, y) = \begin{cases} \dfrac{e^{-1/x^2} y}{e^{-2/x^2} + y^2} & \text{if } x \neq 0 \\ 0 & \text{if } x = 0 \end{cases}$$

Prove that f is discontinuous at the origin.

69. For the function of Exercise 68, prove that $D_1 f(0, 0)$ and $D_2 f(0, 0)$ both exist.

70. If f is a differentiable function of x and y and $u = f(x, y)$, $x = r \cos\theta$, and $y = r \sin\theta$, show that

$$\left(\frac{\partial u}{\partial r}\right)^2 + \frac{1}{r^2}\left(\frac{\partial u}{\partial \theta}\right)^2 = \left(\frac{\partial u}{\partial x}\right)^2 + \left(\frac{\partial u}{\partial y}\right)^2$$

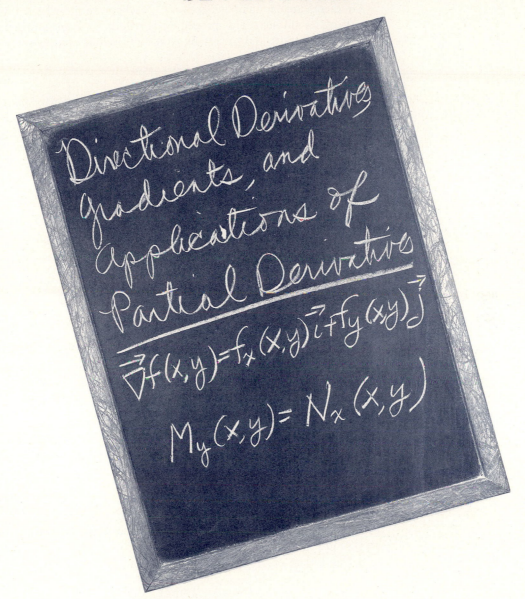

SEVENTEEN

Directional Derivatives, gradients, and Applications of Partial Derivatives

$$\vec{\nabla} f(x,y) = f_x(x,y)\,\vec{i} + f_y(xy)\,\vec{j}$$

$$M_y(x,y) = N_x(x,y)$$

The partial derivatives $f_1(x, y)$ and $f_2(x, y)$ measure the rates of change of the function values $f(x, y)$ in the direction of the x and y axes, respectively. The *directional derivatives*, introduced in Section 17.1, give the rates of change of these function values in any direction. The *gradient*, also introduced in Section 17.1, gives the direction in which the function has its greatest rate of change. This concept is applied in Section 17.2 in our discussion of *tangent planes* and *normals to surfaces*.

Just as we use first and second derivatives to determine maxima and minima of functions of a single variable, we show in Section 17.3 how partial derivatives are applied to find extreme values of functions of two variables. The applications in this section include the *method of least squares*. In Section 17.4 we introduce *Lagrange multipliers*, which are used to compute extrema of a function subject to a constraint.

Gradients appear again in Section 17.5, where we show how to obtain a function from its gradient. This procedure is related to determining if a differential expression is *exact* and to solving *exact differential equations*.

17.1 DIRECTIONAL DERIVATIVES AND GRADIENTS

We now generalize the definition of a partial derivative to obtain the rate of change of a function with respect to any direction. This leads to the concept of a *directional derivative*.

Let f be a function of the two variables x and y and let $P(x, y)$ be a point in the xy plane. Suppose that $\mathbf{U}$ is the unit vector making an angle of radian measure θ with the positive side of the x axis. Then

$$\mathbf{U} = \cos \theta \mathbf{i} + \sin \theta \mathbf{j}$$

Figure 1 shows the representation of $\mathbf{U}$ having its initial point at $P(x, y)$.

17.1.1 DEFINITION

Let f be a function of two variables x and y. If $\mathbf{U}$ is the unit vector $\cos \theta \mathbf{i} + \sin \theta \mathbf{j}$, then the **directional derivative** of f in the direction of $\mathbf{U}$, denoted by $D_{\mathbf{U}}f$, is given by

$$D_{\mathbf{U}}f(x, y) = \lim_{h \to 0} \frac{f(x + h \cos \theta, y + h \sin \theta) - f(x, y)}{h}$$

if this limit exists.

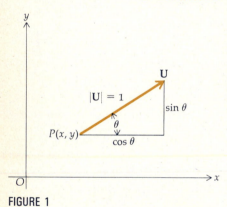

FIGURE 1

FIGURE 2

$Q(x_0 + h \cos \theta, y_0 + h \sin \theta, 0)$

The directional derivative gives the rate of change of the function values $f(x, y)$ with respect to the direction of the unit vector $\mathbf{U}$. This is illustrated in Figure 2. An equation of the surface S in the figure is $z = f(x, y)$. $P_0(x_0, y_0, z_0)$ is a point on the surface, and $R(x_0, y_0, 0)$ and $Q(x_0 + h \cos \theta, y_0 + h \sin \theta, 0)$ are points in the xy plane. The plane through R and Q, parallel to the z axis, makes an angle of θ radians with the positive direction on the x axis. This plane intersects the surface S in the curve C. The directional derivative $D_{\mathbf{U}}f$, evaluated at P_0, is the slope of the tangent line to the curve C at P_0 in the plane of R, Q, and P_0.

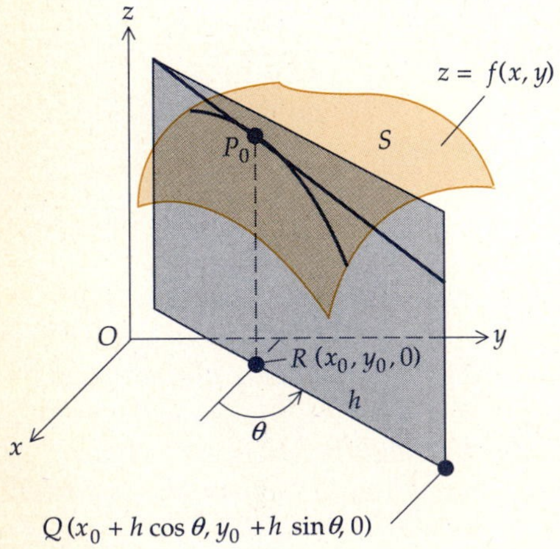

If $\mathbf{U} = \mathbf{i}$, then $\cos \theta = 1$ and $\sin \theta = 0$, and from Definition 17.1.1,

$$D_{\mathbf{i}}f(x, y) = \lim_{h \to 0} \frac{f(x + h, y) - f(x, y)}{h}$$

which is the partial derivative of f with respect to x.

If $\mathbf{U} = \mathbf{j}$, then $\cos \theta = 0$ and $\sin \theta = 1$, and

$$D_{\mathbf{j}}f(x, y) = \lim_{h \to 0} \frac{f(x, y + h) - f(x, y)}{h}$$

which is the partial derivative of f with respect to y.

So f_x and f_y are special cases of the directional derivative in the directions of the unit vectors $\mathbf{i}$ and $\mathbf{j}$, respectively.

▶ **ILLUSTRATION 1**   We apply Definition 17.1.1 to find $D_{\mathbf{U}}f$ if

$$f(x, y) = 3x^2 - y^2 + 4x$$

and $\mathbf{U}$ is the unit vector in the direction $\frac{1}{6}\pi$. Then $\mathbf{U} = \cos \frac{1}{6}\pi \mathbf{i} + \sin \frac{1}{6}\pi \mathbf{j}$; that is, $\mathbf{U} = \frac{1}{2}\sqrt{3}\mathbf{i} + \frac{1}{2}\mathbf{j}$. Thus from Definition 17.1.1,

$$D_{\mathbf{U}}f(x, y) = \lim_{h \to 0} \frac{f(x + \frac{1}{2}\sqrt{3}h, y + \frac{1}{2}h) - f(x, y)}{h}$$

$$= \lim_{h \to 0} \frac{3(x + \frac{1}{2}\sqrt{3}h)^2 - (y + \frac{1}{2}h)^2 + 4(x + \frac{1}{2}\sqrt{3}h) - (3x^2 - y^2 + 4x)}{h}$$

$$= \lim_{h \to 0} \frac{3x^2 + 3\sqrt{3}hx + \frac{9}{4}h^2 - y^2 - hy - \frac{1}{4}h^2 + 4x + 2\sqrt{3}h - 3x^2 + y^2 - 4x}{h}$$

$$= \lim_{h \to 0} \frac{3\sqrt{3}hx + \frac{9}{4}h^2 - hy - \frac{1}{4}h^2 + 2\sqrt{3}h}{h}$$

$$= \lim_{h \to 0} (3\sqrt{3}x + \frac{9}{4}h - y - \frac{1}{4}h + 2\sqrt{3})$$

$$= 3\sqrt{3}x - y + 2\sqrt{3} \qquad ◀$$

We now proceed to obtain a formula that will enable us to calculate a directional derivative in a shorter way than using the definition. Let g be the function of the single variable t, with x, y, and θ fixed, such that

$$g(t) = f(x + t \cos \theta, y + t \sin \theta) \tag{1}$$

and let $\mathbf{U} = \cos \theta \mathbf{i} + \sin \theta \mathbf{j}$. Then by the definition of an ordinary derivative,

$$g'(0) = \lim_{h \to 0} \frac{f(x + (0 + h) \cos \theta, y + (0 + h) \sin \theta) - f(x + 0 \cos \theta, y + 0 \sin \theta)}{h}$$

$$g'(0) = \lim_{h \to 0} \frac{f(x + h \cos \theta, y + h \sin \theta) - f(x, y)}{h}$$

Because the right side of the above is $D_{\mathbf{U}}f(x, y)$,

$$g'(0) = D_{\mathbf{U}}f(x, y) \tag{2}$$

We now find $g'(t)$ by applying the chain rule to the right side of (1), which gives

$$g'(t) = f_1(x + t \cos \theta, y + t \sin \theta) \frac{\partial(x + t \cos \theta)}{\partial t} + f_2(x + t \cos \theta, y + t \sin \theta) \frac{\partial(y + t \sin \theta)}{\partial t}$$

$$= f_1(x + t \cos \theta, y + t \sin \theta) \cos \theta + f_2(x + t \cos \theta, y + t \sin \theta) \sin \theta$$

Therefore

$$g'(0) = f_x(x, y) \cos \theta + f_y(x, y) \sin \theta$$

From this equation and (2) we have the following theorem.

17.1.2 THEOREM

If f is a differentiable function of x and y, and $\mathbf{U} = \cos\theta\mathbf{i} + \sin\theta\mathbf{j}$, then

$$D_{\mathbf{U}}f(x, y) = f_x(x, y)\cos\theta + f_y(x, y)\sin\theta$$

▶ **ILLUSTRATION 2** We apply Definition 17.1.2 to compute $D_{\mathbf{U}}f$ for the function f and the unit vector $\mathbf{U}$ of Illustration 1:

$$f(x, y) = 3x^2 - y^2 + 4x \qquad \mathbf{U} = \cos\tfrac{1}{6}\pi\mathbf{i} + \sin\tfrac{1}{6}\pi\mathbf{j}$$

$$\begin{aligned}
D_{\mathbf{U}}f(x, y) &= f_x(x, y)\cos\tfrac{1}{6}\pi + f_y(x, y)\sin\tfrac{1}{6}\pi \\
&= (6x + 4)\tfrac{1}{2}\sqrt{3} + (-2y)\tfrac{1}{2} \\
&= 3\sqrt{3}x - y + 2\sqrt{3}
\end{aligned}$$

which agrees with the result in Illustration 1. ◀

The directional derivative can be written as the dot product of two vectors. Because

$$f_x(x, y)\cos\theta + f_y(x, y)\sin\theta = (\cos\theta\mathbf{i} + \sin\theta\mathbf{j}) \cdot [f_x(x, y)\mathbf{i} + f_y(x, y)\mathbf{j}]$$

it follows from Theorem 17.1.2 that

$$D_{\mathbf{U}}f(x, y) = (\cos\theta\mathbf{i} + \sin\theta\mathbf{j}) \cdot [f_x(x, y)\mathbf{i} + f_y(x, y)\mathbf{j}] \tag{3}$$

The second vector on the right side of (3) is a very important one, and it is called the *gradient* of the function f. The symbol used for the gradient of f is ∇f, where ∇ is an inverted capital delta and is read "del." Sometimes the abbreviation *grad f* is used.

17.1.3 DEFINITION

If f is a function of two variables x and y, and f_x and f_y exist, then the **gradient** of f, denoted by ∇f (read "del f"), is defined by

$$\nabla f(x, y) = f_x(x, y)\mathbf{i} + f_y(x, y)\mathbf{j}$$

From Definition 17.1.3, Equation (3) can be written as

$$D_{\mathbf{U}}f(x, y) = \mathbf{U} \cdot \nabla f(x, y) \tag{4}$$

Therefore any directional derivative of a differentiable function can be obtained by dot-multiplying the gradient by the unit vector in the desired direction.

EXAMPLE 1   If

$$f(x, y) = \tfrac{1}{16}x^2 + \tfrac{1}{9}y^2$$

find the gradient of f at the point $(4, 3)$. Also find the rate of change of $f(x, y)$ in the direction $\tfrac{1}{4}\pi$ at $(4, 3)$.

Solution   Because $f_x(x, y) = \tfrac{1}{8}x$ and $f_y(x, y) = \tfrac{2}{9}y$,

$$\nabla f(x, y) = \tfrac{1}{8}x\mathbf{i} + \tfrac{2}{9}y\mathbf{j} \qquad \nabla f(4, 3) = \tfrac{1}{2}\mathbf{i} + \tfrac{2}{3}\mathbf{j}$$

The rate of change of $f(x, y)$ in the direction $\tfrac{1}{4}\pi$ at $(4, 3)$ is $D_{\mathbf{U}}f(4, 3)$, where

$$\mathbf{U} = \frac{1}{\sqrt{2}}\mathbf{i} + \frac{1}{\sqrt{2}}\mathbf{j}$$

We find $D_{\mathbf{U}}f(4, 3)$ by dot-multiplying $\nabla f(4, 3)$ by $\mathbf{U}$.

$$D_{\mathbf{U}}f(4, 3) = \left(\frac{1}{\sqrt{2}}\mathbf{i} + \frac{1}{\sqrt{2}}\mathbf{j}\right) \cdot \left(\frac{1}{2}\mathbf{i} + \frac{2}{3}\mathbf{j}\right)$$

$$= \frac{7}{6\sqrt{2}}$$

If α is the radian measure of the angle between the two vectors $\mathbf{U}$ and ∇f, then

$$\mathbf{U} \cdot \nabla f(x, y) = \|\mathbf{U}\| \|\nabla f(x, y)\| \cos \alpha$$

From this equation and (4) it follows that

$$D_{\mathbf{U}}f(x, y) = \|\mathbf{U}\| \|\nabla f(x, y)\| \cos \alpha \qquad (5)$$

We see from (5) that $D_{\mathbf{U}}f$ will be a maximum when $\cos \alpha = 1$, that is, when $\mathbf{U}$ is in the direction of ∇f; and in this case, $D_{\mathbf{U}}f = \|\nabla f\|$. Hence the gradient of a function is in the direction in which the function has its maximum rate of change. In particular, on a two-dimensional topographical map of a landscape where z units is the elevation at a point (x, y) and $z = f(x, y)$, the direction in which the rate of change of z is the greatest is given by $\nabla f(x, y)$; that is, $\nabla f(x, y)$ points in the direction of steepest ascent. This accounts for the name *gradient* (the grade is steepest in the direction of the gradient).

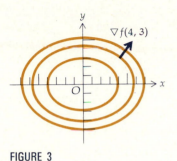

FIGURE 3

▶ **ILLUSTRATION 3**   In Figure 3 there is a contour map showing the level curves of the function of Example 1 at 1, 2, and 3. The level curves are ellipses. The figure also shows the representation of $\nabla f(4, 3)$ having its initial point at $(4, 3)$.   ◀

EXAMPLE 2   Given

$$f(x, y) = 2x^2 - y^2 + 3x - y$$

find the maximum value of $D_{\mathbf{U}}f$ at the point where $x = 1$ and $y = -2$.

Solution   Because $f_x(x, y) = 4x + 3$ and $f_y(x, y) = -2y - 1$,

$$\nabla f(x, y) = (4x + 3)\mathbf{i} + (-2y - 1)\mathbf{j} \qquad \nabla f(1, -2) = 7\mathbf{i} + 3\mathbf{j}$$

Thus the maximum value of $D_{\mathbf{U}}f$ at the point $(1, -2)$ is

$$\|\nabla f(1, -2)\| = \sqrt{49 + 9}$$
$$= \sqrt{58}$$

EXAMPLE 3   The temperature at any point (x, y) of a rectangular plate lying in the xy plane is determined by

$$T(x, y) = x^2 + y^2$$

(a) Find the rate of change of the temperature at the point $(3, 4)$ in the direction making an angle of radian measure $\frac{1}{3}\pi$ with the positive x direction; (b) find the direction for which the rate of change of the temperature at the point $(-3, 1)$ is a maximum.

Solution

(a) We wish to find $D_{\mathbf{U}}T(x, y)$, where

$$\mathbf{U} = \cos \tfrac{1}{3}\pi \mathbf{i} + \sin \tfrac{1}{3}\pi \mathbf{j} \qquad \nabla T(x, y) = T_x(x, y)\mathbf{i} + T_y(x, y)\mathbf{j}$$
$$= \tfrac{1}{2}\mathbf{i} + \tfrac{1}{2}\sqrt{3}\mathbf{j} \qquad\qquad\qquad = 2x\mathbf{i} + 2y\mathbf{j}$$

Therefore

$$D_{\mathbf{U}}T(x, y) = \mathbf{U} \cdot \nabla T(x, y)$$
$$= (\tfrac{1}{2}\mathbf{i} + \tfrac{1}{2}\sqrt{3}\mathbf{j}) \cdot (2x\mathbf{i} + 2y\mathbf{j})$$
$$= x + \sqrt{3}y$$

Thus

$$D_{\mathbf{U}}T(3, 4) = 3 + 4\sqrt{3}$$
$$\approx 9.93$$

So at $(3, 4)$ the temperature is increasing at the rate of approximately 9.93 units per unit change in the distance measured in the direction of $\mathbf{U}$.

(b) $D_{\mathbf{U}}T(-3, 1)$ is a maximum when $\mathbf{U}$ is in the direction of $\nabla T(-3, 1)$. Because $\nabla T(-3, 1) = -6\mathbf{i} + 2\mathbf{j}$, the radian measure of the angle giving the direction of $\nabla T(-3, 1)$ is θ, where $\tan \theta = -\tfrac{1}{3}$. Thus $\theta = \pi - \tan^{-1} \tfrac{1}{3}$. Therefore the rate of change of the temperature at the point $(-3, 1)$ is a maximum in the direction making an angle of radian measure $\pi - \tan^{-1}\tfrac{1}{3}$ with the positive side of the x axis.

We extend the definition of a directional derivative to a function of three variables. In three-dimensional space the direction of a vector is determined by its direction cosines. If $\cos \alpha$, $\cos \beta$, and $\cos \gamma$ are the direction cosines of the unit vector $\mathbf{U}$, then $\mathbf{U} = \cos \alpha \mathbf{i} + \cos \beta \mathbf{j} + \cos \gamma \mathbf{k}$.

17.1.4 DEFINITION   Suppose that f is a function of three variables x, y, and z. If $\mathbf{U}$ is the unit vector $\cos \alpha \mathbf{i} + \cos \beta \mathbf{j} + \cos \gamma \mathbf{k}$, then the **directional derivative** of f in the direction of $\mathbf{U}$, denoted by $D_{\mathbf{U}}f$, is given by

$$D_{\mathbf{U}}f(x, y, z) = \lim_{h \to 0} \frac{f(x + h \cos \alpha, y + h \cos \beta, z + h \cos \gamma) - f(x, y, z)}{h}$$

if this limit exists.

The directional derivative of a function of three variables gives the rate of change of the function values $f(x, y, z)$ with respect to distance in three-dimensional space measured in the direction of the unit vector $\mathbf{U}$.

The following theorem, which provides a method for calculating a directional derivative for a function of three variables, is proved in a manner similar to the proof of Theorem 17.1.2.

17.1.5 THEOREM   If f is a differentiable function of x, y, and z and

$$\mathbf{U} = \cos \alpha \mathbf{i} + \cos \beta \mathbf{j} + \cos \gamma \mathbf{k}$$

then

$$D_{\mathbf{U}}f(x, y, z) = f_x(x, y, z) \cos \alpha + f_y(x, y, z) \cos \beta + f_z(x, y, z) \cos \gamma$$

EXAMPLE 4   Given

$$f(x, y, z) = 3x^2 + xy - 2y^2 - yz + z^2$$

find the rate of change of $f(x, y, z)$ at $(1, -2, -1)$ in the direction of the vector $2\mathbf{i} - 2\mathbf{j} - \mathbf{k}$.

Solution   The unit vector in the direction of $2\mathbf{i} - 2\mathbf{j} - \mathbf{k}$ is

$$\mathbf{U} = \tfrac{2}{3}\mathbf{i} - \tfrac{2}{3}\mathbf{j} - \tfrac{1}{3}\mathbf{k}$$

So from Theorem 17.1.5

$$D_{\mathbf{U}}f(x, y, z) = \tfrac{2}{3}(6x + y) - \tfrac{2}{3}(x - 4y - z) - \tfrac{1}{3}(-y + 2z)$$

Therefore the rate of change of $f(x, y, z)$ at $(1, -2, -1)$ in the direction of $\mathbf{U}$ is given by

$$D_{\mathbf{U}}f(1, -2, -1) = \tfrac{2}{3}(4) - \tfrac{2}{3}(10) - \tfrac{1}{3}(0)$$
$$= -4$$

17.1.6 DEFINITION   If f is a function of three variables x, y, and z and the first partial derivatives f_x, f_y, and f_z exist, then the **gradient** of f, denoted by ∇f, is defined by

$$\nabla f(x, y, z) = f_x(x, y, z)\mathbf{i} + f_y(x, y, z)\mathbf{j} + f_z(x, y, z)\mathbf{k}$$

Just as for functions of two variables, it follows from Theorem 17.1.5 and Definition 17.1.6 that if $\mathbf{U} = \cos \alpha\, \mathbf{i} + \cos \beta\, \mathbf{j} + \cos \gamma\, \mathbf{k}$, then

$$D_{\mathbf{U}}f(x, y, z) = \mathbf{U} \cdot \nabla f(x, y, z)$$

Also, the directional derivative is a maximum when $\mathbf{U}$ is in the direction of the gradient, and the maximum directional derivative is the magnitude of the gradient.

Applications of the gradient occur in physics in problems in heat conduction and electricity. Suppose that $w = f(x, y, z)$. A level surface of this function f at the constant k is given by

$$f(x, y, z) = k \tag{6}$$

If w degrees is the temperature at point (x, y, z), then all points on the surface of Equation (6) have the same temperature of k degrees, and the surface is called an **isothermal surface**. If w is the number of volts in the electric potential at point (x, y, z), then all points on the surface are at the same potential, and the surface is called an **equipotential surface**. The gradient vector at a point gives the direction of greatest rate of change of w. So if the level surface of Equation (6) is an isothermal surface, $\nabla f(x, y, z)$ gives the direction of the greatest rate of change of temperature at (x, y, z). If (6) is an equation of an equipotential surface, then $\nabla f(x, y, z)$ gives the direction of the greatest rate of change of potential at (x, y, z).

EXAMPLE 5   If $V(x, y, z)$ volts is the electric potential at any point (x, y, z) in three-dimensional space and

$$V(x, y, z) = \frac{1}{\sqrt{x^2 + y^2 + z^2}}$$

find: (a) the rate of change of V at the point $(2, 2, -1)$ in the direction of the vector $2\mathbf{i} - 3\mathbf{j} + 6\mathbf{k}$; (b) the direction of the greatest rate of change of V at $(2, 2, -1)$.

Solution

(a) A unit vector in the direction of $2\mathbf{i} - 3\mathbf{j} + 6\mathbf{k}$ is

$$\mathbf{U} = \tfrac{2}{7}\mathbf{i} - \tfrac{3}{7}\mathbf{j} + \tfrac{6}{7}\mathbf{k}$$

We wish to find $D_{\mathbf{U}}V(2, 2, -1)$.

$$\nabla V(x, y, z) = V_x(x, y, z)\mathbf{i} + V_y(x, y, z)\mathbf{j} + V_z(x, y, z)\mathbf{k}$$

$$= \frac{-x}{(x^2 + y^2 + z^2)^{3/2}}\mathbf{i} + \frac{-y}{(x^2 + y^2 + z^2)^{3/2}}\mathbf{j} + \frac{-z}{(x^2 + y^2 + z^2)^{3/2}}\mathbf{k}$$

Then

$$D_{\mathbf{U}}V(2, 2, -1) = \mathbf{U} \cdot \nabla V(2, 2, -1)$$

$$= (\tfrac{2}{7}\mathbf{i} - \tfrac{3}{7}\mathbf{j} + \tfrac{6}{7}\mathbf{k}) \cdot (-\tfrac{2}{27}\mathbf{i} - \tfrac{2}{27}\mathbf{j} + \tfrac{1}{27}\mathbf{k})$$

$$= -\tfrac{4}{189} + \tfrac{6}{189} + \tfrac{6}{189}$$

$$= \tfrac{8}{189}$$

$$\approx 0.042$$

Therefore at $(2, 2, -1)$ the potential is increasing at the rate of approximately 0.042 volt per unit change in the distance measured in the direction of $\mathbf{U}$.

(b) $\nabla V(2, 2, -1) = -\tfrac{2}{27}\mathbf{i} - \tfrac{2}{27}\mathbf{j} + \tfrac{1}{27}\mathbf{k}$. A unit vector in the direction of $\nabla V(2, 2, -1)$ is

$$\frac{\nabla V(2, 2, -1)}{\|\nabla V(2, 2, -1)\|} = \frac{-\tfrac{2}{27}\mathbf{i} - \tfrac{2}{27}\mathbf{j} + \tfrac{1}{27}\mathbf{k}}{\tfrac{3}{27}}$$

$$= -\tfrac{2}{3}\mathbf{i} - \tfrac{2}{3}\mathbf{j} + \tfrac{1}{3}\mathbf{k}$$

The direction cosines of this vector are $-\tfrac{2}{3}$, $-\tfrac{2}{3}$, and $\tfrac{1}{3}$, which give the direction of the greatest rate of change of V at $(2, 2, -1)$.

EXERCISES 17.1

In Exercises 1 through 6, find the directional derivative of the function in the direction of the unit vector $\mathbf{U}$ *by using either Definition 17.1.1 or Definition 17.1.4, and then verify your result by applying either Theorem 17.1.2 or Theorem 17.1.5, whichever one applies.*

1. $f(x, y) = 2x^2 + 5y^2$; $\mathbf{U} = \cos\tfrac{1}{4}\pi\mathbf{i} + \sin\tfrac{1}{4}\pi\mathbf{j}$
2. $g(x, y) = 3x^2 - 4y^2$; $\mathbf{U} = \cos\tfrac{1}{3}\pi\mathbf{i} + \sin\tfrac{1}{3}\pi\mathbf{j}$
3. $h(x, y, z) = 3x^2 + y^2 - 4z^2$; $\mathbf{U} = \cos\tfrac{1}{3}\pi\mathbf{i} + \cos\tfrac{1}{4}\pi\mathbf{j} + \cos\tfrac{2}{3}\pi\mathbf{k}$
4. $f(x, y, z) = 6x^2 - 2xy + yz$; $\mathbf{U} = \tfrac{3}{7}\mathbf{i} + \tfrac{2}{7}\mathbf{j} + \tfrac{6}{7}\mathbf{k}$

5. $g(x, y) = \dfrac{1}{x - y}$; $\mathbf{U} = -\tfrac{12}{13}\mathbf{i} + \tfrac{5}{13}\mathbf{j}$

6. $f(x, y) = \dfrac{1}{x^2 + y^2}$; $\mathbf{U} = \tfrac{3}{5}\mathbf{i} - \tfrac{4}{5}\mathbf{j}$

In Exercises 7 through 14, find the gradient of the function.

7. $f(x, y) = 4x^2 - 3xy + y^2$ 8. $g(x, y) = \dfrac{xy}{x^2 + y^2}$

9. $g(x, y) = \ln\sqrt{x^2 + y^2}$
10. $f(x, y) = e^y \tan 2x$
11. $f(x, y, z) = \dfrac{x - y}{x + z}$
12. $f(x, y, z) = 3z \ln(x + y)$
13. $g(x, y, z) = xe^{-2y} \sec z$
14. $g(x, y, z) = e^{2z}(\sin x - \cos y)$

In Exercises 15 through 22, find the value of the directional derivative at the point P_0 *for the function in the direction of* $\mathbf{U}$.

15. $f(x, y) = x^2 - 2xy^2$; $\mathbf{U} = \cos\pi\mathbf{i} + \sin\pi\mathbf{j}$; $P_0 = (1, -2)$
16. $g(x, y) = 3x^3y + 4y^2 - xy$; $\mathbf{U} = \cos\tfrac{1}{4}\pi\mathbf{i} + \sin\tfrac{1}{4}\pi\mathbf{j}$; $P_0 = (0, 3)$
17. $g(x, y) = y^2 \tan^2 x$; $\mathbf{U} = -\tfrac{1}{2}\sqrt{3}\mathbf{i} + \tfrac{1}{2}\mathbf{j}$; $P_0 = (\tfrac{1}{3}\pi, 2)$
18. $f(x, y) = xe^{2y}$; $\mathbf{U} = \tfrac{1}{2}\mathbf{i} + \tfrac{1}{2}\sqrt{3}\mathbf{j}$; $P_0 = (2, 0)$
19. $h(x, y, z) = \cos(xy) + \sin(yz)$; $\mathbf{U} = -\tfrac{1}{3}\mathbf{i} + \tfrac{2}{3}\mathbf{j} + \tfrac{2}{3}\mathbf{k}$; $P_0 = (2, 0, -3)$

20. $f(x, y, z) = \ln(x^2 + y^2 + z^2)$; $U = \frac{1}{\sqrt{3}}i - \frac{1}{\sqrt{3}}j - \frac{1}{\sqrt{3}}k$;

$P_0 = (1, 3, 2)$

21. $f(x, y) = e^{-3x} \cos 3y$; $U = \cos(-\frac{1}{12}\pi)i + \sin(-\frac{1}{12}\pi)j$;
$P_0 = (-\frac{1}{12}\pi, 0)$

22. $g(x, y, z) = \cos 2x \cos 3y \sinh 4z$;

$U = \frac{1}{\sqrt{3}}i - \frac{1}{\sqrt{3}}j + \frac{1}{\sqrt{3}}k$; $P_0 = (\frac{1}{2}\pi, 0, 0)$

In Exercises 23 through 26, find (a) the gradient of f at P and (b) the rate of change of the function value in the direction of U at P.

23. $f(x, y) = x^2 - 4y$; $P = (-2, 2)$; $U = \cos \frac{1}{3}\pi i + \sin \frac{1}{3}\pi j$
24. $f(x, y) = e^{2xy}$; $P = (2, 1)$; $U = \frac{4}{5}i - \frac{3}{5}j$
25. $f(x, y, z) = y^2 + z^2 - 4xz$; $P = (-2, 1, 3)$; $U = \frac{2}{7}i - \frac{6}{7}j + \frac{3}{7}k$
26. $f(x, y, z) = 2x^3 + xy^2 + xz^2$; $P = (1, 1, 1)$;
$U = \frac{1}{7}\sqrt{21}j - \frac{2}{7}\sqrt{7}k$

27. Draw a contour map showing the level curves of the function of Exercise 23 at 8, 4, 0, -4, and -8. Also show the representation of $\nabla f(-2, 2)$ having its initial point at $(-2, 2)$.
28. Draw a contour map showing the level curves of the function of Exercise 24, at e^8, e^4, 1, e^{-4}, and e^{-8}. Also show the representation of $\nabla f(2, 1)$ having its initial point at $(2, 1)$.

In Exercises 29 through 32, find $D_U f$ at the point P for which U is a unit vector in the direction of $\overrightarrow{PQ}$. Also at P find $D_U f$, if U is a unit vector for which $D_U f$ is a maximum.

29. $f(x, y) = e^x \tan^{-1} y$; $P(0, 1)$, $Q(3, 5)$
30. $f(x, y) = e^x \cos y + e^y \sin x$; $P(1, 0)$, $Q(-3, 3)$
31. $f(x, y, z) = x - 2y + z^2$; $P(3, 1, -2)$, $Q(10, 7, 4)$
32. $f(x, y, z) = x^2 + y^2 - 4xz$; $P(3, 1, -2)$, $Q(-6, 3, 4)$

33. Find the direction from the point $(1, 3)$ for which the value

of f does not change if $f(x, y) = e^{2y} \tan^{-1} \frac{y}{3x}$.

34. The density is $\rho(x, y)$ kilograms per square meter at any point of a rectangular plate in the xy plane and

$$\rho(x, y) = \frac{1}{\sqrt{x^2 + y^2 + 3}}$$

(a) Find the rate of change of the density at the point $(3, 2)$ in the direction of the unit vector $\cos \frac{2}{3}\pi i + \sin \frac{2}{3}\pi j$. (b) Find the direction and magnitude of the greatest rate of change of ρ at $(3, 2)$.

35. The temperature is $T(x, y)$ degrees at any point of a rectangular plate lying in the xy plane, and $T(x, y) = 3x^2 + 2xy$. Distance is measured in meters. (a) Find the maximum rate of change of the temperature at the point $(3, -6)$ on the plate. (b) Find the direction for which this maximum rate of change at $(3, -6)$ occurs.

36. The temperature is $T(x, y, z)$ degrees at any point of a solid in three-dimensional space, and

$$T(x, y, z) = \frac{60}{x^2 + y^2 + z^2 + 3}$$

Distance is measured in inches. (a) Find the rate of change of the temperature at the point $(3, -2, 2)$ in the direction of the vector $-2i + 3j - 6k$. (b) Find the direction and magnitude of the greatest rate of change of T at $(3, -2, 2)$.

37. The electric potential is $V(x, y)$ volts at any point in the xy plane, and $V(x, y) = e^{-2x} \cos 2y$. Distance is measured in feet. (a) Find the rate of change of the potential at the point $(0, \frac{1}{4}\pi)$ in the direction of the unit vector $\cos \frac{1}{6}\pi i + \sin \frac{1}{6}\pi j$. (b) Find the direction and magnitude of the greatest rate of change of V at $(0, \frac{1}{4}\pi)$.

38. An equation of the surface of a mountain is

$$z = 1200 - 3x^2 - 2y^2$$

where distance is measured in meters, the x axis points to the east, and the y axis points to the north. A mountain climber is at the point corresponding to $(-10, 5, 850)$. (a) What is the direction of steepest ascent? (b) If the climber moves in the east direction, is she ascending or descending, and what is her rate? (c) If the climber moves in the southwest direction, is she ascending or descending, and what is her rate? (d) In what direction is she traveling a level path?

17.2 TANGENT PLANES AND NORMALS TO SURFACES

Let S be the surface having the equation

$$F(x, y, z) = 0 \tag{1}$$

and suppose that $P_0(x_0, y_0, z_0)$ is a point on S. Then $F(x_0, y_0, z_0) = 0$. Suppose further that C is a curve on S through P_0 and that a set of parametric equations of C is

$$x = f(t) \qquad y = g(t) \qquad z = h(t) \tag{2}$$

where the value of the parameter t at P_0 is t_0. A vector equation of C is

$$\mathbf{R}(t) = f(t)\mathbf{i} + g(t)\mathbf{j} + h(t)\mathbf{k}$$

Because curve C is on surface S, we have, upon substituting from (2) in (1),

$$F(f(t), g(t), h(t)) = 0 \tag{3}$$

Let $G(t) = F(f(t), g(t), h(t))$. If F_x, F_y, and F_z are continuous and not all zero at P_0, and if $f'(t_0)$, $g'(t_0)$, and $h'(t_0)$ exist, then the total derivative of F with respect to t at P_0 is given by

$$G'(t_0) = F_x(x_0, y_0, z_0)f'(t_0) + F_y(x_0, y_0, z_0)g'(t_0) + F_z(x_0, y_0, z_0)h'(t_0)$$

The right side of this equation can be written as

$$[F_x(x_0, y_0, z_0)\mathbf{i} + F_y(x_0, y_0, z_0)\mathbf{j} + F_z(x_0, y_0, z_0)\mathbf{k}] \cdot [f'(t_0)\mathbf{i} + g'(t_0)\mathbf{j} + h'(t_0)\mathbf{k}]$$

Thus

$$G'(t_0) = \nabla F(x_0, y_0, z_0) \cdot D_t\mathbf{R}(t_0)$$

Since $G'(t) = 0$ for all t under consideration (because of (3)), $G'(t_0) = 0$; so it follows from the above that

$$\nabla F(x_0, y_0, z_0) \cdot D_t\mathbf{R}(t_0) = 0 \qquad (4)$$

From Section 15.8 we know that $D_t\mathbf{R}(t_0)$ has the same direction as a tangent vector to curve C at P_0. Therefore, from (4) we can conclude that the gradient vector of F at P_0 is orthogonal to a tangent vector of every curve C on S through the point P_0. We are led, then, to the following definition.

17.2.1 DEFINITION   A vector that is orthogonal to a tangent vector of every curve C through a point P_0 on a surface S is called a **normal vector** to S at P_0.

From this definition and the preceding discussion we have the following theorem.

17.2.2 THEOREM   If an equation of a surface S is $F(x, y, z) = 0$, and F_x, F_y, and F_z are continuous and not all zero at the point $P_0(x_0, y_0, z_0)$ on S, then $\nabla F(x_0, y_0, z_0)$ is a normal vector to S at P_0.

The concept of a normal vector is used to define the *tangent plane* to a surface at a point.

17.2.3 DEFINITION   If an equation of a surface S is $F(x, y, z) = 0$, then the **tangent plane** of S at a point $P_0(x_0, y_0, z_0)$ is the plane through P_0 having $\nabla F(x_0, y_0, z_0)$ as a normal vector.

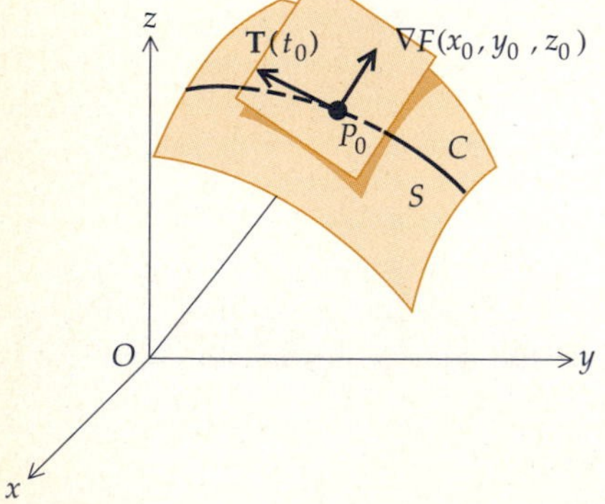

An equation of the tangent plane of the above definition is

$$F_x(x_0, y_0, z_0)(x - x_0) + F_y(x_0, y_0, z_0)(y - y_0) + F_z(x_0, y_0, z_0)(z - z_0) = 0 \quad (5)$$

Refer to Figure 1, which shows the tangent plane to the surface S at P_0 and the representation of the gradient vector having its initial point at P_0.

A vector equation of the tangent plane given by (5) is

$$\nabla F(x_0, y_0, z_0) \cdot [(x - x_0)\mathbf{i} + (y - y_0)\mathbf{j} + (z - z_0)\mathbf{k}] = 0 \qquad (6)$$

FIGURE 1

EXAMPLE 1   Find an equation of the tangent plane to the elliptic paraboloid

$$4x^2 + y^2 - 16z = 0$$

at the point $(2, 4, 2)$.

Solution   Let $F(x, y, z) = 4x^2 + y^2 - 16z$. Then

$$\nabla F(x, y, z) = 8x\mathbf{i} + 2y\mathbf{j} - 16\mathbf{k} \qquad \nabla F(2, 4, 2) = 16\mathbf{i} + 8\mathbf{j} - 16\mathbf{k}$$

From (6) it follows that an equation of the tangent plane is

$$16(x - 2) + 8(y - 4) - 16(z - 2) = 0$$

$$2x + y - 2z - 4 = 0$$

17.2.4 DEFINITION   The **normal line** to a surface S at a point P_0 on S is the line through P_0 having as a set of direction numbers the components of any normal vector to S at P_0.

If an equation of a surface S is $F(x, y, z) = 0$, symmetric equations of the normal line to S at $P_0(x_0, y_0, z_0)$ are

$$\frac{x - x_0}{F_x(x_0, y_0, z_0)} = \frac{y - y_0}{F_y(x_0, y_0, z_0)} = \frac{z - z_0}{F_z(x_0, y_0, z_0)}$$

These symmetric equations follow from Definition 17.2.4 because the denominators are components of $\nabla F(x_0, y_0, z_0)$, which is a normal vector to S at P_0. The normal line at a point on a surface is perpendicular to the tangent plane there.

EXAMPLE 2   Find symmetric equations of the normal line to the surface of Example 1 at $(2, 4, 2)$.

Solution   Because $\nabla F(2, 4, 2) = 16\mathbf{i} + 8\mathbf{j} - 16\mathbf{k}$, it follows that symmetric equations of the required normal line are

$$\frac{x - 2}{2} = \frac{y - 4}{1} = \frac{z - 2}{-2}$$

17.2.5 DEFINITION   The **tangent line** to a curve C at a point P_0 is the line through P_0 having as a set of direction numbers the components of the unit tangent vector to C at P_0.

From Definitions 17.2.3 and 17.2.5 all the tangent lines at the point P_0 to the curves lying on a given surface lie in the tangent plane to the surface at P_0. Refer to Figure 2, showing sketches of a surface and the tangent plane at P_0. Some of the curves through P_0 and their tangent lines are also sketched in the figure.

Consider a curve C that is the intersection of two surfaces having equations

$$F(x, y, z) = 0 \quad \text{and} \quad G(x, y, z) = 0$$

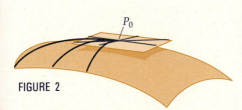

FIGURE 2

respectively. We shall show how to obtain equations of the tangent line to C at a point $P_0(x_0, y_0, z_0)$. Because this tangent line lies in each of the tangent planes to the given surfaces at P_0, it is the line of intersection of the two tangent planes. Let $\mathbf{N}_1$ be a normal vector at P_0 to the surface having the equation $F(x, y, z) = 0$, and let $\mathbf{N}_2$ be a normal vector at P_0 to the surface having the equation $G(x, y, z) = 0$. Then

$$\mathbf{N}_1 = \nabla F(x_0, y_0, z_0) \quad \text{and} \quad \mathbf{N}_2 = \nabla G(x_0, y_0, z_0)$$

Both $\mathbf{N}_1$ and $\mathbf{N}_2$ are orthogonal to the unit tangent vector to C at P_0; so if $\mathbf{N}_1$ and $\mathbf{N}_2$ are not parallel, it follows from Theorem 15.5.10 that the unit tangent vector has the direction that is the same as, or opposite to, the direction of $\mathbf{N}_1 \times \mathbf{N}_2$. Therefore the components of $\mathbf{N}_1 \times \mathbf{N}_2$ serve as a set of direction numbers of the tangent line. From this set of direction numbers and the coordinates of P_0 we can obtain symmetric equations of the required tangent line. This is illustrated in the following example.

EXAMPLE 3   Find symmetric equations of the tangent line to the curve of intersection of the surfaces

$$3x^2 + 2y^2 + z^2 = 49 \quad \text{and} \quad x^2 + y^2 - 2z^2 = 10$$

at the point $(3, -3, 2)$.

Solution   Let

$$F(x, y, z) = 3x^2 + 2y^2 + z^2 - 49 \quad \text{and} \quad G(x, y, z) = x^2 + y^2 - 2z^2 - 10$$

Then $\nabla F(x, y, z) = 6x\mathbf{i} + 4y\mathbf{j} + 2z\mathbf{k}$ and $\nabla G(x, y, z) = 2x\mathbf{i} + 2y\mathbf{j} - 4z\mathbf{k}$. Therefore

$$\begin{aligned}
\mathbf{N}_2 &= \nabla F(3, -3, 2) & \mathbf{N}_2 &= \nabla G(3, -3, 2) \\
&= 18\mathbf{i} - 12\mathbf{j} + 4\mathbf{k} & &= 6\mathbf{i} - 6\mathbf{j} - 8\mathbf{k} \\
&= 2(9\mathbf{i} - 6\mathbf{j} + 2\mathbf{k}) & &= 2(3\mathbf{i} - 3\mathbf{j} - 4\mathbf{k})
\end{aligned}$$

$$\begin{aligned}
\mathbf{N}_1 \times \mathbf{N}_2 &= 4(9\mathbf{i} - 6\mathbf{j} + 2\mathbf{k}) \times (3\mathbf{i} - 3\mathbf{j} - 4\mathbf{k}) \\
&= 4(30\mathbf{i} + 42\mathbf{j} - 9\mathbf{k}) \\
&= 12(10\mathbf{i} + 14\mathbf{j} - 3\mathbf{k})
\end{aligned}$$

Therefore a set of direction numbers of the tangent line is $[10, 14, -3]$. Symmetric equations of the line are, then,

$$\frac{x - 3}{10} = \frac{y + 3}{14} = \frac{z - 2}{-3}$$

If two surfaces have a common tangent plane at a point, the two surfaces are said to be **tangent** at that point. It follows from Definition 17.2.3 that two surfaces S_1 and S_2, whose equations are $F(x, y, z) = 0$ and $G(x, y, z) = 0$, respectively, are tangent at the point $P_0(x_0, y_0, z_0)$ if for some constant k

$$\nabla F(x_0, y_0, z_0) = k\nabla G(x_0, y_0, z_0)$$

EXERCISES 17.2

In Exercises 1 through 12, find an equation of the tangent plane and equations of the normal line to the surface at the indicated point.

1. $x^2 + y^2 + z^2 = 17;\ (2, -2, 3)$
2. $4x^2 + y^2 + 2z^2 = 26;\ (1, -2, 3)$
3. $x^2 + y^2 - 3z = 2;\ (-2, -4, 6)$
4. $x^2 + y^2 - z^2 = 6;\ (3, -1, 2)$
5. $y = e^x \cos z;\ (1, e, 0)$
6. $z = e^{3x} \sin 3y;\ (0, \frac{1}{6}\pi, 1)$
7. $x^2 = 12y;\ (6, 3, 3)$
8. $z = x^{1/2} + y^{1/2};\ (1, 1, 2)$
9. $x^{1/2} + y^{1/2} + z^{1/2} = 4;\ (4, 1, 1)$
10. $zx^2 - xy^2 - yz^2 = 18;\ (0, -2, 3)$
11. $x^{2/3} + y^{2/3} + z^{2/3} = 14;\ (-8, 27, 1)$
12. $x^{1/2} + z^{1/2} = 8;\ (25, 2, 9)$

In Exercises 13 through 20, if the two surfaces intersect in a curve, find equations of the tangent line to the curve of intersection at the given point; if the two surfaces are tangent at the given point, prove it.

13. $x^2 + y^2 - z = 8,\ x - y^2 + z^2 = -2;\ (2, -2, 0)$
14. $x^2 + y^2 - 2z + 1 = 0,\ x^2 + y^2 - z^2 = 0;\ (0, 1, 1)$
15. $y = x^2,\ y = 16 - z^2;\ (4, 16, 0)$

16. $x = 2 + \cos \pi yz$, $y = 1 + \sin \pi xz$; $(3, 1, 2)$
17. $y = e^x \sin 2\pi z + 2$, $z = y^2 - \ln(x + 1) - 3$; $(0, 2, 1)$
18. $x^2 - 3xy + y^2 = z$, $2x^2 + y^2 - 3z + 27 = 0$; $(1, -2, 11)$
19. $x^2 + z^2 + 4y = 0$, $x^2 + y^2 + z^2 - 6z + 7 = 0$; $(0, -1, 2)$
20. $x^2 + y^2 + z^2 = 8$, $yz = 4$; $(0, 2, 2)$
21. Show that the spheres

$$x^2 + y^2 + z^2 = a^2 \qquad \text{and} \qquad (x - b)^2 + y^2 + z^2 = (b - a)^2$$

are tangent at the point $(a, 0, 0)$.

22. Show that the surfaces $xyz = 36$ and $4x^2 + y^2 + 9z^2 = 108$ are tangent at the point $(3, 6, 2)$.

23. Two surfaces are said to be *perpendicular* at a point P_0 of intersection if the normal vectors to the surfaces at P_0 are orthogonal. Show that the surface $x^2 - 2yz + y^3 = 4$ is perpendicular to every member of the family of surfaces $x^2 + (4c - 2)y^2 - cz^2 + 1 = 0$ at the point $(1, -1, 2)$.

24. Prove that every normal line to the sphere $x^2 + y^2 + z^2 = a^2$ passes through the center of the sphere.

17.3 EXTREMA OF FUNCTIONS OF TWO VARIABLES

An important application of the derivative of a function of a single variable is in the study of extreme values of a function, which leads to a variety of problems involving maximum and minimum. This was discussed in Chapter 4, where we proved theorems involving the first and second derivatives, from which relative maximum and minimum values of a function of one variable were determined. In extending the theory to functions of two variables, you will see that it is similar to the one-variable case; however, more complications arise.

17.3.1 DEFINITION

The function f of two variables is said to have a **relative maximum value** at the point (x_0, y_0) if there exists an open disk $B((x_0, y_0); r)$ such that $f(x_0, y_0) \geq f(x, y)$ for all (x, y) in B.

17.3.2 DEFINITION

The function f of two variables is said to have a **relative minimum value** at the point (x_0, y_0) if there exists an open disk $B((x_0, y_0); r)$ such that $f(x_0, y_0) \leq f(x, y)$ for all (x, y) in B.

▶ **ILLUSTRATION 1**   In Figure 1 there is the graph of the function f defined by

$$f(x, y) = \sqrt{25 - x^2 - y^2}$$

Let B be any open disk $((0, 0); r)$ for which $r \leq 5$. From Definition 17.3.1 it follows that f has a relative maximum value of 5 at the point where $x = 0$ and $y = 0$.

In Figure 2 there appears a sketch of the graph of the function g for which

$$g(x, y) = x^2 + y^2$$

Let B be any open disk $((0, 0); r)$. Then from Definition 17.3.2, g has a relative minimum value of 0 at the origin.   ◀

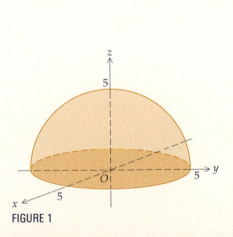

FIGURE 1

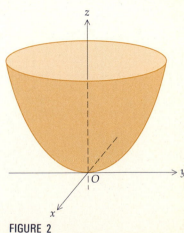

FIGURE 2

Analogous to Theorem 4.1.3 for functions of a single variable is the following one for functions of two variables.

17.3.3 THEOREM

If $f(x, y)$ exists at all points in some open disk $B((x_0, y_0); r)$ and if f has a relative extremum at (x_0, y_0), then if $f_x(x_0, y_0)$ and $f_y(x_0, y_0)$ exist,

$$f_x(x_0, y_0) = 0 \quad \text{and} \quad f_y(x_0, y_0) = 0$$

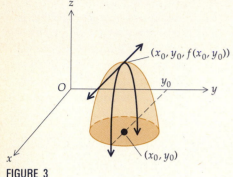

FIGURE 3

Before proving Theorem 17.3.3 we give an informal geometric argument. Let f be a function satisfying the hypothesis, and let f have a relative maximum value at (x_0, y_0). Consider the curve of intersection of the plane $y = y_0$ with the surface $z = f(x, y)$ (refer to Figure 3). This curve is represented by the equations

$$y = y_0 \quad \text{and} \quad z = f(x, y)$$

Because f has a relative maximum value at the point where $x = x_0$, $y = y_0$, it follows that this curve has a horizontal tangent line in the plane $y = y_0$ at $(x_0, y_0, f(x_0, y_0))$. The slope of this tangent line is $f_x(x_0, y_0)$; so $f_x(x_0, y_0) = 0$. In a similar way we can consider the curve of intersection of the plane $x = x_0$ with the surface $z = f(x, y)$ and obtain $f_y(x_0, y_0) = 0$. A similar discussion can be given if f has a relative minimum value at (x_0, y_0). Following is the formal proof.

Proof of Theorem 17.3.3 We prove that if f has a relative maximum value at (x_0, y_0) and if $f_x(x_0, y_0)$ exists, then $f_x(x_0, y_0) = 0$. By the definition of a partial derivative,

$$f_x(x_0, y_0) = \lim_{\Delta x \to 0} \frac{f(x_0 + \Delta x, y_0) - f(x_0, y_0)}{\Delta x}$$

Because f has a relative maximum value at (x_0, y_0), by Definition 17.3.1,

$$f(x_0 + \Delta x, y_0) - f(x_0, y_0) \leq 0$$

whenever Δx is sufficiently small so that $(x_0 + \Delta x, y_0)$ is in B. If Δx approaches zero from the right, $\Delta x > 0$; therefore

$$\frac{f(x_0 + \Delta x, y_0) - f(x_0, y_0)}{\Delta x} \leq 0$$

Hence, by Theorem 2.10.3, if $f_x(x_0, y_0)$ exists, $f_x(x_0, y_0) \leq 0$.
Similarly, if Δx approaches zero from the left, $\Delta x < 0$; so

$$\frac{f(x_0 + \Delta x, y_0) - f(x_0, y_0)}{\Delta x} \geq 0$$

Therefore, by Theorem 2.10.4, if $f_x(x_0, y_0)$ exists, $f_x(x_0, y_0) \geq 0$. We conclude, then, that because $f_x(x_0, y_0)$ exists, both inequalities, $f_x(x_0, y_0) \leq 0$ and $f_x(x_0, y_0) \geq 0$, must hold. Consequently $f_x(x_0, y_0) = 0$.
The proof that $f_y(x_0, y_0) = 0$, if $f_y(x_0, y_0)$ exists and f has a relative maximum value at (x_0, y_0), is analogous and is left as an exercise (see Exercise 37). The proof of the theorem when $f(x_0, y_0)$ is a relative minimum value is also left as an exercise (see Exercise 38). ■

17.3.4 DEFINITION A point (x_0, y_0) for which both $f_x(x_0, y_0) = 0$ and $f_y(x_0, y_0) = 0$ is called a **critical point**.

Theorem 17.3.3 states that a necessary condition for a function of two variables to have a relative extremum at a point, where its first partial derivatives exist, is that this point be a critical point. It is possible for a function of two variables to have a relative extremum at a point at which the partial derivatives do not exist, but we do not consider this situation in this book. Furthermore, the vanishing of the first partial derivatives of a function of two variables is not a sufficient condition for the function to have a relative extremum at the point. Such a situation occurs in the following illustration.

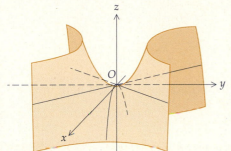

▶ **ILLUSTRATION 2** Let the function f be defined by

$$f(x, y) = y^2 - x^2$$

Then

$$f_x(x, y) = -2x \qquad f_y(x, y) = 2y$$

Both $f_x(0, 0)$ and $f_y(0, 0)$ equal zero. A sketch of the graph of f appears in Figure 4; it is saddle-shaped at points close to the origin. It is apparent that f does not satisfy either Definition 17.3.1 or 17.3.2 when $(x_0, y_0) = (0, 0)$. ◀

FIGURE 4

In Illustration 2, the point $(0, 0)$ is called a *saddle point* of the function f.

There is a second-derivative test that gives conditions that guarantee a function to have a relative extremum at a point where the first partial derivatives equal zero. However, sometimes it is possible to determine relative extrema of a function by Definitions 17.3.1 and 17.3.2, as shown in the following illustration.

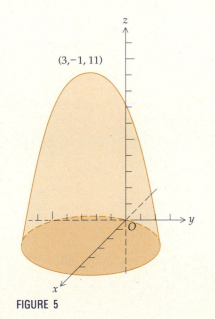

▶ **ILLUSTRATION 3** Let f be the function defined by

$$f(x, y) = 6x - 4y - x^2 - 2y^2$$

We determine if f has any relative extrema.

Because f and its first partial derivatives exist at all (x, y) in R^2, Theorem 17.3.3 is applicable. Differentiating we obtain

$$f_x(x, y) = 6 - 2x \quad \text{and} \quad f_y(x, y) = -4 - 4y$$

Setting $f_x(x, y)$ and $f_y(x, y)$ equal to zero we get $x = 3$ and $y = -1$. See Figure 5 for a sketch of the graph of the equation

$$z = 6x - 4y - x^2 - 2y^2$$

It is a paraboloid having a vertical axis, with vertex at $(3, -1, 11)$ and opening downward. We can conclude that $f(x, y) \le f(3, -1)$ for all (x, y); hence, by Definition 15.3.1, $f(3, -1) = 11$ is a relative maximum function value. ◀

(3, −1, 11)

FIGURE 5

The basic test for determining relative maxima and minima for functions of two variables is the second-derivative test, which is given in the next theorem.

17.3.5 THEOREM
Second-Derivative Test

Let f be a function of two variables such that f and its first- and second-order partial derivatives are continuous on some open disk $B((a, b); r)$. Suppose further that $f_x(a, b) = 0$ and $f_y(a, b) = 0$. Then

(i) f has a relative minimum value at (a, b) if

$$f_{xx}(a, b)f_{yy}(a, b) - f_{xy}{}^2(a, b) > 0 \quad \text{and} \quad f_{xx}(a, b) > 0 \ (\text{or } f_{yy}(a, b) > 0)$$

(ii) f has a relative maximum value at (a, b) if

$$f_{xx}(a, b)f_{yy}(a, b) - f_{xy}{}^2(a, b) > 0 \quad \text{and} \quad f_{xx}(a, b) < 0 \ (\text{or } f_{yy}(a, b) < 0)$$

(iii) $f(a, b)$ is not a relative extremum if

$$f_{xx}(a, b)f_{yy}(a, b) - f_{xy}{}^2(a, b) < 0$$

(iv) We can make no conclusion if

$$f_{xx}(a, b)f_{yy}(a, b) - f_{xy}{}^2(a, b) = 0$$

We defer a discussion of the proof of the second-derivative test until the end of this section where we prove part (i).

EXAMPLE 1   If

$$f(x, y) = 2x^4 + y^2 - x^2 - 2y$$

determine the relative extrema of f if there are any.

Solution   To apply the second-derivative test, we find the first and second partial derivatives of f.

$$f_x(x, y) = 8x^3 - 2x \qquad f_y(x, y) = 2y - 2$$

$$f_{xx}(x, y) = 24x^2 - 2 \qquad f_{yy}(x, y) = 2 \qquad f_{xy}(x, y) = 0$$

Setting $f_x(x, y) = 0$ we get $x = -\frac{1}{2}$, $x = 0$, and $x = \frac{1}{2}$. Setting $f_y(x, y) = 0$ we obtain $y = 1$. Therefore f_x and f_y are both 0 at the points $(-\frac{1}{2}, 1)$, $(0, 1)$, and $(\frac{1}{2}, 1)$, and these are the critical points of f. The results of applying the second-derivative test at these points are summarized in Table 1.

Table 1

Critical point	f_{xx}	f_{yy}	f_{xy}	$f_{xx}f_{yy} - f_{xy}{}^2$	Conclusion
$(-\frac{1}{2}, 1)$	4	2	0	8	f has a relative minimum value
$(0, 1)$	-2	2	0	-4	f does not have a relative extremum
$(\frac{1}{2}, 1)$	4	2	0	8	f has a relative minimum value

At the point $(-\frac{1}{2}, 1)$, $f_{xx} > 0$ and $f_{xx}f_{yy} - f_{xy}{}^2 > 0$; thus from Theorem 17.3.5(i), f has a relative minimum value at $(-\frac{1}{2}, 1)$. At $(0, 1)$, $f_{xx}f_{yy} - f_{xy}{}^2 < 0$; so from Theorem 17.3.5(iii), f does not have a relative extremum at $(0, 1)$. Because $f_{xx} > 0$ and $f_{xx}f_{yy} - f_{xy}{}^2 > 0$ at $(\frac{1}{2}, 1)$, f has a relative minimum value there by Theorem 17.3.5(i).

Because $f(-\frac{1}{2}, 1) = -\frac{9}{8}$ and $f(\frac{1}{2}, 1) = -\frac{9}{8}$, we conclude that f has a relative minimum value of $-\frac{9}{8}$ at each of the points $(-\frac{1}{2}, 1)$ and $(\frac{1}{2}, 1)$.

We now discuss absolute extrema of functions of two variables.

17.3.6 DEFINITION   The function f of two variables is said to have an **absolute maximum value** on its domain D in the xy plane if there is some point (x_0, y_0) in D such that $f(x_0, y_0) \geq f(x, y)$ for all points (x, y) in D. In such a case, $f(x_0, y_0)$ is the absolute maximum value of f on D.

17.3.7 DEFINITION   The function f of two variables is said to have an **absolute minimum value** on its domain D in the xy plane if there is some point (x_0, y_0) in D such that $f(x_0, y_0) \leq f(x, y)$ for all (x, y) in D. In such a case, $f(x_0, y_0)$ is the absolute minimum value of f on D.

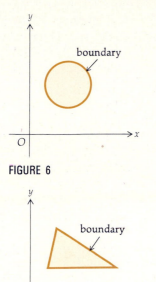

FIGURE 6

For functions of a single variable we had the extreme-value theorem: If the function f is continuous on the closed interval $[a, b]$, then f has an absolute maximum value and an absolute minimum value on $[a, b]$. We learned that an absolute extremum of a function continuous on a closed interval must be either a relative extremum or a function value at an endpoint of the interval. We have a corresponding situation for functions of two variables. In the statement of the extreme-value theorem for functions of two variables we refer to a *closed region* in the xy plane. By a **closed region** we mean that the region includes its *boundary*. In the following illustration we give some closed regions and identify the boundary of each region.

▶ **ILLUSTRATION 4**   (a) A closed disk is a closed region. The boundary is the circumference of the disk. See Figure 6.

(b) The sides of a triangle together with the region enclosed by the triangle is a closed region. The boundary consists of the sides of the triangle. See Figure 7.

(c) The edges of a rectangle together with the region enclosed by the rectangle is a closed region. The boundary consists of the edges of the rectangle. See Figure 8.   ◀

FIGURE 7

17.3.8 THEOREM
The Extreme-Value Theorem
for Functions of Two Variables

Let R be a closed region in the xy plane, and let f be a function of two variables that is continuous on R. Then there is at least one point in R where f has an absolute maximum value and at least one point in R where f has an absolute minimum value.

The proof of this theorem is omitted because it is beyond the scope of the book.

If f is a function satisfying Theorem 17.3.8 and if both $f_x(x, y)$ and $f_y(x, y)$ exist at all points of R, then the absolute extrema of f occur at either a point (x_0, y_0), where $f_x(x_0, y_0) = 0$ and $f_y(x_0, y_0) = 0$, or at a point on the boundary of R.

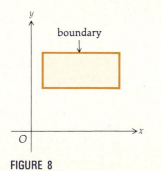

FIGURE 8

EXAMPLE 2   A manufacturer who is a monopolist makes two types of lamps. From experience the manufacturer has determined that if x lamps of the first type and y lamps of the second type are made, they can be sold for $(100 - 2x)$ dollars each and $(125 - 3y)$ dollars each, respectively. The cost of manufacturing x lamps of the first type and y lamps of the second type is $(12x + 11y + 4xy)$

dollars. How many lamps of each type should be produced to realize the greatest profit, and what is the greatest profit?

Solution   The revenue received from the lamps of the first type is $x(100 - 2x)$ dollars, and the revenue received from the lamps of the second type is $y(125 - 3y)$ dollars. Hence, if $f(x, y)$ dollars is the manufacturer's profit,

$$f(x, y) = x(100 - 2x) + y(125 - 3y) - (12x + 11y + 4xy)$$
$$= 88x + 114y - 2x^2 - 3y^2 - 4xy \tag{1}$$

Because x and y both represent the number of lamps, we require that $x \geq 0$ and $y \geq 0$ and allow x and y to be any nonnegative real numbers. Furthermore, $(100 - 2x)$ dollars is the selling price of lamps of the first type. Thus we require that $100 - 2x \geq 0$ or, equivalently, $x \leq 50$. Similarly, because $(125 - 3y)$ dollars is the selling price of lamps of the second type, we require that $y \leq \frac{125}{3}$. Therefore the domain of f is the closed region defined by the set

$$\{(x, y) | 0 \leq x \leq 50 \text{ and } 0 \leq y \leq \tfrac{125}{3}\}$$

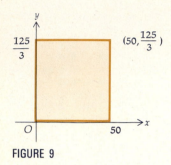

FIGURE 9

This region is rectangular and is shown in Figure 9. The boundary of the region consists of the edges of the rectangle. Because f is a polynomial function, it is continuous everywhere. Hence f is continuous on its domain; so the extreme-value theorem can be applied. The critical points of f are found by determining where $f_x(x, y) = 0$ and $f_y(x, y) = 0$.

$$f_x(x, y) = 88 - 4x - 4y \qquad f_y(x, y) = 114 - 6y - 4x$$

Setting $f_x(x, y) = 0$ and $f_y(x, y) = 0$ we have

$$x + y = 22$$
$$2x + 3y = 57$$

Solving these two equations simultaneously we obtain $x = 9$ and $y = 13$. To apply the second-derivative test we find the second partial derivatives.

$$f_{xx}(x, y) = -4 \qquad f_{yy}(x, y) = -6 \qquad f_{xy}(x, y) = -4$$

At the point $(9, 13)$,

$$f_{xx}(9, 13) = -4 < 0$$

$$f_{xx}(9, 13)f_{yy}(9, 13) - f_{xy}{}^2(9, 13) = (-4)(-6) - (-4)^2$$
$$= 8 > 0$$

It follows, then, by Theorem 17.3.5(ii) that f will have a relative maximum value at $(9, 13)$.

From (1),

$$f(x, y) = x(88 - 2x) + y(114 - 3y) - 4xy \tag{2}$$

Thus

$$f(9, 13) = 9(70) + 13(75) - 468$$
$$= 1137$$

The absolute maximum value of f must occur at either $(9, 13)$ or on the boundary of the domain of f. Let us compare $f(9, 13)$ with the function values on the boundary.

For that part of the boundary on the x axis with $x \in [0, 50]$ we have the function values computed from (2) as follows:

$$f(x, 0) = 88x - 2x^2$$

Let

$$g(x) = 88x - 2x^2 \qquad x \in [0, 50]$$

Then

$$g'(x) = 88 - 4x \quad \text{and} \quad g''(x) = -4$$

Because $g'(22) = 0$ and $g''(22) < 0$, g has a relative maximum value of 968 at $x = 22$. Furthermore, $g(0) = 0$ and $g(50) < 0$. Because $f(9, 13) = 1137 > 968$, the absolute maximum value of f does not occur on the x axis.

For that part of the boundary on the y axis with $y \in [0, \frac{125}{3}]$, from (2),

$$f(0, y) = 114y - 3y^2$$

Let

$$h(y) = 114y - 3y^2 \qquad y \in [0, \tfrac{125}{3}]$$

Then

$$h'(y) = 114 - 6y \quad \text{and} \quad h''(y) = -6$$

Because $h'(19) = 0$ and $h''(19) < 0$, h has a relative maximum value of 1083 at $y = 19$. Moreover, $h(0) = 0$ and $h(\frac{125}{3}) < 0$. Because $f(9, 13) = 1137 > 1083$, the absolute maximum value of f does not occur on the y axis.

We now consider the part of the boundary on the line $x = 50$ with $y \in [0, \frac{125}{3}]$. From (2),

$$f(50, y) = y(114 - 3y) - 600 - 200y \qquad f(0, y) = y(114 - 3y)$$

By comparing these two equations,

$$f(50, y) < f(0, y)$$

Because $f(9, 13) > f(0, y)$ for all y in $[0, \frac{125}{3}]$, then, from the above inequality,

$$f(9, 13) > f(50, y) \qquad \text{for } y \in [0, \tfrac{125}{3}]$$

Hence the absolute maximum value of f does not occur on the line $x = 50$.

Finally, we have that part of the boundary on the line $y = \frac{125}{3}$ with $x \in [0, 50]$. From (2)

$$f(x, \tfrac{125}{3}) = x(88 - 2x) - \tfrac{1375}{3} - \tfrac{500}{3}x \qquad f(x, 0) = x(88 - 2x)$$

From these two equations it follows that $f(x, \frac{125}{3}) < f(x, 0)$. Therefore, because $f(9, 13) > f(x, 0)$ for all x in $[0, 50]$, we can conclude that it is also greater than $f(x, \frac{125}{3})$ for all x in $[0, 50]$. Thus the absolute maximum value cannot occur on the line $y = \frac{125}{3}$.

Therefore the absolute maximum value of f is not on the boundary, and thus it is at the point $(9, 13)$. We conclude then that 9 lamps of the first type and 13 lamps of the second type should be produced for the greatest profit of $1137.

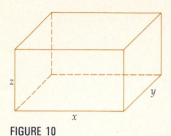

FIGURE 10

EXAMPLE 3   Determine the relative dimensions of a rectangular box, without a top and having a specific volume, if the least amount of material is to be used in its manufacture.

Solution   Let x units be the length of the base of the box, y units be the width of the base of the box, z units be the depth of the box, and S square units be the surface area of the box. The box has a specific volume; therefore, if V cubic units is the volume of the box, V is a constant. Figure 10 shows the box.

Each of the variables x, y, and z is in the interval $(0, +\infty)$. We have the equations

$$S = xy + 2xz + 2yz \quad \text{and} \quad V = xyz$$

Solving the second equation for z in terms of x, y, and the constant V, we get $z = \dfrac{V}{xy}$, and substituting this into the first equation gives

$$S = xy + \frac{2V}{y} + \frac{2V}{x} \tag{3}$$

Differentiating we get

$$\frac{\partial S}{\partial x} = y - \frac{2V}{x^2} \qquad \frac{\partial S}{\partial y} = x - \frac{2V}{y^2}$$

$$\frac{\partial^2 S}{\partial x^2} = \frac{4V}{x^3} \qquad \frac{\partial^2 S}{dy\, \partial x} = 1 \qquad \frac{\partial^2 S}{\partial y^2} = \frac{4V}{y^3}$$

Setting $\dfrac{\partial S}{\partial x} = 0$ and $\dfrac{\partial S}{\partial y} = 0$ we have

$$x^2 y - 2V = 0$$

$$xy^2 - 2V = 0$$

Solving these two equations simultaneously, we obtain $x = \sqrt[3]{2V}$ and $y = \sqrt[3]{2V}$. For these values of x and y,

$$\frac{\partial^2 S}{\partial x^2} = \frac{4V}{(\sqrt[3]{2V})^3} \qquad \frac{\partial^2 S}{\partial x^2} \cdot \frac{\partial^2 S}{\partial y^2} - \left(\frac{\partial^2 S}{\partial y\, \partial x}\right)^2 = \frac{4V}{(\sqrt[3]{2V})^3} \cdot \frac{4V}{(\sqrt[3]{2V})^3} - 1$$

$$= 2 > 0 \qquad\qquad\qquad = 3 > 0$$

From Theorem 17.3.5(i) it follows that S has a relative minimum value when $x = \sqrt[3]{2V}$ and $y = \sqrt[3]{2V}$. Recall that x and y are both in the interval $(0, +\infty)$, and notice from Equation (3) that S is very large when x and y are either close to zero or very large. We therefore conclude that the relative minimum value of S is an absolute minimum value of S.

Because $z = V/(xy)$, then when $x = \sqrt[3]{2V}$ and $y = \sqrt[3]{2V}$,

$$z = \frac{V}{\sqrt[3]{4V^2}}$$

$$= \frac{\sqrt[3]{2V}}{2}$$

Hence the box should have a square base and a depth that is one-half the length of a side of the base.

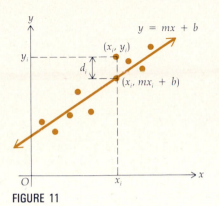

FIGURE 11

An application of extrema of functions of two variables is to obtain the line that *best fits* a set of data points. For instance, suppose that we wish to find a mathematical model for some data given by a set of points $(x_1, y_1), (x_2, y_2), \ldots, (x_n, y_n)$. In particular, y_i may be the number of dollars in a manufacturer's weekly profit when x_i is the number of units sold in a week, or y_i could be a company's total annual sales when x_i years have elapsed since the start of the company. The number of new cases of a certain disease could be y_i when x_i is the number of days since the outbreak of an epidemic of the disease. The desired model is a relationship involving x and y that can be used to make future predictions. Such a relationship is afforded by a line that "fits" the data.

To arrive at a suitable definition for the line of best fit we first indicate how well a particular line fits a set of data points by measuring the vertical distances from the points to the line. For instance, in Figure 11 there are n data points, and the line $y = mx + b$. The point (x_i, y_i) is the ith data point, and corresponding to it on the line is the point $(x_i, mx_i + b)$. The **deviation** (or **error**) between the ith data point and the line is defined to be d_i, where

$$d_i = y_i - (mx_i + b)$$

The sum of the squares of the deviations is

$$\sum_{i=1}^{n} d_i^2 = \sum_{i=1}^{n} [y_i - (mx_i + b)]^2$$

which is never negative and is zero only if each d_i is zero, in which case all the data points lie on the line. We shall take as the line of best fit the one for which $\sum_{i=1}^{n} d_i^2$ is an absolute minimum. This line is called the **regression line** of y on x, and the process for finding it is called the **method of least squares**.

We now give the procedure for using the method of least squares to find the regression line $y = mx + b$ for a set of n data points. Because x_i and y_i are constants and m and b are variables, $\sum_{i=1}^{n} d_i^2$ is a function of m and b. We denote this function by f, so that

$$f(m, b) = \sum_{i=1}^{n} (y_i - mx_i - b)^2$$

We wish to find the values of m and b that make $f(m, b)$ an absolute minimum. We first find the partial derivatives $f_m(m, b)$ and $f_b(m, b)$.

$$f_m(m, b) = \sum_{i=1}^{n} \frac{\partial}{\partial m} [(y_i - mx_i - b)^2]$$

$$= \sum_{i=1}^{n} 2(y_i - mx_i - b)(-x_i)$$

$$= 2 \sum_{i=1}^{n} (-x_i y_i + mx_i^2 + bx_i)$$

$$= 2\left[-\sum_{i=1}^{n} x_i y_i + m \sum_{i=1}^{n} x_i^2 + b \sum_{i=1}^{n} x_i \right]$$

$$f_b(m, b) = \sum_{i=1}^{n} \frac{\partial}{\partial b} \left[(y_i - mx_i - b)^2 \right]$$

$$= \sum_{i=1}^{n} 2(y_i - mx_i - b)(-1)$$

$$= 2 \sum_{i=1}^{n} (-y_i + mx_i + b)$$

$$= 2 \left[-\sum_{i=1}^{n} y_i + m \sum_{i=1}^{n} x_i + nb \right]$$

Setting $f_m(m, b) = 0$ and $f_b(m, b) = 0$ we obtain

$$\left(\sum_{i=1}^{n} x_i^2 \right) m + \left(\sum_{i=1}^{n} x_i \right) b = \sum_{i=1}^{n} x_i y_i \tag{4}$$

$$\left(\sum_{i=1}^{n} x_i \right) m + nb = \sum_{i=1}^{n} y_i$$

These are two simultaneous equations in m and b. Solving the second equation for b we have

$$b = \frac{1}{n} \left[\sum_{i=1}^{n} y_i - m \sum_{i=1}^{n} x_i \right] \tag{5}$$

Substituting this value of b in (4) we get

$$m = \frac{n \sum_{i=1}^{n} x_i y_i - \sum_{i=1}^{n} x_i \sum_{i=1}^{n} y_i}{n \sum_{i=1}^{n} x_i^2 - \left(\sum_{i=1}^{n} x_i \right)^2} \tag{6}$$

In Exercise 41 you are asked to supply the details involved to obtain Equation (6) from (4) and (5). In Exercise 42 you are asked to use the second-derivative test to show that f has a relative minimum value for the values of m and b in (5) and (6). You will see that there is only one relative extremum for f. Also, m and b are both in the interval $(-\infty, +\infty)$ and $f(m, b)$ is large when either the absolute value of m or the absolute value of b is large. Thus we can conclude that the relative minimum value of f is an absolute minimum value.

Observe that there are four distinct summations appearing in formulas (5) and (6). These formulas can be evaluated on a computer or programmable calculator. When a small amount of data is involved, a convenient way of computing the summations is shown in the following examples.

EXAMPLE 4   A rare antique was purchased in 1970 for $1200. Its value was $1800 in 1975, $2500 in 1980, and $3100 in 1985. If the value of the antique were to appreciate according to the same pattern through 1995, estimate the value of the antique in 1995 by the method of least squares.

Solution   To find a regression line $y = mx + b$, we let x be the number of 5-year periods since 1970 and let y dollars be the value of the antique $5x$ years since 1970. Thus we have the data points given in Table 2.

Table 2

x	0	1	2	3
y	1200	1800	2500	3100

Table 3

x_i	y_i	$x_i{}^2$	$x_i y_i$
0	1200	0	0
1	1800	1	1800
2	2500	4	5000
3	3100	9	9300
$\sum$ 6	8600	14	16,100

Table 3 shows the computation of the four summations appearing in Equations (5) and (6). From the table,

$$\sum_{i=1}^{4} x_i = 6 \qquad \sum_{i=1}^{4} y_i = 8600 \qquad \sum_{i=1}^{4} x_i{}^2 = 14 \qquad \sum_{i=1}^{4} x_i y_i = 16{,}100$$

With these values and $n = 4$ we obtain, from (6) and (5),

$$m = \frac{4(16{,}100) - 6(8600)}{4(14) - 6(6)} \qquad b = \tfrac{1}{4}[8600 - 640(6)]$$
$$= 640 \qquad\qquad\qquad\qquad = 1190$$

Hence the regression line has the equation

$$y = 640x + 1190$$

For the year 1995, $x = 5$. For this value of x we have

$$y = 640(5) + 1190$$
$$= 4390$$

Thus in 1995, the value of the antique is estimated to be \$4390.

Table 4

x	1	2	3	4	5
y	20	24	30	35	42

EXAMPLE 5   In Table 4, x days have elapsed since the outbreak of a particular disease, and y is the number of new cases of the disease on the xth day. (a) Find the regression line for the data points (x_i, y_i). (b) Use the regression line to estimate the number of new cases of the disease on the sixth day.

Solution
(a) The required line has the equation $y = mx + b$. To determine m and b we first find the summations in Equations (5) and (6) from the computation in Table 5. From the table,

Table 5

x_i	y_i	$x_i{}^2$	$x_i y_i$
1	20	1	20
2	24	4	48
3	30	9	90
4	35	16	140
5	42	25	210
$\sum$ 15	151	55	508

$$\sum_{i=1}^{5} x_i = 15 \qquad \sum_{i=1}^{5} y_i = 151 \qquad \sum_{i=1}^{5} x_i{}^2 = 55 \qquad \sum_{i=1}^{5} x_i y_i = 508$$

From (6) and (5) with these values and $n = 5$ we obtain

$$m = \frac{5(508) - (15)(151)}{5(55) - (15)(15)} \qquad b = \tfrac{1}{5}[151 - 5.5(15)]$$
$$= 5.5 \qquad\qquad\qquad\qquad = 13.7$$

Therefore the regression line has the equation

$$y = 5.5x + 13.7$$

(b) From the equation of the regression line, when $x = 6$, then $y = 46.7$. Therefore, on the sixth day of the epidemic 47 new cases are estimated.

We conclude this section with the proof of the first part of the second-derivative test.

Proof of Theorem 17.3.5(i)   For simplicity of notation let us define

$$\phi(x, y) = f_{xx}(x, y)f_{yy}(x, y) - f_{xy}{}^2(x, y)$$

We are given $\phi(a, b) > 0$ and $f_{xx}(a, b) > 0$, and we wish to prove that $f(a, b)$ is a relative minimum function value. Because f_{xx}, f_{xy}, and f_{yy} are continuous on $B((a, b); r)$, it follows that ϕ is also continuous on B. Hence there exists an open disk $B'((a, b); r')$, where $r' \leq r$, such that $\phi(x, y) > 0$ and $f_{xx}(x, y) > 0$ for every point (x, y) in B'. Let h and k be constants, not both zero, such that the point $(a + h, b + k)$ is in B'. Then the two equations

$$x = a + ht \quad \text{and} \quad y = b + kt \qquad 0 \leq t \leq 1$$

define all the points on the line segment from (a, b) to $(a + h, b + k)$, and all these points are in B'. Let F be the function of one variable defined by

$$F(t) = f(a + ht, b + kt) \tag{7}$$

By Taylor's formula (formula (2), Section 11.5),

$$F(t) = F(0) + F'(0)t + \frac{F''(\xi)}{2!} t^2$$

where ξ is between 0 and t. If $t = 1$ in this equation, we get

$$F(1) = F(0) + F'(0) + \tfrac{1}{2}F''(\xi) \tag{8}$$

where $0 < \xi < 1$. Because $F(0) = f(a, b)$ and $F(1) = f(a + h, b + k)$, it follows from (8) that

$$f(a + h, b + k) = f(a, b) + F'(0) + \tfrac{1}{2}F''(\xi) \tag{9}$$

where $0 < \xi < 1$.

To find $F'(t)$ and $F''(t)$ from (7) we use the chain rule and obtain

$$F'(t) = hf_x(a + ht, b + kt) + kf_y(a + ht, b + kt) \tag{10}$$

and

$$F''(t) = h^2 f_{xx} + hk f_{yx} + hk f_{xy} + k^2 f_{yy}$$

where each second partial derivative is evaluated at $(a + ht, b + kt)$. From Theorem 16.7.1 it follows that $f_{xy}(x, y) = f_{yx}(x, y)$ for all (x, y) in B'. So

$$F''(t) = h^2 f_{xx} + 2hk f_{xy} + k^2 f_{yy} \tag{11}$$

where each second partial derivative is evaluated at $(a + ht, b + kt)$. Substituting 0 for t in (10) and ξ for t in (11) we get

$$\begin{aligned} F'(0) &= hf_x(a, b) + kf_y(a, b) \\ &= 0 \end{aligned}$$

and

$$F''(\xi) = h^2 f_{xx} + 2hk f_{xy} + k^2 f_{yy}$$

where each second partial derivative is evaluated at $(a + h\xi, b + k\xi)$, where $0 < \xi < 1$. Substituting these values of $F'(0)$ and $F''(\xi)$ into (9) we obtain

$$f(a + h, b + k) - f(a, b) = \tfrac{1}{2}(h^2 f_{xx} + 2hk f_{xy} + k^2 f_{yy}) \tag{12}$$

The terms in parentheses on the right side of (12) can be written as

$$h^2 f_{xx} + 2hk f_{xy} + k^2 f_{yy} = f_{xx}\left[h^2 + 2hk \frac{f_{xy}}{f_{xx}} + \left(k \frac{f_{xy}}{f_{xx}} \right)^2 - \left(k \frac{f_{xy}}{f_{xx}} \right)^2 + k^2 \frac{f_{yy}}{f_{xx}} \right]$$

So from (12),

$$f(a + h, b + k) - f(a, b) = \frac{f_{xx}}{2}\left[\left(h + \frac{f_{xy}}{f_{xx}}k\right)^2 + \frac{f_{xx}f_{yy} - f_{xy}{}^2}{f_{xx}{}^2}k^2\right] \qquad (13)$$

Because $f_{xx}f_{yy} - f_{xy}{}^2$ evaluated at $(a + h\xi, b + k\xi)$ equals

$$\phi(a + h\xi, b + k\xi) > 0$$

it follows that the expression in brackets on the right side of (13) is positive. Furthermore, because $f_{xx}(a + h\xi, b + k\xi) > 0$, it follows from (13) that $f(a + h, b + k) - f(a, b) > 0$. Hence we have proved that

$$f(a + h, b + k) > f(a, b)$$

for every point $(a + h, b + k) \neq (a, b)$ in B'. Therefore, by Definition 17.3.2, $f(a, b)$ is a relative minimum value of f. ■

The proof of part (ii) of Theorem 17.3.5 is similar and is left as an exercise (see Exercise 39). The proof of part (iii) is also left as an exercise (see Exercise 40). Part (iv) is included to cover all possible cases.

EXERCISES 17.3

In Exercises 1 through 12, determine the relative extrema of f, if there are any.

1. $f(x, y) = x^3 + y^2 - 6x^2 + y - 1$
2. $f(x, y) = x^2 - 4xy + y^3 + 4y$
3. $f(x, y) = \frac{1}{x} - \frac{64}{y} + xy$
4. $f(x, y) = 18x^2 - 32y^2 - 36x - 128y - 110$
5. $f(x, y) = e^{xy}$
6. $f(x, y) = x^3 + y^3 - 18xy$
7. $f(x, y) = 4xy^2 - 2x^2y - x$
8. $f(x, y) = \frac{2x + 2y + 1}{x^2 + y^2 + 1}$
9. $f(x, y) = x^3 + y^3 + 3y^2 - 3x - 9y + 2$
10. $f(x, y) = \sin x + \sin y; 0 \le x \le \pi; 0 \le y \le \pi$
11. $f(x, y) = \sin(x + y) + \sin x + \sin y; 0 \le x \le 2\pi; 0 \le y \le 2\pi$
12. $f(x, y) = e^x \sin y$

13. Find the three positive numbers whose sum is 24 such that their product is as great as possible.
14. Find the three positive numbers whose product is 24 such that their sum is as small as possible.
15. Find the point in the plane $3x + 2y - z = 5$ that is closest to the point $(1, -2, 3)$, and find the minimum distance.
16. Find the points on the surface $y^2 - xz = 4$ that are closest to the origin, and find the minimum distance.
17. Find the points on the curve of intersection of the ellipsoid $x^2 + 4y^2 + 4z^2 = 4$ and the plane $x - 4y - z = 0$ that are closest to the origin, and find the minimum distance.
18. A manufacturing plant has two classifications for its workers, A and B. Class A workers earn $14 per run, and class B

workers earn $13 per run. For a certain production run it is determined that in addition to the salaries of the workers, if x class A workers and y class B workers are used, the number of dollars in the cost of the run is $y^3 + x^2 - 8xy + 600$. How many workers of each class should be used so that the cost of the run is a minimum if at least three workers of each class are required for a run?

19. An injection of x milligrams of drug A and y milligrams of drug B causes a response of R units, and $R = x^2y^3(c - x - y)$, where c is a positive constant. What quantity of each drug will cause the maximum response?
20. Suppose that t hours after the injection of x milligrams of adrenalin the response is R units, and $R = te^{-t}(c - x)x$, where c is a positive constant. What values of x and t will cause the maximum response?
21. Find the volume of the largest rectangular parallelepiped that can be inscribed in the ellipsoid $36x^2 + 9y^2 + 4z^2 = 36$ if the edges are parallel to the coordinate axes.
22. A rectangular box without a top is to be made at a cost of $10 for the material. If the material for the bottom of the box costs $0.15 per square foot and the material for the sides costs $0.30 per square foot, find the dimensions of the box of greatest volume that can be made.
23. A closed rectangular box to contain 16 ft³ is to be made of three kinds of materials. The cost of the material for the top and the bottom is $0.18 per square foot, the cost of the material for the front and the back is $0.16 per square foot, and the cost of the material for the other two sides is $0.12 per square foot. Find the dimensions of the box such that the cost of the materials is a minimum.
24. Suppose that T degrees is the temperature at any point (x, y, z) on the sphere $x^2 + y^2 + z^2 = 4$, and $T = 100xy^2z$.

Find the points on the sphere where the temperature is the greatest and also the points where the temperature is the least. Also find the temperature at these points.

25. Suppose that when the production of a particular commodity requires x machine-hours and y person-hours, the cost of production is given by $f(x, y)$, where

$$f(x, y) = 2x^3 - 6xy + y^2 + 500$$

Determine the number of machine-hours and the number of person-hours needed to produce the commodity at the least cost.

26. A clothing store sells two kinds of shirts that are similar but are made by different manufacturers. The cost to the store of the first kind is $40 and the cost of the second kind is $50. It has been determined by experience that if the selling price of the first kind is x dollars and the selling price of the second kind is y dollars, then the number sold monthly of the first kind is $3200 - 50x + 25y$, and the number sold monthly of the second kind is $25x - 25y$. What should be the selling price of each kind of shirt for the greatest gross profit?

27. An early abstract painting was sold by the artist in 1915 for $100. Because of its historical importance its value has increased over the years. Its value was $4600 in 1935, $11,000 in 1955, and $20,000 in 1975. With the assumption that the value of the painting will appreciate according to the same pattern through 1995, use the method of least squares to estimate its value in 1995.

28. A 1985 model car was sold as a used car in 1986 for $6800. Its value was $6200 in 1987, $5700 in 1988, and $4800 in 1990. Use the method of least squares to estimate what its value was in 1989.

29. A motion picture has been playing at Cinema One for 5 weeks, and the weekly attendance (to the nearest 100) for each week is given in the following table.

Week Number	1	2	3	4	5
Attendance	5000	4500	4100	3900	3500

Assume that the weekly attendance will continue to decline according to the same pattern until it reaches 1500. (a) Use the regression line for the data in the table to determine the expected attendance for the sixth week. (b) The film will move over to the smaller Cinema Two when the weekly attendance drops below 2250. How many weeks is the film expected to play in Cinema One?

30. Five trees had their sap analyzed for the amount of a plant hormone that causes the detachment of leaves. For the trees in the following table, when x micrograms (μg) of plant hormone were released, y leaves were detached.

	Oak tree	Maple tree	Birch tree	Pine tree	Locust tree
x	28	57	38	75	82
y	208	350	300	620	719

(a) Find an equation of the regression line for the data in the table. (b) Use the regression line to estimate the number of leaves detached from another kind of tree when 100 μg of plant hormone are released.

31. Five joggers were given examinations to determine their maximum oxygen uptake, a measure used to denote a person's cardiovascular fitness. The results are given in the following table, where x seconds is the jogger's best time for a mile run and y milliliters per minute per kilogram of body weight is the jogger's maximum oxygen uptake.

	Jogger A	Jogger B	Jogger C	Jogger D	Jogger E
x	300.5	350.6	407.3	326.2	512.8
y	350.2	325.8	375.6	418.5	400.2

(a) Find an equation of the regression line for the data in the table. (b) Use the regression line to estimate a jogger's maximum oxygen uptake if the jogger's best time for a mile run is 340.4 sec.

32. The score on a student's entrance examination was used to predict the student's grade-point average at the end of the freshman year. The following table gives the data for six students, where x is the test score and y is the grade-point average.

	Student A	Student B	Student C	Student D	Student E	Student F
x	92	81	73	98	79	85
y	3.4	2.7	3.1	3.8	2.2	3.0

(a) Find an equation of the regression line for the data in the table. (b) Use the regression line to estimate a student's grade-point average at the end of the freshman year if the student had a score of 88 on the college entrance examination.

33. A monopolist produces staplers and staples having demand equations $x = 10/(pq)$ and $y = 20/(pq)$, where $1000x$ staplers are demanded if the price is p dollars per stapler and $1000y$ boxes of staples are demanded if the price per box of staples is q dollars. It costs $2 to produce each stapler and $1 to produce each box of staples. Determine the price of each commodity in order to have the greatest total profit.

34. If the demand equations iin Exercise 33 are $x = 11 - 2p - 2q$ and $y = 19 - 2p - 3q$, show that to have the greatest total profit the staplers should be free and the staples should be expensive.

35. Determine the relative dimensions of a rectangular box, without a top, to be made from a given amount of material for the box to have the greatest possible volume.

36. Prove that the box having the largest volume that can be placed inside a sphere is in the shape of a cube.

37. Prove that $f_y(x_0, y_0) = 0$ if $f_y(x_0, y_0)$ exists and f has a relative maximum value at (x_0, y_0).

38. Prove Theorem 17.3.3 when $f(x_0, y_0)$ is a relative minimum value.

39. Prove Theorem 17.3.5(ii).   40. Prove Theorem 17.3.5(iii).

41. Obtain Equation (6) by substituting from (5) in (4).

42. If $f(m, b) = \sum\limits_{i=1}^{n} (y_i - mx_i - b)^2$, use the second-derivative test to prove that the values of m and b in (5) and (6) give a relative minimum value of f. (*Hint:* First show that $f_{mm}(m, b) > 0$. To show that $f_{mm}(m, b) \cdot f_{bb}(m, b) - f_{mb}^2(m, b) > 0$, you must

prove that $\sum\limits_{i=1}^{n} x_i^2 > \left(\sum\limits_{i=1}^{n} x_i \right)^2$. To prove this, let $\bar{x} = \dfrac{1}{n} \sum\limits_{i=1}^{n} x_i$ and apply properties of the sigma notation to the inequality $\sum\limits_{i=1}^{n} (\bar{x} - x_i)^2 > 0$.)

17.4 LAGRANGE MULTIPLIERS

In the solution of Example 3 of Section 17.3 we minimized the function having function values $xy + 2xz + 2yz$, subject to the condition that x, y, and z satisfy the equation $xyz = V$. Compare this with Example 1 of Section 17.3, in which we found the relative extrema of f for which $f(x, y) = 2x^4 + y^2 - x^2 - 2y$. These are essentially two different kinds of problems because in the first case we had an additional condition, called a *constraint* (or *side condition*). Such a problem is called one in *constrained extrema*, whereas that of the second type is called a problem in *free extrema*.

The solution of Example 3 of Section 17.3 involved obtaining a function of the two variables x and y by replacing z in the first equation by its value from the second equation. Because it is not always feasible to solve the constraint for one of the variables in terms of the others, there is another procedure that can be used to find the critical points to solve a problem in constrained extrema. It is due to Joseph L. Lagrange (1736–1813). Before discussing the theory, we outline the procedure and illustrate it by an example.

Suppose that we wish to find the relative extrema of a function f of the three variables x, y, and z, subject to the constraint $g(x, y, z) = 0$. We introduce a new variable λ, called a **Lagrange multiplier**, and form the auxiliary function F for which

$$F(x, y, z, \lambda) = f(x, y, z) + \lambda g(x, y, z)$$

The problem, then, becomes one of finding the critical points of the function F of the four variables x, y, z, and λ. The values of x, y, and z that give the relative extrema of f are among these critical points. The critical points of F are the values of x, y, z, and λ for which the four first partial derivatives of F are zero:

$$F_x = 0 \qquad F_y = 0 \qquad F_z = 0 \qquad F_\lambda = 0$$

EXAMPLE 1   Solve Example 3 of Section 17.3 by Lagrange multipliers.

Solution   The variables x, y, and z and the constant V are as defined in the solution of Example 3 of Section 17.3. Let

$$\begin{aligned} S &= f(x, y, z) \qquad \text{and} \quad g(x, y, z) = xyz - V \\ &= xy + 2xz + 2yz \end{aligned}$$

We wish to minimize the function f subject to the constraint that

$$g(x, y, z) = 0$$

We form the function F for which

$$\begin{aligned} F(x, y, z, \lambda) &= f(x, y, z) + \lambda g(x, y, z) \\ &= xy + 2xz + 2yz + \lambda(xyz - V) \end{aligned}$$

To find the critical points of F we compute the four partial derivatives F_x, F_y, F_z, and F_λ and set the function values equal to zero.

$$F_x(x, y, z, \lambda): y + 2z + \lambda yz = 0 \tag{1}$$

$$F_y(x, y, z, \lambda): x + 2z + \lambda xz = 0 \tag{2}$$

$$F_z(x, y, z, \lambda): 2x + 2y + \lambda xy = 0 \tag{3}$$

$$F_\lambda(x, y, z, \lambda): xyz - V = 0 \tag{4}$$

Subtracting corresponding members of (2) from those of (1) we obtain

$$y - x + \lambda z(y - x) = 0$$

$$(y - x)(1 + \lambda z) = 0$$

giving the two equations

$$y = x \tag{5}$$

and because $z \neq 0$,

$$\lambda = -\frac{1}{z}$$

Substituting $\lambda = -1/z$ into (2) we get $x + 2z - x = 0$, giving $z = 0$, which is impossible because z is in the interval $(0, +\infty)$. Substituting from (5) into (3) gives

$$2x + 2x + \lambda x^2 = 0$$

$$x(4 + \lambda x) = 0$$

$$\lambda = -\frac{4}{x} \qquad \text{(because } x \neq 0)$$

If in (2), $\lambda = -4/x$, then

$$x + 2z - \frac{4}{x}(xz) = 0$$

$$x + 2z - 4z = 0$$

$$z = \frac{x}{2} \tag{6}$$

Substituting from (5) and (6) into (4) we obtain $\frac{1}{2}x^3 - V = 0$, from which $x = \sqrt[3]{2V}$. From (5) and (6) it follows that $y = \sqrt[3]{2V}$ and $z = \frac{1}{2}\sqrt[3]{2V}$. Therefore $(\sqrt[3]{2V}, \sqrt[3]{2V}, \frac{1}{2}\sqrt[3]{2V})$ is a critical point of the function F, and, as shown in Example 3 of Section 17.3, f has an absolute minimum value at this point.

Note in the solution that the equation $F_\lambda(x, y, z, \lambda) = 0$ is equivalent to the constraint given by the equation $V = xyz$.

The validity of the method of Lagrange multipliers can be shown by considering the general problem of constrained extrema. Suppose we wish to find the relative extrema of a function f of three variables x, y, and z, subject to the constraint

$$g(x, y, z) = 0 \tag{7}$$

We assume that (7) can be solved for z to obtain

$$z = h(x, y)$$

where h is defined on an open disk $B((x_0, y_0); r)$ and $f(x, y, h(x, y))$ has a relative extremum at $(x_0, y_0, h(x_0, y_0))$. We also assume that the first partial derivatives of f, g, and h exist on B and $g_3(x, y, h(x, y)) \neq 0$ on B. Because f has a relative extremum at $(x_0, y_0, h(x_0, y_0))$, the first partial derivatives of f are zero there. We compute these first partial derivatives by the chain rule,

$$\text{at } (x_0, y_0, h(x_0, y_0)) \qquad f_1 + f_3 \frac{\partial h}{\partial x} = 0 \quad \text{and} \quad f_2 + f_3 \frac{\partial h}{\partial y} = 0 \tag{8}$$

If in (7) we differentiate implicitly with respect to x and then y, and consider z as the differentiable function h of x and y, then at a point (x, y) in the open disk B

$$g_1 + g_3 \frac{\partial h}{\partial x} = 0 \quad \text{and} \quad g_2 + g_3 \frac{\partial h}{\partial y} = 0$$

or, equivalently, because $g_3 \neq 0$ on the open disk B,

$$\text{at } (x, y) \text{ in } B \qquad \frac{\partial h}{\partial x} = -\frac{g_1}{g_3} \quad \text{and} \quad \frac{\partial h}{\partial y} = -\frac{g_2}{g_3}$$

where the function values of g_1, g_2, and g_3 are at $(x, y, h(x, y))$. If the above values of $\dfrac{\partial h}{\partial x}$ and $\dfrac{\partial h}{\partial y}$ are substituted in Equations (8), then at the point $(x_0, y_0, h(x_0, y_0))$,

$$f_1 + f_3 \left(-\frac{g_1}{g_3} \right) = 0 \quad \text{and} \quad f_2 + f_3 \left(-\frac{g_2}{g_3} \right) = 0$$

Furthermore,

$$f_3 - g_3 \left(\frac{f_3}{g_3} \right) = 0$$

everywhere $g_3 \neq 0$. Thus at $(x_0, y_0, h(x_0, y_0))$,

$$f_1 + g_1 \left(-\frac{f_3}{g_3} \right) = 0 \qquad f_2 + g_2 \left(-\frac{f_3}{g_3} \right) = 0 \qquad f_3 + g_3 \left(-\frac{f_3}{g_3} \right) = 0$$

If $\lambda = -f_3/g_3$ then these equations can be written as

$$f_1 + \lambda g_1 = 0 \qquad f_2 + \lambda g_2 = 0 \qquad f_3 + \lambda g_3 = 0 \tag{9}$$

Furthermore, because f has a relative extremum at $(x_0, y_0, h(x_0, y_0))$ and this extremum is subject to the constraint $g(x, y, z) = 0$, then

$$g(x_0, y_0, h(x_0, y_0)) = 0 \tag{10}$$

If

$$F(x, y, z, \lambda) = f(x, y, z) + \lambda g(x, y, z) \tag{11}$$

and if $z_0 = h(x_0, y_0)$, then (9) and (10) are equivalent to the equations

$$F_x = 0 \qquad F_y = 0 \qquad F_z = 0 \qquad F_\lambda = 0 \qquad \text{at } (x_0, y_0, z_0) \tag{12}$$

Therefore we can conclude that a point (x_0, y_0, z_0) at which the function f has a relative extremum is among the critical points of the function F defined by (11).

Observe that Equations (9) can be written as the vector equation

$$\nabla f + \lambda \nabla g = \mathbf{0} \qquad \text{at } (x_0, y_0, z_0) \qquad \text{where } \nabla g \neq \mathbf{0}$$

This vector equation together with the equation $g(x_0, y_0, z_0) = 0$ give another form of Equations (12).

EXAMPLE 2 Use Lagrange multipliers to find the shortest distance from the origin to the plane $Ax + By + Cz = D$.

Solution Let w units be the distance from the origin to a point (x, y, z) in the plane. Then

$$w = \sqrt{x^2 + y^2 + z^2}$$

Because w will be a minimum when w^2 is a minimum, we form the function f for which

$$f(x, y, z) = x^2 + y^2 + z^2$$

We wish to find the minimum value of f subject to the constraint

$$Ax + By + Cz - D = 0$$

With the assumption that there is such a minimum value it will occur at a critical point of the function F such that

$$F(x, y, z, \lambda) = x^2 + y^2 + z^2 + \lambda(Ax + By + Cz - D)$$

To find the critical points of F we compute the partial derivatives of F and set them equal to zero.

$$F_x(x, y, z, \lambda): \quad 2x + \lambda A = 0$$

$$F_y(x, y, z, \lambda): \quad 2y + \lambda B = 0$$

$$F_z(x, y, z, \lambda): \quad 2z + \lambda C = 0$$

$$F_\lambda(x, y, z, \lambda): \quad Ax + By + Cz - D = 0 \tag{13}$$

From the first three of these equations

$$x = -\tfrac{1}{2}\lambda A \qquad y = -\tfrac{1}{2}\lambda B \qquad z = -\tfrac{1}{2}\lambda C \tag{14}$$

Substituting these values of x, y, and z into (13) we get

$$-\tfrac{1}{2}\lambda(A^2 + B^2 + C^2) = D$$

$$-\tfrac{1}{2}\lambda = \frac{D}{A^2 + B^2 + C^2}$$

We replace $-\tfrac{1}{2}\lambda$ by this value in Equations (14) and obtain

$$x = \frac{AD}{A^2 + B^2 + C^2} \qquad y = \frac{BD}{A^2 + B^2 + C^2} \qquad z = \frac{CD}{A^2 + B^2 + C^2} \tag{15}$$

The point having these coordinates is the one and only critical point of F. Therefore the minimum distance from the origin to the plane is the distance from the origin to the point (x_0, y_0, z_0), where x_0, y_0, and z_0 are the values of

x, y, and z in Equations (15). The minimum distance then is

$$\sqrt{x_0{}^2 + y_0{}^2 + z_0{}^2} = \sqrt{\frac{A^2 D^2}{(A^2 + B^2 + C^2)^2} + \frac{B^2 D^2}{(A^2 + B^2 + C^2)^2} + \frac{C^2 D^2}{(A^2 + B^2 + C^2)^2}}$$

$$= \frac{|D|}{\sqrt{A^2 + B^2 + C^2}}$$

If several constraints are imposed, the method of Lagrange multipliers can be extended by using several multipliers. In particular, if we wish to find critical points of the function having values $f(x, y, z)$ subject to the two side conditions $g(x, y, z) = 0$ and $h(x, y, z) = 0$, we find the critical points of the function F of the five variables x, y, z, λ, and μ for which

$$F(x, y, z, \lambda, \mu) = f(x, y, z) + \lambda g(x, y, z) + \mu h(x, y, z)$$

The following example illustrates the method.

EXAMPLE 3   Find the relative extrema of the function f if

$$f(x, y, z) = xz + yz$$

and the point (x, y, z) lies on the intersection of the surfaces $x^2 + z^2 = 2$ and $yz = 2$.

Solution   We form the function F for which

$$F(x, y, z, \lambda, \mu) = xz + yz + \lambda(x^2 + z^2 - 2) + \mu(yz - 2)$$

Finding the five partial derivatives and setting them equal to zero we have

$$F_x(x, y, z, \lambda, \mu): \quad z + 2\lambda x = 0 \tag{16}$$

$$F_y(x, y, z, \lambda, \mu): \quad z + \mu z = 0 \tag{17}$$

$$F_z(x, y, z, \lambda, \mu): \quad x + y + 2\lambda z + \mu y = 0 \tag{18}$$

$$F_\lambda(x, y, z, \lambda, \mu): \quad x^2 + z^2 - 2 = 0 \tag{19}$$

$$F_\mu(x, y, z, \lambda, \mu): \quad yz - 2 = 0 \tag{20}$$

From (17) we obtain $\mu = -1$ and $z = 0$. We reject $z = 0$ because this contradicts (20). From (16) we obtain, if $x \neq 0$,

$$\lambda = -\frac{z}{2x}$$

Substituting this value of λ and $\mu = -1$ into (18) we get

$$x + y - \frac{z^2}{x} - y = 0$$

$$x^2 = z^2 \tag{21}$$

Substituting from (21) into (19) we have $2x^2 - 2 = 0$, or $x^2 = 1$. This gives two values for x, namely 1 and -1; and for each of these values of x we get, from

(21), the two values 1 and -1 for z. Obtaining the corresponding values for y from (20) we have four sets of solutions for the five Equations (16) through (20). These solutions are

$$x = 1 \qquad y = 2 \qquad z = 1 \qquad \lambda = -\tfrac{1}{2} \qquad \mu = -1$$

$$x = 1 \qquad y = -2 \qquad z = -1 \qquad \lambda = \tfrac{1}{2} \qquad \mu = -1$$

$$x = -1 \qquad y = 2 \qquad z = 1 \qquad \lambda = \tfrac{1}{2} \qquad \mu = -1$$

$$x = -1 \qquad y = -2 \qquad z = -1 \qquad \lambda = -\tfrac{1}{2} \qquad \mu = -1$$

The first and fourth sets of solutions give $f(x, y, z) = 3$, and the second and third sets of solutions give $f(x, y, z) = 1$. Hence f has a relative maximum function value of 3 and a relative minimum function value of 1.

EXERCISES 17.4

In Exercises 1 through 4, use Lagrange multipliers to find the critical points of the function subject to the constraint.

1. $f(x, y) = 25 - x^2 - y^2$ with constraint $x^2 + y^2 - 4y = 0$
2. $f(x, y) = 4x^2 + 2y^2 + 5$ with constraint $x^2 + y^2 - 2y = 0$
3. $f(x, y, z) = x^2 + y^2 + z^2$ with constraint $3x - 2y + z - 4 = 0$
4. $f(x, y, z) = x^2 + y^2 + z^2$ with constraint $y^2 - x^2 = 1$

In Exercises 5 through 8, use Lagrange multipliers to find the relative extrema of f subject to the constraint. Also find the points at which the extrema occur. Assume that the relative extrema exist.

5. $f(x, y) = x^2 + y$ with constraint $x^2 + y^2 = 9$
6. $f(x, y) = x^2 y$ with constraint $x^2 + 8y^2 = 24$
7. $f(x, y, z) = xyz$ with constraint $x^2 + 2y^2 + 4z^2 = 4$
8. $f(x, y, z) = y^3 + xz^2$ with constraint $x^2 + y^2 + z^2 = 1$

In Exercises 9 and 10, find a relative minimum value of f subject to the constraint. Assume that a relative minimum value exists.

9. $f(x, y, z) = x^2 + y^2 + z^2$ with constraint $xyz = 1$
10. $f(x, y, z) = xyz$ with constraint $x^2 + y^2 + z^2 = 1$

In Exercises 11 and 12, find a relative maximum value of f subject to the constraint. Assume that a relative maximum value exists.

11. $f(x, y, z) = x + y + z$ with constraint $x^2 + y^2 + z^2 = 9$
12. $f(x, y, z) = xyz$ with constraint $2xy + 3xz + yz = 72$
13. Find a relative minimum value of the function f for which $f(x, y, z) = x^2 + 4y^2 + 16z^2$ with the constraint (a) $xyz = 1$; (b) $xy = 1$; (c) $x = 1$.
14. Use Lagrange multipliers to find the shortest distance from the point $(1, 3, 0)$ to the plane $4x + 2y - z = 5$.
15. Use Lagrange multipliers to find the shortest distance from the point $(1, -1, -1)$ to the plane $x + 4y + 3z = 2$.
16. Find the least and greatest distances from the origin to a point on the ellipse $x^2 + 4y^2 = 16$.
17. Find the least and greatest distances from the origin to a point on the ellipsoid $9x^2 + 4y^2 + z^2 = 36$.

18. If $f(x, y, z) = 2x^2 + 3y^2 + z^2$, use Lagrange multipliers to find the point in the plane $x + y + z = 5$ at which $f(x, y, z)$ is least.
19. Use Lagrange multipliers to find a relative minimum function value of f if $f(x, y, z) = x^2 + y^2 + z^2$ with the two constraints $x + 2y + 3z = 6$ and $x - y - z = -1$.
20. Use Lagrange multipliers to find a relative minimum function value of f if $f(x, y, z) = x^2 + y^2 + z^2$ with the two constraints $x + y + 2z = 1$ and $3x - 2y + z = -4$.
21. Use Lagrange multipliers to find a relative maximum function value of f if $f(x, y, z) = xyz$ with the two constraints $x + y + z = 4$ and $x - y - z = 3$.
22. Use Lagrange multipliers to find a relative maximum function value of f if $f(x, y, z) = x^3 + y^3 + z^3$ with the two constraints $x + y + z = 1$ and $x + y - z = 0$.

In Exercises 23 through 32, use Lagrange multipliers to solve the indicated exercise of Exercises 17.3.

23. Exercise 13   **24.** Exercise 14   **25.** Exercise 15
26. Exercise 16   **27.** Exercise 17   **28.** Exercise 22
29. Exercise 23   **30.** Exercise 24   **31.** Exercise 35
32. Exercise 36

33. A circular disk is in the shape of the region bounded by the circle $x^2 + y^2 = 1$. If T degrees is the temperature at any point (x, y) of the disk and $T = 2x^2 + y^2 - y$, find the hottest and coldest points on the disk.
34. A company has three factories, each manufacturing the same product. If factory A produces x units, factory B produces y units, and factory C produces z units, their respective manufacturing costs are $(3x^2 + 200)$ dollars, $(y^2 + 400)$ dollars, and $(2z^2 + 300)$ dollars. If an order for 1100 units is to be filled, use Lagrange multipliers to determine how the production should be distributed among the three factories to minimize the total manufacturing cost.

17.5   OBTAINING A FUNCTION FROM ITS GRADIENT AND EXACT DIFFERENTIALS

In Chapter 19, when studying vector fields, we shall wish to determine if a given vector-valued function is the gradient of some real-valued function f, and if it is, we will want to find such a function f. First let us consider the problem of how to obtain f if its gradient is known. That is, we are given

$$\nabla f(x, y) = f_x(x, y)\mathbf{i} + f_y(x, y)\mathbf{j} \tag{1}$$

and we wish to find $f(x, y)$.

▶ **ILLUSTRATION 1**   Suppose

$$\nabla f(x, y) = (y^2 + 2x + 4)\mathbf{i} + (2xy + 4y - 5)\mathbf{j} \tag{2}$$

Then because Equation (1) must be satisfied, it follows that

$$f_x(x, y) = y^2 + 2x + 4 \tag{3}$$

$$f_y(x, y) = 2xy + 4y - 5 \tag{4}$$

By integrating both sides of (3) with respect to x,

$$f(x, y) = y^2x + x^2 + 4x + g(y) \tag{5}$$

Observe that the "constant" of integration is a function of y and independent of x because we are integrating with respect to x. If we now differentiate both sides of (5) partially with respect to y, we obtain

$$f_y(x, y) = 2xy + g'(y) \tag{6}$$

Equations (4) and (6) give two expressions for $f_y(x, y)$. Hence

$$2xy + 4y - 5 = 2xy + g'(y)$$

Therefore

$$g'(y) = 4y - 5$$

$$g(y) = 2y^2 - 5y + C$$

Substituting this value of $g(y)$ into (5) we have

$$f(x, y) = y^2x + x^2 + 4x + 2y^2 - 5y + C \qquad\qquad ◀$$

EXAMPLE 1   Find $f(x, y)$ if

$$\nabla f(x, y) = e^{y^2} \cos x\mathbf{i} + 2ye^{y^2} \sin x\mathbf{j}$$

Solution   Because Equation (1) must hold,

$$f_x(x, y) = e^{y^2} \cos x \tag{7}$$

$$f_y(x, y) = 2ye^{y^2} \sin x \tag{8}$$

Integrating both sides of (8) with respect to y we obtain

$$f(x, y) = e^{y^2} \sin x + g(x) \tag{9}$$

where $g(x)$ is independent of y. We now partially differentiate both sides of (9) with respect to x and get

$$f_x(x, y) = e^{y^2} \cos x + g'(x) \tag{10}$$

By equating the right sides of (7) and (10),

$$e^{y^2} \cos x = e^{y^2} \cos x + g'(x)$$

$$g'(x) = 0$$

$$g(x) = C$$

We substitute this value of $g(x)$ into (9) and obtain

$$f(x, y) = e^{y^2} \sin x + C$$

All vectors of the form $M(x, y)\mathbf{i} + N(x, y)\mathbf{j}$ are not necessarily gradients, as shown in the next illustration.

▶ **ILLUSTRATION 2**   We show that there is no function f such that

$$\nabla f(x, y) = 3y\mathbf{i} - 2x\mathbf{j} \tag{11}$$

Assume that there is such a function. Then it follows that

$$f_x(x, y) = 3y \tag{12}$$

$$f_y(x, y) = -2x \tag{13}$$

We integrate both sides of (12) with respect to x and obtain

$$f(x, y) = 3xy + g(y)$$

We partially differentiate both sides of this equation with respect to y and have

$$f_y(x, y) = 3x + g'(y)$$

Equating the right sides of this equation and (13) we obtain

$$3x + g'(y) = -2x$$

$$g'(y) = -5x$$

If both sides of this equation are differentiated with respect to x, it must follow that

$$0 = -5$$

which, of course, is not true. Thus the assumption that $3y\mathbf{i} - 2x\mathbf{j}$ is a gradient leads to a contradiction.   ◀

We now investigate a condition that must be satisfied for a vector to be a gradient.

Suppose that M_y and N_x are continuous on an open disk B in R^2. If

$$M(x, y)\mathbf{i} + N(x, y)\mathbf{j} \tag{14}$$

is a gradient on B, then there is a function f such that

$$f_x(x, y) = M(x, y) \tag{15}$$

$$f_y(x, y) = N(x, y) \tag{16}$$

for all (x, y) in B. Because $M_y(x, y)$ exists on B, then, from (15),

$$M_y(x, y) = f_{xy}(x, y) \tag{17}$$

Furthermore, because $N_x(x, y)$ exists on B, it follows from (16) that

$$N_x(x, y) = f_{yx}(x, y) \tag{18}$$

Because M_y and N_x are continuous on B, their equivalents f_{xy} and f_{yx} are also continuous on B. Thus, from Theorem 16.7.1, $f_{xy}(x, y) = f_{yx}(x, y)$ at all points in B. Therefore the left sides of (17) and (18) are equal at all points in B. We have proved that if M_y and N_x are continuous on an open disk B in R^2, a necessary condition for vector (14) to be a gradient on B is that

$$M_y(x, y) = N_x(x, y) \tag{19}$$

Equation (19) is also a sufficient condition for vector (14) to be a gradient on B. If (19) holds, we can show how to find a function f such that vector (14) is a gradient. However, the proof that, whenever (19) holds, such a function exists belongs to a course in advanced calculus. The method for finding f is a generalization of that used in Illustration 1 and Example 1 and in Example 2 below. We have the following theorem.

17.5.1 THEOREM Suppose that M and N are functions of two variables x and y defined on an open disk $B((x_0, y_0); r)$ in R^2, and M_y and N_x are continuous on B. Then the vector

$$M(x, y)\mathbf{i} + N(x, y)\mathbf{j}$$

is a gradient on B if and only if

$$M_y(x, y) = N_x(x, y)$$

at all points in B.

▶ **ILLUSTRATION 3** (a) We apply Theorem 17.5.1 to the vector on the right side of Equation (2) in Illustration 1. Let

$$M(x, y) = y^2 + 2x + 4 \qquad N(x, y) = 2xy + 4y - 5$$
$$M_y(x, y) = 2y \qquad N_x(x, y) = 2y$$

Thus $M_y(x, y) = N_x(x, y)$, and therefore the vector is a gradient.

(b) If we apply Theorem 17.5.1 to the vector on the right side of Equation (11) in Illustration 2, with $M(x, y) = 3y$ and $N(x, y) = -2x$, we obtain

$$M_y(x, y) = 3 \qquad N_x(x, y) = -2$$

Hence $M_y(x, y) \neq N_x(x, y)$; thus the vector is not a gradient. ◀

EXAMPLE 2 Determine if the vector

$$(e^{-y} - 2x)\mathbf{i} - (xe^{-y} + \sin y)\mathbf{j}$$

is a gradient $\nabla f(x, y)$, and if it is, then find $f(x, y)$.

Solution We apply Theorem 17.5.1. Let

$$M(x, y) = e^{-y} - 2x \qquad N(x, y) = -xe^{-y} - \sin y$$
$$M_y(x, y) = -e^{-y} \qquad N_x(x, y) = -e^{-y}$$

Therefore $M_y(x, y) = N_x(x, y)$; so the given vector is a gradient $\nabla f(x, y)$. Furthermore,

$$f_x(x, y) = e^{-y} - 2x \tag{20}$$

$$f_y(x, y) = -xe^{-y} - \sin y \tag{21}$$

Integrating both sides of (20) with respect to x we obtain

$$f(x, y) = xe^{-y} - x^2 + g(y) \tag{22}$$

where $g(y)$ is independent of x. We now partially differentiate both sides of (22) with respect to y and have

$$f_y(x, y) = -xe^{-y} + g'(y)$$

We equate the right members of this equation and (21) and get

$$-xe^{-y} + g'(y) = -xe^{-y} - \sin y$$

$$g'(y) = -\sin y$$

$$g(y) = \cos y + C$$

We substitute this expression for $g(y)$ into (22) and have

$$f(x, y) = xe^{-y} - x^2 + \cos y + C$$

Related to the concept of determining if a given vector-valued function is a gradient of a real-valued function f is that of determining if an expression of the form $M(x, y)\, dx + N(x, y)\, dy$ is the total differential of a function f. Such an expression is said to be an *exact differential*.

17.5.2 DEFINITION   The differential expression

$$M(x, y)\, dx + N(x, y)\, dy$$

is called **exact** on an open disk B in R^2 if there exists a function f such that

$$f_x(x, y) = M(x, y) \quad \text{and} \quad f_y(x, y) = N(x, y)$$

at all points (x, y) in B.

▶ **ILLUSTRATION 4**   The expression

$$xy^4\, dx + 2x^2y^3\, dy \tag{23}$$

is an exact differential because if

$$f(x, y) = \tfrac{1}{2}x^2y^4$$

then $f_x(x, y) = xy^4$ and $f_y(x, y) = 2x^2y^3$. Observe that (23) is the total differential of f.   ◀

The following theorem, which provides a test for determining if an expression is exact, follows immediately from Definition 17.5.2 and Theorem 17.5.1.

18.1.1 DEFINITION Let f be a function defined on a closed rectangular region R. The number L is said to be the **limit** of sums of the form $\sum_{i=1}^{n} f(\xi_i, \gamma_i) \Delta_i A$ if L satisfies the property that for any $\epsilon > 0$ there exists a $\delta > 0$ such that for every partition Δ for which $\|\Delta\| < \delta$ and for all possible selections of the point (ξ_i, γ_i) in the ith rectangle, $i = 1, 2, \ldots, n$,

$$\left| \sum_{i=1}^{n} f(\xi_i, \gamma_i) \Delta_i A - L \right| < \epsilon$$

If such a number L exists, we write

$$\lim_{\|\Delta\| \to 0} \sum_{i=1}^{n} f(\xi_i, \gamma_i) \Delta_i A = L$$

If there is a number L satisfying Definition 18.1.1, it can be shown that it is unique. The proof is similar to the proof of the theorem (2.1.2) regarding the uniqueness of the limit of a function.

18.1.2 DEFINITION A function f of two variables is said to be **integrable** on a closed rectangular region R if f is defined on R and the number L of Definition 18.1.1 exists. The number L is called the **double integral** of f on R, and we write

$$\lim_{\|\Delta\| \to 0} \sum_{i=1}^{n} f(\xi_i, \gamma_i) \Delta_i A = \iint_R f(x, y) \, dA$$

Other symbols for the double integral of f on R are

$$\iint_R f(x, y) \, dx \, dy \quad \text{and} \quad \iint_R f(x, y) \, dy \, dx$$

The following theorem, stated without proof, gives a sufficient condition for which a function of two variables is integrable.

18.1.3 THEOREM If a function of two variables is continuous on a closed rectangular region R, then it is integrable on R.

The approximation of the value of a double integral is shown in the following example.

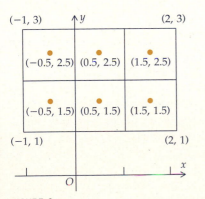

(−1, 3) y (2, 3)

(−0.5, 2.5) (0.5, 2.5) (1.5, 2.5)

(−0.5, 1.5) (0.5, 1.5) (1.5, 1.5)

(−1, 1)   (2, 1)

x

O

FIGURE 2

EXAMPLE 1   Find an approximate value of the double integral

$$\iint_R (3y - 2x^2) \, dA$$

where R is the rectangular region having vertices $(-1, 1)$ and $(2, 3)$. Take a partition of R formed by the lines $x = 0$, $x = 1$, and $y = 2$, and take (ξ_i, γ_i) at the center of the ith subregion.

Solution   Refer to Figure 2, which shows the region R partitioned into six subregions that are squares having sides one unit in length. So for each i, $\Delta_i A = 1$. In each of the subregions the point (ξ_i, γ_i) is at the center of the square. With $f(x, y) = 3y - 2x^2$, an approximation to the given double integral is given by

$$\iint\limits_R (3y - 2x^2)\, dA \approx f(-0.5, 1.5) \cdot 1 + f(0.5, 1.5) \cdot 1 + f(1.5, 1.5) \cdot 1$$
$$+ f(1.5, 2.5) \cdot 1 + f(0.5, 2.5) \cdot 1 + f(-0.5, 2.5) \cdot 1$$
$$= 4 \cdot 1 + 4 \cdot 1 + 0 \cdot 1 + 3 \cdot 1 + 7 \cdot 1 + 7 \cdot 1$$
$$= 25$$

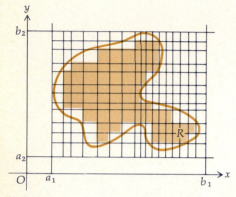

FIGURE 3

The exact value of the double integral in Example 1 is 24, as we will show in Example 1 of Section 18.2.

We now consider the double integral of a function over a more general region. In Section 6.3 a smooth function was defined as one that has a continuous derivative, and a smooth curve is the graph of a smooth function. Let R be a closed region whose boundary consists of a finite number of arcs of smooth curves that are joined together to form a closed curve. As we did with a rectangular region, draw lines parallel to the coordinate axes, which gives a rectangular partition of the region R. Discard the subregions that contain points not in R and consider only those that lie entirely in R (these are shaded in Figure 3). Let the number of these shaded subregions be n and proceed in a manner analogous to the procedure used for a rectangular region. Definitions 18.1.1 and 18.1.2 apply when the region R is the more general one described above. You should intuitively realize that as the norm of the partition approaches zero, n increases without bound, and the area of the region omitted (i.e., the discarded rectangles) approaches zero. Actually, it can be proved that if a function is integrable on a region R, the limit of the approximating sums of the form (1) is the same no matter how we subdivide R, as long as each subregion has a shape to which an area can be assigned.

Just as the integral of a function of a single variable is interpreted geometrically in terms of the area of a plane region, the double integral can be interpreted geometrically in terms of the volume of a three-dimensional solid. Suppose that the function f is continuous on a closed region R in R^2. Furthermore, for simplicity in this discussion, assume that $f(x, y)$ is nonnegative on R. The graph of the equation $z = f(x, y)$ is a surface lying above the xy plane, as shown in Figure 4. The figure shows a particular rectangular subregion of R, having dimensions of measures $\Delta_i x$ and $\Delta_i y$. The figure also shows a rectangular solid having this subregion as a base and $f(\xi_i, \gamma_i)$ as the measure of the altitude, where (ξ_i, γ_i) is a point in the ith subregion. The volume of the rectangular solid is determined by

$$\Delta_i V = f(\xi_i, \gamma_i)\, \Delta_i A$$
$$= f(\xi_i, \gamma_i)\, \Delta_i x\, \Delta_i y$$

The number $\Delta_i V$ is the measure of the volume of the thin rectangular solid shown in Figure 4; thus the sum given in (1) is the sum of the measures of the volumes of n such solids. This sum approximates the measure of the volume of the three-dimensional solid appearing in Figure 4. The solid is bounded above by the graph of f and below by the region R in the xy plane. The sum in (1) also approximates the number given by the double integral

$$\iint\limits_R f(x, y)\, dA$$

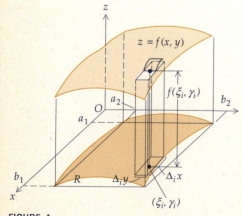

FIGURE 4

It can be proved that the volume of the three-dimensional solid of Figure 4 is the value of the double integral. This fact is stated in the following theorem, for which a formal proof is not given.

18.1.4 THEOREM Let f be a function of two variables that is continuous on a closed region R in the xy plane and $f(x, y) \geq 0$ for all (x, y) in R. If V cubic units is the volume of the solid S having the region R as its base and having an altitude of $f(x, y)$ units at the point (x, y) in R, then

$$V = \lim_{\|\Delta\| \to 0} \sum_{i=1}^{n} f(\xi_i, \gamma_i)\, \Delta_i A$$

$$= \iint\limits_{R} f(x, y)\, dA$$

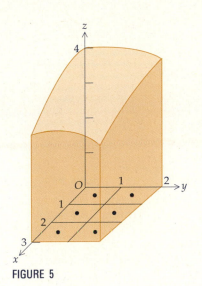

FIGURE 5

EXAMPLE 2   Approximate the volume of the solid bounded by the surface

$$f(x, y) = 4 - \tfrac{1}{9}x^2 - \tfrac{1}{16}y^2$$

the planes $x = 3$ and $y = 2$, and the three coordinate planes. To find an approximate value of the double integral take a partition of the region in the xy plane by drawing the lines $x = 1$, $x = 2$, and $y = 1$, and take (ξ_i, γ_i) at the center of the ith subregion.

Solution   The solid is shown in Figure 5. The rectangular region R is the rectangle in the xy plane bounded by the coordinate axes and the lines $x = 3$ and $y = 2$. From Theorem 18.1.4, if V cubic units is the volume of the solid,

$$V = \iint\limits_{R} (4 - \tfrac{1}{9}x^2 - \tfrac{1}{16}y^2)\, dA$$

Figure 5 shows R partitioned into six subregions that are squares having sides of length one unit. Therefore, for each i, $\Delta_i A = 1$. The point (ξ_i, γ_i) in each subregion is at the center of the square. Then an approximation of V is given by an approximation of the double integral. Therefore

$$V \approx f(0.5, 0.5) \cdot 1 + f(1.5, 0.5) \cdot 1 + f(2.5, 0.5) \cdot 1 + f(0.5, 1.5) \cdot 1 + f(1.5, 1.5) \cdot 1 + f(2.5, 1.5) \cdot 1$$

Using a calculator to compute the function values, we obtain

$$V \approx 3.957 + 3.734 + 3.290 + 3.832 + 3.609 + 3.165$$

$$\approx 21.59$$

Thus the volume is approximately 21.59 cubic units.

In Example 2 of Section 18.2 we show that the exact volume in the above example is 21.5 cubic units.

Analogous to properties of the definite integral of a function of a single variable are several properties of the double integral, and the most important ones are given in the following theorems.

18.1.5 THEOREM If c is a constant and the function f is integrable on a closed region R, then cf is integrable on R and

$$\iint\limits_{R} cf(x, y)\, dA = c \iint\limits_{R} f(x, y)\, dA$$

18.1.6 THEOREM　If the functions f and g are integrable on a closed region R, then the function $f + g$ is integrable on R and

$$\iint_R [f(x, y) + g(x, y)] \, dA = \iint_R f(x, y) \, dA + \iint_R g(x, y) \, dA$$

The result of Theorem 18.1.6 can be extended to any finite number of functions that are integrable. The proofs of Theorems 18.1.5 and 18.1.6 follow from the definition of a double integral.

18.1.7 THEOREM　If the functions f and g are integrable on the closed region R and furthermore $f(x, y) \geq g(x, y)$ for all (x, y) in R, then

$$\iint_R f(x, y) \, dA \geq \iint_R g(x, y) \, dA$$

Theorem 18.1.7 is analogous to Theorem 5.6.8 for the definite integral of a function of a single variable. The proof is similar.

18.1.8 THEOREM　Let the function f be integrable on a closed region R, and suppose that m and M are two numbers such that $m \leq f(x, y) \leq M$ for all (x, y) in R. Then if A is the measure of the area of region R,

$$mA \leq \iint_R f(x, y) \, dA \leq MA$$

The proof of Theorem 18.1.8 is similar to that of Theorem 5.6.9 and is based on Theorem 18.1.7.

18.1.9 THEOREM　Suppose that the function f is continuous on the closed region R and that region R is composed of the two subregions R_1 and R_2 that have no points in common except for points on parts of their boundaries. Then

$$\iint_R f(x, y) \, dA = \iint_{R_1} f(x, y) \, dA + \iint_{R_2} f(x, y) \, dA$$

The proof of Theorem 18.1.9 depends on the definition of a double integral and limit theorems.

EXERCISES 18.1

1. Find an approximate value of the double integral

$$\iint_R (3x - 2y + 1) \, dA$$

where R is the rectangular region having vertices $(0, -2)$ and $(3, 0)$. Take a partition of R formed by the lines $x = 1$, $x = 2$, and $y = -1$, and take (ξ_i, γ_i) at the center of the ith subregion.

2. Find an approximate value of the double integral

$$\iint_R (y^2 - 4x) \, dA$$

where R is the rectangular region having vertices $(-1, 0)$ and $(1, 3)$. Take a partition of R formed by the lines $x = 0$, $y = 1$, and $y = 2$, and take (ξ_i, γ_i) at the center of the ith subregion.

In Exercises 3 through 8, find an approximate value of the double integral, where R is the rectangular region having the vertices P and Q, Δ is a partition of R, and (ξ_i, γ_i) is at the center of each subregion.

3. $\iint_R (x^2 + y) \, dA$; $P(0, 0)$; $Q(4, 2)$; Δ: $x_1 = 0$, $x_2 = 1$, $x_3 = 2$, $x_4 = 3$, $y_1 = 0$, $y_2 = 1$.

4. $\iint\limits_{R} (2 - x - y)\, dA$; $P(0, 0)$; $Q(6, 4)$; Δ: $x_1 = 0$, $x_2 = 2$, $x_3 = 4$, $y_1 = 0$, $y_2 = 2$.

5. $\iint\limits_{R} (xy + 3y^2)\, dA$; $P(-2, 0)$; $Q(4, 6)$; Δ: $x_1 = -2$, $x_2 = 0$, $x_3 = 2$, $y_1 = 0$, $y_2 = 2$, $y_3 = 4$.

6. $\iint\limits_{R} (xy + 3y^2)\, dA$; $P(0, -2)$; $Q(6, 4)$; Δ: $x_1 = 0$, $x_2 = 2$, $x_3 = 4$, $y_1 = -2$, $y_2 = 0$, $y_3 = 2$.

7. $\iint\limits_{R} (x^2 y - 2xy^2)\, dA$; $P(-3, -2)$; $Q(1, 6)$; Δ: $x_1 = -3$, $x_2 = -1$, $y_1 = -2$, $y_2 = 0$, $y_3 = 2$, $y_4 = 4$.

8. $\iint\limits_{R} (x^2 y - 2xy^2)\, dA$; $P(-3, -2)$; $Q(1, 6)$; Δ: $x_1 = -3$, $x_2 = -2$, $x_3 = -1$, $x_4 = 0$, $y_1 = -2$, $y_2 = -1$, $y_3 = 0$, $y_4 = 1$, $y_5 = 2$, $y_6 = 3$, $y_7 = 4$, $y_8 = 5$.

In Exercises 9 through 12, find an approximate value of the double integral, where R is the rectangular region having the vertices P and Q, Δ is a partition of R, and (ξ_i, γ_i) is an arbitrary point in each subregion.

9. The double integral, P, Q, and Δ are the same as in Exercise 3; $(\xi_1, \gamma_1) = (0.25, 0.5)$; $(\xi_2, \gamma_2) = (1.75, 0)$; $(\xi_3, \gamma_3) = (2.5, 0.25)$; $(\xi_4, \gamma_4) = (4, 1)$; $(\xi_5, \gamma_5) = (0.75, 1.75)$; $(\xi_6, \gamma_6) = (1.25, 1.5)$; $(\xi_7, \gamma_7) = (2.5, 2)$; $(\xi_8, \gamma_8) = (3, 1)$.

10. The double integral, P, Q, and Δ are the same as in Exercise 4; $(\xi_1, \gamma_1) = (0.5, 1.5)$; $(\xi_2, \gamma_2) = (3, 1)$; $(\xi_3, \gamma_3) = (5.5, 0.5)$; $(\xi_4, \gamma_4) = (2, 2)$; $(\xi_5, \gamma_5) = (2, 2)$; $(\xi_6, \gamma_6) = (5, 3)$.

11. The double integral, P, Q, and Δ are the same as in Exercise 5; $(\xi_1, \gamma_1) = (-0.5, 0.5)$; $(\xi_2, \gamma_2) = (1, 1.5)$; $(\xi_3, \gamma_3) = (2.5, 2)$; $(\xi_4, \gamma_4) = (-1.5, 3.5)$; $(\xi_5, \gamma_5) = (0, 3)$; $(\xi_6, \gamma_6) = (4, 4)$; $(\xi_7, \gamma_7) = (-1, 4.5)$; $(\xi_8, \gamma_8) = (1, 4.5)$; $(\xi_9, \gamma_9) = (3, 4.5)$.

12. The double integral, P, Q, and Δ are the same as in Exercise 5; $(\xi_1, \gamma_1) = (-2, 0)$; $(\xi_2, \gamma_2) = (0, 0)$; $(\xi_3, \gamma_3) = (2, 0)$; $(\xi_4, \gamma_4) = (-2, 2)$; $(\xi_5, \gamma_5) = (0, 2)$; $(\xi_6, \gamma_6) = (2, 2)$; $(\xi_7, \gamma_7) = (-2, 4)$; $(\xi_8, \gamma_8) = (0, 4)$; $(\xi_9, \gamma_9) = (2, 4)$.

13. Approximate the volume of the solid in the first octant bounded by the sphere $x^2 + y^2 + z^2 = 64$, the planes $x = 3$, $y = 3$, and the three coordinate planes. To find an approximate value of the double integral take a partition of the region in the xy plane formed by the lines $x = 1$, $x = 2$, $y = 1$, and $y = 2$, and take (ξ_i, γ_i) at the center of the ith subregion.

14. Approximate the volume of the solid bounded by the planes $z = 2x + y + 4$, $x = 2$, $y = 3$, and the three coordinate planes. To find an approximate value of the double integral take a partition of the region in the xy plane formed by the lines $x = 1$, $y = 1$, and $y = 2$, and take (ξ_i, γ_i) at the center of the ith subregion.

15. Approximate the volume of the solid bounded by the surface $z = 10 - \frac{1}{4}x^2 - \frac{1}{9}y^2$, the planes $x = 2$, $y = 2$, and the three coordinate planes. To find an approximate value of the double integral take a partition of the region in the xy plane formed by the lines $x = 1$ and $y = 1$, and take (ξ_i, γ_i) at the center of the ith subregion.

16. Approximate the volume of the solid bounded by the surface $100z = 300 - 25x^2 - 4y^2$, the planes $x = -1$, $x = 3$, $y = -3$, $y = 5$ and the xy plane. To find an approximate value of the double integral take a partition of the region in the xy plane formed by the lines $x = 1$, $y = -1$, $y = 1$, and $y = 3$, and take (ξ_i, γ_i) at the center of the ith subregion.

In Exercises 17 through 22, apply Theorem 18.1.8 to find a closed interval containing the value of the double integral.

17. $\iint\limits_{R} (2x + 5y)\, dA$, where R is the rectangular region having vertices $(0, 0)$, $(1, 0)$, $(1, 2)$, and $(0, 2)$.

18. $\iint\limits_{R} (x^2 + y^2)\, dA$, where R is the rectangular region having vertices $(0, 0)$, $(1, 0)$, $(1, 1)$, and $(0, 1)$.

19. $\iint\limits_{R} e^{xy}\, dA$, where R is the rectangular region having vertices $(0, 0)$, $(1, 0)$, $(1, 1)$, and $(0, 1)$.

20. $\iint\limits_{R} (\sin x + \sin y)\, dA$, where R is the rectangular region having vertices $(0, 0)$, $(\pi, 0)$, (π, π), and $(0, \pi)$. (*Hint:* Use the result of Exercise 10 in Exercises 17.3.)

21. $\iint\limits_{R} [\sin(x + y) + \sin x + \sin y]\, dA$, where R is the rectangular region having vertices $(0, 0)$, $(\pi, 0)$, (π, π), and $(0, \pi)$. (*Hint:* Use the result of Exercise 11 in Exercises 17.3.)

22. $\iint\limits_{R} \dfrac{2x + 2y + 1}{x^2 + y^2 + 1}\, dA$, where R is the rectangular region having vertices $(-1, -1)$, $(1, -1)$, $(1, 1)$, and $(-1, 1)$. (*Hint:* Use the result of Exercise 8 in Exercises 17.3.)

18.2 EVALUATION OF DOUBLE INTEGRALS AND ITERATED INTEGRALS

For functions of a single variable, the second fundamental theorem of the calculus provides a method for evaluating a definite integral by finding an antiderivative (or indefinite integral) of the integrand. There is a corresponding method for evaluating a double integral that involves performing successive single integrations. A rigorous development of this method belongs to a course in advanced calculus. Our discussion is an intuitive one, and we use the geometric interpretation of the double integral as the measure of a volume. We first develop the method for the double integral on a rectangular region.

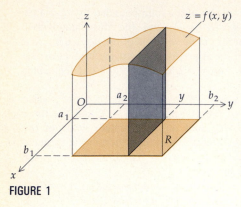

FIGURE 1

Let f be a function that is integrable on a closed rectangular region R in the xy plane bounded by the lines $x = a_1$, $x = b_1$, $y = a_2$, and $y = b_2$. Assume that $f(x, y) \geq 0$ for all (x, y) in R. Refer to Figure 1, showing a sketch of the graph of the equation $z = f(x, y)$ when (x, y) is in R. The number that represents the value of the double integral

$$\iint\limits_{R} f(x, y)\, dA$$

is the measure of the volume of the solid between the surface and the region R. This number can be found by slicing, discussed in Section 6.1, as we now do.

Let y be a number in $[a_2, b_2]$. Consider the plane parallel to the xz plane through the point $(0, y, 0)$. Let $A(y)$ square units be the area of the plane region of intersection of this plane with the solid. The measure of the volume of the solid is expressed by

$$\int_{a_2}^{b_2} A(y)\, dy$$

Because the volume of the solid also is determined by the double integral,

$$\iint\limits_{R} f(x, y)\, dA = \int_{a_2}^{b_2} A(y)\, dy \tag{1}$$

Thus we can find the value of the double integral of the function f on R by evaluating a single integral of $A(y)$. We now must find $A(y)$ when y is given. Because $A(y)$ square units is the area of a plane region, we can find it by integration. In Figure 1, notice that the upper boundary of the plane region is the graph of the equation $z = f(x, y)$ when x is in $[a_1, b_1]$. Therefore $A(y) = \int_{a_1}^{b_1} f(x, y)\, dx$. Substituting from this equation into (1) we obtain

$$\iint\limits_{R} f(x, y)\, dA = \int_{a_2}^{b_2} \left[\int_{a_1}^{b_1} f(x, y)\, dx \right] dy \tag{2}$$

The integral on the right side of (2) is called an **iterated integral**. Because the brackets are usually omitted when writing an iterated integral, (2) can be written as

$$\iint\limits_{R} f(x, y)\, dA = \int_{a_2}^{b_2} \int_{a_1}^{b_1} f(x, y)\, dx\, dy \tag{3}$$

When evaluating the "inner integral" in (3), remember that x is the variable of integration and y is considered a constant. This is comparable to considering y as a constant when finding the partial derivative of $f(x, y)$ with respect to x.

By considering plane sections parallel to the yz plane we obtain an iterated integral that interchanges the order of integration; we have

$$\iint\limits_{R} f(x, y)\, dA = \int_{a_1}^{b_1} \int_{a_2}^{b_2} f(x, y)\, dy\, dx \tag{4}$$

A sufficient condition for (3) and (4) to be valid is that the function be continuous on the rectangular region R.

EXAMPLE 1   Evaluate the double integral

$$\iint_R (3y - 2x^2)\, dA$$

if R is the region consisting of all points (x, y) for which $-1 \le x \le 2$ and $1 \le y \le 3$.

Solution   With $a_1 = -1$, $b_1 = 2$, $a_2 = 1$, and $b_2 = 3$, we have from (3)

$$\iint_R (3y - 2x^2)\, dA = \int_1^3 \int_{-1}^2 (3y - 2x^2)\, dx\, dy$$

$$= \int_1^3 \left[\int_{-1}^2 (3y - 2x^2)\, dx \right] dy$$

$$= \int_1^3 \left[3xy - \tfrac{2}{3}x^3 \right]_{-1}^2 dy$$

$$= \int_1^3 (9y - 6)\, dy$$

$$= \tfrac{9}{2}y^2 - 6y \Big]_1^3$$

$$= 24$$

In Example 1 of Section 18.1 we found an approximate value of the double integral in the above example to be 25.

EXAMPLE 2   Find the volume of the solid bounded by the surface

$$f(x, y) = 4 - \tfrac{1}{9}x^2 - \tfrac{1}{16}y^2$$

the planes $x = 3$ and $y = 2$, and the three coordinate planes.

Solution   Figure 2 shows the graph of the equation $z = f(x, y)$ in the first octant and the given solid. If V cubic units is the volume of the solid, then from Theorem 18.1.4,

$$V = \lim_{\|\Delta\| \to 0} \sum_{i=1}^n f(\xi_i, \gamma_i)\, \Delta_i A$$

$$= \iint_R f(x, y)\, dA$$

$$= \int_0^3 \int_0^2 (4 - \tfrac{1}{9}x^2 - \tfrac{1}{16}y^2)\, dy\, dx$$

$$= \int_0^3 \left[4y - \tfrac{1}{9}x^2 y - \tfrac{1}{48}y^3 \right]_0^2 dx$$

$$= \int_0^3 (\tfrac{47}{6} - \tfrac{2}{9}x^2)\, dx$$

$$= \tfrac{47}{6}x - \tfrac{2}{27}x^3 \Big]_0^3$$

$$= 21.5$$

The volume is therefore 21.5 cubic units.

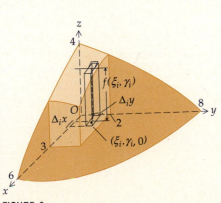

FIGURE 2

In Example 2 of Section 18.1 we found an approximate value of the volume in the above example to be 21.59 cubic units.

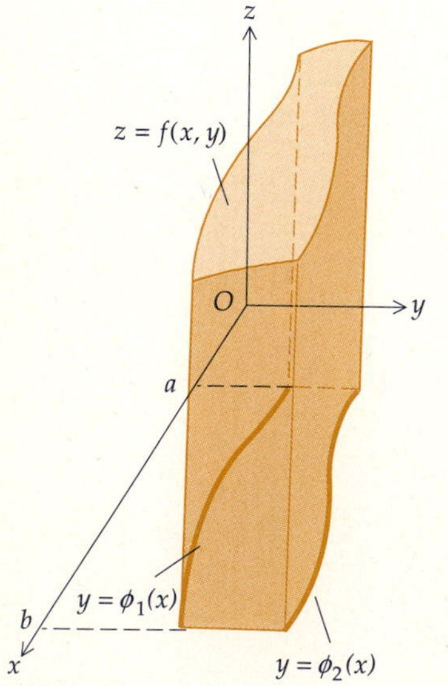

Suppose now that R is the region in the xy plane bounded by the lines $x = a$ and $x = b$, where $a < b$, and by the curves $y = \phi_1(x)$ and $y = \phi_2(x)$, where ϕ_1 and ϕ_2 are functions continuous on the closed interval $[a, b]$. Furthermore, $\phi_1(x) \leq \phi_2(x)$ whenever $a \leq x \leq b$ (see Figure 3). Let Δ be a partition of the interval $[a, b]$ defined by Δ: $a = x_0 < x_1 < \ldots < x_n = b$. Consider the region R of Figure 3 to be divided into vertical strips with widths of $\Delta_i x$ units. A particular strip is shown in the figure. The intersection of the surface $z = f(x, y)$ and a plane $x = \xi_i$, where $x_{i-1} \leq \xi_i \leq x_i$, is a curve. A segment of this curve is over the ith vertical strip. The region under this curve segment and above the xy plane is shown in Figure 4, and the measure of the area of this region is given by

$$\int_{\phi_1(\xi_i)}^{\phi_2(\xi_i)} f(\xi_i, y)\, dy$$

The measure of the volume of the solid bounded above by the surface $z = f(x, y)$ and below by the ith vertical strip is approximately equal to

$$\left[\int_{\phi_1(\xi_i)}^{\phi_2(\xi_i)} f(\xi_i, y)\, dy \right] \Delta_i x$$

If we take the limit, as the norm of Δ approaches zero, of the sum of these measures of volume for n vertical strips of R from $x = a$ to $x = b$, we obtain the measure of the volume of the solid bounded above by the surface $z = f(x, y)$ and below by the region R in the xy plane. (See Figure 5.) This is the double integral of f on R; that is,

$$\lim_{\|\Delta\| \to 0} \sum_{i=1}^{n} \left[\int_{\phi_1(\xi_i)}^{\phi_2(\xi_i)} f(\xi_i, y)\, dy \right] \Delta_i x = \int_a^b \int_{\phi_1(x)}^{\phi_2(x)} f(x, y)\, dy\, dx$$

$$= \iint\limits_R f(x, y)\, dy\, dx \qquad (5)$$

Sufficient conditions for (5) to be valid are that f be continuous on the closed region R and that ϕ_1 and ϕ_2 be smooth functions.

EXAMPLE 3   Express as both a double integral and an iterated integral the measure of the volume of the solid above the xy plane bounded by the elliptic paraboloid $z = x^2 + 4y^2$ and the cylinder $x^2 + 4y^2 = 4$. Evaluate the iterated integral to find the volume of the solid.

Solution   Figure 6 shows the solid. We find the volume of the portion of the solid in the first octant, which from properties of symmetry is one-fourth of the required volume. The region R in the xy plane is that bounded by the x and y axes and the ellipse $x^2 + 4y^2 = 4$. This region appears in Figure 7, which also shows the ith subregion of a rectangular partition of R, where (ξ_i, γ_i) is any point in this ith subregion. If V cubic units is the volume of the given solid, then by Theorem 18.1.4,

$$V = 4 \lim_{\|\Delta\| \to 0} \sum_{i=1}^{n} (\xi_i{}^2 + 4\gamma_i{}^2)\, \Delta_i A$$

$$= 4 \iint\limits_R (x^2 + 4y^2)\, dA$$

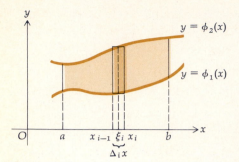

FIGURE 3

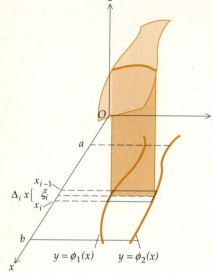

FIGURE 4

FIGURE 5

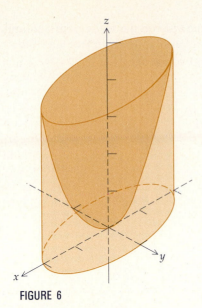

FIGURE 6

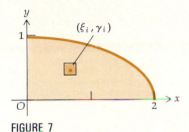

FIGURE 7

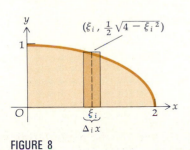

FIGURE 8

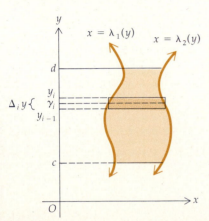

FIGURE 9

To express V as an iterated integral we divide the region R into n vertical strips. Figure 8 shows the region R and the ith vertical strip having width of $\Delta_i x$ units and length of $\frac{1}{2}\sqrt{4 - \xi_i^2}$ units, where $x_{i-1} \le \xi_i \le x_i$. From (5),

$$V = 4 \lim_{\|\Delta\| \to 0} \sum_{i=1}^{n} \left[\int_0^{\sqrt{4 - \xi_i^2}/2} (\xi_i^2 + 4y^2)\, dy \right] \Delta_i x$$

$$= 4 \int_0^2 \int_0^{\sqrt{4 - x^2}/2} (x^2 + 4y^2)\, dy\, dx$$

$$= 4 \int_0^2 \left[x^2 y + \tfrac{4}{3}y^3 \right]_0^{\sqrt{4 - x^2}/2} dx$$

$$= 4 \int_0^2 \left[\tfrac{1}{2}x^2 \sqrt{4 - x^2} + \tfrac{1}{6}(4 - x^2)^{3/2} \right] dx$$

$$= \tfrac{4}{3} \int_0^2 (x^2 + 2)\sqrt{4 - x^2}\, dx$$

$$= -\tfrac{1}{3}x(4 - x^2)^{3/2} + 2x\sqrt{4 - x^2} + 8 \sin^{-1} \tfrac{1}{2}x \Big]_0^2$$

$$= 4\pi$$

Therefore the volume is 4π cubic units.

Suppose the region R is bounded by the curves $x = \lambda_1(y)$ and $x = \lambda_2(y)$ and the lines $y = c$ and $y = d$, where $c < d$, and λ_1 and λ_2 are two functions continuous on the closed interval $[c, d]$ for which $\lambda_1(y) \le \lambda_2(y)$ whenever $c \le y \le d$. Consider a partition Δ of the interval $[c, d]$ and divide the region into horizontal strips, whose widths are $\Delta_i y$ units. See Figure 9, showing the ith horizontal strip. The intersection of the surface $z = f(x, y)$ and a plane $y = \gamma_1$, where $y_{i-1} \le \gamma_i \le y_i$ is a curve, and a segment of this curve is over the ith horizontal strip. Then, as in the derivation of (5), the measure of the volume of the solid bounded above by the surface $z = f(x, y)$ and below by the ith vertical strip is approximately equal to

$$\left[\int_{\lambda_1(\gamma_i)}^{\lambda_2(\gamma_i)} f(x, \gamma_i)\, dx \right] \Delta_i y$$

Taking the limit, as $\|\Delta\|$ approaches zero, of the sum of these measures of volume for n horizontal strips of R from $y = c$ to $y = d$, we obtain the measure of the volume of the solid bounded above by the surface $z = f(x, y)$ and below by the region R in the xy plane. This measure of volume is the double integral of f on R. Hence

$$\lim_{\|\Delta\| \to 0} \sum_{i=1}^{n} \left[\int_{\lambda_1(\gamma_i)}^{\lambda_2(\gamma_i)} f(x, \gamma_i)\, dx \right] \Delta_i y = \int_c^d \int_{\lambda_1(y)}^{\lambda_2(y)} f(x, y)\, dx\, dy$$

$$= \iint\limits_{R} f(x, y)\, dx\, dy \qquad (6)$$

Sufficient conditions for (6) to be valid are that λ_1 and λ_2 be smooth functions and f be continuous on R. In applying both (5) and (6), sometimes it may be necessary to subdivide a region R into subregions on which these sufficient conditions hold.

EXAMPLE 4   Express the volume of the solid of Example 3 by an iterated integral in which the order of integration is the reverse of that of Example 3. Compute the volume.

Solution   Again we find the volume of the solid in the first octant and multiply the result by 4. Figure 10 shows the region R in the xy plane and the ith horizontal strip whose width is $\Delta_i y$ units and whose length is $2\sqrt{1 - \gamma_i^2}$ units. Then by (6),

$$V = 4 \lim_{||\Delta|| \to 0} \sum_{i=1}^{n} \left[\int_0^{2\sqrt{1 - \gamma_i^2}} (x^2 + 4\gamma_i^2)\, dx \right] \Delta_i y$$

$$= 4 \int_0^1 \int_0^{2\sqrt{1 - y^2}} (x^2 + 4y^2)\, dx\, dy$$

$$= 4 \int_0^1 \left[\tfrac{1}{3}x^3 + 4y^2 x \right]_0^{2\sqrt{1 - y^2}} dy$$

$$= 4 \int_0^1 \left[\tfrac{8}{3}(1 - y^2)^{3/2} + 8y^2\sqrt{1 - y^2} \right] dy$$

$$= \tfrac{32}{3} \int_0^1 (2y^2 + 1)\sqrt{1 - y^2}\, dy$$

$$= -\tfrac{16}{3}y(1 - y^2)^{3/2} + 8y\sqrt{1 - y^2} + 8 \sin^{-1} y \Big]_0^1$$

$$= 4\pi$$

Hence the volume is 4π cubic units, which agrees with the answer of Example 3.

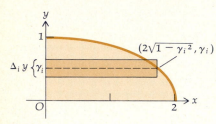

$\Delta_i y \left\{ \gamma_i \right.$ $(2\sqrt{1 - \gamma_i^2}, \gamma_i)$

FIGURE 10

From the solutions of Examples 3 and 4 we see that the double integral $\iint_R (x^2 + 4y^2)\, dA$ can be evaluated by either of the iterated integrals

$$\int_0^2 \int_0^{\sqrt{4 - x^2}/2} (x^2 + 4y^2)\, dy\, dx \quad \text{or} \quad \int_0^1 \int_0^{2\sqrt{1 - y^2}} (x^2 + 4y^2)\, dx\, dy$$

If in either (5) or (6), $f(x, y) = 1$ for all x and y, then the measure A of the area of a region R is expressed as a double integral. We have

$$A = \iint_R dy\, dx \quad \Leftrightarrow \quad A = \iint_R dx\, dy \tag{7}$$

EXAMPLE 5   Find by double integration the area of the region in the xy plane bounded by the curves $y = x^2$ and $y = 4x - x^2$.

Solution   The region is shown in Figure 11. From (7),

$$A = \iint_R dy\, dx$$

$$= \int_0^2 \int_{x^2}^{4x - x^2} dy\, dx$$

$$= \int_0^2 (4x - x^2 - x^2)\, dx$$

$$= 2x^2 - \tfrac{2}{3}x^3 \Big]_0^2$$

$$= \tfrac{8}{3}$$

Hence the area of the region is $\tfrac{8}{3}$ square units.

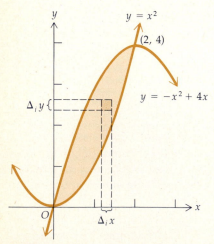

$y = x^2$

$(2, 4)$

$y = -x^2 + 4x$

$\Delta_i y \{$

$\Delta_i x$

FIGURE 11

EXERCISES 18.2

In Exercises 1 through 10, evaluate the iterated integral.

1. $\int_1^2 \int_0^{2x} xy^3 \, dy \, dx$

2. $\int_0^4 \int_0^y dx \, dy$

3. $\int_0^4 \int_0^y \sqrt{9 + y^2} \, dx \, dy$

4. $\int_{-1}^1 \int_1^{e^x} \frac{x}{y} \, dy \, dx$

5. $\int_1^4 \int_{y^2}^y \sqrt{\frac{y}{x}} \, dx \, dy$

6. $\int_1^4 \int_{x^2}^x \sqrt{\frac{y}{x}} \, dy \, dx$

7. $\int_0^1 \int_0^1 |x - y| \, dy \, dx$

8. $\int_0^3 \int_0^x x^2 e^{xy} \, dy \, dx$

9. $\int_{\pi/2}^\pi \int_0^x \sin(4x - y) \, dy \, dx$

10. $\int_{\pi/2}^\pi \int_0^{y^2} \sin \frac{x}{y} \, dx \, dy$

In Exercises 11 through 18, find the exact value of the double integral.

11. The double integral is the same as in Exercise 1 in Exercises 18.1.

12. The double integral is the same as in Exercise 2 in Exercises 18.1.

13. The double integral is the same as in Exercise 3 in Exercises 18.1.

14. The double integral is the same as in Exercise 6 in Exercises 18.1.

15. $\iint\limits_R \sin x \, dA$; R is the region bounded by the lines $y = 2x$, $y = \frac{1}{2}x$, and $x = \pi$.

16. $\iint\limits_R \cos(x + y) \, dA$; R is the region bounded by the lines $y = x$ and $x = \pi$, and the x axis.

17. $\iint\limits_R x^2 \sqrt{9 - y^2} \, dA$; R is the region bounded by the circle $x^2 + y^2 = 9$.

18. $\iint\limits_R \frac{y^2}{x^2} \, dA$; R is the region bounded by the lines $y = x$ and $y = 2$, and the hyperbola $xy = 1$.

19. Find the volume of the solid under the plane $z = 4x$ and above the circle $x^2 + y^2 = 16$ in the xy plane. Draw a sketch of the solid.

20. Find the volume of the solid bounded by the planes $x = y + 2z + 1$, $x = 0$, $y = 0$, $z = 0$, and $3y + z - 3 = 0$. Draw a sketch of the solid.

21. Find the volume of the solid in the first octant bounded by the two cylinders $x^2 + y^2 = 4$ and $x^2 + z^2 = 4$. Draw a sketch of the solid.

22. Find the volume of the solid in the first octant bounded by the paraboloid $z = 9 - x^2 - 3y^2$. Draw a sketch of the solid.

23. Find the volume of the solid in the first octant bounded by the surfaces $x + z^2 = 1$, $x = y$, and $x = y^2$. Draw a sketch of the solid.

24. Find by double integration the volume of the portion of the solid bounded by the sphere $x^2 + y^2 + z^2 = 16$ that lies in the first octant. Draw a sketch of the solid.

In Exercises 25 through 28, use double integrals to find the area of the region bounded by the curves in the xy plane. Draw a sketch of the region.

25. $y = x^3$ and $y = x^2$

26. $y^2 = 4x$ and $x^2 = 4y$

27. $y = x^2 - 9$ and $y = 9 - x^2$

28. $x^2 + y^2 = 16$ and $y^2 = 6x$

29. Express as an iterated integral the measure of the volume of the solid bounded by the ellipsoid

$$\frac{x^2}{a^2} + \frac{y^2}{b^2} + \frac{z^2}{c^2} = 1$$

30. Use double integration to find the area of the region in the first quadrant bounded by the parabola $y^2 = 4x$, the circle $x^2 + y^2 = 5$, and the x axis by two methods: (a) Integrate first with respect to x; (b) integrate first with respect to y. Compare the two methods of solution.

31. Find, by two methods, the volume of the solid below the plane $3x + 8y + 6z = 24$ and above the region in the first quadrant of the xy plane bounded by the parabola $y^2 = 2x$, the line $2x + 3y = 10$, and the x axis: (a) Integrate first with respect to x; (b) integrate first with respect to y. Compare the two methods of solution.

32. Given the iterated integral $\int_0^a \int_0^x \sqrt{a^2 - x^2} \, dy \, dx$. (a) Draw a sketch of the solid the measure of whose volume is represented by the given iterated integral; (b) evaluate the iterated integral; (c) write the iterated integral that gives the measure of the volume of the same solid with the order of integration reversed.

33. Given the iterated integral $\frac{2}{3} \int_0^a \int_0^{\sqrt{a^2 - x^2}} (2x + y) \, dy \, dx$. The instructions are the same as for Exercise 32.

34. Use double integration to find the volume of the solid common to two right-circular cylinders of radius r units, whose axes intersect at right angles. (See Exercise 48 in Exercises 6.1.)

In Exercises 35 and 36, the iterated integral cannot be evaluated exactly in terms of elementary functions by the given order of integration. Reverse the order of integration and perform the computation.

35. $\int_0^4 \int_{\sqrt{x}}^2 \sin \pi y^3 \, dy \, dx$

36. $\int_0^1 \int_y^1 e^{x^2} \, dx \, dy$

18.3 CENTER OF MASS AND MOMENTS OF INERTIA

In Chapter 6 we used single integrals to find the center of mass of a homogeneous lamina. In using single integrals we can consider only laminae of constant area density (except in special cases); however, with double integrals we can find the center of mass of either a homogeneous or a nonhomogeneous lamina.

Suppose we are given a lamina having the shape of a closed region R in the xy plane. Let $\rho(x, y)$ be the measure of the area density of the lamina at any point (x, y) of R where ρ is continuous on R. To find the total mass of the lamina we proceed as follows. Let Δ be a partition of R into n rectangles. If (ξ_i, γ_i) is any point in the ith rectangle having an area of $\Delta_i A$ square units, then an approximation to the measure of the mass of the ith rectangle is given by $\rho(\xi_i, \gamma_i) \Delta_i A$, and the measure of the total mass of the lamina is approximated by

$$\sum_{i=1}^{n} \rho(\xi_i, \gamma_i) \Delta_i A$$

Taking the limit of the above sum as the norm of Δ approaches zero, we express the measure M of the mass of the lamina by

$$M = \lim_{\|\Delta\| \to 0} \sum_{i=1}^{n} \rho(\xi_i, \gamma_i) \Delta_i A$$
$$= \iint\limits_{R} \rho(x, y) \, dA \tag{1}$$

The measure of the moment of mass of the ith rectangle with respect to the x axis is approximated by $\gamma_i \rho(\xi_i, \gamma_i) \Delta_i A$. The sum of the measures of the moments of mass of the n rectangles with respect to the x axis is then approximated by the sum of n such terms. The measure M_x of the moment of mass with respect to the x axis of the entire lamina is given by

$$M_x = \lim_{\|\Delta\| \to 0} \sum_{i=1}^{n} \gamma_i \rho(\xi_i, \gamma_i) \Delta_i A$$
$$= \iint\limits_{R} y \rho(x, y) \, dA$$

Analogously, the measure M_y of its moment of mass with respect to the y axis is given by

$$M_y = \lim_{\|\Delta\| \to 0} \sum_{i=1}^{n} \xi_i \rho(\xi_i, \gamma_i) \Delta_i A$$
$$= \iint\limits_{R} x \rho(x, y) \, dA \tag{2}$$

The center of mass of the lamina is denoted by the point $(\bar{x}, \bar{y})$ and

$$\bar{x} = \frac{M_y}{M} \qquad \bar{y} = \frac{M_x}{M}$$

EXAMPLE 1   A lamina in the shape of an isosceles right triangle has an area density that varies as the square of the distance from the vertex of the right angle. If mass is measured in kilograms and distance is measured in meters, find the mass and the center of mass of the lamina.

Solution   Choose the coordinate axes so that the vertex of the right triangle is at the origin and the sides of length a meters of the triangle are along the coordinate axes (see Figure 1). Let $\rho(x, y)$ kilograms per square meter be the area density of the lamina at the point (x, y). Then $\rho(x, y) = k(x^2 + y^2)$, where k is a constant. Therefore, if M kilograms is the mass of the lamina, we have, from (1),

$$M = \lim_{\|\Delta\| \to 0} \sum_{i=1}^{n} k(\xi_i^2 + \gamma_i^2)\, \Delta_i A$$

$$= k \iint_R (x^2 + y^2)\, dA$$

$$= \int_0^a \int_0^{a-x} (x^2 + y^2)\, dy\, dx$$

$$= k \int_0^a \left[yx^2 + \tfrac{1}{3}y^3 \right]_0^{a-x} dx$$

$$= k \int_0^a \left(\tfrac{1}{3}a^3 - a^2x + 2ax^2 - \tfrac{4}{3}x^3 \right) dx$$

$$= k(\tfrac{1}{3}a^4 - \tfrac{1}{2}a^4 + \tfrac{2}{3}a^4 - \tfrac{1}{3}a^4)$$

$$= \tfrac{1}{6}ka^4$$

To find the center of mass, observe that because of symmetry it must lie on the line $y = x$. Therefore, if we find $\bar{x}$, we also have $\bar{y}$. From (2),

$$M_y = \lim_{\|\Delta\| \to 0} \sum_{i=1}^{n} k\xi_i(\xi_i^2 + \gamma_i^2)\, \Delta_i A$$

$$= k \iint_R x(x^2 + y^2)\, dA$$

$$= k \int_0^a \int_0^{a-x} x(x^2 + y^2)\, dy\, dx$$

$$= k \int_0^a \left[x^3 y + \tfrac{1}{3}xy^3 \right]_0^{a-x} dx$$

$$= k \int_0^a \left(\tfrac{1}{3}a^3 x - a^2 x^2 + 2ax^3 - \tfrac{4}{3}x^4 \right) dx$$

$$= k(\tfrac{1}{6}a^5 - \tfrac{1}{3}a^5 + \tfrac{1}{2}a^5 - \tfrac{4}{15}a^5)$$

$$= \tfrac{1}{15}ka^5$$

Because $M\bar{x} = M_y$, then $M\bar{x} = \tfrac{1}{15}ka^5$; and because $M = \tfrac{1}{6}ka^4$, we get $\bar{x} = \tfrac{2}{5}a$. Therefore the center of mass is at the point $(\tfrac{2}{5}a, \tfrac{2}{5}a)$.

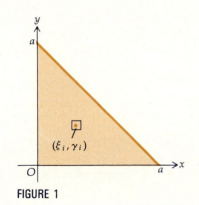

FIGURE 1

18.3.1 DEFINITION   The **moment of inertia** of a particle, whose mass is m kilograms, about an axis is defined to be mr^2 kg-m^2, where r meters is the perpendicular distance from the particle to the axis.

If we have a system of n particles, the moment of inertia of the system is defined as the sum of the moments of inertia of all the particles. That is, if the ith particle has a mass of m_i kilograms and is at a distance of r_i meters from the axis, then I kg-m^2 is the moment of inertia of the system, where

$$I = \sum_{i=1}^{n} m_i r_i^2$$

Extending this concept of moment of inertia to a continuous distribution of mass in a plane such as rods or laminae by processes similar to those previously used, we have the following definition.

18.3.2 DEFINITION   Suppose that we are given a continuous distribution of mass occupying a region R in the xy plane, and suppose that the area density of this distribution at the point (x, y) is $\rho(x, y)$ kilograms per square meter, where ρ is continuous on R. Then the **moment of inertia I_x kg-m^2 about the x axis** of this distribution of mass is determined by

$$I_x = \lim_{||\Delta|| \to 0} \sum_{i=1}^{n} \gamma_i{}^2 \rho(\xi_i, \gamma_i) \, \Delta_i A$$

$$= \iint\limits_{R} y^2 \rho(x, y) \, dA$$

Similarly, the **moment of inertia I_y kg-m^2 about the y axis** is given by

$$I_y = \lim_{||\Delta|| \to 0} \sum_{i=1}^{n} \xi_i{}^2 \rho(\xi_i, \gamma_i) \, \Delta_i A$$

$$= \iint\limits_{R} x^2 \rho(x, y) \, dA$$

and the **moment of inertia I_0 kg-m^2 about the origin**, or the z axis, is given by

$$I_0 = \lim_{||\Delta|| \to 0} \sum_{i=1}^{n} (\xi_i{}^2 + \gamma_i{}^2) \rho(\xi_i, \gamma_i) \, \Delta_i A$$

$$= \iint\limits_{R} (x^2 + y^2) \rho(x, y) \, dA$$

The number I_0 is the measure of the **polar moment of inertia**.

EXAMPLE 2   A homogeneous straight wire has a constant linear density of k kilograms per meter. Find the moment of inertia of the wire about an axis perpendicular to the wire and passing through one end.

Solution   Let the wire be of length a meters, and suppose that it extends along the x axis from the origin. We find its moment of inertia about the y axis. Divide the wire into n segments; the length of the ith segment is $\Delta_i x$ meters. The mass of the ith segment is then $k \, \Delta_i x$ kilograms. Assume that the mass of the ith segment is concentrated at a single point ξ_i, where $x_{i-1} \leq \xi_i \leq x_i$. The moment of inertia of the ith segment about the y axis lies between $k x_{i-1}{}^2 \, \Delta_i x$ kg-m^2 and $k x_i{}^2 \, \Delta_i x$ kg-m^2 and is approximated by $k \xi_i{}^2 \, \Delta_i x$ kg-m^2, where $x_{i-1} \leq \xi_i \leq x_i$. If the moment of inertia of the wire about the y axis is I_y kg-m^2, then

$$I_y = \lim_{||\Delta|| \to 0} \sum_{i=1}^{n} k \xi_i{}^2 \, \Delta_i x$$

$$= \int_0^a k x^2 \, dx$$

$$= \tfrac{1}{3} k a^3$$

Therefore the moment of inertia is $\tfrac{1}{3} k a^3$ kg-m^2.

EXAMPLE 3   A homogeneous rectangular lamina has constant area density of k slugs per square foot. Find the moment of inertia of the lamina about one corner.

Solution   Suppose that the lamina is bounded by the lines $x = a$, $y = b$, the x axis, and the y axis. See Figure 2. If I_0 slug-ft^2 is the moment of inertia about the origin, then

$$I_0 = \lim_{\|\Delta\| \to 0} \sum_{i=1}^{n} k(\xi_i^2 + \gamma_i^2)\, \Delta_i A$$

$$= \iint_R k(x^2 + y^2)\, dA$$

$$= k \int_0^b \int_0^a (x^2 + y^2)\, dx\, dy$$

$$= k \int_0^b \left[\tfrac{1}{3}x^3 + xy^2 \right]_0^a dy$$

$$= k \int_0^b (\tfrac{1}{3}a^3 + ay^2)\, dy$$

$$= \tfrac{1}{3}kab(a^2 + b^2)$$

The moment of inertia is therefore $\tfrac{1}{3}kab(a^2 + b^2)$ slug-ft^2.

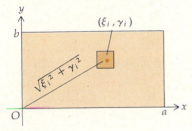

FIGURE 2

It is possible to find the distance from any axis L at which the mass of a lamina can be concentrated without affecting the moment of inertia of the lamina about L. The measure of this distance, denoted by r, is called the *radius of gyration* of the lamina about L. That is, if the mass M kilograms of a lamina is concentrated at a point r meters from L, the moment of inertia of the lamina about L is the same as that of a particle of mass M kilograms at a distance of r meters from L; this moment of inertia is Mr^2 kg-m^2. Thus we have the following definition.

18.3.3 DEFINITION   If I is the measure of the moment of inertia about an axis L of a distribution of mass in a plane and M is the measure of the total mass of the distribution, then the **radius of gyration** of the distribution about L has measure r, where

$$r^2 = \frac{I}{M}$$

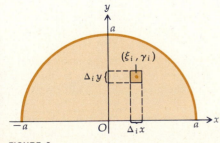

FIGURE 3

EXAMPLE 4   Suppose that a lamina is in the shape of a semicircle and the measure of the area density of the lamina at any point is proportional to the measure of the distance of the point from the diameter. If mass is measured in kilograms and distance is measured in meters, find the radius of gyration of the lamina about the x axis.

Solution   Choose the x and y axes such that the semicircle is the top half of the circle $x^2 + y^2 = a^2$. See Figure 3. The area density of the lamina at the point (x, y) is then ky kilograms per square meter. So if M kilograms is the mass of the lamina, we have

$$M = \lim_{\|\Delta\| \to 0} \sum_{i=1}^{n} k\gamma_i \, \Delta_i A$$

$$= \iint_R ky \, dA$$

$$= \int_0^a \int_{-\sqrt{a^2-y^2}}^{\sqrt{a^2-y^2}} ky \, dx \, dy$$

$$= k \int_0^a \Big[yx\Big]_{-\sqrt{a^2-y^2}}^{\sqrt{a^2-y^2}} dy$$

$$= 2k \int_0^a y\sqrt{a^2 - y^2} \, dy$$

$$= -\tfrac{2}{3}k(a^2 - y^2)^{3/2}\Big]_0^a$$

$$= \tfrac{2}{3}ka^3$$

If I_x kg-m^2 is the moment of inertia of the lamina about the x axis, then

$$I_x = \lim_{\|\Delta\| \to 0} \sum_{i=1}^{n} \gamma_i^2(k\gamma_i) \, \Delta_i A$$

$$= \iint_R ky^3 \, dy \, dx$$

$$= \int_{-a}^a \int_0^{\sqrt{a^2-x^2}} ky^3 \, dy \, dx$$

$$= k \int_{-a}^a \Big[\tfrac{1}{4}y^4\Big]_0^{\sqrt{a^2-x^2}} dx$$

$$= \tfrac{1}{4}k \int_{-a}^a (a^4 - 2a^2x^2 + x^4) \, dx$$

$$= \tfrac{1}{4}k(2a^5 - \tfrac{4}{3}a^5 + \tfrac{2}{5}a^5)$$

$$= \tfrac{4}{15}ka^5$$

Therefore, if r meters is the radius of gyration,

$$r^2 = \frac{\tfrac{4}{15}ka^5}{\tfrac{2}{3}ka^3}$$

$$= \tfrac{2}{5}a^2$$

Thus $r = \tfrac{1}{5}\sqrt{10}\,a$. The radius of gyration is therefore $\tfrac{1}{5}\sqrt{10}\,a$ meters.

EXERCISES 18.3

In Exercises 1 through 12, find the mass and center of mass of the lamina if the area density is as indicated. Mass is measured in kilograms and distance is measured in meters.

1. A lamina in the shape of the rectangular region bounded by the lines $x = 3$ and $y = 2$ and the coordinate axes. The area density at any point is xy^2 kilograms per square meter.

2. A lamina in the shape of the rectangular region bounded by the lines $x = 4$ and $y = 5$ and the coordinate axes. The area density at any point is $(x^2 + y)$ kilograms per square meter.

3. A lamina in the shape of the triangular region whose sides are segments of the coordinate axes and the line $x + 2y = 6$. The area density at any point is y^2 kilograms per square meter.

4. A lamina in the shape of the region in the first quadrant bounded by the parabola $y = x^2$, the line $y = 1$, and the y axis. The area density at any point is $(x + y)$ kilograms per square meter.

5. A lamina in the shape of the region in the first quadrant bounded by the parabola $x^2 = 8y$, the line $y = 2$, and the y axis. The area density varies as the distance from the line $y = -1$.

6. A lamina in the shape of the region bounded by the curve $y = e^x$, the line $x = 1$, and the coordinate axes. The area density varies as the distance from the x axis.

7. A lamina in the shape of the region in the first quadrant bounded by the circle $x^2 + y^2 = a^2$ and the coordinate axes. The area density varies as the sum of the distances from the two straight edges.

8. A lamina in the shape of the region bounded by the triangle whose sides are segments of the coordinate axes and the line $3x + 2y = 18$. The area density varies as the product of the distances from the coordinate axes.

9. A lamina in the shape of the region bounded by the curve $y = \sin x$ and the x axis from $x = 0$ to $x = \pi$. The area density varies as the distance from the x axis.

10. A lamina in the shape of the region bounded by the curve $y = \sqrt{x}$ and the line $y = x$. The area density varies as the distance from the y axis.

11. A lamina in the shape of the region in the first quadrant bounded by the circle $x^2 + y^2 = 4$ and the line $x + y = 2$. The area density at any point is xy kilograms per square meter.

12. A lamina in the shape of the region bounded by the circle $x^2 + y^2 = 1$ and the lines $x = 1$ and $y = 1$. The area density at any point is xy kilograms per square meter.

In Exercises 13 through 18, find the moment of inertia of the homogeneous lamina about the indicated axis if the area density is k kilograms per square meter and distance is measured in meters.

13. A lamina in the shape of the region bounded by $4y = 3x$, $x = 4$, and the x axis; about the x axis.

14. The lamina of Exercise 13; about the line $x = 4$.

15. A lamina in the shape of the region bounded by a circle of radius a meters; about its center.

16. A lamina in the shape of the region bounded by the parabola $x^2 = 4 - 4y$ and the x axis; about the x axis.

17. The lamina of Exercise 16; about the origin.

18. A lamina in the shape of the region bounded by a triangle of sides of lengths a meters, b meters, and c meters; about the side of length a meters.

In Exercises 19 through 22, find for the lamina each of the following: (a) the moment of inertia about the x axis; (b) the moment of inertia about the y axis, (c) the radius of gyration about the x axis, (d) the polar moment of inertia.

19. The lamina of Exercise 1.   20. The lamina of Exercise 4.

21. The lamina of Exercise 9.   22. The lamina of Exercise 10.

23. A homogeneous lamina of area density k slugs per square foot is in the shape of the region bounded by an isosceles triangle having a base of length b feet and an altitude of length h feet. Find the radius of gyration of the lamina about its line of symmetry.

24. A homogeneous lamina of area density k slugs per square foot is in the shape of the region bounded by the curve $x = \sqrt{y}$, the x axis, and the line $x = a$, where $a > 0$. Find the moment of inertia of the lamina about the line $x = a$.

25. A lamina is in the shape of the region enclosed by the parabola $y = 2x - x^2$ and the x axis. Find the moment of inertia of the lamina about the line $y = 4$ if the area density varies as its distance from the line $y = 4$. Mass is measured in kilograms and distance is measured in meters.

18.4   THE DOUBLE INTEGRAL IN POLAR COORDINATES

We now show how the double integral of a function on a closed region in the polar coordinate plane can be defined. We begin by considering the simplest kind of region. Let R be the region bounded by the rays $\theta = \alpha$ and $\theta = \beta$ and by the circles $r = a$ and $r = b$. Then let Δ be a *partition* of this region obtained by drawing rays through the pole and circles having centers at the pole. This is shown in Figure 1. We obtain a network of subregions that we call "curved" rectangles. The norm $\|\Delta\|$ of the partition is the length of the longest of the diagonals of the "curved" rectangles. Let the number of subregions be n, and let $\Delta_i A$ square units be the area of the ith "curved" rectangle. Because the area of the ith subregion is the difference of the areas of two circular sectors,

$$\Delta_i A = \tfrac{1}{2}r_i^2(\theta_i - \theta_{i-1}) - \tfrac{1}{2}r_{i-1}^2(\theta_i - \theta_{i-1})$$
$$= \tfrac{1}{2}(r_i - r_{i-1})(r_i + r_{i-1})(\theta_i - \theta_{i-1})$$

Let $\bar{r}_i = \tfrac{1}{2}(r_i + r_{i-1})$, $\Delta_i r = r_i - r_{i-1}$, and $\Delta_i\theta = \theta_i - \theta_{i-1}$. Then

$$\Delta_i A = \bar{r}_i \Delta_i r \, \Delta_i\theta$$

Take the point $(\bar{r}_i, \bar{\theta}_i)$ in the ith subregion, where $\theta_{i-1} \leq \bar{\theta}_i \leq \theta_i$, and form the sum

$$\sum_{i=1}^{n} f(\bar{r}_i, \bar{\theta}_i) \, \Delta_i A = \sum_{i=1}^{n} f(\bar{r}_i, \bar{\theta}_i)\bar{r}_i \, \Delta_i r \, \Delta_i\theta$$

It can be shown that if f is continuous on the region R, then the limit of this sum, as $\|\Delta\|$ approaches zero, exists and this limit will be the double integral

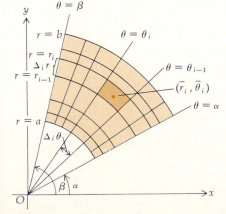

FIGURE 1

of f on R. We write

$$\lim_{\|\Delta\| \to 0} \sum_{i=1}^{n} f(\bar{r}_i, \bar{\theta}_i) \, \Delta_i A = \iint_R f(r, \theta) \, dA$$

$$\Leftrightarrow \quad \lim_{\|\Delta\| \to 0} \sum_{i=1}^{n} f(\bar{r}_i, \bar{\theta}_i) \bar{r}_i \, \Delta_i r \, \Delta_i \theta = \iint_R f(r, \theta) r \, dr \, d\theta$$

Observe that in polar coordinates, $dA = r \, dr \, d\theta$.

The double integral can be shown to be equal to an iterated integral having one of two possible forms:

$$\iint_R f(r, \theta) \, dA = \int_\alpha^\beta \int_a^b f(r, \theta) r \, dr \, d\theta$$

$$= \int_a^b \int_\alpha^\beta f(r, \theta) r \, d\theta \, dr$$

We can define the double integral of a continuous function f of two variables on closed regions of the polar coordinate plane other than the one previously considered. For example, consider the region R bounded by the curves $r = \phi_1(\theta)$ and $r = \phi_2(\theta)$, where ϕ_1 and ϕ_2 are smooth functions, and by the lines $\theta = \alpha$ and $\theta = \beta$. See Figure 2. In the figure, $\phi_1(\theta) \le \phi_2(\theta)$ for all θ in the closed interval $[\alpha, \beta]$. Then it can be shown that the double integral of f on R exists and equals an iterated integral, and we have

$$\iint_R f(r, \theta) \, dA = \int_\alpha^\beta \int_{\phi_1(\theta)}^{\phi_2(\theta)} f(r, \theta) r \, dr \, d\theta$$

If the region R is bounded by the curves $\theta = \chi_1(r)$ and $\theta = \chi_2(r)$, where χ_1 and χ_2 are smooth functions, and by the circles $r = a$ and $r = b$, as shown in Figure 3, where $\chi_1(r) \le \chi_2(r)$ for all r in the closed interval $[a, b]$, then

$$\iint_R f(r, \theta) \, dA = \int_a^b \int_{\chi_1(r)}^{\chi_2(r)} f(r, \theta) r \, d\theta \, dr$$

We can interpret the double integral of a function on a closed region in the polar coordinate plane as the measure of the volume of a solid by using cylindrical coordinates. Figure 4 shows a solid having as its base a region R in the polar coordinate plane and bounded above by the surface $z = f(r, \theta)$, where f is continuous on R and $f(r, \theta) \ge 0$ on R. Take a partition of R giving a network of n "curved" rectangles. Construct the n solids for which the ith solid has as its base the ith "curved" rectangle and as its altitude $f(\bar{r}_i, \bar{\theta}_i)$ units, where $(\bar{r}_i, \bar{\theta}_i)$ is in the ith subregion. Figure 4 shows the ith solid. The measure of the volume of the ith solid is

$$f(\bar{r}_i, \bar{\theta}_i) \, \Delta_i A = f(\bar{r}_i, \bar{\theta}_i) \bar{r}_i \, \Delta_i r \, \Delta_i \theta$$

The sum of the measures of the volumes of the n solids is

$$\sum_{i=1}^{n} f(\bar{r}_i, \bar{\theta}_i) \bar{r}_i \, \Delta_i r \, \Delta_i \theta$$

If V cubic units is the volume of the given solid, then

$$V = \lim_{\|\Delta\| \to 0} \sum_{i=1}^{n} f(\bar{r}_i, \bar{\theta}_i) \bar{r}_i \, \Delta_i r \, \Delta_i \theta$$

$$= \iint_R f(r, \theta) r \, dr \, d\theta \tag{1}$$

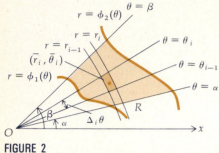

FIGURE 2

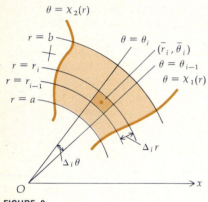

FIGURE 3

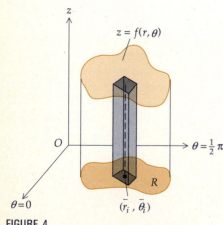

FIGURE 4

EXAMPLE 1   Find the volume of the solid in the first octant bounded by the cone $z = r$ and the cylinder $r = 3 \sin \theta$.

Solution   The solid and the ith element are shown in Figure 5. Using (1) with $f(r, \theta) = r$, we have, where V cubic units is the volume of the given solid,

$$V = \lim_{\|\Delta\| \to 0} \sum_{i=1}^{n} \bar{r}_i \cdot \bar{r}_i \, \Delta_i r \, \Delta_i \theta$$

$$= \iint_{R} r^2 \, dr \, d\theta$$

$$= \int_{0}^{\pi/2} \int_{0}^{3 \sin \theta} r^2 \, dr \, d\theta$$

$$= \int_{0}^{\pi/2} \left[\tfrac{1}{3} r^3 \right]_{0}^{3 \sin \theta} d\theta$$

$$= 9 \int_{0}^{\pi/2} \sin^3 \theta \, d\theta$$

$$= -9 \cos \theta + 3 \cos^3 \theta \Big]_{0}^{\pi/2}$$

$$= 6$$

The volume is therefore 6 cubic units.

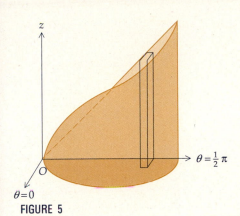

FIGURE 5

EXAMPLE 2   Find the mass of the lamina in the shape of the region inside the semicircle $r = a \cos \theta$, $0 \leq \theta \leq \tfrac{1}{2}\pi$, and whose measure of area density at any point is proportional to the measure of its distance from the pole. The mass is measured in kilograms and distance is measured in meters.

Solution   Figure 6 shows a sketch of the lamina and the ith "curved" rectangle. The area density at the point (r, θ) is kr kilograms per square meter, where k is a constant. If M kilograms is the mass of the lamina, then

$$M = \lim_{\|\Delta\| \to 0} \sum_{i=1}^{n} (k\bar{r}_i)\bar{r}_i \, \Delta_i r \, \Delta_i \theta$$

$$= \iint_{R} kr^2 \, dr \, d\theta$$

$$= k \int_{0}^{\pi/2} \int_{0}^{a \cos \theta} r^2 \, dr \, d\theta$$

$$= \tfrac{1}{3} ka^3 \int_{0}^{\pi/2} \cos^3 \theta \, d\theta$$

$$= \tfrac{1}{3} ka^3 \left[\sin \theta - \tfrac{1}{3} \sin^3 \theta \right]_{0}^{\pi/2}$$

$$= \tfrac{2}{9} ka^3$$

Therefore the mass is $\tfrac{2}{9} ka^3$ kilograms.

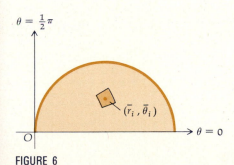

FIGURE 6

EXAMPLE 3   Find the center of mass of the lamina in Example 2.

Solution   Let the cartesian coordinates of the center of mass of the lamina be $\bar{x}$ and $\bar{y}$, where, as is customary, the x axis is along the polar axis and the y axis is along the $\frac{1}{2}\pi$ axis. Let the cartesian coordinate representation of the point $(\bar{r}_i, \bar{\theta}_i)$ be $(\bar{x}_i, \bar{y}_i)$. Then if M_x kg-m is the moment of mass of the lamina with respect to the x axis,

$$M_x = \lim_{||\Delta|| \to 0} \sum_{i=1}^{n} \bar{y}_i (k\bar{r}_i)\bar{r}_i \, \Delta_i r \, \Delta_i \theta$$

Replacing $\bar{y}_i$ by $\bar{r}_i \sin \bar{\theta}_i$ we get

$$\begin{aligned} M_x &= \lim_{||\Delta|| \to 0} \sum_{i=1}^{n} k\bar{r}_i^3 \sin \bar{\theta}_i \, \Delta_i r \, \Delta_i \theta \\ &= \iint\limits_{R} kr^3 \sin \theta \, dr \, d\theta \\ &= k \int_0^{\pi/2} \int_0^{a\cos\theta} r^3 \sin \theta \, dr \, d\theta \\ &= \tfrac{1}{4}ka^4 \int_0^{\pi/2} \cos^4 \theta \sin \theta \, d\theta \\ &= -\tfrac{1}{20}ka^4 \cos^5 \theta \Big]_0^{\pi/2} \\ &= \tfrac{1}{20}ka^4 \end{aligned}$$

If M_y kg-m is the moment of mass of the lamina with respect to the y axis, then

$$M_y = \lim_{||\Delta|| \to 0} \sum_{i=1}^{n} \bar{x}_i (k\bar{r}_i)\bar{r}_i \, \Delta_i r \, \Delta_i \theta$$

Replacing $\bar{x}_i$ by $\bar{r}_i \cos \bar{\theta}_i$ we have

$$\begin{aligned} M_y &= \lim_{||\Delta|| \to 0} \sum_{i=1}^{n} k\bar{r}_i^3 \cos \bar{\theta}_i \, \Delta_i r \, \Delta_i \theta \\ &= \iint\limits_{R} kr^3 \cos \theta \, dr \, d\theta \\ &= k \int_0^{\pi/2} \int_0^{a\cos\theta} r^3 \cos \theta \, dr \, d\theta \\ &= \tfrac{1}{4}ka^4 \int_0^{\pi/2} \cos^5 \theta \, d\theta \\ &= \tfrac{1}{4}ka^4 \left[\sin \theta - \tfrac{2}{3} \sin^3 \theta + \tfrac{1}{5} \sin^5 \theta \right]_0^{\pi/2} \\ &= \tfrac{2}{15}ka^4 \end{aligned}$$

Therefore

$$\begin{aligned} \bar{x} &= \frac{M_y}{M} & \bar{y} &= \frac{M_x}{M} \\ &= \frac{\tfrac{2}{15}ka^4}{\tfrac{2}{9}ka^3} & &= \frac{\tfrac{1}{20}ka^4}{\tfrac{2}{9}ka^3} \\ &= \tfrac{3}{5}a & &= \tfrac{9}{40}a \end{aligned}$$

Hence the center of mass is at the point $(\tfrac{3}{5}a, \tfrac{9}{40}a)$.

In the following example, we show how the area of a region in the polar plane can be found by double integration.

EXAMPLE 4   Find by double integration the area of the region enclosed by one leaf of the rose $r = \sin 3\theta$.

Solution   The region and the ith "curved" rectangle appear in Figure 7. If A square units is the area of the region, then

$$A = \lim_{||\Delta|| \to 0} \sum_{i=1}^{n} \Delta_i A$$

$$= \lim_{||\Delta|| \to 0} \sum_{i=1}^{n} \bar{r}_i \, \Delta_i r \, \Delta_i \theta$$

$$= \iint_R r \, dr \, d\theta$$

$$= \int_0^{\pi/3} \int_0^{\sin 3\theta} r \, dr \, d\theta$$

$$= \tfrac{1}{2} \int_0^{\pi/3} \sin^2 3\theta \, d\theta$$

$$= \tfrac{1}{4}\theta - \tfrac{1}{24} \sin 6\theta \bigg]_0^{\pi/3}$$

$$= \tfrac{1}{12}\pi$$

Hence the area is $\tfrac{1}{12}\pi$ square units.

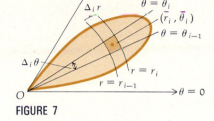

FIGURE 7

Sometimes it is easier to evaluate a double integral by using polar coordinates instead of cartesian coordinates as shown in the following example.

EXAMPLE 5   Evaluate the double integral

$$\iint_R e^{-(x^2+y^2)} \, dA$$

where the region R is in the first quadrant and bounded by the circle $x^2 + y^2 = a^2$ and the coordinate axes.

Solution   Because $x^2 + y^2 = r^2$, and $dA = r \, dr \, d\theta$,

$$\iint_R e^{-(x^2+y^2)} \, dA = \iint_R e^{-r^2} r \, dr \, d\theta$$

$$= \int_0^{\pi/2} \int_0^a e^{-r^2} r \, dr \, d\theta$$

$$= -\tfrac{1}{2} \int_0^{\pi/2} \left[e^{-r^2} \right]_0^a d\theta$$

$$= -\tfrac{1}{2} \int_0^{\pi/2} (e^{-a^2} - 1) \, d\theta$$

$$= \tfrac{1}{4}\pi(1 - e^{-a^2})$$

EXERCISES 18.4

In Exercises 1 through 6, use double integrals to find the area of the region.

1. The region inside the cardioid $r = 2(1 + \sin \theta)$.
2. One leaf of the rose $r = a \cos 2\theta$.
3. The region inside the cardioid $r = a(1 + \cos \theta)$ and outside the circle $r = a$.
4. The region inside the circle $r = 1$ and outside the lemniscate $r^2 = \cos 2\theta$.
5. The region inside the large loop of the limaçon

$$r = 2 - 4 \sin \theta$$

and outside the small loop.
6. The region inside the limacon $r = 3 - \cos \theta$ and outside the circle $r = 5 \cos \theta$.

In Exercises 7 through 12, find the volume of the solid.

7. The solid bounded by the ellipsoid $z^2 + 9r^2 = 9$.
8. The solid cut out of the sphere $z^2 + r^2 = 4$ by the cylinder $r = 1$.
9. The solid cut out of the sphere $z^2 + r^2 = 16$ by the cylinder $r = 4 \cos \theta$.
10. The solid above the polar plane bounded by the cone $z = 2r$ and the cylinder $r = 1 - \cos \theta$.
11. The solid bounded by the paraboloid $z = 4 - r^2$, the cylinder $r = 1$, and the polar plane.
12. The solid above the paraboloid $z = r^2$ and below the plane $z = 2r \sin \theta$.

In Exercises 13 through 19, find the mass and center of mass of the lamina if the area density is as indicated. Mass is measured in kilograms and distance is measured in kilograms and distance is measured in meters.

13. A lamina in the shape of the region of Exercise 1. The area density varies as the distance from the pole.
14. A lamina in the shape of the region of Exercise 2. The area density varies as the distance from the pole.
15. A lamina in the shape of the region inside the limaçon $r = 2 - \cos \theta$. The area density varies as the distance from the pole.
16. A lamina in the shape of the region bounded by the limaçon $r = 2 + \cos \theta$, $0 \le \theta \le \pi$, and the polar axis. The area density at any point is $k \sin \theta$ kilograms per square meter.
17. The lamina of Exercise 16. The area density at any point is $kr \sin \theta$ kilograms per square meter.
18. A lamina in the shape of the region of Exercise 6. The area density varies as the distance from the pole.

19. A lamina in the shape of the region inside the small loop of the limaçon of Exercise 5. The area density varies as the distance from the pole.

In Exercises 20 through 24, find the moment of inertia of the lamina about the indicated axis or point if the area density is as indicated. Mass is measured in kilograms and distance is measured in meters.

20. A lamina in the shape of the region enclosed by the circle $r = \sin \theta$; about the $\frac{1}{2}\pi$ axis. The area density at any point is k kilograms per square meter.
21. The lamina of Exercise 20; about the polar axis. The area density at any point is k kilograms per square meter.
22. A lamina in the shape of the region bounded by the cardioid $r = a(1 - \cos \theta)$; about the pole. The area density at any point is k kilograms per square meter.
23. A lamina in the shape of the region bounded by the cardioid $r = a(1 + \cos \theta)$ and the circle $r = 2a \cos \theta$; about the pole. The area density at any point is k kilograms per square meter.
24. A lamina in the shape of the region enclosed by the lemniscate $r^2 = a^2 \cos 2\theta$; about the polar axis. The area density at any point is k kilograms per square meter.
25. A homogeneous lamina is in the shape of the region enclosed by one loop of the lemniscate $r^2 = \cos 2\theta$. Find the radius of gyration of the lamina about an axis perpendicular to the polar plane at the pole.
26. A lamina is in the shape of the region enclosed by the circle $r = 4$, and the area density varies as the distance from the pole. Find the radius of gyration of the lamina about an axis perpendicular to the polar plane at the pole.
27. Evaluate by polar coordinates the double integral

$$\iint\limits_R e^{x^2 + y^2} \, dA$$

where R is the region bounded by the circles $x^2 + y^2 = 1$ and $x^2 + y^2 = 9$.
28. Evaluate by polar coordinates the double integral

$$\iint\limits_R \frac{x}{\sqrt{x^2 + y^2}} \, dA$$

where R is the region in the first quadrant bounded by the circle $x^2 + y^2 = 1$ and the coordinate axes.

18.5 AREA OF A SURFACE

The double integral can be used to determine the area of the portion of the surface $z = f(x, y)$ that lies over a closed region R in the xy plane. To show this we must first define what we mean by the measure of this area and then obtain a formula for computing it. Assume that f and its first partial derivatives are continuous on R, and suppose that $f(x, y) > 0$ on R. Let Δ be a partition

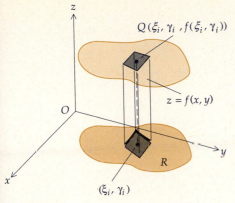

$Q(\xi_i, \gamma_i, f(\xi_i, \gamma_i))$

$z = f(x, y)$

R

(ξ_i, γ_i)

FIGURE 1

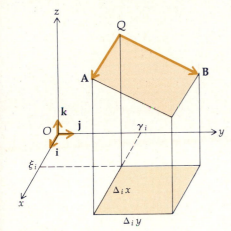

FIGURE 2

of R into n rectangular subregions. The ith rectangle has dimensions $\Delta_i x$ units and $\Delta_i y$ units and an area of $\Delta_i A$ square units. Let (ξ_i, γ_i) be any point in the ith rectangle, and at the point $Q(\xi_i, \gamma_i, f(\xi_i, \gamma_i))$ on the surface consider the tangent plane to the surface. Project vertically upward the ith rectangle onto the tangent plane and let $\Delta_i \sigma$ square units be the area of this projection. Figure 1 shows the region R, the portion of the surface above R, the ith rectangular subregion of R, and the projection of the ith rectangle onto the tangent plane to the surface at Q. The number $\Delta_i \sigma$ is an approximation to the measure of the area of the piece of the surface that lies above the ith rectangle. Because there are n such pieces, the summation

$$\sum_{i=1}^{n} \Delta_i \sigma$$

is an approximation to the measure σ of the area of the portion of the surface that lies above R. This leads to defining σ as follows:

$$\sigma = \lim_{\|\Delta\| \to 0} \sum_{i=1}^{n} \Delta_i \sigma \qquad (1)$$

We now need to obtain a formula for computing this limit. To do this we find a formula for computing $\Delta_i \sigma$ as the measure of the area of a parallelogram. For simplicity in computation take the point (ξ_i, γ_i) in the ith rectangle at the corner (x_{i-1}, y_{i-1}). Let $\mathbf{A}$ and $\mathbf{B}$ be vectors having as representations the directed line segments having initial points at Q and forming the two adjacent sides of the parallelogram whose area is $\Delta_i \sigma$ square units. See Figure 2. Then $\Delta_i \sigma = \|\mathbf{A} \times \mathbf{B}\|$. Because

$$\mathbf{A} = \Delta_i x \mathbf{i} + f_x(\xi_i, \gamma_i) \Delta_i x \mathbf{k} \quad \text{and} \quad \mathbf{B} = \Delta_i y \mathbf{j} + f_y(\xi_i, \gamma_i) \Delta_i y \mathbf{k}$$

it follows that

$$\mathbf{A} \times \mathbf{B} = \begin{vmatrix} \mathbf{i} & \mathbf{j} & \mathbf{k} \\ \Delta_i x & 0 & f_x(\xi_i, \gamma_i) \Delta_i x \\ 0 & \Delta_i y & f_y(\xi_i, \gamma_i) \Delta_i y \end{vmatrix}$$

$$= -\Delta_i x \Delta_i y f_x(\xi_i, \gamma_i) \mathbf{i} - \Delta_i x \Delta_i y f_y(\xi_i, \gamma_i) \mathbf{j} + \Delta_i x \Delta_i y \mathbf{k}$$

Therefore

$$\Delta_i \sigma = \|\mathbf{A} \times \mathbf{B}\|$$
$$= \sqrt{f_x^2(\xi_i, \gamma_i) + f_y^2(\xi_i, \gamma_i) + 1} \, \Delta_i x \, \Delta_i y$$

Substituting this expression for $\Delta_i \sigma$ into (1) we get

$$\sigma = \lim_{\|\Delta\| \to 0} \sum_{i=1}^{n} \sqrt{f_x^2(\xi_i, \gamma_i) + f_y^2(\xi_i, \gamma_i) + 1} \, \Delta_i x \, \Delta_i y$$

This limit is a double integral that exists on R because of the continuity of f_x and f_y on R. We have, then, the following theorem.

18.5.1 THEOREM Suppose that f and its first partial derivatives are continuous on the closed region R in the xy plane. Then if σ square units is the area of the surface $z = f(x, y)$ that lies over R,

$$\sigma = \iint_R \sqrt{f_x^2(x, y) + f_y^2(x, y) + 1} \, dx \, dy$$

EXAMPLE 1   Find the area of the surface that is cut from the cylinder $x^2 + z^2 = 16$ by the planes $x = 0$, $x = 2$, $y = 0$, and $y = 3$.

Solution   The given surface is shown in Figure 3. The region R is the rectangle in the first quadrant of the xy plane bounded by the lines $x = 2$ and $y = 3$. The surface has the equation $x^2 + z^2 = 16$. Solving for z we get $z = \sqrt{16 - x^2}$. Hence $f(x, y) = \sqrt{16 - x^2}$. So if σ square units is the area of the surface, then from Theorem 18.5.1,

$$\sigma = \iint_R \sqrt{f_x{}^2(x, y) + f_y{}^2(x, y) + 1} \; dx \, dy$$

$$= \int_0^3 \int_0^2 \sqrt{\left(\frac{-x}{\sqrt{16 - x^2}}\right)^2 + 0 + 1} \; dx \, dy$$

$$= \int_0^3 \int_0^2 \frac{4}{\sqrt{16 - x^2}} \; dx \, dy$$

$$= 4 \int_0^3 \left[\sin^{-1} \tfrac{1}{4}x\right]_0^2 dy$$

$$= 4 \int_0^3 \tfrac{1}{6}\pi \; dy$$

$$= 2\pi$$

The surface area is therefore 2π square units.

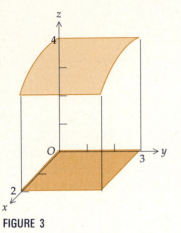

FIGURE 3

EXAMPLE 2   Find the area of the paraboloid $z = x^2 + y^2$ below the plane $z = 4$.

Solution   Figure 4 shows the given surface. From the equation of the paraboloid we see that $f(x, y) = x^2 + y^2$. The closed region in the xy plane bounded by the circle $x^2 + y^2 = 4$ is the region R. If σ square units is the required surface area, then from Theorem 18.5.1,

$$\sigma = \iint_R \sqrt{f_x{}^2(x, y) + f_y{}^2(x, y) + 1} \; dx \, dy$$

$$= \iint_R \sqrt{4(x^2 + y^2) + 1} \; dx \, dy$$

Because the integrand contains the terms $4(x^2 + y^2)$, the evaluation of the double integral is simplified by using polar coordinates. Then $x^2 + y^2 = r^2$. Because $dx \, dy = dA$, then $dx \, dy = r \, dr \, d\theta$. Furthermore, the limits for r are from 0 to 2 and the limits for θ are from 0 to 2π. Therefore

$$\sigma = \iint_R \sqrt{4r^2 + 1}\, r \, dr \, d\theta$$

$$= \int_0^{2\pi} \int_0^2 \sqrt{4r^2 + 1}\, r \, dr \, d\theta$$

$$= \int_0^{2\pi} \left[\tfrac{1}{12}(4r^2 + 1)^{3/2}\right]_0^2 d\theta$$

$$= \tfrac{1}{6}\pi(17\sqrt{17} - 1)$$

Hence the area of the paraboloid below the given plane is $\tfrac{1}{6}\pi(17\sqrt{17} - 1)$ square units.

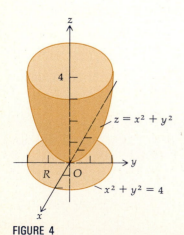

FIGURE 4

parallelepipeds that completely cover S. The parallelepipeds that are entirely inside S or on the boundary of S form a **partition** Δ of S. Choose some system of numbering so that they are numbered from 1 to n. The norm $\|\Delta\|$ of this partition of S is the length of the longest diagonal of any parallelepiped belonging to the partition. The volume of the ith parallelepiped is $\Delta_i V$ cubic units. Let f be a function of three variables that is continuous on S, and let (ξ_i, γ_i, μ_i) be an arbitrary point in the ith parallelepiped. Form the sum

$$\sum_{i=1}^{n} f(\xi_i, \gamma_i, \mu_i)\, \Delta_i V$$

If this sum has a limit as $\|\Delta\|$ approaches zero, and if the limit is independent of the choice of the partitioning planes and the choices of the arbitrary points (ξ_i, γ_i, μ_i) in each parallelepiped, then the limit is called the **triple integral** of f on S, and we write

$$\lim_{\|\Delta\| \to 0} \sum_{i=1}^{n} f(\xi_i, \gamma_i, \mu_i)\, \Delta_i V = \iiint\limits_{S} f(x, y, z)\, dV \qquad (2)$$

It can be proved in advanced calculus that a sufficient condition for the limit in (2) to exist is that f be continuous on S. Furthermore, under the condition imposed on the functions $\phi_1, \phi_2, F_1,$ and F_2 that they be smooth, it can also be proved that the triple integral can be evaluated by the iterated integral

$$\int_a^b \int_{\phi_1(x)}^{\phi_2(x)} \int_{F_1(x,\,y)}^{F_2(x,\,y)} f(x, y, z)\, dz\, dy\, dx$$

Just as the double integral can be interpreted as the measure of the area of a plane region when $f(x, y) = 1$ on R, the triple integral can be interpreted as the measure of the volume of a three-dimensional region. If $f(x, y, z) = 1$ on S, then (2) becomes

$$\lim_{\|\Delta\| \to 0} \sum_{i=1}^{n} \Delta_i V = \iiint\limits_{S} dV$$

and the triple integral is the measure of the volume of the region S.

EXAMPLE 2   Find by triple integration the volume of the solid of Example 3 in Section 18.2.

Solution   The solid lies above the xy plane bounded by the elliptic paraboloid $z = x^2 + 4y^2$ and the cylinder $x^2 + 4y^2 = 4$. See Figure 3. If V cubic units is the volume of the solid, then

$$V = \lim_{\|\Delta\| \to 0} \sum_{i=1}^{n} \Delta_i V$$

$$= \iiint\limits_{S} dV$$

where S is the region bounded by the solid. The z limits are from 0 (the value of z on the xy plane) to $x^2 + 4y^2$ (the value of z on the elliptic paraboloid). The y limits for one-fourth of the volume are from 0 (the value of y on the xz plane) to $\frac{1}{2}\sqrt{4 - x^2}$ (the value of y on the cylinder). The x limits for the first octant are

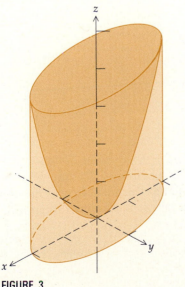

FIGURE 3

from 0 to 2. We evaluate the triple integral by an iterated integral and obtain

$$V = 4 \int_0^2 \int_0^{\sqrt{4-x^2}/2} \int_0^{x^2+4y^2} dz \, dy \, dx$$

$$= 4 \int_0^2 \int_0^{\sqrt{4-x^2}/2} (x^2 + 4y^2) \, dy \, dx$$

This is the same twice-iterated integral that we obtained in Example 3 in Section 18.2, and so the remainder of the solution is the same.

EXAMPLE 3   Find the volume of the solid bounded by the cylinder $x^2 + y^2 = 25$, the plane $x + y + z = 8$ and the xy plane.

Solution   The solid is shown in Figure 4. The z limits for the iterated integral are from 0 to $8 - x - y$ (the value of z on the plane). The y limits are obtained from the boundary region in the xy plane, which is the circle $x^2 + y^2 = 25$. Hence the y limits are from $-\sqrt{25 - x^2}$ to $\sqrt{25 - x^2}$. The x limits are from -5 to 5. If V cubic units is the required volume,

$$V = \lim_{\|\Delta\| \to 0} \sum_{i=1}^n \Delta_i V$$

$$= \iiint_S dV$$

$$= \int_{-5}^5 \int_{-\sqrt{25-x^2}}^{\sqrt{25-x^2}} \int_0^{8-x-y} dz \, dy \, dx$$

$$= \int_{-5}^5 \int_{-\sqrt{25-x^2}}^{\sqrt{25-x^2}} (8 - x - y) \, dy \, dx$$

$$= \int_{-5}^5 \left[(8 - x)y - \tfrac{1}{2}y^2 \right]_{-\sqrt{25-x^2}}^{\sqrt{25-x^2}} dx$$

$$= 2 \int_{-5}^5 (8 - x)\sqrt{25 - x^2} \, dx$$

$$= 16 \int_{-5}^5 \sqrt{25 - x^2} \, dx + \int_{-5}^5 \sqrt{25 - x^2}(-2x) \, dx$$

$$= 16(\tfrac{1}{2}x\sqrt{25 - x^2} + \tfrac{25}{2}\sin^{-1}\tfrac{1}{5}x) + \tfrac{2}{3}(25 - x^2)^{3/2} \Big]_{-5}^5$$

$$= 200\pi$$

The volume is therefore 200π cubic units.

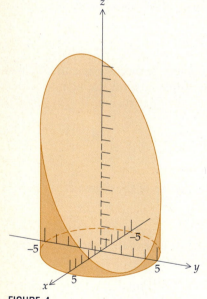

FIGURE 4

EXAMPLE 4   Find the mass of the solid above the xy plane bounded by the cone $9x^2 + z^2 = y^2$ and the plane $y = 9$ if the measure of the volume density at any point (x, y, z) in the solid is proportional to the measure of the distance of the point from the xy plane. The volume density is measured in kilograms per cubic meter.

Solution   Figure 5 shows the solid. Let M kilograms be the mass of the solid. The volume density at any point (x, y, z) in the solid is kz kilograms per cubic meter, where k is a constant. Then if (ξ_i, γ_i, μ_i) is any point in the ith rectangular

parallelepiped of the partition,

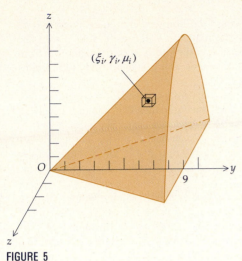

$$M = \lim_{||\Delta|| \to 0} \sum_{i=1}^{n} k\mu_i \, \Delta_i \, V$$

$$= \iiint_S kz \, dV$$

$$= 2k \int_0^9 \int_0^{y/3} \int_0^{\sqrt{y^2 - 9x^2}} z \, dz \, dx \, dy$$

$$= 2k \int_0^9 \int_0^{y/3} \left[\tfrac{1}{2}z^2 \right]_0^{\sqrt{y^2 - 9x^2}} dx \, dy$$

$$= k \int_0^9 \int_0^{y/3} (y^2 - 9x^2) \, dx \, dy$$

$$= \tfrac{2}{9}k \int_0^9 y^3 \, dy$$

$$= \tfrac{729}{2}k$$

The mass is therefore $\tfrac{729}{2}k$ kilograms.

FIGURE 5

EXERCISES 18.6

In Exercises 1 through 8, evaluate the iterated integral.

1. $\int_0^1 \int_0^{1-x} \int_{2y}^{1+y^2} x \, dz \, dy \, dx$

2. $\int_1^2 \int_0^x \int_1^{x+xy} xy \, dz \, dy \, dx$

3. $\int_0^1 \int_0^x \int_0^{x+y} (x + y + z) \, dz \, dy \, dx$

4. $\int_0^2 \int_0^{\sqrt{4-y^2}} \int_0^{2-y} z \, dx \, dz \, dy$

5. $\int_{-1}^0 \int_e^{2e} \int_0^{\pi/3} y \ln z \tan x \, dx \, dz \, dy$

6. $\int_1^2 \int_y^{y^2} \int_0^{\ln x} ye^z \, dz \, dx \, dy$

7. $\int_0^{\pi/2} \int_z^{\pi/2} \int_0^{xz} \cos \frac{y}{z} \, dy \, dx \, dz$

8. $\int_0^2 \int_0^y \int_0^{\sqrt{3}z} \frac{z}{x^2 + z^2} \, dx \, dz \, dy$

In Exercises 9 through 18, evaluate the triple integral.

9. $\iiint_S y \, dV$ if S is the region bounded by the tetrahedron formed by the plane $12x + 20y + 15z = 60$ and the coordinate planes.

10. $\iiint_S (x^2 + z^2) \, dV$ if S is the same region as in Exercise 9.

11. $\iiint_S z \, dV$ if S is the region bounded by the tetrahedron having vertices $(0, 0, 0)$, $(1, 1, 0)$, $(1, 0, 0)$, and $(1, 0, 1)$.

12. $\iiint_S yz \, dV$ if S is the same region as in Exercise 11.

13. $\iiint_S xy \, dV$ if S is the rectangular parallelepiped in the first octant bounded by the coordinate planes and the planes $x = 2$, $y = 3$, and $z = 4$.

14. $\iiint_S x \, dV$ if S is the tetrahedron bounded by the planes $x + 2y + 3z = 6$, $x = 0$, $y = 0$, and $z = 0$.

15. $\iiint_S dV$ if S is the region bounded by the surfaces $z = x^2 + y^2$ and $z = 27 - 2x^2 - 2y^2$.

16. $\iiint_S y^2 \, dV$ if S is the region bounded by the cylinders $x^2 + y = 1$ and $z^2 + y = 1$ and the plane $y = 0$.

17. $\iiint_S (xz + 3z) \, dV$ if S is the region bounded by the cylinder $x^2 + z^2 = 9$ and the planes $x + y = 3$, $z = 0$, and $y = 0$, above the xy plane.

18. $\iiint_S xyz \, dV$ if S is the region bounded by the cylinders $x^2 + y^2 = 4$ and $x^2 + z^2 = 4$.

In Exercises 19 through 32, use triple integration.

19. Find the volume of the solid in the first octant bounded below by the xy plane, above by the plane $z = y$, and laterally by the cylinder $y^2 = x$ and the plane $x = 1$.

20. Find the volume of the solid in the first octant bounded by the cylinder $x^2 + z^2 = 16$, the plane $x + y = 2$, and the three coordinate planes.

21. Find the volume of the solid in the first octant bounded by the cylinders $x^2 + y^2 = 4$ and $x^2 + 2z = 4$ and the three coordinate planes.

22. Find the volume of the solid bounded by the elliptic cone $4x^2 + 9y^2 - 36z^2 = 0$ and the plane $z = 1$.

23. Find the volume of the solid above the elliptic paraboloid $3x^2 + y^2 = z$ and below the cylinder $x^2 + z = 4$.

24. Find the volume of the solid enclosed by the sphere $x^2 + y^2 + z^2 = a^2$.

25. Find the volume of the solid enclosed by the ellipsoid

$$\frac{x^2}{a^2} + \frac{y^2}{b^2} + \frac{z^2}{c^2} = 1$$

26. Find the volume of the solid bounded by the cylinders $z = 5x^2$ and $z = 3 - x^2$, the plane $y + z = 4$, and the xz plane.

27. Find the mass of the homogeneous solid bounded by the cylinder $z = 4 - x^2$, the plane $y = 5$, and the coordinate planes if the volume density at any point is k kilograms per cubic meter.

28. Find the mass of the solid enclosed by the tetrahedron formed by the plane $100x + 25y + 16z = 400$ and the coordinate planes if the volume density varies as the distance from the yz plane. The volume density is measured in kilograms per cubic meter.

29. Find the mass of the solid bounded by the cylinders $x = z^2$ and $y = x^2$, and the planes $x = 1$, $y = 0$, and $z = 0$. The volume density varies as the product of the distances from the three coordinate planes, and it is measured in kilograms per cubic meter.

30. Find the mass of the solid bounded by the surface $z = 4 - 4x^2 - y^2$ and the xy plane. The volume density at any point of the solid is $3z|x|$ kilograms per cubic meter.

31. Find the mass of the solid bounded by the surface $z = xy$ and the planes $x = 1$, $y = 1$ and $z = 0$. The volume density at any point of the solid is $3\sqrt{x^2 + y^2}$ kilograms per cubic meter.

32. A solid has the shape of a right-circular cylinder of base radius r meters and height h meters. Find the mass of the solid if the volume density varies as the distance from one of the bases. The volume density is measured in kilograms per cubic meter.

18.7 THE TRIPLE INTEGRAL IN CYLINDRICAL AND SPHERICAL COORDINATES

If a region S in R^3 has an axis of symmetry, triple integrals on S are easier to evaluate if cylindrical coordinates are used. If there is symmetry with respect to a point, it is convenient to choose that point as the origin and to use spherical coordinates. In this section we discuss the triple integral in these coordinates and apply them to physical problems.

To define the triple integral in cylindrical coordinates we construct a partition of the region S by drawing planes through the z axis, planes perpendicular to the z axis, and right-circular cylinders having the z axis as axis. Figure 1 shows a typical subregion. The elements of the constructed partition lie entirely in S. We call this partition a **cylindrical partition**. The measure of the length of the longest "diagonal" of any of the subregions is the **norm** of the partition. Let n be the number of subregions of the partition and $\Delta_i V$ cubic units be the volume of the ith subregion. The area of the base is $\bar{r}_i \Delta_i r \, \Delta_i \theta$ square units, where $\bar{r}_i = \frac{1}{2}(r_i + r_{i-1})$. Hence if $\Delta_i z$ units is the altitude of the ith subregion,

$$\Delta_i V = \bar{r}_i \, \Delta_i r \, \Delta_i \theta \, \Delta_i z$$

Let f be a function of r, θ, and z, and suppose that f is continuous on S. Choose a point $(\bar{r}_i, \bar{\theta}_i, \bar{z}_i)$ in the ith subregion such that $\theta_{i-1} \le \bar{\theta}_i \le \theta_i$, and $z_{i-1} \le \bar{z}_i \le z_i$. Form the sum

$$\sum_{i=1}^{n} f(\bar{r}_i, \bar{\theta}_i, \bar{z}_i) \, \Delta_i V = \sum_{i=1}^{n} f(\bar{r}_i, \bar{\theta}_i, \bar{z}_i) \bar{r}_i \, \Delta_i r \, \Delta_i \theta \, \Delta_i z \tag{1}$$

As the norm of Δ approaches zero, it can be shown, under suitable conditions on S, that the limit of this sum exists. The limit is called the **triple integral in cylindrical coordinates** of the function f on S, and we write

$$\lim_{\|\Delta\| \to 0} \sum_{i=1}^{n} f(\bar{r}_i, \bar{\theta}_i, \bar{z}_i) \, \Delta_i V = \iiint\limits_{S} f(r, \theta, z) \, dV$$

$$\Leftrightarrow \quad \lim_{\|\Delta\| \to 0} \sum_{i=1}^{n} f(\bar{r}_i, \bar{\theta}_i, \bar{z}_i) \bar{r}_i \, \Delta_i r \, \Delta_i \theta \, \Delta_i z = \iiint\limits_{R} f(r, \theta, z) r \, dr \, d\theta \, dz$$

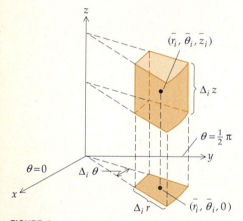

FIGURE 1

Note that in cylindrical coordinates, $dV = r\,dr\,d\theta\,dz$. We can evaluate the triple integral by an iterated integral. For instance, suppose that the region S in R^3 is bounded by the planes $\theta = \alpha$ and $\theta = \beta$, with $\alpha < \beta$, by the cylinders $r = \lambda_1(\theta)$ and $r = \lambda_2(\theta)$, where λ_1 and λ_2 are smooth on $[\alpha, \beta]$ and $\lambda_1(\theta) \leq \lambda_2(\theta)$ for $\alpha \leq \theta \leq \beta$, and by the surfaces $z = F_1(r, \theta)$ and $z = F_2(r, \theta)$, where F_1 and F_2 are functions of two variables that are smooth on some region R in the polar plane bounded by the curves $r = \lambda_1(\theta)$, $r = \lambda_2(\theta)$, $\theta = \alpha$, and $\theta = \beta$. Also, suppose that $F_1(r, \theta) < F_2(r, \theta)$ for every point (r, θ) in R. Then the triple integral can be evaluated by an iterated integral by the formula

$$\iiint\limits_S f(r, \theta, z)r\,dr\,d\theta\,dz = \int_\alpha^\beta \int_{\lambda_1(\theta)}^{\lambda_2(\theta)} \int_{F_1(r,\theta)}^{F_2(r,\theta)} f(r, \theta, z)r\,dz\,dr\,d\theta$$

There are five other iterated integrals that can be used to evaluate the triple integral because there are six possible permutations of the three variables r, θ, and z.

Triple integrals and cylindrical coordinates are especially useful in finding the moment of inertia of a solid with respect to the z axis because the distance from the z axis to a point in the solid is determined by the coordinate r.

EXAMPLE 1   A homogeneous solid in the shape of a right-circular cylinder has a radius of 2 m and an altitude of 4 m. Find the moment of inertia of the solid with respect to its axis.

Solution   Choose the coordinate planes so that the xy plane is the plane of the base of the solid and the z axis is the axis of the solid. Figure 2 shows the portion of the solid in the first octant together with the ith subregion of a cylindrical partition. Using cylindrical coordinates and taking the point $(\bar{r}_i, \bar{\theta}_i, \bar{z}_i)$ in the ith subregion with k kilograms per cubic meter as the volume density at any point, then if I_z kg-m^2 is the moment of inertia of the solid with respect to the z axis,

$$I_z = \lim_{\|\Delta\| \to 0} \sum_{i=1}^n \bar{r}_i^2 k\,\Delta_i V$$

$$= \iiint\limits_S kr^2\,dV$$

There are six different possible orders of integration. Figure 2 shows the order $dz\,dr\,d\theta$. Using this order, we have

$$I_z = \iiint\limits_S kr^2\,dz\,r\,dr\,d\theta$$

$$= 4k \int_0^{\pi/2} \int_0^2 \int_0^4 r^3\,dz\,dr\,d\theta$$

In the first integration the blocks are summed from $z = 0$ to $z = 4$; the blocks become a column. In the second integration the columns are summed from $r = 0$ to $r = 2$; the columns become a wedge-shaped slice of the cylinder. In the third integration the wedge-shaped slice is rotated from $\theta = 0$ to $\theta = \frac{1}{2}\pi$; this sweeps the wedge about the entire three-dimensional region in the first octant.

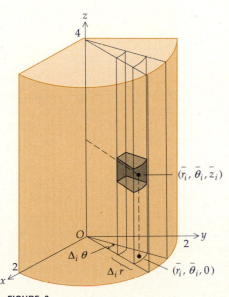

FIGURE 2

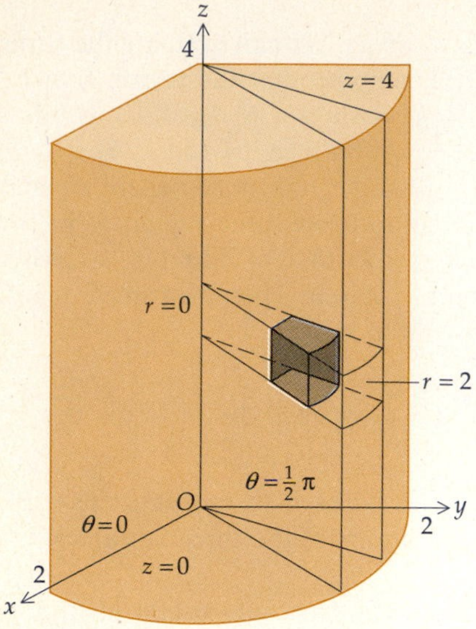

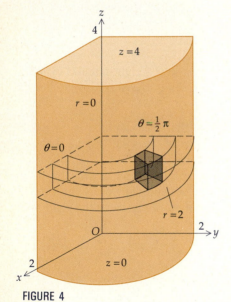

FIGURE 3

FIGURE 4

We multiply by 4 to obtain the entire volume. Performing the integration we obtain

$$I_z = 16k \int_0^{\pi/2} \int_0^2 r^3 \, dr \, d\theta$$

$$= 64k \int_0^{\pi/2} d\theta$$

$$= 32k\pi$$

Hence the moment of inertia is $32k\pi$ kg-m^2.

EXAMPLE 2   Solve Example 1 by taking the order of integration as (a) $dr \, dz \, d\theta$; (b) $d\theta \, dr \, dz$.

Solution

(a) Figure 3 represents the order $dr \, dz \, d\theta$. It shows the block summed from $r = 0$ to $r = 2$ to give a wedge-shaped sector. We then sum from $z = 0$ to $z = 4$ to give a wedge-shaped slice. The slice is rotated from $\theta = 0$ to $\theta = \frac{1}{2}\pi$ to cover the first octant. Then

$$I_z = 4k \int_0^{\pi/2} \int_0^4 \int_0^2 r^3 \, dr \, dz \, d\theta$$

$$= 32k\pi$$

(b) Figure 4 represents the order $d\theta \, dr \, dz$. It shows the blocks summed from $\theta = 0$ to $\theta = \frac{1}{2}\pi$ to give a hollow ring inside the cylinder. These hollow rings are summed from $r = 0$ to $r = 2$ to give a horizontal slice of the cylinder. The horizontal slices are summed from $z = 0$ to $z = 4$. Therefore

$$I_z = 4k \int_0^4 \int_0^2 \int_0^{\pi/2} r^3 \, d\theta \, dr \, dz$$

$$= 32k\pi$$

EXAMPLE 3   Find the mass of a solid hemisphere of radius a meters if the volume density at any point is proportional to the distance of the point from the axis of the solid and is measured in kilograms per cubic meter.

Solution   If we choose the coordinate planes so that the origin is at the center of the sphere and the z axis is the axis of the solid, then an equation of the hemispherical surface above the xy plane is $z = \sqrt{a^2 - x^2 - y^2}$. Figure 5 shows this surface and the solid together with the ith subregion of a cylindrical partition. An equation of the hemisphere in cylindrical coordinates is $z = \sqrt{a^2 - r^2}$. If $(\bar{r}_i, \bar{\theta}_i, \bar{z}_i)$ is a point in the ith subregion, the volume density at this point is $k\bar{r}_i$ kilograms per cubic meter, where k is a constant; and if M kilograms is the mass of the solid, then

$$M = \lim_{||\Delta|| \to 0} \sum_{i=1}^{n} k\bar{r}_i \, \Delta_i V$$

$$= \iiint_S kr \, dV$$

$$= k \int_0^{2\pi} \int_0^a \int_0^{\sqrt{a^2-r^2}} r^2 \, dz \, dr \, d\theta$$

$$= k \int_0^{2\pi} \int_0^a r^2 \sqrt{a^2-r^2} \, dr \, d\theta$$

$$= k \int_0^{2\pi} \left[-\tfrac{1}{4}r(a^2-r^2)^{3/2} + \tfrac{1}{8}a^2 r \sqrt{a^2-r^2} + \tfrac{1}{8}a^4 \sin^{-1}\frac{r}{a} \right]_0^a \, d\theta$$

$$= \tfrac{1}{16}ka^4\pi \int_0^{2\pi} d\theta$$

$$= \tfrac{1}{8}ka^4\pi^2$$

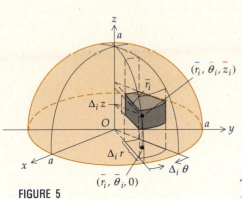

FIGURE 5

The mass of the solid hemisphere is therefore $\tfrac{1}{8}ka^4\pi^2$ kilograms.

EXAMPLE 4   Find the center of mass of the solid of Example 3.

Solution   Let the cartesian coordinate representation of the center of mass be $(\bar{x}, \bar{y}, \bar{z})$. Because of symmetry, $\bar{x} = 0$ and $\bar{y} = 0$. We need to calculate $\bar{z}$. If M_{xy} kg-m is the moment of mass of the solid with respect to the xy plane, then

$$M_{xy} = \lim_{||\Delta|| \to 0} \sum_{i=1}^{n} \bar{z}_i (k\bar{r}_i) \, \Delta_i V$$

$$= \iiint_S kzr \, dV$$

$$= k \int_0^{2\pi} \int_0^a \int_0^{\sqrt{a^2-r^2}} zr^2 \, dz \, dr \, d\theta$$

$$= \tfrac{1}{2}k \int_0^{2\pi} \int_0^a (a^2-r^2)r^2 \, dr \, d\theta$$

$$= \tfrac{1}{15}ka^5 \int_0^{2\pi} d\theta$$

$$= \tfrac{2}{15}ka^5\pi$$

Because $M\bar{z} = M_{xy}$, we get $\bar{z} = M_{xy}/M$; thus

$$\bar{z} = \frac{\tfrac{2}{15}ka^5\pi}{\tfrac{1}{8}ka^4\pi^2}$$

$$= \frac{16}{15\pi} a$$

The center of mass is therefore on the axis of the solid at a distance of $\dfrac{16a}{15\pi}$ meters from the plane of the base.

We now proceed to define the triple integral in spherical coordinates. A spherical partition of the three-dimensional region S is formed by planes containing the z axis, spheres with centers at the origin, and circular cones having vertices

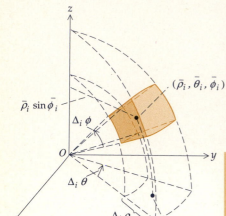

FIGURE 6

at the origin and the z axis as the axis. Figure 6 shows a typical subregion of the partition. If $\Delta_i V$ cubic units is the volume of the ith subregion, and $(\bar{\rho}_i, \bar{\theta}_i, \bar{\phi}_i)$ is a point in it, we can get an approximation to $\Delta_i V$ by considering the region as if it were a rectangular parallelepiped and taking the product of the measures of the three dimensions. These measures are $\bar{\rho}_i \sin \bar{\phi}_i \Delta_i \theta$, $\bar{\rho}_i \Delta_i \phi$, and $\Delta_i \rho$. Figures 7 and 8 show how the first two measures are obtained, and Figure 6 shows the dimension of measure $\Delta_i \rho$. Hence

$$\Delta_i V = \bar{\rho}_i{}^2 \sin \bar{\phi}_i \Delta_i \rho \, \Delta_i \theta \, \Delta_i \phi$$

The **triple integral in spherical coordinates** of a function f on S is given by

$$\lim_{\|\Delta\| \to 0} \sum_{i=1}^{n} f(\bar{\rho}_i, \bar{\theta}_i, \bar{\phi}_i) \Delta_i V = \iiint_S f(\rho, \theta, \phi) \, dV$$

$$\Leftrightarrow \lim_{\|\Delta\| \to 0} \sum_{i=1}^{n} f(\bar{\rho}_i, \bar{\theta}_i, \bar{\phi}_i)\bar{\rho}_i{}^2 \sin \bar{\phi}_i \, \Delta_i \rho \, \Delta_i \theta \, \Delta_i \phi = \iiint_S f(\rho, \theta, \phi)\rho^2 \sin \phi \, d\rho \, d\theta \, d\phi$$

The triple integral can be evaluated by an iterated integral. Observe that in spherical coordinates, $dV = \rho^2 \sin \phi \, d\rho \, d\theta \, d\phi$.

Spherical coordinates are especially useful in some problems involving spheres, as in the following example.

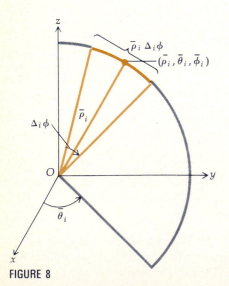

FIGURE 7

EXAMPLE 5   Find the mass of the solid hemisphere of Example 3 if the volume density at any point is proportional to the distance of the point from the center of the base.

Solution   If $(\bar{\rho}_i, \bar{\theta}_i, \bar{\phi}_i)$ is a point in the ith subregion of a spherical partition, the volume density at this point is $k\bar{\rho}_i$ kilograms per cubic meter, where k is a constant. If M kilograms is the mass of the solid, then

$$M = \lim_{\|\Delta\| \to 0} \sum_{i=1}^{n} k\bar{\rho}_i \, \Delta_i V$$

$$= \iiint_S k\rho \, dV$$

$$= 4k \int_0^{\pi/2} \int_0^{\pi/2} \int_0^a \rho^3 \sin \phi \, d\rho \, d\phi \, d\theta$$

$$= a^4 k \int_0^{\pi/2} \int_0^{\pi/2} \sin \phi \, d\phi \, d\theta$$

$$= a^4 k \int_0^{\pi/2} \left[-\cos \phi \right]_0^{\pi/2} d\theta$$

$$= a^4 k \int_0^{\pi/2} d\theta$$

$$= \tfrac{1}{2} a^4 k \pi$$

Hence the mass of the solid hemisphere is $\tfrac{1}{2} a^4 k \pi$ kilograms.

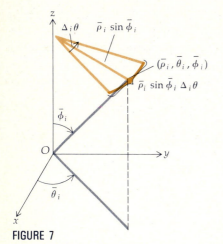

FIGURE 8

It is interesting to compare the solution of Example 5, which uses spherical coordinates, with what is entailed when using cartesian coordinates. By the latter method, a partition of S is formed by dividing S into rectangular boxes by drawing planes parallel to the coordinate planes. If (ξ_i, γ_i, μ_i) is any point in the ith subregion, and because $\rho = \sqrt{x^2 + y^2 + z^2}$, then

$$M = \lim_{||\Delta|| \to 0} \sum_{i=1}^{n} k\sqrt{\xi_i^2 + \gamma_i^2 + \mu_i^2}\, \Delta_i V$$

$$= \iiint_S k\sqrt{x^2 + y^2 + z^2}\, dV$$

$$= 4k \int_0^a \int_0^{\sqrt{a^2 - z^2}} \int_0^{\sqrt{a^2 - y^2 - z^2}} \sqrt{x^2 + y^2 + z^2}\, dx\, dy\, dz$$

The computation involved in evaluating this integral is obviously much more complicated than that using spherical coordinates.

EXAMPLE 6   A homogeneous solid is bounded above by the sphere $\rho = a$ and below by the cone $\phi = \alpha$, where $0 < \alpha < \frac{1}{2}\pi$. Find the moment of inertia of the solid about the z axis. The volume density at any point is k kilograms per cubic meter.

Solution   The solid is shown in Figure 9. Form a spherical partition of the solid and let $(\bar{\rho}_i, \bar{\theta}_i, \bar{\phi}_i)$ be a point in the ith subregion. The measure of the distance of the point $(\bar{\rho}_i, \bar{\theta}_i, \bar{\phi}_i)$ from the z axis is $\bar{\rho}_i \sin \bar{\phi}_i$. Hence if I_z kg-m^2 is the moment of inertia of the given solid about the z axis, then

$$I_z = \lim_{||\Delta|| \to 0} \sum_{i=1}^{n} (\bar{\rho}_i \sin \bar{\phi}_i)^2 k\, \Delta_i V$$

$$= \iiint_S k\rho^2 \sin^2 \phi\, dV$$

$$= k \int_0^\alpha \int_0^{2\pi} \int_0^a (\rho^2 \sin^2 \phi)\rho^2 \sin \phi\, d\rho\, d\theta\, d\phi$$

$$= \tfrac{1}{5}ka^5 \int_0^\alpha \int_0^{2\pi} \sin^3 \phi\, d\theta\, d\phi$$

$$= \tfrac{2}{5}ka^5 \pi \int_0^\alpha \sin^3 \phi\, d\phi$$

$$= \tfrac{2}{5}ka^5\pi \left[-\cos \phi + \tfrac{1}{3} \cos^3 \phi \right]_0^\alpha$$

$$= \tfrac{2}{15}ka^5\pi(\cos^3 \alpha - 3 \cos \alpha + 2)$$

Therefore, $\tfrac{2}{15}ka^5\pi(\cos^3 \alpha - 3 \cos \alpha + 2)$ kg-m^2 is the moment of inertia of the solid about the z axis.

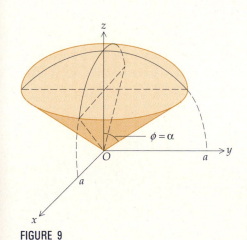

FIGURE 9

EXERCISES 18.7

In Exercises 1 through 6, evaluate the iterated integral.

1. $\int_0^{\pi/4} \int_0^a \int_0^{r \cos \theta} r \sec^3 \theta\, dz\, dr\, d\theta$

2. $\int_0^{\pi/4} \int_{2 \sin \theta}^{2 \cos \theta} \int_0^{r \sin \theta} r^2 \cos \theta\, dz\, dr\, d\theta$

3. $\int_0^\pi \int_2^4 \int_0^1 re^z\, dz\, dr\, d\theta$

4. $\int_0^{2\pi} \int_0^\pi \int_0^2 \rho^3 \sin \phi\, d\rho\, d\phi\, d\theta$

5. $\int_0^{\pi/4} \int_0^{2a \cos \phi} \int_0^{2\pi} \rho^2 \sin \phi\, d\theta\, d\rho\, d\phi$

6. $\int_{\pi/4}^{\pi/2} \int_{\pi/4}^\phi \int_0^{a \csc \theta} \rho^3 \sin^2 \theta \sin \phi\, d\rho\, d\theta\, d\phi$

7. Find the volume of the solid enclosed by the sphere $x^2 + y^2 + z^2 = a^2$ by using (a) cylindrical coordinates and (b) spherical coordinates.

8. If S is the solid in the first octant bounded by the sphere $x^2 + y^2 + z^2 = 16$ and the coordinate planes, evaluate the triple integral $\iiint\limits_{S} xyz \, dV$ by three methods: (a) using spherical coordinates; (b) using rectangular coordinates; (c) using cylindrical coordinates.

In Exercises 9 through 16, use cylindrical coordinates.

9. Find the volume of the solid in the first octant bounded by the cylinder $x^2 + y^2 = 1$ and the plane $z = x$.

10. Find the volume of the solid bounded by the paraboloid $x^2 + y^2 + z = 1$ and the xy plane.

11. Find the volume of the solid bounded by the paraboloid $x^2 + y^2 + z = 12$ and the plane $z = 8$.

12. Find the volume of the solid bounded by the cylinder $x^2 + y^2 = 2y$, the paraboloid $x^2 + y^2 = 2z$ and the xy plane.

13. Find the mass of the solid bounded by a sphere of radius a meters if the volume density varies as the square of the distance from the center. The volume density is measured in kilograms per cubic meter.

14. Find the mass of the solid in the first octant inside the cylinder $x^2 + y^2 = 4x$ and under the sphere $x^2 + y^2 + z^2 = 16$. The volume density varies as the distance from the xy plane, and it is measured in kilograms per cubic meter.

15. Find the moment of inertia with respect to the z axis of the homogeneous solid bounded by the cylinder $r = 5$, the cone $z = r$, and the xy plane. The volume density at any point is k slugs per cubic foot.

16. Find the moment of inertia of the solid bounded by a right-circular cylinder of altitude h meters and radius a meters, with respect to the axis of the cylinder. The volume density varies as the distance from the axis of the cylinder, and it is measured in kilograms per cubic meter.

17. Find the mass of the solid in Exercise 13 by using spherical coordinates.

18. Use spherical coordinates to find the center of mass of the solid bounded by the hemisphere of Example 5. The volume density is the same as that in Example 5.

In Exercises 19 through 22, use spherical coordinates.

19. Find the volume of the solid inside the sphere $x^2 + y^2 + z^2 = 4z$ and above the cone $x^2 + y^2 = z^2$.

20. Find the volume of the solid inside the sphere $x^2 + y^2 + z^2 = 2z$ and above the paraboloid $x^2 + y^2 = z$.

21. Find the moment of inertia with respect to the z axis of the homogeneous solid inside the cylinder $x^2 + y^2 - 2x = 0$, below the cone $x^2 + y^2 = z^2$, and above the xy plane. The volume density at any point is k kilograms per cubic meter.

22. Find the moment of inertia with respect to the z axis of the homogeneous solid bounded by the sphere $x^2 + y^2 + z^2 = 4$. The volume density at any point is k slugs per cubic foot.

In Exercises 23 through 28, use the coordinate system that you decide is best for the problem.

23. Find the mass of a solid hemisphere of radius 2 m if the volume density varies as the distance from the center of the base and is measured in kilograms per cubic meter.

24. Find the mass of the homogeneous solid inside the paraboloid $3x^2 + 3y^2 = z$ and outside the cone $x^2 + y^2 = z^2$ if the constant volume density is k kilograms per cubic meter.

25. Find the moment of inertia about a diameter of the solid between two concentric spheres having radii a feet and $2a$ feet. The volume density varies inversely as the square of the distance from the center, and it is measured in slugs per cubic foot.

26. Find the mass of the solid of Exercise 25. The volume density is the same as that in Exercise 25.

27. Find the center of mass of the solid inside the paraboloid $x^2 + y^2 = z$ and outside the cone $x^2 + y^2 = z^2$. The constant volume density is k kilograms per cubic meter.

28. Find the moment of inertia with respect to the z axis of the homogeneous solid of Exercise 27.

In Exercises 29 through 32, evaluate the iterated integral by using either cylindrical or spherical coordinates.

29. $\int_0^4 \int_0^3 \int_0^{\sqrt{9-x^2}} \sqrt{x^2 + y^2} \, dy \, dx \, dz$

30. $\int_0^1 \int_0^{\sqrt{1-x^2}} \int_0^{\sqrt{1-x^2-y^2}} \frac{z}{\sqrt{x^2+y^2}} \, dz \, dy \, dx$

31. $\int_0^1 \int_0^{\sqrt{1-y^2}} \int_{\sqrt{x^2+y^2}}^{\sqrt{2-x^2-y^2}} z^2 \, dz \, dx \, dy$

32. $\int_0^2 \int_0^{\sqrt{4-y^2}} \int_0^{\sqrt{4-x^2-y^2}} \frac{1}{x^2+y^2+z^2} \, dz \, dx \, dy$

REVIEW EXERCISES FOR CHAPTER 18

In Exercises 1 through 8, evaluate the iterated integral.

1. $\int_0^1 \int_x^{\sqrt{x}} x^2 y \, dy \, dx$

2. $\int_{-2}^2 \int_{-\sqrt{4-y^2}}^{\sqrt{4-y^2}} xy \, dx \, dy$

3. $\int_0^{\pi/2} \int_0^{2 \sin \theta} r \cos^2 \theta \, dr \, d\theta$

4. $\int_0^{\pi} \int_0^{3(1 + \cos \theta)} r^2 \sin \theta \, dr \, d\theta$

5. $\int_0^1 \int_0^z \int_0^{y+z} e^x e^y e^z \, dx \, dy \, dz$

6. $\int_1^2 \int_3^x \int_0^{\sqrt{3}y} \frac{y}{y^2 + z^2} \, dz \, dy \, dx$

7. $\int_0^{\pi/2} \int_{\pi/6}^{\pi/2} \int_0^2 \rho^3 \sin \phi \cos \phi \, d\rho \, d\phi \, d\theta$

8. $\int_0^a \int_0^{\pi/2} \int_0^{\sqrt{a^2-z^2}} zre^{-r^2} \, dr \, d\theta \, dz$

In Exercises 9 through 12, evaluate the multiple integral.

9. $\iint\limits_{R} xy\, dA$; R is the region in the first quadrant bounded by the circle $x^2 + y^2 = 1$ and the coordinate axes.

10. $\iint\limits_{R} (x + y)\, dA$; R is the region bounded by the curve $y = \cos x$ and the x axis from $x = -\frac{1}{2}\pi$ to $x = \frac{1}{2}\pi$.

11. $\iiint\limits_{S} z^2\, dV$; S is the region bounded by the cylinders $x^2 + z = 1$ and $y^2 + z = 1$ and the xy plane.

12. $\iiint\limits_{S} y\cos(x + z)\, dV$; S is the region bounded by the cylinder $x = y^2$ and the planes $x + z = \frac{1}{2}\pi$, $y = 0$, and $z = 0$.

13. Evaluate by polar coordinates the double integral

$$\iint\limits_{R} \frac{1}{x^2 + y^2}\, dA$$

where R is the region in the first quadrant bounded by the two circles $x^2 + y^2 = 1$ and $x^2 + y^2 = 4$.

14. Evaluate by polar coordinates the iterated integral

$$\int_0^1 \int_{\sqrt{3}y}^{\sqrt{4 - y^2}} \ln(x^2 + y^2)\, dx\, dy.$$

In Exercises 15 and 16, evaluate the iterated integral by reversing the order of integration.

15. $\int_0^1 \int_x^1 \sin y^2\, dy\, dx$

16. $\int_0^1 \int_0^{\cos^{-1}y} e^{\sin x}\, dx\, dy$

In Exercises 17 and 18, use double integrals to find the area of the region bounded by the curves in the xy plane. Draw a sketch of the region.

17. $y = x^2$ and $y = x^4$

18. $y = \sqrt{x}$ and $y = x^3$

In Exercises 19 and 20, evaluate the iterated integral by changing to either cylindrical or spherical coordinates.

19. $\int_0^3 \int_0^{\sqrt{9 - x^2}} \int_0^2 \sqrt{x^2 + y^2}\, dz\, dy\, dx$

20. $\int_0^2 \int_0^{\sqrt{4 - x^2}} \int_0^{\sqrt{4 - x^2 - y^2}} z\sqrt{4 - x^2 - y^2}\, dz\, dy\, dx$

21. Use double integration to find the area of the region in the first quadrant bounded by the parabolas $x^2 = 4y$ and $x^2 = 8 - 4y$. Integrate first with respect to x.

22. Use double integration to find the area of the region in the xy plane bounded by the parabolas $y = 9 - x^2$ and $y = x^2 + 1$. Integrate first with respect to x.

23. Use double integration to find the area of the region in Exercise 21 by integrating first with respect to y.

24. Use double integration to find the area of the region in Exercise 22 by integrating first with respect to y.

25. Use double integration to find the volume of the solid bounded by the planes $x = y$, $y = 0$, $z = 0$, $x = 1$, and $z = 1$. Integrate first with respect to x.

26. Use double integration to find the volume of the solid above the xy plane bounded by the cylinder $x^2 + y^2 = 16$ and the plane $z = 2y$. Integrate first with respect to x.

27. Use double integration to find the volume of the solid in Exercise 25 by integrating first with respect to y.

28. Use double integration to find the volume of the solid in Exercise 26 by integrating first with respect to y.

29. Find the volume of the solid above the xy plane bounded by the surfaces $x^2 = 4y$, $y^2 = 4x$, and $x^2 = z - y$.

30. Find the mass of the lamina in the shape of the region bounded by the parabola $y = x^2$ and the line $x - y + 2 = 0$ if the area density at any point is x^2y^2 kilograms per square meter.

31. Find the area of the surface of the cylinder $x^2 + y^2 = 9$ lying in the first octant and between the planes $x = z$ and $3x = z$.

32. Find the area of the surface of the part of the cylinder $x^2 + y^2 = a^2$ that lies inside the cylinder $y^2 + z^2 = a^2$.

33. Use double integration to find the area of the region inside the circle $r = 1$ and to the right of the parabola $r(1 + \cos\theta) = 1$.

34. Find the mass of the lamina in the shape of the region exterior to limaçon $r = 3 - \cos\theta$ and interior to the circle $r = 5\cos\theta$ if the area density at any point is $2|\sin\theta|$ kilograms per square meter.

35. Find the center of mass of the rectangular lamina bounded by the lines $x = 3$ and $y = 2$ and the coordinate axes if the area density at any point is xy^2 kilograms per square meter.

36. Find the center of mass of the lamina in the shape of the region bounded by the parabolas $x^2 = 4 + 4y$ and $x^2 = 4 - 8y$ if the area density at any point is kx^2 kilograms per square meter.

37. Find the mass of the lamina in the shape of the region bounded by the polar axis and the curve $r = \cos 2\theta$, where $0 \leq \theta \leq \frac{1}{4}\pi$. The area density at any point is $r\theta$ kilograms per square meter.

38. Find the moment of inertia about the x axis of the lamina in the shape of the region bounded by the circle $x^2 + y^2 = a^2$ if the area density at any point is $k\sqrt{x^2 + y^2}$ kilograms per square meter.

39. Use cylindrical coordinates to find the volume of the solid bounded by the paraboloid $x^2 + y^2 = 4z$, the cylinder $x^2 + y^2 = 4ay$, and the plane $z = 0$.

40. Use spherical coordinates to find the mass of a spherical solid of radius a meters if the volume density at each point is proportional to the distance of the point from the center of the sphere. The volume density is measured in kilograms per cubic meter.

41. Use triple integration to find the volume of the solid bounded by the plane $z = 1$ and the smaller segment of the sphere $x^2 + y^2 + z^2 = 4$ cut off by this plane.

42. Use triple integration to find the volume of the solid in the first octant bounded by the plane $y + z = 8$, the cylinder $y = 2x^2$, the xy plane, and the yz plane.

43. Find the moment of inertia about the x axis of the lamina in the shape of the region bounded by the curve $y = e^x$, the

line $x = 2$, and the coordinate axes if the area density at any point is xy kilograms per square meter.

44. Find the moment of inertia of the lamina of Exercise 43 about the y axis.

45. Find the moment of inertia with respect to the $\frac{1}{2}\pi$ axis of the homogeneous lamina in the shape of the region bounded by the curve $r^2 = 4 \cos 2\theta$ if the area density at any point is k kilograms per square meter.

46. Find the mass of the lamina of Exercise 45.

47. Find the polar moment of inertia and the corresponding radius of gyration of the lamina of Exercise 45.

48. Find the moment of inertia about the y axis of the lamina in the shape of the region bounded by the parabola $y = x - x^2$ and the line $x + y = 0$ if the area density at any point is $(x + y)$ kilograms per square meter.

49. Find the mass of the solid bounded by the spheres $x^2 + y^2 + z^2 = 4$ and $x^2 + y^2 + z^2 = 9$ if the volume density at any point is $k\sqrt{x^2 + y^2 + z^2}$ kilograms per cubic meter.

50. Find the moment of inertia about the z axis of the solid of Exercise 49.

51. The homogeneous solid bounded by the cone $z^2 = 4x^2 + 4y^2$ between the planes $z = 0$ and $z = 4$ has a volume density at any point of k kilograms per cubic meter. Find the moment of inertia about the z axis for this solid.

52. Find the center of mass of the solid bounded by the sphere $x^2 + y^2 + z^2 - 6z = 0$ and the cone $x^2 + y^2 = z^2$, and above the cone, if the volume density at any point is kz kilograms per cubic meter.

NINETEEN

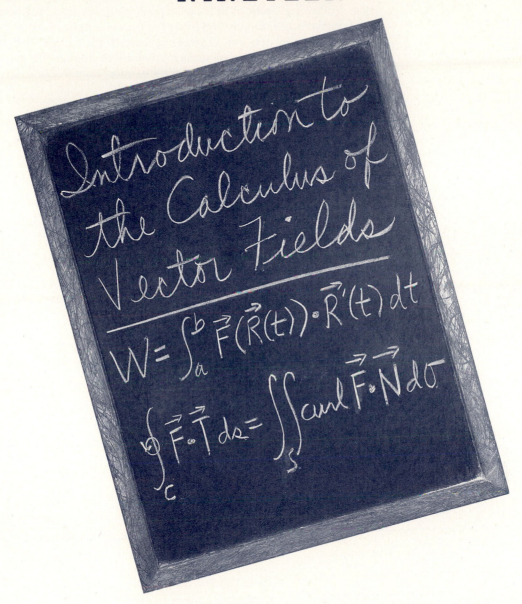

Introduction to
the Calculus of
Vector Fields

$$W = \int_a^b \vec{F}(\vec{R}(t)) \cdot \vec{R}'(t)\, dt$$

$$\oint_C \vec{F} \cdot \vec{T}\, ds = \iint_S \text{curl}\, \vec{F} \cdot \vec{N}\, d\sigma$$

The treatment of the material in this chapter is a brief presentation of topics developed fully in advanced calculus. In Section 19.1 we discuss *vector fields*, which are functions that associate vectors with points in space, and the *divergence* and *curl* of a vector field are introduced. *Line integrals* are defined in Section 19.2 and are applied to find the work done by a force field in moving a particle along a curve. Line integrals independent of the path are then discussed in Section 19.3, where an analogue of the second fundamental theorem of the calculus is presented for line integrals. Also in Section 19.3, we prove the *law of conservation of energy*, a major concept in physics.

Three important theorems in vector calculus are named for the following mathematicians and scientists: George Green, Karl Gauss, and George Stokes. *Green's theorem* on line integrals over curves forms the subject matter of Section 19.4. Following Section 19.5 on *surface integrals*, the final section presents *Gauss's divergence theorem* and *Stokes's theorem*. Applications of these theorems in physics, chemistry, and engineering belong to courses in those fields. We include here applications of surface integrals to finding the mass of a surface and the flux of a velocity field across a surface.

19.1 VECTOR FIELDS

A *vector field* associates a vector with a point in space. For instance, if **F** is a vector-valued function defined on some open ball B in R^3 such that

$$\mathbf{F}(x, y, z) = M(x, y, z)\mathbf{i} + N(x, y, z)\mathbf{j} + R(x, y, z)\mathbf{k} \qquad (1)$$

then **F** associates with each point (x, y, z) in B a vector, and **F** is called a **vector field**. This vector field has as its domain a subset of R^3 and as its range a subset of V_3. If the domain of a vector field is a set of points in a plane and its range is a set of vectors in V_2, then the vector field has an equation of the form

$$\mathbf{F}(x, y) = M(x, y)\mathbf{i} + N(x, y)\mathbf{j}$$

If, instead of a vector, a scalar is associated with each point in space, we have a **scalar field**; thus a scalar field is a real-valued function. An example of a scalar field is obtained by expressing the temperature at a point as a function of the coordinates of the point.

For an example of a vector field, consider the flow of a fluid, such as water through a pipe or blood through an artery. Assume that the fluid consists of infinitely many particles and that the velocity of a particle depends only on its position; thus the velocity is independent of time, and because of this fact the fluid flow is designated as a *steady-state* fluid flow. At a point (x, y, z) the velocity of the fluid is given by $\mathbf{F}(x, y, z)$ defined by an equation of the form (1). Thus **F** is a vector field called the **velocity field** of the fluid. Velocity fields can describe other motions, such as that of a wind or the rotation of a wheel. The vector fields that occur in this book will all be independent of time; they are termed **steady-state vector fields**.

We cannot show in a figure representations of all the vectors of a particular vector field. However, by drawing representations of some of the vectors we may get a visual description of the vector field, as indicated in the following example.

Table 1

(x, y)	$\mathbf{F}(x, y)$
$(1, 1)$	$-\mathbf{i} + \mathbf{j}$
$(1, -1)$	$\mathbf{i} + \mathbf{j}$
$(-1, 1)$	$-\mathbf{i} - \mathbf{j}$
$(-1, -1)$	$\mathbf{i} - \mathbf{j}$
$(1, 2)$	$-2\mathbf{i} + \mathbf{j}$
$(1, -2)$	$2\mathbf{i} + \mathbf{j}$
$(-1, 2)$	$-2\mathbf{i} - \mathbf{j}$
$(-1, -2)$	$2\mathbf{i} - \mathbf{j}$
$(2, 1)$	$-\mathbf{i} + 2\mathbf{j}$
$(2, -1)$	$\mathbf{i} + 2\mathbf{j}$
$(-2, 1)$	$-\mathbf{i} - 2\mathbf{j}$
$(-2, -1)$	$\mathbf{i} - 2\mathbf{j}$
$(2, 2)$	$-2\mathbf{i} + 2\mathbf{j}$
$(2, -2)$	$2\mathbf{i} + 2\mathbf{j}$
$(-2, 2)$	$-2\mathbf{i} - 2\mathbf{j}$
$(-2, -2)$	$2\mathbf{i} - 2\mathbf{j}$

FIGURE 1

EXAMPLE 1 (a) Show on a figure the representations, having initial point at (x, y), of the vectors in the vector field

$$\mathbf{F}(x, y) = -y\mathbf{i} + x\mathbf{j}$$

where x is ± 1 or ± 2 and y is ± 1 or ± 2. (b) Prove that each representation is tangent to a circle having its center at the origin and has a length equal to the radius of the circle.

Solution
(a) Table 1 gives the vectors $\mathbf{F}(x, y)$ associated with the sixteen points (x, y). Representations of these vectors appear in Figure 1.

(b) Let

$$\mathbf{R}(x, y) = x\mathbf{i} + y\mathbf{j}$$

be the position vector whose terminal point is at (x, y). Then

$$\mathbf{R}(x, y) \cdot \mathbf{F}(x, y) = (x\mathbf{i} + y\mathbf{i}) \cdot (-y\mathbf{i} + x\mathbf{j})$$
$$= -xy + xy$$
$$= 0$$

Therefore $\mathbf{R}$ and $\mathbf{F}$ are orthogonal. Thus the representation of $\mathbf{F}$ whose initial point is at (x, y) is tangent to the circle having its center at the origin and radius $\|\mathbf{R}(x, y)\|$. Because

$$\|\mathbf{F}(x, y)\| = \sqrt{(-y)^2 + x^2}$$
$$= \|\mathbf{R}(x, y)\|$$

the length of each representation is equal to the radius of the circle.

The vector field in Example 1 is similar to a velocity field determined by a wheel rotating at the origin.

An example of a vector field in V_3 arises from Newton's inverse square law of gravitational attraction. This law states that the measure of the magnitude of the gravitational force between two particles of mass M units and m units, respectively, is

$$\frac{GMm}{d^2}$$

where d units is the distance between the particles and G is a gravitational constant. Thus if a particle of mass M units is at the origin and a particle of mass 1 unit $(m = 1)$ is at the point $P(x, y, z)$, then if $\mathbf{F}(x, y, z)$ is the gravitational force exerted by the particle at the origin on the particle at P,

$$\|\mathbf{F}(x, y, z)\| = \frac{GM(1)}{\|\mathbf{R}(x, y, z)\|^2}$$

where $\mathbf{R}(x, y, z) = x\mathbf{i} + y\mathbf{j} + z\mathbf{k}$. To obtain the force vector $\mathbf{F}(x, y, z)$, we also need the direction of $\mathbf{F}$. Because this direction is toward the origin, it is the same as the direction of the unit vector $-\dfrac{1}{\|\mathbf{R}\|} \mathbf{R}$. With this direction and the magnitude given above, we have

$$\mathbf{F}(x, y, z) = \frac{GM}{\|\mathbf{R}(x, y, z)\|^2} \left(-\frac{\mathbf{R}(x, y, z)}{\|\mathbf{R}(x, y, z)\|} \right)$$

Because $\|\mathbf{R}(x, y, z)\| = \sqrt{x^2 + y^2 + z^2}$, we obtain

$$\mathbf{F}(x, y, z) = \frac{-GM}{(x^2 + y^2 + z^2)^{3/2}} (x\mathbf{i} + y\mathbf{j} + z\mathbf{k}) \tag{2}$$

The vector field defined by (2) is called a **force field**. Figure 2 shows some of the representations of the vectors in this force field where the object at the origin is a sphere (for instance, the earth) and $\|\mathbf{R}\|$ is greater than the radius of the sphere. Each representation points toward the origin. Representations of

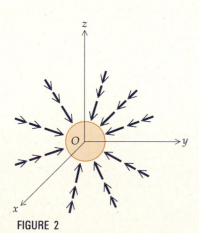

FIGURE 2

vectors at points near the origin are longer than those at points farther away from the origin, and the lengths are the same at points having the same distance from the origin. With these properties, the force field defined by (2) is designated as a *central force field*.

The gradient of a scalar field is a vector field. If ϕ is a scalar field and $\mathbf{F}$ is the vector field defined by $\mathbf{F} = \mathbf{V}\phi$, then $\mathbf{F}$ is called a **gradient vector field**, and ϕ is termed a **potential function** for $\mathbf{F}$. A gradient vector field is also called a **conservative vector field**. The terminology *conservative* will be apparent after you read Section 19.3.

▶ **ILLUSTRATION 1**   Consider the vector field defined by.

$$\mathbf{F}(x, y) = (y^2 + 2x + 4)\mathbf{i} + (2xy + 4y - 5)\mathbf{j}$$

From Illustration 1 of Section 17.5 it follows that if

$$\phi(x, y) = y^2 x + x^2 + 4x + 2y^2 - 5y$$

then

$$\mathbf{F}(x, y) = \mathbf{V}\phi(x, y)$$

Thus $\mathbf{F}$ is a conservative vector field and ϕ is a potential function for $\mathbf{F}$.   ◀

The following illustration shows that the gravitational force field defined by (2) is conservative.

▶ **ILLUSTRATION 2**   In Example 5 of Section 17.1 we showed that if

$$V(x, y, z) = \frac{1}{\sqrt{x^2 + y^2 + z^2}}$$

then

$$\mathbf{V}V(x, y, z) = \frac{-1}{(x^2 + y^2 + z^2)^{3/2}} (x\mathbf{i} + y\mathbf{j} + z\mathbf{k})$$

Thus if

$$\phi(x, y, z) = \frac{GM}{\sqrt{x^2 + y^2 + z^2}}$$

$$\mathbf{V}\phi(x, y, z) = \frac{-GM}{(x^2 + y^2 + z^2)^{3/2}} (x\mathbf{i} + y\mathbf{j} + z\mathbf{k})$$

Comparing this equation and (2) we observe that

$$\mathbf{F}(x, y, z) = \mathbf{V}\phi(x, y, z)$$

Therefore $\mathbf{F}$ is conservative and ϕ is a potential function for $\mathbf{F}$.   ◀

In the above two illustrations it is a simple matter to prove that the vector field is conservative because we know a function ϕ for which $\mathbf{F}$ is its gradient. A more difficult problem is to decide if a given vector field is conservative, and if it is, to find a potential function. We apply Theorems 17.5.1 and 17.5.4 for this purpose, as shown in the following two examples.

EXAMPLE 2   If **F** is the vector field defined by

$$\mathbf{F}(x, y) = \frac{1}{y}\mathbf{i} - \frac{x}{y^2}\mathbf{j}$$

prove that **F** is conservative, and find a potential function for **F**.

Solution   Let

$$M(x, y) = \frac{1}{y} \qquad\qquad N(x, y) = -\frac{x}{y^2}$$

$$M_y(x, y) = -\frac{1}{y^2} \qquad N_x(x, y) = -\frac{1}{y^2}$$

Because $M_y(x, y) = N_x(x, y)$, it follows from Theorem 17.5.1 that **F** is a gradient and hence a conservative vector field. A potential function ϕ satisfies the equation

$$\mathbf{F}(x, y) = \mathbf{\nabla}\phi(x, y)$$

Therefore

$$\phi_x(x, y) = \frac{1}{y} \qquad \phi_y(x, y) = -\frac{x}{y^2} \tag{3}$$

Integrating with respect to x both sides of the first of Equations (3) we have

$$\phi(x, y) = \frac{x}{y} + g(y) \tag{4}$$

where $g(y)$ is independent of x. Partial differentiating both sides of (4) with respect to y gives

$$\phi_y(x, y) = -\frac{x}{y^2} + g'(y)$$

We equate the right sides of this equation and the second of Equations (3) and get

$$-\frac{x}{y^2} + g'(y) = -\frac{x}{y^2}$$

$$g'(y) = 0$$

$$g(y) = C$$

With this value of $g(y)$ in (4) we have

$$\phi(x, y) = \frac{x}{y} + C$$

which is the required potential function.

EXAMPLE 3   If **F** is the vector field defined by

$$\mathbf{F}(x, y, z) = (z^2 + 1)\mathbf{i} + 2yz\mathbf{j} + (2xz + y^2)\mathbf{k}$$

prove that **F** is conservative, and find a potential function for **F**.

Solution   To determine if **F** is conservative we apply Theorem 17.5.4 to ascertain if **F** is a gradient. Let

$$M(x, y, z) = z^2 + 1 \qquad N(x, y, z) = 2yz \qquad R(x, y, z) = 2xz + y^2$$

$$M_y(x, y, z) = 0 \qquad N_x(x, y, z) = 0 \qquad R_x(x, y, z) = 2z$$

$$M_z(x, y, z) = 2z \qquad N_z(x, y, z) = 2y \qquad R_y(x, y, z) = 2y$$

Therefore

$$M_y(x, y, z) = N_x(x, y, z) \quad M_z(x, y, z) = R_x(x, y, z) \quad N_z(x, y, z) = R_y(x, y, z)$$

Thus by Theorem 17.5.4, **F** is a gradient and hence a conservative vector field. We now find a potential function ϕ such that $\mathbf{F}(x, y, z) = \nabla\phi(x, y, z)$; therefore

$$\phi_x(x, y, z) = z^2 + 1 \qquad \phi_y(x, y, z) = 2yz \qquad \phi_z(x, y, z) = 2xz + y^2 \qquad (5)$$

Integrating with respect to x both sides of the first of Equations (5) we have

$$\phi(x, y, z) = xz^2 + x + g(y, z) \tag{6}$$

We partial differentiate with respect to y both sides of (6) and obtain

$$\phi_y(x, y, z) = g_y(y, z)$$

Equating the right sides of this equation and the second of Equations (5) we have

$$g_y(y, z) = 2yz$$

We now integrate with respect to y both sides of this equation and get

$$g(y, z) = y^2z + h(z)$$

Substituting from this equation into (6) we obtain

$$\phi(x, y, z) = xz^2 + x + y^2z + h(z) \tag{7}$$

We partial differentiate with respect to z both sides of (7) and get

$$\phi_z(x, y, z) = 2xz + y^2 + h'(z)$$

We equate the right sides of this equation and the third of Equations (5) and have

$$2xz + y^2 + h'(z) = 2xz + y^2$$

$$h'(z) = 0$$

$$h(z) = C$$

Substituting C for $h(z)$ in (7) we obtain the required potential function ϕ defined by

$$\phi(x, y, z) = xz^2 + x + y^2z + C$$

There are two fields involving derivatives that are associated with a vector field **F**. One is called the *curl* of **F**, which is a vector field, and the other is called the *divergence* of **F**, which is a scalar field. Before giving their definitions, we show how the symbol ∇ is used as an operator. Recall that if f is a scalar function of three variables x, y, and z, the gradient of f is

$$\nabla f(x, y, z) = f_x(x, y, z)\mathbf{i} + f_y(x, y, z)\mathbf{j} + f_z(x, y, z)\mathbf{k} \tag{8}$$

We shall now let the del operator in three dimensions be used to denote

$$\mathbf{i}\frac{\partial}{\partial x} + \mathbf{j}\frac{\partial}{\partial y} + \mathbf{k}\frac{\partial}{\partial z}$$

Therefore, ∇ operating on the scalar function f means

$$\nabla f = \frac{\partial f}{\partial x}\mathbf{i} + \frac{\partial f}{\partial y}\mathbf{j} + \frac{\partial f}{\partial z}\mathbf{k}$$

which agrees with (8).

19.1.1 DEFINITION

Let $\mathbf{F}$ be a vector field on some open ball B in R^3 such that

$$\mathbf{F}(x, y, z) = M(x, y, z)\mathbf{i} + N(x, y, z)\mathbf{j} + R(x, y, z)\mathbf{k}$$

Then the **curl** of $\mathbf{F}$ is defined by

$$\text{curl } \mathbf{F}(x, y, z) = \left(\frac{\partial R}{\partial y} - \frac{\partial N}{\partial z}\right)\mathbf{i} + \left(\frac{\partial M}{\partial z} - \frac{\partial R}{\partial x}\right)\mathbf{j} + \left(\frac{\partial N}{\partial x} - \frac{\partial M}{\partial y}\right)\mathbf{k}$$

if these partial derivatives exist.

A mnemonic device for computing curl $\mathbf{F}$ is to extend the notation for the cross product of two vectors to the "cross product" of the operator ∇ and the vector field $\mathbf{F}$ and write

$$\text{curl } \mathbf{F} = \nabla \times \mathbf{F}$$

$$= \begin{vmatrix} \mathbf{i} & \mathbf{j} & \mathbf{k} \\ \dfrac{\partial}{\partial x} & \dfrac{\partial}{\partial y} & \dfrac{\partial}{\partial z} \\ M & N & R \end{vmatrix}$$

As we indicated when determinant notation was first used for the cross product of two vectors, the elements of the determinant were not all real numbers as is customary. In the above "determinant" the first row contains vectors, the second row contains partial derivative operators, and the third row consists of scalar functions.

EXAMPLE 4   Find curl $\mathbf{F}$ if $\mathbf{F}$ is the vector field defined by

$$\mathbf{F}(x, y, z) = e^{2x}\mathbf{i} + 3x^2yz\mathbf{j} + (2y^2z + x)\mathbf{k}$$

Solution

$$\text{curl } \mathbf{F}(x, y, z) = \begin{vmatrix} \mathbf{i} & \mathbf{j} & \mathbf{k} \\ \dfrac{\partial}{\partial x} & \dfrac{\partial}{\partial y} & \dfrac{\partial}{\partial z} \\ e^{2x} & 3x^2yz & 2y^2z + x \end{vmatrix}$$

$$= (4yz - 3x^2y)\mathbf{i} + (0 - 1)\mathbf{j} + (6xyz - 0)\mathbf{k}$$

$$= (4yz - 3x^2y)\mathbf{i} - \mathbf{j} + 6xyz\mathbf{k}$$

19.1.2 DEFINITION   Let $\mathbf{F}$ be a vector field on some open ball B in R^3 such that

$$\mathbf{F}(x, y, z) = M(x, y, z)\mathbf{i} + N(x, y, z)\mathbf{j} + R(x, y, z)\mathbf{k}$$

Then the **divergence** of $\mathbf{F}$, denoted by div $\mathbf{F}$, is defined by

$$\text{div } \mathbf{F}(x, y, z) = \frac{\partial M}{\partial x} + \frac{\partial N}{\partial y} + \frac{\partial R}{\partial z}$$

if these partial derivatives exist.

We extend the notation for the dot product of two vectors to the "dot product" of the operator $\mathbf{V}$ and the vector field $\mathbf{F}$ to compute the divergence of $\mathbf{F}$, and we write

$$\text{div } \mathbf{F} = \mathbf{V} \cdot \mathbf{F}$$

$$= \left(\mathbf{i}\frac{\partial}{\partial x} + \mathbf{j}\frac{\partial}{\partial y} + \mathbf{k}\frac{\partial}{\partial z} \right) \cdot (M\mathbf{i} + N\mathbf{j} + R\mathbf{k})$$

$$= \frac{\partial M}{\partial x} + \frac{\partial N}{\partial y} + \frac{\partial R}{\partial z}$$

EXAMPLE 5   Find div $\mathbf{F}$ if $\mathbf{F}$ is the vector field of Example 4.

Solution

$$\text{div } \mathbf{F}(x, y, z) = \mathbf{V} \cdot \mathbf{F}(x, y, z)$$

$$= \frac{\partial}{\partial x}(e^{2x}) + \frac{\partial}{\partial y}(3x^2yz) + \frac{\partial}{\partial z}(2y^2z + x)$$

$$= 2e^{2x} + 3x^2z + 2y^2$$

A physical significance of curl $\mathbf{F}$ and div $\mathbf{F}$ in the study of fluid motion will be discussed in Sections 19.4–19.6. In this section we are concerned with learning how to compute them and to prove some properties involving them. Two such properties are given in the following two theorems. You are asked to prove these theorems in Exercises 47 and 48.

19.1.3 THEOREM   Suppose $\mathbf{F}$ is a vector field on an open ball B in R^3 such that

$$\mathbf{F}(x, y, z) = M(x, y, z)\mathbf{i} + N(x, y, z)\mathbf{j} + R(x, y, z)\mathbf{k}$$

If the second partial derivatives of M, N, and R are continuous on B, then

$$\text{div(curl } \mathbf{F}) = 0$$

19.1.4 THEOREM   If f is a scalar field on an open ball B in R^3 and the second partial derivatives of f are continuous on B, then

$$\text{curl}(\mathbf{V}f) = \mathbf{0}$$

The equation in Theorem 19.1.4 states that the curl of the gradient of f equals the zero vector. Consider now the divergence of the gradient of f, that is, $\mathbf{V} \cdot (\mathbf{V}f)$, which can also be written as $\mathbf{V} \cdot \mathbf{V}f$ or $\mathbf{V}^2 f$. By definition,

$$\mathbf{V}^2 f(x, y, z) = \left(\mathbf{i} \frac{\partial}{\partial x} + \mathbf{j} \frac{\partial}{\partial y} + \mathbf{k} \frac{\partial}{\partial z} \right) \cdot \left(\frac{\partial f}{\partial x} \mathbf{i} + \frac{\partial f}{\partial y} \mathbf{j} + \frac{\partial f}{\partial z} \mathbf{k} \right)$$

$$\mathbf{V}^2 f(x, y, z) = \frac{\partial^2 f}{\partial x^2} + \frac{\partial^2 f}{\partial y^2} + \frac{\partial^2 f}{\partial z^2}$$

The expression on the right side of this equation is called the **Laplacian** of f. The following equation obtained by setting the Laplacian equal to zero is called **Laplace's equation**:

$$\frac{\partial^2 f}{\partial x^2} + \frac{\partial^2 f}{\partial y^2} + \frac{\partial^2 f}{\partial z^2} = 0$$

A scalar function satisfying Laplace's equation is said to be **harmonic**. These functions have important applications in physics in the study of heat transfer, electromagnetic radiation, acoustics, and so on.

If $\mathbf{F}$ is a vector field on some open disk B in R^2 such that $\mathbf{F}(x, y) = M(x, y)\mathbf{i} + N(x, y)\mathbf{j}$, then the curl of $\mathbf{F}$ and divergence of $\mathbf{F}$ in two dimensions are defined by

$$\text{curl } \mathbf{F}(x, y) = \left(\frac{\partial N}{\partial x} - \frac{\partial M}{\partial y} \right) \mathbf{k} \qquad \text{div } \mathbf{F}(x, y) = \frac{\partial M}{\partial x} + \frac{\partial N}{\partial y}$$

if these partial derivatives exist. The Laplacian in two dimensions is defined by

$$\mathbf{V}^2 f(x, y) = \frac{\partial^2 f}{\partial x^2} + \frac{\partial^2 f}{\partial y^2}$$

EXAMPLE 6 If $\mathbf{F}(x, y) = 3x^2 y \mathbf{i} - 2xy^3 \mathbf{j}$, find (a) curl $\mathbf{F}(x, y)$ and (b) div $\mathbf{F}(x, y)$.

Solution Since $\mathbf{F}(x, y) = 3x^2 y \mathbf{i} - 2xy^3 \mathbf{j}$, $M(x, y) = 3x^2 y$ and $N(x, y) = -2xy^3$. Therefore,

(a) curl $\mathbf{F}(x, y) = \left(\dfrac{\partial N}{\partial x} - \dfrac{\partial M}{\partial y} \right) \mathbf{k}$ (b) div $\mathbf{F}(x, y) = \dfrac{\partial M}{\partial x} + \dfrac{\partial N}{\partial y}$

$$= (-2y^3 - 3x^2)\mathbf{k} \qquad\qquad\qquad = 6xy - 6xy^2$$

EXERCISES 19.1

In Exercises 1 through 6, show on a figure the representations having initial point at (x, y) of the vectors in the vector field, where x is ± 1 or ± 2 and y is ± 1 or ± 2.

1. $\mathbf{F}(x, y) = x\mathbf{i} - y\mathbf{j}$

2. $\mathbf{F}(x, y) = -x\mathbf{i} + y\mathbf{j}$

3. $\mathbf{F}(x, y) = 4y\mathbf{i} + 3x\mathbf{j}$

4. $\mathbf{F}(x, y) = -3y\mathbf{i} + 4x\mathbf{j}$

5. $\mathbf{F}(x, y) = \dfrac{1}{\sqrt{x^2 + y^2}} (x\mathbf{i} + y\mathbf{j})$

6. $\mathbf{F}(x, y) = y\mathbf{i} + 2\mathbf{j}$

In Exercises 7 through 14, find a conservative vector field having the given potential function.

7. $f(x, y) = 3x^2 + 2y^3$

8. $f(x, y) = 2x^4 - 5x^2 y^2 + 4y^4$

9. $f(x, y) = \tan^{-1} x^2 y$

10. $f(x, y) = ye^x - xe^y$

11. $f(x, y, z) = 2x^3 - 3x^2 y + xy^2 - 4y^3$

12. $f(x, y, z) = \sqrt{x^2 + y^2 + z^2}$

13. $f(x, y, z) = x^2 y e^{-4z}$
14. $f(x, y, z) = z \sin(x^2 - y)$

In Exercises 15 through 20, determine if the vector field is conservative.

15. $\mathbf{F}(x, y) = (3x^2 - 2y^2)\mathbf{i} + (3 - 4xy)\mathbf{j}$
16. $\mathbf{F}(x, y) = (e^x e^y + 6e^{2x})\mathbf{i} + (e^x e^y - 2e^y)\mathbf{j}$
17. $\mathbf{F}(x, y) = y\cos(x + y)\mathbf{i} - x\sin(x + y)\mathbf{j}$
18. $\mathbf{F}(x, y, z) = (3x^2 + 2yz)\mathbf{i} + (2xz + 6yz)\mathbf{j} + (2xy + 3y^2 - 2z)\mathbf{k}$
19. $\mathbf{F}(x, y, z) = (2ye^{2x} + e^z)\mathbf{i} + (3ze^{3y} + e^{2x})\mathbf{j} + (xe^z + e^{3y})\mathbf{k}$
20. $\mathbf{F}(x, y, z) = y\sec^2 x\mathbf{i} + (\tan x - z\sec^2 y)\mathbf{j} + x\sec z\tan z\mathbf{k}$

In Exercises 21 through 32, prove that the vector field is conservative, and find a potential function.

21. $\mathbf{F}(x, y) = y\mathbf{i} + x\mathbf{j}$ **22.** $\mathbf{F}(x, y) = x\mathbf{i} + y\mathbf{j}$
23. $\mathbf{F}(x, y) = e^x \sin y\mathbf{i} + e^x \cos y\mathbf{j}$
24. $\mathbf{F}(x, y) = (\sin y \sinh x + \cos y \cosh x)\mathbf{i}$
$+ (\cos y \cosh x - \sin y \sinh x)\mathbf{j}$
25. $\mathbf{F}(x, y) = (2xy^2 - y^3)\mathbf{i} + (2x^2 y - 3xy^2 + 2)\mathbf{j}$
26. $\mathbf{F}(x, y) = (3x^2 + 2y - y^2 e^x)\mathbf{i} + (2x - 2ye^x)\mathbf{j}$
27. $\mathbf{F}(x, y, z) = (x^2 - y)\mathbf{i} - (x - 3z)\mathbf{j} + (z + 3y)\mathbf{k}$
28. $\mathbf{F}(x, y, z) = yz\mathbf{i} + xz\mathbf{j} + xy\mathbf{k}$
29. $\mathbf{F}(x, y, z) = (ze^x + e^y)\mathbf{i} + (xe^y - e^z)\mathbf{j} + (-ye^z + e^x)\mathbf{k}$
30. $\mathbf{F}(x, y, z) = (\tan y + 2xy\sec z)\mathbf{i} + (x\sec^2 y + x^2 \sec z)\mathbf{j}$
$+ \sec z(x^2 y \tan z - \sec z)\mathbf{k}$

31. $\mathbf{F}(x, y, z) = (2x\cos y - 3)\mathbf{i} - (x^2 \sin y + z^2)\mathbf{j} - (2yz - 2)\mathbf{k}$
32. $\mathbf{F}(x, y, z) = (2y^3 - 8xz^2)\mathbf{i} + (6xy^2 + 1)\mathbf{j} - (8x^2 z + 3z^2)\mathbf{k}$

In Exercises 33 through 42, find curl $\mathbf{F}$ and div $\mathbf{F}$ for the vector field.

33. $\mathbf{F}(x, y) = 2x\mathbf{i} + 3y\mathbf{j}$ **34.** $\mathbf{F}(x, y) = \cos x\mathbf{i} - \sin y\mathbf{j}$
35. $\mathbf{F}(x, y) = e^x \cos y\mathbf{i} + e^x \sin y\mathbf{j}$
36. $\mathbf{F}(x, y) = -\dfrac{y}{x}\mathbf{i} + \dfrac{1}{x}\mathbf{j}$
37. $\mathbf{F}(x, y, z) = x^2\mathbf{i} + y^2\mathbf{j} + z^2\mathbf{k}$
38. $\mathbf{F}(x, y, z) = xz^2\mathbf{i} + y^2\mathbf{j} + x^2 z\mathbf{k}$
39. $\mathbf{F}(x, y, z) = \cos y\mathbf{i} + \cos z\mathbf{j} + \cos x\mathbf{k}$
40. $\mathbf{F}(x, y, z) = (y^2 + z^2)\mathbf{i} + xe^y \cos z\mathbf{j} - xe^y \cos z\mathbf{k}$
41. $\mathbf{F}(x, y, z) = \sqrt{x^2 + y^2 + 1}\mathbf{i} + \sqrt{x^2 + y^2 + 1}\mathbf{j} + z^2\mathbf{k}$
42. $\mathbf{F}(x, y, z) = \dfrac{x}{(x^2 + y^2)^{3/2}}\mathbf{i} + \dfrac{y}{(x^2 + y^2)^{3/2}}\mathbf{j} + \mathbf{k}$

In Exercises 43 through 46, prove that the scalar function is harmonic by showing that its Laplacian is zero.

43. $f(x, y) = e^y \sin x + e^x \cos y$
44. $f(x, y) = \ln(\sqrt{x^2 + y^2})$
45. $f(x, y, z) = 2x^2 + 3y^2 - 5z^2$
46. $f(x, y, z) = (x^2 + y^2 + z^2)^{-1/2}$
47. Prove Theorem 19.1.3.
48. Prove Theorem 19.1.4.

19.2 LINE INTEGRALS

In Chapter 5 the geometric concept of area was used to motivate the definition of the definite integral. To motivate the definition of an integral of a vector field, we use the physical concept of work.

We showed in Section 14.2 that if a constant force of vector measure $\mathbf{F}$ moves a particle along a straight line from a point A to a point B, then if $\mathbf{W}$ is the measure of the work done,

$$\mathbf{W} = \mathbf{F} \cdot \mathbf{V}(\overrightarrow{AB}) \tag{1}$$

Suppose now that the force vector is not constant, and instead of the motion being along a straight line, it is along a curve. Let the force that is exerted on the particle at the point (x, y) in some open disk B in R^2 be given by the force field

$$\mathbf{F}(x, y) = M(x, y)\mathbf{i} + N(x, y)\mathbf{j}$$

where M and N are continuous on B. Let C be a curve lying in B and having the vector equation

$$\mathbf{R}(t) = f(t)\mathbf{i} + g(t)\mathbf{j} \qquad a \le t \le b$$

We require the functions f and g be such that f' and g' are continuous on $[a, b]$. We wish to define the work done by the variable force of vector measure $\mathbf{F}$ in moving the particle along C from the point $(f(a), g(a))$ to $(f(b), g(b))$.

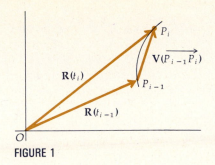

FIGURE 1

At a point $(f(t), g(t))$ on C the force vector is

$$\mathbf{F}(f(t), g(t)) = M(f(t), g(t))\mathbf{i} + N(f(t), g(t))\mathbf{j} \qquad (2)$$

Let Δ be a partition of the interval $[a, b]$:

$$a = t_0 < t_1 < t_2 < \ldots < t_{n-1} < t_n = b$$

On C let P_i be the point $(x_i, y_i) = (f(t_i), g(t_i))$. Refer to Figure 1. The vector $\mathbf{V}(\overrightarrow{P_{i-1}P_i}) = \mathbf{R}(t_i) - \mathbf{R}(t_{i-1})$; therefore

$$\mathbf{V}(\overrightarrow{P_{i-1}P_i}) = f(t_i)\mathbf{i} + g(t_i)\mathbf{j} - [f(t_{i-1})\mathbf{i} + g(t_{i-1})\mathbf{j}]$$

$$\mathbf{V}(\overrightarrow{P_{i-1}P_i}) = [f(t_i) - f(t_{i-1})]\mathbf{i} + [g(t_i) - g(t_{i-1})]\mathbf{j} \qquad (3)$$

Because f' and g' are continuous on $[a, b]$, it follows from the mean-value theorem that there are numbers c_i and d_i in the open interval (t_{i-1}, t_i) such that

$$f(t_i) - f(t_{i-1}) = f'(c_i)(t_i - t_{i-1})$$

$$g(t_i) - g(t_{i-1}) = g'(d_i)(t_i - t_{i-1})$$

Letting $\Delta_i t = t_i - t_{i-1}$, and substituting from the above two equations into (3), we obtain

$$\mathbf{V}(\overrightarrow{P_{i-1}P_i}) = [f'(c_i)\mathbf{i} + g'(d_i)\mathbf{j}]\,\Delta_i t \qquad (4)$$

For each i consider the vector

$$\mathbf{F}_i = M(f(c_i), g(c_i))\mathbf{i} + N(f(d_i), g(d_i))\mathbf{j} \qquad (5)$$

Each of the vectors $\mathbf{F}_i$ $(i = 1, 2, \ldots, n)$ is an approximation to the force vector $\mathbf{F}(f(t), g(t))$, given by (2), along the arc of C from P_{i-1} to P_i. Observe that even though c_i and d_i are in general different numbers in the open interval (t_{i-1}, t_i), the values of the vectors $\mathbf{F}(f(t), g(t))$ are close to the vector $\mathbf{F}_i$. Furthermore, we approximate the arc of C from P_{i-1} to P_i by the line segment $\overrightarrow{P_{i-1}P_i}$. Thus we apply formula (1) and obtain an approximation for the work done by the vector $\mathbf{F}(f(t), g(t))$ in moving a particle along the arc of C from P_{i-1} to P_i. Denoting this approximation by $\Delta_i W$, we have, from formula (1) and Equations (5) and (4)

$$\Delta_i W = [M(f(c_i), g(c_i))\mathbf{i} + N(f(d_i), g(d_i))\mathbf{j}] \cdot [f'(c_i)\mathbf{i} + g'(d_i)\mathbf{j}]\,\Delta_i t$$

$$\Leftrightarrow \quad \Delta_i W = [M(f(c_i), g(c_i))f'(c_i)]\,\Delta_i t + [N(f(d_i), g(d_i))g'(d_i)]\,\Delta_i t$$

An approximation of the measure of the work done by $F(f(t), g(t))$ along C is $\sum_{i=1}^{n} \Delta_i W$ or, equivalently,

$$\sum_{i=1}^{n} [M(f(c_i), g(c_i))f'(c_i)]\,\Delta_i t + \sum_{i=1}^{n} [N(f(d_i), g(d_i))g'(d_i)]\,\Delta_i t$$

Each of these sums is a Riemann sum. The first is a Riemann sum for the function having values $M(f(t), g(t))f'(t)$, and the second is a Riemann sum for the function having values $N(f(t), g(t))g'(t)$. If n increases without bound, these two sums approach the definite integral:

$$\int_a^b [M(f(t), g(t))f'(t) + N(f(t), g(t))g'(t)]\,dt$$

We therefore have the following definition. In the definition we use the notation $\mathbf{F}(\mathbf{R}(t))$ in place of $\mathbf{F}(f(t), g(t))$.

19.2.1 DEFINITION Let C be a curve lying in an open disk B in R^2 for which a vector equation of C is $\mathbf{R}(t) = f(t)\mathbf{i} + g(t)\mathbf{j}$, where f' and g' are continuous on $[a, b]$. Furthermore, let a force field on B be defined by $\mathbf{F}(x, y) = M(x, y)\mathbf{i} + N(x, y)\mathbf{j}$, where M and N are continuous on B. Then if W is the measure of the **work** done by a force of vector measure $\mathbf{F}$ in moving a particle along C from $(f(a), g(a))$ to $(f(b), g(b))$,

$$W = \int_a^b \left[M(f(t), g(t))f'(t) + N(f(t), g(t))g'(t) \right] dt \tag{6}$$

or, equivalently, by using vector notation,

$$W = \int_a^b \langle M(f(t), g(t)), N(f(t), g(t)) \rangle \cdot \langle f'(t), g'(t) \rangle \, dt$$

$$\Leftrightarrow \quad W = \int_a^b \mathbf{F}(\mathbf{R}(t)) \cdot \mathbf{R}'(t) \, dt \tag{7}$$

EXAMPLE 1 Suppose a particle moves along the parabola $y = x^2$ from the point $(-1, 1)$ to the point $(2, 4)$. Find the total work done if the motion is caused by the force field $\mathbf{F}(x, y) = (x^2 + y^2)\mathbf{i} + 3x^2 y\mathbf{j}$. Assume the arc is measured in meters and the force is measured in newtons.

Solution Parametric equations of the parabola are

$$x = t \quad \text{and} \quad y = t^2 \qquad -1 \leq t \leq 2$$

Thus a vector equation of the parabola is

$$\mathbf{R}(t) = t\mathbf{i} + t^2\mathbf{j} \quad \text{and} \quad \mathbf{R}'(t) = \mathbf{i} + 2t\mathbf{j}$$

Because $\mathbf{F}(x, y) = \langle x^2 + y^2, 3x^2 y \rangle$, then

$$\mathbf{F}(\mathbf{R}(t)) = \mathbf{F}(t, t^2)$$
$$= \langle t^2 + t^4, 3t^4 \rangle$$

If W joules is the work done, then, from (7),

$$W = \int_{-1}^2 \mathbf{F}(\mathbf{R}(t)) \cdot \mathbf{R}'(t) \, dt \tag{8}$$

$$= \int_{-1}^2 \langle t^2 + t^4, 3t^4 \rangle \cdot \langle 1, 2t \rangle \, dt$$

$$= \int_{-1}^2 (t^2 + t^4 + 6t^5) \, dt$$

$$= \frac{t^3}{3} + \frac{t^5}{5} + t^6 \Big]_{-1}^2$$

$$= \tfrac{8}{3} + \tfrac{32}{5} + 64 - (-\tfrac{1}{3} - \tfrac{1}{5} + 1)$$

$$= \tfrac{363}{5}$$

Therefore the work done is $\frac{363}{5}$ joules.

The integrals in Equations (6) and (7) are called *line integrals*. For the line integral of Equation (6), a common notation involving the differential form $M(x, y)\, dx + N(x, y)\, dy$ is

$$\int_C M(x, y)\, dx + N(x, y)\, dy$$

This notation is suggested by the fact that because parametric equations of C are $x = f(t)$ and $y = g(t)$, then $dx = f'(t) \, dt$ and $dy = g'(t) \, dt$. A vector notation for the line integral of Equation (7) is

$$\int_C \mathbf{F} \cdot d\mathbf{R}$$

This notation is suggested by considering the vector equation of C, which is $\mathbf{R}(t) = f(t)\mathbf{i} + g(t)\mathbf{j}$, and letting $d\mathbf{R} = \mathbf{R}'(t) \, dt$. Then

$$\mathbf{F}(\mathbf{R}(t)) \cdot d\mathbf{R} = \mathbf{F}(\mathbf{R}(t)) \cdot \mathbf{R}'(t) \, dt$$

We have, then, the following formal definition.

19.2.2 DEFINITION Let C be a curve lying in an open disk B in R^2 and having the vector equation

$$\mathbf{R}(t) = f(t)\mathbf{i} + g(t)\mathbf{j} \qquad a \le t \le b$$

such that f' and g' are continuous on $[a, b]$. Let $\mathbf{F}$ be a vector field on B defined by

$$\mathbf{F}(x, y) = M(x, y)\mathbf{i} + N(x, y)\mathbf{j}$$

where M and N are continuous on B. Then by using differential form notation the **line integral** of $M(x, y) \, dx + N(x, y) \, dy$ over C is given by

$$\int_C M(x, y) \, dx + N(x, y) \, dy = \int_a^b \left[M(f(t), g(t))f'(t) + N(f(t), g(t))g'(t) \right] dt$$

or, equivalently, by using vector notation, the **line integral** of $\mathbf{F}$ over C is given by

$$\int_C \mathbf{F} \cdot d\mathbf{R} = \int_C \mathbf{F}(\mathbf{R}(t)) \cdot \mathbf{R}'(t) \, dt$$

Both the differential form and vector notations are used for line integrals. Consequently, we shall apply the two notations.

▶ **ILLUSTRATION 1**   In Example 1, the integral in Equation (8) that defines W is a line integral. With vector notation, this line integral can be denoted by

$$\int_C \mathbf{F} \cdot d\mathbf{R}$$

where $\mathbf{F}(x, y) = (x^2 + y^2)\mathbf{i} + 3x^2y\mathbf{j}$ and $\mathbf{R}(t) = t\mathbf{i} + t^2\mathbf{j}$. With differential form notation this line integral is written as

$$\int_C (x^2 + y^2) \, dx + 3x^2y \, dy \tag{9}$$

◀

If an equation of C is of the form $y = F(x)$, then x may be used as a parameter in place of t. In a similar manner, if an equation of C is of the form $x = G(y)$, then y may be used as a parameter in place of t.

▶ **ILLUSTRATION 2**   In Example 1 and Illustration 1, the equation of C is $y = x^2$, which is of the form $y = F(x)$. Therefore we can use x as a parameter instead of t. Thus, in integral (9) of Illustration 1 we can replace y by x^2 and

dy by $2x\,dx$, and we have

$$W = \int_{-1}^{2} (x^2 + x^4)\,dx + 3x^2 x^2 (2x\,dx)$$

$$= \int_{-1}^{2} (x^2 + x^4 + 6x^5)\,dx$$

This integral is the same as the third one appearing in the solution of Example 1, except that the variable is x instead of t. ◀

If the curve C in the definition of the line integral is the closed interval $[a, b]$ on the x axis, then $y = 0$ and $dy = 0$. Thus

$$\int_C M(x, y)\,dx + N(x, y)\,dy = \int_a^b M(x, 0)\,dx$$

Therefore, in such a case, the line integral reduces to a definite integral.
In the definition of a line integral, if C has the vector equation

$$\mathbf{R}(t) = f(t)\mathbf{i} + g(t)\mathbf{j} \qquad a \le t \le b$$

we required that f' and g' be continuous on $[a, b]$. If, in addition to this continuity requirement, $f'(t)$ and $g'(t)$ are not both zero at every point in the open interval (a, b), then C is said to be **smooth** on $[a, b]$. If an interval I can be partitioned into a finite number of subintervals on which C is smooth, then C is said to be **sectionally smooth** on I.

▶ **ILLUSTRATION 3**   (a) The curve defined by the vector equation

$$\mathbf{R}(t) = 2\cos t\,\mathbf{i} + 2\sin t\,\mathbf{j} \qquad 0 \le t \le 2\pi$$

is a circle with center at the origin and radius 2. See Figure 2. For this curve,

$$f(t) = 2\cos t \qquad g(t) = 2\sin t$$

$$f'(t) = -2\sin t \qquad g'(t) = 2\cos t$$

Because f' and g' are continuous for all t and $f'(t)$ and $g'(t)$ are not both zero anywhere, the circle is a smooth curve.
(b) The curve defined by the vector equation

$$\mathbf{R}(t) = a(t - \sin t)\mathbf{i} + a(1 - \cos t)\mathbf{j} \qquad t \in (-\infty, +\infty)$$

is a cycloid. This curve was discussed in Section 14.3, and a sketch of its graph on the interval $[-2\pi a, 4\pi a]$ is shown in Figure 3. For the cycloid,

$$f(t) = a(t - \sin t) \qquad g(t) = a(1 - \cos t)$$

$$f'(t) = a(1 - \cos t) \qquad g'(t) = a\sin t$$

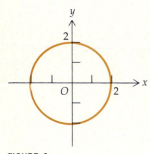

FIGURE 2

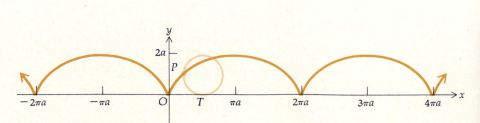

FIGURE 3

The functions f' and g' are continuous for all t, but $f'(t)$ and $g'(t)$ are both zero if $t = 2\pi n$, where n is any integer. Therefore the cycloid is not smooth. However, the cycloid is sectionally smooth, because it is smooth in each subinterval $[2\pi n, 2\pi(n + 1)]$, where n is any integer. ◀

Figures 4 and 5 show two other sectionally smooth curves. We can extend the concept of a line integral to include curves that are sectionally smooth.

19.2.3 DEFINITION   Let the curve C consist of the smooth arcs $C_1, C_2, \ldots, C_n$. Then the **line integral** of $M(x, y)\, dx + N(x, y)\, dy$ over C is defined by

$$\int_C M(x, y)\, dx + N(x, y)\, dy = \sum_{i=1}^{n} \left(\int_{C_i} M(x, y)\, dx + N(x, y)\, dy \right)$$

or, equivalently, by using vector notation, the line integral of **F** over C is defined by

$$\int_C \mathbf{F} \cdot d\mathbf{R} = \sum_{i=1}^{n} \left(\int_{C_i} \mathbf{F}(\mathbf{R}(t)) \cdot \mathbf{R}'(t)\, dt \right)$$

FIGURE 4

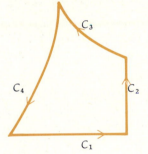

FIGURE 5

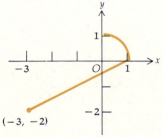

FIGURE 6

EXAMPLE 2   Evaluate the line integral

$$\int_C 4xy\, dx + (2x^2 - 3xy)\, dy$$

if the curve C consists of the line segment from $(-3, -2)$ to $(1, 0)$ and the first quadrant arc of the circle $x^2 + y^2 = 1$ from $(1, 0)$ to $(0, 1)$, traversed in the counterclockwise direction.

Solution   Figure 6 shows the curve C composed of arcs C_1 and C_2. The arc C_1 is the line segment. An equation of the line through $(-3, -2)$ and $(1, 0)$ is $x - 2y = 1$. Therefore C_1 can be represented parametrically by

$$x = 1 + 2t \qquad y = t \qquad -2 \le t \le 0$$

The arc C_2, which is the first quadrant arc of the circle $x^2 + y^2 = 1$, can be represented parametrically by

$$x = \cos t \qquad y = \sin t \qquad 0 \le t \le \tfrac{1}{2}\pi$$

Applying Definition 19.2.2 for each of the arcs C_1 and C_2 we have

$$\int_{C_1} 4xy\, dx + (2x^2 - 3xy)\, dy$$

$$= \int_{-2}^{0} 4(1 + 2t)t(2\, dt) + \left[2(1 + 2t)^2 - 3(1 + 2t)t \right] dt$$

$$= \int_{-2}^{0} (8t + 16t^2 + 2 + 8t + 8t^2 - 3t - 6t^2)\, dt$$

$$= \int_{-2}^{0} (18t^2 + 13t + 2)\, dt$$

$$= 6t^3 + \tfrac{13}{2}t^2 + 2t \Big]_{-2}^{0}$$

$$= -(-48 + 26 - 4)$$

$$= 26$$

and

$$\int_{C_2} 4xy \, dx + (2x^2 - 3xy) \, dy$$

$$= \int_0^{\pi/2} 4 \cos t \sin t(-\sin t \, dt) + [2 \cos^2 t - 3 \cos t \sin t](\cos t \, dt)$$

$$= \int_0^{\pi/2} (-4 \cos t \sin^2 t + 2 \cos^3 t - 3 \cos^2 t \sin t) \, dt$$

$$= \int_0^{\pi/2} [-4 \cos t \sin^2 t + 2 \cos t(1 - \sin^2 t) - 3 \cos^2 t \sin t] \, dt$$

$$= \int_0^{\pi/2} (2 \cos t - 6 \cos t \sin^2 t - 3 \cos^2 t \sin t) \, dt$$

$$= 2 \sin t - 2 \sin^3 t + \cos^3 t \Big]_0^{\pi/2}$$

$$= 2 - 2 - 1$$

$$= -1$$

Therefore, from Definition 19.2.3,

$$\int_C 4xy \, dx + (2x^2 - 3xy) \, dy = 26 + (-1)$$
$$= 25$$

We now extend the concept of a line integral to three dimensions.

19.2.4 DEFINITION   Let C be a curve lying in an open ball B in R^3 and having the vector equation

$$\mathbf{R}(t) = f(t)\mathbf{i} + g(t)\mathbf{j} + h(t)\mathbf{k} \qquad a \le t \le b$$

such that f', g', and h' are continuous on $[a, b]$. Let $\mathbf{F}$ be a vector field on B defined by

$$\mathbf{F}(x, y, z) = M(x, y, z)\mathbf{i} + N(x, y, z)\mathbf{j} + R(x, y, z)\mathbf{k}$$

where M, N, and R are continuous on B. Then by using differential form notation the **line integral** of $M(x, y, z) \, dx + N(x, y, z) \, dy + R(x, y, z) \, dz$ over C is given by

$$\int_C M(x, y, z) \, dx + N(x, y, z) \, dy + R(x, y, z) \, dz$$

$$= \int_a^b [M(f(t), g(t), h(t))f'(t) + N(f(t), g(t), h(t))g'(t) + R(f(t), g(t), h(t))h'(t)] \, dt$$

or, equivalently, by using vector notation, the **line integral** of $\mathbf{F}$ over C is given by

$$\int_C \mathbf{F} \cdot d\mathbf{R} = \int_C \mathbf{F}(\mathbf{R}(t)) \cdot \mathbf{R}'(t) \, dt$$

We can define the work done by a force field in moving a particle along a curve in R^3 as a line integral just as we did in Definition 19.2.1 for a curve in R^2. Such a definition is applied in the following example.

EXAMPLE 3   A particle traverses the twisted cubic

$$\mathbf{R}(t) = t\mathbf{i} + t^2\mathbf{j} + t^3\mathbf{k} \qquad 0 \le t \le 1$$

Find the total work done if the motion is caused by the force field

$$\mathbf{F}(x, y, z) = e^x\mathbf{i} + xe^z\mathbf{j} + x \sin \pi y^2\mathbf{k}$$

Assume that the arc is measured in meters and the force is measured in newtons.

Solution

$$\mathbf{R}(t) = t\mathbf{i} + t^2\mathbf{j} + t^3\mathbf{k} \qquad \mathbf{R}'(t) = \mathbf{i} + 2t\mathbf{j} + 3t^2\mathbf{k}$$

Because $\mathbf{F}(x, y, z) = \langle e^x, xe^z, x \sin \pi y^2\rangle$, then

$$\mathbf{F}(\mathbf{R}(t)) = \mathbf{F}(t, t^2, t^3)$$
$$= \langle e^t, te^{t^3}, t \sin \pi t^4\rangle$$

If W joules is the work done, then from the vector notation for a line integral in Definition 19.2.4

$$W = \int_C \mathbf{F} \cdot d\mathbf{R}$$

$$= \int_0^1 \mathbf{F}(\mathbf{R}(t)) \cdot \mathbf{R}'(t)\, dt$$

$$= \int_0^1 \langle e^t, te^{t^3}, t \sin \pi t^4\rangle \cdot \langle 1, 2t, 3t^2\rangle\, dt$$

$$= \int_0^1 (e^t + 2t^2e^{t^3} + 3t^3 \sin \pi t^4)\, dt$$

$$= e^t + \frac{2}{3}e^{t^3} - \frac{3}{4\pi} \cos \pi t^4 \Big]_0^1$$

$$= e + \frac{2}{3}e - \frac{3}{4\pi} \cos \pi - 1 - \frac{2}{3} + \frac{3}{4\pi} \cos 0$$

$$= \frac{5}{3}e + \frac{3}{2\pi} - \frac{5}{3}$$

Therefore the work done is $\left[\dfrac{5}{3}(e - 1) + \dfrac{3}{2\pi}\right]$ joules.

EXAMPLE 4   Evaluate the line integral

$$\int_C 3x\, dx + 2xy\, dy + z\, dz$$

if the curve C is the circular helix defined by the parametric equations

$$x = \cos t \qquad y = \sin t \qquad z = t \qquad 0 \le t \le 2\pi$$

Solution   From the differential form notation for a line integral in Definition 19.2.4

$$\int_C 3x\, dx + 2xy\, dy + z\, dz$$

$$= \int_0^{2\pi} 3 \cos t(-\sin t\, dt) + 2(\cos t)(\sin t)(\cos t\, dt) + t\, dt$$

$$= \int_0^{2\pi} (-3 \sin t \cos t + 2 \cos^2 t \sin t + t)\, dt$$

$$= -\tfrac{3}{2} \sin^2 t - \tfrac{2}{3} \cos^3 t + \tfrac{1}{2}t^2 \Big]_0^{2\pi}$$

$$= -\tfrac{3}{2}(0) - \tfrac{2}{3}(1) + \tfrac{1}{2}(4\pi^2) + \tfrac{3}{2}(0) + \tfrac{2}{3}(1) + \tfrac{1}{2}(0)$$

$$= 2\pi^2$$

EXERCISES 19.2

In Exercises 1 through 22, evaluate the line integral over the curve C.

1. $\int_C \mathbf{F} \cdot d\mathbf{R}$; $\mathbf{F}(x, y) = y\mathbf{i} + x\mathbf{j}$; C: $\mathbf{R}(t) = t\mathbf{i} + t^2\mathbf{j}$, $0 \leq t \leq 1$.

2. $\int_C \mathbf{F} \cdot d\mathbf{R}$; $\mathbf{F}(x, y) = 2xy\mathbf{i} - 3x\mathbf{j}$; C: $\mathbf{R}(t) = 3t^2\mathbf{i} - t\mathbf{j}$, $0 \leq t \leq 1$.

3. $\int_C \mathbf{F} \cdot d\mathbf{R}$; $\mathbf{F}(x, y) = 2xy\mathbf{i} + (x - 2y)\mathbf{j}$;
C: $\mathbf{R}(t) = \sin t\mathbf{i} - 2 \cos t\mathbf{j}$, $0 \leq t \leq \pi$.

4. $\int_C \mathbf{F} \cdot d\mathbf{R}$; $\mathbf{F}(x, y) = xy\mathbf{i} - y^2\mathbf{j}$; C: $\mathbf{R}(t) = t^2\mathbf{i} + t^3\mathbf{j}$, from the point $(1, 1)$ to the point $(4, -8)$.

5. $\int_C \mathbf{F} \cdot d\mathbf{R}$; $\mathbf{F}(x, y) = (x - y)\mathbf{i} + (y + x)\mathbf{j}$; C: the circle $x^2 + y^2 = 4$ from the point $(2, 0)$ in the counterclockwise direction.

6. $\int_C \mathbf{F} \cdot d\mathbf{R}$; $\mathbf{F}(x, y) = (x - 2y)\mathbf{i} + xy\mathbf{j}$;
C: $\mathbf{R}(t) = 3 \cos t\mathbf{i} + 2 \sin t\mathbf{j}$, $0 \leq t \leq \frac{1}{2}\pi$.

7. $\int_C \mathbf{F} \cdot d\mathbf{R}$; $\mathbf{F}(x, y) = y \sin x\mathbf{i} - \cos x\mathbf{j}$; C: the line segment from $(\frac{1}{2}\pi, 0)$ to $(\pi, 1)$.

8. $\int_C \mathbf{F} \cdot d\mathbf{R}$; $\mathbf{F}(x, y) = 9x^2y\mathbf{i} + (5x^2 - y)\mathbf{j}$;
C: the curve $y = x^3 + 1$ from $(1, 2)$ to $(3, 28)$.

9. $\int_C (x^2 + xy) \, dx + (y^2 - xy) \, dy$; C: the line $y = x$ from the origin to the point $(2, 2)$.

10. The line integral of Exercise 9; C: the parabola $x^2 = 2y$ from the origin to the point $(2, 2)$.

11. The line integral of Exercise 9; C: the x axis from the origin to $(2, 0)$ and then the line $x = 2$ from $(2, 0)$ to $(2, 2)$.

12. $\int_C yx^2 \, dx + (x + y) \, dy$; C: the line $y = -x$ from the origin to the point $(1, -1)$.

13. The line integral of Exercise 12; C: the curve $y = -x^3$ from the origin to the point $(1, -1)$.

14. The line integral of Exercise 12; C: the y axis from the origin to $(0, -1)$ and then the line $y = -1$ from $(0, -1)$ to $(1, -1)$.

15. $\int_C 3xy \, dx + (4x^2 - 3y) \, dy$; C: the line $y = 2x + 3$ from $(0, 3)$ to $(3, 9)$ and then the parabola $y = x^2$ from $(3, 9)$ to $(5, 25)$.

16. $\int_C (xy - z) \, dx + e^x \, dy + y \, dz$; C: the line segment from $(1, 0, 0)$ to $(3, 4, 8)$.

17. $\int_C (x + y) \, dx + (y + z) \, dy + (x + z) \, dz$; C: the line segment from the origin to the point $(1, 2, 4)$.

18. The line integral of Exercise 16; C: $\mathbf{R}(t) = (t + 1)\mathbf{i} + t^2\mathbf{j} + t^3\mathbf{k}$, $0 \leq t \leq 2$.

19. $\int_C \mathbf{F} \cdot d\mathbf{R}$; $\mathbf{F}(x, y, z) = z\mathbf{i} + x\mathbf{j} + y\mathbf{k}$; C: the circular helix $\mathbf{R}(t) = a \cos t\mathbf{i} + a \sin t\mathbf{j} + t\mathbf{k}$, $0 \leq t \leq 2\pi$.

20. $\int_C \mathbf{F} \cdot d\mathbf{R}$; $\mathbf{F}(x, y, z) = 2xy\mathbf{i} + (6y^2 - xz)\mathbf{j} + 10z\mathbf{k}$;
C: the twisted cubic $\mathbf{R}(t) = t\mathbf{i} + t^2\mathbf{j} + t^3\mathbf{k}$, $0 \leq t \leq 1$.

21. The line integral of Exercise 20; C: the line segment from the origin to the point $(0, 0, 1)$; then the line segment from $(0, 0, 1)$ to $(0, 1, 1)$; then the line segment from $(0, 1, 1)$ to $(1, 1, 1)$.

22. The line integral of Exercise 20; C: the line segment from the origin to the point $(1, 1, 1)$.

In Exercises 23 through 36, find the total work done in moving a particle along arc C if the motion is caused by the force field F. Assume the arc is measured in meters and the force is measured in newtons.

23. $\mathbf{F}(x, y) = 2xy\mathbf{i} + (x^2 + y^2)\mathbf{j}$; C: the line segment from the origin to the point $(1, 1)$.

24. The force field of Exercise 23; C: the arc of the parabola $y^2 = x$ from the origin to the point $(1, 1)$.

25. $\mathbf{F}(x, y) = (y - x)\mathbf{i} + x^2y\mathbf{j}$; C: the line segment from the point $(1, 1)$ to $(2, 4)$.

26. The force field of Exercise 25; C: the arc of the parabola $y = x^2$ from the point $(1, 1)$ to $(2, 4)$.

27. The force field of Exercise 25; C: the line segment from $(1, 1)$ to $(2, 2)$ and then the line segment from $(2, 2)$ to $(2, 4)$.

28. $\mathbf{F}(x, y) = -x^2y\mathbf{i} + 2y\mathbf{j}$; C: the line segment from $(a, 0)$ to $(0, a)$.

29. The force field of Exercise 28; C: $\mathbf{R}(t) = a \cos t\mathbf{i} + a \sin t\mathbf{j}$, $0 \leq t \leq \frac{1}{2}\pi$.

30. The force field of Exercise 28; C: the line segment from $(a, 0)$ to (a, a) and then the line segment from (a, a) to $(0, a)$.

31. $\mathbf{F}(x, y, z) = (y + z)\mathbf{i} + (x + z)\mathbf{j} + (x + y)\mathbf{k}$; C: the line segment from the origin to the point $(1, 1, 1)$.

32. $\mathbf{F}(x, y, z) = z^2\mathbf{i} + y^2\mathbf{j} + xz\mathbf{k}$; C: the line segment from the origin to the point $(4, 0, 3)$.

33. $\mathbf{F}(x, y, z) = e^x\mathbf{i} + e^y\mathbf{j} + e^z\mathbf{k}$;
C: $\mathbf{R}(t) = t\mathbf{i} + t^2\mathbf{j} + t^3\mathbf{k}$, $0 \leq t \leq 2$.

34. $\mathbf{F}(x, y, z) = (xyz + x)\mathbf{i} + (x^2z + y)\mathbf{j} + (x^2y + z)\mathbf{k}$;
C: the arc of Exercise 33.

35. The force field of Exercise 34; C: the line segment from the origin to the point $(1, 0, 0)$; then the line segment from $(1, 0, 0)$ to $(1, 1, 0)$; then the line segment from $(1, 1, 0)$ to $(1, 1, 1)$.

36. $\mathbf{F}(x, y, z) = x\mathbf{i} + y\mathbf{j} + (yz - x)\mathbf{k}$; C: $\mathbf{R}(t) = 2t\mathbf{i} + t^2\mathbf{j} + 4t^3\mathbf{k}$, $0 \leq t \leq 1$.

19.3 LINE INTEGRALS INDEPENDENT OF THE PATH

We learned in Section 19.2 that the value of a line integral is determined by the integrand and a curve C between two points P_1 and P_2. However, under certain conditions the value of a line integral depends only on the integrand

and the points P_1 and P_2 and not on the path from P_1 to P_2. Such a line integral is said to be **independent of the path**.

▶ **ILLUSTRATION 1** Suppose a force field

$$\mathbf{F}(x, y) = (y^2 + 2x + 4)\mathbf{i} + (2xy + 4y - 5)\mathbf{j}$$

moves a particle from the origin to the point $(1, 1)$. We show that the total work done is the same if the path is along (a) the line segment from the origin to $(1, 1)$; (b) the segment of the parabola $y = x^2$ from the origin to $(1, 1)$; and (c) the segment of the curve $x = y^3$ from the origin to $(1, 1)$.

If W is the measure of the work done, then

$$W = \int_C (y^2 + 2x + 4)\, dx + (2xy + 4y - 5)\, dy \tag{1}$$

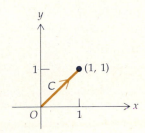

FIGURE 1

(a) See Figure 1. An equation of C is $y = x$. We use x as the parameter and let $y = x$ and $dy = dx$ in (1). Then

$$W = \int_0^1 (x^2 + 2x + 4)\, dx + (2x^2 + 4x - 5)\, dx$$

$$= \int_0^1 (3x^2 + 6x - 1)\, dx$$

$$= x^3 + 3x^2 - x \Big]_0^1$$

$$= 3$$

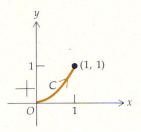

FIGURE 2

(b) See Figure 2. An equation of C is $y = x^2$. Again taking x as the parameter and in (1) letting $y = x^2$ and $dy = 2x\, dx$, we have

$$W = \int_0^1 (x^4 + 2x + 4)\, dx + (2x^3 + 4x^2 - 5)2x\, dx$$

$$= \int_0^1 (5x^4 + 8x^3 - 8x + 4)\, dx$$

$$= x^5 + 2x^4 - 4x^2 + 4x \Big]_0^1$$

$$= 3$$

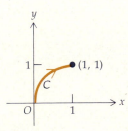

FIGURE 3

(c) See Figure 3. An equation of C is $x = y^3$. We take y as the parameter and in (1) let $x = y^3$ and $dx = 3y^2\, dy$. Then

$$W = \int_0^1 (y^2 + 2y^3 + 4)3y^2\, dy + (2y^4 + 4y - 5)\, dy$$

$$= \int_0^1 (6y^5 + 5y^4 + 12y^2 + 4y - 5)\, dy$$

$$= y^6 + y^5 + 4y^3 + 2y^2 - 5y \Big]_0^1$$

$$= 3 \qquad\qquad ◀$$

In Illustration 1 we see that the value of the line integral is the same over three different paths from $(0, 0)$ to $(1, 1)$. Actually the value of the line integral is the same over any sectionally smooth curve from the origin to $(1, 1)$; so this line integral is independent of the path. (This fact is proved in Illustration 2.)

We now state and prove a theorem which not only gives conditions for which the value of a line integral is independent of the path but also gives a formula for finding the value of such a line integral.

19.3.1 THEOREM   Let C be any sectionally smooth curve lying in an open disk B in R^2 from the point (x_1, y_1) to the point (x_2, y_2). If $\mathbf{F}$ is a conservative vector field continuous on B and ϕ is a potential function for $\mathbf{F}$, then the line integral

$$\int_C \mathbf{F} \cdot d\mathbf{R}$$

is independent of the path C, and

$$\int_C \mathbf{F} \cdot d\mathbf{R} = \phi(x_2, y_2) - \phi(x_1, y_1)$$

Proof   We give the proof if C is smooth. If C is only sectionally smooth, then consider each piece separately; the following proof applies to each smooth piece.

Let parametric equations of C be

$$x = f(t) \qquad y = g(t) \qquad t_1 \leq t \leq t_2$$

Thus a vector equation of C is

$$\mathbf{R}(t) = f(t)\mathbf{i} + g(t)\mathbf{j} \qquad t_1 \leq t \leq t_2$$

Furthermore, the point (x_1, y_1) is $(f(t_1), g(t_1))$ and the point (x_2, y_2) is $(f(t_2), g(t_2))$. Because ϕ is a potential function for $\mathbf{F}$, $\nabla\phi(x, y) = \mathbf{F}(x, y)$ where $\mathbf{F}(x, y) = M(x, y)\mathbf{i} + N(x, y)\mathbf{j}$. Hence

$$\int_C \mathbf{F} \cdot d\mathbf{R} = \int_C \nabla\phi \cdot d\mathbf{R}$$

$$= \int_{t_1}^{t_2} \nabla\phi(\mathbf{R}(t)) \cdot \mathbf{R}'(t)\, dt$$

$$= \int_{t_1}^{t_2} \nabla\phi(f(t), g(t)) \cdot \mathbf{R}'(t)\, dt$$

$$= \int_{t_1}^{t_2} \langle M(f(t), g(t)), N(f(t), g(t)) \rangle \cdot \langle f'(t), g'(t) \rangle\, dt$$

$$= \int_{t_1}^{t_2} [M(f(t), g(t))f'(t)\, dt + N(f(t), g(t))g'(t)\, dt] \tag{2}$$

Observe that because $M(x, y)\, dx + N(x, y)\, dy = d\phi(x, y)$, then

$$M(f(t), g(t))f'(t)\, dt + N(f(t), g(t))g'(t)\, dt = d\phi(f(t), g(t))$$

Substituting from this equation in (2) and then applying the second fundamental theorem of the calculus (Theorem 5.8.2) we get

$$\int_C \mathbf{F} \cdot d\mathbf{R} = \int_{t_1}^{t_2} d\phi(f(t), g(t))$$

$$= \phi(f(t), g(t))\Big]_{t_1}^{t_2}$$

$$= \phi(f(t_2), g(t_2)) - \phi(f(t_1), g(t_1))$$

$$= \phi(x_2, y_2) - \phi(x_1, y_1)$$

which is what we wished to prove.   ∎

Recall from Section 17.5 that stating that a vector field $M(x, y)\mathbf{i} + N(x, y)\mathbf{j}$ is conservative is equivalent to stating that the differential form $M(x, y)\, dx + N(x, y)\, dy$ is exact. Thus we conclude from Theorem 19.3.1 that the line integral $\int_C M(x, y)\, dx + N(x, y)\, dy$ is independent of the path C if the integrand is an exact differential.

Because of the resemblance of Theorem 19.3.1 to the second fundamental theorem of the calculus, it is sometimes called the **fundamental theorem for line integrals**.

▶ **ILLUSTRATION 2**   We use Theorem 19.3.1 to evaluate the line integral in Illustration 1:

$$\int_C (y^2 + 2x + 4)\, dx + (2xy + 4y - 5)\, dy$$

With vector notation this line integral is

$$\int_C \mathbf{F} \cdot d\mathbf{R}$$

where

$$\mathbf{F}(x, y) = (y^2 + 2x + 4)\mathbf{i} + (2xy + 4y - 5)\mathbf{j}$$

In Illustration 1 of Section 19.1 we showed that $\mathbf{F}$ is a conservative vector field having the potential function

$$\phi(x, y) = y^2 x + x^2 + 4x + 2y^2 - 5y$$

Therefore, from Theorem 19.3.1, the line integral is independent of the path, and C can be any sectionally smooth curve from $(0, 0)$ to $(1, 1)$. Furthermore, from Theorem 19.3.1,

$$\int_C (y^2 + 2x + 4)\, dx + (2xy + 4y - 5)\, dy = \phi(1, 1) - \phi(0, 0)$$
$$= 3 - 0$$
$$= 3$$

This result agrees with that of Illustration 1.

Observe that the integrand of this line integral is an exact differential because

$$d\phi(x, y) = (y^2 + 2x + 4)\, dx + (2xy + 4y - 5)\, dy \qquad \blacktriangleleft$$

EXAMPLE 1   Evaluate the line integral

$$\int_C \mathbf{F} \cdot d\mathbf{R}$$

if $\mathbf{F}(x, y) = (e^{-y} - 2x)\mathbf{i} - (xe^{-y} + \sin y)\mathbf{j}$ and C is the first quadrant arc of the circle

$$\mathbf{R}(t) = \pi \cos t\mathbf{i} + \pi \sin t\mathbf{j} \qquad 0 \le t < \tfrac{1}{2}\pi$$

Solution   In Example 2 of Section 17.5 we showed that

$$\nabla(xe^{-y} - x^2 + \cos y) = (e^{-y} - 2x)\mathbf{i} - (xe^{-y} + \sin y)\mathbf{j}$$

Therefore $\mathbf{F}$ is a conservative vector field, and we apply Theorem 19.3.1 with $\phi(x, y) = xe^{-y} - x^2 + \cos y$. The point where $t = 0$ is $(\pi, 0)$ and the point where $t = \frac{1}{2}\pi$ is $(0, \pi)$.

$$\int_C \mathbf{F} \cdot d\mathbf{R} = \phi(0, \pi) - \phi(\pi, 0)$$
$$= \cos \pi - (\pi - \pi^2 + 1)$$
$$= \pi^2 - \pi - 2$$

If the value of a line integral is independent of the path, it is not necessary to find a potential function ϕ. We show the procedure in the next example.

EXAMPLE 2   If $\mathbf{F}(x, y) = \dfrac{1}{y}\mathbf{i} - \dfrac{x}{y^2}\mathbf{j}$ and C is any sectionally smooth curve from the point $(5, -1)$ to the point $(9, -3)$, show that the value of the line integral

$$\int_C \mathbf{F} \cdot d\mathbf{R}$$

is independent of the path, and evaluate it.

Solution   We showed in Example 2 of Section 19.1 that $\mathbf{F}$ is a conservative vector field. Therefore the line integral is independent of the path. We take for the path the line segment from $(5, -1)$ to $(9, -3)$. An equation of the line is $x + 2y = 3$. By letting $y = -t$ and $x = 3 + 2t$, a vector equation of the line is

$$\mathbf{R}(t) = (3 + 2t)\mathbf{i} - t\mathbf{j} \qquad 1 \le t \le 3$$

We compute the value of the line integral by applying Definition 19.2.2.

$$\int_C \mathbf{F} \cdot d\mathbf{R} = \int_C \mathbf{F}(\mathbf{R}(t)) \cdot \mathbf{R}'(t)\, dt$$
$$= \int_1^3 \mathbf{F}(3 + 2t, -t) \cdot \langle 2, -1 \rangle\, dt$$
$$= \int_1^3 \left\langle -\frac{1}{t}, -\frac{3 + 2t}{t^2} \right\rangle \cdot \langle 2, -1 \rangle\, dt$$
$$= \int_1^3 \left(-\frac{2}{t} + \frac{3 + 2t}{t^2} \right) dt$$
$$= \int_1^3 \frac{3}{t^2}\, dt$$
$$= -\frac{3}{t} \Big]_1^3$$
$$= 2$$

▶ **ILLUSTRATION 3**   In Example 2 of Section 19.1 we obtained the following potential function for the conservative vector field $\mathbf{F}(x, y) = \dfrac{1}{y}\mathbf{i} - \dfrac{x}{y^2}\mathbf{j}$:

$$\phi(x, y) = \frac{x}{y} + K$$

where K is an arbitrary constant. Thus if Theorem 19.3.1 is used to evaluate the line integral of Example 2 above, we have

$$\int_C \mathbf{F} \cdot d\mathbf{R} = \phi(9, -3) - \phi(5, -1)$$

$$= \left(\frac{9}{-3} + K \right) - \left(\frac{5}{-1} + K \right)$$

$$= 2$$

which agrees with our previous result.   ◀

Observe in the above illustration that the arbitrary constant K does not appear because $K - K = 0$. Hereafter when applying Theorem 19.3.1 we shall omit the arbitrary constant for the potential function; what we are doing then is choosing as our potential function the one for which $K = 0$.

If for a curve C defined by the vector equation

$$\mathbf{R}(t) = f(t)\mathbf{i} + g(t)\mathbf{j} \qquad a \le t \le b$$

$A = B$

FIGURE 4

the initial point $A(f(a), g(a))$ and the point $B(f(b), g(b))$ coincide, then curve C is said to be **closed**. Figure 4 shows a closed curve where the points A and B coincide. The following theorem regarding the line integral of a conservative vector field over a sectionally smooth closed curve follows immediately from Theorem 19.3.1

19.3.2 THEOREM   If C is any sectionally smooth closed curve lying in some open disk B in R^2 and $\mathbf{F}$ is a conservative vector field on B, then

$$\int_C \mathbf{F} \cdot d\mathbf{R} = 0$$

Proof   We apply Theorem 19.3.1, and because C is closed, the point (x_1, y_1) coincides with the point (x_2, y_2). Therefore

$$\int_C \mathbf{F} \cdot d\mathbf{R} = \phi(x_2, y_2) - \phi(x_1, y_1)$$

$$= 0$$ ■

EXAMPLE 3   A particle moves on the circle

$$\mathbf{R}(t) = 2 \cos t\,\mathbf{i} + 2 \sin t\,\mathbf{j} \qquad 0 \le t \le 2\pi$$

Find the total work done if the motion is caused by the force field

$$\mathbf{F}(x, y) = \left(4 \ln 3y + \frac{1}{x} \right)\mathbf{i} + \frac{4x}{y}\,\mathbf{j}$$

Solution   Let

$$M(x, y) = 4 \ln 3y + \frac{1}{x} \qquad N(x, y) = \frac{4x}{y}$$

$$M_y(x, y) = \frac{4}{y} \qquad N_x(x, y) = \frac{4}{y}$$

Because $M_y(x, y) = N_x(x, y)$, **F** is conservative. Furthermore, C is a closed curve. Therefore, if W is the measure of the work done, we have, from Theorem 19.3.2,

$$W = \int_C \mathbf{F} \cdot d\mathbf{R}$$

$$= 0$$

We now extend our discussion to functions of three variables. The statement of the following theorem and its proof are analogous to Theorem 19.3.1. The proof is left as an exercise (see Exercise 32).

19.3.3 THEOREM   Let C be any sectionally smooth curve lying in an open ball B in R^3 from the point (x_1, y_1, z_1) to (x_2, y_2, z_2). If **F** is a conservative vector field continuous on B and ϕ is a potential function for **F**, then the line integral

$$\int_C \mathbf{F} \cdot d\mathbf{R}$$

is independent of the path C, and

$$\int_C \mathbf{F} \cdot d\mathbf{R} = \phi(x_2, y_2, z_2) - \phi(x_1, y_1, z_1)$$

▶ **ILLUSTRATION 4**   In Example 5 of Section 17.5 we showed that the vector field defined by

$$\mathbf{F}(x, y, z) = (e^x \sin z + 2yz)\mathbf{i} + (2xz + 2y)\mathbf{j} + (e^x \cos z + 2xy + 3z^2)\mathbf{k}$$

is a gradient $\nabla f(x, y, z)$, and

$$f(x, y, z) = e^x \sin z + 2xyz + y^2 + z^3$$

Thus **F** is a conservative vector field. Therefore, if C is any sectionally smooth curve from $(0, 0, 0)$ to $(1, -2, \pi)$, it follows from Theorem 19.3.3 that the line integral

$$\int_C \mathbf{F} \cdot d\mathbf{R}$$

is independent of the path and its value is

$$f(1, -2, \pi) - f(0, 0, 0) = (e \sin \pi - 4\pi + 4 + \pi^3) - 0$$

$$= \pi^3 - 4\pi + 4 \qquad \blacktriangleleft$$

EXAMPLE 4   Show that the line integral

$$\int_C (4x + 2y - z)\, dx + (2x - 2y + z)\, dy + (-x + y + 2z)\, dz$$

is independent of the path, and evaluate the integral if C is any sectionally smooth curve from $(4, -2, 1)$ to $(-1, 2, 0)$.

Solution   The line integral is independent of the path if the integrand is an exact differential. Let

$M(x, y, z) = 4x + 2y - z$	$N(x, y, z) = 2x - 2y + z$	$R(x, y, z) = -x + y + 2z$
$M_y(x, y, z) = 2$	$N_x(x, y, z) = 2$	$R_x(x, y, z) = -1$
$M_z(x, y, z) = -1$	$N_z(x, y, z) = 1$	$R_y(x, y, z) = 1$

Because

$$M_y(x, y, z) = N_x(x, y, z) \qquad M_z(x, y, z) = R_x(x, y, z) \qquad N_z(x, y, z) = R_y(x, y, z)$$

the integrand is an exact differential and the line integral is independent of the path. We take for the path the line segment from $(4, -2, 1)$ to $(-1, 2, 0)$. A set of direction numbers of this line is $[5, -4, 1]$. Therefore equations of the line are

$$\frac{x+1}{5} = \frac{y-2}{-4} = \frac{z}{1}$$

Parametric equations of the line segment are

$$x = -5t - 1 \qquad y = 4t + 2 \qquad z = -t \qquad -1 \le t \le 0$$

Therefore

$$\int_C (4x + 2y - z)\, dx + (2x - 2y + z)\, dy + (-x + y + 2z)\, dz$$

$$= \int_{-1}^{0} [4(-5t - 1) + 2(4t + 2) - (-t)](-5\, dt)$$

$$+ \int_{-1}^{0} [2(-5t - 1) - 2(4t + 2) + (-t)](4\, dt)$$

$$+ \int_{-1}^{0} [-(-5t - 1) + (4t + 2) + 2(-t)](-dt)$$

$$= \int_{-1}^{0} (-28t - 27)\, dt$$

$$= -14t^2 - 27t \Big]_{-1}^{0}$$

$$= -13$$

The line integral in Example 4 can be computed by finding a potential function for the conservative vector field $(4x + 2y - z)\mathbf{i} + (2x - 2y + z)\mathbf{j} + (-x + y + 2z)\mathbf{k}$ and applying Theorem 19.3.3. You are asked to do this in Exercise 31.

EXAMPLE 5   Suppose $\mathbf{F}$ is the gravitational force field exerted by a particle of mass M units at the origin on a particle of mass 1 unit at the point $P(x, y, z)$. Then from Section 19.1,

$$\mathbf{F}(x, y, z) = \frac{-GM}{(x^2 + y^2 + z^2)^{3/2}} (x\mathbf{i} + y\mathbf{j} + z\mathbf{k})$$

Find the work done by the force $\mathbf{F}$ in moving a particle of mass 1 unit along a smooth curve C from $(0, 3, 4)$ to $(2, 2, 1)$.

Solution   In Illustration 2 of Section 19.1 we showed that $\mathbf{F}$ is conservative and that a potential function for $\mathbf{F}$ is given by

$$\phi(x, y, z) = \frac{GM}{\sqrt{x^2 + y^2 + z^2}}$$

If W is the measure of the work done in moving a particle of mass 1 unit along C,

$$W = \int_C \mathbf{F} \cdot d\mathbf{R}$$

From Theorem 19.3.3, the line integral is independent of the path, and

$$W = \phi(2, 2, 1) - \phi(0, 3, 4)$$

$$= \frac{GM}{\sqrt{2^2 + 2^2 + 1^2}} - \frac{GM}{\sqrt{0^2 + 3^2 + 4^2}}$$

$$= \frac{GM}{3} - \frac{GM}{5}$$

$$= \tfrac{2}{15}GM$$

We now show how the results of this section lead to an important conclusion in physics. If the motion of a particle is caused by a conservative force field $\mathbf{F}$, the **potential energy** of the particle at the point (x, y, z) is defined to be a scalar field E such that

$$\mathbf{F}(x, y, z) = -\nabla E(x, y, z)$$

That is, $-E$ is a potential function of $\mathbf{F}$. We shall use the notation $E(P)$ to denote the potential energy of the particle at the point P. If W is the measure of the work done by $\mathbf{F}$ in moving a particle along a sectionally smooth curve C from point A to point B, then from Theorem 19.3.3,

$$W = \int_C \mathbf{F} \cdot d\mathbf{R}$$

$$W = -E(x, y, z)\Big]_A^B$$

$$W = -[E(B) - E(A)]$$

$$W = E(A) - E(B) \tag{3}$$

Thus W is the difference in the potential energies of the particle at A and B.

Now suppose that the particle is at point A at time t_1 and at point B at time t_2 and the curve C has the vector equation

$$\mathbf{R}(t) = f(t)\mathbf{i} + g(t)\mathbf{j} + h(t)\mathbf{k} \qquad t_1 \leq t \leq t_2$$

Then the velocity and acceleration vectors at t are $\mathbf{V}(t)$ and $\mathbf{A}(t)$ defined by

$$\mathbf{V}(t) = \mathbf{R}'(t) \qquad \mathbf{A}(t) = \mathbf{V}'(t)$$

The speed of the particle at t is denoted by $v(t)$, where $v(t) = \|\mathbf{V}(t)\|$. Then another formula for computing W is given by

$$W = \int_C \mathbf{F} \cdot d\mathbf{R}$$

$$W = \int_{t_1}^{t_2} \mathbf{F}(\mathbf{R}(t)) \cdot \mathbf{R}'(t)\, dt$$

$$W = \int_{t_1}^{t_2} \mathbf{F}(\mathbf{R}(t)) \cdot \mathbf{V}(t)\, dt \tag{4}$$

Newton's second law of motion states that if a force $\mathbf{F}$ is acting on a particle of mass m units, then

$$\mathbf{F}(\mathbf{R}(t)) = m\mathbf{A}(t)$$

$$\Leftrightarrow \quad \mathbf{F}(\mathbf{R}(t)) = m\mathbf{V}'(t)$$

Substituting from this equation in (4) we have

$$W = \int_{t_1}^{t_2} m[\mathbf{V}'(t) \cdot \mathbf{V}(t)] \, dt$$

Because $D_t[\mathbf{V}(t) \cdot \mathbf{V}(t)] = 2\mathbf{V}'(t) \cdot \mathbf{V}(t)$ and $\mathbf{V}(t) \cdot \mathbf{V}(t) = [v(t)]^2$, we get

$$W = \tfrac{1}{2}m \int_{t_1}^{t_2} D_t[\mathbf{V}(t) \cdot \mathbf{V}(t)] \, dt$$

$$W = \tfrac{1}{2}m \int_{t_1}^{t_2} D_t[v(t)]^2 \, dt$$

$$W = \tfrac{1}{2}m[v(t)]^2 \Big]_{t_1}^{t_2}$$

$$W = \tfrac{1}{2}m[v(t_2)]^2 - \tfrac{1}{2}m[v(t_1)]^2 \qquad (5)$$

In physics the **kinetic energy** of a particle is defined to be $\tfrac{1}{2}mv^2$. Therefore Equation (5) states that the work done in moving a particle along C from point A to point B is the change in kinetic energy of the particle. If we use the notation $K(P)$ to indicate the kinetic energy of a particle at point P, (5) can be written as

$$W = K(B) - K(A)$$

Equating the values of W from (3) and this equation, we have

$$E(A) - E(B) = K(B) - K(A)$$

$$E(A) + K(A) = E(B) + K(B)$$

The above equation states that the sums of the potential and kinetic energies are equal at the initial point A and the terminal point B. Because A and B can be any points on C, the sum of the two energies is constant along C; that is, the total energy of the particle remains unchanged during the motion. This fact is a major concept in physics called the **law of conservation of energy**. It is for this reason that the terminology *conservative* is used for a force field that is a gradient.

EXERCISES 19.3

In Exercises 1 through 12, use the result of the indicated exercise in Exercises 19.1 to prove that the value of the line integral is independent of the path. Then evaluate the line integral by applying either Theorem 19.3.1 or 19.3.3 and using the potential function found in the indicated exercise. In each exercise, C is any sectionally smooth curve from point A to point B.

1. $\int_C y \, dx + x \, dy$; A is $(1, 4)$ and B is $(3, 2)$; Exercise 21.

2. $\int_C x \, dx + y \, dy$; A is $(-5, 2)$ and B is $(1, 3)$; Exercise 22.

3. $\int_C e^x \sin y \, dx + e^x \cos y \, dy$; A is $(0, 0)$ and B is $(2, \tfrac{1}{2}\pi)$; Exercise 23.

4. $\int_C (\sin y \sinh x + \cos y \cosh x) \, dx$
$\qquad\qquad\qquad + (\cos y \cosh x - \sin y \sinh x) \, dy$
A is $(1, 0)$ and B is $(2, \pi)$; Exercise 24.

5. $\int_C (2xy^2 - y^3) \, dx + (2x^2y - 3xy^2 + 2) \, dy$; A is $(-3, -1)$ and B is $(1, 2)$; Exercise 25.

6. $\int_C (3x^2 + 2y - y^2 e^x) \, dx + (2x - 2ye^x) \, dy$; A is $(0, 2)$ and B is $(1, -3)$; Exercise 26.

7. $\int_C (x^2 - y) \, dx - (x - 3z) \, dy + (z + 3y) \, dz$; A is $(-3, 1, 2)$ and B is $(3, 0, 4)$; Exercise 27.

8. $\int_C yz \, dx + xz \, dy + xy \, dz$; A is $(0, -2, 5)$ and B is $(4, 1, -3)$; Exercise 28.

9. $\int_C (ze^x + e^y) \, dx + (xe^y - e^z) \, dy + (-ye^z + e^x) \, dz$; A is $(1, 0, 2)$ and B is $(0, 2, 1)$; Exercise 29.

10. $\int_C (\tan y + 2xy \sec z) \, dx + (x \sec^2 y + x^2 \sec z) \, dy$
$\qquad\qquad\qquad + \sec z(x^2 y \tan z - \sec z) \, dz$
A is $(2, \tfrac{1}{4}\pi, 0)$ and B is $(3, \pi, \pi)$; Exercise 30.

11. $\int_C (2x \cos y - 3) \, dx - (x^2 \sin y + z^2) \, dy - (2yz - 2) \, dz$; A is $(-1, 0, 3)$ and B is $(1, \pi, 0)$; Exercise 31.

12. $\int_C (2y^3 - 8xz^2)\, dx + (6xy^2 + 1)\, dy - (8x^2z + 3z^2)\, dz$; A is $(2, 0, 0)$ and B is $(3, 2, 1)$; Exercise 32.

In Exercises 13 through 20, show that the value of the line integral $\int_C \mathbf{F} \cdot d\mathbf{R}$ for the given $\mathbf{F}$ and C is independent of the path, and evaluate the line integral.

13. $\mathbf{F}(x, y) = 2(x - y)\mathbf{i} + 2(3y - x)\mathbf{j}$; C is the first quadrant arc of the circle $x^2 + y^2 = 9$ from the point on the x axis to the point on the y axis.

14. $\mathbf{F}(x, y) = (3x^2 + 6xy - 2y^2)\mathbf{i} + (3x^2 - 4xy + 3y^2)\mathbf{j}$; C is the first quadrant arc of the ellipse $4x^2 + 9y^2 = 36$ from the point on the x axis to the point on the y axis.

15. $\mathbf{F}(x, y) = (4e^{2x} - 3e^x e^y)\mathbf{i} + (2e^{2y} - 3e^x e^y)\mathbf{j}$; C is the arc of the parabola $y^2 = 4x$ from the vertex to the endpoint of the latus rectum in the first quadrant.

16. $\mathbf{F}(x, y) = e^x \cos y\,\mathbf{i} - e^x \sin y\,\mathbf{j}$; C is the segment of the line $3x + 4y = 12$ from the point where it intersects the y axis to the point where it intersects the x axis.

17. $\mathbf{F}(x, y, z) = 2x\mathbf{i} + 3y^2\mathbf{j} + \mathbf{k}$; C is the trace of the ellipsoid $4x^2 + 4y^2 + z^2 = 9$ in the xz plane from the positive x axis to the positive z axis.

18. $\mathbf{F}(x, y, z) = (2xy + z^2)\mathbf{i} + (x^2 - 2yz)\mathbf{j} + (2xz - y^2)\mathbf{k}$; C is the trace of the sphere $x^2 + y^2 + z^2 = 1$ in the yz plane from the positive y axis to the positive z axis.

19. $\mathbf{F}(x, y, z) = 2ye^{2x}\mathbf{i} + e^{2x}\mathbf{j} + 3z^2\mathbf{k}$; C is any sectionally smooth curve from the point $(\ln 2, 1, 1)$ to the point $(\ln 2, 2, 2)$.

20. $\mathbf{F}(x, y, z) = \left(\dfrac{1}{z} - \dfrac{y}{x^2}\right)\mathbf{i} + \left(\dfrac{1}{x} + \dfrac{z}{y^2}\right)\mathbf{j} - \left(\dfrac{1}{y} + \dfrac{x}{z^2}\right)\mathbf{k}$; C is any sectionally smooth curve from the point $(1, 2, -1)$ to the point $(2, -4, -2)$.

In Exercises 21 through 30, show that the value of the line integral is independent of the path, and compute the value in any convenient manner. In each exercise, C is any sectionally smooth curve from point A to point B.

21. $\int_C (2y - x)\, dx + (y^2 + 2x)\, dy$; A is $(0, -1)$ and B is $(1, 2)$.

22. $\int_C (\ln x + 2y)\, dx + (e^y + 2x)\, dy$; A is $(3, 1)$ and B is $(1, 3)$.

23. $\int_C \tan y\, dx + x \sec^2 y\, dy$; A is $(-2, 0)$ and B is $(4, \frac{1}{4}\pi)$.

24. $\int_C \sin y\, dx + (\sin y + x \cos y)\, dy$; A is $(-2, 0)$ and B is $(2, \frac{1}{6}\pi)$.

25. $\int_C \dfrac{2y}{(xy + 1)^2}\, dx + \dfrac{2x}{(xy + 1)^2}\, dy$; A is $(0, 2)$ and B is $(1, 0)$.

26. $\int_C \dfrac{x}{x^2 + y^2 + z^2}\, dx + \dfrac{y}{x^2 + y^2 + z^2}\, dy + \dfrac{z}{x^2 + y^2 + z^2}\, dz$; A is $(1, 0, 0)$ and B is $(1, 2, 3)$.

27. $\int_C (y + z)\, dx + (x + z)\, dy + (x + y)\, dz$; A is $(0, 0, 0)$ and B is $(1, 1, 1)$.

28. $\int_C (yz + x)\, dx + (xz + y)\, dy + (xy + z)\, dz$; A is $(0, 0, 0)$ and B is $(1, 1, 1)$.

29. $\int_C (e^x \sin y + yz)\, dx + (e^x \cos y + z \sin y + xz)\, dy + (xy - \cos y)\, dz$; A is $(2, 0, 1)$ and B is $(0, \pi, 3)$.

30. $\int_C (2x \ln yz - 5ye^x)\, dx - (5e^x - x^2 y^{-1})\, dy + (x^2 z^{-1} + 2z)\, dz$; A is $(2, 1, 1)$ and B is $(3, 1, e)$.

31. Evaluate the line integral in Example 4 by finding a potential function for the conservative vector field

$$(4x + 2y - z)\mathbf{i} + (2x - 2y + z)\mathbf{j} + (-x + y + 2z)\mathbf{k}$$

and applying Theorem 19.3.3.

32. Prove Theorem 19.3.3.

In Exercises 33 through 36, find the total work done in moving a particle along arc C if the motion is caused by the force field $\mathbf{F}$. Assume the arc is measured in meters and the force is measured in newtons. (Hint: First show that $\mathbf{F}$ is conservative.)

33. $\mathbf{F}(x, y) = 3(x + y)^2\mathbf{i} + 3(x + y)^2\mathbf{j}$; C: the arc of the parabola $y = x^2$ from the vertex to the point $(2, 4)$.

34. $\mathbf{F}(x, y) = (2xy - 5y + 2y^2)\mathbf{i} + (x^2 - 5x + 4xy)\mathbf{j}$; C: the quarter circle $\mathbf{R}(t) = 2 \cos t\mathbf{i} + 2 \sin t\mathbf{j}$, $0 \le t \le \frac{1}{2}\pi$.

35. $\mathbf{F}(x, y, z) = 2y^2z^3\mathbf{i} + 4xyz^3\mathbf{j} + 6xy^2z^2\mathbf{k}$; C: the arc of the twisted cubic $\mathbf{R}(t) = t\mathbf{i} + t^2\mathbf{j} + t^3\mathbf{k}$ from $t = 1$ to $t = 2$.

36. $\mathbf{F}(x, y, z) = 4y^2z\mathbf{i} + 8xyz\mathbf{j} + 4(3z^3 + xy^2)\mathbf{k}$; C: the arc of the circular helix $\mathbf{R}(t) = 3 \cos t\mathbf{i} + 3 \sin t\mathbf{j} + \mathbf{k}$ from $t = 0$ to $t = \frac{1}{3}\pi$.

37. If $\mathbf{F}$ is the inverse-square force field defined by

$$\mathbf{F}(x, y, z) = \dfrac{k(x\mathbf{i} + y\mathbf{j} + z\mathbf{k})}{(x^2 + y^2 + z^2)^{3/2}}$$

find the work done by $\mathbf{F}$ in moving a particle along the segment of a straight line from the point $(3, 0, 0)$ to $(3, 0, 4)$. Evaluate the line integral by two methods: (a) Use a potential function for $\mathbf{F}$; (b) do not use a potential function for $\mathbf{F}$.

19.4 GREEN'S THEOREM

There is a theorem that expresses a double integral over a plane region R in terms of a line integral around a curve that is a boundary of R. This theorem is named after the English mathematician and physicist George Green (1793–1841), who presented it in a paper on the applications of mathematics to electricity and magnetism. Before stating the theorem, it is necessary to review and introduce some terminology pertaining to plane curves. A *smooth* curve was defined in Section 19.2 and a *closed* curve was defined in Section 19.3. A curve C is said to be **simple** if it does not intersect itself. That is, if a vector equation of C is

$$\mathbf{R}(t) = f(t)\mathbf{i} + g(t)\mathbf{j}$$

and if A is the point $(f(a), g(a))$ and B is the point $(f(b), g(b))$, then C is **simple** between A and B if $(f(t_1), g(t_1))$ is not the same point as $(f(t_2), g(t_2))$ for all t_1 and t_2 in the open interval (a, b).

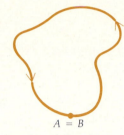

(a) Simple and closed

(b) Simple but not closed

(c) Closed but not simple

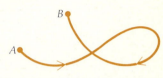

(d) Neither simple nor closed

FIGURE 1

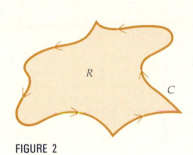

FIGURE 2

The circle and the ellipse are examples of smooth simple closed curves. In Figure 1(a)–(d) there are further examples of smooth curves that may or may not be simple and closed. In (a) the curve is both simple and closed; in (b) the curve is simple but not closed; in (c) the curve is closed but not simple; and in (d) the curve is neither simple nor closed.

In the statement of Green's theorem we refer to a line integral around a sectionally smooth simple closed curve C that forms the boundary of a region R in the plane, and the direction along C is counterclockwise. In Figure 2 such a region R is shown with the required boundary curve C. The line integral around C in the counterclockwise direction is denoted by $\oint_C$.

19.4.1 THEOREM
Green's Theorem

Let M and N be functions of two variables x and y such that they have continuous first partial derivatives on an open disk B in R^2. If C is a sectionally smooth simple closed curve lying entirely in B, and if R is the region bounded by C, then

$$\oint_C M(x, y)\, dx + N(x, y)\, dy = \iint_R \left(\frac{\partial N}{\partial x} - \frac{\partial M}{\partial y} \right) dA$$

The proof of Green's theorem for all regions bounded by curves that are sectionally smooth, simple, and closed belongs to a course in advanced calculus. However, we shall prove the theorem for a particular kind of region, one for which each horizontal line and each vertical line intersect it in at most two points. The proof follows.

Proof   Let R be a region in the xy plane that can be defined by either

$$R = \{(x, y) \mid a \le x \le b, f_1(x) \le y \le f_2(x)\} \qquad (1)$$

or

$$R = \{(x, y) \mid c \le y \le d, g_1(y) \le x \le g_2(y)\} \qquad (2)$$

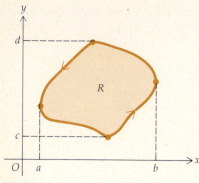

FIGURE 3

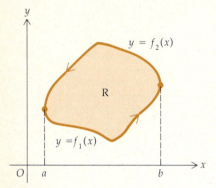

FIGURE 4

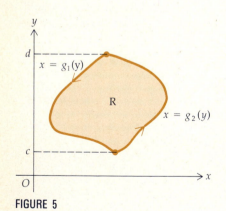

FIGURE 5

where the functions f_1, f_2, g_1, and g_2 are smooth. Figure 3 shows such a region R, regarded as being defined by (1) in Figure 4 and by (2) in Figure 5.

The proof will consist of showing that

$$\oint_C M(x, y)\, dx = -\iint_R \frac{\partial M}{\partial y}\, dA \tag{3}$$

and

$$\oint_C N(x, y)\, dy = \iint_R \frac{\partial N}{\partial x}\, dA \tag{4}$$

To prove (3) we treat R as a region defined by (1). Refer to Figure 4. Let C_1 be the graph of $y = f_1(x)$ from $x = a$ to $x = b$; that is, C_1 is the lower part of the boundary curve C going from left to right. Let C_2 be the graph of $y = f_2(x)$ from $x = b$ to $x = a$; that is, C_2 is the upper part of the boundary curve C going from right to left. Consider the line integral $\oint_C M(x, y)\, dx$.

$$\oint_C M(x, y)\, dx = \int_{C_1} M(x, y)\, dx + \int_{C_2} M(x, y)\, dx$$

$$= \int_a^b M(x, f_1(x))\, dx + \int_b^a M(x, f_2(x))\, dx$$

$$= \int_a^b M(x, f_1(x))\, dx - \int_a^b M(x, f_2(x))\, dx$$

$$= \int_a^b \left[M(x, f_1(x)) - M(x, f_2(x)) \right]\, dx \tag{5}$$

We now deal with the double integral $\iint_R \frac{\partial M}{\partial y}\, dA$, where R is still considered to be defined by (1). Then

$$\iint_R \frac{\partial M}{\partial y}\, dA = \int_a^b \int_{f_1(x)}^{f_2(x)} \frac{\partial M}{\partial y}\, dy\, dx$$

$$= \int_a^b \left(\int_{f_1(x)}^{f_2(x)} \frac{\partial M}{\partial y}\, dy \right) dx$$

$$= \int_a^b M(x, y) \Big]_{f_1(x)}^{f_2(x)}\, dx$$

$$= \int_a^b \left[M(x, f_2(x)) - M(x, f_1(x)) \right]\, dx \tag{6}$$

By comparing (5) and (6) it follows that (3) holds.

To prove (4), R is regarded as a region defined by (2), as in Figure 5. The details of the proof are left as an exercise (see Exercise 43).

By adding corresponding members of Equations (3) and (4) we obtain Green's theorem for this region R. ∎

▶ **ILLUSTRATION 1** We apply Green's theorem to evaluate the line integral $\oint_C y^2\, dx + 4xy\, dy$, where C is the closed curve consisting of the arc of the pa-

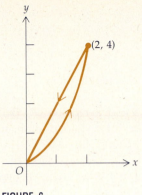

FIGURE 6

rabola $y = x^2$ from the origin to the point $(2, 4)$ and the line segment from $(2, 4)$ to the origin. The region R with the boundary C is shown in Figure 6. From Green's theorem,

$$\oint_C y^2\, dx + 4xy\, dy = \iint_R \left[\frac{\partial}{\partial x}(4xy) - \frac{\partial}{\partial y}(y^2)\right] dA$$

$$= \int_0^2 \int_{x^2}^{2x} (4y - 2y)\, dy\, dx$$

$$= \int_0^2 y^2 \Big]_{x^2}^{2x}\, dx$$

$$= \int_0^2 (4x^2 - x^4)\, dx$$

$$= \tfrac{4}{3}x^3 - \tfrac{1}{5}x^5 \Big]_0^2$$

$$= \tfrac{64}{15}$$

To show the advantage of using Green's theorem we evaluate the same line integral by the method of Section 19.2. If C_1 is the arc of the parabola $y = x^2$ from $(0, 0)$ to $(2, 4)$ and C_2 is the line segment from $(2, 4)$ to $(0, 0)$, then

$$\oint_C y^2\, dx + 4xy\, dy = \oint_{C_1} y^2\, dx + 4xy\, dy + \oint_{C_2} y^2\, dx + 4xy\, dy$$

Parametric equations for C_1 are

$$x = t \qquad y = t^2 \qquad 0 \le t \le 2$$

Therefore

$$\oint_{C_1} y^2\, dx + 4xy\, dy = \int_0^2 (t^2)^2\, dt + 4(t)(t^2)(2t\, dt)$$

$$= \int_0^2 9t^4\, dt$$

$$= \tfrac{9}{5}t^5 \Big]_0^2$$

$$= \tfrac{288}{5}$$

Arc C_2 can be represented parametrically by

$$x = t \qquad y = 2t \qquad \text{from } t = 2 \text{ to } t = 0$$

Thus

$$\oint_{C_2} y^2\, dx + 4xy\, dy = \int_2^0 (2t)^2\, dt + 4(t)(2t)(2\, dt)$$

$$= \int_2^0 20t^2\, dt$$

$$= \tfrac{20}{3}t^3 \Big]_2^0$$

$$= -\tfrac{160}{3}$$

Hence

$$\oint_C y^2 \, dx + 4xy \, dy = \frac{288}{5} - \frac{160}{3}$$
$$= \frac{64}{15}$$

which agrees with the result obtained by using Green's theorem. ◀

EXAMPLE 1 Use Green's theorem to find the total work done in moving an object in the counterclockwise direction once around the circle $x^2 + y^2 = a^2$ if the motion is caused by the force field $\mathbf{F}(x, y) = (\sin x - y)\mathbf{i} + (e^y - x^2)\mathbf{j}$. Assume the arc is measured in meters and the force is measured in newtons.

Solution If W joules is the work done, then

$$W = \oint_C (\sin x - y) \, dx + (e^y - x^2) \, dy$$

where C is the circle $x^2 + y^2 = a^2$. From Green's theorem,

$$W = \iint_R \left[\frac{\partial}{\partial x} (e^y - x^2) - \frac{\partial}{\partial y} (\sin x - y) \right] dA$$

$$= \iint_R (-2x + 1) \, dA$$

We use polar coordinates to evaluate the double integral, with $x = r \cos \theta$ and $dA = r \, dr \, d\theta$. Then

$$W = \int_0^{2\pi} \int_0^a (-2r \cos \theta + 1)r \, dr \, d\theta$$

$$= \int_0^{2\pi} \int_0^a (-2r^2 \cos \theta + r) \, dr \, d\theta$$

$$= \int_0^{2\pi} \left. -\frac{2}{3} r^3 \cos \theta + \frac{r^2}{2} \right]_0^a d\theta$$

$$= \int_0^{2\pi} \left(-\frac{2}{3} a^3 \cos \theta + \frac{a^2}{2} \right) d\theta$$

$$= \left. -\frac{2}{3} a^3 \sin \theta + \frac{a^2}{2} \theta \right]_0^{2\pi}$$

$$= \pi a^2$$

Therefore the work done is πa^2 joules.

The following theorem, which is a consequence of Green's theorem, gives a useful method for computing the area of a region bounded by a sectionally smooth simple closed curve.

19.4.2 THEOREM If R is a region having as its boundary a sectionally smooth simple closed curve C, and A square units is the area of R, then

$$A = \frac{1}{2} \oint_C x \, dy - y \, dx$$

Proof   In the statement of Green's theorem, let $M(x, y) = -\frac{1}{2}y$ and $N(x, y) = \frac{1}{2}x$. Then

$$\oint_C -\frac{1}{2}y\,dx + \frac{1}{2}x\,dy = \iint_R \left[\frac{\partial}{\partial x}\left(\frac{1}{2}x\right) - \frac{\partial}{\partial y}\left(-\frac{1}{2}y\right)\right]dA$$

$$= \iint_R (\tfrac{1}{2} + \tfrac{1}{2})\,dA$$

$$= \iint_R dA$$

Because $\iint_R dA$ is the measure of the area of R,

$$\frac{1}{2}\oint_C x\,dy - y\,dx = A$$ ∎

EXAMPLE 2   Use Theorem 19.4.2 to find the area of the region enclosed by the ellipse

$$\frac{x^2}{a^2} + \frac{y^2}{b^2} = 1$$

Solution   Parametric equations for the ellipse are

$$x = a\cos t \qquad y = b\sin t \qquad 0 \le t \le 2\pi$$

Then $dx = -a\sin t\,dt$ and $dy = b\cos t\,dt$. If C is the ellipse and A square units is the area of the region enclosed by C, then, from Theorem 19.4.2,

$$A = \frac{1}{2}\oint_C x\,dy - y\,dx$$

$$= \frac{1}{2}\int_0^{2\pi} [(a\cos t)(b\cos t\,dt) - (b\sin t)(-a\sin t\,dt)]$$

$$= \frac{1}{2}\int_0^{2\pi} ab(\cos^2 t + \sin^2 t)\,dt$$

$$= \tfrac{1}{2}ab\int_0^{2\pi} dt$$

$$= \pi ab$$

Therefore the area is πab square units.

EXAMPLE 3   Use Green's theorem to evaluate the line integral

$$\oint_C (x^4 - 3y)\,dx + (2y^3 + 4x)\,dy$$

if C is the ellipse $\dfrac{x^2}{9} + \dfrac{y^2}{4} = 1$.

Solution   From Green's theorem,

$$\oint_C (x^4 - 3y)\, dx + (2y^3 + 4x)\, dy = \iint_R \left[\frac{\partial}{\partial x}(2y^3 + 4x) - \frac{\partial}{\partial y}(x^4 - 3y) \right] dA$$

$$= \iint_R (4 + 3)\, dA$$

$$= 7 \iint_R dA$$

The double integral $\iint_R dA$ is the measure of the area of the region enclosed by the ellipse. From Example 2 with $a = 3$ and $b = 2$, the area of the region enclosed by the ellipse is 6π square units. Therefore

$$\oint_C (x^4 - 3y)\, dx + (2y^3 + 4x)\, dy = 42\pi$$

There are two vector forms of Green's theorem that we shall proceed to obtain. Let C be a sectionally smooth simple closed curve in the xy plane. Suppose a vector equation of C is

$$\mathbf{R}(s) = x\mathbf{i} + y\mathbf{j}$$

and $x = f(s)$ and $y = g(s)$, where s units is the length of arc measured in the counterclockwise direction from a particular point P_0 on C to the point P on C. Then if $\mathbf{T}(s)$ is the unit tangent vector of C at P, from Theorem 14.6.3 we have $\mathbf{T}(s) = D_s\mathbf{R}(s)$. Thus

$$\mathbf{T}(s) = \frac{dx}{ds}\mathbf{i} + \frac{dy}{ds}\mathbf{j} \tag{7}$$

The vector $\mathbf{N}(s)$ defined by

$$\mathbf{N}(s) = \frac{dy}{ds}\mathbf{i} - \frac{dx}{ds}\mathbf{j} \tag{8}$$

is a unit normal vector of C at P. To verify this fact observe that $\mathbf{T}(s) \cdot \mathbf{N}(s) = 0$ and the magnitudes of $\mathbf{T}(s)$ and $\mathbf{N}(s)$ are equal. The unit normal vector defined by (8) is selected rather than its negative so that when the direction along C is counterclockwise, $\mathbf{N}(s)$ will point outward from the region R bounded by C, and it is called the **unit outward normal**. See Figure 7. Let

$$\mathbf{F}(x, y) = M(x, y)\mathbf{i} + N(x, y)\mathbf{j}$$

where M and N satisfy the hypothesis of Green's theorem. Because

$$\mathbf{F}(x, y) \cdot \mathbf{N}(s)\, ds = [M(x, y)\mathbf{i} + N(x, y)\mathbf{j}] \cdot \left(\frac{dy}{ds}\mathbf{i} - \frac{dx}{ds}\mathbf{j} \right) ds$$

$$= M(x, y)\, dy - N(x, y)\, dx$$

then

$$\oint_C \mathbf{F}(x, y) \cdot \mathbf{N}(s)\, ds = \oint_C -N(x, y)\, dx + M(x, y)\, dy$$

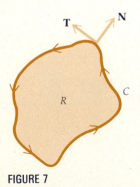

FIGURE 7

We apply Green's theorem to the line integral on the right side of this equation and we have

$$\oint_C \mathbf{F}(x, y) \cdot \mathbf{N}(s)\, ds = \iint_R \left[\frac{\partial M}{\partial x} - \frac{\partial}{\partial y}(-N) \right] dA$$

$$= \iint_R \left(\frac{\partial M}{\partial x} + \frac{\partial N}{\partial y} \right) dA$$

$$= \iint_R \operatorname{div} \mathbf{F}\, dA$$

This vector form of Green's theorem is stated formally as the following theorem, named after the German mathematician and scientist Karl Gauss (1777–1855).

19.4.3 THEOREM
Gauss's Divergence Theorem in the Plane

Let the functions M and N, the curve C, and the region R be as defined in Green's theorem. If $\mathbf{F}(x, y) = M(x, y)\mathbf{i} + N(x, y)\mathbf{j}$ and $\mathbf{N}(s)$ is the unit outward normal vector of C at P, where s units is the length of arc measured in the counterclockwise direction from a particular point P_0 on C to P, then

$$\oint_C \mathbf{F} \cdot \mathbf{N}\, ds = \iint_R \operatorname{div} \mathbf{F}\, dA$$

EXAMPLE 4   Verify Gauss's divergence theorem in the plane if

$$\mathbf{F}(x, y) = 2y\mathbf{i} + 5x\mathbf{j}$$

and R is the region bounded by the circle $x^2 + y^2 = 1$.

Solution   The boundary of R is the unit circle that can be represented parametrically by

$$x = \cos s \qquad y = \sin s \qquad 0 \leq s \leq 2\pi$$

where s units is the length of arc from the point where $s = 0$ to the point P on C. Then a vector equation of C is

$$\mathbf{R}(s) = \cos s\,\mathbf{i} + \sin s\,\mathbf{j} \qquad 0 \leq s \leq 2\pi$$

From (8) the unit outward normal is

$$\mathbf{N}(s) = \cos s\,\mathbf{i} + \sin s\,\mathbf{j}$$

At a point $P(\cos s, \sin s)$ on C, $\mathbf{F}$ is $2\sin s\,\mathbf{i} + 5\cos s\,\mathbf{j}$. Therefore

$$\oint_C \mathbf{F} \cdot \mathbf{N}\, ds = \int_0^{2\pi} (2\sin s\,\mathbf{i} + 5\cos s\,\mathbf{j}) \cdot (\cos s\,\mathbf{i} + \sin s\,\mathbf{j})\, ds$$

$$= \int_0^{2\pi} (2\sin s \cos s + 5\sin s \cos s)\, ds$$

$$= 7\int_0^{2\pi} \sin s \cos s\, ds$$

$$= \tfrac{7}{2}\sin^2 s \Big]_0^{2\pi}$$

$$= 0$$

Because $M = 2y$, $\dfrac{\partial M}{\partial x} = 0$, and because $N = 5x$, $\dfrac{\partial N}{\partial y} = 0$. Thus

$$\iint\limits_{R} \operatorname{div} \mathbf{F} \, dA = \iint\limits_{R} \left(\frac{\partial M}{\partial x} + \frac{\partial N}{\partial y} \right) dA$$

$$= 0$$

We have therefore verified Gauss's divergence theorem in the plane for this $\mathbf{F}$ and R.

Observe in Example 4 that $\displaystyle\iint\limits_{R} \operatorname{div} \mathbf{F} \, dA$ is easier to compute than

$$\oint_{C} \mathbf{F} \cdot \mathbf{N} \, ds.$$

If $\mathbf{F}$ is a vector field and div $\mathbf{F} = 0$, then $\mathbf{F}$ is said to be **divergence free**. The vector field in Example 4 is divergence free. In the study of hydrodynamics (fluid motion), if the velocity field of a fluid is divergence free, the fluid is called **incompressible**. In the theory of electricity and magnetism, a vector field that is divergence free is said to be **solenoidal**.

We now use Gauss's divergence theorem in the plane to give a physical interpretation of the divergence of a vector field. Let the functions M and N, the region R, and the curve C be as defined in Green's theorem. Suppose $\mathbf{F}$ is the velocity field of a two-dimensional fluid (constant depth) and $\mathbf{F}$ is defined by $\mathbf{F}(x, y) = M(x, y)\mathbf{i} + N(x, y)\mathbf{j}$. Suppose that the fluid flows through region R having curve C as its boundary, for which the direction along C is counterclockwise. We assume that the fluid has a constant density in R, and for convenience let the density have unit measure. The flow of the velocity field $\mathbf{F}$ across C is the rate at which the fluid crosses C in a direction perpendicular to C. We shall show how this flow can be expressed as a line integral.

Let s denote the length of arc of curve C measured from a particular point P_0 to a point P. Divide the curve C into n arcs and let $\Delta_i s$ be the length of the ith arc containing the point $P_i(x_i, y_i)$, where s_i is the length of arc of C from P_0 to P_i. Because $\mathbf{F}$ is continuous, an approximation of the velocity of the fluid at each point of the ith arc is $\mathbf{F}(x_i, y_i)$. The amount of fluid that crosses the arc per unit of time is given approximately by the area of a parallelogram having one pair of opposite sides of length $\Delta_i s$ units and an altitude of length $\mathbf{F}(x_i, y_i) \cdot \mathbf{N}(s_i)$ units, where $\mathbf{N}(s_i)$ is the unit outward normal vector of C at $P_i(x_i, y_i)$. See Figure 8. The area of the parallelogram is $\mathbf{F}(x_i, y_i) \cdot \mathbf{N}(s_i) \, \Delta_i s$ square units. The total amount of fluid that crosses C per unit of time is given approximately by

$$\sum_{i=1}^{n} \mathbf{F}(x_i, y_i) \cdot \mathbf{N}(s_i) \, \Delta_i s$$

Taking the limit of this summation as n increases without bound and as each $\Delta_i s$ approaches zero we obtain the line integral

$$\oint_{C} \mathbf{F}(x, y) \cdot \mathbf{N}(s) \, ds$$

which is called the **flux** of $\mathbf{F}$ across C.

Now let $\bar{P}(\bar{x}, \bar{y})$ be a particular point in region R. Consider a circle having its center at $\bar{P}$ and having a small radius δ, and denote this circle by C_δ. Let

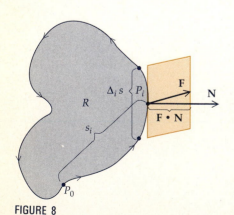

FIGURE 8

R_δ be the region enclosed by C_δ. Then

$$\text{flux of } \mathbf{F} \text{ across } C_\delta = \oint_{C_\delta} \mathbf{F}(x, y) \cdot \mathbf{N} \, ds$$

Applying Theorem 19.4.3 we have

$$\text{flux of } \mathbf{F} \text{ across } C_\delta = \iint_{R_\delta} \text{div } \mathbf{F} \, dA$$

If M_x and N_y are continuous on R_δ, then div $\mathbf{F}$ is continuous there, and for small δ, div $\mathbf{F}$ on R_δ is approximately div $\mathbf{F}(\bar{x}, \bar{y})$. Thus

$$\text{flux of } \mathbf{F} \text{ across } C_\delta \approx \iint_{R} \text{div } \mathbf{F}(\bar{x}, \bar{y}) \, dA$$

Because div $\mathbf{F}(\bar{x}, \bar{y})$ is constant and $\displaystyle\iint_{R_\delta} dA$ is the measure of the area of a circle of radius δ, we have

$$\text{flux of } \mathbf{F} \text{ across } C_\delta \approx \text{div } \mathbf{F}(\bar{x}, \bar{y})(\pi\delta^2) \tag{9}$$

Remember that the flux of $\mathbf{F}$ across C_δ is the total amount of fluid that crosses C_δ per unit of time. Therefore, from (9), div $\mathbf{F}(x, y)$ can be interpreted as the approximate measure of the rate of flow of the fluid per unit area away from the point $(\bar{x}, \bar{y})$. If div $\mathbf{F}(\bar{x}, \bar{y}) > 0$, the fluid is said to have a **source** at $(\bar{x}, \bar{y})$. If div $\mathbf{F}(\bar{x}, \bar{y}) < 0$, the fluid has a **sink** at $(\bar{x}, \bar{y})$. If $\mathbf{F}$ is divergence free at all points in a region, then there are no sources or sinks in the region. As mentioned above, in such a case the fluid is incompressible.

The word *flux* normally means flow; however, the terminology *flux* is applied to vector fields in general, not just to those associated with the velocity of a fluid. Thus if $\mathbf{F}$ is a vector field

$$\text{flux of } \mathbf{F} \text{ across } C = \oint_{C} \mathbf{F} \cdot \mathbf{N} \, ds \tag{10}$$

EXAMPLE 5   The velocity field of a fluid is defined by

$$\mathbf{F}(x, y) = (5x - y)\mathbf{i} + (x^2 - 3y)\mathbf{j}$$

Find the rate of flow of the fluid out of a region R bounded by a smooth closed curve C and whose area is 150 cm^2.

Solution   The rate of flow of the fluid is given by the flux of $\mathbf{F}$ across C. From (10) and Gauss's divergence theorem in the plane,

$$\text{flux} = \oint_{C} \mathbf{F} \cdot \mathbf{N} \, ds$$

$$= \iint_{R} \text{div } \mathbf{F} \, dA$$

$$= \iint_{R} \left[\frac{\partial}{\partial x}(5x - y) + \frac{\partial}{\partial y}(x^2 - 3y) \right] dA$$

$$= \iint_{R} (5 - 3) \, dA$$

$$= 2 \iint_{R} dA$$

Because the area of R is 150 cm², $\iint\limits_R dA = 150$. Thus

flux $= 300$

Hence the rate of flow of the fluid out of the region is 300 cm² per unit of time.

To obtain the second vector form of Green's theorem we consider the dot product of $\mathbf{F}(x, y)$ and the unit tangent vector $\mathbf{T}(s)$ defined by Equation (7). We have

$$\mathbf{F}(x, y) \cdot \mathbf{T}(s)\, ds = [M(x, y)\mathbf{i} + N(x, y)\mathbf{j}] \cdot \left(\frac{dx}{ds}\mathbf{i} + \frac{dy}{ds}\mathbf{j} \right) ds$$

$$= M(x, y)\, dx + N(x, y)\, dy$$

Thus

$$\oint_C \mathbf{F}(x, y) \cdot \mathbf{T}(s)\, ds = \oint_C M(x, y)\, dx + N(x, y)\, dy \qquad (11)$$

The curl of $\mathbf{F}$ in two dimensions was defined in Section 19.1 as

$$\text{curl } \mathbf{F}(x, y) = \left(\frac{\partial N}{\partial x} - \frac{\partial M}{\partial y} \right) \mathbf{k}$$

Therefore

$$\text{curl } \mathbf{F}(x, y) \cdot \mathbf{k} = \left(\frac{\partial N}{\partial x} - \frac{\partial M}{\partial y} \right) \mathbf{k} \cdot \mathbf{k}$$

$$\text{curl } \mathbf{F}(x, y) \cdot \mathbf{k} = \frac{\partial N}{\partial x} - \frac{\partial M}{\partial y}$$

Hence from this equation and (11), the equation of Green's theorem can be written

$$\oint_C \mathbf{F}(x, y) \cdot \mathbf{T}(s)\, ds = \iint\limits_R \text{curl } \mathbf{F}(x, y) \cdot \mathbf{k}\, dA$$

This vector form of Green's theorem is stated formally as the following theorem named after the Irish mathematician and physicist George Stokes (1819–1903).

19.4.4 THEOREM
Stokes's Theorem in
the Plane

Let the functions M and N, the curve C, and the region R be as defined in Green's theorem. If $\mathbf{F}(x, y) = M(x, y)\mathbf{i} + N(x, y)\mathbf{j}$ and $\mathbf{T}(s)$ is the unit tangent vector of C at P, where s units is the length of arc measured from a particular point P_0 on C to P, then

$$\oint_C \mathbf{F} \cdot \mathbf{T}\, ds = \iint\limits_R \text{curl } \mathbf{F} \cdot \mathbf{k}\, dA$$

EXAMPLE 6 Verify Stokes's theorem in the plane for $\mathbf{F}$ and R of Example 4.

Solution As in Example 4, the vector field $\mathbf{F}$ is defined by

$$\mathbf{F}(x, y) = 2y\mathbf{i} + 5x\mathbf{j}$$

and a vector equation of C is

$$\mathbf{R}(s) = \cos s\mathbf{i} + \sin s\mathbf{j} \qquad 0 \le s \le 2\pi$$

Because $\mathbf{T}(s) = D_s\mathbf{R}(s)$,

$$\mathbf{T}(s) = -\sin s\mathbf{i} + \cos s\mathbf{j}$$

At a point $P(\cos s, \sin s)$ on C, $\mathbf{F}$ is $2 \sin s\mathbf{i} + 5 \cos s\mathbf{j}$. Therefore

$$\oint_C \mathbf{F} \cdot \mathbf{T}\, ds = \int_0^{2\pi} (2 \sin s\mathbf{i} + 5 \cos s\mathbf{j}) \cdot (-\sin s\mathbf{i} + \cos s\mathbf{j})\, ds$$

$$= \int_0^{2\pi} (-2 \sin^2 s + 5 \cos^2 s)\, ds$$

$$= -2 \int_0^{2\pi} \frac{1 - \cos 2s}{2}\, ds + 5 \int_0^{2\pi} \frac{1 + \cos 2s}{2}\, ds$$

$$= -s + \tfrac{1}{2} \sin 2s + \tfrac{5}{2}s + \tfrac{5}{4} \sin 2s \Big]_0^{2\pi}$$

$$= \tfrac{3}{2}s + \tfrac{7}{4} \sin 2s \Big]_0^{2\pi}$$

$$= 3\pi$$

Because $N = 5x$, $\dfrac{\partial N}{\partial x} = 5$, and because $M = 2y$, $\dfrac{\partial M}{\partial y} = 2$. Thus

$$\iint_R \operatorname{curl} \mathbf{F} \cdot \mathbf{k}\, dA = \iint_R \left(\frac{\partial N}{\partial x} - \frac{\partial M}{\partial y} \right) dA$$

$$= \iint_R (5 - 2)\, dA$$

$$= 3 \iint_R dA$$

$$= 3\pi$$

Therefore Stokes's theorem in the plane is verified for this $\mathbf{F}$ and R.

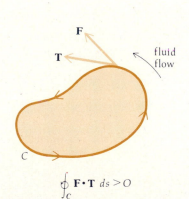

$$\oint_C \mathbf{F} \cdot \mathbf{T}\, ds > 0$$

FIGURE 9

If $\mathbf{F}$ is the velocity field of a fluid, the dot product $\mathbf{F} \cdot \mathbf{T}$ is the tangential component of $\mathbf{F}$ and the line integral $\oint_C \mathbf{F} \cdot \mathbf{T}\, ds$ is called the **circulation** of $\mathbf{F}$ around the closed curve C. In an intuitive sense, we can think of the circulation as being the sum of the tangential components of $\mathbf{F}$ around C. If motion around C is in the counterclockwise direction and $\oint_C \mathbf{F} \cdot \mathbf{T}\, ds > 0$, then the fluid is circulating counterclockwise; see Figure 9. If $\oint_C \mathbf{F} \cdot \mathbf{T}\, ds < 0$, the circulation of the fluid is clockwise; see Figure 10.

Let $\bar{P}(\bar{x}, \bar{y})$ be a particular point in region R and let C_δ be the circle having its center at $\bar{P}$ and having a small radius δ. If R_δ is the region enclosed by C_δ,

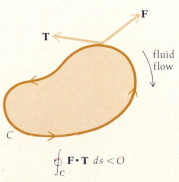

$$\oint_{C} \mathbf{F} \cdot \mathbf{T}\, ds < 0$$

FIGURE 10

$$\oint_{C_\delta} \mathbf{F} \cdot \mathbf{T}\, ds = \iint_{R_\delta} \operatorname{curl} \mathbf{F} \cdot \mathbf{k}\, dA$$

If M_y and N_x are continuous on R_δ, then curl $\mathbf{F} \cdot \mathbf{k}$ is continuous there and for small δ, curl $\mathbf{F} \cdot \mathbf{k}$ on R_δ is approximately curl $\mathbf{F}(\bar{x}, \bar{y}) \cdot \mathbf{k}$. Therefore

$$\oint_{C_\delta} \mathbf{F} \cdot \mathbf{T} \, ds \approx \text{curl } \mathbf{F}(\bar{x}, \bar{y}) \cdot \mathbf{k} \iint_{R_\delta} dA$$

$$\oint_{C_\delta} \mathbf{F} \cdot \mathbf{T} \, ds \approx \text{curl } \mathbf{F}(\bar{x}, \bar{y}) \cdot \mathbf{k}(\pi\delta^2)$$

Thus we interpret curl $\mathbf{F}(\bar{x}, \bar{y}) \cdot \mathbf{k}$ as the approximate measure of the rate of circulation per unit area in the counterclockwise direction at the point $\bar{P}$. When $\mathbf{F}$ and $\mathbf{T}$ are orthogonal vectors, $\mathbf{F} \cdot \mathbf{T} = 0$ and then curl $\mathbf{F} = \mathbf{0}$. In such a case $\mathbf{F}$ is said to be **irrotational**. This terminology is used even if $\mathbf{F}$ is not the velocity field of a fluid.

EXERCISES 19.4

In Exercises 1 through 8, evaluate the line integral by Green's theorem. Then verify the result by the method of Section 19.2.

1. $\oint_C 4y \, dx + 3x \, dy$, where C is the square with vertices at $(0, 0)$, $(1, 0)$, $(1, 1)$, and $(0, 1)$.

2. $\oint_C y^2 \, dx + x^2 \, dy$, where C is the square of Exercise 1.

3. $\oint_C 2xy \, dx - x^2 y \, dy$, where C is the triangle with vertices at $(0, 0)$, $(1, 0)$, and $(0, 1)$.

4. The line integral of Exercise 3, where C is the triangle with vertices at $(0, 0)$, $(1, 0)$, and $(1, 1)$.

5. $\oint_C x^2 y \, dx - y^2 x \, dy$, where C is the circle $x^2 + y^2 = 1$.

6. $\oint_C (x^2 - y^2) \, dx + 2xy \, dy$, where C is the circle $x^2 + y^2 = 1$.

7. The line integral of Exercise 5, where C is the closed curve consisting of the arc of $4y = x^3$ from $(0, 0)$ to $(2, 2)$ and the line segment from $(2, 2)$ to $(0, 0)$.

8. The line integral of Exercise 6, where C is the closed curve of Exercise 7.

In Exercises 9 through 20, use Green's theorem to evaluate the line integral.

9. $\oint_C (x + y) \, dx + xy \, dy$, where C is the closed curve determined by the x axis, the line $x = 2$, and the curve $4y = x^3$.

10. $\oint_C y^2 \, dx + x^2 \, dy$, where C is the closed curve determined by the x axis, the line $x = 1$, and the curve $y = x^2$.

11. $\oint_C (-x^2 + x) \, dy$, where C is the closed curve determined by the line $x - 2y = 0$ and the parabola $x = 2y^2$.

12. $\oint_C (x^2 + y) \, dx$, where C is the closed curve determined by the x axis and the parabola $y = 4 - x^2$.

13. $\oint_C \cos y \, dx + \cos x \, dy$, where C is the rectangle with vertices at $(0, 0)$, $(\frac{1}{3}\pi, 0)$, $(\frac{1}{3}\pi, \frac{1}{4}\pi)$, and $(0, \frac{1}{4}\pi)$.

14. $\oint_C e^{x+y} \, dx + e^{x+y} \, dy$, where C is the circle $x^2 + y^2 = 4$.

15. $\oint_C (\sin^4 x + e^{2x}) \, dx + (\cos^3 y - e^y) \, dy$, where C is the curve $x^4 + y^4 = 16$.

16. $\oint_C x \sin y \, dx - y \cos x \, dy$, where C is the rectangle with vertices at $(0, 0)$, $(\frac{1}{2}\pi, 0)$, $(\frac{1}{4}\pi, \frac{1}{2}\pi)$, and $(0, \frac{1}{4}\pi)$.

17. $\oint_C \frac{x^2 y}{x^2 + 1} \, dx - \tan^{-1} x \, dy$, where C is the ellipse $4x^2 + 25y^2 = 100$.

18. $\oint_C e^y \cos x \, dx + e^y \sin x \, dy$, where C is the curve $x^6 + y^4 = 10$.

19. $\oint_C (e^x - x^2 y) \, dx + 3x^2 y \, dy$, where C is the closed curve determined by $y = x^2$ and $x = y^2$.

20. $\oint_C \tan y \, dx - x \tan^2 y \, dy$, where C is the ellipse $x^2 + 4y^2 = 1$.

In Exercises 21 through 26, use Theorem 19.4.2 to find the area of the region.

21. The region having as its boundary the quadrilateral with vertices at $(0, 0)$, $(4, 0)$, $(3, 2)$, and $(1, 1)$.

22. The region having as its boundary the circle $x^2 + y^2 = a^2$.

23. The region bounded by the graphs of $y = x^2$ and $y = \sqrt{x}$.

24. The region bounded by the parabola $y = 2x^2$ and the line $y = 8x$.

25. The region bounded by the hypocycloid having parametric equations

$$x = a \cos^3 t \qquad y = a \sin^3 t \qquad a > 0 \qquad 0 \le t \le 2\pi$$

26. The region bounded below by the x axis and above by one arch of the cycloid having parametric equations

$$x = t - \sin t \qquad y = 1 - \cos t \qquad 0 \le t \le 2\pi$$

In Exercises 27 through 30, verify Gauss's divergence theorem in the plane and Stokes's theorem in the plane for $\mathbf{F}$ and R.

27. $\mathbf{F}(x, y) = 3x\mathbf{i} + 2y\mathbf{j}$ and R is the region bounded by the circle $x^2 + y^2 = 1$.

28. $\mathbf{F}(x, y) = 3y\mathbf{i} - 2x\mathbf{j}$ and R is the region bounded by $x^{2/3} + y^{2/3} = 1$.

29. $F(x, y) = x^2\mathbf{i} + y^2\mathbf{j}$ and R is the region bounded by the ellipse $4x^2 + 25y^2 = 100$.

30. $F(x, y) = y^2\mathbf{i} + x^2\mathbf{j}$ and R is the region bounded by the circle $x^2 + y^2 = 4$.

In Exercises 31 through 34, use Green's theorem to find the total work done in moving an object in the counterclockwise direction once around curve C if the motion is caused by the force field $\mathbf{F}(x, y)$. Assume the arc is measured in meters and the force is measured in newtons.

31. C is the ellipse $x^2 + 4y^2 = 16$; $F(x, y) = (3x + y)\mathbf{i} + (4x - 5y)\mathbf{j}$.

32. C is the circle $x^2 + y^2 = 25$;
$F(x, y) = (e^x + y^2)\mathbf{i} + (x^2y + \cos y)\mathbf{j}$.

33. C is the triangle with vertices at $(0, 0)$, $(2, 0)$, and $(0, 2)$;
$F(x, y) = (e^{x^2} + y^2)\mathbf{i} + (e^{y^2} + x^2)\mathbf{j}$.

34. C consists of the top half of the ellipse $9x^2 + 4y^2 = 36$ and the interval $[-2, 2]$ on the x axis; $F(x, y) = (xy + y^2)\mathbf{i} + xy\mathbf{j}$.

In Exercises 35 through 38, find the rate of flow of the fluid out of a region R bounded by curve C if $\mathbf{F}$ is the velocity field of the fluid. Assume the velocity is measured in centimeters per second and the area of R is measured in square centimeters.

35. $F(x, y) = (y^2 + 6x)\mathbf{i} + (2y - x^2)\mathbf{j}$; C is the ellipse $x^2 + 4y^2 = 4$.

36. $F(x, y) = (5x - y^2)\mathbf{i} + (3x - 2y)\mathbf{j}$; C is the right triangle having vertices at $(1, 2)$, $(4, 2)$, and $(4, 6)$.

37. $F(x, y) = x^3\mathbf{i} + y^3\mathbf{j}$; C is the circle $x^2 + y^2 = 1$.

38. $F(x, y) = xy^2\mathbf{i} + yx^2\mathbf{j}$; C is the circle $x^2 + y^2 = 9$.

In Exercises 39 through 42, $\mathbf{F}$ is the velocity field of a fluid around the closed curve C, where motion around C is in the counterclockwise direction. Use Stokes's theorem in the plane to compute $\oint_C \mathbf{F} \cdot \mathbf{T}\, ds$, and from the result determine which of the following applies: (i) the circulation of the fluid is counterclockwise; (ii) the circulation of the fluid is clockwise; (iii) $\mathbf{F}$ is irrotational.

39. $F(x, y) = 4y\mathbf{i} + 6x\mathbf{j}$; C is the triangle having vertices at $(0, 0)$, $(3, 0)$, and $(3, 5)$.

40. $F(x, y) = 8y\mathbf{i} + 3x\mathbf{j}$; C is the ellipse $4x^2 + 9y^2 = 1$.

41. $F(x, y) = \sin^2 x\mathbf{i} + \cos^2 y\mathbf{j}$; C is the ellipse $9x^2 + y^2 = 9$.

42. $F(x, y) = y^3\mathbf{i} + x^3\mathbf{j}$; C is the circle $x^2 + y^2 = 25$.

43. Prove $\oint_C N(x, y)\, dy = \iint\limits_R \frac{\partial N}{\partial x}\, dA$ if R is the region defined by $R = \{(x, y) | c \leq y \leq d, g_1(y) \leq x \leq g_2(y)\}$, where g_1 and g_2 are smooth.

19.5 SURFACE INTEGRALS

We now extend the concept of a line integral to that of an integral defined on a surface. We begin by considering a closed region in the xy plane. Let us denote this region by D, instead of R, to avoid confusion with the function defined by $R(x, y, z)$ used later in the discussion. Suppose S is a surface lying over D and having the equation $z = f(x, y)$, where f and its first partial derivatives are continuous on D. Then if σ is the measure of the area of the surface S, we have, from Theorem 18.5.1,

$$\sigma = \iint\limits_D \sqrt{f_x{}^2(x, y) + f_y{}^2(x, y) + 1}\; dx\, dy \qquad (1)$$

We can generalize the integral in (1) by considering a function G of the three variables x, y, and z, where G is continuous on S. We proceed in a manner similar to the discussion in Section 18.5 that precedes the statement of Theorem 18.5.1. Let Δ be a partition of region D into n rectangular subregions, where the ith rectangle has dimensions of measures $\Delta_i x$ and $\Delta_i y$ and an area of measure $\Delta_i A$. Let (ξ_i, γ_i) be any point in the ith rectangle, and at the point $Q(\xi_i, \gamma_i, f(\xi_i, \gamma_i))$ on the surface S consider the tangent plane to the surface. Project vertically upward the ith rectangle onto the tangent plane, and let $\Delta_i \sigma$ be the measure of the area of this projection. See Figure 1. The number $\Delta_i \sigma$ is an approximation to the measure of the area of the piece of the surface that lies above the ith rectangle. We showed in Section 18.5 that

$$\Delta_i \sigma = \sqrt{f_x{}^2(\xi_i, \gamma_i) + f_y{}^2(\xi_i, \gamma_i) + 1}\; \Delta_i A \qquad (2)$$

If we form the sum

$$\sum_{i=1}^n G(\xi_i, \gamma_i, f(\xi_i, \gamma_i))\, \Delta_i \sigma$$

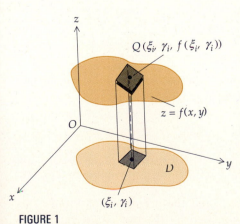

FIGURE 1

and take the limit of this sum as the norm of the partition approaches zero, we have

$$\lim_{||\Delta|| \to 0} \sum_{i=1}^{n} G(\xi_i, \gamma_i, f(\xi_i, \gamma_i)) \, \Delta_i \sigma \tag{3}$$

This limit is called the **surface integral** of G over S and is denoted by

$$\iint_S G(x, y, z) \, d\sigma$$

To obtain a formula for evaluating this surface integral, we substitute from (2) into (3) and we have

$$\lim_{||\Delta|| \to 0} \sum_{i=1}^{n} G(\xi_i, \gamma_i, f(\xi_i, \gamma_i)) \sqrt{f_x^2(\xi_i, \gamma_i) + f_y^2(\xi_i, \gamma_i) + 1} \; \Delta_i A$$

This limit is a double integral over the region D in the xy plane. Thus

$$\iint_S G(x, y, z) \, d\sigma = \iint_D G(x, y, f(x, y)) \sqrt{f_x^2(x, y) + f_y^2(x, y) + 1} \; dA \tag{4}$$

If $G(x, y, z) = 1$, then (4) becomes

$$\iint_S d\sigma = \iint_D \sqrt{f_x^2(x, y) + f_y^2(x, y) + 1} \; dA$$

Comparing this equation with (1), we observe that for this G the surface integral of G over S gives the measure of the area of surface S.

For the surface integral in (4), $z = f(x, y)$ is an equation of the surface S that is projected onto the region D in the xy plane. If an equation of surface S is of the form $y = g(x, z)$ and S is projected onto a region D in the xz plane, and g and its first partial derivatives are continuous on D, then

$$\iint_S G(x, y, z) \, d\sigma = \iint_D G(x, g,(x, z), z) \sqrt{g_x^2(x, z) + g_z^2(x, z) + 1} \; dA \tag{5}$$

Furthermore, if an equation of surface S is of the form $x = h(y, z)$, and S is projected onto a region D in the yz plane, and h and its first partial derivatives are continuous on D, then

$$\iint_S G(x, y, z) \, d\sigma = \iint_D G(h(y, z), y, z) \sqrt{h_y^2(y, z) + h_z^2(y, z) + 1} \; dA \tag{6}$$

EXAMPLE 1   Evaluate the surface integral

$$\iint_S x^2 z^2 \, d\sigma$$

where S is the portion of the cone $x^2 + y^2 = z^2$ between the planes $z = 1$ and $z = 2$.

Solution   Figure 2 shows the surface S and the projection of S onto the region D in the xy plane. Region D is bounded by the two circles of radii 1

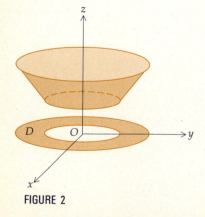

FIGURE 2

and 2 whose centers are at the origin. We solve the equation of S for z, where $z \geq 0$, and we obtain $z = \sqrt{x^2 + y^2}$. Therefore

$$f(x, y) = \sqrt{x^2 + y^2} \qquad f_x(x, y) = \frac{x}{\sqrt{x^2 + y^2}} \qquad f_y(x, y) = \frac{y}{\sqrt{x^2 + y^2}}$$

From (4), with $G(x, y, z) = x^2 z^2$, we obtain

$$\iint_S x^2 z^2 \, d\sigma = \iint_D x^2 (x^2 + y^2) \sqrt{\frac{x^2}{x^2 + y^2} + \frac{y^2}{x^2 + y^2} + 1} \, dA$$

$$= \iint_D x^2 (x^2 + y^2) \sqrt{2} \, dA$$

We evaluate the double integral by using polar coordinates, where $x = r \cos \theta$, $x^2 + y^2 = r^2$, and $dA = r \, dr \, d\theta$. Therefore

$$\iint_S x^2 z^2 \, d\sigma = \sqrt{2} \int_0^{2\pi} \int_1^2 (r^2 \cos^2 \theta) r^2 (r \, dr \, d\theta)$$

$$= \sqrt{2} \int_0^{2\pi} \int_1^2 \cos^2 \theta r^5 \, dr \, d\theta$$

$$= \sqrt{2} \int_0^{2\pi} \left[\cos^2 \theta \frac{r^6}{6} \right]_1^2 \, d\theta$$

$$= \frac{21 \sqrt{2}}{2} \int_0^{2\pi} \frac{1 + \cos 2\theta}{2} \, d\theta$$

$$= \frac{21 \sqrt{2}}{4} \left[\theta + \frac{\sin 2\theta}{2} \right]_0^{2\pi}$$

$$= \frac{21 \pi}{\sqrt{2}}$$

If the measure of the area density at the point (x, y, z) on a surface S is $\rho(x, y, z)$, and if M is the measure of the mass of S, then

$$M = \iint_S \rho(x, y, z) \, d\sigma \qquad (7)$$

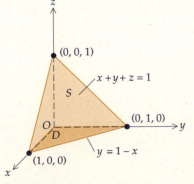

FIGURE 3

EXAMPLE 2   Find the mass of the portion of the plane $x + y + z = 1$ in the first octant if the area density at any point (x, y, z) on the surface is kx^2 kilograms per square meter, where k is a constant.

Solution   Figure 3 shows S, which is the surface of the given plane in the first octant, and the region D, which is the projection of S onto the xy plane. We solve the equation of the plane for z and obtain $z = 1 - x - y$. Therefore

$$f(x, y) = 1 - x - y \qquad f_x(x, y) = -1 \qquad f_y(x, y) = -1$$

From (7), with $\rho(x, y, z) = kx^2$, if M kilograms is the mass of the surface,

$$
\begin{aligned}
M &= \iint\limits_{S} kx^2 \, d\sigma \\
&= \iint\limits_{D} kx^2 \, \sqrt{f_x^2(x, y) + f_y^2(x, y) + 1} \, dA \\
&= \iint\limits_{D} kx^2 \, \sqrt{(-1)^2 + (-1)^2 + 1} \, dA \\
&= \sqrt{3}k \int_0^1 \int_0^{1-x} x^2 \, dy \, dx \\
&= \sqrt{3}k \int_0^1 \left[x^2 y \right]_0^{1-x} dx \\
&= \sqrt{3}k \int_0^1 (x^2 - x^3) \, dx \\
&= \sqrt{3}k \left[\frac{x^3}{3} - \frac{x^4}{4} \right]_0^1 \\
&= \tfrac{1}{12}\sqrt{3}k
\end{aligned}
$$

Thus the mass is $\tfrac{1}{12}\sqrt{3}k$ kilograms.

We now give an application of surface integrals to fluid flow. Let $\mathbf{F}$ be the velocity field of a fluid defined by

$$\mathbf{F}(x, y, z) = M(x, y, z)\mathbf{i} + N(x, y, z)\mathbf{j} + R(x, y, z)\mathbf{k}$$

Furthermore, suppose that the fluid flows through a surface S having the equation $z = f(x, y)$, which lies over a closed region D in the xy plane. Assume that f and its first partial derivatives are continuous on D. At each point of S there are two unit normal vectors to S. The unit normal having a positive $\mathbf{k}$ component is called the **unit upper normal** and the one having a negative $\mathbf{k}$ component is called the **unit lower normal**.

As in our discussion preceding Equation (2), take a partition of D into n rectangular subregions. Choose a point (ξ_i, γ_i) in the ith rectangle. Project vertically upward the ith rectangle onto the tangent plane at the point $Q(\xi_i, \gamma_i, f(\xi_i, \gamma_i))$ on S and let $\Delta_i\sigma$, given by (2), be an approximation to the measure of the area of this projection. Again refer to Figure 1. Now let $\mathbf{N}_i$ be the unit upper normal to S at point Q and let $\mathbf{F}_i$ be the velocity vector of the fluid at Q. The amount of fluid that crosses the projection per unit of time is given approximately by the volume of the parallelepiped having a base of area $\Delta_i\sigma$ square units and an altitude of length $\mathbf{F}_i \cdot \mathbf{N}_i$ units. See Figure 4. The measure of the volume of the parallelepiped is $\mathbf{F}_i \cdot \mathbf{N}_i \, \Delta_i\sigma$. The total amount of fluid that crosses S per unit of time is given approximately by

$$\sum_{i=1}^{n} \mathbf{F}_i \cdot \mathbf{N}_i \, \Delta_i\sigma$$

Taking the limit of this summation as n increases without bound and each $\Delta_i\sigma$ approaches zero, we obtain the surface integral

$$\iint\limits_{S} \mathbf{F} \cdot \mathbf{N} \, d\sigma \tag{8}$$

which is called the **flux** of $\mathbf{F}$ across S.

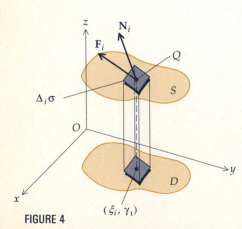

FIGURE 4

To evaluate surface integral (8), write the equation of S in the form $g(x, y, z) = 0$, where

$$g(x, y, z) = z - f(x, y)$$

From Theorem 17.2.2, a unit normal vector of the surface defined by $g(x, y, z) = 0$ is

$$\mathbf{N} = \frac{\nabla g}{\|\nabla g\|}$$

$$= \frac{-f_x(x, y)\mathbf{i} - f_y(x, y)\mathbf{j} + \mathbf{k}}{\sqrt{f_x^2(x, y) + f_y^2(x, y) + 1}}$$

Thus

$$\iint_S \mathbf{F} \cdot \mathbf{N}\, d\sigma = \iint_S (M\mathbf{i} + N\mathbf{j} + R\mathbf{k}) \cdot \left(\frac{-f_x\mathbf{i} - f_y\mathbf{j} + \mathbf{k}}{\sqrt{f_x^2 + f_y^2 + 1}}\right) d\sigma$$

$$= \iint_D \frac{-Mf_x - Nf_y + R}{\sqrt{f_x^2 + f_y^2 + 1}} (\sqrt{f_x^2 + f_y^2 + 1})\, dA$$

Therefore, we conclude that

$$\iint_S \mathbf{F} \cdot \mathbf{N}\, d\sigma = \iint_D (-Mf_x - Nf_y + R)\, dA \qquad (9)$$

where $\mathbf{N}$ is a unit upper normal. If $\mathbf{N}$ is a unit lower normal (where the component of $\mathbf{k}$ is negative)

$$\iint_S \mathbf{F} \cdot \mathbf{N}\, d\sigma = \iint_D (Mf_x + Nf_y - R)\, dA \qquad (10)$$

This formula is proved in a manner similar to that used to prove (9).

EXAMPLE 3   The velocity field of a fluid is given by

$$\mathbf{F}(x, y, z) = y\mathbf{i} - x\mathbf{j} + 8\mathbf{k}$$

and surface S is that part of the sphere $x^2 + y^2 + z^2 = 9$ that is above the region D in the xy plane enclosed by the circle $x^2 + y^2 = 4$. Find the flux of $\mathbf{F}$ across S.

Solution   Figure 5 shows the surface S and the region D in the xy plane. We solve the equation of the sphere for z, with $z > 0$, and obtain $z = \sqrt{9 - x^2 - y^2}$. Therefore

$$f(x, y) = \sqrt{9 - x^2 - y^2} \qquad f_x = \frac{-x}{\sqrt{9 - x^2 - y^2}} \qquad f_y = \frac{-y}{\sqrt{9 - x^2 - y^2}}$$

$$f_x = -\frac{x}{z} \qquad\qquad f_y = -\frac{y}{z}$$

From the definition of flux we have

$$\text{flux of } \mathbf{F} \text{ across } S = \iint_S \mathbf{F} \cdot \mathbf{N}\, d\sigma$$

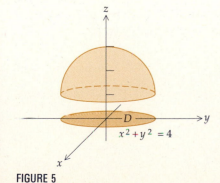

$x^2 + y^2 = 4$

FIGURE 5

From the given velocity field, $M = y$, $N = -x$, and $R = 8$. Therefore, from (9),

$$\text{flux of } \mathbf{F} \text{ across } S = \iint_D (-Mf_x - Nf_y + R) \, dA$$

$$= \iint_D \left[-y \left(-\frac{x}{z} \right) - (-x) \left(-\frac{y}{z} \right) + 8 \right] dA$$

$$= 8 \iint_D dA$$

Because D is the region enclosed by the circle $x^2 + y^2 = 4$, $A = 4\pi$. Thus

$$\text{flux of } \mathbf{F} \text{ across } S = 8(4\pi)$$

$$= 32\pi$$

We conclude, then, that the rate of flow of the fluid across S is 32π cubic units of length per unit of time.

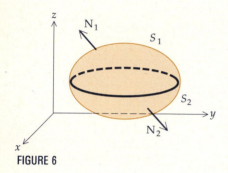

FIGURE 6

Suppose S is a closed surface, examples of which are rectangular parallelepipeds, spheres, and ellipsoids. When using (8) to compute the flux of $\mathbf{F}$ across a closed surface, we select $\mathbf{N}$ as a **unit outward normal**, which is a normal whose direction is away from the solid bounded by the surface. In particular, if S is an ellipsoid, as shown in Figure 6, we consider S as consisting of an upper surface S_1 and a lower surface S_2, as indicated in the figure. In such a case the flux of $\mathbf{F}$ across S is

$$\iint_S \mathbf{F} \cdot \mathbf{N} \, d\sigma = \iint_{S_1} \mathbf{F} \cdot \mathbf{N}_1 \, d\sigma + \iint_{S_2} \mathbf{F} \cdot \mathbf{N}_2 \, d\sigma$$

For the surface integral across S_1, $\mathbf{N}_1$ is a unit upper normal, and for the surface integral across S_2, $\mathbf{N}_2$ is a unit lower normal.

EXAMPLE 4   The velocity field of a fluid is given by $\mathbf{F}(x, y, z) = 5z\mathbf{k}$, and S is the sphere $x^2 + y^2 + z^2 = 16$. Find the flux of $\mathbf{F}$ across S if length is measured in centimeters and time is measured in hours.

Solution   Figure 7 shows the sphere and the region D in the xy plane, which is the circle $x^2 + y^2 = 16$. Because $\mathbf{F}(x, y, z) = 5z\mathbf{k}$, $M = 0$, $N = 0$, and $R = 5z$. The flux of $\mathbf{F}$ across S is

$$\iint_S \mathbf{F} \cdot \mathbf{N} \, d\sigma = \iint_{S_1} \mathbf{F} \cdot \mathbf{N}_1 \, d\sigma + \iint_{S_2} \mathbf{F} \cdot \mathbf{N}_2 \, d\sigma \qquad (11)$$

where S_1 is the top half of the sphere and S_2 is the bottom half. For S_1, $\mathbf{N}_1$ is a unit upper normal and an equation of S_1 is $z = \sqrt{16 - x^2 - y^2}$. Thus $f(x, y) = \sqrt{16 - x^2 - y^2}$. From (9),

$$\iint_{S_1} \mathbf{F} \cdot \mathbf{N}_1 \, d\sigma = \iint_D [-Mf_x - Nf_y + R] \, dA$$

$$= \iint_D 5z \, dA$$

$$= 5 \iint_D \sqrt{16 - x^2 - y^2} \, dA$$

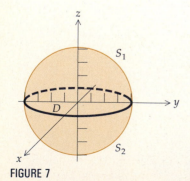

FIGURE 7

$$= 5 \int_0^{2\pi} \int_0^4 \sqrt{16 - r^2} \, r \, dr \, d\theta$$

$$= 5 \int_0^{2\pi} -\frac{1}{3}(16 - r^2)^{3/2} \Big]_0^4 \, d\theta$$

$$= \frac{320}{3} \int_0^{2\pi} d\theta$$

$$= \frac{640}{3}\pi$$

For S_2, $\mathbf{N}_2$ is a unit lower normal and as an equation of S_2 we have $z = -\sqrt{16 - x^2 - y^2}$. Therefore $f(x, y) = -\sqrt{16 - x^2 - y^2}$. From (10),

$$\iint_{S_2} \mathbf{F} \cdot \mathbf{N}_2 \, d\sigma = \iint_D [Mf_x + Nf_y - R] \, dA$$

$$= \iint_D -5z \, dA$$

$$= 5 \iint_D \sqrt{16 - x^2 - y^2} \, dA$$

As in the computation for the flux of $\mathbf{F}$ across S_1 we get

$$\iint_{S_2} \mathbf{F} \cdot \mathbf{N} \, d\sigma = \frac{640}{3}\pi$$

Hence from (11),

$$\iint_S \mathbf{F} \cdot \mathbf{N} \, d\sigma = \frac{640}{3}\pi + \frac{640}{3}\pi$$

$$= \frac{1280}{3}\pi$$

Therefore the rate of flow of the fluid across the sphere is $\frac{1280}{3}\pi$ cm^3/hr.

The concept of flux is not limited to velocity fields of fluids. For instance, if $\mathbf{F}$ is an electric field, then surface integral (8) is an electric flux, and if $\mathbf{F}$ is a magnetic field, the surface integral is a magnetic flux. Surface integral (8) could also represent a flux of heat.

EXERCISES 19.5

In Exercises 1 through 14, evaluate the surface integral $\iint_S G(x, y, z) \, d\sigma$ *for G and S.*

1. $G(x, y, z) = z$; S is the hemisphere $x^2 + y^2 + z^2 = 4$ above the xy plane.
2. $G(x, y, z) = x$; S is the portion of the plane $x + y + z = 1$ in the first octant.
3. $G(x, y, z) = x + 2y - z$; S is the portion of the plane $x + y + z = 2$ in the first octant.
4. $G(x, y, z) = z$; S is the portion of the plane $2x + 3y + z = 6$ in the first octant.
5. $G(x, y, z) = xyz$; S is the same as in Exercise 4.
6. $G(x, y, z) = x^2$; S is the portion of the cylinder $x^2 + y^2 = 1$ between the xy plane and the plane $z = 1$ in the first octant.
7. $G(x, y, z) = x$; S is the portion of the cylinder $z = x^2$ in the

first octant bounded by the coordinate planes and the planes $x = 1$ and $y = 2$.

8. $G(x, y, z) = y$; S is the portion of the cylinder $z = 4 - y^2$ in the first octant bounded by the coordinate planes and the plane $x = 3$.
9. $G(x, y, z) = z^2$; S is the portion of the cone $x^2 + y^2 = z^2$ between the planes $z = 1$ and $z = 2$.
10. $G(x, y, z) = xyz$; S is the portion of the cone $x^2 + y^2 = z^2$ between the planes $z = 1$ and $z = 2$.
11. $G(x, y, z) = x + y$; S is the portion of the plane

$$4x + 3y + 6z = 12$$

in the first octant.

12. $G(x, y, z) = \sqrt{x^2 + y^2 + z^2}$; S is the portion of the cone $x^2 + y^2 = z^2$ between the xy plane and the plane $z = 2$.

13. $G(x, y, z) = xyz$; S is the portion of the cylinder $x^2 + z^2 = 4$ between the planes $y = 1$ and $y = 3$.
14. $G(x, y, z) = x^2$; S is the hemisphere $x^2 + y^2 + z^2 = 9$ above the xy plane.

In Exercises 15 through 20, find the mass of the surface S if the area density at any point (x, y, z) on the surface is $\rho(x, y, z)$ kilograms per square meter.

15. S is the hemisphere $x^2 + y^2 + z^2 = 4$ above the xy plane; $\rho(x, y, z) = k\sqrt{x^2 + y^2 + z^2}$, where k is a constant.
16. S is the portion of the plane $3x + 2y + z = 6$ in the first octant; $\rho(x, y, z) = y + 2z$.
17. S is the portion of the paraboloid $z = 9 - x^2 - y^2$ above the xy plane; $\rho(x, y, z) = 1/\sqrt{4x^2 + 4y^2 + 1}$.
18. S is the hemisphere $x^2 + y^2 + z^2 = 1$ below the xy plane; $\rho(x, y, z) = x^2 + y^2$.
19. S is the portion of the cone $x^2 + y^2 = z^2$ between the planes $z = 2$ and $z = 3$; $\rho(x, y, z) = y^2z^2$.
20. S is the portion of the sphere $x^2 + y^2 + z^2 = 16$ in the first octant; $\rho(x, y, z) = kz^2$, where k is a constant.

In Exercises 21 through 24, find the flux of $\mathbf{F}$ across the surface S where $\mathbf{F}(x, y, z)$ gives the velocity field of a fluid.

21. $\mathbf{F}(x, y, z) = x\mathbf{i} + y\mathbf{j} + z\mathbf{k}$; S is the portion of the plane $3x + 2y + z = 6$ in the first octant.
22. $\mathbf{F}(x, y, z)$ is the same as in Exercise 21; S is the hemisphere $x^2 + y^2 + z^2 = 1$ above the xy plane.
23. $\mathbf{F}(x, y, z) = -2y\mathbf{i} + 2x\mathbf{j} + 5\mathbf{k}$; S is that part of the sphere $x^2 + y^2 + z^2 = 16$ that is above the region in the xy plane enclosed by the circle $x^2 + y^2 = 9$.
24. $\mathbf{F}(x, y, z) = 3x\mathbf{i} + 3y\mathbf{j} + 6z\mathbf{k}$; S is the portion of the paraboloid $z = 4 - x^2 - y^2$ above the xy plane.
25. Suppose $\mathbf{F}(x, y, z) = x^2\mathbf{i} + xy\mathbf{j} + 2z\mathbf{k}$, and S is the cube in the first octant bounded by the coordinate planes and the planes $x = 1$, $y = 1$, and $z = 1$. Find the flux of $\mathbf{F}$ across S by evaluating six surface integrals, one for each face of the cube.
26. If $\mathbf{F}(x, y, z) = 3x\mathbf{i} + y^2\mathbf{j} + yz\mathbf{k}$ and S is the cube of Exercise 25, find the flux of $\mathbf{F}$ across S by evaluating six surface integrals, one for each face of the cube.

19.6 GAUSS'S DIVERGENCE THEOREM AND STOKES'S THEOREM

The discussion in Section 19.4 was concerned with a treatment of Green's theorem and its two vector forms, Gauss's divergence theorem in the plane and Stokes's theorem in the plane. The two vector forms of Green's theorem can be generalized to Gauss's divergence theorem and Stokes's theorem in three dimensions. A rigorous presentation of these theorems belongs to a course in advanced calculus; however, in this section we give a brief intuitive introduction to them.

Gauss's divergence theorem in the plane (Theorem 19.4.3) is a special case of the following theorem in three-dimensional space.

19.6.1 THEOREM
Gauss's Divergence Theorem

Let M, N, and R be functions of three variables x, y, and z such that they have continuous first partial derivatives on an open ball B in R^3. Let S be a sectionally smooth closed surface lying in B, and let E be the region bounded by S. If $\mathbf{F}(x, y, z) = M(x, y, z)\mathbf{i} + N(x, y, z)\mathbf{j} + R(x, y, z)\mathbf{k}$, and $\mathbf{N}$ is a unit outward normal vector of S, then

$$\iint_S \mathbf{F} \cdot \mathbf{N} \, d\sigma = \iiint_E \text{div } \mathbf{F} \, dV$$

This theorem states that the flux of $\mathbf{F}$ across the boundary S of a region E in R^3 is the triple integral of the divergence of $\mathbf{F}$ over E. Its proof is beyond the scope of this book. The following example verifies it for a particular $\mathbf{F}$ and S.

EXAMPLE 1   Use Gauss's divergence theorem to solve Example 4 of Section 19.5.

Solution $\mathbf{F}(x, y, z) = 5z\mathbf{k}$ and S is the sphere $x^2 + y^2 + z^2 = 16$. From Gauss's divergence theorem, the flux of $\mathbf{F}$ across S is

$$\iint_S \mathbf{F} \cdot \mathbf{N} \, d\sigma = \iiint_E \text{div } \mathbf{F} \, dV$$

Since $\mathbf{F}(x, y, z) = 5z\mathbf{k}$, div $\mathbf{F} = \dfrac{\partial}{\partial z}(5z)$; that is, div $\mathbf{F} = 5$. Thus

$$\iint_S \mathbf{F} \cdot \mathbf{N} \, d\sigma = 5 \iiint_E dV$$

Because the volume of E is the volume of a sphere of radius 4, we have

$$\iint_S \mathbf{F} \cdot \mathbf{N} \, d\sigma = 5\left[\tfrac{4}{3}\pi(4)^3\right]$$

$$= \tfrac{1280}{3}\pi$$

By comparing the solution of the above example with that of Example 4 in Section 19.5, observe how Gauss's divergence theorem can simplify the computation of a surface integral.

EXAMPLE 2 If $\mathbf{F}(x, y, z) = x^2 y\mathbf{i} + y^2\mathbf{j} + xz\mathbf{k}$, and S is the cube in the first octant bounded by the planes $x = 1$, $y = 1$, $z = 1$, and the coordinate planes, find the flux of $\mathbf{F}$ across S.

Solution The cube appears in Figure 1. The flux of $\mathbf{F}$ across S is

$$\iint_S \mathbf{F} \cdot \mathbf{N} \, d\sigma$$

To compute this surface integral directly we would have to evaluate six surface integrals, one for each face of the cube. By applying Gauss's divergence theorem with

$$\text{div } \mathbf{F} = \frac{\partial}{\partial x}(x^2 y) + \frac{\partial}{\partial y}(y^2) + \frac{\partial}{\partial z}(xz)$$

$$= 2xy + 2y + x$$

we have

$$\iint_S \mathbf{F} \cdot \mathbf{N} \, d\sigma = \iiint_E \text{div } \mathbf{F} \, dV$$

$$= \int_0^1 \int_0^1 \int_0^1 (2xy + 2y + x) \, dz \, dy \, dx$$

$$= \int_0^1 \int_0^1 (2xy + 2y + x) \, dy \, dx$$

$$= \int_0^1 \left[xy^2 + y^2 + xy \right]_0^1 dx$$

$$= \int_0^1 (2x + 1) \, dx$$

$$= x^2 + x \Big]_0^1$$

$$= 2$$

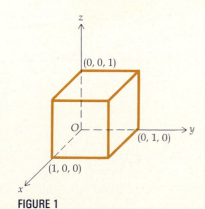

FIGURE 1

In Section 19.4 we had a second vector form of Green's theorem, known as Stokes's theorem in the plane (Theorem 19.4.4):

$$\oint_C \mathbf{F} \cdot \mathbf{T} \, ds = \iint_D \text{curl } \mathbf{F} \cdot \mathbf{k} \, dA$$

where C is a sectionally smooth simple closed curve in R^2 and D is the region bounded by C. We now extend this theorem to three-dimensional space.

19.6.2 THEOREM
Stokes's Theorem

Let M, N, and R be functions of three variables x, y, and z such that they have continuous first partial derivatives on an open ball B in R^3. Let S be a sectionally smooth surface lying in B and let C be a sectionally smooth simple closed curve that is the boundary of S. If $\mathbf{F}(x, y, z) = M(x, y, z)\mathbf{i} + N(x, y, z)\mathbf{j} + R(x, y, z)\mathbf{k}$, $\mathbf{N}$ is a unit upward normal vector of S and $\mathbf{T}$ is a unit tangent vector to C where s units is the length of arc measured from a particular point P_0 on C to P, then

$$\oint_C \mathbf{F} \cdot \mathbf{T} \, ds = \iint_S \text{curl } \mathbf{F} \cdot \mathbf{N} \, d\sigma$$

Stokes's theorem states that the line integral of the tangential component of a vector field $\mathbf{F}$ around the boundary C of a surface S can be computed by evaluating the surface integral of the normal component of the curl of $\mathbf{F}$ over S.

Theorem 19.6.2 is restricted to surfaces for which N is an upward normal of S. A complete statement of Stokes's theorem, involving surfaces having an orientation and for which a unit normal $\mathbf{N}$ can be adequately defined, can be found in an advanced calculus text. The proof of the theorem can also be found there.

Figure 2 shows a surface S with boundary curve C for which Theorem 19.6.2 applies. An equation of S is of the form $z = f(x, y)$, where f has continuous first partial derivatives on region D that is the projection of S onto the xy plane. Curve $\bar{C}$ is the projection of C onto the xy plane, and D and $\bar{C}$ are defined as in Green's theorem (Theorem 19.4.1). The positive direction along C is the same as the positive direction along $\bar{C}$, which is counterclockwise. Figure 2 also shows representations of vectors $\mathbf{N}$ and $\mathbf{T}$.

Another form of the equation of Stokes's theorem is obtained by writing $d\mathbf{R}$ in place of $\mathbf{T} \, ds$ in the line integral on the left. We then have

$$\oint_C \mathbf{F} \cdot d\mathbf{R} = \iint_S \text{curl } \mathbf{F} \cdot \mathbf{N} \, d\sigma \qquad (1)$$

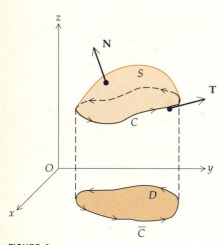

FIGURE 2

EXAMPLE 3   Let the force field $\mathbf{F}$ be defined by

$$\mathbf{F}(x, y, z) = -4y\mathbf{i} + 2z\mathbf{j} + 3x\mathbf{k}$$

and suppose S is the portion of the paraboloid $z = 10 - x^2 - y^2$ above the plane $z = 1$. Verify Stokes's theorem for this $\mathbf{F}$ and S by computing each of the following:

(a) $\oint_C \mathbf{F} \cdot d\mathbf{R}$ where a vector equation of C is $\mathbf{R}(t) = 3 \cos t\mathbf{i} + 3 \sin t\mathbf{j} + \mathbf{k}$;

(b) $\oint_C \mathbf{F} \cdot \mathbf{T} \, ds$;

(c) $\iint_S \text{curl } \mathbf{F} \cdot \mathbf{N} \, d\sigma$.

Solution   Figure 3 shows the surface S and the region D, which is the projection of S onto the xy plane. Region D is bounded by the circle $x^2 + y^2 = 9$. The curve C, which is the boundary of S, is the circle having its center at $(0, 0, 1)$ and radius 3, in the plane $z = 1$.

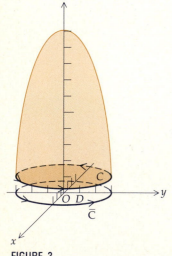

FIGURE 3

(a) We are given the following vector equation of C:

$$\mathbf{R}(t) = 3 \cos t\mathbf{i} + 3 \sin t\mathbf{j} + \mathbf{k} \qquad 0 \leq t \leq 2\pi \tag{2}$$

Thus

$$\mathbf{R}'(t) = -3 \sin t\mathbf{i} + 3 \cos t\mathbf{j} \tag{3}$$

$$\oint_C \mathbf{F} \cdot d\mathbf{R} = \int_C \mathbf{F}(\mathbf{R}(t)) \cdot \mathbf{R}'(t)\, dt$$

$$= \int_0^{2\pi} (-12 \sin t\mathbf{i} + 2\mathbf{j} + 9 \cos t\mathbf{k}) \cdot (-3 \sin t\mathbf{i} + 3 \cos t\mathbf{j})\, dt$$

$$= \int_0^{2\pi} (36 \sin^2 t + 6 \cos t)\, dt$$

$$= 36 \int_0^{2\pi} \frac{1 - \cos 2t}{2}\, dt + 6 \int_0^{2\pi} \cos t\, dt$$

$$= 18t - 9 \sin 2t + 6 \sin t \Big]_0^{2\pi}$$

$$= 36\pi$$

(b) To compute $\oint_C \mathbf{F} \cdot \mathbf{T}\, ds$, we obtain a vector equation of C having s as a parameter, where s units is the arc length measured from the point where $t = 0$. Because $\dfrac{ds}{dt} = \|R'(t)\|$ we have from (3)

$$\frac{ds}{dt} = \sqrt{9 \sin^2 t + 9 \cos^2 t}$$

$$= 3\sqrt{\sin^2 t + \cos^2 t}$$

$$= 3$$

Therefore, $s = 3t + C$, and since $s = 0$ when $t = 0$, $C = 0$. Thus

$$s = 3t$$

From (2) with $t = \tfrac{1}{3}s$, we obtain

$$\mathbf{R}(s) = 3 \cos \tfrac{1}{3}s\mathbf{i} + 3 \sin \tfrac{1}{3}s\mathbf{j} + \mathbf{k} \qquad 0 \leq s \leq 6\pi$$

Because $\mathbf{T}(s) = D_s \mathbf{R}(s)$, we have

$$\mathbf{T}(s) = -\sin \tfrac{1}{3}s\mathbf{i} + \cos \tfrac{1}{3}s\mathbf{j} \qquad 0 \leq s \leq 6\pi$$

We have then

$$\oint_C \mathbf{F} \cdot \mathbf{T}\, ds = \int_C \mathbf{F}(\mathbf{R}(s)) \cdot \mathbf{T}(s)\, ds$$

$$= \int_0^{6\pi} (-12 \sin \tfrac{1}{3}s\mathbf{i} + 2\mathbf{j} + 9 \cos \tfrac{1}{3}s\mathbf{k}) \cdot (-\sin \tfrac{1}{3}s\mathbf{i} + \cos \tfrac{1}{3}s\mathbf{j})\, ds$$

$$= \int_0^{6\pi} (12 \sin^2 \tfrac{1}{3}s + 2 \cos \tfrac{1}{3}s)\, ds$$

$$= 12 \int_0^{6\pi} \frac{1 - \cos \tfrac{2}{3}s}{2}\, ds + 2 \int_0^{6\pi} \cos \tfrac{1}{3}s\, ds$$

$$= 6s - 9 \sin \tfrac{2}{3}s + 6 \sin \tfrac{1}{3}s \Big]_0^{6\pi}$$

$$= 36\pi$$

(c)   We first compute curl **F**.

$$\text{curl } \mathbf{F} = \begin{vmatrix} \mathbf{i} & \mathbf{j} & \mathbf{k} \\ \dfrac{\partial}{\partial x} & \dfrac{\partial}{\partial y} & \dfrac{\partial}{\partial z} \\ -4y & 2z & 3x \end{vmatrix}$$

$$= -2\mathbf{i} - 3\mathbf{j} + 4\mathbf{k}$$

Thus

$$\iint\limits_{S} \text{curl } \mathbf{F} \cdot \mathbf{N} \, d\sigma = \iint\limits_{S} (-2\mathbf{i} - 3\mathbf{j} + 4\mathbf{k}) \cdot \mathbf{N} \, d\sigma$$

To evaluate this surface integral we apply (9) of Section 19.5 because **N** is a unit upper normal. The vector field is $-2\mathbf{i} - 3\mathbf{j} + 4\mathbf{k}$; thus $M = -2$, $N = -3$, and $R = 4$. Because an equation of the surface is $z = 10 - x^2 - y^2$,

$$f(x, y) = 10 - x^2 - y^2 \qquad f_x(x, y) = -2x \qquad f_y(x, y) = -2y$$

Therefore,

$$\iint\limits_{S} \text{curl } \mathbf{F} \cdot \mathbf{N} \, d\sigma = \iint\limits_{D} [-(-2)(-2x) - (-3)(-2y) + 4] \, dA$$

$$= \iint\limits_{D} (-4x - 6y + 4) \, dA$$

$$= \int_{0}^{2\pi} \int_{0}^{3} (-4r \cos \theta - 6r \sin \theta + 4) r \, dr \, d\theta$$

$$= \int_{0}^{2\pi} \left[-\frac{4}{3} r^3 \cos \theta - 2r^3 \sin \theta + 2r^2 \right]_{0}^{3} d\theta$$

$$= \int_{0}^{2\pi} (-36 \cos \theta - 54 \sin \theta + 18) \, d\theta$$

$$= -36 \sin \theta + 54 \cos \theta + 18\theta \Big]_{0}^{2\pi}$$

$$= 36\pi$$

The results of parts (a), (b), and (c) are all 36π. We have therefore verified Stokes's theorem for this **F** and S.

EXAMPLE 4   Use Stokes's theorem to evaluate the line integral

$$\oint_{C} \mathbf{F} \cdot \mathbf{T} \, ds$$

if $\mathbf{F}(x, y, z) = xz\mathbf{i} + xy\mathbf{j} + y^2\mathbf{k}$ and C is the boundary of the surface consisting of the portion of the cylinder $z = 4 - x^2$ in the first octant that is cut off by the coordinate planes and the plane $y = 3$.

Solution   Figure 4 shows the surface S and the boundary curve C that is composed of the four arcs C_1, C_2, C_3, and C_4. From Stokes's theorem,

$$\oint_{C} \mathbf{F} \cdot \mathbf{T} \, ds = \iint\limits_{S} \text{curl } \mathbf{F} \cdot \mathbf{N} \, d\sigma$$

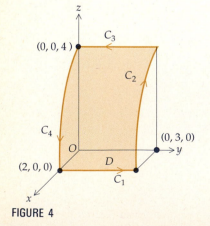

FIGURE 4

$$\text{curl } \mathbf{F} = \begin{vmatrix} \mathbf{i} & \mathbf{j} & \mathbf{k} \\ \dfrac{\partial}{\partial x} & \dfrac{\partial}{\partial y} & \dfrac{\partial}{\partial z} \\ xz & xy & y^2 \end{vmatrix}$$

$$= 2y\mathbf{i} + x\mathbf{j} + y\mathbf{k}$$

Thus

$$\oint_C \mathbf{F} \cdot \mathbf{T} \, ds = \iint_S (2y\mathbf{i} + x\mathbf{j} + y\mathbf{k}) \cdot \mathbf{N} \, d\sigma$$

Because $\mathbf{N}$ is a unit upper normal vector, we compute the value of the surface integral by applying (9) of Section 19.5. Because the vector field is $2y\mathbf{i} + x\mathbf{j} + y\mathbf{k}$, then $M = 2y$, $N = x$, and $R = y$. An equation of S is $z = 4 - x^2$. Therefore

$$f(x, y) = 4 - x^2 \qquad f_x(x, y) = -2x \qquad f_y(x, y) = 0$$

Hence, we have

$$\oint_C \mathbf{F} \cdot \mathbf{T} \, ds = \iint_D \left[-(2y)(-2x) - x(0) + y \right] dA$$

$$= \iint_D (4xy + y) \, dA$$

Region D is enclosed by the rectangle in the xy plane bounded by the x and y axes and the lines $x = 2$ and $y = 3$. Therefore

$$\oint_C \mathbf{F} \cdot \mathbf{T} \, ds = \int_0^2 \int_0^3 (4xy + y) \, dy \, dx$$

$$= \int_0^2 \left[2xy^2 + \frac{1}{2} y^2 \right]_0^3 dx$$

$$= \int_0^2 (18x + \tfrac{9}{2}) \, dx$$

$$= 9x^2 + \tfrac{9}{2}x \Big]_0^2$$

$$= 45$$

EXERCISES 19.6

In Exercises 1 through 4, verify Gauss's divergence theorem for $\mathbf{F}$ and S.

1. $\mathbf{F}(x, y, z) = 2z\mathbf{k}$; S is the sphere $x^2 + y^2 + z^2 = 1$.
2. $\mathbf{F}(x, y, z) = x\mathbf{i} + y\mathbf{j} + z\mathbf{k}$; S is the sphere $x^2 + y^2 + z^2 = 9$.
3. $\mathbf{F}(x, y, z) = xy\mathbf{i} + yz\mathbf{j}$; S is the cube bounded by the coordinate planes and the planes $x = 1$, $y = 1$, and $z = 1$.
4. $\mathbf{F}(x, y, z) = 4x\mathbf{i} - 2y\mathbf{j} + z\mathbf{k}$; S is the portion of the paraboloid $z = x^2 + y^2$ below the plane $z = 4$.

In Exercises 5 through 8, for the $\mathbf{F}$ and S of the indicated exercise in Exercise 19.5, find the flux of $\mathbf{F}$ across S by Gauss's divergence theorem.

5. Exercise 21
6. Exercise 22
7. Exercise 25
8. Exercise 26

In Exercises 9 through 16, use Gauss's divergence theorem to evaluate the surface integral $\iint_S \mathbf{F} \cdot \mathbf{N} \, d\sigma$ for $\mathbf{F}$ and S.

9. $\mathbf{F}(x, y, z) = x^2yz\mathbf{i} + xy^2z\mathbf{j} + xyz^2\mathbf{k}$; S is the cube in the first octant bounded by the coordinate planes and the planes $x = 1$, $y = 2$, and $z = 3$.
10. $\mathbf{F}(x, y, z) = 6x\mathbf{i} + 3y\mathbf{j} + 2z\mathbf{k}$; S is the tetrahedron having vertices at the origin and the points $(3, 0, 0,)$, $(0, 1, 0)$, and $(0, 0, 2)$.
11. $\mathbf{F}(x, y, z) = x\mathbf{i} + y\mathbf{j} + z\mathbf{k}$; S is the sphere $x^2 + y^2 + z^2 = 4$.
12. $\mathbf{F}(x, y, z) = x\mathbf{i} + y\mathbf{j} + z\mathbf{k}$; S is the boundary of the region enclosed on the side by the cylinder $x^2 + y^2 = 9$, below by the xy plane, and above by the plane $z = 4$.
13. $\mathbf{F}(x, y, z) = x^2\mathbf{i} + y^2\mathbf{j} + z^2\mathbf{k}$; S is the surface of Exercise 12.
14. $\mathbf{F}(x, y, z) = x^2\mathbf{i} + y^2\mathbf{j} + z^2\mathbf{k}$; S is the surface of Exercise 11.

15. $\mathbf{F}(x, y, z) = 2x\mathbf{i} + 2yz\mathbf{j} + 3z\mathbf{k}$; S is the boundary of the region enclosed by the coordinate planes and the plane $x + y + z = 1$.

16. $\mathbf{F}(x, y, z) = \dfrac{x\mathbf{i} + y\mathbf{j} + z\mathbf{k}}{x^2 + y^2 + z^2}$; S is the boundary of the region outside the sphere $x^2 + y^2 + z^2 = 1$ and inside the sphere $x^2 + y^2 + z^2 = 4$.

In Exercises 17 through 22, verify Stokes's theorem for $\mathbf{F}$ and S.

17. $\mathbf{F}(x, y, z) = y^2\mathbf{i} + x^2\mathbf{j} + z^2\mathbf{k}$;
S is the hemisphere $x^2 + y^2 + z^2 = 1$ above the xy plane.

18. $\mathbf{F}(x, y, z) = x^2\mathbf{i} + y^2\mathbf{j} + z^2\mathbf{k}$;
S is the hemisphere $x^2 + y^2 + z^2 = 1$ below the xy plane.

19. $\mathbf{F}(x, y, z) = y^2\mathbf{i} + x\mathbf{j} + z^2\mathbf{k}$; S is the portion of the paraboloid $z = x^2 + y^2$ below the plane $z = 1$.

20. $\mathbf{F}(x, y, z) = xy\mathbf{i} + y^2\mathbf{j} + 2\mathbf{k}$; S is the surface of Exercise 19.

21. $\mathbf{F}(x, y, z) = -3y\mathbf{i} + 3x\mathbf{j} + 2\mathbf{k}$; S is the portion of the plane $z = 1$ inside the cylinder $x^2 + y^2 = 9$.

22. $\mathbf{F}(x, y, z) = 2z\mathbf{i} + 3x\mathbf{j} + 4z\mathbf{k}$; S is the portion of the paraboloid $z = 4 - x^2 - y^2$ above the xy plane.

In Exercises 23 through 28, use Stokes's theorem to evaluate the line integral $\oint_C \mathbf{F} \cdot \mathbf{T}\, ds$ for $\mathbf{F}$ and C.

23. $\mathbf{F}(x, y, z) = 4y\mathbf{i} - 3z\mathbf{j} + x\mathbf{k}$; C is the triangle having vertices at $(1, 0, 0)$, $(0, 1, 0)$, and $(0, 0, 1)$.

24. $\mathbf{F}(x, y, z) = (y - x)\mathbf{i} + (x - z)\mathbf{j} + (x - y)\mathbf{k}$; C is the triangle having vertices at $(2, 0, 0)$, $(0, 2, 0)$, and $(0, 0, 1)$.

25. $\mathbf{F}(x, y, z) = -y\mathbf{i} + x\mathbf{j} + z\mathbf{k}$; C is the circle $x^2 + y^2 = 4$ in the xy plane.

26. $\mathbf{F}(x, y, z) = yz\mathbf{i} + xy\mathbf{j} + xz\mathbf{k}$; C is the square having vertices at $(0, 0, 0)$, $(2, 0, 0)$, $(0, 2, 0)$, and $(2, 2, 0)$.

27. $\mathbf{F}(x, y, z) = (2y + \sin^{-1} x)\mathbf{i} + e^{y^2}\mathbf{j} + (x + \ln(z^2 + 4))\mathbf{k}$; C is the triangle having vertices at $(1, 0, 0)$, $(0, 1, 0)$, and $(0, 0, 2)$.

28. $\mathbf{F}(x, y, z) = (2z - e^x)\mathbf{i} + (x^3 + \sin y)\mathbf{j} + (y^2 - \tan z)\mathbf{k}$; C has the vector equation $\mathbf{R}(t) = \cos t\mathbf{i} + \sin t\mathbf{j} + \mathbf{k}$, $0 \le t \le 2\pi$.

REVIEW EXERCISES FOR CHAPTER 19

In Exercises 1 and 2, find a conservative vector field having the potential function f.

1. (a) $f(x, y) = 2x^2y + 3xy^3$; (b) $f(x, y, z) = xe^y - yze^y$

2. (a) $f(x, y) = e^x \cos y + x \sin y$; (b) $f(x, y, z) = \dfrac{1}{x^2 + y^2 + z^2}$

In Exercises 3 through 6, prove that the vector field $\mathbf{F}$ is conservative, and find a potential function.

3. $\mathbf{F}(x, y) = \dfrac{2y^2}{1 + 4x^2y^4}\mathbf{i} + \dfrac{4xy}{1 + 4x^2y^4}\mathbf{j}$

4. $\mathbf{F}(x, y, z) = (6x - 4y)\mathbf{i} + (z - 4x)\mathbf{j} + (y - 8z)\mathbf{k}$

5. $\mathbf{F}(x, y, z) = z^2 \sec^2 x\mathbf{i} + 2ye^{3z}\mathbf{j} + (3y^2e^{3z} + 2z \tan x)\mathbf{k}$

6. $\mathbf{F}(x, y) = (y \sin x - \sin y)\mathbf{i} - (x \cos y + \cos x)\mathbf{j}$

In Exercises 7 through 10, find curl $\mathbf{F}$ and div $\mathbf{F}$.

7. $\mathbf{F}(x, y, z) = e^{yz}\mathbf{i} + e^{xz}\mathbf{j} + e^{xy}\mathbf{k}$ **8.** $\mathbf{F}(x, y) = \sin y\mathbf{i} + \sin x\mathbf{k}$

9. $\mathbf{F}(x, y) = \dfrac{1}{y}\mathbf{i} - \dfrac{2x}{y}\mathbf{j}$ **10.** $\mathbf{F}(x, y, z) = \dfrac{x}{y}\mathbf{i} + \dfrac{y}{z}\mathbf{j} + \dfrac{z}{x}\mathbf{k}$

In Exercises 11 through 18, evaluate the line integral over curve C.

11. $\int_C \mathbf{F} \cdot d\mathbf{R}$; $\mathbf{F}(x, y) = 3y\mathbf{i} - 4x\mathbf{j}$; C: $\mathbf{R}(t) = 2t^2\mathbf{i} - t\mathbf{j}$, $0 \le t \le 1$.

12. $\int_C \mathbf{F} \cdot d\mathbf{R}$; $\mathbf{F}(x, y) = (x + y)\mathbf{i} + (y - x)\mathbf{j}$; C: $\mathbf{R}(t) = t^3\mathbf{i} + t^2\mathbf{j}$ from the point $(8, 4)$ to the point $(1, 1)$.

13. $\int_C (2x + 3y)\, dx + xy\, dy$; C: $\mathbf{R}(t) = 4 \sin t\mathbf{i} - \cos t\mathbf{j}$, $0 \le t \le \frac{1}{2}\pi$.

14. $\int_C (2x + y)\, dx + (x - 2y)\, dy$; C: $x^2 + y^2 = 9$.

15. $\int_C y^2\, dx + z^2\, dy + x^2\, dz$; C: $\mathbf{R}(t) = (t - 1)\mathbf{i} + (t + 1)\mathbf{j} + t^2\mathbf{k}$, $0 \le t \le 1$.

16. $\int_C xe^y\, dx - xe^z\, dy + e^z\, dz$; C: $\mathbf{R}(t) = t\mathbf{i} + t^2\mathbf{j} + t^3\mathbf{k}$, $0 \le t \le 1$.

17. $\int_C \mathbf{F} \cdot d\mathbf{R}$; $\mathbf{F}(x, y, z) = 3xy\mathbf{i} + (4y^2 - xz)\mathbf{j} + 6z\mathbf{k}$; C: the twisted cubic $\mathbf{R}(t) = t\mathbf{i} + t^2\mathbf{j} + t^3\mathbf{k}$, $0 \le t \le 1$.

18. $\int_C \mathbf{F} \cdot d\mathbf{R}$; $\mathbf{F}(x, y, z) = 2x\mathbf{i} + 3y\mathbf{j} + z\mathbf{k}$; C: the circular helix $\mathbf{R}(t) = 2 \cos t\mathbf{i} + 2 \sin t\mathbf{j} + t\mathbf{k}$, $0 \le t \le 2\pi$.

In Exercises 19 through 26, prove that the value of the line integral is independent of the path, and compute the value in any convenient manner. In each exercise C is any sectionally smooth curve from point A to point B.

19. $\int_C 2xe^y\, dx + x^2e^y\, dy$; A is $(1, 0)$ and B is $(3, 2)$.

20. $\int_C \left(\dfrac{1}{y} - y\right) dx + \left(-\dfrac{x}{y^2} - x\right) dy$; A is $(0, 1)$ and B is $(6, 3)$.

21. $\int_C \mathbf{F} \cdot d\mathbf{R}$; $\mathbf{F}(x, y) = (\cos y - y \cos x)\mathbf{i} - (\sin x + x \sin y)\mathbf{j}$; A is $(0, \frac{1}{2}\pi)$ and B is $(\pi, 0)$.

22. $\int_C \mathbf{F} \cdot d\mathbf{R}$; $\mathbf{F}(x, y) = (2xy - 2y)\mathbf{i} + (x^2 - 2x + 3y^2)\mathbf{j}$; A is $(2, -1)$ and B is $(3, 2)$.

23. $\int_C 3y\, dx + (3x + 4y)\, dy - 2z\, dz$; A is $(0, 1, -1)$ and B is $(1, 2, 0)$.

24. $\int_C z \sin y\, dx + xz \cos y\, dy + x \sin y\, dz$; A is $(0, 0, 0)$ and B is $(2, 3, \frac{1}{2}\pi)$.

25. $\int_C \mathbf{F} \cdot d\mathbf{R}$; $\mathbf{F}(x, y, z) = \left(\dfrac{1}{y} - \dfrac{2z}{x^2}\right)\mathbf{i} - \left(\dfrac{1}{z} + \dfrac{x}{y^2}\right)\mathbf{j} + \left(\dfrac{2}{x} + \dfrac{y}{z^2}\right)\mathbf{k}$; A is $(2, -1, 1)$ and B is $(4, 2, -2)$.

26. $\int_C \mathbf{F} \cdot d\mathbf{R}$; $\mathbf{F}(x, y, z) = (2xy + 3yz)\mathbf{i} + (x^2 - 4yz + 3xz)\mathbf{j} + (3xy - 2y^2)\mathbf{k}$; A is $(0, 2, 1)$ and B is $(1, -1, 4)$.

In Exercises 27 through 30, use Green's theorem to evaluate the line integral.

27. $\oint_C (3x + 2y)\, dx + (3x + y^2)\, dy$, where C is the ellipse $16x^2 + 9y^2 = 144$.

28. $\oint_C \ln(y + 1)\, dx - \dfrac{xy}{y + 1}\, dy$, where C is the closed curve determined by the curve $\sqrt{x} + \sqrt{y} = 2$ and the intervals $[0, 4]$ on the x and y axes.

29. $\oint_C e^x \sin y\, dx + e^x \cos y\, dy$, where C is any smooth closed curve.

30. $\oint_C (x^2 - y^3)\, dx + (y^2 + x^3)\, dy$; where C is the circle $x^2 + y^2 = 1$.

In Exercises 31 and 32, use Theorem 19.4.2 to find the area of the region.

31. The region enclosed by the parabola $y = x^2$ and the line $y = x + 2$.

32. The region enclosed by the two parabolas $y = x^2$ and $x^2 = 18 - y$.

In Exercises 33 through 36, find the total work done in moving an object along C if the motion is caused by the force field. Assume the arc is measured in meters and the force is measured in newtons.

33. $\mathbf{F}(x, y) = 2x^2 y\mathbf{i} + (x^2 + 3y)\mathbf{j}$; C: the arc of the parabola $y = 3x^2 + 2x + 4$ from $(0, 4)$ to $(1, 9)$.

34. $\mathbf{F}(x, y) = xy^2\mathbf{i} - x^2 y\mathbf{j}$; C: the arc of the circle $x^2 + y^2 = 4$ from $(2, 0)$ to $(0, 2)$.

35. $\mathbf{F}(x, y, z) = (xy - z)\mathbf{i} + y\mathbf{j} + z\mathbf{k}$; C: the line segment from the origin to the point $(4, 1, 2)$.

36. $\mathbf{F}(x, y, z) = xyz\mathbf{i} + e^y\mathbf{j} + (x + z)\mathbf{k}$; C: $\mathbf{R}(t) = 3t\mathbf{i} + t^2\mathbf{j} + 2t\mathbf{k}$; $0 \le t \le 3$.

In Exercises 37 and 38, verify Gauss's divergence theorem in the plane and Stokes's theorem in the plane for $\mathbf{F}$ and R.

37. $\mathbf{F}(x, y) = 4y\mathbf{i} + 3x\mathbf{j}$, and R is the region bounded by $x^{2/3} + y^{2/3} = 1$.

38. $\mathbf{F}(x, y) = 3x^2\mathbf{i} + 4y^2\mathbf{j}$ and R is the region bounded by the ellipse $9x^2 + 16y^2 = 144$.

In Exercises 39 and 40, use Green's theorem to find the total work done in moving an object in the counterclockwise direction once around C if the motion is caused by the force field $\mathbf{F}(x, y)$. Assume the arc is measured in meters and the force is measured in newtons.

39. C is the circle $x^2 + y^2 = 4$;
$\mathbf{F}(x, y) = (xy^2 + \cos x)\mathbf{i} + (x^2 + e^y)\mathbf{j}$.

40. C is the ellipse $9x^2 + y^2 = 9$; $\mathbf{F}(x, y) = (2x - 3y)\mathbf{i} + (x + 2y)\mathbf{j}$.

In Exercises 41 and 42, find the rate of flow of the fluid out of the region R bounded by C if $\mathbf{F}(x, y)$ is the velocity field of the fluid. Assume the velocity is measured in centimeters per second and the area of R is measured in square centimeters.

41. $\mathbf{F}(x, y) = (4x - 3y)\mathbf{i} + (5y - 4x^2)\mathbf{j}$; C is the right triangle with vertices at $(0, 1)$, $(0, 4)$, and $(4, 4)$.

42. $\mathbf{F}(x, y) = (y^2 + 12x)\mathbf{i} + (4y - x^2)\mathbf{j}$; C is the ellipse $x^2 + 4y^2 = 16$.

43. Find the value of the line integral
$$\int_C \frac{-y}{x^2 + y^2}\, dx + \frac{x}{x^2 + y^2}\, dy$$
if C is the arc of the circle $x^2 + y^2 = 4$ from $(\sqrt{2}, \sqrt{2})$ to $(-\sqrt{2}, \sqrt{2})$.

44. Apply Green's theorem to compute the area of the quadrilateral having vertices at the points $(0, 0)$, $(3, 2)$, $(1, 5)$, and $(-2, 1)$.

45. Evaluate the surface integral $\iint_S xy\, d\sigma$, where S is the portion of the plane $3x + 2y - z = 0$ in the first octant below the plane $z = 6$.

46. Evaluate the surface integral $\iint_S x^2\, d\sigma$, where S is the portion of the cylinder $x^2 + y^2 = 1$ in the first octant bounded by the xy plane and the plane $z = 1$.

47. Evaluate the surface integral $\iint_S x\, d\sigma$, where S is the portion of the cylinder $z = 9 - x^2$ in the first octant bounded by the coordinate planes and the plane $y = 2$.

48. Evaluate the surface integral $\iint_S xyz\, d\sigma$, where S is the portion of the cylinder $y^2 + z^2 = 9$ between the planes $x = 1$ and $x = 4$.

49. Find the mass of the portion of the sphere $x^2 + y^2 + z^2 = 4$ in the first octant if the area density at any point (x, y, z) on the surface is kz^2 kilograms per square meter, where k is a constant.

50. Find the mass of the hemisphere $x^2 + y^2 + z^2 = 4$ above the xy plane if the area density at any point (x, y, z) on the surface is $(4 - z)$ kilograms per square meter.

51. A funnel is in the shape of the portion of the cone $x^2 + y^2 = z^2$ between the planes $z = 1$ and $z = 4$. If the area density at any point (x, y, z) on the surface is $(10 - z)$ kilograms per square meter, find the mass of the funnel.

52. Suppose surface S is that part of the sphere $x^2 + y^2 + z^2 = 9$ that is above the region D in the xy plane enclosed by the circle $x^2 + y^2 = 1$. If the velocity field of a fluid is given by $\mathbf{F}(x, y, z) = -y\mathbf{i} + x\mathbf{j} + 3\mathbf{k}$, find the flux of $\mathbf{F}$ across S.

53. The velocity field of a fluid is given by
$$\mathbf{F}(x, y, z) = 2x\mathbf{i} + 2y\mathbf{j} + 3z\mathbf{k}$$
and surface S is that portion of the paraboloid $z = 4 - x^2 - y^2$ above the xy plane. Find the flux of $\mathbf{F}$ across S.

54. Verify Gauss's divergence theorem if $\mathbf{F}(x, y, z) = \frac{1}{2}z\mathbf{i}$ and S is the sphere $x^2 + y^2 + z^2 = 4$.

In Exercises 55 and 56, use Gauss's divergence theorem to evaluate the surface integral $\iint_S \mathbf{F} \cdot \mathbf{N}\, d\sigma$ for $\mathbf{F}$ and S.

55. $\mathbf{F}(x, y, z) = 2x\mathbf{i} + y\mathbf{j} + 2z\mathbf{k}$; S is the boundary of the region enclosed on the side by the cylinder $x^2 + y^2 = 16$, below by the xy plane and above by the plane $z = 2$.

56. $\mathbf{F}(x, y, z) = x^2\mathbf{i} + y^2\mathbf{j} + z^2\mathbf{k}$; S is the boundary of the region enclosed by the cone $z = \sqrt{x^2 + y^2}$ and the plane $z = 1$.

In Exercises 57 and 58, verify Stokes's theorem for $\mathbf{F}$ and S.

57. $\mathbf{F}(x, y, z) = z\mathbf{i} + 4x\mathbf{j} + 2z\mathbf{k}$; S is the portion of the paraboloid $z = 9 - x^2 - y^2$ above the xy plane.

58. $\mathbf{F}(x, y, z) = xy\mathbf{i} + yz\mathbf{j} + xz\mathbf{k}$; S is the hemisphere $x^2 + y^2 + z^2 = 16$ above the xy plane.

In Exercises 59 and 60, use Stokes's theorem to evaluate the line integral $\oint_C \mathbf{F} \cdot \mathbf{T} \, ds$ for $\mathbf{F}$ and C.

59. $\mathbf{F}(x, y, z) = (z + \ln(x^2 + 1))\mathbf{i} + (\cos y - x^2)\mathbf{j} + (3y^2 - e^z)\mathbf{k}$; C has the vector equation $\mathbf{R}(t) = \cos t\mathbf{i} + \sin t\mathbf{j} + \mathbf{k}$, where $0 \le t \le 2\pi$.

60. $\mathbf{F}(x, y, z) = -2y\mathbf{i} + 3x\mathbf{j} + z\mathbf{k}$; C is the circle $x^2 + y^2 = 1$ in the xy plane.

APPENDIX

A.1 USE OF A TABLE OF INTEGRALS

We have presented various techniques of integration, and you have seen how they are useful for evaluating many integrals. However, there may be occasions when these procedures are either not sufficient or else lead to a complicated integration. In such cases you may wish to use a *table of integrals*. Fairly complete tables of integrals appear in mathematics handbooks, and shorter tables are found in most calculus textbooks. You should be cautioned not to rely too heavily on tables when evaluating integrals. A mastery of integration skills is essential because, as mentioned in Chapter 9, it may be necessary to employ some of the techniques to express the integrand in a form that is found in a table.

A short table of integrals appears on the endpapers of this book. The formulas used in the examples and exercises of this section appear in this table. Observe that in the table there are various headings indicating the form of the integrand. The first heading is *Some Elementary Forms*, and the five formulas listed here are included in those given in the introduction to Chapter 9. The second heading is *Rational Forms Containing a + bu*. The first example utilizes one of these formulas.

EXAMPLE 1   Evaluate

$$\int \frac{x\,dx}{(4-x)^3}$$

Solution   Formula 10 in the table of integrals is

$$\int \frac{u\,du}{(a+bu)^3} = \frac{1}{b^2}\left[\frac{a}{2(a+bu)^2} - \frac{1}{a+bu}\right] + C$$

Using this formula with $u = x$, $a = 4$, and $b = -1$ we have

$$\int \frac{x\,dx}{(4-x)^3} = \frac{1}{(-1)^2}\left[\frac{4}{2(4-x)^2} - \frac{1}{4-x}\right] + C$$

$$= \frac{2}{(4-x)^2} - \frac{1}{4-x} + C$$

EXAMPLE 2   Evaluate

$$\int \frac{dx}{6-2x^2}$$

Solution   Formula 25 in the table is

$$\int \frac{du}{a^2 - u^2} = \frac{1}{2a} \ln \left| \frac{u + a}{u - a} \right| + C \tag{1}$$

Observe that this formula can be used if the coefficient of x^2 in the given integral is 1 instead of 2. Thus we write

$$\int \frac{dx}{6 - 2x^2} = \frac{1}{2} \int \frac{dx}{3 - x^2}$$

To the integral on the right we apply (1) with $u = x$ and $a = \sqrt{3}$, and we have

$$\int \frac{dx}{6 - 2x^2} = \frac{1}{2} \cdot \frac{1}{2\sqrt{3}} \ln \left| \frac{x + \sqrt{3}}{x - \sqrt{3}} \right| + C$$

$$= \frac{\sqrt{3}}{12} \ln \left| \frac{x + \sqrt{3}}{x - \sqrt{3}} \right| + C$$

EXAMPLE 3   Evaluate

$$\int \frac{dx}{8x^2 + 4x}$$

Solution

$$\int \frac{dx}{8x^2 + 4x} = \frac{1}{4} \int \frac{dx}{x(2x + 1)}$$

The integral on the right side is of the form

$$\int \frac{du}{u(a + bu)}$$

where $u = x$, $a = 1$, and $b = 2$. Formula 11 in the table is

$$\int \frac{du}{u(a + bu)} = \frac{1}{a} \ln \left| \frac{u}{a + bu} \right| + C$$

Using this formula we have

$$\frac{1}{4} \int \frac{dx}{x(2x + 1)} = \frac{1}{4} \cdot \frac{1}{1} \ln \left| \frac{x}{1 + 2x} \right| + C$$

$$= \frac{1}{4} \ln \left| \frac{x}{2x + 1} \right| + C$$

EXAMPLE 4   Evaluate

$$\int \frac{dx}{\sqrt{x^2 + 2x - 3}}$$

Solution   Formula 27 in the table is

$$\int \frac{du}{\sqrt{u^2 \pm a^2}} = \ln|u + \sqrt{u^2 \pm a^2}| + C \tag{2}$$

We may be able to apply this formula to the given integral if by completing the square under the radical sign we obtain an expression of the form $u^2 \pm a^2$. To complete the square of $x^2 + 2x$ we add 1, and so we also subtract 1. Thus we write

$$x^2 + 2x - 3 = (x^2 + 2x + 1) - 1 - 3$$
$$= (x + 1)^2 - 4$$

Therefore

$$\int \frac{dx}{\sqrt{x^2 + 2x - 3}} = \int \frac{dx}{\sqrt{(x + 1)^2 - 4}}$$

The integral is of the form

$$\int \frac{du}{\sqrt{u^2 - a^2}}$$

where $u = x + 1$ and $a = 2$. Hence, from (2),

$$\int \frac{dx}{\sqrt{(x + 1)^2 - 4}} = \ln|(x + 1) + \sqrt{(x + 1)^2 - 4}| + C$$
$$= \ln|x + 1 + \sqrt{x^2 + 2x - 3}| + C$$

EXAMPLE 5   Evaluate

$$\int x^2 \sqrt{4x^2 + 1} \, dx$$

Solution   Formula 29 in the table, with the plus sign, is

$$\int u^2 \sqrt{u^2 + a^2} \, du = \frac{u}{8} (2u^2 + a^2)\sqrt{u^2 + a^2} - \frac{a^4}{8} \ln|u + \sqrt{u^2 + a^2}| + C \qquad (3)$$

We can apply this formula by writing the given integral as follows:

$$\int x^2 \sqrt{4x^2 + 1} \, dx = \int x^2 \sqrt{4\left(x^2 + \frac{1}{4}\right)} \, dx$$
$$= 2 \int x^2 \sqrt{x^2 + \frac{1}{4}} \, dx$$

From (3), with $u = x$ and $a = \frac{1}{2}$, we have

$$2 \int x^2 \sqrt{x^2 + \frac{1}{4}} \, dx = 2\left[\frac{x}{8}\left(2x^2 + \frac{1}{4}\right)\sqrt{x^2 + \frac{1}{4}} - \frac{\frac{1}{16}}{8} \ln\left|x + \sqrt{x^2 + \frac{1}{4}}\right|\right] + C$$
$$= \frac{x}{16}(8x^2 + 1)\sqrt{x^2 + \frac{1}{4}} - \frac{1}{64} \ln\left|x + \sqrt{x^2 + \frac{1}{4}}\right| + C$$

Formulas 16, 19, 21, 73, 77, 86, and 98, among others, express one integral in terms of a simpler integral of the same form. Such formulas are called **reduction formulas**. The next example shows how they are applied.

EXAMPLE 6   Evaluate

$$\int \sec^5 x \, dx$$

Solution   Formula 77 is

$$\int \sec^n u \, du = \frac{1}{n-1} \sec^{n-2} u \tan u + \frac{n-2}{n-1} \int \sec^{n-2} u \, du \tag{4}$$

We apply this formula with $u = x$ and $n = 5$, and we have

$$\int \sec^5 x \, dx = \tfrac{1}{4} \sec^3 x \tan x + \tfrac{3}{4} \int \sec^3 x \, dx$$

We now use (4) to the integral on the right side of the above with $n = 3$, and we obtain

$$\int \sec^5 x \, dx = \tfrac{1}{4} \sec^3 x \tan x + \tfrac{3}{4} \left(\tfrac{1}{2} \sec x \tan x + \tfrac{1}{2} \int \sec x \, dx \right)$$

$$= \tfrac{1}{4} \sec^3 x \tan x + \tfrac{3}{8} \sec x \tan x + \tfrac{3}{8} \ln|\sec x + \tan x| + C$$

Because the integral in Example 6 is an odd power of the secant, it can be evaluated by using integration by parts as explained in Section 9.3. As a matter of fact, this integral appears as Exercise 12 in Exercises 9.3.

EXERCISES A.1

In Exercises 1 through 36, use the table of integrals on the end-papers to evaluate the integral. In Exercises 1 through 4, the integrand is a rational form containing $a + bu$. Use one of the formulas 6 through 13.

1. $\displaystyle\int \frac{x \, dx}{2 + 3x}$ **2.** $\displaystyle\int \frac{x \, dx}{(5 - 2x)^3}$

3. $\displaystyle\int \frac{x^2 \, dx}{(6 - x)^2}$ **4.** $\displaystyle\int \frac{dx}{x(7 + 3x)}$

In Exercises 5 through 8, the integrand is a form containing $\sqrt{a + bu}$. Use one of the formulas 14 through 23.

5. $\displaystyle\int x\sqrt{1 + 2x} \, dx$ **6.** $\displaystyle\int x^2\sqrt{1 + 2x} \, dx$

7. $\displaystyle\int \frac{\sqrt{1 + 2x}}{x} \, dx$ **8.** $\displaystyle\int \frac{dx}{x^2\sqrt{1 + 2x}}$

In Exercises 9 and 10, the integrand is a form containing $a^2 \pm u^2$. Use one of the formulas 24 through 26.

9. $\displaystyle\int \frac{dx}{4 - x^2}$ **10.** $\displaystyle\int \frac{dx}{x^2 - 25}$

In Exercises 11 through 14, the integrand is a form containing $\sqrt{u^2 \pm a^2}$. Use one of the formulas 27 through 38.

11. $\displaystyle\int \frac{dx}{\sqrt{x^2 + 6x}}$ **12.** $\displaystyle\int \sqrt{4x^2 + 1} \, dx$

13. $\displaystyle\int \frac{\sqrt{9x^2 + 4}}{x} \, dx$ **14.** $\displaystyle\int \frac{dx}{(x - 1)^2\sqrt{x^2 - 2x - 3}}$

In Exercises 15 and 16, the integrand is a form containing $\sqrt{a^2 - u^2}$. Use one of the formulas 39 through 48.

15. $\displaystyle\int \frac{\sqrt{9 - 4x^2}}{x} \, dx$ **16.** $\displaystyle\int \frac{dx}{x^2\sqrt{25 - 9x^2}}$

In Exercises 17 and 18, the integrand is a form containing $2au - u^2$. Use one of the formulas 49 through 58.

17. $\displaystyle\int x\sqrt{4x - x^2} \, dx$ **18.** $\displaystyle\int \frac{x^2 \, dx}{\sqrt{4x - x^2}}$

In Exercises 19 through 24, the integrand is a form containing trigonometric functions. Use one of the formulas 59 through 88.

19. $\displaystyle\int \sin^5 x \, dx$ **20.** $\displaystyle\int \cos^8 x \, dx$

21. $\displaystyle\int \csc^7 x \, dx$ **22.** $\displaystyle\int \sin 3x \cos 5x \, dx$

23. $\displaystyle\int t^4 \cos t \, dt$ **24.** $\displaystyle\int \sin^3 x \cos^5 x \, dx$

In Exercises 25 and 26, the integrand is a form containing an inverse trigonometric function. Use one of the formulas 89 through 94.

25. $\displaystyle\int \sec^{-1} 3x \, dx$ **26.** $\displaystyle\int \tan^{-1} 4t \, dt$

In Exercises 27 through 34, the integrand is a form containing an exponential or logarithmic function. Use one of the formulas 95 through 106.

27. $\int x^4 e^x \, dx$ **28.** $\int x^3 2^x \, dx$ **29.** $\int x^2 e^{4x} \, dx$

30. $\int x^2 \ln x \, dx$ **31.** $\int x^3 \ln (3x) \, dx$ **32.** $\int 5x^2 e^{-2x} \, dx$

33. $\int e^{2x} \sin 5x \, dx$ **34.** $\int e^{3t} \cos 4t \, dt$

In Exercises 35 and 36, the integrand is a form containing a hyperbolic function. Use one of the formulas 107 through 124.

35. $\int 3y \sinh 5y \, dy$ **36.** $\int e^x \operatorname{sech} e^x \, dx$

In Exercises 37 through 52, use the table of integrals on the end-papers to evaluate the definite integral.

37. $\int_1^2 \dfrac{dx}{x(5-x)^2}$ **38.** $\int_0^3 \dfrac{x \, dx}{(1+x)^2}$

39. $\int_0^3 \dfrac{x^2 \, dx}{\sqrt{x^2+16}}$ **40.** $\int_0^2 \dfrac{dx}{(9+4x^2)^{3/2}}$

41. $\int_1^2 x^4 \ln x \, dx$ **42.** $\int_0^1 x^2 e^{-x} \, dx$

43. $\int_3^4 \sqrt{x^2+2x-15} \, dx$ **44.** $\int_3^5 x^2 \sqrt{x^2-9} \, dx$

45. $\int_1^2 \sqrt{4w-w^2} \, dw$ **46.** $\int_0^{\pi/3} \sec^5 x \, dx$

47. $\int_{\pi/8}^{\pi/4} \sin 3t \sin 5t \, dt$ **48.** $\int_0^{\pi/4} \tan^6 \theta \, d\theta$

49. $\int_0^{\pi/2} \sin^3 2x \cos^3 2x \, dx$ **50.** $\int_4^5 w^2 \sqrt{w^2-16} \, dw$

51. $\int_0^1 x^3 e^{2x} \, dx$ **52.** $\int_0^{\pi/6} e^{2t} \sin 3t \, dt$

THE GREEK ALPHABET

α	alpha	ι	iota	ρ	rho
β	beta	κ	kappa	σ	sigma
γ	gamma	λ	lambda	τ	tau
δ	delta	μ	mu	υ	upsilon
ϵ	epsilon	ν	nu	φ	phi
ζ	zeta	ξ	xi	χ	chi
η	eta	o	omicron	ψ	psi
θ	theta	π	pi	ω	omega

FORMULAS FROM GEOMETRY

The following symbols are used for the measure:
r: radius h: altitude b: base a: base C: circumference A: area S: surface area
B: area of base V: volume

Circle: $A = \pi r^2$; $C = 2\pi r$

Triangle: $A = \frac{1}{2}bh$

Rectangle and parallelogram: $A = bh$

Trapezoid: $A = \frac{1}{2}(a + b)h$

Right circular cylinder: $V = \pi r^2 h$; $S = 2\pi rh$

Right circular cone: $V = \frac{1}{3}\pi r^2 h$; $S = \pi r\sqrt{r^2 + h^2}$

Sphere: $V = \frac{4}{3}\pi r^3$; $S = 4\pi r^2$

Prism (with parallel bases): $V = Bh$

Pyramid: $V = \frac{1}{3}Bh$

FORMULAS FROM TRIGONOMETRY

The Eight Fundamental Trigonometric Identities

$$\sin x \csc x = 1 \qquad \cos x \sec x = 1 \qquad \tan x \cot x = 1 \qquad \tan x = \frac{\sin x}{\cos x} \qquad \cot x = \frac{\cos x}{\sin x}$$

$$\sin^2 x + \cos^2 x = 1 \qquad 1 + \tan^2 x = \sec^2 x \qquad 1 + \cot^2 x = \csc^2 x$$

Sum and Difference Identities

$$\sin(u + v) = \sin u \cos v + \cos u \sin v \qquad \sin(u - v) = \sin u \cos v - \cos u \sin v$$
$$\cos(u + v) = \cos u \cos v - \sin u \sin v \qquad \cos(u - v) = \cos u \cos v + \sin u \sin v$$

$$\tan(u + v) = \frac{\tan u + \tan v}{1 - \tan u \tan v} \qquad \tan(u - v) = \frac{\tan u - \tan v}{1 + \tan u \tan v}$$

Multiple-Measure Identities

$$\sin 2u = 2 \sin u \cos u$$
$$\cos 2u = \cos^2 u - \sin^2 u \qquad \cos 2u = 1 - 2 \sin^2 u \qquad \cos 2u = 2 \cos^2 u - 1$$

$$\tan 2u = \frac{2 \tan u}{1 - \tan^2 u}$$

$$\sin^2 u = \frac{1 - \cos 2u}{2} \qquad \cos^2 u = \frac{1 + \cos 2u}{2} \qquad \tan^2 u = \frac{1 - \cos 2u}{1 + \cos 2u}$$

$$\sin^2 \tfrac{1}{2}t = \frac{1 - \cos t}{2} \qquad \cos^2 \tfrac{1}{2}t = \frac{1 + \cos t}{2}$$

$$\tan \tfrac{1}{2}t = \frac{1 - \cos t}{\sin t} \qquad \tan \tfrac{1}{2}t = \frac{\sin t}{1 + \cos t}$$

Identities for the Product, Sum and Difference of Sine and Cosine

$$\sin u \cos v = \tfrac{1}{2}[\sin(u + v) + \sin(u - v)] \qquad \cos u \sin v = \tfrac{1}{2}[\sin(u + v) - \sin(u - v)]$$
$$\cos u \cos v = \tfrac{1}{2}[\cos(u + v) + \cos(u - v)] \qquad \sin u \sin v = \tfrac{1}{2}[\cos(u - v) - \cos(u + v)]$$

$$\sin s + \sin t = 2 \sin\left(\frac{s + t}{2}\right) \cos\left(\frac{s - t}{2}\right) \qquad \sin s - \sin t = 2 \cos\left(\frac{s + t}{2}\right) \sin\left(\frac{s - t}{2}\right)$$

$$\cos s + \cos t = 2 \cos\left(\frac{s + t}{2}\right) \cos\left(\frac{s - t}{2}\right) \qquad \cos s - \cos t = -2 \sin\left(\frac{s + t}{2}\right) \sin\left(\frac{s - t}{2}\right)$$

Some Reduction Formulas

$$\sin(-x) = -\sin x \qquad \cos(-x) = \cos x \qquad \tan(-x) = -\tan x$$
$$\sin(\tfrac{1}{2}\pi - x) = \cos x \qquad \cos(\tfrac{1}{2}\pi - x) = \sin x \qquad \tan(\tfrac{1}{2}\pi - x) = \cot x$$
$$\sin(\tfrac{1}{2}\pi + x) = \cos x \qquad \cos(\tfrac{1}{2}\pi + x) = -\sin x \qquad \tan(\tfrac{1}{2}\pi + x) = -\cot x$$
$$\sin(\pi - x) = \sin x \qquad \cos(\pi - x) = -\cos x \qquad \tan(\pi - x) = -\tan x$$

Law of Sines and Law of Cosines

a, b, and c represent the measures of the sides of a triangle: α, β, and γ represent the measures of the angles opposite the sides of measures a, b, and c, respectively.

$$\frac{a}{\sin \alpha} = \frac{b}{\sin \beta} = \frac{c}{\sin \gamma} \qquad c^2 = a^2 + b^2 - 2ab \cos \gamma$$

Answers to Odd-Numbered Exercises

EXERCISES 1.1 (Page 12)

1. $(-2, +\infty)$ **3.** $(-\infty, \frac{3}{4}]$ **5.** $[4, 8]$ **7.** $(-\frac{5}{3}, \frac{4}{3}]$ **9.** $(-\infty, -\frac{1}{2}) \cup (0, +\infty)$ **11.** $(-\infty, -1) \cup (\frac{1}{3}, 3)$
13. $(-\infty, -2) \cup (2, +\infty)$ **15.** $(-\infty, -5) \cup (3, +\infty)$ **17.** $[-1, \frac{1}{2}]$ **19.** $(-3, \frac{3}{4})$ **21.** $(\frac{3}{2}, \frac{31}{14}] \cup (\frac{7}{3}, +\infty)$
23. $\{-\frac{5}{2}, 1\}$ **25.** $\{-\frac{1}{4}, 4\}$ **27.** $\{-\frac{2}{3}, \frac{1}{2}\}$ **29.** $\{\frac{4}{3}, 3\}$ **31.** $[\frac{5}{8}, +\infty)$ **33.** $(-\infty, -2] \cup [5, +\infty)$
35. $(-\infty, 1] \cup [4, +\infty)$ **37.** $(-11, 3)$ **39.** $[\frac{2}{3}, 2]$ **41.** $(-\infty, -2) \cup (12, +\infty)$ **43.** $[-\frac{1}{2}, 4]$
45. $(-\infty, 1) \cup (4, +\infty)$ **47.** $(1, +\infty)$ **49.** $[-\frac{9}{2}, \frac{3}{2}]$ **51.** $(-\infty, \frac{10}{9}) \cup (2, +\infty)$ **53.** $|x| > |a|$ **55.** $|x - 2| > 2$

EXERCISES 1.2 (Page 23)

(Sketches of the graphs for Exercises 1 through 5 appear in Figs. EX 1.2-1 through EX 1.2-5.)
1. (a) $(1, 2)$; (b) $(-1, -2)$; (c) $(-1, 2)$; (d) $(-2, 1)$ **3.** (a) $(2, -2)$; (b) $(-2, 2)$; (c) $(-2, -2)$; (d) does not apply
5. (a) $(-1, 3)$; (b) $(1, -3)$; (c) $(1, 3)$; (d) $(-3, -1)$

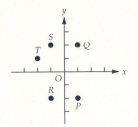

FIGURE EX 1.2-1

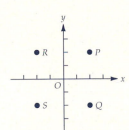

FIGURE EX 1.2-3

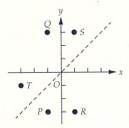

FIGURE EX 1.2-5

9. $\sqrt{26}; \frac{1}{2}\sqrt{89}; \frac{1}{2}\sqrt{53}$ **11.** $|\overline{AB}| = \sqrt{41}, |\overline{AC}| = \sqrt{41}, |\overline{BC}| = \sqrt{82}$, and $|\overline{AB}|^2 + |\overline{AC}|^2 = |\overline{BC}|^2; \frac{41}{2}$ **13.** $\sqrt{578}$ **15.** $(-8, 12)$
19. $(-2 + \frac{3}{2}\sqrt{3}, \frac{3}{2} + 2\sqrt{3})$ and $(-2 - \frac{3}{2}\sqrt{3}, \frac{3}{2} - 2\sqrt{3})$ **21.** -1 **23.** $-\frac{1}{7}$ **25.** $4x - y = 11$ **27.** $2x + y + 1 = 0$
29. $13x - 4y - 28 = 0$ **31.** $4x - 3y + 12 = 0$ **33.** $x + 2y - 4 = 0$ **35.** $y = -7$ **37.** $\sqrt{3}x - y + (2\sqrt{3} - 5) = 0$
39. (a) $-\frac{1}{3}$; (b) 0 **41.** (a) collinear; (b) not collinear   **43.** (a) $y = 0$; (b) $x = 0$; (c) $xy = 0$ **47.** (a) $2x + 3y + 7 = 0$; (b) $\sqrt{13}$

51. $(-\frac{9}{4}, \frac{17}{4}), (\frac{1}{2}, \frac{11}{2}), (\frac{13}{4}, \frac{27}{4})$ **53.** (a) $-\dfrac{A}{B}$; (b) $-\dfrac{C}{B}$; (c) $-\dfrac{C}{A}$; (d) $Bx - Ay = 0$

55. $9x - 4y - 11 = 0$; $y = 1$; $9x + 4y - 19 = 0$. They intersect at $(\frac{5}{3}, 1)$.   **59.** (a) $x = 1$; (b) $y = 1$; (c) $2x + y - 3 = 0$

EXERCISES 1.3 (Page 30)

1. $x^2 + y^2 - 8x + 6y = 0$ **3.** $x^2 + y^2 + 10x + 24y + 160 = 0$ **5.** $x^2 + y^2 - 2x - 4y - 8 = 0$ **7.** $(3, 4); 4$ **9.** $(0, -\frac{2}{3}); \frac{5}{3}$
(Sketches of the graphs for Exercises 7 and 9 appear in Figs. EX 1.3-7 and EX 1.3-9.)
11. circle   **13.** the empty set   **15.** circle
(Sketches of the graphs for Exercises 17 through 43 appear in Figs. EX 1.3-17 through EX 1.3-43.)
45. $x^2 + y^2 + 6x + 10y + 9 = 0$ **47.** $x^2 + y^2 - 4x - 4y - 2 = 0$ **49.** $3x + 4y - 19 = 0$

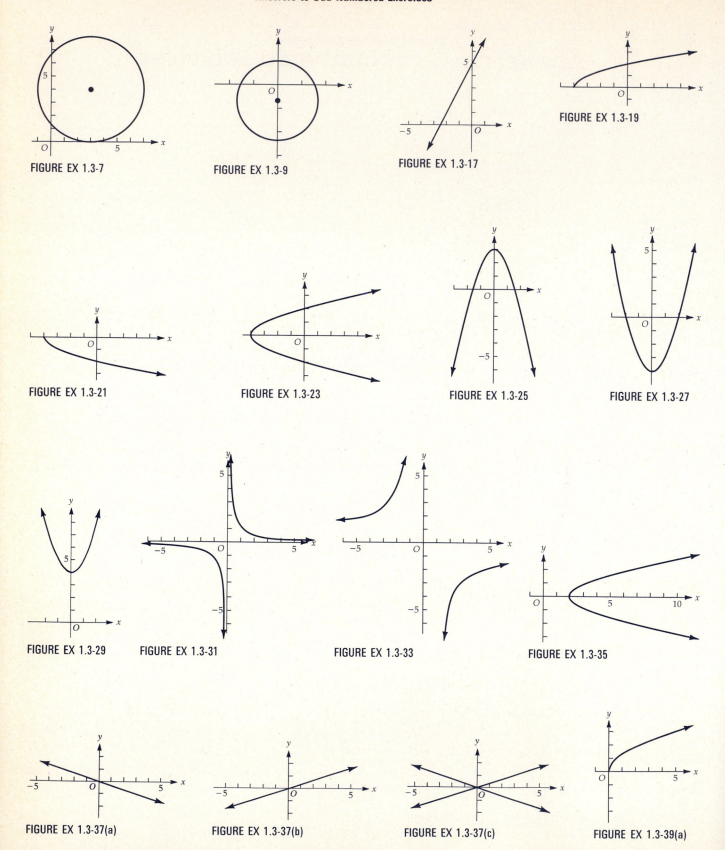

FIGURE EX 1.3-7

FIGURE EX 1.3-9

FIGURE EX 1.3-17

FIGURE EX 1.3-19

FIGURE EX 1.3-21

FIGURE EX 1.3-23

FIGURE EX 1.3-25

FIGURE EX 1.3-27

FIGURE EX 1.3-29

FIGURE EX 1.3-31

FIGURE EX 1.3-33

FIGURE EX 1.3-35

FIGURE EX 1.3-37(a)

FIGURE EX 1.3-37(b)

FIGURE EX 1.3-37(c)

FIGURE EX 1.3-39(a)

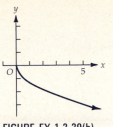

FIGURE EX 1.3-39(b)

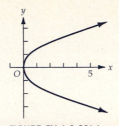

FIGURE EX 1.3-39(c)

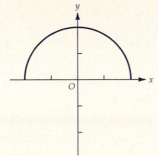

FIGURE EX 1.3-41(a)

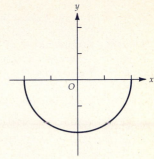

FIGURE EX 1.3-41(b)

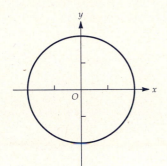

FIGURE EX 1.3-41(c)

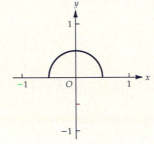

FIGURE EX 1.3-43(a)

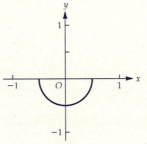

FIGURE EX 1.3-43(b)

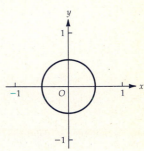

FIGURE EX 1.3-43(c)

EXERCISES 1.4 (Page 39)

1. (a) yes, $x \geq 4$; (b) yes, $|x| \geq 2$; (c) yes, $|x| \leq 2$; (d) no　　**3.** (a) yes, $(-\infty, +\infty)$; (b) no; (c) yes, $(-\infty, +\infty)$; (d) yes, $(-\infty, +\infty)$

5. (a) 5; (b) -5; (c) -1; (d) $2a + 1$; (e) $2x + 1$; (f) $4x - 1$; (g) $4x - 2$; (h) $2x + 2h - 1$; (i) $2x + 2h - 2$; (j) 2

7. (a) -5; (b) -6; (c) -3; (d) 30; (e) $2h^2 + 9h + 4$; (f) $8x^4 + 10x^2 - 3$; (g) $2x^4 - 7x^2$; (h) $2x^2 + (4h + 5)x + (2h^2 + 5h - 3)$;

(i) $2x^2 + 5x + (2h^2 + 5h - 6)$; (j) $4x + 2h + 5$　　**9.** (a) 1; (b) $\sqrt{11}$; (c) 2; (d) 5; (e) $\sqrt{4x + 9}$; (f) $\dfrac{2}{\sqrt{2x + 2h + 3} + \sqrt{2x + 3}}$

11. (a) $x^2 + x - 6$, domain: $(-\infty, +\infty)$; (b) $-x^2 + x - 4$, domain: $(-\infty, +\infty)$; (c) $x^3 - 5x^2 - x + 5$, domain: $(-\infty, +\infty)$;

(d) $\dfrac{x - 5}{x^2 - 1}$, domain: $\{x \mid x \neq -1, x \neq 1\}$; (e) $\dfrac{x^2 - 1}{x - 5}$, domain: $\{x \mid x \neq 5\}$　　**13.** (a) $\dfrac{x^2 + 2x - 1}{x^2 - x}$, domain: $\{x \mid x \neq 0, x \neq 1\}$;

(b) $\dfrac{x^2 + 1}{x^2 - x}$, domain: $\{x \mid x \neq 0, x \neq 1\}$; (c) $\dfrac{x + 1}{x^2 - x}$, domain: $\{x \mid x \neq 0, x \neq 1\}$; (d) $\dfrac{x^2 + x}{x - 1}$, domain: $\{x \mid x \neq 0, x \neq 1\}$;

(e) $\dfrac{x - 1}{x^2 + x}$, domain: $\{x \mid x \neq -1, x \neq 0, x \neq 1\}$　　**15.** (a) $\sqrt{x} + x^2 - 1$, domain: $[0, +\infty)$;

(b) $\sqrt{x} - x^2 + 1$, domain: $[0, +\infty)$; (c) $\sqrt{x}(x^2 - 1)$, domain: $[0, +\infty)$;

(d) $\dfrac{\sqrt{x}}{x^2 - 1}$, domain: $[0, 1) \cup (1, +\infty)$; (e) $\dfrac{x^2 - 1}{\sqrt{x}}$, domain: $(0, +\infty)$　　**17.** (a) $x^2 + 3x - 1$, domain: $(-\infty, +\infty)$;

(b) $x^2 - 3x + 3$, domain: $(-\infty, +\infty)$; (c) $3x^3 - 2x^2 + 3x - 2$, domain: $(-\infty, +\infty)$; (d) $\dfrac{x^2 + 1}{3x - 2}$, domain: $\{x \mid x \neq \frac{2}{3}\}$;

(e) $\dfrac{3x - 2}{x^2 + 1}$, domain: $(-\infty, +\infty)$　　**19.** (a) $\dfrac{x^2 + 2x - 2}{x^2 - x - 2}$, domain: $\{x \mid x \neq -1, x \neq 2\}$;

(b) $\dfrac{-x^2 - 2}{x^2 - x - 2}$, domain: $\{x \mid x \neq -1, x \neq 2\}$; (c) $\dfrac{x}{x^2 - x - 2}$, domain: $\{x \mid x \neq -1, x \neq 2\}$;

(d) $\dfrac{x-2}{x^2+x}$, domain: $\{x \mid x \neq -1, x \neq 0, x \neq 2\}$; (e) $\dfrac{x^2+x}{x-2}$, domain: $\{x \mid x \neq -1, x \neq 2\}$

21. (a) $x+5$, domain: $(-\infty, +\infty)$: (b) $x+5$, domain: $(-\infty, +\infty)$; (c) $x-4$, domain: $(-\infty, +\infty)$; (d) $x+14$, domain: $(-\infty, +\infty)$
23. (a) x^2-6, domain: $(-\infty, +\infty)$; (b) $x^2-10x+24$, domain: $(-\infty, +\infty)$; (c) $x-10$, domain: $(-\infty, +\infty)$;
(d) x^4-2x^2, domain: $(-\infty, +\infty)$ **25.** (a) $\sqrt{x^2-4}$, domain: $(-\infty, -2] \cup [2, +\infty)$; (b) $x-4$, domain: $[2, +\infty)$;
(c) $\sqrt{\sqrt{x-2}-2}$, domain: $[6, +\infty)$ (d) x^4-4x^2+2, domain: $(-\infty, +\infty)$ **27.** (a) $\dfrac{1}{\sqrt{x}}$, domain: $(0, +\infty)$;
(b) $\dfrac{1}{\sqrt{x}}$, domain: $(0, +\infty)$; (c) x, domain: $\{x \mid x \neq 0\}$; (d) $\sqrt[4]{x}$, domain: $[0, +\infty)$ **29.** (a) $|x+2|$, domain: $(-\infty, +\infty)$;
(b) $|x|+2$, domain: $(-\infty, +\infty)$; (c) $|x|$, domain: $(-\infty, +\infty)$; (d) $|x+2|+2$, domain: $(-\infty, +\infty)$
31. (a) $2x^2-3$, domain: $(-\infty, +\infty)$; (b) $4x^2-12x+9$, domain: $(-\infty, +\infty)$; (c) $4x-9$, domain: $(-\infty, +\infty)$
33. (a) 0; (b) 4; (c) $4-2x$ **35.** (a) 1; (b) -1; (c) 1; (d) -1; (e) 1 if $x \leq 0$, -1 if $x > 0$; (f) 1 if $x \geq -1$, -1 if $x < -1$; (g) 1;
(h) -1 if $x \neq 0$, 1 if $x = 0$ **37.** (a) even; (b) neither; (c) odd; (d) even; (e) odd; (f) odd; (g) neither; (h) even; (i) even; (j) odd
39. (a) even; (b) odd; (c) even; (d) even

EXERCISES 1.5 (Page 44)

(Sketches of the graphs appear in Figs. EX 1.5-1 through EX 1.5-43.)
1. domain: $(-\infty, +\infty)$; range: $(-\infty, +\infty)$ **3.** domain: $(-\infty, +\infty)$; range: $[-1, +\infty)$
5. domain: $[-1, +\infty)$; range: $[0, +\infty)$ **7.** domain: $(-\infty, 2]$; range: $[0, +\infty)$ **9.** domain: $(-\infty, 0]$; range: $[0, +\infty)$
11. domain: $(-\infty, +\infty)$; range: $[0, +\infty)$ **13.** domain: $(-\infty, +\infty)$; range: $(-\infty, 4]$
15. domain: $(-\infty, +\infty)$; range: $[4, +\infty)$ **17.** domain: $\{x \mid x \neq 1\}$; range: $\{y \mid y \neq -2\}$
19. domain: $(-\infty, +\infty)$; range: $\{-2, 2\}$ **21.** domain: $(-\infty, +\infty)$; range: $\{y \mid y \neq 3\}$
23. domain: $(-\infty, +\infty)$; range: $[-4, +\infty)$ **25.** domain: $(-\infty, +\infty)$; range: $(-\infty, 4]$
27. domain: $(-\infty, +\infty)$; range: $(-\infty, -2) \cup \{0\} \cup (1, +\infty)$
29. domain: $\{x \mid x \neq -5 \text{ and } x \neq -1\}$; range: $\{y \mid y \neq -7 \text{ and } y \neq -3\}$
31. domain: $(-\infty, -1] \cup [4, +\infty)$; range: $[0, +\infty)$ **33.** domain: $\{x \mid x \neq 2\}$; range: $[0, +\infty)$
35. domain: $\{x \mid x \neq -5\}$; range: $[-6, +\infty)$ **37.** domain: $(-\infty, +\infty)$; range: $[1, +\infty)$
39. domain: $(-\infty, +\infty)$; range: $\{\text{integers}\}$ **41.** domain: $(-\infty, +\infty)$; range: $[0, 1)$
43. domain: $\{x \mid x \neq 0\}$; range: $(-\infty, -1] \cup \{0\} \cup (\tfrac{1}{2}, 1]$

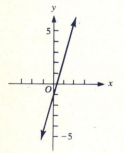

FIGURE EX 1.5-1

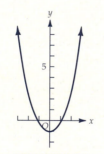

FIGURE EX 1.5-3

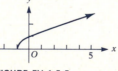

FIGURE EX 1.5-5

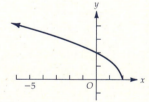

FIGURE EX 1.5-7

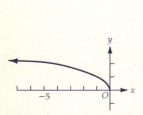

FIGURE EX 1.5-9

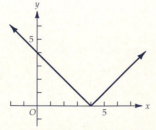

FIGURE EX 1.5-11

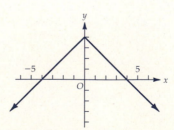

FIGURE EX 1.5-13

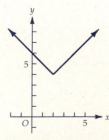

FIGURE EX 1.5-15

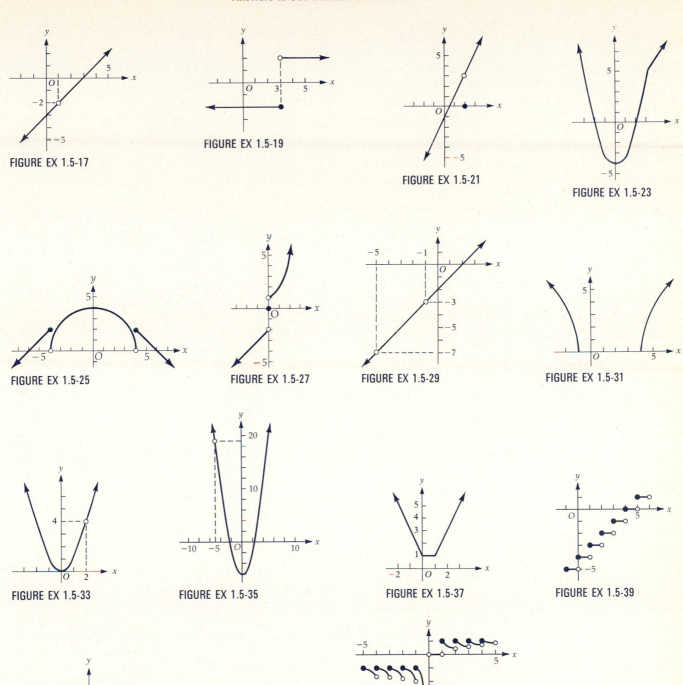

FIGURE EX 1.5-17

FIGURE EX 1.5-19

FIGURE EX 1.5-21

FIGURE EX 1.5-23

FIGURE EX 1.5-25

FIGURE EX 1.5-27

FIGURE EX 1.5-29

FIGURE EX 1.5-31

FIGURE EX 1.5-33

FIGURE EX 1.5-35

FIGURE EX 1.5-37

FIGURE EX 1.5-39

FIGURE EX 1.5-41

FIGURE EX 1.5-43

EXERCISES 1.6 (Page 51)

1. (a) $\frac{1}{3}\pi$; (b) $\frac{3}{4}\pi$; (c) $\frac{7}{6}\pi$; (d) $-\frac{5}{6}\pi$; (e) $\frac{1}{9}\pi$; (f) $\frac{5}{2}\pi$; (g) $-\frac{5}{12}\pi$; (h) $\frac{5}{9}\pi$

3. (a) $45°$; (b) $120°$; (c) $330°$; (d) $-90°$; (e) $28°39'$; (f) $540°$; (g) $-114°36'$; (h) $15°$

5. (a) $\frac{1}{2}$; (b) $\frac{1}{2}\sqrt{2}$; (c) 1; (d) $\frac{1}{2}$ **7.** (a) $-\frac{1}{2}\sqrt{3}$; (b) $\frac{1}{2}\sqrt{2}$; (c) -1; (d) 0   **9.** (a) $\sqrt{3}$; (b) 1; (c) -1; (d) 1

11. (a) $\frac{2}{3}\sqrt{3}$; (b) $\sqrt{2}$; (c) $-\frac{1}{3}\sqrt{3}$; (d) 1   **13.** (a) $\frac{1}{2}\sqrt{2}$; (b) $\frac{1}{2}\sqrt{2}$; (c) $\sqrt{2}$; (d) $\sqrt{2}$ **15.** (a) $-\frac{1}{2}\sqrt{3}$; (b) $-\frac{1}{2}$; (c) -2; (d) $-\frac{2}{3}\sqrt{3}$

17. (a) 0; (b) 1; (c) -1; (d) undefined   **19.** (a) 1; (b) 0; (c) undefined; (d) 1   **21.** (a) -1; (b) -1; (c) $\frac{1}{3}\sqrt{3}$; (d) $\sqrt{3}$

23. (a) $-\sqrt{3}$; (b) $-\frac{1}{3}\sqrt{3}$; (c) 0; (d) 0   **25.** (a) $\frac{1}{2}\pi$; (b) π; (c) $\frac{1}{4}\pi, \frac{5}{4}\pi$; (d) 0   **27.** (a) 0, π; (b) $\frac{1}{2}\pi, \frac{3}{2}\pi$; (c) 0, π; (d) $\frac{1}{2}\pi, \frac{3}{2}\pi$

29. (a) $\frac{7}{6}\pi, \frac{11}{6}\pi$; (b) $\frac{1}{3}\pi, \frac{5}{3}\pi$; (c) $\frac{3}{4}\pi, \frac{7}{4}\pi$; (d) $\frac{1}{6}\pi, \frac{5}{6}\pi$ **31.** (a) $\frac{1}{2}\pi, \frac{3}{2}\pi$; (b) 0, π **33.** (a) π; (b) $\frac{3}{2}\pi$ **35.** (a) $\frac{3}{5}$; (b) 1

37. (a) $-\frac{7}{4}$; (b) $\frac{3}{55}$ **39.** (a) 63°; (b) 19°   **41.** 27°, 45°, 108°   **43.** $3x - y + 7 = 0$; $x + 3y - 11 = 0$

45. (a) $\sin 3x$, $(-\infty, +\infty)$; (b) $\tan \dfrac{x}{2}$, $\{x \mid x \neq (2k + 1)\pi\}$, where k is any integer

47. (a) $\cot \dfrac{1}{x}$, $\left\{x \mid x \neq 0 \text{ and } x \neq \dfrac{1}{k\pi}\right\}$, where k is any integer; (b) $\sec(x - \pi)$, $\{x \mid x \neq \pi \text{ and } x \neq (k + \frac{1}{2})\pi\}$, where k is any integer

---

REVIEW EXERCISES FOR CHAPTER 1 (Page 52)

1. $[5, +\infty)$ **3.** $(-\infty, -\frac{1}{3}) \cup (1, +\infty)$ **5.** $(-\frac{3}{2}, 1)$ **7.** $[-3, \frac{3}{2}]$ **9.** $(-\frac{4}{3}, 4)$ **11.** $(-\infty, -6) \cup (-1, +\infty)$

13. $\{-3, \frac{9}{2}\}$ **15.** $\{-2, 2\}$ **17.** $-4, 14$ **19.** $x^2 + y^2 + 6x - 8y - 75 = 0$ **21.** See Fig. REV 1-21

23. See Fig. REV 1-23   **25.** See Fig. REV 1-25   **29.** 15   **31.** $(-2, -5)$; $(3, -6)$ **35.** $x^2 + y^2 - 2x - 8y - 3 = 0$

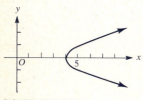

FIGURE REV 1-21

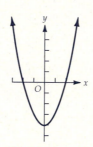

FIGURE REV 1-23

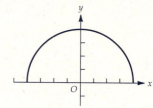

FIGURE REV 1-25

37. $(\frac{3}{2}, -1)$; 1   **39.** $3x - y + 9 = 0$ **41.** $5x + 2y - 19 = 0$ **43.** $7x^2 + 7y^2 + 11x - 19y - 6 = 0$

45. $12x + 3y - 2 = 0$ **47.** $0 \leq k \leq 2$ **49.** (a) 35; (b) $3x^4 + x^2 + 5$; (c) $6x + 3h - 1$

51. (a) odd; (b) even; (c) neither; (d) odd

53. (a) $x^2 + 4x - 7$, domain: $(-\infty, +\infty)$; (b) $x^2 - 4x - 1$, domain: $(-\infty, +\infty)$; (c) $4x^3 - 3x^2 - 16x + 12$, domain: $(-\infty, +\infty)$;

(d) $\dfrac{x^2 - 4}{4x - 3}$, domain: $\{x \mid x \neq \frac{3}{4}\}$; (e) $\dfrac{4x - 3}{x^2 - 4}$, domain: $\{x \mid x \neq \pm 2\}$; (f) $16x^2 - 24x + 5$, domain: $(-\infty, +\infty)$;

(g) $4x^2 - 19$, domain: $(-\infty, +\infty)$

55. (a) $\dfrac{x^2 - 2x + 1}{x^2 - 2x - 3}$, domain: $\{x \mid x \neq -1, x \neq 3\}$; (b) $\dfrac{-x^2 + 4x + 1}{x^2 - 2x - 3}$, domain: $\{x \mid x \neq -1, x \neq 3\}$;

(c) $\dfrac{x}{x^2 - 2x - 3}$, domain: $\{x \mid x \neq -1, x \neq 3\}$; (d) $\dfrac{x + 1}{x^2 - 3x}$, domain: $\{x \mid x \neq -1, x \neq 0, x \neq 3\}$; (e) $\dfrac{x^2 - 3x}{x + 1}$, $\{x \mid x \neq -1, x \neq 3\}$;

(f) $-\dfrac{x + 1}{2x + 3}$, domain: $\{x \mid x \neq -\frac{3}{2}, x \neq -1\}$; (g) $\dfrac{1}{x - 2}$, domain: $\{x \mid x \neq 2, x \neq 3\}$

57. domain: $(-\infty, +\infty)$; range: $[0, +\infty)$; see Fig. REV 1-57   **59.** domain: $\{x \mid x \neq 2\}$; range: $\{y \mid y \neq 6\}$; see Fig. REV 1-59

61. domain: $(-\infty, +\infty)$; range: $[-1, +\infty)$; see Fig. REV 1-61   **63.** domain: $(-\infty, +\infty)$; range: $(-\infty, 4]$; see Fig. REV 1-63

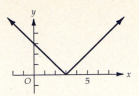

FIGURE REV 1-57

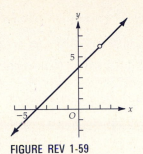

FIGURE REV 1-59

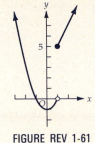

FIGURE REV 1-61

FIGURE REV 1-63

65. (a) $\frac{1}{2}\sqrt{3}$; (b) $-\frac{1}{2}\sqrt{2}$; (c) $\frac{1}{3}\sqrt{3}$; (d) $2 + \sqrt{3}$; (e) -1; (f) $-\dfrac{2}{\sqrt{2-\sqrt{2}}}$ **67.** (a) $\frac{1}{6}\pi, \frac{5}{6}\pi$; (b) 0; (c) $\frac{3}{4}\pi, \frac{7}{4}\pi$; (d) $\frac{1}{6}\pi, \frac{7}{6}\pi$; (e) $\frac{2}{3}\pi, \frac{4}{3}\pi$; (f) $\frac{1}{4}\pi, \frac{3}{4}\pi$

69. $52°, 90°, 106°, 112°$ **71.** $x = 0; 3x + 4y = 0$

75. medians intersect at $(\frac{13}{3}, \frac{13}{3})$; altitudes intersect at $(9, 7)$; center of circumscribed circle is at $(2, 3)$

EXERCISES 2.1 (Page 63)

1. (b) 0.2   **3.** (b) 0.005   **5.** (b) 0.01   **7.** (b) 0.005   **9.** (b) $\frac{1}{1400}$ **11.** (b) 0.0006   **13.** (b) $\frac{1}{3000}$ **15.** (b) 0.0075

17. (b) 0.0008   **19.** (b) 0.01   **21.** (b) $\frac{1}{3000}$ **23.** $\delta = \epsilon$ **25.** $\delta = \frac{1}{2}\epsilon$ **27.** $\delta = \frac{1}{5}\epsilon$ **29.** $\delta = \frac{1}{3}\epsilon$ **31.** $\delta = \frac{1}{3}\epsilon$

33. $\delta = \epsilon$ **35.** $\delta = \min(1, \frac{1}{3}\epsilon)$ **37.** $\delta = \min(1, \frac{1}{8}\epsilon)$ **39.** $\delta = \min(1, \frac{1}{6}\epsilon)$ **41.** $\delta = \min(1, \frac{1}{17}\epsilon)$

EXERCISES 2.2 (Page 72)

1. 8   **3.** 7   **5.** 0   **7.** $\frac{1}{2}$ **9.** $-\frac{1}{22}$ **11.** $\frac{3}{2}$ **13.** $\frac{2}{3}$

15. (a) 0.3333, 0.2857, 0.2564, 0.2506, 0.2501; 0.2000, 0.2222, 0.2439, 0.2494, 0.2499; (b) $\frac{1}{4}$

17. (a) 0.2500, 0.2000, 0.1549, 0.1441, 0.1430, 0.1429; 0, 0.0769, 0.1304, 0.1416, 0.1427, 0.1428; (b) $\frac{1}{7}$

19. (a) 0.1716, 0.1690, 0.1671, 0.1667, 0.1667; 0.1623, 0.1644, 0.1662, 0.1666, 0.1667; (b) $\frac{1}{6}$ **21.** 14   **23.** -6 **25.** $\frac{16}{7}$

27. 12   **29.** $\sqrt{\frac{6}{5}}$ **31.** $\frac{1}{2}$ **33.** $\frac{1}{4}\sqrt{2}$ **35.** $\frac{1}{3}$ **37.** -1 **39.** $\frac{11}{17}$ **45.** (a) 0; (b) See Fig. EX 2.2-45

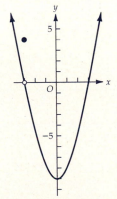

FIGURE EX 2.2-45

EXERCISES 2.3 (Page 76)

(Sketches of the graphs for Exercises 1 through 21 appear in Figs. EX 2.3-1 through EX 2.3-21.)

1. (a) -3; (b) 2; (c) does not exist because $\lim\limits_{x \to 1^+} f(x) \neq \lim\limits_{x \to 1^-} f(x)$ **3.** (a) 8; (b) 0;

(c) does not exist because $\lim\limits_{t \to -4^+} f(t) \neq \lim\limits_{t \to -4^-} f(t)$ **5.** (a) 4; (b) 4; (c) 4   **7.** (a) 5; (b) 5; (c) 5   **9.** (a) 0; (b) 0; (c) 0

11. (a) 0; (b) 0; (c) 0   **13.** (a) -4; (b) -4; (c) -4 **15.** (a) 1; (b) -1; (c) does not exist because $\lim\limits_{x \to 0^+} f(x) \neq \lim\limits_{x \to 0^-} f(x)$

17. (a) 2; (b) 0; (c) does not exist because $\lim\limits_{x \to -2^-} f(x) \neq \lim\limits_{x \to -2^+} f(x)$; (d) 0; (e) -2; (f) does not exist because $\lim\limits_{x \to 2^-} f(x) \neq \lim\limits_{x \to 2^+} f(x)$

19. (a) 0; (b) 0; (c) 0   **21.** (a) 0; (b) 0; (c) 0; (d) 0; (e) 0; (f) 0   **23.** (a) -2; (b) 2; (c) does not exist   **25.** (a) 2; (b) 1;
(c) does not exist   **27.** See Fig. EX 2.3-27 (a) -1; (b) 1; (c) does not exist   **29.** -6 **31.** $a = -\frac{3}{2}$, $b = 1$
35. (a) See Fig. EX 2.3-35 (b) 40; (c) 35; (d) 140; (e) 130

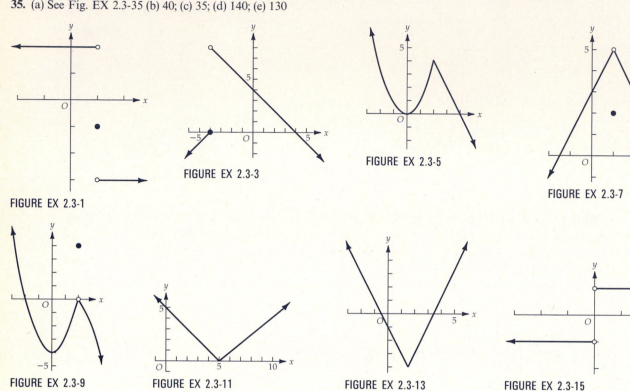

FIGURE EX 2.3-1

FIGURE EX 2.3-3

FIGURE EX 2.3-5

FIGURE EX 2.3-7

FIGURE EX 2.3-9

FIGURE EX 2.3-11

FIGURE EX 2.3-13

FIGURE EX 2.3-15

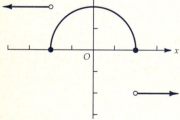

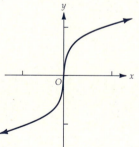

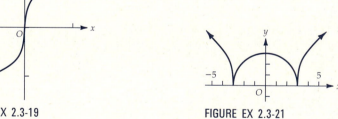

FIGURE EX 2.3-17

FIGURE EX 2.3-19

FIGURE EX 2.3-21

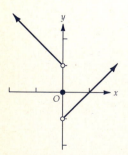

FIGURE EX 2.3-27

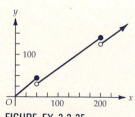

FIGURE EX 2.3-35

EXERCISES 2.4 (Page 87)

1. (a) 1, 2, 10, 100, 1000, 10000; (b) $+\infty$
3. (a) 1, 4, 100, 10000, 1000000, 100000000; 1, 4, 100, 10000, 1000000, 100000000; (b) $+\infty$
5. (a) $-4, -7, -31, -301, -3001, -30001$; (b) $-\infty$ **7.** (a) $-2, -5, -29, -299, -2999, -29999$; (b) $-\infty$
9. (a) 5, 9, 41, 401, 4001, 40001; (b) $+\infty$ **11.** (a) 2.3, 4.3, 20.3, 200.3, 2000.5, 20037; (b) $+\infty$ **13.** $+\infty$ **15.** $-\infty$
17. $-\infty$ **19.** $+\infty$ **21.** $-\infty$ **23.** $+\infty$ **25.** $+\infty$ **27.** $-\infty$ **29.** $-\infty$ **31.** $-\infty$
(Sketches of the graphs for Exercises 33 through 41 appear in Figs EX 2.4-33(a) through EX 2.4-41.)
33. (a) $x = 0$; (b) $x = 0$; (c) $x = 0$; (d) $x = 0$ **35.** $x = 4$ **37.** $x = -3$ **39.** $x = -3$ **41.** $x = -5, x = -3$

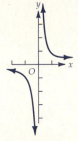

FIGURE EX 2.4-33(a)

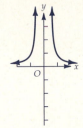

FIGURE EX 2.4-33(b)

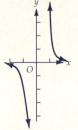

FIGURE EX 2.4-33(c)

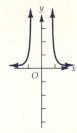

FIGURE EX 2.4-33(d)

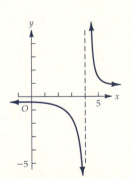

FIGURE EX 2.4-35

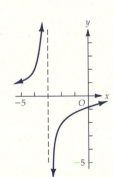

FIGURE EX 2.4-37

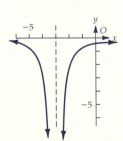

FIGURE EX 2.4-39

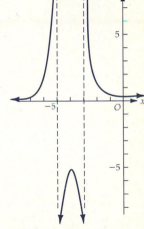

FIGURE EX 2.4-41

EXERCISES 2.5 (Page 97)

1. 4, 1, 0.25, 0.1111, 0.0625, 0.0400, 0.0004, 0.000004; 4, 1, 0.25, 0.1111, 0.0625, 0.0400, 0.0004, 0.000004; (c) 0; (d) 0
3. 1, 0.1250, 0.0156, 0.0046, 0.0020, 0.0010, 10^{-6}, 10^{-9}; $-1, -0.1250, -0.0156, -0.0046, -0.0020, -0.0010, -10^{-6}, -10^{-9}$;
(c) 0; (d) 0   **5.** 0, $-1.5, -2.4, -2.823, -2.919, -2.953, -2.970, -2.9997, -2.999997$; 0, $-1.5, -2.4, -2.823, -2.919,$
$-2.953, -2.970, -2.9997, -2.999997$; (c) -3; (d) -3
7. 3, 2.273, 2.158, 2.015, 2.0015, 2.00015, 2.000015; 1.4, 1.769, 1.857, 1.985, 1.9985, 1.99985, 1.999985; (c) 2; (d) 2
9. 0.75, 0.1944, 0.1100, 0.0101, 0.001001, 0.0001, 0.00001; $-0.25, -0.1389, -0.0900, -0.0099, -0.0010, -0.0001, -0.00001$;
(c) 0; (d) 0   **11.** $\frac{2}{5}$ **13.** $-\frac{2}{5}$ **15.** $\frac{7}{3}$ **17.** 0   **19.** $+\infty$ **21.** $\frac{1}{2}$ **23.** $+\infty$ **25.** $-\infty$ **27.** 1   **29.** -1
31. 0   **33.** $-\infty$ **35.** 0
(Sketches of the graphs for Exercises 37 through 55 appear in Figs. EX 2.5-37 through EX 2.5-55.)
37. $y = 2, x = 3$ **39.** $y = 1, x = 0$ **41.** $y = 0, x = -2, x = 2$ **43.** $y = 4, x = -3, x = 3$ **45.** $y = 0, x = \frac{2}{3}, x = -\frac{5}{2}$
47. $x = -\sqrt{2}, x = \sqrt{2}$ **49.** $y = \frac{2}{3}, x = \frac{4}{3}$ **51.** $y = 1, y = -1$ **53.** $y = -1, y = 1, x = 3$ **55.** $y = 2, x = -1, x = 1$

57. take $N = 1 + \dfrac{1}{\epsilon}$ **59.** take $N = \sqrt{\dfrac{2}{\epsilon} - 1}$ **61.** take $N = \dfrac{\epsilon - 7}{2\epsilon}$

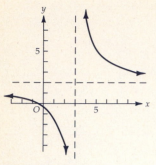

FIGURE EX 2.5-37

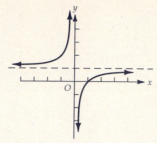

FIGURE EX 2.5-39

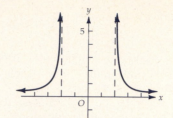

FIGURE EX 2.5-41

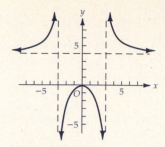

FIGURE EX 2.5-43

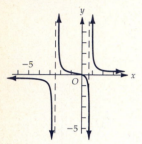

FIGURE EX 2.5-45

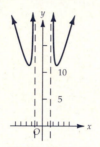

FIGURE EX 2.5-47

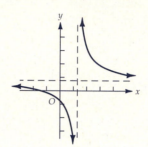

FIGURE EX 2.5-49

FIGURE EX 2.5-51

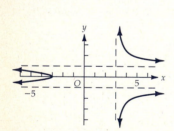

FIGURE EX 2.5-53

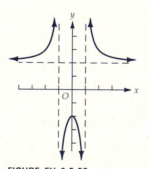

FIGURE EX 2.5-55

EXERCISES 2.6 (Page 105)

(Sketches of the graphs for Exercises 1 through 13 appear in Figs. EX 2.6-1 through EX 2.6-13.)

1. -3; $f(-3)$ does not exist   **3.** -3; $\lim\limits_{x \to -3} g(x) \neq g(-3)$ **5.** 4; $h(4)$ does not exist

7. 4; $\lim\limits_{x \to 4} f(x)$ does not exist   **9.** $-2, 2$; $F(-2)$ and $F(2)$ do not exist   **11.** $-2, 2$; $G(-2)$ and $G(2)$ do not exist

13. 0; $\lim\limits_{x \to 0} f(x)$ does not exist   **15.** (See Fig. EX 2.3-9) 2, $\lim\limits_{t \to 2} g(t) \neq g(2)$ **17.** See Fig. EX 2.6-17, 0; $\lim\limits_{x \to 0} g(x)$ does not exist

19. See Fig. EX 2.3-15, 0; $f(0)$ does not exist   **21.** all integers; $\lim\limits_{x \to k} [\![x]\!]$ does not exist if k is any integer

23. removable; 4   **25.** essential   **27.** removable; 0   **29.** essential   **31.** removable; $-\frac{1}{6}$ **33.** all real numbers

35. all real numbers except 3   **37.** all real numbers except -2 and 2   **39.** all real numbers except 2

41. all real numbers except 1   **43.** (b) 50 and 200   **45.** See Fig. EX 2.6-45, f is continuous everywhere

47. Let $g(0) = \dfrac{1}{3a^2}$ **51.** $f(x) = \begin{cases} 0 \text{ if } x < a \\ 1 \text{ if } a \leq x \end{cases}$; $g(x) = \begin{cases} 1 \text{ if } x < a \\ 0 \text{ if } a \leq x \end{cases}$

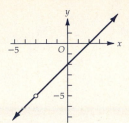

FIGURE EX 2.6-1

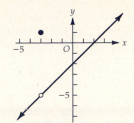

FIGURE EX 2.6-3

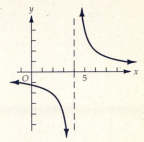

FIGURE EX 2.6-5

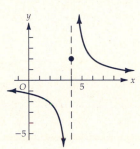

FIGURE EX 2.6-7

FIGURE EX 2.6-9

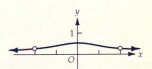

FIGURE EX 2.6-11

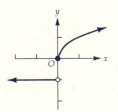

FIGURE EX 2.6-13

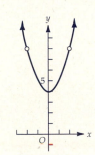

FIGURE EX 2.6-17

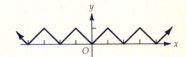

FIGURE EX 2.6-45

EXERCISES 2.7 (Page 112)

1. $(f \circ g)(x) = \sqrt{9 - x^2}$; continuous at all numbers in $(-3, 3)$ **3.** $(f \circ g)(x) = \sqrt{x^2 - 16}$; continuous at all numbers in $(-\infty, -4) \cup (4, +\infty)$ **5.** $(f \circ g)(x) = x^{3/2}$; continuous at all positive numbers

7. $(f \circ g)(x) = \dfrac{1}{x - 2}$; continuous at all numbers except 2   **9.** $(f \circ g)(x) = \dfrac{1}{\sqrt{x - 2}}$; continuous at all numbers in $(2, +\infty)$

11. $(f \circ g)(x) = \dfrac{1}{\sqrt{x} - 2}$; continuous at all positive numbers except 4

13. $(f \circ g)(x) = \dfrac{\sqrt{4 - x^2}}{\sqrt{|x| - 1}}$; continuous at all numbers in $(-2, -1) \cup (1, 2)$

15. continuous; discontinuous; discontinuous; continuous; discontinuous; continuous
17. continuous; continuous; discontinuous; continuous; discontinuous; continuous
19. continuous; continuous; continuous; continuous; discontinuous
21. continuous; discontinuous; discontinuous; discontinuous; continuous
23. continuous; continuous; continuous; continuous; discontinuous; discontinuous   **25.** $[-3, 3]$ **27.** $(-\infty, -4] \cup (4, +\infty)$

29. $(2, +\infty)$ **31.** $[0, 4) \cup (4, +\infty)$ **33.** $[-2, -1] \cup (1, 2]$ **35.** See Fig. EX 2.7-35   **37.** See Fig. EX 2.7-37
39. $(-\infty, -2) \cup [-2, 2] \cup (2, +\infty)$ **41.** (a) $V(x) = x(8 - 2x)(15 - 2x)$; (b) $[0, 4]$ **43.** (a) $A(x) = x(120 - x)$; (b) $[0, 120]$
(Sketches of the graphs for Exercises 45 through 55 appear in Figs. EX 2.7-45 through EX 2.7-55.)
45. $k = 5$ **47.** $c = -3, k = 4$ **49.** $c = \frac{1}{2}(1 + \sqrt{5})$ **51.** $c = -4$ **53.** f is discontinuous at -2
55. f is discontinuous at 1   **57.** no; f will be continuous on $[a, c]$ if $\lim_{x \to b} g(x)$ exists and is equal to $h(b)$ **59.** $\sqrt{10} - 3$

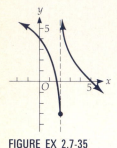

FIGURE EX 2.7-35

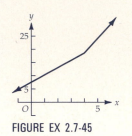

FIGURE EX 2.7-37

FIGURE EX 2.7-45

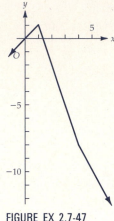

FIGURE EX 2.7-47

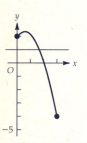

FIGURE EX 2.7-49

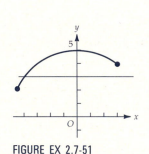

FIGURE EX 2.7-51

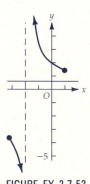

FIGURE EX 2.7-53

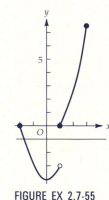

FIGURE EX 2.7-55

EXERCISES 2.8 (Page 122)

1. 4   **3.** $\frac{9}{7}$ **5.** $\frac{3}{5}$ **7.** $\frac{1}{9}$ **9.** 0   **11.** 0   **13.** 12   **15.** $\frac{1}{2}$ **17.** $+\infty$ **19.** 0   **21.** 0   **23.** -1 **25.** 3
27. 0   **29.** -4 **31.** 1   **33.** 0

EXERCISES 2.9 (Page 130)

1. 1   **3.** -1 **5.** 1   **7.** 2   **9.** -2 **11.** $\sqrt{3}$ **13.** no

EXERCISES 2.10 (Page 134)

1. any interval (a, b) where $-2 < a < 3$ and $b > 3$ **3.** any interval (a, b) where $0 \le a < 1$ and $b > 1$
5. any interval (a, b) where $a < -3$ and $-3 < b < 3$ **7.** any interval (a, b) where $-4 \le a < 0$ and $b > 0$
9. (a) $(-\infty, -1) \cup (0, +\infty)$; (b) (a, b) where $a < k < b \le -1$ or $0 \le a < k < b$

11. (a) all numbers in the interval $(-\frac{7}{2}, \frac{3}{2})$; (b) (a, b) where $-\frac{7}{2} \le a < b \le -2$ or $-2 \le a < b \le \frac{3}{2}$ **13.** See Fig. EX 2.10-13
15. See Fig. EX 2.10-15

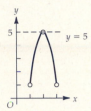

FIGURE EX 2.10-13

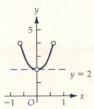

FIGURE EX 2.10-15

REVIEW EXERCISES FOR CHAPTER 2 (Page 135)

1. 9   **3.** -6 **5.** $\sqrt[3]{2}$ **7.** $-\frac{1}{6}$ **9.** $\delta = \frac{1}{2}\epsilon$ **11.** $\delta = \frac{1}{3}\epsilon$ **13.** $\delta = \min(1, \frac{1}{9}\epsilon)$ **15.** $\delta = \min(1, \frac{1}{2}\epsilon)$ **17.** $\delta = \frac{1}{4}\epsilon$
19. $-\frac{5}{2}$ **21.** 8   **23.** $\frac{1}{3}$ **25.** $-\infty$ **27.** $+\infty$ **29.** $-\infty$ **31.** 3   **33.** $-\infty$ **35.** 0   **37.** $\frac{1}{3}$ **39.** $\frac{5}{2}$ **41.** 0

43. $\frac{1}{3}$ **45.** $\dfrac{2}{3a^{1/3}}$ **47.** $-\frac{1}{2}$

(Sketches of the graphs for Exercises 49 through 61 appear in Figs. REV 2-49 through REV 2-61.)
49. $y = 1, x = 4$ **51.** $y = 1, x = 0$ **53.** $y = 5, x = 2, x = -2$ **55.** $y = 3, x = -4$
57. $-2, 1; f(-2)$ and $f(1)$ do not exist   **59.** $-2; \lim\limits_{x \to -2} g(x)$ does not exist   **61.** $0, 1; h(0)$ does not exist, $\lim\limits_{x \to 1} h(x)$ does not exist
63. $(f \circ g)(x) = \sqrt{25 - x^2}$; continuous at all numbers in $(-5, 5)$

65. $(f \circ g)(x) = \dfrac{\sqrt{x^2 - 4}}{\sqrt{3 - |x|}}$; continuous at all numbers in $(-3, -2) \cup (2, 3)$

67. $(f \circ g)(x) = \begin{cases} 1 & \text{if} \quad x < -1 \\ 0 & \text{if} \quad x = -1 \\ -1 & \text{if} \; -1 < x < 1 \\ 0 & \text{if} \quad x = 1 \\ 1 & \text{if} \quad x > 1 \end{cases}$; continuous at all real numbers except -1 and 1

69. $[-5, 5]$ **71.** $(-3, -2] \cup [2, 3)$ **73.** $(-\infty, -4) \cup [-4, 4) \cup [4, +\infty)$ **75.** removable; $\frac{6}{5}$ **77.** essential   **79.** 0
81. See Fig. REV 2-81, $a = 10, b = -23$ **83.** (a) See Fig. REV 2-83; (b) all values of a; (c) all nonintegers
85. See Fig. REV 2-85. (a) yes; (b) no   **87.** the signum function   **89.** See Fig. REV 2-89, $c = \sqrt{7}$

91. $f(x) = \begin{cases} 1 & \text{if } x \ne 0 \\ 0 & \text{if } x = 0 \end{cases}$ **95.** 2   **97.** 3   **99.** (b) any interval (a, b) where $-2 \le a < -\frac{3}{2} < b \le \frac{3}{2}$

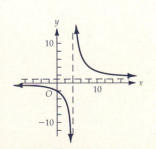

FIGURE REV 2-49

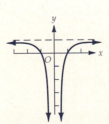

FIGURE REV 2-51

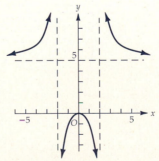

FIGURE REV 2-53

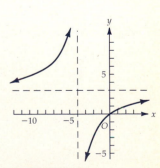

FIGURE REV 2-55

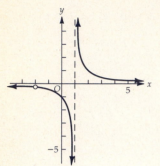

FIGURE REV 2-57

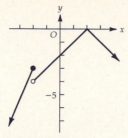

FIGURE REV 2-59

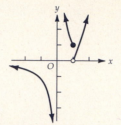

FIGURE REV 2-61

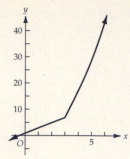

FIGURE REV 2-81

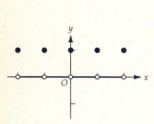

FIGURE REV 2-83

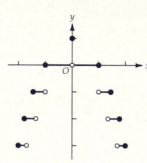

FIGURE REV 2-85

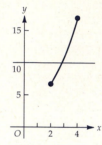

FIGURE REV 2-89

EXERCISES 3.1 (Page 147)

(Sketches of the graphs for Exercises 1 through 11 appear in Figs. EX 3.1-1 through EX 3.1-11.)

1. $-2x_1$ **3.** $-4x_1 + 4$ **5.** $3x_1^2$ **7.** $6x_1 - 12$ **9.** $-\dfrac{1}{2\sqrt{4 - x_1}}$ **11.** $3x_1^2 - 12x_1 + 9$

13. $8x + y + 9 = 0; x - 8y + 58 = 0$ **15.** $6x - y - 16 = 0; x + 6y - 52 = 0$ **17.** $2x + 3y - 12 = 0; 3x - 2y - 5 = 0$

19. $4x + y = 0; x - 4y = 0$ **21.** $8x - y - 5 = 0$ **23.** $4x + 4y - 11 = 0$ **25.** 7   **27.** 0   **29.** $-4x$ **31.** $-3x^2$

33. $\dfrac{1}{2\sqrt{x}}$ **35.** $\dfrac{-13}{(3x - 2)^2}$ **37.** $-\dfrac{2}{x^3} - 1$ **39.** (a) and (c) 7, 6.5, 6.1, 6.01, 6.001; 5, 5.5, 5.9, 5.99, 5.999; (b) and (d) 6

41. (a) and (c) 0.2361, 0.2426, 0.2485, 0.2498, 0.2499; 0.2679, 0.2583, 0.2515, 0.2502, 0.2500; (b) and (d) $\frac{1}{4}$ **43.** -10 **45.** $-\frac{3}{128}$

47. -12 **49.** $-\frac{1}{27}$ **51.** $-\dfrac{8}{x^3} + 3$ **53.** $-\dfrac{7}{2\sqrt{2 - 7x}}$ **55.** $\dfrac{1}{3x^{2/3}}$ **57.** $g(a)$ **61.** $2a$

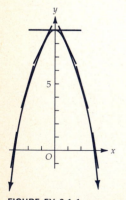

FIGURE EX 3.1-1

FIGURE EX 3.1-3

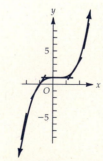

FIGURE EX 3.1-5

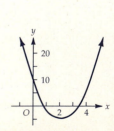

FIGURE EX 3.1-7

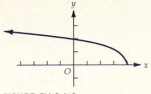

FIGURE EX 3.1-9

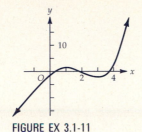

FIGURE EX 3.1-11

EXERCISES 3.2 (Page 155)

(Sketches of the graphs for Exercises 1 through 27 appear in Figs. EX 3.2-1 through EX 3.2-27.)

1. (b) yes; (c) 1, −1; (d) no   **3.** (b) yes; (c) −1, 1; (d) no   **5.** (b) yes; (c) 0, 1; (d) no   **7.** (b) yes; (c) 0, 0; (d) yes   **9.** (b) yes;
(c) does not exist, 0; (d) no   **11.** (b) yes; (c) 8, 8; (d) yes   **13.** (b) yes; (c) neither exists; (d) no   **15.** (b) yes; (c) −6, −6; (d) yes
17. (b) no; (c) 1, 0; (d) no   **19.** (b) no; (c) 12, 12; (d) no   **27.** (b) 0   **29.** (a) 3; (b) no   **31.** $a = 8, b = -9$

33. See Fig. EX 3.2-33. (a) 0; (b) 1; (c) does not exist   **39.** (a) $I(x) = \begin{cases} 15x & \text{if } 0 \le x \le 150 \\ 22.5x - 0.05x^2 & \text{if } 150 < x \le 250 \end{cases}$ **41.** (a) $n > 1$; (b) $n > 1$

FIGURE EX 3.2-1

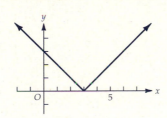

FIGURE EX 3.2-3

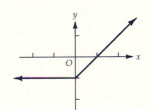

FIGURE EX 3.2-5

FIGURE EX 3.2-7

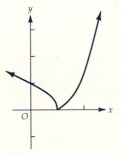

FIGURE EX 3.2-9

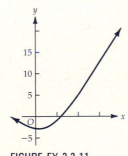

FIGURE EX 3.2-11

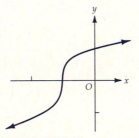

FIGURE EX 3.2-13

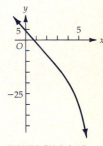

FIGURE EX 3.2-15

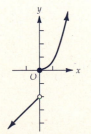

FIGURE EX 3.2-17

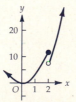

FIGURE EX 3.2-19

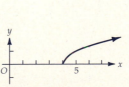

FIGURE EX 3.2-21

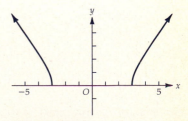

FIGURE EX 3.2-23

FIGURE EX 3.2-25

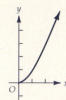

FIGURE EX 3.2-27

FIGURE EX 3.2-33

EXERCISES 3.3 (Page 162)

1. 7   **3.** $-2 - 2x$ **5.** $3x^2 - 6x + 5$ **7.** $x^7 - 4x^3$ **9.** $t^3 - t$ **11.** $4\pi r^2$ **13.** $2x + 3 - \dfrac{2}{x^3}$ **15.** $16x^3 + \dfrac{1}{x^5}$

17. $-\dfrac{6}{x^3} - \dfrac{20}{x^5}$ **19.** $3\sqrt{3}s^2 - 2\sqrt{3}s$ **21.** $70x^6 + 60x^4 - 15x^2 - 6$ **23.** $-18y^2(7 - 3y^3)$

25. $10x^4 - 24x^3 + 12x^2 + 2x - 3$ **27.** $-\dfrac{1}{(x-1)^2}$ **29.** $-\dfrac{4(x+1)}{(x-1)^3}$ **31.** $\dfrac{5(1 - 2t^2)}{(1 + 2t^2)^2}$ **33.** $\dfrac{48y^2}{(y^3 + 8)^2}$

35. $\dfrac{6(x^2 + 10x + 1)}{(x+5)^2}$ **37.** $12x - y = 20$ **39.** $x + 20y + 96 = 0$ **41.** $2x - y = 3$ **43.** $x + 8y + 2 = 0;\ x + 8y - 2 = 0$

45. $28x - y = 99;\ 4x - y = 3$ **49.** $2(3x + 2)(6x^2 + 2x - 3)$ **51.** $3(2x^2 + x + 1)^2(4x + 1)$

EXERCISES 3.4 (Page 171)

1. $v(t) = 6t;\ 18$ **3.** $v(t) = -\dfrac{1}{4t^2};\ -1$ **5.** $v(t) = 6t^2 - 2t;\ 8$ **7.** $v(t) = \dfrac{8}{(4 + t)^2};\ \frac{1}{2}$ **9.** $t < -3$, moving to right; $-3 < t < 1$,

moving to left; $t > 1$ moving to right; changes direction when $t = -3$ and $t = 1$ **11.** $t < -2$, moving to right; $-2 < t < \frac{1}{2}$,
moving to left; $t > \frac{1}{2}$, moving to right; changes direction when $t = -2$ and $t = \frac{1}{2}$ **13.** $t < -3$, moving to left; $-3 < t < 3$,
moving to right; $t > 3$, moving to left; changes direction when $t = -3$ and $t = 3$ **15.** (a) -32 ft/sec; (b) -64 ft/sec; (c) 4 sec;
(d) -128 ft/sec   **17.** 160 cm/sec   **19.** (a) $(20t_1 + 24)$ ft/sec; (b) $\frac{6}{5}$ sec   **21.** (a) 8.600; (b) 8.300; (c) 8.100; (d) 8.050; (e) 8
25. (a) -2.9 degrees per hour; (b) -3 degrees per hour   **27.** (a) 18,750 liters per minute; (b) 17,500 liters per minute
29. (a) $C'(x) = 3 + 2x$; (b) \$83; (c) \$84   **31.** (a) $R'(x) - 600 - \frac{3}{20}x^2$; (b) \$540; (c) \$536.95   **33.** (a) \$3.6 million per year;
(b) 23.1 percent; (c) \$6.8 million per year; (d) 18.7 percent   **35.** (a) 920 people per year; (b) 6.1 percent; (c) 1400 people per year;
(d) 6.4 percent   **37.** (a) profitable; (b) not profitable; (c) 90

EXERCISES 3.5 (Page 180)

3. $3\cos x$ **5.** $\sec^2 x - \csc^2 x$ **7.** $2(\cos t - t \sin t)$ **9.** $x \cos x$ **11.** $4 \cos 2x$ **13.** $-x^2 \sin x$

15. $3 \sec x(2 \tan^2 x + 1)$ **17.** $-\csc y(2 \cot^2 y + 1)$ **19.** $-\dfrac{2(z + 1)\sin z + 2\cos z}{(z + 1)^2}$ **21.** $\dfrac{1}{\cos x - 1}$

23. $\dfrac{1 - 4\sec t + \sin^2 t}{\cos t(\cos t - 4)^2}$ **25.** $\dfrac{2\cos y}{(1 - \sin y)^2}$ **27.** $(1 - \cos x)(x + \cos x) + (1 - \sin x)(x - \sin x)$ **29.** $-\dfrac{5\csc t \cot t}{(\csc t + 2)^2}$

31. 1   **33.** $-\dfrac{2}{\pi}$ **35.** π^2 **37.** 2   **39.** $\sqrt{2}$ **41.** $-\frac{10}{3}$ **43.** (a) 0.0226, 0.2674, 0.4559, 0.4956, 0.4995; 0.8188, 0.6915,

0.5424, 0.5043, 0.5006; (b) $\frac{1}{2}$ **45.** (a) 2.2305, 2.0203, 2.0020, 2.0002, 2.0000; 1.8237, 1.9803, 1.9980, 1.9998, 2.0000; (b) 2
47. (a) $-0.4771, -0.4886, -0.4977, -0.4989, -0.4998; -0.5224, -0.5113, -0.5023, -0.5011, -0.5002$; (b) $-\frac{1}{2}$
49. (a) 0.4929, 0.5736, 0.6468, 0.6567, 0.6647; 0.9116, 0.7770, 0.6872, 0.6768, 0.6687; (b) $\frac{2}{3}$ **51.** (a) $x - y = 0$;
(b) $x - 2y + \sqrt{3} - \frac{1}{3}\pi = 0$; (c) $x + y - \pi = 0$ **53.** (a) $x - y = 0$; (b) $4x - 2y + 2 - \pi = 0$; (c) $4x - 2y - 2 + \pi = 0$

55. (a) $4 \cos t$; (b) $v(0) = 4$, $v(\frac{1}{3}\pi) = 2$, $v(\frac{1}{2}\pi) = 0$, $v(\frac{2}{3}\pi) = -2$, $v(\pi) = -4$ **57.** (a) $3 \sin t$; (b) $v(0) = 0$, $v(\frac{1}{6}\pi) = \frac{3}{2}$, $v(\frac{1}{3}\pi) = \frac{3}{2}\sqrt{3}$,

$v(\frac{1}{2}\pi) = 3$, $v(\frac{2}{3}\pi) = \frac{3}{2}\sqrt{3}$, $v(\frac{5}{6}\pi) = \frac{3}{2}$, $v(\pi) = 0$ **59.** $\frac{\sqrt{2}}{9} W$; (b) $2W$

EXERCISES 3.6 (Page 189)

1. $6(2x + 1)^2$ **3.** $8(x + 2)(x^2 + 4x - 5)^3$ **5.** $2(2t^4 - 7t^3 + 2t - 1)(8t^3 - 21t^2 + 2)$ **7.** $\dfrac{-4x}{(x^2 + 4)^3}$

9. $-12(\sin 3x + \cos 4x)$ **11.** $2 \sec 2x \tan^3 2x$ **13.** $2 \sec^2 x \tan x(2 \tan^2 x + 1)$ **15.** $4 \cot t \csc^2 t$
17. $6(3u^2 + 5)^2(3u - 1)(12u^2 - 3u + 5)$ **19.** $-2(2x - 5)^{-2}(4x + 3)^{-3}(12x - 17)$

21. $10(r^2 + 1)^2(2r - 1)(r + 3)(2r^3 + 4r^2 - r + 1)$ **23.** $\dfrac{18(y - 7)}{(y + 2)^3}$ **25.** $-\dfrac{9(2x - 1)^2(2x^2 - 2x + 1)}{(3x^2 + x - 2)^4}$

27. $\dfrac{2z(z^2 - 5)^2(z^2 + 22)}{(z^2 + 4)^3}$ **29.** $6t \sin(6t^2 - 2)$ **31.** $6(\tan^2 x - x^2)^2(\tan x \sec^2 x - x)$ **33.** $\dfrac{6 \cos 2y(\sin^2 2y + 2)}{(\cos^2 2y + 1)^2}$

35. $-12 \cos 3x \sin(\sin 3x)$ **37.** $24x + y + 39 = 0$; $y = 0$; $y = 1$; $y = 0$; $24x - y - 39 = 0$. See Fig. EX 3.6-37

39. (a) $\dfrac{8t(t^2 - 1)}{(t^2 + 1)^3}$ cm/sec; (b) 0 cm/sec; $\frac{48}{125}$ cm/sec

41. (a) $(5\pi \cos \pi t - 3\pi \sin \pi t)$ cm/sec; (b) $\sqrt{2}\pi$ cm/sec, 3π cm/sec   **43.** (a) $6000 \sin \frac{12}{5}\pi \approx 1854$; (b) 6000 **45.** (a) $5\sqrt{3}$; (b) 10
47. 329 predators/week   **49.** (a) $3x^4$; (b) $6x^5$ **55.** See Figure 9 in Section 1.5.   **57.** See Fig. EX 1.5-37

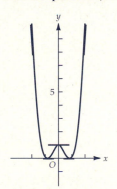

FIGURE EX 3.6-37

EXERCISES 3.7 (Page 194)

1. $x^{-1/2}(2 - \frac{5}{2}x^{-1})$ **3.** $\dfrac{4x}{\sqrt{1 + 4x^2}}$ **5.** $\dfrac{-2}{(5 - 3x)^{1/3}}$ **7.** $\dfrac{y}{(25 - y^2)^{3/2}}$ **9.** $-\dfrac{\sin \sqrt{t}}{\sqrt{t}}$ **11.** $-\dfrac{\sqrt{3}}{2\sqrt{r}} \csc^2 \sqrt{3r}$

13. $-\dfrac{3 \cos 3x}{2(\sin 3x)^{3/2}}$ **15.** $\dfrac{x}{\sqrt{x^2 + 1}} \sec^2 \sqrt{x^2 + 1}$ **17.** $\dfrac{17}{2[(2x - 5)(3x + 1)^3]^{1/2}}$ **19.** $\dfrac{6x^2 - 10x + 1}{3(2x^3 - 5x^2 + x)^{2/3}}$ **21.** $\dfrac{1}{\sqrt{2t}}\left(1 - \dfrac{1}{t}\right)$

23. $\frac{1}{4}x(5 - x^2)^{-1/2}(x^3 + 1)^{-3/4}(-7x^3 + 15x - 4)$ **25.** $\dfrac{1}{x^2\sqrt{x^2 - 1}}$ **27.** $\dfrac{\cos t}{(1 - \sin t)^{3/2}(1 + \sin t)^{1/2}}$

29. $\dfrac{\sec \sqrt{y}}{2\sqrt{y}}(\tan^2 \sqrt{y} + \sec^2 \sqrt{y})$ **31.** $\dfrac{x + 5}{6\sqrt{x - 1}\sqrt[3]{(x + 1)^4}}$ **33.** $\dfrac{-1}{4\sqrt{9 + \sqrt{9 - x}}\sqrt{9 - x}}$ **35.** $\dfrac{\sin 4z}{(1 + \cos^2 2z)^{3/2}}$

37. $2x + 3y = 6$; $2\sqrt[3]{4}x + 3y = 3\sqrt[3]{4}$; $x = 1$; $2\sqrt[3]{4}x - 3y = \sqrt[3]{4}$; $2x - 3y + 2 = 0$ **39.** $4x - 5y + 9 = 0$ **41.** $x + 4y = 0$

43. (a) 0; (b) $\frac{1}{2}$; (c) no value of t **45.** (a) 50 cents per liter; (b) 25 **47.** 100 **49.** 2.7 km/min   **51.** $\dfrac{2x(x^2 - 4)}{|x^2 - 4|}$ **53.** $3x|x|$

EXERCISES 3.8 (Page 198)

1. $-\dfrac{x}{y}$　　**3.** $\dfrac{8y - 3x^2}{3y^2 - 8x}$　　**5.** $-\dfrac{y^2}{x^2}$　　**7.** $-\dfrac{\sqrt{y}}{\sqrt{x}}$　　**9.** $\dfrac{x - xy^2}{x^2y - y}$　　**11.** $\dfrac{3x^2 - 4xy - 1}{2x^2 + 2}$　　**13.** $\dfrac{y^{2/3} + y}{24x^{2/3}y^{5/3} - x}$　　**15.** $\dfrac{y + 4\sqrt{xy}}{\sqrt{x} - x}$

17. $\dfrac{2 + 5x^2 - 4x^{3/2}y}{2x^{1/2}(x^2 + 3)}$　　**19.** $\dfrac{\sin(x - y)}{\sin(x - y) - 1}$　　**21.** $\dfrac{\tan x \sec^2 x}{\cot y \csc^2 y}$　　**23.** $\dfrac{y \sin x - \sin y}{x \cos y + \cos x}$　　**25.** $\dfrac{2 \tan x \sec^2 x + \csc^2(x - y)}{2 \tan y \sec^2 y + \csc^2(x - y)}$

27. $\dfrac{3x^2 - 4y}{4x - 3y^2}$　　**29.** $\dfrac{3x^2 - y^3}{x^3 - 6xy}$　　**31.** $\dfrac{x^3 + 8y^3}{4x^3 - 3x^2y}$　　**33.** $2x + y = 4$　　**35.** $5x - 7y + 11 = 0$　　**37.** -1　　**39.** $(1, 0)$, $(\frac{1}{3}, \frac{4}{3})$

41. (a) $f_1(x) = 2\sqrt{x - 2}$, domain: $x \geq 2$; $f_2(x) = -2\sqrt{x - 2}$, domain: $x \geq 2$; (b)–(c) See Figs. EX 3.8-41(a)–(c);

(d) $f_1'(x) = (x - 2)^{-1/2}$, domain: $x > 2$; $f_2'(x) = -(x - 2)^{-1/2}$, domain: $x > 2$; (e) $\dfrac{2}{y}$; (f) $x - y - 1 = 0$; $x + y - 1 = 0$

43. (a) $f_1(x) = \sqrt{x^2 - 9}$, domain: $|x| \geq 3$; $f_2(x) = -\sqrt{x^2 - 9}$, domain: $|x| \geq 3$; (b)–(c) See Figs. EX 3.8-43(a)–(c);

(d) $f_1'(x) = x(x^2 - 9)^{-1/2}$, domain: $|x| > 3$; $f_2'(x) = -x(x^2 - 9)^{-1/2}$, domain: $|x| > 3$; (e) $\dfrac{x}{y}$; (f) $5x + 4y + 9 = 0$; $5x - 4y + 9 = 0$

45. (a) $f_1(x) = \sqrt{8 - x^2 + 2x} + 2$, domain: $-2 \leq x \leq 4$; $f_2(x) = -\sqrt{8 - x^2 + 2x} + 2$, domain: $-2 \leq x \leq 4$;
(b)–(c) See Figs. EX 3.8-45(a)–(c); (d) $f_1'(x) = (1 - x)(8 - x^2 + 2x)^{-1/2}$, domain: $-2 < x < 4$; $f_2'(x) = (x - 1)(8 - x^2 + 2x)^{-1/2}$,

domain: $-2 < x < 4$; (e) $\dfrac{1 - x}{y - 2}$; (f) $y - 5 = 0$, $y + 1 = 0$　　**47.** (a) decreasing at 11.2 knots; (b) increasing at 34.9 knots

49. $\sqrt{3}x - y + \frac{1}{2}\sqrt{3} = 0$; $\sqrt{3}x + y + \frac{1}{2}\sqrt{3} = 0$

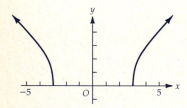

FIGURE EX 3.8-41(a)

FIGURE EX 3.8-41(b)

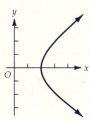

FIGURE EX 3.8-41(c)

FIGURE EX 3.8-43(a)

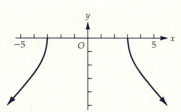

FIGURE EX 3.8-43(b)

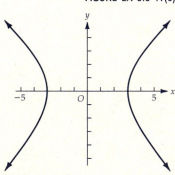

FIGURE EX 3.8-43(c)

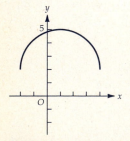

FIGURE EX 3.8-45(a)

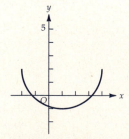

FIGURE EX 3.8-45(b)

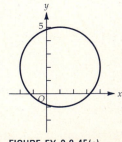

FIGURE EX 3.8-45(c)

EXERCISES 3.9 (Page 204)

1. -3 **3.** -2 **5.** $-\dfrac{\sqrt{3}}{2}$ **7.** $-\dfrac{3}{4}$ **9.** $\dfrac{9}{5}$ ft/sec **11.** $\dfrac{1}{2\pi}$ ft/min **13.** $\dfrac{5}{8\pi}$ m/min **15.** $\dfrac{25}{3}$ ft/sec

17. 0.001π cm^3/day **19.** 0.004π cm^2/day **21.** $\dfrac{6}{25\pi}$ m/min **23.** 1800 lb/ft^2 per min **25.** 128π cm^2/sec **27.** 14 ft/sec

29. $\$1020$ per week **31.** 875 units per month **33.** decreasing at the rate of 55 shirts per week **37.** 22 m^3/min

39. $\frac{1}{194}(3\sqrt{97} + 97)$ ft/sec ≈ 0.65 ft/sec **41.** $\dfrac{2000}{9}$ ft/sec **43.** decreasing at the rate of $\frac{1}{48}$ rad/sec

EXERCISES 3.10 (Page 211)

1. $f'(x) = 5x^4 - 6x^2 + 1$; $f''(x) = 20x^3 - 12x$ **3.** $g'(s) = 8s^3 - 12s^2 + 7$; $g''(s) = 24s^2 - 24s$ **5.** $f'(x) = \frac{5}{2}x^{3/2} - 5$;
$f''(x) = \frac{15}{4}x^{1/2}$ **7.** $f'(x) = x(x^2 + 1)^{-1/2}$; $f''(x) = (x^2 + 1)^{-3/2}$ **9.** $f'(t) = -8t \sin t^2$; $f''(t) = -16t^2 \cos t^2 - 8 \sin t^2$

11. $G'(x) = -2 \cot x \csc^2 x$; $G''(x) = 2 \csc^2 x(3 \cot^2 x + 1)$ **13.** $g'(x) = \dfrac{8x}{(x^2 + 4)^2}$; $g''(x) = \dfrac{32 - 24x^2}{(x^2 + 4)^3}$

15. $f'(x) = \dfrac{\cos x}{2\sqrt{\sin x + 1}}$; $f''(x) = -\frac{1}{4}\sqrt{\sin x + 1}$ **17.** $24x$ **19.** $1152(2x - 1)^{-5}$ **21.** $108 \sec^2 3x(3 \tan^2 3x + 1)$

23. $-32(\sin 2x + \cos 2x)$ **29.** $-\dfrac{3a^4x^2}{y^7}$ **31.** $-\frac{7}{4}$; 5

33. $v = 3t^2 - 18t + 15$; $a = 6t - 18$; when $0 < t < 1$, the particle is at the right of the origin, it is moving to the right, the velocity is decreasing, and the speed is decreasing; when $1 < t < \frac{1}{2}(9 - \sqrt{21})$, the particle is at the right of the origin, it is moving to the left, the velocity is decreasing, and the speed is increasing; when $\frac{1}{2}(9 - \sqrt{21}) < t < 3$, the particle is at the left of the origin, it is moving to the left, the velocity is decreasing, and the speed is increasing; when $3 < t < 5$, the particle is at the left of the origin, it is moving to the left, the velocity is increasing, and the speed is decreasing; when $5 < t < \frac{1}{2}(9 + \sqrt{21})$, the particle is at the left of the origin, it is moving to the right, the velocity is increasing, and the speed is increasing; when $\frac{1}{2}(9 + \sqrt{21}) < t$, the particle is at the right of the origin, it is moving to the right, the velocity is increasing, and the speed is increasing

35. $\frac{3}{2}$ sec; $\frac{7}{4}$ ft; $-\frac{1}{4}$ ft/sec **37.** $\frac{1}{2}$ sec; $\frac{249}{80}$ ft; $-\frac{11}{8}$ ft/sec **39.** $\frac{3}{2}$ sec; $\frac{4}{3}\sqrt{6}$ ft; $\frac{2}{3}\sqrt{6}$ ft/sec

47. $f'(x) = 2|x|$, domain: $(-\infty, +\infty)$; $f''(x) = 2\dfrac{|x|}{x}$, domain: $(-\infty, 0) \cup (0, +\infty)$ **49.** $f'(x) = 4|x^3|$, domain: $(-\infty, +\infty)$;

$f''(x) = 12x|x|$, domain: $(-\infty, +\infty)$ **51.** $f'''(x) = 24|x|$ **53.** $h''(x) = (f'' \circ g)(x)(g'(x))^2 + (f' \circ g)(x)g''(x)$

REVIEW EXERCISES FOR CHAPTER 3 (Page 212)

1. $15x^2 - 14x + 2$ **3.** $\dfrac{x}{2} - \dfrac{8}{x^3}$ **5.** $x^{-1/2} + \frac{1}{4}x^{-3/2}$ **7.** $60t^4 - 39t^2 - 6t - 4$ **9.** $\dfrac{-6x^2}{(x^3 - 1)^2}$

11. $4(2s^3 - 3s + 7)^3(6s^2 - 3)$ **13.** $\dfrac{1 - 2x^3}{3x^{2/3}(x^3 + 1)^{4/3}}$ **15.** $x(4x^2 - 13)(x^2 - 1)^{1/2}(x^2 - 4)^{-1/2}$ **17.** $(x + 1) \sin x + x \cos x$

19. $\tan\dfrac{1}{x} - \dfrac{1}{x}\sec^2\dfrac{1}{x}$ **21.** $\dfrac{2(1 + t^2)\sec^2 t \tan t - 2t \sec^2 t}{(1 + t^2)^2}$ **23.** $-3 \sin 3w \cos(\cos 3w) + 6 \sin 4w$ **25.** $\dfrac{8x}{3y^2 - 8y}$

27. $\dfrac{1}{\sqrt{x^2 - 1}(x - \sqrt{x^2 - 1})}$ **29.** $\dfrac{\cos x \cos y}{\sin x \sin y}$ **31.** $2x + 3[x^3 + (x^4 + x)^2]^2[3x^2 + 2(x^4 + x)(4x^3 + 1)]$ **33.** $\dfrac{y - \sec^2 x}{\sec^2 y - x}$

35. $x - 2y + 9 = 0$; $27x - 54y - 7 = 0$ **37.** $5x - 4y - 6 = 0$; $4x + 5y - 13 = 0$ **39.** $(-1, 0)$ **41.** $-3(3 - 2x)^{-5/2}$
43. $x < -3$ or $x > -1$ **45.** (a) $t = 3$, $t = 8$; (b) when $t = 0$, $v = 24$, and the particle is moving to the right; when $t = 3$, $v = -15$, and the particle is moving to the left; when $t = 8$, $v = 40$, and the particle is moving to the right
47. $v^2 = 4(5 - s)(s - 3)$; $a = 16 - 4s$ **49.** (a) 32 ft/sec; (b) 256 ft; (c) 7 sec; (d) -128 ft/sec **51.** (a) 8.005; (b) 8

53. (a) 3312.2; (b) 3212.5   **55.** $-\frac{1}{3}$ **57.** $\dfrac{2}{\sqrt{4x-3}}$ **59.** $\frac{1}{2}$ **61.** $2(|x+1|-|x|)\left(\dfrac{x+1}{|x+1|}-\dfrac{x}{|x|}\right)$

63. (a) See Fig. REV 3-63; (b) continuous at 3; (c) not differentiable at 3   **65.** (a) See Fig. REV 3-65; (b) 0; (c) 0
67. $a=-\frac{1}{2}$, $b=\frac{3}{2}$ **69.** (a) $C'(x)=2x+40$; (b) \$80; (c) \$81   **71.** 648 fish/week   **77.** 10.6 knots

79. $\dfrac{512}{625\pi}$ in./sec ≈ 0.26 in./sec   **81.** 9.6 ft/sec   **85.** $f(x)=|x|$, $g(x)=x^2$

89. no; if $f(x)=\dfrac{1}{x-1}$ and $g(x)=x+1$, then f and g are differentiable at 0; however, $(f\circ g)(x)=\dfrac{1}{x}$ and $f\circ g$ is not differentiable at 0

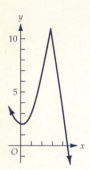

FIGURE REV 3-63

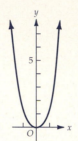

FIGURE REV 3-65

EXERCISES 4.1 (Page 223)

(Sketches of the graphs for Exercises 21 through 57 appear in Figs. EX 4.1-21 through EX 4.1-57.)
1. $-5, \frac{1}{3}$ **3.** $-3, -1, 1$ **5.** $0, 2$ **7.** $-2, 0, 2$ **9.** no critical numbers   **11.** no critical numbers
13. $\frac{1}{6}k\pi$, where k is any integer   **15.** $\frac{1}{8}(2k+1)\pi$, where k is any integer   **17.** $\frac{1}{4}k\pi$, where k is any integer
19. $-1, \frac{1}{5}, 2$ **21.** abs min: $f(2)=-2$ **23.** no absolute extrema   **25.** abs min: $f(-\frac{2}{3}\pi)=-1$; abs max: $f(0)=2$
27. abs min: $f(-3)=0$ **29.** abs min: $h(5)=1$ **31.** abs min: $F(4)=1$ **33.** abs min: $g(0)=2$ **35.** abs max: $f(5)=2$
37. abs min: $f(2)=0$ **39.** abs min: $g(0)=1$ **41.** abs min: $g(-3)=-46$; abs max: $g(-1)=-10$
43. abs min: $f(-2)=0$; abs max: $f(-4)=144$ **45.** abs min: $f(2)=0$; abs max: $f(3)=25$
47. abs min: $f(-\frac{1}{2}\pi)=-2$; abs max: $f(\frac{1}{2}\pi)=2$ **49.** no absolute extrema   **51.** abs min: $f(-1)=-1$; abs max: $f(2)=\frac{1}{2}$
53. abs min: $f(1)=-2$; abs max: $f(0)=-\frac{1}{3}$ **55.** abs min: $F(-3)=-13$; abs max: $F(3)=7$
57. abs min: $f(-1)=0$; abs max: $f(1)=\sqrt[3]{4}$

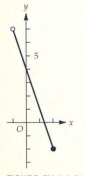

FIGURE EX 4.1-21

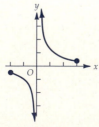

FIGURE EX 4.1-23

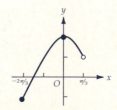

FIGURE EX 4.1-25

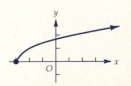

FIGURE EX 4.1-27

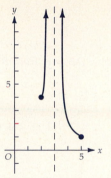

FIGURE EX 4.1-29

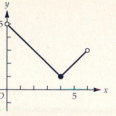

FIGURE EX 4.1-31

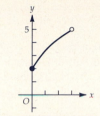

FIGURE EX 4.1-33

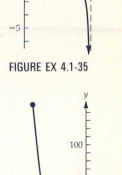

FIGURE EX 4.1-35

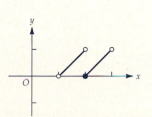

FIGURE EX 4.1-37

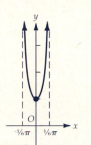

FIGURE EX 4.1-39

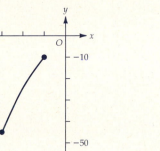

FIGURE EX 4.1-41

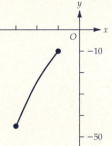

FIGURE EX 4.1-43

FIGURE EX 4.1-45

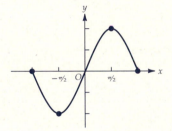

FIGURE EX 4.1-47

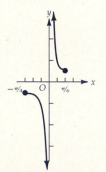

FIGURE EX 4.1-49

FIGURE EX 4.1-51

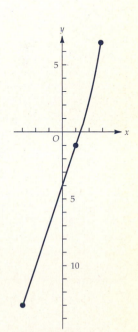

FIGURE EX 4.1-55

FIGURE EX 4.1-53

FIGURE EX 4.1-57

EXERCISES 4.2 (Page 229)

1. $\frac{5}{3}$ in.   **3.** 60 m by 60 m   **5.** 60 m by 120 m   **7.** $\frac{1}{2}$ **9.** $12\sqrt{3}$ sq units

11. from A to P to C, where P is 8 km down the beach from B **13.** from A to P to C, where P is 4 km down the river from B

15. $\frac{1}{4}\pi$ **17.** 225   **19.** 400   **21.** 30   **23.** radius is $3\sqrt{2}$ in., height is $6\sqrt{2}$ in.   **25.** $\frac{50}{3}$ in. by $\frac{50}{3}$ in. by $\frac{100}{3}$ in.

27. (a) $\sqrt{50 - 6\sqrt{41}}$ units $= (\sqrt{41} - 3)$ units ≈ 3.4 units; (b) $\sqrt{50 + 6\sqrt{41}}$ units $= (\sqrt{41} + 3)$ units ≈ 9.4 units   **29.** $\frac{2}{3}k$

33. breadth is $48\sqrt{3}$ cm; depth is $48\sqrt{6}$ cm   **35.** (a) radius of circle is $\dfrac{5}{\pi + 4}$ ft and length of side of square is $\dfrac{10}{\pi + 4}$ ft;

(b) radius of circle is $\dfrac{5}{\pi}$ ft and there is no square

EXERCISES 4.3 (Page 235)

1. 2   **3.** $\frac{1}{4}\pi$ **5.** $\frac{1}{2}$ **7.** $\frac{8}{27}$ **9.** 0

11. (b) (i), (ii), (iii) satisfied; (c) $(\frac{3}{4}, -\frac{9}{8}\sqrt[3]{6})$ **13.** (b) (i) not satisfied   **15.** (b) (ii) not satisfied   **17.** 4

19. $\cos c = \dfrac{2}{\pi}$; $c \approx 0.8807$ **21.** (i) not satisfied   **23.** (ii) not satisfied   **29.** $\frac{3}{2}$

EXERCISES 4.4 (Page 240)

(Sketches of the graphs for Exercises 1 through 39 appear in Figs. EX 4.4-1 through EX 4.4-39.)

1. (a) and (b) $f(2) = -5$, rel min; (c) $[2, +\infty)$; (d) $(-\infty, 2]$

3. (a) and (b) $f(-\frac{1}{3}) = \frac{5}{27}$, rel max; $f(1) = -1$, rel min; (c) $(-\infty, -\frac{1}{3}]$, $[1, +\infty)$; (d) $[-\frac{1}{3}, 1]$

5. (a) and (b) $f(0) = 2$, rel max; $f(3) = -25$, rel min; (c) $(-\infty, 0]$, $[3, +\infty)$; (d) $[0, 3]$

7. (a) and (b) f has a rel max value of 4 if $x = (4k + 1)\pi$, where k is any integer; f has a rel min value of -4 if $x = (4k + 3)\pi$ where k is any integer; (c) $[(4k - 1)\pi, (4k + 1)\pi]$, where k is any integer; (d) $[(4k + 1)\pi, (4k + 3)\pi]$, where k is any integer

9. (a) and (b) $f(0) = 0$, rel min; $f(1) = \frac{1}{4}$, rel max; $f(2) = 0$, rel min; (c) $[0, 1]$, $[2, +\infty)$; (d) $(-\infty, 0]$, $[1, 2]$

11. (a) and (b) $f(-2) = -\frac{1}{15}$, rel max; $f(-1) = -\frac{23}{15}$, rel min; $f(1) = \frac{53}{15}$, rel max; $f(2) = \frac{31}{15}$, rel min; (c) $(-\infty, -2]$, $[-1, 1]$, $[2, +\infty)$; $[-2, -1]$, $[1, 2]$ **13.** (a) and (b) no relative extrema; (c) $(0, +\infty)$; (d) nowhere

15. (a) and (b) $f(\sqrt[3]{2}) = \frac{3}{2}\sqrt[3]{2}$, rel min; (c) $(-\infty, 0)$, $[\sqrt[3]{2}, +\infty)$; (d) $(0, \sqrt[3]{2}]$

17. (a) and (b) $f(\frac{1}{5}) = \frac{3456}{3125}$, rel max; $f(1) = 0$, rel min; (c) $(-\infty, \frac{1}{5}]$, $[1, +\infty)$; (d) $[\frac{1}{5}, 1]$

19. (a) and (b) $f(2) = 4$, rel max; (c) $(-\infty, 2]$; (d) $[2, 3]$

21. (a) and (b) $f(-1) = 2$, rel max; $f(1) = -2$, rel min; (c) $(-\infty, -1]$, $[1, +\infty)$; (d) $[-1, 1]$

23. (a) and (b) $f(4) = 2$, rel max; (c) $(-\infty, 4]$; (d) $[4, +\infty)$ **25.** (a) and (b) $f(\frac{1}{8}) = -\frac{1}{4}$, rel min; (c) $[\frac{1}{8}, +\infty)$; (d) $(-\infty, \frac{1}{8}]$

27. (a) and (b) f has a rel max value of $-\frac{1}{2}$ if $x = \frac{1}{4}\pi + \frac{1}{2}k\pi$, where k is any integer; f has a rel min value of $\frac{1}{2}$ if $x = \frac{1}{2}k\pi$, where k is any integer; (c) $[\frac{1}{2}k\pi, \frac{1}{8}\pi + \frac{1}{2}k\pi)$, $(\frac{1}{8}\pi + \frac{1}{2}k\pi, \frac{1}{4}\pi + \frac{1}{2}k\pi]$, where k is any integer; (d) $[\frac{1}{4}\pi + \frac{1}{2}k\pi, \frac{3}{8}\pi + \frac{1}{2}k\pi)$, $(\frac{3}{8}\pi + \frac{1}{2}k\pi, \frac{1}{2}\pi + \frac{1}{2}k\pi]$, where k is any integer   **29.** (a) and (b) $f(-1) = 0$, rel max; $f(1) = -\sqrt[3]{4}$, rel min; (c) $(-\infty, -1]$, $[1, +\infty)$; (d) $[-1, 1]$

31. (a) and (b) $f(-2) = 5$, rel max; $f(0) = 1$, rel min; (c) $(-\infty, -2]$, $[0, +\infty)$; (d) $[-2, 0]$

33. (a) and (b) $f(-1) = 2$, rel max; $f(0) = 1$, rel min; $f(2) = 5$, rel max; (c) $(-\infty, -1]$, $[0, 2]$; (d) $[-1, 0]$, $[2, +\infty)$

35. (a) and (b) $f(-9) = -8$, rel min; $f(-7) = -4$, rel max; $f(-4) = -5$, rel min; $f(0) = -3$, rel max; $f(2) = -7$, rel min; (c) $[-9, -7]$, $[-4, 0]$, $[2, +\infty)$; (d) $(-\infty, -9]$, $[-7, -4]$, $[0, 2]$ **37.** (a) and (b) no relative extrema; (c) $[0, +\infty)$; (d) nowhere

39. (a) and (b) $f(4) = \frac{1}{4}\sqrt[3]{4}$, rel max; (c) $(-4, 4]$; (d) $(-\infty, -4)$, $[4, +\infty)$ **43.** $a = -3$, $b = 7$

45. $a = -2$, $b = 9$, $c = -12$, $d = 7$

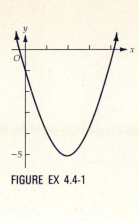

FIGURE EX 4.4-1

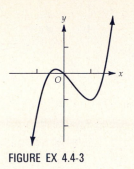

FIGURE EX 4.4-3

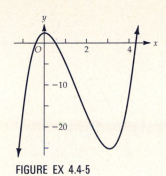

FIGURE EX 4.4-5

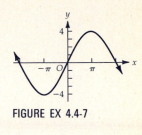

FIGURE EX 4.4-7

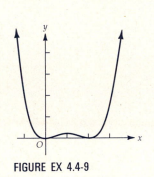

FIGURE EX 4.4-9

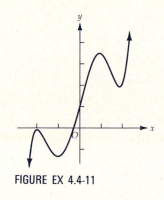

FIGURE EX 4.4-11

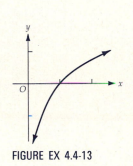

FIGURE EX 4.4-13

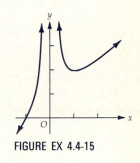

FIGURE EX 4.4-15

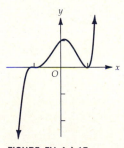

FIGURE EX 4.4-17

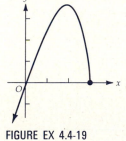

FIGURE EX 4.4-19

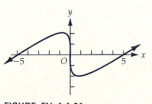

FIGURE EX 4.4-21

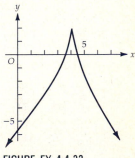

FIGURE EX 4.4-23

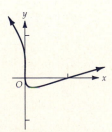

FIGURE EX 4.4-25

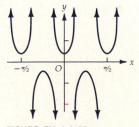

FIGURE EX 4.4-27

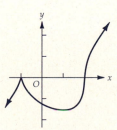

FIGURE EX 4.4-29

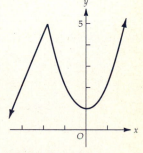

FIGURE EX 4.4-31

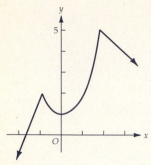

FIGURE EX 4.4-33

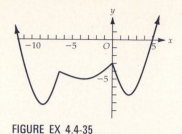

FIGURE EX 4.4-35

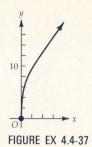

FIGURE EX 4.4-37

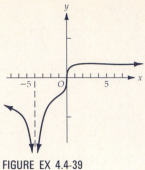

FIGURE EX 4.4-39

EXERCISES 4.5 (Page 248)

(Sketches of the graphs for Exercises 1 through 23 appear in Figs. EX 4.5-1 through 4.5-23.)

1. $(0, 0)$; concave downward for $x < 0$; concave upward for $x > 0$

3. $(-\frac{1}{2}, \frac{15}{2})$; concave downward for $x < -\frac{1}{2}$; concave upward for $x > -\frac{1}{2}$

5. $(0, 0)$, $(4, -256)$; concave upward for $x < 0$ and $x > 4$; concave downward for $0 < x < 4$

7. $(1, 0)$; concave downward for $x < 1$; concave upward for $x > 1$

9. $(-2, 0)$; concave upward for $x < -2$; concave downward for $x > -2$

11. $(-1, \frac{1}{2})$, $(1, \frac{1}{2})$; concave upward for $x < -1$ and $x > 1$; concave downward for $-1 < x < 1$

13. $(-\frac{2}{3}\pi, 0)$, $(-\frac{1}{3}\pi, 0)$, $(0, 0)$, $(\frac{1}{3}\pi, 0)$, $(\frac{2}{3}\pi, 0)$; concave upward for $-\pi < x < -\frac{2}{3}\pi$, $-\frac{1}{3}\pi < x < 0$, and $\frac{1}{3}\pi < x < \frac{2}{3}\pi$; concave downward for $-\frac{2}{3}\pi < x < -\frac{1}{3}\pi$, $0 < x < \frac{1}{3}\pi$, and $\frac{2}{3}\pi < x < \pi$

15. $(0, 0)$; concave downward for $-\pi < x < 0$; concave upward for $0 < x < \pi$

17. no pt. of infl.; concave upward for $x < 2$; concave downward for $x > 2$

19. $(0, 0)$; concave upward for $x < 0$; concave downward for $x > 0$

21. $(0, 0)$; concave downward for $x < 0$; concave upward for $x > 0$

23. $(2, 3)$; concave upward for $x < 2$; concave downward for $x > 2$ **25.** $a = -1$, $b = 3$ **27.** $a = 2$, $b = -6$, $c = 0$, $d = 3$

29. (a) $(\frac{1}{2}\pi + k\pi, 0)$ where k is any integer; (b) -1 at $(\frac{1}{2}\pi + k\pi, 0)$ if k is an even integer, 1 at $(\frac{1}{2}\pi + k\pi, 0)$ if k is an odd integer

31. $(\frac{1}{2}\pi + k\pi, 0)$, where k is any integer; -1 at all pts. of infl.

(Sketches of the graphs for Exercises 33 through 47 appear in Figs. EX 4.5-33 through 4.5-47.)

49. yes   **51.** 10:40 A.M.

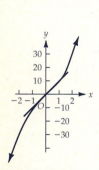

FIGURE EX 4.5-1

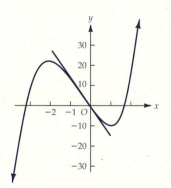

FIGURE EX 4.5-3

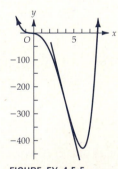

FIGURE EX 4.5-5

FIGURE EX 4.5-7

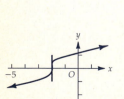

FIGURE EX 4.5-9

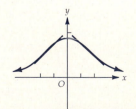

FIGURE EX 4.5-11

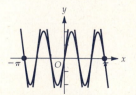

FIGURE EX 4.5-13

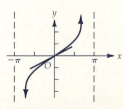

FIGURE EX 4.5-15

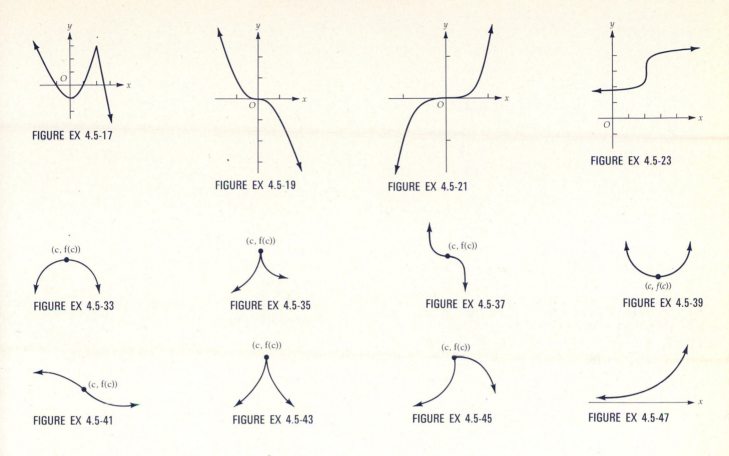

FIGURE EX 4.5-17

FIGURE EX 4.5-19

FIGURE EX 4.5-21

FIGURE EX 4.5-23

FIGURE EX 4.5-33

FIGURE EX 4.5-35

FIGURE EX 4.5-37

FIGURE EX 4.5-39

FIGURE EX 4.5-41

FIGURE EX 4.5-43

FIGURE EX 4.5-45

FIGURE EX 4.5-47

EXERCISES 4.6 (Page 253)

(Sketches of the graphs for Exercises 1 through 25 appear in Figs. EX 4.6-1 through EX 4.6-25.)

1. $f(\frac{1}{3}) = \frac{2}{3}$, rel min; concave upward everywhere

3. $f(\frac{3}{2}) = \frac{81}{4}$, rel max; $f(-1) = -11$, rel min; $(\frac{1}{4}, \frac{37}{8})$, pt. of infl.; concave upward for $x < \frac{1}{4}$; concave downward for $x > \frac{1}{4}$

5. $g(0) = 3$, rel max; $g(2) = \frac{5}{3}$, rel min; $(1, \frac{7}{3})$, pt. of infl.; concave downward for $x < 1$; concave upward for $x > 1$

7. $f(4) = 0$, rel min; concave upward everywhere

9. $h(-\frac{3}{4}) = -\frac{99}{256}$, rel min; $h(0) = 0$, rel max; $h(1) = -\frac{5}{6}$, rel min; pts. of infl. at $x = \frac{1}{12}(1 \pm \sqrt{37})$; concave upward for $x < \frac{1}{12}(1 - \sqrt{37})$ and $x > \frac{1}{12}(1 + \sqrt{37})$; concave downward for $\frac{1}{12}(1 - \sqrt{37}) < x < \frac{1}{12}(1 + \sqrt{37})$

11. $F(\frac{1}{3}\pi) = -1$, rel min; $F(0) = 1$, rel max; $(\frac{1}{6}\pi, 0)$, pt. of infl.; concave downward for $-\frac{1}{6}\pi < x < \frac{1}{6}\pi$; concave upward for $\frac{1}{6}\pi < x < \frac{1}{2}\pi$

13. $f(-\frac{1}{2}) = -\frac{27}{16}$, rel min; $(-2, 0)$, $(-1, -1)$, pts. of infl.; concave upward for $x < -2$ and $x > -1$; concave downward for $-2 < x < -1$

15. $f(1) = 8$, rel min; $(3, \frac{16}{3}\sqrt{3})$, pt. of infl.; concave upward for $0 < x < 3$; concave downward for $x > 3$

17. $h(-2) = -2$, rel min; concave upward for $x > -3$

19. $F(27) = 9$, rel max; $(0, 0)$, $(216, 0)$, pts. of infl.; concave upward for $x < 0$ and $x > 216$; concave downward for $0 < x < 216$

21. $f(-1) = -2$, rel min; $f(1) = 2$, rel max; $(0, 0)$, $(-\frac{1}{2}\sqrt{2}, -\frac{7}{8}\sqrt{2})$, $(\frac{1}{2}\sqrt{2}, \frac{7}{8}\sqrt{2})$, pts. of infl.; concave upward for $x < -\frac{1}{2}\sqrt{2}$ and $0 < x < \frac{1}{2}\sqrt{2}$; concave downward for $-\frac{1}{2}\sqrt{2} < x < 0$ and $x > \frac{1}{2}\sqrt{2}$ **23.** $f(0) = 0$, rel min; concave upward everywhere

25. no relative extrema; $(-\frac{3}{2}\pi, -\frac{3}{2}\pi)$, $(-\frac{1}{2}\pi, -\frac{1}{2}\pi)$, $(\frac{1}{2}\pi, \frac{1}{2}\pi)$, $(\frac{3}{2}\pi, \frac{3}{2}\pi)$, pts. of infl.; concave downward for $-2\pi < x < -\frac{3}{2}\pi$, $-\frac{1}{2}\pi < x < \frac{1}{2}\pi$ and $\frac{3}{2}\pi < x < 2\pi$; concave upward for $-\frac{3}{2}\pi < x < -\frac{1}{2}\pi$ and $\frac{1}{2}\pi < x < \frac{3}{2}\pi$

27. $\cos k\pi = 1$, where k is any even integer, rel max; $\cos k\pi = -1$, where k is any odd integer, rel min

29. $\csc(\frac{1}{2}\pi + 2k\pi) = 1$, where k is any integer, rel min; $\csc(\frac{3}{2}\pi + 2k\pi) = -1$, where k is any integer, rel max

(Sketches of the graphs for Exercises 31 through 35 appear in Figs. EX 4.6-31 through EX 4.6-41.)

43. f has a rel min at $x = \frac{1}{2}\sqrt{2}$ and a rel max at $x = -\frac{1}{2}\sqrt{3}$

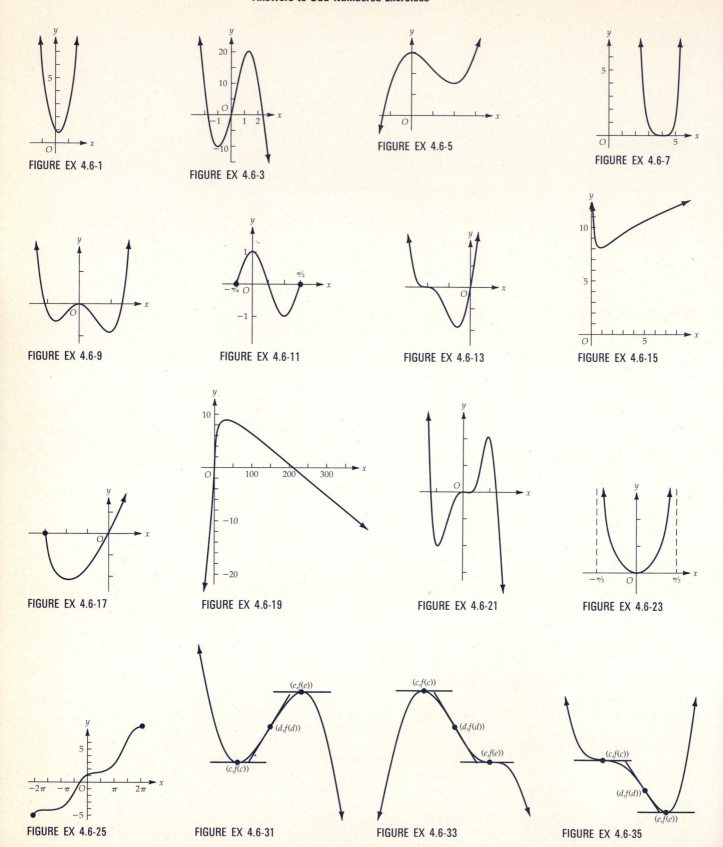

FIGURE EX 4.6-1

FIGURE EX 4.6-3

FIGURE EX 4.6-5

FIGURE EX 4.6-7

FIGURE EX 4.6-9

FIGURE EX 4.6-11

FIGURE EX 4.6-13

FIGURE EX 4.6-15

FIGURE EX 4.6-17

FIGURE EX 4.6-19

FIGURE EX 4.6-21

FIGURE EX 4.6-23

FIGURE EX 4.6-25

FIGURE EX 4.6-31

FIGURE EX 4.6-33

FIGURE EX 4.6-35

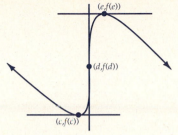

FIGURE EX 4.6-37

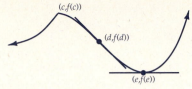

FIGURE EX 4.6-39

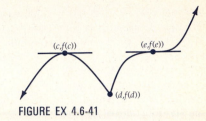

FIGURE EX 4.6-41

EXERCISES 4.7 (Page 259)

(Sketches of the graphs for Exercises 1 through 57 appear in Figs. EX 4.7-1 through EX 4.7-57.)

1. $x = 1$, $y = x + 1$　　**3.** $x = 3$, $y = x + 3$　　**5.** $x = -2$, $y = x - 6$　　**7.** $x = 0$, $y = x + 2$

9. $f(-1) = 5$, rel max; $f(1) = -3$, rel min; $(0, 1)$, pt. of infl.; increasing on $(-\infty, -1]$ and $[1, +\infty)$ decreasing on $[-1, 1]$; concave downward for $x < 0$; concave upward for $x > 0$

11. $f(\frac{3}{2}) = -\frac{27}{16}$, rel min; $(0, 0)$, $(1, -1)$, pts. of infl.; increasing on $[\frac{3}{2}, +\infty)$; decreasing on $(-\infty, \frac{3}{2}]$ concave upward for $x < 0$ and $x > 1$; concave downward for $0 < x < 1$

13. $f(-3) = 5$, rel max; $f(-\frac{1}{3}) = -\frac{121}{27}$, rel min; $(-\frac{5}{3}, \frac{7}{27})$, pt. of infl.; increasing on $(-\infty, -3]$ and $[-\frac{1}{3}, +\infty)$; decreasing on $[-3, -\frac{1}{3}]$; concave downward for $x < -\frac{5}{3}$; concave upward for $x > -\frac{5}{3}$

15. $f(0) = 1$, rel min; $(\frac{1}{2}, \frac{23}{16})$, $(1, 2)$, pts. of infl.; decreasing on $(-\infty, 0]$; increasing on $[0, +\infty)$; concave upward for $x < \frac{1}{2}$ and $x > 1$; concave downward for $\frac{1}{2} < x < 1$

17. $f(-1) = \frac{7}{12}$, rel min; $f(0) = 1$, rel max; $f(2) = -\frac{5}{3}$, rel min; pts. of infl. at $x = \frac{1}{3}(1 \pm \sqrt{7})$; decreasing on $(-\infty, -1]$ and $[0, 2]$; increasing on $[-1, 0]$ and $[2, +\infty)$; concave upward for $x < \frac{1}{3}(1 - \sqrt{7})$ and $x > \frac{1}{3}(1 + \sqrt{7})$; concave downward for $\frac{1}{3}(1 - \sqrt{7}) < x < \frac{1}{3}(1 + \sqrt{7})$

19. $f(0) = 2$, rel min; no pts. of infl.; decreasing on $(-\infty, 0]$; increasing on $[0, +\infty)$; concave upward everywhere

21. $f(0) = 0$, rel min; no pts. of infl.; decreasing on $(-\infty, 0]$; increasing on $[0, +\infty)$; concave upward everywhere

23. no relative extrema; $(0, 0)$ pt. of infl. with horizontal tangent; increasing on $(-\infty, +\infty)$; concave downward on $(-\infty, 0)$; concave upward on $(0, +\infty)$

25. no relative extrema; $(2, 0)$ pt. of infl.; decreasing on $(-\infty, +\infty)$; concave upward for $x < 2$; concave downward for $x > 2$

27. $f(\frac{4}{5}) = \frac{26,244}{3,125}$, rel max; $f(2) = 0$, rel min; pts. of infl. at $(-1, 0)$ and $x = \frac{1}{10}(8 \pm 3\sqrt{6})$; increasing on $(-\infty, \frac{4}{5}]$ and $[2, +\infty)$; decreasing on $[\frac{4}{5}, 2]$; concave downward for $x < -1$ and $\frac{1}{10}(8 - 3\sqrt{6}) < x < \frac{1}{10}(8 + 3\sqrt{6})$; concave upward for $-1 < x < \frac{1}{10}(8 - 3\sqrt{6})$ and $x > \frac{1}{10}(8 + 3\sqrt{6})$

29. $f(-\frac{4}{3}) = \frac{256}{81}$, rel max; $f(0) = 0$, rel min; $(-1, 2)$, pt. of infl.; increasing on $(-\infty, -\frac{4}{3}]$ and $[0, +\infty)$; decreasing on $[-\frac{4}{3}, 0]$; concave downward for $x < -1$; concave upward for $x > -1$

31. $f(-\frac{1}{2}\pi) = -3$, rel min; $f(0) = 3$, rel max; $f(\frac{1}{2}\pi) = -3$, rel min; $(-\frac{3}{4}\pi, 0)$, $(-\frac{1}{4}\pi, 0)$, $(\frac{1}{4}\pi, 0)$, $(\frac{3}{4}\pi, 0)$, pts. of infl.; decreasing on $[-\pi, -\frac{1}{2}\pi]$ and $[0, \frac{1}{2}\pi]$; increasing on $[-\frac{1}{2}\pi, 0]$ and $[\frac{1}{2}\pi, \pi]$; concave downward for $-\pi < x < -\frac{3}{4}\pi$, $-\frac{1}{4}\pi < x < \frac{1}{4}\pi$, and $\frac{3}{4}\pi < x < \pi$; concave upward for $-\frac{3}{4}\pi < x < -\frac{1}{4}\pi$ and $\frac{1}{4}\pi < x < \frac{3}{4}\pi$

33. $(\frac{1}{2}\pi, 0)$, rel min; no pts. of infl.; increasing on $[0, \frac{1}{2}\pi]$ and $[\frac{1}{2}\pi, \pi]$; concave downward for $0 < x < \frac{1}{2}\pi$ and $\frac{1}{2}\pi < x < \pi$

35. no relative extrema; $(0, 0)$, pt. of infl.; increasing on $(-\pi, \pi)$; concave downward for $-\pi < x < 0$; concave upward for $0 < x < \pi$; $x = -\pi$ and $x = \pi$ are asymptotes

37. $f(-\frac{1}{4}\pi) = \sqrt{2}$, rel max; $f(-\frac{3}{4}\pi) = -\sqrt{2}$, rel min; $f(\frac{1}{4}\pi) = \sqrt{2}$, rel max; $f(\frac{3}{4}\pi) = -\sqrt{2}$, rel min; $(-\frac{5}{4}\pi, 0)$, $(-\frac{1}{4}\pi, 0)$, $(\frac{3}{4}\pi, 0)$, $(\frac{7}{4}\pi, 0)$, pts. of infl.; increasing on $[-2\pi, -\frac{7}{4}\pi]$, $[-\frac{3}{4}\pi, \frac{1}{4}\pi]$, and $[\frac{3}{4}\pi, 2\pi]$; decreasing on $[-\frac{7}{4}\pi, -\frac{3}{4}\pi]$, and $[\frac{1}{4}\pi, \frac{3}{4}\pi]$; concave downward for $-2\pi < x < -\frac{5}{4}\pi$, $-\frac{1}{4}\pi < x < \frac{3}{4}\pi$, and $\frac{7}{4}\pi < x < 2\pi$; concave upward for $-\frac{5}{4}\pi < x < -\frac{1}{4}\pi$ and $\frac{3}{4}\pi < x < \frac{7}{4}\pi$

39. $f(0) = 0$, rel max; $f(2) = 4$, rel min; no pts. of infl.; increasing on $(-\infty, 0]$ and $[2, +\infty)$; decreasing on $[0, 1)$ and $(1, 2]$; concave downward for $x < 1$; concave upward for $x > 1$; $x = 1$ and $y = x + 1$ are asymptotes

41. $f(0) = -1$, rel max; no pts. of infl.; increasing on $(-\infty, -1)$ and $(-1, 0]$; decreasing on $[0, 1)$ and $(1, +\infty)$; concave upward for $x < -1$ and $x > 1$; concave downward for $-1 < x < 1$; $y = 1$, $x = -1$, and $x = 1$ are asymptotes

43. $f(-1) = -1$, rel min; $f(1) = 1$, rel max; $(-\sqrt{3}, -\frac{1}{2}\sqrt{3})$, $(0, 0)$ and $(\sqrt{3}, \frac{1}{2}\sqrt{3})$, pts. of infl.; decreasing on $(-\infty, -1]$ and $[1, +\infty)$; increasing on $[-1, 1]$; concave downward for $x < -\sqrt{3}$ and $0 < x < \sqrt{3}$; concave upward for $-\sqrt{3} < x < 0$ and $x > \sqrt{3}$; $y = 0$ is an asymptote

45. $f(0) = 0$, rel min; $f(1) = 1$, rel max; no pts. of infl.; decreasing on $(-\infty, 0]$ and $[1, +\infty)$; increasing on $[0, 1]$; concave downward for $x < 0$ and $x > 0$

47. $f(1) = -1$, rel min; no pts. of infl.; decreasing on $(-\infty, 1]$; increasing $[1, +\infty)$; concave upward for all x

49. no relative extrema; $(3, 2)$, pt. of infl.; increasing on $(-\infty, +\infty)$; concave downward for $x > 3$; concave upward for $x < 3$

51. no relative extrema; $(3, 2)$, pt. of infl. with horizontal tangent; increasing on $(-\infty, +\infty)$; concave downward for $x < 3$; concave upward for $x > 3$

53. $f(0) = 0$, rel min; $f(\frac{16}{5}) = \frac{512}{125}\sqrt{5}$, rel max; pt. of infl. at $x = \frac{1}{15}(48 - 8\sqrt{6})$; decreasing on $(-\infty, 0]$ and $[\frac{16}{5}, 4]$; increasing on $[0, \frac{16}{5}]$; concave upward for $x < \frac{1}{15}(48 - 8\sqrt{6})$; concave downward for $\frac{1}{15}(48 - 8\sqrt{6}) < x < 4$

55. $f(-\frac{2}{3}) = \frac{4}{9}\sqrt{6}$, rel max; no pts. of infl.; increasing on $(-\infty, -\frac{2}{3}]$; decreasing on $[-\frac{2}{3}, 0]$; concave downward for $x < 0$

57. $f(-1) = 0$, rel max; $f(1) = -\sqrt[3]{4}$, rel min; $(2, 0)$, pt. of infl.; increasing on $(-\infty, -1]$ and $[1, +\infty)$; decreasing on $[-1, 1]$; concave upward for $x < -1$ and $-1 < x < 2$; concave downward for $x > 2$

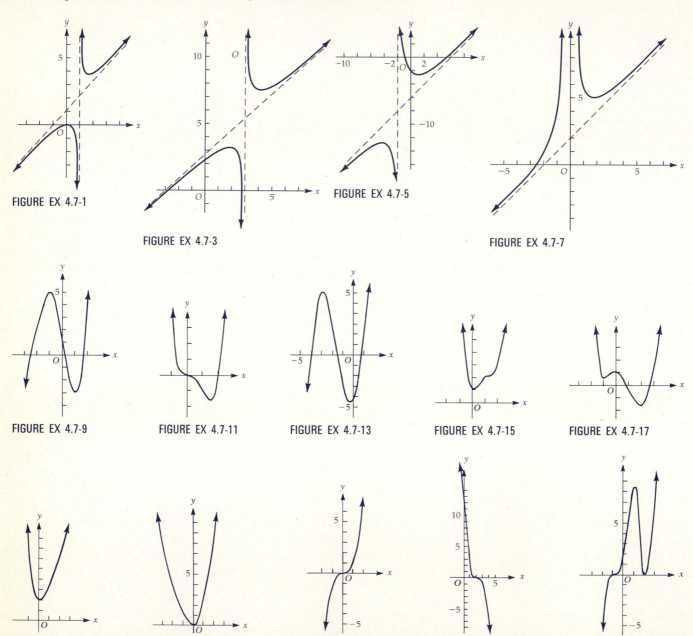

FIGURE EX 4.7-1

FIGURE EX 4.7-3

FIGURE EX 4.7-5

FIGURE EX 4.7-7

FIGURE EX 4.7-9

FIGURE EX 4.7-11

FIGURE EX 4.7-13

FIGURE EX 4.7-15

FIGURE EX 4.7-17

FIGURE EX 4.7-19

FIGURE EX 4.7-21

FIGURE EX 4.7-23

FIGURE EX 4.7-25

FIGURE EX 4.7-27

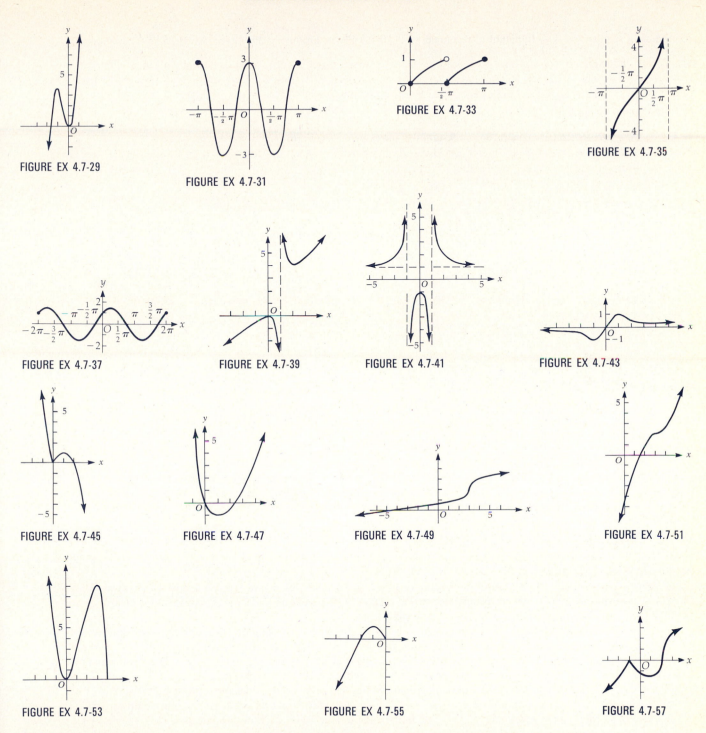

FIGURE EX 4.7-29

FIGURE EX 4.7-31

FIGURE EX 4.7-33

FIGURE EX 4.7-35

FIGURE EX 4.7-37

FIGURE EX 4.7-39

FIGURE EX 4.7-41

FIGURE EX 4.7-43

FIGURE EX 4.7-45

FIGURE EX 4.7-47

FIGURE EX 4.7-49

FIGURE EX 4.7-51

FIGURE EX 4.7-53

FIGURE EX 4.7-55

FIGURE 4.7-57

EXERCISES 4.8 (Page 267)

1. $f(0) = 0$, abs min   **3.** no absolute extrema   **5.** $g(\frac{1}{4}) = \frac{3}{4}$, abs min   **7.** no absolute extrema   **9.** $f(-2) = -13$, abs min
11. $g(-1) = -2$, abs max   **13.** $f(0) = 0$, abs min; $f(\sqrt{2}) = \frac{1}{18}\sqrt{3}$, abs max   **15.** $f(-\frac{1}{6}\pi) = \sqrt{3}$, abs min
17. 45 m by 60 m   **19.** 6 in. by 9 in.   **21.** 12 in. by 4 in. by 6 in.   **23.** $2x - y + 1 = 0$ **25.** 1 month; 7.5 percent
27. height, $4\sqrt[3]{4}$ in.; radius, $\sqrt[3]{4}$ in.   **29.** 40 units   **31.** $1500   **33.** $(\frac{9}{5}, \frac{3}{5})$

35. radius of semicircle, $\dfrac{32}{4 + \pi}$ ft; height of rectangle, $\dfrac{32}{4 + \pi}$ ft   **37.** 90 km/hr   **39.** $\frac{36}{25}$ sec   **41.** $5\sqrt{5}$ ft   **43.** $\sqrt{2}$

45. $2\sqrt{2}$ **47.** $\frac{4}{3}a$

EXERCISES 4.9 (Page 276)

1. 4; 5   **3.** $\frac{1}{6} \approx 0.167$; $\sqrt[3]{10} - 2 \approx 0.154$ **5.** (a) $(-3 - 4x)\,\Delta x - 2(\Delta x)^2$; (b) $(-3 - 4x)\,\Delta x$; (c) $-2(\Delta x)^2$

7. (a) $(3x^2 - 2x)\,\Delta x + (3x - 1)(\Delta x)^2 + (\Delta x)^3$; (b) $(3x^2 - 2x)\,\Delta x$; (c) $(3x - 1)(\Delta x)^2 + (\Delta x)^3$

9. (a) $\dfrac{-2\Delta x}{(x - 1)(x + \Delta x - 1)}$; (b) $\dfrac{-2\Delta x}{(x - 1)^2}$; (c) $\dfrac{2(\Delta x)^2}{(x - 1)^2(x + \Delta x - 1)}$ **11.** (a) 0.0309; (b) 0.03; (c) 0.0009

13. (a) $\frac{1}{42} \approx 0.0238$; (b) $\frac{1}{40} = 0.025$; (c) $-\frac{1}{840} \approx -0.0012$ **15.** (a) -0.875; (b) -1.5; (c) 0.625   **17.** $3(3x^2 - 2x + 1)^2(6x - 2)\,dx$

19. $\dfrac{(14x^2 + 18x)\,dx}{3(2x + 3)^{2/3}}$ **21.** $\dfrac{(1 - 2\sin x + 2\cos x)\,dx}{(2 - \sin x)^2}$ **23.** $2\tan x \sec^2 x(2\tan^2 x + 1)\,dx$ **25.** $-\dfrac{3x}{4y}$ **27.** $-\dfrac{\sqrt{y}}{\sqrt{x}}$

29. 1   **31.** (a) $dx = 4t\,dt$, $dy = 3\,dt$; (b) $\dfrac{3}{4t}$; (c) $\dfrac{9}{4y}$ **33.** (a) $dx = -2\sin t\,dt$, $dy = 3\cos t\,dt$; (b) $-\dfrac{3\cos t}{2\sin t}$; (c) $-\dfrac{9x}{4y}$

35. (a) 6.75 cm³; (b) 0.3 cm²   **37.** $\frac{12}{5}\pi$ m³   **39.** 0.4π cm²   **41.** 0.9π cm³   **43.** 4 percent   **45.** 10 ft³

EXERCISES 4.10 (Page 281)

1. 4.1179   **3.** -1.1673 **5.** 2.649   **7.** 0.507   **9.** -1.128 **11.** 1.73205   **13.** 1.81712   **15.** 0.7391   **17.** 0.8767

19. 2.0288, 4.9132   **21.** 3.14159

REVIEW EXERCISES FOR CHAPTER 4 (Page 282)

(Sketches of the graphs for Exercises 1 through 31 appear in Figs. REV 4-1 through REV 4-31.)

1. abs min: $f(-5) = 0$ **3.** abs min: $f(1) = -\frac{1}{2}$; abs max: $f(2) = 64$ **5.** abs min: $f(3) = 0$; abs max: $f(0) = 9$

7. abs min: $f(\sqrt{6}) = 0$; abs max: $f(5) = 361$ **9.** abs min: $f(-\frac{1}{6}\pi) = -2$; abs max: $f(\frac{1}{6}\pi) = 2$ **11.** abs max: $f(\frac{1}{12}) = \sqrt{3}$

13. $f(-2) = 0$, rel max; $f(0) = -4$, rel min; $(-1, -2)$, pt. of infl.; increasing on $(-\infty, -2]$ and $[0, +\infty)$; decreasing on $[-2, 0]$; concave downward for $x < -1$; concave upward for $x > -1$

15. $f(\frac{8}{5}) = \frac{839,808}{3125}$, rel max; $f(4) = 0$, rel min; pts. of infl. at $x = -2$ and $x = \frac{1}{5}(8 \pm 3\sqrt{6})$; increasing on $(-\infty, \frac{8}{5}]$ and $[4, +\infty)$; decreasing on $[\frac{8}{5}, 4]$; concave upward for $-2 < x < \frac{1}{5}(8 - 3\sqrt{6})$ and $x > \frac{1}{5}(8 + 3\sqrt{6})$; concave downward for $x < -2$ and $\frac{1}{5}(8 - 3\sqrt{6}) < x < \frac{1}{5}(8 + 3\sqrt{6})$

17. no relative extrema; $(4, -3)$, pt. of infl.; increasing on $(-\infty, +\infty)$; concave upward for $x < 4$; concave downward for $x > 4$

19. $f(0) = 0$, rel min; $(\frac{1}{3}, \frac{1}{3})$, $(-\frac{1}{3}, \frac{1}{3})$, pts. of infl.; decreasing on $(-\infty, 0]$; increasing on $[0, +\infty)$; concave downward for $x < -\frac{1}{3}$ and $x > \frac{1}{3}$; concave upward for $-\frac{1}{3} < x < \frac{1}{3}$; $y = \frac{4}{3}$ is an asymptote

21. $f(0) = 0$, rel max; no pts. of infl.; increasing on $(-\infty, -2)$ and $(-2, 0]$; decreasing on $[0, 2)$ and $(2, +\infty)$; concave upward for $x < -2$ and $x > 2$; concave downward for $-2 < x < 2$; $y = 5$, $x = -2$, and $x = 2$ are asymptotes

23. $f(0) = 0$, rel max; $f(6) = 12$, rel min; no pts. of infl.; increasing on $(-\infty, 0]$ and $[6, +\infty)$; decreasing on $[0, 3)$ and $(3, 6]$; concave downward for $x < 3$; concave upward for $x > 3$; $x = 3$ and $y = x + 3$ are asymptotes

25. $f(1) = 0$, rel min; no pts. of infl.; decreasing on $(-\infty, 1]$; increasing on $[1, +\infty)$; concave upward for all x

27. no relative extrema; $(3, 1)$, pt. of infl.; increasing on $(-\infty, +\infty)$; concave downward for $x < 3$; concave upward for $x > 3$

29. $f(-1) = 0$, rel min; $f(0) = 9$, rel max; $f(3) = 0$, rel min; pts. of infl. at $x = \pm\frac{3}{5}\sqrt{5}$; decreasing on $(-\infty, -1]$ and $[0, 3]$; increasing on $[-1, 0]$ and $[3, +\infty)$; concave upward for $x < -\frac{3}{5}\sqrt{5}$ and $x > \frac{3}{5}\sqrt{5}$; concave downward for $-\frac{3}{5}\sqrt{5} < x < \frac{3}{5}\sqrt{5}$

31. $f(-\frac{1}{8}\pi) = -\sqrt{2}$, rel min; $f(\frac{3}{8}\pi) = \sqrt{2}$, rel max; $(\frac{1}{8}\pi, 0)$, pt. of infl.; decreasing on $[-\frac{3}{8}\pi, -\frac{1}{8}\pi]$ and $[\frac{3}{8}\pi, \frac{5}{8}\pi]$; increasing on $[-\frac{1}{8}\pi, \frac{3}{8}\pi]$; concave upward for $-\frac{3}{8}\pi < x < \frac{1}{8}\pi$; concave downward for $\frac{1}{8}\pi < x < \frac{5}{8}\pi$ **33.** $\frac{1}{3}(1 - \sqrt{13})$ **35.** $-\frac{13}{4}$ **41.** $\sqrt{A^2 + B^2}$

43. pt. of infl. at $(0, 2)$; concave downward for $x < 0$; concave upward for $x > 0$

45. $a = 2$, $b = -6$, $c = 3$ **51.** 6, 6   **55.** 20 in. by 20 in. by 10 in.   **57.** $\sqrt[3]{2k}$ in. by $\sqrt[3]{2k}$ in. by $\frac{1}{2}\sqrt[3]{2k}$ in.

59. 1500   **61.** 12 km from point on bank nearest A **65.** $1800   **67.** 25 radios; $525

69. $\frac{3}{2}$ sec; velocity of particle moving horizontally is 1 cm/sec; velocity of particle moving vertically is 3 cm/sec

71. $(h^{2/3} + w^{2/3})^{3/2}$ meters   **73.** 9 m by 18 m   **75.** radius is $\frac{3}{2}r$ cm and altitude is $3h$ cm   **77.** the wire should be cut in half

79. 1000; $11   **81.** (a) -0.16; (b) -0.64 **83.** 2π in.^{3} **85.** an error of approximately $\dfrac{\pi^2}{1610t}$ sec   **91.** 1.168   **93.** 1.9622

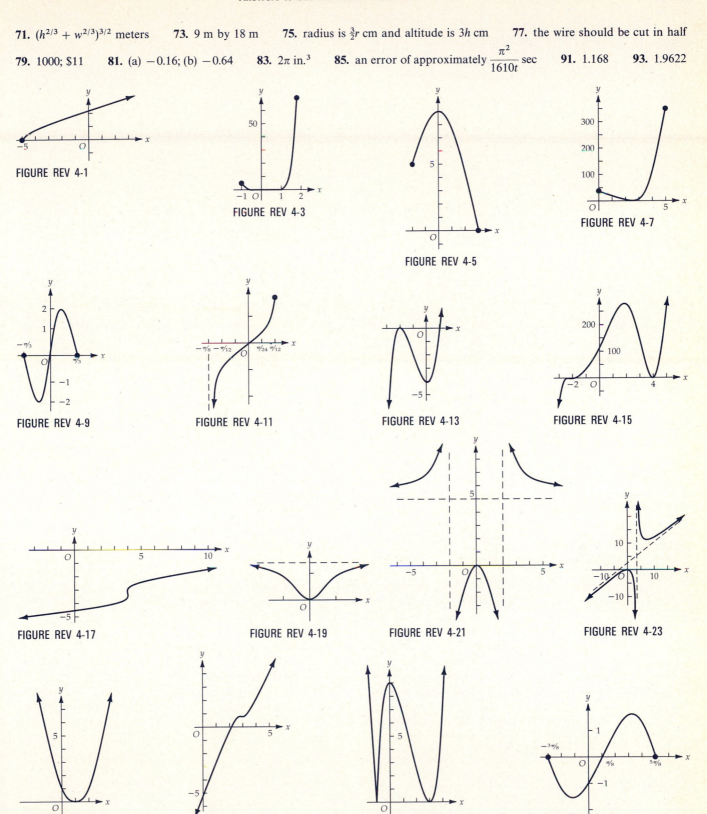

FIGURE REV 4-1

FIGURE REV 4-3

FIGURE REV 4-5

FIGURE REV 4-7

FIGURE REV 4-9

FIGURE REV 4-11

FIGURE REV 4-13

FIGURE REV 4-15

FIGURE REV 4-17

FIGURE REV 4-19

FIGURE REV 4-21

FIGURE REV 4-23

FIGURE REV 4-25

FIGURE REV 4-27

FIGURE REV 4-29

FIGURE REV 4-31

EXERCISES 5.1 (Page 294)

1. $\frac{3}{5}x^5 + C$ **3.** $-\frac{1}{2x^2} + C$ **5.** $2u^{5/2} + C$ **7.** $3x^{2/3} + C$ **9.** $\frac{9}{5}t^{10/3} + C$ **11.** $x^4 + \frac{1}{3}x^3 + C$ **13.** $\frac{1}{3}y^6 - \frac{3}{4}y^4 + C$

15. $3t - t^2 + \frac{1}{3}t^3 + C$ **17.** $\frac{8}{5}x^5 + x^4 - 2x^3 - 2x^2 + 5x + C$ **19.** $\frac{2}{5}x^{5/2} + \frac{2}{3}x^{3/2} + C$ **21.** $\frac{2}{5}x^{5/2} - \frac{1}{2}x^2 + C$

23. $-\frac{1}{x^2} - \frac{3}{x} + 5x + C$ **25.** $\frac{2}{5}x^{5/2} + \frac{8}{3}x^{3/2} - 8x^{1/2} + C$ **27.** $\frac{3}{4}x^{4/3} + \frac{3}{2}x^{2/3} + C$ **29.** $-3\cos t - 2\sin t + C$

31. $\sec x + C$ **33.** $-4\csc x + 2\tan x + C$ **35.** $-2\cot\theta - 3\tan\theta + \theta + C$ **37.** $y = x^2 - 3x + 2$

39. $3y = -2x^3 + 3x^2 + 2x + 6$ **41.** $12y = -x^4 + 6x^2 - 20x + 27$ **43.** $C(x) = x^3 + 4x^2 + 4x + 6$

45. (a) $C(x) = 3x^2 + 8$; (b) $\$800$ **47.** (a) $R(x) = 15x - 2x^2$; (b) $p = 15 - 2x$ **49.** 117π m^3

51. g is not differentiable on $(-1, 1)$

EXERCISES 5.2 (Page 302)

1. $-\frac{1}{6}(1 - 4y)^{3/2} + C$ **3.** $-\frac{3}{8}(6 - 2x)^{4/3} + C$ **5.** $\frac{1}{3}(x^2 - 9)^{3/2} + C$ **7.** $\frac{1}{33}(x^3 - 1)^{11} + C$ **9.** $-\frac{3}{8}(9 - 4x^2)^{5/3} + C$

11. $\frac{1}{32(1 - 2y^4)^4} + C$ **13.** $\frac{3}{11}(x - 2)^{11/3} + C$ **15.** $\frac{2}{5}(x + 2)^{5/2} - \frac{4}{3}(x + 2)^{3/2} + C$ **17.** $-\frac{2}{5}(1 - r)^{-5} + \frac{1}{3}(1 - r)^{-6} + C$

19. $-\frac{3}{4}(3 - 2x)^{3/2} + \frac{3}{10}(3 - 2x)^{5/2} - \frac{1}{28}(3 - 2x)^{7/2} + C$ **21.** $\frac{1}{4}\sin 4\theta + C$ **23.** $-2\cos x^3 + C$ **25.** $\frac{1}{5}\tan 5x + C$

27. $-\frac{1}{6}\csc 3y^2 + C$ **29.** $\frac{1}{6}(2 + \sin x)^6 + C$ **31.** $-2\left(1 + \frac{1}{3x}\right)^{3/2} + C$ **33.** $-\frac{3}{2}(1 + \cos x)^{4/3} + C$ **35.** $-\frac{1}{3}\cos^3 t + C$

37. $\frac{1}{2}\tan 2x - \frac{1}{2}\cot 2x + C$ **39.** $-\frac{1}{3}\sqrt{1 - 2\sin 3x} + C$ **41.** $\frac{2}{3}\sqrt{x^3 + 3x^2 + 1} + C$ **43.** $-\frac{1}{4(3x^4 + 2x^2 + 1)} + C$

45. $\frac{3}{4}(3 - y)^{4/3} - 18(3 - y)^{1/3} + C$ **47.** $\frac{3}{5}(r^{1/3} + 2)^5 + C$ **49.** $\sqrt{x^2 + 4} + \frac{4}{\sqrt{x^2 + 4}} + C$ **51.** $\cos(\cos x) + C$

53. (a) $2x^4 + 4x^3 + 3x^2 + x + C$; (b) $\frac{1}{8}(2x + 1)^4 + C$ **55.** (a) $\frac{2}{3}x^{3/2} - 2x + 2x^{1/2} + C$; (b) $\frac{2}{3}(\sqrt{x} - 1)^3 + C$

57. (a) $\sin^2 x + C$; (b) $-\cos^2 x + C$; (c) $-\frac{1}{2}\cos 2x + C$ **59.** $C(x) = \frac{6}{5}\sqrt{5x + 4} + \frac{38}{5}$ **61.** $p = \frac{4x + 22}{x + 5}$ **63.** $\frac{1}{6}$ coulombs

65. $\$325$ **67.** 3.1 μm^3

EXERCISES 5.3 (Page 310)

1. $y = 2x^2 - 5x + C$ **3.** $y = x^3 + x^2 - 7x + C$ **5.** $y = \frac{-2}{3x^2 + C}$ **7.** $2\sqrt{1 + u^2} = 3v^2 + C$ **9.** $\tan x - \tan y + y = C$

11. $y = \frac{5x^4}{12} + \frac{x^2}{2} + C_1x + C_2$ **13.** $s = -\frac{1}{9}(\sin 3t + \cos 3t) + C_1t + C_2$ **15.** $y = \frac{1}{3}x^3 - x^2 - 4x + 6$

17. $4\sin 3x + 6\cos 2y + 7 = 0$ **19.** $u = 3v^4 + 4v^3 + 2v^2 + 2v$ **21.** $s = \frac{1}{3}(2t + 4)^{3/2} - \frac{8}{3}$

23. $v = 2 + 5t - t^2$, $s = 2t + \frac{5}{2}t^2 - \frac{1}{3}t^3$ **25.** $v = \frac{1}{3}t^3 + t^2 - 4$, $s = \frac{1}{12}t^4 + \frac{1}{3}t^3 - 4t + 1$

27. $v = -2\sqrt{2}\sin(2t - \frac{1}{4}\pi)$, $s = \sqrt{2}\cos(2t - \frac{1}{4}\pi)$ **29.** $1600s = v^2 + 1200$ **31.** $5s^2 + 4s = v^2 + 12$ **33.** (a) $\frac{5}{4}$ sec;

(b) 20 ft/sec; (c) $\frac{5}{8}$ sec; (d) $\frac{25}{4}$ ft   **35.** (a) $\frac{1}{4}\sqrt{555}$ sec; (b) $8\sqrt{555}$ ft/sec   **37.** (a) 3.4 sec; (b) 99 ft/sec

39. (a) $v^2 = -64s + 1600$; (b) 24 ft/sec   **41.** $\frac{3}{2\pi}$ cm to the right of the origin   **43.** 1.62 m/sec^2

45. (a) 3.47 sec; (b) 48.22 m   **47.** 20 m/sec or 72 km/hr   **49.** $x^2 + 2y^2 = C$

EXERCISES 5.4 (Page 322)

1. 51   **3.** 147   **5.** 2025   **7.** $\frac{73}{12}$ **9.** $\frac{63}{4}$ **11.** $\frac{7}{12}$ **13.** 10,400   **15.** $2^n - 1$ **17.** $\frac{100}{101}$ **19.** $n^4 - \frac{2}{3}n^3 - 3n^2 - \frac{4}{3}n$

21. $\frac{8}{3}$ sq units   **23.** 15 sq units   **25.** $\frac{5}{3}$ sq units   **27.** 9 sq units   **29.** $\frac{3}{5}$ sq units   **31.** $\frac{17}{4}$ sq units   **33.** $\frac{27}{4}$ sq units

35. $\frac{1}{2}m(b^2 - a^2)$ sq units   **37.** $\frac{1}{2}h(b_1 + b_2)$ sq units   **39.** 9 sq units   **41.** 15 sq units   **43.** 9 sq units
45. 1.0349 sq units   **47.** 1.1682 sq units   **49.** 1.8530 sq units   **51.** 1.5912 sq units

EXERCISES 5.5 (Page 330)

1. $\frac{247}{32}$ **3.** $\frac{1469}{1320}$ **5.** 0.835   **7.** $\frac{1}{24}(10 + \sqrt{2} + 3\sqrt{3})\pi$ **9.** 9   **11.** $\frac{8}{3}$ **13.** $\frac{15}{4}$ **15.** 66   **17.** 4   **19.** 20 sq units
21. 44 sq units   **23.** $\frac{5}{3}$ sq units   **25.** $\frac{305}{6}$ sq units   **27.** $\frac{31}{4}$ sq units   **29.** 0.2672   **31.** 2.6725

33. $\int_0^2 x^2\, dx$ **35.** $\int_0^1 \frac{1}{x^2}\, dx$

EXERCISES 5.6 (Page 340)

1. 12   **3.** $4\sqrt{5}$ **5.** -5 **7.** 15   **9.** 0   **11.** -21 **13.** $-\frac{3}{2}$ **15.** $4 + \pi$ **17.** $\frac{33}{2}\pi$ **19.** $[24, 56]$ **21.** $[0, 64]$
23. $[0, 27]$ **25.** $[0, 576]$ **27.** $[\frac{1}{12}\pi, \frac{1}{12}\sqrt{3}\pi]$ **29.** $[0, 6]$ **31.** $[-3, \frac{3}{2}]$ **33.** $[-3\sqrt{3}\pi, 0]$ **35.** $\geq$ **37.** $\leq$

EXERCISES 5.7 (Page 343)

1. $\frac{2}{3}\sqrt{3}$ **3.** $\frac{1}{2}\sqrt[3]{30}$ **5.** $-2 + \sqrt{21}$ **7.** 0   **17.** average value is $\frac{1}{2}$ at $x = \frac{1}{2}$ **19.** $\frac{2}{\pi}$; 0.69   **21.** $v = 32t$; 32   **23.** π

EXERCISES 5.8 (Page 351)

1. 12   **3.** 36   **5.** $\frac{3}{2}$ **7.** $\frac{3}{16}$ **9.** $\frac{134}{3}$ **11.** -8 **13.** 1   **15.** $\frac{2}{9}(27 - 2\sqrt{2})$ **17.** $2 - \sqrt[3]{2}$ **19.** $\frac{104}{5}$
21. $\frac{29}{2}$ **23.** $\frac{2}{3}\sqrt{2}$ **25.** $\frac{256}{15}$ **27.** $\frac{5}{6}$ **29.** $\frac{6215}{12}$ **31.** $\frac{11}{6}$ **33.** 0   **35.** $\frac{3}{2}$ **37.** $\sqrt{4 + x^6}$ **39.** $-\sqrt{\sin x}$
41. $\frac{2}{3 + x^2}$ **43.** $3x^2\sqrt[3]{x^6 + 1}$ **45.** 1   **47.** average value is 6 at $x = \sqrt{3}$ **49.** 27   **51.** $\frac{4}{\pi}$ **55.** $\frac{188}{3}$

EXERCISES 5.9 (Page 359)

1. $\frac{32}{3}$ sq units   **3.** $\frac{22}{3}$ sq units   **5.** $\frac{52}{3}$ sq units   **7.** $\frac{343}{6}$ sq units   **9.** 1 sq unit   **11.** 1 sq unit   **13.** $\frac{32}{3}$ sq units
15. $\frac{32}{3}$ sq units   **17.** $\frac{1}{6}$ sq units   **19.** $\frac{12}{5}$ sq units   **21.** $\frac{9}{2}$ sq units   **23.** $\frac{8}{3}\sqrt{2}$ sq units   **25.** $\frac{5}{12}$ sq units   **27.** $\frac{27}{10}$ sq units
29. $\frac{64}{3}$ sq units   **31.** $\frac{253}{12}$ sq units   **33.** $\frac{37}{12}$ sq units   **35.** $(\sqrt{2} - 1)$ sq units   **37.** $\frac{7}{3}$ sq units   **39.** 12 sq units
41. $\frac{128}{5}$ sq units   **43.** 64 sq units   **45.** $(\frac{1}{2}\pi - 1)$ sq units   **47.** $(1 - \frac{1}{4}\pi)$ sq units   **49.** $\frac{16}{3}p^2$ sq units   **51.** 32   **53.** $\frac{3}{2}K$

EXERCISES 5.10 (Page 368)

1. approx: 4.250; exact: 4   **3.** approx: 0; exact: 0   **5.** approx: 0.696   **7.** 0.880   **9.** 0.248   **11.** 3.689
13. $-0.5 \leq \epsilon_T \leq 0$ **15.** $-0.161 \leq \epsilon_T \leq 0.161$ **17.** $-0.007 \leq \epsilon_T \leq -0.001$ **19.** 4.000   **21.** 0.693   **23.** 0.6045
25. 0   **27.** $-0.0005 \leq \epsilon_S \leq 0$ **29.** 0.237   **31.** 1.569   **33.** 1.402   **35.** 3.090   **37.** (a) 15.95; (b) 16.03
39. 26.6 sq units   **41.** 5.9 mi   **43.** 4.109 sq units   **45.** 56   **47.** 222

REVIEW EXERCISES FOR CHAPTER 5 (Page 369)

1. $\frac{1}{2}x^4 - \frac{1}{3}x^3 + 3x + C$ **3.** $2y^2 + 4y^{3/2} + C$ **5.** $-\frac{1}{3}\cos 3t + C$ **7.** $-\frac{1}{3}\cos^3 x + C$ **9.** $-\frac{2}{3x^3} + \frac{5}{x} + C$

11. $\frac{5}{54}(2 + 3x^2)^9 + C$ **13.** $\frac{2}{3}x\sqrt{3x} + \frac{2}{5}\sqrt{5x} + C$ **15.** $-\frac{1}{24(x^4 + 2x^2)^6} + C$ **17.** $\frac{1}{6}(2s + 3)^{3/2} - \frac{3}{2}(2s + 3)^{1/2} + C$

19. $\frac{1}{3}\tan 3\theta - \theta + C$　　**21.** $5\sin x - 3\sec x + C$　　**23.** $\frac{1}{420}(4x+3)^{3/2}(30x^2 - 18x + 79) + C$　　**25.** 6　　**27.** $-\frac{27}{4}$　　**29.** 0

31. $\frac{5}{4}$　　**33.** $\frac{1}{2}$　　**35.** $\frac{652}{15}$　　**37.** $4 - \frac{1}{2}\pi$　　**39.** $y^2 = \dfrac{1}{2x^{-1}+C} + 1$　　**41.** $y = \frac{1}{15}(2x-1)^{5/2} + C_1 x + C_2$

43. (a) $\frac{1}{9}(x^3+1)^3 + C$; (b) $\frac{1}{9}x^9 + \frac{1}{3}x^6 + \frac{1}{3}x^3 + C$　　**45.** $y = 10x - 2x^2 - 9$　　**47.** (a) $R(x) = \frac{1}{4}x^3 - 5x^2 + 12x$;
(b) $p = \frac{1}{4}x^2 - 5x + 12$　　**49.** (a) $V = \frac{2}{3}(t+1)^{3/2} + \frac{1}{3}t^2 + \frac{74}{3}$; (b) 64 cm^3　　**51.** 1.46 cm^3　　**53.** \$5

55. $v = 3\sin 2t + 3$, $s = -\frac{3}{2}\cos 2t + 3t + \frac{1}{2}(5 - 3\pi)$　　**57.** $\dfrac{25\sqrt{817} - 625}{8}$ sec ≈ 11.2 sec　　**59.** (a) 1 sec; (b) -80 ft/sec

61. 4,100,656,560　　**67.** $[0, \pi]$　　**69.** $\frac{313}{2}$　　**71.** $-(3x^2-4)^{3/2}$　　**73.** $\dfrac{1}{x}$　　**75.** 0　　**77.** $\frac{42,304}{175}$　　**81.** 18 sq units

83. $\frac{224}{3}$ sq units　　**85.** $\frac{1}{3}(40\sqrt{5} - 20)$ sq units　　**87.** 36 sq units　　**89.** $\frac{1}{12}$ sq units　　**91.** $2\sqrt{2}$ sq units

93. $(\frac{1}{2}\pi - 1)$ sq units　　**95.** (b) 47°; (c) 60°; (d) 73°; (e) 75°; (f) 67.5°; (g) $60 + \dfrac{9}{\pi}(\sqrt{3} + 2)$ degrees $\approx 70.7°$　　**97.** 2.977

99. 2.958　　**101.** (a) 1.624; (b) 1.563　　**103.** $\sqrt{3} + \frac{1}{6}\pi$　　**105.** 12; $\sqrt{3}$　　**111.** (b) 27

EXERCISES 6.1 (Page 381)

1. $\frac{4}{3}\pi r^3$ cu units　　**3.** $\frac{127}{7}\pi$ cu units　　**5.** 64π cu units　　**7.** $\frac{704}{5}\pi$ cu units　　**9.** $\frac{384}{7}\pi$ cu units　　**11.** $\frac{3456}{35}\pi$ cu units
13. $\frac{256}{15}\pi$ cu units　　**15.** $\frac{128}{5}\pi$ cu units　　**17.** $\frac{4}{3}\pi r^3$ cu units　　**19.** $\frac{1}{3}\pi h(a^2 + ab + b^2)$ cu units　　**21.** π cu units
23. $\frac{1}{2}\pi^2$ cu units　　**25.** $(4\pi - \frac{1}{2}\pi^2)$ cu units　　**27.** $(\sqrt{3}\pi - \frac{1}{3}\pi^2)$ cu units　　**29.** $\frac{1250}{3}\pi$ cu units　　**31.** $\frac{64}{5}\pi$ cu units
33. $\frac{261}{32}\pi$ cu units　　**35.** $\frac{16}{3}\pi$ cu units　　**37.** 180π cm^3　　**39.** $(\frac{8}{3}\pi^2 - 2\sqrt{3}\pi)$ cu units　　**41.** 2　　**43.** $\frac{1372}{3}\sqrt{3}$ cm^3
45. $\frac{686}{3}$ cm^3　　**47.** $\frac{8}{3}r^3$ cu units　　**49.** $\frac{2}{3}r^3$ cm^3

EXERCISES 6.2 (Page 387)

13. $\frac{1}{2}\pi$ cu units　　**15.** $\frac{3}{10}\pi$ cu units　　**17.** $\frac{5}{6}\pi$ cu units　　**19.** $\frac{49}{30}\pi$ cu units　　**21.** 16π cu units　　**23.** $\frac{512}{15}\pi$ cu units
25. $\frac{32}{15}\pi p^3$ cu units　　**27.** $\frac{8}{5}\pi$ cu units　　**29.** $\frac{11}{10}\pi$ cu units　　**31.** $\frac{38}{15}\pi$ cu units　　**33.** $\frac{16}{3}\pi$ cu units　　**35.** $\frac{32}{15}\pi$ cu units
37. $\frac{1}{2}(2 + \sqrt{2})\pi$ cu units　　**39.** π cu units　　**41.** $\frac{224}{3}\pi$ cu units　　**43.** $\sqrt[3]{2744}$

EXERCISES 6.3 (Page 393)

1. $\sqrt{10}$　　**3.** $\sqrt{97}$　　**5.** $\frac{14}{3}$　　**7.** $\frac{33}{16}$　　**9.** $\frac{1}{27}(97^{3/2} - 125)$　　**11.** 12　　**13.** $\frac{22}{3}$　　**15.** $\frac{9}{8}$
17. $\dfrac{8a^3 - (a^2 + 3b^2)^{3/2}}{8(a^2 - b^2)}$ if $b \neq a$; $\frac{9}{8}a$ if $b = a$　　**19.** $2\sqrt{3} - \frac{4}{3}$　　**21.** 2　　**23.** 3.8203　　**25.** 1.0894

EXERCISES 6.4 (Page 399)

1. 250 lb　　**3.** 4000 dynes　　**5.** $\frac{3}{2}$ m/sec^2　　**7.** $\frac{8}{3}$ slugs　　**9.** 4　　**11.** 6　　**13.** 54 kg; $\frac{11}{3}$ m from one end
15. 171 slugs; 5.92 in. from one end　　**17.** 42 g; $\frac{44}{7}$ cm from one end
19. 31.5 kg; $\frac{18}{7}$ m from the end having the greater density　　**21.** 16 slugs; $\frac{16}{5}$ ft from one end

23. 1.2 m from the end having the greater density　　**25.** $\dfrac{20}{L^2}x^2$ slugs/ft　　**27.** 12 kg/m

---

EXERCISES 6.5 (Page 406)

1. $(2, \frac{1}{3})$　　**3.** $\frac{29}{7}$　　**5.** $(\frac{2}{3}, 1)$　　**7.** $(0, \frac{8}{5})$　　**9.** $(0, \frac{12}{5})$　　**11.** $(\frac{16}{15}, \frac{64}{21})$　　**13.** $(\frac{1}{2}, -\frac{3}{2})$　　**15.** $(2, 0)$　　**17.** $\frac{5}{3}p$

21. the point on the bisecting radial line whose distance from the center of the circle is $\dfrac{4}{3\pi}$ times the radius

23. $\frac{1}{3}\pi r^2 h$ cu units　　**25.** $\frac{1}{6}\sqrt{2}(4 + 3\pi)\pi r^3$ cu units

EXERCISES 6.6 (Page 412)

1. $\frac{158}{3}$ ft-lb   **3.** $\frac{1076}{15}$ joules   **5.** $\frac{1}{2}(3 + \sqrt{393})$ **7.** 180 in.-lb   **9.** 8 joules   **11.** 1350 ergs   **13.** 6562.5w ft-lb
15. $256\pi w$ ft-lb   **17.** 100,000 ft-lb   **19.** 5500 ft-lb   **21.** 9,196,875π joules   **23.** 1,017,938π joules

25. $\dfrac{144w}{55}$ sec   **27.** $2\sqrt{3}$ ft

EXERCISES 6.7 (Page 417)

1. $320\rho g$ lb   **3.** $64\rho g$ lb   **5.** $2.25\rho g$ lb   **7.** 941,760 nt   **9.** 4,087,500 nt   **11.** $\sqrt[3]{16.31}$ m ≈ 2.54 m   **15.** $14,000\rho g$ lb
17. $100,000\rho g$ ft-lb   **19.** 756 lb   **21.** $250\sqrt{409}\rho g$ lb   **23.** $11,250\sqrt{3}\rho g$ lb

REVIEW EXERCISES FOR CHAPTER 6 (Page 418)

1. $\frac{1}{9}\pi$ cu units   **3.** π cu units   **5.** π cu units   **7.** 250π cu units   **9.** 1024 cu units   **11.** 3π cu units   **13.** $\frac{25}{6}\pi$ cu units
15. $\frac{16}{3}\sqrt{3}$ **17.** 558π cm^3 **19.** $\frac{1}{5320}\left[\frac{1}{8}(10999)^{3/2} - (2251)^{3/2}\right] \approx 7.03$ **21.** $\frac{2}{13}$ **23.** $(\frac{3}{2}, 2)$
25. $\frac{104}{3}$ slugs; $\frac{298}{65}$ in. from the left end   **27.** $(\frac{9}{8}, \frac{18}{5})$ **29.** $(\frac{9}{20}, \frac{9}{20})$ **31.** $\frac{256}{3}\pi$ m^3 **33.** $\frac{128}{15}\pi$ cu units   **35.** $\frac{1539}{20}\pi$ cu units
37. 6000 ergs   **39.** 400 ft-lb   **41.** 44,145,000 joules   **43.** $\frac{2752}{3}w\pi$ ft-lb   **45.** $\frac{13,600}{3}w\pi$ ft-lb   **47.** 90 ft^3
49. $\frac{832}{3}\pi$ cu units   **51.** $\frac{1}{4}\pi^2$ cu units   **53.** $\frac{18}{5}\rho g$ lb   **55.** $\frac{5120}{3}$ lb

EXERCISES 7.1 (Page 430)

1. one-to-one   **3.** not one-to one   **5.** one-to-one   **7.** one-to-one   **9.** one-to-one   **11.** one-to-one   **13.** one-to-one
15. one-to-one   **17.** not one-to-one   **19.** $f^{-1}(x) = \frac{1}{5}(x + 7)$; domain: $(-\infty, +\infty)$, range: $(-\infty, +\infty)$ **21.** no inverse
23. $f^{-1}(x) = 4 - \sqrt[3]{x}$; domain: $(-\infty, +\infty)$, range: $(-\infty, +\infty)$ **25.** $h^{-1}(x) = \frac{1}{2}x^2 + 3$; domain: $[0, +\infty)$, range: $[3, +\infty)$
27. $F^{-1}(x) = x^3 - 1$; domain: $(-\infty, +\infty)$, range: $(-\infty, +\infty)$ **29.** no inverse

31. $f^{-1}(x) = \frac{1}{32}x^5$; domain: $(-\infty, +\infty)$, range: $(-\infty, +\infty)$ **33.** $f^{-1}(x) = \dfrac{x + 3}{1 - x}$; domain: $\{x \,|\, x \neq 1\}$, range: $\{y \,|\, y \neq -1\}$

35. $g^{-1}(x) = \sqrt{x - 5}$; domain: $[5, +\infty)$, range: $[0, +\infty)$ **37.** $f^{-1}(x) = \frac{1}{2}(\sqrt[3]{x} - 1)$; domain: $[0, 8]$, range: $[-\frac{1}{2}, \frac{1}{2}]$
39. $F^{-1}(x) = \sqrt{9 - x^2}$; domain: $[0, 3]$, range: $[0, 3]$ **41.** (b) $f^{-1}(x) = \frac{1}{4}(x + 3)$ **43.** (b) $f^{-1}(x) = \sqrt[3]{x} - 2$

45. (b) $f^{-1}(x) = \dfrac{4x - 1}{3 - 2x}$ **47.** $f^{-1}(x) = \frac{5}{9}(x - 32)$ **51.** (b) $f_1(x) = x^2 + 4, x \geq 0$; $f_2(x) = x^2 + 4, x \leq 0$;

(c) $f_1^{-1}(x) = \sqrt{x - 4}$; domain: $[4, +\infty)$; $f_2^{-1}(x) = -\sqrt{x - 4}$; domain: $[4, +\infty)$
53. (b) $f_1(x) = \sqrt{9 - x^2}, 0 \leq x \leq 3$; $f_2(x) = \sqrt{9 - x^2}, -3 \leq x \leq 0$;

(c) $f_1^{-1}(x) = \sqrt{9 - x^2}$, domain: $[0, 3]$; $f_2^{-1}(x) = -\sqrt{9 - x^2}$, domain: $[0, 3]$ **55.** $f^{-1}(x) = \begin{cases} x & \text{if } x < 1 \\ \sqrt{x} & \text{if } 1 \leq x \leq 81 \\ \frac{1}{729}x^2 & \text{if } x > 81 \end{cases}$

EXERCISES 7.2 (Page 438)

11. $\frac{2}{3}$ **13.** $\frac{1}{10}$ **15.** $\frac{1}{12}$ **17.** $\frac{1}{21}$ **19.** -2 **21.** $-\frac{1}{4}$ **23.** $-\dfrac{1}{\sqrt{3}}$ **25.** $-\frac{1}{4}$ **27.** $\frac{1}{3}$

29. (a) $f_1(x) = \sqrt{9 - x^2}$, $f_2(x) = -\sqrt{9 - x^2}$; (b) neither has an inverse; (c) $\dfrac{dy}{dx} = -\dfrac{x}{y}, \dfrac{dx}{dy} = -\dfrac{y}{x}$

31. (a) $f(x) = \dfrac{4}{x}$; (b) $f^{-1}(x) = \dfrac{4}{x}$, domain: $\{x \mid x \neq 0\}$; (c) $\dfrac{dy}{dx} = -\dfrac{y}{x}, \dfrac{dx}{dy} = -\dfrac{x}{y}$

33. (a) $f(x) = \dfrac{2x^2 + 1}{3x}$; (b) no inverse; (c) $\dfrac{dy}{dx} = \dfrac{4x - 3y}{3x}, \dfrac{dx}{dy} = \dfrac{3x}{4x - 3y}$ **35.** (b) 6; (c) $\frac{1}{6}$ **37.** $\dfrac{1}{\sqrt{15}}$ **39.** $\dfrac{1}{\sqrt{2}}$

EXERCISES 7.3 (Page 448)

1. $\dfrac{5}{4 + 5x}$ **3.** $\dfrac{5}{8 + 10x}$ **5.** $\dfrac{6}{3t + 1}$ **7.** $\dfrac{6 \ln(3t + 1)}{3t + 1}$ **9.** $-\dfrac{2x}{12 - 3x^2}$ **11.** $\dfrac{5 \cos 5y}{\sin 5y}$ **13.** $-\dfrac{\sin(\ln x)}{x}$ **15.** $2 \sec 2x$

17. $\dfrac{\sec^2 x}{2 \tan x} = \csc 2x$ **19.** $-\dfrac{17}{2(2w - 5)(3w + 1)}$ **21.** $\dfrac{\ln x - 1}{(\ln x)^2}$ **23.** $\dfrac{1 - 2x - x^2}{3(x + 1)(x^2 + 1)}$ **25.** $\dfrac{1}{2(1 + \sqrt{x + 1})}$

27. $-\dfrac{xy + y}{xy + x}$ **29.** $x + y$ **31.** $\dfrac{4x^2y - xy - 2y}{6xy^2 + x}$

(Sketches of the graphs for Exercises 35 through 41 appear in Figs. EX 7.3-35 through EX 7.3-41.)

43. $y = -\frac{1}{2}x + \frac{1}{4} - \ln 2$ **45.** $y = 40x - 40$ **47.** $-\frac{1}{2}$ **49.** (a) $5 per $1 change in budget; (b) $688

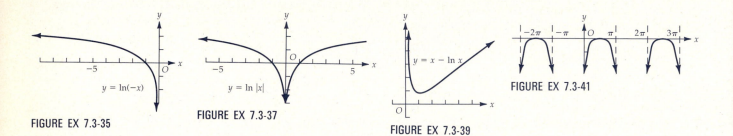

FIGURE EX 7.3-35

$y = \ln(-x)$

FIGURE EX 7.3-37

$y = \ln|x|$

FIGURE EX 7.3-39

$y = x - \ln x$

FIGURE EX 7.3-41

EXERCISES 7.4 (Page 454)

1. $\dfrac{3x^2}{x^3 + 1}$ **3.** $-\dfrac{3 \sin 3x}{\cos 3x}$ **5.** $4 \sec 4x$ **7.** $\dfrac{4 - x^2}{x(x^2 + 4)}$ **9.** $2x(x + 1)^6(x - 1)^2(6x^2 - 2x - 1)$

11. $\dfrac{x(x - 1)(x + 2)^2}{(x - 4)^6}(2x^3 - 30x^2 - 6x + 16)$ **13.** $\dfrac{8x^9 - 4x^7 + 15x^2 + 10}{5(x^7 + 1)^{6/5}}$ **15.** $-\frac{1}{2}\ln|3 - 2x| + C$ **17.** $\frac{3}{2}\ln(x^2 + 4) + C$

19. $\frac{1}{5}\ln|5x^3 - 1| + C$ **21.** $\frac{1}{2}\ln|1 + 2 \sin t| + C$ **23.** $\frac{1}{5}\ln(1 - \cos 5x) + C$ **25.** $\ln(1 + \sin 2x) + \frac{1}{2}\ln|\cos 2x| + C$

27. $x^2 + 4 \ln|x^2 - 4| + C$ **29.** $\ln|\ln x| + C$ **31.** $\frac{1}{3}\ln^3 3x + C$ **33.** $\ln|(\ln x)^2 + \ln x| + C$ **35.** $\ln|\sec(\ln x)| + C$

37. $\ln 5$ **39.** $4 + \ln 2$ **41.** $\frac{1}{2}\ln(4 + 2\sqrt{3})$ **43.** $\dfrac{1}{\ln 4}$ **47.** $\frac{1}{4}\ln 5 \approx 0.402$ **49.** $2000 \ln 2 \text{ lb/ft}^2 \approx 1386 \text{ lb/ft}^2$

51. $\ln 4$ sq units ≈ 1.386 sq units   **53.** $\pi(11 + 8 \ln 2)$ cu units ≈ 51.978 cu units   **55.** $\ln(2 + \sqrt{3})$

EXERCISES 7.5 (Page 462)

1. $5e^{5x}$ **3.** $-6xe^{-3x^2}$ **5.** $-e^{\cos x} \sin x$ **7.** $e^{2x} \cos e^x + e^x \sin e^x$ **9.** $\dfrac{e^{\sqrt{x}} \sec^2 e^{\sqrt{x}}}{2\sqrt{x}}$ **11.** $\dfrac{4}{(e^x + e^{-x})^2}$ **13.** $2x$

15. $2e^{2x} \sec e^{2x} \tan e^{2x} + 2e^{2 \sec x} \sec x \tan x$ **17.** $-e^{y-x}$ **19.** $-\dfrac{y^2 + 2ye^{2x}}{2e^{2x} + 3xy}$ **21.** $-\frac{1}{5}e^{2-5x} + C$ **23.** $e^x - e^{-x} + C$

25. $\dfrac{1}{6(1 - 2e^{3x})} + C$ **27.** $e^x - 3\ln(e^x + 3) + C$ **29.** e^2 **31.** 2   **33.** $\frac{1}{2}$ **35.** $\frac{1}{2}(e^4 - 1)$ **37.** 2.67   **39.** 2.57

41. See Fig EX 7.5-41   **43.** $(e^2 - 1)$ sq units   **45.** $y = -\frac{1}{2}x + \frac{1}{2} + \frac{1}{2}\ln 2$ **47.** $(e^3 + \frac{1}{2})$ ft ≈ 20.586 ft   **49.** -9.17 lb/ft^2/sec

51. 0.006   **53.** $\frac{1}{2}\pi w(e^{-2} - e^{-8})$ ft-lb   **55.** (b) 2.7181459; 2.7184177; 2.7182818   **61.** $f(1) = \dfrac{1}{e}$, rel max; $(2, 2e^{-2})$, pt. of infl.;

increasing on $(-\infty, 1]$; decreasing on $[1, +\infty)$; concave downward for $x < 2$; concave upward for $x > 2$. See Fig. EX 7.5-61

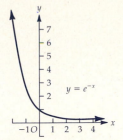

FIGURE EX 7.5-41

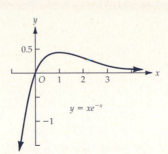

FIGURE EX 7.5-61

EXERCISES 7.6 (Page 468)

1. $(5\ln 3)3^{5x}$ **3.** $4^{3t^2}(\ln 4)\,6t$ **5.** $4^{\sin 2x}(2\ln 4)\cos 2x$ **7.** $2^{5x}3^{4x^2}(5\ln 2 + 8x\ln 3)$ **9.** $\dfrac{1}{x^2}\log_{10}\dfrac{e}{x}$ **11.** $\dfrac{\log_a e}{2x\sqrt{\log_a x}}$

13. $\dfrac{(\log_{10} e)^2}{(x + 1)\log_{10}(x + 1)}$ **15.** $3^{t^2}\sec 3^{t^2}\tan 3^{t^2}(2t\ln 3)$ **17.** $x^{\sqrt{x} - (1/2)}(1 + \frac{1}{2}\ln x)$ **19.** $z^{\cos z - 1}[\cos z - z(\ln z)\sin z]$

21. $(\sin x)^{\tan x}[1 + (\ln\sin x)\sec^2 x]$ **23.** $x^{e^x - 1}e^x(x\ln x + 1)$ **25.** $\dfrac{3^{2x}}{2\ln 3} + C$ **27.** $\dfrac{a^t e^t}{1 + \ln a} + C$ **29.** $\dfrac{10^{x^3}}{3\ln 10} + C$

31. $\dfrac{6^{ey}}{\ln 6} + C$ **33.** 0.621   **35.** 2.999   **45.** (a) 61 sales per day; (b) 2.26 sales per day   **49.** (a) $y = 200 \cdot 2^{t/10}$; (b) \$12,800;

(c) \$877 per year   **51.** $\left(e - 1 - \dfrac{1}{\ln 2}\right)$ sq units ≈ 0.276 sq units   **53.** $\left(\dfrac{4}{\ln 3} - 1\right)\pi$ cu units ≈ 8.297 cu units

---

EXERCISES 7.7 (Page 480)

1. 68.4 years   **3.** 38,720   **5.** \$2734   **7.** 16,000   **9.** (a) 96 percent; (b) 66 percent   **11.** 1389
13. (a) \$52.59; (b) 10.52 percent   **15.** 15.9 years   **17.** 8.7 years   **19.** 43.9 g   **21.** 11.6 kg   **23.** 6451 years ago
25. (a) 34.7; (b) 55.5   **27.** (a) 0.3401; (b) 0.3414   **29.** 0.8427

EXERCISES 7.8 (Page 491)

1. $y = \dfrac{C}{x}$ **3.** $y = C\cos x$ **5.** $y = (x + C)e^x$ **7.** $y = [\ln(1 + e^x) + C]e^{-x}$ **9.** $x = y + Cy^{-2}$

11. $y = \frac{1}{3}\sin x + C\csc^2 x$ **13.** $y = \dfrac{x + C}{1 + x^2}$ **15.** $x = \ln y(\ln|\ln y| + C)$ **17.** $y = x^3 + \frac{1}{2}x^2 - \frac{1}{2}$ **19.** $y = x(e^{x-1} + 1)$

21. $y = 5\sin^2 x - \sin x\cos x$ **23.** $y = 8\sin^2 x - 2$ **25.** 70 min   **27.** (a) 1 min 42 sec; (b) 42.1°
29. (a) $i = 3 - 3e^{-5t}$; (b) 3 amperes   **31.** $i = \frac{1}{45}(e^t - e^{-8t})$ **33.** (a) $q = 8 - 4(1 + 0.01t)^{-1000}$; (b) $i = 40(1 + 0.01t)^{-1001}$;
(c) 8 coulombs   **35.** $y = (\frac{1}{3}x^{-3} + Cx^6)^{-1/3}$ **37.** $y = \pm(\frac{1}{3}e^{-x^2} + Ce^{2x^2})^{-1/2}$

REVIEW EXERCISES FOR CHAPTER 7 (Page 492)

1. (a) $f^{-1}(x) = \sqrt[3]{x+4}$; domain: $(-\infty, +\infty)$; range: $(-\infty, +\infty)$ **3.** no inverse

5. $f^{-1}(x) = \dfrac{4}{3-x}$; domain: $\{x \mid x \neq 3\}$; range: $\{y \mid y \neq 0\}$ **7.** (b) $f^{-1}(x) = x^3 - 1$ **9.** $\frac{1}{4}$ **11.** $\frac{1}{12}$ **13.** $\dfrac{8 \ln x}{x}$

15. $(-4 \ln 2)2^{\cos 4x} \sin 4x$ **17.** $\dfrac{(4 - x^2)e^{x/(4+x^2)}}{(4 + x^2)^2}$ **19.** $\dfrac{\log_{10} e}{(1 - x^2)\sqrt{\log_{10} \dfrac{1+x}{1-x}}}$ **21.** 0

23. $x^{xe^x + e^x - 1}(xe^x \ln^2 x + e^x \ln x + 1)$ **25.** $\frac{3}{2} \ln(1 + e^{2x}) + C$ **27.** $\frac{1}{3}\left(e^{3x} + \dfrac{2^{3x}}{\ln 2} + C\right)$ **29.** $\frac{1}{6}\sqrt{1 + e^{6x^2}} + C$

31. $\dfrac{2}{3 \ln 2}\sqrt{3 \cdot 2^x + 4} + C$ **33.** $\frac{1}{3}(e^8 - 1)$ **35.** $\frac{3}{2} \ln 2$ **37.** $1 + 5 \ln \frac{3}{4}$ **39.** $-\dfrac{ye^x + e^y + 1}{e^x + xe^y + 1}$ **41.** 5.004

43. $v = e^t - e^{-t} + 1$; $s = e^t + e^{-t} + t$ **45.** $\frac{1}{2}\pi(1 - e^{-2b})$; $\frac{1}{2}\pi$ **47.** 8212 years   **49.** $g(x) = -e^x$; domain: $(-\infty, +\infty)$
53. $3000 \ln \frac{3}{2}$ in.-lb   **55.** 8.66 years   **57.** 187,500   **59.** (a) 32; (b) 43.5; (c) 3.83   **61.** 8.63 min   **63.** $(1, e)$
71. $\operatorname{sgn} t(1 - e^{-|t|})$ **73.** (a) $\ln \frac{4}{5} > \ln \frac{1}{2}$; (b) $\ln 2 > -\ln 2$; (c) $\ln 2 > \ln \frac{5}{4}$ **75.** $y = x + x^{-1} + Cx^{-2}$
77. $y = (x + 2)(\sec x - \tan x)$ **79.** (a) $i = 10 - 10(1 + \frac{1}{3}t)^{-1000}$; (b) 10 amperes   **81.** 73.7°

EXERCISES 8.1 (Page 502)

1. (a) $\frac{1}{6}\pi$; (b) $-\frac{1}{6}\pi$; (c) $\frac{1}{3}\pi$; (d) $\frac{2}{3}\pi$ **3.** (a) $\frac{1}{6}\pi$; (b) $-\frac{1}{3}\pi$; (c) $\frac{1}{6}\pi$; (d) $\frac{7}{6}\pi$ **5.** (a) $\frac{1}{2}\pi$; (b) $-\frac{1}{2}\pi$; (c) $\frac{1}{2}\pi$; (d) $-\frac{1}{2}\pi$; (e) 0

7. (a) $\frac{2}{3}\sqrt{2}$; (b) $\frac{1}{4}\sqrt{2}$; (c) $2\sqrt{2}$; (d) $\frac{3}{4}\sqrt{2}$; (e) 3   **9.** (a) $\frac{2}{3}\sqrt{2}$; (b) $-\frac{1}{4}\sqrt{2}$; (c) $-2\sqrt{2}$; (d) $\frac{3}{4}\sqrt{2}$; (e) -3

11. (a) $-\frac{2}{5}\sqrt{5}$; (b) $\frac{1}{5}\sqrt{5}$; (c) $-\frac{1}{2}$; (d) $\sqrt{5}$; (e) $-\frac{1}{2}\sqrt{5}$ **13.** (a) $-\frac{2}{3}$; (b) $-\frac{1}{3}\sqrt{5}$; (c) $\frac{2}{5}\sqrt{5}$; (d) $\frac{1}{2}\sqrt{5}$; (e) $-\frac{3}{5}\sqrt{5}$

15. (a) $\frac{1}{6}\pi$; (b) $-\frac{1}{6}\pi$; (c) $\frac{1}{6}\pi$; (d) $-\frac{1}{6}\pi$ **17.** (a) $\frac{1}{3}\pi$; (b) $\frac{1}{3}\pi$; (c) $\frac{2}{3}\pi$; (d) $\frac{2}{3}\pi$ **19.** (a) $\frac{1}{6}\pi$; (b) $-\frac{1}{3}\pi$; (c) $\frac{1}{6}\pi$; (d) $-\frac{1}{3}\pi$

21. (a) $\frac{1}{6}\pi$; (b) $\frac{2}{3}\pi$; (c) $\frac{1}{6}\pi$; (d) $\frac{2}{3}\pi$ **23.** (a) $\frac{1}{3}\pi$; (b) $\frac{1}{3}\pi$; (c) $\frac{4}{3}\pi$; (d) $\frac{4}{3}\pi$ **25.** (a) $\frac{1}{6}\pi$; (b) $-\frac{5}{6}\pi$; (c) $\frac{1}{6}\pi$; (d) $-\frac{5}{6}\pi$

27. (a) $\sqrt{3}$; (b) $\frac{1}{7}\sqrt{21}$ **29.** (a) $\frac{1}{2}\sqrt{3}$; (b) $\frac{1}{2}\sqrt{3}$ **31.** $\frac{119}{169}$ **33.** $\frac{2}{9}(1 + \sqrt{10})$ **35.** $\frac{1}{27}(7\sqrt{5} + 8\sqrt{2})$ **37.** $\frac{1}{39}(48 - 25\sqrt{3})$

39. $\frac{1}{15}(4\sqrt{10} + \sqrt{5})$ **43.** See Fig EX 8.1-43   **45.** See Fig EX 8.1-45   **47.** See Fig EX 8.1-47   **49.** See Fig EX 8.1-49

51. (a) $\left\{t \,\middle|\, t = \dfrac{1}{4\pi}\sin^{-1}\dfrac{y}{2} - \dfrac{1}{8} + \dfrac{k}{2}\right\} \cup \left\{t \,\middle|\, t = \dfrac{1}{8} - \dfrac{1}{4\pi}\sin^{-1}\dfrac{y}{2} + \dfrac{k}{2}\right\}$, where k is any integer; (b) $\frac{1}{12}, \frac{5}{12}, \frac{7}{12}$

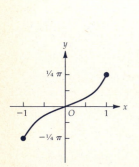

FIGURE EX 8.1-43

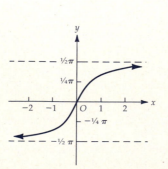

FIGURE EX 8.1-45

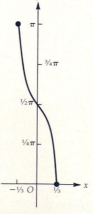

FIGURE EX 8.1-47

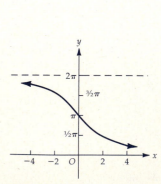

FIGURE EX 8.1-49

EXERCISES 8.2 (Page 509)

1. $\dfrac{1}{\sqrt{4-x^2}}$ **3.** $\dfrac{2}{1+4x^2}$ **5.** $-\dfrac{1}{\sqrt{x-x^2}}$ **7.** 0 **9.** $-\dfrac{x}{|x|\sqrt{1-x^2}}$ **11.** $\dfrac{4}{4+x^2}$ **13.** $\sin^{-1}2y+\dfrac{2y}{\sqrt{1-4y^2}}$

15. $2x\sec^{-1}\dfrac{1}{x}-\dfrac{x|x|}{\sqrt{1-x^2}}$ **17.** $-\dfrac{\cos x}{|\cos x|}$ **19.** $\dfrac{2x}{(1+x^4)\tan^{-1}x^2}$ **21.** $2\sqrt{4-x^2}$ **23.** $-\dfrac{3}{\sqrt{4e^{6x}-1}}$ **25.** $\cot^{-1}x$

27. $-1-\dfrac{1}{\cos^{-1}x\sqrt{1-x^2}}$ **29.** $\dfrac{(1+y^2)(3x^2+\sin y)}{1-x\cos y(1+y^2)}$ **31.** $2\sqrt{3}x-6y+2\pi-\sqrt{3}=0;\ 6\sqrt{3}x+6y-2\pi-3\sqrt{3}=0$

33. $\sqrt{10}$ ft **35.** 0.078 rad/sec **37.** $\frac{52}{3}\pi$ km/min **39.** 8 ft/sec **41.** $\dfrac{6}{x\sqrt{x^2-64}}$

EXERCISES 8.3 (Page 513)

1. $\frac{1}{2}\sin^{-1}2x+C$ **3.** $\dfrac{1}{12}\tan^{-1}\dfrac{3x}{4}+C$ **5.** $\dfrac{1}{2}\tan^{-1}\dfrac{x-1}{2}+C$ **7.** $\dfrac{1}{16}\sec^{-1}\dfrac{x}{4}+C$ **9.** $\dfrac{\sqrt{5}}{5}\sin^{-1}\dfrac{\sqrt{10}}{2}x+C$

11. $\dfrac{1}{6}\sin^{-1}\dfrac{3r^2}{4}+C$ **13.** $\dfrac{1}{\sqrt{7}}\tan^{-1}\dfrac{e^x}{\sqrt{7}}+C$ **15.** $2\tan^{-1}\sqrt{x}+C$ **17.** $\dfrac{2}{\sqrt{7}}\tan^{-1}\dfrac{2x-1}{\sqrt{7}}+C$ **19.** $\cos^{-1}\dfrac{1-x}{4}+C$

21. $\cos^{-1}\dfrac{1+x}{2}-\sqrt{3-2x-x^2}+C$ **23.** $\sin^{-1}\dfrac{1+x}{\sqrt{5}}-\sqrt{4-2x-x^2}+C$

25. $\frac{1}{2}x^2+2x+\frac{5}{4}\ln(2x^2-4x+3)-\frac{1}{2}\sqrt{2}\tan^{-1}\sqrt{2}(x-1)+C$ **27.** $\frac{1}{4}\pi+\frac{1}{2}\ln 2$ **29.** $\frac{1}{3}\pi$ **31.** $\tan^{-1}e-\frac{1}{4}\pi$

33. $\frac{1}{4}\pi$ **35.** π sq units **37.** $\frac{1}{3}\pi$ sq units

EXERCISES 8.4 (Page 522)

17. $\dfrac{4}{5}\operatorname{sech}^2\dfrac{4x+1}{5}$ **19.** $2e^{2y}\cosh e^{2y}$ **21.** $-8\operatorname{sech}^2 4w\tanh 4w$ **23.** e^{2x} **25.** $2\operatorname{csch}2t$ **27.** $2x\operatorname{sech}x^2$

29. $x^{\sinh x-1}(x\cosh x\ln x+\sinh x)$ **35.** $\frac{1}{2}\cosh^2(e^t)+C$ **37.** $2\cosh\sqrt{x}+C$ **39.** $x-\frac{1}{3}\coth 3x+C$

41. $\frac{1}{6}\tanh^6 x+C$ **43.** $\frac{1}{2}\ln^2\cosh x+C$ **45.** $\frac{4}{5}$ **47.** $\frac{1}{4}\sinh^4 2$

49. $\operatorname{sech}(0)=1$, rel max; pts. of infl. at $x=\pm\ln(1+\sqrt{2})$; increasing on $(-\infty,0]$; decreasing on $[0,+\infty)$; concave upward for $x<-\ln(1+\sqrt{2})$ and $x>\ln(1+\sqrt{2})$; concave downward for $-\ln(1+\sqrt{2})<x<\ln(1+\sqrt{2})$

51. $a^2\sinh\dfrac{x_1}{a}$ sq units **53.** $v=e^{-ct/2}[(B-\frac{1}{2}cA)\sinh t+(A-\frac{1}{2}cB)\cosh t]$;

$a=e^{-ct/2}[(A-cB+\frac{1}{4}c^2A)\sinh t+(B-cA+\frac{1}{4}c^2B)\cosh t]$; $a=K_1s+K_2v$, where $K_1=1-\frac{1}{4}c^2$ and $K_2=-c$

EXERCISES 8.5 (Page 527)

9. (a) $\ln\frac{1}{4}(1+\sqrt{17})$; (b) $\frac{1}{2}\ln 3$ **11.** $\dfrac{2x}{\sqrt{x^4+1}}$ **13.** $\dfrac{4}{1-16x^2}$ **15.** $-\dfrac{1}{2x+3x^2}$ **17.** $2x\left(\cosh^{-1}x^2+\dfrac{x^2}{\sqrt{x^4-1}}\right)$

19. $|\sec x|$ **21.** $-\dfrac{\csc x\cot x}{|\cot x|}$ **23.** $\dfrac{6z(\coth^{-1}z^2)^2}{1-z^4}$ **25.** $-e^x\csc e^x$ **27.** $\sinh^{-1}x$

REVIEW EXERCISES FOR CHAPTER 8 (Page 527)

1. (a) $\frac{4}{5}$; (b) $\frac{3}{5}$; (c) $\frac{3}{4}$; (d) $-\frac{3}{4}$ **3.** (a) $-\frac{120}{169}$; (b) $\frac{3}{4}$ **5.** $\dfrac{2^x\ln 2}{1+2^{2x}}$ **7.** $6\sinh^2 2x\cosh 2x$ **9.** $\dfrac{x}{|x|(x^2+1)}$ **11.** $\dfrac{\operatorname{sech}\sqrt{x}}{2\sqrt{x}}$

13. $-4e^{2x} \operatorname{csch}^2 e^{2x} \coth e^{2x}$ **15.** $(\cosh x)^{1/x} \left(\dfrac{x \tanh x - \ln \cosh x}{x^2} \right)$ **19.** $\dfrac{6xy - 9y^4 - x^4y^2}{3x^2 + 18xy^3 + 2x^5y}$ **21.** $\frac{3}{2} \sin^{-1} x^2 + C$

23. $x - \frac{1}{3} \tanh 3x + C$ **25.** $\dfrac{2}{\sqrt{31}} \tan^{-1} \dfrac{4t + 3}{\sqrt{31}} + C$ **27.** $\frac{1}{4}\sqrt{2} \sec^{-1} \frac{1}{4}\sqrt{2}e^x + C$ **29.** $\frac{2}{3}\pi + \sqrt{3} - 2$ **31.** $2\sqrt{2} \sinh \frac{1}{2}$

33. (a) $\left\{ t \mid t = \dfrac{k}{60} + \dfrac{1}{120\pi} \cos^{-1} \dfrac{E}{20} \right\} \cup \left\{ t \mid t = \dfrac{k}{60} - \dfrac{1}{120\pi} \cos^{-1} \dfrac{E}{20} \right\}$, where k is any integer; (b) $\frac{1}{360}$; (c) 0.0035; (d) $\frac{1}{180}$; (e) 0.0048

37. $9 \sin^{-1} \frac{2}{3}\sqrt{2}$ sq units **39.** $\frac{1}{4}\pi$ cu units **41.** (a) 120 rad/hr; (b) 60 rad/hr **43.** decreasing at the rate of 60 rad/hr

45. $\frac{1}{10}\pi$ hr; he walks all the way **47.** $\ln(2 + \sqrt{3})$ **49.** $2w \sinh^{-1} 2w + \dfrac{2w^2}{\sqrt{4w^2 + 1}}$ **51.** $\dfrac{2e^{2x}}{\sqrt{e^{4x} - 1}}$

EXERCISES 9.1 (Page 536)

1. $\frac{1}{3}xe^{3x} - \frac{1}{9}e^{3x} + C$ **3.** $x \sec x - \ln|\sec x + \tan x| + C$ **5.** $x \ln x - x + C$ **7.** $x \ln^2 x - 2x \ln x + 2x + C$

9. $\frac{1}{2} \tan^{-1} x(x^2 + 1) - \frac{1}{2}x + C$ **11.** $\dfrac{e^x}{x + 1} + C$ **13.** $-\cos x \ln(\cos x) + \cos x + C$ **15.** $\frac{1}{2}e^x(\cos x + \sin x) + C$

17. $-x^2\sqrt{1 - x^2} - \frac{2}{3}(1 - x^2)^{3/2} + C$ **19.** $x^2 \cosh x - 2x \sinh x + 2 \cosh x + C$ **21.** $2\sqrt{z} \cot^{-1} \sqrt{z} + \ln(1 + z) + C$

23. $2\sqrt{x} \sin \sqrt{x} + 2 \cos \sqrt{x} + C$ **25.** $\dfrac{36}{\ln 3} - \dfrac{36}{(\ln 3)^2} + \dfrac{16}{(\ln 3)^3}$ **27.** $\frac{9}{16}$ **29.** $\frac{1}{4}(3e^4 + 1)$ **31.** $\frac{4}{25}(e^{3\pi/4} + 1)$

33. $\frac{5}{6}\pi - \sqrt{3} + 1$ **35.** $(e^2 + 1)$ sq units **37.** $\frac{1}{2}\pi(3e^4 + 1)$ cu units **39.** $(8 - 24e^{-2})$ sq units

41. $2(1 - e^{-6})$ kg; $\dfrac{e^6 - 7}{e^6 - 1}$ m from one end **43.** $(0.267, 0.604)$ **45.** $\frac{1}{4}(1 - 9e^{-8}) w\pi$ ft-lb **47.** $C(x) = x \ln x - x + 6$

49. (b) $\frac{1}{4}x^4 \ln x - \frac{1}{16}x^4 + C$ **51.** (b) $\frac{1}{5} \sec^4 x \tan x + \frac{4}{15} \sec^2 x \tan x + \frac{8}{15} \tan x + C$ **55.** (a) 0.4156; (b) -0.4133

57. (a) $i = \dfrac{3}{625 + 9\pi^2} (250 \sin 120\pi t - 30\pi \cos 120\pi t) + \dfrac{90\pi}{625 + 9\pi^2} e^{-1000t}$ (b) $i \approx \dfrac{3}{625 + 9\pi^2} (250 \sin 120\pi t - 30\pi \cos 120\pi t)$

EXERCISES 9.2 (Page 541)

1. $\frac{1}{5} \sin^5 x + C$ **3.** $-\frac{1}{16} \cos^4 4x + C$ **5.** $\frac{1}{3} \cos^3 x - \cos x + C$ **7.** $\frac{3}{8}z - \frac{1}{4} \sin 2z + \frac{1}{32} \sin 4z + C$

9. $\frac{1}{2}x + \frac{1}{2} \sin x + C$ **11.** $\frac{1}{3} \sin^3 x - \frac{1}{5} \sin^5 x + C$ **13.** $-\frac{1}{3} \cos^3 x + \frac{2}{5} \cos^5 x - \frac{1}{7} \cos^7 x + C$ **15.** $\frac{1}{8}t - \frac{1}{96} \sin 12t + C$

17. $\frac{1}{2} \sin^{2/3} 3x - \frac{1}{8} \sin^{8/3} 3x + C$ **19.** $\frac{1}{14} \sin 7x + \frac{1}{2} \sin x + C$ **21.** $-\frac{1}{16} \cos 8y + \frac{1}{4} \cos 2y + C$

23. $t - \sin t - \frac{1}{8} \sin 4t + \frac{1}{5} \sin 5t - \frac{1}{12} \sin 6t + C$ **25.** $\frac{2}{3}$ **27.** $\frac{3}{8}$ **29.** $\frac{1}{8}$ **31.** $\frac{1}{8}(\sqrt{2} - 1)$

33. (a) $\sin^2 x + C$; (b) $-\cos^2 x + C$; (c) $-\frac{1}{2} \cos 2x + C$ **39.** $\frac{1}{2}\pi$ sq units **41.** $\frac{3}{8}\pi^2$ cu units **43.** $\frac{5}{8}\pi^2$ cu units

45. $(\frac{1}{2}\pi - 1, \frac{1}{8}\pi)$ **49.** 31,250,000 lb

---

EXERCISES 9.3 (Page 545)

1. $\frac{1}{5} \tan 5x - x + C$ **3.** $-\frac{1}{4} \cot 2x^2 - \frac{1}{2}x^2 + C$ **5.** $-\frac{1}{2} \cot^2 t - \ln|\sin t| + C$

7. $\frac{1}{15} \tan^5 3x - \frac{1}{9} \tan^3 3x + \frac{1}{3} \tan 3x - x + C$ **9.** $\frac{1}{3} \tan^3 x + \tan x + C$ **11.** $-\frac{1}{2} \csc x \cot x + \frac{1}{2} \ln|\csc x - \cot x| + C$

13. $\frac{1}{3} \tan^3 e^x - \tan e^x + e^x + C$ **15.** $\frac{1}{9} \tan^9 x + \frac{1}{7} \tan^7 x + C$ **17.** $-\frac{1}{15} \cot^5 3x - \frac{1}{9} \cot^3 3x + C$

19. $\frac{1}{2}(\tan 2x - \cot 2x) + C$ **21.** $2 \sec w - \tan w + C$ **23.** $\frac{1}{12} \tan^4 3x - \frac{1}{6} \tan^2 3x + \frac{1}{3} \ln|\sec 3x| + C$

25. $u - 2 \tan \frac{1}{4}u + C$ **27.** $-\frac{1}{3} \csc^3 x + C$ **29.** $\frac{1}{4} \tan^4(\ln x) + \frac{1}{3} \tan^6(\ln x) + \frac{1}{8} \tan^8(\ln x) + C$ **31.** $\frac{1}{4} - \frac{1}{8} \ln 2$ **33.** $\frac{56}{15}$

35. $\frac{1}{5}$ **37.** $(1 - \frac{1}{4}\pi)$ sq units **39.** $\frac{4}{3}\pi$ cu units

---

EXERCISES 9.4 (Page 550)

1. $-\dfrac{\sqrt{4 - x^2}}{4x} + C$ **3.** $\frac{1}{2} \ln \left| \dfrac{\sqrt{x^2 + 4} - 2}{x} \right| + C$ **5.** $\frac{1}{5} \ln \left| \dfrac{5 - \sqrt{25 - x^2}}{x} \right| + C$ **7.** $\ln|x + \sqrt{x^2 - a^2}| + C$

9. $\frac{1}{4}\tan^{-1}\frac{1}{2}x - \frac{x}{2(x^2+4)} + C$ **11.** $-\frac{x}{9\sqrt{4x^2-9}} + C$ **13.** $\frac{1}{5}\ln\left(\frac{\sqrt{t^4+25}-5}{t^2}\right) + C$ **15.** $\ln|x+2+\sqrt{4x+x^2}| + C$

17. $\frac{x+2}{9\sqrt{5-4x-x^2}} + C$ **19.** $\frac{\tan x}{4\sqrt{4-\tan^2 x}} + C$ **21.** $\frac{1}{3}\sqrt{\ln^2 w - 4}(8 + \ln^2 w) + C$ **23.** $-\frac{e^t+4}{9\sqrt{e^{2t}+8e^t+7}} + C$

25. $\frac{128}{3} - 24\sqrt{3}$ **27.** $\frac{1}{27}(6-2\sqrt{3})$ **29.** $\frac{1}{2}\cos^{-1}\frac{1}{3} - \frac{1}{6}\pi$ **31.** $\frac{625}{16}\pi$ **33.** (a) $\sec^{-1}\frac{2}{3}x + C$; (b) $-\frac{5}{2}\sqrt{3-2x^2} + C$

35. $\ln\left(\frac{\sqrt{10}-1}{3\sqrt{2}-3}\right) + \sqrt{10} - \sqrt{2}$ **37.** $\frac{81}{16}\pi^2$ cu units   **39.** $\frac{392}{60+27\ln 3}$ cm from the left end

41. $\left(\frac{20-15\cos^{-1}\frac{3}{5}}{5\ln 3 - 4}, \frac{26}{225(5\ln 3 - 4)}\right)$ **43.** $(\frac{8}{3}\pi + 3\sqrt{3})\rho g$ lb   **45.** $(\frac{512}{3}\pi + 192\sqrt{3})\rho$ oz

---

EXERCISES 9.5 (Page 560)

1. $\frac{1}{4}\ln\left|\frac{x-2}{x+2}\right| + C$ **3.** $\ln|C(x-2)^2(x+2)^3|$ **5.** $\ln\left|\frac{C(w+4)^3}{2w-1}\right|$ **7.** $\frac{1}{4}\ln\left|\frac{Cx^4(2x+1)^3}{2x-1}\right|$ **9.** $\frac{1}{9}\ln\left|\frac{x+3}{x}\right| - \frac{1}{3x} + C$

11. $2\ln\left|\frac{x+1}{x}\right| - \frac{1}{x} - \frac{1}{x+1} + C$ **13.** $\frac{3}{x+1} + \ln|x+1| - \frac{1}{2}\ln|2x+3| + C$ **15.** $\frac{5}{16(z+2)} - \frac{7}{16(z-2)} + \frac{1}{32}\ln\left|\frac{z+2}{z-2}\right| + C$

17. $\frac{1}{2}x^2 + 2x - \frac{3}{x-1} - \ln|x^2+2x-3| + C$ **19.** $-\ln[(3x+2)^{2/3}(x-1)^2] - \frac{1}{3(3x+2)} - \frac{3}{x-1} + C$

21. $4\ln\frac{4}{3} - \frac{3}{2}$ **23.** $\ln\frac{27}{4} - 2$ **25.** $13\ln 2 - 4\ln 5$ **27.** $\ln\frac{7}{2} - \frac{5}{3}$ **29.** $\ln 4.5$ sq units

31. $2\pi(2+6\ln 3 - 2\ln 2)$ cu units   **33.** $\left(\frac{6\ln 3 - 2\ln 2 + 2}{2\ln 3 - \ln 2}, \frac{48\ln 2 - 48\ln 3 + 35}{24(2\ln 3 - \ln 2)}\right)$

35. (a) $f(t) = \frac{5000}{1+4999e^{-0.5t}}$; (b) 1328; (c) 4075; (d) 5000   **37.** 50 percent   **39.** \$11,201,100   **41.** $\frac{31}{19}$ **43.** 7.4 lb

---

EXERCISES 9.6 (Page 565)

1. $\frac{1}{2}\ln\left|\frac{Cx^2}{2x^2+1}\right|$ **3.** $\frac{1}{8}\ln\left|\frac{2x-1}{2x+1}\right| - \frac{1}{4}\tan^{-1}2x + C$ **5.** $\frac{1}{10}\ln|(t^2+1)(2t+1)^3| + \frac{2}{5}\tan^{-1}t + C$

7. $\ln|x-1| + \tan^{-1}x + C$ **9.** $\frac{1}{2}\ln\left(\frac{Cx^2}{x^2+x+1}\right) - \frac{1}{\sqrt{3}}\tan^{-1}\left(\frac{2x+1}{\sqrt{3}}\right)$ **11.** $\ln\left(\frac{Cx^2}{x^2+1}\right) - \frac{1}{2}\tan^{-1}x - \frac{x}{2(x^2+1)}$

13. $\frac{5}{2}\ln(z^2-2z+5) + \frac{65}{16}\tan^{-1}\left(\frac{z-1}{2}\right) - \frac{47z-15}{8(z^2-2z+5)} + C$

15. $\frac{5}{162}\ln|9x^2+3x+1| - \frac{2}{81}\ln|3x-1| + \frac{5}{9\sqrt{3}}\tan^{-1}\left(\frac{6x+1}{\sqrt{3}}\right) + C$ **17.** $\frac{1}{6}\tan^{-1}\frac{2}{3}x + \frac{x}{4x^2+9} + C$

19. $\ln|\tan x + 1| + \frac{2}{\sqrt{3}}\tan^{-1}\left(\frac{2\tan x - 1}{\sqrt{3}}\right) + C$ **21.** $6\ln 2$ **23.** $\ln\frac{12}{5} + \frac{3}{2}\ln\frac{20}{13}$ **25.** $\frac{1}{4}\pi$ **27.** $\frac{3}{4}\ln 2 + \frac{5}{8}\pi$

29. $\frac{3}{8}\ln\frac{125}{41}$ **31.** $\frac{1}{16}$ sq units   **33.** $(\frac{2}{9}\sqrt{3}\pi^2 - \frac{2}{3}\pi\ln 3)$ cu units   **35.** $\frac{3}{50}\ln\frac{(t_1+2)^2}{4(t_1^2+1)} - \frac{7}{5(t_1+2)} - \frac{4}{25}\tan^{-1}t_1 + \frac{7}{10}$

EXERCISES 9.7 (Page 569)

1. $\frac{2}{3}x^{3/2} - 3x + 18\sqrt{x} - 54\ln(3+\sqrt{x}) + C$ **3.** $\ln\left|\frac{\sqrt{1+4x}-1}{\sqrt{1+4x}+1}\right| + C$ **5.** $-2\sqrt{1+x} + \sqrt{2}\ln\left|\frac{\sqrt{1+x}+\sqrt{2}}{\sqrt{1+x}-\sqrt{2}}\right| + C$

7. $\frac{3}{2}(x-2)^{2/3} - 3(x-2)^{1/3} + 3\ln|1+(x-2)^{1/3}| + C$ **9.** $2\sqrt{2x} + 2\sqrt{x+4} + 4\sqrt{2}\ln\left|\frac{\sqrt{x+4}-2\sqrt{2}}{\sqrt{x}+2}\right| + C$

11. $\frac{1}{9}\sqrt{1 + 2x^3}(2x^3 + 7) + C$ **13.** $\frac{6}{\sqrt{15}}\tan^{-1}\left(\frac{1}{\sqrt{15}}\tan\frac{x}{2}\right) + C$ **15.** $\frac{\sqrt{15}}{5}\ln\left|\frac{\sqrt{15} + \tan\frac{1}{2}x}{\sqrt{15} - \tan\frac{1}{2}x}\right| + C$

17. $\frac{1}{2}\ln|\tan\frac{1}{2}x| - \frac{1}{4}\tan^2\frac{1}{2}x + C$ **19.** $\frac{1}{4}\ln\left|\frac{\tan\frac{1}{2}x - 3}{\tan\frac{1}{2}x - \frac{1}{3}}\right| + C$ **21.** $\frac{1}{5}\ln\left|\frac{\tan\frac{1}{2}x - \frac{1}{3}}{\tan\frac{1}{2}x + 3}\right| + C$ **23.** $\sqrt{2}\ln\left|\frac{\tan x + \sqrt{2}}{\tan x - \sqrt{2}}\right| + C$

25. $\frac{1}{2}x + \frac{\sqrt{3}}{6}\ln\left|\frac{\sqrt{3} - \tan\frac{1}{2}x}{\sqrt{3} + \tan\frac{1}{2}x}\right| + C$ **27.** $-\frac{1}{2}\ln|13 - \tan^2 x| + C$ **29.** $2\tan^{-1}(2 + \tan\frac{1}{2}x) + C$

31. $3\tan^{-1}\sqrt[6]{x} + \frac{3\sqrt[6]{x}}{1 + \sqrt[3]{x}} + C$ **33.** $4 - 2\ln 3$ **35.** $\ln\frac{11}{10}$ **37.** $\frac{1}{4}\ln 3$ **39.** $\frac{1}{2}\sqrt{3}\ln(1 + \frac{1}{2}\sqrt{3})$

41. $2\sqrt{3}\ln(1 + \sqrt{3})$ **43.** $\frac{3}{2}\pi - \frac{152}{35}$ **45.** $2\ln|\sqrt{x} - 1| + C$

EXERCISES 9.8 (Page 574)

1. $\sinh^{-1}\frac{1}{2}x + C = \ln\frac{1}{2}(x + \sqrt{x^2 + 4}) + C$ **3.** $\frac{1}{2}\cosh^{-1}x^2 + C = \frac{1}{2}\ln(x^2 + \sqrt{x^4 - 1}) + C$

5. $\begin{cases} -\frac{1}{12}\tanh^{-1}(\frac{3}{4}x) + C \text{ if } |x| < \frac{4}{3} \\ -\frac{1}{12}\coth^{-1}(\frac{3}{4}x) + C \text{ if } |x| > \frac{4}{3} \end{cases} = \frac{1}{24}\ln\left|\frac{4 - 3x}{4 + 3x}\right| + C$ **7.** $\sinh^{-1}\frac{\sin x}{\sqrt{3}} + C = \ln(\sin x + \sqrt{\sin^2 x + 3}) + C$

9. $\frac{1}{\sqrt{5}}\cosh^{-1}(\sqrt{5}e^t) + C = \frac{1}{\sqrt{5}}\ln(\sqrt{5}e^t + \sqrt{5e^{2t} - 1}) + C$

11. $\begin{cases} \frac{1}{\sqrt{6}}\tanh^{-1}\left(\frac{x + 2}{\sqrt{6}}\right) + C \text{ if } |x + 2| < \sqrt{6} \\ \frac{1}{\sqrt{6}}\coth^{-1}\left(\frac{x + 2}{\sqrt{6}}\right) + C \text{ if } |x + 2| > \sqrt{6} \end{cases} = \frac{1}{2\sqrt{6}}\ln\left|\frac{\sqrt{6} + 2 + x}{\sqrt{6} - 2 - x}\right| + C$

13. $\begin{cases} -\tanh^{-1}(x + 5) + C \text{ if } |x + 5| < 1 \\ -\coth^{-1}(x + 5) + C \text{ if } |x + 5| > 1 \end{cases} = -\frac{1}{2}\ln\left|\frac{x + 6}{x + 4}\right| + C$

15. $\frac{3}{2}\sinh^{-1}\left(\frac{2\ln w}{3}\right) + C = \frac{3}{2}\ln(2\ln w + \sqrt{4\ln^2 w + 9}) + C$

17. $\ln\frac{5 + \sqrt{21}}{3 + \sqrt{5}}$ **19.** $\ln 3$ **21.** $\frac{1}{3}\ln\frac{7 + \sqrt{40}}{4 + \sqrt{7}}$

REVIEW EXERCISES FOR CHAPTER 9 (Page 575)

1. $\frac{1}{8}x - \frac{1}{128}\sin 16x + C$ **3.** $-2\sqrt{4 - e^x} + C$ **5.** $(x + 1)\tan^{-1}\sqrt{x} - \sqrt{x} + C$ **7.** $\frac{1}{2}x + \frac{3}{4}\sin\frac{2}{3}x + C$

9. $\ln|x - 1| - 2(x - 1)^{-1} - (x - 1)^{-2} + C$ **11.** $\frac{1}{4}\sin 2x - \frac{1}{8}\sin 4x + C$ **13.** $3\ln\left|\frac{x^{1/3}}{1 + x^{1/3}}\right| + C$

15. $\frac{1}{3}\tan 3x - \frac{1}{3}\cot 3x + \frac{2}{3}\ln|\tan 3x| + C$ **17.** $2t + \ln\frac{t^2}{(t + 2)^{10}} - \frac{15}{t + 2} + C$ **19.** $x - \tan^{-1}x + \frac{1}{2}\ln\left|\frac{x - 1}{x + 1}\right| + C$

21. $\frac{1}{16}x - \frac{1}{192}\sin 12x - \frac{1}{144}\sin^3 6x + C$ **23.** $\sin^{-1}\left(\frac{r + 2}{\sqrt{7}}\right) + C$ **25.** $\frac{1}{2}x^2\sin x^2 + \frac{1}{2}\cos x^2 + C$

27. $\frac{2}{17}e^{t/2}(4\sin 2t + \cos 2t) + C$ **29.** $\frac{1}{4}\tan^{-1}(\frac{1}{2}\sin^2 x) + C$ **31.** $\begin{cases} \frac{1}{n}\left(-\cos nx + \frac{2}{3}\cos^3 nx - \frac{1}{5}\cos^5 nx\right) + C \text{ if } n \neq 0 \\ C \hspace{5cm} \text{if } n = 0 \end{cases}$

33. $-\frac{1}{4}\csc^3 x \cot x - \frac{3}{8}\csc x \cot x + \frac{3}{8}\ln|\csc x - \cot x| + C$ **35.** $2\ln|y - 2| - 8(y - 2)^{-1} - \frac{9}{2}(y - 2)^{-2} + C$

37. $-\tan^{-1}(\cos x) + C$ **39.** $2\sin^{-1}\left(\frac{t - 2}{2}\right) + \frac{1}{2}(t - 2)\sqrt{4t - t^2} + C$ **41.** $\frac{1}{3}\ln|1 - x^{-3}| + C$ **43.** $\frac{1}{3}\sin^{-1}(\frac{3}{2}e^x) + C$

45. $-\frac{1}{15}\cot^5 3x - \frac{1}{9}\cot^3 3x + C$ **47.** $\frac{1}{3}x^3\sin^{-1}x + \frac{1}{9}(x^2 + 2)\sqrt{1 - x^2} + C$ **49.** $\tan^{-1}(\cos x) + C$

51. $\frac{2}{3}\sec^{-1}(2\sin 3t) + C$ **53.** $-\dfrac{(x^2 + a^2)^{3/2}}{3a^2x^3} + C$ **55.** $\sqrt{2t} - \sqrt{1 - 2t}\,\sin^{-1}\sqrt{2t} + C$

57. $\frac{4}{3}\sqrt{2 + \sqrt{x-1}}\,(\sqrt{x-1} - 4) + C$

59. $\dfrac{\sqrt{2}}{4}\ln\left|\dfrac{\tan x - \sqrt{2\tan x} + 1}{\tan x + \sqrt{2\tan x} + 1}\right| + \dfrac{\sqrt{2}}{2}\tan^{-1}(\sqrt{2\tan x} - 1) + \dfrac{\sqrt{2}}{2}\tan^{-1}(\sqrt{2\tan x} + 1) + C$

61. $\begin{cases}\dfrac{x^{n+1}\ln x}{n+1} - \dfrac{x^{n+1}}{(n+1)^2} + C & \text{if } n \neq -1 \\ \frac{1}{2}\ln^2 x + C & \text{if } n = -1\end{cases}$ **63.** 4   **65.** $\frac{1}{2} + 2\ln\frac{6}{5}$ **67.** $\frac{16}{3} - \frac{8}{3}\sqrt{2}$ **69.** $\frac{4}{3}\sqrt{3} - \frac{1}{2}\pi$ **71.** $\frac{4}{3}$

73. $\frac{1}{2} - \frac{1}{4}\ln 2$ **75.** $-a^2(\frac{9}{8}\sqrt{3} - \frac{1}{2}\pi)$ **77.** $\frac{1}{2}\ln\frac{9}{2} - \frac{1}{6}\pi$ **79.** 5   **81.** $\frac{1}{6} + \ln\frac{3}{2}$ **83.** $\frac{4}{3}$ **85.** $1 - \frac{1}{2}\ln 3$ **87.** $\frac{1}{24}\pi$

89. $\sqrt{3} - \frac{1}{2}\ln(2 + \sqrt{3})$ **91.** $\frac{1}{5}\ln\frac{3}{2}$ **93.** $\frac{256}{15}$ **95.** $\frac{1}{3}k(1 - e^{-9})$ kg; $\dfrac{e^9 - 10}{3(e^9 - 1)}$ m from one end

97. $9\sqrt{2} - 3\sqrt{5} + \frac{3}{2}\ln\left(\dfrac{3 + 2\sqrt{2}}{\sqrt{5} + 2}\right)$ **99.** $\frac{1}{8}\pi$ sq units   **101.** $\pi(e^2\ln 2 + \frac{1}{2}e^2 + \frac{1}{8})$ cu units

103. (a) $x = 300\left(\dfrac{18^t - 17^t}{3\cdot 18^t - 2\cdot 17^t}\right)$; (b) 35.94 lb   **105.** $\left(0, \dfrac{32}{15\pi}\right)$ **107.** $(\frac{1}{2}\pi - 1, \frac{1}{2})$

109. (a) 4824; (b) 8.18 weeks   **111.** $3\rho g$ lb

EXERCISES 10.1 (Page 585)

(Sketches of the graphs for Exercises 1 through 7 appear in Figs. EX 10.1-1 through EX 10.1-7.)
1. $(0, 1)$; $y = -1$; 4   **3.** $(-2, 0)$; $x = 2$; 8   **5.** $(0, -\frac{1}{4})$; $y = \frac{1}{4}$; 1   **7.** $(\frac{9}{8}, 0)$; $x = -\frac{9}{8}$; $\frac{9}{2}$ **9.** $y^2 = 20x$
11. $x^2 = -8y$ **13.** $y^2 = 2x$ **15.** $y^2 = -6x$ **17.** $y^2 = 10x$ **19.** $x^2 = -y$ **21.** $x'^2 + y'^2 = 13$ **23.** $y'^2 = 6x'$
25. $y' = 2x'^3$ **27.** $(-3, \frac{1}{4})$; $(-3, -\frac{3}{4})$; $x = -3$; $y = \frac{5}{4}$ **29.** $(1, -5)$; $(-\frac{1}{2}, -5)$; $y = -5$; $x = \frac{5}{2}$
31. $(\frac{4}{3}, 1)$; $(\frac{7}{24}, 1)$; $y = 1$; $x = \frac{25}{24}$ **33.** $y^2 + 20x - 8y - 24 = 0$ **35.** $x^2 + 2x - 8y + 41 = 0$
37. $x^2 - 6x - 6y - 3 = 0$; $x^2 - 6x + 6y + 21 = 0$ **39.** $y^2 - 4x - 4y - 12 = 0$ **41.** (a) $-\frac{5}{2}$; (b) $x^2 = -10y$ **43.** 16.6 m
47. $\frac{32}{45}$ in.   **49.** (a) $y = x$; (b) $(\frac{1}{2}, \frac{1}{2})$; (c) $2\sqrt{2}$ **53.** translate the axes so that the new origin is $(\frac{1}{4}\pi, -3)$

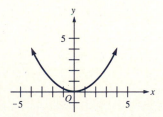

FIGURE EX 10.1-1

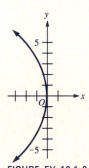

FIGURE EX 10.1-3

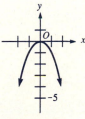

FIGURE EX 10.1-5

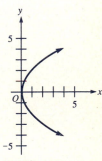

FIGURE EX 10.1-7

EXERCISES 10.2 (Page 593)

1. center: $(0, 0)$; vertices: $(\pm 3, 0)$; foci: $(\pm\sqrt{5}, 0)$; extremities of minor axis: $(0, \pm 2)$
3. center: $(0, 0)$; vertices: $(0, \pm 5)$; foci: $(0, \pm\sqrt{21})$; extremities of minor axis: $(\pm 2, 0)$
5. center: $(0, 0)$; vertices: $(\pm 3, 0)$; foci: $(\pm\sqrt{3}, 0)$; extremities of minor axis: $(0, \pm\sqrt{6})$
7. center: $(0, 0)$; vertices: $(0, \pm\frac{1}{2})$; foci: $(0, \pm\frac{1}{4}\sqrt{3})$; extremities of minor axis: $(\pm\frac{1}{4}, 0)$
9. center: $(2, 3)$; vertices: $(-1, 3)$, $(5, 3)$; foci: $(2 \pm \sqrt{3}, 3)$; extremities of minor axis: $(2, 3 \pm \sqrt{6})$
11. center: $(0, \frac{1}{2})$; vertices: $(0, \frac{1}{2} \pm \frac{1}{2}\sqrt{17})$; foci: $(0, \frac{1}{2} \pm \frac{1}{10}\sqrt{170})$; extremities of minor axis: $(\pm\frac{1}{10}\sqrt{255}, \frac{1}{2})$ **13.** point $(-\frac{5}{2}, 4)$
15. center: $(1, 0)$; vertices: $(1 \pm \sqrt{5}, 0)$; foci: $(1 \pm \sqrt{2}, 0)$; extremities of minor axis: $(1, \pm\sqrt{3})$ **17.** point $(1, -1)$

(Sketches of the graphs for Exercises 19 through 25 appear in Figs. EX 10.2-19 through EX 10.2-25.)

19. $16x^2 + 25y^2 = 100$ **21.** $3x^2 + 2y^2 = 54$ **23.** $x^2 + 4y^2 = 4$ **25.** $\dfrac{(x-4)^2}{25} + \dfrac{(y+2)^2}{9} = 1$

27. $\dfrac{(y-3)^2}{64} + \dfrac{(x+1)^2}{48} = 1$ **29.** $2x + 3y - 12 = 0$ **33.** $\frac{42}{5}$ m   **35.** $9x^2 + 25y^2 = 56{,}250{,}000$ **37.** 98π in.^{3}

39. $(h \pm \sqrt{a^2 - b^2}, k)$ **41.** $18\pi \rho g$ lb

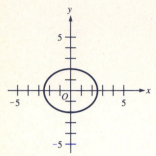

FIGURE EX 10.2-19

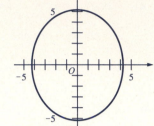

FIGURE EX 10.2-21

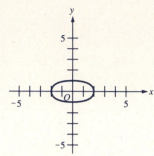

FIGURE EX 10.2-23

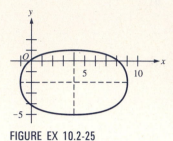

FIGURE EX 10.2-25

EXERCISES 10.3 (Page 603)

1. center: $(0, 0)$; vertices: $(\pm 2, 0)$; foci: $(\pm\sqrt{13}, 0)$; asymptotes: $y = \pm\frac{3}{2}x$

3. center: $(0, 0)$; vertices: $(\pm 5, 0)$; foci: $(\pm\sqrt{29}, 0)$; asymptotes: $y = \pm\frac{2}{5}x$

5. center: $(0, 0)$; vertices: $(0, \pm 2)$; foci: $(0, \pm 2\sqrt{5})$; asymptotes: $y = \pm\frac{1}{2}x$

7. center: $(0, 0)$; vertices: $(0, \pm\frac{1}{3})$; foci: $(0, \pm\frac{5}{12})$; asymptotes: $y = \pm\frac{4}{3}x$

9. center: $(-4, -1)$; vertices: $(-10, -1)$, $(2, -1)$; foci: $(-4 \pm 6\sqrt{2}, -1)$; asymptotes: $x - y + 3 = 0$, $x + y + 5 = 0$

11. center: $(-3, -1)$; vertices: $(-3, -1 \pm \frac{2}{3}\sqrt{2})$; foci: $(-3, -1 \pm \frac{2}{3}\sqrt{6})$; asymptotes: $\pm x + \sqrt{2}y + \sqrt{2} \pm 3 = 0$

13. center: $(-1, 4)$; vertices: $(-1, 4 \pm 2\sqrt{7})$; foci: $(-1, -3)$, $(-1, 11)$; asymptotes: $\pm 2x + \sqrt{3}y \pm 2 - 4\sqrt{3} = 0$

15. center: $(1, -2)$; vertices: $(1, -5)$, $(1, 1)$; foci: $(1, -2 \pm \sqrt{13})$; asymptotes: $3x + 2y + 1 = 0$, $3x - 2y - 7 = 0$

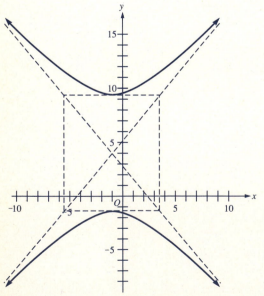

FIGURE EX 10.3-13

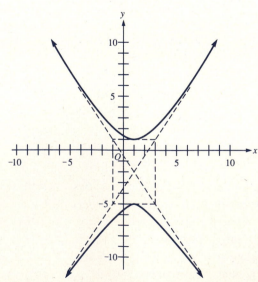

FIGURE EX 10.3-15

(Sketches of the graphs for Exercises 17 through 25 appear in Figs. EX 10.3-17 through EX 10.3-25.)

17. $9x^2 - 4y^2 = 36$ **19.** $32y^2 - 33x^2 = 380$ **21.** $25x^2 - 144y^2 = 14,400$ **23.** $\dfrac{(y+1)^2}{144} - \dfrac{(x+2)^2}{81} = 1$

25. $72(x-3)^2 - 9(y-4)^2 = 128$ **27.** (a) $\dfrac{(x+2)^2}{1} - \dfrac{(y+1)^2}{4} = 1$; (b) $2x - y + 3 = 0, 2x + y + 5 = 0$

29. $2x - \sqrt{2}y + 1 = 0$ **31.** $7x^2 - 4y^2 = 28$ **33.** the right branch of the hyperbola $16x^2 - 9y^2 = 14,400$
35. $\frac{4}{3}\pi ab^2$ cu units **37.** (3, 4)

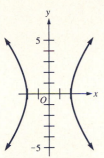

FIGURE EX 10.3-17

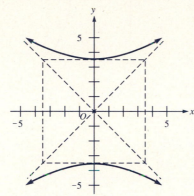

FIGURE EX 10.3-19

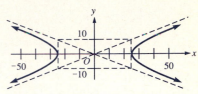

FIGURE EX 10.3-21

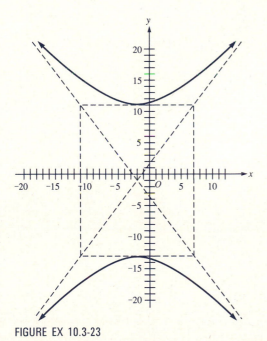

FIGURE EX 10.3-23

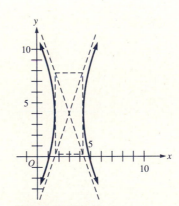

FIGURE EX 10.3-25

EXERCISES 10.4 (Page 608)

(Sketches of the graphs for Exercises 1 through 21 appear in Figs. EX 10.4-1 through EX 10.4-21.)
1. (a) $\bar{x}^2 - \bar{y}^2 = 16$ **3.** (a) $\bar{x}\bar{y} = -4$ **5.** $16\bar{y}^2 - 9\bar{x}^2 = 36$ **7.** $\bar{x}^2 + 4\sqrt{2}\bar{y} = 0$ **9.** $\bar{y}^2 - \bar{x}^2 = 32$
11. $9\bar{x}^2 + 4\bar{y}^2 = 36$ **13.** $3\bar{x}'^2 + \bar{y}'^2 = 18$ **15.** $\bar{x}'^2 + 4\bar{y}'^2 = 16$ **17.** $\sqrt{2}\bar{x}'^2 = \bar{y}'$
19. $\bar{x}'^2 - 4\bar{y}'^2 = 16$ **21.** $5\bar{x}^2 + 6\sqrt{5}\bar{y} = 0$

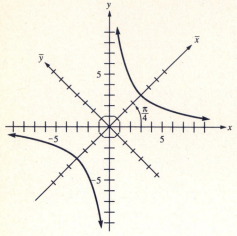

FIGURE EX 10.4-1

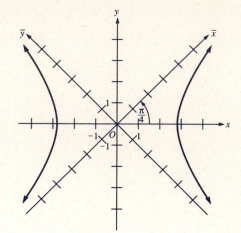

FIGURE EX 10.4-3

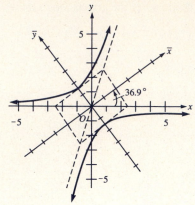

FIGURE EX 10.4-5

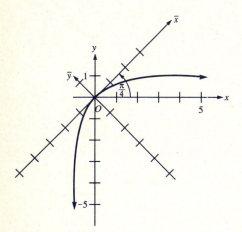

FIGURE EX 10.4-7

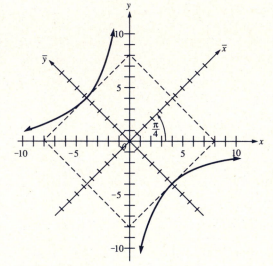

FIGURE EX 10.4-9

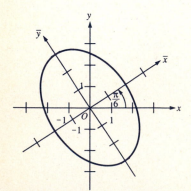

FIGURE EX 10.4-11

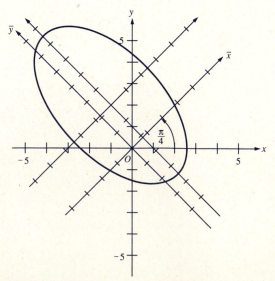

FIGURE EX 10.4-13

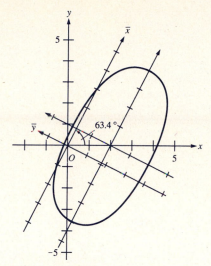

FIGURE EX 10.4-15

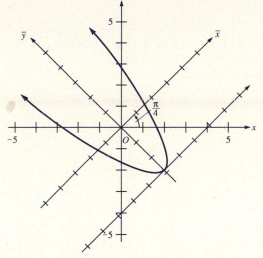

FIGURE EX 10.4-17

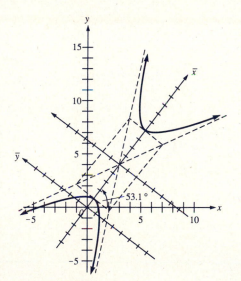

FIGURE EX 10.4-19

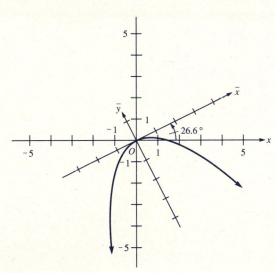

FIGURE EX 10.4-21

EXERCISES 10.5 (Page 613)

5. (a) $(-4, \frac{5}{4}\pi)$; (b) $(4, -\frac{7}{4}\pi)$; (c) $(-4, -\frac{3}{4}\pi)$　　**7.** (a) $(-2, \frac{3}{2}\pi)$; (b) $(2, -\frac{3}{2}\pi)$; (c) $(-2, -\frac{1}{2}\pi)$　　**9.** (a) $(-\sqrt{2}, \frac{3}{4}\pi)$; (b) $(\sqrt{2}, -\frac{1}{4}\pi)$;
(c) $(-\sqrt{2}, -\frac{5}{4}\pi)$　　**11.** (a) $(-2, \frac{3}{4}\pi)$; (b) $(-2, -\frac{5}{4}\pi)$; (c) $(2, \frac{15}{4}\pi)$　　**13.** $(3, \frac{4}{3}\pi)$; $(-3, \frac{1}{3}\pi)$　　**15.** $(-4, -\frac{7}{6}\pi)$; $(4, -\frac{1}{6}\pi)$

17. $(-2, \frac{3}{4}\pi)$; $(2, \frac{7}{4}\pi)$　　**19.** $(2, 2\pi + 6)$; $(-2, 6 - \pi)$　　**21.** (a) $(-3, 0)$; (b) $(-1, -1)$; (c) $(2, -2\sqrt{3})$; (d) $(\frac{1}{2}\sqrt{3}, -\frac{1}{2})$

23. (a) $(\sqrt{2}, \frac{7}{4}\pi)$; (b) $(2, \frac{5}{6}\pi)$; (c) $(2\sqrt{2}, \frac{1}{4}\pi)$; (d) $(5, \pi)$　　**25.** $r = |a|$　　**27.** $r = \dfrac{2}{1 - \cos\theta}$　　**29.** $r = 6\sin\theta$

31. $r^2 = 4\cos 2\theta$　　**33.** $r = \dfrac{3a\sin 2\theta}{2(\sin^3\theta + \cos^3\theta)}$　　**35.** $(x^2 + y^2)^2 = 4xy$　　**37.** $(x^2 + y^2)^3 = x^2$　　**39.** $y = x\tan(x^2 + y^2)$

41. $x = -1$　　**43.** $4x^2 - 5y^2 - 36y - 36 = 0$

EXERCISES 10.6 (Page 624)

1. (a) line through the pole with slope $\sqrt{3}$; (b) circle with center at the pole and radius $\frac{1}{3}\pi$

3. (a) line through the pole with slope $\tan^{-1} 2$; (b) circle with center at the pole and radius 2

5. (a) line parallel to the $\frac{1}{2}\pi$ axis and 4 units to the right of it; (b) circle tangent to the $\frac{1}{2}\pi$ axis with center at the point (0, 2)

7. (a) line parallel to the polar axis and 4 units below it; (b) circle tangent to the polar axis with center at the point $(2, \frac{3}{2}\pi)$

(Sketches of the graphs for Exercises 9 through 23 appear in Figs. EX 10.6-9 through EX 10.6-23.)

25. logarithmic spiral, containing the points (r, θ) given in the following table.

r	1	$e^{\pi/2} \approx 5$	$e^{\pi} \approx 23$	$e^{3\pi/2} \approx 111$	$e^{2\pi} \approx 535$	$e^{5\pi/2} = 2576$	$e^{3\pi} \approx 12{,}392$
θ	0	$\frac{1}{2}\pi$	π	$\frac{3}{2}\pi$	2π	$\frac{5}{2}\pi$	3π

27. reciprocal spiral, containing the points (r, θ) given in the following tables.

r	$\frac{6}{\pi} \approx 1.9$	$\frac{3}{\pi} \approx 0.95$	$\frac{2}{\pi} \approx 0.63$	$\frac{1}{\pi} \approx 0.32$	$\frac{1}{2\pi} \approx 0.16$	$\frac{1}{3\pi} \approx 0.12$	$\frac{1}{4\pi} \approx 0.08$	$\frac{1}{6\pi} \approx 0.05$
θ	$\frac{1}{6}\pi$	$\frac{1}{3}\pi$	$\frac{1}{2}\pi$	π	2π	3π	4π	6π

(Sketches of the graphs for Exercises 29 through 35 appear in Figs. EX 10.6-29 through EX 10.6-35.)

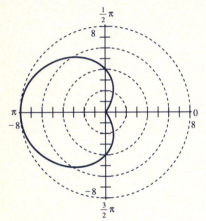

FIGURE EX 10.6-9 Cardioid

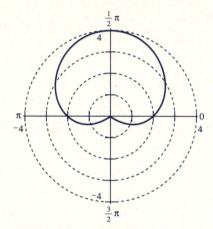

FIGURE EX 10.6-11 Cardioid

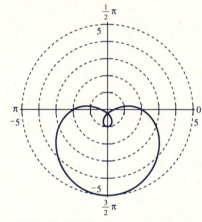

FIGURE EX 10.6-13 Limaçon with a Loop

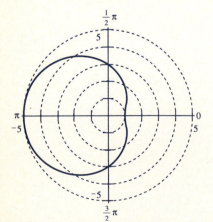

FIGURE EX 10.6-15 Limaçon with a Dent

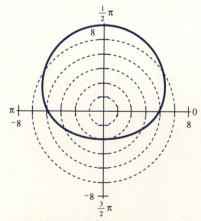

FIGURE EX 10.6-17 Convex Limaçon

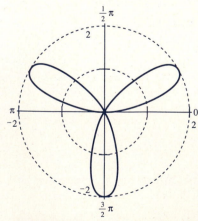

FIGURE EX 10.6-19 Three-Leafed Rose

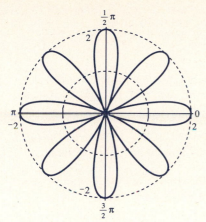

FIGURE EX 10.6-21 Eight-Leafed Rose

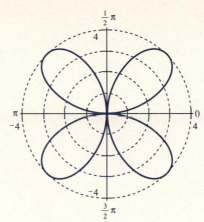

FIGURE EX 10.6-23 Four-Leafed Rose

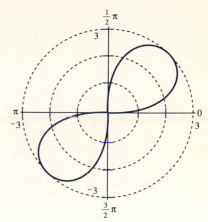

FIGURE EX 10.6-29 Lemniscate

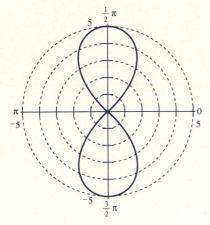

FIGURE EX 10.6-31 Lemniscate

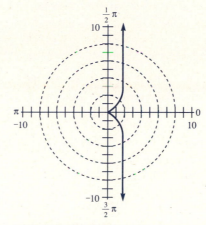

FIGURE EX 10.6-33 Cissoid

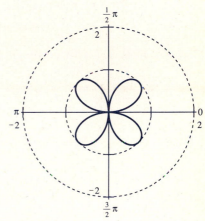

FIGURE EX 10.6-35 Four-Leafed Rose

37. $(\frac{3}{2}, \frac{1}{6}\pi)$; $(\frac{3}{2}, \frac{5}{6}\pi)$ **39.** pole; $(\sqrt{2}, \frac{1}{4}\pi)$ **41.** $(\frac{1}{2}\pi, \frac{1}{8}\pi)$ **43.** pole; $(\sqrt{15}, \cos^{-1}\frac{1}{4})$; $(\sqrt{15}, \pi - \cos^{-1}\frac{1}{4})$
45. $(3, \frac{1}{3}\pi)$; $(3, -\frac{1}{3}\pi)$ **47.** $(2\sqrt{2}, \frac{1}{4}\pi)$ **49.** $(1, \frac{3}{2}\pi)$; $(1, \frac{1}{2}\pi)$; $(\frac{1}{2}, \frac{4}{3}\pi)$; $(\frac{1}{2}, \frac{2}{3}\pi)$; pole; $(0.22, 2.47)$; $(0.22, 3.82)$
51. pole; $(\frac{1}{2}\sqrt{2}, \frac{1}{8}(2n+1)\pi)$, where n is $0, 1, \ldots, 7$ **55.** pole; $(\frac{1}{2}\sqrt{2}, \frac{1}{6}(2n+1)\pi)$, where n is $0, 1, 2, 3, 4, 5$

EXERCISES 10.7 (Page 628)

1. $\frac{9}{4}\pi$ sq units   **3.** 4π sq units   **5.** 4 sq units   **7.** $\frac{9}{16}\pi^3$ sq units   **9.** $\frac{9}{8}\pi$ sq units
11. $(\frac{11}{4}\pi - \frac{11}{2}\sin^{-1}\frac{1}{3} - 3\sqrt{2})$ sq units   **13.** $(\frac{19}{3}\pi - \frac{11}{2}\sqrt{3})$ sq units   **15.** $(\frac{9}{2}\pi - 9)$ sq units   **17.** $a^2(2 - \frac{1}{4}\pi)$ sq units
19. $\frac{1}{2}(\pi + 1)$ sq units   **21.** $\frac{1}{2}\pi a^2$ sq units   **23.** 4 sq units   **25.** $16a^2\pi^3$ sq units

---

EXERCISES 10.8 (Page 638)

1. (a) $e = \frac{1}{3}\sqrt{5}$; foci: $(\pm\sqrt{5}, 0)$; directrices: $x = \pm\frac{9}{5}\sqrt{5}$ **3.** (a) $e = \frac{1}{5}\sqrt{21}$; foci: $(0, \pm\sqrt{21})$; directrices: $y = \pm\frac{25}{21}\sqrt{21}$
5. (a) $e = \frac{1}{5}\sqrt{29}$; foci: $(\pm\sqrt{29}, 0)$; directrices: $x = \pm\frac{25}{29}\sqrt{29}$ **7.** (a) $e = \frac{5}{3}$; foci: $(\pm 5, 0)$; directrices: $x = \pm\frac{9}{5}$ **9.** (a) parabola;
(b) hyperbola; (c) ellipse; (d) circle   **11.** (a) 1; (b) parabola; (c) $r\cos\theta = -2$ **13.** (a) $\frac{1}{2}$; (b) ellipse; (c) $r\sin\theta = 5$
15. (a) $\frac{2}{3}$; (b) ellipse; (c) $r\cos\theta = -3$ **17.** (a) $\frac{6}{5}$; (b) hyperbola; (c) $r\sin\theta = -\frac{3}{2}$ **19.** (a) $\frac{2}{7}$; (b) ellipse; (c) $r\sin\theta = -5$

21. (a) $\frac{5}{4}$; (b) hyperbola; (c) $r\cos\theta = 2$ **23.** $r = \dfrac{8}{1 - \sin\theta}$ **25.** $r = \dfrac{36}{3 + 4\cos\theta}$ **27.** $r = \dfrac{3}{2 - \cos\theta}$

29. (a) $r = \dfrac{5}{1 - 3 \cos \theta}$; (b) $r \cos \theta = -\frac{5}{3}$ **31.** $\frac{16}{3}\sqrt{3}\pi$ sq units   **37.** (a) $r = \dfrac{40,000,000}{1 - \cos \theta}$; (b) 20,000,000 miles

39. $r^2 = \dfrac{a^2(1 - e^2)}{1 - e^2 \cos^2 \theta}$

EXERCISES 10.9 (Page 646)

1. $\sqrt{3}$ **3.** $\sqrt{2} + 1$ **5.** $2\sqrt{3}$ **7.** $-\frac{7}{3}\sqrt{3}$ **9.** horizontal tangent lines at $(7, \frac{1}{2}\pi)$, $(1, \frac{3}{2}\pi)$, $(2, 3.87)$, $(2, 5.55)$; vertical tangent lines at $(5.34, 0.46)$, $(5.34, 2.68)$ **11.** horizontal tangent lines at $(4.73, 1.95)$, $(4.73, 4.34)$; vertical tangent lines at $(2, 0)$, $(6, \pi)$

13. horizontal tangent lines at $(2, \frac{3}{2}\pi)$, $(\frac{1}{2}, \frac{1}{6}\pi)$, $(\frac{1}{2}, \frac{5}{6}\pi)$; vertical tangent lines at $(0, \frac{1}{2}\pi)$, $(\frac{3}{2}, \frac{7}{6}\pi)$, $(\frac{3}{2}, \frac{11}{6}\pi)$

15. horizontal tangent lines at $(3.68, 0.98)$, $(3.68, 5.31)$, $(-0.68, 2.68)$, $(-0.68, 3.61)$; vertical tangent lines at $(5, 0)$, $(-1, \pi)$, $(1, 1.91)$, $(1, 4.37)$ **17.** horizontal tangent lines at $(-1, \frac{1}{2}\pi)$, $(-1, \frac{3}{2}\pi)$, $(\frac{2}{3}, 0.42)$, $(\frac{2}{3}, 2.72)$, $(\frac{2}{3}, 3.56)$, $(\frac{2}{3}, 5.86)$; vertical tangent lines at $(1, 0)$, $(1, \pi)$, $(-\frac{2}{3}, 1.15)$, $(-\frac{2}{3}, 1.99)$, $(-\frac{2}{3}, 4.29)$, $(-\frac{2}{3}, 5.13)$

19. horizontal tangent lines at $(0, 0)$, $(\sqrt[4]{12}, \frac{1}{3}\pi)$, $(-\sqrt[4]{12}, \frac{1}{3}\pi)$; vertical tangent lines at $(0, \frac{1}{2}\pi)$, $(\sqrt{2}, \frac{1}{6}\pi)$, $(-\sqrt{2}, \frac{1}{6}\pi)$

21. $\chi = \frac{3}{4}\pi$, $\alpha = 1 + \frac{3}{4}\pi$ **23.** $\chi = 2.68$; $\alpha = 2.29$ **25.** $\chi = 0.67$; $\alpha = 2.25$ **27.** $\chi = \frac{5}{6}\pi$; $\alpha = 0$ **29.** $90°$ **31.** $60°$

33. $0°$ at pole; $90°$ at $(1, 0)$; $90°$ at $(1, \pi)$ **35.** $0°$ at $(0, \frac{1}{2}\pi)$; $90°$ at $(0, 0)$; $79°6'$ at $(\frac{1}{2}\sqrt{3}, \frac{1}{6}\pi)$; $79°6'$ at $(-\frac{1}{2}\sqrt{3}, \frac{5}{6}\pi)$

REVIEW EXERCISES FOR CHAPTER 10 (Page 647)

1. $x^2 = -12y$; (b) 12   **3.** (a) $y^2 = 4x$; (b) 4   **5.** (a) $(0, \frac{3}{2})$; (b) $y = -\frac{3}{2}$ **7.** (a) $(2, 3)$; (b) $y = 3$; (c) $(0, 3)$; (d) $x = 4$

9. $(x - 5)^2 = 8(y - 1)$ **11.** 9π cm   **13.** center: $(0, 0)$; vertices: $(\pm 5, 0)$; foci: $(\pm \sqrt{21}, 0)$; extremities of minor axis: $(0, \pm 2)$

15. center: $(0, 0)$; vertices: $(\pm 3, 0)$; foci: $(\pm 5, 0)$; asymptotes: $y = \pm \frac{4}{3}x$

(Sketches of the graphs for Exercises 17 through 23 appear in Figs. REV 10-17 through REV 10-23.)

17. center: $(-3, 8)$; vertices: $(-3, 4)$, $(-3, 12)$; foci: $(-3, 8 \pm 2\sqrt{3})$; extremities of minor axis: $(-5, 8)$, $(-1, 8)$

19. center: $(-1, 3)$; vertices: $(-2, 3)$, $(0, 3)$; foci: $(-1 \pm \sqrt{26}, 3)$; asymptotes: $5x + y + 2 = 0$, $5x - y + 8 = 0$

21. $\dfrac{x^2}{20} + \dfrac{(y - 2)^2}{36} = 1$ **23.** $\dfrac{(x + 2)^2}{4} - \dfrac{(y - 1)^2}{5} = 1$ **25.** $r^2(4 \cos^2 \theta - 9 \sin^2 \theta) = 36$ **27.** $r = 9 \cos \theta - 8 \sin \theta$

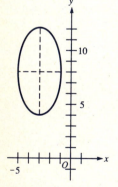

FIGURE REV 10-17

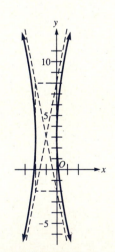

FIGURE REV 10-19

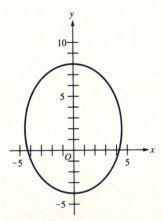

FIGURE REV 10-21

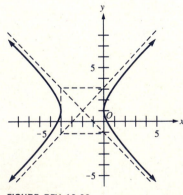

FIGURE REV 10-23

29. $4x^4 + 8x^2y^2 + 4y^4 + 36x^3 + 36xy^2 - 81y^2 = 0$ **31.** $(x^2 + y^2)^2 = 4x^3 - 12xy^2$ **39.** the pole

41. (a) $(16\pi - 24\sqrt{3})$ sq units; (b) $(32\pi + 24\sqrt{3})$ sq units   **43.** $a^2(\frac{1}{3}\pi + \frac{1}{2}\sqrt{3})$ sq units   **45.** $\dfrac{e^{4k\pi} - 1}{4k}$ sq units

49. $r^2 - 2r_0 r \cos (\theta - \theta_0) + r_0^2 = a^2$ **51.** (a) $e = \frac{1}{2}$; (b) ellipse; (c) $r \sin \theta = -2$ **53.** (a) $e = \frac{3}{2}$; (b) hyperbola; (c) $3r \cos \theta = 4$

55. $r = \dfrac{8}{1 - 3\cos\theta}$ **57.** $r = \dfrac{6}{1 - \sin\theta}$ **59.** $21\bar{x}'^2 - 49\bar{y}'^2 = 72$ **61.** (a) $e = \frac{1}{3}\sqrt{34}$; foci $(\pm\sqrt{34}, 0)$; directrices: $x = \pm\frac{9}{34}\sqrt{34}$

63. $4\sqrt{3}\pi\, a^2 b$ cu units   **65.** 600 mi   **67.** $(3 \pm \frac{3}{4}\sqrt{6})$ million miles   **69.** $r = \dfrac{1}{1 - \cos\theta}$

71. the left branch of the hyperbola $16x^2 - 9y^2 = 1{,}440{,}000$ **77.** first quadrant parabola

79. horizontal tangent lines at $(6, \frac{3}{2}\pi)$, $(\frac{3}{2}, \frac{1}{6}\pi)$, $(\frac{3}{2}, \frac{5}{6}\pi)$; vertical tangent lines at $(0, \frac{1}{2}\pi)$, $(\frac{9}{2}, \frac{7}{6}\pi)$, $(\frac{9}{2}, \frac{11}{6}\pi)$ **81.** $3\sqrt{3}$

83. $\frac{1}{2}\pi$ at pole; $\frac{1}{6}\pi$ at $(3, \frac{2}{3}\pi)$, $\frac{1}{6}\pi$ at $(3, \frac{4}{3}\pi)$ **85.** $\chi = 1.98$, $\alpha = 0.93$

EXERCISES 11.1 (Page 659)

1. $\frac{4}{3}$ **3.** $\frac{1}{2}\pi$ **5.** $-\frac{11}{4}$ **7.** 1   **9.** 2   **11.** $+\infty$ **13.** $-\frac{1}{8}$ **15.** $\ln\frac{2}{3}$ **17.** $\frac{1}{6}$ **19.** $\frac{1}{3}$ **21.** $\frac{1}{2}$ **23.** 1

25. $\frac{3}{5}$ **27.** 2   **29.** -1 **31.** $\dfrac{Et}{L}$ **33.** $a = -3$, $b = \frac{9}{2}$ **37.** (b) $f'(0) = -\frac{1}{2}$

EXERCISES 11.2 (Page 664)

1. 0   **3.** 0   **5.** 1   **7.** 0   **9.** 0   **11.** $\frac{1}{2}$ **13.** 1   **15.** $\frac{1}{2}$ **17.** 1   **19.** e^2 **21.** e^a **23.** 1   **25.** e^2

27. $e^{-1/3}$ **29.** $\frac{1}{2}$ **31.** 0   **33.** 1   **35.** 1   **37.** $\dfrac{1}{\ln 3}$ **41.** $f(e) = e^{1/e}$, rel max; $y = 1$ is an asymptote

43. (a) 0; (b) 0; no

EXERCISES 11.3 (Page 672)

1. 1   **3.** $-\dfrac{1}{2\ln 5}$ **5.** $\dfrac{1}{(\ln 2)^2}$ **7.** divergent   **9.** divergent   **11.** $\frac{1}{3}\pi$ **13.** 2   **15.** 1   **17.** divergent

19. (a) divergent; (b) 0   **21.** (a) 0   **23.** $\frac{1}{2}\pi$ **25.** $\frac{1}{2}\pi$ **27.** (a) 0.565; (b) 0.287   **29.** (a) 0.203; (b) 0.188   **31.** $\dfrac{1}{k}$

33. $\dfrac{1000}{0.08 + \ln 2}$ dollars $\approx \$1293.41$ **37.** $n > 1$ **39.** $\frac{1}{3}$; $\frac{1}{9}\ln\frac{32}{27}$

EXERCISES 11.4 (Page 676)

1. 2   **3.** -4 **5.** $\frac{1}{3}\pi$ **7.** divergent   **9.** divergent   **11.** divergent   **13.** divergent   **15.** divergent   **17.** 0

19. divergent   **21.** 0   **23.** $\frac{1}{3}\pi$ **25.** 0   **27.** $n > -1$; $\dfrac{1}{n + 1}$ **29.** $n > -1$; $\dfrac{2}{(n + 1)^3}$ **31.** yes; 6π

EXERCISES 11.5 (Page 684)

1. $P_3(x) = -1 - (x - 1) - (x - 1)^2 - (x - 1)^3$; $R_3(x) = \dfrac{(x - 1)^4}{(\xi - 2)^5}$, ξ between 1 and x

3. $P_3(x) = 8 + 3(x - 4) + \frac{3}{16}(x - 4)^2 - \frac{1}{128}(x - 4)^3$; $R_3(x) = \dfrac{3(x - 4)^4}{128\xi^{5/2}}$, ξ between 4 and x

5. $P_3(x) = \frac{1}{2} + \frac{1}{2}\sqrt{3}(x - \frac{1}{6}\pi) - \frac{1}{4}(x - \frac{1}{6}\pi)^2 - \frac{1}{12}\sqrt{3}(x - \frac{1}{6}\pi)^3$; $R_3(x) = \frac{1}{24}\sin\xi(x - \frac{1}{6}\pi)^4$, ξ between $\frac{1}{6}\pi$ and x

7. $P_4(x) = x + \frac{1}{6}x^3$; $R_4(x) = \frac{1}{120}(\cosh \xi)x^5$, ξ between 0 and x **9.** $P_3(x) = x - 1 - \frac{1}{2}(x-1)^2 + \frac{1}{3}(x-1)^3$;
$R_3(x) = -\frac{1}{4}\xi^{-4}(x-1)^4$; ξ between 1 and x **11.** $P_3(x) = -\ln 2 - \sqrt{3}(x - \frac{1}{3}\pi) - 2(x - \frac{1}{3}\pi)^2 - \frac{4}{3}\sqrt{3}(x - \frac{1}{3}\pi)^3$;
$R_3(x) = -\frac{1}{12}(3\sec^4 \xi - 2\sec^2 \xi)(x - \frac{1}{3}\pi)^4$, ξ between $\frac{1}{3}\pi$ and x **13.** $P_3(x) = 1 + \frac{3}{2}x + \frac{3}{8}x^2 - \frac{1}{16}x^3$;

$R_3(x) = \frac{3}{128}(1 + \xi)^{-5/2}x^4$, ξ between 0 and x **15.** 2.71828 **17.** $|\text{error}| < \dfrac{(0.1)^4}{24} < 0.000005$ **19.** 0.515 **21.** 0.1823

23. $\dfrac{55\sqrt{2}}{672}$, $|\text{error}| < \frac{1}{7680}\sqrt{2}$ **27.** $x \approx \dfrac{\pi}{2(1 + m)}$ **29.** $2(x - 1) + 5(x - 1)^2 + 3(x - 1)^3 + (x - 1)^4$

REVIEW EXERCISES FOR CHAPTER 11 (Page 684)

1. $+\infty$ **3.** $\frac{1}{3}$ **5.** $+\infty$ **7.** 1 **9.** $-\frac{1}{2}$ **11.** e^{12} **13.** 0 **15.** 1 **17.** e **19.** divergent **21.** $\frac{1}{2}$

23. divergent **25.** $\dfrac{32}{\ln 2}$ **27.** divergent **29.** $\frac{1}{4}\pi$ **31.** divergent **33.** (b) $f'(0) = -1$ **35.** (a) no; (b) 0

37. (a) divergent; (b) 0 **39.** $P_6(x) = 1 - \frac{1}{2}x^2 + \frac{1}{24}x^4 - \frac{1}{720}x^6$; $R_6(x) = \frac{1}{5040}(\sin \xi)x^7$, ξ between 0 and x
41. $P_4(x) = \frac{1}{3} - \frac{1}{54}(x - 9) + \frac{1}{648}(x - 9)^2 - \frac{5}{34,992}(x - 9)^3 + \frac{35}{2,519,424}(x - 9)^4$; $R_4(x) = -\frac{63}{256}\xi^{-11/2}(x - 9)^5$, ξ between 9 and x
43. $+\infty$ **45.** $\frac{64}{3}$ **47.** 1 **49.** (a) 0.148; (b) 0.699; (c) 0.018 **51.** \$152,500

EXERCISES 12.1 (Page 693)

1. $\frac{1}{2}$ **3.** divergent **5.** -2 **7.** 0 **9.** 1 **11.** divergent **13.** divergent **15.** $e^{1/3}$ **17.** 1 **19.** 0

21. choose $N = 1 + \dfrac{3}{\epsilon}$ **23.** choose $N = \dfrac{6}{\epsilon} - \dfrac{3}{2}$ **25.** choose $N = \dfrac{1}{5}\sqrt{\dfrac{2 - 5\epsilon}{\epsilon}}$

EXERCISES 12.2 (Page 699)

1. increasing **3.** decreasing **5.** not monotonic **7.** not monotonic **9.** decreasing

11. increasing after the first two terms **13.** increasing **15.** decreasing **19.** not bounded **31.** $\left\{ \dfrac{(-1)^{n+1}}{n} \right\}$

EXERCISES 12.3 (Page 708)

1. $s_n = \dfrac{n}{2n + 1}$; $\frac{1}{2}$ **3.** $s_n = \dfrac{5n}{3n + 1}$; $\frac{5}{3}$ **5.** $s_n = -\ln(n + 1)$; divergent **7.** $s_n = \dfrac{5}{2}\left(1 - \dfrac{1}{5^n}\right)$; $\frac{5}{2}$ **9.** $\displaystyle\sum_{n=1}^{+\infty} \dfrac{2}{(3n - 2)(3n + 1)}$; $\frac{2}{3}$

11. $\dfrac{1}{3} - \displaystyle\sum_{n=2}^{+\infty} \dfrac{2}{3^n}$; 0 **13.** $\displaystyle\sum_{n=1}^{+\infty} \ln\left(\dfrac{2n + 1}{2n - 1}\right)$; divergent **15.** divergent **17.** 2 **19.** divergent **21.** 1 **23.** $\dfrac{1}{e - 1}$
25. divergent **27.** $\frac{3}{11}$ **29.** $\frac{137}{111}$ **31.** 8 m **33.** $\frac{3100}{9}$ m **35.** 1000

EXERCISES 12.4 (Page 713)

1. divergent **3.** divergent **5.** 3 **7.** $\frac{10}{3}$ **9.** $\dfrac{63 \cdot 2^{10} + 1}{2^{16}}$ **11.** divergent **13.** $\frac{3}{2}$ **15.** divergent

17. divergent **19.** 2 **21.** divergent **23.** $\displaystyle\sum_{n=1}^{+\infty} \dfrac{1}{n}$ and $\displaystyle\sum_{n=1}^{+\infty} \dfrac{1}{n + 1}$ are both divergent; $\displaystyle\sum_{n=1}^{+\infty} \dfrac{1}{n^2 + n}$ is convergent

EXERCISES 12.5 (Page 722)

1. convergent **3.** convergent **5.** divergent **7.** convergent **9.** divergent **11.** convergent **13.** divergent
15. convergent **17.** divergent **19.** convergent **21.** divergent **23.** convergent **25.** convergent

EXERCISES 12.6 (Page 726)

1. divergent **3.** convergent **5.** convergent **7.** convergent **9.** divergent **11.** convergent **13.** convergent

15. convergent **17.** convergent **19.** convergent **21.** divergent **27.** $0.7032 < \sum\limits_{m=50}^{100} \frac{1}{m} < 0.7134$

EXERCISES 12.7 (Page 730)

1. convergent **3.** convergent **5.** convergent **7.** convergent **9.** convergent **11.** divergent **13.** convergent
15. $|R_4| < \frac{1}{5}$ **17.** $|R_4| < \frac{1}{81}$ **19.** $|R_4| < \frac{1}{25}$ **21.** $|R_4| < \dfrac{1}{6 \ln 6}$ **23.** 0.333 **25.** 0.632 **27.** 0.113 **29.** 0.406

EXERCISES 12.8 (Page 738)

1. absolutely convergent **3.** absolutely convergent **5.** absolutely convergent **7.** divergent **9.** absolutely convergent
11. absolutely convergent **13.** absolutely convergent **15.** absolutely convergent **17.** absolutely convergent
19. divergent **25.** (b) convergent

REVIEW EXERCISES FOR CHAPTER 12 (Page 740)

1. $1, \frac{3}{2}, \frac{9}{5}, 2; 3$ **3.** $0, \frac{3}{5}, \frac{4}{5}, \frac{15}{17}; 1$ **5.** $1, 3, 1, 3$; no limit **7.** $4, \frac{81}{16}, \frac{4096}{729}, \frac{390,625}{65,536}; e^2$ **9.** $\frac{3}{16}, \frac{15}{64}, \frac{63}{256}, \frac{255}{1024}; s_n = \dfrac{4^n - 1}{4^{n+1}}; \frac{1}{4}$

11. convergent; 3 **13.** divergent **15.** convergent; $4 + 2\sqrt{3}$ **17.** convergent; $\frac{1}{6}$ **19.** $\frac{649}{729}$ **21.** convergent
23. divergent **25.** convergent **27.** divergent **29.** convergent **31.** divergent **33.** divergent **35.** convergent
37. convergent **39.** absolutely convergent **41.** conditionally convergent **43.** divergent **45.** absolutely convergent
47. absolutely convergent **49.** $\frac{437}{330}$ **51.** 90 in.

EXERCISES 13.1 (Page 750)

1. $[-1, 1)$ **3.** $[-1, 1]$ **5.** $[-\frac{1}{2}, \frac{1}{2}]$ **7.** $(-3, 3)$ **9.** $(-\infty, +\infty)$ **11.** $(-5, -1)$ **13.** $(-9, 9]$ **15.** $(0, 2]$
17. $\left(-\dfrac{1}{e^2}, \dfrac{1}{e^2}\right)$ **19.** $[-1, 1]$ **21.** $(-4, 6)$ **23.** $[4, 6)$ **25.** $[-1, 1]$ **27.** $(-e, e)$ **29.** $+\infty$

EXERCISES 13.2 (Page 759)

1. (a) $R = 1; [-1, 1]$; (b) $\sum\limits_{n=1}^{+\infty} \dfrac{x^{n-1}}{n}; R = 1$; (c) $[-1, 1)$ **3.** (a) $R = 1; [-1, 1)$; (b) $\sum\limits_{n=1}^{+\infty} \sqrt{n}x^{n-1}; R = 1$; (c) $(-1, 1)$

5. (a) $R = +\infty; (-\infty, +\infty)$; (b) $\sum\limits_{n=1}^{+\infty} (-1)^{n-1} \dfrac{x^{2n-2}}{(2n-2)!}; R = +\infty$; (c) $(-\infty, +\infty)$

7. (a) $R = \frac{1}{3}; (0, \frac{2}{3})$; (b) $\sum\limits_{n=1}^{+\infty} 3n(n+1)(3x-1)^n; R = \frac{1}{3}$; (c) $(0, \frac{2}{3})$ **9.** (a) $R = 3; [-2, 4)$; (b) $\sum\limits_{n=1}^{+\infty} \dfrac{(x-1)^{n-1}}{3^n}; R = 3$; (c) $(-2, 4)$

11. $\dfrac{1}{2}\sum\limits_{n=2}^{+\infty} n(n-1)x^{n-2}$ **13.** $\sum\limits_{n=0}^{+\infty} (-1)^n(n+1)x^n$ **15.** (a) $\sum\limits_{n=0}^{+\infty} 2^n x^n$ if $|x| < \tfrac{1}{2}$; (b) $\sum\limits_{n=1}^{+\infty} n2^n x^{n-1}$ if $|x| < \tfrac{1}{2}$

17. (a) $\sum\limits_{n=0}^{+\infty} \dfrac{x^{2n}}{n!}$; (b) $\sum\limits_{n=0}^{+\infty} \dfrac{x^{2n+1}}{n!}$ **19.** 0.60653   **21.** (a) $\sum\limits_{n=0}^{+\infty} \dfrac{x^{2n+1}}{(2n+1)!}$; (b) $\sum\limits_{n=0}^{+\infty} \dfrac{x^{2n}}{(2n)!}$ **23.** 2   **25.** (a) $\sum\limits_{n=0}^{+\infty} (-1)^n \dfrac{x^{n+2}}{n!}$

27. $\sum\limits_{n=0}^{+\infty} \dfrac{x^n}{n!}$

EXERCISES 13.3 (Page 766)

1. $\sum\limits_{n=0}^{+\infty} \dfrac{x^{n+1}}{(n+1)!}$; $R = +\infty$ **3.** $\sum\limits_{n=0}^{+\infty} \dfrac{x^{n+1} - 2^{n+1}}{4^{n+1}(n+1)}$; $R = 4$ **5.** 1.718   **7.** 0.693   **9.** $\sum\limits_{n=1}^{+\infty} \dfrac{x^n}{n(n!)}$; $R = +\infty$

11. $\sum\limits_{n=0}^{+\infty} \dfrac{x^{2n+1}}{(2n+1)(2n+1)!}$; $+\infty$ **13.** 1.318   **15.** 1.057   **17.** 0.485   **19.** 0.747   **21.** 0.041   **23.** 0.450

25. $\sum\limits_{n=0}^{+\infty} \dfrac{x^{2n+1}}{(2n+1)!}$ **27.** 0.2450   **29.** 0.221   **31.** $\sum\limits_{n=0}^{+\infty} \dfrac{x^{2n+1}}{2n+1}$ **37.** $\sum\limits_{n=0}^{+\infty} \dfrac{(-1)^n x^{2n+1}}{(2n+1)!}$; $R = +\infty$

EXERCISES 13.4 (Page 775)

7. $\ln 2 + \sum\limits_{n=1}^{+\infty} (-1)^{n-1} \dfrac{(x-2)^n}{n2^n}$ **9.** $\ln 2 + \sum\limits_{n=1}^{+\infty} (-1)^{n-1} \dfrac{(x-1)^n}{n2^n}$; $R = 2$

11. $2 + \tfrac{1}{4}(x-4) + 2\sum\limits_{n=2}^{+\infty} (-1)^{n-1} \dfrac{1 \cdot 3 \cdot 5 \cdot \ldots \cdot (2n-3)(x-4)^n}{2 \cdot 4 \cdot 6 \cdot \ldots \cdot (2n) \cdot 4^n}$; $R = 4$

13. $\tfrac{1}{2} - \tfrac{1}{2}\sqrt{3}(x - \tfrac{1}{3}\pi) - \tfrac{1}{4}(x - \tfrac{1}{3}\pi)^2 + \tfrac{1}{12}\sqrt{3}(x - \tfrac{1}{3}\pi)^3 + \tfrac{1}{48}(x - \tfrac{1}{3}\pi)^4 - \ldots$; $R = +\infty$ **15.** $\dfrac{1}{2}\sum\limits_{n=1}^{+\infty} \dfrac{(-1)^{n-1}(2x)^{2n}}{(2n)!}$

17. (a) $x + \tfrac{1}{3}x^3 + \tfrac{2}{15}x^5$; (b) $1 + x^2 + \tfrac{2}{3}x^4$; (c) $\tfrac{1}{2}x^2 + \tfrac{1}{12}x^4 + \tfrac{1}{45}x^6$ **19.** 0.5299   **21.** 1.97435   **23.** -0.2231

25. 2.7182818   **27.** 0.0415   **29.** 0.0048   **31.** 0.2398   **33.** $a_4 = 3$; $a_3 = -5$; $a_2 = 2$; $a_1 = -1$; $a_0 = 6$

EXERCISES 13.5 (Page 780)

1. $1 + \dfrac{x}{2} + \sum\limits_{n=2}^{+\infty} \dfrac{(-1)^{n+1} \cdot 1 \cdot 3 \cdot \ldots \cdot (2n-3)}{2^n n!} x^n$; $R = 1$ **3.** $\dfrac{1}{2} + \dfrac{1}{2}\sum\limits_{n=1}^{+\infty} \dfrac{(-1)^n \cdot 1 \cdot 3 \cdot 5 \cdot \ldots \cdot (2n-1)x^n}{8^n n!}$; $R = 4$

5. $1 + \sum\limits_{n=1}^{+\infty} -\dfrac{2 \cdot 5 \cdot 8 \cdot \ldots \cdot (3n-4)}{3^n n!} x^{3n}$; $R = 1$ **7.** $\dfrac{1}{3} + \dfrac{1}{3}\sum\limits_{n=1}^{+\infty} \dfrac{(-1)^n \cdot 1 \cdot 3 \cdot 5 \cdot \ldots \cdot (2n-1)x^{4n}}{18^n n!}$; $R = \sqrt{3}$

9. $x^2 + \sum\limits_{n=1}^{+\infty} \dfrac{(-1)^n \cdot 1 \cdot 3 \cdot 5 \cdot \ldots \cdot (2n-1)x^{n+2}}{2^n n!}$; $R = 1$ **11.** 4.899   **13.** 5.010   **15.** 0.198

17. $x + \sum\limits_{n=1}^{+\infty} \dfrac{(-1)^n \cdot 1 \cdot 3 \cdot 5 \cdot \ldots \cdot (2n-1)x^{2n+1}}{2^n n!(2n+1)}$; $R = 1$ **19.** (a) $1 + \dfrac{1}{4}x + \sum\limits_{n=2}^{+\infty} \dfrac{(-1)^{n+1} \cdot 3 \cdot 7 \cdot 11 \cdot \ldots \cdot (4n-5)}{4^n n!} x^n$;

(b) $1 + \dfrac{1}{4}x^2 + \sum\limits_{n=2}^{+\infty} \dfrac{(-1)^{n+1} \cdot 3 \cdot 7 \cdot 11 \cdot \ldots \cdot (4n-5)}{4^n n!} x^{2n}$; (c) 0.510   **21.** 0.335   **23.** 2.0271   **25.** 0.4970

27. 0.5082   **29.** $\dfrac{1}{p+1}x^{p+1} + \sum\limits_{n=1}^{+\infty} \dfrac{1 \cdot 3 \cdot 5 \cdot \ldots \cdot (2n-1)x^{2n+p+1}}{2^n n!(2n+p+1)}$; $R = 1$

REVIEW EXERCISES FOR CHAPTER 13 (Page 780)

1. $[-1, 1)$ **3.** $[-3, 3]$ **5.** $x = 3$ **7.** $(-7, 5)$ **9.** $(-1, 3]$ **11.** $(-1, 1)$ **13.** (a) $R = 1$, $[-1, 1]$;

(b) $\sum\limits_{n=1}^{+\infty} (-1)^n x^{2n-1}$, $R = 1$; (c) $(-1, 1)$ **15.** (a) $R = +\infty$, $(-\infty, +\infty)$; (b) $\sum\limits_{n=1}^{+\infty} \dfrac{nx^{n-1}}{(n!)^2}$, $R = +\infty$; (c) $(-\infty, +\infty)$

17. $\sum_{n=0}^{+\infty} \dfrac{(-1)^n x^{2n+1}}{2^{4n+4}(2n+1)}$; $R = 4$ **19.** $\sum_{n=0}^{+\infty} \dfrac{(-1)^n x^{2n+1}}{(2n+1)(2n+1)!}$; $R = +\infty$ **21.** 0.161 **23.** 0.493 **25.** 0.1974 **27.** 5.0658

29. 0.9986 **31.** 1.6094 **33.** 0.0124 **35.** $\sum_{n=0}^{+\infty} \dfrac{(\ln a)^n x^n}{n!}$; $(-\infty, +\infty)$ **37.** $\dfrac{3}{4} \sum_{n=1}^{+\infty} \dfrac{(-1)^{n-1}(3^{2n}-1)x^{2n+1}}{(2n+1)!}$; $(-\infty, +\infty)$

39. $\sum_{n=1}^{+\infty} \dfrac{(-1)^n(3x+\pi)^{2n-1}}{(2n-1)!}$ **41.** $\sum_{n=1}^{+\infty} -\dfrac{(x+1)^n}{n}$

EXERCISES 14.1 (Page 793)

1. (b) 5 **3.** (b) $\frac{1}{2}\sqrt{1+4e^2}$ **5.** (a) $\frac{7}{4}\pi$; (b) π; (c) $\tan^{-1} 0.4 \approx 0.38$ **7.** $\langle 2, -3 \rangle$ **9.** $\langle 5, 6 \rangle$ **11.** $(-4, 3)$ **13.** $(12, -5)$
15. (a) $\langle -1, 9 \rangle$; (b) $\langle 1, -5 \rangle$ **17.** (a) $\langle -9, -4 \rangle$; (b) $\langle 4, -e \rangle$ **19.** (a) $\langle 6, 1 \rangle$; (b) $\sqrt{74}$; (c) $\sqrt{1061}$
21. (a) $10\mathbf{i} + 15\mathbf{j}$; (b) $-24\mathbf{i} + 6\mathbf{j}$; (c) $6\mathbf{i} + 2\mathbf{j}$; (d) $2\sqrt{10}$ **23.** (a) $\sqrt{13} + \sqrt{17}$; (b) $-14\mathbf{i} + 21\mathbf{j}$; (c) $7\sqrt{13}$; (d) $5\sqrt{13} - 6\sqrt{17}$

25. (a) $-28\mathbf{i} + 6\mathbf{j}$; (b) $2\sqrt{205}$ **27.** $\dfrac{11}{\sqrt{137}}\mathbf{i} + \dfrac{4}{\sqrt{137}}\mathbf{j}$ **29.** (a) $5(\frac{3}{5}\mathbf{i} - \frac{4}{5}\mathbf{j})$, $\frac{3}{5}\mathbf{i} - \frac{4}{5}\mathbf{j}$; (b) $2\sqrt{2}(\cos \frac{1}{4}\pi\mathbf{i} + \sin \frac{1}{4}\pi\mathbf{j})$, $\frac{1}{2}\sqrt{2}\mathbf{i} + \frac{1}{2}\sqrt{2}\mathbf{j}$

31. (a) $8(\cos \frac{2}{3}\pi\mathbf{i} + \sin \frac{2}{3}\pi\mathbf{j})$, $-\frac{1}{2}\mathbf{i} + \frac{1}{2}\sqrt{3}\mathbf{j}$; (b) $16(\cos \pi\mathbf{i} + \sin \pi\mathbf{j})$, $-\mathbf{i}$ **33.** $h = 2, k = 3$ **37.** (a) 135 lb; (b) 17°
39. 29.0° **41.** 346.1°; (b) 243 mi/hr **43.** (a) 28.1°; (b) 1.7 mi/hr; (c) 0.53 mi **51.** (a) $\langle 1, -2 \rangle$; (b) $\langle 1, -2 \rangle$

EXERCISES 14.2 (Page 800)

1. 10 **3.** -1 **11.** $\frac{1}{10}\sqrt{2}$ **13.** $-\frac{16}{65}$ **15.** $10, -\frac{2}{5}$ **17.** (a) 0; (b) no k **19.** $\dfrac{-240 + \sqrt{85,683}}{407}$

21. $-\frac{4}{17}\sqrt{17}\mathbf{i} + \frac{1}{17}\sqrt{17}\mathbf{j}$; $\frac{4}{17}\sqrt{17}\mathbf{i} - \frac{1}{17}\sqrt{17}\mathbf{j}$ **23.** $\pm\dfrac{1}{\|\mathbf{A}\|}(a_2\mathbf{i} - a_1\mathbf{j})$ **25.** (a) $-\dfrac{80}{\sqrt{85}}$; (b) $\dfrac{-112}{17}\mathbf{i} + \dfrac{96}{17}\mathbf{j}$ **27.** $\frac{29}{50}\sqrt{50}$

29. (a) 24 ft-lb; (b) $24\sqrt{3}$ ft-lb **31.** $(18 - 9\sqrt{3})$ ft-lb ≈ 2.41 ft-lb

---

EXERCISES 14.3 (Page 807)

1. $(-\infty, 0) \cup (0, 4]$ **3.** $[-1, 1]$ **5.** all real numbers not in $(-4, 3)$ **7.** $\frac{4}{3}t; \frac{4}{9}$

9. $\dfrac{1 + \ln t}{te^t(2 + t)}$; $\dfrac{(2 + t) - (1 + \ln t)(2 + 4t + t^2)}{t^3 e^{2t}(2 + t)^3}$ **11.** $-\dfrac{b}{a}\cot t$; $-\dfrac{b}{a^2}\csc^3 t$ **13.** $(y - 1)^2 = x$ **15.** $25x^2 - 9y^2 = 225$

17. $x^2 - y^2 = 1, x \geq 1$ **19.** $9x^2 + 16y^2 = 144$ **21.** $5\sqrt{3}x + 2y = 20$ **23.** $y = 1; x = -1$

27. $\dfrac{dy}{dx} = 0; \dfrac{d^2y}{dx^2} = -\dfrac{1}{4a}; \dfrac{d^3y}{dx^3} = 0$ **31.** $x^{2/3} + y^{2/3} = a^{2/3}$ **35.** $3\pi a^2$ sq units

---

EXERCISES 14.4 (Page 814)

1. $4\mathbf{i} + 4\mathbf{j}$ **3.** $2\mathbf{i}$ **5.** $\mathbf{i}$ **7.** $\mathbf{R}'(t) = 2e^{2t}\mathbf{i} + \dfrac{1}{t}\mathbf{j}$; $\mathbf{R}''(t) = 4e^{2t}\mathbf{i} - \dfrac{1}{t^2}\mathbf{j}$ **9.** $\mathbf{R}'(t) = (1 + t^2)^{-1}\mathbf{i} + 2^t \ln 2\mathbf{j}$;

$\mathbf{R}''(t) = -2t(1 + t^2)^{-2}\mathbf{i} + 2^t(\ln 2)^2\mathbf{j}$ **11.** $\mathbf{R}'(t) = -2t(t^2 + 4)^{-2}\mathbf{i} - \frac{5}{2}(1 - 5t)^{-1/2}\mathbf{j}$; $\mathbf{R}''(t) = (6t^2 - 8)(t^2 + 4)^{-3}\mathbf{i} - \frac{25}{4}(1 - 5t)^{-3/2}\mathbf{j}$
13. $\mathbf{R}'(t) = 10 \cos 2t\mathbf{i} - 4 \sec 4t \tan 4t\mathbf{j}$; $\mathbf{R}''(t) = -20 \sin 2t\mathbf{i} + (16 \sec 4t - 32 \sec^3 4t)\mathbf{j}$ **15.** $(2t - 3)(2t^2 - 6t + 5)^{-1/2}$
21. $20t$ **23.** $8e^{4t} - 8e^{-4t}$ **29.** $\ln|\sec t|\mathbf{i} - \ln|t|\mathbf{j} + \mathbf{C}$ **31.** $\frac{1}{3}e^{3t}\mathbf{i} + \ln|t - 1|\mathbf{j} + \mathbf{C}$ **33.** $(t \ln t - t)\mathbf{i} + \frac{1}{3}t^3\mathbf{j} + \mathbf{C}$
35. $\frac{1}{2}(t - \frac{1}{2}\sin 2t - \pi)\mathbf{i} + (t + \frac{1}{2}\sin 2t - \pi)\mathbf{j}$ **37.** $x^2 + y^2 = 1; 0$ **39.** 0

EXERCISES 14.5 (Page 820)

1. $1 + \frac{1}{2}\sqrt{2}\ln(1 + \sqrt{2})$ **3.** $2\sqrt{10} + \sqrt{2}\ln(2 + \sqrt{5})$ **5.** $\frac{2}{27}[(40)^{3/2} - (13)^{3/2}]$ **7.** 120 **9.** $\sqrt{2}(e - 1)$
11. $\ln(1 + \sqrt{2})$ **13.** $6a$ **15.** 8π **17.** $a[\ln \cosh 2 + \ln \cosh 1]$ **19.** 5π **21.** $2\pi a$ **23.** 12 **25.** $\frac{1}{2}\sqrt{5}(e^8 - 1)$
27. $\frac{1}{2}a(\theta_1 - \frac{3}{2}\sin \frac{2}{3}\theta_1)$ **29.** 200 m

EXERCISES 14.6 (Page 824)

1. $\mathbf{T}(t) = \dfrac{t^2 - 1}{t^2 + 1}\mathbf{i} + \dfrac{2t}{t^2 + 1}\mathbf{j}$; $\mathbf{N}(t) = \dfrac{2t}{t^2 + 1}\mathbf{i} + \dfrac{1 - t^2}{t^2 + 1}\mathbf{j}$ **3.** $\mathbf{T}(t) = \dfrac{e^t}{\sqrt{e^{2t} + e^{-2t}}}\mathbf{i} - \dfrac{e^{-t}}{\sqrt{e^{2t} + e^{-2t}}}\mathbf{j}$;

$\mathbf{N}(t) = \dfrac{e^{-t}}{\sqrt{e^{2t} + e^{-2t}}}\mathbf{i} + \dfrac{e^t}{\sqrt{e^{2t} + e^{-2t}}}\mathbf{j}$ **5.** $\mathbf{T}(t) = -\sin t\mathbf{i} + \cos t\mathbf{j}$; $\mathbf{N}(t) = -\cos t\mathbf{i} - \sin t\mathbf{j}$ **7.** $\mathbf{T}(t) = -\sin kt\mathbf{i} + \cos kt\mathbf{j}$;

$\mathbf{N}(t) = -\cos kt\mathbf{i} - \sin kt\mathbf{j}$ **9.** $\mathbf{T}(t) = -(1 + \cot^4 t)^{-1/2}\mathbf{i} + (1 + \tan^4 t)^{-1/2}\mathbf{j}$; $\mathbf{N}(t) = \dfrac{-\cos^2 t}{\sqrt{\sin^4 t + \cos^4 t}}\mathbf{i} - \dfrac{\sin^2 t}{\sqrt{\sin^4 t + \cos^4 t}}\mathbf{j}$

11. $\mathbf{T}(t) = \dfrac{2t}{\sqrt{4t^2 + e^{2t}}}\mathbf{i} + \dfrac{e^t}{\sqrt{4t^2 + e^{2t}}}\mathbf{j}$; $\mathbf{N}(t) = \dfrac{e^t(1 - t)}{|t - 1|\sqrt{4t^2 + e^{2t}}}\mathbf{i} + \dfrac{2t(t - 1)}{|t - 1|\sqrt{4t^2 + e^{2t}}}\mathbf{j}$ **13.** $\mathbf{T}(t) = \sin t\mathbf{i} - \cos t\mathbf{j}$;

$\mathbf{N}(t) = \cos t\mathbf{i} + \sin t\mathbf{j}$ **15.** $\mathbf{T}(t) = \dfrac{1}{\sqrt{1 + t^2}}\mathbf{i} + \dfrac{t}{\sqrt{1 + t^2}}\mathbf{j}$; $\mathbf{N}(t) = -\dfrac{t}{\sqrt{1 + t^2}}\mathbf{i} + \dfrac{1}{\sqrt{1 + t^2}}\mathbf{j}$ **19.** $x = 2 + \cos s$, $y = 3 + \sin s$

21. $x = \frac{1}{9}[(27s + 8)^{2/3} - 4]$, $y = \frac{1}{27}[(27s + 8)^{2/3} - 4]^{3/2}$ **23.** $x = a\left(\dfrac{3a - 2s}{3a}\right)^{3/2}$, $y = a\left(\dfrac{2s}{3a}\right)^{3/2}$ **25.** $s = \sqrt{2}(e^t - 1)$

27. $\frac{27}{185}\sqrt{37}$

EXERCISES 14.7 (Page 831)

1. $\frac{1}{8}\sqrt{2}$; $4\sqrt{2}$ **3.** $\frac{1}{4}\sqrt{2}$; $2\sqrt{2}$ **5.** $\dfrac{\sqrt{2}|1 - t^2|^3}{(1 + 6t^2 + t^4)^{3/2}}$; $\sqrt{2}$; $\frac{1}{2}\sqrt{2}$ **7.** $\frac{1}{2}$; 2 **9.** $\frac{1}{4}\sqrt{2}$; $2\sqrt{2}$ **11.** $\frac{4}{49}\sqrt{7}$; $\frac{7}{4}\sqrt{7}$

13. $\frac{2}{289}\sqrt{17}$; $\frac{17}{2}\sqrt{17}$ **15.** $\dfrac{(2 - x^2)^{3/2}}{|x|}$ **17.** $\frac{1}{576}(16x^2 + 81y^2)^{3/2}$ **19.** $\dfrac{2(x + y)^{3/2}}{a^{1/2}}$ **21.** $4|a \sin \frac{1}{2}t|$ **25.** $(3, 9)$

27. $(-3, -1)$ **29.** $x^2 + y^2 + 4x - 6y + 5 = 0$ **31.** $\frac{23}{98}\sqrt{7}$; $\frac{14}{23}\sqrt{7}$ **33.** $\dfrac{1}{16|a|}$; $16|a|$ **37.** 2; $\frac{1}{2}$; $(0, -\frac{1}{2})$

39. $\left(3x + 2p, -\dfrac{y^3}{4p^2}\right)$ **41.** $\left(\dfrac{a^2 - b^2}{a}\cos^3 t, \dfrac{b^2 - a^2}{b}\sin^3 t\right)$

EXERCISES 14.8 (Page 837)

1. (a) $2t\mathbf{i} + \mathbf{j}$; (b) $2\mathbf{i}$; (c) $\sqrt{37}$; (d) 2 **3.** (a) $-10 \sin 2t\mathbf{i} + 6 \cos 2t\mathbf{j}$; (b) $-20 \cos 2t\mathbf{i} - 12 \sin 2t\mathbf{j}$; (c) 10; (d) 12
5. (a) $\mathbf{i} + \tan t\mathbf{j}$; (b) $\sec^2 t\mathbf{j}$; (c) $\sqrt{2}$; (d) 2 **7.** (a) $\cos t\mathbf{i} + \sec^2 t\mathbf{j}$; (b) $-\sin t\mathbf{i} + 2 \sec^2 t \tan t\mathbf{j}$; (c) $\frac{1}{6}\sqrt{91}$; (d) $\frac{1}{18}\sqrt{849}$
9. (a) $2\mathbf{i} + 6\mathbf{j}$; (b) $2\mathbf{j}$; (c) $2\sqrt{10}$; (d) 2 **11.** (a) $2\mathbf{i} + 8\mathbf{j}$; (b) $2\mathbf{i} + 16\mathbf{j}$; (c) $2\sqrt{17}$; (d) $2\sqrt{65}$ **13.** (a) $3\mathbf{j}$; (b) $-4\mathbf{i}$; (c) 3; (d) 4

15. (a) $\mathbf{i} + \sqrt{3}\mathbf{j}$; (b) $-\sqrt{3}\mathbf{i} + \mathbf{j}$; (c) 2; (d) 2 **17.** $\dfrac{2t - 3}{t - 1}\mathbf{i} + \dfrac{4 - 2t - t^2}{2}\mathbf{j}$ **19.** $(e^{-t} + 3t - 1)\mathbf{i} + (\frac{1}{2}e^{2t} + \frac{5}{2})\mathbf{j}$

21. (a) $\dfrac{390{,}625}{2}$ ft; (b) $\dfrac{390{,}625}{8}$ ft; (c) $1250\sqrt{2}\mathbf{i} - 1250\sqrt{2}\mathbf{j}$ **23.** $(25 + \sqrt{631})$ sec; $(20{,}000\sqrt{3} + 800\sqrt{1893})$ ft   **25.** 283 m/sec

27. $40°8'$ **29.** 29 ft

EXERCISES 14.9 (Page 842)

1. $\mathbf{V}(t) = \mathbf{i} + 2t\mathbf{j}$; $\mathbf{A}(t) = 2\mathbf{j}$; $A_T(t) = \dfrac{4t}{\sqrt{1 + 4t^2}}$; $A_N(t) = \dfrac{2}{\sqrt{1 + 4t^2}}$ **3.** $\mathbf{V}(t) = t \cos t\mathbf{i} + \sin t\mathbf{j}$;

$\mathbf{A}(t) = (\cos t - t \sin t)\mathbf{i} + (\sin t + t \cos t)\mathbf{j}$; $A_T(t) = 1$; $A_N(t) = t$ **5.** $\mathbf{V}(0) = -\mathbf{i} + \mathbf{j}$; $\mathbf{A}(0) = \mathbf{i} + \mathbf{j}$; $A_T(0) = 0$; $A_N(0) = \sqrt{2}$

7. $\mathbf{V}(\frac{1}{4}\pi) = \frac{3}{4}\sqrt{2}\mathbf{i} - \frac{3}{4}\sqrt{2}\mathbf{j}$; $\mathbf{A}(\frac{1}{4}\pi) = \frac{3}{4}\sqrt{2}\mathbf{i} + \frac{3}{4}\sqrt{2}\mathbf{j}$; $A_T(\frac{1}{4}\pi) = 0$; $A_N(\frac{1}{4}\pi) = \frac{3}{2}$ **9.** $\|\mathbf{V}(t)\| = t\sqrt{4 + 9t^2}$; $A_T(t) = \dfrac{4 + 18t^2}{\sqrt{4 + 9t^2}}$;

$A_N(t) = \dfrac{6t}{\sqrt{4 + 9t^2}}$; $K(t) = \dfrac{6}{t(4 + 9t^2)^{3/2}}$ **11.** $\mathbf{V}(t) = 2\mathbf{i} + 2t\mathbf{j}$; $\mathbf{A}(t) = 2\mathbf{j}$; $\mathbf{T}(t) = \dfrac{1}{\sqrt{1 + t^2}}\mathbf{i} + \dfrac{t}{\sqrt{1 + t^2}}\mathbf{j}$;

$\mathbf{N}(t) = \dfrac{-t}{\sqrt{1 + t^2}}\mathbf{i} + \dfrac{1}{\sqrt{1 + t^2}}\mathbf{j}$; $\|\mathbf{V}(t)\| = 2\sqrt{1 + t^2}$; $A_T(t) = \dfrac{2t}{\sqrt{1 + t^2}}$; $A_N(t) = \dfrac{2}{\sqrt{1 + t^2}}$; $K(t) = \dfrac{1}{2(1 + t^2)^{3/2}}$; $\mathbf{V}(2) = 2\mathbf{i} + 4\mathbf{j}$;

$\mathbf{T}(2) = \dfrac{1}{\sqrt{5}}\mathbf{i} + \dfrac{2}{\sqrt{5}}\mathbf{j}$; $\mathbf{A}(2) = 2\mathbf{j}$; $\mathbf{N}(2) = \dfrac{-2}{\sqrt{5}}\mathbf{i} + \dfrac{1}{\sqrt{5}}\mathbf{j}$; $\|\mathbf{V}(2)\| = 2\sqrt{5}$; $A_T(2) = \dfrac{4}{\sqrt{5}}$; $A_N(2) = \dfrac{2}{\sqrt{5}}$; $K(2) = \dfrac{1}{10\sqrt{5}}$

13. $\mathbf{V}(t) = -15 \sin 3t\mathbf{i} + 15 \cos 3t\mathbf{j}$; $\mathbf{A}(t) = -45 \cos 3t\mathbf{i} - 45 \sin 3t\mathbf{j}$; $\mathbf{T}(t) = -\sin 3t\mathbf{i} + \cos 3t\mathbf{j}$; $\mathbf{N}(t) = -\cos 3t\mathbf{i} - \sin 3t\mathbf{j}$;
$\|\mathbf{V}(t)\| = 15$; $A_T(t) = 0$; $A_N(t) = 45$; $K(t) = \frac{1}{5}$; $\mathbf{V}(\frac{1}{3}\pi) = -15\mathbf{j}$; $\mathbf{A}(\frac{1}{3}\pi) = 45\mathbf{i}$; $\mathbf{T}(\frac{1}{3}\pi) = -\mathbf{j}$; $\mathbf{N}(\frac{1}{3}\pi) = \mathbf{i}$; $\|\mathbf{V}(\frac{1}{3}\pi)\| = 15$

15. $\mathbf{V}(t) = e^t\mathbf{i} - e^{-t}\mathbf{j}$; $\mathbf{A}(t) = e^t\mathbf{i} + e^{-t}\mathbf{j}$; $\mathbf{T}(t) = \dfrac{e^{2t}}{\sqrt{e^{4t} + 1}}\mathbf{i} - \dfrac{1}{\sqrt{e^{4t} + 1}}\mathbf{j}$; $\mathbf{N}(t) = \dfrac{1}{\sqrt{e^{4t} + 1}}\mathbf{i} + \dfrac{e^{2t}}{\sqrt{e^{4t} + 1}}\mathbf{j}$;

$\|\mathbf{V}(t)\| = \dfrac{\sqrt{e^{4t} + 1}}{e^t}$; $A_T(t) = \dfrac{e^{4t} - 1}{e^t\sqrt{e^{4t} + 1}}$; $A_N(t) = \dfrac{2e^t}{\sqrt{e^{4t} + 1}}$; $K(t) = \dfrac{2e^{3t}}{(e^{4t} + 1)^{3/2}}$; $\mathbf{V}(0) = \mathbf{i} - \mathbf{j}$; $\mathbf{A}(0) = \mathbf{i} + \mathbf{j}$; $\mathbf{T}(0) = \dfrac{1}{\sqrt{2}}\mathbf{i} - \dfrac{1}{\sqrt{2}}\mathbf{j}$;

$\mathbf{N}(0) = \dfrac{1}{\sqrt{2}}\mathbf{i} + \dfrac{1}{\sqrt{2}}\mathbf{j}$; $\|\mathbf{V}(0)\| = \sqrt{2}$; $A_T(0) = 0$; $A_N(0) = \sqrt{2}$; $K(0) = \dfrac{1}{\sqrt{2}}$ **17.** (a) $\mathbf{i} + 4\mathbf{j}$; (b) $\mathbf{i} + 8\mathbf{j}$; (c) $8\mathbf{j}$; (d) $\frac{64}{65}\sqrt{65}$; (e) $\frac{8}{65}\sqrt{65}$

19. $\mathbf{R} = 2\mathbf{i} + 4\mathbf{j}$; if k is constant speed, $\mathbf{V} = \dfrac{k}{\sqrt{2}}\mathbf{i} + \dfrac{k}{\sqrt{2}}\mathbf{j}$; $\mathbf{A} = \dfrac{k^2}{16}\mathbf{i} - \dfrac{k^2}{16}\mathbf{j}$; $\mathbf{T} = \dfrac{1}{\sqrt{2}}\mathbf{i} + \dfrac{1}{\sqrt{2}}\mathbf{j}$; $\mathbf{N} = \dfrac{1}{\sqrt{2}}\mathbf{i} - \dfrac{1}{\sqrt{2}}\mathbf{j}$;

$A_T = 0$; $A_N = \dfrac{k^2\sqrt{2}}{16}$

REVIEW EXERCISES FOR CHAPTER 14 (Page 842)

1. $-25\mathbf{i} + 63\mathbf{j}$ **3.** $\sqrt{4594}$ **5.** $15\sqrt{2} - 14\sqrt{13}$ **7.** 92 **9.** $\dfrac{9}{\sqrt{106}}\mathbf{i} - \dfrac{5}{\sqrt{106}}\mathbf{j}$ **11.** $h = -\frac{1}{2}$; $k = \frac{1}{2}$ **13.** $-\frac{19}{5}\sqrt{2}$

15. $-\frac{19}{25}\mathbf{i} - \frac{133}{25}\mathbf{j}$ **17.** $-\frac{19}{26}\sqrt{13}$ **19.** (a) 104.4 lb; (b) 35.5°   **21.** -89.30 ft-lb

23. (a) all real numbers in $[0, +\infty)$ except 1; (b) $\frac{1}{2}\mathbf{i} + \frac{1}{2}\mathbf{j}$; (c) $\dfrac{-1}{(t + 1)^2}\mathbf{i} + \dfrac{2t^{1/2} - t - 1}{2t^{1/2}(t - 1)^2}\mathbf{j}$ **25.** $\dfrac{dy}{dx} = \dfrac{1}{6t}$; $\dfrac{d^2y}{dx^2} = -\dfrac{1}{108t^3}$

27. $x = 12$; $y = 16$; $y = -16$ **29.** $-\dfrac{4(t^{12} + t^{10}) + 144(t^2 - 1)^3}{t^9(t^2 - 1)^3}$ **31.** $\frac{3}{2}\sqrt{37} + \frac{1}{4}\ln(6 + \sqrt{37})$ **37.** $|at|$ **39.** $\frac{2}{75}$; $\frac{75}{2}$

41. (a) $9y = 12x - x^2$; (b) $\mathbf{V}(t) = 3\mathbf{i} + (4 - 2t)\mathbf{j}$, $\mathbf{A}(t) = -2\mathbf{j}$; (c) $\mathbf{V}(1) = 3\mathbf{i} + 2\mathbf{j}$, $\mathbf{A}(1) = -2\mathbf{j}$

43. $\dfrac{dy}{dx} = -\tan t$; $\dfrac{d^2y}{dx^2} = \dfrac{1}{3a}\sec^4 t \csc t$ **45.** $\mathbf{T}(t) = \dfrac{e^t - e^{-t}}{e^t + e^{-t}}\mathbf{i} + \dfrac{2}{e^t + e^{-t}}\mathbf{j}$; $\mathbf{N}(t) = \dfrac{2}{e^t + e^{-t}}\mathbf{i} - \dfrac{e^t - e^{-t}}{e^t + e^{-t}}\mathbf{j}$

47. $x = 2(3s + 17\sqrt{17})^{2/3} - 34$; $y = \frac{1}{3}[(3s + 17\sqrt{17})^{2/3} - 16]^{3/2}$ **49.** $h = \dfrac{1}{2g}(v_0 \sin \alpha)^2$ **55.** $\mathbf{V}(t) = 2 \sinh 2t\mathbf{i} + 2 \cosh 2t\mathbf{j}$;

$\mathbf{A}(t) = 4 \cosh 2t\mathbf{i} + 4 \sinh 2t\mathbf{j}$; $\|\mathbf{V}(t)\| = 2\sqrt{\cosh 4t}$; $A_T(t) = \dfrac{4 \sinh 4t}{\sqrt{\cosh 4t}}$; $A_N(t) = \dfrac{4}{\sqrt{\cosh 4t}}$ **57.** $A_T = \dfrac{4t - 8}{\sqrt{4t^2 - 16t + 25}}$;

$A_N = \dfrac{6}{\sqrt{4t^2 - 16t + 25}}$

EXERCISES 15.1 (Page 851)

1. (b) (7, 2, 0), (0, 0, 3), (0, 2, 0), (0, 2, 3), (7, 0, 3), (7, 0, 0); (c) $\sqrt{62}$ **3.** (b) (2, 1, 2), (−1, 3, 2), (−1, 1, 5), (2, 3, 2), (−1, 3, 5), (2, 1, 5);
(c) $\sqrt{22}$ **5.** (b) (3, −1, 0), (3, 3, 0), (1, 3, 0), (1, 3, 5), (1, −1, 5), (3, −1, 5); (c) $3\sqrt{5}$ **7.** (a) 3; (b) $(2, 5, \frac{5}{2})$ **9.** (a) $\frac{13}{2}$; (b) $(\frac{5}{4}, -1, 2)$

11. (a) $7\sqrt{2}$; (b) $(-1, \frac{9}{2}, -\frac{1}{2})$ **13.** $(\pm 4\sqrt{6}, 4, 2)$ **17.** (a) $|\overline{AB}| = 9\sqrt{2}$; $|\overline{AC}| = 2\sqrt{62}$; $|\overline{BC}| = \sqrt{62}$; (b) midpoint of AB: $(\frac{1}{2}, 1, \frac{3}{2})$; midpoint of AC: $(-1, 2, 5)$; midpoint of BC: $(-\frac{5}{2}, 8, \frac{7}{2})$ **21.** sphere with center at $(4, -2, -1)$ and $r = 5$
23. the point $(0, 0, 3)$ **25.** the empty set   **27.** $x^2 + (y - 1)^2 + (z + 4)^2 = 9$

EXERCISES 15.2 (Page 859)

1. (a) $\langle 21, -13, -2 \rangle$; (b) $\langle -25, -26, 5 \rangle$; (c) $7\sqrt{59} - 5\sqrt{41}$; (d) $\sqrt{1326}$

3. (a) $\langle -19, -16, -1 \rangle$; (b) $\langle -6\sqrt{91}, -8\sqrt{91}, -2\sqrt{91} \rangle$ **5.** $a = 0, b = 0$ **7.** $\frac{4}{\sqrt{89}}; \frac{3}{\sqrt{89}}; \frac{8}{\sqrt{89}}$ **9.** $-\frac{6}{\sqrt{86}}; -\frac{1}{\sqrt{86}}; -\frac{7}{\sqrt{86}}$

11. $(\frac{13}{3}, 0, -\frac{4}{3})$ **13.** $(\frac{17}{3}, \frac{4}{3}, -6)$ **15.** (a) $7(-\frac{6}{7}\mathbf{i} + \frac{2}{7}\mathbf{j} + \frac{3}{7}\mathbf{k})$; (b) $\sqrt{14}\left(-\frac{2}{\sqrt{14}}\mathbf{i} + \frac{1}{\sqrt{14}}\mathbf{j} - \frac{3}{\sqrt{14}}\mathbf{k}\right)$

17. (a) $\langle \frac{1}{9}, \frac{8}{9}, \frac{4}{9} \rangle$; (b) $\left\langle -\frac{1}{\sqrt{3}}, \frac{1}{\sqrt{3}}, \frac{1}{\sqrt{3}} \right\rangle$ **21.** (a) -44; (b) -468; (c) -31; (d) $\langle -84, 198, 124 \rangle$

23. (a) $-\frac{1}{5}\sqrt{5}$; (b) -3; (c) $\langle 2, 1, -2 \rangle$ **25.** (a) $-\frac{13}{9}\sqrt{6}$; (b) $\langle -\frac{26}{27}, -\frac{91}{27}, \frac{13}{27} \rangle$ **29.** $\frac{1}{11}\sqrt{4422}$ **33.** $\frac{7}{2}\sqrt{3}$ sq units

35. $-\frac{43}{30}\sqrt{6}$ **37.** 25 ft-lb   **39.** $\frac{5}{3}\sqrt{6}$ ft-lb   **41.** $\frac{2}{5}$ **45.** $\cos^{-1}\frac{1}{\sqrt{3}}$ or $\cos^{-1}\left(-\frac{1}{\sqrt{3}}\right)$ **49.** $r = 5, s = 9, t = -12$

EXERCISES 15.3 (Page 867)

1. $x + 2y - 3z + 1 = 0$ **3.** $y - z + 3 = 0$ **5.** $x - 3y - 4z - 3 = 0$ **7.** $3x + 2y + 6z = 23$

9. $\langle \frac{2}{3}, -\frac{1}{3}, \frac{2}{3} \rangle; \langle -\frac{2}{3}, \frac{1}{3}, -\frac{2}{3} \rangle$ **11.** $\langle \frac{4}{13}, \frac{3}{13}, -\frac{12}{13} \rangle; \langle -\frac{4}{13}, -\frac{3}{13}, \frac{12}{13} \rangle$ **13.** $\left\langle \frac{3}{\sqrt{13}}, 0, \frac{2}{\sqrt{13}} \right\rangle; \left\langle -\frac{3}{\sqrt{13}}, 0, -\frac{2}{\sqrt{13}} \right\rangle$

15. $5x - 3y + 7z + 14 = 0$ **17.** $2x - y - z + 1 = 0$ **19.** $4y - 3z - 1 = 0$ and $z = 1$ **21.** $67.6°$ **23.** $69.2°$
25. $\frac{16}{15}\sqrt{6}$ **27.** $\frac{3}{2}$

EXERCISES 15.4 (Page 872)

1. $x = 1 + 4t, y = 2 - 3t, z = 1; \frac{x - 1}{4} = \frac{y - 2}{-3}, z = 1$ **3.** $x = 13t, y = -12t, z = -8t; \frac{x}{13} = \frac{y}{-12} = \frac{z}{-8}$

5. $x = 2t - 2, y = -16t; z = 13t + 3; \frac{x + 2}{2} = \frac{y}{-16} = \frac{z - 3}{13}$ **7.** $x = 4 + t, y = -5 + 3t, z = 20 - 6t; \frac{x - 4}{1} = \frac{y + 5}{3} = \frac{z - 20}{-6}$

9. $\frac{x - \frac{1}{7}}{2} = \frac{y}{7} = \frac{z - \frac{10}{7}}{13}$ **13.** $8x - y - 66 = 0; 13x - 5z - 102 = 0; 13y - 40z + 42 = 0$ **15.** $4x + y + 3 = 0; 3x - z + 4 = 0;$

$3y + 4z - 7 = 0$ **17.** $\frac{5}{18}\sqrt{6}$ **19.** $4x + 7y - 3z + 7 = 0$ **21.** $4x + 2y - 3z + 5 = 0$ **23.** $(\frac{5}{3}, -\frac{17}{6}, \frac{5}{12})$

25. $\frac{x - 3}{1} = \frac{y - 6}{2} = \frac{z - 4}{1}$ **27.** $\frac{2}{5}\sqrt{70}$ **31.** $x = x_0, y = y_0$

EXERCISES 15.5 (Page 882)

1. $\langle 7, 13, -11 \rangle$ **3.** -490 **11.** $\langle 9, -1, -23 \rangle$ **15.** $\frac{2}{3}\sqrt{2}$ **17.** $\sqrt{89}$ sq units   **19.** $9\sqrt{29}$ sq units

21. $5x - 2y + 7z = 0$ **23.** $x + 2y + z - 2 = 0$ **25.** $\pm\frac{1}{\sqrt{3}}(\mathbf{i} + \mathbf{j} - \mathbf{k})$ **27.** $\pm\frac{1}{\sqrt{6}}(\mathbf{i} + 2\mathbf{j} + \mathbf{k})$ **29.** 20 cu units

31. $\frac{38}{3\sqrt{78}}$

EXERCISES 15.6 (Page 887)

13. $x^2 + z^2 = 4y$　　**15.** $x^2 + 4y^2 + 4z^2 = 16$　　**17.** $y^2 = 9x^2 + 9z^2$　　**19.** $y^2 + z^2 = \sin^2 x$　　**21.** $x^2 + z^2 = 16$; x axis
23. $x^2 - z^2 = 4$; z axis　　**25.** $z = \sqrt{|y|}$; y axis　　**27.** $y^2 = 9x^2$; y axis

EXERCISES 15.7 (Page 894)

1. elliptic hyperboloid of one sheet　　**3.** hyperbolic paraboloid　　**5.** hyperbolic cylinder　　**7.** ellipsoid
9. elliptic hyperboloid of one sheet　　**11.** elliptic cone　　**13.** elliptic paraboloid　　**15.** hyperbolic paraboloid
17. elliptic hyperboloid of two sheets　　**19.** (a) $1 < |k| < \sqrt{2}$; (b) $|k| < 1$　　**21.** vertex: $(1, -\frac{1}{3}, 0)$; focus: $(1, 0, 0)$　　**25.** 8π cu units
27. $\dfrac{abh^2}{2c} \pi$ cu units

EXERCISES 15.8 (Page 900)

1. $\dfrac{1}{\sqrt{4t^2 + 5}}(\mathbf{i} - 2t\mathbf{j} - 2\mathbf{k})$　　**3.** $\mathbf{T}(t) = \frac{1}{3}\sqrt{3}[(\cos t - \sin t)\mathbf{i} + (\cos t + \sin t)\mathbf{j} + \mathbf{k}]$

5. $\dfrac{1}{\sqrt{4t^2 + 29}}[2(\cos t - t\sin t)\mathbf{i} + 5\mathbf{j} + 2(\sin t + t\cos t)\mathbf{k}]$　　**7.** $\sqrt{21} + \frac{3}{2} + \frac{5}{4}\ln(4 + \sqrt{21})$　　**9.** $\sqrt{3}(e^3 - 1)$　　**11.** 13

15. $\mathbf{R}(t) = t\mathbf{i} + e^t\mathbf{j} + te^t\mathbf{k}$　　**17.** $-\frac{3}{5}$　　**19.** $\mathbf{T}(1) = \frac{1}{14}\sqrt{14}\mathbf{i} + \frac{1}{7}\sqrt{14}\mathbf{j} + \frac{3}{14}\sqrt{14}\mathbf{k}$; $\mathbf{N}(1) = -\frac{11}{266}\sqrt{266}\mathbf{i} - \frac{4}{133}\sqrt{266}\mathbf{j} + \frac{9}{266}\sqrt{266}\mathbf{k}$;
$\mathbf{B}(1) = \dfrac{\sqrt{19}}{19}(3\mathbf{i} - 3\mathbf{j} + \mathbf{k})$; $K(1) = \frac{1}{98}\sqrt{266}$　　**21.** $\mathbf{T}(-1) = \frac{1}{3}(\mathbf{i} + 2\mathbf{j} - 2\mathbf{k})$; $\mathbf{N}(-1) = \frac{2}{15}\sqrt{5}\mathbf{i} - \frac{1}{3}\sqrt{5}\mathbf{j} - \frac{4}{15}\sqrt{5}\mathbf{k}$;
$\mathbf{B}(-1) = -\frac{2}{5}\sqrt{5}\mathbf{i} - \frac{1}{5}\sqrt{5}\mathbf{k}$; $K(-1) = \frac{2}{27}\sqrt{5}$　　**23.** $\mathbf{T}(0) = \frac{1}{3}\sqrt{3}(\mathbf{i} + \mathbf{j} + \mathbf{k})$; $\mathbf{N}(0) = -\frac{1}{2}\sqrt{2}(\mathbf{i} - \mathbf{j})$; $\mathbf{B}(0) = -\frac{1}{6}\sqrt{6}(\mathbf{i} + \mathbf{j} - 2\mathbf{k})$; $K(0) = \frac{1}{3}\sqrt{2}$
25. $\mathbf{V}(\frac{1}{2}\pi) = -a\mathbf{i} + \mathbf{k}$; $\mathbf{A}(\frac{1}{2}\pi) = -a\mathbf{j}$; $\|\mathbf{V}(\frac{1}{2}\pi)\| = \sqrt{a^2 + 1}$　　**27.** $\mathbf{V}(1) = 2e^2\mathbf{i} - 2e^{-2}\mathbf{j} + 3e^2\mathbf{k}$;
$\mathbf{A}(1) = 4e^2\mathbf{i} + 4e^{-2}\mathbf{j} + 8e^2\mathbf{k}$; $\|\mathbf{V}(1)\| = \sqrt{13e^4 + 4e^{-4}}$　　**31.** 2

EXERCISES 15.9 (Page 904)

1. (a) $(0, 3, 5)$; (b) $(-\frac{7}{2}, \frac{7}{2}\sqrt{3}, -4)$; (c) $(\cos 1, \sin 1, 1)$　　**3.** (a) $(\sqrt{6}, \sqrt{2}, 2\sqrt{2})$; (b) $(0, 2\sqrt{3}, 2)$; (c) $(\frac{1}{2}\sqrt{3}, \frac{3}{2}, -\sqrt{3})$
5. (a) $(2, \frac{2}{3}\pi, -2\sqrt{3})$; (b) $(0, \frac{3}{4}\pi, -\sqrt{2})$; (c) $(\sqrt{6}, \frac{1}{3}\pi, \sqrt{6})$　　**7.** ellipsoid; $r^2 + 4z^2 = 16$　　**9.** elliptic paraboloid; $r^2 = 3z$
11. elliptic cone; $r^2\cos 2\theta = 3z^2$　　**13.** sphere; $\rho = 9\cos\phi$　　**15.** right circular cylinder; $\rho\sin\phi = 3$
17. sphere; $\rho = 8\sin\phi\cos\theta$　　**19.** (a) right circular cylinder; $x^2 + y^2 = 16$; (b) plane through z axis; $y = x$　　**21.** $x^2 - y^2 = z^3$
23. (a) sphere; $x^2 + y^2 + z^2 = 81$; (b) plane through z axis; $x = y$; (c) half of a cone with vertex at origin; $z = \sqrt{x^2 + y^2}$
25. right circular cylinder; $x^2 + y^2 = 36$　　**27.** $x\sqrt{x^2 + y^2 + z^2} = 2y$　　**31.** (b) $2\pi\sqrt{a^2 + 1}$

REVIEW EXERCISES FOR CHAPTER 15 (Page 905)

1. a point on the x axis in R, a line parallel to the y axis in R^2, a plane parallel to the yz plane in R^3　　**3.** the x axis
5. the circle in the xz plane with center at the origin and radius 2　　**7.** the plane perpendicular to the xy plane and
intersecting the xy plane in the line $y = x$　　**9.** the paraboloid of revolution generated by revolving $y^2 = 9z$ about the z axis
11. the right circular cone generated by revolving $y = x$ about the x axis　　**13.** $\mathbf{i} + 26\mathbf{j} - 16\mathbf{k}$　　**15.** -3　　**17.** $7\sqrt{1270}$
19. $-\frac{1}{3}\sqrt{21}$　　**21.** $\frac{1}{9}(\mathbf{i} + 2\mathbf{j} - 2\mathbf{k})$　　**23.** $\langle 60, -40, 80\rangle$　　**25.** 16　　**27.** 295　　**29.** $(-3, \sqrt{167}, 1), (-3, -\sqrt{167}, 1)$
31. $(x + 2)^2 + (y + 1)^2 + (z - 3)^2 = 17$　　**33.** $z^2 = e^{4y}$ or $x^2 = e^{4y}$; the y axis　　**35.** 3

37. (a) $\cos\alpha = -\dfrac{7}{\sqrt{78}}$, $\cos\beta = -\dfrac{5}{\sqrt{78}}$, $\cos\gamma = -\dfrac{2}{\sqrt{78}}$; (b) $-\dfrac{7}{\sqrt{78}}\mathbf{i} - \dfrac{5}{\sqrt{78}}\mathbf{j} - \dfrac{2}{\sqrt{78}}\mathbf{k}$　　**39.** (a) $-\frac{4}{5}\sqrt{3}$; (b) $-\frac{28}{25}\mathbf{i} + \frac{4}{25}\mathbf{j} - \frac{4}{5}\mathbf{k}$

41. $x - 6y - 10z + 23 = 0$ **43.** $\frac{47}{10}\sqrt{2}$ **45.** 3   **47.** $\frac{1}{3}\sqrt{3}$ **49.** $\frac{x}{4} = \frac{y}{-3} = \frac{z}{1}$, $x = 4t$, $y = -3t$, $z = t$ **53.** $\frac{54}{25}\pi$ sq units

55. 24 cu units   **57.** $\frac{25}{4}[\sqrt{2} + \ln(\sqrt{2} + 1)]$ **59.** $\mathbf{V}(\frac{1}{2}\pi) = -\frac{1}{2}\pi\mathbf{i} + \mathbf{j} + \mathbf{k}$; $\mathbf{A}(\frac{1}{2}\pi) = -2\mathbf{i} - \frac{1}{2}\pi\mathbf{j}$; $\|\mathbf{V}(\frac{1}{2}\pi)\| = \frac{1}{2}\sqrt{8 + \pi^2}$

61. $\mathbf{T}(\frac{1}{2}\pi) = -\dfrac{3}{\sqrt{13}}\mathbf{i} - \dfrac{2}{\sqrt{13}}\mathbf{j}$; $\mathbf{N}(\frac{1}{2}\pi) = -\mathbf{k}$; $\mathbf{B}(\frac{1}{2}\pi) = \dfrac{2}{\sqrt{13}}\mathbf{i} - \dfrac{3}{\sqrt{13}}\mathbf{j}$ **63.** $(\frac{3}{2}\sqrt{3}, \pi, \frac{3}{2})$

65. (a) $z = r^2(1 + \sin 2\theta) + 1$; (b) $r^2(25\cos^2\theta + 4\sin^2\theta) = 100$

EXERCISES 16.1 (Page 916)

1. (a) $-\frac{1}{7}$; (b) 5; (c) $\dfrac{x + y}{x - y + 2}$; (d) 0   **3.** (a) 1; (b) $\frac{1}{2}\sqrt{2}$; (c) $\frac{1}{2}\sqrt{16 - x^2 - y^2 - z^2}$; (d) $4x + 4y + 8$ **5.** $\{(x, y)\,|\,x^2 + y^2 \neq 1\}$

7. $\{(x, y)\,|\,x^2 + y^2 \leq 1\}$ **9.** $\{(x, y)\,|\,x^2 - y^2 \geq 1\}$ **11.** $\{(x, y)\,|\,x^2 + y^2 \geq 1\}$ **13.** $\{(x, y)\,|\,x^2 + y^2 < 1\}$

15. $\{(x, y)\,|\,y \neq \pm x\}$ **17.** $\{(x, y)\,|\,-1 \leq x - y \leq 1\}$ **19.** $\{(x, y)\,|\,xy > 1\}$ **21.** $\{(x, y, z)\,|\,x - y - z \neq 0\}$

23. $\{(x, y, z)\,|\,x^2 + 4y^2 + z^2 \leq 16\}$ **25.** $\{(x, y, z)\,|\,|x| \leq 1, |y| \leq 1, |z| \leq 1\}$ **27.** $\{(x, y, z)\,|\,x^2 + y^2 < 4\}$

29. $\{(x, y)\,|\,x^2 + y^2 \leq 16\}$ **31.** R^2 **33.** R^2 **35.** R^2 **37.** circles of radius $\sqrt{16 - k^2}$, $k = 0, 1, 2, 3, 4$

39. circles of radius $\sqrt{16 - k}$, $k = 16, 12, 7, 0, -9, -20$ **41.** hyperbolas $x^2 - y^2 = k$, $k = 16, 9, 4, 0, -4, -9, -16$

43. circles of radius $\sqrt{2k}$, $k = 8, 6, 4, 2, 0$ **45.** $h(x, y) = \sin^{-1}\sqrt{1 - x^2 - y^2}$; domain: $\{(x, y)\,|\,x^2 + y^2 \leq 1\}$

47. (a) 2; (b) 6; (c) $\sqrt{x - y^2}$; (d) $|x - y|$; (e) $|x - y|$

EXERCISES 16.2 (Page 926)

1. 0   **3.** -4 **5.** 2   **7.** 0   **9.** $\delta = \frac{1}{7}\epsilon$ **11.** $\delta = \frac{1}{5}\epsilon$ **13.** $\delta = \min(1, \frac{1}{6}\epsilon)$ **15.** $\delta = \min(1, \frac{1}{8}\epsilon)$ **23.** 0   **25.** 0

27. limit exists and is 0; take $\delta = \sqrt{\epsilon}$ **29.** limit does not exist   **33.** $\frac{1}{4}\pi$ **35.** $\frac{1}{2}$ **39.** $\frac{1}{2}$ **41.** 3   **47.** 0; take $\delta = \frac{1}{3}\epsilon$

EXERCISES 16.3 (Page 930)

1. continuous at every point (x, y) in R^2 that is not on the line $y = 1$

3. continuous at every point (x, y) in R^2 that is not on the y axis

5. continuous at every point (x, y) in R^2 that is not on the line $y = 2x$

7. continuous at every point (x, y) in R^2 that is interior to the circle $x^2 + y^2 = 25$ **9.** continuous at every point in R^2

11. continuous at every point $(x, y) \neq (0, 0)$ in R^2 **13.** continuous at every point in R^2

15. all points (x, y) in R^2 that are interior to the circle $x^2 + y^2 = 16$

17. all points (x, y) in R^2 that are exterior to the ellipse $4x^2 + 9y^2 = 36$ **19.** all points (x, y) in R^2 for which $|xy| \geq 1$

21. all points (x, y) in R^2 in the first and third quadrants for which $|x + y| < 1$ **23.** all points in R^2 **25.** essential

27. removable; $f(0, 0) = 0$ **29.** essential   **31.** removable; $f(0, 0) = 0$

35. continuous at every point (x, y, z) in R^3 that is exterior to the sphere $x^2 + y^2 + z^2 = 1$ **37.** continuous at all points in R^3

EXERCISES 16.4 (Page 938)

1. 6   **3.** $3x - 2y$ **5.** $\dfrac{x}{\sqrt{x^2 + y^2}}$ **7.** $x^2 - 6xy + 2z$ **9.** $xy + yt + zt$ **11.** 4   **13.** $\dfrac{x}{\sqrt{x^2 + y^2}}$

15. $-2\sin 3\theta \sin 2\phi$ **17.** $\dfrac{e^{y/x}}{xy}\left(y \ln \dfrac{x^2}{y} - x\right)$ **19.** $\dfrac{-z}{(x^2 + y^2 + z^2)^{3/2}}$ **21.** $4xy + \dfrac{1}{z}$ **23.** $xze^{xyz} + \dfrac{3xz^2}{z^4 + 9x^2y^2}$

25. (a) -1; (b) 12   **27.** $-\ln \sin x$; $\ln \sin y$ **31.** (a) 1; (b) 1   **33.** (a) -2; (b) 0   **35.** 4   **37.** -4 deg/cm; -8 deg/cm

39. (a) $\dfrac{100}{i^2}\left[\dfrac{9i + 1}{(1 + i)^9} - 1\right]$; (b) $\dfrac{1}{0.0036}\left[\dfrac{1.54}{(1.06)^9} - 1\right] \approx -24.4$; (c) $\dfrac{5000 \ln 1.06}{3(1.06)^t}$; (d) $-\dfrac{5000 \ln 1.06}{3(1.06)^8} \approx -61$

41. 0.0943 m²/kg; 6.42 m²/m

EXERCISES 16.5 (page 947)

1. (a) $3(\Delta x)^2 + 2(\Delta x)(\Delta y) - (\Delta y)^2 + 14\Delta x - 6\Delta y$; (b) 0.5411; (c) $14\Delta x - 6\Delta y$; (d) 0.54

3. (a) $(2 + \Delta x)(-4 + \Delta y)e^{(2+\Delta x)(-4+\Delta y)} + 8e^{-8}$; (b) -0.0026; (c) $28e^{-8}\,\Delta x - 14e^{-8}\,\Delta y$; (d) -0.0019

5. (a) $\Delta x + 4\Delta y + (\Delta x)(\Delta y) + \ln(1 + \Delta y) + \ln(5 + \Delta z) - \ln 5$; (b) 0.2141; (c) $\Delta x + 5\Delta y + \frac{1}{5}\Delta z$; (d) 0.214

7. $(12x^2 - y^2)\,dx + (3 - 2xy)\,dy$ **9.** $(\cos y - y \cos x)\,dx + (-x \sin y - \sin x)\,dy$ **11.** $\dfrac{2x\,dx + 2y\,dy + 2z\,dz}{x^2 + y^2 + z^2}$

13. $\tan^{-1} z\,dx - \dfrac{2y}{z}\,dy + \left(\dfrac{x}{1+z^2} + \dfrac{y^2}{z^2}\right) dz$

15. (a) $2(x_0 y_0 - y_0)\,\Delta x + (x_0{}^2 - 2x_0)\,\Delta y + (y_0\,\Delta x + \Delta x\,\Delta y)\,\Delta x + 2(x_0\,\Delta x - \Delta x)\,\Delta y$; (b) $\epsilon_1 = y_0\,\Delta x + \Delta x\,\Delta y$; $\epsilon_2 = 2(x_0\,\Delta x - \Delta x)$

17. (a) $\dfrac{2x_0 y_0\,\Delta x + y_0(\Delta x)^2 - x_0{}^2\,\Delta y}{y_0{}^2 + y_0\,\Delta y}$; (b) $\epsilon_1 = \dfrac{y_0{}^2\,\Delta x - 2x_0 y_0\,\Delta y}{y_0{}^3 + y_0{}^2\,\Delta y}$; $\epsilon_2 = \dfrac{x_0{}^2\,\Delta y}{y_0{}^3 + y_0{}^2\,\Delta y}$

31. (a) $(y_0 - z_0)\,\Delta x + x_0\,\Delta y + (2z_0 - x_0)\,\Delta z - (\Delta z)(\Delta x) + (\Delta x)(\Delta y) + (\Delta z)(\Delta z)$; (b) $\epsilon_1 = -\Delta z, \epsilon_2 = \Delta x, \epsilon_3 = \Delta z$ **35.** 7.36 m^3

37. 0.14 cm; 1.4 percent   **39.** $\frac{13}{1600}$; 0.325 percent   **41.** \$7200

43. $D_1 f(x, y) = \begin{cases} 2x \sin \dfrac{1}{\sqrt{x^2+y^2}} - \dfrac{x}{\sqrt{x^2+y^2}} \cos \dfrac{1}{\sqrt{x^2+y^2}} & \text{if } (x, y) \neq (0, 0) \\ 0 & \text{if } xy = (0, 0) \end{cases}$;

$D_2 f(x, y) = \begin{cases} 2y \sin \dfrac{1}{\sqrt{x^2+y}} - \dfrac{y}{\sqrt{x^2+y^2}} \cos \dfrac{1}{\sqrt{x^2+y^2}} & \text{if } (x, y) \neq (0, 0) \\ 0 & \text{if } (x, y) = (0, 0) \end{cases}$

EXERCISES 16.6 (Page 955)

1. $\dfrac{\partial u}{\partial r}$: (a) $6x - 2y$; (b) $16r - 10s$; $\dfrac{\partial u}{\partial s}$: (a) $-2x - 4y$; (b) $-10r - 6s$ **3.** $\dfrac{\partial u}{\partial r}$: (a) $13x - 2y + 5$; (b) $24r - 41s + 5$;

$\dfrac{\partial u}{\partial s}$: (a) $-17x - 7y - 10$; (b) $-41r + 44s - 10$ **5.** $\dfrac{\partial u}{\partial r}$: (a) $\dfrac{2e^{y/x}}{x^2}(2x \sin t - y \cos t)$; (b) 0; $\dfrac{\partial u}{\partial t}$: (a) $\dfrac{2re^{y/x}}{x^2}(y \sin t + 2x \cos t)$;

(b) $2e^{2 \tan t} \sec^2 t$ **7.** $\dfrac{\partial u}{\partial r} = 2r(2x + y) + 3x$; $\dfrac{\partial u}{\partial s} = 2s(2x + y) - 2x$ **9.** $\dfrac{\partial u}{\partial r} = \dfrac{6re^s + s \cos rs}{\sqrt{1 - (3x+y)^2}}$; $\dfrac{\partial u}{\partial s} = \dfrac{3r^2 e^s + r \cos rs}{\sqrt{1 - (3x+y)^2}}$

11. $\dfrac{\partial u}{\partial r} = \dfrac{6s}{x^2} \sinh \dfrac{y}{x}(xe^r - ry)$; $\dfrac{\partial u}{\partial s} = \dfrac{3}{x^2} \sinh \dfrac{y}{x}(2xe^r - yr^2) = 0$

13. $\dfrac{\partial u}{\partial r} = 2x \sin \phi \cos \theta + 2y \sin \phi \sin \theta + 2z \cos \phi$; $\dfrac{\partial u}{\partial \phi} = 2xr \cos \phi \cos \theta + 2yr \cos \phi \sin \theta - 2zr \sin \phi$;

$\dfrac{\partial u}{\partial \theta} = -2xr \sin \phi \sin \theta + 2yr \sin \phi \cos \theta$ **15.** (a) $e^x(\cos t - y \sin t) + e^y(x \cos t - \sin t)$;

(b) $e^{\cos t}(\cos t - \sin^2 t) + e^{\sin t}(\cos^2 t - \sin t)$ **17.** (a) $\dfrac{x \sec^2 t - y \sin t + z \cos t}{\sqrt{x^2 + y^2 + z^2}}$; (b) $\tan t \sec t$ **19.** $\dfrac{txe^t - y}{t(x^2 + y^2)}$

21. $\dfrac{x + y + 2t + ty - tx}{t(y + t)^2}$ **23.** $\dfrac{\partial z}{\partial x} = \dfrac{3y - 6x - 4z}{2z + 4x}$; $\dfrac{\partial z}{\partial y} = \dfrac{3x - 2y}{2z + 4x}$ **25.** $\dfrac{\partial z}{\partial x} = -\dfrac{z}{x}$; $\dfrac{\partial z}{\partial y} = \dfrac{xyz + 1}{3xy \tan 3xz - xy^2}$

29. decreasing at a rate of $\frac{8}{61}$ rad/min   **31.** increasing at a rate of 3840π cm^3/min   **33.** increasing at a rate of 1.6 liters/min

35. $\dfrac{\partial u}{\partial v} = \cos w \sinh v \dfrac{\partial u}{\partial x} + \sin w \cosh v \dfrac{\partial u}{\partial y}$; $\dfrac{\partial u}{\partial w} = -\sin w \cosh v \dfrac{\partial u}{\partial x} + \cos w \sinh v \dfrac{\partial u}{\partial y}$

EXERCISES 16.7 (Page 962)

1. (a) $\dfrac{2}{y} - \dfrac{6y}{x^4}$; (b) $\dfrac{2x^2}{y^3}$ **3.** (a) $4e^{2x} \sin y$; (b) $-e^{2x} \sin y$ **5.** (a) $2 \tan^{-1} \dfrac{y}{x} - \dfrac{2xy}{x^2 + y^2}$; (b) $2 \tan^{-1} \dfrac{y}{x} + \dfrac{2xy}{x^2 + y^2}$

7. (a) $3y \cosh x$; (b) $4x \sinh y$ **9.** (a) $e^x \cos y - \dfrac{2x \ln y}{(1 + x^2)^2}$; (b) $-e^x \cos y - \dfrac{\tan^{-1} x}{y^2}$ **11.** (a) $12x + 20y$; (b) $12x + 20y$

13. (a) 0; (b) e^y **15.** (a) $-2e^z \sin e^z$; (b) $-2we^z(\sin e^z + e^z \cos e^z)$ **17.** (a) $\dfrac{-320rst}{(r^2 + 4s^2 - 5t^2)^3}$; (b) $\dfrac{16r(5t^2 + 12s^2 - r^2)}{(r^2 + 4s^2 - 5t^2)^3}$

25. neither exist   **27.** $f_{12}(0, 0) = -1$; $f_{21}(0, 0) = 1$ **31.** $6se^{r-s}(2 + r) - 8e^{-2s}$ **33.** $10 \cos^2 \theta + 8$ **35.** $-10r \sin 2\theta$

EXERCISES 16.8 (Page 967)

3. $\xi_1 = 2$ **5.** $\xi_1 = 6 - 2\sqrt{3}$ **7.** 0   **9.** 0   **11.** $\xi_2 = 2\sqrt{3} - 4$ **13.** $\xi_2 = 3 + \ln\left(\dfrac{e^2 - 1}{2}\right)$

REVIEW EXERCISES FOR CHAPTER 16 (Page 968)

1. $\{(x, y) \,|\, x^2 + 4y^2 \geq 16\}$ **3.** $\{(x, y) \,|\, y > x^2\}$ **5.** $\{(x, y, z) \,|\, y \neq \pm z\}$; the set of all points in R^3 except those on plane $y = \pm z$

7. $\{(x, y) \,|\, 4x^2 + 9y^2 \leq 36\}$; the upper half of the ellipsoid $\dfrac{x^2}{9} + \dfrac{y^2}{4} + \dfrac{z^2}{36} = 1$ **9.** $y = \dfrac{k}{4x^{1/2}}$, $k = 16, 8, 4, 2$

11. (a) $4xy - 3y^2 + 4$; (b) $2x^2 - 6xy - 2$; (c) $4y$; (d) $-6x$; (e) $4x - 6y$; (f) $4x - 6y$ **13.** (a) $\dfrac{2x}{3y^2}$; (b) $\dfrac{y - 2x^2}{3y^3}$; (c) $-\dfrac{4x}{3y^3}$; (d) $-\dfrac{4x}{3y^3}$

15. (a) $t^2 \cos st^2 + te^s$; (b) $2st \cos st^2 + e^s$; (c) $2t(\cos st^2 - st^2 \sin st^2) + e^s$; (d) $2t(\cos st^2 - st^2 \sin st^2) + e^s$ **17.** (a) $\dfrac{1}{y} e^{x/y} + \dfrac{1}{x}$;

(b) $-\dfrac{x}{y^2} e^{x/y} - \dfrac{1}{y}$; (c) $\dfrac{1}{y^2} e^{x/y} - \dfrac{1}{x^2}$; (d) $\dfrac{2x}{y^3} e^{x/y} + \dfrac{x^2}{y^4} e^{x/y} + \dfrac{1}{y^2}$ **19.** (a) $\dfrac{y^2 + z^2 - x^2}{(x^2 + y^2 + z^2)^2}$; (b) $\dfrac{-2xy}{(x^2 + y^2 + z^2)^2}$; (c) $\dfrac{-2xz}{(x^2 + y^2 + z^2)^2}$

21. (a) $\dfrac{-320uvw}{(u^2 + 4v^2 - 5w^2)^3}$; (b) $\dfrac{16u(12v^2 + 5w^2 - u^2)}{(u^2 + 4v^2 - 5w^2)^3}$ **23.** (a) $\dfrac{1}{rt^2}$; (b) $\dfrac{-2}{rt^3}$; (c) $\dfrac{2}{r^2t^3}$ **27.** (a) $\dfrac{\partial u}{\partial t} = \dfrac{6y(x + y)}{x^2 + y^2} + 3 \ln(x^2 + y^2)$;

$\dfrac{\partial u}{\partial s} = \dfrac{4y(x - y)}{x^2 + y^2} - 2 \ln(x^2 + y^2)$; (b) $\dfrac{\partial u}{\partial t} = (3t - 2s)\dfrac{18t}{4s^2 + 9t^2} + 3 \ln(8s^2 + 18t^2)$; $\dfrac{\partial u}{\partial s} = (3t - 2s)\dfrac{8s}{4s^2 + 9t^2} - 2 \ln(8s^2 + 18t^2)$

29. (a) $18xyse^{3rs} + 6yse^{3rs} + 9x^2r^2s^2 + 6xr^2s^2 - 9zr^2s^2$; (b) $[9(1 + 2rs)e^{6rs} + 6(1 + rs)e^{3rs} - 9 \ln 4]r^2s^2$

31. (a) $3x \cos t - 4(y + 2x) \sin t$; (b) $12 \cos 2t - 16 \sin 2t$; $\dfrac{du}{dt}\bigg|_{t = \pi/4} = -16$

33. (a) $3(-1 + \Delta x)(3 + \Delta y)^2 - 5(-1 + \Delta x)(2 + \Delta z)^2 - 2(-1 + \Delta x)(3 + \Delta y)(2 + \Delta z) - 5$; (b) -0.48; (c) $-5\Delta x - 14\Delta y + 26\Delta z$;

(d) -0.48 **35.** 2   **37.** $\frac{1}{6}\pi$ **39.** $\delta = \min(1, \frac{1}{35}\epsilon)$ **41.** limit exists and is 0; take $\delta = \sqrt[4]{\epsilon}$ **43.** limit does not exist

45. continuous at all points (x, y) in R^2 not on the lines $x = \pm 2y$

47. continuous at all points (x, y) in R^2 except $(x, y) = (2n + 1, 2m + 1)$, where n and m are any integers

49. continuous at all points in R^2 **51.** 39¢   **53.** -3200π cm^3/min   **55.** decreasing at the rate of 0.44 atm/min

59. (a) 12; (b) -3

EXERCISES 17.1 (Page 978)

1. $2\sqrt{2}x + 5\sqrt{2}y$ **3.** $3x + \sqrt{2}y + 4z$ **5.** $\dfrac{17}{13(x - y)^2}$ **7.** $(8x - 3y)\mathbf{i} + (2y - 3x)\mathbf{j}$ **9.** $\dfrac{x}{x^2 + y^2}\mathbf{i} + \dfrac{y}{x^2 + y^2}\mathbf{j}$

11. $\dfrac{y + z}{(x + z)^2}\mathbf{i} - \dfrac{1}{x + z}\mathbf{j} - \dfrac{x - y}{(x + z)^2}\mathbf{k}$ **13.** $e^{-2y} \sec z(\mathbf{i} - 2x\mathbf{j} + x \tan z\mathbf{k})$ **15.** 6   **17.** -42 **19.** -2

21. $-3e^{\pi/4} \cos \frac{1}{12}\pi$ **23.** (a) $\langle -4, -4 \rangle$; (b) $-2 - 2\sqrt{3}$ **25.** (a) $\langle -12, 2, 14 \rangle$; (b) $\frac{6}{7}$ **29.** $\frac{3}{20}\pi + \frac{2}{5}$; $\frac{1}{4}\sqrt{\pi^2 + 4}$ **31.** $-\frac{29}{11}$; $\sqrt{21}$

33. $\theta = \tan^{-1} \dfrac{3}{3\pi + 1}$ and $\theta = \tan^{-1} \dfrac{3}{3\pi + 1} - \pi$ **35.** (a) $6\sqrt{2}$ degrees per meter; (b) direction of the vector $\dfrac{1}{\sqrt{2}}\mathbf{i} + \dfrac{1}{\sqrt{2}}\mathbf{j}$

37. (a) -1; (b) direction of the vector $-\mathbf{j}$ and magnitude 2

EXERCISES 17.2 (Page 982)

1. $2x - 2y + 3z = 17$; $\dfrac{x-2}{2} = \dfrac{y+2}{-2} = \dfrac{z-3}{3}$ **3.** $4x + 8y + 3z + 22 = 0$; $\dfrac{x+2}{4} = \dfrac{y+4}{8} = \dfrac{z-6}{3}$ **5.** $ex - y = 0$;

$\dfrac{x-1}{-e} = \dfrac{y-e}{1}$, $z = 0$ **7.** $x - y - 3 = 0$; $\dfrac{x-6}{1} = \dfrac{y-3}{-1}$, $z = 3$ **9.** $x + 2y + 2z - 8 = 0$; $\dfrac{x-4}{1} = \dfrac{y-1}{2} = \dfrac{z-1}{2}$

11. $3x - 2y - 6z + 84 = 0$; $\dfrac{x+8}{-3} = \dfrac{y-27}{2} = \dfrac{z-1}{6}$ **13.** $\dfrac{x-2}{4} = \dfrac{y+2}{-1} = \dfrac{z}{20}$ **15.** $x = 4$, $y = 16$

17. $\dfrac{x}{1-8\pi} = \dfrac{y-2}{-2\pi} = \dfrac{z-1}{-1}$ **19.** surfaces are tangent

EXERCISES 17.3 (Page 995)

1. $f(4, -\frac{1}{2}) = -\frac{133}{4}$, rel min   **3.** $f(-\frac{1}{4}, 16) = -12$, rel max   **5.** no relative extrema   **7.** no relative extrema
9. $f(1, 1) = -5$, rel min; $f(-1, -3) = 31$, rel max   **11.** $f(\frac{1}{3}\pi, \frac{1}{3}\pi) = \frac{3}{2}\sqrt{3}$, rel max; $f(\frac{5}{3}\pi, \frac{5}{3}\pi) = -\frac{3}{2}\sqrt{3}$, rel min

13. 8, 8, 8   **15.** $(\frac{41}{14}, -\frac{5}{7}, \frac{33}{14})$; $\frac{9}{14}\sqrt{14}$ **17.** $\left(0, \dfrac{1}{\sqrt{17}}, -\dfrac{4}{\sqrt{17}}\right)$ and $\left(0, -\dfrac{1}{\sqrt{17}}, \dfrac{4}{\sqrt{17}}\right)$; 1

19. $\frac{1}{3}c$ mg of drug A and $\frac{1}{2}c$ mg of drug B **21.** $\frac{16}{3}\sqrt{3}$ cu units   **23.** length of the base is $2\frac{2}{3}$ ft; width of the base is 2 ft;
depth is 3 ft   **25.** 3 machine-hours and 9 person-hours   **27.** $25,400   **29.** (a) 3120; (b) 9   **31.** (a) $y = 0.1560x + 314.9$;
(b) 368.0 milliliters per minute per kilogram   **33.** there is no greatest total profit   **35.** l:w:h = $1:1:\frac{1}{2}$

EXERCISES 17.4 (Page 1002)

1. $(0, 0)$ and $(0, 4)$ **3.** $(\frac{6}{7}, -\frac{4}{7}, \frac{2}{7})$ **5.** $f(\pm\frac{1}{2}\sqrt{35}, \frac{1}{2}) = \frac{37}{4}$, rel max; $f(0, 3) = 3$, rel min; $f(0, -3) = -3$, rel min
7. $f(-\frac{2}{3}\sqrt{3}, -\frac{1}{3}\sqrt{6}, -\frac{1}{3}\sqrt{3}) = f(-\frac{2}{3}\sqrt{3}, \frac{1}{3}\sqrt{6}, \frac{1}{3}\sqrt{3}) = f(\frac{2}{3}\sqrt{3}, -\frac{1}{3}\sqrt{6}, \frac{1}{3}\sqrt{3}) = f(\frac{2}{3}\sqrt{3}, \frac{1}{3}\sqrt{6}, -\frac{1}{3}\sqrt{3}) = -\frac{2}{9}\sqrt{6}$, rel min;
$f(\frac{2}{3}\sqrt{3}, \frac{1}{3}\sqrt{6}, \frac{1}{3}\sqrt{3}) = f(-\frac{2}{3}\sqrt{3}, -\frac{1}{3}\sqrt{6}, \frac{1}{3}\sqrt{3}) = f(\frac{2}{3}\sqrt{3}, -\frac{1}{3}\sqrt{6}, -\frac{1}{3}\sqrt{3}) = f(-\frac{2}{3}\sqrt{3}, \frac{1}{3}\sqrt{6}, -\frac{1}{3}\sqrt{3}) = \frac{2}{9}\sqrt{6}$, rel max
9. 3   **11.** $3\sqrt{3}$ **13.** (a) 12; (b) 4; (c) 1   **15.** $\frac{4}{13}\sqrt{26}$ **17.** 2, 6   **19.** $\frac{37}{13}$ **21.** $\frac{7}{32}$
33. hottest at $(\pm\frac{1}{2}\sqrt{3}, -\frac{1}{2})$; coldest at $(0, \frac{1}{2})$

EXERCISES 17.5 (Page 1010)

1. $f(x, y) = 2x^2 - \frac{3}{2}y^2 + C$ **3.** $f(x, y) = 3x^2 - 5xy + 2y^3 + C$ **5.** $f(x, y) = 2x^3y^2 - 7x^2y + 3x - 8y + C$

7. $f(x, y) = \dfrac{2x^2 - 2y^2 - x}{2xy^2} + C$ **9.** $f(x, y) = x^2 \sec 2y + C$ **11.** $f(x, y) = x^2 \cos y - x + C$ **13.** not an exact differential

15. $f(x, y) = xe^y - x^2y + C$ **17.** $\frac{1}{2}x^2 + y^2 + xy = C$ **19.** $ye^x - x^2 = C$ **21.** not exact   **23.** $x \sin\dfrac{x}{y} + \sin x = C$

25. $f(x, y, z) = 2xy - 5xz + 8yz + C$ **27.** $f(x, y, z) = xe^y \sin z + C$ **29.** $f(x, y, z) = e^{x+z} + e^y \ln z - e^x \ln y + C$
31. $f(x, y, z) = 2x^2y + 3xyz - 5yz^2 - 2x + z + C$ **33.** $f(x, y, z) = xz \tan y + C$

REVIEW EXERCISES FOR CHAPTER 17 (Page 1011)

1. 14   **3.** $-\frac{5}{2}\sqrt{2}$ **5.** $-\frac{8}{3}$ **7.** (a) $\frac{1}{2}\mathbf{i} + \frac{1}{2}\mathbf{j}$; (b) $\frac{1}{4}(1 + \sqrt{3})$ **9.** (a) $12\mathbf{i} + 3\mathbf{j} - 4\mathbf{k}$; (b) $\frac{13}{7}\sqrt{14}$ **11.** $f(x, y) = e^{x^2} \ln y + C$

13. $f(x, y, z) = \dfrac{y}{x+z} - \dfrac{1}{x} - \dfrac{2}{z} + C$ **15.** $4x + 2y + z - 12 = 0$; $\dfrac{x-2}{4} = \dfrac{y-1}{2} = \dfrac{z-2}{1}$ **17.** $\dfrac{x-1}{17} = \dfrac{y+2}{20} = \dfrac{z-11}{-4}$

19. (a) the direction of the vector $-12\mathbf{i} - 150\mathbf{j}$; (b) ascending;

(c) the direction of either of the vectors: $\dfrac{25}{\sqrt{629}}\mathbf{i} - \dfrac{2}{\sqrt{629}}\mathbf{j}$ or $-\dfrac{25}{\sqrt{629}}\mathbf{i} + \dfrac{2}{\sqrt{629}}\mathbf{j}$ **21.** $f(-1, -1) = 1$, rel max

23. $f(\pm\sqrt{5}, 0) = 10$, rel min **25.** $f(9, 11, 15) = -362$, rel max **27.** $\frac{7}{6}\sqrt{6}$ **29.** $f(x, y) = 2x^3y + \frac{5}{2}x^2 + \frac{3}{2}y^2 + C$

31. $f(x, y, z) = e^x \tan y + z \cot y + C$ **33.** $x^3y^2 + 2xy^3 = C$ **35.** $ye^{-x} + y^2 - \cos x = C$ **37.** $\frac{100}{3}, \frac{100}{3}, \frac{100}{3}$

39. $2\sqrt{3}$ by $\frac{2}{3}\sqrt{3}$ by $2\sqrt{3}$ **41.** (a) $-\dfrac{3\sqrt{3}+2}{11}$ degrees per cm; (b) $\frac{2}{11}\sqrt{13}$ degrees per cm in the direction of the vector

$-\dfrac{3}{\sqrt{13}}\mathbf{i} - \dfrac{2}{\sqrt{13}}\mathbf{j}$ **43.** 18 ft by 18 ft by 18 ft **45.** square base and a depth that is one-half the length of a side of the base

47. (a) $81x - 284y - 270 = 0$; (b) 11 days **49.** (a) $y = 9.628x - 1488.2$; (b) 2740 kilograms per hectare

---

EXERCISES 18.1 (Page 1018)

1. 45 **3.** 50 **5.** 1368 **7.** 704 **9.** 50.75 **11.** 1376 **13.** 68.6 **15.** 38.2 **17.** $[0, 24]$ **19.** $[1, e]$
21. $[0, \frac{3}{2}\sqrt{3\pi^2}]$

EXERCISES 18.2 (Page 1025)

1. 42 **3.** $\frac{98}{3}$ **5.** $-\frac{49}{5}$ **7.** $\frac{1}{3}$ **9.** $\frac{1}{3}$ **11.** 45 **13.** $\frac{152}{3}$ **15.** $\frac{3}{2}\pi$ **17.** $\frac{864}{5}$ **19.** $\frac{512}{3}$ cu units **21.** $\frac{16}{3}$ cu units

23. $\dfrac{15\pi - 32}{120}$ cu units **25.** $\frac{1}{12}$ sq units **27.** 72 sq units **29.** $c\displaystyle\int_{-a}^{a}\int_{-b\sqrt{1-(x/a)^2}}^{b\sqrt{1-(x/a)^2}} \sqrt{1 - \dfrac{x^2}{a^2} - \dfrac{y^2}{b^2}}\, dy\, dx$

31. $\frac{337}{30}$ cu units **33.** (b) $\frac{2}{3}a^3$; (c) $\frac{2}{3}\displaystyle\int_0^a \int_0^{\sqrt{a^2-x^2}} (2x + y)\, dx\, dy$ **35.** 0

EXERCISES 18.3 (Page 1030)

1. 12 kg; $(2, \frac{3}{2})$ **3.** $\frac{27}{2}$ kg; $(\frac{6}{5}, \frac{9}{5})$ **5.** $\frac{176}{15}k$ kg; $(\frac{35}{22}, \frac{102}{77})$ **7.** $\frac{2}{3}ka^3$ kg; $(\frac{3}{32}a(2 + \pi), \frac{3}{32}a(2 + \pi))$ **9.** $\frac{1}{4}k\pi$ kg; $\left(\dfrac{\pi}{2}, \dfrac{16}{9\pi}\right)$

11. $\frac{4}{3}$ kg; $(\frac{6}{5}, \frac{6}{5})$ **13.** $9k$ kg-m^2 **15.** $\frac{1}{2}\pi ka^4$ kg-m^2 **17.** $\frac{96}{35}k$ kg-m^2 **19.** (a) $\frac{144}{5}$ kg-m^2; (b) 54 kg-m^2; (c) $\frac{2}{5}\sqrt{15}$ m;

(d) $\frac{414}{5}$ kg-m^2 **21.** (a) $\frac{3}{32}\pi k$ kg-m^2; (b) $\frac{1}{24}\pi(2\pi^2 - 3)k$ kg-m^2; (c) $\frac{1}{4}\sqrt{6}$ m; (d) $(\frac{1}{12}\pi^3 - \frac{1}{32}\pi)k$ kg-m^2 **23.** $\frac{1}{12}b\sqrt{6}$ ft

25. $\dfrac{19,904}{315}\, k$ kg-m^2

EXERCISES 18.4 (Page 1036)

1. 6π sq units **3.** $\frac{1}{4}a^2(8 + \pi)$ sq units **5.** $(4\pi + 12\sqrt{3})$ sq units **7.** 4π cu units **9.** $\frac{128}{9}(3\pi - 4)$ cu units
11. $\frac{7}{2}\pi$ cu units **13.** $\frac{40}{3}\pi k$ kg; $(0, \frac{21}{10})$ **15.** $\frac{22}{3}k\pi$ kg; $(-\frac{57}{44}, 0)$ **17.** $\frac{20}{3}k$ kg; $(\frac{23}{25}, \frac{531}{1280}\pi)$

19. $\frac{8}{9}(27\sqrt{3} - 14\pi)k$ kg; $\left(0, \dfrac{3}{10}\cdot\dfrac{297\sqrt{3} - 160\pi}{27\sqrt{3} - 14\pi}\right)$ **21.** $\frac{5}{64}k\pi$ kg-m^2 **23.** $\frac{11}{16}k\pi a^4$ kg-m^2 **25.** $\frac{1}{4}\sqrt{2\pi}$ m **27.** $\pi e(e^8 - 1)$

EXERCISES 18.5 (Page 1041)

1. $\sqrt{6}$ sq units **3.** $2\sqrt{1633}$ sq units **5.** 9 sq units **7.** $\frac{1}{12}(135\sqrt{10} - 13\sqrt{26})$ sq units **9.** 8π sq units
11. 12π sq units **13.** 32 sq units **15.** $9\sqrt{2}$ sq units **17.** $\pi b\sqrt{a^2 + b^2}$ sq units **19.** $2\pi a^2(1 - e^{-1})$ sq units
21. 12π sq units **23.** $\frac{72}{5}[2 + \sqrt{2}\ln(1 + \sqrt{2})]$ sq units

EXERCISES 18.6 (Page 1045)

1. $\frac{1}{10}$ **3.** $\frac{7}{8}$ **5.** $-e(\ln 2)^2$ **7.** $\frac{1}{2}\pi - 1$ **9.** $\frac{15}{2}$ **11.** $\frac{1}{24}$ **13.** 36   **15.** $\frac{243}{2}\pi$ **17.** $\frac{648}{5}$ **19.** $\frac{1}{4}$ cu units
21. $\frac{3}{2}\pi$ cu units   **23.** 4π cu units   **25.** $\frac{4}{3}\pi abc$ cu units   **27.** $\frac{80}{3}k$ kg   **29.** $\frac{1}{28}k$ kg   **31.** $\frac{2}{5}(2\sqrt{2} - 1)$ kg

EXERCISES 18.7 (Page 1051)

1. $\frac{1}{3}a^3$ **3.** $6\pi(e - 1)$ **5.** πa^3 **7.** $\frac{4}{3}\pi a^3$ **9.** $\frac{1}{3}$ **11.** 8π **13.** $\frac{4}{5}a^5\pi k$ kg   **15.** $1250\pi k$ slug-ft^2 **17.** $\frac{4}{5}a^5\pi k$ kg
19. 8π **21.** $\frac{512}{75}k$ kg-m^2 **23.** $8k\pi$ kg   **25.** $\frac{56}{9}\pi a^3 k$ slug-ft^2 **27.** $(0, 0, \frac{1}{2})$ **29.** 18π **31.** $\frac{1}{15}\pi(2\sqrt{2} - 1)$

REVIEW EXERCISES FOR CHAPTER 18 (Page 1052)

1. $\frac{1}{40}$ **3.** $\frac{1}{8}\pi$ **5.** $\frac{1}{8}e^4 - \frac{3}{4}e^2 + e - \frac{3}{8}$ **7.** $\frac{3}{4}\pi$ **9.** $\frac{1}{8}$ **11.** $\frac{1}{3}$ **13.** $\frac{1}{2}\pi \ln 2$ **15.** $\frac{1}{2}(1 - \cos 1)$ **17.** $\frac{4}{15}$ sq units
19. 9π **21.** $\frac{8}{3}$ sq units   **23.** $\frac{8}{3}$ sq units   **25.** $\frac{1}{2}$ cu units   **27.** $\frac{1}{2}$ cu units   **29.** $\frac{1104}{35}$ cu units   **31.** 18 sq units
33. $(\frac{1}{2}\pi - \frac{2}{3})$ sq units   **35.** $(2, \frac{3}{2})$ **37.** $\frac{1}{108}(3\pi - 7)$ kg   **39.** $6\pi a^4$ cu units   **41.** $\frac{5}{3}\pi$ cu units   **43.** $\frac{1}{64}(7e^8 + 1)$ kg-m^2
45. $k(\pi + \frac{8}{3})$ kg-m^2 **47.** $2k\pi$ kg-m^2; $\frac{1}{2}\sqrt{2\pi}$ m   **49.** $65k\pi$ kg   **51.** $\frac{32}{5}k\pi$ kg-m^2

EXERCISES 19.1 (Page 1063)

7. $6x\mathbf{i} + 6y^2\mathbf{j}$ **9.** $\dfrac{2xy}{1 + x^4y^2}\mathbf{i} + \dfrac{x^2}{1 + x^4y^2}\mathbf{j}$ **11.** $(6x^2 - 6xy + y^2)\mathbf{i} + (-3x^2 + 2xy - 12y^2)\mathbf{j}$

13. $2xye^{-4z}\mathbf{i} + x^2e^{-4z}\mathbf{j} - 4x^2ye^{-4z}\mathbf{k}$ **15.** conservative   **17.** not conservative   **19.** conservative
21. $\phi(x, y) = xy + C$ **23.** $\phi(x, y) = e^x \sin y + C$ **25.** $\phi(x, y) = x^2y^2 - xy^3 + 2y + C$
27. $\phi(x, y, z) = \frac{1}{3}x^3 + \frac{1}{2}z^2 - xy + 3yz + C$ **29.** $\phi(x, y, z) = ze^x + xe^y - ye^z + C$
31. $\phi(x, y, z) = x^2 \cos y - yz^2 - 3x + 2z + C$ **33.** 0; 5 **35.** $2e^x \sin y\mathbf{k}$; $2e^x \cos y$ **37.** 0; $2x + 2y + 2z$

39. $\sin z\mathbf{i} + \sin x\mathbf{j} + \sin y\mathbf{k}$; 0 **41.** $\dfrac{x - y}{\sqrt{x^2 + y^2 + 1}}\mathbf{k}$; $\dfrac{x + y}{\sqrt{x^2 + y^2 + 1}} + 2z$

EXERCISES 19.2 (Page 1072)

1. 1   **3.** $\pi - \frac{8}{3}$ **5.** 8π **7.** 1   **9.** $\frac{16}{3}$ **11.** $\frac{4}{3}$ **13.** $-\frac{5}{12}$ **15.** $\frac{1477}{2}$ **17.** $\frac{35}{2}$ **19.** $\pi(a^2 + 2a)$ **21.** 8
23. $\frac{4}{3}$ joules   **25.** $20\frac{3}{4}$ joules   **27.** $27\frac{3}{4}$ joules   **29.** $(\frac{1}{16}\pi a^4 + a^2)$ joules   **31.** 3 joules   **33.** $(e^2 + e^4 + e^8 - 3)$ joules
35. $2\frac{1}{2}$ joules

EXERCISES 19.3 (Page 1081)

1. 2   **3.** e^2 **5.** -4 **7.** 15   **9.** $-4e$ **11.** -14 **13.** 18   **15.** $2e^2 - 3e^3 + e^4$ **17.** $\frac{3}{4}$ **19.** 11   **21.** $\frac{13}{2}$
23. 4   **25.** 0   **27.** 3   **29.** 4   **31.** -13 **33.** 216 joules   **35.** 32,766 joules   **37.** $\frac{2}{15}k$ joules

EXERCISES 19.4 (Page 1094)

1. -1 **3.** $-\frac{5}{12}$ **5.** $-\frac{1}{2}\pi$ **7.** $-\frac{32}{15}$ **9.** $-\frac{3}{7}$ **11.** $\frac{1}{5}$ **13.** $\frac{1}{24}(5 - 4\sqrt{2})\pi$ **15.** 0   **17.** -10π **19.** $\frac{41}{70}$
21. $\frac{9}{2}$ sq units   **23.** $\frac{1}{3}$ sq units   **25.** $\frac{3}{8}\pi a^2$ sq units   **31.** 24π joules   **33.** 0   **35.** 16π cm^2/sec   **37.** $\frac{3}{2}\pi$ cm^2/sec
39. 15; (i)   **41.** 0; (iii)

EXERCISES 19.5 (Page 1101)

1. 8π **3.** $\frac{8}{3}\sqrt{3}$ **5.** $\frac{9}{5}\sqrt{14}$ **7.** $\frac{1}{6}(5\sqrt{5}-1)$ **9.** $\frac{15}{2}\sqrt{2}\pi$ **11.** $\frac{7}{3}\sqrt{61}$ **13.** 0 **15.** $16k\pi$ kg   **17.** 9π kg
19. $\frac{665}{6}\sqrt{2}\pi$ kg   **21.** 18 **23.** 45π **25.** $\frac{7}{2}$

EXERCISES 19.6 (Page 1107)

1. $\frac{8}{3}\pi$ **3.** 1 **5.** 18 **7.** $\frac{7}{2}$ **9.** 27 **11.** 32π **13.** 144π **15.** $\frac{11}{12}$ **17.** 0 **19.** π **21.** 54π **23.** -1
25. 8π **27.** -2

REVIEW EXERCISES FOR CHAPTER 19 (Page 1108)

1. (a) $(4xy+3y^3)\mathbf{i}+(2x^2+9xy^2)\mathbf{j}$; (b) $e^y\mathbf{i}+(xe^y-ze^y-yze^y)\mathbf{j}-ye^y\mathbf{k}$ **3.** $\phi(x,y)=\tan^{-1}2xy^2+C$

5. $\phi(x,y,z)=z^2\tan x+y^2e^{3z}+C$ **7.** $x(e^{xy}-e^{xz})\mathbf{i}+y(e^{yz}-e^{xy})\mathbf{j}+z(e^{xz}-e^{yz})\mathbf{k}$; 0 **9.** $\dfrac{1-2y}{y^2}\mathbf{k}; \dfrac{2x}{y^2}$ **11.** $-\frac{4}{3}$

13. $\frac{44}{3}-3\pi$ **15.** $\frac{27}{10}$ **17.** $\frac{19}{4}$ **19.** $9e^2-1$ **21.** π **23.** 13 **25.** 4 **27.** 12π **29.** 0
31. $\frac{9}{2}$ sq units   **33.** $\frac{1568}{15}$ joules   **35.** $\frac{23}{6}$ joules   **39.** 0 **41.** 54 cm²/sec   **43.** $\frac{1}{2}\pi$ **45.** $\frac{3}{2}\sqrt{14}$
47. $\frac{1}{6}(37\sqrt{37}-1)$ **49.** $\frac{8}{3}k\pi$ kg   **51.** $108\sqrt{2}\pi$ kg   **53.** 56π **55.** 160π **59.** 0

EXERCISES A.1 (Page A-4)

1. $\frac{1}{3}x-\frac{2}{9}\ln|2+3x|+C$ **3.** $x+\dfrac{36}{6-x}+12\ln|6-x|+C$ **5.** $\frac{1}{30}(3x-1)(1+2x)^{3/2}+C$

7. $2\sqrt{1+2x}+\ln\left|\dfrac{\sqrt{1+2x}-1}{\sqrt{1+2x}+1}\right|+C$ **9.** $\dfrac{1}{4}\ln\left|\dfrac{x+2}{x-2}\right|+C$ **11.** $\ln|x+3+\sqrt{x^2+6x}|+C$

13. $\sqrt{9x^2+4}-2\ln\left|\dfrac{2+\sqrt{9x^2+4}}{3x}\right|+C$ **15.** $\sqrt{9-4x^2}-3\ln\left|\dfrac{3+\sqrt{9-4x^2}}{2x}\right|+C$

17. $\dfrac{1}{3}(x^2-x-6)\sqrt{4x-x^2}+4\cos^{-1}\left(\dfrac{2-x}{2}\right)+C$ **19.** $-\frac{1}{5}\sin^4 x\cos x-\frac{4}{15}\sin^2 x\cos x-\frac{8}{15}\cos x+C$

21. $-\frac{1}{6}\csc^5 x\cot x-\frac{5}{24}\csc^3 x\cot x-\frac{5}{16}\csc x\cot x+\frac{5}{16}\ln|\csc x-\cot x|+C$

23. $t^4\sin t+4t^3\cos t-12t^2\sin t-24t\cos t+24\sin t+C$ **25.** $x\sec^{-1}3x-\frac{1}{3}\ln|3x+\sqrt{9x^2-1}|+C$

27. $e^x(x^4-4x^3+12x^2-24x+24)+C$ **29.** $\dfrac{e^{4x}}{32}(8x^2-4x+1)+C$ **31.** $\dfrac{x^4}{16}(4\ln 3x-1)+C$

33. $\frac{2}{29}e^{2x}\sin 5x-\frac{5}{29}e^{2x}\cos 5x+C$ **35.** $\frac{3}{5}y\cosh 5y-\frac{3}{25}\sinh 5y+C$ **37.** $\frac{1}{60}+\frac{1}{25}\ln\frac{8}{3}$ **39.** $\frac{15}{2}-8\ln 2$
41. $\frac{32}{5}\ln 2-\frac{31}{25}$ **43.** $\frac{15}{2}-8\ln 2$ **45.** $\frac{1}{3}\pi+\frac{1}{2}\sqrt{3}$ **47.** $\frac{1}{4}-\frac{1}{8}\sqrt{2}$ **49.** 0 **51.** $\frac{1}{8}(e^2+3)$

Index